COMPREHENSIVE MEDICINAL CHEMISTRY II

Editors-in-Chief

Dr John B Taylor

Former Senior Vice-President for Drug Discovery, Rhône-Poulenc Rorer, Worldwide, UK

Professor David J Triggle

State University of New York, Buffalo, NY, USA

Volume 7

THERAPEUTIC AREAS II: CANCER, INFECTIOUS DISEASES, INFLAMMATION & IMMUNOLOGY AND DERMATOLOGY

Volume Editors

Dr Jacob J Plattner

Anacor Pharmaceuticals, Palo Alto, CA, USA

Dr Manoj C Desai

Gilead Sciences, Inc., Foster City, CA, USA

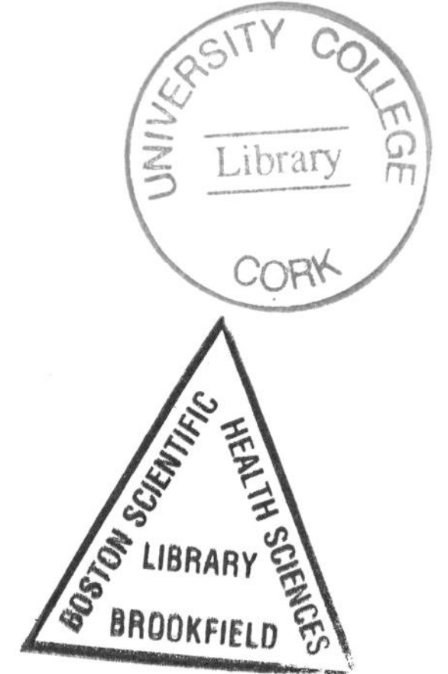

ELSEVIER

AMSTERDAM BOSTON HEIDELBERG LONDON NEW YORK OXFORD
PARIS SAN DIEGO SAN FRANCISCO SINGAPORE SYDNEY TOKYO

Elsevier Ltd.
The Boulevard, Langford Lane, Kidlington, Oxford OX5 1GB, UK

First edition 2007

British Library Cataloguing in Publication Data
A catalogue record for this book is available from the British Library

Library of Congress Catalog Number: 2006936669

ISBN-13: 978-0-08-044513-7
ISBN-10: 0-08-044513-6

Printed and bound in Spain

06 07 08 09 10 10 9 8 7 6 5 4 3 2 1

COMPREHENSIVE MEDICINAL CHEMISTRY II

Disclaimers

Contents

Anti Viral

Anti Fungal

Anti Bacterials

[✠]Deceased.

Contents of all Volumes

Volume 7 Therapeutic Areas II: Cancer, Infectious Diseases, Inflammation & Immunology and Dermatology

Volume 8 Case Histories and Cumulative Subject Index

Preface

The first edition of *Comprehensive Medicinal Chemistry* was published in 1990 and was intended to present an integrated and comprehensive overview of the then rapidly developing science of medicinal chemistry from its origins in organic chemistry. In the last two decades, the field has grown to embrace not only all the sophisticated synthetic and technological advances in organic chemistry but also major advances in the biological sciences. The mapping of the human genome has resulted in the provision of a multitude of new biological targets for the medicinal chemist with the prospect of more rational drug design (CADD). In addition, the development of sophisticated in silico technologies for structure–property relationships (ADMET) enables a much better understanding of the fate of potential new drugs in the body with the subsequent development of better new medicines.

It was our ambitious aim for this second edition, published 16 years after the first edition, to provide both scientists and research managers in all relevant fields with a comprehensive treatise covering all aspects of current medicinal chemistry, a science that has been transformed in the twenty-first century. The second edition is a complete reference source, published in eight volumes, encompassing all aspects of modern drug discovery from its mechanistic basis, through the underlying general principles and exemplified with comprehensive therapeutic applications. The broad scope and coverage of *Comprehensive Medicinal Chemistry II* would not have been possible without our panel of authoritative Volume Editors whose international recognition in their respective fields has been of paramount importance in the enlistment of the world-class scientists who have provided their individual 'state of the science' contributions. Their collective contributions have been invaluable.

Volume 1 (edited by Peter D Kennewell) overviews the general socioeconomic and political factors influencing modern R&D in both the developed and developing worlds. Volume 2 (edited by Walter H Moos) addresses the various strategic and organizational aspects of modern R&D. Volume 3 (edited by Hugo Kubinyi) critically reviews the multitude of modern technologies that underpin current discovery and development activities. Volume 4 (edited by Jonathan S Mason) highlights the historical progress, current status, and future potential in the field of computer-assisted drug design (CADD). Volume 5 (edited by Bernard Testa and Han van de Waterbeemd) reviews the fate of drugs in the body (ADMET), including the most recent progress in the application of 'in silico' tools. Volume 6 (edited by Michael Williams) and Volume 7 (edited by Jacob J Plattner and Manoj C Desai) cover the pivotal roles undertaken by the medicinal chemist and pharmacologist in integrating all the preceding scientific input into the design and synthesis of viable new medicines. Volume 8 (edited by John B Taylor and David J Triggle) illustrates the evolution of modern medicinal chemistry with a selection of personal accounts by eminent scientists describing their lifetime experiences in the field, together with some illustrative case histories of successful drug discovery and development.

We believe that this major work will serve as the single most authoritative reference source for all aspects of medicinal chemistry for the next decade and it is intended to maintain its ongoing value by systematic electronic upgrades. We hope that the material provided here will serve to fulfill the words of Antoine de Saint-Exupery (1900–44) and allow future generations of medicinal chemists to discover the future.

'As for the future, your task is not to foresee it but to enable it'
Citadelle (1948)

John B Taylor and David J Triggle

Preface to Volume 7

Volume 7 of *Comprehensive Medicinal Chemistry II* follows Volume 6 in format and presents principles of medicinal chemistry and drug discovery in a disease-oriented format. This format is subdivided into four main sections that include cancer, infectious diseases, immunological diseases, and dermatological-based diseases. Within these sections are 34 individual chapters.

The first section addresses the medicinal chemistry of cancer chemotherapy. This section is introduced by two chapters covering first, a general discussion of cancer biology, followed by a chapter that describes the basic principles of chemotherapy and pharmacology as they relate to cancer treatment. The next six chapters describe therapeutic approaches based upon biological target or mechanism-based approaches. The final chapter in this section describes a number of the newer mechanism-based approaches that are currently being explored in clinical or preclinical studies.

The disease category of infectious diseases comprises four sections. These include viral diseases, fungal diseases, bacterial diseases, and parasitic infections. The viral diseases section starts with an introductory chapter overviewing the field of antiviral chemotherapy and is followed by three chapters covering drug approaches for several DNA viruses (herpes viruses and hepatitis B virus). Next, a chapter focusing exclusively on HIV is given. The last chapter describes some of the most common RNA viruses (influenza viruses, respiratory syncytial virus and hepatitis C virus).

A section delineating medicinal chemistry treatment options for fungal infections includes two chapters, one describing the medically important fungi and their resultant diseases and a second chapter covering both new and existing chemotherapeutic modalities for the principal human fungal infections.

A 10-chapter section on bacterial diseases starts with a general overview of the field and is followed by specific chapters organized by chemical class (e.g., β-lactams, macrolides, and fluoroquinolones). The section concludes with a chapter focusing on the pathology and treatment options for mycobacterial infections and a final chapter describing the prospects for bacterial genomics impacting antibacterial drug discovery.

Human parasitic diseases are covered by three chapters. An introductory chapter reviews the primary parasites causing the disease, their geographic distribution along with clinical manifestations of disease. Two chapters then follow, with the first focusing on the medicinal chemistry approaches to trypanosomatic parasites and the second one covering drug discovery approaches for malaria.

The remaining sections include immunological diseases and dermatological diseases. Three chapters each are devoted to these two disease areas.

For immunological diseases, the first chapter provides an overview of inflammatory and immunological diseases and then transitions to a focus on new therapeutic advances for arthritis. Two successive chapters then follow, with the first dealing with drug discovery approaches for asthma and chronic obstructive pulmonary disease while the latter describes treatment options for transplant rejection and multiple sclerosis.

The final section includes three chapters dealing with dermatological disorders. An introductory chapter reviews some of the more common afflictions treated by dermatologists and is followed by a chapter describing acne pathogenesis and treatments while the last chapter describes drug discovery approaches for atopic dermatitis and psoriasis.

Over the course of editing this volume, it has been our pleasure to work with highly professional chapter authors, whose critical contributions comprise Volume 7.

Jacob J Plattner and Manoj C Desai

Editors-in-Chief

John B Taylor, DSc, was formerly Senior Vice President for Drug Discovery at Rhône-Poulenc Rorer. He obtained his BSc in chemistry from the University of Nottingham in 1956 and his PhD in organic chemistry at the Imperial College of Science and Technology with Nobel Laureate Professor Sir Derek Barton in 1962. He subsequently undertook postdoctoral research fellowships at the Research Institute for Medicine and Chemistry in Cambridge (US) with Sir Derek and at the University of Liverpool (UK), before entering the pharmaceutical industry.

During his career in the pharmaceutical industry Dr Taylor spent more than 30 years covering all aspects of research and development in an international environment. From 1970 to 1985 he held a number of positions in the Hoechst Roussel organization, ultimately as research director for Roussel Uclaf (France). In 1985 he joined Rhône-Poulenc Rorer holding various management positions in the research groups worldwide before becoming Senior Vice President for Drug Discovery in Rhône-Poulenc Rorer.

Dr Taylor is the co-author of two books on medicinal chemistry and has more than 50 publications and patents in medicinal chemistry. He was joint executive editor for the first edition of Comprehensive Medicinal Chemistry, a visiting professor for medicinal chemistry at the City University (London) from 1974 to 1984 and was awarded a DSc in medicinal chemistry from the University of London in 1991.

David J Triggle, PhD, is the University Professor and a Distinguished Professor in the School of Pharmacy and Pharmaceutical Sciences at the State University of New York at Buffalo. Professor Triggle received his education in the UK with a BSc degree in chemistry at the University of Southampton and a PhD degree in chemistry at the University of Hull working with Professor Norman Chapman. Following postdoctoral fellowships at the University of Ottawa (Canada) with Bernard Belleau and the University of London (UK) with Peter de la Mare he assumed a position in the School of Pharmacy at the University at Buffalo. He served as Chairman of the Department of Biochemical Pharmacology from 1971 to 1985 and as Dean of the School of Pharmacy from 1985 to 1995. From 1996 to 2001 he served as Dean of the Graduate School and from 1999 to 2001 was also the University Provost. He is currently the University Professor, in which capacity he teaches bioethics and science policy, and is President of the Center for Inquiry Institute, a secular think tank located in Amherst, New York.

Professor Triggle is the author of three books dealing with the autonomic nervous system and drug–receptor interactions, the editor of a further dozen books, some 280 papers, some 150 chapters and reviews, and has presented over 1000 invited lectures worldwide. The Institute for Scientific Information lists him as one of the 100 most highly cited scientists in the field of pharmacology. His principal research interests have been in the areas of drug–receptor interactions, the chemical pharmacology of drugs active at ion channels, and issues of graduate education and scientific research policy.

Editors of Volume 7

Jacob J Plattner received his BS degree in chemistry from the University of Illinois. He completed his PhD degree in organic chemistry in 1972, working with Prof Henry Rapoport at the University of California, Berkeley. In 1972, Plattner joined Pfizer as a research scientist working in the Medicinal Chemistry Department with a focus on analgesics, antipsychotics, and prostaglandins. In 1977, Dr Plattner joined Abbott Laboratories as a Chemistry Group Leader for the antihypertensive project. In the Pharmaceutical Division at Abbott, he progressed through increasing levels of responsibility to become Vice President of the Anti-infective Research Division in 1992. In this capacity, he led fully integrated project teams focusing on antibacterials, antifungals, natural products, anticancer agents, and antivirals. In 1998, Dr Plattner joined Chiron Corporation as Vice President of Small Molecule Discovery Research. At Chiron, he focused on building a small molecule discovery platform that involved anti-infective and anticancer therapeutics. Dr Plattner joined Anacor Pharmaceuticals in February 2004 as Senior Vice President of R&D.

Manoj C Desai, is Vice President of Medicinal Chemistry at Gilead Sciences (2003–current). He received his PhD in Natural Products from Dr Sukh Dev at the M.S. University of Baroda. Desai was a postdoctoral fellow in the laboratory of Prof Herbert C. Brown (1981–83) and Prof E. J. Corey (1983–86). He joined Pfizer Inc.'s Central Research Division, Groton, in 1986, where he worked for seven years as a medicinal chemist. His key contributions were in the discovery of nonpeptidic substance P antagonists and in the implementation of highly automated nonpeptidic library synthesis protocol. In 1994, he joined Chiron Corporation as Director of the chemistry department and rose to be a Divisional Vice President, Research. In this capacity, he was a functional head of chemistry department that included medicinal chemistry, chemical informatics, computational chemistry, protein x-ray crystallography, combinatorial chemistry, analytical chemistry, and bioorganic chemistry.

In addition, he is a section editor for *Annual Report in Medicinal Chemistry* (2003–present) and coeditor of 'The chemistry of Drug Design and Lead Discovery' issues of the *Current Opinion in Drug Discovery and Development* (1998–present). Personal research interest includes developing a pipeline of small-molecule clinical candidates, medicinal chemistry, implementation of strategies for acquiring diverse corporate collection for high-throughput screening (HTS) and optimization of screening derived leads.

Contributors to Volume 7

N E Allen
Indiana University School of Medicine, Indianapolis, IN, USA

T Arrhenius
Del Mar, CA, USA

M A Avery
University of Mississippi, University, MS, USA

S J Baker
Anacor Pharmaceuticals, Palo Alto, CA, USA

A J Barker
AstraZeneca, Macclesfield, UK

J F Barrett*
Merck Research Laboratories, Rahway, NJ, USA

A S Bell
Pfizer Global Research and Development, Sandwich, UK

S J Brickner
Pfizer Inc., Groton, CT, USA

E De Clercq
Rega Institute for Medical Research, Leuven, Belgium

W A Denny
University of Auckland, Auckland, New Zealand

J A Dodge
Lilly Research Laboratories, Indianapolis, IN, USA

C K Donawho
Abbott Laboratories, Abbott Park, IL, USA

P Dorr
Pfizer Global Research and Development, Sandwich, UK

T J Dougherty
Pfizer Global Research and Development, Groton, CT, USA

M J Elices
PharmaMar USA, Cambridge, MA, USA

*Deceased.

D Fabbro
Novartis Institutes for BioMedical Research, Basel, Switzerland

S Fidanze
Abbott Laboratories, Abbott Park, IL, USA

C García-Echeverría
Novartis Institutes for BioMedical Research, Basel, Switzerland

A M Ginsberg
Global Alliance for TB Drug Development, New York, NY, USA

D W Green
Amgen Inc., Cambridge, MA, USA

R Griffith
Genelabs Technologies, Inc., Redwood City, CA, USA

B R Hearn
Kosan Biosciences, Hayward, CA, USA

J R Henry
Lilly Research Laboratories, Indianapolis, IN, USA

R D Hubbard
Abbott Laboratories, Abbott Park, IL, USA

C Hubschwerlen
Actelion Pharmaceuticals Ltd, Allschwil, Switzerland

D M Huryn
University of Pennsylvania, Philadelphia, PA, USA

M Y Ismail
Paratek Pharmaceuticals, Inc., Boston, MA, USA

H R Jalian
David Geffen School of Medicine at UCLA, Los Angeles, CA, USA

T Kaneko
Pfizer Global Research and Development, Groton, CT, USA

T Kelly
Boehringer-Ingelheim Inc., Ridgefield, CT, USA

J Kim
David Geffen School of Medicine at UCLA, Los Angeles, CA, USA

H A Kirst
Consultant, Indianapolis, IN, USA

V J Lee
Adesis, Inc., New Castle, DE, USA

E Littler
MEDIVIR UK Ltd, Little Chesterford, UK

L Lou
Genelabs Technologies, Inc., Redwood City, CA, USA

T A Lyle
Merck Research Laboratories, West Point, PA, USA

Z Ma
Global Alliance for TB Drug Development, New York, NY, USA

M M Mader
Lilly Research Laboratories, Indianapolis, IN, USA

T V Magee
Pfizer Global Research and Development, Groton, CT, USA

R Magolda
Wyeth Research, Princeton, NJ, USA

M-S Maira
Novartis Institutes for BioMedical Research, Basel, Switzerland

A A Mortlock
AstraZeneca, Macclesfield, UK

K M Muraleedharan
University of Mississippi, University, MS, USA

D C Myles
Kosan Biosciences, Hayward, CA, USA

M L Nelson
Paratek Pharmaceuticals, Inc., Boston, MA, USA

R Newton
Incyte Corporation, Wilmington, DE, USA

E Ottow
Schering AG, Berlin, Germany

J P Palma
Abbott Laboratories, Abbott Park, IL, USA

M A Pearson
Novartis Institutes for BioMedical Research, Basel, Switzerland

T I Richardson
Lilly Research Laboratories, Indianapolis, IN, USA

C Roberts
Genelabs Technologies, Inc., Redwood City, CA, USA

P J Rosenthal
University of California, San Francisco, CA, USA

G P Roth
Abbott Bioresearch Center, Worcester, MA, USA

U Schmitz
Genelabs Technologies, Inc., Redwood City, CA, USA

S J Shaw
Kosan Biosciences, Hayward, CA, USA

A R Shoemaker
Abbott Laboratories, Abbott Park, IL, USA

J S Skotnicki
Wyeth Research, Pearl River, NY, USA

M Spigelman
Global Alliance for TB Drug Development, New York, NY, USA

S Takahashi
David Geffen School of Medicine at UCLA, Los Angeles, CA, USA

J Trzaskos
Bristol-Myers Squibb Corporation, Lawrenceville, NJ, USA

A S Wagman
Novartis Institutes for BioMedical Research, Inc., Emeryville, CA, USA

O B Wallace
Lilly Research Laboratories, Indianapolis, IN, USA

H Weinmann
Schering AG, Berlin, Germany

M P Wentland
Rensselaer Polytechnic Institute, Troy, NY, USA

P M Woster
Wayne State University, Detroit, MI, USA

L S Young
California Pacific Medical Center Research Institute, San Francisco, CA, USA

X-X Zhou
Medivir AB, Huddinge, Sweden

7.01 Cancer Biology

M-S Maira, M A Pearson, D Fabbro, and C García-Echeverría, Novartis Institutes for BioMedical Research, Basel, Switzerland

7.01.1 Introduction

Cancer can be considered a general term that covers a plethora of different malignancies. These pathogenic conditions are characterized by uncontrolled cellular proliferation and growth, and under special conditions, tumor cell migration, invasion, and spreading to other organs and tissues. Different factors and conditions can transform normal cells into cancerous ones by altering the normal function of a wide spectrum of regulatory, apoptotic, and signal transduction pathways. The complexity of these biochemical processes and networks represent a major challenge in the elucidation of the changes that must occur in order to initiate and maintain cancer progression. Although our understanding of the biology of human cancer is not fully complete, numerous genes and proteins that are causally involved in the initiation and progression of cancer have been identified in the past few years, but due to space constraints and the broad scope of this topic, we have emphasized here the most recent advances in our understanding of the principles underlying tumor cell transformation, growth, survival, invasion, and metastasis. To start with, we briefly cover the epidemiology of cancers and provide information about the available treatment modalities for the four major types of cancer in adults (breast, lung, colorectal, and prostate). After this, different biological processes and signal transduction pathways that are altered in cancer cells or used by the malignant cells to spread to other organs and tissues are reviewed. Throughout the chapter, several review articles on individual topics have been included, and the reader is referred to the most recent publications in the text.

7.01.2 Cancer Epidemiology

Each year cancer is newly diagnosed in about 10 million of people worldwide, and it causes 5 million deaths. Cancer is second to cardiovascular disease as a cause of death in developed countries, and overall causes 10% of all deaths in the world. In most developed countries, one person in three will develop cancer during their lifetime, and this will increase to one in two by the year 2010 provided that the age of the population will increase as predicted.[1,2]

Cancer comprises over 200 distinct entities differing in their genetic basis, etiology, clinical characteristics, patterns of progression, and final outcome. In broad terms cancer can be classified into carcinomas and sarcomas according to the fetal germ layer from which tumors arise. Carcinomas arise within tissues derived from the fetal ectoderm or endoderm and include most of the common cancers in adults. Sarcomas are seen more frequently in children and arise from tissues originating from the fetal mesoderm which generate tumors of the bone, muscle, connective tissues, and blood vessels. In developed countries, about 50% of all cancer cases are carcinomas of the lung, colon, prostate, and breast while hematological cancers (leukemias and lymphomas) account for about 8–10% of all cancer cases.[1,2]

The genetic basis for cancer is well established through studies with tumor viruses, carcinogenesis models, molecular biology, somatic cell genetics, and genetic epidemiology. Cancer is the result of multiple mutations that occur in oncogenes, tumor suppressors and/or DNA repair genes of somatic cells.[3–14] Although in some cases these mutations can occur in the germ line and are the cause for genetic predisposition, the majority of cancers are sporadic and arise due to somatic mutations in the tumor cells.[3–14] Cancer follows a characteristic natural history.[1–6] First, normal cells become dysplastic showing subtle morphological abnormalities indicative of the initiation of transformation. Characteristic abnormalities of both form and proliferation then lead to a carcinoma in situ that does not invade the underlying basement membrane of the tissue of origin. These early phases are highly curable and may be detected with screening programs. Localized cancer is a stage I disease when the tumor exhibits invasion and disruption of local tissues to form a primary lesion. Then tumor cells invade local lymphatics and spread to the regional (stage II) or distant regional (stage III) draining lymph nodes as secondary tumors. Finally, tumor cells invade the bloodstream initiating the characteristic patterns of bloodborne metastasis characteristic of stage IV disease. Staging correlates with survival and provides an essential guide both to prognosis and to the design of treatment plans.[1] Tumors vary in the extent to which they follow these phases. Melanoma usually has locoregional phase, while breast cancer is systemic from the beginning.[1]

7.01.2.1 Molecular Cancer Epidemiology

Molecular and genetic epidemiology represent two separate branches of epidemiology whose boundaries are overlapping. While molecular epidemiology evaluates the association of variations in known genes with risk of cancer, genetic epidemiology aims to identify the unknown genes that influence risk of malignancies.[3–14] Moreover, molecular epidemiology also uses molecular markers to link exposures to cancer.[7–13] The two terms are very often used interchangeably and are referred to in the following as molecular cancer epidemiology.

Molecular cancer epidemiology identifies risk factors for the different cancer types at the population level. These studies are usually carried out as either case-control, cohort cross-sectional, or ecological studies.[14] Case-control studies investigate patients with a specific cancer in relation to a healthy group (which serves as the control) while the cohort study assesses distribution of exposures to certain risk factors in a subset of a population and correlates it with the future occurrence of cancer. In contrast cross-sectional studies survey the distribution of cancer and its putative risk factors in a subset of the population while ecological studies attempt to demonstrate the relation between risk factors and cancer in a single population.[14]

Remarkable progress in our understanding of the molecular pathogenesis of cancer has been made and is beginning to have positive impact across the whole field of oncology: from prevention through screening and diagnosis to the development of molecularly targeted therapies. Major advances in molecular (or genetic) markers have been made which have a great impact on health risk assessment. For example, rather than treating all cases of breast cancer as the same disease, an epidemiologist can use tumor markers to identify potentially more heterogeneous subsets.

7.01.2.2 Risk Factors and Cancer

Cancer is a multifactorial disease, in which both environmental and genetic factors play a role.[10,15] Part of the environmental factors are conditions of life that result in exposures to carcinogens depending on where people live and work as well as changes that people make in the world. Cancer shows both geographic and temporal variability, and there are different patterns of cancer at different places and different times which depend both on lifestyle and habits as well as environmental hazards.[15] Risk factors in cancer etiology comprise four classes of external agents in carcinogenesis (carcinogens): physical, chemical and biological agents, and diet.[12,13,15]

The increasing knowledge of the process of carcinogenesis induced by chemical agents provides a major basis for cancer control and has unraveled the association between smoking and lung cancer. About 30% of all cancer deaths in the USA are due to the use of tobacco, and this death toll is still increasing reflecting smoking habits among young women since the 1950s.[1,15] Smoking causes more than 90% of all cases of lung cancer and is the main cause of cancers of the larynx, mouth, esophagus, bladder, kidney, and pancreas, while about 25% of colon cancer and polyps can be attributed to smoking. Five years after stopping smoking, the risk of cancer decreases to half and to the level of lifelong nonsmokers after 10–15 years. Reducing the epidemic of tobacco smoking is currently the most effective means of cancer prevention.[15–17]

Diet and obesity in adults account for 30% of all cancer deaths in the USA. Diet has been shown to play a significant role in the causation of cancer but little is known about how it plays its role as a carcinogen.[15,18] Excessive fat in the diet raises the risk of colorectal and breast cancer and possibly prostate cancer. Adult obesity is associated with endometrial cancer, postmenopausal breast cancer, and cancers of the colon, rectum, and kidney.[15,18] Obesity in concert with other risk factors such as low activity level, menopausal status, and predisposition to insulin resistance significantly increase the risk of cancer. While some methods of food preparation and preservation have been shown to increase the risk of various forms of cancers, certain classes of foods appear to contain protective substances against cancer including vegetables, whole grain products (fiber), and citrus fruits.[12,15,18,20] Salt intake has been associated with risk of stomach cancer, but no other food additive or contaminant (except for aflatoxins)[21] has been linked conclusively to cancer. Therefore, a diet that reduces cancer risk has been proposed.[12,15,18,20]

Occupational factors may account for 5% of all cancer deaths, and these include mostly cancers of the lung, bladder, and bone marrow.[1,19] Workplace exposure, the most important carcinogen, which can increase the rate of cancer by 10 to 100 times the rate in unexposed people, is second to tobacco smoke.[16,17,19] About 15% of lung cancers and about 10% of skin and bladder cancers are caused by workplace exposure.

A family history of cancer is also an important risk factor and can increase the risk of cancers of the breast, colon, prostate, and lung by 1.5–3-fold accounting for perhaps 5% of all cancer deaths in the USA.[1,7,9,22,23] Susceptibility to cancer is due to genetic mutations of key regulatory genes that occur in the germline.[9,23] About 5–10% of most types of cancer are due to defects in single genes that run in families. Cancer incidence also depends on genetic polymorphisms that affect the absorption, transport, metabolic activation, or detoxification of environmental carcinogens and may act as cancer-facilitating influences.[19,24]

Both DNA and RNA viruses are responsible for about 5% of human cancer but are a more common causes of cancer in animals where they play a central role in the identification of oncogenes.[25,26] About 5% of adult T cell leukemias/lymphomas are due to human lymphotropic virus type I[27] while Epstein–Barr virus accounts for 10–15% of non-Hodgkin's lymphoma, Burkitt's lymphoma (almost all children in central Africa and 20% of cases occurring elsewhere), 35–50% of Hodgkin's disease, and 40–70% of nasopharyngeal carcinoma (especially in southern Chinese).[28] Hepatitis B virus accounts for 40–60% of hepatocellular carcinoma and hepatitis C virus for 20–30%.[29] Some subtypes of the human papilloma viruses (HPV16 and HPV18) account for 90% of cervical cancer. HPV infection is now recognized to be a sexually transmitted disease with special risk from early sexual exposure before the cervix is fully mature. HPV viruses cause benign warts and may be involved in cancers of the oral cavity and upper respiratory tract.[30] The developing countries bear the greatest impact from these cancers from very early infections with these agents.

Establishment of latent and chronic infections with associated higher levels of viral replication[31–33] increases the probability of causing secondary genetic damage to target tissues with associated immune dysfunction contributing to carcinogenesis.

Reproductive factors associated with hormonal changes such as early age at menarche, late age at first birth, late age at menopause, and nulliparity each increase the risk of breast cancer. Nulliparity is also associated with endometrial and ovarian cancers. Early age at first intercourse increases the risk of cervical cancer, more likely through HPV infections.[34,35]

Alcohol and tobacco smoking are the main cause of cancers of the upper respiratory and gastrointestinal tracts. Alcohol by itself plays a role in liver cancer (cirrhosis) and possibly in a proportion of colon and breast cancer.[36] Poverty is associated with increased exposure to tobacco smoke, alcoholism, poor nutrition, and certain infectious agents. Thus, poverty can act as a carcinogen, which suggests that fighting cancer also requires fighting poverty.

Surprisingly, the contribution of air pollution in causing lung cancer is less than anticipated (1% of lung cancer deaths yearly in the USA with a higher risk for urban smokers compared with rural smokers).[37] Risks associated with environmental exposures are several orders of magnitude less than that associated with smoking. Aromatic organochlorines dichlorodiphenyltrichloroethane (DDT) and polychlorinated biphenyls (PCBs) accumulate along the food chain in fatty tissues and can bind to estrogen receptors causing estrogenic or antiestrogenic effects in animals.

Over 90% of melanomas, basal cell carcinomas, and squamous cell carcinomas are due to sunlight exposure.[38] Most skin cancers are nonmelanoma skin cancers and account for 40% of all new cancers in the USA yearly. Although the major DNA-damaging spectrum is ultraviolet (UV)-B radiation the excessive exposure to the UVA-ray as used in sun lamps and sun bed can also cause DNA damage.[19,24,38]

Ionizing radiation is a universal but weak carcinogen.[1,39] However, cumulative exposures from medical diagnostic and treatment procedures, commercial, occupational sources, or waste increase the risk of cancer. Leukemias and cancers of the breast, lung, and thyroid are typical but cancers of the stomach, colon, and bladder, and potentially any human tumor may be seen. Radiation can cause most types of cancer, especially myelogenous leukemia and cancers of the breast, thyroid, and lung. Some cancers that have not been linked to radiation include chronic lymphocytic leukemia, non-Hodgkin's lymphoma, Hodgkin's disease, and cancers of the cervix, testis, prostate, and pancreas. Despite the massive radiation contamination resulting from the nuclear reactor accidents at Chernobyl and Chelyabinsk thus far the only well-documented increase in cancer is childhood thyroid cancer.[40] Finally, genetic predisposition can increase the risk of developing cancer by exposure to radiation, as in the cases of inherited retinoblastoma (bone tumors) and hereditary ataxia telangiectasis (lymphoid tumors).[41–43]

Cancer chemotherapy, especially alkylating agents, increases the risk of leukemia and other secondary cancers.[41] Immunosuppressive agents increase the risk of some cancers, especially lymphoma.[44] While hormones and hormone antagonists increase the risk of some cancers, they decrease it for others. Hormone replacement therapy with estrogens for the treatment of menopausal symptoms increases the risk of endometrial cancer and breast cancer.[34] Oral contraceptives reduce by half the risk of ovarian and endometrial cancers; but they may increase the risk of liver cancer.[34] It should be emphasized that overall the risk of cancer associated with conventional medicines is very much smaller than the direct benefits of therapy.

Although changes in lifestyle can reduce cancer incidence, such changes may be very difficult to achieve, as anyone who has tried to stop smoking can attest. As the last two decades have brought little gain in terms of survival benefit to patients with most types of cancer, there is now a major research emphasis on the application of behavioral science in health promotion and prevention programs to create lifestyle change at the population level that reduce the incidence of the disease in the population.

7.01.2.3 Cancer Screening, Diagnosis, and Prevention

7.01.2.3.1 Genetic markers

Molecular biomarkers may be used to identify early responses to DNA damage and to assess cancer risk.[45–47] Unscheduled DNA synthesis, chromosome aberrations, sister chromatid exchanges, and the micronucleus tests are some examples of these tumor markers. These markers are also referred to as genetic markers and can be scored in advance of preneoplastic lesions that are usually detected by histological methods.[45–47] Genetic markers can also be used to show that the vast majority of cancers originate from a single stem cell. In addition to the clonogenic nature of cancer, there are numerous epigenetic events that create an environment for abnormal cell division, such as chronic inflammation or persistent stimulation of the immune system.

Solid tumors and a majority of hematological malignancies display various degrees of abnormality in the chromosomal karyotype including translocation, deletion, and duplications of chromosomes as well as more subtle aberrations such as rearrangements, deletions, and amplifications.[48,49] All of these aberrations may have diagnostic as well as prognostic significance. Chromosome aberrations may be primarily related to the formation of a specific tumor and may be the only genetic abnormality present, as in case of chronic myeloid leukemia, chronic monocytic myeloid leukemia, and other malignancies.[48,49] Since these types of translocation are never seen in normal cells, the chromosomal abnormalities may serve as useful tumor markers not only for diagnosis and assessment of disease stage, but also as indicators for a successful treatment or relapse.[48–50]

A number of chromosomal abnormalities are associated with predisposition to cancer.[48,49,51] The molecular pathology of the chromosome rearrangement are well understood and have been shown to inactivate tumor suppressors like the Wilms' tumor gene[52] and adenomatous polyposis coli. Other situations include disorders characterized by genomic instability as in xeroderma pigmentosa, in which there is extreme sensitivity to the effects of sunlight and carcinogens due to a reduced ability to repair damaged DNA, or as in Bloom's syndrome, in which there is a defect in a gene encoding a helicase leading to inefficient joining of nucleotides.[53,54] The genes involved in these syndromes have been identified and tests have been devised to identify heterozygotes, thus making genetic counseling more precise.[53,54]

Certain carcinogens increase the risk of developing cancer by increasing the frequency of mutations or by interfering with chromosome organization.[55] Genetic markers where the germline mutation or polymorphism is present at birth have been widely used in epidemiological studies, mainly to identify high-risk subjects. Much emphasis has been placed on the so-called major cancer genes that usually show a high penetrance, like *BRCA1* and *BRAC2*.[56] The prevalence of these mutations is very low but the risk of cancer associated with mutations in these genes is very high. As a consequence screening a whole population to detect these relatively rare genes would make no sense. Screening for polymorphism in so-called minor cancer genes such as metabolic and DNA repair genes makes more sense.[57,58] Correct identification of the population frequency of a polymorphism including its function in noncoding regions as well as its linkage to other functional polymorphisms is key to each correlation or association study. For most of the known polymorphisms, the presence of an environmental cue may exacerbate the risk associated with such a polymorphism. Although polymorphisms in susceptibility genes are diffuse in the healthy population, their absolute frequency varies with ethnicity.[57,58]

7.01.2.3.2 Diagnostic and prognostic biomarkers

Localized tumors can be removed effectively by surgery or radiotherapy, while metastatic disease requires relatively ineffective systemic therapy in particular in patients with solid tumors. The identification of patients with early stage disease is imperative to improve the chances of local control and cure. Therefore, candidate markers and genes may have a great potential for cancer screening to facilitate diagnosis, to monitor disease progression, and to assess the risk of recurrence after local therapy.

Thus far, biological markers (biomarkers) have been used to diagnose cancer, and to monitor disease progression and recurrence after therapy.[59–62] Biomarkers have substantially improved the understanding of the molecular mechanism of action of carcinogenesis and risk, although they are more useful for monitoring the consequences of the disease and therapy rather than assessing the effects of risk factors on the onset of the disease. For example the serum-based markers calcitonin, prostate-specific antigen, and CA-125 are all elevated in medullary thyroid, prostate cancer, and a small subset of ovarian cancers, respectively.[63–65] Urine-based biomarkers like bladder tumor antigen, survivin, and calreticulin have been recently used as diagnostic marker for bladder cancer.[66]

More recently the marriage between multiple biomarkers and bioinformatics has allowed the discovery and clinical validation of gene and protein profiles resulting in signatures that are characteristic of particular cancers. Cancer signatures are useful for predicting the outcome of cancer (association with cancer stages and prognosis) and/or cancer therapy, and to distinguish invasive versus noninvasive, metastatic versus nonmetastatic, and indolent versus life-threatening cancer types. This is best exemplified for prostate and breast cancers which are diagnosed relatively early. Using biomarker analysis performed on the primary tumor recommendations can be made as to which patients will benefit from no treatment or from localized treatment such as surgery and/or radiation or early systemic therapy, which should likely lead to improved therapeutic success rates and quality of life. Several potential prognostic biomarkers have been proposed and are being used for different tumors.[58–66]

The diagnosis of cancer is usually based on the recognition by a histopathologist of aberrant cellular patterns in a biopsy sample. More and more frequently, histological examination of fine needle aspirates is being used not only for making the diagnosis but also for evaluating the sites of spread, a process called staging. This examination is supported by DNA or protein-based assays that can be performed on the same biopsy. The use of the polymerase chain reaction (PCR) to detect mutant *RAS* genes in a fine needle aspirate from a pancreatic biopsy is one example of attempts to use our knowledge of oncogenes for clinical purposes.[67]

Unfortunately many of the current screening programs to detect early stages of solid tumors are not very efficient, as for example mammographic screening for breast cancer.[68] Regular screening is relatively nonspecific and identifying benign lesions requires further investigation and biopsy. Large studies have shown that mammographic screening has only a small effect on overall breast cancer mortality and only in the 50–64 age group.[68]

Another example of relatively ineffective screening is the use of the biochemical tumor marker prostate-specific antigen to identify men with asymptomatic prostate cancer.[65] This is because a large percentage of men over 50 years old actually have prostate cancer which remains confined to the prostate gland and has no significant impact on the individual's health. Removal of these localized tumors may not have any beneficial effect on overall survival as death is mainly caused by the effects of metastatic disease.

Therefore improved modern molecular screening approaches may identify: (1) high-risk patient groups who might benefit from more intensive screening as well as inclusion in trials of preventive agents; and (2) indicators of metastatic potential of a cancer providing a rationale for more aggressive therapy. Understanding the progressive accumulation of genetic damage of breast, pancreatic, thyroid, and colorectal cancer could lead to novel precancer detection systems. An example of this type of novel approach would be the identification by PCR of mutant *RAS* in fecal material to identify early colorectal cancer.[69,70]

7.01.3 Available Treatment Modalities

Anticancer treatment often includes more than one approach, and the treatment modality adopted by the oncologist is largely dependent on the type of tumor and how far it has progressed in the patient. Surgery, chemotherapy, radiotherapy, and photodynamic therapy are among the most common and broadly used treatment strategies available today. These treatment modalities are briefly discussed in this section.

7.01.3.1 Surgery

Cancer surgery attempts to remove localized, well-defined tumors or precancerous conditions. Additional treatment with chemotherapy or radiotherapy is often required due to the impossibility of removing the tumor completely or of eliminating malignant cells that may linger after surgery or metastasize to other tissues or organs.

7.01.3.2 Chemotherapy

Clinical cancer chemotherapy uses low-molecular-mass agents to destroy rapidly dividing cells (*see* 7.02 Principles of Chemotherapy and Pharmacology). This strategy was initially dominated by genotoxic drugs that target the integrity of the cell's genetic material. Currently, more than 100 chemotherapeutic drugs are used either alone or in combination regimes. The mechanism of action of these anticancer drugs is very broad and expands from the initial DNA interactive agents, which interfere with the growth of cancer cells by reacting with DNA to block its replication, to the most recent targeted therapeutic drugs, which block components of deregulated signal transduction pathways. (Other types of chemotherapy medications include antimetabolites (*see* 7.03 Antimetabolites), antitumor antibiotics, mitotic inhibitors, hormonal therapies, nitrosoureas.) Cancer chemotherapy offers a unique advantage over the other treatment modalities: it can treat the entire body, even the cells that may have escaped from the primary tumor(s). Chemotherapy is often used in conjunction with other treatments to improve its efficiency, and in this case, patients may receive neoadjuvant chemotherapy to reduce the size of the tumor before surgery or radiation, or adjuvant chemotherapy to eliminate cancer cells that might linger following earlier treatment. Normal cells that divide quickly (e.g., those of bone marrow or the reproductive system) are also affected by chemotherapy and substantial drug discovery efforts have been devoted in the past few years to the identification of more effective agents with better side effect profiles. One of the main issues with this approach is the development of drug resistance, and in order to overcome this problem, combinations of different agents and dosage regimes are often used to maximize benefit for the patient (*see* 7.02 Principles of Chemotherapy and Pharmacology).

7.01.3.3 **Radiotherapy**

Radiotherapy (also called radiation therapy) uses high-energy radiation, e.g., photon or particle radiation, to eliminate or shrink localized tumors.[71] The radiation damages the DNA of cancer cells, blocking their ability to divide and proliferate. Unfortunately, the surrounding normal cells are also affected and, to protect them, radiation treatments are spread over time. This strategy (termed fractionation) tries to minimize the damage to normal tissue by allowing cells to be repaired during the time between treatments. The different repair rate between normal and cancer cells means that only a small fraction of malignant cells, which have a lower recovery rate, will have been repaired by the time of the next treatment. Ideally the vast majority of (or indeed all) cancer cells will be dead after the last treatment session. As an alternative to external beam therapy, until now the most traditional way to deliver radiation, radioisotopes (e.g., cesium-137 or iodine-125) can be implanted near the tumor allowing the delivery of radiation to localized areas. This mode of therapy is particularly useful for cancers where surgery or beam therapy would be detrimental to tissues surrounding the tumor (e.g., prostate or cervical cancer).

7.01.3.4 **Photodynamic Therapy**

Photodynamic therapy combines a photosensitizer agent (e.g., porfimer sodium or temoporfin) with a specific type of light in an oxygen-rich environment.[72] The photosensitizer agent is injected into the bloodstream, and few hours later, when most of the agent is cleared from normal tissue but localized in cancer cells, the malignant tissue is irradiated with a light source. The photosensitizer agent in the tumor absorbs the light and raises the agent to an excited single state. The excited dye transfers energy to the available oxygen in the tumor, and the molecular oxygen is transformed into highly reactive singlet oxygen that oxidizes biological targets (e.g., cell membranes) killing nearby cancer cells. In addition to its direct effect on tumor cells, the photosensitizer may activate the immune system to attack tumor cells and also damage tumor blood vessels, thereby preventing cancer cells from receiving oxygen and nutrients or removing waste. Photodynamic therapy is currently used to treat or relieve symptoms of esophageal cancer and non-small cell lung cancer.

7.01.4 **Treatment Options for Major Cancers and Future Directions**

According to the most recent statistics, one in three people will be diagnosed with cancer during their lifetime. Increased life expectancy (cancer can be considered a disease of advanced years), environmental factors (e.g., exposure to carcinogens), and social habits (e.g., diet and/or smoking) have dramatically increased the risk of cancer in Western populations. Breast, lung, colorectal, and prostate are the four major types of cancer in adults, and they account for over half of all cases diagnosed. The treatments of these major cancers depend upon a variety of factors, but the most common ones involved surgery combined with radiation and/or chemotherapy. These treatments are briefly discussed in this section. **Table 1** shows representative examples of anticancer agents and their mechanism of action.

Depending on the tumor size and the stage of the disease, treatment for breast cancer may involved surgery, radiation therapy, chemotherapy, hormone therapy, or a combination of two or more of the preceding methods.[73] The most common chemotherapeutic agents used to treat this malignancy are cyclophosphamide, doxorubicin, methotrexate, and fluorouracil. In addition to these agents, paclitaxel is often prescribed when breast cancer cells have metastasized to the lymph nodes or after breast cancer surgery. Other chemotherapy drugs for breast cancer include, among others, capecitabine and mitomycin. In addition to these agents, selective estrogen-receptor modulators (SERMs) (e.g., tamoxifen or raloxifen), aromatase inhibitors (e.g., anastozole, letrozole), and luteinizing hormone-releasing hormone (LHRH) analogs (e.g, goserelin) are used for the adjuvant treatment of estrogen receptor-positive breast cancers. Recently, new treatment options have been available to control the progression of breast tumors. Thus, trastuzumab, a humanized monoclonal antibody that targets the extracellular domain of the $p185^{HER2}$ cell-surface receptor, is an option for the treatment of erbB-2-positive metastatic breast cancer.[74]

Treatment options for lung cancer are determined by the type: small cell lung carcinoma (SCLC) or non-small cell lung carcinoma (NSCC), and by stage of the tumor.[75] Small cell lung carcinoma, which accounts for approximately 20% of all primary lung cancers and tends to be particularly aggressive, is often widespread by the time of diagnosis and treatment is often limited to chemotherapy and/or chest radiation therapy. Representative examples of combination therapies for SCLC are: etoposide and cisplatin (EC); etoposide, cisplatin, and vincristine sulfate (ECV);

Table 1 Representative examples of the most commonly used anticancer agents and their mechanism of action

Generic name	*Type of drug*
Cyclophosphamide, cisplatin	Alkylating agent[a]
Doxorubicin, mitomycin	Antibiotic[a]
Methotrexate, fluorouracil, capecitabine, gemcitabine	Antimetabolite[b]
Paclitaxel	Taxane[c]
Tamoxifen, raloxifene	Antiestrogen[d]
Anastrozole, letrozole, examestane	Aromatase inhibitor[d]
Goserelin, leuprolide	Luteinizing hormone-releasing hormone analog[d]
	Gonadotropin-releasing hormone analog[d]
Trastuzumab, cetuximab, becacizumab	Monoclonal antibody
Etoposide	Topoisomerase II inhibitor[e]
Vincristine, vinorelbine	Vinca alkaloid[c]
Leucovorin	Vitamin derivative
Irinotecan	Topoisomerase I inhibitor[e]
Flutamide, bicalutamide	Antiandrogen[d]
Zoledronic acid	Bisphosphonic acid
Imatinib, erlonitib	Kinase inhibitor[f]

[a] *See* 7.06 Alkylating and Platinum Antitumor Compounds.
[b] *See* 7.03 Antimetabolites.
[c] *See* 7.04 Microtubule Targeting Agents.
[d] *See* 7.07 Endocrine Modulating Agents.
[e] *See* 7.05 Deoxyribonucleic Acid Topoisomerase Inhibitors.
[f] *See* 7.08 Kinase Inhibitors for Cancer.

cyclophosphamide, doxorubicin, and vincristine sulfate (CAV); cyclophosphamide, doxorubicin, and etoposide (CAE); etoposide and cisplatin (EP); and cyclophosphamide, etoposide, and vincristine (CEV). Depending on the extent of the disease, non-small cell lung carcinoma can be removed by surgical resection or treated with radiation therapy in combination with chemotherapy (e.g., gemcitabine hydrochloride together with cisplatin and vinorelbine). In addition to the preceding treatment modalities, photodynamic therapy is use for the care of patients with inoperable lung cancer or with distant metastasis.

Surgery is the treatment of choice for colorectal cancer and, depending on the stage of the disease, chemotherapy and radiation are used as adjuvant treatment.[76] For example, a cocktail of different agents (fluorouracil, leucovorin, and irinotecan) is used for metastatic colorectal cancer. An important advance in the treatment of colorectal cancer has been reported recently with bevacizumab, which is a humanized vascular endothelial growth factor (VEGF) antagonist. This protein blocks VEGF-induced signaling in endothelial cells and partially inhibits VEGF-driven angiogenesis. Bevacizumib in combination with intravenous 4-fluorouracial was approved in February 2004 by the US Food and Drug Administration (FDA) for the treatment of patients with first-line or previously untreated metastatic colorectal cancer.[77] The antibody showed improved survival benefit compared to chemotherapy alone. Additional clinical trials in breast cancer (phase II/III, E-2100; treatment of metastatic breast cancer) and non-small cell lung cancer (phase II/III, E-4599; advanced non-squamous NSCLC in combination with carboplatin and paclitaxel) are ongoing and results from these studies are expected in late 2005 or early 2006.

Treatment options for prostate cancer include surgery and radiotherapy (e.g., external beam or radioactive seed implants, called brachytherapy; see above).[78] Hormonal treatment, chemotherapy, and radiation (or a combination of them) are used for metastatic diseases and as supplemental/additional therapies for early stage tumors. Hormonal treatment, which controls the progression of the disease by shrinking the size of the primary tumor and reduces symptoms (e.g., pain), involves the use of antiandrogens to block production of testosterone, a hormone that prostate cancer cells use to grow. Examples of drugs used for this specific treatment include flutamide, leuprolide acetate, bicalutamide, and goserelin acetate. Bone metastasis is often associated with prostate cancer. The secondary tumor causes the breakdown of bone tissue releasing substantial amounts of calcium into the bloodstream. In order to treat the excess calcium in the blood (hypercalcemia), bisphosphonates (e.g., zoledronic acid), are used to decrease bone loss, increase bone density, and reduce the risk of bone fractures.

A deeper understanding of the molecular events leading to tumor formation, invasion, angiogenesis, and metastasis has provided a new mechanistic basis for oncology drug discovery: targeted anticancer therapy.[79] The rationale behind this approach is relatively simple: specific inhibitors of proteins involved in aberrant signaling mechanisms would interfere with cancer progression, altering the natural course of the disease while sparing normal tissues. Although numerous disappointments have been harvested, we start to see how a new generation of targeted cancer agents – both low-molecular-mass inhibitors (e.g., imatinib, gefitinib, erlotinib)[80] and humanized monoclonal antibodies (e.g., trastuzumab, cetuximab, becacizumab)[77,81,82] – can provide incremental improvements over existing drug treatments. In addition to the identification of more effective anticancer agents, other areas under development include gene therapy (administration of tumor suppressor genes), immunotherapy (to boost the body's immune system to better fight off or destroy cancer cells), and improved radiation therapy approaches (the use of neutrons to cause cell damage in an oxygen-independent manner).[83] Of special interest is the recent results obtained with CP-675206, a fully human immunoglobulin G2 (IgG2) antibody that is directed against human cytotoxic T lymphocyte-associated antigen 4 (CTLA4). This antibody enlists the immune system to attack malignant tumor cells responsible for the disease. CP-675206 is currently in clinical trials for the treatment of prostate cancer (phase II in combination with neoadjuvant androgen ablation) and advanced melanoma (phase I/II). In this last indication, midtrial results have shown a 43% response rate, a major breakthrough in comparison with the current response rates for the two FDA-approved therapies for metastatic melanoma (from 6% to 15%).

7.01.5 Cell Proliferation, Differentiation, and Apoptosis

7.01.5.1 The Cell-Cycle and Cancer

A unique feature of cancer cells is high, indefinite cellular proliferation that continues without regard for the surrounding environment and the regulatory mechanisms that exist in normal cells. In normal tissues, cell division occurs in the context of a series of sequential biological events known as the cell cycle, which is composed of four different phases (**Figure 1**). DNA replication occurs in the so-called S (synthesis) phase, and cell division occurs in the M (mitosis) phase. These two main activities are preceded by two gap phases: G1 to prepare cells for DNA synthesis, and G2 to allow cells the opportunity to check the integrity of the newly synthesize DNA and to prepare cell division. Phase transitions are controlled by a series of serine/threonine kinases called cyclin-dependent kinases (CDKs) and their regulatory subunits named cyclins. Misregulation the cell cycle is a general event in diseases of uncontrolled cell growth and some of the most common alterations of this pathway are discussed in the following sections.

7.01.5.1.1 The G1/S transition and the increase of activity of cyclin-dependent kinases

Cells entering the G1 phase will actively prepare to divide (newly synthesized proteins are produced, the cell size increases), until a certain point called the G1 checkpoint. This crucial checkpoint is controlled by the retinoblastoma tumor suppressor gene product (Rb), which is a transcriptional regulator. This protein imposes a block on G1 progression that is released by its phosphorylation by Cdk4/cyclin D1, Cdk6/cyclin D3, and Cdk2/cyclin E. Hyperphosphorylated Rb (pRb) dissociates from the transcription factors of the E2F-DP family (principally E2F1, 2, and 3) making these proteins available to direct the expression of proteins essential for DNA synthesis such as cyclin A, dihydrofolate reductase, thymidine kinase, thymidylate synthase, or DNA polymerase-α. The activity of the CDKs is tightly regulated by endogenous inhibitors of Cdk/cyclin complexes. Thus, the $p21^{cip1}$, $p27^{kip1}$, and $p57^{kip2}$ gene products can bind and inhibit all Cdk/cyclin D, E, and A complexes, and the products of the *INK4* gene ($p16^{Ink4a}$, $p15^{Ink4b}$, $p18^{Ink4c}$, and $p16^{Ink4d}$) have the ability to interact specifically with CDK4 and 6.

The G1 checkpoint is often deficient in human tumors, often due to deregulation or absence of the Rb protein. Although germline mutations in the *RB* gene cause the highly penetrant hereditary retinoblastoma,[84] the frequency of *RB* mutation is low among the sporadic cancers; however, it has been reported in osteosarcomas, small cell lung carcinomas, and breast carcinomas. Rb protein inactivation, found in a wide variety of human cancers,[85] may be the result of three possible causes. First, the Rb protein can be sequestered from its physiological partners, when bound to viral oncoproteins, such as the SV40 T antigen, the adenovirus E1A protein, or the papilloma E7 protein.[86] These events are frequently observed in human cervical tumors. The second cause, and probably the most common one, is the

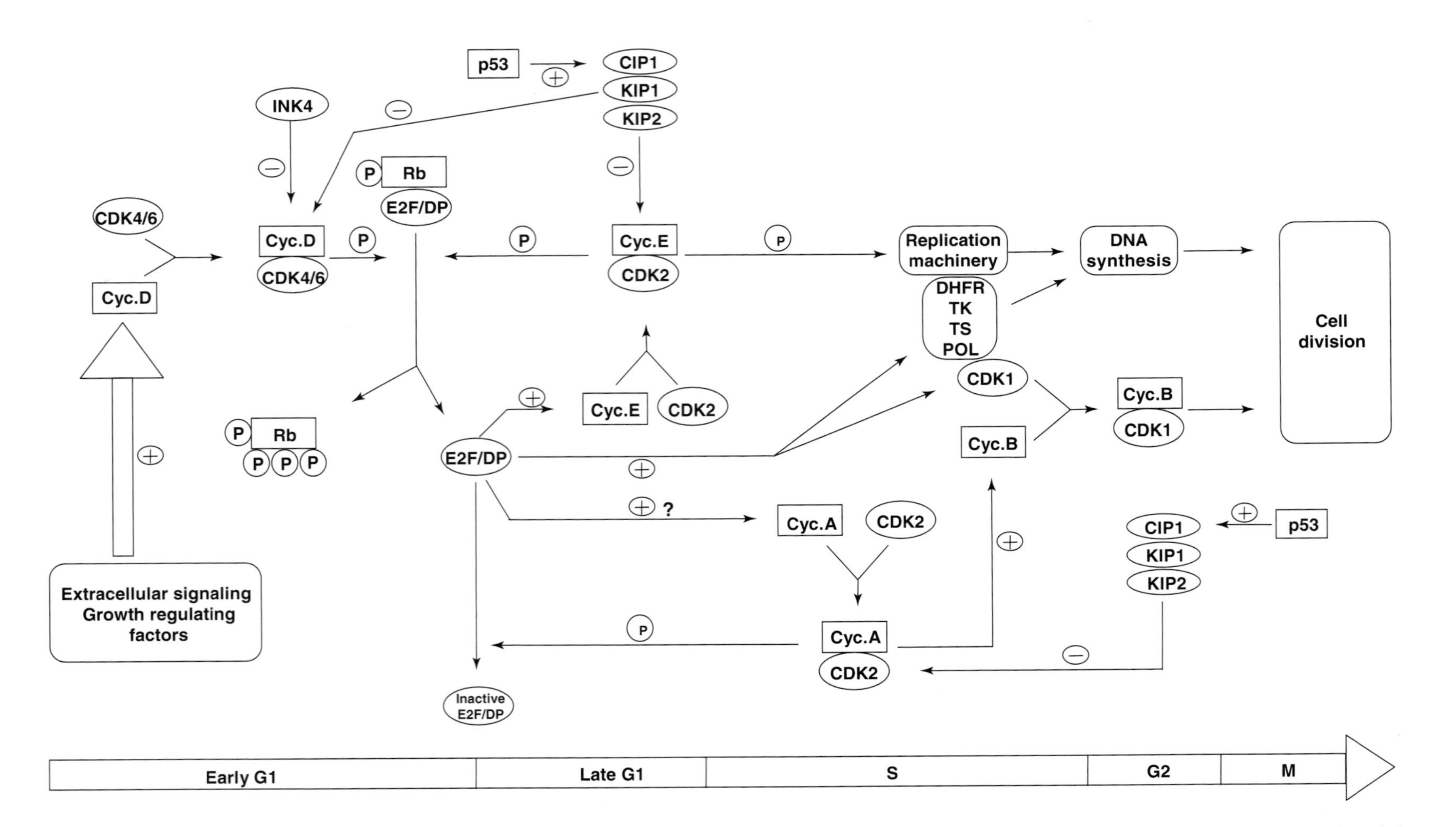

Figure 1 Molecular mechanisms governing the cell cycle. Symbols: + and – symbols represents the effects on the activity of the targeted protein as well as transcriptional regulation; P represent a phosphorylation event; ? represent a probable effect, but not fully demonstrated. DHFR, dihydrofolate reductase; TK, thymidine kinase; TS, thymidilate synthase; POL, DNA polymerase-α.

loss of Rb function through permanent hyperphosphorylation, leading to accumulation of active E2F factors. This can occur by deregulated expression of cyclin D or CDK4, as a result of amplification or translocation of the respective genes. For example, CDK4 is amplified in gliomas and sarcomas.[87] Alternatively, point mutations abrogating $p16^{Ink4a}$ binding have also been identified in CDK4.[88] Third, CKI genes such as the *INK4* gene are often deleted [89] or silenced by hypermethylation of the gene promoter[90] in human tumors.

Key observations made in different biological systems have also identified Rb as an important player in cell fate determination (i.e., the differentiation process), by inducing apoptosis. This can be accomplished by two different mechanisms: (1) through regulation of apoptosis either in a E2F1, $p19^{ARF}$ (the sixth and last product of the INK4A/ARF locus) and p53-dependent fashion,[91,91–95] or (2) in a E2F1-independent manner, through c-Jun N-terminal kinase (JNK), nuclear c-Abl, and p84N5.[96–100] Because for a cell to become tumorigenic, it has to turn out to be resistant to apoptosis (*see* Section 7.01.5.3) and acquire properties leading to a strong blockade of the cell death machinery.

7.01.5.1.2 The G2/M transition, control of deoxyribonucleic acid (DNA) damage, and the spindle checkpoint

Historically identified as mitosis promoting factor (MPF), the Cdk1/cyclin B complex is the key element responsible for the onset of mitosis.[101] Regulation of the activity of this complex is performed at multiple levels. First, the cyclin B protein is expressed uniquely at the end of the S phase and during the M phase, being rapidly degraded as cells enter the next G1 phase.[102] Second, Cdk1 (also known as Cdc2) kinase activity is tightly regulated by an antagonistic posttranslation modification: phosphorylation by the Wee1 and myt1 kinases[103] and dephosphorylation by the Cdc25B/C phosphatases.[104] Phosphorylated Cdk1 is kept inactive during the G2 phase and its dephosphorylation is in fact the rate-limiting step for entry into mitosis.[105] Finally, because the primary sequence of cyclin B contains a nuclear exclusion signal, the dephosphorylated and active Cdk1/cyclinB complex is actively exported from the nucleus into the cytosol until the beginning of prophase.[106] Similarly, Cdc25C is also regulated by proteolysis and nuclear exclusion.[107]

Exact duplication of the DNA during the S phase dictates the entry of the dividing cells into the M phase. The DNA contained in every mammalian cells is under constant attack by agents (e.g., UV radiation or chemicals) that can either break the phosphodiester bonds on the backbone of the DNA helix or damage its bases. Defects in the repair of double-strand breaks can lead to chromosomal instability, a phenomenon intrinsically linked to carcinogenesis,[108] as all malignant tumor types have been shown to contain chromosomal aberrations.[109] Different mechanisms are required to repair the DNA damages or mismatches and the challenge to fix the problems can vary in the different phases of the cell cycle (**Figure 2**). Halting or slowing DNA replication until the error or damage has been repaired limits heritable mutations in daughter cells and controls the viability of the damaged cells. Initiation of the activity of ATR and ATM, which are members of the (phosphatidylinositol-3-OH kinase)-like kinase (PIKK) family, is the first step in the DNA damage/repair pathways that inhibit cell cycle progression.[110] ATR kinase seems to be critical for DNA damages located in DNA being replicated (leading to replication fork arrest) whereas ATM seems to be primarily activated following

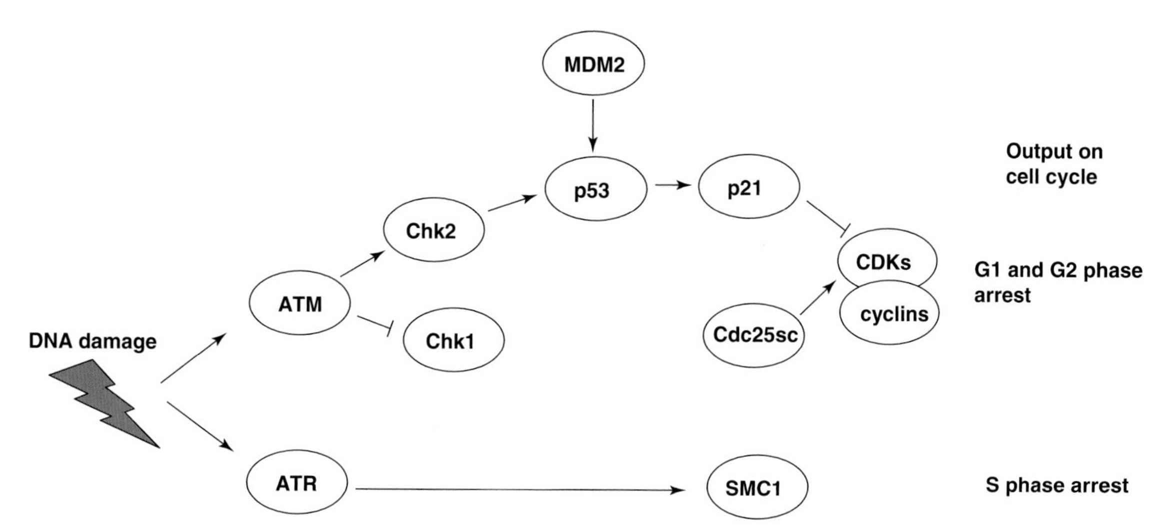

Figure 2 DNA damage and cell cycle checkpoints. This representative scheme shows the activity of the proteins involved in the DNA damage response and the final output on cell cycle.

DNA damage in nonreplicating DNA.[111] ATR and ATM work closely with the downstream transducer kinases Chk1 and Chk2, and the members of the polo-like kinases family (Plks) to control cell progression and the cellular responses to DNA repair, transcription, chromatin assembly, and cell death.

As described above, the DNA repair and cell cycle checkpoint pathways facilitate cellular responses to endogenous and exogenous sources of DNA damage, and it is reasonable to assume that alterations in these pathways increase the risk of cancer developing by permitting the survival or the continued growth of cells with genomic abnormalities. The dysfunction of some of the components of these pathway in cancer cells is briefly discussed in the following paragraphs.

The *TP53* tumor suppressor gene encodes the p53 protein, also called the guardian of the human genome. This protein is activated upon DNA damage[112] leading to G1 cell cycle arrest[113,114] and eventually to apoptosis.[115] It can also affect the G2/M transition through cyclin B downregulation, or disruption of the Cdk1/Cyclin B complex.[116] *TP53* is found mutated in around 50% of all human cancers,[117] and the p53 inhibitory partner Mdm2, which targets p53 for degradation by the ubiquitin/proteasome pathway, is amplified in 10–20% of bone and soft tissue tumors.[118] Mice carrying a heterozygous deletion of *TP53* are highly prone to spontaneous tumor development. Individuals with the Li–Fraumeni syndrome, carrying a germline deletion of only one p53 allele, are highly susceptible to cancer development. [119] This underlines the critical role of p53 as a tumor suppressor.

Although the ATM gene is not considered to be a tumor suppressor gene, its loss of function is frequently observed in patients with ataxia telangiectasia, a tumor-prone neurodegenerative disorder.[120] Moreover, somatic mutations in ATM have been identified in some sporadic cancers, particularly leukemia.[121] Typical cytogenetic changes seen in tumors from individuals with ataxia telangiectasia often involve aberrant oncogenic rearrangements at the T cell receptor loci,[122] underlining a link between normal DNA damage repair (the immunoglobulin-gene recombination) and the physiological function of the ATM protein.[123]

Chk2 is a stable protein expressed at constant levels throughout the cell cycle, but aberrant low levels of Chk2 have been found in subsets of human breast carcinomas, testicular tumors, and lymphomas. Furthermore, germline mutations in the Chk2 gene have been identified in some Li–Fraumeni syndrome patients with the wild-type p53 allele.[124]

7.01.5.2 Differentiation: Tumor versus Normal Cells

Embryonic development and metastasizing tumors are characterized by parallel processes of invasion, migration, and colonization so that cancer cells can be considered to behave as undifferentiated or inappropriately differentiated embryonic cells. Probably the best example to illustrate this point is the highly malignant teratocarcinoma tumors that contain a variety of differentiated tissues, including muscle, bone, teeth, and hair, showing an erroneous execution of the development program.[125]

There are two hypotheses that could explain why a cancer cell lacks all the features of a fully differentiated cell. First, the development process may have been stopped at some point. This seems to be the case for some hematological malignancies such as acute promyelocytic leukemia.[126] For this special type of leukemia, therapies using all-*trans* retinoic acid are particularly efficacious due to the ability of the drug to resume the differentiation program of the tumor cells.[127] Second, and this may apply more generally to solid tumors, cancer cells may have undergo dedifferentiation, which does not lead to a more primitive cell type, but to a new, nonstandard, and abnormal cell type. Probably the best example of this is in colorectal cancers. Inactivation of the tumor suppressor gene *APC* is the most important event for sporadic initiation of colorectal cancers and progression toward adenocarcinoma.[128] *APC* mutations were found to cause dysplasia, and this change in cell character is critical for initiation of colorectal cancers.[129] Moreover, spontaneous tumors in *APC* knockout mice were formed by cells retaining crypt-cell-like features instead of continuing their differentiation.[130] The biochemical function of the APC protein is to downregulate beta-catenin levels to reduce the transcription activity of the TCF4/beta-catenin complex.[131] Abrogation of TCF4 expression in mice results in early lethality, due to a complete absence of stem cells of the gut.[132] Beta-catenin/TCF4 signaling seems to be necessary for the maintenance of the stem cell population, and one of the main functions of *APC* is to control the transition of stem cell to daughter. Last but not least, it was recently shown that epigenetic changes in specific gene promoters alter the balance of differentiated versus undifferentiated cells. The loss of imprinting in the IGF-2 gene is present in about 30% of patients with colorectal cancer. Reconstitution of this loss of imprinting in heterozygous $APC^{+/-}$ mice resulted in a higher incidence (twofold) of intestinal tumors. The heterozygous animals also showed a shift toward a less differentiated normal intestinal epithelium.[133]

It is also important to mention that a lack of differentiation of nontumor cells may contribute to tumor growth. Thus, a defective host antitumor immune response is an important mechanism allowing tumor cells to evade the control of the immune system. Induction of an effective antitumor response requires the active participation of host bone-marrow-derived antigen presenting cells (APCs). Dendritic cells are the most potent APCs, and those cells are sometimes defective in cancer patients due to a fault in their maturation (e.g., lack of expression of costimulatory molecules). Different factors are at the origin of defective DC differentiation, including cytokines and angiogenic factors.[134] However, the link between production of those factors and tumor growth remains to be elucidated.

7.01.5.3 Apoptosis

Apoptosis is the mechanism that orchestrates programed cell death when a cell cannot overcome certain type of insults coming from its environement. Apoptosis is a multistep process involving many proteins, and also the participation of the mitochondria. The resulting apoptotic phenotype is characterized by cell shrinking, condensed chromatin with nuclear margination and DNA fragmentation, and membrane blebbing.

7.01.5.3.1 Intrinsic and extrinsic apoptosis pathways

Originally thought to be exclusively devoted to energy production, the role of mitochondria was recently revisited as these organelles actively participate in the intrinsic (or stress-induced) apoptosis machinery (**Figure 3**). Upon activation of apoptotic stimuli (e.g., p53 stabilization), proapoptotic members of the Bcl-2 superfamily proteins such as Bax or Bak can translocate into the membrane of the mitochondria. They are thought to disrupt the outer membrane of mitochondria, thereby leading to the release of proapoptotic mediators (cytochrome-*c*, Sma/Diablo, Omi/HtrA2, endonuclease G, flavoprotein AIF), normally located in the mitochondria, into the cytosol. This chain of events, termed mitochondrial membrane permeabilization, is nonreversible and is considered to be the apoptosis checkpoint.[135] The exact mechanism by which Bax/Bak induces mitochondrial membrane permeabilization is still highly controversial, but it has been shown that Bax oligomers can form in liposomes pores that are large enough for cytochrome-*c* release.[136] Released cytochrome-*c*, interacts with Apaf-1 and the resulting complex (the apoptosome complex), will lead to the maturation of zymogen procaspase 9 to active caspase 9, which in turn will activate the executioner procaspases 3, 6, and 7 into their active forms.[137]

In parallel to the intrinsic pathway, the extrinsic apoptosis pathway (**Figure 3**) is activated upon ligands tumor necrosis factor-alpha (TNF-α), Fas-L, and TNF-related apoptosis-inducing ligand (TRAIL) binding to their cognate membrane-bound receptors of the tumor necrosis family (TNF) super family of receptor proteins (death receptors TNFR-I and TNFR-II, Fas, and DR4/5). Activation of the receptors leads to the maturation of procaspases 8 and 10 which will then activate the executioner caspases 3, 6, and 7, without any participation of the mitochondria.

7.01.5.3.2 Defects of the apoptosis machinery in cancer cells

Cancer cells exhibit enhanced glycolytic ATP generation and decreased respiratory phosphorylation, even under normal oxygen tension (the Warburg effect). This metabolic alteration might be linked to a change in the composition or the regulation of the mitrochondria organization, which would prevent or reduce mitochondrial apoptosis. Indeed, mitochondra with resistance to mitochondrial membrane permeabilization have been observed in leukemia cell lines.[138]

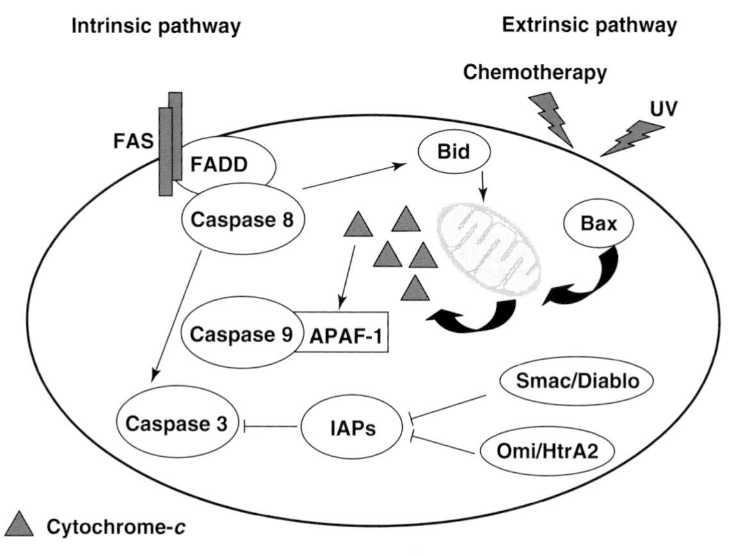

Figure 3 The apoptosis machinery. The different players involved in either intrinsic or extrinsic apoptosis cascades discussed in the text are shown in this simplified scheme.

Antiapoptotic members of the Bcl-2 superfamily all presumably have oncogenic potential due to their ability to block cytochrome-*c* release. This is indeed the case for Bcl-2 itself and for Bcl-xL. The Bcl-2 gene is often translocated in human follicular lymphoma, where it is constitutively expressed.[139] It is now well accepted that high levels of Bcl2 or Bcl-xL may impede the oligomerization of the proapoptotic members.[140] However, in transgenic mouse models mimicking the Bcl-2 translocation, spontaneous lymphomas take many months to develop, and the penetrance of the disease is rather low, suggesting that Bcl-2 overexpression on its own is not highly oncogenic and that progression of malignancy requires synergistic mutations. The oncogenic potential of Bcl-2 is more effective when the *Myc* oncogene is coexpressed, either in vitro or in vivo. Consistent with this, lymphoid malignancies with elevated Bcl-2 expression also contains *Myc* translocation[141] and p53 mutations.[142]

In contrast to Bcl-2-like proteins, proapoptotic members could be considered to be tumor suppressor proteins. Since Bax and Bak have redundant functions, tumor promotion would require inactivation of both genes. Some human colorectal and hematopoietic tumors exhibit mutations in the Bax or Bak genes. Reduced expression of the Bax proapoptotic protein has also been reported in colorectal carcinoma, and there is often a good correlation between the expression levels of the various Bcl-2 members and clinical outcome.[143]

Human tumor cells may be defective in the assembly of the apoptosome. This is often the result of a loss of expression of the Apaf-1 protein, as observed in melanomas, leukemias, glioblastoma, and ovarian cancers. Apoptosome assembly may also be blocked by the overexpression of certain chaperones or oncogenes.[144] Thus, hsp70 overexpression in breast cancer is correlated with a shorter disease-free survival interval, increased frequency of metastasis, and decreased responsiveness to chemotherapy.[145] Expression of the ProT oncogene suppresses apotosome formation,[146] and increased ProT expression have been found in human malignancies including lung, hepatocellular, breast, and colon cancers.

Finally, caspase activation is a step in the apoptosis pathway that could be blocked by direct interaction with the inhibitors of apoptosis (IAPs). From the IAPs family members (NAIP, c-AIP1, c-IAP2, XIAP, survivin, livin, ILP-2, c-FLIPL, ARC, and BAR), survivin appears to have the most clinically relavant link to cancer, as this protein has been shown to be overexpressed uniquely in cancer cells but not in the normal surrounding tissues.[147] A correlation has been established between survivin overexpression and poor prognosis for a wide variety of tumor types.[148] XIAP and ML-IAP are also overexpressed in acute myelogenous leukemia and melanoma, respectively.

7.01.6 Invasion and Metastasis

7.01.6.1 Invasive Growth and the Tumor Microenvironment

The spread of primary tumors to other sites in the body to form metastasis is responsible for the majority of cancer-related deaths. Growth of occult metastasis is dependent on cancer cells invading the basement membrane surrounding the primary tumor and entering the lymphatic or blood system. Within the circulation, cells must be able to resist anoikis (programed cell death due to loss of anchorage), evade the immune response, and overcome shear stress before arresting within the capillary bed of the distant site. At this site, the cell must either proliferate within the vasculature or extravasate into the surrounding tissue, survive, divide, and coopt the extant vascular system. These multiple, essential, and interrelated steps explain why very few clinically relevant metastasis occur despite the fact that tumors can shed millions of cancer cells daily into the vasculature.

Although much remains unclear, it is apparent that during invasion and metastasis fundamental changes occur in the manner in which cells respond to and modify their local environment. Invasive growth is characterized by unregulated tissue remodeling in which normal tissue is replaced by tumor tissue. The actions of malignant and neighboring cells, responding to soluble factors released by the tumor, result in characteristic changes in the surrounding stroma. The tumor stroma microenvironment is rich in endothelial cells, smooth muscle cells, and myelofibroblasts.[149] Under the influence of the tumor, these cells proliferate, migrate, and differentiate, releasing enzymes that degrade the surrounding extracellular matrix as well as growth factors and cytokines that can act in a paracrine fashion, promoting tumor growth and invasion. In addition, the release of chemotactic factors leads to the accumulation and activation of dendritic and inflammatory cells, thereby increasing levels of local cytokines. Defining the changes in the manner with which tumors interact with their microenvironment is therefore key to understanding the mechanisms underlying invasive growth.

7.01.6.2 Adhesion Molecules

Adhesion molecules play a major role in the interchange of signals between the cell and its environment. Differential regulation of several classes of adhesion molecules has been implicated in invasion and metastasis of which perhaps the best characterized are cadherins and integrins.

7.01.6.3 Cadherins and epithelial–mesenchymal transition (EMT)

Cadherins play an essential role in maintaining the integrity of the epithelial layer. Cadherins mediate tight cell–cell junctions and are integral components of adherens junctions and desmosomes.[150] The large extracellular domain forms a calcium-dependent homophilic interaction with cadherin molecules on neighboring cells while the cytoplasmic domain associates with alpha-, beta-, and gamma-catenins. These molecules interact, in turn, with the actin cytoskeleton and intracellular signaling components. As described above, free beta-catenin can translocate to the nucleus, associate with the TCF/LEF transcription factor complex, and drive expression of proproliferative genes, such as c-Myc and cyclin D1. Disruption of cadherin complexes would therefore affect cell–cell interaction and regulation of the actin cytoskeleton, and potentially increase the levels of available beta-catenin. In fact, accumulating evidence suggests that alterations in cadherin expression, or 'cadherin switching', is a critical event during tumor progression. Loss of E-cadherin, predominantly expressed in epithelial cells, has been associated with more aggressive, invasive, and poorly differentiated tumors.[151] Various mechanisms of E-cadherin downregulation have been characterized in tumors including loss-of-function mutations, upregulation of transcriptional repressors such as TWIST and members of the SLUG/SNAIL family, aberrant promoter methylation, proteolytic degradation as well as phosphorylation by both receptor (RTK) and Src family tyrosine kinases (SFK).[152] Tyrosine phosphorylation and recruitment of Hakai, a Cbl-like E3-ubiquitin ligase, leads to internalization of E-cadherin by endocytosis. In agreement with a seminal role in tumor invasion, forced overexpression of E-cadherin impairs invasiveness in an experimental tumor model while its downregulation, via targeted proteolytic degradation, leads to the development of breast carcinoma.[153]

Changes in expression of other cadherins accompanies downregulation of E-cadherin, in a situation reminiscent of the epithelial–mesenchymal transition that occurs during embryonic development. This transition is characterized by a loss of epithelial cell polarity and cell–cell contact, gain of mesenchymal markers such as N-cadherin, and an increased migratory phenotype.[154] Increased levels of N-cadherin, normally expressed in neuronal cells and fibroblasts in adults, are observed in breast and prostate cancer and invasive melanoma.[155] N-cadherin is thought to mediate a less stable and more dynamic form of cell–cell adhesion, and its overexpression enhances invasion and formation of metastasis. These changes occurred despite the presence of E-cadherin, suggesting that N-cadherin may play a dominant role. N-cadherin invasive activity occurs, at least partly, via association with the fibroblast growth factor receptor-1 (FGFR-1). Interaction results in stabilization of FGFR-1 leading to sustained downstream signaling and upregulation of proteolytic activity.[156]

Other cadherins have also been implicated in tumor progression. Aberrant methylation of the H-cadherin promoter results in its in silencing in breast, colorectal, and lung cancers.[157] Loss of H-cadherin, normally expressed in the ductal epithelium, was observed in breast carcinomas before invasion was apparent.[158] Furthermore, expression of H-cadherin, but not E-cadherin, was sufficient to suppress the invasiveness and tumorigenicity of an aggressive breast cancer cell line, MDA-MB-435.[159] Therefore alterations in H-cadherin may represent a necessary earlier event in a cancer's progression to a more invasive phenotype.

7.01.6.4 Integrins and Downstream Signaling Pathways

Cells secrete insoluble molecules that are assembled and remodeled to form the extracellular matrix (ECM). The ECM provides a scaffold for the appropriate organization of groups of cells into tissues, but also controls multiple aspects of cellular behavior. The majority of the effects of the ECM on cells are mediated via integrins that directly interact with ECM components, such as collagen, vitronectin, laminins, and fibrinogen. Each integrin is composed of a heterodimer formed between an alpha and a beta transmembrane subunit. Over 18 alpha and eight beta subunits have been characterized, and different heterodimers have been shown to bind to distinct but overlapping ECM epitopes.[160,161] The intracellular tail of the receptor is found within focal adhesion plaques that are sites of interaction with the actin cytoskeleton and contain numerous signaling molecules, including focal adhesion kinase (FAK), integrin-linked kinase (ILK), Src family kinases (SFKs) and protein kinase C.[162] Upregulation of FAK has been observed in multiple invasive human cancers, and FAK-null fibroblasts display defects in migration, particularly in integration of epidermal (EGF) and platelet-derived growth factor (PDGF) mediated promigratory signals.[163,164] Integrin mediated activation of FAK, leading to recruitment and activation of SFKs such as c-Src, c-Yes, and c-Fyn, has been extensively studied in tumors. Increased Src activity has been observed in several tumor types, including mammary, colon and pancreatic carcinoma.[165] A correlation between Src activity and disease progression led to investigation of the role of Src in tumor invasion and metastasis. Activated c-Src is observed at focal adhesions and genetic ablation of Src interferes with cellular adhesion and spreading. Transformation by Src leads to changes in cell–cell and cell–ECM adhesion and protects cells from anoikis, facilitating anchorage-independent growth in vitro and enhanced invasive behavior of implanted tumors. Activation of Src is dependent on dephosphorylation of a negative regulatory tyrosine residue within the C-terminal domain of the kinase. A number of phosphatases have implicated in activating Src, including PTPα, PTPB1, and Shp2. Deletion of PTPB1 or Shp2 is associated with reduced Src activity and perturbation of cell adhesion and spreading.[165]

Multiple signaling components are activated by FAK and SFKs including ETK, an intracellular tyrosine kinase found at high levels in metastatic carcinoma cells and the ERK/MAPK and JUN kinase cascades, which, in addition to modifying gene expression, can affect motility by direct phosphorylation of cytoskeletal components.[166] Activated SFKs phosphorylate and initiate signaling from paxillin and p130CAS.[165] Modulation of motility and the actin cytoskeleton occurs principally via the consequent activation of the Rho GTPase family members Rho, Rac, and cdc42. Both Rac and Cdc42 are required for carcinoma motility and invasion, promoting actin polymerization at the leading edge of migrating cells.[167] RhoA and RhoC are upregulated in metastatic carcinomas, and RhoC overexpression favors colonization of the lung in an experimental melanoma metastasis model.[168,169] Rho, acting via two effectors, ROCK and mDIA, regulates actomyosin fiber assembly and contraction, contributing toward pulling forward the trailing edge of cells during migration. Rho–ROCK signaling appears to regulate aspects of carcinoma dissemination and invasion.[162] Finally, and importantly, FAK activation can protect tumor cells from anoikis by activating PI3K and PKB and upregulating antiapoptotic protein such as BCL2.[162,165]

Binding of integrins to the tissue ECM, therefore, governs cellular polarity and migration and provides positional cues that are integrated into the cells response to other stimuli, such as growth factors or cytokines. Although the malignant process is characterized by a partial loss of dependency on ECM-mediated signals, it is equally apparent that it is accompanied by specific changes in integrin expression: upregulation of integrins that facilitate invasion and metastasis and downregulation of those integrins that oppose these processes. Expression of particular oncogenes has been demonstrated to directly modulate integrin expression; however, it appears more likely that these changes occur as a consequence of the selective pressure on cells to adapt to the different environments encountered during invasion and metastasis.[170]

7.01.6.5 Integrins in Extracellular Matrix Remodeling and Invasion

Integrin signaling is implicated in multiple aspects of matrix remodeling by cancer cells. Experimentally, activation of JNK by FAK in v-Src transformed fibroblasts increases expression of ECM degrading proteases.[171] In particular, $\alpha_v\beta_3$ integrin, upregulated in melanomas and glioblastomas,[172,173] has been shown to combine with diverse RTKs to promote cell migration.[174] In addition, $\alpha_v\beta_3$ appears to regulate the recruitment and activation of various enzymes involved in ECM degradation at the leading edge of cells, a necessary part of the invasive process.[175,176]

In normal tissues, expression of $\alpha_6\beta_4$ integrin is associated with cellular compartments that contain populations of rapidly dividing cells. In keratinocytes, binding of $\alpha_6\beta_4$ to laminin-5 permits progression from G1 to S phase in response to EGF and is required for Ras-mediated transformation in culture.[162] Upregulation of $\alpha_v\beta_4$ is observed in carcinomas, and introduction of β4 is sufficient to enhance the invasive potential of breast carcinoma cells.[177] These effects appear to be due to both enhanced activation of integrin-mediated signals and interaction between $\alpha_6\beta_4$ and RTK, involved in invasion. Recruitment of $\alpha_6\beta_4$ by RTKs, such as c-Met, c-Ron, EGFR, and ErbB2, induces tyrosine phosphorylation of the β4 subunit, increasing downstream signaling and enhancing migration and invasion. In the case of Met, this appears to be independent of laminin-5–integrin interaction, whereas ECM binding transactivates EGFR and ErbB2 in an-SFK-dependent manner indicating that a two-way passage of information can occur.[178,179]

7.01.6.6 Integrins and Angiogenesis

As described below, neovascularisation is an essential event in the growth of tumors. Tumor angiogenesis occurs as a result of the interaction between positive and negative regulators, and integrins have been shown to play an intimate role in these processes. In endothelial cells, several integrins are upregulated and associate with FAK in response to proangiogenic factors, such as VEGF and bFGF. Knockout studies have implicated a role for $\alpha_5\beta_1$ and αv integrins in angiogenesis, and ECM-derived fragments with potent antiangiogenic effects, such as endostatin and tumstatin, have been shown to act by blocking the action of integrins $\alpha_5\beta_1$ and $\alpha_v\beta_3$, respectively.[180]

7.01.6.7 Proteolytic Enzymes and the Invasive Process

Proteolytic enzymes have been demonstrated to have a role not only in the degradation of the ECM, but also in the release of factors that regulate the growth and behavior of cells. Under normal conditions, proteases and their inhibitors are finely balanced, allowing appropriate remodeling of the ECM to maintain normal tissue architecture.[181] Breakdown of this balance is a fundamental step in the progression of a tumor from a benign to a malignant state. Of the various classes of proteases capable of degrading components of the ECM, the matrix metalloproteinases (MMPs) and plasminogen activator family are most strongly implicated in cancer progression.

7.01.6.8 Metalloproteinase Family

MMPs can degrade all components of the basement membrane and ECM and play important physiological roles in tissue repair, morphogenesis, and angiogenesis. Currently, over 24 MMPs have been identified and classified according to their domain structures: archetypal MMPs, matrilysins, gelatinases, and convertase-activatable MMPs.[181] Although MMPs occur predominantly as secreted proteins, a subgroup is found on the cell surface, attached via a transmembrane domain or a glycosylphosphatidylinositol anchor. MMP activity is regulated at multiple levels although transcriptional control and posttranslational activation/inhibition appear to be the most important. Several growth factors and cytokines have been demonstrated to upregulate levels and activity of MMPs. Most MMPs are secreted as inactive zymogens, remaining latent until cleaved by another MMP or other protease. Refolding follows and typically a second cleavage, occuring either autocatalytically or via another MMP, is required for full activity.[182] Endogenous MMP inhibitors, tissue inhibitors of metalloproteinases (TIMPs), provide a further level of regulation of protease activity. TIMPs are expressed within the extracellular space, either as soluble or membrane-anchored proteins, that directly bind MMPs with high affinity but little specificity within different MMP subclasses.[183]

Cleavage of MMPs at the tumor cell–matrix interface is highly regulated by cell surface MT-MMP, as well as by adhesion molecules that have been implicated in proangiogenic and proinvasive activities (see above). Malignant cells within primary tumors have been reported to express MMPs; MMP-7 is commonly expressed by epithelial tumors, but it is apparent that stromal cells within the vicinity of tumors a represent a major, if not the principal, source of MMP activity.[184] Stromal myofibroblasts and inflammatory cells, such as mast cells, appear to be responsible for the production of MMPs at the tumor–matrix interface. In agreement with this, mice deficient in particular MMPs (MMP-2 and MMP-9) display resistance to tumor development in response to carcinogenic insult and delayed growth of experimentally implanted tumors.[184,185]

A number of MMP targets relevant for cancer invasion have been identified (*see* 7.09 Recent Development in Novel Anticancer Therapies). Cleavage products of laminin-5 and elastin, frequently observed at the invasive front of tumors, possess chemotactic properties. MMPs can also target the extracellular domains of adhesion molecules; MMP3 has been proposed to shed the ectodomain of E-cadherin, thereby disrupting cell–cell attachment. Administration of TIMP1, or pharmacological inhibitors, blocked this process, restoring cell membrane localization of E-cadherin and beta-catenin. MMPs can additionally increase local growth factor concentrations and bioavailablity by releasing them from the extracellular matrix, e.g., fibroblast growth factor (FGF) and transforming growth factor beta (TGF-β), or inhibitory proteins, e.g., insulin-like growth factor binding proteins 1. Consistently, transgenic mice with reduced MMP activity, due to high TIMP-1 expression levels, display decreases in IGF1 availability and activation of IGF1R.[186,187]

Epidemiological studies have clearly demonstrated that cancer progression is associated with the upregulation of multiple MMPs, redundancy rather than specificity being the rule. MMP levels correlate with tumor aggressiveness, and increased expression of MMPs has been linked to poorer prognosis in locally invasive cancers, such as gastric and esophageal carcinoma.[188,189] MMP activity has been implicated in positive and negative regulation of tumor angiogenesis. MMP-9 is upregulated in highly vascularized tumors, such as lung, colon and renal carcinoma, and was shown experimentally to mediate angiogenesis and intravascular colonization of the lung by implanted tumor cells. In contrast, as described above, the action of MMP on type IV, type XVIII, and type XV collagen releases peptides with potent antiangiogenic activity. Expression of MMP-12 is associated with increased survival in colon carcinoma patients, presumably due to its inhibitory actions on angiogenesis and invasion, mediated by cleaving the urokinase-type plasminogen activator receptor (uPAR).[190] Extensive studies, using antisense RNA, small interfering RNA (siRNA), and low-molecular-weight inhibitors, have demonstrated that decreasing MMP activity can reduce tumor growth, invasion, and vascularization. Several MMP inhibitors, with broad specificity, have been tested in clinical trials. However, the observed patient response has not replicated the effectiveness seen in preclinical studies, potentially due to dose-limiting toxicities and to redundancy of action that occurs between MMPs and plasminogen activation (see below).[181]

7.01.6.9 Plasminogen Activation and Cancer

Plasminogen is present in the plasma and interstitial fluids as a zymogen that is locally converted to an active serine protease by the action of either of two plasminogen activators, urokinase-type PA (uPA) and tissue-type PA. It seems that uPA-dependent activation mediates physiological and pathological tissue remodeling processes.[191] Both uPA and plasmin are inhibited by the action of specific inhibitors, plasminogen activator inhibitor (PAI-1) and α2-antiplasmin, respectively; uPA is secreted as a proenzyme that is reciprocally activated by the action of plasmin. Activation of the cascade is locally confined by the high-affinity binding between uPA or pro-uPA and the cell surface receptor uPAR. Mice lacking uPA display defects in extravascular fibrin degradation while cooperation is observed in transgenic animals that coexpress uPA and uPAR.[192] In a similar manner to MMPs, stromal cells appear to be the major sources of plasminogen activity in most tumors. The stromal cell type producing uPA depends on the origin of the tumor, myofibroblasts being responsible in breast and colon and macrophages in prostate cancer. Elevated levels of uPA and uPAR in the blood and tissue are associated with poor prognosis in many types of cancer, regardless of their respective sources.[193,194] Manipulation of uPA levels by overexpression, treatment with anti-uPA antibodies, or using uPA-deficient mice has demonstrated a key role in cancer growth, invasion, and metastasis.[191,195]

7.01.6.10 Growth of Tumor Cells at Secondary Sites

As described above, a cancer cell needs to overcome a series of challenges in order to successfully metastasize. Once it has gained the capacity to invade and migrate, allowing it to leave the primary tumor and enter the vasculature, the cell must survive transit within the vasculature and arrest at the secondary site. The latter steps have been proposed to be rate-limiting in the metastatic process, as a correlation between metastatic efficiency and resistance to apoptosis at the site of extravasation has been experimentally observed.[196]

7.01.6.11 Tumor Necrosis Family: Apoptosis and Immune Surveillance

It is clear that tumor cells can potentially utilize multiple mechanisms to evade apoptosis; however, several frequently employed pathways have been identified. Perturbation of the action of ligands of the tumor necrosis family (TNF), FasL, and TRAIL have been particularly studied in this context. FasL is a transmembrane protein that can trigger apoptosis in cells expressing its receptor Fas (CD95/APO-1). De novo expression of FasL by tumors, including colon, gastric, and lung carcinomas, is proposed to mediate immune privilege by inducing apoptosis in antitumor lymphocytes that express Fas.[197–200] FasL-expressing colorectal and gastric tumors were characterized by increased numbers of apoptotic tumor-infiltrating lymphocytes (TIL).[199,201] The ability of tumor-expressed FasL to suppress the immune response has been extensively verified experimentally.[202] In addition, increased FasL expression accompanies disease progression in colorectal carcinoma and has been significantly associated with the probability of breast and cervical carcinomas to form lymph node metastasis.[197,203] In an analogous fashion, upregulation of TRAIL has been observed in hepatocellular carcinoma, and its expression in gastric carcinoma is correlated with metastasis and the presence of apoptotic TIL.[204,205]

One necessary consequence of increased FasL or TRAIL levels is that tumors themselves need to develop resistance to their apoptotic effects. FasL resistance commonly occurs via decreased cell surface levels of Fas expression or expression of an antagonistic decoy receptor.[197] Alternatively, upregulation of intracellular inhibitors of the proapoptotic pathway, such as cFLIP (FLICE inhibitory protein), IAP, or Bcl-2 and Bcl-XL has been documented (see above). Evading apoptosis is also relevant at the site of arrest, particularly in tissues that express high levels of TNF ligands, such as the lung and lymph nodes.[198,206,207] In animal models, Fas expression inversely correlates with the capacity of cells to metastasize to the lung, as normally nonmetastatic cells (with high Fas levels) induce as many metastases as subclones that express lower levels of Fas when injected into mice lacking FasL.[162,208] In patients, mutations in the Fas pathway, identified in NSCLC, significantly correlated with development of lymph node metastases.[209] TRAIL knockout mice have also been shown to be more susceptible to development of metastases, while administration of TRAIL, or agonistic antibodies that activate its cognate receptors DR4 and DR5 suppress the formation of metastases.[210,211–213] Consistently, mutations of DR4 and DR5 were found to be significantly associated with highly metastatic breast cancers.[214] Mutations appear to be relatively rare, however, and other mechanisms, such as deficient DR transport to the cell surface or abrogation of the intracellular apoptosis inducing pathways (described above), may be more common. DAP kinase (DAPK), a serine threonine kinase involved in mediating FasL and TRAIL proapoptotic signals, has been linked to metastases in several studies. Aberrant methylation of DAPK has been associated with clinical aggressiveness of NSCLC, colon, bladder, and breast carcinoma and with development of lymph node metastasis in head and neck squamous cell carcinoma (HNSCC).[215] Restoration of DAPK activity to highly metastatic cell lines reduced their capacity to metastasize to mouse lungs.[216,217]

Cells that do not die either remain dormant or begin to proliferate to form microscopic metastasis extra- or intravascularly. Dormant solitary cancer cells have been observed in experimental models and cancer patients and may be responsible for the recurrence of cancer due to their resistance to chemotherapeutic agents.[216,218] In order to form occult metastasis, dormant cells must start to proliferate, maintain growth, and become vascularized. Experimental studies have suggested that each of these steps is inefficient, with only a fraction of cells surviving each stage.

Studies undertaken to establish the factors that maintain tumor cells in a dormant state have led to the identification of metastasis suppressors. The first metastasis suppressor, NM23, was identified in 1988; however, eight more have since been identified.[216,219] Most were identified based on differential expression in metastatic and nonmetastatic cell lines and appear to act at discrete stages in the process of metastatic colonization. Exogenous expression of NM23 impairs motility and the metastatic potential of diverse cell lines in vivo but does not affect proliferation or primary tumor size.[220] NM23 possesses histidine kinase activity that is essential for its metastasis suppressor activity.[221] Kinase suppressor of RAS(KSR), a ERK-MAPK scaffold protein, binds and acts as a substrate for NM23. More recently, histidine phosphorylation has been shown to destabilize KSR leading to reduced extracellular signal regulated kinase (ERK) activation which was correlated with a reduction in metastatic capacity of a breast cancer cell line.[222]

KiSS-1 encodes a neuropeptide ligand for a GPCR hGPR54(AXOR12), identified as a gatekeeper of the reproductive cascade.[223] KiSS-1 and hGPR54 display complementary expression patterns in the brain and in the placenta where they have been identified as regulators of trophoblast invasion, inducing migration and protease expression.[224] KiSS-1 appears to play a similar role in tumor invasion and metastasis. Forced expression of KiSS-1 suppresses metastasis of melanomas and human breast carcinomas without affecting tumorigenicity.[225–227] Although the action of Kiss-1 may depend on tumor type, loss of KiSS-1 expression was correlated with tumor recurrence in gastric cancer and venous invasion and distant metastasis in invasive bladder cancer patients. Additionally, loss of KiSS-1 and/or hGPR54 gene expression was found to be a significant predictor of lymph node metastasis in esophageal squamous cell carcinoma (ESCC).[217]

Upon entering the vasculature, mechanical forces contribute to the delivery of tumor cells to specific target organs. The relatively large size of solid tumor cells will tend to lead to their arrest in the first capillary bed they encounter. However, it apparent that tumors arising in particular organs have preferential secondary sites of metastasis, indicating that interaction between the cancer cells (seed) and the organ microenvironment (soil) can influence the fate of the cells.[228] In addition to negative regulators, such as TNF ligand expression as described above, diverse factors that can promote the formation of tissue-specific metastasis have been postulated. Chemokines have been identified as key regulators of this process. Chemokines are soluble ligands that are involved in recognition and homing of multiple cell types, including hematopoietic cells, lymphocytes, and germ cells.[229] Breast cancer cell lines, as well as carcinomas and metastasis, express high levels of the chemokine receptors CXCR4 and chemokine receptor 7(CCR7).[230] The ligands for these receptors, stromal cell-derived factor-1, CCL19, and CCL21, are expressed in those tissues where breast cancers most often metastasize, such as lymph nodes, lung, liver, and bone marrow.[231] Binding of SDF-1 to CXCR4 upregulates expression of proteases and adhesion molecules and enhances invasion and migration in vitro. Disruption of the receptor–ligand interaction was sufficient to inhibit spontaneous metastasis to the lung and lymph nodes.[232–234] Similar studies have implicated CXCR4, as well as other chemokine receptor and ligand family members, in the metastatic behavior of other tumor types, including melanoma (CXCR4, CCR7, and CCR10), colorectal cancer (CXCR4 and CCR7), and prostate cancer (CXCR4 and CCR9).[235] Notably, these agents frequently affect both tumor growth and invasiveness, suggesting CXCR4 may also function at the site of the primary tumor. A more recent publication has demonstrated that breast carcinoma-associated fibroblasts (CAF) express high levels of SDF-1. Expression of SDF-1 by CAF enhanced growth of tumors, via activation of CXCR4, and recruited endothelial cell precursors, promoting vascularization.[236]

7.01.7 Tumor Angiogenesis

Cells cannot survive in the absence of oxygen and nutrients. As tumors grow beyond a certain size simple diffusion is no longer sufficient, necessitating the complementary development of a vascular network, a process described as neovascularization or angiogenesis. Control of physiological angiogenesis is highly complex, involving both positive and negative stimuli. Vascularization of tumors is associated with imbalance of these regulatory mechanisms, and increased vessel density and upregulation of proangiogenic factors correlates with more aggressive cancer and poorer prognosis (*see* 7.08 Kinase Inhibitors for Cancer).

7.01.7.1 Vascular Endothelial Growth Factor

The majority of studies have focused on sprouting angiogenesis, the proliferation and migration of endothelial cells from pre-existing blood vessels to form vascular structures. Sprouting angiogenesis appears to be mainly controlled by VEGF family members. First identified by Dvorak and colleagues due to its ability to induce vascular leaking and permeability in ascites, VEGF-A plays a central role in angiogenesis, inducing endothelial cell proliferation, migration, and survival.[237] In addition, VEGF-A induces changes in endothelial cell morphology and motility, and upregulates expression of proteases, integrins, and a variety of mitogens. These changes facilitate endothelial cell migration and degradation of the basement membrane, necessary steps in the angiogenic process.

Homozygous or heterozygous deletion of VEGF-A results in embryonic lethality due to cardiovascular abnormalities and defects in vasculogenesis.[238] VEGF-A is also required postnatally in physiological vasculogenic processes such as wound healing, ovulation and pregnancy.[239] In common with other VEGF-A family members, VEGF-A occurs both in freely secreted and ECM-bound forms that can be released by the action of plasmin or MMPs.[240] Binding to either VEGFR-1 (Flt-1) or VEGFR-2 (KDR or flk-1) is sufficient and necessary for the majority of VEGF's known functions.

Originally identified on endothelial cells, both VEGFR-1 and VEGFR-2 are also expressed within the hematopoietic compartment.[241,242] In contrast, expression of a third family member, VEGFR-3 (Flt-4), is restricted to the lymphatic vasculature where it has been demonstrated to play a significant role in lymphangiogenesis.[243] All three receptors possess an intrinsic intracellular kinase domain that is activated upon ligand binding. VEGF-A is bound by both VEGFR-1 and VEGFR-2 whereas receptor specific interactions are observed between other VEGF-A family members. Signaling through VEGFR-2 appears to mediate most of the actions of VEGF-A on endothelial cell biology. Similarly to the VEGF-A knockout phenotype, VEGFR-2 deficient mice die in utero due to defects in vascular development.[244] In contrast, the precise role of VEGFR-1 in physiological and pathological angiogenesis remains to be fully clarified. Initial studies suggested that the normal physiological function of VEGFR-1 was to act as a negative regulator of VEGFR-2.[245,246] More recent findings have indicated that VEGFR-1 has positive functions in certain cell types and can heterodimerize and transphosphorylate VEGFR-2. In particular, the positive actions of VEGFR-1 appear to be coopted in pathological conditions including tumor angiogenesis.[239]

Increased expression of VEGF-A within the tumor has been associated with disease progression, development of metastasis, and poor prognosis in multiple cancers.[239,247,248] VEGF-A levels correlate with increased tumor vascularization and microvessel density, independent prognostic factors for breast carcinoma and NSCLC.[239,249] Antagonistic antibodies and low molecular-weight inhibitors of VEGF-A signaling display antiangiogenic and antitumor effects in vivo, appearing to target mainly neovascularization rather than established vessels (see below). Despite the role of VEGF-A in physiological angiogenesis, several therapeutic entities have demonstrated clinical effectiveness while retaining an acceptable safety profile. The most advanced, Avastin, a VEGF-A-binding antibody, has been approved for treatment of patients with metastatic colorectal cancer.[239,250,251]

VEGF-A is one of the most potent inducers of vascular permeability known, an important function considering the extreme leakiness of tumor vessels. Increased intracellular trafficking (vesicovascular organelles), nitrous oxide (NO) production, and induction of endothelial fenestrations have been proposed as mechanisms by which VEGF-A induces increased permeability. Inhibition of VEGF-A signaling has been shown to be effective in preventing pathological conditions associated with increased permeability, such as pleural effusion and the formation of malignant ascites.[239,252,253]

7.01.7.2 Regulators of Vascular Endothelial Growth Factor-A and Vascular Endothelial Growth Factor Receptor Expression

VEGF-A expression by tumor cells is upregulated by multiple stimuli, including cytokines, growth factors, hypoxia, and hypoglycemia in addition to activation of oncogenes and mutation of tumor suppressors. The mechanisms by which VEGF-A levels are induced by hypoxia have been extensively elaborated. In normoxic conditions the transcription factor HIF-1α is maintained at low levels, due to the action of the von Hippel–Lindau tumor suppressor (VHL) which directs its ubiquitinylation and proteosome mediated degradation. Hypoxia results in stabilization of HIF-1α, permitting activation of target genes that includes VEGF-A, as well as proteases and adhesion molecules.[238,239] Many cancers are characterized by areas of hypoxia, and an association between hypoxia or HIF-1α levels in primary tumors and probability of metastasis has been clinically established.[254] In agreement, experimental manipulation of VHL or HIF-1α activity is associated with changes in tumor growth, vascularization, and metastatic potential.[255]

Multiple studies have demonstrated that growth factor-mediated stimulation of tumor cells can lead to release of VEGF-A. In particular, activation of both EGFR and HER2 has been demonstrated to play a key role in breast, colon, and pancreatic tumor angiogenesis. Other growth factors that induce VEGF-A release, including IGF1 and hepatocyte growth factor (HGF), are frequently found to be overexpressed, together with their respective receptors, in tumors. VEGF-A upregulation occurs via both HIF-1α dependent and independent mechanisms and blockade of receptor activation is associated with decreased VEGF-A levels and compromised angiogenesis in experimental tumors.[256]

7.01.7.3 Fibroblast Growth Factor and Angiogenesis

Similarly to VEGF-A, FGF signaling has been implicated in multiple aspects of tumor angiogenesis.[235] In addition to promoting endothelial cell proliferation, survival and chemotaxis, FGF has been shown to upregulate protease activity and direct expression of uPAR to the leading edge of the endothelial cell migration front.[256] FGF2 stimulation induces the reorganization of endothelial cells into capillary-like structures when plated in type 1 collagen; a process that involves upregulation of cell–matrix (integrins $\alpha_v\beta_3$) and cell–cell (cadherins) adhesion molecules.[257,258] Whereas mice lacking FGFR1 die during gastrulation, prior to the appearance of blood vessels, expression of a dominant negative FGFR1 impairs vascular development and stabilization.[259,260]

In contrast to VEGF-A, endothelial cells coexpress both FGF ligands and receptors, permitting autocrine and paracrine pathway activation. Autocrine activation of the FGFR has been implicated in pathogenic endothelial lesions such as Kaposi's sarcoma and hemangioma; however, the role of FGFs in the normal angiogenenic process remains uncertain, as mice lacking the predominantly expressed forms do not display vascular abnormalities.[261] One explanation for this apparent lack of an overt phenotype is the redundancy and cross-talk that occurs between the FGF and VEGF-A signaling pathways. FGFs stimulate VEGF-A and VEGFR expression, and inhibitors of VEGFR-1 and VEGFR-2 antagonize FGF-induced angiogenesis.[262] Conversely, there is evidence that some of VEGFs effects on angiogenesis are dependent on expression of endogenous FGF2.[263–265]

Multiple studies have documented expression of FGF ligands, particularly FGF2, within tumors. Abrogation of FGF signaling has been shown to inhibit neovascularization and growth of experimental tumors, while a synergistic effect on tumor vessel density is observed upon administration of VEGF-A and FGF.[266,267] However, whereas a consistent correlation between tumor microvessel density and VEGF-A expression has been documented, there appears to be marked heterogeneity when FGF levels are examined. One notable exception is melanoma, in which FGF2 levels and microvessel density are clearly correlated.[268]

7.01.7.4 Platelet-Derived Growth Factor

Whereas VEGF-A is predominantly required during the initial steps of angiogenesis (formation of the endothelial plexus) subsequent steps, involving recruitment and vessel coverage by pericytes and smooth muscle cells, appear to be dependent on other soluble factors, e.g., PDGF. PDGF-B acts as a pericyte and smooth muscle cell chemoattractant whereas TGF-β has been implicated in vessel stabilization.[269,270] Microscopic examination indicates that tumor capillaries possess mural cells; however, unlike normal vessels in which pericytes are well organized and closely associated with the endothelial compartment, tumor pericytes were present at a lower density and appeared more loosely attached.[271] Positive immunohistochemical staining for PDGFR-B, originating from associated pericytes, was observed in tumors.[194] Experimental manipulation of PDGF levels has demonstrated that release of PDGF from both endothelial and tumor cells can impinge on pericyte recruitment to tumor vessels.[272] Activation of the intrinsic PDGF-R kinase domain, as well as the phosphatidylinositol-3-kinase (PI3K) signaling pathway, has been shown to be important in this process. Increased pericyte abundance, observed upon overexpression of PDGF, leads to increased tumor growth but does not increase microvessel density in a murine melanoma model, suggesting that it modulates the efficiency rather than number of vessels.[273]

Investigation of the clinically proven activity of VEGF-A antagonists suggest that they target immature vessels and that pericyte coverage renders vessels resistant to their actions. The hypothesis that a combination of PDGF and VEGF-A inhibitors would provide therapeutic benefit has been experimentally validated. Administration of SU6668, a joint PDGFR/VEGFR inhibitor, resulted in blocking a tumor growth, while SU5416, a more specific VEGFR inhibitor, was not effective.[274]

Glivec has been used to demonstrate the functionality of PDGFR expression on endothelial cells. Normally occurring at low levels, PDGFR was observed to be specifically unregulated in hyperproliferative endothelium and endothelial cells found within bone metastasis.[275] Enhanced antitumor activity, observed upon combination of Glivec with chemotherapeutic agents, was associated with increased endothelial cell apoptosis, indicating that PDGF signaling may transmit prosurvival signals in these tumors.[276,277]

In addition to endothelial cells, PDGFR expression has been documented in other tumor-associated stromal cells. Manipulation of PDGF levels enhances stroma recruitment, decreases tumor necrosis, and reduces the latency time to formation of tumors in vivo. One functional consequence of stromal recruitment is the high interstitial fluid pressure that characterizes many tumors. Reduction of this pressure would be therapeutically beneficial as it would increase the penetration and effectiveness of pharmacological agents. A role for PDGF in control of interstitial fluid pressure in normal tissue has been documented, while decreased tumor interstitial fluid pressure, observed upon administration of Glivec, is associated with enhanced uptake and antitumor activity of coadministered chemotherapeutic agents.

7.01.8 Growth Factor Signal Transduction in Cancer

7.01.8.1 Deregulated Receptor Tyrosine Kinases

Receptor protein tyrosine kinases (RPTKs) are a subclass of transmembrane-spanning proteins with ligand-stimulatable kinase activity. These enzymes are important regulators of intracellular signal transduction pathways involved in a number of cell functions, such as cell differentiation and proliferation. The activation of these kinases by their cognate ligands involves a complex and multistep process. Binding of the growth factor to the extracellular domain of the protein results in receptor dimerization and subsequent autotransphosphorylation of each intracellular domain at specific tyrosine residues.[278] The phosphorylated tyrosine residues act as docking sites for protein adaptors that initiate a cascade of intracellular biological events.[279] The activity of RPTK is tightly controlled under normal physiological conditions,[280] but many different tumor types have been shown to have dysfunctional RPTKs as a consequence of mutations or genetic alterations. Irrespective of the cause, this leads to enhanced or constitutive kinase activity and, in turn, to aberrant and inappropriate post receptor cellular signaling within the tumor cell. The RPTKs involved in oncogenic transformation and the intracellular components of the pathways initiated by these receptors have become attractive targets for cancer drug discovery programs, and many efforts have focused in the last few years on preventing receptor activation or blocking deregulated signal transduction pathways (*see* 7.08 Kinase Inhibitors for Cancer). Two representative examples of deregulated receptor tyrosine kinases in cancer cells are given below.

7.01.8.1.1 Epidermal growth factor receptors

The EGF family of type I receptor tyrosine kinases comprises four structurally related proteins: EGFR (erbB-1, HER1); erbB-2 (HER2, Neu), erbB-3 (HER3), and erbB-4 (HER4). The biological activities of these transmembrane proteins are intimately interrelated, and their activation by a broad spectrum of growth factors (e.g., EGF, TGF-α amphiregulin, betacellulin, heparin-binding EGF, and, epiregulin) triggers the initiation of signal transduction pathways that in the main result in cellular proliferation, apoptosis, differentiation, angiogenesis, motility, and invasion. This family of receptors was first implicated in cancer when the avian erythroblastosis tumor virus was found to encode an aberrant form of the human EGFR. Additional studies have supported an important role for this family of RTKs in the development and progression of numerous human tumors. Overexpression of EGFR or coexpression of both receptor and ligand(s) is a frequent event in a large variety of epithelial cancers and is associated with advanced disease and poor prognosis. In a significant proportion of these tumors, gene amplification is accompanied by rearrangements that result in constitutively active receptors. Thus, overexpression of erbB-2 occurs in around 30% of breast cancers[281] and coexpression of elevated levels of EGFR and erbB-2 has been observed in ovarian cancer patients. Interestingly, although erbB-2 is an orphan receptor, it participates in EGF receptor signaling by heterodimerization with other members of this receptor family. Dimerization causes activation of the kinase domain, leading to initiation of signal transduction pathways linked to cell survival and division. Constitutive activation has also been observed in mutated forms of the EGFR. The most common mutation (EGFRvIII), which is found in gliomas, NSCLC, and breast cancer, lacks domains I and II of the extracellular domain, and, despite being unable to bind to the ligands, displays constitutive kinase activity.

7.01.8.1.2 Oncogenic activation of c-Kit

The proto-oncogenic c-Kit receptor responds specifically to its physiological ligand, the stem cell factor (SCF), but its activation in tumor cells can proceed by alternative mechanisms. Thus, more than 30 gain-of-function mutations in the extracellular dimerization domain, the intracellular juxtamembrane domain, and the kinase domain have been identified in the c-Kit receptor.[282,283] The transforming mechanism of these mutations (single or multiple amino acid changes) involves dimer formation resulting in constitutive SCF-independent kinase activation.

Upregulation of the kinase activity of c-Kit by somatic mutations has been documented in a number of human malignancies, particularly in gastrointestinal stromal tumors (GISTs) (around 85–90% of all diagnosed cases). The constitutive c-Kit kinase activity observed in GISTs was hypothesized to be crucial to the pathogenesis of this disease,[284,285] and this premise has been recently confirmed with the objective responses observed with c-Kit kinase inhibitors (e.g., imatinib) in GISTs clinical trails.[286] Thus, clinical responses in GIST patients following treatment with imatinib appear to be associated with the presence of activating mutations of c-Kit, as patients expressing wild-type c-Kit, had a significantly lower response.

In addition to GISTs, activation of the kinase activity of c-Kit by somatic mutations has been documented in seminoma, acute myelogenous leukaemia, and mastocytosis, and paracrine or autocrine activation of this receptor has also been postulated for SCLC and ovarian cancer.

7.01.8.2 Deregulated Cytosolic Signaling Pathways

For additional information on this topic, see Chapter 7.08.

7.01.8.2.1 The Ras–Raf–MEK–MAPK pathway

Different growth factors and cytokines transduce their growth-promoting signals through the activation of the small G protein Ras (**Figure 4**). This leads to activation of members of the Raf family of serine/threonine kinases (c-Raf, A-Raf, and B-Raf) and its downstream effector, the mitogen-activated protein kinase (MAPK) kinase (MEK). This protein then phosphorylates and activates MAPK, the so-called extracellular signal-regulated kinase (ERK). The Ras–Raf–MEK–MAPK pathway controls the growth and survival of a broad spectrum of cancers. Thus, it has been shown by the expression of dominant negative or activated forms of MEK that the expression of cyclin D1, which leads to transition from G1 to S phase, is controlled by MEK signaling.[287] Activating mutations in Ras and Raf are present in a large percentage of solid tumors. For example, the Ras oncogene is found mutated in its oncogenic form in 15% of human cancers, with subsequent activation of the MAPK pathway.[288] The MEK and ERK kinases are rarely found constitutively activated in human tumors; however, B-Raf is often mutated and activated in certain human tumors such as melanoma (66%), ovarian (30%), or papillary thyroid (35–70%). All the mutations are within the kinase domain, either in the P loop or in the activation segment adjacent to the Asp-Phe-Gly (DFG) motif. Interestingly, roughly 90% of the activating mutations involve the replacement of the hydrophobic valine residue in the activation loop by glutamic acid, which due to its negative charge seems to mimic the phosphorylation of activation loop.[289] Ras and B-Raf activating mutations are generally mutually exclusive in human, showing that activation of the MAPK pathway at the Ras or the B-Raf level is probably equivalent.[290]

7.01.8.2.2 The phosphatidylinositol-3-kinase–PDK1–protein kinase B pathway

The phosphatidylinositol-3-kinase (PI3K)–protein kinase B (PKB) pathway (**Figure 5**) regulates fundamental cellular functions (e.g., transcription, translation, proliferation, growth or survival), and it is often deregulated in a wide range of tumor types.[291,292] Activation of the PI3K–PKB survival pathway can begin with the stimulation of growth factor receptor tyrosine kinases, followed by the recruitment of PI3K to the plasma membrane. Once localized to the plasma membrane, PI3K phosphorylates phosphatidylinositol-4,5-bisphosphate (PtdIns(4,5)P_2) at the 3′-OH position of the

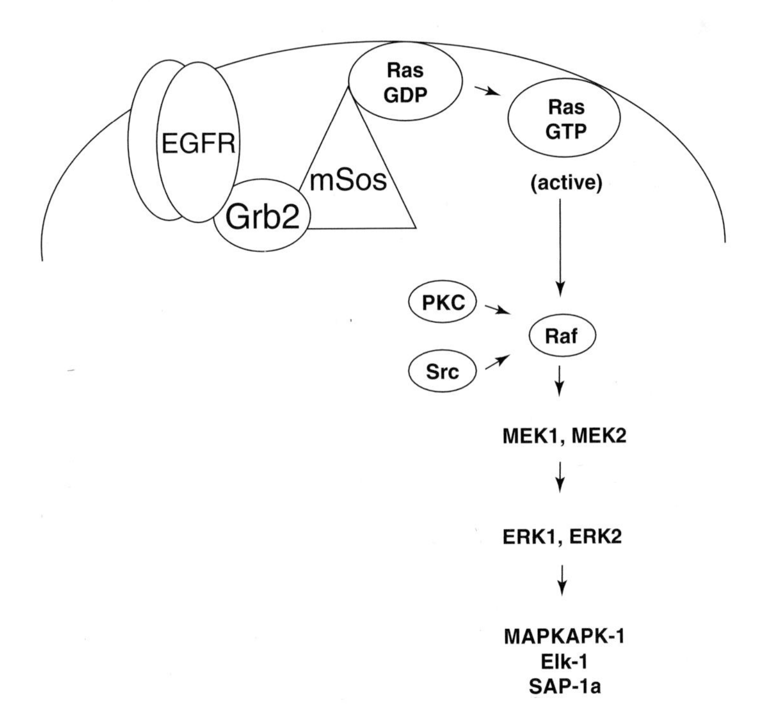

Figure 4 Activation of the Ras–Raf–MAPK pathway. Upon cell stimulation with EGF, the EGFR recruits Grb2 that in turn will engage mSoS, the Ras-specific GEF protein (left panel). The GTP-bound Ras protein will then recruit effector molecules such as Raf kinase, subsequent activation of the MAPK cascade down to transcription factors Elk-1 and SAP-1a (right panel).

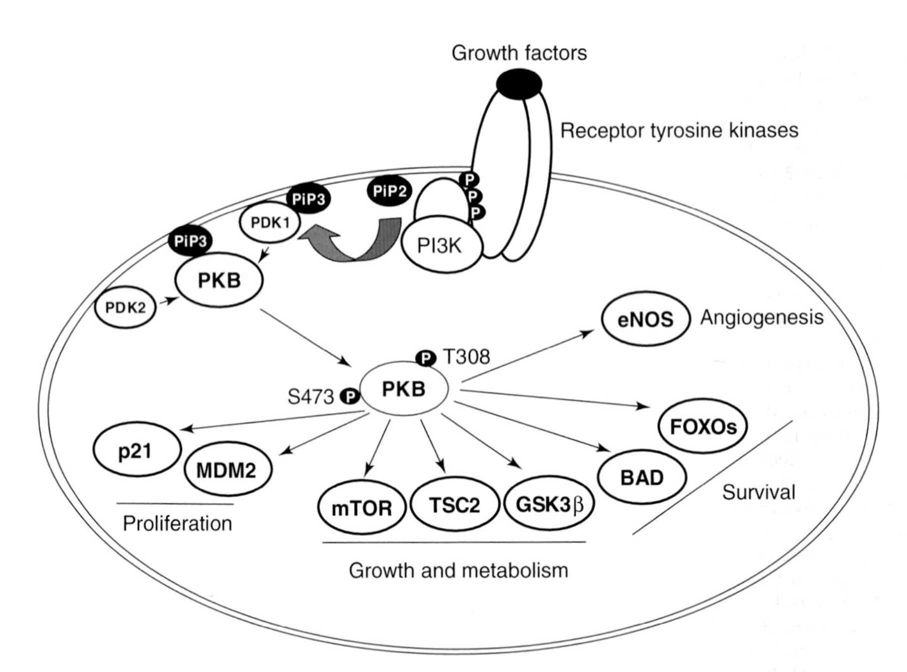

Figure 5 The PI3K pathway. This simplified scheme represent the molecular events leading to PKB phosphorylation and activation, and the subsequent activation of its downstream effectors.

inositol ring to generate the second messenger phosphatidylinositol-3,4,5-trisphosphate (PtdIns(3,4,5)P_3). PKB and its upstream activating kinase PDK1 are then recruited at the plasma membrane, an event accomplished by direct interaction between its PH domains and the generated PtdIns(3,4,5)P_3 molecules. Colocalization of PDK1 at the plasma membrane allows the phosphorylation of Thr308 in the activation loop by this enzyme, which is necessary and sufficient for PKB activation.[293,294] However, maximal enzymatic activation of PKB requires phosphorylation of Ser473, a residue located in its C-terminal regulatory domain.[295] The molecular identity(ies) of the kinase(s) responsible for the phosphorylation of the Ser473 residue, often refered as PDK2, is still controversial and has been hypothesized, among other proteins, to be DNA-PK [296,297] or the mTOR/rictor complex.[298]

The activated PKB phosphorylates and modulates the function of important regulators of cell proliferation (e.g., $p21^{Cip1}$, $p27^{Kip1}$), survival (e.g., Bad, caspase 9, CREB), and metabolism (e.g., GSK3β). As a direct consequence of this broad spectrum of activities,[299] the biological consequences of uncontrolled PI3K–PKB activation in cancer cells are critical in inhibiting apoptosis and in promoting tumor cell growth, proliferation and angiogenesis.

Activation of the PI3K–PKB pathway in tumor cells is accomplished by various mechanisms that reflect the importance of this signaling cascade in the biology of human cancers. In addition to persistent recruitment and activation of PI3K by deregulated RPTKs (e.g., erbB-2), it has been reported that one of the catalytic subunits of PI3K (the catalytic p110αsubunit of class IA PI3Ks) is amplified and overexpressed in tumors. Moreover, activating mutations in PIK3CA[300–304] have been identified in different types of cancers: 74 of 199 colorectal cancers (32%), 4 of 15 glioblastomas (27%), 3 of 12 gastric cancers (25%), 1 of 12 breast cancers (8%), and 1 of 24 lung cancers (4%). Other biological alterations can also affect the correct regulation of the PtdIns(3,4,5)P_3 signal transducer.[305] Loss of the dual-specificity phosphatase and tensin (PTEN) homolog deleted on chromosome 10, which modulates the activation state of the pathway by removing the 3′-phosphate group from PtdIns(3,4,5)P_3, has been found in a large fraction of advanced human cancers,[306–310] including glioblastomas, endometrial, breast, thyroid, and prostate cancers, and melanoma. Germline mutation of PTEN also results in autosomal dominant cancer syndromes such as Cowden disease, a heritable cancer risk syndrome, and several related conditions (e.g., Bannayan–Riley–Ruvalcaba syndrome or Proteus syndrome). In addition to the preceding alterations, PKB isoforms have been found to be amplified or overexpressed to different degrees in various human tumors, including ovarian, breast, prostate, and pancreatic cancers.[311–313]

References

1. DeVita, V. T., Jr.; Hellman, S.; Rosenberg, S. A. *Cancer: Principles and Practice of Oncology*; Lippincott-Raven: Philadelphia, PA, 1997.
2. American Cancer Society. *Cancer Facts and Figures*. www.cancer.org (accessed Aug 2006).

3. Hanahan, D.; Weinberg, R. A. *Cell* **2000**, *100*, 57–70.
4. Vogelstein, B.; Kinzler, K. W. *Nat. Med.* **2004**, *10*, 789–799.
5. Fearon, E. R.; Vogelstein, B. *Cell* **1990**, *61*, 759–767.
6. Vogelstein, B.; Fearon, E. R.; Hamilton, S. R.; Kern, S. E.; Preisinger, A. C.; Leppert, M. et al. *N. Engl. J. Med.* **1988**, *319*, 525–535.
7. Goldgar, D. E. *Biochimie* **2002**, *84*, 19–25.
8. Ponder, B. A. *Nature* **2001**, *411*, 336–341.
9. Hemminki, K.; Rawal, R.; Chen, B.; Bermejo, J. L. *Int. J. Cancer* **2004**, *111*, 944–950.
10. Bocchetta, M.; Carbone, M. *Oncogene* **2004**, *23*, 6484–6491.
11. Potter, J. D. *Trends Genet.* **2003**, *19*, 690–695.
12. Wild, C. P. *Handbook of Experimental Pharmacology*, 2003; Vol. 156, pp 307–321.
13. Hussain, S. P.; Harris, C. C. *Toxicol. Lett.* **1998**, *102*, 219–225.
14. Franco, E. L. Epidemiology in the Study of Cancer. In *Encyclopedia of Cancer*; Bertino, J., Ed.; Academic Press: New York, 1997; Vol. 1, pp 621–641.
15. Luch, A. *Nat. Rev. Cancer* **2005**, *5*, 113–125.
16. Peto, R.; Lopez, A. D.; Boreham, J.; Thun, M.; Heath, C. *Lancet* **1992**, *339*, 1268–1278.
17. Sasco, A. J.; Secretan, M. B.; Straif, K. *Lung Cancer* **2004**, *45*, S3–S9.
18. Go, V. L. W.; Wong, D. A.; Wang, Y.; Butrum, R. R.; Norman, H. A.; Wilkerson, L. A. *J. Nutr.* **2004**, *134*, 3513S–3516S.
19. Boffetta, P. *Oncogene* **2004**, *23*, 6392–6403.
20. Prior, R. L.; Joseph, J. Berries and Fruits in Cancer Chemoprevention. In *Phytopharmaceuticals in Cancer Chemoprevention*; Bagchi, D., Preuss, H. G., Eds.; CRC Press: Boca Raton, FL, 2005, pp 465–479.
21. Wang, J.-S.; Tang, L. *J. Toxicol. Toxin Rev.* **2004**, *23*, 249–271.
22. Stiller, C. A. *Oncogene* **2004**, *23*, 6429–6444.
23. Couch, F. J. *Cancer Biol. Ther.* **2004**, *3*, 509–514.
24. Carbone, M.; Pass, H. I. *Semin. Cancer Biol.* **2004**, *14*, 399–405.
25. Herrera, L. A.; Benitez-Bribiesca, L.; Mohar, A.; Ostrosky-Wegman, P. *Environ. Mol. Mutagen.* **2005**, *45*, 284–303.
26. Nakachi, K.; Hayashi, T.; Imai, K.; Kusunoki, Y. *Cancer Sci.* **2004**, *95*, 921–929.
27. Yoshida, M.; Seiki, M. *Haematol. Blood Transfus.* **1985**, *29*, 331–334.
28. Baumforth, K. R. N.; Young, L. S.; Flavell, K. J.; Constandinou, C.; Murray, P. G. *Mol. Pathol.* **1999**, *52*, 307–322.
29. Blumberg, B. S. *Proc. Natl. Acad. Sci. USA* **1997**, *94*, 7121–7125.
30. McCance, D. *J. Biochim. Biophys. Acta* **1986**, *823*, 195–205.
31. Gallo, R. C.; Shaw, G.; Hahn, B.; Wong-Staal, F.; Popovic, M.; Schupbach, J.; Sarngadharan, M. G.; Arya, S.; Salahuddin, S. Z.; Reitz, M. S. *Dev. Oncol.* **1985**, *28*, 191–205.
32. Miehlke, S. *Falk Sympos* **2003**, *132*, 100–108.
33. Sur, M.; Cooper, K. *Infect. Causes Cancer* **2000**, *23*, 435–448.
34. Brewster, A. B. M. Clinical and Molecular Epidemiology of Breast Cancer. In *Molecular Oncology of Breast Cancer*; Ross, J. S., Hortobagyi, G. N., Eds.; Jones and Bartlett: Sudbury, MA, 2005, pp 34–47.
35. Coe, K. The Epidemiology of Ovarian Cancer: The Role of Reproductive Factors and Environmental Chemical Exposure. In *Ovarian Toxicology*; Hoyer, P. B., Ed.; CRC Press: Boca Raton, FL, 2004, pp 171–201.
36. Poeschl, G.; Seitz, H. K. *Alcohol Alcohol.* **2004**, *39*, 155–165.
37. Hrelia, P.; Maffei, F.; Angelini, S.; Forti, G. C. *Toxicol. Lett.* **2004**, *149*, 261–267.
38. Saladi, R. N.; Persaud, A. N. *Drugs Today* **2005**, *41*, 37–53.
39. Duport, P. *Review of Cancer Risk in Animals Exposed to Low Doses of Ionizing Radiation*. Proceedings of the 8th International Conference on Nuclear Engineering, Baltimore, MD, Apr 2–6, 2000.
40. Reiners, C. *Nuklearmedizin* **1994**, *33*, 229–234.
41. Iki, S.; Urabe, A. *Cancer Chemother* **2000**, *27*, 1635–1640.
42. Bellet, D. *J. Natl. Cancer Inst. Monogr.* **1992**, *12*, 115–121.
43. Stoppa-Lyonnet, D.; Aurias, A. *Bull. Cancer* **1992**, *79*, 645–650.
44. Asten, P.; Barrett, J.; Symmons, D. *J. Rheumatol.* **1999**, *26*, 1705–1714.
45. Preston, R. *J. Environ. Mol. Mutagen.* **2005**, *45*, 214–221.
46. Goncalves, A.; Borg, J.-P.; Pouyssegur, J. *Drug Disc. Today* **2004**, *1*, 305–311.
47. Verma, M. *Techno. Cancer Res. Treat.* **2004**, *3*, 505–514.
48. Radford, I. R. *Int. J. Radiat. Biol.* **2004**, *80*, 543–557.
49. Mitelman, F.; Johansson, B.; Mertens, F. *Nat. Genet.* **2004**, *36*, 331–334.
50. Hughes, T.; Branford, S. *Semin. Hematol.* **2003**, *40*, 62–68.
51. Dutrillaux, B. *Adv. Cancer Res.* **1995**, *67*, 59–82.
52. Huff, V.; Miwa, H.; Haber, D. A.; Call, K. M.; Housman, D.; Strong, L. C. et al. *Am. J. Hum. Genet* **1991**, *48*, 997–1002.
53. Daya-Grosjean, L.; Sarasin, A. *Mutat. Res.* **2005**, *571*, 43–56.
54. Kaneko, H.; Kondo, N. *Expert Rev. Mol. Diagn.* **2004**, *4*, 393–401.
55. Ayesh, R.; Idle, J. R.; Ritchie, J. C.; Crothers, M. J.; Hetzel, M. R. *Nature* **1984**, *312*, 169–173.
56. Calderon-Margalit, R.; Paltiel, O. *Int. J. Cancer* **2004**, *112*, 357–364.
57. Thomas, F. J.; McLeod, H. L.; Watters, J. W. *Curr. Top. Med. Chem.* **2004**, *4*, 1399–1409.
58. Stoehlmacher, J.; Iqbal, S.; Lenz, H.-J. Polymorphisms in Genes of Drug Targets and Metabolism, and DNA Repair. In *Handbook of Anticancer Pharmacokinetics and Pharmacodynamics*, 1st ed.; Figg, W., McLeod, H., Eds.; Human Press: Totowa, NJ, 2003, pp 231–243.
59. Preston, R. J. *Environ. Mol. Mutagen.* **2005**, *45*, 214–221.
60. Davies, A. M.; Mack, P. C.; Lara, P. N., Jr.; Lau, D. H.; Danenberg, K.; Gumerlock, P. H.; Gandara, D. R. *J. Natl. Comp. Cancer Network* **2004**, *2*, 125–131.
61. Stoss, O.; Henkel, T. *Drug Disc. Today* **2004**, *3*, 228–237.
62. Duffy, M. J. *Crit. Revi. Clin. Lab. Sci.* **2001**, *38*, 225–262.
63. Lips, C. J. M.; Hoppener, J. W. M.; van der Luijt, R. B. Medullary Thyroid Carcinoma. *In Encyclopedia of Endocrine Diseases*; Martini, L., Ed.; Elsevier, 2004; Vol. 3, pp 218–227.

64. Rustin, G. J. S.; Bast, R. C.; Kelloff, G. J.; Barrett, J. C.; Carter, S. K.; Nisen, P. D.; Sigman, C. C.; Parkinson, D. R.; Ruddon, R. W. *Clin. Cancer Res.* **2004**, *10*, 3919–3926.
65. Catalona, W. J. *Prostate* **1996**, 7, 64–69.
66. Han, K.; Pantuck, A. J.; Belldegrun, A. S.; Rao, J.-Y. *Front. Biosci.* **2002**, 7, E19–E26.
67. Leung, H. Y.; Lemoine, N. R. Pancreatic Cancer. In *Cancer: A Molecular Approach*; Lemoine, N. R., Neoptolomos, J., Cooke, T., Eds.; Blackwell Scientific: Oxford, UK, 1994, pp 105–115.
68. Wald, N.; Cuckle, H.; Frost, C. *Br. Med. J.* **1991**, *302*, 845.
69. Sidransky, D.; Tokino, T.; Hamilton, S. R.; Kinzler, K. W.; Levin, B.; Frost, P. et al. *Science* **1992**, *256*, 102.
70. Nollau, P.; Moser, C.; Weinland, G.; Wagener, C. *Int. J. Cancer* **1996**, *66*, 332–336.
71. Corn, B. W. *Drug News Perspect.* **2004**, *17*, 469–475.
72. Rodriguez, E.; Baas, P.; Friedberg, J. S. *Thorac. Surg. Clin.* **2004**, *14*, 557–566.
73. Bickell, N. A.; Mendez, J.; Gutt, A. A. *Surg. Oncol. Clin. N. Am.* **2005**, *14*, 103–117.
74. Ocana, A.; Rodriguez, C. A.; Cruz, J. J. *Clin. Trans. Oncol.* **2005**, 7, 99–100.
75. Chen, Y.; Okunieff, P. *Hematol. Oncol. Clin. N. Am.* **2004**, *18*, 55–80.
76. Etzioni, S.; Rosenfeld, K. *N. Engl. J. Med.* **2005**, *352*, 1820–1822.
77. Ellis, L. M. *Nat. Rev.* **2005**, *4*, S8–S9.
78. Assikis, V. J.; Simon, J. W. *Semin. Oncol.* **2004**, *31*, 26–32.
79. Abou-Jawde, R.; Choueiri, T.; Alemany, C.; Mekhail, T. *Clin. Ther.* **2003**, *25*, 2121–2133.
80. Garcia-Echeverria, C.; Fabbro, D. *Mini-Rev. Medi. Chem.* **2005**, *4*, 273–283.
81. Bicknell, R. *Br. J. Cancer* **2005**, *92*, S2–S5.
82. Goldberg, R. *Nat. Rev.* **2005**, *4*, S10–S11.
83. Leaf, C. *Fortune* **2005**, 76.
84. Friend, S. H.; Bernards, R.; Rogelj, S.; Weinberg, R. A.; Rapaport, J. M.; Albert, D. M.; Dryja, T. P. *Nature* **1986**, *323*, 643–646.
85. Hunter, T. *Cell* **1997**, *88*, 333–346.
86. Scheffner, M.; Munger, K.; Byrne, J. C.; Howley, P. M. *Proc. Natl. Acad. Sci. USA* **1991**, *88*, 5523–5527.
87. Sherr, C. J. *Science* **1996**, *274*, 1672–1677.
88. Wolfel, T.; Hauer, M.; Schneider, J.; Serrano, M.; Wolfel, C.; Klehmann-Hieb, E.; De Plaen, E.; Hankeln, T.; Meyer zum Buschenfelde, K. H.; Beach, D. *Science* **1995**, *269*, 1281–1284.
89. Ruas, M.; Peters, G. *Biochim. Biophys. Acta* **1998**, *1378*, F115–F177.
90. Merlo, A.; Herman, J. G.; Mao, L.; Lee, D. J.; Gabrielson, E.; Burger, P. C.; Baylin, S. B.; Sidransky, D. *Nat. Med.* **1995**, *1*, 686–692.
91. Bates, S.; Phillips, A. C.; Clark, P. A.; Stott, F.; Peters, G.; Ludwig, R. L.; Vousden, K. H. *Nature* **1998**, *395*, 124–125.
92. Lowe, S. W.; Ruley, H. E. *Genes Dev.* **1993**, 7, 535–545.
93. Wu, X.; Levine, A. J. *Proc. Natl. Acad. Sci. USA* **1994**, *91*, 3602–3606.
94. Kowalik, T. F.; DeGregori, J.; Leone, G.; Jakoi, L.; Nevins, J. R. *Cell Growth Differ.* **1998**, *9*, 113–118.
95. DeGregori, J.; Leone, G.; Miron, A.; Jakoi, L.; Nevins, J. R. *Proc. Natl. Acad. Sci. USA* **1997**, *94*, 7245–7250.
96. Wang, J. Y. *Oncogene* **1991**, *49*, 5643–5650.
97. Welch, P. J.; Wang, J. Y. A. *Cell* **1993**, *75*, 779–790.
98. Shim, J.; Park, H. S.; Kim, M. J.; Park, J.; Park, E.; Cho, S. G.; Eom, S. J.; Lee, H. W.; Joe, C. O.; Choi, E. J. *J. Biol. Chem.* **2000**, *275*, 14107–14111.
99. Doostzadeh-Cizeron, J.; Evans, R.; Yin, S.; Goodrich, D. W. *Mol. Biol. Cell* **1999**, *10*, 3251–3261.
100. Doostzadeh-Cizeron, J.; Yin, S.; Goodrich, D. W. *J. Biol. Chem.* **2000**, *275*, 25336–25341.
101. Pines, J. *Nat. Cell Biol.* **1999**, *1*, E73–E79.
102. Gallant, P.; Nigg, E. A. *J. Cell Biol.* **1992**, *117*, 213–224.
103. Heald, R.; McLoughlin, M.; McKeon, F. *Cell* **1993**, *74*, 463–474.
104. Krek, W.; Nigg, E. A. *EMBO J.* **1991**, *10*, 305–316.
105. Draetta, G.; Eckstein, J. *Biochim. Biophys. Acta* **1997**, *1332*, M53–M63.
106. Hagting, A.; Karlsson, C.; Clute, P.; Jackman, M.; Pines, J. *EMBO J.* **1998**, *17*, 4127–4138.
107. Gabrielli, B. G.; Clark, J. M.; McCormack, A. K.; Ellem, K. A. *J. Biol. Chem.* **1997**, *272*, 28607–28614.
108. Nowell, P. C. *Science* **1994**, *4260*, 23–28.
109. Heim, S.; Mandahl, N.; Mitelman, F. *Cancer Res.* **1988**, *48*, 5911–5916.
110. Sancar, A.; Lindsey-Boltz, L. A.; Unsal-Kacmaz, K.; Linn, S. *Annu. Rev. Biochem.* **2004**, *73*, 39–85.
111. Kastan, M. B.; Bartek, J. *Nature* **2004**, *432*, 316–323.
112. Bunz, F.; Dutriaux, A.; Lengauer, C.; Waldman, T.; Zhou, S.; Brown, J. P.; Sedivy, J. M.; Kinzler, K. W.; Vogelstein, B. *Science* **1998**, *282*, 1497–1501.
113. El-Deiry, W. S.; Tokino, T.; Velculescu, V. E.; Levy, D. B.; Parsons, R.; Trent, J. M.; Lin, D.; Mercer, W. E.; Kinzler, K. W.; Vogelstein, B. *Cell* **1993**, *75*, 817–825.
114. Harper, J. W.; Adami, G. R.; Wei, N.; Keyomarsi, K.; Elledge, S. J. *Cell* **1993**, *75*, 805–816.
115. Lowe, S. W.; Schmitt, E. M.; Smith, S. W.; Osborne, B. A.; Jacks, T. *Nature* **1993**, *362*, 847–849.
116. Taylor, W. R.; DePrimo, S. E.; Agarwal, A.; Agarwal, M. L.; Schonthal, A. H.; Katula, K. S.; Stark, G. R. *Mol. Biol. Cell* **1999**, *10*, 3607–3622.
117. Hollstein, M.; Hergenhahn, M.; Yang, Q.; Bartsch, H.; Wang, Z. Q.; Hainaut, P. *Mutat. Res./Fundament. Mol. Mech. Mutag.* **1999**, *431*, 199–209.
118. Lonardo, F.; Ueda, T.; Huvos, A. G.; Healey, J.; Ladanyi, M. *Cancer* **1997**, *79*, 1541–1547.
119. Donehower, L. A.; Harvey, M.; Slagle, B. L.; McArthur, M. J.; Montgomery, C. A.; Butel, J. S.; Bradley, A. *Nature* **1992**, *356*, 215–221.
120. van Gent, D. C.; Hoeijmakers, J. H.; Nat. Kanaar, R. *Rev. Genet.* **2001**, *2*, 196–206.
121. Thorstenson, Y. R.; Roxas, A.; Kroiss, R.; Jenkins, M. A.; Yu, K. M.; Bachrich, T.; Muhr, D.; Wayne, T. L.; Chu, G.; Davis, R. W. et al. *Cancer Res.* **2003**, *63*, 3325–3333.
122. Gumy-Pause, F.; Wacker, P.; Sappino, A. P. *Leukemia* **2004**, *18*, 238–242.
123. Liao, M. J.; Van Dyke, T. *Genes Dev.* **1999**, *13*, 1246–1250.
124. Sullivan, A.; Yuille, M.; Repellin, C.; Reddy, A.; Reelfs, O.; Bell, A.; Dunne, B.; Gusterson, B. A.; Osin, P.; Farrell, P. J. et al. *Oncogene* **2002**, *21*, 1316–1324.
125. da Costa, L. F. *Curr. Opin. Oncol.* **2001**, *13*, 58–62.

126. Lutterbach, B.; Hiebert, S. W. *Gene* **2000**, *245*, 223–235.
127. Bruserud, O.; Tjonnfjord, G.; Gjertsen, B. T.; Foss, B.; Ernst, P. *Stem Cells* **2000**, *18*, 343–351.
128. Kinzler, K. W.; Nilbert, M. C.; Su, L. K.; Vogelstein, B.; Bryan, T. M.; Levy, D. B.; Smith, K. J.; Preisinger, A. C.; Hedge, P.; McKechnie, D. *Science* **1991**, *253*, 661–665.
129. Roncucci, L.; Pedroni, M.; Vaccina, F.; Benatti, P.; Marzona, L.; De Pol, A. *Cell Prolif.* **2000**, *33*, 1–18.
130. Oshima, H.; Oshima, M.; Kobayashi, M.; Tsutsumi, M.; Taketo, M. M. *Cancer Res.* **1997**, *57*, 1644–1649.
131. Barker, N.; Morin, P. J.; Clevers, H. *Adv. Cancer Res.* **2000**, *77*, 1–24.
132. Korinek, V.; Barker, N.; Moerer, P.; van Donselaar, E.; Huls, G.; Peters, P. J.; Clevers, H. *Nat. Genet.* **1998**, *4*, 379–383.
133. Sakatani, T.; Kaneda, A.; Iacobuzio-Donahue, C. A.; Carter, M. G.; Boom Witzel, S.; Okano, H.; Ko, M. S. H.; Ohlsson, R.; Longo, D. L.; Feinberg, A. P. *Science* **2005**, *307*, 1976–1978.
134. Almand, B.; Resser, J. R.; Lindman, B.; Nadaf, S.; Clark, J. I.; Kwon, E. D.; Carbone, D. P.; Gabrilovich, D. I. *Clin. Cancer Res.* **2000**, *6*, 1755–1766.
135. Debatin, K. M.; Poncet, D.; Kroemer, G. *Oncogene* **2002**, *21*, 8786–8803.
136. Saito, M.; Korsmeyer, S. J.; Schlesinger, P. H. *Nat. Cell Biol.* **2000**, *2*, 553–555.
137. Philchenkov, A. *J. Cell. Mol. Med.* **2004**, *8*, 432–444.
138. Chandra, J.; Mansson, E.; Gogvadze, V.; Kaufmann, S. H.; Albertioni, F.; Orrenius, S. *Blood* **2002**, *99*, 655–663.
139. Yunis, J. J.; Frizzera, G.; Oken, M. M.; McKenna, J.; Theologides, A.; Arnesen, M. *N. Eng. J. Med.* **1987**, *316*, 79–84.
140. Cory, S.; Huang, D. C.; Adams, J. M. *Oncogene* **2003**, *22*, 8590–8607.
141. Yano, T.; Jaffe, E. S.; Longo, D. L.; Raffeld, M. *Blood* **1992**, *80*, 758–767.
142. Sander, C. A.; Yano, T.; Clark, H. M.; Harris, C.; Longo, D. L.; Jaffe, E. S.; Raffeld, M. *Blood* **1993**, *82*, 1994–2004.
143. Kroemer, G. *Nat. Med.* **1997**, *3*, 614–620.
144. Jolly, C.; Morimoto, R. I. *J. Natl. Cancer Inst.* **2000**, *92*, 1564–1572.
145. Vargas-Roig, L. M.; Gago, F. E.; Tello, O.; Aznar, J. C.; Ciocca, D. R. *Int. J. Cancer.* **1998**, *79*, 468–475.
146. Jiang, X.; Kim, H. E.; Shu, H.; Zhao, Y.; Zhang, H.; Kofron, J.; Donnelly, J.; Burns, D.; Ng, S. C.; Rosenberg, S. et al. *Science* **2003**, *299*, 223–226.
147. Ambrosini, G.; Adida, C.; Sirugo, G.; Altieri, D. C. *J. Biol. Chem.* **1998**, *273*, 11177–11182.
148. Magdalena, C.; Dominguez, F.; Loidi, L.; Puente, J. L. *Br. J. Cancer* **2000**, *82*, 584–590.
149. Micke, P.; Ostman, A. *Lung Cancer* **2004**, *45*, S163–S175.
150. Troyanovsky, S. *Eur. J. Cell Bio.* **2005**, *84*, 225–233.
151. Berx, G.; Van, R. F. *Breast Cancer Res.* **2001**, *3*, 289–293.
152. Cavallaro, U.; Christofori, G. *Ann. NY Acad. Sci.* **2004**, *1014*, 58–66.
153. Fujita, Y.; Krause, G.; Scheffner, M.; Zechner, D.; Leddy, H. E.; Behrens, J.; Sommer, T.; Birchmeier, W. Hakai, A. *Nat. Cell Biol.* **2002**, *4*, 222–231.
154. Pla, P.; Moore, R.; Morali, O. G.; Grille, S.; Martinozzi, S.; Delmas, V.; Larue, L. *J. Cell. Physiol.* **2001**, *189*, 121–132.
155. Hazan, R. B.; Qiao, R.; Keren, R.; Badano, I.; Suyama, K. *Ann. NY Acad. Sci.* **2004**, *1014*, 155–163.
156. Nieman, M. T.; Prudoff, R. S.; Johnson, K. R.; Wheelock, M. J. *J. Cell Bio.* **1999**, *147*, 631–644.
157. Toyooka, S.; Toyooka, K. O.; Harada, K.; Miyajima, K.; Makarla, P.; Sathyanarayana, U. G.; Yin, J.; Sato, F.; Shivapurkar, N.; Meltzer, S. J. et al. *Cancer Res.* **2002**, *62*, 3382–3386.
158. Zhong, Y.; Delgado, Y.; Gomez, J.; Lee, S. W.; Perez-Soler, R. *Clin. Cancer Res.* **2001**, *7*, 1683–1687.
159. Lee, S. W.; Reimer, C. L.; Campbell, D. B.; Cheresh, P.; Duda, R. B.; Kocher, O. *Carcinogenesis* **1919**, *6*, 1157–1159.
160. Hynes, R. O. *Cell* **2002**, *110*, 673–687.
161. Giancotti, F. G.; Ruoslahti, E. *Science* **1999**, *285*, 1028–1033.
162. Guo, W.; Giancotti, F. G. *Nat. Rev. Mol. Cell Biol.* **2004**, *5*, 816–826.
163. Gabarra-Niecko, V.; Schaller, M. D.; Dunty, J. M. *Cancer Metastasis Rev.* **2003**, *22*, 359–374.
164. Ilic, D.; Furuta, Y.; Kanazawa, S.; Takeda, N.; Sobue, K.; Nakatsuji, N.; Nomura, S.; Fujimoto, J.; Okada, M.; Yamamoto, T. *Nature* **1995**, *377*, 539–544.
165. Playford, M. P.; Schaller, M. D. *Oncogene* **2004**, *23*, 7928–7946.
166. Klemke, R. L.; Cai, S.; Giannini, A. L.; Gallagher, P. J.; de Lanerolle, P.; Cheresh, D. A. *J. Cell Biol.* **1997**, *137*, 481–492.
167. Pollard, T. D.; Borisy, G. G. *Cell* **2003**, *112*, 453–465.
168. Keely, P. J.; Westwick, J. K.; Whitehead, I. P.; Der, C. J.; Parise, L. V. *Nature* **1997**, *390*, 632–636.
169. Clark, E. A.; Golub, T. R.; Lander, E. S.; Hynes, R. O. *Nature* **2000**, *406*, 532–535.
170. Plantefaber, L. C.; Hynes, R. O. *Cell* **1989**, *56*, 281–290.
171. Hsia, D. A.; Mitra, S. K.; Hauck, C. R.; Streblow, D. N.; Nelson, J. A.; Ilic, D.; Huang, S.; Li, E.; Nemerow, G. R.; Leng, J. et al. *J. Cell Biol.* **2003**, *160*, 753–767.
172. Gladson, C. L.; Cheresh, D. A. *J. Clin. Invest.* **1991**, *88*, 1924–1932.
173. Albelda, S. M.; Mette, S. A.; Elder, D. E.; Stewart, R.; Damjanovich, L.; Herlyn, M.; Buck, C. A. *Cancer Res.* **1990**, *50*, 6757–6764.
174. Giancotti, F. G.; Tarone, G. *Annu. Revi. Cell Dev. Biol.* **1991**, *7*, 173–206.
175. Brooks, P. C.; Stromblad, S.; Sanders, L. C.; von Schalscha, T. L.; Aimes, R. T.; Stetler-Stevenson, W. G.; Quigley, J. P.; Cheresh, D. A. *Cell* **1996**, *85*, 683–693.
176. Chapman, H. A.; Wei, Y. *Thrombos. Haemostas.* **2001**, *86*, 124–129.
177. Shaw, L. M.; Rabinovitz, I.; Wang, H. H.; Toker, A.; Mercurio, A. M. *Cell* **1997**, *91*, 949–960.
178. Trusolino, L.; Bertotti, A.; Comoglio, P. M. *Cell* **2001**, *107*, 643–654.
179. Mariotti, A.; Kedeshian, P. A.; Dans, M.; Curatola, A. M.; Gagnoux-Palacios, L.; Giancotti, F. G. *J. Cell Bio.* **2001**, *155*, 447–458.
180. Eliceiri, B. P.; Cheresh, D. A. *Curr. Opin. Cell Biol.* **2001**, *13*, 563–568.
181. Vihinen, P.; Ala-aho, R.; Kahari, V. M. *Curr. Cancer Drug Targets* **2005**, *5*, 203–220.
182. Westermarck, J.; Kahari, V. M. *FASEB J.* **1999**, *13*, 781–792.
183. Brew, K.; Dinakarpandian, D.; Nagase, H. *Biochimi. Biophys. Acta* **2000**, *1477*, 267–283.
184. Powell, W. C.; Matrisian, L. M. *Curr. Topics Microbiol. Immunol.* **1996**, *213*, 1–21.
185. Boulay, A.; Masson, R.; Chenard, M. P.; El, F. M.; Cassard, L.; Bellocq, J. P.; Sautes-Fridman, C.; Basset, P.; Rio, M. C. *Cancer Res.* **2001**, *61*, 2189–2193.
186. Cairns, R. A.; Khokha, R.; Hill, R. P. *Curr. Mol. Med.* **2003**, *3*, 659–671.

187. Martin, D. C.; Fowlkes, J. L.; Babic, B.; Khokha, R. *J. Cell Biol.* **1999**, *146*, 881–892.
188. Yasui, W.; Oue, N.; Aung, P. P.; Matsumura, S.; Shutoh, M.; Nakayama, H. *Gastric Cancer* **2005**, *8*, 86–94.
189. Yamashita, K.; Tanaka, Y.; Mimori, K.; Inoue, H.; Mori, M. *Int. J. Cancer* **2004**, *110*, 201–207.
190. Handsley, M. M.; Edwards, D. R. *Int. J. Cancer* **2005**, *115*, 849–860.
191. Dano, K.; Behrendt, N.; Hoyer-Hansen, G.; Johnsen, M.; Lund, L. R.; Ploug, M.; Romer, J. *Thrombos. Haemostas.* **2005**, *93*, 676–681.
192. Almholt, K.; Lund, L. R.; Rygaard, J.; Nielsen, B. S.; Dano, K.; Romer, J.; Johnsen, M. *Int. J. Cancer* **2005**, *113*, 525–532.
193. Grondahl-Hansen, J.; Peters, H. A.; van Putten, W. L.; Look, M. P.; Pappot, H.; Ronne, E.; Dano, K.; Klijn, J. G.; Brunner, N.; Foekens, J. A. *Clin. Cancer Res.* **1995**, *1*, 1079–1087.
194. Furuhashi, M.; Sjoblom, T.; Abramsson, A.; Ellingsen, J.; Micke, P.; Li, H.; Bergsten-Folestad, E.; Eriksson, U.; Heuchel, R.; Betsholtz, C. et al. *Cancer Res.* **2004**, *64*, 2725–2733.
195. Almholt, K.; Lund, L. R.; Rygaard, J.; Nielsen, B. S.; Dano, K.; Romer, J.; Johnsen, M. *Int. J. Cancer* **2005**, *113*, 525–532.
196. Wong, C. W.; Lee, A.; Shientag, L.; Yu, J.; Dong, Y.; Kao, G.; Al-Mehdi, A. B.; Bernhard, E. J.; Muschel, R. J. *Cancer Res.* **2001**, *61*, 333–338.
197. Houston, A.; O'Connell, J. *Curr. Opin. Pharmacol.* **2004**, *4*, 321–326.
198. Bennett, M. W.; O'Connell, J.; O'Sullivan, G. C.; Roche, D.; Brady, C.; Collins, J. K.; Shanahan, F. *Dis. Esophagus* **1999**, *12*, 90–98.
199. Bennett, M. W.; O'Connell, J.; O'Sullivan, G. C.; Roche, D.; Brady, C.; Kelly, J.; Collins, J. K.; Shanahan, F. *Gut* **1999**, *44*, 156–162.
200. Elsasser-Beile, U.; Gierschner, D.; Welchner, T.; Wetterauer, U. *Anticancer Res.* **2003**, *23*, 433–437.
201. Bennett, M. W.; O'Connell, J.; O'Sullivan, G. C.; Brady, C.; Roche, D.; Collins, J. K.; Shanahan, F. *J. Immunol.* **1998**, *160*, 5669–5675.
202. Zheng, H. C.; Sun, J. M.; Wei, Z. L.; Yang, X. F.; Zhang, Y. C.; Xin, Y. *World J. Gastroenterol.* **2003**, *9*, 1415–1420.
203. Kase, H.; Aoki, Y.; Tanaka, K. *Gynecol. Oncol.* **2003**, *90*, 70–74.
204. Shiraki, K.; Yamanaka, T.; Inoue, H.; Kawakita, T.; Enokimura, N.; Okano, H.; Sugimoto, K.; Murata, K.; Nakano, T. *Int. J. Oncol.* **2005**, *26*, 1273–1281.
205. Koyama, S.; Koike, N.; Adachi, S. *J. Cancer Res. Clin. Oncol.* **2002**, *128*, 73–79.
206. Kokkonen, T. S.; Augustin, M. T.; Makinen, J. M.; Kokkonen, J.; Karttunen, T. J. *J. Histochem. Cytochem.* **2004**, *52*, 693–699.
207. Gochuico, B. R.; Miranda, K. M.; Hessel, E. M.; De Bie, J. J.; Van Oosterhout, A. J.; Cruikshank, W. W.; Fine, A. *Am. J. Physiol.* **1998**, *274*, L444–L449.
208. Shin, M. S.; Kim, H. S.; Lee, S. H.; Lee, J. W.; Song, Y. H.; Kim, Y. S.; Park, W. S.; Kim, S. Y.; Lee, S. N.; Park, J. Y. et al. *Oncogene* **2002**, *21*, 4129–4136.
209. Koomagi, R.; Volm, M. *Int. J. Cancer* **1999**, *84*, 239–243.
210. Cretney, E.; Takeda, K.; Yagita, H.; Glaccum, M.; Peschon, J. J.; Smyth, M. J. *J. Immunol.* **2002**, *168*, 1356–1361.
211. Seki, N.; Hayakawa, Y.; Brooks, A. D.; Wine, J.; Wiltrout, R. H.; Yagita, H.; Tanner, J. E.; Smyth, M. J.; Sayers, T. J. *Cancer Res.* **2003**, *63*, 207–213.
212. Pukac, L.; Kanakaraj, P.; Humphreys, R.; Alderson, R.; Bloom, M.; Sung, C.; Riccobene, T.; Johnson, R.; Fiscella, M.; Mahoney, A. et al. *Br. J. Cancer* **2005**, *92*, 1430–1441.
213. Lee, S. H.; Shin, M. S.; Kim, H. S.; Lee, H. K.; Park, W. S.; Kim, S. Y.; Lee, J. H.; Han, S. Y.; Park, J. Y.; Oh, R. R. et al. *Cancer Res.* **1999**, *59*, 5683–5686.
214. Toyooka, S.; Toyooka, K. O.; Miyajima, K.; Reddy, J. L.; Toyota, M.; Sathyanarayana, U. G.; Padar, A.; Tockman, M. S.; Lam, S.; Shivapurkar, N. et al. *Clin. Cancer Res.* **2003**, *9*, 3034–3041.
215. Iqbal, B.; Cohen, O.; Polak-Charcon, S.; Kopolovic, J.; Vadai, E.; Eisenbach, L.; Kimchi, A. *Nature* **1997**, *390*, 180–184.
216. Pantel, K.; Brakenhoff, R. H. *Nat. Rev. Cancer.* **2004**, *4*, 448–456.
217. Paget, S. *Cancer Metastasis Rev.* **1989**, *8*, 98–101.
218. Steeg, P. S.; Ouatas, T.; Halverson, D.; Palmieri, D.; Salerno, M. *Clin. Breast Cancer* **2003**, *4*, 51–62.
219. Steeg, P. S.; Bevilacqua, G.; Kopper, L.; Thorgeirsson, U. P.; Talmadge, J. E.; Liotta, L. A.; Sobel, M. E. *J. Natl. Cancer Inst.* **1988**, *80*, 200–204.
220. Freije, J. M.; Blay, P.; MacDonald, N. J.; Manrow, R. E.; Steeg, P. S. *J. Biol. Chem.* **1997**, *272*, 5525–5532.
221. Hartsough, M. T.; Morrison, D. K.; Salerno, M.; Palmieri, D.; Ouatas, T.; Mair, M.; Patrick, J.; Steeg, P. S. *J. Biol. Chem.* **2002**, *277*, 32389–32399.
222. Seminara, S. B.; Kaiser, U. B. *Endocrinology* **2005**, *146*, 1686–1688.
223. Bilban, M.; Ghaffari-Tabrizi, N.; Hintermann, E.; Bauer, S.; Molzer, S.; Zoratti, C.; Malli, R.; Sharabi, A.; Hiden, U.; Graier, W. et al. *J. Cell Sci.* **2004**, *117*, 1319–1328.
224. Ohtaki, T.; Shintani, Y.; Honda, S.; Matsumoto, H.; Hori, A.; Kanehashi, K.; Terao, Y.; Kumano, S.; Takatsu, Y.; Masuda, Y. et al. *Nature* **2001**, *411*, 613–617.
225. Dhar, D. K.; Naora, H.; Kubota, H.; Maruyama, R.; Yoshimura, H.; Tonomoto, Y.; Tachibana, M.; Ono, T.; Otani, H.; Nagasue, N. *Int. J. Cancer* **2004**, *111*, 868–872.
226. Ikeguchi, M.; Yamaguchi, K. I.; Kaibara, N. *Clin. Cancer Res.* **2004**, *10*, 1379–1383.
227. Sanchez-Carbayo, M.; Capodieci, P.; Cordon-Cardo, C. *Am. J. Pathol.* **2003**, *162*, 609–617.
228. Kulbe, H.; Levinson, N. R.; Balkwill, F.; Wilson, J. L. *Int. J. Dev. Biol.* **2004**, *48*, 489–496.
229. Walser, T. C.; Fulton, A. M. *Breast Disease* **1920**, 137-143.
230. Kucia, M.; Jankowski, K.; Reca, R.; Wysoczynski, M.; Bandura, L.; Allendorf, D. J.; Zhang, J.; Ratajczak, J.; Ratajczak, M. Z. *J. Mol. Histol.* **2004**, *35*, 233–245.
231. Muller, A.; Homey, B.; Soto, H.; Ge, N.; Catron, D.; Buchanan, M. E.; McClanahan, T.; Murphy, E.; Yuan, W.; Wagner, S. N. et al. *Nature 410*, 50-56.
232. Singh, S.; Singh, U. P.; Stiles, J. K.; Grizzle, W. E.; Lillard, J. W. *Clin. Cancer Res.* **2004**, *10*, 8743–8750.
233. Murakami, T.; Cardones, A. R.; Hwang, S. T. *J. Dermatol. Sci.* **2004**, *36*, 71–78.
234. Schimanski, C. C.; Schwald, S.; Simiantonaki, N.; Jayasinghe, C.; Gonner, U.; Wilsberg, V.; Junginger, T.; Berger, M. R.; Galle, P. R.; Moehler, M. *Clin. Cancer Res.* **2005**, *11*, 1743–1750.
235. Orimo, A.; Gupta, P. B.; Sgroi, D. C.; Renzana-Seisdedos, F.; Delaunay, T.; Naeem, R.; Carey, V. J.; Richardson, A. L.; Weinberg, R. A. *Cell* **2005**, *121*, 335–348.
236. Senger, D. R.; Galli, S. J.; Dvorak, A. M.; Perruzzi, C. A.; Harvey, V. S.; Dvorak, H. F. *Science* **1983**, *219*, 983–985.
237. Carmeliet, P.; Ferreira, V.; Breier, G.; Pollefeyt, S.; Kieckens, L.; Gertsenstein, M.; Fahrig, M.; Vandenhoeck, A.; Harpal, K.; Eberhardt, C. *Nature* **1996**, *380*, 435–439.

238. Ferrara, N.; Gerber, H. P.; LeCouter, J. *Nat. Med.* **2003**, *9*, 669–676.
239. Dvorak, H. F. *J. Clin. Oncol.* **2002**, *20*, 4368–4380.
240. Matthews, W.; Jordan, C. T.; Gavin, M.; Jenkins, N. A.; Copeland, N. G.; Lemischka, I. R. *Proc. Natl. Acad. Sci. USA* **1991**, *88*, 9026–9030.
241. Kaipainen, A.; Korhonen, J.; Mustonen, T.; van Hinsbergh, V. W. M.; Fang, G.; Dumont, D.; Breitman, M.; Alitalo, K. *Proc. Natl. Acad. Sci. USA* **1995**, *92*, 3566–3570.
242. Paavonen, K.; Puolakkainen, P.; Jussila, L.; Jahkola, T.; Alitalo, K. *Am. J. Pathol.* **2000**, *156*, 1499–1504.
243. Shalaby, F.; Rossant, J.; Yamaguchi, T. P.; Gertsenstein, M.; Wu, X. F.; Breitman, M. L.; Schuh, A. C. *Nature* **1995**, *376*, 62–66.
244. Fong, G. H.; Rossant, J.; Gertsenstein, M.; Breitman, M. L. *Nature* **1995**, *376*, 66–70.
245. LeCouter, J.; Moritz, D. R.; Li, B.; Phillips, G. L.; Liang, X. H.; Gerber, H. P.; Hillan, K. J.; Ferrara, N. *Science* **2003**, *299*, 890–893.
246. Hiratsuka, S.; Maru, Y.; Okada, A.; Seiki, M.; Noda, T.; Shibuya, M. *Cancer Res.* **2001**, *61*, 1207–1213.
247. De, P. F.; Granato, A. M.; Scarpi, E.; Monti, F.; Medri, L.; Bianchi, S.; Amadori, D.; Volpi, A. *Int. J. Cancer* **2002**, *98*, 228–233.
248. O'Byrne, K. J.; Koukourakis, M. I.; Giatromanolaki, A.; Cox, G.; Turley, H.; Steward, W. P.; Gatter, K.; Harris, A. L. *Br. J. Cancer* **2000**, *82*, 1427–1432.
249. Hurwitz, H.; Fehrenbacher, L.; Novotny, W.; Cartwright, T.; Hainsworth, J.; Heim, W.; Berlin, J.; Baron, A.; Griffing, S.; Holmgren, E. et al. *N. Engl. J. Med.* **2004**, *350*, 2335–2342.
250. Yoshiji, H.; Kuriyama, S.; Hicklin, D. J.; Huber, J.; Yoshii, J.; Ikenaka, Y.; Noguchi, R.; Nakatani, T.; Tsujinoue, H.; Fukui, H. *Hepatology* **2001**, *33*, 841–847.
251. Yuan, F.; Chen, Y.; Dellian, M.; Safabakhsh, N.; Ferrara, N.; Jain, R. K. *Proc. Natl. Acad. Sci. USA* **1996**, *93*, 14765–14770.
252. Iliopoulos, O.; Levy, A. P.; Jiang, C.; Kaelin, W. G.; Goldberg, M. A. *Proc. Natl. Acad. Sci. USA* **1996**, *93*, 10595–10599.
253. Maxwell, P. H.; Wiesener, M. S.; Chang, G. W.; Clifford, S. C.; Vaux, E. C.; Cockman, M. E.; Wykoff, C. C.; Pugh, C. W.; Maher, E. R.; Ratcliffe, P. J. *Nature* **1999**, *399*, 271–275.
254. Carroll, V. A.; Ashcroft, M. *Expert Rev. Mol. Med.* **2005**, *7*, 1–16.
255. Hicklin, D. J.; Ellis, L. M. *J. Clin. Oncol.* **2005**, *23*, 1011–1027.
256. Presta, M.; Dell'era, P.; Mitola, S.; Moroni, E.; Ronca, R.; Rusnati, M. *Cytokine Growth Factor Rev.* **2005**, *16*, 159–178.
257. Yamaguchi, T. P.; Harpal, K.; Henkemeyer, M.; Rossant, J. *Genes Dev.* **1994**, *8*, 3032–3044.
258. Lee, S. H.; Schloss, D. J.; Swain, J. L. *J. Biol. Chem.* **2000**, *275*, 33679–33687.
259. Ensoli, B.; Gendelman, R.; Markham, P.; Fiorelli, V.; Colombini, S.; Raffeld, M.; Cafaro, A.; Chang, H. K.; Brady, J. N.; Gallo, R. C. *Nature* **1994**, *371*, 674–680.
260. Takahashi, K.; Mulliken, J. B.; Kozakewich, H. P.; Rogers, R. A.; Folkman, J.; Ezekowitz, R. A. J. *Clin. Invest.* **1994**, *93*, 2357–2364.
261. Tille, J. C.; Wood, J.; Mandriota, S. J.; Schnell, C.; Ferrari, S.; Mestan, J.; Zhu, Z.; Witte, L.; Pepper, M. S. *J. Pharmacol. Exp. Ther.* **2001**, *299*, 1073–1085.
262. Tomanek, R. J.; Sandra, A.; Zheng, W.; Brock, T.; Bjercke, R. J.; Holifield, J. S. *Circ. Res.* **2001**, *88*, 1135–1141.
263. Konerding, M. A.; Fait, E.; Dimitropoulou, C.; Malkusch, W.; Ferri, C.; Giavazzi, R.; Coltrini, D.; Presta, M. *Am. J. Pathol.* **1998**, *152*, 1607–1616.
264. Wang, Y.; Becker, D. *Med. Nat.* **1997**, *3*, 887–893.
265. Polnaszek, N.; Kwabi-Addo, B.; Peterson, L. E.; Ozen, M.; Greenberg, N. M.; Ortega, S.; Basilico, C.; Ittmann, M. *Cancer Res.* **2003**, *63*, 5754–5760.
266. Straume, O.; Akslen, L. A. *Am. J. Pathol.* **2002**, *160*, 1009–1019.
267. Nishishita, T.; Lin, P. C. *J. Cell. Biochem.* **2004**, *91*, 584–593.
268. Morikawa, S.; Baluk, P.; Kaidoh, T.; Haskell, A.; Jain, R. K.; McDonald, D. M. *Am. J. Pathol.* **2002**, *160*, 985–1000.
269. Sundberg, C.; Ljungstrom, M.; Lindmark, G.; Gerdin, B.; Rubin, K. *Am. J. Pathol.* **1993**, *143*, 1377–1388.
270. Apte, S. M.; Fan, D.; Killion, J. J.; Fidler, I. J. M. *Clin. Cancer Res.* **2004**, *10*, 897–908.
271. Abramsson, A.; Berlin, O.; Papayan, H.; Paulin, D.; Shani, M.; Betsholtz, C. *Circulation* **2002**, *105*, 12–117.
272. Bergers, G.; Song, S.; Meyer-Morse, N.; Bergsland, E.; Hanahan, D. *J. Clin. Invest.* **2003**, *111*, 1287–1295.
273. Hermansson, M.; Nister, M.; Betsholtz, C.; Heldin, C. H.; Westermark, B.; Funa, K. *Proc. Natl. Acad. Sci. USA* **1988**, *85*, 7748–7752.
274. Uehara, H.; Kim, S. J.; Karashima, T.; Shepherd, D. L.; Fan, D.; Tsan, R.; Killion, J. J.; Logothetis, C.; Mathew, P.; Fidler, I. J. *J. Natl. Cancer Inst.* **2003**, *95*, 458–470.
275. Forsberg, K.; Valyi-Nagy, I.; Heldin, C. H.; Herlyn, M.; Westermark, B. *Proc. Natl. Acad. Sci. USA* **1993**, *90*, 393–397.
276. Pietras, K.; Ostman, A.; Sjoquist, M.; Buchdunger, E.; Reed, R. K.; Heldin, C. H.; Rubin, K. *Cancer Res.* **2001**, *61*, 2929–2934.
277. Pietras, K.; Rubin, K.; Sjoblom, T.; Buchdunger, E.; Sjoquist, M.; Heldin, C. H.; Ostman, A. *Cancer Res.* **2002**, *62*, 5476–5484.
278. Schlessinger, J. *Cell* **2000**, *103*, 211–225.
279. Schlessinger, J.; Lemmon, M. A. *Signal Transduct. Knowl. Environ.* **2003**, *191*, RE12.
280. Schlessinger, J. *Science* **2003**, *300*, 750–752.
281. Hynes, N. E.; Stern, D. F. *Biochim. Biophys. Acta* **1994**, *1198*, 165–184.
282. Vliagoftis, H.; Worobec, A. S.; Metcalfe, D. D. *J. Allergy Clin. Immunol.* **1997**, *100*, 435–440.
283. Longley, B. J.; Reguera, M. J.; Ma, Y. *Leukemia Res.* **2001**, *25*, 571–576.
284. Huizinga, J. D.; Thuneberg, L.; Kluppel, M.; Malysz, J.; Mikkelsen, H. B.; Bernstein, A. *Nature* **1995**, *373*, 347–349.
285. Hirota, S.; Isozaki, K.; Moriyama, Y.; Hashimoto, K.; Nishida, T.; Ishiguro, S.; Kawano, K.; Hanada, M.; Kurata, A.; Takeda, M. et al. *Science* **1998**, *279*, 577–580.
286. Joensuu, H.; Roberts, P. J.; Sarlomo-Rikala, M.; Andersson, L. C.; Tervahartiala, P.; Tuveson, D.; Silberman, S.; Capdeville, R.; Dimitrijevic, S.; Druker, B. et al. *N. Engl. J. Med.* **2001**, *344*, 1052–1056.
287. Cheng, M.; Sexl, V.; Sherr, C. J.; Roussel, M. F. *Proc. Natl. Acad. Sci. USA* **1998**, *95*, 1091–1096.
288. Sebolt-Leopold, J. S.; Herrera, R. *Nat. Rev. Cancer* **2004**, *4*, 937–947.
289. Wan, P. T.; Garnett, M. J.; Roe, S. M.; Lee, S.; Niculescu-Duvaz, D.; Good, V. M.; Jones, C. M.; Marshall, C. J.; Springer, C. J.; Barford, D. et al. *Cell* **2004**, *116*, 855–867.
290. Davies, H.; Bignell, G. R.; Cox, C.; Stephens, P.; Edkins, S.; Clegg, S.; Teague, J.; Woffendin, H.; Garnett, M. J.; Bottomley, W. et al. *Nature* **2002**, *417*, 949–954.
291. Katso, R.; Okkenhaug, K.; Ahmadi, K.; White, S.; Timms, J.; Waterfield, M. D. *Annu. Rev. Cell Dev. Biol.* **2001**, *17*, 615–675.
292. Wymann, M. P.; Zvelebil, M.; Laffargue, M. *Trends Pharmacol. Sci.* **2003**, *24*, 366–376.
293. Alessi, D. R.; James, S. R.; Downes, C. P.; Holmes, A. B.; Gaffney, P. R.; Reese, C. B.; Cohen, P. *Curr. Biol.* **1997**, *7*, 261–269.
294. Stokoe, D.; Stephens, L. R.; Copeland, T.; Gaffney, P. R.; Reese, C. B.; Painter, G. F.; Holmes, A. B.; McCormick, F.; Hawkins, P. T. *Science* **1997**, *277*, 567–570.

295. Brazil, D. P.; Yang, Z. Z.; Hemmings, B. A. *Trends Biochem. Sci.* **2004**, *29*, 233–242.
296. Feng, J.; Park, J.; Cron, P.; Hess, D.; Hemmings, B. A. *J. Biol. Chem.* **2004**, *279*, 41189–41196.
297. Dragoi, A. M.; Fu, X.; Ivanov, S.; Zhang, P.; Sheng, L.; Wu, D.; Li, G. C.; Chu, W. M. *EMBO J.* **2005**, *24*, 779–789.
298. Sarbassov, D. D.; Guertin, D. A.; Ali, S. M.; Sabatini, D. M. *Science* **2005**, *307*, 1098–1101.
299. Yang, Z. Z.; Tschopp, O.; Baudry, A.; Dummler, B.; Hynx, D.; Hemmings, B. A. *Biochem. Soc. Trans.* **2004**, *32*, 350–354.
300. Samuels, Y.; Wang, Z.; Bardelli, A.; Silliman, N.; Ptak, J.; Szabo, S.; Yan, H.; Gazdar, A.; Powell, S. M.; Riggins, G. J. *Science* **2004**, *304*, 554.
301. Broderick, D. K.; Di, C.; Parrett, T. J.; Samuels, Y. R.; Cummins, J. M.; McLendon, R. E.; Fults, D. W.; Velculescu, V. E.; Bigner, D. D.; Yan, H. *Cancer Res.* **2004**, *64*, 5048–5050.
302. Bachman, K. E.; Argani, P.; Samuels, Y.; Silliman, N.; Ptak, J.; Szabo, S.; Konishi, H.; Karakas, B.; Blair, B. G.; Lin, C. et al. *Cancer Biol. Ther.* **2004**, *3*, 772–775.
303. Lee, J. W.; Soung, Y. H.; Kim, S. Y.; Lee, H. W.; Park, W. S.; Nam, S. W.; Kim, S. H.; Lee, J. Y.; Yoo, N. J.; Lee, S. H. *Oncogene* **2005**, *24*, 1477–1480.
304. Saal, L. H.; Holm, K.; Maurer, M.; Memeo, L.; Su, T.; Wang, X.; Yu, J. S.; Malmstrom, P. O.; Mansukhani, M.; Enoksson, J. et al. *Cancer Res.* **2005**, *65*, 2554–2559.
305. Maehama, T.; Taylor, G. S.; Dixon, J. E. *Annu. Rev. Biochem.* **2001**, *70*, 247–279.
306. Myers, M. P.; Pass, I.; Batty, I. H.; Van der Kaay, J.; Stolarov, J. P.; Hemmings, B. A.; Wigler, M. H.; Downes, C. P.; Tonks, N. K. *Proc. Natl. Acad. Sci. USA* **1998**, *95*, 13513–13518.
307. Li, D. M.; Sun, H. *Proc. Natl. Acad. Sci. USA* **1998**, *95*, 15406–15411.
308. Furnari, F. B.; Huang, H. J.; Cavenee, W. K. *Cancer Res.* **1998**, *58*, 5002–5008.
309. Haas-Kogan, D.; Shalev, N.; Wong, M.; Mills, G.; Yount, G.; Stokoe, D. *Curr. Biol.* **1998**, *8*, 1195–1198.
310. Wu, X.; Senechal, K.; Neshat, M. S.; Whang, Y. E.; Sawyers, C. L. *Proc. Natl. Acad. Sci. USA* **1998**, *95*, 15587–15591.
311. Bellacosa, A.; de Feo, D.; Godwin, A. K.; Bell, D. W.; Cheng, J. Q.; Altomare, D. A.; Wan, M.; Dubeau, L.; Scambia, G.; Masciullo, V. *Int. J. Cancer* **1995**, *64*, 280–285.
312. Cheng, J. Q.; Ruggeri, B.; Klein, W. M.; Sonoda, G.; Altomare, D. A.; Watson, D. K.; Testa, J. R. *Proc. Natl. Acad. Sci. USA* **1996**, *93*, 3636–3641.
313. Ruggeri, B. A.; Huang, L.; Wood, M.; Cheng, J. Q.; Testa, J. R. *Mol. Carcinogen.* **1998**, *21*, 81–86.

Biographies

Michel-Saveur Maira did his PhD in molecular and cellular biology in the laboratory of Dr B Wasylyk (IGBMC, ULP, Strasbourg, France). After a first postdoctoral post in the Research and Development group of Rhone-Poulenc-Rorer (CRVA, Paris, France), he joined the group of Dr B Hemmings at the Friedrich Miescher Institute (FMI, Basel, Switzerland), sponsored by a postdoctoral EMBO fellowship. Since 2001, when he joined Novartis Pharma (NIBR Oncology, Basel, Switzerland) as a laboratory head, his work has mostly focused on targeted therapy drug discovery, most notably against signaling cascade kinases.

Mark A Pearson investigated the molecular mechanisms underlying the synergistic interaction of stem cell factor (SCF) and other cytokines in primitive hematopoietic progenitors in the laboratory of Dr Claire Heyworth Paterson Institute Manchester. After taking his PhD in the Paterson Institute of Cancer Research, Manchester University, he moved to the laboratory of Prof Pier Guiseppe Pelicci, European Institute of Oncology, Milan, assessing the role of the promyeloytic leukemia (PML) protein in the pathophysiology of acute promyelocytic leukemia, in particular the effect of the fusion protein PML-RAR-alpha on regulation of the tumor suppressor p53 by PML. Dr Pearson joined Novartis Oncology in February 2002.

Doriano Fabbro, PhD, is currently Head of the Kinase Biology of the Expertise platform kinases (EPK) at Novartis Institute of Biomedical Research (NIBR). Dr Fabbro's career includes 12 years as Group Leader in the Molecular Tumor Biology Unit of the University of Basel working mainly on the mechanisms of activation of PKC as well as on prognostic markers for breast cancer; in 1985 he started a collaboration with Prof Alex Matter (Ciba-Geigy) on PKC inhibitors which subsequently led to the successful discovery of Gleevec. In 1991, Dr Fabbro joined the Pharmaceuticals Research of Ciba-Geigy Basel first as a Group Leader and Indication Area Head in Oncology and then as Unit Head in Novartis Oncology Research (after the merger of Ciba-Geigy with Sandoz to form Novartis) where he was member of the integration team oncology research. In this role he was responsible for the drug discovery efforts on ATP-dependent enzymes. More recently, Dr Fabbro has become Head of the EPK in NIBR. Over the year he has contributed significantly to the discovery and development of various protein kinase inhibitors for the treatment of cancer including among others Gleevec, PKC412, AEE788, AEW541, PTK787, and RAD001. Dr Fabbro is a member of numerous professional societies and several journal editorial boards as well as author of over 170 publications and numerous patents in the areas of protein kinase regulation, structure, screening, and drug discover.

Carlos García-Echeverría earned his PhD degree in organic chemistry at the University of Barcelona under the supervision of Dr Fernando Albericio and Dr Miquel Pons. After a three-year postdoctoral stay at the University of Madison-Wisconsin with Prof Daniel Rich, he joined the Exploratory Research Unit of Ciba-Geigy (now Novartis Institutes for BioMedical Research) in 1993. Since his incorporation at the Oncology Research Group in 1995, he has been the medicinal chemistry sponsor and team head of different programs linked to tumor cell growth control and apoptosis. More recently, he has become Head of Oncology Research in Basel. His research activities have been mainly focused on the identification and development of antagonists of intramolecular protein–protein interactions inhibitors of protein kinases and proteolytic enzymes. He is an inventor on 20 patents (issued or pending), and has published 7 book chapters and over 100 articles and review papers.

Comprehensive Medicinal Chemistry II
ISBN (set): 0-08-044513-6

ISBN (Volume 7) 0-08-044520-9; pp. 1–31

7.02 Principles of Chemotherapy and Pharmacology

C K Donawho, A R Shoemaker, and J P Palma, Abbott Laboratories, Abbott Park, IL, USA

7.02.1 Introduction

Cancer remains one of the major causes of death in the world and it has been estimated that there will be 15 million new cases and 10 million deaths in 2020.[1] Since mustine was first used to treat an acute lymphoblastic leukemia patient in 1943,[2] approximately 100 cancer drugs (and counting) have been approved by the US Food and Drug Administration (FDA). During the last 63 years, survival in some cancers has improved but the urgent need to develop new and improved drugs has continued. Initially, therapies were directly cytotoxic, consisting of alkylating agents, antimetabolites, and antitumor antibiotics. As biological understanding of cancer cell growth kinetics continued to increase, other successful strategies were employed, i.e., interruption of tubulin/microtubule formation, disruption of normal nucleotide incorporation, interference with nuclear hormone receptors or inhibition of topoisomerases I and II,[3] thereby increasing the cytotoxic arsenal for killing or slowing the growth of tumor cells. While these drugs continue to be refined, resistance reduced, or drugs combined to improve survival and/or reduce toxicities in cancer patients, novel molecular targets continue to emerge as additional opportunities to increase effectiveness of standard therapeutics.

The important characteristics differentiating cancer cells from normal cells are: (1) limitless replicative potential; (2) self-sufficiency in growth signals; (3) nonresponsiveness to normal growth-inhibitory signals; (4) escape from senescence and apoptotic death; (5) ability to invade normal tissues and metastasize to distant sites; and (6) support angiogenesis within the growing tumor – all characteristics targeted by recent therapies. In addition, the modulation of resistance with new drugs to potentate existing chemotherapeutic agents and/or the chemoprevention of primary tumors provide exciting new targets for cancer therapeutics.

7.02.2 Pharmacology

Pharmacology (study of drugs) comprises two arms: pharmacokinetics (PK) (drug movement) and pharmacodynamics (PD) (drug power). These two aspects dictate all testing done both in vitro and in vivo for the screening for lead agents, evaluation of these leads, determination of therapeutic window/index, and preclinical support of the aspects of dose, schedule, and potential combination therapy for first-in-humans clinical trials.

7.02.2.1 Pharmacokinetics: The In Vivo Movement of Drugs

PK is the time course studies of drug absorption, distribution, metabolism, and elimination in vivo. Drug absorption is a highly variable process dependent on inherent drug physiochemical properties. Passive diffusion is by far the most important process by which drugs move across cell membranes. Both the thickness of the cell membrane and presence of drug efflux pumps (e.g., ATP-binding cassette (ABC) transporters such as P-glycoprotein) will determine the rate and extent of drug absorption. While oral (enteral) drug administration is convenient, safe, and economical, the majority of cancer drugs often have a high first-pass effect and this results in low bioavailability (F), describing the fractional extent of the dose reaching the systemic circulation. This first-pass effect results from poor absorption in the gastrointestinal tract, or significant metabolism or excretion by gastrointestinal mucosa and/or liver prior to entering the systemic circulation.

7.02.2.2 Pharmacodynamics: The Mechanisms of Drug Action

PD is the study of drug action on the biologic and physiologic processes of both cells and biologic systems. The targets of drug interaction are usually specific macromolecules that induce a physiologic or biochemical change; DNA, RNA, or other macromolecules involved in cell division (e.g., microtubules); an enzyme found either intracellularly or in the plasma; an ion channel protein or structural protein.[4] Many newer drugs bind to receptors that normally bind an endogenous regulatory ligand (e.g., growth factors, neurotransmitters, hormones) and drug binding to this receptor alters its function and/or its downstream signaling responses. Agonists are drugs that mimic the endogenous ligand for the receptor (e.g., opioids, granulocyte colony stimulating factor (G-CSF), recombinant human erythropoietin (rhEPO). Antagonists are drugs that block the effects of the endogenous ligand (e.g., trastuzumab (Herceptin) monoclonal antibody against HER-2/neu; flutamide, an androgen receptor antagonist). Partial agonists are agents that possess both agonist/antagonist activities (e.g., tamoxifen-mixed estrogen receptor agonist/antagonist). Enzyme inhibitors bind directly to endogenous enzyme and inhibit their function (e.g., methotrexate, dihydrofolate reductase inhibitors; epidermal growth factor receptor (EGFR)-associated tyrosine kinase I inhibitors, i.e., Tarceva; topoisomerase I inhibitors, i.e., camptothecins).

7.02.3 Evaluation of New Drugs or Agents

7.02.3.1 Assay Development (High-Throughput) or In Vitro Cell Panels

High-volume screening in vitro is a critical aspect of the lead selection and drug development process (**Figures 1** and **2**) initially started with the development of nitrogen mustard derivatives in 1946, as reviewed by Dendy and Hill.[5] Until 1985, for the most part screening was performed in vivo. Around this time the National Cancer Institute (NCI) initiated a program for improving anticancer drug discovery[6]; they deployed an empirical scheme based on an initial three-cell line prescreen panel, followed by a 60-cell line screen. For compounds with significant (cytotoxic) activity, the drug would probably be evaluated in the hollow-fiber (HF) assay (in vivo model) developed by Hollingshead *et al.*[7] This approach was or is predominantly used for the discovery of antiproliferative and cytotoxic drugs, but for the drugs that target signal transduction pathway, small molecules, static (antiangiogenic) drugs, and antibodies, the need for both biochemical 'target-driven' screening assays and tailored cellular screens became evident. For these agents, the use of cell-free target-specific, high-throughput screening (HTS) using miniaturization, robot-aided automation, and data management allowed large libraries of compounds to be tested efficiently.[8] An indepth review of HTS and current assay strategies can be found in the *Anticancer Drug Development Guide*.[9]

7.02.4 Preclinical Tumor Models in Cancer Drug Discovery

For the last 50 years, animal tumor model systems have been utilized as a key component for the evaluation of antitumor efficacy of novel cancer therapeutic agents prior to initiation of clinical trials.[10,11] The earliest models were

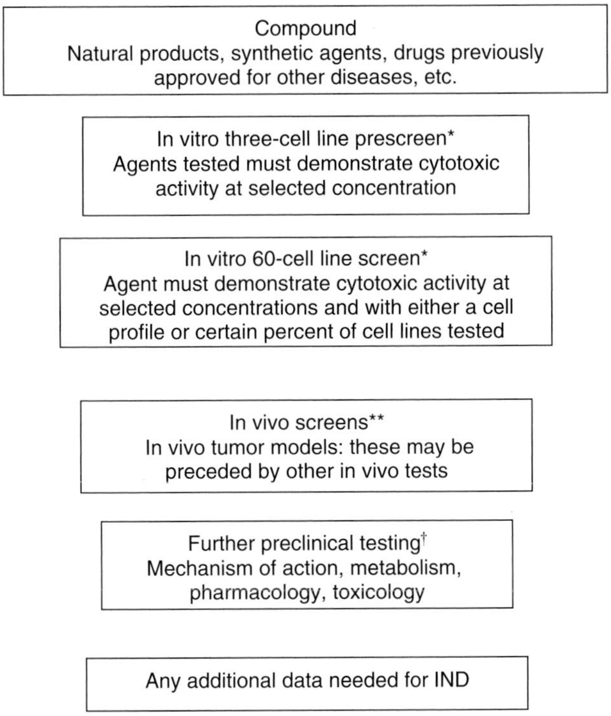

Compound
Natural products, synthetic agents, drugs previously approved for other diseases, etc.

In vitro three-cell line prescreen*
Agents tested must demonstrate cytotoxic activity at selected concentration

In vitro 60-cell line screen*
Agent must demonstrate cytotoxic activity at selected concentrations and with either a cell profile or certain percent of cell lines tested

In vivo screens**
In vivo tumor models: these may be preceded by other in vivo tests

Further preclinical testing†
Mechanism of action, metabolism, pharmacology, toxicology

Any additional data needed for IND

Figure 1 National Institutes of Health/classical cytotoxic preclinical drug screening process. *Cellular screens are usually performed on permanent cell lines using some type of growth inhibition or cytotoxicity assay (e.g., methylthiazolidiphenyl tetrazolium (MTT), sulforhodamine B (SRB)). **In vivo models can include xenografts/immuno-suppressed mice, murine tumors/mice, other species (e.g., rats) with syngeneic or spontaneous tumors in companion animals (e.g., dogs), hollow fiber assays, etc. †Aspects of mechanism of action, metabolism, PK/PD, and toxicology can be addressed at any point or concurrently with earlier testing. (*Source*: http://dtp.nci.nih.gov.)

syngeneic leukemia/lymphoma models that were typically conducted as an intraperitoneal inoculation of mouse tumor cells, with subsequent development of ascitic disease. In their most simplistic forms, efficacy was evaluated as a treatment-associated delay in the development of morbidity.[12] These model systems were used quite successfully to develop effective therapies for the treatment of childhood leukemia.[13,14] With the utilization of methods to suppress immune function in mice and, more importantly, the discovery and characterization of immunocompromised strains of mice (e.g., *Foxn1*$^{nu/nu}$ (nude) and *Prkdc*$^{scid/scid}$ (scid)), numerous solid tumor lines of human origin were added to the tumor model repertoire (xenograft models).[15,16] These early solid tumor models were usually rapidly growing, subcutaneously implanted tumors that often required in vivo propagation to maintain the line. Today, dozens of human tumor cell lines are commonly used to evaluate efficacy of experimental agents.[17] Although some of these lines still require in vivo propagation, many can be easily maintained in cell culture, thus allowing for potentially more direct in vitro/in vivo correlates of activity.

7.02.4.1 Common/Classical Preclinical In Vivo Approaches

7.02.4.1.1 Subcutaneous or flank tumor models

In a typical murine tumor experiment, the cell line of interest is expanded in vitro and harvested at a point when the cells are in log growth. The number of inoculated cells needed for robust growth in vivo must be determined empirically and can vary from approximately 5×10^4 to 1×10^7, depending on the cell line. Important factors to consider when determining the appropriate cell titer include rate of tumor growth, variability in growth, and degree of tumor ulceration and necrosis. The strain of mice used (e.g., *nude, scid, scid*-bg for human or syngeneic for murine cell lines) can also significantly impact these parameters. Growth factors (usually in the form of Matrigel) are also often added to the cellular inoculum to enhance the initial in vivo growth phase. An *n* of 5 mice per treatment group is generally considered to be a minimum sample size, with a group size of 10 being preferred. Efficacy experiments are conducted either as early treatment studies (sometimes referred to as tumor growth inhibition) in which therapy is

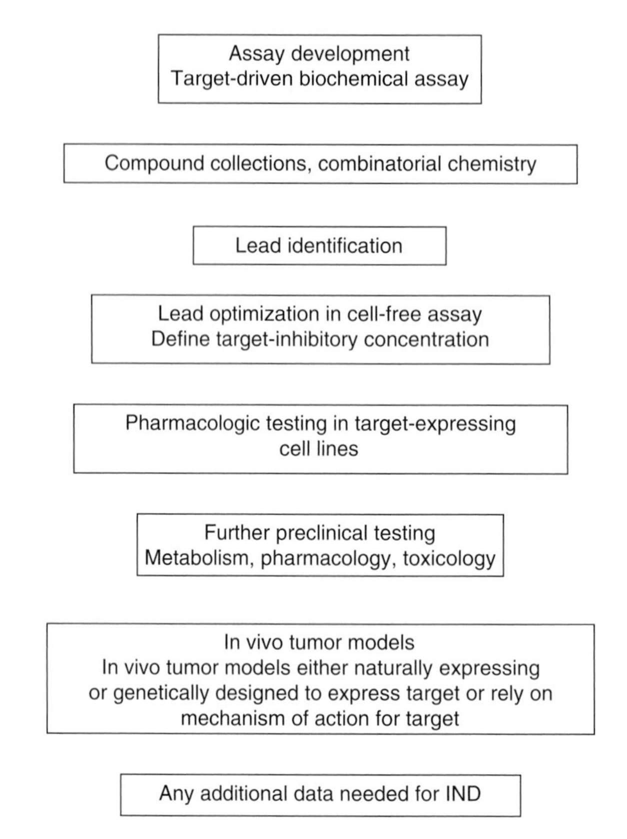

Figure 2 Contemporary preclinical drug development process.

initiated simultaneously or shortly after tumor cell inoculation (but prior to the development of palpable tumors) or as staged tumor studies (tumor growth delay), where animals are assigned to treatment groups after tumors have reached a defined size.

Once therapy is initiated, tumor size is determined by caliper measurements conducted with defined frequency (weekly to daily, depending on the tumor growth rate). One common representation of tumor size is the calculated tumor volume determined by the formula $V=$ (larger diameter $\times$ shorter diameter^2) $\times$ 0.5. The tumor growth inhibition (TGI) properties of an experimental compound can then be examined by determining the average tumor volume of the treated (T) group relative to the control (C) group (%T/C or %TGI, calculated as 100% – %T/C).[18] While this is generally determined at the end of the dosing period, it can also be valuable to analyze %T/C values throughout the study (early in the treatment cycle and posttreatment). This additional information about efficacy can be useful if the %T/C changes dramatically through the duration of a chronically dosed trial, as it may be indicative of either tumor resistance or changes in the PK or PD of the drug within the animal. This is especially informative with the novel targeted agents. An additional efficacy parameter, tumor growth delay, is another important measure of activity. This value is determined by calculating the mean or median number of days required for the treated versus vehicle groups to reach a specific endpoint (e.g., a tumor volume of 1000 mm^3), statistically compared using parametric or nonparametric tests and reported as percent increased lifespan (%ILS).

7.02.4.1.2 Liquid/hematologic tumor models

Net log cell kill is an additional parameter of efficacy evaluation that is commonly used for the evaluation of cytotoxic agents in leukemia models but is less frequently determined in solid tumor models.[18]

7.02.4.1.3 Orthotopic/metastatic tumor models

In addition to these flank models, tremendous progress has also been made in the development of more sophisticated orthotopic and metastatic tumor models that provide an opportunity for the potential evaluation of some stroma–tumor interactions and extravasation/intravasation processes.[19–21] The expression of molecular targets can be variable depending on the organ or stromal environment of the growing tumor and this can be helpful when working with targeted agents. In addition, the PK/PD relationships in different tumor-containing organs can be evaluated. However, relative to flank models, orthotopic and metastatic model systems often have increased complications associated with staging of trials and, without sensitive imaging technologies and/or genetic alteration of tumors cells, the inability to monitor tumor sizes remains a drawback. Furthermore, without tumor staging, these models can require large cohorts of tumor-bearing animals for convincing statistical analysis. These models also often require a high degree of technical skill and can be quite time-consuming, thus potentially restricting the size of trials that can be conducted. However, the evaluation of biomarkers in selected orthotopic or metastatic models, such as prostate-specific antigen (PSA) in the LuCap 23.1 (human prostrate carcinoma) or CA15-3 in the MDA-MB-231 (human breast carcinoma) models can greatly aid staged studies in these xenograft models (*see* Section 7.02.8). In addition, advances in small-animal imaging technologies (for instance, dynamic contrast-enhanced magnetic resonance imaging (DCE-MRI), computed tomography (μCT), etc.) will allow these approaches to be an increasingly important component of the preclinical evaluation of experimental cancer agents.

7.02.4.1.4 **Transgenics, knockouts, and in vivo gene targeting**

With the advent of transgenic and gene targeting technologies in mice, there are now available an increasingly sophisticated variety of genetic tumor models.[22] While scientifically allowing the precise control of disease-associated genes, these genetic model systems often have increased complications associated with phenotypic analysis. Further, these investigations require large cohorts of tumor-bearing animals for efficacy studies that require analysis of multiple agents, doses, and schedules. Currently, given the time and expense of these models, only very specific targets or dedicated biology-driven projects are likely to find this approach helpful. However, as our ability to generate these models and analyze their phenotypes increases, these approaches will be an increasingly important component of the preclinical evaluation of experimental cancer agents.

7.02.4.1.5 **Predictive value of murine in vivo models**

During the last several years, there has been much argument and counterargument regarding the validity of current tumor model systems (particularly subcutaneous or flank models) as predictors of clinical activity.[23–27] A recent review demonstrated that, for 39 anticancer agents, those with activity in at least one-third of xenograft models tested also had significant activity in phase II trials.[28] Further, all currently registered cancer drugs have shown activity in preclinical animal models. However, antitumor activity in a xenograft model representing a specific histological type within a given organ (e.g., colon adenocarcinoma) does not necessarily translate into clinical activity in that organ type.[28] While animal models are most important in showing the spectrum of activity, they have limitations and may not be fully predictive due to many confounding factors, outlined below.

It should come as no surprise that immortalized tumor cell lines grown for many generations on a plastic surface (or even in vivo) are not likely to reflect perfectly the complex biology of spontaneously arising tumors in humans. Clearly there are limitations with how extensively xenograft models can mimic factors such as tumor blood supply, tumor–stromal interactions, genomic integrity, and other processes that influence the ability of any potential drug to impact tumor growth. Another problem unique to model development for cancer drug discovery is that the disease is an outgrowth from normal cells and tissues, yet effective therapy requires killing (or at least significant inhibition of growth) of cancer cells without dramatically compromising the well-being of the host cells. Part of the tremendous challenge of developing neoplastic biological model systems is that cancer represents more than 100 diseases, each with biology unique to the site from which it arose. For experimental agents where there is precise knowledge of the tumor type(s) that are likely to be highly dependent on the therapeutic target, it is possible (in some cases) to develop quite sophisticated tumor models that would be more likely to reflect the clinical setting.[29] However, in the vast majority of cases the prediction of tumor types likely to be dependent on the target of interest is associated with a high degree of uncertainty. This problem, in fact, may be addressed by one of the significant benefits of xenograft model systems in that literally hundreds of cell lines representing most major and minor tumor types exist for potential study.

It is worth discussing in some detail the factors that contribute to the perception that existing tumor models have limited value as predictors of clinical activity. Part of the problem lies with the lack of rigor with which efficacy studies are often conducted. It is not uncommon to see a report claiming significant activity in a tumor model, yet examination of the data shows that, at best, no conclusion about activity can be drawn.[10] In addition to the factors described above (appropriate sample size, accounting for tumor variability and ulceration), it is important that the evaluation of efficacy is conducted in a biologically relevant fashion. Reporting a difference in tumor volume between treated and control groups when the control groups have only grown to a volume of a few hundred cubic millimeters or less is unlikely to be informative. The method of bilateral inoculation of tumors in order to increase sample size should also be avoided, as it appears that the growth of one tumor can sometimes influence the growth of the other (perhaps due to effects on tumor angiogenesis), in ways that are not predictable.[30] For many tumor lines there now also exist a variety of cytotoxic agents that are known to show significant antitumor activity in a particular xenograft model (**Table 1**).[31] Use of these compounds side by side with the experimental agent not only serves as a powerful benchmark to judge activity, but they are also important positive controls for the numerous variables that can affect tumor growth from one trial to the next.

Another important criterion is defining the concept of 'significant activity' for an experimental agent in a tumor model. Depending on the treatment group sample size and the variability of growth in a given model, one could expect that a treatment-induced *%T/C* of 50–60% (40–50% TGI) would register as statistically significant. If this level of TGI is achieved at only a single measurement during the course of the trial, the effect should be considered of limited value. However, if this level of inhibition is observed throughout the treatment period and can be accurately reproduced from one trial to the next, then a significant biological effect can be argued more convincingly. Does this level of inhibition signify a level of efficacy that could translate to a clinical response? For a purely cytostatic agent that has limited toxicological side effects and can therefore be administered for long duration, the answer may be 'yes,' although the ultimate benefit of such agents in clinic may be in combination with cytotoxic drugs. Numerous clinical trials are currently ongoing to test this hypothesis.[32,33] However, in the context of the more traditional way in which clinical activity is measured for cytotoxic agents, the answer is almost certainly 'no.' Given that a 50% reduction in tumor mass is generally considered the clinical standard for partial response (PR), then a 40–50% inhibition of tumor growth rate would almost certainly translate to progressive disease.[10,26] A more robust measure of significant efficacy for a bona fide drug candidate in xenograft models is tumor regression (or at least complete tumor stasis), observed at the end of the treatment period.[18] It seems logical to argue that the more pronounced the regression and the more durable the response, the more likely the activity is to translate to clinical effect. Of course, even when tumor regression is observed, a reasonable therapeutic window must exist and the translation of activity for agents used in combination can be much more complex. Toxicological evaluations of experimental anticancer agents are typically quite rudimentary during the early preclinical stages. Dose-dependent antitumor activity is usually reported relative to the frequency of deaths or significant weight loss (typically 15–20% of total body weight) within the efficacy trial. However, in some cases where modulation of target activity is expected to have a specific (target-related) toxicological liability, it is worth examining these potential limitations extensively early in the discovery process.[34,35]

Although 40–50% inhibition of tumor growth is generally not sufficient efficacy to declare an experimental agent ready for clinical trial, it nonetheless can be quite informative to identify appropriate model(s) that are critical for subsequent medicinal chemistry refinement of the lead. Identifying an appropriate model system is vital for efficient evaluation of potential improvements in potency and PK that ultimately result in a drug candidate with more robust efficacy.

In addition to defining the significance of the level of activity in a given model, one must also consider how many models need to be examined in order to try to extrapolate preclinical efficacy to clinical response. This question, of course, needs to be considered in the context of the pathway that is being targeted. Thus, molecules designed to inhibit receptor tyrosine kinase (RTK) signaling critical in angiogenesis are likely to be active in a variety of tumor types, as these pathways have been shown to be important in a broad range of neoplasms.[36,37] Indeed, a number of these RTK inhibitors have been described that do show broad activity in xenograft tumor models. Several of these inhibitors have entered clinical evaluation and are showing promising activity as monotherapy and/or in combination with other drugs.[36,37] For other targets, efficacy may be expected to be limited to more specific type(s) of tumors. In some cases, however, it has been shown that significant activity in one model that expresses the target is likely to be predictive of clinical response for that class of cancer (e.g., Herceptin).

As has already been mentioned, identification of the appropriate xenograft tumor type(s) for efficacy evaluation is critical for effective lead identification and refinement. This process can be greatly assisted by learning as much as possible about the biology and genetics of the xenograft tumors. The pediatric oncology community has recently established a consortium for evaluation of promising therapeutic agents in fairly large panels of xenograft models representing a variety of tumor types as a key component for selection of drugs for clinical evaluation.[26] Although it is not always possible to trace the clinical origin of each line, it is often helpful to know if the line was isolated from a primary or metastatic lesion (and if so, from what site), and the treatment history prior to isolation. For example, in some cases multiple tumor lines have been established from the same patient: this is an important factor if one wishes to evaluate efficacy in a panel of models of a certain tumor type.[38,39] In addition, long-term culture of cell lines also carries with it the inherent risk that cells may develop changes in phenotype or may become cross-contaminated over time.[40]

Table 1 Examples of cytotoxic drug efficacy in xenograft and syngeneic tumor models

Tumor	*Cell line/drug/schedule*	*TGI*[a]	*Time to 1 cc (days)*[b]
Bladder			
	EJ-1		21.4
	Paclitaxel		
	$25\,mg\,kg^{-1}\,day^{-1}$, i.p., q4 days × 3 ET	85.3	
	$25\,mg\,kg^{-1}\,day^{-1}$, i.p., q4 days × 3 ST	73.1	
	Gemcitabine		
	$120\,mg\,kg^{-1}\,day^{-1}$, i.p., q3 days × 4 ET	93.7	
	FTI		
	$6.25\,mg\,kg^{-1}\,day^{-1}$, p.o., b.i.d. × 21 ET	77.7	
Brain			
	U-87 MG		33.9
	Paclitaxel		
	$40\,mg\,kg^{-1}\,day^{-1}$, i.p., qd × 1 ST	48.3	
	BCNU		
	$20\,mg\,kg^{-1}\,day^{-1}$, i.p., q4 days × 3 ST	70.7	
	Temozolomide		
	$10\,mg\,kg^{-1}\,day^{-1}$, p.o., qd × 1 ST	73.9	
Breast			
	MDA-435-LM		30.9
	Paclitaxel		
	$25\,mg\,kg^{-1}\,day^{-1}$, i.p., q4 days × 3 ET	74.0	
	$25\,mg\,kg^{-1}\,day^{-1}$, i.p., q4 days × 3 ST	35.4	
	MDA-MB-231		50.2
	Paclitaxel		
	$25\,mg\,kg^{-1}\,day^{-1}$, i.p., q4 days × 3 ET	78.7	
	$25\,mg\,kg^{-1}\,day^{-1}$, i.p., q4 days × 3 ST	53.1	
Colon			
	DLD-1		26.7
	Cyclophosphamide		
	$100\,mg\,kg^{-1}\,day^{-1}$, i.p., q4 days × 3 ET	72.8	
	Gemcitabine		
	$120\,mg\,kg^{-1}\,day^{-1}$, i.p., q3 days × 4 ET	51.3	
	Irinotecan		
	$100\,mg\,kg^{-1}\,day^{-1}$, i.p., q4 days × 4 ET	82.5	
	$160\,mg\,kg^{-1}\,day^{-1}$, i.p., q4 days × 4 ST	81.6	
	$10\,mg\,kg^{-1}\,day^{-1}$, i.p., q4 days × 4 ST	5.7	
	$10\,mg\,kg^{-1}\,day^{-1}$, i.p., qd × 21 ST	92.9	
	5-Fluorouracil		
	$50\,mg\,kg^{-1}\,day^{-1}$, i.p., qd × 5 ET	43.0	
	Doxorubicin		
	$1\,mg\,kg^{-1}\,day^{-1}$, i.p., qd × 5 ET	24.5	
	Oxaliplatin		
	$14\,mg\,kg^{-1}\,day^{-1}$, i.p., q7 days × 3 ST	−10.1	
	RTKI		
	$25\,mg\,kg^{-1}\,day^{-1}$, p.o., b.i.d. × 21 ST	71.4	
	HCT-116		25.0

continued

Table 1 Continued

Tumor	*Cell line/drug/schedule*	*TGI*[a]	*Time to 1 cc (days)*[b]
	Gemcitabine		
	120 mg kg^{-1} day^{-1}, i.p., q3 days × 4 ET	86.2	
	Paclitaxel		
	30 mg kg^{-1} day^{-1}, i.p., q4 days × 3 ET	85.5	
	Irinotecan		
	50 mg kg^{-1} day^{-1}, i.p., q4 days × 4 ET	72.8	
	50 mg kg^{-1} day^{-1}, i.p., qd × 21 ST	90.0	
	5-Fluorouracil		
	20 mg kg^{-1} day^{-1}, i.p., qd × 5 ET	18.1	
	Cisplatin		
	2 mg kg^{-1} day^{-1}, i.p., q4 days × 3 ST	18.2	
	RTKI		
	25 mg kg^{-1} day^{-1}, p.o., b.i.d. × 21 ST	81.5	
	SW-620		29.8
	Temozolomide		
	100 mg kg^{-1} day^{-1}, p.o., qd × 1 ST	63.8	
	Irinotecan		
	40 mg kg^{-1} day^{-1}, i.p., qd × 5 ST	91.3	
	2.5 mg kg^{-1} day^{-1}, i.p. (5 days on/3 days off)3	79.0	
	Etoposide		
	30 mg kg^{-1} day^{-1}, i.p., q3 days × 3 ST	18.4	
	Oxaliplatin		
	15 mg kg^{-1} day^{-1}, i.p., qd × 1 ST	26.6	
	5-Fluorouracil		
	25 mg kg^{-1} day^{-1}, i.p., q2 days × 7 ST	2.7	
Lung			
	LX-1 (squamous cell carcinoma)		21.1
	Paclitaxel		
	25 mg kg^{-1} day^{-1}, i.p., q4 days × 3 ET	87.5	
	25 mg kg^{-1} day^{-1}, i.p., q4 days × 3 ST	70.7	
	Irinotecan		
	50 mg kg^{-1} day^{-1}, i.p., q4 days × 4 ST	84.2	
	1 mg kg^{-1} day^{-1}, i.p. (5 days on/2 days off)3 ST	58.6	
	Carboplatin		
	50 mg kg^{-1} day^{-1}, i.p., q4 days × 4 ST	10.5	
	Etoposide		
	40 mg kg^{-1} day^{-1}, i.p., q3 days × 3 ST	26.9	
	FTI		
	25 mg kg^{-1} day^{-1}, p.o., b.i.d. × 21 ST	50.1	
	RTKI		
	25 mg kg^{-1} day^{-1}, p.o., b.i.d. × 21 ST	76.9	
	A549 (NSCLC)		41.1
	Paclitaxel		
	30 mg kg^{-1} day^{-1}, i.p., q4 days × 3 ET	57.1	
	30 mg kg^{-1} day^{-1}, i.p., q4 days × 3 ST	57.1	
	Docetaxel		
	30 mg kg^{-1} day^{-1}, i.p., q4 days × 3 ST	64.0	
	Doxorubicin		
	1 mg kg^{-1} day^{-1}, i.p., qd × 5 ET	3.3	
	Cisplatin		
	3 mg kg^{-1} day^{-1}, i.p., q4 days × 3 ET	4.6	

Table 1 Continued

Tumor	*Cell line/drug/schedule*	*TGI*[a]	*Time to 1 cc (days)*[b]
	Calu-6 (NSCLC)		28.1
	Etoposide 30 mg kg^{-1} day^{-1}, i.p., q3 days × 3 ST	35.9	
	Carboplatin 50 mg kg^{-1} day^{-1}, i.p., q4 days × 4 ST	55.5	
	Cisplatin 10 mg kg^{-1} day^{-1}, i.p., qd × 1 ST	54.7	
	Oxaliplatin 7 mg kg^{-1} day^{-1}, i.p., q7 days × 3 ST	–4.3	
	H146 (SCLC)		49.5
	Paclitaxel 30 mg kg^{-1} day^{-1}, i.p., q4 days × 3 ST	90.5	
	Etoposide 25 mg kg^{-1} day^{-1}, i.p., q4 days × 3 ST	22.8	
	Vincristine 0.5 mg kg^{-1} day^{-1}, i.v., q7 days × 4 ST	83.2	
	H526 (SCLC)		26.0
	Carboplatin 50 mg kg^{-1} day^{-1}, i.p., q4 days × 3 ST	98.9	
	RTKI 25 mg kg^{-1} day^{-1}, p.o., b.i.d. × 21 ST	62.0	
Lymphoma	**DoHH-2**		21.4
	Doxorubicin 3.3 mg kg^{-1} day^{-1}, i.v., qd × 1 ST	51.3	
	Rituximab 10 mg kg^{-1} day^{-1}, i.v., qd × 1 ST	81.7	
	CHOP 25/3/0.25/0.5, i.p./i.v./i.v./p.o., qd × 1 ST	77.1	
	WSU-DLCL2		29.8
	Rituximab 10 mg kg^{-1} day^{-1}, i.v., qd × 1 ST	76.0	
	CHOP 25/3/0.25/0.5, i.p./i.v./i.v./p.o., qd × 1 ST	45.9	
Melanoma			
	A-375		38.0
	Temozolomide 140 mg kg^{-1} day^{-1}, p.o., qd × 5 ST	8.0	
	B16F10		12.1
	Paclitaxel 25 mg kg^{-1} day^{-1}, i.p., q4 days × 3 ET	46.6	
	Temozolomide 100 mg kg^{-1} day^{-1}, p.o., qd × 5 ET	81.8	
	5-Fluorouracil 30 mg kg^{-1} day^{-1}, i.p., qd × 5 ET	20.6	
	RTKI 25 mg kg^{-1} day^{-1}, p.o., b.i.d., d1-end ET	79.0	

continued

Table 1 Continued

Tumor	*Cell line/drug/schedule*	*TGI*[a]	*Time to 1 cc (days)*[b]
Pancreas			
	MiaPaCa-2		27.8
	Gemcitabine		
	120 mg kg^{-1} day^{-1}, i.p., q3 days × 4 ET	64.0	
	120 mg kg^{-1} day^{-1}, i.p., q3 days × 4 ST	48.8	
	Cisplatin		
	3 mg kg^{-1} day^{-1}, i.p., q4 days × 3 ET	37.0	
	5-Fluorouracil		
	25 mg kg^{-1} day^{-1}, i.p., q4 days × 3 ET	20.7	
	Rapamycin		
	20 mg kg^{-1} day^{-1}, i.p., qd × 21 ST	59.0	
	FTI		
	50 mg kg^{-1} day^{-1}, p.o., b.i.d. × 21 ET	50.9	
Prostate			
	PC-3		38.1
	Paclitaxel		
	25 mg kg^{-1} day^{-1}, i.p., q4 days × 3 ET	73.5	
	25 mg kg^{-1} day^{-1}, i.p., q4 days × 3 ST	73.6	
	Rapamycin		
	20 mg kg^{-1} day^{-1}, i.p., q4 days × 3 ST	89.9	
	FTI		
	25 mg kg^{-1} day^{-1}, s.c., qd × 21 ST	58.8	

All experiments were conducted by inoculation of cells or tumor brei into the flanks of immunocompromised mice (*nude, scid,* or *scid*-bg). Trials were conducted either as early treatment (ET) studies with therapy beginning prior to the development of measurable tumors, or as staged tumor (ST) studies, with therapy initiated after tumors attained a specific size (typically 150–300 mm^3). Tumor volumes were determined by two or three times weekly measurements using digital calipers.

[a] Tumor growth inhibition. Calculated as 100 − %*T/C* based on the average determined from between one and more than 10 trials.

[b] Time for untreated or vehicle-treated tumors to reach an average tumor volume of 1000 mm^3. Determined from an analysis of between one and more than 10 trials per tumor type.

Immunohistochemical (IHC) markers and/or DNA fingerprinting are methods that can be used to ensure cell line integrity. Most xenograft tumor lines used to establish models have been maintained in culture for many years and have highly aneuploidic genomes.[41] Utilization of methods such as comparative genomic hybridization (CGH) can be informative for determining chromosomal regions of interest that may be amplified or deleted in the tumor lines. Mutational analysis of specific genes of interest may also be warranted, such as for examination of specific mutations in the epidermal growth factor receptor that confer higher degrees of sensitivity to inhibitors of this protein.[42–44] Expression profiling of tumors and the cell lines from which they originated can help elucidate expression of target protein pathways, subclassify lines both within a given tumor type and between different tumor types, and identify changes in gene expression between cells grown in vitro versus in vivo that may help explain differences in compound activity between these two settings. In cases where differential sensitivity (in vitro and/or in vivo) is observed for a particular agent (or class of agents) between different tumor lines, expression profiling can be used to identify genes that may be potentially used as predictive markers of activity in a patient population in the clinical setting.[45] Histological analysis of basic morphologic characteristics such as degree of necrosis at various tumor sizes, relative extent of tumor angiogenesis (e.g., microvessel density or MVD), and proliferative/apoptotic index provides a more sophisticated assessment of tumor growth properties. Furthermore, IHC methods can be used to quantify, at the cellular level, drug-mediated effects on proliferation and/or apoptosis and MVD.[46] IHC can also be used to examine target gene expression within a tumor as well as study downstream proteins that can be used as pharmacodynamic markers of activity in vivo. The histological and immunohistochemical analysis of tumors has been greatly aided by the development of tissue microarray technology.[47] With this method, small cores of multiple (as many as several hundred) tumors are collected and reassembled in a single paraffin block. In this way many different tumor samples can be analyzed on a single slide.

Concern about the predictive ability of xenograft models has been reinforced by some notable examples of agents that have shown (to date) relatively limited clinical activity despite quite significant antitumor activity (including tumor regression) in xenograft models.[48,49] In preclinical xenograft models, several farnesyl transferase inhibitors (FTIs) have demonstrated significant preclinical activity in colon, bladder, lung, breast, and other tumors but, for the most part, have shown limited clinical activity.[50] Although these results certainly raise concerns about interpretations of xenograft efficacy (note that at least one FTI also showed significant activity in a transgenic model expressing oncogenic *ras*), one must be cautious to dismiss the efficacy observed in the preclinical models as uninformative. As discussed above, analysis of a broader set of models of each tumor type may help elucidate determinants that contribute to clinical resistance. In addition, the stratification of patients based on these selective criteria could significantly improve chances of a clinical response.

In the case of EGFR inhibitors, clinical evaluation of a very large set of patients revealed that specific mutations in the target were likely to underlie sensitivity.[51] In other cases the disconnect between preclinical activity and poor clinical response may not be reflective of the model but rather due to differences in PK and/or metabolism of the experimental agent in humans compared to rodents. Peterson and Houghton[26] describe several examples of agents (e.g., irofulven/MGI-114: MGI Pharma) that showed significant activity in xenograft models, yet failed in the clinic for the simple reason that it was not possible to obtain drug exposures in humans equivalent to exposures in mice that were required for activity. PK and metabolic interactions must also be considered when evaluating combination therapies. As it has now become standard practice to combine experimental agents with standard therapies early in both preclinical and clinical development, it is critical to demonstrate that potentiation or synergistic activity is not simply the result of an indirect enhanced exposure of the standard therapy.

7.02.5 Analysis of Cytotoxic Agents in Xenograft Models

Retrospective evaluation of activity of established cytotoxic agents in clinical trials versus preclinical models shows that, in general, activity in xenograft models is predictive of some level of clinical response. Several reports have detailed good correlations between xenograft response and clinical response for rhabdomyosarcoma, colon cancer, lung cancer (particularly small-cell lung cancer), breast cancer, and myeloma.[26,52,53] **Table 1** shows a partial compilation of data generated in our department examining the activity of numerous standard cytotoxic agents as well as selected novel targeted agents in a variety of xenograft and syngeneic flank tumor models. Both early treatment (ET) and staged tumor (ST) trials are included and it can be seen that in some cases the magnitude of response is quite similar in the two cases (e.g., paclitaxel in the A549 model), while in others there are significant differences in response (e.g., paclitaxel in breast cancer).

The data include several examples of cytotoxic agents that show significant xenograft activity that is reflective of clinical activity: irinotecan is highly active in several models of colon cancer, temozolomide is efficacious in a model of glioma and one of two models of melanoma, paclitaxel is active in models of bladder, breast, lung, and prostate cancer, rituximab is highly active in two models of B-cell lymphoma, and gemcitabine is active in a model of pancreatic cancer. Note, however, that this activity can be quite dependent on the dose and schedule employed (note irinotecan in colon cancer models). Correlating dose and schedule with PK (e.g., efficacy driven by C_{max}, area under the curve (AUC), time over threshold, etc.) is a vital aspect of preclinical evaluation in order to help define the optimum clinical schedule (once- versus twice-daily administration, continuous versus intermittent therapy, etc.). **Table 1** also highlights some of the differences in activity that can be observed for a given agent in different models of the same tumor type (note carboplatin and cisplatin in different models of lung cancer). There are also examples of drugs that are known to be clinically active yet show little to no activity in the representative models shown here (for example, 5-fluorouracil in colon cancer). While these results indicate a reasonable correlation between activity, it is much more difficult to correlate the extent of activity. For example, complete regression and/or curative activity can be attained with agents such as irinotecan in some models of colon cancer, but this extent of response is difficult to achieve clinically, resulting in part from toxicity-associated complications (algorithmic dosing/differential drug metabolism).[54]

On the other hand, there are also published reports suggesting that xenograft models are not predictive of cytotoxic activity, or that they are not predictive for a given tumor type.[23,28] To some extent these studies suffered from limited critical review of the legitimacy of the claims of preclinical activity and/or failure to consider issues such as whether optimal drug levels were achieved clinically. It is interesting to note, however, that there seems to be a trend toward more favorable correlations between preclinical and clinical activity when the xenografts used were from early passage tumor lines compared to lines that have been maintained in culture for extended periods of time.[28,52,53] This is in agreement with the assertion made by Peterson and Houghton that early passage lines are preferable as xenograft models.[26]

There is little debate that xenograft models (and probably all tumor models) have significant limitations as perfect predictors of clinical activity for experimental anticancer agents. However, it also appears quite clear that, when used properly and augmented by appropriate biological validation, these models can be very useful tools for the development of active agents, especially for broadly acting cytotoxic agents. The current challenge is to determine whether these models, or any existing tumor models, will be predictive of activity for the newer generation of 'targeted' therapies that are being developed. The first decade of attempts to develop more targeted therapies has resulted in a few wonderful success stories but also some great disappointments.[50,55–58] The tremendous unmet need of effective therapies for cancer patients has undoubtedly increased the pressure to initiate clinical trials with agents where significant preclinical efficacy was somewhat lacking.

Our ability to make significant advances in the development of cancer therapeutics is also hindered by the cumbersome and overly rigid methods of clinical evaluation. Lessons can be learnt from the antiviral field, where the utilization of cocktails of experimental agents led to profound improvements in clinical response.[59]

7.02.6 Drug Resistance and its Clinical Circumvention

Drug resistance is the most significant reason for treatment failure in cancer and there are several underlying causes for this phenomenon. Perhaps the most well-defined and researched factor is the presence of ABC transporters, including: P-glycoprotein (P-gp), the product of the *MDR1* gene; the MRP or ABCC transporter subfamily; ABCG2, a mitoxantrone transporter (also known as the breast cancer resistance protein (BCRP); the ABC transporter in placenta (ABCP); and mitoxantrone-resistance gene (*MXR*) (**Table 2**).

An understanding of other methods of tumor escape from sensitivity to cytotoxicity has resulted in opportunities for new drugs such as inhibitors of the Bcl-2 family proteins, inhibitors of p53/mdm2 interactions, proteasome family proteins, and poly(ADP-ribose) polymerase (PARP).[46,60–62] Bortezomib (Velcade: Millennium Pharmaceuticals) is an approved proteasome inhibitor that blocks nuclear factor κB (NFκB) by inhibiting the 26S proteasome and preventing IκB degradation, thereby preventing the activation of NFκB. Velcade has shown a significant effect on myeloma in clinical trials both in survival and the significant reduction of the M protein, a biomarker for myeloma. Velcade is also being tested in clinical trials to overcome cisplatin resistance in ovarian cancer.[63]

7.02.7 Toxicology by Organ Systems

Toxicities of therapeutic agents are summarized in **Table 3**.

7.02.8 Biological Markers and Clinical Development

Biological markers (biomarkers) are objectively measured characteristics or analytes evaluated as an indicator of normal and pathogenic processes as well as pharmacologic responses to therapeutic intervention.[64,65] Biomarkers can be used to diagnose, monitor disease, or predict a response (**Figure 3**). It is the predictive value of a biomarker that has great potential to accelerate and improve both drug discovery and development and ultimately increase patient benefit. As rather low success rates in translating preclinical to clinical activity and druggable target hit rates have resulted in a number of drugs being withdrawn from clinical trials/market,[66] the need for predictive biomarkers is apparent.

7.02.8.1 Types of Biomarkers

The concept of biological markers of disease is both old and new. Historically, markers for disease have been used to diagnose and monitor disease risk, progression, improvement, and severity.[67] Examples are clinical chemistry for many disease states, histological markers such as Papanicolaou smears for cervical cancer, magnetic resonance imaging (MRI) for multiple sclerosis, as well as serum markers associated with cancer states, i.e., prostate-specific antigen (PSA), carcinoembryonic antigen (CEA), and cancer antigen-125 (CA-125).[29,65,68–73] Markers that can be assessed from blood and/or urine samples are very reliable in determining the clinical status of the disease. In most references, the definitions of biomarkers and its categories are often used interchangeably, but for clarity, the functional categories of biomarkers will be used in this discussion (**Figure 3**).[64,65,67,74] Biomarkers can be categorized as disease biomarkers

Table 2 Chemotherapy substrates of 11 ABC transporters

HUGO name[a]	*ABCA2*	*ABCB1*	*ABCB4*	*ABCB11*	*ABCC1*	*ABCC2*	*ABCC3*	*ABCC4*	*ABCC5*	*ABCC6*	*ABCG2*
Common name	*ABC2*	*MDR1*	*MDR2 MDR3*	*SPgp BSEP*	*MRP1*	*MRP2 cMOAT*	*MRP3*	*MRP4*	*MRP5*	*MRP6*	*BCRP MXR ABCP*
Adriamycin		+			+	+					
Paclitaxel		+	+	+							
Vinblastine		+									
Vincristine		+			+	+					
Etoposide		+			+	+	+				
Cisplatin						+					
Organic anions					+	+	+	+	+	+	
Methotrexate					+	+	+	+		+	+
17β-E2G					+	+	+	+		+	
6-Mercaptopurine								+	+		
6Thioguanine								+	+		
Topotecan		+									+
SN-38						+					+
Mitoxantrone											+
Estramustine	+	+									

[a] 17β-E2G, 17β-Estradiol glucuronide, SN38, active metabolite of irinotecan.[103]

Table 3 Related toxicities of therapeutic agents

Therapeutic agent (generic names)	*Mode of action/drug classification*	*Some related toxicities*
Dactinomycin, mitoxantrone, bleomycin, mitomycin C, daunorubicin, doxorubicin, epirubicin, idarubicin	Antitumor antibiotics	Myelosuppression, cardiotoxicity, nausea/vomiting/diarrhea, hepatic, renal, pulmonary, mucositis
Cisplatin, carboplatin, oxaliplatin	Intercalating agents	Nephrotoxicity, myelosuppression, neurotoxicity, neuropathy, nausea/vomiting, anaphylactic reaction
Busulfan, chlorambucil, carmustine, lomustine, mechlorethamine, cyclophosphamide, thiotepa, ifosfamide, streptozocin melphalan, L-PAM	Alkylating agents	Myelosuppression, hypersensitivity, neurotoxicity, nausea/vomiting, skin, renal, hepatotoxicity, pulmonary skin, renal, hepatotoxicity, pulmonary
Procarbazine, temozolomide	Nonclassic alkylating agents	Myelosuppression, nausea/vomiting, flu-like symptoms, infection, headache/fatigue
Cytarabine, floxuridine, fludarabine, thioguanine, gemcitabine, methotrexate, pemetrexed, mercaptopurine, 5-fluorouracil, cladribine, capecitabine, hydroxyurea, trimetrexate	Antimetabolites	Myelosuppression, nausea/vomiting, immunosuppression, gastrointestinal, hepatotoxicity, pulmonary, skin, renal, cardiovascular
Leucovorin	Biochemical modulating agent	Leukopenia, gastrointestinal, skin, nausea/vomiting
Estramustine, vinorelbine, Vinblastine, vincristine	Antimicrotubule agents	Cardiovascular, respiratory, thrombosis, nausea/vomiting, gastrointestinal, neurologic, myelosuppression
Docetaxel, paclitaxel	Taxanes, antimitotic agents	Myelosuppression, hypersensitivity, respiratory, fluid retention, skin, neurotoxicity, diarrhea, cardiotoxicity
Etoposide, teniposide	Topoisomerase II inhibitors	Myelosuppression, nausea/vomiting, gastrointestinal, hypersensitivity, alopecia, mucositis, hypotension
Irinotecan, topotecan	Topoisomerase I inhibitors	Gastrointestinal, myelosuppression, diarrhea, central nervous system, transient hepatotoxicity, skin, respiratory
Aldesleukin, interferon-α	Biologic response modifiers	Flu-like symptoms (fever, chills, malaise), vascular leak syndrome, myelosuppression, hepatoxicity
Tamoxifen	Antiestrogen	Menopausal symptoms, fluid retention, tumor flare-associated toxicities
Megestrol acetate	Hormonal agent Estrogen receptor-modulating agent	Weight gain, thromboembolic events, nausea/vomiting, affects menses, heart failure
Thalidomide	Antiangiogenic, immunomodulatory	Teratogenic, neurologic, neutropenia, skin, sedation/fatigue, thromboembolic, constipation immunomodulatory
Gefitinib, erlotinib	Antiangiogenic EGFR/tyrosine kinase inhibitor	Pulmonary, potential embryotoxicity, diarrhea, skin, mild hepatotoxicity, asthenia/anorexia, nausea/vomiting
Bevacizumab	Antiangiogenic Monoclonal antibody to VEGF	Gastrointestinal, hemorrhage, wound healing, hypertension, proteinuria
Cetuximab	Antiangiogenic	

Table 3 Continued

Therapeutic agent (generic names)	*Mode of action/drug classification*	*Some related toxicities*
		Infusion-related (airway obstruction, urticaria, hypotension), pulmonary, cardiovascular, immunogenic, dermatologic
	Monoclonal antibody to EGFR	
Alemtuzumab	Monoclonal antibody to CD52	Infusion-related (fever, chills, fatigue, headache, etc.), immunosuppressive agent, myelosuppression
Trastuzumab	*Her2/neu* monoclonal antibody	Infusion-related (fever, chills, fatigue, headache, etc.), nausea/vomiting, cardiotoxicity, myelosuppression, respiratory, skin
Ibritumomab tiuxetan	Immunoconjugate of monoconal antibody to CD20 and indium-111 and yttrium-90	Infusion-related (fever, chills, urticaria, fatigue, headache), respiratory myelosuppression, asthenia, cardiovascular, gastrointestinal, edema
Rituximab	Monoclonal antibody to CD20	Infusion-related (fever, chills, fatigue, headache, etc.), tumor lysis syndrome, mucocutaneous reactions, arrhythmia/chest pain, renal
Tositumomab	Radioimmunotherapeutic, monoclonal antibody to CD20 linked to iodine-131	Infusion-related (fever, chills, urticaria, fatigue, headache, etc.), hyper-sensitivity/anaphylaxis, cytopenia, asthenia, infections, gastrointestinal, cardiovascular, central nervous system
Imatinib mesylate	*bcr-abl* tyrosine kinase inhibitor	Nausea/vomiting, transient edema, occasional myalgias, diarrhea, hemorrhage
Bortezomib	Reversible proteasome inhibitor	Fatigue/malaise, gastrointestinal, myelosuppression, peripheral neuropathy, nausea/vomiting, diarrhea, respiratory infection

Sources: http://www.fda.gov/cder/drug
http://www.thomsonhc.com/pdrel/librarian (requires PDR subscription)
http://www.nlm.nih.gov/medlineplus/aboutmedlineplus.html
http://www.cancercare.on.ca/index_drugformularydrugsbygenericandtradename.htm
http://www.accessdata.fda.gov/scripts/cder/onctools/druglist.cfm
http://www.cancersourcemd.com/drugdb3/index.cfm
***Disclaimer: Not for patient use. To be used as a research guide only.

(monitors and predictors of disease risk, progression, improvement, and severity); surrogate endpoints (substitutes or correlates with measurable clinical endpoints such as how a patient feels, functions, or survives, e.g., viral titer, serum cholesterol, blood pressure); pharmacogenomic biomarker (monitors or predicts drug response, e.g., *Her2/neu*, EGFR single nucleotide polymorphisms (SNPs), CYP2D6 variants); pharmacodynamic biomarkers (demonstrates that a drug is on target and correlates with drug plasma levels in vivo, e.g., receptor occupancy, kinase phosphorylation) and functional response biomarkers (demonstrates a functional effect of the drug on a relevant pathway or mechanism of action, e.g., apoptosis markers).

Disease biomarkers are usually based on large population studies that monitor diseases such as cancer (CEA, PSA),[69] diabetes (hemoglobin A1c), and autoimmunity (rheumatoid factor).[73] These markers not only allow the possibility for early detection that may influence success of treatment, but also serial analysis to potentially indicate whether disease

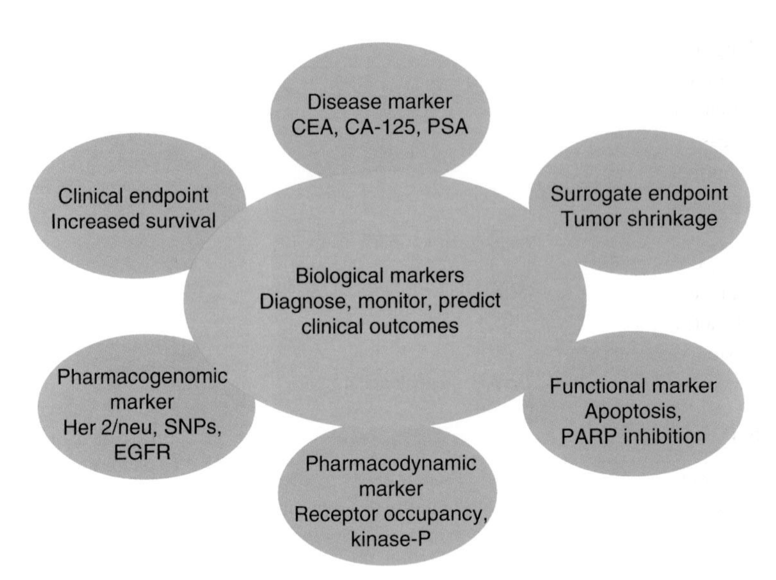

Figure 3 Types of biomarkers. CEA, carcinoembryonic antigen; CA-125, cancer antigen-125; PSA, prostate-specific antigen; SNPs, single nucleotide polymorphisms; EGFR, epidermal growth factor receptor; kinase-P, kinase phosphorylation; PARP inhibition, poly(ADP-ribose) polymerase inhibition.

is stable, progressing, or in remission. However, both inter- and/or intrapatient variability can complicate the interpretation of results.

Clinical endpoints are direct variables attributed to a response to therapeutic intervention and the gold standard in cancer trials is increase in long-term survival.[75] However, this takes time to achieve and can only be evaluated at the end-stages of a study. Consequently, surrogate endpoints are attractive since they could provide early signals of therapeutic benefit and aid in treatment decisions that could save time and suffering. However, not all surrogate endpoints may accurately predict clinical outcomes because only one parameter or effect of the drug is evaluated.[64] Although this may be of benefit in monitoring therapeutic treatment in devastating diseases such as cancer, it may be less useful in chronic diseases when drugs are taken over a long duration or there is no certainty of clinical benefit. Therefore, validation of surrogate markers must reflect the mechanism of action of the drug, pathophysiology of disease, or treatment benefit that correlates with clinical outcomes. Evaluating other drugs with similar mechanisms is important and can better ensure predictability. With the acceptance of imaging (DCE-MRI, high-resolution CT, positron emission tomography (PET), and multimodal PET-CT) that more accurately measures lesion size and/or metabolic assessment (use of tracers) for tumor growth,[76] the evaluation of efficacy of treatment regimens has increased. However, further validation of these technologies is still necessary. Owing to the difficulties associated with demonstrating definite predictive value, few biomarkers have attained US FDA approval as surrogate endpoints. Exceptions include blood pressure monitoring and cholesterol screening for cardiology; tumor shrinkage for oncology; and the $CD4^{+}$ cell count for human immunodeficiency virus (HIV)/acquired immune deficiency syndrome (AIDS).[77]

Recently, molecular profiles that can potentially predict drug response have been possible with advances in genomics. Examples of pharmacogenomic biomarkers are polymorphisms for enzymes involved in drug metabolism (i.e., CYP2D6 of the cytochrome P450 system and thiopurine *S*-methyltransferase or TPMT).[78] In December 2004, the FDA approved the first DNA microarray test (Amplichip Cytochrome P450 Genotyping Test: Roche Molecular Systems) for genetic variations in the *2D6* and *2C19* genes that have been associated with the predictive phenotypes of drug metabolism that affect drug efficacy and potential adverse reactions with certain drugs.[79]

Other genomic markers, such as *Her2/neu* expression in tumors are being screened to select patients more likely to respond to trastuzumab, the therapeutic monoclonal antibody to the (*Her2/neu*)/ErbB2 receptor (Herceptin: Genentech/Roche).[80,81] Similarly, imatinib (Gleevac: Novartis), a small-molecule inhibitor of the mutant in *bcr-abl* kinase, is an example where a subset of patients with a translocation in *bcr-abl* kinase predicts response to therapy.[58] EGFR/ErbB1/*Her1* mutations appear to correlate with a response to the small-molecule EGFR tyrosine kinase inhibitor, erlotinib (Tarceva: Genentech/OSI Oncology).[82–84] Additionally, the Oncotype Dx test (Genomic Health),[85] profiling 21 genes from tumor samples predicting recurrence in certain breast cancers and response to chemotherapy, is now on the market. These examples and many others in development have and will impact disease outcomes.

PD markers have long been used in phase I studies to determine the optimal dose by assessing serum/plasma levels of the drug in conjunction with safety and toxicology parameters. Other analytes in either blood or urine associated with disease can also be evaluated if the pathogenesis of disease is understood and it is in this context where discovery of novel biomarkers of disease may be greatly aided by genomic, proteomics and metabolomics.[67,86] Functional biomarkers, an extension of a pharmacodynamic type of biomarker, are biochemical measures that indicate the drug is reaching its target and affecting its proposed mechanism of action. Evaluating cell death in cancer therapies may soon be possible using an enzyme-linked immunosorbent assay (ELISA)-based assay measuring caspase 3-mediated apoptosis in solid tumors (Cyclacel, UK),[75] but it will be limited to accessible tumor types. Recently, functional biomarker assays for PARP inhibition were demonstrated in metastatic melanoma from patients in a phase I study.[87] Potentially, the correlation of PK and functional biomarkers could shed light on cancer treatments where variable efficacy is observed and thereby influence treatment decisions earlier in the clinical development process.

7.02.9 Tools for Biomarker Discovery

With the advent of new technologies and targeted therapies (described in the previous sections), there is continued interest in biomarker discovery for clinical development. Recent advances in genomic, proteomic, metabolomic, and imaging technologies continue to provide important tools in biomarker identification.[64,79] In addition, the assessment of potential biomarkers using antibody arrays that capitalize on humoral responses to tumor antigens has also garnered interest.[88]

7.02.9.1 Genomics

Most diseases are complex and the incomplete understanding of their pathogenic mechanisms hampers identification of useful biomarkers. Genomic analysis (SNPs, microarrays) may offer a solution to discover novel biomarkers such as identification of variants in the P450 system which are important in drug metabolism.[78,79] There are an estimated 10 million SNPs in the human genome[89,90] that could be a potential treasure trove of disease biomarkers. As previously discussed, the expression profiling of tumor types can aid clinical decisions (Oncotype Dx, Genomic Health).[85] While these molecular signatures have great potential, the translation of a genotype to distinct phenotypes can be challenging in light of many other confounding factors.[91] This is because diseases are multigenic in nature; therefore, multiple SNPs or a panel of biomarkers may prove more correlative to disease status.[92,93]

7.02.9.2 Proteomics and Metabolomics

There are an estimated 100 000 protein transcripts and 1000–10 000 metabolites that may similarly offer proteomic (protein analytes) or metabolomic (nonprotein metabolite) profiles of disease in tissue or biological fluids.[67] The potential advantage of proteomics and metabolomics is the ability to detect subclinical parameters.[94] The past two decades have produced serum-based proteins that monitor general disease conditions, for instance, CA-125 for ovarian cancer, CA-15-3 and 27.29 for breast cancer, CEA for ovarian, lung, breast, pancreas, and gastrointestinal cancers, and PSA for prostrate cancer.[95] Further, clinical chemistry has likewise proven useful in providing metabolic profiles that can aid in diagnosis of a variety of diseased conditions.[96] However, the complexity of the sample analyzed can greatly impact its predictive value. While the number of transcripts and metabolites increases possibilities, a disadvantage is that relevant biomarkers may be difficult to identify if obscured by more abundant proteins. Nevertheless, using advances such as laser scanning cytometry,[64] combined liquid chromatography/mass spectroscopy (LC/MS),[97] multiplex immunoassays (xMAP technology/Luminex),[79,98,99] and array platforms (Ciphergen, Compugen, Icoria)[79,100,101] to identify novel biomarkers in the proteome and metabolome holds much promise.

7.02.10 Challenges in Biomarker Development

While many challenges remain, the important role of biomarkers in diagnosis, monitoring, and predicting disease is unquestionable.[65,73,86] Technology has advanced but the identification of novel biomarkers remains a challenge due to the samples being analyzed containing an abundance of analytes (known/unknown genes, protein/protein fragments, metabolic factors) expressed in a wide dynamic range.[102] Moreover, differences between normal and patient populations at this level may not be readily apparent due to lack of comprehensive data. It can still be difficult to correlate a genotypic, proteomic, metabolomic, or biochemical profile to a distinct phenotype. Factors such as variable penetrance, variations in disease phenotype, and/or environmental factors resulting in varied clinical manifestations make biomarkers less likely to be fully predictive of disease outcome. Therefore, extensive characterization and validation of the biomarker are crucial.

Rigorous evaluation of biomarkers in preclinical models is important whenever possible, but it has limitations. Physiological differences between humans and animal models can hinder translation to the clinic. Though costly, incorporation of biomarkers in clinical trials is necessary to achieve full validation of a particular biomarker. The consideration of scale and population studied (including the proper controls) must also be carefully considered. In the initial Cardiac Arrhythmic Suppression Trial (CAST), the surrogate marker used (electrocardiograms) did not demonstrate a similar benefit when a large, multiyear follow-up study with proper controls was conducted.[79]

Expectedly, technical and regulatory hurdles also pose significant challenges for the biomarker field.[64,65,73] The measures necessary (reproducibility, statistical performance, and predictive value) prior to classification as a bona fide biomarker are not completely defined. Apart from assay validation (sensitivity and specificity), reagent availability, standardization, and information management (medical records privacy, intellectual property) across institutions have yet to be universally implemented.[86] Rectifying this problem will require a concerted effort from academic, government, research, and pharmaceutical institutions. A consortium has recently been established by the NCI (and several programs by the National Institutes of Health) to accelerate biomarker discovery and clinical application.[65]

The need not only to diagnose and monitor disease, but also to predict clinical benefit/outcome has become increasingly important in drug development. Clinical outcome is most often assessed by clinical endpoints as well as surrogate endpoints that may only be apparent later in development (clinical trials).[64,74] Attempts have been made to co-develop biomarkers early in drug discovery/development rather than investigate them retrospectively, to gain maximum measurable benefit to the patient as well as streamline drug development, minimizing costs by early identification of nonviable drug candidates. The predictive value of a validated biomarker may also reduce sample size by identification of patients most likely to have a positive response to therapy (*Her2/neu* and Herceptin) and abbreviate the duration of studies accelerating go/no-go decisions.

Despite existing challenges that may preclude attempts to incorporate biomarkers in clinical practice, remarkable strides to date have demonstrated proof of principle. Altogether, biomarkers have immense potential to impact not only the initial phases of drug discovery/later development but also in patient care, monitoring, and the early evaluation of successful therapies.

References

1. Parkin, D. M. *Lancet Oncol.* **2001**, *2*, 533–543.
2. Karnofsky, D. A. *Ann. NY Acad. Sci.* **1958**, *68*, 899–914.
3. Banerji, U.; Judson, I.; Workman, P. Molecular Targets. In *Handbook of Anticancer Pharmacokinetics and Pharmacodynamics*; Figg, W. D., Mcleod, H. L., Eds.; Humana Press: New Jersey, 2004, pp 1–27.
4. Beelen, A.; Lewis, L. D. Clinical Pharmacology Overview. In *Handbook of Anticancer Pharmacokinetics and Pharmacodynamics*; Figg, W. D., Mcleod, H. L., Eds.; Humana Press: New Jersey, 2004, pp 111–127.
5. Dendy, P.; Hill, B. T. *Human Tumor Drug Sensitivity Testing In Vitro: Techniques and Clinical Applications*; Academic Press: New York, 1983.
6. Shoemaker, R.; McLemore, T. L.; Abbott, B. J.; Fine, D. L.; Gorelik, E.; Mayo, J. G.; Fodstad, O.; Boyd, M. R. Human Tumor Xenograft Models for Use with an In Vitro-Based Disease-Oriented Anti-Tumor Drug Screening Program. In *Human Tumor Xenografts in Anticancer Drug Development*; Winograd, B., Peckham, M. J., Pinedo, H. M., Eds.; Springer-Verlag: Berlin, 1988, pp 115–120.
7. Hollingshead, M.; Plowman, J.; Alley, M. C.; Mayo, J.; Sausville, E. The Hollow Fiber Assay. In *Contributions to Oncology: Relevance of Tumor Models for Anticancer Drug Development*; Fiebig, H. H., Burger, A. M., Eds.; Karger: Basel, 1999, pp 109–120.
8. Aherne, W.; Garret, M.; McDonald, T.; Workman, P. Mechanism-Based High-Throughput Screening for Novel Anticancer Drug Discovery. In *Anticancer Drug Development*; Baguley, B. C., Kerr, D. J., Eds.; Academic Press: London, UK, 2002, pp 249–267.
9. Teischer, B.; Andrews, P. A. *Anticancer Drug Development Guide*, 2nd ed.; Humana Press: Tarrytown, NY, 2004.
10. Harrison, S. Perspectives on the History of Tumor Models. In *Tumor Models in Cancer Research*; Humana Press: Totowa, NJ, 2002; pp 3–19.
11. Suggitt, M.; Bibby, M. C. *Clin. Cancer Res.* **2005**, *11*, 971–981.
12. Schabel, F. M. J.; Griswold, D. P., Jr.; Laster, W. R., Jr.; Corbett, T. H.; Lloyd, H. H. *Pharm. Ther.* **1977**, *1*, 411–435.
13. Frei, E., III *Cancer* **1984**, *53*, 2013–2025.
14. Dykes, D. J.; Waud, W. R. Murine L1210 and P388 Leukemias. In *Tumor Models in Cancer Research*; Teicher, B. A., Ed.; Humana Press: Totowa, NJ, 2002, pp 23–40.
15. Giovanella, B. C.; Fogh, J. *Adv. Cancer Res.* **1985**, *44*, 69–120.
16. Bosma, M. J.; Carroll, A. M. *Annu. Rev. Immunol.* **1991**, *9*, 323–350.
17. Fogh, J.; Fogh, J. M.; Orfeo, T. *J. Natl. Cancer Inst.* **1977**, *59*, 221–226.
18. Teicher, B. A. In Vivo Tumor Response End Points. In *Tumor Models in Cancer Research*, 1st ed.; Teicher, B. A., Ed.; Humana Press: Totowa, NJ, 2002; pp 593–616.
19. Pocard, M.; Tsukui, H.; Salmon, R. J.; Dutrillaux, B.; Poupon, M. F. *In Vivo* **1996**, *10*, 463–469.
20. Burgos, J. S.; Rosol, M.; Moats, R. A.; Khankaldyyan, V.; Kohn, D. B.; Nelson, M. D., Jr.; Laug, W. E. *Biotechniques* **2003**, *34*, 1184–1188.
21. Campostrini, N.; Pascali, J.; Hamdan, M.; Astner, H.; Marimpietri, D.; Pastorino, F.; Ponzoni, M.; Righetti, P. G. *J Chromatogr. B. Analyt. Technol. Biomed. Life Sci.* **2004**, *808*, 279–286.
22. Zambrowicz, B. P.; Sands, A. T. *Nat. Rev. Drug Disc.* **2003**, *2*, 38–51.

23. Voskoglou-Nomikos, T.; Pater, J. L.; Seymour, L. *Clin. Cancer Res.* **2003**, *9*, 4227–4239.
24. Kelland, L. R. *Eur. J. Cancer* **2004**, *40*, 827–836.
25. Leaf, C. *Fortune* **2004**, *149*, 76–97.
26. Peterson, J. K.; Houghton, P. J. *Eur. J. Cancer* **2004**, *40*, 837–844.
27. Kamb, A. *Nat. Rev. Drug Disc.* **2005**, *4*, 161–165.
28. Johnson, J. I.; Decker, S.; Zaharevitz, D.; Rubinstein, L. V.; Venditti, J. M.; Schepartz, S.; Kalyandrug, S.; Christian, M.; Arbuck, S.; Hollingshed, M. et al. *Br. J. Cancer* **2001**, *84*, 1424–1431.
29. Holland, E. *Mouse Models of Human Cancer*; A. John Wiley: New Jersey, 2004.
30. Kisker, O.; Onizuka, S.; Banyard, J.; Komiyama, T.; Becker, C. M.; Achilles, E. G.; Barnes, C. M.; O'Reilly, M. S.; Folkman, J.; Pirie-Shepherd, S. R. *Cancer Res.* **2001**, *61*, 7298–7304.
31. Alley, M. C.; Hollingshead, M. G.; Dykes, D. J.; Waud, W. R. Human Tumor Xenograft Models in NCI Drug Development. In *Anticancer Drug Development Guide*, 2nd ed.; Teicher, B. A., Andrews, P. A., Eds.; Humana Press: Totowa, NJ, 2004, pp 125–152.
32. Nelson, N. J. *J. Natl. Cancer Inst.* **1998**, *90*, 960–963.
33. Thomas, J. P.; Arzoomanian, R. Z.; Alberti, D.; Marnocha, R.; Lee, F.; Friedl, A.; Tutsch, K.; Dresen, A.; Geiger, P.; Pluda, J. et al. *J. Clin. Oncol.* **2003**, *21*, 223–231.
34. Goetz, M. P.; Toft, D.; Reid, J.; Ames, M.; Stensgard, B.; Safgren, S.; Adjei, A. A.; Sloan, J.; Atherton, P.; Vasile, V. et al. *J. Clin. Oncol.* **2005**, *23*, 1078–1087.
35. Price, J. T.; Quinn, J. M.; Sims, N. A.; Vieusseux, J.; Waldeck, K.; Docherty, S. E.; Myers, D.; Nakamura, A.; Waltham, M. C.; Gillespie, M. T. et al. *Cancer Res.* **2005**, *65*, 4929–4938.
36. Zondor, S. D.; Medina, P. J. *Ann. Pharmacother.* **2004**, *38*, 1258–1264.
37. Rosen, L. S. *Oncologist* **2005**, *10*, 382–391.
38. Vermeulen, S. J.; Chen, T. R.; Speleman, F.; Nollet, F.; Van Roy, F. M.; Mareel, M. M. *Cancer Genet. Cytogenet.* **1998**, *107*, 76–79.
39. Chen, T. R.; Dorotinsky, C. S.; McGuire, L. J.; Macy, M. L.; Hay, R. J. *Cancer Genet. Cytogenet.* **1995**, *81*, 103–108.
40. Drexler, H. G.; Dirks, W. G.; Matsuo, Y.; MacLeod, R. A. *Leukemia* **2003**, *17*, 416–426.
41. Roschke, A. V.; Tonon, G.; Gehlhaus, K. S.; McTyre, N.; Bussey, K. J.; Lababidi, S.; Scudiero, D. A.; Weinstein, J. N.; Kirsch, I. R. *Cancer Res.* **2003**, *63*, 8634–8647.
42. Learn, C. A.; Hartzell, T. L.; Wikstrand, C. J.; Archer, G. E.; Rich, J. N.; Friedman, A. H.; Friedman, H. S.; Bigner, D. D.; Sampson, J. H. *Clin. Cancer Res.* **2004**, *10*, 3216–3224.
43. Perea, S.; Hidalgo, M. *Clin. Lung Cancer* **2004**, *6*, S30–S34.
44. Pao, W.; Miller, V. A.; Politi, K. A.; Riely, G. J.; Somwar, R.; Zakowski, M. F.; Kris, M. G.; Varmus, H. *PLoS Med.* **2005**, *2*, e73.
45. Staunton, J. E.; Slonim, D. K.; Coller, H. A.; Tamayo, P.; Angelo, M. J.; Park, J.; Scherf, U.; Lee, J. K.; Reinhold, W. O.; Weinstein, J. N. et al. *Proc. Natl. Acad. Sci.* **2001**, *98*, 10787–10792.
46. Oltersdorf, T.; Elmore, S. W.; Shoemaker, A. R.; Armstrong, R. C.; Augeri, D. J.; Belli, B. A.; Bruncko, M.; Deckwerth, T. L.; Dinges, J.; Hajduk, P. J. et al. *Nature* **2005**, *435*, 677–681.
47. Bubendorf, L.; Nocito, A.; Moch, H.; Sauter, G. *J. Pathol.* **2001**, *195*, 72–79.
48. Cohen, S. J.; Ho, L.; Ranganathan, S.; Abbruzzese, J. L.; Alpaugh, R. K.; Beard, M.; Lewis, N. L.; McLaughlin, S.; Rogatko, A.; Perez-Ruixo, J. J. et al. *J. Clin. Oncol.* **2003**, *21*, 1301–1306.
49. Sawyers, C. L. *Cancer Cell* **2003**, *4*, 343–348.
50. Zhu, K.; Hamilton, A. D.; Sebti, S. M. *Curr. Opin. Invest. Drugs* **2003**, *4*, 1428–1435.
51. Shih, J. Y.; Gow, C. H.; Yang, P. C. *N. Engl. J. Med.* **2005**, *353*, 207–208.
52. Steel, G. G.; Courtenay, V. D.; Peckham, M. J. *Br. J. Cancer* **1983**, *47*, 1–13.
53. Fiebig, H. H.; Maier, A.; Burger, A. M. *Eur. J. Cancer* **2004**, *40*, 802–820.
54. Seiter, K. *Exp. Opin. Drug Safe.* **2005**, *4*, 45–53.
55. Twombly, R. *J. Natl. Cancer Inst.* **2002**, *94*, 1520–1521.
56. Avivi, I.; Robinson, S.; Goldstone, A. *Br. J. Cancer* **2003**, *89*, 1389–1394.
57. Vogel, C. L.; Franco, S. X. *Breast J.* **2003**, *9*, 452–462.
58. Druker, B. J. *Adv. Cancer Res.* **2004**, *91*, 1–30.
59. Komarova, N. L.; Wodarz, D. *Proc. Natl. Acad. Sci.* **2005**, *102*, 9714–9719.
60. Kane, R. C.; Bross, P. F.; Farrell, A. T.; Pazdur, R. *Oncologist* **2003**, *8*, 508–513.
61. Vassilev, L. T.; Vu, B. T.; Graves, B.; Carvajal, D.; Podlaski, F.; Filipovic, Z.; Kong, N.; Kammlott, U.; Lukacs, C.; Klein, C. et al. *Science* **2004**, *303*, 844–848.
62. Graziani, G.; Szabo, C. *Pharmacol. Res.* **2005**, *52*, 109–118.
63. Dizon, D.; Aghajanian, C.; Jun Yan, X.; Spriggs, D. Targeting NF-kappaB to Increase the Activity of Cisplatin in Solid Tumors. In *Combination Cancer Therapy*; Humana Press: New Jersey, 2005, pp 197–208.
64. Frank, R.; Hargreaves, R. *Nat. Rev. Drug Disc.* **2003**, *2*, 566–580.
65. Srivastava, S.; Srivastava, R. G. *J. Proteome Res.* **2005**, *4*, 1098–1103.
66. Need, A. C.; Motulsky, A. G.; Goldstein, D. B. *Nat. Genet.* **2005**, *37*, 671–681.
67. Seo, D.; Ginsburg, G. S. *Curr. Opin. Chem. Biol.* **2005**, *9*, 381–386.
68. Shingleton, H. M.; Patrick, R. L.; Johnston, W. W.; Smith, R. A. *CA Cancer J. Clin.* **1995**, *45*, 305–320.
69. Bast, R. C., Jr.; Ravdin, P.; Hayes, D. F.; Bates, S.; Fritsche, H., Jr.; Jessup, J. M.; Kemeny, N.; Locker, G. Y.; Mennel, R. G.; Somerfield, M. R. *J. Clin. Oncol.* **2001**, *19*, 1865–1878.
70. Dhanasekaran, S. M.; Barrette, T. R.; Ghosh, D.; Shah, R.; Varambally, S.; Kurachi, K.; Pienta, K. J.; Rubin, M. A.; Chinnaiyan, A. M. *Nature* **2001**, *412*, 822–826.
71. Duffy, M. *J. Clin. Chem.* **2001**, *47*, 624–630.
72. Grizzle, W. E.; Manne, U.; Jhala, N. C.; Weiss, H. L. *Arch. Pathol. Lab. Med.* **2001**, *125*, 91–98.
73. LaBaer, J. *J. Proteome Res.* **2005**, *4*, 1053–1059.
74. Group, B. D. W. *Clin. Pharmacol. Ther.* **2001**, *69*, 89–95.
75. Bouchie, A. *Nat. Biotechnol.* **2004**, *22*, 6–7.
76. Therasse, P.; Arbuck, S. G.; Eisenhauer, E. A.; Wanders, J.; Kaplan, R. S.; Rubinstein, L.; Verweij, J.; Van Glabbeke, M.; van Oosterom, A. T.; Christian, M. C. et al. *J. Natl. Cancer Inst.* **2000**, *92*, 205–216.

77. Rolan, P.; Atkinson, A. J., Jr.; Lesko, L. J. *Clin. Pharmacol. Ther.* **2003**, *73*, 284–291.
78. Weinshilboum, R.; Wang, L. *Nat. Rev. Drug Disc.* **2004**, *3*, 739–748.
79. Baker, M. *Nat. Biotechnol.* **2005**, *23*, 297–304.
80. Baselga, J.; Gianni, L.; Geyer, C.; Perez, E. A.; Riva, A.; Jackisch, C. *Semin. Oncol.* **2004**, *31*, 51–57.
81. Ross, J. S.; Fletcher, J. A.; Bloom, K. J.; Linette, G. P.; Stec, J.; Symmans, W. F.; Pusztai, L.; Hortobagyi, G. N. *Mol. Cell Proteomics* **2004**, *3*, 379–398.
82. Lynch, T. J.; Bell, D. W.; Sordella, R.; Gurubhagavatula, S.; Okimoto, R. A.; Brannigan, B. W.; Harris, P. L.; Haserlat, S. M.; Supko, J. G.; Haluska, F. G. et al. *N. Engl. J. Med.* **2004**, *350*, 2129–2139.
83. Paez, J. G.; Janne, P. A.; Lee, J. C.; Tracy, S.; Greulich, H.; Gabriel, S.; Herman, P.; Kaye, F. J.; Lindeman, N.; Boggon, T. J. et al. *Science* **2004**, *304*, 1497–1500.
84. Tsao, M. S.; Sakurada, A.; Cutz, J. C.; Zhu, C. Q.; Kamel-Reid, S.; Squire, J.; Lorimer, I.; Zhang, T.; Liu, N.; Daneshmand, M. et al. *N. Engl. J. Med.* **2005**, *353*, 133–144.
85. Paik, S.; Shak, S.; Tang, G.; Kim, C.; Baker, J.; Cronin, M.; Baehner, F. L.; Walker, M. G.; Watson, D.; Park, T. et al. *N. Engl. J. Med.* **2004**, *351*, 2817–2826.
86. Aebersold, R.; Anderson, L.; Caprioli, R.; Druker, B.; Hartwell, L.; Smith, R. *J. Proteome Res.* **2005**, *4*, 1104–1109.
87. Plummer, E. R.; Middleton, M. R.; Jones, C.; Olsen, A.; Hickson, I.; McHugh, P.; Margison, G. P.; McGown, G.; Thorncroft, M.; Watson, A. J. et al. *Clin. Cancer Res.* **2005**, *11*, 3402–3409.
88. Anderson, K. S.; LaBaer, J. *J. Proteome Res.* **2005**, *4*, 1123–1133.
89. Kruglyak, L.; Nickerson, D. A. *Nat. Genet.* **2001**, *27*, 234–236.
90. Lai, E. *Genome Res.* **2001**, *11*, 927–929.
91. Weiss, S. T.; Silverman, E. K.; Palmer, L. J. *Pharmacogenomics J.* **2001**, *1*, 157–158.
92. Bell, J. *Nature* **2004**, *429*, 453–456.
93. Seo, D.; Wang, T.; Dressman, H.; Herderick, E. E.; Iversen, E. S.; Dong, C.; Vata, K.; Milano, C. A.; Rigat, F.; Pittman, J. et al. *Arterioscler. Thromb. Vasc. Biol.* **2004**, *24*, 1922–1927.
94. McDonald, W. H.; Yates, J. R. *Dis. Markers* **2002**, *18*, 99–105.
95. Perkins, G. L.; Slater, E. D.; Sanders, G. K.; Prichard, J. G. *Am. Fam. Phys.* **2003**, *68*, 1075–1082.
96. Anderson, N. L.; Anderson, N. G. *Mol. Cell Proteomics* **2002**, *1*, 845–867.
97. Wang, W.; Zhou, H.; Lin, H.; Roy, S.; Shaler, T. A.; Hill, L. R.; Norton, S.; Kumar, P.; Anderle, M.; Becker, C. H. *Anal. Chem.* **2003**, *75*, 4818–4826.
98. Dunbar, S. A. *Clin. Chim. Acta* **2006**, *363*, 71–82.
99. Pang, S.; Smith, J.; Onley, D.; Reeve, J.; Walker, M.; Foy, C. *J. Immunol. Methods* **2005**, *302*, 1–12.
100. Tang, N.; Tornatore, P.; Weinberger, S. R. *Mass. Spectrom. Rev.* **2004**, *23*, 34–44.
101. Wiesner, A. *Curr. Pharm. Biotechnol.* **2004**, *5*, 45–67.
102. Liotta, L. A.; Ferrari, M.; Petricoin, E. *Nature* **2003**, *425*, 905.
103. Bates, S. E.; Fojo, T. ABC Transporters: Involvement in Multidrug Resistance and Drug Disposition. In *Handbook of Anticancer Pharmacokinetics and Pharmacodynamics*; Figg, W. D., McLeod, H. L., Eds.; Humana Press: New Jersey, 2004, pp 267–288.

Biographies

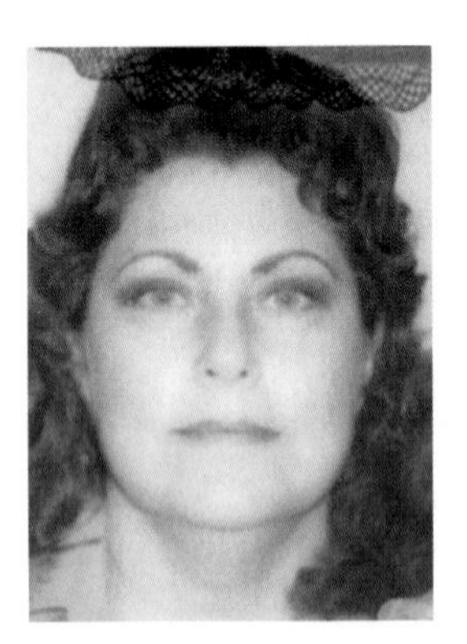

Cherrie K Donawho received her BA in Zoology from the University of Texas at Austin in 1973, and Medical Technology (ASCP) in 1974. She worked in microbiology and blood banking areas of the hospital laboratory and received her Specialty in Blood Banking (ASCP) in 1980. She was a research manager for the development of biological laboratory reagents with Gamma Biologicals, Houston, Texas, until 1987. From 1987 to 1990 she was a doctoral student at the University of Texas Graduate School of Biomedical Sciences at Houston, where she received her PhD degree in 1990 in Biomedical Sciences. Her doctoral work was carried out at the University of Texas M D Anderson Cancer Center, focusing on isolation and cell culture of primary melanoma tumor lines recovered from chemical/UV-irradiation carcinogenesis in mice. She used these cell lines as useful tumor immunology/cancer models to examine the role of UV-irradiation in the increased incidence and progression of murine melanoma. After receiving her PhD, she remained at the University of Texas M D Anderson Cancer Center, where she continued to work on the tumor immunology and biology of melanoma using in vivo tumor models and immunohistochemistry to examine changes in the host–tumor microenvironment. In 2002, as an Assistant Professor in the Cancer Biology Department, she left to join the In Vivo Tumor Biology Department at Abbott Laboratories where she heads an in vivo tumor biology group supporting the discovery and development of multitargeted kinease inhibitors, antiangiogenic agents, and monoclonal antibody evaluation.

Alexander R Shoemaker received his BS and PhD degrees in genetics from the University of Wisconsin-Madison. His doctoral work was carried out at the McArdle Laboratory for Cancer Research and focused on genetic factors that influence intestinal tumorigenesis in mice. After completing his thesis work in 1997, Dr Shoemaker conducted post-doctoral research at the Ludwig Institute for Cancer Research in La Jolla, CA, where he studied the genetics of DNA repair and chromosomal instability. Dr Shoemaker joined the In Vivo Tumor Biology Department at Abbott Laboratories in 2000, where he heads an in vivo tumor biology group focusing on chemotherapeutic regulators of apoptotic processes.

Joann P Palma finished her PhD graduate work at Michigan State University studying immune mechanisms affected by the platinum drugs cisplatin and carboplatin. This led to post graduate work delineating the immune-mediated pathogenic mechanisms of a virally-induced autoimmune mouse model for human multiple sclerosis at Northwestern University Medical School. Currently, she holds a senior scientist position in cancer drug discovery in the In Vivo Tumor Biology Department at Abbott Laboratories working on various projects, including development of biological markers.

Comprehensive Medicinal Chemistry II
ISBN (set): 0-08-044513-6

ISBN (Volume 7) 0-08-044520-9; pp. 33–53

7.03 Antimetabolites

M M Mader and J R Henry, Lilly Research Laboratories, Indianapolis, IN, USA

7.03.1 Disease State

Cancer broadly describes diseases in which abnormal cells divide without control. Cancer cells can invade nearby tissues and can spread through the bloodstream and lymphatic system to other parts of the body. The US National Institutes of Health (NIH) defines four general types of cancer.[1] Carcinoma is cancer that begins in the skin or in tissues that line or cover internal organs. Sarcoma is cancer that begins in bone, cartilage, fat, muscle, blood vessels, or other connective or supportive tissue. Leukemia is cancer that starts in blood-forming tissue such as the bone marrow, and causes large numbers of abnormal blood cells to be produced and enter the bloodstream. Lymphoma and multiple myeloma are cancers that begin in the cells of the immune system. The antimetabolite chemotherapeutics described in this review have generally been used to treat leukemias, lymphomas, and carcinomas.

Cancer of all types is the second leading cause of death in the USA, with 22.9% of deaths attributed to cancer in 2001, and it is estimated that almost 1.4 million new cases will be diagnosed in 2005 in the USA.[2] Cancer is a more significant health issue in the developed countries than in less developed nations, with incidence for all cancers almost twofold higher in the former. The annual incidence per 100 000 is 271 cases in developed nations such as the USA, Canada, and the countries of the European Union in comparison to 144 cases in less developed nations.[3] The US *Annual Report to the Nation on the Status of Cancer, 1975–2001*[4] shows overall cancer incidence rates increased at a rate of 1.5% per year over the

Table 1 Cancer incidence, mortality, and 5-year relative survival rates in the USA

Sites	*Incidence*[a] *(per 100 000)*	*5-Year relative survival (%)*	*Study period*[b]
All sites	470.3	64.1	Current
	252.9 (est.)	35	Historic
Leukemias	12.1	46.4	Current
	9.4	10	Historic
Lymphomas	21.7	63.3	Current
	8.5	30	Historic
Lung	61.7	15.2	Current
	17.0	6	Historic
Colon and rectum	53.7	63.4	Current
	45.9	37	Historic
Pancreas	11.0	4.2	Current
	7.9	1.0	Historic

[a] Incidence rates are age-adjusted to the 2000 US standard population by 5-year age groups.
[b] 'Current' data for incidence and survival are for the period 1997-2001 as reported by Ries[4]; 'Historic' data for incidence is for the period 1947–50 as reported by Devesa[122]; data for 5-year survival encompass 1950–54.[4]

period 1950–2001, for a net increase of 85.9% from approx 253 cases to 470 cases per 100 000, but 5-year survival rates improved from 35% to 65% over the same period for all cancers (**Table 1**). The increased incidence in the USA over time, and in developed nations, can be attributed largely to increased life expectancy in the populations. The impact of antimetabolite oncolytics is reflected in significantly improved survival rates for the indications for which these drugs are used, most notably in leukemias and lymphomas. Antimetabolites have only more recently been approved for use in solid tumors of the lung, pancreas, and colon/rectum and although survival rates for these cancers have doubled over 50 years, the rates remain low relative to leukemia and lymphoma.

7.03.2 Disease Basis

Cancers are characterized by abnormal cell proliferation, but the causes for this are myriad, even for cancers in the same tissue. Antimetabolite oncology agents work by interfering with the formation or utilization of a normal cellular metabolite. Most antimetabolites interfere with the enzymes involved in the synthesis of new DNA, are incorporated into the newly formed DNA, or in some cases both processes are important to an agent's efficacy. As a result, many antimetabolites are derivatives of the building blocks of DNA itself, such as the nucleoside based inhibitors, or analogs of critical cofactors such as the antifolates. A variety of key cellular pathways can be disrupted with antimetabolite therapy, via inhibition of the thymidine and purine nucleotide biosynthesis pathway and inhibition of ribonucleoside reductase. Given their mechanisms of action, it is not surprising that the observed benefits of antimetabolites are often accompanied by significant toxicity, due to the fact that the affected cellular metabolites are critical to both normal and cancer cells. Single antimetabolite agents can act on a single pathway, or on multiple pathways at once, but in either instance, they are often used in combination with other therapies in the clinic.

7.03.2.1 Thymidine Biosynthesis

Critical to the cell's process of replication is its ability to synthesize thymidine.[5,6] This process involves several key enzymes including thymidylate synthase (TS), dihydrofolate reductase (DHFR), and serine hydroxymethyl transferase (SHMT) (**Figure 1**). The methylation of deoxyuridine 5′-monophosphate (dUMP) to produce deoxythymidine 5′-monophosphate (dTMP) is mediated by TS.[7] The methyl group for dTMP is provided by N5,N10-methylene tetrahydrofolate (N5,N10-CH_2-THF) through its conversion to 7,8-dihydrofolate (7,8-DHF). The 7,8-DHF must then

Figure 1 Key steps in thymidine biosynthesis. (Reprinted from Henry, J. R.; Mader, M. M. Recent Advances in Antimetabolite Cancer Chemotherapies. In *Annual Reports in Medicinal Chemistry*; Doherty, A. M., Ed.; Elsevier: Oxford, UK, 2004, pp 161–172, with permission from Elsevier.)

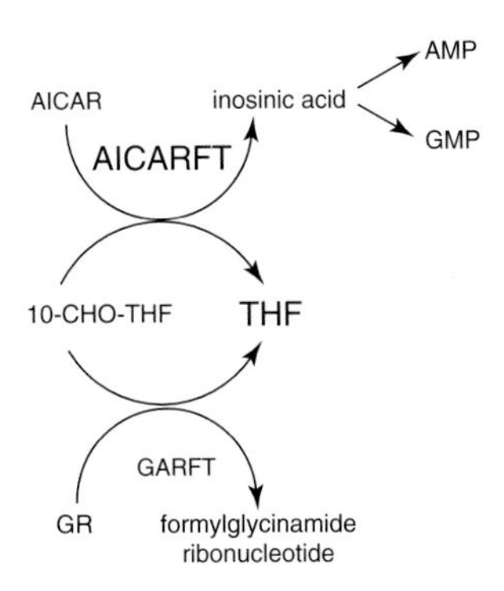

Figure 2 Key steps in purine biosynthesis. GMP, guanine monophosphate. (Reprinted from Henry, J. R.; Mader, M. M. Recent Advances in Antimetabolite Cancer Chemotherapies. In *Annual Reports in Medicinal Chemistry*; Doherty, A. M., Ed.; Elsevier: Oxford, UK, 2004, pp 161–172, with permission from Elsevier.)

be converted to tetrahydrofolate (THF) by DHFR,[8] followed by further transformation back to N5,N10-CH_2-THF through the action of SHMT.[9] Therefore, inhibition of TS, DHFR, or SHMT with an appropriate agent would interrupt the process of thymidine biosynthesis. Low thymidine levels cause defects in DNA that in turn activate stress response elements, such as the Fas ligand/Fas death pathway leading to apoptosis.[10] It has also been proposed that defects in this Fas-dependent apoptotic signaling pathway are one cause of cellular resistance to drugs.

7.03.2.2 Purine Nucleotide Synthesis

The cell's ability to provide the needed purine nucleotides for DNA and RNA synthesis is also critical to its survival. The de novo biosynthesis of purine nucleotides involves 10 separate enzyme catalyzed reactions starting with 5-phosphoribosyl-1-pyrophosphate and leading to inosinic acid.[11] Both adenosine monophosphate (AMP) and guanine monophosphate are then derived from inosinic acid (**Figure 2**). The third step in this process is the biosynthesis of formylglycinamide ribonucleotide from glycinamide ribonucleotide (GR) via glycinamide ribonucleotide formyltransferase (GARFT). The last two steps in the synthesis of inosinic acid occur via a bifunctional enzyme having both aminoimidazolecarboxamide ribonucleotide formyltransferase (AICARFT) and inosine monophosphate cyclohydrolase (IMPCH) activity. This enzyme is composed of a 39 kDa C-terminal AICARFT active fragment along with a 25 kDa N-terminal IMPCH active fragment.[12] Both GARFT and AICARFT catalyze the transfer of a formyl group from 10-CHO-tetrahydrofolate (10-CHO-THF) to GR or aminoimidazolecarboxamide ribonucleotide (AICAR) respectively, returning THF as the second product of the reaction.

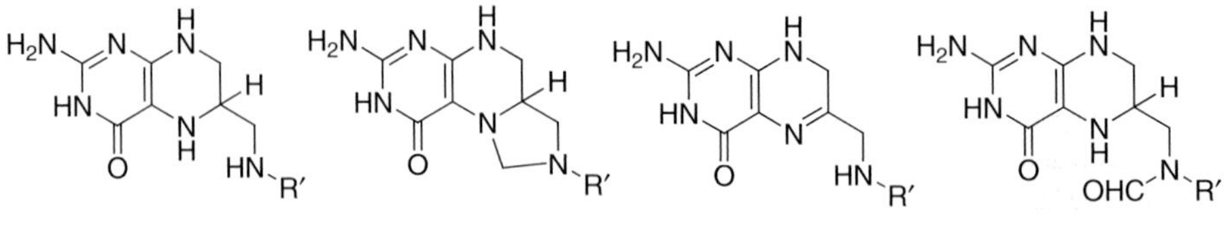

THF N5, N10-CH_2-THF 7, 8-DHF 10-CHO-THF

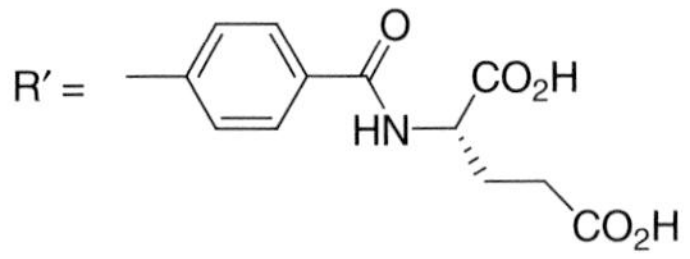

R′ =

7.03.2.3 Ribonucleotide Reductase

The synthesis of new DNA within a cell requires the production of deoxynucleotides. The four required deoxynucleotides (adenosine, guanosine, cytidine, and thymidine) are all produced by reduction of the appropriate ribonucleotide substrate with ribonucleotide reductase, also referred to a nucleoside diphosphate reductase (NDPR) (**Figure 3**).[13] The resulting oxidized form of NDPR can then be reduced back to NDPR by the action of glutaredoxin, which is in turn oxidized to thioredoxin.[14]

NDPR is a dimer, with each monomer made up of two subunits: a larger (M1) and a smaller (M2) subunit. The M1 subunit contains two allosteric sites involved in regulation of the overall activity of the enzyme and the enzyme's substrate specificity. The deoxynucleotide triphosphates bind to this allosteric site, and regulate their own synthesis. The M2 subunit is responsible for the key reduction reaction, carrying a tightly bound iron atom that stabilizes the tyrosyl free radical critical to reduction. Deoxynucleotide pools in proliferating cells are sufficient for only a few minutes of DNA synthesis without regeneration, thus making NDPR inhibition an attractive candidate for cancer chemotherapy.[15]

7.03.2.4 Active Transport of Antifolates

Classical antifolates such as methotrexate (MTX, **1**), an antifolate agent targeting DHFR, and pemetrexed (Alimta, **2**), a multitargeted antifolate, are actively transported into cells by the reduced folate carrier (RFC). Antifolates can also be transported via the membrane folate receptors (MFRs) through either endocytosis or potocytosis.[16] The RFC is a high affinity transporter for reduced folate cofactors such as 5-methyl-THF, 5-formyl-THF as well as MTX.[17] Folic acid, on the other hand, is a poor substrate for the RFC. The MFRs are linked to the outer plasma membrane via a glycosylphosphatidylinositol anchor, and have a high affinity for folic acid. The affinity of the MFRs for MTX is about 100-fold less than for folic acid, but other antifolates such as pemetrexed show enhanced affinity over folic acid.

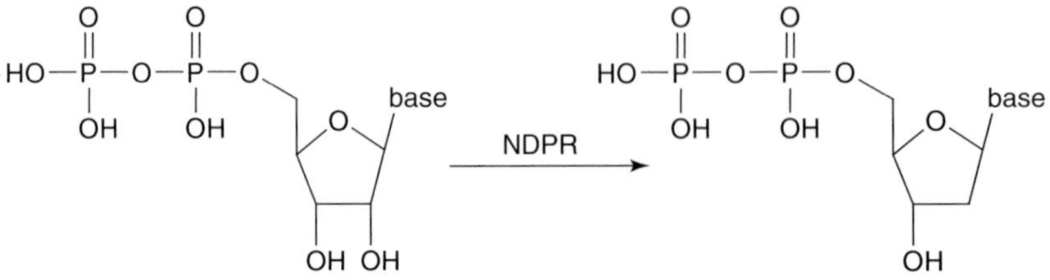

NDPR

Figure 3 Reversible oxidation of nucleoside diphosphate reductase (NDPR). (Reprinted from Henry, J. R.; Mader, M. M. Recent Advances in Antimetabolite Cancer Chemotherapies. In *Annual Reports in Medicinal Chemistry*; Doherty, A. M., Ed.; Elsevier: Oxford, UK, 2004, pp 161–172, with permission from Elsevier.)

1 MTX

site of polyglutamate formation

2 Pemetrexed

Defective RFC transport has been recognized as a common mechanism of resistance of cell lines to the effects of antifolate drugs.[17] Resistance is manifested as either decreased affinity for drug by RFC, or reduced expression of the RFC. A third mechanism of resistance has also recently been identified wherein an altered RFC exhibits increased affinity for folic acid. The increased level of folic acid in the cell then results in a loss of polyglutamation and therefore activity of the antifolate agent.[18]

7.03.2.5 Intracellular Transformations of Antimetabolites

Many antimetabolites must undergo modification within the cell before they are active agents, and so in essence are prodrugs. MTX and pemetrexed exert much of their pharmacological effects as polyglutamates. Classical antifolate drugs usually require polyglutamation within the cell to achieve maximum efficacy.[19] This transformation is carried out by the enzyme folylpolyglutamate synthetase (FPGS). Formation of the polyglutamate of antifolate drugs can cause a dramatic increase in the activity of the agent toward its intended target. Further, polyglutamates (above diglutamate) are less susceptible to cellular efflux, thus providing a long-lived pool of drug within the cell.[20] Most natural folate cofactors exist as polyglutamates, and so the beneficial action of FPGS on antifolate drugs is not surprising. It also follows that any cellular change leading to decreased FPGS activity could lead to antifolate resistance. Nonclassical or lipophilic antifolates do not require or are incapable of polyglutamation through the action of FPGS, and are thus not susceptible to resistance related to decreased FPGS activity.

Most nucleoside derived antimetabolite analogs also undergo intracellular transformations to become active agents (**Figure 4**). The earliest of these agents, 5-fluorouracil (5-FU, **3**) is converted into three major metabolites that are responsible for its activity. 5-FU is changed into 5-fluoro-2′-deoxyuridine monophosphate (FdUMP, **4**) which acts as a mimic

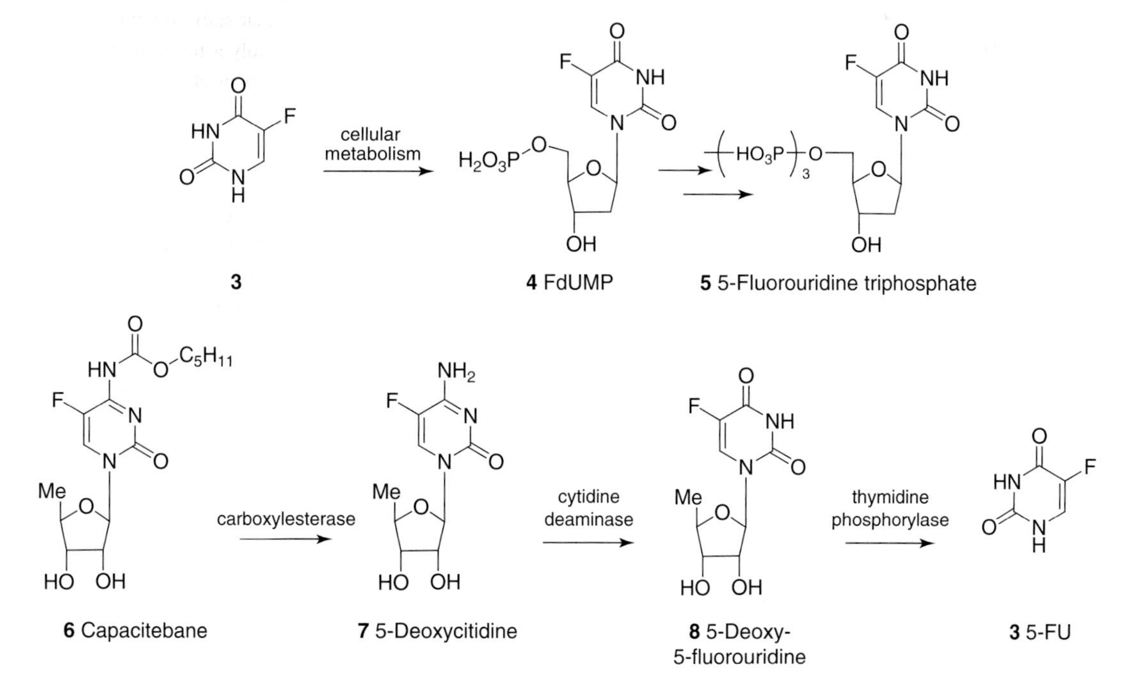

Figure 4 Intracellular transformations of antimetabolites. (Reprinted from Henry, J. R.; Mader, M. M. Recent Advances in Antimetabolite Cancer Chemotherapies. In *Annual Reports in Medicinal Chemistry*; Doherty, A. M., Ed.; Elsevier: Oxford, UK, 2004, pp 161–172, with permission from Elsevier.)

for the natural substrate of TS, dUMP, thus inhibiting TS activity. 5-FU can also be transformed to 5-fluoro-2′-deoxyuridine triphosphate, which is eventually incorporated into DNA causing DNA damage, and finally to 5-fluorouridine triphosphate (**5**), which is incorporated into RNA leading to impaired RNA function.[21] Capecitabine (**6**) is a prodrug of 5-FU that is metabolized in vivo to **3**. It is readily absorbed in the gastrointestinal tract, then passes intact through the intestinal mucosa. It is subsequently converted to 5′-deoxycitidine (**7**) by carboxylesterase in the liver, 5′-deoxy-5-fluorouridine (**8**) by cytidine deaminase, and finally to 5-FU (**3**) by thymidine phosphorylase.

7.03.3 Experimental Disease Models

Preclinical development of antimetabolites typically progresses through a sequence of assays to determine activity in vitro in isolated enzymes and cells, and in vivo in a variety of animal models. In the past, antimetabolites were evaluated initially for their effect in cell proliferation, employing an immortalized, cultured human tumor cell line relevant to the targeted indication. Elucidation of the biochemical pathways of purine, pyrimidine, and folate biosynthesis has enabled approaches in which the targeted enzyme is isolated and purified, and the small molecules are assayed to determine their inhibitory effect. Activity may also be determined in homologous enzymes or others in the biochemical pathway. Subsequent to this screening, the molecules are evaluated in tumor cells in vitro and then in animal models.[22]

The in vivo models employed to predict clinical efficacy for carcinomas include rodent xenografts of human tumor cell lines implanted and grown subcutaneously in immunocompromised mice.[23] The complications inherent in this strategy are numerous.[24] The tumor cells are derived from human sources (the HeLa cell line originated from the cervical tumors of Henrietta Lacks, in 1951), but are then passaged through multiple generations in vitro. Consequently, the cells that survive over generations may mutate to possess different characteristics from the parent cells. In addition, the tumor types preferred for xenografts grow rapidly, in days or weeks, to enable a timely assessment of the compound's efficacy, but such rapidly growing tumors are not reflective of all human cancers. Finally, differences in rodent physiology relative to humans impact metabolism, exposure, and, ultimately, efficacy of the drugs. General toxicity is evaluated through decreases in the animals' total body weight during the study period, and nausea can be reflected in decreased appetite and weight loss. However, several of the side effects frequently experienced by humans cannot be experienced or reported by rodents, including hair loss, vomiting, and hand and foot syndrome. Immunocompromised mice lack hair, a vomiting reflex, and the ability to report numbness or tingling in their paws. Syngeneic tumors (allografts) in rodents are also used, and these have an identical or closely similar genetic makeup to the rodent, allowing the transplantation of tissue without provoking an immune response. Such tumors address the problem of using immunocompromised mice, but retain many of the previously described complications.

It is not unheard of for drugs to be evaluated in the laboratory against certain tumor types and then be found to have activity in different tumors in the clinic.[25] Analyses of the correlation between activity in human tumor cell lines in vitro and in vivo to phase II clinical activity demonstrate that some tumor types correlate more positively than others, and that the correlations are also dependent on the drug. Johnson *et al.* found that only non-small cell lung cancer (NSCLC) xenografts were predictive of clinical activity in the same histology. Breast xenograft models were most useful in predicting activity in NSCLC, melanoma, and ovarian cancer, but not clinical breast cancer.[26] The analysis also showed that compounds found to be active in at least one-third of broader panels of xenografts were more likely to show activity in patients in phase II trials although, again, the correlation of specific in vitro cell type to clinical histology was not good. An extension of this analysis to selected syngeneic tumors in mice supported the previous finding and additionally found that the allografts are even less predictive than xenografts.[27] Thus, a drug that will be successful in the clinic generally must show preclinical activity against a variety of xenografts, and the phase II trials should also investigate multiple tumor types. In light of these challenges, it is not surprising to find that the clinical failure rate for oncology drugs is twofold higher than that of chronic indications such as cardiovascular disease and arthritis.[24,28]

7.03.4 Clinical Trial Issues

In addition to the difficulties encountered in translating preclinical models to success in the clinic, toxicity and or no significant improvement in efficacy over existing treatments is a significant factor in nonapproval of drugs. The attrition of oncology drugs in the clinic is especially disappointing in light of the fact that oncology chemotherapies are intended for acute dosing of potentially fatal conditions, and typically are allowed smaller therapeutic indices in comparison to chemotherapies directed at chronic, non-life-threatening indications. Some antimetabolites marketed outside the USA, but not approved by the US Food and Drug Administration (FDA), such as doxiflurabine and raltitrexed,[29] may fall into this category.

A selected set of commonly reported adverse events from phase III trials for several antimetabolites is found in **Table 2**. Others events, such as liver or pulmonary toxicity, may be observed as well. Comparisons are made for the

Table 2 Percentage of patients reporting representative adverse events in phase III trials

	5-FU[a]	*Capecitabine*[b]	*Gemcitabine*[c]	*Pemetrexed*[d]	*Paclitaxel*[e]
Non-laboratory					
Diarrhea	31	55	24	21	84
Nausea	58	43	64	39	31
Vomiting[f]	(58)	27	(64)	25	15
Alopecia	16	6	18	<5	92
Laboratory					
Anemia	45	80	65	33	51
Leukopenia	15	<5	71	13	12
Neutropenia	18	13	62	11	31
Percent of phase III patients who discontinued due to adverse events	4.8	13	14.3	<5	5

Reprinted from Henry, J. R.; Mader, M. M. Recent Advances in Antimetabolite Cancer Chemotherapies. In *Annual Reports in Medicinal Chemistry*; Doherty, A. M., Ed.; Elsevier: Oxford, UK, 2004, pp 161–172, with permission from Elsevier.
[a] 5-FU (once weekly, intravenously, 600 mg m^{-2} over 30 min) reported in control arm of gemcitabine phase III pancreatic cancer trial.[40]
[b] Capecitabine (1250 mg m^{-2} twice daily, orally, for 2 weeks followed by 1 week rest period) reported in phase III trial for breast cancer.[123]
[c] Gemcitabine (once weekly, intravenously, 1000 mg m^{-2} over 30 min) reported in phase III pancreatic cancer trial versus 5-FU.[40]
[d] Pemetrexed (day 1 of 21 day cycle, intravenously, 500 mg m^{-2} over 10 min plus folic acid pretreatment 400 μg × 5 days) reported in phase III trial for NSCLC.[124]
[e] Paclitaxel (intravenously, 175 mg m^{-2} on days 1 and 8 of 21 day cycle) reported in control arm of phase III breast cancer trial versus gemcitabine/paclitaxel combination.[40]
[f] Nausea and vomiting reported as a combined adverse event.

benchmark nucleoside antimetabolite 5-FU, its prodrug capecitabine, the nucleoside analog gemcitabine, which has supplanted 5-FU as the standard of care in pancreatic cancer, and the antifolate pemetrexed. Also for comparison, representative adverse events for paclitaxel, which acts by inhibiting microtubule formation, are included. The antimetabolite drugs act by inhibiting processes in rapidly dividing cells, in both tumors and normal tissue. Thus, the observed toxicities tend to be associated with tissues that undergo relatively frequent cell division, such as bone marrow, gastrointestinal tract, and hair. A significant number of patients in the trials discontinued treatment due to the adverse events, indicating another challenge in the clinical use of such drugs.

An additional hurdle for all chemotherapies is adapting the preclinical dosing regimen to the clinic. Efficacy doses for oncology drugs are determined in rodents (usually mice), and maximum tolerated doses are determined in mice, rats, dogs, or other species as warranted by the absorption, distribution, metabolism, excretion, and toxicity (ADMET) profile of the drug in discovery phase. Again, due to differences in metabolism between species, the predictive ability of these models is limited. For example, preclinical evaluation of gemcitabine found it to have antitumor activity in several tumor xenografts, including breast (MX-1), colon (CX-1, HC-1, and GC3), NSCLC (Calu-6, LX-1, and NCI-H460), and pancreas (HS766T, PaCa-2, and PANC-1).[30] The dosing regimens in these mouse models required 80–160 mg kg^{-1} of gemcitabine administered intraperitoneally every third day for 4 cycles. In contrast, the optimal dosing in humans for gemcitabine as a single agent in phase III trials for pancreatic cancer was found to be 1000 mg m^{-2} administered intravenously over 30 min once weekly for up to 7 weeks.[31]

7.03.5 Current Treatments

The first approved antimetabolites, 5-FU and MTX, have been in clinical use since for over 40 years. Both 5-FU and MTX were initially developed to treat carcinomas, but MTX and later other nucleoside analogs were found to be useful agents in hematopoietic therapy. However, in recent years, more antimetabolites have been developed to be effective in the treatment of solid tumors (**Tables 3** and **4**). Five antimetabolites have been approved for clinical use since 1996: gemcitabine, capecitabine, pemetrexed, azacitidine, and clofarabine. New methods for delivery of antimetabolites have

Table 3 Selected nucleoside analogs approved or in clinical trials as anticancer agents

Clinical agent	*Trade name*	*Year approved (USA)*	*Indication*	*Reference*
5-Fluorouracil (5-FU)	Adrucil	1962	Colorectal cancer	125
Cytarabine (ara-C)	Cytosar-U	1969	Leukemia	126
Fludarabine	Fludara	1991	Leukemia	127
Pentostatin	Nipent	1991	Leukemia	128
Cladribine	Leustatin	1993	Leukemia	129
Gemcitabine	Gemzar	1996	Pancreatic cancer	31
		1998	NSCLC	41
		2004	Breast cancer	130
Cytarabine Lyposomal	DepoCyt	1999	Lymphomatous meningitis	72
Capecitabine	Xeloda	1998	Breast cancer	131
		2001	Colorectal cancer	132
Clofarabine	Clolar	2004	Leukemia (pediatric)	133
Azacitidine	Vidaza	2004	Myelodysplastic syndrome	56
Decitabine	Dacogen	2006	Myelodysplastic syndrome	33
Nelarabine	Arranon	2005	Leukemia	64
Thymectacin		Phase II	Colorectal cancer	68
Troxacitabine	Troxatyl	Phase II	Leukemia, solid tumors	69
DAVANAT-1		Phase II	Colorectal cancer	76

Table 4 Selected folate antagonists in clinical use

Clinical agent	*Trade name*	*Year approved (USA)*	*Indication*	*Reference*
Methotrexate (MTX)		1953, 1959, 1971	Leukemia	79
Trimetrexate (TMTX)	NeuTrexin	(Launched outside USA)	(Infections)	88
Raltitrexed (RTX)	Tomudex	1996 (outside USA)	Colorectal cancer	87
Pemetrexed	Alimta	2004	Mesothelioma	84
		2004	NSCLC	134, 135
Nolatrexed (NTX)	Thymitaq	Phase III	Liver cancer	136
Plevitrexed (PTX)		Phase II/III	NCSLC	101
Pralatrexate (PDX)		Phase II	NSCLC	92
Pelitrexol (PTO)		Phase I	Solid tumors	94
PT-523		Phase I/II	NSCLC	105
MDAM		Phase I	Solid tumors	103

been developed including DepoCyt, which was given accelerated approval in 1999. Additional oncolytic indications for some of the approved antimetabolites are in clinical trials, and several other antimetabolites or improved delivery methods are currently undergoing phase III or phase II study. The antimetabolites can be broadly categorized as nucleoside mimics and folate antagonists, and we will describe each in turn.[32]

7.03.5.1 Nucleoside Mimics

The nucleoside mimics tend to have a structural modification of the sugar residue, such as di-fluorination of the 2′-position (e.g., gemcitabine, **11**) or inversion of hydroxyl configuration at 2′-position (cytarabine, **9**). Others possess modified base residues, such as azacitidine (**13**) and decitabine (**14**). Some structures combine modification of the sugar with modification of the base, such as clofarabine (**12**) and nelarabine (**15**). The nucleoside mimics inhibit a variety of enzymes that are important in nucleoside synthesis or DNA replication, including thymidylate synthase, adenosine deaminase, ribonucleotide reductase, and DNA polymerase.

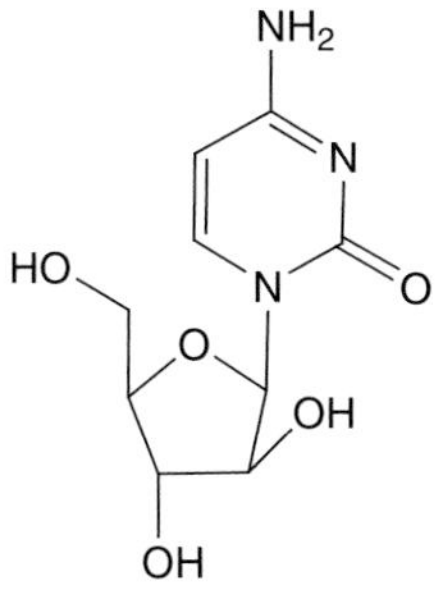

9 Cytarabine

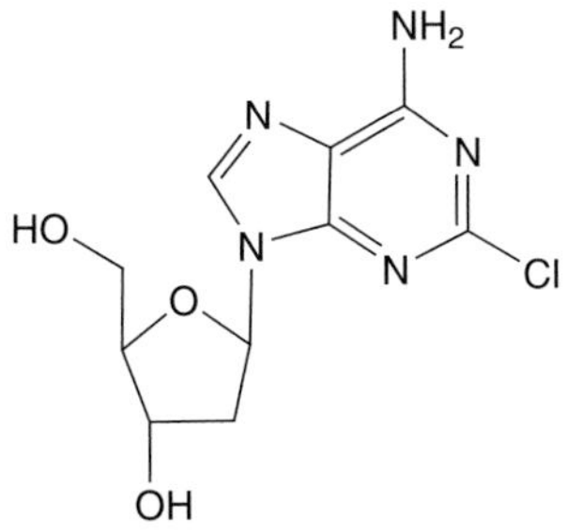

10 Cladribine

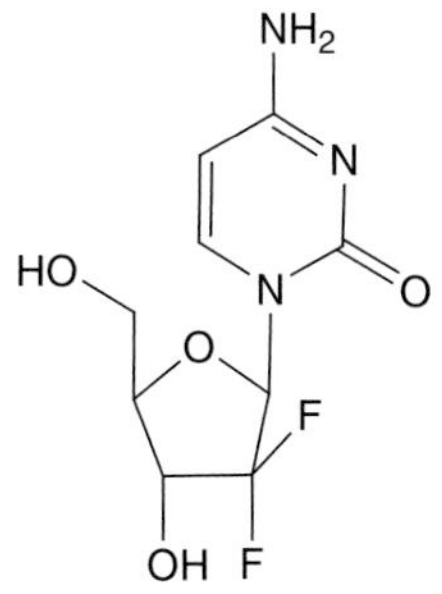

11 Gemcitabine

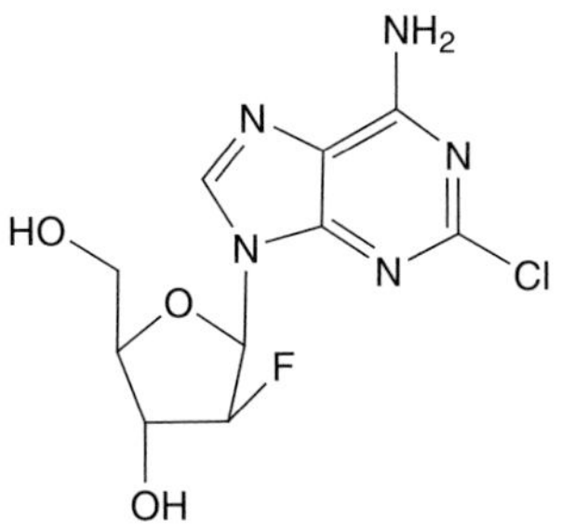

12 Clofarabine

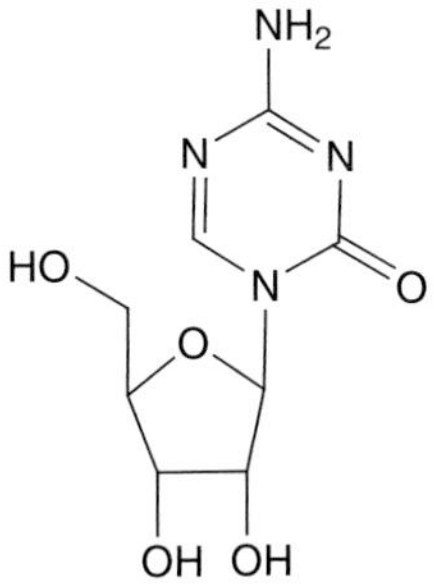

13 Azacitidine

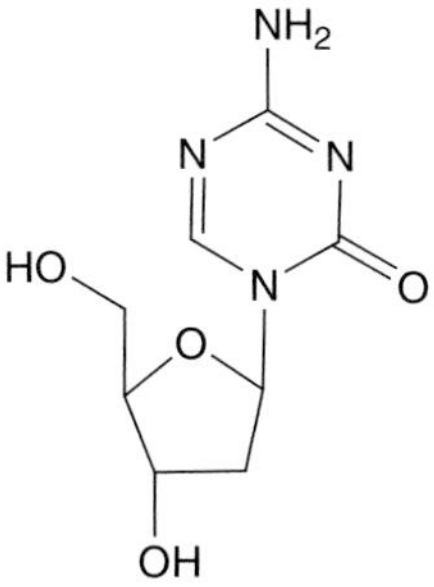

14 Decitabine

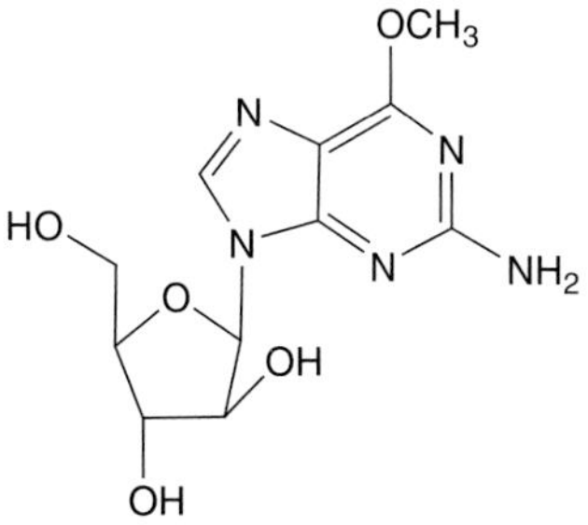

15 Nelarabine

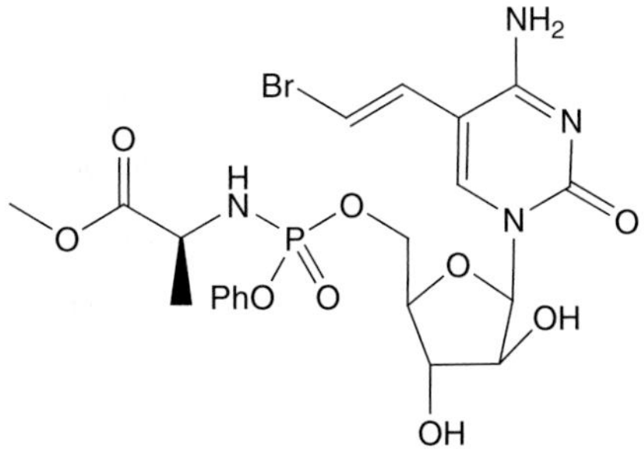

16 Thymectacin

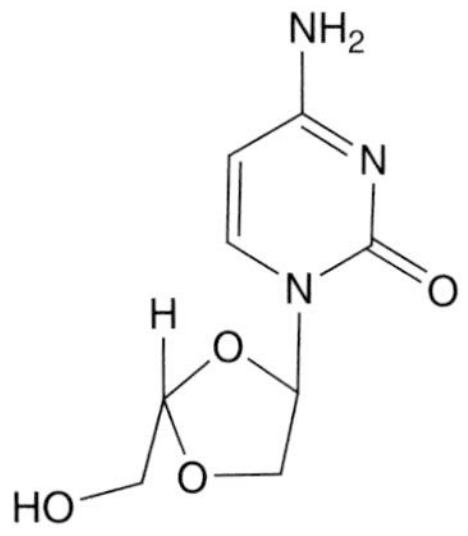

17 Troxacitabine

Generally the drugs act not through inhibition of a single enzyme, but through interaction as substrate or inhibitor of several enzymes; the primary mechanism is shown in **Table 5**. For example, decitabine (**14**) competes with deoxycytidine for conversion to the monophosphate by deoxycytidine kinase.[33] In doing so, intracellular levels of cytidine monophosphate are decreased. In addition, decitabine monophosphate is converted via its diphosphate to the triphosphate, decitabine triphosphate, which is then incorporated into DNA. It acts as a cytosine residue until it covalently inhibits DNA methyltransferase. Hypermethylation silences signal transduction pathways in tumor cells, and thus decitabine inhibits tumor cell proliferation.

Table 5 Summary of mechanism of action and active form of drug for selected nucleoside analogs

Clinical agent	*Analog of which nucleoside?*	*Mechanism of action*	*Active form of drug*
5-FU	Pyrimidine	Inhibits thymidylate synthase	5-Fluorodeoxyuridine monophosphate
Cytarabine	Pyrimidine	Inhibits DNA polymerases	Triphosphate
Cladribine	Purine	Inhibits adenosine deaminase	Monophosphate
Gemcitabine	Pyrimidine	Inhibits ribonucleotide reductase	Diphosphate
		Substrate for dCTP/triphosphate	Triphosphate
Capecitabine	Pyrimidine	(prodrug of 5-FU)	—
Clofarabine	Purine	Inhibits ribonucleotide reductase	Triphosphate
Azacitidine	Pyrimidine	(prodrug of decitabine)	—
Decitabine	Pyrimidine	Inhibits DNA methyltransferase	Triphosphate
Nelarabine	Purine	(prodrug of guanine arabinoside)	Triphosphate of guanine arabinoside
Thymectacin	Pyrimidine	Inhibits thymidylate synthase	Monophosphate
Troxacitabine	Pyrimidine	Inhibits DNA polymerases	Triphosphate

7.03.5.1.1 5-Fluorouracil/Leucovorin

The standard of care for many carcinomas for several years was 5-FU. Although 5-FU is widely used, its pharmacokinetic (PK) profile is not ideal. Its optimal method of delivery is by continuous intravenous infusion, as its bioavailability after oral administration is variable. 5-FU is rapidly metabolized, with a mean half-life ($t_{1/2}$) of elimination of approximately 16 min. Within 3 h, no intact drug can be detected in plasma. 5-FU is more effective when coadministered with folinic acid (also referred to as LV or leucovorin), a prodrug of 5,10-CH_2-THF. Inhibition of TS by the 5-FU metabolite **4** is dependent on the cofactor 5,10-CH_2-THF, which combines with TS and **4** to form a covalent ternary complex. Excess cofactor decreases the dissociation rate of this complex, and consequently addition of leucovorin increases the cytotoxicity of 5-FU. The major toxicities of 5-FU are to bone marrow and mucous membranes.[34]

7.03.5.1.2 Capecitabine

Capecitabine (**6**) is an orally administered fluoropyrimidine carbamate that is metabolized in vivo to 5-FU (**3**).[35] PK studies in patients showed rapid gastrointestinal absorption of capecitabine, followed by extensive conversion to **8**, with only low systemic levels of 5-FU.[36] In preclinical animal models, capecitabine appeared to deliver drug selectively to tumors. Analysis of the tumor: plasma area under curve (AUC) ratios of capecitabine versus 5-FU in four human tumor xenografts in mice (HCT116, CXF280, COLO205, and WiDr) at the maximum tolerated dose (oral) showed that although the $t_{1/2}$ of 5-FU was similar in all four tumors, the tumor: plasma AUC ratio of 5-FU was significantly higher for animals dosed with capecitabine. For example, in HCT116, 5-FU exposure was 127-fold higher in tumor than plasma in animals treated with capecitabine, and 209-fold higher in CXF280. In addition, the AUC_{inf} of 5-FU in tumor via capecitabine dosing was 18- to 32-fold higher than via direct 5-FU dosing, demonstrating that capecitabine is efficiently converted to 5-FU in the tumors.[37] In the clinic, the efficacy of capecitabine equals or exceeds 5-FU, and some differences in tumor versus adjacent tissue exposure are observed, although not to the same extent as in the murine models.[38] In colorectal tumors, the concentration of 5-FU was found to be 3.2-fold higher in the tumor than in the adjacent healthy tissues. In liver metastasis, however, no difference in exposure was found between the diseased and healthy tissues. The tumor-preferential activation of capecitabine to 5-FU is explained by tissue differences in the activity of cytidine deaminase and thymidine phosphorylase, key enzymes in the capecitabine → 5-FU conversion. The interpatient variability in AUC and C_{max} (from 27% to 89%) is also attributed to the different levels of expression of the same enzymes across the patient population.

The elimination $t_{1/2}$ of parent capecitabine in the phase III studies ranged from 0.55 to 0.89 h, and it reached its peak plasma concentration (t_{max}) in 2 h. Thus, the $t_{1/2}$ of capecitabine after oral administration of 1250 mg m^{-2} is comparable to that of 5-FU, but its ease of dosing by mouth offers an advantage to patients. Capecitabine is approved for the treatment of breast and colorectal cancer.

7.03.5.1.3 Gemcitabine

Gemcitabine (**11**) is a nucleoside analog that exhibits cell phase specificity, primarily killing cells undergoing DNA synthesis (S phase) and also blocking the progression of cells through the G1/S phase boundary. The cytotoxic effect of gemcitabine is attributed to the actions of both its diphosphate and the triphosphate nucleosides, which leads to inhibition of DNA synthesis. First, gemcitabine diphosphate inhibits NDPR, which causes a reduction in the concentrations of deoxynucleotides, including deoxycytidine triphosphate (dCTP).[39] Second, gemcitabine triphosphate competes with dCTP for incorporation into DNA. The reduction in the intracellular concentration of dCTP (by the action of the diphosphate) enhances the incorporation of gemcitabine triphosphate into DNA (self-potentiation). Following gemcitabine nucleotide incorporation into DNA, only one additional nucleotide is added to the growing DNA strand, followed by inhibition of further DNA synthesis. Because of the addition of this final nucleotide, DNA polymerase epsilon is unable to remove the gemcitabine nucleotide and repair the growing DNA strands (masked chain termination).[40]

Gemcitabine has been approved in the USA for use either as a single agent or in combination for three carcinomas: pancreatic, NSCLC, and breast (**Table 6**). First approved by the FDA in 1996, gemcitabine was demonstrated to have a significant clinical benefit response in advanced pancreatic cancer patients compared to 5-FU, with a survival advantage of 5.6 months versus 4.4 months in the 5-FU treated patients.[31] The clinical benefit was measured as improvement in three symptoms present in most pancreatic cancer patients: pain, functional impairment, and weight loss. Gemcitabine has become accepted as the standard of care for the treatment of advanced pancreatic cancer, and in 1998 the FDA approved its combination with cisplatin for the treatment of NSCLC.[41] The gemcitabine/cisplatin combination demonstrated a survival advantage of 9.0 months versus 7.6 months for cisplatin alone, and increased the median time to disease progression by 1.5 months. In 2004, gemcitabine was approved for use in breast cancer in combination with paclitaxel, with a significant increase in time to disease progression from 2.9 months for paclitaxel alone to 5.2 months for the gemcitabine/paclitaxel doublet.

The PK parameter of elimination $t_{1/2}$ for gemcitabine is similar to that of 5-FU.[42] In a population study of 353 patients, $t_{1/2}$ ranged from 42 to 92 min, with significant age and gender differences. The longest $t_{1/2}$ of 92 min was observed for women over age 75, and the shortest was for men under age 30. Clearance decreased with age and gender and was approximately 25% less for women than men, regardless of age. The average clearance value of 87.5 $\mathrm{L\,h^{-1}\,kg^{-1}}$ indicates a two-compartment model of metabolism in which tissues in addition to liver contribute to metabolism.

Clinical trials are ongoing for other indications, including ovarian, bladder, and non-Hodgkin's lymphoma (NHL),[43] employing gemcitabine as a single agent or in combination with other drugs. Trials with gemcitabine monotherapy as a

Table 6 Gemcitabine activity in solid tumors (approved indications in USA)

Cancer	*Gemcitabine or gemcitabine + SOC*[40]	*Standard of care (SOC)*
Pancreatic cancer	Gemcitabine	5-FU
Dosing schedule (once weekly, up to 7 weeks)	1000 $\mathrm{mg\,m^{-2}}$ over 30 min	600 $\mathrm{mg\,m^{-2}}$ over 30 min
Median survival, months	5.7	4.2
Median time to disease progression, months	2.1	0.9
NSCLC	Gemcitabine/cisplatin	Cisplatin
Dosing schedule (28 day cycle)	Gemcitabine: 1000 $\mathrm{mg\,m^{-2}}$ over 30 min on days 1, 8, 15 + cisplatin: 100 $\mathrm{mg\,m^{-2}}$ on day 1	100 $\mathrm{mg\,m^{-2}}$ on day 1
Median survival, months	9.0	7.6
Median time to disease progression, months	5.2	3.7
Breast cancer	Gemcitabine/paclitaxel	Paclitaxel
Dosing schedule (21 day cycle)	Gemcitabine: 1000 $\mathrm{mg\,m^{-2}}$ over 30 min, days 1, 8 + paclitaxel: 175 $\mathrm{mg\,m^{-2}}$ on day 1	175 $\mathrm{mg\,m^{-2}}$ on day 1
Median time to disease progression, months	5.2	2.9
Overall response rate	40.8%	22.1%

second-line treatment for ovarian cancer show it to be well tolerated, although the objective response rate of ~15% in four studies was assessed as 'modest but definite.'[44] In combination with cisplatin as a first-line treatment, overall response rates (ORRs) of 70% have been reported in two studies.[45] The gemcitabine/carboplatin combination was approved for the treatment of ovarian cancer in Europe in July 2004 following a phase III study of 356 patients previously treated with platinum-based therapy. The gemcitabine/carboplatin study group showed an increased median time to progression of 8.6 months versus 5.8 months for patients treated with carboplatin alone, and ORR of 47% in the treatment group versus 31% in the carboplatin control group.[46] The triplet combination of gemcitabine/carboplatin/paclitaxel as a first-line therapy has shown even greater efficacy, and the ORR was determined to be 89.7–100%, with 35 of 53 patients having complete remission.[47] In bladder cancer, gemcitabine monotherapy is effective with ORR ranging from 36% to 45%,[48,49] as are doublet combinations gemcitabine with cisplatin, paclitaxel, or docetaxel (ORR 60%, 40%, and 33%, respectively).[50–52]

7.03.5.1.4 **Clofarabine**

Clofarabine (**12**) was given accelerated approval in late December 2004 for use in pediatric acute lymphoblastic leukemia (ALL), based on induction of complete responses in 6 of 49 patients, and 9 additional partial responses or complete responses without platelet recovery. Clinical studies demonstrating increased survival or other clinical benefit have not been conducted. The synthesis of clofarabine reported in 1992[53] stemmed from attempts to modify the structure of fludarabine (**18**) and cladribine (**10**) and retain desirable cytotoxic effects while avoiding neurotoxicities associated with fludarabine. Cladribine replaces the 2-fluoro substituent with chlorine, and clofarabine additionally replaces the 2′-hydroxyl with fluorine.

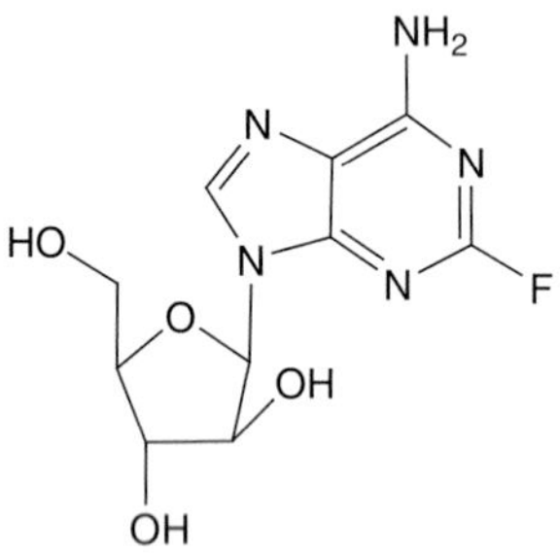

NH_2 N N HO N N F O OH OH

18 Fludarabine

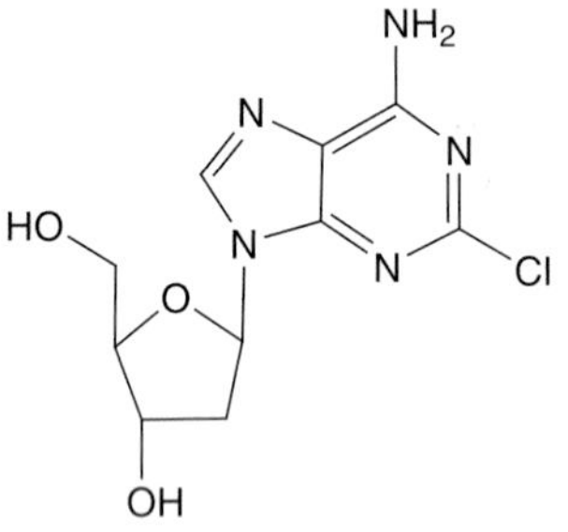

NH_2 N N HO N N Cl O OH

10 Cladribine

NH_2 N N HO N N Cl O F OH

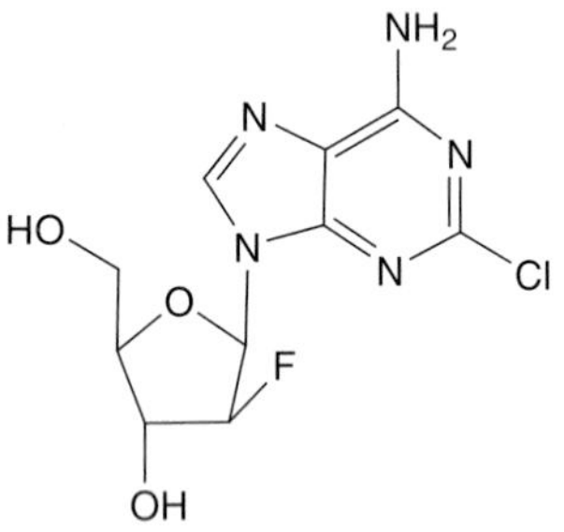

12 Clofarabine

Like the other antimetabolites, clofarabine disrupts several pathways. It is metabolized intracellularly to its active form, the 5′-triphosphate. As the triphosphate, it inhibits NDPR, thereby decreasing cellular deoxynucleotide triphosphate pools and preventing DNA synthesis. It is also incorporated into DNA chains, but terminates chain elongations and inhibits DNA repair by inhibition of DNA polymerases.

Although activity in a preclinical model of leukemia (murine p388 cells) was successfully translated into the approved indication, the previously described issues of dosing, exposure, and adverse events profile recur for this compound as well. In the murine model, the optimal dosage was determined to be $20\,\mathrm{mg\,kg^{-1}}$, intraperitoneally q3 h × 8 on days 1, 5, and 9 of treatment.[53] In the clinic, the recommended pediatric dose for ALL is $52\,\mathrm{mg\,m^{-2}}$, intravenously, daily for 5 days, repeated every 2–6 weeks. However, in early clinical examinations in patients with solid tumors, the dose had to be reduced from 15 to $2\,\mathrm{mg\,m^{-2}}$ due to toxicities,[54] and development of the drug for the solid tumor indications was ultimately discontinued. In rodents, the half-life was estimated to be 1.84 h when dosed at $25\,\mathrm{mg\,kg^{-1}}$,[55] but in humans it was estimated at 5.2 h at $52\,\mathrm{mg\,m^{-2}}$. Typical cytotoxic adverse events were reported (nausea, vomiting, etc.), but cardiac and hepatic toxicity were also noted. In particular, four of 96 patients evidenced capillary leak syndrome (pleural and pericardial effusions) or systemic inflammation response syndrome, which is characterized by tachycardia, hypotension, and pulmonary edema. Clofarabine administration was discontinued for these patients.

Clofarabine is in preregistration phase for acute myelogenous leukemia (AML) in the USA, and is in trials in Europe for ALL and AML.

7.03.5.1.5 **Azacitidine**

The FDA approved azacitidine (**13**) in May 2004 as an orphan drug to be used in the treatment of myelodysplastic syndrome (MDS). MDS is a collection of bone marrow disorders in which not enough red blood cells are made. The condition may develop following treatment with drugs or radiation therapy for other diseases, and some of the forms of MDS can progress to AML. At the time of its approval, the primary treatment for MDS was supportive care in the form of blood transfusions or administration of hematopoietic factors, but no curative treatment was available for the 7000–12 000 new cases diagnosed each year in the USA.[56]

Upjohn originally developed the compound, an N-5 aza analog of cytidine, as a cytotoxic agent over 30 years ago. As such, its enzymatic target was not known during its early development history, but it is now understood to act as a prodrug for decitabine (**14**). Metabolic dehydroxylation at C-2′ generates **14**, which is incorporated into DNA and inhibits methylation by DNA methyltransferases on cytosine polyguanine (CpG) rich regions of DNA. This inhibition is of consequence in tumor cells because they possess CpG regions which are rare in normal cells, but which must be methylated to activate signal transduction pathways resulting in cell proliferation. In the presence of decitabine, hypermethylation is shut down, slowing the cell cycle progression from S phase.[57] Azacitidine is also incorporated into RNA, but very little is known about the biological effects of this.

The ORR of 15.7% for patients treated with 75–100 $mg\,m^{-2}$ of azacitidine (suspension injected subcutaneously daily for 7 days, with the cycle repeating every 4 weeks) is remarkable in light of the 0% ORR for the observation group. Patients in the observation group who crossed over into the treatment group had an ORR of 12.7%. The adverse events reported for azacitidine were those typical of cytotoxics of this type, including nausea, vomiting, and anemia.[58]

7.03.5.1.6 Nucleoside mimics in phase II and phase III clinical trials

Decitabine (**14**) is a pyrimidine nucleoside analog differing from azacitidine in that the hydroxyl group at C-2′ is absent. Like cytarabine and clofarabine, which treat acute AML and ALL, decitabine is being developed as a therapy for hematological malignancies. Like azacitidine, its synthesis has been known in the literature for some time,[59] but clinical trials in solid tumors were complicated by its significant marrow toxicity. Interest in it was renewed with the recognition of its inhibitory activity in DNA methyltransferases, and that reversal of hypermethylation may lead to differentiation and apoptosis in malignant cells.

Preclinical studies found activity in xenografts of breast, colon, and lung tumors in mice,[33] but decitabine showed only modest clinical activity in solid tumors. Thus, ongoing focus has been on its development as a therapy for hematological malignancies. In the phase III trial for MDS, dosing with decitabine by 3 h infusion of 15 $mg\,m^{-2}\,h^{-1}$ every 8 h on 3 consecutive days every 6 weeks resulted in an ORR of 17% in the decitabine treatment arm, with 0% ORR for those receiving supportive care.[60] Decitabine is in preregistration phase with the FDA for MDS, with a review date anticipated on 1 Sept 2005. Initiation of a phase III trial in AML was reported in January 2006.

Decitabine has presented challenges in its solution and metabolic stability. It decomposes rapidly in both acidic and basic solution, and significantly (7%) in 1 h at pH7 and 37 °C.[61] While the mean $t_{1/2}$ of azacitidine and its associated metabolites is about 4 h by both intravenous and subcutaneous administration,[58] the $t_{1/2}$ of decitabine ranges from 7 to 35 min.[62] The high clearance and total urinary excretion of $<1\%$ indicate that decitabine is rapidly metabolized through phosphorylation and deamination.[62] Further clinical experience with both azacitidine and decitabine will be required to determine which compound offers greatest benefit in MDS, ALL and AML.

Nelarabine (**15**) (also referred to as 506U78) is a purine nucleoside analog that is a prodrug of 9-β-D-arabinofuranosyl guanine (ara-G). Ara-G was first synthesized in 1964, but its clinical development was slowed by a difficult synthesis and poor solubility. Nelarabine is a 6-methoxy derivative that is approximately 10 times more soluble in water than ara-G,[63] and is rapidly demethylated in the presence of adenosine deaminase to generate ara-G. Ara-G is cytotoxic to T cells through accumulation and retention of its triphosphate, ara-GTP. It has been observed that patients with T cell lymphoblasts generally have higher levels of ara-GTP than those having myeloblasts or B cell lymphoblasts. As such, nelarabine chemotherapy could be indicated for patients suffering T-cell acute lymphoblastic leukemia (T-ALL) rather than those with B cell leukemias.[64]

In a phase I study, the harmonic mean $t_{1/2}$ over a variety of doses of nelarabine in pediatric and adult patients was found to be 14.1 min and 16.5 min, respectively, with a maximum tolerated dose (MTD) of 60 $mg\,m^{-2}$ over 45–120 min intravenously daily for 5 days. Interestingly, the clearance of ara-G was higher in pediatric patients (0.312 $L\,h^{-1}\,kg^{-1}$) as compared with adult patients (0.213 $L\,h^{-1}\,kg^{-1}$), and the $t_{1/2}$ of ara-G was shorter in pediatric patients as compared with adult patients (2.1 h versus 3.0 h). These studies demonstrated that nelarabine is an effective prodrug of ara-G, allowing systemic concentrations of ara-G that result in clinical activity.[65] Phase II studies showed greater efficacy and lesser toxicity in pediatric than adult patients,[66] and based on these Phase II studies, in October 2005 the FDA granted accelerated approval for use of nelarabine in both adults and children with T-ALL and T-cell lymphoblastic lymphoma (T-LBL) who had relapsed or not responded to two chemotherapy regimes. The principal dose-limiting side effect in pediatric patients was neurogical, with 10 of 84 patients in a phase II study reporting grade 3 or 4 events such as peripheral neuropathy (six patients). The structural similarity of nelarabine to the purine analogs fludarabine (**18**) and clofarabine (**12**) should be noted.

Thymectacin (**16**, NB 1011), a nucleosidephosphoramidate derivative, was designed as an enzyme catalyzed therapeutic agent (ECTA). In this strategy, the molecule acts as a substrate for an enzyme that is overexpressed in abnormal cells, such as TS. Upon reaction with the enzyme catalyst, the compound is, theoretically, released as a toxic product that can inhibit cell proliferation through interaction with other enzymes or proteins in the cell. This variation on the targeted therapy approach has the potential of having an improved therapeutic index, as the cytotoxic would be delivered in situ to the tumor. Initial studies confirm that thymectacin is indeed a substrate in vitro for TS, and is cytotoxic to tumor cell lines that overexpress TS.[67] Subsequent work revealed that the product of the conversion with TS, BVdUMP (**19**), modifies cellular proteins, not DNA, despite the fact that it is a nucleoside analog.[68] The compound is currently in phase II clinical trials for colorectal cancer.

thymidylate synthase

16 Thymectacin

19 BrVdUMP

Troxacitabine[69] (**17**) is a pyrimidine nucleoside analog possessing two unique structural features: L-configuration at the C-4′ carbon and replacement of the C-3′ ring carbon with oxygen, resulting in a dioxolane ring. It is active as its triphosphate metabolite and inhibits DNA polymerases. However, due to the inversion of configuration at C-4′, it is not a substrate for deoxycytidine deaminase, unlike other pyrimidine nucleoside analogs such as cytarabine and gemcitabine. In addition, unlike cytarabine and gemcitabine, radiolabeling studies show that the major route of cellular uptake is passive diffusion. Consequently, it may be active in patients who are refractory to cytarabine and gemcitabine when resistance is linked to active transport mechanisms.

This compound also demonstrates significant PK differences between rats and humans. In rats, the PK was independent of dosage concentration, the elimination $t_{1/2}$ is 1.65 h and its clearance is 1.38 L h^{-1} kg^{-1} (single dose at 10, 25, or 50 mg kg^{-1} administered intravenously or orally).[70] After 5 days of treatment in humans, (intravenous infusion over 30 min, daily × 5 at 0.12–1.8 mg m^{-2}) the $t_{1/2}$ was found to be 39 h and clearance was 127 mL min^{-1}.[71] The longer $t_{1/2}$ relative to other nucleoside analogs is attributed to its inactivity in deoxycytidine deaminase, so the drug is not cleared rapidly.

Troxacitabine was investigated in phase I trials in patients with either solid or hematological malignancies. Phase II investigations were completed in a variety of leukemias including AML and ALL, and in pancreatic cancer.

7.03.5.1.7 Advances in delivery of nucleoside mimics

The activity of capecitabine in colorectal cancer demonstrates an effective strategy to enhance the activity of nucleoside mimics: increase the exposure of the active drug by delivery as a prodrug. Other delivery strategies have been effective as well, including formulations of drug in liposomes, in implanted microspheres, or in polysaccharides.

DepoCyt is an injectable, sustained release form of cytarabine (ara-C; **9**), for the treatment of antineoplastic meningitis (NM) arising from lymphoma (lymphomatous meningitis).[72] Ara-C acts by inhibiting DNA polymerase as well as through incorporation of its triphosphate into DNA. A phase III trial of DepoCyt in lymphomatous NM showed it to be more convenient and associated with a higher positive response rate than ara-C. The DepoCyt formulation of ara-C is encapsulated in the aqueous chambers of a spherical 20-μm matrix comprising of lipids biochemically similar to normal human cell membranes (phospholipids, triglycerides, and cholesterol). When injected into the cerebrospinal fluid (CSF) at room temperature, the particles spread throughout the neuroaxis and slowly release ara-C. A single injection of free unencapsulated ara-C maintains cytotoxic concentrations in the CSF for <24 h,[73] whereas a single injection of 50 mg of DepoCyt maintains cytotoxic concentrations of ara-C in most patients in the CSF for >14 days.[74] As cytotoxicity is a function of both drug concentration and duration of exposure, this formulation maintains high concentrations of ara-C in the cancer cell for prolonged periods of time and increases the efficacy of the agent.

5-FU has been investigated in implanted microspheres for treatment of brain tumors (glioblastoma).[75] The microspheres are biodegradable poly(lactide-co-glycolide) with a mean size of 40 μm and drug loading of $21.6\% \pm 1.3\%$. The microspheres allow for controlled release of 5-FU into the tumor due to a combination of diffusion and degradation. In this indication, 5-FU acts as a local radiosensitizer, improving the efficacy of radiation therapy. Phase I results from a 10 patient trial showed an overall survival time of 40 weeks, and 5-FU was detected at very low levels (0–25 ng mL^{-1}) in the blood and CSF of some, but not all, patients. The results were sufficiently encouraging for a phase II study to be conducted, and the results are pending analysis.

Release of 5-FU from a polysaccharide matrix is currently under phase I investigation in DAVANAT (1,4-β-D-galactomannan). The rationale for this delivery system is threefold. First, sugars on the tumor cells can differ from those of normal cells, allowing for tumor-cell specific delivery of drug. In addition, galactoside-containing carbohydrates disturb cell association, thus preventing new tumor cells from adhering to host tissue, and disrupting metastasis. Lastly, entrapment of hydrophobic drugs can improve their solubility and distribution to the target tumors. DAVANAT is a soluble polysaccharide, derived from plant sources, which is co-administered with 5-FU by intravenous injection. In preclinical studies, coadministration of DAVANAT with 5-FU in mouse xenografts of COLO205 and HT-29 tumors resulted in a decrease of median tumor volume and increase in mean survival time.[76] Interim analysis of five of six cohorts in the phase I study showed mean exposure parameters of AUC and C_{max} increased with repeated doses of 5-FU and with escalating doses of DAVANAT.[77] Details regarding the compound have not been widely disclosed in peer-reviewed literature, and thus the full promise of the strategy remains to be understood.

7.03.5.2 Antifolates

The biosynthesis of nucleic acid precursors depends on folate metabolism. Tetrahydrofolate, along with its cofactors, are the main carriers of one carbon units in the synthesis of thymidine and purine nucleosides. Agents interfering with the metabolism of folate are termed antifolates, and can be divided into three general subtypes. Classical antifolates are generally highly polar molecules that are actively transported by the RFC and in some cases the MFR. They are also substrates for FPGS and exert much of their effects through polyglutamates. Antifolates that are not substrates for FPGS have been termed nonclassical antifolates, and can be broken into two further subgroups, type A and type B. The type A nonclassical antifolate is generally a more lipophilic drug that is passively transported and does not utilize the RFC. The type B antifolate can be thought of as a hybrid between the classical and nonclassical antifolates. While still being nonpolyglutamatable like their type A counterparts, type B antifolates are generally polar, water-soluble inhibitors that utilize the RFC for cell entry. Type B antifolates typically cannot form polyglutamates because the gamma-carboxyl of the glutamic acid moiety has been blocked or otherwise modified.[78]

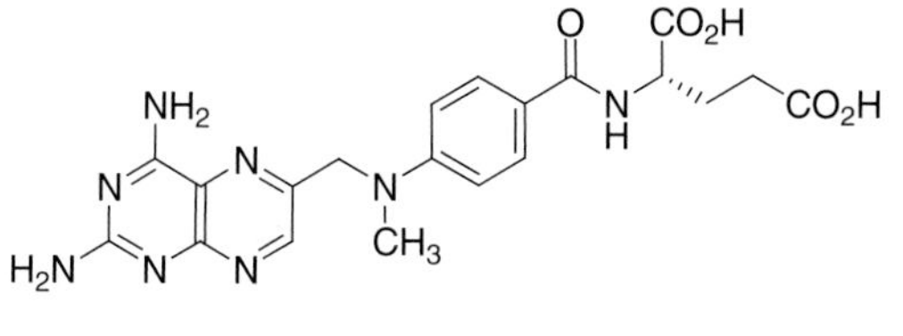

1 Methotrexate

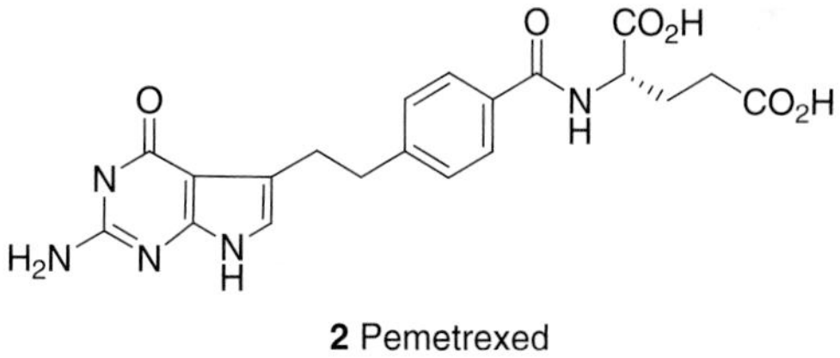

2 Pemetrexed

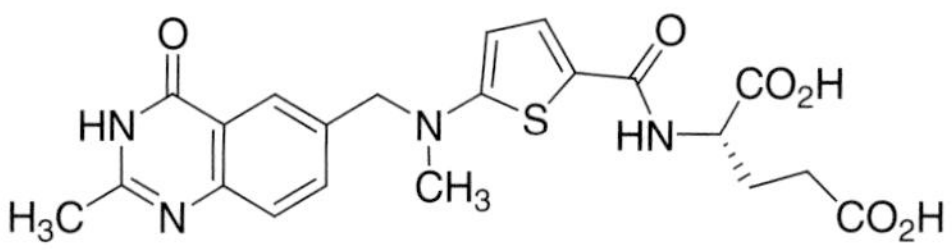

20 Raltitrexed (RTX)

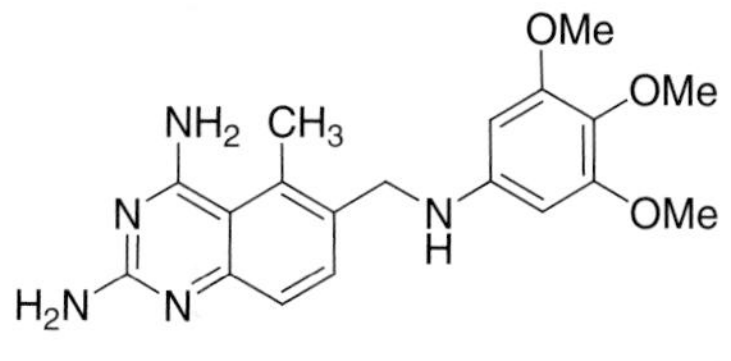

21 Trimetrexate (TMTX)

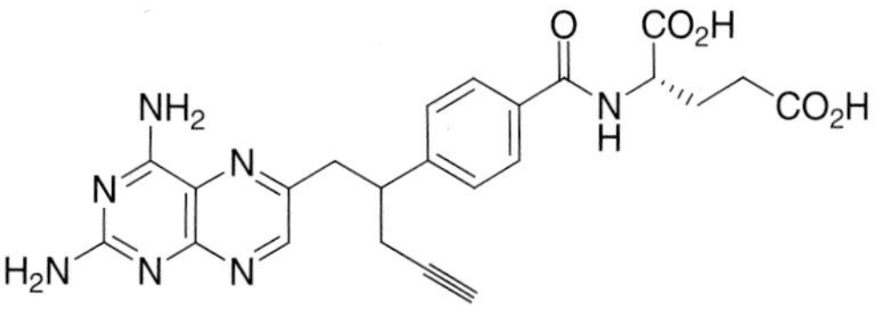

22 Pralatrexate (PDX)

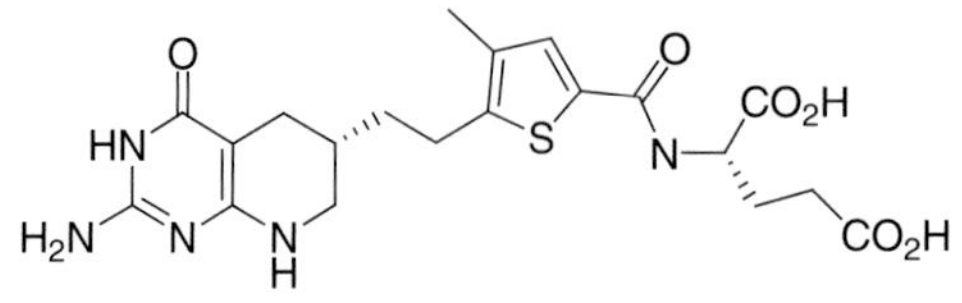

23 Pelitrexol (PTO)

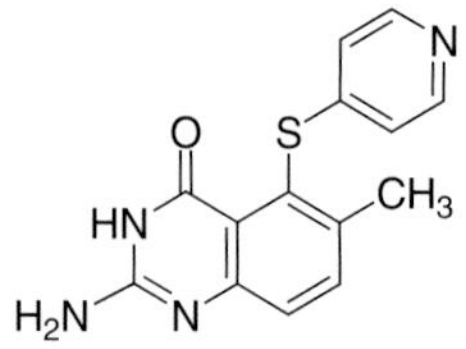

24 Nolatrexed (NTX)

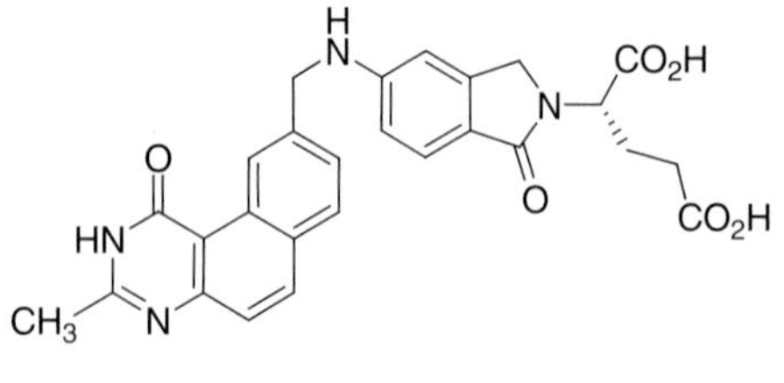

25 OSI-7904L (OSI)

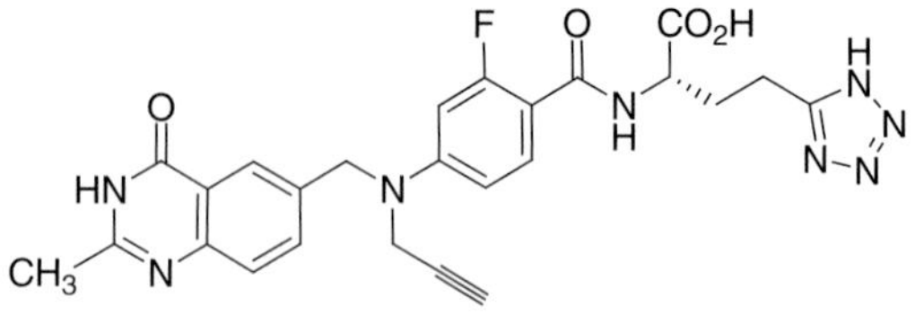

26 Plevitrexed (PTX)

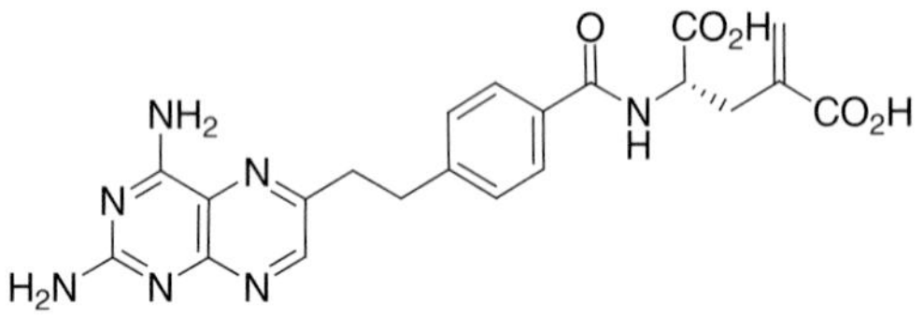

27 MDAM

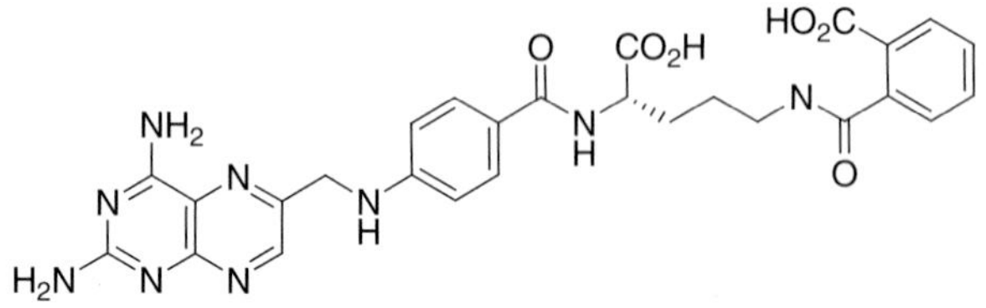

28 PT-523

7.03.5.2.1 Methotrexate

MTX (**1**) is a classical antifolate that has been approved for over 50 years and is widely used for the treatment of a variety of cancers and autoimmune diseases. As mentioned previously, MTX is an inhibitor of DHFR, thus inhibiting the production of THF and ultimately thymidine biosynthesis. MTX has also been shown to inhibit purine biosynthesis due to reductions in THF levels as well as through inhibition of GARFT and AICARFT.[79]

MTX is indicated for the treatment of gestational choriocarcinoma, chorioadenoma destruens, and hydatidiform mole. MTX is used alone or in combination with other agents for the treatment of NHL, mycosis fungoides, breast cancer, epidermoid cancers of the head and neck, and lung cancer. MTX is also used to treat psoriasis and rheumatoid arthritis.

Because of its mechanism of action, malignant cells as well as actively proliferating cells of the host such as bone marrow, fetal cells, buccal and intestinal mucosa, and cells of the urinary bladder are sensitive to MTX. The primary toxicities of MTX are myelosuppression and gastrointestinal toxicity, particularly mucositis. Renal toxicity caused by MTX precipitation in the renal tubules has been observed, as well as a direct drug effect. Leucovorin rescue is often used in conjunction with MTX treatment. LV treatment helps restore folate levels in patients treated with MTX and other antimetabolic agents. MTX can be delivered orally, parenterally, and intrathecally.[79]

Clinical resistance to MTX has been described in the excellent review of Gorlick.[79] Intrinsic resistance of leukemic blast cells appears to stem predominately from impaired polyglutamation. In a study looking at acquired resistance to MTX, 70% of patients with relapsed ALL showed evidence of impaired MTX transport versus only 13% of patients who were untreated. Further analysis showed that six of nine transport defective ALL samples showed reduced or no RFC mRNA expression. Finally, resistance has been noted due to increased enzymatic activity of DHFR due to gene amplification.

7.03.5.2.2 Pemetrexed

Pemetrexed (**2**),[80] is a classical antifolate that inhibits multiple folate-requiring enzymes including TS, DHFR, GARFT, and to a lesser extent AICARFT.[81] Pemetrexed was discovered through structure–activity relationship (SAR) studies revolving around the antifolate lometrexol (**29**). The key change was the replacement of the tetrahydropyridine ring of lometrexol with a pyrrole moiety to give a pyrrolopyrimidine nucleus. This change caused a shift in activity from the inhibition of purine biosynthesis through lometrexol's effects on GARFT, to the inhibition of components of thymidylate biosynthesis. While pemetrexed is a weak GARFT inhibitor (9300 nM), its polyglutamates derived through the action of FPGS are much more active (glu$_5$ 65 nM).[82]

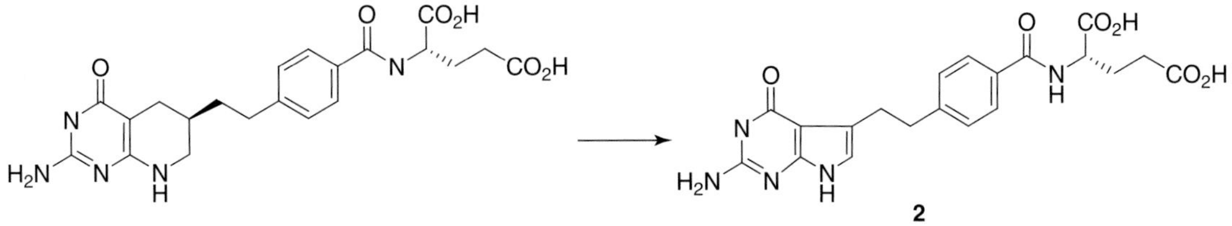

29 Lometrexol

2

Having multiple sites of inhibition results in an activity profile that differs from the TS inhibitor, 5-FU, or the DHFR inhibitor, MTX. Folic acid and vitamin B_{12} supplementation modulate pemetrexed's overall toxicity while enhancing its cytotoxic effects, and pretreatment with folic acid is a component of the clinical regimen.[83] In a phase II study of pemetrexed as a single agent in patients with malignant pleural mesothelioma (MPM), a 17% response rate was observed (nine of 64 patients). In combination with cisplatin, however, a phase III trial found a response rate of 41.3% and a median survival of 12.3 months.[84] The control arm of the study received cisplatin monotherapy, and the response rate in these patients was 16.7% with median survival of 9.3 months.[85] Based on these findings, the pemetrexed/cisplatin combination was approved by the FDA in February 2004 as a treatment for MPM. In August 2004, the FDA approved pemetrexed for the treatment of locally advanced or metastatic NSCLC in previously treated patients. In January 2005, pemetrexed was launched in the UK for the treatment of mesothelioma and NSCLC. Clinical trials of pemetrexed are under way as a therapy for solid tumors including NSCLC, pancreatic, metastatic breast, colorectal, and gastric cancers.[86]

7.03.5.2.3 Raltitrexed

Raltitrexed (**20**, RTX) is a classical antifolate and was originally launched in the UK in 1996 for the first-line palliative treatment of advanced colorectal cancer. The compound has since been launched in 40 other countries, but not in the USA.

RTX was designed to be more water-soluble than the leading TS inhibitor at the time, CB3717 (**30**). Key modifications included replacement of the 2-amino group with a methyl, and the incorporation of the thiophene moiety. In this heterocyclic series, it was also found that the propargyl group of CB3717 could be replaced with a methyl without loss of activity.

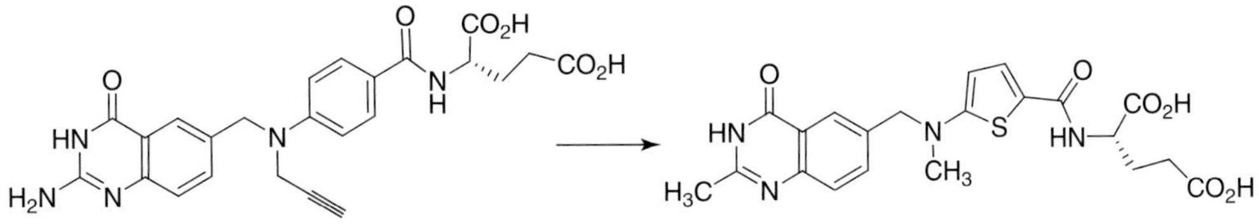

30 CB3717

20 Raltitrexed (RTX)

Like MTX, RTX utilizes the RFC and is an excellent substrate for FPGS, and much of its activity arises from its polyglutamates. Unlike MTX however, RTX is a relatively weak inhibitor of DHFR, instead exerting its efficacy through inhibition of TS. Because the polyglutamates are retained within cells for prolonged periods, RTX has a convenient dosing regimen of once every 3 weeks. As with other drugs dependent on the RFC and or FPGS, mutations or reduced expression of these proteins leads to resistance to RTX.[87]

7.03.5.2.4 Trimetrexate

Trimetrexate (**21**, TMTX) is a nonclassical type A lipophilic DHFR inhibitor. Unlike MTX and RTX, TMTX does not utilize the RFC and cannot be polyglutamated. Thus, from the standpoint of RFC or FPGS related resistance, TMTX should have some advantage. A substituent at the C5 position of the quinazoline ring lends improved activity to the scaffold. Chloro or methyl substituents are an order of magnitude more potent against L1210 leukemia cells than the C5-H analogs.[88] TMTX has been launched for the treatment of pneumocystis carinii infections as well as neoplasms in many countries. The TMTX/LV combination is both effective and nontoxic in the treatment of pneumocystis carinii, as

LV cannot rescue this organism due to the lack of a LV transporter, while normal host cells are protected by LV. With LV, TMTX has also shown promise in combination therapy. In particular, the combination of TMTX, 5FU, and LV has shown a 20% response in previously treated gastrointestinal cancer and a 50% response in previously untreated patients with colorectal cancer.[89] Further it has been shown that leukemic cells resistant to MTX due to aberrant transport mechanisms are sensitive to TMTX.[90]

7.03.5.2.5 Antifolates in clinical trials

Pralatrexate (**22**, PDX) is a classical DHFR inhibitor, currently in phase II single agent trials against NSCLC. Phase I trials in combination with docetaxel for the same indication are also under way. A new phase I trial started in 2005 in combination with vitamin B_{12} and folic acid also in NSCLC. Synthesis of PDX was based on previous observations that 10-N-propargyl-5,8-dideazafolic acid showed excellent TS inhibition. Replacement of the 10-nitrogen group in the typical *para*-aminobenzoic acid side chain with a carbon resulted in PDX, which showed a 10-fold advantage over MTX in L1210 cells.[91] The primary side effect of PDX in phase II trials with previously treated NSCLC patients was stomatitis with no significant myelosuppression observed. Of the 38 evaluable patients, four had major objective responses and 12 had stable disease. Reverse transcription-PCR was performed to measure the relative expression of FPGS and RFC-1 in patients, with the data showing no trend with respect to FPGS. The polymerase chain reaction (PCR) data did suggest that tumors expressing high levels of RFC might be more sensitive to PDX.[92]

Pelitrexol (**23**, PTO) is a GARFT inhibitor currently in phase I clinical trials. A closely related analog, AG-2034 (**31**), is a classical antifolate showing affinity for FPGS. Glutamation (glu-5) results in a 28-fold increase in affinity for GARFT.[93] Mid-stage development was reported in June 2003.[94]

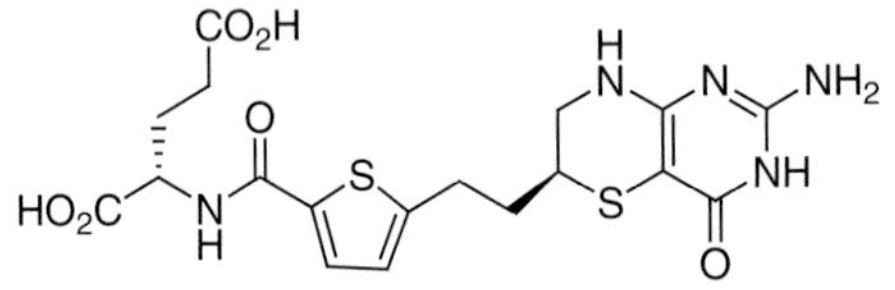

CO2H HO2C N H O S H N N NH2 NH S O

31 AG-2034

OSI-7904L (**25**, OSI) is being developed as a liposomal formation of GW1843 with phase II trials under way against a variety of tumor types including colorectal, gastric, gallbladder and biliary. Combination studies have also initiated with cisplatin and oxoplatin.

OSI behaves as a classical antifolate, but is an unusual tricyclic inhibitor of TS. X-ray analysis of OSI bound to *Escherichia coli* TS showed the tricyclic unit bound in a fashion similar to other inhibitors and the folate cofactor, but the enzyme distorts to accommodate the side chain resulting in dislocations of active-site amino acids. The folate-binding site is nearly identical in *E. coli* and human TS.

OSI evolved from an original series of simple benzoquinazolines (**32**). While the original compounds were highly potent TS inhibitors (20 nM), they had poor cellular activity. Using the benzoquinazoline as a pterin substitute, the group prepared *para*-aminobenzoylglutamate analogs attached to the 9-position. Cyclization of the *para*-aminobenzoylglutamate of typical folate inhibitors to the lactam of OSI had little effect on TS inhibition, but it resulted in improved transportability and a nearly 200-fold increase in FPGS activity. Enzyme kinetic analysis has shown OSI inhibition of TS is noncompetitive with respect to both TS substrates. OSI is actively transported into cells by the RFC, and is a substrate for FPGS. Its polyglutamation properties are unique, however, in that the diglutamate is the primary product. It has been shown that polyglutamation is not required for potent TS inhibition.[95]

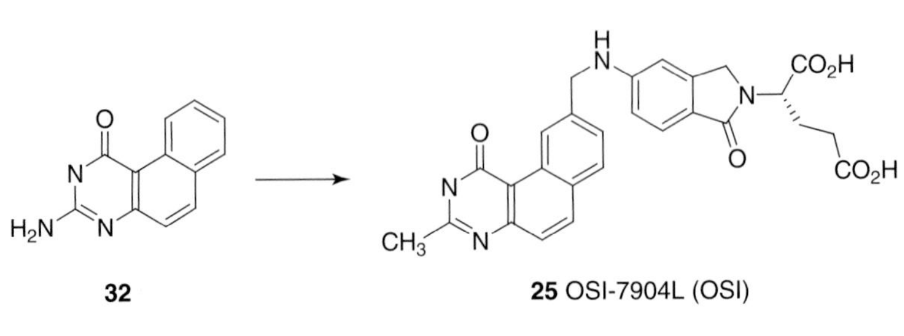

O N H2N N → H N CO2H N O CO2H O N CH3 N

32 **25** OSI-7904L (OSI)

Nolatrexed (**24**, NTX) is a TS inhibitor targeted to the folate-binding site of the TS enzyme. Like TMTX, NTX is a nonclassical type A antifolate that does not contain a glutamate substrate for FPGS and does not require active transport into cells through the RFC, thus potentially overcoming resistance associated with both mechanisms. NTX is currently in phase III clinical trials for the treatment of hepatocellular carcinoma. NTX is also in phase II studies in combination with docetaxel for the treatment of breast, lung, head, and neck cancers. NTX was designed with the aid of x-ray crystallographic data using *E. coli* TS. The 6-methyl group of NTX is a key substituent, aligning the thiopyridine in an optimal configuration, giving a 10-fold increase in human TS activity over the des-methyl analog **33**.[96] Conversion of the 2-methyl group to an amine resulted in a further six-fold increase in potency resulting in the 15 nM TS inhibitor NTX.[97] Cell lines resistant to NTX have shown increased TS activity as well as structural changes in the enzyme.[98] Also P-glycoprotein-associated multidrug resistance (P-gp-MDR) has been noted.[99] Dose-limiting toxicities of NTX included myelosuppression and mucositis.

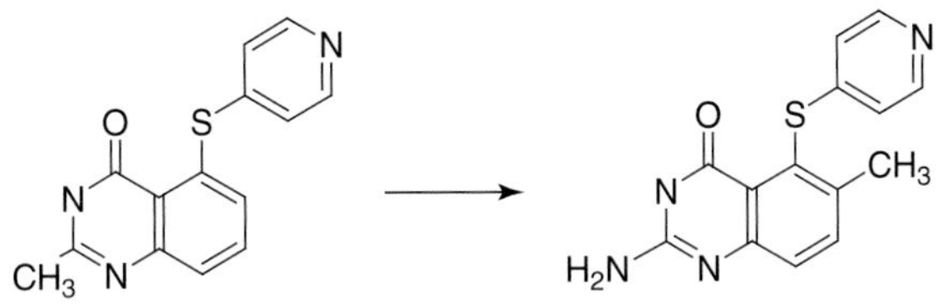

33

24 Nolatrexed (NTX)

Plevitrexed (**26**, PTX) is a potent TS inhibitor containing a tetrazole as a glutamic acid mimic. As such, PTX is a substrate for the RFC, but is not polyglutamated by FPGS, and would be considered a nonclassical type B antifolate. PTX is a noncompetitive inhibitor of TS, with respect to N5,N10-CH_2-THF. PTX has been shown to be effective in cell lines with low FPGS activity. PTX was designed to be a nonpolyglutamatable analog of ICI198583 (**34**). Incorporation of the 2′F substituent provides increased binding, presumably by reinforcement of the preferred near-planar conformation of the benzamide through a hydrogen bond between the 2′F and the glutamate NH. The tetrazole proved to be a highly effective gamma-acid mimic, providing a highly potent, 0.001 nM TS inhibitor.[100]

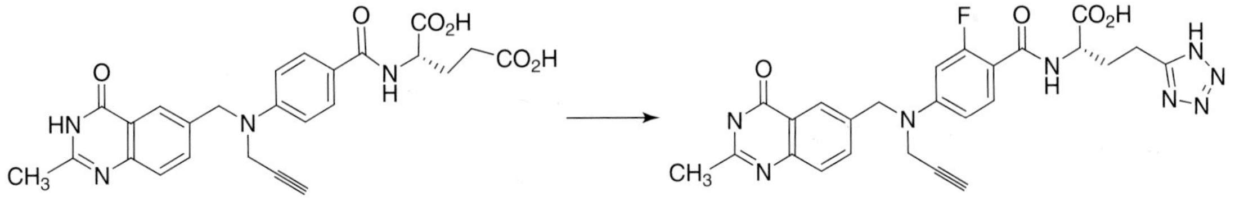

34 ICI198583

26 Plevitrexed (PTX)

PTX is currently in phase I/II clinical trials for the treatment of gastric and other solid tumors. A previous phase II/III trial in metastatic pancreatic cancer was terminated due to the severity of adverse events. Phase II data for patients with recurrent NSCLC has been reported, and showed 43% achieving stable disease following a 3 week cycle of treatment on days 1 and 8.[101]

MDAM (**27**) is a nonpolyglutamatable type B nonclassical antifolate DHFR inhibitor. The gamma-methylene group completely blocks polyglutamation via FPGS, but MDAM does not inhibit the action of FPGS.[102] MDAM is currently in phase I clinical trials in patients with advanced solid tumors. Of 17 evaluable patients in the study, three demonstrated stable disease. No complete or partial responses were observed.[103] MDAM is a DHFR inhibitor, with superior activity relative to MTX in rodent models, perhaps partially due to a combination of enhanced transport to tumor cells and slower deactivation by aldehyde oxidase.[104]

PT-523 (**28**) is a nonpolyglutamatable type B nonclassical antifolate currently in phase I clinical trials. It has been shown to be a potent inhibitor of DHFR, and efficiently utilizes the RFC for cellular influx.[105] PT-523 was designed as an analog of APA-L-ORN, which was shown to be a potent inhibitor of DHFR and FPGS, but had poor cellular activity. It was felt that inhibition of DHFR and FPGS could lead to more efficacious compounds and an effort ensued to attach potentially labile groups to the terminal NH_2 of APA-L-ORN that may deliver the agent more efficiently into cells.[106]

PT-523 came out of this effort as a very potent DHFR inhibitor with cell growth inhibition substantially greater than MTX. However, detailed mechanistic studies reveal that this increased activity is not due to cleavage to APA-L-ORN and inhibition of FPGS.[107] PT-523 is currently in phase I/II trials in NSCLC, as well as phase I trials in patients with advanced solid tumors who had failed curative or survival-prolonging therapy.

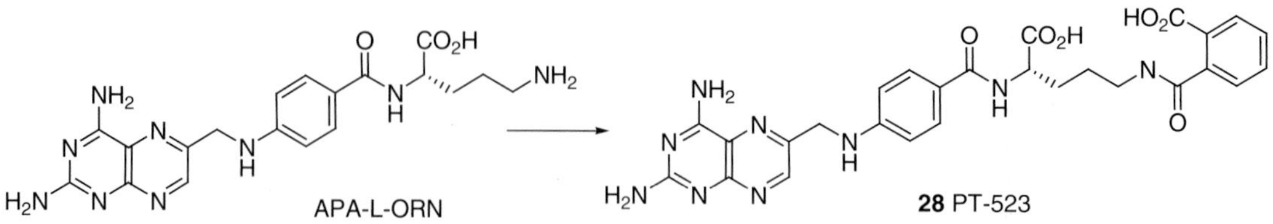

APA-L-ORN → **28** PT-523

7.03.6 Unmet Medical Needs

As the understanding of clinical resistance to current antifolate therapy increases, new molecules designed to overcome these challenges have and will continue to emerge. Antifolates that do not utilize the RFC, and are not substrates for FPGS will potentially meet this need. Also, compounds that do not induce the synthesis of their target enzymes will be valuable tools. Recent work directed at finding molecules that use the MFR rather than the RFC has been reported. The ubiquitous expression of the RFC in normal tissue reduces patient tolerability of antifolates utilizing this transporter. MFR-α, however, is overexpressed on some tumors, and has limited distribution in normal tissue. CB300638 (**35**) has proven to be a potent inhibitor of TS and has shown high affinity for MFR-α while at the same time having low affinity for the RFC.[108] One challenge to developing an agent targeted towards tumors overexpressing MFR-α would be the need to stratify patients based on an individual's particular tumor MFR-α expression. It is known however that 90% of ovarian cancers overexpress MFR-α.

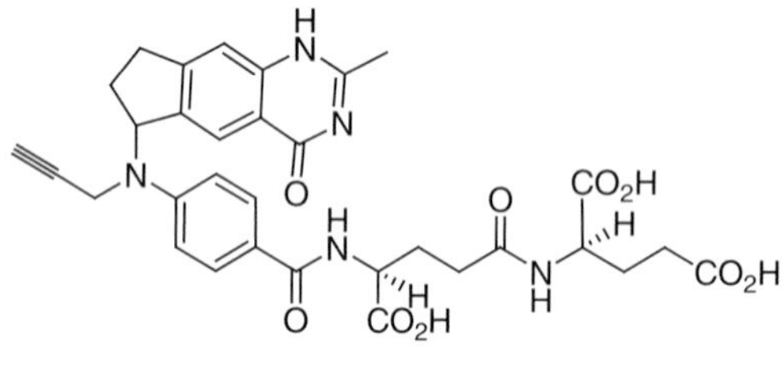

35 CB300638

7.03.7 New Research Areas

A central focus in the development of any oncology therapy at this time is balancing efficacy with toxicity. Preclinical assessment of efficacy and toxicity has been poorly predictive (*see* Section 7.03.3), but movement away from animal models is not likely. However, optimizing the therapeutic window of oncology drugs (increasing efficacy while simultaneously decreasing or minimizing toxicity) is a goal that is achievable. Efforts under way include investigating combinations of antimetabolites that may be less toxic and expression profiling of tumors to identify patients who are more likely to benefit from a particular drug or combination.

Anticancer agents are rarely given singly, as combinations of drugs have proven to be far superior to single agent therapy for a variety of cancers. Antimetabolites are no exception; they have been combined with other antimetabolites and with other chemotherapeutic agents. The goal of combination therapy is to find agents whose activities are synergistic, i.e., a regime where the combined effect is greater than what would be expected from the sum of the two individual agents' activities, and have nonoverlapping toxicities. Since most antimetabolites interfere with the process of DNA synthesis or growth, many combinations with drugs that react with DNA have been used (e.g., the gemcitabine/cisplatin combination for NSCLC).[109] Other potential combinations would be with compounds targeted towards inducing apoptosis, preventing angiogenesis, or with antimetabolites targeting different enzymes in the same pathway. An example is the combination of cyclophosphamide (a DNA alkylating agent), MTX, and 5-FU. This regimen, referred to as CMF, was an early standard of care in the treatment of metastatic breast cancer.[110] 5-FU targets the thymidine biosynthesis pathway by inhibiting TS, while MTX targets the same general pathway by inhibiting DHFR.

Thymidylate synthase inhibitors including 5-FU, capecitabine, and raltitrexed have been tested in the clinic in combination with gemcitabine. Gemcitabine and the TS inhibitors inhibit DNA and RNA synthesis by different mechanisms and possess almost no overlapping toxicity. Gemcitabine inhibits NDPR, depleting cellular dUMP pools, thereby decreasing the dUMP competition with 5-FdUMP at TS. Raltitrexed prevents binding of the folate cofactor on TS, and also has little overlapping toxicity with gemcitabine. However, although several phase III trials in pancreatic cancer have found gemcitabine/5-FU regimens to be tolerable in terms of toxicities, an optimal dosing schedule has yet to be found which improves the median survival of patients with advanced pancreatic carcinoma compared with single-agent gemcitabine.[111] Likewise, phase II trials of the combinations of gemcitabine/raltitrexed[112] and gemcitabine/capecitabine[113,114] were found to have little symptomatic toxicity, but no more efficacious in terms of survival benefit over gemcitabine monotherapy.

Gemcitabine has also been investigated with pemetrexed in pancreatic cancer and NSCLC. Pemetrexed depletes the intracellular supply of both purine and thymidine deoxynucleosides, while gemcitabine is incorporated into nascent DNA strands ultimately resulting in strand termination. Thus, the two agents together would interfere with DNA replication at both the nucleoside and strand synthesis level. Early in vitro cell assays and tumor xenograft models indicated that gemcitabine/pemetrexed would show synergism in vivo, but the degree of activity was dependent on cell type and dosing schedule.[83,115,116] Phase II results for the gemcitabine/pemetrexed combination have been reported, with patients showing a 15% partial response rate, with 29% of the evaluable patients surviving for 12 months. As the three measures (partial response, median survival, and 1-year survival) for the combination showed improvement relative to therapy with gemcitabine or pemetrexed alone (**Table 7**), a phase III trial was initiated. Enrollment for the phase III trial has concluded, and but a final data analysis has not been reported at this writing.

Expression profiling allows for identification of the over- or underexpression of key metabolic enzymes in tumors to enable the chemotherapeutic regimen to be tailored to the metabolic pathway. Examples have already been noted (nelarabine versus PNP expression in T-ALL; CB300638 versus MFR-α in vitro), and the strategy has been applied to the nucleoside analogs as well. Elevated TS expression has been linked to sensitivity to 5-FU treatment in patients with breast,[117] kidney,[118] colorectal,[119] and pancreatic[120] cancers. Low TS levels in normal colonic mucosa have been associated with increased toxicity (grade 2–3 diarrhea and stomatitis) of 5-FU patients receiving 5-FU as adjuvant therapy.[121] Profiling of DNA methylation patterns or chromosomal analysis may allow for identification of the patients who are most likely to respond to azacitidine or decitabine, drugs that work through inhibition of DNA methyl transferases and consequent reactivation of genes silenced by hypermethylation. However, successful implementation of such profiling has not yet been reported.[56] Nonetheless, pretreatment diagnosis of patients whose tumors show high susceptibility to the therapy but whose physiology has low susceptibility to that therapy's toxicity offers the hope of making oncology chemotherapy a less intimidating and disagreeable experience.

Table 7 Comparisons of selected antimetabolite single agent and combination clinical trials

Study	*Patients (total/evaluated)*	*Treatment*	*Agent*	*Partial response (n (%))*	*Median survival (mo)*	*1-year survival (%)*
Advanced pancreatic cancer						
Burris[31]	63/56	Phase III	Gemcitabine	3 (5.4)	5.7	18
Miller[137]	42/35	Phase II	Pemetrexed	2 (5.7)	6.5	28
Kindler[138]	42/40	Phase II	Gemcitabine/pemetrexed	6 (15)	6.5	29
Non-Small Cell Lung Cancer						
Schiller[139]	301/288	Phase III	Gemcitabine/cisplatin	60 (21)	8.1	36
Rusthoven[140]	33/30	Phase II	Pemetrexed	7 (23)	9.2	25
Monnerat[141]	60/54	Phase II	Gemcitabine/pemetrexed	9 (17)[a]	11.3	46

[a] Objective response.

7.03.8 Conclusions

Much progress has been made in the last decade in the development and clinical use of antimetabolites as chemotherapeutics for the treatment of solid tumors. Both mono- and combination therapies have been found to be efficacious, and clinical trials are under way to determine efficacies against a greater variety of tumor types, and of regimens involving two-, three-, and four-drug combinations. Toxicities associated with the therapies have spurred investigation into methods of diagnosing tumors by expression levels of targeted enzymes, to increase the likelihood of efficacy in specific patient populations. As for continued discovery of antimetabolite oncolytics, new chemical entities in this class are not reported as frequently as in the period 1985–95 in the peer-reviewed medicinal chemistry literature, as interest appears to have shifted to the development of kinase inhibitors rather than cytotoxic agents as 'targeted therapies' for carcinomas, lymphomas, and leukemias. Nonetheless, many antimetabolites are the standards of care in oncology, and are likely to remain so for several years to come.

References

1. *NCI Dictionary of Cancer Terms*; National Institutes of Health. National Cancer Institute: Bethesda, MD, 2006. Available at: http://www.nci.nih.gov/dictionary (accessed Aug 2006).
2. Jemal, A.; Murray, T.; Ward, E.; Samuels, A.; Tiwari, R. C.; Ghafoor, A.; Feuer, E. J.; Thun, M. J. *Cancer J. Clin.* **2005**, *55*, 10–30.
3. Ferlay, J.; Bray, F.; Pisani, P.; Parkin, D. M. *GLOBOCAN 2002: Cancer Incidence, Mortality and Prevalence Worldwide*. Available at: http://www-depdb.iarc.fr/globocan/GLOBOframe.htm (accessed Aug 2006).
4. Ries, L. A. G.; Eisner, M. P.; Kosary, C. L.; Hankey, B. F.; Miller, B. A.; Clegg, L.; Mariotto, A.; Feuer, E. J.; Edwards, B. K. *SEER Cancer Statistics Review, 1975–2001*; National Institutes of Health. National Cancer Institute: Bethesda, MD, 2006. Available at: http://seer.cancer.gov/csr/1975_2001/ (accessed Aug 2006).
5. Foye, W. O. *Cancer Chemotherapeutic Agents*; American Chemical Society: Washington, DC, 1995.
6. Jackman, A. L. *Antifolate Drugs in Cancer Therapy*; Humana Press: Totowa, NJ, 1999.
7. Santi, D. V.; Danenberg, P. V. Folates in Pyrimidine Nucleotide Biosynthesis. In *Folates and Pterins*; Blakely, R. L., Benkovic, S. J., Eds.; John Wiley: New York, 1984, pp 345–398.
8. Blakely, R. L. Dihydrofolate Reductase. In *Folates and Pterins*; Blakely, R. L., Benkovic, S. J., Eds.; John Wiley: New York, 1984, pp 191–254.
9. Schirch, V. Folates in Serine and Glycine Metabolism. In *Folates and Pterins*; Blakely, R. L., Benkovic, S. J., Eds.; John Wiley: New York, 1984, pp 399–432.
10. Houghton, J. A.; Harwood, F. G.; Tillman, D. M. *Proc. Natl. Acad. Sci. USA* **1997**, *94*, 8144–8149.
11. Rowe, P. B. Folates in the Biosynthesis and Degradation of Purines. In *Folates and Pterins*; Blakely, R. L., Benkovic, S. J., Eds.; John Wiley: New York, 1984, pp 329–344.
12. Kan, J. L.; Moran, R. G. *Nucleic Acids Res.* **1997**, *25*, 3118–3123.
13. Cory, J. G.; Cory, A. H.; *Inhibitors of Ribonucleoside Diphosphate Reductase Activity*; Pergamon: New York, 1989.
14. Holmgren, A. *Annu. Rev. Biochem.* **1985**, *54*, 237–271.
15. Cory, J. G.; Chiba, P. *Pharmacol. Ther.* **1985**, *29*, 111–127.
16. Anderson, R. D.; Kamen, B. A.; Rothberg, K. G.; Lacey, S. W. *Science* **1992**, *255*, 410–411.
17. Jansen, G. Receptor- and Carrier-Mediated Transport Systems for Folates and Antifolates. In *Antifolate Drugs in Cancer Therapy*; Jackman, A. L., Ed.; Humana Press: Totowa, NJ, 1999, pp 293–321.
18. Tse, A.; Brigle, K. E.; Moran, R. G. *Proc. Am. Assoc. Cancer Res.* **1997**, *38*, 162.
19. Matherly, L. H.; Seither, R. L.; Goldman, I. D. *Pharmacol. Ther.* **1987**, *35*, 27–56.
20. McGuire, J. J. Antifolate Polyglutamation in Preclinical and Clinical Antifolate Resistance. In *Antifolate Drugs in Cancer Therapy*; Jackman, A. L., Ed.; Humana Press: Totowa, NJ, 1999, pp 339–363.
21. Montgomery, J. A. Antimetabolites. In *Cancer Chemotherapeutic Agents*; Foye, W. O., Ed.; American Chemical Society: Washington, DC, 1995, pp 49–58.
22. Pearce, H. L. *Ann. Oncol.* **1995**, *6*, S55–S62.
23. Hirst, G. L.; Balmain, A. *Eur. J. Cancer* **2004**, *40*, 1974–1980.
24. Kamb, A. *Nat. Rev. Drug Disc.* **2005**, *4*, 161–164.
25. Kelland, L. R. *Eur. J. Cancer* **2004**, *40*, 827–836.
26. Johnson, J. I.; Decker, S.; Zaharevitz, D.; Rubinstein, L. V.; Venditti, J. M.; Schepartz, S.; Kalyandrug, S.; Christian, M.; Arbuck, S.; Hollingshead, M. et al. *Br. J. Cancer* **2001**, *84*, 1424–1431.
27. Voskoglou-Nomikos, T.; Pater, J. L.; Seymour, L. *Clin. Cancer Res.* **2003**, *9*, 4227–4239.
28. Booth, B.; Glassman, R.; Ma, P. *Nat. Rev. Drug Disc.* **2003**, *2*, 609–610.
29. Cunningham, D.; Zalcberg, J.; Maroun, J.; James, R.; Clarke, S.; Maughan, T. S.; Vincent, M.; Schulz, J.; Gonzalez Baron, M.; Facchini, T. *Eur. J. Cancer* **2002**, *38*, 478–486.
30. Merriman, R. L.; Hertel, L. W.; Schultz, R. M.; Houghton, P. J.; Houghton, J. A.; Rutherford, P. G.; Tanzer, L. P.; Boder, G. B.; Grindey, G. B. *Invest. New Drugs* **1996**, *14*, 243–247.
31. Burris, H. A., III; Moore, M. J.; Andersen, J.; Green, M. R.; Rothenberg, M. L.; Modiano, M. R.; Cripps, M. C.; Portenoy, R. K.; Storniolo, A. M.; Tarassoff, P. et al. *J. Clin. Oncol.* **1997**, *15*, 2403–2413.
32. Henry, J. R.; Mader, M. M. Recent Advances in Antimetabolite Cancer Chemotherapies. In *Annual Reports in Medicinal Chemistry*; Doherty, A. M., Ed.; Elsevier: Oxford, UK, 2004, pp 161–172.
33. Lyons, J.; Bayar, E.; Fine, G.; McCullar, M.; Rolens, R.; Rubinfield, J.; Rosenfield, C. *Curr. Opin. Investig. Drugs* **2003**, *4*, 1442–1450.
34. Moore, M. J.; Erlichman, C. Pharmacology of Anticancer Drugs. In *The Basic Science of Oncology*, 3rd ed.; Tannock, I. F., Hill, R. P., Eds.; McGraw-Hill: New York, 1998, pp 370–391.
35. Shimma, N.; Umeda, I.; Arasaki, M.; Murasaki, C.; Masubuchi, K.; Kohchi, Y.; Miwa, M.; Ura, M.; Sawada, N.; Tahara, H. *Bioorg. Med. Chem.* **2000**, *8*, 1697–1706.

36. Mackean, M.; Planting, A.; Twelves, C.; Schellens, J.; Allman, D.; Osterwalder, B.; Reigner, B.; Griffin, T.; Kaye, S.; Verweij, J. *J. Clin. Oncol.* **1998**, *16*, 2977–2985.
37. Ishikawa, T.; Utoh, M.; Sawada, N.; Nishida, M.; Fukase, Y.; Sekiguchi, F.; Ishitsuka, H. *Biochem. Pharmacol.* **1998**, *55*, 1091–1097.
38. Reigner, B.; Blesch, K.; Weidekamm, E. *Clin. Pharmacokin.* **2001**, *40*, 85–104.
39. Baker, C. H.; Banzon, J.; Bollinger, J. M.; Stubbe, J.; Samano, V.; Robins, M. J.; Lippert, B.; Jarvi, E.; Resvick, R. J. *Med. Chem.* **1991**, *34*, 1879–1884.
40. *Gemzar (Gemcitabine HCl for Injection) Prescribing Information*, US Food and Drug Administration. US Government Printing Office: Washington, DC, 2005. Available at: http://www.fda.gov/cder/foi/label/2004/020509s029lbl.pdf (accessed Aug 2006).
41. Sandler, A.; Ettinger, D. S. *Oncologist* **1999**, *4*, 241–251.
42. Storniolo, A. M.; Allerheiligen, S. R. B.; Pearce, H. L. *Semin. Oncol.* **1997**, *24*, S7-2–S7-7.
43. Chau, I. W.; Cunningham, D. *Clin. Lymphoma* **2002**, *3*, 97–104.
44. Markham, M. *Semin. Oncol.* **2002**, *29*, 9–10.
45. Thigpen, T. *Semin. Oncol.* **2002**, *29*, 11–16.
46. Pfisterer, J.; Plante, M.; Vergote, I.; Du Bois, A.; Wagner, U.; Hirte, H.; Lacave, A. J.; Stähle, A.; Kimmig, R.; Eisenhauer, E. *J. Clin. Oncol.* **2004**, *22*, 5005.
47. Mutch, D. G. *Gynec. Oncol.* **2003**, *90*, S16–S20.
48. Castagneto, B.; Zai, S.; Marenco, D.; Bertetto, O.; Repetto, L.; Scaltriti, L.; Mencoboni, M.; Ferraris, V.; Botta, M. *Oncology* **2004**, *67*, 27–32.
49. Tsavaris, N.; Kosmas, C.; Gouveris, P.; Gennatas, K.; Polyzos, A.; Mouratidou, D.; Tsipras, H.; Margaris, H.; Papastratis, G.; Tzima, E. et al. *Invest. New Drugs* **2004**, *22*, 193–198.
50. Kaufman, D. S.; Carducci, M. A.; Kuzel, T. M.; Todd, M. B.; Oh, W. K.; Smith, M. R.; Ye, Z.; Nicol, S. J.; Stadler, W. M. *Urol. Oncol.: Semin. Orig. Invest.* **2004**, *22*, 393–397.
51. Doval, D. C.; Sekhon, J. S.; Gupta, S. K.; Fuloria, J.; Shukla, V. K.; Gupta, S.; Awasthy, B. S. *Br. J. Cancer* **2004**, *90*, 1516–1520.
52. Gitlitz, B. J.; Baker, C.; Chapman, Y.; Allen, H. J.; Bosserman, L. D.; Patel, R.; Sanchez, J. D.; Shapiro, R. M.; Figlin, R. A. *Cancer* **2003**, *98*, 1863–1869.
53. Montgomery, J. A.; Shortnacy-Fowler, A. T.; Clayton, S. D.; Riordan, J. M.; Secrist, J. A., III. *J. Med. Chem.* **1992**, *35*, 397–401.
54. Kantarjian, H. M.; Gandhi, V.; Kozuch, P.; Faderl, S.; Giles, F.; Cortes, J.; O'Brien, S.; Ibrahim, N.; Khuri, F.; Du, M. et al. *J. Clin. Oncol.* **2003**, *21*, 1167–1173.
55. Qian, M.; Wang, X.; Shanmuganathan, K.; Chu, C. K.; Gallo, J. M. *Cancer Chemother. Pharmacol.* **1994**, *33*, 484–488.
56. Issa, J.-P.; Kantarjian, H. M.; Kirkpatrick, P. *Nat. Rev. Drug Disc.* **2005**, *4*, 275–276.
57. Paz, M. F.; Fraga, M. F.; Avila, S.; Guo, M.; Pollan, M.; Herman, J. G.; Esteller, M. *Cancer Res.* **2003**, *63*, 1121–1141.
58. *Vidaza (Azacitidine for injectionable suspension) Prescribing Information*, US Food and Drug Administration. US Government Printing Office: Washington, DC, 2004. Available at: www.fda.gov/cder/foi/label/2004/050794lbl.pdf (accessed Aug 2006).
59. Pliml, J.; Sorm, F. *Coll. Czech. Chem. Commun.* **1964**, *29*, 2576–2577.
60. Saba, H.; Rosenfeld, C.; Issa, J.-P.; DiPersio, J.; Raza, A.; Klimek, V.; Slack, J.; de Castro, C.; Mettinger, K.; Kantarjian, H. *Blood* **2004**, *104*, Abstr. 67.
61. Cihak, A.; Vesely, J.; Hynie, S. *Biochem. Pharm.* **2004**, *29*, 2929–2932.
62. van Groeningen, C. J.; Leyva, A.; O'Brien, A. M.; Gall, H. E.; Pinedo, H. M. *Cancer Res.* **1986**, *46*, 4831–4836.
63. Lambe, C. U.; Averett, D. R.; Paff, M. T.; Reardon, J. E.; Wilson, J. G.; Krenitsky, T. A. *Cancer Res.* **1995**, *55*, 3352–3356.
64. Gandhi, V.; Plunkett, W.; Weller, S.; Du, M.; Ayres, M.; Rodriguez, C. O., Jr.; Ramakrishna, P.; Rosner, G. L.; Hodge, J. P.; O'Brien, S. et al. *J. Clin. Oncol.* **2001**, *19*, 2142–2152.
65. Kisor, D. F.; Plunkett, W.; Kurtzberg, J.; Mitchell, B.; Hodge, J. P.; Ernst, T.; Keating, M. J.; Gandhi, V. J. *Clin. Oncol.* **2000**, *18*, 995–1003.
66. Czuczman, M. S.; Porcu, P.; Johnson, J.; Niedzwiecki, D.; Canellos, G. P.; Cheson, B. D. *Blood* **2004**, *104*, Abstr. 2486.
67. Lackey, D. B.; Groziak, M. P.; Sergeeva, M.; Beryt, M.; Boyer, C.; Stroud, R. M.; Sayre, P.; Park, J. W.; Johnston, P.; Slamon, D. et al. *Biochem. Pharmac.* **2001**, *61*, 179–189.
68. Sergeeva, M. V.; Cathers, B. E. *Biochem. Pharmacol.* **2003**, *65*, 823–831.
69. Ecker, G. *Curr. Opin. Investig. Drugs* **2002**, *3*, 1533–1538.
70. Moore, L. E.; Boudinot, F. D.; Chu, C. K. *Cancer Chemother. Pharmacol.* **1997**, *39*, 532–536.
71. de Bono, J. S.; Stephenson, J. J.; Baker, S. D.; Hidalgo, M.; Patnaik, A.; Hammond, L. A.; Weiss, G.; Goetz, A.; Siu, L.; Simmons, C. et al. *J. Clin. Oncol.* **2002**, *20*, 96–109.
72. Jaeckle, K. A.; Phuphanich, S.; van den Bent, M. J.; Aiken, R.; Batchelor, T.; Campbell, T.; Fulton, D.; Gilbert, M.; Heros, D.; O'Day, S. J. et al. *Br. J. Cancer* **2001**, *84*, 157–163.
73. Zimm, S.; Collins, J. M.; Miser, J.; Chatterji, D.; Poplack, D. G. *Clin. Pharmacol. Ther.* **1984**, *35*, 826–830.
74. Kim, S.; Chatelut, E.; Kim, J. C.; Howell, S. B.; Cates, C.; Kormanik, P. A.; Chamberlain, M. C. *J. Clin. Oncol.* **1993**, *11*, 2186–2193.
75. Menei, P.; Jadaud, E.; Faisant, N.; Boisdron-Celle, M.; Michalak, S.; Fournier, D.; Delhaye, M.; Benoit, J.-P. *Cancer* **2004**, *100*, 405–410.
76. Klyosov, A. A.; Platt, D.; Zomer, E. *Preclinica* **2003**, *1*, 175–183.
77. Squeglia, A. *Press Release: Pro-Pharmaceuticals reports Phase I for Davanat-1 in Refractory cancer patients is now in its sixth and final cohort*. Available at: http://www.pro-pharmaceuticals.com/press/pr-06-23-04.pdf (accessed Aug 2006).
78. Rosowsky, A. Development of Nonpolyglutamatable DHFR Inhibitors. In *Antifolate Drugs in Cancer Therapy*; Jackman, A. L., Ed.; Humana Press: Totowa, NJ, 1999, pp 59–100.
79. Gorlick, R.; Bertino, J. R. Clinical Pharmacology and Resistance to Dihydrofolate Reductase Inhibitors. In *Antifolate Drugs in Cancer Therapy*; Jackman, A. L., Ed.; Humana Press: Totowa, NJ, 1999, pp 37–57.
80. Taylor, E. C.; Kuhnt, D.; Shih, C.; Rinzel, S. M.; Grindey, G. B.; Barredo, J.; Jannatipour, M.; Moran, R. G. *J. Med. Chem.* **1992**, *35*, 4450–4454.
81. Shih, C.; Chen, V. J.; Gossett, L. S.; Gates, S. B.; Mackellar, W. C.; Habeck, L. L.; Shackelford, K. A.; Mendelsohn, L. G.; Soose, D. J.; Patel, V. F. et al. *Cancer Res.* **1997**, *57*, 1116–1123.
82. Shih, C.; Thornton, D. E. Preclinical Pharmacology Studies and the Clinical Development of a Novel Multitargeted Antifolate, MTA (LY231514). In *Antifolate Drugs in Cancer Therapy*; Jackman, A. L., Ed.; Humana Press: Totowa, NJ, 1999, pp 183–202.
83. Worzalla, J. F.; Shih, C.; Schultz, R. M. *Anticancer Res.* **1998**, *18*, 3235–4240.
84. Vogelzang, N. J.; Rusthoven, J. J.; Symanowski, J.; Denham, C.; Kaukel, E.; Ruffie, P.; Gatzemeier, U.; Boyer, M.; Emri, S.; Manegold, C. et al. *J. Clin. Oncol.* **2003**, *21*, 2636–2644.
85. Scagliotti, G. V.; Shin, D.-M.; Kindler, H. L.; Vasconcelles, M. J.; Keppler, U.; Manegold, C.; Burris, H.; Gatzemeier, U.; Blatter, J.; Symanowski, J. T. et al. *J. Clin. Oncol.* **2003**, *21*, 1556–1561.

86. Paz-Ares, L.; Bezares, S.; Tabernero, J. M.; Castellanos, D.; Cortes-Funes, H. *Cancer* **2003**, *97*, 2056–2063.
87. Hughes, L. R.; Stephens, T. C.; Boyle, F. T.; Jackman, A. L. Ralitrexed (Tomudex), a Highly Polyglutamatable Antifolate Thymidylate Synthase Inhibitor. In *Antifolate Drugs in Cancer Therapy*; Jackman, A. L., Ed.; Humana Press: Totowa, NJ, 1999, pp 147–165.
88. Elslager, E. F.; Johnson, J. L.; Werbel, L. M. *J. Med. Chem.* **1983**, *26*, 1753–1760.
89. Blanke, C. D.; Kasimis, B.; Schein, P.; Capizzi, R.; Kurman, M. *J. Clin. Oncol.* **1997**, *15*, 915–920.
90. Jackson, R. C.; Fry, D. W.; Bortizki, T. J.; Besserer, J. A.; Leopold, W. R.; Sloan, B. J.; Elslager, E. F. *Adv. Enzym. Reg.* **1984**, *22*, 187–206.
91. DeGraw, J. L.; Colwell, W. T.; Piper, J. R.; Sirotnak, F. M. *J. Med. Chem.* **1993**, *36*, 2228–2231.
92. Krug, L. M.; Azzoli, C. G.; Kris, M. G.; Miller, V. A.; Khokhar, N. Z.; Tong, W.; Ginsberg, M. S.; Venkatraman, E.; Tyson, L.; Pizzo, B. et al. *Clin. Cancer Res.* **2003**, *9*, 2072–2078.
93. Boritzki, T. J.; Zhang, C.; Bartlett, C. A.; Jackson, R. C. AG2034, a GARFT Inhibitor with Selective Cytotoxicity to Cells that Lack a G1 Checkpoint. In *Antifolate Drugs in Cancer Therapy*; Jackman, A. L., Ed.; Humana Press: Totowa, NJ, 1999, pp 281–292.
94. Robert, F.; Garrett, C.; Dinwoodie, W. R.; Sullivan, D. M.; Bishop, M.; Amantea, M.; Zhang, M.; Reich, S. D. *J. Clin. Oncol.* **2004**, *22*, 3075.
95. Smith, G. K.; Bigley, J. W.; Dev, I. K.; Duch, D. S.; Ferone, R.; Pendergast, W. A Potent, Noncompetitive Thymidylate Synthase Inhibitor–Preclinical and Preliminary Clinical Studies. In *Antifolate Drugs in Cancer Therapy*; Jackman, A. L., Ed.; Humana Press: Totowa, NJ, 1999, pp 203–227.
96. Webber, S. E.; Bleckman, T. M.; Attard, J.; Deal, J. G.; Kathardekar, V.; Welsh, K. M.; Webber, S.; Janson, C. A.; Mathews, D. A.; Smith, W. W. et al. *J. Med. Chem.* **1993**, *36*, 733–746.
97. Hughes, A.; Calvert, A. H. Preclinical and Clinical Studies with the Novel Thymidylate Synthase Inhibitor Nolatrexed Dihydrochloride (Thymitaq^{TM}, AG337). In *Antifolate Drugs in Cancer Therapy*; Jackman, A. L., Ed.; Humana Press: Totowa, NJ, 1999, pp 229–241.
98. Tong, Y.; Banerjee, D.; Bertino, J. R. *Proc. Am. Assoc. Cancer Res.* **1996**, *37*, 384.
99. van Triest, B.; Pinedo, H. M.; Telleman, F.; van der Wilt, C. L.; Jansen, G.; Peters, G. J. *Biochem. Pharm.* **1997**, *53*, 1855–1866.
100. Boyle, F. T.; Stephens, T. C.; Averbuch, S. D.; Jackman, A. L. ZD9331 – Preclinical and Clinical Studies. In *Antifolate Drugs in Cancer Therapy*; Jackman, A. L., Ed.; Humana Press: Totowa, NJ, 1999, pp 243–260.
101. Kahanic, S.; Hainsworth, J. D.; Garcia-Vargas, J. E.; Skinner, M.; Garnet, S.; Riddell, P. *Proceedings of the American Society for Clinical Oncology*, Orlando, FL, May 18–21, 2002; Abstr. 2682.
102. Abraham, A.; McGuire, J. J.; Galivan, J.; Nimec, Z.; Kisliuk, R. L.; Gaumont, Y.; Nair, M. G. *J. Med. Chem.* **1991**, *34*, 222–227.
103. Johansen, M.; Zukowski, T.; Hoff, P. M.; Newman, R. A.; Ni, D.; Hutto, T.; Abbruzzeese, J.; Berghorn, E.; Hausheer, F.; Madden, T. *Cancer Chemother. Pharmacol.* **2004**, *53*, 370–376.
104. Cao, S.; Abraham, A.; Nair, M. G.; Pati, R.; Galivan, J. H.; Hausheer, F. H.; Rustum, Y. M. *Clin. Cancer Res.* **1996**, *2*, 707–712.
105. Rosowsky, A. *Curr. Med. Chem.* **1999**, *6*, 329–352.
106. Rosowsky, A.; Bader, H.; Cucchi, C. A.; Moran, R. G.; Kohler, W.; Freisheim, J. H. *J. Med. Chem.* **1988**, *31*, 1332–1337.
107. Rosowsky, A.; Bader, H.; Wright, J. E.; Keyomarsi, K.; Matherly, L. H. *J. Med. Chem.* **1994**, *37*, 2167–2174.
108. Jackman, A. L.; Theti, D. S.; Gibbs, D. D. *Adv. Drug. Del. Rev.* **2004**, *56*, 1111–1125.
109. Ackland, S. P.; Kimbell, R. Antifolates in Combination Therapy. In *Antifolate Drugs in Cancer Therapy*; Jackman, A. L., Ed.; Humana Press: Totowa, NJ, 1999, pp 365–382.
110. Bonadonna, G.; Valagussa, P.; Rossi, A.; Tancini, G.; Brambilla, C.; Zambetti, M.; Veronesi, U. *Breast Cancer Res. Treat.* **1985**, *5*, 95–115.
111. Heinemann, V. *Semin. Oncol.* **2002**, *29*, 25–35.
112. Kralidis, E.; Aebi, S.; Friess, H.; Buchler, M. W.; Borner, M. M. *Ann. Oncol.* **2003**, *14*, 574–579.
113. Hess, V.; Salzberg, M.; Borner, M.; Morant, R.; Roth, A. D.; Ludwig, C.; Herrmann, R. *J. Clin. Oncol.* **2003**, *21*, 66–68.
114. Scheithauer, W.; Schull, B.; Ulrich-Pur, H.; Schmid, K.; Raderer, M.; Haider, K.; Kwasny, W.; Depisch, D.; Schneeweiss, B.; Lang, F. et al. *Ann. Oncol.* **2003**, *14*, 97–104.
115. Teicher, B. A.; Chen, V.; Shih, C.; Menon, K.; Forler, P. A.; Phares, V. G.; Amsrud, T. *Clin. Cancer Res.* **2000**, *6*, 1016–1023.
116. Giovannetti, E.; Mey, V.; Danesi, R.; Mosca, I.; Del Tacca, M. *Clin. Cancer Res.* **2004**, *10*, 2936–2943.
117. Foekens, J. A.; Romain, S.; Look, M. P.; Martin, P.-M.; Klijn, J. G. M. *Cancer Res.* **2001**, *61*, 1421–1425.
118. Mizutani, Y.; Wada, H.; Yoshida, O.; Fukushima, M.; Nonomura, M.; Nakao, M.; Miki, T. *Clin. Cancer Res.* **2003**, *9*, 1453–1460.
119. Hosokawa, A.; Yamada, Y.; Shimada, Y.; Muro, K.; Hamaguchi, T.; Morita, H.; Araake, M.; Orita, H.; Shirao, K. *Int. J. Clin. Oncol.* **2004**, *9*, 388–392.
120. Hu, Y. C.; Komorowski, R. A.; Graewin, S.; Hostetter, G.; Kallioniemi, O.-P.; Pitt, H. A.; Ahrendt, S. A. *Clin. Cancer Res.* **2003**, *9*, 4165–4171.
121. Santini, D.; Vincenzi, B.; Perrone, G.; Rabitti, C.; Borzomati, D.; Caricato, M.; La Cesa, A.; Grilli, C.; Verzi, A.; Coppola, R. et al. *Oncology* **2004**, *67*, 135–142.
122. Devesa, S. S.; Silverman, D. T.; Young, J. L., Jr.; Pollack, E. S.; Brown, C. C.; Horm, J. W.; Percy, C. L.; Myers, M. H.; McKay, F. W.; Fraumeni, J. F., Jr. *J. Nat. Cancer Inst.* **1987**, *79*, 701–770.
123. *Xeloda (Capecitabine Tablets) Prescribing Information*, US Food and Drug Administration. US Government Printing Office: Washington, DC, 2000. Available at: http://www.fda/cder/foi/label/2000/20896lbl.pdf (accessed Aug 2006).
124. *Alimta (Pemetrexed for Injection) Prescribing Information*. US Food and Drug Administration. US Government Printing Office: Washington, DC, 2004 Available at: http://www.fda.gov/cder/foi/label/2004/021677lbl.pdf (accessed Aug 2006).
125. Ananthan, S. Fluoropyrimidines. In *Cancer Chemotherapeutic Agents*; Foye, W. O., Ed.; American Chemical Society: Washington, DC, 1995, pp 49–57.
126. Secrist, J. A., III. 2′-Deoxyribonucleoside Analogs. In *Cancer Chemotherapeutic Agents*; Foye, W. O., Ed.; American Chemical Society: Washington, DC, 1995, pp 71–82.
127. Adkins, J. C.; Peters, D. H.; Markham, A. *Drugs* **1997**, *53*, 1005–1037.
128. Dearden, C. E.; Matutes, E.; Catovsky, D. *Semin. Oncol.* **2000**, *27*, 22–26.
129. Tallman, M. S.; Hakimian, D. *Semin. Hematol.* **1996**, *33*, 23–27.
130. Delfino, C.; Caccia, G.; Gonzalez, L. R.; Mickiewicz, E.; Rodger, J.; Balbiani, L.; Morales, D. F.; Comba, A. Z.; Brosio, C. *Oncology* **2004**, *66*, 18–23.
131. Hoff, P. M.; Ansari, R.; Batist, G.; Cox, J.; Kocha, W.; Kuperminc, M.; Maroun, J.; Walde, D.; Weaver, C.; Harrison, E. et al. *J. Clin. Oncol.* **2001**, *19*, 2282–2292.
132. Twelves, C. *Eur. J. Cancer* **2002**, *38*, 15–20.
133. Sternberg, A. *Curr. Opin. Investig. Drugs* **2003**, *4*, 1479–1487.
134. Adjei, A. A. *Clin. Lung Cancer* **2003**, *4*, S64–S67.
135. Le Chevalier, T. *Semin. Oncol.* **2003**, *30*, 37–44.
136. Stuart, K.; Tessitore, J.; Rudy, J.; Clendennin, N.; Johnston, A. *Cancer* **1999**, *86*, 410–414.
137. Miller, K. D.; Picus, J.; Blanke, C.; John, W.; Clark, J.; Shulman, L. N.; Thornton, D.; Rowinsky, E.; Loehrer, P. J. S. *Ann. Oncol.* **2000**, *11*, 101–103.
138. Kindler, H. L.; Dugan, W. M.; Hochster, H.; Strickland, D.; Jacobs, A.; Hayden, A. M.; Leipa, A. M.; John, W. J. *Am. J. Cancer* **2005**, *4*, 185–191.

139. Schiller, J. H.; Harrington, D.; Belani, C. P.; Langer, C.; Sandler, A.; Krook, J.; Zhu, J.; Johnson, D. H. N. *Engl. J. Med.* **2002**, *346*, 92–98.
140. Rusthoven, J. J.; Eisenhauer, E.; Butts, C.; Gregg, R.; Dancey, J.; Fisher, B.; Iglesias, J. *J. Clin. Oncol.* **1999**, *17*, 1194–1199.
141. Monnerat, C.; Le Chevalier, T.; Kelly, K.; Obasaju, C. K.; Brahmer, J.; Novello, S.; Nakamura, T.; Liepa, A. M.; Bozec, L.; Bunn, P. A. et al. *Clin. Cancer Res.* **2004**, *10*, 5439–5446.

Biographies

Mary M Mader, PhD, earned her bachelor's degree in chemistry in 1985 from the Ohio State University and then worked for a year as a structure editor at Chemical Abstracts Service. She received her doctorate in organic chemistry from the University of Notre Dame in 1991. As a NIH postdoctoral fellow at the University of California–Berkeley, she pursued the computer-aided design and synthesis of peptidic thioacetal inhibitors of cysteine proteases. From 1993 to 1999 she served as a faculty member at Grinnell College and was promoted to the rank of associate professor in 1999. Her research with undergraduate scientists centered on synthetic methods and computational analysis of organosilane oxidations. Dr Mader joined Eli Lilly in 2000 and has been active in the discovery of kinase inhibitors for the cardiovascular, inflammation, and oncology therapeutic areas.

J R Henry received a bachelor's degree from Virginia Tech in 1989, and a PhD in organic chemistry in 1994 from The Pennsylvania State University under the direction of Prof Steven Weinreb. After an NIH postdoctoral fellowship with Prof Andrew Kende at The University of Rochester, he joined the Drug Discovery group at the R W Johnson Pharmaceutical Research Institute in Raritan, NJ as a medicinal chemist. During this time, his efforts were focused on the design of novel p38 MAP kinase inhibitors for the treatment of inflammatory diseases. In 2000, he moved to his current position at Eli Lilly where he is a Principal Research Scientist. His research continues to focus on the design and synthesis of novel kinase inhibitors for a variety of diseases, primarily cancer.

Comprehensive Medicinal Chemistry II
ISBN (set): 0-08-044513-6

ISBN (Volume 7) 0-08-044520-9; pp. 55–79

7.04 Microtubule Targeting Agents

B R Hearn, S J Shaw, and D C Myles, Kosan Biosciences, Hayward, CA, USA

7.04.1 Introduction: Microtubule Targeting Agents

7.04.1.1 Microtubule Dynamics and the Cell Cycle

Microtubules and their associated proteins are cytoskeletal components that play central roles in a number of cellular processes, including mitosis, motility, and cell shape. They are formed by a guanidine triphosphate (GTP)-driven self-assembly process that is initiated by the association of alpha- and beta-tubulin. These heterodimers assemble head to tail, resulting in a spiral protofiliment. In the final step of the process the protofiliments aggregate into sheets that curl up to form the microtubule. The complex equilibria that govern the creation and destruction of microtubules are influenced by a number of cellular factors, including GTP concentration, ionic strength, temperature, and the cellular concentration of certain microtubule-associated proteins. The intricacy of this assembly process and its essential nature make the microtubules and associated proteins ideal targets for the development of cancer chemotherapies.

Microtubule inhibitors target the mitotic spindle (in contrast to other mitotic inhibitors that target nucleic acids) and arrest the cell cycle during the metaphase phase, specifically arresting mitosis at the transition from metaphase to anaphase (**Figure 1**).

Microtubule-targeting agents can be grouped into two classes: (1) those that inhibit the polymerization process (microtubule-destabilizing agents); and (2) those that promote the polymerization of tubulin (microtubule-stabilizing agents). Each of these two major groups will be discussed in following sections of this review. The effects of microtubule interacting agents can be observed in vitro using mammalian alpha- and beta-tubulin. Thus, data from in vitro cytotoxicity, microtubule bundling, and flow cytometry can all be used to drive medicinal chemistry efforts directed toward identifying novel potent analogs of existing microtubule interacting agents.

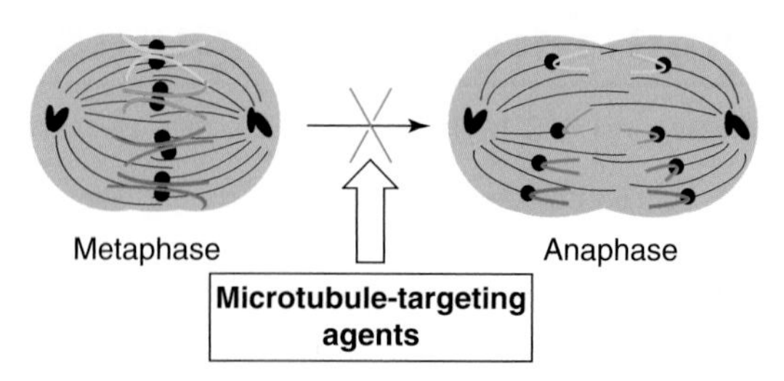

Figure 1 Microtubule-targeting agents' effects on cell cycle.

7.04.2 Inhibitors of Microtubule Assembly

7.04.2.1 Inhibitors Targeting the Vinca Alkaloid Binding Domain

In this well-known class, tubulin binding agents can be divided broadly into two categories, peptidic and nonpeptidic.[1] The nonpeptide compounds bind within the vinca domain and can be either competitive or noncompetitive inhibitors of vinca alkaloid binding to tubulin.[2] The peptides and depsipeptides, however, are exclusively noncompetitive inhibitors of vinca-tubulin binding, and recent evidence suggests that these compounds bind tubulin at the so-called peptide binding site, which is distinct from and yet proximal to the vinca site.[3] Rather than simply destabilizing microtubules, the vinca alkaloids alter the dynamic equilibrium of tubulin addition and loss at mitotoic spindles.[4–6]

Many of these compounds have generated clinical interest, and currently three vinca alkaloids are in clinical use. A number of vinca domain inhibitors, including analogs of halichondrin, dolastatin, and hemiasterlin, are currently under clinical evaluation, and several excellent reviews covering their clinical progress have recently been published.[7–10]

7.04.2.1.1 Nonpeptide vinca site inhibitors

7.04.2.1.1.1 Vinca alkaloids

The vinca *bis*-indole alkaloids, isolated from *Catharanthus roseus*, include naturally occurring vinblastine **1** and vincristine **2**. These compounds have demonstrated clinical utility against leukemias and lymphomas since the early 1960s.[11,12] Vinblastine, primarily used in the treatment of Hodgkin's disease,[13] and vincristine, part of combination therapy for non-Hodgkin's lymphoma, are currently the subjects of several human clinical trials.[14] The mode of action of the vinca alkaloids involves inhibition of tubulin dynamics by binding at the vinca site.[15] More recently, it has been suggested that antiangiogenesis may play a role in the antiproliferative activity of these compounds.[16]

In addition to the naturally occurring vinca alkaloids (**Figure 2**), a semisynthetic analog, vinorelbine **4**,[17,18] has been approved by the US Food and Drug Administration for treatment of nonsmall-cell lung cancer either as a single agent or in combination with cisplatin.[19] Further evaluation of the vinca structure–activity relationship (SAR)[20–23] has led to the identification of several analogs, including vinflunine **5** and KAR-2 **6**, that have progressed to advanced preclinical and clinical trials.[24–26]

7.04.2.1.1.2 Maytansines

The isolation and structure of maytansine **7** from the East African shrubs *Maytenus serrata* and *M. buchananii* was reported by Kupchan *et al.* in 1972.[27] This compound inhibits tubulin polymerization by binding within the vinca domain adjacent to the vinca site. Thus, binding of maytansine at this site inhibits vinblastine binding and prevents association of GTP with tubulin. This binding site is shared by rhizoxin **8**.[28,29] In addition to several naturally occurring maytansinoids, a number of semisynthetic analogs have been prepared to aid in establishing an SAR, and the resulting data are summarized in **Figure 3**.[30–33] Although the presence of a C-3 ester is required for activity, structural variability of the side chain is allowed; furthermore, the presence of the C-9 carbinolamide is essential for antiproliferative activity. Although no maytansinoids are currently in clinical use, an immunoconjugate of DM1, a C-3 side-chain analog of maytansine, was recently evaluated in phase I human clinical trials as part of an antibody-directed therapy.[34] The immunoconjugate was delivered intravenously over 30 min at doses ranging from 22 to 296 mg m^{-2}. The most common toxicity involved elevation of hematologic transaminases. Nausea, vomiting, fatigue, and diarrhea were observed.[35]

7.04.2.1.1.3 Rhizoxin

Rhizoxin **8**, a 16-membered macrocyclic lactone isolated in 1984 from the fermentation broth of *Rhizopus chinensis*, was originally identified as the fungal phytotoxin responsible for rice seedling blight.[36,37] It is biosynthesized by an

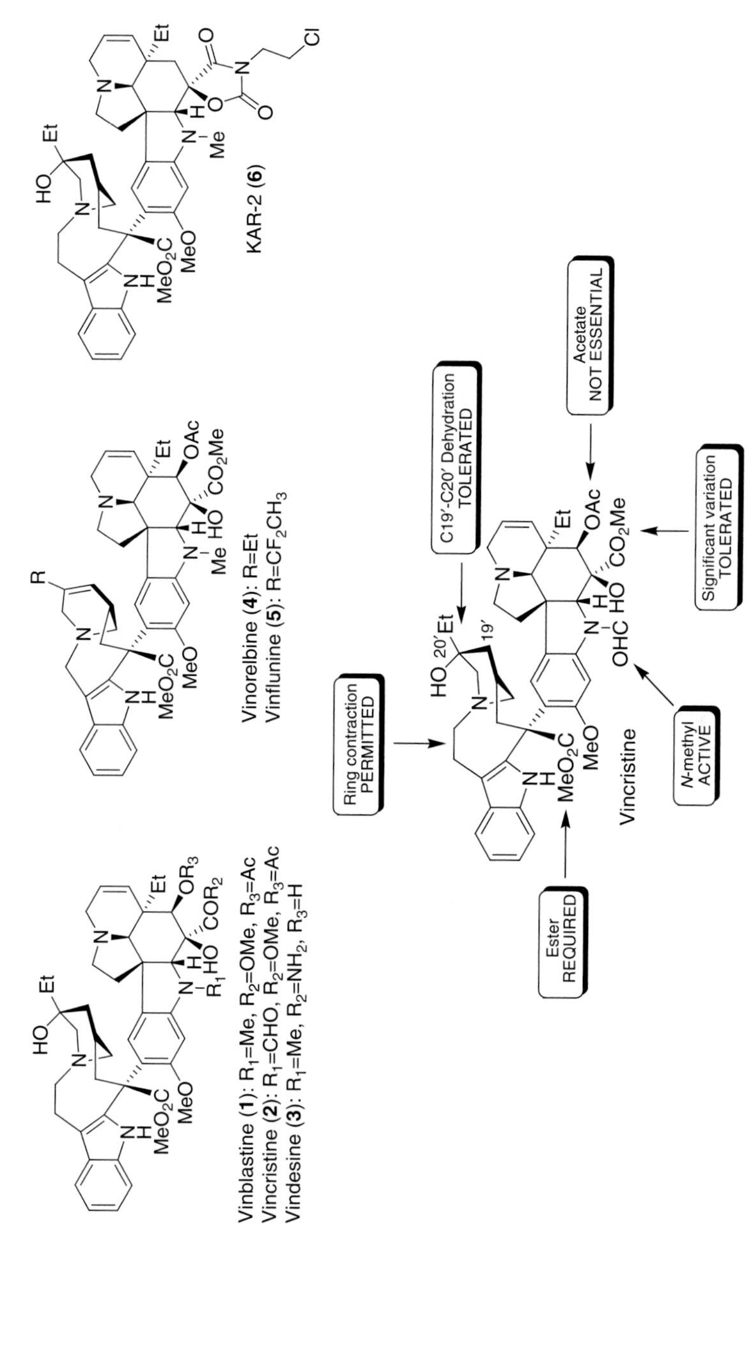

Figure 2 Vinca alkaloids and SAR.

Figure 3 Maytansine SAR.

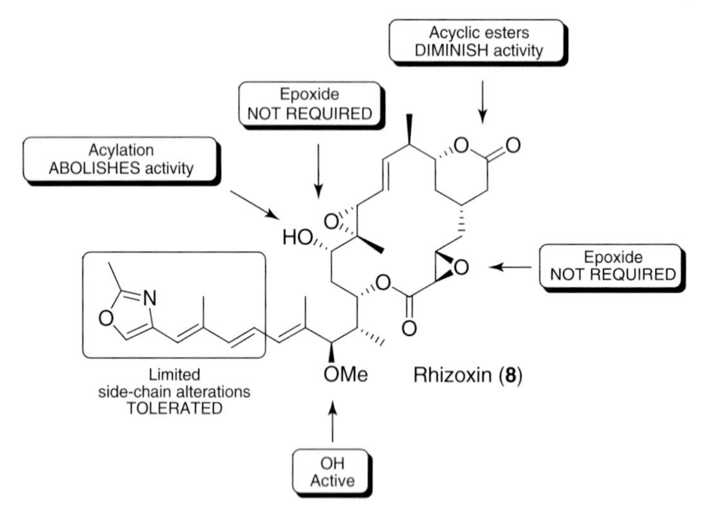

Figure 4 Rhizoxin SAR.

endosymbiotic bacterium from the genus *Burkholderia*.[38] The absolute structure of rhizoxin, unknown at the time of its discovery, was determined in 1986 by single-crystal x-ray crystallographic analysis.[39] Rhizoxin was found to inhibit tubulin polymerization and to destabilize microtubules by rapid, reversible tubulin binding within the vinca domain at the maytansine binding site.[40] Several semisynthetic analogs of rhizoxin have been prepared, and the resulting SAR data are detailed in **Figure 4**.[41,42] In addition, a number of total chemical syntheses of rhizoxin have been reported. The chemistry and biology of this compound have recently been reviewed.[43]

Rhizoxin has been advanced through preclinical and clinical evaluation to phase II. Preclinically, xenograft models showed a strong schedule dependence with antitumor activity evident only with prolonged exposure. In human evaluation of this compound, a 5-min bolus injection showed only modest antitumor activity, possibly due to rapid systemic clearance. To address this pharmacokinetic issue and to exploit the potential for schedule-dependent behavior of rhizoxin, a dose and infusion time-escalating study was initiated. This study reached a dose of 1.2 mg m^{-2} infused over 72 h every 3 weeks. Dose-limiting toxicities were mucositis and neutropenia. No antitumor activity was observed.[44]

7.04.2.1.1.4 Halichondrins

The halichondrins are polyether macrolides isolated from a variety of marine sponges.[45–49] Halichondrin B **9** (**Figure 5**), the most potent member of this family, is a noncompetitive inhibitor of vinblastine binding to tubulin and likely inhibits microtubule formation through binding at a unique site within the vinca domain.[50,51] Analysis of the halichondrin SAR revealed that significant structural variations of the C-30 to C-54 side chain have only minimal effects on potency.

Figure 5 Structures of halichondrin B and analogs.

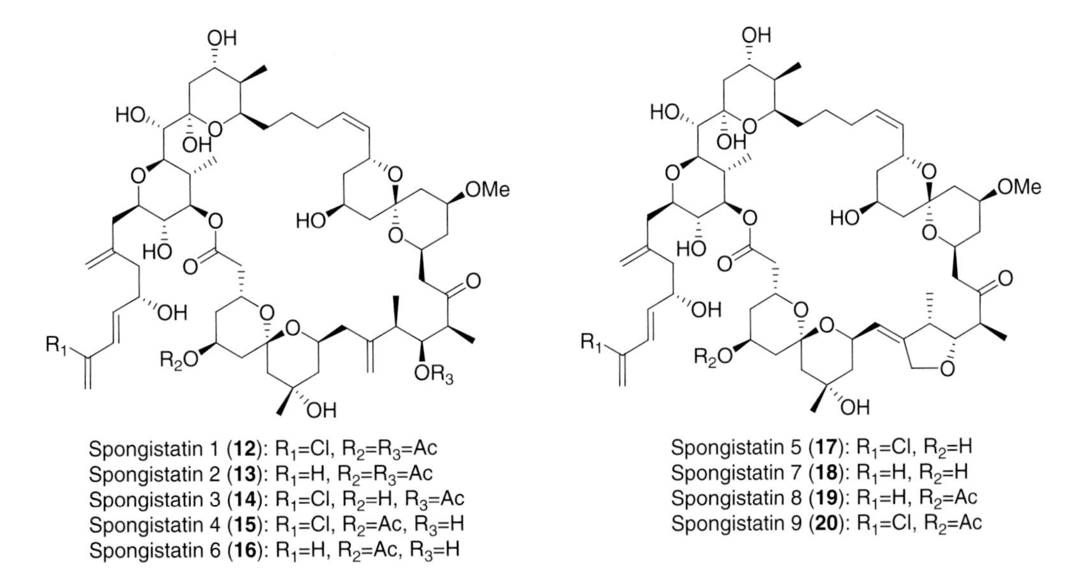

Figure 6 Structures of spongistatins.

Eisai has subsequently developed substantially simplified analogs, ER-086526 **10** and ER-076349 **11**, that demonstrate subnanomolar IC_{50} values versus a variety of human cancer cell lines.[52–54] ER-086526 (E7389) is currently undergoing phase I human clinical trials.

7.04.2.1.1.5 Spongistatins

The spongistatins are macrocyclic polyethers of marine origin (**Figure 6**).[55] Spongistatin 1 **12**, the most highly studied member of this class of compounds, shows potent cytotoxicity against a range of tumor cell types with IC_{50} values in the low picomolar range. Its antiproliferative activity results from its interaction with microtubules by binding within the vinca domain but not at the vinca binding site.[56] Initially, spongistatin 1 was thought to inhibit microtubule assembly, but more recently, an unusual mechanism involving microtubule severing, not simply altering the microtubule dynamics, was proposed.[57] This compound has not advanced beyond the early preclinical stage due to a scarcity of material. The potency and scarcity of these compounds have made them highly attractive targets for total chemical synthesis. Several elegant strategies to this class of compounds have appeared in the scientific literature.[58–64] Not only does this synthetic work have the potential to address the supply problem, but it also provides the opportunity for the expansion of the spongistatin SAR outside the scope of naturally occurring analogs.

7.04.2.1.1.6 Pironetin

Pironetin **21**, an unsaturated lactone originally isolated from a *Streptomyces* sp. fermentation broth, possesses both plant growth-regulatory and immunosuppressant activities.[65–67] In addition, pironetin and structurally related analogs inhibit microtubule assembly by binding to tubulin near the vinca site (**Figure 7**). Replacement of the olefin with an epoxide,

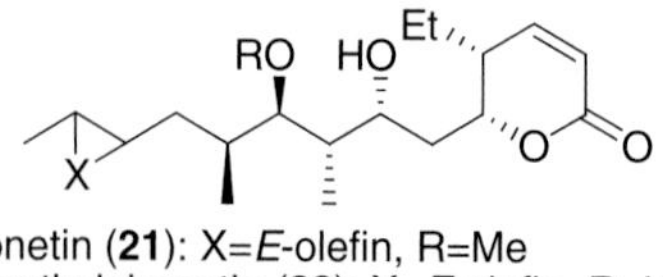

Pironetin (**21**): X=*E*-olefin, R=Me
Demethylpironetin (**22**): X=*E*-olefin, R=H
Epoxypironetin (**23**): X=O, R=Me

Figure 7 Structures of pironetins.

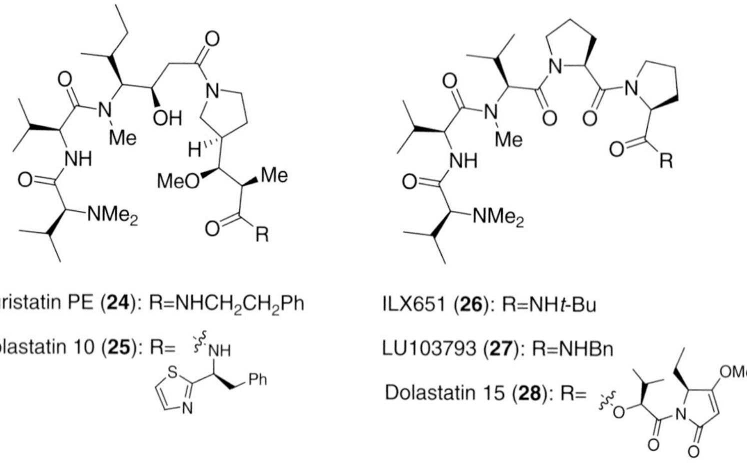

Auristatin PE (**24**): R=$NHCH_2CH_2Ph$
Dolastatin 10 (**25**): R=
ILX651 (**26**): R=NH*t*-Bu
LU103793 (**27**): R=NHBn
Dolastatin 15 (**28**): R=

Figure 8 Structures of dolastatins and analogs.

as in epoxypironetin **23**, was found to result in a substantially weaker antimitotic potency.[68,69] It has been shown by systematic alanine scanning that pironetin binds covalently at Lys352 of alpha-tubulin. This lysine residue is positioned near the opening of a small pocket in alpha-tubulin facing the beta-tubulin of the next dimer, perhaps destabilizing the protein–protein interaction.[70]

7.04.2.1.2 **Peptides and depsipeptides**

7.04.2.1.2.1 Dolastatins

Dolastatin 10 **25** (**Figure 8**), a linear pentapeptide, was originally isolated from the extracts of the sea hare *Dolabella auriculari*. It has since been isolated directly from a dietary cyanobacterium, *Symploca* species.[71–73] With ED_{50} values as low as 4.6×10^{-5} ng mL^{-1} dolastatin 10 was the most potent antiproliferative compound known at the time of its isolation. The antimitotic activity of this compound results from both its inhibition of microtubule formation and its inhibition of tubulin-dependent GTP hydrolysis.[74] It was these findings, as well as the noncompetitive inhibition of vincristine binding to tubulin, that led to a distinct tubulin binding site (peptide site) being proposed. This site is proximal to both the vinca and exchangeable GTP sites and exists for dolastatin 10 and related peptide agents.[75,76] Promising preclinical data led to clinical evaluation of dolastatin 10.[77,78] Although dolastatin 10 was ultimately not successful in human clinical trials, it did progress to phase II and served as the parent compound for a number of structurally related clinical candidates.[79,80] These include auristatin PE **24**,[81,82] ILX651 **26**,[83] LU103793 **27**,[84–86] and dolastatin 15 **28**.[87]

7.04.2.1.2.2 Cryptophycins

The cryptophycins are a family of potent antimitotic depsipeptides isolated in the early 1990s from both *Nostoc* sp. ATCC 53789[88] and *N.* sp. GSV 224.[89] The correct structures of the cryptophycins were determined via total synthesis.[90] Although initially identified as antifungal agents, these depsipeptides also inhibit mitosis by binding at the peptide site within the vinca domain and are active against a number of human cancer cell lines, including cell lines resistant to other agents, with IC_{50} values in the low picomolar range.[91–94] The extraordinary potency of the cryptophycins resulted in intense efforts to generate therapeutic analogs from both total and semisynthesis.[95–100] An overview of the resulting SAR is depicted in **Figure 9**.

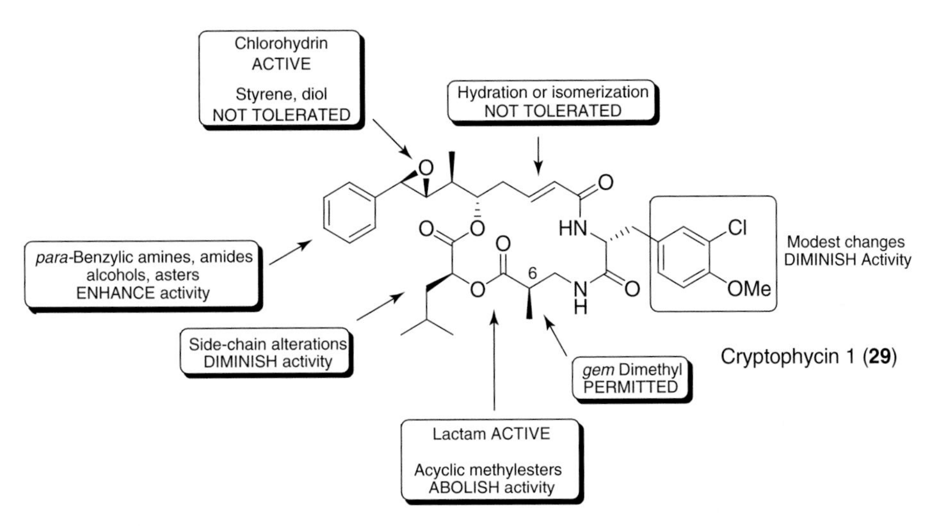

Figure 9 Cryptophycin SAR.

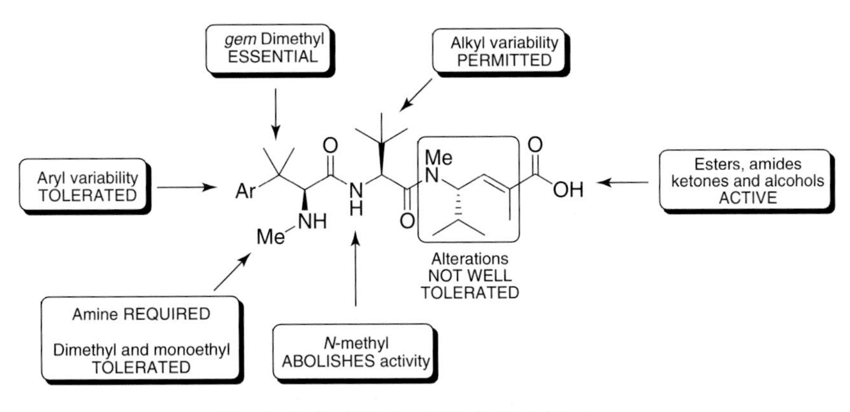

Hemiasterlin (**30**): Ar=*N*-Methylindol-2-yl
HTI-286 (**31**): Ar=Ph

Figure 10 Hemiasterlin SAR.

Cryptophycin 52, the C-6 *gem* dimethyl analog of cryptophycin 1 **29**, was initially the most clinically promising member of this family.[101] Preclinical studies indicated that cryptophycin 52 was 40–400 times more potent than paclitaxel and broad-spectrum in vivo activity was observed in several xenograft models, including multidrug resistant (MDR)-expressing tumors.[102] A phase I dose-escalating study (0.1–$1.92\,mg\,m^{-2}$) monitored toxicity in patients who received 2-h infusions once every 3 weeks. The most common toxicities observed included acute peripheral neuropathy and myalgia.[103] This totally synthetic cryptophycin analog was withdrawn from clinical trials in 2002 due to limited efficacy; however, a number of other cryptophycin analogs are presently in preclinical studies.[104]

7.04.2.1.2.3 Hemiasterlin tripeptides

Hemiasterlin **30**, a potent cytotoxic peptide that was originally isolated from the sea sponge, *Hemiasterella minor*, noncompetitively inhibits vincristine–tubulin binding through interaction at the vinca peptide site (**Figure 10**).[105] Its inhibition of tubulin polymerization resulted in IC_{50} values as low as $0.0014\,\mu g\,mL^{-1}$ in human cancer cells.[106,107] Consequently, hemiasterlin has generated substantial interest as a lead compound in the search for new cancer chemotherapies.[108–110] SAR data reveal that structural diversification of either the aryl substituent or the carboxylic acid may provide analogs that maintain antiproliferative activity.[111,112]

HTI-286 **31** (**Figure 10**) is a synthetic derivative of hemiasterlin that has the advantage of maintaining potency in MDR cell lines that overexpress P-glycoprotein (P-gp). Although in vitro resistance of the KB-3-1 epidermoid carcinoma cell line has been observed, this phenomenon results from a point mutation in alpha-tubulin and from a novel efflux pump, not P-gp.[113] Preclinical mouse xenograft studies, dosing of HTI-286 at the maximum tolerated dose ($1.6\,mg\,kg^{-1}$ intravenously) resulted in up to 98% growth inhibition of both paclitaxel-sensitive and paclitaxel-resistant tumor cells.[114] Phase I human clinical trials involved administering this molecule intravenously over 30 min every 21 days at doses ranging from 0.06 to $2.0\,mg\,m^{-2}$ and generated toxicities including neutropenia, nausea, alopecia, and pain. The resulting recommended phase II dose of HIT-286 was near $1.5\,mg\,m^{-2}$.[115]

7.04.2.2 Inhibitors Targeting the Colchicine Alkaloid Binding Domain

Binders to the colchicine domain of tubulin are inhibitors of mitosis, although precise details of the mechanism are unclear. For example, binding of the drug molecule to microtubule ends, which destabilizes the microtubules leading to depolymerization, only occurs at drug concentrations well above those required to show activity. At lower concentrations these compounds bind to the alpha, beta-tubulin dimer, which is subsequently incorporated into the growing polymer. Although this binding does not stop the ability of the microtubule to grow, it does appear to disrupt the lattice, impairing the delicately balanced microtubule dynamics required for progression of mitosis.[116] Early labeling experiments had localized the binding of colchicines to alpha-tubulin.[117,118] More recently, these results were confirmed through a crystal structure of colchicine itself bound to a tubulin heterodimer,[119] which shows that the heterodimer undergoes a conformational change to a more bent form in the presence of the drug. This conformational change results in a loss of lateral contacts between the heterodimers, which it is believed results in the observed lower microtubule dynamic stability. It is inferred from these results that compounds that bind to this site achieve the antimitotic effect via a similar mechanism.

7.04.2.2.1 Colchicine and analogs

Colchicine **32** is a naturally occurring alkaloid from the plant meadow saffron (*Colchium autumnnale* L.). It was used as a poison in Roman times but was not isolated in pure form until 1820.[120] The structure caused much debate and was finally confirmed by x-ray crystallography in the early 1950s (**Figure 11**).[121] While the molecule was not successful as an anticancer agent, due primarily to its toxicity, it has been used clinically for the treatment of gout,[122] familial Mediterranean fever,[123] and liver cirrhosis.[124]

Colchicine was first produced synthetically in 1959 and, while there are now many syntheses employing a range of strategies in the literature,[125] it still remains a synthetic challenge to integrate all the complex structural features into this molecule. Using total synthesis and semisynthetic approaches, many derivatives of colchicine have been synthesized. From these compounds it has been shown that the B ring is not involved in binding, but does hold the molecule in the active conformation with respect to the A and C rings. Indeed the unnatural 7*R*-colchicine **33**, which exists as the other atropisomer, does not bind tubulin. Furthermore, the 7-acetamide is not required for tubulin binding. The naturally occurring 7-ketone, colchicone,[126] is similar to colchicine, while introducing a double bond into the B ring **34** improves binding to tubulin.[127]

In the tropolone C ring, swapping the keto and methoxy groups leads to inactive compounds. The 10-methoxy group can be varied without affecting tubulin binding (**Figures 12** and **13**). Halides, alkyl, alkoxy, and amino groups have all been synthesized without affecting activity.[128] However, activity is decreased as the steric bulk increases at this position. Moving from a seven-membered tropolone to the aromatic phenol compound, androbiphenylene **35**, a natural product, does not impact tubulin binding. In a study of aromatic biphenyls,[129] tubulin binding could be maintained upon replacement of the C ring by a 4-methylketone, but was lost with a 4-methoxy group. This study also investigated the effect of methoxy groups on the A ring. It was found that the 4-methoxy was required for strong binding. By contrast, the 2- and 3-methoxy groups are not critical for strong binding, although the 2-methoxy group might be important in setting the correct conformation of the molecule.

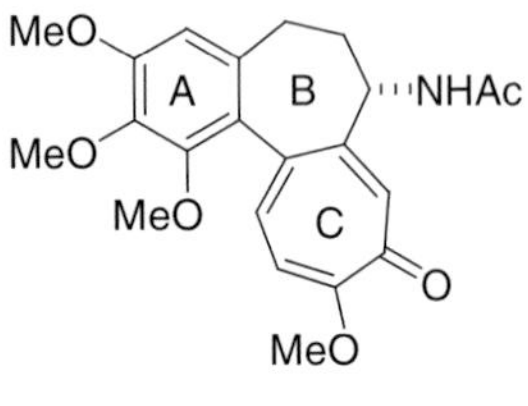

MeO A B NHAc MeO MeO C O MeO

Colchicine (**32**)

Figure 11 Structure of colchicine.

(33) (34) (35)

Figure 12 Structures of colchicine analogs.

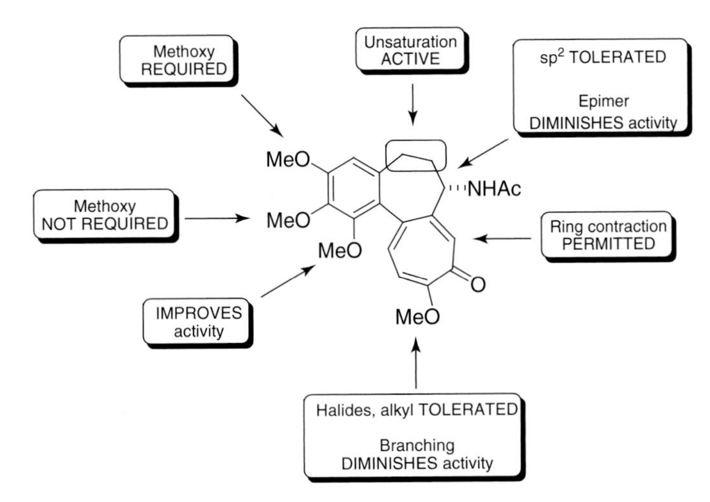

Figure 13 Colchicine SAR.

7.04.2.2.2 Podophyllotoxin and steganacin

Podophyllotoxin **36** is found in a number of plants together with a number of related compounds (**Figure 14**). The chemistry and biology of this compound and related plant natural products have been reviewed.[130] The biology and SAR for this class of compounds have been complicated by the presence of a second potential mechanism for cytotoxicity, inhibition of topoisomerase II (topo II). The SAR described in this section is, unless otherwise noted, for tubulin inhibition.

Podophyllotoxin itself is a potent antimitotic agent, which binds in the colchicine-binding site of tubulin at submicromolar concentrations. In cell culture experiments, the subnanomolar cytotoxicity of this compound led to its investigation as an anticancer agent. This compound failed clinically due in part to its toxicity; however, podophyllotoxin has been used topically for the treatment of venereal warts for several decades.

From the many related compounds, a good understanding of the SAR has been obtained.[131] The stereochemistry of the C-1 position is important for positioning the E ring, for the enantiomer is inactive. Replacing the methyl of the 4′-methoxy with bulky groups lowers the ability of the molecule to bind tubulin; however, its removal has little effect. Removal of the 3′-methoxy in combination with the methyl of the 4′-methoxy lowers the activity of the compound. The configuration of the *trans* fused D-ring lactone is crucial to activity, with the C-2 epimer, obtained upon treatment with mild base, being inactive. The lactone moiety itself is not necessary. Indeed, opening of the lactone with hydrazines yields the podophyllic acid hydrazides, which are weaker, but the ethyl hydrizide **37** has been used clinically as an anticancer agent. The cyclic ether **38** is only twofold less active than podophyllotoxin, but increasing the size of the heteroatom lowers the activity, suggesting a steric constraint in this portion of the molecule.

Aromatization of the C ring or ring opening leads to compounds with low or no activity; presumably this is due to the E ring no longer being positioned correctly. However, dehydration to the apopicropodophyllotoxin **39** has improved activity. Inversion of the 4-hydroxyl leads to a set of naturally occurring and semisynthetic compounds that are usually glycosylated. These are characterized by etoposide **40**, which is used clinically for the treatment of a range of cancers. However, etoposide does not bind to tubulin; it is a topoisomerase II inhibitor. All analogs appear to exert their activity by this mechanism.[132]

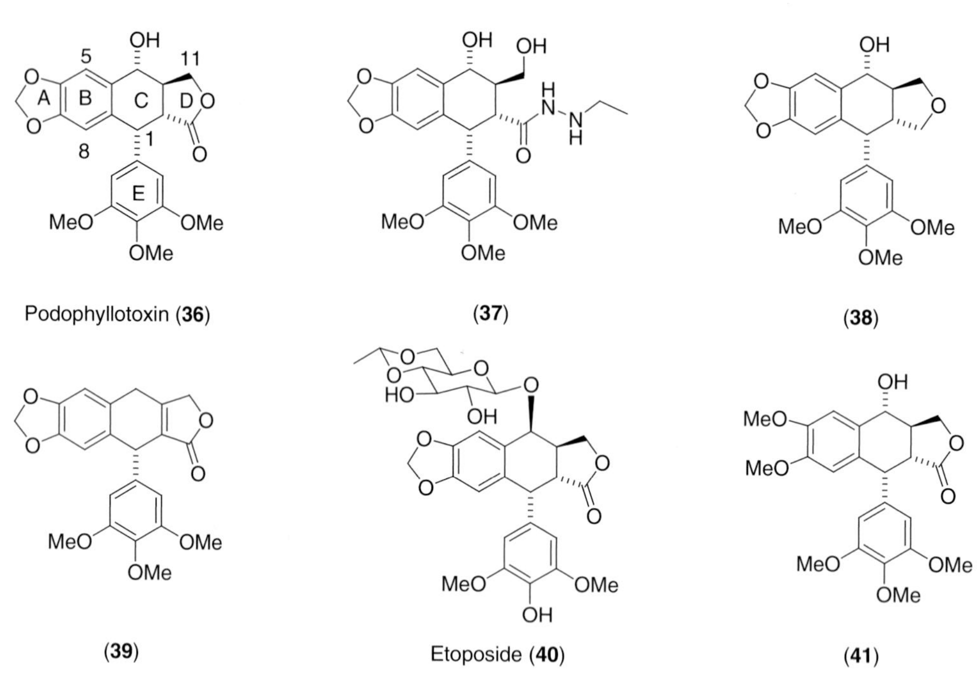

Figure 14 Structures of podophyllotoxin and analogs.

(42)

(43)

Figure 15 Structures of quinoline analogs of podophyllotoxin.

The introduction of a hydroxyl on the B ring at the C-5 position is found in several related natural products as well as semisynthetic analogs. In general this is well tolerated, with a range of alkyl ethers and esters actually increasing the activity of the molecule. Replacement of the A ring with two methoxy groups, as found in sikkimotoxin **41**, results in a 10-fold loss in cytotoxicity.

A number of quinoline and dihydroquinoline derivatives of podophyllotoxin have been synthesized (**Figure 15**).[133] The trends for these compounds are similar to those in the podophyllotoxin, with the corresponding quinoline **42** and analogs having weak cytotoxicity. The dihydroquinolines are significantly more potent. In this series a number of A-ring compounds have been synthesized. The indane and the six-membered dioxane **43** rings are very potent, while the dimethoxy derivative is weaker. These results suggest that the requirements in this sector of the molecule may be due more to steric requirements than electronic. In the E ring there is a 10-fold loss of activity on removal of the 3′-methoxy compared with the 4′-methoxy in podophyllotoxin.

Azatoxin **44** was designed as a topoisomerase II inhibitor through rational design-based modeling from preexisting inhibitors (**Figure 16**).[134] In screening against the National Cancer Institute (NCI) 60-cell-line panel, it was observed that this molecule showed a similar profile to colchicine-site tubulin destabilizers. Indeed, at low concentrations this was found to be the prevailing mechanism of action.[135] It has been shown that the tubulin binding can be engineered out by the introduction of bulky groups at the C-11 position, while the topoisomerase II activity is removed either by methylation of the 4′-position to form methylazatoxin **45**[136] or by appending an aromatic ring to the indole functionality to form benz[e]aztoxin **46**.[137]

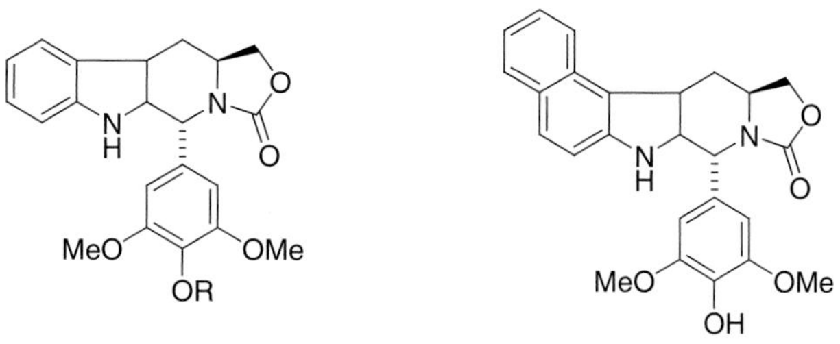

Azatoxin (**44**): R=H
(**45**): R=Me

(**46**)

Figure 16 Structures of azatoxin and analogs.

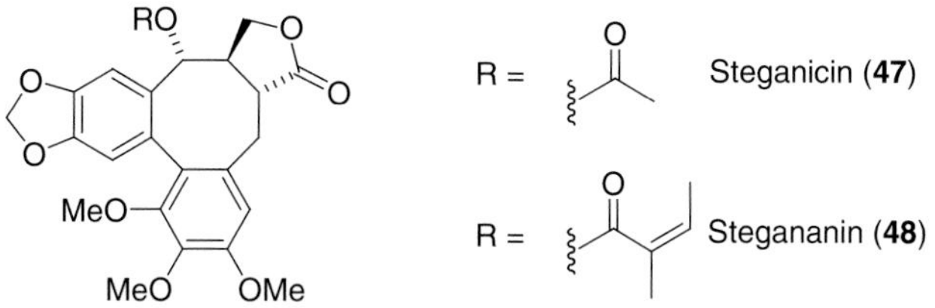

Steganicin (**47**)

Steganacin (**48**)

Figure 17 Structures of steganicin and steganacin.

A structural relation to podophyllotoxin is the steganacin family. These were isolated from the extract of stems and stem bark of the east African tree *Steganataenia araliacea* in the early 1970s. In this series, steganicin **47** and steganacin **48** (**Figure 17**) are of similar potency, both being slightly weaker inhibitors of tubulin polymerization than podophyllotoxin (single-digit nanomolar). It appears that correct positioning of the trimethoxy ring is essential for activity. This is achieved through the eight-membered ring holding the biaryl moiety as the correct atropisomer. In this way inversion of the C-5 stereocenter or ring-opening results in an inactive molecule.[138,139]

In general, the SAR of this series (**Figure 18**) is dominated by the ability of the scaffold to position the trimethoxy aryl group correctly. For this reason, the stereochemistry about the C ring, be it a six- or eight-membered ring, is critical.

7.04.2.2.3 Combretastatins

The combretastatins **49** consist of a group of compounds isolated from the root bark or stem of the South African tree *Combretum caffrum*. The first molecules in this series were characterized in 1982 (**Figure 19**).[140] Since that time many related compounds have been discovered, of which combretastatin A-4 **50**, with single-digit nanomolar cytotoxicity in cell culture, is the most potent.[141] Some of the compounds in this series have been shown to be among the most potent antitubulin agents known. Due to their structural simplicity many analogs have also been synthesized.

In the A ring, replacing the 4-methoxy with a phenol reduced activity significantly, although inhibition of tubulin polymerization is not affected. Increasing the size of the ether at this position destroys both cytotoxicity and tubulin polymerization inhibition. The 5-methoxy group appears to be critical for activity.[142] The strict requirements for activity in this portion of the molecule mirror the A ring of colchicine.

In the B ring, removal of the 3-hydroxyl reduces activity, but the compounds maintain potency. However, moving this methoxy around the ring or replacing it with a halide-, alkyl-, or nitrogen-containing group lowers activity severely.[143] Inversion of the methoxy and hydroxy groups as in isocombretastatin **51** (**Figure 20**) results in an inactive compound in a similar fashion to that seen in the colchicine series. Glycosylation of this position results in low or no activity.[144]

In investigating the use of combretastatin as a potential therapeutic, the low solubility of the compound made formulation difficult. To address this issue a number of prodrugs of the B-ring phenol have been synthesized.[145] The sodium phosphate compound **52** was stable and increased the water solubility significantly. This prodrug showed similar nanomolar activity to the parent compound. An interesting characteristic of this molecule is its ability to target the vasculature. It has been shown to lower blood flow to tumors, leading to hemorrhage. Although the reason for this unusual activity is unclear, it is thought that cytoskeletal changes take place in the endothelial cells of the new blood vessels due to tubulin binding which results in disruption of the blood flow. This activity was observed in a phase I

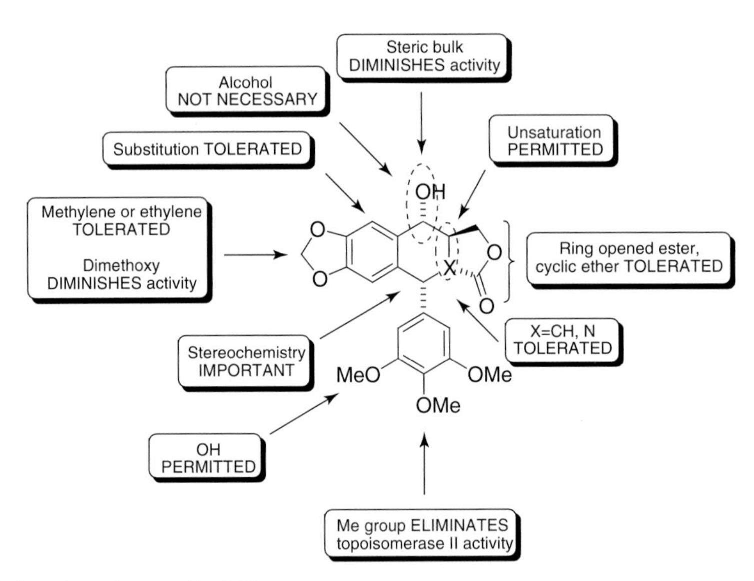

Figure 18 Podophyllotoxin and steganicin SAR.

Combrestatin (**49**) (**50**)

Figure 19 Structures of combrestatins.

Isocombrestatin (**51**)

Figure 20 Structure of isocombrestatin.

study, in which the molecule showed signs of efficacy, including one complete response.[146] Replacing the phenol by a bioisosteric boronic acid is an alternative approach to increasing the aqueous solubility of the molecule. Molecule **53** was designed based on modeling studies and is expected to displace a water molecule, which is thought to make a hydrogen bond to the combretastatin phenol. It shows similar binding to tubulin as the natural product and nanomolar cytotoxicity (**Figure 21**).[147]

The linkage between the two aromatic rings has been varied both by nature and by chemical synthesis. This linkage helps to hold the two aromatic rings in the correct conformation for potency, which appears from crystal structures to be 66°.[148] The saturated analogs of combretastatin A-4 are weaker, while the *trans* isomer has no cytotoxicity. Introduction of a carboxylic acid, ester, or amide off the alkene lowers the potency by at least two orders of magnitude.[149] The ketone analog of combretastatin **54** is only weakly cytotoxic; however, when the linkage is shortened to a single carbon with a ketone, as in phenstatin **55**,[150] a potent molecule is obtained. The introduction of a further hydroxyl at the

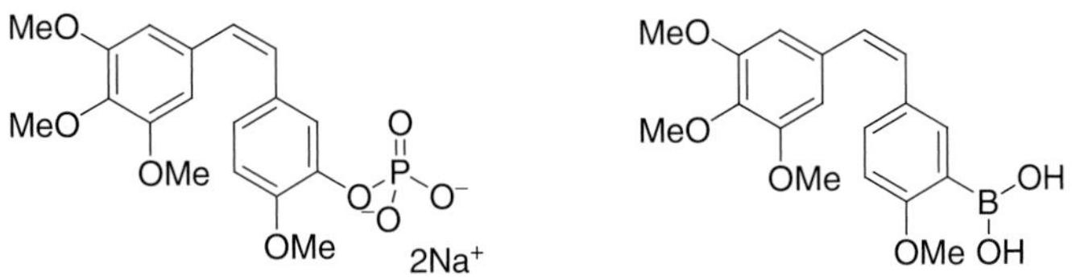

(52) (53)

Figure 21 Structures of water-soluble combrestatins.

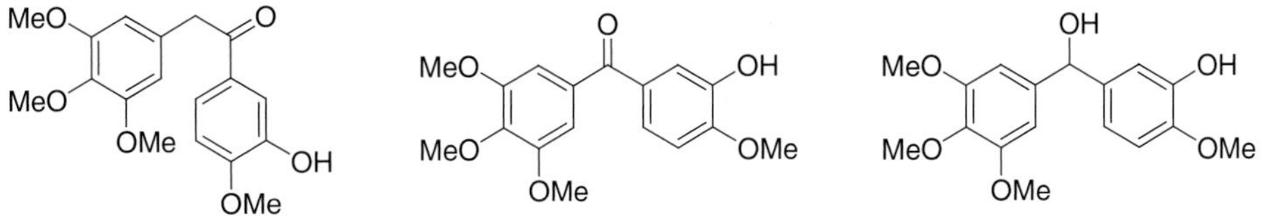

(54) (55) (56)

Figure 22 Structures of additional combrestatin analogs.

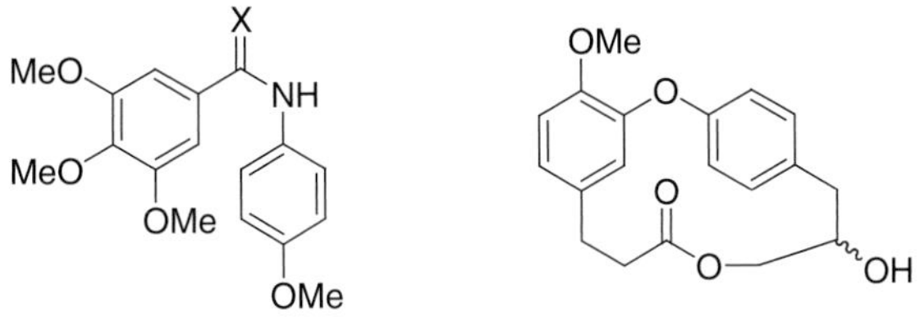

X = H_2 (57)
X = O (58)
(59)

Figure 23 Structures of weakly cytotoxic combrestatin analogs.

2-position of the B ring lowers the potency of the molecule (**Figures 22–24**).[151] The prodrug of this molecule is significantly poorer than the corresponding combretastatin A-4 **52** prodrug, while reduction of the ketone, hydroxyphenestatin **56**, results in only weak activity. Replacement of the ketone with a sulfur or selenium results in significantly poorer compounds.[152]

The introduction of a heteroatom into the bridge in the form of a benzamine **57** is inactive while the benzamide **58** shows lower cytotoxicity.[153] The cyclization to a biaryl ether system **59** results in weak tubulin binders; however, the binding site is maintained.[154]

7.04.2.2.4 Flavinoids

Screening the plant extract *Polymnia fruticosa* against the NCI panel of tumor cell lines showed high correlations with the other microtubule-interacting compounds. The active compound, the flavanol centaureidin **60** (**Figures 25** and **26**), was isolated and found to compete with colchicine and podophyllotoxin for tubulin, suggesting that it bound to a similar site.[155] Many more compounds of this class have since been isolated for a variety of plants with varying substitution about the rings.[156,157]

In this series of compounds, a twofold loss in activity is observed if the hydroxy and methoxy of the C ring are reversed. Introducing a further oxidation on to this ring lowers the activity. Compounds with a hydroxyl at the 3-position are generally better than the corresponding methoxy compounds, while a trimethoxy arrangement of the A ring results in superior activity, suggesting that it is binding in a similar place to that of the A ring of colchicine.

7.04.2.2.5 Other natural products

Curacin A **61** was isolated from the marine cyanobacterium *Lyngbya majuscula* and shows nanomolar cytotoxicity against a range of cell lines (**Figure 27**).[158] Although the molecule is very different in structure from other tubulin binders at the

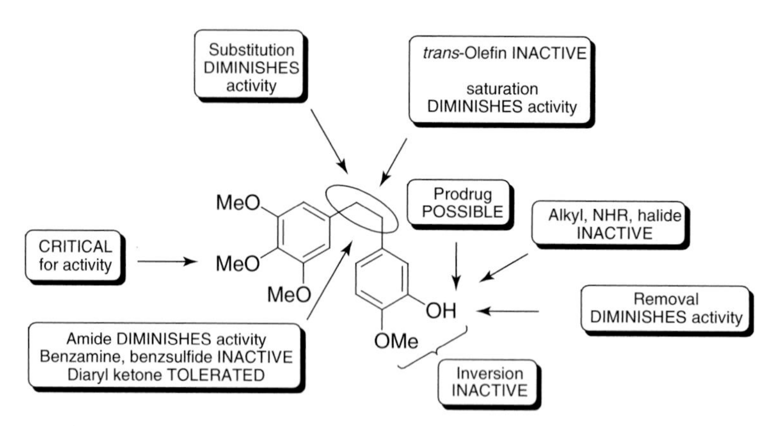

Figure 24 Combrestatin SAR.

(60)

Figure 25 Structure of centaureidin.

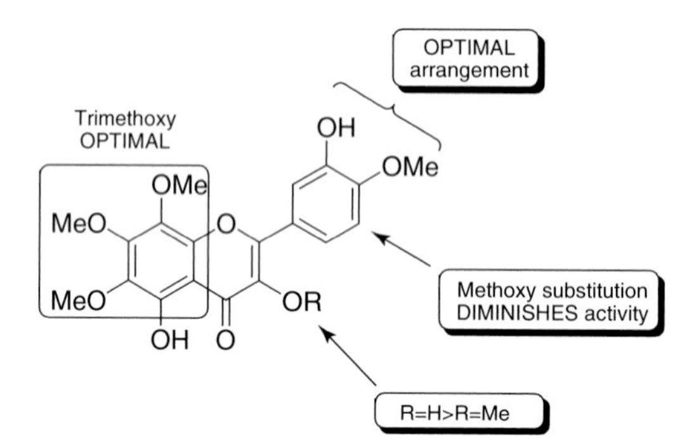

Figure 26 Centaureidin SAR.

colchicine site, it has been shown to be a competitive inhibitor of colchicine binding, suggesting that it binds to the same site.[159] It is assumed that in solution it adopts a conformation that allows it to interact with tubulin. Significantly, screening a range of fatty-acid derivatives resulted in no active compounds. The stereochemistry at the 2-position of the thiazoline ring is important, for the epimer is inactive.[160] However, removal of the cyclopropane, **62**, or epimerization of the cyclopropane only modestly reduces potency.

The secondary metabolite of estrogen, 2-methoxyestradiol **63**, was found while screening for angiogenesis. The molecule is a potent inhibitor of endothelial cell proliferation and migration as well as angiogenesis.[161] Further studies showed that this molecule inhibited angiogenesis through its interaction with tubulin and was an effective inhibitor of colchicine binding.[162] 2-Methoxyestradiol has been shown to suppress tumor growth in vivo without apparent signs of toxicity. Moreover, analyses showed that there was a reduction in tumor vessel density and proliferation rate, suggesting that there is direct inhibition of both endothelial and tumor cell compartments.[163] Initial screening showed that 2-hydroxyestriol **64** and 2-methoxyestradiol-3-methyl ether **65** were 50–100-fold weaker in activity (**Figure 28**).

These results suggested that the A ring was playing a similar role in binding tubulin to that of the C ring of colchicine. To test this theory a collection of tropolone A-ring analogs were synthesized, in which the tropolone ring was

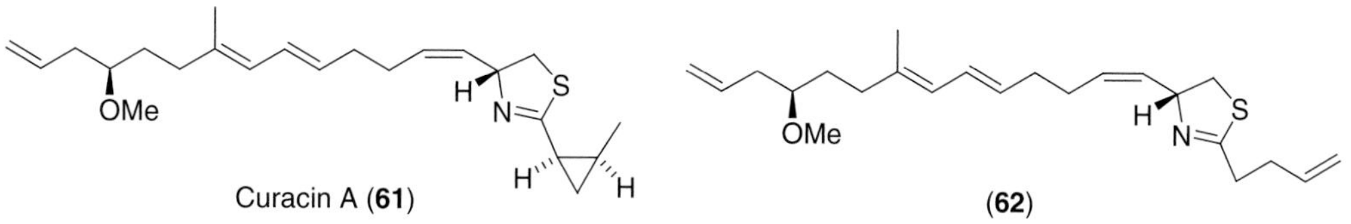

Curacin A (**61**)

(**62**)

Figure 27 Structures of curacin A and analog.

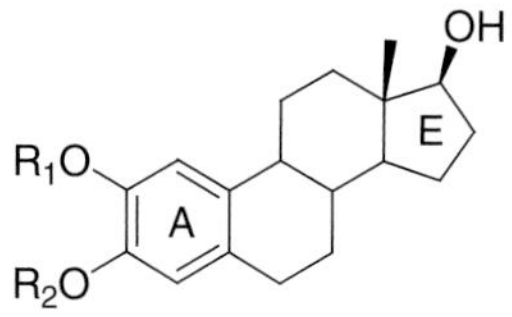

(**63**): R_1=H, R_2=Me

(**64**): R_1=H, R_2=H

(**65**): R_1=Me, R_2=Me

Figure 28 Structures of estradiols.

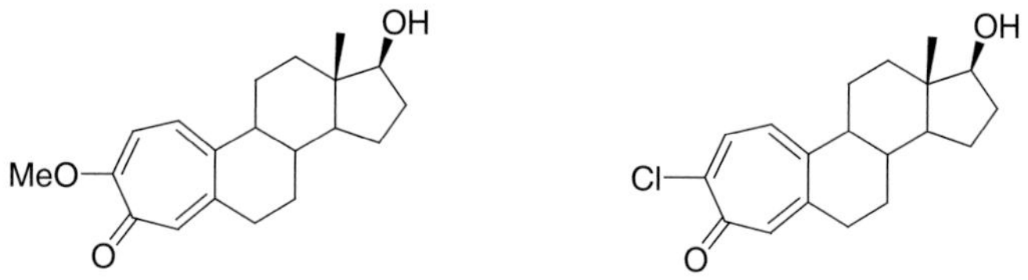

(**66**) (**67**)

Figure 29 Structures of tropolone estradiol analogs.

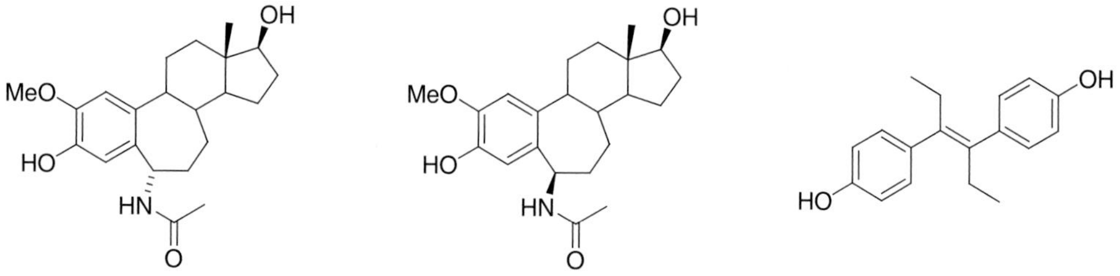

(**68**) (**69**) (**70**)

Figure 30 Structures of additional steroid-derived analogs.

varied (**Figures 29** and **30**). The molecule **66** most resembling colchicine showed the best binding to tubulin. A further improvement was obtained when the methoxy was substituted with a chloride **67**.[164]

Expansion of the B ring and introduction of an acetamide similar to that found in colchicine resulted in a slightly weaker compound **68**; however, the epimer **69** was significantly poorer.[165] Interestingly, some estrogenic drugs, for example E-diethylstilbestrol **70**, also appear to bind to tubulin at the colchicine-binding site.[166]

7.04.2.2.6 Synthetic compounds

Two classes of synthetic compounds that function by destabilizing tubulin have entered human clinical trials. These classes are the sulfonamides and the aryl carbamates. Sulfonamides are a class of compounds that shows a range of biological activity (**Figure 31**). This made them an interesting class of compounds to screen for antimitotic activity. The sulfonamide **71** was found through the screening.[167] This molecule was optimized using in vitro and in vivo assays. It was shown that the methoxy group improved activity over a methyl group, while the in vivo activity was improved by introducing a nitrogen atom into the ring and introducing a hydroxyl group into the anilino ring to generate ABT-751 (formally E7010) **72**. The molecule showed good activity against a range of human tumors in vivo, including MDR expressing resistant cell lines,[168] and was subsequently shown to inhibit tubulin polymerization by binding to the

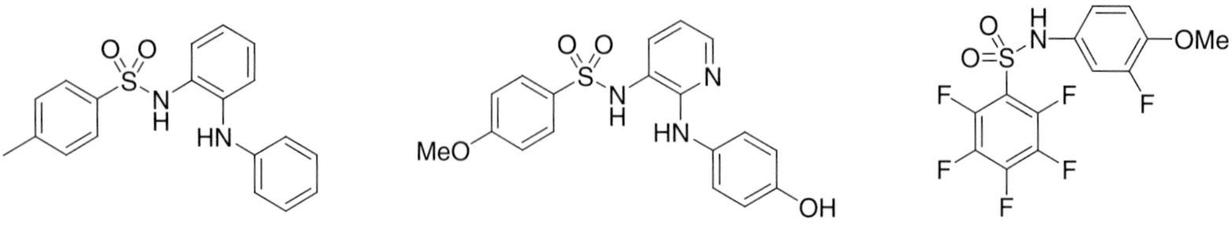

(71) (72) (73)

Figure 31 Structures of aryl sulfonamides.

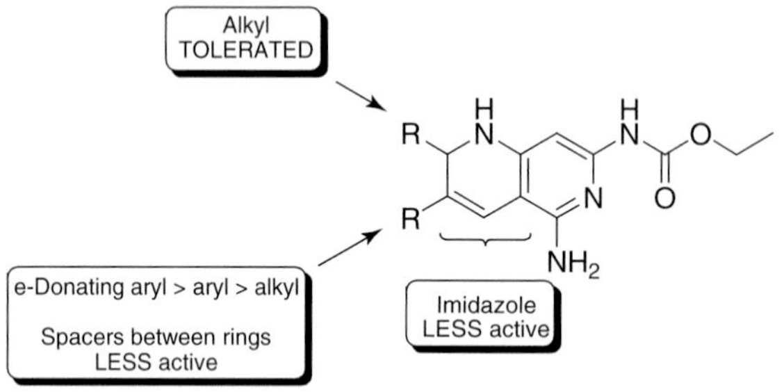

Figure 32 Pyridopyrazine SAR.

colchicine binding site.[169] This molecule is orally active with good dose-dependent bioavailability and is currently undergoing both phase I and phase II clinical trials for nonsmall-cell lung, breast, and colorectal cancers.[170]

Further modification of this scaffold resulted in a group of *N*-(7-indole)-1,4-benzenesulfonamides, which inhibited cell cycle at the G1 phase and did not inhibit tubulin polymerization.[171] One of these compounds (E7070) is currently in phase II trials. A pentafluorophenylsulfonamide (T138067) **73** has been reported to bind irreversibly to the Cys-239 of tubulin, which is found in the colchicine-binding site.[172] It is thought that sulfhydryl group displaces the 4-fluoro group of the pentafluorophenyl ring. The molecule has undergone a phase II trial.

Since the beginning of the 1980s, carbamates of aromatic aminoaryl compounds have been reported to possess antimitotic activity. The pyridopyrazine core of NSC-330770 has been the subject of an SAR study in which the phenyl group at the 2-position was varied along with the substituent at the 3-position (**Figure 32**).[173] It was found that the electron-donating groups on the aryl ring maintained or improved activity, while bulky groups diminished potency. The introduction of an alkyl group at the 2-position was generally beneficial. The pyrazinyl ring is significantly superior to the imidazole ring: this is thought to be due to the better positioning of the aromatic ring for binding (**Figure 33**).[174]

A set of aromatic ureas have also been shown to bind tubulin at the colchicine site. Molecules containing an *N*-(2-chloroethyl)urea were shown to bind covalently to tubulin thought to be through reaction of the Cys-239.[175] Further structural modification is needed to improve the activity of these molecules; however, they represent another class of compounds that are able to bind tubulin. The most clinically advanced members of this class are nocadazole **74**, NSC-330770 **75**, and tubulozole **76**, all of which have been shown to inhibit tubulin polymerization by binding competitively with colchicine. Nocadozole and tubazole and the more water-soluble erbulozole **77** have undergone clinical evaluation, without reaching the market. Erbulozole showed acute central nervous system toxicity with symptoms similar to Wernicke's syndrome.[176]

The chalcones were a rationally designed group of compounds based on the A ring of colchicine and the presence of a sulfhydryl residue in the colchicine–tubulin binding site that could be used to generate an irreversible inhibitor (**Figures 34** and **35**). In this way a molecule containing a trimethoxy aryl group and a Michael acceptor enone **78** was produced. The molecule was found to display good antimitotic activity in vitro and an SAR study was carried out based on this scaffold.[177]

It was found that the trimethoxy aryl ring tolerated alkyl groups, halides, and amides; however, these did not increase the activity. Substitution at the 3-position lowered activity in all cases. Replacement of the amido substituent with a dialkyl amine improved the activity by 1000-fold. Again, moving this substituent to the 3-position destroyed activity. Substitution of the double bond at the alpha-position enhanced activity while at the beta-position, activity was lost. Inversion of the enone lowered potency. These molecules have been found to bind to tubulin and inhibit polymerization. They are inhibited by colchicine, suggesting that they share a similar binding site.[178]

Figure 33 Structures of additional synthetic aromatics.

Figure 34 Structure of a substituted chalcone.

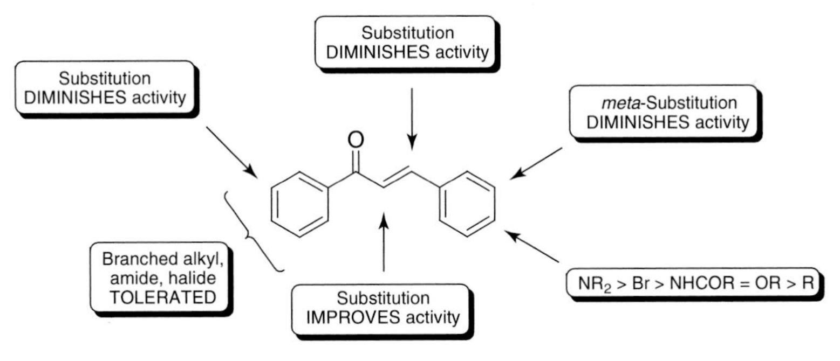

Figure 35 Chalcone SAR.

A series of quinolones was synthesized, inspired by the activity of flavoloids. In an initial screen a collection of 2-phenyl quinolones were synthesized with substitutions at the 6-, 7-, and 8-positions of the ring (**Figures 36** and **37**).[179] Substitutions at the 8-position resulted in poor cytotoxicity; however, substituents at the 6- and 7-positions were well tolerated. Activity was destroyed by both methylation of the quinolone nitrogen and aromatization to the corresponding quinoline. The more potent molecules were tested in the NCI 60-cell-line panel where they showed a similar profile to colchicine. Subsequently it was established that these molecules inhibited tubulin polymerization at similar concentrations to colchicine; however, they were only competitive inhibitors of colchicine at high concentrations. Further studies showed that the 6-methoxy and 6,7-methylenedioxy substitution led to submicromolar cytotoxicity and compounds that could inhibit colchicine-tubulin binding.[180] Replacement of the phenyl ring with heteroaromatic rings or constraining it through an extra ring resulted in no activity. Substitution of the phenyl ring at the *para* position with substituents larger than a methyl group resulted in no activity, suggesting a steric requirement. By contrast, *meta* halo, amino, and alkyl groups gave compounds with high nanomolar activity.[181]

By replacing the methylenedioxy group with a heterocyclic group at the 6-position, low-nanomolar compounds were obtained, with the pyrrolinyl compound **79** being the most potent. This compound is a strong inhibitor of tubulin

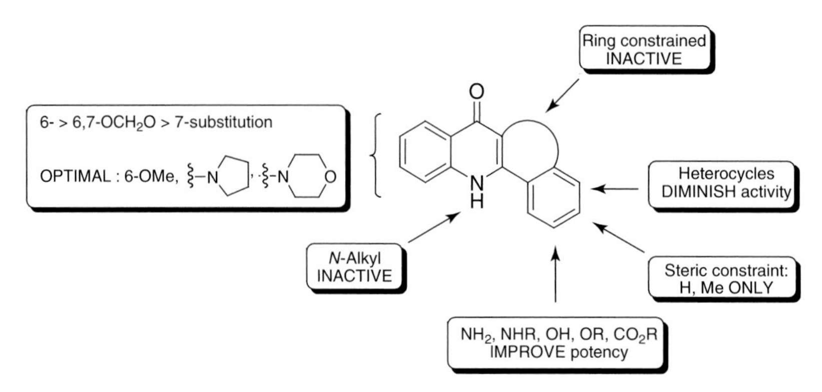

Figure 36 Quinolone SAR.

(**79**) (**80**)

(**81**) (**82**)

Figure 37 Structures of additional aromatics.

4 steps

Baccatin III (**83**) Paclitaxel (**84**)

Figure 38 Conversion of baccatin III to paclitaxel.

binding. The analogous dihydroquinazoline **80** shows a similar ability to inhibit tubulin polymerization through binding to the colchicine-binding site.[182] The corresponding strylyquinazolinones have also been shown to be potent molecules, which appear to show a similar SAR to the quinolones.[183] A number of alternative scaffolds have been tested, including naphthyridinones (e.g., **81**)[184] and 3-formyl-2-phenylindoles **82**.[185] The indoles again show similar SAR elements to that of the quinolone series.

7.04.2.2.7 Conclusion

Binding to tubulin, either in the vinca domain or at the colchicine site, is an effective method of disrupting microtubule dynamics. This results in cell cycle arrest in the G2/M phase and ultimately apoptosis. Within the vinca domain, several overlapping sites are targeted by a structurally diverse set of peptide, depsipeptide, and polyketide compounds that inhibit the binding of vinca alkaloids to tubulin. These vinca domain inhibitors have received significant clinical interest; however, only the vinca *bis*-indole alkaloids and closely related structural analogs are currently in clinical use. Human clinical trials involving analogs of halichondrin, dolastatin, and hemiasterlin are presently under way. Although tubulin binding at the colchicines site was first observed with colchicine, it is not limited to this molecule. Indeed, there are many compounds with a variety of structures that bind in this region. Since the exact nature of each interaction is unknown, it is impossible to know whether they bind in exactly the same area and it is probably the case that they do not; however, there appears to be a considerable overlap between the structures. Most classes are characterized by a di- or trimethoxyaryl group which appears to be analogous to the A ring of colchicine and has been determined to make the initial binding contact. The binding pocket in this region is often restricted with specific steric requirements. The remainder of the molecules varies more widely. However, there are many examples of molecules with a second aromatic group that mimics the tropolone functionality found in colchicine. A number of molecules have Michael acceptors, which are used to take advantage of the Cys-239 residue that appears to be in the colchicine-binding site.

One disadvantage of the colchicines site inhibitors appears to be their associated toxicity. While there are a number of examples that have been taken into trials, there are few examples of molecules that have made it to market. However, the more recent finding that these molecules are very effective at preventing angiogenesis, a requirement for tumors,[186] may spur further investigation in this field.

7.04.3 Microtubule Stabilizers and Promotors of Microtubule Assembly

7.04.3.1 Paclitaxel and Related Taxanes

Paclitaxel **84**, the first member of this class to be identified, was isolated in 1971 from the bark of the pacific yew (*Taxus brevifolia*) (**Figure 38**).[187] This compound has been the object of a tremendous amount of interest over the years.[188] The taxanes and related microtubule-stabilizing agents target the dimer of alpha- and beta-tubulin, binding in a hydrophobic cleft on the beta form (**Figures 39** and **40**). This binding promotes the formation of, and imparts unusual stability to, the microtubule.[189] The chemistry and structural biology of this interaction have been objects of considerable study using a range of techniques, particularly photoaffinity labeling.[190] Not surprisingly, these experiments show that the site targeted by microtubule-stabilizing drugs is distinct and well separated from the site at which other mechanistic classes of microtubule-interacting agents bind to the target. More recently, an electron diffraction structure of oligomeric alpha-beta-tubulin prepared in the presence of paclitaxel confirmed the presence of the drug in the site suggested by the earlier labeling experiments.[191] The low resolution (*c*. 4 Å) of the structure means that structure-based drug design based on it is not yet possible.

The discovery of the potent anticancer properties and subsequent elucidation of the mechanism of action of paclitaxel firmly established this compound as a leader in the treatment of difficult cancers. The last decade has seen paclitaxel and related taxanes emerge as aggressive frontline therapies for advanced tumors, including breast, lung, and ovarian carcinomas. Initially the development of paclitaxel for chemotherapy was slowed by poor availability of the natural product. However, the discovery that paclitaxel could be prepared semisynthetically from a more abundant natural product, baccatin III **83**, alleviated this problem.[192]

Hundreds of synthetic and semisynthetic analogs in the taxane class have been prepared. From this effort, an understanding of the structural components required for antimitotic activity has been developed. It is clear from this SAR that the core structure of the taxanes is required for the maintenance of potency. Furthermore, much of the

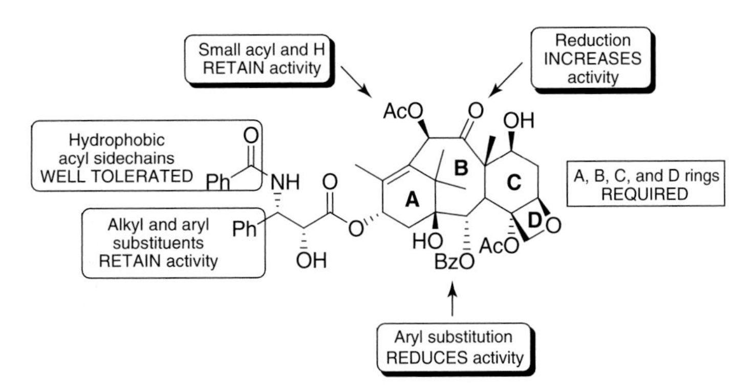

Figure 39 Taxane SAR.

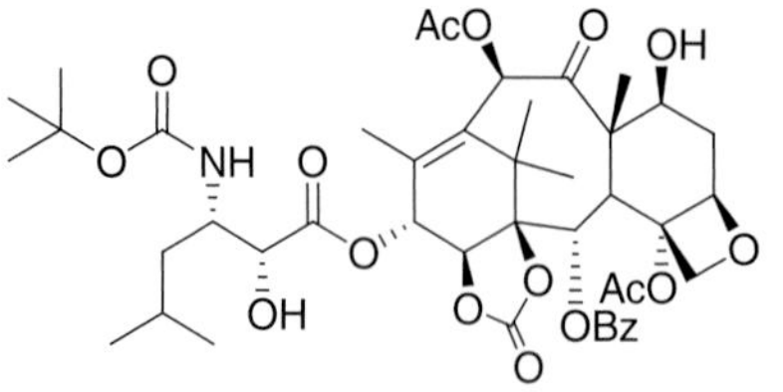

(85) 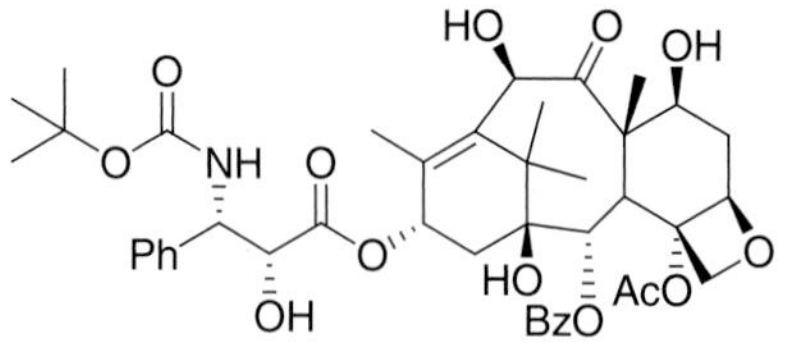 (86)

Figure 40 Structures of clinically relevant taxanes.

peripheral functionality is required. Notable exceptions to this rule are the aryl side chains, in which some variability is permitted, and the carbonyl, which may be reduced.

With the growing clinical use of paclitaxel, P-gp-based MDR emerged as a factor that could limit the clinical utility of the drug. Furthermore, the formulated drug product showed both drug- and formulation-related toxicities, including neuropathy and allergic reactions. Analog work to address these issues resulted in the identification of orataxel **85**, an orally active taxane, structurally similar to docetaxel **86**, that shows potency not only against wild-type cancers, but also P-gp expressing cancers. This compound showed excellent pharmacological properties preclinically and has advanced to clinical trials.[193] In addition to its improved activity against taxol-resistant tumors, because of its high level of oral bioavailability, this compound does not require the clinically cumbersome cremaphore-based formulation used by other taxanes.

Although there is considerable use of taxanes for a range of indications, commanding the lion's share of the market, and generating combined yearly sales of nearly 2 billion dollars, the high toxicity and poor solubility of these drugs may ultimately limit their utility. In addition, these compounds are substrates for P-gp, a transmembrane efflux pump that serves to limit the intracellular concentration of drug substrates in cells. Thus, tumor cells that express P-gp are resistant to taxanes and other hydrophobic drugs that are substrates for the efflux pump. These issues combine to create an opportunity for new taxanes and other drugs, free of these liabilities, to enter the market and gain acceptance.[194]

7.04.3.2 Epothilones

Of the new microtubule stabilizers, the epothilone class is by far the most advanced clinically. Epothilones A **87** and B **88** were isolated from the myxobacterium *Sorangium cellulosum* and originally identified due to their antifungal activity.[195] The potent antitumor properties and the elucidation of the microtubule-stabilizing effect of these compounds heightened interest in the epothilones.[196] As well as epothilones A and B, the culture of the producing organism also yields a wide variety of other epothilone-related structures, including epothilones C, D **89**, and F **90**.[197] In addition to these major natural products, a number of totally synthetic and semisynthetic analogs have been prepared and studied in vitro.[198–202]

The SARs that emerge from this body of work are shown in **Figure 41**. In general, the macrocycle can be divided into two halves for the purposes of discussing the SAR. The C-1 to C-8 region of the molecule is highly sensitive to modification, with even modest changes abrogating activity. In contrast, the C-9 to C-17 portion affords a greater degree of flexibility. In various forms, epoxide isosteres, small heterocycles, as well as alkyl substituents, may be substituted in the C-12 to C-13 region with modest or no reduction in activity. Likewise, compounds containing a range of side chains emanating from C-15 have been found to maintain varying degrees of activity (**Figure 42**).

To date, seven epothilone analogs have entered human clinical trials. Compound **88** entered phase II clinical trials in 2002. Likewise, the semisynthetic lactam analog of epothilone B, BMS-247550 **91**, was advanced to phase III in 2004. The synthesis of this material, prepared in a three-step, one-pot procedure from **88**, was presumably motivated by its more secure patent position and the possibility for it to show reduced hydrolytic instability as compared to the natural product. In 2004, epothilone D (named KOS-862) **89**, entered phase II human clinical trials. Interestingly, in a head-to-head comparison with **88**, the in vivo toxicity of **89** in mice was found to be as much as 40-fold lower.[203] A recent epothilone analog to enter the clinic is BMS-310705, the 21-amino analog of epothilone B **88**. Two epothilone analogs prepared by total chemical synthesis have entered human clinical trials. KOS-1584 **94**, the 9,10-dehydro analog of KOS-862, and ZK-epo **95** entered phase I human clinical trials in late 2004. Interestingly, the C-6 allyl moiety present in ZK-epo is an exception to the otherwise tight SAR in this region of the molecule. This compound does, however, take advantage of the observation that the side-chain substitution is well tolerated. Methylthio analog **93** was also recently advanced into clinical study.

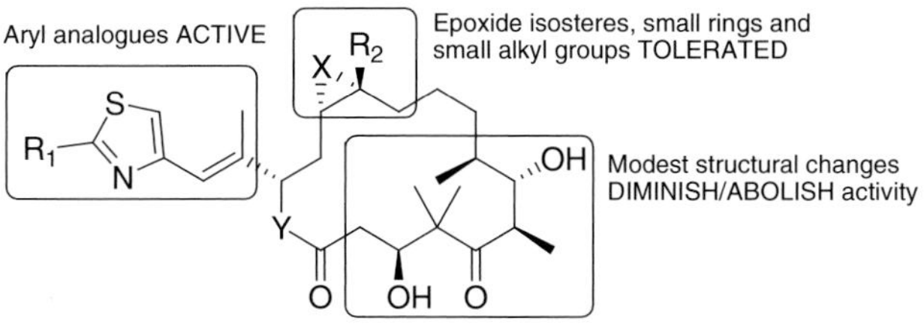

Aryl analogues ACTIVE

Epoxide isosteres, small rings and small alkyl groups TOLERATED

Modest structural changes DIMINISH/ABOLISH activity

(**87**): R_1=CH_3, Y=O, X=O, R_2=H
(**88**): R_1=CH_3, Y=O, X=O, R_2=CH_3
(**89**): R_1=CH_3, Y=O, X=bond, R_2=CH_3
(**90**): R_1=CH_2OH, Y=O, X=O, R_2=CH_3
(**91**): R_1=CH_3, Y=NH, X=O, R_2=CH_3
(**92**): R_1=CH_2OH, Y=O, X=O, R_2=CH_3
(**93**): R_1=SCH_3, Y=O, X=O, R_2=CH_3

Figure 41 Epothilone SAR.

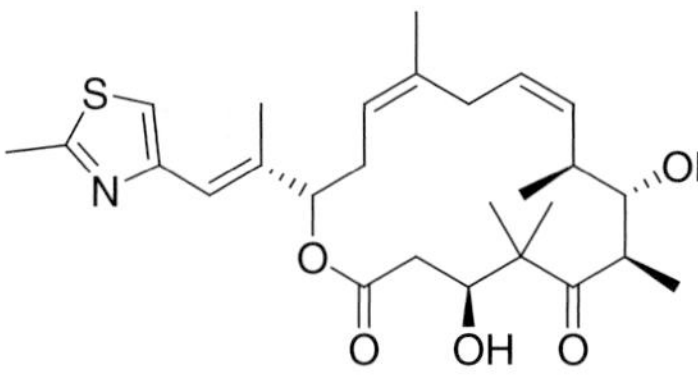

(**94**) 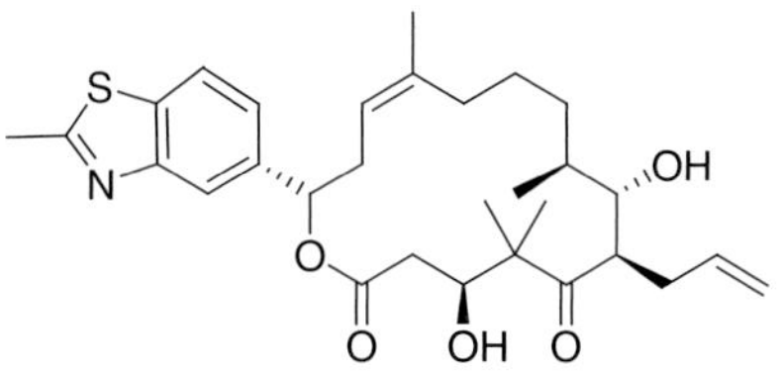 (**95**)

Figure 42 Structures of additional epothilones.

The epothilones that have advanced to the clinic have undergone extensive preclinical evaluation.[204] The compounds all show excellent potency in cell culture against a range of susceptible and MDR-expressing resistant cell lines. Although much of the information regarding the clinical behavior of this class of compounds has only appeared in poster or abstract form, the clinical prospects of the class appear good.[205]

7.04.3.3 Discodermolide

Discodermolide **96** (**Figures 43** and **44**) was isolated from the deep-water sponge *Discodermia dissoluta* by Gunasekera and co-workers.[206] Initial biological tests indicated that discodermolide has immunosuppressive activity, while recent interest has focused on its potent antimitotic activity. Notably, **96** has been reported to show potent activity against MDR carcinoma cell lines, due to its reduced affinity for the P-gp efflux pump.[207] It is thought to be *c.* 100-fold more soluble in water than paclitaxel and thus is likely to be more easily formulated.

The natural source of disocdermolide does not provide sufficient material for advanced evaluation, with the reported isolated yield of 7 mg from 434 g of frozen sponge. At this time, the only reliable source of meaningful quantities of the natural product is total chemical synthesis. Total syntheses of **96** have been reported by several groups.[208–213] A detailed discussion of these syntheses has been published.[214]

This synthetic effort has resulted in the production of sufficient material to serve as the foundation for a phase I dose escalation clinical trial of discodermolide, which began in 2002. This clinical trial was suspended when 3 patients receiving 14.4–19.2 mg m^{-2} q3w experienced, after the fourth cycle of therapy, interstitial pneumonitis that proved fatal.[215] In this clinical study, it was also noted that the pharmacokinetics of discodermolide suggest that it is subject to hepatobiliary recirculation.

From analogs obtained by synthesis and semisynthesis, an in vitro discodermolide SAR is now emerging. Polyacetylated versions of the compound show a range of activity.[216] Totally synthetic analogs, in which methyl groups at either C-16[217] or C-14 have been deleted, retain activity, as does the 2,3-dehydro analog.[218] Analogs of **96** in which the C-1 lactone cabonyl has been reduced and stored as the thiophenyl acetal show activity similar to the natural product. The retention of activity by analogs in which the lactone ring has been modified suggest that this region of the molecule may be a fruitful area for further analog work. A recent series of publications describes the replacement of the

Figure 43 Structures of discodermolides.

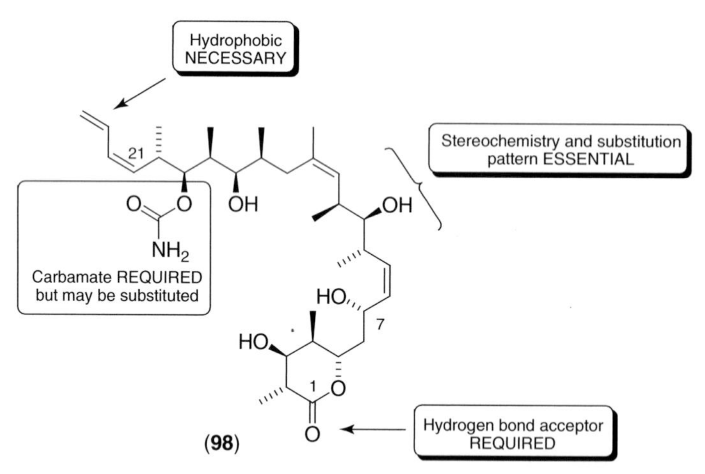

Figure 44 Discodermolide SAR.

C-1 to C-8 region of the molecule with dramatically simplified structures, for example 97, resulting in compounds that are in some cases fivefold more potent in vitro.[219–222] One group has prepared in excess of 50 discodermolide analogs by total synthesis and has indicated that certain isosteres of the C-21 to C-24 diene maintain potency while substitution of the C-13 to C-15 carbon atoms with an *N*-methyl amide abolish activity.[223]

7.04.3.4 Other Natural Products

Eleuthrobin **99** (**Figure 45**) and the structurally related sarcodictyins **100** and **101** are marine natural products that were identified as cytotoxins.[224,225] The unique structure of these compounds and their antiproliferative activity initially made them quite attractive to the synthetic and medicinal chemistry community. These compounds have been synthesized and a number of analogs have been prepared and studied in vitro as potential anticancer agents. Like the taxanes, these compounds were found to be substrates for P-gp and hence show dramatically reduced cytotoxicity against MDR-expressing cell lines. The SAR of the eleuthrobin/sarcodictyin class of microtubule-stabilizing agents has been explored in a combinatorial fashion.[226] This effort led to the conclusion that the conserved urocanic acid side chain was required for activity and that acetal and ester variants (R_1 and R_2 in **100** and **101**) were more easily tolerated. Although these compounds showed early promise as novel antiproliferative agents, the failure of this class to address adequately resistant tumor lines suggests that it is unlikely that it will yield clinically useful cytotoxins.

The microtubule-stabilizing macrocyclic lactone laulimalide **102** (**Figure 46**), also known as fijianolide, has been reported in small quantities from a number of different sponge species.[227–229] This compound is quite unstable at low pH and undergoes a rapid conversion to isolaulimalide **103** via displacement of the epoxide by the C-20 alcohol with concomitant loss in potency. Despite its promise as an antiproliferative agent, the in vivo evaluation of laulimalide has not yet been described, presumably due to lack of material for such experiments. Not surprisingly, laulimalide has been the

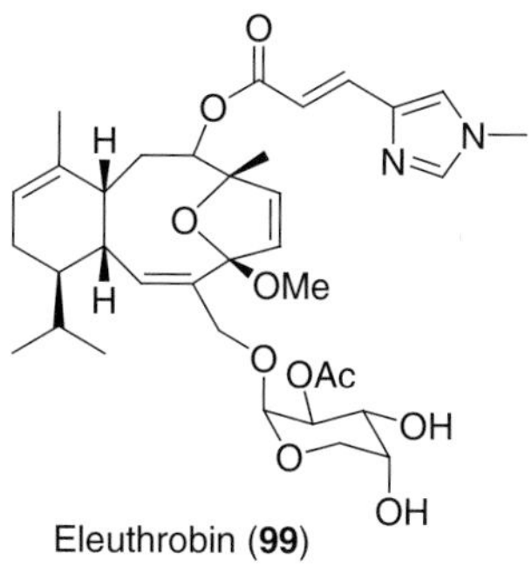

Eleuthrobin (**99**)

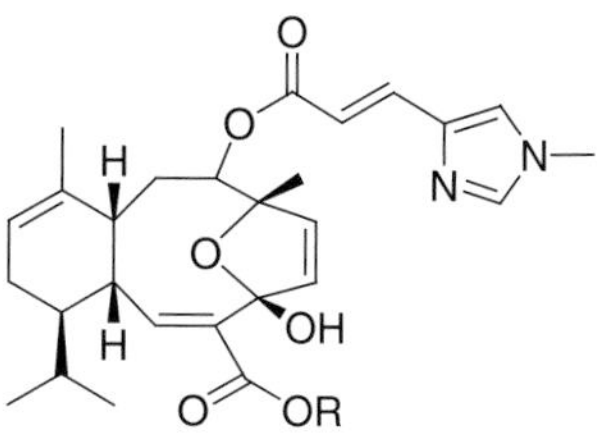

R=Me (**100**)

R=Et (**101**)

Figure 45 Structures of eleuthrobin and sarcodictyins.

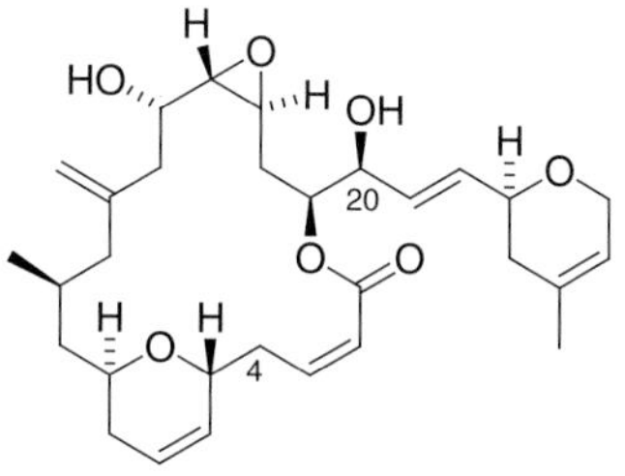

Laulimalide (**102**)

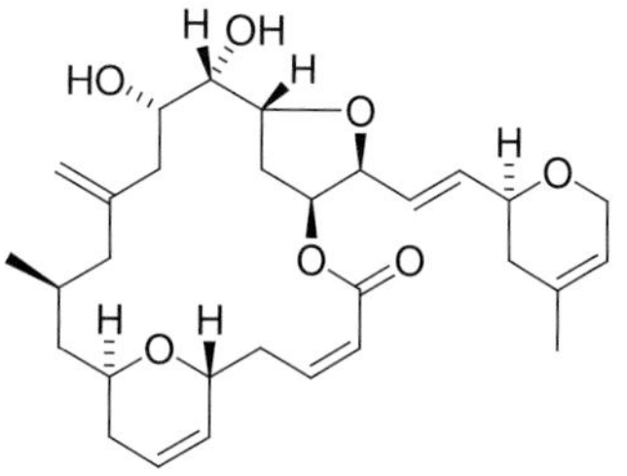

Isolaulimalide (**103**)

Figure 46 Structures of laulimalides.

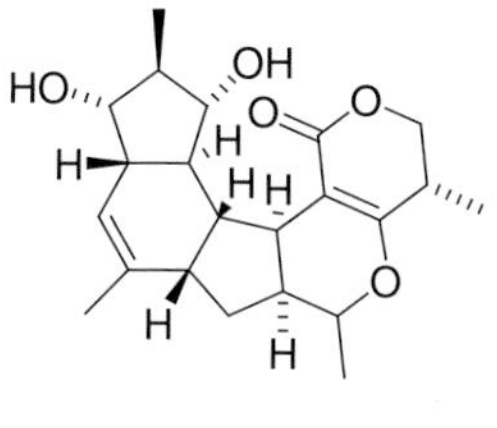

(**104**)

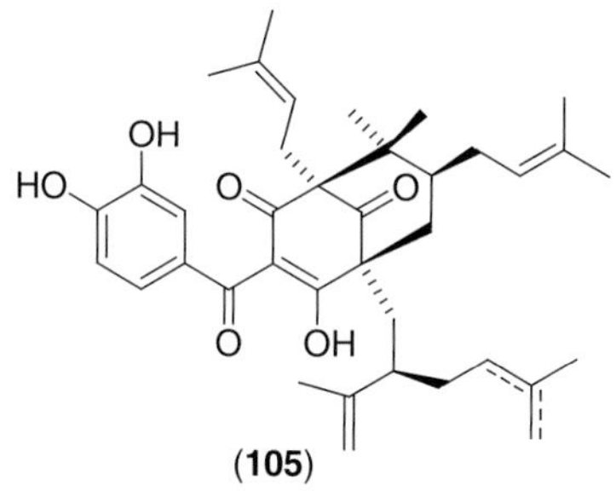

(**105**)

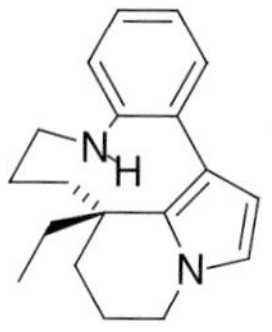

(**106**)

Figure 47 Structures of other naturally occurring microtubule stabilizers.

subject of considerable synthetic effort. Three total syntheses of the compound have recently appeared.[230–232] These and later total syntheses will create additional material with which to study this interesting molecule and its analogs.

Several other molecules of natural origin have been reported to be stabilizers of microtubules (**Figure 47**). The isolation and characterization of WS9885B **104** have been reported.[233] This material was found to possess cytotoxicity similar to paclitaxel.[234] A family of polyisoprenyl benzophenones (e.g., **105**) has been isolated from the fruit of the Malaysian plant *Garcinia pyrifera* and has shown low micromolar cytotoxicities.[235] Although **105** is only weakly cytotoxic and structurally complex, it is isolated in relatively large quantities (27 g of **105** from 1.8 kg of dried fruit) and has served as the starting point for an analog program based on semisynthesis. The polycyclic alkaloid (–)-rhainilum **106** was isolated from *Melodinus australia* and other species.[236] Although this compound and its congeners are only weak (micromolar) cytotoxins, the possibility remains that more potent analogs may be identified.

7.04.3.5 Synthetic Microtubule-Stabilizing Agents

Several synthetic microtubule stabilizers have also been identified (**Figure 48**). Although the best of these compounds are significantly less potent than paclitaxel and other clinically useful microtubule-stabilizing agents, they are as a group

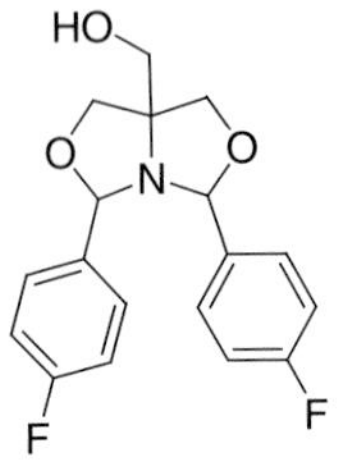

Figure 48 Structures of synthetic microtubule stabilizers.

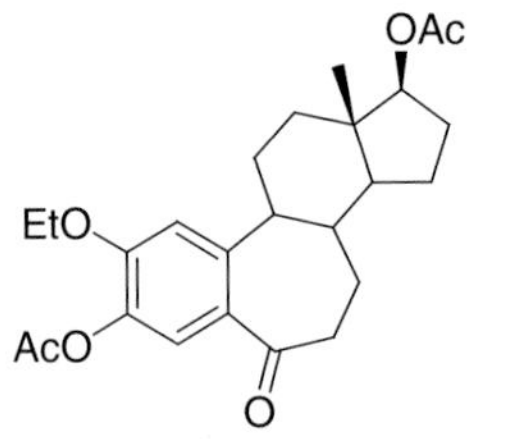

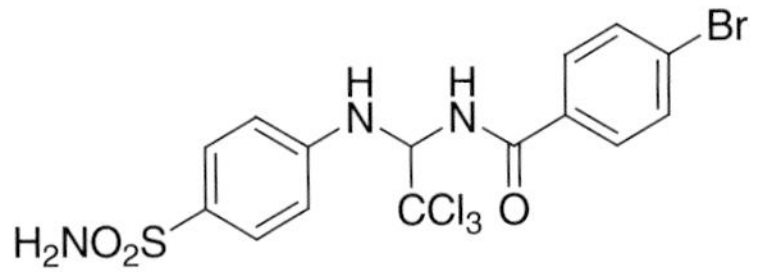

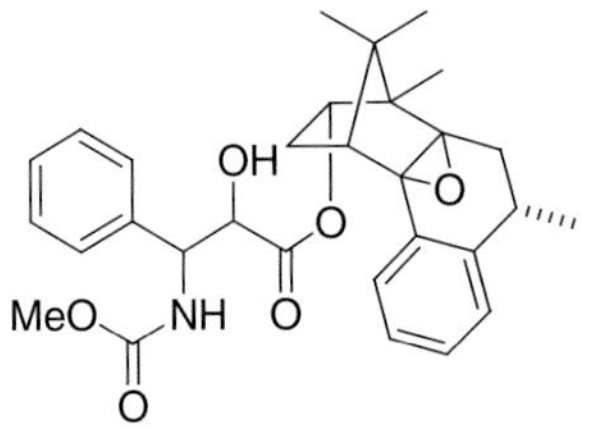

considerably less complex than their natural-product counterparts. Thus, it is likely that these compounds can serve as more amenable lead structures for medicinal chemistry. GS-64 **107**, a small heterobicycle that shows all the mechanistic hallmarks expected of an microtubule stabilizer, has recently been described.[237] This compound was found to possess micromolar cytotoxicity, roughly 1000 times less potent than paclitaxel. Homologated estradiol **108** was likewise found to be a promoter of microtubule assembly at the micromolar level.[238] Other workers have identified sulfonamide **109** via a high-throughput screening strategy. This compound has been reported to stabilize microtubules in vitro with potency similar to paclitaxel; however, it is far less potent as a cytotoxin.[239] A similar divergence between in vitro and cell-based potency was observed for synthetic borneol esters such as **110**.[240]

7.04.3.6 Conclusion

The clinical effectiveness of the taxanes has paved the way for the more recently discovered nontaxane microtubule-stabilizing agents. These new compounds, most discovered in the last 10 years, have provided new avenues for research in cancer chemotherapy, as well as significant challenges to the synthetic community to supply sufficient quantities of these frequently complex structures. The early clinical success of the epothilones is an encouraging first step toward the introduction of a nontaxane microtubule stabilizer into widespread clinical use. The advancement of discodermolide toward clinical evaluation is further cause for optimism. Total synthesis has played a key role in the advancement of discodermolide through preclinical evaluation. Furthermore, totally synthetic material will be used in clinical trials. The complexity of this undertaking illustrates what may be the single biggest challenge for further evaluation of other microtubule-stabilizing natural products, that being the lack of adequate supplies of material. Biotechnology and heterologous expression of biosynthetic machinery of these natural products can alleviate this issue, as is the case with the epothilones. The discoveries of a small number of simple, synthetic materials that show cytotoxiciy and function as microtubule stabilizers may serve to promote additional medicinal chemistry efforts on these structures.

References

1. Gupta, G.; Bhattacharyya, B. *Mol. Cell. Biochem.* **2003**, *253*, 41–47.
2. Sackett, D. L. *Biochemistry* **1995**, *34*, 7010–7019.
3. Mitra, A.; Sept, D. *Biochemistry* **2004**, *43*, 13955–13962.
4. Jordan, M. A.; Margolis, R. L.; Himes, R. H.; Wilson, L. *J. Mol. Biol.* **1986**, *187*, 61–73.
5. Jordan, M. A.; Wilson, L. *Biochemistry* **1990**, *29*, 2730–2739.
6. Jordan, M. A.; Thrower, D.; Wilson, L. *Cancer Res.* **1991**, *51*, 2212–2222.
7. Jordan, M. A.; Wilson, L. *Nat. Rev.* **2004**, *4*, 253–265.

8. Cragg, G. M.; Newman, D. J. *J. Nat. Prod.* **2004**, *67*, 232–244.
9. Newman, D. J.; Cragg, G. M. *J. Nat. Prod.* **2004**, *67*, 1216–1238.
10. Simmons, T. L.; Andrianasolo, E.; McPhail, K.; Flatt, P.; Gerwick, W. H. *Mol. Cancer. Ther.* **2005**, *4*, 333–342.
11. Neuss, N.; Gorman, M.; Hargrove, W.; Cone, N. J.; Biemann, K.; Büchi, G.; Manning, R. E. *J. Am. Chem. Soc.* **1964**, *86*, 1440–1442.
12. Moncrief, J. W.; Lipscomb, W. N. *J. Am. Chem. Soc.* **1965**, *87*, 4963–4964.
13. Duflos, A.; Kruczynski, A.; Barret, J.-M. *Curr. Med. Chem. – Anti-Cancer Agents* **2002**, *2*, 55–70.
14. Armitage, J. O. *Clin. Lymphoma* **2002**, *3*, S5–S11.
15. Jordan, M. A.; Wilson, L. *Biochemistry* **1990**, *29*, 2730–2739.
16. Vacca, A.; Iurlaro, M.; Ribatti, D.; Minischetti, M.; Nico, B.; Ria, R.; Pellegrino, A.; Dammacco, F. *Blood* **1999**, *12*, 4143–4155.
17. Bruno, S.; Puerto, V. L.; Mickiewicz, E.; Hegg, R.; Texeira, L. C.; Gaitan, L.; Martinez, L.; Fernandez, O.; Otero, J.; Kesselring, G. et al. *Am. J. Clin. Oncol.* **1995**, *18*, 392–396.
18. Budman, D. R. *Cancer Invest.* **1997**, *15*, 475–490.
19. http://www.fda.gov/cder/cancer/druglistframe.htm (accessed Sept 2006).
20. Barnett, C. J.; Cullinan, G. J.; Gerzon, K.; Hoying, R. C.; Jones, W. E.; Newlon, W. M.; Poore, G. A.; Robison, R. L.; Sweeney, M. J.; Todd, G. C. *J. Med. Chem.* **1978**, *21*, 88–96.
21. Mangeney, P.; Andriamialisoa, R. Z.; Lallemand, J.-Y.; Langlois, N.; Langlois, Y.; Potier, P. *Tetrahedron* **1979**, *35*, 2175–2179.
22. Lavielle, G.; Hautefaye, P.; Schaeffer, C.; Boutin, J. A.; Cudennec, C. A.; Pierré, A. *J. Med. Chem.* **1991**, *34*, 1998–2003.
23. Orosz, F.; Kovács, J.; Löw, P.; Vértessy, B. G.; Urbányi, Z.; Ács, T.; Keve, T.; Ovádi, J. *Br. J. Pharmacol.* **1997**, *121*, 947–954.
24. Adenis, A.; Pion, J.-M.; Fumoleau, P.; Pouillart, P.; Marty, M.; Giroux, B.; Bonneterre, J. *Cancer Chemother. Pharmacol.* **1995**, *35*, 527–528.
25. Orosz, F.; Comin, B.; Raïs, B.; Puigjaner, J.; Kovács, J.; Tárkányi, B.; Ács, T.; Keve, T.; Cascante, M.; Ovádi, J. *Br. J. Cancer* **1999**, *79*, 1356–1365.
26. Kruczynski, A.; Colpaert, F.; Tarayre, J.-P.; Mouillard, P.; Fahy, J.; Hill, B. T. *Cancer Chemother. Pharmacol.* **1998**, *41*, 437–447.
27. Kupchan, S. M.; Komoda, Y.; Court, W. A.; Thomas, G. J.; Smith, R. M.; Karim, A.; Gilmore, C. J.; Haltiwanger, R. C.; Bryan, R. F. *J. Am. Chem. Soc.* **1972**, *94*, 1354–1356.
28. Huang, A. B.; Lim, C. M.; Hamel, E. *Biochem. Biophys. Res. Commun.* **1985**, *128*, 1239–1246.
29. Takahashi, M.; Iwasaki, S.; Kobayashi, H.; Okuda, S.; Murai, T.; Sato, Y. *Biochim. Biophys. Acta* **1987**, *926*, 215–223.
30. Kupchan, S. M.; Komoda, Y.; Branfman, A. R.; Dailey, R. G., Jr.; Zimmerly, V. A. *J. Am. Chem. Soc.* **1974**, *96*, 3706–3708.
31. Kupchan, S. M.; Branfman, A. R.; Sneden, A. T.; Verma, A. K.; Dailey, R. G., Jr.; Komoda, Y.; Nagao, Y. *J. Am. Chem. Soc.* **1975**, *97*, 5294–5295.
32. Kupchan, S. M.; Sneden, A. T.; Branfman, A. R.; Howie, G. A.; Rebhun, L. I.; McIvor, W. E.; Wang, R. W.; Schnaitman, T. C. *J. Med. Chem.* **1978**, *21*, 31–37.
33. Kawai, A.; Akimoto, H.; Kozai, Y.; Ootsu, K.; Tanida, S.; Hashimoto, N.; Nomura, H. *Chem. Pharm. Bull.* **1984**, *32*, 3441–3451.
34. Liu, C.; Tadayoni, B. M.; Bourret, L. A.; Mattocks, K. M.; Derr, S. M.; Widdison, W. C.; Kedersha, N. L.; Ariniello, P. D.; Goldmacher, V. S.; Lambert, J. M. et al. *Proc. Natl. Acad. Sci. USA* **1996**, *93*, 8618–8623.
35. Tolcher, A. W.; Ochoa, L.; Hammond, L. A.; Patnaik, A.; Edwards, T.; Takimoto, C.; Smith, L.; de Bono, J.; Schwartz, G.; Mays, T. et al. *J. Clin. Oncol.* **2003**, *21*, 211–222.
36. Iwasaki, S.; Kobayashi, H.; Furukawa, J.; Namikoshi, M.; Okuda, S.; Sato, Z.; Matsuda, I.; Noda, T. *J. Antibiot.* **1984**, *37*, 354–362.
37. Kiyoto, S.; Kawai, Y.; Kawakita, T.; Kino, E.; Okuhara, M.; Uchida, I.; Tanaka, H.; Hashimoto, M.; Terano, H.; Kohsaka, M. et al. *J. Antibiot.* **1986**, *39*, 762–772.
38. Partida-Martinez, L. P.; Hertweck, C. *Nature* **2005**, *437*, 884–888.
39. Iwasaki, S.; Namikoshi, M.; Kobayashi, H.; Furukawa, J.; Okuda, S.; Itai, A.; Kasuya, A.; Iitaka, Y.; Sato, Z. *J. Antibiot.* **1986**, *39*, 424–429.
40. Takahashi, M.; Iwasaki, S.; Kobayashi, H.; Okuda, S.; Murai, T.; Sato, Y. *Biochim. Biophys. Acta* **1987**, *926*, 215–223.
41. Takahashi, M.; Iwasaki, S.; Kobayashi, H.; Okuda, S.; Murai, T.; Sato, Y.; Haraguchi-Hiraoka, T.; Nagano, H. *J. Antibiot.* **1987**, *40*, 66–72.
42. Kato, Y.; Ogawa, Y.; Imada, T.; Iwasaki, S.; Shimazaki, N.; Kobayashi, T.; Komai, T. *J. Antibiot.* **1991**, *44*, 66–75.
43. Hong, J.; White, J. D. *Tetrahedron* **2004**, *60*, 5653–5681.
44. Tolcher, A. W.; Aylesworth, C.; Rizzo, J.; Izbicka, E.; Campbell, E.; Kuhn, J.; Weiss, G.; Von Hoff, D. D.; Rowinsky, E. K. *Ann. Oncol.* **2000**, *11*, 333–338.
45. Uemura, D.; Takahashi, K.; Yamamoto, T.; Katayama, C.; Tanaka, J.; Okumura, Y.; Hirata, Y. *J. Am. Chem. Soc.* **1985**, *107*, 4796–4798.
46. Hirata, Y.; Uemura, D. *Pure Appl. Chem.* **1986**, *58*, 701–710.
47. Pettit, G. R.; Herald, C. L.; Boyd, M. R.; Leet, J. E.; Dufresne, C.; Doubek, D. L.; Schmidt, J. M.; Cerny, R. L.; Hooper, J. N. A.; Rützler, K. C. *J. Med. Chem.* **1991**, *34*, 3340–3342.
48. Pettit, G. R.; Tan, R.; Gao, F.; Williams, M. D.; Doubek, D. L.; Boyd, M. R.; Schmidt, J. M.; Chaupis, J.-C.; Hamel, E.; Bai, R. et al. *J. Org. Chem.* **1993**, *58*, 2538–2543.
49. Litaudon, M.; Hickford, S. J. H.; Lill, R. E.; Lake, R. J.; Blunt, J. W.; Munro, M. H. G. *J. Org. Chem.* **1997**, *62*, 1868–1871.
50. Bai, R.; Paull, K. D.; Herald, C. L.; Malspeis, L.; Pettit, G. R.; Hamel, E. *J. Biol. Chem.* **1991**, *266*, 15882–15889.
51. Ludue, R. F.; Roach, M. C.; Prasad, V.; Pettit, G. R. *Biochem. Pharmacol.* **1993**, *45*, 421–427.
52. Littlefield, B. A., Palme, M. H., Seletsky, B. M., Towle, M. J., Yu, M. J., Zheng, W. US Patent 6,214,865, 2001.
53. Towle, M. J.; Salvato, K. A.; Budrow, J.; Wels, B. F.; Kuznetsov, G.; Aalfs, K. K.; Welsh, S.; Zheng, W.; Seletsky, B. M.; Palme, M. H. *Cancer Res.* **2001**, *61*, 1013–1021.
54. Jordan, M. A.; Kamath, K.; Manna, T.; Okouneva, T.; Miller, H. P.; Davis, C.; Littlefield, B. A.; Wilson, L. *Mol. Cancer Ther.* **2005**, *4*, 1086–1095.
55. Pettit, G. R. *J. Nat. Prod.* **1996**, *59*, 812–821.
56. Bai, R.; Cichacz, Z. A.; Herald, C. L.; Pettit, G. R.; Hamel, E. *Mol. Pharmacol.* **1993**, *44*, 757–766.
57. Ovechkina, Y. Y.; Pettit, R. K.; Cichacz, Z. A.; Pettit, G. R.; Oakley, B. R. *Antimicrob. Agents Chemother.* **1999**, *43*, 1993–1999.
58. Evans, D. A.; Trotter, B. W.; Cote, B.; Coleman, P. J.; Dias, L. C.; Tyler, A. N. *Angew. Chem. Int. Ed. Engl.* **1997**, *36*, 2744–2747.
59. Guo, J.; Duffy, K. J.; Stevens, K. L.; Dalko, P. I.; Roth, R. M.; Hayward, M. M.; Kishi, Y. *Angew. Chem. Int. Ed. Engl.* **1998**, *37*, 187–192.
60. Hayward, M. M.; Roth, R. M.; Duffy, K. J.; Dalko, P. I.; Stevens, K. L.; Guo, J.; Kishi, Y. *Angew. Chem. Int. Ed. Engl.* **1998**, *37*, 192–196.
61. Paterson, I.; Chen, D. Y.-K.; Coster, M. J.; Acena, J. L.; Bach, J.; Gibson, K. R.; Keown, L. E.; Oballa, R. M.; Trieselmann, T.; Wallace, D. J. et al. *Angew. Chem. Int. Ed. Engl.* **2001**, *40*, 4055–4060.
62. Crimmins, M. T.; Katz, J. D.; Washburn, D. G.; Allwein, S. P.; McAtee, L. F. *J. Am. Chem. Soc.* **2002**, *124*, 5661–5663.
63. Smith, A. B., III; Zhu, W.; Shirakami, S.; Sfouggatakis, C.; Doughty, V. A.; Bennett, C. S.; Sakamoto, Y. *Org. Lett.* **2003**, *5*, 761–764.
64. Heathcock, C. H.; McLaughlin, M.; Medina, J.; Hubbs, J. L.; Wallace, G. A.; Scott, R.; Claffey, M. M.; Hayes, C. J.; Ott, G. R. *J. Am. Chem. Soc.* **2003**, *125*, 12844–12849.

65. Kobayashi, S.; Tsuchiya, K.; Harada, T.; Nishide, M.; Kurokawa, T.; Nakagawa, T.; Shimada, N.; Kobayashi, K. *J. Antibiot.* **1994**, *47*, 697–702.
66. Yasui, K.; Tamura, Y.; Nakatani, T.; Horibe, I.; Kawada, K.; Koizumi, K.; Suzuki, R.; Ohtani, M. *J. Antibiot.* **1996**, *49*, 173–180.
67. Tsuchiya, K.; Kobayashi, S.; Nishikiori, T.; Nakagawa, T.; Tatsuta, K. *J. Antibiot.* **1997**, *50*, 259–260.
68. Kondoh, M.; Usui, T.; Nishikiori, T.; Mayumi, T.; Osada, H. *Biochem. J.* **1999**, *340*, 411–416.
69. Kondoh, M.; Usui, T.; Kobayashi, S.; Tsuchiya, K.; Nishikawa, K.; Nishikiori, T.; Mayumi, T.; Osada, H. *Cancer Lett.* **1998**, *126*, 29–32.
70. Usui, T.; Watanabe, H.; Nakayama, H.; Tada, Y.; Kanoh, N.; Kondoh, M.; Asao, T.; Takio, K.; Watanabe, H.; Nishikawa, K. et al. *Chem. Biol.* **2004**, *11*, 799–806.
71. Pettit, G. R.; Kamano, Y.; Herald, C. L.; Tuinman, A. A.; Boettner, F. E.; Kizu, H.; Schmidt, J. M.; Baczynskyj, L.; Tomer, K. B.; Bontems, R. J. *J. Am. Chem. Soc.* **1987**, *109*, 6883–6885.
72. Pettit, G. R.; Singh, S. B.; Hogan, F.; Lloyd-Williams, P.; Herald, D. L.; Burkett, D. D.; Clewlow, P. J. *J. Am. Chem. Soc.* **1989**, *111*, 5463–5465.
73. Leusch, H.; Moore, R. E.; Paul, V. J.; Mooberry, S. L.; Corbett, T. H. *J. Nat. Prod.* **2001**, *64*, 907–910.
74. Bai, R.; Pettit, G. R.; Hamel, E. *Biochem. Pharmacol.* **1990**, *39*, 1941–1949.
75. Bai, R.; Pettit, G. R.; Hamel, E. *J. Biol. Chem.* **1990**, *265*, 17141–17149.
76. Bai, R.; Pettit, G. R.; Hamel, E. *Biochem. Pharmacol.* **1992**, *43*, 2637–2645.
77. Beckwith, M.; Urba, W. J.; Longo, D. L. *J. Natl. Cancer Inst.* **1993**, *85*, 483–488.
78. Aherne, G. W.; Hardcastle, A.; Valenti, M.; Bryant, A.; Rogers, P.; Pettit, G. R.; Srirangam, J. K.; Kelland, L. R. *Cancer Chemother. Pharmacol.* **1996**, *38*, 225–232.
79. Pitot, H. C.; McElroy, E. A.; Reid, J. M.; Windebank, A. J.; Sloan, J. A.; Erlichman, C.; Bagniewski, P. G.; Walker, D. L.; Rubin, J.; Goldberg, R. M. et al. *Clin. Cancer Res.* **1999**, *5*, 525–531.
80. Vaishampayan, U.; Glode, M.; Du, W.; Kraft, A.; Hudes, G.; Wright, J.; Hussain, M. *Clin. Cancer Res.* **2000**, *6*, 4205–4208.
81. Miyazaki, K.; Kobayashi, M.; Natsume, T.; Gondo, M.; Mikami, T.; Sakakibara, K.; Tsukagoshi, S. *Chem. Pharm. Bull.* **1995**, *43*, 1706–1718.
82. Yamamoto, N. A.; Andoh, M.; Kawahara, M.; Fukuoka, M.; Niitani, H. *Proc. Am. Soc. Clin. Oncol.* **2002**, *21*, 420 (abstract).
83. Newman, D. J.; Cragg, G. M. *J. Nat. Prod.* **2004**, *67*, 1216–1238.
84. de Arruda, M.; Cocchiaro, C. A.; Nelson, C. M.; Grinnell, C. M.; Janssen, B.; Haupt, A.; Barlozzari, T. *Cancer Res.* **1995**, *55*, 3085–3092.
85. Mross, K.; Berdel, W. E.; Fiebig, H. H.; Velagapudi, R.; von Broen, I. M.; Unger, C. *Ann. Oncol.* **1998**, *9*, 1323–1330.
86. Villalona-Calero, M. A.; Baker, S.; Hammond, L.; Aylesworth, C.; Eckhardt, S. G.; Kraynak, M.; Fram, R.; Fischkoff, S.; Velagapudi, R.; Toppmeyer, D. et al. *J. Clin. Oncol.* **1998**, *16*, 2770–2779.
87. Pettit, G. R.; Kamano, Y.; Dufresne, C.; Cerny, R. L.; Herald, C. L.; Schmidt, J. M. *J. Org. Chem.* **1989**, *54*, 6005–6006.
88. Schwartz, R. E.; Hirsch, C. F.; Sesin, D. F.; Flor, J. E.; Chartrain, M.; Fromtling, R. E.; Harris, G. H.; Salvatore, M. J.; Liesch, J. M.; Yudin, K. *J. Ind. Microbiol.* **1990**, *5*, 113–124.
89. Trimurtulu, G.; Ohtani, I.; Patterson, G. M. L.; Moore, R. E.; Corbett, T. H.; Valeriote, F. A.; Demchik, L. *J. Am. Chem. Soc.* **1994**, *116*, 4729–4737.
90. Barrow, R. A.; Hemscheidt, T.; Liang, J.; Paik, S.; Moore, R. E.; Tius, M. A. *J. Am. Chem. Soc.* **1995**, *117*, 2479–2490.
91. Smith, C. D.; Zhang, X.; Mooberry, S. L.; Patterson, G. M. L.; Moore, R. E. *Cancer Res.* **1994**, *54*, 3779–3784.
92. Bai, R.; Schwartz, R. E.; Kepler, J. A.; Pettit, G. R.; Hamel, E. *Cancer Res.* **1996**, *56*, 4398–4406.
93. Smith, C. D.; Zhang, X. *J. Biol. Chem.* **1996**, *271*, 6192–6198.
94. Mooberry, S. L.; Busquets, L.; Tien, G. *Int. J. Cancer* **1997**, *73*, 440–448.
95. Golakoti, T.; Ogino, J.; Heltzel, C. E.; Husebo, T. L.; Jensen, C. M.; Larsen, L. K.; Patterson, G. M. L.; Moore, R. E.; Mooberry, S. L.; Corbett, T. H. et al. *J. Am. Chem. Soc.* **1995**, *117*, 12030–12049.
96. Al-awar, R. S.; Ray, J. E.; Schultz, R. M.; Andis, S. L.; Kennedy, J. H.; Moore, R. E.; Liang, J.; Golakoti, T.; Subbaraju, G. V.; Corbett, T. H. *J. Med. Chem.* **2003**, *46*, 2985–3007.
97. Chaganty, S.; Golakoti, T.; Heltzel, C.; Moore, R. E.; Yoshida, W. Y. *J. Nat. Prod.* **2004**, *67*, 1403–1406.
98. Buck, S. B.; Huff, J. K.; Himes, R. H.; Georg, G. I. *J. Med. Chem.* **2004**, *47*, 696–702.
99. Buck, S. B.; Huff, J. K.; Himes, R. H.; Georg, G. I. *J. Med. Chem.* **2004**, *47*, 3697–3699.
100. Larie, V. D.; Shih, C.; Hay, D. A.; Andis, S. L.; Corbett, T. H.; Gossett, L. S.; Janisse, S. K.; Martinelli, M. J.; Moher, E. D.; Schultz, R. M. et al. *Bioorg. Med. Chem. Lett.* **1999**, *9*, 369–374.
101. Panda, D.; DeLuca, K.; Williams, D.; Jordan, M. A.; Wilson, L. *Proc. Natl. Acad. Sci. USA* **1998**, *95*, 9313–9318.
102. Wagner, M. M.; Paul, D. C.; Shih, C.; Jordan, M. A.; Wilson, L.; Williams, D. C. *Cancer Chemother. Pharmacol.* **1999**, *43*, 115–125.
103. Sessa, C.; Weigang-Kohler, K.; Pagani, O.; Greim, G.; Mora, O.; De Pas, T.; Burgess, M.; Weimar, I.; Johnson, B. *Eur. J. Cancer* **2002**, *38*, 2388–2396.
104. Liang, J.; Moore, R. E.; Moher, E. D.; Munroe, J. E.; Alawar, R. S.; Hay, D. A.; Varie, D. L.; Zhang, T. Y.; Aikins, J. A.; Martinelli, M. J. et al. *Invest. New Drugs* **2005**, *23*, 213–224.
105. Talpir, R.; Benayahu, Y.; Kashman, Y.; Pannell, L.; Schleyer, M. *Tetrahedron Lett.* **1994**, *35*, 4453–4456.
106. Anderson, H. J.; Coleman, J. E.; Andersen, R. J.; Roberge, M. *Cancer Chemother. Pharmacol.* **1996**, *39*, 223–226.
107. Bai, R.; Durso, N. A.; Sackett, D. L.; Hamel, E. *Biochemistry* **1999**, *38*, 14302–14310.
108. Coleman, J. E.; de Silva, E. D.; Kong, F.; Andersen, R. J.; Allen, T. M. *Tetrahedron* **1995**, *51*, 10653–10662.
109. Nieman, J. A.; Coleman, J. E.; Wallace, D. J.; Piers, E.; Lim, L. Y.; Roberge, M.; Andersen, R. J. *J. Nat. Prod.* **2003**, *66*, 183–199.
110. Zask, A.; Birnberg, G.; Cheung, K.; Kaplan, J.; Niu, C.; Norton, E.; Suayan, R.; Yamashita, A.; Cole, D.; Tang, Z. et al. *J. Med. Chem.* **2004**, *47*, 4774–4786.
111. Yamashita, A.; Norton, E. B.; Kaplan, J. A.; Niu, C.; Loganzo, F.; Hernandez, R.; Beyer, C. F.; Annable, T.; Musto, S.; Discafini, C. et al. *Bioorg. Med. Chem. Lett.* **2004**, *14*, 5317–5322.
112. Zask, A.; Birnberg, G.; Cheung, K.; Kaplan, J.; Niu, C.; Norton, E.; Yamashita, A.; Beyer, C.; Krishnamurthy, G.; Greenberger, L. M. et al. *Bioorg. Med. Chem. Lett.* **2004**, *14*, 4353–4358.
113. Loganzo, F.; Hari, M.; Annable, T.; Tan, X.; Morilla, D. B.; Musto, S.; Zask, A.; Kaplan, J.; Minnick, A. A., Jr.; May, M. K. et al. *Mol. Cancer Ther.* **2004**, *3*, 1319–1327.
114. Loganzo, F.; Discafani, C.; Annable, T.; Beyer, C.; Musto, S.; Hari, M.; Tan, X.; Hardy, C.; Hernandez, R.; Baxter, M. et al. *Cancer Res.* **2003**, *63*, 1838–1845.
115. Ratain, M. J.; Undevia, S.; Janisch, L.; Roman, S.; Mayer, P.; Buckwalter, M.; Foss, D.; Hamilton, B. L.; Fischer, J.; Bukowski, R. M. *Proc. Am. Soc. Clin. Oncol.* **2003**, *22*, 516 (abstract).
116. Jordan, M. A. *Curr. Med. Chem.* **2002**, *9*, 1–17.

117. Uppuluri, S.; Knipling, L.; Sackett, D. L.; Wolff, J. *Proc. Natl. Acad. Sci. USA* **1993**, *90*, 11598–11602.
118. Bai, R.; Covell, D. G.; Pei, X.-F.; Ewell, J. B.; Nguyen, N. Y.; Brossi, A.; Hamel, E. *J. Biol. Chem.* **2000**, *275*, 40443–40452.
119. Ravelli, R. B. G.; Gigant, B.; Curmi, P. A.; Jourdain, I.; Lachkar, S.; Sobel, A.; Knossow, M. *Nature* **2004**, *428*, 198–202.
120. Pelletier, J. P.; Caventou, J. B. *Ann. Chim. Phys.* **1820**, *14*, 69.
121. King, M. V.; De Vries, J. L.; Pepinsky, R. *Acta Crystallogr. Sect. B* **1952**, *5*, 437.
122. Moreland, L. W.; Ball, G. V. *Arthritis Rheum.* **1991**, *34*, 782–786.
123. Zemer, D.; Livneh, A.; Pras, M.; Sohar, E. *Am. J. Med. Genet.* **1993**, *45*, 340–344.
124. Brossi, A.; Yeh, H. J.; Chrzanowska, M.; Wolf, J.; Hamel, E.; Lin, C. M.; Quin, F.; Suffness, M.; Silverton, J. *J. Med. Res. Rev.* **1988**, *8*, 77.
125. Graening, T.; Schmalz, H.-G. *Angew. Chem. Int. Ed. Engl.* **2004**, *43*, 3230–3256.
126. Al-Tel, T. H.; AbuZarga, M. H.; Sabri, S. S.; Freyer, A. J.; Shamma, M. *J. Nat. Prod.* **1990**, *53*, 623–629.
127. Banwell, M. G.; Peters, S. C.; Greenwood, R. J.; Mackay, M. F.; Hamel, E.; Lin, C. M. *Aust. J. Chem.* **1992**, *45*, 1577–1588.
128. Staretz, M. E.; Hastie, S. B. *J. Med. Chem.* **1993**, *36*, 758–764.
129. Andrew, J. M.; Ramirez, P. B.; Gorbunoff, M. J.; Alaya, D.; Timasheff, N. *Biochemistry* **1998**, *37*, 8356–8368.
130. Srivastava, V.; Negi, A. S.; Kumar, J. K.; Gupta, M. M.; Khanuja, S. P. S. *Bioorg. Med. Chem.* **2005**, *13*, 5892–5908.
131. Sackett, D. L. *Pharmacol. Ther.* **1993**, *59*, 163–228.
132. Ji, Z.; Wang, H.-K.; Bastow, K. F.; Zhu, X.-K.; Cho, S. J.; Cheng, Y.-C.; Lee, K.-H. *Bioorg. Med. Chem. Lett.* **1997**, 7, 607–612.
133. Hitotsuyanagi, Y.; Fukuyo, M.; Tsuda, K.; Kobayashi, M.; Ozeki, A.; Itokawa, H.; Takeya, K. *Bioorg. Med. Chem. Lett.* **2000**, *10*, 315–317.
134. Leteurtre, F.; Madalengoitia, J.; Orr, A.; Guzi, T. J.; Lehnert, E.; Macdonald, T.; Pommier, Y. *Cancer Res.* **1992**, *52*, 4478–4483.
135. Solary, E.; Leteurtre, F.; Paull, K. D.; Scudiero, D.; Hamel, E.; Pommier, Y. *Biochem. Pharmacol.* **1993**, *45*, 2449–2456.
136. Leteurtre, F.; Sackett, D. L.; Madalengoitia, J.; Kohlhagen, G.; Macdonald, T.; Hamel, E.; Paull, K. D.; Pommier, Y. *Biochem. Pharmacol.* **1995**, *49*, 1283–1290.
137. Miller, T. A.; Vachaspati, P. R.; Labroli, M. A.; Thompson, C. D.; Bulman, A. L.; Macdonald, T. L. *Bioorg. Med. Chem. Lett.* **1998**, *8*, 1065–1070.
138. Kupchan, M. S.; Bitton, R. W.; Zeilger, M. F.; Gilmore, C. J.; Restivo, R. J.; Bryan, R. F. *J. Am. Chem. Soc.* **1973**, *95*, 1335–1336.
139. Zavala, F.; Guenard, D.; Robin, J.-P.; Brown, E. *J. Med. Chem.* **1980**, *23*, 546–549.
140. Petit, G. R.; Cragg, G. M.; Herald, D. L.; Schmidt, J. M.; Lohavanuaya, P. *Can. J. Chem.* **1982**, *60*, 1374–1376.
141. Pettit, G. R.; Singh, S. B.; Boyd, M. R.; Hamel, E.; Pettit, R. K.; Schmidt, J. M.; Hogan, F. *J. Med. Chem.* **1995**, *38*, 1666–1672.
142. Cushman, M.; Nagarathnam, D.; Gopal, D.; Chakraborti, A. K.; Lin, C. M.; Hamel, E. *J. Med. Chem.* **1991**, *34*, 2579–2588.
143. Cushman, M.; Nagarathnam, D.; Gopal, D.; He, H.-M.; Lin, C. M.; Hamel, E. *J. Med. Chem.* **1992**, *35*, 2293–2306.
144. Brown, R. T.; Fox, B. W.; Hadfield, J. A.; McGown, A. T.; Mayalarp, S. P.; Pettit, G. R.; Woods, J. A. *J. Chem. Soc., Perkin Trans.* **1995**, *1*, 577–581.
145. Pettit, G. R.; Temple, C., Jr.; Narayanan, V. L.; Varma, R.; Simpson, M. J.; Boyd, M. R.; Rener, G. A.; Bansal, N. *Anti-Cancer Drug Design* **1995**, *10*, 299–309.
146. Dowlati, A.; Robertson, K.; Cooney, M.; Petros, W. P.; Stratford, M.; Jesberger, J.; Rafie, N.; Overmoyer, B.; Makkar, V.; Stambler, B. et al. *Cancer Res.* **2002**, *62*, 3408–3416.
147. Kong, Y.; Grembecka, J.; Edler, M. C.; Hamel, E.; Mooberry, S. L.; Sabat, M.; Rieger, J.; Brown, M. L. *Chem. Biol.* **2005**, *12*, 1007–1014.
148. Pettit, G. R.; Singh, S. B.; Niven, M. L.; Hamel, E.; Schmidt, J. M. *J. Nat. Prod.* **1987**, *50*, 119–131.
149. Cushman, M.; Nagarathnam, D.; Gopal, D.; He, H.-M.; Lin, C. M.; Hamel, E. *J. Med. Chem.* **1992**, *35*, 2293–2306.
150. Pettit, G. R.; Toki, B.; Herald, D. L.; Verdier-Pinard, P.; Boyd, M. R.; Hamel, E.; Pettit, R. K. *J. Med. Chem.* **1998**, *41*, 1688–1695.
151. Pettit, G. R.; Grealish, M. P.; Herald, D. L.; Boyd, M. R.; Hamel, E.; Pettit, R. K. *J. Med. Chem.* **2000**, *43*, 2731–2737.
152. Woods, J. A.; Hadfield, J. A.; McGown, A. T.; Fox, B. W. *Bioorg. Med. Chem. Lett.* **1993**, *1*, 333–340.
153. Cushman, M.; Nagarathnam, D.; Gopal, D.; He, H.-M.; Lin, C. M.; Hamel, E. *J. Med. Chem.* **1992**, *35*, 2293–2306.
154. Couladouros, E. A.; Li, T.; Moutsos, V. I.; Pitsinos, E. N.; Soufli, I. C. *Bioorg. Med. Chem. Lett.* **1999**, *9*, 2928–2929.
155. Beutler, J. A.; Cardellina, J. H., II; Lin, C. M.; Hamel, E.; Cragg, G. M.; Boyd, M. R. *Bioorg. Med. Chem. Lett.* **1993**, *3*, 581–584.
156. Lichius, J. J.; Thoison, O.; Montagnac, A.; Pias, M.; Gueritte-Voegelen, F.; Sevenet, T.; Cosson, J.-P.; Hadi, A. H. A. *J. Nat. Prod.* **1994**, *57*, 1012–1016.
157. Shi, Q.; Chen, K.; Li, L.; Chang, J.-J.; Autry, C.; Kozuka, M.; Konoshima, T.; Estes, J. R.; Lin, C. M.; Hamel, E. et al. *J. Nat. Prod.* **1995**, *58*, 475–482.
158. Gerwick, W. H.; Proteau, P. J.; Nagle, D. G.; Hamel, E.; Blokhin, A.; Slate, D. L. *J. Org. Chem.* **1994**, *59*, 1243–1245.
159. Blokhin, A. V.; Yoo, H.-D.; Geralds, R. S.; Nagle, D. G.; Gerwick, W. H.; Hamel, E. *Mol. Pharmacol.* **1995**, *48*, 523–531.
160. Verdier-Pinard, P.; Lai, J.-Y.; Yoo, H.-D.; Yu, J.; Marquez, B.; Nagle, D. G.; Nambu, M.; White, J. D.; Falck, J. R.; Gerwick, W. H. et al. *Mol. Pharmacol.* **1998**, *53*, 62–76.
161. Fotsis, T.; Zhang, Y.; Pepper, M. S.; Adlercreutz, H.; Montesano, R.; Nawroth, P. P.; Scweigerer, L. *Nature* **1994**, *368*, 237–239.
162. A'Amato, R. J.; Lin, C. M.; Flynn, E.; Folkman, J.; Hamel, E. *Proc. Natl. Acad. Sci. USA* **1994**, *91*, 3964–3968.
163. Klauber, N.; Parangi, S.; Flynn, E.; Hamel, E.; D'Amato, R. *J. Cancer Res.* **1997**, *57*, 81–86.
164. Miller, T. A.; Bulman, A. L.; Thompson, C. D.; Garest, M. E.; Macdonald, T. L. *J. Med. Chem.* **1997**, *40*, 3836–3841.
165. Wang, Z.; Yang, D.; Mohnakrishnan, A. K.; Fanwick, P. E.; Nampoothiri, P.; Hamel, E.; Cushman, M. *J. Med. Chem.* **2000**, *43*, 2419–2429.
166. Chaudoreille, M. M.; Peyrot, V.; Braguer, D.; Codaccioni, F.; Crevat, A. *Biochem. Pharmacol.* **1991**, *41*, 685–693.
167. Yoshino, H.; Ueda, N.; Niijima, J.; Sugumi, H.; Kotake, Y.; Koyanagi, N.; Yoshimatsu, K.; Asada, M.; Watanabe, T.; Nagasu, T. et al. *J. Med. Chem.* **1992**, *35*, 2496–2497.
168. Koyanagi, N.; Nagasu, T.; Fujita, F.; Watanabe, T.; Tsukahara, K.; Funahashi, Y.; Fujita, M.; Taguchi, T.; Yoshino, H.; Kitoh, K. *Cancer Res.* **1994**, *54*, 1702–1706.
169. Yoshimatsu, K.; Yamaguchi, A.; Yoshino, H.; Koyanagi, N.; Kitoh, K. *Cancer Res.* **1997**, *57*, 3208–3213.
170. Yamamoto, K.; Noda, K.; Yoshimura, A.; Fukuoka, M.; Furuse, K.; Niitani, H. *Cancer Chemother. Pharmacol.* **1998**, *42*, 127–134.
171. Owa, T.; Yoshino, H.; Okauchi, T.; Yoshimastsu, K.; Ozawa, Y.; Sugi, N. H.; Nagasu, T.; Koyanagi, N.; Kitoh, K. *J. Med. Chem.* **1999**, *42*, 3789–3799.
172. Shan, B.; Medina, J. C.; Santha, E.; Frankmoelle, W. P.; Chou, T.-C.; Learned, R. M.; Narbut, M. R.; Stott, D.; Wu, P.; Jaen, J. C. et al. *Proc. Natl. Acad. Sci. USA* **1999**, *96*, 5686–5691.
173. Temple, C., Jr.; Rener, G. A.; Comber, R. N.; Waud, W. R. *J. Med. Chem.* **1991**, *34*, 3176–3181.
174. Temple, C., Jr. *J. Med. Chem.* **1990**, *33*, 656–661.
175. Legault, J.; Gaulin, J.-F.; Mounetou, E.; Bolduc, S.; Lacroix, J.; Poyet, P.; Gaudreault, R. C. *Cancer Res.* **2000**, *60*, 985–992.
176. Van Belle, S. J.; Distelmans, W.; Vandebroek, J.; Bruynseels, J.; Van Ginckel, R.; Storme, G. A. *Anticancer Res.* **1993**, *13*, 2389–2391.

177. Edwards, M. L.; Stemerick, D. M.; Sunkara, P. S. *J. Med. Chem.* **1990**, *33*, 1948–1954.
178. Peyrot, V.; Leynadier, D.; Sarrazin, M.; Briand, C.; Menendez, M.; Laynez, J.; Andreu, J. M. *Biochemistry* **1992**, *31*, 11125–11132.
179. Kuo, S. C.; Lee, H.-Z.; Juang, J.-P.; Lin, Y.-T.; Wu, T.-S.; Chang, J.-J.; Lednicer, D.; Paull, K. D.; Lin, C. M.; Hamel, E. et al. *J. Med. Chem.* **1993**, *36*, 1146–1156.
180. Li, L.; Wang, H.-K.; Kuo, S.-C.; Wu, T.-S.; Lednicer, D.; Lin, C. M.; Hamel, E.; Lee, K.-H. *J. Med. Chem.* **1994**, *37*, 1126–1135.
181. Li, L.; Wang, H.-K.; Kuo, S.-C.; Wu, T.-S.; Mauger, A.; Lin, C. M.; Hamel; Lee, K.-H. *J. Med. Chem.* **1994**, *37*, 3400–3407.
182. Hamel, E.; Lin, C. M.; Plowman, J.; Wang, H.-K.; Lee, K.-H.; Paull, K. D. *Biochem. Pharmacol.* **1996**, *51*, 53–59.
183. Jiang, J. B.; Hesson, D. P.; Dusak, B. A.; Dexter, D. L.; Kang, G. J.; Hamel, E. *J. Med. Chem.* **1990**, *33*, 1721–1728.
184. Zhang, S.-X.; Bastow, K. F.; Tachibana, Y.; Kuo, S.-C.; Hamel, E.; Mauger, A.; Narayanan, V. L.; Lee, K.-H. *J. Med. Chem.* **1999**, *42*, 4081–4087.
185. Gastpar, R.; Goldbrunner, M.; Marko, D.; von Angerer, E. *J. Med. Chem.* **1998**, *41*, 4965–4972.
186. Hanahan, D.; Weinberg, R. A. *Cell* **2000**, *100*, 57–70.
187. Wani, M. C.; Taylor, H. L.; Wall, M. E.; Coggon, P.; McPhail, A. T. *J. Am. Chem. Soc.* **1971**, *93*, 2325–2327.
188. Suffness, M., Ed. *Taxol: Science and Applications*; CRC Press: New York, 1995.
189. He, L.; Orr, G. A.; Horowitz, S. B. *Drug Disc. Today* **2001**, *6*, 1153–1164.
190. Jimenez-Barbero, J.; Amat-Guerri, F.; Snyder, J. P. *Curr. Med. Chem.* **2002**, *2*, 91–122.
191. Nogales, E.; Wolf, S. G.; Downing, K. H. *Nature* **1998**, *391*, 199–203.
192. Della Casa de Marcano, D. P.; Halsall, T. G. *J. Chem. Soc., Chem. Commun.* **1975**, 365–366.
193. Ojima, I.; Geney, R.; Ungureanu, I. M.; Li, D. *Life* **2002**, *53*, 269–274.
194. Stachel, S. J.; Biswas, K.; Danishefsky, S. *J. Curr. Pharm. Design* **2001**, 7, 1277–1290.
195. Hofle, G.; Bedorf, N.; Gerth, H.; Reichenbach, H. *Angew. Chem. Int. Ed. Engl.* **1996**, *35*, 1567–1569.
196. Bolag, D. M.; McQueney, P. A.; Zhu, J.; Hensens, O.; Koupal, L.; Liesch, J.; Goetz, M.; Lazarides, E.; Woods, C. M. *Cancer Res.* **1995**, *55*, 2325–2333.
197. Hardt, I. H.; Steinmetz, H.; Gerth, K.; Sasse, F.; Reichenbach, H.; Hofle, G. *J. Nat. Prod.* **2001**, *64*, 847–856.
198. Florsheimer, A.; Altmann, K.-H. *Expert Opin. Ther. Patent* **2001**, *11*, 951–968.
199. Nicolaou, K. C.; Roschanger, F.; Vouloumis, D. *Angew. Chem. Int. Ed. Engl.* **1998**, *37*, 2014–2045.
200. Rivkin, A.; Yoshimura, F.; Gabarda, A. E.; Chou, T.-C.; Dong, H.; Tong, W. P.; Danishefsky, S. J. *J. Am. Chem. Soc.* **2003**, *125*, 2899–2901.
201. Chou, T.-C.; Dong, H.; Rivkin, A.; Yoshimura, F.; Gabarda, A. E.; Cho, Y. S.; Tong, W. P.; Danishefsky, S. J. *Angew. Chem. Int. Ed. Engl.* **2003**, *42*, 4761–4767.
202. Rivkin, A.; Cho, Y. S.; Gabarda, A. E.; Yoshimura, F.; Danishefsky, S. J. *J. Nat. Prod.* **2004**, *67*, 139–143.
203. Chou, T.-C.; Zhang, X.-G.; Balog, A.; Su, D.-S.; Meng, D.; Savin, K.; Bertino, J. R.; Danishefsky, S. *J. Proc. Natl. Acad. Sci. USA* **1998**, *95*, 9642–9647.
204. Kolmar, A. *Curr. Opin. Investig. Drugs* **2005**, *6*, 616–622.
205. Altman, K.-H. *Curr. Pharm. Design* **2005**, *11*, 1595–1613.
206. Gunasekera, S. P.; Gunasekera, M.; Longley, R. E.; Shulte, G. K. *J. Org. Chem.* **1991**, *55*, 4912–4915 (corrigendum: *J. Org. Chem.* **1991**, *56*, 1346).
207. Kowalski, J. R.; Giannakakou, P.; Gunasekera, S. P.; Longley, R. E.; Day, B. W.; Hamel, E. *Mol. Pharmacol.* **1997**, *52*, 613–622.
208. Nerenberg, J. B.; Hung, D. T.; Somers, P. K.; Schreiber, S. L. *J. Am. Chem. Soc.* **1993**, *115*, 12621–12622.
209. Harried, S. S.; Yang, G.; Strawn, M. A.; Myles, D. C. *J. Org. Chem.* **1997**, *62*, 6098–6099.
210. Smith, A. B., III; Beauchamp, T. J.; LaMarche, M. J.; Arimoto, H. *Org. Lett.* **1999**, *1*, 1823–1826.
211. Marshall, J. A.; Johns, B. A. *J. Org. Chem.* **1999**, *63*, 7885–7892.
212. Paterson, I.; Florence, G. J.; Gerlach, K.; Scott, J. P. *Angew. Chem. Int. Ed. Engl.* **2000**, *39*, 377–380.
213. Mickel, S. J.; Sedelmeier, G. H.; Seeger-Weibel, M.; Berold, B.; Schaer, K.; Gamboni, R.; Chen, S.; Chen, W.; Jagoe, C. T.; Kinder, F. R., Jr. et al. *Org. Proc. Res. Dev.* **2004**, *8*, 92–130.
214. Kalesse, M. *ChemBioChem* **2000**, *1*, 171–175.
215. Mita, A.; Lockhart, A. C.; Chen, T.-L.; Bochinski, K.; Curtright, J.; Cooper, W.; Hammond, L.; Rothenberg, M.; Rowinsky, E.; Sharma, S. *Proc. Am. Soc. Clin. Oncol.* **2004**, *22*, 2025 (abstract).
216. Insbruker, R. A.; Gunasekera, S. P.; Longley, R. E. *Cancer Chemother. Pharmacol.* **2001**, *48*, 29–36.
217. Hung, D. T.; Nerenberg, J. B.; Schreiber, S. L. *J. Am. Chem. Soc.* **1996**, *118*, 11054–11080.
218. Martello, L. A.; LeMarche, M. J.; He, L.; Beuachamp, T. J.; Smith, A. B., III; Horowitz, S. B. *Chem. Biol.* **2001**, *8*, 843–855.
219. Burlingame, M. A.; Shaw, S. J.; Sundermann, K. F.; Zhang, D.; Petryka, J.; Mendoza, E.; Lui, F.; Myles, D. C.; LaMarche, M. J.; Hirose, T. *Bioorg. Med. Chem. Lett.* **2004**, *14*, 2335–2338.
220. Shaw, S. J.; Sundermann, K. F.; Burlingame, M. A.; Myles, D. C.; Freeze, B. S.; Xain, M.; Brouard, I.; Smith, A. B., Jr. *J. Am. Chem. Soc.* **2005**, *127*, 6532–6533.
221. Smith, A. B., III; Freeze, B. S.; LaMarche, M. J.; Hirose, T.; Brouard, I.; Xian, M.; Sundermann, K. F.; Shaw, S. J.; Burlingame, M. A.; Horwitz, S. B. et al. *Org. Lett.* **2005**, 7, 315–318.
222. Smith, A. B., III; Xian, M. *Org. Lett.* **2005**, *6*, asap.
223. Kinder, F. R., Jr.; Bair, K. W.; Chen, W.; Florence, G.; Francavilla, C.; Geng, P.; Gunasekera, S.; Lassota, P. T.; Longley, R. E.; Palermo, M. et al. Abstract number 3650 AACR, 2002.
224. Long, B. H.; Carboni, J. M.; Wasserman, A. J.; Cornell, L. A.; Casazza, A. M.; Jensen, P. R.; Lindel, T.; Fenical, W. H.; Firchild, C. R. *Cancer Res.* **1998**, *58*, 1111–1115.
225. D'Ambrosio, M.; Guerriero, A.; Pietra, F. *Helv. Chim. Acta* **1987**, *70*, 2019–2027.
226. Nicolaou, K. C.; Wissinger, D.; Vouloumis, D.; Oshima, T.; Kim, S.; Pfefferkorn, J.; Xu, J.-Y.; Li, T. *J. Am. Chem. Soc.* **1998**, *120*, 10814–10826.
227. Quinoa, E.; Kakou, Y.; Crews, P. *J. Org. Chem.* **1988**, *53*, 3642–3644.
228. Corley, D. G.; Herb, R.; Moore, R. E.; Scheuer, P. J.; Paul, V. J. *J. Org. Chem.* **1988**, *53*, 3644–3646.
229. Tanaka, J. I.; Higa, T.; Bernadinelli, G.; Jefford, C. W. *Chem. Lett.* **1996**, *25*, 255–256.
230. Mulzer, J.; Ohler, E. *Angew. Chem. Int. Ed. Engl.* **2001**, *40*, 3842–3846.
231. Paterson, I.; De Savi, C.; Tudge, M. *Org. Lett.* **2001**, *3*, 3149–3152.
232. Ghosh, A. K.; Wang, Y. *J. Am. Chem. Soc.* **2000**, *122*, 11027–11028.
233. Muramatsu, H.; Miyauchi, M.; Sato, B.; Yoshimura, S. 40th Symposium on the Chemistry of Natural Products, Fukuoka Japan, 1998, abstract 83.

234. Vanderwal, C. D.; Vosberg, D. A.; Weiler, S.; Sorensen, E. *J. Org. Lett.* **1999**, *1*, 645–648.
235. Roux, D.; Hadi, H. A.; Thoret, S.; Guenard, D.; Thoison, O.; Pais, M.; Sevenet, T. *J. Nat. Prod.* **2000**, *63*, 1070–1076.
236. Kam, T. S.; Tee, Y. M.; Subramanium, G. *Nat. Prod. Lett.* **1998**, *12*, 307–310.
237. Shintani, Y.; Tanaka, T.; Nozaki, Y. *Cancer Chemother. Pharmacol.* **1997**, *40*, 513–520.
238. Wang, Z.; Yang, D.; Mohanakrishnan, A.; Fanwick, P. E.; Nampoothiri, P.; Hamel, E.; Cushman, M. *J. Med. Chem.* **2000**, *43*, 2419–2429.
239. Haggarty, S. J.; Mayer, T. U.; Miyamoto, D. T.; Fathi, R.; King, R. W.; Mitchesonand, T. J.; Schreiber, S. L. *Chem. Biol.* **2000**, 7, 275–286.
240. Klar, U.; Graf, H.; Schenk, O.; Rohr, B.; Schulz, H. *Bioorg. Med. Chem. Lett.* **1998**, *8*, 1397–1402.

Biographies

Brian R Hearn, PhD, is a scientist in the chemistry department at Kosan Biosciences. He joined Kosan in 2003 and worked on the total synthesis and semisynthesis of a number of polyketide natural products and their analogs. Prior to Kosan, he completed an NIH postdoctoral fellowship with Prof Paul A Wender at Stanford University, CA, USA. He obtained his PhD from the University of Notre Dame under the supervision of Prof Richard E Taylor.

Simon J Shaw, PhD, is a scientist in the chemistry department of Kosan Biosciences. Since joining the department in 2001, he has been involved in research in the areas of infectious disease and cancer. Prior to joining Kosan, he spent 2 years in the laboratories of Prof Dan Kahne as a postdoctoral scholar. He received a PhD from Imperial College, London, under the direction of Prof Donald Craig.

David C Myles, PhD, is Executive Director, Chemistry of Kosan Biosciences. Prior to joining Kosan in May 2001, he was Associate Director, Organic and Medicinal Chemistry at Chiron Corporation where he was involved in research directed toward the development of new antiviral, antibacterial, and anticancer agents. From 1991 to 1998, he held a faculty position in the Department of Chemistry and Biochemistry at the University of California, Los Angeles, CA. From 1989 through 1991, he held the NIH/Lawton Chiles Biotechnology Fellowship in the Department of Chemistry at Harvard University, MA, where he studied organic synthesis using biocatalysts. He earned his PhD in 1989 from Yale University.

Comprehensive Medicinal Chemistry II
ISBN (set): 0-08-044513-6

ISBN (Volume 7) 0-08-044520-9; pp. 81–110

7.05 Deoxyribonucleic Acid Topoisomerase Inhibitors

W A Denny, University of Auckland, Auckland, New Zealand

7.05.1 Introduction

7.05.1.1 The Topoisomerase Enzymes

The mammalian topoisomerase enzymes topo I, topo IIα, and topo IIβ are regulatory homodimeric enzymes that catalyze the breakage and religation of DNA. They provide swivel points for relaxing the supercoils generated during the transcription and replication of DNA, and by the segregation of chromatids prior to mitosis.[1] The topo II enzyme has major isozymes coded by two separate genes.[2] The IIα isozyme (170 kDa) maps[3] to chromosome 17, is regulated during the cell cycle, and is the target of virtually all of the DNA intercalating topo inhibitors. The IIβ form (180 kDa) maps[4] to chromosome 3, and becomes the predominant isozyme in both noncycling cells and in cells resistant to 'classical' topo II agents. The gene for the topo I enzyme gene is located on chromosome 20, with the gene copy number varying from two to eight in a panel of colorectal cancer cell lines.[5]

There are three main steps in the action of these enzymes. The first is binding to recognition sequences on the DNA. These binding sites for topo I and II involve 15–19 and 30 contiguous nucleotide pairs respectively,[6,7] although initial binding is probably to a smaller segment of about five base pairs, followed by conformational changes to allow binding to the full site. Enzyme-promoted hydrolysis of a phosphodiester bond then occurs, with the protein becoming covalently bound via a tyrosine hydroxyl group to the 5′-end of the broken DNA chain.[8] Topo II generally effects cuts on each strand about five base pairs apart, whereas topo I generates a single-strand cut. The resulting complex, termed the cleavable complex, is held together by the interactions between the enzyme domains, while a second strand (in the case of topo II) is passed through the gap to relieve torsional stress.[8] The process is regulated so that minimum time is spent in this fragile state before the breaks are religated in an ATP-driven process.[9]

7.05.1.2 Mechanism of Topoisomerase Inhibition by Drugs

A large number of drugs, generally referred to as topo inhibitors, interact with topo enzymes and affect one of these actions. Collectively they are a major class of anticancer agents, and one of the mainstays of current cancer chemotherapy. At high concentrations they can bind to the DNA or the protein, and compete with the formation of the initial ternary enzyme–DNA–drug complex, although this is unlikely to be the primary mechanism of clinical activity for the majority of the drugs. More important is their subsequent interference with the DNA cleavage, DNA strand passing (in the case of topo IIα and IIβ), or DNA re-ligation steps. DNA intercalating agents in particular can unwind the double helix and distort DNA structure at the enzyme cleavage sites. This is thought to hinder registration between the cleaved ends of the DNA during re-ligation, leading to accumulation of the cleavable complex and an increase in strand breaks. Such drugs essentially convert the topo enzyme into a DNA damaging agent, and are also termed topo poisons.[1] The selective killing of tumor cells by topo inhibitors arises because many tumor cells overexpress these enzymes to enhance cellular proliferation, and the degree of poisoning is a function of the amount of the enzyme present.

7.05.1.3 Classification of Topoisomerase Inhibitors

The primary classification of topoisomerase inhibitors is a functional one; inhibitors of topo I, topo II, or dual topo I/II inhibitors.[10] While it would be preferable to fully classify these compounds by their detailed mechanism of action there is in general not enough known to allow this, and the next level of classification is by their mode of DNA interaction. The largest class of drugs by far (and the majority of the topo II and dual topo I/II inhibitors) are the DNA intercalators. Intercalation as a mode of reversible binding of ligands to DNA was first described by Lerman,[11] and involves insertion of the ligand between the base pairs. This is now understood to be the major DNA binding mode of virtually any flat polyaromatic ligand of sufficiently large surface area and suitable steric properties, and is driven primarily by stacking (charge-transfer and dipole-induced dipole) and electrostatic interactions, with entropy (dislodgement of ordered water around the DNA) of lesser and variable importance.[12] A great deal of work has been done delineating the ligand structural properties that favor intercalation, the geometry, kinetics, and DNA sequence-selectivity of the binding process, and the effect of such binding on the structure of the DNA substrate.[13] A much smaller class of compounds (no clinical agents) bind to DNA in the minor groove, and there is a substantial and diverse class of non-DNA binders (compounds with little direct DNA affinity but which nevertheless form ternary complexes). Finally, a number of unrelated compounds affect topoisomerase function by virtue of their effects on the multiple signaling pathways that interact with the topoisomerases, but these compounds are not primarily known as topo inhibitors and will not be covered here. This review notes the important clinical inhibitors of the topo I and/or topo II enzymes, as well as some less clinically successful examples that have been of scientific importance. It focuses on the more recent emerging drugs, which demonstrate how the major drawbacks of these classes of compounds are being addressed.

7.05.2 Topoisomerase I Inhibitors

7.05.2.1 Camptothecin and Analogs

The drug camptothecin (**1**) is a natural product, first isolated and characterized from the plant *Camptotheca acuminate*.[14] The successful clinical introduction was substantially delayed, largely due to problems with insolubility and instability, particularly of the ring E lactone. This is absolutely required for activity, yet is unstable in aqueous solution, being in equilibrium with the corresponding, and possibly toxic, carboxylate at physiological pH. Following the clinical success of camptothecin,[15] the major driving forces behind analog development were to increase water solubility and bioavailability, after extensive experimental structure–activity relationship (SAR) studies showed that only substitution at positions 9–11 of ring A, and position 7 of ring B were permitted. Of a large number of analogs evaluated, topotecan (**2**) and the carbamate-based prodrug irinotecan (**3**) have established themselves,[16] but several others are in early clinical development.

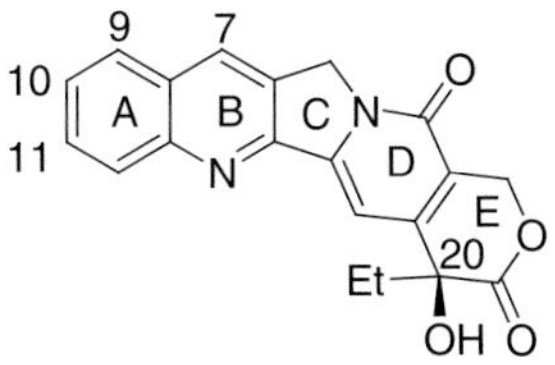

1 Camptothecin

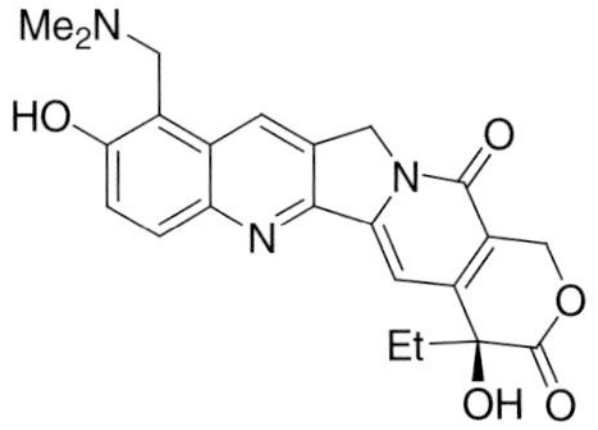

2 Topotecan

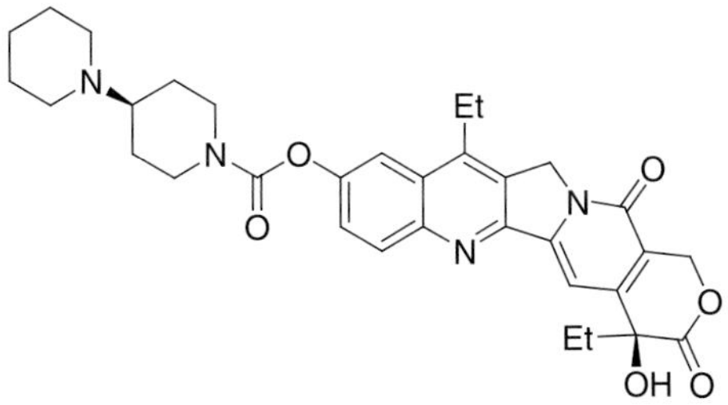

3 Irinotecan

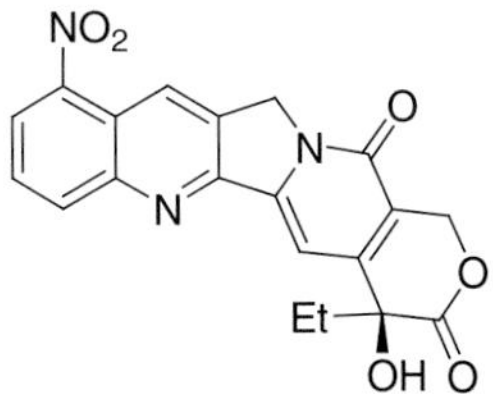

4 Rubitecan

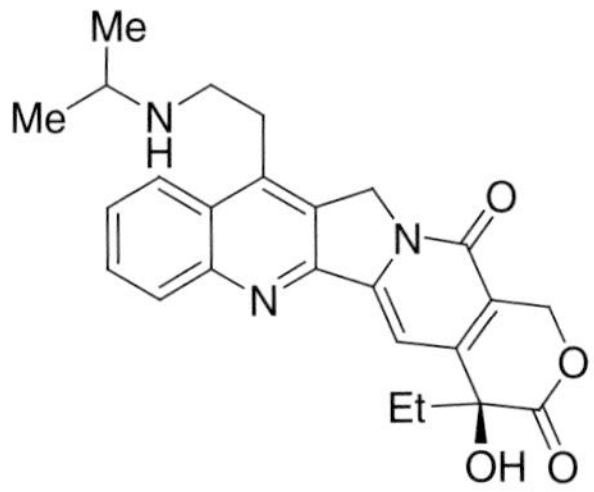

5 CKD-602

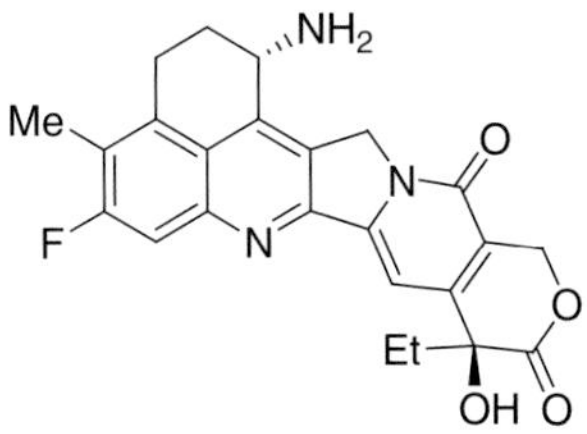

6 Exatecan

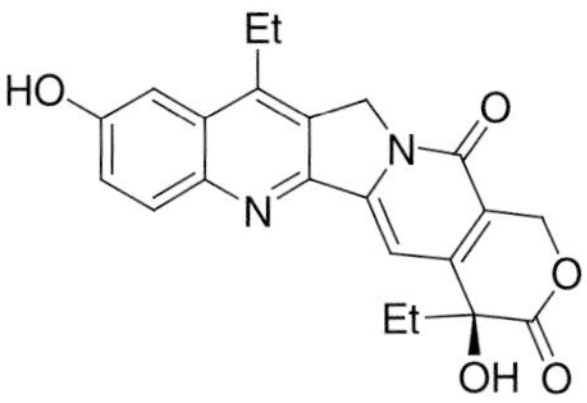

7 SN38

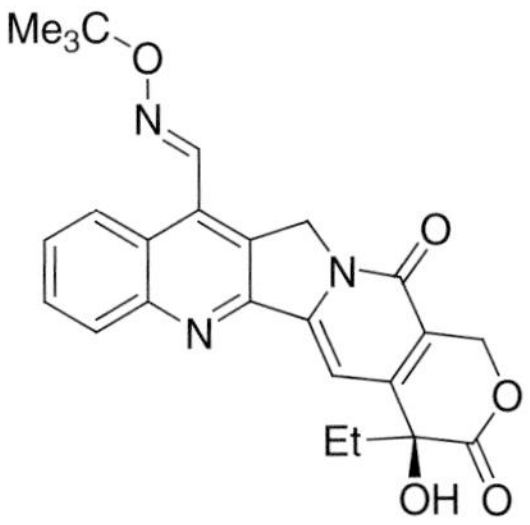

8 Gimatecan

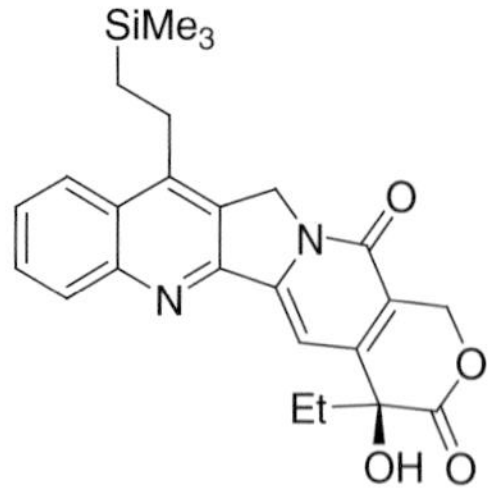

SiMe3 N N O Et OH O O

9 Karenatecin

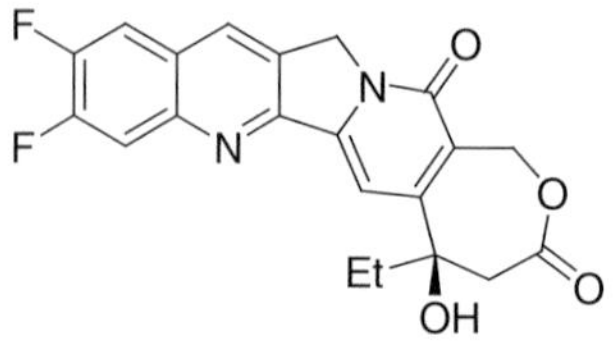

F F N N O O Et OH O

10 Diflomotecan

OH HN OH O N O HO N H N OH O OH HO HO OH

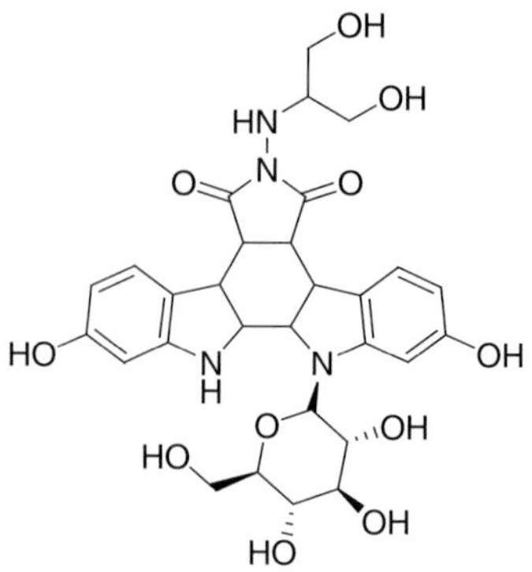

11 Edotecarin

7.05.2.1.1 Rubitecan

Rubitecan (9-nitrocamptothecin) (**4**) is a semisynthetic compound, derived from camptothecin by direct nitration, and the most advanced of all the new camptothecin analogs. Early syntheses using classic nitration conditions gave poor yields (5–10%),[17] but a later process[18] using urea-mediated nitration gives yields of up to 40%. Rubitecan undergoes rapid in vivo reduction to the 9-amino analog in mice,[19] rats,[20] and humans.[21] The latter compound (NSC-603071) is also an active drug that has been evaluated clinically, albeit without much success.[22] Rubitecan itself induces apoptosis associated with the upregulation of the cytokines cyclin B1 and cdc2, but not of cyclins A, E, and cdk2.[23] Overexpression of P-glycoprotein (P-gp) or multidrug resistance protein types 1 and 2 or the breast cancer resistance protein (BCRP) have little effect on the cytotoxicity of rubitecan, the latter activity distinguishing it from camptothecin.[24] It has broad-spectrum activity in human tumor xenografts in nude mice, showing growth inhibition in all 30 lines tested, with complete tumor abrogation in 24 of these, at its maximum tolerated dose of 1 mg $kg^{-1}day^{-1}$.[25] A number of Phase II clinical trials have been reported, mostly using oral dosing at 1.5 mg m^{-2} per day for five consecutive days (5 days on – 2 days off) on a continuous basis. These suggested limited or no activity of the drug as a single agent in a number of tumor types, including metastatic breast cancer[26] advanced small cell lung cancer,[27] advanced soft-tissue sarcomas,[28] glioblastoma,[29] colon cancer,[30] and nonsmall cell lung cancer[31] (the latter at up to 2 mg $m^{-2}day^{-1}$). The drug was well tolerated, suggesting a possible escalation of dose. A study using 1.5 mg $m^{-2}day^{-1}$ on the same schedule in previously treated pancreatic cancer patients did show activity, with 3/43 partial responses and 7/43 stable disease in this very refractory cancer,[32] and Phase III trials are in progress. Because of the insolubility of rubitecan, there has been interest in developing liposomal formulation to improve delivery[33] and a recent Phase II trial of aerosolized liposomal drug in 25 patients, using a dose equivalent to 0.5 mg $m^{-2}day^{-1}$, resulted in two partial remissions in patients with uterine cancer, and stable disease in three patients with primary lung cancer.[34]

7.05.2.1.2 CKD-602

CKD-602 (**5**) represents an attempt to improve the solubility of camptothecin, by adding an *N*-aminoethyl group at the 7-position of ring B where there is some bulk tolerance, and was the best of a number of such analogs evaluated.[35] It is equipotent to camptothecin at inhibiting human topo I in vitro, measured by the cleavable complex assay (IC_{50} 0.27 and 0.26 μM, respectively), and has higher solubility and comparable cytotoxicity to camptothecin in a wide range of human tumor cell lines (IC_{50} values from 4.6 nM to 4.6 μM), but is susceptible (like camptothecin) to P-gp-mediated efflux.[36] It showed good activity in a range of human tumor xenografts in nude mice (e.g., SKOV-3 ovarian, MX-1, LX-1, and HT29, WIDR, and CX-1 colon lines), with activity broadly comparable to topotecan at the same dose.[36] It was not mutagenic in *Salmonella typhimurium* TA 98, TA 100, TA 1535, and TA 1537 strains with and without metabolic activation, but did show an increased incidence over controls of chromosome aberrations (micronuclei) in rats.[37] A preliminary report of a Phase II trial of CKD-602, at a dose of 0.6 mg $m^{-2}day^{-1}$ in patients with ovarian cancer, said that 20% showed partial responses, with the major toxicity being neutropenia.[38]

7.05.2.1.3 Exatecan

Exatecan (**6**) is a more extensively modified derivative of camptothecin, with an additional alicyclic ring fused to rings A and B that bears a solubilising primary amine (equivalent to a 7-CH_2NH_2 substituent on camptothecin). There are also lipophilic substituents at positions 10 and 11 on ring A that may help to enhance membrane permeability.[39] It is about 2.5-fold more potent than SN38 (**7**, the active metabolite of irinotecan) at stabilizing the topo I/DNA complex, and as a cytotoxin in a variety of cell lines.[40] It is also fully active in P-gp-overexpressing cell lines,[41] possibly due to its enhanced cell penetration ability.[42] In one model, resistance to exatecan appears to involve changes in topoisomerase expression.[43] Extensive Phase II trial studies suggest the drug has little activity as a single agent in gastric, biliary tract, and colon cancer, but may have some utility in the treatment of refractory ovarian and pancreatic cancer.[44]

7.05.2.1.4 Gimatecan

Gimatecan (**8**) has an unusual lipophilic butyloxyliminomethyl substituent at the 7-position, and was the best of a series of similar compounds evaluated. Quantitative SARs for these indicated that overall lipophilicity was the property most highly correlated with both cytotoxicity and the potency of inhibition of topo I.[45] Fluorescence studies suggest a quite different pattern of cellular disposition of this lipophilic drug compared with topotecan (lysosomal rather than mitochondrial).[46] It is approximately 2–5-fold more potent than SN38 in cell culture, causing G2/M arrest.[45] Despite lower levels of intracellular accumulation than other camptothecin analogs, gimatecan was the most active against neuroblastoma cell lines, suggested due to an ability to cause high levels of DNA breaks.[47] In Phase I clinical studies, oral gimatecan had an acceptable toxicity profile, with myelotoxicity being dose-limiting. It showed a very long terminal half-life and favorable pharmacokinetics, with a significant incidence of tumor responses.[48]

7.05.2.1.5 Karenitecin

Karenitecin (BNP1350) (**9**) is another semisynthetic, lipophilic camptothecin derivative, designed to have higher oral bioavailability and greater lactone stability than camptothecin by employing an ethyltrimethylsilyl derivative at the 7-position. Cell culture studies in a variety of tumor cell lines showed that karenitecin was significantly more effective than SN38, and not affected by overexpression of P-gp.[49] Karenitecin proved to be a much poorer substrate than topotecan for BCRP, which may be a major reason for its broader-spectrum activity.[50] Karenitecin-resistant A2780 variants overexpressed BCRP, had reduced catalytic activity of topo I, and showed cross-resistance to many other camptothecins. Later studies in karenitecin-resistant cell lines suggest that resistance is related to upregulation of the chk1 signaling pathways that mediate the G2/M cell cycle checkpoint.[51] In vitro studies[52] with cloned cytochrome P450 (CYP) isoenzymes suggested that karenitecin is metabolized by CYP3A4, CYP2C8, and CYP2D6, and is an inhibitor of CYP3A4 and CYP2C8. Karenitecin was also active in a wide range of adult and pediatric central nervous system malignancies growing in athymic nude mice, suggesting possible clinical utility in patients with such tumors.[53] A Phase II trial of karenitecan in patients with metastatic melanoma (which expresses high levels of topo I), using an intravenous infusion of 1 mg m^{-2} karenitecin daily for 5 days every 3 weeks, showed 1/43 complete response, and 33% with some degree of stable disease. The major toxicity was reversible noncumulative myelosuppression.[54]

7.05.2.1.6 Diflomotecan

Diflomotecan (**10**) is prepared by a totally synthetic and stereospecific route,[55] and was selected for development from a series of E ring modified homocamptothecin analogs possessing a 7- rather than a 6-membered hydroxylactone ring.[56] On the basis of its high activity, and the improved stability of the lactone ring (half-life about 2 h in human plasma at 37°C, compared with about 5 min for camptothecin), it was the first homocamptothecin taken to clinical trial.[57] Diflomotecan is more effective than camptothecin in inducing topo I-mediated DNA cleavage, due to a greater ability to stabilize the topo I–DNA cleavage complexes.[57] It is also a very potent cytotoxin in cell line assays, with higher overall antiproliferative activity than camptothecin, topotecan, or SN38 against a series of 43 early-passage human colon cancer cell lines in culture.[58] It has broad activity in mouse models, including HT29 human colon adenocarcinoma xenografts (tumor growth delay of 25 days compared with a 4-day growth delay with camptothecin[57]), and with oral dosing against a PC-3 prostate model (27 days compared with 15 days[59]). In vitro studies[60] suggest it is metabolized in humans primarily by CYP3A4. A Phase I clinical trial of oral diflomotecan in adult patients with solid malignant tumors suggested a maximum tolerated dose of 0.27 mg day^{-1} daily for 5 days, repeated every 3 weeks. Pharmacokinetics were linear over the dose range studied, and the dose-limiting toxicity was thrombocytopenia.[61]

7.05.2.2 Indolocarbazoles

7.05.2.2.1 Edotecarin

Edotecarin (**11**) is an indolocarbazole-based derivative that was developed from the related NB-506 by moving the hydroxyl groups to the 2,10 positions, and replacing the 6-*N*-formyl group with a more bulky diol.[62] A synthetic route utilizes nontoxic protected intermediates, followed by a last step deprotection.[63] It is a potent inducer of single-strand

DNA cleavage by topo I,[64] and forms a very stable DNA–topo I cleavable complex.[62] Resistant cell lines show overexpression of BCRP, and enhanced drug efflux.[65] Edotecarin is more resistant than NB-506 to both phthalimide ring opening[66] and glucuronidation, suggested[67] due to the larger size of the N-6 group. Edotecarin was active against human tumor xenografts of LX-1 and PC-3 cells in mice, and in cell lines that overexpressed P-gp.[68] Activity has been shown in Phase I trials in patients with refractory metastatic colon and gastric cancers, and with breast and esophageal cancer. Phase II trials in patients with irinotecan- or 5-FU-refractory colorectal cancer produced objective responses in irinotecan-naive, 5-FU refractory patients, but not in irinotecan-refractory patients.[69]

7.05.3 Topoisomerase II Inhibitors

7.05.3.1 Anthracyclines

The great clinical success of the anthracycline doxorubin (**12**)[70] has ensured that a large effort has been put into other anthracycline analogs. Among the goals of this research has been to lower the cumulative cardiotoxicity seen with this drug, and also to develop semisynthetic or wholly synthetic analogs. Several such new anthracycline derivatives are reported on here.

Another, broader goal for improving all classes of DNA-intercalating topo II agents is to decrease their susceptibility to cell efflux pumps such as P-gp, and this is a second broad theme. However, the search for new topo II agents is not restricted to anthracyclines, or even to DNA intercalators, with a wide variety of both synthetic and natural product structures under development.

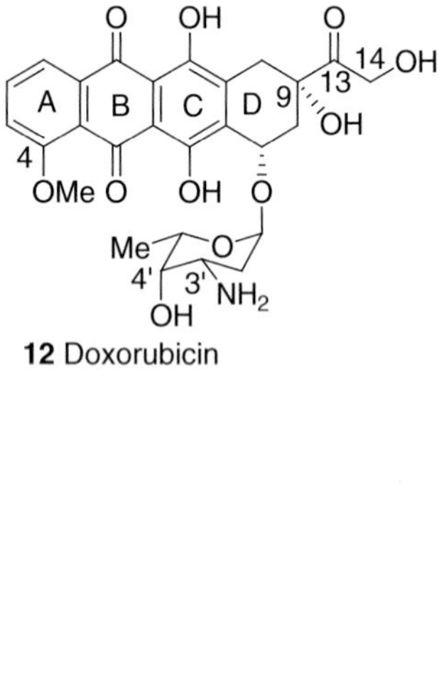

12 Doxorubicin

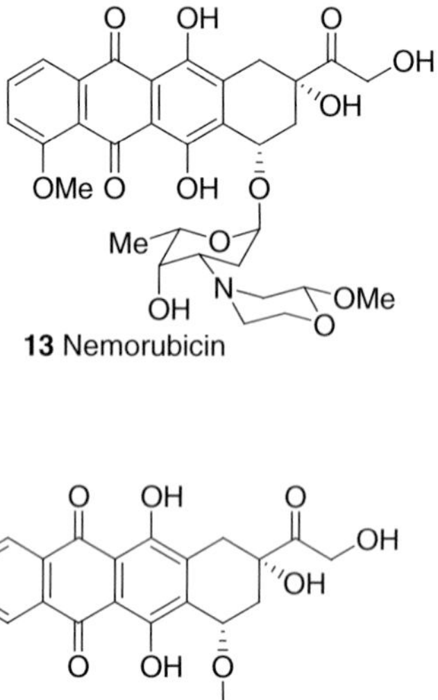

13 Nemorubicin

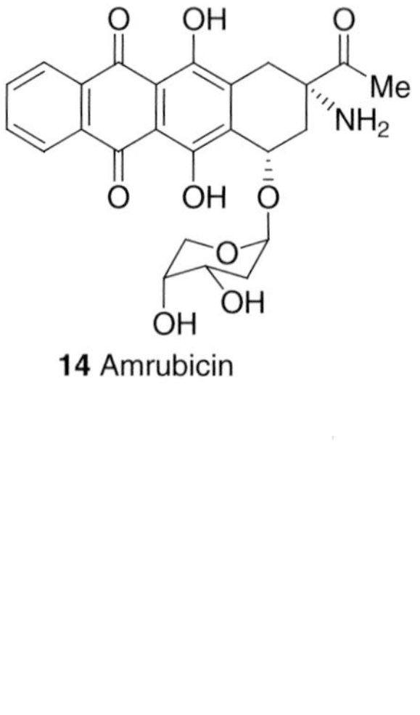

14 Amrubicin

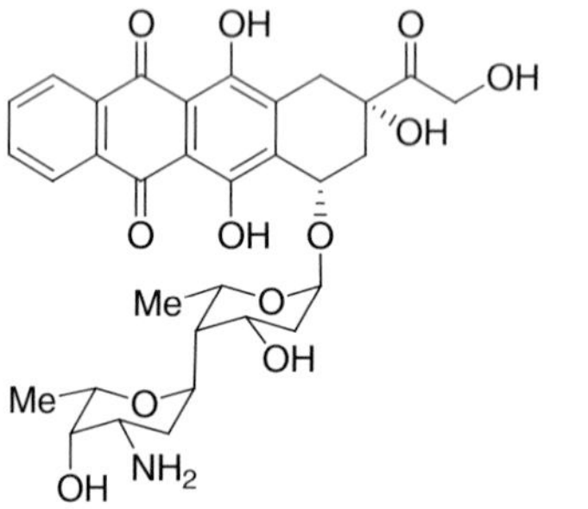

15 MEN-10755

7.05.3.1.1 Nemorubicin

Nemorubicin (PNU-152243) (**13**) is a lipophilic analog of doxorubicin, where the amino group has been modified by a lipophilic and base-weakening 2-methoxymorpholide substituent. Both of these effects contribute to a higher lipophilicity than for doxorubicin, likely providing more rapid extravascular diffusion and cell uptake, and are considered to contribute to its observed lower cardiotoxicity. Nemorubicin acts as a prodrug, being converted in the liver, primarily by CYP3A, to a series of metabolites,[71] one of which (PNU-159682) is considerably more potent than the parent compound, and is also active in in vivo tumor models.[72] The potency and toxicity of nemorubicin can both be modulated by stimulators or inhibitors of CYP3A.[71] It showed good in vitro activity against murine L1210 cells resistant to cis-platin[73] and melphalan,[74] and in vivo against P388 or LoVo cells and MX-1 mammary tumor xenografts,[75] especially human hepatocellular carcinoma xenografts.[76] In patients, nemorubicin was cleared rapidly from the circulation by extensive tissue distribution, with little renal excretion.[77] In a Phase II trial as second-line therapy in soft-tissue sarcoma, using 1.23 $mg\,m^{-2}$ every 4 weeks, nemorubicin gave 1/28 partial responses and 6/28 stable disease for at least 2 months, with no significant cardiotoxicity seen.[78] It is currently in Phase II/III evaluation in hepatocellular carcinoma.[72]

7.05.3.1.2 Amrubicin

Amrubicin (**14**) is a totally synthetic anthracycline, and differs more substantially than nemorubicin from the parent doxorubicin. It lacks the 4-methoxy and 14-hydroxy groups on the chromophore and the 3′-amino group in the sugar unit, but has an amino group at the 9-position on the chromophore. It was designed as a less cardiotoxic anthracycline[79] and is a prodrug, with the major metabolite (the 13-alcohol amrubicinol) being 3–8-fold more cytotoxic in cell culture than amrubicin itself.[80] Both compounds act via topo II-generated double-strand DNA breaks.[81] In human tumor xenografts the metabolite accumulates to higher levels than that of doxorubicin in the tumors but at lower levels in normal tissues, suggesting that its selective distribution plays a large part in the observed more potent therapeutic activity.[82] Amrubicin was less active than doxorubicin against a panel of human lung tumor cell lines in culture,[83] but was superior to doxorubicin in vivo in both murine tumor models and human tumor xenografts.[84] In a series of human tumor xenografts in nude mice, there was a positive correlation between activity and the levels of the 13-alcohol metabolite in the tumors.[85] Several Phase I trials, using different protocols, indicated maximum tolerated doses of 50–100 $mg\,m^{-2}$, with leukocytopenia and thrombocytopenia as the major toxicities.[86] A recent Phase I/II study of amrubicin and irinotecan in patients showed maximum tolerated doses of 45 $mg\,m^{-2}$ and 100 $mg\,m^{-2}$ respectively, with responses seen in about 10% of patients.[87]

7.05.3.2.3 MEN-10755

MEN-10755 (**15**) is a des-4-methoxy derivative of doxorubicin, but with a disaccharide side chain; it was the first anthracycline disaccharide to be used clinically. The crystal structure of MEN-10755 bound to the oligodeoxynucleotide $d(CGATGG)_2$ is broadly similar to other anthracycline–DNA complexes, except that two different binding sites exist; in one both sugar rings lie in the minor groove, while in the other the second sugar H-bonds to a guanine of a second DNA helix.[88] MEN-10755 is a more potent topo II poison than doxorubicin (suggested due to replacement of the amino group on the first sugar), while having lower cardiotoxicity. The latter is possibly related to its slower rate of cell uptake, which leads to a higher cytoplasmic/nuclear ratio than for doxorubicin.[89] In cumulative dose studies in rats comparing a range of anthracyclines, cardiotoxicity correlated with myocardial levels of the corresponding alcohol metabolites.[90] Studies with human cardiac cytosol also showed that MEN-10755 had a lower rate of formation of the cardiotoxic alcohol metabolite than did doxorubicin, and that this alcohol has lesser ability to react at the active 4Fe–4S site of cytoplasmic aconitase.[91] MEN-10755 is more effective than doxorubicin in human tumor xenografts in nude mice,[92] but was not effective in cells expressing transport-mediated resistance.[93] A Phase I trial of MEN-10755 established the maximum tolerated dose as 45 $mg\,m^{-2}$, with the dose-limiting toxicity being neutropenia.[94] No antitumor responses were seen in 25 patients. A second Phase I study[92] recommended a 3-weekly regimen for Phase II.

7.05.3.2 Anthraquinones

These were developed in a search for intercalators with lower cardiotoxicity than doxorubicin, with the major early success being mitoxantrone (**16**).[95] This compound, like the anthracyclines, binds tightly to DNA by intercalation, but the exact mode of this binding, and of the other anthraquinones (parallel or perpendicular to the base pair axis) has not been exactly established.

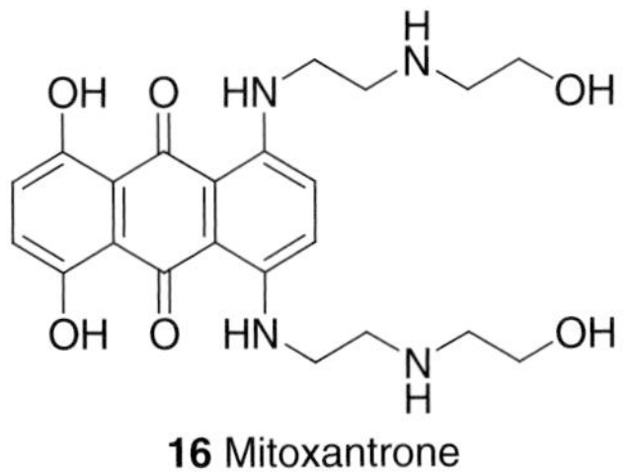

16 Mitoxantrone

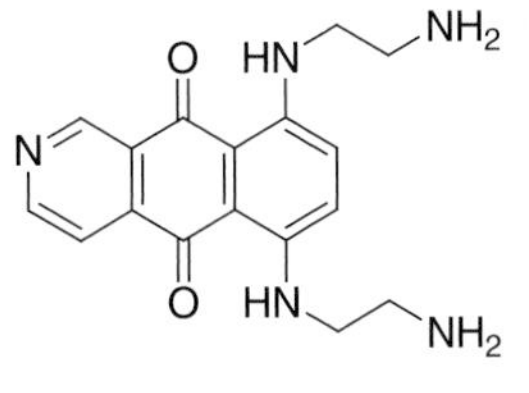

17 Pixantrone

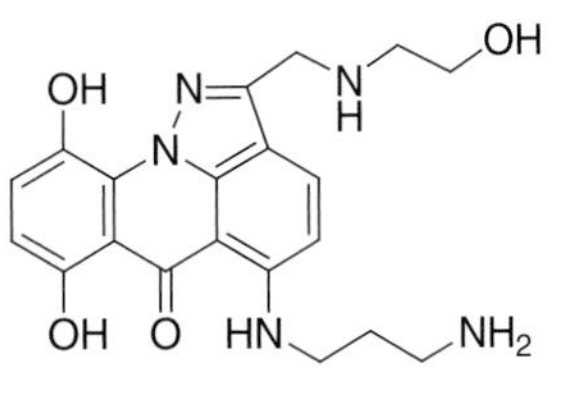

18 KW-2170

7.05.3.2.1 Pixantrone

Pixantrone (**17**) is an azaanthraquinone, evolved from mitoxantrone in a search for drugs with lower cardiotoxicity.[96] It was selected from a series of analogs on the basis of high cytotoxic activity, significant in vivo antitumor efficacy (especially against lymphomas and leukemias) over a wide dose range, and a lack of delayed cardiotoxicity compared with other anthracenediones.[97] It is a DNA intercalating agent and topo II inhibitor,[98] but it is not clear if this is its only mechanism of action, since its DNA damaging effects do not correlate with cytotoxicity.[99] While pixantrone is not as potent as mitoxantrone it is also less cardiotoxic at equieffective doses, less myelosuppressive, and shows better in vivo activity in leukemia and lymphoma models.[96] The drug appears to be most useful in the treatment of non-Hodgkin's lymphoma[100]; a recent multicenter Phase II trial used a dose of 85 $mg\,m^{-2}$ in a 3-weekly protocol, gave 9/33 remissions with a median relapse time of 17+ months, and recommended Phase III trials.[101] Its low cardiotoxicity has led to proposals for its use instead of mitoxantrone in the long-term treatment of multiple sclerosis.[102]

7.05.3.2.2 KW-2170

KW-2170 (**18**) is a synthetic pyrazoloacridone derivative, prepared in a 13-step synthesis in 12% overall yield.[103] KW-2170 is a topo II inhibitor with lower cardiotoxicity and lower cross-resistance to other topo II agents than doxorubicin. It was superior to doxorubicin against a variety of subcutaneous syngeneic and human tumor xenografts in mice (sarcoma 180, breast carcinoma MM102, fibrosarcoma Meth A, the doxorubicin-resistant human ovary carcinoma A2780/ADM, and nasopharynx carcinoma KB-A1).[104] Several Phase I studies have been reported,[105] with maximum tolerated dose suggested between 25 and 50 $mg\,m^{-2}$. A recent Phase I study using a 30-min intravenous infusion every 4 weeks showed a maximum tolerated dose of 53 $mg\,m^{-2}$, with the major toxicity being neutropenia. There were no objective responses, but no cardiotoxicity.[106]

7.05.3.3 Acridines and Related Intercalators

The acridines have been widely used as drugs, primarily as antibacterial and antiprotozoal agents, but increasing also as anticancer drugs.

7.05.3.3.1 Amsacrine

The 9-anilinoacridine amsacrine (**19**)[107] is a relatively weak DNA binder, with an association constant of $1.8 \times 10^5\,M^{-1}$ for calf thymus DNA in 0.01 M salt, via reversible, enthalpy-driven[108] intercalation of the acridine chromophore. It was one of the drugs first used to show that topo II was the target of most DNA intercalators,[109] and has been widely used since then as a probe to study topoisomerase action. It is postulated to bind with the anilino ring lodged in the minor groove, with the 1′-substituent pointing tangentially away from the helix.[110] Amsacrine appears unique among topo II poisons in that its ability to trap both topo IIα- and topo IIβ-induced lesions is only modestly reduced in ATP-depleted cells, implying that it produces mainly pre-strand passage DNA lesions, whereas other topo II poisons stabilize post-strand passage DNA lesions in intact cells.[111] It was one of the first synthetic topo II inhibitors to reach clinical trial.[112] The clinical use of amsacrine is mainly in acute myeloid leukemia,[113] although successful use in various adult leukemias has also been reported; a recent Phase II trial[114] of amsacrine with ara-C and etoposide in refractory acute leukemia achieved a complete remission rate of 55%.

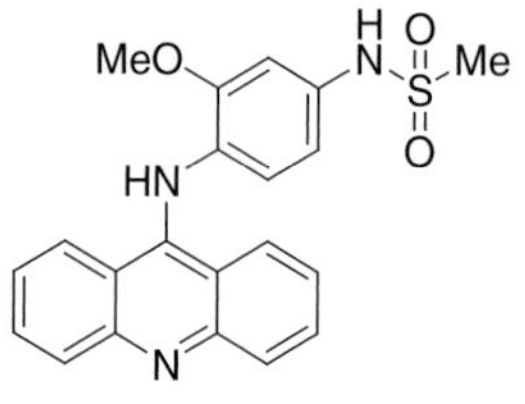

19 Amsacrine 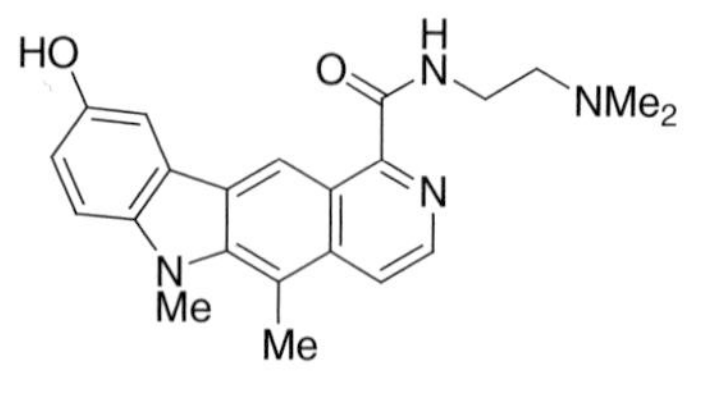 **20** S-16020-2 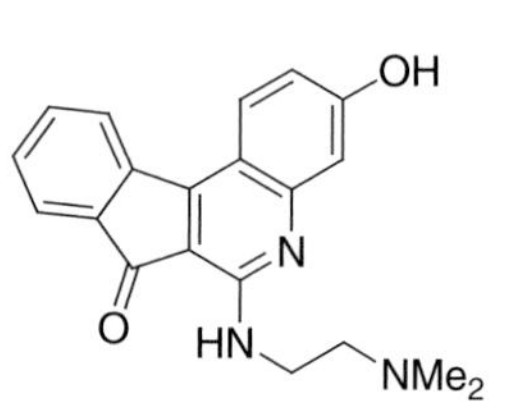 **21** TAS-103

7.05.3.3.2 S-16020-2

S-16020-2 (**20**) is a derivative of the olivacine class of pyridocarbazoles,[115] containing a (dimethylamino)ethylcarboxamide side chain that confers tight DNA intercalative binding properties. The drug is a potent stimulator of topo II-mediated DNA cleavage.[116] It appears to be little affected by P-gp-mediated multidrug resistance, both in cell lines in culture,[117] and in vivo.[118] S-16020-2 was superior to doxorubicin in 5/13 human tumor xenografts in mice

(NCI-H460, A549, NCI-H69, SCLC6, and NIH:OVCAR-3) in terms of growth delay when given on a weekly schedule.[119] It was superior to doxorubicin and cyclophosphamide against chemoresistant A549 nonsmall cell lung carcinomas in nude mice, A549 tumor cells metastases in severe combined immunodeficiency (SCID) mice, and intravenously implanted Lewis lung carcinomas.[120] Preliminary conference reports of Phase I trials suggest a maximum tolerated dose of about 150 mg m^{-2}, with the major toxicity being acne-like lesions.

7.05.3.3.3 TAS-103

TAS-103 (**21**) is a synthetic indenoquinoline that intercalates into DNA with a moderate binding constant ($2.2 \times 10^{-6} M^{-1}$).[121] While TAS-103 was initially reported as a dual topo I/II inhibitor, later studies suggested that its cytotoxicity is due to topo IIα poisoning via blockage of DNA religation.[122] TAS-103 induces cellular apoptosis[123] by a pathway that involves interleukin-1 converting enzyme (ICE)-like proteases, and which is blocked by Bcl-2. It is a potent cytotoxin, with IC_{50} values in the low nM range and a broad spectrum of activity in human lung, colon, stomach, breast, and pancreatic tumor xenografts.[124] It was not affected by the overexpression of P-gp or other transporters, or by mutations in the topo I enzyme.[123] In a series of three cell lines made resistant to TAS-103 by chronic exposure to increasing concentrations of drug, topo II levels decreased by $>75\%$, with lower expression of both mRNA and protein than in the wild-type line.[125] The drug is extensively metabolized in humans, primarily to glucuronides.[126] A Phase I trial of a weekly $\times$ 3 protocol suggested a maximum tolerated dose of 130–160 mg m^{-2}, with the immediate toxicity being neutropenia,[127] but a prolonged QTc interval was seen in some patients, suggestive of cardiotoxicity. Probably because of this, no Phase II trials have been reported.

7.05.3.4 Etoposide and Analogs

The epipodophyllin etoposide (**22**) is a semisynthetic drug, prepared from the natural product podophyllotoxin, which is isolated from the plant *Podophyllum peliatum*.[128] It is been widely used in cancer therapy, and has sparked intense research into other derivatives, of which teniposide (**23**) is also a well-established cancer drug,[129] and other analogs continue to be developed.

22 Etoposide: R = Me

23 Teniposide: R = 2-thienyl

24 TOP-53

25 XL-119

26 IST-622: R = $CO(CH_2)_2OEt$

27 A-132: R = H

28 Elinafide

7.05.3.4.1 TOP-53

TOP-53 (**24**) is a epipodophyllotoxin derivative with a basic aminoalkyl side chain. This not only improves drug solubility, but is also suggested to confer binding of the drug to phospholipids (especially phosphatidylserine, association constant $K_a = 5.6 \times 10^{-6}\,M^{-1}$), resulting in its selective accumulation in lung tissue.[130] TOP-53 stabilizes topo II–DNA cleavable-complex formation by interfering with the DNA religation activity of the enzyme, with a DNA cleavage site specificity that is identical to that of etoposide.[131] It is a potent cytotoxin (IC_{50} 0.07–0.7 μM in a range of murine tumor cell lines), with broad in vivo activity against human tumor xenografts (NL-22 and NL-17 colon cancer, UV2237M fibrosarcoma, and K1735M2 melanoma), especially lung metastases.[132] Preliminary conference reports of Phase I trials suggest a maximum tolerated dose of about 110 mg m^{-2}, with the major toxicity being leukopenia.

7.05.3.5 Miscellaneous Compounds

7.05.3.5.1 XL-119

XL-119 (NSC 655649) (**25**) is a more water-soluble (about 4 mg mL^{-1} as the HCl salt), monoethylated derivative of the natural antitumor antibiotic rebeccamycin,[133] and can be prepared from it by direct alkylation.[134] It evolved from a synthesis program aimed at developing analogs with better in vivo distributive properties than rebeccamycin.[135] In studies using 14 established cell lines and 20 early-passage clinical isolates of pediatric solid tumors in a growth inhibition assay, XL-119 had IC_{50} values in the range 0.5–1 μM, similar to plasma levels achieved during adult Phase I clinical trials.[136] In Phase I trials, XL-119 showed dose-dependent pharmacokinetics that fitted a three-compartment model, and a long terminal half-life (49 h and 154 h in the two studies respectively).[137,138] A Phase II study in advanced renal cell cancer patients showed that a dose of 165 mg m^{-2} daily intravenously for 5 days was well tolerated and had modest antitumor activity, with 8% partial responses and 46% stable disease. The major toxicity was myelosuppression.[139] A Phase II trial in patients with minimally treated metastatic colorectal cancer given 500 mg m^{-2} once every 3 weeks showed no activity.[140]

7.05.3.5.2 IST-622

IST-622 (**26**) is a semisynthetic analog of the *Actinomycete*-derived antitumor antibiotic chartreusin, and is a potent topo II inhibitor.[141] It has excellent activity in a wide range of in vivo murine tumor models, including Colon 26 and Colon 38 adenocarcinomas, and M5076 reticulum cell sarcoma, and was also active in two of seven human tumor xenografts. In vivo, IST-622 acts as a prodrug, being rapidly metabolized into the exo-benzylidene metabolite A-132 (**27**) but not into chartreusin.[142,143] A Phase I study gave the drug for 5 consecutive days, and derived a maximum tolerated dose of 525–700 mg m^{-2}. A follow-up Phase II trial in 18 patients with breast cancer used oral dosing at 280 or 525 mg m^{-2} once daily for 5 days, and measured the plasma concentrations of the drug and its main metabolites. The major toxicity was myelosuppression, but therapeutic effects were not reported.[144]

7.05.3.5.3 Elinafide

Elinafide (LU-79553) (**28**) is a symmetric dimeric bis(naphthalimide),[145] and an example of a broader class of compounds comprised of neutral chromophores joined by cationic linker chains.[146] Elinafide is reported as a DNA bis-intercalator, suggested to bind via the major groove.[147] A nuclear magnetic resonance (NMR) structure of elinafide bound to the oligodeoxynucleotide duplex $d(ATGCAT)_2$ showed that the two naphthalimide chromophores bis-intercalate at the TpG and CpA steps of the DNA. The linker chain lies in the major groove, with the two amino groups H-bonded to the guanine bases. The naphthalimide rings exchange by rotational ring flipping (at 1800 s^{-1} at 36 °C), without affecting the binding of the linker chain region.[148] Elinafide proved highly cytotoxic in vitro (IC_{50} from 2×10^{-7} to 5×10^{-10} M), and very active in a series of human xenograft models in nude mice, achieving complete or partial regressions in LX-1, CX-1 (colon), DLD (colon), and LOX (melanoma) xenografts.[149] A Phase I trial of elinafide given intravenously for 5 days every 3 weeks established the maximum tolerated dose as 18 mg m^{-2}, with the major toxicity being a cumulative muscular toxicity. The cumulative nature of the principal toxicities suggested that rigorous, long-term toxicological monitoring is required.[150] A second Phase I trial also found dose-limiting neuromuscular toxicity,[151] which may limit further development.

7.05.4 Dual Topo I/II Inhibitors

A number of drugs have been identified that simultaneously target both topo I and topo II.[10] Interest in these compounds is based on observations that the timing of expression of topo I and topo II differs, with topo II levels at their highest during S phase, while levels of topo I remain relatively constant through the cell cycle.[152] Cellular resistance to topo I inhibitors is often accompanied by a concomitant rise in the level of topo II and vice versa, since expression of either enzyme appears to be sufficient to support cell division.[153] While it is still not clear how viable this concept is, some of these compounds show high activity in preclinical tests, and in some cases suggestions of novel mechanisms of action, although it is difficult to decide whether both activities contribute to the cytotoxicity.[10]

7.05.4.1 Camptothecins

7.05.4.1.1 BN-80927

BN-80927 (**29**) is a homocamptothecin and a potent topo I poison, but also a catalytic inhibitor of topo II.[154,155] It has excellent cytotoxicity against a number of human tumor cell lines in culture (e.g., IC_{50} of 6.6 and 3 nM in HT29 colon and DU-145 prostate cell lines respectively).[156] BN-80927 was more potent than SN38 (**6**) in both wild-type and camptothecin-resistant tumor cell lines, and was highly effective in tumor xenografts of the human androgen-independent prostate tumors PC-3 and DU-145; clinical trials have been recommended[155] but are not yet reported.

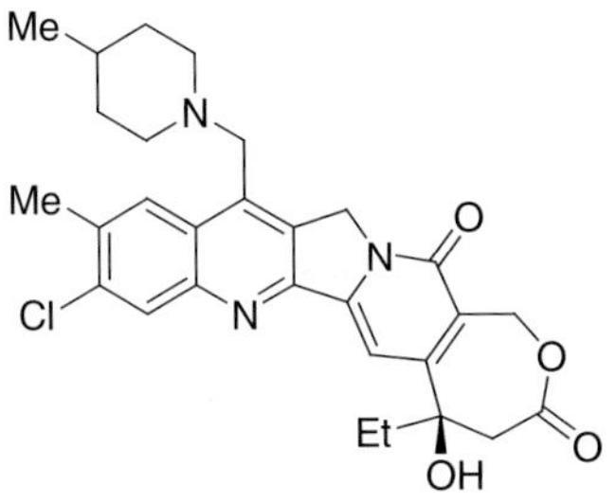

29 BN-80927

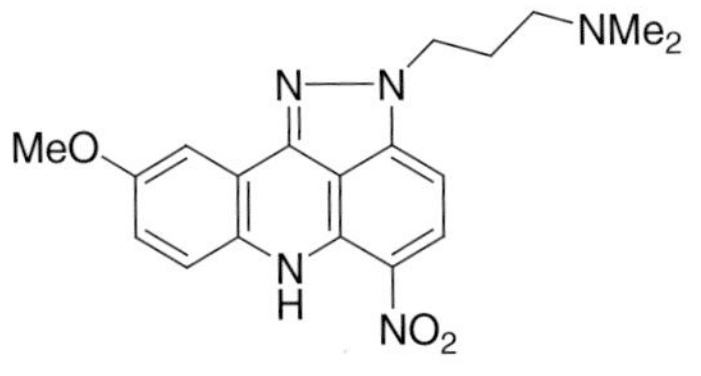

30 Pyrazoloacridine

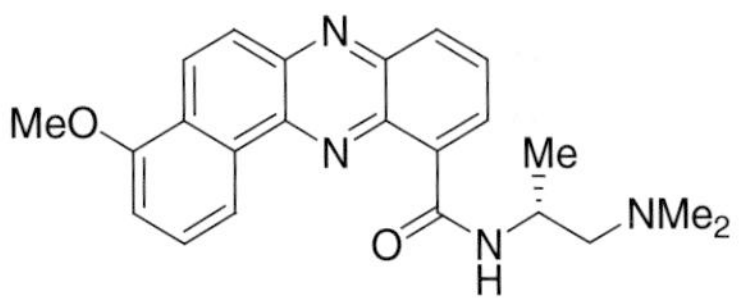

31 XR-11576

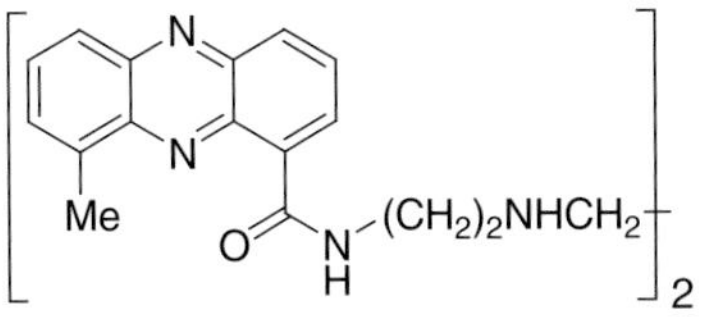

32 MLN-944

7.05.4.2 Deoxyribonucleic Acid Intercalators

7.05.4.2.1 Pyrazoloacridine

Pyrazoloacridine (PZA, PD-115934) (**30**) came from a program evaluating polycyclic heterocycles as DNA intercalating agents, and was selected for its outstanding activity in solid tumor compared with leukemia models.[157] It is a potent inhibitor of the catalytic activity of both topo I and topo II enzymes,[158] but without stabilization of the topo–DNA cleavable complexes,[159] suggesting an unusual mechanism of action. It is also a hypoxia-selective agent,[160] with bioreduction of the 5-nitro group likely to form transient alkylating species. In support of the latter point, the corresponding 5-amino derivative is a major metabolite in mice.[161] The three main oxidative metabolites of PZA in mice were 9-desmethyl-PZA, *N*-demethyl-PZA, and PZA *N*-oxide; studies with a panel of cloned enzymes and inhibitors showed that 9-desmethyl-PZA was largely produced by CYP1A2, *N*-demethyl-PZA formation by CYP3A4, and PZA *N*-oxide by flavin monooxygenase. PZA *N*-oxide and *N*-demethyl-PZA were detected in urine from patients after PZA administration.[162] Pyrazoloacridine has received extensive clinical trial because of its broad-spectrum activity in animal solid tumor models,[160] and its novel mechanism of action, but was inactive in many Phase II trials.[163] More recently, PZA was shown to be highly cytotoxic in a series of multidrug-resistant neuroblastoma cell lines, suggesting it may be effective clinically against neuroblastoma.[164] However, a recent Phase II study of a PZA/carboplatin combination in patients with recurrent glioma showed only modest activity, with short-term disease stabilization in about 38% of the patients.[165]

7.05.4.2.2 XR-11576

XR-11576 (**31**) is a lipophilic benzo[*a*]phenazine derivative, developed from earlier work on benzophenazinecarboxamide intercalators.[166] Detailed SAR studies[167] established the 4-methoxybenz[*a*]phenazine-11-carboxamide series as the most potent, and led to selection of the *R*-methyl enantiomer XR-11576 for clinical development (the first chiral synthetic DNA intercalating agent to be brought to clinical trial). It is a dual inhibitor of topo I and IIα in enzyme assays, inducing cleavable complex formation by both at concentrations as low as 30 nM.[168] XR-11576 is a potent cytotoxin in a range of murine and human cell lines, with the activity being unaffected by either transport resistance mechanisms or by atypical drug resistance generated by low topo IIα levels.[168] It showed significant and comparable activity against subcutaneous HT29 human colon tumor xenografts in mice using either intravenous (52.5 mg kg^{-1}) or oral (75 mg kg^{-1}) dosing on days 1–5, repeating every 21 days.[168] A Phase I trial giving drug orally on days 1–5 every 3 weeks identified the maximum tolerated dose as 180 mg per day, with diarrhea and fatigue as the major toxicities, although extensive support antiemetic therapy was also needed.[169]

7.05.4.2.3 MLN-944

MLN-944 (**32**) is in the same broad structural class as elinafide (see above). It binds tightly ($K_a = 1.6 \times 10^9\,M^{-1}$) to DNA by intercalation, preferentially to GC-rich regions, suggesting a binding mode with the linker chain lying in the major groove.[170] This was confirmed by an NMR structure of MLN-944 complexed with the DNA duplex d(ATGCAT)$_2$ showing the two phenazine rings intercalate at the 5′-TpG sites, with the linker chain lying in the major groove of DNA.[171] It was the most potent (IC_{50} of 0.08 nM in human leukemia cells) of a series of compounds reported to be dual topo I/II inhibitors and potent cytotoxins,[172] and was shown to stabilize cleavable complex formation between DNA and both topo I and human topo IIα, with fragmentation patterns different to that generated by the specific inhibitors etoposide and camptothecin.[173] Later electrophoretic gel mobility shift studies showed that MLN-944 significantly inhibited c-Jun DNA binding to the AP-1 site.[171] Studies in synchronized human HCT 116 cells showed MLN-944 induced G1 and G2 arrest, unlike the typical G2–M arrest noted with known topoisomerase poisons. Transcriptional profiling analysis of treated xenografts showed clusters of regulated genes distinct from those observed in irinotecan-treated tumors. This suggested that the primary mechanism does involve DNA binding, but not topoisomerase inhibition.[174] More generally, MLN-944 inhibited transcription initiation of all RNA polymerases, as well as inhibiting transcription elongation at higher concentrations, suggesting that transcription is the primary target, and is the reason for its high cytotoxicity.[175] In vivo, MLN-944 induced regressions of both HT29 and H69/P xenografts, inducing 100% cures in the latter model at an intravenous dose of 5 mg kg^{-1} on a daily schedule.[173] Phase I clinical trials are in progress, but have not been formally reported on.

7.05.4.3 Epipodophyllotoxins

7.05.4.3.1 Tafluposide

Tafluposide (F11872) (**33**) is a novel phosphate prodrug of a lipophilic, perfluorinated epipodophyllotoxin, and is an example where significant modification of a topo II inhibitor has altered its spectrum of enzyme activity. Tafluposide has superior antitumor activity in vivo compared to etoposide.[176] Although it does not inhibit the religation step of the catalytic cycle of either topo I or topo II, it is a potent inhibitor of the catalytic activities of both enzymes. It does not bind to DNA, but inhibits the binding of the enzymes to DNA in a drug- and enzyme-dependent manner[177] and possibly represents a new class of topoisomerase agent. A resistant P388 leukemia subline retained marked collateral sensitivity to cisplatin, topotecan, colchicine, and Vinca alkaloids, with no overexpression of resistance-related proteins or modification of the glutathione-mediated detoxification process. However, nucleotide excision repair activity was decreased threefold, suggesting that both topoisomerase IIα and DNA repair enzymes are major targets.[178] Tafluposide showed exceptionally high in vivo activity across a variety of tumor models, including P388 leukemia and B16 melanoma, and MX-1 and LX-1 xenografts, with an unusually broad therapeutic range.[179] It shows synergistic effects in the A549 human nonsmall cell lung cancer line in culture with cisplatin, etoposide, doxorubicin, and mitomycin C,[180] and activity against early passage cell lines from cancer patients.[181] The drug has been advanced to clinical trial, but results have not yet been reported.

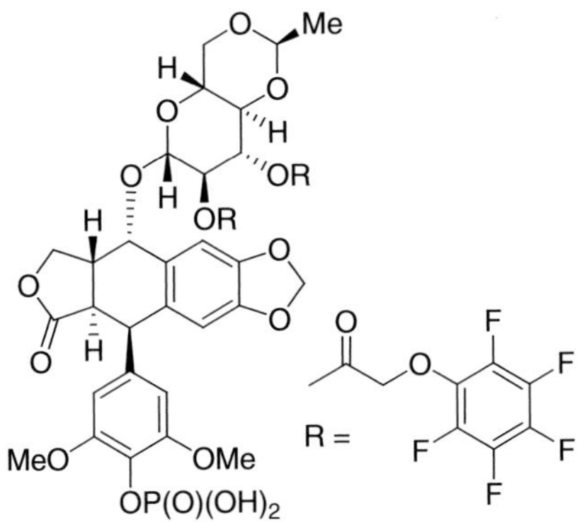

33 Tafluposide

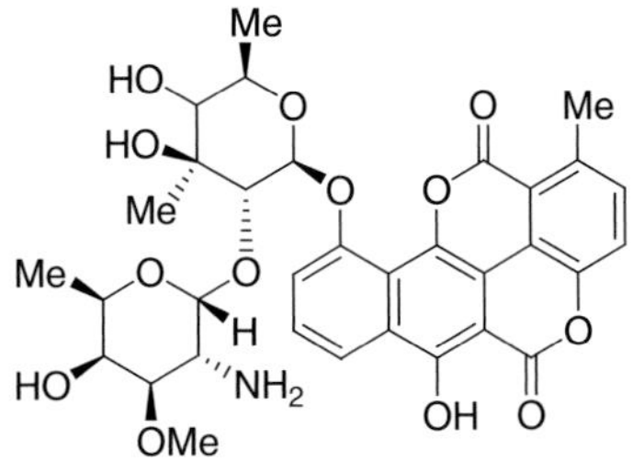

34 Elsamitrucin

7.05.4.4 Chartreusin Analogues

7.05.4.4.1 Elsamitrucin

Elsamitrucin (elsamicin A) (**34**) is an antitumor antibiotic isolated from *Actinomycete* strain J-1907-21. It is related to chartreusin and to the topo II inhibitor XL-119 (*see* Section 7.05.3.5.1), but has a water-solubilizing amino sugar as part of the disaccharide side chain. Elsamitrucin generates single strand breaks at 5′-GG sites by hydroxyl radical-induced oxidative cleavage.[182] It is a potent inhibitor of topo II (IC_{50} 0.4 μM),[183] and while it does not inhibit topo I by trapping of covalent DNA–enzyme cleavage complexes, it may do so simply by its tight binding to DNA.[184] It also binds tightly to the P1 and P2 promoter regions of the *c-myc* oncogene and inhibits the binding of the Sp1 transcription factor.[165] As a cytotoxin it is 10–15 times more potent than chartreusin,[185] and has in vivo antitumor activity in a wide range of in vivo models, both murine and human.[186] In early Phase I trials the major toxicity of elsamitrucin was established as reversible disruption of liver enzyme function, rather than the more usual myelosuppression. In Phase II trials employing a dose of 25 mg m^{-2} weekly, in patients with advanced solid tumors[187] and with nonsmall cell lung cancer,[188] no responses were seen.

7.05.5 Conclusions

Although the overall mechanism of action of topoisomerase inhibitors is well understood, there are yet few molecular-level details of the critical enzyme–drug–DNA ternary complexes that are formed. It is thus difficult to apply modern structural biology tools to the design of better agents. Nevertheless, this review shows that there is a vigorous research program into the development of new topoisomerase agents. This takes two general approaches.

In the absence of detailed structural biology, a major effort has gone into improving the broad deficiencies of established agents. In the topo I inhibitor class, several camptothecin analogs (exatecan, gimatecan, karenitecin) have lipophilic 7-substituents, providing compounds with enhanced membrane permeability as a way of counteracting cell efflux mechanisms. This may also provide better lactone stability (a major deficiency of camptothecin), as seen with karenitecin, by altering drug disposition. A more direct approach to the latter issue is seen in the homocamptothecin diflomotecan, where the 7-membered lactone ring is chemically more stable. All of these compounds show equal or superior potency to camptothecin, together with broad-spectrum activity and improved resistance to cell efflux mechanisms, and constitute an exciting set of potentially improved agents. For topo II inhibitors, a major drug design target is the cumulative cardiotoxicity shown by the key drug doxorubicin. The anthracycline/anthraquinone analogs discussed here address this in different ways, through more rapid extravascular diffusion and cell uptake (nemorubicin), differential tissue accumulation of the cardiotoxic alcohol metabolite (amrubicin), or lesser ability of this metabolite to bind to the critical enzyme (MEN-10755). Another issue, especially for the DNA-intercalating topo II agents, is to decrease their susceptibility to cell efflux pumps. This is not addressed as specifically by these compounds, with MEN-10755 in particular still susceptible to transport-mediated resistance.

The second major approach is to seek compounds with novel structures and mechanisms of action, although this is less apparent in the topo I field, where only the indolocarbazole edotecarin is reported on. This compound is more stable than previous indolocarbazole analogs, and active in P-gp-expressing cell lines. Overall, there is a much wider range of structures in the topo II inhibitor class, and many of these have the advantage of being generally not cardiotoxic, and usually lipophilic enough to avoid transport-mediated resistance (e.g., the pyridocarbazole S-16020-2, the pyrazoloacridone KW-2170, the indenoquinoline TAS-103). Many of these compounds appear to inhibit topo II in different ways to the anthracyclines, and some may act by additional mechanisms (e.g., selective accumulation in lung tissue via phosphatidylserine binding by the podophyllotoxin derivative TOP-53, hydroxyl radical-induced oxidative cleavage of DNA by the chartreusin analog IST-622). In general, more work on the mechanisms of action of these compounds would be beneficial.

The dual topo I/II inhibitors represent the newest class of topoisomerase-active drugs. While it is still not clear how viable the original rationale (resistance to one class of inhibitors resulting in an increase in level of the other enzyme) is for seeking such compounds, the class has given rise to some interesting examples. The homocamptothecin BN-80927, the etoposide tafluposide, and the chartreusin analog elsamitrucin pose interesting questions about the properties needed to convert a topo I or II inhibitor to a dual one, and vice versa. More work is needed to explore this area. Some of the compounds (e.g., tafluposide, elsamitrucin, XR-11576) appear to inhibit topoisomerase in a novel manner, and some (e.g., MLN-944) may in fact work primarily by nontopo mechanisms.

Overall, the work reviewed here indicates that the field is still developing vigorously, with a variety of specific new drug design approaches being followed.

References

1. Liu, L. F. *Annu. Rev. Biochem.* **1989**, *58*, 351–375.
2. Chung, T. D.; Drake, F. H.; Tan, K. B.; Per, S. R.; Crooke, S. T.; Mirabelli, C. K. *Proc. Natl. Acad. Sci. USA* **1989**, *86*, 9431–9435.
3. Tan, K. B.; Dorman, T. F.; Falls, K. M.; Chung, T. D. Y.; Mirabelli, C. K.; Crooke, S. T.; Mao, J. *Cancer Res.* **1992**, *52*, 231–234.
4. Drake, F. H.; Hofmann, G. A.; Bartus, H. F.; Mattern, M. R.; Crooke, S. T.; Mirabelli, C. K. *Biochemistry* **1989**, *28*, 8154–8160.
5. Boonsong, A.; Marsh, S.; Rooney, P. H.; Stevenson, D. A. J.; Cassidy, J.; McLeod, H. L. *Cancer Genet. Cytogenet.* **2000**, *121*, 56–60.
6. Stevnsner, T.; Mortensen, U. H.; Westergaard, O.; Bonven, B. J. *J. Biol. Chem.* **1989**, *264*, 10110–10113.
7. Berger, J. M.; Gamblin, S. J.; Harrison, S. C.; Wang, J. C. *Nature* **1996**, *379*, 225–232.
8. Osheroff, N. *J. Biol. Chem.* **1986**, *261*, 9944–9950.
9. Lindsley, J. E.; Wang, J. C. *Proc. Natl. Acad. Sci. USA* **1991**, *88*, 10485–10489.
10. Denny, W. A.; Baguley, B. C. *Curr. Top. Med. Chem.* **2003**, *3*, 339–353.
11. Lerman, L. S. *J. Mol. Biol.* **1961**, *3*, 18–30.
12. Wadkins, R. M.; Graves, D. E. *Biochemistry* **1991**, *30*, 4277–4283.
13. Denny, W. A. *Anti-Cancer Drug Des.* **1989**, *4*, 241–263.
14. Wall, M. E. *Med. Res. Rev.* **1998**, *18*, 299–314.
15. Thomas, C. J.; Rahier, N. J.; Hecht, S. M. *Bioorg. Med. Chem. Lett.* **2004**, *12*, 1585–1604.
16. Pizzolato, J. F.; Saltz, L. B. *Lancet* **2003**, *361*, 2235–2242.
17. Wani, M. C.; Nicholas, A. W.; Wall, M. E. *J. Med. Chem.* **1986**, *29*, 2358–2363.
18. Puri, S. C.; Handa, G.; Suri, O. P.; Qazi, G. N. *Synth. Commun.* **2004**, *34*, 3443–3448.
19. Zamboni, W. C.; Jung, L. L.; Egorin, M. J.; Hamburger, D. R.; Joseph, E.; Jin, R.; Strychor, S.; Ramanathan, R. K.; Eiseman, J. L. *Clin. Cancer Res.* **2005**, *11*, 4867–4874.
20. Li, K.; Chen, X.; Zhong, D.; Li, Y. *Drug Metab. Dispos.* **2003**, *31*, 792–797.
21. Jung, L. L.; Ramanathan, R. K.; Egorin, M. J.; Jin, R.; Belani, C. P.; Potter, D. M.; Strychor, S.; Trump, D. L.; Walko, C.; Fakih, M. et al. *Cancer Chemother. Pharmacol.* **2004**, *54*, 487–496.
22. Takimoto, C. H.; Thomas, R. *Ann. NY Acad. Sci.* **2000**, *922*, 224–236.
23. Chatterjee, D.; Wyche, J. H.; Pantazis, P. *Anticancer Res.* **2000**, *20*, 4477–4482.
24. Rajendra, R.; Gounder, M. K.; Saleem, A.; Schellens, J. H. M.; Ross, D. D.; Bates, S. E.; Sinko, P.; Rubin, E. H. *Cancer Res.* **2003**, *63*, 3228–3233.
25. Giovanella, B. C.; Stehlin, J. S.; Hinz, H. R.; Kozielski, A. J.; Harris, N. J.; Vardeman, D. M. *Int. J. Oncol.* **2002**, *20*, 81–88.
26. Miller, K. D.; Soule, S. E.; Haney, L. G.; Guiney, P.; Murry, D. J.; Lenaz, L.; Sledge, S.-L.; Sun, G. W. *Invest. New Drugs* **2004**, *22*, 69–73.
27. Punt, C. J. A.; de Jonge, M. J. A.; Monfardini, S.; Daugaard, G.; Fiedler, W.; Baron, B.; Lacombe, D.; Fumoleau, P. *Eur. J. Cancer* **2004**, *40*, 1332–1334.
28. Patel, S. R.; Beach, J.; Papadopoulos, N.; Burgess, M. A.; Trent, J.; Jenkins, J.; Benjamin, R. S. *Cancer* **2003**, *97*, 2848–2852.
29. Raymond, E.; Campone, M.; Stupp, R.; Menten, J.; Chollet, P.; Lesimple, T.; Fety-Deporte, R.; Lacombe, D.; Paoletti, X.; Fumoleau, P. *Eur. J. Cancer* **2002**, *38*, 1348–1350.
30. Schoffski, P.; Herr, A.; Vermorken, J. B.; Van den Brande, J.; Beijnen, J. H.; Rosing, H.; Volk, J.; Ganser, A.; Adank, S.; Botma, H. J. et al. *Eur. J. Cancer* **2002**, *38*, 807–813.
31. Baka, S.; Ranson, M.; Lorigan, P.; Danson, S.; Linton, K.; Hoogendam, I.; Mettinger, K.; Thatcher, N. *Eur. J. Cancer* **2005**, *41*, 1547–1550.
32. Burris, H. A.; Rivkin, S.; Reynolds, R.; Harris, J.; Wax, A.; Gerstein, H.; Mettinger, K. L.; Staddon, A. *Oncologist* **2005**, *10*, 183–190.
33. Koshkina, N. V.; Kleinerman, E. S.; Waldrep, C.; Jia, S.-F.; Worth, L. L.; Gilbert, B. E.; Knight, V. *Clin. Cancer Res.* **2000**, *6*, 2876–2880.
34. Verschraegen, C. F.; Gilbert, B. E.; Loyer, E.; Huaringa, A.; Walsh, G.; Newman, R. A.; Knight, V. *Clin. Cancer Res.* **2004**, *10*, 2319–2326.
35. Jew, S.-S.; Kim, M. G.; Kim, H.-J.; Rho, E.-Y.; Park, H.-G.; Kim, J.-K.; Han, H.-J.; Lee, H. *Bioorg. Med. Chem. Lett.* **1998**, *8*, 1797–1800.
36. Lee, J. H.; Lee, J. M.; Kim, J. K.; Ahn, S. K.; Lee, S. J.; Kim, M. Y.; Jew, S. S.; Park, J. G.; Hong, C. I. *Arch. Pharm. Res.* **1998**, *21*, 581–590.
37. Ha, K. W.; Oh, H. Y.; Heo, O. S.; Park, C. H.; Sohn, S. J.; Han, E. S.; Kim, J. W.; Kang, I. H.; Kang, H. J.; Lee, S. J. et al. *Env. Mutagens Carcinogens* **1998**, *18*, 129–134.

38. Song, Y.; Seo, S.-S.; Bang, Y.-J.; Kang, S.-B.; Nam, J.-H.; Ryu, S.-Y.; Lee, K.-H.; Park, S.-Y.; Hong, C.-I.; Lee, H.-P. *Proc. Am. Soc. Clin. Oncol.* **2003**, *22*, 467, Abstr. 1877.
39. Bom, D.; Curran, D. P.; Chavan, A. J.; Kruszewski, S.; Zimmer, S. G.; Fraley, K. A.; Burke, T. G. *J. Med. Chem.* **1999**, *42*, 3018–3022.
40. Kumazawa, E.; Tohgo, A. *Expert Opin. Investig. Drugs* **1998**, *7*, 625–632.
41. Kumazawa, E.; Jimbo, T.; Ochi, Y.; Tohgo, A. *Cancer Chemother. Pharmacol.* **1998**, *42*, 210–220.
42. Joto, N.; Ishii, M.; Minami, M.; Kuga, H.; Mitsui, I.; Tohgo, A. *Int. J. Cancer* **1997**, *72*, 680–686.
43. Nomoto, T.; Nishio, K.; Ishida, T.; Mori, M.; Saijo, N. *Jpn. J. Cancer Res.* **1998**, *89*, 1179–1186.
44. Chilman-Blair, K.; Mealy, N. E.; Castaner, J.; Bayes, M. *Drugs Future* **2004**, *29*, 9–22.
45. Dallavalle, S.; Ferrari, A.; Merlini, L.; Penco, S.; Carenini, N.; De Cesare, M.; Perego, P.; Pratesi, G.; Zunino, F. *Bioorg. Med. Chem. Lett.* **2001**, *11*, 291–294.
46. Croce, A. C.; Bottiroli, G.; Supino, R.; Favini, E.; Zuco, V.; Zunino, F. *Biochem. Pharm.* **2004**, *67*, 1035–1045.
47. Di Francesco, A. M.; Riccardi, A. S.; Barone, G.; Rutella, S.; Meco, D.; Frapolli, R.; Zucchetti, M.; D́Incalci, M.; Pisano, C.; Carminati, P. et al. *Biochem. Pharmacol.* **2005**, *70*, 1125–1136.
48. Pratesi, G.; Beretta, G. L.; Zunino, F. *Anti-Cancer Drugs* **2004**, *15*, 545–552.
49. Van Hattum, A. H.; Pinedo, H. M.; Schluper, H. M.; Hausheer, F. H.; Boven, E. *Int. J. Cancer* **2000**, *88*, 260–266.
50. Van Hattum, A. H.; Schluper, H. M.; Hausheer, F. H.; Pinedo, H. M.; Boven, E. *Int. J. Cancer* **2002**, *100*, 22–29.
51. Yin Ming, B.; Hapke, G.; Wu, J.; Azrak, R.; Frank, C.; Wrzosek, C.; Rustum, Y. M. *Biochem. Biophys. Res. Commun.* **2002**, *295*, 435–444.
52. Smith, J. A.; Newman, R. A.; Hausheer, F. H.; Madden, T. *J. Clin. Pharmacol.* **2003**, *43*, 1008–1014.
53. Keir, S. T.; Hausheer, F.; Lawless, A. A.; Bigner, D. D.; Friedman, H. S. *Cancer Chemother. Pharmacol.* **2001**, *48*, 83–87.
54. Daud, A.; Valkov, N.; Centeno, B.; Derderian, J.; Sullivan, P.; Munster, P.; Urbas, P.; DeConti, R. C.; Berghorn, E.; Liu, Z. et al. *Clin. Cancer Res.* **2005**, *11*, 3009–3016.
55. Gabarda, A. E.; Du, W.; Isarno, T.; Tangirala, R. S.; Curran, D. P. *Tetrahedron* **2002**, *58*, 6329–6341.
56. Lavergne, O.; Demarquay, D.; Bailly, C.; Lanco, C.; Rolland, A.; Huchet, M.; Coulomb, H.; Muller, N.; Baroggi, N.; Camara, J. et al. *J. Med. Chem.* **2000**, *43*, 2285–2289.
57. Lesueur, G. L.; Demarquay, D.; Kiss, R.; Kaspryzk, P. G.; Lavergne, O. *Cancer Res.* **1999**, *59*, 2939–2943.
58. Philippart, P.; Harper, L.; Chaboteaux, C.; Decaestecker, C.; Bronckart, Y.; Gordover, L.; Lesueur-Ginot, L.; Malonne, H.; Lavergne, O.; Bigg, D. C. H. et al. *Clin. Cancer Res.* **2000**, *6*, 1557–1562.
59. Demarquay, D.; Huchet, M.; Coulomb, H.; Lesueur-Ginot, L.; Lavergne, O.; Kasprzyk, P. G.; Bailly, C.; Camara, J.; Bigg, D. C. H. *Anti-Cancer Drugs* **2001**, *12*, 9–19.
60. Sola, J.; Gay-Feutry, C.; Massiere, F.; Peraire, C.; Obach, R.; Principe, P. *Drug Metab. Rev.* **2003**, *35*, 51.
61. Gelderblom, H.; Salazar, R.; Verweij, J.; Pentheroudakis, G.; de Jonge, M. J. A.; Devlin, M.; van Hooije, C.; Seguy, F.; Obach, R.; Prunonosa, J. et al. *Clin. Cancer Res.* **2003**, *9*, 4101–4107.
62. Ohkubo, M.; Nishimura, T.; Honma, T.; Nishimura, I.; Ito, S.; Yoshinari, T.; Suda, H. A.; Morishima, H.; Nishimura, S. *Bioorg. Med. Chem. Lett.* **1999**, *9*, 3307–3312.
63. Akao, A.; Hiraga, S.; Iida, T.; Kamatani, A.; Kawasaki, M.; Mase, T.; Nemoto, T.; Satake, N.; Weissman, S. A.; Tschaen, D. M. et al. *Tetrahedron* **2001**, *57*, 8917–8923.
64. Yoshinari, T.; Ohkubo, M.; Fukasawa, K.; Egashira, S. I.; Hara, Y. *Cancer Res.* **1999**, *59*, 4271–4275.
65. Komatani, H.; Kotani, H.; Hara, Y.; Nakagawa, R.; Matsumoto, M.; Arakawa, H.; Nishimura, S. *Cancer Res.* **2001**, *61*, 2827–2832.
66. Goossens, J.-F.; Kluza, J.; Vezin, H.; Kouach, M.; Briand, G.; Baldeyrou, B.; Wattez, N.; Bailly, C. *Biochem. Pharmacol.* **2003**, *65*, 25–34.
67. Takenaga, N.; Ishii, M.; Kamei, T.; Yasumori, T. *Drug Metab. Dispos.* **2002**, *30*, 494–497.
68. Arakawa, H.; Morita, M.; Kodera, T.; Okura, A.; Ohkubo, M.; Morishima, H.; Nishimura, S. *Jpn. J. Cancer Res.* **1999**, *90*, 1163–1170.
69. Saif, M. W.; Diasio, R. B. *Clin. Colorectal Cancer* **2005**, *5*, 27–36.
70. Minotti, G.; Menna, P.; Salvatorelli, E.; Cairo, G.; Gianni, L. *Pharmacol. Rev.* **2004**, *56*, 185–229.
71. Quintieri, L.; Rosato, A.; Napoli, E.; Sola, F.; Geroni, C.; Floreani, M.; Zanovello, P. *Cancer Res.* **2000**, *60*, 3232–3238.
72. Quintieri, L.; Geroni, C.; Fantin, M.; Battaglia, R.; Rosato, A.; Speed, W.; Zanovello, P.; Floreani, M. *Clin. Cancer Res.* **2005**, *11*, 1608–1617.
73. Grandi, M.; Ballinari, D.; Capolongo, L.; Pastori, A.; Ripamonti, M.; Suarato, A.; Spreafico, F. *Haematologica* **1991**, *76*, 181–183.
74. Ripamonti, M.; Pezzoni, G.; Pesenti, E.; Pastori, A.; Farao, M.; Bargiotti, A.; Suarato, A.; Spreafico, F.; Grandi, M. *Br. J. Cancer* **1992**, *65*, 703–707.
75. Grandi, M.; Pezzoni, G.; Ballinari, D.; Capolongo, L.; Suarato, A.; Bargiotti, A.; Faiardi, D.; Spreafico, F. *Cancer Treat. Rev.* **1990**, *17*, 133–138.
76. Yuan, S.; Zhang, X.; Lu, L.; Xu, C.; Yang, W.; Ding, J. *Anti-Cancer Drugs* **2004**, *15*, 641–646.
77. Vasey, P. A.; Bissett, D.; Strolin-Benedetti, M.; Poggesi, I.; Breda, M.; Adams, L.; Wilson, P.; Pacciarini, M. A.; Kaye, S. B.; Cassidy, J. *Cancer Res.* **1995**, *55*, 2090–2096.
78. Graul, A.; Leeson, P. A.; Castaner, J. *Drugs Future* **1997**, *22*, 1319–1324.
79. Tsujimoto, S.; Satoh, E.; Sugimoto, S.; Katoh, T.; Kaneko, M.; Takada, H.; Katoh, T. *Pharmacometrics* **1996**, *52*, 351–370.
80. Yamaoka, T.; Hanada, M.; Ichii, S.; Morisada, S.; Noguchi, T.; Yanagi, Y. *Jpn. J. Cancer Res.* **1998**, *89*, 1067–1073.
81. Hanada, M.; Mizuno, S.; Fukushima, A.; Saito, Y.; Noguchi, T.; Yamaoka, T. *Jpn. J. Cancer Res.* **1998**, *89*, 1229–1238.
82. Hanada, M.; Noguchi, T.; Murayama, T. *Nippon Yakurigaku Zasshi* **2003**, *122*, 141–150.
83. Ohe, Y.; Nakagawa, K.; Fujiwara, Y.; Sasaki, Y.; Minato, K.; Bungo, M.; Niimi, S.; Horichi, N.; Fukuda, M.; Saijo, N. *Cancer Res.* **1989**, *49*, 4098–4102.
84. Morisada, S.; Yanagi, Y.; Noguchi, T.; Kashiwazaki, Y.; Fukui, M. *Jpn. J. Cancer Res.* **1989**, *80*, 69–76.
85. Noguchi, T.; Ichii, S.; Morisada, S.; Yamaoka, T.; Yanagi, Y. *Jpn. J. Cancer Res.* **1998**, *89*, 1055–1060.
86. Inoue, K.; Ogawa, M.; Horikoshi, N.; Mukaiyama, T.; Itoh, Y.; Imajoh, K.; Ozeki, H.; Nagamine, D.; Shinagawa, K. *Cancer Chemother.* **1988**, *15*, 1771–1776.
87. Sugiura, T.; Ariyoshi, Y.; Negoro, S.; Nakamura, S.; Ikegami, H.; Takada, M.; Yana, T.; Fukuoka, M. *Invest. New Drugs* **2005**, *23*, 331–337.
88. Temperini, C.; Messori, L.; Orioli, P.; Di Bugno, C.; Animati, F.; Ughetto, G. *Nucleic Acids Res.* **2003**, *31*, 1464–1469.
89. Bigioni, M.; Salvatore, C.; Bullo, A.; Bellarosa, D.; Iafrate, E.; Animati, F.; Capranico, G.; Goso, C.; Maggi, C. A. et al. *Biochem. Pharmacol.* **2001**, *62*, 63–70.
90. Sacco, G.; Giampietro, R.; Salvatorelli, E.; Menna, P.; Bertani, N.; Graiani, G.; Animati, F.; Goso, C.; Maggi, C. A.; Manzini, S. et al. *Br. J. Pharmacol.* **2003**, *139*, 641–651.

91. Mintti, G.; Licata, S.; Saponiero, A.; Menna, P.; Calafiore, A. M.; Di Giammarco, G.; Liberi, G.; Animati, F.; Cipollone, A.; Manzini, S. et al. *Chem. Res. Toxicol.* **2000**, *13*, 1336–1341.
92. Bos, A. M. E.; de Vries, E. G. E.; Dombernovsky, P.; Aamdal, S.; Uges, D. R. A.; Schrijvers, D.; Wanders, J.; Roelvink, M. W. J.; Hanauske, A. R.; Bortini, S. et al. *Cancer Chemother. Pharmacol.* **2001**, *48*, 361–369.
93. Pratesi, G.; De Cesare, M.; Caserini, C.; Perego, P.; Dal Bo, L.; Polizzi, D.; Supino, R.; Bigioni, M.; Manzini, S.; Iafrate, E. et al. *Clin. Cancer Res.* **1998**, *4*, 2833–2839.
94. Schrijvers, D.; Bos, A. M. E.; Dyck, J.; de Vries, E. G. E.; Wanders, J.; Roelvink, M.; Fumoleau, P.; Bortini, S.; Vermorken, J. B. *Ann. Oncol.* **2002**, *13*, 385–391.
95. Van der Graaf, W. T. A.; De Vries, E. G. E. *Anti-Cancer Drugs* **1990**, *1*, 109–125.
96. Krapcho, A. P.; Petry, M. E.; Getahun, Z.; Landi, J. J.; Stallman, J.; Polsenberg, J. F.; Gallagher, C. E.; Maresch, M. J.; Hacker, M. P.; Giuliani, F. C. *J. Med. Chem.* **1994**, *37*, 828–837.
97. Beggiolin, G.; Crippa, L.; Menta, E.; Manzotti, C.; Cavalletti, E.; Pezzoni, G.; Torriani, D.; Randisi, E.; Cavagnoli, R.; Sala, F. et al. *Tumori* **2001**, *87*, 407–416.
98. Andoh, T.; Ishida, R. *Biochim. Biophys. Acta–Gene Struct. Expr.* **1998**, *1400*, 155–171.
99. De Isabella, P.; Palumbo, M.; Sissi, C.; Capranico, G.; Carenini, N.; Menta, E.; Oliva, A.; Spinelli, S.; Krapcho, A. P. *Mol. Pharmacol.* **1995**, *48*, 30–38.
100. Borchmann, P.; Schnell, R. *Expert Opin. Investig. Drugs* **2005**, *14*, 1055–1061.
101. Borchmann, P.; Morschhauser, F.; Parry, A.; Schnell, R.; Harousseau, J. L.; Gisselbrecht, C.; Rudolph, C.; Wilhelm, M.; Gunther, H.; Pfreundschuh, D. M. et al. *Haematologica* **2003**, *88*, 888–894.
102. Gonsette, R. E. *J. Neurol. Sci.* **2004**, *223*, 81–86.
103. Mimura, T.; Kato, N.; Sugaya, T.; Ikuta, M.; Kato, S.; Kuge, Y.; Tomioka, S.; Kasai, M. *Synthesis* **1999**, *6*, 947–952.
104. Ashizawa, T.; Shimizu, M.; Gomi, K.; Okabe, M. *Anti-Cancer Drugs* **1998**, *9*, 263–271.
105. Verschraegen, C. F. *IDrugs* **2002**, *5*, 1000–1003.
106. Saeki, T.; Eguchi, K.; Takashima, S.; Sugiura, T.; Hida, T.; Horikoshi, N.; Aiba, K.; Kuwabara, T.; Ogawa, M. *Cancer Chemother. Pharmacol.* **2004**, *54*, 459–468.
107. Baguley, B. C. *Drugs Today* **1984**, *20*, 237–245.
108. Wadkins, R. M.; Graves, D. E. *Nucleic Acids Res.* **1989**, *17*, 9933–9946.
109. Nelson, E. M.; Tewey, K. M.; Liu, L. F. *Proc. Natl. Acad. Sci. USA* **1984**, *81*, 1361–1365.
110. Wilson, W. R.; Baguley, B. C.; Wakelin, L. P. G.; Waring, M. J. *Mol. Pharmacol.* **1981**, *20*, 404–414.
111. Sorensen, M.; Sehested, M.; Jensen, P. B. *Mol. Pharmacol.* **1999**, *55*, 424–431.
112. Grove, W. R.; Fortner, C. L.; Wiernik, P. H. *Clin. Pharm.* **1982**, *1*, 320–326.
113. Steuber, C. P.; Krischer, J.; Holbrook, T.; Camitta, B.; Land, V.; Sexauer, C.; Mahoney, D.; Weinstein, H. *J. Clin. Oncol.* **1996**, *14*, 1521–1525.
114. Sung, W. J.; Kim, D. H.; Sohn, S. K.; Kim, J. G.; Baek, J. H.; Jeon, S. B.; Moon, J. H.; Ahn, B. M.; Lee, K. B. *Jpn. J. Clin. Oncol.* **2005**, *35*, 612–616.
115. Jaszfold-Howorko, R.; Landras, C.; Pierre, A.; Atassi, G.; Guilbaud, N.; Kraus-Berthier, L.; Leonce, S.; Rolland, Y.; Prost, J.-F.; Bisagni, E. *J. Med. Chem.* **1994**, *37*, 2445–2452.
116. Le Mees, P. A.; Markovits, J.; Atassi, G.; Jacquemin-Sablon, A.; Saucier, J. M. *Mol. Pharmacol.* **1998**, *53*, 213–220.
117. Pierre, A.; Leonce, S.; Perez, V.; Atassi, G. *Cancer Chemother. Pharmacol.* **1998**, *42*, 454–460.
118. Guilbaud, N.; Kraus-Berthier, L.; Saint-Dizier, D. *Cancer Chemother. Pharmacol.* **1996**, *38*, 513–521.
119. Kraus-Berthier, L.; Guilbaud, N.; Jan, M.; Saint-Dizier, D.; Rouillon, M. H.; Burbridge, M. F.; Pierre, A.; Atassi, G. *Eur. J. Cancer* **1997**, *33*, 1881–1887.
120. Guilbaud, N.; Kraus-Berthier, L.; Saint-Dizier, D.; Rouillon, M.-H.; Jan, M.; Burbridge, M.; Pierre, A.; Atassi, G. *Anti-Cancer Drugs* **1997**, *8*, 276–282.
121. Fortune, J. M.; Velea, L.; Graves, D. E.; Utsugi, T.; Yamada, Y.; Osheroff, N. *Biochemistry* **1999**, *38*, 15580–15586.
122. Byl, J. A.; Fortune, J. M.; Burden, D. A.; Nitiss, J. L.; Utsugi, T.; Yamada, Y.; Osheroff, N. *Biochemistry* **1999**, *38*, 15573–15579.
123. Ohyama, T.; Li, Y.; Utsugi, T.; Irie, S.; Yamada, Y.; Sato, T.-A. *Jpn. J. Cancer Res.* **1999**, *90*, 691–698.
124. Utsugi, T.; Aoyagi, K.; Asao, T.; Okazaki, S.; Aoyagi, Y.; Sano, M.; Wierzba, K.; Yamada, Y. *Jpn. J. Cancer Res.* **1997**, *88*, 992–1002.
125. Aoyagi, Y.; Kobunai, T.; Utsugi, T.; Wierzba, K.; Yamada, Y. *Jpn. J. Cancer Res.* **2000**, *91*, 543–550.
126. Azuma, R.; Saeki, M.; Yamamoto, Y.; Hagiwara, Y.; Growchow, L. B.; Donehower, R. C. *Xenobiotica* **2002**, *32*, 63–72.
127. Ewesuedo, R. B.; Iyer, L.; Das, S.; Koenig, A.; Mani, S.; Vogelzang, N. J.; Schilsky, R. L.; Brenckman, W.; Ratain, M. J. *J. Clin. Oncol.* **2001**, *19*, 2084–2090.
128. Baldwin, E. L.; Osheroff, N. *Curr. Med. Chem. Anti-Cancer Agents* **2005**, *5*, 363–372.
129. Kettenes-Van den Bosch, J. J.; Holthuis, J. J. M.; Bult, A. *Anal. Profiles Drug Subst.* **1990**, *19*, 575–600.
130. Yoshida, M.; Kobunai, T.; Aoyagi, K.; Saito, H.; Utsugi, T.; Wierzba, K.; Yamada, Y. *Clin. Cancer Res.* **2000**, *6*, 4396–4401.
131. Byl, J.; Cline, S. D.; Utsugi, T.; Kobanai, T.; Yamada, Y.; Osheroff, N. *Biochemistry* **2001**, *40*, 712–718.
132. Utsugi, T.; Shibata, J.; Sugimoto, Y. *Cancer Res.* **1996**, *56*, 2809–2814.
133. Prudhomme, M. *Curr. Med. Chem.* **2000**, 7, 1189–1212.
134. Kaneko, T.; Wong, H.; Utzig, J.; Schuring, J.; Doyle, T. W. *J. Antibiot.* **1990**, *43*, 125–127.
135. Long, B. H.; Rose, W. C.; Vyas, D. M.; Matson, J. A.; Forenza, S. *Curr. Med. Chem. Anti-Cancer Agents* **2002**, *2*, 255–266.
136. Weitman, S.; Moore, R.; Barrera, H.; Cheung, N. K.; Izbicka, E.; Von Hoff, D. D. *J. Pediatr. Hematol. Oncol.* **1998**, *20*, 136–139.
137. Tolcher, A. W.; Eckhardt, S. G.; Kuhn, J.; Hammond, L.; Weiss, G.; Rizzo, J.; Aylesworth, C.; Hidalgo, M.; Patnaik, A.; Schwartz, G. et al. *J. Clin. Oncol.* **2001**, *19*, 2937–2947.
138. Dowlati, A.; Hoppel, C. L.; Ingalls, S. T.; Majka, S.; Li, X.; Sedransk, N.; Spiro, T.; Gerson, S. L.; Ivy, P.; Remick, S. C. *J. Clin. Oncol.* **2001**, *19*, 2309–2318.
139. Hussain, M.; Vaishampayan, U.; Heilbrun, L. K.; Jain, V.; LoRusso, P. M.; Ivy, P.; Flaherty, L. *Invest. New Drugs* **2003**, *21*, 465–471.
140. Goel, S.; Wadler, S.; Hoffman, A.; Volterra, F.; Baker, C.; Nazario, E.; Ivy, P.; Silverman, A.; Mani, S. *Invest. New Drugs* **2003**, *21*, 103–107.

141. Kon, K.; Sugi, H.; Tamai, K.; Ueda, Y.; Yamada, N. *J. Antibiot.* **1990**, *43*, 372–382.
142. Tashiro, T.; Kon, K.; Yamamoto, M.; Yamada, N.; Tsuruo, T.; Tsukagoshi, S. *Cancer Chemother. Pharmacol.* **1994**, *34*, 287–292.
143. Portugal, J. *Curr. Med. Chem. Anti-Cancer Agents* **2003**, *3*, 411–420.
144. Asai, G.; Yamamoto, N.; Toi, M.; Shin, E.; Nishiyama, K.; Sekine, T.; Nomura, Y.; Takashima, S.; Kimura, M.; Tominaga, T. *Cancer Chemother. Pharmacol.* **2002**, *49*, 468–472.
145. Brana, M. F.; Ramos, A. *Curr. Med. Chem. Anti-Cancer Agents* **2001**, *1*, 237–255.
146. Spicer, J. A.; Gamage, S. A.; Atwell, G. J.; Finlay, G. J.; Baguley, B. C.; Denny, W. A. *Anti-Cancer Drug Des.* **1999**, *14*, 281–289.
147. Brana, M.; Waring, M. J.; Bailly, C. *Eur. J. Biochem.* **1996**, *240*, 195–208.
148. Gallego, J.; Reid, B. R. *Biochemistry* **1999**, *38*, 15104–15115.
149. Bousquet, P. F.; Brana, M. F.; Conlon, D.; Fitzgerald, K. M.; Perron, D.; Cocchiaro, C.; Miller, R.; Moran, M.; George, J. *Cancer Res.* **1995**, *55*, 1176–1180.
150. Villalona-Calero, M. A.; Eder, J. P.; Toppmeyer, D. L.; Allen, L. F.; Fram, R.; Velagapudi, R.; Myers, M.; Amato, A.; Kagen-Hallet, K.; Razvillas, B. et al. *J. Clin. Oncol.* **2001**, *19*, 857–869.
151. Awada, A.; Thoedtmann, R.; Piccart, M. J.; Wanders, J.; Schrijvers, A. H. G. J.; Von Broen, I.-M.; Hanauske, A. R. *Eur. J. Cancer* **2003**, *39*, 742–747.
152. Heck, M. M. S.; Hittelman, W. N.; Earnshaw, W. C. V. *Proc. Natl. Acad. Sci. USA* **1988**, *85*, 1086–1090.
153. Whitacre, C. M.; Zborowska, E.; Gordon, N. H.; Mackay, W.; Berger, N. A. *Cancer Res.* **1997**, *57*, 1425–1428.
154. Huchet, M.; Demarquay, D.; Coulomb, H.; Kasprzyk, P.; Carlson, M.; Lauer, J.; Lavergne, O.; Bigg, D. *Ann. NY Acad. Sci.* **2000**, *922*, 303–305.
155. Demarquay, D.; Huchet, M.; Coulomb, H.; Lesueur-Ginot, L.; Lavergne, O.; Camara, J.; Kasprzyk, P. G.; Prevost, G.; Bigg, D. C. H. *Cancer Res.* **2004**, *64*, 4942–4949.
156. Lavergne, O.; Harnett, J.; Rolland, A.; Lanco, C.; Lesueur-Ginot, L.; Demarquay, D.; Huchet, M.; Coulomb, H.; Bigg, D. C. *Bioorg. Med. Chem. Lett.* **1999**, *9*, 2599–2602.
157. Sebolt, J. S.; Scavone, S. V.; Pinter, C. D.; Hamelehle, K. L.; Von Hoff, D. D.; Jackson, R. C. *Cancer Res.* **1987**, *47*, 4299–4304.
158. Adjei, A. A.; Charron, M.; Rowinsky, E. K.; Svingen, P. A.; Miller, J.; Reid, J. M.; Sebolt-Leopold, J.; Ames, M. M.; Kaufmann, S. H. *Clin. Cancer Res.* **1998**, *4*, 683–691.
159. Grem, J. L.; Politi, P. M.; Balis, F. M.; Sinha, B. K.; Dahut, W.; Allegra, C. J. *Biochem. Pharmacol.* **1996**, *51*, 1649–1659.
160. Capps, D. B.; Dunbar, J.; Kesten, S. R.; Shillis, J.; Werbel, L. M.; Plowman, J.; Ward, D. L. *J. Med. Chem.* **1992**, *35*, 4770–4778.
161. Palomino, E.; Foster, B.; Kempff, M.; Corbett, T.; Wiegand, R.; Horwitz, J.; Baker, L. *Cancer Chemother. Pharmacol.* **1996**, *38*, 453–458.
162. Reid, J. M.; Walker, D. L.; Miller, J. K.; Benson, L. M.; Tomlinson, A. J.; Naylor, S.; Blajeski, A. L.; LoRusso, P. M.; Ames, M. M. *Clin. Cancer Res.* **2004**, *10*, 1471–1480.
163. Berg, S. L.; Blaney, S. M.; Sullivan, J.; Bernstein, M.; Dubowy, R.; Harris, M. *J. Pediatr. Hematol. Oncol.* **2000**, *22*, 506–509.
164. Keshelava, N.; Tsao-Wei, D.; Reynolds, C. P. *Clin. Cancer Res.* **2003**, *9*, 3492–3502.
165. Galanis, E.; Buckner, J. C.; Maurer, M. J.; Reid, J. M.; Kuffel, M. J.; Ames, M. M.; Scheithauer, B. W.; Hammack, J. E.; Pipoly, G.; Kuross, S. A. *Invest. New Drugs* **2005**, *23*, 495–503.
166. Vicker, N.; Burgess, L.; Chuckowree, I. S.; Dodd, R.; Folkes, A.; Hardick, D. J.; Hancox, T. C.; Miller, W.; Milton, J.; Sohal, S. et al. *J. Med. Chem.* **2002**, *45*, 721–739.
167. Wang, S.; Miller, W.; Milton, J.; Vicker, N.; Stewart, A.; Charlton, P.; Mistry, P.; Hardick, D. J.; Denny, W. A. *Bioorg. Med. Chem. Lett.* **2002**, *12*, 415–418.
168. Mistry, P.; Stewart, A. J.; Dangerfield, W.; Baker, M.; Liddle, C.; Bootle, D.; Kofler, B.; Laurie, D.; Denny, W. A.; Baguley, B. C. et al. *Anticancer Drugs* **2002**, *13*, 15–28.
169. De Jonge, M. J. A.; Kaye, S.; Verweij, J.; Brock, C.; Reade, S.; Scurr, M.; van Doorn, L.; Verheij, C.; Loos, W.; Brindley, C. et al. *Br. J. Cancer* **2004**, *91*, 1459–1465.
170. Gamage, S. A.; Spicer, J. A.; Finlay, G. J.; Stewart, A. J.; Charlton, P.; Baguley, B. C.; Denny, W. A. *J. Med. Chem.* **2001**, *44*, 1407–1415.
171. Dai, J.; Punchihewa, C.; Mistry, P.; Ooi, A. T.; Yang, D. *J. Biol. Chem.* **2004**, *279*, 46096–46103.
172. Spicer, J. A.; Gamage, S. A.; Rewcastle, G. W.; Finlay, G. J.; Bridewell, D. J. A.; Baguley, B. C.; Denny, W. A. *J. Med. Chem.* **2000**, *43*, 1350–1358.
173. Stewart, A. J.; Mistry, P.; Dangerfield, W.; Bootle, D.; Baker, M.; Kofler, B.; Okiji, S.; Baguley, B. C.; Denny, W. A.; Charlton, P. *Anti-Cancer Drugs* **2001**, *12*, 359–367.
174. Sappal, D. S.; McClendon, A. K.; Fleming, J. A.; Thoroddsen, V.; Connolly, K.; Reimer, C.; Blackman, R. K.; Bulawa, C. E.; Osheroff, N.; Charlton, P. et al. *Mol. Cancer Ther.* **2004**, *3*, 47–58.
175. Byers, S. A.; Schafer, B.; Sappal, D. S.; Brown, J.; Price, D. H. *Mol. Cancer Ther.* **2005**, *4*, 1260–1267.
176. Barret, J.-M.; Montaudon, D.; Etievant, C.; Perrin, D.; Kruczynski, A.; Robert, J.; Hill, B. T. *Anticancer Res.* **2000**, *20*, 4557–4562.
177. Perrin, D.; van Hille, B.; Barret, J.-M.; Kruczynski, A.; Etievant, C.; Imbert, T.; Hill, B. T. *Biochem. Pharmacol.* **2000**, *59*, 807–819.
178. Kruczynski, A.; Barret, J.-M.; Van Hille, B.; Chansard, N.; Astruc, J.; Menon, Y.; Duchier, C.; Creancier, L.; Hill, B. T. *Clin. Cancer Res.* **2004**, *10*, 3156–3168.
179. Kruczynski, A.; Etievant, C.; Perrin, D.; Imbert, T.; Colpaert, F.; Hill, B. T. *Br. J. Cancer* **2000**, *83*, 1516–1524.
180. Barret, J.-M.; Kruczynski, A.; Hill, B. T.; Etievant, C. *Cancer Chemother. Pharmacol.* **2002**, *49*, 479–486.
181. Sargent, J. M.; Elgie, A. W.; Williamson, C. J.; Hill, B. T. *Anti-Cancer Drugs* **2003**, *16*, 467–473.
182. Uesugi, M.; Sekida, T.; Matsuki, S.; Sugiura, Y. *Biochemistry* **1991**, *30*, 6711–6715.
183. Lorico, A.; Long, B. H. *Eur. J. Cancer* **1993**, *29A*, 1985–1991.
184. Rodriguez, C.; Azorin, F.; Portugal, J. *Biochemistry* **1996**, *35*, 11177–11182.
185. Gaver, R. C.; Deeb, G.; George, A. M. *Cancer Chemother. Pharmacol.* **1989**, *25*, 195–201.
186. Schurig, J. E.; Bradner, W. T.; Basler, G. A.; Rose, W. C. *Invest. New Drugs* **1989**, 7, 173–178.
187. Verweij, J.; Wanders, J.; Nielsen, A. L.; Pavlidis, N.; Calabresi, F.; ten Bokkel Huinink, W.; Bruntsch, U.; Piccart, M.; Franklin, H.; Kaye, S. B. *Ann. Oncol.* **1994**, *54*, 375–376.
188. Goss, G.; Letendre, F.; Stewart, D.; Shepherd, F.; Schacter, L.; Hoogendoorn, P.; Eisenhauer, E. *Invest. New Drugs* **1994**, *12*, 315–317.

Biography

William A Denny was trained at the universities of Auckland and Oxford as a medicinal chemist, and is the Director of the Auckland Cancer Society Research Centre in the Medical School of the University of Auckland. He is a founding scientist of Proacta Therapeutics Ltd, a biotechnology company focused around the development of hypoxia-activated drugs for cancer therapy. He is a past President of both the NZ Society for Oncology and the NZ Institute of Chemistry, and was the 2005 Adrien Albert Medallist of the UK Royal Society of Chemistry. He has been closely involved in the design and development of nine new cancer drugs to clinical trials in NZ, the USA, and Europe, and is author of about 530 scientific papers and 75 patent applications.

Comprehensive Medicinal Chemistry II
ISBN (set): 0-08-044513-6

ISBN (Volume 7) 0-08-044520-9; pp. 111–128

7.06 Alkylating and Platinum Antitumor Compounds

R D Hubbard and S Fidanze, Abbott Laboratories, Abbott Park, IL, USA

7.06.1 Introduction

The prolific use of chemical weapons during World War I and II, principally sulfur mustard (**1**, **Figure 1**), highlighted the powerful vesicant properties of this class of compounds, particularly toward the skin, eyes, and respiratory tract. A subsequent autopsy of soldiers revealed that exposure to **1** was characterized by the following: leukopenia, ulceration of the gastrointestinal tract, bone marrow aplasia, and dissolution of lymphoid tissue.[1] The profound cytotoxic effect of **1** indicated that **1** or related compounds might be potentially efficacious chemotherapeutics. Unfortunately, the

Cl–CH2CH2–S–CH2CH2–Cl (1)   Cl–CH2CH2–N(CH3)–CH2CH2–Cl (2)

Figure 1 Chemical structures of sulfur mustard (**1**) and mechlorethamine (**2**).

systematic toxicity of **1** was unacceptable, even upon direct injection into the tumor tissue.[2] Therefore, a less toxic mustard was sought, and eventually yielded the isoelectronic mustard mechlorethamine (Mustargen, **2**). The antitumor properties of **2** had been determined as early as 1942; however, the information was not published until 1946.[3] In that seminal paper, **2** was disclosed as being most effective against Hodgkin's disease. This disclosure has been widely recognized as the advent of modern cancer chemotherapy.

During the past 60 years, the focus of the medicinal chemist has been not only to understand better the mechanism of cytotoxicity of **1** and **2**, but also the discovery of safer, more selective chemotherapeutics. As a result, a variety of structurally distinct chemotypes have emerged that possess the desired cytotoxic properties, yet in some cases have dramatically reduced toxicity. The structural classes that will be the focus of this discussion are: (1) nitrogen mustards, (2) oxazaphosphorine mustards, (3) triazenes, (4) nitrosoureas, (5) mitomycins, and (6) platinum complexes. A key feature of these classes of compounds is their ability to alkylate key endogenous nucleophiles, principally DNA.

7.06.2 Nitrogen Mustards

Gillman and Philips' initial disclosure of **1** was accompanied by a mechanistic rationale that the sulfur mustard exerted its antiproliferative effects by the irreversible alkylation of 'phosphokinases.' The electrophilicity of **1** and the related nitrogen-based mustards react, via the intermediacy of either a thiarinium or an aziridinium cation, with the following exogenous nucleophiles: sulfides, alcohols, amines, phosphates, imidazoles, and carboxylates. In doing so, these agents possess the ability to alkylate proteins; however, their antiproliferative effects appear to be a function of their ability to covalently modify DNA and subsequent formation of DNA interstrand cross-links.[4] The order of the relative nucleophilicity of positions on DNA for electrophiles has been determined to be *N*-7 of guanine >*N*-3 of adenine >*N*-7 of adenine >*N*-3 of guanine >*N*-1 of adenine >*N*-1 of cytosine[5]; see **Figure 2** for numbering of the purines (A and G) and pyrimidines (T and C).

Mustards covalently modify DNA by the mechanism shown in **Figure 3**. Ionization of one of the chlorides of the beta-chloroethylamine-substituent of **3**, via an S_N1 pathway, yields the highly reactive aziridinium cation **4** (**Figure 3**). Nucleophilic capture of **4** by one of the aforementioned sites on DNA yields mono-adduct **5**. At this point, reformation of an aziridinium cation generates intermediate **6**, which has the potential to be attacked by exogenous nucleophiles (X), for example, water, proteins, or other nucleic acids yielding **7**. In addition, **6** can undergo a second alkylation by the same piece of DNA generating either the intrastrand cross-link **8** or the interstrand cross-links **9**.[6]

The ability of a mustard agent, for example **2**, to alkylate DNA begins a series of events that have a profound cumulative effect on the cellular machinery, ultimately leading to cell death. Treating cells with **2** results in formation of several alkylated moieties. Of particular interest are the interstrand cross-links formed between the *N*-7 of two different guanines. The resulting imidazolium species increases the enolic character of the residue, emulating the hydrogen bond accepting/donating array of adenine. During DNA replication, the alkylated base pairs with thymine resulting in a net GC to AT substitution.[7,7a] The increased imidazolium character of the modified guanine allows for possible hydrolysis of the heterocycle, or depurination of the guanine residue, both of which lead to DNA lesions,[8] and ultimately strand scission.

The reactivity of the beta-chloroethylamine group is highly dependent on the nucleophilicity of the mustard nitrogen. For example, mechlorethamine reacts within minutes in the body upon intravenous dosing, thereby limiting its clinical utility. In fact, the chemical instability of **2** translates to a rather nonselective alkylating agent in vivo, with a variety of toxicities elicited in rapidly dividing tissues and systems. In an effort to find analogs that have increased chemical stability, as well as reduced toxicity, a variety of substituted arenes have been conjugated to the nitrogen to moderate the chemical reactivity of the resulting mustard. The most successful analogs have employed conjugation of the mustard group to an appropriately substituted arene, yielding chlorambucil (Leukeran, **10**; **Figure 4**) and melphalan (Alkeran, **11**). The conjugation of the arene onto the mustard nitrogen significantly retards the rate of aziridinium cation formation allowing for increased chemical stability, yielding orally active drugs. Unfortunately, the increased chemical stability of **10** and **11** does not translate into greater selectivity for malignant cells versus nontransformed cells. As a result, the chemical modifications of **2** to afford **10** and **11** have greatly impacted the physical properties of this class of compounds, while yielding only marginally decreased toxicity to normal tissue.

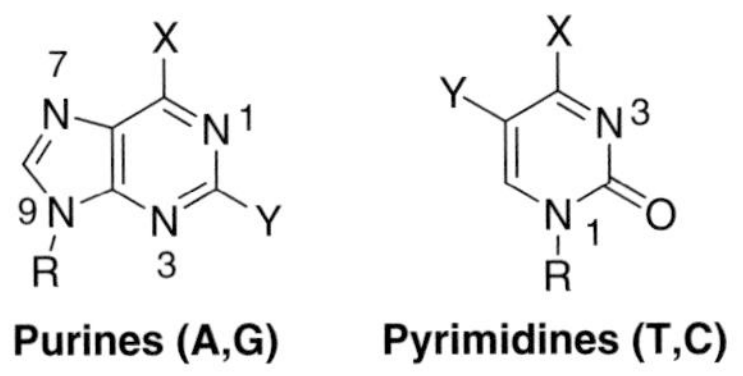

Purines (A,G) Pyrimidines (T,C)

Figure 2 Numbering of the purine and pyrimidine bases of DNA.

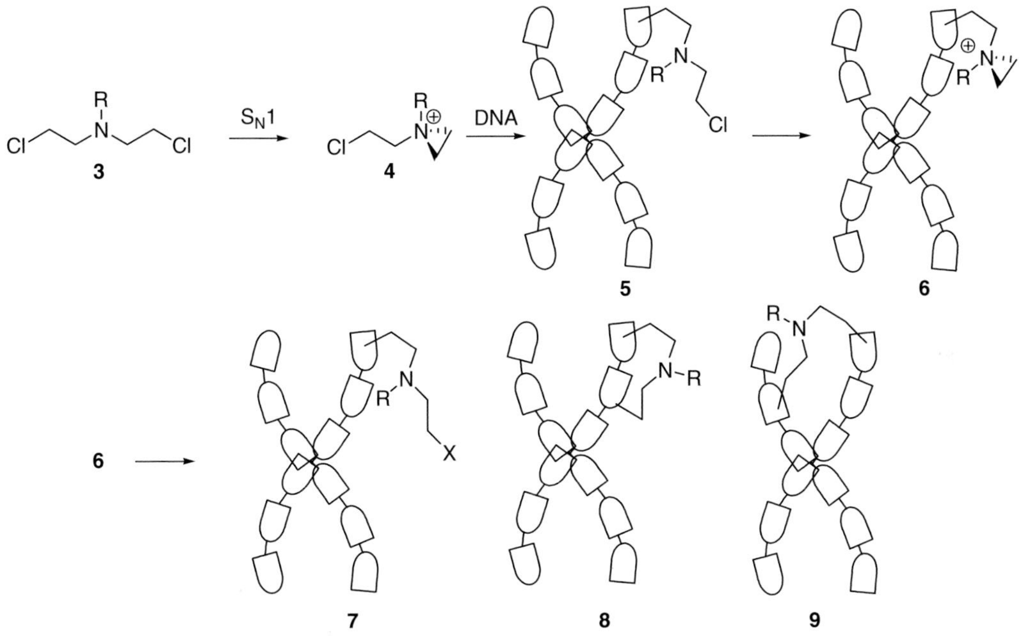

Figure 3 Schematic representation of a generic mustard **3** reaction with DNA.

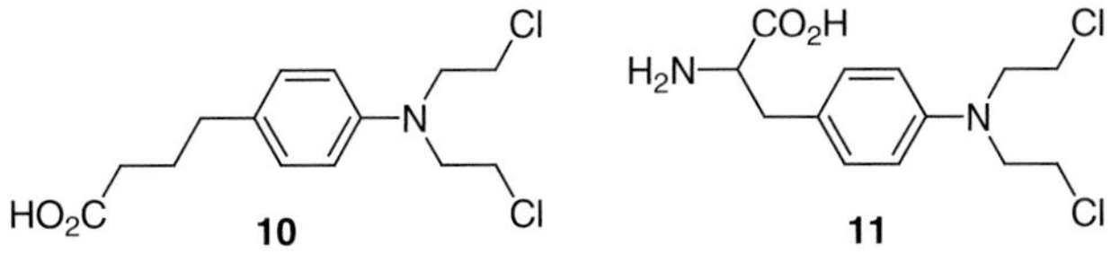

Figure 4 Structures of chlorambucil (**10**) and melphalan (**11**).

7.06.2.1 Clinical Use

7.06.2.1.1 Mechlorethamine (2)

The clinical utility of **2** has not only defined but is also somewhat limited by its intrinsic chemical instability. For example, mechlorethamine, the most reactive mustard, reacts rapidly with water or other exogenous nucleophiles, such that the intravenous (i.v.) solution that is used for dosing must be prepared immediately before use. Mechlorethamine is primarily used to treat advanced Hodgkin's disease as a component of the MOPP regimen, which is a combination of vincristine, mechlorethamine, procarbazine, and prednisone.[9] In the clinic, the toxic side effects observed upon administration of **2** include nausea, anemia, lymphocytopenia, leukopenia, and myelosuppression. In addition, care must be taken to avoid contact with the eyes or skin due to the vesicant properties of **2**. The powerful vesicant activity of **2** also increases the severity of side reactions resulting from extravasation. As mentioned previously, the narrow therapeutic window of **2** has hampered its clinical utility and has been largely supplanted by more chemically stable alkylating agents, like melphalan and the oxazaphosphorines (vide infra).

7.06.2.1.2 Chlorambucil (10)

The slightly enhanced chemical stability of **10** versus **2** has yielded a larger spectrum of use. The clinical indications for **10** are the following: palliative treatment of chronic lymphocytic leukemia, malignant lymphomas, and Hodgkin's disease. Upon oral administration of **10**, the compound is highly metabolized to phenylacetic acid mustard, which also possesses antineoplastic activity.[10] The pharmacokinetic properties of **10** and phenylacetic acid mustard are remarkably similar with respect to C_{max} and AUC. The standard dose is 4–10 mg day^{-1}, over a 3–6 week treatment period. During the course of therapy, slow and progressive lymphopenia is normal, and reversible after completion of therapy.[11] However, should an abrupt reduction in white blood cell count occur, the dosage is reduced to minimize possible permanent marrow damage.

7.06.2.1.3 Melphalan (11)

The ability to use melphalan as an orally active cytotoxic agent has proven useful for the palliation of multiple myeloma, its primary clinical indication.[11] The typical oral dose is 6 mg day^{-1} for 2–3 weeks, followed by a 4-week hiatus to enable the reestablishment of adequate leukocyte and platelet counts. Once sufficient bone marrow function has been determined, the maintenance portion of chemotherapy begins, which consists of 2–4 mg doses of **11** daily. High-dose intravenous melphalan has been approved for the treatment of both neuroblastoma (100–240 mg m^{-2}) and multiple myeloma (100–200 mg m^{-2}), which will usually necessitate autologous stem cell rescue.[12] The mustard functionality's reactivity toward hydrolysis or indiscriminate reaction with other exogenous nucleophiles besides DNA coupled with incomplete 'first-pass' hepatic clearance has led to highly variable plasma levels, which can be moderated by appropriate dose individualization.

The toxicity of **11** is similar to chlorambucil and mechlorethamine in that the primary dose-limiting toxicities are hematologic in origin. In practice, the dose of **11** is escalated until moderate myelosuppression is detected; the degree of myelosuppression has become a clinical pharmacodynamic marker of efficacy. However, oral usage of melphalan does not tend to have negative effects on the gastrointestinal system and vesicant properties that mechlorethamine displays.

7.06.2.2 Carcinogenicity of the Nitrogen-Based Mustards

Given the history of the mustard-based chemotherapeutics, one would expect the therapeutic window to be small. Indeed, all of the above agents have profound dose-limiting hematological toxicities. Historically, through the continued use of **2**, **10**, or **11** we have gained a better understanding of how to manage toxic side effects and the 'curative' potential of these agents are stressed. However, over time these agents have been reported to be both mutagenic[13,14] and carcinogenic. The extent to which the nitrogen mustards contribute to mutagenicity/carcinogenicity is difficult to determine, given that most of the regimens that employ the mustards are used in combination with other cytotoxics and/or ionizing radiation. However, a seminal study by Reimer and colleagues revealed that in a review of 5455 patients treated for ovarian cancer, 13 cases of acute nonlymphocytic leukemia were reported. All of the patients that developed leukemia had been previously treated with an alkylating agent, such as melphalan, chlorambucil, or cyclophosphamide.[15] As stated above, melphalan is a mainstay for myeloma treatment; unfortunately, the treatment increases the risk of developing acute leukemia by 200 times within 4 years of therapy.[16]

7.06.2.3 Resistance to Mustard-Based Chemotherapy

The difficulty in effectively treating resistant malignancies in patients is a function of first defining which mechanism(s) enables resistance, then having an agent(s) that will overcome the resistant malignancy. However, as will be discussed, a given therapeutic agent can give rise to differing mechanisms of resistance depending on the type of tumor being treated, thereby hampering the development of a universal approach to render resistant cells sensitive to therapy. The cellular mechanisms responsible for evasion of the cytotoxic effects of chemotherapy can be roughly categorized into the following: (1) decreased cellular transport of the agent; (2) increased expression of the multigene family of glutathione *S*-transferases (GSTs) that catalyze the conjugation of glutathione and electrophilic alkylating agents; (3) increased expression of key DNA repair enzymes; and (4) increased metabolism of key detoxifying intermediates.

The depletion of glutathione (GSH) levels in melphalan-resistant pancreatic cancer cell lines with buthionine sulfoximine resulted in sensitive cell lines, comparable to the responsiveness found in control lines.[17] However, for chlorambucil- or melphalan-resistant chronic lymphocytic leukemia patients, resistance did not correlate with the expression levels of GST, metabolism mediated by GSH, or altered kinetics of cellular transport.

The pioneering work of Goldenberg and colleagues demonstrated that melphalan requires active, carrier-mediated transport to the cell,[18] which is different from the passive cellular uptake employed by the structurally related chlorambucil.[19] The use of carrier transport, which is analogous to the system employed by leucine,[20] has prompted speculation that impairment of the transporter protein could lead to resistance to melphalan-based therapy. Redwood and Colvin reported that the melphalan resistance in the L1210 cell line was a result of impaired transport.[21] However, Parsons determined that there was no difference in transport between sensitive and resistant MM253 cell lines.[22] The differential mechanisms of resistance employed by the two tumor types (melanoma versus leukemia) might explain the inconsistencies between the two reports.

7.06.3 Oxazaphosphorines

The historical significance of the mustard class of cytotoxics cannot be underestimated in that this class of compounds revealed that organic synthesis could provide chemotherapeutically valuable materials. Although the exact mechanism of cytotoxicity of the mustards was not fully appreciated, medicinal chemists appreciated the necessity of the beta-chloroethylamine group for in vivo activity. As a result, the desire to generate compounds that would either mask the reactivity of the mustard group and/or use properties specific to the cancerous cells to increase the selectivity of these compounds versus normal tissue became the desired profile for potential chemotherapeutics. The oxazaphosphorine class of cytotoxics was originally devised with both of the above properties in mind.

Initial profiling of cancerous cells to determine potential differences versus nontransformed cells revealed that cancer cells possessed increased expression of phosphoamidases. Therefore, a compound that conjugated the mustard nitrogen to a phosphate group would not only yield a more chemically robust analog, but also afforded a substrate for cleavage by phosphoamidases. The net result was the discovery of cyclophosphamide (Cytoxan, **12**) shown in **Figure 5**, and the structurally related ifosfamide (Ifex, **13**). The synthesis of **12** begins with condensation of phosphoryl chloride ($POCl_3$) with bis(2-chloroethyl)amine hydrochloride (**14**) providing **15**. Subsequent treatment of the intermediate phosphoramide with 3-aminopropanol (**16**) affords **12**.

The development of **12** would have been severely hampered in the current paradigm of drug discovery in that only compounds that are highly active (cytotoxic, antiproliferative, induce apoptosis, etc.) versus transformed cell lines in vitro are further progressed to in vivo xenograft murine models of cancer. Cyclophosphamide is only weakly cytotoxic versus transformed cells in vitro. Nevertheless, administration of **12** in vivo to either animals, or subjects with sensitive tumors, provides dramatic antineoplastic effects. The initial rationalization for the in vivo activity of **12** was based on phosphatase or phosphoamidase cleavage of the P–N bond, which would release the mustard group. This hypothesis has been subsequently proven incorrect (for a summary of the development of oxazaphosphorine-based agents see [23]). The accepted mechanism for cyclophosphamide's impressive in vivo activity is shown in **Figure 6**. Upon administration of **12**, metabolic activation occurs via the cytochrome P450 monooxygenase system of the liver to afford 4-hydroxycyclophosphamide (**17**), a common oxidative precursor in all of the oxazaphosphorine therapeutics. The specific P450 isoforms responsible for the generation of **17** are species dependent. The human P450 isoforms involved in the metabolism of **12** are CYP2B6, CYP3A4, and CYP2C9, with CYP2B6 being primarily responsible for cyclophosphamide's metabolism in vivo.[24] The intermediate **17** simultaneously affords the desired cytotoxic metabolites (vide infra) and allows for cell permeability.[25] The highly unstable aminal of **17** undergoes spontaneous and reversible ring opening to afford aldophosphamide **18**, which affords either inactive or cytotoxic metabolites depending on subsequent chemical or enzymatic modifications. The generation of the inactive metabolites occurs if **18** is oxidized by either alcohol dehydrogenase yielding **19**, or by aldehyde dehydrogenase (ALDH1) affording the corresponding acid **20**. Alternatively, nonenzymatic elimination of the phosphoryl group from the aldehyde simultaneously generates phosphoramide mustard (**21**) and acrolein (**22**) The former is responsible for the generation of interstrand DNA cross-links, and the latter results in additional DNA lesions,[26] and has been implicated in the urotoxicity, principally hemorrhagic cystitis, witnessed with the use of **12**.

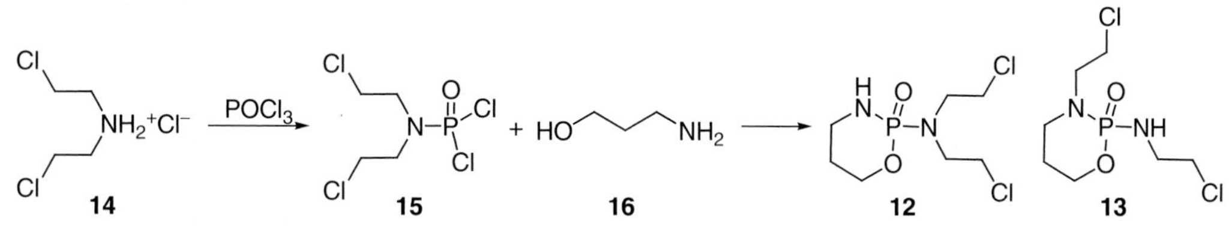

$POCl_3$

14 15 16 12 13

Figure 5 Synthesis of cyclophosphamide (**12**), and the structure of ifosfamide (**13**).

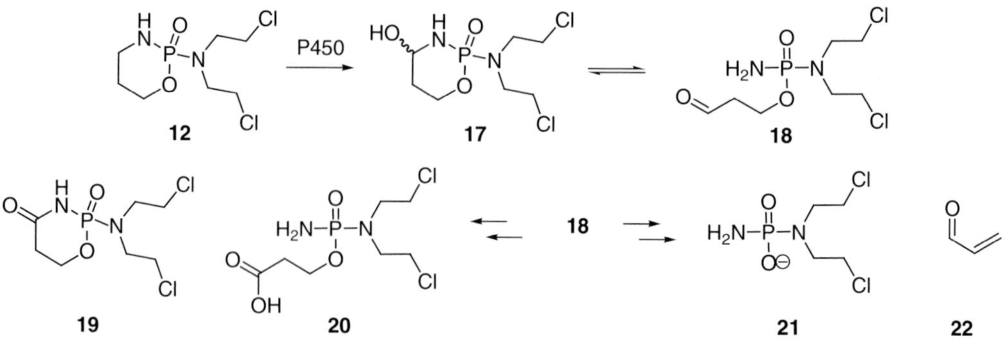

Figure 6 Metabolism of cyclophosphamide (**12**).

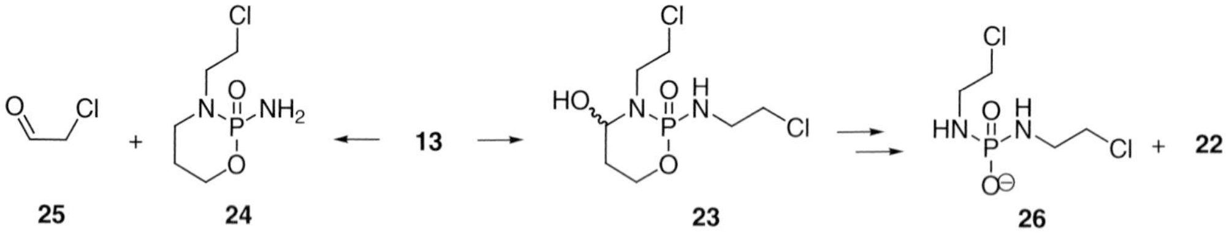

Figure 7 Metabolism of ifosfamide (**13**).

The structurally related oxazaphosphorine ifosfamide undergoes metabolic activation by CYP3A4 to form 4-hydroxyifosfamide (**23**, **Figure 7**).[27] In contrast, the rate of the initial oxidation of **13** to **23** is slow, relative to **12**, allowing side chain metabolism to occur resulting in dechloroethylated product **24** and equimolar amounts of chloroacetaldehyde (**25**). For some patients, the dechloroethylation of **13** is the major metabolic pathway accounting for as much as 60% of the metabolism of ifosfamide.[28] Nevertheless, 4-hydroxyifosfamide can be eventually processed to **22** and isophosphoramide mustard **26**, the latter being the desired cytotoxic component resulting from **13**. The overall process by which **13** exerts its cytotoxic effects is quite similar to **12**. However, key metabolites differ in not only their composition but also their amounts. For example, ALDH1-mediated detoxification of **12** is one of the major metabolites, but for **13** this process is only minor, compared to dechloroethylation. The differential rates of oxidation of **12** compared to **13** coupled with differing tether lengths of the beta-chloroethyl groups yield variations in not only their clinical use but also their toxicology (vide infra).

7.06.3.1 Clinical Use

7.06.3.1.1 Cyclophosphamide (12)

Cyclophosphamide is one of the most frequently employed cytotoxic agents, either used alone or in combination with other chemotherapeutics.[29] The ability to use **12** either orally or intravenously has led to a wide variety of dosing regimens and protocols of usage. As a single agent for induction therapy, cyclophosphamide is typically given as a dose of $100\,\mathrm{mg\,m^{-2}}$ orally for 14 days for either leukemias or lymphomas, or up to $500\,\mathrm{mg\,m^{-2}}$ is given intravenously for treatment of breast cancer or other types of lymphoma. Cyclophosphamide is a component of the CHOP regimen (cyclophosphamide, doxorubicin, vincristine, and prednisone) that is commonly used for treating non-Hodgkin's lymphoma.[30,30a] Another common combination therapy employing cyclophosphamide is the CMF regimen for breast cancer, which consists of cyclophosphamide, methotrexate, and 5-fluorouracil.[31,31a] According to the product insert from the manufacturer, cyclophosphamide has been approved to treat the following malignancies: lymphoma, leukemia, multiple myeloma, mycosis fungoides, neuroblastoma, adenocarcinoma of the ovary, retinoblastoma, and breast carcinoma.

The reliance of cyclophosphamide upon metabolism in order to generate the desired cytotoxic species **21** creates unique pharmacokinetic challenges in using **12** in the clinic. Depending on the concentration and the activity of CYP2B6, the key human isozyme believed to be responsible for bioactivation of **12**, the half-life of **12** can vary from 6 to 9 h, with 5–25% of the dose excreted unchanged in urine. During the course of repeated administration of **12**, increased rates of metabolism of cyclophosphamide simultaneously lead to decreased half-life and increased clearance,[32] which result from increased expression of key CYP isozymes.[33] As a result, interpatient variance makes predicting the efficacious dose of cyclophosphamide difficult. For this reason, **12** is typically dosed until the total white blood cell count falls to 2000–3000 cells per mL.[34]

In contrast to the aforementioned nitrogen mustard-based agents, cyclophosphamide shows reduced reports of nonhematopoetic toxicity. The reduced toxicity has allowed the development of a high-dose regimen of cyclophosphamide to be used in combination with busulfan or with total body irradiation, for allogenic or autologous bone marrow transplantation.[35] The toxicity observed with cyclophosphamide includes nausea, myelosuppression, and alopecia. In addition, the generation of equimolar quantities of acrolein can lead to dose-limiting urotoxicity, principally hemorrhagic cystitis, which is an issue at elevated doses of **12** ($>1\,\mathrm{g\,m^{-2}}$). Fortunately, the urotoxicity from **22** can be minimized by coadministration of the sulfhydryl-based nucleophile mesna (2-mercaptoethanesulfonate).[36,36a]

7.06.3.1.2 **Ifosfamide (13)**

The relatively minor structural disposition of the beta-chloroethyl group of **13** versus **12** has profound effects on not only the spectrum of anticancer activity and chemical reactivity, but also pharmacokinetics and toxicity. The early clinical trials of **13** revealed superior activity compared with **12** in some tumor populations. However, the clinical development of **13** was initially hampered by dose-limiting toxicity (hemorrhagic cystitis), which can be effectively managed by scavenging the acrolein released from the metabolism of **13** by co-dosing with mesna.[37] The dosing of the two agents in combination has allowed **13** to be used to treat both pediatric and adult malignancies. In addition to ifosfamide's use as a monotherapy, **13** has been used in combination with a wide variety of other chemotherapeutics: platinum agents, antimitotics, etoposide, dactinomycin, and mitomycin.[38] The combination of these agents and **13** has led to either synergistic cytotoxicity or has enabled activity against previously resistant tumors. The net result is that ifosfamide-based regimens have shown activity against cyclophosphamide-resistant tumors[39,39a] and activity that includes, but is not limited to, the following malignancies: Ewing's sarcoma, cervical and lung carcinoma, soft-tissue sarcoma, germ cell carcinoma, breast cancer, lymphoma, melanoma, prostate cancer, and neuroblastoma.[40]

The typical dosing regimen of ifosfamide for adults consists of an intravenous dose of $1.2\,\mathrm{g\,m^{-2}}$ for 5 days, which is approximately 3–10 times greater than the amount of **12** needed for an equicytotoxic effect. The extensive dechloroethylation of ifosfamide is the reason why such large amounts of **13** are needed for clinical efficacy. This regimen is typically repeated for 3 weeks. The compound is usually dosed intravenously to minimize neurotoxicity observed upon oral dosing.[41] In addition, mesna is given as either a bolus dose equivalent to the dose of **13**, or fractionated over 3 doses, each 20% of the initial ifosfamide dose. Adequate hydration (>2 L of liquid) either orally or intravenously is given daily during treatment to minimize nephrotoxicity.

Ifosfamide causes additional side effects besides urotoxicity, some being quite serious. Ifosfamide has shown dose-related and dose-limiting myelosuppression, as determined by the incidence of leukopenia, and to a lesser degree thrombocytopenia,[42] which is reversible reaching maximal suppression of the white blood cells between days 8 and 13 of use. The generation of equimolar amounts of chloroacetaldehyde has been implicated in the neurotoxicity witnessed in both high-dose and oral regimens of ifosfamide, and in the nephrotoxicity cited with pediatric patients.[43,43a] The neurotoxic side effects can range from confusion to coma.[44] Chloroacetaldehyde has been shown to cause rapid glutathione depletion in vitro, thereby diminishing the cell's ability to remove potentially toxic moieties.[45,45a] Cerebral levels of glutathione in mice were reduced upon administration of ifosfamide in the presence of CYP3A4 inducers, which correlated with increased neurotoxicity.[46] Chloroacetaldehyde-induced neurotoxicity can be managed by treating with methylene blue.[47,47a]

Comparison of the principal cytotoxic agents derived from cyclophosphamide and ifosfamide, namely **21** and **26**, reveal that **21** is structurally similar to **3**. Therefore, the mechanism by which **21** forms interstrand DNA cross-links is likely analogous to **3**. However, the presence of two secondary nitrogens in **26** compared to the tertiary mustard nitrogen in **21** provides a very different chemical reactivity profile (**Figure 8**). For example, the initial S_N1 reaction of isophosphoramide mustard affords the aziridine intermediate **31**, which is roughly four times less reactive than the analogous intermediate generated from phosphoramide mustard **27** ($t_{1/2}=80$ min versus 20 min). Alkylation by *N*-7 of guanine, or another exogenous nucleophile, affords **32**. A subsequent S_N1 reaction occurs and, after proton transfer, yields **33**.[48] This intermediate is much longer lived (170 min for reaction of **33** with an additional nucleophile to afford **34**) as compared to the second alkylation of the monoadduct from phosphoramide mustard **29** (20 min).[49] In the examples in which both nucleophiles are from the *N*-7 of guanine, the resulting cross-linked product, from cyclophosphamide, is **35**, and likewise **36** is derived from ifosfamide. The striking differences in the rates of reactivity of **26** and **21** have led to the assertion that the increased spectrum of cytotoxicity of **13** might be due, in part, to elevated concentrations of **26** reaching the nucleus, generating increased numbers of DNA interstrand cross-links.

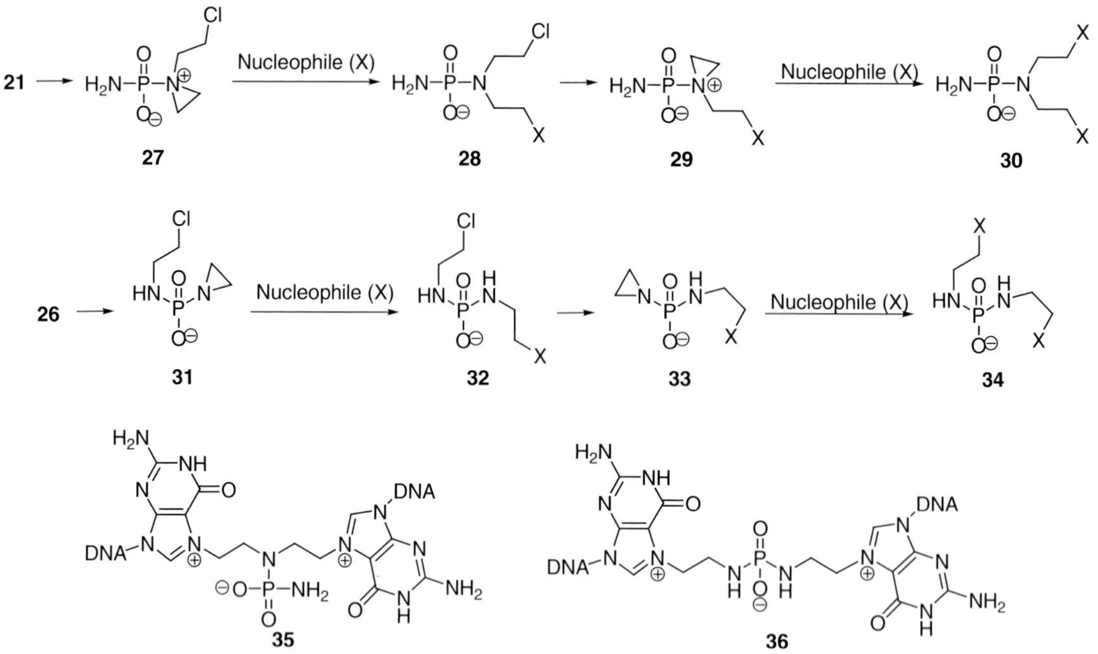

Figure 8 Mechanism of alkylation of **21** and **26**.

7.06.3.2 Resistance to Oxazaphosphorine-Based Chemotherapy

The primary mode of resistance to oxazaphosphorine-based therapies is the increased expression of aldehyde dehydrogenase activity, thereby increasing the concentration of inactive metabolite **20** that is generated from cyclophosphamide.[50] To underscore the complexity of how a cell might become resistant, the characterization of a cyclophosphamide-resistant breast cell line (MCF-7) revealed not only overexpression of ALDH1, but also increased levels of GSH and increased GST activity.[51,51a,51b] Recent reports have indicated that the activity of the O^6-alkylguanine alkyltransferase (AGT) and defective DNA repair can lead to resistance.[52] The role of AGT is to irreversibly remove an alkyl group from the O^6-position of guanine. For the triazene and nitrosourea class of therapeutics, the inability of the cell to repair these lesions has been correlated to their cytotoxicity and mutagenicity.[53] However, these types of lesions have not been isolated from cells treated with cyclophosphamide. As a result, the exact manner in which the expression of AGT confers resistance to **12** is unknown. A possible explanation is that AGT acts to remove acrolein-derived lesions from the O^6-guanine,[54] and/or acts as a molecular scavenger for acrolein.[55]

7.06.4 Triazenes

The development of the triazene class of therapeutics can trace its beginnings to the mid-1950s, where early predecessors of the triazenes (aryl-dimethyl triazenes) were metabolic precursors to cytotoxic agents. Contemporaneously, the pyrazolo-triazines were reported to have modest activity in a mouse model of sarcoma. Over nearly 60 years of research, three compounds have been widely studied for their anticancer properties: temozolomide (**37**, TMZ, Temodar), mitozolomide (**38**), and dacarbazine (**42**, DTIC-Dome). Preclinical evaluation of **38** in mice demonstrated promising cytotoxic activity. However, mitozolomide exhibited dose-limiting thrombocytopenia and hematological toxicities in phase I studies, even after dose reduction. The unpredictable toxicity of mitozolomide rendered further development impractical (for a discussion of the historical development of the triazene class of therapeutics see [56,56a]); consequently, the following discussion is restricted to **37** and **42**.

7.06.4.1 Mechanism of Action

The minor structural difference between the bicyclic agents (**37**, **38**) and the monocyclic agent **42** yields major differences in the manner in which the agents exert their cytotoxic effects. For example, TMZ possesses good aqueous stability at low pH, making oral administration of TMZ possible. However, at pH >7, base promoted attack of the C-4 carbonyl generates the intermediate **39** and subsequent decomposition yields **40** (**Figure 9**). The decomposition of

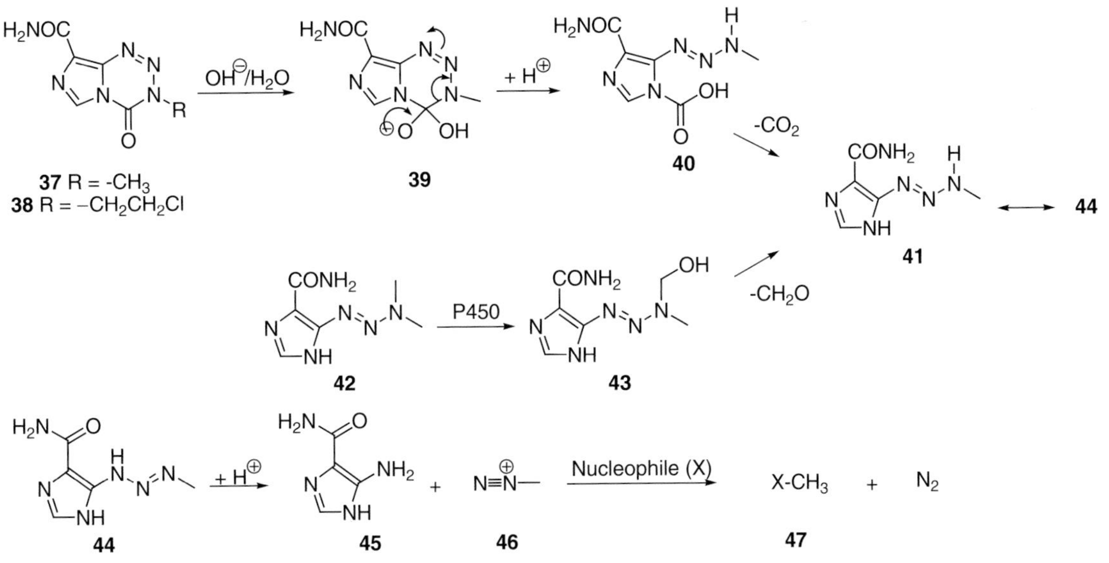

Figure 9 Bioactivation of the triazenes.

TMZ has been discussed in detail.[57] Decarboxylation of **40** affords a common intermediate of TMZ and dacarbazine: 5-(3-methyltriazen-1-yl)imidazole-4-carboxamide (**41** or **44**, MTIC). For dacarbazine metabolic activation via cytochrome P450 oxidation yields **43**,[58,59] which upon formal loss of formaldehyde affords MTIC. Tautomerization of **41** to **44**, followed by protonation yields 5-aminoimidazole-4-carboxamide (**45**, AIC) and methyl diazonium cation (**46**), the agent responsible for the cytotoxic activity of the triazenes. A key difference between the triazenes (TMZ, DTIC) and the mustard-based cytotoxics is that these agents are incapable of creating DNA cross-links. Nucleophilic capture of **46** by DNA yields lesions at the following positions and percentages: N^7 (70%) and O^6 (5%) positions of guanine and the N^3 (9%) of adenine.[60] The DNA-base selectivity preference toward guanine from MTIC has been argued for both steric and electronic reasons.[60] Although the carboxamide (C_8, for TMZ) does not participate in either the metabolism of **42** or the decomposition of **38**, this functionality has been postulated to play a key role in aligning the prodrugs along runs of guanines in DNA.[61]

The methylation of O^6 guanine, although only formed in minor amounts relative to the N^7 lesion, has been proposed to be the key lesion responsible for cytotoxicity. Gerson has indicated that about 6000 of these lesions are needed for cell death versus only 5–10 cross-links derived from nitrosourea treatment.[62] During DNA replication, the preferred base pair to O^6-methylguanine is thymine,[63] not cytosine, creating a net mismatch. During the course of mismatch repair (MMR), long single-stranded gaps are generated[64]; in doing so, double-stranded breaks can occur if another lesion is present on the opposite strand.[65,65a] The presence of multiple DNA single or double-stranded gaps/breaks eventually leads to apoptosis.[66] Tumor lines with either increased levels of AGT[67] or deficient mismatch repair[68] tend to be resistant to the triazene class of cytotoxics.

7.06.4.2 Clinical Use

7.06.4.2.1 Dacarbazine (42)

The clinical use of dacarbazine is primarily for metastatic melanoma[69] and as a component of the AVBD (doxorubicin, vinblastine, bleomycin, dacarbazine) regimen for treating Hodgkin's lymphoma.[70,70a] The dosing regimen for metastatic melanoma is 250 mg m^{-2} for 5 days, repeated every 3 weeks. For Hodgkin's disease, DTIC is dosed at 150 mg m^{-2} for 5 days, repeated every 4 weeks. An alternative dosing schedule for DTIC in the AVBD regimen is to administer **42** at 350–375 mg m^{-2} on the first day of dosing and repeat every 15 days.[71] Intravenous administration of DTIC results in biphasic elimination of the parent compound with the initial half-life of 19 min, and the terminal half-life of 5 h, although for patients with renal and hepatic impairment, significantly increased half-lives are witnessed.[72] As discussed previously, metabolic activation of DTIC is required for cytotoxic activity, however, up to 40% of the drug is excreted unchanged in the urine. The inability of the body to effectively metabolize the entire dose of DTIC has hampered the effectiveness of this agent. In addition, the use of dacarbazine for melanoma can actually select a more aggressive melanoma, by promoting the increased expression of both interleukin-8 and vascular endothelial growth factor.[73]

7.06.4.2.2 Temozolomide (37)

Temozolomide is an orally active agent that is currently used for the treatment of adult anaplastic astrocytoma and glioblastoma multiforme.[74] The dose and duration of drug used are highly dependent on the identity of the tumor, the expression levels of AGT, and on MMR functional status.[75] TMZ has also been used, off label, for the treatment of metastatic melanoma. The results from a randomized phase III clinical trail revealed that TMZ was superior to DTIC in median progression-free survival time, and improvement in quality of life parameters with respect to reduced fatigue and insomnia.[76]

TMZ is readily absorbed, although a pronounced food effect can occur, and is highly metabolized upon oral administration. The lipophilic nature of TMZ allows for a large volume of distribution and blood–brain barrier penetration, achieving 20% of the concentration found in plasma.[77] The ring-opened intermediates have been shown to preferentially partition into tissue versus plasma, and higher levels of TMZ were found in tumor tissue as compared to normal tissue.[78]

The major dose-limiting toxicity of TMZ is myelosuppression. In contrast to the toxicity witnessed with **38**, the hematological-based toxicity of TMZ is predictable with the nadir occurring between 21 and 28 days after administration. Temozolomide's myelosuppressive toxicity does not appear to be cumulative and its nonhematological toxicities range from mild to moderate and include the following: nausea/vomiting, fatigue, constipation, and headache.[79]

7.06.4.3 Resistance to the Triazenes

For the triazenes, principally TMZ, the combination therapies that have been developed to overcome resistance mediated through DNA repair processes encompass: AGT, MMR, and a newer strategy that involves inhibition of poly(ADP-ribose) synthetase (PARP).[80] For AGT, detection of an O^6-alkyl group on guanine causes the enzyme to utilize the conserved active-site cysteine to irreversibly remove the alkyl group; thereby, inactivating the protein.[81,81a] For maximal cytotoxicity, TMZ requires a large amount of these types of lesions to be present; therefore, repair of these lesions via elevated concentrations of AGT allows for decreased cell death. This has been shown experimentally, in that high concentrations of AGT in vivo predicts for resistance to TMZ,[82] and the related chloroethylating agents, like **52**.[83] Fortunately, most cells have fully functional MMR, which is crucial for TMZ-derived lesions to eventually lead to cell death. However, cells that are MMR deficient are highly resistant to TMZ, regardless of the levels of AGT. Recent research has demonstrated that both the effectiveness of MMR and the levels of AGT together affect the sensitivity of a given cell line.[84] These results suggest that maximal TMZ sensitivity may be associated with low levels of AGT and fully functional MMR.

The methods that have been employed to sensitize previously resistant cells to TMZ involve either dose modification of TMZ, or use of an AGT inhibitor, like O^6-benzylguanine (BG, **48**). TMZ has been shown to act as an indirect inhibitor of AGT, which can lead to AGT depletion and increased cellular sensitivity to chemotherapy. Acceptable dosing of TMZ for glioblastoma multiforme was either 150 mg m^{-2} day^{-1} or 200 mg m^{-2} day^{-1} for 5 days in every 28 days.[85] The higher dose was used for chemotherapy-pretreated patients. Unfortunately, increasing the dose of TMZ, as a mechanism for AGT depletion, leads to increased toxicity. However, an alternative dosing schedule of 75 mg m^{-2} day^{-1}, once daily for 6–7 weeks, for a group of 24 patients, 17 with malignant glioma, yielded an overall response rate of 33%. A net twofold increase in drug exposure was observed compared to the aforementioned 200 mg m^{-2} day^{-1} regimen, without an increase of dose-limiting toxicity.[86] Tolcher has demonstrated that prolonged administration of TMZ leads to prolonged depletion of AGT, which could lead to increased therapeutic activity.[87]

The second strategy for overcoming AGT-mediated resistance has been to use the specific and potent inhibitor of AGT, BG. The mechanism by which BG depletes AGT, O^6-benzylguanine (**48**, BG) is shown in **Figure 10**. The nucleophilic active-site sulfur of AGT reacts with BG causing transfer of the O^6-benzyl group liberating the masked guanine **49**. Irreversible alkylation of AGT results in net sensitization of cells toward TMZ-based therapy.[88] For example, treating TMZ-resistant T98G glioblastoma cells with BG prior to TMZ results in a significant decrease in cell viability.[89] Friedman and co-workers have demonstrated in a murine model of malignant glioma that a triple combination of TMZ, BG, and irinotecan yielded growth delays greater than the combination of TMZ and irinotecan alone (150 versus 43 days). The impressive activity in vivo of the combination of TMZ, BG, and irinotecan has prompted the design of a phase I clinical trial to develop an optimal dosing regimen.[90]

7.06.5 Nitrosoureas

The development of nitrosourea-based agents (**Figure 11**) was the culmination of extensive medicinal chemistry and a fortuitous choice of in vivo screening model.[91] The preferred in vivo screening model during the early to mid-1950s was

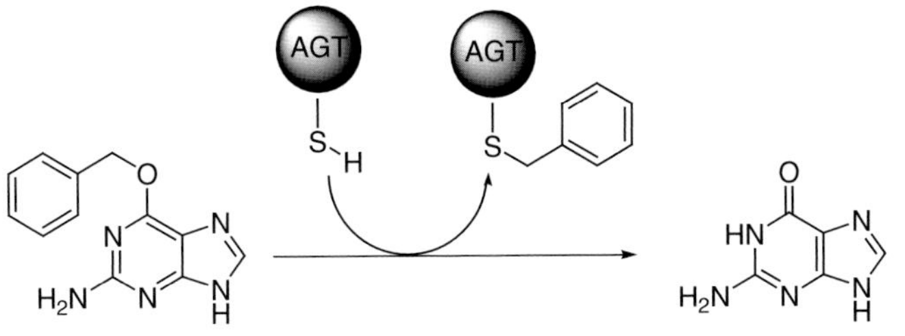

Figure 10 Mechanism for AGT depletion by O^6-benzylguanine (**48**).

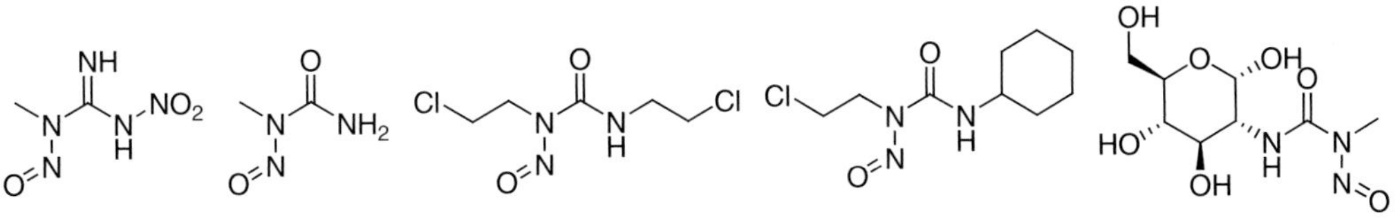

Figure 11 Chemical structures of MNNG (**50**), MNU (**51**), BCNU (**52**), CCNU (**53**), and STZ (**54**).

a murine model of leukemia, L1210. The elevated expression of murine AGT relative to human AGT afforded a higher therapeutic window in this model than what would be witnessed in the clinic. Nevertheless, the predecessors of the nitrosoureas, namely *N*-methyl-*N′*-nitro-*N*-nitrosoguanidine (**50**, MNNG) and *N*-methyl-*N*-nitrosourea (**51**, MNU), showed moderate activity in the L1210 tumor model, the latter being more efficacious. The promising activity of **51** prompted Montgomery's group to synthesize >200 analogs, before finding one with superior activity to **51**, which was *N*,*N′*-bis(2-chloroethyl)-*N*-nitrosourea (**52**, BCNU, BiCNU, Carmustine).[92,92a] Once again, the presence of 2-chloroethyl groups proved to be instrumental in providing increased efficacy for this class of therapeutics relative to **51**. The preliminary in vivo results indicated that BCNU had cured mice from either intraperitoneally or intracerebrally implanted L1210 leukemia cells. Multiple compounds have been optimized for anticancer activity within this class, two such examples being Lomustine (**53**, CCNU) and streptozocin (**54**, STZ, Zanosar). For leading references regarding the chemical and biological properties of STZ, see[93]. The discussion will focus on the chemical and biological properties of the prototypical nitrosourea, BCNU.

7.06.5.1 Mechanism of Action of BCNU

Upon intravenous administration, **52** undergoes base catalyzed degradation, within as little as 5 min, to afford **55** and **56**, the former being responsible for the cytotoxic activity of **52**, and the latter acting as a carbamoylating agent (**Figure 12**).[94] The mechanism of how BCNU generates interstrand cross-links, shown below, needs to address two seemingly conflicting pieces of data. First, research had shown that treating cells with BCNU results in formation of the G–C crosslink **61**,[95] which was formed by attack of the poorly nucleophilic *N*-1 position of guanine and the *N*-3 position of cytosine. Second, Erickson had demonstrated that the in vitro cytotoxic effects of the nitrosoureas were diminished with increased levels of AGT, which would imply that BCNU alkylates the O^6 position of guanine.[96] To address both of these issues Ludlum has proposed the following reaction mechanism, which begins with the bis-electrophilic ethylating agent **55** reacting initially with the O^6 position of guanine **57**, yielding **58**. The *N*-1 position of guanine displaces the chloride, yielding the highly electrophilic intermediate **59**, which is trapped by *N*-3 of the Watson–Crick paired cytosine to yield the interstrand cross-linked base pair **61**.[97] The cytotoxic activity of **52** is strongly correlated to its ability to generate **61**.

7.06.5.2 Clinical Use and Administration of BCNU (52)

The dramatic activity of BCNU in preclinical rodent models has not translated into equally dramatic responses in man. Nevertheless, BCNU has been used clinically either alone or in combination for treating the following malignancies: primary or metastatic melanoma, multiple myeloma, Hodgkin's disease, and non-Hodgkin's lymphoma. The compound is highly metabolized, with a half-life of 15 min, after intravenous administration. The majority of the metabolites are

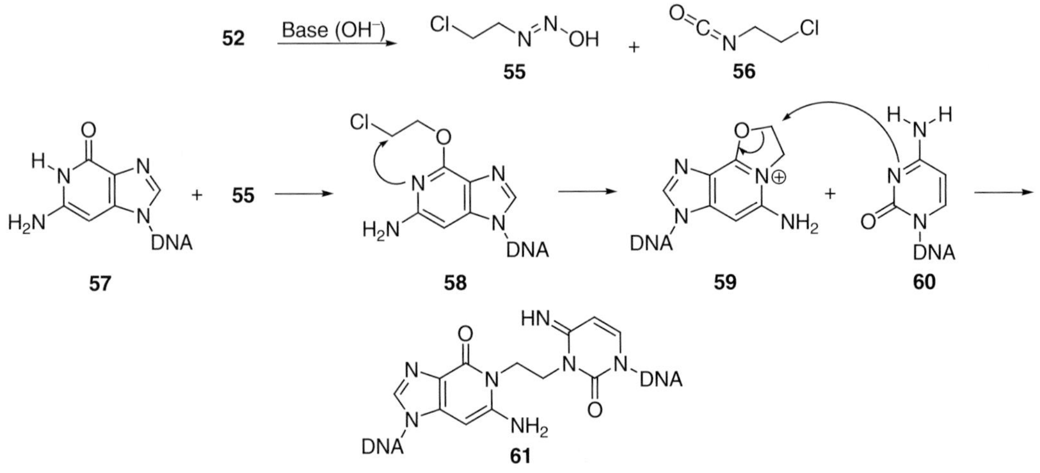

Figure 12 Proposed mechanism of cytotoxicity of **52**.

excreted in the urine. The high lipophilicity and lack of highly charged metabolites allows the carmustine-derived agent to penetrate the blood–brain barrier. As a single agent, the compound is given i.v. over 2 h, every 6 weeks.[98] The major dose-limiting toxicity of BCNU is delayed and there is cumulative bone marrow suppression, which necessitates dose individualization if the patient is currently taking other myelosuppressive agents.

7.06.6 Mitomycin C

The mitomycins are antitumor antibiotics first isolated from *Streptomyces* spp. bacteria in the 1950s.[99] Mitomycin C (**62**, MC, Mutamycin; **Figure 13**) contains an unusual molecular architecture including an aziridine, a quinone, and a primary carbamate. This structure allows for a unique mechanism of bioactivation of MC.[100] The quinone is reduced in vivo by one of a number of enzymatic reductases, followed by the elimination of methanol to give the mitocene **63** (**Figure 13**).[101] The 2-amino group of a guanine residue of DNA opens the aziridine yielding **64**. Nucleophilic displacement of the carbamate forms a 1,2-interstrand adduct to a guanine of the complementary strand **65**. MC cross-linking of DNA is specific to dCpG CpG duplex sites in DNA.[102,103]

7.06.6.1 Clinical Use of Mitomycin C (62)

MC was initially approved for clinical use in 1974 for treatment of stomach and pancreatic cancers. It has since been used in several other cancer types, including bladder, breast, cervical, colorectal, head and neck, and nonsmall cell lung cancer.[104] For example, although typically used in combination, MC has shown single-agent activity in breast cancer.[105] In the FILM combination for first-line treatment of advanced breast cancer, 5-fluorouracil (750 mg m^{-2}), ifosfamide (1 g m^{-2}), leucovorin (200 mg m^{-2}), and mitomycin C (6 mg m^{-2}, alternate cycles) are given in 3-week cycles on an outpatient basis. From a trial cohort of 90 patients on the FILM regimen, 70% were in remission and 86% were alive after 5 years.[106]

One common side effect of MC is dose-dependent myelosuppression.[107] This is evidenced by delayed leucopenia and thrombocytopenia. The dose-limiting toxicity is hemolytic uremic syndrome. The end result can be irreversible renal damage, which is usually lethal. This side effect can generally be avoided by maintaining a cumulative dose of <30 mg m^{-2}.

7.06.6.2 Resistance to Mitomycin C

Resistance to MC can be caused by a lack of activation. Single electron reduction of MC by several enzymes can be reversed by molecular oxygen. However, under normoxic conditions, MC can be reduced in a two-electron process by DT-diaphorase (DTD) or xanthine dehydrogenase to an oxygen-insensitive intermediate.[108] Thus, resistance is less pronounced under hypoxic conditions, but under normoxic conditions, downregulation of DTD can lead to resistance.[109]

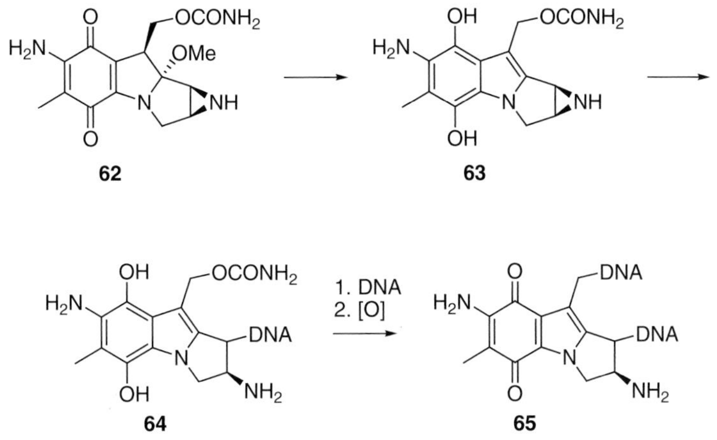

Figure 13 Structure and activation of mitomycin C (**62**).

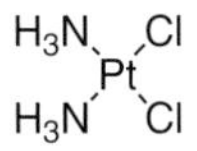

Figure 14 Chemical structure of cisplatin (**66**).

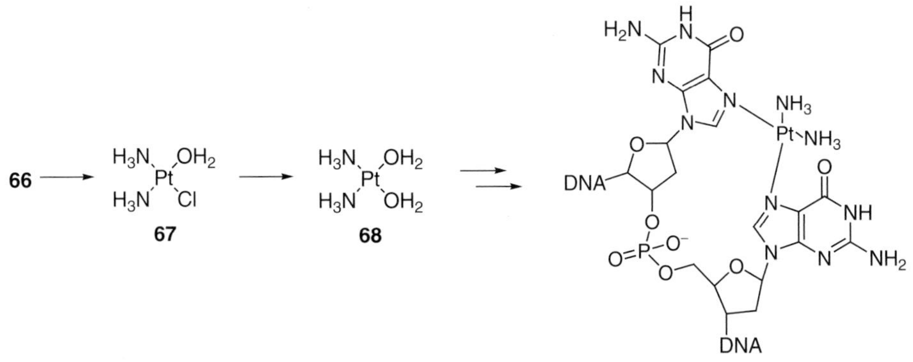

Figure 15 Activation of cisplatin.

7.06.7 Platinum Complexes

The parent compound of all platinum-based chemotherapy is cisplatin (Platinol, **66**; **Figure 14**). Rosenberg and co-workers serendipitously discovered the antiproliferative effects of cisplatin in 1965.[110] This study demonstrated that cisplatin inhibited cell division, but not growth, of *Escherichia coli*. Later studies by Rosenberg showed that cisplatin was active in vivo versus Sarcoma 180 and L1210 xenograft mouse models of cancer.[111] Cisplatin was approved as an antitumor agent in 1978.[112]

7.06.7.1 Cisplatin Mechanism of Action

Cisplatin is activated by intracellular hydrolysis of the chlorides to either hydroxyl or aqua ligands (**Figure 15**).[113] Initially, **66** was assumed to enter the cell via passive diffusion; however, recent investigations have demonstrated that cisplatin enters the cell via a copper transporter.[114] The fully activated platinum complex, **68**, generates intrastrand purine cross-links in DNA, 65% of which are dGpG (**69**).[115,116] These lesions are recognized by a number of proteins to trigger several cellular processes. The downstream effect is generally cytotoxicity via apoptosis, although cisplatin-mediated necrosis has also been documented.[117]

7.06.7.2 Clinical Uses of Cisplatin (66)

Cisplatin has found widespread use in a variety of solid tumor types. It is not orally available, and is administered via intravenous infusion. Instead of using body surface area, dosing can be calculated using projected AUC to account for renal function in an effort to help minimize potential side effects.[118]

Cisplatin has been a particularly effective drug in the treatment of testicular and ovarian cancers. For example, cisplatin is used in combination with bleomycin and etoposide (BEP regimen) for treating testicular cancer, which consists of three cycles, each lasting 3 days. Etoposide is given at a dose of $165\,mg\,m^{-2}\,day^{-1}$ for 3 days, bleomycin $50\,mg\,m^{-2}$ for 1 day, and cisplatin $50\,mg\,m^{-2}\,day^{-1}$ for 2 days yielding ~90% progression-free survival rate after 2 years of completing this therapy.[119]

7.06.7.3 Cisplatin Toxicity

There are a number of toxicities associated with cisplatin use. The major dose-limiting toxicity is neuropathy. This may begin as a mild peripheral tingling sensation, but can progress to impairment of fine motor function. While this may gradually improve over time after cessation of therapy, in some cases it can be permanent. The presence of this effect is related to the current dose given, and to the overall cumulative dose. High-dose regimens ($>120\,mg\,m^{-2}$) may cause symptoms from a single treatment, while at more typical levels ($50–75\,mg\,m^{-2}$), symptoms are less likely until a cumulative dose of $>300\,mg\,m^{-2}$ has been achieved.[118]

Cisplatin's clinical use also results in acute renal toxicity. The mechanism of this toxicity is not yet understood, but several strategies have been developed to help alleviate kidney toxicity. These include delivering the drug over a longer time period, either by increasing the infusion time, or by administering a course of treatment over several days instead of a 1-day treatment.[120] The use of hypertonic saline or a nephroprotective agent is also effective.[118]

One of the most pervasive cisplatin toxicities is emesis, which occurs in >90% of patients that are not treated with a prophylactic antiemetic.[121] The onset of emesis can be acute or delayed by >24 h. Fortunately, this can often be prevented with antiemetic prophylactic treatment, including use of a 5-hydroxytryptamine-3 ($5HT_3$) antagonist and, more recently, a neurokinin 1 (NK_1) antagonist.[122]

Cisplatin is known to lack bone marrow suppression, thereby enhancing its clinical utility by allowing combinations with other cytotoxics that do have bone marrow suppressive effects. However, this is not true of all platinum complexes, as carboplatin possesses significant hematological toxicity.[118]

7.06.7.4 Resistance to Cisplatin

Extensive investigations focused on determining which cellular mechanisms have led to resistance to platinum-based cytotoxics have yielded several possible pathways. As the primary mechanism of action of cisplatin involves damaging DNA, increased levels of DNA repair machinery within the cell can also increase resistance to cisplatin.[123] In particular, tumors that are clinically resistant to cisplatin overexpress the DNA-repair gene ERCC1.[124] Tumor cells that are deficient in MMR are also typically resistant to cisplatin.[125]

About 90% of cisplatin in blood is protein bound, mostly to albumin, and activated cisplatin reacts with many cytosolic nucleophiles other than DNA,[117] for example, glutathione and metallothionin.[126] The glutathione–cisplatin complex is then actively exported from the cell by the ATP-dependent glutathione *S*-conjugate pump.[127]

Treatment with cisplatin also activates several survival pathways, including the ERK and PI3K/Akt pathways. In ovarian cancer cell lines, inhibition of either of these pathways sensitizes resistant cells to cisplatin.[128]

7.06.7.5 Other Platinum-Based Cytotoxics

The search for newer analogs of cisplatin has been focused to address the following issues. First, one goal has been to affect tumors that are resistant to cisplatin. Another goal has been the alleviation of some of the more serious side effects of cisplatin. Finally, more recent compounds have sought to achieve oral bioavailability. Indeed, new compounds have been introduced that have addressed all of these issues. There are currently two other platinum compounds marketed in the US: carboplatin (Paraplatin, **70**, US approval in 1989; **Figure 16**) and oxaliplatin (Eloxatin, **71**, European approval in 1996, US in 2002; **Figure 16**).

Carboplatin and oxaliplatin both require activation via intracellular hydrolysis, with retention of the amine ligands. For these compounds, hydrolysis of the dicarboxylate groups is much slower than for the chlorides in cisplatin. The slower hydrolysis comes with a number of differences in vivo, particularly in side effect profiles. The two newer

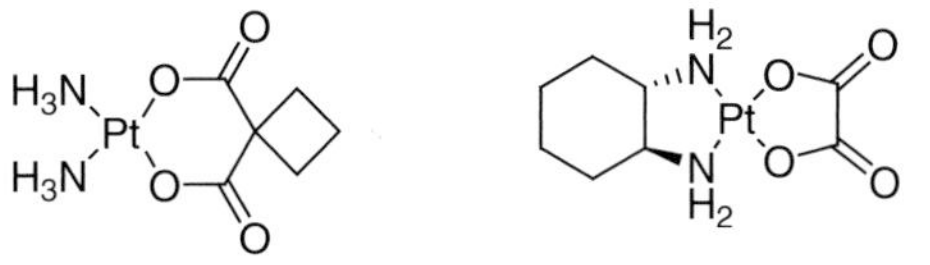

70 71

Figure 16 Structures of carboplatin **70** and oxaliplatin **71**.

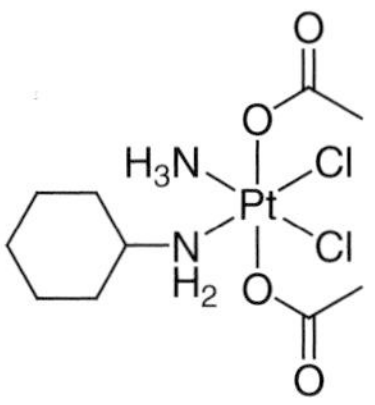

72

Figure 17 Structure of satraplatin **72**.

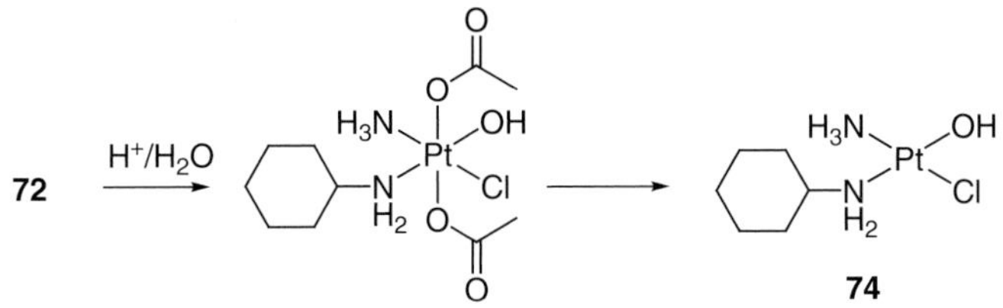

72 H^+/H_2O 73 74

Figure 18 Active species from satraplatin.

compounds are not considered to be nephrotoxic, whereas this toxicity can be severe with cisplatin.[118] Carboplatin is also less neurotoxic than cisplatin and oxaliplatin.

One negative aspect of carboplatin is that it demonstrates pronounced bone marrow suppression, in contrast to cisplatin and oxaliplatin. This toxicity can be controlled to some extent by dosing based on projected AUC, which accounts for renal function. Clinically, typical treatment programs include carboplatin at an AUC of 4–7 ($mg\,mL^{-1}$) min.[129] Although both carboplatin and paclitaxel suppress bone marrow, the combination of carboplatin with paclitaxel results in significantly lower hematological toxicity than either agent alone, through a currently unknown mechanism.[130]

Cisplatin and carboplatin form similar DNA cross-links, and are effective against similar tumors. They are also cross-resistant with each other; thus, they are affected by the same resistance mechanisms. Oxaliplatin can circumvent some forms of cisplatin resistance, particularly when mediated by cellular mismatch repair mechanisms.[131] Indeed, a key component in MMR has been shown to be twofold selective to bind to cisplatin over oxaliplatin modified DNA.[132]

A common factor among cisplatin, carboplatin, and oxaliplatin is that they are all square planar Pt(II) complexes. All lack oral bioavailability, and are thus dosed by intravenous infusion. An important quality of life consideration for the patient has been the development of orally active platinum-based therapeutics. The use of an octahedral Pt(IV) complex has provided several orally active candidates for clinical trials. One member of this class that has reached phase III trials is satraplatin **72** (JM216, BMS-182751; **Figure 17**).[133]

Satraplatin must be reduced and hydrated in vivo to the active platinum (II) species **74**, via the intermediacy of **73** (**Figure 18**). The use of the cyclohexylamine ligand does confer some activity to tumors that are resistant to cisplatin. Satraplatin-DNA adducts are not recognized by mismatch repair proteins.[134] In addition, satraplatin also has different transport properties from cisplatin, and in vitro experiments suggest that it can overcome cisplatin resistance caused by reduced platinum transport.[135]

Satraplatin has shown some promise in a phase II trial against small-cell lung cancer.[136] The regimen of treatment in this trial was an oral dose of $120\,mg\,m^{-2}\,day^{-1}$ for 5 consecutive days, with a course given every 3 weeks.

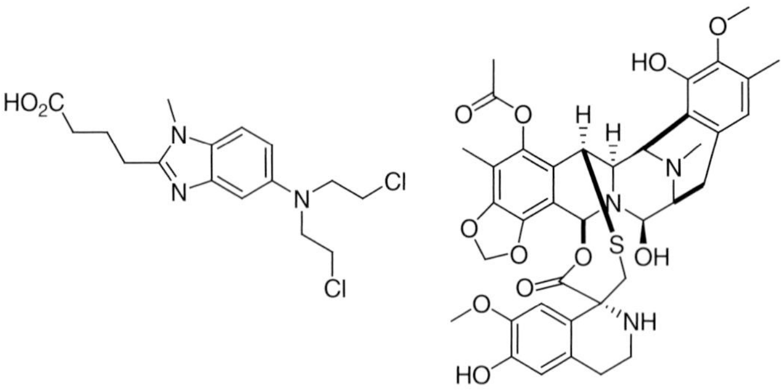

Figure 19 Chemical structures of bendamustine (75) and ET-743 (76).

Unfortunately, the drug suffers from nonlinear pharmacokinetics due to saturable absorption.[137] However, for this trial the dosing level of **72** was within the range of linear absorption. Satraplatin showed a response rate of 38%, which was similar to cisplatin. The major dose-limiting toxicity associated with satraplatin is myelosuppression.

7.06.8 Future Directions

Collectively, agents that covalently modify DNA constitute the most widely used agents in cancer chemotherapy. Agents that were profoundly toxic have found utility by either appropriate dosing regimens or by use in combination with other agents, like **48**. As the science has evolved, chemists have been searching for molecules with increased therapeutic utility. In doing so, the resulting compounds range from known chemotypes appended to a novel scaffold, for example bendamustine (**75**),[138] to the complex, like ET-743 (**76**)[139,139a] (**Figure 19**). Further clinical evaluation will determine if either of these or as yet undiscovered agents will afford truly selective, efficacious, and safe cancer chemotherapeutics.

References

1. Chabner, B. A.; Ryan, D. P.; Paz-Ares, L.; Carbonero, R. G.; Calabresi, P. Antineoplastic Agents. In *Goodman & Gilman's The Pharmalogical Basis of Therapeutics*, 10th ed.; Hardman, J. G.; Limbard, L. E.; Gilman, A. G., Eds.; McGraw-Hill: New York, 2001, Chapter 52, pp 1389–1459.
2. Silverman, R. B. *The Organic Chemistry of Drug Design and Drug Action*, 2nd ed.; Elsevier: Boston, MA, 2004; Chapter 6, pp 353–366.
3. Gilman, A.; Philips, F. S. *Science* **1946**, *103*, 409–415.
4. Rajski, S. R.; Williams, R. M. *Chem. Rev.* **1998**, *98*, 2723–2795.
5. Beranek, D. T. *Mutat. Res.* **1990**, *231*, 11–30.
6. Hopkins, P. B.; Millard, J. T.; Woo, J.; Weidner, M. F. *Tetrahedron* **1991**, *47*, 2475–2489.
7. Kim, M.-S.; Guengerich, F. P. *Chem. Res. Toxicol.* **1998**, *11*, 311–316.
7a. Oida, T.; Humphreys, G.; Guengerich, F. P. *Biochemistry* **1991**, *30*, 10513–10522.
8. Gates, K. S.; Nooner, T.; Dutta, S. *Chem. Res. Toxicol.* **2004**, *17*, 839–856.
9. Longo, D. L.; Young, R. C.; Wesley, M.; Hubbard, S. M.; Duffey, P. L.; Jaffe, E. S.; DeVita, V. T., Jr. *J. Clin. Oncol.* **1986**, *4*, 1295–1306.
10. Lee, F. Y.; Coe, P.; Workman, P. *Cancer Chem. Pharm.* **1986**, *17*, 21–29.
11. GlaxoSmithKline. For information regarding the clinical use and toxicity, the product information insert is particularly informative. http://www.gsk.com (accessed May 2006).
12. Samuels, B. L.; Bitran, J. D. *J. Clin. Oncol.* (Offical Journal of ASCO) **1995**, *13*, 1786–1799.
13. Olsen, L. S.; Korsolm, B.; NexØ, B. A.; Wasserman, K. *Arch. Toxicol.* **1997**, *71*, 198–201.
14. Sanderson, B. J. S.; Shield, A. J. *Mutat. Res.* **1996**, *355*, 41–57.
15. Reimer, R. R.; Hoover, R.; Fraumeni, J. F., Jr.; Young, R. C. *New Engl. J. Med.* **1977**, *297*, 177–181.
16. Seiber, S. M.; Adamson, R. H. *Adv. Cancer Res.* **1975**, *22*, 57–155.
17. Ripple, M.; Mulcahy, R. T.; Wilding, G. *J. Urol.* **1993**, *150*, 209–214.
18. Goldenberg, G. J.; Lee, M.; Lam, H. Y. P.; Begleiter, A. *Cancer Res.* **1977**, *37*, 755–760.
19. Begleiter, A.; Goldenberg, G. J. *Biochem. Pharm.* **1983**, *32*, 535–539.
20. Vistica, D. T.; Rabon, A.; Rabinovitz, M. *Cancer Lett.* **1979**, *6*, 7–13.
21. Redwood, W. R.; Colvin, M. *Cancer Res.* **1980**, *40*, 1144–1149.
22. Parsons, P. G.; Carter, F. B.; Morrison, L.; Mary, Sister R. *Cancer Res.* **1981** *41*, 1525–1534.
23. Brock, N. *Cancer Res.* **1989**, *49*, 1–7.
24. Boddy, A. V.; Yule, S. M. *Clin. Pharmokinet.* **2000**, *38*, 291–304.
25. Draeger, U.; Peter, G.; Hohorst, H. J. *Cancer Treat. Rep.* **1976**, *60*, 355–359.

26. Colvin, M. E.; Quong, J. N. DNA-Alkylating Events Associated with Nitrogen Mustard Based Anticancer Drugs and the Metabolic Byproduct Acrolein. In *Advances in DNA Sequence-Specific Agents*, Vol. 4; Chapman, B. J., Ed.; Elsevier Sciences: AE Amsterdam, Netherlands, 2002, pp 29–46.
27. Walker, D.; Flinois, J.-P.; Monkman, S. C.; Beloc, C.; Boddy, A. V.; Cholerton, S.; Daly, A. K.; Lind, M. J.; Pearson, A. D. J.; Beaune, P. H.; Idle, J. R. *Biochem. Pharmacol.* **1994**, *47*, 1157–1163.
28. Kaijser, G. P.; Beijnen, J. H.; Bult, A.; Underberg, W. J. *Anticancer Res.* **1994**, *14*, 517–531.
29. Colvin, O. M. *Curr. Pharm. Design* **1999**, *5*, 555–560.
30. Marcus, R. *Leukemia Lymphoma* **2003**, *44*, S15–S27.
30a. Fisher, R. I.; Shah, P. *Leukemia* **2003**, *17*, 1948–1960.
31. Nabholtz, J.-M.; Reese, D.; Lindsay, M.-A.; Riva, A. *Int. J. Clin. Oncol.* **2002**, 7, 245–253.
31a. Fossati, R.; Confalonieri, C.; Torri, V.; Ghislandi, E.; Penna, A.; Pistotti, V.; Tinazzi, A.; Liberati, A. *J. Clin. Oncol.* **1998**, *16*, 3439–3460.
32. D'Incalci, M.; Bolis, G.; Facchinetti, T.; Mangioni, C.; Marasca, L.; Morazzoni, P.; Salmona, M. *Eur. J. Cancer* **1979**, *15*, 7–10.
33. Chang, T. K. H.; Yu, L.; Maurel, P.; Waxman, D. J. *Cancer Res.* **1997**, *57*, 1946–1954.
34. Bristol-Myers Squibb. For information regarding the use of white blood cell count as a surrogate pharmacodynamic marker http://www.bms.com (accessed May 2006).
35. Ferry, C.; Socié, G. *Exp. Hematol.* **2003**, *31*, 1182–1186.
36. Brock, N. *Recent Results Cancer Res.* **1980**, *74*, 270–278.
36a. Olver, I.; Keefe, D.; Myers, M.; Caruso, D. *Chemotherapy* **2005**, *51*, 142–146.
37. Bryant, B. M.; Jarman, M.; Ford, H. T. *Lancet* **1980**, *2*, 657–659.
38. Brade, W. P.; Herdrich, K.; Klein, H. O. *Contrib. Oncol.* **1987**, *26*, 22–52.
39. Brade, W.; Seeber, S.; Herdrich, K. *Cancer Chemother. Pharmacol.* **1986**, *18*, S1–S9.
39a. Bramwell, V. H. C.; Mouridsen, H. T.; Santoro, A.; Blackledge, G.; Somers, R.; Verweij, J.; Dombernowsky, P.; Onsrud, M.; Thomas, D.; Sylvester, R. *Cancer Chemother. Pharmacol.* **1993**, *31*, S180–S184.
40. Kerbush, T.; de Kraker, J.; Keizer, H. J.; van Putten, J. W. G.; Groen, H. J. M.; Jansen, R. L. H.; Schellens, J. H. M.; Beijnen, J. H. *Clin. Pharmakinet.* **2001**, *40*, 41–62.
41. Nicolao, P.; Giometto, B. *Oncology* **2003**, *65*, 11–16.
42. Cormier, J. N.; Patel, S. R.; Herzog, C. E.; Ballo, M. T.; Burgess, M. A.; Feig, B. W.; Hunt, K. K.; Raney, R. B.; Zagars, G. K.; Benjamin, R. S. et al. *Cancer* **2001**, *92*, 1550–1555.
43. Loebstein, R.; Atanackovic, G.; Bishai, R.; Wolpin, J.; Khattak, S.; Hashemi, G.; Gorbrial, M.; Baruchel, B.; Ito, S.; Koren, G. *J. Clin. Pharm.* **1999**, *39*, 454–461.
43a. Skinner, R.; Sharkey, I. M.; Pearson, A. D. J.; Craft, A. W. *J. Clin. Oncol.* **1993**, *11*, 173–190.
44. Cerny, T.; Kupfer, A. *Ann. Oncol.* **1992**, *3*, 678–681.
45. Binotto, G.; Trentin, L.; Semenzato, G. *Oncology* **2003**, *65*, 17–20.
45a. Issels, R. D.; Meier, T. H.; Mueller, E.; Multhoff, G.; Wilmanns, W. *Mol. Aspects Med.* **1993**, *14*, 281–286.
46. Sood, C.; O'Brien, P. J. *Br. J. Cancer* **1994**, *74*, S287–S293.
47. Donegan, S. *J. Oncol. Pharm. Prac.* **2000**, *6*, 153–165.
47a. Pelgrims, J.; De Vos, F.; Van Den Brande, J.; Schrijvers, D.; Prove, A.; Vermorken, J. B. *Brit. J. Cancer* **2000**, *82*, 291–294.
48. Springer, J. B.; Colvin, M. E.; Colvin, O. M.; Ludeman, S. M. *J. Org. Chem.* **1998**, *63*, 7218–7222.
49. Boal, J. H.; Deamond, S. F.; Callahan, D.; Bruce, S. A.; Ts'o, P. O.; Kan, L. S. *J. Med. Chem.* **1989**, *32*, 1768–1773.
50. Sládek, N. E. *Curr. Pharm. Design* **1999**, *5*, 607–625.
51. Sreerama, L.; Sládek, N. E. *Biochem. Pharmacol.* **1993**, *45*, 2487–2505.
51a. Chen, G. A.; Waxman, D. J. *Biochem. Pharmacol.* **1995**, *49*, 1691–1701.
51b. Gamcsik, M. P.; Mills, K. K.; Colvin, O. M. *Cancer Res.* **1995**, *55*, 2012–2016.
52. Gamcsik, M. P.; Dolan, M. E.; Andersson, B. S.; Murray, D. *Curr. Pharm. Design* **1999**, *5*, 587–605.
53. Gerson, S. L. *Nat. Cancer Rev.* **2004**, *4*, 296–307.
54. Freidman, H. S.; Pegg, A. E.; Johnson, S. P.; Loktionova, N. A.; Dolan, M. E.; Modrich, P.; Moschel, R. C.; Struck, R.; Brent, T. P.; Ludeman, S. et al. *Cancer Chemother. Pharmacol.* **1999**, *43*, 80–85.
55. Krokan, H.; Graftstrom, R. C.; Sundqvist, K.; Esterbauer, H.; Harris, C. C. *Carcinogenesis* **1985**, *6*, 1755–1759.
56. Newlands, E. S.; Stevens, M. F. G.; Wedge, S. R.; Wheelhouse, R. T.; Brock, C. *Cancer Treat. Rev.* **1997**, *23*, 35–61.
56a. Stevens, M. F. G.; Newlands, E. S. *Eur. J. Cancer* **1993**, *29A*, 1045–1047.
57. Denny, B. J.; Wheelhouse, R. T.; Stevens, M. F. G.; Tsang, L. L. H.; Slack, J. A. *Biochemistry* **1994**, *33*, 9045–9051.
58. Yamagata, S.-I.; Ohmori, S.; Suzuki, N.; Hino, M.; Ishii, I.; Kitada, M. *Drug Metab. Dispos.* **1998**, *26*, 379–382.
59. Gersher, A.; Threadgill, M. D. *Pharmacol. Ther.* **1987**, *32*, 191–205.
60. Denny, B. J.; Wheelhouse, R. T.; Stevens, M. F. G.; Tsang, L. L. H.; Slack, J. A. *Biochemistry* **1994**, *33*, 9045–9051.
61. Lowe, P. R.; Sansom, C. E.; Schwalbe, C. H.; Stevens, M. F. G.; Clark, A. S. *J. Med. Chem.* **1992**, *35*, 3377–3382.
62. Gerson, S. L. *J. Clin. Oncol.* **2002**, *20*, 2388–2399.
63. Pauly, G. T.; Hughes, S. H.; Moschel, R. C. *Biochemistry* **1994**, *33*, 9169–9177.
64. Schärer, O. D. *Angew. Chem. Int. Ed.* **2003**, *42*, 2946–2974.
65. Gerson, S. L. *J. Clin. Oncol.* **2002**, *20*, 2388–2399.
65a. Armstrong, M. J.; Galloway, S. M. *Mutat. Res.* **1997**, *373*, 167–178.
66. Hickman, M. J.; Samson, L. D. *Proc. Natl. Acad. Sci. USA* **1999**, *96*, 10764–10769.
67. Friedman, H. S.; Dolan, M. E.; Pegg, A. E.; Marcelli, S.; Keir, S.; Catino, J. J.; Bigner, D. D.; Schold, S. C., Jr. *Cancer Res.* **1995**, *55*, 2853–2857.
68. Liu, L.; Markowitz, S.; Gerson, S. L. *Cancer Res.* **1996**, *56*, 5375–5379.
69. Eggermont, A. M. M.; Kirkwood, J. M. *Eur. J. Cancer* **2004**, *40*, 1825–1836.
70. Volker, D. *J. Clin. Oncol.* **2004**, *22*, 15–18.
70a. Canellos, G. P. *Sem. Hematol.* **2004**, *41*, 26–31.
71. Bayer. Information obtained from the product insert from the manufacturer http://www.bayer.com (accessed May 2006).
72. Loo, T. L.; Housholder, G. E.; Gerulath, A. H.; Saunders, P. H.; Farquhar, D. *Cancer Treat. Rep.* **1976**, *60*, 149–152.
73. Lev, D. C.; Omn, A.; Melinkova, W. O.; Miller, C.; Stone, V.; Ruiz, M.; McGary, E. C.; Ananthaswamy, H. N.; Price, J. E.; Bar-Eli, M. *J. Clin. Oncol.* **2004**, *22*, 2092–2100.

74. See: http://www.schering-plough.com (accessed May 2006).
75. Nagasubramanian, R.; Dolan, M. E. *Curr. Opin. Oncol.* **2003**, *15*, 412–418.
76. Middleton, M. R.; Grob, J. J.; Aaronson, N.; Fierlbeck, G.; Tilgen, W.; Seiter, S.; Gore, M.; Aamdal, S.; Cebon, J.; Coates, A. et al. *J. Clin. Oncol.* **2000**, *18*, 158–166.
77. Ostermann, S.; Csajka, C.; Buclin, T.; Leyvraz, S.; Ferdy, L.; Decosterd, L. A.; Stupp, R. *Clin. Cancer Res.* **2004**, *10*, 3728–3736.
78. Saleem, A.; Brown, G. D.; Brady, F.; Aboagye, E. O.; Osman, S.; Luthra, S. K.; Ranicar, A. S. O.; Brock, C. S.; Stevens, M. F. G.; Newlands, E. et al. *Cancer Res.* **2003**, *63*, 2409–2415.
79. Danson, S. J.; Middleton, M. R. *Exp. Rev. Anticancer Ther.* **2001**, *1*, 13–19.
80. Curtin, N. J.; Wang, L.-Z.; Yiakouvaki, A.; Kyle, S.; Arris, C. A.; Canan-Koch, S.; Webber, S. E.; Durkacz, B. W.; Calvert, H. A.; Hostomsky, Z. et al. *Clin. Cancer Res.* **2004**, *10*, 881–889.
81. Moore, M. H.; Dodson, E. J.; Demple, B.; Moody, P. C. *EMBO J.* **1994**, *13*, 1495–1501.
81a. Wibley, J. E.; Pegg, A. E.; Moody, P. C. *Nucleic Acids Res.* **2000**, *28*, 383–401.
82. Freidman, H. S.; Dolan, M. E.; Pegg, A. E.; Marcelli, S.; Keir, S.; Catino, J. J.; Bigner, D. D.; Schold, S. C., Jr. *Cancer Res.* **1995**, *55*, 2853–2857.
83. Brent, T. P.; Houghton, P. J.; Houghton, J. A. *Proc. Natl. Acad. Sci. USA* **1985**, *82*, 2985–2989.
84. Nagasubramanian, R.; Dolan, M. E. *Curr. Opin. Oncol.* **2003**, *15*, 412–418.
85. Gaya, A.; Rees, J.; Greenstein, A.; Stebbing, J. *Cancer Treat. Rev.* **2002**, *28*, 115–120.
86. Brock, C. S.; Newlands, E. S.; Wedge, S. R.; Bower, M.; Evans, H.; Colquhoun, I.; Roddie, M.; Glaser, M.; Brampton, M. H.; Rustin, G. J. *Cancer Res.* **1998**, *58*, 4363–4367.
87. Tolcher, A. W.; Gerson, S. L.; Denis, L.; Geyer, C.; Hammond, L. A.; Patnaik, A.; Goetz, A. D.; Schwartz, G.; Edwards, T.; Reyderman, L. et al. *Br. J. Cancer* **2003**, *88*, 1004–1011.
88. Dolan, M. E.; Pegg, A. E. *Clin. Cancer. Res.* **1997**, *3*, 837–847.
89. Kanzawa, T.; Bedwell, J.; Kondo, Y.; Kondo, S.; Germano, I. M. *J. Neurosurg.* **2003**, *99*, 1047–1052.
90. Friedman, H. S.; Keir, S.; Pegg, A. E.; Houghton, P. J.; Colvin, M. O.; Moschel, R. C.; Bigner, D. D.; Dolan, M. E. *Mol. Cancer. Therapeut.* **2002**, *1*, 943–948.
91. McCormick, J. E.; McElhinney, R. S. *Eur. J. Cancer* **1990**, *26*, 207–221.
92. Johnston, T. P.; McCaleb, G. S.; Montgomery, J. A. *J. Med. Chem.* **1963**, *6*, 669–681.
92a. Johnston, T. P.; McCaleb, G. S.; Oplinger, P. S.; Montgomery, J. A. *J. Med. Chem.* **1966**, *9*, 892–911.
93. Weiss, R. B. *Cancer Treat. Rep.* **1982**, *66*, 427–438.
94. Panasci, L. C.; Green, D.; Nagourney, R.; Fox, P.; Schein, P. S. *Cancer Res.* **1977**, *37*, 2615–2618.
95. Kohn, K. W. *Cancer Res.* **1977**, *37*, 1450–1454.
96. Erickson, L. C.; Laurent, G.; Sharkey, N. A.; Kohn, K. W. *Nature* **1980**, *288*, 727–729.
97. Tong, W. P.; Kirk, M. C.; Ludlum, D. B. *Cancer Res.* **1982**, *42*, 3102–3105.
98. Bristol-Myers Squibb. For clinical use and pharmacokinetic data, see the product insert. http://www.bms.com (accessed May 2006).
99. Hata, T.; Sano, Y.; Sugawara, R.; Matsumae, A.; Kanamori, K.; Shima, T.; Hoshi, T. *J. Antibiotics Ser. A* **1956**, *9*, 141–146.
100. Iyer, V. N.; Szybalski, W. *Science* **1964**, *145*, 55–58.
101. Kumar, G. S.; Lipman, R.; Cummings, J.; Tomasz, M. *Biochemistry* **1997**, *36*, 14128–14136.
102. Tomasz, M.; Das, A.; Tang, K. S.; Ford, M. G. J.; Minnock, A.; Musser, S. M.; Waring, M. J. *J. Am. Chem. Soc.* **1998**, *120*, 11581–11593.
103. Hopkins, P. B.; Millard, J. T.; Woo, J.; Weidner, M. F.; Kirchner, J. J.; Sigurdsson, S. T.; Raucher, S. *Tetrahedron* **1991**, *47*, 2475–2489.
104. Bradner, W. T. *Cancer Treat. Rev.* **2001**, *27*, 35–50.
105. Walters, R. S.; Frye, D.; Buzdar, A. U.; Holmes, F. A.; Hortobagyi, G. N. *Cancer* **1992**, *69*, 476–481.
106. Davidson, N. G. P.; Davis, A. S.; Woods, J.; Snooks, S.; Cheverton, P. D. *Cancer Chemother. Pharmacol.* **1999**, *44*, S18 – S23.
107. Moertel, C. G.; Reitmeier, R. J.; Hahn, R. G. *JAMA* **1968**, *204*, 1045–1048.
108. Cummings, J. *Drug Resist. Updates* **2000**, *3*, 143–148.
109. Cummings, J.; Spanswick, V. J.; Smyth, J. F. *Eur. J. Cancer* **1995**, *31A*, 1928–1933.
110. Rosenberg, B.; Van Camp, L.; Krigas, T. *Nature* **1965**, *205*, 698–699.
111. Rosenberg, B.; Van Camp, L.; Trosko, J. E.; Mansour, V. H. *Nature* **1969**, *222*, 385–386.
112. Chabner, B. A.; Roberts, T. G. *Nature Rev: Cancer* **2005**, *5*, 65–72.
113. Judson, I.; Kelland, L. R. *Drugs* **2000**, *59*, 29–36.
114. Ishida, S.; Lee, J.; Thiele, D. J.; Herskowitz, I. *Proc. Natl. Acad. Sci. USA* **2002**, *99*, 14298–14302.
115. Cohen, G. L.; Ledner, J. A.; Bauer, W. R.; Ushay, H. M.; Caravana, C.; Lippard, S. J. *J. Am. Chem. Soc.* **1980**, *102*, 2487–2488.
116. Caradonna, J. P.; Lippard, S. J.; Gait, M. J.; Singh, M. *J. Am. Chem. Soc.* **1982**, *104*, 5793–5795.
117. Fuertes, M. A.; Castilla, J.; Alonso, C.; Perez, J. M. *Curr. Med. Chem.* **2003**, *10*, 257–266.
118. Markman, M. *Exp. Opin. Drug Saf.* **2003**, *2*, 597–607.
119. De Wit, R.; Roberts, J. T.; Wilkinson, P. M.; de Mulder, P. H. M.; Mead, G. M.; Fossa, S. D.; Cook, P.; de Prijck, L.; Stenning, S.; Collette, L. *J. Clin. Oncol.* **2001**, *19*, 1629–1640.
120. Stewart, D. J.; Dulberg, C. S.; Mikhael, N. Z.; Redmond, M. D.; Montpetit, V. A. J.; Goel, R. *Cancer Chemother. Pharmacol.* **1997**, *40*, 293–308.
121. Laszlo, J.; Lucas, V. S., Jr. *New Engl. J. Med.* **1981**, *305*, 948–949.
122. Chawla, S. P.; Grunberg, S. M.; Gralla, R. J.; Hesketh, P. J.; Rittenberg, C.; Elmer, M. E.; Schmidt, C.; Taylor, A.; Carides, A. D.; Evans, J. K. et al. *Cancer* **2003**, *97*, 2290–2300.
123. Chu, G. J. *Biol. Chem.* **1994**, *269*, 787–790.
124. Dabholkar, M.; Bostick-Bruton, F.; Weber, C.; Bohr, V. A.; Egwuagu, C.; Reed, E. *J. Natl. Cancer Inst.* **1992**, *84*, 1512–1517.
125. Fuertes, M. A.; Alonso, C.; Perez, J. M. *Chem. Rev.* **2003**, *103*, 645–662.
126. Hector, S.; Bolanowska-Higdon, W.; Zdanowicz, J.; Hitt, S.; Pendyala, L. *Cancer Chemother. Pharmacol.* **2001**, *48*, 398–406.
127. Ishikawa, T.; Ali-Osman, F. J. *Biol. Chem.* **1993**, *268*, 20116–20125.
128. Hayakawa, J.; Ohmichi, M.; Kurachi, H.; Kanda, Y.; Hisamoto, K.; Nishio, Y.; Adachi, K.; Tasaka, K.; Kanzaki, T.; Murata, Y. *Cancer Res.* **2000**, *60*, 5988–5994.
129. Calvert, A. H.; Newell, D. R.; Gumbrell, L. A.; O'Reilly, S.; Burnell, M.; Boxall, F. E.; Siddik, Z. H.; Judson, I. R.; Gore, M. E.; Wiltshaw, E. *J. Clin. Oncol.* **1989**, 7, 1748–1956.

130. Pertussini, E.; Ratajczak, J.; Majka, M.; Vaughn, D.; Ratajczak, M. Z.; Gewirtz, A. M. *Blood* **2001**, *97*, 638–644.
131. Fuertes, M. A.; Castilla, J.; Nguewa, P. A.; Alonso, C.; Perez, J. M. *Med. Chem. Rev.* [Online] **2004**, *1*, 187–198.
132. Zdraveski, Z. Z.; Mello, J. A.; Farinelli, C. K.; Essigmann, J. M.; Marinus, M. G. *J. Biol. Chem.* **2002**, *277*, 1255–1260.
133. Kelland, L. R. *Exp. Opin. Investig. Drugs* **2000**, *9*, 1373–1382.
134. Fink, D.; Nebel, S.; Aebi, S.; Zheng, H.; Cenni, B.; Nehme, A.; Christen, R. B.; Howell, S. B. *Cancer Res.* **1996**, *56*, 4881–4886.
135. Sharp, S. Y.; Rogers, P. M.; Kelland, L. R. *Clin. Cancer Res.* **1995**, *1*, 981.
136. Fokkema, E.; Groen, H. H. M.; Bauer, J.; Uges, D. R. A.; Weil, C.; Smith, I. E. *J. Clin. Oncol.* **1999**, *17*, 3822.
137. McKeage, M. J.; Mistry, P.; Ward, J.; Boxall, F. E.; Loh, S.; O'Neill, C.; Ellis, P.; Kelland, L. E.; Morgan, S. E.; Murrer, B. et al. *Cancer Chemother. Pharmacol.* **1995**, *36*, 451–458.
138. Balfour, J. A. B.; Goa, K. L. *Drugs* **2001**, *61*, 631–638.
139. Cvetkovic, R. S.; Figgitt, D. P.; Plosker, G. L. *Drugs* **2002**, *62*, 185–1192.
139a. Elzbieta, I.; Tolcher, A. W. *Curr. Opin Investig. Drugs* **2004**, *5*, 587–591.

Biographies

Robert D Hubbard was born in Franklin, Indiana in 1973. He attended Indiana University at Bloomington, IN, where he received his BS degree in chemistry in 1995. In the fall of 1995, he began to pursue his doctoral degree at the University of Rochester in Rochester, New York, eventually working in the laboratories of Prof Benjamin L Miller. His doctoral thesis focused on the synthesis and biological properties of a class of novel ter-cyclopentane scaffolds. Upon graduation in 2000, he accepted a postdoctoral assistant position in the laboratories of Prof Paul A Wender at Stanford University, Stanford, CA. His work focused on the total synthesis of the natural product Laulimalide, and the development of rhodium-mediated multi component reactions. In 2002, he joined the oncology medicinal chemistry group in GlaxoSmithKline, at Research Triangle Park, NC. In October 2004, he joined the oncology discovery group at Abbott Laboratories, Abbott Park, IL.

Steve Fidanze was born in Chicago, IL in 1969. He attended the University of Illinois at Urbana-Champaign, where he received his BS degree in chemistry in 1992. After working for 3 years in Quality Assurance at Griffith Laboratories in Alsip, IL, he began to pursue his doctoral degree in the fall of 1995 at the University of Illinois at Chicago. He graduated in 2000 from the laboratories of Prof Arun Ghosh, where his work focused on the asymmetric titanium-mediated aldol reaction of aminoindanol-based ester enolates. In 2000, he accepted a postdoctoral position in the laboratories of Prof Yoshito Kishi at Harvard University. His work there focused on the use of NMR databases to assign the stereochemistry of the natural products Mycolactone A and Mycolactone B, as well as their total synthesis. In 2002, he joined the oncology discovery group at Abbott Laboratories, Abbott Park, IL.

Comprehensive Medicinal Chemistry II
ISBN (set): 0-08-044513-6

ISBN (Volume 7) 0-08-044520-9; pp. 129–148

7.07 Endocrine Modulating Agents

J A Dodge, T I Richardson, and O B Wallace, Lilly Research Laboratories, Indianapolis, IN, USA

7.07.1 Disease State

Modulation of endocrine hormones has been a long-standing approach to the treatment of cancers that are dependent on hormones for tumor growth. While early treatment focused on surgical ablation to impact tumor regression, the discovery of pathways that control the biosynthesis and binding of steroid hormones, along with their underlying mechanisms, has provided important strategies for identifying novel anticancer therapies. The treatment of breast and prostate cancers has benefited from the advent of therapeutics that target the steroid hormones, estrogen and androgen, respectively, that can control the growth of these tumors. In particular, research into breast cancer treatment has pioneered the modulation of endocrine hormones for therapeutic benefit. In this chapter, we will primarily focus on the evolution of breast cancer treatment from a drug discovery perspective as well as the implications of this approach to other forms of hormone-dependent cancer.

Breast cancer is the most common malignancy in women and the second leading cause of cancer death (exceeded by lung cancer in 1985).[1] Breast cancer is three times more common than all gynecologic malignancies put together. The incidence of breast cancer has been increasing steadily from an incidence of 1:20 in 1960 to 1:7 women today. The American Cancer Society estimates that 211 000 new cases of invasive breast cancer will be diagnosed in 2006 in the US and 43 300 patients will die from the disease.

The incidence of breast cancer is very low for women in their 20s, then gradually increases to reach a plateau at the age of 45, and then increases dramatically after the age of 50. Fifty percent of breast cancer is diagnosed in women over the age of 65, indicating the ongoing necessity of yearly screening throughout a woman's life. Generally, breast cancer is a much more aggressive disease in younger women. Autopsy studies show that 2% of the population has undiagnosed breast cancer at the time of death. Older women typically have much less aggressive disease than younger women.

7.07.1.1 Types of Breast Cancer[1]

7.07.1.1.1 Ductal carcinoma in situ (DCIS)

DCIS is generally divided into two types, comedo (blackhead: the cut surface of the tumor demonstrates extrusion of dead and necrotic tumor cells similar to a blackhead) and noncomedo. DCIS is early breast cancer confined to the inside of the ductal system. The distinction between comedo and noncomedo types is important as comedocarcinoma in situ generally behaves more aggressively and may show areas of microinvasion (small areas of invasion through the ductal wall into surrounding tissue).

7.07.1.1.2 Infiltrating ductal carcinoma

Infiltrating ductal carcinoma is the most common type of breast cancer, representing 78% of all malignancies. These lesions can be stellate (starlike in appearance on mammography) or well circumscribed (rounded). Patients with stellate lesions generally have a poorer prognosis.

7.07.1.1.3 Medullary carcinoma

Medullary carcinomas comprise 15% of breast cancers. These lesions are generally well circumscribed and may be difficult to distinguish from fibroadenoma by mammography or sonography. Medullary carcinoma is estrogen and progesterone receptor-negative 90% of the time. Medullary carcinoma usually has a better prognosis than ordinary breast cancer.

7.07.1.1.4 Infiltrating lobular carcinoma

Representing 15% of breast cancer, these lesions generally present in the upper outer quadrant of the breast as a subtle thickening and are difficult to diagnose by mammography. Infiltrating lobular carcinoma can be bilateral (involving both breasts). Microscopically, these tumors exhibit a linear array of cells and grow around the ducts and lobules.

7.07.1.1.5 Tubular carcinoma

Tubular carcinoma is an orderly or well-differentiated carcinoma of the breast. These lesions make up about 2% of breast cancers. They have a favorable prognosis, with almost 95% 10-year survival.

7.07.1.1.6 Mucinous carcinoma

Mucinous carcinoma represents 1–2% of carcinoma of the breast and has a favorable prognosis. These lesions are usually well circumscribed (rounded).

7.07.1.1.7 Inflammatory breast cancer

A particularly aggressive type of breast cancer, the presentation is usually noted in changes in the skin of the breast, including redness, thickening of the skin, and prominence of the hair follicles. The diagnosis is made by a skin biopsy, which reveals tumor in the lymphatic and vascular channels 50% of the time.

7.07.2 Disease Basis[1]

Risk factors for breast cancer include the following:

7.07.2.1 Early Onset of Menses and Late Menopause

Onset of the menstrual cycle before the age of 12 and menopause after 50 are associated with increased risk of developing breast cancer.

7.07.2.2 Diets High in Saturated Fat

The types of fat are important. Monounsaturated fats such as canola oil and olive oil do not appear to increase the risk of developing breast cancer in the same way as polyunsaturated fats, corn oil, and meat.

7.07.2.3 Family History of Breast Cancer

Patients with a positive family history of breast cancer are at increased risk for developing the disease. However, 85% of women with breast cancer have a negative family history. Family history includes immediate relatives – mother, sisters, and daughters. If a family member was postmenopausal (50 or older) when she was diagnosed with breast cancer, the lifetime risk is only increased by 5%. If the family member was premenopausal, the lifetime risk is 18.6%. If the family member was premenopausal and had bilateral breast cancer, the lifetime risk is 50%.

7.07.2.4 Late or no Pregnancies

Pregnancies before the age of 26 are somewhat protective.

7.07.2.5 Moderate Alcohol Intake

More than two alcoholic beverages increases the risk for breast cancer.

7.07.2.6 Estrogen Replacement Therapy

Most studies indicate that taking estrogen for more than 10 years may lead to a slight increase in risk for developing breast cancer. However, these studies indicate that, depending on the patient, the positive benefits of taking estrogen as far as reducing the risk for osteoporosis and colon cancer may outweigh the increase in risk that may be associated with estrogen replacement therapy.

7.07.2.7 History of Previous Breast Cancer

Patients with a history of previous breast cancer are at increased risk for developing breast cancer in the other breast. This risk is 1% per year or a lifetime risk of 10%.

7.07.2.8 Therapeutic Irradiation to Chest Wall, i.e., for Hodgkin's Disease (Cancer of Lymph Nodes)

Patients who have had therapeutic irradiation to the chest are at increased risk for developing breast cancer.

7.07.2.9 Moderate Obesity

The relationship of breast cancer to obesity is more complex but is associated with an increased risk.

7.07.3 Experimental Disease Models

The experimental disease models for estrogen-responsive tissues will be discussed in the section on current treatment, below.

7.07.4 Clinical Trial Issues

While the estrogen receptor is expressed in most breast cancer, endocrine therapy by treating with tamoxifen is not effective in all estrogen receptor-positive breast cancer patients, i.e., in a random population of patients with metastatic breast cancer, 30–35% will respond positively, whereas 55–60% of patients with tumors known to be estrogen receptor-positive will respond positively. In addition to this intrinsic form of resistance to endocrine therapy, there is an acquired resistance that occurs following a tumor response when the disease subsequently progresses despite continued treatment.

One of the most significant current clinical trials involving the endocrine modulation of breast cancer is the Study of Raloxifene and Tamoxifen, or STAR. This is a large phase III, double-blind trial in which postmenopausal women are assigned to take tamoxifen ($20\,\text{mg}\,\text{day}^{-1}$) or raloxifene ($60\,\text{mg}\,\text{day}^{-1}$) for 5 years. The primary aim of this trial is to compare these two selective estrogen receptor modulators (SERMs) directly for efficacy and safety parameters with respect to breast cancer, coronary heart disease, and osteoporosis. In particular, the effects of long-term raloxifene therapy on preventing the occurrence of invasive breast cancer in postmenopausal women who are identified as being at risk for the disease will be investigated.

Other trials include the study of SERMs combined with aromatase inhibitors to determine if health outcomes can be improved by combining these two therapies.

7.07.5 Current Treatment

Endocrine therapy has been practiced since the turn of the twentieth century. Pioneering studies by George Beatson in 1896 showed that surgical removal of the ovaries, the primary source of endogenous estrogen, from premenopausal women with metastatic breast cancer could cause tumor regression and improve clinical outcomes.[2] Beatson's landmark study provided the critical link between a steroid hormone, estrogen, and cancer (breast) regulation. Remarkably, Beatson's findings significantly predate the discovery of both estrogen and the estrogen receptor by decades.

7.07.5.1 The Estrogen Receptors and Endocrine Modulation of Breast Cancer

The interaction of the estrogen receptor with its natural ligand, 17β-estradiol (E_2), mediates a number of fundamental physiological processes, including regulation of the female reproductive system and the maintenance of skeletal and cardiovascular health. Pharmacological modulation of the estrogen receptor with the estrogens found in hormone replacement therapy (HRT) provides important clinical benefits in women for the treatment of hot flashes and osteoporosis. However, the increased risk of breast, uterine, and cardiovascular side effects that is associated with estrogen therapies has led to the development of ligands with improved risk-to-benefit ratios. These SERMs, such as tamoxifen and raloxifene, have demonstrated the ability to mimic estrogen in some tissues (bone, liver, and the cardiovascular system) while suppressing the effects of estrogen in other tissues (breast, uterus: **Figure 1**). This unique tissue-selective profile has proven beneficial for the prevention and treatment of breast cancer as well as osteoporosis.

7.07.5.1.1 Tamoxifen and other triphenylethylenes

Tamoxifen is currently approved by the US Food and Drug Administration for the treatment of breast cancer in postmenopausal women, estrogen receptor-positive, metastatic breast cancer in premenopausal women, and metastatic breast cancer in men. Tamoxifen is also approved for adjuvant therapy, either alone or following chemotherapy for early-stage, estrogen receptor-positive, breast cancer in pre- and postmenopausal women. Adjuvant therapy with tamoxifen is approved for 5 years of treatment.

Tamoxifen belongs to a class of nonsteroidal estrogen receptor modulators that are termed triphenylethylenes (TPEs). In recent years, the estrogen agonist activities of these compounds in the skeletal and cardiovascular systems have been described. Although they partially antagonize the effects of estrogen on the uterus, evidence to date suggests that, in the absence of estrogen, the members of this structural class also tend to induce some level of uterine stimulation, hence their classification here as partial agonists.

One of the first TPEs to achieve clinical significance was clomiphene,[3] available as a one-to-one mixture of double-bond isomers. Paradoxically, although developed as a contraceptive, it has been mainly utilized for the induction of ovulation in anovulatory women.[4] Estrogen agonist effects of clomiphene in skeletal tissue have been reported in an ovariectomized (OVX) rat model of postmenopausal osteoporosis.[5,6] Clomiphene has also been reported to inhibit bone resorption in vitro and to decrease serum markers of bone resorption in menopausal/castrated women.[7,8] In the rat

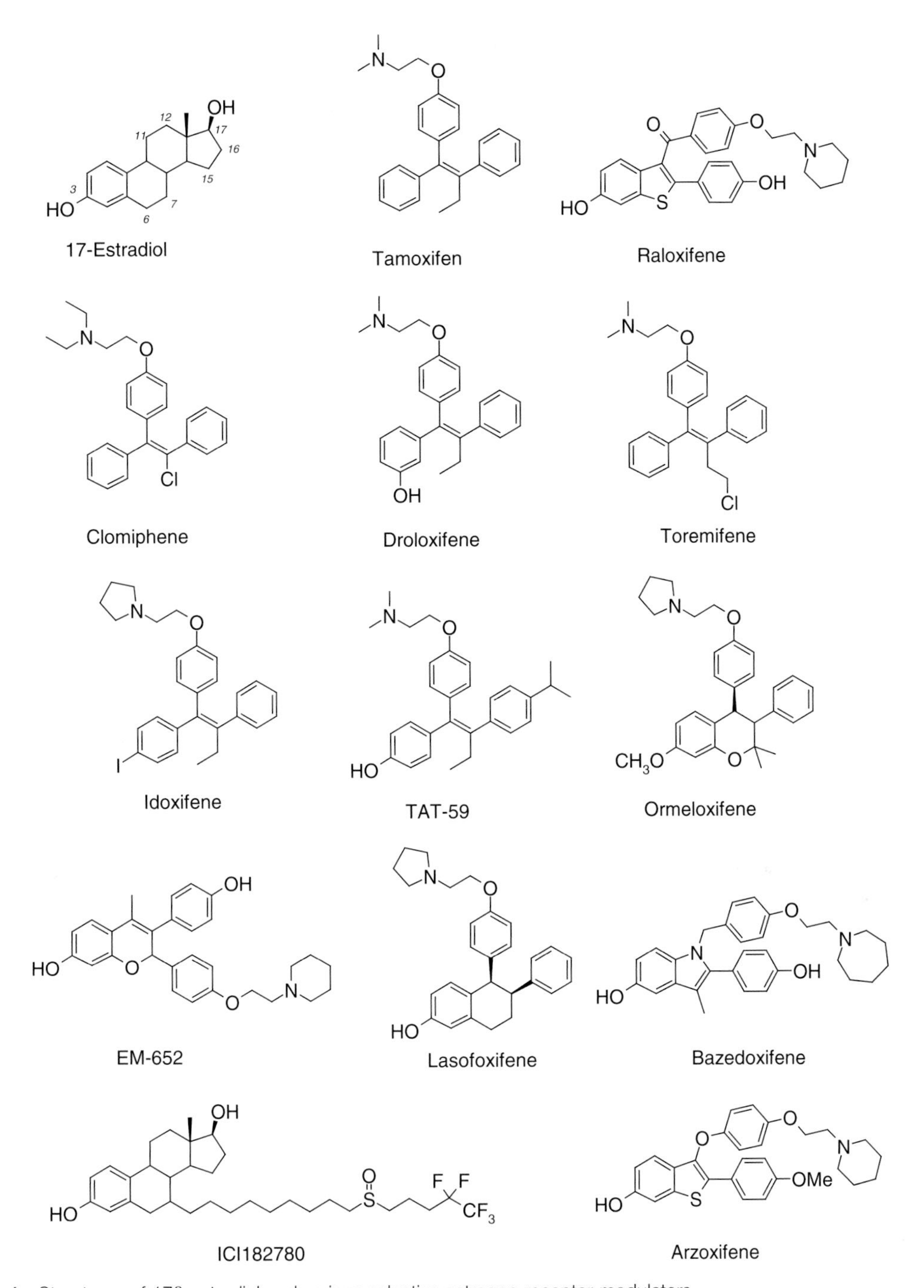

Figure 1 Structures of 17β-estradiol and various selective estrogen receptor modulators.

uterus, clomiphene and each of its components have been shown to stimulate the epithelium potently, and to stimulate a variety of estrogenic effects in OVX animals.[9–11]

The widespread clinical use of tamoxifen for the treatment of breast cancer has resulted in a large body of evidence with respect to its effects in other tissues. In several clinical trials, tamoxifen has demonstrated effectiveness in the preservation of bone mineral density at the lumbar spine, femoral neck, and forearm in postmenopausal women.[12–15] These effects on bone density are accompanied by an estrogen-like reduction of serum markers of bone turnover.[16]

Likewise, significant decreases in risk factors for cardiovascular disease have been observed in postmenopausal women treated with tamoxifen.[17] In general, significant decreases in low-density lipoprotein cholesterol and lipoprotein

(a) have been observed, with modest increases in triglycerides and little or no change in high-density lipoprotein cholesterol.[18–20] These effects have coincided with a reduction in mortality due to cardiovascular disease in patients treated with tamoxifen.[21–23] Whether the cardiovascular benefits of tamoxifen are mediated via its interaction with the estrogen receptor have not been fully established. Alternative mechanisms, including inhibition of cholesterol synthesis,[24] inhibition of lipid peroxidation,[25] and decreases in membrane fluidity, have been proposed.[26] In a primate model, tamoxifen has also been shown to inhibit the progression of coronary artery atherosclerosis significantly.[27]

Although clearly an antagonist of estrogen action in breast tissue, tamoxifen has been shown to induce endometrial hyperplasia in the OVX rat model.[28] Notwithstanding the considerable positive effects described above, there continues to be concern about the increased risk of endometrial cancer associated with tamoxifen use.[29–32] Tamoxifen has also been shown to induce DNA adduct formation and liver cancer in rats.[33,34]

Concern over this potential carcinogenicity in the uterus and liver has led to the development of tamoxifen analogs. It has been speculated that blocking the metabolic hydroxylation of tamoxifen, which occurs primarily at the 4-position, might lead to agents in which these unwanted side effects are reduced. Several of these drugs, including droloxifene, toremifene, and idoxifene, are being evaluated for the treatment of breast cancer.[35–38]

Toremifene has been reported to reduce serum cholesterol in breast cancer patients and to be less uterotrophic than tamoxifen in the rat.[39,40] In postmenopausal women, however, its estrogenic effects on the uterus have been reported to be comparable to tamoxifen.[41]

In OVX rats, droloxifene has been reported to reduce serum cholesterol by 40–46% and to protect against loss of bone mineral density, but with less uterine stimulation than tamoxifen.[42,43] Estrogenic effects on the skeleton have also been observed by histomorphometry in both cortical and cancellous bone.[44]

Ormeloxifene is a structurally distinct estrogen receptor modulator that has been reported to have tissue-selective effects on bone in OVX rats.[45] Other authors, however, have reported potent uterine stimulation with this compound in OVX rats.[46] Ormeloxifene has also been found to inhibit osteoclastic bone resorption in vitro.[47]

The discovery that some compounds are able to mimic the effects of estrogen in the skeletal and cardiovascular system, yet produce almost complete antagonism in the breast and uterus, has led the term selective estrogen receptor modulator to identify more appropriately the pharmacology of this class of compounds. One of the most investigated of these compounds in this regard is raloxifene (keoxifene, LY139481; HCl, LY156758). Raloxifene was originally developed as a therapeutic agent for breast cancer.[48] Raloxifene is currently approved for the prevention and treatment of osteoporosis in postmenopausal women. Long-term raloxifene therapy to women who are postmenopausal and osteoporotic reduced the incidence of invasive breast cancer by 66% and invasive estrogen receptor-positive breast cancer by 76% compared to placebo.[49]

Uterine effects observed with raloxifene have been qualitatively different from those observed with TPEs. Although a modest increase in uterine weight has been observed in OVX rats treated with raloxifene, this increase was not dose-dependent and was not accompanied by similar changes in epithelial cell height or other estrogen-sensitive parameters of uterine histology.[50] In OVX rats, the TPEs tamoxifen, droloxifene, and idoxifene were found to induce a larger maximal stimulation of uterine weight and to induce uterine eosinophilia, while raloxifene did not.[51] Furthermore, raloxifene has been shown to antagonize the uterotrophic effects effectively of tamoxifen in OVX rats.[52] In postmenopausal women, raloxifene has been shown to exert a significant suppression of estrogen effects on the uterine epithelium relative to placebo-treated controls.[53]

Benzopyrans based on EM 652 have been reported to have an activity profile similar to raloxifene with respect to bone protection and serum cholesterol reduction in OVX rats.[54] Other compounds in clinical development for the prevention and/or treatment of osteoporosis include arzoxifene, lasofoxifene, and bazedoxifene.[55]

7.07.5.1.2 Pure antiestrogens

While SERMs can mimic the pharmacology of the natural hormone, pure antiestrogens represent a class of therapeutic agents that are devoid of estrogen agonism, regardless of the target tissue. Initially introduced by Wakeling in 1988, these compounds exhibit no estrogenic activity in the rat uterus, vagina, and hypothalamic–pituitary axis, as well as effectively antagonizing the stimulatory effects of estrogen.[56] ICI 182,780 (Faslodex) is being investigated as first-line therapy for metastatic breast cancer and as second-line therapy in combination with aromatase inhibitors. The drug has also proven effective against tamoxifen-resistant and tamoxifen-stimulated tumors in vitro and in vivo.

The clinical development of SERMs such as tamoxifen, raloxifene, and ICI 182,780 has stemmed the discovery of novel estrogen receptor ligands for ERα and ERβ. The remainder of this section focuses on these new ligands, their structure–activity relationships, and biological activities.

7.07.5.1.3 Diaryls, biphenyls, and related structures

Substituted biphenyls, which were assembled using a Suzuki reaction, have been demonstrated to have very high affinity for the estrogen receptor.[57] Substituted 4-hydroxy-4′-hydroxymethyl-biphenyls were extensively explored following a report on the estrogenic activity of related molecules.[58] *Ortho*-disubstituted analogs such as biphenyl 1 were shown to have the highest binding affinity for human recombinant ERα (relative binding affinity (RBA) = 106%; **Figure 2**). Affinity was reduced when either or both bromines were replaced, although the *ortho*-dichloro analog still possessed good affinity (RBA = 34%). Unfortunately no data on the transcriptional activity of these compounds were reported.

Relatively simple bisphenolic amides have also been shown to act as estrogen receptor ligands (**Figure 3**). Stauffer and co-workers have prepared a series of *N*-phenyl benzamides and *N*-phenyl acetamides, together with their thioamide analogs in some cases.[59] In general the thioamide derivatives were higher affinity ligands compared to their amide congeners. The *N*-phenyl benzamides and thioamides, particularly those with a benzylic-CF_3 substitution pattern, had the highest affinity when assayed using purified estrogen receptor preparations at 0 °C, as exemplified by thioamide (**2**: RBA ERα = 25%; RBA ERβ = 8.9%). The enantiomers of **2** were subsequently separated and tested. One enantiomer displayed about eightfold selectivity for ERα, while the other enantiomer was considerably less ERα-selective (ERα/β ratio = 1.6). In a cellular transcription assay using human endometrial cancer (HEC-1) cells, the benzamides in general were full or almost full agonists through ERα and most were partial antagonists through ERβ. The highest-affinity carboxamide reported, benzamide (**3**: RBA ERα = 27%; RBA ERβ = 1.3%) was found to be a full agonist through ERα and ERβ in transfected HEC-1 cells. However, it was 500-fold more potent at activating transcription through ERα than ERβ. Interestingly, this 500-fold transactivation selectivity was considerably greater than the 21-fold binding selectivity. Computational, x-ray, and nuclear magnetic resonance studies suggest that the compounds exhibited a preference for the *s-cis* conformation where the aromatic rings are *syn*.

1,2-Diphenylenediamine platinum complexes such as **4** (**Figure 4**) have previously been shown to have estrogenic and antitumor properties.[60] The aryl rings are forced into a synclinal orientation, unlike traditional estrogen receptor

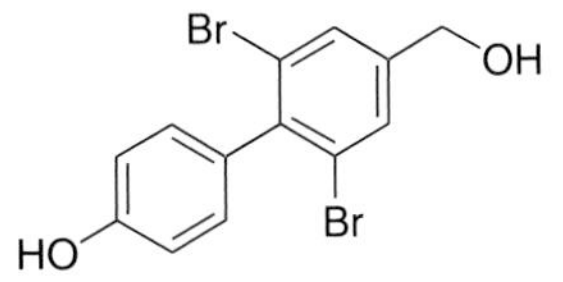

Br OH HO Br
1

Figure 2 Representative biphenyl estrogen receptor ligand.

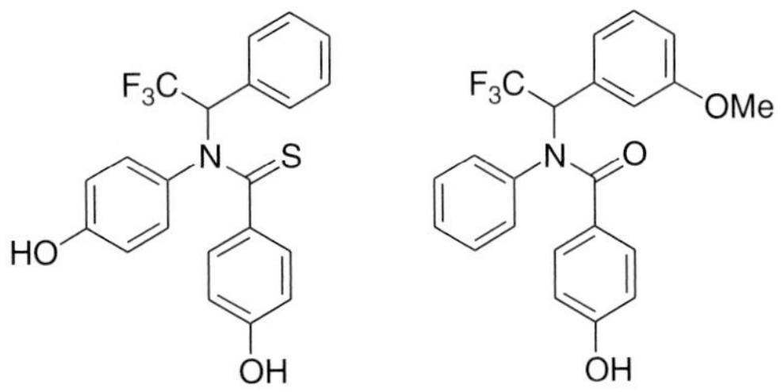

F3C N S HO OH
2
F3C OMe N O OH
3

Figure 3 Representative bisphenolic thioamides and amides.

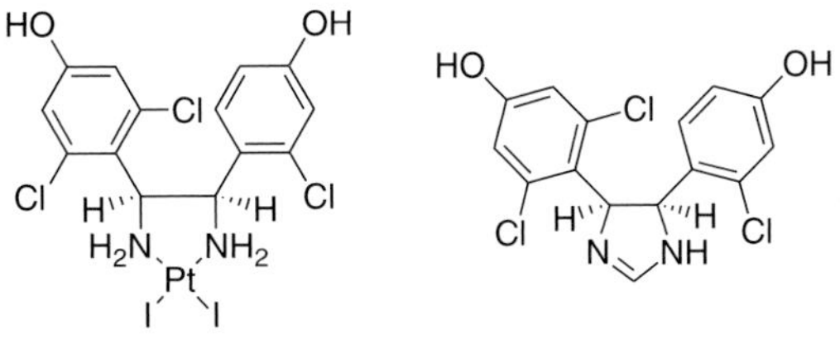

HO OH Cl Cl H H Cl H2N NH2 Pt I I
4
HO OH Cl Cl H H Cl N NH
5

Figure 4 1,2-Diphenylenediamine derivatives.

ligands such as diethylstilbestrol, which consists of an antiperiplanar *E*-stilbene. In an effort to explore the estrogenic behavior of related structures further, Gust and co-workers prepared a series of conformationally constrained analogs of **4**, where the orientation of the aryl rings is defined using a heterocyclic ring.[61] Piperazine, imidazoline, and imidazole cores were examined, with the most interesting compounds emanating from the imidazoline series.

The compounds were tested in competition experiments using calf uterine cytosol. All showed very weak binding (RBA$\leq$0.1%) in this assay. However, in spite of the weak binding affinity, the compounds were capable of activating transcription in a human breast cancer cell line (MCF-7). For example, imidazoline (**5**) showed maximal efficacy equivalent to that of E_2 at 5×10^{-7} mol L^{-1}. The authors postulate that one phenol of **5** forms contacts with the receptor at Glu-353 and Arg-394, while the other phenol interacts with Asp-351. Compounds such as **5** which do not displace E_2 from its binding site but do demonstrate gene activation have been classified as type II estrogens, as opposed to type I estrogens, such as diethylstilbestrol, which do bind in a similar mode to E_2.[62]

Gust *et al.* further optimized the 2,3-diaryl piperazine series by exploring substitution of the aryl rings and the effect of *N*-alkylation (**Figure 5**).[63] Analogs lacking the *N*-alkyl substituent, such as **6**, were essentially devoid of estrogen receptor-binding activity (RBA<0.02%, calf uterine cytosol, 4 °C) while the *N*-ethyl derivatives displayed marginally higher affinity (e.g., **7**, RBA = 0.42%). The compounds were then tested for their ability to activate gene expression in stably transfected MCF-7–2a cells. Both **6** and **7** were shown to activate gene expression at a 1 μmol L^{-1} concentration to the level of 73% and 74%, respectively, relative to E_2. Methylation of the phenols significantly abrogated gene activation (<5% relative activation).

The pyrroloindolizine ring structure (**Figure 6**), has recently been used as a core in the preparation of new estrogen receptor ligands.[64] The compounds were assessed using estrogen receptor-rich cytosol from rabbit uterine tissue. Not surprisingly, *para*-hydroxylation on the phenyl ring was found to be necessary for potent binding affinity. An additional hydroxy group on the pyrroloindolizine core improved estrogen receptor affinity significantly, as demonstrated by **8** (IC_{50} = 0.9 nmol L^{-1}) and **9** (IC_{50} = 0.9 nmol L^{-1}). Given the lack of a basic side chain, which is a common feature of estrogen receptor antagonists, these compounds are expected to be agonists in uterine tissue. Indeed, NNC 45–0095 (**10**, IC_{50} = 9.5 nmol L^{-1}) was evaluated in an Ishikawa human endometrial adenocarcinoma cell line and was shown to be a full agonist (EC_{50} = 13 nmol L^{-1}) with a maximal efficacy of 105% relative to the agonist moxestrol.[65] The in vivo estrogenic nature of **10** was examined in immature mice, where it was shown to be a potent agonist on the uterus (EC_{50} = 0.96 nmol g^{-1}). The uterotrophic nature of pyrroloindolizine (**10**) was also seen in mature OVX mice at doses of 0.1–10 nmol g^{-1}. The ability of **10** to prevent loss of bone mineral density of the distal femur completely was demonstrated at a 5 nmol g^{-1} dose, administered subcutaneously five times per week for 5 weeks.

Using the known estrogen receptor agonist coumestrol (**11**) as a conceptual lead, Jacquot *et al.* prepared a series of benzothiazinone-based estrogen receptor ligands (**Figure 7**).[66] The compounds were tested in MCF-7 proliferation and binding assays. Both methyl ether (**12**) and phenol (**13**) were shown to be high-affinity estrogen receptor ligands

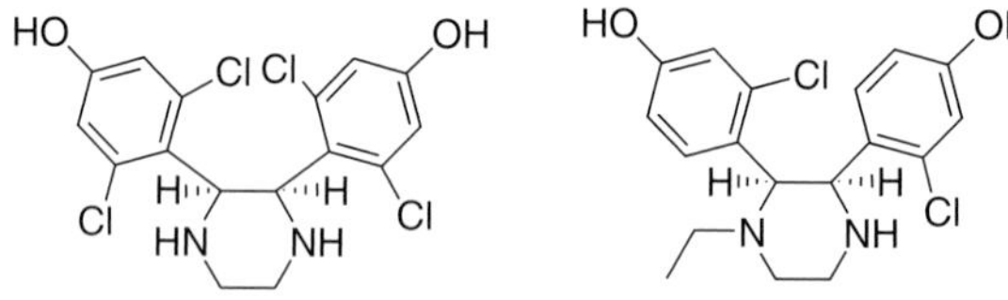

6 **7**

Figure 5 Estrogenic piperazine derivatives.

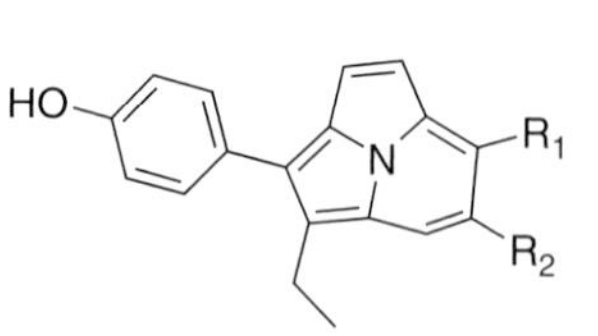

8: R_1=OH; R_2=H
9: R_1=H; R_2=OH
10: R_1=R_2=H (NNC 45-0095)

Figure 6 Pyrroloindolizine-based estrogen receptor ligands.

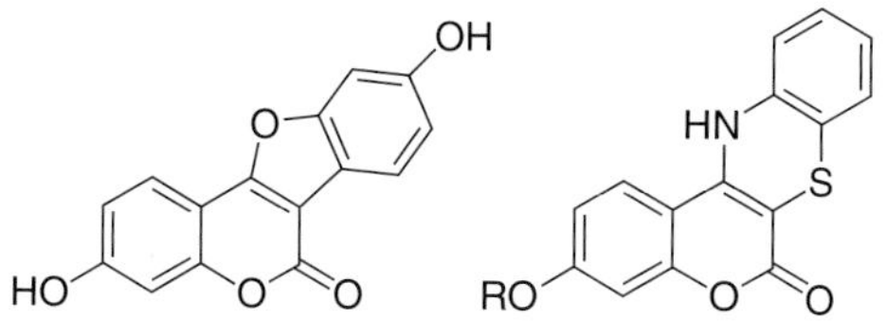

Figure 7 Benzothiazinone-based estrogen receptor ligands.

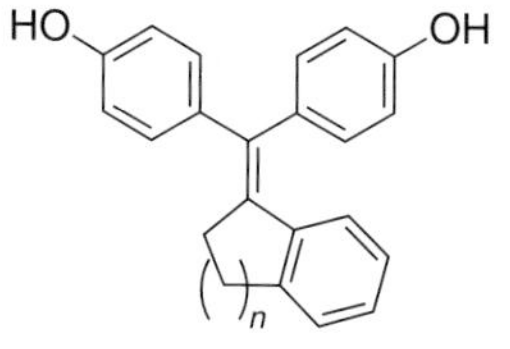

Figure 8 Cyclic tamoxifen derivatives.

($K_d = 2.54$ and $1.60\,\text{nmol}\,\text{L}^{-1}$, respectively) and both compounds stimulated MCF-7 proliferation, with methyl ether (**12**) displaying greater activity at a $1\,\mu\text{mol}\,\text{L}^{-1}$ concentration; the EC_{50} of **12** was calculated to be $2.4 \times 10^{-8}\,\text{mol}\,\text{L}^{-1}$. The activity was inhibited by the antiestrogen ICI 182,780, thus confirming that the effect was estrogen receptor-mediated. In addition, molecular modeling studies comparing the benzothiazinones to E_2 were consistent with the compounds acting as estrogen receptor ligands. In transient transfection studies using MCF-7 cells, **12** and **13** showed a modest capacity to activate transcription.

7.07.5.1.4 Triphenylethylenes and related structures

Tamoxifen and related structures containing a triphenylethene backbone continue to generate interest in the search for new estrogen receptor ligands with unique structural or biological properties. Many modifications to the backbone have been investigated, with heterocyclic replacements predominating. The process of designing new molecules which modulate estrogen receptor activity often begins by first synthesizing high-affinity ligands which possess agonist activity, and subsequently appending a basic side chain in an effort to produce a true SERM, with tissue-selective antagonism.

7.07.5.1.4.1 Triphenylethylenes without antagonist side chains

Kim and Katzenellenbogen have reported a series of ring-constrained tamoxifen analogs where the alkyl substituent is tethered via a five-, six-, or seven-membered ring to the distal phenyl ring (**Figure 8**).[67] The compounds generally bound well to both ERα and ERβ, with the six- and seven-membered ring being favored (14, $n = 2$ or 3). Interestingly, 14 ($n = 2$) displayed modest binding selectivity for ERβ (RBA = 240%, purified commercial estrogen receptor) compared to ERα (RBA = 132%), unlike 4-hydroxytamoxifen, which exhibited a twofold selectivity for ERα, and **14** ($n = 3$), which exhibited approximately threefold selectivity for ERα. Reduction of the olefin which afforded the saturated analogs led to compounds with lower binding affinity and no ERα/β-selectivity. The affinities measured with purified receptor were much higher than those measured in a uterine cytosol preparation, an observation attributed to the high lipophilicity of the compounds and the likelihood of nonspecific protein binding in the cytosol preparation.

The construction of estrogen receptor ligands using solid-phase chemistry has been somewhat limited in scope, since nonsteroidal ligands tend to be relatively complex structures synthesized using a series of carbon–carbon bond-forming reactions which can be lowyielding and difficult to generalize for library preparation. Stauffer and Katzenellenbogen[68] have successfully demonstrated the use of solid-phase, split–split chemistry to synthesize a series of tetrasubstituted pyrazole derivatives targeted at the estrogen receptor. Modifications at the N1 and C5 positions were explored most extensively with limited alkyl substitution at C4 (**Figure 9**). A 12-member pilot library was initially prepared, followed by a more comprehensive 96-member library, with the average purity of the compounds being 50%. The crude compounds were tested in a radioligand binding assay using lamb uterine cytosol at 0 °C and selected compounds were purified by chromatography before retesting. Hydroxy substitution was preferred at

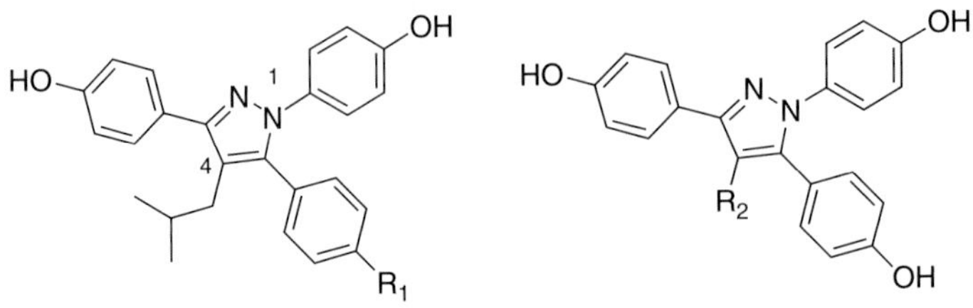

Figure 9 Pyrazole estrogen receptor ligands.

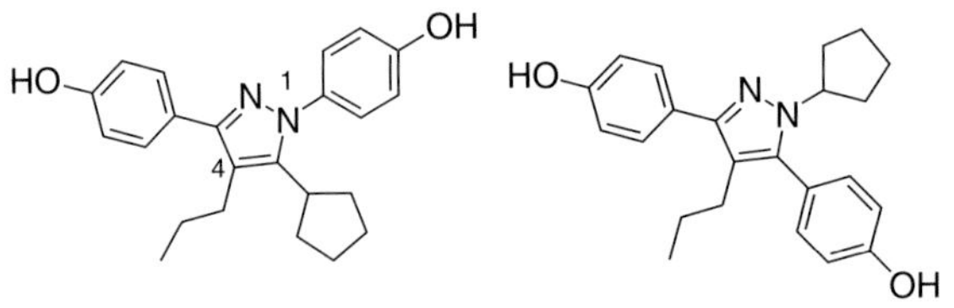

Figure 10 4,5- and 1,4-dialkylpyrazole estrogen receptor ligands.

R_1 (**15**, RBA = 7.6%), although the brominated derivative (**16**) also displayed reasonable binding affinity (RBA = 3.3%. In subsequent work, Stauffer and co-workers[69] explored the nature of the optimal pyrazole C4 alkyl group (**Figure 9**). Modifications at this position (R_2 = Me, Et, Pr, Bu) demonstrated that binding affinity increased with intermediate-size substituents (Et, *n*-Pr and *i*-Bu). Interestingly, propylpyrazole triol (**17**), termed PPT, showed not only a high affinity for ERα (RBA = 49%, purified, full-length, human estrogen receptor) but was also 410-fold selective for ERα (RBA ERβ = 0.12%). PPT was shown to activate transcription via ERα in transfected HEC-1 cells, but had no effect on ERβ-mediated transcription, making PPT the first reported ERα-selective agonist.

Both molecular modeling and chemical studies were used in an attempt to elucidate the binding mode of PPT and related analogs. Systematic deletion of the phenol hydroxyls suggested a binding preference for a mode where the C3-phenol acts as the estrogen A-ring mimic. Docking studies suggested that the N1-phenyl occupies space corresponding to the E_2 C- and D-ring binding pocket. In an effort to rationalize the significant ERα-selectivity, PPT was modeled using an x-ray crystal structure of diethylstilbestrol bound to ERα.[70] The pyrazole core and C4-propyl group were predicted to have contacts to Leu-384 in ERα, which corresponds to Met-384 in ERβ. Because of the increased size of the methionine residue in ERβ relative to the leucine of ERα, the pyrazole is displaced, causing a significant shift in the position of the N1-phenol, the C5-phenol, and the propyl group. This shift in position is thought to account for the considerably weaker binding to the ERβ subtype.

The same group of authors further expanded their work on pyrazole-based SERMs by exploring tetrasubstituted pyrazoles with new substitution patterns.[71] The 4,5-dialkyl pyrazoles (**Figure 10**), as exemplified by **18**, displayed relatively weak binding affinity to both human, purified ERα (RBA = 3.6%) and ERβ (RBA = 0.6%). Compounds with fused 4,5-cycloalkyl substitution were also prepared and likewise were found to have weak binding affinity. The isomeric 1,4-dialkyl pyrazoles were considerably more interesting and were among the highest-affinity pyrazole ligands reported. For example, pyrazole (**19**) is a high-affinity, nonselective ligand (RBA ERα = 74%; RBA ERβ = 71%). It is interesting to note the lack of selectivity in this series since the previously discussed pyrazoles (e.g., PPT) show a distinct preference for ERα.

Having successfully demonstrated the use of pyrazole-based structures as high-affinity estrogen receptor ligands, the researchers working with Katzenellenbogen subsequently explored the synthesis and biological evaluation of ligands with single heteroatom, five-membered heterocyclic cores, such as furans, pyrazoles, and thiophenes.[72] Because of synthetic difficulties, the desired tetrasubstituted thiophenes could not be prepared. However, two trisubstituted thiophenes were investigated but these proved to have very weak estrogen receptor binding (the compounds were initially tested using estrogen receptor from lamb uterine cytosol and the higher affinity ligands were further investigated with human purified receptor). This was not unexpected since trisubstituted pyrazoles were previously shown to have weak affinity.[73] The one pyrrole prepared (**20**) also displayed unexpectedly low affinity, in spite of its

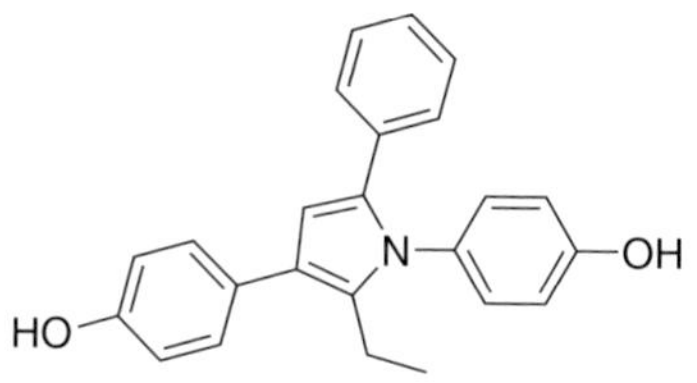

20

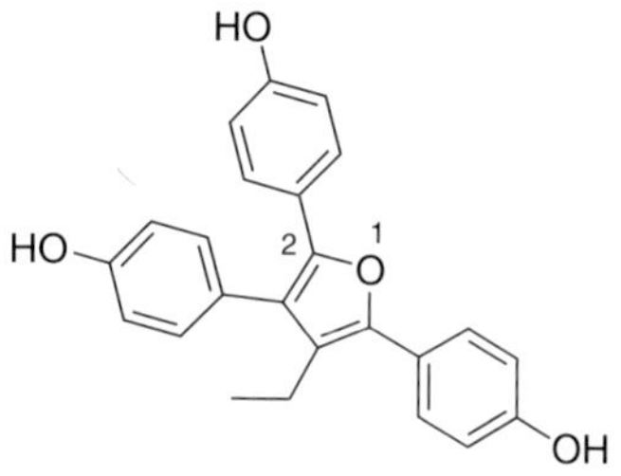

21

Figure 11 Heterocyclic estrogen receptor ligands.

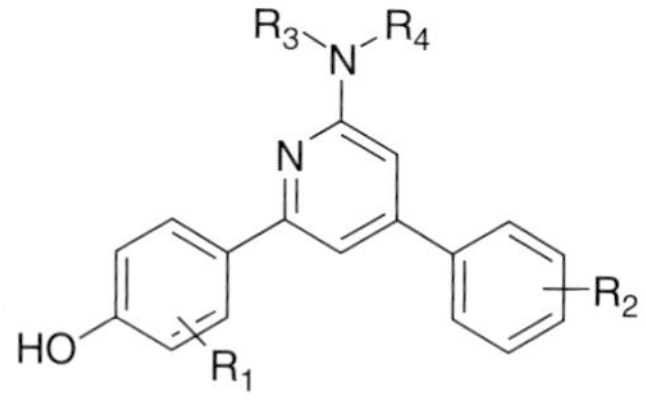

22: R_1=*m*-Me; R_2=H;
R_3=Me; R_4=CH_2CH_2Ph

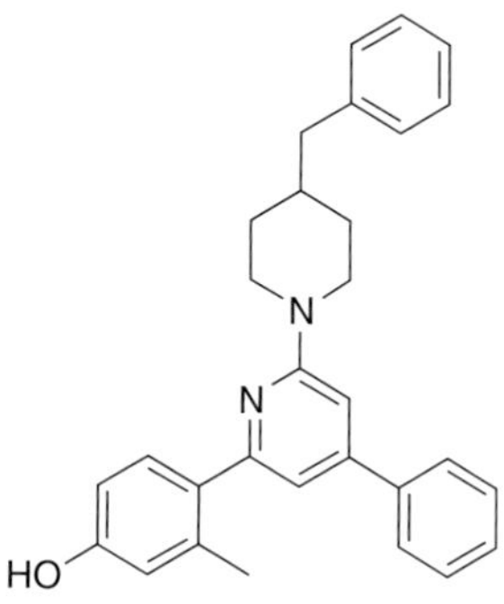

23

Figure 12 Aminopyridine-based estrogen receptor ligands.

tetrasubstitution (**Figure 11**). The higher polarity of the pyrrole relative to the furans and pyrazoles was postulated to account for the lower affinity. The tetrasubstituted furan derivatives, as represented by **21**, proved to be the highest-affinity ligands. While all the furans examined exhibited binding selectivity for ERα, the highest selectivity was achieved with tris-phenolic structures. The optimal C3 alkyl group was shown to be either ethyl or propyl. Furan (**21**) is a high-affinity ERα-selective ligand (RBA ERα = 140%; RBA ERβ = 2.9%). Transcriptional activation studies in HEC-1 cells demonstrated that furan (**21**) was a potent agonist through ERα (EC_{50} = 0.33 nmol L^{-1}) but had no activity on ERβ at concentrations up to 1 μmol L^{-1}. Comparison of **21** with its monophenol and bisphenol analogs suggested that it binds in an orientation where the C2 phenol mimics the estrogen A-ring. Molecular modeling studies were consistent with this hypothesis.

Solution-phase, parallel synthesis techniques were used to prepare a series of 2-amino-4,6-diarylpyridines as estrogen receptor ligands (**Figure 12**).[74] Substitution of the two pendant aryl rings and alkylation of the 2-amino group were investigated. Although most of the compounds displayed modest binding (K_i > 100 nmol L^{-1}), a significant increase in potency was observed for analogs containing a methyl group *meta* to the phenol. For example, *meta*-methyl derivative (**22**) was the most potent compound prepared in this series (K_i ERα: 20 nmol L^{-1}; K_i ERβ: 110 nmol L^{-1}) (compounds were assayed in a scintillation proximity assay using bacterial lysate with overexpressed human estrogen receptor ligand binding domain). Large hydrophobic groups on the 2-amino nitrogen were necessary for increased binding affinity. Unlike many estrogen receptor ligands, the addition of a second *para*- or *meta*-phenol led to a dramatic loss of potency (i.e., R_1 = H; R_2 = *p*-OH). This was rationalized based on molecular modeling studies which suggested that the 4- and 6-phenyl rings span a greater distance than the aryl rings of traditional estrogen receptor ligands such as raloxifene; thus introduction of substitution may cause adverse steric interactions. Compounds of this class display interesting functional activity. In a transient transfection assay using human breast carcinoma cells (T47D), aminopyridine **23** was shown to be a full antagonist at ERβ (IC_{50} = 160 nmol L^{-1}), but a weak partial agonist at ERα (EC_{50} = 30 nmol L^{-1}, 16% efficacy relative to E_2).

A key feature of natural and synthetic estrogen receptor ligands is the presence of an A-ring phenolic group. Tamoxifen, which appears to lack this feature, is in fact metabolized to 4-hydroxytamoxifen, which then regulates transcription via the estrogen receptor. The interaction of the A-ring hydroxyl with Glu-353 and Arg-394 of the receptor is thought to be a key contact for high-affinity ligands. In an effort to identify nonphenolic A-ring isosteres, Minutolo and

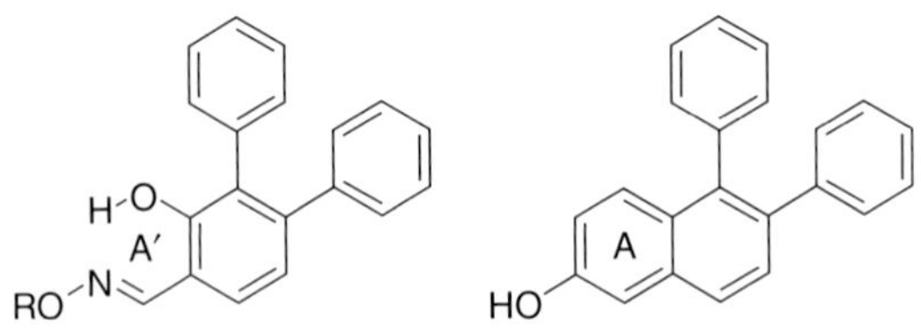

24: R=H
25: R=Me

26

Figure 13 Salicylaldoxime A-ring isosteres.

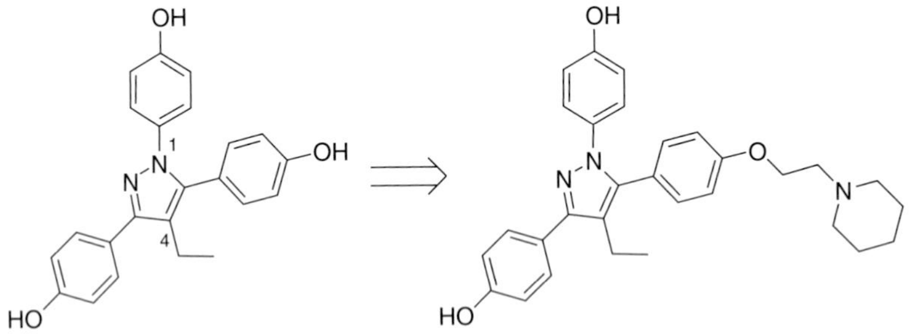

27

28

Figure 14 Pyrazole agonists to antagonists.

co-workers have investigated salicylaldoxime-based estrogen receptor ligands.[75] In such a system, the oxime hydroxy was designed to mimic the phenol (**Figure 13**). This proposition appeared attractive since the estrogen A-ring and the aldoxime A-ring have approximately the same size and geometry, the pK_a of the oxime OH is roughly the same as a phenol, and the oxime OH would be positioned in the same space as the phenolic OH of E_2.

Salicylaldoxime (**24**) was demonstrated to bind to both receptor subtypes (RBA ERα = 1.13%; RBA ERβ = 1.71%, purified, full-length human estrogen receptor) but had 30- to 100-fold lower affinity than the naphthalene analog (**26**). However, methylation of the oxime hydroxyl (i.e., **25**) significantly decreased binding affinity (RBA <0.015% for both receptors), as did the aldehyde precursor to the oxime, suggesting that the oxime hydroxyl may indeed be acting as an A-ring phenol isostere.

7.07.5.1.4.2 Triphenylethylenes with antagonist side chains

The group of Katzenellenbogen has attempted to convert pyrazole-based estrogen receptor agonists such as **27** (**Figure 14**), into antagonists by appending a piperidinyl-ethyl side chain found in many SERMs such as raloxifene.[76] Four pyrazole derivatives were prepared with the basic side chain capping each phenol sequentially or appended to the pyrazole C4 position via an alkyl chain. Of the compounds tested, the C5-piperidinyl-ethoxy-substituted pyrazole (**28**) displayed the highest affinity for both ERα (RBA = 11.5%) and ERβ (RBA = 0.65%) (purified recombinant estrogen receptor) and was ~18-fold selective for ERα. Not surprisingly, the addition of the basic side chain had a significant effect on transcriptional activity. The compound showed no stimulation of transcription in transfected HEC-1 cells, and was found to be a full antagonist on ERα (IC_{50} ~20 nmol L^{-1}) and ERβ (IC_{50} ~160 nmol L^{-1}). The selectivity in the transfection assay parallels the binding selectivity, in that **28** displays ERα-selectivity in both assays. Molecular modeling studies based on the x-ray crystal structure of raloxifene bound to ERα suggested that the N1 phenol occupies the E_2 A-ring binding pocket, and that the C4-ethyl occupies a subpocket normally occupied by the 18-methyl of E_2. This binding mode contrasts with that predicted for pyrazoles such as PPT (**Figure 9**), which lack the basic side chain where the C3-phenol acts as the estrogen A-ring mimic.

In a similar effort to convert triaryl furan estrogen receptor agonists into antagonists, a series of compounds containing the ethoxypiperidine side chain has been examined.[77] Furan (**29**) (**Figure 15**), was shown to have the highest affinity and was almost 25-fold selective for ERα (RBA ERα = 75%; RBA ERβ = 3.1%, recombinant human, full-length estrogen receptor). Derivatives lacking the phenolic OH group at various positions indicated that compound **29**

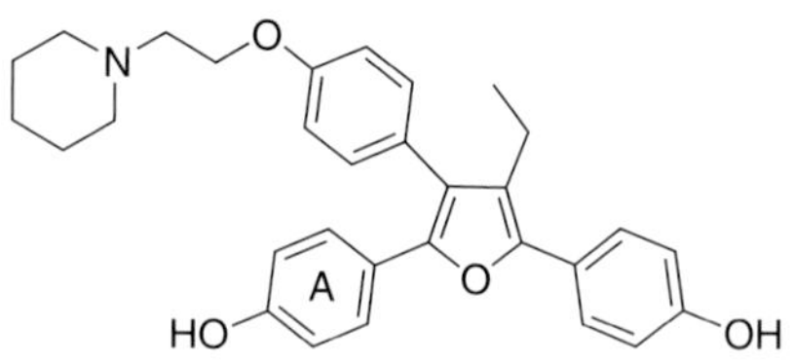

Figure 15 Furan-based SERMs.

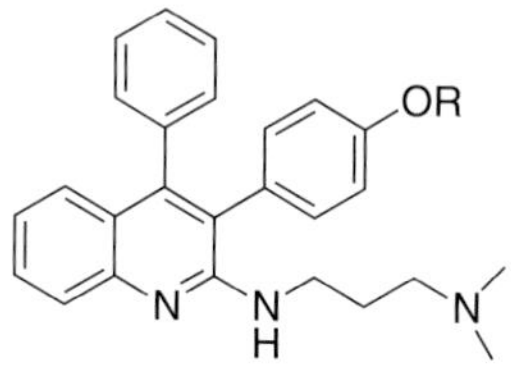

Figure 16 Diphenylquinoline estrogen receptor ligands.

binds to estrogen receptor in a manner where the C5 aryl ring mimics the A-ring of E_2. A similar binding mode was seen previously for the pyrazole series.

Interestingly, both the pyrazole and furan ligands substituted with the basic side chain appear to bind in a different mode to the unsubstituted, parent molecules. In the case of furan and pyrazole agonists such as **17**, it appears that the C2 phenol corresponds to the E_2 A-ring. Compound **29** was subsequently examined in a cotransfection assay using HEC-1 cells and was found to be a full antagonist at ERα at 0.1 μmol L^{-1} but had no effect on transcription through ERβ at this concentration.

Two series of diphenyl quinolines and isoquinolines were prepared where the aminoalkyl basic side chain was designed to occupy a different spatial location compared to typical SERMs such as tamoxifen and raloxifene (**Figure 16**).[78] Binding to ERα was determined with cytosolic estrogen receptor from MCF-7 cells. The compounds, in general, proved to have weak affinity and, as a class, tended to be cytotoxic. Interestingly, compounds that possessed a phenolic group (e.g., **30**) failed to exhibit any binding affinity. Molecular modeling studies attributed this to the fact that the 3-phenyl ring most likely does not occupy the same binding pocket as the E_2 A-ring and hence lacks the electrostatic interaction with Glu-353 and Arg-394. Estrogenicity and antiestrogenicity were measured by progesterone receptor induction in MCF-7 cells. Compound **31**, which was shown to bind ERα weakly, was demonstrated to possess weak antiestrogenic activity, and antiproliferative and cytotoxic effects in MCF-7 cells (IC_{50} = 7.1 μmol L^{-1}). Compounds lacking the basic side chain displayed minimal estrogen receptor-mediated activity.

Although the 2-phenylindole moiety has been known for some time to be a suitable core for the generation of estrogen receptor ligands,[79] recent progress has resulted in the identification of two SERMs which are currently in clinical trials.[80] The compounds (**32** and **33**), which differ only in the nature of the basic side chain, possess a more rigid side chain than the previously disclosed structures (**Figure 17**).

Both compounds display modest α-selectivity and have reasonable binding affinity for ERα (IC_{50} <25 nmol L^{-1}) and ERβ (IC_{50} <90 nmol L^{-1}). The more rigid side chain clearly plays a positive role in blocking transcriptional activity. The compounds were shown to be potent antagonists in MCF-7 cells (IC_{50} <5 nmol L^{-1}). The indoles were also examined for uterotrophic activity in the 3-day immature rat model. Both compounds showed no significant increase in uterine wet weight when dosed alone subcutaneously for 3 days and, in antagonist mode, completely inhibited E_2-induced uterine wet weight gain. In a 6-week OVX rat bone assay, azepine derivative (**33**) was shown to prevent bone loss in the proximal tibia at a dose of 0.3 mg kg^{-1} per day. Total cholesterol was also reduced in this assay at a 0.1 mg kg^{-1} per day dose. Indole (**33**) was also tested in a rat hot-flash assay to assess the ability of the compound to agonize or antagonize the vasomotor response.[81] As was observed for other SERMs such as raloxifene, **33** is an estrogen antagonist in this assay. However, the antagonism was only significant above a 1 mg kg^{-1} dose, which is threefold higher than the bone-sparing dose of 0.3 mg kg^{-1}, thus suggesting that a therapeutic window may exist for treating osteoporosis without inducing hot flashes. Indole (**33**), which has been named bazedoxifene, is currently in phase III trials for the treatment of osteoporosis and pipindoxifene (**32**) has been in phase II trials for the treatment of

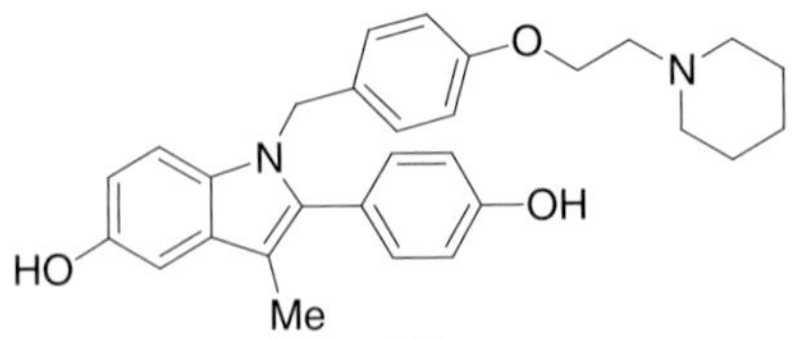

Figure 17 Indole-based SERMs.

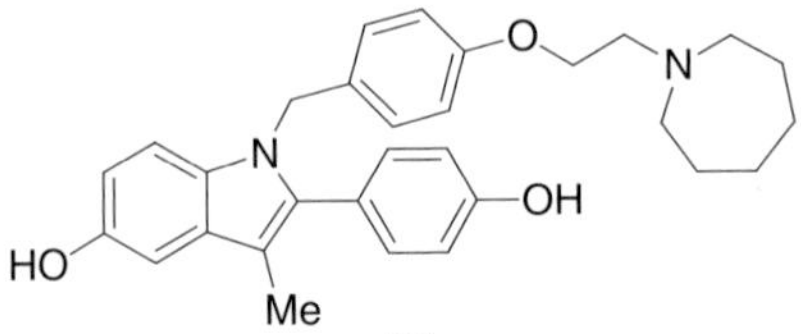

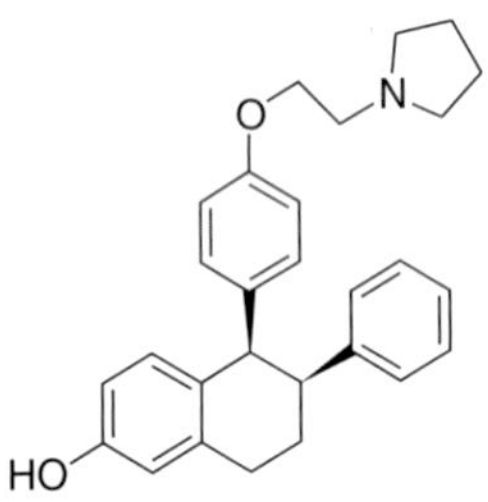

Figure 18 Tetrhydronaphthlyl (lasofoxifene) and chromane SERMs.

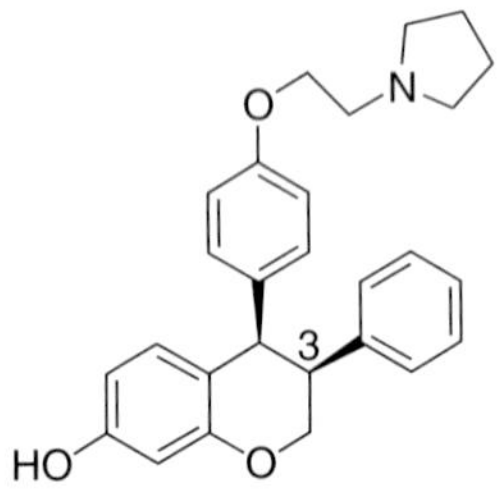

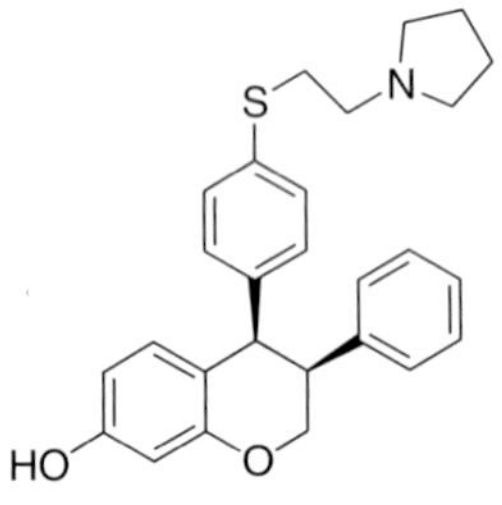

metastatic breast cancer. There have been no reports of an increase in the incidence of hot flashes with **33** during phase I or II trials.

Bury and co-workers have recently reported chromane-derived compounds which act as SERMs (**Figure 18**).[82] The lead compound in the series, NNC 45–0781 (**35**), is structurally related to lasofoxifene (**34**), which is currently in clinical development.[83]

Structure–activity relationship studies indicated that the C3 aryl ring is tolerant of ubstitution, and increased affinity was seen with both *para*- and *meta*-hydroxyl groups. The enantiomers were then separated, and for each compound examined, one enantiomer demonstrated significantly higher affinity. The parent compound, (−) **35**, which was a high-affinity estrogen receptor ligand ($IC_{50} = 2\,nmol\,L^{-1}$, rabbit uterine cytosol), was further evaluated in a series of in vivo assays. In an OVX rat bone assay, chromane (**35**) was shown fully to prevent volumetric bone mineral density loss from the proximal tibia at oral doses of $10\,nmol\,g^{-1}$ per day. Additionally, **35** prevented OVX-induced increases in osteoclast numbers, as evaluated by changes in tartrate-resistant acid phosphatase. Total serum cholesterol was also lowered with compound **35** at doses of $10\,nmol\,g^{-1}$ and higher compared to OVX animals. Positive effects were also observed on levels of endothelial nitric oxide synthase expression, which is speculated to play a role in the atheroprotective effect of estrogen. Beneficial effects were also observed with the maturation of vaginal mucosa; these were similar to the effects of E_2 treatment. In OVX rats, the compound caused a reduction in uterine wet weight and had a nonproliferative effect on uterine epithelia.

In related work, two thio-substituted chromanes were prepared which contained either a pyrrolidine or piperidine basic side chain (**Figure 18**).[84] The compounds were shown to have similar estrogen receptor binding affinity to other, nonsteroidal ligands such as raloxifene, and were partial agonists in Ishikawa human endometrial adenocarcinoma cells. For example, chromane **36** (estrogen receptor binding $IC_{50} = 6.5\,nmol\,L^{-1}$, rabbit uterine cytosol) had a maximal agonist efficacy of 5% relative to moxestrol, with an EC_{50} of $0.5\,nmol\,L^{-1}$. The oxy-substituted analog (±) **35** bound to estrogen receptor with lower affinity ($IC_{50} = 19\,nmol\,L^{-1}$) and was a more efficacious agonist in the Ishikawa cellular assay ($EC_{50} = 0.9\,nmol\,L^{-1}$; $E_{max} = 36\%$).

The introduction of flexibility into the rigid tamoxifen TPE core has been recently investigated.[85] A series of compounds which contained a methylene spacer between the ethylene group and phenyl rings was prepared in an

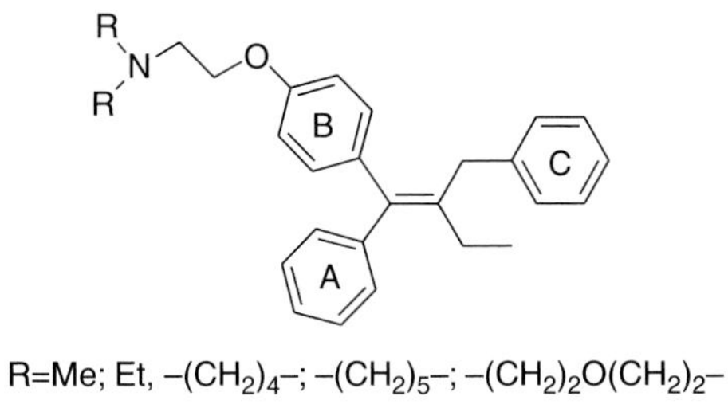

R=Me; Et, $-(CH_2)_4-$; $-(CH_2)_5-$; $-(CH_2)_2O(CH_2)_2-$

37

Figure 19 Tamoxifen analogs.

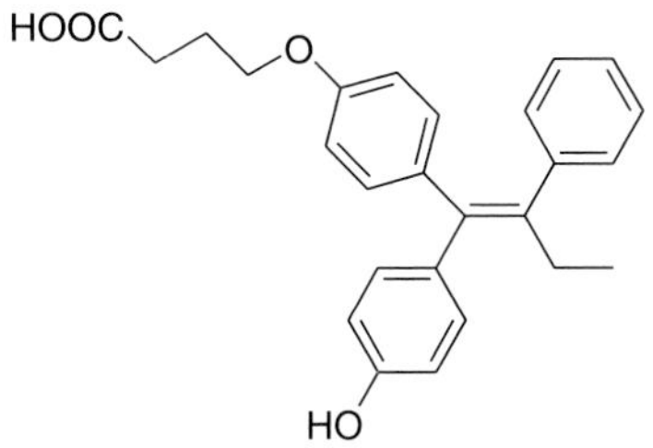

38

Figure 20 Tamoxifen analog containing an acid side chain.

effort to explore the effect on antiestrogenic activity (**Figure 19**). The nature of the basic side chain was also varied, with cyclic (e.g., piperidine) and acyclic (e.g., *N,N*-dimethyl-) amines being investigated. The compounds were tested as mixtures of *E*/*Z* isomers. In general, the compounds where the methylene was inserted between the C-ring and ethylene group displayed the highest activity in MCF-7 cells.

For example, compound **37** ($R = -(CH_2)_4-$) is a $3.54\,\mu mol\,L^{-1}$ inhibitor of MCF-7 proliferation which is approximately threefold more potent than tamoxifen in this assay ($IC_{50} = 11.28\,\mu mol\,L^{-1}$). Other flexible analogs displayed reduced potency, i.e., the introduction of a methylene spacer between the ethylene moiety and the phenyl ring bearing the basic side chain was detrimental to activity, no matter what side chain was used. The compounds were tested in a cytoxicity assay using lactate dehydrogenase as a marker in MCF-7 cells, and were shown to be cytostatic rather than cytotoxic. In general, the flexible analogs were shown to be less cytotoxic than tamoxifen itself.

Although the basic side chain of SERMs is a ubiquitous feature that appears to be critical for conferring tissue selectivity, it has become apparent that the amine functionality can be replaced with a carboxylic acid without losing selective estrogenic activity. Tamoxifen derivatives containing acrylic acid[86] and oxyacetic acid[87] side chains have been previously identified as selective estrogen receptor ligands. More recently, in an effort to reduce the polarity and ionizability of the molecules, Rubin and co-workers have explored oxybutyric acid tamoxifen derivatives (**Figure 20**).[88] For example, oxybutyric acid (**38**: RBA = 10.3%, human recombinant ERα), dosed either orally or subcutaneously, suppressed osteocalcin and deoxypyridinoline bone markers to a degree similar to that of E_2, but had significantly less uterotrophic activity. However, **38** was assayed as a mixture of *E*/*Z* isomers, and it is not clear how each component contributes to the overall profile of the mixture. The synthesis of pure geometrical isomers of tamoxifen analogs continues to generate interest, and a recent report has been published that describes the synthesis of (*Z*)-4-hydroxytamoxifen and an analog containing an acetic acid side chain.[89]

7.07.5.1.5 Conformationally constrained selective estrogen receptor modulators

One of the key structural features that differentiates tamoxifen (a partial agonist on the uterus) from raloxifene (a uterine antagonist) is the position of the basic side chain, which adopts a nonplanar orientation in the case of raloxifene. Grese and co-workers previously described a novel series of SERMs where the side chain is locked in a similar, nonplanar orientation by a trisubstituted, asymmetric carbon, as depicted by 39 (**Figure 21**).[90] In an effort to remove the asymmetric center, a series of related phenanthridine derivatives was subsequently investigated.[91] Compound 40 (RBA = 24%, MCF-7 cell lysate) had the highest affinity of the compounds examined. The corresponding amide (41) was dramatically less potent (RBA = 1%), which suggests that this region of the binding pocket is

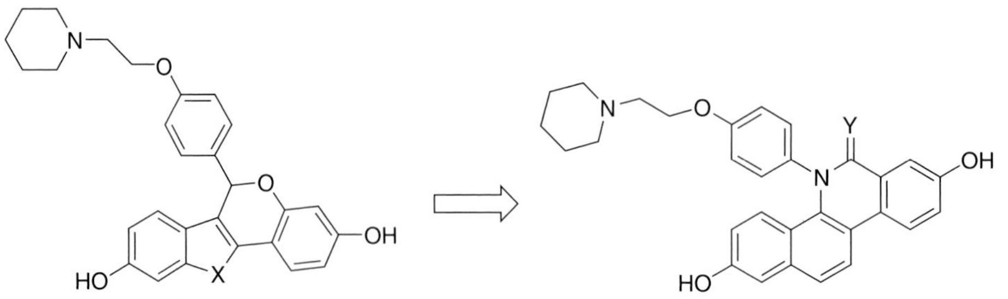

X=S, O, CH=CH

39

40: Y=H, H
41: Y=O

Figure 21 Conformationally constrained raloxifene analogs.

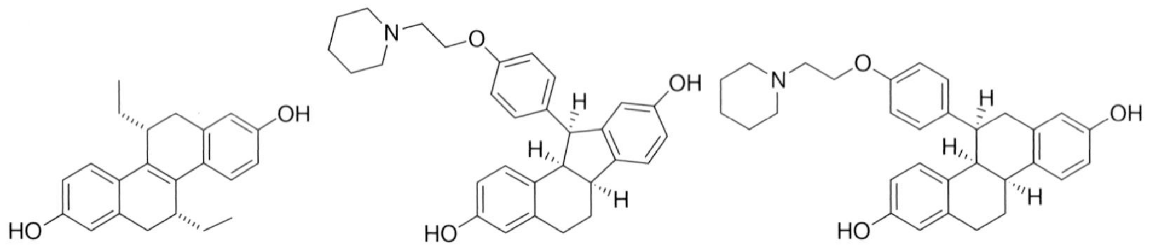

42 **43** **44**

Figure 22 Tetrahydrochrysenes and related structures.

intolerant to hydrophilic substitution. Phenanthridine (**40**) was determined to be an antagonist of MCF-7 proliferation in vitro ($IC_{50} = 1\,nmol\,L^{-1}$) and a partial estrogen antagonist in the immature rat model at oral doses of 1 and $10\,mg\,kg^{-1}$, which is similar to the profile of tamoxifen. The compound also lowered serum cholesterol at oral doses of $0.1–10\,mg\,kg^{-1}$.

Tetrahydrochrysene (THC) derivatives have previously been described as estrogen receptor ligands with interesting biological properties.[92] For example, *cis*-diethyl THC derivative (**42**: **Figure 22**), has been shown to be an agonist through ERα but an antagonist on ERβ.[93] Related saturated analogs containing a *cis*-fused ring junction have recently been described.[94] Interestingly, compounds lacking the basic side chain show some selectivity for binding to ERβ, whereas compounds possessing the basic side chain are ERα-selective ligands. For example, tetrahydrobenzofluorene (**43**: RBA ERα = 38%; RBA ERβ = 10%) and hexahydrochrysene (**44**: RBA ERα = 1.1%; RBA ERβ = 0.06%) have increased affinity for ERα relative to ERβ. Both **43** and **44** have been shown to possess interesting transactivation profiles. Compound **43** was demonstrated to be an antagonist on both ERα and ERβ, whereas **44** was an antagonist on ERβ but was a mixed agonist–antagonist through ERα. The higher affinity of **43** compared to **44** and its higher antagonistic character may be due to a more favorable orientation of the basic side, thus facilitating key contacts between the piperidine nitrogen and the receptor (Asp-354).

Recently, a SERM with improved selectivity for uterus and ovaries has been identified.[95] This naphthalene sulfone-derived SERM (**Figure 23**) binds with high affinity to both estrogen receptors and is a potent inhibitor of MCF-7 cell proliferation ($IC_{50} = 0.86\,nmol\,L^{-1}$). The effects on uterine tissue were assessed at the in vitro level in Ishikawa cells in the presence (antagonism) and absence (agonism) of E_2. In the antagonist mode, this SERM blocks the effects of $1\,nmol\,L^{-1}$ E_2 by >90% with an IC_{50} of $10.7\,nmol\,L^{-1}$. The agonist activity was similar to that of 1 ($29.0 \pm 3.7\%$ over control versus $28.6 \pm 8.50\%$ for 1) and significantly less than that of 4-hydroxytamoxifen ($123 \pm 24\%$), a known uterine agonist. When tested in rodents, this compound proved to be a highly potent, orally active uterine antagonist with an ED_{50} of $0.07\,mg\,kg^{-1}$ at blocking estrogen-induced uterine hypertrophy in immature, ovary-intact rats. Significantly, this analog is greater than fivefold more potent than raloxifene as a uterine antagonist. In addition, it does not have agonist properties in the uterus when administered to OVX rats for 4 days at all doses examined ($0.01–10.0\,mg\,kg^{-1}$, data not shown) based on eosinophil peroxidase activity, a sensitive marker for determining the uterine agonist properties of SERMs.[96] Long-term treatment (42 days) to OVX rats does not cause uterine weight gain.[97] Taken together, the uterine data indicate that this SERM is a potent estrogen antagonist without significant agonist properties

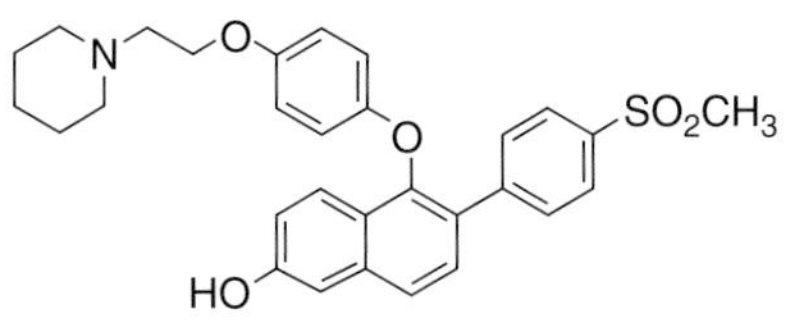

Naphthalene sulfone

Figure 23 An SERM designed for uterine fibroids.

in the uterus. The effects on the uterus and ovaries were studied in 6-month-old ovary-intact female rats.[17] Oral administration of naphthalene sulfone for 35 days at doses of 0.05, 0.5, 1.0, and 3.0 mg kg^{-1} results in a dose-dependent decrease in uterine weight with a maximal inhibitory dose of 1 mg kg^{-1} (51% reduction in uterine wet weight) and an ED_{50} of 0.15 mg kg^{-1}. Morphometric measurements at the 0.5 mg kg^{-1} dose group compared to the control group show that the total area of the uterine wall was significantly reduced by 53% with nearly equal contributions in reduction of endometrial (54%) and myometrial (51%) compartments. These data confirm substantial antagonist activity in both compartments of the uterus in the presence of circulating levels of E_2 over several estrous cycles in the rat. The effects on the ovaries in this study were determined by measuring serum E_2 levels and histologic evaluation of ovarian cross-sections. Treatment with the naphthalene sulfone results in serum E_2 levels that are similar to vehicle-treated animals at doses that exceed the inhibitory effects on the uterus by greater than 60-fold. Statistically significant increases in E_2 are observed at the high dose but are within the normal proestrus range for Sprague–Dawley rats. Histological evaluation of the ovaries of the rats treated with the naphthalene sulfone SERM indicates minimal ovarian stimulation relative to untreated controls, i.e., ovarian weights are decreased at doses >0.5 mg kg^{-1}, there are no ovarian cysts, and granulosa cell hyperplasia is observed in only a few animals and generally at a mild level. These data collectively indicate that this naphthalene sulfone SERM is a potent uterine antagonist with minimal ovarian stimulation in rats.

7.07.5.1.6 ERα-selective ligands

SERMs selective for ERα have recently been reported, including the flavanone,[98] dihydrobenzoxathiin,[99] chromane[100] scaffolds (**Figure 24**). Members of the chromane family have been shown to be highly receptor subtype-selective, while demonstrating potent in vivo antagonism of E_2 in an immature rat uterine weight assay, effectively inhibited ovariectomy-induced bone resorption in a 42-day treatment paradigm, and lowered serum cholesterol levels in OVX adult rat models. The dihydrobenzoxathiins class has typically exhibited low to subnanomolar binding to ERα, with 50- to 100-fold selectivity, and as a result of further study, a derivative has been targeted for development as a potential agent for the treatment of osteoporosis.

7.07.5.1.7 ERβ-selective ligands

The discovery in 1996 of another mediator of estrogen activity has further complicated our understanding of estrogen physiology. This novel receptor, now known as ERβ to distinguish it from the classical ERα, is highly homologous in the DNA-binding domain (95%), but shares only 55% sequence identity in the C-terminal region responsible for ligand binding, nuclear localization, and ligand-dependent transactivation properties. Indeed, knockout models of each subtype (αERKO and βERKO) have demonstrated some of the unique biological roles played by each receptor.[101] Differences in the C-terminal region make possible the discovery and design of ligands with divergent binding affinities and potencies of both agonist and antagonist activity.

Although the overall sequence identity of the C-terminal region is modest between ERα and ERβ, the overall structural differences in the ligand-binding pocket are subtle. Three-dimensional structures of human ERβ bound to genistein and rat ERβ bound to raloxifene have been described.[102] In the structure of genistein (**45**) bound to ERβ, the phenolic hydroxyl of genistein interacts with the Glu-Arg-water triad through a hydrogen bond, while the flavone hydroxy at C7 interacts with the distal His at the end of the cavity (**Figure 25**). The remaining hydroxy at C5, which is presumably hydrogen-bonded to the adjacent carbonyl group, does not interact with the protein. The overall pocket size is slightly smaller in ERβ (390 Å^3 for ERβ-genistein versus 490 Å^3 for ERα-E_2). This reduction in size is mainly due to the substitution of Leu-384 in ERα with Met-336 in ERβ. Of the 22 hydrophobic residues that line the pocket, there is only one other amino acid substitution: Met-421 in ERα is Ile-373 in ERβ.

Krege and co-workers have described the ERβ-selective activity of several naturally occurring phytoestrogens (**Figure 26**).[101] Daidzein (**46**) and genistein (**45**) are closely related, differing only by the addition of a hydroxyl group

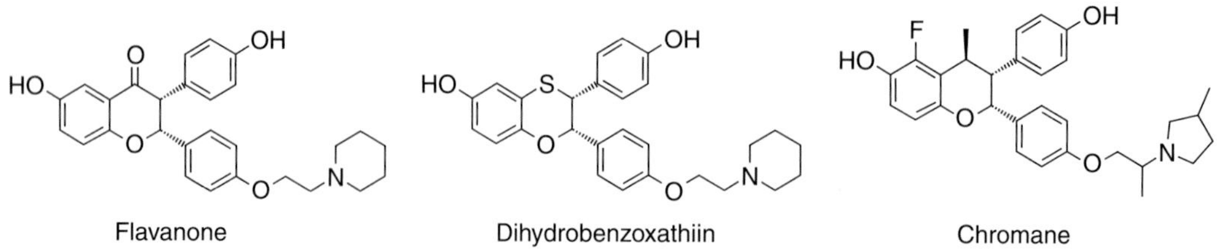

Flavanone

Dihydrobenzoxathiin

Chromane

Figure 24 ERα-selective SERMs.

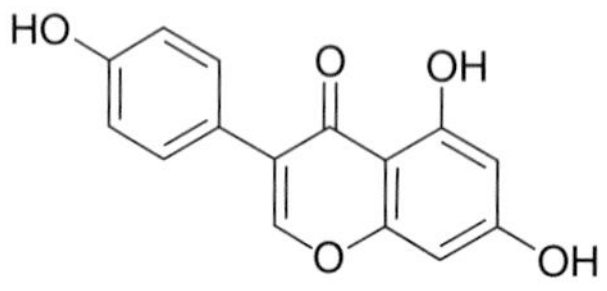

45

Figure 25 Genistein.

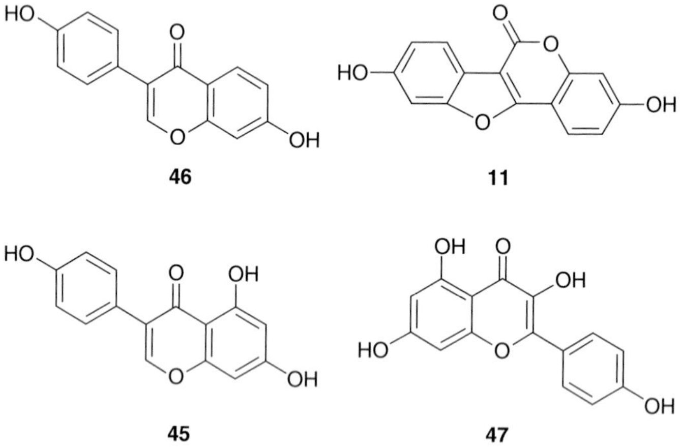

46

11

45

47

Compound	RBA[a] ERβ	RBA[a] ERα	Selectivity β/α
Daidzein (**46**)	0.1	0.5	5
Coumestrol (**11**)	20	140	7
Genistein (**45**)	4	87	20
Kaempferol (**47**)	0.1	3	30

[a]Ratio of IC_{50} values competitive to 17β-estradiol.

Figure 26 ERβ-selective phytoestrogens.

at C5 on the isoflavone ring system of genistein. Although this additional hydroxyl has no specific interaction with the protein, its presence confers a significant increase in binding selectivity for ERβ (20-fold versus fivefold). Indeed, coumestrol (**11**) and kaempferol (**47**) also appear to have additional polar functionalities that are more tolerated by ERβ than ERα and therefore confer selectivity for ERβ. The relative transactivational activity of each compound was also determined at a concentration of 1000 nmol L^{-1}. Although they were all less potent than E_2, some (genistein, zearalenone, and coumestrol) were able to generate a response of the same or greater magnitude than E_2. In addition, the rankings of the estrogenic potency of the phytoestrogens differed for the two subtypes.

Schopfer and co-workers have explored the potential of aryl benzothiophenes as lead structures for the generation of selective ERβ agonists.[104] A series of aryl benzothiophene compounds (**Figure 27**), was synthesized via a Suzuki

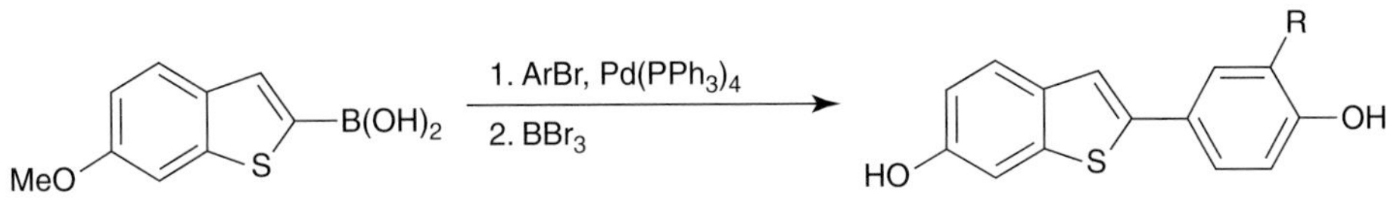

Compd	–R	Binding: EC50 ERβ (nmol L^{-1})	Binding: EC50 ERα (nmol L^{-1})	Function (HeLa): EC50 ERβ (nmol L^{-1})	Function (HeLa): EC50 ERα (nmol L^{-1})	Function (neuronal): EC50 ERβ (nmol L^{-1})
48	–H	114	1410	17	50	10
49	–F	115	1030	32	138	13
50	–Me	1830	3650	302	148	NA
51	–Et	4710	7440	257	191	NA
52	–iPr	3940	>10 000	186	162	257
53	–OH	249	4980	525	363	NA

Figure 27 Benzothiophene compounds.

coupling reaction of benzothiophene boronic acids with aryl bromides and then screened for binding affinity and selectivity at ERβ and ERα. The compounds were also evaluated for their ability to activate an estrogen response element in HeLa cells stably transfected with human ERα or ERβ receptors using a luciferase reporter gene assay. The binding selectivities for ERβ reached 20-fold (R = OH) with binding affinities for ERβ of 100 nmol L^{-1} (R = H or F) to 4 μmol L^{-1} (R = Et). The reported agonist activities are good (EC_{50} = 17 nmol L^{-1}, R = H; EC_{50} = 32 nmol L^{-1}, R = F) but the selectivities are more modest (nonselective to fourfold selective for R = F). The compounds were also assayed for agonist activity in a human neuronal cell line (neuroblastoma SH-SY5Y transfected with human ERβ and the luciferase reporter gene bearing an estrogen response element). All compounds retained agonism with an EC_{50} comparable to that observed in HeLa cells.

Katzenellenbogen's group discovered ERβ-selective diarylpropionitrile (**54**) by screening a select group of compounds for transcriptional activity by ERα and ERβ in human endometrial cancer cells (**Figure 28**).[105] The RBA of diarylpropionitrile (**54**) is 70-fold selective for ERβ and is a full agonist with 170-fold higher relative potency in transcription assays.

A series of analogs were prepared in which the ligand core and aromatic rings of diarylpropionitrile (**54**) were modified (**Figure 28**). The aromatic ring analogs lacking either phenolic hydroxy group lost substantial affinity (30- to 60-fold for ERα and 250- to 370-fold for ERβ). Repositioning of the phenolic hydroxy groups demonstrated a preference for *para*-substitution. Affinity for ERα is lowered twofold if R_2 is *meta*-OH and 10-fold if R_1 is *meta*-OH. For ERβ a *meta*-OH group resulted in a six- to eightfold drop in affinity. When both R_1 and R_2 are *p*-OH, the addition of a methyl group *ortho* to R_2 is tolerated; but interestingly, the addition of a methyl group *ortho* to R_1 resulted in a threefold boost in binding affinity for both ERα and ERβ. Extension of the nitrile by a single methylene resulted in moderate loss of affinity and a significant decrease in selectivity. Addition of a methyl group α to the nitrile is more tolerated, while addition of an α-ethyl group resulted in a 7- to 10-fold increase in ERβ affinity but concurrently diminished ERβ selectivity. The preference of the receptor pocket for the antiperiplanar conformation with respect to the aryl rings is demonstrated by a pair of dinitrile diastereomers. The overall affinity of the meso diastereomer, which prefers an antiperiplanar conformation, is 10-fold more than the DL-diastereomer, which prefers the synclinal conformation. ERβ selectivity is maintained by these dinitriles. The unsaturated analog of diarylpropionitrile (**54**) is identical to the parent while the unsaturated dinitrile has 32-fold more affinity for ERα but only eightfold selectivity for ERβ. The steric analogs in which the nitrile group is replaced by acetylene or propyne have a 10- to 12-fold increase in binding affinity for ERα and a two- to fourfold increase for ERβ relative to the parent nitrile. Thus, the acetylene analogs have better affinity but lower selectivity. The perfluoroalkyl analogs have slightly better affinity but lower selectivity. The polar analogs lose overall binding affinity and ERβ selectivity.

The analogs were examined for transcriptional activities in human endometrial cancer cells. All of the nitrile analogs examined were agonists at ERα and ERβ, although some low-affinity analogs did not reach full efficacy. In general, nitrile analogs with the highest affinities also had the highest potencies. Compounds **54**, **60**, **62**, **64**, and **65** were all more potent and selective than genistein. The steric, perfluoroalkyl, and polar analogs all had only modest transcriptional potency selectivity.

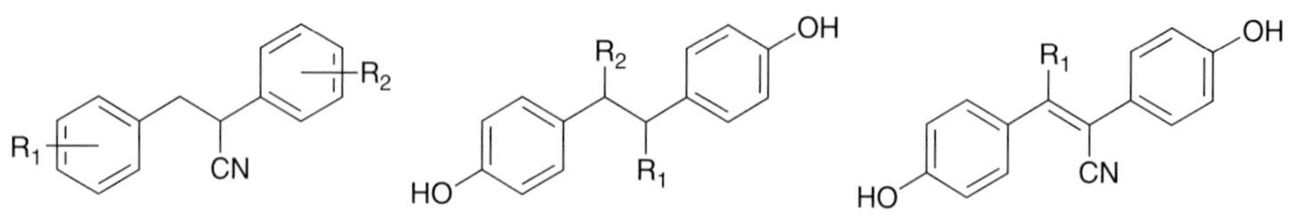

54–60 **61–65, 68–77** **66,67**

Compd	–R_1	–R_2	RBA (%) ERα	RBA (%) ERβ	β/α
54	*p*-OH	*p*-OH	0.25	18	72
55	*p*-OH	–H	0.005	0.048	10
56	–H	*p*-OH	0.01	0.071	7
57	*p*-OH	*m*-OH	0.17	2.9	17
58	*m*-OH	*p*-OH	0.03	2.2	73
59	*p*-OH, *o*-Me	*p*-OH	0.87	60	69
60	*p*-OH	*p*-OH, *o*-Me	0.41	18	44
61	CH_2CN	–H	0.16	2.2	14
62	CN	–Me	1.7	48	27
63	CN	–Et	17	75	4
64	meso-CN	meso-CN	0.55	29	53
65	dl-CN	dl-CN	0.04	2.8	70
66	–H		0.30	17	57
67	–CN		9.8	82	8
68	–C≡CH	–H	3.3	78	24
69	–C≡CMe	–H	3.8	43	11
70	–C≡CH	–C≡CH	0.48	14	29
71	–CF_3	–H	0.71	22	31
72	–CF_2CF_3	–H	10	35	4
73	–$COCF_3$	–H	0.11	1.3	12
74	–COMe	–H	0.24	2.3	10
75	–$CONEt_2$	–H	0.012	0.040	3
76	–CO_2Me	–H	0.76	8.3	11
77	–CH_2OH	–H	0.045	0.071	2

Figure 28 Nitrile analogs.

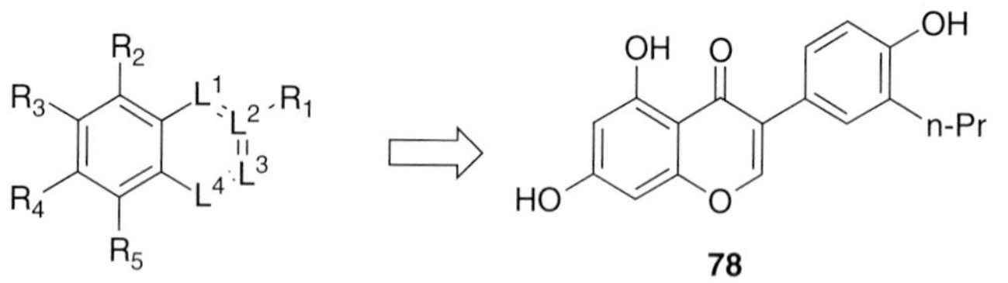

78

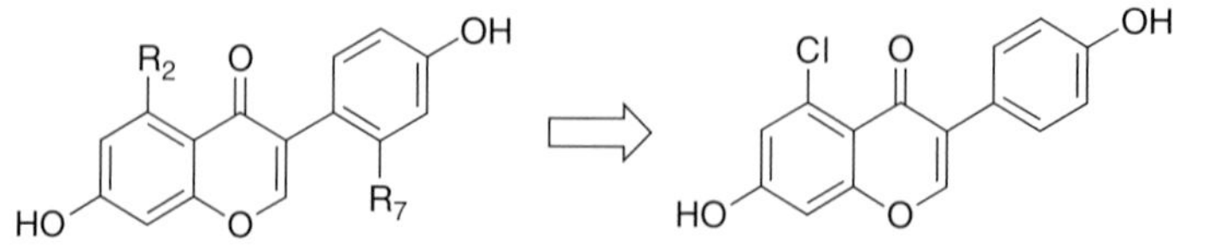

79

Figure 29 Genistein derivatives.

Several groups have disclosed ERβ-selective ligands in the patent literature. A group at AstraZeneca has claimed a series of genistein analogs with impressive binding selectivity for ERβ (**Figure 29**).[106] For example, the genistein analog (**78**) with an *n*-propyl substituent at the C3′ position of the isoflavone ring system has a reported binding K_i of $0.75\ \text{nmol}\ L^{-1}$ at ERα and $>3000\ \text{nmol}\ L^{-1}$ at ERα. Another disclosure from the same group claims a series of analogs

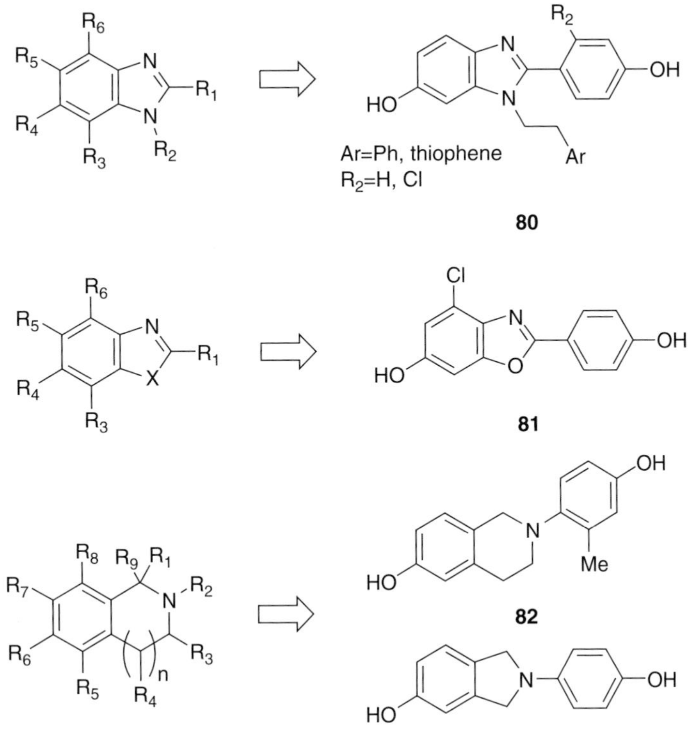

Ar=Ph, thiophene
R_2=H, Cl

80

81

82

83

Figure 30 Heterocyclic ERβ-selective ligands.

in which the C5 hydroxyl of the genistein isoflavone ring system has been replaced by halogens, nitrile, and thiomethyl.[107] The chloro-analog (**79**) has a binding affinity of 4.3 nmol L^{-1} for ERβ with a selectivity of 139-fold.

Researchers at AstraZeneca have also disclosed a series of benzimidazoles[108] of generic structure **80** as selective ERβ ligands (**Figure 30**). The compounds are reported to have binding affinity of 15–2000 nmol L^{-1} with selectivity for ERβ of 50–0.5. The compounds were also assayed for functional activity in human embryonic kidney cells (HEK 293) transiently transfected with either ERα or ERβ and a β-galactosidase reporter. They have reported EC_{50} =1–1200 nmol L^{-1} with selectivity for ERβ of 250–0.005. Related series of benzoxazoles and benzothiazoles[109] have also been disclosed (**Figure 30**). Benzoxazole (**81**) is reported to have a binding affinity of 1.7 nmol L^{-1} with a selectivity for ERβ of 10-fold and is a full agonist (EC_{50} = 1.2 nmol L^{-1}) with selectivity for ERβ of 52-fold. A related series of isoquinolines and isoindolines[110] has also been disclosed (**Figure 30**). Isoquinoline (**82**) is reported to have a binding affinity of 8 nmol L^{-1}, with a selectivity for ERβ of sevenfold and is a full agonist (EC_{50} = 36 nmol L^{-1} for ERβ and 28 nmol L^{-1} for ERα). Isoindoline (**83**) is reported to have a binding affinity of 55 nmol L^{-1} with a selectivity for ERβ of 3.5-fold but is only a partial agonist (EC_{50} = 161 nmol L^{-1}, 77% relative efficiency for ERβ and EC_{50} = 192 nmol L^{-1}, 60% relative efficiency for ERα).

A group from Akzo Nobel have disclosed a series of chromane derivatives of general structure (**84**: **Figure 31**).[111] Chromane (**85**) is reported to have a potency ratio > 10% relative to E_2 and a selectivity for ERβ > 30-fold. The same group has also disclosed a series of tetracyclic compounds of general structure **86**.[112] Tetracycle (**87**) has a potency transactivation of ERβ > 10% relative to E_2 and a selectivity for ERβ > 30-fold. In a related filing the same tetracyclic compounds (**88**) are adorned with a basic side chain to provide ERβ antagonists.[113] Tetracycle (**89**) is reported to have good ERβ antagonist activity (> 100% compared to ICI 164,384).

Merck has disclosed tetrahydrofluorenones of general structures **90** and **92** (**Figure 32**).[114] Compounds of general structure **90** have binding affinities to ERα in the range of IC_{50} = 2.8–5625 nmol L^{-1} and to ERβ in the range of IC_{50} = 0.6–126 nmol L^{-1}. Compounds of general structure **92** have binding affinities to ERα in the range of IC_{50} = 36–> 10 000 nmol L^{-1} and to ERβ in the range of IC_{50} = 1.4–283 nmol L^{-1}.

Recent advances in ERβ scaffolds include 2-phenylnaphthalenes[115] (WAY-202196) and benzisoxazoles[116] (ERB-041), which are being investigated for chronic inflammatory diseases (**Figure 33**).

Figure 31 ERβ-selective ligands from Akzo Nobel.

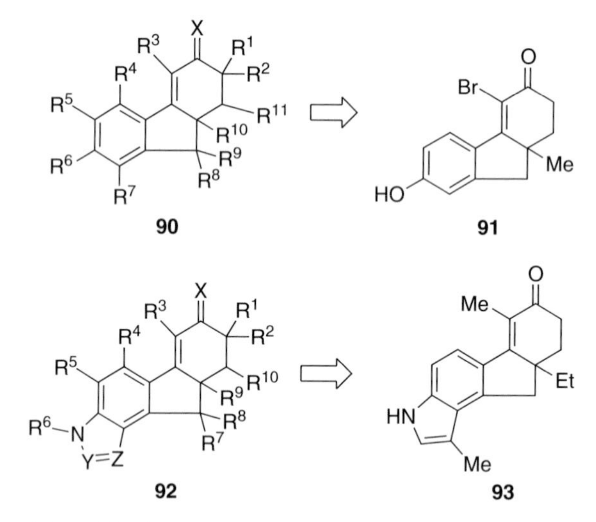

Figure 32 ERβ-selective ligands from Merck.

7.07.5.1.8 Steroid-derived modulators

The estrogen platform has been used extensively to study the structure and function of the estrogen receptor. In fact, the attachment of the appropriate functionality on to the steroid backbone has led to the development of SERMs as well as pure antagonists, affinity labels, and other important tools for understanding estrogen-mediated pharmacology. The effects of substitution on the estrogen nucleus have been reviewed in depth recently, leading to the derivation of a binding site model.[117]

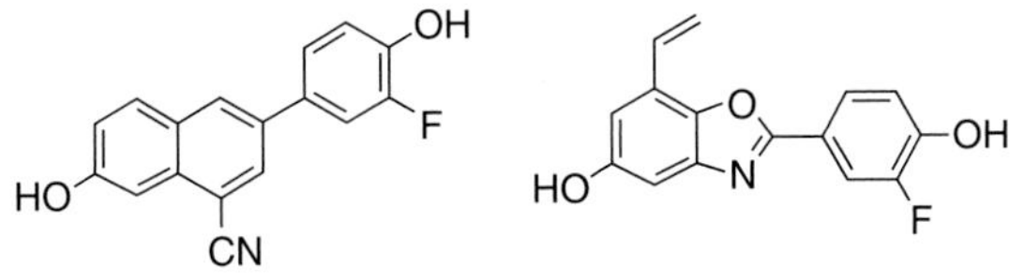

WAY-202196

ERb-041

Figure 33 Wyeth ERβ ligands.

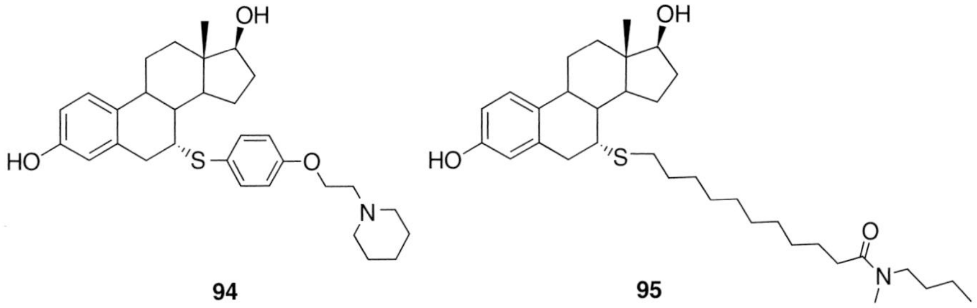

	94	95
hERα IC_{50}	Not active	9 nmol L^{-1}
hERβ IC_{50}	Not active	9 nmol L^{-1}

Figure 34 7α-Thioestratriene derivatives.

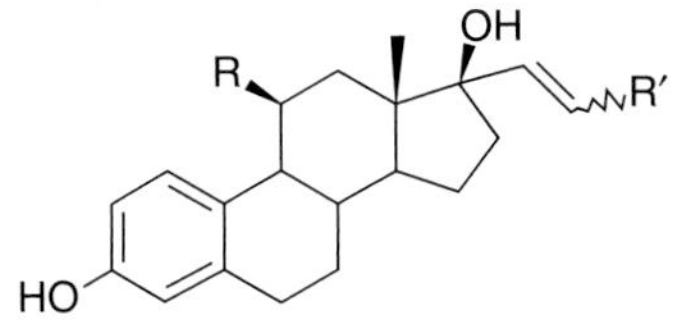

Cmpd	R	X	*E*-isomer[a]	*Z*-isomer[a]
96	H	Cl	89	199
97	H	I	62	776
98	OMe	Cl	65	195
99	OMe	I	65	413
100	$CH{=}CH_2$	Cl	724	704
101	$CH{=}CH_2$	I	447	1202

[a]Relative binding affinities from lamb uterine cytosol at 25 °C.

Figure 35 Estradiol-based ligands with substitution at the 11β and 17α positions.

Miller and co-workers have prepared a series of 7α-thioestratrienes represented by compounds **94** and **95** (**Figure 34**).[118] Substitution of the steroid nucleus at 7α with a basic aryloxypiperidine side chain (**94**) common to many SERMs resulted in no affinity to ERα or ERβ. However, 7α-substitution with an amide side chain (**95**) resulted in a high-affinity ligand ($IC_{50} = 9\,nmol\,L^{-1}$) at ERα or ERβ that was also an inhibitor of E_2-stimulated growth in vitro (ERE/MCF-7) and in vivo (uterine weight gain in the immature rat). Substitution at the 6-position with carbonyl or alcohol functionality resulted in a three- to fourfold reduction in binding affinity.

Hanson and co-workers have prepared E_2-based ligands in which substitution at the 11β and 17α positions has been investigated.[119] In this series, the binding behavior is dependent on the substituents and geometry of the alkene as well as the nature of the 11β position (**Figure 35**). Optimal receptor recognition is achieved by small lipophilic groups in the 11β position, i.e., H or vinylhalo > MeO. The presence of 17α-*Z*-halovinyl substitution enhances RBA values relative to E_2 (**96–99**) while 17α-*E* substitution has a less dramatic effect on receptor affinity.

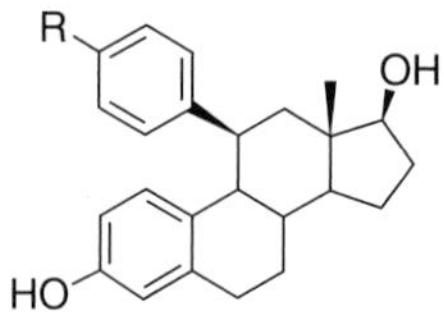

Cmpd	R	RAC[a]	Covalent binding
Estradiol		100	–
102	OCH_2CH_2OMs	39.9	No
103	OCH_2CH_2I	10.6	No
104	$OCH_2CH_2NHCOCH_2Cl$	3.3	Yes
105	$OCH_2CH_2N(Me)COCH_2Br$	4.2	No
106	$OCH_2CH_2N(Me)COCH_2Cl$	7.0	Yes
107	$OCH_2CH_2CH_2CH_2CH_2I$	8.2	Yes
108	$OCH_2CH_2CH_2CH_2CH_2NHCOCH_2Br$	17	Yes
109	$OCH_2CH_2CH_2CH_2CH_2N(Me)COCH_2Br$	9.7	No
110	$CCCH_2NHCOCH_2Cl$	14	Yes
111	$CCCH_2N(Me)COCH_2Cl$	19	Yes

[a]RAC = relative affinity constants for the cytosolic lamb uterine ER at 20 °C.

Figure 36 11β-Substituted estradiol derivatives.

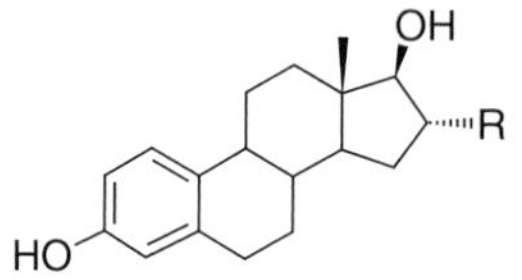

Cmpd	R	RBA[a]	Cmpd	R	RBA[a]
	H	100		$CH_2CO_2CH_3$	5
	CO_2CH_3	35		$CH_2CO_2CH_2CH_3$	5
	$CO_2CH_2CH_3$	40		CH_2CO_2H	0
	$CO_2CH_2CH_2CH_3$	34		$CH_2CH_2CO_2CH_3$	1
	$CO_2CH(CH_3)_2$	7		$CH_2CH_2CO_2H$	0.1
	$CO_2CH_2CH_2CH_2CH_3$	38			
	$CO_2CH_2CH_2F$	35			
	$CO_2CH_2CHF_2$	10			
	$CO_2CH_2CF_3$	7			
	CO_2H	0			

[a]Relative binding affinities (RBAs) from rat uterine cytosol at approximately 4 °C.

Figure 37 16α-Carboxylic esters and acids.

Aliau and co-workers have prepared a series of steroidal affinity labels by attaching electrophilic groups (**Figure 36**).[120] Evaluation of the ability of these compounds to inhibit the binding of ^{3}H-estradiol irreversibly indicates that nucleophilic sites in the receptor are relatively remote from the steroid backbone. In addition, cysteine residues were determined to play a role in the covalent attachment sites, as determined by modification of the estrogen receptor with methyl methanthiosulfonate.

Steroid analogs have been used by Labaree and co-workers to evaluate the estrogenic effects of local versus systemic administration of 16α-carboxylic esters/acids (**Figure 37**).[121] Small, aliphatic ester derivatives were shown to bind well to the estrogen receptor and have estrogen-like activity in a human endometrial carcinoma cell line (Ishikawa). In all cases, the corresponding carboxylic acids are substantially less active, as are bulkier esters.[116] Increasing the distance of the ester functionality from the steroid ring by homologation of the methylene tether reduces the binding affinity. Receptor affinity was also determined for human ERα and ERβ. In general, the affinity of the esters to ERβ was significantly diminished relative to ERα. In rodent models for determining estrogenic activity, esters exhibited good efficacy when administered locally but poor activity when administered systemically. The authors postulate that these esters act as 'soft estrogens' that are rapidly inactivated by conversion to the carboxylic acid by hydrolytic esterases found near the site of administration.

Hillisch and co-workers have designed highly selective estrogen receptor isotype-selective and potent ligands using the crystal structure of the ERα LBD and a homology model of the ERβ LBD (**Figure 38**).[151] These ligands are approximately 200-fold selective. In rats, the ERα agonist induced uterine growth and was bone-protective whereas the ERβ agonist did not have an effect on uterus or bone. Simultaneous administration of both agonists to rats did not lead to an attenuation of the ERα-mediated effects on the uterus.

Conjugated equine estrogens (Premarin) have served as starting points for identifying molecules that have the positive attributes of estrogen replacement therapy (ERT) but with diminished estrogenic activity. Conjugated equine estrogens contain sulfate esters of two structural classes of estrogen that include ring B saturated steroids such as E_2 as well as ring B unsaturated estrogens (**Figure 39**) such as equilin, equilenin, 17β-dihydroequilenin, and 17α-dihydroequilenin (17α-DHEqn), among others. In 1991, Bhavnani and Woolover examined a number of ring B unsaturated steroids in their unconjugated form in order to determine their RBAs for the estrogen receptor and their in vivo effects on uterine hypertrophy in ovary-intact rats.[122] All of the individual equine components caused an increase in uterine wet weight, with the exception of 17α-DHEqn, which lacked a significant effect at the highest dose tested. More recently, the sulfate ester conjugate of 17β-DHEqn has been shown to lower serum cholesterol,[123] increase hippocampal dendritic spine density in rats,[124] and improve arterial vasomotor function in macaques.[125] Neuroprotective effects for this compound have also been observed.[126] Oral treatment with 17α-DHEqn for 5 weeks prevents bone loss in the proximal metaphysis of excised femurs in a dose-dependent manner. In addition, 17α-DHEqn efficaciously lowers total serum cholesterol with an ED_{50} of <0.1 mpk. Uterine wet weight is significantly increased by 100% and 142% relative to OVX controls at doses of 1 and 10 mpk, respectively. Thus, the efficacy demonstrated by 17α-DHEqn as an estrogen agonist in preventing bone loss and lowering cholesterol is only observed at doses at which agonist effects are also observed in the uterus, as determined by induction of uterine weight gain.

Benzothiophene (Bt) analogs of the equilenins have been evaluated as SERMS (**Figure 40**). These compound bind to ERα with K_i ranging from 4 to 24 nmol L^{-1}.[127] Weaker relative affinity to ERβ is observed in each case. The

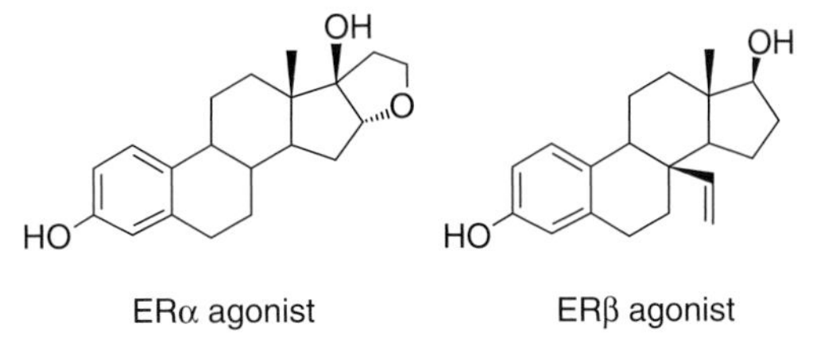

ERα agonist

ERβ agonist

Figure 38 Steroidal subtype selective agonists.

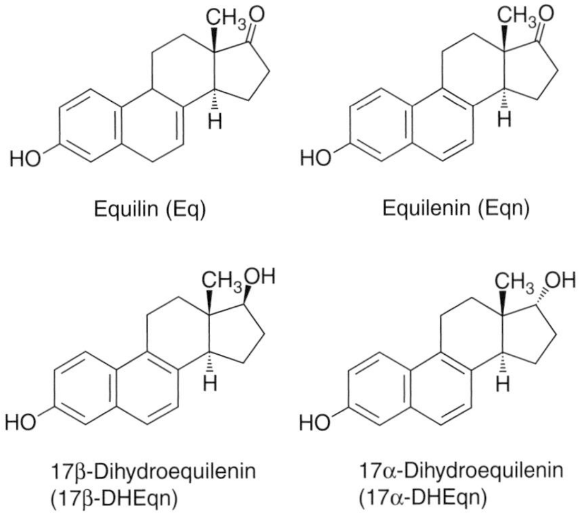

Equilin (Eq)

Equilenin (Eqn)

17β-Dihydroequilenin (17β-DHEqn)

17α-Dihydroequilenin (17α-DHEqn)

Figure 39 Equine estrogen components.

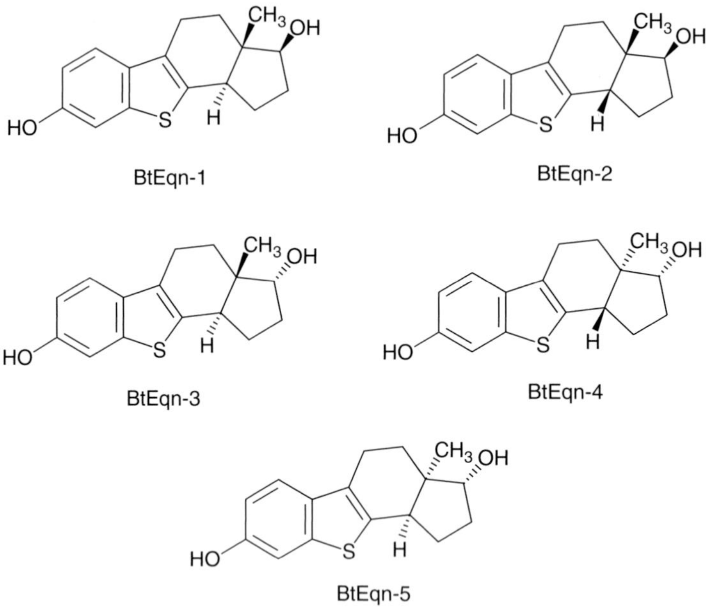

BtEqn-1

BtEqn-2

BtEqn-3

BtEqn-4

BtEqn-5

hERα and hERβ binding affinity for 17α-DHEqn and analogs

Compound	ERα K_i nmol L^{-1}	ERβ K_i nmol L^{-1}
17α-DHEqn	8.0	24
BtEqn-1	4.0	11
BtEqn-2	10	47
BtEqn-3	23	73
BtEqn-4	13	17
BtEqn-5	24	120

Figure 40 Thiaequilenins.

ligand with the highest affinity to ERα is BtEqn-1 ($K_i = 4\,\text{nmol}\,L^{-1}$), while its enantiomer (BtEqn-4) has threefold weaker affinity ($K_i = 13\,\text{nmol}\,L^{-1}$). Inversion of the 17β-OH to 17α-OH results in a greater than fivefold loss in potency (see BtEqn-4 and BtEqn-5), data which are consistent with known trends for E_2 and analogs.[128] While the effects of modifying the C17 position in the steroid backbone are well documented, structure–activity studies examining the variations in the C/D ring juncture are limited.[16] In the thiaequilenin series, the C/D *trans*-17β-OH analog BtEqn-1 binds with higher affinity to ERα than its *cis* counterpart BtEqn-5. A similar trend is observed for the analogous compounds in the 17α-OH series (BtEqn-4 and BtEqn-5). The orientation of the angular methyl group has little influence on binding in the 17α-OH series, as shown by analogs BtEqn-3 and BtEqn-5 that differ only by the orientation of the methyl group and have identical affinity. When comparing ligands that are enantiomeric pairs, BtEqn-1 and BtEqn4 or BtEqn-2 and BtEqn-5, the binding affinity of the enantiomer with the 17β-OH is more potent than its 17α counterpart. In general, compounds with the 17β-OH functionality have higher affinity to ERα regardless of whether the ring C/D ring juncture is *cis* or *trans*. However, the overall differences in affinity between ligands are relatively small, indicating that the receptor can readily accommodate changes in the spatial arrangement of the C/D ring juncture without dramatic consequences to the binding affinity. Overall, these data support the hypothesis that there is a flexible binding pocket in the estrogen receptor that allows for D-ring modifications in the ligand.[16]

7.07.5.1.9 Mechanism of action

The origin(s) of the tissue selectivity demonstrated by SERMs is complex.[129] Two subtypes of the estrogen receptor have been described, ERα and ERβ, and some of the observed selectivity may be derived from differential binding to these subtypes.[130,131] Crystal structures of the estrogen receptor bound to different ligands, such as estrogen and raloxifene, indicate that compounds of different sizes and shapes induce a spectrum of receptor conformations.[132]

In turn, these conformational states interact uniquely with co-regulators as well as target gene promoters. Thus, SERM pharmacology is dependent on interaction of the ligand with the receptor and the presence of gene promoter elements and co-regulatory proteins. Regardless of the origin, a key molecular determinant of receptor conformation, and thus tissue selectivity, is the interaction of the ligand with the nuclear receptor. The estrogen receptor contains two transcriptional activation functions, AF-1 and AF-2 which operate upon genes containing an estrogen response element to initiate transcription.[133] Of these activation functions, only AF-2 requires ligand binding. It has been demonstrated that tamoxifen binding inhibits AF-2 function and therefore gene transcription in some cell types.[134] In other cell types, however, AF-1 activation is sufficient to induce gene transcription and in these tissues tamoxifen functions as an estrogen agonist. Whether or not AF-1 is sufficient to activate transcription is dependent upon the cell type and the individual gene promoter involved.[135] This mechanistic picture has been further complicated by the discovery that AP-1 transcriptional sites within DNA can also respond to the estrogen receptor–ligand complex.[136] Once again, activation by a particular ligand is dependent upon cell type, as tamoxifen has shown agonist activity at AP-1 sites in uterine but not breast tissue.[137] Altered transcriptional profiles have been observed for E_2, TPEs, raloxifene, and ICI 164,384; thus, four distinct classes of estrogen receptor modulators have been identified to date.[138] The crystallization of the ligand-binding domain of ERα with E_2 and SERMs has provided direct insight into the shape of the ligand–protein complex.[139,140] The interaction between raloxifene and the LBD of the receptor differs significantly from that of the natural ligand, E_2, in that the tertiary amine of raloxifene interacts with Asp-351 of the protein to reorient helix 12 such that coactivators cannot bind in an effective manner. Clearly, the agonist–antagonist profile for any estrogen receptor modulator must be determined for each target tissue in question. It is expected that, as additional target tissues are explored and new estrogen-regulated genes are uncovered, new classes of estrogen receptor modulators will be developed with unique profiles of selectivity.

7.07.5.2 Endocrine Modulation by Inhibiting Estrogen Synthesis

Two primary approaches have been developed to reduce the growth-stimulatory effects of estrogens in breast cancer: (1) interfering with the ability of estrogen to bind to its receptor; and (2) decreasing circulating levels of estrogen. The first approach has been discussed and is represented by the effects of tamoxifen treatment on breast cancer. The second approach consists primarily of aromatase inhibitors and gonadotropin-releasing hormone (GnRH) agonists/antagonists. Along these lines, effective aromatase inhibitors have been developed as therapeutic agents for controlling estrogen-dependent breast cancer. Investigations on the development of aromatase inhibitors began in the 1970s and have expanded greatly since then. GnRH agonists have been used in the treatment of prostate cancer in men and estrogen-dependent uterine leiomyoma tumors in women.

7.07.5.2.1 Aromatase inhibitors[141]

E_2 is biosynthesized from androgens by the cytochrome P450 enzyme complex, called aromatase. Aromatase is present in breast tissue and is the source of local estrogen production in breast cancer tissues. Inhibition of aromatase is an important approach for reducing growth-stimulatory effects of estrogens in estrogen-dependent breast cancer. Steroidal inhibitors that have been developed to date build upon the basic androstenedione nucleus and incorporate chemical substituents at varying positions on the steroid.[142,143] Aromatase inhibitors are of two general types: competitive and suicide inhibitors (**Figure 41**). Suicide inhibitors include exemestane (Lentaron) and formestane (Aromasin). Formestane is approved outside the US for the treatment of advanced breast cancer in postmenopausal women. It has poor oral bioavailability and is generally administered as an intramuscular injection. Exemestane is orally bioavailable and is effective at suppressing estrogen levels in women. Competitive aromatase inhibitors include aminoglutethimide, anastrazole (Arimidex), and letrozole (Femera). Aminoglutethimide has been used broadly for the treatment of advanced breast cancer. Administration leads to the inhibition of adrenal steroid synthesis, including cortisol. Anasatrazole is approved for the treatment of metastatic breast cancer in postmenopausal women who do not respond to tamoxifen. It does not affect cortisol or aldosterone secretion. Letrozole was recently approved for the treatment of tamoxifen-resistant, advanced breast cancer in postmenopausal women.

7.07.5.2.2 Gonadotropin-releasing hormone agonists and antagonists

GnRH is a hypothalamic decapeptide that binds to GnRH receptors on pituitary gonadotrope cells to modulate the synthesis and secretion of the gonadotropins, luteinizing hormone and follicle-stimulating hormone. These gonadotropins in turn regulate gonadal steroidogenesis and gametogenesis. Structure–activity studies of GnRH analogs have led to the discovery of peptidic GnRH agonists (**Figure 42**) and antagonists (**Figure 43**). These

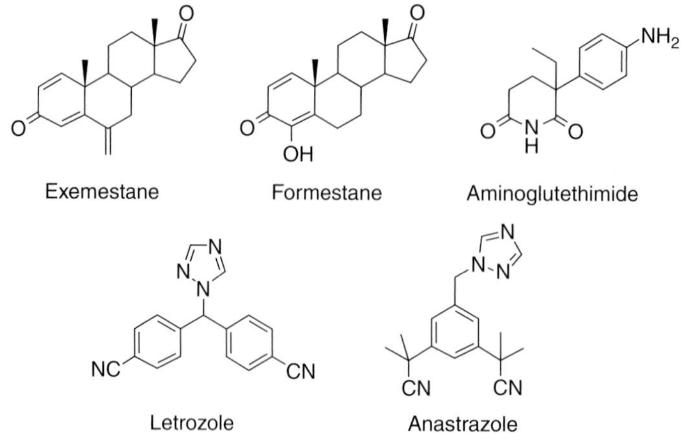

Exemestane Formestane Aminoglutethimide

Letrozole Anastrazole

Figure 41 Aromatase inhibitors.

GnRH	pGlu1-His2-Trp3-Ser4-Tyr5-Glu6-Leu7-Arg8-Pro9-Gly10-NH_2
Triptorelin	pGlu1-His2-Trp3-Ser4-Tyr5-D-Tryp6-Leu7-Arg8-Pro9-Gly10-NH_2
Leuprolide	pGlu1-His2-Trp3-Ser4-Tyr5-D-Leu6-Leu7-Arg8-Pro9-NHC_2H_5
Buserelin	pGlu1-His2-Trp3-Ser4-Tyr5-D-Ser(tBu)6-Leu7-Arg8-Pro9-NHC_2H_5
Goserelin	pGlu1-His2-Trp3-Ser4-Tyr5- Ser(tBu)6-Leu7-Arg8-Pro9-AzGly10-NH_2
Nafarelin	pPyr1-His2-Trp3-Ser4-Tyr5-D-2-Nal6-Leu7-Arg8-Pro9-Gly10-NH_2

Figure 42 Peptide GnRH agonists.

GnRH	pGlu1-His2-Trp3-Ser4-Tyr5-Glu6-Leu7-Arg8-Pro9-Gly10-NH_2
Nal-Glu-GnRH	Ac-D-Nal1-D-Cpa2-D-Pal3-Ser4-Arg5-D-Glu6(AA)-Leu7-Arg8-Pro9-D-Ala10-NH_2
Antide	Ac-D-Nal1-D-Cpa2-D-Pal3-Ser4-NicLys5-D-NicLys6-Leu7-ILys8-Pro9-D-Ala10-NH_2
Cetrorelix	Ac-D-Nal1-D-Cpa2-D-Pal3-Ser4-Tyr5-D-Cit6-Leu7-Arg8-Pro9-D-Ala10-NH_2
Ganirelix	Ac-D-Nal1-D-Cpa2-D-Pal3-Ser4-Tyr5-D-hArg(Et_2)6-Leu7-hArg(Et_2)8-Pro9-D-Ala10-NH_2

Figure 43 Peptide GnRH antagonists.

peptides have been widely used in a clinical setting in which modulation of the production of sex steroid hormones is beneficial to prevent the development or progression of benign conditions (e.g., endometriosis, uterine fibroids) or malignant tumors (e.g., breast, ovarian, endometrial, and prostate carcinoma). GnRH agonists have increased potency for the short-term release of gonadotropins. However, they show paradoxical action in that chronic treatment results in inhibition of GnRH production as a result of desensitization of the gonadotropins and downregulation of its receptor. In contrast, peptidic GnRH antagonists produce a rapid and dose-dependent suppression of gonadotropin release by competitive blockade of the GnRH receptors without any initial stimulatory effect, as seen with agonists. Zoladex is the only GnRH agonist approved by the US Food and Drug Administration for the treatment of metastatic breast cancer in pre- and postmenopausal women. It is administered on a monthly basis as an implant. Response rates range from 30% to 60% in premenopausal women. Major side effects include hot flushes and osteoporosis.

Small-molecule GnRH antagonists have been reported (**Figure 44**). The first high-affinity nonpeptidic antagonist of the human GnRH receptor T-98475 was reported in 1998.[144] Since this initial disclosure, several small molecules from different chemical classes have appeared in the literature. Arylquinolones are potent antagonists of the human ($IC_{50} = 0.44\,\mathrm{nmol\,L^{-1}}$) and rat ($IC_{50} = 4\,\mathrm{nmol\,L^{-1}}$) GnRH receptor.[145] Tryptamine GnRH antagonists have demonstrated oral bioavailability in rats and dogs.[146] Recently, TAK-013 has been reported to be clinically efficacious in humans in suppressing luteinizing hormone when given orally.[147] A series of pyrazolopyrimidones,[148] imidazolopyrimidones,[149] and uracils[150] have also been reported.

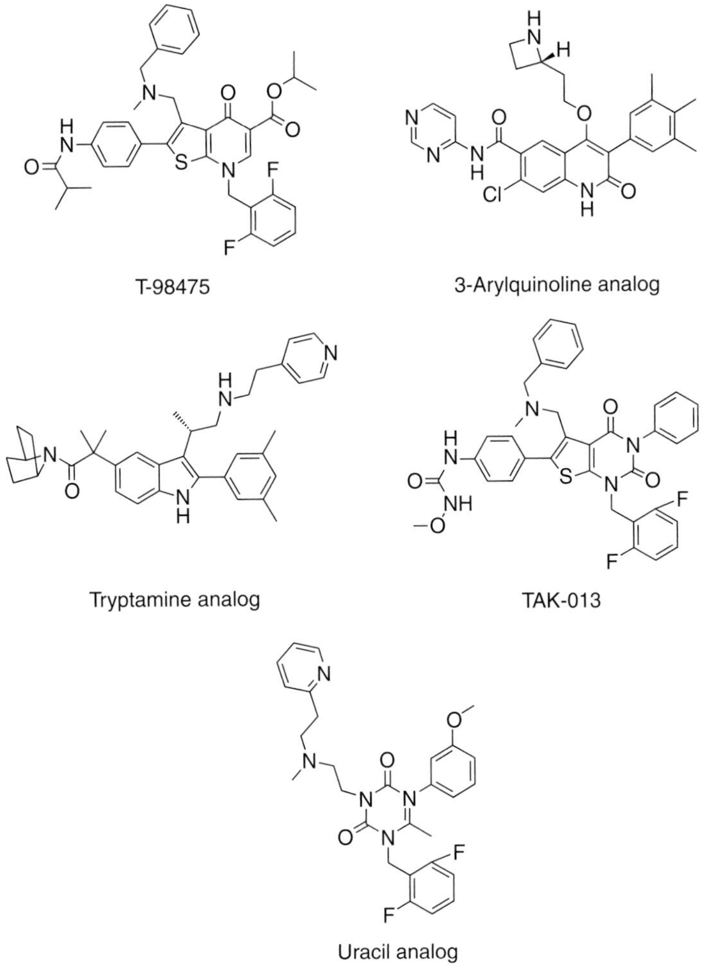

T-98475

3-Arylquinoline analog

Tryptamine analog

TAK-013

Uracil analog

Figure 44 Small-molecule GnRH antagonists.

7.07.6 Unmet Medical Needs and New Research Areas

The advent of endocrine therapy has led to significant advances in the treatment of cancer. In particular, modulation of the estrogen receptor and the proteins that regulate estrogen synthesis have led to improved treatments for breast cancer. The clinical manifestation of tissue-selective compounds such as tamoxifen has led to the identification of SERMs as viable treatments for therapies ranging from breast cancer to osteoporosis. Moreover, this discovery has laid the groundwork for the identification of tissue-selective agents for other steroid receptors such as the androgen, glucocorticoid, and progesterone receptors, i.e., SARMs, SGRMs, and SPRMs, respectively. Future research will lie in obtaining the appropriate tissue selectivity for the disease state in question. An in depth understanding of ligand–protein interactions, including co-regulators, will be important for fine-tuning tissue-selective pharmacology. In the area of the regulation of estrogen synthesis, small-molecule, orally active GnRH antagonists have been investigated and are being clinically validated for the treatment of gynecological disorders and prostate cancer. Combination studies of inhibitors of hormone action with inhibitors of hormone synthesis are underway and may provide benefits in clinical outcomes.

References

1. National Breast Cancer Foundation Inc. http://www.nationalbreastcancer.org/signs_and_symptoms/index.html (accessed April 2006).
2. Beatson, G. *Lancet* **1896**, *2*, 104–107.
3. Holtkamp, D. E.; Greslin, S. C.; Root, C. A.; Lerner, L. J. *Proc. Soc. Exp. Biol. Med.* **1960**, *105*, 197–201.
4. Huppert, L. C. *Fertil. Steril.* **1979**, *31*, 1.
5. Beall, P. T.; Misra, L. K.; Yound, R. L.; Spjut, H. J.; Evans, H. J.; Leblanc, A. *Calcif. Tissue Int.* **1984**, *36*, 123–127.

6. Chakraborty, P. K.; Brown, J. L.; Ruff, C. B.; Nelson, M. F; Mitchell, A. S. *J. Steroid Biochem. Mol. Biol.* **1991**, *40*, 725–731.
7. Stewart, P. J.; Stern, P. H. *Endocrinology* **1986**, *118*, 125–131.
8. Young, R. L.; Goldzieher, J. W.; Elkind-Hirsch, K.; Hickox, P. G.; Chakraborty, P. K. *Int. J. Fertil.* **1991**, *36*, 167–171.
9. Clark, J. H.; Markaverich, B. M. *Pharm. Ther.* **1982**, *15*, 467–519.
10. Clark, J. H.; Guthrie, S. C. *Biol. Reprod.* **1981**, *25*, 667–672.
11. Young, R. L.; Goldzieher, J. W.; Chakraborty, P. K.; Panko, W. B.; Bridges, C. N. *Int. J. Fertil.* **1991**, *36*, 291–295.
12. Grey, A. B.; Stapleton, J. P.; Evans, M. C.; Tatnell, M. A.; Ames, R. W.; Reid, I. R. *Am. J. Med.* **1995**, *99*, 636–641.
13. Love, R. R.; Barden, H. S.; Mazess, R. B.; Epstein, S.; Chappell, R. J. *Arch. Intern. Med.* **1994**, *154*, 2585–2588.
14. Ward, R. L.; Morgan, G.; Dalley, D.; Kelly, P. J. *Bone Miner.* **1993**, *22*, 87–94.
15. Kristensen, B.; Ejlertsen, B.; Dalgaard, P.; Larsen, L.; Holmegaard, S. N.; Transbøl, I.; Mouridsen, H. T. *J. Clin. Oncol.* **1994**, *12*, 992–997.
16. Kenny, A. M.; Prestwood, K. M.; Pilbeam, C. C.; Raisz, L. G. *J. Clin. Endocrinol. Metab.* **1995**, *80*, 3287–3291.
17. Sismondi, P.; Biglia, N.; Giai, M.; Sgro, L.; Campagnoli, C. *Anticancer Res.* **1994**, *14*, 2237–2244.
18. Grey, A. B.; Stapleton, J. P.; Evans, M. C.; Reid, I. R. *J. Clin. Endocrinol. Metab.* **1995**, *80*, 3191–3195.
19. Shewmon, D. A.; Stock, J. L.; Rosen, C. J.; Heiniluoma, K. M.; Hogue, M. M.; Morrison, A.; Doyle, E. M.; Ukena, T.; Weale, V.; Baker, S. *Arterioscler. Thromb.* **1994**, *14*, 1586–1593.
20. Love, R. R.; Wiebe, D. A.; Feyzi, J. M.; Newcomb, P. A.; Chappell, R. J. *J. Natl. Cancer Inst.* **1994**, *86*, 1534–1539.
21. Rutqvist, L. E.; Mattson, A. *J. Natl. Cancer Inst.* **1993**, *85*, 1398–1401.
22. Fisher, B.; Constantino, J. P.; Redmond, C. K.; Fisher, E. R.; Wickerham, D. L.; Cronin, W. M. *J. Natl. Cancer Inst.* **1994**, *86*, 527–531.
23. Early Breast Cancer Trialists' Collaborative Group. *Lancet* 1992, *339*, 1–15.
24. Gylling, H.; Antylä, A. M.; Miettinen, T. A. *Atherosclerosis* **1992**, *96*, 245–250.
25. Guetta, V.; Cannon, R. O. *Circulation* **1995**, *92*, 164.
26. Wiseman, H.; Quinn, P.; Halliwell, B. *FEBS* **1993**, *330*, 53–55.
27. Williams, J. K.; Adams, M. R. *Circulation* **1995**, *92*, 627.
28. Fuchs-Young, R.; Glasebrook, A. L.; Short, L. L.; Draper, M. W.; Rippy, M. K.; Cole, H. W.; Magee, D. W.; Termine, J. D.; Bryant, H. U. *Ann. NY Acad. Sci.* **1995**, *761*, 355–360.
29. Kedar, R. P.; Bourne, T. H.; Powles, T. J.; Collins, W. P.; Ashley, S. E.; Cosgrove, D. O.; Campbell, S. *Lancet* **1994**, *343*, 1318–1321.
30. Assikis, V. J.; Jordan, V. C. *Endocrinol. Relat. Cancer* **1995**, *2*, 235–241.
31. Kuo, D. Y.-S.; Runowicz, C. D. *Med. Oncol.* **1995**, *12*, 87–94.
32. Robinson, D. C.; Bloss, J. D.; Schiano, M. A. *Gynecol. Oncol.* **1995**, *59*, 186–190.
33. Osborne, M. R.; Hewer, A.; Hardcastle, I. R.; Carmichael, P. L.; Phillips, D. H. *Cancer Res.* **1996**, *56*, 66–71.
34. Williams, G. M.; Iatropoulos, M. J.; Djordjevic, M. V.; Kaltenberg, O. P. *Carcinogenesis* **1993**, *14*, 315–317.
35. Rauschning, W.; Pritchard, K. I. *Breast Cancer Res. Treat.* **1994**, *31*, 83–94.
36. Sterbygaard, L. E.; Hernstedt, J.; Thomsen, J. F. *Breast Cancer Res. Treat.* **1993**, *25*, 57–63.
37. Coombes, R. C.; Haynes, B. P.; Dowsett, M.; Quigley, M.; English, J.; Judson, I. R.; Griggs, L. J.; Potter, G. A.; McCague, R.; Jarman, M. *Cancer Res.* **1995**, *55*, 1070–1074.
38. Koh, J.-I.; Kubota, T.; Asanuma, F.; Yamada, Y.; Kawamura, E.; Hosoda, Y.; Hashimoto, M.; Yamamoto, O.; Sakai, S.; Maeda, K. et al. *J. Surg. Oncol.* **1992**, *51*, 254–260.
39. Gylling, H.; Pyrhönen, S.; Mäntylä, E.; Mäqenpää, H.; Kangas, L.; Miettinen, T. A. *J. Clin. Oncol.* **1995**, *13*, 2900–2905.
40. Di Salle, E.; Zaccheo, T.; Ornati, G. *J. Steroid Biochem. Mol. Biol.* **1990**, *36*, 203–206.
41. Tomás, E.; Kauppila, A.; Blanco, G.; Apaja-Sarkkinen, M.; Laatikainen, T. *Gynecol. Oncol.* **1995**, *59*, 261–264.
42. Löser, R.; Seibel, K.; Roos, W.; Eppenberger, U. *J. Cancer Clin. Oncol.* **1985**, *21*, 985–987.
43. Ke, H. Z.; Simmons, H. A.; Pirie, C. M.; Crawford, D. T.; Thompson, D. D. *Endocrinology* **1995**, *136*, 2435–2441.
44. Ke, H. E.; Chen, H. K.; Qi, H.; Pirie, C. M.; Simmons, H. A.; Ma, Y. F.; Jee, W. S. S.; Thompson, D. D. *Bone* **1995**, *17*, 491–496.
45. Bain, S. D.; Celino, D. L.; Bailey, M. C.; Strachan, M. J.; Piggott, J. R.; Labroo, V. M. *Calcif. Tiss.* **1994**, *55*, 338–341.
46. Trivedi, R. N.; Chauhan, S. C.; Dwivedi, A.; Kamboj, V. P.; Singh, M. M. *Contraception* **1995**, *51*, 367–379.
47. Hall, T. J.; Nyugen, H.; Schaueblin, M.; Fournier, B. *Biochem. Biophys. Res. Commun.* **1995**, *216*, 662–668.
48. Gradishar, W.; Glusman, J.; Lu, Y.; Vogel, C.; Cohen, F. J.; Sledge, G. W. *Cancer* **2000**, *88*, 2047–2053.
49. Martino, S.; Cauley, J. A.; Barrett-Connor, E.; Powles, T. J.; Mershon, J.; Disch, D.; Secrest, R. J.; Cummings, S. R. *J. Natl. Cancer Inst.* **2004**, *96*, 1751–1761.
50. Black, L. J.; Sato, M.; Rowley, E. R.; Magee, D. E.; Bekele, A.; Williams, D. C.; Cullinan, G. J.; Bendele, R.; Kauffman, R. F.; Bensch, W. R. et al. *J. Clin. Invest.* **1994**, *93*, 63–73.
51. Washburn, S. A.; Adams, M. R.; Clarkson, T. B.; Adelman, S. J. *Am. J. Obstet. Gynecol.* **1993**, *169*, 251–256.
52. Fuchs-Young, R.; Magee, D. E.; Cole, H. W.; Short, L.; Glasebrook, A. L.; Rippy, M. K.; Termine, J. D.; Bryant, H. U. *Endocrinology* **1995**, *136*, 57.
53. Boss, S. M.; Huster, W. J.; Neild, J. A.; Glant, M. D.; Eisenhut, C. C.; Draper, M. W. *Am. J. Obstet. Gynecol.* **1997**, *177*, 1458–1464.
54. Grese, T. A.; Sluka, J. P.; Bryant, H. U.; Cole, H. W.; Kim, J. R.; Magee, D. E.; Rowley, E. R.; Sato, M. *Bioorg. Med. Chem. Lett.* **1996**, *6*, 2683–2686.
55. Wallace, O. B.; Richardson, T. I.; Dodge, J. A. *Curr. Top. Med. Chem.* **2003**, *3*, 1663–1680.
56. Wakeling, A. E.; Bowler, J. *J. Steroid Biochem. Mol. Biol.* **1992**, *43*, 173–177.
57. Lesuisse, D.; Albert, E.; Bouchoux, F.; Cérède, E.; Lefrancois, J.-M.; Levif, M.-O.; Tessier, S.; Tric, B.; Teutsch, G. *Bioorg. Med. Chem. Lett.* **2001**, *11*, 1709–1712.
58. Korach, K. S.; Sarver, P.; Chae, K.; McLachlan, J. A.; McKinney, J. D. *Mol. Pharmacacol.* **1998**, *33*, 120–126.
59. Stauffer, S. R.; Sun, J.; Katzenellenbogen, B. S.; Katzenellenbogen, J. A. *Bioorg. Med. Chem.* **2000**, *8*, 1293–1316.
60. Gust, R.; Burgemeister, T.; Mannschreck, A.; Schönenberger, H. *J. Med. Chem.* **1990**, *33*, 2535–2544.
61. Gust, R.; Keilitz, R.; Schmidt, K. *J. Med. Chem.* **2001**, *44*, 1963–1970.
62. Jordan, V. C.; Schafer, J.; Levenson, A. S.; Liu, H.; Pease, K. M.; Simons, L. A.; Zapf, J. W. *Cancer Res.* **2001**, *61*, 6619–6623.
63. Gust, R.; Keilitz, R.; Schmidt, K. *J. Med. Chem.* **2001**, *44*, 1670–1673.
64. Jørgensen, A. S.; Jacobsen, P.; Christiansen, L. B.; Bury, P. S.; Kanstrup, A.; Thorpe, S. M.; Nćrum, L.; Wassermann, K. *Bioorg. Med. Chem. Lett.* **2000**, *10*, 2383–2386.
65. Jørgensen, A. S.; Jacobsen, P.; Christiansen, L. B.; Bury, P. S.; Kanstrup, A.; Thorpe, S. M.; Bain, S.; Nćrum, L.; Wassermann, K. *Bioorg. Med. Chem. Lett.* **2000**, *10*, 399–402.
66. Jacquot, Y.; Bermont, L.; Giorgi, H.; Refouvelet, B.; Adessi, G. L.; Daubrosse, E.; Xicluna, A. *Eur. J. Med. Chem.* **2001**, *36*, 127–136.

67. Kim, S.-H.; Katzenellenbogen, J. A. *Bioorg. Med. Chem.* **2000**, *8*, 785–793.
68. Stauffer, S. R.; Katzenellenbogen, J. A. *J. Comb. Chem.* **2000**, *2*, 318–329.
69. Stauffer, S. R.; Coletta, C. J.; Tedesco, R.; Nishiguchi, G.; Carlson, K.; Sun, J.; Katzenellenbogen, B. S.; Katzenellenbogen, J. A. *J. Med. Chem.* **2000**, *43*, 4934–4947.
70. Shiau, A. K.; Barstad, D.; Loria, P. M.; Cheng, L.; Kushner, P. J.; Agard, D. A.; Greene, G. L. *Cell* **1998**, *95*, 927–937.
71. Nishiguchi, G. A.; Rodriguez, A. L.; Katzenellenbogen, J. A. *Bioorg. Med. Chem. Lett.* **2002**, *12*, 947–950.
72. Mortensen, D. S.; Rodriguez, A. L.; Carlson, K. E.; Sun, J.; Katzenellenbogen, B. S.; Katzenellenbogen, J. A. *J. Med. Chem.* **2001**, *44*, 3838–3848.
73. Fink, B. E.; Mortensen, D. S.; Stauffer, S. R.; Aron, Z. D.; Katzenellenbogen, J. A. *Chem. Biol.* **1999**, *6*, 205–219.
74. Henke, B. R.; Drewry, D. H.; Jones, S. A.; Stewart, E. L.; Weaver, S. L.; Wiethe, R. W. *Bioorg. Med. Chem. Lett.* **2001**, *11*, 1939–1942.
75. Minutolo, F.; Bertini, S.; Papi, C.; Carlson, K. E.; Katzenellenbogen, J. A.; Macchia, M. *J. Med. Chem.* **2001**, *44*, 4288–4291.
76. Stauffer, S. R.; Huang, Y. R.; Aron, Z. D.; Coletta, C. J.; Sun, J.; Katzenellenbogen, B. S.; Katzenellenbogen, J. A. *Bioorg. Med. Chem.* **2001**, *9*, 151–161.
77. Mortensen, D. S.; Rodriguez, A. L.; Sun, J.; Katzenellenbogen, B. S.; Katzenellenbogen, J. A. *Bioorg. Med. Chem. Lett.* **2001**, *11*, 2521–2524.
78. Croisy-Delcey, M.; Croisy, A.; Carrez, D.; Huel, C.; Chiaroni, P.; Ducrot, P.; Bisagni, E.; Jin, L.; Leclercq, G. *Bioorg. Med. Chem.* **2000**, *8*, 2629–2641.
79. Von Angerer, E.; Prekajac, J.; Strohmeier, J. *J. Med. Chem.* **1984**, *27*, 1439–1447.
80. Miller, C. P.; Collini, M. D.; Tran, B. D.; Harris, H. A.; Kharode, Y. P.; Marzolf, J. T.; Moran, R. A.; Henderson, R. A.; Bender, R. H. W.; Unwalla, R. J. et al. *J. Med. Chem.* **2001**, *44*, 1654–1657.
81. Komm, B. S.; Lyttle, C. R. *Ann. NY Acad. Sci.* **2001**, *949*, 317–326.
82. Bury, P. S.; Christiansen, L. B.; Jacobsen, P.; Jørgensen, A. K.; Kanstrup, A.; Nérum, L.; Bain, S.; Fledelius, C.; Gissel, B.; Hansen, B. S. et al. *Bioorg. Med. Chem.* **2002**, *10*, 125–145.
83. Rosati, R. L.; Jardine, P. D.; Cameron, K. O.; Thompson, D. D.; Ke, H. Z.; Toler, S. M.; Brown, T. A.; Pan, L. C.; Ebbinghaus, C. F.; Reinhold, A. R. et al. *J. Med. Chem.* **1998**, *41*, 2928–2931.
84. Christiansen, L. B.; Wenckens, M.; Bury, P. S.; Gissel, B.; Hansen, B. S.; Thorpe, S. M.; Jacobsen, P.; Kanstrup, A.; Jørgensen, A. S.; Nérum, L. et al. *Bioorg. Med. Chem. Lett.* **2002**, *12*, 17–19.
85. Meegan, M. J.; Hughes, R. B.; Lloyd, D. G.; Williams, D. C.; Zisterer, D. M. *J. Med. Chem.* **2001**, *44*, 1072–1084.
86. Willson, T. M.; Henke, B. R.; Momtahen, T. M.; Charifson, P. S.; Batchelor, K. W.; Lubahn, D. B.; Moore, L. B.; Oliver, B. B.; Sauls, H. R.; Triantafillou, J. A. et al. *J. Med. Chem.* **1994**, *37*, 1550–1552.
87. Ruenitz, P. C.; Shen, Y.; Li, M.; Whitehead, R. D., Jr.; Pun, S.; Wronski, T. J. *Bone* **1998**, *23*, 537–542.
88. Rubin, V. N.; Ruenitz, P. C.; Boudinot, F. D.; Boyd, J. L. *Bioorg. Med. Chem.* **2001**, *9*, 1579–1587.
89. Detsi, A.; Koufaki, M.; Calogeropoulou, T. *J. Org. Chem.* **2002**, *67*, 4608–4611.
90. Grese, T. A.; Pennington, L. D.; Sluka, J. P.; Adrian, M. D.; Cole, H. W.; Fuson, T. R.; Magee, D. E.; Phillips, D. L.; Rowley, E. R.; Shetler, P. K. et al. *J. Med. Chem.* **1998**, *41*, 1272–1283.
91. Grese, T. A.; Adrian, M. D.; Phillips, D. L.; Shetler, P. K.; Short, L. L.; Glasebrook, A. L.; Bryant, H. U. *J. Med. Chem.* **2001**, *44*, 2857–2860.
92. Hwang, K. J.; O'Neil, J. P.; Katzenellenbogen, J. A. *J. Org. Chem.* **1992**, *57*, 1262–1271.
93. Meyers, M. J.; Sun, J.; Carlson, K. E.; Katzenellenbogen, B. S.; Katzenellenbogen, J. A. *J. Med. Chem.* **1999**, *42*, 2456–2468.
94. Tedesco, R.; Youngman, M. K.; Wilson, S. R.; Katzenellenbogen, J. A. *Bioorg. Med. Chem. Lett.* **2001**, *11*, 1281–1284.
95. Hummel, C. W.; Geiser, A. G.; Bryant, H. U.; Cohen, I. R.; Dally, R. D.; Fong, K. C.; Frank, S. A.; Hinklin, R.; Jones, S. A.; Lewis, G. et al. *J. Med. Chem.* **2005**, *48*, 6772–6775.
96. Lyttle, C. R.; DeSombre, E. R. *Proc. Natl. Acad. Sci. USA* 1977, *74*, 3162–3166.
97. Geiser, A. G.; Hummel, C. W.; Draper, M. W.; Henck, J. W.; Cohen, I. R.; Rudmann, D. G.; Donnelly, K. B.; Adrian, M. D.; Shepherd, T. A.; Wallace, O. B. et al. *Endocrinology* **2005**, *146*, 4524–4535.
98. Chen, H. Y.; Dykstra, K. D.; Birzin, E. T.; Frisch, K.; Chan, W.; Yang, Y. T.; Mosley, R. T.; DiNinno, F.; Rohrer, S. P.; Schaeffer, J. M. et al. *Bioorg. Med. Chem. Lett.* **2004**, *14*, 1417–1421.
99. Blizzard, T. A.; DiNinno, F.; Morgan, J. D.; Chen, H. Y.; Wu, J. Y.; Kim, S.; Chan, W.; Birzin, E. T.; Yang, Y. T.; Pai, L. Y. et al. *Bioorg. Med. Chem. Lett.* **2005**, *15*, 107.
100. Tan, Q.; Blizzard, T. A.; Morgan, J. D.; Birzin, E. T.; Chan, W.; Yang, Y. T.; Pai, L. H.; Hayes, E. C.; DaSilva, C. A.; Warrier, S. et al. *Bioorg. Med. Chem. Lett.* **2005**, *15*, 1675–1681.
101. Krege, J. H.; Hodgin, J. B.; Couse, J. F.; Enmark, E.; Warner, M. *Proc. Natl. Acad. Sci. USA* **1998**, *95*, 15677–15682.
102. Pike, A. C. W.; Brzozowski, A. M.; Hubbard, R. E.; Bonn, T.; Thorsell, A.-G. *EMBO J.* **1999**, *18*, 4608–4618.
104. Schopfer, U.; Schoeffter, P.; Bischoff, S. F.; Nozulak, J.; Feuerbach, D. *J. Med. Chem.* **2002**, *45*, 1399–1401.
105. Meyers, M. J.; Sun, J.; Carlson, K. E.; Marriner, G. A.; Katzenellenbogen, B. S. *J. Med. Chem.* **2001**, *44*, 4230–4251.
106. Barlaam, B. C.; Piser, T. M. *PCT Int. Appl.* WO 0062765, 2000.
107. Barlaam, B.; Folmer, J. J.; Piser, T. M. *PCT Int. Appl.* WO 0230407, 2002.
108. Barlaam, B.; Dock, S.; Folmer, J. *PCT Int. Appl.* WO0246168, 2002.
109. Barlaam, B.; Bernstein, P.; Dantzman, C.; Warwick, P. *PCT Int. Appl.* WO 0251821, 2002.
110. Barlaam, B.; Dantzman, C. *PCT Int. Appl.* WO 0246164, 2002.
111. Veeneman, G. H.; Teerhuis, N. M. *PCT Int. Appl.* WO 0164665, 2001.
112. Veeneman, G.; De Zwart, E.; Loozen, H.; Mestres, J. *PCT Int. Appl.* WO 0172713, 2001.
113. Veeneman, G. H.; Loozen, H. J.; Mestres, J.; De Zwart, E. W. *PCT Int. Appl.* WO 0216316, 2002.
114. Parker, D. L.; Ratcliffe, R. W.; Wilkening, R. R.; Wildonger, K. J. *PCT Int. Appl.* WO 0182923, 2001; Wilkening, R. R.; Parker, D. L., Jr.; Wildonger, K. J.; Meng, D.; Ratcliffe, R. W. *PCT Int. Appl.* WO 0241835, 2002.
115. Mewshaw, R. E.; Edsall, R. J.; Yang, C.; Manas, E. S.; Xu, Z. B.; Henderson, R. A.; Keith, J. C.; Harris, H. A. *J. Med. Chem.* **2005**, *48*, 3953–3979.
116. Manas, E. S.; Unwalla, R. J.; Xu, Z. B.; Malamas, M. S.; Miller, C. P.; Harris, H. A.; Hsiao, C.; Akopian, T.; Hum, W.; Malakian, K. *J. Am. Chem Soc.* **2004**, *126*, 15106–15119.
117. Anstead, G. M.; Carlson, K. E.; Katzenellenbogen, J. A. *Steroids* **1997**, *62*, 268–303.
118. Miller, C. P.; Jirkovsky, I.; Tran, B. D.; Harris, H. A.; Moran, R. A.; Kromm, B. S. *Bioorg. Med. Chem. Lett.* **2000**, *10*, 147–151.
119. Hanson, R. N.; Napolitano, E.; Fiaschi, R. *J. Med. Chem.* **1998**, *41*, 4686–4692.
120. Aliau, S.; Delettre, G.; Mattras, H.; Garrouj, D. E.; Nique, F.; Teutsch, G.; Borgna, J.-L. *J. Med. Chem.* **2000**, *43*, 613–628.
121. Labaree, D. C.; Reynolds, T. Y.; Hochberg, R. B. *J. Med. Chem.* **2001**, *44*, 1802–1814.
122. Bhavnani, B. R.; Woolover, C. A. *Steroids* **1991**, *56*, 201.

123. Washburn, S. A.; Adams, M. R.; Clarkson, T. C.; Adelman, S. J. *Am. J. Obstet. Gynecol.* **1993**, *169*, 251.
124. Washburn, S. A.; Lewis, C. E.; Johnson, J. E.; Voytko, M. L.; Shively, C. A. *Brain Res.* **1997**, *758*, 241.
125. Washburn, S. A.; Honore, E. K.; Cline, J. M.; Helman, M.; Wagner, J. D.; Adelman, S. J.; Clarkson, T. B. *Am. J. Obstet. Gynecol.* **1996**, *175*, 341.
126. Berco, B. M.; Bhavnani, B. R. *J. Soc. Obstet. Gynecol.* **2001**, *8*, 245.
127. Wallace, O. B.; Lauwers, K. S.; Jones, S. A.; Dodge, J. A. *Bioorg. Med. Chem. Lett.* **2003**, *13*, 1907.
128. Anstead, G. M; Carlson, K. A.; Katzenellenbogen, J. A. *Steroids* **1997**, *62*, 268–303.
129. McDonnell, D. P.; Norris, J. D. *Science* **2002**, *296*, 1642–1644; Katzenellenbogen, B. S.; Katzenellenbogen, J. A. *Science* **2002**, *295*, 2380–2381.
130. Kuiper, G. G. J. M.; Enmark, E.; Pelto-Huikko, M.; Nilsson, S.; Gustafsson, J.-A. *Proc. Natl. Acad. Sci. USA* **1996**, *93*, 5925–5930.
131. Mosselman, S.; Polman, J.; Dijkema, R. *FEBS Lett.* **1996**, *93*, 5925–5930.
132. Brzozowski, A. M.; Pike, A. C.; Dauter, Z.; Hubbard, R. E.; Bonn, T.; Engstrom, O.; Ohman, L.; Greene, G. L.; Gustafsson, J. A.; Carlquist, M. *Nature* **1997**, *389*, 753–758.
133. Kumar, V.; Green, S.; Stack, G.; Berry, M.; Jin, J. R.; Chambon, P. *Cell* **1988**, *54*, 199–951.
134. Berry, M.; Metzger, D.; Chambon, P. *EMBO J.* **1990**, *9*, 2811–2818.
135. Tzukerman, M. T.; Esty, A.; Santiso-Mere, D.; Danielian, P.; Parker, M. G.; Stein, R. B.; Pike, J. W.; McDonnell, D. P. *Mol. Endocrinol.* **1994**, *8*, 21–30.
136. Umayahara, Y.; Kawamori, R.; Watada, H.; Imano, E.; Iwams, N.; Morishima, T.; Yamasaki, Y.; Kajimoto, Y.; Kamada, T. *J. Biol. Chem.* **1994**, *269*, 16433–16441.
137. Webb, P.; Lopez, G. N.; Uht, R. M.; Kushner, P. J. *Mol. Endocrinol.* **1995**, *9*, 443–456.
138. McDonnell, D. P.; Dana, S. L.; Hoener, P. A.; Lieberman, B. A.; Imhof, M. O.; Stein, R. B. *Ann. NY Acad. Sci.* **1995**, *761*, 121–137.
139. Brzozowski, A. M.; Pike, A. C. W.; Dauter, Z.; Hubbard, R. E.; Bonn, T.; Engstrom, O.; Ohman, L.; Greene, G. L.; Gustafsson, J.-A.; Carlquist, M. *Nature* **1997**, *389*, 753–758.
140. Shiau, A. K.; Barstad, D.; Loria, P. M.; Cheng, L.; Kushner, P. J.; Agard, D. A.; Greene, G. L. *Cell* **1998**, *95*, 927–937.
141. Recanatini, M.; Cavalli, A.; Valenti, P. *Med. Res. Rev.* **2002**, *22*, 282–304.
142. Brodie, A.; Lu, Q.; Long, B. J. *Steroid Biochem. Mol. Biol.* **1999**, *69*, 205–210.
143. Narashimamurthy, J.; Rao, A. R. R.; Narahari, S. G. *Curr. Med. Chem. Anti-Cancer Agents* **2004**, *4*, 523–534.
144. Armer, R. E.; Smelt, K. H. *Curr. Med. chem.* **2004**, *11*, 3017–3028; Tucci, F. C.; Chen, C. *Curr. Opin. Drug Disc. Dev.* **2004**, 7, 832–847; Zhu, Y. F.; Chen, C.; Struthers, R. S. *Annu. Rep. Med. Chem.* **2004**, *39*, 99–110.
145. DeVita, R. J.; Walsh, T. F.; Young, J. R.; Jiang, J.; Ujjainwalla, F.; Toupence, R. B.; Parikh, M.; Huang, S. X.; Fair, J. A. D.; Goulet, M. T. *J. Med. Chem.* **2001**, *44*, 917–922; Jiang, J.; DeVita, R. J.; Goulet, M. T.; Wyvratt, M. J.; Lo, J.-L.; Ren, N.; Yudkovitz, J. B.; Cui, J.; Yang, Y. T.; Cheng, K. et al. *Bioorg. Med. Chem. Lett.* **2004**, *14*, 1795–1798.
146. Ashton, W. T.; Sisco, R. M.; Kieczykowski, G. R.; Yang, Y. T.; Yudkovitz, C. J.; Mount, G. R.; Ren, R. N.; Wu, T.-J.; Shen, X.; Lyons, K. A. et al. *Bioorg. Med. Chem. Lett.* **2001**, *11*, 2597–2602.
147. Clark, E.; Boyce, M.; Warrington, S.; Johnston, A.; Suzuki, N.; Cho, N.; Dote, N.; Nishizawa, A.; Furuya, S. *Br. J. Clin. Pharmacol.* 2003, January 7–10, 443.
148. Zhu, Y.-F.; Struthers, R. S.; Connors, P. J., Jr.; Gao, Y.; Gross, T. D.; Saunders, J.; Wilcoxen, K.; Reinhart, G. J.; Ling, N.; Chen, C. *Bioorg. Med. Chem. Lett.* **2002**, *12*, 399–402; Zhu, Y.-F.; Wilcoxen, K.; Struthers, R. S.; Connors, P., Jr.; Saunders, J.; Gross, T.; Gao, Y.; Reinhart, G.; Chen, C. *Bioorg. Med. Chem. Lett.* **2002**, *12*, 403–406; Tucci, F. C.; Zhu, Y.-F.; Guo, Z.; Gross, T. D.; Connors, P. J., Jr.; Struthers, R. S.; Reinhart, G. J.; Wang, X.; Saunders, J.; Chen, C. *Bioorg. Med. Chem. Lett.* **2002**, *12*, 3491–3495.
149. Wilcoxen, K. M.; Zhu, Y.-F.; Connors, P. J., Jr.; Saunders, J.; Gross, T. G.; Gao, Y.; Reinhart, G. J.; Struthers, R. S.; Chen, C. *Bioorg. Med. Chem. Lett.* **2002**, *12*, 2179–2183; Gross, T. D.; Zhu, Y.-F.; Saunders, J.; Wilcoxen, K. M.; Gao, Y.; Connors, P. J.; Guo, Z.; Struthers, R. S.; Reinhart, G. J.; Chen, C. *Bioorg. Med. Chem. Lett.* **2002**, *12*, 2183–2189; Zhu, Y.-F.; Guo, Z.; Gross, T. D.; Gao, Y.; Connors, P. J., Jr.; Struthers, R. S.; Xie, Q.; Tucci, F. C.; Reinhart, G. J.; Wu, D. et al. *J. Med. Chem.* **2003**, *46*, 1769–1772.
150. Tucci, F. C.; Zhu, Y.-F.; Guo, Z.; Gross, T. D.; Connors, P. J., Jr.; Gao, Y.; Rowbottom, M. W.; Struthers, R. S.; Reinhart, G. J.; Xie, Q. et al. *J. Med. Chem.* **2004**, *47*, 3486–3487.
151. Hillisch, A.; Peters, O.; Kosemund, D.; Mueller, G.; Walter, A.; Schneider, B.; Reddersen, G.; Elger, W.; Fritzemeier, K.-H. *Mol. Endocrinal.* **2004**, *18*, 1599–1609.

Biographies

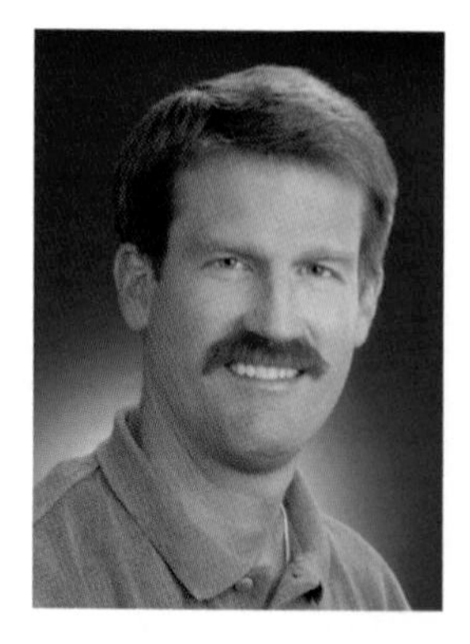

Jeffrey A Dodge obtained his BS in Chemistry in 1984 at Eckerd College in St Petersburg, FL. In 1989, he completed his PhD with Prof Dick Chamberlin at the University of California, Irvine in the field of molecular recognition. He joined Eli Lilly and Company in 1991 after postdoctoral studes with Prof Steve Martin at the University of Texas,

Austin in natural products synthesis. Dr Dodge has extensive expertise in nuclear receptors, particularly the area of estrogen receptor medicinal chemistry and pharmacology. He is currently a Research Fellow at Lilly Research Laboratories.

Timothy I Richardson graduated from the University of Minnesota in 1992 with a BS in Biochemistry. He then joined the laboratories of Scott Rychnovsky in the Chemistry Department at Minnesota where he worked on the stereochemical assignment and synthesis of polyene macrolide antibiotics. He received his doctorate in 1997 and then joined the laboratories of David Evans at Harvard University where he helped complete the total synthesis of vancomycin aglycone. Dr Richardson joined Lilly as a senior organic chemist in 1999 where he has contributed to the development of clinical candidates for endocrine-related diseases. He is currently a Principle Research Scientist.

Owen B Wallace completed his undergraduate education at University College, Cork, Ireland, before moving to the US. He completed his PhD at Yale University on natural product total synthesis under direction of Prof Frederick E Ziegler. He then joined Bristol Myers Squibb in 1995, where he worked on projects directed toward the treatment of Alzheimer's disease and HIV. He moved to Eli Lilly and Company in 2000 where he has contributed to several endocrine-related projects. He is currently Director, Discovery Chemistry Research and Technologies.

Comprehensive Medicinal Chemistry II
ISBN (set): 0-08-044513-6

ISBN (Volume 7) 0-08-044520-9; pp. 149–181

7.08 Kinase Inhibitors for Cancer

A A Mortlock and A J Barker, AstraZeneca, Macclesfield, UK

7.08.1 The Challenge of Cancer

Cancer is fast becoming the leading cause of death in developed countries, causing untold suffering and misery for both patients and their families. Real progress has been achieved in treating certain cancers through the use of novel surgical approaches and pharmaceuticals. While cancers are still typically treated according to their anatomical origin, as our understanding of the underlying genetic causes increases, tumors will be increasingly typed according to the molecular mechanisms driving growth. Allied to the progress in treatment options from drugs (or combinations of drugs) targeting these mechanisms, the future treatment of cancer is likely to change substantially, and offers the promise of major improvements in the treatment of the disease. However, at present, many tumor types remain almost untreatable while factors such as an aging population,[1] environmental pollution,[2] tobacco smoking,[3] and poor diet[4] continue to impact negatively on the occurrence of cancer.

7.08.2 The Basis of the Disease and the Role of Kinases

7.08.2.1 Signal Transduction and Cancer

Cancer is a complex family of diseases characterized by host of cells that have acquired a malignant phenotype, rendering them more genetically unstable and less responsive to external stimuli and their own environment. In a landmark paper, Hanahan and Weinberg have described six 'hallmarks of cancer' that collectively lead to malignant growth[5]: self-sufficiency in growth signals, insensitivity to antigrowth signals, evasion of apoptosis, limitless replicative potential, sustained angiogenesis, and tissue invasion/metastasis. One common underlying mechanism that can operate in many of these areas is aberrant intracellular signal transduction, brought about by the increased genetic instability of the cell.[6] There are numerous examples where upregulation, or constitutive activation, of key signaling pathways allows the cell to proliferate in the absence of extracellular mitogenic signals[7,8] or where downregulation of a death-signaling pathway allows the cell to evade apoptosis.[9]

For primary neoplasia to develop into carcinoma, cellular invasion into adjacent tissue through the basement membrane is necessary, followed by vascularization of the growing tumor mass.[10] Invasion and metastasis result from changes in intracellular signaling pathways, which allow the cell to become insensitive to the normal regulatory effects of cell–cell and cell–matrix contacts, and acquisition of a motile phenotype.[11] Progression of a tumor beyond a certain size requires it to develop a new vasculature from the untransformed endothelial cells of the host. Endothelial cells do not typically acquire intracellular signaling defects, and angiogenesis is often driven through overexpression of extracellular growth factors.[12]

Many of the features of cancer cells described above are mediated through intracellular signal transduction pathways, and involve reversible protein phosphorylation[13] (phosphorylation of nonproteins, such as lipids, can also mediate signal transduction,[14] but this is not considered in this review). These pathways are critical for normal cells to function, but are often inappropriately activated in many tumors. Selective inhibition of such inappropriate signaling offers new treatment opportunities for cancer patients and the potential to find well-tolerated and effective drugs targeted at the specific molecular lesions that drive tumor cell growth.

7.08.2.2 Receptor Tyrosine Kinases (RTKs)

RTKs typically comprise an extracellular N-terminal ligand-binding domain (ECD) (often glycosylated), a hydrophobic transmembrane domain, an intracellular juxtamembrane domain (often autoinhibitory), and a cytoplasmic kinase domain containing multiple tyrosine residues capable of phosphorylation.[15] Most RTKs are monomeric single-chain peptides in the unactivated form, although the Met/Ron family comprise α and β subunits (linked by disulfide bridges),[16] and the insulin receptor family consists of two extracellular α chains disulfide-linked to two transmembrane β subunits.[17] Ligand binding to the ECD causes receptor dimerization/oligomerization, which allows intracellular auto- and/or transphosphorylation within the newly formed signaling complex.[18] Typically, homodimers are formed, although some kinases can form heterodimers, and these may include partners that do not bind external ligands (e.g., ErbB2)[19] or which lack catalytic kinase activity (e.g., ErbB3).[20] These signaling complexes may also include chaperone proteins (e.g., HSP90),[21] which confer conformational stability and are essential for signal transduction. The resultant phosphotyrosine residues within the kinase signaling complex can recruit downstream signaling molecules that contain phosphotyrosine recognition elements such as SH2 or PTB domains.[22,23] Some 30 families of RTKs have been described, on the basis of structural homology, and crystallographic data on 10 different RTK kinase domains have recently been reviewed.[24]

RTK ligand upregulation at both the RNA and protein level can result in uncontrolled proliferative drive, as can direct effects on kinase protein levels.[25] In a small number of well-documented instances, germline mutations can result in the expression of activated kinase.[26] However, activating somatic mutations are a much more significant cause of kinase activation in cancer, and can occur either through direct mutation in the kinase domain or via remote mutations that drive conformational change or limit the influence of internal regulatory domains or external regulatory proteins.[27] A number of mechanisms for therapeutic intervention have been identified for RTKs; those that are common to cytoplasmic kinases will be discussed later. Inhibitors that target the extracellular domains of RTKs, such as antibodies, can block ligand-induced activation of the intracellular kinase domain.[28] An alternative approach is the administration of an agent (typically a solubilized fragment of the extracellular domain of the kinase itself), which sequesters the ligand.[29]

7.08.2.3 Cytoplasmic and Nuclear Kinases

The majority of kinases are intracellular and do not have extracellular domains for ligand binding, although, as with RTKs, they can often localize to membranes (e.g., Fak)[30] and form multimeric signaling complexes with other proteins (e.g., Aurora B).[31] Cytoplasmic kinases are typically expressed in inactive forms that can be activated in a number of ways, including (auto)phosphorylation.[32] Compounds such as imatinib (**1**) (see **Figure 1**) have been shown to bind preferentially to inactive forms of the kinase, preventing adoption of an inactive conformation and resulting in

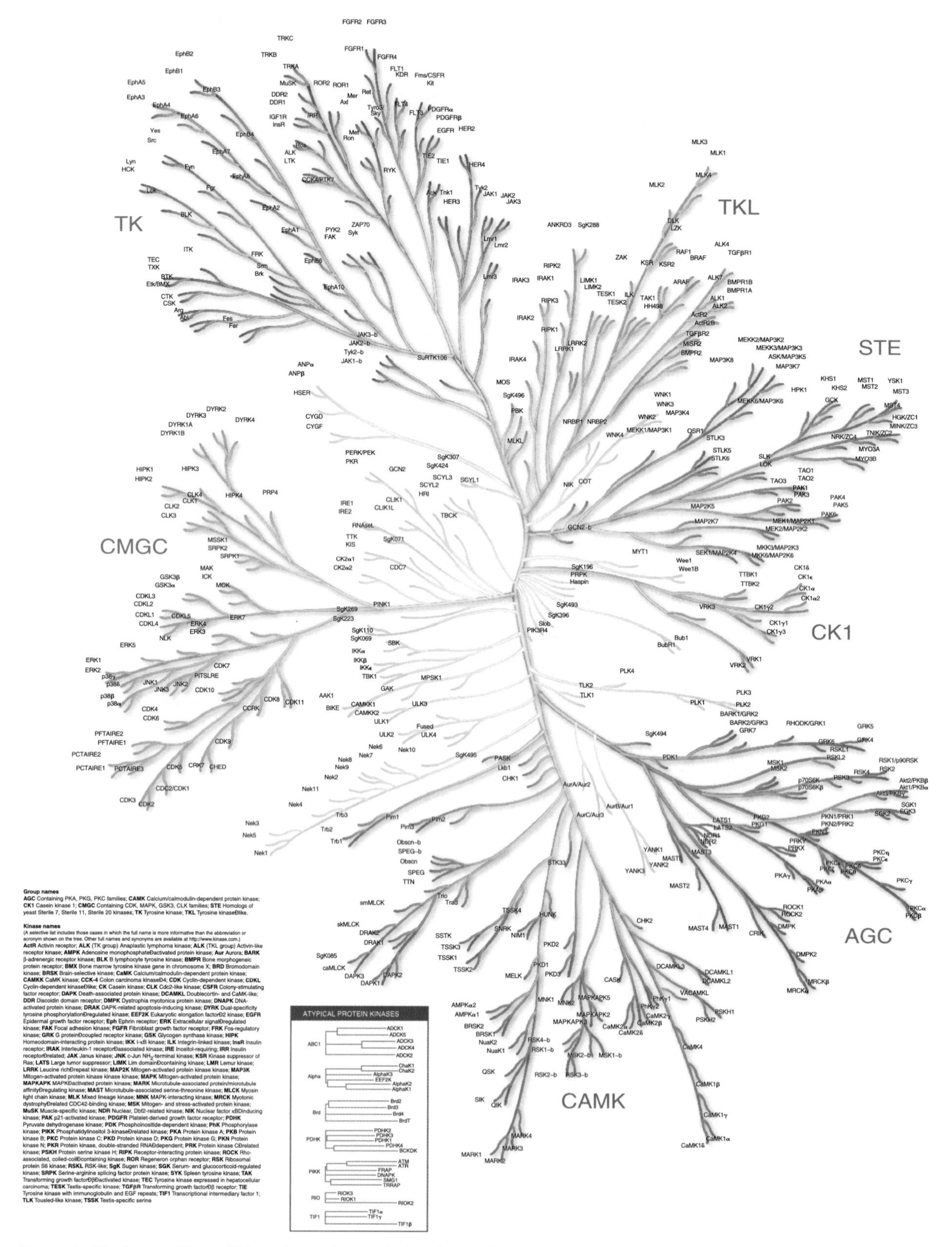

Figure 1 The human kinome.This phylogenetic tree depicts the relationships between members of the complete superfamily of human protein kinases. The main diagram illustrates the similarity between the protein sequences of these catalytic domains. Each kinase is at the tip of a branch, and the similarity between various kinases is inversely related to the distance between their positions on the tree diagram. Most kinases fall into small families of highly related sequences, and most families are part of larger groups. The seven major groups are labeled and colored distinctly. Other kinases are shown in the center of the tree, colored gray. The inset diagram shows trees for seven atypical protein kinase families. These proteins have verified or strongly predicted kinase activity, but have little or no sequence similarity to members of the protein kinase superfamily. A further eight atypical protein kinases in small families of one or two genes are not shown. (Courtesy of Cell Signaling Technology, Inc. (www.cellsignal.com).)

downregulation of the downstream signaling cascade.[33] However, the most widely exploited mechanism of inhibition is to deliver compounds that target the ATP-binding site in the activated kinase.[34,35] Even here, it is possible to subdivide compounds into those that compete directly with ATP for the adenine-binding site, those that occupy an adjacent site (e.g., the 'selectivity pocket'), and those inhibitors that, by virtue of a chemically reactive moiety, compete with ATP for its binding site, but bind irreversibly. Thus, gefitinib (**27**) and erlotinib (**28**) (see **Figure 6**) are epidermal growth factor receptor (EGFR) inhibitors that compete reversibly with ATP for the adenine-binding site within the hinge region,[36,37] while CI-1033 (**31**) (see **Figure 7**) is an irreversible EGFR inhibitor that forms a covalent interaction with a specific cysteine residue within the hinge region.[38] In contrast, the MEK (mitogen-activated protein (MAP) kinase/extracellular signal-related kinase kinase) inhibitor CI-1040 (**78**) (see **Figure 15**) occupies an allosteric site adjacent to the ATP-binding site.[39]

Compounds that target substrate binding can also be identified,[40] and inhibitors of p38 kinase, which seem to target the binding site used by regulatory proteins, have also been described recently.[41] Both mechanisms are potentially attractive as they may offer ways of improving kinase selectivity. Nature has a clear need to regulate kinase activity, and a variety of phosphatases[42] and other regulatory proteins have evolved to this purpose, including many well-established tumor-suppressors such as phosphatase and tensin homologue deleted on chromosome ten (PTEN), which is lost in ~50% of human tumors.[43]

7.08.3 The Human Kinome

Following analysis of the human genome, a recent paper has identified 518 human protein kinases comprising 478 eukaryotic protein kinases (ePKs) and 40 atypical protein kinases (aPKs).[44,45] Of these 518 human kinases, 510 have orthologs in the mouse kinome, which has recently been described,[46] whereas the homology with worm, fly, and yeast kinomes shows clear evolutionary development, underlying the importance of kinases in regulating cellular function.[47]

The 518 human kinases have been classified into seven groups (TK, tyrosine kinase;TKL, tyrosine kinase-like; STE, homolog to yeast sterile 7, 11, and 20 kinases; CK1, containing casein kinase; AGC, containing protein kinase A (PKA), G (PKG), and C (PKC); CAMK, calcium/calmodulin-dependent-like kinase; CMGC, containing cyclin-dependent kinase (CDK), mitogen-activated protein kinase (MAPK), glycogen synthase kinase 3 (GSK-3), and Cdc2-like kinase (CLK) kinases), which are further divided into 134 families and 201 subfamilies (**Figure 1**).[44,48] Hierarchical clustering based on the sequence of the catalytic domain is useful in defining kinase families, but does not cluster kinases by biochemical role or by inhibition profile. The latter can be achieved using selectivity data, and a recent paper describes this approach, using published data on 43 kinases.[24]

7.08.4 Clinical Trial Issues

The strategy for clinical development of anticancer agents has, to date, been dominated by the specific needs of nontargeted cytotoxic agents.[49] Thus, dose escalation studies in Phase I trials have sought to establish a maximum tolerated dose (MTD), which is the dose used in subsequent Phase II trials where the response rate (the percentage of patients whose tumors shrink on receiving treatment) is measured. Phase III studies will evaluate the benefit (typically with survival as the primary endpoint) of the novel agent in combination with, or as a replacement for, the accepted best treatment for the particular tumor type.

The clinical development of imatinib (**1**) has rightly been hailed as a 'paradigm shift' in the way novel anticancer agents are clinically evaluated.[50] Only patients who had been shown to have the activated kinase target (chronic myelogenous leukemia (CML), Philadelphia chromosome positive (Ph+)) were included in the Phase I studies. As MTD was not reached, a biologically effective dose was identified; the side effect profile was examined at that dose, and subsequent studies were carried out to establish the clinical efficacy of the compound at doses below the MTD. Lessons from the development of imatinib are guiding the development of other kinase inhibitors in cancer; patient selection (based on molecular diagnosis as well as tumor origin) and the availability of a validated biomarker in an accessible tissue are critical to minimize the size of the patient cohort, and to provide clear data to evaluate the potential of the agent.[51–53] While survival data in large randomized Phase III studies will remain a long-term regulatory requirement in advanced disease, a number of kinase inhibitors have achieved 'fast-track' status on the basis of early biomarker and safety data, providing an opportunity for terminally ill patients to seek benefit from novel drugs. As many kinase inhibitors may have less severe side effects (compared with current chemotherapy), they should be of value in treating earlier, less advanced disease and this will undoubtedly raise new challenges in how these agents can be evaluated against endpoints other than survival.[54,55]

7.08.5 Tyrosine Kinases and their Inhibitors

7.08.5.1 Nonreceptor Tyrosine Kinases

Even though the non-RTKs represent less than 10% of the human kinome, they have already provided a number of valuable anticancer drug targets such as the Abl and Src families. While this review will attempt to cover the latter in detail, it is also important to highlight other kinases that are emerging as potential targets.

The Jak (Janus) and Fak (focal adhesion) kinase families are both characterized by containing an integrin-binding domain. Jak tyrosine kinases[56] are activated by interleukins, and are often found to be constitutively active in hematopoietic malignancies,[57] with a recent report describing a link to breast cancer.[58] Fak[59] is a substrate and binding partner of Src, and there is an emerging literature highlighting the ability of Fak to interact with downstream signaling pathways, causing both proliferation and evasion of apoptosis.[60] In contrast, Csk (C-terminal Src kinase) is a negative regulator for Src, and inhibition of the kinase results in increased cell migration and in vitro invasiveness.[61] While less studied to date, Pyk2 (proline-rich tyrosine kinase 2) has been shown to regulate apoptosis in multiple myeloma cells,[62] and may also act downstream from fibroblast growth factor receptor (FGFR-3).[63] In breast carcinomas there is an emerging link between Brk (breast tumor kinase) activity and disease progression,[64] while Syk (spleen tyrosine kinase) protein levels have an inverse correlation with survival and metastasis in this disease, suggesting that Syk acts as a tumor suppressor gene in breast epithelium.[65] It is likely that the elucidation of the roles of other non-RTKs (including Btk (Bruton's tyrosine kinase),[66] Fes tyrosine kinase,[67] and Fer tyrosine kinase[68]) may provide additional targets for anticancer drugs.

7.08.5.2 Bcr-Abl

7.08.5.2.1 Bcr-Abl and cancer

Imatinib (Gleevec, Glivec, or STI571, **1**) (**Figure 2**),[50,69–71] an inhibitor of the Bcr-Abl tyrosine kinase, was one of the first small-molecule kinase inhibitors to reach the market, being approved for the treatment of CML in May 2001.

CML is a clonal disease of hematopoietic progenitor cells, and is characterized by progressive granulocytosis, marrow hypercellularity, and splenomegaly.[72–75] The link between CML and Ph+, caused by a reciprocal translocation (between chromosomes 9 and 12) was made as long ago as 1960.[76] Of the two fusion genes created by this transformation,[77,78] the Bcr-Abl gene on the Philadelphia chromosome (22q–) encodes a 210 kDa protein with upregulated tyrosine kinase activity.[79] By 1990, a transgenic mouse model had been developed that demonstrated that Bcr-Abl was a leukemic oncogene.[80] Subsequently, Bcr-Abl has been shown to be present in 95% of patients diagnosed with CML, and is also found in 10–15% of patients with acute lymphoblastic leukemia (ALL).[81] The clinical utility of imatinib in CML is now well established, and demonstrates that inhibition of a primary disease driver using a kinase inhibitor can result in significant clinical activity.

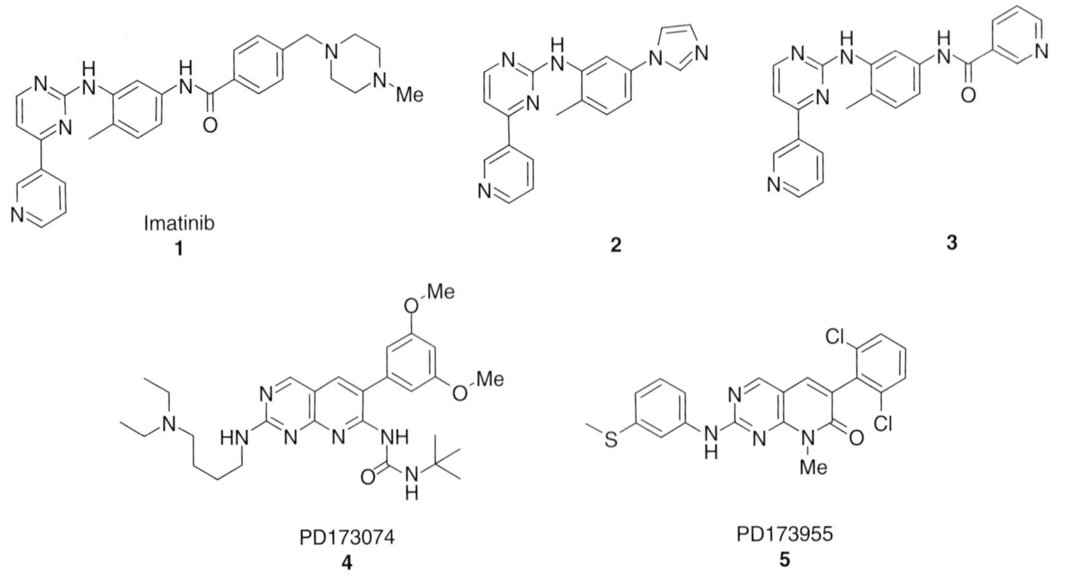

Imatinib
1

2

3

PD173074
4

PD173955
5

Figure 2 First-generation Bcr-Abl inhibitors.

7.08.5.2.2 Inhibitors of Bcr-Abl

The first Bcr-Abl inhibitor to enter clinical development, and the only marketed drug in this area is Novartis's imatinib (**1**). The discovery of imatinib was a result of judicious optimization of a series of phenylaminopyrimidines that had been identified as inhibitors of the serine/threonine kinase PKC-α.[82] Further profiling of compound **2** showed that it inhibited a number of other serine/threonine and tyrosine kinases, including Bcr-Abl, platelet-derived growth factor receptor β (PDGFR-β), and Src.[83,84] Refinement of the substituent at the 4-position of the pyrimidine ring, and introduction of the 'flag methyl group' to the 6-position of the phenyl ring gave compounds with significantly improved potency against both Bcr-Abl and PDGFR-β.[85] Compounds such as **3** also showed no inhibitory activity toward a range of serine/threonine kinases, such as PKC-α, but were typically insoluble and lacked oral bioavailability. Exploiting an observation made in the quinolinone antibiotic area,[86] an *N*-methyl piperazinyl group was introduced to the benzoic acid moiety, and gave imatinib (**1**), which was selected for clinical development.

Imatinib is an ATP-competitive inhibitor of Bcr-Abl (IC_{50} = 188 nM, K_i = 85 nM).[70,87] When tested against a panel of kinases (recombinant kinase domains), it also showed activity against c-Kit (IC_{50} = 400 nM) and PDGFR-β (IC_{50} = 400 nM), but had no activity against Src, PKC-α, and eight other kinases.[85] Autophosphorylation of Bcr-Abl in cells was inhibited (250 nM),[88] as was signal output due to PDGFR, c-Kit, and MAP kinase activation.[89,90] Imatinib blocks proliferation of Bcr-Abl-positive CML and ALL cell lines, causing apoptosis.[91,92] In vivo activity was demonstrated after three-times daily oral dosing (160 mg kg^{-1}) to a KU812 (Bcr-Abl-positive) xenograft in nude mice, causing complete inhibition of tumor growth over the duration of dosing.[93]

7.08.5.2.3 Mechanistic understanding of imatinib

To rationalize the kinase selectivity of imatinib, retrospective modeling studies were carried out,[85] building on the published structure of PD173074 (**4**) bound to the tyrosine kinase domain of FGFR1.[94] This analysis suggested that the aminopyrimidine group bound to the nucleotide binding site, and that the terminal *N*-piperazine lay outside the ATP-binding site in a solvent-accessible region.[95] In a key paper, Kuriyan reported the crystal structure of the imatinib analog **3**, with the inhibitor bound to an inactive form of Bcr-Abl.[96] In this inactive from, Tyr393 in the activation loop is not phosphorylated, leaving it free to hydrogen bond to Asp363, a conserved residue that is essential for kinase activity.

It appears that imatinib specifically recognizes this particular inactive conformation of Bcr-Abl, which is presumed to be in a dynamic equilibrium with a catalytically active conformation. While imatinib specifically targets an inactive from of Bcr-Abl, PD173955 (**5**) has been demonstrated to bind to a conformation of Bcr-Abl in which the activation loop adopts an orientation similar to that found in many active kinases.[97] PD173955 is significantly more potent against Bcr-Abl than imatinib, and is also active against the Src family of kinases.[98] Thus, the different selectivity profile between imatinib and PD173955 can be rationalized in terms of the former targeting an inactive form of Bcr-Abl that is structurally distinct from the inactive conformations of the Src kinases. As has been discussed earlier, this detailed understanding of the molecular mechanism of the action of imatinib has stimulated considerable interest in examining whether 'activation inhibitors' can be found for other kinases.

7.08.5.2.4 Bcr-Abl inhibitors in the clinic

Phase I trials with imatinib (**1**) began in June 1998 in CML patients who had failed interferon therapy with dosing in the range 25–1000 mg day^{-1}.[99] The compound was well absorbed, showed dose-proportional exposure, and had a plasma half-life of between 13 and 16 h. In general, imaitinib was well tolerated; the MTD was not reached, and the most common adverse events were nausea, myalgias, edema, and diarrhea.[100] Encouragingly, hematological responses were seen in all patients receiving doses of 140 mg day^{-1} or more, and 53 of 54 patients who received doses of 300 mg day^{-1} or greater (for a period of at least 4 weeks) showed complete hematological response. Of these 54 patients, 54% showed cytogenic responses, with 13% having a complete cytogenic response. Phase II studies, starting in 1999, focused on CML patients with either interferon refractory disease or with myeloid blast crisis, and also patients with Ph+ ALL.[101] Imatinib was approved by the US Food and Drugs Administration (FDA), following accelerated review, in May 2001 for the treatment of CML refractory to treatment with interferon.

Clinical experience with imatinib has shown that the outcome is critically dependent on which phase of CML (chronic, accelerated, or blast crisis) the patient is in.[102] While 95% of patients in the chronic phase show complete hematological remission, the majority of patients in blast crisis quickly relapse.[99,103] Resistance to imatinib in blast phase CML is believed to be due to secondary mutations in the kinase rather than Bcr-Abl amplification or targeted drug efflux.[104,105] An analysis of 11 patients who relapsed during imatinib treatment showed that all 11 had re-activated Bcr-Abl signaling rather than acquiring Bcr-Abl-independent growth.[106]

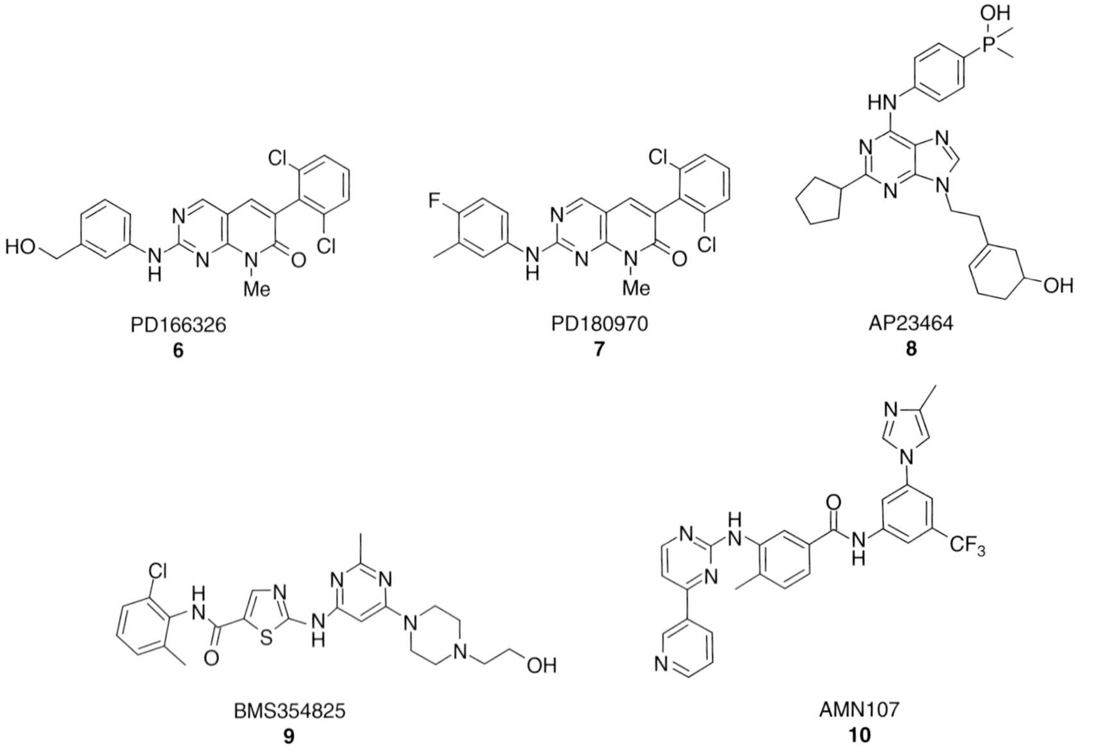

PD166326
6

PD180970
7

AP23464
8

BMS354825
9

AMN107
10

Figure 3 Second-generation Bcr-Abl kinase inhibitors.

Further analysis of clinical material led to the discovery of a number of mutations in Bcr-Abl, including three (T351I and Y255F/H) that are not significantly inhibited by imatinib.[107] Of the known mutations, six (T315I, Y235F/H, E255V/K, and M351T) occur in 60% of blast phase patients, with a mutational frequency of 30–90%. The functional consequence of these mutations is highly variable; many mutated kinases remain effectively inhibited by imatinib (e.g., E355G and M244V). The significance of mutations such as T315I and E255V is that they cause a greater than 200-fold decrease in biochemical sensitivity to imatinib.

7.08.5.2.5 Future directions

While imatinib has revolutionized the treatment of CML, there remains a real need to identify second-generation compounds that target key Bcr-Abl mutations. As many of the mutations seem to prevent the kinase adopting the specific inactive form that imatinib binds to, there is renewed interest in inhibitors that target activated Bcr-Abl.[108] A number of mixed Bcr-Abl and Src kinase inhibitors are in early clinical development, and are showing promising results, although inhibition of the T315I mutated form (the 'gate keeper' residue) remains challenging.[108] Recent publications from various groups have described the activity of compounds such as PD166326 (**6**)[109,110] (**Figure 3**), PD180970 (**7**),[111] AP23464 (**8**),[112] and AG1024 (**39**)[113] (see **Figure 9**) against imatinib-resistant cell lines. BMS354825 (**9**) (Sprycell, Dasatanib) is a mixed Src:Brc-Abl kinase inhibitor which shows good in vivo activity against all Bcr-Abl mutants other than T315I and has recently been approved for the treatment of CML.[114,115] Recent publications from the Novartis group have described the activity of AMN107 (**10**), which is structurally related to imatinib (**1**) but has activity against a number of the mutated forms of Bcr-Abl, although it too is inactive against the T315I form.[116,117]

7.08.5.3 Src Family Kinases

7.08.5.3.1 Src family kinases in cancer

Identification of the Rous sarcoma virus (RSV) in chickens demonstrated that some cancers could be of infectious rather than endogenous in origin.[118–121] In 1978, v-Src (the protein product of the v-*src* gene)[122,123] was shown to be a protein tyrosine kinase.[124–126] In parallel, a normal gene homolog, c-*src*, had been identified,[127] suggesting that the c-*src* sequence had been captured by the virus through recombination at both sides of the c-*src* gene.[128] It is now known that c-*src* is a proto-oncogene,[129] and that its reduced transforming ability, compared with v-*src*, is due to the presence of a C-terminal regulatory domain (absent from v-*src*).[130] Phosphorylation of Tyr527 within the regulatory domain of c-Src, by Csk,[131] acts to modulate Src activity.[132]

The Src family of non-RTKs currently consists of nine members (Src, Yes, Fgr, Yrk, Fyn, Lyn, Hck, Lck, and Blk), all of which share a common structural motif. The molecules comprise an N-terminal myristoylation sequence, a proline-rich linker region, SH3 and SH2 protein interaction domains, a kinase domain, and a C-terminal regulatory domain.[133] The crystal structures of human Src,[134] chicken Src,[135] and Hck[136,137] have been published, and show that the SH2 and SH3 domains lie at the back of the kinase domain, with the SH2 domain interacting with phosphotyrosine 527. This interaction, along with an interaction between the SH3 domain and the polyproline linker (which forms a type II helix), maintains the kinase in an inactive form, and explains how the SH2 and SH3 domain ligands activate Src family kinases.[138]

Increased Src activity has been demonstrated in prostate,[139] ovarian,[140] breast,[141] and colorectal cancers,[142] with some studies suggesting that Src plays a significant role in invasion and metastasis, with the kinase often highly expressed in metastatic tissue.[143] The only normal tissues with high levels of Src are platelets, neural cells, and osteoclasts; gene knockout studies in mice are characterized by osteopetrosis.[144] In cancer cells, Src kinase activity is localized to the cell membrane, and is associated with adhesion and cytoskeletal changes, which promote a motile, invasive phenotype.[145] Src activity is critical in the turnover of focal adhesions,[146] again promoting cell motility. With clear links between Src and the well-known growth factor signaling pathways,[147] it may be surprising that a clear link between elevated Src kinase activity and a proliferative cellular phenotype has not yet been made.[129]

7.08.5.3.2 Src family kinase inhibitors

Among the earliest Src family inhibitors were the pyrrolo[2,3-d]pyrimidines,[148] exemplified by Pfizer's PP1 (**11**) and PP2 (**12**), which show nanomolar potency against Lck and Fyn (**Figure 4**).[149] Starting with the related

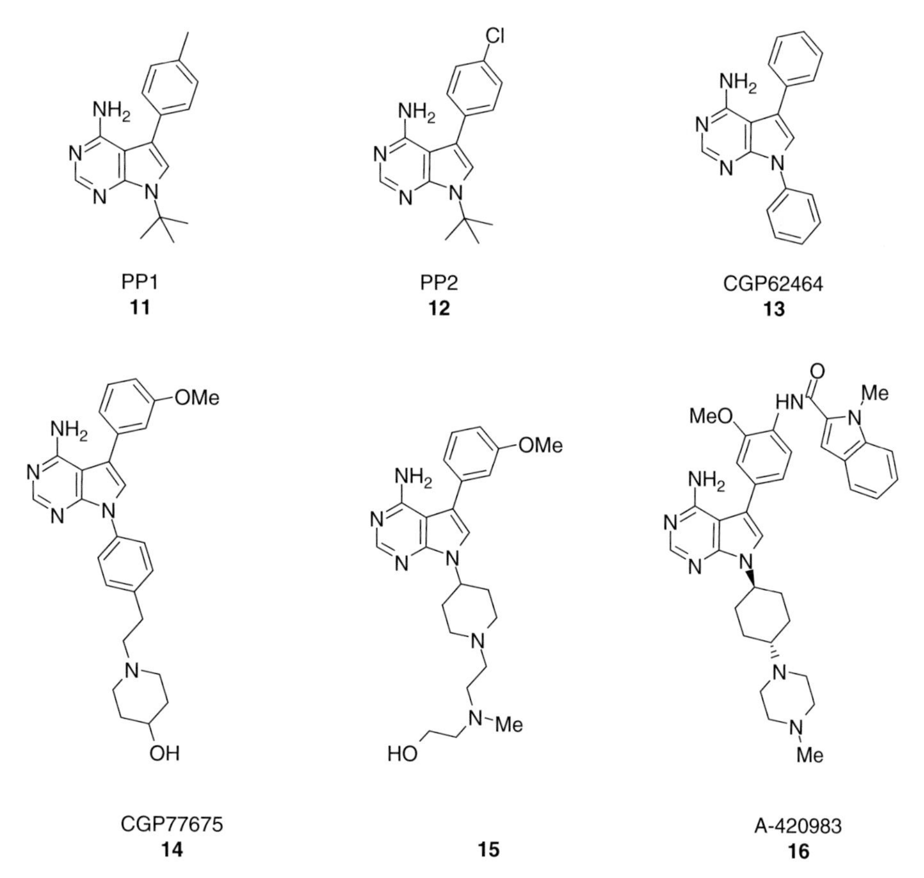

Figure 4 First-generation Src family kinase inhibitors.

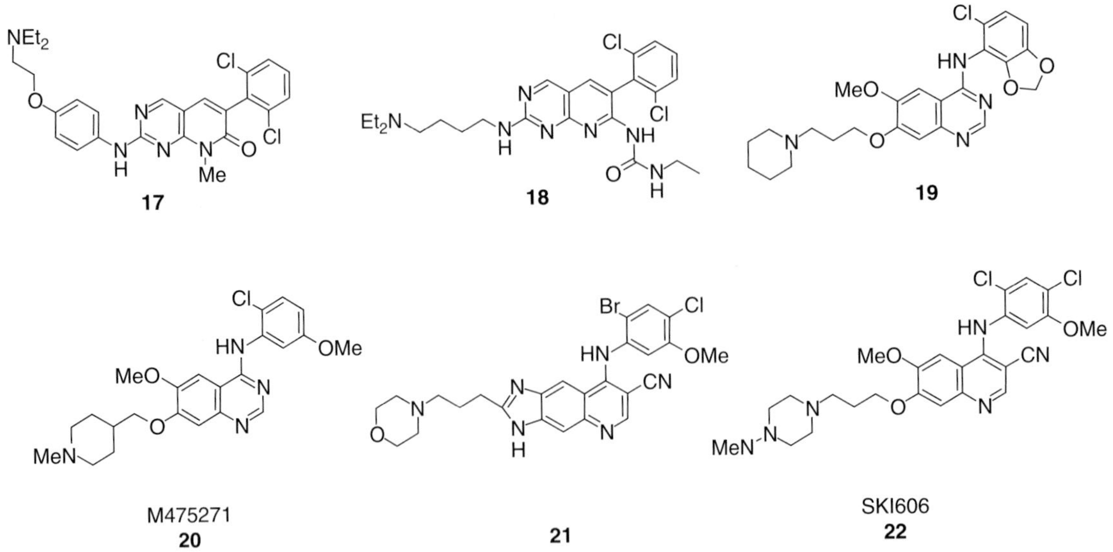

Figure 5 Second-generation Src family kinase inhibitors.

pyrrolo[2,3-d]pyrimidine CGP62464 (**13**),[150] researchers at Novartis were able to introduce a *meta*-methoxy group to the 5-phenyl ring, to improve its interaction with the hydrophobic pocket. Introduction of polar substituents to the 7-phenyl ring further improved potency, and gave compounds such as CGP77675 (**14**), which is selective for Src over other kinases such as EGFR and vascular endothelial growth factor receptor 2 (VEGFR2).[151] The 7-aryl ring is not essential for activity in this series, and compounds such as **15**, where this ring has been replaced with alicyclic groups, also show good activity.[152,153] While compound **14** shows a modest selectivity for Src over other family members such as Lck and Yes, this selectivity can be reversed by substituting the part of the inhibitor that interacts with the hydrophobic pocket; BASF/Abbott has described compounds, including A-420983 (**16**), a potent, orally active Lck inhibitor that has a modest selectivity (2–5-fold) over Src.[154,155]

Two closely related series based on a common pyridopyrimidine core have been developed at Parke-Davis (Pfizer). Compound **17** is very potent against Src ($IC_{50}=0.009\,\mu M$), and retains some activity against EGFR, PDGFR, and FGFR (IC_{50} $<0.1\,\mu M$ for all three kinases) (**Figure 5**).[156] A further refinement led to compound **18**, which shows a similar potency and selectivity profile to compound **17**, but which has been demonstrated to show significant antitumor growth in vivo when given daily at a dose of $26\,mg\,kg^{-1}$ to nude mice bearing Colo-205 tumors.[157]

Starting with an anilinoquinazoline lead, the group at AstraZeneca demonstrated that changes in the substitution pattern around the aniline ring could give compounds such as **19**, which is highly selective for the Src family and inhibits Src with IC_{50} $<0.004\,\mu M$.[158] Another compound from this series, M475271 (**20**), shows good in vivo antitumor activity against a human pancreatic cancer grown orthotopically in a nude mouse, and sensitizes tumor cells to gemcitabine.[159] The substitution pattern around the aniline rings in these quinazolines is similar to that in the corresponding 4-phenylamino-3-cyanoquinolines (**21** and **22**), developed at Wyeth.[160,161] SKI-606 (**22**) is a potent inhibitor of Src ($IC_{50}=0.004\,\mu M$), shows good activity against Src-dependent cellular proliferation, and has been demonstrated to be a potent ($IC_{50}=0.001\,\mu M$) inhibitor of Bcr-Abl.[162]

The antiresorptive activity of Src inhibitors demonstrated in vivo (Src plays a key in osteoclast signaling) has generated an interest in compounds that achieve tissue selectivity via preferential distribution to the bone surface.[163] Researchers at Ariad Pharmaceuticals have shown that dual Src:Bcr-Abl dual inhibitors, such as AP23464 (**8**), retain excellent potency against both kinases and display a high affinity for hydroxyapatite, a major bone constituent.[164,165] The clinical evaluation of Src inhibitors is at an early stage, and it could be many years before the true value of these compounds in treating human disease can be judged.

7.08.5.4 Receptor Tyrosine Kinases

Of the 20 structurally distinct subfamilies of RTKs, 13 contain enzymes that have been associated with human malignancies.[13] Although space limitations allow detailed discussion of only the most significant five classes here, it is almost certain that inhibitors of some of these other targets will be important in the future treatment of certain tumors. For example, Tie and Tek are strongly linked to angiogenesis in endothelial cells,[166] and have been associated with gastric cancer[167] and head and neck squamous cell carcinoma.[168] Overexpression of c-Met and its ligand, hepatocyte growth factor, is strongly linked to the acquisition of an invasive phenotype[169] and disease progression, notably in cervical cancer and hepatocellular carcinoma.[170,171] Fusion proteins containing activated anaplastic lymphoma kinase (ALK) have been found in anaplastic large cell lymphoma.[172] The Ret oncogene causes both papillary and medullary thyroid carcinomas, depending on the exact nature of the germline point mutation.[173,174] Axl (with its ligand GAS6) was first linked to cancer through its discovery in DNA from patients suffering from CML[175] but it has subsequently been associated with prostate, colon, and endometrial cancer as well as acute myeloid leukemia (AML).[176,177] The three Trk subfamily kinases are receptors for nerve growth factor, and have been linked to the progression of neuroblastoma and prostate cancer.[178]

7.08.5.5 Erb Family Tyrosine Kinase

7.08.5.5.1 Epidermal growth factor receptor tyrosine kinase in cancer

The EGFR (erbB1) tyrosine kinase is involved in the regulation of normal cell growth and differentiation,[179] following interaction with the ligands EGF and transforming growth factor α (TGFα). A wide range of human tumors show high expression levels of EGFR,[180] express mutated and/or constitutively activated forms of the kinase,[181] or overexpress the ligand.[182] The resulting inappropriate or amplified signaling through the erbB1 signaling cascade can be transforming, and lead to malignancies in a wide variety of epithelial tissues.[183]

7.08.5.5.2 Inhibitors of epidermal growth factor receptor tyrosine kinase

Among the first compounds to show activity against the isolated EGFR were flavones and isoflavones, exemplified by genistein (**23**) (**Figure 6**).[184] Genistein (**23**) was competitive with ATP in binding to the enzyme, but relatively unselective, and shows only modest activity in an EGF-driven cell proliferation assay. Erbstatin (**24**),[185] a natural product inhibitor, was sufficiently simple, in chemical terms, to provide scope for an investigation of the srtucture–activity relationships surrounding kinase inhibition.[186] Modification of the erbstatin structure, and removal of the potentially unstable enamide, resulted in a series of molecules termed tyrphostins, such as RG-13022 (**25**).[187] Levitzki and his group went on to develop this series into a range of compounds showing some selective inhibition of a variety of kinases, including EGFR.[188] Some had submicromolar activity against the isolated kinase, with cellular activity at similar levels. An important breakthrough was the demonstration of in vivo activity in mouse models of human tumor disease driven by EGF.[189] RG-13022 (**25**) was active when given intraperitoneally to mice bearing an MH85 squamous cell carcinoma at a dose of $10\,\mathrm{mg\,kg^{-1}\,day^{-1}}$, and, in contrast to conventional cytotoxic therapy, appeared well tolerated. The lack of selectivity and potential for metabolism made RG-13022 unattractive as a drug. Many groups began to screen compounds for small-molecule leads as start points for EGFR

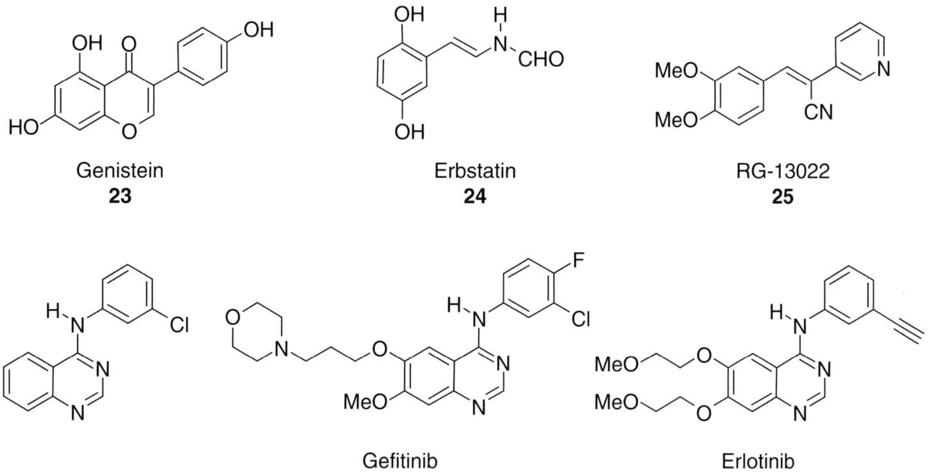

Genistein **23**; Erbstatin **24**; RG-13022 **25**; **26**; Gefitinib **27**; Erlotinib **28**

Figure 6 First-generation EGFR kinase inhibitors.

inhibitor programs, and a large number of pharmacophore types have been identified, which have been comprehensively reviewed.[190–194]

AstraZeneca identified the simple anilinoquinazoline lead **26**.[195] The compound had good activity against the kinase (IC_{50} = 0.04 μM), and inhibited EGF-driven cellular growth (IC_{50} = 1.2 μM) as well as showing selectivity over a variety of other kinases. Its potency was increased by adding electronegative substituents to the 6- and 7-positions of the quinazoline, and selectivity and metabolic stability resulted from changing the aniline substitution. Finally, modifying the physicochemical properties of the molecule by incorporating a weak base in the alkoxy side chain resulted in gefitinib (Iressa, ZD1839, **27**),[196] the first small-molecule EGFR tyrosine kinase inhibitor to reach the clinic. A group at Parke Davis was also very active in investigating the anilinoquinazoline pharmacophore.[197] Roche/OSI has also taken Erlotinib (Tarceva, OSI-774, **28**) to the clinic,[198] and a wide variety of other small molecules have now been identified as EGFR kinase inhibitors and selected for clinical evaluation.

7.08.5.5.3 Structural and mechanistic understanding of epidermal growth factor receptor

Most ATP-competitive EGFR tyrosine kinase inhibitors interact with the enzyme through up to three hydrogen bond donor–acceptor interactions with backbone amide functions used to bind ATP.[199] Interactions of the small molecule with nonconserved amino acids not required for binding ATP can provide selectivity. In particular, a hydrophobic pocket (the 'selectivity' pocket), not used in binding ATP, is believed to be one of the principal reasons for the ability to achieve selectivity between different kinases with anilinoquinazolines.[200] The aniline portion of these molecules interacts with the variable residues in this region. The aniline substituents and the dihedral angle at which the aniline is held relative to the quinazoline then dictate how energetically favorable such interactions in the 'selectivity' pocket will be. The crystal structures of erlotinib (**28**) bound in the kinase domain of EGFR have confirmed this binding mode.[199] As with many other kinases, the selectivity pocket is able to take part in 'adaptive binding' and change conformation to accommodate large groups if the energetics of the interaction are favorable.

7.08.5.5.4 Epidermal growth factor receptor mutations

Mutations of the extracellular domain of the EGFR tyrosine kinase have been known for some time, the commonest being the EGFRvIII mutation,[201] which results in constitutive activity of the kinase. Recent evidence has shown that mutations of the EGFR kinase domain are present in some non-small cell lung cancer (NSCLC) tumors, and can result in a more intense and prolonged activation of the kinase by the ligand and increased sensitivity to inhibition by some inhibitors.[202–205] Recent results suggest that most of the NSCLC patients in a subgroup responding to gefitinib treatment had these latter mutations that stabilize the ATP–kinase interaction.[206] These mutations also stabilize the interaction between the kinase and gefitinib, and so these variants are particularly sensitive to the action of the drug. While these observations do not explain all the responses to the drug, they may open up the possibility of therapies targeted toward patients with specific EGFR mutations.

7.08.5.5.5 Epidermal growth factor receptor tyrosine kinase inhibitors in the clinic

A chimeric mouse–human monoclonal antibody, cetuximab (C-225, Erbitux), which prevents activation of EGFR by inhibiting ligand binding to the extracellular ligand-binding domain of the receptor,[207] has been approved for the treatment of metastatic colorectal cancer. Side effects related to the mechanism of action include an acneiform skin rash, which appears to be typical of EGFR signaling inhibition. The first small-molecule EGFR tyrosine kinase inhibitors to reach the clinic and obtain regulatory approval have been gefitinib (**27**)[208–212] and erlotinib (**28**).[213–215] Both compounds act by binding reversibly at the ATP-binding site of the activated kinase domain of the protein.

Gefitinib has an IC_{50} for inhibition of the EGFR kinase of 23 nM (ATP concentration = 20 μM), and shows good selectivity over the erbB2 kinase. Inhibition of EGFR-driven cellular growth of KB cells is seen at 0.08 μM. The compound has good pharmacokinetic properties, and is active in vivo against a range of human tumor xenografts in mice at daily oral doses of 10–200 mg kg^{-1}. Gefitinib was progressed to Phase I trials, where it was well tolerated at a dose of 750 mg day^{-1}.[216] Diarrhea and an acneiform skin rash (both of which are believed to be related to the pharmacology of the agent) were observed as toxicities, and are clearly different from the typical side effects seen with conventional cytotoxic agents. Further trials established that biomarkers could be used to assess the degree of inhibition of the kinase in patients and help design later-stage trials. Two large Phase II trials of gefitinib in patients with advanced NSCLC who had relapsed following treatment with platinum-based chemotherapy were carried out at 250 mg kg^{-1} and 500 mg day^{-1} continuous dosing.[217,218] Responses were seen in 12–19%, and disease control in 36–54%, of patients receiving the drug. Based on these results, gefitinib was approved for the treatment of inoperable NSCLC in Japan in

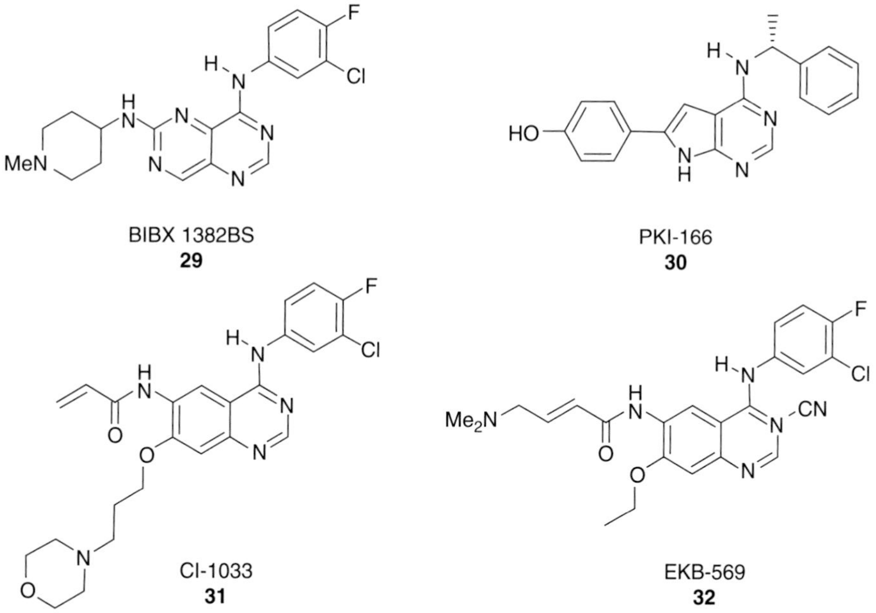

BIBX 1382BS
29

PKI-166
30

CI-1033
31

EKB-569
32

Figure 7 Second-generation EGFR kinase inhibitors.

2002, and other countries rapidly followed. A recent Phase III trial in NSCLC using a continuous daily dose of 250 mg failed to achieve the primary endpoint (survival).

Erlotinib (**28**) is closely related to gefitinib; it is a reversible ATP-competitive inhibitor of the kinase ($IC_{50} = 2$ nM), inhibits EGF-driven cellular growth ($IC_{50} = 0.1\ \mu M$), and is orally active in disease models at 10–100 mg kg^{-1} daily doses.[219] Phase I trials achieved target blood levels dosing at 100 mg day^{-1} or greater.[220] The MTD is 150 mg, and a familiar pattern of toxicity including diarrhea and skin rash was seen. The human half-life is about 24 h (approximately 48 h for gefitinib), and some antitumor activity was observed in these trials. Recent data from a Phase III trial, dosing erlotinib at 150 mg day^{-1}, shows a statistically significant increase in life expectancy compared with controls in patients with NSCLC, and the drug has been approved in the USA for this condition.[221]

These two compounds have resulted in many other EGFR inhibitors progressing to clinical studies, including those having modifications to the basic quinazoline pharmacophore, such as the pyrimidopyrimidine BIBX 1382BS (**29**)[222] and the pyrrolopyrimidine PKI-166 (**30**) (**Figure 7**).[223] Of interest is CI-1033 (canertinib, **31**),[224] which differs from gefitinib and erlotinib in that it both binds irreversibly at the ATP site of the kinase and has activity at the erbB2 kinase through a similar mechanism. The acrylamide residue at the 6-position of the quinazoline is not a particularly good Michael acceptor, but in binding at the ATP site is brought into close contact with Cys773 of the EGFR (and Cys805 in the erbB2), and results in alkylation.[225] This approach may have the advantage that less frequent dosing is required for inhibition of kinase function – for instance, 72 h after a single 40 mg kg^{-1} oral dose to mice there is still 75% inhibition of the EGFR kinase. However, the approach carries risks in that alkylated proteins can often cause in-host immune responses, which could limit further dosing. CI-1033 (**31**) has progressed to Phase II trials in patients and shown modest activity, but with more severe toxicity than was seen with the reversible agents. A second EGFR/erbB2 irreversible inhibitor built on the 3-cyanoquinoline pharmacophore EKB-569 (**32**)[226] is in Phase II trials.

7.08.5.5.6 ErbB2 and cancer

A large amount of evidence is available that highlights the important role that the erbB2 RTK plays in mediating signaling by ligand-dependent activation of other erbB family members.[227,228] While a natural ligand is unknown, this activation is achieved through heterodimerization with the other erbB family members.[229] Overexpression of the kinase is transforming and tumorigenic.[230] It is overexpressed in many other solid tumors,[231] and the link with cancer is best characterized in breast cancer, where approximately 30% of tumors overexpress the kinase,[232] and correlates with poor prognosis.

7.08.5.5.7 Inhibitors of ErbB2

The only erbB2 inhibitor presently approved for treatment is a humanized inactivating monoclonal antibody, trastuzumab (Herceptin).[233,234] It produces a 15% response rate in erbB2-overexpressing breast cancer as a

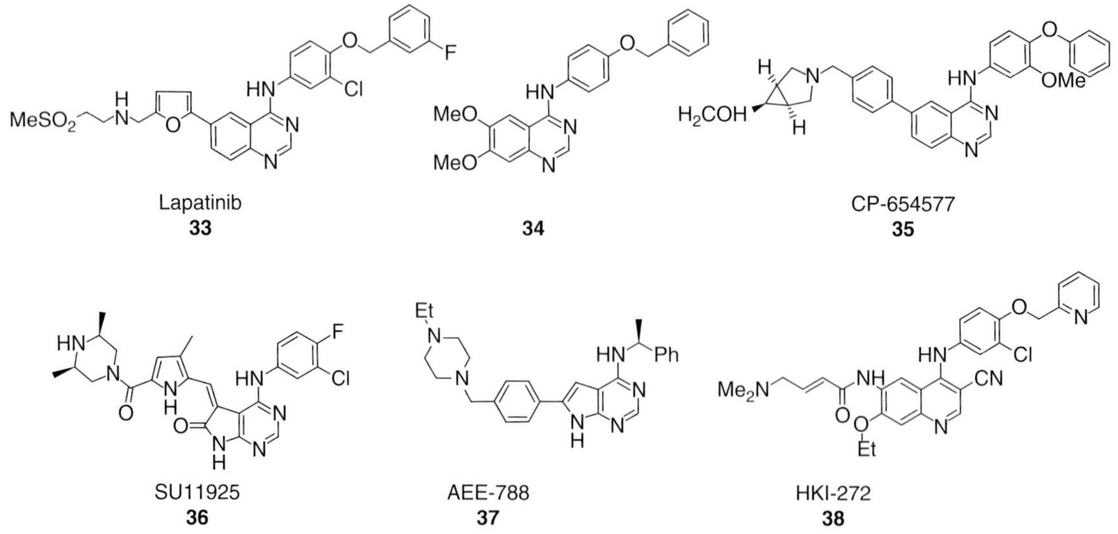

Lapatinib **33** **34** CP-654577 **35**

SU11925 **36** AEE-788 **37** HKI-272 **38**

Figure 8 ErbB2 kinase inhibitors.

monotherapy, and a 49% response rate in combination with paclitaxel.[235] As it binds at the extracellular ligand-binding site, it is not an inhibitor of the kinase domain itself; nevertheless, it has encouraged others to adopt a similar approach, and a range of such monoclonal antibodies are in trial, including pentuzumab (2C4),[236] which targets erbB2 dimerization.

The leading small-molecule erbB2 inhibitor is lapatinib (GW572016, **33**)[237,238] a 6-furylquinazoline that reversibly binds at the ATP site of the erbB2 kinase (IC_{50} = 9 nM) (**Figure 8**); the compound also inhibits EGFR (IC_{50} = 11 nM), and so should truly be considered a dual inhibitor. The compound was developed from the lead compound **34**.[239] Large aniline substituents were found to enhance erbB2 activity relative to EGFR, and the extended furan side chain provides an improved therapeutic index and better pharmacokinetic properties, presumably through modification of the physicochemical properties of the molecule. Lapatinib (**33**) was shown to be selective for the inhibition of the growth of tumor cells over normal proliferating cells, and demonstrated good in vivo activity against BT474 tumor xenografts in mice when given as an oral dose of 10–30 mg kg twice daily. The compound is in Phase III trials in a variety of solid tumor types, and clinical activity has been reported on a single daily dosing schedule.[240] Lapatinib has been granted fast-track status by the US FDA for patients with refractory breast cancer.

CP-654577 (**35**),[241] a 6-phenyl substituted quinazoline from Pfizer, is selective for inhibition of erbB2 (IC_{50} = 11 nM); the IC_{50} for the EGFR is 670 nM, the erbB2 selectivity being provided by the extended aniline substituent. The compound has antiproliferative activity against cells whose growth is driven by the erbB2 receptor, and is active in vivo against FRE-erbB2 tumors implanted in nude mice when given intraperitoneally twice daily at doses of between 12.5 and 50 mg kg^{-1}.

A variety of other small molecules have been identified as erbB2 kinase inhibitors, many of which also have activity against EGFR, and are exemplified by **32**,[226] SU-11925 (**36**),[242] and AEE-788 (**37**),[243] which in addition have significant activity against the VEGFR family. Selective irreversible inhibition has also been an approach to erbB2 (in addition to the relatively unselective agents mentioned in Section 7.08.5.5.5). HKI-272 (**38**) irreversibly binds to both erbB2 and EGFR.[244] Cellular effects are seen at much lower doses, and the agent is orally active in vivo at continuous daily doses of between 20 and 80 mg kg^{-1}.

7.08.5.5.8 Structural and mechanistic understanding of ErbB2

While many of the agents described above bind to erbB2 as described for EGFR, lapatinib appears to be different. The crystal structure of lapitinib (**33**) bound to the EGFR kinase domain reveals that the compound binds to a distinct inactive-like conformation,[245] different to the structures seen with erlotinib bound in the same kinase. Lapatinib appears to 'lock' EGFR (and, by implication, the erbB2 homolog) into an inactive conformation with the extended aniline buried deep within the protein fold. A slow rate of change to the conformational state required for lapitinb association/dissociation may account for its slow off-rate from the kinase, and the observation of inhibition long after extracellular compound has been cleared.

7.08.5.5.9 Future directions

The EGFR/erbB2 inhibitors are some of the first new agents for the treatment of cancer to reach the clinic and receive approval. They differ from classical treatments, and new methods of development have been required. As neither gefitinib nor erlotinib has yet shown benefit in combination with cytotoxics, it may be that the most effective use of these agents remains to be explored. The class may well offer a significant opportunity to progress personalized medicine in which specific tumor lesions are typed and targeted with drugs having a high chance of efficacy in the patient population identified as likely responders.

First-generation agents have shown clear differences in binding kinetics, and while the leading compounds are competitive with ATP, there may be benefits in exploiting compounds with different binding modes and slower dissociation rates from the kinase. Furthermore, the benefit of pan-erB inhibitors compared with more selective agents remains to be evaluated.

7.08.5.6 Insulin-Like Growth Factor-1 Receptor (IGF-1R)

7.08.5.6.1 Insulin-like growth factor-1 receptor and cancer

IGF-1R plays an important role in proliferation, survival, and transformation, and is widely expressed in human tissues.[246] Overexpression and increased activity have been demonstrated in many types of cancer, including breast, colon, and prostate, and there is good evidence that signaling through IGF-1R protects cells from apoptosis and confers a growth advantage.[247] Antibodies to the ligand-binding domain have demonstrated antitumor properties in experimental models, both alone and in combination with cytotoxics, and have progressed to Phase I clinical studies. A key issue for clinical development is to achieve good selectivity over the structurally similar insulin receptor (INSR), as this plays a critical role in the metabolic processes of normal cell function.

7.08.5.6.2 Inhibitors of insulin-like growth factor-1 receptor

Relatively few selective inhibitors of the IGF-1R have been identified. Early work by Levitski's group identified several tyrphostins such as AG1024 (**39**) that had modest activity against the IGF-1R kinase but also inhibit INSR (**Figure 9**).[248] The most advanced small-molecule kinase inhibitor is AEW-541 (**40**), from the Novartis group, that was developed from an high-throughput screening hit, and has now advanced to clinical development.[249] This pyrrolo[2,3-d]pyrimidine is competitive with ATP, shows approximately 27-fold selectivity over insulin signaling in cells, and is selective against a range of other kinases, including EGFR, c-Kit, PDGFR, and Bcr-Abl. The compound inhibits IGF-1-mediated survival of MCF-7 cells, and is active in vivo against an NWT-21 fibrosarcoma in nude mice, dosed orally twice daily at 20–50 mg kg^{-1}. The compound appears to be well tolerated, and no significant differences in plasma glucose or insulin levels were observed at efficacious doses, suggesting minimal inhibition of INSR occurs.

7.08.5.7 Type III Receptor Tyrosine Kinases (Platelet-Derived Growth Factor Receptor α, Platelet-Derived Growth Factor Receptor β, Colony Stimulating Factor 1R (CSF-1R), c-Kit, and Fms-Like Tyrosine Kinase (Flt3))

7.08.5.7.1 Type III receptor tyrosine kinases and cancer

The five members of the type III subfamily of RTKs are characterized by an extracellular domain that contains five immunoglobulin-like repeats and a split cytoplasmic kinase domain.[13,250] Abnormal proliferation induced by PDGF-α

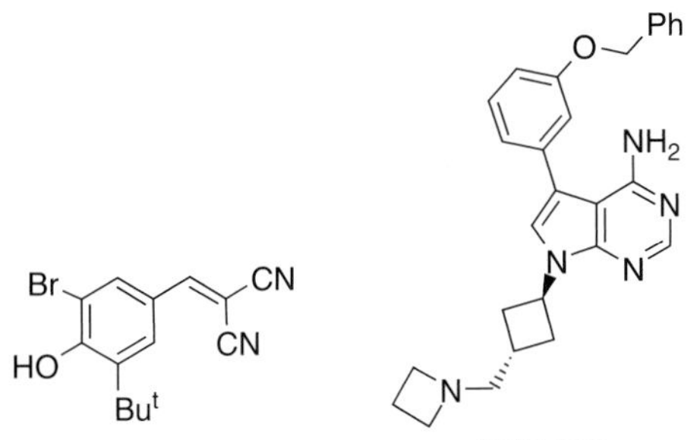

AG1024
39

AEW-541
40

Figure 9 IGF-1R kinase inhibitors.

and PDGR-β[251,252] has been linked to atherosclerosis, pulmonary fibrosis, and restenosis,[253,254] as well as breast, colon, lung, prostate, and ovarian cancer.[255] PDGFR-β promotes endothelial cell and pericyte migration and proliferation, and preclinical data show that PDGFR inhibitors have an antiangiogenic phenotype.[256] Fusion proteins of both receptors have been identified that, as exemplified in the recent discovery of a PDGFR-β:rabaptin 5 fusion protein in chronic myelomonocytic leukemia (CMML), cause constitutive activation of PDGFR-β.[257]

c-Kit is the receptor for stem cell factor (SCF),[258,259] and plays a critical role in normal hematopoiesis and mast cell development.[260] More than 30 activating mutations of c-Kit have been identified that separate into two clusters; those in the juxtamembrane domain are associated with GIST (gastrointestinal stromal tumor),[261] whereas mutations in the kinase domain (notably D816V)[262,263] are linked with mast cell and myeloid leukemias.[264] Flt3 (fms-like tyrosine kinase-3)[265] has an important role in hematopoietic precursor expansion,[266] but is also strongly expressed by the leukemic blasts in AML.[267] As with c-Kit and other kinases,[268,269] most activating mutations of Flt3 are in the juxtamembrane region, and illustrate that this domain has evolved a negative regulatory role.[270] Macrophage colony stimulating factor 1 (CSF-1)[271] is the ligand for the receptor CSF-1R, which is encoded by the c-*fms* proto-oncogene.[272] Mice that are nullizygous for CSF-1R have skeletal abnormalities and developmental defects in a number of tissues caused by a reduction in the number of tissue macrophages and osteoclasts.[273] CSF-1 and CSF-1R are upregulated in ovarian, endometrial, and breast cancer,[274] potentially indicating a role for tumor-associated macrophages in tumor invasion and metastasis.[275]

7.08.5.7.2 Type III receptor tyrosine kinase inhibitors

Imatinib (**1**) (*see* Section 7.08.5.2) has significant activity against both PDGFR and c-Kit, and has already allowed the potential of type III RTK inhibitors to be tested in the clinic. A close structural analog of Imatinib, CGP53716 (**41**) (**Figure 10**), has also demonstrated good activity when given orally to a transfected BALB/c3T3 xenograft at a dose of 3–25 $\mathrm{mg\,kg^{-1}\,day^{-1}}$.[276] Among the earliest classes of PDGFR inhibitor were a series of potent 3-substituted quinolines described by RPR.[277] Development of this series gave the quinoxaline RPR101511A (**42**), which shows good selectivity for PDGFR and has been demonstrated to show in vivo activity in a porcine model of restenosis dosed at $2 \times 30\,\mathrm{mg\,kg^{-1}\,day^{-1}}$ orally.[278] AG1296 (**43**)[279] is a structurally related quinoxaline that formed the basis of a series of tricyclic quinoxalines that have been recently described as inhibitors of PDGFR, Kit, and Flt3.[280] Kirin has described a series of 4-piperazinyl quinazolines based on an early lead, KN1022 (**44**), that show good potency toward PDGFR-β, c-Kit, and Flt3 but that are selective over CSF-1R.[281,282] Two other notable early classes of PDGFR inhibitors with novel structures are the *N*-phenyl benzimidazoles (typified by compound **45**)[283] and SU101 (**46**).[284]

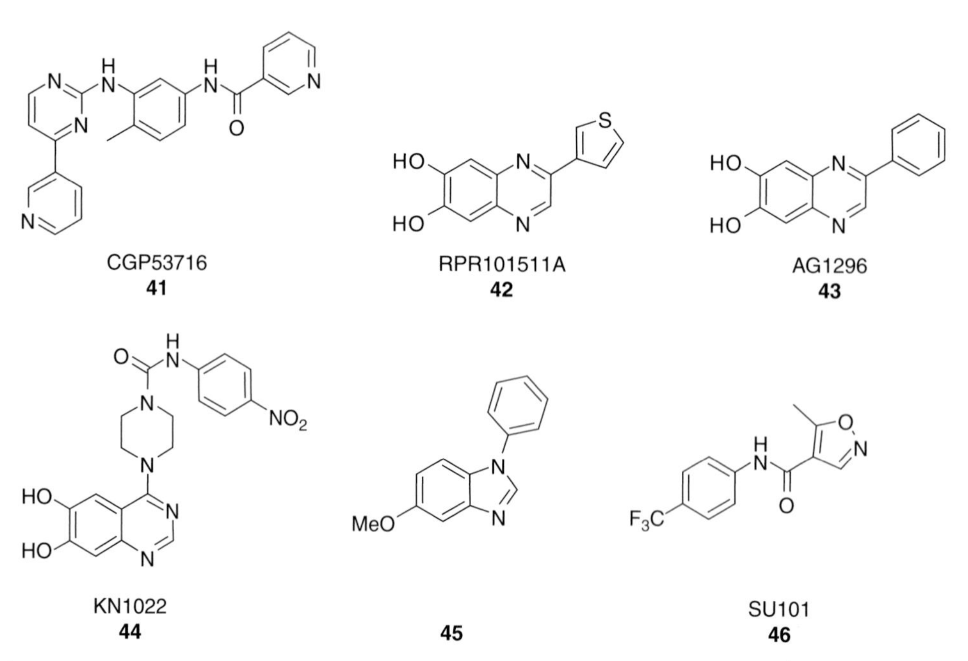

Figure 10 First-generation PDGFR kinase inhibitors.

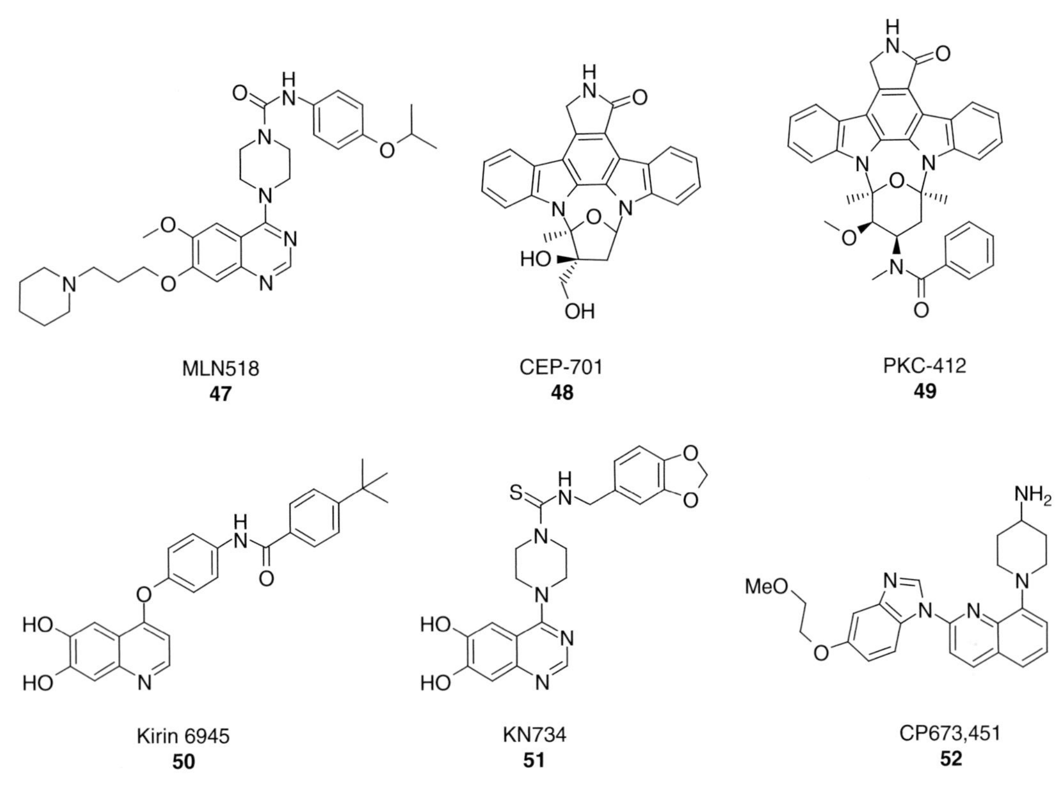

Figure 11 Second-generation type III RTK inhibitors.

Five Flt3 inhibitors have entered clinical trials,[265] including Sugen's indolinones **53** and **55** (see later) and Millennium's MLN518/CT53518 (**47**)[285,286] (**Figure 11**), which is probably the most advanced compound. MLN518 is a potent inhibitor of wild-type Flt3 (IC_{50} = 220 nM), and also inhibits some, but not all, of the AML-derived Flt3 activation loop mutatations.[287] CEP-701 (**48**)[288] from Cephalon was originally identified as a TrkA inhibitor (IC_{50} = 2.9 nM),[289] but has recently been shown to be a potent inhibitor of Flt3 (IC_{50} = 3 nM).[290] Novartis has described PKC-412 (**49**),[291] a staurosporine derivative originally identified as an inhibitor of PKC, but more recently shown to inhibit Flt3 (IC_{50} = 3 nM) as well.[292]

7.08.5.7.3 Type III receptor tyrosine kinases in the clinic

Following registration for CML, imatinib (**1**) has been exploited in the treatment of a number of cancers that are not linked to Bcr-Abl. Imatinib, as a dual c-Kit and PDGFR-α inhibitor, is registered for use in GIST at a dose of 400 mg day^{-1}.[293] This compound has also been used successfully in the treatment of CMML[294] and dermatofibrosarcoma protuberans,[295] which are both linked to constitutive activation of PDGFR-β, although the genetic causes of activation are specific and different.

PKC412 (**49**) has been given orally to AML/myelodysplastic syndrome patients at a dose of 75 mg three times daily. Two patients suffered fatal pulmonary events of unknown etiology, but 50–70% of patients experienced significant reductions in blasts, and in the majority of these patients a reduction in Flt3 phosphorylation could be detected.[296] Phase I trials with CEP-701 (**48**) established that a dose of 40 mg twice daily is tolerated, and a small number of patients showed signs of disease stabilization.[297] Early clinical studies with the dual VEGFR/PDGFR inhibitor SU11248 (**54**)[298] are described in Section 7.08.5.8.4.

7.08.5.7.4 Future directions

Novel PDGFR kinase inhibitors showing improved potency and selectivity continue to be reported. Kirin's Ki6945 (**50**) shows good selectivity over EGFR,[299] while further optimization of the earlier KN1022 (**44**) has allowed the development of CTG52923/KN734 (**51**), which has improved physicochemical properties.[300] Pfizer has described CP673,451 (**52**), which is extremely potent in both enzyme (PDGFR-β IC_{50} = 1 nM) and cellular phosphorylation assays, and differs from earlier compounds in having little activity against c-Kit.[301] The compound has good physical properties and low clearance, although the oral bioavailability in the rat is only 35%, and decreases to 21% in the dog, and to <3% in the monkey. When given orally to a Colo205 mouse xenograft model, CP673,451 (**52**) showed significant growth inhibition at doses of 10–100 $mg\,kg^{-1}day^{-1}$.

7.08.5.8 Vascular Endothelial Growth Factor Receptor

7.08.5.8.1 Vascular endothelial growth factor receptor and cancer

In the past decade it has become clear that tumor progression beyond a limited physical size requires the development of tumor vasculature resulting from tumor-induced angiogenesis.[302] Antiangiogenic therapy is a potentially attractive approach to cancer treatment, in that many key targets are expressed on genetically stable vascular endothelial cells, and the approach offers the potential for treating a broad range of solid tumors.[303,304] VEGF-A has been shown to be a primary driver of the angiogenic cascade, promoting local invasion, inducing migration and mitosis, and also increasing vascular permeability.[305,306] Two high-affinity RTKs, Flt-1 (VEGFR-1) and KDR (VEGFR-2), expressed on vascular endothelial cells, can bind VEGF-A, with KDR representing the more attractive antitumor target.[307,308] A third RTK, Flt-4 (VEGFR-3) is expressed on adult lymphatic endothelial cells; it does not bind VEGF-A but promotes lymphangiogenesis when stimulated with VEGF-C or VEGF-D.[309,310]

7.08.5.8.2 Biopharmaceutical vascular endothelial growth factor inhibitors in the clinic[311]

Bevacizumab (Avastin) is a recombinant humanized monoclonal anti-VEGF antibody developed by Genentech, recently approved for the treatment of metastatic colorectal cancer. An excellent review of the clinical development of Bevacizumab has been published, and highlights work in colorectal, renal cell, and non-small cell lung cancer.[312] In a Phase III trial in metastatic colorectal cancer, bevacizumab was administered as a 5 $mg\,kg^{-1}$ i.v. infusion every 2 weeks in combination with standard IFL (5-fluorouracil, leucovorin, and irinotecan) therapy. Demonstrable clinical benefit was observed in survival, progression-free survival, and objective response, while the side effects included hypertension, gastrointestinal perforation, and epistaxis.[313]

ImClone is developing the anti-KDR antibody IMC-1C11 (derived from DC101), which has completed Phase I trials.[314–316] Ribozyme and Chiron have Angiozyme (RPI-4610), which is a hammerhead ribozyme that targets the mRNA for VEGFR-1.[317] Angiozyme is currently in Phase II trials for breast, non-small cell lung (in combination with carboplatin/paclitaxel), and colorectal cancers (in combination with IFL). Regeneron has pioneered the 'VEGF trap,' which is a chimeric protein containing binding domains from both Flt-1 and KDR that has high affinity for all forms of VEGF-A.[318] Inhibition of tumor angiogenesis has been demonstrated in murine models dosed subcutaneously with 25 $mg\,kg^{-1}$. VEGF trap RIR2 every other day.[319] The protein is now in Phase I trials in patients with solid tumors and non-Hodgkin's lymphoma.

7.08.5.8.3 Small-molecule inhibitors of vascular endothelial growth factor receptor

Alongside a number of biopharmaceutical approaches, three classes of 'first-generation' small-molecule inhibitors have entered clinical development.[320,321] SU5416 (semaxanib, **53**), the first of three indolinones developed by Sugen (now Pfizer) (**Figure 12**),[322,323] inhibits both KDR and PDGFR-β (not FGFR1), but suffers from low aqueous solubility and a plasma half-life of <30 min in the rat and dog.[324] Despite this, the compound has demonstrated good activity (23–85% inhibition) in a broad range of xenograft models when given intraperitoneally at a dose of 25 $mg\,kg^{-1}day^{-1}$. SU6668 (**54**), bearing a carboxylic acid group, is a more soluble indolinone with modest potency for KDR and FGFR1 that has shown 81% inhibition of a C6 glioma xenograft dosed orally at 200 $mg\,kg^{-1}day^{-1}$.[325,326] Significant inhibition, coupled with a reduction in tumor microvessel density and induction of apoptosis, was observed in a range of lung, colon, melanoma, and ovarian tumors, and daily doses of 60 $mg\,kg^{-1}$ of SU6668 (**54**) increased the survival times of mice with CT-26 liver metastases by 58%.[327] SU11248 (**55**) is a much more potent inhibitor of KDR, PDGFR, c-Kit, and Flt3 kinases that inhibits the growth in a range of xenograft and metastatic tumor models when given at a dose of 20–80 $mg\,kg^{-1}day^{-1}$.[328] SU11248 (**55**) has acceptable protein binding and pharmacokinetics, such that plasma drug levels can exceed the biologically effective concentration for 12 h following a single oral dose of 80 $mg\,kg^{-1}$.[329]

PTK787 (Novartis/Schering AG – Valatinib/ZK222584, **56**) is an orally active phthalazine discovered through screening and found to be suitable for clinical development without structural modification.[330,331] In contrast to the indolinones, PTK787 has a narrower range of kinase activity and much improved physical properties, which contribute to it having an attractive pharmacokinetic profile after oral dosing. PTK787 (**56**) has no activity against EGFR, and is equipotent against the three VEGF receptors, in contrast to AstraZeneca's ZD6474 (**57**), which

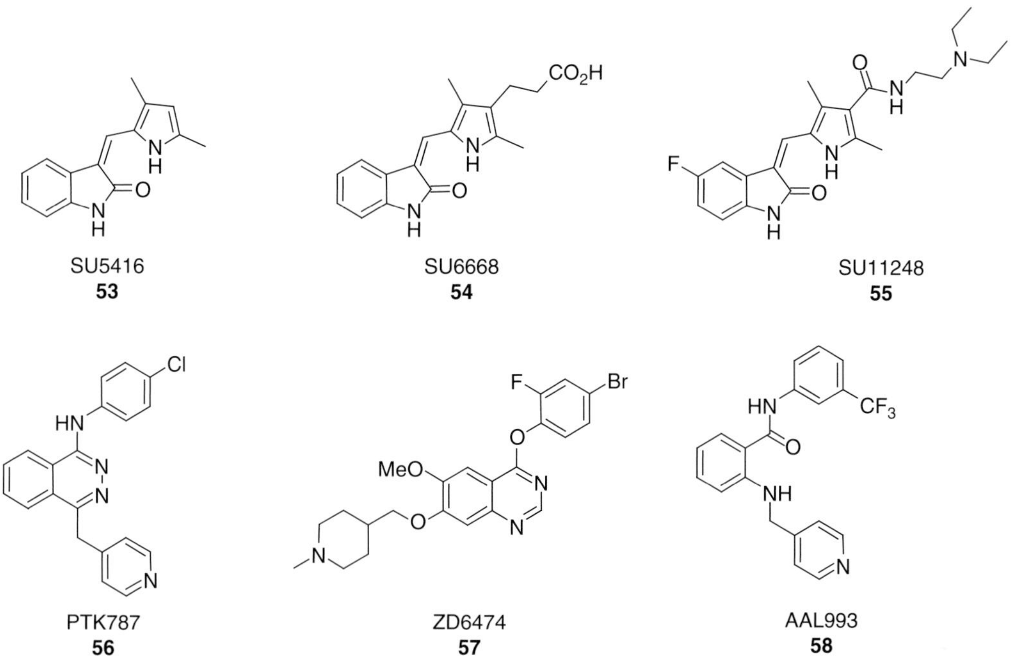

SU5416
53

SU6668
54

SU11248
55

PTK787
56

ZD6474
57

AAL993
58

Figure 12 First-generation VEGFR kinase inhibitors.

is equipotent against KDR and EGFR but has less activity against Flt-1.[332] ZD6474 (**57**) has good biopharam-aceutical and pharmacokinetic properties, and has been shown to inhibit a Calu-6 xenograft growth at doses of 12.5–100 mg kg^{-1} day^{-1}.[332]

7.08.5.8.4 Small-molecule vascular endothelial growth factor receptor inhibitors in the clinic

In Phase I trials, SU5416 (semaxanib, **53**) had an MTD of 145 mg m^{-2} when provided as intravenous doses twice weekly in Cremophor. However, the MTD in Phase I/II trials in combination with gemcitabine and cisplatin was reduced to 85 mg m^{-2} due to observation of thromboembolic events at the higher dose.[333,334] A Phase III trial with SU5416 (**53**) (in combination with IFL) was terminated in 2002 after interim analysis showed a lack of clinical benefit.[321] SU6668 (**54**) has a low volume of distribution, and, in Phase I trials, the plasma half-life of the compound was 1 h in fasted patients, and 2 h in fed patients.[321] Little toxicity was observed when the compound was administered once per day (100–2400 mg m^{-2}), but twice daily dosing caused pain, fatigue, thrombocytopenia, and gastrointestinal toxicity.[335] In Phase I trials, SU11248 (**55**) (Sutent, Sunitinib) achieved the target drug plasma levels when dosed at 50 mg day^{-1} orally and has recently been approved for treatment of metastatic renal cell carcinoma on a schedule of 50 mg day^{-1} for 4 weeks followed by a 2 week recovery period. In this Phase I study, the dose-limiting toxicities were severe fatigue, hypertension, nausea, and vomiting.[321]

PTK787 (**56**) was well tolerated in Phase I trials (in metastatic renal cell carcinoma) at 1200 mg day^{-1}, with 19% of patients showing measurable responses, and 64% with stable disease. The compound has been used in combination studies with standard cytotoxic regimens (colorectal cancer and glioblastoma), with toxicities reported to include hypertension, nausea, vomiting, transaminase elevation, and fatigue. PTK787 (**56**) is currently in Phase III trials in metastatic colorectal cancer at a dose of 1250 mg day^{-1} in combination with FOLFOX (folinic acid, 5-fluorouracil, and oxiplatin). ZD6474 (**57**) is well tolerated at doses up to 500 mg day^{-1}, with dose-limiting toxicities including diarrhea, hypertension, thrombocytopenia, and QT prolongation. Acneiform skin rash is also seen as a less serious side effect, and may be attributable to EGFR inhibition. ZD6474 (**57**) is currently in Phase II trials.

7.08.5.8.5 Future directions

A number of compounds with novel structures have recently been described, some of which are poised to enter clinical evaluation. Working from the structure of PTK787 (**56**), the Novartis group has developed a series of anthranilamides such as AAL993 (**58**), which has been demonstrated to bind selectively to an inactive (DFG out) conformation of KDR.[320,336]

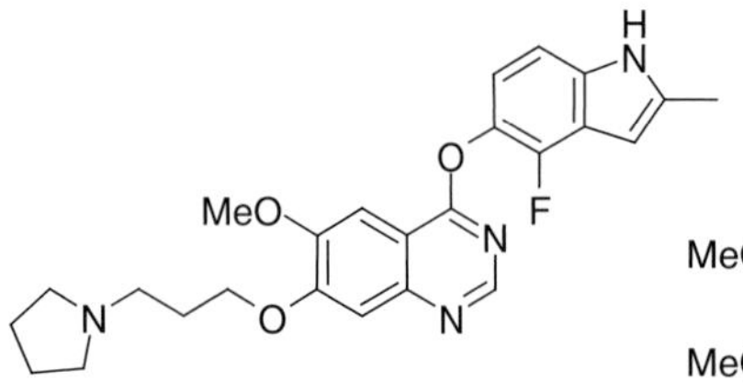

AZD2171
59

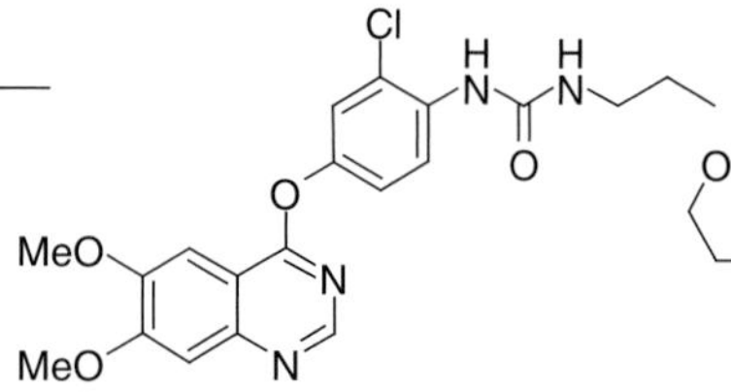

KRN633
60

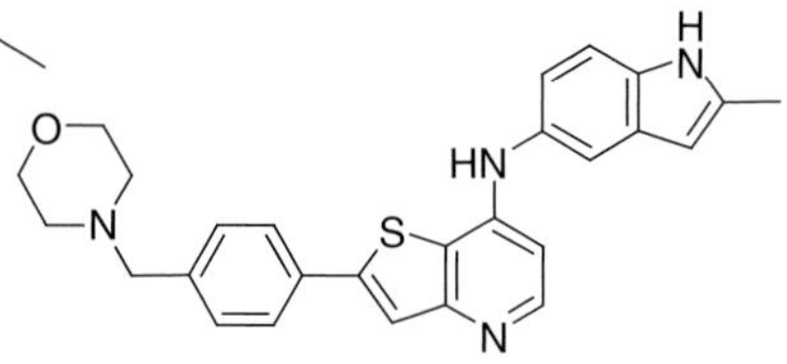

61

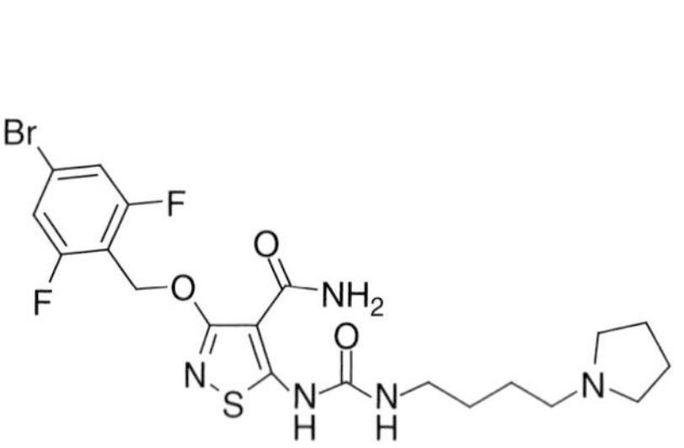

CP547632
62

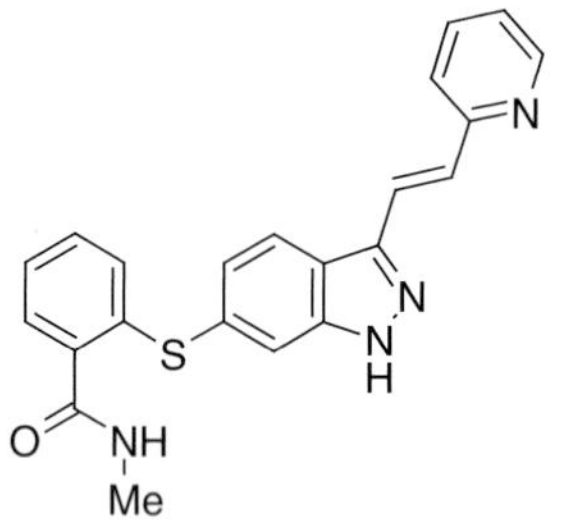

AG013736
63

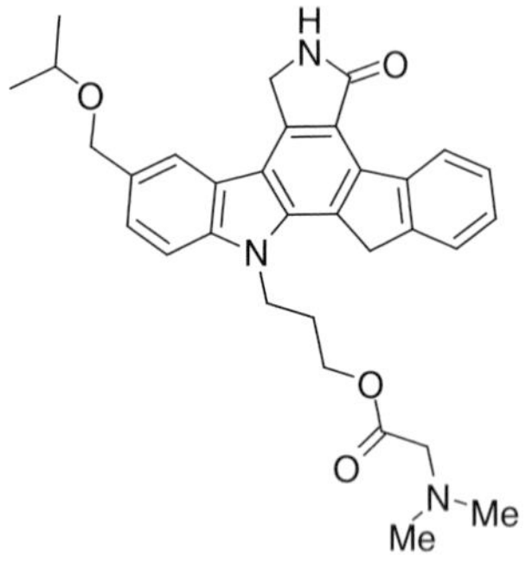

CEP7033
64

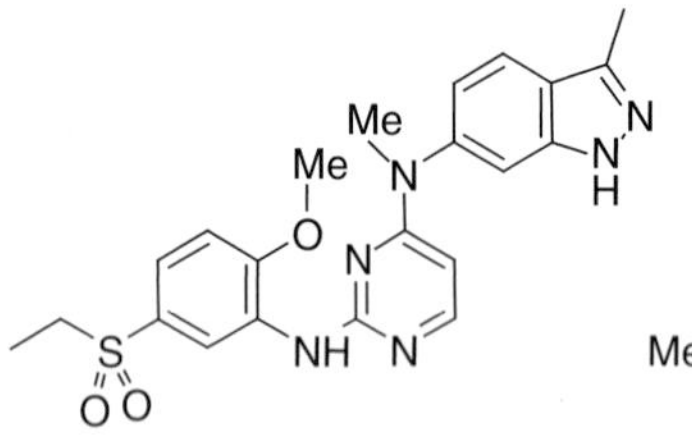

GW654652
65

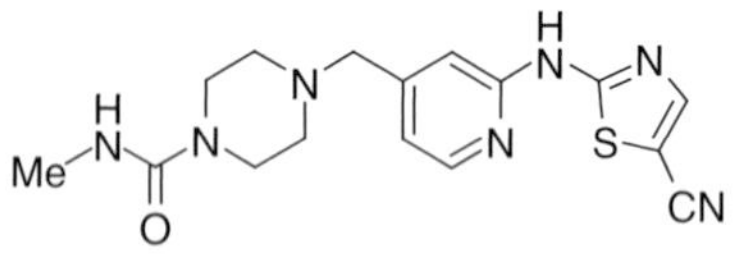

66

Figure 13 Second-generation VEGFR kinase inhibitors.

Further refinement of ZD6474 has resulted in ZD2171 (**59**) (**Figure 13**), which is significantly more potent than the earlier compound, and has lost the activity against EGFR.[337] Two structurally related analogs of these quinazolines are Kirin's KRN633 (**60**) and Pfizer's thienylpyridine (**61**).[338,339] Pfizer has also described CP547,632 (**62**), which has entered Phase I trials, as has AG013736 (**63**).[340,341] CEP-7055 (**64**) is an *N,N*-dimethylglycine ester prodrug of CEP-5214 that Cephalon is partnering with Sanofi-Synthelabo.[342] GlaxoSmithKline has described preclinical studies with GW654652 (**65**),[343–345] and a group from Merck has recently published work with *N*-(1,3-thiazol-2-yl)pyridine-2-amine **66**.[346]

7.08.5.9 Fibroblast Growth Factor Receptor (FGFR)

7.08.5.9.1 Fibroblast growth factor receptor and cancer

To date, 23 different fibroblast growth factors have been identified, including FGF-1 (acidic FGF) and FGF-2 (basic FGF or bFGF).[347–349] Four RTKs (FGFR1–4) have been identified for these growth factors, and are characterized by having an extracellular domain that contains three different immunoglobulin domains, with the first two separated by an acidic region.[350] Due to the existence of a large number of splice variants in the ligand-binding domain, it is possible for the four FGFR genes to encode a much larger number of receptor proteins.[351] There is a clear link between various FGF family members (including FGF-2) and prostate cancer, where the growth factors are implicated in both proliferation and angiogenesis/invasion.[352,353] The involvement of FGF family members and their receptors is also emerging in bladder (FGFR3),[354] renal cell (FGF-2 and FGF-5),[355] and testicular (FGF-4)[356] cancers.

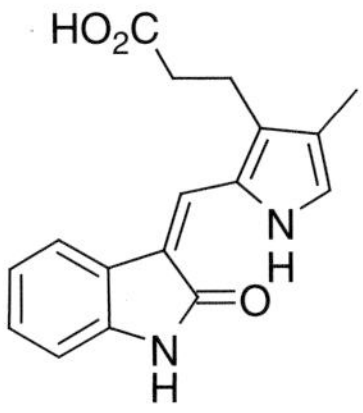

SU5402 **67** **68** 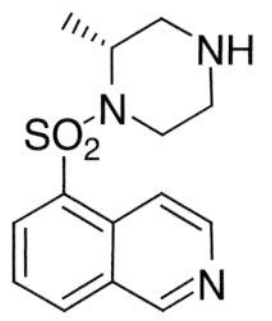 **69**

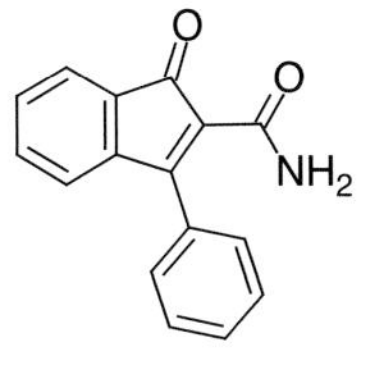

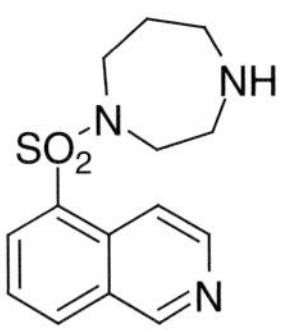

Fasudil **70** Y-27632 **71**

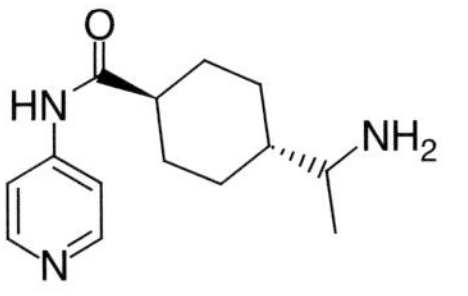

Figure 14 FGFR and ROCK inhibitors.

7.08.5.9.2 Fibroblast growth factor receptor inhibitors[357]

Curiously, only a few FGFR inhibitors have been described, and achieving good selectivity over VEGF inhibition looks to be a significant challenge. SU5402 (**67**) (**Figure 14**) from the ubiquitous indolinone series has good potency toward FGFR1 ($IC_{50}=20\,nM$) but is equiactive against KDR and has relatively poor physical properties.[358] PD173074 (**4**) is also from a well-known chemical class of inhibitors, but achieves good selectivity over PDGFR-β, Src, EGFR, and IGF-1R, with an IC_{50} of 26 nM against FGFR1.[156,359] Finally, 1-oxo-3-phenyl-1*H*-indene-2-carboxamide (**68**) is a relatively weak FGFR1 inhibitor ($IC_{50}=4.7\,\mu M$), but may represent a novel starting point for future work.[360]

7.08.6 Serine/Threonine Kinases

7.08.6.1 Serine/Threonine Kinases and Cancer

There are considerably more serine/threonine kinases than tyrosine kinases in the human kinome, but despite this numerical advantage, no serine/threonine kinase inhibitor is currently approved for use in cancer. Promising data in the MEK and cell cycle (Aurora/CDK) areas has been reported, and these kinases will be discussed in detail alongside the less well-studied proteins Raf, Akt, PDK1, and PKC. Other kinases with potential value as oncology targets include ROCK (rho kinase),[361] Plk (polo-like kinase),[362] P70S6 kinase,[363] TGF-β receptor-associated kinases (ALK1–ALK7),[364] CK1e,[365] PAK4,[366] and Nek family kinases.[367]

Some progress in finding inhibitors for these kinases has been made, with some encouraging results against ROCK. Thus, H-7 (**69**) was identified as an ATP-competitive inhibitor,[368] and modification resulted in HA-1077 (Fasudil, **70**),[369] which has modest activity for the kinase; recent work in the isoquinoline series has provided more potent inhibitors.[370] Another ROCK inhibitor is Y-27632 (**71**),[371] which has an IC_{50} of 0.25 μM; the structure of this inhibitor bound to PKA[372] should aid the design of more potent inhibitors.

7.08.6.2 Mitogen-Activated Protein Kinase/Extracellular Signal-Related Kinase Kinase (MEK) (MEK1 and MEK2)

7.08.6.2.1 Mitogen-activated protein kinase/extracellular signal-related kinase kinase and cancer

MEK (MAP kinase kinase) is a key signal amplification point in the Ras-MEK-ERK (extracellular signal-related kinase) transduction pathway driving growth and survival signaling in mammalian cells.[373] Two closely related isoforms, MEK1 and MEK2, are ubiquitously expressed, and are unusual in being dual specificity kinases, phosphorylating both serine/threonine and tyrosine residues. MEK phosphorylation of its substrates, ERK1 and ERK2, results in a large amplification of growth factor signals derived from RTKs and other sources. MEK (MEK1 and MEK2) has not been identified as an oncogene product[374] but constitutive activation does result in cellular transformation and this is implicated in the progression of a broad range of human tumors.[375,376] Recent clinical trials with MEK inhibitors have demonstrated activity in human tumors and confirm these enzymes as important targets for treatment of human tumor disease.[377]

7.08.6.2.2 Inhibitors of mitogen-activated protein kinase/extracellular signal-related kinase kinase

The first inhibitor of MEK identified was the flavone PD98059 (**72**) (**Figure 15**).[378] The molecule was identified from library screening, and, although not potent ($IC_{50}=2–7\,\mu M$), it prevents activation of MEK.[379] The compound does not compete directly with ATP, and this led to the hypothesis that the compound inhibited the enzyme by binding to an unphosphorylated site on MEK, preventing activation. Physicochemical limitations and activity against other enzymes prevented further development of **72**.

A variety of natural products have been identified as MEK inhibitors, including (10*E*)-hymenialdisine (**73**),[380] lactone Ro-09-2210 (**74**),[381] and the closely related lactone L-783277 (**75**),[382] the last being the most potent ($IC_{50}=4\,nM$). DuPont identified U-0126 (**76**) as a potent MEK inhibitor ($IC_{50}=72\,nM$);[383] this agent is not competitive with ATP, and is more potent against unactivated MEK. Analogs of **76** have been synthesized, and the (*Z,Z*) isomers usually possess most potency,[384] although the area is complex, since such molecules often cyclize when dissolved in dimethyl sulfoxide.

More conventional inhibitors of MEK are represented by 3-cyanoquinoline (**77**).[385] This compound is ATP-competitive and highly potent ($IC_{50}=2\,nM$) and with antiproliferative activity against LoVo cells at 5 nM, although it is unclear if this activity is solely the result of MEK inhibition. The most advanced MEK inhibitors are those represented by Parke-Davis's CI-1040 (PD184352, **78**).[373,386] This agent is not competitive with ATP, and acts as an allosteric inhibitor of the enzyme, which prevents activation of the kinase by blocking phosphorylation. Because of this mechanism, the compound is highly selective (MEK1 $IC_{50}=17\,nM$) against the enzyme, which translates into antiproliferative activity in cells at concentrations of $\sim 0.1\,\mu M$. The activity in cells correlates well with inhibition of

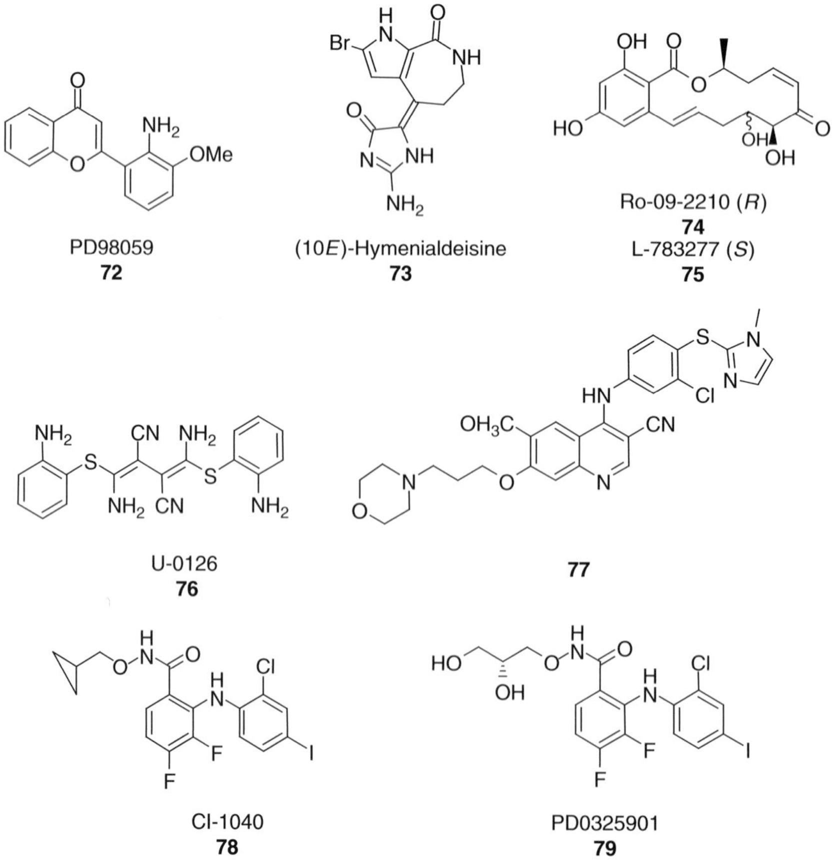

PD98059
72

(10*E*)-Hymenialdeisine
73

Ro-09-2210 (*R*)
74
L-783277 (*S*)
75

U-0126
76

77

CI-1040
78

PD0325901
79

Figure 15 MEK inhibitors.

phosphorylation of ERK1, a downstream substrate of MEK. The compound is also active in vivo, where at twice daily oral doses of 48–200 mg kg^{-1} to mice bearing human tumor xenografts, it shows significant inhibition of tumor growth and is well tolerated.[386]

Compound **78** has progressed to the clinic,[387] where it was given at a dose of 1600 mg once daily or 800 mg twice daily, with the latter regime resulting in better exposure. Side effects seen included fatigue, skin rash, and diarrhea, but target therapeutic plasma levels were achieved, and stable disease was seen in ~30% of patients. However, Phase II trials in patients with advanced NSCLC, breast, colon, and pancreatic cancers, dosing at 800 mg twice daily, no objective responses were seen.[377] This may be due to the relatively poor pharmacokinetic and physicochemical properties of the compound, leading to variable absorption and blood levels. A second-generation inhibitor, PD0325901 (**79**), which has a number of advantages including improved aqueous solubility, achieved by introducing a hydrophilic side chain, is already in Phase I studies.[388]

7.08.6.2.3 Structural studies

The unique mode of action of the allosteric inhibitors of MEK has stimulated a great deal of work to understand how they interact with the enzyme, and, recently, structures of a bromo derivative of inhibitor **79**, bound to both the MEK1 and MEK2 isoforms, have been solved.[389] Both enzymes have what appear to be unique binding pockets close to the ATP-binding site to which molecules such as **79** bind. When these molecules bind, they induce the same type of conformational changes that activating phosphorylations produce, closing the active site, but resulting in a catalytically inactive kinase through disruption of the competent orientation of catalytic residues.

7.08.6.2.4 Future directions

Just as the high selectivity of imatinib (**1**) is attributable to its distinct mechanism of inhibition, the high selectivity of inhibitors such as **78** and **79** appears to be linked to their unique mechanism of action. If these compounds prove active at tolerated doses, this approach could provide impetus for looking at similar inhibition mechanisms for other clinically relevant kinases. The reduced potential for nontarget-based pharmacology may also make such an approach of interest in other disease areas.

7.08.6.3 Cyclin-Dependent Kinases

7.08.6.3.1 Cyclin-dependent kinases and cancer

The CDK family of serine/threonine kinases is an essential component of the cell cycle machinery of every mammalian cell, and is critical to effective cell proliferation.[390,391] CDK activity is tightly controlled throughout the cell cycle by a group of activating proteins termed cyclins,[392] which are transiently expressed throughout the replication process, and activate the required CDK. While the CDKs are rarely overexpressed in tumors, there is clear evidence that the activating cyclins can be amplified[393] and that natural inhibitors of the CDKs, such as p27, can be downregulated.[393] The consequent growth advantage obtained by cells under these circumstances is a common feature in human cancers, and has generated great interest in CDKs as potential selective targets for the treatment of the disease.[394,395] While apoptosis is commonly the result of cell cycle inhibition of tumor cells, normal cycling cells simply arrest. CDK1, CDK2, and CDK4 are directly involved in cell cycle progression; CDK5, CDK7, and CDK9 are less directly involved, but may still be worthwhile targets.

7.08.6.3.2 Cyclin-dependent kinase inhibitors

CDK inhibitors typically fall into two broad classes; those which inhibit CDK1, CDK2, and CDK5, and those inhibiting CDK4 and CDK6.[390] That said, UCN-01 (**101**)[396] and flavopiridol (**80**) (**Figure 16**),[397] both derived from natural product leads and among the first agents to be identified, inhibit the majority of CDKs, in addition to several other kinases. More recently, the availability of structural information from CDK2–inhibitor complexes[398,399] (and the detailed interactions of small molecules within the ATP-binding site) have resulted in the development of relatively selective inhibitors.

The purine derivatives olomucine (**81**)[400] and purvalanol A (**82**)[401] and seliciclib ((*R*)-roscovitine, CYC202, **83**)[402] are all more selective inhibitors of CDK2. Seliciclib inhibits a range of CDKs, including CDK1, CDK2, CDK7, and CDK9 at micromolar levels, and this translates into cell cycle blocks at both the G_1/S and G_2/M checkpoints, consistent with the inhibition profile of the enzymes.

The Pharmacia group has identified a range of 3-aminopyrazole derivatives, including PNU-292137 (**84**), as CDK2 inhibitors[403] (IC_{50} = 37 nM) that also inhibit CDK1 and CDK5. Inhibition of HCT116 cell growth is observed

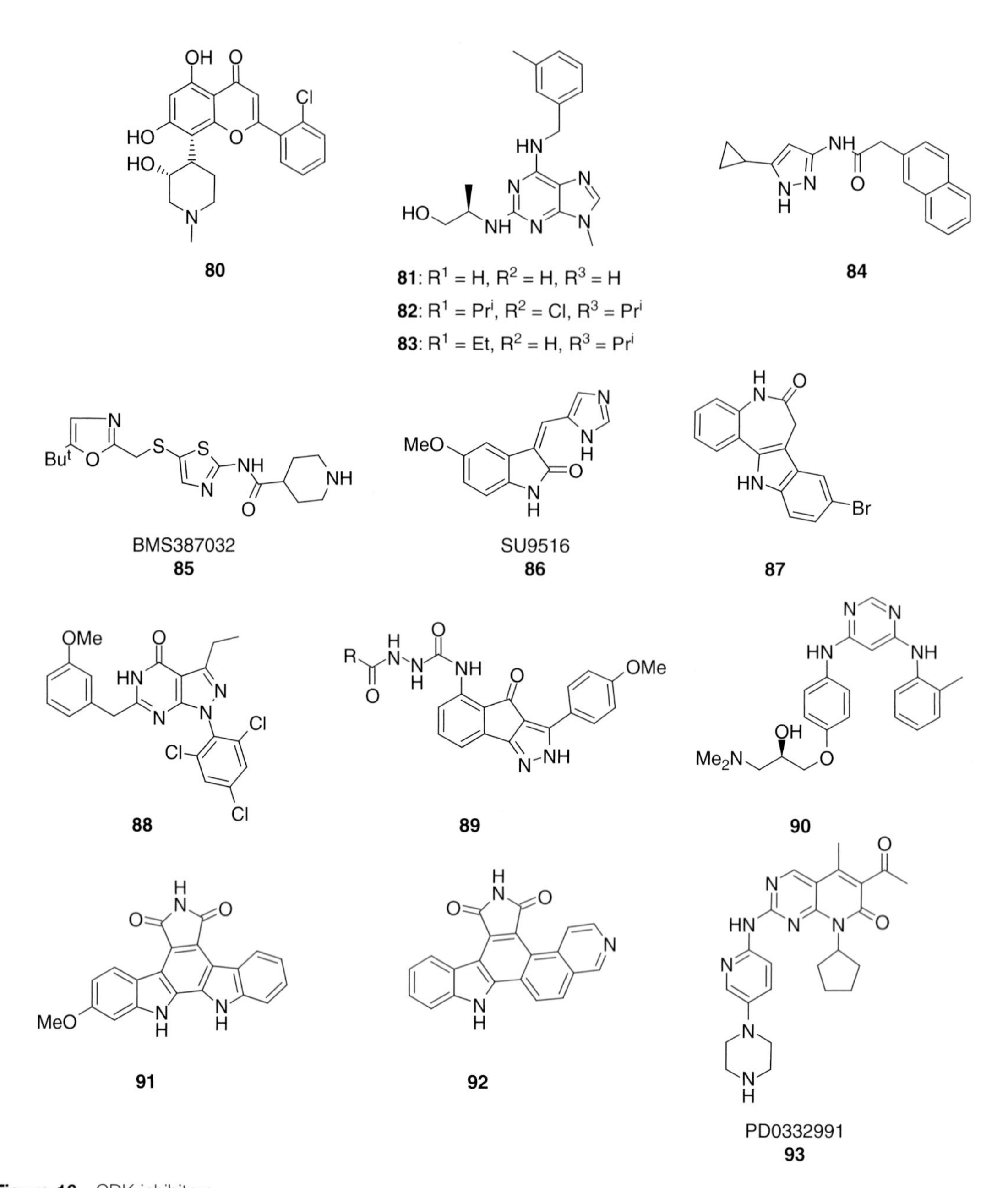

Figure 16 CDK inhibitors.

($IC_{50} = 73$ nM), and the compound is active in vivo at an oral dose of 7.5 mg kg^{-1} twice daily against a human ovarian cancer xenograft (A2780) grown in nude mice, and is well tolerated. The BMS group has also developed potent and selective aminothiazole CDK1 and CDK2 inhibitors,[404,405] culminating in the identification of BMS387032 (**85**), which has good in vitro potency and in vivo activity. A wide variety of other structures have been identified as CDK2 inhibitors, including SU9516 (**86**),[406] paullones (**87**),[407] pyrazolopyrimidines (**88**),[408] and indenopyrazoles (**89**).[409]

Inhibitors of CDK4 are structurally related to those of CDK2, but selectivity is achievable. AstraZeneca has highlighted a series of bis-anilinopyrimidines (**90**),[410] and used structural information derived from CDK2–inhibitor complexes to enhance selectivity for CDK4. Other pharmacophores offering selectivity for CDK4 include the carbazoles (**91**)[411] and (**92**).[412] Of major interest is PD0332991 (**93**), developed by the Pfizer group;[413] this compound inhibits CDK4 and CDK6 ($IC_{50} = 10$ and 15 nM, respectively), and has good selectivity against a range of other kinases. Inhibition of cellular growth is seen at 0.1–0.2 μM, and the compound causes growth inhibition in in vivo models of human tumor disease.

7.08.6.3.3 Cyclin-dependent kinase inhibitors in the clinic

Flavopiridol (**80**) has progressed to multiple Phase II trials,[399] and subsets of patients have shown stable disease and partial responses. Toxicities include diarrhea, nausea, vomiting, and some neutropenia. Development in solid tumor disease has stopped, but work in combination with other agents in lymphomas and leukemias continues. Seliciclib (**83**) has reached Phase II,[414] and appears to be well tolerated up to a dose of 2.5 g day^{-1}. Disease stabilization has been seen in some patients, and the use of the drug in combination with cytotoxic agents is being studied. The rapid metabolism and modest potency of **83** coupled with high doses may complicate further development. Some patients in the Phase I trials of aminothiazole BMS387032 (**85**) demonstrated stable disease and regression when **85** was periodically infused at a dose of 5–23 mg m^{-2}. PD0332991 (**93**), the CDK4 and CDK6 inhibitor, has also entered Phase I studies.

7.08.6.3.4 Future directions

CDK inhibitors appear to have the promise of clinical utility based on early results, but they may have to be used in combination to reach their full potential, and this will need careful scheduling. It is clear that further work will be required to capitalize on the different clinical profiles and therapeutic margins for this group of kinase inhibitors.

7.08.6.4 Aurora Family Kinases

7.08.6.4.1 Aurora kinases and cancer

Aurora A and B (and C) are serine/threonine protein kinases that are critical for the proper regulation of mitosis;[415,416] they are commonly overexpressed in human tumors,[417] and regulate key mitotic events.[418] Aurora A is expressed during mitosis, and localizes to the centrosomes and the poles of the mitotic spindle.[419] Repression of Aurora A activity in cells delays their entry into mitosis,[420] while overexpression of Aurora A can inhibit cytokinesis,[421] and compromise the spindle checkpoint. Aurora B is a 'chromosomal passenger protein'[422] that plays a key role in kinetochore function, and is required for accurate chromosome alignment and segregation during mitosis.[423] Crystal structures of the N-terminal truncated catalytic domain of Aurora A show a relatively large hydrophobic pocket, which is exploited by many of the small-molecule inhibitors to achieve kinase selectivity.[424–426]

7.08.6.4.2 Aurora kinase inhibitors

Published work on Aurora kinase inhibitors has focused on compounds **94–96** (**Figure 17**).[427] VX-680 (**94**) is a potent inhibitor of Aurora A, B, and C (IC_{50} = 0.0006, 0.018, and 0.0046 μM, respectively) and shows good selectivity over a broad kinase panel, with some activity against Flt3 (IC_{50} = 0.03 μM).[428] In MCF7 cells, VX-680 (**94**) inhibits phosphorylation of histone H3 (IC_{50} = 0.003–0.300 μM), and induces accumulation of cells with 4n DNA content. VX-680 (**94**) blocks tumor cell proliferation in a panel of tumor cell lines (EC_{50} = 0.015–0.113 μM) after 96 h of treatment. Tumor regression was observed in leukemic (AML: HL-60), pancreatic (MIA PaCa-2), and colon (HCT-116) xenografts following either intraperitoneal or intravenous dosing. As with other Aurora kinase inhibitors, the main dose-limiting toxicity in rodents is mechanism-related neutropenia, which is reversible on drug withdrawal.

Hesperadin (**95**) is a member of a series of indolinone compounds prepared as cell proliferation inhibitors that inhibit Aurora B kinase (IC_{50} = 0.25 μM).[429] While the compound inhibits a number of other kinases, hesperadin is a potent inhibitor of histone H3 phosphorylation in cells, and has a phenotype similar to VX-680 (**94**). ZM447439 (**96**) is a quinazoline that is equiactive against Aurora A and B (IC_{50} = 0.11 and 0.13 μM, respectively), and which shows good

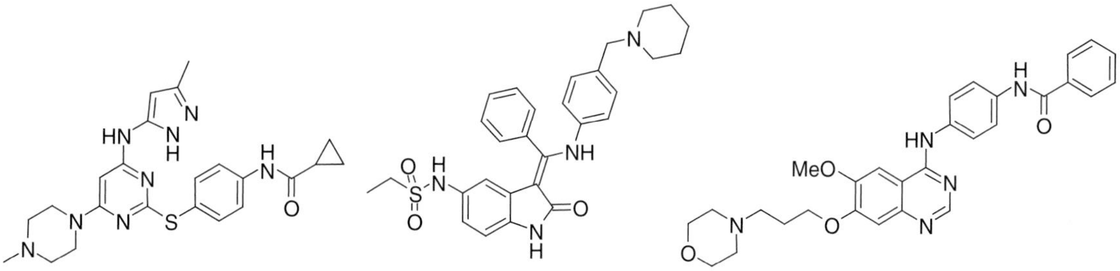

VX-680
94

Hesperadin
95

ZM447439
96

Figure 17 Aurora kinase inhibitors.

selectivity over a range of other kinases.[423] Published studies with compound **96** shows that it causes a dose-dependent increase in 4n DNA, due to a failure of cytokinesis, that is consistent with the proposed mechanism of action. Furthermore, compound **96** selectively induced apoptosis in cycling cells but had relatively little effect on G_1-arrested MCF-7 cells when profiled in in vitro clonigenicity assays.[423]

7.08.6.5 Raf Kinase

7.08.6.5.1 Raf and cancer

The three closely related Raf kinases, A-raf, B-raf and C-raf, are serine/threonine kinases that are activated by ras-GTP and/or PKC-α. Raf, in turn, activates MEK by phosphorylation of S218 and S222, transducing the signal from upstream Ras down the Ras-Raf-MEK-ERK pathway.[430] Since about 30% of human tumors possess activating *ras* mutations, the downstream effector Raf is an attractive target for anticancer therapy. It is now recognized that activating somatic mutations to B-raf are common in some human tumors such as melanoma.[431]

7.08.6.5.2 Inhibitors of Raf

A variety of small-molecule approaches to the inhibition of Raf have been reported,[430,432] the most advanced being BAY 43-9006 (**97**) (**Figure 18**).[433,434] The compound was developed from a bis-aryl lead series, and interacts at the ATP-binding site of the kinase (B-raf $IC_{50} = 22$ nM). Compound **97** has additional kinase activity, notably against c-Kit, Flt3, and KDR, and also inhibits the common V599E mutant of B-raf. It shows oral activity in mice bearing a range of human tumor xenografts with mutant K-ras, and entered Phase I clinical studies in 2000.[435] These studies showed it to be well tolerated, and, at a dose of 400 mg twice daily, side effects were diarrhea, skin rash, and fatigue. Some disease stabilization was seen and BAY 43-9006 (**97**) (Nexavar, Sorafenib) has recently been approved for the treatment of renal cell carcinoma.[436] Further work in this series has been undertaken to address the poor aqueous solubility of **97**.[437]

7.08.6.6 Akt/3-Phosphoinositide-Dependent Kinase 1 (PDK1)

7.08.6.6.1 Akt/3-phosphoinositide-dependent kinase 1 and cancer

Akt (PKB) is a serine/threonine kinase within the AGC kinase group having a high homology with both PKA and PKC. Three closely related isoforms (Akt1, Akt2, and Akt3) are known, and constitute part of the PI3′ kinase signaling pathway, frequently inappropriately activated in human cancers.[438] A compelling piece of evidence, linking increased Akt signaling with cancer, is that loss of PTEN activity, a tumor suppressor encoding an Akt-deactivating phosphatase, is observed in 30–50% of human tumors.[439,440]

PDK1 (3-phosphoinositide-dependent kinase 1), another serine/threonine kinase member of the AGC group, is also part of the PI3′ kinase signaling pathway, but is upstream of Akt. Phosphoinositides produced by PI3′ kinase recruit the

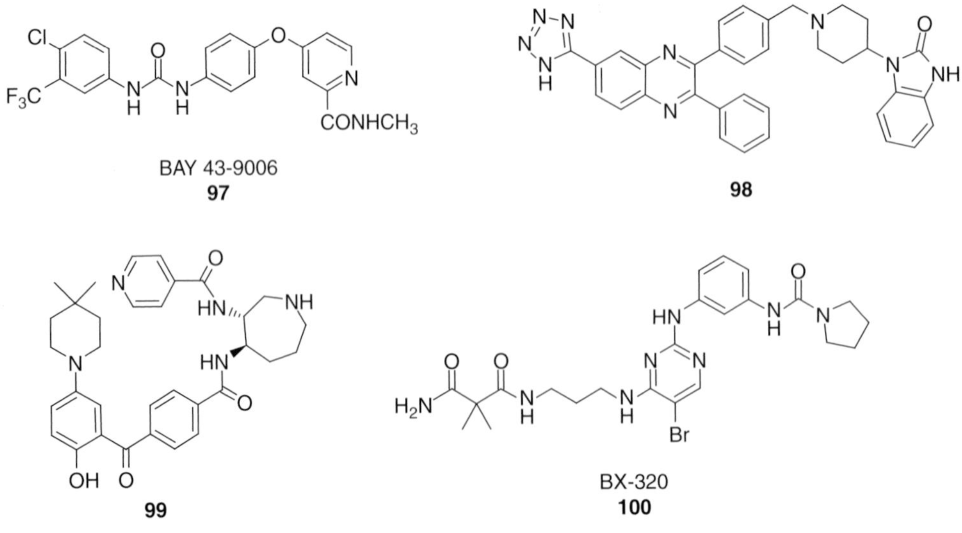

BAY 43-9006
97

98

99

BX-320
100

Figure 18 Raf, Akt, and PDK1 inhibitors.

plectrin homology domains of Akt, causing conformational changes in the Akt activation loop, which can then be phosphorylated and activated by PDK1. There is compelling evidence for the role of all three enzymes (PI3′ kinase, PDK1,[441] and Akt) in aberrant signaling in cancer, which makes all these enzymes attractive oncology targets.

7.08.6.6.2 Inhibitors of Akt/3-phosphoinositide-dependent kinase 1

While Akt/PDK1 are attractive targets for cancer chemotherapy, progress in finding potent and selective inhibitors has been relatively slow. A series of substituted quinoxalines (**98**)[442] have been identified that have some selectivity for the individual isoforms of Akt and that may bind at an allosteric site. Analogs of **98** suggest a requirement to inhibit Akt1 and Akt2 to achieve maximal pro-apoptotic response in vitro. Recent work has identified a series of azepanes (**99**) that were designed using crystal structures of probe compounds, bound to the related kinase PKA,[443] and the structure of the natural product kinase inhibitor (−)-balanol. A recent crystal structure of Akt[444] supports conclusions drawn from this work.

Selective inhibitors of PDK1 have been described by a Berlex group, and include compounds such as BX-320 (**100**).[445] This compound inhibits PDK1 ($IC_{50} = 40$ nM) and inhibits cell growth at 0.1–0.4 μM, appearing to promote apoptosis in tumor cells but not in normal cells. The compound is also active in vivo when given orally at 200 mg kg^{-1} twice daily in a mouse model of metastatic melanoma, and binds to the ATP-binding site of the kinase according to an x-ray crystal structure.[445]

7.08.6.7 Protein Kinase C

7.08.6.7.1 Protein kinase C and cancer

The three major isoform classes of PKC, known as conventional (PKC-α, PKC-β, and PKC-γ), novel (PKC-δ, PKC-ε, PKC-η, and PKC-θ), and atypical (PKC-ζ and PKC-ι), constitute a family of serine/threonine kinase involved in multiple signal transduction pathways.[446] The enzymes are involved in signal transduction from classical RTKs and G-protein-coupled receptors, and play a role in nuclear hormone signaling with much of the specificity of the activated isoforms appearing to derive from their subcellular localization. The various isoforms require other co-factors for activation (e.g., 1,2-*sn*-diacylglycerol, calcium, phosphatidyl-L-serine, or phorbol esters), in addition to phosphorylation of serine or threonine residues and membrane translocation to be fully active.[446] There is a limited understanding of the role of PKC in tumor initiation and growth,[447] and the complexity of PKC function and activity has made the rational choice of isoform targets difficult.

7.08.6.7.2 Inhibitors of protein kinase C

A number of natural products have been identified as inhibitors of the PKC family,[448] and most synthetic work has used these agents as starting points. A series of staurosporine derivatives, such as UCN-01 (**101**),[449] midostaurin (CGP41251, **102**),[450] Ro 31-8220 (**103**),[451] and LY 333531 (**104**)[452] (**Figure 19**), have been described as PKC inhibitors, and have progressed to clinical evaluation. UCN-01 (**101**) inhibits PKC-α ($IC_{50} = 7$ nM), but also has activity against a wide range of other kinases such as the CDKs.[453] It was administered by intravenous infusion to patients in a Phase I trial in cancer,[398] and shown to have a very long half-life (400–1500 h), which may be related to its

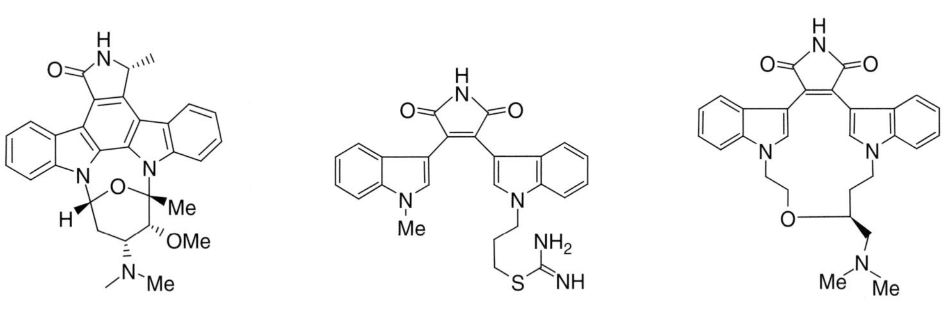

UCN-01: R^1 = OH, R^2 = H
101
CGP41251: R^1 = H, R^2 = COPh
102

Ro-31-8220
103

LY333531
104

Figure 19 Inhibitors of PKC.

very high binding to the α-acid glycoprotein component of human blood. Midostaurin (PKC-α $IC_{50} = 3\,nM$) inhibits cancer cell growth at $\sim 0.5\,\mu M$, and is active in animal models of metastatic melanoma when given at a dose of $75\,mg\,kg^{-1}$ three times daily by the intraperitoneal route.[454] It too is highly bound to α-acid glycoprotein, has a human half-life of 40 h, and has progressed to Phase II trials,[455] principally in leukemias. The compound has demonstrated some antitumor effects, with side effects of nausea, diarrhea and neutropenia. Ro 31-8220 (**103**) has also entered clinical trials,[446] although the true effect of PKC inhibition may be masked by the broad kinase inhibitory profile of the compound.

Bryostatin, a marine-derived natural product, inhibits PKC by interacting with the regulatory domain of the kinase,[456] acting as an analog of diacylglycerol. It is unusual in that it initially activates PKC, which leads to a rapid downregulation of the enzyme. Phase II studies against a variety of tumor types gave no conclusive activity, but did result in some disease stabilization.[457] Lipid analogs such as imofosine[458] and perifosine,[459] which also target the regulatory domain of PKC, have also progressed to the clinic for cancer treatment.

7.08.7 Summary and Future Directions

The first kinase inhibitors have established themselves alongside existing agents in the treatment of many cancers. This position has only been reached through a huge global investment by scientists and clinicians, companies, and research institutes, and has challenged the way in which novel anticancer drugs are developed. The search for novel protein kinase inhibitors may constitute more than half of the current medicinal chemistry effort in oncology. The lives of millions of current and future cancer patients will be touched by the ability of medicinal chemists, and their partners in drug discovery, to learn lessons from the compounds described in this review. Future progress in the field requires exploitation of the burgeoning structural and bioinformatic data, the conquering of drug-induced resistance, and the opening up of novel inhibitory paradigms, outside the ATP-binding site of active kinases.

References

1. Anisimov, V. N. *Crit. Rev. Oncol. Hematol.* **2003**, *45*, 277–304.
2. Rushton, L. *Occup. Environ. Med.* **2003**, *60*, 150–156.
3. Doll, R.; Peto, R. *J. Natl. Cancer Inst.* **1981**, *66*, 1191–1308.
4. Calle, E. E.; Thun, M. J. *Oncogene* **2004**, *23*, 6365–6378.
5. Hanahan, D.; Weinberg, R. A. *Cell* **2000**, *100*, 57–70.
6. Vogelstein, B.; Kinzler, K. W. *Trends Genet.* **1993**, *9*, 138–141.
7. Mitsiades, C. S.; Mitsiades, N.; Koutsilieris, M. *Curr. Cancer Drug Targets* **2004**, *4*, 235–256.
8. Chang, F.; Steelman, L. S.; Lee, J. T.; Shelton, J. G.; Navolanic, P. M.; Blalock, W. L.; Franklin, R. A.; McCubrey, J. A. *Leukemia* **2003**, *17*, 1263–1293.
9. Yang, J. Y.; Michod, D.; Walicki, J.; Widmann, C. *Biochem. Pharmacol.* **2004**, *68*, 1027–1031.
10. Jain, R. K. *Science* **2005**, *307*, 58–62.
11. Woodhouse, E. C.; Chuaqui, R. F.; Liotta, L. A. *Cancer* **1997**, *80*, 1529–1537.
12. Tammela, T.; Enholm, B.; Alitalo, K.; Paavonen, K. *Cardiovasc. Res.* **2005**, *65*, 550–563.
13. Blume-Jensen, P.; Hunter, T. *Nature* **2001**, *411*, 355–365.
14. Leevers, S. J.; Vanhaesebroeck, B.; Waterfield, M. D. *Curr. Opin. Cell Biol.* **1999**, *11*, 219–225.
15. Hubbard, S. R.; Till, J. H. *Annu. Rev. Biochem.* **2000**, *69*, 373–398.
16. Maulik, G.; Shrikhande, A.; Kijima, T.; Ma, P. C.; Morrison, P. T.; Salgia, R. *Cytokine Growth Factor Rev.* **2002**, *13*, 41–59.
17. Marino-Buslje, C.; Martin-Martinez, M.; Mizuguchi, K.; Siddle, K.; Blundell, T. L. *Biochem. Soc. Trans.* **1999**, *27*, 715–726.
18. Schlessinger, J. *Cell* **2002**, *110*, 669–672.
19. Peneul, E.; Schafer, G.; Akita, R. W.; Sliwkowski, M. X. *Semin. Oncol.* **2001**, *28*, S36–S42.
20. Kani, K.; Warren, C. M.; Kaddis, C. S.; Loo, J. A.; Landgraf, R. *J. Biol. Chem.* **2005**, *280*, 8238–8247.
21. Citri, A.; Gan, J.; Mosesson, Y.; Vereb, G.; Szollosi, J.; Yarden, Y. *EMBO Rep.* **2004**, *5*, 1165–1170.
22. Pawson, T.; Gish, G. D.; Nash, P. *Trends Cell. Biol.* **2001**, *11*, 504–511.
23. Uhlik, M. T.; Temple, B.; Bencharit, S.; Kimple, A. J.; Siderovski, D. P.; Johnson, G. L. *J. Mol. Biol.* **2005**, *345*, 1–20.
24. Vieth, M.; Higgs, R. E.; Robertson, D. H.; Shapiro, M.; Gragg, E. A.; Hemmerle, H. *Biochem. Biophys. Acta* **2004**, *1697*, 243–257.
25. Menard, S.; Casalini, P.; Campiglio, M.; Pupa, S.; Agresti, R.; Tagliabue, E. *Ann. Oncol.* **2001**, *12*, S15–S19.
26. Lynch, H. T.; Fusaro, R. M.; Lynch, J. F. *Ann. NY Acad. Sci.* **1997**, *833*, 1–28.
27. Vogelstein, B.; Kinzler, K. W. *Nat. Med.* **2004**, *10*, 789–799.
28. Albanell, J.; Codony, J.; Rovira, A.; Mellado, B.; Gascon, P. *Adv. Exp. Med. Biol.* **2003**, *532*, 253–268.
29. Konner, J.; Dupont, J. *Clin. Colorectal Cancer* **2004**, *4*, S81–S85.
30. Wozniak, M. A.; Modzelewska, K.; Kwong, L.; Keely, P. J. *Biochim. Biophys. Acta* **2004**, *1692*, 103–119.
31. Sampath, S. C.; Ohi, R.; Leismann, O.; Salic, A.; Pozniakovski, A.; Funabiki, H. *Cell* **2004**, *118*, 187–202.
32. Mohammadi, M.; Dikic, I.; Sorokin, A.; Burgess, W. H.; Jaye, M.; Schlessinger, J. *Mol. Cell. Biol.* **1996**, *16*, 977–989.
33. Huse, M.; Kuriyan, J. *Cell* **2002**, *109*, 275–282.
34. Cherry, M.; Williams, D. H. *Curr. Med. Chem.* **2004**, *11*, 663–673.
35. Madhusudan, S.; Ganesan, T. S. *Clin. Biochem.* **2004**, *37*, 618–635.

36. Ranson, M.; Wardell, S. *J. Clin. Pharm. Ther.* **2004**, *29*, 95–103.
37. Norman, P. *Curr. Opin. Investig. Drugs* **2001**, *2*, 298–304.
38. Allen, L. F.; Lenehan, P. F.; Eiseman, I. A.; Elliott, W. L.; Fry, D. W. *Semin. Oncol.* **2002**, *29*, S11–S21.
39. Allen, L. F.; Sebolt-Leopold, J.; Meyer, M. B. *Semin. Oncol.* **2003**, *30*, 105–116.
40. Lawrence, D. S.; Niu, J. *Pharmacol. Ther.* **1998**, *77*, 81–114.
41. Davidson, W.; Frego, L.; Peet, G. W.; Kroe, R. R.; Labadia, M. E.; Lukas, S. M.; Snow, R. J.; Jakes, S.; Grygon, C. A.; Pargellis, C. et al. *Biochemistry* **2004**, *43*, 11658–11671.
42. Zhang, Z.-Y.; Zhou, B.; Xie, L. *Pharmacol. Ther.* **2002**, *93*, 307–317.
43. Steelman, L. S.; Bertrand, F. E.; McCubrey, J. A. *Expert Opin. Ther. Targets* **2004**, *8*, 537–550.
44. Manning, G.; Whyte, D. B.; Martinez, R.; Hunter, T.; Sudarsanam, S. *Science* **2002**, *298*, 1912–1934.
45. The human kinome. http://www.kinase.com/human/kinome (accessed Aug 2006).
46. Caenepeel, S.; Charydczak, G.; Sudarsanam, S.; Hunter, T.; Manning, G. *Proc. Natl. Acad. Sci. USA* **2004**, *101*, 11707–11712.
47. Manning, G.; Plowman, G. D.; Hunter, T.; Sudarsanam, S. *Trends Biochem. Sci.* **2002**, *27*, 514–520.
48. Hanks, S. K.; Hunter, T. *FASEB J.* **1995**, *9*, 576–596.
49. Newell, D. R. *Eur. J. Cancer* **2005**, *41*, 676–682.
50. Lydon, N. B.; Druker, B. J. *Leukemia Res.* **2004**, *28S1*, S29–S38.
51. Sawyers, C. L. *Genes Dev.* **2003**, *17*, 2998–3010.
52. Dancey, J.; Sausville, E. A. *Nat. Rev. Drug Disc.* **2003**, *2*, 296–313.
53. Hanke, J. H.; Webster, K. R.; Ronco, L. V. *Eur. J. Cancer Prev.* **2004**, *13*, 297–305.
54. Williams, G.; Pazdur, R.; Temple, R. *J. Biopharm. Stat.* **2004**, *14*, 5–21.
55. Connors, T. *Oncologist* **1996**, *1*, 180–181.
56. Verma, A.; Kambhampati, S.; Parmar, S.; Platanias, L. C. *Cancer Metastasis Rev.* **2003**, *22*, 423–434.
57. Opdam, F. J. M.; Kamp, M.; de Bruijn, R.; Roos, E. *Oncogene* **2004**, *23*, 6647–6653.
58. Garcia, R.; Bowman, T. L.; Niu, G.; Yu, H.; Minton, S.; Muro-Cacho, C. A.; Cox, C. E.; Falcone, R.; Fairclough, R.; Parsons, S. et al. *Oncogene* **2001**, *20*, 2499–2513.
59. Gabarra-Niecko, V.; Schaller, M. D.; Dunty, J. M. *Cancer Metastasis Rev.* **2003**, *22*, 359–374.
60. Schlaepfer, M. D.; Hauck, C. R.; Sieg, D. J. *Prog. Biophys. Mol. Biol.* **1999**, *71*, 435–478.
61. Rengifo-Cam, W.; Konishi, A.; Morishita, N.; Matsuoka, H.; Yamori, T.; Nada, S.; Okada, M. *Oncogene* **2004**, *23*, 289–297.
62. Chauhan, D.; Pandey, P.; Hideshima, T.; Treon, S.; Raje, N.; Davies, F. E.; Shima, Y.; Tai, Y. T.; Rosen, S.; Avraham, S. et al. *J. Biol. Chem.* **2000**, *275*, 27845–27850.
63. Meyer, A. N.; Gastwirt, R. F.; Schlaepfer, D. D.; Donoghue, D. J. *J. Biol. Chem.* **2004**, *279*, 28450–28457.
64. Harvery, A. J.; Crompton, M. R. *Anti-Cancer Drugs* **2004**, *15*, 107–111.
65. Moroni, M.; Soldatenkv, V.; Zhang, L.; Zhang, Y.; Stoica, G.; Gehan, E.; Rashidi, B.; Singh, B.; Ozdemirli, M.; Mueller, S. C. *Cancer Res.* **2004**, *64*, 7346–7354.
66. Vassilev, A. O.; Uckun, F. M. *Curr. Pharm. Des.* **2004**, *10*, 1757–1766.
67. Lionberger, J. M.; Smithgall, T. E. *Cancer Res.* **2000**, *60*, 1097–1103.
68. Greer, P. *Nat. Rev. Mol. Cell Biol.* **2002**, *3*, 278–289.
69. O'Dwyer, M. E.; Druker, B. J. *Curr. Cancer Drug Targets* **2001**, *1*, 49–57.
70. Traxler, P.; Bold, G.; Buchdunger, E.; Caravatti, G.; Furet, P.; Manley, P.; O'Reilly, T.; Wood, J.; Zimmermann, J. *Med. Res. Rev.* **2001**, *21*, 499–512.
71. Savage, D. G.; Antman, K. H. *N. Engl. J. Med.* **2002**, *346*, 683–693.
72. Lee, S. J. *Br. J. Haematol.* **2000**, *111*, 993–1009.
73. Sawyers, C. L. *N. Engl. J. Med.* **1999**, *340*, 1330–1340.
74. Faderl, S.; Talpaz, M.; Estrov, Z.; O'Brien, S.; Kurzrock, R.; Kantarjian, H. M. *N. Engl. J. Med.* **1999**, *341*, 164–172.
75. Sawyers, C. L.; Druker, B. J. *Cancer J. Sci. Am.* **1999**, *5*, 63–69.
76. Nowell, P. C.; Hungerford, D. A. *Science* **1960**, *132*, 1497.
77. Rowley, J. D. *Nature* **1973**, *243*, 290–293.
78. Heisterkamp, N.; Stam, K.; Groffen, J.; de Klein, A.; Grosveld, G. *Nature* **1985**, *315*, 758–761.
79. Ben-Neriah, Y.; Daley, G. Q.; Mes-Masson, A.-M.; Witte, O. N.; Baltimore, D. *Science* **1986**, *233*, 212–214.
80. Daley, G. Q.; Van Etten, R. A.; Baltimore, D. *Science* **1990**, *247*, 824–830.
81. Druker, B. J. *J. Clin. Oncol.* **2003**, *21*, 239s–245s.
82. Paul, R.; Hallett, W. A.; Reich, M. F.; Johnson, B. D.; Lenhard, R. H.; Dusza, J. P.; Kerwar, S. S.; Lin, Y.; Pickett, W. C. *J. Med. Chem.* **1993**, *36*, 2716–2730.
83. Zimmermann, J.; Buchdunger, E.; Mett, H.; Meyer, T.; Lydon, N. B.; Traxler, P. *Bioorg. Med. Chem. Lett.* **1996**, *6*, 1221–1226.
84. Zimmermann, J.; Buchdunger, E.; Mett, H.; Meyer, T.; Lydon, N. B. *Bioorg. Med. Chem. Lett.* **1997**, *7*, 187–192.
85. Buchdunger, E.; Matter, A.; Druker, B. J. *Biochim. Biophys. Acta* **2001**, *1551*, M11–M18.
86. Matsumoto, J.; Minami, S. *J. Med. Chem.* **1975**, *18*, 74–79.
87. Druker, B. J.; Tamura, S.; Buchdunger, E.; Ohno, S.; Segal, G. M.; Fanning, S.; Zimmermann, J.; Lydon, N. B. *Nat. Med.* **1996**, *2*, 561–566.
88. Buchdunger, E.; Cioffi, C. L.; Law, N.; Stover, D.; Ohno-Jones, S.; Druker, B. J.; Lydon, N. B. J. *Pharmacol. Exp. Ther.* **2000**, *295*, 139–145.
89. Heinrich, M. C.; Griffith, D. J.; Druker, B. J.; Wait, C. L.; Ott, K. A.; Zigler, A. J. *Blood* **2000**, *96*, 925–932.
90. Beran, M.; Cao, X.; Estrov, Z.; Jeha, S.; Jin, G.; O'Brien, S.; Talpaz, M.; Arlinghaus, R. B.; Lydon, N. B.; Kantarjian, H. *Clin. Cancer Res.* **1998**, *4*, 1661–1672.
91. Gambacorti-Passerini, C.; le Coutre, P.; Mologni, L.; Fanelli, M.; Bertazzoli, C.; Marchesi, E.; Di Nicola, M.; Biondi, A.; Corneo, G. M.; Belotti, D. et al. *Blood Cell Mol. Dis.* **1997**, *23*, 380–394.
92. Wang, J. Y. J. *Oncogene* **2000**, *19*, 5643–5650.
93. Le Coutre, P.; Mologni, L.; Cleris, L.; Marchesi, E.; Buchdunger, E.; Giardini, R.; Formelli, F.; Gambacorti-Passerini, C. *J. Natl. Cancer Inst.* **1999**, *91*, 163–168.
94. Mohammadi, M.; Froum, S.; Hamby, J. M.; Schroeder, M. C.; Panek, R. L.; Lu, G. H.; Eliseenkova, A. V.; Green, D.; Schlessinger, J.; Hubbard, S. R. *EMBO J.* **1998**, *17*, 5896–5904.
95. Corbin, A. S.; Buchdunger, E.; Pascal, F.; Druker, B. J. *J. Biol. Chem.* **2002**, *277*, 32214–32219.

96. Schindler, T.; Bornmann, W.; Pellicena, P.; Miller, W. T.; Clarkson, B.; Kuriyan, J. *Science* **2000**, *289*, 1938–1942.
97. Nagar, B.; Bornmann, W. G.; Pellicena, P.; Schindler, T.; Veach, D. R.; Miller, W. T.; Clarkson, B.; Kuriyan, J. *Cancer Res.* **2002**, *62*, 4236–4243.
98. Moasser, M. M.; Srethapakdi, M.; Sachar, K. S.; Kraker, A. J.; Rosen, N. *Cancer Res.* **1999**, *59*, 6145–6152.
99. Druker, B. J.; Sawyers, C. L.; Kantarjian, H.; Resta, D. J.; Fernandes-Reese, S.; Ford, J. M.; Capdeville, R.; Talpaz, M. *N. Engl. J. Med.* **2001**, *344*, 1038–1042.
100. Druker, B. J.; Talpaz, M.; Resta, D. J.; Peng, B.; Buchdunger, E.; Ford, J. M.; Lydon, N. B.; Kantarjian, H.; Capdeville, R.; Ohno-Jones, S. et al. *N. Engl. J. Med.* **2001**, *344*, 1031–1037.
101. Ottmann, O. G.; Druker, B. J.; Sawyers, C. L.; Goldman, J. M.; Reiffers, J.; Silver, R. T.; Tura, S.; Fischer, T.; Deininger, M. W.; Schiffer, C. A. et al. *Blood* **2002**, *100*, 1965–1971.
102. Hochhaus, A.; Kreil, S.; Corbin, A. S.; La Rosee, P.; Muller, M. C.; Lahaye, T.; Hanfstein, B.; Schoch, C.; Cross, N. C.; Berger, U. et al. *Leukemia* **2002**, *16*, 2190–2196.
103. Gorre, M. E.; Mohammed, M.; Ellwood, K.; Hsu, N.; Paquette, R.; Rao, P. N.; Sawyers, C. L. *Science* **2001**, *293*, 876–880.
104. Corbin, A. S.; La Rosee, P.; Stoffregen, E. P.; Druker, B. J.; Deininger, M. W. *Blood*, **2003**, *101*, 4611–4614.
105. Branford, S.; Rudzki, Z.; Walsh, S.; Grigg, A.; Arthur, C.; Taylor, K.; Herrmann, R.; Lynch, K. P.; Hughes, T. P. *Blood* **2002**, *99*, 3472–3475.
106. Roumiantsev, S.; Shah, N. P.; Gorre, M. E.; Nicoll, J.; Brasher, B. B.; Sawyers, C. L.; Van Etten, R. A. *Proc. Natl. Acad. Sci. USA* **2002**, *99*, 10700–10705.
107. Shah, N. P.; Nicoll, J. M.; Nagar, B.; Gorre, M. E.; Paquette, R. L.; Kuriyan, J.; Sawyers, C. L. *Cancer Cell* **2002**, *2*, 117–125.
108. Travis, J. *Science* **2004**, *305*, 319–320.
109. Wisniewski, D.; Lambek, C. L.; Liu, C.; Strife, A.; Veach, D. A.; Nagar, B.; Young, M. A.; Schindler, T.; Bornmann, W. G.; Bertino, J. R. et al. *Cancer Res.* **2002**, *62*, 4244–4255.
110. Huron, D. R.; Gorre, M. E.; Kraker, A. J.; Sawyers, C. L.; Rosen, N.; Moasser, M. M. *Clin. Cancer Res.* **2003**, *9*, 1267–1273.
111. La Rosee, P.; Corbin, A. S.; Stoffregen, E. P.; Deininger, M. W.; Druker, B. J. *Cancer Res.* **2002**, *62*, 7149–7153.
112. O'Hare, T.; Pollock, R.; Stoffregen, E. P.; Keats, J.; Abdullah, O. M.; Moseson, E. M.; Rivera, V. M.; Tang, H.; Metcalf, C. A.; Bohacek, R. S. et al. *Blood* **2004**, *104*, 2532–2539.
113. Deutsch, E.; Maggiorella, L.; Wen, B.; Bonnet, M. L.; Khanfir, K.; Frascogna, V.; Turhan, A. G.; Bourhis, J. *Br. J. Cancer* **2004**, *91*, 1735–1741.
114. Shah, N. P.; Tran, C.; Lee, F. Y.; Chen, P.; Norris, D.; Sawyers, C. L. *Science* **2004**, *305*, 399–401.
115. Lombardo, L. J.; Lee, F. Y.; Chen, P.; Norris, D.; Barrish, J. C.; Behnia, K.; Castaneda, S.; Cornelius, L. A. M.; Das, J.; Doweyko, A. M. et al. *J. Med. Chem.* **2004**, *47*, 6658–6661.
116. Weisberg, E.; Manley, P. W.; Breitenstein, W.; Bruggen, J.; Cowan-Jacob, S. W.; Ray, A.; Huntly, B.; Fabbro, D.; Fendrich, G.; Hall-Meyers, E. et al. *Cancer Cell* **2005**, *7*, 129–141.
117. O'Hare, T.; Walters, D. K.; Deininger, M. W.; Druker, B. J. *Cancer Cell* **2005**, *7*, 117–119.
118. Martin, G. S. *Oncogene* **2004**, *23*, 7910–7917.
119. Martin, G. S. *Nat. Rev. Mol. Cell Biol.* **2001**, *2*, 467–475.
120. Rous, P. *J. Exp. Med.* **1911**, *13*, 397–411.
121. Gross, L. *Proc. Soc. Exp. Biol. Med.* **1957**, *94*, 767–771.
122. Wang, L. H.; Duesberg, P. H.; Kawai, S.; Hanafusa, H. *Proc. Natl. Acad. Sci. USA* **1976**, *73*, 447–451.
123. Brugge, J. S.; Erikson, R. L. *Nature* **1977**, *269*, 346–348.
124. Collett, M. S.; Erikson, R. L. *Proc. Natl. Acad. Sci. USA* **1978**, *75*, 2021–2024.
125. Levinson, A. D.; Oppermann, H.; Levintow, L.; Varmus, H. E.; Bishop, J. M. *Cell* **1978**, *15*, 561–572.
126. Hunter, T.; Sefton, B. M. *Proc. Natl. Acad. Sci. USA* **1980**, *77*, 1311–1315.
127. Stehelin, D.; Varmus, H. E.; Bishop, J. M.; Vogt, P. K. *Nature* **1976**, *260*, 170–173.
128. Takeya, T.; Hanasafura, H. *Cell* **1983**, *32*, 881–890.
129. Frame, M. C. *Biochim. Biophys. Acta* **2002**, *1602*, 114–130.
130. Courtneidge, S. A. *EMBO J.* **1985**, *4*, 1471–1477.
131. Okada, M.; Nakagawa, H. *J. Biol. Chem.* **1989**, *264*, 20286–20893.
132. Cooper, J. A.; Gould, K. L.; Cartwright, C. A.; Hunter, T. *Science* **1986**, *231*, 1431–1434.
133. Williams, J. C.; Wierenga, R. W.; Saraste, M. *Trends Biochem. Sci.* **1998**, *23*, 179–184.
134. Xu, W.; Harrison, S. C.; Eck, M. J. *Nature* **1997**, *385*, 595–602.
135. Williams, J. C.; Weijland, A.; Gonfloni, S.; Thompson, A.; Courtneidge, S. A.; Superti-Furga, G.; Wierenga, R. K. *J. Mol. Biol.* **1997**, *274*, 757–775.
136. Sicheri, F.; Moarefi, I.; Kuriyan, J. *Nature* **1997**, *385*, 602–609.
137. Schindler, T.; Sicheri, F.; Pico, A.; Gazit, A.; Levitzki, A.; Kuriyan, J. *Mol. Cell* **1999**, *3*, 639–648.
138. Porter, M.; Schindler, T.; Kuriyan, J.; Miller, W. T. *J. Biol. Chem.* **2000**, *275*, 2721–2726.
139. Lee, L.-F.; Louie, M. C.; Desai, S. J.; Yang, J.; Chen, H.-W.; Evans, C. P.; Kung, H.-J. *Oncogene* **2004**, *23*, 2197–2205.
140. Wiener, J. R.; Widham, T. C.; Estrella, V. C.; Parikh, N. U.; Thall, P. F.; Deavers, M. T.; Bast, R. C.; Mills, G. B.; Gallick, G. E. *Gynecol. Oncol.* **2003**, *88*, 73–79.
141. Reissig, D.; Clement, J.; Sanger, J.; Berndt, A.; Kosmeshi, H.; Bohmer, F. D. *J. Cancer Res. Clin. Oncol.* **2001**, *127*, 226–230.
142. Allgayer, H.; Boyd, D. D.; Heiss, M. M.; Abdalla, E. K.; Curley, S. A.; Gallick, G. E. *Cancer* **2002**, *94*, 344–351.
143. Talmonti, M. S.; Roh, M. S.; Curley, S. A.; Gallick, G. E. *J. Clin. Invest.* **1993**, *91*, 3–60.
144. Lowe, C.; Yoneda, T.; Boyce, B. F.; Chen, H.; Mundy, G. R.; Sonano, P. *Proc. Natl. Acad. Sci. USA* **1993**, *90*, 4485–4489.
145. Boyer, B.; Vallas, A. M.; Edmen, N. *Biochem. Pharmacol.* **2000**, *60*, 1091–1099.
146. Fincham, V. J.; Frame, M. C. *EMBO J.* **1998**, *17*, 81–92.
147. Roche, S.; Kogel, M.; Braone, M. V.; Roussel, M. F.; Courtneidge, S. A. *Mol. Cell. Biol.* **1995**, *15*, 1102–1109.
148. Dave, C. G.; Shah, P. R.; Upadhyaya, S. P. *Indian J. Chem.* **1988**, *27B*, 778–780.
149. Hanke, J. H.; Gardner, J. P.; Dow, R. L.; Chnagelian, P. S.; Brissette, W. H.; Weringer, E. J.; Pollok, B. A.; Connelly, P. A. *J. Biol. Chem.* **1996**, *271*, 695–701.
150. Missbach, M.; Altmann, E.; Widler, L.; Susa, M.; Buchdunger, E.; Mett, H.; Meyer, T.; Green, J. *Bioorg. Med. Chem. Lett.* **2000**, *10*, 945–949.
151. Missbach, M.; Jeschke, M.; Feyen, J.; Muller, K.; Glatt, M.; Green, J.; Susa, M. *Bone* **1999**, *24*, 437–449.
152. Widler, L.; Green, J.; Missbach, M.; Susa, M.; Altmann, E. *Bioorg. Med. Chem. Lett.* **2001**, *11*, 849–852.
153. Altmann, E.; Missbach, M.; Green, J.; Susa, M.; Wagenknecht, H.-A.; Widler, L. *Bioorg. Med. Chem. Lett.* **2001**, *11*, 853–856.

154. Borhani, D. W.; Calderwood, D. J.; Friedman, M. M.; Hirst, G. C.; Li, N.; Leung, A. K. W.; McRae, B.; Ratnofsky, S.; Ritter, K.; Waegell, W. *Bioorg. Med. Chem. Lett.* **2004**, *14*, 2613–2616.
155. Burchat, A. F.; Calderwood, D. J.; Friedman, M. M.; Hirst, G. C.; Li, N.; Rafferty, P.; Ritter, K.; Skinner, B. S. *Bioorg. Med. Chem. Lett.* **2002**, *12*, 1687–1690.
156. Klutchko, S. R.; Hamby, J. M.; Boschelli, D. H.; Wu, Z.; Kraker, A. J.; Amar, A. M.; Hartl, B. G.; Shen, C.; Klohs, W. D.; Steinkampf, R. W. et al. *J. Med. Chem.* **1998**, *41*, 3276–3292.
157. Schroeder, M. C.; Hamby, J. M.; Connolly, C. J. C.; Grohar, P. J.; Winters, R. T.; Barvian, M. R.; Moore, C. W.; Boushelle, S. L.; Crean, S. M.; Kraker, A. J. et al. *J. Med. Chem.* **2001**, *44*, 1915–1926.
158. Ple, P. A.; Green, T. P.; Hennequin, L. F.; Curwen, J.; Fennell, M.; Allen, J.; Lambert-van der Brempt, C.; Costello, G. *J. Med. Chem.* **2004**, *47*, 871–887.
159. Yezhelyev, M. V.; Koehl, G.; Guba, M.; Brabletz, T.; Jauch, K.-W.; Ryan, A.; Brage, A.; Green, T.; Fennell, M.; Bruns, C. J. *Clin. Cancer Res.* **2004**, *10*, 8028–8036.
160. Berger, D.; Dutia, M.; Powell, D.; Wu, B.; Wissner, A.; DeMorin, F.; Weber, J.; Boschelli, F. *Bioorg. Med. Chem. Lett.* **2002**, *12*, 2761–2765.
161. Boschelli, D. H.; Ye, F.; Wang, Y. D.; Dutia, M.; Johnson, S. L.; Wu, B.; Miller, K.; Powell, D. W.; Yaczko, D.; Young, M. et al. *J. Med. Chem.* **2001**, *44*, 3965–3977.
162. Boschelli, D. H.; Wu, B.; Barrios Sosa, A. C.; Durutlic, H.; Ye, F.; Raifeld, Y.; Golas, J. M.; Boschelli, F. *J. Med. Chem.* **2004**, *47*, 6666–6668.
163. Susa, M.; Missbach, M.; Green, J. *Trends Pharmacol. Sci.* **2000**, *21*, 489–495.
164. Vu, C. B.; Luke, G. P.; Kawahata, N.; Shakespeare, W. C.; Wang, Y.; Sundaramoorthi, R.; Metcalf, C. A., III; Keenan, T. P.; Pradeepan, S.; Corpuz, E. et al. *Bioorg. Med. Chem. Lett.* **2003**, *13*, 3071–3074.
165. Dalgarno, D.; Stehle, T.; Narula, S.; Schelling, P.; van Schravendijk, M. R.; Adams, S.; Andrade, L.; Keats, J.; Ram, M.; Jin, L. et al. *Chem. Biol. Drug Des.* **2006**, *67*, 46–57.
166. Jones, N.; Dumont, D. J. *Cancer Metastasis Rev.* **2000**, *19*, 13–17.
167. Nakayama, T.; Yoshizaki, A.; Kawahara, N.; Ohtsuru, A.; Wen, C. Y.; Fukuda, E.; Nakashima, M.; Sekine, I. *Histopathology* **2004**, *44*, 232–239.
168. Homer, J. J.; Greenman, J.; Drevs, J.; Marme, D.; Stafford, N. D. *Head Neck* **2002**, *24*, 773–778.
169. Ma, P. C.; Maulik, G.; Christensen, J.; Salgia, R. *Cancer Metastasis Rev.* **2003**, *22*, 309–325.
170. Baykal, C.; Ayhan, A.; Al, A.; Yuce, K. *Gynecol. Oncol.* **2003**, *88*, 123–129.
171. Ueki, T.; Fujimoto, J.; Suzuki, T.; Yamamoto, H.; Okamoto, E. *Hepatology* **1997**, *25*, 862–866.
172. Gascoyne, R. D.; Aoun, P.; Wu, D.; Chhanabhai, M.; Skinnider, B. F.; Greiner, T. C.; Morris, S. W.; Connors, J. M.; Vose, J. M.; Viswanatha, D. S. et al. *Blood* **1999**, *93*, 3913–3921.
173. Putzer, B. M.; Drosten, M. *Trends Mol. Med.* **2004**, *10*, 351–357.
174. Hansford, J. R.; Mulligan, L. M. *J. Med. Genet.* **2000**, *37*, 817–827.
175. O'Bryan, J. P.; Frye, R. A.; Cogswell, P. C.; Neubauer, A.; Kitch, B.; Prokop, C.; Espinosa, R., III; Le Beau, M. M.; Earp, H. S.; Liu, E. T. *Mol. Cell. Biol.* **1991**, *11*, 5016–5031.
176. Jacob, A. N.; Kalapurakal, J.; Davidson, W. R.; Kandpal, G.; Dunson, N.; Prashar, Y.; Kandpal, R. P. *Cancer Detect. Prev.* **1999**, *23*, 325–332.
177. Rochlitz, C.; Lohri, A.; Bacchi, M.; Schmidt, M.; Nagel, S; Fopp, M.; Fey, M. F.; Herrmann, R.; Neubauer, A. *Leukemia* **1999**, *13*, 1352–1358.
178. Nakagawara, A. *Cancer Lett.* **2001**, *169*, 107–114.
179. Yarden, Y.; Slikowski, M. X. *Nat. Rev. Mol. Cell Biol.* **2001**, *2*, 127–137.
180. Hong, W. K.; Ullrich, A. U. *Oncol. Ther.* **2000**, *1*, 1–30.
181. Sordella, R.; Bell, D. W.; Haber, D. A.; Settleman, J. *Science* **2004**, *305*, 1163–1167.
182. Ciardiello, F.; Tortora, G. *Clin. Cancer Res.* **2001**, *7*, 2958–2970.
183. Arteaga, C. L. *Oncologist* **2002**, *7*, 31–39.
184. Ogawara, H.; Akiyama, T.; Wtanabe, S. *J. Antibiot. (Tokyo)* **1989**, *42*, 340–343.
185. Umezawa, H.; Imoto, M.; Sawa, T.; Isshiki, K.; Matsuda, N.; Uchida, T.; Iinuma, H.; Hamada, M.; Takeuchi, T. *J. Antibiot. (Tokyo)* **1986**, *39*, 170–173.
186. McLeod, H. L.; Brunton, V. G.; Eckhardt, N.; Lear, M. J.; Robins, D. J.; Workman, P.; Graham, M. A. *Br. J. Cancer* **1996**, *74*, 1714–1718.
187. Yaish, P.; Gazit, A.; Gilon, C.; Levitzki, A. *J. Biol. Chem.* **1988**, *242*, 933–935.
188. Levitzki, A.; Gazit, A. *Science* **1995**, *267*, 1782–1788.
189. Yoneda, T.; Lyall, R. M.; Alsina, M. M.; Persons, P. E.; Spada, A. P.; Levitzki, A. Z.; Mundy, G. R. *Cancer Res.* **1991**, *51*, 4430–4435.
190. Boschelli, D. H. *Drugs Future* **1999**, *24*, 515–537.
191. Bridges, A. J. *Chem. Rev.* **2001**, *101*, 2541–2571.
192. Traxler, P.; Lydon, N. *Drugs Future* **1995**, *20*, 1261–1274.
193. Fry, D. W. *Exp. Opin. Invest. Drugs* **1994**, *3*, 577–595.
194. Palmer, B. D.; Trumpp-Kallmeyer, S.; Fry, D. W.; Nelson, J. M.; Showalter, H. D.; Denny, W. A. *J. Med. Chem.* **1997**, *40*, 1519–1529.
195. Ward, W. H. J.; Cook, P. N.; Slater, A. M.; Davies, D. H.; Holdgate, G. A.; Green, L. R. *Biochem. Pharmacol.* **1994**, *48*, 659–666.
196. Woodburn, J. R.; Barker, A. J.; Gibson, K. H.; Ashton, S. E.; Wakeling, A. E.; Curry, B. J.; Scarlett, L.; Henthorn, L. R. 88th Annual Meeting of the American Assocation fpr. Cancer Research, San Diego, CA, 1997; Abst 4251.
197. Fry, D. W.; Kraker, A. J.; McMichael, A.; Ambroso, L. A.; Nelson, J. M.; Leoplod, W. R.; Connors, R. W.; Bridges, A. J. *Science* **1994**, *265*, 1093–1095.
198. Moyer, J. D.; Barbacci, E. G.; Iwata, K. K.; Arnold, L.; Boman, B.; Cunningham, A.; DiOrio, C.; Doty, J.; Morin, M. J.; Moyer, M. P. et al. *Cancer Res.* **1997**, *57*, 4838–4848.
199. Stamos, J.; Slikowski, M. X.; Eigenbrot, C. *J. Biol. Chem.* **2002**, *277*, 46265–46272.
200. Bhattacharya, S. K.; Cox, E. D.; Kath, J. C.; Mathiowetz, A. M.; Morris, J.; Moyer, J. D.; Pustilnik, L. R.; Rafidi, K.; Richter, D. T.; Su, C. et al. *Biochem. Biophys. Res. Commun.* **2003**, *307*, 267–273.
201. Lorimer, I. A. J. *Curr. Cancer Drug Targets* **2002**, *2*, 91–102.
202. Kuan, C.-T.; Wikstrand, C. J.; Bigner, D. D. *Endocr. Relat. Cancer* **2001**, *8*, 83–96.
203. Amann, J.; Kalyankrishna, S.; Massion, P. P.; Ohm, J. E.; Girard, L.; Shigematsu, H.; Peyton, M.; Juroske, D.; Huang, Y.; Salmon, J. S. et al. *Cancer Res.* **2005**, *65*, 226–235.
204. Bell, D. W.; Gore, I.; Okimoto, R. A.; Godin-Heymann, N.; Sordella, R.; Mulloy, R.; Sharma, S. V.; Brannigan, B. W.; Mohapatra, G.; Settleman, J. et al. *Nat. Genet.* **2005**, *37*, 1315–1316.

205. Paez, J. G.; Janne, P. A.; Lee, J. C.; Tracy, S.; Greulich, H.; Gabriel, S.; Herman, P.; Kaye, F. J.; Lindeman, N.; Boggon, T. J. et al. *Science* **2004**, *304*, 1497–1500.
206. Lynch, T. J.; Bell, D. W.; Sordella, R.; Gurubhagavatula, S.; Okimoto, R. A.; Brannigan, B. W.; Harris, P. L.; Haserlat, S. M.; Supko, J. G.; Huluska, F. G. et al. *N. Engl. J. Med.* **2004**, *350*, 2129–2139.
207. Herbst, R. S.; Langer, C. J. *Semin. Oncol.* **2002**, *29*, 27–36.
208. Baselga, J.; Averbuch, S. D. *Drugs* **2000**, *60*, 33–40.
209. Herbst, R. S. *Semin. Oncol.* **2003**, *30*, 34–46.
210. Blackledge, G.; Averbuch, S. D. *Br. J. Cancer* **2004**, *90*, 566–572.
211. Arteaga, C. L.; Johnson, D. H. *Curr. Opin. Oncol.* **2001**, *13*, 491–498.
212. Agelaki, S.; Georgoulias, V. *Expert Opin. Emerg. Drugs* **2005**, *10*, 855–874.
213. Perez-Soler, R.; Chachoua, A.; Huberman, M.; Karp, D.; Rigas, J.; Hammond, L.; Rowinsky, E.; Preston, G.; Ferrante, K. J.; Allen, L. F. et al. *Proc. Am. Soc. Clin. Oncol.* **2001**, *20*, 310.
214. Grunwald, V.; Hidalgo, M. *Adv. Exp. Med. Biol.* **2003**, *532*, 235–246.
215. Akita, R. W.; Sliwkowski, M. X. *Semin. Oncol.* **2003**, *30*, 15–24.
216. Kris, M.; Ranson, M.; Ferry, D.; Hammond, L.; Averbush, S.; Ochs, J.; Rowinsky, J. *Clin. Cancer Res.* **1999**, *5*, 3749S.
217. Fukuoka, M.; Yano, S.; Giaccone, G.; Tamura, T.; Nakagawa, K.; Douillard, J.-Y.; Nishiwaki, Y.; Vansteenkiste, J.; Kudoh, S.; Rischin, D. et al. *J. Clin. Oncol.* **2003**, *21*, 2237–2246.
218. Kris, M. G.; Natale, R. B.; Herbst, R. S.; Lynch, T. J.; Prager, D.; Belini, C. P.; Schiller, J. H.; Kelly, K.; Spiridonidis, H.; Sandler, A. et al. *JAMA* **2003**, *290*, 2149–2158.
219. Pollack, V. A.; Savane, D. M.; Baker, D. A.; Tsaparikos, K. E.; Sloan, D. E.; Moyer, J. D.; Barbacci, E. G.; Pustilnik, L. R.; Smolarek, T. A.; Davis, J. A. et al. *J. Pharmacol. Exp. Ther.* **1999**, *291*, 739–748.
220. Hidalgo, M.; Siu, L. L.; Nemunaitis, J.; Rizzo, J.; Hammond, L. A.; Takimoto, C.; Eckhardt, G.; Tolcher, A.; Britten, C. D.; Denis, L. et al. *J. Clin. Oncol.* **2001**, *13*, 3267–3279.
221. Herbst, R. S.; Prager, D.; Hermann, R.; Miller, V.; Fehrenbacher, L.; Hoffman, P.; Johnson, B.; Sandler, A. B.; Mass, R.; Johnson, D. H. *Proc. Am. Soc. Clin. Oncol.* **2004**, *23*, 617.
222. Dittrich, C.; Greim, G.; Bomer, M.; Weigang-Kohler, K.; Huisman, H.; Amelsberg, A.; Ehret, A.; Wanders, J.; Hanauske, A.; Fomelau, P. *Eur. J. Cancer* **2002**, *38*, 1072–1080.
223. Hoekstra, R.; Dumez, H.; van Oosterom, A. T.; Sizer, K. C.; Ravera, C.; Vaidyanathan, S.; Verweij, J.; Eskens, F. A. *Proc. Am. Soc. Clin. Oncol.* **2002**, *21*, 86.
224. Rowinsky, E. K.; Garrison, M.; Lorusso, P.; Patnaik, A.; Hammond, L.; DeBono, J.; McCreery, H.; Eiseman, I.; Lenehan, P.; Tolcher, A. *Proc. Am. Soc. Clin. Oncol.* **2003**, *22*, 201.
225. Slichenmeyer, W. J.; Elliott, W. L.; Fry, D. W. *Semin. Oncol.* **2001**, *28*, 80–85.
226. Hidalgo, M.; Erhlichman, C.; Rowinsky, E. K.; Koepp-Norris, J.; Jensen, K.; Boni, J.; Korth-Bradley, D.; Zacharchuk, C. *Proc. Am. Soc. Clin. Oncol.* **2002**, *21*, 17.
227. Riese, D. J.; Stern, D. F. *Bioessays* **1998**, *20*, 41–48.
228. Alroy, I.; Yarden, Y. *FEBS Lett.* **1997**, *410*, 83–86.
229. Hynes, N. E.; Stern, D. F. *Biochem. Biophys. Acta* **1994**, *1198*, 165–184.
230. Slamon, D. J.; Godolphin, W.; Jones, L. A.; Holt, J. A.; Wong, S. G.; Keith, D. E.; Levin, W. J.; Stuart, S. G.; Udove, J.; Ullrich, A. et al. *Science* **1989**, *244*, 707–712.
231. Rowinsky, E. K. *Annu. Rev. Med.* **2004**, *55*, 433–457.
232. Slamon, D. J.; Clark, G. M.; Wong, S. G.; Levin, W. J.; Ullrich, A.; McGuire, W. L. *Science* **1987**, *235*, 177–182.
233. Vogel, C. L.; Cobleigh, M. A.; Tripathy, D.; Guthiel, J. C.; Harris, L. N.; Fehrenbacher, L.; Slamon, D. J.; Murphy, M.; Novotny, W. F.; Burchmore, S. et al. *Oncology* **2001**, *61*, 37–42.
234. Slamon, D. J.; Leyland-Jones, B.; Shak, S.; Fuchs, H.; Paton, V.; Bajamonde, A.; Fleming, T.; Eierman, W.; Wolter, J.; Pegram, M. et al. *N. Engl. J. Med.* **2001**, *344*, 783–792.
235. Cobleigh, M. A.; Vogel, C. L.; Tripathy, D.; Robert, N. J.; Scholl, S.; Fehrenbacher, L.; Wolter, J. M.; Paton, V.; Shak, S.; Lieberman, G. et al. *J. Clin. Oncol.* **1999**, *17*, 2639–2648.
236. Vogel, C. L.; Cobleigh, M. A.; Tripathy, D.; Guthiel, J. C.; Harris, L. N.; Fehrenbacher, L.; Slamon, D. J.; Murphy, M.; Novotny, W. F.; Burchmore, M. et al. *J. Clin. Oncol.* **2002**, *20*, 719–726.
237. Gaul, M. D.; Guo, Y.; Affleck, K.; Cockerill, G. S.; Gilmer, T. M.; Griffin, R. J.; Guntrip, S.; Keith, B. R.; Knight, W. B.; Mullin, R. J. et al. *Bioorg. Med. Chem. Lett.* **2003**, *13*, 637–640.
238. Xia, W.; Mullin, R. J.; Keith, B. R.; Liu, L.-H.; Ma, H.; Rusnak, D. W.; Owens, G.; Alligood, K. J.; Spector, N. L. *Oncogene* **2002**, *21*, 6255–6263.
239. Cockerill, G. S.; Lackey, K. E. *Curr. Top. Med. Chem.* **2002**, *2*, 1001–1010.
240. Spector, N.; Raefsky, E.; Hurwitz, H.; Hensing, T.; Dowlati, A.; Dees, C.; O'Neil, B.; Smith, A.; Mangum, S.; Burris, H. A. *Proc. Am. Soc. Clin. Oncol.* **2003**, *22*, 193.
241. Barbacci, E. G.; Pustilnik, L. R.; Rossi, A. M. K.; Emerson, E.; Miller, P. E.; Boscoe, B. P.; Cox, E. D.; Iwata, K. K.; Jani, J. P.; Provoncha, K. et al. *Cancer Res.* **2003**, *63*, 4450–4459.
242. Sun, L.; Cui, J.; Liang, C.; Zhou, Y.; Nematalla, A.; Wang, X.; Chen, H.; Tang, C.; Wei, J. *Bioorg. Med. Chem. Lett.* **2002**, *12*, 2153–2157.
243. Traxler, P.; Allegrini, P. R.; Brandt, R.; Brueggen, J.; Cozens, R.; Fabbro, D.; Grossios, K.; Lane, H. A.; McSheehy, P.; Mestan, J. et al. *Cancer Res.* **2004**, *64*, 4931–4941.
244. Rabindran, S. K.; Discafani, C. M.; Rosfjord, E. C.; Baxter, M.; Floyd, M. B.; Golas, J.; Hallett, W. A.; Johnson, B. D.; Nilakantan, R.; Overbeek, E. et al. *Cancer Res.* **2004**, *64*, 3958–3965.
245. Wood, E. R.; Truesdale, A. T.; McDonald, O. B.; Yuan, D.; Hassell, A.; Dickerson, S. H.; Ellis, B.; Pennisi, C.; Horne, E.; Lackey, K. et al. *Cancer Res.* **2004**, *64*, 6652–6659.
246. LeRoith, D.; Roberts, C. T. *Cancer Lett.* **2003**, *195*, 127–137.
247. Zhang, H.; Yee, D. *Expert Opin. Investig. Drugs* **2004**, *13*, 1569–1577.
248. Parrizas, M.; Gazit, A.; Levitzki, A.; Wertheimer, E.; LeRoith, D. *Endocrinology* **1997**, *138*, 1427–1433.
249. Garcia-Echeverria, C.; Pearson, M. A.; Marti, A.; Meyer, T.; Mestan, J.; Zimmermann, J.; Gao, J.; Brueggen, J.; Capraro, H.-G.; Cozens, R. et al. *Cancer Cell* **2004**, *5*, 231–239.
250. Cheetham, G. M. T. *Curr. Opin. Struct. Biol.* **2004**, *14*, 700–705.

251. Pietras, K.; Sjoblom, T.; Rubin, K.; Heldin, C.-H.; Ostman, A. *Cancer Cell* **2003**, *3*, 439–443.
252. Claesson-Welsh, L. *J. Biol. Chem.* **1994**, *269*, 32023–32026.
253. Levitzki, A. *Cardiovasc. Res.* **2005**, *65*, 581–586.
254. Levitzki, A. *Cytokine Growth Factor Rev.* **2004**, *15*, 229–235.
255. Schmandt, R. E.; Broaddus, R.; Lu, K. H.; Shvartsman, H.; Thornton, A.; Malpica, A.; Sun, C.; Bodurka, D. C.; Gershenson, D. M. *Cancer* **2003**, *98*, 758–764.
256. Li, H.; Frederiksson, L.; Li, X.; Eriksson, U. *Oncogene* **2003**, *22*, 1501–1510.
257. Magnusson, M. K.; Meade, K. E.; Brown, K. E.; Arthur, D. C.; Krueger, L. A.; Barrett, A. J.; Dunbar, C. E. *Blood* **2002**, *100*, 1088–1091.
258. Ashman, L. K. *Int. J. Biochem. Cell Biol.* **1999**, *31*, 1037–1051.
259. Broudy, V. C. *Blood* **1997**, *90*, 1345–1364.
260. Lyman, S. D.; Jacobsen, S. E. W. *Blood* **1998**, *91*, 1101–1134.
261. Corless, C. L.; Fletcher, J. A.; Heinrich, M. C. *J. Clin. Oncol.* **2004**, *22*, 3813–3825.
262. Moriyama, Y.; Tsujimura, T.; Hashimoto, K.; Morimoto, M.; Kitayama, H.; Matsuzawa, Y.; Kitamura, Y.; Kanakura, Y. *J. Biol. Chem.* **1996**, *271*, 3347–3350.
263. Longley, B. J.; Reguera, M. J.; Ma, Y. *Leuk. Res.* **2001**, *25*, 571–576.
264. Furitsu, T.; Tsujimura, T.; Tono, T.; Ikeda, H.; Kitayama, H.; Koshimizu, U.; Sugahara, H.; Butterfield, J. H.; Ashman, L. K.; Kanayama, Y. et al. *J. Clin. Invest.* **1993**, *92*, 1736–1744.
265. Levis, M.; Small, D. *Expert Opin. Investig. Drugs* **2003**, *12*, 1951–1962.
266. Gilliland, D. G.; Griffin, J. D. *Leukemia* **2003**, *17*, 1532–1542.
267. Sawyers, C. L. *Cancer Cell* **2002**, *1*, 413–415.
268. Hubbard, S. R.; Mohammadi, M.; Schlessinger, J. *J. Biol. Chem.* **1998**, *273*, 11987–11990.
269. Wybenga-Groot, L. E.; Baskin, B.; Ong, S. H.; Tong, J.; Pawson, T.; Sicheri, F. *Cell* **2001**, *106*, 745–757.
270. Hubbard, S. R. *Mol. Cell* **2001**, *8*, 481–482.
271. Pixley, F. J.; Stanley, E. R. *Trends Cell Biol.* **2004**, *14*, 628–638.
272. Roussel, M. F.; Downing, J. R.; Ashmun, R. A.; Rettenmier, C. W.; Scherr, C. J. *Proc. Natl. Acad. Sci. USA* **1988**, *85*, 5903–5907.
273. Dai, X.-M.; Ryan, G. R.; Hapel, A. J.; Dominguez, M. G.; Russell, R. G.; Kapp, S.; Sylvestre, V.; Stanley, E. R. *Blood* **2002**, *99*, 111–120.
274. Kacinski, B. M. *Ann. Med.* **1995**, *27*, 79–85.
275. Wrobel, C. N.; Debnath, J.; Lin, E.; Beausoleil, S.; Roussel, M. F.; Brugge, J. S. *J. Cell Biol.* **2004**, *165*, 263–273.
276. Buchdunger, E.; Zimmermann, J.; Mett, H.; Meyer, T.; Muller, M.; Regenass, U.; Lydon, N. B. *Proc. Natl. Acad. Sci. USA* **1995**, *92*, 2558–2562.
277. Maguire, M. P.; Sheets, K. R.; McVety, K.; Spada, A. P.; Zilberstein, A. *J. Med. Chem.* **1994**, *37*, 2129–2137.
278. Bilder, G.; Wentz, T.; Leadley, R.; Amin, D.; Byan, L.; O'Conner, B.; Needle, S.; Galczenski, H.; Bostwick, J.; Kasiewski, C. et al. *Circulation* **1999**, *99*, 3292–3299.
279. Kovalenko, M.; Ronnstrand, L.; Heldin, C. H.; Loubtchenkov, M.; Gazit, A.; Levitzki, A.; Bohmer, F. D. *Biochemistry* **1997**, *36*, 6260–6269.
280. Gazit, A.; Yee, K.; Uecker, A.; Bohmer, F. D.; Sjoblom, T.; Ostman, A.; Waltenberger, J.; Golomb, G.; Banai, S.; Heinrich, M. C. et al. *Bioorg. Med. Chem.* **2003**, *11*, 2007–2018.
281. Matsuno, K.; Ichimura, M.; Nakajima, T.; Tahara, K.; Fujiwara, S.; Kase, H.; Ushiki, J.; Giese, N. A.; Pandey, A.; Scarborough, R. M. et al. *J. Med. Chem.* **2002**, *45*, 3057–3066.
282. Pandey, A.; Volkots, D. L.; Seroogy, J. M.; Rose, J. W.; Yu, J. C.; Lambing, J. L.; Hutchaleelaha, A.; Hollenbach, S. J.; Abe, K.; Giese, N. A. et al. *J. Med. Chem.* **2002**, *45*, 3772–3793.
283. Palmer, B. D.; Smaill, J. B.; Boyd, M.; Boschelli, D. H.; Doherty, A. M.; Hamby, J. M.; Khatana, S. S.; Kramer, J. B.; Kraker, A. J.; Panek, R. L. et al. *J. Med. Chem.* **1998**, *41*, 5457–5465.
284. Shawver, L. K.; Schwartz, D. P.; Mann, E.; Chen, H.; Tsai, J.; Chu, L.; Taylorson, L.; Longhi, M.; Meredith, S.; Germain, L. et al. *Clin. Cancer Res.* **1997**, *3*, 1167–1177.
285. Kelly, L. M.; Yu, J.-C.; Boulton, C. L.; Apatira, M.; Li, J.; Sullivan, C. M.; Williams, I.; Amaral, S. M.; Curley, D. P.; Duclos, N. et al. *Cancer Cell* **2002**, *1*, 421–432.
286. Corbin, A. S.; Griswold, I. J.; La Rosee, P.; Yee, K. W. H.; Heinrich, M. C.; Reimer, C. L.; Druker, B. J.; Deininger, M. W. N. *Blood* **2004**, *104*, 3754–3757.
287. Clark, J. J.; Cools, J.; Curley, D. P.; Yu, J.-C.; Lokker, N. A.; Giese, N. A.; Gilliland, D. G. *Blood* **2004**, *104*, 2867–2872.
288. Miknyoczki, S. J.; Chang, H.; Klein-Szanto, A.; Dionne, C. A.; Ruggeri, B. A. *Clin. Cancer Res.* **1999**, *5*, 2205–2212.
289. Camoratto, A. M.; Jani, J. P.; Angeles, T. S.; Maroney, A. C.; Sanders, C. Y.; Murakata, C.; Neff, N. T.; Vaught, J. L.; Isaacs, J. T.; Dionne, C. A. *Int. J. Cancer* **1997**, *72*, 673–679.
290. Smith, B. D.; Levis, M.; Beran, M.; Giles, F.; Kantarjian, H.; Berg, K.; Murphy, K. M.; Dauses, T.; Allebach, J.; Small, D. *Blood* **2004**, *103*, 3669–3676.
291. Weisberg, E.; Boulton, C.; Kelly, L. M.; Manley, P.; Fabbro, D.; Meyer, T.; Gilliland, D. G.; Griffin, J. D. *Cancer Cell* **2002**, *1*, 433–443.
292. Levis, M.; Allebach, J.; Tse, K. F.; Zheng, R.; Baldwin, B. R.; Smith, B. D.; Jones-Bolin, S.; Ruggeri, B.; Dionne, C.; Small, D. *Blood* **2002**, *99*, 3885–3891.
293. Demetri, G. D.; von Mehren, M.; Blanke, C. D.; Van den Abbeele, A. D.; Eisenberg, B.; Roberts, P. J.; Heinrich, M. C.; Tuveson, D. A.; Singer, S.; Janicek, M. et al. *N. Engl. J. Med.* **2002**, *347*, 472–480.
294. Gunby, R. H.; Cazzaniga, G.; Tassi, E.; Le Coutre, P.; Pogliani, E.; Specchia, G.; Biondi, A.; Gambacorti-Passerini, C. *Haematologica* **2003**, *88*, 408–415.
295. Labropoulos, S. V.; Fletcher, J. A.; Oliveira, A. M.; Papadopoulos, S.; Razis, E. D. *Anticancer Drugs* **2005**, *16*, 461–466.
296. Stone, R. M.; DeAngelo, D. J.; Klimek, V.; Galinsky, I.; Estey, E.; Nimer, S. D.; Grandin, W.; Lebwohl, D.; Wang, Y.; Cohen, P. et al. *Blood* **2005**, *105*, 54–60.
297. Marshall, J. L.; Kindler, H.; Deeken, J.; Bhargava, P.; Vogelzang, N. J.; Rizvi, N.; Luhtala, T.; Boylan, S.; Dordal, M.; Robertson, P. et al. *Invest. New Drugs* **2005**, *23*, 31–37.
298. Fiedler, W.; Serve, H.; Dohner, H.; Schwittay, M.; Ottmann, O. G.; O'Farrell, A.-M.; Bello, C. L.; Allred, R.; Manning, W. C.; Cherrington, J. M. et al. *Blood* **2005**, *105*, 986–993.
299. Kubo, K.; Ohyama, S.; Shimizu, T.; Takami, A.; Murooka, H.; Nishitoba, T.; Kato, S.; Yagi, M.; Kobayashi, Y.; Iinuma, N. et al. *Bioorg. Med. Chem. Lett.* **2003**, *11*, 5117–5133.

300. Matsuno, K.; Nakajima, T.; Ichimura, M.; Giese, N. A.; Yu, J.-C.; Lokker, N. A.; Ushiki, J.; Ide, S.; Oda, S.; Nomoto, Y. *J. Med. Chem.* **2002**, *45*, 4513–4523.
301. Roberts, W. G.; Whalen, P. M.; Soderstrom, E.; Moraski, G.; Lyssikatos, J. P.; Wang, H.-F.; Cooper, B.; Baker, D. A.; Savage, D.; Dalvie, D. et al. *Cancer Res.* **2005**, *65*, 957–966.
302. Bergers, G.; Benjamin, L. E. *Nat. Rev. Cancer* **2003**, *3*, 401–410.
303. Kerbel, R.; Folkman, J. *Nat. Rev. Cancer* **2002**, *2*, 727–739.
304. Kerbel, R. S. *Carcinogenesis* **2000**, *21*, 505–515.
305. Glade-Bender, J.; Kandel, J. J.; Yamashiro, D. J. *Expert Opin. Biol. Ther.* **2003**, *3*, 263–276.
306. Ferrara, N. *Am. J. Physiol. Cell Physiol.* **2001**, *280*, C1358–C1366.
307. Gille, H.; Kowalski, J.; Li, B.; LeCouter, J.; Moffat, B.; Zioncheck, T. F.; Pelletier, N.; Ferrara, N. *J. Biol. Chem.* **2001**, *276*, 3222–3230.
308. Inoue, M.; Hager, J. H.; Ferrara, N.; Gerber, H. P.; Hanahan, D. *Cancer Cell.* **2002**, *1*, 193–202.
309. Veikkola, T.; Jussila, L.; Makinen, T.; Karpanen, T.; Jeltsch, M.; Petrova, T. V.; Kubo, H.; Thurston, G.; McDonald, D. M.; Achen, M. G. et al. *EMBO J.* **2001**, *20*, 1223–1231.
310. Stacker, S. A.; Achen, M. G.; Jussila, L.; Baldwin, M. E.; Alitalo, K. *Nat. Rev. Cancer* **2002**, *2*, 573–583.
311. Hicklin, D. J.; Witte, L.; Zhu, Z.; Liao, F.; Wu, Y.; Li, Y.; Bohlen, P. *Drug Disc. Today* **2001**, *6*, 517–528.
312. Ferrara, N.; Hillan, K. J.; Gerber, H.-P.; Novotny, W. *Nat. Rev. Drug Disc.* **2004**, *3*, 391–400.
313. Hurwitz, H.; Fehrenbacher, L.; Novotny, W.; Cartwright, T.; Hainsworth, J.; Heim, W.; Berlin, J.; Baron, A.; Griffing, S.; Holmgren, E. et al. *N. Engl. J. Med.* **2004**, *350*, 2335–2342.
314. Zhu, Z.; Rockwell, P.; Lu, D.; Kotanides, H.; Pytowski, B.; Hicklin, D. J.; Bohlen, P.; Witte, L. *Cancer Res.* **1998**, *58*, 3209–3214.
315. Witte, L.; Hicklin, D. J.; Zhu, Z.; Pytowski, B.; Kotanides, H.; Rockwell, P.; Bohlen, P. *Cancer Metastasis Rev.* **1998**, *17*, 155–161.
316. Prewett, M.; Huber, J.; Li, Y.; Santiago, A.; O'Connor, W.; King, K.; Overholser, J.; Hooper, A.; Pytowski, B.; Witte, L. et al. *Cancer Res.* **1999**, *59*, 5209–5218.
317. Pavco, P. A.; Bouhana, K. S.; Gallegos, A. M.; Agrawal, A.; Blanchard, K. S.; Grimm, S. L.; Jensen, K. L.; Andrews, L. E.; Wincott, F. E.; Pitot, P. A. et al. *Clin. Cancer Res.* **2000**, *6*, 2094–2103.
318. Holash, J.; Davis, S.; Papadopoulos, N.; Croll, S. D.; Ho, L.; Russell, M.; Boland, P.; Leidich, R.; Hylton, D.; Burova, E. et al. *Proc. Natl. Acad. Sci. USA* **2002**, *99*, 11393–11398.
319. Wulff, C.; Wilson, H.; Wiegand, S. J.; Rudge, J. S.; Fraser, H. M. *Endocrinology* **2002**, *143*, 2797–2807.
320. Manley, P. W.; Bold, G.; Bruggen, J.; Fendrich, G.; Furet, P.; Mestan, J.; Schnell, C.; Stolz, B.; Meyer, T.; Meyhack, B. et al. *Biochim. Biophys. Acta* **2004**, *1697*, 17–27.
321. Sepp-Lorenzino, L.; Thomas, K. A. *Expert Opin. Investig. Drugs* **2002**, *11*, 1447–1465.
322. Mendel, D. B.; Laird, D.; Smolich, B. D.; Blake, R. A.; Liang, C.; Hannah, A. L.; Shaheen, R. M.; Ellis, L. M.; Weitman, S.; Shawyer, L. K. et al. *Anti-Cancer Drug Des.* **2000**, *15*, 29–41.
323. Fong, T. A. T.; Shawyer, L. K.; Sun, L.; Tang, C.; App, H.; Powell, T. J.; Kim, Y. H.; Schreck, R.; Wang, X.; Risau, W. et al. *Cancer Res.* **1999**, *59*, 99–106.
324. Mendel, D. B.; Schreck, R. E.; West, D. C.; Li, G.; Strawn, L. M.; Tanciongco, S. S.; Vasile, S.; Shawyer, L. K.; Cherrington, J. M. *Clin. Cancer Res.* **2000**, *6*, 4848–4858.
325. Hoekman, K.; Hoekman, K. *Cancer J.* **2001**, 7, S134–S138.
326. Laird, A. D.; Christensen, J. G.; Li, G.; Carver, J.; Smith, K.; Xin, X.; Moss, K. G.; Louie, S. G.; Mendel, D. B.; Cherrington, J. M. *FASEB J.* **2002**, *16*, 681–690.
327. Shaheen, R. M.; Tseng, W. W.; Davis, D. W.; Liu, W.; Reinmuth, N.; Vellagas, R.; Wieczorek, A. A.; Ogura, Y.; McConkey, D. J.; Drazan, K. E. et al. *Cancer Res.* **2001**, *61*, 1464–1468.
328. Sun, Li.; Liang, C.; Shirazian, S.; Zhou, Y.; Miller, T.; Cui, J.; Fukuda, J. Y.; Chu, J.-Y.; Nematalla, A.; Wang, X. et al. *J. Med. Chem.* **2003**, *46*, 1116–1119.
329. Mendel, D. B.; Laird, A. D.; Xin, X.; Louie, S. G.; Christensen, J. G.; Li, G.; Schreck, R. E.; Abrams, T. J.; Ngai, T. J.; Lee, L. B. et al. *Clin. Cancer Res.* **2003**, *9*, 327–337.
330. Bold, G.; Altmann, K.-H.; Frei, J.; Lang, M.; Manley, P. W.; Traxler, P.; Wietfeld, B.; Bruggen, J.; Buchdunger, E.; Cozens, R. et al. *J. Med. Chem.* **2000**, *43*, 2310–2323.
331. Wood, J. M.; Bold, G.; Buchdunger, E.; Cozens, R.; Ferrari, S.; Frei, J.; Hofmann, F.; Mestan, J.; Mett, H.; O'Reilly, T. et al. *Cancer Res.* **2000**, *60*, 2178–2189.
332. Hennequin, L. F.; Stokes, E. S. E.; Thomas, A. P.; Johnstone, C.; Ple, P. A.; Ogilvie, D. J.; Dukes, M.; Wedge, S. R.; Kendrew, J.; Curwen, J. O. *J. Med. Chem.* **2002**, *45*, 1300–1312.
333. Zangari, M.; Anaissie, E.; Stopeck, A.; Morimoto, A.; Tan, N.; Lancet, J.; Cooper, M.; Hannah, A.; Garcia-Manero, G.; Faderl, S. et al. *Clin Cancer Res.* **2004**, *10*, 88–95.
334. Kuenen, B. C.; Rosen, L.; Smit, E. F.; Parson, M. R.; Levi, M.; Ruijter, R.; Huisman, H.; Kedde, M. A.; Noordhuis, P.; van der Vijgh, W. J. et al. *J. Clin. Oncol.* **2002**, *20*, 1657–1667.
335. Xiong, H. Q.; Herbst, R.; Faria, S. C.; Scholz, C.; Davis, D.; Jackson, E. F.; Madden, T.; McConkey, D.; Hicks, M.; Hess, K. et al. *Invest. New Drugs* **2004**, *22*, 459–466.
336. Manley, P. W.; Furet, P.; Bold, G.; Bruggen, J.; Mestan, J.; Meyer, T.; Schnell, C. R.; Wood, J.; Haberey, M.; Huth, A. et al. *J. Med. Chem.* **2002**, *45*, 5687–5693.
337. Wedge, S. R.; Kendrew, J.; Hennequin, L. F. A.; Valentine, P. J.; Barry, S. T.; Brave, S. R.; Smith, N. R.; James, N. H.; Dukes, M.; Curwen, J. O. et al. *Cancer Res.* **2005**, *65*, 4389–4400.
338. Nakamura, K.; Yamamoto, A.; Kamishohara, M.; Takahashi, K.; Taguchi, E.; Miura, T.; Kubo, K.; Shibuya, M.; Isoe, T. *Mol. Cancer Ther.* **2004**, *3*, 1639–1649.
339. Munchhof, M. J.; Beebe, J. S.; Casavant, J. M.; Cooper, B. A.; Doty, J. L.; Higdon, R. C.; Hillerman, S. M.; Soderstrom, C. I.; Knauth, E. A.; Marx, M. A. et al. *Bioorg. Med. Chem. Lett.* **2004**, *14*, 21–24.
340. Beebe, J. S.; Jani, J. P.; Knauth, E.; Goodwin, P.; Higson, C.; Rossi, A. M.; Emerson, E.; Finkelstein, M.; Floyd, E.; Harriman, S. et al. *Cancer Res.* **2003**, *63*, 7301–7309.
341. Inai, T.; Mancuso, M.; Hashizume, H.; Baffert, F.; Haskell, A.; Baluk, P.; Hu-Lowe, D. D.; Shalinsky, D. R.; Thurston, G.; Yancopoulos, G. D. et al. *Am. J. Pathol.* **2004**, *165*, 35–52.

342. Gingrich, D. E.; Reddy, D. R.; Iqbal, M. A.; Singh, J.; Aimone, L. D.; Angeles, T. S.; Albom, M.; Yang, S.; Ator, M. A.; Meyer, S. L. et al. *J. Med. Chem.* **2003**, *46*, 5375–5388.
343. Podar, K.; Catley, L. P.; Tai, Y.-T.; Shringarpure, R.; Carvalho, P.; Hayashi, T.; Burger, R.; Sclossman, R. L.; Richardson, P. G.; Pandite, L. N. et al. *Blood* **2004**, *103*, 3474–3479.
344. Dev, I. K.; Dornsife, R. E.; Hopper, T. M.; Onori, J. A.; Miller, C. G.; Harrington, L. E.; Dold, K. M.; Mullin, R. J.; Johnson, J. H.; Crosby, R. M. et al. *Br. J. Cancer* **2004**, *91*, 1391–1398.
345. Huh, J.-I.; Calvo, A.; Stafford, J.; Cheung, M.; Kumar, R.; Philp, D.; Kleinman, H. K.; Green, J. E. *Oncogene* **2005**, *24*, 790–800.
346. Bilodeau, M. T.; Balitza, A. E.; Koester, T. J.; Manley, P. J.; Rodman, L. D.; Buser-Doepner, C.; Coll, K. E.; Fernandes, C.; Gibbs, J. B.; Heimbrook, D. C. et al. *J. Med. Chem.* **2004**, *47*, 6363–6372.
347. Cronauer, M. V.; Schulz, W. A.; Seifert, H.-H.; Ackermann, R.; Burchardt, M. *Eur. Urol.* **2003**, *43*, 309–319.
348. Jeffers, M.; LaRochelle, W. J.; Lichenstein, H. S. *Expert Opin. Ther. Targets* **2002**, *6*, 469–482.
349. Ornitz, D. M.; Itoh, N. *Genome Biol.* **2001**, *2*, 1–12.
350. Plotnikov, A. N.; Schlessinger, J.; Hubbard, S. R.; Mohammadi, M. *Cell* **1999**, *98*, 641–650.
351. Ornitz, D. M.; Xu, J.; Colvin, J. S.; McEwen, D. G.; MacArthur, C. A.; Coulier, F.; Gao, G.; Goldfarb, M. *J. Biol. Chem.* **1996**, *271*, 15292–15297.
352. Cronauer, M. V.; Hittmair, A.; Eder, I. E.; Hobisch, A.; Culig, Z.; Ramoner, R.; Zhang, J.; Bartsch, G.; Reissigl, A.; Radmayr, C. et al. *Prostate* **1997**, *31*, 223–233.
353. Ozen, M.; Giri, D.; Ropiquet, F.; Mansukhani, A.; Ittman, M. *J. Natl. Cancer Inst.* **2001**, *93*, 1790–1793.
354. Sibley, K.; Stern, P.; Knowles, M. A. *Oncogene* **2001**, *20*, 4416–4418.
355. Yoshimura, K.; Eto, H.; Miyake, H.; Hara, I.; Arakawa, S.; Kamidono, S. *Cancer Lett.* **1996**, *103*, 91–97.
356. Strohmeyer, T.; Peter, S.; Hartmann, M.; Munemitsu, S.; Ackermann, R.; Ullrich, A.; Slamon, D. J. *Cancer Res.* **1991**, *51*, 1811–1816.
357. Manetti, F.; Botta, M. *Curr. Pharm. Des.* **2003**, *9*, 567–581.
358. Sun, L.; Tran, N.; Liang, C.; Tang, F.; Rice, A.; Schreck, R.; Waltz, K.; Shawver, L. K.; McMahon, G.; Tang, C. *J. Med. Chem.* **1999**, *42*, 5120–5130.
359. Dimitroff, C. J.; Klohs, W.; Sharma, A.; Pera, P.; Driscoll, D.; Veith, J.; Steinkampf, R.; Schroeder, M.; Klutchko, S.; Sumlin, A. et al. *Invest. New Drugs* **1999**, *17*, 121–135.
360. Barvian, M. R.; Panck, R. L.; Lu, G. H.; Kraker, A. J.; Amar, A.; Hartl, B.; Hamby, J. M.; Showalter, H. D. H. *Bioorg. Med. Chem. Lett.* **1997**, *7*, 2903–2908.
361. Croft, D. R.; Sahai, E.; Mavria, G.; Li, S.; Tsai, J.; Lee, W. M. F.; Marshall, C. J.; Olson, M. F. *Cancer Res.* **2004**, *64*, 8994–9001.
362. Ahmad, N. *FASEB J.* **2004**, *18*, 5–7.
363. Lane, H. A.; Fernandez, A.; Lamb, N. J.; Thomas, G. *Nature* **1993**, *363*, 170–172.
364. Lebrin, F.; Goumans, M.-J.; Jonker, L.; Carvalho, R. L. C.; Valdimarsdottir, G.; Thorikay, M.; Mummery, C.; Arthur, H. M.; ten Dijke, P. *EMBO J.* **2004**, *23*, 4018–4028.
365. Knippschild, U.; Gocht, A.; Wolff, S.; Huber, N.; Lohler, J.; Stoter, M. *Cell. Signal.* **2005**, *17*, 675–689.
366. Callow, M. G.; Clairvoyant, F.; Zhu, S.; Schryver, B.; Whyte, B. D.; Bischoff, J. R.; Jallal, B.; Smeal, T. *J. Biol. Chem.* **2002**, *277*, 550–558.
367. Bowers, A. J.; Boylan, J. F. *Gene* **2004**, *328*, 135–142.
368. Felipe, V.; Minana, M. D.; Cabedo, H.; Perez-Minguez, F.; Llombart-Bosch, A.; Grisolia, S. *Eur. J. Cancer* **1994**, *30A*, 252–257.
369. Suzuki, Y.; Yamamoto, M.; Wada, H.; Ito, M.; Nakano, T.; Sasaki, Y.; Narumaya, S.; Shiku, H.; Nishikawa, M. *Blood* **1999**, *93*, 3408–3417.
370. Takami, A.; Iwabuko, M.; Okada, Y.; Kawata, T.; Odai, H.; Takahashi, N.; Shindo, K.; Kimura, K.; Tagami, Y.; Miyake, M. et al. *Bioorg. Med. Chem.* **2004**, *12*, 2115–2137.
371. Ishizaki, T.; Uehata, M.; Tamechika, I.; Keel, J.; Nonomura, K.; Maekawa, M.; Narumiya, S. *Mol. Pharmacol.* **2000**, *57*, 976–983.
372. Breitenlechner, C.; Gassel, M.; Hidaka, H.; Kinzel, V.; Huber, R.; Engh, R. A.; Bossemeyer, D. *Structure* **2003**, *11*, 1595–1607.
373. Sebolt-Leopold, J. S. *Curr. Pharm. Des.* **2004**, *10*, 1907–1914.
374. Sebolt-Leopold, J. S. *Oncogene* **2000**, *19*, 6594–6599.
375. Sebolt-Leopold, J. S.; Van Beceleare, K.; Dudley, D.; Herrera, R.; Gowan, R.; Tecle, H.; Barrett, S.; Bridges, A.; Przybranowski, S.; Leopold, W. R. et al. R. 90th Annual Meeting of the American Association for Cancer Research, Philadelphia, PA, 1999; Abst 785.
376. Hoshino, R.; Chatani, Y.; Yamori, T.; Tsuruo, T.; Oka, H.; Yoshida, O.; Shimada, Y.; Ari-I, S.; Wada, H.; Fujimoto, J. et al. *Oncogene* **1999**, *18*, 813–822.
377. Rinehart, J.; Adjei, A. A.; LoRusso, P. M.; Waterhouse, D.; Hecht, J. R.; Natale, R. B.; Hamid, O.; Varterasian, M.; Asbury, P.; Kaldjian, E. P. et al. *J. Clin. Oncol.* **2004**, *22*, 4456–4462.
378. Dudley, D. T.; Pang, L.; Decker, S. J.; Bridges, A. J.; Saltiel, A. R. *Proc. Natl. Acad. Sci. USA* **1995**, *92*, 7686–7689.
379. Alessi, D. R.; Cuenda, A.; Cohen, P.; Dudley, D. T.; Saltiel, A. R. *J. Biol. Chem.* **1995**, *270*, 27489–27494.
380. Tasdemir, D.; Mallon, R.; Greenstein, M.; Feldberg, L. R.; Kim, S. C.; Collins, K.; Wojciechowicz, D.; Mangalindan, G. C.; Concepcion, G. P.; Harper, M. K. et al. *J. Med. Chem.* **2002**, *45*, 529–532.
381. Williams, D. H.; Wilkinson, S. E.; Purton, T.; Lamont, A.; Flotow, H.; Murray, E. J. *Biochemistry* **1998**, *37*, 9579–9585.
382. Zhao, A.; Lee, S. H.; Mojena, M.; Jenkins, R. G.; Patrick, D. R.; Huber, H. E.; Goetz, M. A.; Hensens, O. D.; Zink, D. L.; Vilella, D. et al. *J. Antibiot. (Tokyo)* **1999**, *52*, 1086–1094.
383. Favata, M. F.; Horiuchi, K. Y.; Manos, E. J.; Daulerio, A. J.; Stradley, D. A.; Feeser, W. S.; Van Dyk, D. E.; Pitts, W. J.; Earl, R. A.; Hobbs, F. et al. *J. Biol. Chem.* **1998**, *273*, 18623–18632.
384. Duncia, J. V.; Santella, J. B.; Higley, C. A.; Pitts, W. J.; Wityak, J.; Frietze, W. E.; Rankin, F. W.; Sun, J.-H.; Earl, R. A.; Tabaka, C. et al. *Bioorg. Med. Chem. Lett.* **1998**, *8*, 2839–2844.
385. Berger, D.; Dutia, M.; Powell, D. W.; Wu, B.; Wissner, A.; Boschelli, D. H.; Floyd, M. B.; Zhang, N.; Torres, N.; Levin, J. et al. *Bioorg. Med. Chem. Lett.* **2003**, *13*, 3031–3034.
386. Sebolt-Leopold, J. S.; Dudley, D. T.; Herrera, R.; Van Beceleare, K.; Wiland, A.; Gowan, R. C.; Tecle, H.; Barrett, S. D.; Bridges, A.; Przybranowski, S. et al. *Nat. Med.* **1999**, *5*, 810–816.
387. Allen, L. F.; Sebolt-Leopold, J. S.; Meyer, M. B. *Semin. Oncol.* **2003**, *30*, 105–116.
388. Wallace, E. M.; Lyssikatos, J. P.; Yeh, T.; Winkler, J. D.; Koch, K. *Curr. Topics Med. Chem.* **2005**, *5*, 215–229.
389. Ohren, J. F.; Chen, H.; Pavlovsky, A.; Whitehead, C.; Zhang, E.; Kuffa, P.; Yan, C.; McConnell, P.; Spessard, C.; Banotai, C. et al. *Nat. Struct. Mol. Biol.* **2004**, *11*, 1192–1197.
390. Senderowicz, A. M. *Curr. Opin. Cell Biol.* **2004**, *16*, 670–678.

391. Nigg, E. A. *Bioessays* **1995**, *17*, 471–480.
392. Johnson, D. G.; Walker, C. L. *Annu. Rev. Pharmacol. Toxicol.* **1999**, *39*, 295–312.
393. Sherr, C. J. *Science* **1996**, *274*, 1672–1677.
394. Senderowicz, A. M. *Oncogene* **2003**, *22*, 6609–6620.
395. Vermeulen, K.; Van Bockstaele, D. R.; Berneman, Z. N. *Cell Prolif.* **2003**, *36*, 131–149.
396. Sausville, E. A.; Arbuck, S. G.; Messmann, R.; Headlee, D.; Bauer, K. S.; Lush, R. M.; Murgo, A.; Figg, W. D.; Lahusen, T.; Jaken, S. et al. *J. Clin. Oncol.* **2001**, *19*, 2319–2333.
397. Shapiro, G. I. *Clin. Cancer Res.* **2004**, *10*, 4270–4275.
398. Jeffrey, P. D.; Russo, A. A.; Polyak, K.; Gibbs, E.; Hurwitz, J.; Massagne, J.; Pavletich, N. P. *Nature* **1995**, *376*, 313–320.
399. Rosenblatt, J.; De Bondt, H.; Jancarik, J.; Morgan, D. O.; Kim, S. H. *J. Mol. Biol.* **1993**, *230*, 1317–1319.
400. Gray, N.; Detivaud, L.; Doerig, C.; Meijer, L. *Curr. Med. Chem.* **1999**, *6*, 859–875.
401. Rosania, G. R.; Merlie, J.; Gray, N.; Chang, Y. T.; Schultz, P. G.; Heald, R. *Proc. Natl. Acad. Sci. USA* **1999**, *96*, 4797–4802.
402. McClue, S. J.; Blake, D.; Clarke, R.; Cowan, A.; Cumming, L.; Fischer, P. M.; Mackenzie, M.; Melville, J.; Stewart, K.; Wang, S. et al. *Int. J. Cancer* **2002**, *102*, 463–468.
403. Pavarello, P.; Brasca, M. G.; Amici, R.; Orsini, P.; Traquandi, G.; Corti, L.; Piutti, C.; Sansonna, D.; Villa, M.; Pierce, B. S. et al. *J. Med. Chem.* **2004**, *47*, 3367–3380.
404. Misra, R. N.; Xiao, H.; Williams, D. K.; Kim, K. S.; Lu, S.; Keller, K. A.; Mulheron, J. G.; Batorsky, R.; Tokarski, J. S.; Sack, J. S. et al. *Bioorg. Med. Chem. Lett.* **2004**, *14*, 2973–2977.
405. Kim, K. S.; Kimball, S. D.; Misra, R. N.; Rawlins, D. B.; Hunt, J. T.; Xiao, H. Y.; Lu, S.; Qian, L.; Han, W. C.; Shan, W. et al. *J. Med. Chem.* **2002**, *45*, 3905–3927.
406. Lane, M. E.; Yu, B.; Rice, A.; Lipson, K. E.; Liang, C.; Sun, L.; Tang, C.; McMahon, G.; Pestell, R. G.; Wadler, S. *Cancer Res.* **2001**, *61*, 6170–6177.
407. Schultz, C.; Link, A.; Leost, M.; Zaharevitz, D. W.; Gussion, R.; Sausville, E. A.; Meijer, L.; Kunick, C. *J. Med. Chem.* **1999**, *42*, 2909–2919.
408. Markwalder, J. A.; Arnone, M. R.; Benfield, P. A.; Boisclair, M.; Burton, C. R.; Chang, C.-H.; Cox, S. S.; Czerniak, P. M.; Dean, C. L.; Doleniak, D. et al. *J. Med. Chem.* **2004**, *47*, 5894–5911.
409. Nugiel, D. A.; Vidwans, A.; Dzierba, C. D. *Bioorg. Med. Chem. Lett.* **2004**, *14*, 5489–5491.
410. Beattie, J. F.; Breault, G. A.; Ellston, R. P.; Green, S.; Jewsbury, P. J.; Midgley, C. J.; Naven, R. T.; Pauptit, R. A.; Tucker, J. A.; Pease, J. E. *Bioorg. Med. Chem. Lett.* **2003**, *13*, 2955–2960.
411. Zhu, G.; Conner, S.; Zhou, X.; Shih, C.; Li, T.; Anderson, B. D.; Brooks, H. B.; Campbell, R. M.; Considine, E.; Dempsey, J. A. et al. *J. Med. Chem.* **2003**, *46*, 2027–2030.
412. Zhu, G.; Conner, S.; Zhou, X.; Shih, C.; Brooks, H. B.; Considine, E.; Dempsey, J. A.; Ogg, C.; Patel, B.; Schultz, R. M. et al. *Bioorg. Med. Chem. Lett.* **2003**, *13*, 1231–1255.
413. Fry, D. W.; Harvey, P. J.; Keller, P. R.; Elliott, W. L.; Meade, M.; Trachet, E.; Albassam, M.; Zheng, X.; Leopold, W. R.; Pryer, N. K. et al. *Mol. Cancer Ther.* **2004**, *3*, 1427–1437.
414. Laurence, V.; Faivre, S.; Vera, K.; Pierga, J.; Delbaldo, C.; Bekradda, M.; Armand, J.; Gianella-Borradori, A.; Diera, S. V.; Raymond, E. *Eur. J. Cancer,* **2002**, *38*, S50, poster 150.
415. Keen, N. J.; Taylor, S. S. *Nat. Rev. Cancer* **2004**, *4*, 927–936.
416. Carmena, M.; Earnshaw, W. C. *Nat. Rev. Mol. Cell Biol.* **2003**, *4*, 842–854.
417. Bischoff, J. R.; Anderson, L.; Zhu, Y.; Mossie, K.; Ng, L.; Souzza, B.; Schryver, B.; Flanagan, P.; Clairvoyant, F.; Ginther, C. et al. *EMBO J.* **1998**, *17*, 3052–3065.
418. Nigg, E. A. *Nat. Rev. Mol. Cell Biol.* **2001**, *2*, 21–32.
419. Marumoto, T.; Zhang, D.; Saya, H. *Nat. Rev. Cancer* **2005**, *5*, 42–50.
420. Hirota, T.; Kunitoku, N.; Sasayama, T.; Marumoto, T.; Zhang, D.; Nitta, M.; Hatakeyama, K.; Saya, H. *Cell* **2003**, *114*, 585–598.
421. Anand, S.; PenrhynLowe, S.; Venkitaraman, A. R. *Cancer Cell.* **2003**, *3*, 51–62.
422. Murata-Hori, M.; Tatsuka, M.; Wang, Y.-L. *Mol. Biol. Cell* **2002**, *13*, 1099–1108.
423. Ditchfield, C.; Johnson, V. L.; Tighe, A.; Ellston, R.; Haworth, C.; Johnson, T.; Mortlock, A.; Keen, N.; Taylor, S. S. *J. Cell. Biol.* **2003**, *161*, 267–280.
424. Cheetham, G. M. T.; Knegtel, R. M. A.; Coll, J. T.; Renwick, S. B.; Swenson, L.; Weber, P.; Lippke, J. A.; Austen, D. A. *J. Biol. Chem.* **2002**, *277*, 42419–42422.
425. Nowakowski, J.; Cronin, C. N.; McRee, D. E.; Knuth, M. W.; Nelson, C. G.; Pavletich, N. P.; Rogers, J.; Sang, B.-C.; Scheibe, D. N.; Swanson, R. V. et al. *Structure* **2002**, *10*, 1659–1667.
426. Bayliss, R.; Sardon, T.; Vernos, I.; Conti, E. *Mol. Cell.* **2003**, *12*, 851–862.
427. Mortlock, A. A.; Keen, N. J.; Jung, F. H.; Heron, N. M.; Foote, N. M.; Wilkinson, R.; Green, S. *Curr. Top. Med. Chem.* **2005**, *5*, 199–213.
428. Harrington, E. A.; Bebbington, D.; Moore, J.; Rasmussen, R. K.; Ajose-Adeogun, A. O.; Nakayama, T.; Graham, J. A.; Demur, C.; Hercend, T.; Diu-Hercend, A. et al. *Nat. Med.* **2004**, *10*, 262–267.
429. Hauf, S.; Cole, R. W.; LaTerra, S.; Zimmer, C.; Schnapp, G.; Walter, R.; Heckel, A.; van Meel, J.; Rieder, C. L.; Peters, J.-M. *J. Cell. Biol.* **2003**, *161*, 281–294.
430. Strumberg, D.; Seeber, S. *Onkologie* **2005**, *28*, 101–107.
431. Sharma, A.; Trivedi, N. R.; Zimmerman, M. A.; Tuveson, D. A.; Smith, C. D.; Robertson, G. P. *Cancer Res.* **2005**, *65*, 2412–2421.
432. Bollag, G.; Freeman, S.; Lyons, J. F.; Post, L. E. *Curr. Opin. Investig. Drugs* **2003**, *4*, 1436–1441.
433. Wilhelm, S.; Chien, D.-S. *Curr. Pharm. Des.* **2002**, *8*, 2255–2257.
434. Lyons, J. F.; Wilhelm, S.; Hibner, B.; Bollag, G. *Endocr. Relat. Cancer* **2001**, *8*, 219–225.
435. Strumberg, D.; Richly, H.; Hilger, R. A.; Schleucher, N.; Korfee, S.; Tewes, M.; Faghih, M.; Brendel, E.; Voliotis, D.; Haase, C. G. et al. *J. Clin. Oncol.* **2005**, *23*, 965–972.
436. Lee, J. T.; McCubrey, J. A. *Curr. Opin. Investig. Drugs* **2003**, *4*, 757–763.
437. Khire, U. R.; Bankston, D.; Barbosa, J.; Brittelli, D. R.; Caringal, Y.; Carlson, R.; Dumas, J.; Heald, S. L.; Hibner, B.; Johnson, J. S. et al. *Bioorg. Med. Chem. Lett.* **2004**, *14*, 783–786.
438. Gills, J.; Dennis, P. A. *Expert Opin. Investig. Drugs* **2004**, *13*, 787–797.
439. Vivanco, I.; Sawyers, C. L. *Nat. Rev. Cancer* **2002**, *2*, 489–501.
440. Hsu, J. H.; Shi, Y.; Hu, L.; Fisher, M.; Franke, T. F.; Lichtenstein, A. *Oncogene* **2002**, *21*, 1391–1400.

441. Stein, R. C. *Endocr. Relat. Cancer* **2001**, *8*, 237–248.
442. Lindsley, C. W.; Zhao, Z.; Leister, W. H.; Robinson, R. G.; Barnett, S. F.; DeFeo-Jones, D.; Hartman, G. D.; Huff, J. R.; Huber, H. E.; Duggan, M. E. *Bioorg. Med. Chem. Lett.* **2005**, *15*, 761–764.
443. Breitenlechner, C. B.; Friebe, W.-G.; Brunet, E.; Werner, G.; Graul, K.; Thomas, U.; Kunkele, K.-P.; Schafer, W.; Gassel, M.; Bossemeyer, D. et al. *J. Med. Chem.* **2005**, *48*, 163–170.
444. Yang, J.; Cron, P.; Good, V. M.; Thompson, V.; Hemmings, B. A.; Barford, D. *Nat. Struct. Biol.* **2002**, *9*, 940–944.
445. Feldman, R. I.; Wu, J. M.; Polokoff, M. A.; Kochanny, M. J.; Dinter, H.; Zhu, D.; Biroc, S. L.; Alicke, B.; Bryant, J.; Yuan, S. et al. *J. Biol. Chem.* **2005**, *280*, 19867–19874.
446. Goekjian, P. G.; Jirousek, M. R. *Expert Opin. Investig. Drugs* **2001**, *10*, 2117–2140.
447. Carter, C. A. *Curr. Drug Targets* **2000**, *1*, 163–183.
448. Carter, C. A.; Kane, C. J. M. *Curr. Med. Chem.* **2004**, *11*, 2883–2902.
449. Akinaga, S.; Sugiyama, K.; Akiyama, T. *Anti-Cancer Drug Des.* **2000**, *15*, 43–52.
450. Fabbro, D; Reutz, S.; Bodis, S.; Pruschy, M.; Csermak, K.; Man, A.; Campochiaro, P.; Wood, J.; O'Reilly, T.; Meyer, T. *Anti-Cancer Drug Des.* **2000**, *15*, 17–28.
451. Davis, P. D.; Hill, C. H.; Keech, E.; Lawton, G.; Nixon, J. S.; Sedgwick, A. D.; Wadsworth, J.; Westmacott, D.; Wilkinson, S. E. *FEBS Lett.* **1989**, *259*, 61–63.
452. Teicher, B. A.; Menon, K.; Alvarez, E.; Shih, P. L. C.; Faul, M. M. *In Vivo* **2001**, *15*, 185–193.
453. Davies, S. P.; Reddy, H.; Caivano, M.; Cohen, P. *Biochem. J.* **2000**, *351*, 95–105.
454. Meyer, T.; Regenass, U.; Fabbro, D.; Alteri, E.; Rosel, J.; Muller, M.; Caravatti, G.; Matter, A. *Int. J. Cancer* **1989**, *43*, 851–856.
455. Virchis, A.; Ganeshaguru, K.; Hart, S.; Jones, D.; Fletcher, L.; Wright, F.; Wickremasinghe, R.; Man, A.; Csermak, K.; Meyer, T. et al. *Haematol. J.* **2002**, *3*, 131–136.
456. Smith, J. B.; Smith, L.; Pettit, G. R. *Biochem. Biophys. Res. Commun.* **1985**, *132*, 939–945.
457. Varterasian, M. L.; Mohammad, R. M.; Shurafa, M. S.; Hulburd, K.; Pemberton, P. A.; Rodriguez, D. H.; Spadoni, V.; Eilender, D. S.; Murgo, A.; Wall, N. et al. *Clin. Cancer Res.* **2000**, *6*, 825–828.
458. Herrmann, D. B.; Bicker, U. *Drugs Future* **1988**, *13*, 543–554.
459. Van Ummersen, L.; Binger, K.; Volkman, J.; Marnocha, R.; Tutsch, K.; Kolesar, J.; Arzoomanian, R.; Alberti, D.; Wilding, G. *Clin. Cancer Res.* **2004**, *10*, 7450–7456.

Biographies

Andrew A Mortlock was educated at the University of Oxford, UK, completing his BA in chemistry in 1988 and his PhD in 1991, under the supervision of Prof Stephen Davies. Following postdoctoral studies at the University of California at Berkeley, CA, in the group of Prof Clayton Heathcock, he joined ICI Pharmaceuticals (later Zeneca Pharmaceuticals and now AstraZeneca) at Alderley Park, UK, in 1992.

Between 1992 and 1995, he worked on the endothelin antagonist program, which led to the selection of three ETa-selective inhibitors for clinical evaluation. In 1995 he transferred to the cancer research group, where he has been involved in anticancer projects in the areas of antiinvasives, cell cycle inhibitors, antiangiogenics, and signal transduction inhibitors. These projects have involved targets within the kinase, protease, GPCR, nuclear hormone receptor, and protein–protein interaction areas. In recent years, he led the medicinal chemistry group that developed AstraZeneca's novel Aurora kinase inhibitors.

In 2003, he was appointed a director of medicinal chemistry, and currently leads a group of 50 chemists involved across a broad range of anticancer projects while having particular responsibility for the portfolio of lead generation projects. In the course of his career, he has been the author of some 40 scientific publications and patents.

Andy J Barker obtained his first degree and PhD from Nottingham University, UK, under the supervision of Prof Gerry Pattenden, with whom he also completed a 2-year postdoctoral study. After working for Beecham Pharmaceuticals and Hoechst, he joined ICI (later Zeneca Pharmaceuticals, and now AstraZeneca) at Alderley Park, UK, in 1987.

He is currently a director of medicinal chemistry at AstraZeneca, leading a group of over 50 chemists working on a portfolio of lead optimisation projects within oncology. His publication record includes over 70 papers, patents, and oral presentations at national and international meetings.

Barker has worked in a range of therapeutic areas including obesity, bacterial infection, antimetabolites, inflammation, analgesia, and oncology, and been involved in delivering at least nine drug candidates into development. His work in oncology has covered antimetabolites, cell cycle, and signal transduction, and has had a particular focus on kinases as targets; he was the lead chemist in the program that delivered gefitinib (Iressa).

Comprehensive Medicinal Chemistry II
ISBN (set): 0-08-044513-6

ISBN (Volume 7) 0-08-044520-9; pp. 183–220

7.09 Recent Development in Novel Anticancer Therapies

H Weinmann and E Ottow, Schering AG, Berlin, Germany

7.09.1 Disease State

Cancer forms a group of different diseases which are characterized by uncontrolled growth of a highly heterogeneous malignant cell population. The estimated worldwide incidence of different types of cancer is around 10 million, roughly half of which is in the developed countries.[1] Although impressive progress in diagnosis, surgery, and therapy has been achieved over the last decades, overall cancer mortality is still very high. In the USA, the cancer deaths (in thousands) in 2000 compared to 1985 dropped from 160 to 120 for lung, 110 to 60 for colon–rectum, 41 to 38 for breast, 32 to 25 for prostate, 24 to 19 for urinary tract, and 14 to 12 for ovary.[1,2] The 5-year survival rate for the average of the more common cancers is about 60%, but it is still as low as 14% for lung cancer. Overall the 5-year survival rate is increasing; however, this more a reflection of the great progress in early diagnosis than a sign of the real possibilities of therapeutic control.[3]

In spite of the large number of currently available chemotherapeutic drugs, the medical need is still largely unmet. The main reasons are: the toxicity of conventional drugs due to lack of selectivity; the spreading of metastases, resulting in early tumor implantation in organs other than original site; the heterogeneity of the disease, which comprises about 100 types of cancer; resistance to chemotherapy developed after few therapeutic cycles.

Therefore, great efforts have been initiated for the identification of novel anticancer targets and the discovery of anticancer drugs. Furthermore, the global effort of sequencing the human genome has delivered an enormous number of potential targets associated with cancer therapy. The main focus of targets for anticancer drugs is now on targets and signaling pathways involved in the growth and metastasis of cancer cells as well as tumor angiogenesis.[4] A number of innovative strategies that address novel targets to overcome malignant abnormalities of tumor cells with higher selectivity are in development, and are beginning to give important results.

7.09.2 Disease Basis

The basic facts of cancer with special focus of the underlying biochemical, genetical, and environmental origins as well as epidemiological status have been described in the preceding chapters of this volume.

7.09.3 Experimental Disease Models

The standard models for testing cancer drugs include cultured human tumor cell lines and rodent xenografts that comprise many of the same human cell lines grown subcutaneously in immunocompromised animals.[5] The artificial nature of tumor cell lines which are typically passaged for many generations in culture and may not be representative of the tumor in its native state is one problem related to these models. Cells in culture lack the architectural and cellular complexity of real tumors, which contain inflammatory, vasculature, and other stromal components. Additionally, the xenograft–host interactions under the skin might be different from those in the tissue of tumor origin. Whereas normal human tumors develop over a number of years, mouse xenografts are chosen so that they can be assayed in the timeframe of days or weeks. Even if a xenograft represents significant aspects of the tumor from which it was derived, it probably captures only a fraction of the total genetic and epigenetic heterogeneity of a given tumor subtype. Scientists in oncological drug development have historically advanced compounds that have a major effect on the growing xenograft.[6] Efficacy in these models must be associated with minimal toxicity (e.g., a maximum of 10% loss in animal body weight during the course of 2-week dosing). Pharmacodynamic measurements are increasingly used in culture and in animals and could improve the current ability to predict efficacy and toxicity.[7] The described complexities and limitations of the currently used models explain why it is still difficult to predict the outcome of clinical trials from preclinical experimental data and why despite the rational approaches followed in cancer research many investigational anticancer drugs eventually fail in the clinics.

7.09.4 Clinical Trial Issues

The oncology market has grown remarkably over the last 5 years. The total oncology market is expected to reach more than $32 billion by 2005, driven by factors including a continued trend toward aggressive chemotherapy, patient populations that demand access to the latest therapies, and continued high prices due to strong unmet medical need.[8]

In spite of all this positive momentum, pharmaceutical and biotech companies that do research in this area face the challenges of late-stage clinical attrition. Even with more than 500 oncology compounds in development, only a few achieve regulatory approval each year and there are only approximately 90 approved oncology drugs in the USA today. Gefitinib (Iressa) and Cetuximab (Erbitux) are only two examples of recent setbacks which highlight the challenges that oncology drug candidates are facing. Although oncology projects usually have higher average success rates than other therapeutic areas in early-stage trials (clinical phase I and II), these projects tend to have a lower success rate than other therapeutic areas at phase III. Drug development projects in oncology have a higher overall risk compared to those in other therapeutic areas, precisely at the stage of clinical development in which costs are highest.

The high rate of attrition of oncology compounds is a result of several key factors: market pressures, scientific challenges, and changes in the regulatory environment. On the market side, companies are under significant pressure to reach for the broadest possible label on first regulatory approval to capture a broad market. Therefore, many initial label applications target one or several of the four major tumor types: breast, prostate, lung, and colon cancer. The challenge in doing this is that the standard clinical practice for these tumors often involves multiple drug cocktails of cytotoxic chemotherapy agents. Therefore clinical trials need to show incremental improvement over these existing chemotherapy regimens, which can be difficult to demonstrate. The basic science of oncology can also give rise to high attrition rates for anticancer compounds. With the rapid accumulation of new knowledge, there are more and more novel approaches to anticancer drug development. More than 40% of the compounds that are currently in development for cancer are directed against novel mechanisms and almost 70% of the drug targets that are being investigated in discovery are unprecedented. Therefore a thorough target validation at the beginning of a new drug discovery project is an absolutely crucial prerequisite for decreasing the risk of late stage failure.

7.09.5 Current Treatment

Cancer is a collection of many diseases, and currently anticancer treatments are often applied as if each tissue was affected by only one type of cancer. A problem is to identify the subclasses of diseases that make up cancer or even one type of cancer. A number of different approaches to classifying tumors (e.g., by using microarrays) are being tested. In the following sections of this chapter, several innovative concepts for possible treatments with new compounds that are currently in development are discussed.

7.09.5.1 Histone Deacetylase Inhibitors

7.09.5.1.1 Basic facts

Histone deacetylase (HDAC) inhibitors have become the subject of considerable attention over the last decade due to a strong increase in understanding of the role that acetylation of histone and nonhistone proteins plays in the regulation of normal and cancer cell behavior. Histone deacetylases (HDACs) are a family of enzymes that compete with histone acetyltransferases (HATs) to modulate chromatin structure and to regulate gene transcriptional activity by changing the acetylation status of lysines of nucleosomal histones. Several forms of cancer are characterized by an altered expression or mutation of genes that encode HATs or HDACs. Furthermore, aberrant repression of genes mediated by HDACs is associated with the pathogenesis of various types of solid tumors and hematological malignancies. HDAC inhibitors induce histone hyperacetylation, reactivate suppressed genes, and consequently inhibit the cell cycle, activate differentiation programs, or induce apoptosis. Several HDAC inhibitors of various structural families have now advanced into phase I and II clinical trials. Recently several reviews describing HDAC as a new target for cancer chemotherapy[9–12] and the current status of HDAC inhibitors in clinical development have appeared.[13–16]

Mammalian HDACs form three distinct subclasses.[17] Class I deacetylases (HDACs 1, 2, 3, and 8) share homology in their catalytic sites with molecular weights of 42–55 kDa. Class II deacetylases includes HDACs 4, 5, 6, 7, 9, and 10. They have molecular weights between 120 and 130 kDa. HDACs 4, 5, 7, and 9 share homology in two regions, the C-terminal catalytic domain and an N-terminal regulatory domain.[18] Recently, HDAC11 was cloned and characterized.[19] It contains conserved residues in the catalytic core regions shared by both Class I and Class II mammalian enzymes. HDAC6 and HDAC10 have two regions of homology with the Class II catalytic site. The third class of deacetylases is the conserved Sir2 family of proteins which are dependent on NAD^+ for activity, whereas Class I and II HDACs operate by zinc-dependent mechanisms.

A structural rationale for HDAC inhibition is suggested from the x-ray crystal structure of trichostatin A-bound HDAC-like protein (HDLP), a homolog of Class I/II HDAC with about 35% sequence identity.[20] In this structure, the hydroxamic acid of trichostatin A (TSA), **1**, penetrates a narrow, hydrophobic channel and chelates a buried zinc ion. Despite the variety of their structural characteristics, most HDAC inhibitors can be rationalized in the light of the HDLP structure. Inhibitors typically possess a metal-binding moiety and a surface recognition part which interacts with amino acids at the entrance of the *N*-acetyl lysine binding channel. These two parts of the molecule are connected by a linker which is often a 5–6 hydrocarbon chain. Besides binding to the zinc ion, the hydroxamic acid functionality interacts via hydrogen bonding with two imidazole groups from histidines and another tyrosine in the active site which provides further rationale for the potent inhibitory activity of these agents.

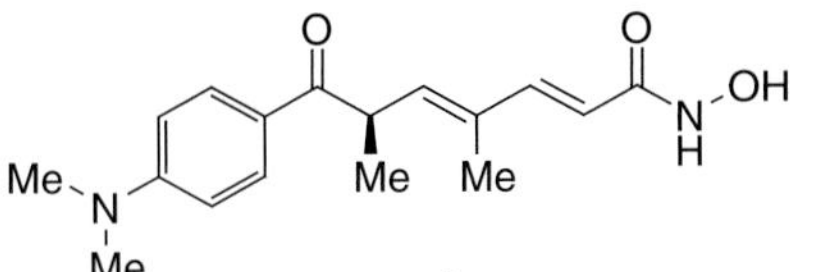

1

Surface recognition | Linker | Metal binding

Recently the first crystal structure of a human HDAC isoenzyme was published.[21] The structures of human HDAC8 complexed with four structurally diverse hydroxamate inhibitors was determined. HDAC8 inhibition leads to the hyperacetylation of histones H3 and H4. Recent evidence suggests that HDAC8 may play a role in one of the most frequent types of acute myeloid leukemia (AML). This crystal structure sheds light on the catalytic mechanism of the HDACs, and on differences in substrate specificity across the HDAC family. A comparison of the structures of the four HDAC8 inhibitor complexes demonstrated considerable structural differences in the protein surface in the vicinity of the opening to the active site. These differences suggest that this region is highly flexible and able to structure changes to accommodate binding to a variety of different ligands. From a physiological point of view, this flexibility suggests that HDAC8 might be able to bind to acetylated lysines that are presented in a variety of structural contexts. Another independent crystal structure elucidation study of HDAC8 came to similar results.[22] In addition, this group demonstrated that knockdown of HDAC8 by RNA interference inhibits growth of human lung, colon, and cervical cancer cell lines, highlighting the importance of this HDAC subtype for tumor cell proliferation and therefore as a target for possible antitumor agents.

7.09.5.1.2 Histone deacetylase inhibitors in development

HDAC inhibitors contain great potential as new drugs because of their ability to modulate transcription, and are endowed with cytodifferentiating, antiproliferative, and apoptogenic properties.[23] Furthermore the anticancer activity of HDAC inhibitors may be mediated in part by the inhibition of angiogenesis, since it was shown recently that TSA, **1**, specifically inhibited hypoxia-induced angiogenesis by reducing the expression of genes required for angiogenesis.[24] Several structural classes of compounds have been described as HDAC inhibitors.[25–30] The most important of these compound classes are short-chain fatty acids, hydroxamic acids, benzamides, and cyclic tetrapeptides.

7.09.5.1.2.1 Short-chain fatty acids

This compound class with butyrate, phenyl butyrate, or valproic acid as most important examples exhibits the least potency, with IC_{50} in the millimolar range.[31–33] In a recent structural optimization study, valproate, butyrate, phenylacetate, and phenylbutyrate were coupled with zinc-chelating motifs (hydroxamic acid and *o*-phenylenediamine) through aromatic amino acid linkers.[34] This strategy led to a novel class of short-chain fatty acid derivatives that exhibited varying degrees of HDAC inhibitory potency with the best compounds in the nanomolar range.

7.09.5.1.2.2 Hydroxamic acids

Hydroxamic acids constitute the largest class of HDAC inhibitors and are still rapidly increasing in number. Solution- and solid-phase synthesis methods as well as potential therapeutic applications of hydroxamic acids[35–37] have been reviewed recently. A fully automated multistep solution-phase synthesis method using polymer supported reagents for the preparation of hydroxamic acid HDAC inhibitors has been described which might have the potential to further accelerate the output of medicinal and combinatorial chemistry groups in this field.[38] Besides TSA, **1**, other first-generation hydroxamic acid HDAC inhibitors like suberoyl anilide hydroxamic acid (SAHA), **2**, Pyroxamide, **3**, CBHA, **4**, Oxamflatin, **5**, and Scriptaid, **6**, have been playing important roles as pharmacological tool compounds with some of them undergoing clinical trials.[39,40] SAHA is one of the most advanced HDAC inhibitors and is currently undergoing clinical phase II trials.

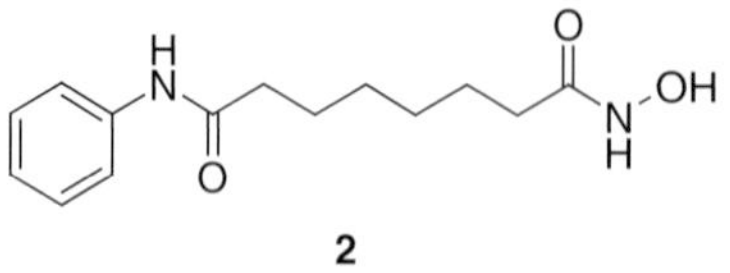

2

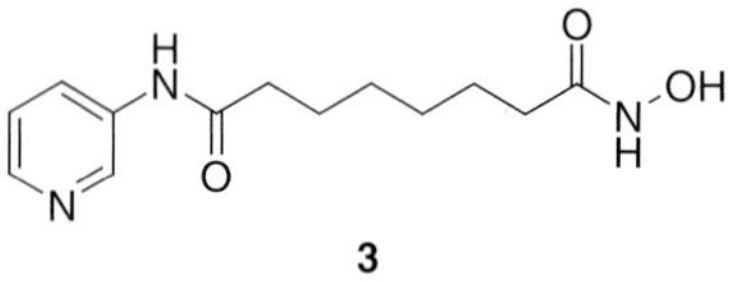

3

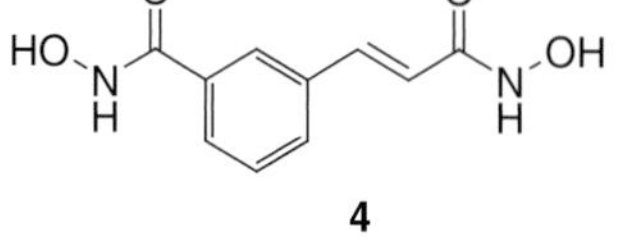

4

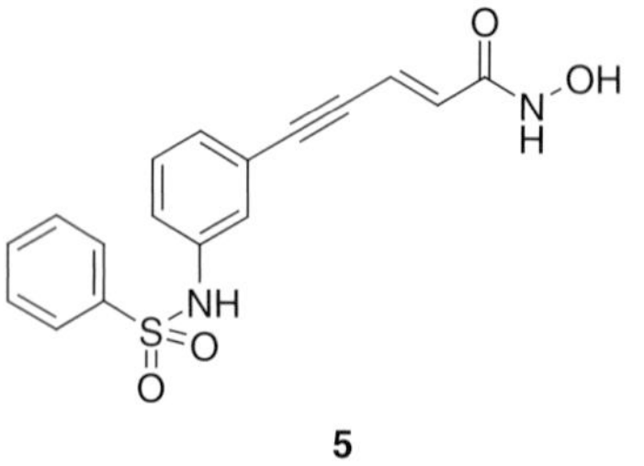

5

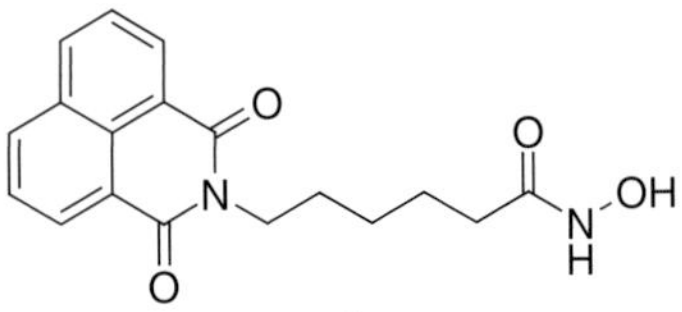

6

The cinnamyl hydroxamic acid NVP-LAK974, **7**, was found as a HTS hit. This compound had good enzyme and cellular potency, but poor efficacy in vivo. A systematic structural exploration of cinnamyl hydroxamates based on NVP-LAK974 was undertaken with the goal of finding a novel, well-tolerated and efficacious HDAC inhibitor.[41,42] NVP-LAQ824, **8**, showed activity both in vitro (HDAC $IC_{50} = 32$ nM, HCT116 $GI_{50} = 10$ nM) and in vivo, and is currently undergoing phase I clinical trials against both solid tumors and leukemia.[43]

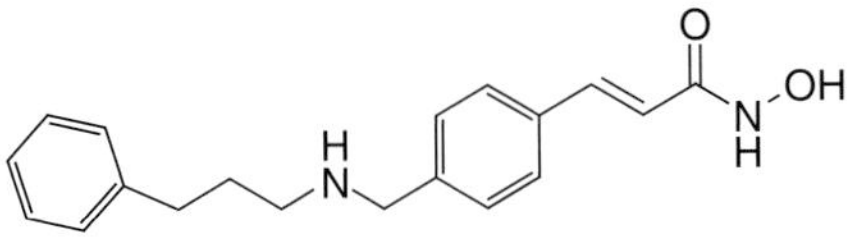

7

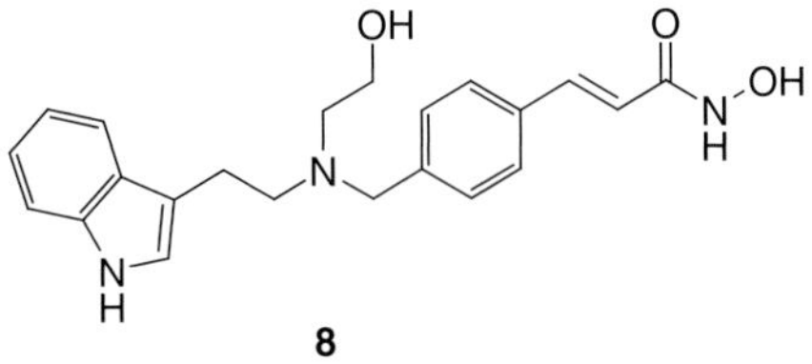

8

A new set of sulfonamide hydroxamic acids, **9**, and anilides have been synthesized.[44,45] One of the best compounds was **10** which had an IC_{50} in the HDAC1 in vitro assay of 90 nM and showed 57% tumor growth inhibition against A549 tumors. The reversal of the sulfonamide functionality resulted in similar activity.

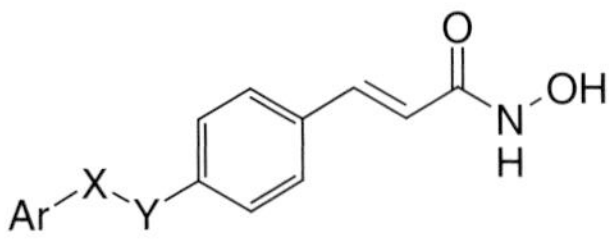

X=SO_2, Y=NH **9**
X=NH,Y=SO_2

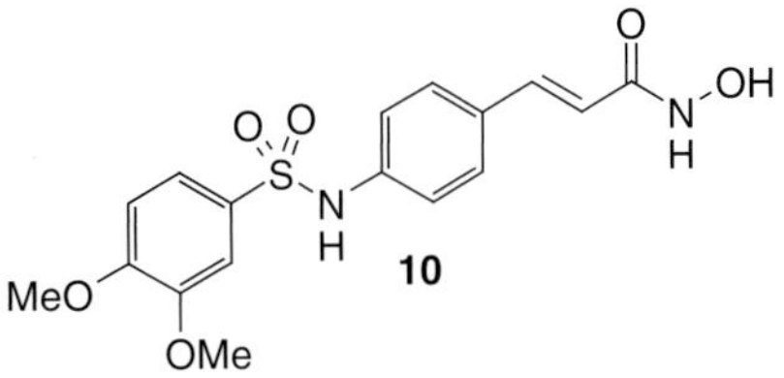

10

PXD101, **11**, is a new hydroxamate-type HDAC inhibitor which inhibits histone deacetylase activity with an IC_{50} of 27 nM.[46] Replacement of the cinnamic acid moiety with an alkyl chain leads to a reduction in activity and substitution of the sulfonamide nitrogen also reduces activity. During further optimization of this series, piperazine sulfonamide derivatives (e.g., **12**) have been prepared.[47] Most of these compounds inhibit HDAC with $IC_{50} = 20–200$ nM.

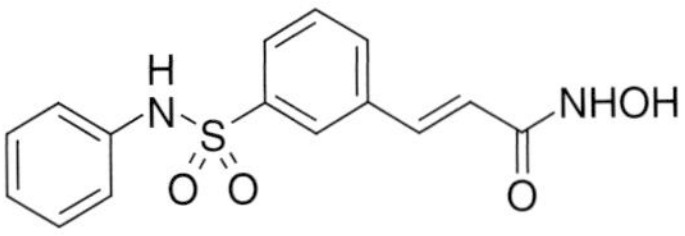

11 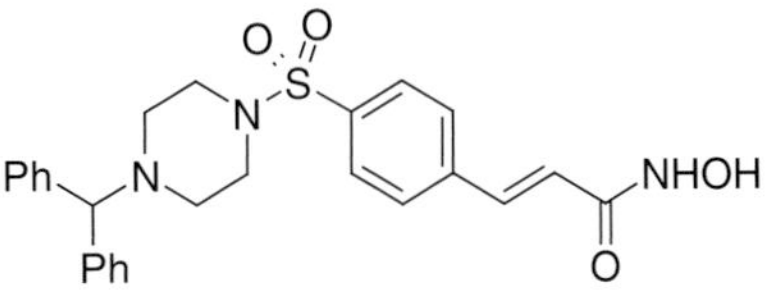 12

Novel 3-(4-substituted-phenyl)-*N*-hydroxy-2-propenamides have been prepared recently.[48,49] Incorporation of a 1,4-phenylene carboxamide linker and a 4-(dimethylamino)phenyl or 4-(1-pyrrolidinyl)phenyl group as a cap substructure generated highly potent hydroxamic acid-based HDAC inhibitors.

Aroyl-pyrrole-hydroxy-amides (APHAs) are another important class of α,β-unsaturated hydroxamic acid inhibitors in which the benzene ring was replaced by a pyrrole ring.[50] Pyrrole N-substitution with groups larger than methyl gave a reduction in HDAC inhibiting activity, and replacement of hydroxamate function with various nonhydroxamate, metal ion-complexing groups yielded poorly active or totally inactive compounds. On the contrary, proper substitution at the pyrrole carbonyl side chain favorably affected enzyme inhibiting potency, leading to **14** which was 38-fold more potent than **13** in in vitro anti-HD2 assay. Such enhancement of inhibitory activity can be explained by the higher flexibility of the carbonyl side chain substituent of **14** which accounts for a considerably better fit into the HDAC1 pocket compared to **13**. The enhanced fit allows a closer positioning of the hydroxamate moiety to the zinc ion. These findings were supported by docking studies performed on both APHAs and reference drugs TSA and SAHA.[51]

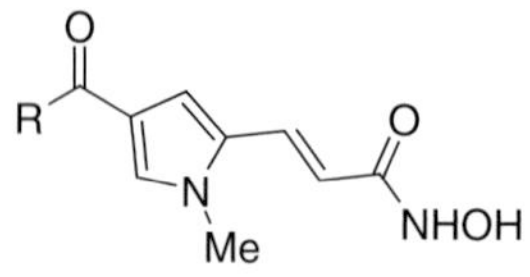

13: R = –Ph
14: R = $-CH_2-Ph$

A new class of pyrimidine-derived hydroxamic acid HDAC inhibitors was prepared recently.[52–55] JNJ-16241199, **15**, was found to be a potent ($IC_{50} = 6\,nM$) and orally bioavailable HDAC inhibitor. Orally administered JNJ-16241199 was associated with strong inhibition of the growth of ovarian A2780, human non-small cell lung carcinoma NCI-H460, and human colon carcinoma HCT 116 xenografts in immunodeficient mice. Urea derivatives of these pyrimidine hydroxamic acids (e.g., **16**) showed potency in the HDAC inhibitor assay as well as compounds in which the sulfonamide group in **15** was replaced by an amide or an alkyl or aryl moiety.[56–58]

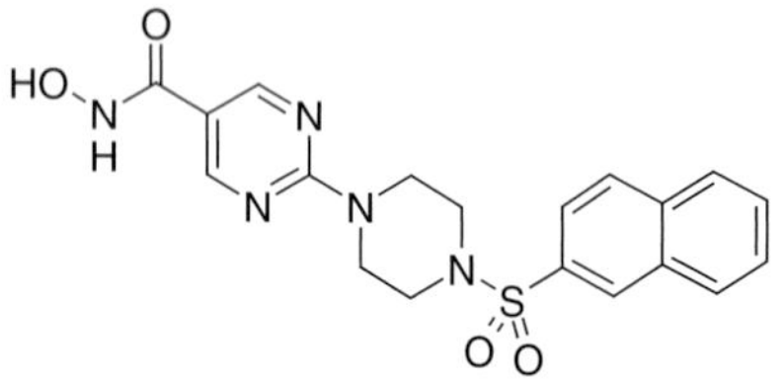

15 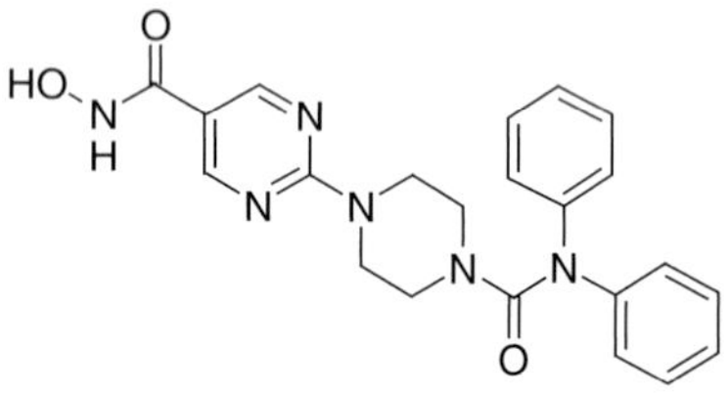 **16**

Novel aromatic dicarboxylic acid derivatives have been prepared and tested as HDAC inhibitors.[59] The in vitro cytotoxic activity of these compounds against HT29 human colon carcinoma cells was determined (**17**, $IC_{50} = 0.03\,\mu M$). The synthesis of a series of heterocyclic-amide hydroxamic acids demonstrated the highly potent HDAC inhibitory activity of indole-amides.[60] Derivatives of 2-indole amides were found to be the most active inhibitors among the different regioisomers. Introduction of substituents on the indole ring further improved the potency and generated a series of low nanomolar HDAC inhibitors (e.g., **18**, $IC_{50} = 3.1\,nM$) with strong antiproliferative activity (HT1080 $IC_{50} = 120\,nM$).

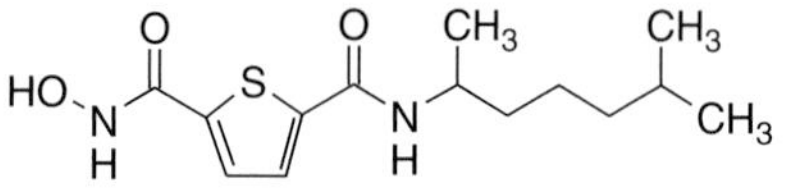

17

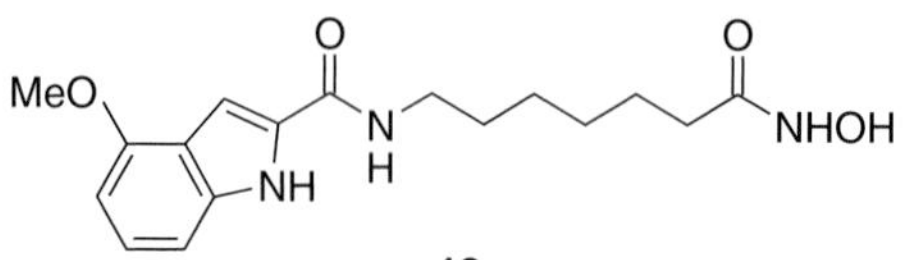

18 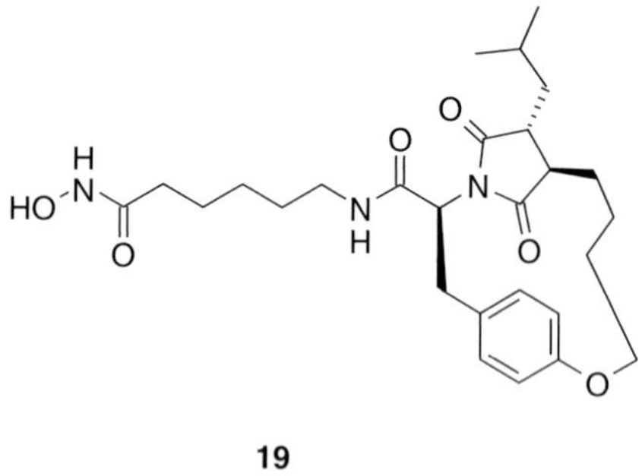 **19**

A series of succinimide hydroxamic acids which contain a macrocyclic surface recognition domain was described.[61] The best compound, **19**, gave an HDAC IC_{50} of 38 nM and in the antiproliferation assay using HT1080 fibrosarcoma cells an IC_{50} of 250 nM was measured. The number, identity, and disposition of macrocycle substituents appears to be critical for activity. The removal of the succinimide substituents or the phenylalanine side chain depresses activity. Replacement of the succinimide with a lactam or phthalimide also led to reduced activity. The length of the linker domain alkyl group was also found to be critical, with the 5-methylene analog showing maximal activity. The pharmacokinetic and in vivo tumor study data for **19** were rather disappointing. In mouse or monkey, **19** had a very short half-life of approximately 20 min and little or no measurable exposure after oral or intraperitoneal dosing, due to rapid hydrolysis to the inactive carboxylic acid.[62] In the HT1080 mouse model **19** had only marginal antitumor activity.

7.09.5.1.2.3 Nonhydroxamate histone deacetylase inhibitors

Although hydroxamic acids are frequently employed as zinc-binding groups, they often present metabolic and pharmacokinetic problems such as glucuronidation and sulfation that result in a short in vivo half-life. Many hydroxamates are unstable in vivo and are prone to hydrolysis. Concerns about the metabolic stability and the toxicity[63] associated with hydroxamic acids have triggered various research activities to find replacement groups with strong HDAC inhibitory potency.

Trifluoromethyl ketones were found to be active as HDAC inhibitors.[64] It is likely that in these and other electrophilic ketones the hydrated form of the carbonyl group probably acts in a similar way as a transition state analog and coordinates the zinc ion in the active site. Optimization of this series led to the identification of submicromolar inhibitors which showed antiproliferative effects.

In another series various heterocyclic ketones were tested for their HDAC inhibitory potency.[65] α-Keto oxazoles were found to be submicromolar inhibitors. One of the most potent compounds was **20** with HDAC $IC_{50} = 30$ nM and antiproliferative activity in MDA 435 cells ($IC_{50} = 2.3$ μmol). However, both types of electrophilic ketones exhibited short half-lives in vivo and in vitro due to rapid reduction to the corresponding inactive alcohols. Recently α-keto acids, α-keto esters, and α-keto amides were studied as hydroxamic acid substitutes.[66,67] Compound **21** is an HDAC inhibitor ($IC_{50} = 9$ nM) which shows antiproliferative activity in vitro, as well as significant efficacy in an HT1080 mouse tumor model. However, the α-keto amide **21** was also rapidly metabolized to the inactive α-hydroxy amide. Overall exposure was low with only transient concentrations of inhibitor above the cellular proliferation IC_{50} value. Despite the poor pharmacokinetics of the α-keto amides, they exhibit significant antitumor effects in xenograft models, suggesting that transient exposure to HDAC inhibitors may be sufficient for antitumor effects.

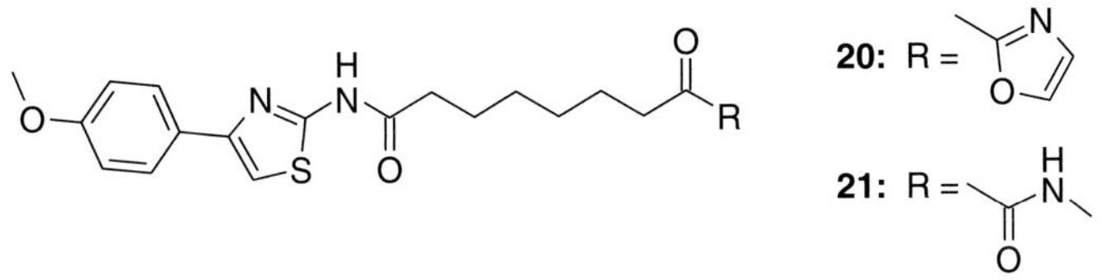

20: R = (oxazol-2-yl)

21: R = C(O)NHCH3

In another attempt to substitute the hydroxamic acid moiety three analogs of SAHA with phosphorus metal-chelating functionalities were synthesized as inhibitors of histone deacetylases.[68] The compounds showed only weak HDAC inhibitory potency in the millimolar range, suggesting that the transition state of HDAC is not analogous to zinc proteases. Also derived from SAHA HDAC inhibitors with an *N*-formyl hydroxylamine head group have been prepared.[69] Replacing the hydroxamic acid group by this moiety led to a 50-fold drop in potency. This loss in activity could be offset by increasing the size of the hydrophobic region to afford HDAC inhibitors with low micromolar activity in cellular histone hyperacetylation assay. Further studies will be necessary to determine whether these compounds offer enhanced pharmacological and pharmacokinetic properties relative to SAHA.

Novel series of mercaptoacetamide derivatives of SAHA were described independently by two research groups as alternatives to hydroxamic acids.[70,71] Mercaptoacetamide, **22**, had an HDAC IC_{50} of 390 nM and it exhibits strong competitive inhibition versus acetylated lysine substrate. The mercaptoacetamide group likely interacts with the zinc in the active site. Some compounds showed better HDAC inhibitory activity than SAHA.

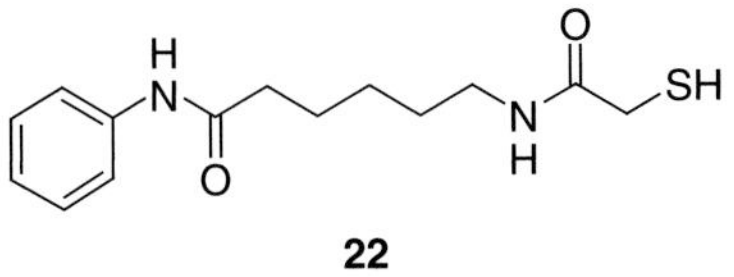

22

7.09.5.1.2.4 Benzamides

The benzamide class of HDAC inhibitors, which is in general less potent than the corresponding hydroxamic acid and tetrapeptide classes, includes MS-275,[72] **23**, and CI-994, **24**. The structure–activity relationship (SAR) study of MS-275 revealed that a 2′-amino or 2′-hydroxyl moiety is critical for inhibitory activity.

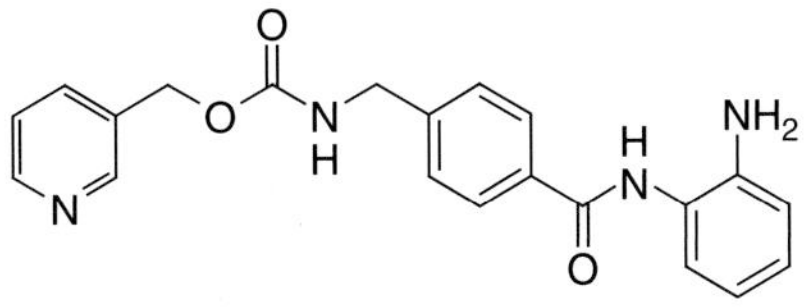

23 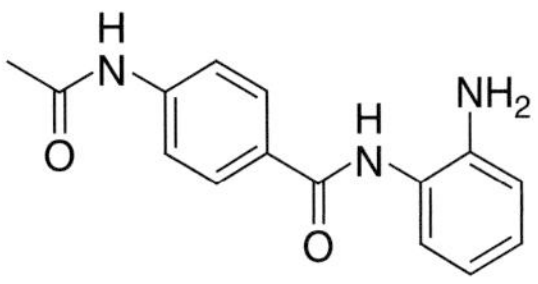 **24**

MS-275 is currently under clinical evaluation. A variety of ω-substituted alkanoic acid (2-amino-phenyl)-amides were prepared. They inhibit recombinant human HDACs with IC_{50} values in the low micromolar range.[73] Compounds in this class (e.g., **25**) showed efficacy in human tumor xenograft models. However, the activity was lower than that of MS-275. A heterocyclic series of benzamides, **26**, derived from *o*-phenylene diamine has been reported recently and found to be active in the in vitro HDAC assay in the micromolar range.[74]

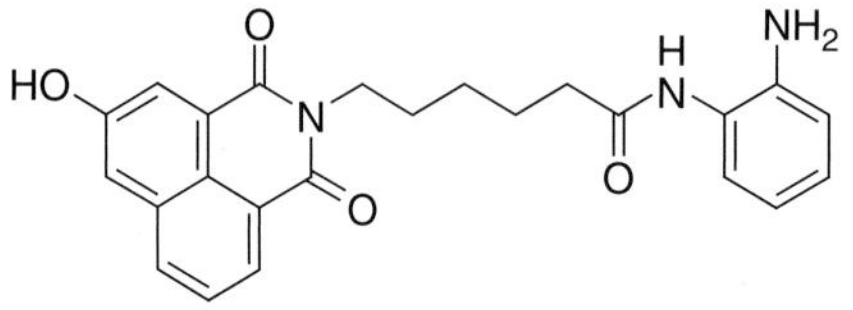

25 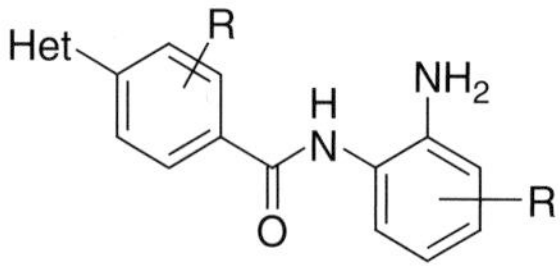 **26**

7.09.5.1.2.5 Cyclic peptides

Compounds containing cyclic peptide structures constitute the structurally most complex class of HDAC inhibitors. Well-known examples are the natural product FK-288 (Depsipeptide, **27**) which is currently undergoing clinical evaluation, CHAP31, and Trapoxin A and B and derivatives.

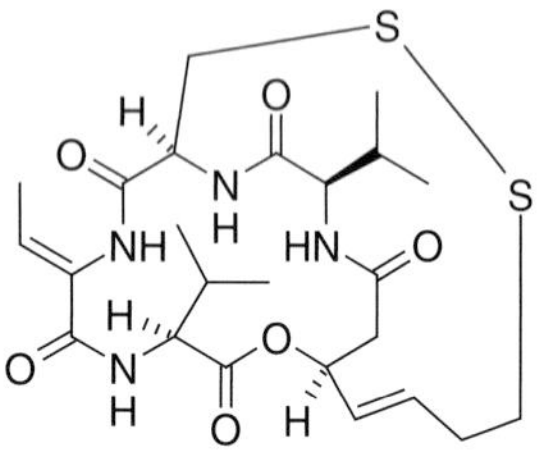

27 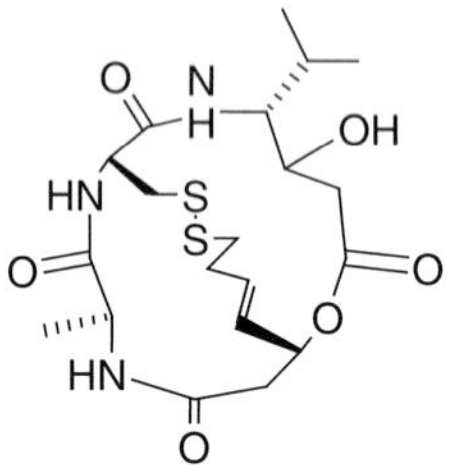 **28**

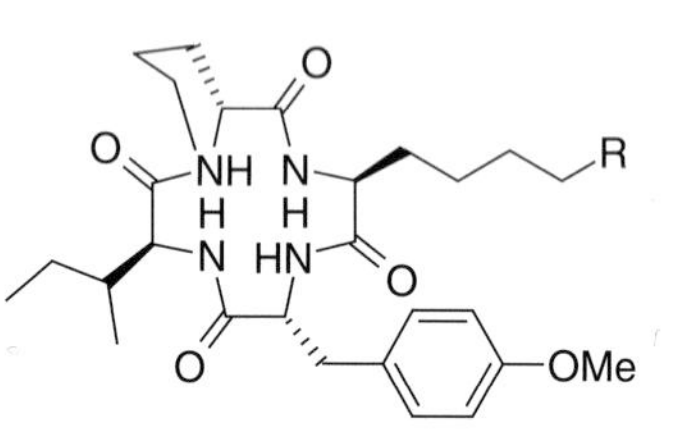

29: R= CH_2–S–S–(4-pyridyl)

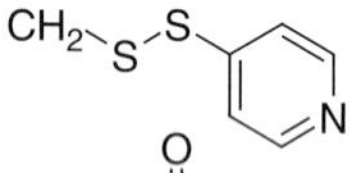

30: R= CH_2–C(=O)–CF_3

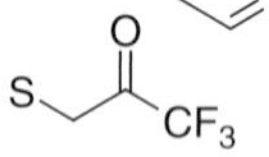

Recently the total synthesis of Spiruchostatin A, **28**, was accomplished, unambiguously confirming its structure.[75] Spiruchostatin A is shown to have biological activity similar to that of FK-228. The Spiruchostatin A analog, epimeric at the β-hydroxy acid, is inactive, highlighting the importance of stereochemistry at this position for interactions with HDACs. New inhibitors of HDAC containing a sulfhydryl group (e.g., **29**) were designed based on the CHAP31 skeleton (i.e., cyclic hydroxamic acid-containing peptide) and the HDAC-binding functional group of FK-228.[76] HDAC inhibitor potency was in the nanomolar range. Based on the same cyclic tetrapeptide scaffold, another series of potent HDAC inhibitors with trifluoromethyl and pentafluoroethyl ketones as alternative zinc-binding groups was designed. The IC_{50} for HDAC1 inhibition of **30** was 47 nM.[77]

7.09.5.1.3 **Inhibitors selective for special histone deactylase isoforms**

Despite the progress made in the last few years, the role of the various individual HDAC isoforms is still under investigation. Therefore new isoform selective HDAC inhibitors will be useful as tools for probing the biology of these enzymes and eventually as new anticancer agents with hopefully decreased toxicity. Class II HDACs differ from Class I HDACs depending on their tissue expression, subcellular localization, and biological roles. Class I HDACs are ubiquitously expressed, whereas Class II enzymes display tissue-specific expression in humans and mice.[78]

There is evidence for differences in sensitivity of different members of Class I and II HDACs to different inhibitors. Few HDAC inhibitors (e.g., Trapoxin A, Trapoxin B, CHAP1, and FK-228) are known to be selective for class I HDACs. To probe the steric requirements for deacetylation, lysine-derived small-molecule substrates were synthesized as tool compounds and their SARs were tested with various histone deacetylases.[79] A benzyloxycarbonyl substituent on the α-amino group yielded the highest conversion rates. Replacing the ε-acetyl group with larger lipophilic acyl substituents led to a decrease in conversion by Class I and II enzymes, whereas the Class III enzyme displayed a greater tolerance. Incubations with recombinant human HDACs 1, 3, and 6 showed a distinct subtype selectivity among small-molecule substrates. This information could be useful for refining homology models of human HDACs and could be used as an starting point for rational drug design.

Tubacin, **31**, which was discovered through a multidimensional, chemical genetics screen of 7392 molecules, has been described as specific α-tubulin deacetylation inhibitor in mammalian cells.[80–82] Between the two HDAC6 catalytic domains, Tubacin selectively binds only that with tubulin deacetylase activity, without affecting histone acetylation, gene expression, or cell cycle progression.[83] These results suggest that small molecules that selectively inhibit HDAC 6-mediated α-tubulin deacetylation might have therapeutic applications as antimetastatic and antiangiogenic agents. Interestingly, Histacin, a closely related *o*-aminoanilide derivative of **31**, is inactive toward HDAC6 while apparently inhibiting deacetylases that act upon histone substrates.[84]

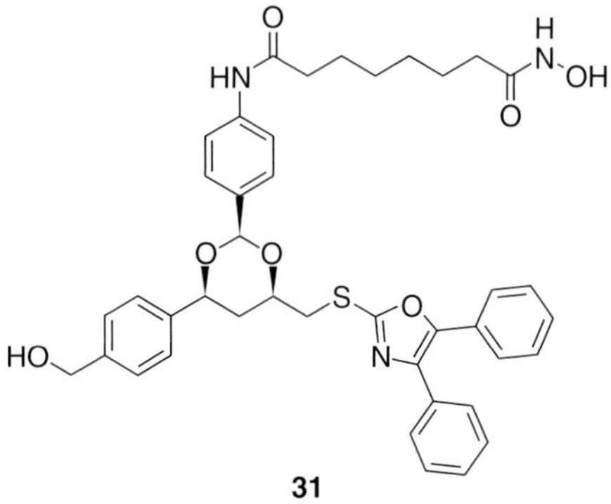

31 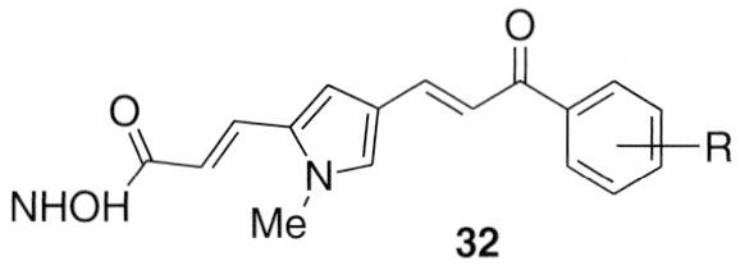 **32** 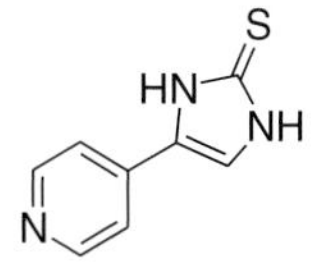 **33**

A series of (aryloxopropenyl)pyrrolyl hydroxyamides **32** (R = H, 2-Cl, 3-Cl, 4-Cl, 2-Me, 3-Me, 4-Me) were designed as the first representatives of selective inhibitors of Class IIa histone deacetylase (HDAC4, HDAC5, HDAC 7, and HDAC9).[85] **32** showed better inhibitory activity against maize HD1-A than HD1-B (two homologs of mammalian Class IIa and I HDACs) in the submicromolar range with 7- to 78-fold selectivity. The unsubstituted compound showed good inhibitory activity against both HD1-B and HD1-A, lacking in class selectivity. Insertion of a chlorine atom or a methyl group at any position of the benzene ring caused a dramatic change in the inhibiting effect on the two deacetylases: 2-substituted compounds were the most potent against both enzymes whereas 3-substitution gave rise to the highest selectivities.

In an HDAC8 enzyme-based HTS several Class I HDAC selective hits were found.[86] SB-42920 preferentially inhibited HDAC1 ($IC_{50} \sim 1.5\ \mu M$) whereas SB-379872, **33**, only inhibited HDAC8 ($IC_{50} \sim 0.5\ \mu M$). Recently a method for selectively inhibiting HDAC7 and HDAC8 was described by either inhibiting expression at the nucleic acid level using antisense oligonucleotides or by inhibiting enzymatic activity at the protein level with small-molecule inhibitors derived from the phenylenediamine benzamide series.[87]

7.09.5.1.4 Conclusions

HDAC inhibitors represent a prototype of molecularly targeted agents that perturb signal transduction, cell cycle regulatory, and survival-related pathways. Newer generation HDAC inhibitors have been introduced into the clinics that are considerably more potent than their predecessors and are beginning to show early evidence of activity, particularly in hematopoietic malignancies. It seems that each HDAC enzyme has a particular role in controlling transcription, the cell cycle, cell motility, DNA damage response, and senescence by deacetylating histone and nonhistone proteins. Therefore enzyme subtype specific inhibitors will pave a new way to therapeutic drugs that control the specific function and the downstream pathway of HDACs.

7.09.5.2 Hsp90 Inhibitors

7.09.5.2.1 Basic facts

Heat shock proteins have been investigated in the study of cellular biology for a long time. Originally they were identified as proteins whose levels increase dramatically when cells are cultivated at elevated temperatures. Newer experiments gave insights that heat shock proteins are molecular chaperones, or proteins that help other proteins fold correctly while they are being synthesized, or to protect proteins that might unfold and thereby lose their active conformation during a stress event, such as thermal denaturation. It has been demonstrated that heat shock proteins and especially Hsp90 are overexpressed in human tumors, and therefore these chaperone proteins are recognized as an exciting new target for the treatment of cancer. In addition to playing an important role in response to proteotoxic heat shock and others stresses, Hsp90 is also critical for maintaining normal cellular homeostasis. Hsp90 isresponsible for ensuring the conformational stability, shape, and function of a selected range of key proteins, including many kinases and transcription factors. Hsp90 is involved in dealing with the cellular stress associated with malignancy, as well as being essential for a range of important oncogenic proteins, like ErbB2, Raf-1, Akt/PKB, and mutant p53. A major attraction of Hsp90 inhibitors is their potential to inhibit many cancer signaling pathways in parallel and thereby exhibiting broad-spectrum antitumor activity.[88,89] First experimental evidence as well as a few pioneering clinical trials suggest that blocking Hsp90 activity results in a therapeutic advantage in the treatment of cancer.[90] However, of major concern to investigators has been that, despite the fact that cancer cells can produce high levels of Hsp90, it is also abundant in normal cells. This could mean that drugs targeting Hsp90 might be unacceptably toxic. However, the first Hsp90 inhibitor, 17-allylaminogeldanamycin (17-AAG) has entered clinical trials with promising early results and has been well tolerated by patients. The drug seems to be targeting tumor cells in preference to normal cells. Hsp90 derived from clinical cancer biopsies has a 100-fold higher binding affinity for 17-AAG than Hsp90 from normal tissues.[91] This binding affinity change is likely induced when Hsp90 associates with other chaperone proteins because Hsp90 is present in multichaperone complexes with high ATPase activity in tumors, whereas in normal tissues Hsp90 is present in a free state with low ATPase activity. The changes in affinity of Hsp90 in cells could be reconstituted in vitro. These biochemical changes detected in different cell types correlated with the cytotoxic activity of 17-AAG in cellular assays. As tumor cells gradually accumulate mutant and overexpressed signaling proteins, Hsp90 becomes engaged in stabilizing these proteins. As tumors become dependent upon these signaling proteins for their survival, they become dependent upon Hsp90 activity as well. This dependence, which is similar in principle to oncogene dependency, could make Hsp90 a specific target in tumors, despite its broad expression in all tissues.

7.09.5.2.2 Inhibitors of Hsp90 function

Besides 17-AAG a range of other promising drug candidates is under investigation and preclinical development.

7.09.5.2.2.1 Binders to the N-terminal ATP pocket of Hsp90

The N-terminus of the chaperone has a regulatory pocket which binds and hydrolyzes ATP.[92] While bound to Hsp90, the nucleotide adopts a bent shape which is found only in ATPases belonging to the GHKL family (G = DNA gyrase subunit B, H = Hsp90, K = histidine kinases, and L = MutL). These enzymes share the same left-handed β–α–β fold, called the Bergerat fold, which is observed neither in the binding sites of kinases nor in other chaperones like Hsp70.[93] Therefore, it is highly probable that Hsp90 inhibitors are in a bent conformation when inside the pocket to achieve high-affinity binding. Furthermore it might be possible to discover new inhibitors with a high degree of selectivity by identifying those that specifically bind to Hsp90 via the N-terminal ATPase pocket.

7.09.5.2.2.2 Ansamycins

It was demonstrated that the ansamycin antibiotics herbimycin, **34**, and geldanamycin, **35**, exert their activity by binding to the regulatory pocket in the N-terminal domain of Hsp90.[94,95] Thereby the ansamycins alter chaperone function by preventing the dissociation of Hsp90 client proteins from the chaperone complex. The trapped proteins cannot achieve their mature functional conformation and are degraded by the proteasome.[96] Although these natural products show cellular potency, their use as clinical agents has been limited by their associated hepatotoxicity and their instability. The liver toxic effects of the ansamycin class are believed to be caused by the benzoquinone functionality because radicicol, another natural product that does not contain a benzoquinone moiety, shows biological activity similar to that of ansamycins without hepatotoxicity.

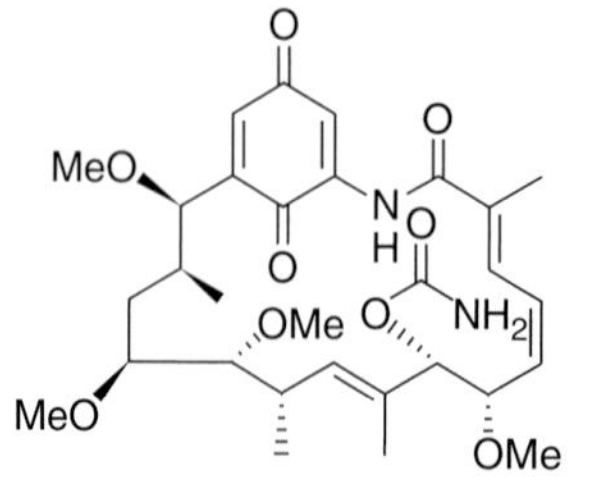

34

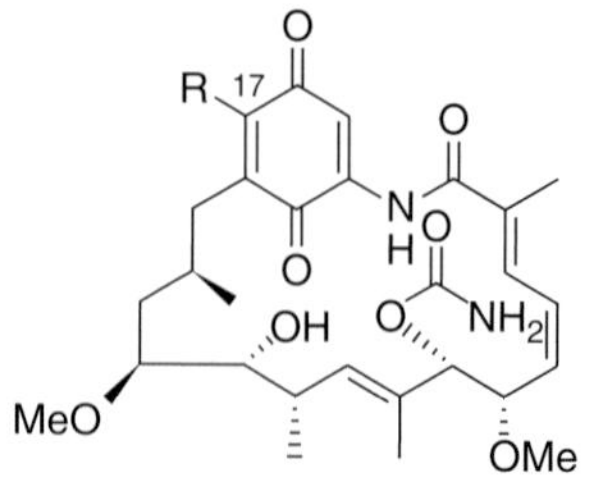

35: R = OMe
36: R = $NHCH_2CH{=}CH_2$
37: R = $NHCH_2CH_2N(CH_3)_2$

17-Allylaminogeldanamycin (17-AAG, **36**), which is a derivative of geldanamycin, has similar cellular effects but lower hepatotoxicity compared to the parent compound.[97] The compound shows activity at nontoxic doses in a subset of breast, prostate, colon, and non-small cell lung cancer (NSCLC) animal models.[98,99] 17-AAG has entered clinical trials in cancer patients in the USA and UK. Toxicity of 17-AAG was found to be dependent on the dosing schedule. With five times daily or three times daily dosing schedules, hepatic toxicity was found to be the limiting factor. This side effect was less severe with intermittent dosing schedules. Clinical phase II studies are ongoing in melanoma. Downregulation of client proteins including Raf-1, cdk4, and Akt in lymphocytes has been found at well-tolerated doses in patients treated in phase I clinical trials.[100,101] Early evidence of therapeutic activity in melanoma patients was reported where stable disease has been seen for over 2 years as well as prolonged stable disease in patients with renal cancer and tumor regressions with the combination of 17-AAG and docetaxel. Despite these early promising results and its important role as proof of principle Hsp90 inhibitor, 17-AAG has several potential limitations like limited solubility, difficult formulation, and extensive metabolization, which could result in the generation of toxic species.[102] Efforts to improve the solubility and bioavailability of 17-AAG have resulted in 17-DMAG, **37**; this has comparable in vivo and in vitro activity to 17-AAG but is water soluble, potentially orally bioavailable, and metabolically more stable.[103,104] This compound has entered phase I clinical trials in patients with advanced cancers.

A new geldanamycin analog, KOSN1559, which binds to Hsp90 with a fourfold greater affinity than that of 17-AAG, was produced by using a newly developed special method for engineered biosynthesis.[105] This analog lacked the quinone moiety which is believed to lead to hepatotoxicity of 17-AAG. Application of this biosynthesis method led to the efficient production of unique geldanamycin analogs that would be very difficult to produce by conventional chemical modification.

Synthesis and biological evaluation of a new class of geldanamycin derivatives containing special amide side chains with improved pharmacological properties was also reported recently.[106]

Recently, the identification of a geldanamycin dimer (EC4) with an extended duration of action has been reported. This derivative was more potent in killing cells with defects in apoptosis pathways and exhibited increased in vivo activity compared to 17-AAG.[107]

Despite the advantages of these novel ansamycins over 17-AAG, it was not possible to eliminate the benzoquinone moiety without affecting activity. Therefore, hepatotoxicity is likely to remain a limiting factor in the clinical use of this compound class. For this reason, the identification of synthetic small-molecule inhibitors of Hsp90 with novel structural features is a major focus of research in the field.

7.09.5.2.2.3 Radicicol derivatives

Radicicol, **38**, is a natural product which was isolated from the fungus *Monosporium bonorden*. The compound also potently inhibits Hsp90 function by binding to its N-terminal ATP pocket.[108] The crystal structure of radicicol bound to Hsp90 demonstrates that it interacts differently with Hsp90 compared to geldanamycin, but it also adopts the bent conformation.[109] Despite promising in vitro activity, radicicol is inactive in vivo due to its instability in serum. Research efforts to modify its structure have resulted in oxime derivatives (e.g., KF55823, **39**). These derivatives have potent activity in vitro and in vivo. These oximes do not cause serious liver toxicity, which indicates that the liver toxicity observed with the ansamycins is not caused by Hsp90 inhibition.[110] Despite their promising activity, these compounds have not advanced further into clinical development.

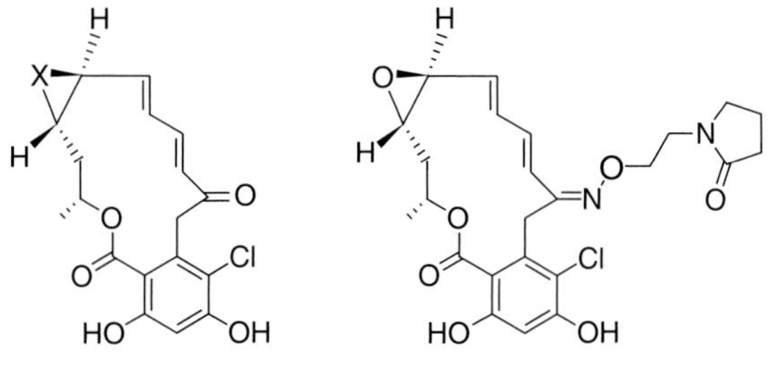

38: X = O
40: X = CH_2

39

Several successful total syntheses of the Hsp90 inhibitor radicicol have been reported. Further studies in the radicicol skeleton have resulted in cyclopropylradicicol, **40**, as a highly promising preclinical anticancer agent targeting Hsp90.[111] A novel second-generation approach utilizing a key Diels–Alder cycloaddition was designed through which cycloproparadicicol may be synthesized efficiently in large-scale quantities.[112]

7.09.5.2.2.4 Purine-scaffold derivatives

To overcome the limitations of natural product derived Hsp90 inhibitors, novel pharmacological scaffolds have been studied. The first described member of the purine-based family of Hsp90 inhibitors is PU3, **41**, which inhibits the binding of Hsp90 to immobilized geldanamycin and mimics the cellular effects of ansamycins, although with modest potency.[113] Preparation of a small library of 70 purine compounds resulted in the identification of PU24FCl, **42**, a compound with an affinity for Hsp90 close to that of geldanamycin.[114,115] PU24FCl has 10–50 times higher affinity for Hsp90 from transformed cells compared to normal tissue.

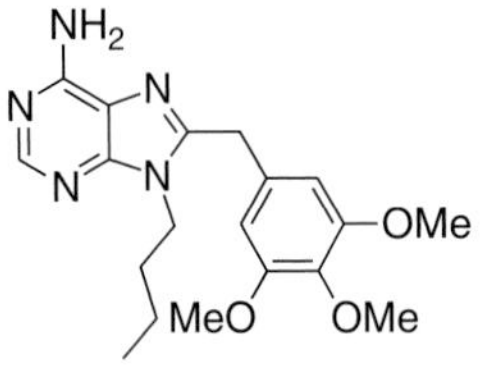

NH2 N N N N OMe MeO OMe

41

NH2 N N F N N Cl OMe MeO OMe

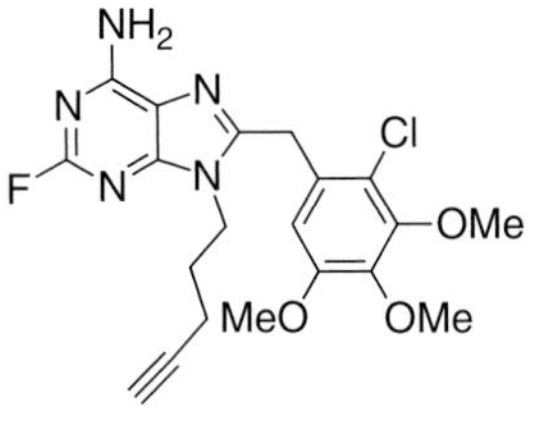

42

According to crystal structures of PU3 and PU24FCl in complex with the N-terminal region of human Hsp90α, these compounds induce a conformational change in the Hsp90 ATP pocket.[116] PU24FCl adopts during binding the bent shape characteristic of ligands specific to this pocket. PU24FCl has wide-ranging anticancer activities that occur at doses of 3–6 μM in all tested tumor cell lines, including those resistant to 17-AAG. In concordance with its higher affinity for tumor Hsp90, PU24FCl accumulates in vivo in tumors while being rapidly cleared from normal tissues. In MCF-7 tumor xenografts, one dose of PU24FCl causes a significant depletion of receptor tyrosine kinases (Her2, Her3, and Her4) as well as degradation and inactivation of Akt and Raf-1. In a 30-day study of antitumor efficacy in the MCF-7 xenograft model, treatment with PU24FCl resulted in a 72% reduction in tumor burden compared with a control group.[117] Due to their broad antitumor activity, the purine class of Hsp90 inhibitors could have clinical use in a wide range of tumor types.

7.09.5.2.2.5 Pyrazoles and other heterocyclic derivatives

Another novel scaffold Hsp90 inhibitor based upon a pyrazole core structure (CCT018159, **43**) was identified by HTS of a library of 60 000 compounds.[118] The assay was designed to identify inhibitors of the intrinsic and biologically essential ATPase activity of yeast Hsp90. CCT018159 inhibits the growth of human colon, ovarian, and melanoma tumor cells at concentrations similar to those that inhibit human Hsp90 ATPase activity (~8 μM). Analogs of CCT018159 (e.g., **44**) with improved activity have been reported.[119–121] In further studies, isoxazole derivatives[122] and 8-heteroaryl-6-phenyl-imidazo[1,2-a]pyrazines[123] have been described as new scaffolds for Hsp90 inhibitors.

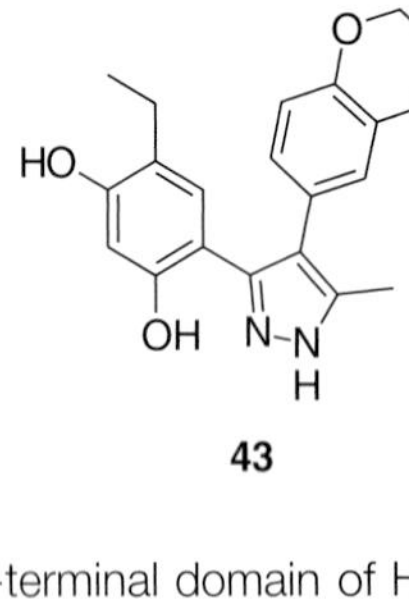

O O HO OH N-N H

43

H N Cl HO N OH N-N H

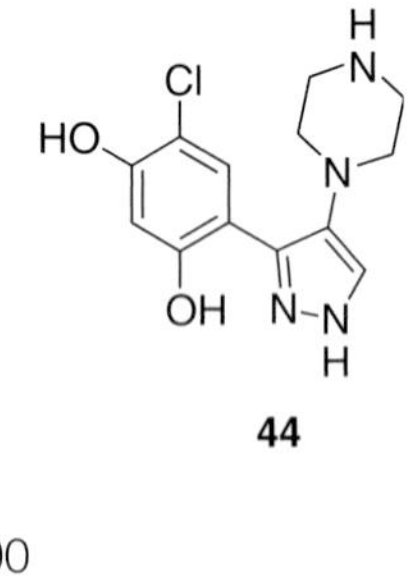

44

7.09.5.2.2.6 Binders to the C-terminal domain of Hsp90

Many Hsp90 inhibitors function by binding to the N-terminal ATP pocket, but the chaperone has several other possibilities for inactivation. Compounds that interact with its C-terminus or modify its posttranslational status represent additional ways of interfering with Hsp90 activity. In contrast to the well-studied N-terminus, the crystal structure of the C-terminal region has not been solved so far. This region has been implicated in the binding of a second ATP molecule. Studies suggest that the site becomes available to ATP only after the N-terminal ATP pocket is occupied by either ATP or an inhibitor such as geldanamycin. Although the contribution of this site to the function of Hsp90 remains unclear, there is a hypothesis that it might regulate the ATPase activity of the N-terminal region.[124] Agents that interact with this region of the chaperone might also impair Hsp90 function and result in anticancer effects.

Novobiocin, an antibiotic which is known as an inhibitor of DNA gyrase subunit B, was reported to interact with Hsp90.[125] Cells exposed to high micromolar concentrations of novobiocin revealed destabilization of various Hsp90 client proteins, including Her2, Raf-1, mutant p53, and v-src.[126] The binding site of novobiocin has been mapped to a region in the C-terminus of the chaperone.[127]

It was found that cisplatin binds also to Hsp90 and induces a conformational change in the structure of the chaperone.[128] Cisplatin interacts with the C-terminal region of Hsp90 and interferes with nucleotide binding in the region.

7.09.5.2.2.7 Inactivators of Hsp90 function by posttranslational modifications

The importance of posttranslational modifications in regulating Hsp90 function is still under investigation. It has been discovered that Hsp90 phosphorylation leads to the release of the chaperone from the target protein. This process can be inhibited by geldanamycin.[129] Acetylation and ubiquitinylation of Hsp90 have also been found to change its activity although it is not yet elucidated which sites are responsible for such posttranslational regulation. HDAC inhibitors can induce growth arrest and apoptosis in a variety of human cancer cells by mechanisms that cannot be attributed solely to histone acetylation. It was found that Hsp90 is a downstream target of the HDAC inhibitors. These compounds induce acetylation of Hsp90. This results in inhibition of the ATP binding to Hsp90 which impairs association of Hsp90 with its client proteins.[130] Treatment of NSCLC cells with the HDAC inhibitor FK-228, **27**, results in reduced expression of mutant, but not wild-type, p53, depletion of Her1, Her2, and Raf-1 proteins, and lower ERK1/2 activity. These effects are similar to the inhibition of Hsp90 ATPase activity by binders of its N-terminal ATP pocket. Similar effects and induction of Hsp70 have been detected with the HDAC inhibitor LAQ824 in Her2-overexpressing breast cancer cells and in lymphocytes from patients who had been treated with this compound.[131]

Hsp90 function can also be changed by ubiquitinylation.[132] Treatment of cells with hypericin enhances chaperone ubiquitinylation resulting in the proteasome-independent degradation of Hsp90 client proteins, such as mutant p53, Cdk4, Raf-1, and Plk. Treated cells exhibit retardation at the G2/M checkpoint, increased cell volume, and multinucleation. These findings suggest that, together with cochaperones and ATP hydrolysis, the posttranslational status of Hsp90 might be responsible for regulating the activation state of the Hsp90 superchaperone complexes.

7.09.5.2.3 Conclusion

Early clinical results with 17-AAG, the first Hsp90 inhibitor to enter clinical trials, suggest that doses of drug that are sufficient to inhibit Hsp90 function can be administered alone and in combination with cytotoxic chemotherapy with surprisingly little target-associated toxicity. These findings have prompted a search for novel classes of chaperone inhibitors including natural products and synthetic small-molecule inhibitors of Hsp90 and regulators of Hsp90 acetylation and ubiquitinylation.

These structurally different inhibitors might have better selectivity and improved pharmacological profiles over ansamycins. Several of these agents are in late preclinical development or have already entered clinical trials.

7.09.5.3 Farnesyltransferase Inhibitors

7.09.5.3.1 Basic facts

Ras protein, which is a low-molecular-weight GDP/GTP-binding guanine triphosphatase encoded by the *Ras* gene, plays a critical role in signal transduction of cell growth and differentiation.[133,134] In normal process of signal transduction, Ras performs its function in a GTP-binding form. However, Ras itself has no ability to bind to the membrane due to its low hydrophobicity. It must be modified by enzymes, with a lipid modification called farnesylation, which enhances its hydrophobicity, binding to the cell inner membrane. After performing its function, Ras protein is hydrolyzed into the GDP-binding form. Ras mutations have been described in malignant transformation, invasion, and spread of cancer. This results in continuous cell growth signals that are out of control, which causes excessive cell differentiation and proliferation and finally leads to tumorigenesis.

The *Ras* gene is overexpressed in 30–40% of cases of thyroid carcinoma, over 50% of colon carcinoma, and 90% of pancreatic carcinoma, which supports the principal role of Ras in cell signal transduction and tumorigenesis. Therefore, inhibiting the activity of the Ras protein can prevent the cell signal transduction, which is one of the most important targets for anticancer drug design.[135]

7.09.5.3.2 Farnesyltransferase inhibitors in development

Based on the theoretical assumption that preventing Ras farnesylation might result in the inhibition of Ras functions, a range of farnesyltransferase (FTase) inhibitors have been synthesized.[136,137] Their biology is fascinating since after substantial investigation and their use in several phase II studies and at least two phase III trials, the exact mechanism of action still remains unclear.[138] Farnesyltransferase inhibitors can block the farnesylation of several additional proteins, such as RhoB, prelamins A and B, and centromere proteins (CENP-E, CENP-F). While the farnesyltransferase inhibitors clearly do not or only partly target Ras, these agents appear to have clinical activity in leukemia[139] and in some solid tumors regardless of their Ras mutational status.

In all members of the Ras family there is a specific C-terminal sequence known as CAAX, where C is cysteine, AA is aliphatic amino acids, and X is any amino acid, preferably methionine or serine. Farnesylation is catalyzed by farnesyltransferase, with a farnesyl in farnesylpyrophosphate bound with the thiol group of Cys, anchoring Ras to the cell membrane, which is a required step of the cancer-causing activity of Ras candidates. Early attempts to discover inhibitors of farnesyltransferase focused on modifications of the isoprenoid and CAAX polypeptide substrates of the enzyme. While potent farnesylpyrophosphate-derived inhibitors have been discovered, most attention has been given to analogs of the CAAX peptide. Medicinal chemistry optimization of initial peptidomimetics led to L-778,123, **45**, with an FTase $IC_{50} = 2$ nM. The compound inhibited the growth of H-ras-transformed cells with an IC_{50} value of 15 nM. A phase I study examined the administration of **45** to patients with advanced solid malignancies.[140] The observed dose-limiting toxicities were myelosuppression, prolongation of the QTc interval, and fatigue. No objective responses were seen, and the clinical development of this inhibitor has been discontinued.

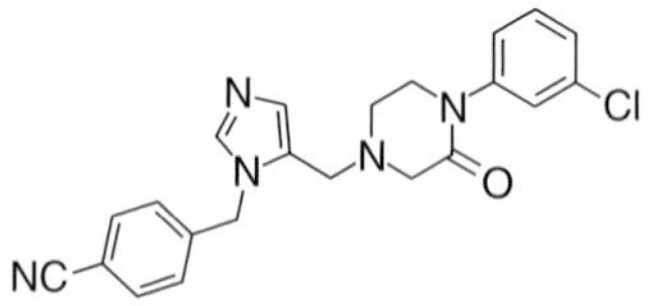

45 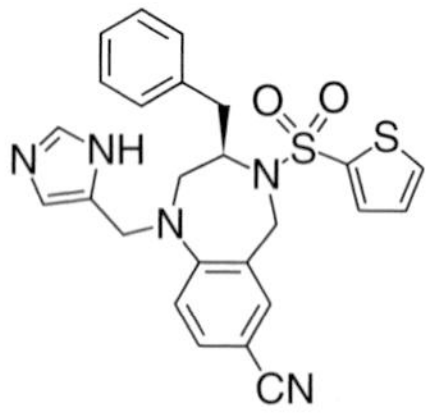 **46**

Another example of an imidazole-based farnesyltransferase inhibitor is BMS-214662, **46**, which has an FTase IC50 of 1.4 nM.[141] It is notable that these researchers independently discovered the potency-enhancing effects of a cyanophenyl group. It seems likely that this moiety is taking advantage of similar binding interactions to the cyanobenzyl as in **45**. Compound **46** is a potent and selective farnesyltransferase inhibitor that is reported to promote apoptosis to a greater extent than other farnesyltransferase inhibitors of comparable potency, suggesting that the apoptosis might result from a mechanism unrelated to farnesyltransferase. Compound **46** has been investigated clinically and there have been several reports of minor clinical responses in solid tumors.[142] In a phase I study, treatment with **46** led to objective responses in 24% of patients with advanced hematologic malignancies.

In contrast to the gradual evolution from CAAX peptides to small molecules such as **45** and **46**, other research groups obtained attractive leads for the development of farnesyltransferase inhibitors from screening of compound libraries. SCH37370 (**47**, FTase $IC_{50} = 27\,000$ nM), a close analog of the H1 receptor antagonist loratadine was found as a screening hit.[143] SAR exploration (e.g., addition of bromo substituents and saturation of the vinyl bond) were found to improve potency, ultimately leading to the clinical candidate **48** (SCH66336, lonafarnib, Sarasar) with an FTase IC_{50} of 1.9 nM.[144] This tricyclic inhibitor is distinguished from the other farnesyltransferase inhibitors tested in humans by its lack of a ligand for the active site zinc ion in FTase. In nude mouse xenograft experiments, orally administered **48** demonstrated good activity and achieved complete growth inhibition for some tumor types. SCH66336 is currently in phase II clinical trials. Efficiency of SCH66336 with a combination of Temozolomide for recurrent glioblastoma multiforme (GBM) has been investigated.[145] Two out of three human GBM xenografts demonstrated substantial growth inhibition in response to SCH66336, with up to 69% growth inhibition after 21 days of treatment.

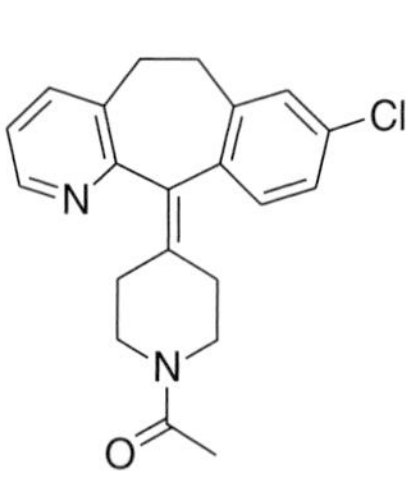

47 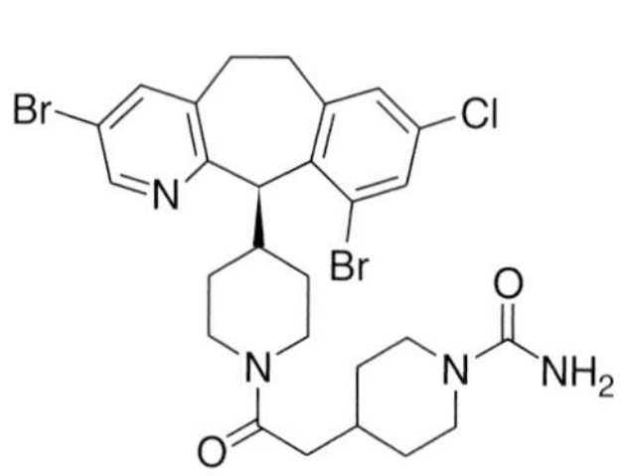 **48** 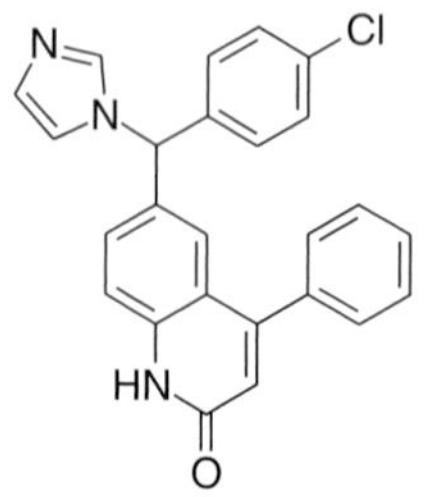 **49** 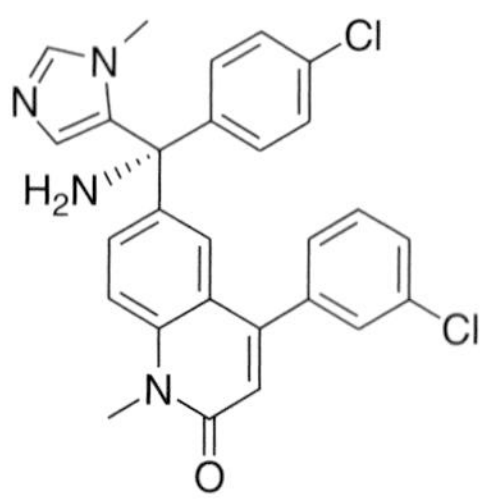 **50**

Screening for farnesyltransferase inhibitors afforded lead quinolinone compounds such as **49** (FTase $IC_{50} = 180\,nM$). It was found that attachment of the imidazole moiety via the 5-position, combined with *N*-methylation of the imidazole, increased potency against FTase and selectivity against cytochrome P450-dependent enzymes. Incorporation of a benzylic amino group was shown to improve cell potency, and further optimization produced **50** (R115777, tipifarnib, Zarnestra; FTase $IC_{50} = 0.86\,nM$), the first farnesyltransferase inhibitor to advance to human clinical trials.[146] Zarnestra inhibited the growth of H-ras-transformed NIH 3T3 cells with an impressive IC_{50} value of 1.7 nM. Clinical trials for breast cancer (phase II)[147] and pancreatic carcinoma (phase II/III)[148] are ongoing. Although inhibition of farnesyltransferase by these compounds has also been well documented in normal tissues, their toxic effects seem to be manageable. The main dose-limiting toxicities that have been reported are myelosuppression, fatigue, and neurotoxicity with R115777. Two other phase III trials of R115777 in colorectal (versus placebo) and pancreatic (with gemcitabine versus placebo) cancers have failed to show a survival benefit.

7.09.5.3.3 Conclusion

Overall preliminary results of early phase II/III studies suggest that the activity of farnesyltransferase inhibitors, as a single agent, is modest and generally lower than that obtained by standard cytotoxic drugs.[149] Ongoing clinical studies are assessing the role of farnesyltransferase inhibitors for early stage disease or in combination with cytotoxic agents or with other molecular targeted therapies for advanced stage tumors. It is likely that the future clinical direction of farnesyltransferase inhibitors will be as a combination therapy.[150] Synergies have been seen in a couple of preclinical studies especially with taxanes.[151] Further insights in the molecular mechanism of action of farnesyltransferase inhibitors might help in better defining their optimal use in combination with standard therapies in the treatment of cancer patients.

7.09.5.4 Proteasome Inhibitors

7.09.5.4.1 Basic facts

The ubiquitin proteasome pathway constitutes the most important system for protein degradation in eukaryotic cells.[152] It prevents the accumulation of nonfunctional, potentially toxic proteins and allows elimination of normal proteins that are no longer required. This pathway depends on the central role of a complex macromolecular structure, the 26S proteasome, which is composed of one 20S and two 19S subunits. The 20S proteasome has multiple proteolytic activities. Three major proteolytic activities can be distinguished: trypsin-like, chymotrypsin-like, and peptidyl-glutamyl peptide hydrolase activities. Peptide bonds are cleaved on the carboxyl side of basic, hydrophobic, and acidic amino acid residues.[153] These proteolytic activities rely on an unusual mechanism involving the N-terminal threonine residue of particular subunits as the catalytic nucleophile.[154] The exact substrate specificity of the different catalytic sites is still not well understood. Although the distinct activities are generally defined by the amino acid in the P1 position of a synthetic peptide substrate, the specificity determinants go well beyond the P1 position. Therefore, characterization of the effects of novel proteasome inhibitors is useful in unraveling the specificity of the different catalytic sites. Selective inhibitors of the proteasome are also important tools to study proteasome function in cells. Proteasome inhibitors have a potential for treating cancer and neurodegenerative diseases and can also be used to prevent cancer- or AIDS-associated muscle cachexia.[155]

7.09.5.4.2 Proteasome inhibitors in development

Several groups of proteasome inhibitors have been widely used to study the role of the ubiquin proteasome pathway in various cellular processes or as anticancer drugs.[156] Peptidomimetics have been developed to circumvent problems inherent in peptides such as poor bioavailability and protease-mediated degradation, while retaining biological activity.

7.09.5.4.2.1 Peptide aldehyde inhibitors and peptidomimetics

Peptide aldehyde inhibitors (e.g., ALLN, **51**, and MG132, **52**) are still commonly used as tool compounds but they are relatively nonspecific.[157,158] Furthermore, these peptide aldehydes are not configurationally stable. The activity of peptide-based proteasome inhibitors is often characterized by poor bioavailability and protease-mediated degradation. Therefore efforts were made to overcome these disadvantages, while retaining the biological activity of the peptide. These experiments have led to the design of peptidomimetics. A particular interesting class of peptidomimetics is formed by peptoids[159–161] and analogs such as azapeptides and azapeptoids,[162–164] ureapeptoids,[165] amino-oxypeptoids,[166] β-peptoids,[167] and hydrazino-azapeptoids,[168,169] in which the side chains are linked to nitrogen atoms. Recently, a series of retro hydrazinoazapeptoids, which contain NHNRCO bonds instead of CONRNH bonds, was synthesized and screened for proteasome inhibition and cell cytotoxicity.[170] Only relatively weak proteasome inhibition was found for these compounds.

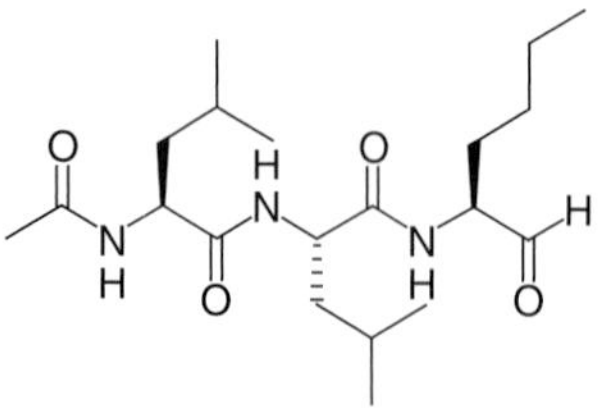

51 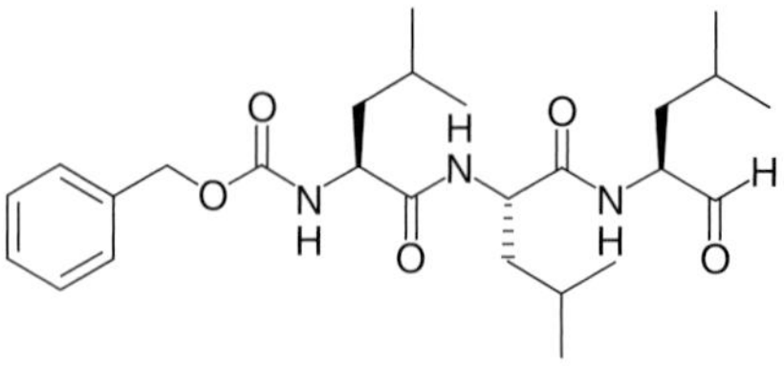 **52**

Novel peptidic compounds (e.g., **53**) and methods of their use for the treatment of disorders mediated by a proteasome, including cancer, were claimed recently.[171] This patent also described methods of using the compounds in combination with radiation or chemotherapy for the treatment of cancer. Compound **53** is stated to be useful for the treatment of multiple myeloma, leukaemia, and colorectal, prostate, breast, and lung cancers.

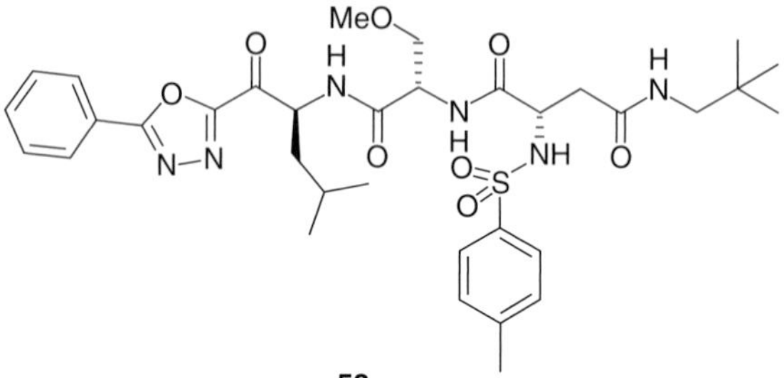

53

7.09.5.4.2.2 Dipeptidyl boronic acids

Novel dipeptidyl boronic acids have been shown to be stable and highly potent reversible inhibitors of the proteasome.[172–174] Boronic acids act as transition-state analogs for serine proteases because the boron can accept the oxygen lone pair of the active site serine residue. It seems likely that these compounds react similarly with the catalytic N-terminal threonine residue of the proteasome catalytic subunits.[175,176]

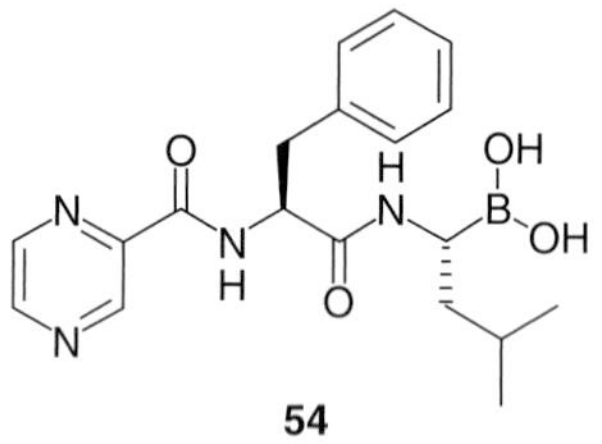

54

In 2003 Bortezomib (PS341, Velcade), **54**, became the first proteasome inhibitor to be approved by the FDA.[177] Bortezomib was approved for multiple myeloma, a blood cancer that affects two to three people per 100 000, in the record time of 4 months on the basis of positive outcomes in early phase trials. Multiple myeloma is a malignant B cell tumor characterized by osteolytic bone lesions. It is the second most common hematological cancer (after non-Hodgkin's lymphoma), and treatment has relied predominantly on glucocorticoids as well as alkylating agents. Although these treatments improve the survival rate (5-year survival is ~29%), the disease remains incurable. Proteasomes selectively destroy potentially harmful abnormal or unfolded proteins, which might be produced in increased amounts in various forms of cancers and especially in multiple myeloma. In vitro, myeloma cells are up to 1000-fold more sensitive to apoptosis induced by bortezomib than normal plasma cells.[178] A key factor in this differential response seems to be the ability of proteasome inhibitors to block the activation of the transcription factor nuclear factor κB (NFκB), which is constitutively expressed in myeloma cells, some leukemias, and solid tumors. In normal cells, NFκB is bound to the inhibitory protein IκB, which maintains it in the inactive form in the cytosol. Certain tumors have activated forms of NFκB, and the proteasome is essential for this activation, as it catalyzes the proteolytic generation of the NFκB subunit p50 from the inactive p105 precursor and the destruction of the inhibitory IκB.[179] The activated NFκB can then enter the nucleus, which allows it to carry out many functions in the tumor cell that help the cell to survive and proliferate. NFκB also has antiapoptotic effects and therefore proteasome inhibitors can sensitize cells to other cancer treatments. It was demonstrated that a combination of bortezomib and a histone deacetylase inhibitor (e.g., SAHA) sensitizes NSCLC cells to HDAC inhibitor-induced apoptosis.[180]

Bortezomib was specifically designed to fit active sites of the proteasome. The proteasome has three types of catalytic activity: chymotryptic-like, tryptic-like, and caspase-like. The boronic acid group forms a complex with the threonine hydroxyl group in the chymotrypsin-like active site and acts as a reversible inhibitor of the chymotryptic-like activity of the proteasome, which is sufficient to inhibit proteolysis. A phase II study was carried out in 202 patients with relapsed and refractory multiple myeloma who had received at least two prior therapies. Out of the 193 patients evaluated, 35% showed a response to bortezomib. The most commonly reported adverse events were gastrointestinal symptoms, thrombocytopenia, and peripheral sensory neuropathy.[181]

7.09.5.4.2.3 β-lactones

The streptomyces metabolite lactacystin, **55**, is a specific and irreversible inhibitor of the proteasome.[182] The cell membrane is impermeable to lactacystin, and lactacystin itself does not react with the proteasome. However, in cell culture media lactacystin is spontaneously converted to a reactive β-lactone that easily traverses the plasma membrane. The β-lactone inhibits the proteasome through reaction with the hydroxyl group on the active site threonine to form an acyl enzyme conjugate. Under intracellular conditions the active β-lactone is a highly unstable molecule that is rapidly inactivated. Despite this drawback, lactacystin is more selective than the peptide aldehydes.

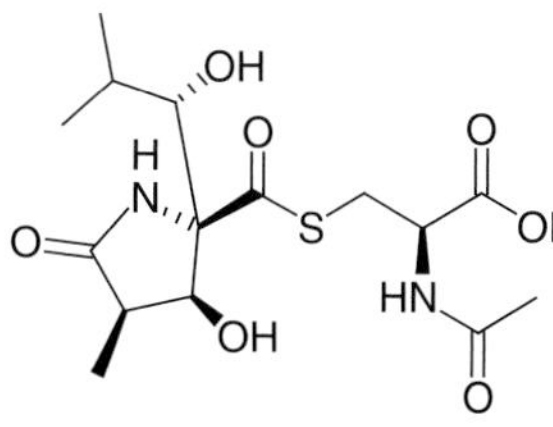

55 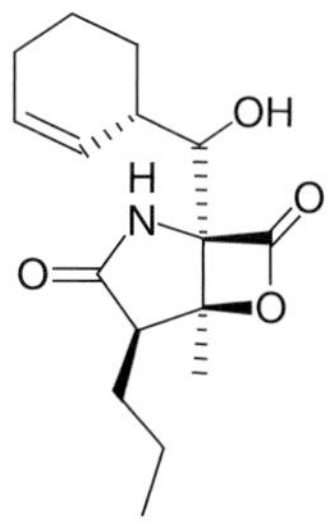 **56** 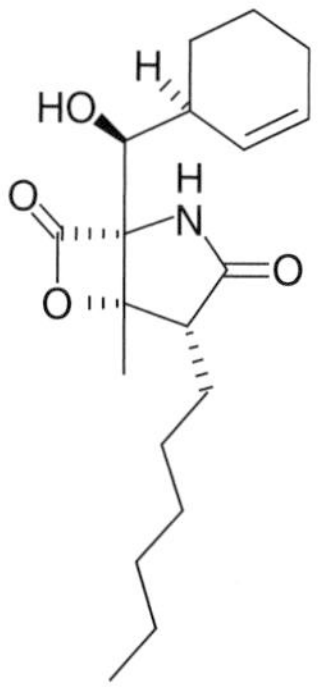 **57**

NPI-0052, **56**, is an orally active proteasome inhibitor which is currently under development for the treatment of cancer. It was discovered during the fermentation of *Salinospora* sp., a new marine Gram-positive actinomycete. In leukemia cell lines NPI-0052 induced apoptosis. In human cancer cell lines (breast, colorectal, lung, ovarian, prostate, and T cell leukemia), murine melanoma NPI-0052 produced apoptosis at lower concentrations (EC_{50} of 10–100 nM). NPI-0052 is a potent inhibitor of human proteasomes and is more potent and selective than bortezomib in in vitro studies. NPI-0052 is active against multiple myeloma cells that are resistant to bortezomib, steroid therapy, and thalidomide. The compound has also shown efficacy in animal models of myeloma, colon, pancreatic, and lung cancer when administered orally or intravenously and is progressing toward phase I clinical trials.[183]

In a further variation, 2-pyrrolidinone derivatives like **57** have been claimed recently as potent inhibitors of the proteasome. Fermentation processes for their preparation and methods of their use for the treatment of inflammation or cancer have been also described.[184]

7.09.5.4.2.4 Epoxyketones

Eponemycin and epoxomicin, natural epoxyketone peptides produced by various actinomycetes, are potent antitumor agents with powerful antiangiogenic activity. They both share the proteasome as a common intracellular target. ER-807446, **58**, is a representative compound in a series of epoxyketone derivatives based on the natural product eponemycin, under investigation as proteasome inhibitors for the treatment of cancer. ER-807446 showed good antitumor activity against MBA-MD-435 human xenografts in mice. Compounds in this series had 100 times improved proteasome inhibitory activity compared to the initial lead ER-804191. Pharmacodynamic studies using murine splenocyte proteasome activity as a surrogate marker revealed that ER-807446 demonstrated proteasome inhibition for

72 h and this inhibition was reversible. Combination of ER-807446 plus CPT-11 generated tumor regression; the combination was tolerated well with reversible toxicity. Most interesting of all, this effect is more pronounced with larger tumors: treating mice with small tumors (~200 μL) led to 25% regression, while treating mice with large tumors (~500 μL) led to ~60% tumor regression.[185]

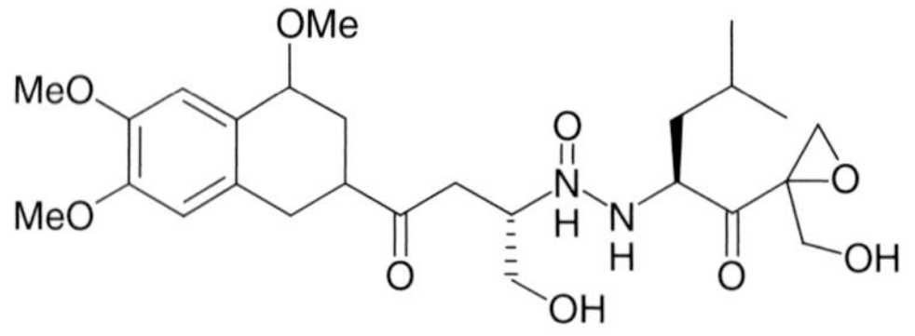

58

7.09.5.4.3 Conclusion

The success of bortezomib in the treatment of multiple myeloma provides evidence that proteasome inhibitors can be effective clinically. However, resistance to bortezomib and issues of toxicity provide significant opportunities for the development of second-generation proteasome inhibitors.

7.09.5.5 Inhibitors of p53–MDM2 Interaction

7.09.5.5.1 Basic facts

MDM2 (often referred to as HDM2 in human) is a zinc finger oncoprotein[186] that has been well characterized as the principal negative regulator of the p53 tumor suppressor protein.[187] p53 is a transcription factor that is pivotal to cellular responses to genotoxic and other stress.[188] Wild-type p53 function is critical for maintaining the genomic integrity of cells. The tumor suppressor functions of p53 stem from transcription-dependent or -independent induction of apoptosis, cell cycle arrest, differentiation, and senescence, as well as its involvement in DNA repair.[189] Under physiological conditions, p53 seems to be inactive, and its levels are kept very low. This is caused mainly by destabilization by MDM2. This allows for normal balance in cell growth, proliferation, and survival. However, under genotoxic (DNA damage) or other stress conditions such as oncogenic activation, oxidation, hypoxia, ribonucleotide depletion, or mitotic spindle damage, p53 becomes activated and stabilized, resulting in cell cycle arrest or apoptosis.

In approximately half of human cancers[190] loss-of-function p53 gene mutations occur which result in more aggressive and drug-resistant tumor phenotypes.[191] In the other half of human cancers, which posses a wild-type p53, its tumor suppressor functions can still be compromised by the overexpression or deregulation of MDM2 which also confers tumor aggressiveness and drug resistance. Therefore the inhibition of MDM2 function is seen as an attractive means of triggering or enhancing cancer cell death by promoting p53-induced cell cycle arrest and apoptosis.[192–195]

7.09.5.5.2 Inhibitors of p53–MDM2 interaction in development

7.09.5.5.2.1 Targeting p53–MDM2 binding with small molecules

The recent successes achieved in the therapeutic exploitation of molecular targets in cancer cell proliferation and survival promoting signaling pathways, as exemplified by small-molecule drugs Gleevec (imatinib),[196] Iressa (gefitinib),[197] and Velcade (bortezomib),[198] also triggered significant efforts to target p53 pathways with small molecules for cancer therapy.[199]

The p53–MDM2 interaction is the best characterized of all MDM2 interactions and offers a logical target for cancer therapy. Targeting p53–MDM2 interaction with small molecules is one of several approaches to activating p53 for cancer treatment.[200] The selection of a target for possible anticancer therapy includes the criterion that a target must be 'druggable.' In the pharmaceutical industry this term means that the target is usually a protein with a catalytic activity, and especially a catalytic site that is accessible to inhibition by small-molecular-weight chemicals. Protein–protein interactions are generally believed to be difficult to inhibit with small molecules due to factors like large and shallow interaction interfaces often lacking deep binding pockets, which are difficult to cover with small molecules. Fortunately, in the case of p53–MDM2 binding at the N-terminal of MDM2, the interface is relatively small, and has a well-defined deep basinlike cleft with well-defined hydrophobic pockets that small molecules can bind with high affinity and effectively compete against p53 binding.[201]

Early successful studies with p53-derived peptides demonstrating that the MDM–p53 interaction could be effectively disrupted gave first hints that this could be also useful for therapeutic purposes.[202] The assays that were developed through peptide studies, e.g., enzyme-linked immunosorbent assay (ELISA) and gel mobility shift assays enabled the search for inhibitors. Thus, several classes of compounds have been reported to exhibit high-affinity competitive binding to the p53 binding site on MDM2, and thereby cause p53 accumulation and enhance cell cycle arrest, and apoptosis in cancer cells.[203–205] The current increase in discovery of small molecule p53–MDM2 inhibitors is built on more sophisticated structure-based drug design tools and better assays.

7.09.5.5.2.2 Screening approaches to the discovery of small-molecule MDM2 antagonists

The discovery of chalcone derivatives that bound to MDM2 at the p53 transactivation domain binding site using an ELISA that employed a p53 peptide was reported in 2001.[206] These inhibitors were of low affinity, with IC_{50} values in the range of 50–250 mM. The general low potency of these chalcones possibly reflects a failure of the groups to adequately occupy the hydrophobic pocket.

Through the screening of a diverse library of synthetic compounds, a series of *cis*-imidazoline derivatives has been identified with suitable hydrophobic substituents well positioned to occupy the hydrophobic subpockets of the critical triad of amino acid residues that have been found to be necessary for p53 binding.[207] The compounds were named Nutlins-1, -2, and -3 (**59–61**), and exhibited IC_{50} values in the nanomolar range (100–300 nM) in displacing p53 from its complex with MDM2. Testing the imidazoline analogs in a range of cancer cell based assays provided strong evidence that they activated the p53 pathway, leading to cell cycle arrest and apoptosis. Oral administration of one of the compounds in mice xenograft models was well tolerated and resulted in 90% inhibition of tumor growth relative to vehicle controls, compared with 81% inhibition using intravenous administration of the maximal tolerated dose of the traditional cytotoxic drug doxorubicin. This study was the first real proof of principle that p53–MDM2 interaction is a viable target for targeted cancer therapy. Although approximately 50% of human tumors have lost wild-type p53 and therefore would not be expected to be affected by inhibitors of the p53–MDM2 interaction, these results indicate that activating the tumor suppressor capability of p53 with such compounds might be beneficial in the other cancer types in which the wild-type form of p53 is retained.[208] Furthermore the demonstration that a protein–protein interaction can be successfully targeted by small-molecule inhibitors provides encouragement for the growing number of research programs pursuing this challenging goal.

One of the compounds, Nutlin-2, **60**, was cocrystallized with MDM2 and the structure solved by x-ray diffraction. The complex obtained showed that the compound bound with the imidazole ring as a scaffold projecting three hydrophobic groups into the three hydrophobic pockets in an analogous manner to p53. The two vicinal 4- and 5-position *para*-bromophenyl substituents occupy the same positions as p53 residues Leu26 and Trp23, while the *ortho*-ethoxy substituent on the 2-position phenyl substituent occupies the same pocket as p53 residue Phe19. However, compared to the bound p53 peptide, there is an induced fit in the Nutlin-MDM2 complex involving MDM2 residue Tyr100 that is different from that in the p53–MDM2 complex. In the p53–MDM2 complex, this tyrosine residue of MDM2 hydrogen bonds with Asn29 of the p53 peptide, whereas in the Nutlin–MDM2 complex, this tyrosine moves toward the cleft to bond with the *para*-bromo substituent on the imidazoline 4-position bromophenyl. This demonstrates a flexibility of this residue that could be used to design further classes of inhibitors. The Nutlin–MDM2 structure has an induced fit at the receptor site that will be suitable for structure-based design of small-molecule inhibitors, whereas the p53–MDM2 complex site will better serve peptide ligand discovery.

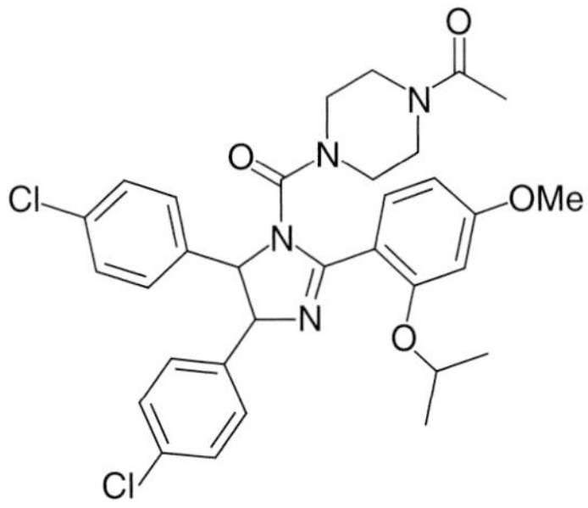

59 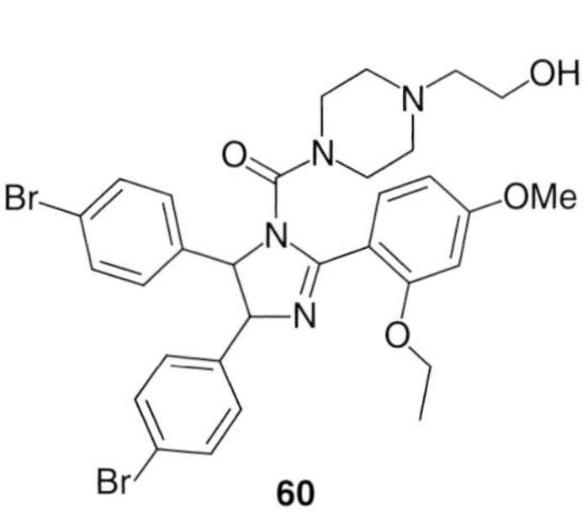 **60** 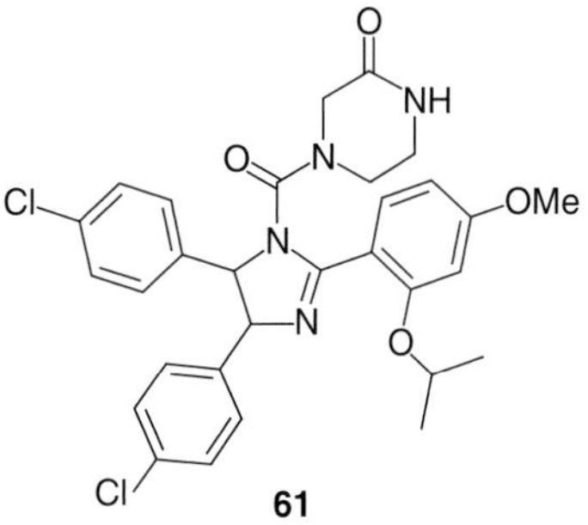 **61**

Recently, a library of 22 000 1,4-benzodiazepine-2,5-diones was screened for binding to the p53-binding domain of MDM2.[209] The hits obtained were further shown to bind to MDM2 in the p53-binding pocket using a fluorescence polarization (FP) peptide displacement assay. This series was further optimized for potency to obtain submicromolar antagonists of the p53–MDM2 interaction. The most potent antagonists (e.g., **62**, $IC_{50} = 420$ nM) possessed two chlorophenyl substituents that occupy two of the three hydrophobic pockets of the MDM2 cleft, while an iodophenyl or chlorophenyl group (iodo better than chloro) occupies the third hydrophobic pocket. The antitumor activities of these compounds have not yet been published.

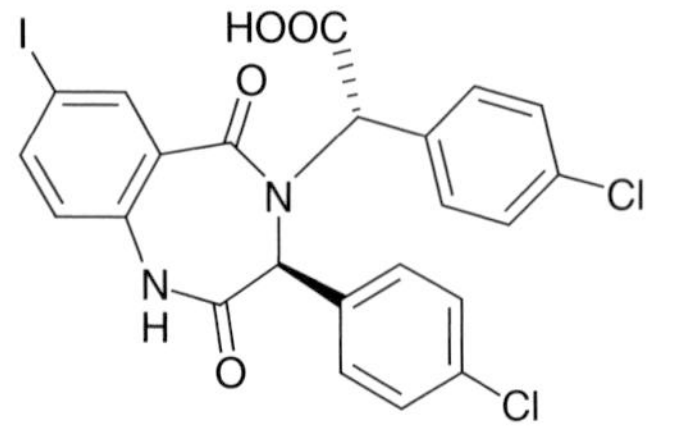

62

7.09.5.5.2.3 Structure-based design of p53–MDM2 inhibitors

The discovery of various classes of ligands that competitively bind at the N-terminal p53 binding site on MDM2, and the x-ray crystallographic elucidation of the 3D structure of the p53–MDM2 binding interface,[210] have triggered attempts for structure-based drug design approaches to the discovery of small-molecule inhibitors. The binding interaction between MDM2 and p53 at the p53 transactivation domain and the MDM2 N-terminal domain has been well characterized.[211] The N-terminal domain of MDM2 binds to the p53 transactivation domain. The MDM2 N-terminal domain has an unique fold with an extensive hydrophobic cleft at its center. There are three prominent hydrophobic pockets within the p53 binding site on MDM2 that are occupied by the hydrophobic side chains of p53 residues Phe19, Tryp23, and Leu26. Available information suggests that occupation of all three pockets is required for highly potent inhibitors. Therefore, the structure-based design of small molecules has focused on these three hydrophobic pockets. Several hydrophobic groups, especially phenyl, *para*-chlorophenyl, *para*-bromophenyl, and *para*-iodophenyl groups, have been shown to be suitable occupants. The specific moieties may vary depending on the scaffold employed to anchor the groups.

A de novo structure-based design approach based on the human p53–MDM2 complex structure was used to synthesize and test a series of new norbornane derivatives designated as syc compounds.[212] These compounds had hydrophobic groups attached to the norbornane scaffold, and were intended to take advantage of the hydrophobic pockets occupied by p53 residues Phe19, Trp23, and Leu26. Five of 23 synthetic compounds that were tested possessed concentration-dependent affinity for MDM2. These compounds (syc-7, syc-8, syc-11, syc-12, and syc-13) also showed moderate growth inhibitory activity against MCF-7 (breast), NCI-H446 (small cell lung), HCT-8 (colon), and HeLa (cervical) cancer cell lines. The compound designated syc-7 (2-benzoyloxy-3-hydroxy-5,6-(*N*,*N*′-diphenyl)-carboxamide-dicyclo[2.2.1]-heptane, **63**) showed about fivefold selectivity between MCF7 cells and NEC normal cells, stimulated p53 and p21 accumulation, and induced apoptosis.

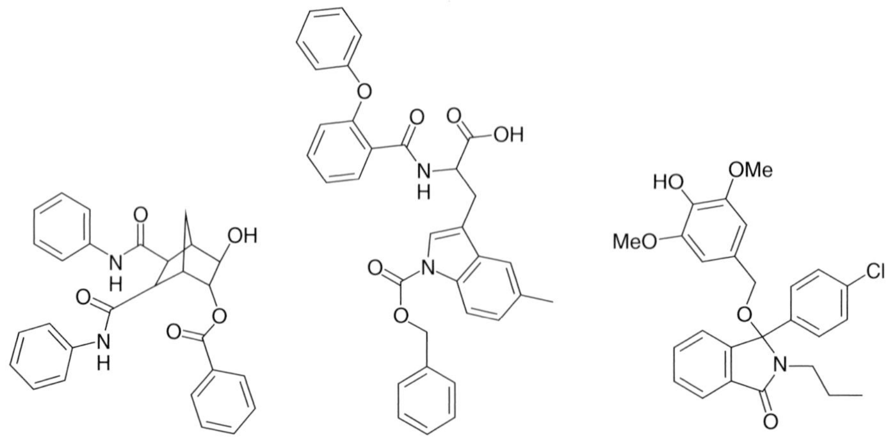

63 **64** **65**

High-affinity binding tryptophan derivatives, **64**, were also recently designed based on the binding site of p53's Trp23 on MDM2.[213] Anticancer activity of this series has not been published so far. Another series of structurally novel isoindolinone inhibitors was designed by using the published structure of the MDM2–p53 binding site in combination with focused library synthesis.[214] The most potent compound, **65**, has an IC_{50} of 5.3 μM in a cell-free binding assay and shows dose-dependent induction of MDM2 and p21 when used to treat an intact MDM2 amplified human sarcoma cell line.

7.09.5.5.2.4 Targeting other domains of p53–MDM2 interactions

Besides the binding of p53 to MDM2 at the N-terminal domain, other domains of the MDM2 oncoprotein interact with p53. For example the RING finger domain of MDM2 interacts with p53 as an E3 ligase to ubiquitinate p53 and target it for proteasomal destruction. This interaction has also been shown to be amenable to inhibition by small molecules.[215] Three small-molecule compounds, an anilidosulfonamide, **66**, a bis-(amidinophenyl)-urea, **67**, and a benzoylimidazolone, **68**, can inhibit the E3 ligase activity of MDM2, and antagonize p53 ubiquitination. This area provides an opportunity for identifying compounds that may synergize with the p53–MDM2 binding inhibitors in multipronged approaches to targeting MDM2 for cancer therapy.

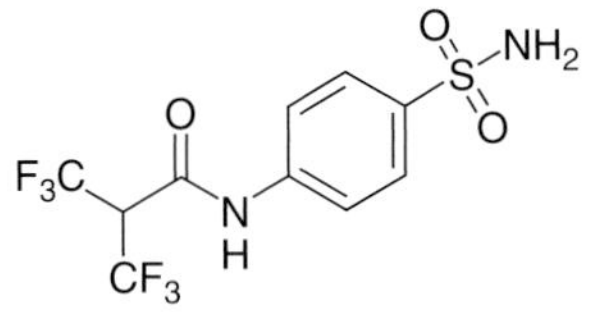

66 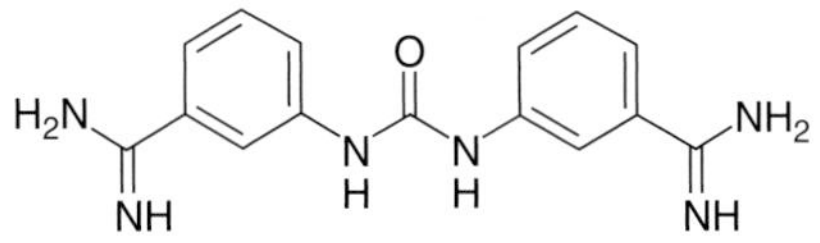 **67** 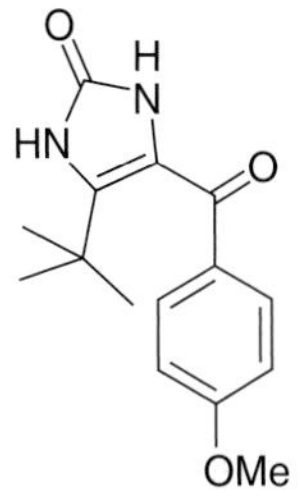 **68**

7.09.5.5.3 Conclusion

The disruption of the p53–MDM2 interaction with small-molecule inhibitors has been validated as a potentially viable and attractive cancer therapeutic target. MDM2 amplification and overexpression occurs in a wide variety of human cancers, several of which can be treated experimentally with MDM2 antagonists. Inhibition of the p53–MDM2 interaction has been shown to cause selective cancer cell death, as well as sensitize cancer cells to chemotherapy or radiation effects.

Current drug discovery efforts in this area are focused on the binding of p53 to the N-terminal domain of MDM2, which has been extensively studied at the molecular level. There is very little information available so far on small-molecules targeting other regions of MDM2, like the RING finger domain. There are still many opportunities to discover small-molecule inhibitors of MDM2 as there is a plethora of other proteins besides p53 that interact with MDM2.[216] Many of those interactions enhance the ability of MDM2 to downregulate p53, and therefore these interactions with MDM2 might be potential novel anticancer targets.

The drug development process for small-molecule MDM2 inhibitors is still at an early preclinical stage and no compounds have advanced into phase I clinical trials. Proof of principle studies have been conducted in rodents but primate studies have not yet been reported. Despite this rather early stage, targeting MDM2 holds a lot of promise for future clinical applications in multiple cancer therapies.[217] MDM2 antagonists should be effective chemo- and radiation sensitizing agents in treatment of tumors expressing functional p53, which constitute approximately half of all human cancers.[218] It should be noted that although MDM2 overexpression is oncogenic and indicates poor prognosis in most situations, there are certain cases where overexpression of MDM2 may be associated with good prognosis. Therefore there may be some caveats to using MDM2 inhibitors in cancer therapy, but in most cases they may prove to be very useful novel anticancer agents.

7.09.5.6 Matrix Metalloproteinase Inhibitors

7.09.5.6.1 Basic facts

It has been known for many years that proteolytic enzymes play a role in tumor invasion. This process is initiated via the degradation of extracellular matrix (ECM), basement membranes, basal laminae, and interstitial stroma by the proteolytic enzymes to allow invasive cells to migrate into adjacent tissues. Matrix metalloproteinases (MMPs) are a family of calcium- and zinc-containing endopeptidases involved in the degradation and remodeling of ECM proteins, tissue remodelling, and wound healing.[219–222] Extensive biochemical and molecular biological studies have revealed that they play an essential role in tumor growth and angiogenesis.[223] The activity of MMPs is regulated by endogenous tissue inhibitors of matrix metalloproteinases (TIMPs). In the presence of specific stimuli like cytokines and growth factors, MMPs are upregulated and thereby the balance between MMPs and TIMPs is destroyed. This results in the chronic activation of MMPs and an excessive degradation of ECM components, which are believed to contribute to numerous pathological conditions, e.g., cancer, osteoarthritis, rheumatoid arthritis, angiogenesis, periodontal disease, pulmonary emphysema, skin ulceration, atherosclerosis, and central nervous system diseases.[224] So far at least 24 mammalian MMPs have been identified, and they can be subdivided into five classes: collagenases, gelatinases, stromelysins, membrane-type MMPs (MT-MMPs), and other enzymes. These enzymes have considerable overlap in their substrate specificity and their domain structure has a high degree of similarity. All family members contain a pre-domain involved in enzyme secretion, a pro-domain that is autoinhibitory, a catalytic domain responsible for enzyme activity and, with the exception of matrilysins (MMP-7 and MMP-26) and MMP-23, a C-terminal domain involved in substrate recognition. The sixth member of MT-MMP (MMP-25) has been identified. It is expressed at high levels in brain tumors and colon carcinoma cells but not in normal brain or colon.[225,226] In contrast, MMP-26 is widely expressed in cancer cells of epithelial origin.[227] The most recently discovered member of MMPs is MMP-28. Structurally it is relatively similar to MMP-19 and is overexpressed in testis and keratinocytes as well as in carcinomas.[228] Despite these findings the pathological role of these MMPs is still under evaluation. Under normal physiological conditions, the enzymatic activity of MMPs is generally very low and is well regulated by TIMPs. Expression of these enzymes has been shown to participate in various normal biological processes such as embryonic development, wound healing, angiogenesis, ovulation, and nerve growth. However, chronic or stimulated activation of MMPs may result in an imbalance between the activity of these proteolytic enzymes and TIMPs. As a consequence pathological conditions may occur due to an excessive degradation of ECM or related components by the MMPs. Therefore MMP inhibitors may be beneficial for the treatment of these disorders and several reviews on the medicinal chemistry of MMP inhibitors have appeared recently.[229–231]

7.09.5.6.2 Matrix metalloproteinase inhibitors in development

Several MMP inhibitors are in various developmental stages for different symptoms, mostly in cancer[232,233] and rheumatoid arthritis. Compounds tested in clinical trials as MMP inhibitors include marimastat, **69**, tanomastat (Bay-129566), **70**, prinomastat (AG3340), **71**, and batimastat, **72**. All these compounds have been applied to treat different types of cancer, such as ovarian cancer, breast cancer, malignant glioma, pancreatic cancer, NSCLC, and advanced bladder carcinoma.[234] Although these compounds possess different inhibitory potencies toward the various MMPs, none of them was found to be selective for a particular enzyme. Unfortunately most of these clinical trials of MMPIs have yielded disappointing results so far. Positive results have been achieved in gastric cancer with marimastat, **69**.[235]

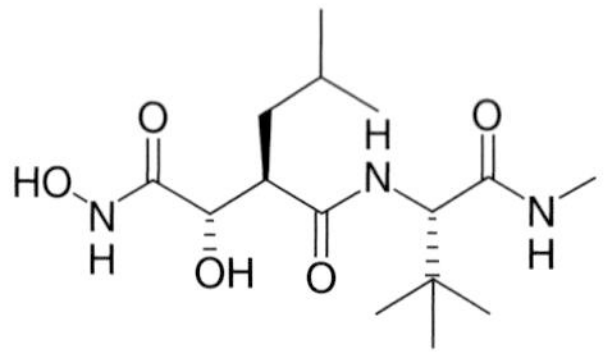

69

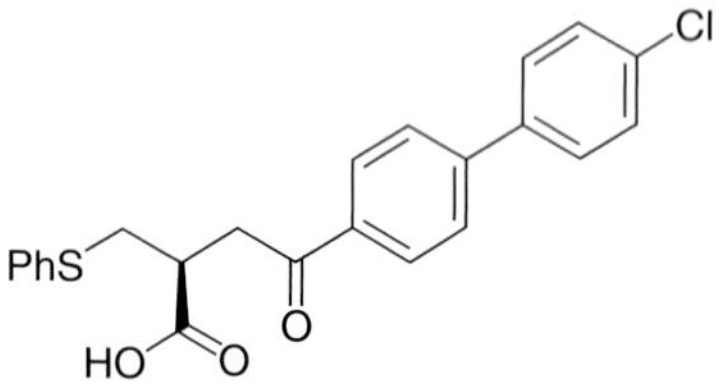

70

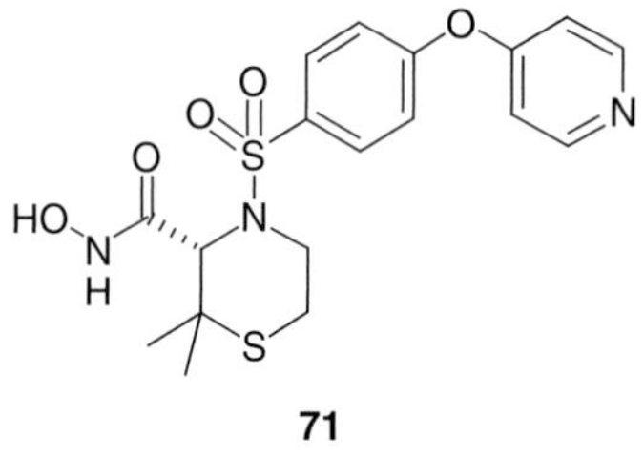

71

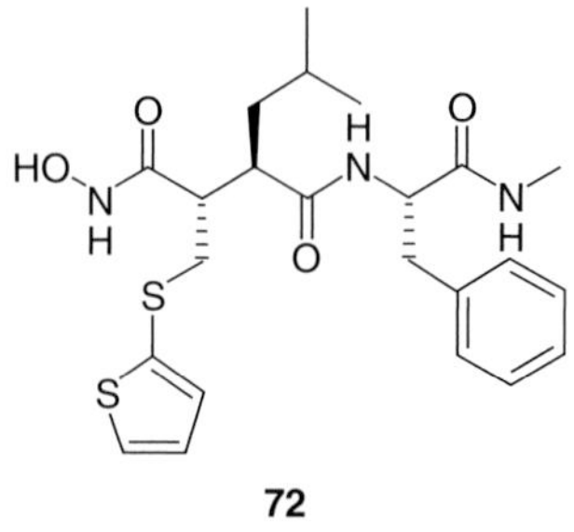

72

7.09.5.6.2.1 Batimastatat

Batimastat, **72**, was the first MMP inhibitor of the hydroxamic acid series which was tested in clinical trials for treatment of cancer. In 2002, development of batimastat for all indications, including cardiac failure and cerebral infarction, was discontinued.

7.09.5.6.2.2 Marimastat

The results of three clinical studies of marimastat (hydroxamic acid series), **69**, for lung cancer have been reported. In 2001, it was reported that marimastat did not show any beneficial effect over placebo in two phase III clinical trials in patients with small cell lung cancer. Severe musculoskeletal pain was noted in 18% of patients treated with marimastat and quality of life was significantly worse for these patients. Marimastat did not improve the time to progression or overall survival in this study.[236]

Glioblastoma multiforme is the most commonly diagnosed malignant primary brain tumor in adults. Oral treatment with temozolomide was found to be beneficial for patients with recurrent glioblastoma multiforme and therefore a combination of temozolomide and marimastat was evaluated. Despite promising results in a phase II study a phase III clinical trial involving 162 patients with glioblastoma multiforme or gliosarcoma following surgery and radiotherapy showed no significant difference between the marimastat- and placebo-treated groups in terms of survival.[237]

Another phase III clinical study was conducted in patients with advanced gastric cancer. The patients received 10 mg marimastat. Overall survival in this group of patients was 18% compared with 5% in the placebo group.[238] These data support the utility of marimastat as a maintenance treatment in gastric cancer patients following chemotherapy. In other phase III clinical trials of marimastat in patients with metastatic breast cancer,[239] advanced ovarian cancer, and inoperable colorectal cancer liver metastases no significant advantage for treatment with marimastat in overall survival rate was found.[240]

7.09.5.6.2.3 Prinomastat

Prinomastat, **71**, is a member of the sulfonamide class of MMP inhibitors. It was tested in a number of phase II clinical trials to evaluate its effects in esophageal adenocarcinoma,[241] glioblastoma multiforme,[242] metastatic melanoma,[243] and progressive breast cancer.[244] The study in esophageal adenocarcimoma was prematurely terminated due to a high number of patients suffering unexpected life-threatening thromboembolic events. The glioblastoma multiforme and metastatic melanoma clinical trials failed to show clinical efficacy. However, promising results were obtained in the metastatic breast cancer study. Phase III clinical studies involving patients with metastatic, hormone-refractory prostate cancer and in patients with NSCLC did not show significant differences in progression-free survival or overall survival.[245]

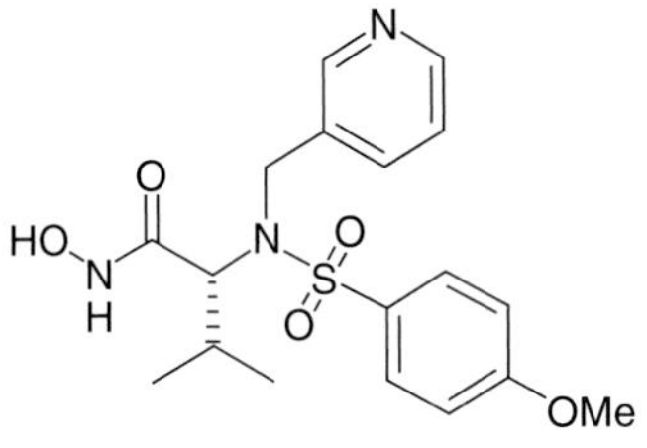

73 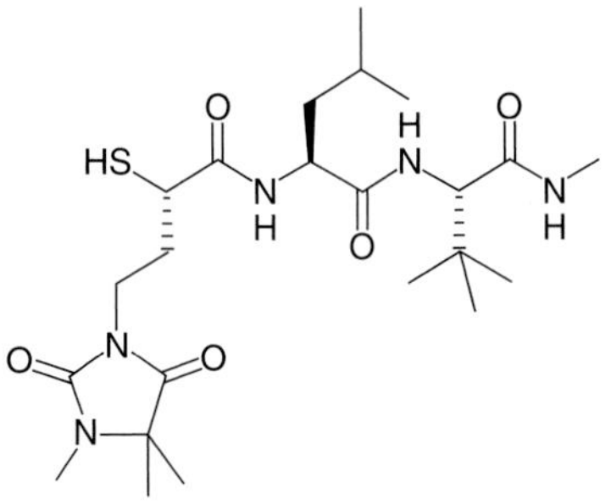 **74**

7.09.5.6.2.4 MMI-270

A phase I clinical study of MMI-270 (hydroxamic acid derivative), **73**, in advanced solid cancer was performed. The primary aims were to evaluate toxicity and pharmacokinetics, while the secondary aims were to measure tumor response and changes in biological markers. Stable disease for 90 days or longer was seen in 19 patients, but no tumor regression was found.[246]

7.09.5.6.2.5 BMS-275291 (D-2163)

In a phase II clinical study, 75 patients with stage IIIb or IV NSCLC were given paclitaxel and carboplatin in combination with BMS-275291 (hydantoin series), **74**. BMS-275291 was generally well tolerated when given in combination with paclitaxel/carboplatin. It was not associated with dose-limiting arthrotoxicity and did not appear to impact adversely on early tumor shrinkage with chemotherapy. The study has progressed to phase III to evaluate the effect of BMS-275291 in combination with paclitaxel/carboplatin on progression-free survival and overall survival.[247]

7.09.5.6.2.6 Hydroxyproline-derivatives

Hydroxyproline is known as one of the specific amino acids of collagens, which are the substrates of MMPs. Therefore it was assumed that the derivatives of hydroxyproline might specifically interact with MMPs in a competitive manner.

Caffeic acid or gallic acid have proved to inhibit MMP-2 and MMP-9, and therefore they were linked with hydroxyproline to find potent compounds with inhibiting activity against MMP-2 and MMP-9. It was demonstrated that pyrrolidine peptidomimetic inhibitors **75** and **76** have high inhibitory activity against MMP-2 and -9 with IC_{50} of 11.5 and 7.7 nM.[248] These derivatives also displayed favorable inhibitory potency to metastasis of tumor cells with the metastasis inhibition rate of H22 mouse liver carcinoma model higher than 92%.

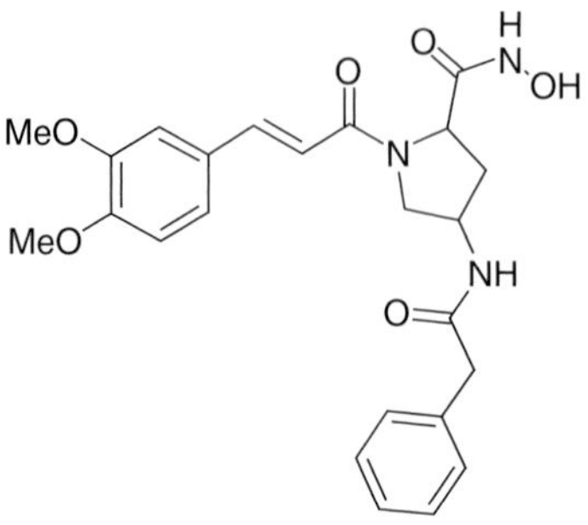

75 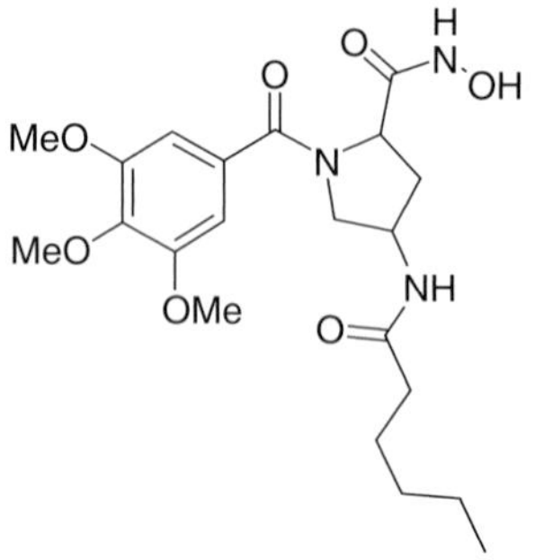 **76**

7.09.5.6.3 Conclusion

Preclinical studies during the past two decades have documented the pathogenic role of MMPs in various diseases, including cancer and arthritis. The possibility that anticancer effects might be realized by limiting angiogenesis using MMP inhibitors in combination with cytotoxic drugs or radiation therapy to reduce tumor growth and metastasis has generated considerable interest in this compound class. Therefore a number of lower molecular-weight MMP inhibitors have entered clinical trials. Unfortunately, the results of these trials have been extremely disappointing and have led many investigators to conclude that MMP inhibitors have no therapeutic benefit in human cancer. The first-generation MMP inhibitors exhibited poor bioavailability while second-generation compounds revealed that prolonged treatment caused musculoskeletal pain and inflammation or had a lack of efficacy. The fact that MMP inhibitors failed to demonstrate efficacy in clinical studies, in contrast to the results observed in preclinical experiments, gives rise to the question whether the animal models have been really adequate. For many human diseases and particularly in various types of advanced cancers, it is still difficult to find the right animal models. Therefore, positive effects of MMP inhibitors obtained in preclinical studies are not necessarily reproduced in human clinical trials. In order to improve the clinical results, care must be taken to make rational decisions on the appropriate enzymes based on correlation with clinical outcome. As it is now known that, e.g., the expression of MMP-11 and/or MMP-14, but not MMP-2, is a negative prognostic marker for small cell lung cancer in human, compounds such as tanomastat that possess potent inhibitory activity against MMP-2 but not MMP-11 are unlikely to demonstrate efficacy in this disease and should therefore be used to treat other more appropriate cancer type.[249] It is not unexpected that many MMP inhibitors display broad-spectrum inhibition of the MMP family of enzymes, as the majority of compounds have been designed by incorporating a zinc-binding ligand such as a hydroxamic acid, a thiol, a carboxylic acid, or a phosphorus-containing group to chelate the zinc ion at the catalytic center of the enzyme. Although this approach has successfully generated potent MMP inhibitors, the selectivity of these compounds toward various MMP members is still not predictable. Therefore the discovery of potent and selective MMP-13 inhibitors, (e.g., **77**, MMP-13 IC_{50} 6.2 nM, **78**, MMP-13 IC_{50} 5.6 nM, **79**, MMP-13 IC_{50} 8 nM) which do not contain any of the traditional zinc-binding groups represents an advance in structure-based inhibitor design.[250,251]

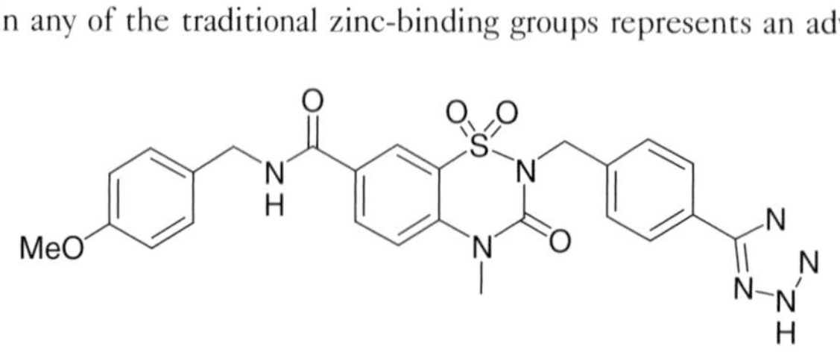

77 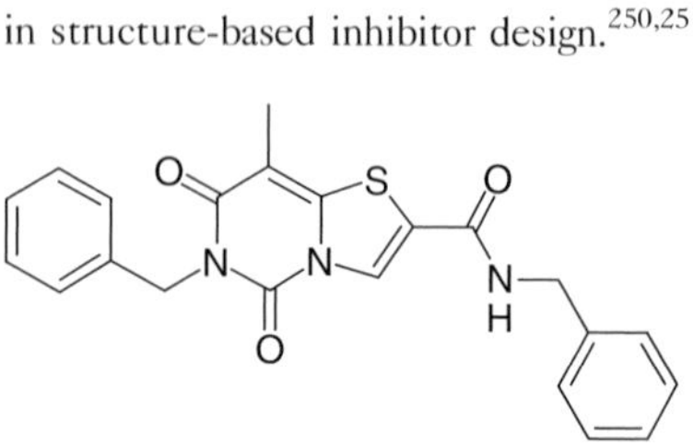 **78**

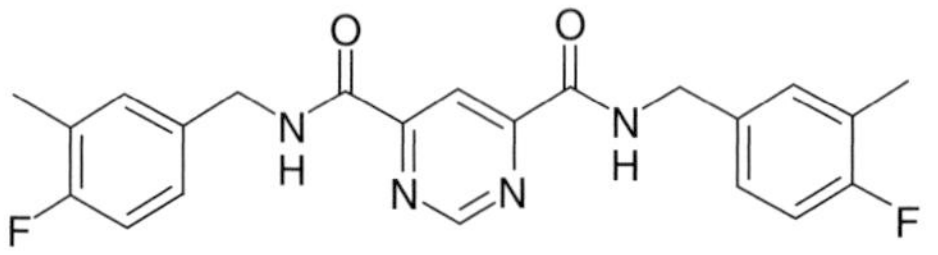

79

Overall the development of MMP inhibitors for the treatment of cancer and arthritis has been more challenging than originally expected. The first-generation MMP inhibitors have shown poor bioavailability, while the second-generation compounds suffer from a lack of clinical efficacy and serious side effects. To achieve effectiveness in the future, third-generation MMP inhibitors will require a better understanding of the role of MMPs involved in human diseases together with structure-based design of more selective inhibitors of the relevant enzymes.

7.09.6 Unmet Medical Needs

Despite the fact that chemotherapy has become an important component of cancer treatment over the past 60 years, this medical treatment option still has many unmet needs that limit its utility for cancer treatment and satisfactory results can be obtained in a small range of cancers only.[252] Chemotherapeutic regimens have yielded cures in several childhood cancers and certain adult malignancies such as lymphoma, leukemia, or testicular cancer. Long-lasting remission and improved survival can also result from adjuvant drug treatment of breast and colorectal cancer. Sometimes, mechanistically based chemotherapies are clinically effective, but only for a short period of time. For example, antihormonal treatments of prostate and breast cancer can initially shrink tumors but fail when tumor cell growth becomes hormone-independent. For many others, there are drugs with only a limited efficacy and existing chemotherapeutic treatments are largely palliative, in particular for those tumors that are in their advanced metastatic form. Another problem related to chemotherapy is that efficacy is highly variable with only limited ability to predict the outcome of individual patients. Genetic variability exists not only between different tumor types, but also between tumors with apparently similar pathological features in different patients. Investigation of tumor genotype of each patient might provide useful information related to a more effective therapy. Furthermore the toxic invariably associated with drug treatment limits the dose of drug that can be used and the low therapeutic index prevents the clinical development of potentially effective agents. A further problem is that after a first line of successful treatment, the majority of tumors that are sensitive to chemotherapy relapse and develop resistance mechanisms not only toward the same class of drugs, but also against structurally different compounds ('multidrug resistance' or MDR). The identification of biological differences between tumor and normal cells which will allow the development of innovative selective therapies designed to specifically block the mechanisms that account for malignant transformation is crucial for further progress in cancer chemotherapy.

7.09.7 New Research Areas

A significant fraction of current research in this area is now moving away from relatively nonselective cytotoxic drugs toward the generation of new therapeutic agents that target the key molecular abnormalities that drive malignant transformation and progression and which therefore should have a fundamental impact on cancer cell survival.[253] The current view of anticancer drug discovery and development is based on the concept that more selective and effective therapies will exhibit less harmful effects to normal cells than traditional cytotoxic agents and will emerge by identifying the genetic defects that create and drive the malignant phenotype. These attempts are also supported by the recent advances of new technologies and the progress made in the field of molecular cell biology.[254] A couple of these promising new research areas have been summarized in this article and there is still a plethora of promising potential targets for cancer drugs to be explored by future research efforts.

References

1. Greenlee, R. T.; Murray, T.; Bolden, S.; Wingo, P. A. *Cancer J. Clin.* **2000**, *50*, 7–33.
2. Silverberg, E. *Cancer J. Clin.* **1985**, *35*, 19–35.
3. Welch, H. G.; Schwartz, L. M.; Woloshin, S. *JAMA* **2000**, *284*, 2053–2055.
4. Li, Q.; Xu, W. *Curr. Med. Chem. Anti-Cancer Agents* **2005**, *5*, 53–63.
5. Kamb, A. *Nat. Rev. Drug Disc.* **2005**, *4*, 161–165.
6. Oskoglou-Nomikos, T.; Pater, J. L.; Seymour, L. *Clin. Cancer Res.* **2003**, *9*, 4227–4239.
7. Peterson, J. K.; Houghton, P. J. *Eur. J. Cancer* **2004**, *40*, 837–844.
8. Booth, B.; Glassman, R.; Ma, P. *Nat. Rev. Drug Disc.* **2003**, *2*, 609–610.
9. Mai, A.; Massa, A.; Rotili, D.; Cerbara, I.; Valente, S.; Pezzi, R.; Simeoni, S.; Ragno, R. *Med. Res. Rev.* **2005**, *25*, 261–309.
10. Marks, P. A.; Richon, V. M.; Miller, T.; Kelly, W. K. *Adv. Cancer Res.* **2004**, *91*, 137–168.
11. McLaughlin, F.; La Thangue, N. *Biochem. Pharmacol.* **2004**, *68*, 1139–1144.
12. Mei, S.; Ho, A. D.; Mahlknecht, U. *Int. J. Oncology* **2004**, *25*, 1509–1519.
13. Rosato, R.; Grant, S. *Expert Opin. Invest. Drugs* **2004**, *13*, 21–38.

14. Marks, P. A.; Richon, V. M.; Breslow, R.; Rifkind, R. A. *Curr. Opin. Oncol.* **2001**, *13*, 477–483.
15. Kelly, W. K.; O'Connor, O. A.; Marks, P. A. *Expert Opin. Invest. Drugs* **2002**, *11*, 1695–1713.
16. Yoshida, M.; Furumai, R.; Nishiyama, M.; Komatsu, Y.; Nishino, N.; Horinouchi, S. *Cancer Chemother. Pharmacol.* **2001**, *48*, S20–S26.
17. De Ruijter, A. J. M.; van Gennip, A. H.; Caron, H. N.; Kemp, S.; van Kuilenburg, A. B. P. *Biochem. J.* **2003**, *370*, 737–749.
18. Gray, G. G.; Ekstrom, T. J. *Exp. Cell Res.* **2001**, *262*, 75–83.
19. Gao, L.; Cueto, M. A.; Asselbergs, F.; Ataoja, P. *J. Biol. Chem.* **2002**, *277*, 25748–25755.
20. Finnin, M. S.; Donigian, J. R.; Cohen, A.; Richon, V. M.; Rifkind, R. A.; Marks, P.; Breslow, R.; Pavletich, N. P. *Nature* **1999**, *401*, 188–193.
21. Somoza, J. R.; Skene, R. J.; Katz, B. A.; Mol, C.; Ho, J. D.; Jennings, A. J.; Luong, C.; Arvai, A.; Buggy, J. J.; Chi, E. et al. *Structure* **2004**, *12*, 1325–1334.
22. Vannini, A.; Volpari, C.; Filocamo, G.; Casavola, E. C.; Brunetti, M.; Renzoni, D.; Chakravarty, P.; Paolini, C.; De Francesco, R.; Gallinari, P. et al. *Proc. Natl. Acad. Sci. USA* **2004**, *101*, 15064–15069.
23. Yoshida, M.; Matsuyama, A.; Komatsu, Y.; Nishino, N. *Curr. Med. Chem.* **2003**, *10*, 2351–2358.
24. Kim, M. S.; Kwon, H. J.; Lee, Y. M.; Baek, J. H.; Jang, J. E.; Lee, S. W.; Moon, E. J.; Kim, H. S.; Lee, S. K.; Chung, H. Y. et al. *Nat. Med.* **2001**, 7, 437–443.
25. Monneret, C. *Eur. J. Med. Chem.* **2005**, *40*, 1–13.
26. Miller, T. A.; Witter, D. J.; Belvedere, S. *J. Med. Chem.* **2003**, *46*, 5097–5117.
27. Gomez-Vidal, J. A.; Campos, J.; Marchal, J. A.; Boulaiz, H.; Gallo, M. A.; Carrillo, E.; Espinosa, A.; Aranega, A. *Curr. Topics Med. Chem.* **2004**, *4*, 175–202.
28. Remiszewski, S. W. *Curr. Opin. Drug Disc. Dev.* **2002**, *5*, 487–499.
29. Marks, P. A.; Miller, T.; Richon, V. M. *Curr. Opin. Pharmacol.* **2003**, *3*, 344–351.
30. Weinmann, H.; Ottow, E. *Annu. Rep. Med. Chem.* **2004**, *39*, 185–196.
31. Newmark, H. L.; Lupton, J. R.; Young, C. W. *Cancer Lett.* **2001**, *78*, 1–9.
32. DiGiuseppe, J. A.; Wenig, L. J.; Yu, K. H.; Fu, S.; Kastan, M. B.; Samid, D.; Gore, S. D. *Leukemia* **1999**, *13*, 1243–1253.
33. Göttlicher, M.; Minucci, S.; Zhu, P.; Krämer, O. H.; Schimpf, A.; Giavara, S.; Sleeman, J. P.; Lo Coco, F.; Nervi, C.; Pelicci, P. G. et al. *EMBO J.* **2001**, *20*, 6969–6978.
34. Lu, Q.; Yang, Y.-T.; Chen, C.-S.; Davis, M.; Byrd, J. C.; Etherton, M. R.; Umar, A.; Chen, C.-S. *J. Med. Chem.* **2004**, *47*, 467–474.
35. Yang, K.; Lou, B. *Mini Rev. Med. Chem.* **2003**, *3*, 349–360.
36. Lou, B.; Yang, K. *Mini Rev. Med. Chem.* **2003**, *3*, 609–620.
37. Curtin, M. L. *Curr. Opin. Drug Disc. Dev.* **2004**, 7, 848–868.
38. Vickerstaffe, E.; Warrington, B. H.; Ladlow, M.; Ley, S. V. *Org. Biomol. Chem.* **2003**, *1*, 2419–2422.
39. Shabbeer, S.; Carducci, M. A. *Invest. Drugs* **2005**, *8*, 144–154.
40. McLaughlin, F.; La Thangue, N. B. *Biochem. Pharmacol.* **2004**, *68*, 1139–1144.
41. Remiszewski, S. W. *Curr. Med. Chem.* **2003**, *10*, 2393–2402.
42. Remiszewski, S.; Sambucetti, L.; Bair, K.; Bontempo, J.; Cesarz, D.; Chandramouli, N.; Chen, R.; Cheung, M.; Cornell-Kennon, S.; Dean, K. et al. *J. Med. Chem.* **2003**, *46*, 4609–4624.
43. Atadja, P.; Gao, L.; Kwon, P.; Trogani, N.; Walker, H.; Hsu, M.; Yeleswarapu, L.; Chandramouli, N.; Perez, L.; Versace, R. et al. *Cancer Res.* **2004**, *64*, 689–695.
44. Bouchain, G.; Leit, S.; Frechette, S.; Khalil, E. A.; Lavoie, R.; Moradei, O.; Woo, S. H.; Fournel, M.; Yan, P. T.; Kalita, A. et al. *J. Med. Chem.* **2003**, *46*, 820–830.
45. Bouchain, G.; Delorme, D. *Curr. Med. Chem.* **2003**, *10*, 2359–2372.
46. Plumb, J. A.; Finn, P. W.; Williams, R. J.; Bandara, M. J.; Romero, M. R.; Watkins, C. J.; La Thangue, N. B.; Brown, R. *Mol. Cancer Ther.* **2003**, *2*, 721–728.
47. Watkins, C. J.; Romero-Martin, M.-R.; Ritchie, J.; Finn, P. W.; Kalvinsh, I.; Loza, E.; Dikovska, K.; Starchenkov, I.; Lolya, D.; Gailite, V. PCT Int. Patent Appl., WO2003082288-A1, 2003.
48. Kim, D.-K.; Lee, J. Y.; Kim, J.-S.; Ryu, J.-H.; Choi, J.-Y.; Lee, J. W.; Im, G.-J.; Kim, T.-K.; Seo, J. W.; Park, H.-J. *J. Med. Chem.* **2003**, *46*, 5745–5751.
49. Kim, D.-K.; Lee, J. Y.; Lee, N. K.; Kim, J.-S.; Ryu, J.-H.; Lee, J. W.; Lee, S. H.; Choi, J. Y.; Kim, N. H.; Im, G.-J. PCT Int. Patent Appl., WO2003087066-A1, 2003.
50. Mai, A.; Massa, S.; Ragno, R.; Cerbara, I.; Jesacher, F.; Loidl, P.; Brosch, G. *J. Med. Chem.* **2003**, *46*, 512–524.
51. Mai, A.; Massa, S.; Cerbara, I.; Valente, S.; Ragno, R.; Bottoni, P.; Scatena, R.; Loidl, P.; Brosch, G. *J. Med. Chem.* **2004**, *47*, 1098–1109.
52. Arts, J. American Association for Cancer Research (AACR)/National Cancer Institute (NCI)/European Organization for Research and Treatment of Cancer (EORTC) Intl. Conf. Mol. Targets Cancer Ther., Boston, **2003**, Abstr. A153.
53. Van Emelen, K.; Arts, J.; Backx, L. J. J.; De Winter, H. L. J.; Van Brandt, S. F. A.; Verdonck, M. G. C.; Meerpoel, L.; Pilatte, I. N. C.; Poncelet, V. S.; Dyatkin, A. B. PCT Int. Patent Appl., WO2003076422-A1, 2003.
54. Van Emelen, K.; PCT Int. Patent Appl., WO2003076438-A1, 2003.
55. Van Emelen, K.; Backx, L. J. J.; Van Brandt, S. F. A.; Angibaud, P. R.; Pilatte, I. N. C.; Verdonck, M. G. C.; De Winter, H. L. J. PCT Int. Patent Appl., WO2003076401-A1, 2003.
56. Van Emelen, K.; De Winter, H. L. J.; Dyatkin, A. B.; Verdonck, M. G. C.; Meerpoel, L. PCT Int. Patent Appl., WO2003076421-A1, 2003.
57. Van Emelen, K.; Verdonck, M. G. C.; Van Brandt, S. F. A.; Angibaud, P. R.; Meerpoel, L.; Dyatkin, A. B. PCT Int. Patent Appl., WO2003075929-A1, 2003.
58. Angibaud, P. R.; Pilatte, I. N. C.; Van Brandt, S. F. A.; Roux, B.; Ten Holte, P.; Verdonck, M. G. C.; Meerpoel, L.; Dyatkin, A. B. PCT Int. Patent Appl., WO2003076400-A1, 2003.
59. Leser-Reiff, U.; Sattelkau, T.; Zimmermann, G. PCT Int. Patent, WO2003011851-A2, 2003.
60. Dai, Y.; Guo, Y.; Guo, J.; Pease, L. J.; Li, J.; Marcotte, P. A.; Glaser, K. B.; Tapang, P.; Albert, D. H.; Richardson, P. L. et al. *Bioorg. Med. Chem. Lett.* **2003**, *13*, 1897–1901.
61. Curtin, M. L.; Garland, R. B.; Heyman, H. R.; Frey, R. R.; Michaelides, M. R.; Li, J.; Pease, L. J.; Glaser, K. B.; Marcotte, P. A.; Davidsen, S. K. *Bioorg. Med. Chem. Lett.* **2002**, *12*, 2919–2923.
62. Curtin, M.; Glaser, K. *Curr. Med. Chem.* **2003**, *10*, 2373–2392.
63. Vanhaecke, T.; Papeleu, P.; Elaut, G.; Rogiers, V. *Curr. Med. Chem.* **2004**, *11*, 1629–1643.

64. Frey, R. R.; Wada, C. K.; Garland, R. B.; Curtin, M.; Michaelides, M. R.; Li, J.; Pease, L. J.; Glaser, K. B.; Marcotte, P. A.; Bouska, J. J. et al. *Bioorg. Med. Chem. Lett.* **2002**, *12*, 3443–3447.
65. Vasudevan, A.; Ji, Z.; Frey, R. R.; Wada, C. K.; Steinman, D.; Heyman, H. R.; Guo, Y.; Curtin, M. L.; Guo, J.; Li, J. et al. *Bioorg. Med. Chem. Lett.* **2003**, *13*, 3909–3913.
66. Wada, C. K.; Frey, R. R.; Ji, Z.; Curtin, M. L.; Garland, R. B.; Holms, J. H.; Li, J.; Pease, L. J.; Guo, J.; Glaser, K. B. et al. *Bioorg. Med. Chem. Lett.* **2003**, *13*, 3331–3335.
67. Dai, Y.; Guo, Y.; Curtin, M. L.; Li, J.; Pease, L. J.; Guo, J.; Marcotte, P. A.; Glaser, K. B.; Davidsen, S. K.; Michaelides, M. R. *Bioorg. Med. Chem. Lett.* **2003**, *13*, 3817–3820.
68. Kapustin, G. V.; Fejer, G.; Gronlund, J. L.; McCafferty, D. G.; Seto, E.; Etzkorn, F. A. *Org. Lett.* **2003**, *5*, 3053–3056.
69. Wu, T. Y. H.; Hassig, C.; Wu, Y.; Ding, S.; Schultz, P. G. *Bioorg. Med. Chem. Lett.* **2004**, *14*, 449–453.
70. Suzuki, T.; Matsuura, A.; Kouketsu, A.; Nakagawa, H.; Miyata, N. *Bioorg. Med. Chem. Lett.* **2005**, *15*, 331–335.
71. Chen, B.; Petukhov, P. A.; Jung, M.; Velena, A.; Eliseeva, E.; Dritschilo, A.; Kozikowski, A. P. *Bioorg. Med. Chem. Lett.* **2005**, *15*, 1389–1392.
72. Suzuki, T.; Ando, T.; Tsuchiya, K.; Fukazawa, N.; Saito, A. *J. Med. Chem.* **1999**, *42*, 3001–3003.
73. Wang, J.; Woo, S. H.; Fournel, M.; Yan, P.; Trachy-Bourget, M. C.; Kalita, A.; Beaulieu, C.; Li, Z.; MacLeod, A. R.; Besterman, J. M. et al. *Bioorg. Med. Chem. Lett.* **2004**, *14*, 283–287.
74. Stokes, E. S. E.; Roberts, C. A.; Waring, M. J. PCT Int. Patent, WO2003087057- A1, 2003.
75. Yurek-George, A.; Habens, F.; Brimmell, M.; Packham, G.; Ganesan, A. *J. Am. Chem. Soc.* **2004**, *126*, 1030–1031.
76. Nishino, N.; Jose, B.; Okamura, S.; Ebisusaki, S.; Kato, T.; Sumida, Y.; Yoshida, M. *Org. Lett.* **2003**, *5*, 5079–5082.
77. Jose, B.; Oniki, Y.; Kato, T.; Nishino, N.; Sumida, Y.; Yoshida, M. *Bioorg. Med. Chem. Lett.* **2004**, *14*, 5343–5346.
78. Bertos, N. R.; Wang, A. H.; Yang, X. J. *Biochem. Cell Biol.* **2001**, *79*, 243–252.
79. Heltweg, B.; Dequiedt, F.; Marshall, B. L.; Brauch, C.; Yoshida, M.; Nishino, N.; Verdin, E.; Jung, M. *J. Med. Chem.* **2004**, *47*, 5235–5243.
80. Sternson, S. M.; Wong, J. C.; Grozinger, C. M.; Schreiber, S. L. *Org. Lett.* **2001**, *3*, 4239–4342.
81. Koeller, K. M.; Haggarty, S. J.; Perkins, B. D.; Leykin, I.; Wong, J. C.; Kao, M. J.; Schreiber, S. L. *Chem. Biol.* **2003**, *10*, 397–410.
82. Wong, J. C.; Hong, R.; Schreiber, S. L. *J. Am. Chem. Soc.* **2003**, *125*, 5586–5587.
83. Haggarty, S. J.; Koeller, K. M.; Wong, J. C.; Grozinger, C. M.; Schreiber, S. L. *Proc. Natl. Acad. Sci. USA* **2003**, *100*, 4389–4394.
84. Haggarty, S. J.; Koeller, K. M.; Wong, J. C.; Butcher, R. A.; Schreiber, S. L. *Chem. Biol.* **2003**, *10*, 383–396.
85. Mai, A.; Massa, S.; Pezzi, R.; Rotili, D.; Loidl, P.; Brosch, G. *J. Med. Chem.* **2003**, *46*, 4826–4829.
86. Hu, E.; Dul, E.; Sung, C.; Chen, Z.; Kirkpatrick, R.; Zhang, G.-F.; Johanson, K.; Liu, R.; Lago, A.; Hofmann, G. et al. *J. Pharmacol. Exp. Ther.* **2003**, *307*, 720–728.
87. Besterman, J. M.; Li, Z.; Delorme, D.; Bonfils, C. PCT Int. Patent Appl., WO2004005513-A2, 2004.
88. Workman, P. *Cancer Lett.* **2004**, *206*, 149–157.
89. Chiosis, G.; Vilenchik, M.; Kim, J.; Solit, D. *Drug Disc. Today* **2004**, *9*, 881–888.
90. Golsteyn, R. M. *Drug News Perspect.* **2004**, *17*, 405–416.
91. Kamal, A.; Thao, L.; Sensitaffar, J.; Zhang, L.; Boehm, M. F.; Fritz, L. C.; Burrows, F. L. *Nature* **2003**, *425*, 407–410.
92. Prodromou, C.; Roe, S. M.; O'Brien, R.; Ladbury, J. E.; Piper, P. W.; Pearl, L. H. *Cell* **1997**, *90*, 65–75.
93. Chene, P. *Nat. Rev. Drug Disc.* **2002**, *1*, 665–673.
94. Stebbins, C. E.; Russo, A. A.; Schneider, C.; Rosen, N.; Hartl, F. U.; Pavletich, N. P. *Cell* **1997**, *89*, 239–250.
95. Neckers, L.; Schulte, T. W.; Mimnaugh, E. *Invest. New Drugs* **1999**, *17*, 361–373.
96. Schulte, T. W.; Neckers, L.; An, W. G. *Biochem. Biophys. Res. Commun.* **1997**, *239*, 655–659.
97. Schulte, T. W.; Neckers, L. M. *Cancer Chemother. Pharmacol.* **1998**, *42*, 273–279.
98. Solit, D. B.; Zheng, F. F.; Drobnjak, M.; Munster, P. N.; Higgins, B.; Verbel, D.; Heller, G.; Tong, W.; Cordon-Cardo, C.; Agus, D. B. et al. *Clin. Cancer Res.* **2002**, *8*, 986–993.
99. Kelland, L. R.; Sharp, S. Y.; Rogers, P. M.; Myers, T. G.; Workman, P. *J. Natl. Cancer Inst.* **1999**, *91*, 1940–1949.
100. Banerji, U. *Proc. Am. Assoc. Cancer Res.* **2003**, *44*, 677–678.
101. Sausville, E. A.; Tomaszewski, J. E.; Ivy, P. *Curr. Cancer Drug Targets* **2003**, *3*, 377–383.
102. Egorin, M. J.; Lagattuta, T. F.; Hamburger, D. R.; Covey, J. M.; White, K. D.; Musser, S. M.; Eiseman, J. L. *Cancer Chemother. Pharmacol.* **2002**, *49*, 17–19.
103. Kaur, G.; Belotti, D.; Burger, A. M.; Fisher-Nielson, K.; Borsotti, P.; Riccardi, E.; Thillainathan, J.; Hollingshead, M.; Sausville, E. A.; Giavazzi, R. *Clin. Cancer Res.* **2004**, *10*, 4813–4821.
104. Burger, A. M.; Fiebig, H. H.; Stinson, S. F.; Sausville, E. A. *Anticancer Drugs* **2004**, *15*, 377–387.
105. Patel, K.; Piagentini, M.; Rascher, A.; Tian, Z. Q.; Buchanan, G. O.; Regentin, R.; Hu, Z.; Hutchinson, C. R.; McDaniel, R. *Chem. Biol.* **2004**, *11*, 1625–1633.
106. Le Brazidec, J.-Y.; Kamal, A.; Busch, D.; Thao, L.; Zhang, L.; Timony, G.; Grecko, R.; Trent, K.; Lough, R.; Salazar, T. et al. *J. Med. Chem.* **2004**, *47*, 3865–3873.
107. Kim, G. *94th Annual Meeting of AACR*, Washington, DC, July **2003**.
108. Schulte, T. W.; Akinaga, S.; Soga, S.; Sullivan, W.; Stensgard, B.; Toft, D.; Neckers, L. M. *Cell Stress Chaperones* **1998**, *3*, 100–108.
109. Roe, S. M.; Prodromou, C.; O'Brien, R.; Ladbury, J. E.; Piper, P. W.; Pearl, L. H. *J. Med. Chem.* **1999**, *42*, 260–266.
110. Soga, S.; Shiotsu, Y.; Akinaga, S.; Sharma, S. V. *Curr. Cancer Drug Targets* **2003**, *3*, 359–369.
111. Yamamoto, K.; Garbaccio, R. M.; Stachel, S. J.; Solit, D. B.; Chiosis, G.; Rosen, N.; Danishefsky, S. J. *Angew. Chem. Int. Ed. Engl.* **2003**, *42*, 1280–1284.
112. Geng, X.; Yang, Z.-Q.; Danishefsky, S. J. *Synlett*, **2004**, 1325–1333.
113. Chiosis, G.; Timaul, M. N.; Lucas, B.; Munster, P. N.; Zheng, F. F.; Sepp-Lorenzino, L.; Rosen, N. *Chem. Biol.* **2001**, *8*, 289–299.
114. Chiosis, G.; Lucas, B.; Shtil, A.; Huezo, H.; Rosen, N. *Bioorg. Med. Chem.* **2002**, *10*, 3555–3564.
115. Chiosis, G.; Lucas, B.; Huezo, H.; Solit, D.; Basso, A.; Rosen, N. *Curr. Cancer Drug Targets* **2003**, *3*, 371–376.
116. Wright, L.; Barril, X.; Dymock, B.; Sheridan, L.; Surgenor, A.; Beswick, M.; Drysdale, M.; Collier, A.; Massey, A.; Davies, N. et al. *Chem. Biol.* **2004**, *11*, 775–785.
117. Vilenchik, M.; Solit, D.; Basso, A.; Huezo, H.; Lucas, B.; He, H.; Rosen, N.; Spampinato, C.; Modrich, P.; Chiosis, G. *Chem. Biol.* **2004**, *11*, 787–797.
118. Rowlands, M. G.; Newbatt, Y. M.; Prodromou, C.; Pearl, L. H.; Workman, P.; Aherne, W. *Anal. Biochem.* **2004**, *327*, 176–183.

119. Sharp, S. et al. In *95th Annual Meeting of AACR*, 27–31 March 2004, Orlando, FL, USA.
120. Beswick, M. C.; Drysdale, M. J.; Dymock, B. W.; McDonald E. PCT Int. Patent Appl., WO2004056782, 2004.
121. Barril-Alonso, X.; Dymock, B. W.; Drysdale, M. J. PCT Int. Patent Appl., WO2004096212, 2004.
122. Drysdale, M. J.; Dymock, B. W.; Finch, H.; PCT Int. Patent Appl., WO2004072051, 2004.
123. Currie, K. S.; Desimone, R. W.; Pippin, D. A.; Darrow, J. W.; Mitchell, S. A. PCT Int. Patent Appl., WO2004072080, 2004.
124. Marcu, M. G.; Neckers, L. M. *Curr. Cancer Drug Targets* **2003**, *3*, 343–347.
125. Marcu, M. G.; Chadli, A.; Bouhouche, I.; Catelli, M.; Neckers, L. M. *J. Biol. Chem.* **2000**, *275*, 37181–37186.
126. Marcu, M. G.; Schulte, T. W.; Neckers, L. *J. Natl. Cancer Inst.* **2000**, *92*, 242–248.
127. Soti, C.; Racz, A.; Csermely, P. *J. Biol. Chem.* **2002**, *277*, 7066–7075.
128. Itoh, H.; Ogura, M.; Komatsuda, A.; Wakui, H.; Miura, A. B.; Tashima, Y. *Biochem. J.* **1999**, *343*, 697–703.
129. Zhao, Y. G.; Gilmore, R.; Leone, G.; Coffey, M. C.; Weber, B.; Lee, P. W. *J. Biol. Chem.* **2001**, *276*, 32822–32827.
130. Yu, X.; Sheng Guo, Z.; Marcu, M. L.; Neckers, L.; Nguyen, D. M.; Aaron, G.; Schrump, D. S. *J. Natl. Cancer Inst.* **2002**, *94*, 504–513.
131. Atadja, P.; Hsu, M.; Kwon, P.; Trogani, N.; Bhalla, K.; Remiszewski, S. *Novartis Found. Symp.* **2004**, *259*, 249–266.
132. Blank, M.; Mandel, M.; Keisari, Y.; Meruelo, D.; Lavie, G. *Cancer Res.* **2003**, *63*, 8241–8247.
133. Tremont-Lukats, I. W.; Gilbert, M. R. *Cancer Control* **2003**, *10*, 125–137.
134. Gibbs, R. A.; Zahn, T. J.; Sebolt-Leopold, J. S. *Curr. Med. Chem.* **2001**, *8*, 1437–1465.
135. Singh, S. B.; Lingham, R. B. *Curr. Opin. Drug. Disc. Dev.* **2002**, *5*, 225–244.
136. Dinsmore, C. J.; Bell, I. M. *Curr. Top. Med. Chem.* **2003**, *3*, 1075–1093.
137. Bell, I. M. *J. Med. Chem.* **2004**, *47*, 1869–1878.
138. Russo, P.; Loprevite, M.; Cesario, A.; Ardizzoni, A. *Curr. Med. Chem. Anti-Cancer Agents* **2004**, *4*, 123–138.
139. Santos, E. S.; Rosenblatt, J. D.; Goodman, M. *Expert Rev. Anticancer Ther.* **2004**, *4*, 843–856.
140. Britten, C. D.; Rowinsky, E. K.; Soignet, S.; Patnaik, A.; Yao, S.-L.; Deutsch, P.; Lee, Y.; Lobell, R. B.; Mazina, K. E.; McCreery, H. et al. *Clin. Cancer Res.* **2001**, 7, 3894–3903.
141. Hunt, J. T.; Ding, C. Z.; Batorsky, R.; Bednarz, M.; Bhide, R.; Cho, Y.; Chong, S.; Chao, S.; Gullo-Brown, J.; Guo, P. et al. *J. Med. Chem.* **2000**, *43*, 3587–3595.
142. Haluska, P.; Dy, G. K.; Adjei, A. A. *Eur. J. Cancer* **2002**, *38*, 1685–1700.
143. Taveras, A. G.; Kirschmeier, P.; Baum, C. M. *Curr. Top. Med. Chem.* **2003**, *3*, 1103–1114.
144. Njoroge, F. G.; Taveras, A. G.; Kelly, J.; Remiszewski, S.; Mallams, A. K.; Wolin, R.; Afonso, A.; Cooper, A. B.; Rane, D. F.; Liu, Y.-T. et al. *J. Med. Chem.* **1999**, *42*, 2125–2135.
145. Feldkamp, M. M.; Lau, N.; Roncari, L.; Guha, A. *Cancer Res.* **2001**, *61*, 4425–4431.
146. Venet, M.; End, D.; Angibaud, P. *Curr. Top. Med. Chem.* **2003**, *3*, 1095–1102.
147. Johnston, S. R.; Hickish, T.; Ellis, P.; Houston, S.; Kelland, L.; Dowsett, M.; Salter, J.; Michiels, B.; Perez-Ruixo, J. J.; Palmer, P. et al. *J. Clin. Oncol.* **2003**, *21*, 2492–2499.
148. Van Cutsem, E.; van de Velde, H.; Karasek, P.; Oettle, H.; Vervenne, W. L.; Szawlowski, A.; Schoffski, P.; Post, S.; Verslype, C.; Neumann, H. et al. *J. Clin. Oncol.* **2004**, *22*, 1430–1438.
149. Zhu, K.; Hamilton, A. D.; Sebti, S. M. *Curr. Opin. Invest. Drugs* **2003**, *4*, 1428–1435.
150. Mazieres, J.; Pradines, A.; Favre, G. *Cancer Lett.* **2004**, *206*, 159–167.
151. Kelland, L. R. *Expert Opin. Invest. Drugs* **2003**, *12*, 413–421.
152. Hendil, K. B.; Hartmann-Petersen, R. *Curr. Protein Pept. Sci.* **2004**, *5*, 135–151.
153. Cardozo, C. *Enzyme Protein* **1993**, *47*, 296–305.
154. Lowe, J.; Stock, D.; Jap, B.; Zwickl, P.; Baumeister, W.; Huber, R. *Science* **1995**, *268*, 533–539.
155. Delcros, J. G.; Floc'h, M. B.; Prigent, C.; Arlot-Bonnemains, Y. *Curr. Med. Chem.* **2003**, *10*, 479–503.
156. Adams, J. *Drug Disc. Today* **2003**, *8*, 307–315.
157. Rock, K. L.; Gramm, C.; Rothstein, L.; Clark, K.; Stein, R.; Dick, L.; Hwang, D.; Goldberg, A. L. *Cell* **1994**, *78*, 761–771.
158. Iqbal, M.; Chatterjee, S.; Kauer, J. C.; Das, M.; Messina, P.; Freed, B.; Biazzo, W.; Siman, R. J. *Med. Chem.* **1995**, *38*, 2276–2277.
159. Simon, R. J.; Kania, R. S.; Zuckermann, R. N.; Huebner, V. D.; Jewell, D. A.; Bandville, S.; Ng, S.; Wang, L.; Rosenberg, S.; Marlowe, C. K. et al. *Proc. Natl. Acad. Sci. USA* **1992**, *89*, 9367–9371.
160. Zuckermann, R. N.; Kerr, J. M.; Kent, S. B. H.; Moos, W. H. J. *J. Am. Chem. Soc.* **1992**, *114*, 10646–10647.
161. Kessler, H. *Angew. Chem. Int. Ed. Engl.* **1993**, *32*, 543–544.
162. Gante, J. *Synthesis* **1989**, *6*, 405–408.
163. Han, H.; Janda, K. D. *J. Am. Chem. Soc.* **1996**, *118*, 2539–2544.
164. Gibson, C.; Goodman, S. L.; Hahn, D.; Holzemann, G.; Kessler, H. *J. Org. Chem.* **1999**, *64*, 7388–7394.
165. Kruijtzer, J. A. W.; Lefeber, D. J.; Liskamp, R. M. J. *Tetrahedron Lett.* **1997**, *38*, 5335–5338.
166. Shin, I.; Park, K. *Org. Lett.* **2002**, *4*, 869–872.
167. Hamper, B. C.; Kolodziej, S. A.; Scates, A. M.; Smith, R. G.; Cortez, E. *J. Org. Chem.* **1998**, *63*, 708–718.
168. Cheguillaume, A.; Lehardy, F.; Bouget, K.; Baudy-Floch, M.; Le Grel, P. *J. Org. Chem.* **1999**, *64*, 2924–2927.
169. Bouget, K.; Aubin, S.; Delcros, J. G.; Arlot-Bonnemains, Y.; Baudy-Floch, M. *Bioorg. Med. Chem.* **2003**, *11*, 4881–4889.
170. Aubin, S.; Martin, B.; Delcros, J.-G.; Arlot-Bonnemains, Y.; Baudy-Floch, M. *J. Med. Chem.* **2005**, *48*, 330–334.
171. Burrill, L. C.; Mendonca, R. V.; Palmer, J. T.; Rydzewski, R. M. PCT Int. Patent Appl., WO2004014882, 2004.
172. Iqbal, M.; Chatterjee, S.; Kauer, J. C.; Mallamo, J. P.; Messina, P. A.; Reibolt, A. *Biorg. Med. Chem. Lett.* **1996**, *6*, 287–290.
173. Adams, J.; Behnke, M.; Chen, S.; Cruickshank, A. A.; Dick, L. R.; Grenier, L.; Klunder, J. M.; Ma, Y. T.; Plamondon, L.; Stein, R. L. *Bioorg. Med. Chem. Lett.* **1998**, *8*, 333–338.
174. Adams, J.; Palombella, V. J.; Sausville, E. A.; Johnson, J.; Destree, A.; Lazarus, D. D.; Maas, J.; Pien, C. S.; Prakash, S.; Elliott, P. J. *Cancer Res* **1999**, *59*, 2615–2622.
175. Teicher, B. A.; Ara, G.; Herbst, R.; Palombella, V. J.; Adams, J. *Clin. Cancer Res.* **1999**, *5*, 2638–2645.
176. Lightcap, E. S.; Pien, T. A.; McCormack, C. S.; Chau, V.; Adams, J.; Elliott, P. *Clin. Chem.* **2000**, *46*, 673–683.
177. Paramore, A.; Frantz, S. *Nat. Rev. Drug Disc.* **2003**, *2*, 611–612.
178. Hideshima, T.; Richardson, P.; Chauhan, D.; Palombella, V. J.; Elliott, P. J.; Adams, J.; Anderson, K. C. *Cancer Res.* **2001**, *61*, 3071–3076.
179. Palombella, V. J.; Rando, O. J.; Goldberg, A. L.; Maniatis, T. *Cell* **1994**, *78*, 773–785.
180. Denlinger, C. E.; Rundall, B. K.; Jones, D. R. *J. Thorac. Cardiovasc. Surg.* **2004**, *128*, 740–748.

181. Anderson, P. G. *N. Engl. J. Med.* **2003**, *348*, 2609–2613.
182. Kisselev, A. F.; Goldberg, A. L. *Chem. Biol.* **2001**, *8*, 739–758.
183. Information from the company web page. http://www.nereuspharm.com (accessed April 2006).
184. Stadler, M.; Seip, S.; Müller, H.; Mayer-Bartschmid, A.; Brüning, M.-A.; Benet-Buchholz, J.; Togame, H.; Dodo, R.; Reinemer, P.; Bacon, K. et al. PCT Int. Patent Appl., WO2004071382, 2004.
185. Agoulnik, S.; Akasaka, K.; Fang, F.; Harmange, J.-C.; Hawkins, L.; Jiang, Y.; Johannes, C.; Li, X.-Y.; Mcguinness, P.; Murphy, E. et al. PCT Int. Patent Appl., WO2003059898, 2003.
186. Momand, J.; Zambetti, G. P.; Olson, D. C.; George, D.; Levine, A. J. *Cell* **1992**, *69*, 1237–1245.
187. Finlay, C. A.; Hinds, P. W.; Levin, A. J. *Cell* **1989**, *57*, 1083–1093.
188. Chene, P. *Expert Opin. Ther. Patents* **2001**, *11*, 923–935.
189. Selivanova, G. *Curr. Cancer Drug Targets* **2004**, *4*, 385–402.
190. Lane, D. P.; Lain, S. *Trends Mol. Med.* **2002**, *8*, S38–S42.
191. Buttitta, F.; Marchetti, A.; Gadducci, A.; Pellegrini, S.; Morganti, M.; Carnicelli, V.; Cosio, S.; Gagetti, O.; Genazzani, A. R.; Bevilacqua, G. *Br. J. Cancer* **1997**, *75*, 230–235.
192. Zhang, R.; Wang, H. *Curr. Pharm. Design* **2000**, *6*, 393–416.
193. Chene, P. *Nat. Rev. Drug Disc.* **2003**, *3*, 102–109.
194. Klein, C.; Vassilev, L. T. *Br. J. Cancer* **2004**, *91*, 1415–1419.
195. Zheleva, D. I.; Lane, D. P.; Fischer, P. M. *Mini Rev. Med. Chem.* **2003**, *3*, 257–270.
196. Druker, B. J.; Talpaz, M.; Resta, D. J.; Peng, B.; Buchdunger, E.; Ford, J. M.; Lydon, N. B.; Kantarjian, H.; Capdeville, R.; Ohno-Jones, S. et al. *N. Engl. J. Med.* **2001**, *344*, 1031–1037.
197. Wakeling, A. E.; Guy, S. P.; Woodburn, J. R.; Ashton, S. E.; Curry, B. J.; Barker, A. J.; Gibson, K. H. *Cancer Res.* **2002**, *62*, 5749–5754.
198. Richardson, P. G.; Barlogie, B.; Berenson, J.; Singhal, S.; Jagannath, S.; Irwin, D.; Rajkumar, S. V.; Srkalovic, G.; Alsina, M.; Alexanian, R. et al. *N. Engl. J. Med.* **2003**, *348*, 2609–2617.
199. Buolamwini, J. K.; Addo, J.; Kamath, S.; Patil, S.; Mason, D.; Ores, M. *Curr. Cancer Drug Targets* **2005**, *5*, 57–68.
200. Hupp, T. R.; Lane, D. P.; Ball, K. L. *Biochem. J.* **2000**, *352*, 1–17.
201. Kussie, P. H.; Gorina, S.; Marechal, V.; Elenbaas, B.; Moreau, J.; Levine, A. J.; Pavletich, N. P. *Science* **1996**, *274*, 948–953.
202. Bottger, A.; Bottger, V.; Sparks, A.; Liu, W.-L.; Howard, S. F.; Lane, D. P. *Curr. Biol.* **1997**, *7*, 860–869.
203. Chene, P. *Curr. Med. Chem. Anti-Cancer Agents* **2001**, *1*, 151–161.
204. Chene, P. *Mol. Cancer Res.* **2004**, *2*, 20–28.
205. Vassilev, L. T. *Cell Cycle* **2004**, *3*, 419–421.
206. Stoll, R.; Renner, C.; Hansen, S.; Palme, S.; Klein, C.; Belling, A.; Zeslawski, W.; Kamionka, M.; Rehm, T.; Muhlhahn, P. et al. *Biochemistry* **2001**, *40*, 336–344.
207. Vassilev, L. T.; Vu, B. T.; Graves, B.; Carvajal, D.; Podlaski, F.; Filipovic, Z.; Kong, N.; Kammlott, U.; Lukacs, C.; Klein, C. et al. *Science* **2004**, *303*, 844–848.
208. Kirkpatrick, P. *Nat. Rev. Drug Disc.* **2004**, *3*, 111.
209. Parks, D. J.; LaFrance, L. V.; Calvo, R. R.; Milkiewicz, K. L.; Gupta, V.; Lattanze, J.; Ramachandren, K.; Carver, T. E.; Petrella, E. C.; Cummings, M. D. et al. *Bioorg. Med. Chem. Lett.* **2005**, *15*, 765–770.
210. Kussie, P. H.; Gorina, S.; Marechal, V.; Elenbaas, B.; Moreau, J.; Levine, A. J.; Pavletich, N. P. *Science* **1996**, *274*, 948–953.
211. Schon, O.; Friedler, A.; Freund, S.; Fersht, A. R. *J. Mol. Biol.* **2004**, *336*, 197–202.
212. Zhao, J.; Wang, M.; Chen, J.; Luo, A.; Wang, X.; Wu, M.; Yin, D.; Liu, Z. *Cancer Lett.* **2002**, *183*, 69–77.
213. Zhang, R.; Mayhood, T.; Lipari, P.; Wang, Y.; Durkin, J.; Syto, R.; Gesell, J.; McNemar, C.; Windsor, W. *Anal. Biochem.* **2004**, *331*, 138–146.
214. Hardcastle, I. R.; Ahmed, S. U.; Atkins, H.; Calvert, A. H.; Curtin, N. J.; Farnie, G.; Golding, B. T.; Griffin, R. J.; Guyenne, S.; Hutton, C. et al. *Bioorg. Med. Chem. Lett.* **2005**, *15*, 1515–1520.
215. Lai, Z.; Yang, T.; Kim, Y. B.; Sielecki, T. M.; Diamond, M. A.; Strack, P.; Rolfe, M.; Caligiuri, M.; Benfield, P. A.; Auger, K. R. et al. *Proc. Natl. Acad. Sci. USA* **2002**, *99*, 14734–14739.
216. Zhang, Z.; Zhang, R. *Curr. Cancer Drug Targets* **2005**, *5*, 9–20.
217. Rayburn, E.; Zhang, R.; He, J.; Wang, H. *Curr. Cancer Drug Targets* **2005**, *5*, 27–41.
218. Wang, H.; Oliver, P.; Zhang, Z.; Agrawal, S.; Zhang, R. *Ann. NY Acad. Sci.* **2003**, *1002*, 217–235.
219. Hidalgo, M.; Eckhardt, S. G. *J. Natl. Cancer. Inst.* **2001**, *93*, 178–193.
220. Skiles, J. W.; Gonnella, N. C.; Jeng, A. Y. *Curr. Med. Chem.* **2001**, *8*, 425–474.
221. Stamenkovic, I. *Semin. Cancer Biol.* **2000**, *10*, 415–433.
222. Ramnath, N.; Creaven, P. J. *Curr. Oncol. Rep.* **2004**, *6*, 96–102.
223. Summers, J. B.; Davidsen, S. K. *Annu. Rep. Med. Chem.* **1998**, *33*, 131–140.
224. Yong, V. W. *Expert Opin. Invest. Drugs* **1999**, *8*, 255–268.
225. Velasco, G.; Cal, S.; Merlos-Suárez, A.; Ferrando, A. A.; Alvarez, S.; Nakano, A.; Arribas, J.; López-Otín, C. *Cancer Res.* **2000**, *60*, 877–882.
226. Kojima, S.; Itoh, Y.; Matsumoto, S.; Masuho, Y.; Seiki, M. *FEBS Lett.* **2000**, *480*, 142–146.
227. Marchenko, N. D.; Marchenko, G. N.; Strongin, A. Y. *J. Biol. Chem.* **2002**, *277*, 18967–18972.
228. Lohi, J.; Wilson, C. L.; Roby, J. D.; Parks, W. C. *J. Biol. Chem.* **2001**, *276*, 10134–10144.
229. Skiles, J. W.; Gonnella, N. C.; Jeng, A. Y. *Curr. Med. Chem.* **2004**, *11*, 2911–2977.
230. Wada, C. K. *Curr. Top. Med. Chem.* **2004**, *4*, 1255–1267.
231. Kontogiorgis, C. A.; Papaioannou, P.; Hadjipavlou-Litina, D. J. *Curr. Med. Chem.* **2005**, *12*, 339–355.
232. Egeblad, M.; Werb, Z. *Nat. Rev. Cancer* **2002**, *2*, 161–174.
233. Overall, C.; Lopez-Otin, C. *Nat. Rev. Cancer* **2002**, *2*, 657–672.
234. Sridhar, S. S.; Shepherd, F. A. *Lung Cancer* **2003**, *42*, S81–S91.
235. Folgueras, A. R.; Pendas, A. M.; Sanchez, L. M.; Lopez-Otin, C. *Int. J. Dev. Biol.* **2004**, *48*, 411–424.
236. Giaccone, G.; Shepherd, F.; Debruyne, C.; Hirsh, V.; Smylie, M.; Rubin, S.; Martins, H.; Lamont, A.; Krzakowski, M.; Zee, B. *Eur. J. Cancer* **2001**, *37*, S152.
237. Phuphanich, S.; Levin, V. A.; Yung, W.; Forsyth, P.; Maestro, R.; Perry, J.; Elliott, M.; Baillet, M. *Proc. Am. Soc. Clin. Oncol.* **2001**, *20*, 52a.

238. Bramhall, S. R.; Hallissey, M. T.; Whiting, J.; Scholefield, J.; Tierney, G.; Stuart, R. C.; Hawkins, R. E.; McCulloch, P.; Maughan, T.; Brown, P. D. et al. *Br. J. Cancer* **2002**, *86*, 1864–1870.
239. Sparano, J. A.; Bernardo, P.; Gradishar, W. J.; Ingle, J. N.; Zucker, S.; Davidson, N. E. *Proc. Am. Soc. Clin. Oncol.* **2002**, *21*, 44a.
240. King, J.; Clingan, P.; Morris, D. L. *Proc. Am. Soc. Clin. Oncol.* **2002**, *21*, 135a.
241. Heath, E. I.; Burtness, B. A.; Kleinberg, L.; Salem, R.; Yang, S. C.; Heitmiller, R. F.; Canto, M. I.; Knisely, J. P. S.; Topazian, M.; Rohmiller, B. et al. *Proc. Am. Soc. Clin. Oncol.* **2002**, *21*, 173a.
242. Levin, V.; Phuphanich, S.; Glantz, M. J.; Mason, W. P.; Groves, M.; Recht, L.; Shaffrey, M.; Puduvalli, V.; Roeck, B.; Zhang, M. et al. *Proc. Am. Soc. Clin. Oncol.* **2002**, *21*, 26a.
243. Collier, M. A.; Gonzalez, R.; Smylie, M.; O'Day, S.; Anderson, C.; Bedikian, A. Y.; Apodaca, D.; Pithavala, Y. *Proc. Am. Soc. Clin. Oncol.* **2002**, *21*, 5b.
244. Rugo, H. S.; Budman, D.; Vogel, C.; Baidas, S.; Fleming, G.; Collier, M.; Dixon, M.; Pithavala, Y.; Clendeninn, N. J.; Tripathy, D. et al. *Proc. Am. Soc. Clin. Oncol.* **2001**, *20*, 48a.
245. Smylie, M.; Mercier, R.; Aboulafia, D.; Tucker, R.; Bonomi, P.; Collier, M.; Keller, M.; Stuart-Smith, J.; Knowles, M.; Clendeninn, N. J. et al. *Proc. Am. Soc. Clin. Oncol.* **2001**, *20*, 307a.
246. Levitt, N. C.; Eskens, F. A. L. M.; O'Byrne, K. J.; Propper, D. J.; Denis, L. J.; Owen, S. J.; Choi, L.; Foekens, J. A.; Wilner, S.; Wood, J. M. et al. *Clin. Cancer Res.* **2001**, *7*, 1912–1922.
247. Douillard, J.; Petersen, V.; Shepherd, F.; Paz-Ares, L.; Amold, A.; Tonato, M.; Ottaway, J.; Davis, M.; Van Vreckem, A.; Humphrey, J. et al. *Eur. J. Cancer* **2001**, *37*, S19–S20.
248. Xu, W. F. *226th ACS National Meeting* **2003**.
249. Coussens, L. M.; Fingleton, B.; Matrisian, L. M. *Science* **2002**, *295*, 2387–2392.
250. Breuer, E.; Frant, J.; Reich, R. *Expert Opin. Ther. Patents* **2005**, *15*, 253–269.
251. Engel, C. K.; Pirard, B.; Schimanski, S.; Kirsch, R.; Habermann, J.; Klingler, O.; Schlotte, V.; Weithmann, K. U.; Wendt, K. U. *Chem. Biol.* **2005**, *12*, 181–189.
252. Boyer, M. J.; Tannock, I. F. In *The Basic Science of Oncology*, 3rd ed.; Tannock, I. F., Hill, R. P., Eds.; McGraw Hill: New York, 1998, pp 349–376.
253. Marchini, S.; D'Incalci, M.; Broggini, M. *Curr. Med. Chem. Anti-Cancer Agents* **2004**, *4*, 247–262.
254. Marton, M. J.; DeRisi, J. L.; Bennett, H. A.; Iyer, V. R.; Meyer, M. R.; Roberts, C. J.; Stoughton, R.; Burchard, J.; Slade, D.; Dai, H. et al. *Nat. Med.* **1998**, *4*, 1293–1301.

Biographies

Hilmar Weinmann studied chemistry at the Universities of Tuebingen and Hannover. After receiving his Dr.rer.nat. degree under the supervision of Ekkehard Winterfeldt he joined Schering AG in 1995. Initially he worked in Process Research where he was responsible as a team leader for the successful scale-up of several clinical candidates. In 2000, he became group leader of the Automated Process Optimization group which is dedicated to the application of synthesis robots to high-throughput reaction screening and optimization. In 2003, he joined Medicinal Chemistry at Schering AG and since 2004 he has been department head of Medicinal Chemistry I. The major topics in Hilmar Weinmann's department are lead discovery and optimization studies in the fields of oncology, gender healthcare, and dermatology. He is a member of the Schering Hit-to-Lead team providing guidance and support for project teams in the lead discovery process to rapidly generate high-quality lead structures.

Eckhard Ottow studied chemistry at the University of Hannover and received his Dr.rer.nat. degree under the supervision of Ekkehard Winterfeldt in 1982. He continued his chemical education taking a postdoctoral position in Paul Bartlett's group before he joined Schering AG in 1983. After working for 9 years on antihormones in the field of endocrinology he took over a head of department position in 1992 leading oncology and CNS projects. In 2000, he switched to Process Research in Chemical Development for 1 1/2 years and became member of the Strategic Task Force. Thereafter he returned to research, and his former area of responsibility was extended by heading the microbiological chemistry group as well. In 2002, he became the head of Medicinal Chemistry at Schering, being responsible for seven MedChem departments in Berlin and Jena. Since 2004 Eckhard has been a honorary professor at the Technical University of Berlin.

Comprehensive Medicinal Chemistry II
ISBN (set): 0-08-044513-6

ISBN (Volume 7) 0-08-044520-9; pp. 221–251

7.10 Viruses and Viral Diseases

E De Clercq, Rega Institute for Medical Research, Leuven, Belgium

7.10.1 Introduction

There are at present some 40 antiviral drugs that have been formally licensed for clinical use in the treatment of viral infections (**Table 1**).[1] These are mainly used in the treatment of infections caused by human immunodeficiency virus (HIV), hepatitis B virus (HBV), herpes viruses (herpes simplex virus (HSV), varicella-zoster virus (VZV), cytomegalovirus (CMV)), orthomyxoviruses (influenza), paramyxoviruses (respiratory syncytial virus (RSV)), and hepaciviruses (hepatitis C virus (HCV)). As these are the viruses that are most in demand of antiviral therapy, they have prompted the search for new antiviral strategies and drugs directed toward either the same molecular targets as the approved antiviral drugs or to other targets.

Most of the newly described antiviral compounds (that are currently in development) are targeted at HIV, HBV, or HCV. Some are targeted at HSV, VZV, or CMV, but, there are in addition many other important viral pathogens for which medical intervention, either prophylactic or therapeutic, is urgently needed, and, these are, among the DNA viruses, the papillomaviruses (human papilloma virus (HPV)), adenoviruses, poxviruses (variola, vaccinia, monkeypox,

Table 1 The past, present, and future of antiviral drugs

Virus	*Compound*		
	Approved for medical use	*In clinical development*	*In preclinical evaluation*
Parvo (B19)	—	—	—
Polyoma (JC, BK)	Cidofovir (off label)	—	—
Papillomas (HPV)	Cidofovir (off label)	—	cPr PMEDAP and other acyclic nucleoside phosphonates Biphenylsulfonacetic acid derivatives
Adeno	Cidofovir (off label)	—	HPMPO-DAPy
Alpha-herpes (HSV-1, HSV-2, VZV)	Acyclovir Valaciclovir Penciclovir (topical) Famciclovir Brivudin Idoxuridine (topical) Trifluridine (topical)	H2G prodrug	A-5021 Synguanol Cyclopropavir (ZSM-I-62) BAY 57-1293 BCNA Cf 1743
Beta-herpes (CMV, HHV-6, HHV-7)	Ganciclovir Valganciclovir Cidofovir Foscarnet Fomivirsen	Maribavir	CMV 423
Gamma-herpes (EBV, HHV-8)	Cidofovir (off label)	—	*North*-methanocarbathymidine (N-MCT)
Pox (variola, vaccinia, monkeypox, molluscum contagiosum, orf, etc.)	Cidofovir (off label)	—	HPMPO-DAPy HDP-CDV, ODE-CDV CI-1033
Hepadna (HBV)	Lamivudine Adefovir dipivoxil Entecavir Pegylated interferon-α	Telbivudine Valtorcitabine Clevudine	3′-Fluoro-2′,3′-dideoxyguanosine
Picorna (entero, rhino)	—	Pleconaril Ruprintrivir	Pyrrolidinyl pentenoic acid (ethyl ester) Mycophenolic acid (MPA) mofetil Interferon (inducers)
Flavi (yellow fever, dengue, West Nile, etc.)	—	—	Interferon (inducers)
Hepaci (HCV)	Pegylated interferon-α combined with ribavirin	BILN 2061 (Ciluprevir) VX-950 NM 283 (Valopicitabine) Viramidine SCH 503034	2′-*C*-methylcytidine 2′-*O*-methylcytidine 2′-*C*-methyladenosine 7-deaza-2′-*C*-methyladenosine 2′-*C*-methylguanosine DKA compound 30 Benzimidazole derivative Indole-*N*-acetamide derivative Benzothiadiazine derivative Phenylalanine derivative

Table 1 Continued

Virus	*Compound*		
	Approved for medical use	*In clinical development*	*In preclinical evaluation*
			Thiophene 2-carboxylic acid derivative Dihydropyranone derivative Tetrahydropyranoindolyl acetic acid derivative HCV-371 *N*-1-Aza-4-hydroxyquinolone benzothiadiazine A-782759
Corona (SARS)	—	Pegylated interferon-α (off label)	Calpain inhibitors (III, VI) Niclosamide anilide Phe–Phe dipeptide Bananin Valinomycin Glycyrrhizin Chloroquine Niclosamide Nelfinavir
Orthomyxo (influenza)	Amantadine Rimantadine Zanamivir Oseltamivir Ribavirin (off label)	—	RWJ-270201 A-192558 A-315675 T-705 Flutimide
Paramyxo (parainfluenza, measles, mumps, RSV, hMPV, etc.)	Ribavirin (approved for RSV only)	—	VP-14637 JNJ-2408068 BCX 2798 & BCX-2855 BMS-433771
Arena (Lassa, etc.)	Ribavirin (off label)	—	—
Bunya (Crimean-Congo, Rift Valley, etc.)	Ribavirin (off label)	—	—
Rhabdo (rabies)	—	—	—
Filo (Ebola, Marburg)	—	—	3-Deazaneplancin A
Retro (HIV)	Zidovudine Didanosine Zalcitabine Stavudine Lamivudine Abacavir Emtricitabine Tenofovir disoproxil fumarate Nevirapine Delavirdine Efavirenz Saquinavir	BMS-378806 BMS-488043 AMD-3100 SCH-C Vicriviroc Aplaviroc Maraviroc Racivir AVX-754 Reverset Elvucitabine Alovudine	Cyclotriazadisulfonamide Cyanovirin N KRH-1636 TAK-779 TAK-220 TAK-652 MIV-210 DOT 4′-Ed4T PMEO-DAPy PMPO-DAPy PMDTA

continued

Table 1 Continued

Virus	*Compound*		
	Approved for medical use	*In clinical development*	*In preclinical evaluation*
	Ritonavir	Amdoxovir	PMDTT
	Indinavir	Capravirine	Thiocarboxanilide UC-781
	Nelfinavir	Etravirine	Dapivirine
	Amprenavir	TMC-114	Rilpivirine
	Lopinavir	PA-457	L-870810
	Atazanavir	GW678248	L-870812
	Fosamprenavir		Dihydroxytropolone
	Tipranavir		Pyrimidinyl diketo acid
	Enfuvirtide		Indolyl aryl sulfone

etc.) and the herpesviruses Epstein–Barr (EBV) and human herpesvirus type 6 (HHV-6), and, among the RNA viruses, enteroviruses (e.g., Coxsackie B and Echo), coronaviruses (e.g., severe acute respiratory syndrome (SARS)-associated coronavirus), flaviviruses (e.g., dengue, yellow fever), and other RNA viruses associated with hemorrhagic fever (arenaviruses (e.g., Lassa fever), bunyaviruses (e.g., Rift Valley fever, Crimean–Congo fever), and filoviruses (e.g., Ebola and Marburg)).

Here I will describe, for each viral family (1) which are the antiviral drugs that have been formally approved, (2) which are the compounds that are under clinical development and thus may be considered as antiviral drug candidates, and (3) which compounds are in the preclinical stage of development and still have a long route ahead before they could qualify as antiviral drugs. The virus families to be addressed are the following: parvo-, polyoma-, papilloma-, adeno-, herpes-, pox-, hepadna-, picorna-, flavi-, hepaci-, corona-, orthomyxo-, paramyxo-, arena-, bunya-, rhabdo-, filo-, reo-, and retroviruses.

7.10.2 Parvoviruses

No significant attempts have been made to develop compounds with potential activity against B19, the only parvovirus that is pathogenic for humans, and which is responsible for erythema infectiosum, the so-called fifth disease, in children.

7.10.3 Polyomaviruses

No antiviral drugs have been formally approved for the treatment of polyomavirus (JC and BK)-associated diseases such as progressive multifocal leukoencephalopathy (PML) and hemorrhagic cystitis in patients with acquired immune deficiency syndrome (AIDS). There are, however, anecdotal case reports pointing to the efficacy of cidofovir (*S*)-1-(3-hydroxy-2-phosphonylmethoxypropyl)cytosine, HPMPC, **1**), which has been licensed under the trademark name Vistide for the intravenous treatment of CMV retinitis in AIDS patients, in the treatment of polyoma (JC and BK) virus infections, particularly PML, in AIDS patients.[2]

7.10.4 Papillomaviruses

As for polyomaviruses, no antivirals have been licensed for the treatment of human papillomavirus-associated diseases, including warts, condylomata acuminata, papillomas, and cervical, vulvar, penile, and (peri)anal dysplasia (evolving to carcinoma). Cidofovir has been used 'off label,' with success, in the topical and, occasionally, systemic treatment of HPV-associated papillomatous lesions.[2] In many instances, a virtually complete and durable resolution of the lesions was achieved following topical application of cidofovir as a 1% gel or cream. In addition to cidofovir, other acyclic nucleoside phosphonates, such as cPrPMEDAP (N^6-cyclopropyl-9-(2-phosphonylmethoxyethyl)-2,6-diaminopurine, **2**), are being explored for their potential in the treatment of HPV-associated papillomas and dysplasias.[2,3] These compounds have been shown to specifically induce apoptosis in HPV-infected cells, which, in turn, may be related to their ability to restore the function of the tumor suppressor proteins p53 and pRb (which are neutralized by the oncoproteins E6 and E7, respectively, in HPV-infected cells).

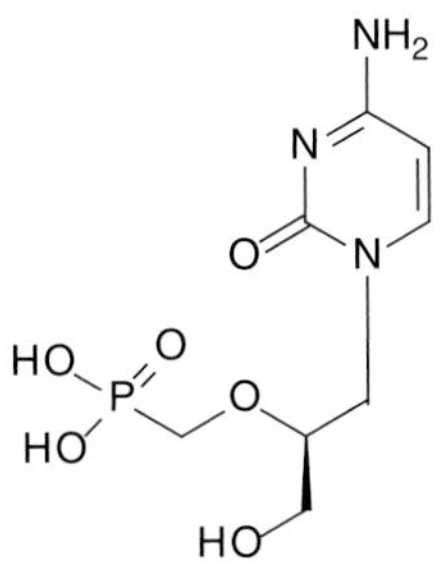

1 HPMPC Cidofovir
Vistide

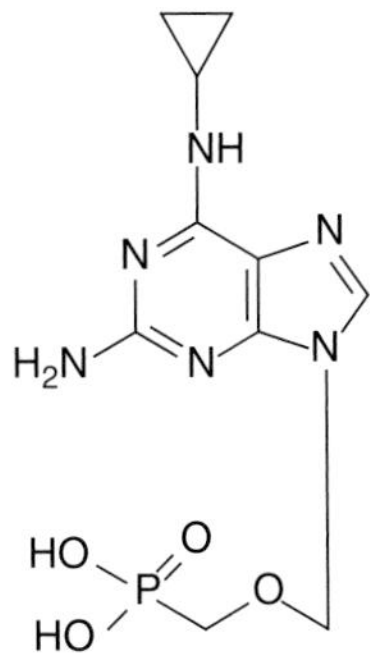

2 cPrPMEDAP

Recently, biphenylsulfonacetic acid derivatives have been described as inhibitors of HPV E1 helicase-associated ATP hydrolysis.[4,5] Although these novel ATPase inhibitors can hardly be considered to be good drug candidates, they may serve as leads for further optimization as potential antiviral agents active against multiple HPV types.[5]

7.10.5 Adenoviruses

For the treatment of adenovirus infections, which can be quite severe in immunocompromised patients (e.g., allogeneic hematopoietic stem-cell transplant recipients), no antiviral drugs have been officially approved. Anecdotal reports have pointed to the efficacy of cidofovir against adenovirus infections in such patients.[2] Among the novel compounds that could be further explored for the treatment of adenovirus infections are (*S*)-2,4-diamino-6-[3-hydroxy-2-phosphono-methoxy)propoxy]pyrimidine (HPMPO-DAPy), **3**),[3] which akin to some 'older' compounds like (*S*)-9-(3-hydroxy-2-phosphono-methoxypropyl)adenine (HPMPA), the N7-substituted acyclic nucleoside 2-amino-7-(1,3-dihydroxy-2-propoxymethyl)purine S-2242, the 2′,3′-dideoxynucleosides zalcitabine (ddC) and alovudine (FddT, FLT) have been found to inhibit adenovirus replication in vitro.[6] Also, ether lipid-ester (hexadecyloxypropyl (HDP) and octadecyloxyethyl (ODE)) prodrugs of HPMPC and HPMPA have been designed that inhibit adenovirus replication in vitro at significantly lower concentrations than the parent compounds.[7]

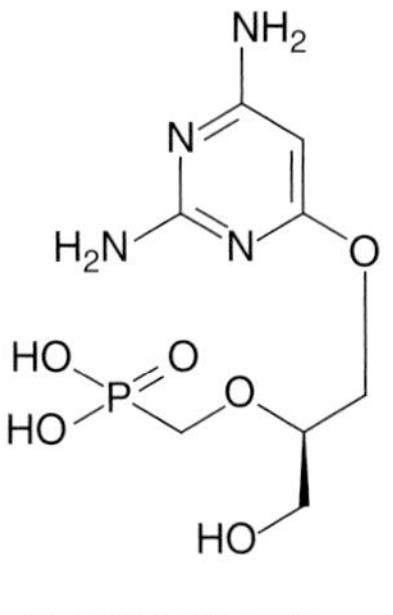

3 HPMPO-DAPy

7.10.6 Herpesviruses

7.10.6.1 Alpha-Herpesviruses (HSV-1, HSV-2, VZV)

For the treatment of HSV-1, HSV-2, and VZV, a number of compounds have been approved: acyclovir (**4**) and its oral prodrug valaciclovir (**5**); penciclovir (**6**) and its oral prodrug famciclovir (**7**); idoxuridine (**8**), trifluridine (**9**), and brivudin (**10**). Penciclovir, idoxuridine, and trifluridine are used topically, primarily in the treatment of herpes labialis (penciclovir) and herpetic keratitis (idoxuridine, trifluridine). Acyclovir can be used orally, intravenously, or topically, whereas valaciclovir and famciclovir are administered orally, in the treatment of both HSV and VZV infections. Brivudin (available in some European countries) is used orally for the treatment of herpes zoster, but is also effective against HSV-1 infections.

4 Acyclovir (ACV)
Zovirax

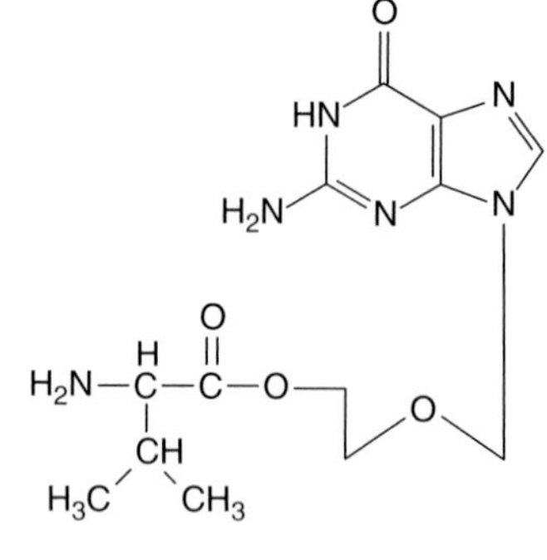

5 Valaciclovir (VACV)
Zelitrex, Valtrex

6 Penciclovir (PCV)
Denavir, Vectavir

7 Famciclovir (FCV)
Famvir

8 Idoxuridine
Herpid, Stoxil

9 Trifluridine
Viroptic

10 (*E*)-5-(2-Bromovinyl)-2′-deoxyuridine
BVDU
Brivudin
Zostex, Brivirac

While acyclovir (and its oral prodrug valaciclovir) have remained the gold standard for the treatment of HSV and VZV infections, few attempts have been made to bring other anti-HSV (or anti-VZV) agents into the clinic, with the exception of the H2G prodrug (**11**), which after quite a number of years is still in clinical development for the treatment of herpes zoster.[8] Worth considering for clinical development as anti-HSV (and anti-VZV) agents are a number of carbocyclic guanosine analogs, such as A-5021 (**12**), cyclohexenylguanine, and the methylene cyclopropane synguanol (**13**).[9] All these compounds owe their selective antiviral activity to a specific phosphorylation by the HSV- or VZV-encoded thymidine kinase (TK); upon phosphorylation to their triphosphate form, they act as chain terminators in the DNA polymerization reaction. In the (rare) circumstances that HSV or VZV becomes resistant to the acyclic (or carbocyclic) guanosine analogs due to TK deficiency (TK^-), the pyrophosphate analog foscarnet (**14**) could be useful for treating TK^- HSV or TK^- VZV infections (in immunocompromised patients).

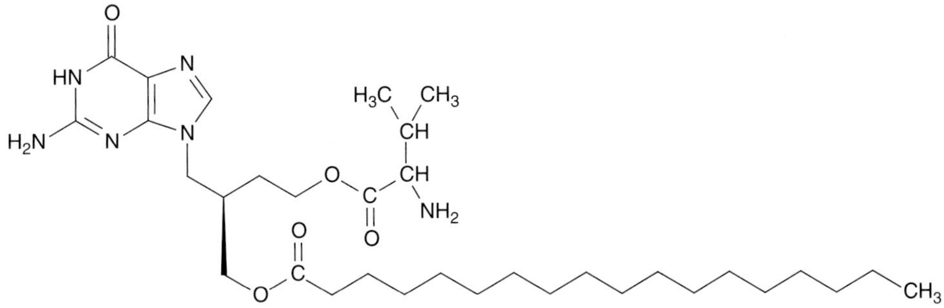

11 H2G prodrug
MIV-606
Valomaciclovir stearate

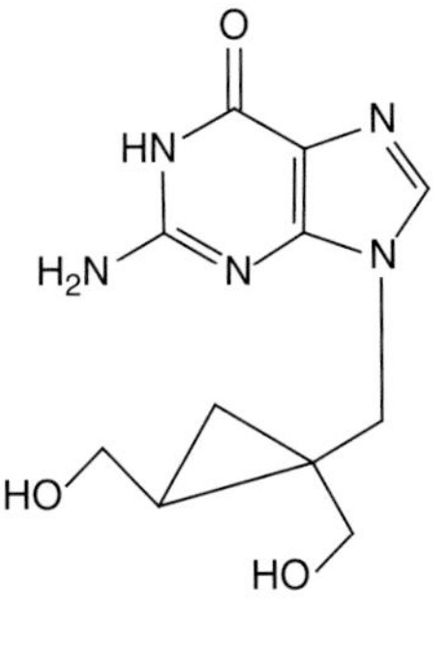

12 A-5021

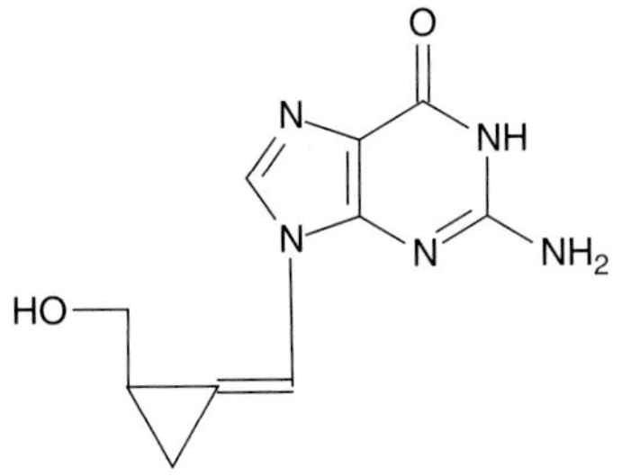

13 Synguanol

Recently, a second generation of methylene cyclopropane analogs, the 2,2-bishydroxymethyl derivatives, has been synthesized.[10] These compounds may have potential, not only for the treatment of HSV-1, HSV-2, and VZV, but also beta-herpes (CMV, HHV-6, HHV-7) and gamma-herpes (EBV, HHV-8) infections.[11] In particular, ZSM-I-62 (Cyclopropavir) (**15**) has been reported to be very effective in reducing mortality of mice infected with murine CMV.[12]

New anti-HSV agents targeting the viral helicase–primase complex, the thiazolylphenyl derivatives BILS 179BS and BAY 57-1293 (**16**), were recently reported to have in vivo efficacy in animal models of HSV-1 and HSV-2 infections.[13,14] These compounds seem to function by diminishing the affinity of the helicase–primase complex for the HSV DNA. The antiviral potency of BAY 57-1293 was claimed to be superior to all compounds that are currently used to treat HSV infections.[15] If so, this lead should be further pursued from a clinical viewpoint.

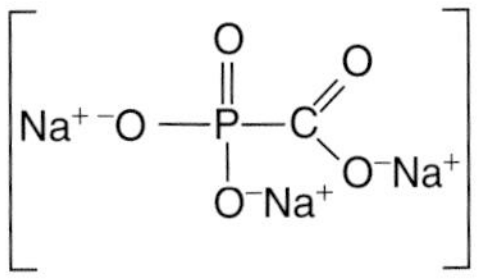

14 Foscarnet
Foscavir

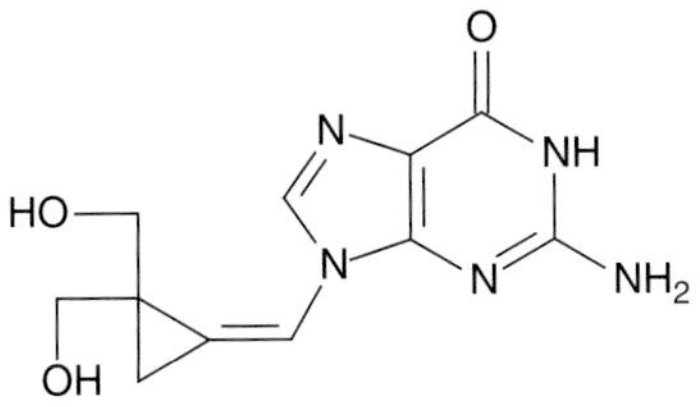

15 Cyclopropavir
(ZSM-I-62)

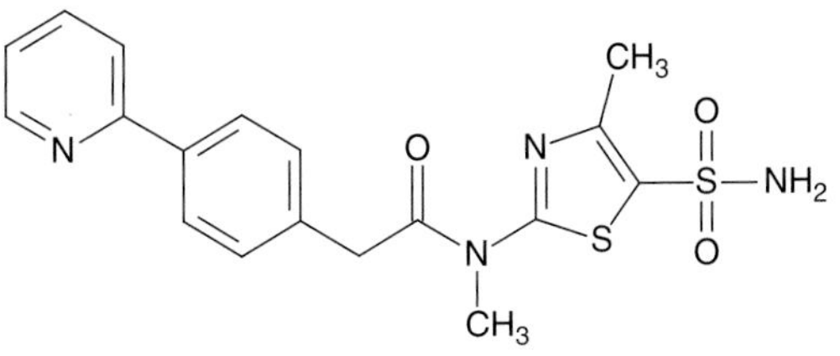

16 BAY 57-1293

A new class of anti-VZV compounds are the bicyclic furo (2,3-*d*)pyrimidine nucleoside analogs (BCNAs), represented by Cf 1742 and Cf 1743 (**17**). These compounds are exquisitely active against VZV.[16] They inhibit the replication of VZV, but not that of other viruses (including HSV), at subnanomolar concentrations, with a selectivity index in excess of 100 000.[17] Given the extremely high potency and selectivity of the BCNAs they warrant to be further developed toward clinical use, e.g., against herpes zoster.

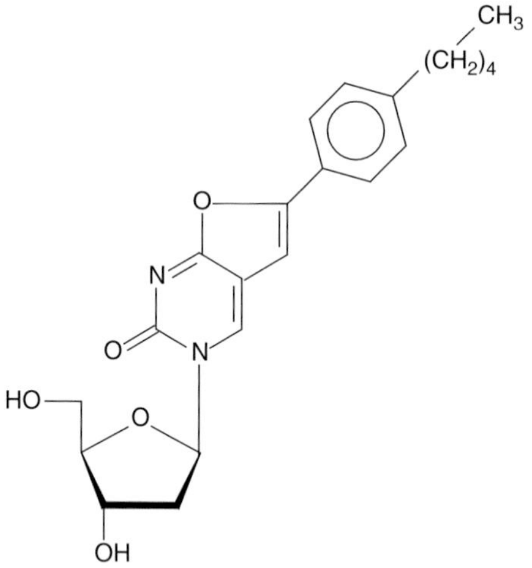

CH_3
$(CH_2)_4$
O
N
O
N
HO
O
OH

17 BCNA Cf 1743

7.10.6.2 Beta-Herpesviruses (CMV, HHV-6, HHV-7)

Five compounds have been licensed to treat CMV infections: ganciclovir (**18**), its oral prodrug valganciclovir (**19**), foscarnet (**14**), cidofovir (**1**), and fomivirsen (**20**). With the exception of fomivirsen (an antisense oligonucleotide) which targets the CMV immediate–early mRNA, all other licensed anti-CMV drugs target the viral DNA polymerase. Ganciclovir must first be phosphorylated by the CMV-encoded protein kinase (the UL97 gene product) which is also the principal site for mutations engendering resistance toward this compound. Toxic side effects (i.e., bone-marrow suppression for ganciclovir, nephrotoxicity for foscarnet and cidofovir) have prompted the search for new inhibitors of CMV.[18]

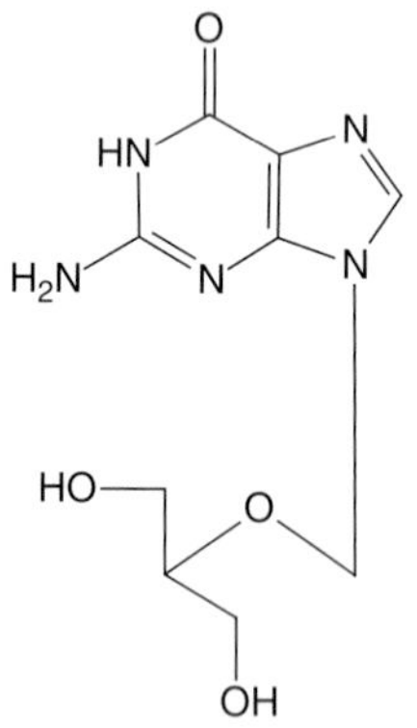

O
HN
N
H_2N
N
N
HO
O
OH

18 Ganciclovir
Cymevene, Cytovene

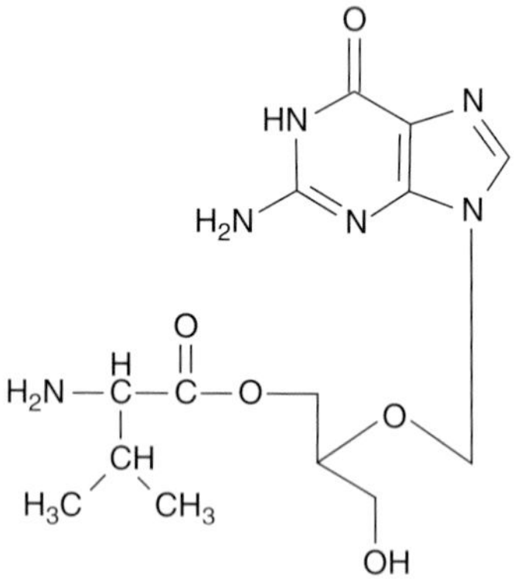

O
HN
N
H_2N
N
N
O
H
$H_2N—C—C—O$
CH
H_3C CH_3
O
OH

19 Valganciclovir
Valcyte

5′-d-[G*C*G*T*T*T*G*C*T*C*T*T*C*T*T*C*T*T*G*C*G]-3′

* = racemic phosphorothioate

20 Fomivirsen
Vitravene

Several benzimidazole ribonucleosides, including maribavir (previously also known as 1263W94) (**21**), have been accredited with specific activity against human CMV. Maribavir seems to target the UL97 protein kinase,[19] and, as the

UL97 gene product has been shown to account for the release of CMV nucleocapsids from the nucleus,[20] maribavir may be assumed to target a stage in the viral life cycle that follows viral DNA maturation and packaging. Preclinical pharmacokinetic and toxicological studies have shown that maribavir has a favorable safety profile and excellent oral bioavailability.[21] Phase I/II dose-escalation trials in HIV-infected men with asymptomatic CMV shedding further indicated that maribavir is rapidly absorbed following oral dosing and reduces CMV titers in semen.[22]

While maribavir is primarily active against CMV, 2-chloro-3-pyridin-3-yl-5,6,7,8-tetrahydroindolizine-1-carboxamide (CMV 423) (**22**) has potent and selective in vitro activity against all three human beta-herpesviruses, CMV, HHV-6, and HHV-7.[23] As compared to ganciclovir and foscarnet, CMV 423 has higher antiviral potency and lower cytotoxicity. It is targeted at an early stage of the viral replication cycle (following viral entry but preceding viral DNA replication), which is regulated by a cellular process that may involve protein tyrosine kinase activity. The in vitro antiviral action profile of CMV 423 is such that it deserves to further explored for its in vivo potential in the treatment of CMV and HHV-6 infections.

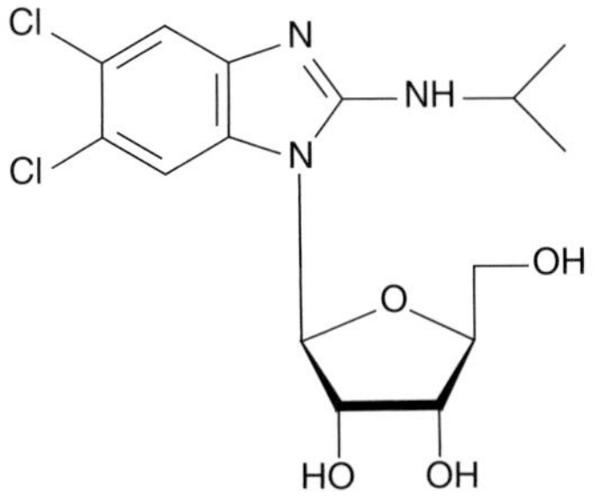

Cl Cl N N NH O OH HO OH

21 1263W94
Maribavir

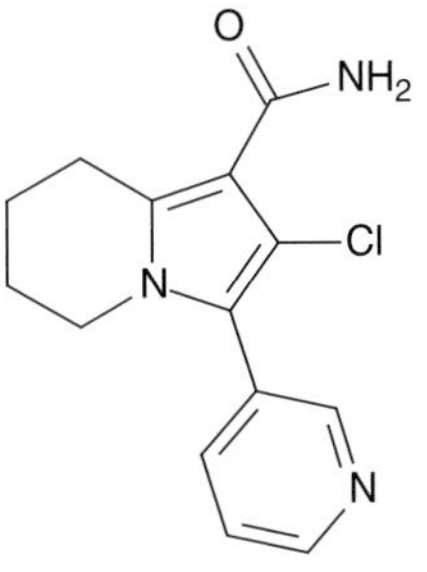

O NH_2 Cl N N

22 CMV 423

There is, at present, no standardized antiviral treatment for HHV-6 infections. From a comparative study, A-5021, foscarnet, S2242, and cidofovir emerged as the most potent compounds with the highest antiviral selectivity against HHV-6. The latter three also proved to be the most potent against HHV-7.[24] However, indications for the clinical use of anti-HHV-6 and anti-HHV-7 agents remain ill-defined.

7.10.6.3 Gamma-Herpesviruses (EBV, HHV-8)

Although a number of the aforementioned approved antiherpetic drugs, such as acyclovir, ganciclovir, brivudin, and cidofovir, have proven to be effective against the in vitro replication of EBV and HHV-8,[24] none of these (or any other) antiviral drugs has been formally approved for the treatment of diseases associated with EBV (e.g., mononucleosis infectiosa, B-cell lymphoma, lymphoproliferative syndrome, Burkitt's lymphoma, nasopharyngeal carcinoma) or HHV-8 (e.g., Kaposi's sarcoma, primary effusion lymphoma, multicentric Castleman's disease). It would seem appealing to further examine established antiherpetic drugs, such as cidofovir, and other acyclic nucleoside phosphonates such as HPMPA, or prodrugs thereof, for their potential in the therapy of EBV- and HHV-8-associated malignancies. Also, new nucleoside analogs, such as *north*-methanocarbathymidine (N-MCT) (**23**),[25] which have been previously shown to block the replication of HSV-1 and HSV-2, should be further explored for their potential in the prevention and treatment of HHV-8-associated malignancies: in particular, N-MCT, which is specifically triphosphorylated in HHV-8-infected cells undergoing lytic replication, efficiently blocks HHV-8 DNA replication in these cells.[25]

7.10.7 Poxviruses (Variola, Vaccinia, Monkeypox, Molluscum Contagiosum, Orf, etc.)

Several nucleoside analogs (e.g., S2242, 8-methyladenosine, idoxuridine) and nucleotide analogs (e.g., cidofovir, HPMPO-DAPy) have proven to be effective in various animal models of poxvirus infections.[26] In particular, cidofovir has shown high efficacy, even after administration of a single systemic (intraperitoneal) or intranasal (aerosolized) dose, in protecting mice from a lethal respiratory infection with either vaccinia or cowpox. Cidofovir has demonstrated high effectiveness in the treatment of disseminated progressive vaccinia in athymic-nude mice.[27] In humans, cidofovir has been used successfully, by both the topical and intravenous route, in the treatment of orf and recalcitrant molluscum contagiosum in immunocompromised patients.[28] Given the in vitro activity of cidofovir against variola (smallpox), and the in vivo efficacy of cidofovir against various poxvirus infections in animal models and humans, it can be reasonably assumed that cidofovir should be effective in the therapy and/or prophylaxis of smallpox in case of an inadvertent outbreak or biological attack with the variola virus.

Being a phosphonate analog, cidofovir only has limited oral bioavailability. In case of an outbreak of smallpox, it would be useful to have an orally active drug at hand.[29] To this end, hexadecyloxypropyl-cidofovir (HDP-CDV, **24**) and octadecyloxyethyl-cidofovir (ODE-CDV, **25**) were designed as potential oral prodrugs of cidofovir. These alkyloxyalkyl esters of cidofovir were found to significantly enhance inhibition of the replication of orthopoxviruses (e.g., vaccinia, cowpox) in vitro.[30] HDP-CDV and ODE-CDV given orally were as effective as cidofovir given parenterally for the treatment of vaccinia and cowpox infections.[31] HDP-CDV has also proven effective in the treatment of a lethal vaccinia virus respiratory infection in mice.[32] Furthermore, HDP-CDV and ODE-CDV, when given orally, proved highly efficacious in a lethal (aerosol) mousepox (ectromelia) virus model,[33] further attesting as to the potential usefulness of the alkyloxyalkyl esters of cidofovir in the oral therapy and prophylaxis of poxvirus infections.

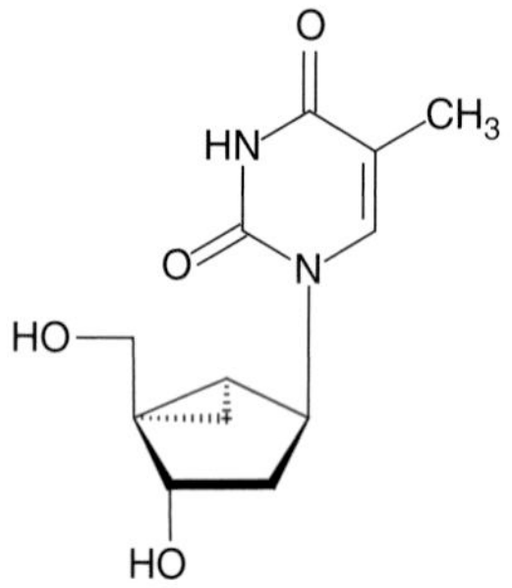

23 N-MCT
North-Methanocarbathymidine

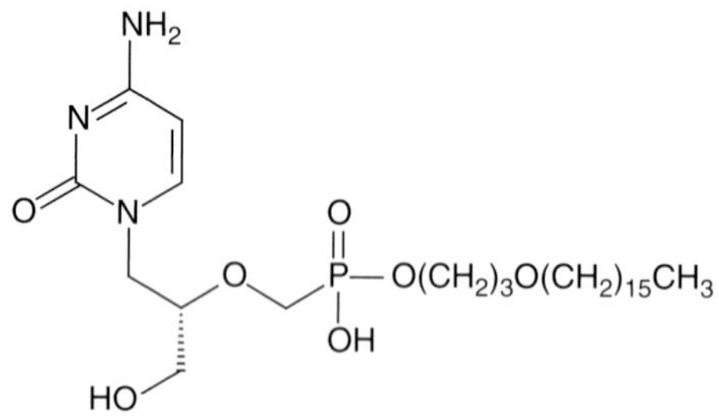

24 HDP-CDV

In fact, as attested by the most recent findings with cidofovir in mice infected with ectromelia (mousepox) virus encoding interleukin-4,[34] and monkeys infected with monkeypox,[35] cidofovir (CDV) (and HDP-CDV and/or ODE-CDV) still provides the best current hope for effective control of virulent poxvirus infections.

In addition to the nucleotide analog cidofovir, which primarily acts as a viral DNA chain terminator (for vaccinia virus DNA polymerase after it has been incorporated at the penultimate position),[36] antiviral strategies for poxvirus infections may also be based on inhibitors of cellular processes, i.e., signal transduction pathways. In this respect, the 4-anilinoquinazoline CI-1033 (**26**), an ErbB tyrosine kinase inhibitor, was found to block variola virus replication in vitro and vaccinia virus infection in vivo.[37]

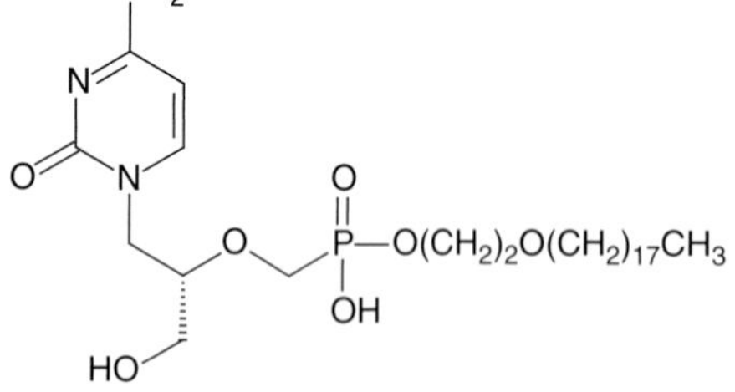

25 ODE-CDV

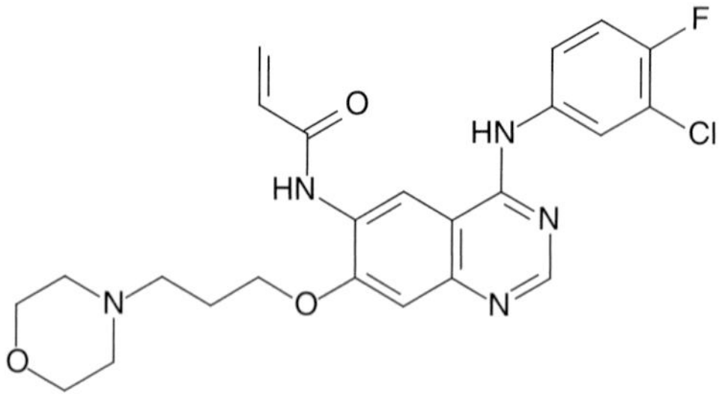

26 Anilinoquinazoline
CI-1033

Likewise, Gleevec (STI-571, Imatimib), an Abl-family kinase inhibitor used to treat chronic myelogenous leukemia in humans, was shown to suppress poxviral dissemination in vivo by several orders of magnitude and to promote survival in infected mice,[38] suggesting possible use for this drug in treating smallpox or complications associated with vaccination against smallpox. Because the drug targets host rather than viral molecules, it is less likely to engender resistance compared to more specific antiviral agents. Collectively,[37,38] inhibitors of host-signaling pathways exploited by poxviral pathogens may represent potential antiviral therapies.

Recently, a new antipoxvirus compound (ST-246, **27**) has been described, which is orally bioavailable, acts according to a novel mechanism of action, targeting a specific viral product (i.e., vaccinia virus F13L) required for extracellular virus particle formation, and protecting mice from a lethal orthopoxvirus challenge.[39] These properties make ST-246 an attractive candidate for development as a smallpox antiviral drug that could be stockpiled for use in the treatment and prevention of smallpox virus infection in the event of a bioterrorist threat.

7.10.8 Hepadnaviruses (HBV)

An estimated 400 million people worldwide are chronically infected with the hepadnavirus HBV; approximately 1 million die each year from complications of infection, including cirrhosis, hepatocellular carcinoma, and end-stage liver disease. Formally approved for the treatment of chronic hepatitis B are lamivudine (**28**), adefovir dipivoxil (**29**), (pegylated) interferon-α2, and entecavir (**30**). Whereas lamivudine, adefovir, and entacavir (and other nucleoside analogs that are still in (pre)clinical development) act as genuine antiviral agents at the HBV-associated reverse transcriptase, interferon, in the chronic hepatitis B setting, primarily acts as an immunomodulator. Pegylated interferon-α2b is effective in the treatment of hepatitis B e-antigen (HBeAg)-positive chronic hepatitis B, but no additional benefit is achieved if it is combined with lamivudine.[40]

Whereas interferon therapy, also because of its unavoidable side effects (influenza-like symptoms) is not recommended for treatment lasting longer than 1 year, the nucleos(t)ide analogs can, in principle, be administered for quite a number of years.

For lamivudine (3TC), however, this prolonged treatment is compounded by the emergence of both virological and clinical resistance at an accumulating rate of approximately 20% of patients per year. Resistance to adefovir dipivoxil may also emerge, but much less frequently (not more than 6% after 3 years).[41] Adefovir dipivoxil is the oral prodrug of adefovir (PMEA, 9-(2-phosphonylmethoxyethyl)adenine), which, after intracellular conversion to the diphosphate form, acts as a competitive inhibitor or alternative substrate for the HBV reverse transcriptase, and, when incorporated into the DNA, acts as a chain terminator, thereby preventing DNA chain elongation.[42]

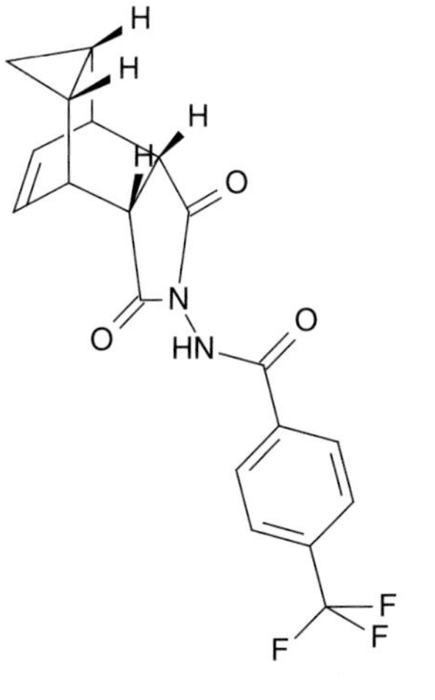

27 ST-246

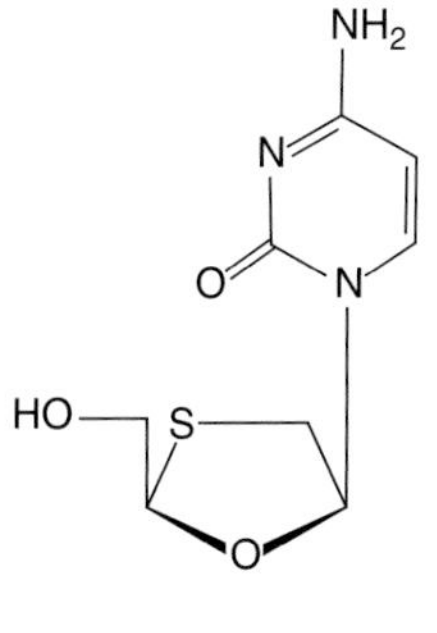

28 Lamivudine
(−)-β-L-3′-Thia-2′,3′-dideoxycytidine
(3TC)
Epivir, Zeffix

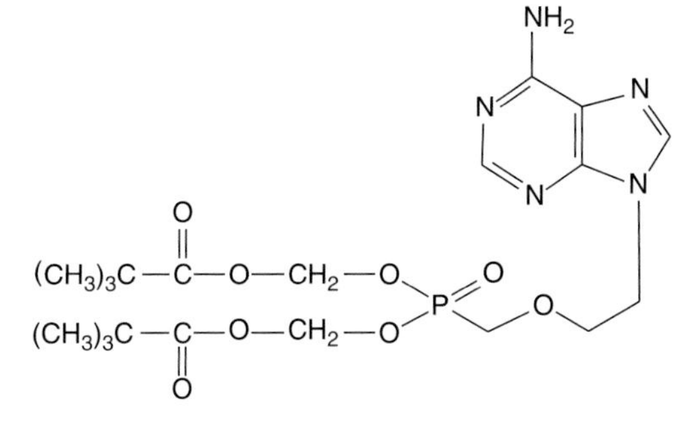

29 Adefovir dipivoxil
Bis(pivaloyloxymethyl)ester of
9-(2-phosphonylmethoxyethyl)adenine,
bis(POM)PMEA
Hepsera

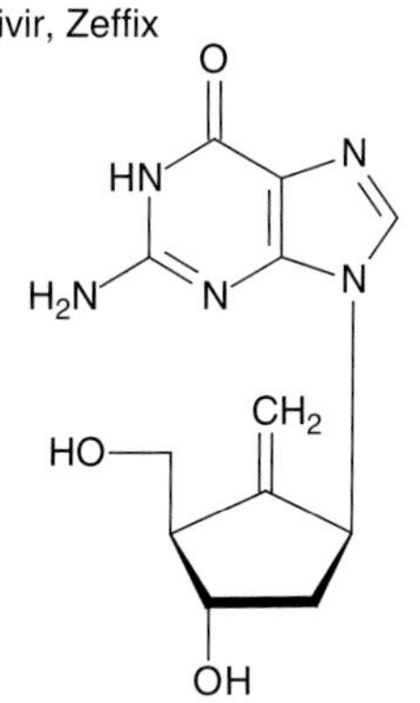

30 Entecavir
Baraclude

In patients with chronic HBV infection who were either positive[43] or negative[44] for HBeAg, 48 weeks of treatment with a dose of adefovir dipivoxil as low as 10 mg day^{-1} resulted in significant improvement of all parameters of the disease (histological liver abnormalities, serum HBV DNA titers, and serum alanine aminotransferase levels). In patients with HBeAg-negative chronic hepatitis B, the benefits achieved from 48 weeks of adefovir dipivoxil were lost when treatment was discontinued, but maintained if treatment was continued through week 144.[41]

Entecavir, one of the most recent antiviral drugs launched for clinical use, has in vitro and in vivo potency that seems to be greater than that of lamivudine: in patients with chronic hepatitis B infection it has proven efficacious at a dose as low as 0.5 mg day^{-1}.[45] The active (triphosphate) metabolite of entecavir would accumulate intracellularly at concentrations that are inhibitory to 3TC-resistant HBV DNA polymerase.[46] How this translates to the clinical efficacy of entecavir the treatment of 3TC-resistant HBV infections remains to be followed up.

In addition to entecavir, a number of L-nucleosides, e.g., β-L-thymidine (L-dT, Telbivudine, **31**), the 3′-valine ester of β-L-2′-deoxycytidine (Val-L-dC, Valtorcitabine, **32**) and 1-(2-fluoro-5-methyl-β-L-arabinosyl)uracil (L-FMAU, Clevudine, **33**) are in clinical development (Clevudine involves the role of deoxythymidylate (dTMP) kinase[47]) for the treatment of chronic hepatitis B. Other compounds in preclinical development include 2′,3′-dideoxy-3′-fluoroguanosine (FLG, **34**),[48] racivir, and L-Fd4C. These compounds are also in preclinical development for HIV (see below).

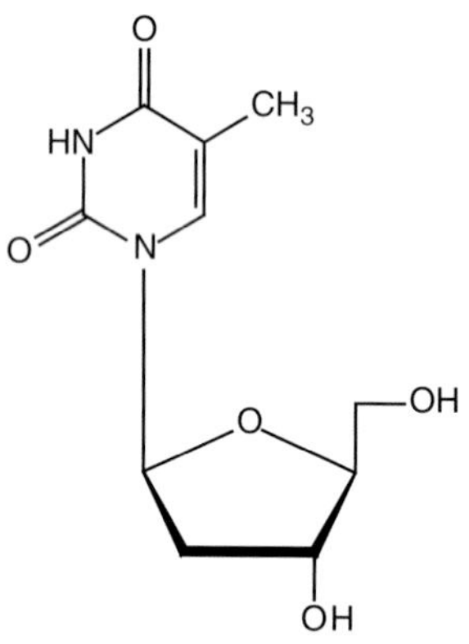

31 L-dT
Telbivudine

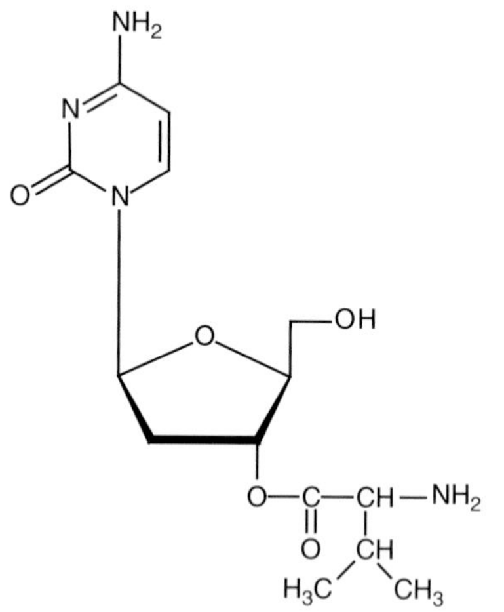

32 3′-Val-L-dC
3′-Valine ester of β-L-deoxycytidine
Valtorcitabine

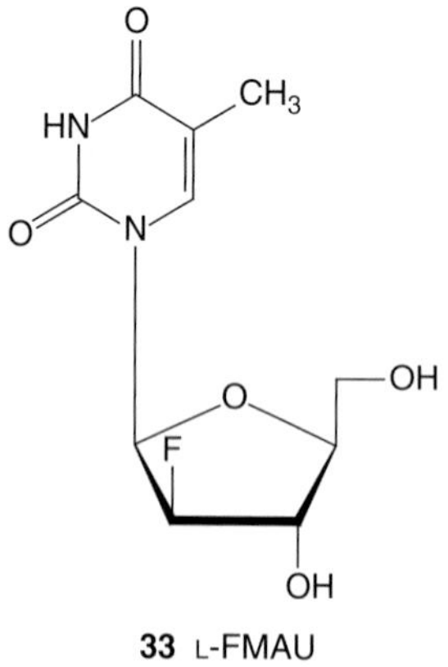

33 L-FMAU
Clevudine

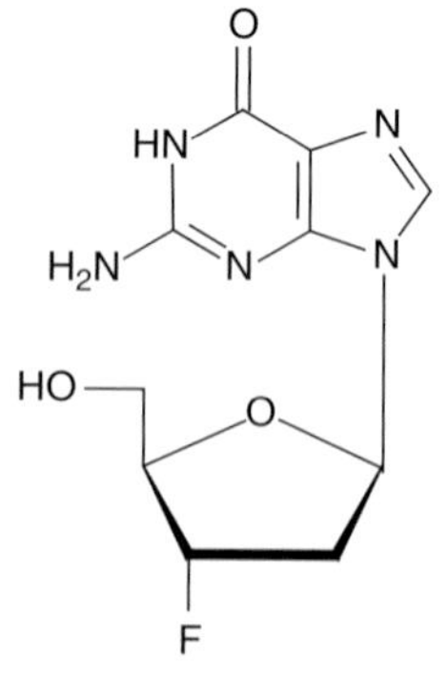

34 FLG
3′-Fluoro-2′,3′-dideoxyguanosine

Moreover, tenofovir disoproxil fumarate (TDF, **35**) and emtricitabine ((-)-FTC, the 5-fluoro-substituted counterpart of lamivudine, **36**)), which have both been licensed, individually and in combination, for the treatment of HIV infections (AIDS), may also be considered and further pursued, most likely in combination, for use in the treatment of chronic hepatitis B. In fact, TDF has been considered an important new therapeutic tool for the induction of complete remission in patients wit lamivudine-resistant HBV infection.[49]

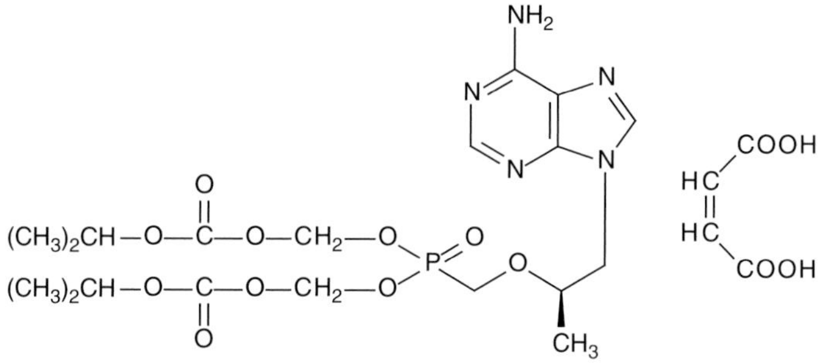

35 Tenofovir disoproxil fumarate (TDF)
Fumarate salt of bis(isopropoxycarbonyloxymethyl) ester of
(*R*)-9-(2-phosphonylmethoxypropyl)adenine
bis(POC)-PMPA
Viread

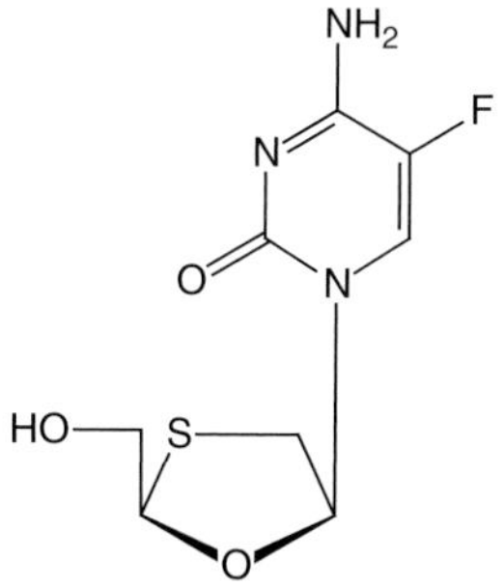

36 Emtricitabine
(−)-β-L-3′-thia-2′,3′-dideoxy-5-fluorocytidine [(−)-FTC]
Emtriva

An interesting recommendation has been proposed for the care of patients with chronic HBV and HIV coinfection.[50] They should be put on the combination of TDF with (−)-FTC, which would cover both the HBV and HIV infection. Only if no antiretroviral therapy is used in these patients, adefovir dipivoxil and/or pegylated interferon could be installed depending on whether they are HBeAg-negative or -positive, respectively.[50]

7.10.9 Picornaviruses (Entero- and Rhinoviruses)

Among the enteroviruses, polio and hepatitis A can be efficiently controlled by vaccination: for polio both a live attenuated and an inactivated ('killed') virus vaccine available, whereas for hepatitis A an inactivated virus vaccine is available. The other enteroviruses (Coxsackie A and B and echoviruses) and the rhinoviruses need to be approached by chemotherapeutic agents. No single antiviral drug has ever been licensed for clinical use against entero- or rhinovirus infections. The most extensively studied for its potential against enteroviruses has been pleconaril (**37**). This compound binds to a hydrophobic pocket beneath the 'canyon floor' of the VP1 capsid protein of picornaviruses,[51] thereby 'freezing' the viral capsid and preventing its dissociation (uncoating) from the viral RNA genome. The clinical efficacy of pleconaril has been evaluated in experimentally induced enterovirus (Coxsackie A21) respiratory infections in adult volunteers[52] and, on a compassionate basis, against potentially life-threatening enterovirus infections.[53] Pleconaril has also been shown to reduce the duration and severity of picornavirus-associated viral respiratory illnesses in adolescents and adults.[54,55]

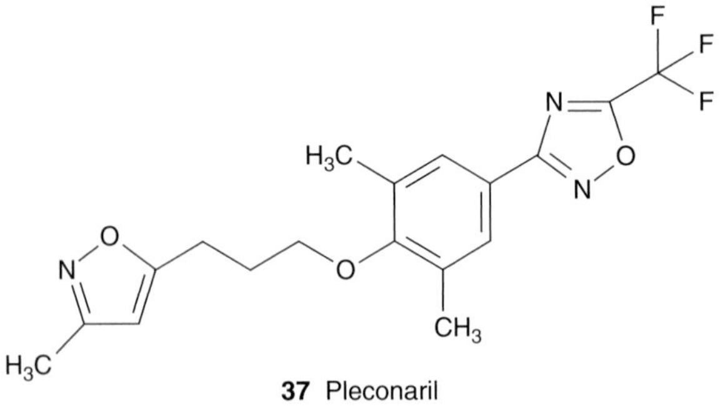

37 Pleconaril

For the prevention and/or treatment of rhinovirus infections (common colds) inhibitors of the human rhinovirus (HRV) 3C protease have been extensively investigated. Ruprintrivir (**38**) is an irreversible 3C protease inhibitor,[56] which, upon intranasal administration in human volunteers, appeared to be safe and well tolerated.[57] In experimentally induced rhinovirus colds in healthy volunteers, ruprintrivir prophylaxis reduced the proportion of subjects with positive viral cultures but did not decrease the frequency of colds.[58] Another, irreversible inhibitor of HRV 3C protease, here referred to as a pyrrolidinyl pentenoic acid ethyl ester (compound 3[59] or compound 1[60]) (**39**), offers the advantage of being orally bioavailable: in healthy volunteers, single oral doses of this compound appeared to be safe and well tolerated, although the compound is currently not progressing toward clinical development.[60]

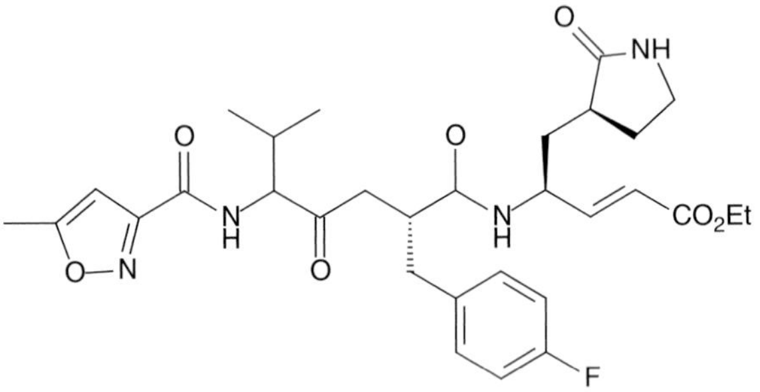

38 Ruprintrivir

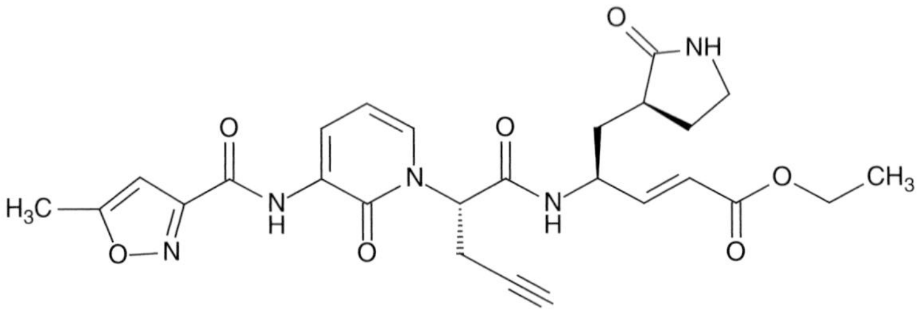

39 Pyrrolidinyl pentenoic acid ethyl ester

In great need of antiviral treatment are the often severe complications of Coxsackie B virus infections, such as myocarditis which may lead to idiopathic dilated cardiomyopathy. In mice, Coxsackie B3 virus-induced myocarditis is inhibited by the immunosuppressive agent mycophenolic acid (MPA) mofetil (**40**).[61] This beneficial outcome must apparently result from the immunosuppressive effect of MPA (through inhibition of IMP dehydrogenase and, hence, GTP supply), since MPA did not reduce the infectious virus titers in the myocarditis. A more pronounced inhibitory effect on Coxsackie B3 virus-induced myocarditis, accompanied by a marked reduction in the virus titers in the heart, was obtained with the interferon inducer poly(I).poly(C) and poly(I).poly(C_{12}U) (also known as Ampligen), and to a lesser extent with (pegylated) interferon-α2b.[62] Combination of an inhibitor of viral replication (such as Ampligen) with an immunosuppressant (such as MPA mofetil) could be an ideal treatment strategy for viral myocarditis, whether due to Coxsackie B or other viruses. How to implement such as treatment regimen in the clinical setting should be further addressed.

7.10.10 Flaviviruses (Yellow Fever, Dengue, West Nile, etc.)

No antivirals are currently available for the treatment of flavivirus infections (although there is a live virus vaccine routinely used for the prophylaxis of yellow fever), and the prospects for an effective therapy of flavivirus infections do not seem encouraging.[63] Antiviral compounds such as ribavirin (**41**) have only weak activity against flaviviruses. Greater hope may be vested in interferon and interferon inducers. Based on infection of hamsters with the murine Modoc virus, an experimental flavivirus encephalitis model has been developed, which is reminiscent of Japanese encephalitis virus infection in humans.[64] In a related model with Modoc virus in SCID mice, both interferon-α2b (whether pegylated or not) and interferon inducers (poly(I).poly(C) and Ampligen) were shown to significantly delay virus-induced morbidity (paralysis) and mortality (due to progressive encephalitis).[65] Ribavirin did not provide any beneficial effect in this model, whether given alone or in combination with interferon.

7.10.11 Hepaciviruses (HCV)

Current, approved therapy for chronic hepatitis C consists of pegylated interferon-α2 (180 μg, parenterally, once weekly) combined with ribavirin (1000 or 1200 orally, daily).[66,67] This treatment regimen is associated with a sustained viral response in at least 50% of the patients infected with HCV genotype 1, and of 80% in patients infected with another genotype (2, 3, or 4) of HCV. Duration of treatment is 48 weeks (or longer) for patients infected with HCV genotype 1, but may be reduced to 24 weeks for patients infected with another genotype. Interferon appears to be targeted at the phosphoprotein encoded by the nonstructural NS5A gene of the HCV genome,[68] whereas ribavirin, akin to MPA, primarily acts as an inhibitor of IMP dehydrogenase, thus reducing the biosynthesis of GTP.

The combination of pegylated interferon α-2a with ribavirin has also been advocated for the therapy of HCV infection in patients with HIV coinfection.[69] However, it should not be forgotten that, should these patients be treated (for their HIV infection) with azidothymidine (zidovudine, ZDV, **42**), the latter may be antagonized by ribavirin.[70] Therefore it was reassuring to note that ribavirin (at 800 mg day^{-1}) administered in combination with pegylated interferon α-2a did not significantly affect the intracellular phosphorylation or plasma pharmacokinetics of ZDV (or other pyrimidine dideoxynucleosides such as 3TC or d4T) in HIV/HCV coinfected patients.[71]

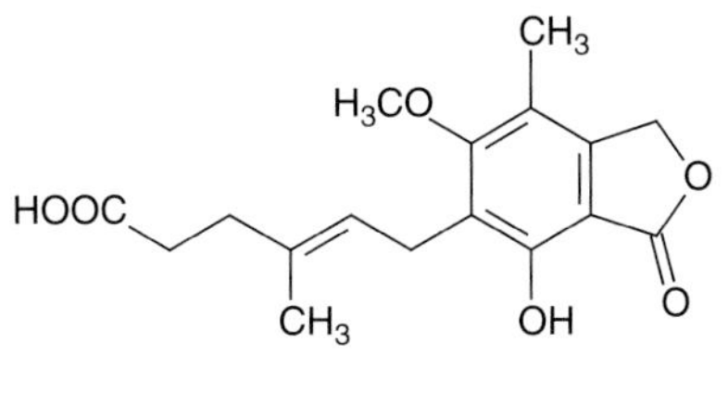

40 Mycophenolic acid (MPA) mofetil

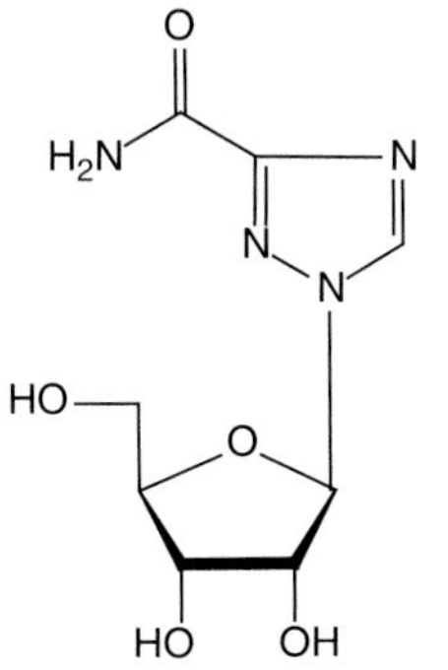

41 Ribavirin

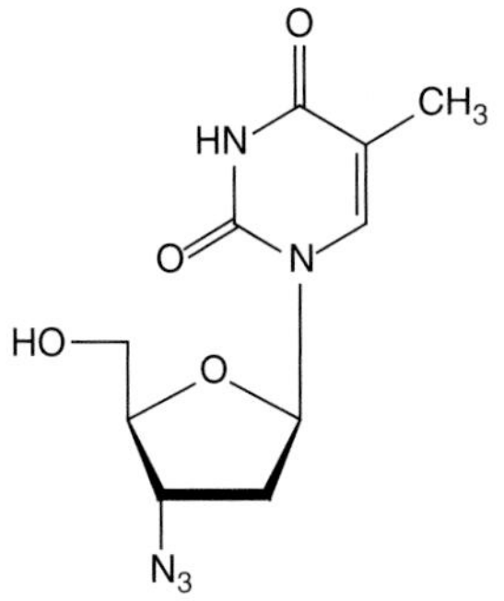

42 Zidovudine
3′-Azido-2′,3′-dideoxythymidine,
azidothymidine (AZT)
Retrovir

In addition to pegylated interferon α-2a and pegylated interferon α-2b, other interferons are already in use or development, e.g., albuferon-α (Interferon α-2b fused to human serum albumin) which allows dosing at intervals of 2–4 weeks compared with 1 week for the pegylated interferons.[72] Consensus interferon (i.e., alfacon-1), when combined with ribavirin, has been shown to achieve a higher sustained response rate in naiive patients with chronic hepatitis C as compared to standard Interferon-α and ribavirin.[72]

In the combination with pegylated interferon, ribavirin may be advantageously replaced by viramidine (**43**), its amidine analog (which is converted, mainly in hepatocytes, by adenosine deaminase, to ribavirin), as the latter has a reduced uptake by, and, therefore, lesser toxicity for red blood cells, as compared to ribavirin.[73] Viramidine would give less anemia as compared with ribavirin. Phase III studies with viramidine combined with pegylated interferon, as compared to ribavirin combined with pegylated interferon, are eagerly awaited to assess which one to choose, ribavirin or viramidine.

Taking into account the duration of the combined interferon plus ribavirin treatment, the therewith associated side effects and costs, and the partial responses observed with this treatment regimen, fierce attempts have been made, rightfully, to develop more selective anti-HCV agents, targeted at specific viral proteins such as the NS3 protease and RNA helicase, and the NS5B RNA replicase (RNA-dependent RNA polymerase). Also the HCV p7 protein, which forms an ion channel and can be blocked by long-alkyl-chain iminosugar derivatives, has been considered as a potential target for antiviral therapy.[74]

Proof of principle that compounds targeted at the NS3 protease could reduce plasma concentrations of HCV RNA has already been delivered with BILN 2061 (**44**) administered orally for no longer than 2 days in patients infected with HCV genotype 1.[75] BILN 2061 (Ciluprevir) was able to reduce HCV RNA levels by 2-3 $\log_{10}$ in patients infected with HCV genotype 1, after 2 days of treatment,[76] but in patients infected with HCV genotypes 2 or 3, it proved less effective, apparently due to a lower affinity of BILN 2061 for the HCV protease of these genotypes.[77]

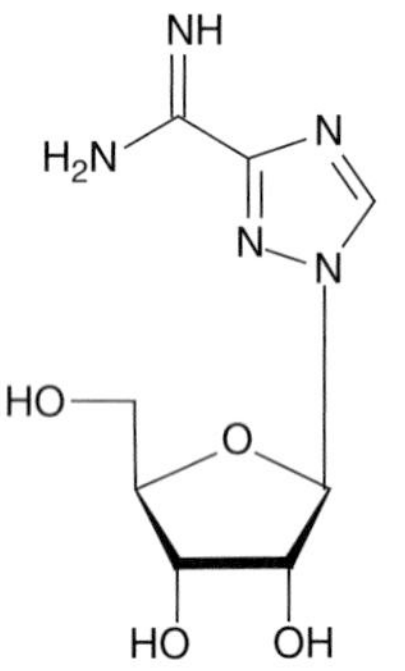

43 Viramidine

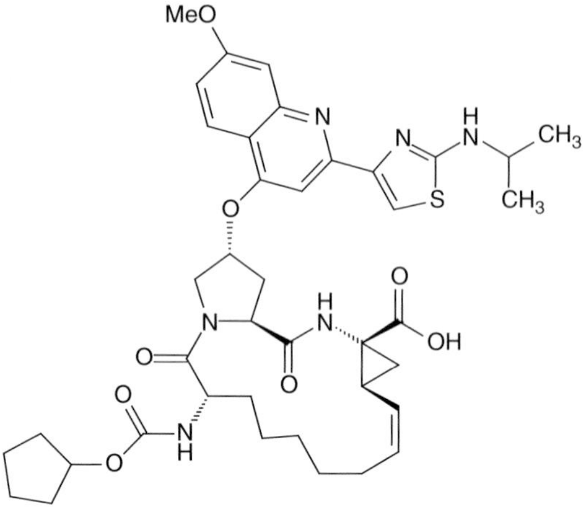

44 BILN 2061
Ciluprevir

Another NS3 protease inhibitor, which differs in its in vitro resistance profile from BILN 2061, is VX-950 (**45**)[78]: i.e., the major BILN 2061-resistant mutations at Asp168 are fully susceptible to VX-950, and similarly, the dominant resistant mutation against VX-950 at Ala156 remains sensitive to BILN 2061.[78] Thus, VX-950 and BILN 2061 elicit resistance to HCV protease (NS3A) by different mechanisms.[79] VX-950 (750 mg every 8 h) was found to achieve, at the end of a 14-day treatment, a main reduction of HCV RNA of 4.4 $\log_{10}$. (In some patients dosed with VX-950, the virus became undetectable at day 14 of dosing.) The overall preclinical profile of VX-950 supports its candidacy as a novel oral therapy against hepatitis C.[80] Other HCV NS3 protease inhibitors may be announced in the future as behaving similarly.[72] Various other 7-hydroxy-1,2,3,4-tetrahydroisoquinoline-3-carboxylic acid-based macrocyclic inhibitors of HCV NS3 protease[81] as well as SCH 503034 (**46**), a mechanism-based inhibitor of HCV NS3 protease,[82] are in preclinical development. In fact, the latter was found to act synergistically with α-interferon in suppressing HCV replicon synthesis.[82]

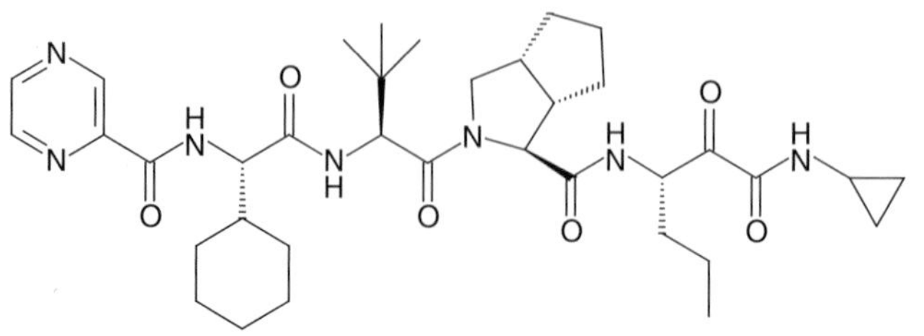

45 VX-950 (Telaprevir)

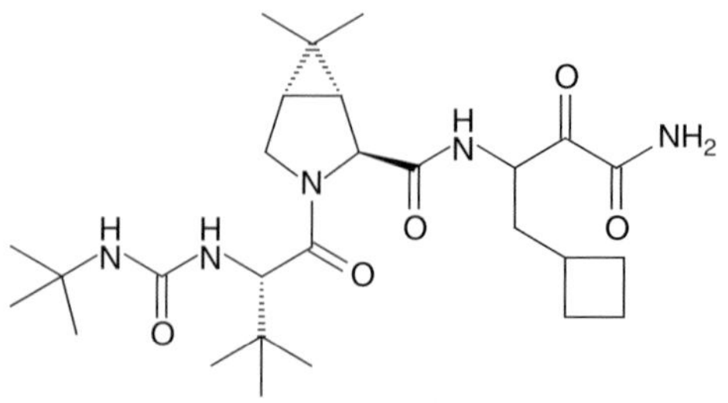

46 SCH 503034

In addition to the NS3 protease, the NS5B RNA replicase has also been perceived as an attractive target for the development of HCV inhibitors. Highly potent and selective antiviral agents targeted at the viral RNA replicase have been described to inhibit the replication of bovine viral diarrhea virus (BVDV), a pestivirus which could be considered as a surrogate virus for HCV.[83,84] We have recently described a novel series of compounds (prototype: 5-[(4-bromophenyl)methyl]-2-phenyl-5*H*-imidazo[4,5-c]pyridine (BPIP, **47**)) that act as 'nonnucleoside' RNA replicase inhibitors (NNRRIs) and effect a highly potent and selective inhibition of the replication of BVDV.[85] From the BPIP class of compounds, new congeners have been derived that act equally efficiently against HCV replication. In future treatment strategies for HCV infections, these NNRRIs may likely to be combined with 'nucleoside' RNA replicase inhibitors (NRRIs), in analogy with the strategy followed for the treatment of HIV infections, where NRTIs are combined with NNRTIs (see below).

The sole NRRI which has already proceeded to phase I/II clinical trials for the therapy of hepatitis C is the 3′-*O*-valine ester of 2′-*C*-methylcytidine (NM-283, valopicitabine, **48**), which can be administered by the oral route[86,87] and shows enhanced antiviral efficacy if combined with pegylated interferon.[88] In addition to 2′-*C*-methylcytidine, several other ribonucleoside analogs (NRRIs) have been reported to inhibit HCV replication[89]: 2′-*O*-methylcytidine (**49**),[90] 2′-*C*-methyladenosine (**50**),[90,91] 7-deaza-2′-*C*-methyladenosine (**51**),[92] and 2′-*C*-methylguanosine (**52**).[93,94] All these compounds act as nonobligate chain-terminating nucleoside analogs. It would seem interesting to prepare and evaluate the corresponding 3′-deoxyribonucleoside analogs for their anti-HCV activity.

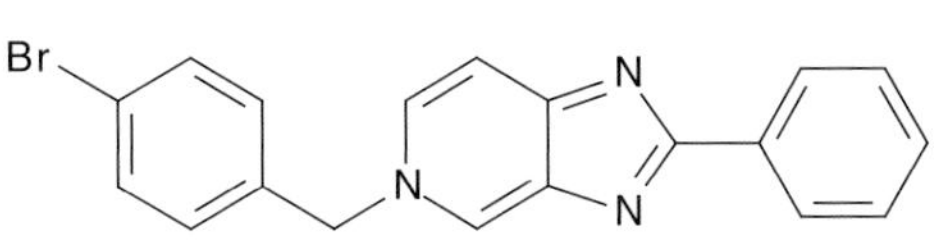

47 BPIP

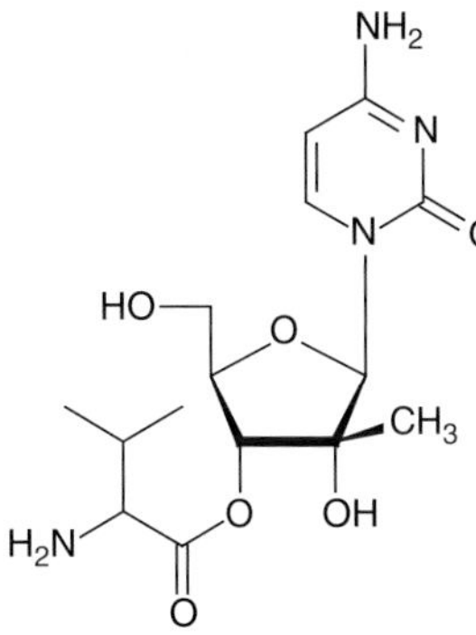

48 NM283
3′-Valine ester of 2′-*C*-methylcytidine
Valopicitabine

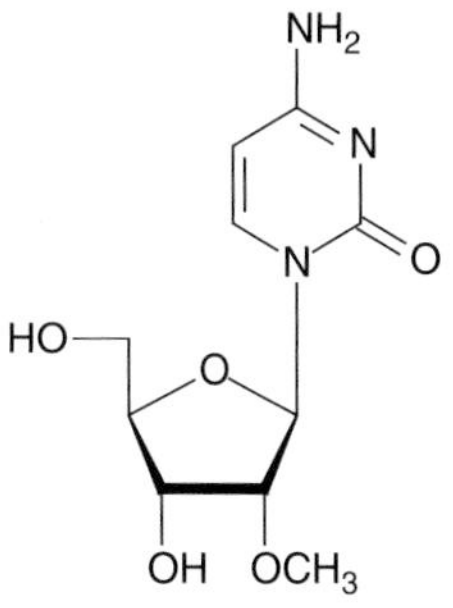

49 2′-*O*-methylcytidine

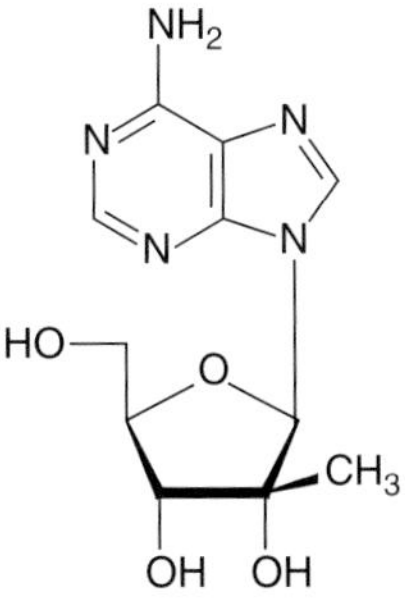

50 2′-*C*-methyladenosine

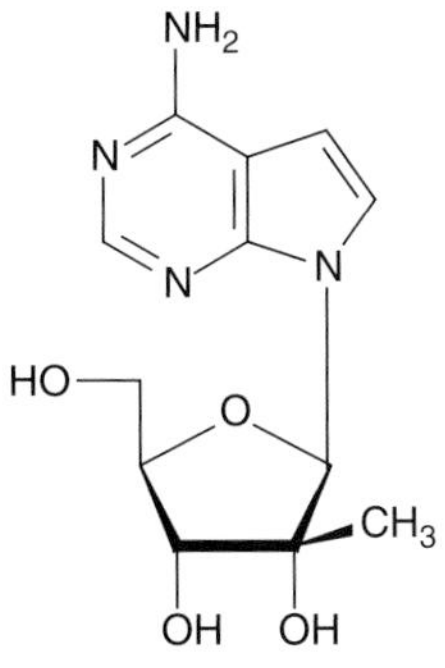

51 7-deaza-2′-*C*-methyladenosine

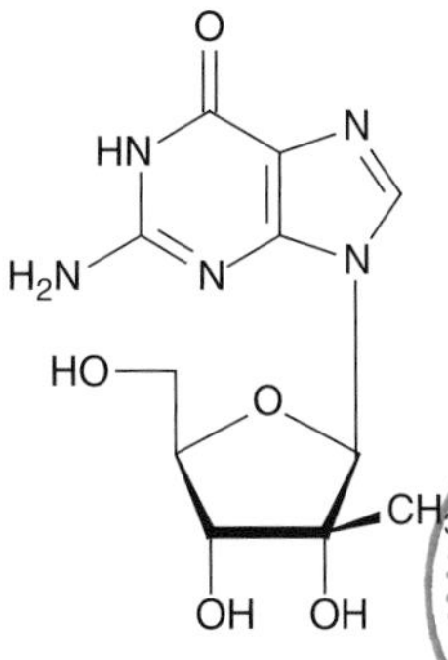

52 2′-*C*-methylguanosine

A new class of HCV NS5B RNA replicase inhibitors is represented by the alpha, gamma-diketo acids (DKAs),[95,96] one of the more active DKAs being DKA compound 30 (**53**).[95] These compounds are reminiscent of the DKA type of HIV integrase inhibitors (see below) and are assumed to inhibit the HCV polymerate activity via chelation of the active site Mg^{2+} ions. In a certain sense they may be considered as 'pyrophosphate mimics,' acting as product-like inhibitors of the polymerase reaction.[89]

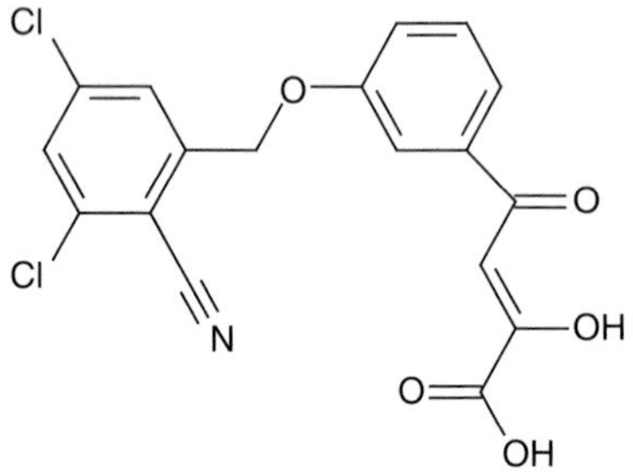

53 DKA compound 30

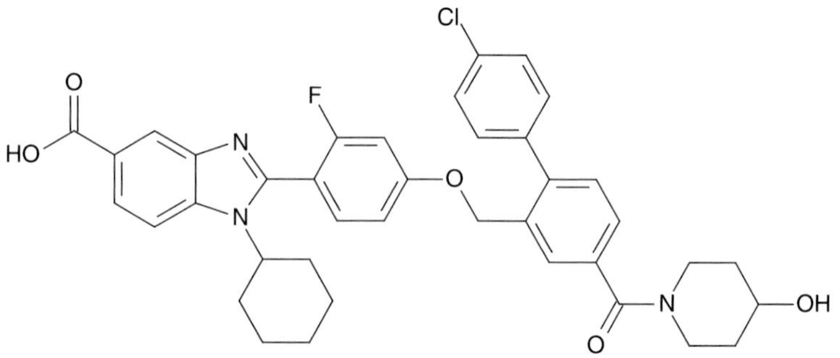

54 Benzimidazole derivative

In recent years, a constellation of NNRRIs has been described as acting in a very similar ('allosteric') fashion with HCV NS5B RNA replicase as NNRTIs do with respect to the HIV reverse transcripase: benzimidazole-based derivatives (**54**),[97] indole-*N*-acetamide derivatives (**55**),[98] benzo-1,2,4thiadiazine derivatives (**56**),[99,100] phenylalanine derivatives (**57**),[101] thiophene 2-carboxylic acid derivatives (**58**),[102] dihydropyranone derivatives (**59**),[103] the tetrahydropyranoindolyl acetic acid derivative HCV-371 (**60**)[104] and the *N*-1-aza-4-hydroxyquinolone benzothiadiazine A-782759 (**61**).[105] The allosteric binding site for some of these compounds have already been identified by crystallographic studies.[101,102] It corresponds to a narrow cleft on the protein's surface in the 'thumb' domain, about 3.0–3.5 nm from the enzyme's catalytic center.[101,102] Curiously, most of the NNRRIs that are active against the HCV NS5B polymerase contain, besides a large hydrophobic region, a carboxylic acid group (or a similar motif) that allows hydrogen bonding with main chain amide nitrogen atoms (i.e., Ser476 and Tyr477, as demonstrated for the phenylalanine derivative).[101]

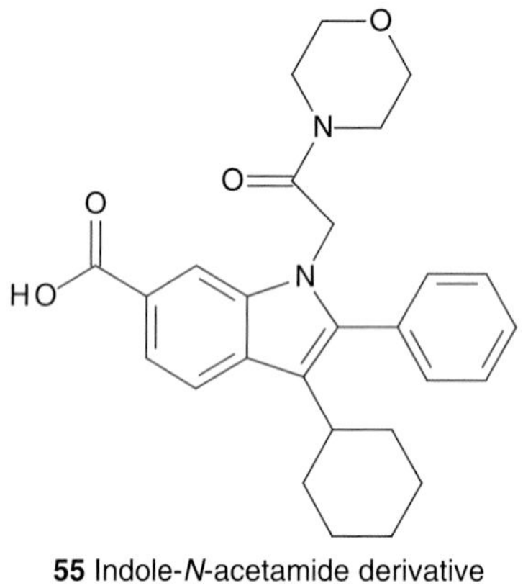

55 Indole-*N*-acetamide derivative

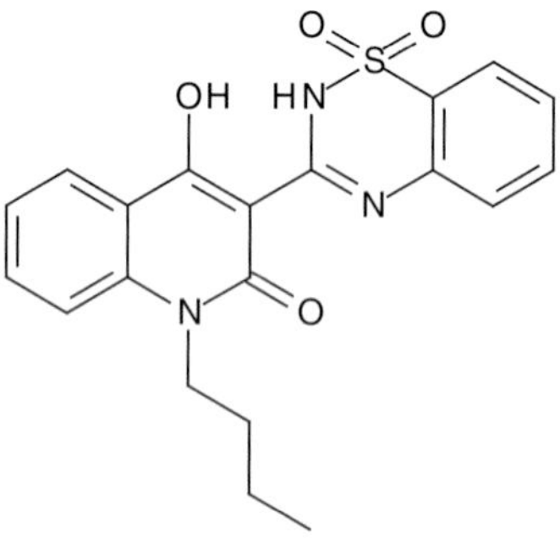

56 Benzothiadiazine derivative

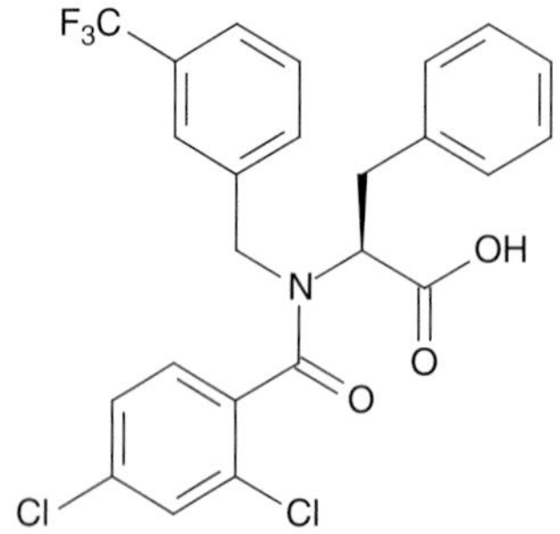

57 Phenylalanine derivative

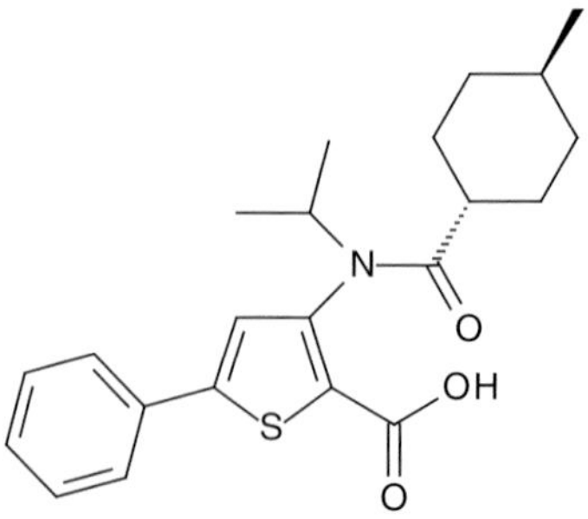

58 Thiophene 2-carboxylic acid derivative

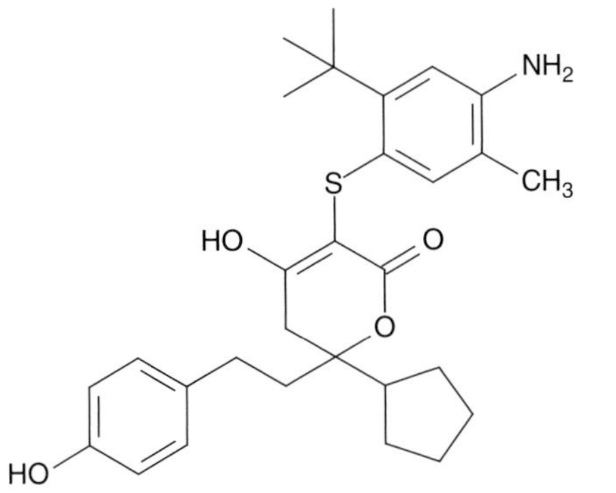

59 Dihydropyranone derivative

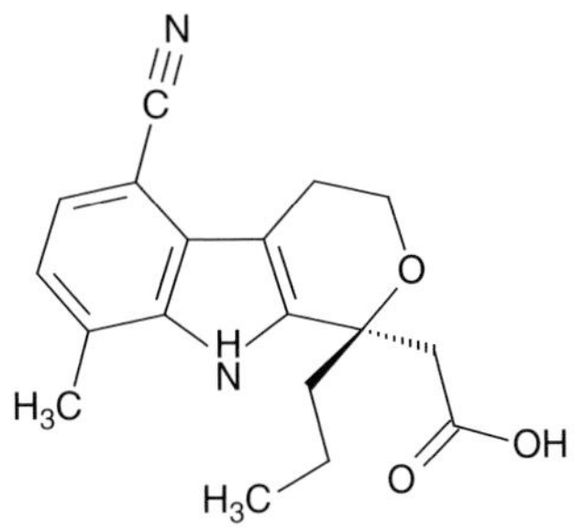

60 Tetrahydropyranoindolyl acetic acid derivative HCV-371

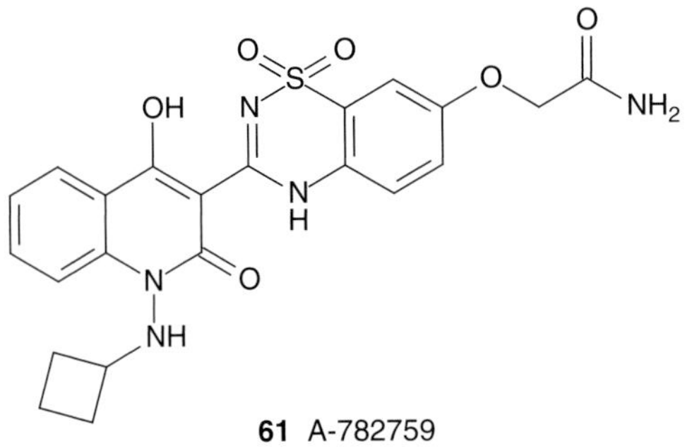

61 A-782759

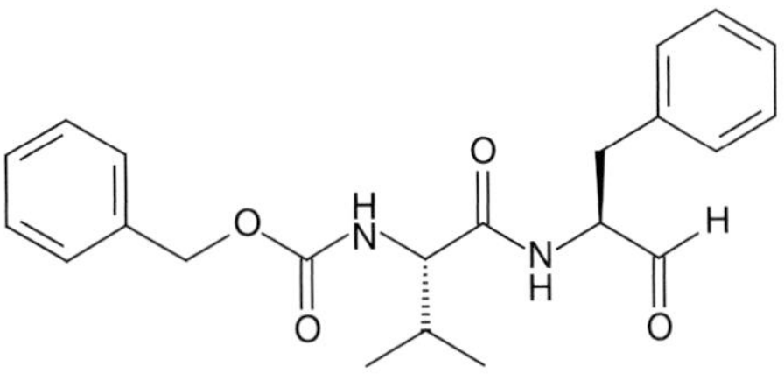

62 Calpain inhibitor III

Mutations conferring resistance to both the HCV NS5B RNA replicase (i.e., H95Q, N411S, M414L, M414T, or Y448H) and NS3 protease (i.e., A156V or D168V) have been identified.[105] These mutations conferred high levels of resistance to A-782759 and BILN 2061, respectively. However, the A-782759-resistant mutants remained susceptible to the NRRIs and other classes of NNRRIs, as well as interferon. In addition, the dually (A-782759- and BILN 2061-) resistant mutants displayed significantly reduced replicative ability as compared to the wild-type. These findings support a rationale for drug combinations in the therapy of HCV infections.[105]

As recently reviewed,[72] several other approaches, including ribozymes, antisense oligonucleotides, and RNA interference (RNAi) based on small interfering (si)RNAs could be envisaged to target the HCV genome. In particular, siRNAs aimed at posttranscriptional gene silencing may be considered an attractive approach.[106]

7.10.12 Coronaviruses (SARS)

As for HCV, there are several proteins encoded by the SARS coronavirus which could be considered as targets for chemotherapeutic intervention: i.e., the spike (S) protein, the 3C-like main protease, the NTPase/helicase, the RNA-dependent RNA polymerase (RNA replicase), and, possibly, other viral (or cellular) protein-mediated processes.[107] The severe acute respiratory syndrome (SARS) coronavirus S protein mediates infection of permissive cells through interaction with its receptor, the angiotensin-converting enzyme 2 (ACE 2),[108] and monoclonal antibody to the S1 domain was found to neutralize the virus by blocking its association with the receptor ACE 2.[109]

Also the fusion of the SARS coronavirus with the cell could be considered an attractive target. To the extent that this fusion process bears resemblance to the fusogenic mechanism of HIV, i.e., with regard to heptad repeat interactions and six-helix bundle formation, it might be feasible to develop SARS coronavirus inhibitors, analogous to the HIV fusion inhibitor enfuvirtide.[110]

Following receptor binding and induced conformational changes in the spike glycoprotein, a third step would be involved in the viral entry process, namely cathepsin L proteolysis within endosomes.[111] The cathepsin-L-specific inhibitor, MDL 28170 (also known as calpain inhibitor III (**62**), or Z-Val-Phe(CHO)), at the same time inhibited cathepsin L activity and S protein-mediated infection (at IC_{50} of 2.5 nM and 0.1 μM, respectively). In addition to calpain inhibitor III, some other calpain inhibitors have been described as inhibitors of SARS coronavirus replication, the most selective (selectivity index >100) being calpain inhibitor VI (4-fluorophenylsulfonyl-Val-Leu(CHO),**63**).[112]

The crystal structure of the SARS coronavirus protease has been revealed.[113,114] This offers a solid basis for the rational drug design of SARS protease inhibitors. For other potential targets such as the NTPase/helicase and the RNA replicase (RNA-dependent RNA polymerase) such structural basis still has to be delineated.

Of a number of peptidomimetic compounds (aziridinyl peptides, keto-glutamine analogs, chymotrypsin-like protease inhibitors, and peptide anilides) that have been reported as inhibitors of the SARS coronavirus main protease, the niclosamide anilide (**64**), with a $K_i = 0.03\,\mu M$ ($IC_{50} = 0.06\,\mu M$), proved to the most potent (competitive) inhibitor.[115] There are only a few cases where the 3C-like protease inhibitors were shown to inhibit both the SARS coronavirus protease activity and virus replication in cell culture. For example, the phe–phe dipeptide inhibitor (**65**) was found to inhibit the 3C-like protease at an IC_{50} of 1 μM and inhibited virus replication in Vero cells at an EC_{50} of 0.18 μM, while not being toxic to the host cells at a concentration of 200 μM (selectivity index >1000).[116] An octapeptide specifically designed for the SARS coronavirus main protease, namely AVLQSGFR, was reported to inhibit SARS coronavirus replication in Vero cells at an EC_{50} of 0.027 $\mu g\,mL^{-1}$, while not being cytotoxic at 100 $\mu g\,mL^{-1}$, thus establishing a selectivity index of >3700.[117] Whether this highly selective antiviral effect was actually mediated by an inhibition of the SARS coronavirus main protease was not ascertained in this study.[117]

The SARS coronavirus NTPase/helicase has been considered a potential target for the development of anti-SARS agents.[118] Bananin (**66**) and three of its derivatives (iodobananin, vanillinbananin, and eubananin) were shown to inhibit both the ATPase and helicase activity of the SARS coronavirus NTPase/helicase, with IC_{50} values (for the ATPase activity) in the range of 0.5–3 μM.[119] Bananin was also found to inhibit SARS-CoV replication in fetal rhesus kidney (FRhK-4) cells at an EC_{50} of less than 10 μM and a CC_{50} of over 300 μM, thus exhibiting a selectivity index of over 30.[119] Wheter the antiviral effect obtained in cell culture was causally linked to inhibition of the NTPase/helicase was not ascertained.

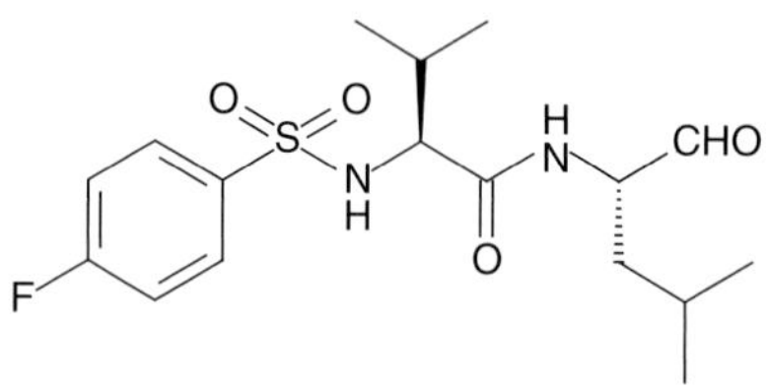

63 Calpain inhibitor VI

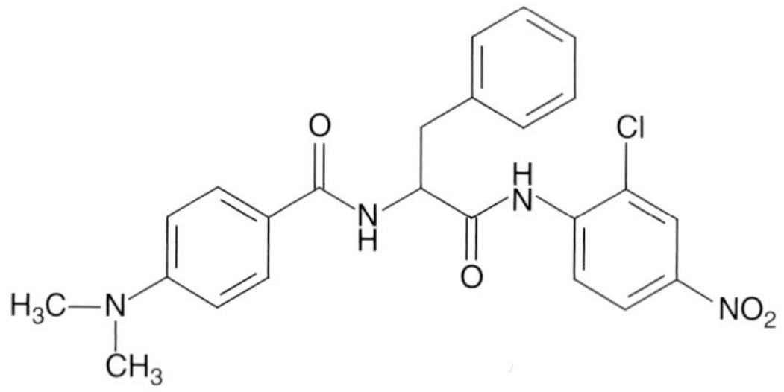

64 Niclosamide anilide (JMF 1507)

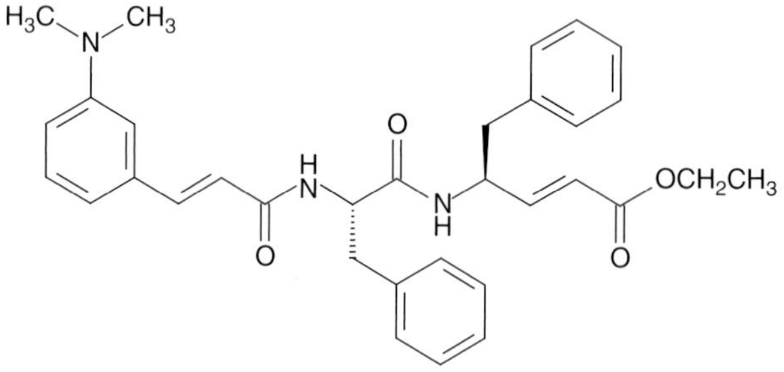

65 Phe–phe dipeptide

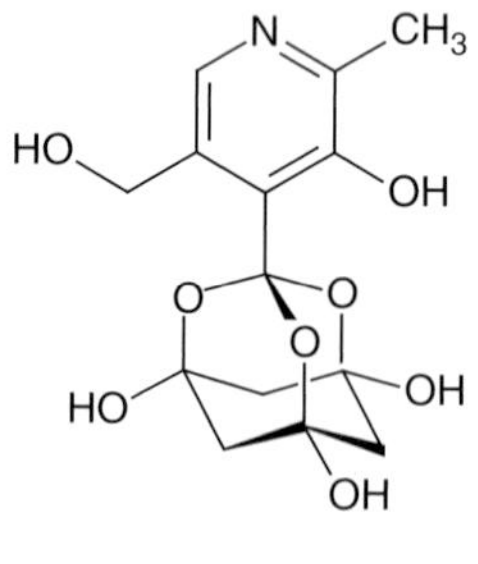

66 Bananin

The SARS coronavirus RNA-dependent RNA polymerase (RdRp), because of its pivotal role in viral replication, represents another potential target for anti-SARS therapy. This enzyme does not contain a hydrophobic pocket for nonnucleoside inhibitors similar to those that have proven effective against the HCV polymerase or HIV-1 reverse transcriptase.[120] In fact, nonnucleoside HIV-1 reverse transcriptase inhibitors were shown to have no evident inhibitory effect on SARS coronavirus RdRp activity.[121] At present, few, if any, nucleoside analogs have been recognized as specific inhibitors of the SARS coronavirus RdRp. There is N^4-hydroxycytidine, which has been accredited with both anti-HCV and anti-SARS coronavirus effects. Against SARS coronavirus it proved active at an EC_{50} of 10 μM (selectivity index ⩾10).[112] Whether this antiviral effect was mediated by an inhibition of the viral RdRp was not ascertained, however.

A wide variety of 'old' and 'new' compounds have been reported to inhibit the in vitro replication of the SARS coronavirus at relatively high concentration (⩾1 μM).[122] There is no shortage of small molecules that inhibit the replication of the SARS virus within the 1–10 μM (or higher) concentration range,[123] but whether any of these molecules would be able to prevent or suppress SARS in vivo, remains to be determined. Typical examples of such miscellaneous compounds, often with an ill-defined mode of action but selectivity indexes up to 100, that have been reported to inhibit SARS coronavirus replication are valinomycin (**67**),[123] glycyrrhizin (**68**),[124] chloroquine (**69**),[125] niclosamide (**70**),[126] and nelfinavir (**71**).[127]

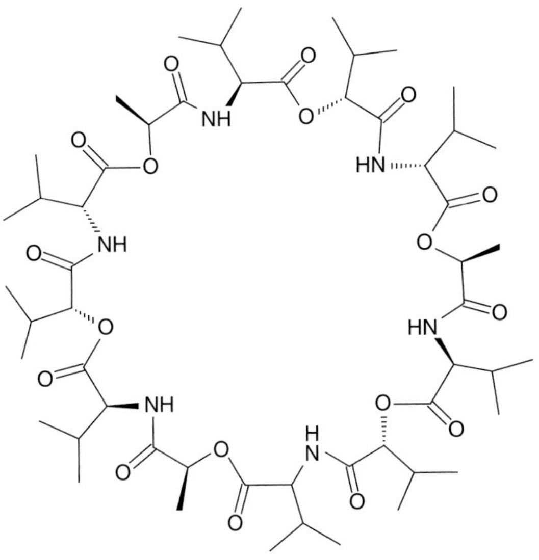

67 Valinomycin

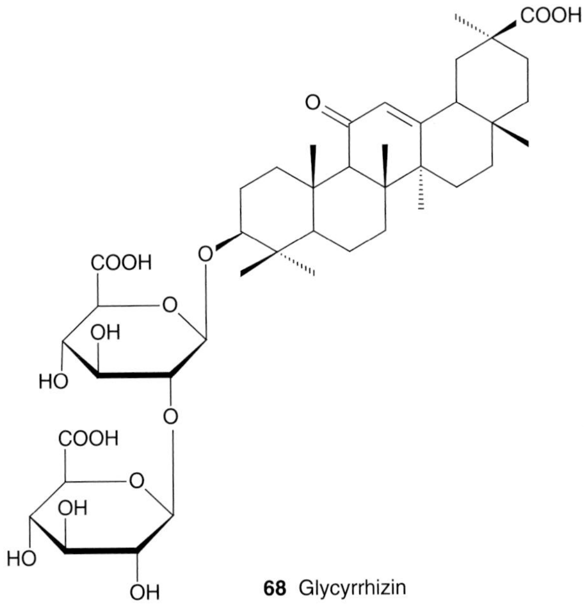

68 Glycyrrhizin

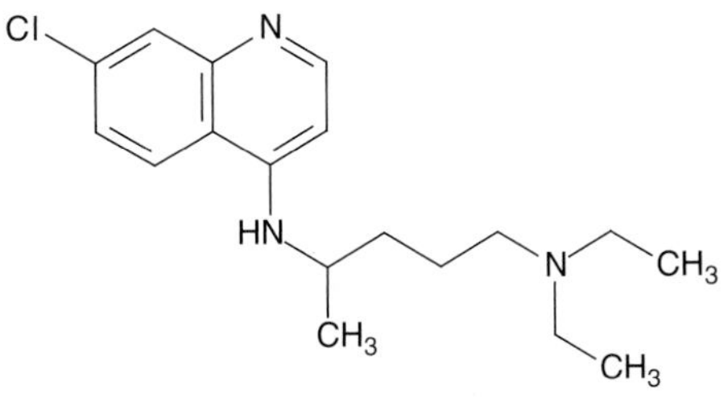

69 Chloroquine
(Nivaquine)
7-chloro-4-(4-diethylamino-1-methylbutylamino)quinoline

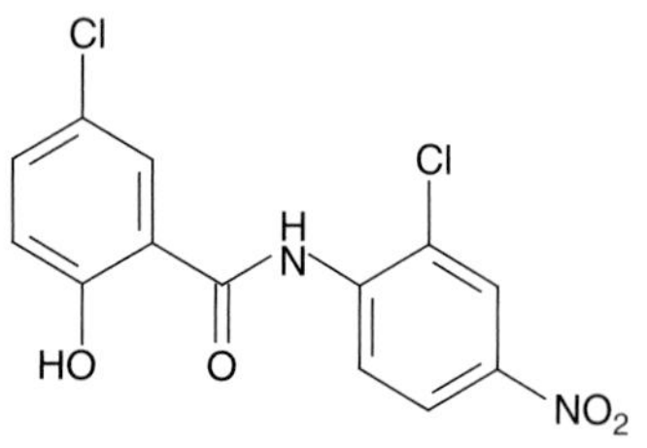

70 Niclosamide

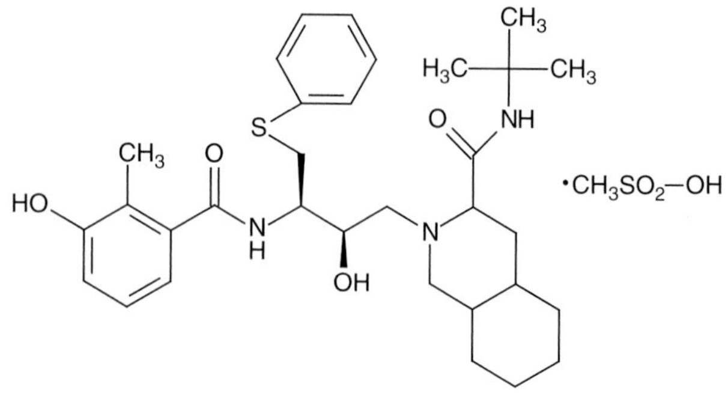

71 Nelfinavir
Viracept

SiRNAs have been developed that target the replicase[128] and spike (S)[129] genes of the SARS coronavirus genome, thereby silencing their expression in cell culture. Potent siRNA inhibitors of SARS coronavirus in vitro (i.e., the siRNA duplexes siSC2 (forward sequence: 5′-GCUCCUAAUUACACUCAACdtdt-3′) and siSC5 (forward sequence: 5′-GGAUGAGGAAGGCAAUUUAdtdt-3′), targeting the SARS coronavirus genome at S protein-coding and nonstructural protein-12-coding regions, respectively) were further evaluated for their efficacy in a rhesus macaque SARS model,[130] and found to provide relief from SARS coronavirus infection-induced fever, diminish SARS-CoV levels, and reduce acute diffuse alveolar damage. Whether SARS can be conquered by the siRNA approach remains to be proven, however.

Shortly after SARS coronavirus was identified as the causative agent of SARS, interferons were shown to inhibit the replication of SARS coronavirus in cell culture in vitro, interferon-β being more potent than either interferon-α or -γ.[131] These observations were subsequently confirmed and extended in several other studies.[132–134] Interferon-β, in conjunction with interferon-γ, was found to synergistically inhibit the replication of SARS coronavirus in Vero cells.[134] Being a prophylactic rather than therapeutic agent, interferon(s) may have their highest utility in the prophylaxis or early postexposure management of SARS. Pegylated interferon-α has been shown to reduce viral replication and excretion, viral antigen expression by type 1 pneumocytes, and the attendant pulmonary damage in cynomolgous macaques that were infected experimentally with SARS coronavirus.[135] Pegylated interferon-α is commercially available for the treatment of hepatitis C (where it is generally used in combination with ribavirin) and hepatitis B. Pegylated interferon-α as well as the other commercially available interferons (interferon-β, alfacon-1, etc.) could be considered for prevention and/or early postexposure treatment of SARS should it reemerge.

7.10.13 Orthomyxoviruses (Influenza A, B)

For many years, amantadine (**72**) and rimantadine (**73**) have been used for the prophylaxis and therapy of influenza A virus infections, but they never gained wide acceptance, primarily because of the risk of rapid emergence of drug-resistant virus mutants. These compounds interact specifically with the matrix M2 protein, which through its function as an hydrogen ion (H^+) channel, helps in the decapsidation (uncoating) of the influenza A virus particles. Influenza A (H1N1) viruses harboring amantadine resistance mutations are as virulent as wild-type virus strains[136] and can be readily transmitted during antiviral pressure in the clinical setting.

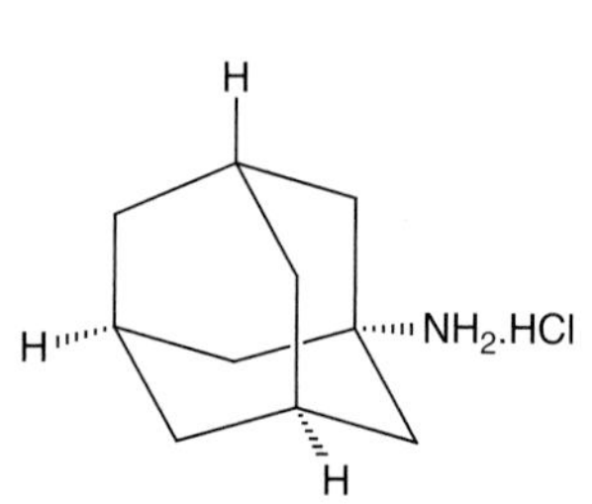

72 Amantadine
Symmetrel, Mantadix

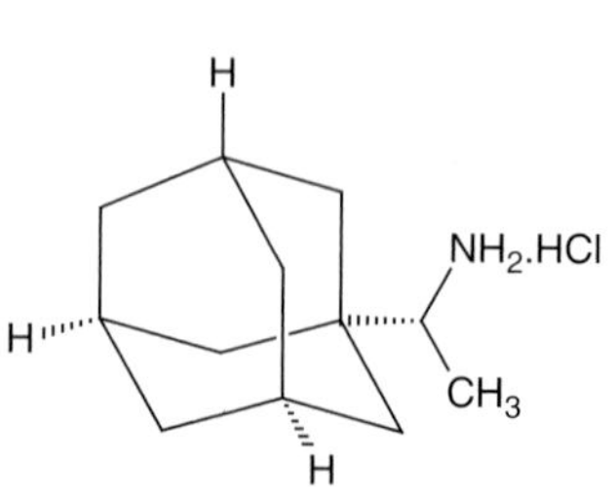

73 Rimantadine
Flumadine

In recent years, the neuraminidase inhibitors zanamivir (**74**)[137] and oseltamivir (**75**)[138] have become available for the therapy and/or prophylaxis of influenza A and B virus infections. Influenza has adopted a unique replication strategy by using one of its surface glycoproteins, hemagglutinin (H), to bind to the target cell receptor (which contains a terminal sialic acid, N-acetylneuraminic acid (NANA)), and another surface glycoprotein neuraminidase (N), to cleave off the terminal sialic acid, thus allowing the virus particles to leave the cells after the viral replicative cycle has been completed.[122] Neuraminidase inhibitors block the release of progeny virus particles from the virus-infected cells, thus preventing virus spread to other host cells.

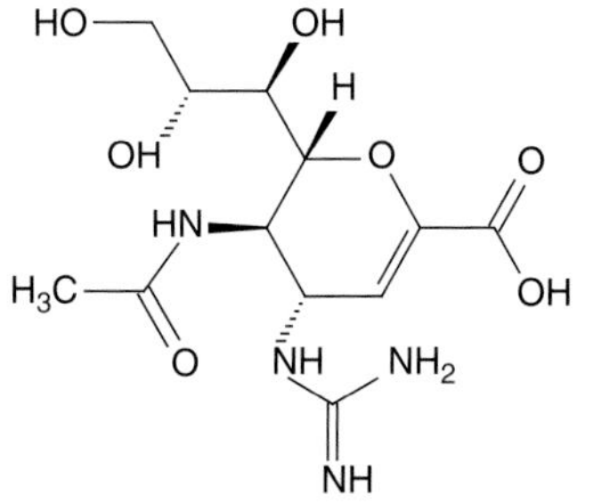

74 Zanamivir
Relenza

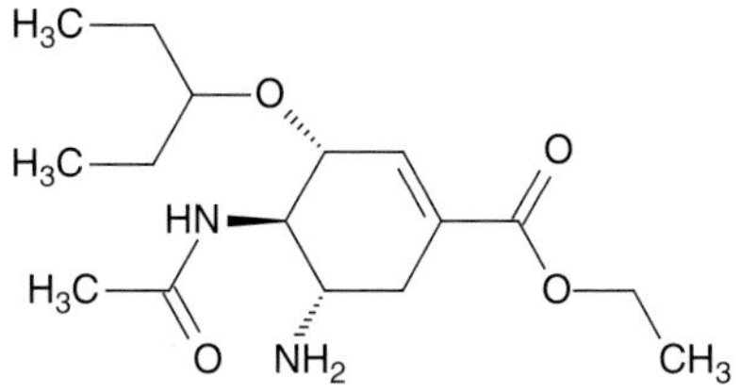

75 Oseltamivir
Tamiflu

When used therapeutically, neuraminidase inhibitors lead to a reduction in illness by 1–2 days, a reduction in virus transmission to household or healthcare contacts, a reduction in the frequency and severity of complications (such as sinusitis and bronchitis) and a diminished use of antibiotics.[139] When used prophylactically, neuraminidase inhibitors significantly reduced the number of new influenza cases.[140] Although resistance of human influenza viruses to neuraminidase inhibitors can develop,[141] there is no evidence of naturally occurring resistance to either zanamivir or oseltamivir.[142] Zanamivir and oseltamivir should be effective against both influenza A and B, and, among influenza A, the prevailing variants H_1N_1, H_3N_2, and also the 'avian flu' H_5N_1. Zanamivir, which must be taken through (oral) inhalation, and, in particular, oseltamivir, which can be more conveniently administered as oral capsules, should be stockpiled to confront a potential influenza pandemic in the future.[143,144] With the increasing threat of the avian flu, the need for a sufficient supply of neuraminidase inhibitors, such as oseltamivir and zanamivir, has become extremely urgent. In comparison with the neuraminidase inhibitors, the existing influenza vaccines are likely to be of limited value against newly emerging influenza virus strains.

Following zanamivir and oseltamivir, similar structure-based neuraminidase inhibitors have been developed, such as the cyclopentane derivative RWJ-270201 (peramivir, **76**)[145,146] and the pyrrolidines A-192558 (**77**)[147] and A-315675 (**78**).[148,149] These novel neuraminidase inhibitors may themselves be considered as potential drug candidates, and, while being amenable to further optimization,[150] lead to the development of yet newer compounds with improved activity, bioavailability and/or resistance profiles.

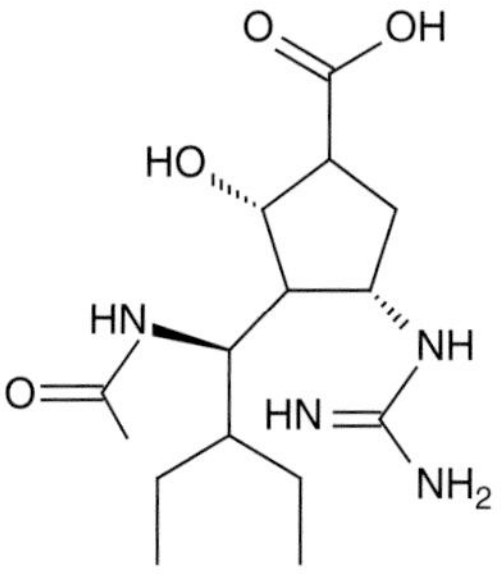

76 RWJ-270201
Peramivir

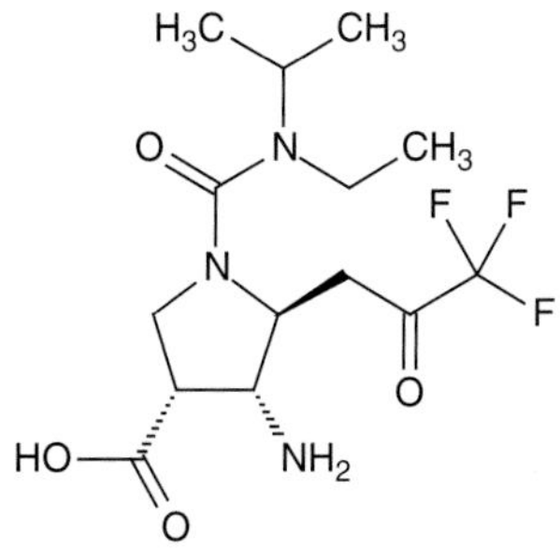

77 A-192558

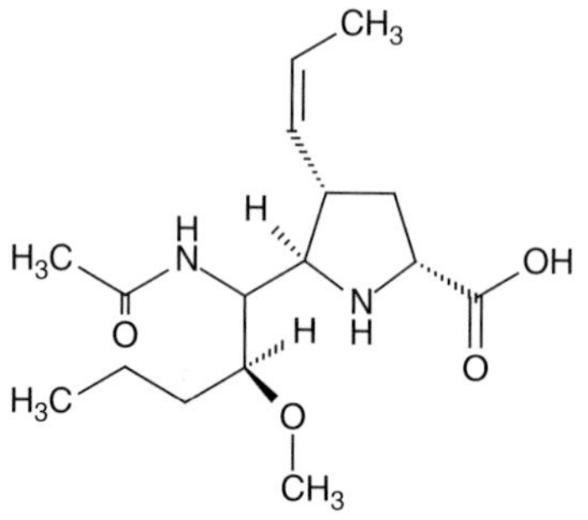

78 A-315675

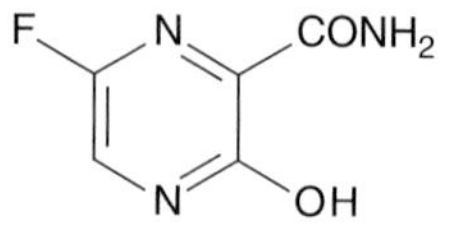

79 T-705

In fact, both peramivir (RWJ-270201) and A-315675 proved effective against a panel of five zanamivir-resistant and six oseltamivir-resistant A and B influenza virus strains.[151] Oseltamivir resistance in clinical isolates of human influenza A has been associated with mutations at positions 119, 198, 274, 292, or 294 of the neuraminidase. Recently, resistance of avian influenza A H_5N_1 against oseltamivir was shown to be caused by the H274Y mutation.[152,153] The H274Y variant still appeared sensitive to zanamivir.[152] Prominent among the other neuraminidase inhibitor-resistant influenza A (H_3N_2) virus mutations (not yet demonstrated for H_5N_1) are E119V and R292K: whereas the R292K mutation was associated with compromised virus growth and transmissibility, the growth and transmissibility of the E119V variant were comparable to those of wild-type virus.[154]

Are there other antiviral agents, besides amantadine, rimantadine, and the neuraminidase inhibitors, that may be considered for their potential, in the prevention and/or therapy, of influenza A virus infections, including avian influenza? Ribavirin has since long been recognized as a broad-spectrum antiviral agent, with particular activity against both ortho- and paramyxoviruses.[155] Recently, viramidine, the carboxamidine analog of ribavirin, was shown to have efficacy similar to ribavirin against influenza virus infections, and considering its lesser toxicity, viramidine may warrant further evaluation as a possible therapy for influenza, including H_5N_1.[156] Yet, other recently described compounds with specific activity against influenza A, B, and C viruses are T-705 (6-fluoro-3-hydroxy-2-pyrazine carboxamide, **79**) and the 2,6-diketopiperazine flutimide (**80**), which would target the viral polymerase[157] and cap-dependent endonuclease,[158] respectively.

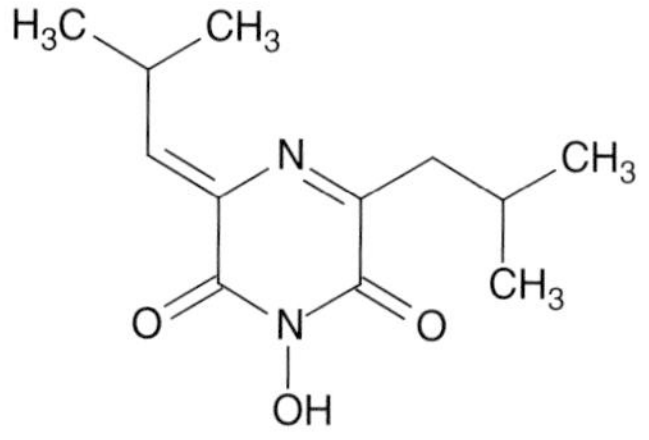

80 Flutimide

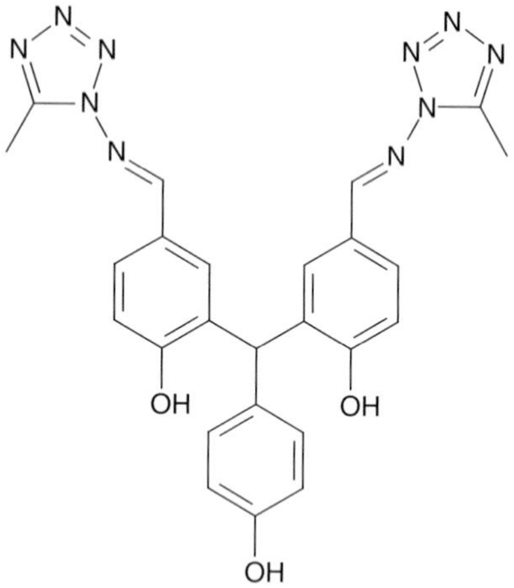

81 VP-14637

7.10.14 Paramyxoviruses (Parainfluenza, Measles, Mumps, RSV, hMPV, Nipah, etc.)

Of the paramyxoviruses, parainfluenza (types 1–5) has received little attention from either a preventative or curative viewpoint; mumps and measles, like the rubellivirus rubella, are now sufficiently contained by vaccination, which makes respiratory syncytial virus (RSV), human metapneumovirus (hMPV), and Nipah the paramyxoviruses with the greatest need for antiviral therapy. For RSV the only approved antiviral therapy is aerosol administration of ribavirin. In practice, however, ribavirin is rarely used owing to the technical burden of delivery by aerosol under the given circumstances (RSV bronchopneumonitis in young infants). Given the high incidence of RSV infections (which are often diagnosed as influenza), there is a high (and as yet unmet) medical need for an appropriate therapy (and prophylaxis) of RSV infections; the same holds for hMPV infections, which usually occur during the same (winter) season as RSV, mainly in young children, elderly people, and immunocompromised individuals. Ribavirin certainly holds promise for the treatment of hMPV infections, as has recently been demonstrated in the mouse model for hMPV.[159] Also, as has been mentioned above for HCV,[106] siRNA may also be applicable, if properly designed and administered (intranasally) in the treatment of respiratory virus infections.[160]

Recently, a number of small molecules, e.g., VP-14637 (**81**) and JNJ-2408068 (**82**) (formerly known as R-170591), although structurally dissimilar, have been shown to fit into a small hydrophobic cavity in the inner core of the RSV fusion (F) protein, thereby interacting with the heptad repeats HR1 and HR2 domains, and to inhibit RSV fusion.[161–164] Although the therapeutic potential of these compounds in the treatment of RSV infections is presently unclear, there is no doubt that further exploration of the mechanism of interaction between these inhibitors and the F protein should facilitate the design of new RSV fusion inhibitors.[161]

BMS-433771 (**83**) was found to be a potent inhibitor of RSV replication in vitro[165]; it exhibited excellent potency against multiple laboratory and clinical isolates of both A and B RSV with an average EC_{50} of 20 nM. BMS-433771 inhibits fusion the (viral and cellular) lipid membranes during both the early and virus entry stage and late-stage syncytium formation.[165] BMS-433771 was shown to be orally active against RSV in BALB/c mice and cotton rats, even if administered as a single oral dose 1 h prior to intranasal RSV inoculation.[166] It could be considered the prototype of low-molecular-weight inhibitors that target the formation of the six helical coiled-coil bundles as a prelude to virus–cell fusion, not only of RSV but also HIV.[167]

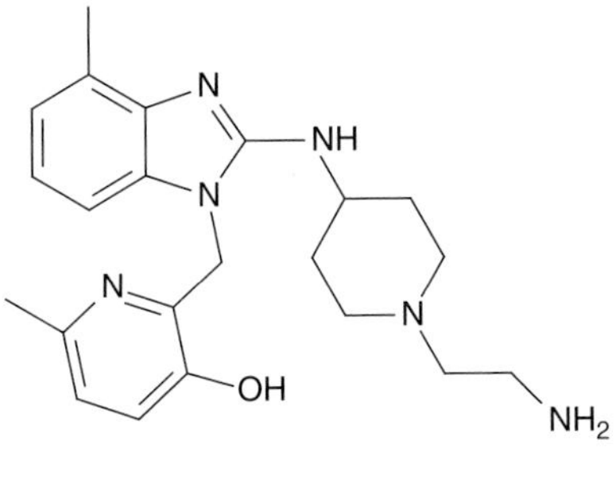

82 JNJ-2408068
R-170591

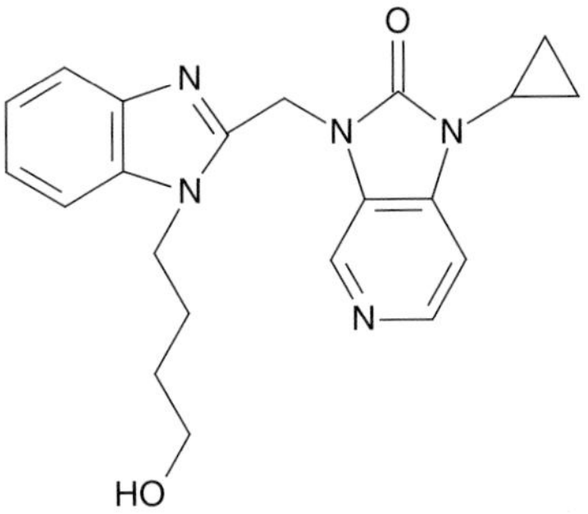

83 BMS-433771

Although human parainfluenza viruses are important respiratory tract pathogens, especially in children, they have received little attention from either prophylactic (vaccine) or therapeutic viewpoint. Yet, they contain a unique target, the major surface glycoprotein hemagglutinin-neuraminidase (HN) that serves, at the same time, for cell attachment and virus spread. The HN inhibitors BCX-2798 (**84**) and BCX-2855 (**85**) were found to inhibit both functions, and to block infection with parainfluenza viruses both in vitro and in vivo.[168] These compounds may limit parainfluenza virus infections in humans. Other compounds that may be further pursued for their activity against parainfluenza viruses, RSV, as well as influenza viruses, include flavonoids,[169] uncinosides,[170] and polyoxotungstates.[171]

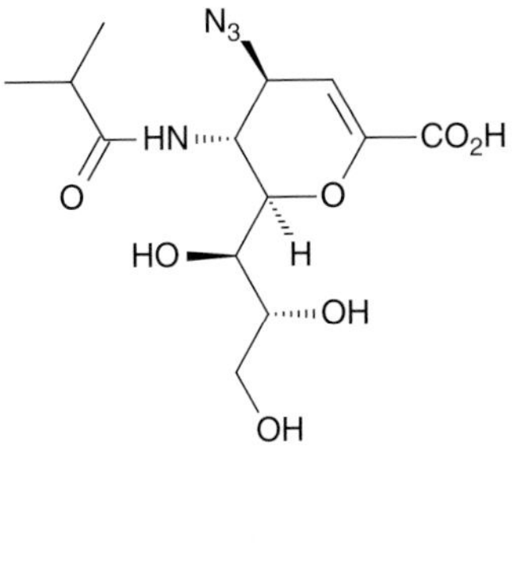

84 BCX-2798

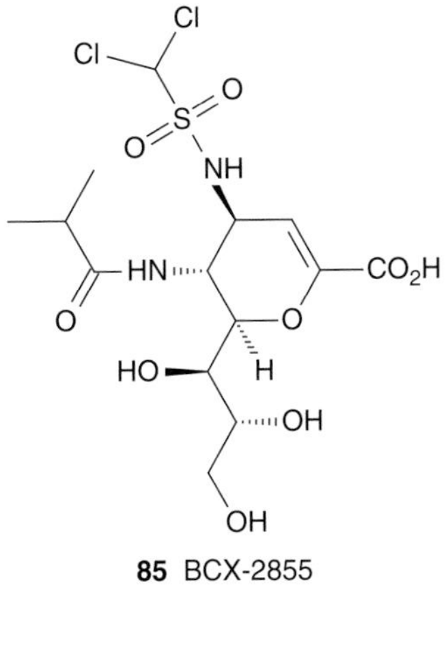

85 BCX-2855

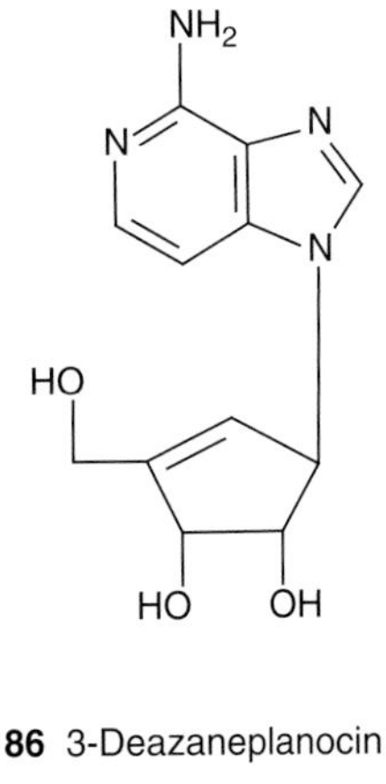

86 3-Deazaneplanocin A

7.10.15 Arena-, Bunya-, Rhabdo-, and Filoviruses

7.10.15.1 Arenaviruses

Of the 23 arenaviruses known, five are associated with viral hemorrhagic fever: Lassa, Junin, Machupo, Guanarito, and Sabia.[172] Ribavirin has proven to be effective in postexposure prophylaxis and therapy of experimental arenavirus infections in animal models, and anecdotal reports suggest that it might also be effective in the treatment of arenavirus infections (i.e., Machupo and Sabia) in humans.[172] The most convincing evidence for the (clinical) efficacy was obtained in the case of Lassa fever, where it was found to reduce the case-fatality rate, irrespective of the time point in the illness when treatment was started.[173]

7.10.15.2 Bunyaviruses

Of the bunyaviruses, one of the most feared (because it is highly infectious, easily transmitted between humans, and associated with a case-fatality rate of approximately 30%) is Crimean–Congo hemorrhagic fever virus.[174] Bunyaviruses are sensitive to ribavirin, and this has also been demonstrated in experimental animal models.[175] Also, interferon and interferon inducers have proved effective in the treatment of experimental bunyavirus infections, and, likewise, interferon-α should be considered for the treatment of arenavirus infections, as warranted by its efficacy in the therapy of Pichindi virus infection in hamsters.[176]

7.10.15.3 Rhabdoviruses

Of the rhabdoviruses, rabies, which is almost invariably fatal if no control measures are taken, can be contained by repeated injections of specific immunoglobulin and/or the inactivated ('killed') rabies vaccine as soon as possible after the infection. For the filovirus infections Ebola and Marburg no vaccine is (yet) available. Specific immunoglobulin or interferon-α may only be of limited value in the treatment of filovirus infections, as indicated by experimental findings in rhesus macaques infected with Ebola (Zaire) virus.[177]

7.10.15.4 Filoviruses

No antiviral drugs that are currently in clinical use, including ribavirin, provide meaningful protection against filoviruses in vivo.[178] A possible therapeutic strategy may be based on the use of *S*-adenosylhomocysteine (SAH) hydrolase inhibitors.[122] SAH hydrolase inhibitors, such as 3-deazaneplanocin A (**86**), interfere with *S*-adenosylmethionine (SAM)-dependent methylation reactions, particularly those involved in the 'capping' of viral mRNA. Some viruses, such as the rhabdovirus vesicular stomatitis virus (VSV), rely heavily on mRNA capping and are particularly sensitive to inhibition by SAH hydrolase inhibitors, including 3-deazaneplanocin A.[179] As, biochemically, filo- and rhabdoviruses are quite similar in their replication machinery, both requiring 5′-capping of their mRNAs, SAH hydrolase inhibitors such as 3-deazaneplanocin A may logically be expected to be effective in the treatment of Ebola virus infections. In fact, when administered as a single dose of 1 $mg\,kg^{-1}$, 3-deazaneplanocin A was found to protect mice against a lethal infection with Ebola virus (Zaire strain).[180,181] This protective effect was accompanied, and probably mediated, by the production of high concentrations of interferon in the Ebola virus-infected mice.[181] It can be hypothesized that, by blocking the 5′-capping of the nascent (+)RNA viral strands (and, hence, their maturation toward mRNAs), 3-deazaneplanocin A stimulated the formation of double-stranded (±)RNA complexes, which have long been known to be excellent inducers of interferon.[182]

Like SAH hydrolase, inosine monophosphate (IMP) dehydrogenase is another cellular enzyme that may be envisaged as a target for antiviral agents. IMP dehydrogenase is a crucial enzyme involved in the biosynthesis of GTP, and, although ribavirin may act against distinct viruses by distinct mechanisms (e.g., IMP dehydrogenase inhibition, immunomodulatory effect, RNA capping interference, polymerase inhibition, lethal mutagenesis),[183] the predominant mechanism by which ribavirin exerts its antiviral activity in vitro against flaviviruses and paramyxoviruses is mediated by inhibition of IMP dehydrogenase.[184]

Recently, a new class of compounds, phosphorodiamidate morpholino oligomers (PMO), conjugated to arginine-rich cell-penetrating peptides (P-PMO) and designed to base pair with the translation start region of Ebolavirus VP35 positive-sense RNA, were reported to inhibit Ebolavirus replication and to protect mice against a lethal Ebolavirus infection.[185]

7.10.16 Reoviruses

Of the reoviruses, rotavirus, which is associated with viral gastrointestinal infections, is by far the most clinically important pathogen. Several attempts have been, and are still being, made to develop an effective vaccine for rotavirus infections. Current treatment for rotavirus diarrhea is mainly based on the administration of fluids to prevent dehydration. There are no attempts to develop an antiviral drug for this disease, although it is worthy to note that the replication of reo- (and rota)viruses is exquisitely sensitive to SAH hydrolase inhibitors such as 3-deazaneplanocin A.[179]

7.10.17 Retroviruses (HIV)

There are at present some 20 compounds available for the treatment of HIV infections.[186] These compounds fall into five categories: (1) nucleoside reverse transcriptase inhibitors (NRTIs): zidovudine, didanosine (**87**), zalcitabine (**88**), stavudine (**89**), lamivudine, abacavir (**90**) and emtricitabine; (2) nucleotide reverse transcriptase inhibitors (NRTIs): tenofovir disoproxil fumarate; (3) nonnucleoside reverse transcriptase inhibitors (NNRTIs): nevirapine (**91**), delavirdine (**92**), and efavirenz (**93**); (4) protease inhibitors (PIs): saquinavir (**94**), ritonavir (**95**), indinavir (**96**), nelfinavir, amprenavir (**97**), lopinavir (**98**) (combined at a 4-to-1 ratio with ritonavir), atazanavir (**99**), fosamprenavir (**100**), and tipranavir (**101**); and (5) fusion inhibitors (FIs): enfuvirtide (**102**). Several of these compounds are also available as fixed dose combinations: zidovudine with lamivudine, lamivudine with abacavir, and emtricitabine with tenofovir disoproxil fumarate. A triple-drug fixed dose combination, containing efavirenz, emtricitabine, and tenofovir disoproxil fumarate is forthcoming.

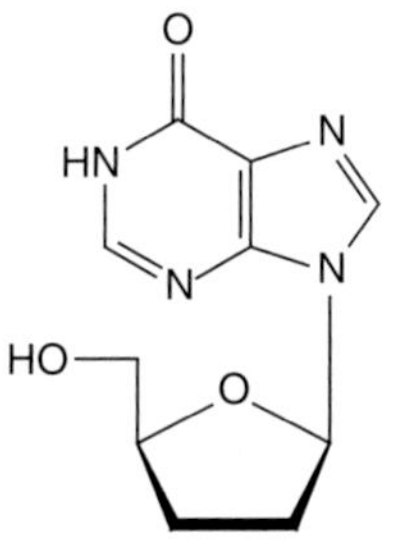

87 Didanosine
2′,3′-Dideoxyinosine (ddI)
Videx, Videx EC

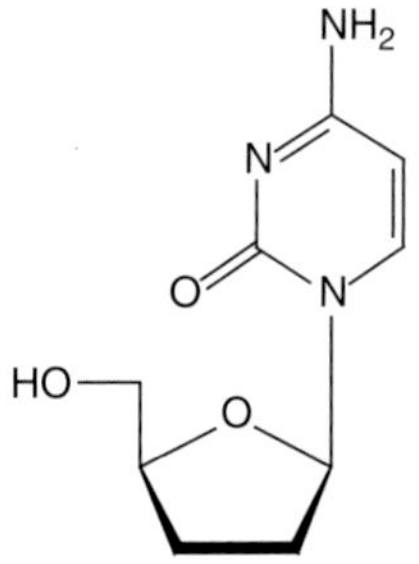

88 Zalcitabine
2′,3′-Dideoxycytidine (ddC)
Hivid

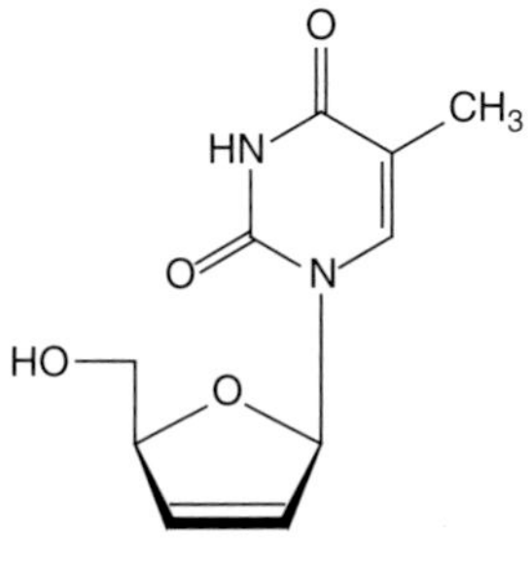

89 Stavudine
2′,3′-Didehydro-2′,3′-dideoxythymidine (d4T)
Zerit

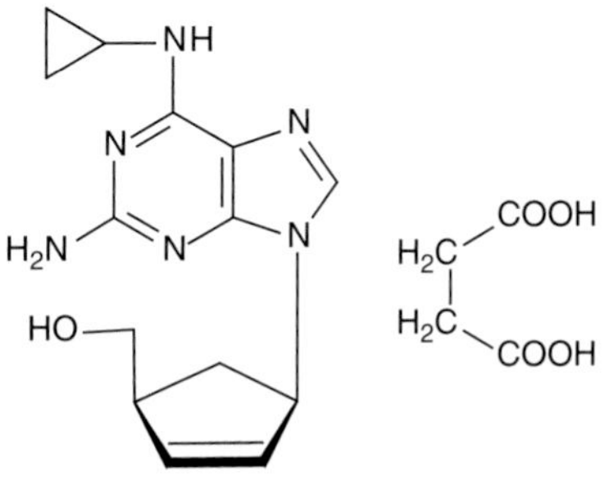

90 Abacavir
(1*S*, 4*R*)-4-[2-amino-6-(cyclopropylamino)-9*H*-purin-9-yl]-2-cyclopentene-1-methanol succinate (ABC)
Ziagen

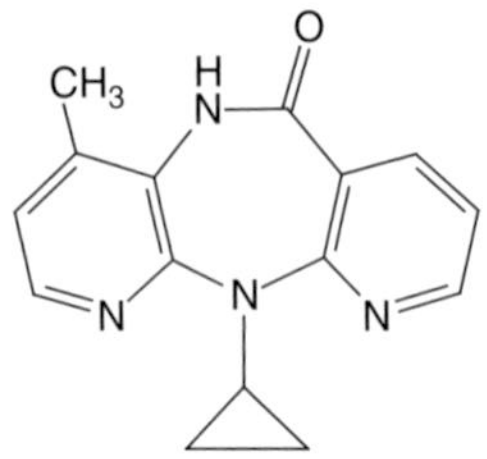

CH_3 H N O N N N

91 Nevirapine
Viramune

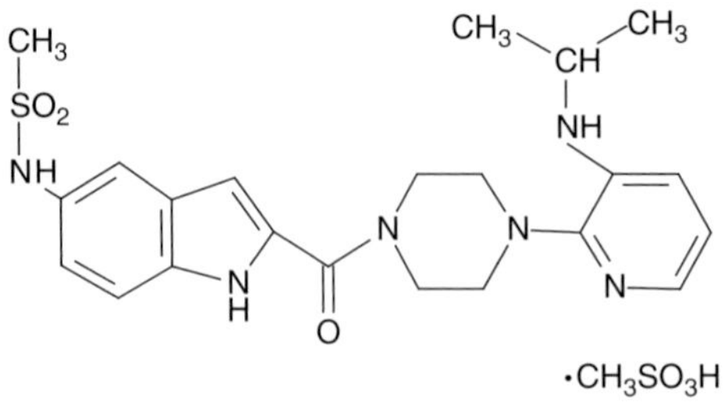

CH_3 SO_2 NH N H O N N CH_3 CH CH_3 NH N

$\cdot CH_3SO_3H$

92 Delavirdine
Rescriptor

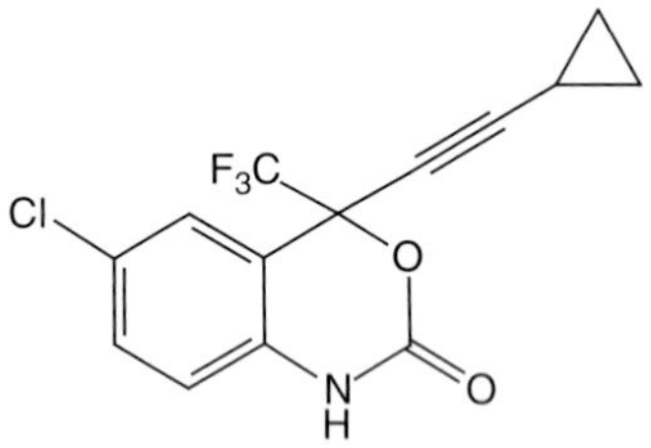

Cl F_3C O N H O

93 Efavirenz
Sustiva, Stocrin

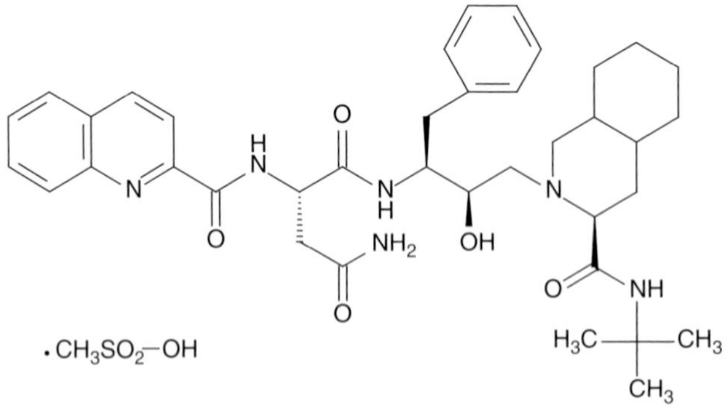

N H N O O N H NH_2 O OH N O NH H_3C CH_3 CH_3

$\cdot CH_3SO_2{-}OH$

94 Saquinavir
hard gel capsules, Invirase
soft gelatin capsules, Fortovase

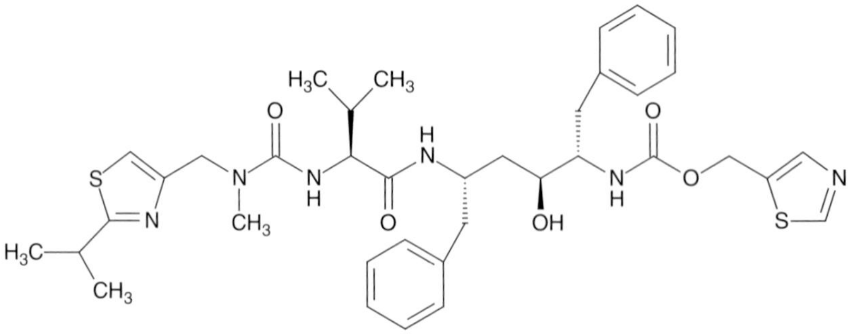

H_3C CH_3 O S N H_3C CH_3 N CH_3 N H O H N OH O N H O N S

95 Ritonavir
Norvir

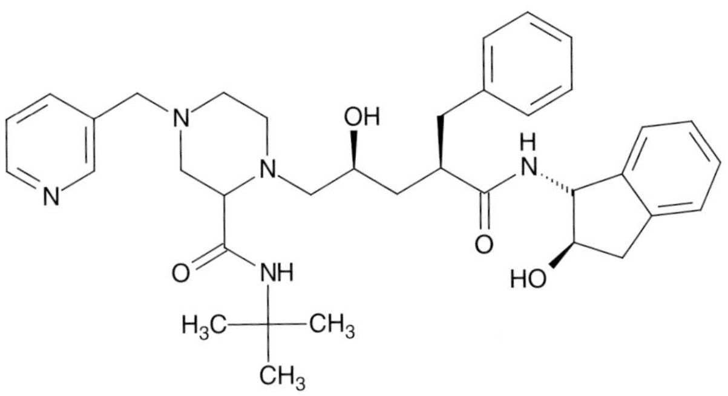

96 Indinavir
Crixivan

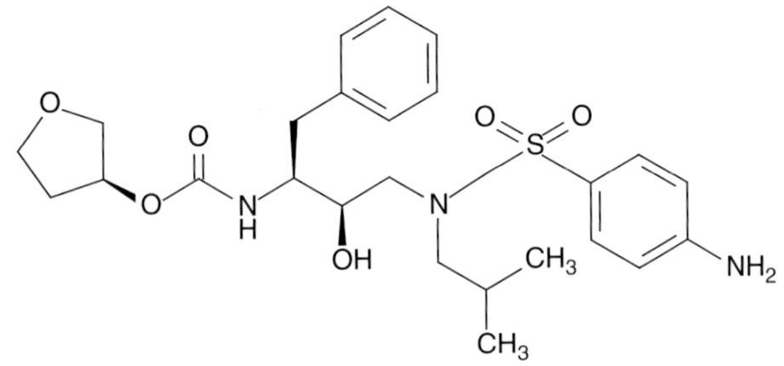

97 Amprenavir
Agenerase, Prozei

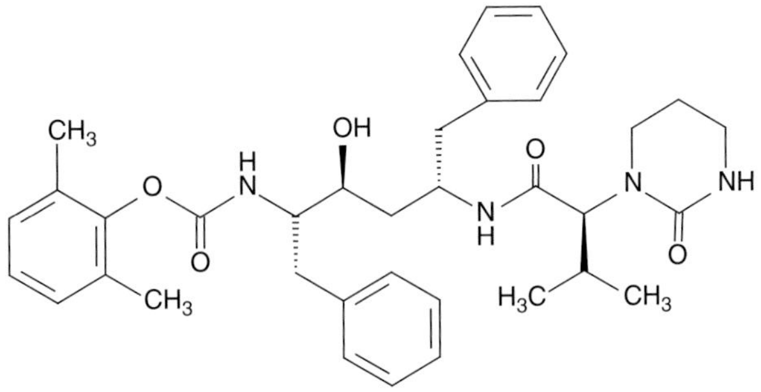

98 Lopinavir
combined with ritonavir at 4/1 ratio
Kaletra

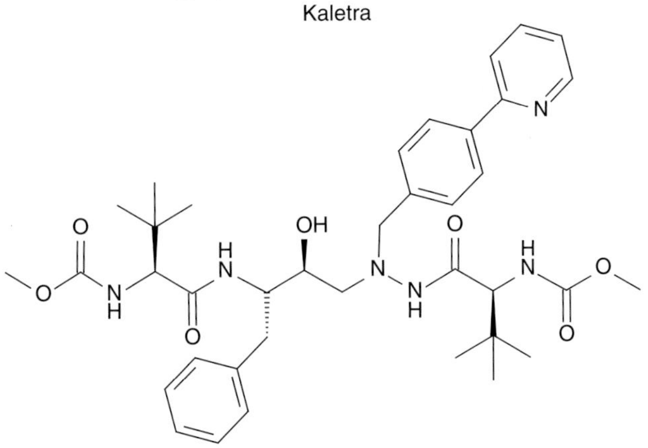

99 Atazanavir
Reyataz

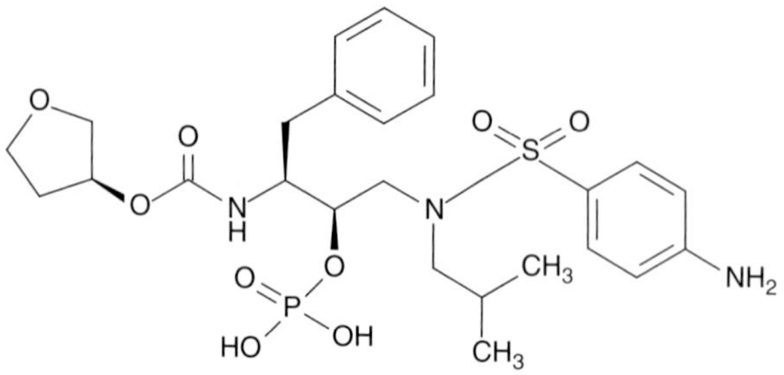

100 Fosamprenavir
Lexiva, Telzir

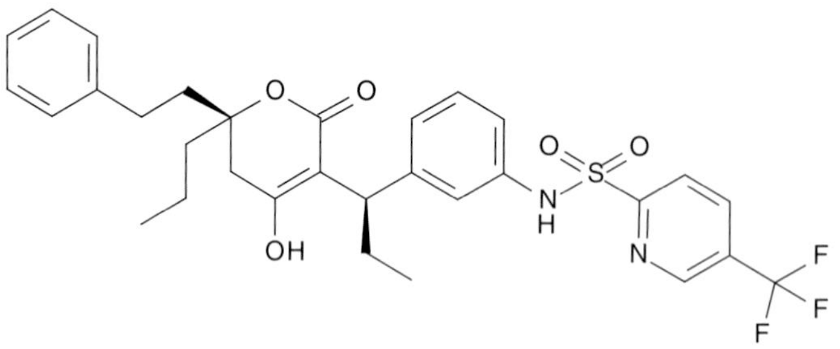

101 Tipranavir (U-140690)
Aptivus

YTSLIHSLIEESQNQQEKNEQELLELDKWASLWNWF

102 Enfuvirtide
DP-178, pentafuside, T20
Fuzeon

In addition to the 21 licensed anti-HIV compounds, various others are (or have been) in clinical (phase II or III) development: the HIV-1 attachment inhibitors BMS-378806 (**103**) and BMS-488043 (**104**),[187] the CXCR4 antagonist AMD-3100 (**105**) (as stem cell mobilizer for stem cell transplantation in patients with non-Hodgkin lymphoma or multiple myeloma),[188] the CCR5 antagonists[189] SCH-C (**106**), vicriviroc (SCH-D, SCH 417690) (**107**),[190] aplaviroc (873140) (**108**),[191] and maraviroc (UK-427857) (**109**),[192] the NRTIs Racivir (**110**), (-)-dOTC (AVX-754 (SPD-754) (**111**), which has been accredited with activity against most other NRTI-resistant HIV-1 strains[193]), Reverset (**112**), elvucitabine (**113**), alovudine (**114**), and amdoxovir (**115**), the NNRTIs capravirine (**116**) and etravirine (**117**), the protease inhibitor TMC-114 (**118**),[186,194] and the gag (p24) maturation inhibitor PA-457 (**119**).[195] Also, a prodrug of the benzophenone GW678248 (**120**) has recently progressed to phase II clinical trials.[196]

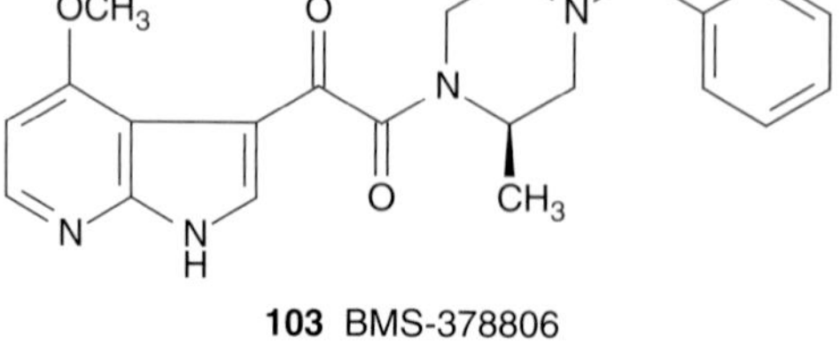

103 BMS-378806

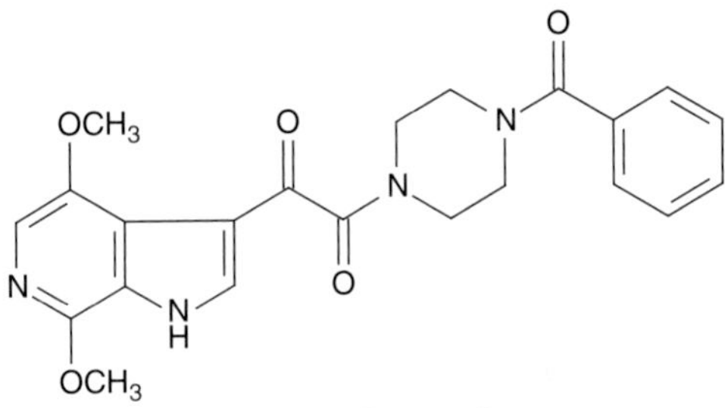

104 BMS-488043

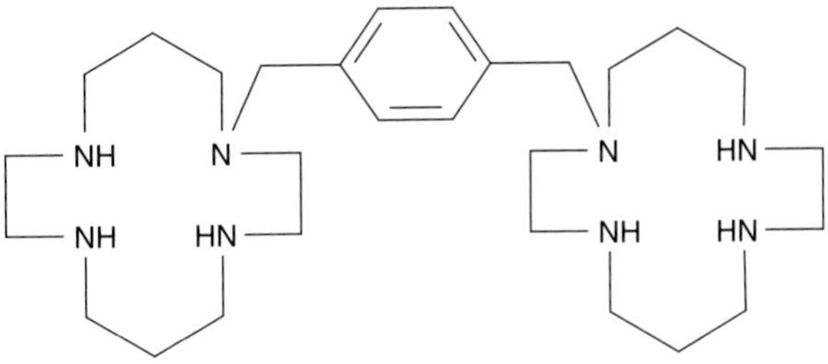

105 AMD 3100
Mozobil™

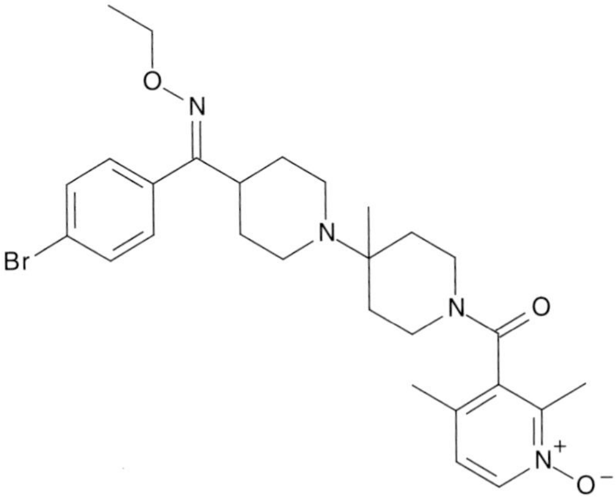

106 SCH-C (SCH-351125)

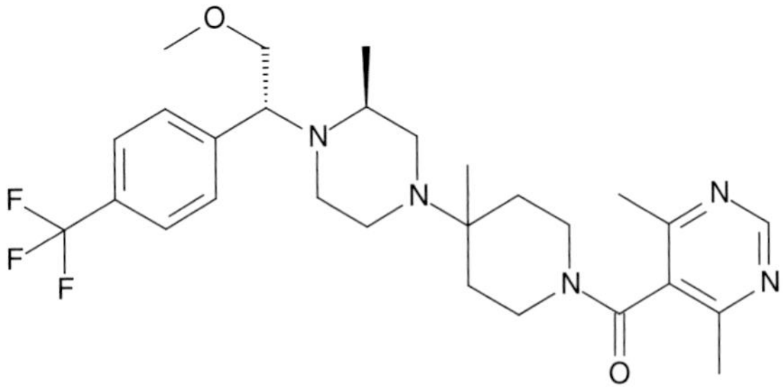

107 SCH-D (SCH-417690)
Vicriviroc

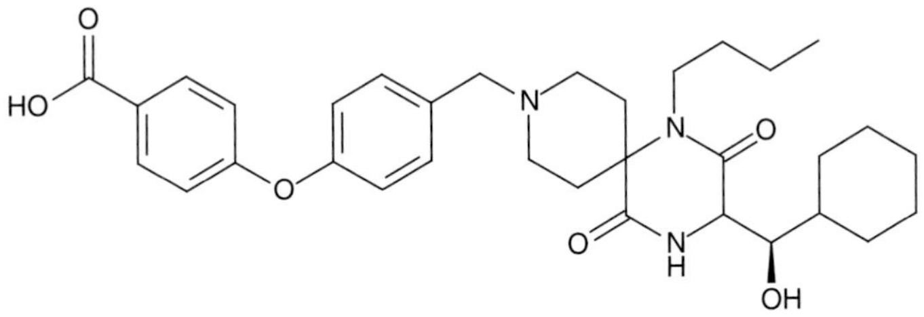

108 873140 (GW-873140, ONO-4128, AK-602)
Aplaviroc

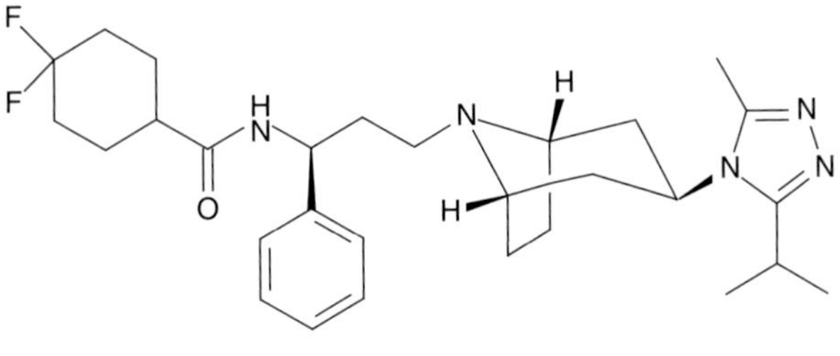

109 UK-427857
Maraviroc

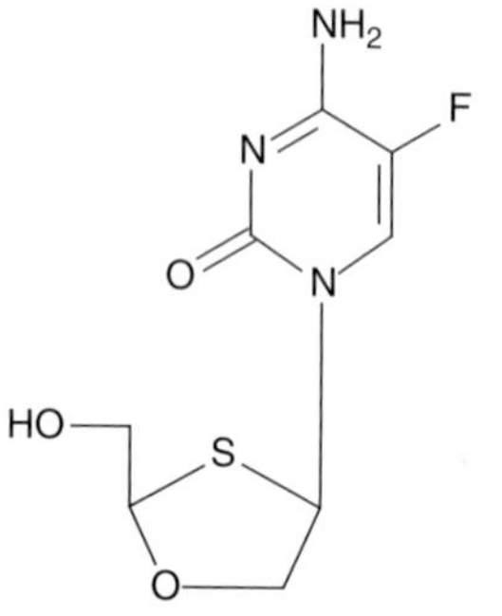

110 Racemic (±)FTC (FdOTC)
Racivir

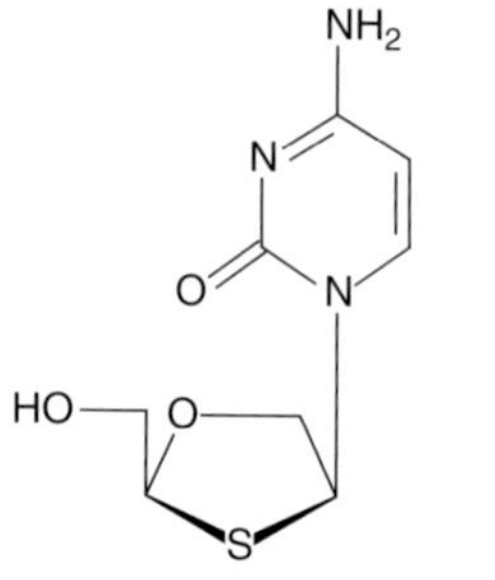

111 AVX-754 ((–)-dOTC)

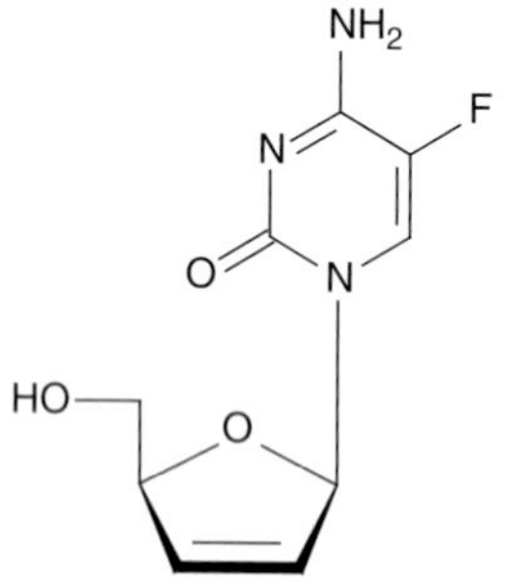

112 DPC-817 (β-D-Fd4C)
Reverset

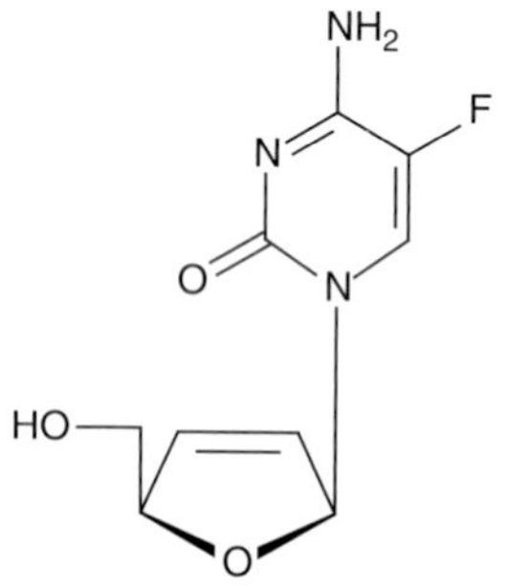

113 ACH-126443 (β-L-Fd4C)
Elvucitabine

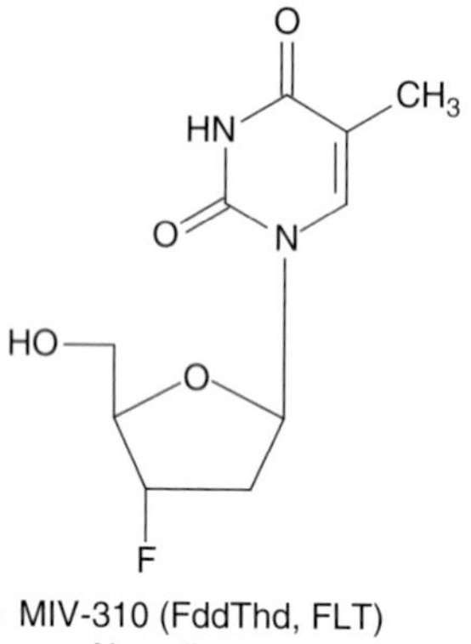

114 MIV-310 (FddThd, FLT)
Alovudine

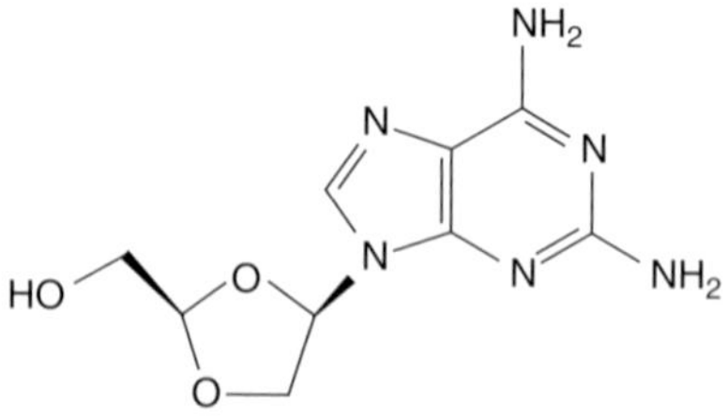

115 Diaminopurine dioxolane (DAPD)
Amdoxovir

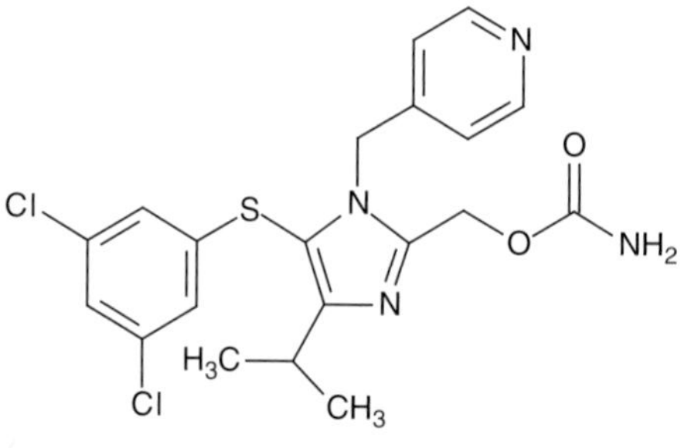

116 Capravirine (S-1153, AG1549)

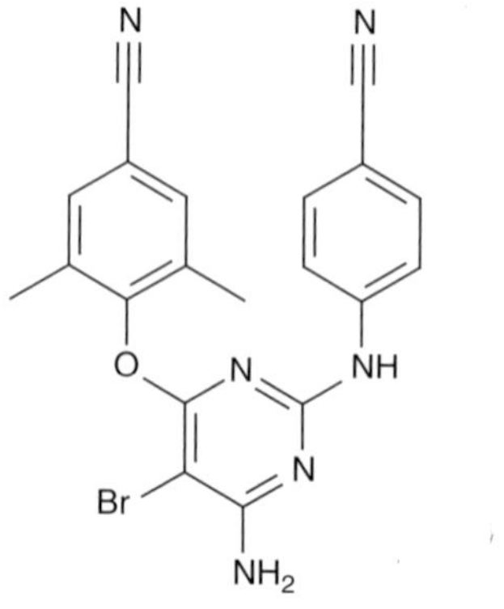

117 Etravirine (TMC-125, R-165335)

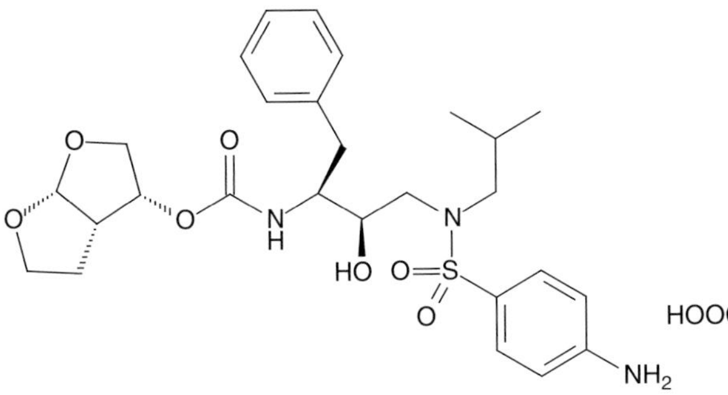

118 TMC-114
Darunavir

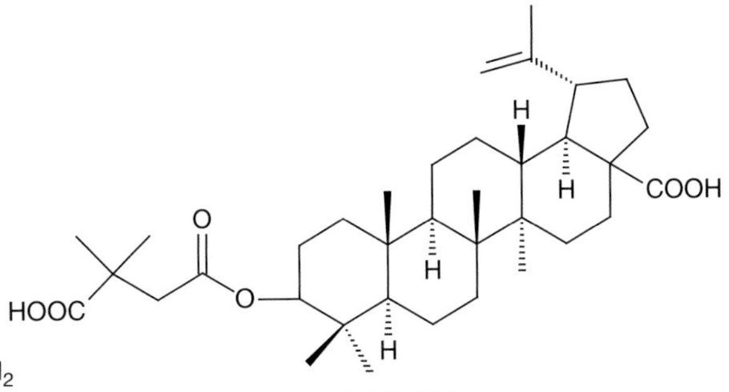

119 PA-457

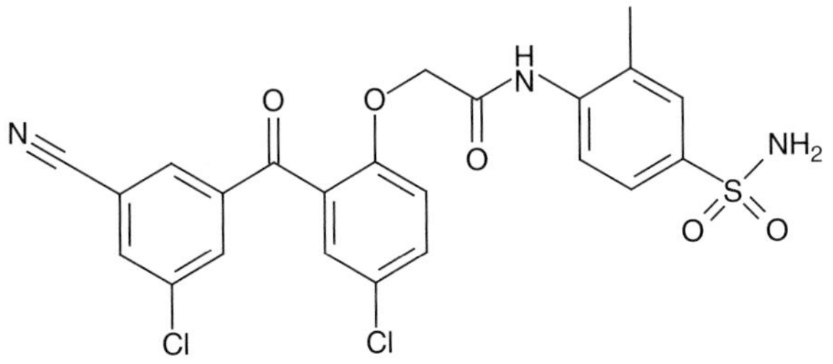

120 GW678248

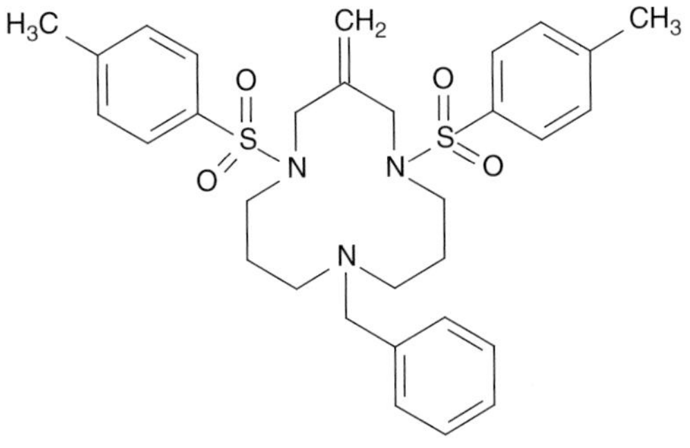

121 Cyclotriazadisulfonamide
CADA

Yet other compounds are in preclinical development and/or may soon proceed to clinical phase I/II clinical trials: the CD4 (HIV receptor) downmodulator cyclotriazadisulfonamide (CADA) (**121**)[197]; the HIV gp120 envelope-binding protein cyanovirin-N (**122**) as a topical microbicide[198]; KRH-2731, a CXCR4 antagonist, structurally related to KRH-1636 (**123**)[199]; the CXCR4 antagonist AMD-070 (a derivative of the bicyclam AMD3100, which is currently being pursued in phase II/III clinical trials, in combination with granulocyte colony-stimulating factor (G-CSF), for the mobilization of autologous hematopoietic progenitor cells)[200]; TAK-220 (**124**), a CCR5 antagonist, structurally related to TAK-779 (**125**),[201] which has proved to be a highly potent (orally bioavailable) inhibitor of CCR5-using (R5) HIV-1 strains,[202,203] and acts synergistically with other antiretrovirals[204]; TAK-652 (**126**), another orally bioavailable inhibitor of CCR5-mediated HIV infection[205]; MIV-210 (**127**), a prodrug of the NRTI 3′-fluoro-2′,3′-dideoxyguanosine; the thymine dioxolane DOT (**128**), another NRTI[206]; 4′-Ed4T (2′,3′-didehydro-3′-deoxy-4′-ethynyl-2′-deoxythymidine) (**129**), which has favorable oral bioavailabity and a unique drug resistance profile, different from the other NRTIs[207]; the NtRTIs 6-[2-(phosphonomethoxy)alkoxy]-2,4-diaminopyrimidines PMPO-DAPy, PMEO-DAPy, and 5-substituted derivatives thereof (**130**)[3]; the deoxythreosyl nucleoside phoshonates phosphonomethyldeoxythreosyladenine (PMDTA, **131**) and -thymine (PMDTT, **132**)[208]; the NNRTIs thiocarboxanilide UC-781 (**133**) and dapivirine (TMC-120) (**134**), both as topical microbicides, and rilpivirine (R-278474) (**135**), one of the most potent anti-HIV agents ever described[209]; GW678248, a novel benzophenone NNRTI,[210,211] which has activity at 1 nM against the K103N and Y181C RT HIV-1 mutants associated with clinical resistance to efavirenz and nevirapine, respectively[196]; and a number of compounds, including the 1,6-naphthyridine-7-carboxamides L-870810 (**136**) and L-870812 (**137**), which are targeted at the HIV-1 integrase.[212,213] The 3,7-dihydroxytropolones (**138**) represent an interesting platform for the design of inhibitors of both the reverse transcriptase (and RNase H) as well as the HIV integrase.[214] Similarly, indolyl aryl sulfone (**139**) may serve as a platform for the design of new NNRTIs effective against K103N HIV-1 variants.[215] Recently, diketo acids bearing a nucleobase scaffold have been described as highly potent HIV integrase inhibitors[216]: the prototype compound, 4-(1,3-dibenzyl-1,2,3,4-tetrahydro-2,4-dioxopyrimidin-5-yl)-2-hydroxy-4-oxo-but-2-enoic acid (**140**), exhibited an anti-HIV selectivity index in cell culture of >4000.

(H_2N)Leu—Gly—Lys—Phe—Ser—Gln—Thr—Cys—Tyr—Asn—Ser—Ala—

—Ile—Gln—Gly—Ser—Val—Leu—Thr—Ser—Thr—Cys—Glu—Arg—Thr—Asn—Gly—Gly—Tyr—Asn—Thr—Ser—

—Ser—Ile—Asp—Leu—Asn—Ser—Val—Ile—Glu—Asn—Val—Asp—Gly—Ser—Leu—Lys—Trp—Gln—Pro—Ser—

—Asn—Phe—Ile—Glu—Thr—Cys—Arg—Asn—Thr—Gln—Leu—Ala—Gly—Ser—Ser—Glu—Leu—Ala—Ala—Glu—

—Cys—Lys—Thr—Arg—Ala—Gln—Gln—Phe—Val—Ser—Thr—Lys—Ile—Asn—Leu—Asp—Asp—His—Ile—Ala—

—Asn—Ile—Asp—Gly—Thr—Leu—Lys—Tyr—Glu(COOH)

122 Cyanovirin-N

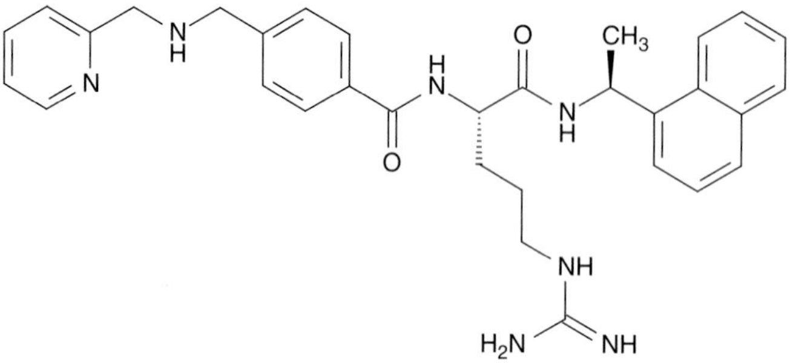

123 KRH-1636

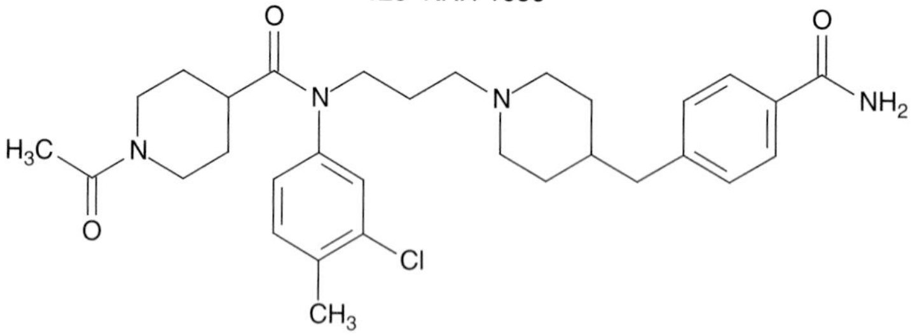

124 TAK-220

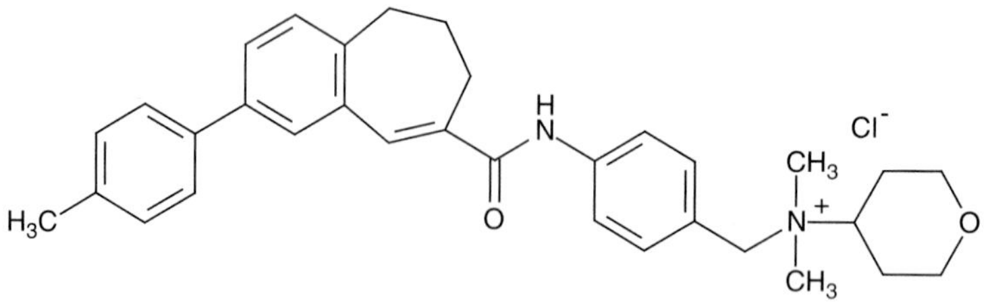

125 TAK-779

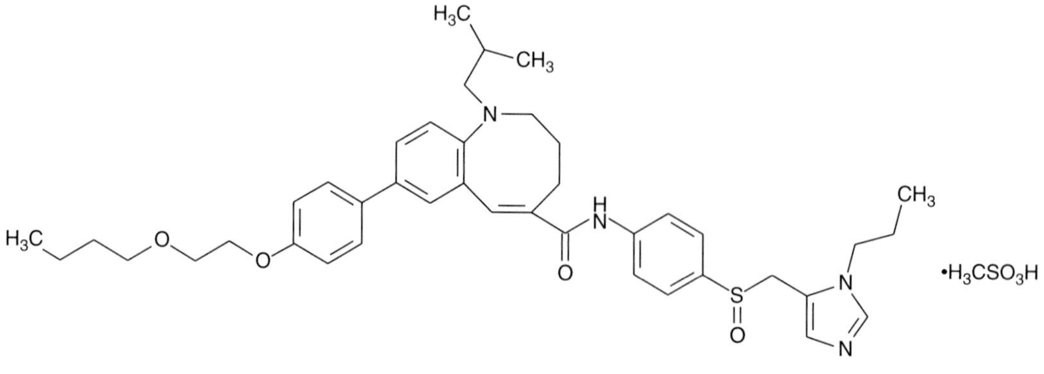

126 TAK-652

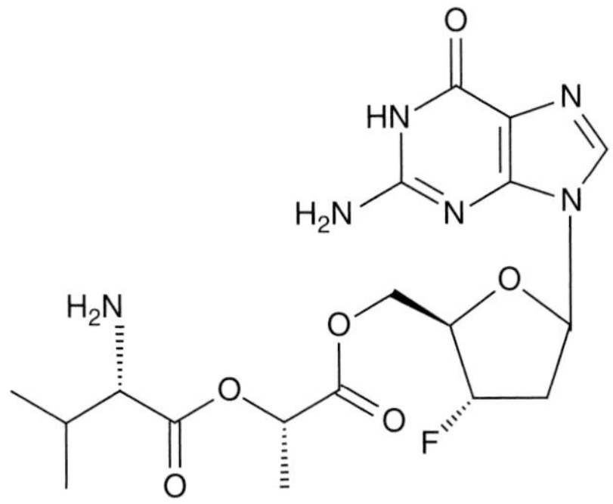

127 MIV-210 [FLG (3′-fluoro-2′,3′-dideoxyguanosine) prodrug]

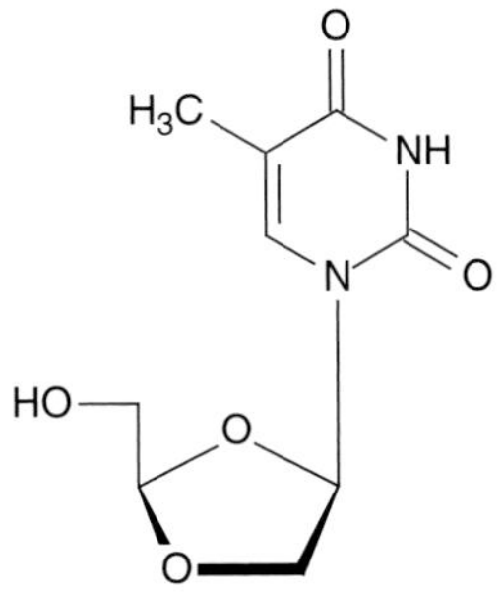

128 1-(β-D-dioxolane)thymine DOT

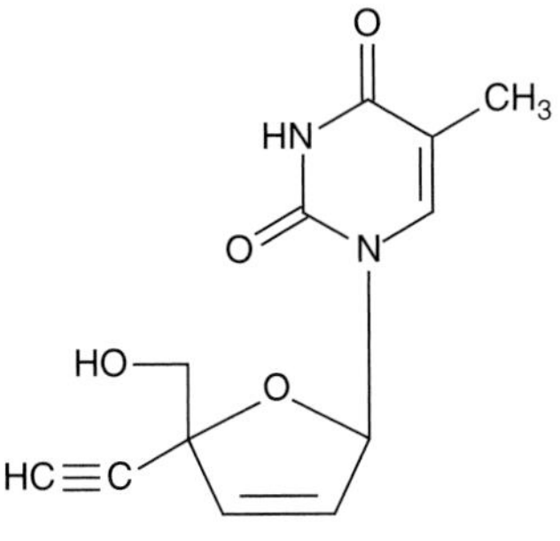

129 4′-Ed4T

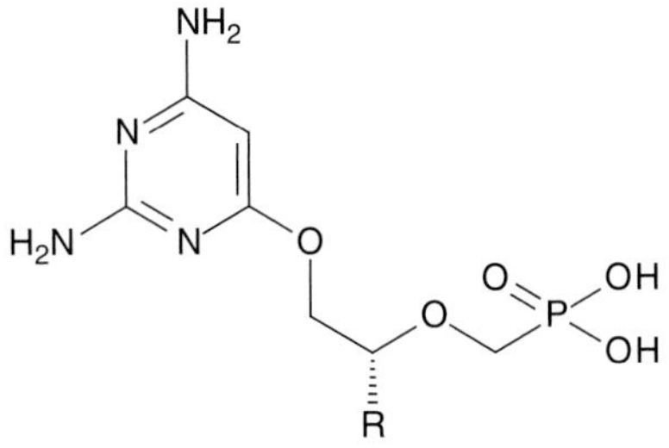

130 R = CH_3 : PMPO-DAPy
R = H : PMEO-DAPy

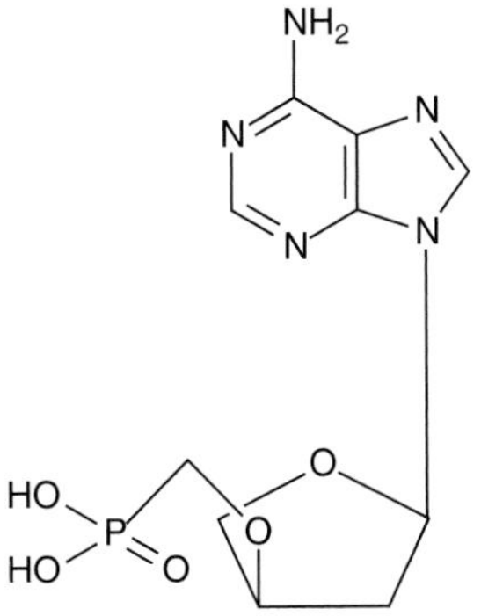

131 PMDTA

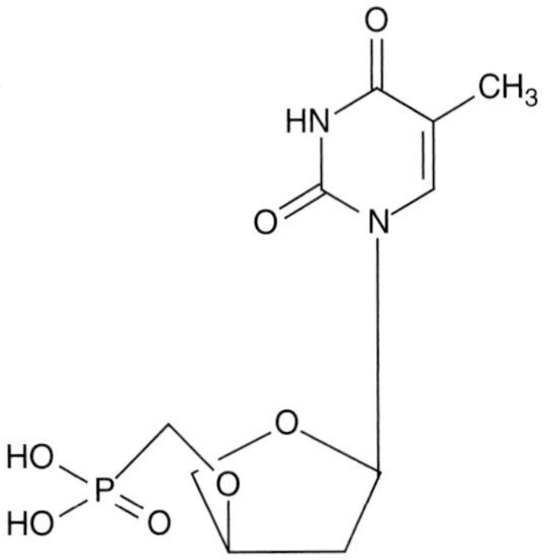

132 PMDTT

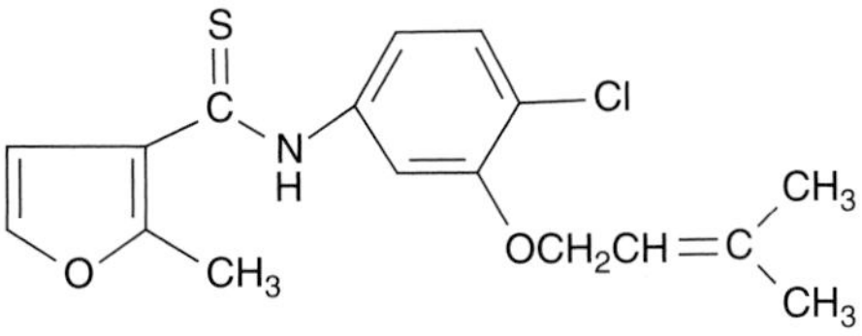

133 Thiocarboxanilide UC-781

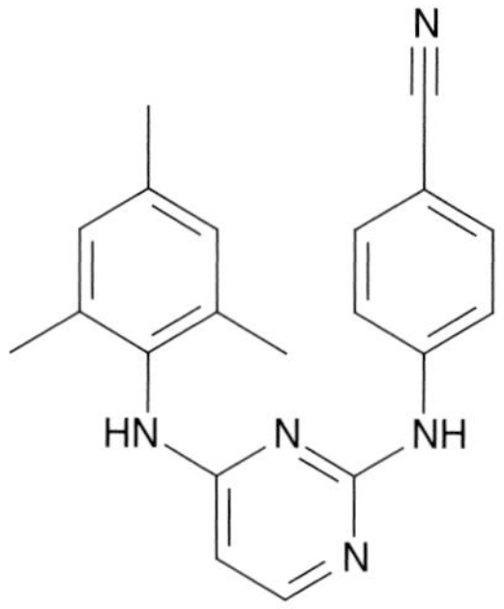

134 Dapivirine (TMC-120, R-147681)

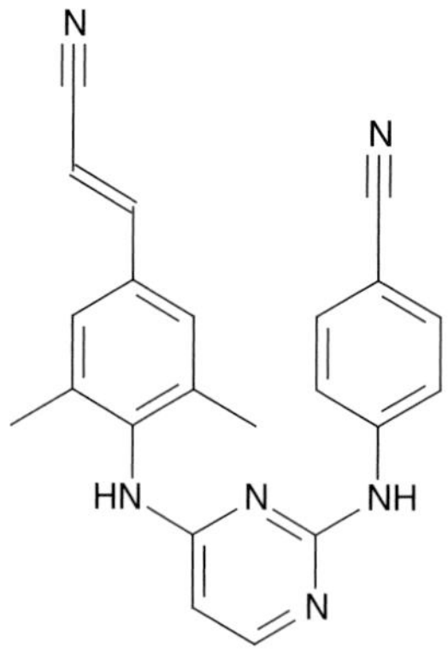

135 Rilpivirine (R-278474)

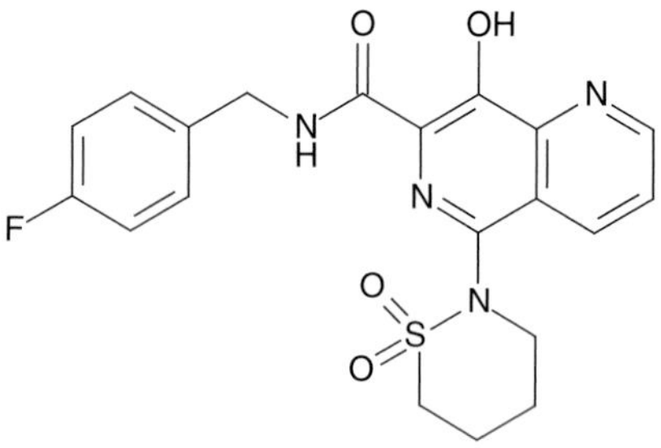

136 L-870810

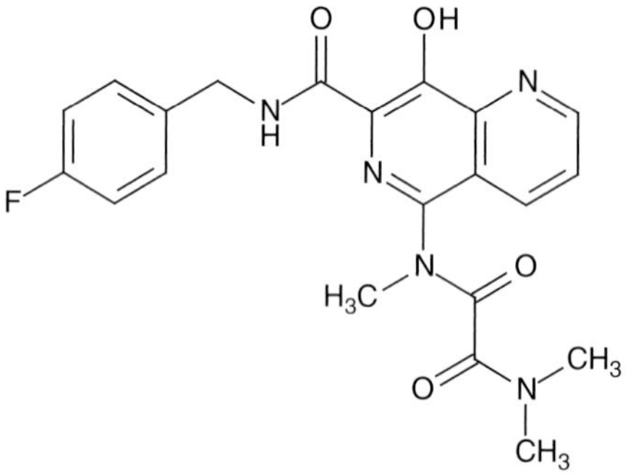

137 L-870812

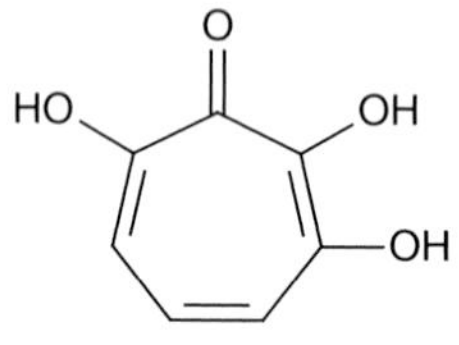

138 3,7-dihydroxytropolone

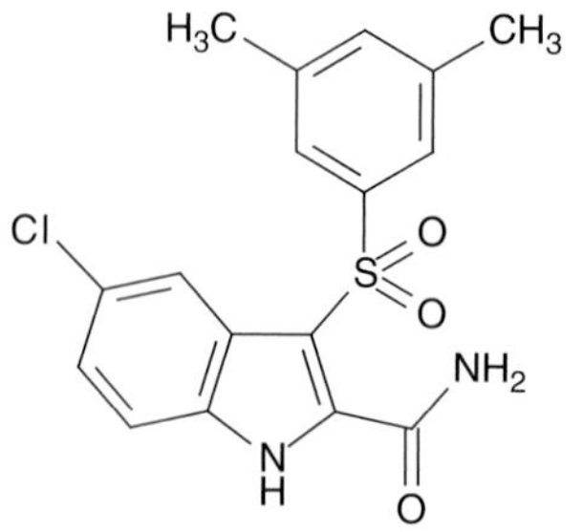

139 Indolyl aryl sulfone RS 1588

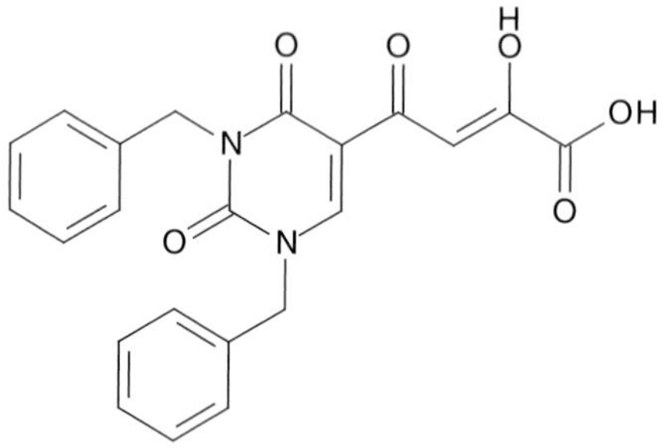

140 4-(1,3-dibenzyl-1,2,3,4-tetrahydro-2,4-dioxopyrimidin-5-yl)-2-hydroxy-4-oxo-but-2-enoic acid

In addition to the aforementioned cyanovirin-N, thiocarboxanilide UC-781, and dapivirine, there are some other compounds that could be further developed as topical (e.g., vaginal) microbicides, namely the aglycons of the glycopeptide antibiotics vancomycin, teicoplanin, and eremomycin which specifically interact with the gp120 glycoprotein.[217] Also the plant lectins, e.g., *Galanthus nivalis* agglutinin (GNA) and *Hippeastrum* hybrid agglutinin (HHA), represent potential candidate anti-HIV microbicides: they show marked stability at relatively low pH and high temperatures for prolonged time periods, they directly interact with the viral envelope and prevent entry of HIV into its target cells.[218] Upon prolonged exposure of HIV in cell culture to HHA or GNA, the virus acquires resistance mutations in the gp120 glycoprotein which are predominantly located at the N-glycosylation (asparagine) sites.[219]

An avenue to be further explored is the combination of different microbicides, such as the NNRTI thiocarboxanilide UC-781 with the cellulose acetate 1,2-benzenedicarboxylate (CAP) viral entry inhibitor, which exhibit synergistic and complementary effects against HIV-1 infection.[220] There is, in addition, no shortage of sulfated and sulfonated polymers (starting off with suramin, the first polysulfonate ever shown to be active against HIV) which could be considered as topical anti-HIV microbicides.[221]

7.10.18 Conclusion

About 40 compounds are registered as antiviral drugs, at least half of which are used to treat HIV infections. An even greater number of compounds are under clinical or preclinical development, with again, as many targeting HIV as all the other viruses taken together. This implies that HIV, since its advent, has remained the main target in antiviral drug development. Antiviral agents can, as guided by the anti-HIV agents as examples, be divided in roughly five categories: (1) nucleoside analogs, (2) nucleotide analogs (or acyclic nucleoside phosphonates), (3) nonnucleoside analogs, (4) protease inhibitors, and (5) virus–cell fusion inhibitors. Molecular targets are for (1) and (2) the viral DNA polymerase (whether DNA-dependent as in the case of herpesviruses, or RNA-dependent as in the case of HIV or HBV); for (3) RNA-dependent DNA polymerase (reverse transcriptase), associated with HIV, or RNA-dependent RNA polymerase (RNA replicase) associated with HCV; for (4) the proteases associated with HIV and HCV; and for (5) the fusion process of HIV (and, potentially, other viruses such as the SARS coronavirus and RSV). Antiviral agents may also exert their antiviral effects through an interaction with cellular targets such as IMP dehydrogenase (ribavirin) and SAH hydrolase (3-deazaneplanocin A). The latter enzymes are essential for viral RNA synthesis (through the supply of GTP) and viral mRNA maturation (through 5′-capping), respectively. Finally, interferons (now generally provided in their pegylated form) may be advocated in the therapy of those viral infections (actually, HBV and HCV; prospectively, Coxsackie B, SARS, ...) that, as yet, cannot be sufficiently curbed by other therapeutic measures.

References

1. De Clercq, E. *J. Clin. Virol.* **2004**, *30*, 115–133.
2. De Clercq, E. *Clin. Microbiol. Rev.* **2003**, *16*, 569–596.
3. De Clercq, E.; Andrei, G.; Balzarini, J. et al. *Nucleos. Nucleot. Nucleic Acids* **2005**, *24*, 331–341.
4. Faucher, A. M.; White, P. W.; Brochu, C.; Grand-Maitre, C.; Rancourt, J.; Fazal, G. *J. Med. Chem.* **2004**, *47*, 18–21.
5. White, P. W.; Faucher, A.-M.; Massariol, M.-J. et al. *Antimicrob. Agents Chemother.* **2005**, *49*, 4834–4842.
6. Naesens, L.; Lenaerts, L.; Andrei, G. et al. *Antimicrob. Agents Chemother.* **2005**, *49*, 1010–1016.
7. Hartline, C. B.; Gustin, K. M.; Wan, W. B. et al. *J. Infect. Dis.* **2005**, *191*, 396–399.
8. De Clercq, E.; Field, H. J. *Br. J. Pharmacol.* **2005**, *147*, 1–11.
9. De Clercq, E.; Andrei, G.; Snoeck, R. et al. *Nucleos. Nucleot. Nucleic Acids* **2001**, *20*, 271–285.
10. Zhou, S.; Breitenbach, J. M.; Borysko, K. Z. et al. *J. Med. Chem.* **2004**, *47*, 566–575.
11. Kern, E. R.; Kushner, N. L.; Hartline, C. B. et al. *Antimicrob. Agents Chemother.* **2005**, *49*, 1039–1045.
12. Kern, E. R.; Bidanset, D. J.; Hartline, C. B.; Yan, Z.; Zemlicka, J.; Quenelle, D. C. *Antimicrob. Agents Chemother.* **2004**, *48*, 4745–4753.
13. Crute, J. J.; Grygon, C. A.; Hargrave, K. D. et al. *Nat. Med.* **2002**, *8*, 386–391.
14. Kleymann, G.; Fischer, R.; Betz, U. A. et al. *Nat. Med.* **2002**, *8*, 392–398.
15. Betz, U. A.; Fischer, R.; Kleymann, G.; Hendrix, M.; Rübsamen-Waigmann, H. *Antimicrob. Agents Chemother.* **2002**, *46*, 1766–1772.
16. De Clercq, E. *Med. Res. Rev.* **2003**, *23*, 253–274.
17. Andrei, G.; Sienaert, R.; McGuigan, C.; De Clercq, E.; Balzarini, J.; Snoeck, R. *Antimicrob. Agents Chemother.* **2005**, *49*, 1081–1086.
18. De Clercq, E. *J. Antimicrob. Chemother.* **2003**, *51*, 1079–1083.
19. Biron, K. K.; Harvey, R. J.; Chamberlain, S. C. et al. *Antimicrob. Agents Chemother.* **2002**, *46*, 2365–2372.
20. Krosky, P. M.; Baek, M. C.; Coen, D. M. *J. Virol.* **2003**, *77*, 905–914.
21. Koszalka, G. W.; Johnson, N. W.; Good, S. S. et al. *Antimicrob. Agents Chemother.* **2002**, *46*, 2373–2380.
22. Lalezari, J. P.; Aberg, J. A.; Wang, L. H. et al. *Antimicrob. Agents Chemother.* **2002**, *46*, 2969–2976.
23. De Bolle, L.; Andrei, G.; Snoeck, R. et al. *Biochem. Pharmacol.* **2004**, *67*, 325–336.
24. De Clercq, E.; Naesens, L.; De Bolle, L.; Schols, D.; Zhang, Y.; Neyts, J. *Rev. Med. Virol.* **2001**, *11*, 381–395.
25. Zhu, W.; Burnette, A.; Dorjsuren, D. et al. *Antimicrob. Agents Chemother.* **2005**, *49*, 4965–4973.
26. De Clercq, E.; Neyts, J. *Rev. Med. Virol.* **2004**, *14*, 289–300.
27. Neyts, J.; Leyssen, P.; Verbeken, E.; De Clercq, E. *Antimicrob. Agents Chemother.* **2004**, *48*, 2267–2273.
28. De Clercq, E. *Antivir. Res.* **2002**, *55*, 1–13.
29. Painter, G. R.; Hostetler, K. Y. *Trends Biotechnol.* **2004**, *22*, 423–427.
30. Kern, E. R.; Hartline, C.; Harden, E. et al. *Antimicrob. Agents Chemother.* **2002**, *46*, 991–995.
31. Quenelle, D. C.; Collins, D. J.; Wan, W. B.; Beadle, J. R.; Hostetler, K. Y.; Kern, E. R. *Antimicrob. Agents Chemother.* **2004**, *48*, 404–412.
32. Smee, D. F.; Wong, M.-H.; Bailey, K. W.; Beadle, J. R.; Hostetler, K. Y.; Sidwell, R. W. *Int. J. Antimicrob. Agents* **2004**, *23*, 430–437.
33. Bulller, R. M.; Owens, G.; Schriewer, J.; Melman, L.; Beadle, J. R.; Hostetler, K. Y. *Virology* **2004**, *318*, 474–481.
34. Robbins, S. J.; Jackson, R. J.; Fenner, F. et al. *Antivir. Res.* **2005**, *66*, 1–7.
35. Stittelaar, K. J.; Neyts, J.; Naesens, L. et al. *Nature* **2006**, *439*, 745–748.
36. Magee, W. C.; Hostetler, K. Y.; Evans, D. H. *Antimicrob. Agents Chemother.* **2005**, *49*, 3153–3162.
37. Yang, H.; Kim, S.-K.; Kim, M. et al. *J. Clin. Invest.* **2005**, *115*, 379–387.
38. Reeves, P. M.; Bommarius, B.; Lebeis, S. et al. *Nat. Med.* **2005**, *11*, 731–739.
39. Yang, G.; Pevear, D. C.; Davies, M. H. et al. *J. Virol.* **2005**, *79*, 13139–13149.
40. Janssen, H. L. A.; van Zonneveld, M.; Senturk, H. et al. *Lancet* **2005**, *365*, 123–129.
41. Hadziyannis, S. J.; Tassopoulos, N. C.; Heathcote, E. J. et al. *N. Engl. J. Med.* **2005**, *352*, 2673–2681.
42. De Clercq, E. *Exp. Rev. Anti-Infect. Ther.* **2003**, *1*, 21–43.
43. Marcellin, P.; Chang, T. T.; Lim, S. G. et al. *N. Engl. J. Med.* **2003**, *348*, 808–816.
44. Hadziyannis, S. J.; Tassopoulos, N. C.; Heathcote, E. J. et al. *N. Engl. J. Med.* **2003**, *348*, 800–807.

45. Lai, C. L.; Rosmawati, M.; Lao, J. et al. *Gastroenterology* **2002**, *123*, 1831–1838.
46. Levine, S.; Hernandez, D.; Yamanaka, G. et al. *Antimicrob. Agents Chemother.* **2002**, *46*, 2525–2532.
47. Hu, R.; Li, L.; Degrève, B.; Dutschman, G. E.; Lam, W.; Cheng, Y.-C. *Antimicrob. Agents Chemother.* **2005**, *49*, 2044–2049.
48. Jacquard, A.-C.; Brunelle, M.-N.; Pichoud, C. et al. *Antimicrob. Agents Chemother.* **2006**, *50*, 955–961.
49. van Bömmel, F.; Wünsche, T.; Mauss, S. et al. *Hepatology* **2004**, *40*, 1421–1425.
50. Soriano, V.; Puoti, M.; Bonacini, M. et al. *AIDS* **2005**, *19*, 221–240.
51. Ledford, R. M.; Patel, N. R.; Demenczuk, T. M. et al. *J. Virol.* **2004**, *78*, 3663–3674.
52. Schiff, G. M.; Sherwood, J. R. *J. Infect. Dis.* **2002**, *181*, 20–26.
53. Rotbart, H. A.; Webster, A. D. *Clin. Infect. Dis.* **2001**, *32*, 228–235.
54. Hayden, F. G.; Coats, T.; Kim, K. et al. *Antiviral Ther.* **2002**, *7*, 53–65.
55. Hayden, F. G.; Herrington, D. T.; Coats, T. L. et al. *Clin. Infect. Dis.* **2003**, *36*, 1523–1532.
56. Dragovich, P. S.; Prins, T. J.; Zhou, R. et al. *J. Med. Chem.* **1999**, *42*, 1213–1224.
57. Hsyu, P.-H.; Pithavala, Y. K.; Gersten, M.; Penning, C. A.; Kerr, B. M. *Antimicrob. Agents Chemother.* **2002**, *46*, 392–397.
58. Hayden, F. G.; Turner, R. B.; Gwaltney, J. M. et al. *Antimicrob. Agents Chemother.* **2003**, *47*, 3907–3916.
59. Dragovich, P. S.; Prins, T. J.; Zhou, R. et al. *J. Med. Chem.* **2003**, *46*, 4572–4585.
60. Patick, A. K.; Brothers, M. A.; Maldonado, F. et al. *Antimicrob. Agents Chemother.* **2005**, *49*, 2267–2275.
61. Padalko, E.; Verbeken, E.; Matthys, P.; Aerts, J. L.; De Clercq, E.; Neyts, J. *BMC Microbiol.* **2003**, *3*, 25.
62. Padalko, E.; Nuyens, D.; De Palma, A. et al. *Antimicrob. Agents Chemother.* **2004**, *48*, 267–274.
63. Leyssen, P.; Charlier, N.; Paeshuyse, J.; De Clercq, E.; Neyts, J. *Adv. Virus Res.* **2003**, *61*, 511–553.
64. Leyssen, P.; Croes, R.; Rau, P. et al. *Brain Pathol.* **2003**, *13*, 279–290.
65. Leyssen, P.; Drosten, C.; Paning, M. et al. *Antimicrob. Agents Chemother.* **2003**, *47*, 777–782.
66. Fried, M. W.; Shiffman, M. L.; Reddy, K. R. et al. *N. Engl. J. Med.* **2002**, *347*, 975–982.
67. McHutchison, J. G.; Manns, M.; Patel, K. et al. *Gastroenterology* **2002**, *123*, 1061–1069.
68. Tan, S. L.; Pause, A.; Shi, Y.; Sonenberg, N. *Nat. Rev. Drug Disc.* **2002**, *1*, 867–881.
69. Torriani, F. J.; Rodriguez-Torres, M.; Rockstroh, J. K. et al. *N. Engl. J. Med.* **2004**, *351*, 438–450.
70. Vogt, M. W.; Hartshorn, K. L.; Furman, P. A. et al. *Science* **1987**, *235*, 1376–1379.
71. Rodriguez-Torres, M.; Torriani, F. J.; Soriano, V. et al. *Antimicrob. Agents Chemother.* **2005**, *49*, 3997–4008.
72. Cornberg, M.; Manns, M. P. *Future Virol.* **2006**, *1*, 99–107.
73. Watson, J. *Curr. Opin. Investig. Drugs* **2002**, *3*, 680–683.
74. Pavlovic, D.; Neville, D. C. A.; Argaud, O. et al. *Proc. Natl. Acad. Sci. USA* **2003**, *100*, 6104–6108.
75. Lamarre, D.; Anderson, P. C.; Bailey, M. et al. *Nature* **2003**, *426*, 186–189.
76. Hinrichsen, H.; Benhamou, Y.; Wedemeyer, H. et al. *Gastroenterology* **2004**, *127*, 1347–1355.
77. Reiser, M.; Hinrichsen, H.; Benhamou, Y. et al. *Hepatology* **2005**, *41*, 832–835.
78. Lin, C.; Lin, K.; Luong, Y.-P. et al. *J. Biol. Chem.* **2004**, *279*, 17508–17514.
79. Lin, C.; Gates, C. A.; Rao, B. G. et al. *J. Biol. Chem.* **2005**, *280*, 36784–36791.
80. Perni, R. B.; Almquist, S. J.; Byrn, R. A. et al. *Antimicrob. Agents Chemother.* **2006**, *50*, 899–909.
81. Chen, K. X.; Njoroge, F. G.; Pichardo, J. et al. *J. Med. Chem.* **2006**, *49*, 567–574.
82. Malcolm, B. A.; Liu, R.; Lahser, F. et al. *Antimicrob. Agents Chemother.* **2006**, *50*, 1013–1020.
83. Baginski, S. G.; Pevear, D. C.; Seipel, M. et al. *Proc. Natl. Acad. Sci. USA* **2000**, *97*, 7981–7986.
84. Sun, J. H.; Lemm, J. A.; O'Boyle, D. R., II; Racela, J.; Colonno, R.; Gao, M. *J. Virol.* **2003**, *77*, 6753–6760.
85. Paeshuyse, J.; Leyssen, P.; Mabery, E. et al. *J. Virol.* **2005**, *80*, 149–160.
86. De Francesco, R.; Migliaccio, G. *Nature* **2005**, *436*, 953–960.
87. Afdhal, N.; Godofsky, E.; Dienstag, J. et al. *Hepatology* **2004**, *40*, 726A.
88. Afdhal, N.; Rodriguez-Torres, M.; Lawitz, E. et al. *J. Hepatol.* **2004**, *42*, 39–40.
89. Tomei, L.; Altamura, S.; Paonessa, G.; De Francesco, R.; Migliaccoi, G. *Antivir. Chem. Chemother.* **2005**, *16*, 225–245.
90. Carroll, S. S.; Tomassini, J. E.; Bosserman, M. et al. *J. Biol. Chem.* **2003**, *278*, 11979–11984.
91. Tomassini, J. E.; Getty, K.; Stahlhut, M. W. et al. *Antimicrob. Agents Chemother.* **2005**, *49*, 2050–2058.
92. Olsen, D. B.; Eldrup, A. B.; Bartholomew, L. et al. *Antimicrob. Agents Chemother.* **2004**, *48*, 3944–3953.
93. Migliaccio, G.; Tomassini, J. E.; Carroll, S. S. et al. *J. Biol. Chem.* **2003**, *278*, 49164–49170.
94. Eldrup, A. B.; Allerson, C. R.; Bennett, C. F. et al. *J. Med. Chem.* **2004**, *47*, 2283–2295.
95. Summa, V.; Petrocchi, A.; Pace, P. et al. *J. Med. Chem.* **2004**, *47*, 14–17.
96. Summa, V.; Petrocchi, A.; Matassa, V. G. et al. *J. Med. Chem.* **2004**, *47*, 5336–5339.
97. Tomei, L.; Altamura, S.; Bartholomew, L. et al. *J. Virol.* **2003**, *77*, 13225–13231.
98. Harper, S.; Pacini, B.; Avolio, S. et al. *J. Med. Chem.* **2005**, *48*, 1314–1317.
99. Dhanak, D.; Duffy, K. J.; Johnston, V. K. et al. *J. Biol. Chem.* **2002**, *277*, 38322–38327.
100. Tomei, L.; Altamura, S.; Bartholomew, L. et al. *J. Virol.* **2004**, *78*, 938–946.
101. Wang, M.; Ng, K. K.; Cherney, M. M. et al. *J. Biol. Chem.* **2003**, *278*, 9489–9495.
102. Chan, L.; Das, S. K.; Reddy, T. J. et al. *Bioorg. Med. Chem. Lett.* **2004**, *14*, 793–796.
103. Love, R. A.; Parge, H. E.; Yu, X. et al. *J. Virol.* **2003**, *77*, 7575–7581.
104. Howe, A. Y. M.; Bloom, J.; Baldick, C. J. et al. *Antimicrob. Agents Chemother.* **2004**, *48*, 4813–4821.
105. Mo, H.; Lu, L.; Pilot-Matias, T. et al. *Antimicrob. Agents Chemother.* **2005**, *49*, 4305–4314.
106. Kronke, J.; Kittler, R.; Buchholz, F. et al. *J. Virol.* **2004**, *78*, 3436–3446.
107. Stadler, K.; Masignani, V.; Eickmann, M. et al. *Nat. Rev. Microbiol.* **2003**, *1*, 209–218.
108. Li, W.; Moore, M. J.; Vasilieva, N. et al. *Nature* **2003**, *426*, 450–454.
109. Siu, J.; Li, W.; Murakami, A. et al. *Proc. Natl. Acad. Sci. USA* **2004**, *101*, 2536–2541.
110. Liu, S.; Xiao, G.; Chen, Y. et al. *Lancet* **2004**, *363*, 938–947.
111. Simmons, G.; Gosalia, D. N.; Rennekamp, A. J.; Reeves, J. D.; Diamond, S. L.; Bates, P. *Proc. Natl. Acad. Sci. USA* **2005**, *102*, 11876–11881.
112. Barnard, D. L.; Hubbard, V. D.; Burton, J. et al. *Antiviral Chem. Chemother.* **2004**, *15*, 15–22.

113. Yang, H.; Yang, M.; Ding, Y. et al. *Proc. Natl. Acad. Sci. USA* **2003**, *100*, 13190–13195.
114. Lee, T.-W.; Cherney, M. M.; Huitema, C. et al. *J. Mol. Biol.* **2005**, *353*, 1137–1151.
115. Shie, J.-J.; Fang, J.-M.; Kuo, C.-J. et al. *J. Med. Chem.* **2005**, *48*, 4469–4473.
116. Shie, J.-J.; Fang, J.-M.; Kuo, T.-H. et al. *Bioorg. Med. Chem.* **2005**, *13*, 5240–5252.
117. Gan, Y.-R.; Huang, H.; Huang, Y.-D. et al. *Peptides* **2006**, *27*, 622–625.
118. Tanner, J. A.; Watt, R. M.; Chai, Y. B. et al. *J. Biol. Chem.* **2003**, *278*, 39578–39582.
119. Tanner, J. A.; Zheng, B.-J.; Zhou, J. et al. *Chem. Biol.* **2005**, *12*, 303–311.
120. Xu, X.; Liu, Y.; Weiss, S.; Arnold, E.; Sarafianos, S. G.; Ding, J. *Nucl. Acids Res.* **2003**, *31*, 7117–7130.
121. Cheng, A.; Zhang, W.; Xie, Y. et al. *Virology* **2005**, *335*, 165–176.
122. De Clercq, E. *Nat. Rev. Microbiol.* **2004**, *2*, 704–720.
123. Wu, C.-Y.; Jan, J.-T.; Ma, S.-H. et al. *Proc. Natl. Acad. Sci. USA* **2004**, *101*, 10012–10017.
124. Cinatl, J.; Morgenstern, B.; Bauer, G.; Chandra, P.; Rabenau, H.; Doerr, H. W. *Lancet* **2003**, *361*, 2045–2046.
125. Keyaerts, E.; Vijgen, L.; Maes, P.; Neyts, J.; Van Ranst, M. *Biochem. Biophys. Res. Commun.* **2004**, *323*, 264–268.
126. Wu, C.-J.; Jan, J.-T.; Chen, C.-M. et al. *Antimicrob. Agents Chemother.* **2004**, *48*, 2693–2696.
127. Yamamoto, N.; Yang, R.; Yoshinaka, Y. et al. *Biochem. Biophys. Res. Commun.* **2004**, *318*, 719–725.
128. He, M.-L.; Zheng, B.; Peng, Y. et al. *J. Am. Med. Assoc.* **2003**, *290*, 2665–2666.
129. Zhang, Y.; Li, T.; Fu, L. et al. *FEBS Lett* **2004**, *560*, 141–146.
130. Li, B.-j.; Tang, Q.; Cheng, D. et al. *Nat. Med.* **2005**, *11*, 944–951.
131. Cinatl, J.; Morgenstern, B.; Bauer, G.; Chandra, P.; Rabenau, H.; Doerr, H. W. *Lancet* **2003**, *362*, 293–294.
132. Hensley, L. E.; Fritz, E. A.; Jahrling, P. B.; Karp, C. L.; Huggins, J. W.; Geisberg, T. W. *Emerg. Infect. Dis.* **2004**, *10*, 317–319.
133. Ströher, U.; DiCaro, A.; Li, Y. et al. *J. Infect. Dis.* **2004**, *189*, 1164–1167.
134. Sainz, B., Jr.; Mossel, E. C.; Peters, C. J.; Garry, R. F. *Virology* **2004**, *329*, 11–17.
135. Haagmans, B. L.; Kuiken, T.; Martina, B. E. et al. *Nat. Med.* **2004**, *10*, 290–293.
136. Abed, Y.; Goyette, N.; Boivin, G. *Antimicrob. Agents Chemother.* **2005**, *49*, 556–559.
137. von Itzstein, M.; Wu, W. Y.; Kok, G. B. et al. *Nature* **1993**, *363*, 418–423.
138. Kim, C. U.; Lew, W.; Williams, M. A. et al. *J. Am. Chem. Soc.* **1997**, *119*, 681–690.
139. Kaiser, L.; Wat, C.; Mills, T.; Mahoney, P.; Ward, P.; Hayden, F. *Arch. Intern. Med.* **2003**, *163*, 1667–1672.
140. Hayden, F. G.; Belshe, R.; Villanueva, C. et al. *J. Infect. Dis.* **2004**, *189*, 440–449.
141. Kiso, M.; Mitamura, K.; Sakai-Tagawa, Y. et al. *Lancet* **2004**, *364*, 759–765.
142. McKimm-Breschkin, J.; Trivedi, T.; Hampson, A. et al. *Antimicrob. Agents Chemother.* **2003**, *47*, 2264–2272.
143. Ward, P.; Small, I.; Smith, J.; Suter, P.; Dutkowski, R. *J. Antimicrob. Chemother.* **2005**, *55*, i5–i21.
144. Oxford, J. *Exp. Opin. Pharmacother.* **2005**, *6*, 2493–2500.
145. Smee, D. F.; Huffman, J. H.; Morrison, A. C.; Barnard, D. L.; Sidwell, R. W. *Antimicrob. Agents Chemother.* **2001**, *45*, 743–748.
146. Sidwell, R. W.; Smee, D. F.; Huffman, J. H. et al. *Antimicrob. Agents Chemother.* **2001**, *45*, 749–757.
147. Wang, G. T.; Chen, Y.; Wang, S. et al. *J. Med. Chem.* **2001**, *44*, 1192–1201.
148. DeGoey, D. A.; Chen, H.-J.; Flosi, W. J. et al. *J. Org. Chem.* **2002**, *67*, 5445–5453.
149. Hanessian, S.; Bayrakdarian, M.; Luo, X. *J. Am. Chem. Soc.* **2002**, *124*, 4716–4721.
150. Maring, C. J.; Stoll, V. S.; Zhao, C. et al. *J. Med. Chem.* **2005**, *48*, 3980–3990.
151. Mishin, V. P.; Hayden, F. G.; Gubareva, L. V. *Antimicrob. Agents Chemother.* **2005**, *49*, 4515–4520.
152. Le, Q. M.; Kiso, M.; Someya, K. et al. *Nature* **2005**, *437*, 1108.
153. de Jong, M. D.; Tran, T. T.; Truong, H. K. et al. *N. Engl. J. Med.* **2005**, *353*, 2667–2672.
154. Yen, H.-L.; Herlocher, L. M.; Hofmann, E. et al. *Antimicrob. Agents Chemother.* **2005**, *49*, 4075–4084.
155. Sidwell, R. W.; Huffman, J. H.; Khare, G. P.; Allen, L. B.; Witkowski, J. T.; Robins, R. K. *Science* **1972**, *177*, 705–706.
156. Sidwell, R. W.; Bailey, K. W.; Wong, M. H.; Barnard, D. L.; Smee, D. F. *Antivir. Res.* **2005**, *68*, 10–17.
157. Furuta, Y.; Takahashi, K.; Kuno-Maekawa, M. et al. *Antimicrob. Agents Chemother.* **2005**, *49*, 981–986.
158. Tomassini, J. E.; Davies, M. E.; Hastings, J. C. et al. *Antimicrob. Agents Chemother.* **1996**, *40*, 1189–1193.
159. Hamelin, M.-E.; Prince, G. A.; Boivin, G. *Antimicrob. Agents Chemother.* **2006**, *50*, 774–777.
160. Bitko, V.; Musiyenko, A.; Shulyayeva, O.; Barik, S. *Nat. Med.* **2005**, *11*, 50–55.
161. Douglas, J. L.; Panis, M. L.; Ho, E. et al. *Antimicrob. Agents Chemother.* **2005**, *49*, 2460–2466.
162. Douglas, J. L.; Panis, M. L.; Ho, E. et al. *J. Virol.* **2003**, 77, 5054–5064.
163. Andries, K.; Moeremans, M.; Gevers, T. et al. *Antivir. Res.* **2003**, *60*, 209–219.
164. Wyde, P. R.; Chetty, S. N.; Timmerman, P.; Gilbert, B. E.; Andries, K. *Antivir. Res.* **2003**, *60*, 221–231.
165. Cianci, C.; Yu, K.-L.; Combrink, K. et al. *Antimicrob. Agents Chemother.* **2004**, *48*, 413–422.
166. Cianci, C.; Genovesi, E. V.; Lamb, L. et al. *Antimicrob. Agents Chemother.* **2004**, *48*, 2448–2454.
167. Cianci, C.; Meanwell, N.; Krystal, M. *J. Antimicrob. Chemother.* **2005**, *55*, 289–292.
168. Alymova, I. V.; Taylor, G.; Takimoto, T. et al. *Antimicrob. Agents Chemother.* **2004**, *48*, 1495–1502.
169. Wei, F.; Ma, S.-C.; Ma, L.-Y.; But, P. P.-H.; Lin, R.-C.; Khan, I. A. *J. Nat. Prod.* **2004**, *67*, 650–653.
170. Ma, L.-Y.; Ma, S.-C.; Wei, F. et al. *Chem. Pharm. Bull.* **2003**, *51*, 1264–1267.
171. Shigeta, S.; Mori, S.; Kodama, E.; Kodama, J.; Takahashi, K.; Yamase, T. *Antivir. Res.* **2003**, *58*, 265–271.
172. Charrel, R. N.; de Lamballerie, X. *Antivir. Res.* **2003**, *57*, 89–100.
173. McCormick, J. B.; King, I. J.; Webb, P. A. et al. *N. Engl. J. Med.* **1986**, *314*, 20–26.
174. Clement, J. P. *Antivir. Res.* **2003**, *57*, 121–127.
175. Sidwell, R. W.; Smee, D. F. *Antivir. Res.* **2003**, *57*, 101–111.
176. Gowen, B. B.; Barnard, D. L.; Smee, D. F. et al. *Antimicrob. Agents Chemother.* **2005**, *49*, 2378–2386.
177. Jahrling, P. B.; Geisbert, T. W.; Geisbert, J. B. et al. *J. Infect. Dis.* **1999**, *179*, S224–S234.
178. Bray, M. *Antivir. Res.* **2003**, *57*, 53–60.
179. De Clercq, E.; Cools, M.; Balzarini, J. et al. *Antimicrob. Agents Chemother.* **1989**, *33*, 1291–1297.
180. Bray, M.; Driscoll, J.; Huggins, J. W. *Antivir. Res.* **2000**, *45*, 135–147.

181. Bray, M.; Raymond, J. L.; Geisbert, T.; Baker, R. O. *Antivir. Res.* **2002**, *55*, 151–159.
182. Carter, W. A.; De Clercq, E. *Science* **1974**, *186*, 1172–1178.
183. Graci, J. D.; Cameron, C. E. *Rev. Med. Virol.* **2006**, *16*, 37–48.
184. Leyssen, P.; Balzarini, J.; De Clercq, E.; Neyts, J. *J. Virol.* **2005**, *79*, 1943–1947.
185. Enterlein, S.; Warfield, K. L.; Swenson, D. L. et al. *Antimicrob. Agents Chemother.* **2006**, *50*, 984–993.
186. De Clercq, E. *Exp. Opin. Emerg. Drugs* **2005**, *10*, 241–274.
187. Madani, N.; Perdigoto, A. L.; Srinivasan, K. et al. *J. Virol.* **2004**, *78*, 3742–3752.
188. De Clercq, E. *Nat. Rev. Drug Disc.* **2003**, *2*, 581–587.
189. Westby, M.; van der Ryst, E. *Antivir. Chem. Chemother.* **2005**, *16*, 339–354.
190. Strizki, J. M.; Tremblay, C.; Xu, S. et al. *Antimicrob. Agents Chemother.* **2005**, *49*, 4911–4919.
191. Watson, C.; Jenkinson, S.; Kazmierski, W.; Kenakin, T. *Mol. Pharmacol.* **2005**, *67*, 1268–1282.
192. Dorr, P.; Westby, M.; Dobbs, S. et al. *Antimicrob. Agents Chemother.* **2005**, *49*, 4721–4732.
193. Gu, Z.; Allard, B.; de Muys, J. M. et al. *Antimicrob. Agents Chemother.* **2006**, *50*, 625–631.
194. De Meyer, S.; Azijn, H.; Surleraux, D. et al. *Antimicrob. Agents Chemother.* **2005**, *49*, 2314–2321.
195. Li, F.; Goila-Gaur, R.; Salzwedel, K. et al. *Proc. Natl. Acad. Sci. USA* **2003**, *100*, 13555–13560.
196. Romines, K. R.; Freeman, G. A.; Schaller, L. T. et al. *J. Med. Chem.* **2006**, *49*, 727–739.
197. Vermeire, K.; Princen, K.; Hatse, S. et al. *AIDS* **2004**, *18*, 2115–2125.
198. Boyd, M. R.; Gustafson, K. R.; McMahon, J. B. et al. *Antimicrob. Agents Chemother.* **1997**, *41*, 1521–1530.
199. Ichiyama, K.; Yokoyama-Kumakura, S.; Tanaka, Y. et al. *Proc. Natl. Acad. Sci. USA* **2003**, *100*, 4185–4190.
200. Flomenberg, N.; Devine, S. M.; DiPersio, J. F. et al. *Blood* **2005**, *106*, 1867–1874.
201. Baba, M.; Nishimura, O.; Kanzaki, N. et al. *Proc. Natl. Acad. Sci. USA* **1999**, *96*, 5698–5703.
202. Takashima, K.; Miyake, H.; Kanzaki, N. et al. *Antimicrob. Agents Chemother.* **2005**, *49*, 3474–3482.
203. Nishikawa, M.; Takashima, K.; Nishi, T. et al. *Antimicrob. Agents Chemother.* **2005**, *49*, 4708–4715.
204. Tremblay, C. L.; Giguel, F.; Guan, Y.; Chou, T.-C.; Takashima, K.; Hirsch, M. S. *Antimicrob. Agents Chemother.* **2005**, *49*, 3483–3485.
205. Baba, M.; Takashima, K.; Miyake, H. et al. *Antimicrob. Agents Chemother.* **2005**, *49*, 4584–4591.
206. Chu, C. K.; Yadav, V.; Chong, Y. H.; Schinazi, R. F. *J. Med. Chem.* **2005**, *48*, 3949–3952.
207. Nitanda, T.; Wang, X.; Kumamoto, H. et al. *Antimicrob. Agents Chemother.* **2005**, *49*, 3355–3360.
208. Wu, T.; Froeyen, M.; Kempeneers, V. et al. *J. Am. Chem. Soc.* **2005**, *127*, 5056–5065.
209. Janssen, P.; Lewi, P. J.; Arnold, E. et al. *J. Med. Chem.* **2005**, *48*, 1901–1909.
210. Ferris, R. G.; Hazen, R. J.; Roberts, G. B. et al. *Antimicrob. Agents Chemother.* **2005**, *49*, 4046–4051.
211. Hazen, R. J.; Harvey, R. J.; St. Clair, M. H. et al. *Antimicrob. Agents Chemother.* **2005**, *49*, 4465–4473.
212. Hazuda, D. J.; Anthony, N. J.; Gomez, R. P. et al. *Proc. Natl. Acad. Sci. USA* **2004**, *101*, 11233–11238.
213. Hazuda, D. J.; Young, S. D.; Guare, J. P. et al. *Science* **2004**, *305*, 528–532.
214. Didierjean, J.; Isel, C.; Querré, F. et al. *Antimicrob. Agents Chemother.* **2005**, *49*, 4884–4894.
215. Cancio, R.; Silvestri, R.; Ragno, R. et al. *Antimicrob. Agents Chemother.* **2005**, *49*, 4546–4554.
216. Nair, V.; Chi, G.; Ptak, R.; Neamati, N. *J. Med. Chem.* **2006**, *49*, 445–447.
217. Printsevskaya, S. S.; Solovieva, S. E.; Olsufyeva, E. N. et al. *J. Med. Chem.* **2005**, *48*, 3885–3890.
218. Balzarini, J.; Hatse, S.; Vermeire, K. et al. *Antimicrob. Agents Chemother.* **2004**, *48*, 3858–3870.
219. Balzarini, J.; Van Laethem, K.; Hatse, S. et al. *J. Virol.* **2004**, *78*, 10617–10627.
220. Liu, S.; Lu, H.; Neurath, A. R.; Jiang, S. *Antimicrob. Agents Chemother.* **2005**, *49*, 1830–1836.
221. Scordi-Bello, I. A.; Mosoian, A.; He, C. et al. *Antimicrob. Agents Chemother.* **2005**, *49*, 3607–3615.
222. Updated overview originally published in *Future Virol.* **2006**, *1*, 19–35.

Biography

Erik De Clercq, MD, PhD is Chairman of the Department of Microbiology and Immunology of the Medical School at the Katholieke Universiteit Leuven and also is the President of the Rega Foundation and Chairman of the Board of the Rega Institute for Medical Research. He is a director of the Belgian Royal Academy of Medicine, a member of the Academia Europaea, and fellow of the American Association for the Advancement of Science. He has also been the titular of the Prof P De Somer Chair for Microbiology. He teaches the courses of Cell Biology, Biochemistry, and Microbiology at the K U Leuven (and Kortrijk) Medical School. Professor De Clercq is the co-inventor of Gilead's nucleotide analogs cidofovir, adefovir, and tenofovir and received the Hoechst Marion Roussel (now called 'Aventis') award, the Maisin Prize for Biomedical Sciences (National Science Foundation, Belgium), R Descartes Prize (European Union Commission), and B Pascal Award (European Academy of Sciences) for his pioneering efforts in the field of antiviral research. His scientific interests are in the antiviral chemotherapy field, and, in particular, the development of new antiviral agents for various viral infections, including HSV, VZV, CMV, HIV, HBV, HPV, and HCV.

Comprehensive Medicinal Chemistry II
ISBN (set): 0-08-044513-6

ISBN (Volume 7) 0-08-044520-9; pp. 253–293

7.11 Deoxyribonucleic Acid Viruses: Antivirals for Herpesviruses and Hepatitis B Virus

E Littler, MEDIVIR UK Ltd, Little Chesterford, UK
X-X Zhou, Medivir AB, Huddinge, Sweden

7.11.1 Treatment of Herpesvirus Diseases

7.11.1.1 Herpesvirus Biology and Clinical Significance

For a comprehensive review of the herpesviruses refer to relevant chapters in *Field's Virology*.[1] The herpesviruses comprise a large family of viruses infecting most, if not all, vertebrates. The virus particle is about 100 nm in diameter and consists of an icosahedral nucleocapsid surrounded by an outer membrane. A critical defining property of the herpesviruses is their ability to form a latent infection as part of their life cycle. The family is divided into three main groups namely the alpha, beta, and gamma herpesviruses. This division was initially based upon aspects of their biology such as cell tropism or disease pathology but more recently deoxyribonucleic acid (DNA) sequence analysis has served to consolidate this classification. In humans there are eight identified viruses, namely herpes simplex type 1 (HSV-1), herpes simplex type 2 (HSV-2), varicella zoster virus (VZV), Epstein–Barr virus (EBV), human cytomegalovirus (HCMV), and human herpesviruses 6, 7, and 8 (HHV-6, -7, and -8). About 70% of the human population is infected by HSV which is responsible for causing cold sores either on the mouth (usually HSV-1) or genitals (usually HSV-2). However, they can also be the cause of central nervous system (CNS) infections including encephalitis. Both HSV-1 and HSV-2 form latent infections in either trigeminal or cervical ganglia respectively.[2]

Primary infection by VZV usually occurs in children and is the cause of chickenpox after which the virus becomes latent in the cervical ganglia. Reactivation of VZV occurs when the host's immune response is reduced either naturally due to old age or due to immune suppression. The reactivated virus causes shingles, a debilitating disease, which in the case of the immunocompromised can be fatal.[3]

EBV is the primary cause of infectious mononucleosis in adolescents (usually of high socioeconomic background who have avoided the infection as children). EBV forms a latent infection in B lymphocytes.[4] Infection by EBV is highly associated with two human cancers: Burkitt's lymphoma, the most common tumor in African children, and nasopharyngeal carcinoma, the most common tumor of Chinese males and the second most common tumor of Chinese females.[5]

HCMV is rarely associated with disease in healthy people, the exception being infection to the unborn baby during pregnancy which may result in moderate to severe damage to the fetal CNS. In immunocompromised patients CMV can cause severe infections such as pneumonia (often the cause of death in acquired immune deficiency syndrome (AIDS) patients), retinitis, or gastrointestinal infections. HCMV forms latent infections in macrophages.[6]

The remaining herpesviruses are not sufficiently linked with significant human disease to warrant the development of their own antivirals but instead depend upon the selection of suitable drugs available for treatment of more commercially attractive herpesviruses. HHV-6 was first isolated from patients with B cell tumors in 1986 [7] and has been shown to be a cause of exanthum subitum in young children.[8] HHV-7 was isoloated in 1970 and is also a cause of exanthum subitum in young children.[9] HHV-6 and HHV-7 are closely related to each other and to HCMV which suggests that the optimal therapeutic agent should be ganciclovir; however, HHV-7 seems to have a slightly different profile than HHV-6.[10] Both HHV-6 and HHV-7 appear to be widespread in the human population with initial infection occurring during childhood (HHV-6 earlier than HHV-7) but their association with serious clinical conditions is limited and sporadic requiring little intervention. The one exception may be CNS disease in children with HHV-6 and varying infections in the immunocompromised.[11] Finally HHV-8 or Karposi's sarcoma (KS) associated herpesvirus is a member of the gamma herpesviruses but resembling primate herpesviruses rather than EBV. It is found in 95% of KS tumors which are commonly found in HIV or other immunocompromised patients but can also be found in geographical clusters in elderly men of Mediterranean or Easter European origin.[12] HHV-8 appears most sensitive to cidofovir and foscarnet with only low sensitivity to ACV.[13]

All herpesviruses have a large double-stranded DNA genome coding for around 100 genes depending upon the virus. Herpesviruses code for several genes that are involved directly or indirectly in the replication of the virus, for example, DNA polymerase, helicase/primase, thymidine kinase, or ribonucleotide reductase.[14] Indeed it is these replication proteins, especially the DNA polymerase, that form the target of most of the inhibitors of virus replication.

7.11.1.2 Antiviral Drugs in Clinical Use

The major antiherpes drugs that are currently in clinical use are inhibitors of the viral DNA polymerase, which block replication of viral DNA. Among the inhibitors of viral DNA polymerase, most are nucleosides, or occasionally nucleotides, that need to undergo phosphorylation to their triphosphates to be active as antivirals. The triphosphates are competitive inhibitors of the viral DNA polymerases; however, they are also incorporated into the DNA chain, preventing the extension of the viral DNA due to their modified structures. It is this latter property that represents the mechanism by which the nucleoside analogs inhibit virus replication.[15]

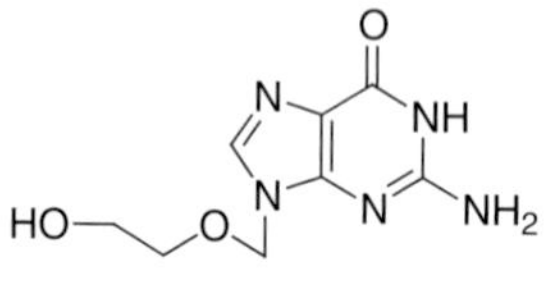

1 ACV

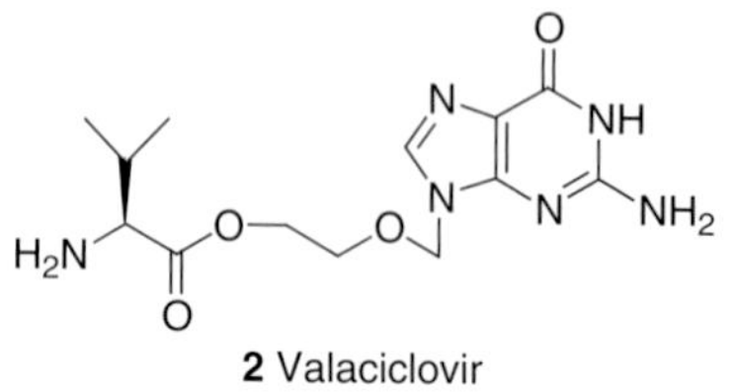

2 Valaciclovir

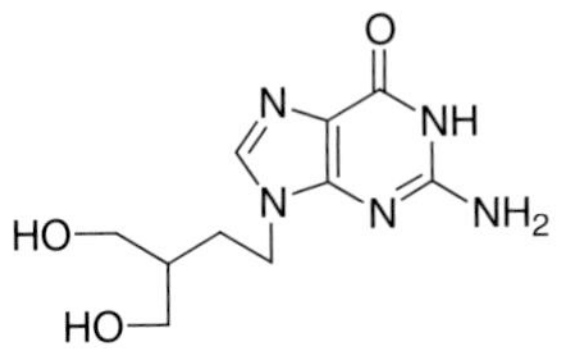

3 PCV

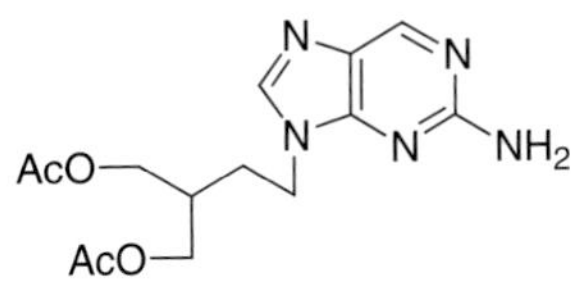

4 FCV

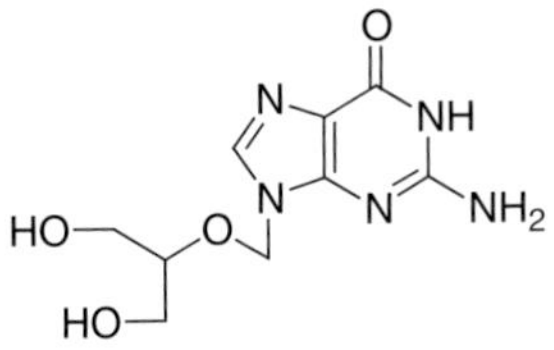

5 Ganciclovir

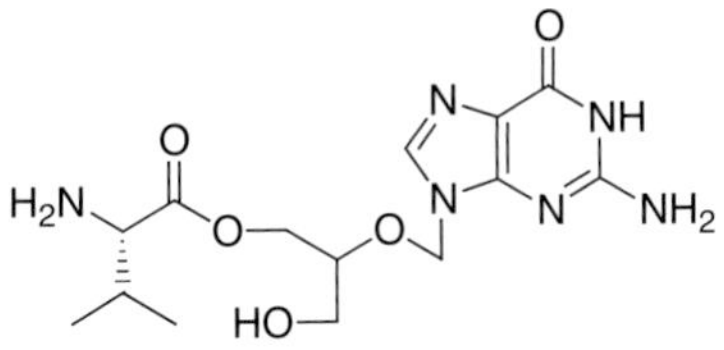

6 Valganciclovir

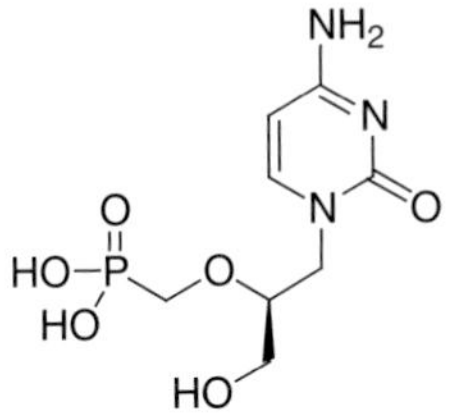

7 Cidofovir

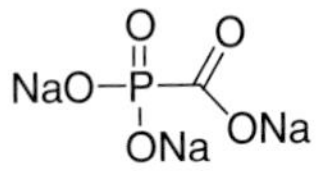

8 Foscarnet

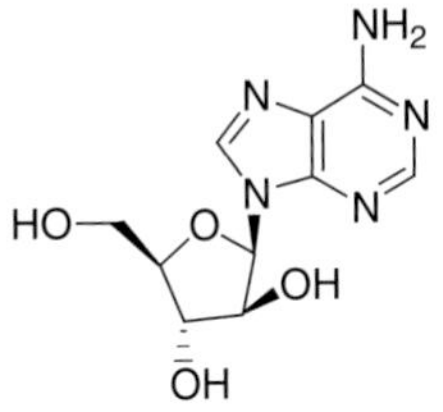

9 Vidarabine

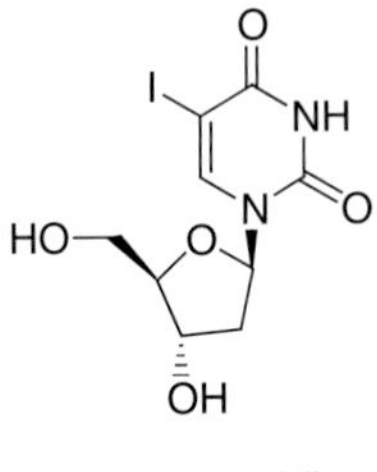

10 Idoxuridine

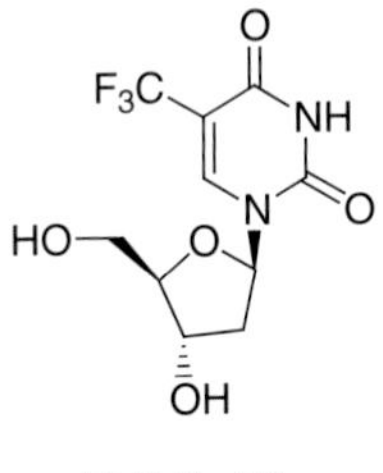

11 Trifluridine

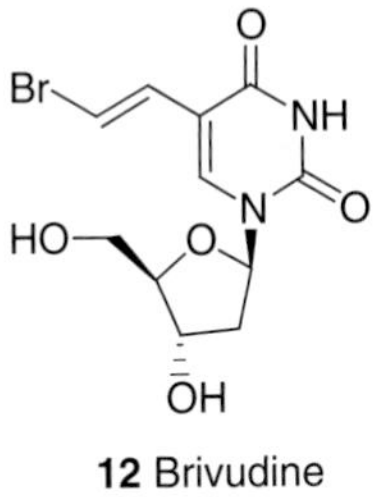

12 Brivudine

7.11.1.2.1 Aciclovir

Aciclovir, 9-(2-hydroxyethoxymethyl)guanine (ACV, **1**), has been the principal drug for the treatment and prophylaxis of HSV. It is potent against HSV-1 and HSV-2 with an EC_{50} of 0.1 and 0.5 μM, respectively,[16] and is also active, although to a lesser extent, against other herpesviruses such as VZV ($EC_{50} = 1.1\,\mu g\,mL^{-1}$), EBV ($EC_{50} = 27\,\mu g\,mL^{-1}$), and CMV ($EC_{50} = 16\,\mu g\,mL^{-1}$).[17] The antiviral efficacy is through the active metabolite, ACV triphosphate. ACV is first phosphorylated by the virus-coded thymidine kinase (TK) to ACV-MP, which is subsequently further phosphorylated by cellular enzymes to ACV diphosphate and ACV triphosphate (**Figure 1**). ACV-TP is incorporated into the nascent viral DNA chain and blocks viral DNA synthesis. So far no cellular kinase has been found to be able to catalyze the synthesis of the ACV-MP, leading to the high selectivity of ACV for cells that are infected by herpesvirus, which explains the excellent safety profile of the compound.

As the recognition of viral/host kinases for phosphorylation and the binding to the viral DNA polymerase for incorporation are essential for the antiviral effect, the favored structures are expected to have certain similarity to the natural substrate nucleosides. The structures of several TKs encoded by herpesviruses have been resolved by x-ray crystallography[18–21] as well as their complex with natural substrate and acyclic nucleoside inhibitors, which sheds light

ACV → ACV-MP → ACV-DP → ACV-TP → Incorporation into virus DNA → Chain termination

Herpesvirus TK Host kinases Host kinases

Figure 1 Mechanism of action of ACV.

on understanding the interaction between ACV and the viral thymidine kinases. The active site of the HSV TK consists of the substrate binding pocket, ATP binding site, and the catalytic residues that coordinate magnesium ions. The sugar moiety of the natural substrate, thymidine, has a close contact with HSV-1 TK. Its 5′-OH is hydrogen bonded to the Arg163 of the TK and its 3′-OH to Tyr101.[18] In the complex of ACV and HSV-1 TK, it was found that the 4-hydroxyl in the side chain forms an H-bond with Arg163 in a similar manner as thymidine. The guanine base also has close interaction with the enzyme. This explains how ACV is so well accepted by the HSV-1 TK, with the phosphorylation rate equivalent to about 27% of thymidine.[22–24]

Herpes polymerase has a DNA-dependent DNA polymerase activity as well as a 3′–5′ exonuclease activity. Similar to other DNA polymerases, it has palm, fingers, and thumb domains which together exercise the function of polymerization. The crystal structure of HSV-1 DNA polymerase has been reported[25]; however no complex with an inhibitor has yet been described.

Extensive work has been done in exploring the structure–activity relationship (SAR) for ACV. For optimal antiherpes activity, the side chain with four atoms bearing a primary hydroxyl group at the 4-position is favored.[26,27] Although compound with an amino group replacing hydroxyl (**17**) in ACV showed slight inhibitory effect against HSV ($IC_{50} = 8\,\mu M$), it is believed that the inhibition is due to the deamination of the compound by a different enzyme to generate a low level of ACV.[27] An increase in the side chain length reduces the activity (**13–15**), while maintaining essentially a similar level of phosphorylation for the analogs with longer chain of five to six atoms. The reduced antiviral activity comes presumably from the lower affinity to the viral DNA polymerase.[27] Replacement of the 2-oxygen with a sulfur (**16**) is tolerable, with only slightly reduced potency. A methylene group replacement abolishes the activity. The change of the nucleobase will affect the activity negatively in most cases. The adenine counterpart (**18**), which is known to be a good substrate of adenosine deaminase, has a very low activity in vivo.[28] However, 2,6-diaminopurine analog (**19**) which can be converted to guanine base by adenosine deaminase maintains a very good activity both in vitro and in vivo.[29]

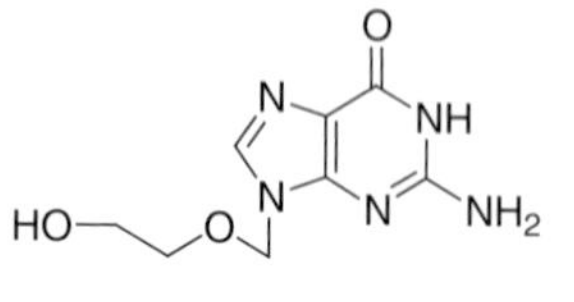

1 ACV 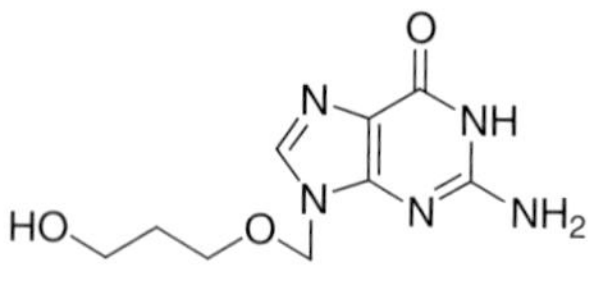 **13** 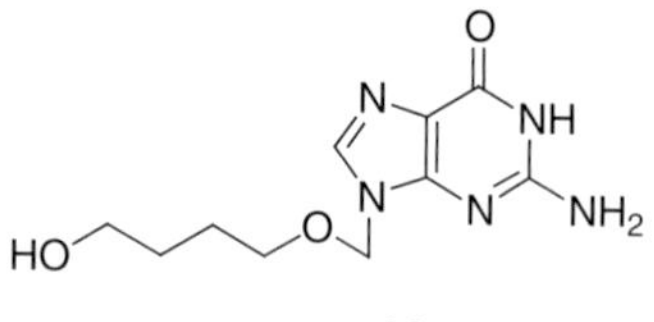 **14**

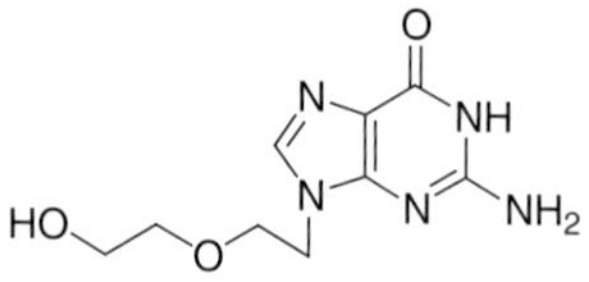

15 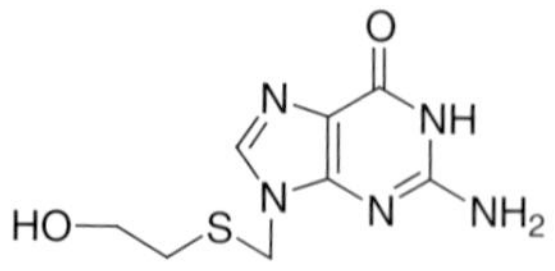 **16** 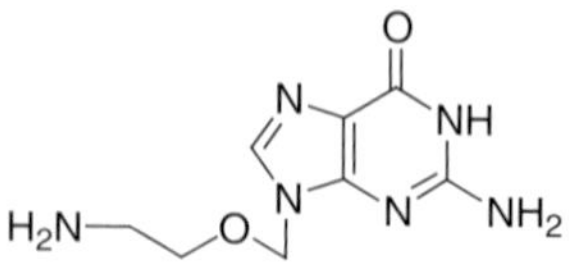 **17**

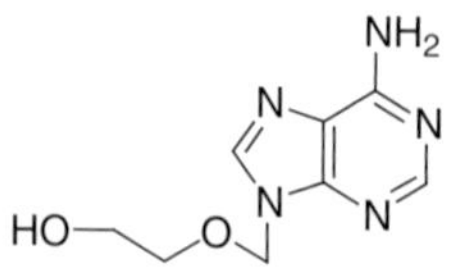

18 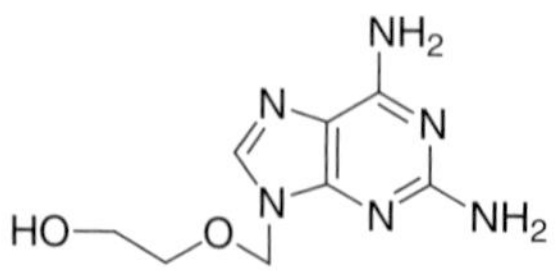 **19**

ACV is widely used both orally and topically for the treatment and prophylaxis of HSV-1, HSV-2, and VZV. Treatment of herpes labialis with ACV marginally decreases the duration of symptoms and the healing time.[31] Topical ACV reduces viral shedding and the length of the episode, but is less effective than oral or intravenous ACV[32] because of poor skin penetration.

ACV has a low oral bioavailability in human (15–21%).[30] Urinary excretion is the major route for the excretion of ACV and only less than 1% of the dose is secreted into the bile.[35] In order to achieve higher plasma concentration, intravenous administration is necessary. The half-life of ACV in humans is about 2–3 h. Much effort has been made to improve the pharmacokinetic properties of ACV through formulation or prodrug approaches.

Mutation in the HSV genome can result in a deficiency or alteration in viral TK activity, preventing ACV from being phosphorylated.[33] Deficinecy in TK expression is the most common mechanism for conferring ACV resistance and the cross-resistance to the other acyclic nucleosides inhibitors which are also dependent on the viral TK for their phosphorylation. However it has been shown by animal models that TK-deficient HSV fails to reactivate efficiently from a latent infection and hence clinical resistance to ACV is rare in all but immunocompromised patients where this selective pressure is negated by the lack of a fully active immune response. An alteration in the substrate specificity of the viral TK is not common but can occur. Far less frequently mutation in the viral DNA polymerase gene can result in a failure to incorporate ACV triphosphate into progeny DNA molecules, thus leading to resistance to ACV.[34]

7.11.1.2.2 Valaciclovir

Because of the limited oral bioavailability of ACV, plasma levels adequate for inhibition of less sensitive viruses can only be achieved by intravenous dosing. Therefore, a highly efficient and safe prodrug of ACV that can be dosed orally and gives a plasma level comparable to intravenous dosing was needed. Valaciclovir (VACV, **2**) is the L-valyl ester of ACV. The orally administered prodrug is absorbed in the gastrointestinal tract through both passive transcellular mechanism and active transporter mechanism. Peptide transporter PEPT1, which is abundant in the intestine, has a high affinity to VACV,[35] and helps to enhance the absorption very efficiently. The parent compound ACV is not recognized as a substrate by PEPT1. The prodrug is rapidly converted to ACV and valine by enzymes in the liver and intestinal walls.[36] Levels of ACV in plasma after oral VACV administration increases almost proportionally to the dose administrated. The oral bioavailability reaches as high as 50% versus 15–21% for ACV. In clinical trials VACV has a similar antiviral activity and safety profile to that of ACV.[37,38]

Various structural classes of prodrugs have been reported for ACV.[39–47] Prodrugs containing a water-soluble ester moiety have been regarded as a general approach to increase the solubility and dissolution rate of parent drugs to be dosed. The most commonly used esters for this purpose are those with an ionic or ionizable group.[48,49] Alpha-amino acids are an ideal selection due to their biocompatible nature. Indeed, the solubility of VACV is about 130-fold higher than that of ACV. The stability of the amino acid esters of ACV vary depending on the structure of the side chains of the amino acid. The glycyl esters under physiological pH and temperature can undergo substantial hydrolysis within 1 h.[50] Branching on the beta-carbon can significantly increase the chemical stability of the prodrug. L-Valyl and L-isoleucyl ACV enhance the oral bioavailability up to 50–60% in rats. The stereochemistry of the amino acid in the prodrug esters has a marked effect on the absorption, with L-amino acid being much preferred over its D-isomer. That can be explained by the contribution of the stereospecific transporter which plays an important role in the absorption.[46] The side chain structures of the amino acids in the prodrug also affect the stability toward esterases in vivo, where steric hindrance and the configuration of the amino acid are determining factors.[30,51]

7.11.1.2.3 Famciclovir

Famciclovir, 2-[2-(2-amino-9H-purin-9-yl)ethyl]-1,3-propanediol diacetate (FCV, **4**), is an oral prodrug of penciclovir, 9-(4-hydroxy-3-hydroxymethyl-but-1-yl)guanine (PCV, **3**). PCV has a broad spectrum antiviral activity against HSV-1, HSV-2, VZV, and EBV. The 50% effective dose versus laboratory and clinical isolates of HSV-1 and HSV-2 for PCV is 0.3–2.0 $\mu g\,mL^{-1}$ compared to 0.1–1.9 $\mu g\,mL^{-1}$ for ACV. Similar to ACV, PCV is phosphorylated in HSV-1 and HSV-2 infected cells by the viral TK. PCV is a good substrate for the viral TK and crystallography studies showed that the two hydroxy groups of the side chain are H-bonded to the Arg163 and Tyr101 in a manner that mimics the natural substrate thymidine.[19] The further phosphorylation of PCV-MP is also favored. In the infected cells the intracellular PCV-TP concentration can reaches an approximately 200-fold higher level than ACV triphosphate (ACV-TP). Moreover, PCV-TP has an extremely long half-life in HSV-1 and HSV-2 infected cells: about 10 and 20 h, respectively, versus 0.7 and 1.0 h for ACV-TP. PCV-TP is a competitive inhibitor of HSV-1 and HSV-2 DNA polymerases with K_i of 11 and 6 μM, respectively.[52] Although PCV-TP is about 80–140-fold less active than ACV-TP (0.07 and 0.08 μM, respectively), the higher intracellular levels of PCV-TP compensates its antiviral effect. Unlike ACV, PCV does not act as an obligate DNA chain terminator, but can permit limited DNA chain elongation, presumably due to the second hydroxyl group. Incorporated PCV is thought to distort the conformation of the DNA and induce a blockade of further elongation.[52]

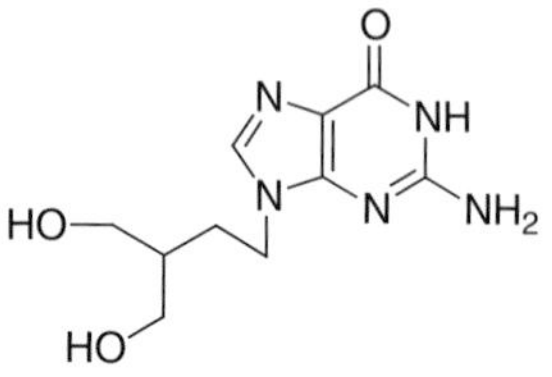

3 PCV 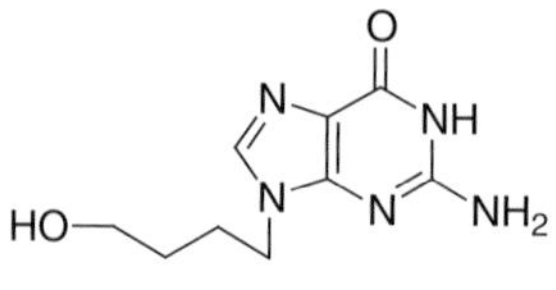 **20** 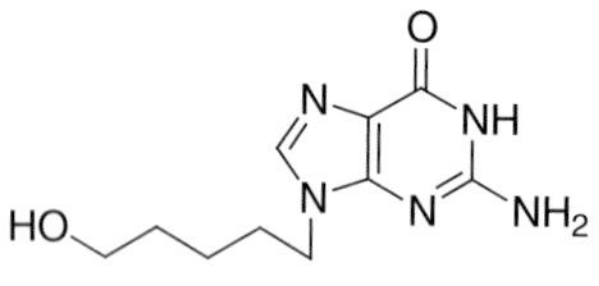 **21** 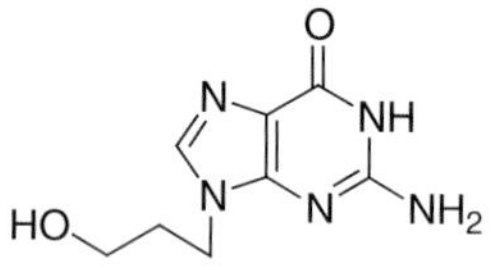 **22** 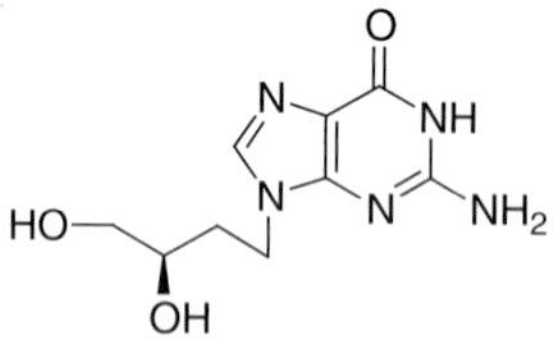 **23** 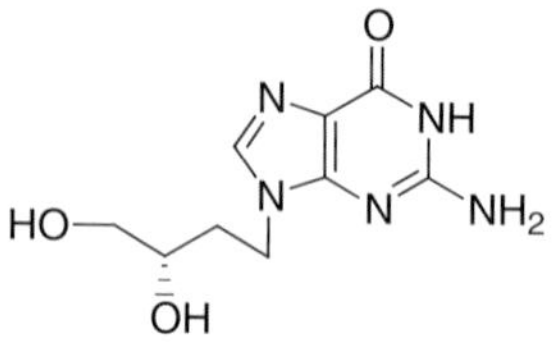 **24** 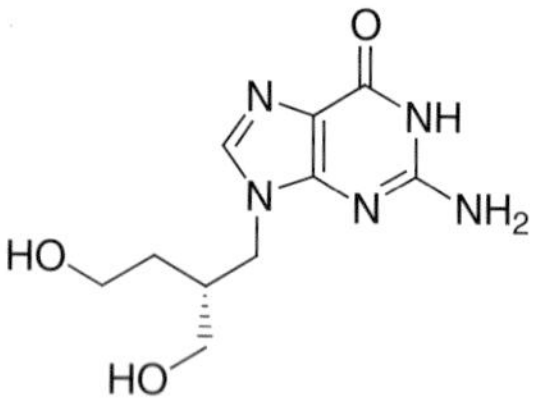 **25** 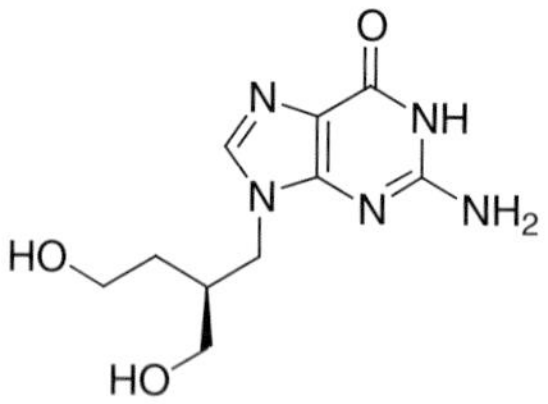 **26** 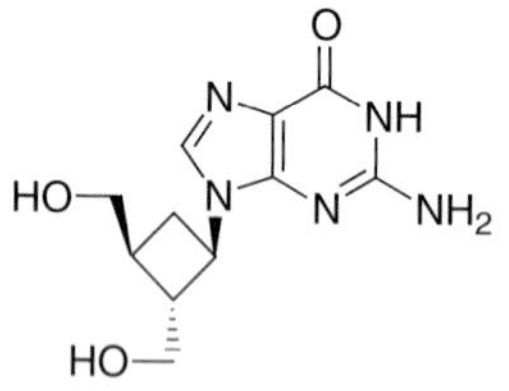 **27** 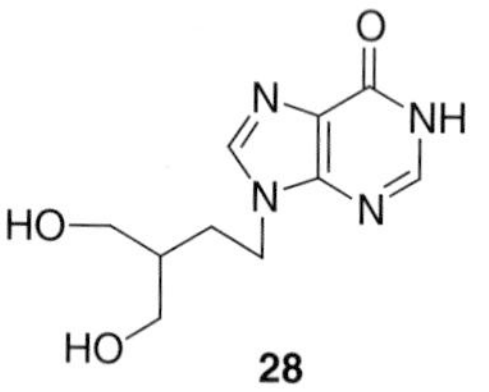 **28** 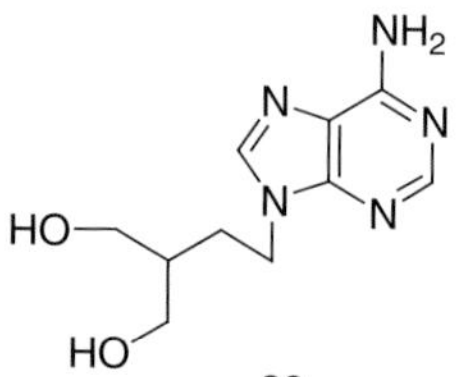 **29**

The SAR studies for PCV showed that the optimal activity comes with the acyclic side chain of four carbon atoms bearing a hydroxyl at the 4-position.[53] Reducing or increasing the length by one atom will abolish most activity (**21**, **22**). The hydroxymethyl substitution on the 3-position is important for the activity, however not essential. 9-(4-Hydroxybut-1-yl)guanine (**20**) maintains good activity ($IC_{50}=3\,\mu M$ against HSV-1 and $8\,\mu M$ against HSV-2). The 9-(3,4-dihydroxybut-1-yl)guanine (bucyclovir, **23**) with a hydroxyl directly attached to the butyl side chain is very potent against HSV-1 and HSV-2 ($IC_{50}=0.4\,\mu M$ against HSV-1 and $0.9\,\mu M$ against HSV-2). The stereochemistry of the 3-position of bucyclovir is crucial in determining the antiherpetic activities. While the *R* configuration is very potent, the *S*-isomer (**24**) is essentially inactive against HSV-1 and HSV-2.[22,54] The position for hydroxymethyl branching can affect the antiherpetic profile. The 2-hydroxylmethyl substitution gives two stereoisomers of 9-(4-hydroxy-2-hydroxymethyl-but-1-yl)guanine (**25**, **26**). The *S* configuration (**26**) is virtually inactive, but the *R*-isomer (H2G, **25**) is very potent against VZV, being about 20-fold more potent than PCV, while its activity against HSV-1 and HSV-2 are of a

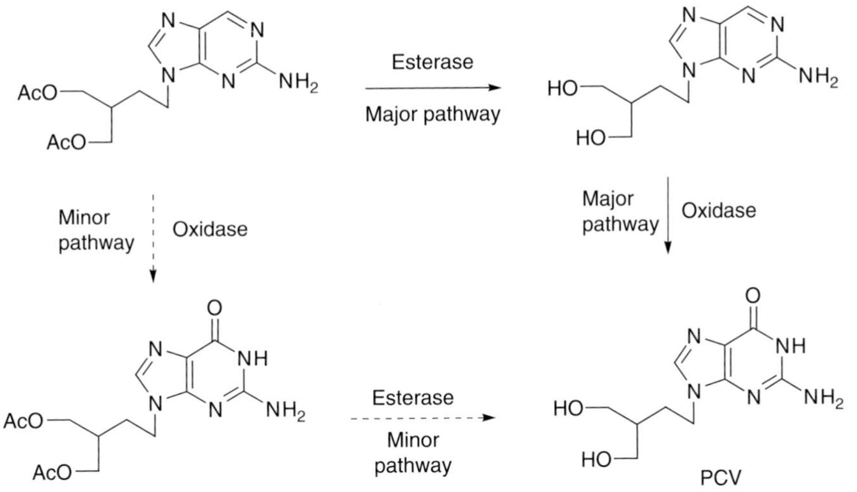

Figure 2 Metabolism of FCV.

similar level to PCV. Lobucavir (**27**), a compound with dihydroxymethylcyclobutane as side chain, maintains a similar antiherpes profile to PCV.[55] The attempts to replace guanine base of PCV mostly lead to inactive compounds or substantially less active compounds (**28**, **29**).[56]

When dosed orally in human, PCV has an extremely low oral bioavailability (2–5%), and thus can only be used in topical form. However, its prodrug FCV (**4**) has substantially higher oral bioavailability (77%). FCV is absorbed rapidly in the intestine and is deacetylated and oxidated by xanthine oxidase to PCV (**Figure 2**). Xanthine oxidase occurs in the intestinal wall and liver in high concentration[57] and has broad substrate specificity. Various prodrugs of PCV have been reported.[58,59] After FCV dosing, approximately 65% of the administrated dose is excreted in urine as PCV and 5% as 6-deoxy PCV.[60]

Since PCV is phosphorylated by viral TK, the TK deficient strains are resistant to PCV in a similar manner to ACV.

7.11.1.2.4 Ganciclovir

Ganciclovir, 9-(1,3-dihydroxy-2-propoxymethyl)guanine (GCV, **5**), is a broad-spectrum antiherpes substance ($EC_{50}=0.1$, 0.1, 2.8, 3.4, and 0.1 μM against HSV-1, HSV-2, VZV, HCMV, and EBV, respectively).[61] Due to the excellent potency against HCMV, it is extensively used for the treatment and prophylaxis of cytomegalovirus (CMV) retinitis in immunocompromised patients and in the recipients of solid organ transplants.

Similar to ACV and PCV, the triphosphate of GCV inhibits viral DNA polymerase. The first phosphorylation of GCV to monophosphate is catalyzed by viral enzymes. In HSV-1, HSV-2, and VZV infected cells, the virus TK are responsible for the phosphorylation of GCV with the two hydroxy groups of the GCV side chain form H-bondings with the viral TK in a similar manner to PCV. However, the HCMV genome lacks the corresponding thymidine kinase gene. Littler *et al.* have reported a protein kinase, a product of UL97 gene, which catalyses the phosphorylation of GCV.[62,63] The dependence on UL97 for phosphorylation is confirmed by the observation that the mutations in the UL97 gene result in resistance to ganciclovir.[64] Similarly, HHV-6, which is sensitive to ganciclovir, encodes a homolog of UL97 responsible for the phosphorylation of ganciclovir.[63,65,66] GCV monophosphate is converted by cellular enzymes to the triphosphate, which blocks viral DNA synthesis.

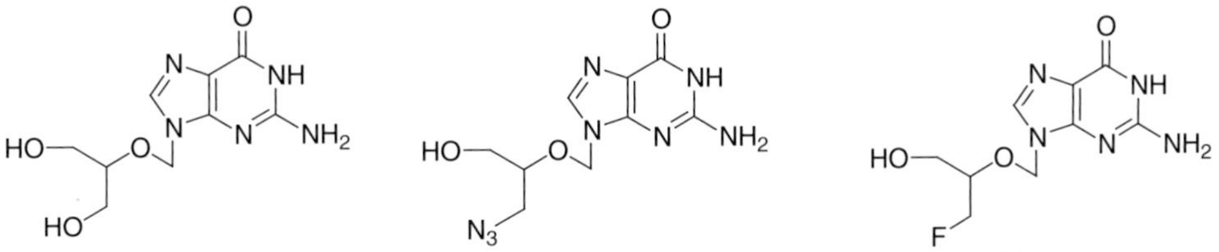

GCV (**5**) **30** **31**

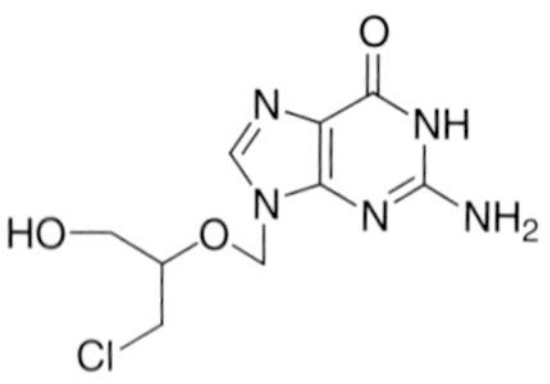

32 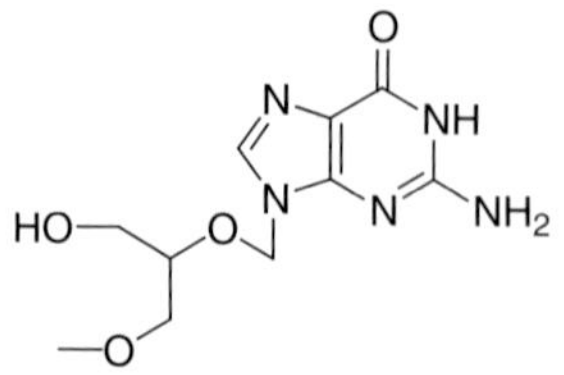 **33** 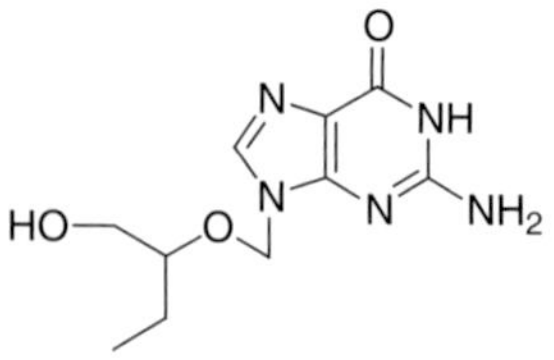 **34**

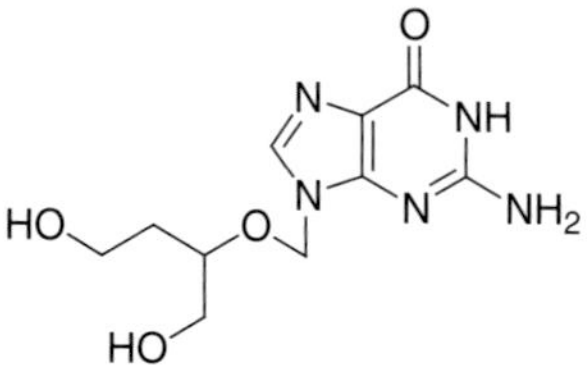

35 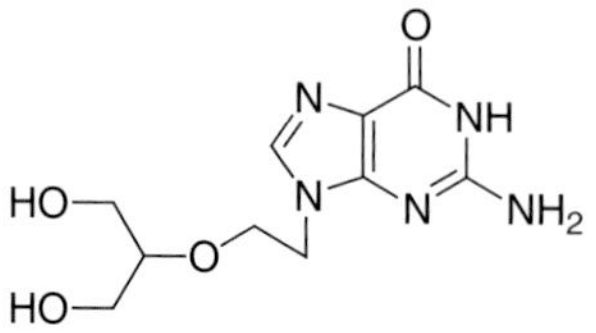 **36** 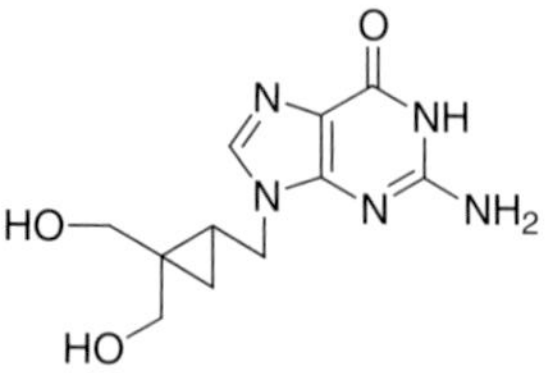 **37**

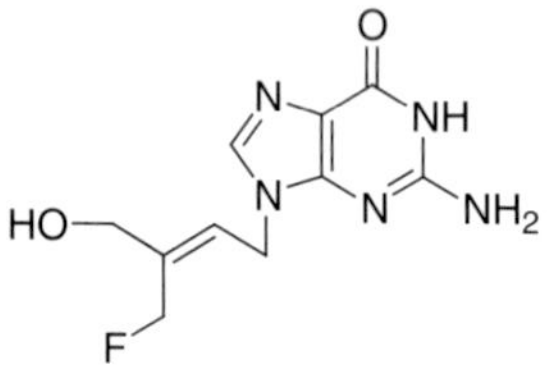

38 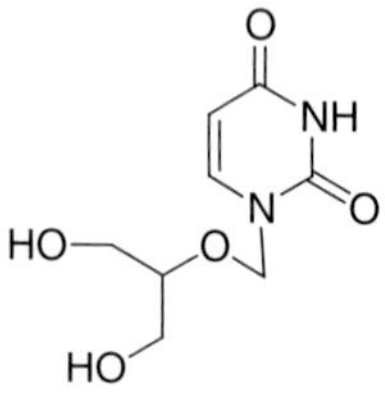 **39** 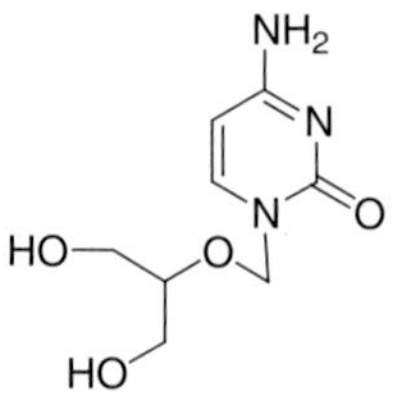 **40** 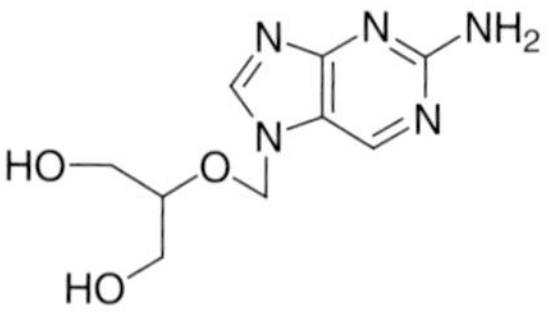 **41** (SS2242)

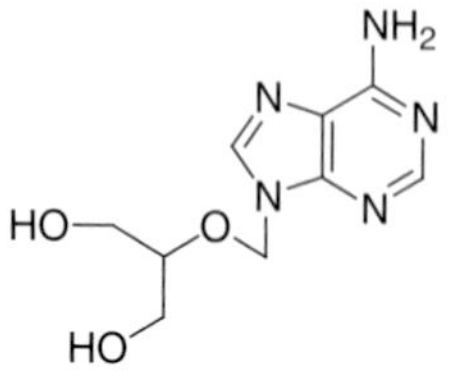

42 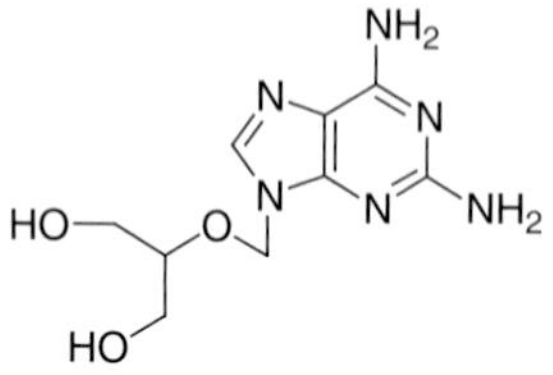 **43** 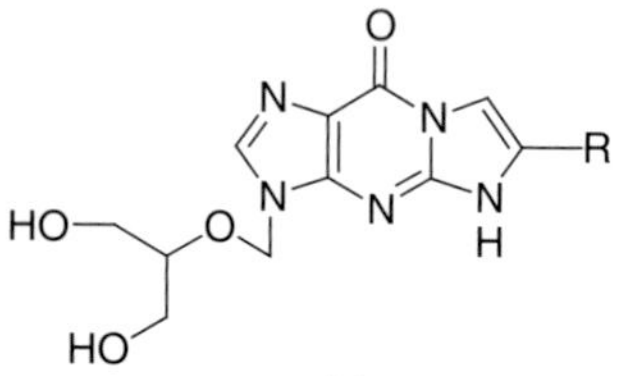 **44**

The structure of GCV is reminiscent of deoxyribonucleoside with two hydroxyls in the position of 5′ and 3′. Replacing the hydroxyl group at the 3′-position by azide, amino, or halogen abolishes the antiherpetic activity (**30**, **31**, **32**),[67] as does blocking the hydroxyl with a methoxy group (**33**).[67] However, 3′-deoxy compound (**34**) shows some activity, though about 10-fold less compared to GCV.[67] Extension of the side chain with one carbon atom at the either side of the oxygen is intolerable (**35**, **36**).[67] Replacing oxygen in the side chain with a cyclopropane or a vinyl loses the activity (**37**, **38**).[68] Modification or replacement of guanine base will affect the activity negatively. Various pyrimidines with 1,3-dihydroxy-2-propoxymethyl side chain are inactive (**39**, **40**).[61,69] Interestingly, though deoxycytidine is a good substrate for viral thymidine kinase, acyclic cytidine analogs are not accepted by the enzyme.[70] Other purine analogs like adenine (**42**), hypoxanthine, or 2-aminopurine have substantially reduced antiherpes effect. 2,6-Diaminopurine analog (**43**) shows good activity, though less than GCV, which is due to the fact that 2,6-diaminopurine analog can be effectively converted to guanine analog by adenosine deaminase. The 7-isomer of the 6-deoxy guanine substitution (**41**) gives a compound with broad antiherpes activity. It is worth mentioning that the compound is extremely active against HHV-6.[70,71] Several analog with tricyclic 3,9-dihydro-9-oxo-5H-imidazo(1,2-α)purine (**44**), formed by linking 2-NH_2 and N^1 position of guanine with an etheno bridge, show antiherpes activities,[72] though with big variation of potency.

GCV has an unfavorable pharmacokinetic profile. When given in oral dose in human, the oral bioavailability is around 2.6–7.3%.[73] The plasma half-life is about 3.0–7.3 h. Intravenously dosed GCV is excreted unchanged in urine in patients with normal renal function, indicating that little or no extra renal elimination occurs.[74] GCV prodrugs have been explored. Valganciclovir (**6**), the monovalyl ester of GCV, shows very promising properties, with the oral bioavailability being raised to over 60%. GCV is not recognized by intestinal PEPT1 and renal PETT2, but the addition of a valyl ester makes it a good substrate of the two peptide transporters, which effectively enhances the absorption.[75] The ester linkage is easily hydrolyzable by intracellular esterases, which explains why valganciclovir is readily converted to GCV, once the prodrug enters the mucosal cells after crossing the brush border membrane.[75] After oral dosing of valganciclovir, the quick appearance of GCV in plasma and the high peak levels of GCV indicated a rapid hydrolysis of valganciclovir in vivo.[76] The half-life of GCV following oral dosing of valganciclovir ranges from 3.7 to 4.6 h, which is nearly identical to that after intravenous GCV.[76,77]

Two genetic loci are found to be involved in resistance to GCV in HCMV,[78] namely *UL97* protein kinase involved in initial phosphorylation and *UL54* encoding for HCMV DNA polymerase. In patients on short-term GCV therapy, resistance is predominately at the UL97 locus, resulting in a 5–12-fold increase in the IC_{50}. However, in patients with prolonged treatment, mutations in the *UL54* gene in addition to existing mutation in *UL97* are observed.[79] As expected *UL54* mutations associated with GCV also show cross-resistance to other antiherpes substances.

7.11.1.2.5 Cidofovir

Cidofovir (*S*-HPMPC), *S*-*N*-(3-hydroxy-2-phosphonylmethoxy)propylcytosine (**7**), is a pyrimidine nucleotide analog and specific inhibitor of viral replication. The compound is active against a variety of herpesviruses (EC_{50} = 5.4, 2.3, and 0.2 μg mL^{-1} against HSV-1, HSV-2, and HCMV in plaque assays).[80,81] It is used in clinic for the treatment of HSV infections and CMV infections in acquired immune deficiency syndrome (AIDS) patients. HPMPC diphosphate is the active species that inhibits the viral DNA polymerase, which is formed by the catalysis of cellular kinases and independent of viral kinases.[82] As shown in **Figure 3**, HPMPC is converted to HPMPC-pp by pyrimidine nucleoside monophosphate kinase and diphosphate kinase.[83–86] HPMPC is also possibly converted to HPMPC-p-cholin, which could function as an intracellular reservoir of HPMPC. This may explain the long intracellular half-life (17–65 h) of cidofovor and the long-lasting antiviral effect. HPMPC is incorporated into viral DNA, inhibiting DNA synthesis of a wide variety of herpesviruses.[87] For full termination of viral DNA synthesis, the incorporation of two consecutive HPMPC molecules is needed.[86] In contrast to ACV, GCV, or PCV, where the selectivity results partly from the preferential phosphorylation by virus-coded TK or protein kinase in infected cells, the selectivity of HPMPC comes mainly from the difference of its affinities between the viral DNA polymerase and the cellular DNA polymerases. Nevertheless, HPMPC has a high selectivity index >1000 (viral DNA synthesis versus cellular DNA synthesis).

HPMPC has a plasma half-life of 3–4 h in human after intravenous dosing and is excreted renally 90% unchanged. A prodrug, cyclized HPMPC (cHPMPC) (**45**), was reported to give comparable in vitro and in vivo activity. This prodrug is converted intracellularly to the active HPMPC and related phosphorylation metabolites and it has less degenerative kidney toxicity in rats,[88,89] which is the main safety concern for HPMPC.

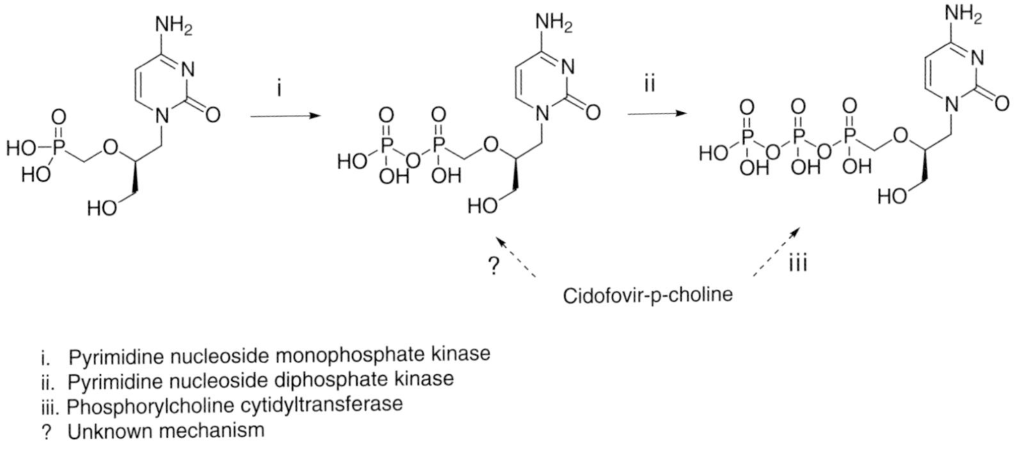

i. Pyrimidine nucleoside monophosphate kinase
ii. Pyrimidine nucleoside diphosphate kinase
iii. Phosphorylcholine cytidyltransferase
? Unknown mechanism

Cidofovir-p-choline

Figure 3 Activation of cidofovir.

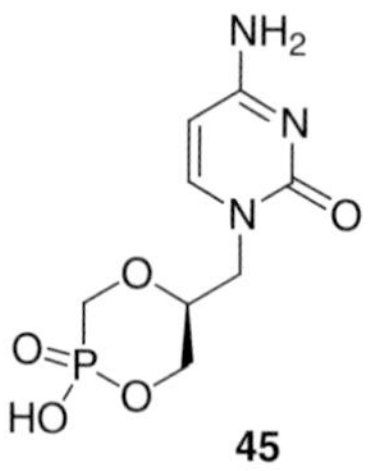

45

The SAR on HPMPC and phosphonate acyclic nucleoside analogs has been the subject of an excellent review.[90] The phosphonate is regarded as an isostere of monophosphate and is stable toward the enzymatic hydrolysis. The nucleotides with HPMP side chain phosphonate are active against all DNA virus, including herpesviruses, hepatitis B virsus (HBV), and even retrovirus HIV. The antiviral activity is linked in all cases to (*S*)-enantiomers at the 2-position (**46**).[91] The enantioselectivity is probably due to the specificity of the nucleotide kinase. Similarly, for 3-deoxy side chain (PMP), the (*R*)-enantiomer retains the activity (**47,48**). PME series, formed by removing the hydroymethyl group from the HPMP side chain (**49**), are also active against herpesviruses, though their activity is less pronounced as compared to HPMP series. Exploration on the length of the side chain between the nucleobase and the phosphonate group shows that four atoms are optimal. Increasing or decreasing the length will affect the activity negatively (**50**). Thus, the regioisomer of phosphonate group, 2-hydroxy-3-phosphonomethoxypropyl analogs (**51**), is inactive against herpesviruses. Replacing the 2-oxygen is intolerable. The carbo or thio version of the HPMPC (**52, 53**) results in the total loss of the activity as well.[92] Introducing a double bond to replace the ether bond (**54**) significantly reduces the activity.[93] Changing 1-carbon to oxygen forming an O–N bond (**55**) will abolish the activity.[93] While HPMPC powerfully inhibits the herpesvirus, an introduction of methyl group on the 5-position of cytosine (**56**) can abolish its antiviral activity. Other pyrimidine base modifications like uracil (**57**) or thymine have no antiviral activity. Purine base replacements are tolerable, especially guanine (**60**), 2-6-diamino purine (**59**), and adenine (**58**). Surprisingly, (*S*)HPMPG (**60**) and (*R*)HPMPG (**61**) are both very potent against various herpesviruses, i.e., the stereospecificity seen in HPMPC structures is not observed.[91]

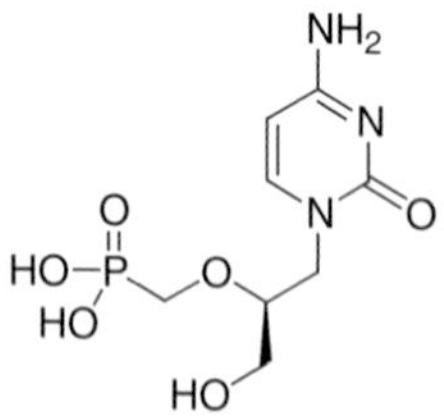

Cidofovir (**7**) 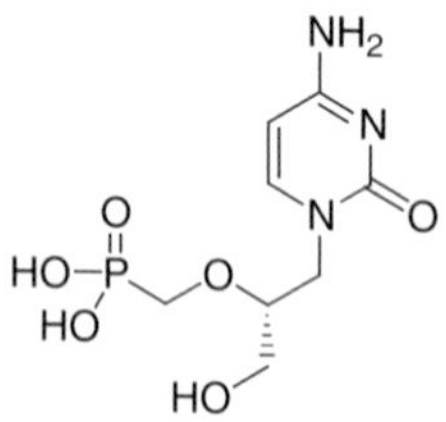 **46** 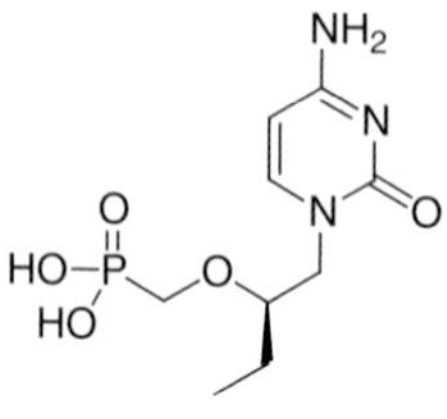 **47** 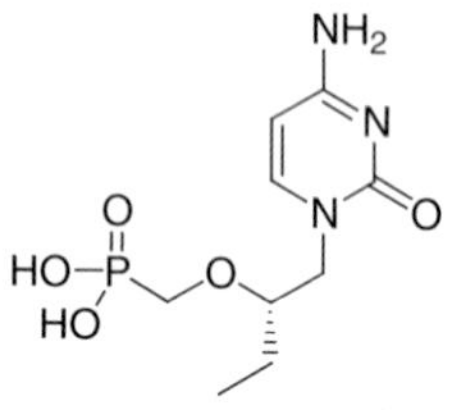 **48**

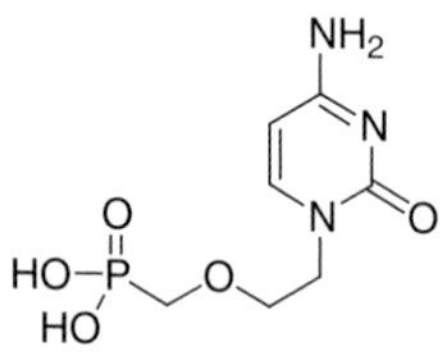

49 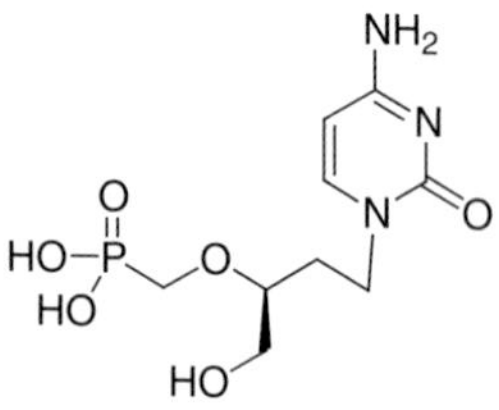 **50** 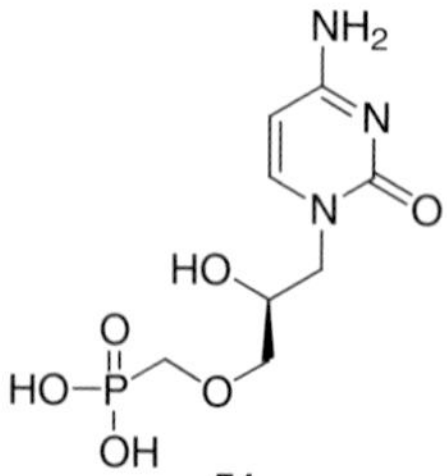 **51** 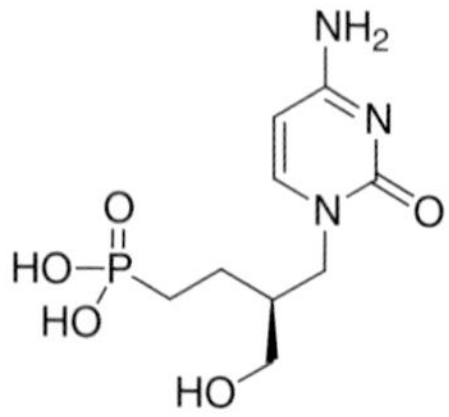 **52**

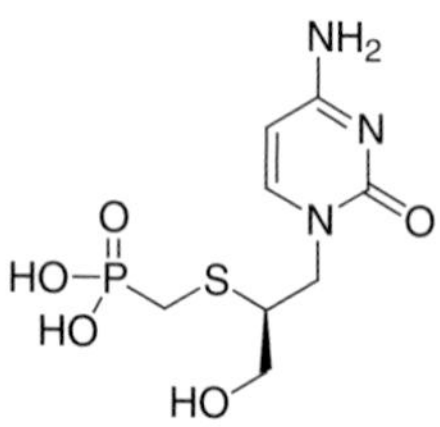

53 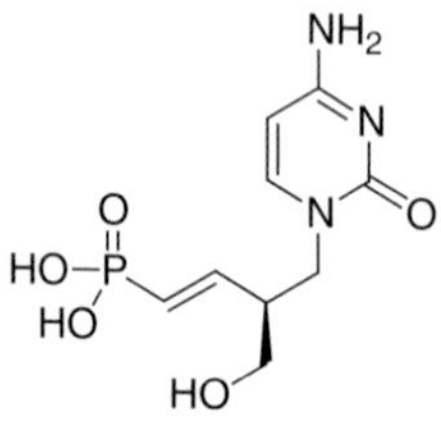 **54** 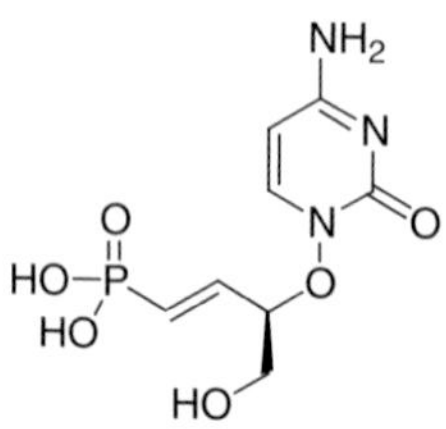 **55** 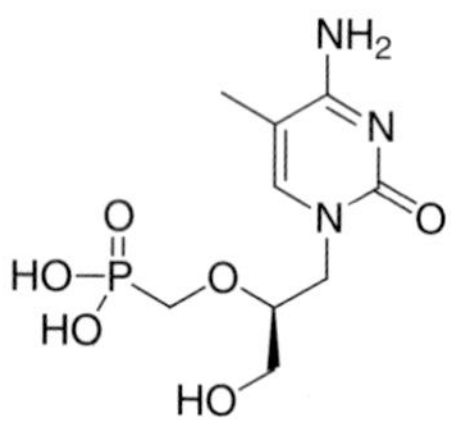 **56**

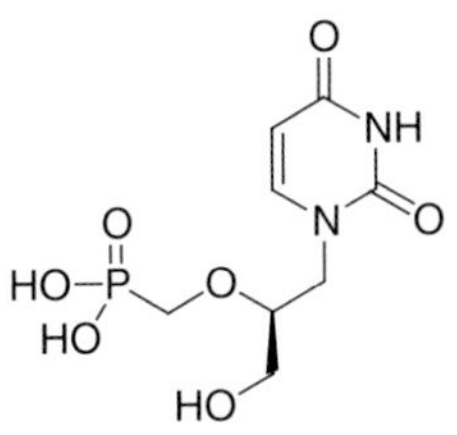

57 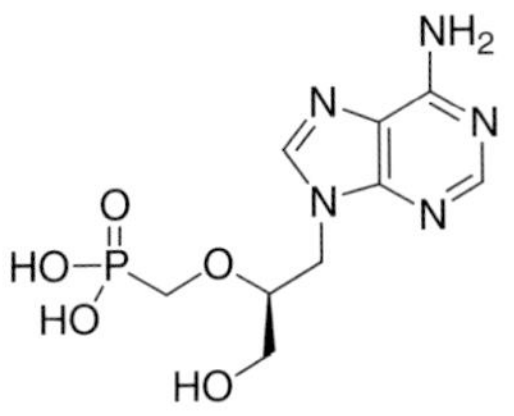 58 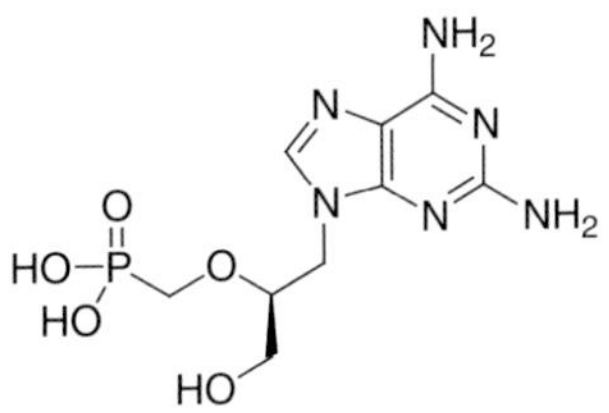 59

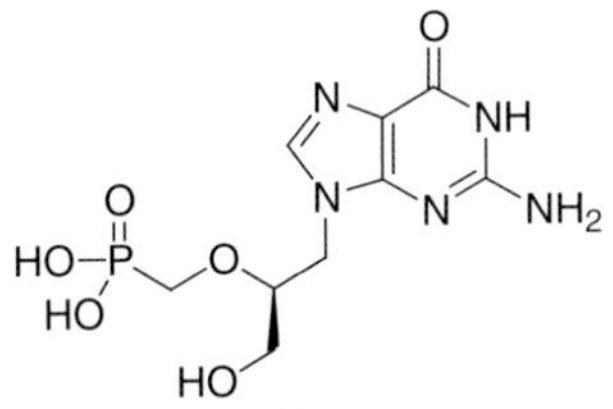

60 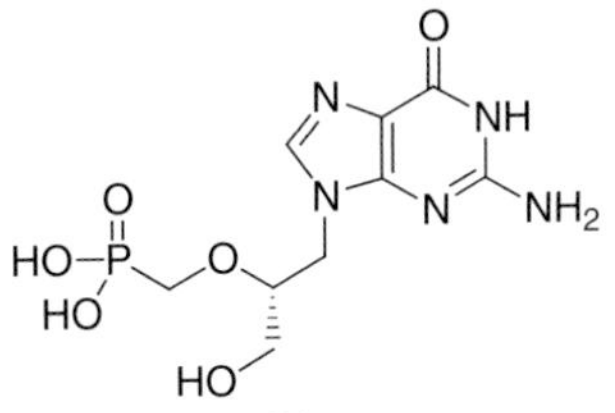 61

Since no viral kinase activity is needed for activation, HPMPC does not show any cross-resistance toward the kinase defective mutant strains. It is reported that HPMPC even enhances susceptibility by 4–20-fold against herpesviruses with altered or deficient viral TK activity.[94] The enhancement of the susceptibility in kinase mutant strain may be due to that the mutant virus induces less dCTP in the infected cells, which competes against HPMPC diphosphate for viral DNA polymerase. Mutant strains that have altered viral DNA polymerase may cause resistance to HPMPC, and even cross-resistance to ACV or GCV.[79]

7.11.1.2.6 Foscarnet

Foscarnet, phosphonoformic acid (**8**), is a potent and broad-spectrum antiherpes agent (IC_{50} = 0.3, 0.5, 0.3, and 0.5 μM for HSV-1, HSV-2, HCMV, and EBV) and is used in the clinic for the treatment of HCMV infection.[95–97] As a pyrophosphate-mimicking analog, it inhibits the viral DNA polymerases with a unique mechanism.[98] Kinetic studies revealed that the inhibition of viral polymerases by foscarnet is noncompetitive with respect to dNTP, uncompetitive with the DNA primer, and competitive with pyrophosphate,[99–102] suggesting that foscarnet blocks the pyrophosphate binding site of the viral DNA polymerase and prevents the cleavage of pyrophosphate from the deoxynucleoside triphosphates. The proposed mechanism is further supported by the observation that foscarnet-resistant HSV-1 DNA polymerase is also less sensitive to the inhibition of pyrophosphate.[103]

Due to its high polarity and low oral absorption, foscarnet is given intravenously, and high plasma concentration can thus be achieved. Foscarnet is not metabolized to any significant extent and is eliminated principally by renal route.[104] Much effort has been made to improve the low oral bioavailability of foscarnet (9%) either by prodrug or formulation approaches;[105–115] however, so far intravenous dosing is still the route used in the clinic.

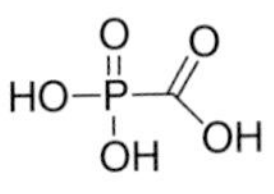

Foscarnet (**8**) 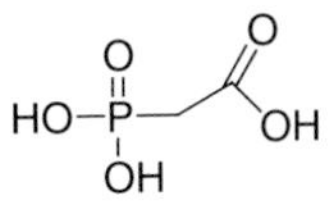 62 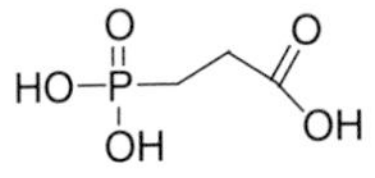 63

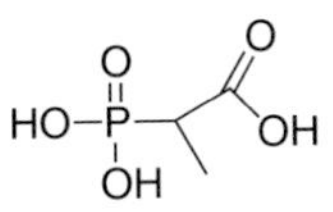

64 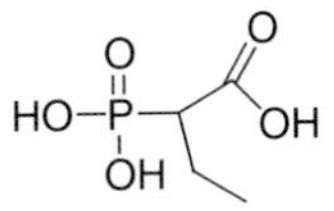 65 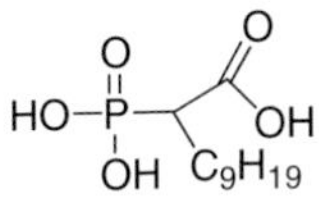 66

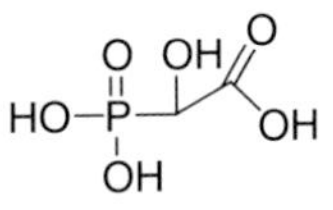

67 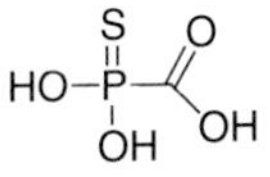 68

Foscarnet is the shortest version of phosphonocarboxylic acids. The modification of the length of the carboxylic acid part has been studied. When the length increases, the antiviral activity diminishes. Phosphonoacetic acid (PAA, **62**) retains similar activity against HSV-1, HSV-2, HCMV, and EBV.[95] Phosphonopropanoic acid (**63**) loses all activity. Substitution on the alpha-carbon of PAA is tolerable. Methyl, ethyl, and even bigger alkyl substitutions are acceptable (**64**, **65**, **66**), though not as active as foscarnet and PAA.[116] Alpha-hydroxy PAA is about 10-fold less active (**67**), but it is still highly active in inhibiting herpes DNA polymerases. Thiophosphonoformate (**68**), where one oxygen on the phosphorus is replaced by sulfur, retains its activity against HSV.[117,118]

In vitro studies shown that serial passage of HSV-1 in the presence of foscarnet will produce drug-resistant mutants.[99,103] As expected the resistance is due to the mutation of the viral DNA polymerase, and several sites have been found to confer this resistance. Clinically, drug resistance of HCMV did not appear to be the cause of a lack of response to foscarnet in immunocompromised patients coinfected with HCMV.[119]

7.11.1.3 Other Substances Used in the Clinic

Beside the main drugs mentioned above, there are other antiherpes drugs being used in the clinic, including Vidarabine (Ara-A) (**9**), Idoxuridine (**10**), Trifluridine (**11**), and Brivudine (**12**). Due to the safety concerns, those compounds are used to a lesser extent. Also used in clinic are antiherpes compounds such as docosanol and formivirsen. Their mechanism of action is not based on the inhibition of viral polymerases. Docosanol is a saturated 22-carbon aliphatic alcohol and is used as a 10% topical cream for the treatment of herpes labialis. It is suggested docosanol can inhibit fusion between host cell plasma membrane and the HSV envelop, thus blocking the viral entry.[120] Formivirsen is an antisense oligonucleotide with phosphothioate linkage in the backbone and is complementary to the mRNA transcript of HCMV encoding proteins responsible for gene expression. Due to this unique mechanism, formivirsen retains activity against ganciclovir- or HPMPC-resistant strains. The substance is used topically for the treatment of CMV retinitis in immunocompromised patients who are unresponsive to GCV, HPMPC, and foscarnet.

7.11.1.4 Substances Under Development

7.11.1.4.1 (*R*)-9-(4-Hydroxy-2-hydroxymethylbutyl)guanine

(*R*)-9-(4-Hydroxy-2-hydroxymethylbutyl)guanine (H2G, **69**) is a potent inhibitor of HSV-1 ($EC_{50}=0.8\,\mu M$), HSV-2 ($EC_{50}=2\,\mu M$), VZV ($EC_{50}=0.9\,\mu M$), and EBV ($EC_{50}=0.9\,\mu M$).[121] The antiviral activity is only seen in the (*R*)-isomer, and the (*S*)-isomer (**70**) is essentially inactive.[121] H2G is mainly phosphorylated by the viral TK, which gives high selectivity for the virus-infected cells. In the uninfected cell, only small amount of H2G triphosphate can be detected, which is due to the action of mitochondrial deoxyguanosine kinase. However, the level of H2G-TP in uninfected cell is about 200–2000-fold less than that in the herpesvirus-infected cells.[122] H2G is not an obligate chain terminator, but it only supports limited DNA chain extension. Compared to ACV, H2G has a superior activity against VZV. In plaque reduction assays, the EC_{50} for H2G is 60–400-fold lower than that of ACV, depending on the virus strains.[123] The substantially enhanced efficacy of H2G can be partly explained by the extremely high H2G-TP concentration compared to ACV-TP concentration (>170-fold). Meanwhile, H2G-TP has a longer half-life in the VZV infected cell (3–7 h) while ACT-TP has a half-life of about 1–2 h under similar assay conditions.[122] Due to the excellent potency against VZV, H2G is in development primarily for the treatment of VZV infection. Similar to the other antiherpesvirus acyclic guanines, H2G has an unfavorable pharmacokinetic profile, with an oral bioavailability in rat below 10%. Various prodrugs of H2G have been studied, and 9-(2-stearoyloxymethyl-4-valyloxybutyl)guanine (valomciclovir) (**71**) showed a high oral bioavailability of 56%. It is interesting to note that its regioisomer 9-(2-valyloxymethyl-4-stearoyloxybutyl)guanine can also reach an oral bioavailability over 50% in rat (**72**).[124,125]

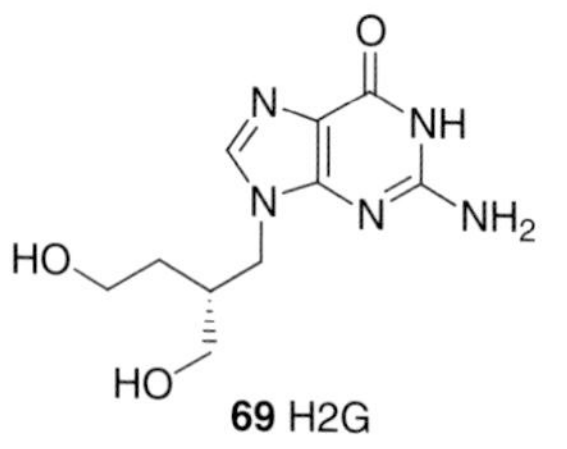

O N NH N N NH2 HO HO

69 H2G

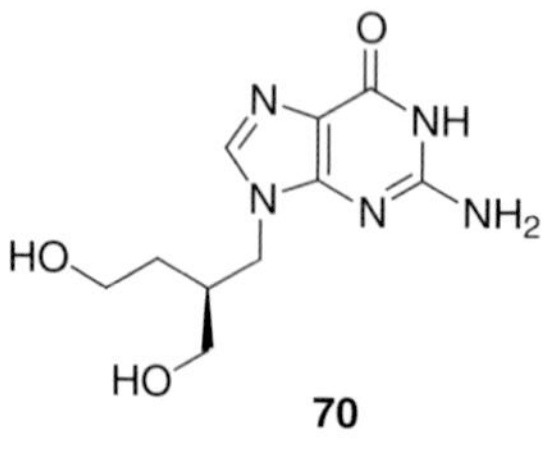

O N NH N N NH2 HO HO

70

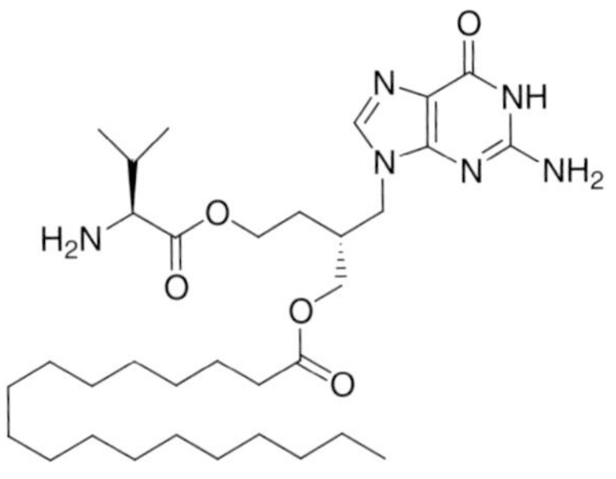

71 Valomciclovir

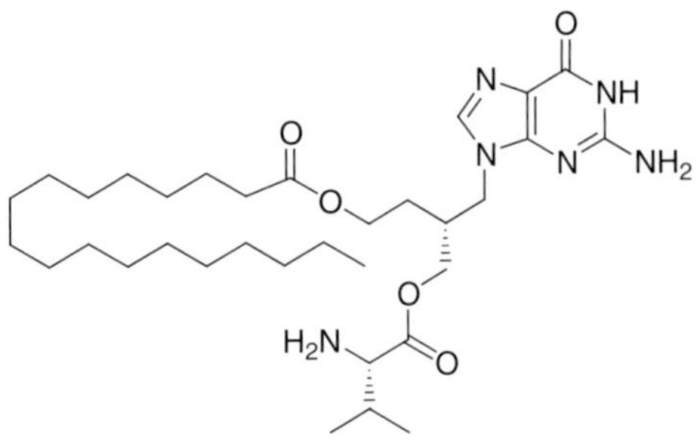

72

Several factors contribute to the success of H2G in treating shingles caused by recurrent VZV infection in clinical phase II studies, namely the high potency of H2G against the VZV, the high plasma concentration achieved by oral dosing of valomciclovir, the high level of phosphorylation, the high concentration of H2G-TP in infected cells, and the long half-life of H2G-TP. The drastic inhibition of the virus leads to the fast healing of the lesion, which may give the possibility of reducing the postherpetic neuralgia.

7.11.1.4.2 Bicyclic pyrimidine nucleoside analogs

A series of bicyclic pyrimidine nucleoside analogs (BCNA) are reported to be very potent anti-VZV agents. The most potent compound, CF1743 (**73**), is reported to have an $EC_{50} = 0.0001\ \mu M$.[126,127] Given its low cytotoxicity, the selective index reaches as high as 10^6. One unique characteristic of the class structure is that BCNA only inhibits VZV and has no inhibitory effect on the other herpesviruses, including the closely related HSV-1 and HSV-2. BCNA is dependent on VZV viral kinases for the phosphorylation and the BCNA monophosphate can be further phosphorylated to diphosphate by the same enzyme.[128] However, BCNA is not recognized by either the HSV-1 or HSV-2 TKs or the cellular kinases TK1 and TK2. Docking BCNA into HSV-1 TK showed that substitution at the 6-position clashes with residues **167** and **168** of the enzyme, preventing the binding to the enzymes.

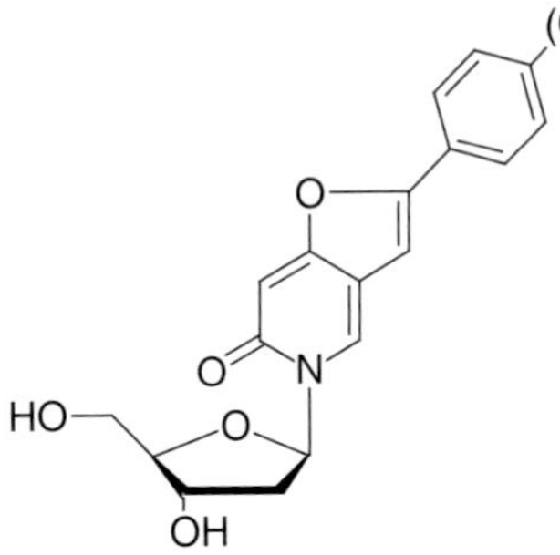

73 CF1743

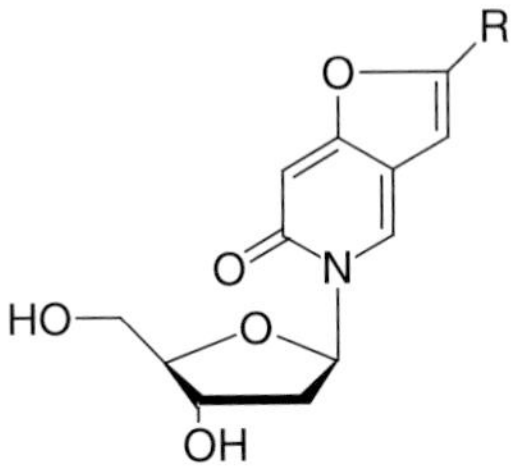

74

R = C8-C10 Alkyl

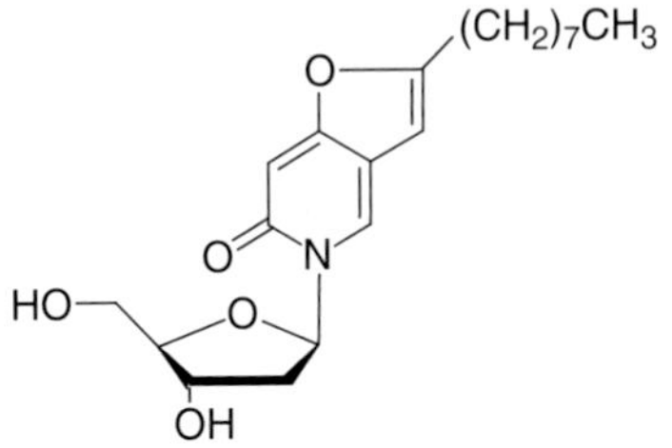

75 CF1368

The substituents on the fused furan ring influence the potency of the BCNA against VZV. Among the alkyl substituents, the C8–C10 long chains give the best activity (**74**). The substitution on the *para*-position of phenyl group (**73**) is important. Both alkyl and alkoxy groups were explored as the *para*-substitution on the phenyl ring.[129–131] Though the long aliphatic chain increases the liphophilicity of the molecule and reduces its water solubility, with proper formulation a high oral bioavailability (>50%) was reported in mice for CF1368 (**75**).[132]

BCNA compounds are currently under further investigation.

7.11.1.4.3 Nonnucleoside viral deoxyribonucleic acid polymerase inhibitors

The success of the HIV nonnucleoside reverse transcriptase inhibitors has certainly its impact on the development of antiherpes agents. A series of quinoline carboxamides was reported. Compound PNU-183792 (**76**) is a broad-spectrum antiherpes agent, with IC_{50} of 3.5, 3.4, 0.34, 0.9, 0.2, and 3.1 μM against HSV-1, HSV-2, VZV, HCVM, EBV, and HHV-8, respectively.[133–135] The antiviral effect is parallel to its inhibition of the viral DNA polymerase, which confirms its mechanism of action. Further evidence of the action on polymerase is provided by the isolation of a resistant strain with a mutation of V823A, located in domain III of those viral polymerases. PNU-183792 is active against all members of herpes family except HHV-6 and HHV-7. In HHV-6, the equivalent residue 823 is an alanine, which can explain to some extent the insensitivity of the HHV-6 toward the compound. PNU-183792 was developed from a lead compound naphthalene amide (**77, 78**). The substitution on the naphthalene ring improves the potency and the hydroxy at *ortho*-position of carboxamide give good result. Using quinoline to replace naphthalene (**79**) improves the pharmacokinetic properties of the compound, presumably due to the enhanced aqueous solubility of the molecules. Those nonnucleoside polymerase inhibitors are not cross-resistance with the nucleoside inhibitors currently in clinical use, their resistant strains are in some cases even hypersensitive to the nucleoside inhibitors,[135] providing a possibility of combination or alternative therapy.

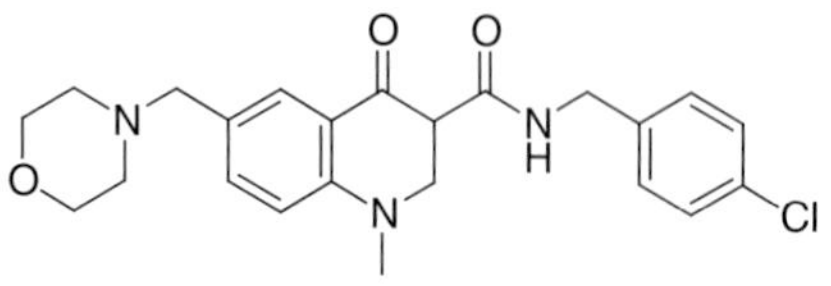

76 PNU-183792

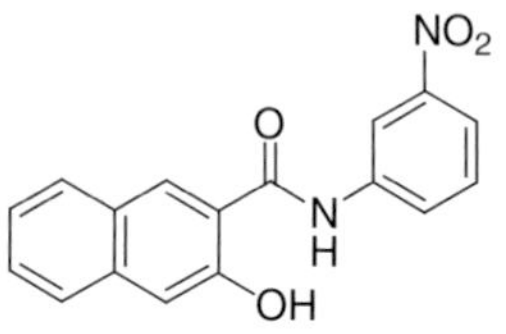

77

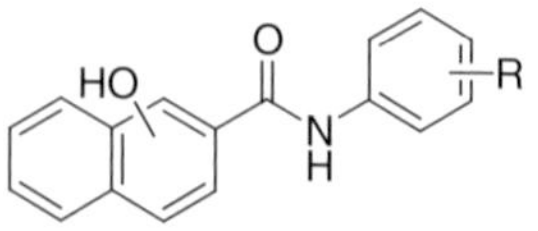

78

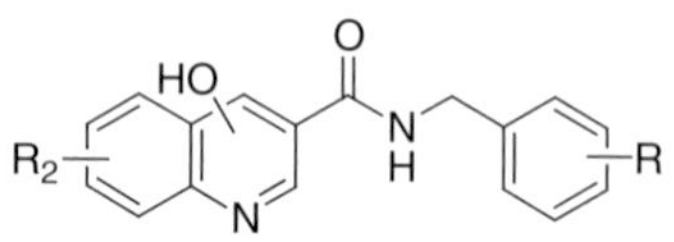

79

7.11.1.4.4 Antiherpes compounds based on the mechanisms other than the inhibition of viral deoxyribonucleic acid polymerases

7.11.1.4.4.1 Inhibition of human cytomegalovirus deoxyribonucleic acid cleavage/packaging

Benzimidazole ribosides were first reported by Drach and Townsend. TCRB, 2,5,6-trichloro-1-(β-D-ribofuranosyl)-benzimidazole (**80**) is a potent inhibitor against HCMV ($EC_{50} = 2.9\,\mu M$).[136,137] Though the structure of the compounds has some similarity to the nucleoside analogs, the mechanism of inhibition is different from the other nucleoside herpesvirus inhibitors. The compound inhibits the cleavage of concatemeric viral DNA in cell culture, thus preventing the packaging process.[138,139] The resistance of HCMV to benzimidazole ribosides maps to two gene products *UL89* and *UL56*, encoding proteins involved in DNA maturation. BDCRB (**81**), another member of the class, was shown to inhibit the nuclease activity of *UL89* gene product and ATPases activity of UL56 gene product.[140,141] Although TCRB shows good inhibition of HCMV in cell culture, it has an unfavorable pharmacokinetic profile, mainly due to that the quick cleavage of glycosidic bond results in a short half-life in plasma.[142] The attempt to improve its pharmacokinetic properties led to compound GW273175X (**82**), which entered clinical studies; but its progression has not been reported lately. Other attempt to reduce the glycosidic bond cleavage of TCRB in vivo is to use indole (**83**) or pyrazolo(3,4-β) indole (**84**) instead of benzimidazole. Compound **84** is several-fold more potent; however, the cytotoxicity was also increased.[143,144]

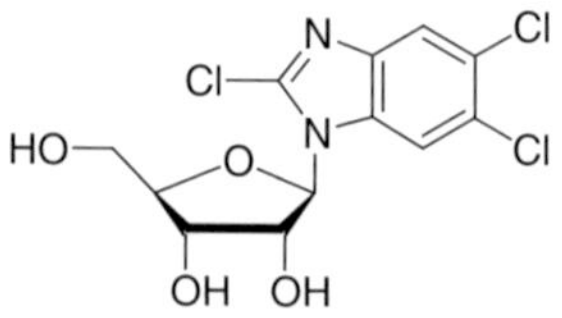

80 TCRB

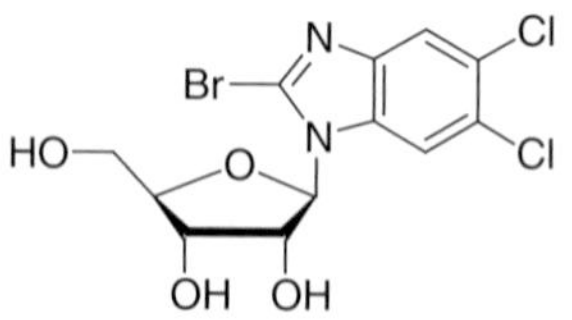

81 BDCRB

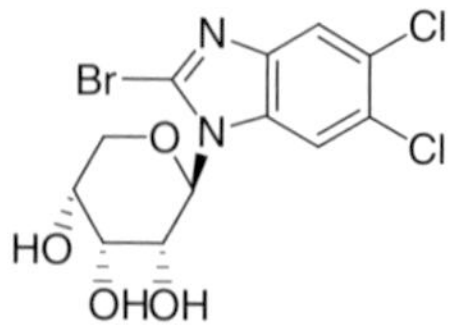

82 GW273175X

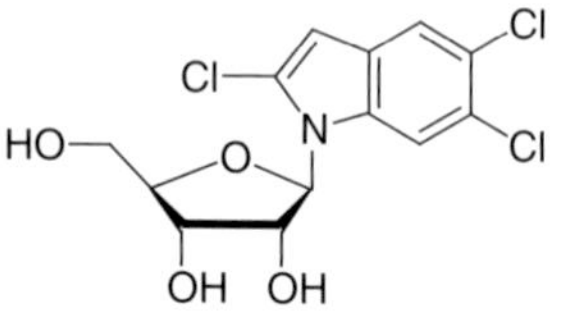

83

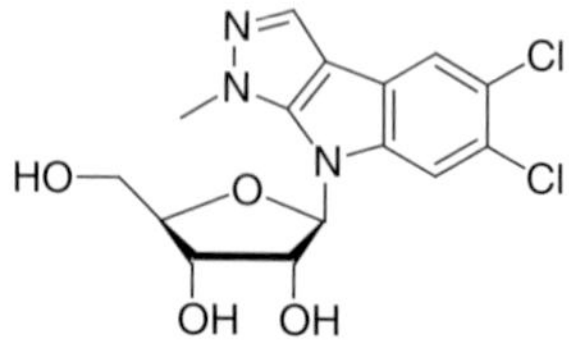

84

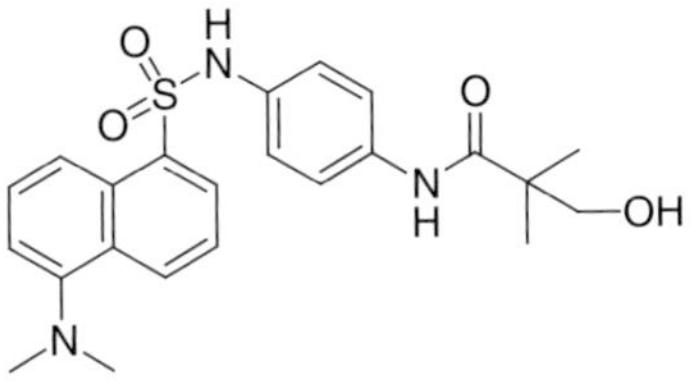

85 BAY38-4766

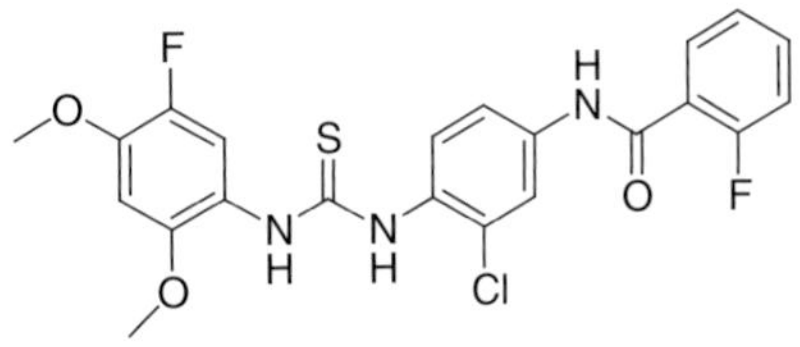

86 WAY-150138

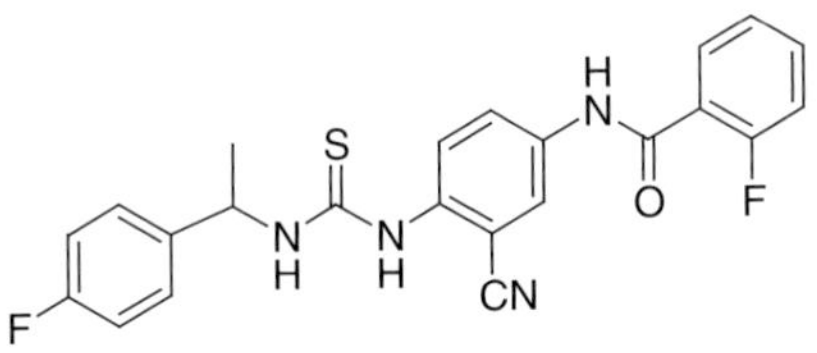

87

Another class of compounds, phenylenediamine sulfonamide (BAY38-4766, **85**) inhibits HCMV by a similar mechanism, namely preventing the cleavage of concatemeric viral DNA.[145] The mechanism is supported by the finding of the resistant mutation in the *UL89* and *UL56* genes. Interestingly, BAY38-4766 does not show any cross-resistance with benzimidazole ribosides.[146] The structure of BAY38-4766 differs considerably from benzimidazole riboside, hence the molecular mode of action may be different between the two classes of compounds in spite that they both act on the same target proteins. The phase I clinical study of BAY38-4766 showed that the compound was well tolerated and high plasma concentration could be reached.

Some bis-substituted thiourea (**86**, **87**) were reported to inhibit the concatemeric virus DNA cleavage in HSV-1 and VZV.[147,148] They are targeting at the portal protein of the viruses, UL6 protein in HSV-1 and ORF 54 in VZV.

7.11.1.4.4.2 Inhibition of the helicase/primase complex

The helicase/primase complex is known to be essential for DNA replication in herpesviruses.[149–151] The helicase activity separates or unwinds duplex viral DNA and the primase activity synthesizes ribonucleic aicd (RNA) primers on the single-stranded DNA upon which DNA synthesis can be initiated by viral DNA polymerases. Inhibition of those activities can block the viral replication.[152–155] The HSV helicase/primase complex consists of three proteins that are the gene products of *UL5*, *UL8*, and *UL52* respectively. The proteins contain intrinsic DNA helicase, RNA primase, and single-stranded DNA stimulated ATPase activities.[151,156] Several structural classes have been reported to be effective inhibitors of the herpesvirus helicase/primase complex.

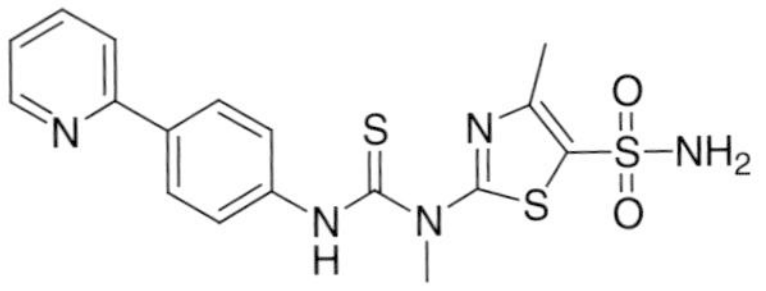

88 BAY57-1293

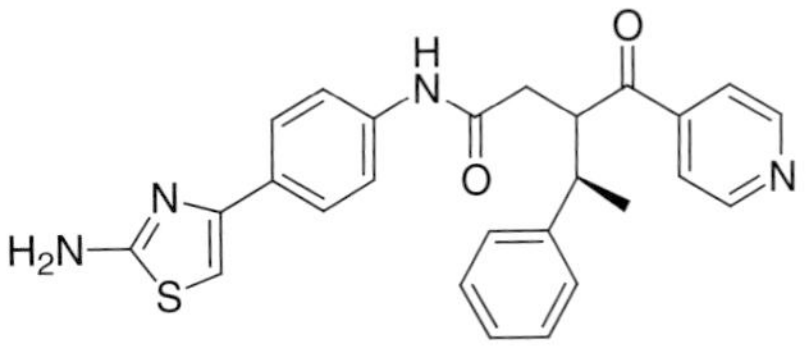

89 BILS 179 BS

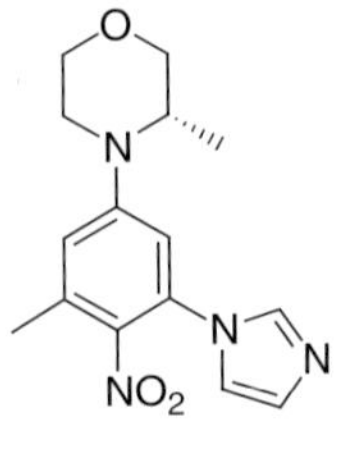

90 T0902611

Kleymann *et al.* reported thiazole urea derivatives that inhibit both HSV-1 and HSV-2. Compound BAY57-1293 (**88**)[157–160] has an EC_{50} of 0.02 and 0.02 μM against HSV-1 and HSV-2 in cell culture. Mechanistically, BAY57-1293 inhibits single-stranded DNA-stimulated ATPase activity. As expected, the thiazole urea compounds are active against

ACV-resistant viruses. Resistant mutants can be selected against BAY57-1293 at a very low frequency, with a major mutation on *UL5* (K356N for HSV-1 and K355N for HSV-2).[157] The SAR study showed that the pharmacophore of this class of compounds consists of sulfonamide, central carbonyl, and the phenyl ring. BAY57-1293 exhibits low toxicity.[161] When given as an oral dosing, the compound reaches an oral bioavailability over 60% in both rats and dogs, and shows high efficacy on the healing of HSV-1 infection in mice and HSV-2 infection in guinea pig.[162]

BILS 179 BS (**89**)[150] acts on the helicase/primase complex of HSV-1 ($EC_{50} = 0.03\,\mu M$) and HSV-2 ($EC_{50} = 0.1\,\mu M$). Differing from the action mode of BAY57-1293, BILS 179 BS enhances the affinity between the helicase/primase complex and DNA, which prevents the propagation of the catalytic cycle by the enzyme complex. BILS 179 BS is active orally. In a murine model for cutaneous HSV-1 infection, it shows a dose-dependent reduction of disease scores. Even dosed 65 h after infection, BILS 179 BS can still substantially reduce the HSV pathology, whereas ACV does not have any significant effect in this case.

T0902611 (**90**) is a helicase/primase inhibitor specific for HCMV. In cell culture it is about 30-fold more potent than GCV.[163,164] T0902611 modifies the HCMV primase component of the helicase/primase complex.[165] The compound entered phase II clinical trials and was discontinued later.

7.11.1.4.4.3 Inhibition of human cytomegalovirus UL97 protein kinase

Maribavir is an L-riboside (**91**) against HCMV ($EC_{50} = 0.12\,\mu M$) which was found in the modification of benzimidazole ribosides. Interestingly, it acts through a different mechanism in inhibiting HCMV as BDCRB (**81**) though it bears a certain structural similarity to BDCRB.[166,167] Maribavir inhibits UL97 viral protein kinase with an IC_{50} of 3 nM, hence the viral replication. The compound is inactive against HCMV DNA polymerase in biochemical assays. As expected, maribavir is effective in inhibiting GCV-resistant HCMV isolates.

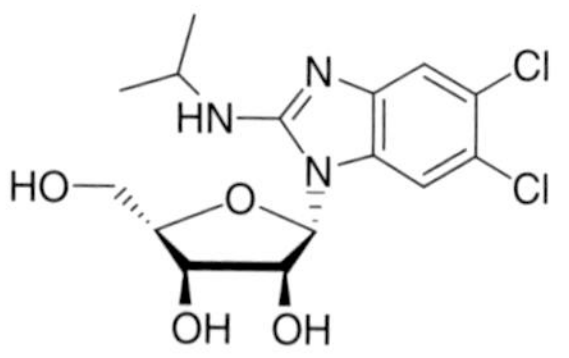

91 Maribavir

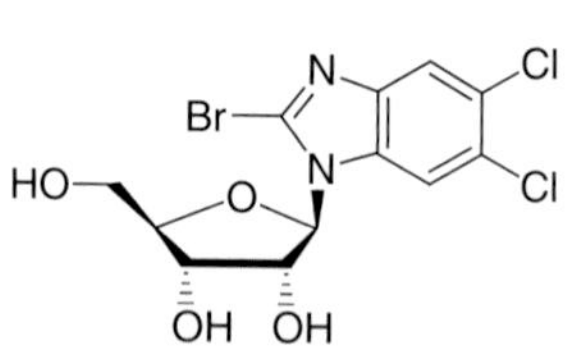

81 BDCRB

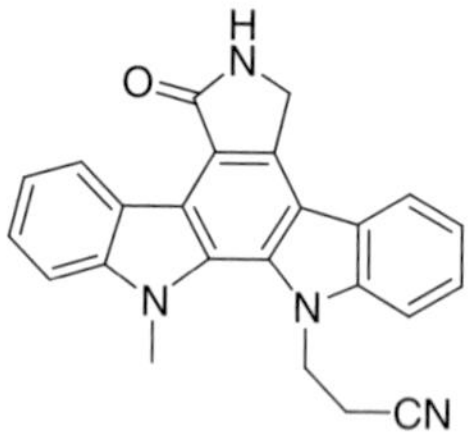

92 (Gö6976)

A phase I clinical trial revealed that maribavir is quickly absorbed after oral dosing and the compound was well tolerated. Although maribavir has a high serum protein binding (over 97%), sufficient free-drug concentration can be achieved. In HCMV patients coinfected with HIV, maribavir showed in vivo anti-HCMV activity and substantially reduced the HCMV viral load.[168]

Indocarbazole derivatives (**92**) were reported to be HCMV protein kinase inhibitor.[169] They strongly inhibit both autophosphorylation of UL97 protein and UL97 protein kinases activity. Gö6967 has an EC_{50} of about $0.05\,\mu M$ against HCMV in cell culture and is essentially inactive against HSV-1.

7.11.1.4.4.4 Inhibitors of the proteases of herpesvirus

Human herpesviruses encode a serine protease which has the function of maturational processing of the assembly protein scaffold. The processing is essential for the packaging of viral DNA, thus for viral replication.[170,171] The herpesvirus proteases represent good targets for the development of antiherpes agent,[172] and three-dimensional structures of the serine proteases of HCMV, HSV-1, HSV-2, and VZV have been reported.[173–179] The proteases are active in dimer form and the active site is composed of a His–His–Ser catalytic triad, which differs from that of most other serine proteases.[180]

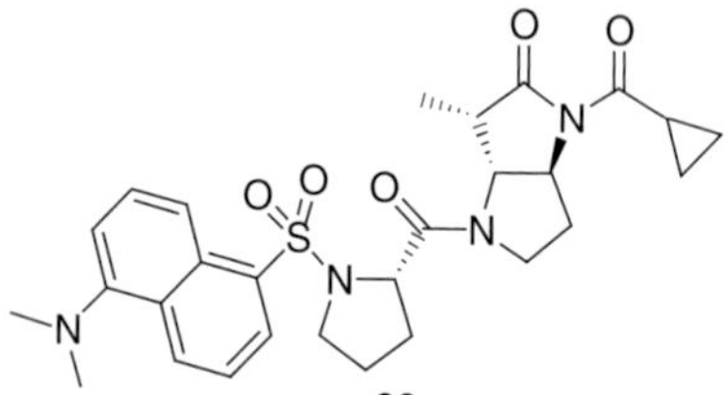

93

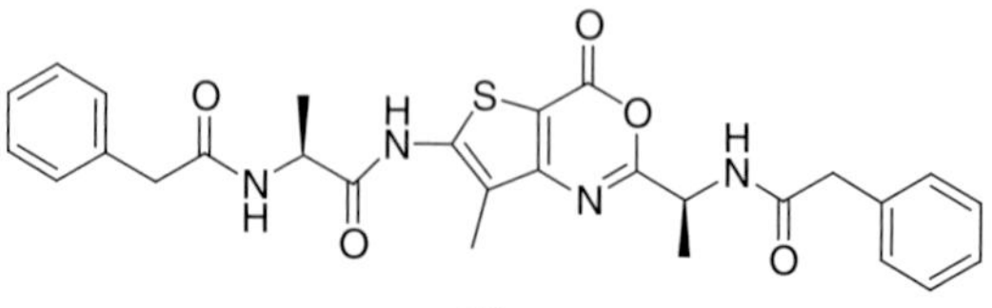

94

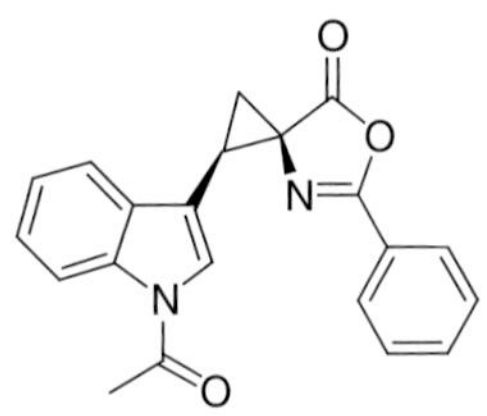

95

A series of pyrrolidine-5, 5-*trans*-lactams (**93**) was reported to be active against HCMV protease. Starting from a natural product extracted from a plant, the optimization leads to a compound which inhibits HCMV protease with a K_i of 34 nM and IC_{50} of 0.5 μM against HCMV in cell assay.[181–183] Meanwhile, it shows a good selectivity against other serine proteases like thrombin and elastase. Other structural classes (**94**, **95**) are also reported for HCMV protease inhibition. Most of those protease inhibitors are under an early stage of the development.

7.11.1.5 Challenges in Herpesvirus Therapy and the Development of New Antiherpes Agents

Drug resistance in viruses is generally regarded as an acquired heritable change in losing the sensitivity to a particular inhibitor. Unlike the situation in retroviruses or RNA viruses, the incidence of clinical resistance for herpesviruses remains relatively rare.[184,185] As the major antiherpes agents in clinical use are polymerase inhibitors, the main resistant isolates identified bear mutations in viral proteins responsible for phosphorylation of the nucleoside analog or in DNA polymerase.[186–189] Resistance in the normal population is around 0.7%; however, in immunocompromised patients, the resistance rate is reported to be higher, at around 3–5%.[190,191] The availability of a selection of antiherpes agents provides the option for treating resistant strains although it has been reported that certain mutant strains in viral DNA polymerase showed cross-resistance to inhibitors with different mechanism of action.[192] Thus there is a need, albeit not a major one, for alternative therapies.

Of more significance is the production of pathologies triggered by virus replication such as inflammation or pain which is often the major need of patients. Thus, it is highly desirable to have a therapy treating recurrent infection which takes into consideration the pathology of recurrent infections.[193] Recently, a novel concept of combining antiviral and anti-inflammatory preparations in the treatment of recurrent HSV infection has been developed.[194]

Despite the medical need for more antiherpes agents, efforts in developing new therapies have been dampened in recent years. Many pharmaceutical companies have discontinued their research in herpes indications, presumably due to the high threshold of efficacy, safety, and pricing set by ACV or similar compounds which have raised the hurdles of market entry and return on investment. The prospects of effective vaccination programs in Western European countries and the USA have brought about uncertainty over the market for a new anti-VZV compound. It may need a strong incentive to revitalize the commitment in herpes research by the pharmaceutical industry.

7.11.2 Treatment for Hepatitis B Virus Infection

7.11.2.1 Hepatitis B Virus Biology

HBV is a member of the Hepadnaviridae which consists of only four well-characterized members including human, woodchuck, duck, and ground squirrel viruses, although several other animals have been identified.[195] HBV is thought to have infected some 2 billion people worldwide of which some 400 million people are chronically infected and some 1 million are expected to die of the disease each year. Infection rates vary considerably, with very high rates in SE Asia and Africa and very low rates in Europe or the USA. Infection is usually either by sexual contact, by transmission from mother to baby, or may be blood-borne.[196]

HBV resembles retroviruses in that the replication cycle includes a phase in which the genomic sequence changes from RNA to DNA using a virus-coded reverse transcriptase. However HBV differs from HIV in that the input nucleic acid in the newly formed virus particle is an RNA copy transcribed by host RNA polymerases. The viral reverse transcriptase is incorporated into the virus particle and subsequently it is within the virus particle that the viral pregenomic RNA is reverse transcribed into a DNA copy. It is this DNA copy that then goes on to infect further hepatocytes.

The viral genome is relatively small (3.2 kb), even for a virus, and it utilizes alternative initiation codons to produce a number of viral gene products including components of the virus particle and the polymerase.[195] The molecular simplicity of HBV suggests that the polymerase is likely to be the only credible target for the development of inhibitors. One crucial question in the use of antiviral drugs for HBV is the choice of patient population. The outcomes of HBV infection can be highly variable from an inapparant disease to a transient hepatitis which self-resolves. However in about 2–10% of infected patients a chronic infection is established which may lead on to induce pathological changes in the liver including inflammation, fibrosis, and cirrhosis, and a small number of patients may progress to hepatocellular carcinoma.[196]

Although not the subject for discussion here HBV has been treated for several years with interferon with all of its adverse events and disadvantages. More recently pegylated interferon, which was developed to address some of these adverse events in the treatment of hepatitis C virus, has been applied in the treatment of HBV and this may change the usage of interferon significantly.[197]

The development of small-molecule inhibitors, which will be discussed in this chapter, has focused upon nucleoside inhibitors of the HBV polymerase. Given the similarity between the HIV and HBV polymerases it is no surprise that some of the HIV inhibitors also have an effect on HBV; however, that is not always the case due to the subtle differences in the recognition site of the two polymerases. More often the inability of hepatocytes to phosphorylate some nucleosides to their active triphosphates, due to differences in their kinases and other metabolic enzymes, can lead to some compounds such as AZT having activity against HIV but not HBV. One of the first compounds to be developed and tested in the clinic was fialuridine (FIAU, **96**) which was shown to have antiviral effect in trials also induced severe adverse events and led to the death of several patients. It is thought that the toxicity of FIAU is due to effects on mitochondria.[198] After this initial setback it was a long time before a second nucleoside analog, lamivudine, being developed for the treatment of HIV, was shown to have activity against HBV.[199] Drug discovery in HBV is severely limited by available in vitro and in vivo systems. In a true sense it is not possible to grow HBV in the laboratory. This is due to the exquisite selectivity of HBV to replicate in liver cells and the corresponding difficulty in growing and maintaining hepatocytes in an undifferentiated state. The breakthrough came when a construct was made consisting of the whole of the HBV genome and a selectable marker. Several groups have utilized this approach but the 2.2.15 clone of HepG2 transfected cells is the most commonly used.[200] These cells can be monitored for the presence of viral nucleic acid, including closed circular DNA, and viral proteins HBsAG and HBeAG. The main systems used to determine the effect of an inhibitor of HBV are either the woodchuck or duck models using the relevant hepatitis virus and not the human HBV.[195] The results generated in these animal models are not always indicators of the potency of inhibitors in the clinic for human disease.

7.11.2.2 Anti-Hepatitis B Virus Agents in Clinical Use

7.11.2.2.1 Lamivudine

Lamivudine, β-L-2′,3′-dideoxy-3′-thiacytidine (**97**), was the first small-molecular inhibitor launched against HBV. It is used for the treatment of HBV infection as a monotherapy and in combination with interferon. Lamivudine is potent against HBV with an EC_{50} of 10 nM against HBV in HepG2.2.15 cell culture.[199,201] The compound is phosphorylated by a cellular kinase to monophosphate and further to triphosphate by pyrimidine nucleotide kinases and nucleoside diphosphate kinase.[202,203] Lamivudine triphosphate acts as a competitive inhibitor of the HBV DNA polymerase and blocks viral replication through inhibiting the viral polymerase by chain termination.[204] Although lamivudine has an L-configuration, it can be recognized by the cellular kinases which have D-nucleosides as the natural substrates and have a low stereospecificity for substrates. The enzyme dCK catalyses the phosphorylation of lamivudine to its monophosphate. The K_m of lamivudine is about 7 μM, which is comparable to that of the natural substrate deoxycytidine (K_m 8 μM).[205] Lamivudine triphosphate inhibits HBV reverse transcriptase with a K_i of about 21 nM.[205] Mechanistically, it acts on both stages of the HBV reverse transcriptase function, namely the RNA-dependent DNA synthesis and the DNA-dependent DNA synthesis. Lamivudine triphosphate has only low inhibitory effects against cellular DNA polymerases alpha, beta, and delta, which leads to a good safety margin. Although lamivudine triphosphate gives a substantial inhibition against human DNA polymerase gamma, which is the major DNA polymerase in mitochondria, its inhibition of mitochondrial DNA synthesis is limited. At a concentration of 50 μM, lamivudine reduces the mitochondrial DNA concentration by 50% in the lymphoblastoid cell line CEM.[206,207] The low toxicity of lamivudine toward mitochondrial DNA synthesis can be partly explained by the observation that lamivudine is not phosphorylated by mitochondrial deoxypyrimidine nucleoside kinase and that the 3′–5′exonuclease repair activity of DNA polymerase gamma can effectively excise the incorporated lamivudine monophosphate from the DNA chain.[208,209]

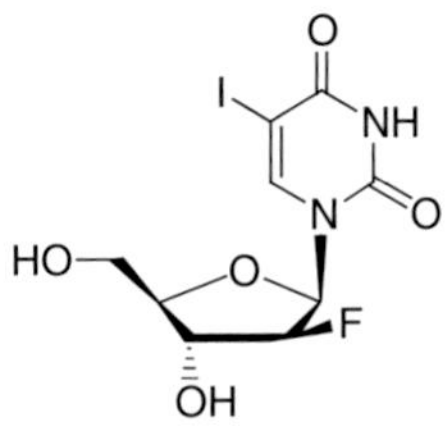

96 FIAU 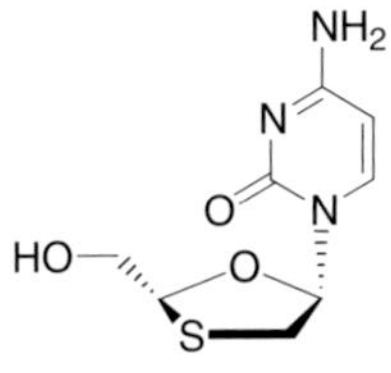 **97** Lamivudine 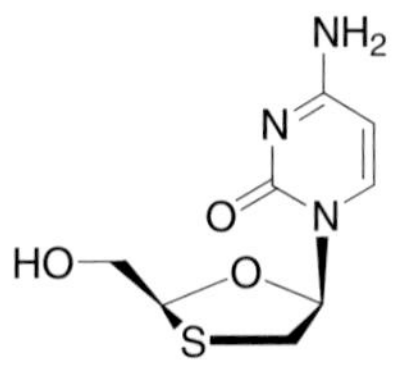 **98** 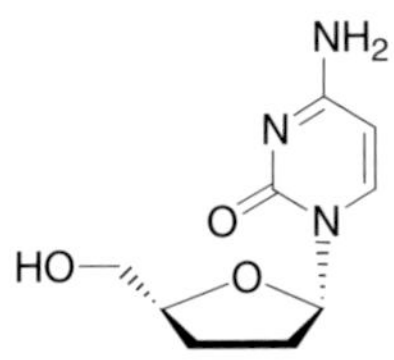 **99** 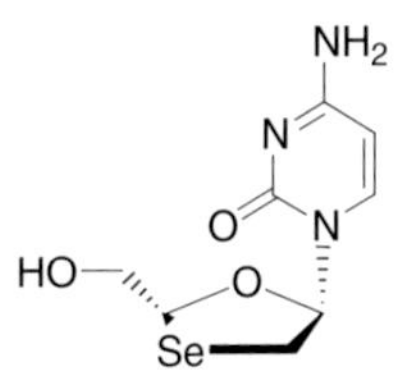 **100**

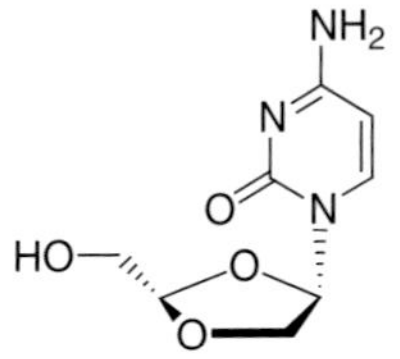

101 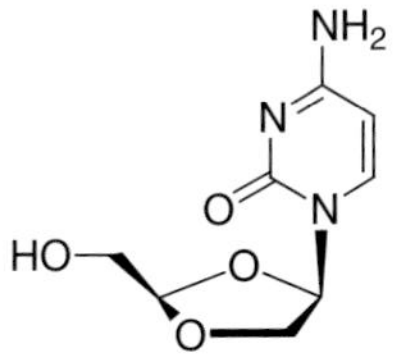 **102** 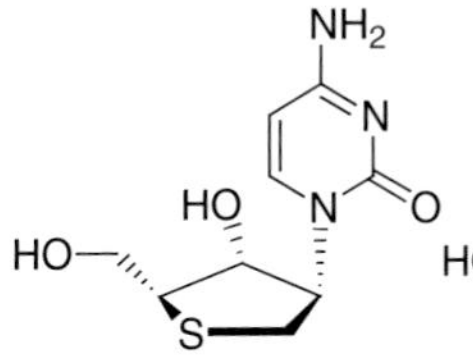 **103** 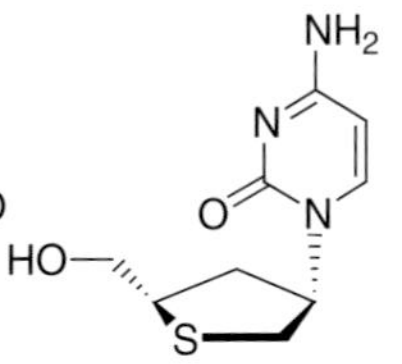 **104**

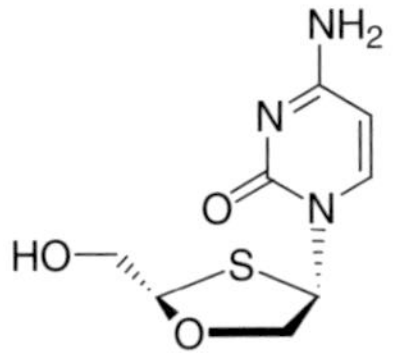

105 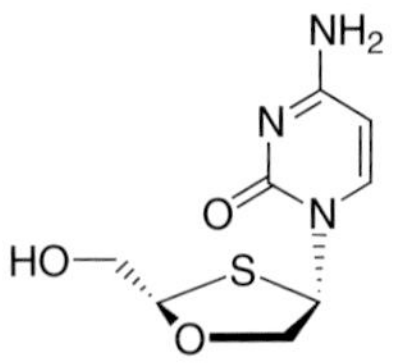 **106** 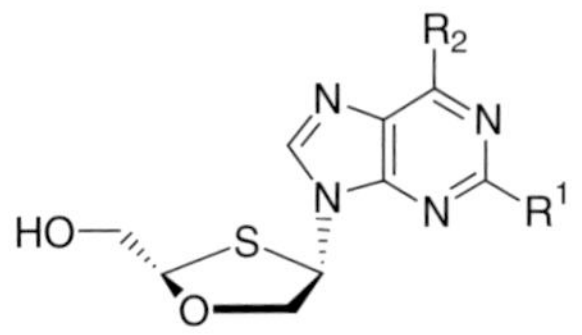 **107**

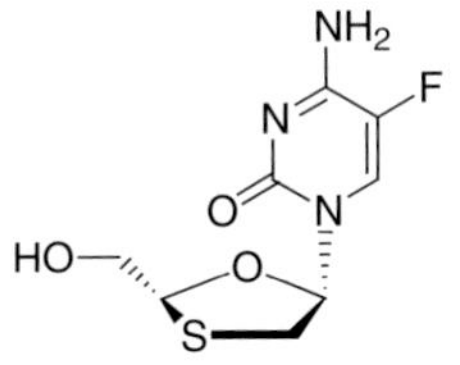

108 Emtricitabine 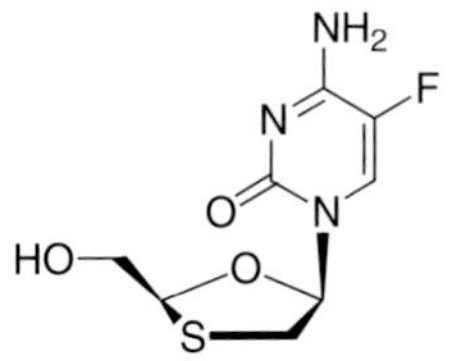 **109** (+)-FTC

Extensive SAR work has been published around lamivudine. The enantiomer of lamivudine, namely β-D-2′, 3′-dideoxy-3′-thiacytidine (**98**), has an inhibitory effect about 50-fold less compared to lamivudine. The D-enantiomer is prone to deamination by deoxycytidine deaminase forming the uracil analog, while lamivudine is completely resistant to the hydrolysis of the enzyme. In CEM cells the D-enantiomer is about 25-fold more toxic for cell growth and about 12-fold more toxic against mitochondrial DNA synthesis.[210] This difference appears to be due to the fact that the D-enantiomer is much more efficiently incorporated into mitochondrial DNA. Lamivudine triphosphate has some threefold higher potency against HBV DNA polymerase compared to the triphosphate of D-analog.[206] The sulfur atom at the 3′-position is important for the anti-HBV activity. Replacing the sulfur with a carbon atom gives L-ddC (**99**) which, while active against HBV, is about 10-fold less potent.[210,211] Chu *et al.* [212] reported an analog of lamivudine with a selenium atom at the 3′-position (**100**), which is isoelectronic with sulfur. The racemic β-oxaselenolane cytidine nucleoside shows an EC_{50} of 1.2 μM against HBV in HepG2.2.15 cells. Using oxygen replacement of the sulfur at the 3′-position, L-dioxolane C (**101**) exhibits extremely high activity against HBV ($EC_{50} = 0.0005$ μM) which is slightly more potent than lamivudine. However, the compound also shows fairly high cytotoxicity in CEM and Vero cell lines. The D-dioxolane C (**102**) has a lower activity against HBV, and, interestingly, the D-isomer is tens of fold less cytotoxic in comparison with its L-isomer.[213] Replacing the oxygen in the oxathiolane ring (**103**, **104**) abolishes the antiviral activity.[214] A series of analogs formed by switching the positions of oxygen and sulfur atoms in the oxathiolane ring

(**105–107**) leads to essentially inactive compounds against HBV, though several compounds in the series have fairly good antiviral effect against HIV.[215] Most purine oxathiolane nucleoside analogs have lower antiviral effect.

Lamivudine has a favorable clinical pharmacokinetic profile. The compound is well absorbed in humans after oral dosing and its oral bioavailablility can reach over 80%.[216] Lamivudine is moderately bound to plasma proteins (36%) and distributes into total body fluid (steady state volume about 1.3 $L\,kg^{-1}$). The half-life is about 8 h. The major route for elimination is urine excretion. About 70% of drug is found to be excreted unchanged in urine and less than 5% of a dose is oxidized to sulfoxide metabolites.[217] The compound has a relatively low CNS penetration. The cerebrospinal fluid concentration is about 6% of the plasma concentration.[218] In human hepatoma cell line (2.2.15) the intracellular lamivudine triphosphate is the major metabolite and the triphosphate has a half-life of about 17–19 h.[202]

7.11.2.2.2 Emtricitabine

Emtricitabine, *cis*-L-5-fluoro-1-[2-(hydroxymethyl)-1,3-oxathiolan-5-yl]cytosine (**108**) ((−)-FTC), is used for the treatment of HIV infection in the clinic. It is also in late stage development for the treatment of HBV infection.[219–221] In the HepG 2.2.15 cell line, emtricitabine was found to be a potent inhibitor of viral replication, with an apparent IC_{50} of 10 nM, while the (+)-isomer (**109**) was considerably less active.[199] Similar to lamivudine, emtricitabine triphosphate is the active form responsible for the antiviral activity. Emtricitabine was found to be a good substrate for human 2′-deoxycytidine kinase with a K_m value of 42.1 μM, with the affinity of the (−)-enantiomer greater than that of the (+)-enantiomer.[199,222]

Emtricitabine is essentially nontoxic to HepG cells and human peripheral blood mononuclear cells (PBMCs). It also shows minimal toxicity in the human bone marrow progenitor cell assay.[199,221] The inhibition against human DNA polymerases alpha, beta, gamma, and epsilon are limited with K_i values of 6.0, 17, 6.0, and 150 μM, respectively.[223] Emtricitabine has a good pharmacokinetic profile. The compound is absorbed rapidly and the oral bioavailability reaches 60–90% in humans. The compound is extensively phosphorylated to the active metabolite triphosphate. The major route of elimination is via renal excretion principally as unchanged drug. It is reported that a small amount of glucuronide ether and a diastereomeric sulfoxide were found in urine.[224] Emtricitabine was not detectably deaminated at either the nucleoside or nucleotide level, whereas (+)-FTC was deaminated by cytidine deaminase at a much faster rate.[225] It is suggested that the difference between the (−)-form and the (+)-form toward the cellular metabolic enzymes is one of the reasons that may account for the different antiviral effects.[226]

7.11.2.2.3 Adefovir dipivoxil

Adefovir dipivoxil, bis(pivaloyloxymethyl)-9-(2-phosphonylmethoxyethyl)adenine or bis-POM PMEA (**110**) is an oral prodrug of adefovir, 9-(2-phosphonylmethoxyethyl)adenine or PMEA (**111**). Adefovir has been shown to have good antiviral activity against several viruses, including HIV and HBV, and herpesviruses, such as HSV, EBV, VZV, and CMV.[227–230] Although the development of the parent compound adefovir has now been discontinued, the prodrug, adefovir dipivoxil, has been approved for the treatment of HBV in the clinic.

Adefovir has an IC_{50} of 0.7 μM against HBV in the HepG2.2.15 cell line.[231] Studies to elucidate the mechanism of action of PMEA against herpesvirus replication revealed that the phosphorylation of the compound occurred intracellularly and was carried out by host cellular enzymes. The diphosphorylated derivatives of PMEA targeted the viral DNA polymerase and also acted as DNA chain terminators.[232] Adenylate kinase was shown to be responsible for the first phosphorylation, which was followed by ADP kinase and creatine kinase, forming adefovir diphosphate.[233,234] Another alternative mechanism was also suggested through the pyrophosphorylation catalyzed by 5-phosphoribosyl 1-pyrophosphate synthetase.[235]

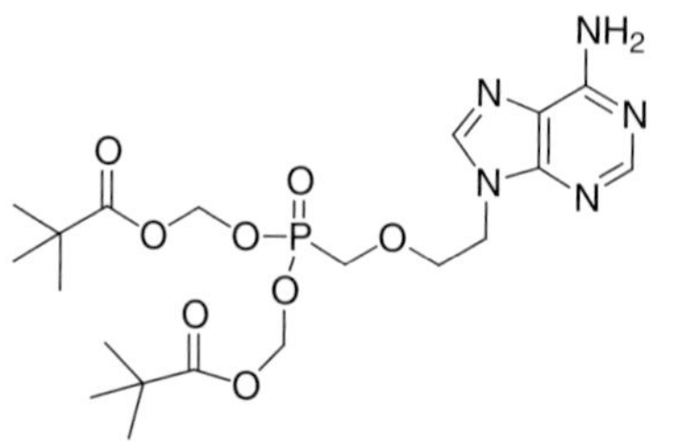

110 Adefovir dipivoxil

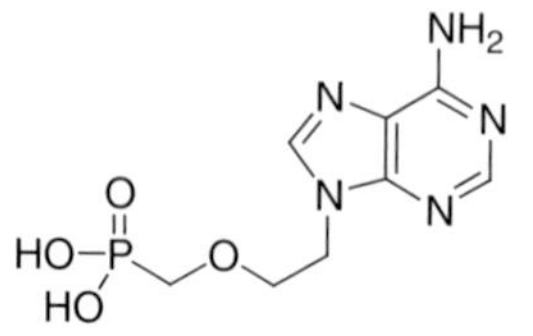

111 Adefovir

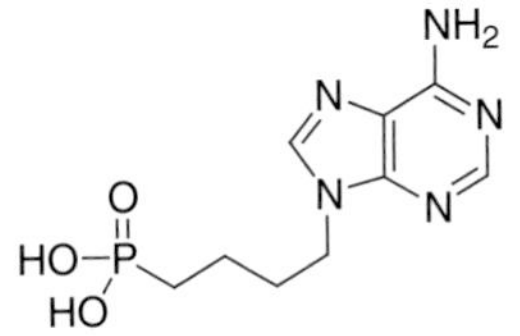

112 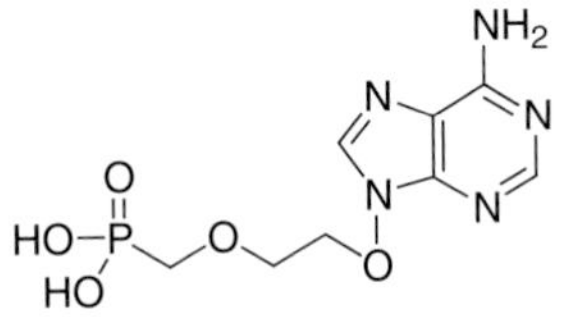 113 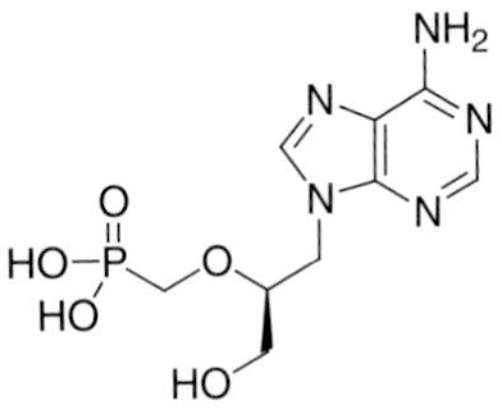 114

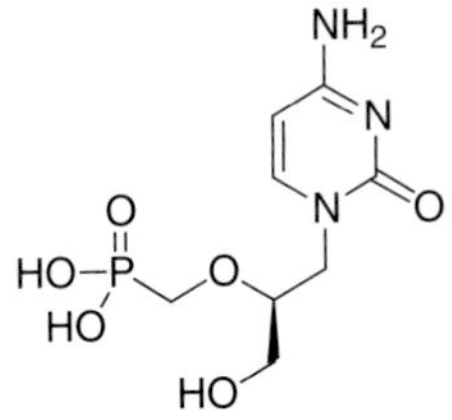

Cidofovir (7) 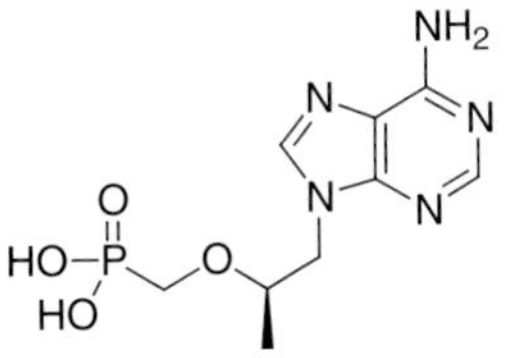 115 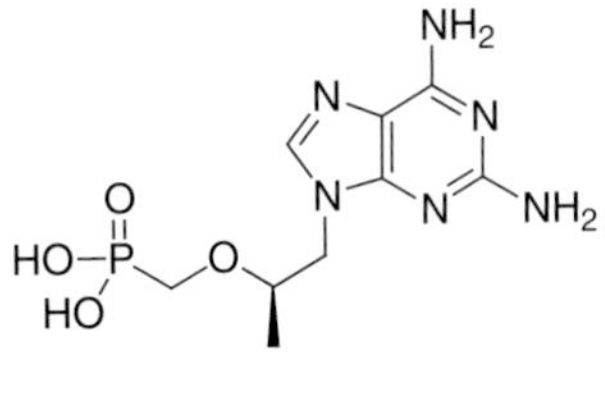 116

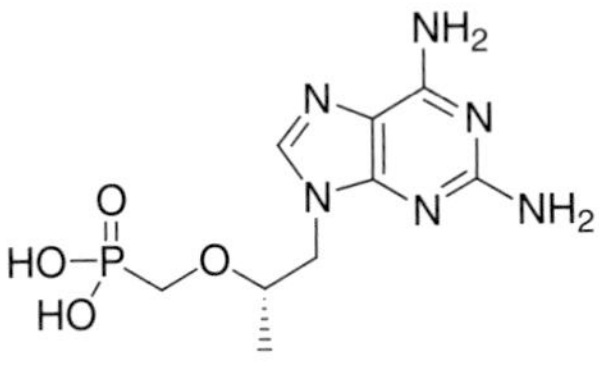

117 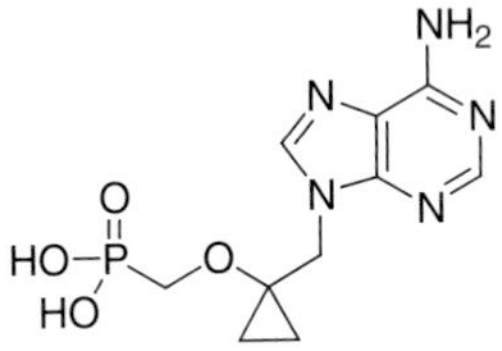 118 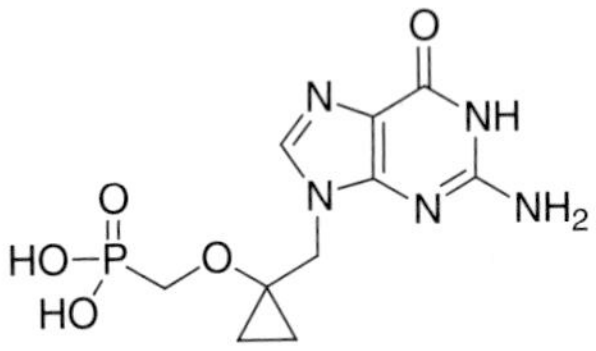 119

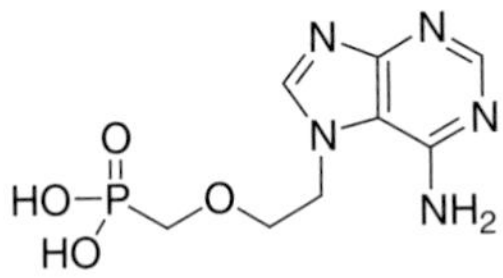

120 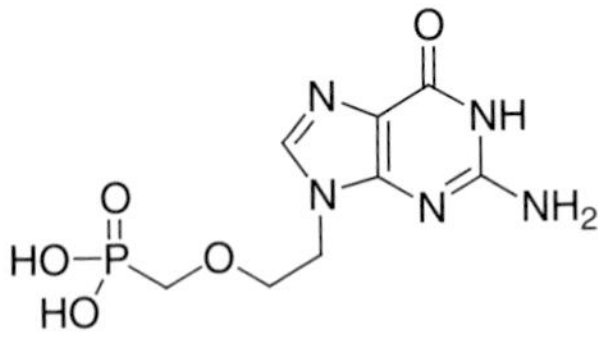 121 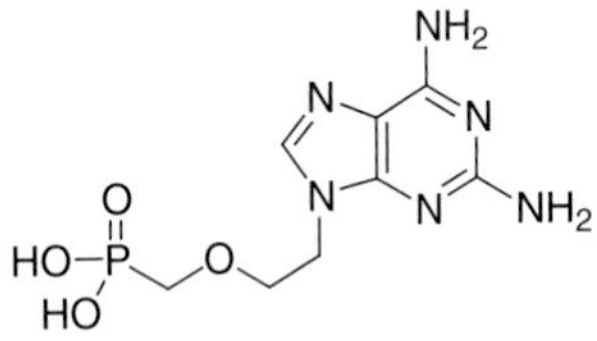 122

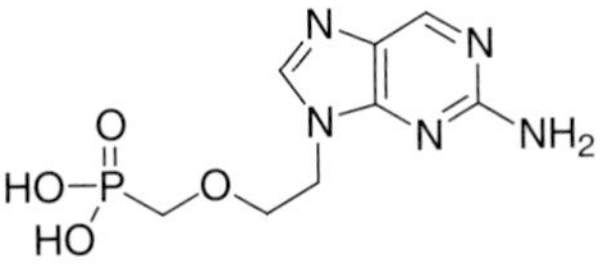

123 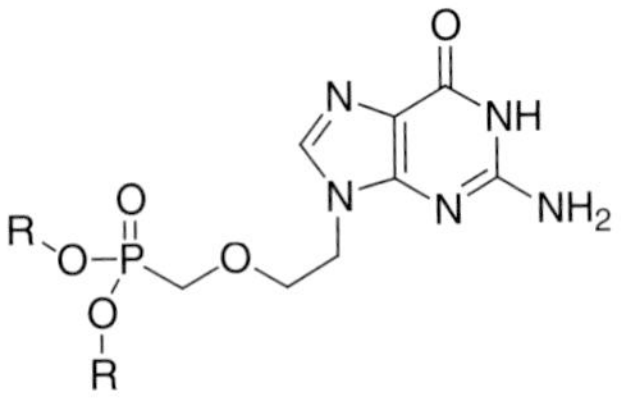 124

Similar to the antiherpetic acyclic nucleotides, the oxygen atom at the beta-position of phosphonate is vital for the antiviral activities and the alkyl side chain modification (**112**) leads to totally inactive compounds.[92,236] Insertion of an extra oxygen atom between the nucleobase and side chain (**113**) substantially reduces the anti-HBV activity of adefovir.[237] (*S*)-HPMPA (**114**) shows good anti-HBV activity, with an EC_{50} of around 0.4 μM in the HB 611 cell line, whereas its congener (*S*)-HPMPC (cidofovir, **7**) is about 20-fold less potent.[228] PMPA (tenofovir, **115**) is very active against HBV with an EC_{50} about twofold higher than that of adefovir. For PMPDAP (**116**), both stereoisomers are active against HBV, with the (*R*)-isomer about 20-fold more potent than the (*S*)-isomer (**117**).[238] A new structural class with phophonomethoxycyclopropylmethyl side chain has been reported to be potent anti-HBV agents. PMCA, 9-[(1-phosphonomethoxycyclopropyl)methyl]adenine (**118**), is slight less active than adefovir. However, the guanine analog, PMCG (**119**), is several-fold more potent than adefovir. The series of compounds is reported to have low cytotoxicity.[239]

For the PME side chain series, the antiviral activity was connected with 9-regioisomers of the purine isomers, and their N-3 or N-7 isomers (**120**) lose their antiviral activity.[240] 9-(2-Phosphonylmethoxyethyl)guanine (PMEG, **121**) is almost equally potent against HBV as adefovir. However, it is about 10 times more cytotoxic.[241] Similarly, 9-(2-phosphonylmethoxyethyl)-2,6-diaminopurine (PMEDAP, **122**) is several-fold more efficacious than adefovir, but its cytotoxicity is higher. In contrast, 9-(2-phosphonylmethoxyethyl)-2-aminopurine (**123**) loses almost all anti-HBV activity.[242]

Due to the polar nature of adefovir, the compound has poor oral bioavailability ($<10\%$) in vivo studies.[243,244] Efforts have been made to develop prodrugs of adefovir in order to improve its absorption. The prodrugs of phosphonate ester structures (**124**) were explored and showed to be rapidly degraded to PMEA.[243] However, their bioavailability remained low as did their stability in physiological condition, therefore they are unlikely to be therapeutically useful. The bis-(pivaloyloxymethyl)ester prodrug of PMEA, adefovir dipivoxil, was found to have an oral bioavailability approximately twice as high as adefovir (17–22%).[245,246] The prodrug is hydrolyzed by esterases, giving the parent compound adefovir.[247]

A clinical pharmacokinetic evaluation of PMEA in HIV patients revealed that, when administered intravenously, 90% of the drug was recovered in the urine within 12 h. The steady-state volume of distribution of adefovir in patients was found to be about $0.4\,L\,kg^{-1}$. It is also known to be unbound to plasma protein. The prodrug had 100-fold greater cellular uptake than the parent compound in MT-2 cells, but was susceptible to hydrolysis in serum.[246] The active metabolite PMEApp has a very long intracellular half-life (16–18 h). Adefovir is not a substrate of adenosine or AMP deaminases.[235]

7.11.2.2.4 Entecavir

Entecavir (Baraclude, 9-[(1*S*,3*R*,4*S*)-4-hydroxy-3-hydroxymethyl-2-methylenecyclopentyl]guanine, **125**) is a carbocyclic analog of 2′-deoxyguanosine in which the furanose ring is replaced by a cyclopentyl with an exocyclic methylene group. It has been approved for the treatment of chronic HBV infection.

Entecavir is a selective and potent inhibitor for HBV, with little activity against HIV and other DNA viruses. In the HepG2.2.15. cell line, entecavir had an EC_{50} of 3.75 nM for HBV as determined by analysis of secreted HBV DNA.[248,249] The cytotoxicity of entecavir is rather low. In HepG2 cells the CD_{50} is about 30 μM, which gives a selective index over 8000. The compound is efficiently converted to its triphosphate by cellular kinases in HepG2 cells.[250] The K_i value for the inhibition of HBV polymerase entecavir triphosphate was 3.2 nM.[251] Binding was competitive with respect to the natural substrate dGTP. Importantly, entecavir triphosphate blocks the three distinct phases of HBV reverse transcriptase activity, i.e., priming, reverse transcription, and DNA-dependent DNA synthesis.[252] Entecavir triphosphate has little inhibitory effect on cellular DNA polymerases with K_i values ranging from 18 to 160 μM. Entecavir triphosphate is also a poor substrate for the human mitochondrial DNA polymerase. Consequently, treatment of HepG2 cells with entecavir resulted in no apparent inhibitory effects on mitochondrial DNA content.[248] It has been reported that entecavir is clastogenic and is a rodent carcinogen and cannot be eliminated as a risk to humans.[315]

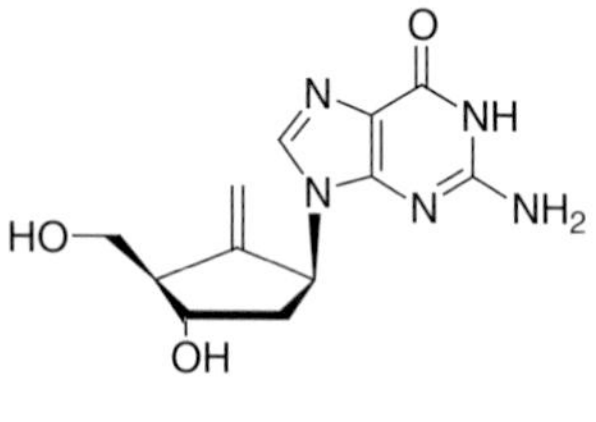

125 Entecavir

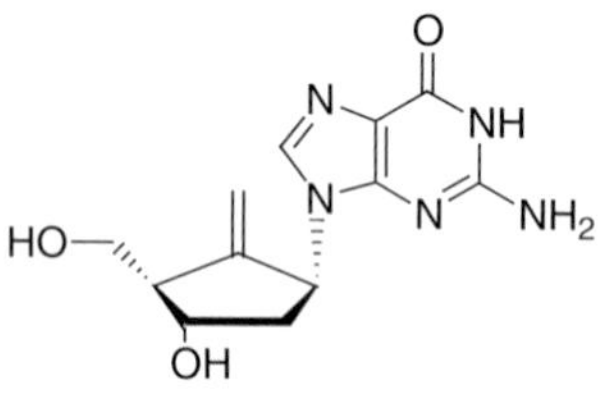

126

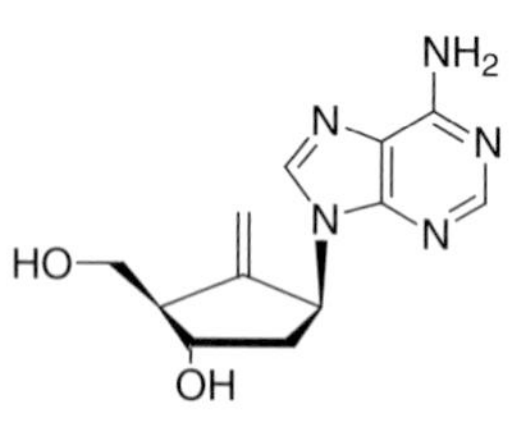

127

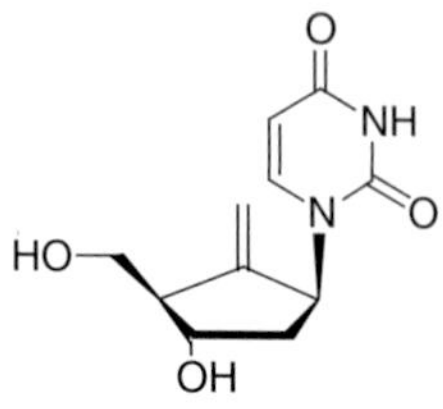

128

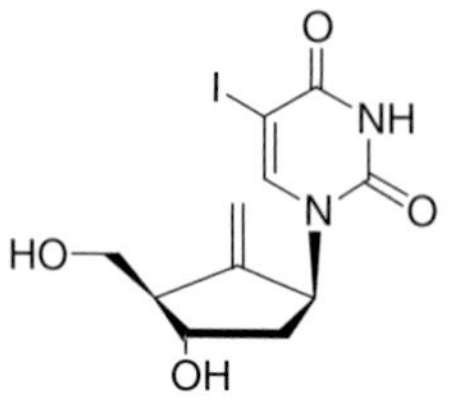

129

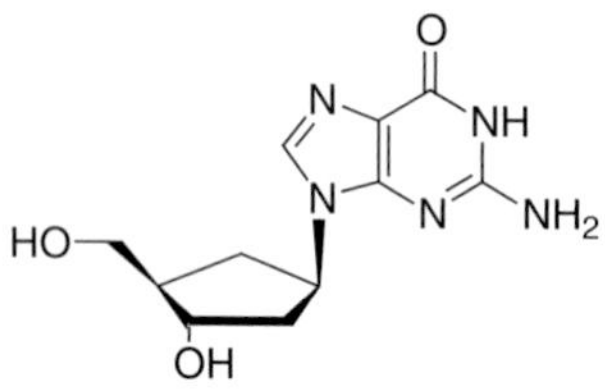

130 CDG

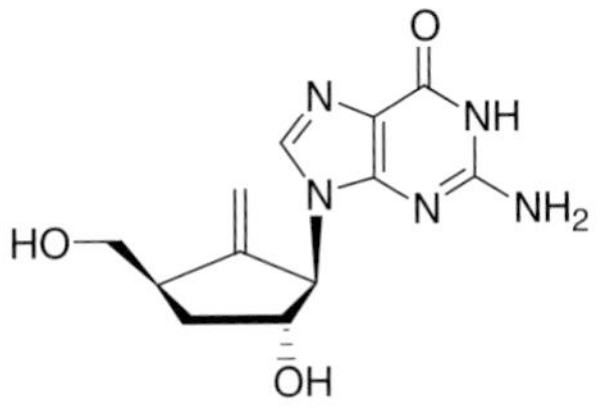

131 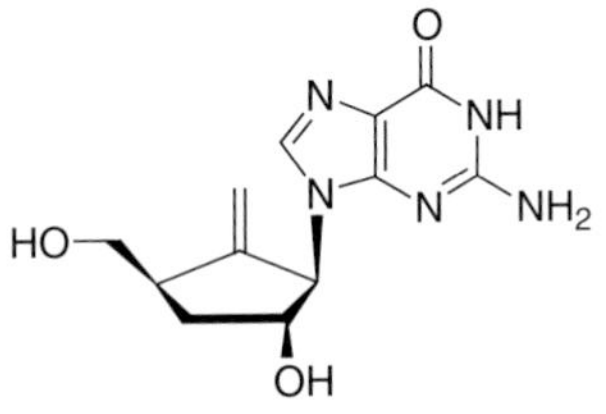 **132** 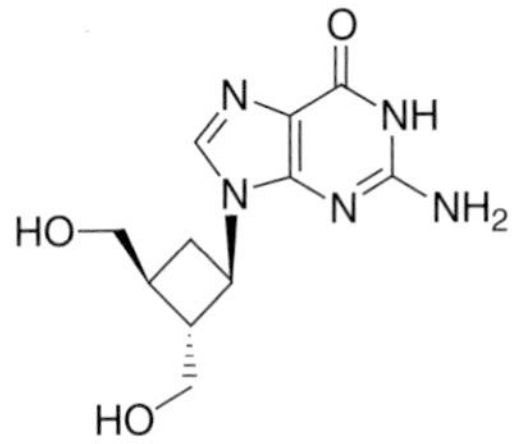 **133** Lobucavir

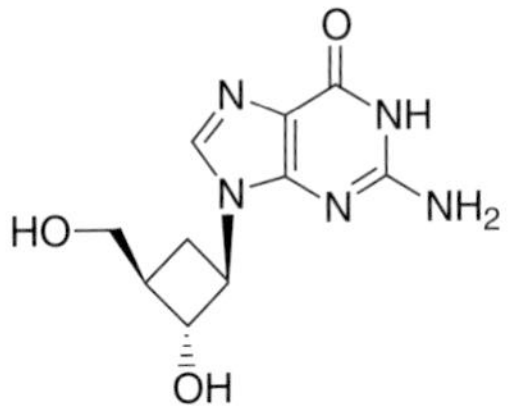

134 Cyclobut G 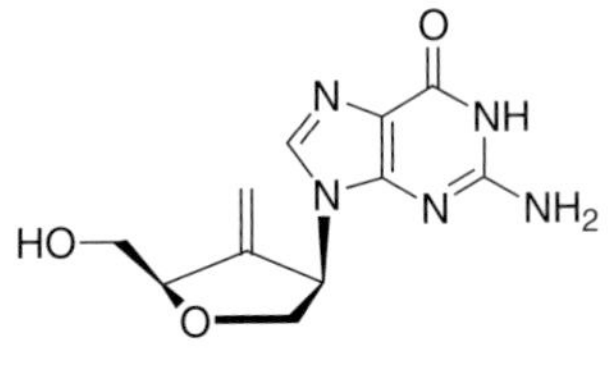 **135** 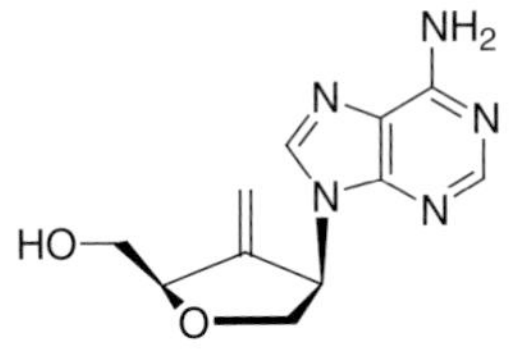 **136**

Though entecavir is an extremely potent anti-HBV agent, its enantiomer (**126**) was totally inactive.[249,252] The base modification adenine analog (**127**) retained significant activity, although with an EC_{50} about 30–40-fold higher.[249] The pyrimidine analogs, thymine (**128**) or iodouracil (**129**) analog, were essentially inactive. The exocyclic double bond is important for the activity. 2′-CDG (**130**), where the exocyclic methylene group is removed, has about a 10-fold lower potency than entecavir, but is still quite active with an EC_{50} of 50 nM in cell culture.[248] Shifting the hydroxy group from the 3′-position to the 2′-position (**131**, **132**) abolishes the activity, which is understandable since the 2′-hydroxyl often steers selectivity between DNA polymerases and RNA polymerases.[249] Decreasing the carbocyclic ring size will reduce the anti-HBV activity; lobucavir (**133**) is more than 100 times less active than entecavir and 10 times less than CDG,[251] although the K_i of lobucavir triphosphate for HBV polymerase is just slightly higher than that of entecavir triphosphate. Under the same conditions, it was found that the lobucavir triphosphate concentration is over 200 times lower than that of entecavir triphosphate in HepG2.2.15 cells, which probably explains the difference of their anti-HBV activity.[251] Cyclobut G (**134**) is just marginally active for HBV.[248] Using an oxygen atom to replace the carbon at the 3′-position will abolish the activity for HBV (**135**, **136**). Isodeoxyguanosine with exocyclic methylene is totally inactive, while its adenine shows modest activity.[253]

7.11.2.3 Resistance Related to Anti-Hepatitis B Virus Treatment

Studies of large numbers of patients receiving anti-HBV drugs show that resistance is generated to all the compounds currently used in the clinic. The most comprehensive information has been obtained for lamivudine but it is worth noticing that direct comparisons between resistance rates with other nucleosides must be taken with some caution as most other compounds have not been in the clinic for very long. In the case of lamivudine, genotypic resistance appears in patients at a rate of 24%, 42%, 53%, and 70% of patients after 1, 2, 3, and 4 years, respectively; key mutations lying in the conserved YMDD motif of the HBV polymerase (the corresponding mutation in the HIV reverse transcriptase is M184V).[254–258] It is thought that changes in this catalytic domain cause steric hindrance preventing binding of lamivudine triphosphate.[259] The pattern of resistance found in the HBV polymerase resembles that found in HIV with mutations with L180 M + M204V/I being the most common (70%) and the additional mutation V173L found in around 20% of cases.[260,261] The emergence of the YMDD mutations is associated with rebounds in serum HBV DNA to levels sufficient to drive disease. In some cases additional compensatory mutations at positions 204 or 128 have been detected.[262]

As expected, emtricitabine (FTC) appears to induce almost identical rates and patterns of resistance to lamivudine.[263] Telbivudine (LdT), active against HBV but not HIV, appears to select for at least the M204I lamivudine-resistant virus, but due perhaps to its greater potency the rate of resistance appears to be lower.[264]

Entecavir appears to give rise to mutations such as I169 T, M250V, T184G, and S202I but at a low rate, and it appears that multiple changes giving rise to lamivudine resistance may be a prerequisite to select for entecavir resistance. Resistance to entecavir was highest when both the T184G and S202I changes where combined with the preexisting lamivudine mutations.[265] However, clinical studies are at an early stage and these observations need to be substantiated in larger numbers of patients over an extended period of time.

Finally of all the current therapies adefovir appears to give rise to a unique resistance pattern with apparently a lower rate.[266,267] It has been speculated that the close molecular similarity to dATP or the flexibility of binding may explain the apparent low rate of resistance. The most common mutation seen so far is N236T which reduces sensitivity by 4–13-fold and A181V which gives rise to a 2–3-fold shift in sensitivity. Due to the differing resistance profile, trials are under way to test the effects of combinations of lamivudine and adefovir in reducing clinical resistance.[267]

7.11.2.4 Anti-Hepatitis B Virus Substances Under Development

7.11.2.4.1 Acyclic nucleosides

Some acyclic nucleosides that are antiherpes drugs show weak activity against HBV. Penciclovir (**3**) was reported to be able to inhibit HBV replication. However, the reported EC_{50} in HepG2.2.15 varied from submicromolar up to 100 μM in different laboratories.[248,268–270] Its prodrug FCV (**4**) was investigated in the clinic either alone or in combination with interferon or lamivudine.[271–275] Although a reduction of viral load was observed, a clinical benefit was not clear and the development of FCV for HBV therapy was terminated.

Aciclovir triphosphate is a potent inhibitor of HBV polymerase with a submicromolar EC_{50}. However, ACV (**1**) is inactive in cell-based assays and did not enhance the effect of interferon in treatment of HBV due to the lack of phosphorylation of ACV in hepatocytes.[276,277] Efforts made to develop ACV phosphate prodrugs to bypass the first phosphorylation step were successful and several prodrugs show anti-HBV effect in both in vitro and in vivo studies.[278–280]

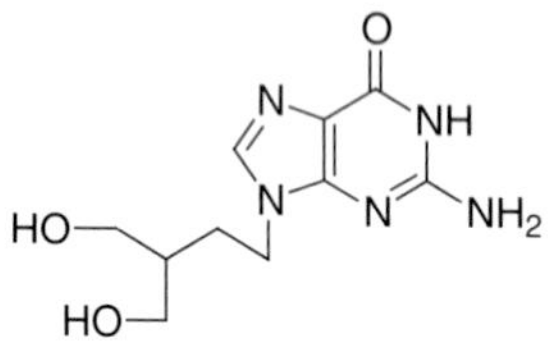

Penciclovir (**3**) 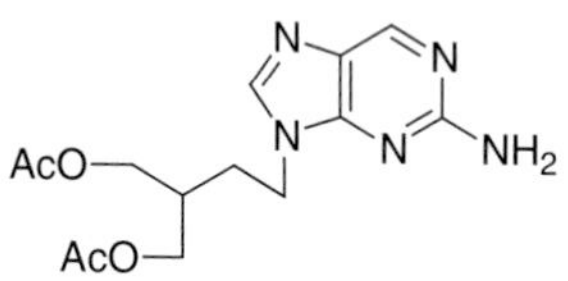 FCV (**4**) 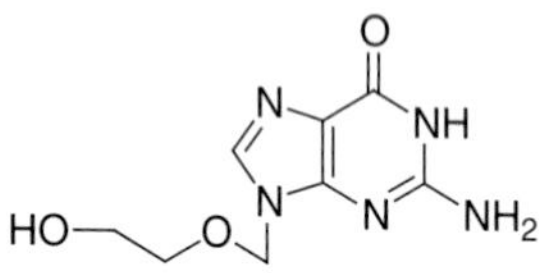 ACV (**1**)

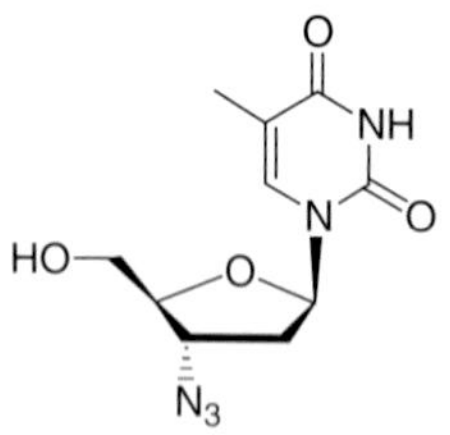

137 AZT 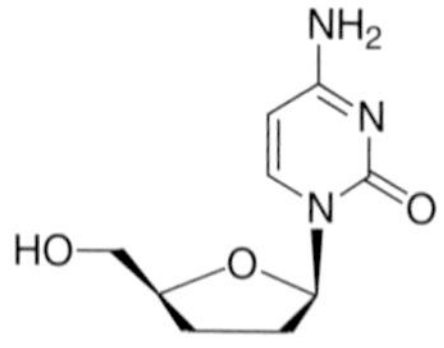 **138** ddC 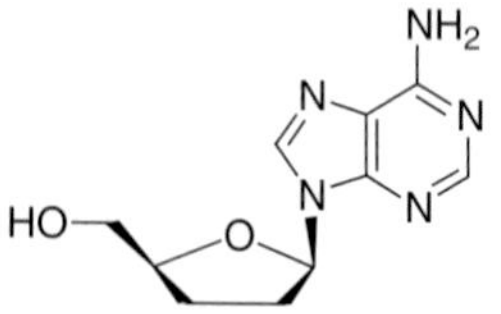 **139** ddA

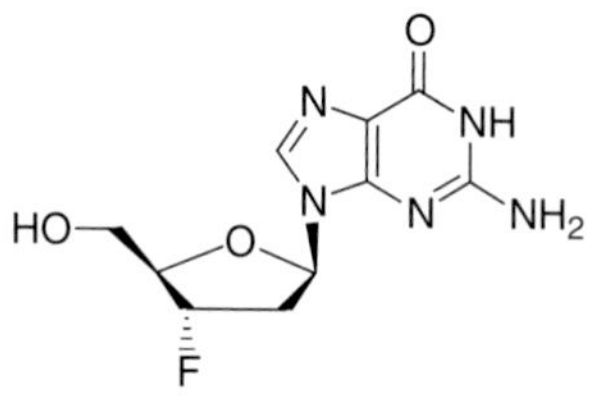

140 FLG 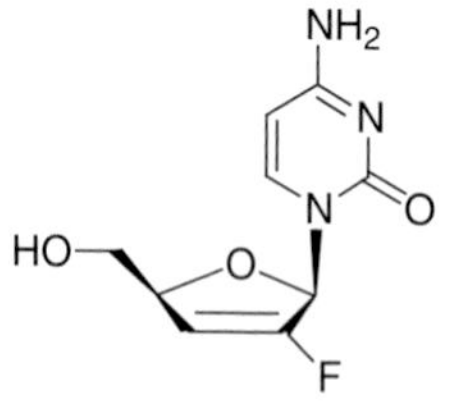 **141** Fd4C 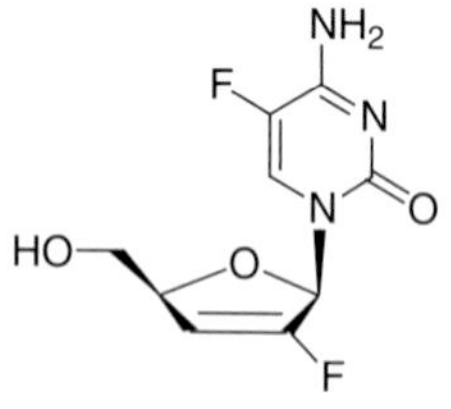 **142** Fd4FC

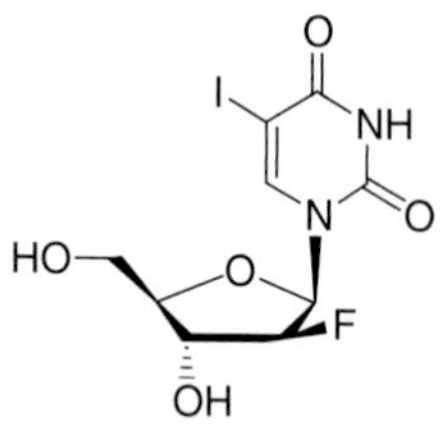

96 FIAU 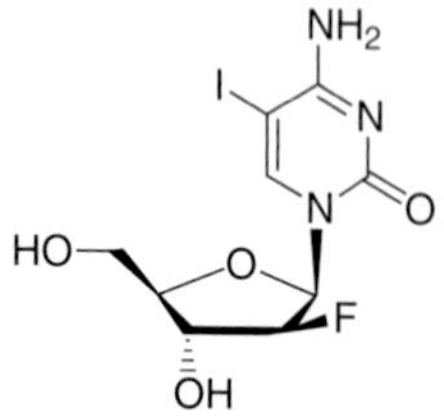 **144** FIAC 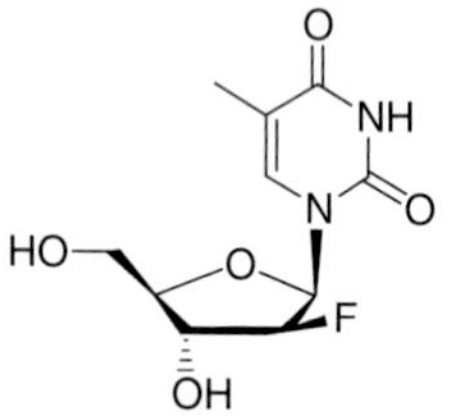 **145** FMAU

7.11.2.4.2 2′,3′-Modified nucleosides

Many 2′,3′-modified nucleosides are active against HBV. Notably, some compounds are active against both HIV and HBV, like AZT (**137**), ddC (**138**), and ddA (**139**).[281] 3′-Fluoro-2′,3′-dideoxyguanosine (FLG, **140**) is a very potent HBV and HIV inhibitor in vitro studies, and it has been shown to be highly efficacious against hepatitis B in both duck and woodchuck animal models.[282,283] Due to the low absorption of FLG, prodrugs have been developed, which enhance the oral bioavailability to more than 70% in human.[284]

Some fluorine-substituted 2′,3′-dideoxydidehydro nucleosides are active HBV inhibitors (**141**, **142**). Fd4FC (**142**) has an EC_{50} of 0.05 μM in HepG2.2.15 cell assay.[285] Several nucleosides with fluorine substitution at the 2′-position with arabino configuration were investigated for their anti-HBV activity, such as FIAU (**96**), FIAC (**144**), and FMAU (**145**).[284–289] However, their strong interaction with cellular mitochondrial DNA polymerase resulted in serious safety concerns.[290–292]

7.11.2.4.3 L-Nucleosides

A number of L-nucleosides are under investigation as anti-HBV treatments. L-dC (valtoricitabine) (**146**), L-dT (Telbivudine) (**147**), and L-dA (**148**) showed potent anti-HBV activity in HepG2.2.15 cell culture.[293,294] Though the configuration differs from the natural D-nucleosides, the compound can be efficiently phosphorylated by cellular kinases, dCK, and/or TK.[295,296] The compounds were selective for HBV, and showed no activity against other viruses like HIV or herpesviruses. However, removal of the 3′-hydroxyl may broaden the antiviral activity,[295] showing activity against both HBV and HIV as with L-ddC (**149**), L-5FddC (**150**), and L-ddAMP (**151**). 2′,3′-Unsaturated L-nucleosides are very potent against HBV (L-d4C (**152**), L-d4FC (**153**), and L-d4A (**154**)), and introducing a fluorine atom on to the unsaturated bond is favored for anti-HBV activity.[297–299] L-Fd4C (**155**) and L-Fd4FC (**156**) are among the most potent anti-HBV agents with EC_{50} of low nanomolar level in HepG2.2.15 cell assay.

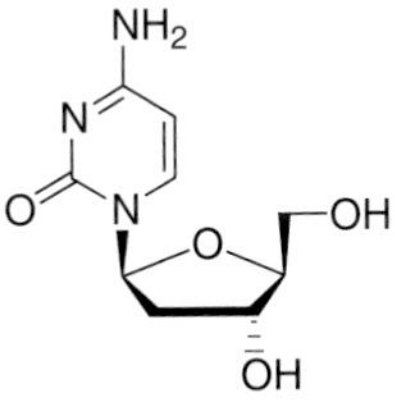

146 L-dC

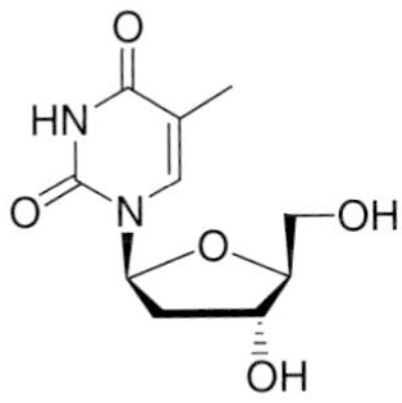

147 L-dT

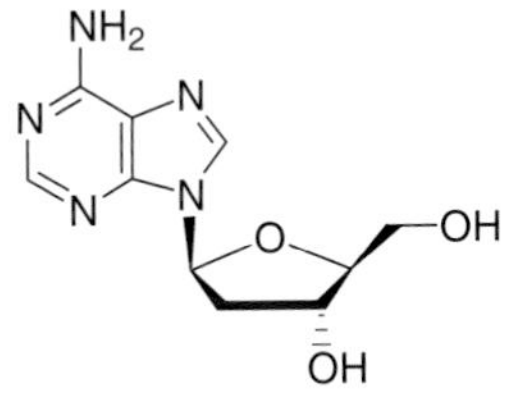

148 L-dA

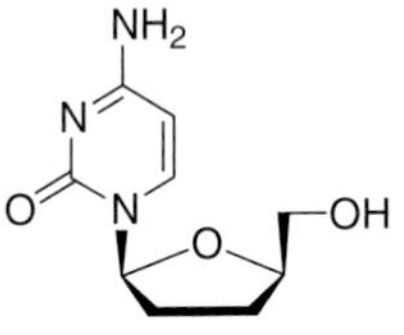

149 L-ddC

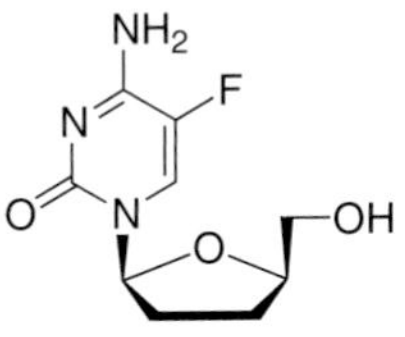

150 L-5FddC

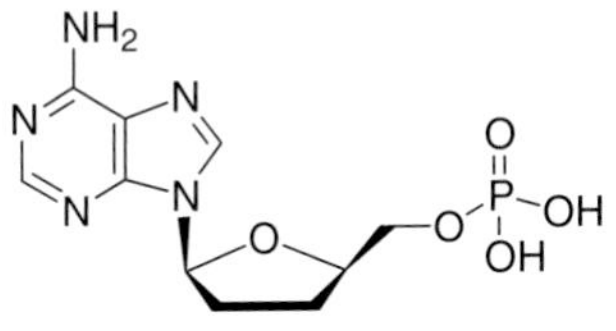

151 L-ddAMP

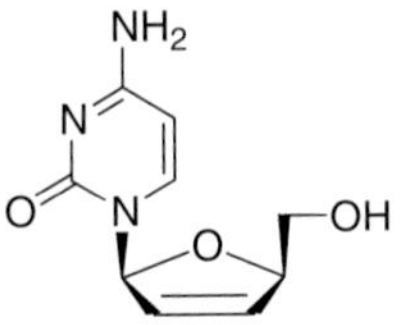

152 L-d4C

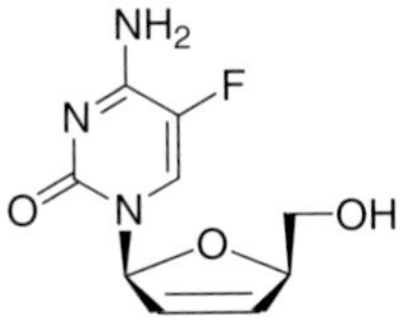

153 L-d4FC

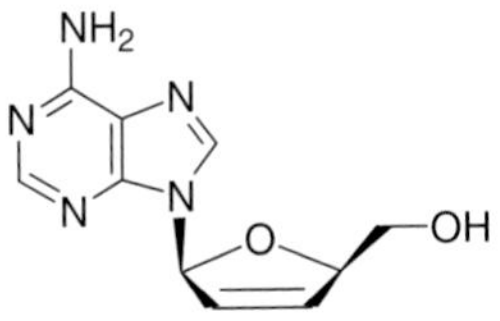

154 L-d4A

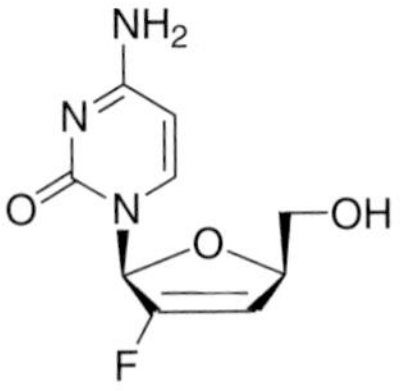

155 L-2′-Fd4C

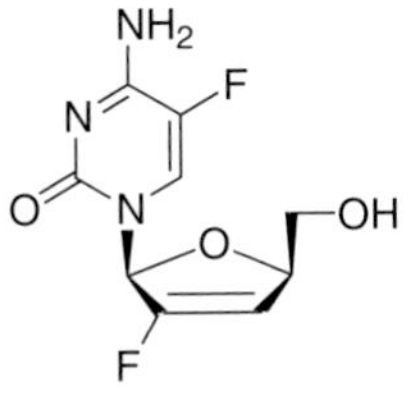

156 L-2′-Fd4FC

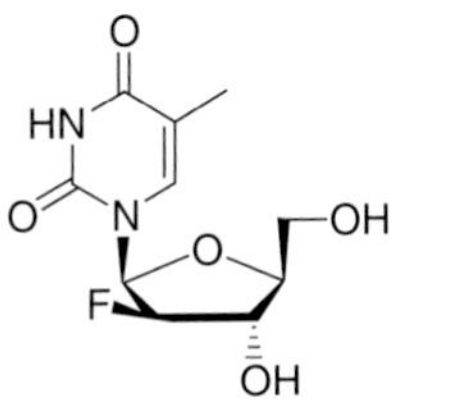

157 L-FMAU

L-Nucleosides are generally less toxic compared to the D-counterparts, due to the fact that L-analogs are not efficient substrates for human cellular DNA polymerases.[295,301] An example is L-FMAU (**157**), which is more potent than FMAU (**145**) and has a selective index of 2000 versus 25 for FMAU.[302,303] Currently L-nucleosides present one of the most important structural classes under investigation as anti-HBV substances.

7.11.2.4.4 Nonnucleoside hepatitis B virus polymerase inhibitors

A series of 2,5-pyridinedicarboxylic acid derivatives (**158**) were reported to be active against HBV.[304] The mechanism of inhibition on HBV reverse transcriptase was established by a cell-free DNA polymerization assay.[305] Depending on the substituents R^1 and R^2, the compounds showed the IC_{50} varying between 0.2 and 0.001 $\mu g\,mL^{-1}$ and, generally, the compounds had very low cytotoxicity. This class of compounds suggests the possibility of combination with nucleoside HBV reverse transcriptase inhibitors. A natural product, robustaflavone (**159**), which inhibits HBV with EC_{50} of 0.25 μM in HepG2.2.15 cells, was also suggested to be active on the HBV polymerase.[306]

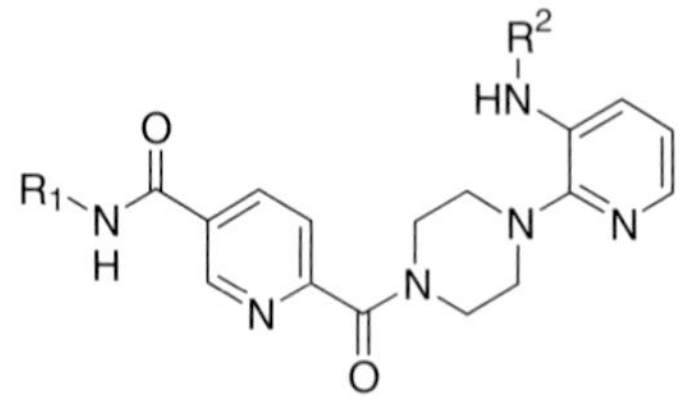

158

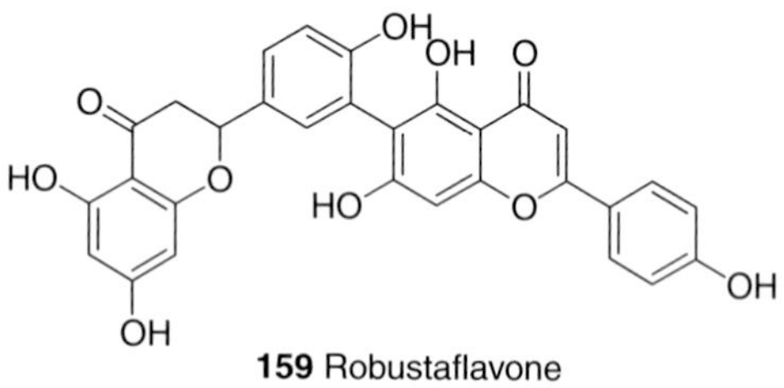

159 Robustaflavone

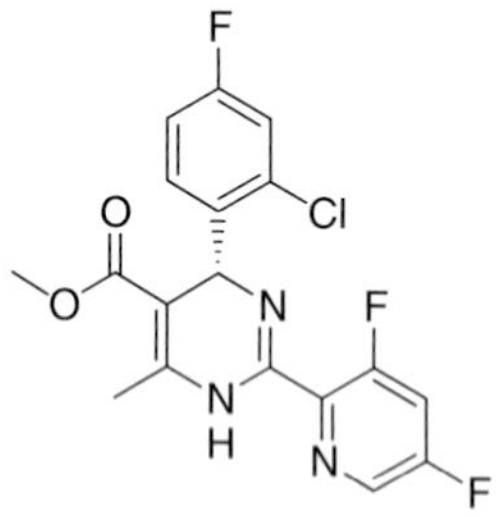

160 BAY41-4109

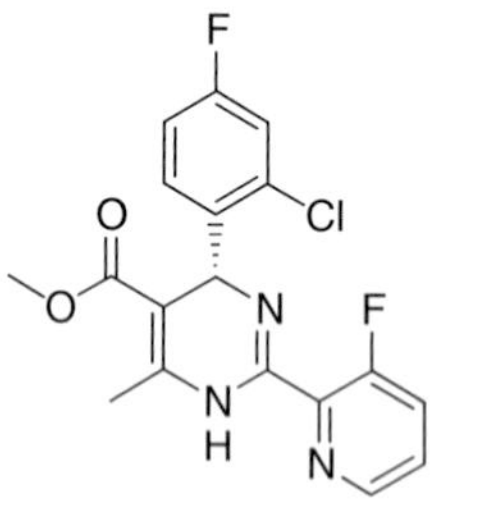

161 BAY38-7690

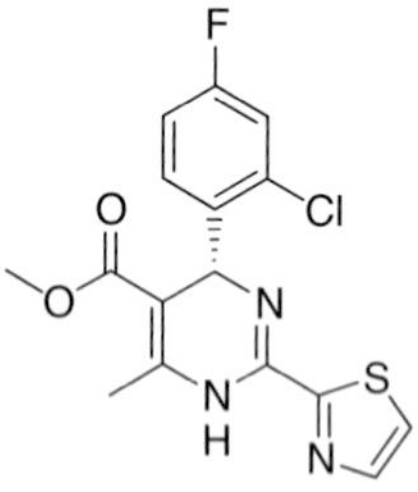

162 BAY39-5493

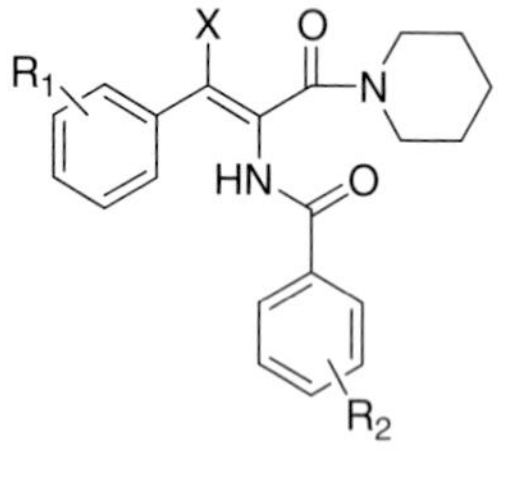

163

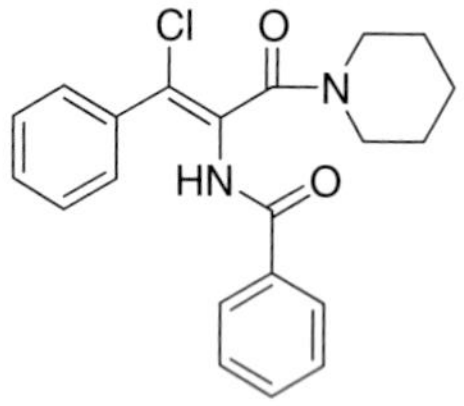

164 AT-61

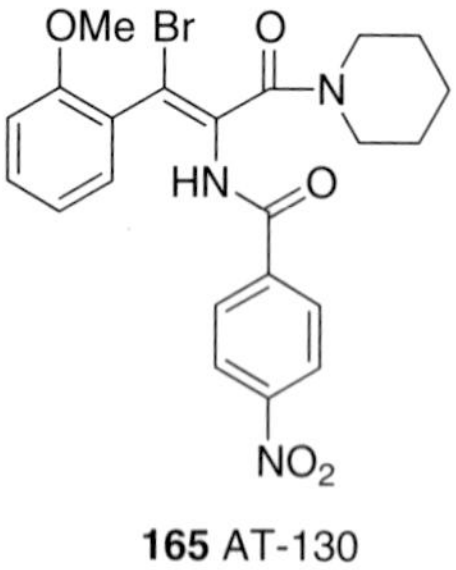

165 AT-130

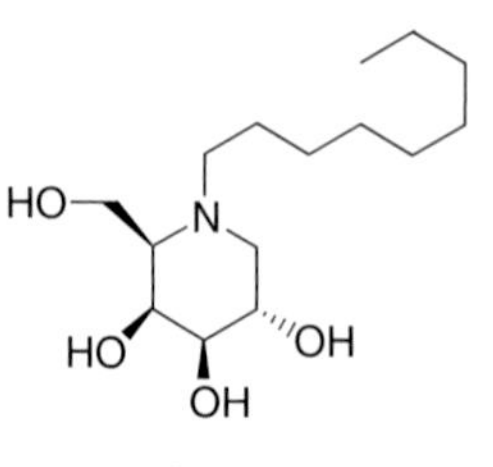

166

7.11.2.4.5 Inhibitors by other mechanisms

HBV viral nucleocapsids are stable aggregates assembled from HBV core proteins, and the correct formation of nucleocapsid is essential for producing a replicable virus.[307,308] A class of biaryl substituted pyrimidine, BAY41-4109 (**160**), BAY38-7690 (**161**), and BAY39-5493 (**162**) was shown to be able to inhibit HBV by affecting the formation of viral nucleocapsid. BAY41-4109 had an EC_{50} of 0.05 μM against HBV in HepG2.2.15, which was more potent than lamivudine in same assay. Anti-HBV efficacy was also seen in an HBV-infected transgenic mice model. The detailed studies pointed to the high affinity of the compounds for the viral core protein which prevented the correct aggregation of core proteins, even though the formation and the concentration of the core proteins were not affected.

A group of phenylpropenamide compounds (**163–165**) showed potent inhibitory effect against HBV in cell assay.[309–311] The representative structures AT-61 (**164**) and AT-130 (**165**) had the EC_{50} of 0.5–1.2 μM in cell assay. The cytotoxicity toward the same cell line was generally very low for this structural class. The compounds inhibit HBV replication by preventing the packaging of pregenomic RNA into the core particle. The unique mechanism suggests that the compounds could be used for inhibiting the resistant strains from the mutation in HBV reverse transcriptase. Indeed, the compounds maintained the same activity against both wild-type and lamivudine resistant strains.

One iminosugar derivative, *N*-nonyl-deoxygalactonojirimycin (**166**), was under clinical investigation for its anti-HBV activity. The compound was suggested to inhibit the proper encapsidation of viral pregenomic RNA.[312–314]

7.11.2.5 Challenges in Hepatitis B Virus Treatment

Effective treatment of chronic HBV infection is a significant challenge. While a small number of anti-HBV nucleoside analogs are in use in the clinic, they have already begun to show the classical issues of resistance and potential toxicity. We are likely to enter a phase in which we will see the emergence of compounds with inherently better resistance profiles or combinations of compounds with such properties. Ultimately we will need to see compounds that are able to effectively clear the covently closed circular HBV DNA which would lead to improved long-term outcomes including potential 'cures.' Finally the major challenge of treatment of HBV will be the target population. Will a compound be developed with the correct profile to encourage the treatment of asymptomatic or early stage chronically infected patients? This is a matter not only of the properties of compounds (cost, resistance, patient compliance) but also of the attitudes of health authorities based upon their ability to measure clinical benefit.

References

1. *Fields Virology*, 4th ed.; Knipe, D. M., Howley, P. M., Eds.-in-Chief; Lippincott Williams & Wilkins: Philadelphia, PA, 2001.
2. Whitley, R. J. In *Fields Virology*, 4th ed.; Knipe, D. M., Howley, P. M., Eds.-in-Chief; Lippincott Williams & Wilkins: Philadelphia, PA, 2001, pp 2461–2511.
3. Arvin, A. M. In *Fields Virology*, 4th ed.; Knipe, D. M., Howley, P. M., Eds.-in-Chief; Lippincott Williams & Wilkins: Philadelphia, PA, 2001, pp 2731–2769.
4. Kieff, E.; Rickinson, A. B. In *Fields Virology*, 4th ed.; Knipe, D. M., Howley, P. M., Eds.-in-Chief; Lippincott Williams & Wilkins: Philadelphia, PA, 2001, pp 2511–2575.
5. Rickinson, A. B.; Kieff, E. In *Fields Virology*, 4th ed.; Knipe, D. M., Howley, P. M., Eds.-in-Chief; Lippincott Williams & Wilkins: Philadelphia, PA, 2001, pp 2575–2629.
6. Pass, R. F. In *Fields Virology*, 4th ed.; Knipe, D. M., Howley, P. M., Eds.-in-Chief; Lippincott Williams & Wilkins: Philadelphia, PA, 2001, pp 2675–2707.
7. Grulich, A. E.; Olsen, S. J.; Luo, K.; Hendry, O.; Cunningham, P.; Cooper, D. A.; Gao, S. J.; Chang, Y.; Moore, P. S.; Kaldor, J. M. *J. Acq. Imm. Def. Syndr. Hum. Retrovirol.* **1999**, *20*, 387–393.
8. Yamanishi, K.; Okuno, T.; Shiraki, K.; Takahashi, M.; Kondo, T.; Asano, Y.; Kurata, T. *Lancet* **1988**, *1*, 1065–1067.
9. Frenkel, N.; Schirmer, E. C.; Wyatt, L. S.; Katsafanas, G.; Roffman, E.; Danovich, R. M.; June, C. H. *Proc. Natl. Acad. Sci. USA* **1990**, *87*, 748–753.
10. Yoshida, M.; Yamada, M.; Chatterjee, S.; Lakeman, F.; Nii, S.; Whitley, R. J. *J. Virol. Methods* **1996**, *58*, 137–143.
11. Jee, S. H.; Long, C. E.; Schnabel, K. C.; Sehgal, N.; Epstein, L. G.; Hall, C. B. *Pedia. Infect. Dis. J.* **1998**, *17*, 43–48.
12. Moore P. S.; Chang, Y. In *Fields Virology*, 4th ed.; Knipe, D. M., Howley, P. M., Eds.-in-Chief; Lippincott Williams & Wilkins: Philadelphia, PA, 2001, pp 2803–2835.
13. Kedes, D. H.; Ganem, D. *J. Clin. Invest.* **1997**, *99*, 2082–2086.
14. Challberg, M. D. *Proc. Natl. Acad. Sci. USA* **1986**, *83*, 9094–9098.
15. Reardon, J. E.; Spector, T. *J. Biol. Chem.* **1989**, *264*, 7405–7411.
16. Keller, P. M.; Fyfe, J. A.; Beauchamp, L.; Lubbers, C. M.; Furman, P. A.; Schaeffer, H. L.; Elion, G. B. *Biochem. Pharmacol.* **1981**, *30*, 3071–3077.
17. de Clercq, E.; Andrei, G.; Snoeck, R.; de Bolle, L.; Naesens, L.; Degreve, B.; Balzarini, J.; Zhang, Y.; Schols, D.; Leyssen, P. et al. *Nucleosides Nucleotides Nucleic Acids* **2001**, *20*, 271–285.
18. Brown, D. G.; Visse, R.; Sandu, G.; Davies, A.; Rizkallah, P. J.; Melitz, C.; Summers, W. C.; Sanderson, M. *Nat. Struct. Biol.* **1995**, *2*, 876–881.
19. Champness, J.; Bennett, M.; Wien, F.; Visse, R.; Summer, W.; Herdewijn, P.; de Clercq, E.; Ostrowski, T.; Jarvest, R.; Sanderson, M. *Prot. Struct. Funct. Genet.* **1998**, *32*, 350–361.

20. Vogt, J.; Perrozzo, R.; Pautsch, A.; Prota, A.; Schellin, P.; Pilger, B.; Folkers, G.; Scapozza, L.; Schultz, G. *Prot. Struct. Funct. Genet.* **2000**, *41*, 545–553.
21. Bird, L.; Ren, J.; Wright, A.; Leslie, K.; Degreve, B.; Balzarini, J.; Stammer, D. *J. Biol. Chem.* **2003**, *278*, 24680–24687.
22. Ericsson, A.; Larsson, A.; Aoki, Y.; Yisak, W.; Johansson, N. G.; Öberg, B.; Datema, R. *Antimicrob. Agents Chemother.* **1985**, *27*, 753–759.
23. Smee, D. F.; Martin, J. C.; Verheyden, J. P.; Matthews, T. R. *Antimicrob. Agents Chemother.* **1983**, *23*, 676–682.
24. Ashton, W. T.; Karkas, J. D.; Field, A. K.; Tolman, R. L. *Biochem. Biophys. Res. Commun.* **1982**, *108*, 1716–1721.
25. Liu, S.; Baldwin, E.; Crystals of HSV-1 DNA polymerase. PCT Int. Patent WO2004108946, 2004.
26. Johansson, N. G. *Adv. Antivir. Drug Design* **1993**, *1*, 87–177.
27. Kelley, J. L.; Krochmal, M. P.; Schaeffer, H. J. *J. Med. Chem.* **1981**, *24*, 1528–1531.
28. Elion, G. B. *Science* **1989**, *244*, 41–47.
29. Elion, G. B.; Furman, P. A.; Fyfe, J. A.; De Miranda, P.; Beauchamp, L.; Schaeffer, H. J. *Proc. Natl. Acad. Sci. USA* **1977**, *74*, 5716–5720.
30. Ahmed, S.; Imai, T.; Yoshigne, Y.; Otagiri, M. *Life Sci.* **1997**, *61*, 1879–1887.
31. Spruance, S. L.; Stewart, J. C.; Rowe, N. H.; McKeough, M. B.; Wenerstrom, G.; Freeman, D. J. *J. Infect. Dis.* **1990**, *161*, 185–190.
32. Whitley, R. J.; Gnann, J. W., Jr. *N. Engl. J. Med.* **1992**, *327*, 782–789.
33. Hill, E. L.; Hunter, G. A.; Ellis, M. N. *Antimicrob. Agents Chemother.* **1991**, *35*, 2322–2328.
34. Sacks, S. L.; Wanklin, R. J.; Reece, D. E. *Ann. Intern. Med.* **1989**, *111*, 893–899.
35. Ganapathy, M. E.; Huang, W.; Wang, H.; Ganapathy, V.; Leibach, F. H. *Biochem. Biophys. Res. Commun.* **1998**, *246*, 470–475.
36. Weller, S.; Blum, M. R.; Doucett, M. *Clin. Pharmacol. Ther.* **1997**, *54*, 595–605.
37. Tyring, S. K.; Baker, D.; Snowden, W. *J. Infect. Dis.* **2002**, *186*, S40–S46.
38. Szczech, G. M. *Clin. Infect. Dis.* **1996**, *22*, 355–360.
39. Colla, L.; De Clercq, E.; Busson, R.; Vanderhaeghe, H. *J. Med. Chem.* **1983**, *26*, 602–604.
40. Selby, P.; Powles, R. L.; Blake, S.; Stolle, K.; Mbidde, E. K.; McElwain, T. J.; Hickmott, E.; Whiteman, P. D.; Fiddian, A. P. *Lancet* **1984**, *11*, 1428–1430.
41. Welch, C. J.; Larsson, A.; Ericsson, A. C.; Oeberg, B.; Datema, R.; Chattopadhyaya, J. *Acta. Chem. Scand. (B)* **1985**, *39*, 47–54.
42. Kumar, S.; Oakes, F. T.; Wilson, S. R.; Leonard, N. J. *Heterocycles* **1988**, *27*, 2891–2901.
43. Bundgaard, H.; Falch, E.; Jensen, E. *J. Med. Chem.* **1989**, *32*, 2503–2507.
44. Bundgaard, H.; Jensen, E.; Falch, E. *Pharm. Res.* **1991**, *8*, 1087–1093.
45. Stimac, A.; Kobe, J. *Synthesis* **1990**, 461–465.
46. Beauchamp, L. M.; Orr, G. F.; deMiranda, P.; Burnett, T.; Krenitsky, T. A. *Antivir. Chem. Chemother.* **1992**, *3*, 157–164.
47. Krenitsky, T. A.; Beauchamp, L. M. Therapeutic Nucleosides. EP308065, 1988.
48. Bundgaard, H.; Falch, E.; Jensen, E. *J. Med. Chem.* **1989**, *32*, 2503–2507.
49. Bundgaard, H. In *Design of Prodrugs*; Bundgaard, H., Ed.; Elsevier: Amsterdam, The Netherlands, 1985, p 1.
50. Kovach, I. M.; Pitman, I. H.; Higuchi, T. *J. Pharm. Sci.* **1981**, *70*, 881–885.
51. Buchwald, P.; Bodor, N. *J. Med. Chem.* **1999**, *42*, 5160–5168.
52. Earnshaw, D. L.; Bacon, T. H.; Darlison, S. J.; Edmonds, K.; Perkins, R. M.; Vere Hodge, R. A. *Antimicrob. Agents Chemother.* **1992**, *36*, 2747–2757.
53. Öberg, B.; Johansson, N. G. *J. Antimicrob. Chemother.* **1984**, *14*, 5–26.
54. Larsson, A.; Öberg, B.; Alenius, S.; Hagberg, E.; Johansson, N. G.; Lindborg, B.; Stening, G. *Antimicrob. Agents Chemother.* **1983**, *23*, 664–670.
55. Norbeck, D.; Kern, E.; Hayashi, S. *J. Med. Chem.* **1990**, *33*, 1281–1285.
56. Harnden, M. R.; Jarvest, R. L. *J. Chem. Soc. Perkin Trans.* **1989**, *1*, 2207–2213.
57. Krenitsky, T. A.; Tuttle, J. V.; Catteau, E. L., Jr.; Wang, P. *Comp. Biochem. Physiol. B* **1974**, *49*, 687–703.
58. Harnden, M. R.; Jarvest, R. J.; Boyd, M. R.; Suttun, D.; Vene Hodge, R. A. *J. Med. Chem.* **1989**, *32*, 1738–1743.
59. Kim, D.; Lee, N.; Jin, G.; Kim, Y.; Chang, K.; Kim, H.; Cho, Y.; Choi, W.; Jung, I.; Kim, K. *Biorg. Med. Chem. Lett.* **1996**, *6*, 1849–1854.
60. Pue, M. A.; Benet, L. Z. *Antivir. Chem. Chemother.* **1993**, *4*, 47–55.
61. Beauchamp, L. M.; Serling, B. L.; Kelsey, J. E.; Biron, K. K.; Collins, P.; Selway, J.; Lin, J. C.; Schaeffer, H. *J. Med. Chem.* **1988**, *31*, 144–149.
62. Littler, E.; Stuart, A. D.; Chee, M. *Nature* **1992**, *358*, 160–162.
63. Sullivan, V.; Talarico, C. L.; Stanat, S. C.; Davis, M.; Coen, D. M.; Biron, K. K. *Nature* **1992**, *358*, 162–164.
64. Chou, S.; Waldemer, R.; Senters, A. E.; Michels, K. S.; Kemble, G. W.; Miner, R. C.; Drew, W. L. *J. Infect. Dis.* **2002**, *185*, 162–169.
65. Di Luca, D.; Katasafanas, G.; Schirmer, E. C.; Balachandran, N.; Frenkel, N. *Virology* **1990**, *175*, 199–210.
66. Fyfe, J. A.; Keller, P. M.; Furman, P. A.; Miller, R. L.; Elion, G. B. *J. Biol. Chem.* **1978**, *253*, 8721–8727.
67. Martin, J. C.; McGee, D. P.; Jeffrey, G. A.; Hobbs, D. W.; Smee, D. F.; Mattews, T. M.; Verheyden, J. P. *J. Med. Chem.* **1986**, *29*, 1384–1389.
68. Ashton, W. T.; Meurer, L. C.; Cantone, C.; Field, A. K.; Hannah, J.; Karkas, J. D.; Liou, R.; Patel, G. F.; Wagner, A.; Walton, E. et al. *J. Med. Chem.* **1988**, *31*, 2304–2315.
69. Martin, J. C.; Jeffrey, G. A.; McGee, D. P.; Tippie, M. A.; Smee, D. F.; Mattews, T. M.; Verheyden, J. P. *J. Med. Chem.* **1985**, *28*, 358–362.
70. Neyts, J.; Andrei, G.; Snoeck, R. *Antimicrob. Agents Chemother.* **1994**, *38*, 2710–2716.
71. Neyts, J.; Jähne, G.; Andrei, G.; Snoeck, R.; Winkler, I.; de Clercq, E. *Antimicrob. Agents Chemother.* **1995**, *39*, 56–60.
72. Golankiewicz, B.; Ostrowski, T.; Golinski, T.; Januszczyk, P.; Zeidler, J.; Baranowski, D.; de Clercq, E. *J. Med. Chem.* **2001**, *44*, 4284–4287.
73. Spector, S. A.; Busch, D. F.; Follansbee, S.; Squires, K.; Lalezari, J. P.; Jacobson, M. A.; Mastre, B.; Buhles, W.; Drew, W. L. *J. Infect. Dis.* **1995**, *171*, 1431–1437.
74. Faulds, D.; Heel, R. *Drugs* **1990**, *39*, 597–638.
75. Sugawara, M.; Huang, W.; Fei, Y.; Leibach, F. H.; Ganapathy, V.; Ganapathy, M. *J. Pharm. Sci.* **2000**, *89*, 781–789.
76. Jung, D.; Dorr, A. *J. Clin. Pharmacol.* **1999**, *39*, 800–804.
77. Brown, F.; Banken, L.; Saywell, K. *Clin. Pharmacokinet.* **1999**, *37*, 167–176.
78. Emery, V. C. *J. Clin. Virol.* **2001**, *21*, 223–228.
79. Smith, I. L.; Cherrington, J. M.; Jiles, R. E.; Fuller, M. D.; Freeman, W. R.; Spector, S. A. *J. Infect. Dis.* **1997**, *176*, 69–77.
80. Wos, J. A.; Bronson, J. J.; Martin, J. C. *Tetrahedron Lett.* **1988**, *29*, 5475–5478.
81. Bronson, J. J.; Ghazzouli, I.; Hitchcock, J. M.; Webb, R. R.; Martin, J. C. *J. Med. Chem.* **1989**, *32*, 1457–1463.
82. Connelly, M. C.; Robbins, B. L.; Fridland, A. *Biochem. Pharmacol.* **1993**, *46*, 1053–1057.
83. Hwang, J.; Choi, J. *Drugs Future* **2004**, *29*, 163–177.
84. De Clercq, E. *Nature Rev. Drug Disc.* **2002**, *1*, 13–25.

85. Neyts, J.; de Clercq, E. *Biochem. Pharmacol.* **1994**, *47*, 39–41.
86. Xiong, X.; Smith, J.; Chen, M. *Antimicrob. Agents Chemother.* **1997**, *41*, 594–599.
87. de Clercq, E. *Biochem. Pharmacol.* **1994**, *47*, 155–169.
88. Alexander, P.; Arimilli, M. N.; Bischofberger, N. W.; Hitchcock, M. Method for Dosing Therapeutic Compounds. PCT Patent WO19949507919, 1994.
89. Bischofberger, N.; Hitchcock, M. J. M.; Chen, M. S. *Antimicrob. Agents Chemother.* **1994**, *38*, 2387–2391.
90. Holy, A. *Curr. Pharm. Des.* **2003**, *9*, 2567–2592.
91. Bazarini, J.; Holy, A.; Jindrich, J.; Naesens, L.; Snoeck, R.; Schols, D. *Antimicrob. Agents Chemother.* **1993**, *37*, 332–338.
92. Kim, C.; Luh, B.; Misco, P.; Bronson, J.; Hitchcock, M.; Ghazzouli, I.; Martin, J. *J. Med. Chem.* **1990**, *33*, 1207–1213.
93. Harnden, M.; Parkin, A.; Parrat, M.; Perkins, R. *J. Med. Chem.* **1993**, *36*, 1343–1355.
94. Mendel, D. B.; Barkhimer, D. B.; Chen, M. S. *Antimicrob. Agents Chemother.* **1995**, *39*, 2120–2122.
95. Öberg, B. *Pharmacol. Therapeut.* **1989**, *40*, 213–285.
96. Chrisp, P.; Clissold, S. *Drugs* **1991**, *41*, 104–129.
97. Öberg, B. *Pharmacol. Therapeut.* **1983**, *19*, 387–415.
98. Öberg, B. In *Human Herpesviruses Infection*; Lopes, C., Roizman, B., Eds.; Raven Press: New York, 1986, p 141.
99. Derse, D.; Bastow, K. F.; Cheng, Y. C. *J. Biol. Chem.* **1982**, *257*, 10251–10260.
100. Eriksson, B.; Larsson, A.; Helgstrand, E.; Johansson, N. G.; Öberg, B. *Biochim. Biophys. Acta* **1980**, *607*, 53–64.
101. Eriksson, B.; Öberg, B.; Wahren, B. *Biochim. Biophys. Acta* **1982**, *696*, 115–123.
102. Ostrander, M.; Cheng, Y. C. *Biochim. Biophys. Acta* **1980**, *609*, 232–245.
103. Eriksson, B.; Öberg, B. *Antimicrob. Agents Chemother.* **1979**, *15*, 758–762.
104. Sjövall, J.; Karlsson, A.; Ogenstad, S.; Sandström, E.; Saarimäki, M. *Clin. Pharmacol. Therapeut.* **1988**, *44*, 65–73.
105. Camaioni Neto, C.; Steim, J. M.; Sarin, P. S.; Sun, D. K.; Bhongle, N. N.; Piratla, R. K.; Turcotte, J. G. *Biochem. Biophys. Res. Commun.* **1990**, *171*, 458–464.
106. Chu, C.; Szoka, F. C. *J. Liposome Res.* **1992**, *2*, 67–92.
107. Walker, I.; Nicholls, D.; Irwin, W. J.; Freeman, S. *Int. J. Pharm.* **1994**, *104*, 157–167.
108. Mitchell, A. G.; Nicholls, D.; Irwin, D.; Freeman, S. *J. Chem. Soc. Perkin Trans.* **1992**, *2*, 1145–1150.
109. Mitchell, A. G.; Nicholls, D.; Walker, I.; Irwin, W. J.; Freeman, S. *J. Chem. Soc. Perkin Trans.* **1991**, *2*, 1297–1303.
110. Noren, J. O.; Helgstrand, E.; Johansson, N. G.; Misiorny, A.; Stening, G. *J. Med. Chem.* **1983**, *26*, 264–270.
111. Iyer, R.; Philips, L. R.; Biddle, J. A.; Thakker, D. R.; Egan, W. *Tetrahedron Lett.* **1989**, *30*, 7141–7144.
112. Vaghefi, A. M.; McKernan, P. A.; Robins, R. K. *J. Med. Chem.* **1986**, *29*, 1389–1393.
113. Glazier, A. Phosphorus Prodrugs. PCT Patent WO19909119721, 1990.
114. Hostetler, K. Lipid Derivatives of Phosphonoacids for Liposomal Incorporation and Method of Use. PCT Patent WO19929413682, 1992.
115. Maloisel, J. L.; Pring, B. G.; Tiden, A. K.; Dypbuck, J.; Liungdahl-Stahle, E. *Antivir. Chem. Chemother.* **1999**, *10*, 333–345.
116. Mao, J. C.; Otis, E. R.; von Esch, A. M.; Herrin, T. R.; Fairgrieve, J. S.; Shipkowitz, N. L.; Duff, R. G. *Antimicrob. Agents Chemother.* **1985**, *27*, 197–202.
117. Hutchinson, D. W.; Masson, S. *IRCS Med. Sci.* **1986**, *14*, 176.
118. McKenna, C. E.; Ye, T. G.; Levy, J. N.; Pham, T.; Wen, T.; Starnes, M.; Cheng, Y. C. *Phosphorus, Sulfur Silicon Relat. Elem.* **1990**, *49–50*, 183–186.
119. Åkesson-Johansson, A.; Lernestedt, J. O.; Ringden, O.; Lönnqvist, B.; Wahren, B. *Bone Marrow Transplant.* **1986**, *1*, 215–220.
120. Pope, L. E.; Marceletti, J. F.; Katz, L. R.; Lin, J. Y.; Katz, D. H.; Parish, M. L.; Spear, P. G. *Antivir. Res.* **1998**, *40*, 85–94.
121. Abele, G.; Cox, S.; Bergman, S.; Lindborg, B.; Vissgården, A.; Karlström, A.; Harmenberg, J.; Wahren, B. *Antivir. Chem. Chemother.* **1991**, *2*, 163–169.
122. Lowe, D. M.; Alderton, W. K.; Ellis, M. R.; Parmar, V.; Miller, W. H.; Roberts, G. B.; Fyfe, J. A.; Gaillard, R.; Ertl, P. *Antimicrob. Agents Chemother.* **1995**, *39*, 1802–1808.
123. Ng, T. I.; Shi, Y.; Huffaker, H.; Kati, W.; Liu, Y.; Chen, C. M.; Lin, Z.; Maring, C.; Kolbrenner, W.; Akhteruzzaman, M. *Antimicrob. Agents Chemother.* **2001**, *45*, 1629–1636.
124. Engelhardt, P.; Högberg, M.; Zhou, X. X.; Lindborg, B.; Johansson, N. G. PCT Patent WO9730051, August 21, 1997.
125. Engelhardt, P.; Högberg, M.; Zhou, X. X.; Lindborg, B.; Johansson, N.G. PCT Patent WO9730052, August 21, 1997.
126. McGuigan, C.; Yarnold, C. J.; Jones, G.; Velazquez, S.; Barucki, H.; Brancale, A. *J. Med. Chem.* **1999**, *42*, 4479–4484.
127. McGuigan, C.; Barucki, H.; Blewett, S.; Carangio, A.; Erichsen, J. T.; Andrei, G. *J. Med. Chem.* **2000**, *43*, 4993–4997.
128. Sienaert, R.; Naesens, L.; Branchale, A.; de Clercq, E.; McGuigan, C.; Balzarini, J. *Mol. Pharmacol.* **2002**, *61*, 249–254.
129. McGuigan, C.; Barucki, H.; Carangio, A.; Blewett, S.; Srinivasan, S.; Andrei, G.; Snoeck, R.; de Clercq, E.; Balzarini, J. *Nucleosides Nucleotides Nucleic Acids* **2001**, *20*, 287–296.
130. Carangio, A.; McGuigan, C.; Cahard, D.; Andrei, G.; Snoeck, R.; de Clercq, E.; Balzarini, J. *Nucleosides Nucleotides Nucleic Acids* **2001**, *20*, 653–656.
131. Blewett, S.; McGuigan, C.; Barucki, H.; Andrei, G.; Snoeck, R.; de Clercq, E.; Balzarini, J. *Nucleosidessides Nucleotidesides Nucleic Acids* **2001**, *20*, 1063–1066.
132. Balzarini, J.; McGuigan, C. *J. Antimicrob. Chemother.* **2002**, *50*, 5–9.
133. Oien, N. L.; Brideau, R. J.; Hopkins, T. A. *Antimicrob. Agents Chemother.* **2002**, *46*, 724–730.
134. Brideau, R. J.; Knechtel, M. L.; Huang, A. *Antivir. Res.* **2002**, *54*, 19–28.
135. Thomsen, D. R.; Orien, N. L.; Hopkinws, T. A.; Knechtel, M. L.; Brideau, R. J.; Wathen, M. W.; Homa, F. L. *J. Virol.* **2003**, 77, 1868–1876.
136. Townsend, L.; Devivare, R. V.; Turk, S. R. *J. Med. Chem.* **1995**, *38*, 4098–4105.
137. Zou, R.; Drach, J. C.; Townsend, L. *J. Med. Chem.* **1997**, *40*, 811–818.
138. Krosky, P. M.; Underwood, M. R.; Turk, S. R. *J. Virol.* **1998**, *72*, 4721–4728.
139. Underwood, M. R.; Harvey, R. J.; Stanat, S. C. *J. Virol.* **1998**, *72*, 717–725.
140. Scheffczik, H.; Savva, C. G.; Holzeenburg, A.; Kolesnikova, L.; Bogner, E. *Nucleic Acids Res.* **2002**, *30*, 1695–1703.
141. Scholz, B.; Rechter, S.; Drach, J.; Townsend, L.; Bogner, E. *Nucleic Acid Res.* **2003**, *31*, 1426–1433.
142. Good, S. S.; Owens, B. S.; Townsend, L.; Drach, J. C. *Antivir. Res.* **1994**, *23*, 103.
143. Chen, J. J.; Wei, Y.; Drach, J.; Townsend, L. *J. Med. Chem.* **2000**, *43*, 2449–2456.
144. Williams, J. D.; Drach, J. C.; Townsend, L. B. *Nucleosides Nucleotides Nucleic Acids* **2004**, *23*, 805–812.
145. Buerger, I.; Reffschlager, J.; Bender, W.; Eckenberg, P.; Popp, A.; Weber, O.; Graeper, S.; Klenk, H. D.; Ruebsamen-Waigmann, H.; Hallenberger, S. *J. Virol.* **2001**, *75*, 9077–9086.

146. Evers, D. L.; Komazin, G.; Shin, D.; Hwang, D. D.; Townsend, L.; Drach, J. C. *Antivir. Res.* **2002**, *56*, 61–72.
147. van Zeijl, M.; Fairhurst, J.; Jones, T. R.; Vernon, S. K.; Morin, J.; LaRocque, J.; Feld, B.; O'Hara, B.; Bloom, J. D.; Johann, S. V. *J. Virol.* **2000**, *74*, 9054–9061.
148. Visali, R. J.; Fairhurst, J.; Srinivas, S.; Hu, W.; Feld, B.; DiGrandi, M.; Curran, K.; Ross, A.; Bloom, J. D.; van Zeijl, M. et al. *J. Virol.* **2003**, *77*, 2349–2358.
149. Crumpacker, C. S.; Schaffer, P. A. *Nat. Med.* **2002**, *8*, 327–328.
150. Crute, J. J.; Grygon, C. A.; Hargrave, K. D.; Simoneau, B.; Faucher, A. M.; Bolger, G.; Kiber, P.; Liuzzi, M.; Cordingley, M. G. *Nat. Med.* **2002**, *8*, 386–391.
151. Crute, J. J.; Lehman, I. R. *J. Biol. Chem.* **1991**, *266*, 4484–4488.
152. Frick, D. *Drug News Perspect.* **2003**, *16*, 355–362.
153. Falkenberg, M.; Elias, P.; Lehman, I. R. *J. Biol. Chem.* **1998**, *273*, 32154–32157.
154. Crute, J. J.; Tsurumi, T.; Zhu, L. A.; Weller, S. K.; Olivo, P. D.; Chalberg, M. D.; Mocarski, E. S.; Lehman, I. R. *Proc. Natl. Acad. Sci. USA* **1989**, *86*, 2186–2189.
155. Kleymann, G. *Drugs Future* **2003**, *28*, 257–265.
156. Zhu, L. A.; Weller, S. K. *J. Virol.* **1992**, *66*, 458–468.
157. Kleymann, G.; Fisher, R.; Betz, U. A.; Hendrix, M.; Bender, W.; Schneider, U.; Handke, K.; Keldenich, J.; Jensen, A.; Kolb, J. et al. *Nat. Med.* **2002**, *8*, 392–398.
158. Fisher, R.; Kleymann, G.; Betz, U. A.; Baumeister, J.; Bender, W.; Eckenberg, P.; Handke, K.; Hendrix, M.; Henniger, K.; Jensen, A. et al. PCT Patent WO0147904.
159. Fisher, R.; Kleymann, G.; Betz, U. A.; Baumeister, J.; Bender, W.; Eckenberg, P.; Handke, K.; Hendrix, M.; Henniger, K.; Jensen, A. et al. Topical Application of Thiazolyl Amides. PCT Patent WO03000259, 2002.
160. Fisher, R.; Kleymann, G.; Betz, U. A.; Baumeister, J.; Bender, W.; Eckenberg, P.; Handke, K.; Hendrix, M.; Henniger, K.; Jensen, A. et al. PCT Patent WO03000260.
161. Kleymann, G. *Curr. Med. Chem.* **2004**, *3*, 69–83.
162. Betz, U. A.; Fischer, R.; Kleymannn, G.; Hendrix, M.; Rubsamen-Waigmann, H. *Antimicrob. Agents Chemother.* **2002**, *46*, 1766–1772.
163. Kearney, P. C.; Sivaraja, M.; Jaen J. C.; Fleygare, J. A.; Medina, J.C. PCT Patent WO9942455, 1998.
164. Spector, F. C.; Liang, L.; Giordano, H.; Sivaraja, M.; Peterson, M. G. *J. Virol.* **1998**, *72*, 6979–6987.
165. Chen, X.; Adrian, J.; Cushing, T.; 41st Interscience Conference on Antimicrob Agents Chemother. (ICAAC), Chicago, IL, 2001, Abstract F-1672.
166. Biron, K. K.; Harvey, R. J.; Chamberlain, S. C.; Good, S. S.; Smith, I. A.; Davis, M. G.; Talarico, C. L.; Townsend, L.; Koszalka, G. W. *Antimicrob. Agents Chemother.* **2002**, *46*, 2365–2372.
167. Williams, S. W.; Hartline, C. B.; Kushner, N. I.; Harden, E. A.; Bidanst, D. J.; Drach, J. C.; Townsend, L.; Undrwood, M. R.; Biron, K. K.; Kern, E. R. *Antimicrob. Agents Chemother.* **2003**, *47*, 2186–2192.
168. Lalezari, J. P.; Aberg, J. A.; Wang, L. H.; Wire, M. B.; Miner, R.; Snowdden, W.; Talarico, C. L.; Shaw, S.; Jacobson, M. A.; Drew, W. L. *Antimicrob. Agents Chemother.* **2002**, *46*, 2969–2976.
169. Zimmermann, A.; Wilts, H.; Lenhardt, M.; Hahn, M.; Mertens, T. *Antivir. Res.* **2000**, *48*, 49–60.
170. Liu, F.; Roizman, B. *J. Virol.* **1991**, *65*, 5149–5156.
171. Welch, A.; Woods, A. S.; MacNally, L. M.; Cotter, R. J.; Gibson, W. *Proc. Natl. Acad. Sci. USA* **1991**, *88*, 10792–10796.
172. Gibson, W.; Welch, A. R.; Hall, M. R. *Prospect. Drug Disc. Des.* **1994**, *2*, 413–426.
173. Tong, L.; Qian, C.; Massariol, M. J.; Bonneau, P. R.; Cordingley, M. G.; Lagace, L. *Nature* **1996**, *383*, 272–275.
174. Qiu, X.; Culp, J. S.; DiLella, A. G.; Hellmig, B.; Hoog, S. S.; Janson, C. A.; Smith, W. W.; Abdel-Meguid, S. S. *Nature* **1996**, *383*, 275–279.
175. Shieh, H. S.; Kurumbail, R. G.; Stevens, A. M.; Stegeman, R. A.; Sturman, E. J.; Pak, J. Y.; Wittwer, A. J.; Palmier, M. O.; Wiegand, R. C.; Holwerda, B. C. et al. *Nature* **1996**, *383*, 279–282.
176. Chen, P.; Tsuge, H.; Almassy, R. J.; Gribskov, C. L.; Katoh, S.; Vaderpool, D. L.; Margosiak, S. A.; Pinko, C.; Matthews, D. A.; Kan, C. C. *Cell* **1996**, *86*, 835–843.
177. Tong, L.; Qian, C.; Massariol, M. J.; Deziel, R.; Yoakim, C.; Lagace, L. *Nat. Struct. Biol.* **1998**, *5*, 819–826.
178. Hoog, S. S.; Smith, W. W.; Qiu, X.; Janson, C. A.; Hellmig, B.; McQueney, M. S.; ÓDonnell, K.; ÓShannessy, D.; DiLella, A. G. et al. *Biochemistry* **1997**, *36*, 14023–14029.
179. Qiu, X.; Janson, C. A.; Culp, J. S.; Richardson, S. B.; Debouk, C.; Smith, W. W.; Abdel-Meguid, S. S. *Proc. Natl. Acad. Sci. USA* **1997**, *94*, 2874–2879.
180. Batra, R.; Khyat, R.; Tong, L. *Protein Pept. Lett.* **2001**, *8*, 333–342.
181. Borthwick, A. D.; Angier, S. J.; Crame, A. J.; Exall, A. M.; Haley, T. M.; Hart, G. J.; Mason, A. M.; Pennell, A. M.; Weingarten, G. W. *J. Med. Chem.* **2000**, *43*, 4452–4464.
182. Borthwick, A. D.; Crame, A. J.; Ertl, P. F.; Exall, A. M.; Haley, T. M.; Hart, G. J.; Mason, A. M.; Pennell, A. M.; Singh, O. M.; Weingarten, G. W. et al. *J. Med. Chem.* **2002**, *45*, 1–18.
183. Borthwick, A. D.; Exall, A. M.; Haley, T. M.; Jackson, D. L.; Mason, A. M.; Weingarten, G. W. *Bioorg. Med. Chem. Lett.* **2002**, *12*, 1719–1722.
184. Field, H. *J. Clinic. Virol.* **2001**, *21*, 261–269.
185. Morfin, F.; Thouvenot, D. *J. Clinic. Virol.* **2003**, *26*, 29–37.
186. Coen, D. M.; Schaffer, P. A. *Proc. Natl. Acad. Sci. USA* **1980**, *77*, 2265–2269.
187. Schnipper, L. E.; Crumpacker, C. S. *Proc. Natl. Acd. Sci. USA* **1980**, *77*, 2270–2273.
188. Christophers, J.; Clayton, J.; Craske, J.; Ward, R.; Collins, P.; Trowbridge, M.; Darby, G. *Antimicrob. Agents Chemother.* **1998**, *42*, 868–872.
189. Boon, R. J.; Bacon, T. H.; Geiringer, K.; Khan, S.; Schultz, M.; Bradford, D. C.; Hodge-Savola, C. *Antivir. Res.* **2000**, *46*, A73.
190. Wade, J. C.; Newton, B.; McLaren, C.; Fluornoy, N.; Keeney, R. E.; Meyers, J. D. *N. Ann. Intern. Med.* **1982**, *96*, 265–269.
191. Englund, J. A.; Zimmerman, M. E.; Swierkosz, E. M.; Goodman, J. L.; Scholl, D. R.; Balfour, H. H. *Ann. Intern. Med.* **1990**, *112*, 416–422.
192. Åkesson-Johansson, A.; Harmenberg, J.; Wahren, B.; Linde, A. *Antimicrob. Agents Chemother.* **1990**, *34*, 2417–2419.
193. Harmenberg, J.; Awan, A.; Alenius, S.; Ståhle, L.; Erlandsson, A.; Lekare, G.; Flink, O.; Augustsson, E.; Larsson, T. *Antiviral Chem. Chemother.* **2003**, *14*, 205–215.
194. Evans, T. G.; Bernstein, D. I.; Raborn, G. W.; Harmenberg, J.; Kowalski, J.; Spruance, S. L. *Antimicrob. Agents Chemother.* **2002**, *46*, 1870–1874.
195. Ganem, D.; Schneider, R. J. In *Fields Virology*, 4th ed.; Knipe, D. M., Howley, P. M., Eds.-in-Chief; Lippincott Williams & Wilkins: Philadelphia, PA, 2001, pp 2923–2971.

196. Hollinger, F. B.; Liang, T. J. In *Fields Virology*, 4th ed.; Knipe, D. M., Howley, P. M., Eds.-in-Chief; Lippincott Williams & Wilkins: Philadelphia, PA, 2001, pp. 2971–3037.
197. Cooksley, W. G. *Semin. Liver. Dis.* **2004**, *24*, 45–53.
198. Cui, L.; Yoon, S.; Schinazi, R. F.; Sommadossi, J. P. *J. Clin. Invest.* **1995**, *95*, 555–563.
199. Furman, P. A.; Davis, M.; Liotta, D.; Paff, M.; Frick, L. W.; Nelson, D. J.; Dornsife, R. E.; Wurster, J. A.; Wilson, L.; Fyfe, J. A. et al. *Antimicrob. Agents Chemother.* **1992**, *36*, 2686–2692.
200. Sells, M. A.; Zelent, A. Z.; Shvartsman, M.; Acs, G. *J. Virol.* **1988**, *62*, 2836–2844.
201. Doong, S. L.; Tsai, C. H.; Schinazi, R. F.; Liotta, D. C.; Cheng, Y. C. *Proc. Natl. Acad. Sci. USA* **1991**, *88*, 8495–8499.
202. Johnson, M. A.; Moore, K. H.; Yuen, G. J. *Clin. Pharmacokinet.* **1999**, *36*, 41–66.
203. Ericsson, S.; Liu, X. Y. In *Recent Advances in Nucleosides: Chemistry and Chemotherapy*; Chu, C. K., Ed.; Elsevier: Amsterdam, The Netherlands, 2002, pp 455–475.
204. Severini, A.; Liu, X. Y.; Wilson, J. S.; Tyrrell, D. L. *Antimicrob. Agents Chemother.* **1995**, *39*, 1430–1435.
205. Chang, C. N.; Skalski, V.; Zhou, J. H.; Cheng, Y. C. *J. Biol. Chem.* **1992**, *267*, 22414–22420.
206. Chang, C. N.; Doong, S. L.; Zhou, J. H.; Beach, J. W.; Jeong, L.; Chu, C. K.; Tsai, C. H.; Cheng, Y. C. *J. Biol. Chem.* **1992**, *267*, 13938.
207. Zhu, Y. L.; Dutchman, G. E.; Liu, S. H.; Bridges, E. G.; Cheng, Y. C. *Antimicrob. Agents Chemother.* **1998**, *42*, 1805–1810.
208. Jarvis, B.; Fauld, D. *Drugs* **1999**, *58*, 101–141.
209. Gray, N. M.; Marr, C. L.; Penn, C. R. *Biochem. Pharmacol.* **1995**, *50*, 1043–1051.
210. Lin, T. S.; Luo, M. Z.; Liu, M. C.; Pai, S. B.; Dutchman, G. E.; Cheng, Y. C. *Biochem. Pharmacol.* **1994**, *47*, 171–174.
211. Bryant, M. L.; Bridges, E. G.; Placidi, L.; Faraj, A.; Loi, A. G.; Pierra, C.; Dukhan, D.; Gosselin, G.; Imbach, J. L.; Hernadez, B. et al. *Antimicrob. Agents Chemother.* **2001**, *45*, 229–235.
212. Chu, C. K.; Ma, L. I.; Olgen, S.; Pierra, C.; Du, J.; Gumina, G.; Gullen, E.; Cheng, Y. C.; Schinazi, R. F. *J. Med. Chem.* **2000**, *43*, 3906–3912.
213. Kim, H. O.; Shanmuganathan, K.; Alves, A. J.; Jeong, L. S.; Beach, J. W.; Schinazi, R. F.; Chang, C. N.; Cheng, Y. C.; Chu, C. K. *Tetrahedron Lett.* **1992**, *33*, 6899–6902.
214. Yamada, K.; Sakata, S.; Yoshimura, Y. *J. Org. Chem.* **1998**, *63*, 6891–6899.
215. Mansour, T. S.; Jin, H.; Wang, W.; Hooker, E. U.; Ashman, C.; Cammack, N.; Salomon, H.; Belmonte, A. R.; Wainberg, M. A. *J. Med. Chem.* **1995**, *38*, 1–4.
216. Yuen, G. J.; Morris, D. M.; Mydlow, P. K. *J. Clin. Pharmacol.* **1995**, *35*, 1174–1180.
217. van Leeruwen, R.; Lange, J. M.; Hussey, E. K. *AIDS* **1992**, *6*, 1471–1475.
218. Perry, C. M.; Faulds, D. *Drugs* **1997**, *53*, 657–680.
219. Chayama, K.; Suzuki, Y.; Kobayashi, M. *Hepatology* **1998**, *27*, 1711–1716.
220. Hoong, L. K.; Strange, L. E.; Liotta, D. C.; Koszalka, G. W.; Burns, C. L.; Schinazi, R. F. *J. Org. Chem.* **1992**, *57*, 5563–5565.
221. Jeong, L. S.; Schinazi, R. F.; Beach, J. W.; Kim, H. O.; Nampalli, S.; Shanmuganathan, K.; Alves, A. J.; McMillan, A.; Chu, C. K.; Mathis, R. *J. Med. Chem.* **1993**, *36*, 181–195.
222. Shewach, D. S.; Liotta, D. C.; Schinazi, R. F. *Biochem. Pharmacol.* **1995**, *45*, 1540–1543.
223. Painter, G. R.; St Clair, M.; Ching, S.; Noblin, J.; Wang, L. H.; Furman, P. A. *Drugs Future* **1995**, *20*, 761–765.
224. Shockcor, J. P.; Wurm, R. M.; Frick, L. W.; Sanderson, P. N.; Farrant, R. D.; Sweatman, B. C.; Lindon, J. C. *Xenobiotica* **1996**, *26*, 189–199.
225. Paff, M. T.; Averett, D. R.; Prus, K. L.; Miller, W. H.; Nelson, D. J. *Antimicrob. Agents Chemother.* **1994**, *38*, 1230–1238.
226. Ma, T.; Yip, Y. Y.; Chen, S. H. *Fron. Biotech. Pharmaceut.* **2004**, *4*, 74–122.
227. Balzarini, J.; Holy, A.; Jindrich, J.; Dvorakova, H.; Hao, Z.; Snoeck, R.; Herdewijn, P.; Johns, D. G.; de Clercq, E. *Proc. Natl. Acad. Sci. USA* **1991**, *88*, 4961–4965.
228. Yokota, T.; Mochizuki, S.; Konno, K.; Mori, S.; Shigeta, S.; de Clercq, E. *Antimicrob. Agents Chemother.* **1991**, *35*, 394–397.
229. Holy, A.; Rosenberg, I. *Collect. Czech. Chem. Commun.* **1987**, *52*, 2801–2829.
230. de Clercq, E.; Holy, A.; Rosenberg, I.; Sakuma, T.; Balzarini, J.; Maudgal, P. C. *Nature* **1986**, *323*, 464–467.
231. Heijtink, R. A.; de Wilde, G. A.; Kruining, J.; Berk, L.; Balzarini, J.; de Clercq, E.; Holy, T. *Antivir. Res.* **1993**, *21*, 141–153.
232. Neyts, J.; de Clerq, E. *Biochem. Pharmacol.* **1994**, *47*, 39–41.
233. Robbins, B. L.; Greenhaw, J.; Connelly, M. C.; Fridland, A. *Antimicrob. Agents Chemother.* **1995**, *39*, 2304–2308.
234. Merta, A.; Votruba, I.; Jindrich, J.; Holy, A.; Cihlar, T.; Rosenberg, I.; Otmar, M.; Herve, T. Y. *Biochem. Pharmacol.* **1992**, *44*, 2067–2077.
235. Balzarini, J.; Hao, Z.; Herdewijn, P.; Johns, D. G.; de Clercq, E. *Proc. Natl. Acad. Sci. USA* **1991**, *88*, 1499–1503.
236. Rosenberg, I.; Holy, A.; Masojidkova, M. *Collect. Czech. Chem. Commun.* **1988**, *53*, 2753–2777.
237. Balzarini, J.; Kruining, J.; Heijtink, R.; de Clercq, E. *Antivir. Chem. Chemother.* **1994**, *5*, 360–365.
238. Ying, C.; Holy, A.; Hockova, D.; Havlas, Z.; de Clercq, E.; Neyts, J. *Antimicrob. Agents Chemother.* **2005**, *49*, 1177–1180.
239. Choi, J. R.; Cho, D. G.; Roh, K. Y.; Hwang, J. T.; Ahn, S.; Jang, H. S.; Cho, W. Y.; Kim, K. W.; Cho, Y. G.; Kim, J. et al. *J. Med. Chem.* **2004**, *47*, 2864–2869.
240. Holy, T. In *Recent Advances in Nucleosides: Chemistry and Chemotherapy*; Chu, C. K., Ed.; Elsevier: Amsterdam, The Netherlands, 2002, pp 167–238.
241. Ying, C.; Holy, A.; Hockova, D.; Havlas, Z.; de Clercq, E.; Neyts, J. *Antimicrob. Agents Chemother.* **2005**, *49*, 1177–1180.
242. Yokota, T.; Konno, K.; Shigeta, S.; Holy, A.; Balzarini, J.; de Clercq, E. *Antivir. Chem. Chemother.* **1994**, *5*, 57–62.
243. Starrett, J. E.; Tortolani, D. R.; Russell, J.; Hitchcock, J. M.; Whiterock, V.; Martin, J. C.; Mansuri, M. M. *J. Med. Chem.* **1994**, *37*, 1857–1864.
244. Cundy, K. C.; Barditch, C. P.; Walker, R. E.; Collier, A. C.; Ebeling, D.; Toole, J.; Jaffe, H. S. *Antimicrob. Agents Chemother.* **1995**, *39*, 2401–2405.
245. Naesens, L.; Neyts, J.; Balzarini, J.; Bischofberger, N.; de Clercq, E. *Antivir. Res.* **1995**, *26*, 3.
246. Srinivas, R. V.; Robbins, B. L.; Connelly, M. C.; Gong, Y. F.; Bischofberger, N.; Fridland, A. *Antimicrob. Agents Chemother.* **1993**, *37*, 2247–2250.
247. Srinivas, R.; Robbins, B. L.; Connelly, M. C.; Gong, Y. F.; Bischofberger, N.; Fridland, A. *Int. Antivir. News* **1994**, *2*, 53–55.
248. Innaimo, S. F.; Seifer, M.; Bisacchi, G. S.; Standring, D. N.; Zahler, R.; Colonno, R. J. *Antimicrob. Agents Chemother.* **1997**, *41*, 1444–1448.
249. Bisacchi, G. S.; Chao, S. T.; Bachard, C.; Daris, J. P.; Innaimo, R.; Colonno, R. J.; Zahler, R. *Bioorg. Med. Chem. Lett.* **1997**, 7, 127–132.
250. Yamanaka, G.; Wilson, T.; Innaimo, S.; Bisacchi, G. S.; Egli, P.; Rinehart, J. K.; Zahler, R.; Colonno, R. J. *Antimicrob. Agents Chemother.* **1999**, *43*, 190–193.
251. Seifer, M.; Hamatake, R. K.; Colonno, R. J.; Standring, D. N. *Antimicrob. Agents Chemother.* **1998**, *42*, 3200–3208.
252. Ruediger, E.; Martel, A.; Meanwell, N.; Solomon, C.; Turmel, B. *Tetrahedron Lett.* **2004**, *45*, 739–742.
253. Yoo, S. J.; Kim, H. O.; Lim, Y.; Kim, J.; Jeong, L. S. *Bioorg. Med. Chem.* **2002**, *10*, 215–226.
254. Dienstag, J. L.; Schiff, E. R.; Wright, T. L.; Perrillo, R. P.; Hann, H. W.; Goodman, Z. *N. Engl. J. Med.* **1999**, *341*, 1256–1263.

255. Lai, C. L.; Chien, R. N.; Leung, N. W.; Chang, T. T.; Guan, R.; Tai, D. I. et al. *N. Engl. J. Med.* **1998**, *339*, 61–68.
256. Schalm, S. W.; Heathcote, J.; Cianciara, J.; Farrell, G.; Sherman, M.; Willems, B. *Gut* **2000**, *46*, 562–568.
257. Liaw, Y. F.; Leung, N. W. Y.; Chang, T. T.; Guan, R.; Tai, D. I.; Ng, K. Y. *Gastroenterology* **2001**, *119*, 172–180.
258. Leung, N. W. Y.; Lai, C. L.; Chang, T. T.; Guan, R.; Lee, C. M.; Ng, K. Y. et al. *Hepatology* **2001**, *33*, 1527–1532.
259. Sarafianos, S. G.; Das, K.; Clark, A. D.; Ding, J.; Boyer, P. L.; Hughes, S. H. *Proc. Natl. Acad. Sci. USA* **1999**, *96*, 10027–10032.
260. Bartholomeusz, A.; Tehan, B. G.; Chalmers, D. K. *Antivir. Ther.* **2004**, *9*, 149–160.
261. Bartholomew, M. M.; Jansen, R. W.; Jeffers, L. J.; Reddy, K. R.; Johnson, L. C.; Bunzendahl, H. *Lancet* **1997**, *349*, 20–22.
262. Li, M. W.; Hou, W.; Wo, J. E.; Liu, K. Z. *Zhejiang Univ. Sci.* **2005**, *6*, 664–667.
263. Gish, R. G.; Trinh, H.; Leung, N.; Chan, F. K.; Fried, M. W.; Wright, T. L.; Wang, C.; Anderson, J.; Mondou, E.; Snow, A. et al. *J. Hepatol.* **2005**, *43*, 60–66.
264. Yuen, M. F.; Lai, C. L. *Exp. Rev. Anti. Infect. Ther.* **2005**, *3*, 489–494.
265. Tenney, D. J.; Levine, S. M.; Rose, R. E.; Walsh, A. W.; Weinheimer, S. P.; Discotto, L.; Plym, M.; Pokornowski, K.; Yu, C. F.; Angus, P. et al. *Antimicrob. Agents Chemother.* **2004**, *48*, 3498–3507.
266. Villeneuve, J. P.; Durantel, D.; Durantel, S.; Westland, C.; Xiong, S.; Brosgart, C. L.; Gibbs, C. S.; Parvaz, P.; Werle, B.; Trepo, C. et al. *J. Hepatol.* **2003**, *39*, 1085–1089.
267. Maynard, M.; Parvaz, P.; Durantel, S.; Chevallier, M.; Chevallier, P.; Lot, M.; Trepo, C.; Zoulim, F. *J. Hepatol.* **2005**, *42*, 279–281.
268. Korba, B. E.; Boyd, M. R. *Antimicrob. Agents Chemother.* **1996**, *40*, 1282–1284.
269. Ying, C.; de Clercq, E.; Nicholson, W.; Furman, P.; Neyts, J. *J. Viral. Hepatol.* **2000**, *7*, 161–165.
270. Davis, M. G.; Wilson, J. E.; van Draanen, N. A.; Miller, W. H.; Freeman, G. A.; Daluge, S. M.; Boyd, F. L.; Aulabaugh, A. E.; Painter, C. R.; Boone, L. R. *Antiviral Res.* **1996**, *30*, 133–145.
271. Rayes, N.; Seehofer, D.; Bechstein, W. O.; Muller, A. R.; Berg, T.; Neuhaus, R.; Neuhaus, P. *Clin. Transplant.* **1999**, *13*, 447–452.
272. Singh, N.; Gayowski, T.; Wannstedt, C. F.; Wagner, M. M.; Marino, I. R. *Transplantation* **1997**, *63*, 1415–1419.
273. Kruger, M.; Boker, K. H.; Zeidler, H.; Manns, M. P. *J. Hepatol.* **1997**, *26*, 935–939.
274. Zoulim, F. *J. Clin. Virol.* **2001**, *21*, 243–253.
275. Mutimer, D. *J. Clin. Virol.* **2001**, *21*, 239–242.
276. Hantz, O.; Allaudeen, H. S.; Ooka, T.; de Clercq, E.; Trepo, C. *Antivir. Res.* **1984**, *4*, 187–199.
277. Berk, L.; Schalm, S. W.; de Man, R. A.; Heytink, R. A.; Berthelot, P.; Brechot, C.; Boboc, B.; Degos, F.; Marcellin, P.; Benhamou, J. P. et al. *J. Hepatol.* **1992**, *14*, 305–309.
278. Perigaud, C.; Gosselin, G.; Girardet, J.-L.; Korba, B. E.; Imbach, J.-L. *Antivir. Res.* **1999**, *40*, 167–178.
279. Hantz, O.; Perigaud, C.; Borel, C.; Jamard, C.; Zoulim, F.; Trepo, C.; Imbach, J.-L.; Gosselin, G. *Antivir. Res.* **1999**, *40*, 179–187.
280. Hostetler, K. Y.; Beadle, J. R.; Hornbuckee, W. E.; Bellezza, C. A.; Tochkov, I. A.; Cote, P.; Gerin, J. L.; Korba, B. E.; Tennant, B. C. *Antimicrob. Agents Chemother.* **2000**, *44*, 1964–1969.
281. Gumina, G. G.; Song, G. Y.; Chu, C. K. *Antivir. Chem. Chemother.* **2001**, *12*, 93–117.
282. Schröder, I.; Holmengren, B.; Öberg, M.; Löfgren, B. *Antivir. Res.* **1998**, *37*, 57–66.
283. Kumar, R.; Agrawal, B. *Curr. Opin. Invest. Drugs* **2004**, *5*, 171–178.
284. Zhou, X.X.; Wähling, H.; Johansson, N.G. PCT Patent WO9909031, 1999.
285. Lee, K.; Choi, Y.; Gullen, E.; Schlueter-Wirtz, S.; Schinazi, R. F.; Cheng, Y. C.; Chu, C. K. *J. Med. Chem.* **1999**, *42*, 1320–1328.
286. Staschke, K. A.; Colacino, J. M.; Mabry, T. E.; Jones, C. D. *Antivir. Res.* **1994**, *23*, 45–61.
287. Korba, B. E.; Gerin, J. L. *Antivir. Res.* **1992**, *19*, 55–70.
288. Fourel, I.; Li, J.; Hantz, O.; Jacquet, C.; Fox, J. J.; Treppo, C. *J. Med. Virol.* **1992**, *37*, 122–126.
289. Fried, M. W.; Di Bisceglie, A. M.; Strauss, S. E.; Savarese, B.; Beames, M. P.; Hoofnagle, J. H. *Hepatology* **1992**, *16*, 861–864.
290. Parker, W. B.; Cheng, Y. C. *J. NIH Res.* **1994**, *6*, 57–61.
291. Lewis, W.; Levine, E. S.; Griniuviene, B.; Tankersley, K. O.; Colacino, J. M.; Sommadossi, J. P.; Watanabe, K. A.; Perrino, F. W. *Proc. Natl. Acad. Sci. USA* **1996**, *93*, 3592–3597.
292. Cui, L.; Yoon, S.; Schinazi, R. F.; Sommadossi, J. P. *J. Clin. Invest.* **1995**, *95*, 555–563.
293. Gosselin, G.; Imbach, J. L.; Bryant, L. PCT Patent WO0009531, 1999.
294. Standring, D. N.; Bridges, E. G.; Placidi, L.; Faraj, A.; Loi, A. G.; Pierra, C.; Dukhan, D.; Gosselin, G.; Imbach, J. L.; Hernandez, B. et al. *Antivir. Chem. Chemother.* **2001**, *12*, 119–129.
295. Bryant, M.; Bridges, E. G.; Placidi, L.; Faraj, A.; Loi, A.; Pierra, C.; Benzaria, S.; Dukhan, D.; Gosselin, G.; Imbach, J. L. et al. In *Frontiers in Viral Hepatitis*; Schinazi, R. F. et al. Eds.; Elsevier: Amsterdam, The Netherlands, 2003, pp 245–261.
296. Placidi, L.; Hernandez, B.; Cretton-Scott, E.; Faraj, A.; Brynat, M.; Imbach, J. L.; Gosselin, G.; Pierra, C.; Dukhan, D.; Sommadossi, J. P. *Antivir. Ther.* **1999**, *4*, A122.
297. Lin, T. S.; Luo, M. Z.; Liu, M. C.; Pai, S. B.; Dutschman, G. E.; Cheng, Y. C. *J. Med. Chem.* **1996**, *39*, 1757–1759.
298. Bryant, M.; Bridges, E. G.; Placidi, L.; Faraj, A.; Loi, A.; Pierra, C.; Dukhan, D.; Gosselin, G.; Imbach, J. L.; Hernandez, B. et al. *Antimicrob. Agents Chemother.* **2001**, *45*, 229–235.
299. Lee, K.; Choi, Y.; Gullen, E.; Schluetter-Wirtz, S.; Schinazi, R. F.; Cheng, Y. C.; Chu, C. K. *J. Med. Chem.* **1999**, *42*, 1320–1328.
301. Semizarov, D. G.; Arzumanov, A. A.; Dyatkina, N. B.; Meyer, A.; Vichier-Guerre, S.; Gosselin, G.; Rayner, B.; Imbach, J. L.; Krayevsky, A. A. *J. Biol. Chem.* **1997**, *272*, 9556–9560.
302. Ma, T.; Lin, T. S.; Newton, M. G.; Cheng, Y. C.; Chu, C. K. *J. Med. Chem.* **1997**, *40*, 2750–2754.
303. Chu, C. K.; Ma, T.; Shanmuganathan, K.; Wang, C.; Xiang, Y.; Pai, S. B.; Yao, G. Q.; Sommaddossi, J. P.; Cheng, Y. C. *Antimicrob. Agents Chemother.* **1995**, *39*, 979–981.
304. Lee, J.; Shim, H. S.; Partk, Y. K.; Park, S. J.; Shin, J. S.; Yang, W. Y.; Lee, H. D.; Park, W. J.; Chung, Y. H.; Lee, S. W. *Bioorg. Med. Chem. Lett.* **2002**, *12*, 2715–2717.
305. Yoon, S. J.; Kim, J. W.; Huh, Y.; Rho, H. M.; Jung, G. H. US Patent 5,968,781, 1999.
306. Lin, Y. M.; Zembower, D. E.; Flavin, M. T.; Schure, R. M.; Anderson, H. M.; Korba, B. E.; Chen, F. C. *Bioorg. Med. Chem. Lett.* **1997**, *7*, 2325–2328.
307. Deres, K.; Schröder, C. H.; Paessens, A.; Goldmann, S.; Hacckedr, H. J.; Weber, O.; Krämer, T.; Niewöhner, U.; Pleiss, U.; Stoltefuss, J. et al. *Science* **2003**, *299*, 893–896.
308. Weber, O.; Schlemmer, K. H.; Hartmann, E.; Hagelschuer, I.; Paessens, A.; Graef, E.; Deres, K.; Goldmann, S.; Niewoehner, U.; Stoltefuss, J. et al. *Antivir. Res.* **2002**, *54*, 69–78.

309. King, R. W.; Ladner, S. K.; Miller, T. J.; Zaifert, K.; Perni, R. B.; Conway, S. C.; Otto, M. J. *Antimicrob. Agent Chemother.* **1998**, *42*, 3179–3186.
310. Ladner, S. K.; Miller, T. J.; King, R. W. *Antimicrob. Agents Chemother.* **1998**, *42*, 2128–2131.
311. Perni, R. B.; Conway, S. C.; Ladner, S. K.; Zaifert, K.; Otto, M. J.; King, R. W. *Bioorg. Med. Chem. Lett.* **2000**, *10*, 2687–2690.
312. Block, T. M.; Jordan, R. *Antivir. Chem. Chemother.* **2001**, *12*, 317–325.
313. Mehta, A.; Carrouee, S.; Conyers, B.; Jordan, R.; Butters, T.; Dwek, R. A.; Block, T. M. *Hepatology* **2001**, *33*, 1488–1495.
314. Mehta, A.; Conyers, B.; Tyrrell, D. L.; Walters, K. A.; Tipples, G. A.; Dwek, R. A.; Block, T. M. *Antimicrob. Agents Chemother.* **2002**, *46*, 4004–4008.
315. Entecavir (BMS-200475), Antiviral Drugs Advisory Committee (AVDAC) Briefing Document NDA 21-797, NDA 21-79, March 11, 2005.

Comprehensive Medicinal Chemistry II
ISBN (set): 0-08-044513-6

ISBN (Volume 7) 0-08-044520-9; pp. 295–327

7.12 Ribonucleic Acid Viruses: Antivirals for Human Immunodeficiency Virus

T A Lyle, Merck Research Laboratories, West Point, PA, USA

7.12.1 Disease State

7.12.1.1 Origins: Manifestation, Symptoms

In the summer of 1981, an unusual number of cases of *Pneumocystis carinii* pneumonia (PCP), primarily in young homosexual male patients, was reported to the Centers for Disease Control and Prevention (CDC) by physicians in Los Angeles, CA.[1] This type of pneumonia was uncommon, and was normally observed in elderly or severely immunosuppressed patients. Subsequently, a number of other clusters of rare opportunistic infections and cancers such as Kaposi's sarcoma (KS) were reported in major metropolitan areas throughout the USA in this same patient population.[2,3] This pattern of diseases in otherwise healthy young adults was named acquired immunodeficiency syndrome (AIDS). An examination of USA medical records indicated that AIDS cases had been occurring since 1979,[4] and that the disease also affected female heterosexuals, intravenous drug users, and hemophiliacs.[5] A common trait among all AIDS patients was a severely depleted number of T helper immune cells,[4] resulting in the host's inability to mount an effective immune response to other infectious agents or carcinogens. Eventually, the cause of AIDS would be traced to a retrovirus (*see* Section 7.12.2.1) now called the human immunodeficiency virus (HIV). A thorough genetic analysis of HIV obtained from individuals infected in the early stages of the epidemic has led to an estimate of 1967 when the virus first entered the USA.[6] Similar genetic analyses and comparison with the related simian immunodeficiency viruses (SIV) have provided evidence that HIV-1 is derived from the SIVcpz strain, and made the jump from chimpanzees to humans as early as the 1930s, and possibly much earlier.[6] Similarly, HIV-2 appears to be related to the SIVsm strain that occurs in sooty mangabeys.

Because the number of reported cases of AIDS appeared to be growing very rapidly, a concerted effort was made in laboratories around the world to identify the causative agent. From epidemiological studies, it soon became apparent that the infection was spread by contact with bodily fluids between individuals.

7.12.1.2 Disease Prevalence

In December 2005, over 40 million people were estimated to be infected with HIV worldwide, including approximately 1.2 million cases in North America, and 720 000 in Western Europe. By far the most affected region in the world is sub-Saharan Africa where it is estimated that over 25 million people are infected with HIV. Another 7.4 million persons are believed to be infected in South and Southeast Asia, a region that includes the highly populated countries of India, China, and Thailand. Approximately, 3.1 million deaths are estimated to have occurred in 2005, including 570 000 children under the age of 15.[7]

7.12.1.3 Symptomatic Classification

Infection with HIV is often followed by the appearance of influenza-like symptoms such as fever, enlarged lymph nodes, or headache within several weeks after initial exposure to the virus. These initial symptoms may last up to 1 month and are followed by a so-called 'latent period' in which the patient may be relatively symptom-free for several years.

Until 1993, a patient was considered to have AIDS when he or she presented with severe symptoms from a serious disease or diseases that were attributable to a compromised immune system caused by infection with HIV.

More recently, a more objective method of classifying the course of infection has been established. The current CDC classifications for HIV infection are based on the number of $CD4^+$ T lymphocytes per microliter of blood, which is inversely correlated with the risk of developing opportunistic infections. Patients with greater than 500 cells per microliter are classified as Category 1, those with 200–499 as Category 2, and those with less than 200 are defined as Category 3. There are three categories for the clinical classification of HIV infection. Category A includes patients who are HIV positive and are asymptomatic, have lymphadenopathy, or have symptoms associated with primary HIV infection such as fever or influenza-like symptoms. Category B includes patients with a spectrum of less severe diseases attributable to HIV infection including thrush, shingles, or peripheral neuropathy. Patients with the most severe diseases caused by a compromised immune system are included in Category C. Such diseases may include PCP, KS, Burkitt's lymphoma, *Mycobacterium tuberculosis*, and cytomegalovirus eye infection.[8,9] A patient is now considered to have AIDS when his or her $CD4^+$ count is either less than 200 or a $CD4^+$ percentage of less than 14, along with having one of the diseases defined in Category C.[9]

7.12.2 Disease Basis

7.12.2.1 Identification of Infectious Agent: Human Immunodeficiency Virus

Because AIDS patients exhibited a wide range of opportunistic infections and rare cancers, it was difficult to pinpoint the causative agent or agents. A number of viral, bacterial, fungal, and noninfectious causes were considered; however, none could explain all of the cases reported at the time. It was known that the disease caused a significant decrease in $CD4^+$ T cells, but the inability to propagate such cells in culture was a major obstacle in the search for a cause. A key discovery was the identification of a new growth factor, now known as interleukin-2 (IL2), which facilitated the growth of T cells in culture.[10] Several laboratories uncovered evidence for the involvement of retroviruses related to human T cell lymphoma virus (HTLV)[11,12] or feline immunodeficiency virus (FIV).[13] Using blood collected from AIDS patients, it was now possible to grow the virus in cell culture, and it was soon isolated and characterized. Initially, the virus was known either as HTLV-III,[14] lymphadenopathy-associated virus (LAV), or immune deficiency associated virus (IDAV).[15] It was soon discovered that the virus belonged to a family of ribonucleic acid (RNA) lentiviruses known as Retroviridae, and it was given the name HIV.

7.12.2.2 Viral Life Cycle

A diagram depicting the various stages of HIV infection of a host cell is shown in **Figure 1**. In the first phase, the HIV viral particle (virion) attaches itself to the host cell membrane. This event involves binding of the viral coat protein GP-120 to specific CD4 receptors on the human T cell. This is accompanied by a binding event that involves a second site on the T cell: either the CXCR-4 or CCR-5 receptor, depending on the strain of the virus.[16] This is followed by a process termed fusion in which the viral membrane fuses with the host cell membrane, allowing the contents of the virus to enter the host cell. The second phase includes uncoating of the viral core, which permits the viral contents to enter the cytoplasm of the host cell.[17] Among these contents are two copies of a single-stranded RNA genome that encodes a number of structural, regulatory, and accessory proteins. In addition, the three critical viral enzymes reverse transcriptase (RT), integrase, and protease are introduced into the host cell. The HIV genome is flanked by two identical sequences called long terminal repeats (LTRs). Just downstream of the 5′ LTR is a region called the primer binding site (PBS), and just upstream of the 3′ LTR is a purine-rich region called the polypurine tract (PPT). The first viral enzyme, RT, is responsible for producing a complementary double-stranded DNA (dsDNA) copy of the genome,[18] called proviral DNA. In **Figure 2**, the x-ray crystal structure[19] of RT shows that the enzyme exists as a heterodimer consisting of a p51 protein (shown in yellow) complexed with a larger p66 protein (shown in cyan). The p66 portion has been likened to an upward-facing right hand with regions corresponding to the fingers, palm, and thumb. This remarkable enzyme functions as an RNA-dependent DNA polymerase, a DNA-dependent DNA polymerase, and a riboendonuclease (RNaseH). The polymerase active site is near the palm domain, and the RNaseH active site is near the thumb region as indicated in **Figure 2**. Starting from a single strand of viral RNA (shown in red) as the template, a primer from the host cell (lysine transfer RNA) binds to the primer binding site on the template as shown in step 2 (**Figure 3**). In step 3, DNA synthesis is initiated by extending the 3′ end of the primer in a stepwise fashion by incorporation of the appropriate deoxyribonucleotide complementary to the ribonucleotide template of the viral RNA genome. The complementary DNA is shown in green. In step 4, the ribonuclease (RNaseH) activity of RT is responsible for removal of the 5′ LTR RNA from the RNA–DNA double-stranded hybrid. This is followed in step 5 by a process called strand transfer in which the newly synthesized DNA with the tRNA primer still attached relocates to the 3′ end of the RNA and binds to the complementary 3′ LTR. Step 6 illustrates further DNA synthesis, followed by the selective action of RNaseH, which removes the remaining viral RNA except for the PPT (step 7). This small segment of RNA remains bound to the new single plus-strand DNA to act as a new primer for DNA synthesis of the complementary minus-strand. As shown in step 8, DNA synthesis proceeds from the PPT toward the tRNA primer. In step 9, RNaseH removes the remaining RNA, followed by circularization driven by the binding of the two complementary DNA PBS regions (step 10). DNA synthesis proceeds from the PBS in a clockwise fashion to the end of the 3′ LTR shown in step 11. In step 12, DNA synthesis of the complementary 5′ LTR completes the formation of the proviral dsDNA. (A more detailed description of reverse transcription is described by Wainberg and co-workers.[18]) The polymerase activity of RT was the target for the development of the first anti-HIV drug, azidothymidine (AZT), from the nucleoside class of RT inhibitors (NRTIs).

In the next step of viral replication (**Figure 1**, phase 3), the proviral DNA is inserted into the host cell genome by the second viral enzyme HIV integrase.[20,21] As shown in **Figure 4**, the first step entails the formation of a preintegration complex (PIC) between the integrase enzyme and the newly synthesized dsDNA. This is followed by integrase-catalyzed cleavage of the two terminal nucleotides from the 3′ ends of the ds-DNA (step 2). This cleavage

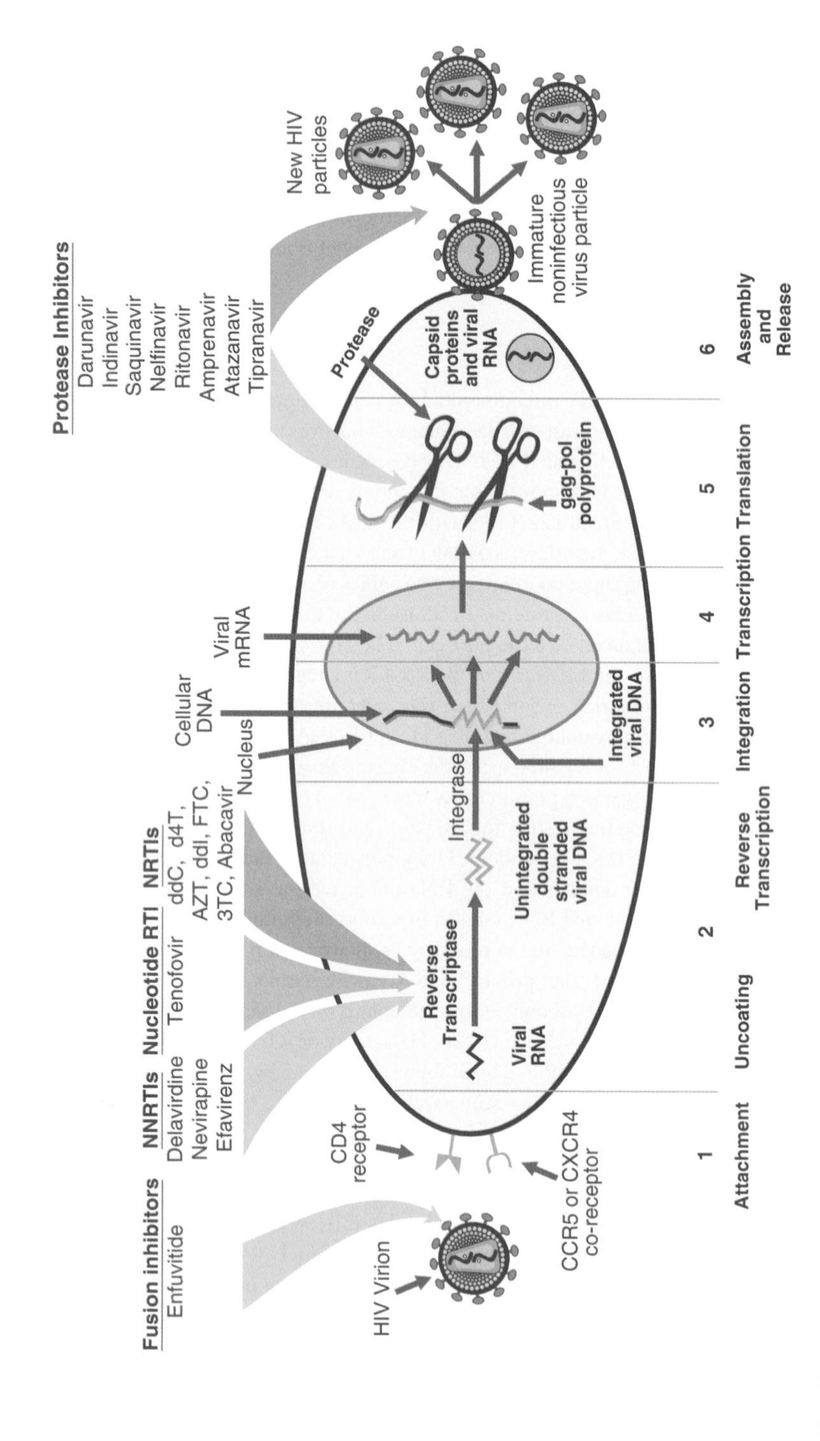

Figure 1 Viral life cycle of human immunodeficiency virus.

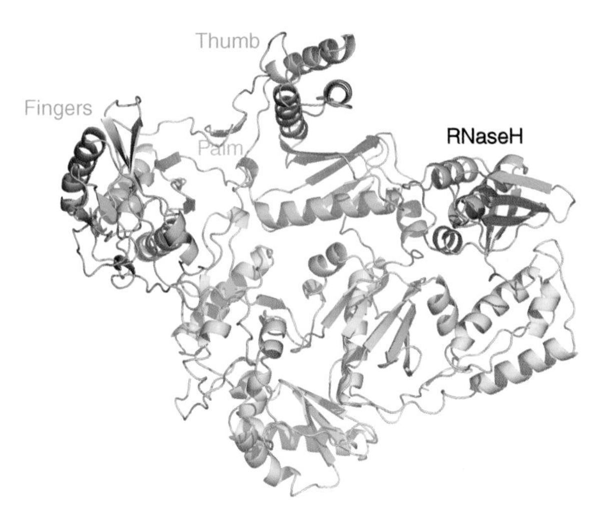

Figure 2 X-ray crystal structure of native reverse transcriptase. (All figures showing crystal structures were created using PyMol: Warren L. DeLano 'The PyMOL Molecular Graphics System.' DeLano Scientific LLC, San Carlos, CA, USA. http://www.pymol.org).

event is specific for CAGT terminal sequences, and cleavage occurs between the A and G sites. This action is termed 3′ processing, and is required for the subsequent insertion of the proviral DNA into the host cell genome. After the PIC enters the nucleus of the cell and binds to the host cell DNA (step 3a), integrase then catalyzes a *trans*-esterification involving the attack of the 3′ OH processed ends of the viral DNA onto the host DNA, in a process called strand transfer (step 3b). This phosphate *trans*-esterification reaction results in cleavage of the host DNA as well as joining the viral DNA to the host DNA. The sites along the host DNA at which strand transfer occurs are apparently random. After cellular enzymes remove the two extra nucleotides on the 5′ ends of the viral DNA, the resulting gaps at the viral DNA–host DNA junctions are repaired to complete the integration process.[21]

As shown in **Figure 1** phase 4, viral transcription factors[22] cause the normal cellular machinery to produce multiple copies of viral messenger RNA, some of which are translated (**Figure 1**, phase 5) into polyproteins containing the sequences of all the viral enzymes and proteins. The final steps in the life cycle of HIV involve the assembly of the polyproteins and viral RNA at the inner surface of the host cell membrane, followed by budding[23] and release of the new viral particles as shown in **Figure 1**, phase 6. During these final steps, the third viral enzyme HIV protease cleaves the polyproteins to the proper length, a process called maturation.[24]

7.12.2.3 Routes of Transmission

Through a variety of epidemiological and virology studies, it has been determined that HIV is spread by intimate contact with bodily fluids.[8] The virus may be transmitted by contact with blood, semen, or genital secretions, and even though the virus can be detected in the saliva of infected individuals, there is very little evidence to suggest that this is a viable route of transmission.[25] In the USA, the most common mode of transmission is via unprotected sexual encounters. Early in the epidemic, blood transfusions and contaminated products derived from blood used by hemophiliacs were also a significant problem. Due to stringent surveillance of the blood supply in the USA, this is no longer the case. In many parts of the world, transmission via contaminated needles used by intravenous drug users is a major route of transmission. Another significant mode of viral infection is from an infected mother to her child, otherwise known as vertical transmission.[26] Such infections can occur prenatally or postnatally via the baby's ingestion of contaminated milk during breastfeeding. Given the intimate relationship between mother and fetus, it is remarkable that a significant number of uninfected children are born to infected mothers. In the absence of treatment, it has been estimated that the vertical transmission rate of HIV is approximately 25%. Treatment of infected mothers with AZT monotherapy during the latter months of pregnancy can reduce the baby's infection rate by more than half.[27]

A potentially serious emerging health crisis involves the spread of drug-resistant HIV by infected individuals who have received suboptimal treatment with antiviral agents. It has been estimated that between 1% and 25% of newly infected individuals carry HIV with at least one mutation known to harbor resistance to one of the currently marketed drugs,[28] depending on the population sampled. It stands to reason that treatment options for dealing with these viruses can be limited.

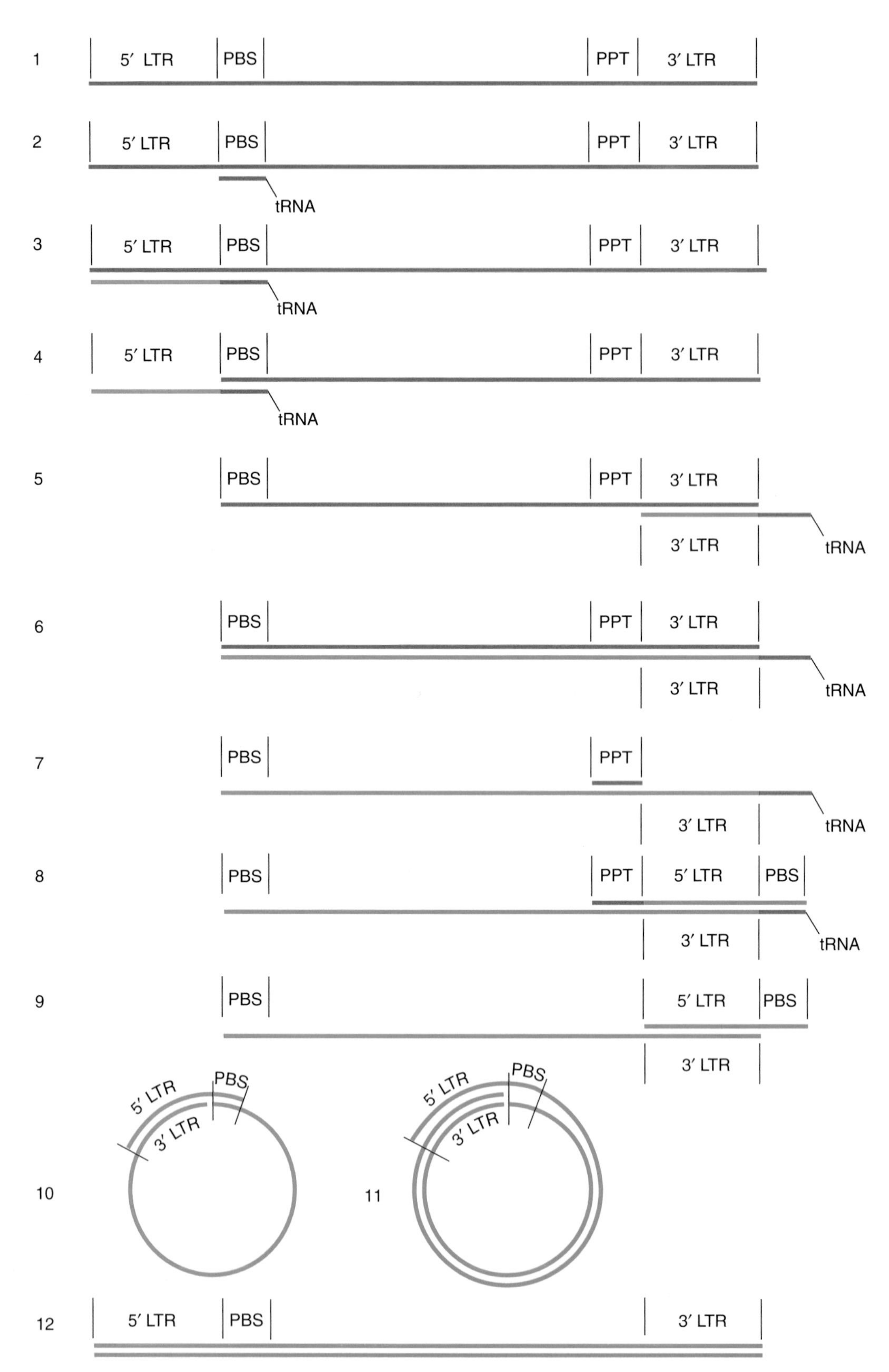

Figure 3 Steps involved in reverse transcription.

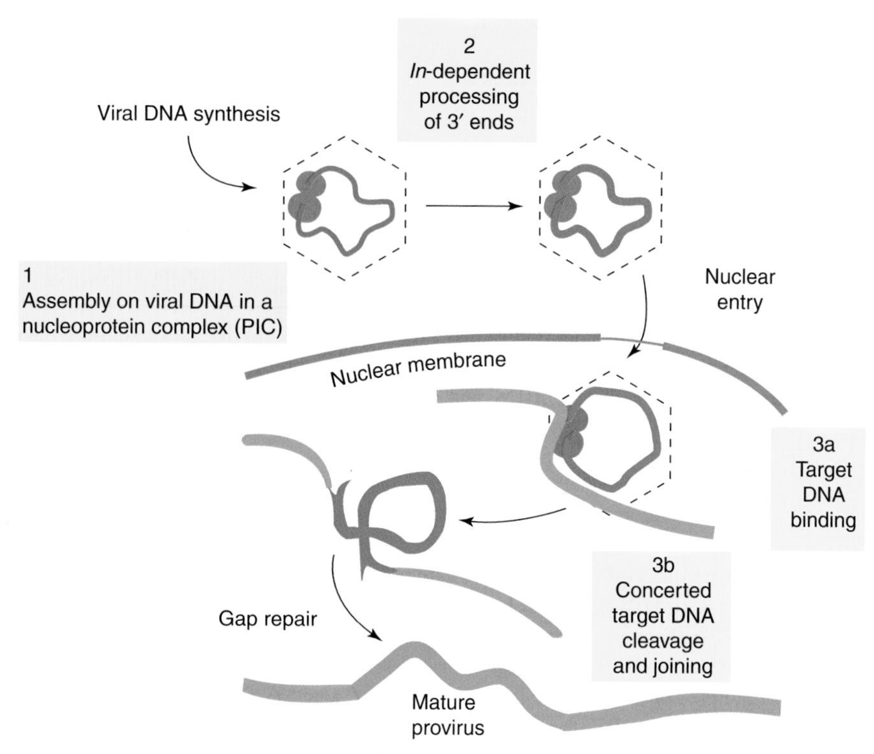

Figure 4 Integration of proviral DNA.

7.12.2.4 Disease Severity

Because AIDS is defined by the manifestation of opportunistic infections and cancers, the severity of AIDS can be defined by the symptoms resulting from these secondary diseases. A more clinically useful definition is based on the severity of HIV infection, and the resulting effect on the immune system of the patient. These definitions have been quite useful in deciding when to initiate antiviral therapy, and what particular treatment regimens should be used. Early in the course of the epidemic, levels of p24 antigen (a major core protein) were used to determine the course of the disease.[29] However, the sensitivity of this assay was shown to be insufficient in some asymptomatic patients, and the measurement of plasma viral RNA load was adopted.[30,31] This is one of two major biochemical markers that are utilized by clinicians to determine the severity of infection in infected patients, and is reported as the number of copies of viral RNA per volume of plasma, typically copies per milliliter. Concentrations of viral RNA as low as 5 copies mL^{-1} can be detected, and viral titers of 10^8 copies mL^{-1} are sometimes observed in the acute phase of infection. Literature reports of 'undetectable' viral RNA commonly refer to assays in which less than either 400 or 50 copies mL^{-1} are observed.

The second biochemical assay used is the number of T cells that express the CD4 receptor on their cell surface per volume of plasma. This assay is an indication of the health of the patient's immune system,[32] and correlates with the propensity to develop opportunistic infections.[33] Typically, these $CD4^+$ cell counts decrease from a normal level of approximately 750 mm^{-3} to less than 50 mm^{-3} as the disease progresses.[8] Current treatment guidelines[34] recommend that initiation of antiviral therapy should be deferred until a patient's $CD4^+$ count drops below 350 mm^{-3}.

7.12.2.5 Viral Classification: Geographical Distribution of Clades

AIDS is caused by two human immunodeficiency viruses, HIV-1 and HIV-2. HIV-2 is found primarily in western Africa, and is significantly less pathogenic[35] and infectious[36] than HIV-1. The vast majority of infections occur worldwide with HIV-1. Genetic sequencing of the HIV-1 viral genome has led to its classification into three major groups: M (main), O (outlier), and N (non M, non O). The M group is responsible for the current worldwide epidemic, and has been further subclassified into clades A through K based on the similarity of their genetic sequences.[6] These genetic subtypes are geographically distributed. For example, the predominant clade found in Western Europe and the USA is clade B, clade C is found in China, India, and South Africa, whereas clades A and D are most commonly found in East Africa.

7.12.3 Experimental Disease Models

7.12.3.1 Viral Replication in Cell Culture

A cornerstone for the preclinical assessment of therapeutic agents directed at HIV has been a cell culture-based assay in which infected cells are added to culture medium containing uninfected cells. In the absence of drug, the viral infection will spread to the uninfected cells over time. After a period of a few days depending upon the type of cells and virus used, the extent of infection is commonly determined by the amount of HIV-specific p24 antigen present. The amount of antigen is conveniently measured by employing an enzyme-linked immunosorbance assay (ELISA).[37] Typically the assay is run in the presence of increasing concentrations of drug, and the concentration of drug that provides 50%, 90%, or 95% reduction in p24 levels is determined, yielding an inhibitory concentration (e.g., IC_{50}). Because the activity of many drugs may be compromised as a result of binding to plasma proteins, the assay may also be run in the presence of human serum.[38] This type of assay has been highly predictive of the efficacy of a variety of drugs that have been approved for treatment of HIV/AIDS, and is used extensively in research.

7.12.3.2 Simian Models

Because exposure to HIV does not infect nonhuman mammals, it has been difficult to develop a reliable animal model for AIDS research. The significant exception to this has been the use of SIV to infect nonhuman primates. Antibodies to SIV can be detected in a wide variety of nonhuman primate populations, with an incidence of less than 1% to greater than 34%.[39,40] In sharp contrast to HIV, SIV often has no significant effect on the health of these animals. However, there are SIV strains that are pathogenic in rhesus macaques that have been used as models for AIDS. In order to reproduce the devastating effects of HIV in nonhuman primates, hybrid viruses have been created that are genetically engineered to contain elements of both SIV and HIV, in which one or more SIV genes are replaced by the corresponding HIV genes.[41] These viruses, termed SHIVs, are designed to retain the ability to infect nonhuman primates, and can often cause lethal AIDS-like symptoms in infected animals.[42] This animal model has been useful in proof of concept studies for both vaccines and small-molecule therapeutics.[43,44]

7.12.4 Clinical Trial Issues

7.12.4.1 Resistance

Early experience with the first drug (AZT) to be studied for the treatment of HIV infection in a clinical setting gave an inkling of the problems that resistance would pose. Viral isolates from patients treated with AZT for longer than 6 months were found to have reduced susceptibility to the drug.[45] This pattern was ultimately attributed to the emergence of a viral population that was resistant to AZT. Analysis of the resistant viral genome revealed three or more mutations in the RT sequence that caused the loss of sensitivity.[46] The emergence of resistance was much more rapid when the first nonnucleoside reverse transcriptase inhibitors (NNRTIs) were clinically evaluated.[47] Virtually, every drug approved for the treatment of HIV has caused the selection of resistant virus in some patients.

7.12.4.2 Compliance

Because the activity of many antiretroviral drugs depends upon sustained plasma concentrations in patients, it is important that these drugs be administered as indicated. Inadequate plasma concentrations of antiretroviral drugs is highly correlated with the development of resistance.[48]

7.12.4.2.1 Tolerability: side effects

Certain drugs can have very unpleasant side effects that can negatively impact compliance. Nausea, vomiting, perioral numbness, neurological effects, and poor palatability are some of the reasons that patients might miss dosing (*see* Section 7.12.5).

7.12.4.2.2 Dosing regimens

The dosing recommendations for several drugs that have been approved for the treatment of HIV/AIDS require patients to take multiple pills or capsules up to three times daily, some with meals, and some before or after meals. Because most patients must take at least three different antiretroviral drugs in addition to other medications they might require, this often leads to impractical and confusing dosing regimens that often result in missed or inappropriate dosing.

7.12.4.3 Combination Chemotherapy

7.12.4.3.1 Cross-resistance

Soon after the first clinical studies with AZT were under way, the issue of cross-resistance was investigated. Viral isolates from patients who had failed on AZT monotherapy were found to also be resistant to other similar drugs within the same class (NRTIs).[49] With the advent of NNRTIs, the issue of cross-resistance was much more serious, because single amino acid changes within RT could confer broad cross-resistance to all of the approved agents within this class.[50] Cross-resistance to protease inhibitors (PIs) has also become a problem; however, multiple amino acid changes within the protease sequence are usually required for high-level resistance.[51]

7.12.4.3.2 Drug interactions

Combination chemotherapy of HIV/AIDS can involve a bewildering number of possible combinations of the numerous approved drugs. Because drugs of the same class are often metabolized by the same metabolic enzymes, drugs that interact with one member of a class will often interact with most other agents within that class.[52] For example, many PIs and NNRTIs are oxidatively metabolized by cytochrome P450 3A4 (Cyp3A4). When administered with a potent inhibitor or inducer of Cyp3A4, dose adjustment is often required. In practice, such drug interactions are somewhat unpredictable and must be determined in clinical studies. In addition to pharmacokinetic drug interactions, the resistance profile of HIV drugs, as well as any preexisting drug resistance mutations in the patient's viral population, must also be considered when prescribing drug combinations.

7.12.4.3.3 Treatment sequencing

Because of the large number of approved drugs, treatment regimens can now be crafted that will give the greatest benefit to the individual patient. Factors such as likelihood of compliance, preexisting resistance, and tolerability can be useful in helping physicians design the most effective drug combinations.[53] Treatment failure on any particular regimen is quite common, so subsequent treatment options must also be considered. For example, some patients will incur treatment failure due to the various side effects associated with PIs, so ideally they should be able to switch to potent regimens that do not include a PI.

7.12.4.4 Structured Treatment Interruptions

Normally, treatment of HIV/AIDS requires continuous daily administration of several antiretroviral drugs for the life of the patient. Because of the severity of drug-induced side effects experienced by some patients, new therapeutic strategies have been evaluated that stop treatment for time periods of days to weeks. These 'drug holidays' allow patients to minimize drug-induced side effects and the inconvenience of taking the drugs on a daily basis.[54] Most of these strategies are based on various fixed time periods when patients are either on or off therapy, and have met with variable success, depending on the design of the trial. More recently, the trigger for going on or off therapy has been tied to either viral load or $CD4^+$ cell count.[55,56]

7.12.4.5 Treatment Naïve versus Treatment Experienced Patients

The requirements for registration and regulatory approval for new antiretroviral agents may include a demonstration of an antiviral effect as monotherapy in previously untreated infected patients. Due to the number of new agents entering clinical trials, these patients are in high demand, and are often difficult to recruit. After a potential drug has demonstrated an antiviral effect and an adequate safety and tolerability profile when administered as monotherapy for 7–10 days, it can then be evaluated as part of a combination. This normally entails a comparative trial where one arm of the trial is a drug combination that has been shown to be effective. The other arm of the trial would normally substitute the investigational drug for one of the drugs in that combination, or add the investigational drug to the combination, giving a direct comparison of the two drug regimens in a multidrug background.[57]

7.12.4.6 Salvage Therapy

For those patients whose HIV infection has not been controlled by various drug combinations, new investigational drugs may offer the only viable treatment option. In such cases, clinical trials have been designed to add the new drug to the patient's existing therapy in hopes of demonstrating a treatment-related reduction in viral load.[58] If such trials are successful, regulatory approval can be obtained under the salvage therapy indication.

7.12.5 Current Treatment

From the time that AIDS was recognized as a disease, therapy has evolved from treatment of the AIDS-defining opportunistic infections and cancers to sophisticated multidrug regimens designed to lower the burden of HIV in the body.

7.12.5.1 Drug Classes

At the present time, there are four classes of approved antiretroviral therapies for the treatment of HIV. They are: NRTIs, NNRTIs, PIs, and entry inhibitors.

7.12.5.1.1 Nucleoside reverse transcriptase inhibitors

7.12.5.1.1.1 Discovery

The first successful clinical trial for the treatment of HIV infection[59] involved the use of zidovudine (AZT, Retrovir), which received regulatory approval in 1987 as monotherapy. The report of the first synthesis of zidovudine[60] (**Figure 5**) occurred in 1964, and its potential as an antiviral agent was published in 1980.[61] Its activity against HIV was shown to be due to inhibition of the activity of the viral enzyme reverse transcriptase when present as its 5′-triphosphate (AZT-TP).[62] Formation of AZT-TP occurs intracellularly and involves two different kinase enzymes.[63,64] Zidovudine is first readily converted to its 5′-monophosphate by the cellular enzyme thymidine kinase, followed by slower conversion to the triphosphate by the enzyme thymidylate kinase. Of critical importance is the substitution of the thymidine 3′-hydroxyl group with an azide function, which cannot be phosphorylated. Because AZT-TP is an acceptable substrate for RT, it is incorporated normally on the growing complementary DNA chain, but due to the lack of the 3′-hydroxyl group, the DNA polymerization reaction is blocked, resulting in the disruption of proviral DNA synthesis and termination of the viral life cycle. The structure of DNA terminated with zidovudine is shown in **Figure 6**. All of the agents in this class share this mechanism of action; however, a wide variety of structural diversity in both the base and sugar portions of the inhibitors is evident in the eight members of this class that have received marketing approval (**Figure 5**). The other drug based on thymidine is stavudine (d4T, Zerit), which incorporates an olefin within the sugar moiety and is therefore a dideoxynucleoside. The first synthesis of stavudine was reported in 1966 as part of an investigation into the biosynthesis of 2′-deoxyribonucleotides,[65] and its activity against HIV was described 21 years later.[66,67] The metabolism of stavudine differs from zidovudine in that the monophosphate form of the drug does not accumulate in cells when compared with the di- and triphosphate forms.[68] Initial clinical studies of stavudine in

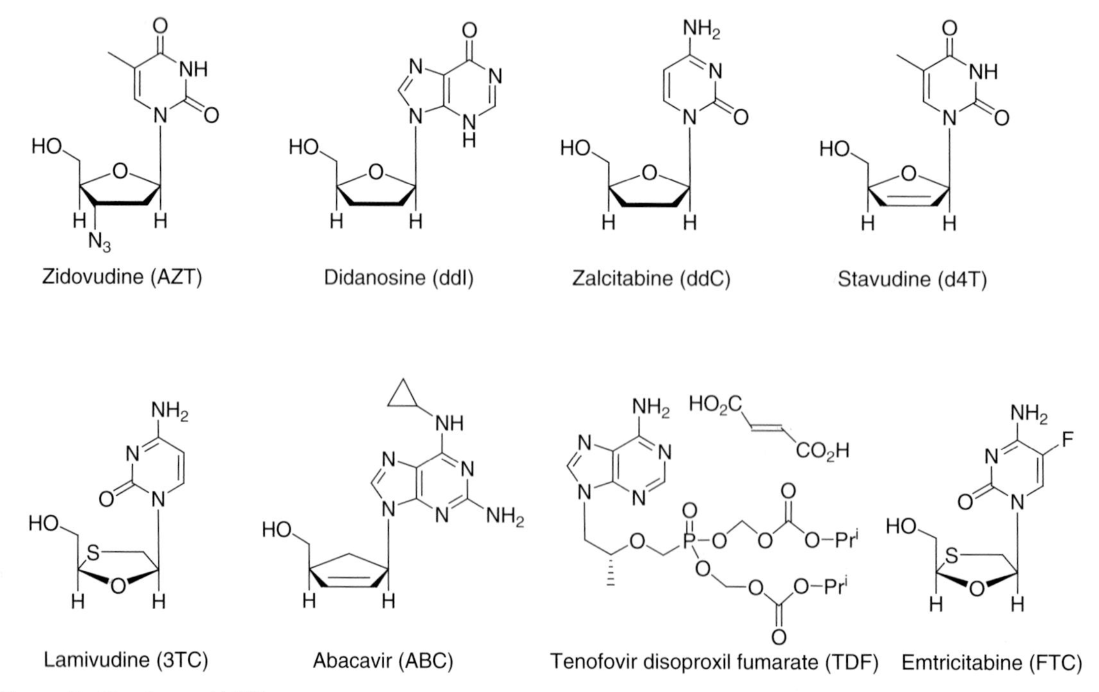

Figure 5 Structures of NRTIs.

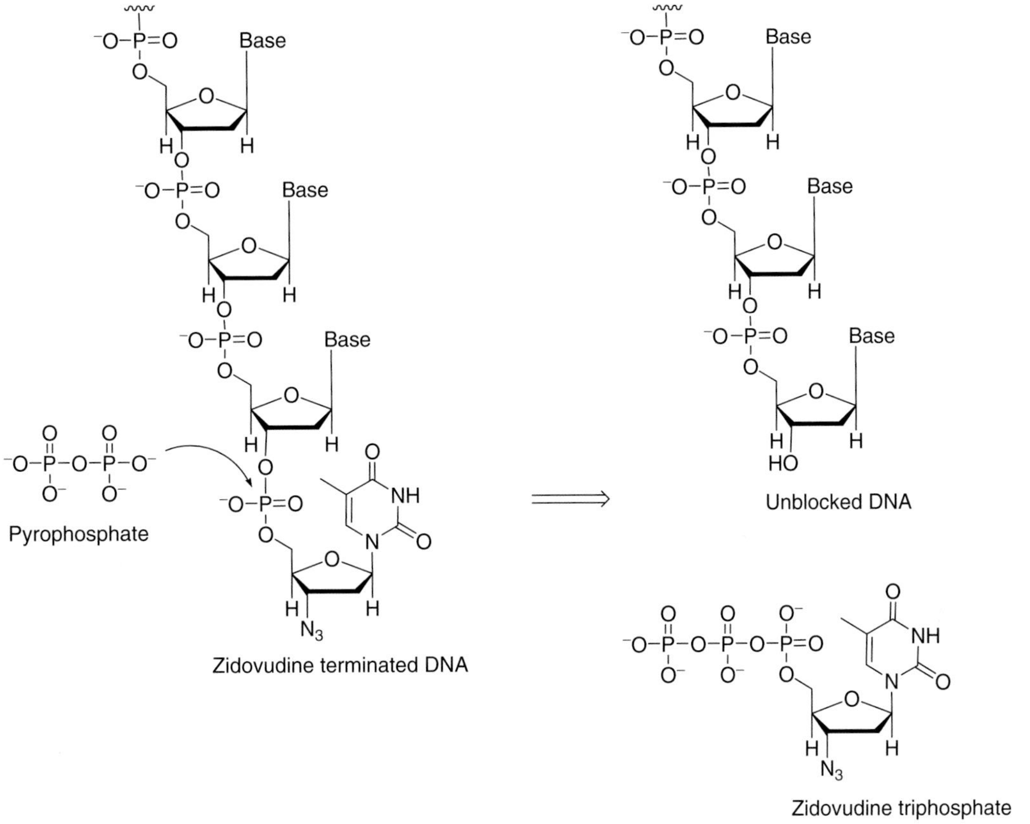

Figure 6 Removal of AZT-TP from blocked DNA.

HIV-infected patients showed a potent antiviral effect with little evidence for resistance after 18 months of therapy.[69] As shown in **Figure 5**, three HIV drugs are based on cytidine, including zalcitabine (ddC, Hivid), which is also a dideoxynucleoside, lamivudine (3TC, Epivir), and emtricitabine (FTC, Emtriva). The first synthesis of zalcitabine was carried out from the corresponding 2′–3′ unsaturated derivative in order to evaluate its activity as a chain-terminator of DNA synthesis.[70] Among all the dideoxynucleosides, zalcitabine was found to be the most potent against HIV in cell culture.[71] The closely related compounds lamivudine and emtricitabine both incorporate an oxathiolane ring replacement for the cytidine tetrahydrofuran ring. The first synthesis of the racemic form[72] of lamivudine was followed by the preparation and detailed descriptions of the individual isomers.[73,74] The activity of the two enantiomers in cell culture against HIV was found to be identical; however, the levorotatory (–) enantiomer (lamivudine) was found to be considerably less toxic than the dextrorotatory (+) enantiomer.[74] Lamivudine was also shown to have potent activity against hepatitis B virus (HBV),[73] and has received approval for that indication. The first reported synthesis of the 5-fluoro derivative of lamivudine (emtricitabine) was published in 1992.[75] As was found for lamivudine, the preferred enantiomer in terms of activity against HIV, HBV, and toxicity profile was found to be the (–) antipode.[76] In an effort to explain the apparent potency advantage in clinical trials, it was demonstrated that the triphosphate form of emtricitabine was more efficiently incorporated into DNA by RT than the triphosphate form of lamivudine.[77] Didanosine (ddI, Videx) is a dideoxyinosine (**Figure 5**) that was first prepared by the enzymatic deamination of dideoxyadenosine.[78] The intracellular metabolism of didanosine to its active form dideoxyadenosine triphosphate form is a complex process in which monophosphorylation is followed by amination by adenylosuccinate synthetase/lyase, followed by diphosphorylation by cellular kinases.[79] In early clinical studies, didanosine was administered for up to 48 weeks in HIV-infected patients, including some patients who could not tolerate the only approved HIV drug at the time, zidovudine. Didanosine was found to have significant antiviral activity and was well tolerated.[80] As illustrated in **Figure 5**, abacavir (ABC, Ziagen) is a prodrug that is converted intracellularly to the triphosphate form of carbovir, a guanosine-based inhibitor containing a cyclopentene sugar replacement. The first step of this conversion is the phosphorylation of abacavir to its monophosphate by adenosine phosphotransferase, followed by conversion of the cyclopropylamine at the 6 position to a hydroxyl group by a novel cellular enzyme distinct from adenine deaminase.[81] This process generates carbovir monophosphate, which is further phosphorylated to carbovir triphosphate by cellular kinases. The most recently approved NRTI is tenofovir disoproxil fumarate (TFV, Viread), which is sometimes classified as a nucleotide RTI. It is structurally unique (**Figure 5**) in that it incorporates an acyclic, hydrolytically stable phosphonate group that mimics a nucleoside monophosphate. The phosphonate is masked as a bis(isopropyloxy-carbonyloxymethyl) ester prodrug to improve its oral bioavailability. After cleavage of the prodrug to release tenofovir, cellular kinases provide the active diphosphorylated form. The rate-limiting step in the formation of NRTI triphosphates is often the formation of the monophosphates, and tenofovir avoids this problematic step.[82]

7.12.5.1.1.2 Extent of use

Because monotherapy of HIV/AIDS with zidovudine was unsatisfactory for many patients due to the emergence of resistant viral populations or problems with toxicity and tolerability, the idea of using combinations of drugs was investigated. The required use of combination chemotherapy for the treatment of certain bacterial infections such as tuberculosis served as a model for the treatment of HIV.[83] One of the first studies to indicate that combinations of HIV drugs might be beneficial was carried out in cell culture using zidovudine plus zalcitabine.[84] In an early clinical study, alternating 7-day regimens of zidovudine and zalcitabine were well tolerated and provided significant improvements in viral load and T cell number.[85] As new single NRTIs were approved for the treatment of HIV, they were invariably tested in combination with the NRTIs that had previously been approved. One of the most effective combination treatments for HIV using NRTIs was the addition of lamivudine to zidovudine-experienced patients, which was shown to be significantly more efficacious than continued zidovudine monotherapy.[86] These drugs were eventually combined in a single capsule marketed as Combivir, which simplified dosing requirements for many patients. Addition of abacavir to this combination has also been approved, and is marketed as Trizivir. The combination of abacavir with lamivudine is marketed as Epzicom. The most recently approved combination of NRTIs includes tenofovir with emtricitabine, sold as Truvada. Various combinations of the eight approved NRTIs have become a cornerstone of HIV therapy. Current treatment guidelines recommend a dual NRTI 'backbone' with the addition of at least one other antiretroviral.[34] In July 2006, a single-tablet combination of tenofovir, emtricitabine and efavirenz was approved by the FDA. Atripla represents the first complete once-daily recommended regimen contained in a single dose.

One of the most important uses of zidovudine has been in the prevention of mother-to-child (vertical) infection (*see* Section 7.12.2.3). It has been shown that treatment of the mother during the latter weeks of pregnancy followed by treatment of the infant for the first 6 weeks of life is effective in preventing infection of the child.[87] Currently, there are simpler alternatives for the prevention of vertical transmission (*see* Section 7.12.5.1.2.4).

7.12.5.1.1.3 Clinical limitations

7.12.5.1.1.3.1 Resistance

During the clinical studies utilizing zidovudine as monotherapy, it was apparent that the antiviral effect diminished over time for many patients, and the loss in activity was ascribed to acquisition of mutations within the RT sequence.[45] The magnitude of activity loss generally corresponded to the number of acquired mutations. The RT that has been sequenced from patients whose virus has become highly resistant to zidovudine contains four or more mutations at M41L, D67N, K70R, L210W, T215, and K219Q.[88] **Figure 7** shows a global representation of the crystal structure of zidovudine-terminated DNA bound to RT.[89] **Figure 8** is a detailed view of the active site showing the interaction of zidovudine with the complementary base on the other DNA strand and the active site magnesium atom. Interestingly, zidovudine triphosphate is essentially able to inhibit these mutant enzymes with similar efficiency as wild-type enzyme. This finding was key to the discovery that the mechanism of resistance to zidovudine involves the ability of RT to unblock the capped DNA by removal of zidovudine monophosphate by RT-catalyzed phosphorolysis, resulting in repair of the DNA allowing further chain elongation,[90] a process shown in **Figure 6**. RT containing more than two zidovudine-associated mutations is more efficient at excision of some NRTI-terminated DNAs, including zidovudine, stavudine, didanosine (dideoxyadenosine), and abacavir (carbovir).[91] Resistance to stavudine follows a similar course, and generally selects for similar mutations, termed thymidine-associated mutations (TAMs). In contrast, resistance to zalcitabine can be largely attributed to a decreased affinity of the mutant enzyme for the triphosphate form of the drug, resulting in less efficient blocking of the viral DNA.[92] Treatment with lamivudine or emtricitabine as monotherapy or in combination with other antiretroviral agents often results in the rapid emergence of viral populations with a single M184V or M184I mutation (**Figure 8**) that results in more than a 1000-fold loss in activity. This loss in activity has also been ascribed to the decreased sensitivity of the mutant RT enzyme to the triphosphate form of the drugs.[93] It is indeed fortunate that RT containing this mutation is also significantly less able to repair NRTI-terminated

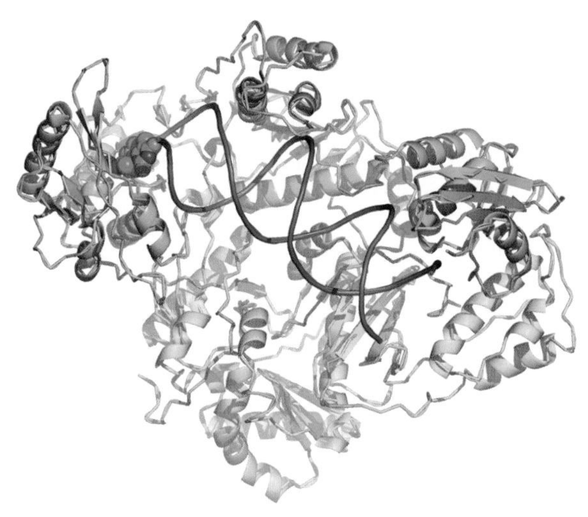

Figure 7 X-ray crystal structure of AZT bound to reverse transcriptase. The dsDNA is represented by the dark blue coiled structures and zidovudine is represented by the spheres color-coded by atom type: green, C; blue, N; red, O; purple, P.

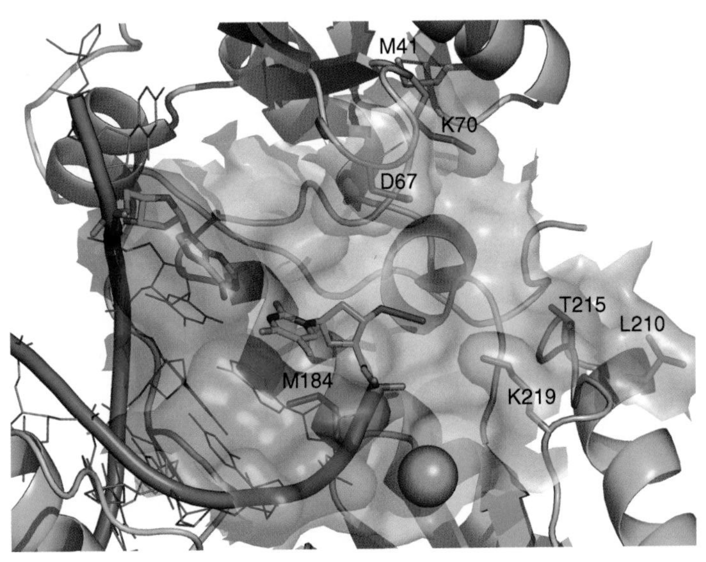

Figure 8 Detail of x-ray crystal structure of AZT bound to reverse transcriptase. The active site magnesium atom is shown as the large orange sphere. The locations of the common NRTI-mutated amino acids are shown in pink.

DNA, providing a mechanistic explanation for the synergy between zidovudine and lamivudine in clinical practice.[94,95] Other important factors in characterizing drug-resistant viruses are the growth characteristics (fitness) relative to wild-type or other mutant viruses as well as the host immune response. For example, although the M184V mutation can appear relatively rapidly in patients taking regimens that include lamivudine or emtricitabine, the relative fitness and immunogenicity of viruses containing this mutation can be significantly different from wild-type virus.[96] This mutation has also been reported to cause increased sensitivity to tenofovir.[97] As the number of possible combinations of NRTIs has increased, the complexity of the mutation profiles generated by treatment with these regimens has also increased dramatically. This has resulted in an empirical approach to determine the appropriate drug combination for each patient. In particular, some triple NRTI regimens (such as tenofovir, lamivudine, and didanosine or abacavir, lamivudine, and tenofovir) have been associated with virologic failure, and should be avoided.[98]

7.12.5.1.1.3.2 Tolerability and compliance Because most of the approved NRTIs also inhibit human DNA polymerases to some extent, mechanism-based toxicity can lead to side effects that cause patients to become ill and stop taking these drugs. A decrease in mitochondrial DNA synthesis due to the inhibition of human DNA polymerase gamma has been implicated[99] in the development of symptoms such as pancreatitis, peripheral neuropathy, lipoatrophy, and liver enzyme elevations.[100] NRTIs have been implicated in rare incidences of serious lactic acidosis that can lead to liver steatosis.[34] Stavudine has been reported to have more toxicity issues than other NRTIs.[101] Abacavir has been associated with a potentially fatal hypersensitivity reaction with symptoms that can include fever, nausea, and rash.[102] Because NRTIs must be combined with other antiretrovirals, it is often difficult to ascribe toxicity to any individual agent.

7.12.5.1.1.3.3 Dosing regimens Initially NRTIs were given either three times daily or twice daily, but recently approved drugs such as lamivudine, emtricitabine, abacavir, and tenofovir are indicated for once-daily dosing.[103] This has had the much desired effect of increasing the probability of patient compliance and improving convenience. With the exception of didanosine, none of the NRTIs is significantly affected by food intake. **Table 1** shows the recommended dosing regimens for Food and Drug Administration (FDA)-approved NRTIs.[101] The combination products Truvada and Epzicom have been approved as single tablets given once daily.

7.12.5.1.1.3.4 Compatibility with other antiretrovirals Significant interactions with other HIV treatments requiring dosage modifications are relatively rare with NRTIs. However, concomitant use of either didanosine or lamivudine with stavudine can exacerbate mitochondrial toxicities or result in diminished antiviral activity, respectively.[101] Combinations of zidovudine with stavudine have also been shown to be antagonistic in clinical studies.[104] Because emtricitabine and lamivudine share a common resistance profile, that combination is not recommended. Triple NRTI regimens are generally not recommended except in special circumstances when zidovudine, lamivudine, and abacavir (Trizavir) can be used. Also, tenofovir can alter the plasma levels of both atazanavir and didanosine requiring some dose modification.[101]

7.12.5.1.1.4 Comparison of commonly used agents

Certain combinations of NRTIs with other HIV treatments have become popular based on dosing convenience, tolerability, and resistance expectations. Drug combinations such as tenofovir, emtricitabine, and efavirenz provide patients with the convenience of once-daily dosing, minimal pill burden, and the option of switching to a PI-based regimen at a later date if necessary.

7.12.5.1.1.5 Outlook

An intensive research effort continues to identify NRTIs that have improved convenience, tolerability, resistance profiles, and compatibility with other HIV treatments. The complexity of pharmacokinetic, resistance, and metabolic interactions between individual NRTIs and other reverse transcriptase inhibitors is increasing as the number of

Table 1 Dosing regimens for NRTIs[a]

Drug	*Dosage (mg)*	*Frequency*[a]
Zidovudine	300	bid
	200	tid
Stavudine	30–40	bid
Zalcitabine	0.75	tid
Didanosine	250–400	qd
	125–200	bid
Lamivudine	300	qd
	150	bid
Emtricitabine	200	qd
Abacavir	600	qd
	300	bid
Tenofovir DF	300	qd

[a] qd, once daily; bid, twice daily; tid, three times daily.

approved treatments proliferates. Such interactions can be difficult to define preclinically, requiring expensive clinical evaluation. Currently, dual NRTIs in combination with at least one other antiretroviral agent is the mainstay of HIV therapy.

7.12.5.1.2 Nonnucleoside reverse transcriptase inhibitors

The second major class of drugs to be used in the battle against HIV are the NNRTIs. Although these compounds are also targeted at RT, their mechanism of action is substantially different than the NRTI class.

7.12.5.1.2.1 Discovery

Early in 1990, the first example of a nonnucleoside inhibitor of HIV-1 RT (NNRTI) was published. This new class of inhibitor was discovered by an interesting screening paradigm in which approximately 600 compounds that were otherwise inactive against all other biochemical targets and were reasonably well tolerated after acute dosing to rodents were evaluated for their activity against HIV-1 and their cytotoxicity in cell culture.[105] This led to the identification of tricyclic benzodiazepine R14458 (**Figure 9**), which was an effective inhibitor of HIV in cell culture. Structure–activity relationship (SAR) studies provided a series of related compounds with improved activity in the cell-based assay. In-depth studies with compound R82150 demonstrated that its mechanism of action was related to inhibition of RT, and that the compound was inactive against closely related viruses such as HIV-2 and SIV. Pharmacokinetics in dogs[105] indicated a plasma half-life of 3.2 h with a volume of distribution of 10.2 L kg^{-1}. In phase I clinical studies, once-daily dosing of up to 200 mg of R82913, a significantly more potent analog, gave trough plasma concentrations that were often below the IC_{90} in cell culture.[106] In phase II clinical studies, where the compound was administered either orally or intravenously to HIV-infected patients, limited efficacy was achieved, probably due to inadequate exposure.[107] Later in 1990 a second group published the structure and antiviral activity of a compound from an unrelated chemical class that inhibited HIV-1 in a similar manner.[108] A series of diaryldiazepinones was evaluated resulting in the identification of the specific dipyridodiazepinone BI-RG-587 as a clinical candidate based on its potency versus HIV-1 RT and in cell culture. Enzyme kinetic studies revealed that the compound was noncompetitive with guanosine triphosphate (GTP) and indicated that binding to RT might be occurring at an allosteric site distinct from the active (GTP-binding) site.[108] This compound underwent many years of both preclinical and clinical testing to ultimately be marketed as nevirapine (Viramune), the first NNRTI of HIV-1 to receive regulatory approval (**Figure 10**). The human metabolism of nevirapine is mediated by Cyp3A4 and involves oxidative hydroxylation at multiple sites, followed by glucuronide conjugation and urinary excretion.[109] Nevirapine has also been found to be a potent inducer of cytochrome P450 metabolism. After dosing with 200 mg per day in healthy volunteers for 14 days, a significant increase in plasma clearance and a corresponding decrease in plasma half-life were noted.[109]

Using a directed screening approach against HIV-1 RT, the pyridinone L-345,516 (**Figure 11**) was identified that exhibited significant inhibitory activity. The compound was found to have suboptimal chemical stability, so several related derivatives were prepared and evaluated. Optimization of this series for potency and pharmacokinetics resulted

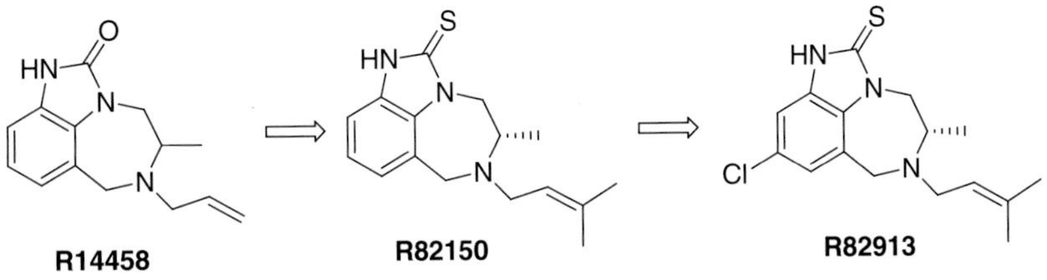

R14458 R82150 R82913

Figure 9 Evolution of R82913.

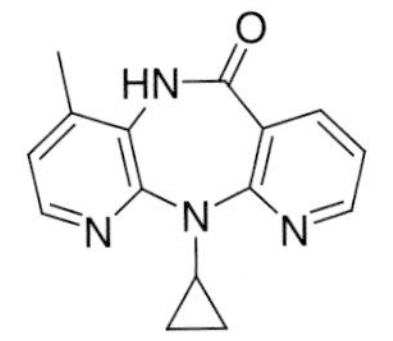

Figure 10 Nevirapine (NVP).

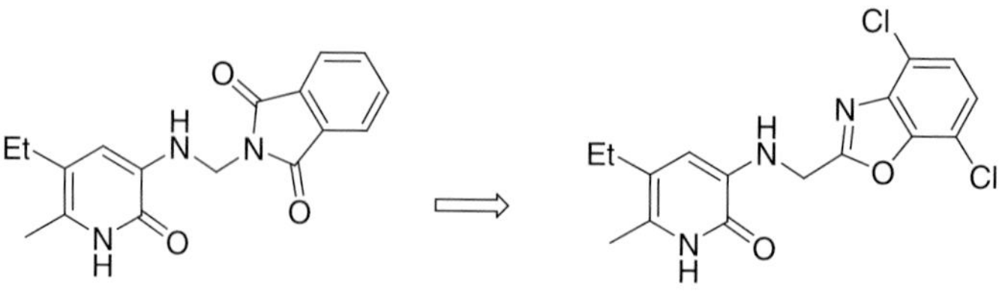

L-345,516 L-697,661

Figure 11 Evolution of L-697,661.

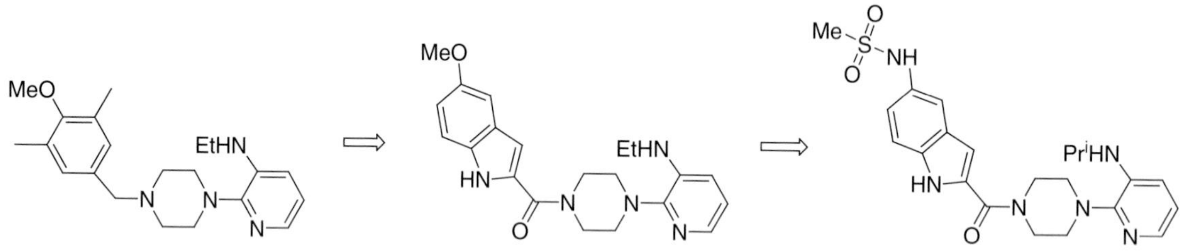

U-80493E Atevirdine Delavirdine (DLV)

Figure 12 Evolution of delavirdine.

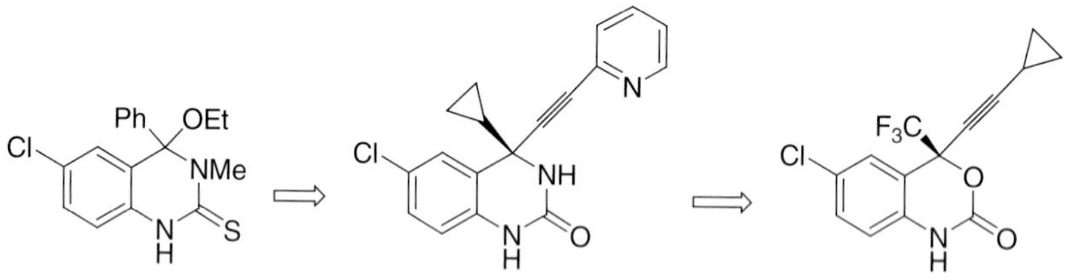

L-608,788 L-738,372 Efavirenz

Figure 13 Evolution of Efavirenz (EFV).

in the discovery of L-697,661.[110] In early clinical trials, monotherapy with this compound was associated with a rapid drop in plasma viremia for up to 8 weeks. However, after 12 weeks, a rebound in viremia was observed that was attributed to the emergence of HIV that was highly resistant to L-697,661.[30]

Before the advent of high-throughput screening, it was not practical to test very large numbers of compounds in a sample collection. Using a technique called computational dissimilarity analysis, a subset consisting of 1500 compounds was selected for testing against HIV-1 RT. This resulted in the selection of 100 compounds for further testing in cell culture for antiviral activity and toxicity. The lead compound that emerged from this approach was the arylpiperazine U-80493E (**Figure 12**). Optimization of this lead structure lead to the first clinical candidate from this class, atevirdine (U-87201E).[111] Unfortunately, no significant antiviral effect was seen in HIV-infected patients at doses of 600 mg three times daily, which may have been due to inadequate drug trough concentrations.[112] Further optimization of substituents on the indole portion of the molecule were guided by potency and metabolic concerns, leading to the identification[113] of U-90152S and advancement of delavirdine (Rescriptor) into clinical trials[114] and regulatory approval.

Because of the relatively rapid selection for resistant viral populations in patients undergoing therapy with NNRTIs, a search for second-generation inhibitors with improved pharmacokinetics and a higher barrier to resistance was under way in several laboratories. Beginning from a second structural class of dihydroquinazolinethiones identified from directed screening, L-608,788 became the lead compound (**Figure 13**). Optimization of this series was initially driven by chemical and metabolic stability issues, as well as potency in cell culture.[115] In particular, oxidative metabolism of either the alkyl group on the nitrogen atom at the 3-position or substituents at the 4-position were thought to be responsible for rapid clearance in animals. Introduction of an acetylene substituent and cyclopropyl group at the 4-position allowed the 3-position to go unsubstituted, thus circumventing the metabolism issues. This led to the identification of L-738,372, which possessed an excellent pharmacokinetic profile in animals[116] and good potency in cell culture. Detailed enzymatic studies demonstrated that the compound was a slow-binding, noncompetitive inhibitor that was synergistic with NRTIs.[117] A second approach entailed the substitution of an oxygen atom at the 3-position to give the benzoxazine ring system. Further optimization was guided by relative activity versus a panel of single and double mutants previously identified as conferring resistance to a variety of NNRTIs. The optimized compound L-743,726 (DMP-266) displayed an attractive combination of good pharmacokinetics and an improved resistance profile.[118] Clinical studies[119] confirmed the preclinical profile, and efavirenz (Sustiva, Stocrin) received regulatory approval as one element of combination therapy for HIV, and the first HIV drug indicated for once-daily dosing.

7.12.5.1.2.2 Extent of use

NNRTIs have become an integral part of currently recommended regimens for the treatment of HIV/AIDS.[34] Because of its overall potency and convenience of once-daily dosing, efavirenz has been recommended as a key component in initial antiretroviral therapy. For patients in whom efavirenz is contraindicated or not well tolerated, nevirapine is the NNRTI of choice.[34] These drugs may be combined with either PIs and/or NRTIs depending upon the profile of the individual being treated.

7.12.5.1.2.3 Clinical limitations

7.12.5.1.2.3.1 Resistance

Preclinical studies of all members of the NNRTI class indicated that these inhibitors were selective for HIV-1, having significantly reduced activity against HIV-2 or SIV.[120–122] Molecular and structural biology studies indicated that upon addition to HIV-1 RT, NNRTIs induce a conformational change in the native enzyme that allows binding of all members of this class to the same site.[123,124] This binding site does not appear in RT from either HIV-2 or SIV, which explains the NNRTIs lack of significant activity against these enzymes or viruses.[123] **Figure 14** shows the x-ray crystal structure of efavirenz bound to HIV-1 RT.[125] In early clinical trials that evaluated NNRTIs as monotherapy, relatively rapid emergence of resistant viral populations occurred within a few weeks from the initiation of treatment. Sequencing of the RT domain of these resistant viruses often led to the identification of single amino acid changes that were responsible for the resistance. However, each NNRTI tended to select for different individual mutations. For example, virus isolated from patients treated with L-697,661 as monotherapy had predominantly the Y181C mutation,[126] whereas delavirdine selected predominantly for the K103N mutant.[127] **Figure 15** shows a detailed view of the crystal structure of efavirenz bound to the active site, along with the locations

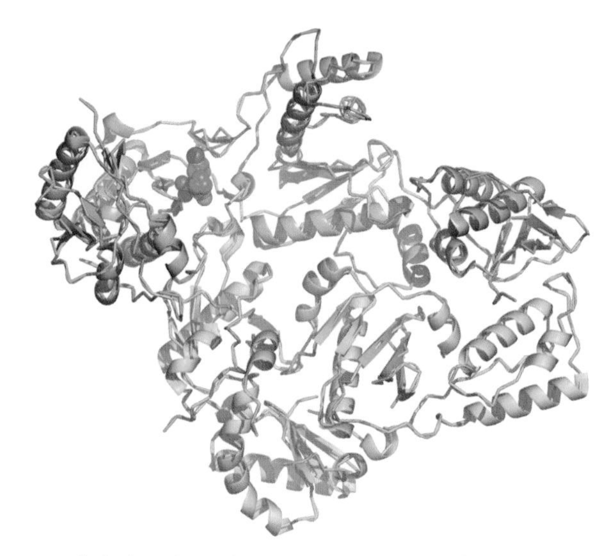

Figure 14 X-ray crystal structure of efavirenz bound to reverse transcriptase. Efavirenz is shown as spheres color-coded by atom type: green, C; red, O; blue, N; and purple, F or Cl. This NNRTI allosteric binding site is approximately 65 angstroms from the polymerase active site.

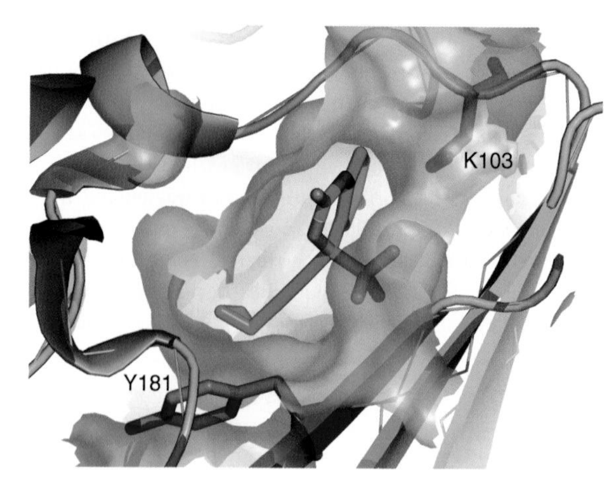

Figure 15 Detail of x-ray crystal structure of efavirenz bound to reverse transcriptase. Locations of the K103 and Y181 mutations are shown in red.

Table 2 Dosing regimens for NNRTIs[a]

Drug	*Dosage (mg)*	*Frequency*[a]
Nevirapine	200	qd 14 days
	200	bid after
Delavirdine	400	tid
Efavirenz	600	qd

[a] qd, once daily; bid, twice daily; tid, three times daily.

of the K103 and Y181 residues.[125] The change of a single tyrosine at position 181 to a cysteine (Y181C) or a lysine at position 103 to asparagine (K103N) can lead to dramatic losses in susceptibility to the inhibitors, and a corresponding loss of activity in cell culture. Although efavirenz is effective against the Y181C mutant, K103N causes a significant loss in activity. More significantly, K103N engenders cross-resistance among all the approved NNRTIs. The inability of these drugs to reposition themselves to allow efficient binding in the mutated binding site could explain such cross-resistance.[125] Under current guidelines, therapy consisting of NNRTIs in combination with NRTIs and/or PIs can give rise to a wide variety of single and multiple mutants within the RT portion of the genome that are highly resistant to NNRTIs.[128]

7.12.5.1.2.3.2 Tolerability and compliance A side effect that is common for all three approved NNRTIs is the appearance of a rash serious enough to cause discontinuation in clinical studies. The incidence was 7% for nevirapine, 4.3% for delavirdine, and 1.7% for efavirenz.[101] Nevirapine has also been associated with Stevens–Johnson syndrome and hepatotoxicity.[129] The most commonly reported side effects with efavirenz are neurological symptoms such as insomnia, dizziness, abnormal dreams, confusion, and euphoria.[130] Efavirenz has also been found to be a teratogen in monkeys, and has recently been reclassified as 'Pregnancy Category D' by the FDA.[131] This category indicates that human safety data show some risk that could be outweighed by benefits.

7.12.5.1.2.3.3 Dosing regimens Efavirenz was one of the first AIDS drugs to be approved for once-daily dosing. Initially, patients took three 200-mg capsules of efavirenz daily; however, a 600-mg tablet has recently become available that reduces the pill burden to one tablet daily. Because nevirapine causes autoinduction, it is recommended that 200-mg tablets be given daily for the first 14 days on therapy, followed by 200-mg twice daily thereafter.[109] Because of the relatively poor exposure and plasma half-life of delavirdine, the recommended dose is two 200-mg tablets three times daily. **Table 2** indicates the recommended dosing regimens for NNRTIs.

7.12.5.1.2.3.4 Compatibility with other antiretrovirals The three approved NNRTIs are metabolized by the cytochrome P450 pathway, and all either cause induction and/or inhibition of Cyp3A4. Because Cyp3A4 is also involved in the metabolism of most of the PIs, pharmacokinetic drug interactions with this class are common and can require dosing adjustments.[132] Due to the relatively common resistance profiles for NNRTIs, combining them in a multidrug regimen is uncommon.

7.12.5.1.2.4 Comparison of commonly used agents

Convenience, tolerability, and durability are factors that determine the choice of NNRTI to be used in multidrug combinations. When tolerability is not a factor, efavirenz is usually the agent of choice due to the convenience of dosing once daily with only one tablet. In one study, efavirenz and nevirapine gave similar results with respect to efficacy and discontinuation rates; however, the reasons for discontinuation were significantly different for the two drugs. Efavirenz discontinuations were associated with CNS and metabolic effects, whereas nevirapine discontinuations were associated with hypersensitivity and liver toxicity.[130] This would suggest that the choice of NNRTI should be driven primarily by individual patient tolerability. Nevirapine has been extensively used for prevention of vertical transmission, and has been particularly useful in the developing world.[133] Efavirenz is contraindicated for use in this patient population due to teratogenicity concerns. Due to the high pill burden and frequency of dosing, delavirdine is less commonly used.

7.12.5.1.2.5 Outlook

Nevirapine and efavirenz have proven invaluable as one element in drug combination therapy for the treatment of HIV/AIDS. However, issues related to tolerability, side effects, and resistance have been problematic for many patients. This has inspired the search to identify NNRTIs that address one or more of these issues. Significant progress has been achieved in several laboratories with respect to drug resistance. Most notably, the bis(arylamino)pyrimidine rilpivirine (R278474, TMC-278) (**Figure 16**) has been reported to be active against a panel of viruses including L100I, K103N, Y181C, Y188L, G190S, and K103N/Y181C, many of which are insensitive to the marketed NNRTIs. It is significantly more potent than efavirenz, has good oral bioavailability, and has been evaluated in 1-month oral toxicity studies in rats and dogs.[134] In a recent clinical study in patients not currently receiving antiviral treatment, oral doses of between 25 and 150 mg of rilpivirine in PEG-400 solution were administered once daily for 7 days. All doses produced a 1.0–1.3 $\log_{10}$ drop in viral load by day 8, with trough plasma levels well above the protein-adjusted IC_{90}. The compound was generally well tolerated, and no evidence of resistance was observed.[135]

7.12.5.1.3 Human immunodeficiency virus protease inhibitors

Until the introduction of HIV PIs into clinical practice, the treatment of HIV with NRTIs alone was often met with limited success in suppressing the virus to undetectable levels. Utilization of a PI drug in combination with NRTIs revolutionized the treatment of HIV/AIDS.

7.12.5.1.3.1 Discovery

The third viral enzyme is HIV protease, and is responsible for the maturation of the newly created viruses during or shortly after the budding process. Because the products from transcription and translation of the proviral DNA are expressed as long polyproteins, the protease enzyme is required to cleave at nine distinct sites to provide the individual functional and structural proteins of the virus.[136,137] This enzyme belongs to the family of aspartic acid proteases (renin, pepsin, etc.), and exists as a homodimer composed of two 99 amino acid proteins. From a number of studies, it became apparent that this enzyme would be an attractive biochemical target for HIV therapeutics.[137,138] A number of drugs have been developed that target this enzyme, and represent the second major mechanism for anti-HIV drugs. A common approach to inhibitor design was implemented by several laboratories in the late 1980s. The strategy evolved from the identification of HIV protease as a member of the aspartic acid proteases and built on the large body of knowledge related to the structure and inhibition of members of this enzyme class. The proper folding of the two identical proteins results in a C2 symmetric tertiary structure that has been solved by x-ray crystallography and that has been a major focus of structure-based drug design in research laboratories. Before the structure of the enzyme was determined experimentally, molecular modeling techniques were used to predict the three-dimensional structure based on sequence homology.[139] The results of this prediction were borne out when the three-dimensional structure of

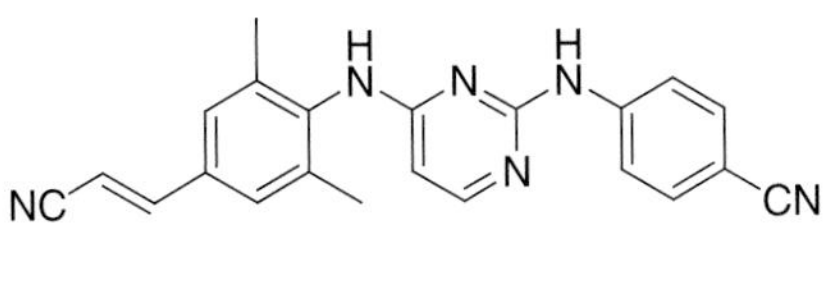

Figure 16 Rilpivirine.

the native enzyme was determined by x-ray crystallography (**Figure 17**).[140,141] When potent inhibitors were later identified, a number of inhibitors were cocrystallized with the enzyme and the structure of the enzyme–inhibitor complex determined crystallographically. **Figure 18** shows the structure of indinavir bound to HIV protease.[142] When compared with the native structure (**Figure 17**) the indinavir-containing enzyme clearly shows the movement of the flap regions toward the inhibitor. Mechanistic studies of this family of aspartyl proteases indicated that two aspartates in the active site catalyze the addition of a water molecule to the amide bond at the cleavage site, leading to a tetrahedral transition state shown in **Figure 19**.[143] Breakdown of this intermediate leads to the formation of the two smaller peptide fragments. The protease enzyme is responsible for cleavage of the gag and gag-pol polyproteins at multiple sites, and the specific amino acids on either side of each cleavage site were determined.[144] From this information, it then became possible to design PIs based on the principle of the transition state isostere model.[145] This approach entails replacement of the scissile amide bond within a peptide substrate with a nonhydrolyzable replacement, termed an isostere. Because the structure of the isostere mimics the geometry and structure of the transition state of amide bond hydrolysis, tight binding to the enzyme results in potent inhibition.[145] This concept had

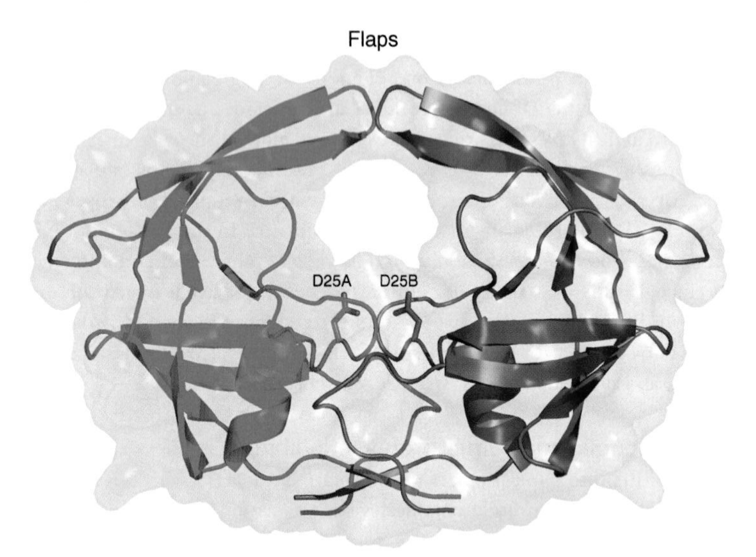

Figure 17 X-ray crystal structure of native HIV protease. One of the monomers is shown in green and the other in blue. The 'floor' of the active site is near the two catalytic aspartate residues D25A and D25B, and the 'ceiling' includes the mobile flap regions that close down on the substrate or inhibitor upon binding.

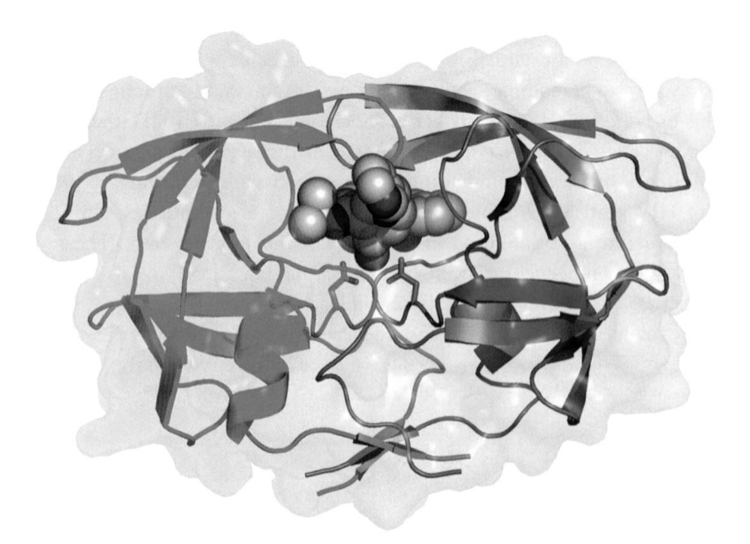

Figure 18 X-ray crystal structure of indinavir bound HIV protease.

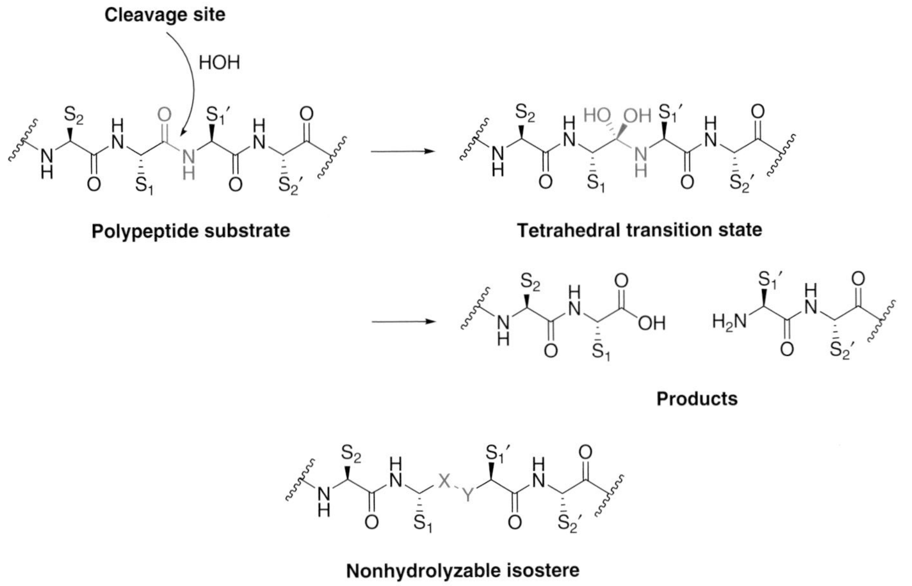

Figure 19 Tetrahedral transition state.

been previously applied to the related aspartyl protease renin.[146] A wide variety of functional groups (X and Y in **Figure 19**) can fulfill the role of transition state isostere, and a number of the known structures have been incorporated into substrates of HIV protease to make inhibitors.[147–152]

7.12.5.1.3.1.1 Discovery of saquinavir The first PI to be approved for the treatment of HIV infection in the USA was saquinavir (Invirase, Fortovase). The medicinal chemistry approach[152] taken for this compound began from one of the key substrates for the enzyme. Although HIV protease is required for cleavage at a variety of amino acid pairs within the viral proteome, one of the more unusual cleavages is between either tyrosine (Tyr) or phenylalanine (Phe) and proline (Pro). Because most mammalian proteases are unable to cleave between these amino acid residues, inhibitors based on the corresponding transition state isosteres were synthesized. Based on the sequence of amino acids 165–169 in the pol polyprotein (Leu-Asn-Phe-Pro-Ile) around the Phe-Pro cleavage site, a series of inhibitors were synthesized that included a stable hydroxyl-containing isostere to replace the amide bond and mimic the presumed tetrahedral intermediate resulting from enzymatic hydrolysis (**Figure 20**). For a series of compounds containing the transition state isostere, it was noted that the stereochemistry of the key hydroxyl group was critical for potency against the enzyme. Optimization of this series included an effort to determine the smallest sequence that preserved tight binding to the enzyme, followed by optimization of the N and C termini, as well as the amino acid side chains. This led to the identification of saquinavir as a development candidate. Regulatory approval of the first HIV PI, saquinavir, by the FDA was granted in December 1995 as part of a combination with at least one other antiviral agent approved for the treatment of HIV.

A few months later, in February 1996, the HIV PIs ritonavir (Norvir) and indinavir (Crixivan) also received regulatory approval.

7.12.5.1.3.1.2 Discovery of ritonavir The medicinal chemistry program that led to the discovery of ritonavir began with an appreciation of the C2 symmetry inherent in the tertiary structure of the enzyme, as well as the substrate sequence.[153] The particular cleavage site used in the initial design was also the Phe-Pro dipeptide transition state (**Figure 21**). Application of a C2 symmetry element to the N-terminal side of the transition state gives a structure that contains a 2,5-diamino-3,4-dihydroxy-1,6-diphenylhexane fragment in place of the dipeptide. Due to the highly lipophilic nature of the substrate sequence containing the transition state mimetic, many of the early inhibitors had low aqueous solubility and displayed very poor oral absorption. It was therefore necessary to include functional groups that would help to make the inhibitors more soluble in aqueous environments. Incorporation of pyridyl groups on both ends of the tetrapeptide mimetic led to sufficient aqueous solubility to allow the selection of A-77003 as an initial clinical candidate, administered by the intravenous route.[154] Further optimization of this series was driven by potency, physical

Leu-Asn-Phe-Pro-Ile substrate sequence

Z-Leu-Asn-Phe[CHOH-CH$_2$]Pro-IleNHBu^t
IC$_{50}$ = 750 nM

Transition state isostere

Z-Asn-Phe[CHOH-CH$_2$]PicNHBu^t
IC$_{50}$ = 18 nM

Saquinavir (SQV)
IC$_{50}$ < 0.4 nM

Figure 20 Evolution of saquinavir.

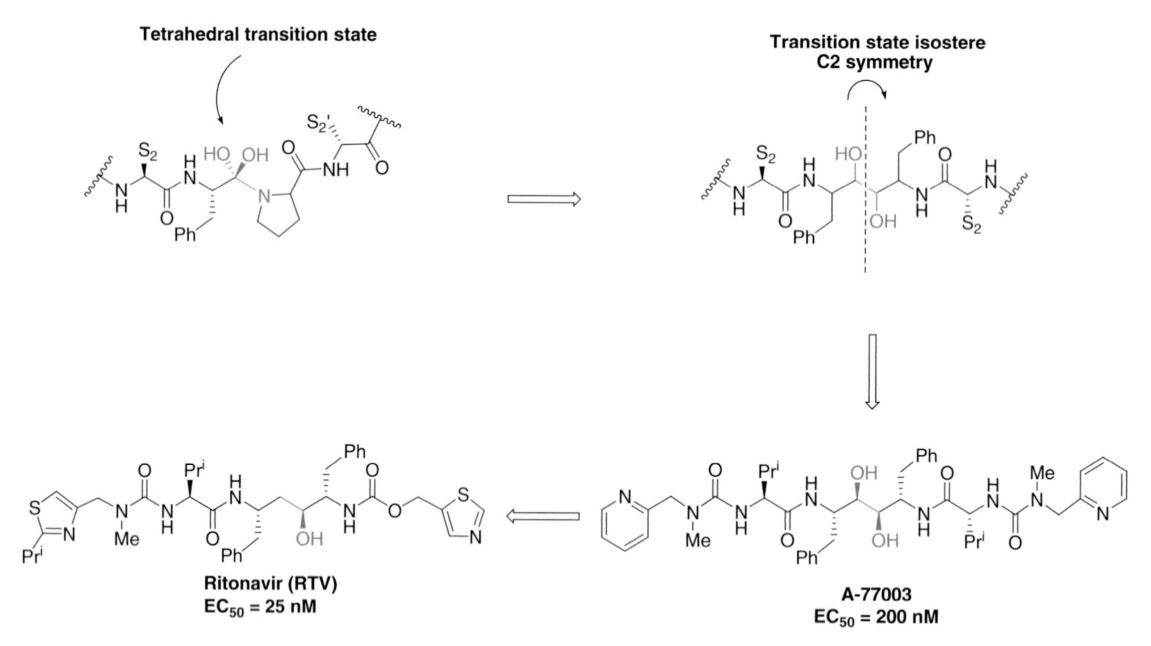

Figure 21 Evolution of ritonavir. The Phe-Pro dipeptide transition state is shown in red.

properties, and pharmacokinetic behavior in rats. Truncation of A-77003 gave inhibitors with good exposure after oral dosing, but limited plasma half-lives. By examining the metabolic fate of these inhibitors, the labile pyridyl groups were replaced with thiazoles, which led to the identification of ritonavir.[155] In clinical trials, ritonavir produced a rapid and significant decrease in the levels of plasma HIV and a corresponding increase in the T cell population.[156] Ritonavir was the first PI to show a survival benefit in patients with AIDS.[157]

7.12.5.1.3.1.3 Discovery of indinavir The design path to indinavir began from the knowledge that HIV protease was also related to the aspartyl protease renin. Renin is a key enzyme in the renin–angiotensin pathway and is

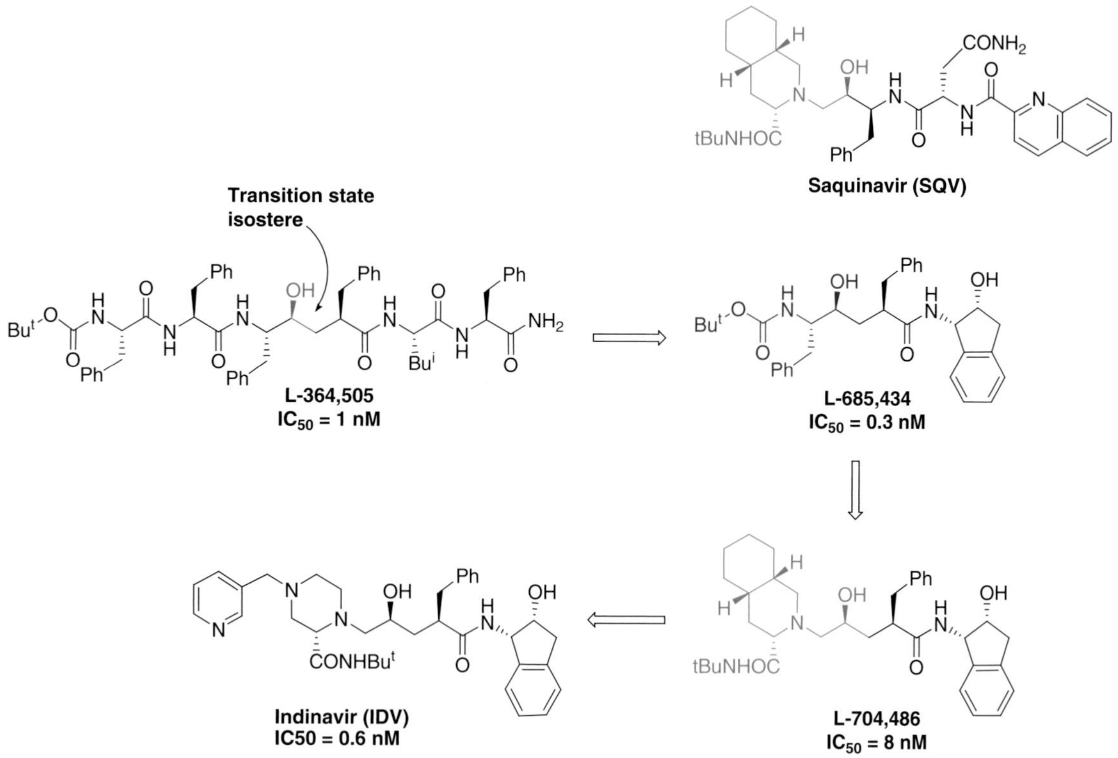

Figure 22 Evolution of indinavir. N-terminus of L-685,434 shown in blue; DIQ region of saquinavir shown in pink.

intimately involved in the control of blood pressure. A relatively large collection of synthetic renin inhibitors were available for counterscreening to identify compounds that would also inhibit HIV protease.[38] However, in order to provide enough HIV protease to conduct such a screen, a chemical synthesis of the enzyme was undertaken.[158] This effort provided sufficient quantities of the enzyme that were crucial for the timely identification of PIs from the renin collection. A hexapeptide-based inhibitor (L-364,505) containing a hydroxyethylene isostere with relatively weak activity against human renin[159] was soon identified as a potent inhibitor of HIV protease (**Figure 22**).[38] This isostere was a replacement for a Phe-Phe dipeptide, and only one of the many possible diastereomers gave potent inhibitors. Deletion of the two N-terminal Phe residues and replacement of the C-terminal dipeptide amide with an aminoindanol amide gave a series of compounds with equivalent potency and significantly diminished molecular weight.[160] Although L-685,434 was of manageable size and lacked any traditional amino acid residues, it had very poor oral bioavailability and limited water solubility. This prompted a new strategy that involved a hybrid design using the published structure of saquinivir.[152] As noted previously, the stereochemistry of the hydroxyl group in the transition state isosteres was critical for both series of inhibitors. This observation led to the hypothesis that the bound conformation of the decahydroisoquinoline (DIQ) portion of saquinavir might occupy the same space as the N-terminal portion of L-685,434. To test this hypothesis, a hybrid structure was synthesized that replaced the N-terminus of the current inhibitors.[161] Importantly, this design also introduced a basic nitrogen atom into the backbone of the inhibitor that improved the solubility characteristics of the compounds. Optimization of this new hybrid inhibitor led to the replacement of the DIQ with a substituted piperidine, and ultimately to the synthesis of indinavir.[161,162]

7.12.5.1.3.1.4 Discovery of nelfinavir The design of the fourth PI to receive regulatory approval was also derived in part from the structure of saquinavir. Extensive utilization of the structure of a series of inhibitors bound in the active site of HIV protease had provided small-molecule inhibitors (molecular weight <600 Da) with relatively weak activity and poor water solubility.[163–165] Using a strategy similar to the approach taken to indinavir, it was appreciated that the dipeptide isostere in saquinavir should bind to the enzyme in a similar manner to the C-terminal portion of compound **1** shown in red in **Figure 23**. Replacement of this C-terminal domain with the DIQ portion of saquinavir shown in blue (**Figure 23**) afforded the potent and orally bioavailable inhibitor **2**, which was further optimized for antiviral potency to

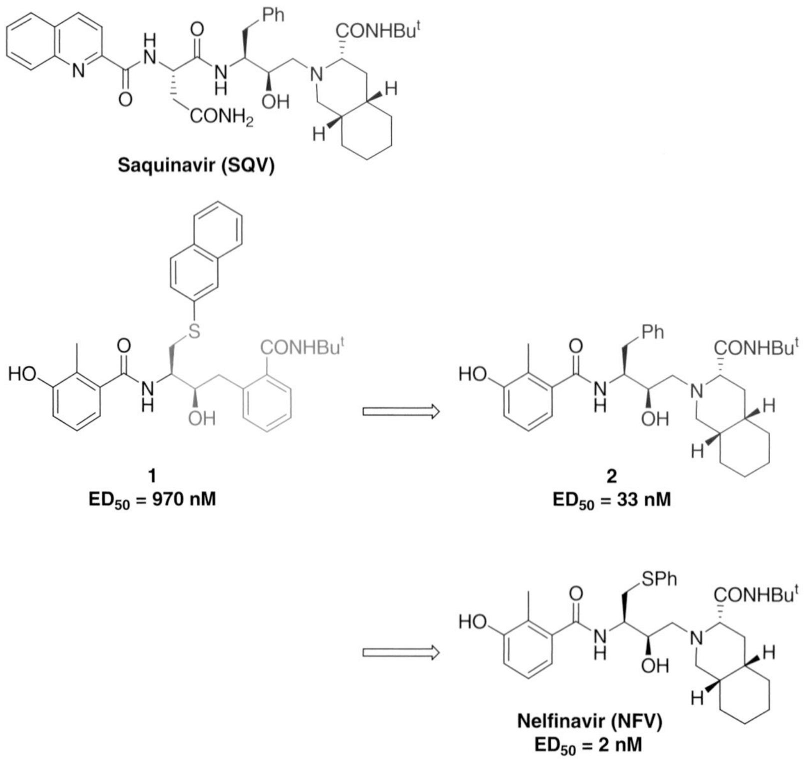

Saquinavir (SQV)

1
ED_{50} = 970 nM

2
ED_{50} = 33 nM

Nelfinavir (NFV)
ED_{50} = 2 nM

Figure 23 Evolution of nelfinavir.

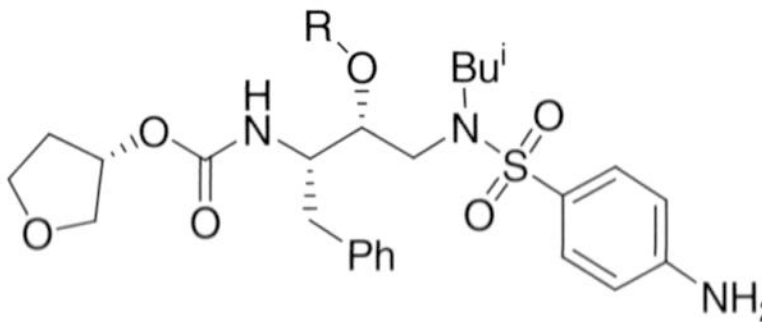

Amprenavir (APV) R = H, K_i = 0.6 nM
Fosamprenavir (f-APV) R = PO_3H_2

Figure 24 Structure of amprenavir and fosamprenavir.

give nelfinavir mesylate (Viracept).[166] Advancement of nelfinavir into clinical development was based on a combination of antiviral activity and pharmacokinetic properties. In clinical studies nelfinavir was shown to be suitable for twice-daily dosing, the first HIV PI to achieve this milestone.[167]

7.12.5.1.3.1.5 Discovery of amprenavir and fosamprenavir In an effort to minimize molecular size and optimize binding contacts with the protease enzyme, extensive use of inhibitor–enzyme crystal structures led to the discovery of amprenavir (Agenerase) as a potent, low-molecular-weight drug that was approved for the treatment of HIV/AIDS in 1999. The structure-based design[168] was driven by an effort to maximize the interaction of the inhibitors with the aspartate residues in the active site, as well as the water molecule that interacts with the flap portion of the enzyme. Inhibitors were also designed to maximize water solubility and minimize the amount of conformational reorganization upon binding. Ultimately, the design included the use of the tetrahydrofuranyl carbamate group (**Figure 24**) that was shown to significantly improve potency in a different series of inhibitors.[169] The C-terminal portion of the molecule included a novel alkyl sulfonamide group that helped to reduce the molecular weight of the inhibitor.[170] Clinical studies of amprenavir indicated that twice-daily dosing gave acceptable exposure in HIV-infected adults.[171] However, problems with formulating amprenavir required patients to take eight 150-mg capsules twice daily. In an effort to

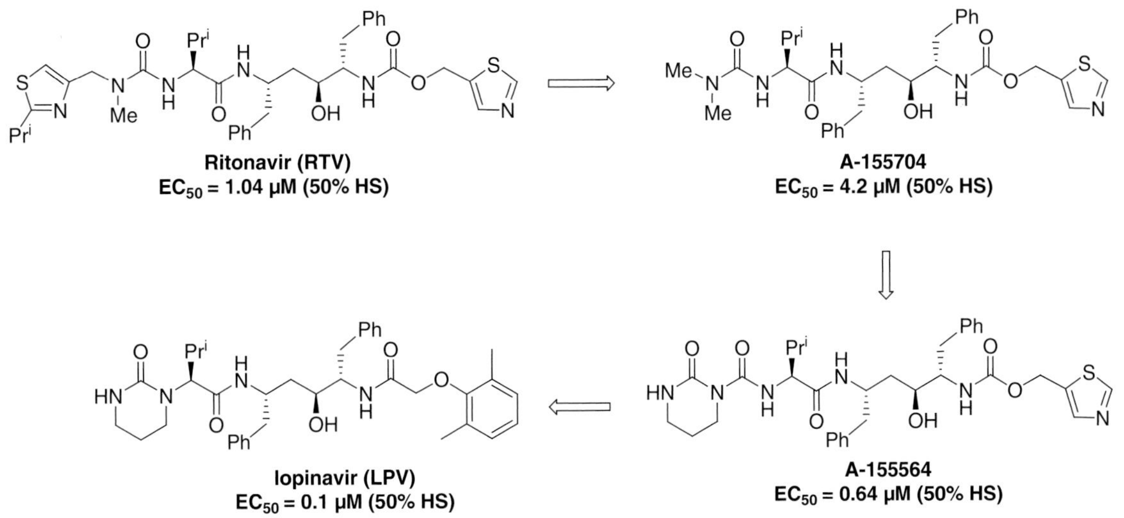

Ritonavir (RTV)
EC_{50} = 1.04 μM (50% HS)

A-155704
EC_{50} = 4.2 μM (50% HS)

lopinavir (LPV)
EC_{50} = 0.1 μM (50% HS)

A-155564
EC_{50} = 0.64 μM (50% HS)

Figure 25 Evolution of lopinavir.

improve the pharmacokinetic profile of amprenavir, the calcium salt of a phosphate prodrug of amprenavir was investigated. This modification resulted in two important improvements, including a lower pill burden and no significant interactions with foods.[172] These improvements led to the approval of fosamprenavir (Lexiva) in 2003.

7.12.5.1.3.1.6 Discovery of lopinavir Based on extensive experience with ritonavir in clinical studies, a variety of shortcomings with ritonavir became apparent. These included the relatively rapid emergence of resistance, suboptimal pharmacokinetics, and tight binding to plasma proteins that compromised potency. Starting from the crystal structure of ritonavir bound to HIV protease, modifications were made to the inhibitor to address these issues.[173] Because protease mutations at V82 were in part responsible for the high-level resistance found in clinical isolates, the structure of ritonavir was modified to minimize the interaction of the inhibitor with the amino acid side chain at this position. Removal of the isopropylthiazole group residing on the N-terminus of ritonavir gave A-155704 (**Figure 25**), which eliminated this unwanted interaction but caused a decrease in antiviral activity in cell culture measured in the presence of 50% human serum. This drop in potency was regained by constraining the dimethyl urea within a six-membered ring to give A-155564. Further optimization led to the replacement of the thiazole carbamate by a phenoxymethyl amide. This afforded the inhibitor lopinavir that has a significantly improved resistance profile and reduced susceptibility to plasma protein binding. Because the metabolism of lopinavir is inhibited in the presence of low concentrations of ritonavir, concomitant dosing of the two agents provides dramatically improved plasma concentrations of lopinavir in animals and humans.[173] Based on these results and other clinical studies, the combination product Kaletra containing a fixed-dose combination of lopinavir and ritonavir received regulatory approval in 2000.

7.12.5.1.3.1.7 Discovery of atazanavir As shown in **Figure 26**, starting from a symmetry-based design similar to that used for ritonavir, an aza-peptide[174,175] was utilized as a dipeptide isostere. By replacing one of the backbone carbon atoms of a ritonavir-type isostere with a nitrogen atom, a series of aza-peptide inhibitors of HIV protease have been discovered.[176,177] Optimization of this series of compounds was based on the crystal structure of CGP 53820 bound to the enzyme. Using this technique, compounds such as **3** (**Figure 26**) with significantly improved antiviral activity were identified by incorporating relatively minor structural changes.[178] Unfortunately, the oral bioavailability of compounds like **3** was relatively modest, and required further optimization. Introduction of a biaryl replacement for the isobutyl group attached to the hydrazine portion of the inhibitor led to atazanavir (Reyataz), a compound that retained the excellent antiviral potency of **2**, but with much-improved animal pharmacokinetics.[179] Atazanavir was active against several drug-resistant HIV strains insensitive to both indinavir and saquinavir.[180] Clinical studies showed that once-daily dosing with atazanavir either as short-term monotherapy or in combination with dual NRTIs caused significant drops in viral load. It became the first HIV PI approved for once-daily dosing in 2003.[181]

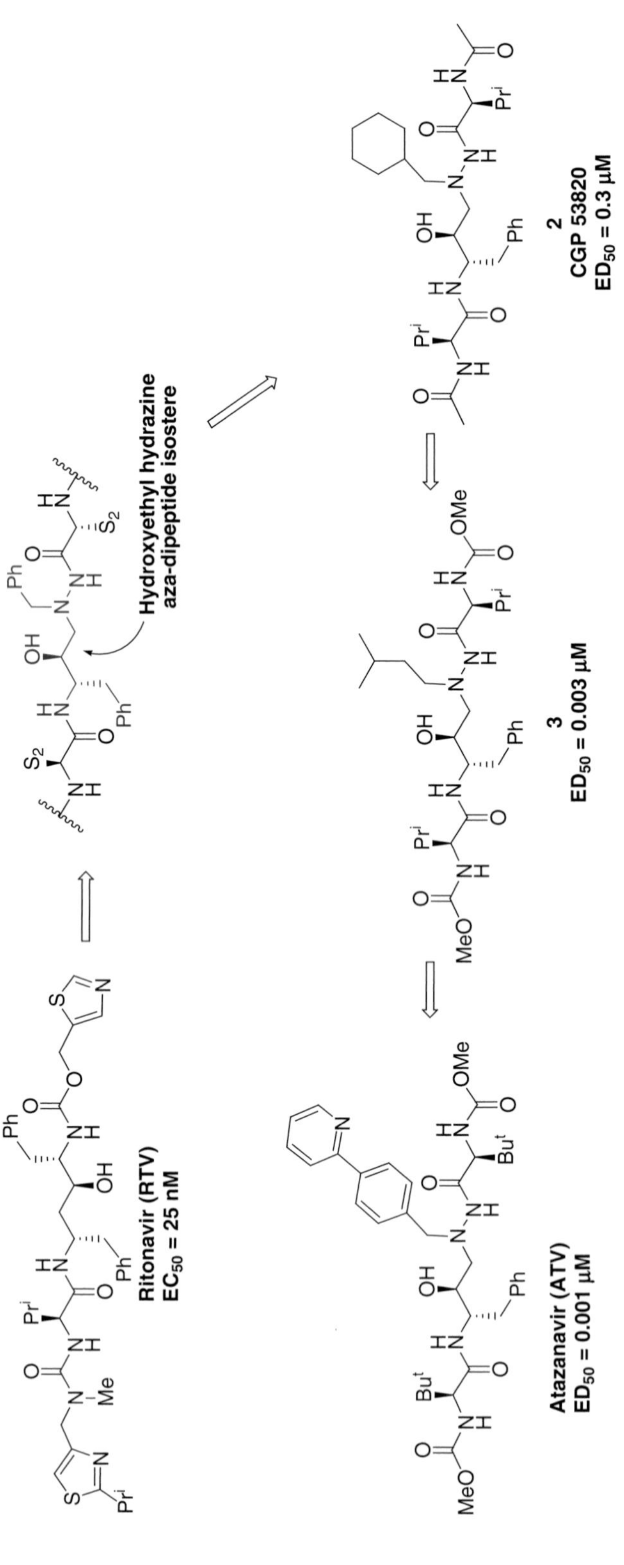

Ritonavir (RTV)
EC_{50} = 25 nM

Hydroxyethyl hydrazine aza-dipeptide isostere

2
CGP 53820
ED_{50} = 0.3 μM

3
ED_{50} = 0.003 μM

Atazanavir (ATV)
ED_{50} = 0.001 μM

Figure 26 Evolution of atazanavir. Aza-peptide (blue) used as a dipeptide isostere.

7.12.5.1.3.1.8 Discovery of tipranavir In 1993 it was reported that the anticoagulant warfarin (**Figure 27**) was able to block the spread of HIV in cell culture.[182] Shortly thereafter, the antiviral mechanism for warfarin was found to be inhibition of HIV protease.[183] Such relatively potent inhibition by a small nonpeptide molecule was unprecedented at the time. In 1994, two laboratories independently reported the discovery of similar compounds from a broad-based screening approach designed to identify novel nonpeptidic HIV PIs.[184,185] Using a combination of a Monte Carlo based automatic docking program and x-ray crystallography, optimization of the lead compound PD099560 (**Figure 27**) led to compound **31** with modest improvements in potency against the enzyme.[186] Further optimization involving major structural modifications led to the very potent inhibitor **13-(S)** that also displayed very good bioavailability in mice and dogs and a broad resistance profile.[187]

At about the same time, the warfarin analog phenprocoumon (**Figure 27**) was also identified through a broad-based screening approach.[184] Through a series of iterative structure-based design modifications utilizing the structures of the inhibitors bound to the enzyme, a significantly more potent compound U-96988 containing a 5,6-dihydro-4-hydroxy-2-pyrone core was identified. This compound exhibited promising oral bioavailability and was advanced into phase I clinical trials.[184] By comparing the enzyme–inhibitor crystal structure of U-96988 to those of unrelated peptidomimetic compounds, modifications were introduced that led to U-103017 (**Figure 27**), a significantly more potent inhibitor[188] and the second clinical candidate from this series.[189] Combining the desirable properties of U-96988 and U-103017 with extensive use of crystallographic data ultimately led to the identification of tipranavir (Aptivus) as a structurally novel inhibitor with a unique resistance profile.[190–192] In vitro studies of tipranavir have shown that it retains significant activity against a wide variety of clinical isolates resistant to multiple PIs.[193,194] In phase I/II clinical trials, the pharmacokinetic profile and efficacy of tipranavir were significantly enhanced when it was coadministered with

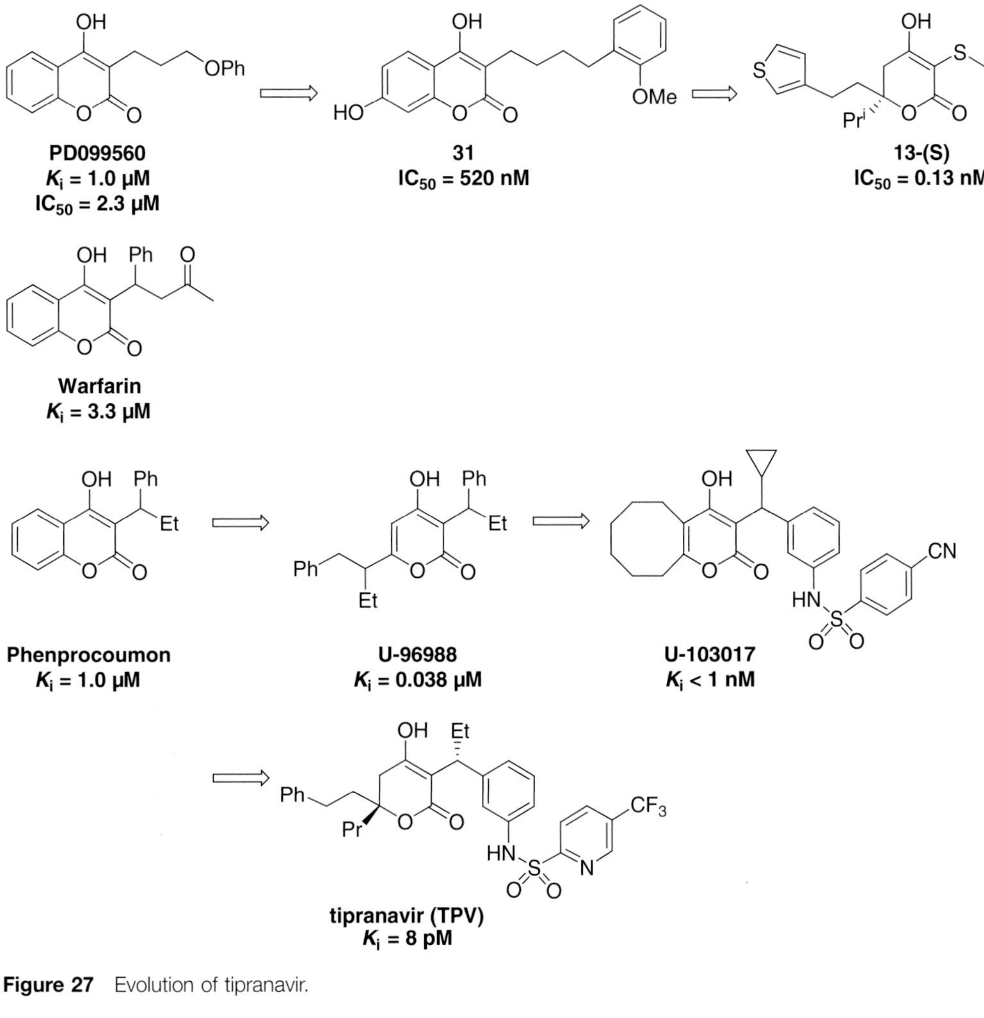

PD099560 K_i = 1.0 μM, IC_{50} = 2.3 μM

31 IC_{50} = 520 nM

13-(S) IC_{50} = 0.13 nM

Warfarin K_i = 3.3 μM

Phenprocoumon K_i = 1.0 μM

U-96988 K_i = 0.038 μM

U-103017 K_i < 1 nM

tipranavir (TPV) K_i = 8 pM

Figure 27 Evolution of tipranavir.

low-dose ritonavir.[195] Longer-term studies in patients experiencing virologic failure after treatment with at least 12 antiretroviral drugs showed that tipranavir coadministered with low-dose ritonavir was superior to the comparator PIs.[196] It is worth noting that the relative stereochemistry of the small and large substituents on the pyrone ring is different for tipranavir and **13-(S)** (**Figure 27**). Tipranavir was approved by the FDA in June 2005.

7.12.5.1.3.2 Extent of use

Until the introduction of PIs into clinical practice, reduction of viral replication to undetectable levels in HIV-infected patients was rare. Combination therapy of a PI with two nucleosides accomplished this goal in a substantial percentage of patients, and was referred to as highly active antiretroviral treatment (HAART).[33,197] As the number of approved drugs to treat HIV/AIDS increases, a large number of combination therapies continue to be evaluated, many that include one or more PIs. In current practice, most PIs are used in combination with low-dose ritonavir, which significantly improves exposure, plasma half-life, and efficacy.[198]

7.12.5.1.3.3 Clinical limitations

7.12.5.1.3.3.1 Resistance Some of the earliest clinical studies of HIV PIs included monotherapy arms, which often resulted in an initial drop in viral load followed by viral rebound that was due to the emergence of resistant viral populations.[199–201] Genetic sequencing provided evidence that multiple mutations within the protease sequence were required for high-level resistance, and that these mutations occurred in a stepwise fashion.[202,203] It is common to observe cross-resistance between several members of the PI class, depending upon the particular set of mutations (**Table 3**). Mutations occurring at 20 or more of the 99 amino acids of the enzyme have been associated with protease resistance; however, the following sites are among those considered to be primary mutants: 10, 30, 46, 48, 54, 82, 84, and 90.[204] Analyses of the location of these many mutations on the three-dimensional structure of the enzyme has revealed that changes occur close to the active site of the enzyme, in peripheral locations within the protease sequence, and changes at the cleavage sites in the polyprotein substrate.[205] Cleavage site mutations alone have not been implicated in PI resistance, rather they appear to increase the efficiency of cleavage by the mutant enzymes.[204,205] **Figure 28** shows indinavir bound to HIV-1 protease. Enzyme crystal structures of a number of mutant enzymes have yielded high-resolution structural data on the effects that various mutations can have on the structure of the enzyme.[206] Such changes can affect enzymatic activity, substrate specificity, and enzyme stability. Generally, mutations within the active site lead to reduced affinity of the drug, resulting in a drop in potency. Mutations at the periphery of the enzyme are often compensatory mutations that can lead to enzymes with increased activity or to conformational effects that cause changes in the active site.

7.12.5.1.3.3.2 Tolerability and compliance Patients taking HIV PIs can experience a wide variety of side effects depending the particular inhibitor and other drugs being taken.[207] Because PIs must normally be coadministered with other antiretroviral agents, it is often difficult to determine the contribution of any individual drug on the side effect profile. However, in phase I or early phase II studies, drugs are typically given to either normal subjects or patients for up to 2 weeks, allowing the identification of side effects specific to that medication. Typical side effects experienced by a significant percentage of patients can include gastrointestinal disturbances,[208,209] nephrolithiasis,[210] and hyperbilirubinemia.[211] Until the approval of atazanavir, PIs were implicated in a number of metabolic complications

Table 3 Mutations closely associated with individual PIs

Drug	*Mutations*
Saquinavir (sgc)[a]	G48V, L90M
Ritonavir	I54V, V82A, L90M
Indinavir	V82A/T/F/S, L90M
Nelfinavir	D30N, L90M
Amprenavir	I50V, I54M, I84V
Atazanavir	I50L, N88S
Tipranavir	L33F

[a] sgc, soft gel capsule, marketed as Fortovase.

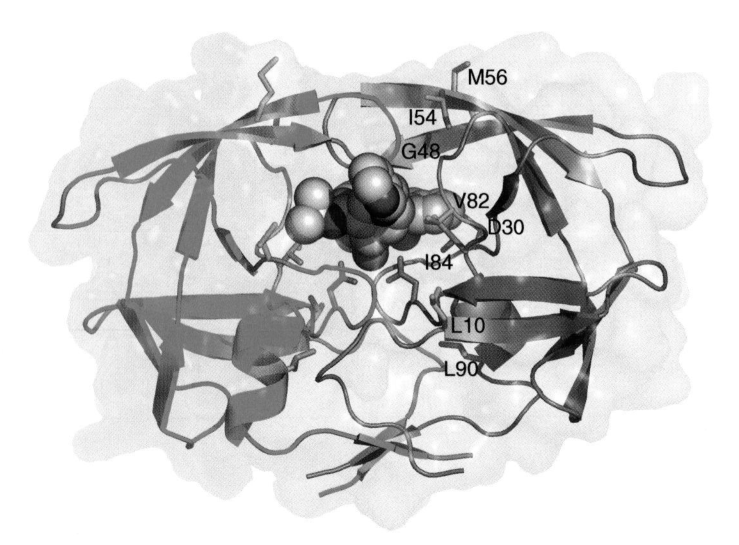

Figure 28 X-ray crystal structure of indinavir-bound HIV protease with mutants; locations of commonly mutated amino acids shown in orange.

Table 4 Dosing regimens for PIs

Drug	*Dosage (mg)*	*Frequency*[a]
Saquinavir (sgc)[b]	1200	tid
Ritonavir	600	bid
Indinavir	800	tid
Nelfinavir	1250	bid
	750	tid
Amprenavir	1400	bid
Atazanavir	400	qd
Fosamprenavir	1400	bid

[a] qd, once daily; bid, twice daily; tid, three times daily.
[b] sgc, soft gel capsule, marketed as Fortovase.

including increased serum cholesterol and triglyceride levels, insulin resistance, and lipodystrophy.[212] Because most of the approved PIs are coadministered with low-dose ritonavir to improve their pharmacokinetic profiles, side effects from ritonavir dosing are common.

7.12.5.1.3.3.3 Dosing regimens Due to the suboptimal pharmacokinetics of the first PIs, dosing regimens were characterized by complicated requirements that could involve dosing up to three times daily, taking up to 16 capsules per day, and dosing either before or after meals (**Table 4**). Because the majority of approved PIs are metabolized via cytochrome P450 pathways, it was discovered that coadministration with lower doses of the potent cytochrome P450 inhibitor ritonavir could dramatically increase the oral exposure of many of these drugs. In many cases this has allowed less frequent dosing and dosing with fewer food restrictions (**Table 5**). This practice is often referred to as 'ritonavir boosting,' and has improved the antiviral effectiveness of saquinavir, indinavir, lopinavir, amprenavir, fosamprenavir, tipranavir, and atazanavir.[198,213] Atazanavir is the only PI currently recommended for once-daily dosing without ritonavir.

7.12.5.1.3.3.4 Compatibility with other antiretrovirals Because the metabolism of most PIs is not affected by NRTIs, compatibility is generally good, and current dosing recommendations normally include two NRTIs. The exception is atazanavir whose plasma levels are decreased when dosed with tenofovir.[214] In the case of NNRTIs that either inhibit or induce cytochrome P450, dose adjustment may be necessary.

Table 5 Dosing regimens for PIs with ritonavir[a]

Drug	*Dosage (mg)*	*Frequency*[a]	*Ritonavir dose (mg)*
Saquinavir	1000	bid	100
Indinavir	800	bid	100 or 200
Lopinavir[b]	400	bid	100
Fosamprenavir	700	bid	100
	1400	qd	200
Atazanavir	300	qd	100
Tipranavir	500	bid	200

[a] qd, once daily; bid, twice daily.
[b] Combination marketed as Kaletra.

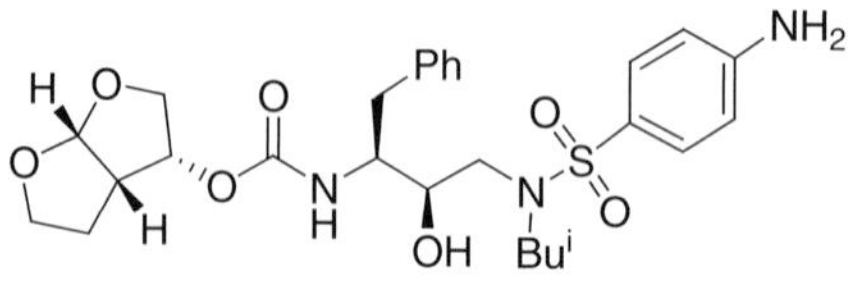

Figure 29 Structure of TMC-114.

7.12.5.1.3.4 Comparison of commonly used agents

Due to its ability to increase exposure of other PIs, ritonavir is commonly used as a boosting agent at doses between 100 and 400 mg d^{-1}. Consequently, all of the other approved PIs except nelfinavir are routinely prescribed with low-dose ritonavir. The combination of ritonavir plus lopinavir (Kaletra) offers the advantage of a lower pill burden. Atazanavir also has been shown to have a negligible effect on serum lipid levels; however, hyperbilirubinemia and electrocardiographic PR-interval increases should be considered.[132] Fosamprenavir also decreased the pill burden and food effects when compared to the earlier drug amprenavir, but rash and gastrointestinal side effects can be problematic in some patients. In boosted indinavir regimens, dosing frequency can be reduced to twice daily, and food effects are minimized. However, a higher rate of nephrolithiasis requires patients to increase their fluid intake to 1.5–2.0 L d^{-1}. Nelfinavir has superior safety and pharmacokinetic profiles in pregnant women when compared to the other PIs, but can have a higher rate of virologic failure when compared to fosamprenavir and Kaletra.

7.12.5.1.3.5 Outlook

PIs have been shown to be a valuable part of many effective antiviral regimens, giving durable reductions in viral load in those patients who tolerate therapy. There remains a need for newer PIs that have improved resistance profiles, more convenient dosing, reduced effects on lipid and glucose metabolism, and fewer drug interactions. **Figure 29** shows the experimental PI TMC-114, which is a close analog of amprenavir.[215] This compound has a significantly improved resistance profile and is currently being evaluated in clinical trials.[216] TMC-114 (darunavir) in combination with ritonavir was approved by the FDA in June 2006.

7.12.5.1.4 **Entry inhibitors**

7.12.5.1.4.1 Discovery of enfuvirtide

The first step in viral replication involves a binding event between specific glycoproteins on the surface of the virus and one or more receptors on the surface of the targeted cell. The viral proteins involved in this event are gp-41 and gp-120, which are derived from the larger glycoprotein gp-160 and are closely bound in a trimeric structure within the membrane of HIV.[217] Subsequent to the binding of gp-120 to a CD4 receptor on the target T cell, a conformational change allows an interaction between a co-receptor (either CXCR4 or CCR5) on the cell surface and gp-120. This second binding event is thought to cause gp-41 to undergo a refolding event that brings the viral membrane into close contact with the cell membrane permitting fusion to occur.[218] This is followed by movement of the viral core into the

CH_3CO-Tyr-Thr-Ser-Leu-Ile-His-Ser-Leu-Ile-Glu-Glu-Ser-
Gln-Asn-Gln-Gln-Glu-Lys-Asn-Glu-Gln-Glu-Leu-Leu-Glu-
Leu-Asp-Lys-Trp-Ala-Ser-Leu-Trp-Asn-Trp-Phe-NH_2

Enfuvirtide (ENF)

Figure 30 Sequence of enfuvirtide.

cell. Several laboratories have shown that peptides of varying length corresponding to regions within the HIV envelope can have antiviral properties.[219–222] In an effort to further elucidate the mechanism of this binding interaction, a series of peptides corresponding to various regions of gp-41 were synthesized and evaluated.[223] An N-terminal acetylated, C-terminal amide peptide (DP-178) (**Figure 30**) corresponding to residues 643–678 of HIV-1 gp-41, just on the extracellular side of the transmembrane region, was found to have remarkable antiviral properties at very low concentrations. DP-178 was found to inhibit both cell-to-cell and cell-free transmission of HIV-1 in cell culture at concentrations between 1 and 80 $ng\,mL^{-1}$. The compound, now known as enfuvirtide (Fuzeon), was more than 100-fold less potent versus HIV-2, but inhibited a variety of HIV-1 isolates.[223] It has been shown that enfuvirtide inhibits the gp-41 refolding event required for viral fusion, and interacts with the HR-1 heptad repeat region.[224] Due to its potent antiviral activity and novel mechanism of action, enfuvirtide was proposed as a candidate for clinical studies. Because it would not likely survive conditions in the intestinal tract, the first clinical studies were carried out using intravenous administration in treatment-naive patients. Over the 2 weeks of treatment, viral load reductions of 1.96 $\log_{10}$ were observed at the highest dose of 100 mg twice daily.[225] In subsequent studies, the more easily administered route of subcutaneous injection was found to provide sufficient exposure and plasma half-life to be given only twice daily.[226,227] Enfuvirtide was shown to give significant reductions in viral load, even in patients who were failing therapy on combinations of RT or PIs.[227] Enfuvirtide received regulatory approval in 2003 for the treatment of HIV-infected patients with evidence of viral replication despite ongoing antiviral therapy.

7.12.5.1.4.2 Extent of use

A significant number of patients have tried a variety of the currently approved drugs to treat HIV in various combinations without success. The reasons for failure of individual drug combinations are usually the development of cross-resistance that limits the effectiveness of an entire class of drugs, or the inability to continue treatment due to intolerable side effects. In these patients, enfuvirtide is one of the last options for antiviral therapy. It has not gained wider acceptance in other patient populations mainly due to the relatively inconvenient and sometimes painful twice-daily injection dosing requirements.

7.12.5.1.4.3 Clinical limitations

7.12.5.1.4.3.1 Resistance In the initial phase I monotherapy trials of enfuvirtide, patients receiving suboptimal doses were observed to have a continuous drop in viral load through the 10th day of dosing, followed by a trend to viral rebound by the end of the 14-day study. Viruses cloned from these patients showed single and multiple mutations within the gp-41 HR-1 region with reduced susceptibility to enfuvirtide.[228] In viral isolates from patients receiving long-term enfuvirtide combination therapy, a similar resistance pattern was noted with the addition of a mutation in the HR-2 heptad repeat region in some isolates. It was hypothesized that this mutation in HR-2 could be a compensatory mutation perhaps involving viral fitness.[229]

7.12.5.1.4.3.2 Tolerability and compliance Injection site reactions of any severity were reported by 98% of patients taking enfuvirtide in phase III clinical studies. Such reactions can include pain or discomfort, induration (hardening), erythema (redness), nodules and cysts, pruritus (itching), and ecchymosis (hemorrhagic spot).[132] Other serious side effects reported in a much smaller proportion ($<1\%$) of patients include bacterial pneumonia and systemic hypersensitivity reactions.

7.12.5.1.4.3.3 Dosing regimens For adults, 90 mg of enfuvirtide should be given by subcutaneous injection twice daily. In HIV-infected pediatric patients a dose of 2 $mg\,kg^{-1}$ twice daily is recommended.[226]

7.12.5.1.4.3.4 Compatibility with other antiretrovirals Enfuvirtide has been shown to have a very low propensity to interact with other drugs. It has little if any effect on drugs metabolized by the cytochrome P450 pathways, and its plasma levels are not significantly affected by either ritonavir or rifampin.[226]

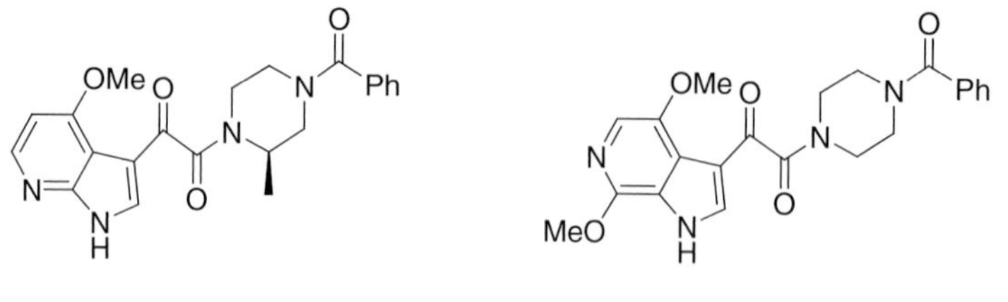

BMS-378806

BMS-488043

Figure 31 Structures of BMS-378806 and BMS-488043.

7.12.5.1.4.4 Outlook

Enfuvirtide has amply demonstrated the proof of concept for treatment of HIV by influencing the entry pathway. Due to the large number of both viral and cellular proteins involved in viral fusion and entry, the prospects for new therapeutics based on these mechanisms are promising. In particular, the compound BMS-378806 (**Figure 31**) was discovered as part of a cell-based screening program to identify novel inhibitors of HIV replication without regard to mechanism. The EC_{50} of BMS-378806 toward a panel of 83 clinical isolates of HIV-1 varied between 1 and $>10\,000$ nM with little if any activity against HIV-2 or a number of other viruses. Mechanistic studies indicated that interference with the binding of gp-120 to CD4 was the mode of action for this new series of compounds.[230,231] In a pilot clinical study of the closely related analog BMS-488043 (**Figure 31**), 8 out of 12 patients experienced a drop in viral load of greater than 1.0 $\log_{10}$ after receiving 1800 mg twice daily.[232]

Several new drug candidates are currently being evaluated that target the two most common cellular co-receptors CCR5 and CXCR4. HIV viral strains that target only CCR5 receptors are primarily macrophage tropic (M-tropic) and constituted approximately 80% of isolates in a sampling of both antiretroviral naive and experienced patients. Viruses that utilize only the CXCR4 receptor are classified as T cell tropic (T-tropic) and constituted less than 1% of the samples isolates. The remaining 20% consisted of dual tropic viruses that utilize both receptors.[233] As the course of HIV infection progresses in AIDS patients, the tropism of their virus often changes from M-tropic to T-tropic. These T-tropic viruses have been associated with higher viral load and increased pathogenesis, so there are some concerns that selective pressure on CCR5 viruses might promote the change from M-tropic to T-tropic virus.[233] These concerns have not yet been adequately addressed in clinical trials. The first report of a small molecule that could block the interaction of gp-120 with CCR5 was TAK-779 (**Figure 32**).[234] Since that disclosure, several laboratories have discovered a wide variety of structurally distinct small-molecule antagonists of CCR5.[235] A proof of concept study using intravenously administered CMPD-167 was carried out in six rhesus macaques infected with either SIVmac, SIVB670, or the CXCR4 tropic SHIV-89.6P. All of the animals showed a rapid decrease in viral load; however, the magnitude and duration of the responses were quite variable. No effect of the drug was noted in the animals infected with the SHIV.[236] Three compounds have advanced into later stages of clinical development (**Figure 32**). Maraviroc (UK-427857) at a dose of 100 mg twice daily was found to lower viral load by 1.42 $\log_{10}$ in a 10-day study in patients whose viruses used only CCR5.[237] A comparable outcome was achieved when GW-873140 was given orally at 100 mg twice daily in patients screened for CCR5 receptor usage, giving a drop in viral load of 1.5 $\log_{10}$.[238] In patients treated for 14 days with 10, 25, or 50 mg twice daily of Sch-417690, average viral load decreases of at least 1.0 $\log_{10}$ were observed.[239]

Compounds have also been discovered that bind selectively to the other major co-receptor CXCR4. From an investigation of the antiviral activities of certain metal-chelating compounds called polyoxometallates, a series of nonmetal containing bicyclic compounds called bicyclams were found to inhibit HIV in cell culture at low concentrations.[240] After many years of research, these compounds, and AMD-3100 (**Figure 33**) in particular, were shown to bind specifically to CXCR4 receptors on T cells. In a patient cohort infected with either dual tropic CXCR4/CCR5 virus or a mixed population of viruses, intravenous administration of AMD-3100 for 10 days resulted in a significant reduction in viral load.[241] The clinical benefit of CXCR4 receptor antagonists will likely depend on the identification of effective, orally bioavailable compounds.

7.12.6 Unmet Medical Needs

Because of the introduction of many antiretroviral drugs into clinical practice, AIDS and infection with HIV have evolved from untreatable terminal conditions to a manageable chronic disease for those with access to adequate medical care.

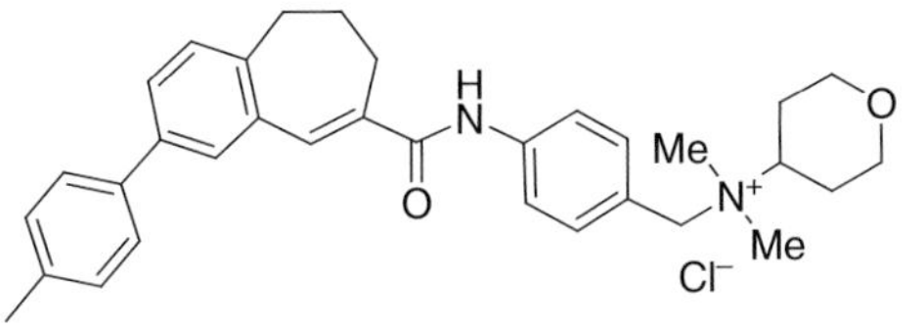

TAK-779

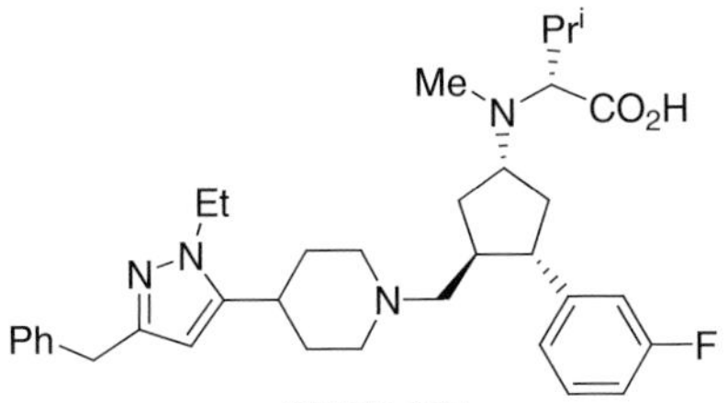

CMPD-167

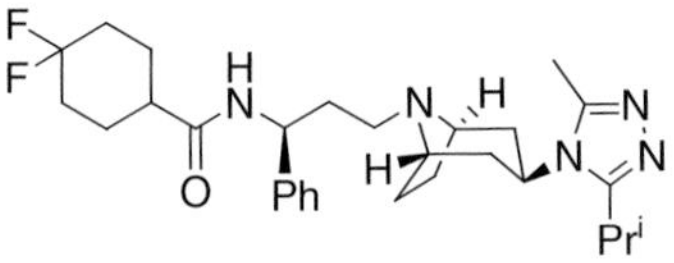

Maraviroc

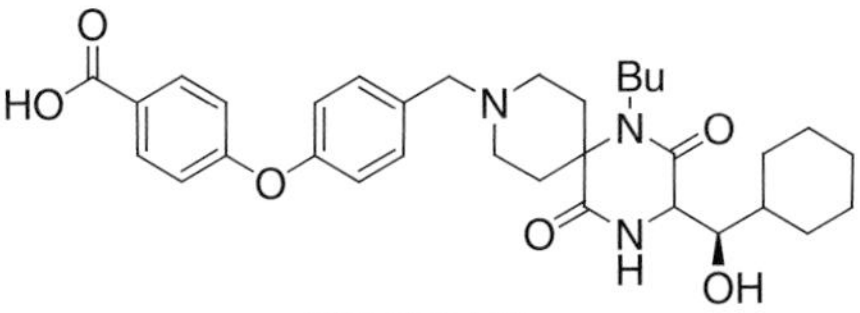

GW-873140

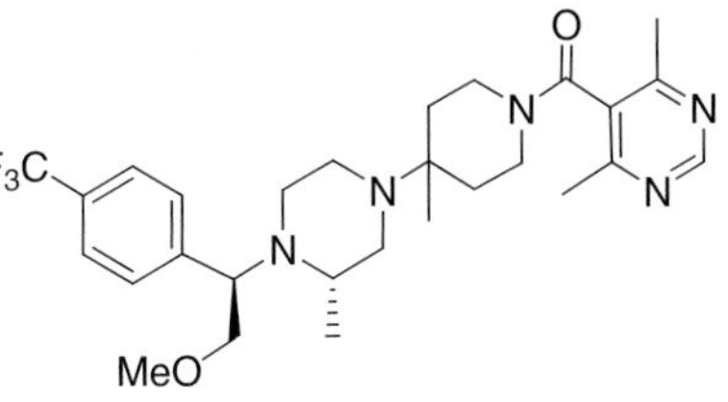

Sch-417690

Figure 32 Structures of CCR5 antagonists.

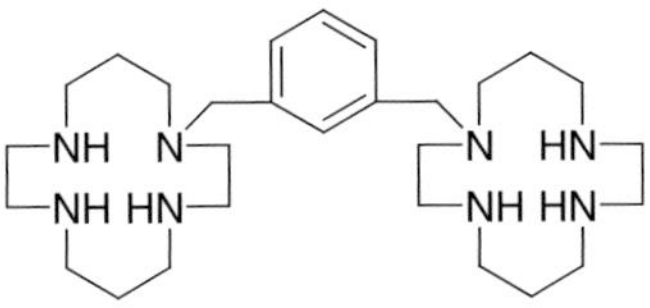

AMD-3100
EC_{50} = 3 nM

Figure 33 Structure of AMD-3100.

7.12.6.1 Developing World Challenges

Some of the poorest nations in the world are at the epicenter of the current AIDS pandemic. Countries such as India and South Africa have enormous numbers of infected individuals and many more at high risk of becoming infected.

Many of these countries have an extremely limited healthcare infrastructure, and lack the means to transport medical equipment or drugs to a significant proportion of their population. The challenges of providing antiretroviral therapy and monitoring on a large scale in such circumstances are formidable.[242] The lack of trained medical personnel and facilities is also an ongoing problem. AIDS is affecting primarily persons under the age of 25 in the prime of life. Consequently, AIDS has had a devastating effect on the economies and social structure of developing world countries. Parents, teachers, social workers, and healthcare workers have been struck down by HIV, magnifying the challenges mentioned above. All of these issues serve to focus attention on the obvious solutions: prevention or an effective vaccine.[243,244] Unfortunately, many of the same issues surrounding healthcare also apply to vaccine delivery, education, and the behavior modifications necessary for prevention of HIV spread. People living in the developing world often lack access to radio, television, print media, or electronic media. Many people living in the developing world are unable to read, even if educational materials were available. Cultural traditions can also be an obstacle to efforts at changing innate behavior.[245] Nevertheless, prevention has become a central concern to those agencies dealing with the AIDS epidemic, and some progress has been reported.[246]

For the millions of people already infected with HIV, it is imperative that antiretroviral therapy be made available at an affordable cost. In many countries this will not be possible until the aforementioned infrastructure problems are solved. A number of organizations have implemented a variety of approaches to deal with this issue. Several pharmaceutical companies have either donated their antiretroviral drugs or made them available at a significant discount to the prices common in the developed world. There are significant intellectual property concerns surrounding the manufacture, marketing, and distribution of antiretroviral drugs, and preventing their export to the developed world. In spite of these problems, significant progress is being made in several countries around the globe.[246]

7.12.6.2 Resistance and Compliance

As has been emphasized in several of the previous sections, resistance to antiretroviral drugs is inevitable in a significant percentage of treated patients. The failure of a given antiviral treatment is related to regimen adherence, viral resistance to that regimen, and the fitness of the resistant virus.[247] Multidrug cross-resistance has severely compromised the efficacy of entire classes of HIV drugs. Every year the percentage of viruses that are transmitted with resistance to at least one of the approved drugs is increasing. Because of these problems, the need for medications with a higher barrier to resistance and new drugs targeted at novel biochemical targets is pressing. Because lack of compliance is a major contributor to resistance, there is also a urgent need for drugs with improved tolerability and dosing convenience.

7.12.6.3 Long-Term Adverse Effects

The present outlook for an outright cure for HIV infection is not promising. It is likely that HIV-infected individuals will face lifelong therapy with antiretroviral drugs and perhaps some type of immunotherapy. The side effects of HIV infection and long-term antiviral therapy can be quite serious, and can include metabolic abnormalities, fat redistribution, lipid and cholesterol elevations, and various organ toxicities. Because many of the biochemical HIV drug targets do not exist in humans, mechanism-based toxicity should not be an issue. However, off-target activity continues to be problematic to some extent for the majority of AIDS drugs.[248] There is significant room for improvement in the side effect profile for many of the approved agents.

7.12.7 New Research Areas

7.12.7.1 Integrase Inhibitors

The identification of effective inhibitors of the viral enzyme integrase has not been as rapid as for the other viral enzymes RT and protease. **Figure 34** shows some of the many structures reported to have integrase inhibitory activity, including the diketoacid derived compounds L-731,988[249,250] and S-1360,[251] and the naphthyridines L-870,810[252] and L-870,812.[43] A comparison of the mutation profiles of viruses obtained by serial passage of HIV-1 in the presence of diketoacids,[252,253] S-1360,[254] or naphthyridines[252] shows striking differences. This suggests that even though both series of compounds inhibit integrase by mechanistically identical means, they bind to the enzyme in different orientations.[252] **Figure 35** shows a representation of the x-ray crystal structure of the integrase core domain.[255] Virus resistant to either L-870,810 or the diketoacids remains fully sensitive to the other. From the many compounds that have been reported to be active against integrase and HIV in cell culture, only the naphthyridines L-870,812 and

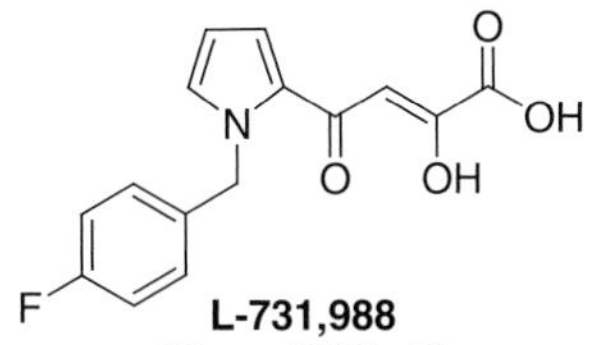

L-731,988
IC_{95} = 9600 nM

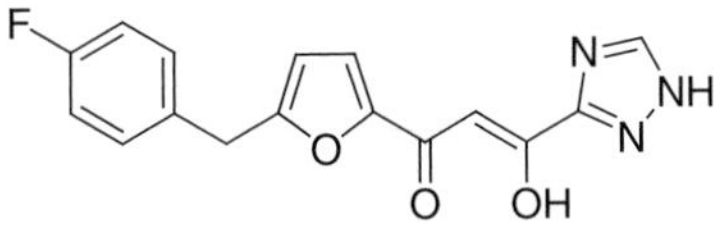

S-1360
EC_{90} = 990 nM

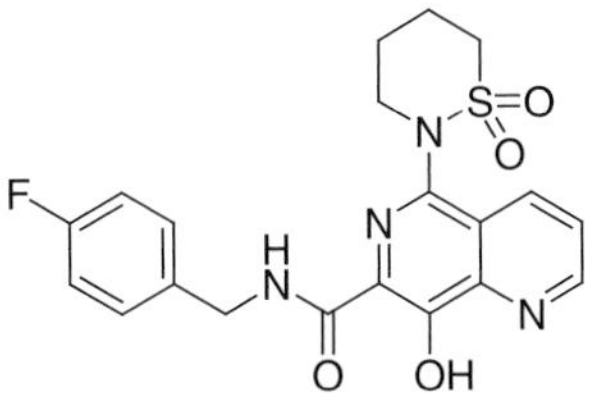

L-870,810
IC_{95} = 15 nM (10% FBS)
IC_{95} = 100 nM (50% NHS)

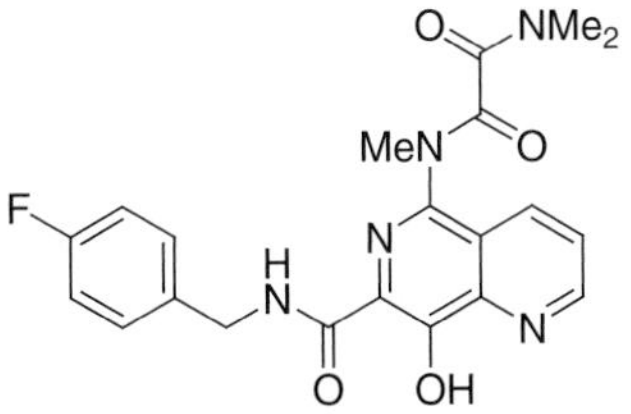

L-870,812
IC_{95} = 250 nM

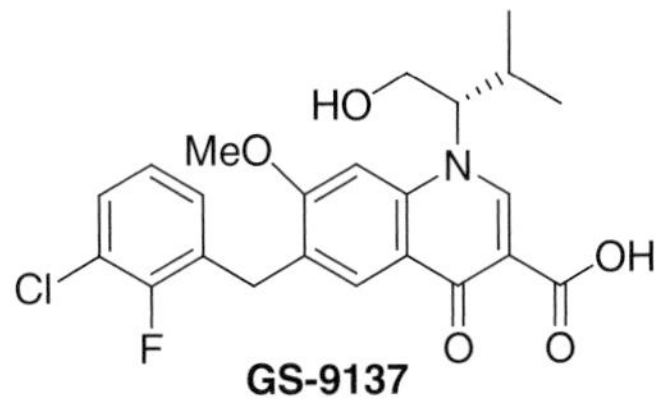

GS-9137
EC_{50} = 0.9 nM

Figure 34 Structures of integrase inhibitors.

L-870,810 (**Figure 34**) have also demonstrated in vivo activity. In particular, oral administration of L-870,812 to rhesus macaques infected with SHIV 89.6P was shown to cause a sustained decrease in viral load and maintenance of CD4 levels.[43] In a phase Ib clinical trial, L-870,810 caused an average reduction in viral load of greater than 1.7 $\log_{10}$ when administered at either 200 or 400 mg twice daily over a 10-day period.[256] This study represents the first clinical proof of concept for the use of integrase inhibitors for the treatment of HIV infection. Although the clinical development of L-870,810 was put on hold due to long-term toxicity observed in dogs,[256] two other integrase inhibitors have also demonstrated clinical proof-of-concept. The quinolone GS-9137[256a] (**Figure 34**) demonstrated viral load decreases of between 1.48–2.03 log 10 over a 10-day period when dosed at 200, 400 or 800 mg twice daily. Similarly, a 50-mg dose of GS-9137 administered once daily with 100 mg of ritonavir gave a 2.03 log10 decrease.[256b] In a study population of treatment-experienced patients with documented resistance to all three drug classes (NRTIs, NNRTIs, and PIs), the integrase inhibitor MK-0518 was shown to drop viral loads below 400 copies mL^{-1} in 85–92% of patients after 8 weeks of dosing 200, 400 or 600 mg twice daily that was added to their existing therapy.[256c]

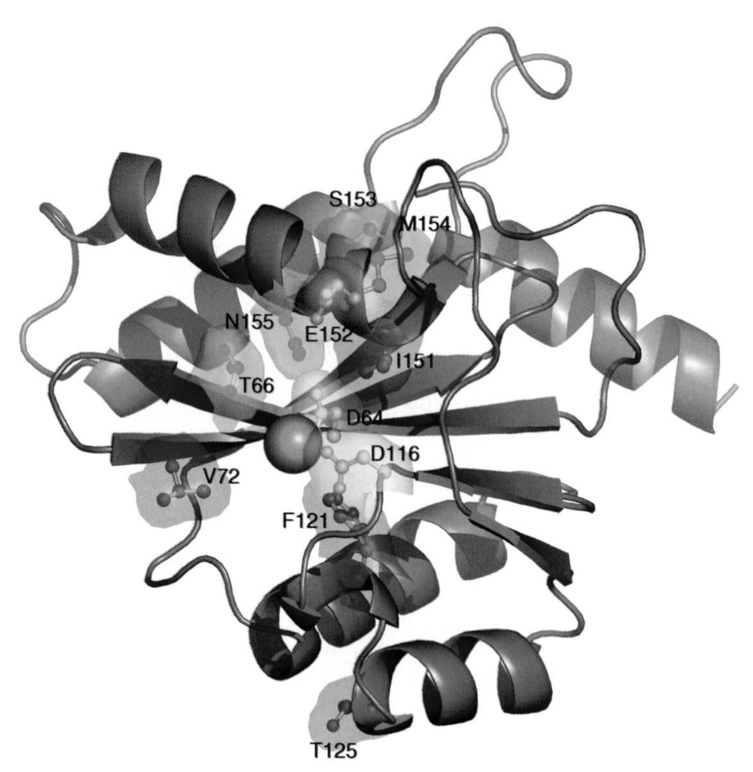

Figure 35 X-ray crystal structure of the integrase core domain. The three catalytic residues D64, D116, and E152 are shown in yellow coordinated with a magnesium atom shown in orange. The residues T66, S153, M154, and N155 highlighted in green are associated with resistance to diketoacids. Serial passage of virus in the presence of increasing concentrations of S-1360 caused mutations at T66 and L74 which also engender cross-resistance to the diketoacids. Mutations at residues V72, F121, T125, and I151 shown in red are associated with resistant virus when passaged in the presence of increasing concentrations of the naphthyridine L-870,810.

Betulinic acid
IC_{50} = 1400 nM

PA-457
IC_{50} = 10 nM

Figure 36 Structures of maturation inhibitors.

7.12.7.2 Maturation Inhibitors

From a screening program directed at compounds that inhibit HIV replication without regard to mechanism of action, betulinic acid (**Figure 36**) was found to be a weak inhibitor.[257] Optimization of this activity afforded the betulinic acid derivative PA-457 that was found to be a potent inhibitor of a variety of HIV clinical isolates and drug-resistant mutant viruses. From a series of experiments it was found that PA-457 inhibits one of the key steps in the latter stage of viral replication, namely the proteolytic processing of the gag p25 capsid precursor (CA-SP1) to mature gag p24 capsid protein (CA). This results in the production of defective viral particles that lack replicative capacity. HIV-1 passaged in the presence of increasing concentrations of PA-458 results in a mutant virus with a single amino acid change (Ala to Val) at the N-terminal side of the CA/SP1 cleavage site.[258] In phase I clinical studies, PA-457 was shown to cause a drop in viral load of 0.7 $\log_{10}$ after single oral doses of 75, 150, or 250 mg.[259]

7.12.7.3 RNaseH Inhibitors

One of the enzymatic activities of RT that has been shown to be strictly required for viral replication is the removal of RNA from the RNA–DNA hybrids during reverse transcription (**Figure 3**).[260] The active site where this endonuclease activity originates is near the thumb region of the p66 domain (**Figure 2**). Several compounds have been reported to specifically inhibit the ribonuclease activity of RT, but none has advanced to clinical status.[261–264] Because inhibition of RT is a well-validated means of treating HIV, this mechanism should provide clinical benefit when optimized compounds are identified, and pre-existing resistance should not be a problem.

7.12.7.4 Nonenzymatic Viral Targets

The HIV proteome includes a number of accessory proteins[265] such as rev, tat, nef, vif, vpu, and vpr that have not yet been successfully targeted by antiviral drugs. Many of these proteins interact with host-cell biomolecules, and are important for viral propagation in vivo. For example, it has been reported that the binding of tat to the host coactivator protein PCAF can be blocked by small molecules.[266] As a better understanding emerges for the role that accessory proteins play in viral replication, it is likely that disruption of some of these interactions will lead to antiviral drugs.

There are also a number of host proteins that play an important role in viral replication. An example of such a protein is TSG101, which plays an important role in viral assembly and release.[267] It may be possible to interfere selectively with the interaction of the virus with host proteins (*see* Section 7.12.5.1.4.4) without disruption of their normal role in cell function. Because resistance will almost certainly continue to arise from the viral targets identified thus far, the need for new drugs acting through novel mechanisms is pressing. The challenge of combating AIDS in the developing world will depend in part on the development of new effective antiviral therapies that are easily administered and are well tolerated.

References

1. *CDC Morbid Mortal Weekly Rep.* **1981**, *30*, 250–252.
2. *CDC Morbid Mortal Weekly Rep.* **1981**, *30*, 305–308.
3. Fauci, A. S. *Ann. Intern. Med.* **1982**, *96*, 777–779.
4. Masur, H.; Michelis, M. A.; Greene, J. B.; Onorato, I.; Stouwe, R. A.; Holzman, R. S.; Wormser, G.; Brettman, L.; Lange, M.; Murray, H. W. et al. *N. Engl. J. Med.* **1981**, *305*, 1431–1438.
5. *CDC Morbid Mortal Weekly Rep.* **1982**, *31*, 300–301.
6. Korber, B.; Muldoon, M.; Theiler, J.; Gao, F.; Gupta, R.; Lapedes, A.; Hahn, B. H.; Wolinsky, S.; Bhattacharya, T. *Science* **2000**, *288*, 1789–1796.
7. AIDS Epidemic Update 2005 – Joint United Nations Programme on HIV/AIDS (UNAIDS) World Health Organization, 2005. Geneva. http://www.unaids.org/epi/2005/doc/report_pdf.asp (accessed Aug 2006).
8. *Human Immunodeficiency Virus Infection*. In *The Merck Manual of Diagnosis and Therapy*, 17th ed.; Beers, M. H., Berkow, R., Eds.; Merck Research Laboratories: Whitehouse Station, NJ, 1999, pp 1312–1323. http://www.merck.com/mrkshared/mmanual/home.jsp (accessed Aug 2006).
9. Castro, K. G.; Ward, J. W.; Slutsker, L.; Buehler, J. W.; Jaffe, H. W.; Berkelman, R. L.; Curran, J. W. *CDC Morbid Mortal Weekly Rep.: Recommendations and Reports* **1992**, *41*, RR–17.
10. Mier, J. W.; Gallo, R. C. *Proc. Natl. Acad. Sci. USA* **1980**, *77*, 6134–6138.
11. Gallo, R. C.; Salahuddin, S. Z.; Popovic, M.; Shearer, G. M.; Kaplan, M.; Haynes, B. F.; Palker, T. J.; Redfield, R.; Oleske, J.; Safai, B. et al. *Science* **1984**, *224*, 500–503.
12. Barre-Sinoussi, F.; Chermann, J. C.; Rey, F.; Nugeyre, M. T.; Chamaret, S.; Gruest, J.; Dauguet, C.; Axler-Blin, C.; Vezinet-Brun, F.; Rouzioux, C. et al. *Science* **1983**, *220*, 868–871.
13. Gallo, R. C. *Immunol. Rev.* **2002**, *185*, 236–265.
14. Popovic, M.; Sarngadharan, M. G.; Read, E.; Gallo, R. C. *Science* **1984**, *224*, 497–500.
15. Rey, M. A.; Spire, B.; Dormont, D.; Barre-Sinoussi, F.; Montagnier, L.; Chermann, J. C. *Biochem. Biophys. Res. Commun.* **1984**, *121*, 126–133.
16. Berger, E. A.; Murphy, P. M.; Farber, J. M. *Annu. Rev. Immunol.* **1999**, *17*, 657–700.
17. Bukrinskaya, A.; Brichacek, B.; Mann, A.; Stevenson, M. *J. Exp. Med.* **1998**, *188*, 2113–2125.
18. Götte, M.; Li, X.; Wainberg, M. A. *Arch. Biochem. Biophys.* **1999**, *365*, 199–210.
19. Esnouf, R.; Ren, J.; Ross, C.; Jones, Y.; Stammers, D.; Stuart, D. *Nat. Struct. Biol.* **1995**, *2*, 303–308.
20. Craigie, R. *J. Biol. Chem.* **2001**, *276*, 23213–23216.
21. Anthony, N. J. *Curr. Top. Med. Chem.* **2004**, *4*, 979–990.
22. Bannwarth, S.; Gatignol, A. *Curr. HIV Res.* **2005**, *3*, 61–71.
23. Morita, E.; Sundquist, W. I. *Annu. Rev. Cell Dev. Biol.* **2004**, *20*, 395–425.
24. Hill, M.; Tachedjian, G.; Mak, J. *Curr. HIV Res.* **2005**, *3*, 73–85.
25. Shugars, D. C.; Sweet, S. P.; Malamud, D.; Kazmi, S. H.; Page-Shafer, K.; Challacombe, S. J. *Oral Dis.* **2002**, *8*, 169–175.
26. Andiman, W. A. *Curr. Opin. Pediatr.* **2002**, *14*, 78–85.
27. Brenner, B. G.; Wainberg, M. A. *Ann. NY Acad. Sci.* **2000**, *918*, 9–15.
28. Tang, J. W.; Pillay, D. *J. Clin. Virol.* **2004**, *30*, 1–10.
29. Caruso, A.; Terlenghi, L.; Ceccarelli, R.; Verardi, R.; Foresti, I.; Scura, G.; Manca, N.; Bonfanti, C.; Turano, A. *J. Virol. Methods* **1987**, *17*, 199–210.
30. Davey, R. T., Jr.; Dewar, R. L.; Reed, G. F.; Vasudevachari, M. B.; Polis, M. A.; Kovacs, J. A.; Falloon, J.; Walker, R. E.; Masur, H.; Haneiwich, S. E. et al. *Proc. Natl. Acad. Sci. USA* **1993**, *90*, 5608–5612.

31. Dewar, R. L.; Sarmiento, M. D.; Lawton, E. S.; Clark, H. M.; Kennedy, P. E.; Shah, A.; Baseler, M.; Metcalf, J. A.; Lane, H. C.; Salzman, N. P. *J. Acquir. Immune Defic. Syndr.* **1992**, *5*, 822–828.
32. Samuelsson, A.; Brostrom, C.; van Dijk, N.; Sonnerborg, A.; Chiodi, F. *Virology* **1997**, *238*, 180–188.
33. Autran, B.; Carcelain, G.; Li, T. S.; Blanc, C.; Mathez, D.; Tubiana, R.; Katlama, C.; Debre, P.; Leibowitch, J. *Science* **1997**, *277*, 112–116.
34. Yeni, P. G.; Hammer, S. M.; Hirsch, M. S.; Saag, M. S.; Schechter, M.; Carpenter, C. C.; Fischl, M. A.; Gatell, J. M.; Gazzard, B. G.; Jacobsen, D. M. et al. *JAMA* **2004**, *292*, 251–265.
35. Popper, S. J.; Sarr, A. D.; Travers, K. U.; Gueye-NDiaye, A.; Mboup, S.; Essex, M. E.; Kanki, P. J. *J. Infect. Dis.* **1999**, *180*, 1116–1121.
36. Gilbert, P. B.; McKeague, I. W.; Eisen, G.; Mullins, C.; Gueye-NDiaye, A.; Mboup, S.; Kanki, P. J. *Stat. Med.* **2003**, *22*, 573–593.
37. Hardy, C. T.; Damrow, T. A.; Villareal, D. B.; Kenny, G. E. *J. Virol. Methods* **1992**, *37*, 259–273.
38. Vacca, J. P.; Guare, J. P.; deSolms, S. J.; Sanders, W. M.; Giuliani, E. A.; Young, S. D.; Darke, P. L.; Zugay, J.; Sigal, I. S.; Schleif, W. A. et al. *J. Med. Chem.* **1991**, *34*, 1225–1228.
39. Otsyula, M.; Yee, J.; Jennings, M.; Suleman, M.; Gettie, A.; Tarara, R.; Isahakia, M.; Marx, P.; Lerche, N. *Ann. Trop. Med. Parasitol.* **1996**, *90*, 65–70.
40. Kodama, T.; Silva, D. P.; Daniel, M. D.; Phillips-Conroy, J. E.; Jolly, C. J.; Rogers, J.; Desrosiers, R. C. *AIDS Res. Hum. Retrovir.* **1989**, *5*, 337–343.
41. Reimann, K. A.; Li, J. T.; Veazey, R.; Halloran, M.; Park, I. W.; Karlsson, G. B.; Sodroski, J.; Letvin, N. L. *J. Virol.* **1996**, *70*, 6922–6928.
42. Reimann, K. A.; Li, J. T.; Voss, G.; Lekutis, C.; Tenner-Racz, K.; Racz, P.; Lin, W.; Montefiori, D. C.; Lee-Parritz, D. E.; Lu, Y. et al. *J. Virol.* **1996**, *70*, 3198–3206.
43. Hazuda, D. J.; Young, S. D.; Guare, J. P.; Anthony, N. J.; Gomez, R. P.; Wai, J. S.; Vacca, J. P.; Handt, L.; Motzel, S. L.; Klein, H. J. et al. *Science* **2004**, *305*, 528–532.
44. Parker, R. A.; Regan, M. M.; Reimann, K. A. *J. Virol.* **2001**, *75*, 11234–11238.
45. Larder, B. A.; Darby, G.; Richman, D. D. *Science* **1989**, *243*, 1731–1734.
46. Larder, B. A.; Kemp, S. D. *Science* **1989**, *246*, 1155–1158.
47. Richman, D. D.; Havlir, D.; Corbeil, J.; Looney, D.; Ignacio, C.; Spector, S. A.; Sullivan, J.; Cheeseman, S.; Barringer, K.; Pauletti, D. *J. Virol.* **1994**, *68*, 1660–1666.
48. Birch, C. *J. HIV Ther.* **1998**, *3*, 63–66.
49. Larder, B. A.; Chesebro, B.; Richman, D. D. *Antimicrob. Agents Chemother.* **1990**, *34*, 436–441.
50. Bacheler, L.; Jeffrey, S.; Hanna, G.; D'Aquila, R.; Wallace, L.; Logue, K.; Cordova, B.; Hertogs, K.; Larder, B.; Buckery, R. et al. *J. Virol.* **2001**, *75*, 4999–5008.
51. Condra, J. H.; Schleif, W. A.; Blahy, O. M.; Gabryelski, L. J.; Graham, D. J.; Quintero, J. C.; Rhodes, A.; Robbins, H. L.; Roth, E.; Shivaprakash, M. et al. *Nature* **1995**, *374*, 569–571.
52. Robertson, S. M.; Penzak, S. R.; Pau, A. K. *Expert Opin. Pharmacother.* **2005**, *6*, 233–253.
53. Keiser, P. *J. Acquir. Immune Defic. Syndr.* **2002**, *29*, S19–S27.
54. Montaner, L. J. *Trends Immunol.* **2001**, *22*, 92–96.
55. Boschi, A.; Tinelli, C.; Ortolani, P.; Moscatelli, G.; Morigi, G.; Arlotti, M. *AIDS* **2004**, *18*, 2381–2389.
56. Ananworanich, J.; Hirschel, B. *Expert Rev. Anti Infect. Ther.* **2005**, *3*, 51–60.
57. Kyriakides, T. C.; Babiker, A.; Singer, J.; Cameron, W.; Schechter, M. T.; Holodniy, M.; Brown, S. T.; Youle, M.; Gazzard, B. *Control Clin. Trials* **2003**, *24*, 481–500.
58. Montaner, J. S.; Mellors, J. W. *Antivir. Ther.* **1999**, *4*, 59–60.
59. Yarchoan, R.; Klecker, R. W.; Weinhold, K. J.; Markham, P. D.; Lyerly, H. K.; Durack, D. T.; Gelmann, E.; Lehrman, S. N.; Blum, R. M.; Barry, D. W. et al. *Lancet* **1986**, *1*, 575–580.
60. Horwitz, J. P.; Chua, J.; Noel, M. *J. Org. Chem.* **1964**, *29*, 2076–2078.
61. De Clercq, E.; Balzarini, J.; Descamps, J.; Eckstein, F. *Biochem. Pharmacol.* **1980**, *29*, 1849–1851.
62. Mitsuya, H.; Weinhold, K. J.; Furman, P. A.; St Clair, M. H.; Lehrman, S. N.; Gallo, R. C.; Bolognesi, D.; Barry, D. W.; Broder, S. *Proc. Natl. Acad. Sci. USA* **1985**, *82*, 7096–7100.
63. Furman, P. A.; Fyfe, J. A.; St Clair, M. H.; Weinhold, K.; Rideout, J. L.; Freeman, G. A.; Lehrman, S. N.; Bolognesi, D. P.; Broder, S.; Mitsuya, H. et al. *Proc. Natl. Acad. Sci. USA* **1986**, *83*, 8333–8337.
64. Lavie, A.; Schlichting, I.; Vetter, I. R.; Konrad, M.; Reinstein, J.; Goody, R. S. *Nat. Med.* **1997**, *3*, 922–924.
65. Horwitz, J. P.; Chua, J.; Da Rooge, M. A.; Noel, M.; Klundt, I. L. *J. Org. Chem.* **1966**, *31*, 205–211.
66. Baba, M.; Pauwels, R.; Herdewijn, P.; De Clercq, E.; Desmyter, J.; Vandeputte, M. *Biochem. Biophys. Res. Commun.* **1987**, *142*, 128–134.
67. Lin, T. S.; Schinazi, R. F.; Prusoff, W. H. *Biochem. Pharmacol.* **1987**, *36*, 2713–2718.
68. Balzarini, J.; Herdewijn, P.; De Clercq, E. *J. Biol. Chem.* **1989**, *264*, 6127–6133.
69. Riddler, S. A.; Anderson, R. E.; Mellors, J. W. *Antiviral Res.* **1995**, *27*, 189–203.
70. Horwitz, J. P.; Chua, J.; Noel, M.; Donatti, J. T. *J. Org. Chem.* **1967**, *32*, 817–818.
71. Mitsuya, H.; Broder, S. *Proc. Natl. Acad. Sci. USA* **1986**, *83*, 1911–1915.
72. Belleau, B.; Belleau, P.; Nguyen, B.-N. Preparation of 2-(hydroxymethyl)-5-cytosinyl-1,3-oxathiolanes and analogs as virucides. JAF Biochem International Inc., 1990, EP 382526 A2.
73. Beach, J. W.; Jeong, L. S.; Alves, A. J.; Pohl, D.; Kim, H. O.; Chang, C. N.; Doong, S. L.; Schinazi, R. F.; Cheng, Y. C.; Chu, C. K. *J. Org. Chem.* **1992**, *57*, 2217–2219.
74. Coates, J. A. V.; Mutton, I. M.; Penn, C. R.; Storer, R.; Williamson, C. Preparation of 1,3-oxathiolane nucleoside analogs and pharmaceutical compositions containing them. IAF Biochem International Inc., 1991, WO 9117159.
75. Liotta, D. C.; Schinazi, R. F.; Choi, W.-B. Antiviral activity and resolution of 2-hydroxymethyl-5-(5-fluorocytosin-1-yl)-1,3-oxathiolane. Emory University, USA, 1992, WO 9214743.
76. Schinazi, R. F.; McMillan, A.; Cannon, D.; Mathis, R.; Lloyd, R. M.; Peck, A.; Sommadossi, J.-P.; St Clair, M.; Wilson, J.; Furman, P. A. et al. *Antimicrob. Agents Chemother.* **1992**, *36*, 2423–2431.
77. Feng, J. Y.; Shi, J.; Schinazi, R. F.; Anderson, K. S. *FASEB J.* **1999**, *13*, 1511–1517.
78. Plunkett, W.; Cohen, S. S. *Cancer Res.* **1975**, *35*, 1547–1554.
79. Perry, C. M.; Noble, S. *Drugs* **1999**, *58*, 1099–1135.
80. Yarchoan, R.; Mitsuya, H.; Thomas, R. V.; Pluda, J. M.; Hartman, N. R.; Perno, C. F.; Marczyk, K. S.; Allain, J. P.; Johns, D. G.; Broder, S. *Science* **1989**, *245*, 412–415.
81. Faletto, M. B.; Miller, W. H.; Garvey, E. P.; St Clair, M. H.; Daluge, S. M.; Good, S. S. *Antimicrob. Agents Chemother.* **1997**, *41*, 1099–1107.

82. Lyseng-Williamson, K. A.; Reynolds, N. A.; Plosker, G. L. *Drugs* **2005**, *65*, 413–432.
83. Broder, S.; Yarchoan, R. *Am. J. Med.* **1990**, *88*, 31S–33S.
84. Spector, S. A.; Ripley, D.; Hsia, K. *Antimicrob. Agents Chemother.* **1989**, *33*, 920–923.
85. Yarchoan, R.; Perno, C. F.; Thomas, R. V.; Klecker, R. W.; Allain, J. P.; Wills, R. J.; McAtee, N.; Fischl, M. A.; Dubinsky, R.; McNeely, M. C. et al. *Lancet* **1988**, *1*, 76–81.
86. Staszewski, S.; Loveday, C.; Picazo, J. J.; Dellarnonica, P.; Skinhoj, P.; Johnson, M. A.; Danner, S. A.; Harrigan, P. R.; Hill, A. M.; Verity, L. et al. *JAMA* **1996**, *276*, 111–117.
87. Connor, E. M.; Sperling, R. S.; Gelber, R.; Kiselev, P.; Scott, G.; O'Sullivan, M. J.; VanDyke, R.; Bey, M.; Shearer, W.; Jacobson, R. L. et al. *N. Engl. J. Med.* **1994**, *331*, 1173–1180.
88. Kellam, P.; Boucher, C. A.; Larder, B. A. *Proc. Natl. Acad. Sci. USA* **1992**, *89*, 1934–1938.
89. Sarafianos, S. G.; Clark, A. D., Jr.; Das, K.; Tuske, S.; Birktoft, J. J.; Ilankumaran, P.; Ramesha, A. R.; Sayer, J. M.; Jerina, D. M.; Boyer, P. L. et al. *EMBO J.* **2002**, *21*, 6614–6624.
90. Arion, D.; Kaushik, N.; McCormick, S.; Borkow, G.; Parniak, M. A. *Biochemistry* **1998**, *37*, 15908–15917.
91. Naeger, L. K.; Margot, N. A.; Miller, M. D. *Antimicrob. Agents Chemother.* **2002**, *46*, 2179–2184.
92. Isel, C.; Ehresmann, C.; Walter, P.; Ehresmann, B.; Marquet, R. *J. Biol. Chem.* **2001**, *276*, 48725–48732.
93. Schinazi, R. F.; Lloyd, R. M., Jr.; Nguyen, M. H.; Cannon, D. L.; McMillan, A.; Ilksoy, N.; Chu, C. K.; Liotta, D. C.; Bazmi, H. Z.; Mellors, J. W. *Antimicrob. Agents Chemother.* **1993**, *37*, 875–881.
94. Götte, M.; Arion, D.; Parniak, M. A.; Wainberg, M. A. *J. Virol.* **2000**, *74*, 3579–3585.
95. Turner, D.; Brenner, B. G.; Routy, J. P.; Petrella, M.; Wainberg, M. A. *New Microbiol.* **2004**, *27*, 31–39.
96. Turner, D.; Brenner, B.; Wainberg, M. A. *Clin. Diagn. Lab. Immunol.* **2003**, *10*, 979–981.
97. Wainberg, M. A.; Miller, M. D.; Quan, Y.; Salomon, H.; Mulato, A. S.; Lamy, P. D.; Margot, N. A.; Anton, K. E.; Cherrington, J. M. *Antivir. Ther.* **1999**, *4*, 87–94.
98. Wainberg, M. A.; Turner, D. *J. Acquir. Immune Defic. Syndr.* **2004**, *37*, S36–S43.
99. Collins, M. L.; Sondel, N.; Cesar, D.; Hellerstein, M. K. *J. Acquir. Immune Defic. Syndr.* **2004**, *37*, 1132–1139.
100. Moyle, G. *Clin. Ther.* **2000**, *22*, 911–936.
101. Guidelines for the Use of Antiretroviral Agents in HIV-1 Infected Adults and Adolescents. 05-04-2006. Developed by the DHHS Panel on Antiretroviral Guidelines for Adults and Adolescents - A Working Group of the Office of AIDS Research Advisory Council. http://aidsinfo.nih.gov/ContentFiles/AdultandAdolescentGL.pdf (accessed Aug 2006).
102. Clay, P. G. *Clin. Ther.* **2002**, *24*, 1502–1514.
103. Piliero, P. J. *J. Acquir. Immune Defic. Syndr.* **2004**, *37*, S2–S12.
104. Havlir, D. V.; Tierney, C.; Friedland, G. H.; Pollard, R. B.; Smeaton, L.; Sommadossi, J. P.; Fox, L.; Kessler, H.; Fife, K. H.; Richman, D. D. *J. Infect. Dis.* **2000**, *182*, 321–325.
105. Pauwels, R.; Andries, K.; Desmyter, J.; Schols, D.; Kukla, M. J.; Breslin, H. J.; Raeymaeckers, A.; Van Gelder, J.; Woestenborghs, R.; Heykants, J. et al. *Nature* **1990**, *343*, 470–-474..
106. De Wit, S.; Hermans, P.; Sommereijns, B.; O'Doherty, E.; Westenborghs, R.; van der, V.; Cauwenbergh, G. F.; Clumeck, N. *Antimicrob. Agents Chemother.* **1992**, *36*, 2661–2663.
107. Pialoux, G.; Youle, M.; Dupont, B.; Gazzard, B.; Cauwenbergh, G. F.; Stoffels, P. A.; Davies, S.; de Saint, M. J.; Janssen, P. A. *Lancet* **1991**, *338*, 140–143.
108. Merluzzi, V. J.; Hargrave, K. D.; Labadia, M.; Grozinger, K.; Skoog, M.; Wu, J. C.; Shih, C. K.; Eckner, K.; Hattox, S.; Adams, J. et al. *Science* **1990**, *250*, 1411–1413.
109. Riska, P.; Lamson, M.; MacGregor, T.; Sabo, J.; Hattox, S.; Pav, J.; Keirns, J. *Drug Metab. Dispos.* **1999**, *27*, 895–901.
110. Goldman, M. E.; Nunberg, J. H.; O'Brien, J. A.; Quintero, J. C.; Schleif, W. A.; Freund, K. F.; Gaul, S. L.; Saari, W. S.; Wai, J. S.; Hoffman, J. M. et al. *Proc. Natl. Acad. Sci. USA* **1991**, *88*, 6863–6867.
111. Romero, D. L.; Morge, R. A.; Biles, C.; Berrios-Pena, N.; May, P. D.; Palmer, J. R.; Johnson, P. D.; Smith, H. W.; Busso, M.; Tan, C. K. et al. *J. Med. Chem.* **1994**, *37*, 999–1014.
112. Been-Tiktak, A. M. M.; Williams, I.; Vrehen, H. M.; Richens, J.; Aldam, D.; van Loon, A. M.; Loveday, C.; Boucher, C. A. B.; Ward, P.; Weller, I. V. D. et al. *Antimicrob. Agents Chemother.* **1996**, *40*, 2664–2668.
113. Romero, D. L.; Morge, R. A.; Genin, M. J.; Biles, C.; Busso, M.; Resnick, L.; Althaus, I. W.; Reusser, F.; Thomas, R. C.; Tarpley, W. G. *J. Med. Chem.* **1993**, *36*, 1505–1508.
114. Davey, R. T., Jr.; Chaitt, D. G.; Reed, G. F.; Freimuth, W. W.; Herpin, B. R.; Metcalf, J. A.; Eastman, P. S.; Falloon, J.; Kovacs, J. A.; Polis, M. A. et al. *Antimicrob. Agents Chemother.* **1996**, *40*, 1657–1664.
115. Tucker, T. J.; Lyle, T. A.; Wiscount, C. M.; Britcher, S. F.; Young, S. D.; Sanders, W. M.; Lumma, W. C.; Goldman, M. E.; O'Brien, J. A.; Ball, R. G. et al. *J. Med. Chem.* **1994**, *37*, 2437–2444.
116. Prueksaritanont, T.; Balani, S. K.; Dwyer, L. M.; Ellis, J. D.; Kauffman, L. R.; Varga, S. L.; Pitzenberger, S. M.; Theoharides, A. D. *Drug Metab. Dispos.* **1995**, *23*, 688–695.
117. Carroll, S. S.; Stahlhut, M.; Geib, J.; Olsen, D. B. *J. Biol. Chem.* **1994**, *269*, 32351–32357.
118. Young, S. D.; Britcher, S. F.; Tran, L. O.; Payne, L. S.; Lumma, W. C.; Lyle, T. A.; Huff, J. R.; Anderson, P. S.; Olsen, D. B.; Carroll, S. S. et al. *Antimicrob. Agents Chemother.* **1995**, *39*, 2602–2605.
119. Staszewski, S.; Morales-Ramirez, J.; Tashima, K. T.; Rachlis, A.; Skiest, D.; Stanford, J.; Stryker, R.; Johnson, P.; Labriola, D. F.; Farina, D. et al. *N. Engl. J. Med.* **1999**, *341*, 1865–1873.
120. Witvrouw, M.; Pannecouque, C.; Van Laethem, K.; Desmyter, J.; De Clercq, E.; Vandamme, A. M. *AIDS* **1999**, *13*, 1477–1483.
121. Emini, E. A.; Byrnes, V. W.; Condra, J. H.; Schleif, W. A.; Sardana, V. V. *Arch. Virol.* **1994**, *9*, 11–17.
122. Condra, J. H.; Emini, E. A.; Gotlib, L.; Graham, D. J.; Schlabach, A. J.; Wolfgang, J. A.; Colonno, R. J.; Sardana, V. V. *Antimicrob. Agents Chemother.* **1992**, *36*, 1441–1446.
123. Tantillo, C.; Ding, J.; Jacobo-Molina, A.; Nanni, R. G.; Boyer, P. L.; Hughes, S. H.; Pauwels, R.; Andries, K.; Janssen, P. A. J.; Arnold, E. *J. Mol. Biol.* **1994**, *243*, 369–387.
124. Das, K.; Lewi, P. J.; Hughes, S. H.; Arnold, E. *Prog. Biophys. Mol. Biol.* **2005**, *88*, 209–231.
125. Ren, J.; Milton, J.; Weaver, K. L.; Short, S. A.; Stuart, D. I.; Stammers, D. K. *Struct. Fold Des.* **2000**, *8*, 1089–1094.
126. Saag, M. S.; Emini, E. A.; Laskin, O. L.; Douglas, J.; Lapidus, W. I.; Schleif, W. A.; Whitley, R. J.; Hildebrand, C.; Byrnes, V. W.; Kappes, J. C. et al. *N. Engl. J. Med.* **1993**, *329*, 1065–1072.

127. Para, M. F.; Meehan, P.; Holden-Wiltse, J.; Fischl, M.; Morse, G.; Shafer, R.; Demeter, L. M.; Wood, K.; Nevin, T.; Virani-Ketter, N. et al. *Antimicrob. Agents Chemother.* **1999**, *43*, 1373–1378.
128. Delaugerre, C.; Rohban, R.; Simon, A.; Mouroux, M.; Tricot, C.; Agher, R.; Huraux, J. M.; Katlama, C.; Calvez, V. *J. Med. Virol.* **2001**, *65*, 445–448.
129. Patel, S. M.; Johnson, S.; Belknap, S. M.; Chan, J.; Sha, B. E.; Bennett, C. *J. Acquir. Immune Defic. Syndr.* **2004**, *35*, 120–125.
130. Manfredi, R.; Calza, L.; Chiodo, F. *J. Acquir. Immune Defic. Syndr.* **2004**, *35*, 492–502.
131. Mofenson, L. M. *AIDS Clin. Care* **2005**, *17*, 17.
132. *Physicians' Desk Reference*, 58th ed.; Thompson PDR: Montvale, NJ, 2004.
133. Sweat, M. D.; O'Reilly, K. R.; Schmid, G. P.; Denison, J.; de Zoysa, I. *AIDS* **2004**, *18*, 1661–1671.
134. Janssen, P. A. J.; Lewi, P. J.; Arnold, E.; Daeyaert, F.; de Jonge, M.; Heeres, J.; Koymans, L.; Vinkers, M.; Guillemont, J.; Pasquier, E. et al. *J. Med. Chem.* **2005**, *48*, 1901–1909.
135. Goebel, F.; Yakovlev, A.; Pozniak, A.; Vinogradova, E.; Lewi, P.; Boogaerts, G.; Hoetelmans, R.; de Béthune, M.-P.; Peeters, M.; Woodfall, B. *Abstracts of the 12th Conference on Retroviruses and Opportunistic Infections, Boston, MA, Abs. 160, Feb. 22–25, 2005,* **2005**.
136. Ratner, L.; Haseltine, W.; Patarca, R.; Livak, K. J.; Starcich, B.; Josephs, S. F.; Doran, E. R.; Rafalski, J. A.; Whitehorn, E. A.; Baumeister, K. et al. *Nature* **1985**, *313*, 277–284.
137. Kohl, N. E.; Emini, E. A.; Schleif, W. A.; Davis, L. J.; Heimbach, J. C.; Dixon, R. A.; Scolnick, E. M.; Sigal, I. S. *Proc. Natl. Acad. Sci. USA* **1988**, *85*, 4686–4690.
138. Kramer, R. A.; Schaber, M. D.; Skalka, A. M.; Ganguly, K.; Wong-Staal, F.; Reddy, E. P. *Science* **1986**, *231*, 1580–1584.
139. Pearl, L. H.; Taylor, W. R. *Nature* **1987**, *329*, 351–354.
140. Navia, M. A.; Fitzgerald, P. M.; McKeever, B. M.; Leu, C. T.; Heimbach, J. C.; Herber, W. K.; Sigal, I. S.; Darke, P. L.; Springer, J. P. *Nature* **1989**, *337*, 615–620.
141. Wlodawer, A.; Miller, M.; Jaskolski, M.; Sathyanarayana, B. K.; Baldwin, E.; Weber, I. T.; Selk, L. M.; Clawson, L.; Schneider, J.; Kent, S. B. *Science* **1989**, *245*, 616–621.
142. Munshi, S.; Chen, Z.; Li, Y.; Olsen, D. B.; Fraley, M. E.; Hungate, R. W.; Kuo, L. C. *Acta Crystallogr. D Biol. Crystallogr.* **1998**, *54*, 1053–1060.
143. Dunn, B. M. *Chem. Rev.* **2002**, *102*, 4431–4458.
144. Tritch, R. J.; Cheng, Y. E.; Yin, F. H.; Erickson-Viitanen, S. *J. Virol.* **1991**, *65*, 922–930.
145. Wolfenden, R. *Annu. Rev. Biophys. Bioeng.* **1976**, *5*, 271–306.
146. Szelke, M.; Leckie, B.; Hallett, A.; Jones, D. M.; Sueiras, J.; Atrash, B.; Lever, A. F. *Nature* **1982**, *299*, 555–557.
147. Billich, S.; Knoop, M. T.; Hansen, J.; Strop, P.; Sedlacek, J.; Mertz, R.; Moelling, K. *J. Biol. Chem.* **1988**, *263*, 17905–17908.
148. McQuade, T. J.; Tomasselli, A. G.; Liu, L.; Karacostas, V.; Moss, B.; Sawyer, T. K.; Heinrikson, R. L.; Tarpley, W. G. *Science* **1990**, *247*, 454–456.
149. Moore, M. L.; Bryan, W. M.; Fakhoury, S. A.; Magaard, V. W.; Huffman, W. F.; Dayton, B. D.; Meek, T. D.; Hyland, L.; Dreyer, G. B.; Metcalf, B. W. et al. *Biochem. Biophys. Res. Commun.* **1989**, *159*, 420–425.
150. Rich, D. H.; Green, J.; Toth, M. V.; Marshall, G. R.; Kent, S. B. *J. Med. Chem.* **1990**, *33*, 1285–1288.
151. Dreyer, G. B.; Metcalf, B. W.; Tomaszek, T. A., Jr.; Carr, T. J.; Chandler, A. C., III; Hyland, L.; Fakhoury, S. A.; Magaard, V. W.; Moore, M. L.; Strickler, J. E. et al. *Proc. Natl. Acad. Sci. USA* **1989**, *86*, 9752–9756.
152. Roberts, N. A.; Martin, J. A.; Kinchington, D.; Broadhurst, A. V.; Craig, J. C.; Duncan, I. B.; Galpin, S. A.; Handa, B. K.; Kay, J.; Kröhn, A. et al. *Science* **1990**, *248*, 358–361.
153. Kempf, D. J.; Norbeck, D. W.; Codacovi, L.; Wang, X. C.; Kohlbrenner, W. E.; Wideburg, N. E.; Paul, D. A.; Knigge, M. F.; Vasavanonda, S.; Craig-Kennard, A. et al. *J Med Chem* **1990**, *33*, 2687–2689.
154. Kempf, D. J.; Marsh, K. C.; Paul, D. A.; Knigge, M. F.; Norbeck, D. W.; Kohlbrenner, W. E.; Codacovi, L.; Vasavanonda, S.; Bryant, P.; Wang, X. C. et al. *Antimicrob. Agents Chemother.* **1991**, *35*, 2209–2214.
155. Kempf, D. J.; Sham, H. L.; Marsh, K. C.; Flentge, C. A.; Betebenner, D.; Green, B. E.; McDonald, E.; Vasavanonda, S.; Saldivar, A.; Wideburg, N. E. et al. *J. Med. Chem.* **1998**, *41*, 602–617.
156. Markowitz, M.; Saag, M.; Powderly, W. G.; Hurley, A. M.; Hsu, A.; Valdes, J. M.; Henry, D.; Sattler, F.; La Marca, A.; Leonard, J. M. et al. *N. Engl. J. Med.* **1995**, *333*, 1534–1539.
157. MacDougall, D. S. *J. Int. Assoc. Physicians AIDS Care* **1996**, *2*, 38–44.
158. Nutt, R. F.; Brady, S. F.; Darke, P. L.; Ciccarone, T. M.; Colton, C. D.; Nutt, E. M.; Rodkey, J. A.; Bennett, C. D.; Waxman, L. H.; Sigal, I. S. et al. *Proc. Natl. Acad. Sci. USA* **1988**, *85*, 7129–7133.
159. Evans, B. E.; Rittle, K. E.; Homnick, C. F.; Springer, J. P.; Hirshfield, J.; Veber, D. F. *J. Org. Chem.* **1985**, *50*, 4615–4625.
160. Lyle, T. A.; Wiscount, C. M.; Guare, J. P.; Thompson, W. J.; Anderson, P. S.; Darke, P. L.; Zugay, Z. A.; Emini, E. A.; Schleif, W. A.; Quintero, J. C. et al. *J. Med. Chem.* **1991**, *34*, 1228–1230.
161. Dorsey, B. D.; Levin, R. B.; McDaniel, S. L.; Vacca, J. P.; Guare, J. P.; Darke, P. L.; Zugay, J. A.; Emini, E. A.; Schleif, W. A.; Quintero, J. C. et al. *J. Med. Chem.* **1994**, *37*, 3443–3451.
162. Vacca, J. P.; Dorsey, B. D.; Schleif, W. A.; Levin, R. B.; McDaniel, S. L.; Darke, P. L.; Zugay, J.; Quintero, J. C.; Blahy, O. M.; Roth, E. et al. *Proc. Natl. Acad. Sci.* USA **1994**, 91, 4096–4100.
163. Kaldor, S. W.; Appelt, K.; Fritz, J.; Hammond, M.; Crowell, T. A.; Baxter, A. J.; Hatch, S. D.; Wiskerchen, M. A.; Muesing, M. A. *Bioorg. Med. Chem. Lett.* **1995**, *5*, 715–720.
164. Kalish, V. J.; Tatlock, J. H.; Davies, J. F.; Kaldor, S. W.; Dressman, B. A.; Reich, S.; Pino, M.; Nyugen, D.; Appelt, K. *Bioorg. Med. Chem. Lett.* **1995**, *5*, 727–732.
165. Kaldor, S. W.; Hammond, M.; Dressman, B. A.; Fritz, J. E.; Crowell, T. A.; Hermann, R. A. *Bioorg. Med. Chem. Lett.* **1994**, *4*, 1385–1390.
166. Kaldor, S. W.; Kalish, V. J.; Davies, J. F.; Shetty, B. V.; Fritz, J. E.; Appelt, K.; Burgess, J. A.; Campanale, K. M.; Chirgadze, N. Y.; Clawson, D. K. et al. *J. Med. Chem.* **1997**, *40*, 3979–3985.
167. Moyle, G. J.; Youle, M.; Higgs, C.; Monaghan, J.; Prince, W.; Chapman, S.; Clendeninn, N.; Nelson, M. R. *J. Clin. Pharmacol.* **1998**, *38*, 736–743.
168. Navia, M. A.; Murcko, M. A. *Curr. Opin. Struct. Biol.* **1992**, *2*, 202–210.
169. Ghosh, A. K.; Thompson, W. J.; McKee, S. P.; Duong, T. T.; Lyle, T. A.; Chen, J. C.; Darke, P. L.; Zugay, J. A.; Emini, E. A.; Schleif, W. A. et al. *J. Med. Chem.* **1993**, *36*, 292–294.
170. Kim, E. E.; Baker, C. T.; Dwyer, M. D.; Murcko, M. A.; Rao, B. G.; Tung, R. D.; Navia, M. A. *J. Am. Chem. Soc.* **1995**, *117*, 1181–1182.
171. Sadler, B. M.; Gillotin, C.; Lou, Y.; Stein, D. S. *Antimicrob. Agents Chemother.* **2001**, *45*, 30–37.
172. Furfine, E. S.; Baker, C. T.; Hale, M. R.; Reynolds, D. J.; Salisbury, J. A.; Searle, A. D.; Studenberg, S. D.; Todd, D.; Tung, R. D.; Spaltenstein, A. *Antimicrob. Agents Chemother.* **2004**, *48*, 791–798.

173. Sham, H. L.; Kempf, D. J.; Molla, A.; Marsh, K. C.; Kumar, G. N.; Chen, C. M.; Kati, W.; Stewart, K.; Lal, R.; Hsu, A. et al. *Antimicrob. Agents Chemother.* **1998**, *42*, 3218–3224.
174. Lecoq, A.; Marraud, M.; Aubry, A. *Tetrahedron Lett.* **1991**, *32*, 2765–2768.
175. Gante, J. *Synthesis* **1989**, 405–413.
176. Fässler, A.; Rösel, J.; Gruetter, M.; Tintelnot-Blomley, M.; Alteri, E.; Bold, G.; Lang, M. *Bioorg. Med. Chem. Lett.* **1993**, *3*, 2837–2842.
177. Sham, H. L.; Betebenner, D. A.; Zhao, C.; Wideburg, N. E.; Saldivar, A.; Kempf, D. J.; Plattner, J. J.; Norbeck, D. W. *J. Chem. Soc. Chem. Commun.* **1993**, 1052–1053.
178. Fässler, A.; Bold, G.; Capraro, H. G.; Cozens, R.; Mestan, J.; Poncioni, B.; Rösel, J.; Tintelnot-Blomley, M.; Lang, M. *J. Med. Chem.* **1996**, *39*, 3203–3216.
179. Bold, G.; Fässler, A.; Capraro, H.-G.; Cozens, R.; Klimkait, T.; Lazdins, J.; Mestan, J.; Poncioni, B.; Rösel, J.; Stover, D. et al. *J. Med. Chem.* **1998**, *41*, 3387–3401.
180. Gong, Y. F.; Robinson, B. S.; Rose, R. E.; Deminie, C.; Spicer, T. P.; Stock, D.; Colonno, R. J.; Lin, P. F. *Antimicrob. Agents Chemother.* **2000**, *44*, 2319–2326.
181. Sanne, I.; Piliero, P.; Squires, K.; Thiry, A.; Schnittman, S. *J. Acquir. Immune Defic. Syndr.* **1999**, *32*, 18–29.
182. Bourinbaiar, A. S.; Tan, X.; Nagorny, R. *AIDS* **1993**, *7*, 129–130.
183. Tummino, P. J.; Ferguson, D.; Hupe, D. *Biochem. Biophys. Res. Commun.* **1994**, *201*, 290–294.
184. Thaisrivongs, S.; Tomich, P. K.; Watenpaugh, K. D.; Chong, K. T.; Howe, W. J.; Yang, C. P.; Strohbach, J. W.; Turner, S. R.; McGrath, J. P.; Bohanon, M. J. et al. *J. Med. Chem.* **1994**, *37*, 3200–3204.
185. Tummino, P. J.; Ferguson, D.; Hupe, L.; Hupe, D. *Biochem. Biophys. Res. Commun.* **1994**, *200*, 1658–1664.
186. Lunney, E. A.; Hagen, S. E.; Domagala, J. M.; Humblet, C.; Kosinski, J.; Tait, B. D.; Warmus, J. S.; Wilson, M.; Ferguson, D.; Hupe, D. et al. *J. Med. Chem.* **1994**, *37*, 2664–2677.
187. Hagen, S. E.; Domagala, J.; Gajda, C.; Lovdahl, M.; Tait, B. D.; Wise, E.; Holler, T.; Hupe, D.; Nouhan, C.; Urumov, A. et al. *J. Med. Chem.* **2001**, *44*, 2319–2332.
188. Skulnick, H. I.; Johnson, P. D.; Howe, W. J.; Tomich, P. K.; Chong, K.-T.; Watenpaugh, K. D.; Janakiraman, M. N.; Dolak, L. A.; McGrath, J. P.; Lynn, J. C. et al. *J. Med. Chem.* **1995**, *38*, 4968–4971.
189. Zhong, W. Z.; Williams, M. G.; Borin, M. T.; Padbury, G. E. *Chirality* **1998**, *10*, 210–216.
190. Turner, S. R.; Strohbach, J. W.; Tommasi, R. A.; Aristoff, P. A.; Johnson, P. D.; Skulnick, H. I.; Dolak, L. A.; Seest, E. P.; Tomich, P. K.; Bohanon, M. J. et al. *J. Med. Chem.* **1998**, *41*, 3467–3476.
191. Thaisrivongs, S.; Strohbach, J. W. *Biopolymers* **1999**, *51*, 51–58.
192. Thaisrivongs, S.; Skulnick, H. I.; Turner, S. R.; Strohbach, J. W.; Tommasi, R. A.; Johnson, P. D.; Aristoff, P. A.; Judge, T. M.; Gammill, R. B.; Morris, J. K. et al. *J. Med. Chem.* **1996**, *39*, 4349–4353.
193. Larder, B. A.; Hertogs, K.; Bloor, S.; van den Eynde, C. H.; DeCian, W.; Wang, Y.; Freimuth, W. W.; Tarpley, G. *AIDS* **2000**, *14*, 1943–1948.
194. Rusconi, S.; La Seta, C. S.; Citterio, P.; Kurtagic, S.; Violin, M.; Balotta, C.; Moroni, M.; Galli, M.; d'Arminio-Monforte, A. *Antimicrob. Agents Chemother.* **2000**, *44*, 1328–1332.
195. McCallister, S.; Valdez, H.; Curry, K.; MacGregor, T.; Borin, M.; Freimuth, W.; Wang, Y.; Mayers, D. L. *J. Acquir. Immune Defic. Syndr.* **2004**, *35*, 376–382.
196. Feinberg, J. *AIDS Clin. Care* **2005**, *17*, 5–6.
197. Gulick, R. M.; Mellors, J. W.; Havlir, D.; Eron, J. J.; Gonzalez, C.; McMahon, D.; Richman, D. D.; Valentine, F. T.; Jonas, L.; Meibohm, A. et al. *N. Engl. J. Med.* **1997**, *337*, 734–739.
198. Zeldin, R. K.; Petruschke, R. A. *J. Antimicrob. Chemother.* **2004**, *53*, 4–9.
199. Condra, J. H.; Holder, D. J.; Schleif, W. A.; Blahy, O. M.; Danovich, R. M.; Gabryelski, L. J.; Graham, D. J.; Laird, D.; Quintero, J. C.; Rhodes, A. et al. *J. Virol.* **1996**, *70*, 8270–8276.
200. Jacobsen, H.; Haenggi, M.; Ott, M.; Duncan, I. B.; Andreoni, M.; Vella, S.; Mous, J. *Antiviral Res.* **1996**, *29*, 95–97.
201. Schmit, J. C.; Ruiz, L.; Clotet, B.; Raventos, A.; Tor, J.; Leonard, J.; Desmyter, J.; De Clercq, E.; Vandamme, A. M. *AIDS* **1996**, *10*, 995–999.
202. Drusano, G. L.; Bilello, J. A.; Stein, D. S.; Nessly, M.; Meibohm, A.; Emini, E. A.; Deutsch, P.; Condra, J.; Chodakewitz, J.; Holder, D. J. *J. Infect. Dis.* **1998**, *178*, 360–367.
203. Molla, A.; Korneyeva, M.; Gao, Q.; Vasavanonda, S.; Schipper, P. J.; Mo, H. M.; Markowitz, M.; Chernyavskiy, T.; Niu, P.; Lyons, N. et al. *Nat. Med.* **1996**, *2*, 760–766.
204. de Mendoza, C.; Soriano, V. *Curr. Drug Metab.* **2004**, *5*, 321–328.
205. Kaufmann, G. R.; Suzuki, K.; Cunningham, P.; Mukaide, M.; Kondo, M.; Imai, M.; Zaunders, J.; Cooper, D. A. *AIDS Res. Hum. Retrovir.* **2001**, *17*, 487–497.
206. Mahalingam, B.; Boross, P.; Wang, Y. F.; Louis, J. M.; Fischer, C. C.; Tozser, J.; Harrison, R. W.; Weber, I. T. *Proteins* **2002**, *48*, 107–116.
207. Sax, P. E.; Kumar, P. *J. Acquir. Immune Defic. Syndr.* **2004**, *37*, 1111–1124.
208. Castiglione, B.; Chan-Tompkins, N. H. *J. Infect. Dis. Pharmacother.* **1999**, *3*, 33–59.
209. Gatti, G.; Di Biagio, A.; Casazza, R.; De Pascalis, C.; Bassetti, M.; Cruciani, M.; Vella, S.; Bassetti, D. *AIDS* **1999**, *13*, 2083–2089.
210. Daudon, M.; Estepa, L.; Viard, J. P.; Joly, D.; Jungers, P. *Lancet* **1997**, *349*, 1294–1295.
211. Busti, A. J.; Hall, R. G.; Margolis, D. M. *Pharmacotherapy* **2004**, *24*, 1732–1747.
212. Hui, D. Y. *Prog. Lipid. Res.* **2003**, *42*, 81–92.
213. Orrick, J. J.; Steinhart, C. R. *Ann. Pharmacother.* **2004**, *38*, 1664–1674.
214. Taburet, A. M.; Piketty, C.; Chazallon, C.; Vincent, I.; Gerard, L.; Calvez, V.; Clavel, F.; Aboulker, J. P.; Girard, P. M. *Antimicrob. Agents Chemother.* **2004**, *48*, 2091–2096.
215. Surleraux, D. L.; Tahri, A.; Verschueren, W. G.; Pille, G. M.; de Kock, H. A.; Jonckers, T. H.; Peeters, A.; De Meyer, S.; Azijn, H.; Pauwels, R. et al. *J. Med. Chem.* **2005**, *48*, 1813–1822.
216. Young, B.; Markowitz, M. *J. Acquir. Immune Defic. Syndr.* **2004**, *35*, S3–S12.
217. Earl, P. L.; Doms, R. W.; Moss, B. *Proc. Natl. Acad. Sci. USA* **1990**, *87*, 648–652.
218. Greenberg, M.; Cammack, N.; Salgo, M.; Smiley, L. *Rev. Med. Virol.* **2004**, *14*, 321–337.
219. Owens, R. J.; Tanner, C. C.; Mulligan, M. J.; Srinivas, R. V.; Compans, R. W. *AIDS Res. Hum. Retrovir.* **1990**, *6*, 1289–1296.
220. Jiang, S.; Lin, K.; Strick, N.; Neurath, A. R. *Biochem. Biophys. Res. Commun.* **1993**, *195*, 533–538.
221. Qureshi, N. M.; Coy, D. H.; Garry, R. F.; Henderson, L. A. *AIDS* **1990**, *4*, 553–558.

222. Nehete, P. N.; Arlinghaus, R. B.; Sastry, K. J. *J. Virol.* **1993**, *67*, 6841–6846.
223. Wild, C. T.; Shugars, D. C.; Greenwell, T. K.; McDanal, C. B.; Matthews, T. J. *Proc. Natl. Acad. Sci. USA* **1994**, *91*, 9770–9774.
224. Kliger, Y.; Shai, Y. *J. Mol. Biol.* **2000**, *295*, 163–168.
225. Kilby, J. M.; Hopkins, S.; Venetta, T. M.; DiMassimo, B.; Cloud, G. A.; Lee, J. Y.; Alldredge, L.; Hunter, E.; Lambert, D.; Bolognesi, D. et al. *Nat. Med.* **1998**, *4*, 1302–1307.
226. Patel, I. H.; Zhang, X.; Nieforth, K.; Salgo, M.; Buss, N. *Clin. Pharmacokinet.* **2005**, *44*, 175–186.
227. Kilby, J. M.; Lalezari, J. P.; Eron, J. J.; Carlson, M.; Cohen, C.; Arduino, R. C.; Goodgame, J. C.; Gallant, J. E.; Volberding, P.; Murphy, R. L. et al. *AIDS Res. Hum. Retrovir.* **2002**, *18*, 685–693.
228. Wei, X.; Decker, J. M.; Liu, H.; Zhang, Z.; Arani, R. B.; Kilby, J. M.; Saag, M. S.; Wu, X.; Shaw, G. M.; Kappes, J. C *Antimicrob. Agents Chemother.* **2002**, *46*, 1896–1905.
229. Xu, L.; Pozniak, A.; Wildfire, A.; Stanfield-Oakley, S. A.; Mosier, S. M.; Ratcliffe, D.; Workman, J.; Joall, A.; Myers, R.; Smit, E. et al. *Antimicrob. Agents Chemother.* **2005**, *49*, 1113–1119.
230. Wang, T.; Zhang, Z.; Wallace, O. B.; Deshpande, M.; Fang, H.; Yang, Z.; Zadjura, L. M.; Tweedie, D. L.; Huang, S.; Zhao, F. et al. *J. Med. Chem.* **2003**, *46*, 4236–4239.
231. Si, Z.; Madani, N.; Cox, J. M.; Chruma, J. J.; Klein, J. C.; Schon, A.; Phan, N.; Wang, L.; Biorn, A. C.; Cocklin, S. et al. *Proc. Natl. Acad. Sci. USA* **2004**, *101*, 5036–5041.
232. Colonno, R. J. *Abstracts of the 13th International Symposium on HIV and Emerging Infectious Diseases,* 2005, OP4.6.
233. Moyle, G. J.; Wildfire, A.; Mandalia, S.; Mayer, H.; Goodrich, J.; Whitcomb, J.; Gazzard, B. G. *J. Infect. Dis.* **2005**, *191*, 866–872.
234. Baba, M.; Nishimura, O.; Kanzaki, N.; Okamoto, M.; Sawada, H.; Iizawa, Y.; Shiraishi, M.; Aramaki, Y.; Okonogi, K.; Ogawa, Y. et al. *Proc. Natl. Acad. Sci. USA* **1999**, *96*, 5698–5703.
235. Mills, S. G.; DeMartino, J. A. *Curr. Top. Med. Chem.* **2004**, *4*, 1017–1033.
236. Veazey, R. S.; Klasse, P. J.; Ketas, T. J.; Reeves, J. D.; Piatak,, M., Jr.; Kunstman, K.; Kuhmann, S. E.; Marx, P. A.; Lifson, J. D.; Dufour, J. et al. *J. Exp. Med.* **2003**, *198*, 1551–1562.
237. Hitchcock, C. A. *Abstracts of the 13th International Symposium on HIV and Emerging Infectious Diseases,* 2005, OP4.5.
238. Lalezari, J.; Thompson, M.; Kumar, P.; Piliero, P.; Davey, R.; Murtaugh, T.; Patterson, K.; Shachoy-Clark, A.; Adkinson, K.; Demerest, J. et al. *Abstracts of the 44th Interscience Conference on Antimicrobial Agents and Chemotherapy,* 2005, Abstr. H-1137B.
239. Schurmann, D.; Rouzier, R.; Nougarede, R.; Reynes, J.; Fatkenheuer, G.; Raffi, F.; Michelet, C.; Tarral, A.; Hoffmann, C.; Kiunke, J. et al. *Abstracts of the 11th Conference on Retroviruses and Opportunistic Infections,* San Francisco, CA, 2004, Abstr. 140LB.
240. De Clercq, E.; Yamamoto, N.; Pauwels, R.; Baba, M.; Schols, D.; Nakashima, H.; Balzarini, J.; Debyser, Z.; Murrer, B. A.; Schwartz, D. et al. *Proc. Natl. Acad. Sci. USA* **1992**, *89*, 5286–5290.
241. De Clercq, E. *Nat. Rev. Drug Disc.* **2003**, *2*, 581–587.
242. Dionisio, D.; Esperti, F.; Messeri, D.; Vivarelli, A. *Curr HIV Res.* **2004**, *2*, 377–393.
243. Klausner, R. D.; Fauci, A. S.; Corey, L.; Nabel, G. J.; Gayle, H.; Berkley, S.; Haynes, B. F.; Baltimore, D.; Collins, C.; Douglas, R. G. et al. *Science* **2003**, *300*, 2036–2039.
244. Emini, E. A.; Weiner, D. B. *Expert Rev. Vaccines* **2004**, *3*, S1–S2.
245. Adler, M. W. *Lancet* **1997**, *349*, 498–500.
246. Piot, P.; Feachem, R. G.; Lee, J. W.; Wolfensohn, J. D. *Science* **2004**, *304*, 1909–1910.
247. Lucas, G. M. *J Antimicrob. Chemother.* **2005**, *55*, 413–416.
248. Nolan, D.; Reiss, P.; Mallal, S. *Expert Opin. Drug. Saf.* **2005**, *4*, 201–218.
249. Wai, J. S.; Egbertson, M. S.; Payne, L. S.; Fisher, T. E.; Embrey, M. W.; Tran, L. O.; Melamed, J. Y.; Langford, H. M.; Guare, J. P., Jr.; Zhuang, L. et al. *J. Med. Chem.* **2000**, *43*, 4923–4926.
250. Hazuda, D. J.; Felock, P.; Witmer, M.; Wolfe, A.; Stillmock, K.; Grobler, J. A.; Espeseth, A.; Gabryelski, L.; Schleif, W.; Blau, C. et al. *Science* **2000**, *287*, 646–650.
251. Billich, A. *Curr. Opin. Investig. Drugs* **2003**, *4*, 206–209.
252. Hazuda, D. J.; Anthony, N. J.; Gomez, R. P.; Jolly, S. M.; Wai, J. S.; Zhuang, L.; Fisher, T. E.; Embrey, M.; Guare, J. P., Jr.; Egbertson, M. S. et al. *Proc. Natl. Acad. Sci. USA* **2004**, *101*, 11233–11238.
253. Fikkert, V.; Van Maele, B.; Vercammen, J.; Hantson, A.; Van Remoortel, B.; Michiels, M.; Gurnari, C.; Pannecouque, C.; De Maeyer, M.; Engelborghs, Y. et al. *J. Virol.* **2003**, 77, 11459–11470.
254. Fikkert, V.; Hombrouck, A.; Van Remoortel, B.; De Maeyer, M.; Pannecouque, C.; De Clercq, E.; Debyser, Z.; Witvrouw, M. *AIDS* **2004**, *18*, 2019–2028.
255. Maignan, S.; Guilloteau, J. P.; Zhou-Liu, Q.; Clement-Mella, C.; Mikol, V. *J. Mol. Biol.* **1998**, *282*, 359–368.
256. Little, S. J.; Drusano, G.; Schooley, R.; Haas, D. W.; Kumar, P.; Hammer, S.; McMahon, D.; Squires, K.; Asfour, R.; Richman, D. et al. *Abstracts of the 12th Conference on Retroviruses and Opportunistic Infections,* Boston, MA, Feb. 22–25, 2005, **2005**, Abs 161.
256a. Sato, M.; Motomura, T.; Aramaki, H.; Matsuda, T.; Yamashita, M.; Ito, Y.; Kawakami, H.; Matsuzaki, Y.; Watanabe, W.; Yamataka, K. et al. *J. Med. Chem.* **2006**, *49*, 1506–1508.
256b. DeJesus, E.; Berger, D.; Markowitz, M.; Cohen, C.; Hawkins, T.; Ruane, P.; Elion, R.; Farthing, C.; Cheng, A.; Kearney, B.; the 183-0101 Study Team. *Abstracts of the 13th Conference on Retroviruses and Opportunistic Infections,* Denver, CO, Feb. 5–8, 2006, Abstr. 160LB.
256c. Grinsztejn, B.; Nguyen, B.-Y.; Katlama, C.; Gatell, J.; Lazzarin, A.; Vittecoq, D.; Gonzalez, C.; Chen, J.; Isaacs, R.; The Protocol 005 Study Team. *Abstracts of the 13th Conference on Retroviruses and Opportunistic Infections,* Denver, CO, Feb. 5–8, 2006, Abstr. 159LB.
257. Kashiwada, Y.; Hashimoto, F.; Cosentino, L. M.; Chen, C. H.; Garrett, P. E.; Lee, K. H. *J. Med. Chem.* **1996**, *39*, 1016–1017.
258. Li, F.; Goila-Gaur, R.; Salzwedel, K.; Kilgore, N. R.; Reddick, M.; Matallana, C.; Castillo, A.; Zoumplis, D.; Martin, D. E.; Orenstein, J. M. et al. *Proc. Natl. Acad. Sci. USA* **2003**, *100*, 13555–13560.
259. Martin, D.; Jacobson, J.; Schurmann, D.; Osswald, E.; Doto, J.; Wild, C.; Allaway, G. *Abstracts of the 12th Conference on Retroviruses and Opportunistic Infections,* Boston, MA, Feb 22–25, 2005, Abstr. 159.
260. Tisdale, M.; Schulze, T.; Larder, B. A.; Moelling, K. *J. Gen. Virol.* **1991**, *72*, 59–66.
261. Budihas, S. R.; Gorshkova, I.; Gaidamakov, S.; Wamiru, A.; Bona, M. K.; Parniak, M. A.; Crouch, R. J.; McMahon, J. B.; Beutler, J. A.; Le Grice, S. F. *Nucleic Acids Res.* **2005**, *33*, 1249–1256.
262. Sluis-Cremer, N.; Arion, D.; Parniak, M. A. *Mol. Pharmacol.* **2002**, *62*, 398–405.
263. Shaw-Reid, C. A.; Munshi, V.; Graham, P.; Wolfe, A.; Witmer, M.; Danzeisen, R.; Olsen, D. B.; Carroll, S. S.; Embrey, M.; Wai, J. S. et al. *J. Biol. Chem.* **2003**, *278*, 2777–2780.

264. Hang, J. Q.; Rajendran, S.; Yang, Y.; Li, Y.; In, P. W.; Overton, H.; Parkes, K. E.; Cammack, N.; Martin, J. A.; Klumpp, K. *Biochem. Biophys. Res. Commun.* **2004**, *317*, 321–329.
265. Bour, S.; Strebel, K. *Adv. Pharmacol.* **2000**, *48*, 75–120.
266. Zeng, L.; Li, J.; Muller, M.; Yan, S.; Mujtaba, S.; Pan, C.; Wang, Z.; Zhou, M. M. *J. Am. Chem. Soc.* **2005**, *127*, 2376–2377.
267. Goff, A.; Ehrlich, L. S.; Cohen, S. N.; Carter, C. A. *J. Virol.* **2003**, *77*, 9173–9182.

Biography

Terry A Lyle is a Senior Director in the Department of Medicinal Chemistry within the Merck Research Laboratories at West Point, PA. He is a project leader for a group of organic chemists working on the discovery of novel treatments for HIV/AIDS. Terry began his career at Merck in late 1981 and has contributed to a number of projects, including: somatostatin agonists, atrial natriuretic peptides, NMDA antagonists (MK-801 and ^{18}F-labeled MK-801 analogs for PET imaging), and thrombin inhibitors. In the HIV field his group made key contributions to the discovery of indinavir and efavirenz. He received the BA, MS, and PhD degrees from the University of New Mexico during 1975–79 under the direction of Prof Guido H Daub. Terry was awarded an NIH postdoctoral fellowship at Stanford University, working in the laboratories of Prof William S Johnson from 1979 to 1980, and then spent a year at the ETH in Zurich, Switzerland with Prof Oskar Jeger.

Honors he has received include a Merck Management Award (1985), the Merck Divisional Scientific Award for contributions to the discovery of indinavir (1998), and election to Phi Beta Kappa and Sigma Xi. Terry is married to Paulette Lyle and lives in Lederach, Pennsylvania.

Comprehensive Medicinal Chemistry II
ISBN (set): 0-08-044513-6
ISBN (Volume 7) 0-08-044520-9; pp. 329–371

7.13 Ribonucleic Acid Viruses: Antivirals for Influenza A and B, Hepatitis C Virus, and Respiratory Syncytial Virus

U Schmitz, L Lou, C Roberts, and R Griffith, Genelabs Technologies, Inc., Redwood City, CA, USA

7.13.1 Influenza A and B

Influenza infection, commonly known as 'the flu,' has plagued mankind for centuries and is a disease that afflicts almost everybody at some point in their life. Influenza infection occurs with a wide range of severity, depending on the particular strain of the virus and the age and health status of the befallen individual. In the vast majority of cases, the infection is not lethal and is overcome within a couple of weeks without treatment. The worst worldwide influenza pandemic on record is the 1918 Spanish flu, which killed more than 20 million people, sometimes within hours after the first symptoms appeared. Other deadly outbreaks occurred in 1957 Asian flu and 1968 Hong Kong flu. Over the past couple of decades a wealth of knowledge[1,2] has been acquired regarding the viral life cycle, virus–host relationships, especially the role of avian hosts as a genetic reservoir, and the virus' genetic escape maneuvers. At the same time, vaccination against influenza has been a great success and several drugs have been developed for treatment and prophylaxis, described in this chapter.

7.13.1.1 Replicative Cycle

Influenza A and B viruses are members of the Orthomyxoviridae family of single-stranded RNA viruses. The negative-strand RNA genome is segmented into eight RNA molecules which are independently encapsidated by a viral RNA polymerase and nucleoproteins forming viral ribonucleoprotein complexes (vRNP). These vRNPs are enclosed by a lipoprotein envelope. A layer of matrix protein (M1) lines the inside of the envelope and is bound to the vRNP.

The viral envelope contains three different membrane proteins, hemagglutinin (H), neuraminidase (N), and ion channel protein M2 in influenza A virus and NB in influenza B virus.[3] Hemagglutinin and neuraminidase are the two major proteins on the surface of all influenza viruses and are important in determining infectivity and pathogenicity. Influenza A viruses are further classified into subtypes based on the differences in their hemagglutinin and neuraminidase sequences, e.g., H1N9, the subtype which was extensively used for structural studies.

Influenza infection is essentially confined to cells of the upper and lower respiratory tract. The replication of influenza viruses (**Figure 1**) is initiated by binding of viral hemagglutinin on the surface of the viral envelope to respiratory epithelial cells via a specific hemagglutinin receptor. This hemagglutinin-mediated attachment of the virus relies heavily on the glycosylation side chains presented by the receptor on the cell surface. Once bound, the virus enters the cell via endocytosis. The acidic pH in the endosome opens an ion channel on the surface of the virus particle permitting hydrogen ions to enter the virion. While for influenza A, this ion channel consists of the viral M2 protein and is well understood, there is no direct functional counterpart in influenza B, but proteins NB and BM2 are thought to be most closely related.[4] The acidification of the virus is necessary for uncoating, a critical event involving dissociation of the M1–vRNP complex so that the vRNP complex may enter the nucleus. Both mRNA synthesis and genomic replication takes place inside the nucleus. These reactions are catalyzed by the vRNP complex, which is composed of nucleoproteins (NP) and the heterotrimeric RNA-dependent RNA polymerase (PB1, PB2, and PA subunits). After export of genomic and mRNA from the nucleus with the help of viral nuclear export protein (NEP), virus proteins are produced in association with cellular membranes. Progeny virions are assembled and then bud from the cell surface. It is the viral neuraminidase on the outside of nascent virions that is responsible for the cleavage of the terminal sialic acid residues from glycosylated macromolecules on the surface in the respiratory epithelial cell and the influenza virus envelope. Removal of the sialic acid enables the release of the virus from the cell surface into the respiratory lumen. In addition, neuraminidase prevents clumping and helps to transport the virus through mucosal secretions in the lungs which are rich in sialic acid-containing macromolecules. Neuraminidase from the two major types of influenza strains A and B exhibits strict conservation at the catalytic site. It is therefore an attractive target for anti-influenza drug discovery.

7.13.1.2 Transmission and Epidemiology

Influenza is a highly contagious respiratory tract infection caused by influenza types A and B viruses. Influenza A viruses are known to infect birds and other animals. Influenza B viruses are known only in humans. Virus-laden respiratory aerosols expelled during coughing and sneezing perpetuate infection throughout the community. Seasonal epidemics affect 10–20% of the global population in an average year. The World Health Organization (WHO) estimates that annual influenza epidemics result in 3–5 million cases of severe respiratory illness and up to 500 000 deaths. In the USA, over 150 000 persons are hospitalized and 20 000 die annually as a result of influenza.

A majority of the infections are caused by influenza type A. All age groups are susceptible. Although rates of infection are highest in children, rates of serious illness and death related to influenza are greatest among persons aged >65 years, those with chronic cardiopulmonary diseases, and immune-compromised individuals. These people are deemed high risk because influenza may exacerbate their underlying condition or predispose them to life-threatening illnesses such as bacterial pneumonia.[5]

Although there are only a few different influenza viruses circulating in a particular flu season, susceptibility in any host is continuous. The reason is that influenza viruses are continually changing. When the H and N proteins of influenza A and B viruses, which are targeted by the vaccines, undergo periodic mutations, slightly different viruses

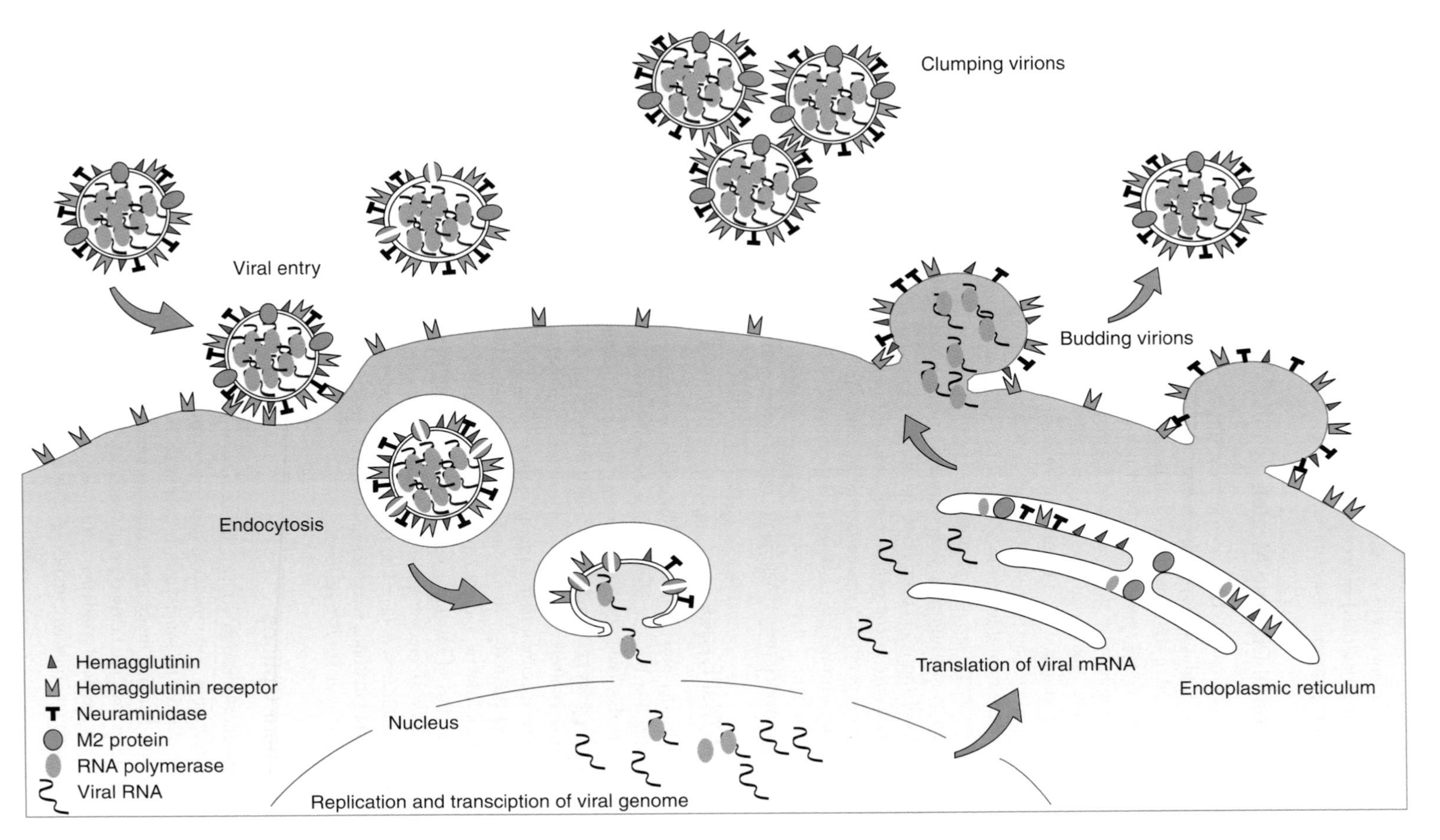

Figure 1 Replication cycle of influenza virus. Distinct phases are labeled in the figure.

emerge that are not recognized by the host immune system. These minor changes (antigenic drift) are responsible for seasonal epidemics. Influenza A viruses, in particular, also go through major changes (antigenic shift). When this happens, a new subtype of virus emerges and this is responsible for occasional pandemics. Antigenic shifts happen because of the segmented arrangement of the influenza genome and its propensity for recombination of entire segments (shuffling of genetic materials) between human and avian viral genomes. With the vast avian reservoir of type A influenza, the threatening potential of a highly virulent pandemic strain cannot be underestimated.[5,6]

7.13.1.3 Experimental Disease Models

Cell-based and animal infection models have been successfully developed to measure antiviral activity of experimental drugs. For cell-based assays, Madin–Darby canine kidney (MDCK) cells serve as the susceptible host to the viruses. When MDCK cells are cultured in the presence of influenza viruses, plaques (clear zones) occur if cells are destroyed by viral growth; suppression of viral replication reduces plaque formation.[7–10] Alternatively, antiviral activity is determined by reduction of virus-induced cytopathic effect by measuring a reduction of dye uptake by living cells.

Mouse and ferret influenza models have been developed.[7–10] In mice, animals are inoculated with 90% to 100% lethal doses of influenza by intranasal inoculation. Dosing regimens can start prior to or after inoculation and continue for designated durations. Antiviral activity is measured by increase in survival rate and reduction in viral titer in lung homogenate. In ferrets, animals are infected by intranasal inoculation and reduction of viral titer from nasal washings and reduction of the febrile responses are measured.

Animal models are used to measure drug penetration to target tissue. Since influenza virus replicates primarily in the surface epithelial cells of the respiratory tract, the ability of a compound to be delivered to the bronchoalveolar lining fluid following an oral dose of compound is an important indicator of its potential efficacy.

The earliest clinical testing of efficacy was performed in experimental influenza infection. Healthy human volunteers were inoculated intranasally with infectious doses of influenza A or B. Treatment started either several hours before inoculation (prophylaxis) or 1 or 2 days afterward (early or delayed treatment).[7,8] The endpoints were laboratory-confirmed infection by positive culture, HA antibody in the serum, and viral titer (viral shedding) from nasal washings.

7.13.1.4 Clinical Trial Issues

Making timely and accurate diagnosis of influenza infections is a major and unresolved problem in treatment. Clinical diagnosis of influenza infection by symptoms cannot be done with confidence because the symptoms of influenza overlap considerably with other cocirculating respiratory viruses. Virological diagnosis by virus identification using cell culture requires 3–5 days, a delay that renders treatment ineffective. Although diagnostic tests detecting viral antigens or RNA are available, testing is often not used because of less than optimum sensitivity and specificity of the tests, and the problems with obtaining specimen from patients' nasal washes. Determining the efficacy of therapeutics in clinical trails, and the application of them to seasonal epidemics is hampered by the difficulty of a correct and early diagnosis.

In order to test the prophylactic and therapeutic effectiveness of investigational drugs, healthy volunteers were inoculated with specific strains of influenza viruses A or B during the initial clinical testing.[11,12] The experimental human influenza studies allowed specific questions such as route of administration, time of treatment initiation, and duration of treatment to be addressed. Based on the information gained from these preliminary studies, pivotal trials to investigate natural influenza infections were designed. Later clinical trials included patients with influenza-like illness which were influenza-positive in 50–70% of cases.

7.13.1.5 Treatment Options

Vaccination against influenza A and B viruses has been and remains the most attractive way to prevent influenza infection. However, protection is often incomplete due to insufficient vaccine coverage but also to a lack of protective immunogenicity in the elderly. In some years a mismatch between vaccine strains and circulating strains renders the vaccine less effective. Antiviral medications serve as an adjunct to immunization in controlling the impact of the disease. [11,12] Antiviral drugs may also be used prophylactically for those patients in close contact to the infected or high-risk populations.[13,14] In the event of pandemic influenza caused by a major antigenic shift, vaccines most likely cannot be adapted quickly enough and only antiviral drugs will be useful.

7.13.1.5.1 Vaccines

Inactivated vaccines have been successfully employed for more than 60 years and new vaccines such as live attenuated intranasal vaccines have recently become available. H protein extracted from recombinant virus forms the basis of today's trivalent vaccines. Influenza vaccine used in the USA contains two type A strains and one type B strain representing the viruses likely to circulate in the USA in the upcoming winter. The vaccine is made from highly purified, egg-grown viruses that have been made noninfectious. The effectiveness of the vaccine depends primarily on the age and immunocompetence of the vaccine recipient and the degree of similarity between the viruses in the vaccine and those in circulation. At best, influenza vaccine can prevent illness in approximately 70–90% of healthy persons aged below 65 years.[5]

7.13.1.5.2 M2 ion channel inhibitors

The transmembrane domain of the influenza A M2 ion channel is the target of the adamantane group of antivirals. Amantadine (Symmetrel), **1**, was the first specific influenza antiviral drug licensed in the USA (1966) for the treatment and prevention of influenza A. Rimantadine (Flumadine), **2**, was licensed in 1993. The adamantylamines block the ion channel activity of M2 through allosteric inhibition. M2 inhibition blocks viral uncoating and RNA release and results in inhibition of viral replication. These drugs are only effective toward influenza type A and not type B because M2 is specific for type A (**Scheme 1**).

Amantadine inhibits influenza type A replication in infected MDCK cells with IC_{50} ranging from 1.1 to $>25\,\mu M$. Rimantadine is 4- to 10-fold more active. Both drugs are orally bioavailable and exposure at the nasal mucus is the same as in circulation. Both are equally effective provided treatment is initiated within 48 h of the onset of symptoms. Most studies show a reduction in symptom scores and fever by 1 to 2 days as well as a decrease in viral shedding. They are also useful as prophylactics, although there is no convincing evidence that these drugs reduce complications of influenza infections.

The use of the amantadines is limited by the adverse effects and early development of drug resistance.[11,12] Both amantadine and rimantadine can cause severe central nervous system and gastrointestinal side effects. Although the incidence is less with rimantadine, usage in elderly patients is still limited to lower doses. Resistance to amantadine and rimantadine may emerge within the first 3–5 days in up to 50% of children, elderly people, and immune-compromised patients. The mechanism of resistance appears to be due to single amino acid changes in the M2 protein. The rapid and extensive emergence of resistance severely limits the use of these drugs as therapeutic and prophylactic regimens especially in close-contact environments. The resistant viruses are cross-resistant with both amantadine and rimantadine.

7.13.1.5.3 Neuraminidase inhibitors

Two heavily glycosylated proteins are presented at the surface of all influenza viruses, hemagglutinin (H) and neuraminidase (N). The enzymatic activity of neuraminidase is essential for the release of new viral particles from host cells. Neuraminidase is a glycohydrolase which cleaves the α-ketosidically linked sialic acid (neuraminic acid), **3**, off the

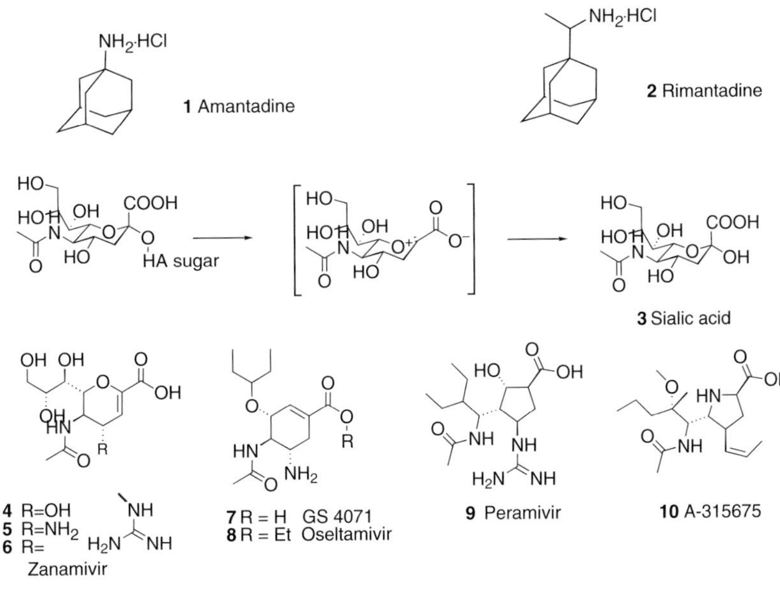

$NH_2 \cdot HCl$ **1** Amantadine

$NH_2 \cdot HCl$ **2** Rimantadine

HA sugar

3 Sialic acid

4 R=OH
5 R=NH_2
6 R= (guanidino)
Zanamivir

7 R = H GS 4071
8 R = Et Oseltamivir

9 Peramivir

10 A-315675

Scheme 1

ends of the glycosidic side chains of hemagglutinin and other glycoproteins and glycolipids. This capacity contributes to the infectious spread in multiple ways. First, budding virions pick up HA receptors from host cell surfaces such that a nascent virus remains tethered to the host cell via HA/HA receptor interactions. New viral particles also stick together since both HA and HA receptors are displayed on the viral surface. The removal of sialic acid by neuraminidase renders the HA receptor impotent, thus allowing viral particles to emerge from host cell surfaces and viral clumps. Second, neuraminidase is capable of digesting various glycosylated components in respiratory mucus involving neuraminic acid, thereby contributing to viral infection.

With a clear understanding of its role and its genetic and structural makeup, the discovery of several clinically efficacious, small-molecule neuraminidase active site inhibitors is a success story of modern structure-based drug design,[2,15,16] which will be reviewed herein. This work led to the approval in 1999 by the US Food and Drug Administration (FDA) of zanamivir, **6**, (for inhalation) and oseltamivir, **8**, an oral prodrug of GS 4071, **7**. On the heels of the latter, peramivir (BCX-1812), **9**, the first compound exploiting a cyclopentane scaffold, showed disappointing phase III results in 2001.[17] Another, pyrrolidine-based compound, A-315675, **10**, is showing promising in vitro results but its developmental status is unclear.

7.13.1.5.3.1 Discovery of neuraminidase inhibitors

The importance of influenza neuraminidase as a determinant of subtype virulence and the fact that neuraminidase 'heads' can be harvested from virus particles in sufficient yield and purity made this enzyme a prime candidate for early crystallographic studies.[15] In 1978, collaborative efforts of the laboratories of G. Laver and P. Colman in Australia yielded the 1.9 Å crystal structure of neuraminidase of influenza A subtype H11N9.[15,18,19] Influenza neuraminidase is a homo-tetramer. The individual subunits fold into the prototypic beta-propeller, with six four-stranded antiparallel beta-sheets arranged like blades of a propeller (**Figure 2**). The active site is a deep cavity, and it is lined entirely by amino acids that are highly conserved in both influenza A and B subtypes. Cocrystal structures with sialic acid,[20] the product of neuraminidase catalyzed cleavage, revealed the molecular details of enzyme substrate recognition (**Figure 2b**) and quickly led to the rational design of potent neuraminidase inhibitors. The neuraminidase active site is highly polar with ten Arg, Asp, and Glu residues and four hydrophobic residues. Five subsites (S1–S5) of interaction can be defined (**Figure 2b**). The pyranose ring itself is found in a boat conformation, and all of the ring substituents make interactions with the five subsites. Most importantly, the carboxylate group makes strong electrostatic and hydrogen bonding contacts with three arginines in S1 (R292, R371, R118). The *N*-acetyl group makes both polar (R152) and nonpolar interactions in S3 (W178, I222, R224). The glycerol moiety makes two hydrogen bonds with E276 in S5 and weak hydrophobic contacts to A248 and R224 in S4. The hydroxyl groups exhibit hydrogen bonds to D151 and E119. The finding of a high-energy pyranose conformation along with the assumption that the neuraminidase enzymatic pathway most likely involves a partially flat transition state[21] (**Scheme 1**) led to the discovery of 2-deoxy-2,3-dehydro-*N*-acetylneuraminic acid (**4**, Neu5Ac2en, also known as DANA). This compound showed micromolar inhibitory activity against a range of viral, bacterial, and mammalian neuraminidases. The structure of DANA bound to neuraminidase (PDB: 1f8b)[22] is very similar to that of sialic acid and it was noted that the most underutilized interaction site was S2. The presence of two acidic amino acids (D151 and E119) prompted the exploration of positively charged substituents on C4 yielding 4-amino and 4-guanidino-Neu5Ac2en, **5** and **6**.[23]

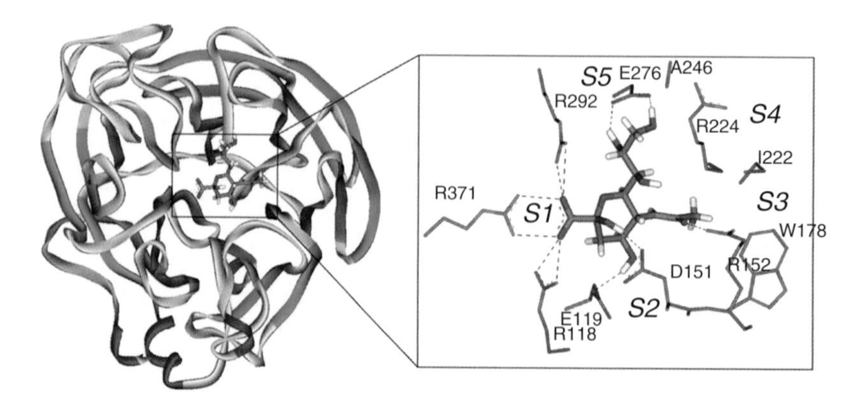

Figure 2 Secondary structure depiction of influenza neuraminidase (PDB: 1mwe). Insert shows the active site complexed with sialic acid. Distinct subsites of the neuraminidase active site are labeled S1–S5.

7.13.1.5.3.1.1 Zanamivir Zanamivir, **6**, and its amino analog, **5**, were very potent in enzyme inhibition assays using intact influenza A virions (N2 A/Tokyo/3/67) exhibiting K_i-values of 0.2 and 50 nM, respectively. The structure of zanamivir bound to neuraminidase did indeed reveal the salt bridge between the 4-guanidino group and the two carboxylate groups of E119 and E227 (PDB: n/a). The fact that this nanomolar compound was much less active against other, nonviral neuraminidases suggested that polar subsite S2 in part determines the specificity of the neuraminidase active site for inhibitors. Zanamivir was also an effective inhibitor of influenza A and B in tissue culture, with IC_{50} values (concentration for 50% reduction of plaque formation in MDCK cells) of 14 and 5 nM, respectively.[23] The highly polar nature of zwitterionic zanamivir proved devastating for its oral bioavailability, but it was successfully developed for inhalation.[7] A crystal structure of zanamivir bound to influenza B neuraminidase (B/BEIJING/1/87) became available (PDB: 1a4g) toward the end of the development process (**Figure 3a**).[24] Compared to the structures of sialic acid and DANA bound to neuraminidase of influenza A, the zanamavir cocrystal structure exhibits some adjustments of the neuraminidase side chains. Interestingly, R153 (156, influenza A numbering) is pulled in closer toward the 4-guanidino group and it clearly participates in the salt bridge network involving E116(119), D148(151), and E225(227). The near-flat conformation of the dehydropyran facilitates tighter interaction of C2-carboxylate with the arginine cluster (R115, R291, R373). Furthermore, the 4-guanidino group expels the two water molecules around the 4-OH seen in the sialic acid and DANA crystal structures.[24] As one would expect from the zanamivir crystal structure, substitutions on the guanidino nitrogens are not well tolerated, but 5-trifluoroacetamido and 5-sulfonamide derivates remained as potent as zanamivir.[25] The 6-glycerol substituent of zanamivir interacts with E274(276), but it does not exploit the hydrophobic nature of subsite S4. However, replacing the glycerol chain with ether, ketone,[26] or carboxamide[24] linked alkyl chains yields active zanamivir derivatives with a strong selectivity for influenza A, and much reduced affinity for influenza B.

From the structures discussed thus far, it is clear that the pyran and dehydropyran rings merely act as a scaffold without engaging in any direct contacts to neuraminidase. A number of other scaffolds have been designed which can direct the substituents properly into some or all of the five subsites.

7.13.1.5.3.1.2 Oseltamivir Oseltamivir, **8**, is the ethyl ester prodrug of GS 4071, **7**, taking advantage of the fact that the cyclohexene scaffold is also a close mimic of the putative oxocarbonium transititon state. The design of **7** by Gilead scientists was driven in part by the need to reduce the highly polar character of zanamivir to achieve oral bioavailability.[27] Besides the replacement of the pyran oxygen by a carbon atom[28] and the utilization of the initially less potent, but also less basic 4-amino group (see above), great improvements in potency were achieved by exploring a wide range of hydrophobic substiutents directed at subsite S4 and S5. Exemplary IC_{50} values for influenza A and B enzymes in **Table 1** demonstrate the tight structure–activity relationship (SAR) for a series of alkoxy substituents targeting this pocket, which surprisingly also mediates influenza A versus B specificity of the ligands. The 3-pentyl side chain clearly offered the best inhibition for both subtypes. The profound differences in activity for different alkoxy substituents are readily explained by the cocrystal structure of **7** (**Figure 3b**).[29,30] **7** binds to influenza A neuraminidase in a similar fashion as zanamivir, although some rearrangements of the amino acids in the binding pocket are evident. Most importantly, the 3-pentyloxy chain of GS 4071 induces a reorganization of the amino acid side chains in S4 and S5 and gains substantial hydrophobic interactions not seen in the zanamivir cocrystal structure. In this reorganization, E276 is turned over toward R224 to form a tight salt bridge, allowing the rightmost branch of the pentyl chain to have close hydrophobic contacts with A246 and I222. The other branch seems to be pushing out the proximal amino acid side chains to induce a strong hydrogen bonding network between the backbone carbonyl of A247, the carboxamide of N294, and the guanidino group of R292. The crystal structure of GS 4071 bound to influenza B has been reported as well.[27] The main difference in GS 4071 binding to type A and B neuraminidase is the degree of displacement of E276. In type B neuraminidase, E276 hardly moves from the original position seen in the sialic acid cocrystal structure, explaining the trend in the SAR of GS 4071 and its alkoxy derivatives seen in **Table 1** where the most potent compounds inhibit type A neuraminidase more efficiently.

The strong preference for the *S*-enantiomers in case of the phenethyl and cyclohexylethyl compounds from **Table 1** is also readily understood in light of the crystal structure. In the *S*-configuration, the phenyl and cyclohexyl group is attached to the rightmost branch of the pentyloxy chain, which will place them in a hydrophobic sandwich between A246 and I222 (see [27] for detail). Comparing the activity data for the phenethyl and cyclohexylethyl derivatives also suggests that this hydrophobic pocket between A246 and I222 is less adaptable in type B neuraminidase, thereby contributing to the specificity of the ligand protein interaction. Beyond those profound changes induced by the 3-substituent of GS 4071, more subtle changes can be seen in subsite S2 related to the repositioning of the double bond relative to zanamivir. Moving the double bond from 5en to the 1en position, more closely mimicking the stereochemistry of the oxocarbonium transition state, allows the C4–NH2 moiety to participate more efficiently in the salt bridge with E119 and D151. This more advantageous positioning of the double bond is also evident from the greatly different

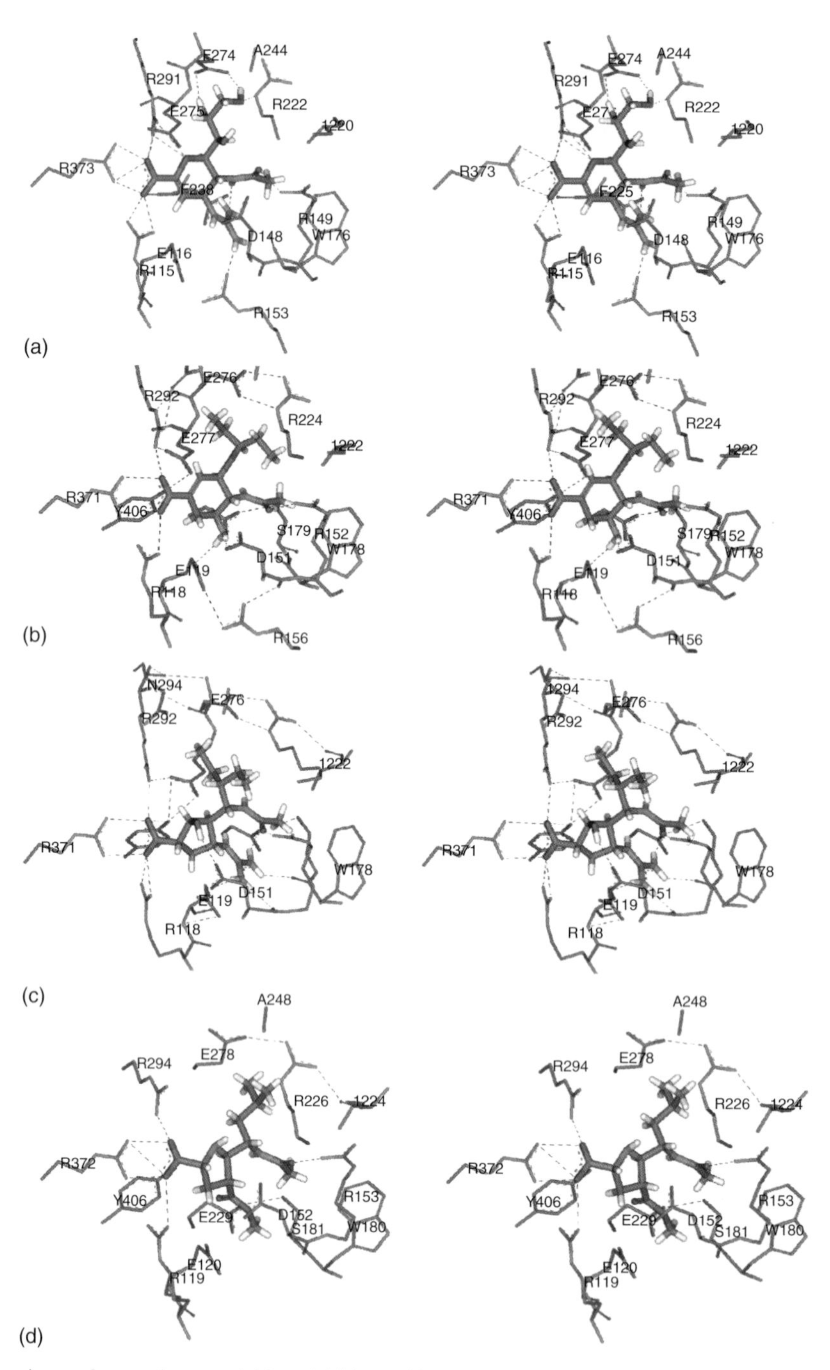

Figure 3 Stereoviews of several neuraminidase inhibitors: (a) zanamivir, **6** (PDB: 1a4g); (b) GS4071, **7** (PDB: 2qwk); (c) peramivir, **9** (PDB: 1l7f); (d) pyrrolidine derivative, **16a** (PDB: 1xoe). Inhibitors are shown in large stick representation. For the neuramin; dase, only active site residues are shown (hydrogens have been omitted for clarity). Residue labeling might differ depending on actual influenza A or B sequence used. Hatched lines indicate possible hydrogen bonds. Color-coding is by atom type.

activities of pairs of isomers with the double bond at either C1–C2 or C5–C1. For the 5en isomer of GS 4071, the activity is reduced 30-fold for type A neuraminidase and over 100-fold for type B neuraminidase.[27,30] Using the crystal structure of GS 4071, one can rationalize most of the SAR described for other GS 4071 analogs: the C3-linker atom does not contact the protein and replacement with CH_2-, S-, NH-, and even NCH_3-linkers results in compounds with potency comparable to that with the O-linker. With respect to the C2-substituents which is near R292, even small groups,

Table 1 Cyclohexene based neuraminidase inhibitors

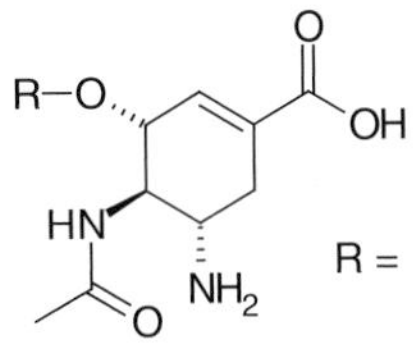 R =	IC_{50} (nM) Infl. A	Infl.B
H	6300	ND
CH3–	300	ND
CH3CH2–	2000	185
CH3(CH2)2–	180	15
CH3(CH2)3–	300	215
CH3(CH2)5–	150	1450
(CH3CH2)2CH–	1	3
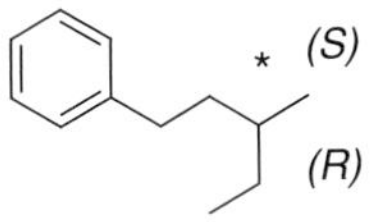 * (S)	0.3	70
(R)	12	35
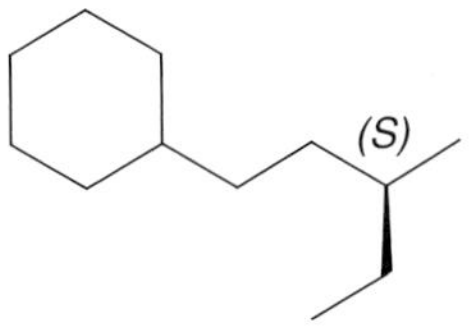 (S)	1	2150

Adapted from Lew, W.; Chen, X.; Kim, C. U. *Curr. Med. Chem.* **2000**, *7*, 663–672.

i.e., Cl and CH_3, destroy activity.[29] Not surprisingly, the F-analog of GS 4071 is roughly as potent as its parent against type A neuraminidase and 30-fold less potent against type B neuraminidase.

For a direct comparison between the zanamivir and oseltamivir chemotypes, it should be noted that GS 4071 is 150-fold more active against type A neuraminidase than its 4-amino zanamivir counterpart, **5** (1 versus 150 nM, respectively). On the other hand, the 4-guanidino analog of GS 4071 is only five fold more potent than parent GS 4071 and zanamivir (0.2 versus 1 nM, respectively). Clearly, the contribution of the 4-guanidino group is larger for the zanamivir series than for the oseltamivir series.[27]

7.13.1.5.3.1.3 Peramivir Peramivir (**9**, RWJ-270201 or BCX-1812) emerged from rational drug design efforts at Biocryst based on crystallographic studies of another natural neuraminidase inhibitor, α/β-6-acetyl-amino-3,6-dideoxy-D-glycero-altro-2-nonulofuranosonic acid, **11**. The latter has micromolar inhibitory activity similar to DANA, **4**, and exhibits a five-membered ring scaffold presenting essentially the same substitutents (carboxylic acid, glycerol, acetamido, and C4-OH groups) to the neuraminidase (**Scheme 2**).[31]

The crystal structure of this natural inhibitor showed that what is essential for activity is not the absolute position of the scaffold but rather the relative position of the interacting substituents. This data suggested that other five-membered rings with the same, one-carbon spacer for the acetamido and glycerol groups should yield potent neuraminidase inhibitors as well. Cyclopentane-based compound **12** was initially synthesized as a mixture of diastereoisomers. Crystallography studies employing soaks of type A and B neuraminidase crystals with the compound mixture ultimately revealed the nature of the active isomer as 1*S*,3*R*,4*R*. It was noted that the butyl chain adopted different binding modes in the type A versus type B neuraminidase structures. In type B neuraminidase, the butyl chain of **12** interacts with the hydrophobic portion of A246, I222, and R224, whereas in type A neuraminidase, the

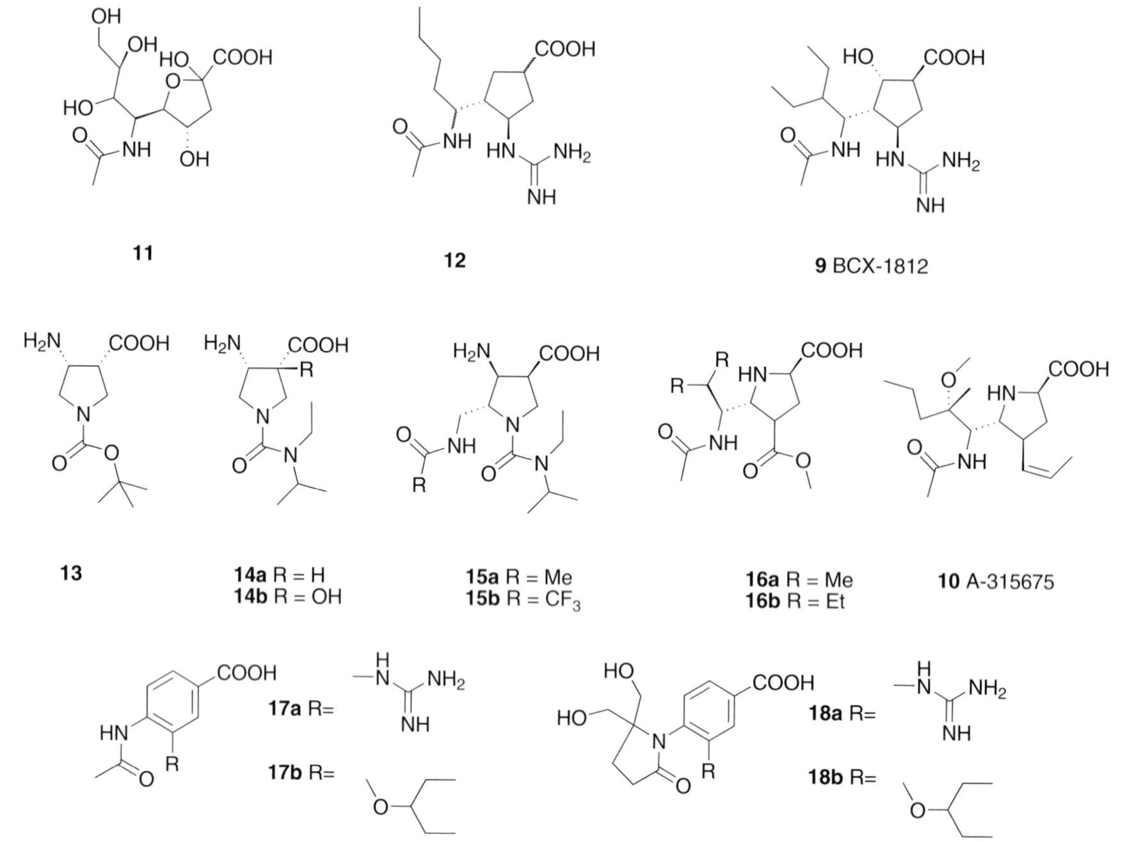

11 **12** **9** BCX-1812

13 **14a** R = H **14b** R = OH **15a** R = Me **15b** R = CF_3 **16a** R = Me **16b** R = Et **10** A-315675

17a R= **17b** R= **18a** R= **18b** R=

Scheme 2

interaction is with the hydrophobic portion of E276. The latter is found in the salt-bridged form with R224, very similar to what has been seen for GS 4071. This observation led to the replacement of the butyl group by the 3-pentyl group, which maximizes the hydrophobic interactions in both areas, as it is evident in the crystal structure of the optimized peramivir, **9** (**Figure 3c**). Peramivir is 100-fold more potent than **12** and inhibits various strains of type A and type B neuraminidase in the low nanomolar range (0.1–1.4 versus 0.6–11 nM, respectively).[31] The structure of peramivir was also determined using soaks with diastereomeric mixtures, revealing the active isomer to exhibit the 1*S*,2*S*,3*R*,4*R*,1′ *S*-configuration. The cyclopentane ring is found in a low-energy envelope conformation with the C4 atom out-of-plane. Interestingly, the *R*-stereochemistry of C4 directs the 4-guanidino group into a position different from that seen for zanamivir. For peramivir, the guanidino group is projected behind the cyclopentane plane, making hydrogen bonds with the backbone carbonyl atoms of S179, W178, and D151. Its electrostatic interactions involve the carboxylic acid groups of D151 and E227, but no longer E119, which was critical for zanamivir binding. Not surprisingly, the zanamivir resistant type A neuraminidase mutant E119G remains fully sensitive to inhibition by peramivir. It should be noted that the 2-hydroxy substituent of peramivir which contributes a hydrogen bond to D151 does not increase the inhibitory activity. However, peramivir is easier to synthesize than its des-hydroxy counterpart.[32]

7.13.1.5.3.1.4 Pyrrolidine-based inhibitors Following the success stories of the other neuraminidase active site inhibitors, researchers at Abbott Laboratories sought to obtain novel scaffolds by screening a small library of α/β-amino carboxylic acids. Their efforts yielded the pyrrolidine-based compound, **13**, a ~50 μM inhibitor of type A neuraminidase.[33,34] Optimization of this particular scaffold using small combinatorial libraries led to compounds **14a** and **14b** which exhibited submicromolar activity (type A neuraminidase $IC_{50} = 500$ and 360 nM, respectively) but failed to reach the potency of the compounds described above. The Abbott team used a combination of rapid crystallography using soaks with racemic mixtures of compounds and modeling to drive the evolution of these pyrrolidine-based inhibitors.[34] Initially, the crystal structure[33] of **14a** bound to type A neuraminidase showed that the C4-amino group makes a hydrogen bond to Y406, which had not been seen with any other scaffold. But even the addition of the C3–OH in compound **14b** which contributes an additional hydrogen bond to the D152 carboxylate group in the crystal structure, did not greatly improve potency. The explanation for this can be seen in the crystal structure, where compounds like **14a** do not occupy subsites S2 and S3. This led to the design of tetra-substituted pyrrolidines like compounds **15a** and **15b**. Unexpectedly the activity of **15a** was worse than that of **14a**. The most potent compound in this series was **15b**, showing enzyme inhibition of type A neuraminidase with an IC_{50} of 280 nM.[35] The crystal structure of **15a** yielded two unexpected findings which explained the undesirable drop in potency: (1) the chirality of the two carbon atoms C3 and C4 switched from 3*R*,4*R* to 3*S*,4*S*, and (2) the pyrrolidine scaffold itself is rotated about 90° with respect to the orientation of compound **14a**.[36] Crystallographic analysis of a number of cyclopentane analogs of compounds like **13–15** ultimately pointed to a 180° rotated binding mode, in which the urethane/urea substituent is interacting with subsite S2, formed by carboxylic acids of E120 and D152. The 4-amino group was found pointing toward subsite S5 without making any strong contribution. The potential of this new binding mode is beautifully realized in compounds **16a** and **16b**, where the 4-amino group has been collapsed into the ring, a methyl ester moiety is directed into subsites 2 and the hydrophobic moiety has been expanded to reach into subsites S4 and S5. Inhibition values for type A neuraminidase are 37 and 0.8 nM for **16a** and **16b**, respectively. The crystal structure of **16a** bound to type A neuraminidase is shown in **Figure 3d**. The pyrrolidine scaffold is found in the same envelope conformation as the cyclopentane ring of peramivir. The most striking feature of the **16a** structure is the unusual placement of the methyl ester moiety partially stacked against the carboxylic acid moiety of E120 mediated by π–π interactions. It is noteworthy that the replacement of the methyl ester moiety by an amino group, which should increase the electrostatic interactions with subsite S2 similar to GS 4071, results in a dramatic, 170-fold loss in activity. For the pyrrolidine scaffold, the optimal ligand substituent targeting subsite S2 is clearly hydrophobic. Following this new, hydrophobic paradigm, the biologically labile methyl ester was replaced by more stable hydrophobic groups, leading to clinical candidate A-315675 (**10**). Compound **10** inhibits neuraminidases of a wide variety of type A and B strains with subnanomolar efficiency ($K_i = 0.024$–0.31 nM),[37] similar or superior to GS 4071, especially for type B neuraminidases. These properties of **10** translated into efficient inhibition in cell culture assays against both laboratory- and clinic-isolated B strain influenza viruses. The oral bioavailability of the ethyl ester prodrug of 9 was ~50% in dogs.

The tetrahydrofuryl analog of **16a** is 10-fold less active against type A and ~20 against type B, indicating that the zwitterionic character is an important electrostatic component of binding.[38]

7.13.1.5.3.1.5 Benzoic acid-based inhibitors The partially flat, six-membered ring of the sialic acid transition state inspired the investigation of the achiral, completely flat benzoic acid scaffold. BANA-113, **17a**, is a micromolar inhibitor of type A and B neuraminidase,[39,40] and it was expected that the guanidino group would be directed into subsite S2 as seen with zanamivir. However, in the crystal structure of **17a**, the benzene scaffold has flipped such that the guanidino group interacts with E276 of subsite S5. When the guanidino group is replaced by 3-pentyloxy, **17b** , the inhibitory activity against type A neuraminidase remains but **17b** is completely inactive against type B neuraminidase.[41] This activity shift is most likely related to the ligand induced reorganization of subsite S4 and S5, involving the formation of the E276 R224 salt bridge. This adjustment seems to be energetically more costly in type B neuraminidase, as we have seen in the SAR discussion of GS 4071 analogs. A significant improvement in the inhibitory activity for this compound series came from the introduction of bis(hydroximethyl)pyrrolidone replacing the methylcarboxamide. While **18a** shows comparable activity to **16a** inhibiting both type A and B neuraminidase, **18b** inhibits type A neuraminidase with a much improved IC_{50} of 48 nM.[42,43] However, **18b** is over 1000-fold less active against type B neuraminidase, very much like **17b**. This lack of dual activity ultimately prevented this scaffold from producing preclinical candidates.

7.13.1.5.3.2 In vitro antiviral activity

The overall superb neuraminidase inhibitory activity of the four most successful NAIs – zanamivr, GS 4071, peramivir, and A-315657 – is best judged from side-by-side comparisons.[37,44–47] Most recently, Abbott researchers profiled all four compounds against four type A and two type B neuraminidases.[37] K_i values were mostly subnanomolar for all compounds (total range is 0.014–2.1 nM) and there is a tendency for all compounds to inhibit type A neuraminidase with slightly more efficiency. Based on the enzyme inhibitory activity, all four compounds should be considered equipotent. Comparative inhibition studies of influenza virus replication in cell cultures (cell culture plaque reduction with MDCK cells) are available as well.[37,46] Sidwell and Smee compared the inhibition of the viral cytopathic effect of 19 type A and 5 type B influenza strains by peramivir, zanamivir, and GS 4071. With the exception of two type A strains and three type B strains, all other strains were sensitive to all the compounds with EC_{50} values largely below 100 nM.[46] In this study, peramivir was slightly better against type A strains than the other two NAIs, and zanamivir showed the best activity against type B strains. In a different cell-based viral protection study employing three type A and two type B strains,[37] A-315675 was compared with GS 4071 and peramivir. All compounds showed potent inhibition of viral replication in all strains. The EC_{50} ranges are 0.1–41 nM, 1.2–79 nM, and 0.4–14 nM for peramivir, GS 4071, and A-315675, respectively. Peramivir and A-315675 displayed a very similar potency pattern and were slightly better than GS 4071. A-315675 was clearly the most potent NAI against a particularly dangerous, difficult to inhibit type B strain (B/HK/5/72) (EC_{50} = 14.1 versus 79 nM for GS 4071).

7.13.1.5.3.3 DMPK and efficacy in animal models

The earliest NAI, zanamivir, exhibits very low oral bioavailability (1–5% in humans) but direct topical delivery to the infection site, predominately the lungs, is possible by inhalation or nasal sprays. Once zanamivir has entered the blood stream, its half-life is fairly short ($t_{1/2}$ after intravenous injection is ~10 min in mice, 50 min in dogs, and 2 h in humans).[7] Zanamivir has very low toxicity.[7,48] Animal efficacy studies were conducted with mice and ferrets using intranasal application of zanamivir prophylatically and for day 0–3 after infection with different type A and B influenza strains.[49] Doses as low as 0.05 $mg\,kg^{-1}$ twice daily reduced mortality and viral titers in lung homogenates. It is important to realize that in typical animal models of influenza infection the compounds are administered several hours before inoculation with a 70–100% lethal dose of a particular strain. In general, less dramatic efficacy results are seen when compounds are administered 24 or 48 h after infection, which would be more typical of influenza infection in humans (see below).[44,46]

Oseltamivir, the ethyl ester of GS 4071, is readily absorbed from the gastrointestinal tract and is rapidly metabolized by hepatic enzymes. Oral bioavailability of oseltamivir is ~30% in mice, ~60% in dogs, and ~80% in humans.[50] It reaches peak concentrations 3–4 h after ingestion, and it is excreted renally with a terminal elimination half-life of 6–10 h.[11] Oseltamivir has very low cytotoxicity[9] and demonstrated strong antiviral activity in mouse and ferret models of influenza infection.[8] For example, in mice, oral oseltamivir (10 $mg\,kg^{-1}$ twice daily for 5 days) reduced lung homogenate viral titers by over 100-fold compared with untreated animals and also protected against the lethal effects of various types (A and B) of influenza viruses used in the study.[9]

Peramivir with its free carboxylate and guanidino group is chemically more similar to zanamivir than to oseltamivir and low oral bioavailability would be expected. No detailed account on peramivir's oral bioavailability has been published to date. However, mouse efficacy studies showed that peramivir performed well when compared directly with oseltamivir and zanamivir, using either oral gavage or intranasal spray, respectively.[44,46,47] In a prophylactic study, oral peramivir gave complete protection from the lethal effects of several type A strains at doses between 1 and 10 $mg\,kg^{-1}$ twice daily dependent on the strain used. In the delayed-treatment model, an oral dose of 10 $mg\,kg^{-1}$ twice daily of peramivir or oseltamivir at 24 h postinfection gave essentially complete protection against lethality. If treatment is started 48 h postinfection, neither drug offered significant protection. In another prophylactic model, all three compounds gave complete protection from lethality when using intranasal application of 0.1 $mg\,kg^{-1}$ once daily. In general, these mouse efficacy studies show that peramivir exhibits a potency similar to zanamivir and oseltamivir.

For the more hydrophobic pyrrolidine NAI A-315675, **10**, good oral bioavailability has been reported,[51] but no animal efficacy data is available. However, it should be noted that ethyl and isopropyl ester prodrugs of **10** were reported to have efficacy similar or slightly better than oseltamivir in prophylactic mouse models of infection with the deadly B/HK/5/72 strain.[52]

7.13.1.5.3.4 Clinical studies

As promising as the NAIs have appeared in the in vitro and in vivo studies above, measuring real life efficacy against influenza infection in humans has its difficulties.[12] As seen in the mouse efficacy studies, NAIs are expected to work best when applied prophylactically or shortly after infection. Therefore, typical phase II studies in humans are conducted with experimentally induced influenza infection in healthy volunteers such that treatment can be applied in a time-controlled manner. In such experimental influenza phase II studies, both of the FDA-approved NAIs, topical zanamivir and oral oseltamivir, as well as oral peramivir, proved to be efficacious in protecting from infection and reduced viral shedding compared to the placebo group in both early and postinfection treatment studies.[7,8,47] For example, oseltamivir (100 mg once or twice daily for 5 days), administered prophylactically, starting 26 h before infection with a type A strain, reduced the number of infections in an 8-day follow-up period from 67% (placebo group) to 38%[8] and the number of individuals shedding virus from 50% to 0%. In a treatment study with oseltamivir (75 or 150 mg twice daily for 5 days, starting 28 h after inoculation), the quantity of viral shedding (71–96% viral titer reduction compared to placebo as measured by area under the curve of the viral titer time-curve) and the time to cessation of shedding (median time to cessation of viral shedding is 18 h for the oseltamivir groups versus 96 h for placebo) were greatly reduced.[8] In a similar influenza A challenge study, peramivir (400 mg once daily for 5 days) reduced viral titers by 73% compared to the placebo group. This data might be an indication of peramivir's somewhat reduced clinical efficacy when compared to oseltamivir.

While phase II studies with experimental infection involve individuals with 100% influenza infection by design and measure the drug's true inherent potency, phase III studies are conducted with the intended target population for the drug, which involves a large percentage of patients with influenza-like symptoms caused by other respiratory viruses. Even with stringent selection criteria based on the presented symptoms (fever, headache, myalgia, cough, etc.), only about two-thirds of a typical trial population actually have influenza. Therefore, phase III results are usually presented for the total trial population, ITT (intention to treat), and the confirmed influenza population. Inhaled zanamivir (10 mg twice daily) reduced the median duration of symptoms by approximately 1 day for the whole ITT group and 30 h for the influenza group.[11,53] Similarly, oseltamivir treatment (75 mg twice daily started within 48 h of onset of symptoms) reduced time to alleviation of symptoms by 1 day in the ITT group and by 33 h in the influenza-positive group and the time to resume normal activity was reduced by 2–3 days. In an early versus late start of-treatment study (not placebo controlled) with over 1400 patients, early treatment with oseltamivir was clearly more advantageous; treatment within 12 h from onset of symptoms reduced the illness duration by 3 days more than treatment initiated 48 h after onset.[11,54] Oseltamivir has also been shown to be efficacious in children and in high-risk patients, i.e., the elderly and those with chronic cardiac and respiratory diseases.[55] Furthermore, there is evidence that oseltamivir can significantly reduce the number of naturally acquired influenza cases compared to placebo when local influenza virus activity is high in a largely unvaccinated community, demonstrating the prophylactic potential of oseltamivir. Although both zanamivir and oseltamivir demonstrated efficacy in the clinical setting, it is largely the ease of use (oral versus inhaler) especially for the children and high-risk patients that made oseltamivir the more successful drug.

No new drug application (NDA) was submitted for peramivir after disappointing, yet unpublished phase III studies in Europe and the USA. Considering how subnanomolar activities for neuraminidase inhibition in vitro translates into a fairly small margin for clinical efficacy, it is imaginable that less than perfect oral bioavailability and pharmacodynamics lead to the erosion of measurable clinical benefits.

7.13.1.5.3.5 Resistance to neuraminidase inhibitors

Both approved NAI inhibitors produce resistant mutants during in vitro serial passages of influenza viruses. The basic mechanisms of resistance involve alterations in hemagglutinin and its receptor binding site and/or changes in the neuraminidase active site, notably residues E119 and R292. The former mutations decrease the affinity of hemagglutinin for sialic acid, thereby reducing the dependence of viral replication on neuraminidase activity.[56,57] The latter mutations in the neuraminidase active site are typically accompanied by reduced enzymatic activity. The R292K mutation which exhibits a disruption of the arginine triad in subsite S1 was observed in vitro for both zanamivir and oseltamivir. However, this mutation led to greatly reduced infectivity and transmissibility in vivo.[58] Overall, the occurrence of in vivo resistance has been rare for both NAIs, although for oseltamivir, the published frequency of posttreatment viruses with resistant neuraminidase mutations is higher. It is conceivable that the way oseltamivir's pentyloxy side chain induces tight hydrophobic packing and rearrangement (see above) makes it more susceptible to small amino acid changes.

The notion that the rationally designed NAIs are generally associated with insignificant in vivo resistance was challenged by a recent Japanese study.[59,60] Of 50 children treated with oseltamivir, nine (18%) harbored viruses with resistant neuraminidase mutations. Besides known mutations at amino acids R293 and E119, a new mutation at N294 was found. This study confirmed that the rate of in vivo resistance is clearly higher in children. Although it is not clear whether these mutants are still highly infectious and transmissible, the Japanese study suggests that with such high mutation rates, oseltamivir resistance is not inconceivable.

7.13.1.5.4 **Other experimental drugs**

Besides targeting the neuraminidase-related viral budding process, at least two other parts of the viral life cycle can be targeted with small molecules: the fusion of viral and host cell membranes in the endosome and the endonuclease-related capping of viral mRNA for transcription and replication (**Scheme 3**).

Endosomal fusion of virus and cell membrane is facilitated through the conformational change of the hemagglutinin trimer, with each monomer containing the two subunits HA1 and HA2, which are linked by a disulfide bridge. Entry of the influenza virus into host cells relies on the interaction of HA1 with the sialic acid-containing HA receptor. In the endosome, the lowered pH leads to a conformational change which exposes the hydrophobic N-terminus of the HA2 subunit. The exposed N-terminus then mediates the fusion process. Researchers at Bristol-Myers Squibb identified a class of salicylamides that protect MDCK cells from the cythopathic effect of influenza infection at submicromolar

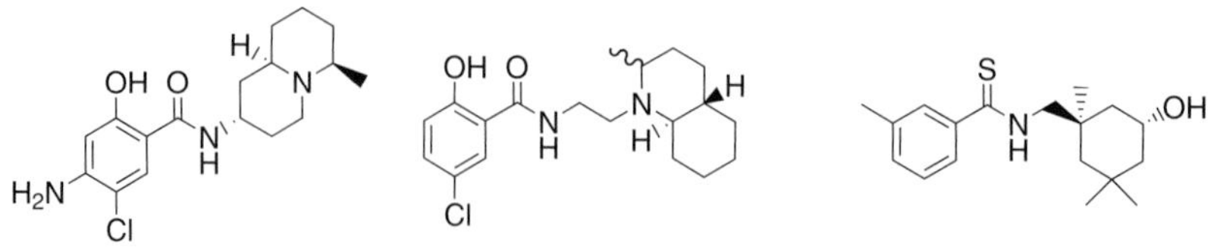

Scheme 3

concentrations.[61–63] The lead compound BMY-27709, **19**, eventually evolved into two different chemotypes, represented by compounds **20** and **21**.[62,63] **20** and **21** exhibited EC_{50} values of $90\,ng\,mL^{-1}$ and $15\,ng\,mL^{-1}$, respectively. The fusion-targeting mode of action of this class of compounds has been elucidated by resistant mutations which were found in the HA1 (M313) and HA2 subunits (F110) and photoaffinity labeling of a region in HA2 involving residues 84–106. This region is part of defining a pocket, that has been proposed as the binding site by molecular modeling studies.[61] The fact that this class of compounds is only active against influenza subtypes H1 and H2, but not H3, is explained by the sequence similarity among subtype H1 and H2 in this particular pocket and limits the utility of this class of inhibitors significantly.

The influenza polymerase complex contains an endonuclease function, which provides the polymerase with the capped RNA primers necessary for the initiation of transcription. Capped host mRNAs bound to the polymerase complex are cleaved by the endonuclease releasing capped RNA primers with free 3′-OH on which the nascent viral RNA chain is assembled. Specific influenza endonuclease inhibitors have been reported,[64–66] but their potency is still in the micromolar range.

7.13.1.6 Future Outlook

Among the three RNA viruses discussed in this chapter, influenza virus is clearly countered most efficiently by both vaccines and antiviral drugs. While one can expect vaccine generation to keep up with the gradual changes of the influenza virus, flares of deadly avian flu infections in Asia should be an important reminder of the fact that deadly flu pandemics are still possible. Fortunately, to date, none of the more lethal strains observed in Asia seemed to have had a high degree of infectivity between humans, which is required to cause a pandemic. In case of a pandemic, vaccine generation would be to slow, and it will be important to combat the flu with antiviral drugs, such as the neuraminidase inhibitors. As the latter are expected to retain their potency against many influenza mutants yet to come, it is not surprising that some physicians have called for 'stockpiling' neuraminidase inhibitors.

7.13.2 Hepatitis C Virus

Chronic infection with hepatitis C virus (HCV) has become a major health problem associated with liver cirrhosis, hepatocellular carcinoma and liver failure. An estimated 170 million chronic carriers worldwide are at risk of developing liver disease.[67,68] In the USA. alone 2.7 million are chronically infected with HCV, and the number of HCV-related deaths in 2000 was estimated between 8000 and 10 000, a number that is expected to increase significantly over the next years.[69] Liver failure resulting from chronic HCV infection is now recognized as the leading cause of liver transplantation in the USA. Considering the slow course of the infection and the nature of the current, insufficient therapy, treatment of HCV-related illness is a heavy burden on the public healthcare system.

7.13.2.1 Replicative Cycle

HCV is a member of the Flaviviridae family of RNA viruses that affect animals and humans. The genome is a single ~9.6-kb strand of RNA, and consists of one open reading frame that encodes for a polyprotein of ~3000 amino acids flanked by untranslated regions (UTRs) at both 5′ and 3′ ends. The polyprotein serves as the precursor to at least 10 separate viral proteins critical for replication and assembly of progeny viral particles. Because the replicative cycle of HCV does not involve any DNA intermediate and the virus is not integrated into the host genome, HCV infection can theoretically be cured. While the pathology of HCV infection affects mainly the liver, the virus is found in other cell types in the body including peripheral blood lymphocytes.[70,71]

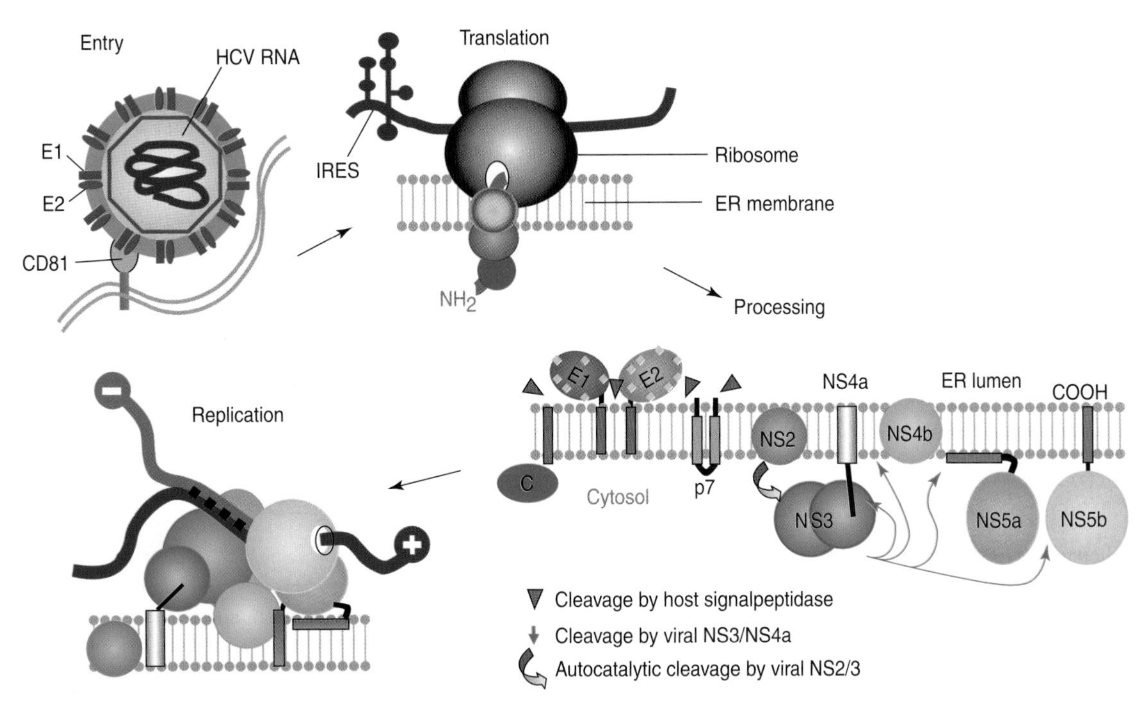

Figure 4 HCV life cycle and genome processing. 5′-UTR contains the internal ribosome entry site (IRES). Both 5′- and 3′-UTR have conserved elements essential for replication. C, core protein for nucleocapsid; E1 and E2 are hypervariable and heteromeric envelope proteins; p7 is a putative ion channel; NS2 is a metalloproteinase; NS3 is a protease/helicase; NS4a is a cofactor for NS3 protease; NS4b is a membrane protein with function unknown; NS5a is a phosphoprotein mediating resistance to interferon; NS5b is an RNA-dependent RNA polymerase.

The complete life cycle of HCV has not been clearly elucidated. This research has been hindered by the lack of an efficient experimental system of infection and virus production. A working model of the life cycle is based mainly on knowledge of other Flaviviridae viruses. The replicative cycle of HCV involves cell attachment and entry, translation and processing of viral proteins, replication of the RNA genome, and then assembly and release of progeny viruses (**Figure 4**).[71–73]

Several host cell surface proteins have been proposed to be involved with viral attachment and entry into host cells via endocytosis. These include a cell surface tetraspanin CD81,[74] the low-density lipoprotein receptor (LDLR),[75] the scavenger receptor B1,[76] and C-type lectins such as DC-SIGN and L-SIGN.[77,78] Some of them selectively bind the HCV envelop E2 glycoprotein but none has been shown to be essential for infectivity by authentic HCV particles. Recently, an experimental pseudotyped virus based on retrovirus modified with HCV envelope proteins was used to demonstrate that CD81 can mediate entry into host cells.[79–81]

Upon entry into the cytoplasm of host cells, uncoating of envelope and nucleocapsid takes place and the viral genome RNA is liberated and translated using host ribosomes.[72,73] Translation initiation is mediated by a highly conserved RNA element within the 5′-UTR, the internal ribosome entry site (IRES). The product of translation, a polyprotein precursor, is processed by both host and viral proteases to yield 10 individual viral proteins. The viral proteins are in the following order from the N- to the C-termini: C–E1–E2–p7–NS2–NS3–NS4a–NS4b–NS5a–NS5b. Host enzymes, signal peptide peptidase and signal peptidase, cleave to separate the C, E1, E2, and p7 proteins, and viral NS2/3 protease releases NS2. Viral NS3/4a protease cleaves all the downstream sites to separate the nonstructural proteins. The core protein interacts with genome RNA to form the nucleocapsid; E1 and E2 are glycosylated envelop proteins that interact with cell surface receptors of permissive cells, and p7 is a putative ion channel. The nonstructural (NS) proteins are the enzymes or associated factors responsible for catalyzing polyprotein processing and RNA replication, as well as anchoring the replication machinery to the endoplasmic reticulum membrane.

Viral replication requires an active 'replicase' complex that is associated with the endoplasmic reticulum in the cytoplasm. This multiprotein replication machinery is believed to consist of all the HCV NS proteins plus yet to be identified proteins of cellular origin. The major enzyme components are NS3/4a protease-helicase and NS5b polymerase. The initial replication is to copy the positive strand of RNA genome to produce the negative strand. Then the negative strand serves as a template to produce multiples copies of the positive strand for packaging into new viral particles. Conserved RNA elements in both the 5′- and 3′-UTR are essential for production of the negative and positive RNA strands, respectively.

The assembly of the nucleocapsid and the viral envelop occurs at the endoplasmic reticulum and the viral particles bud into the lumen of the ER. The release of the virus is thought to utilize the Golgi apparatus and cellular export machinery where the viral particles in vesicles are targeted to the cell plasma membrane for secretion.

In a chronically infected individual, an estimated 10^{10}–10^{12} new virus particles are produced each day.[74,82] It is also estimated that there is one mutation per cycle of genome replication due to the lack of proofreading of the viral polymerase. The high rate of production together with the mutational frequency may be the basis for the heterogeneity of the HCV population. Most of the mutations are localized to a hypervariable region (HVR-1) of the E2 protein, a region that would be exposed to antibody.[83] Such rapid generation of immunologically diverse viral population in an infected individual could render the immune response ineffective thus providing a way to evade host defense. Therefore, the replication of HCV not only produces progeny to spread the infection, it is also a means to enhance persistence of the infection.

7.13.2.2 Transmission and Epidemiology

A hepatitis C epidemic actually occured 30 years ago, but diagnosis and prevention was possible only after the viral genome was sequenced in 1989. Among the infected cases, 14% to 46% clear the virus naturally, and the infections are resolved. For the rest of the patients, the virus persists and the infection is long-lasting; although most of them have no clinical signs or symptoms.[84]

Viral transmission is by direct contact of HCV-infected blood. Presently in developed countries, most infections are due to illegal intravenous drug use. In developing countries, unsafe therapeutic injection practices are responsible for high rates of infection. Prior to screening of blood donors in the 1990s, millions of cases of infection were caused by blood transfusion. Because of blood donor testing, the number of new infections per year has declined from an average of 240 000 in the late 1980s to about 30 000 in 2003.

Certain regions of the HCV genome are more variable among different isolates. Analyzing the variations of these regions between individual isolates has resulted in a system to classify the virus into at least six main groups.[85] These groups are called the genotypes and are further divided into subtypes. Patients can be infected by more than one genotype. In a single infected individual, although there may be one or two predominant subtypes, there are millions of quasispecies. These viral particles are heterogeneous in their genomic sequences, a result of mutations due to high error rates in RNA replication. These variants are immunologically distinct. The diversity of the quasispecies is thought to be related to the rate of progression to liver disease.[86,87]

Genotypes do not appear to be linked to disease severity. Genotype variations can influence the response to the currently available treatments. Patients with genotypes 2 and 3 are almost three times more likely than patients with genotype 1 to respond to therapy with interferon-α or the combination of interferon-α and ribavirin.[88]

7.13.2.3 Experimental Disease Models

The pathogenesis of HCV infection involves acute hepatitis, which can be influenced by the host immune response, and chronic hepatitis, which can lead to serious liver diseases. Currently, the available experimental models deal mainly with HCV infection and replication. There is very limited knowledge in evaluating compounds for clearing the virus or improving liver damage. An HCV-infected cell model has been an exceptionally challenging system to establish. While HCV infectivity of human primary hepatocytes and peripheral blood mononuclear cells has been reported, the efficiency of replication and viral yield is low and difficult to reproduce. Infection models using closely related viruses, such as bovine viral diarrhea, virus infection of bovine kidney cells, and GB Virus-B virus (GVB-B) infection of tamarin and marmoset hepatocytes, have been developed as alternatives.[89,90] However, the ability of these surrogate systems to predict susceptibility of HCV to test compounds has not been convincingly demonstrated. The development of HCV replicons was a major breakthrough in providing a robust cell culture system for examining viral replication. These replicons are HCV RNA molecules capable of replicating autonomously in a human hepatoma cultured cell line Huh-7. The original system was developed in 1999 using a genotype 1b strain of HCV.[91] Since then, a number of replicon constructs have been reported, and they include different strains of genotypes 1a and 1b as well as chimeras of 1b and other genotypes.[92,93] The different systems all consist of the HCV 5′- and 3′-UTRs plus either the entire open reading frame (full-length) or only the NS sequences (subgenomic). Most are also engineered to express antibiotic resistant markers for selection and different reporter genes for convenient assays. Cell cultures that harbor HCV replicons are stable and support replication of both positive and negative strands of HCV RNA and express HCV proteins in a robust manner. However no viral particles are produced from the replicons. Furthermore, the ability of the HCV replicon to propagate in cell culture is inversely related to infectivity in animals. Most recently, a cell-based model of HCV virion production was described.[94] This system involves an RNA construct of two ribozymes flanking an HCV clone that is infectious in animals. When introduced into Huh-7 cells, HCV RNA is liberated from the ribozymes, followed by viral replication and release of virions. Although infectivity of these particles has not been established, this system can be a useful tool for studying virus assembly and secretion. Separately, the viral entry has been investigated using pseudotyped viruses, which are particles displaying HCV envelope proteins E1 and E2 assembled with retroviral core particles.[77,78,80,94] These engineered viruses are infectious in human hepatoma cells, and they may be a useful tools to study cell attachment and uptake of HCV.

A convenient small animal model for HCV infection is not available in spite of intensive efforts. The only primate model susceptible to HCV infection is chimpanzee,[83] whose use for drug discovery is prohibitive by both ethical and cost considerations. As an alternative, a surrogate virus GBV-B has been used to infect marmosets and tamarins.[89] Although GBV-B does not cause infection in humans, the GBV-B protease is similar to that of HCV such that in vivo efficacy of protease inhibitors can be assessed to a certain degree.

A murine HCV infection model with chimeric mouse–human livers may be useful for evaluating animal efficacy of anti-HCV compounds.[95] A breed of transgenic, immune-deficient mice was created that express a hepatotoxic transgene. The survival of the animals is accomplished by repopulating the livers by viable primary human hepatocytes after transplantation. The resulting animals have human and mouse hepatocytes in their livers; they are susceptible to HCV infection and support high level of HCV replication. However, the utility of this model is limited by the laborious nature and the requirement for specialized expertise.

7.13.2.4 Clinical Trial Issues

Most clinical trials have been conducted with interferon with or without ribavirin. Endpoints are biological (ALT, serum alanine aminotransferase) and virological (HCV RNA) responses as well as histopathological (liver damage). An essential criterion to define treatment success is the ability to achieve sustained virologic response (SVR) which is defined as continuous undetectable serum HVC RNA ($<50\,\mathrm{IU\,mL^{-1}}$) 24 weeks after treatment completion. These endpoints are likely to remain the standards for present and future clinical trails since they are consistent with the treatment goals of eradicating the virus, halting progression of fibrosis, reducing the chance of cirrhosis and hepatocellular carcinoma.

Genotype variation is an important determinant for interferon response, and it is likely to be an issue with agents that target HCV proteins at regions that allow genetic variability. Examples are HCV enzyme inhibitors that target mutationally more susceptible allosteric sites. Some factors that determine response to therapy include host race and immune condition, extent of liver disease, and viral load.[96,97] Thus, defining the patient selection criteria is critical to the evaluation of clinical candidates.

7.13.2.5 Treatment Options

Currently, the only approved treatment for HCV is interferon-α and interferon-α in combination with ribavirin, **22**. Interferon-α belongs to a family of low-molecular-weight proteins involved in the antiviral immune response. The antiviral mechanism of interferon-α is twofold. Its direct mechanism involves the stimulation of an RNase that degrades viral (and cellular) RNA and stimulation of PKR (double-stranded RNA dependent protein kinase R) which inhibits protein synthesis. The indirect mechanism is believed to involve the innate immune response and recruitment of cytotoxic T lymphocytes (**Scheme 4**).[82,98,99]

22 Ribavirin **23** Viramidine **24** Merimepodib

Scheme 4

Initially, HCV treatment involved interferon-α monotherapy. Due to a relatively short plasma half-life, interferon is dosed thrice weekly by subcutaneous injections. Treatment for 24 weeks with interferon-α resulted in a sustained virological response or SVR (HCV RNA undetectable 6 months posttreatment) of only ~10%. Doubling the treatment duration to 48 weeks resulted in an improved SVR of ~20%. Subsequent to that, pegylated interferon-α was introduced. This product consists of the protein conjugated to polyethyleneglycol (PEG). The primary advantage was pharmacodynamic; the increased half-life allows for once-weekly dosing. Improved SVR (25–39% depending on HCV genotype) is due to more uniform exposure with the pegylated protein relative to the extreme peak–trough exposures seen with interferon-α. Both interferon products are associated with similar side effects: flulike symptoms, neuropsychiatric manifestations, neutropenia, and thrombocytopenia that make treatment demanding and compliance difficult.

The next incremental improvement was the addition of ribavirin, **22**. Ribavirin is a nucleoside analog discovered in 1970 with a broad range of antiviral activities. This combination therapy was found to increase the SVR's to 44–47% with interferon-α and to 54–56% with PEG-interferon-α.[100,101]

The addition of ribavirin comes with its own side effects: primarily, but not exclusively, anemia. This, coupled with the side effects of interferon, results in 15–25% of patients ceasing treatment. The role of ribavirin in the sustained virologic response is not well understood.[102] Ribavirin is known to inhibit inosine monophosphate dehydrogenase (IMPDH) which results in guanosine nucleotide pool depletion.[103] It has also been speculated that ribavirin is incorporated (as its NTP) by HCV RNA-dependent RNA polymerase with a resulting increase in mutations, some of which will be lethal to HCV, a process termed 'error-catastrophe.'[104]

Regardless of the exact role exerted by ribavirin, the current state of treatment is PEG-interferon in combination with ribavirin. Numerous factors influencing successful outcomes have been identified, including HCV genotype, HCV viral load, and progression of liver disease among others. The serious adverse events and roughly 50% SVR have prompted a significant amount of effort into the discovery of new HCV therapeutic agents.

While most of the effort is being expended in discovering and developing direct antivirals, several potentially important improvements of interferon/ribavirin combinations are under development. Albuferon-α is a recombinant protein containing interferon-α fused to human serum albumin. Phase Ib/IIa clinical trials demonstrated that the median half-life of 142 h would support dosing once every 2–4 weeks. An early virological response (>0.5 log reduction at two or more consecutive time points) in over 50% of the patients has prompted its sponsor to evaluate the compound further in phase II studies.[105]

The ribavirin prodrug viramidine, **23**, is currently in phase III trials as a replacement for ribavirin in PEG-interferon-α/ribavirin combination therapy. Phase II data demonstrated that only 4% of the patients developed anemia in the viramidine arm compared to 27% in the ribavirin arm ($p<0.001$). While there was no statistical difference in SVR between the two arms, these data suggest that this prodrug may supplant ribavirin in future treatment.[106,107]

Furthermore, another inhibitor of IMPDH, mycophenolic acid derivate VX-497, **24**, is currently in phase II clinical trials in combination with PEG-interferon.[108,109]

7.13.2.6 Targets for Hepatitis C Virus Therapy

Two viral proteins have emerged as the prime targets for inhibiting viral replication, the NS3 protease and the NS5b RNA polymerase. In the following, only the most advanced classes of small molecule inhibitors are described. More detailed reviews on these and other targets are available.[110–117]

7.13.2.6.1 Protease inhibitors

Proper proteolytic processing of the HCV translation product is essential for release of the viral structural and nonstructural proteins. The C-terminal portion of the polypeptide with all nonstructural proteins is processed by viral proteases NS2/3 and NS3/4a.[118–120] The cleavage between NS2 and NS3 is an autocatalytic step,[121] releasing NS2 which by itself does not harbor a distinct proteolytic activity. Processing of the remaining polypeptide is carried out by the N-terminal domain of NS3 (180aa), which is a serine-protease of the chymotrypsin family (**Figure 5**). The larger, C-terminal domain (480aa) connected only by a single amino acid chain contains the RNA-dependent helicase and ATPase activity. Both domains have been shown to retain their catalytic activities when separated from each other.[122] On the other hand, both helicase and protease domains modulate the activity of the adjacent domain.[122–125] Another peculiar feature of the NS3 protease is that it requires NS4a (54aa) as a cofactor to achieve full catalytic activity. It is the 12 central amino acids of NS4a that complete the eight-stranded beta-barrel motif typically found in the N-terminal region of chymotrypsin-like proteases. This missing beta-strand can also be supplied as the 12mer peptide or engineered into the NS3 sequence as an addition to the N-terminus (the latter strategy produced the material used for

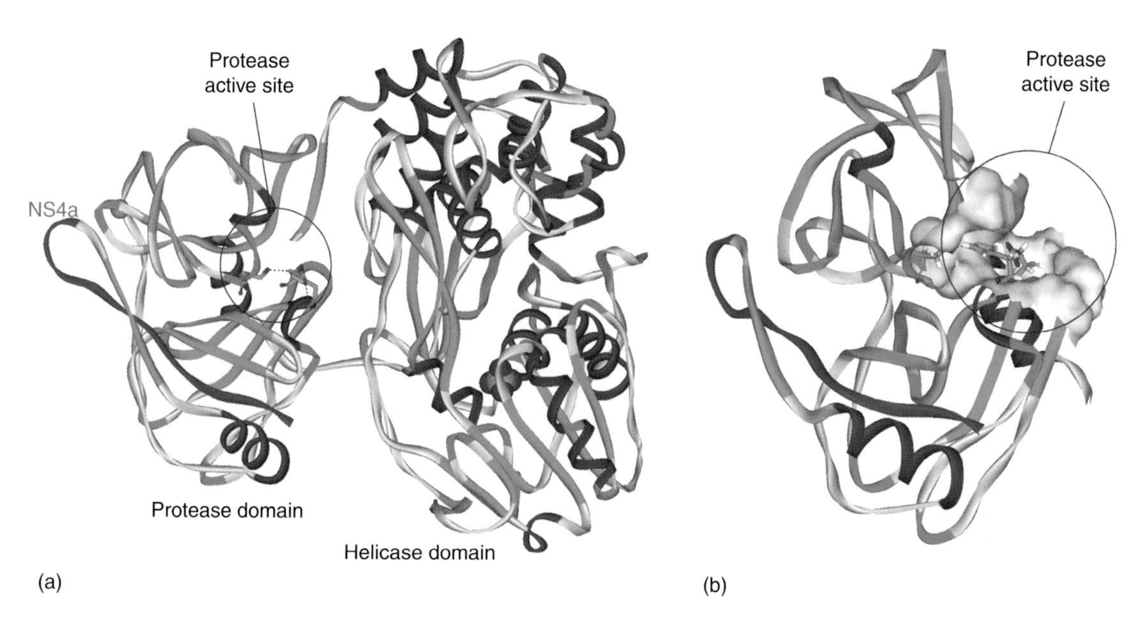

Figure 5 HCV NS3/4a protease (PDB: 1cu1). (a) Secondary structure depiction of the full-length NS3/4a protein. The NS4a cofactor is shown in magenta; active site residues S139, H57, and D81 are shown as sticks. (b)The exposed active site of the NS3 protease domain is shown as a solvent accessible surface with the three active site residues as sticks.

the crystal structure shown in **Figure 5**). Binding of cofactor NS4a reduces the plasticity of the protease and induces a much more tightly organized structure including the active site,[126] which is situated in a shallow, solvent-exposed cleft between the two beta-barrel motifs. Two of the amino acids of the catalytic triad, H57 and A81, reside on the N-terminal subdomain of the NS3 protease, while the third, S139, arises from the C-terminal portion. The proteolytic mechanism of a serine protease[127] is depicted in **Figure 6a**. The NS3 proteolytic cascade is initiated by the binding of cofactor NS4a, followed by the *cis* cleavage at the NS3/4a junction. The NS3/4a complex then carries out three more *trans* cleavage reactions to free NS4b, NS5a, and NS5b viral proteins. Although both proteases, NS2/3 and NS3/4a, are essential for viral replication,[128] NS3/4a has proven to be much more targetable than the elusive NS2/3.[129–131] The NS3/4a inhibitors have now reached the clinic.[132–134] Similar to the development of inhibitors of the HIV protease, early leads toward peptide-based NS3/4a inhibitors arose from the observation that the N-terminal, P-site products of the cleavage reaction had micromolar inhibitory activity, but not the P′-site peptides.[135–138] For example, the *N*-acetyl derivative of hexapeptide NH2-DDIVPC-COOH, based on the P6→P1 consensus for the HCV cleavage sites (**Figure 6c**), inhibits the truncated and full length NS3/4a protease with IC_{50} values of 39 and 800 μM, respectively.[137] These weak peptides ultimately led to the development of potent inhibitors with either a covalent ('serine traps') or noncovalent mode of interaction. The class of covalent inhibitors is largely composed of peptides bearing a reactive center at the cleavage site trapping the hydroxyl of S139, e.g., α-keto, aldehyde, boronic acid, or lactam moieties. The most advanced is an α-ketoamide VX-950, **25**.[139–141] Among the noncovalent inhibitors, the most successful ones to date are the macrocyclic tripeptides, such as BILN 2061, **26**, which arose from linking the P1 and P3 amino acid side chains.[133] Potent linear peptide inhibitors have been described as well, for which the activity is largely driven by extending to include P1′-groups and adding aromatic substituents to the P2-proline.[142,143]

7.13.2.6.1.1 Discovery and SAR of macrocyclic NS3 inhibitors: BILN 2061

Detailed accounts of the optimization of a weak lead into a low nanomolar inhibitor tell a remarkable story in which the nuclear magnetic resonance (NMR)-derived structures of bound, micromolar peptides were pivotal to success.[133,134,144,145] Starting with the hexapeptide, it was found early that the chemically undesirable thiol moiety of the P1-cysteine could be replaced with the *n*-propyl (norvalin) side chain without significant loss in activity.[137,146] However, the first large activity improvements came from substitutions on the P2-proline ring. Hexapeptide **27** exhibits an IC_{50} of 7.6 μM, almost 100-fold better than the lead peptide.[134,146] A similar hexapeptide (phenyl instead of naphtyl) was studied with transferred no observable effect (NOE) experiments,[133,147,148] in which the observed NOE intensities for a weakly binding ligand in the presence of protein reveal traits of the bound conformation. Comparison of

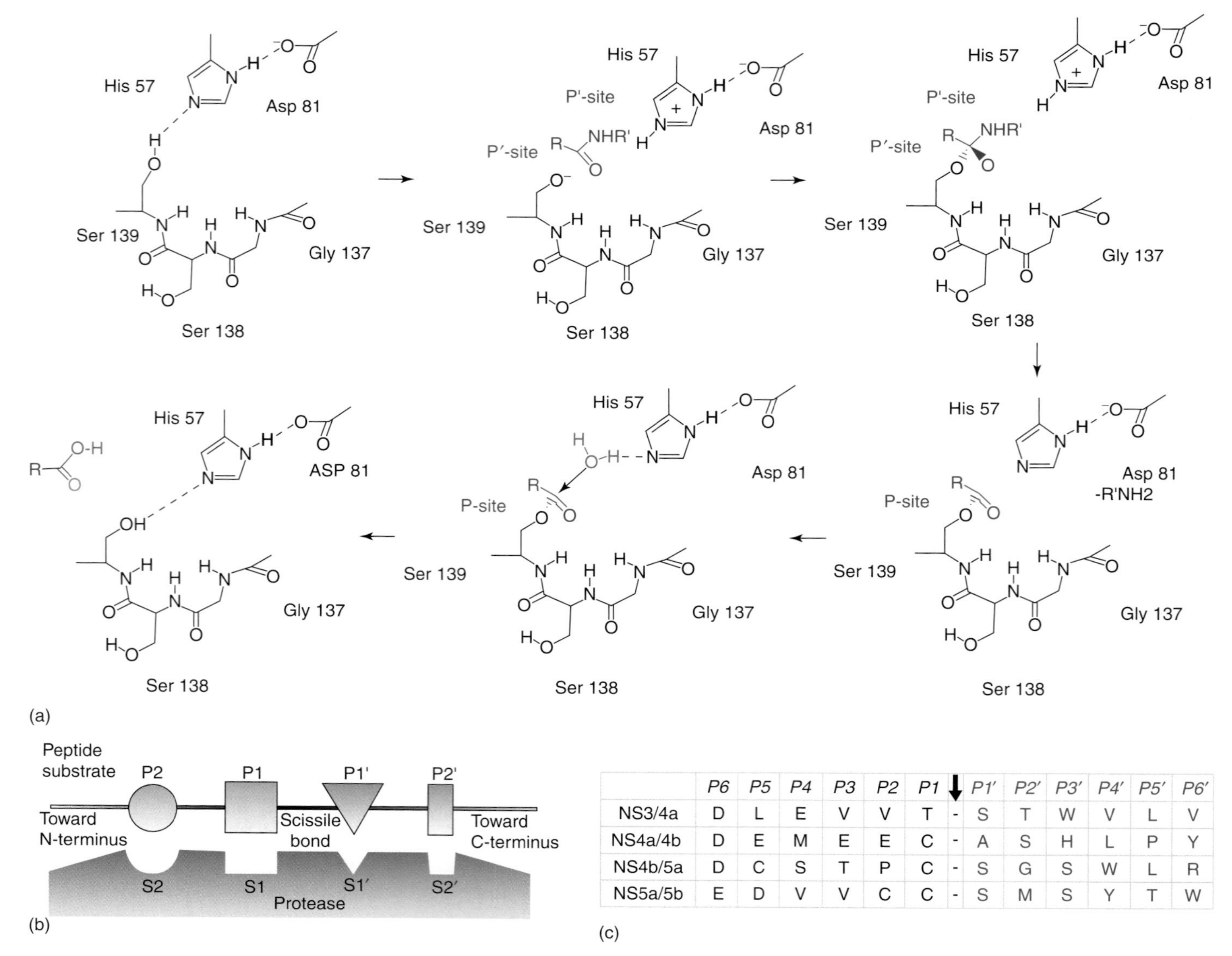

	P6	*P5*	*P4*	*P3*	*P2*	*P1*	⬇	*P1′*	*P2′*	*P3′*	*P4′*	*P5′*	*P6′*
NS3/4a	D	L	E	V	V	T	-	S	T	W	V	L	V
NS4a/4b	D	E	M	E	E	C	-	A	S	H	L	P	Y
NS4b/5a	D	C	S	T	P	C	-	S	G	S	W	L	R
NS5a/5b	E	D	V	V	C	C	-	S	M	S	Y	T	W

Figure 6 (a) Mechanism of a serine protease. (b) Nomenclature for proteases and (c) examples for peptide sequences from HCV that are cleaved by NS3/4a.

the structures in free and bound states revealed that the hexapeptide partially rigidifies upon binding, adopting an extended beta-strand conformation. Protease-induced differential line-broadening in the ^{1}H-NMR spectra indicate that the P5 and P6 side chains are largely solvent exposed, and the P3-valine side chain is solvent-exposed as well. Most importantly, the extended beta-strand conformation positions the P1 and P3 side chains into close proximity to each other. Modeling of the transfer NOE-derived ligand structure into the NS3 protease suggested that the P1 *n*-propyl group is folded inside the S1 pocket, placing the δCH_3 close to the αH of the P1 amino acid and the δCH_3 of the P3 valine (**Figure 7**).[147,148] Similar NOE-derived observations were obtained for P1–P4 tetrapeptides, and it was clear that in order to go from the free to the bound form a 180° rotation around the P1–NHCα bond had to take place. Therefore, a constrained scaffold where this conformation is fixed would avoid the entropic penalty and should be more potent. Compounds **27–31** illustrate the path to BILN 2061. Hexapeptide **27** barely loses potency when P5- and P6-aspartic acids are eliminated. The P1 *n*-propyl side chain can be replaced by the cyclopropyl group without loss of activity. Motivated by the NMR-derived model, small hydrophobic groups were added to P1 to improve interactions with the S1 pocket. This leads to (*1R,2S*)-vinyl aminocyclopropane carboxylic acid as the optimal residue, shown in compound **28** and **29**. For a number of tetrapeptides, the addition of the vinyl group improves activity 20- to 30-fold compared to the unsubstituted cyclopropane.[149] The IC_{50} values for **28** and **29** are 700 and 38 nM (**Scheme 5**), respectively. However, the cell based activity even for more potent **29** is still poor ($EC_{50} > 5\,\mu M$). The difference in potency between **28** and **29** also hints at the importance of the heteroaromatic P2-substituent. As suggested by the NMR-derived structure, connecting P1 and P3 sidechains with an additional 3-carbon chain leads to a dramatic improvement in the replicon activity. P3-capped cyclic tripeptide **32** exhibits an IC_{50} of 11 nM and an EC_{50} of 77 nM.[148] Note, the corresponding open chain tripeptide, **30**, is 36-fold less potent ($IC_{50} = 400$ nM) than **32**. As much as this cyclization is a corollary for potency, the P2-quinoline substituent is a profound modulator of activity, especially replicon activity. Cyclic tripeptide, **31**, not harboring any proline substituent, is barely active ($IC_{50} = 400\,\mu M$). Ultimately, fine-tuning of the P2-quinoline with additional small substituted heterocycles[133,145] and adjusting the P3 capping group yielded clinical candidate BILN 2061, **26** ($IC_{50} = 6.4$ nM, $EC_{50} = 3.0$ nM). The crystal structure of macrocycles **26** and **32** has been reported by several groups but no coordinates have been released.[133,150,151] Nevertheless, key binding interactions of this class of compounds can be gleaned from the NMR-based model structure of a similar compound bound to the NS3 active site (**Figure 8**).[144] The P1 carboxylate group engages in strong polar interactions. One of the carboxylate oxygens binds to the 'oxyanion hole' formed by the NH of G137 and S139, while the other forms hydrogen bonds with the ring nitrogen of H57 and the S139 OH. More consistent hydrogen bonds are observed between the A157 C=O and the P3 amino group and the R155 C=O and P1 NH. The aliphatic linker of the inhibitor clearly interacts with the S1/S3 pocket made up of hydrophobic residues A156, F154, A157, V132, and part of K136. With respect to the contribution of the P2-aromatic substituent, it appears that the quinoline is engaged in π–π stacking interactions with the guanidinium group of R155. In the crystal structure of **32** bound to NS3,[134,152] ligand movement of the R155 side chain leads to the formation of an interaction between the methoxy of the P2 quinoline and the guanidinium group, suggestive of more

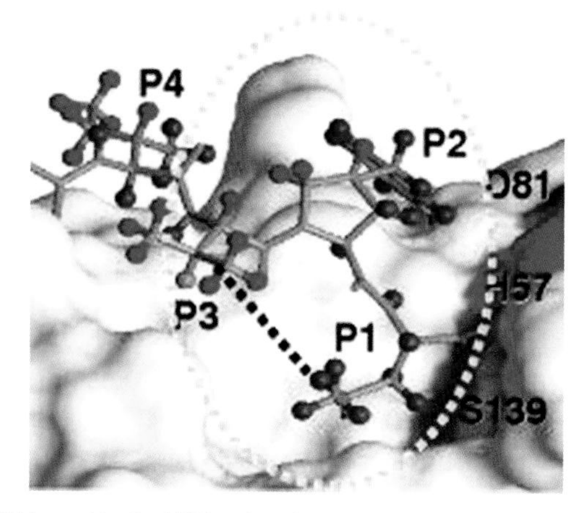

Figure 7 Model of hexapeptide **27** bound to the NS3 active site, generated by docking the NOE-derived bound conformation into the NS3/4a crystal structure (protons in blue indicate negative differential line broadening, protons in red indicate positive differential line broadening). Black dotted line indicates the proximity of P1 and P3 side chains which were linked in macrocycles. (Reprinted from Tsantrizs, Y. S. *Biopolymers* **2004**, *76*, 309–323, with permission of John Wiley & Sons, Inc.)

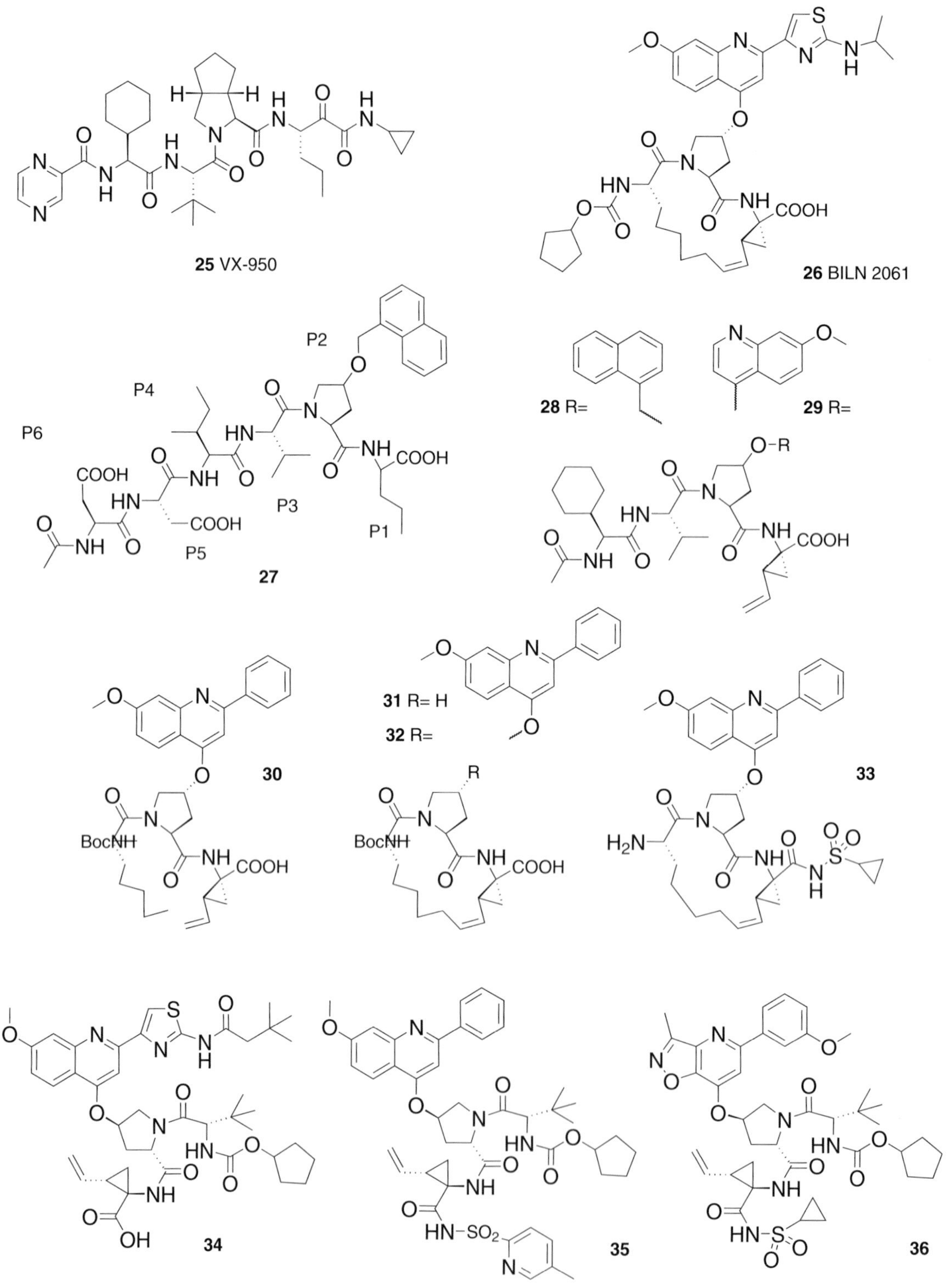

25 VX-950

26 BILN 2061

P2 P4 P6 P3 P5 P1

27

28 R=

29 R=

30

31 R= H

32 R=

33

34

35

36

Scheme 5

complicated electronic interaction (dipole/quadrupole interaction) of the P2 substituent with the protease. But most importantly, the primary interactions of these macrocycles with NS3 seen in the model and the crystal structure involve largely the residues conserved across all HCV subtypes. This explains why the IC_{50} values for BILN 2061 against proteases of HCV subtypes 2a, 2b, and 3a are still below 100 nM.[133]

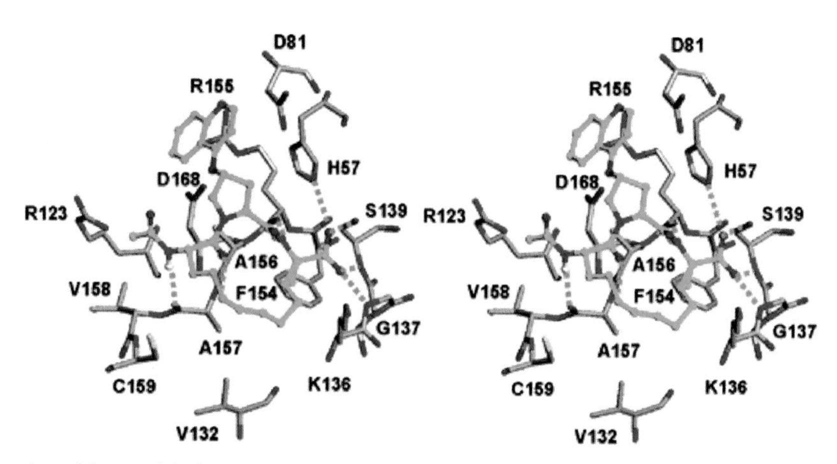

Figure 8 Stereoview of the model of a macrocyclic NS3-bound inhibitor similar to **32** (here, R is a quinoline): key interactions that may be implicated in hydrogen bonds with the active site residues are indicated with dashed orange lines. Carbon atoms of the inhibitor are colored in green, those of the protein in gray. Nitrogens and oxygen atoms are in blue and red, respectively. The sulfur atom of C159 is in yellow. (Reprinted from Goudreau, N.; Brochu, C.; Cameron, D. R.; Duceppe, J. S.; Faucher, A. M.; Ferland, J. M.; Grand-Maitre, C.; Poirier, M.; Simoneau, B.; Tsantrizos, Y. S. *J. Org. Chem.* **2004**, *69*, 6185–6201.Copyright (2004) American Chemical Society.)

The utility of peptidomimetic compounds is often challenged by poor pharmacodynamic parameters, i.e., short plasma half-life and high clearance rates.[153] Not surprisingly, many of the potent macrocycles faced these problems, but the plasma half-life and clearance was clearly modulated by the size and nature of the substituent of the 2-aminothiazole moiety. Using the isopropyl substituent led to acceptable pharmarokinetic parameters.[154] In rats, BILN 2061 exhibits a C_{max} of 2.5 μM, a plasma half-life of 1.3 h, and a clearance rate of 13 mL min^{-1} kg^{-1}. Oral bioavailability was 42% in rats; for dogs, values were as high as 38%.[132]

BILN-2061 yielded clinical proof of concept data as its outstanding in vitro activity was mirrored by a dramatic 2–3 log_{10} reduction of viral loads within 24 h in the majority of patients that had received **26** (200 mg in an oral solution of a PEG 400/ethanol 80 : 20 mixture, twice daily for 2 days).[132,155] While **26** was well tolerated up to 2000 mg and no serious clinical or laboratory findings were obtained in a human safety study, further phase II clinical trials are currently on hold because cardiac lesions were observed in routine chronic safety testing in monkeys at supratherapeutic doses.[152,156]

In terms of alternative designs for noncovalent peptide inhibitiors, it should be noted that the carboxylate group in **26** can be replaced by a cyclopropyl-acylsulfonamide moiety without loss in activity. A potent example of this class of macrocycles is compound **33**, which exhibits IC_{50} and EC_{50} values of 21 and 124 nM, respectively.[157] Note that an unprotected amino group and a shortened P1–P3 linker are compatible with good potency. Also unconstrained P2 proline-substituted tripeptides, **34–36**, retain good replicon activity.[142,143,157,158] Compound **35** shows that the carboxylate group can be replaced by a wide range of alkyl and especially substituted aryl- and heteroarylacylsulfonamides.[157] Compound **36** serves as an example of a large group of compounds with both IC_{50} and EC_{50} below 100 nM in which the heterocyclic proline-substituent assumes a range of substituted condensed ring systems.[157,158]

7.13.2.6.1.2 Discovery and SAR of covalent NS3 inhibitors: VX-950

Several years ago Merck researchers showed that capped P4–P1 peptides bearing an α-keto group inserted before the carboxylic acid can inhibit the NS3 protease by reacting with the S139 hydroxyl. Crystal structures of these ligand-protease adducts were readily obtained (PDB: 1dy8, 1dy9)[136,159] and showed the beta-strand orientation of the inhibitor, similar to what was observed by the Boehringer Ingelheim group. In a similar approach, Vertex researchers used N-capped P4–P1 peptides with an aldehyde group instead of the carboxylic acid as a starting point.[141] The potency of these aldehydes could be increased into the submicromolar range by using larger hydrophobic P2 substituents instead of the benzyl ether, as in compound **37**. Since aldehydes carry a general metabolic liability, a useful alternative was found with the α-ketoamide group.[160] While aldehyde **37** has an IC_{50} value of 12 μM (**Scheme 6**), α-ketoamides **38** and **39** exhibit IC_{50} values of 0.9 and 0. 1 μM, respectively. The respective replicon EC_{50} values of **38** and **39** are 4.8 and 0.9 μM.[140] In an attempt to find a balance between molecular size and potency, systematic optimization of the peptide side chains showed that the large P2 substituent was expendable as well as the P1′ group, resulting in compound **40** (IC_{50} = 0.15 μM). In a collaborative effort between Vertex and Eli Lilly,[161–163] among a series

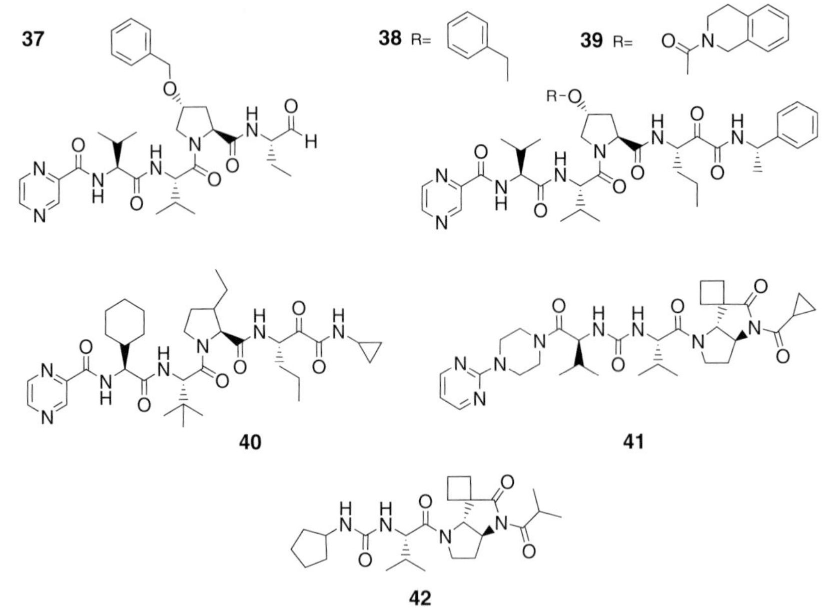

Scheme 6

of bicycloproline derivatives cyclopentylproline was found to be optimal, leading to VX-950, **25**. The latter was shown to bind slowly and covalently, yet reversibly to the NS3 protease ($K_i^* = 3$ nm, $t_{1/2} = 58$ min) with a multistep mechanism. The replicon EC_{50} value is 354 nM,[150] which is relatively high compared to the macrocycles described above, but this might be a consequence of the covalent mechanism. Nevertheless, in a cell-based viral clearance assay, VX-950 showed a ~1000-fold reduction of viral RNA levels at concentrations above 3.5 μM. The cocrystal structure of VX-950 has been reported[150] and shows that the binding area partially overlaps with the region accessed by BILN 2061.This explains why VX-950 retains full activity against NS3 mutants (e.g., D168V) that are more than 1000-fold less sensitive to BILN 2061.[150]

Pharmacokinetic studies showed a plasma half-life of about 1 h and in rats and dogs and the oral bioavailability was described as good.[164] VX-950 yields high liver exposure in animals and humans and showed promising data in a phase Ib clinical trial.[165] VX-950 was well tolerated in a group of patients infected with HCV subtype 1 that were either treatment naive or refractory to standard therapy. In this monotherapy clinical trial,[165] all three dose groups (450 mg 8qd, 750 mg 8qd, and 1250 mg 12qd over a 14-day period) experienced significant reductions in HCV-RNA; within 3 days of treatment, the median reduction in HCV-RNA was greater than 3 $\log_{10}$. For the 750 mg 8qd group, the median of reduction in HCV-RNA after 14 days was 4.4 $\log_{10}$.

A number of modifications that can be applied to the VX-950 scaffold without significant loss of potency have been described.[139,161]

Another class of covalent inhibitors is that of the pyrrolidine-*trans*-lactams, exemplified by **41** and **42**. For this class of inhibitors, the S139 hydroxyl opens the lactam ring by attaching itself to the carbonyl group, forming a hemiketal.[166–168] As older compounds in this series exhibited carbamate moieties at the N-terminus, more potent compounds favor the urea linker. There is crystallographic evidence that both amide hydrogens engage in hydrogen bonding to A157.[169] **41** exhibits replicon EC_{50} values of 300 nM, although it was found to be fivefold more active in the enzyme assay.[169] Due to the general chemical instability of the lactam moiety, compounds like **41** exhibit undesirable, fast clearance rates after intravenous administration. GW0014, **42**, however, although less potent, showed significantly improved in vivo pharmacokinetic properties. **Figure 9** shows the crystal structure of **42** when it is attached to the protease. This compound demonstrated antiviral activity in a surrogate animal model with GBV-B infected marmosets following subcutaneous administration.[166,168]

7.13.2.6.2 Ribonucleic acid polymerase inhibitors

Because HCV is an RNA virus, it must encode its own polymerase to replicate. The HCV nonstructural protein NS5b is an RNA-dependent RNA polymerase that is essential for viral replication and is highly conserved among the different

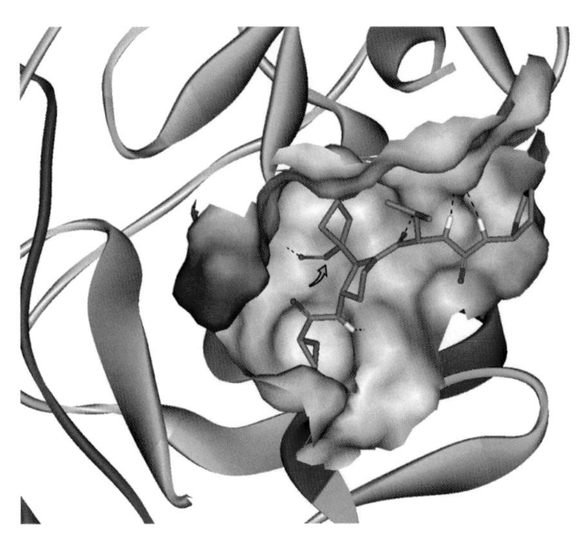

Figure 9 Crystal structure of the covalently bound lactam NS3 inhibitor **42** (PDB: 1rtl). The inhibitor is shown in stick representation while the protease is shown in secondary structure (coloring as in **Figure 5**) depiction with the active site as solvent accessible surface. (Surface coloring is by electrostatic potential: blue shows positive, red indicates negative potentials.) The arrow indicates where the attachment point is.

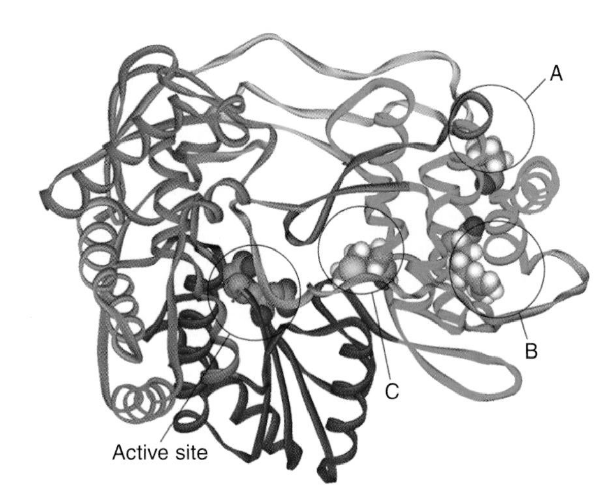

Figure 10 Location of active site and allosteric binding sites in NS5b, shown in secondary structure representation. The color coding emphasizes the 'handlike' domain organization of NS5b with the palm in red, fingers in blue, and the thumb in green. The fingerloops, connecting finger, and thumb domains are in cyan. Circles indicate active site (catalytic triad GDD is shown in spacefilling spheres) and allosteric sites: (A) benzimidazole site (P495 is shown in spacefilling spheres), (B) Shire/Pfizer site (R422 is shown in spacefilling spheres), and (C) benzothiadiazine site (M414 is shown in spacefilling spheres).

HCV genotypes. NS5b uses single-stranded HCV RNA as a template to initiate de novo RNA synthesis, and this process has no homologous mammalian counterpart. NS5b displays the hallmark finger–palm–thumb domain organization, typical for RNA polymerases (**Figure 10**). The unique structure of NS5b, its de novo mechanism of action, and the absence of a homologous mammalian enzyme suggested that this protein is a prime target for the discovery of potent and selective HCV antiviral agents. The medicinal chemistry strategies to target this polymerase are analogous to those employed for HIV reverse transcriptase, with parallel and complementary efforts in both nucleoside and nonnucleoside inhibitor discovery. The sequence of NS5b displays multiple amino acid polymorphism among the different HCV genotypes. The active site retains a high level of conservation among the different genotypes, while much amino acid variability occurs on the surface of the enzyme. The implications on drug discovery are profound. The anti-HCV nucleosides that have been discovered display equivalent activity against most genotypes, not surprisingly as they work in the highly conserved active site. To date however, nonnucleoside inhibitors have been

Table 2 Effect of NS5b enzyme constructs on kinetics and inhibition

	HT[a] -NS5b	*HT-NS5bΔ21*	*NS5bΔ57-HT*	*NS5bΔ21-HT*	*NS5b*
K_m (P-T)[b] (nM)	210	58	34	25	25
K_m (UTP) (nM)	6200	12000	1800	5200	3300
IC_{50} (nM)[c]	54	440	2200	3000	5700

Adapted from McKercher, G. *et al. Nucleic Acids Res.* **2004**, *32*, 422–431.
[a] His_6-tagged.
[b] Primer-template.
[c] IC_{50} of a benzimidazole type inhibitor, similar to **56**.

discovered that target three well-described allosteric inhibitor sites (**Figure 10**). Each site in each genotype has various susceptibilities to inhibition, raising the challenge of developing successful nonnucleosides that are broadly active against all the major HCV genotypes.

The enzyme contains a 21 amino acid hydrophobic C-terminus that localizes and anchors the polymerase to the endoplasmic reticulum. However, under enzymatic screening conditions this hydrophobic C-terminus causes several assay complications. Not only is the solubility of the wild-type protein decreased, but the presence of this tail also affects K_m for both RNA template and nucleotide substrates (**Table 2**). More soluble NS5b constructs are obtained via truncation of the C-terminal 21 or 55 amino acids. N- or C-terminal His_6 tags as purification handles further modulate the affinities. This has the effect, particularly in the case of some nonnucleoside inhibitors, of causing IC_{50} values to vary dramatically (**Table 2**).[170] Another factor influencing an inhibitor's IC_{50} is the choice of the RNA template (i.e., homopolymeric versus heteropolymeric RNA templates, with or without primers). Consequently, one must clearly understand the enzymatic assay systems of reported results to properly interpret SAR.

7.13.2.6.2.1 Anti-NS5b Nucleosides

Antiviral nucleosides are prodrugs that are actively transported into cells and then activated by cellular kinases to the nucleotide triphosphate (NTP). This NTP is now able to competitively inhibit the enzyme or, more commonly, act as a substrate and be incorporated into the nascent RNA chain. Chemical or structural features of the incorporated nucleotide subsequently prevent further replication of the viral genome (chain termination). While nucleoside inhibitors can vary in either (or both) the ribose or base portion of the molecule, the initially reported anti-HCV inhibitors were modified ribose analogs (**Scheme 7**).

In contrast to anti-HIV nucleosides which are primarily 2′-deoxyribose analogs, almost all the nucleosides targeted against HCV have retained the 2′-oxygen of ribose. One of the first nucleosides reported to be active against HCV NS5b was **43**, 2′-*O*-methyl cytidine.[171] Its corresponding triphosphate **44** was found to be active against the polymerase with an IC_{50} of 2.7 μM. Furthermore, incorporation of this nucleotide resulted in chain termination. In vitro the nucleoside **43** inhibited HCV replication with an EC_{50} of 21 μM and no cytotoxicity (CC_{50} > 100 μM). The poor replicon activity was attributable to low intra cellular levels of NTP **44**, (0.12 pmol per 10^6 cells), as well as extensive deamination to the uridine metabolite.

This situation highlights the difficulty of nucleoside drug discovery. As the nucleoside (e.g., **43**) is a prodrug of the active species (e.g., **44**), a successful nucleoside drug must be efficiently localized intracellularly and then be successfully metabolized to the NTP without being metabolized into unproductive (or toxic) species via the many nucleic acid metabolic pathways.

In contrast to the poor in vitro activity of **43**, 2′-methyl-adenosine, **45**, was found to be a potent replicon active compound.[171] The NTP of **45** displayed an IC_{50} of 1.9 μM against NS5b as a result of incorporation and chain termination. The nucleoside itself exhibited an EC_{50} of 0.3 μM with essentially no cytotoxicity (CC_{50} > 100 μM). This excellent activity correlates with the high concentration of NTP achieved intracellularly, 75 pmol per 10^6 cells. Prolonged treatment of replicon cells with **45** resulted in a mutation, S282T that conferred ~40-fold resistance.[172] The NTP of **45** is also ~40-fold less potent against the NS5b containing this point mutation (84 versus 1.9 μM). Kinetic analysis of this mutant revealed that while the S282T mutant is functional, it displays a reduced catalytic efficiency, that is, an increased ATP K_m and a decreased ATP k_{cat} relative to wild-type. Interestingly however, the 2′-*O*-methyl cytidine NTP, **44**, retains activity against this mutant.

Models of the initiation complex built with the apo-form of NS5b and the RNA portion taken from a related phage RNA polymerase[173] shed light into this resistance and the mechanisms of action of the 2′-methyl nucleosides.[172]

Scheme 7

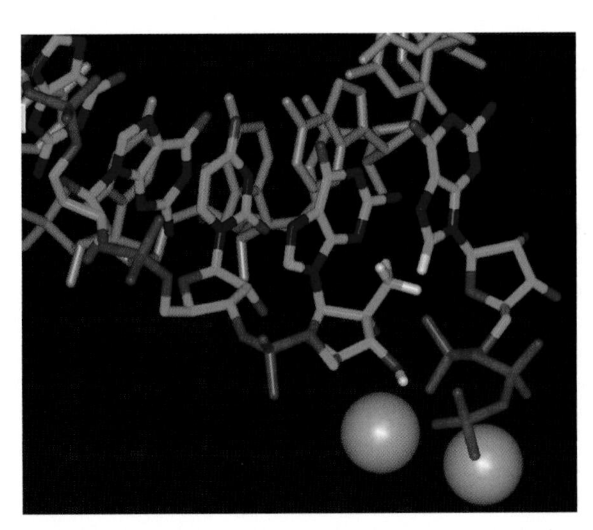

Figure 11 Model for the chain terminating mechanism of 2′-methyl nucleosides. This growing chain model was generated by attaching double-stranded RNA to the first two G:C base pairs found in the phi6 phage RNA polymerase (PDB: 1hi1) and energy minimization of the junction. The template strand is shown in gray. Carbon atoms of the growing RNA strand is shown in green. Only hydrogens of interest are shown. Magnesium atoms involved in catalysis are shown as magenta spheres.

Following the incorporation of a 2′-methyl nucleoside into the growing RNA chain, the subsequent NTP is sterically prevented from incorporation. The incorporated 2′-methyl blocks the approach of this NTP via steric repulsion of O-4′ and/or purine H-8 (**Figure 11**). In context with the polymerase, the model also shows that for the S282T mutation, the additional threonine methyl group occupies, at least to some extent, space otherwise occupied by the 2′methyl of the nucleoside.[172]

While **45** displays excellent potency in vitro its further development has not been pursued due to other liabilities, namely that it is quantitatively converted by adenosine deaminase to inosine as well as having slower glycosidic bond cleavage to adenine by purine nucleoside phosphorylase.[174] A further issue with this compound is its complete lack of in vivo oral bioavailibility (F% = 0) when dosed in rats. The NTP of 2′-methyl guanosine analog **46** was found to be a very potent enzyme inhibitor (EC_{50} = 0.15 μM). This guanosine analog was only moderately active in the replicon (IC_{50} = 3.5 μM), presumably due to the poor accumulation of intracellular NTP (0.2 pmol per 10^6 cells), but had excellent oral bioavailibility in rats (F% = 85). This stands in sharp contrast to the adenosine analog **45** which had potent in vitro anti-HCV activity but no bioavailibility in rats, further demonstrating the difficulty of nucleoside antiviral discovery.

The anti-HCV activity of the 2′-methyl nucleosides has been further explored by others with the pyrimidine (cytidine) analog **47**.[175–177] **47** was investigated, not with HCV, but with bovine diarrheal virus (BVDV), a well-characterized HCV surrogate. The BVDV NS5b polymerase was inhibited by the **47** NTP, **48**, with an IC_{50} = 0.74 μM and a K_i of 0.16 μM and was demonstrated to be a chain terminator as well. Treatment of BVDV infected cells resulted in an EC_{50} = 0.67 μM with no cytotoxicity observed (CC_{50} > 100 μM). Long-term exposure of BVDV infected cells to 16 μM **47** resulted in a 10^6 drop in BVDV RNA levels, and when used in combination with interferon-α, a synergistic antiviral effect was observed. Exposure of human hepatocytes to 10 μM extracellular **47** resulted in 10.7 μM of intracellular NTP with a half-life of 13.9 h, further correlating the observed in vitro activity to high concentrations of chain-terminating triphosphate. Due to the poor bioavailability of **47**, as well as its lack of patentability,[178] the 3′-*O*-valyl ester (**49**, Valopicitabine) was chosen as the prodrug that was moved forward into in vivo evaluation of both pharmacokinetics and efficacy in chimpanzees. As stated in the introduction, chimpanzees are the only nonhuman mammals in which chronic infection occurs. Animals were dosed once daily for a week with **49** at 10 and 20 mg kg^{-1} (8.3 and 16.6 mg kg^{-1} of **47** free base, respectively). Peak serum levels of 2.9–12.1 μM and trough levels of 0.2–0.4 μM were achieved. No adverse safety events were reported and approximately a 1 $\log_{10}$ drop in HCV RNA was observed at the end of the 1 week dosing.[177] Based on this encouraging data, **49** was advanced into clinical trials. Initial human clinical pharmacokinetics showed that the conversion of **49** to **47** in plasma occurred quickly, with a $t_{1/2}$ = 1.8 h. A phase Ib/IIa study in interferon/ribavirin nonresponding patients at 800 mg once daily treatment for 12 weeks showed an approximately 1.7 $\log_{10}$ drop in HCV RNA as monotherapy and a 3.0 $\log_{10}$ drop in HCV RNA when used in combination with interferon-α.[179]

In addition to 2′-methyl pyrimidine nucleoside **47**, several purine 2′-methyl nucleosides have been extensively characterized as well. Nucleoside **50** was discovered to be a potent inhibitor of NS5b both enzymatically and in vitro.[174,180] The NTP of **50** was found to inhibit the enzyme with an IC_{50} = 0.108 μM and a K_i = 0.024 μM. Similarly, **50** showed an EC_{50} = 0.3 μM with the HCV replicon with no cytotoxicity observed (CC_{50} = 800–1000 μM in Jurkat and replicon cells). As with the other 2′-methyl nucleosides, the NTP of **50** functioned as an efficient chain terminator. Additionally, **50** exhibits activity against other Flaviviridae and Picornaviridae viruses, members of RNA virus families that are closely related to HCV, such as BVDV, West Nile, dengue type-2, yellow fever, Rhinoviruses, and type-3 polio viruses, but not the more distantly related Western or Venezuelan equine encephalitis viruses. Similarly, an even more distantly related RNA virus, respiratory syncytial virus (RSV), was not affected. The NTP of **50** was completely inactive against human DNA polymerase α, β, and γ, a common cytotoxic mechanism with 2′-deoxy nucleoside drugs.

Extracellular dosing of the replicon cells with 2 μM of **50** for 23 h resulted in excellent levels of parent and NTP: 5.6 pmol per 10^6 cells of **50** and 3.4 pmol per 10^6 cells of **50** NTP. The 7-deaza modification of **50** protected it from metabolism by adenosine deaminase, as compared to the quantitative metabolism of the adenosine analog **45**. In vivo pharmacokinetic evaluation in rats, dogs, and rhesus monkeys showed $t_{1/2}$ = 1.6, 14, and 9 h, respectively, and oral bioavailability (F%) = 51, 98, and 51%, respectively. Importantly, the disease target organ, the liver, showed approximately 100-fold higher concentrations than the plasma exposures. Finally, in vivo toxicity was assessed by 14 day oral dosing in mice with a LD_{50} > 2000 mg kg^{-1}.

Other ribose analogs have also been discovered. **51** is reported to have an EC_{50} of 1.2 μM inhibitor of the HCV replicon.[181]

More recently, 2′-deoxy-2′-methyl-2′-fluorocytidine **52** has been reported to be active against HCV.[182] The NTP of **52** inhibits NS5b with an IC_{50} = 1.7 μM and the nucleoside inhibits the replicon (EC_{90} = 4.6 μM, as compared with **47** EC_{90} = 21.9 μM) with minimal cytotoxicity (CC_{50} > 100 μM). Interestingly, the NS5b S282T mutant is not resistant to 2′-methyl-2′-fluoro-CTP (IC_{50} = 2.0 μM). In vivo pharmacokinetic evaluation of **52** in rhesus monkeys following an oral dose of 33.3 mg kg^{-1} demonstrated a $t_{1/2}$ = 3.9 h and moderate bioavailibility (F% = 21).

The rapid discovery of multiple nucleosides that are active in vitro against HCV is now evolving to a situation where the first clinical validation is suggesting that nucleosides are likely to play an important role in future anti-HCV therapeutic strategy, as they are currently in anti-HIV therapy.

7.13.2.6.2.2 Anti-NS5b nonnucleosides

As with HIV therapy, anti-HCV nonnucleosides are anticipated to be complementary to nucleoside therapy. Consequently, a tremendous amount of effort has been expended in searching for these inhibitors. As described above, and in contrast to nucleosides, the variability of the amino acid sequences in the allosteric sites among genotypes has proven to be a very important issue with these compounds.

One of the first allosteric inhibitor sites was discovered in the fingerloop/thumb region, which is close to the secondary GTP binding site.[183] Several families of benzimidazole and indole inhibitors of this site (site A, **Figure 10**) have been discovered and optimized (**Scheme 8**).

Compound **53** was one of the first non-nucleoside inhibitors of HCV NS5b polymerase reported.[184] This it is a potent inhibitor of the enzyme ($IC_{50} < 0.006\ \mu M$) and active in the replicon ($EC_{50} = 0.5\ \mu M$).[185] This compound illustrates the inherent variability of the enzyme to assay conditions (described above). The reported value of 6 nM is from an enzyme assay that uses a protein construct with a higher K_m (i.e., lower affinity) toward the RNA templates. Furthermore, this assay uses a homopolymeric template with priming, resulting in a very potent IC_{50}. The same compound assayed by another group using a lower K_m enzyme and heteropolymeric RNA from the 3′-end of HCV resulted in a 20-fold shift in IC_{50} against NS5b ($IC_{50} = 0.12\ \mu M$) while the replicon value was essentially the same ($EC_{50} = 0.3\ \mu M$).[186] Compounds from this class are currently in early clinical trials in Japan.

Another group independently identified similar benzimidazole compounds from high-throughput screening (HTS) and subsequent optimization using ^1H NMR.[187] One of the early lead compounds, **54**, was reported to have an $IC_{50} = 1.6\ \mu M$; optimization of the carboxylic acid led to compound **55**. This compound was almost 100-fold more potent ($IC_{50} = 0.019\ \mu M$) than the initial lead NS5b inhbitor **54**.[188,189] However, this entire series of tryptophan benzimidazoles, while highly potent against the enzyme, were poorly active in the replicon assay, presumably due to the multiple carboxylate groups. However, cinnamic acid derivates without the central carboxyl group also showed some activity in the replicon assay; e.g., **56** exhibits an $IC_{50} = 0.055\ \mu M$ and an $EC_{50} = 1.0\ \mu M$. Most recently, a series of compounds have been described wherein the benzimidazole is replaced by an indole whose N-1 is alkylated with various substituents, e.g., compound **57**.[190] The binding mode (**Figure 12**) of this class of compounds within the allosteric site near P495 close to the secondary GTP binding site has recently been reported.[191] This patent reveals that this class of inhibitors interferes with the interaction between the tip of the fingerloops and the thumb domain, causing a change in protein structure and/or stability. This might lead to an altered RNA-binding of the NS5b polymerase and thus inhibition of initiation.[170]

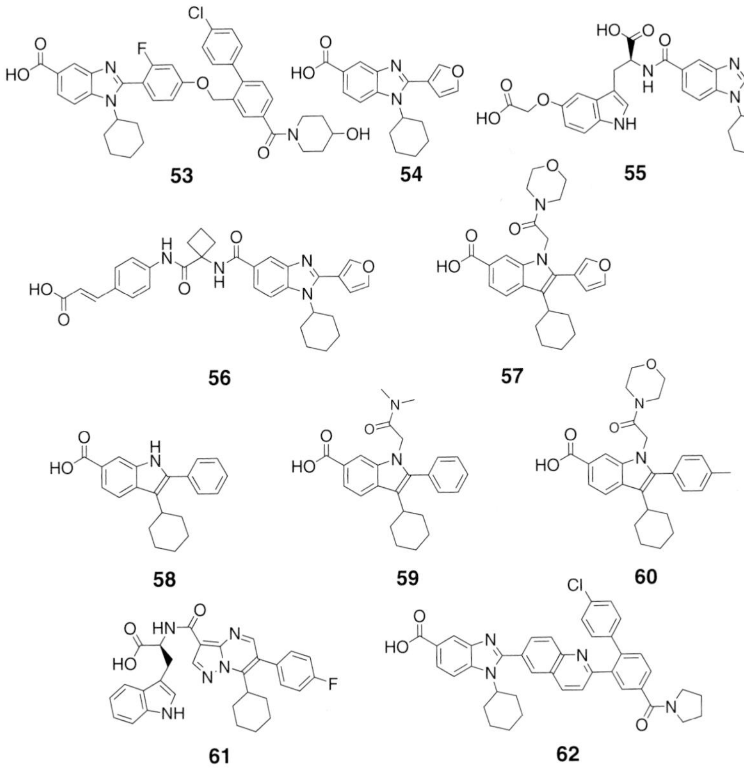

Scheme 8

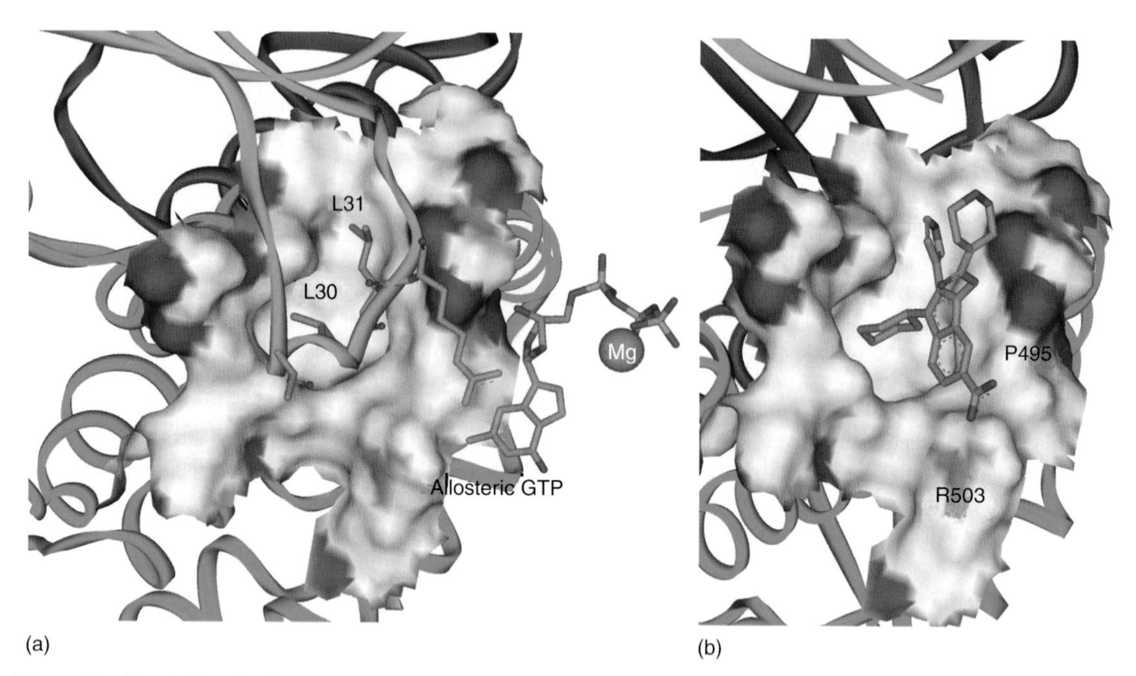

Figure 12 Benzimidazole allosteric binding site in fingerloop/thumb contact region of NS5b. (a) Apo-form of NS5b (PDB: 1gx5) in secondary structure representation (color code as in **Figure 10**). The area on the thumb to which the fingerloop binds (7 Å contact map) is shown as a solvent accessible surface with atom color coding. The fingerloop side chains of Leu30 and 31 are shown which bind in the hydrophobic pocket. For orientation, the location of the allosteric GTP is shown as well. (b) Polymerase inhibitor **57** docked into the hydrophobic pocket of the apo-form of NS5b after fingerloop residues 18 through 35, which are disordered in the reported crystal structure, have been removed. R503 is involved in salt bridge with compound **57**. P495, whose mutation leads to resistance to the benzimidazole class of inhibitors, packs tighlty against the six-membered portion of the indole moiety of **57**.

Indole derivative **58** was remarkably potent against the enzyme ($IC_{50}=0.048\,\mu M$) but only poorly active in the replicon ($EC_{50}=6.7\,\mu M$).[185] Addition at N-1 of a dimethylacetamido group **59** had minimal effect on the enzyme potency ($IC_{50}=0.059\,\mu M$), but improved the replicon activity ($EC_{50}=1.5\,\mu M$).

This compound was further profiled and shown to have no cytotoxicity ($CC_{50}>50\,\mu M$), and not to affect human DNA or RNA polymerases. In vivo pharmacokinetic analysis in rats showed a good half-life ($t_{1/2}=4.5\,h$), but high clearance ($Cl=32\,mL\,min^{-1}\,kg^{-1}$) due to formation of the acyl-glucoronide and moderate bioavailability (F% = 45). Evaluation in dogs resulted in full bioavailability, good half-life ($t_{1/2}=4.8\,h$) and a low clearance ($Cl=3.0\,mL\,min^{-1}\,kg^{-1}$), for in nonrodents glucoronide-mediated clearance is often less prevalent. While **59** showed promising pharmacokinetic properties, related compounds (like **60**) with improved antiviral activity ($IC_{50}=0.011\,\mu M$, $EC_{50}=0.3\,\mu M$) are being investigated further.[185]

Finally, other compounds that bind in this allosteric site have recently been reported. Compounds wherein the benzimidazole 6 + 5 ring system orientation is 'reversed' to 5 + 6, such as **61**, have been reported with excellent enzyme potency ($IC_{50}=0.02\,\mu M$).[192–194] A recently reported benzimidazole compound, **62**, wherein the flexible benzyl ether linkage of **53** is replaced by a constrained quinoline system, has displayed potent enzyme and replicon potency ($IC_{50}=0.095\,\mu M$, $EC_{50}=0.21\,\mu M$, respectively) (**Scheme 9**).[186,195]

A second allosteric binding site in the thumb region has been identified (**Figure 1**, site B) and inhibitors discovered. Using crystallography and structure-based design, compound **63** was synthesized and found to be a potent NS5b inhibitor ($IC_{50}=0.7\,\mu M$) and active against the replicon ($EC_{50}=0.2\,\mu M$). The crystal structure reveals the cyclohexyl ring binding tightly in a hydrophobic pocket near the beginning of the C-terminal tail.[196] Similarly, another group reported compound **64** bound to the same site with the cyclopentyl group binding to the same hydrophobic pocket with a nanomolar IC_{50} against NS5b (**Figure 13**).[197,198]

A third class of compounds, the pyranoindoles, also targeting this allosteric binding site has progressed into clinical trials. Compound **65** (HCV-371) was discovered through the optimization of an HTS hit.[199] It displayed a range of potencies against the NS5b enzymes from different HCV genotypes (**Table 3**).

63 **64** **65**

66 **67**

68 **69** **70**

Scheme 9

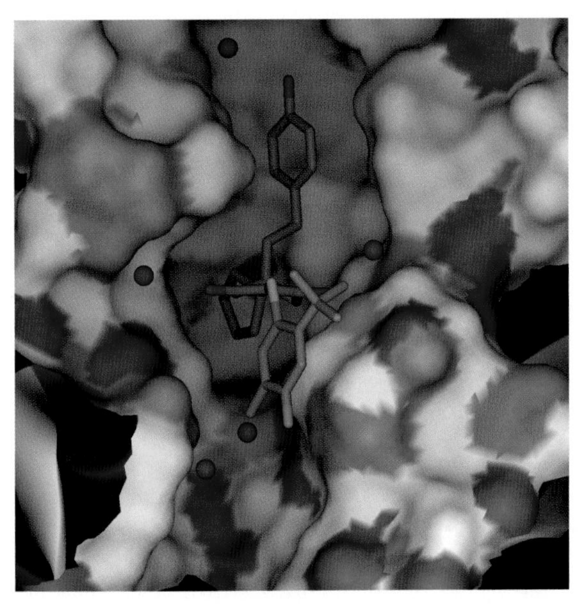

Figure 13 Crystal structure of inhibitor **64** bound to NS5b polymerase (PDB: 1os5). The binding site is shown as a solvent accessible surface. (The surface is colored by atom type. Color coding is as in **Figure 12**; additionally polar hydrogens are in pink.) Water molecules involved in adding to the inhibitor protein binding are shown as red balls.

Table 3 NS5b activity of pyranoindole **65** (HCV-371)

	Genotype 1a	*Genotype 1b*	*Genotype 3a*	*Genotype 4*
NS5b IC_{50}	0.6–4.8 μM ($n=5$)	0.3–0.5 μM ($n=8$)	1.8 μM ($n=1$)	17.8 μM ($n=1$)

The wide range of potency highlights the structural diversity of the different genotypes and the difficulty of discovering a pan-genotype inhibitor, particularly a nonactive site inhibitor. **65** was found to bind tightly to site B with a K_D of 0.15 μM and inhibited a genotype 1b replicon with an $EC_{50}=4.8$ μM and a genotype 1a replicon with $EC_{50}=6.1$ μM. The compound showed minimal cytotoxicity in four cell lines ($CC_{50}>160$ μM). Treatment of a 1b replicon for 20 days with 20 μM **65** ($5\times EC_{50}$) resulted in a 3 $\log_{10}$ drop in HCV RNA. The activity was found to be additive with pegylated interferon-μ.[199,200] This compound was taken into a phase I/II clinical trial where, although the compound was safe, it failed to reduce HCV RNA levels, presumably due to the marginal replicon potency.[201]

Another series of nonnucleoside inhibitors have been reported and characterized, and are being developed. Benzothiadiazine, **66**, was identified by HTS as a very potent enzyme and replicon inhbitor (NS5b $IC_{50}=0.08$ μM, replicon $EC_{50}=0.5$ μM).[202] Mechanistic investigation of this class of inhibitors reveals that they function by preventing the initiation of RNA synthesis.[203] This compound was used to generate resistant replicons wherein the key resistance mutation was mapped to M414T (**Figure 10**, site C).[113] This third allosteric site on NS5b is located in the interior of the protein, near the active site.

Separately, another group has identified and optimized the benzothiadiazine scaffold to potent compounds. Replacing the C-1 of the isopentyl tail of the benzothiadiazine with nitrogen opened up a new class of improved compounds.[204] Extensive modification led to compound **67** which was a potent inhibitor of NS5b. Whereas the early compounds of this scaffold suffered from activity against only 1b genotypes, addition of the acetamide ether functionality led to both 1b and 1a genotype activity (NS5b $IC_{50}=0.008$ and 0.033 μM, respectively). Importantly, when used against both 1b and 1a genotype replicons, the activity observed as the enzyme translated into in vitro activity as well ($EC_{50}=0.015$ and 0.136 μM, respectively). Finally, initial pharmacokinetic evaluation in rats (oral administration, 5 mg kg^{-1}) showed that the compound exhibits encouraging in vivo properties. Moderate bioavailibility (F% = 27.5) and a long half-life ($t_{1/2}=5.1$ h) were seen with a high exposure of drug in the liver observed ($C_{max}=22.8$ μg mL^{-1}). When compared to the plasma C_{max} (0.18 μg mL^{-1}), this greater than 100-fold concentration in the target organ warrants their further investigation.

Other nonnucleoside inhibitors have been investigated as well. An interesting class of active site targeting compounds that mimic the pyrophosphate of the incoming nucleotide triphosphates have been investigated.[205] Dihydroxypyridimine **68** was found to inhibit the enzyme with an $IC_{50}=5.8$ μM in a traditional assay that uses magnesium as an active site cofactor. However, when the divalent cation was changed to manganese, the potency of **68** improved to 0.6 μM. Presumably, this 10-fold shift is due to varying affinities and sensitivities of RNA catalysis by the two metals.

Though not successful to date, the strategy of targeting the active site with nonnucleosides is attractive. If this approach does succeed, the genotype specificity issue that plagues nonnucleosides could be circumvented, as the active site is much more highly conserved between genotypes.

Finally, a potent dichloroacetamide, **69** (R-803), was identified from screening the replicon. This binds to the enzyme with a $K_D=1.4$ μM. Resistant mutations suggest the compound might bind in an area on the thumb domain, which has strong interactions with the C-terminal tail.[206,207] This compound was evaluated in a phase Ib/IIa clinical trial and while safe, it failed to generate any reduction in HCV RNA levels due to no liver exposure.[208]

While NS5b polymerase inhibitors are still in the early stages of development, it is likely that the future of HCV therapy will involve polymerase inhibitors. Scenarios wherein they are initially used in combination with interferon and then with other antivirals, such as protease inhibitors or combinations of nucleosides and nonnucleosides, make this a continually exciting and competitive race to better therapeutics.

7.13.2.6.3 Nonviral targets

Similar to interferon, whose mechanism of action largely involves stimulation of the host immune response, several other compounds with the potential of broad-spectrum antiviral activity through immunomodulatory effects have found their way into the clinic. Most advanced is the 28mer peptide thymosin α1 (Zadaxine),[209] which is currently in a phase III clinical trial in the USA. Another interesting approach is presented by the guanosine derivative isatoribine, **70**,[210] and its prodrugs, which are slated to go into phase II clinical trials.

7.13.2.6.3.1 Thymosin α1

This *N*-acetylated peptide occurs naturally in the thymus (serum concentration in healthy adults $<1\,\mu g\,mL^{-1}$) and has immunomodulatory properties. The basis for using this peptide in chronic HBV and HCV infection is the observation that chronic infection develops only in patients with a weak cytotoxic T cell response, and, thus, cannot clear the virus. A relative deficiency of thymosin α1 is thought to contribute to the inadequate cytotoxic T cell response. The acidic thymosin α1 is chemically synthesized and has the following sequence: Ac–S–D–A–A–V–Asp–T–S–S–E–I–T–T–K–D–L–K–E–K–K–E–V–V–E–E–A–E–N–OH. Thymosin α1 is applied subcutaneously and reaches its fairly high peak concentration in serum within 1.3 h ($47\,\mu g\,mL^{-1}$ when dosed at $900\,\mu g\,m^{-2}$).[211] It clears swiftly with an elimination half-life of ~2 h. In terms of clinical efficacy, thymosin α1 was shown to be not effective as monotherapy in chronic HCV infection.[212] However, several small, yet promising trials showed that combination treatment with interferon yielded responses similar to those obtained with the combination of ribavirin and interferon,[213] the current first-line treatment for HCV. An important consideration for thymosin α1 is that it exhibits fewer and less severe side effects compared to ribavirin. Thymosin α1 has already been approved in several Asian and South American countries for the treatment of chronic HCV,[214,215] but it is more widely used for chronic HBV infection. SciClone and its European partner Sigma-Tau are currently conducting phase III trials using thymosin α1 in combination with pegylated interferon for the treatment of HCV in nonresponders.

7.13.2.6.3.2 Isatoribine

Isatoribine, **70**, was initially discovered in the late 1980s by ICN researchers, and was shown to have immunopotentiating properties in animals.[216–218] this and a few other 7,8-modified guanosine derivates[218,219] lacked direct antiviral activity but induced interferon production. In various animal models of viral infection (not including HCV), **70** demonstrated protection from infection and/or lethality.[216] Lacking a clear mechanism of action or a molecular target, this class of compounds was hard to develop. However, recently, it was demonstrated that **70** interacts with the toll-like receptor 7 (TLR7),[210] which is involved in detecting the presence of viral RNA after a virus has invaded a cell via endocytosis. Recent work has elucidated the role and mechanism of TLR7 mediated antiviral immune response, and it is clear now that TLR7 stimulates production of interferon and other cytokines via a complex signaling cascade.[220–222] Anadys Pharmaceuticals is now developing **70** and two prodrugs (ANA971 and ANA975) with increased oral bioavailability for HCV.[223] **70** was well tolerated in healthy volunteers, and in a separate phase Ib trial, **70** showed a statistically significant decrease in viral load at the end of 1 week in the 12 HCV infected patients in the 800 mg cohort.

7.13.2.7 Future Outlook

Research to discover drugs for the treatment of HCV and specifically to discover drugs that target HCV viral specific proteins is presently at high intensity both in academia and especially in the pharmaceutical industry. The discovery of a cell-based model for HCV viral replication and compound screening by Bartenschlager and coworkers in late 1999 has been pivotal to the rapid expansion and success in discovering new potential HCV drugs. Compounds targeting both the HCV protease and HCV polymerase enzymes have been discovered and optimized in several companies. The first compounds targeting these proteins have entered early phase Ib/II clinical trials and demonstrated proof of principle efficacy in reducing HCV viral titers in patients. There is still a long way to go clinically in the evaluation of these compounds, but optimism is high that an NDA approval for a specific viral targeted small molecule will be achieved before the end of the decade.

7.13.3 Respiratory Syncytial Virus

Human RSV is a single-stranded, negative RNA virus of the Paramyxoviridae family. RSV has been recognized as the major cause of severe lower respiratory tract infections in young infants worldwide, and it is the most common pathogen leading to hospital admission in children under 5 years of age.[224,225] RSV is also considered a widespread pathogen responsible for nosocomial infections, threatening the elderly, immunocompromised patients, and bone-marrow transplant patients.[226] Despite over 50 years of RSV research, there is still no effective treatment nor the prospect of a suitable vaccine.

7.13.3.1 Replicative Cycle and Virus Structure

RSV enters epithelial cells through fusion with the cell membrane and replicates in the cytoplasm of infected host cells using its own viral polymerase for transcribing and replicating its ~15 kb nonsegmented RNA genome. After

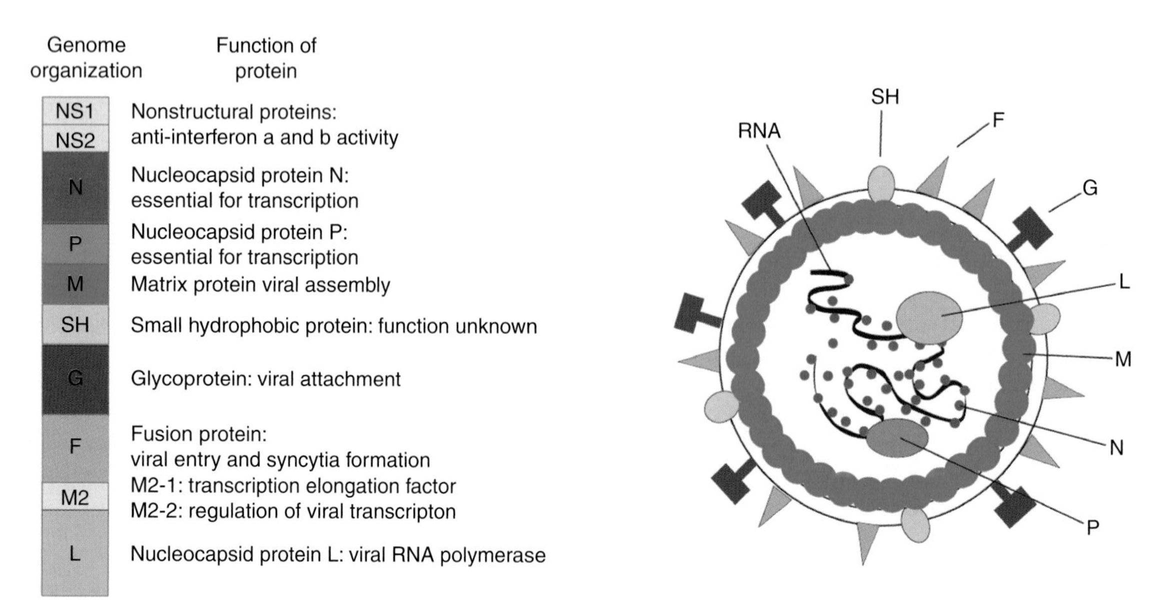

Figure 14 Genome organization of respiratory syncytial virus (RSV), function of viral proteins, and structure of the virus.

production of sufficient viral proteins by the host's translation machinery, viral assembly is coordinated by key viral proteins. RSV budding appears to be a reverse penetration of the cell membrane and occurs through the apical membrane providing the virus with its lipid membrane. In a unique way, RSV budding virions can also fuse immediately with the membrane of neighboring cells, thereby leading to cell-to-cell fusion with cytoplasmic continuity.[224,227] RSV is named after this unique process of syncytia formation.[228]

Figure 14 depicts the genome organization of RSV, the function of the gene products, and the structure of the virus. The genome encodes ten major proteins, three of which are displayed on the viral surface (F, G, and SH proteins), anchored to the lipid membrane. Besides matrix proteins M and M2, three other proteins (nucleocapsid proteins N, L, and P) bound directly to the genomic RNA, are packaged into the virion. Additionally, RSV exhibits two nonstructural proteins (NS1 and NS2) that accumulate in infected cells but are packaged into virions only at trace concentration. The M2 gene contains a second open reading frame, encoding protein M2-2, which governs the transition form transcription of viral RNA to the production of genomic RNA. Matrix protein M is thought to play the role of the coordinator of the assembly of the envelope proteins F and G with the nucleocapsid proteins N, P, and L.

7.13.3.2 Transmission and Epidemiology

RSV infections have occurred worldwide and epidemics can last up to 5 months, typically starting in late fall and lasting through early spring. RSV is highly contagious and can attack up to 50% of infants during an epidemic. The infection is spread through oral and nasal fluids.[228]

The highest rates of RSV illness occur in infants 2 to 6 months old. In older children and adults, RSV causes only mild upper respiratory symptoms, much like the 'common cold,' but for children under 3 years, symptoms much more severely impact the lungs (bronchiolitis or pneumonia) and can lead to respiratory failure. A typical, self-limiting infection lasts 7 to 14 days. Children who are hospitalized with lower respiratory tract infection usually spend 5 to 7 days in the hospital. In the USA, about 0.5 to 2% of infected children (up to 75 000) are hospitalized due to RSV and up to 1900 die each year.[228,229] Most vulnerable to serious RSV infections are children born prematurely, as their lungs are immature. The cost of RSV-related hospitalization is significant considering the vulnerable age of the infected and the fact that mechanical ventilation is often required ($2.6 billion for the years from 1997 to 2000 in the USA alone).[230]

RSV infection generally does not produce lasting immunity. As children age, reinfection with RSV is common but usually less severe. It should be noted that RSV is responsible for 60–80% greater mortality than influenza during a typical winter season.[227,231]

Similar to influenza, preventing RSV infection with vaccines would be most desirable and seems logical. However, vaccine development for RSV has been severely hampered in part by major problems with the initial vaccine trials many years ago.[232] The early trial vaccine did not just fail to protect from infection but it also caused an increase in the severity of the infection ensuing from the next exposure. Despite the difficult immunological situation which is beyond the scope of this chapter, RSV vaccines are still being pursued vigorously and reviews can be found elsewhere.[233–237]

7.13.3.3 Treatment Options

The only medication currently approved by the FDA to treat RSV infection is the broad-spectrum antiviral ribavirin (**22**, virazole), which is administered topically as an aerosol. Similar to the situation with HCV and influenza, the precise mode of antiviral activity of ribavirin is not entirely clear.[102]

Although not a treatment per se, passive immunization with RSV-neutralizing antibodies has become a useful option, especially for the prophylactic treatment of the high-risk patient groups.[227] In 1996, the FDA approved RespiGam, a polyclonal RSV immunoglobulin derived from human plasma, which is administered though intravenous infusion. Two years later, Synagis (palivizumab, MEDI493) was approved, largely replacing the inferior RespiGam. Synagis is a recombinant, humanized mouse immunoglobulin monoclonal antibody directed against the RSV fusion protein and is administered by monthly intramuscular injections.[238] Synagis was fairly potent in the prophylactic cotton rat model of RSV infection and several clinical trials with high-risk pediatric patients showed that the rate of hospitalization is roughly cut in half compared to the placebo group.[239] However, the clinical studies do not suggest that RSV infection was less severe among RSV hospitalized patients who received Synagis. Beyond this prophylactic use, it was shown recently that Synagis also exhibited a therapeutic effect when it was administered to immunosuppressed cotton rats harboring RSV infection.[240,241] Nevertheless, lacking vaccines and a real treatment, RSV infection represents a critical unmet medical need.

7.13.3.4 Experimental Drugs: Fusion Inhibitors

The process of virus to cell membrane fusion has been well elucidated[242,243]; and for RSV, the F protein has emerged as the prime target for therapeutic intervention, both as the epitope for antibodies and vaccines as well as a target for small-molecule antiviral agents.[227] Similar to fusion proteins of other viruses (HIV, influenza), the F protein of RSV is a *N*-glycosylated polypeptide of 67 kDa (F_0) which requires activation to its fusion-competent form. The latter consists of two disulfide-linked subunits, F_1 and F_2, which arise from proteolytic cleavage by host cell endoproteinases. A schematic of the RSV F protein in **Figure 15** shows how this activation exposes the fusion peptide, a short hydrophobic, glycine-rich segment, at the N-terminus of the F_1 subunit. The transmembrane domain is close to the C-terminus of the F_1 subunit. Adjacent to both the fusion peptide and transmembrane domains are two regions containing highly conserved, hydrophobic heptad repeats (HR1 and HR2), that were shown to form trimeric hairpinlike structures [243] similar to gp41 of HIV.[244] **Figure 15** depicts the crystal structure obtained from cocrystallization of the HR1 (57aa) and HR2 (45aa) segments, which together form a stable trimer of heterodimers.[243] The three HR1 45mer peptides pack in an antiparallel manner against the long hydrophobic grooves formed on the surface of the core helix bundle consisting of the three HR2 57mer peptides. Extrapolating back to the biological context of the activated form of the F protein, this observed packaging mode puts the fusion peptide segment, located at the N-terminus of HR1, in close proximity to the transmembrane segment at the C-terminus of HR2. This observed structure confirmed the notion that RSV fusion occurs in a fashion similar to HIV and other viruses.[242,245] After unmasking of the fusion peptide and its insertion into the host cell membrane, the powerful conformational change driven by the helix bundle formation of the two heptad repeat segments accomplishes the apposition of the viral and host cell membranes, and ultimately the merging of viral and host cell membranes occurs. Inhibition of the formation of the coiled coil helix bundle should lead to potent antiviral activitiy; examples are discussed below.

7.13.3.4.1 Peptidic fusion inhibitors

In the case of HIV, potent peptidic fusion inhibitors have been derived from the sequences of the viral envelope glycoprotein, gp41,[244,246–248] which is the equivalent of the RSV protein F. The anti-HIV drug Fuzeon (*see* 7.12 Ribonucleic Acid Viruses: Antivirals for Human Immunodeficiency Virus) is a highly alpha-helical, natural peptide (36aa) that binds to the first heptad repeat (HR1) in the gp41 subunit of the HIV envelope glycoprotein and prevents the formation of the coiled coil helix bundle necessary for the fusion of HIV and cellular membranes.[249] Fueled by the success of Fuzeon, researchers at Trimeris identified peptide sequences from the RSV HR2 segment with potent antiviral activity. The most potent potent one, T118, comprises the HR2 sequence from 488 to 522 and exhibited an IC_{50} of 51 nM in a cell culture assays.[250] T118 has the following sequence: Ac-F-D-A-S-S-I-S-Q-V-N-E-K-I-N-Q-S-L-A-F-I-R-K-S-D-E-L-L-H-N-V-B-A-G-K-S-T.

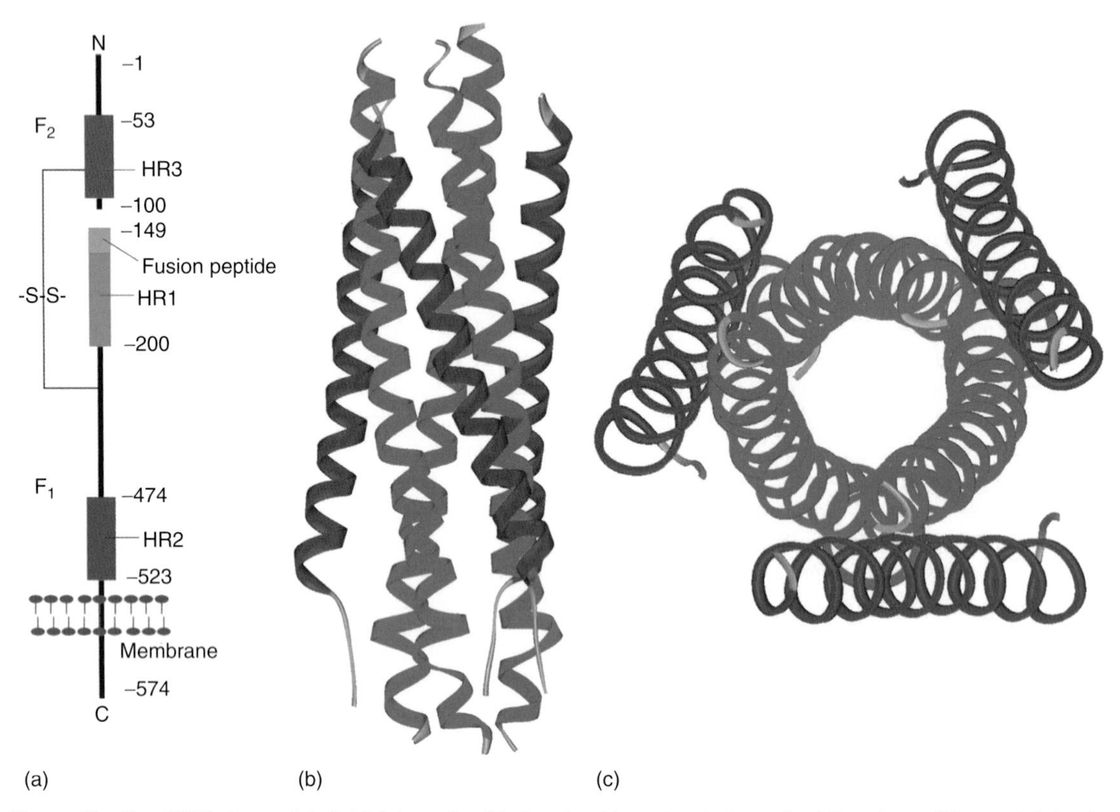

Figure 15 The RSV fusion protein F. (a) Schematic of its functional two-domain form. (b–c) Two views of the crystal structure (PDB:1g2c) of the coiled coil obtained through cocrystallization of a 57mer peptide from HR1 (green) and a 45mer peptide from HR2 (red). In (b) the N-termini of the HR1 57mer helices are pointing toward the top of the page and those of the HR2 45mer helices toward the bottom. (c) View along the threefold symmetry axis of the trimer of hairpins.

In light of above crystal structure, one can assume that T118 blocks the interaction of the HR2 helices with the HR1 trimeric helical core and thereby disables formation of the fusogenic state. However, none of the peptidic fusion inhibitors has yet been pursued in the clinic.

7.13.3.4.2 Small-molecule fusion inhibitors

7.13.3.4.2.1 BMS-433771

Researchers at Bristol-Meyer Squibb identified benzotriazole **71** (BMS-433771) as a potent inhibitor of RSV replication in a high-throughput tissue culture screen. **71** protected HEp-2 cells from the RSV-induced cytopathic effect with an EC_{50} of 0.34 μM, while the CC_{50} cytotoxic concentration was 84 μM (**Scheme 10**).[251,252] The protective effect of **71** was observed only for RSV and not for related viruses, i.e., influenza, parainfluenza-3, or HIV. Initial optimization of this screening hit led to a number of potent compounds ($EC_{50} < 30$ nM), but due to their poor metabolic stability and unsuitable pharmacokinetic parameters (poor oral bioavailability), it was hard to obtain proof of their in vivo antiviral potency.[253] Eventually, proof of concept was obtained using a cotton rat model of RSV infection in which a highly water-soluble representative of this class, **72**, was administered topically. Much of the chemical optimization[254,255] was geared toward improving oral bioavailability and the metabolic instability associated with the benzimidazolone moiety. As a result, azabenzimidazolone BMS-433771, **73**, emerged as a potent (average $EC_{50} = 20.4$ nM), highly selective antiviral with low toxicity ($CC_{50} > 218$ μM) and an excellent pharmacokinetic profile.[253] Mechanism of action studies quickly revealed that **73** interferes with the fusion process. Resistant mutant viruses were generated with this class of compounds and one particular mutation emerged consistently, K394R, in the F_1 subunit. The direct binding site of **73** in the F protein was mapped using the radiolabeled, photoaffinity probe **74**.[253,256] Upon Ultraviolet activation, the diazirine moiety of **74** generates highly reactive carbene species that can insert covalently into proximal amino acids in the binding site. Peptide mapping revealed that **73** labeled specifically residue Y198 in the N-terminal HR1 domain of

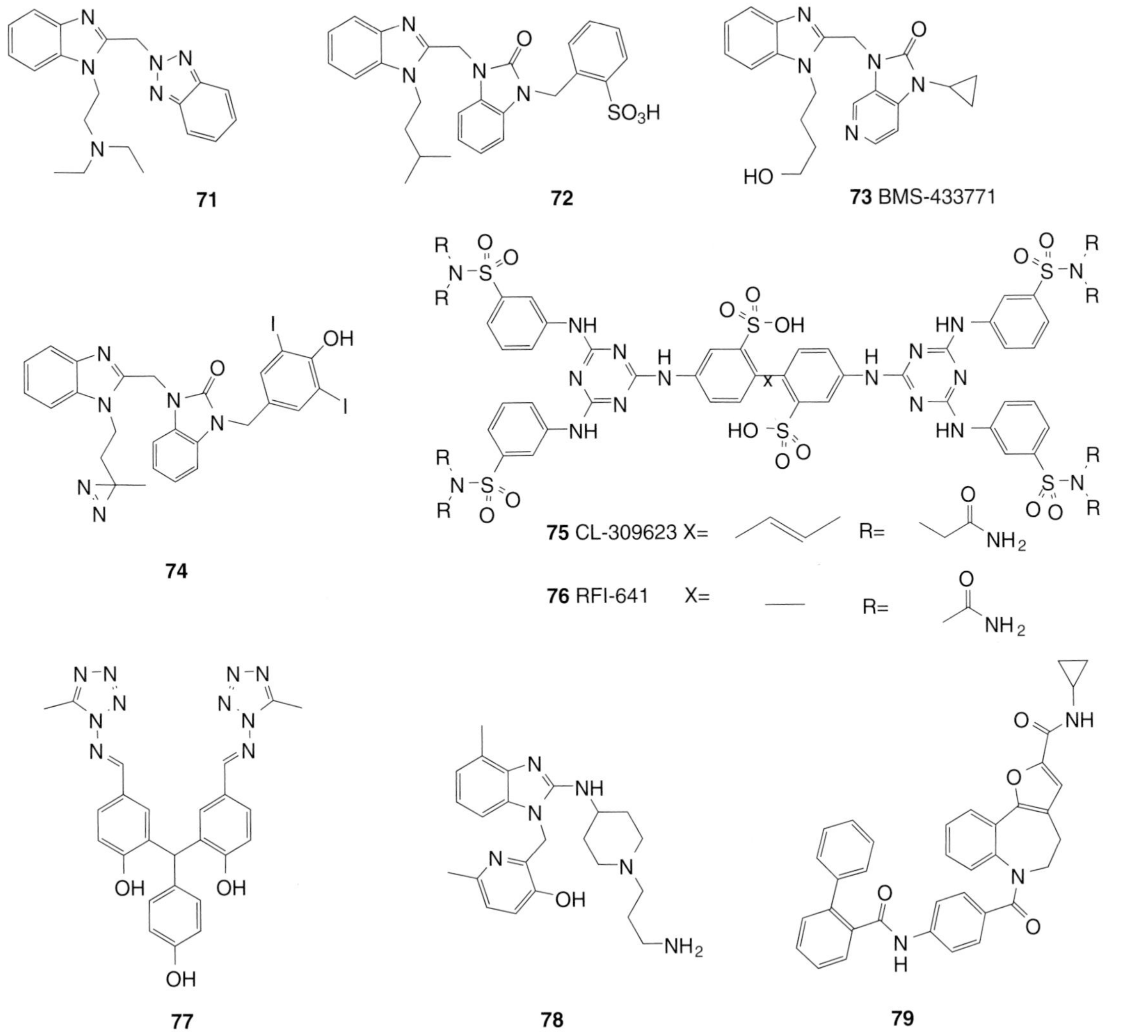

71

72

73 BMS-433771

74

75 CL-309623 X= R=

76 RFI-641 X= R=

77

78

79

Scheme 10

the F_1 subunit (**Figure 15**). This tyrosine residue is found at the edge of a deep hydrophobic pocket formed from the packing of the inner three helices (HR1, green in **Figure 15**). From the crystallographic work discussed above, this pocket had been suggested to be a candidate for small-molecule targeting since it is fairly deep and hosts a number of key interactions between the HR1 core helices and the outer HR2 helices.[243] **Figure 16** shows a model of **73** bound to this pocket and thereby displacing one of the outer HR2 helices.[256] Note that the threefold symmetry of the HR1/HR2 helix bundle leads to the hydrophobic pocket being present three times as well, thereby enabling amplification of the inhibition by **73**.

The in vivo efficacy of **73** was tested in a BALB/c mouse model of RSV infection, by administering the compound at 50 mg kg^{-1} in a single oral dose 1 h or 5 min before inoculation or 1 h postinoculation. Only prophylactic treatment was efficacious, reducing infectious lung titers in most animals to the assay detection limit.[251] Possible explanation for the lack of efficacy in the group in which treatment started 1 h postinoculation could be related to the pathogenesis of the virus in this particular animal model and/or the mechanism of **73** consisting of inhibition of the early step of membrane fusion. A similar discrepancy between prophylactic and treatment efficacy was also seen for other small-molecule fusion inhibitors (see below), which makes it difficult to extrapolate from RSV animal models to humans.

Bristol-Myers Squibb is not taking BMS-433771 forward into the clinic, but it has made the compound available for licensing.[257]

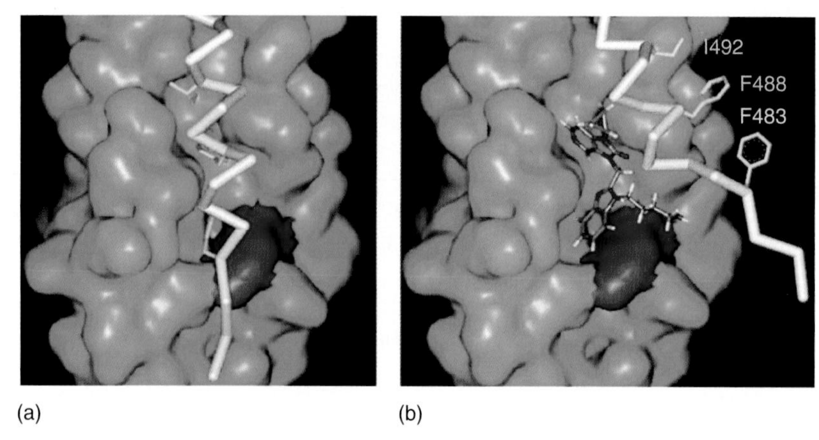

Figure 16 Model for the binding of BMS-433771 and the mechanism of inhbition. (a) Interaction of one HR2 helix (white stick display; F483, F488, and I492 side chains highlighted in yellow) with the hydrophobic pocket on the core three-helix bundle of the HR1 helices (green surface display; Tyr198, the attachment point of the inhibitor, is shown in red). The HR2 amino acids F483 and F488 bind in the HR1 hydrophobic pocket. (b) BMS-433771, docked into the hydrophobic HR1 pocket, blocks the HR2 F483, F488, and I492 associations. (Reproduced from Cianci, C.; Langley, D. R.; Dischino, D. D.; Sun, Y.; Yu, K. L.; Stanley, A.; Roach, J.; Li, Z.; Dalterio, R.; Colonno, R. *et al. Proc. Natl. Acad. Sci. USA* **2004**, *101*, 15046–15051. Copyright (2004) National Academy of Sciences, USA.)

7.13.3.4.2.2 RFI-641

Researchers at Wyeth discovered highly potent CL-309623, **75**, in a cell-based screening campaign for anti-RSV compounds.[258] It had been synthesized initially in the 1950s as a brightener for industrial applications[259] and defies all principles of drug-likeness. The compound was resynthesized and derivatives showed clear SAR leading to RFI-641, **76**. RFI-641 inhibits a variety of clinical and laboratory strains of RSV A and B with average IC_{50} values of 0.055 and 0.018 $\mu g\,mL^{-1}$, respectively, without exhibiting significant cytotoxicity.[260] By means of a fluorescence-quenching assay applied to a wild-type and a mutant virus that contained only the fusion protein on its surface, it was shown that this class of compounds inhibits virus binding and fusion with the host cell membrane[261,262] as well as syncytium formation by direct interaction with the F protein.[260] Cell-based time of addition studies indicated that the compound must be present before the fusion event takes place. RFI-641 showed antiviral activity in three different, prophylactic animal models, utilizing intranasal administration in a mouse model, a cotton rat model, and a African green monkey model of RSV infection. No efficacy was seen when RFI-641 was administered to mice via the intravenous route. Results for therapeutic models where RFI-641 is administered 1 h postinoculation are difficult to interpret. In the cotton rat model, there was no therapeutic effect of RFI-641 when administered postinfection.[263] However, when African green monkeys were treated 12–24 h postinfection via intranasal administration of RFI-641, viral titers were reduced in both nasal and throat washings.[260] It has been noted that the intranasal form of RFI-641 used in the animal models would not provide good exposure of the lungs and therefore an inhaled formulation[264] has been developed for phase II clinical testing.

7.13.3.4.2.3 VP-14637

Researchers at Viropharma discovered the triphenol VP-14637, **77**, as a low nanomolar antiviral agent through HTS.[265,266] This compound also showed in vivo activity in a cotton rat model of RSV infection. More recently a more detailed characterization of this compound and the benzimidazole, **78**, has been reported.[267,268] VP-14637 inhibits RSV replication and RSV fusion with EC_{50} values of 1.4 nM and 5.4 nM, respectively. Furthermore, VP-14637 binds to RSV-infected Hep-2 cells with a K_D of 9.3 nM. To identify the protein target of VP-14637, a recombinant vaccinia virus expressing only the RSV F protein at the surface at levels comparable to wild-type RSV was used to demonstrate that RSV protein F is necessary and sufficient for low nanomolar binding. Further proof of the fusion targeting mechanism involving protein F came from the selection and characterization of VP-14637 resistant mutants. Two independent mutations give rise to over 1000-fold reduced sensitivity to VP-14637; the first, F488Y, located in the HR2 segment of the F_1 subunit (the outer layer of the six-helix bundle), and the second, T400A, in the large linker region between HR1 and HR2 (**Figure 15**). Initially, these results were interpreted such that VP-14637 would bind to protein F in a transient conformation encountered during the fusion process.[267] However, the fact that F488 is an integral part of the intricate network of interactions in the hydrophobic pocket at the bottom of HR1, the binding site of BMS-433771, led to the suggestion that VP-14637 could indeed bind to this particular pocket and that the F488Y mutation in HR2 alters the helical conformation such that the HR2 coils can bind with VP-14637 in the pocket.[268] VP-14637 exhibits poor solubility and oral bioavailability. Nevertheless, phase I clinical studies have been completed, in which the compound was administered by inhalation.[269] However, the FDA has raised issues with the particular formulation which to date have not been resolved.

7.13.3.4.2.4 R170591

The Janssen Research Foundation identified benzimidazole R170591 (**78**, JNJ-2408068), through a cell-based assay as potent inhibitors of fusion of RSV-infected HeLa cells.[268,270] This compound shows in vivo efficacy in the prophylactic version of cotton rat model of RSV infection. Further R170591 inhibits RSV replication and RSV fusion with EC_{50} values of 2.1 and 0.9 nM, respectively, and binds to RSV-infected Hep-2 cells with a K_D of 2.9 nM.[268] Resistant mutations against this compound were not identical to the ones for VP-14637, but very similar; E487D, D486N, and K399I. It has been proposed that the mechanism of action of **78** is the same as that for VP-14637 described above. The oral bioavailability for the highly basic **78** is poor and the developmental status for this compound is unclear.

7.13.3.5 Miscellaneous Other Experimental Drugs

Recently, researchers at Yamanouchi Pharmaceuticals discovered a number of potent benzazapine carboxamides in a large-scale random screening assay based on inhibition of the cytopathic effect of RSV-infection on HeLa cells.[225] YM-53403, **79**, was the most potent exhibiting an EC_{50} of 200 nM and did not inhibit a number of other RNA viruses. Time-of-addition studies demonstrated that **79** inhibits the RSV life cycle at a later phase than the fusion inhibitors described above, implicating an effect on early transcription and/or replication of the RSV genome. Consistent with this notion, **79**-resistant mutants have a single point mutation (Y1631H) in the L nucleocapsid protein, which is the RNA polymerase.

7.13.3.6 Future Outlook

In the current clinical practice, with safe vaccines still unavailable, emphasis is on temporary immunization with monoclonal antibodies of high-risk patient groups. Although there is a clear benefit in the reduction of RSV-related hospitalization and presumably reduction of infant mortality, actual treatment methods with superior effects than the current, cumbersome ribavirin inhalation therapy are urgently needed. The inhibition of virus–host cell fusion has emerged as a great area for antiviral intervention with small molecules, but the vulnerability of the target population as well as the potential impact of future vaccines and improved temporary immunization methods have been slowing the development of small-molecule drug candidates.

References

1. Cunha, B. A. *Infect. Dis. Clin. North Am.* **2004**, *18*, 141–155.
2. Laver, G.; Bischofberger, N.; Webster, R. G. *Sci. Am.* **1998**, *280*, 78–87.
3. Mould, J. A.; Paterson, R. G.; Takeda, M.; Ohigashi, Y.; Venkataraman, P.; Lamb, R. A.; Pinto, L. H. *Dev. Cell* **2003**, *5*, 175–184.
4. Imai, M.; Watanabe, S.; Ninomiya, A.; Obuchi, M.; Odagiri, T. *J. Virol.* **2004**, *78*, 11007–11015.
5. Bridges, C. B.; Harper, S. A.; Fukuda, K.; Uyeki, T. M.; Cox, N. T.; Singleton, J. A. *Morb. Mortal. Wkly Rep.* **2003**, *52*, 1–36.
6. Kaiser, J. *Science* **2004**, *306*, 394–397.
7. Elliott, M. *Phil. Trans. R. Soc. Lond. B* **2001**, *356*, 1885–1893.
8. McClellan, K.; Perry, C. M. *Drugs* **2001**, *61*, 263–283.
9. Mendel, D. B.; Tai, C. Y.; Escarpe, P. A.; Li, W.; Sidwell, R. W.; Huffman, J. H.; Sweet, C.; Jakeman, K. J.; Merson, J.; Lacy, S. A. et al. *Antimicrob. Agents Chemother.* **1998**, *42*, 640–646.
10. Sidwell, R. W.; Huffman, J. H.; Barnard, D. L.; Bailey, K. W.; Wong, M. H.; Morrison, A.; Syndergaard, T.; Kim, C. U. *Antiviral Res.* **1998**, *37*, 107–120.
11. Schmidt, A. C. *Drugs* **2004**, *64*, 2031–2046.
12. Stiver, G. *Can. Med. Assoc. J.* **2003**, *168*, 49–56.
13. Dumyati, G.; Falsey, A. R. *Drugs Aging* **2002**, *19*, 777–786.
14. Fagan, H. B.; Moeller, A. H. *Am. Fam. Physician* **2004**, *70*, 1331–1332.
15. Laver, W. G. *Perspect. Biol. Med.* **2004**, *47*, 590–596.
16. Wade, R. C. *Structure* **1997**, *5*, 1139–1145.
17. Meanwell, N.; Serrano-Wu, M. H.; Snyder, L. B. *Annu. Rep. Med. Chem.* **2003**, *38*, 213–228.
18. Colman, P. M.; Varghese, J. N.; Laver, W. G. *Nature* **1983**, *303*, 41–44.
19. Varghese, J. N.; Laver, W. G.; Colman, P. M. *Nature* **1983**, *303*, 35–40.
20. Burmeister, W. P.; Ruigrok, R. W.; Cusack, S. *EMBO J.* **1992**, *11*, 49–56.
21. Chong, A. K.; Pegg, M. S.; Taylor, N. R.; von Itzstein, M. *Eur. J. Biochem.* **1992**, *207*, 335–343.

22. Smith, B. J.; Colman, P. M.; von Itzstein, M.; Danylec, B.; Varghese, J. N. *Protein Sci.* **2001**, *10*, 689–696.
23. von Itzstein, M.; Wu, W. Y.; Kok, G. B.; Pegg, M. S.; Dyason, J. C.; Jin, B.; Van Phan, T.; Smythe, M. L.; White, H. F.; Oliver, S. W. *Nature* **1993**, *363*, 418–423.
24. Taylor, N. R.; Cleasby, A.; Singh, O.; Skarzynski, T.; Wonacott, A. J.; Smith, P. W.; Sollis, S. L.; Howes, P. D.; Cherry, P. C.; Bethell, R. et al. *J. Med. Chem.* **1998**, *41*, 798–807.
25. Smith, P. W.; Sollis, S. L.; Howes, P. D.; Cherry, P. C.; Starkey, I. D.; Cobley, K. N.; Weston, H.; Scicinski, J.; Merritt, A.; Whittington, A. et al. *J. Med. Chem.* **1998**, *41*, 787–797.
26. Smith, P. W.; Robinson, J. E.; Evans, D. N.; Sollis, S. L.; Howes, P. D.; Trivedi, N.; Bethell, R. C. *Bioorg. Med. Chem. Lett.* **1999**, *9*, 601–604.
27. Lew, W.; Chen, X.; Kim, C. U. *Curr. Med. Chem.* **2000**, *7*, 663–672.
28. Kim, C. U.; McGee, L. R.; Krawczyk, S. H.; Harwood, E.; Harada, Y.; Swaminathan, S.; Bischofberger, N.; Chen, M. S.; Cherrington, J. M.; Xiong, S. F. et al. *J. Med. Chem.* **1996**, *39*, 3431–3434.
29. Kim, C. U.; Lew, W.; Williams, M. A.; Wu, H.; Zhang, L.; Chen, X.; Escarpe, P. A.; Mendel, D. B.; Laver, W. G.; Stevens, R. C. *J. Med. Chem.* **1998**, *41*, 2451–2460.
30. Kim, C. U.; Lew, W.; Williams, M.; Liu, H.; Zhang, L.; Swaminathan, S.; Bischofberger, N.; Chen, M. S.; Mendel, D. B.; Tai, C. et al. *J. Am. Chem. Soc.* **1997**, *119*, 681–690.
31. Babu, Y. S.; Chand, P.; Bantia, S.; Kotian, P.; Dehghani, A.; El Kattan, Y.; Lin, T. H.; Hutchison, T. L.; Elliott, A. J.; Parker, C. D. et al. *J. Med. Chem.* **2000**, *43*, 3482–3486.
32. Chand, P.; Kotian, P. L.; Dehghani, A.; El Kattan, Y.; Lin, T. H.; Hutchison, T. L.; Babu, Y. S.; Bantia, S.; Elliott, A. J.; Montgomery, J. A. *J. Med. Chem.* **2001**, *44*, 4379–4392.
33. Kati, W. M.; Montgomery, D.; Maring, C.; Stoll, V. S.; Giranda, V.; Chen, X.; Laver, W. G.; Kohlbrenner, W.; Norbeck, D. W. *Antimicrob. Agents Chemother.* **2001**, *45*, 2563–2570.
34. Stoll, V.; Stewart, K. D.; Maring, C. J.; Muchmore, S.; Giranda, V.; Gu, Y. G.; Wang, G.; Chen, Y.; Sun, M.; Zhao, C. et al. *Biochemistry* **2003**, *42*, 718–727.
35. Wang, G. T.; Chen, Y.; Wang, S.; Gentles, R.; Sowin, T.; Kati, W.; Muchmore, S.; Giranda, V.; Stewart, K.; Sham, H. et al. *J. Med. Chem.* **2001**, *44*, 1192–1201.
36. Wang, T.; Wade, R. C. *J. Med. Chem.* **2001**, *44*, 961–971.
37. Kati, W. M.; Montgomery, D.; Carrick, R.; Gubareva, L.; Maring, C.; McDaniel, K.; Steffy, K.; Molla, A.; Hayden, F.; Kempf, D. et al. *Antimicrob. Agents Chemother.* **2002**, *46*, 1014–1021.
38. Wang, G. T.; Wang, S.; Gentles, R.; Sowin, T.; Maring, C. J.; Kempf, D. J.; Kati, W. M.; Stoll, V.; Stewart, K. D.; Laver, G. *Bioorg. Med. Chem. Lett.* **2005**, *15*, 125–128.
39. Williams, M.; Bischofberger, N.; Swaminathan, S.; Kim, C. U. *Bioorg. Med. Chem. Lett.* **1995**, *5*, 2251–2254.
40. Singh, S.; Jedrzejas, M. J.; Air, G. M.; Luo, M.; Laver, W. G.; Brouillette, W. J. *J. Med. Chem.* **1995**, *38*, 3217–3225.
41. Atigadda, V. R.; Brouillette, W. J.; Duarte, F.; Babu, Y. S.; Bantia, S.; Chand, P.; Chu, N.; Montgomery, J. A.; Walsh, D. A.; Sudbeck, E. et al. *Bioorg. Med. Chem. Lett.* **1999**, *7*, 2487–2497.
42. Atigadda, V. R.; Brouillette, W. J.; Duarte, F.; Ali, S. M.; Babu, Y. S.; Bantia, S.; Chand, P.; Chu, N.; Montgomery, J. A.; Walsh, D. A. et al. *J. Med. Chem.* **1999**, *42*, 2332–2343.
43. Finley, J. B.; Atigadda, V. R.; Duarte, F.; Zhao, J. J.; Brouillette, W. J.; Air, G. M.; Luo, M. *J. Mol. Biol.* **1999**, *293*, 1107–1119.
44. Bantia, S.; Parker, C. D.; Ananth, S. L.; Horn, L. L.; Andries, K.; Chand, P.; Kotian, P. L.; Dehghani, A.; El Kattan, Y.; Lin, T. et al. *Antimicrob. Agents Chemother.* **2001**, *45*, 1162–1167.
45. Sidwell, R. W.; Smee, D. F.; Huffman, J. H.; Barnard, D. L.; Bailey, K. W.; Morrey, J. D.; Babu, Y. S. *Antimicrob. Agents Chemother.* **2001**, *45*, 749–757.
46. Sidwell, R. W.; Smee, D. F. *Expert Opin. Invest. Drugs* **2002**, *11*, 859–869.
47. Young, D.; Fowler, C.; Bush, K. *Phil. Trans. R. Soc. Lond. B* **2001**, *356*, 1905–1913.
48. Dines, G. D.; Bethell, R.; Daniel, M. *Drug Saf.* **1998**, *19*, 233–241.
49. Ryan, D. M.; Ticehurst, J.; Dempsey, M. H. *Antimicrob. Agents Chemother.* **1995**, *39*, 2583–2584.
50. Gubareva, L. V.; Kaiser, L.; Hayden, F. G. *Lancet* **2000**, *355*, 827–835.
51. Maring, C.; McDaniel, K.; Krueger, A.; Zhao, C.; Sun, M.; Madigan, D. L.; DeGoey, D.; Chen, J.; Yeung, C.; Flosi, W. et al. *Antiviral Res.* **2001**, *50*, A76.
52. Zielinski Mozny, N. A.; Wilson, S. T.; Maring, C.; McDaniel, K.; Kati, W.; Kempf, D.; Carrick, R.; Steffy, K.; Molla, A.; Kohlbrenner, W. et al. *40th Meeting ICAAC*, Washington, DC, **2001**, Abs H1583.
53. Makela, M. J.; Pauksens, K.; Rostila, T.; Fleming, D. M.; Man, C. Y.; Keene, O. N.; Webster, A. *J. Infect.* **2000**, *40*, 42–48.
54. Aoki, F. Y.; Macleod, M. D.; Paggiaro, P.; Carewicz, O.; El Sawy, A.; Wat, C.; Griffiths, M.; Waalberg, E.; Ward, P. *J. Antimicrob. Chemother.* **2003**, *51*, 123–129.
55. Whitley, R. J.; Hayden, F. G.; Reisinger, K. S.; Young, N.; Dutkowski, R.; Ipe, D.; Mills, R. G.; Ward, P. *Pediatr. Infect. Dis. J.* **2001**, *20*, 127–133.
56. Cheer, S. M.; Wagstaff, A. J. *Drugs* **2002**, *62*, 71–106.
57. McKimm-Breschkin, J. L.; Blick, T. J.; Sahasrabudhe, A.; Tiong, T.; Marshall, D.; Hart, G. J.; Bethell, R. C.; Penn, C. R. *Antimicrob. Agents Chemother.* **1996**, *40*, 40–46.
58. Tai, C. Y.; Escarpe, P. A.; Sidwell, R. W.; Williams, M. A.; Lew, W.; Wu, H.; Kim, C. U.; Mendel, D. B. *Antimicrob. Agents Chemother.* **1998**, *42*, 3234–3241.
59. Kiso, M.; Mitamura, K.; Sakai-Tagawa, Y.; Shiraishi, K.; Kawakami, C.; Kimura, K.; Hayden, F. G.; Sugaya, N.; Kawaoka, Y. *Lancet* **2004**, *364*, 759–765.
60. Moscona, A. *Lancet* **2004**, *364*, 733–734.
61. Cianci, C.; Yu, K. L.; Dischino, D. D.; Harte, W.; Deshpande, M.; Luo, G.; Colonno, R. J.; Meanwell, N. A.; Krystal, M. *J. Virol.* **1999**, *73*, 1785–1794.
62. Combrink, K. D.; Gulgeze, H. B.; Yu, K. L.; Pearce, B. C.; Trehan, A. K.; Wei, J.; Deshpande, M.; Krystal, M.; Torri, A.; Luo, G. et al. *Bioorg. Med. Chem. Lett.* **2000**, *10*, 1649–1652.
63. Yu, K. L.; Torri, A. F.; Luo, G.; Cianci, C.; Grant-Young, K.; Danetz, S.; Tiley, L.; Krystal, M.; Meanwell, N. A. *Bioorg. Med. Chem. Lett.* **2002**, *12*, 3379–3382.

64. Hastings, J. C.; Selnick, H.; Wolanski, B.; Tomassini, J. E. *Antimicrob. Agents Chemother.* **1996**, *40*, 1304–1307.
65. Parkes, K. E.; Ermert, P.; Fassler, J.; Ives, J.; Martin, J. A.; Merrett, J. H.; Obrecht, D.; Williams, G.; Klumpp, K. *J. Med. Chem.* **2003**, *46*, 1153–1164.
66. Singh, S. B.; Tomassini, J. E. *J. Org. Chem.* **2001**, *66*, 5504–5516.
67. Szabo, E.; Lotz, G.; Paska, C.; Kiss, A.; Schaff, Z. *Pathol. Oncol. Res.* **2003**, *9*, 215–221.
68. Hoofnagle, J. H. *Hepatology* **1997**, *26*, 15S–20S.
69. Kim, W. R.; Gross, J. B., Jr.; Poterucha, J. J.; Locke, G. R., III; Dickson, E. R. *Hepatology* **2001**, *33*, 201–206.
70. Thomson, B. J.; Finch, R. G. *Clin. Microbiol. Infect.* **2005**, *11*, 86–94.
71. Moriishi, K.; Matsuura, Y. *Antivir. Chem. Chemother.* **2003**, *14*, 285–297.
72. Moradpour, D.; Blum, H. E. *Liver Int.* **2004**, *24*, 519–525.
73. Bartenschlager, R.; Frese, M.; Pietschmann, T. *Adv. Virus Res.* **2004**, *63*, 71–180.
74. Pileri, P.; Uematsu, Y.; Campagnoli, S.; Galli, G.; Falugi, F.; Petracca, R.; Weiner, A. J.; Houghton, M.; Rosa, D.; Grandi, G. et al. *Science* **1998**, *282*, 938–941.
75. Agnello, V.; Abel, G.; Elfahal, M.; Knight, G. B.; Zhang, Q. X. *Proc. Natl. Acad. Sci. USA* **1999**, *96*, 12766–12771.
76. Scarselli, E.; Ansuini, H.; Cerino, R.; Roccasecca, R. M.; Acali, S.; Filocamo, G.; Traboni, C.; Nicosia, A.; Cortese, R.; Vitelli, A. *EMBO J.* **2002**, *21*, 5017–5025.
77. Cormier, E. G.; Durso, R. J.; Tsamis, F.; Boussemart, L.; Manix, C.; Olson, W. C.; Gardner, J. P.; Dragic, T. *Proc. Natl. Acad. Sci. USA* **2004**, *101*, 14067–14072.
78. Lozach, P. Y.; Amara, A.; Bartosch, B.; Virelizier, J. L.; Arenzana-Seisdedos, F.; Cosset, F. L.; Altmeyer, R. *J. Biol. Chem.* **2004**, *279*, 32035–32045.
79. Cormier, E. G.; Tsamis, F.; Kajumo, F.; Durso, R. J.; Gardner, J. P.; Dragic, T. *Proc. Natl. Acad. Sci. USA* **2004**, *101*, 7270–7274.
80. Zhang, J.; Randall, G.; Higginbottom, A.; Monk, P.; Rice, C. M.; McKeating, J. A. *J. Virol.* **2004**, *78*, 1448–1455.
81. Bartosch, B.; Vitelli, A.; Granier, C.; Goujon, C.; Dubuisson, J.; Pascale, S.; Scarselli, E.; Cortese, R.; Nicosia, A.; Cosset, F. L. *J. Biol. Chem.* **2003**, *278*, 41624–41630.
82. Neumann, A. U.; Lam, N. P.; Dahari, H.; Gretch, D. R.; Wiley, T. E.; Layden, T. J.; Perelson, A. S. *Science* **1998**, *282*, 103–107.
83. Bukh, J.; Miller, R. H.; Purcell, R. H. *Clin. Exp. Rheumatol.* **1995**, *13*, S3–S7.
84. Seeff, L. B. *Hepatology* **2002**, *36*, S35–S46.
85. Mahaney, K.; Tedeschi, V.; Maertens, G.; Di Bisceglie, A. M.; Vergalla, J.; Hoofnagle, J. H.; Sallie, R. *Hepatology* **1994**, *20*, 1405–1411.
86. Di Bisceglie, A. M. *Hepatology* **2000**, *31*, 1014–1018.
87. Farci, P.; Shimoda, A.; Coiana, A.; Diaz, G.; Peddis, G.; Melpolder, J. C.; Strazzera, A.; Chien, D. Y.; Munoz, S. J.; Balestrieri, A. et al. *Science* **2000**, *288*, 339–344.
88. Hnatyszyn, H. J. *Antivir. Ther.* **2005**, *10*, 1–11.
89. Bright, H.; Carroll, A. R.; Watts, P. A.; Fenton, R. J. *J. Virol.* **2004**, *78*, 2062–2071.
90. Beames, B.; Chavez, D.; Lanford, R. E. *ILAR J.* **2001**, *42*, 152–160.
91. Bartenschlager, R.; Lohmann, V. *Antiviral Res.* **2001**, *52*, 1–17.
92. Pietschmann, T.; Bartenschlager, R. *Clin. Liver Dis.* **2003**, *7*, 23–43.
93. Guha, C.; Lee, S. W.; Chowdhury, N. R.; Chowdhury, J. R. *Lab. Anim.* **2005**, *34*, 39–47.
94. Heller, T.; Saito, S.; Auerbach, J.; Williams, T.; Moreen, T. R.; Jazwinski, A.; Cruz, B.; Jeurkar, N.; Sapp, R.; Luo, G. et al. *Proc. Natl. Acad. Sci. USA* **2005**, *102*, 2579–2583.
95. Mercer, D. F.; Schiller, D. E.; Elliott, J. F.; Douglas, D. N.; Hao, C.; Rinfret, A.; Addison, W. R.; Fischer, K. P.; Churchill, T. A.; Lakey, J. R. et al. *Nat. Med.* **2001**, *7*, 927–933.
96. Hadziyannis, S. J.; Sette, H., Jr.; Morgan, T. R.; Balan, V.; Diago, M.; Marcellin, P.; Ramadori, G.; Bodenheimer, H., Jr.; Bernstein, D.; Rizzetto, M. et al. *Ann. Intern. Med.* **2004**, *140*, 346–355.
97. Russo, M. W.; Fried, M. W. *Curr. Gastroenterol. Rep.* **2004**, *6*, 17–21.
98. Zein, N. N. *Expert Opin. Invest. Drugs* **2001**, *10*, 1457–1469.
99. Guo, J. T.; Sohn, J. A.; Zhu, Q.; Seeger, C. *Virology* **2004**, *325*, 71–81.
100. Fried, M. W.; Shiffman, M. L.; Reddy, K. R.; Smith, C.; Marinos, G.; Goncales, F. L., Jr.; Haussinger, D.; Diago, M.; Carosi, G.; Dhumeaux, D. et al. *N. Engl. J. Med.* **2002**, *347*, 975–982.
101. Manns, M. P.; McHutchison, J. G.; Gordon, S. C.; Rustgi, V. K.; Shiffman, M.; Reindollar, R.; Goodman, Z. D.; Koury, K.; Ling, M.; Albrecht, J. K. *Lancet* **2001**, *358*, 958–965.
102. Reyes, G. R. *Curr. Opin. Drug Disc. Dev.* **2001**, *4*, 651–656.
103. Zhou, S.; Liu, R.; Baroudy, B. M.; Malcolm, B. A.; Reyes, G. R. *Virology* **2003**, *310*, 333–342.
104. Crotty, S.; Cameron, C. E.; Andino, R. *Proc. Natl. Acad. Sci. USA* **2001**, *98*, 6895–6900.
105. Balan, V.; Sulkowski, M.; Nelson, D.; Everson, G.; Dickson, R.; Lambiase, L.; Post, A.; Redfield, R.; Weisner, R.; Recta, J. *39th Annual Meeting of the European Association for the Study of the Liver*, Berlin, Germany, Apr 15, 2004.
106. Lin, C. C.; Xu, C.; Teng, A.; Yeh, L. T.; Peterson, J. *55th Annual Meeting of AASLD*, Oct 29–Nov 2, 2004, Abs 515.
107. Gish, R. G.; Arora, S.; Nelson, D.; Fried, M. W.; Reddy, K. R.; Xu, Y.; Peterson, J.; Murphy, B. *55th Annual Meeting of AASCD*, Oct 29–Nov 2, 2004, Abs 519.
108. Jain, J.; Almquist, S. J.; Ford, P. J.; Shlyakhter, D.; Wang, Y.; Nimmesgern, E.; Germann, U. A. *Biochem. Pharmacol.* **2004**, *67*, 767–776.
109. Tossing, G. *IDrugs* **2003**, *6*, 372–376.
110. De Francesco, R.; Altamura, S.; Migliaccio, G.; Summa, V.; Tomei, L. *Antiviral Res.* **2003**, *58*, 1–16.
111. Narjes, F.; Koch, U.; Steinkuehler, C. *Expert Opin. Invest. Drugs* **2003**, *12*, 153–163.
112. Ni, Z. J.; Wagman, A. S. *Curr. Opin. Drug Disc. Dev.* **2004**, *7*, 446–459.
113. Sarisky, R. T. *J. Antimicrob. Chemother.* **2004**, *54*, 14–16.
114. Zhang, X. *IDrugs* **2002**, *5*, 154–158.
115. Beaulieu, P. L.; Tsantrizos, Y. S. *Curr. Opin. Invest. Drugs* **2004**, *5*, 838–850.
116. Gordon, C. P.; Keller, P. A. *J. Med. Chem.* **2005**, *48*, 1–20.
117. Griffith, R. C.; Roberts, C. D.; Lou, L.; Schmitz, U. *Annu. Rev. Med. Chem.* **2005**, *2004*, 81–101.
118. Kolykhalov, A. A.; Agapov, E. V.; Rice, C. M. *J. Virol.* **1994**, *68*, 7525–7533.
119. Urbani, A.; Bianchi, E.; Narjes, F.; Tramontano, A.; De Francesco, R.; Steinkuhler, C.; Pessi, A. *J. Biol. Chem.* **1997**, *272*, 9204–9209.
120. Walker, M. P.; Yao, N.; Hong, Z. *Expert Opin. Invest. Drugs* **2003**, *12*, 1269–1280.

121. Pallaoro, M.; Lahm, A.; Biasiol, G.; Brunetti, M.; Nardella, C.; Orsatti, L.; Bonelli, F.; Orru, S.; Narjes, F.; Steinkuhler, C. *J. Virol.* **2001**, *75*, 9939–9946.
122. Kim, D. W.; Gwack, Y.; Han, J. H.; Choe, J. *Biochem. Biophys. Res. Commun.* **1995**, *215*, 160–166.
123. Bartenschlager, R. *J. Viral Hepatol.* **1999**, *6*, 165–181.
124. Gallinari, P.; Brennan, D.; Nardi, C.; Brunetti, M.; Tomei, L.; Steinkuhler, C.; De Francesco, R. *J. Virol.* **1998**, *72*, 6758–6769.
125. Morgenstern, K. A.; Landro, J. A.; Hsiao, K.; Lin, C.; Gu, Y.; Su, M. S.; Thomson, J. A. *J. Virol.* **1997**, *71*, 3767–3775.
126. Love, R. A.; Yu, X.; Diehl, W.; Hickey, M. J.; Parge, H. E.; Gao, J.; Fuhrman Shella. U.S. Patent Appl. 02009387, 2002.
127. Hedstrom, L. *Chem. Rev.* **2002**, *102*, 4501–4524.
128. Kolykhalov, A. A.; Mihalik, K.; Feinstone, S. M.; Rice, C. M. *J. Virol.* **2000**, *74*, 2046–2051.
129. Mak, P.; Palant, O.; Labonte, P.; Plotch, S. *FEBS Lett.* **2001**, *503*, 13–18.
130. Walker, M. P.; Yao, N.; Hong, Z. *Expert Opin. Invest. Drugs* **2003**, *12*, 1269–1280.
131. Whitney, M.; Stack, J. H.; Darke, P. L.; Zheng, W.; Terzo, J.; Inglese, J.; Strulovici, B.; Kuo, L. C.; Pollok, B. A. *J. Biomol. Screen.* **2002**, *7*, 149–154.
132. Lamarre, D.; Anderson, P. C.; Bailey, M.; Beaulieu, P.; Bolger, G.; Bonneau, P.; Boes, M.; Cameron, D. R.; Cartier, M.; Cordingley, M. G. et al. *Nature* **2003**, *426*, 186–189.
133. Tsantrizos, Y. S.; Bolger, G.; Bonneau, P.; Cameron, D. R.; Goudreau, N.; Kukolj, G.; LaPlante, S. R.; Llinas-Brunet, M.; Nar, H.; Lamarre, D. *Angew. Chem. Int. Ed. Engl.* **2003**, *42*, 1356–1360.
134. Tsantrizos, Y. S. *Biopolymers* **2004**, *76*, 309–323.
135. Ingallinella, P.; Altamura, S.; Bianchi, E.; Taliani, M.; Ingenito, R.; Cortese, R.; De Francesco, R.; Steinkuhler, C.; Pessi, A. *Biochemistry* **1998**, *37*, 8906–8914.
136. Lahm, A.; Yagnik, A.; Tramontano, A.; Koch, U. *Curr. Drug Targets* **2002**, *3*, 281–296.
137. Llinas-Brunet, M.; Bailey, M.; Fazal, G.; Goulet, S.; Halmos, T.; LaPlante, S.; Maurice, R.; Poirier, M.; Poupart, M. A.; Thibeault, D. et al. *Bioorg. Med. Chem. Lett.* **1998**, *8*, 1713–1718.
138. Pacini, L.; Vitelli, A.; Filocamo, G.; Bartholomew, L.; Brunetti, M.; Tramontano, A.; Steinkuhler, C.; Migliaccio, G. *J. Virol.* **2000**, *74*, 10563–10570.
139. Perni, R. B.; Britt, S. D.; Court, J. C.; Courtney, L. F.; Deininger, D. D.; Farmer, L. J.; Gates, C. A.; Harbeson, S. L.; Kim, J. L.; Landro, J. A. et al. *Bioorg. Med. Chem. Lett.* **2003**, *13*, 4059–4063.
140. Perni, R. B.; Pitlik, J.; Britt, S. D.; Court, J. C.; Courtney, L. F.; Deininger, D. D.; Farmer, L. J.; Gates, C. A.; Harbeson, S. L.; Levin, R. B. et al. *Bioorg. Med. Chem. Lett.* **2004**, *14*, 1441–1446.
141. Perni, R. B.; Kwong, A. D. *Prog. Med. Chem.* **2002**, *39*, 215–255.
142. Campbell, J. A.; Good, A. C. PCT Patent Appl. WO2003053349, 2003.
143. Llinas-Brunet, M.; Gorys, V. J. PCT Patent Appl. WO2003064456, 2003.
144. Goudreau, N.; Brochu, C.; Cameron, D. R.; Duceppe, J. S.; Faucher, A. M.; Ferland, J. M.; Grand-Maitre, C.; Poirier, M.; Simoneau, B.; Tsantrizos, Y. S. *J. Org. Chem.* **2004**, *69*, 6185–6201.
145. Goudreau, N.; Cameron, D. R.; Bonneau, P.; Gorys, V.; Plouffe, C.; Poirier, M.; Lamarre, D.; Llinas-Brunet, M. *J. Med. Chem.* **2004**, *47*, 123–132.
146. Llinas-Brunet, M.; Bailey, M.; Fazal, G.; Ghiro, E.; Gorys, V.; Goulet, S.; Halmos, T.; Maurice, R.; Poirier, M.; Poupart, M. A. et al. *Bioorg. Med. Chem. Lett.* **2000**, *10*, 2267–2270.
147. LaPlante, S. R.; Cameron, D. R.; Aubry, N.; Lefebvre, S.; Kukolj, G.; Maurice, R.; Thibeault, D.; Lamarre, D.; Llinas-Brunet, M. *J. Biol. Chem.* **1999**, *274*, 18618–18624.
148. LaPlante, S. R.; Aubry, N.; Bonneau, P. R.; Kukolj, G.; Lamarre, D.; Lefebvre, S.; Li, H.; Llinas-Brunet, M.; Plouffe, C.; Cameron, D. R. *Bioorg. Med. Chem. Lett.* **2000**, *10*, 2271–2274.
149. Rancourt, J.; Cameron, D. R.; Gorys, V.; Lamarre, D.; Poirier, M.; Thibeault, D.; Llinas-Brunet, M. *J. Med. Chem.* **2004**, *47*, 2511–2522.
150. Lin, C.; Lin, K.; Luong, Y. P.; Rao, B. G.; Wei, Y. Y.; Brennan, D. L.; Fulghum, J. R.; Hsiao, H. M.; Ma, S.; Maxwell, J. P. et al. *J. Biol. Chem.* **2004**, *279*, 17508–17514.
151. Lu, L.; Pilot-Matias, T. J.; Stewart, K. D.; Randolph, J. T.; Pithawalla, R.; He, W.; Huang, P. P.; Klein, L. L.; Mo, H.; Molla, A. *Antimicrob. Agents Chemother.* **2004**, *48*, 2260–2266.
152. Hinrichsen, H.; Benhamou, Y.; Wedemeyer, H.; Reiser, M.; Sentjens, R. E.; Calleja, J. L.; Forns, X.; Erhardt, A.; Cronlein, J.; Chaves, R. L. et al. *Gastroenterology* **2004**, *127*, 1347–1355.
153. Kempf, D. J. *Methods Enzymol.* **1994**, *241*, 334–354.
154. Llinas-Brunet, M.; Bailey, M. D.; Bolger, G.; Brochu, C.; Faucher, A. M.; Ferland, J. M.; Garneau, M.; Ghiro, E.; Gorys, V.; Grand-Maitre, C. et al. *J. Med. Chem.* **2004**, *47*, 1605–1608.
155. Hinrichsen, H.; Benhamou, Y.; Wedemeyer, H.; Reiser, M.; Sentjens, R. E.; Calleja, J. L.; Forns, X.; Erhardt, A.; Cronlein, J.; Chaves, R. L. et al. *Gastroenterology* **2004**, *127*, 1347–1355.
156. LaPlante, S. R.; Llinas-Brunet, M. *Curr. Med. Chem.* **2005**, *4*, 111–132.
157. Campbell, J. A.; Good, A. C. PCT Patent Appl. WO2003099316, 2003.
158. Wang, X. A.; Sun, L.-Q.; Sit, S.-Y.; Sin, N.; Scola, P. M.; Hewawasam, P.; Good, A. C.; Chen, Y.; Campbell, J. A. PCT Patent Appl. WO2003099274, 2003.
159. Di Marco, S.; Rizzi, M.; Volpari, C.; Walsh, M. A.; Narjes, F.; Colarusso, S.; De Francesco, R.; Matassa, V. G.; Sollazzo, M. *J. Biol. Chem.* **2000**, *275*, 7152–7157.
160. Narjes, F.; Koehler, K. F.; Koch, U.; Gerlach, B.; Colarusso, S.; Steinkuhler, C.; Brunetti, M.; Altamura, S.; De Francesco, R.; Matassa, V. G. *Bioorg. Med. Chem. Lett.* **2002**, *12*, 701–704.
161. Victor, F.; Lamar, J.; Snyder, N.; Yip, Y.; Guo, D.; Yumibe, N.; Johnson, R. B.; Wang, Q. M.; Glass, J. I.; Chen, S. H. *Bioorg. Med. Chem. Lett.* **2004**, *14*, 257–261.
162. Yip, Y.; Victor, F.; Lamar, J.; Johnson, R.; Wang, Q. M.; Barket, D.; Glass, J.; Jin, L.; Liu, L.; Venable, D. et al. *Bioorg. Med. Chem. Lett.* **2004**, *14*, 251–256.
163. Yip, Y.; Victor, F.; Lamar, J.; Johnson, R.; Wang, Q. M.; Glass, J. I.; Yumibe, N.; Wakulchik, M.; Munroe, J.; Chen, S. H. *Bioorg. Med. Chem. Lett.* **2004**, *14*, 5007–5011.
164. Perni, R. B.; Chandorkar, G.; Chaturvedi, P. R.; Courtney, L. F.; Decker, C. J.; Gates, C. A.; Harbeson, S. L.; Kwong, A. D.; Lin, C.; Luong, Y. P. et al. *54th Annual Meeting American Association for the Study. Liver Disease*, Oct 24-28, Boston, MA, **2003**, Abs 972.

165. Vertex Pharmaceuticals, Press Release 2005. http://www.vrtx.com/Pressreleases2005/pr051705.html (accessed Aug 2006).
166. Andrews, D. M.; Barnes, M. C.; Dowle, M. D.; Hind, S. L.; Johnson, M. R.; Jones, P. S.; Mills, G.; Patikis, A.; Pateman, T. J.; Redfern, T. J. et al. *Org. Lett.* **2003**, *5*, 4631–4634.
167. Andrews, D. M.; Borthwick, A. D.; Chaignot, H.; Jones, P. S.; Robinson, J. E.; Shah, P.; Slater, M. J.; Upton, R. J. *Synthesis* **2003**, 1722–1726.
168. Andrews, D. M.; Chaignot, H. M.; Coomber, B. A.; Dowle, M. D.; Lucy Hind, S.; Johnson, M. R.; Jones, P. S.; Mills, G.; Patikis, A.; Pateman, T. J. et al. *Eur. J. Med. Chem.* **2003**, *38*, 339–343.
169. Slater, M. J.; Amphlett, E. M.; Andrews, D. M.; Bamborough, P.; Carey, S. J.; Johnson, M. R.; Jones, P. S.; Mills, G.; Parry, N. R.; Somers, D. O. et al. *Org. Lett.* **2003**, *5*, 4627–4630.
170. McKercher, G.; Beaulieu, P. L.; Lamarre, D.; LaPlante, S.; Lefebvre, S.; Pellerin, C.; Thauvette, L.; Kukolj, G. *Nucleic Acids Res.* **2004**, *32*, 422–431.
171. Carroll, S. S.; Tomassini, J. E.; Bosserman, M.; Getty, K.; Stahlhut, M. W.; Eldrup, A. B.; Bhat, B.; Hall, D.; Simcoe, A. L.; LaFemina, R. et al. *J. Biol. Chem.* **2003**, *278*, 11979–11984.
172. Migliaccio, G.; Tomassini, J. E.; Carroll, S. S.; Tomei, L.; Altamura, S.; Bhat, B.; Bartholomew, L.; Bosserman, M. R.; Ceccacci, A.; Colwell, L. F. et al. *J. Biol. Chem.* **2003**, *278*, 49164–49170.
173. Laurila, M. R.; Makeyev, E. V.; Bamford, D. H. *J. Biol. Chem.* **2002**, *277*, 17117–17124.
174. Eldrup, A. B.; Prhavc, M.; Brooks, J.; Bhat, B.; Prakash, T. P.; Song, Q.; Bera, S.; Bhat, N.; Dande, P.; Cook, P. D. et al. *J. Med. Chem.* **2004**, *47*, 5284–5297.
175. LaColla, P.; Sommadossi, J. P. U.S. Patent 09/863,816, 2004.
176. Sommadossi, J.-P. PCT Patent Appl. WO2004003000, 2004.
177. Sommadossi, J.-P.; La Colla, P. PCT Patent Appl. WO2004002422, 2004.
178. Walton, E. U.S. Patent 3,480,613, 1969.
179. Afdahl, N.; Rodriguez-Torres, M.; Lawitz, E.; Godofsky, E.; Chao, G.; Fielman, B.; Knox, S.; Brown, N. *40th Annual Meeting of the European Association for the Study of the Liver*, Paris, France, Apr 13–17, 2005.
180. Olsen, D. B.; Eldrup, A. B.; Bartholomew, L.; Bhat, B.; Bosserman, M. R.; Ceccacci, A.; Colwell, L. F.; Fay, J. F.; Flores, O. A.; Getty, K. L. et al. *Antimicrob. Agents Chemother.* **2004**, *48*, 3944–3953.
181. Martin, J. A.; Sarma, K.; Smith, D. B.; Smith, M. Roche, P. A. C. U.S. Patent Appl. US 6,846,81, 2005.
182. Clark, J. PCT Patent Appl. WO2005003147, 2005.
183. Bressanelli, S.; Tomei, L.; Rey, F. A.; De Francesco, R. *J. Virol.* **2002**, *76*, 3492.
184. Hashimoto, H.; Mizutani, K.; Yoshida, A. Japan Tobacco Inc. Ed. EP1162196 A1, 2001.
185. Harper, S.; Pacini, B.; Avolio, S.; Di Filippo, M.; Migliaccio, G.; Laufer, R.; De Francesco, R.; Rowley, M.; Narjes, F. *J. Med. Chem.* **2005**, *48*, 1314–1317.
186. Latour, D.; Hancock, C.; Michelotti, E.; Koo-McCoy, S.; Fung, K.; Kirk, M.; Lou, L.; Shi, D.-F.; Dyatkina, N.; Zheng, X. et al. HEP DART 2003, Kauai, HI, Dec 14–18, 2003.
187. LaPlante, S. R.; Jakalian, A.; Aubry, N.; Bousquet, Y.; Ferland, J. M.; Gillard, J.; Lefebvre, S.; Poirier, M.; Tsantrizos, Y. S.; Kukolj, G. et al. *Angew. Chem., Int. Ed. Engl.* **2004**, *43*, 4306–4311.
188. Beaulieu, P. L.; Bos, M.; Bousquet, Y.; Fazal, G.; Gauthier, J.; Gillard, J.; Goulet, S.; LaPlante, S.; Poupart, M. A.; Lefebvre, S. et al. *Bioorg. Med. Chem. Lett.* **2004**, *14*, 119–124.
189. Beaulieu, P. L.; Bos, M.; Bousquet, Y.; DeRoy, P.; Fazal, G.; Gauthier, J.; Gillard, J.; Goulet, S.; McKercher, G.; Poupart, M. A. et al. *Bioorg. Med. Chem. Lett.* **2004**, *14*, 967–971.
190. Beaulieu, P. L.; Brochu, C.; Chabot, C.; Jolicoeur, E.; Kawai, S.; Poupart, M. A.; Tsantrizos, Y. S. PCT Patent Appl. WO2004065367, 2004.
191. Columbe, R. B. P. L.; Jolicoeur, E.; Kukolj, G.; LaPlante, S.; Poupart, M. A. PCT Patent Appl. WO2004099241, 2004.
192. Shipps, G. W., Jr.; Rosner, K. E.; Popovici-Muller, J.; Deng, Y.; Wang, T.; Curran, P. PCT Patent Appl. WO2003101993, 2003.
193. Deng, Y.; Popovici-Muller, J.; Shipps, G. W., Jr.; Rosner, K. E.; Wang, T.; Curran, P.; Girijavallabhan, V.; Butkiewicz, N.; Cable, M. *229th ACS National Meeting*, San Diego, CA, Mar 13–17, 2005.
194. Popovici-Muller, J.; Shipps, G. W., Jr.; Rosner, K. E.; Deng, Y.; Wang, T.; Curran, P.; Brown, M. A.; Cooper, A. B.; Cable, M.; Butkiewicz, N. et al. *229th ACS National Meeting*, San Diego, CA, Mar 13–17, 2005.
195. Schmitz, F. U.; Roberts, C. D.; Griffith, R. C.; Botyanszki, J.; Gezginci, M. H.; Gralapp, J. M.; Shi, D.-F.; Liehr, S. J. PCT Patent Appl. WO2005012288, 2005.
196. Wang, C.; Pflugheber, J.; Sumpter, R., Jr.; Sodora, D. L.; Hui, D.; Sen, G. C.; Gale, M., Jr. *J. Virol.* **2003**, *77*, 3898–3912.
197. Borchardt, A. J.; Goble, M. P. Pfizer, I. PCT Patent Appl. WO2003082848, 2003.
198. Gonzalez, J.; Borchardt, A. J.; Dragovich, P. S.; Jewell, T. M.; Linton, M. A.; Zhou, R.; Li, H.; Tatlock, J. H.; Abreo, M. A.; Prins, T. J. et al. PCT Patent Appl. WO2003095441, 2003.
199. Gopalsamy, A.; Lim, K.; Ciszewski, G.; Park, K.; Ellingboe, J. W.; Bloom, J.; Insaf, S.; Upeslacis, J.; Mansour, T. S.; Krishnamurthy, G. et al. *J. Med. Chem.* **2004**, *47*, 6603–6608.
200. Howe, A. Y.; Bloom, J.; Baldick, C. J.; Benetatos, C. A.; Cheng, H.; Christensen, J. S.; Chunduru, S. K.; Coburn, G. A.; Feld, B.; Gopalsamy, A. et al. *Antimicrob. Agents Chemother.* **2004**, *48*, 4813–4821.
201. Viropharma, Inc., Press Release 2003. http://www.irconnect.com/vphm/pages/news_releases. html?d = 42687 (accessed Aug 2006).
202. Dhanak, D.; Duffy, K. J.; Johnston, V. K.; Lin-Goerke, J.; Darcy, M.; Shaw, A. N.; Gu, B.; Silverman, C.; Gates, A. T.; Nonnemacher, M. R. et al. *J. Biol. Chem.* **2002**, *277*, 38322–38327.
203. Gu, B.; Johnston, V. K.; Gutshall, L. L.; Nguyen, T. T.; Gontarek, R. R.; Darcy, M. G.; Tedesco, R.; Dhanak, D.; Duffy, K. J.; Kao, C. C. et al. *J. Biol. Chem.* **2003**, *278*, 16602–16607.
204. Pratt, J. J.; Beno, D.; Donner, P.; Jiang, W.; Kati, W.; Kempf, D.; Koev, G.; Liu, D.; Liu, Y.; Maring, C. et al. *55th Annual Meeting of the AASLD*, Boston, MA, 2004, Abs 1224-A.
205. Summa, V.; Petrocchi, A.; Pace, P.; Matassa, V. G.; De Francesco, R.; Altamura, S.; Tomei, L.; Koch, U.; Neuner, P. *J. Med. Chem.* **2004**, *47*, 14–17.
206. Goff, D.; Lu, H.; Singh, R.; Sun, T.; Issankani, S. D. PCT Patent Appl. WO2003040112, 2003.
207. Lu, H. PCT Patent Appl. WO2005000308, 2005.
208. Rigel Pharmaceuticals, Press Release 2004, http://www.rigel.com/rigel/pr_1101094254 (accessed Aug 2006).
209. Billich *Curr. Opin. Investig. Drugs* **2002**, *3*, 698–707.

210. Lee, J.; Chuang, T. H.; Redecke, V.; Pitha, P. M.; Carson, D. A.; Raz, E.; Cottam, H. B. *Proc. Natl. Acad. Sci. USA* **2003**, *100*, 6646–6651.
211. Rost, K. L.; Wierich, W.; Masayuki, F.; Tuthill, C. W.; Horwitz, D. L.; Herrmann, W. M. *Int. J. Clin. Pharmacol. Ther.* **1999**, *37*, 51–57.
212. Andreone, P.; Cursaro, C.; Gramenzi, A.; Buzzi, A.; Covarelli, M. G.; Di Giammarino, L.; Miniero, R.; Arienti, V.; Bernardi, M.; Gasbarrini, G. *Liver* **1996**, *16*, 207–210.
213. Andreone, P.; Gramenzi, A.; Cursaro, C.; Felline, F.; Loggi, E.; D'Errico, A.; Spinosa, M.; Lorenzini, S.; Biselli, M.; Bernardi, M. *J. Viral Hepatol.* **2004**, *11*, 69–73.
214. Abbas, Z.; Hamid, S. S.; Tabassum, S.; Jafri, W. *J. Pak. Med. Assoc.* **2004**, *54*, 571–574.
215. Kullavanuaya, P.; Treeprasertsuk, S.; Thong-Ngam, D.; Chaermthai, K.; Gonlachanvit, S.; Suwanagool, P. *J. Med. Assoc. Thai.* **2001**, *84*, S462–S468.
216. Smee, D. F.; Alaghamandan, H. A.; Cottam, H. B.; Sharma, B. S.; Jolley, W. B.; Robins, R. K. *Antimicrob. Agents Chemother.* **1989**, *33*, 1487–1492.
217. Smee, D. F.; Alaghamandan, H. A.; Cottam, H. B.; Jolley, W. B.; Robins, R. K. *J. Biol. Response Mod.* **1990**, *9*, 24–32.
218. Smee, D. F.; Alaghamandan, H. A.; Gilbert, J.; Burger, R. A.; Jin, A.; Sharma, B. S.; Ramasamy, K.; Revankar, G. R.; Cottam, H. B.; Jolley, W. B. *Antimicrob. Agents Chemother.* **1991**, *35*, 152–157.
219. Smee, D. F.; Alaghamandan, H. A.; Ramasamy, K.; Revankar, G. R. *Antiviral Res.* **1995**, *26*, 203–209.
220. Crozat, K.; Beutler, B. *Proc. Natl. Acad. Sci. USA* **2004**, *101*, 6835–6836.
221. Heil, F.; Hemmi, H.; Hochrein, H.; Ampenberger, F.; Kirschning, C.; Akira, S.; Lipford, G.; Wagner, H.; Bauer, S. *Science* **2004**, *303*, 1526–1529.
222. Lund, J. M.; Alexopoulou, L.; Sato, A.; Karow, M.; Adams, N. C.; Gale, N. W.; Iwasaki, A.; Flavell, R. A. *Proc. Natl. Acad. Sci. USA* **2004**, *101*, 5598–5603.
223. Horsmans, Y.; Berg, T.; Desager, J. P.; Mueller, T.; Schott, E.; Fletcher, S. P.; Steffy, K. R.; Bauman, L. A.; Kerr, B. M.; Averett, D. R. *Hepatology* **2005**, *42*, 724–731.
224. Hacking, D.; Hull, J. *J. Infect.* **2002**, *45*, 18–24.
225. Sudo, K.; Miyazaki, Y.; Kokima, N.; Kobayashi, M.; Suzuki, H.; Shintani, M.; Shimizu, Y. *Antiviral Res.* **2005**, *65*, 125–131.
226. Falsey, A. R.; Walsh, E. E. *Clin. Microbiol. Rev.* **2000**, *13*, 371–384.
227. Meanwell, N. A.; Krystal, M. *Drug Disc. Today* **2000**, *5*, 241–252.
228. Tripp, R. A. *Viral Immunol.* **2004**, *17*, 165–181.
229. Maggon, K.; Barik, S. *Rev. Med. Virol.* **2004**, *14*, 149–168.
230. Leader, S.; Kohlhase, K. *J. Pediatr.* **2003**, *143*, S127–S132.
231. Nicholson, K. G. *Epidemiol. Infect.* **1996**, *116*, 51–63.
232. Hall, C. B. *Science* **1994**, *265*, 1393–1394.
233. Domachowske, J. B.; Bonville, C. A.; Rosenberg, H. F. *Pediatr. Infect. Dis. J.* **2004**, *23*, S228–S234.
234. Durbin, J. E.; Durbin, R. K. *Viral Immunol.* **2004**, *17*, 370–380.
235. McIntosh, E. D.; Malinoski, F. J.; Randolph, V. B. *Expert Rev. Vaccines* **2004**, *3*, 353–357.
236. Olszewska, W.; Helson, R.; Openshaw, P. J. *Expert Opin. Investig. Drugs* **2004**, *13*, 681–689.
237. Kneyber, M. C.; Kimpen, J. L. *Curr. Opin. Investig. Drugs* **2004**, *5*, 163–170.
238. Fenton, C.; Scott, L. J.; Plosker, G. L. *Paediatr. Drugs* **2004**, *6*, 177–197.
239. *Physicians' Desk Reference*, 59th ed.; Thomson PDR: Montvale, NJ, 2005.
240. Ottolini, M. G.; Porter, D. D.; Hemming, V. G.; Zimmerman, M. N.; Schwab, N. M.; Prince, G. A. *Bone Marrow Transplant.* **1999**, *24*, 41–45.
241. Ottolini, M. G.; Curtis, S. R.; Mathews, A.; Ottolini, S. R.; Prince, G. A. *Bone Marrow Transplant.* **2002**, *29*, 117–120.
242. Eckert, D. M.; Kim, P. S. *Annu. Rev. Biochem.* **2001**, *70*, 777–810.
243. Zhao, X.; Singh, M.; Malashkevich, V. N.; Kim, P. S. *Proc. Natl. Acad. Sci. USA* **2000**, *97*, 14172–14177.
244. Chan, D. C.; Kim, P. S. *Cell* **1998**, *93*, 681–684.
245. Root, M. J.; Steger, H. K. *Curr. Pharm. Des.* **2004**, *10*, 1805–1825.
246. Dando, T. M.; Perry, C. M. *Drugs* **2003**, *63*, 2755–2766.
247. Este, J. A.; Este, J. A. *Curr. Med. Chem.* **2003**, *10*, 1617–1632.
248. Williams, I. G. *Int. J. Clin Pract.* **2003**, *57*, 890–897.
249. Chen, R. Y.; Kilby, J. M.; Saag, M. S. *Expert Opin. Investig. Drugs* **2002**, *11*, 1837–1843.
250. Lambert, D. M.; Barney, S.; Lambert, A. L.; Guthrie, K.; Medinas, R.; Davis, D. E.; Bucy, T.; Erickson, J.; Merutka, G.; Petteway, S. R., Jr. *Proc. Natl. Acad. Sci. USA* **1996**, *93*, 2186–2191.
251. Cianci, C.; Genovesi, E. V.; Lamb, L.; Medina, I.; Yang, Z.; Zadjura, L.; Yang, H.; D'Arienzo, C.; Sin, N.; Yu, K. L. et al. *Antimicrob. Agents Chemother.* **2004**, *48*, 2448–2454.
252. Cianci, C.; Yu, K. L.; Combrink, K.; Sin, N.; Pearce, B.; Wang, A.; Civiello, R.; Voss, S.; Luo, G.; Kadow, K. et al. *Antimicrob. Agents Chemother.* **2004**, *48*, 413–422.
253. Cianci, C.; Meanwell, N.; Krystal, M. *J. Antimicrob. Chemother.* **2005**, *55*, 289–292.
254. Yu, K. L.; Zhang, Y.; Civiello, R. L.; Kadow, K. F.; Cianci, C.; Krystal, M.; Meanwell, N. A. *Bioorg. Med. Chem. Lett.* **2003**, *13*, 2141–2144.
255. Yu, K. L.; Zhang, Y.; Civiello, R. L.; Trehan, A. K.; Pearce, B. C.; Yin, Z.; Combrink, K. D.; Gulgeze, H. B.; Wang, X. A.; Kadow, K. F. et al. *Bioorg. Med. Chem. Lett.* **2004**, *14*, 1133–1137.
256. Cianci, C.; Langley, D. R.; Dischino, D. D.; Sun, Y.; Yu, K. L.; Stanley, A.; Roach, J.; Li, Z.; Dalterio, R.; Colonno, R. et al. *Proc. Natl. Acad. Sci. USA* **2004**, *101*, 15046–15051.
257. Borman, S. *Chem. Eng. News* **2004**, *82*, 30–32.
258. Nikitenko, A. A.; Raifeld, Y. E.; Wang, T. Z. *Bioorg. Med. Chem. Lett.* **2001**, *11*, 1041–1044.
259. Gazumyan, A.; Mitsner, B.; Ellestad, G. A. *Curr. Pharm. Des.* **2000**, *6*, 525–546.
260. Huntley, C. C.; Weiss, W. J.; Gazumyan, A.; Buklan, A.; Feld, B.; Hu, W.; Jones, T. R.; Murphy, T.; Nikitenko, A. A.; O'Hara, B. et al. *Antimicrob. Agents Chemother.* **2002**, *46*, 841–847.
261. Razinkov, V.; Gazumyan, A.; Nikitenko, A.; Ellestad, G.; Krishnamurthy, G. *Chem. Biol.* **2001**, *8*, 645–659.
262. Razinkov, V.; Huntley, C.; Ellestad, G.; Krishnamurthy, G. *Antiviral Res.* **2002**, *55*, 189–200.
263. Wyde, P. R.; Moore-Poveda, D. K.; O'Hara, B.; Ding, W. D.; Mitsner, B.; Gilbert, B. E. *Antiviral Res.* **1998**, *38*, 31–42.
264. Weiss, W. J.; Murphy, T.; Lynch, M. E.; Frye, J.; Buklan, A.; Gray, B.; Lenoy, E.; Mitelman, S.; O'Connell, J.; Quartuccio, S. et al. *J. Med. Primatol.* **2003**, *32*, 82–88.

265. Nitz, T.; Pevear, D. PCT Patent Appl. WO199938508, 1999.
266. Pevear, D.; Tull, T.; Direnzo, R.; Nitz, T.; Collins, C. L. *Antiviral Res.* **2000**, *46*, A52.
267. Douglas, J. L.; Panis, M. L.; Ho, E.; Lin, K. Y.; Krawczyk, S. H.; Grant, D. M.; Cai, R.; Swaminathan, S.; Cihlar, T. *J. Virol.* **2003**, *77*, 5054–5064.
268. Douglas, J. L.; Panis, M. L.; Ho, E.; Lin, K. Y.; Krawczyk, S. H.; Grant, D. M.; Cai, R.; Swaminathan, S.; Chen, X.; Cihlar, T. *Antimicrob. Agents Chemother.* **2005**, *49*, 2460–2466.
269. McKimm-Breschkin, J. *Curr. Opin. Investig. Drugs* **2000**, *1*, 425–427.
270. Andries, K.; Gevers, T.; Willebrords, R.; Lacrampe, J.; Jannssen, F.; Moeremans, M. *Antiviral Res.* **2001**, *50*, A76.

Comprehensive Medicinal Chemistry II
ISBN (set): 0-08-044513-6

ISBN (Volume 7) 0-08-044520-9; pp. 373–417

7.14 Fungi and Fungal Disease

P Dorr, Pfizer Global Research and Development, Sandwich, UK

7.14.1 Medically Important Fungi and Major Fungal Diseases

7.14.1.1 General Classification and Structure of Medically Important Fungi

All living organisms are classified into five kingdoms, one of which is Fungi. This represents a diverse group of eukaryotic microorganisms (usually with dimensions of 3–50 μm) that are heterotrophic (i.e., require elaborated organic substrate for growth, such as human tissues in the case of pathogenic fungi), and are devoid of chlorophyll. In light of their heterotrophic nature, all fungi exist either as saprophytes or parasites. It is the latter that will be discussed in more detail in this chapter.

The basic classification of fungi is based on their appearance and method (or indeed absence) of sexual reproduction. Their nomenclature is in accordance with the International Code of Botanical Nomenclature, in light of their original identification as plants in Roman times. Most fungi, including fungal pathogens such as *Aspergillus fumigatus* (**Figure 1**), exist in a vegetative phase as microscopic filaments or hyphae. Hyphae usually consist of tubular cells that are surrounded by a rigid, chitin-containing cell wall. The hyphae extend by tip growth, and multiply by branching, creating a fine network called a mycelium, which can be of sufficient biomass to be macroscopic. Consistent with their eukaryotic nature, hyphae contain nuclei, mitochondria, ribosomes, Golgi, and membrane-bound vesicles within a plasma membrane-bound cytoplasm. The subcellular structures are supported and organized by microtubules and endoplasmic reticulum. Under certain conditions, septa develop in hyphae, which initiate the differentiation into nonvegetative resting spore structures. The yeasts represent a large group of fungi that consist of separate round, oval, or elongated cells that propagate by budding out similar cells from their surface. Many fungi, including the most significant fungal pathogen *Candida albicans*, can exist in a yeast or mycelial form (**Figure 1**).

A comprehensive review of the structural biology and taxonomy of pathogenic fungi can be found in the recently published textbook edited by San-Blas and Calderone.[1]

7.14.1.2 Human Pathogenic Fungi and Disease States

It is estimated that about 300 000 different species of fungi are currently in existence, of which about 600 are associated with human disease, with about 20 that cause >99% of human fungal infections (mycoses). The vast majority of fungi are therefore free-living organisms with no dependence on humans for survival. With few exceptions, fungal infections originate from an exogenous source in the environment, and are acquired through inhalation, ingestion, or traumatic implantation.

Fungal diseases are described in many ways, although the most practical is a subdivision into the mycoses groups described below.

7.14.1.2.1 Superficial mycoses

Superficial (or cutaneous) mycoses are fungal diseases that are confined to the outer layers of the skin, nail, or hair, (keratinized layers), rarely invading the deeper tissue or viscera, without inducing a cellular response from the host. The fungi involved are called dermatophytes (listed in **Table 1**). Superficial fungal infections are very common, of worldwide distribution, and are rarely serious or difficult to treat. These infections are often so innocuous that patients are often unaware of their condition. All topical antifungal drugs such as clotrimazole (Canestan) or terbinafine (Lamisil) are highly efficaceous and available where treatment is sought. The epidemiology and therapeutic options for the treatment of these infections has been extensively reviewed by Brandt and Warnock.[2]

7.14.1.2.2 Subcutaneous mycoses

These are rare but chronic localized infections of the skin and subcutaneous tissue following the traumatic implantation of the etiologic agent. The causative fungi are all soil saprophytes of regional epidemiology, whose ability to adapt to the tissue environment and elicit disease is extremely variable (**Table 2**). Management options are limited, difficult, and can vary according to pathogen.[3] Surgical intervention in combination with systemic antifungal chemotherapy is often required, with variable success.[4] Amphotericin B has been generally used for the treatment of these infections, although azole therapy is becoming more widespread as sensitivities to these drugs are being reported.[5,6] Sporotrichosis and chromoblastomycosis have been reported to respond to oral therapy with the azole itraconazole, and, more recently, voriconazole and allylamines, sometimes in combination with amphotericin.[6]

7.14.1.2.3 Cutaneous mycoses

These extremely common and worldwide diseases are superficial fungal infections of the skin, hair, or nails. Although no living tissue is invaded, a variety of pathological changes occur in the host because of the presence of the infectious

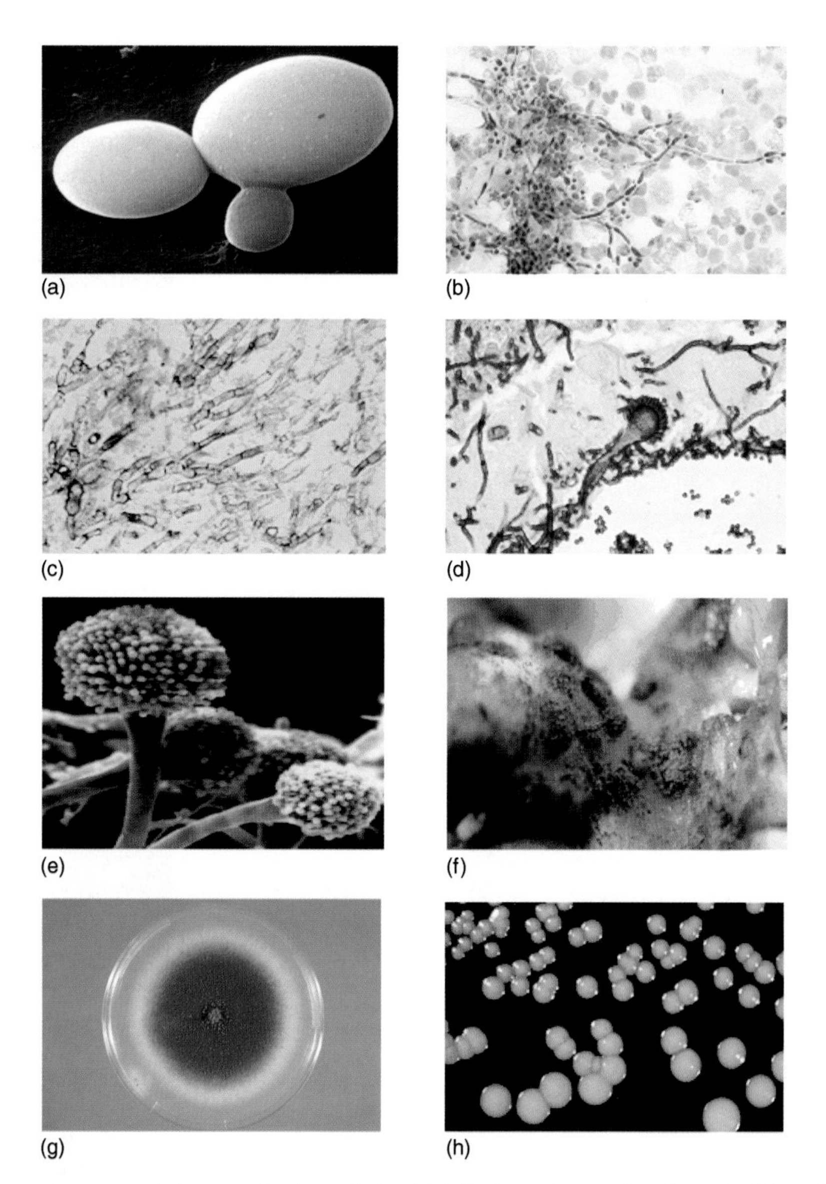

Figure 1 Micro- and macroscopic images of pathogenic fungi. (a) Electron micrograph of *Candida albicans* in yeast form and (b) mixed yeast–mycelium form from a urine sample (per acid stained). (c) Methenamine silver staining of clinical samples highlights the hyphal growth of the filamentous pathogen *Aspergillus fumigatus* invading lung tissue. The spore-forming (conidial head) structure of *A. fumigatus* is also highlighted (d) within invaded lung tissue, and (e) in more detail as an electron micrograph following laboratory culture. Extensive invasion can lead to macroscopic emergence of *Aspergillus* biomass, as shown by (f) visible growth on lung tissue. Colony presentation of (g) *Aspergillus* and (h) *C. albicans* on agar plates. (a–d; f–h: reproduced by permission of Dr David Ellis, School of Molecular & Biomedical Science, University of Adelaide and the Kaminski digital image library; e: reproduced by permission of Dr Jean-Paul Latge of the Pasteur Institute.)

Table 1 Diseases and causative organisms of superficial mycoses

Disease	*Causative organism*
Pityriasis versicolor (**Figure 1a**)	*Malassezia furfur* (**a lipophilic yeast**)
Tinea negra	*Exophiala werneckii*
White piedra	*Trichosporon beigelii*
Black piedra	*Piedraia hortae*

Table 2 Subcutaneous mycoses. The causative pathogen is emboldened for the clinical case highlighted

Disease	*Causative organism*	*Presentation*
Sporotrichosis	***Sporothrix schenckii***	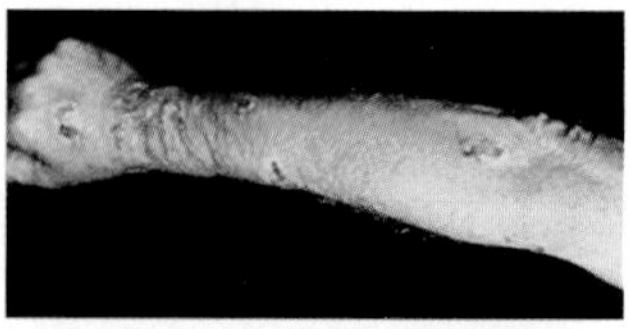
Chromoblastomycosis	*Fonsecaea*, *Phialophora*, ***Cladosporium***	
Phaeohyphomycosis	*Cladosporium*, *Exophiala*, ***Wangiella***, *Bipolaris*, *Exserohilum*, *Curvularia*	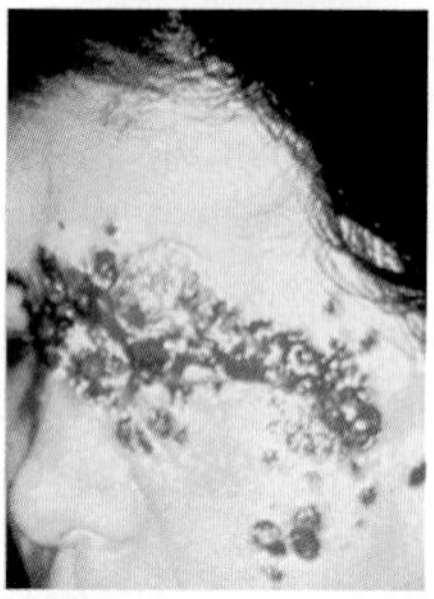
Mycotic mycetoma	***Pseudallescheria***, *Madurella*, *Acremonium*, *Exophiala*, etc.	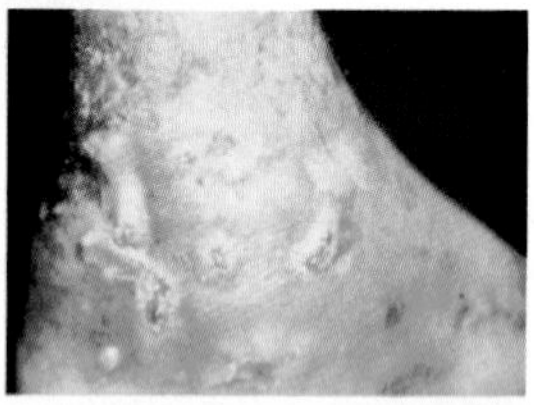
Subcutaneous zygomycosis (including mucoromycosis)	***Basidiobolus*** *ranarum*, *Conidiobolus coronatu*, *Rhizopus*, *Mucor*, *Rhizomucor*	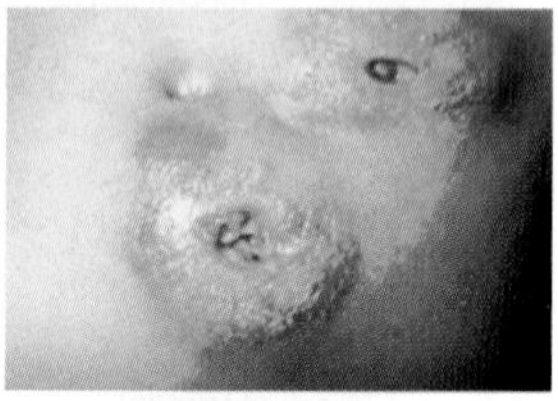
Rhinosporidiosis	***Rhinosporidium seeberi***	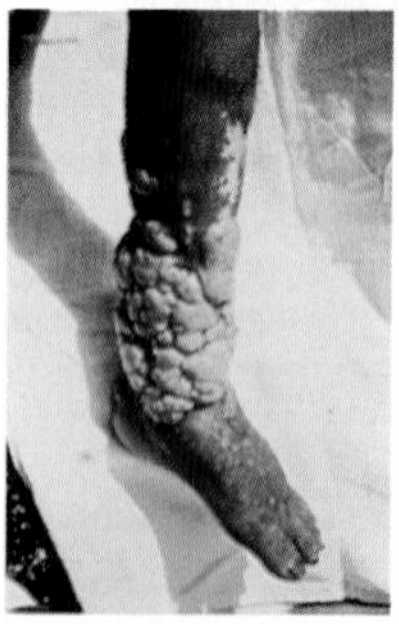

Reproduced by permission of Dr David Ellis, School of Molecular & Biomedical Science, University of Adelaide.

agent and its metabolic products. The causative pathogen is usually a filamentous dermatophyte (**Table 3**), from which *Trychophyton* (usually *T. mentagrophytes* or *T. rubrum*), *Epidermophyton*, and *Microsporium* species are by far the most common pathogens, although superficial candidiasis is not uncommon.[7] These pathogens infect various areas of the body, from which the infection description is derived (e.g., *Tinea capitis* (scalp and/or hair), *Tinea corporis* (a dermatophyte

Table 3 Cutaneous mycoses. The causative pathogen is emboldened for the clinical case highlighted

Disease	*Causative organism*	*Presentation*
Dermatophytosis. Ringworm of the scalp, glabrous skin, and nails	Dermatophytes (***Microsporum***, *Trichophyton*, *Epidermophyton*)	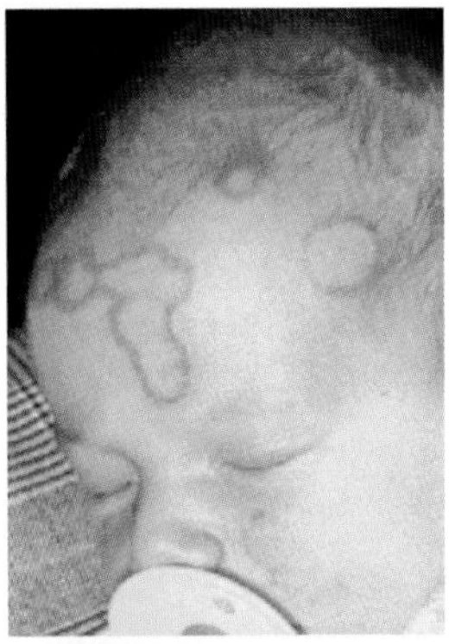
Candidiasis of skin, mucous membranes and nails	***Candida albicans*** and related species	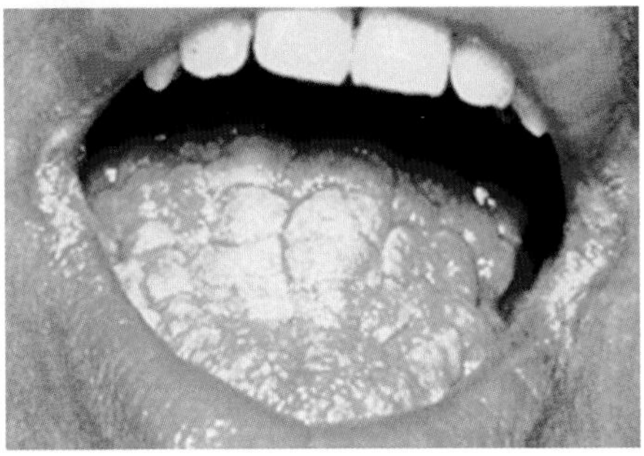
Dermatomycosis	Nondermatophyte moulds (*Hendersonula toruloidea*, *Scytalidium hyalium*, *Scopulariopsis brevicaulis*)	

Reproduced by permission of Dr David Ellis, School of Molecular & Biomedical Science, University of Adelaide.

infection of the trunk, legs, and arms), *T. pedis* (feet), *T. mannum* (hands), *T. unguium* (nails)). The vast majority of infections are mild, and readily diagnosed by microscopy and culture of samples, following symptoms of localized irritation, lesions, and inflammation. These infections respond well to topical therapies such as azole and allylamine drugs (in particular, terbinafine).[8] In cases where the infection is showing signs of deeper and wider invasion into peripheral tissues, oral systemic therapy is required, usually by itraconazole or terbinafine.[9,10] Of these infections, *T. unguium* or onychomycosis is the least well addressed in terms of therapy.[8] Although generally a mild infection due to dermatophytes and also *Candida* species, it can spread to case discomfort and localized disfigurement. Effective therapy is restricted to long-term (3–6 month) oral terbinafine or itraconazole. Topical agents have very limited efficacy,[8,10] but would be preferred to limit systemic exposure of the oral agents, where liver tests are often required to monitor possible side effects.

7.14.1.2.4 Dimorphic systemic mycosis

Dimorphic systemic mycoses are rare fungal infections of the body caused by dimorphic fungal pathogens that can overcome the physiological and cellular defenses of the normal human host by changing their morphological form. They are geographically restricted, and the primary site of infection is usually pulmonary, following the inhalation of conidia (fungal spores from the associated pathogen). The severity of disease is extremely variable, sometimes inapparent (e.g., most cases of histoplasmosis), but can be extreme, leading to gross disfiguration and fatal outcomes[7] (**Table 4**). Most immunocompetent patients have a benign, self-limiting illness, and will recover without antifungal treatment. However, in patients who require treatment, amphotericin B is currently the drug of choice, often with azole therapy for maintenance, although the broad spectrum of voriconazole (Vfend) and recent demonstration of its successful treatment of these infections offer a therapeutic option in the absence of the side effects associated with the polyenes.[11–13]

Table 4 Dimorphic systemic mycoses

Disease	*Causative organism*	*Presentation*
Histoplasmosis	*Histoplasma capsulatum*	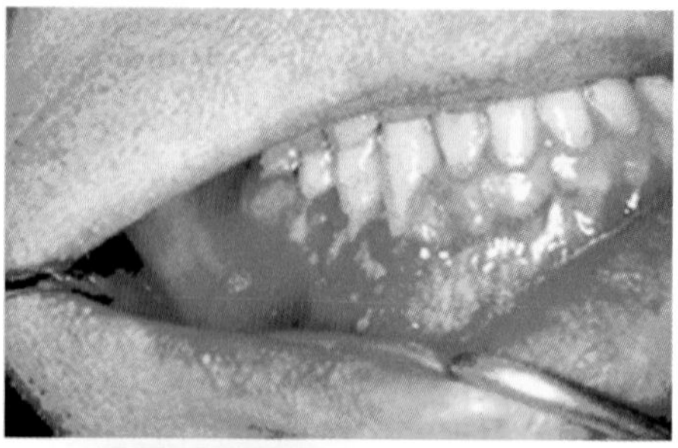
Coccidioidomycosis	*Coccidioides immitis*	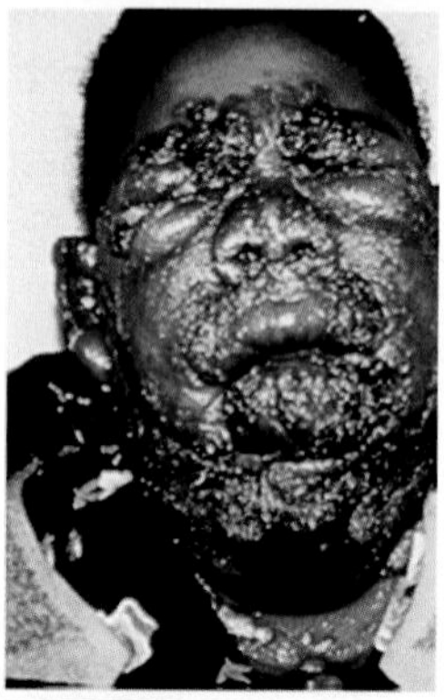
Blastomycosis	*Blastomyces dermatitidis*	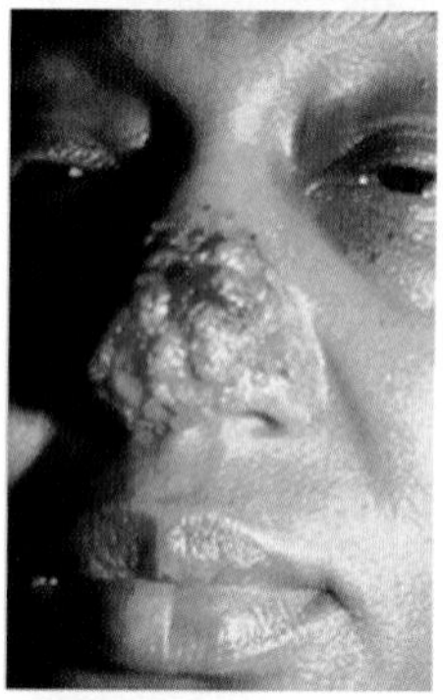
Paracoccidioidomycosis	*Paracoccidioides brasiliensis*	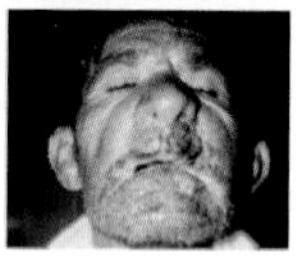

Reproduced by permission of Dr David Ellis, School of Molecular & Biomedical Science, University of Adelaide.

7.14.1.2.5 Opportunistic systemic mycoses

These are fungal infections of the body that occur almost exclusively in debilitated patients whose normal defense mechanisms are impaired. Of all the mycoses, these represent by far the greatest challenge in terms of unmet medical need. The great increase in this patient population has led to a massive increase in the frequency of these diseases, which are associated with high morbidity and mortality. The organisms involved (**Table 5** and **Figure 2**) are cosmopolitan fungi that have a very low inherent virulence. The increased incidence of these infections and the diversity of fungi causing them has paralleled the decrease in host resistance due to the emergence of AIDS, more aggressive cancer and post-transplantation chemotherapy, and the use of antibiotics, cytotoxins, immunosuppressives, and corticosteroids.[14–16] In light of these mycoses, in particular candidiasis and aspergillosis becoming more prevalent, scientists and the pharmaceutical industry have focused considerable effort toward the discovery and development of

Table 5 Opportunistic systemic mycoses

Disease	*Causative organism*
Candidiasis	*Candida albicans* and other *Candida* spp.
Aspergillosis	*Aspergillus fumigatus*
Cryptococcosis	*Cryptococcus neoformans*
Pseudallescheriasis	*Pseudallescheria boydii*
Zygomycosis (mucormycosis)	*Rhizopus*, *Mucor*, *Rhizomucor*, *Absidia*, etc.
Hyalohyphomycosis	*Penicillium*, *Paecilomyces*, *Beauveria*, *Fusarium*, *Scopulariopsis*, etc.
Phaeohyphomycosis	*Cladosporium*, *Exophiala*, *Wangiella*, *Bipolaris*, *Exserohilum*, *Curvularia*

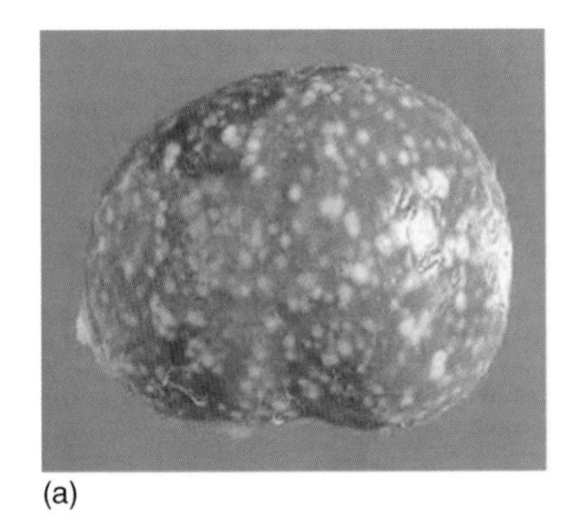

(a)

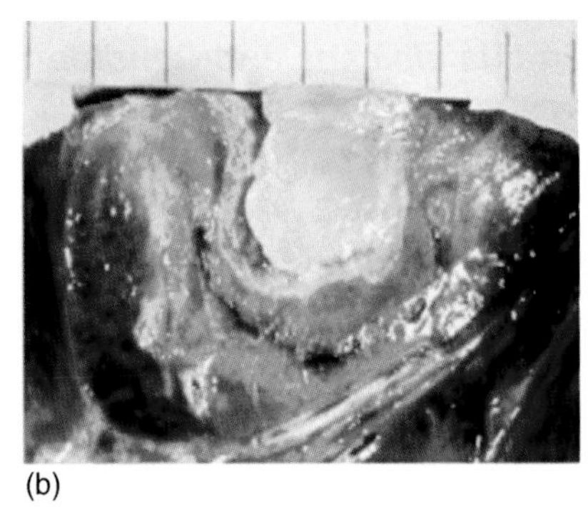

(b)

Figure 2 Tissue invasion by *Candida albicans* in kidney: (a) colony formation on tissue surface highlighted and (b) an apsergilloma (fungal ball) caused by *Aspergillus fumigatus* (excised from lung).

new systemic antifungal agents. It is this area of antifungal drug discovery that is of particular challenge to the medicinal chemist. Diagnosis of fungal infections and treatment are described in this chapter in more detail, together with emerging targets and chemistry starting points for the discovery of new antifungal drugs.

7.14.1.3 Emerging Fungal Diseases

Results of many surveys have shown that yeasts are still the most frequent cause of systemic mycoses, particularly those caused by *Candida* species. Indeed, *Candida* species account for 70–80% of invasive bloodstream fungal infections, and collectively they represent the fourth most common nosocomial bloodstream infection.[17] Among the *Candida* species causing invasive infections, *C. albicans*, *C. parapsilosis*, *C. tropicalis*, and *C. glabrata* account for about 80–90% of fungal isolates encountered in the clinical laboratory.[18] *Aspergillus* species are the predominant molds, with *Scedosporium* species now accounting for 25% of mold infections other than aspergillosis in organ transplant recipients. Following these pathogens, previously uncommon hyaline filamentous fungi (e.g., *Fusarium* species, *Acremonium* species, *Paecilomyces* species, and *Pseudallescheria boydii*), dematiaceous filamentous fungi (e.g., *Bipolaris* species, *Cladophialophora bantiana*, *Dactylaria gallopava*, *Exophiala* species, and *Alternaria* species) and yeast-like pathogens (e.g., *Trichosporon* species, *Blastoschizomyces capitatus*, *Malassezia* species, and *Rhodotorula rubra*) are increasingly encountered as causing life-threatening invasive infections that are often refractory to conventional therapies.[19,20] On the basis of past and current trends, the spectrum of fungal pathogens will continue to evolve in the settings of an expanding population of immunocompromised hosts, selective antifungal pressures, and shifting conditions in hospitals and the environment.[21,22] An expanded and refined drug arsenal combined with the further elucidation of pathogenesis and resistance mechanisms, the establishment of in vitro/in vivo correlations, and, finally, the implementation of combination (and possibly immunotherapies) will offer hope for substantial progress in prevention and treatment of these mycoses.[22,23] These include resistant strains from common infection-causing fungi, emergence of azole-resistant and consequently high mortality-associated *Candida* strains, and new species.[24] Less common and emerging fungal pathogens are often resistant to conventional antifungal therapy, and may cause severe morbidity and mortality in immunocompromised hosts. Emerging pathogens are increasingly reported as causing invasive mycoses refractory to amphotericin B therapy, although the new agents caspofungin (Cancidas) and voriconazole in particular have been shown to have impressive potency against these species in vitro and in the clinic.[25–30] Pfaller and Diekema provide an excellent review of new and emerging fungal infections together with changes in epidemiology and treatment modalities.[25]

7.14.2 Pathogenesis of Major Fungal Diseases

As discussed above, the treatments for superficial and dermatophyte-associated infections are highly effective, well tolerated, readily available, and in general address medical need. However, life-threatening systemic mycoses, especially as caused by *Candida* and *Aspergillus* species, will be reviewed in detail in this chapter in light of their high medical need, the challenges faced in their treatment, and since these are no longer considered as rare infections. To put these specific mycoses in perspective, the crude mortality from invasive aspergillosis is around 85%, while that for *Candida* bloodstream infections is 40%.[14,25,31,32]

7.14.2.1 *Candida albicans* and *Candida* Species

The ability of this otherwise commensal microorganism to cause disease is usually a consequence of host immunocompromisation rather than fungal virulence.[15,33–35] In general, pathogenesis is associated with factors that enable fungal adherence, tissue invasion, and growth. Adherence and persistence are necessary for the initiation of mycosis,[36] for which, in the case of *Candida*, mannoproteins are believed to play a role, and indeed mannose transferase has been identified as a virulence factor.[37] To assist in the invasion of host tissues, proteases and phospholipase secretion by the pathogen is enabled, and a degree of correlation exists between the proteolytic activity of the *Candida* species with pathogenicity. Invasion is also associated with a switch from the generally noninvasive yeast to the invasive hyphal form for *C. albicans*.[34,35,37–41] Detailed reviews of fungal pathogen virulence factors have been recently published.[1,35]

The extreme virulence of *Candida* species reflects the diversity of disease states caused. These are briefly described below. In addition to the text references above, comprehensive descriptions of *Candida* pathogenicity and disease states can be found online.[37a]

C. albicans and, to a lesser extent, non-*albicans* species of *Candida* cause a variety of severe infection, including:

- esophageal candidiasis (frequently associated with AIDS and severe immunosuppression following treatment for leukemia or solid tumors);
- gastrointestinal candidiasis;
- pulmonary candidiasis;
- peritonitis (usually from colonization of indwelling catheters used for peritoneal dialysis or gastrointestinal perforation);
- urinary tract and renal candidiasis (pyelonephritis);
- meningitis;
- hepatic and hepatosplenic candidiasis;
- endocarditis, myocarditis, and pericarditis;
- candidemia (*Candida* septicemia) and disseminated candidiasis.

The complications associated with *Candida* infections is compounded by the emergence of non-*albicans* species as both a colonizer and a pathogen. More than 17 different species of *Candida* have been identified as etiologic agents of bloodstream infections, with approximately 95% of these being caused by four species: *C. albicans*, *C. glabrata*, *C. parapsilosis*, and *C. tropicalis*.[25] The greater efficacy of fluconazole as a prophylactic in transplant patients has indeed led to a shift toward the less susceptible *C. glabrata* species as the predominant pathogen in such immunocompromised patients, as well as more emerging opportunistic pathogens.[25] The inherent resistance of a few non-*albicans* species to commonly used systemic antifungals such as fluconazole and amphotericin B can pose a therapeutic challenge in the management of candidaemia. Encouragingly, newer drugs, including broad-spectrum triazoles such as voriconazole, and candins such as caspofungin, are emerging as therapeutic alternatives for non-*albicans* infection treatment, as discussed elsewhere (*see* 7.15 Major Antifungal Drugs).

7.14.2.2 *Aspergillus*

Aspergillosis is a spectrum of diseases of humans (and animals) caused by members of the genus *Aspergillus*. *Aspergillus fumigatus* is overwhelmingly the predominant pathogenic species. The type of disease and severity depends upon the physiologic state of the host and the species of *Aspergillus* involved. The etiological agents are ubiquitous, and include *A. fumigatus*, *A. flavus*, *A. niger*, *A. nidulans*, and *A. terreus*.

Numerous virulence factors have been associated with aspergillosis, including phospholipases and proteases to facilitate adhesion and invasion,[38,42] and the biosynthesis of factors that enable evasion of limited host defenses in the immunosuppressed patient.[41] As with *Candida* infections, the clinical manifestations resulting from *Aspergillus* virulence are varied, although pulmonary infection and aspergilloma caused by the saprophytic colonization of preformed cavities (**Figure 2b**) is the most common. Acute invasive pulmonary aspergillosis (usually a result of prolonged neutropenia, especially in leukemia patients or in bone marrow transplant recipients) generates symptoms similar to acute bacterial pneumonia. Dissemination can lead to tissue-specific infections. Abscesses may occur in the brain (cerebral aspergillosis), kidney (renal aspergillosis), heart (endocarditis, myocarditis), bone (osteomyelitis), and gastrointestinal tract. Ocular lesions (mycotic keratitis, endophthalmitis, and orbital aspergilloma) manifest themselves as erythematous papules or macules with progressive central necrosis.

7.14.3 Host Defenses

As discussed earlier, host defense against fungal infection usually ensures that mycoses are restricted to minor superficial, nonlife-threatening cases, which are well addressed by current therapies. The rarity of systemic infections prior to the dramatic increase in the immunocompromised patient population[15,21] has highlighted the extensive role that the immune system must play in keeping opportunistic fungi at bay, in particular *C. albicans*, other *Candida* species, and *A. fumigatus*. A number of comprehensive reviews of host defenses against mycoses have been published.[1,43]

The stratified squamous epithelium of the skin normally functions as an effective barrier to *Candida*. Mechanical breakdown is the most important factor for skin infections. However, mucosal surfaces are not so well defended against colonization, although infection in this tissue is greatly influenced by the host's status and factors such as diabetes, pH/hormonal factors (vaginal thrush), and competition with endogenous flora. After mucosal and skin barriers are penetrated by fungi, and in particular *Candida* species, it is the cellular immune system complemented by humoral factors that restrict invasion.[44–46] Indeed, it is the former in particular that is compromised by HIV and many immune-compromising medical advances that has led to the massive increase in fungal infections. Host immune cell interactions and immune responses are extremely complex, and multiple arms of the immune system play roles to combat fungal pathogens, with distinct patterns that are dependent on the fungal species.[47] Humoral factors such as complement have been shown to play a role in restricting opportunistic fungal infection.[45,48] Antibody-mediated defenses are believed to play a relatively small role in host defense, whereas cell-mediated immunity is believed to be the predominant arm of the host defense. This has been shown by animal studies, specific defects in the cell-mediated immunity in patients, and the known selective cytotoxic effects of immune-compromising medications and HIV.[49–54] The high incidence of *Candida* in leukemic patients is mainly attributable to neutropenia and decreased neutrophil functionality.[43,55] Phagocytosis by neutrophils, eosinophils, monocytes, and macrophages specifically represents the most important defense mechanism against the invading pathogenic fungus.[43,49,50,56] In vitro studies have indicated the importance of Toll-like receptor (TLR) signaling in response to the fungal pathogens *C. albicans* and *A. fumigatus*. However, the functional consequences of the complex interplay between fungal morphogenesis and TLR signaling in vivo remain largely undefined.[57]

7.14.4 Diagnosis of Fungal Infection

7.14.4.1 General

The diagnosis of systemic fungal diseases can present a major dilemma for the clinician. With the majority of suspected infections in a given patient being bacterial in origin, relief by appropriate antibiotic use is a predominant course of action. However, with immune-compromised patients in particular, the ever-increasing concern is that the causative pathogen is a fungus. Given the speed of debilitation resulting from systemic mycoses, rapid diagnosis to enable a timely decision is needed, although this is not always possible. In light of the high morbidity and mortality surrounding systemic fungal infections, particularly those of *Candida* and *Aspergillus*, rapid administration even before confirmed diagnosis is often the favored option, even when the administered agent has severe side effects, such as those associated with amphotericin B. However, in recent years, the choice of antifungal drugs for systemic diseases has increased with the more widespread use of fluconazole for *C. albicans* infections (in particular maintenance therapy), to voriconazole (particularly for *Aspergillus* infections, and broad-spectrum use), and caspofungin (which is a better-tolerated infused agent compared with amphotericin).[58] A comprehensive general review of the diagnosis (and management) of fungal infections has been published.[3]

Much has been achieved recently in the improvement of laboratory diagnosis of mycoses, such as advances in blood culture systems, and the development of new biochemical assays, antigen detection assays, and molecular methodologies.[59] More standardized susceptibility testing guidelines provide for better therapeutic interventions. In an era of economic cutbacks in healthcare, future challenges include the development of cost-effective and technically simplified systems, which will provide for the early detection and identification of common and emerging fungal pathogens. The conventional diagnostic methods such as culture are often slow and lack sensitivity and specificity, resulting in additional alternative diagnostic assays. Among the most promising new techniques are the detection of fungal DNA, metabolites (see following sections), and serology. Fungal DNA can be detected with high sensitivity and specificity when performed with specimens from sterile sites such as blood.[60,61] Polymerase chain reaction (PCR) assays can be used to detect a broad range of fungal pathogens, and combined with species identification, although a lack of standardization has restricted its use. The sero-diagnosis of invasive fungal infections has become an important tool in the management of invasive fungal infections. Both serology and PCR can be used to monitor the response to antifungal therapy.[60–62] These newer approaches are generally used to help confirm the diagnosis of systemic infection; however, their impact on mortality may be greatest when they are used to screen high-risk patients.[61] While the mainstay of infectious disease diagnosis remains in microbiological culture with sensitivity follow-up, significant steps in providing rapid diagnosis in commercialized and standardized assays have been achieved.

7.14.4.1.1 Microscopy and culture of pathogenic fungi

Clinical samples for suspected fungal pathogens, in particular *Candida* and *Aspergillus* species, are varied, and include skin and nail scrapings (usually for cutaneous mycoses); urine, sputum and bronchial washings; cerebrospinal fluid, pleural fluid and blood; tissue biopsies from various visceral organs; and indwelling catheter tips (subcutaneous and systemic infections). Direct microscopy of skin and, to a lesser extent, body fluid samples for fungal identification can be achieved with staining (e.g., using 10% KOH and Parker ink or calcofluor white mounts), although direct microscopy of sterile body fluids, such as cerebrospinal fluid, vitreous humor, joint fluid, and peritoneal fluid is relatively insensitive, and positive culture is usually required to make a diagnosis. A positive culture from blood, or other sterile body fluid, or tissue biopsy (usually on blood or Sabourauds agar plates) should be considered significant. *Candida* colonies are usually white to cream colored, with a smooth, glabrous to waxy surface, whereas *Aspergillus* colonies and mycelia are white, yellow, yellow-brown, brown to black, or green in color (see **Figure 1**). Direct microscopy to identify pathogenic fungi in tissue samples is enabled by staining (e.g., cell wall staining using periodic acid, Gram stain, or methenamine silver stain – see **Figure 1**). Most *Candida* species are also characterized by the presence of well-developed pseudohyphae; however, this characteristic may be absent in certain species such as *C. glabrata*.

7.14.4.1.2 Diagnosis of fungal infection by metabolite detection

The fungal infection diagnostic kit Fungitell has been commercialized for hospital/clinical use. This serological assay is for the qualitative detection of β-glucan in the serum of patients with symptoms of, or medical conditions predisposing the patient to, invasive fungal infection, and as an aid in the diagnosis of deep-seated mycoses and fungemias. Fungitell is sensitive to a few trillionths of a gram of (1,3)-β-D-glucan, and has been shown to provide significant advantages in the diagnosis of invasive fungal infection alone and in combination with traditional blood culture methods.[63,64] The Fungitell assay does not detect certain fungal species such as the genus *Cryptococcus*, which produces very low levels of (1,3)-β-D-glucan. This assay also does not detect the Zygomycetes, such as *Absidia*, *Mucor*, and *Rhizopus*, which are not known to produce (1,3)-β-D-glucan.

Galactomannan, a component of the *Aspergillus* cell wall, is released during invasive disease, and can be detected in blood, cerebrospinal fluid, and bronchial fluid[65,66], where it has proved useful as a biomarker for infection. Exploitation of this to enable a commercialized kit for diagnosis has been achieved via enzyme-linked immunosorbent assay (ELISA) technology (Platelia Aspergillus). Many studies have validated this diagnostic kit, and report excellent sensitivities and specificities.[67,68] The degree of galactomannan antigenemia correlates with tissue fungal burden, and the course of antigenemia in patients has been shown to correspond with clinical outcome and to have prognostic value.[69] The utility of such technology has resulted from years of extensive studies on circulating *Candida* antigens, in particular mannan-based metabolites derived from the cell wall of *Candida* species.[70,71] The Platelia Candida sandwich ELISA represents a sensitive serological method, utilizing a monoclonal antibody that recognizes the sequences of linked oligomannoses present in a large number of mannoproteins extracted from different *Candida* species. An issue of concern regarding the Platelia Aspergillus antigen/antibody assay is the variable reactivity of different *Candida* species.[72]

7.14.4.1.3 Diagnosis of fungal infection by gene sequence detection

The amplification of gene sequences unique to fungi is conceptually appealing, offering the potential for rapid, specific, and sensitive diagnosis of systemic mycosis. Experimental models and clinical studies have shown PCR to be more sensitive than culture for detection of candidaemia.[73,74] Several PCR protocols have been developed utilizing different extraction methods, amplification targets, and amplicon detection formats. These have been extensively reviewed.[75] PCR-based detection of pulmonary-based infections may be more sensitive than traditional culture methods, and have been successfully evaluated in bronchoalveolar lavage specimens from patients with pulmonary infection.[76] Further prospective studies have confirmed the utility of screening patients at high risk of systemic fungal infection using PCR. Sensitivity and specificity values are encouraging, and PCR positivity precedes the development of clinical and radiological signs and administration of empirical antifungal therapy in most patients.[77]

The gradual move toward dedicated custom PCR-based kits for the clinical diagnosis of systemic fungal infections highlights the progress being made toward rapid and accurate pathogen identification.[78] Indeed, kits have been commercialized for PCR-based *Candida* detection in research samples. The Lightcycler (Roche) is intended as a research tool, but highlights its potential in the use of PCR technology for fungal diagnostics. DNA preparations from blood culture bottles, isolates from culture plates, and direct preparations from a wide range of specimens, such as urine, swabs, bronchoalveolar lavage, and sputum, can be tested. Such advances may help to develop species-specific antifungal drugs. For broad-spectrum diagnosis by this technology, amplification of a conserved region (e.g., 18S ribosomal DNA), may represent the way forward.

7.14.5 Current Antifungal Drugs: Clinical Pharmacology and Treatment Modalities

7.14.5.1 General

The antifungal drugs that are currently licensed or approaching this milestone are listed in **Table 6**. These have been discovered and developed over many years, with only a few now being used extensively in clinical practice (*see* 7.15 Major Antifungal Drugs). Despite the number of drugs, the mechanisms by which they operate are very limited, with disruption of the ergosterol function (direct or indirect) a dominant theme (see **Figure 3**). The use of the agents varies according to the type of infection for which treatment is sought. Systemic, life-threatening fungal infections are an ever-increasing medical problem faced by the immunosuppressed population. The arsenal of antifungal drugs for treatment of these mycoses is gradually improving, in particular with the emergence of new-generation triazoles and candins. However, limitations exist with the current range of antifungal drugs, in terms of an agent that combines broad-spectrum fungicidality, predictive pharmacokinetics with manageable drug–drug interactions, high safety/tolerance, and oral plus intravenous formulations. New targets and leads to increase future treatment options are needed. This chapter is aimed at describing the causative pathogens of fungal infections (in particular systemic mycoses), the current drugs for their treatment, and future directions for new drugs, for which medicinal chemists will play a key role. In the following sections, the leading therapies will be discussed in more detail.

7.14.5.2 Superficial Mycoses

As described above (*see* Sections 7.14.1.2.1 and 7.14.1.2.3), these infections are generally mild, and are well addressed by current therapies, which are mostly topical. A range of topical therapies that are available has been extensively reviewed,[2,3,15,79] with therapies dominated by azole and allylamine topical drugs, in particular terbinafine. Terbinafine has well-documented activity when used both topically and systemically for infections of nails and skin. It is active in vitro against a wide range of fungi, including dermatophytes, molds, and some yeasts.[80] Terbinafine selectively inhibits fungal squalene epoxidase (and is essentially inactive against human squalene epoxidase[81]), which converts squalene into lanosterol (see **Figure 3**). Lanosterol is eventually converted into the functional sterol ergosterol. Terbinafine leads to ergosterol deficiency and accumulation of intracellular squalene, which leads to a fungicidal endpoint, particularly so for filamentous fungi.[82] Highly effective therapy for skin infections is also achieved with azole

Table 6 Licensed and late-stage development antifungal drugs

Antifungal class (mode of action)	*Drug*	*Route*
Allylamines (squalene epoxidase)	Amorolfine	Topical
	Butenafine	Topical
	Naftifine	Topical
	Terbinafine	Oral, topical
Antimetabolite (DNA)	Flucytosine	Oral
Azoles (cytochrome P450 demethylase)	Fluconazole	Oral, intravenous
	Itraconazole	Oral, intravenous
	Ketoconazole	Oral, topical
	Posaconazole	Oral, intravenous
	Ravuconazole	Oral, intravenous
	Voriconazole	Oral, intravenous
	Clotrimazole	Topical
	Econazole	Topical
	Miconazole	Topical
	Oxiconazole	Topical
	Sulconazole	Topical
	Terconazole	Topical
	Tioconazole	Topical
Candins (glucan synthase inhibitors)	Caspofungin	Intravenous
	Micafungin	Intravenous
	Anidulafungin	Intravenous
Polyenes (ergosterol)	Amphotericin B	Intravenous
	Amphotericin B lipid complex	Intravenous
	Amphotericin B colloidal dispersion	Intravenous
	Liposomal amphotericin B	Intravenous
	Amphotericin B oral suspension	Oral
	Liposomal nystatin	Intravenous
	Topical Nystatin	Topical
	Pimaricin	Ophthalmic
Other systemic drugs	Griseofulvin	Oral
Other topical drugs	Ciclopirox olamine	Topical
	Haloprogin	Topical
	Tolnaftate	Topical
	Undecylenate	Topical

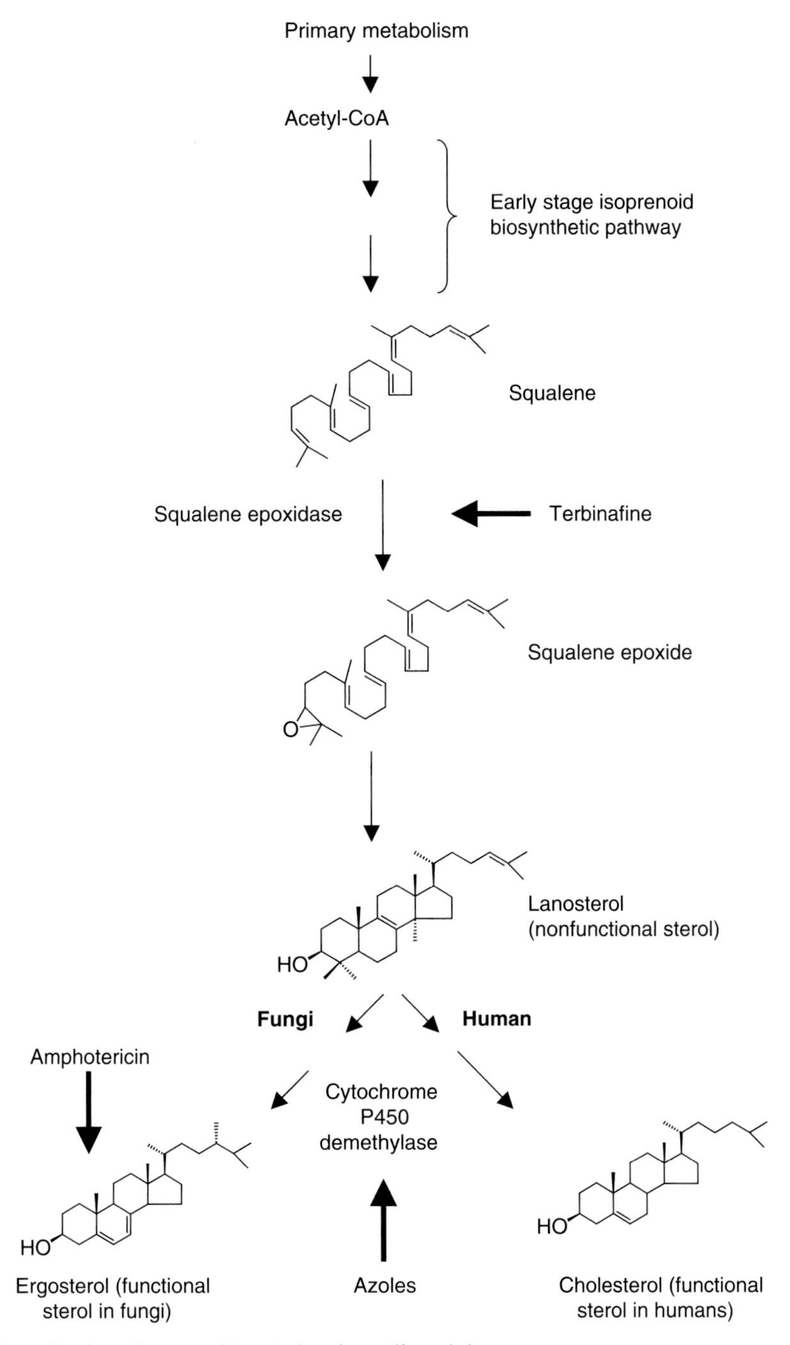

Figure 3 Sterol biosynthesis pathway and target sites for antifungal drugs.

creams and pessaries, and oral fluconazole is effective for vaginal thrush.[3] In light of these infections being in general well addressed, and thereby the need for the involvement of the medicinal chemist to discover new agents being reduced, the greater challenge of finding and prescribing antifungal drugs for the treatment of systemic infections by *Candida*, *Aspergillus*, and emerging pathogens will be focused on here.

7.14.5.3 Systemic Mycoses

Treatment of invasive fungal infections is increasingly complex. Efforts to standardize treatments as far as practicable, and set guidelines, have progressed in light of advances in diagnosis and, moreover, the greater availability, efficacy, and overall safety of newer antifungal drugs. Comprehensive reviews of treatment modalities and clinical pharmacology of currently available antifungal drugs have been published, and are very useful reference points.[82–85]

The treatment of systemic fungal infections has been an evolutionary process as newer agents have become available and comparative trials for efficacy and safety have proceeded. Generally, amphotericin led the way as a broad-spectrum intravenous agent, which has efficacy but very poor tolerance. Azole antifungals represent the next major advance and have added to the antifungal arsenal. Early azoles such as ketoconazole showed significant side effects, and were rapidly supplanted by fluconazole as a highly safe systemic antifungal, with good efficacy against the predominant pathogen *C. albicans*. Further progress was made with broader-spectrum azole itraconazole, although this met with limitations due to reduced tolerance and unpredictable pharmacokinetics relative to fluconazole. In more recent years the introduction of a new generation of azoles and glucan synthase-inhibiting candin antifungals have substantially broadened therapeutic options for clinicians. The profile of the leading antifungals amphotericin and fluconazole and new agents are reviewed in more detail below.

7.14.5.3.1 Amphotericin

Amphotericin was first isolated from *Streptomyces nodosus* in 1955. It is an amphoteric compound composed of a hydrophilic polyhydroxyl chain along one side and a lipophilic polyene hydrocarbon chain on the other (*see* 7.15 Major Antifungal Drugs). This parenternal drug was the gold standard antifungal for neutropenic patients with invasive candidiasis for many years. The three lipid formulations are liposomal amphotericin B (AmBisome), amphotericin B lipid complex (Abelcet), and amphotericin B colloidal dispersion (Amphocil). These formulations are more expensive than conventional amphotericin B, but generally have the advantage of less infusional and renal toxicity, and provide the opportunity to administer higher doses of amphotericin B.[86] Animal studies suggest similar efficacy between liposomal amphotericin B and amphotericin B lipid complex, as do retrospective comparisons of patient outcomes.[82] Amphotericin and polyenes in general operate by inhibiting sterol function in the fungal cell membrane, to exert an antifungal effect (**Figure 3**; for a review, see Polak[87]). Amphotericin achieves functional ergosterol blockade by directly binding this functional sterol, to induce dysfunctional membrane properties, including leakage of intracellular potassium, magnesium, sugars, and metabolites, and then cellular death. The mechanism of action is the same for all the preparations, and is due to the intrinsic antifungal activity of amphotericin B. Target binding is only moderately selective with respect to cholesterol binding, in particular at cholesterol-rich sites such as the kidney tubules, which underpins the mechanistic toxicity of amphotericin B.[15,88] In addition to the poor tolerance, the pharmacokinetic profile is poorly defined, and administration is limited to intravenous infusion. Amphotericin B is not bioavailable following oral administration.[87,88] Generally speaking, amphotericin B has a very broad range of activity, and is active against most pathogenic fungi. Notable exceptions include the emerging pathogens *Trichosporon beigleii* and *Fusarium* species.[89,90] It is testament to the rapid progression and high mortality and morbidity of systemic fungal infections that such a poorly tolerated drug as amphotericin has been the gold standard for antifungal therapy for so many years.

7.14.5.3.2 Fluconazole

Fluconazole is an extremely safe and effective drug for the treatment and prevention of *C. albicans* infections (*see* 7.15 Major Antifungal Drugs). Fluconazole also has the advantages of oral and intravenous administration, and has highly predictable pharmacokinetics.[91,92] However, like all azoles, it is fungistatic and not fungicidal against *C. albicans*, and has a very limited spectrum with no efficacy against molds. Furthermore, some *Candida* species are intrinsically resistant to fluconazole. Almost all *C. krusei* are resistant, and approximately 50% of *C. glabrata* isolates are resistant or have intermediate dose-dependent susceptibility to fluconazole.[3,25] Until recently, oral fluconazole has been the drug of choice for controlling oropharyngeal candidiasis in AIDS patients, and for prophylactic use (predominantly transplant and cancer patients).[93,94] The success of azoles and polyenes in addressing a high medical need has been significant, despite their mutually exclusive limitations. Both drug classes operate by inhibiting sterol function in the fungal cell membrane, to exert an antifungal effect (see **Figure 3**).[87] However, azoles indirectly target ergosterol function via fungal-specific inhibition of 14-sterol demethylase, to block the synthesis of functional ergosterol from nonfunctional lanosterol. No significant effect on human/mammalian cholesterol biosynthesis is achieved in vivo as a result of this activity, which underpins the mechanistic safety of azoles and fluconazole in particular.[95,96]

7.14.5.3.3 New antifungal drugs

The dominance of fluconazole and amphotericin B in the treatment of serious fungal infections has been recently challenged by newer drugs. Indeed, the new antifungal agents voriconazole, caspofungin, and micafungin are rapidly establishing a place in the clinician's antifungal arsenal, and are now available for the treatment of systemic fungal infections. Voriconazole is an extended-spectrum triazole that is fungicidal for many filamentous fungi, including *Aspergillus*, *Scedosporium*, *Fusarium*, and *Paecilomyces*, and is active against all species of *Candida*.[97,98] Clinical trials have shown that treatment with voriconazole cleared *Candida* from the blood as quickly as amphotericin B, with a lower incidence of treatment-related adverse events. Subsequently, it was approved for the first-line treatment of invasive aspergillosis and as salvage therapy for fungal infections caused by the pathogens *Scedosporium apiospermum* and *Fusarium* species. More recently, voriconazole was approved for use in treating esophageal candidiasis.[98] This is a reflection of its impressive activity against filamentous fungi, in particular *Aspergillus*, where it exerts fungicidal activity, as seen in a variety of in vivo and in vitro models.[99,100] Voriconazole is given either by the oral or the intravenous route. Unlike fluconazole, voriconazole is not renally cleared in light of its higher lipophilicity. Clearance is cytochrome P450-mediated. The effects of voriconazole on cytochrome P450-mediated metabolism means that clinicians must be aware of drug–drug interactions. The clinical pharmacokinetics of voriconazole have been described in detail by Purkins *et al.*[101] Voriconazole represents the most advanced and widely used third-generation azole antifungal, with posaconazole and ravuconazole (*see* 7.15 Major Antifungal Drugs) also showing promising broad-spectrum activity.

The most obvious rational starting point for a drug discovery program would be to target the cell wall of fungal pathogens, in light of its essentiality for pathogen viability, and total absence in host cells, despite the commonality of their eukaryotic origin. This approach has been aggressively targeted by drug researchers, as seen by the massive success demonstrated with the penicillin and cephalosporin series of antibacterial antibiotics.[102] Despite the complex nature of fungal cell walls (a highly structured complex of mannan, β-glucan, and *n*-acetyl glucosamine (chitin)-based polymers[103]) and their biosynthesis, highlighting a host of target areas for antifungal drug discovery, it is arguably disappointing that relatively little success has been made (with one exception) in terms of novel drug series. Numerous attempts and screens to find inhibitors of chitin synthases have been undertaken, and inhibitors such as the natural products nikkomycin and poloxin have been discovered.[104] However, these have not led to clinical candidates, and other more amenable inhibitors to a drug discovery program have not been forthcoming. Similarly, the pradamicins, which bind to mannans in the cell wall, have proven to be of limited value as an antifungal drug lead.[102] These are far from being antifungal drugs due to inherent limitations in potency and unfavorable physicochemical properties that limit their systemic bioavailability. Despite the substantial effort of medicinal chemists and biotransformation scientists to improve the pradamicins, little of therapeutic use has emerged from these starting points. One notable exception in the search for new drugs that target the cell wall of fungi is the discovery, development, and successful commercialization of the echinocandin and pneumocandin drugs (reviewed by Denning,[105] and discussed in detail in 7.15 Major Antifungal Drugs). These agents inhibit the glucan synthase enzyme responsible for the β-glucan polymer, a major constituent of the cell wall. Clinical trials have highlighted amphotericin-like efficacy, in empirical-based therapy, with superior tolerance.[86] The most successful of these to date is caspofungin (Cancidas), which is a semisynthetic analog of a pneumocandin lead.[106] Caspofungin has subsequently progressed through clinical development, and met with considerable success for the empirical therapy of fungal infections in febrile neutropenic patients. Approval was based on results from the largest prospective antifungal empirical therapy trial published to date in neutropenic patients, where caspofungin was as effective as amphotericin (Ambisome) but with fewer side effects.[86] This represents a breakthrough in supplanting the gold standard amphotericin B with a safer agent for many deep-seated life-threatening infections. Caspofungin is also active against aspergillosis,[107] and has been used successfully in combination with voriconazole for this life-threatening mycoses.[108] Caspofungin is currently the most widely used glucan synthase inhibitor for the treatment of serious fungal infections. Micafungin (Mycamine), has a very similar profile, and has been approved for prophylaxis against *Candida* infection as well as esophageal candidasis. Other agents in this class are also showing great promise. Clinical studies with anidulafungin (Eraxis) have been encouraging,[109] where recent Phase III studies showed it to be superior to fluconazole against invasive candidiasis, yet with a similar safety profile.

However, a limitation of caspofungin, and the candin class of antifungal drugs in general, is the lack of an oral route of administration. Efforts by medicinal chemists and drug discoverers are ongoing to find new small molecule templates for the inhibition of glucan synthase (e.g., see Onishi[110]), although this has currently met with limited success.

7.14.6 Antifungal Drug Discovery and Development: Future Directions and New Modes of Action

7.14.6.1 Antifungal Compounds in Clinical Development

7.14.6.1.1 Isoleucyl-tRNA synthetase inhibitors (ITRS)

ITRS is the target of icofungipen (formerly PLD-118, see **Figure 4**), which is a highly novel clinical development compound for the treatment of *Candida* infections.[111,112] Icofungipen is a synthetic derivative of the naturally occurring

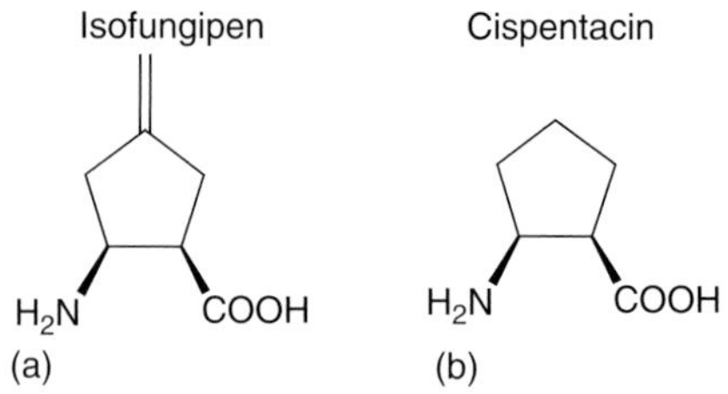

Isofungipen Cispentacin

H_2N COOH H_2N COOH
(a) (b)

Figure 4 (a) Isofungipen and (b) cispentacin.

β-amino acid cispentacin (derived from *Bacillus cereus* and *Streptomyces setonii*) that blocks isoleucyl-tRNA synthetase, resulting in the inhibition of protein synthesis and growth of fungal cells. Icofungipen, like fluconazole, has the advantage of renal clearance to drive highly predictive pharmacokinetics and avoid drug–drug interactions as a result of hepatic cytochrome P450-based clearance. Like fluconazole, it is also a small, highly soluble and orally bioavailable molecule. Although active against a number of *Candida* species, including fluconazole-resistant strains, it has only moderate antifungal potency in vitro (minimum inhibitory concentration (MIC) = 8–64 $\mu g\,mL^{-1}$ with no fungicidal activity). However, this compound showed promising efficacy in models of candidiasis (*C. albicans*), which was similar to that seen with amphotericin B.[112,113] Icofungipen showed a clear pharmacokinetic–pharmacodynamic relationship, with no overt safety alerts, as shown by a well-tolerated profile against numerous toxicity markers. In light of promising preclinical data, icofungipen has progressed to Phase II for oropharyngeal comparison with fluconazole, where efficacy was seen, albeit less than that demonstrated for fluconazole, and with a more frequent dosing regimen.[114]

Unlike most natural-product antifungal leads, the small molecular size and simple structure of icofungipen make it an attractive starting point for the medicinal chemist, since added substituents to increase potency may not compromise the physicochemical properties, and thus hamper the favorable oral bioavailability and phamacokinetics.[115]

7.14.6.1.2 Sphingolipid biosynthesis inhibitors – isophosphorylceremide synthase (IPCS)

Sphingolipid biosynthesis, although not a fungal specific anabolic pathway, has attracted much interest for the discovery of a broad-spectrum fungicidal agent. Blockade of the pathway is associated with rapid cell death, as demonstrated by a number of natural product inhibitors of the pathway.[116,117] Most importantly, one step in the pathway, the conversion of phosphoinositol into the C-1 hydroxy group of ceramide by IPCS, is associated with extremely rapid fungal death when inhibited, and is unique to fungi (**Figure 5**). Aureobasidin A (**Figure 6**), a cyclic depsipeptide produced by *Aureobasidium pullulans*, is a potent inhibitor of IPCS from *Candida* and *Aspergillus*.[118,119] The discovery of Abureobasidin A was prompted by observations of very potent and rapid fungicidal activity (IC_{50} against IPCS in the nanomolar range, and the MIC against *Candida* species in the sub-microgram per milliliter range),[120–122] and subsequent identification and characterization of resistance-encoding genes highlighting IPCS as the molecular target.[123–125] The absence of the target and lack of cytotoxicity prompted interest in this target, and, indeed, aureobasidin as a potential new antifungal agent or starting point. Aureobasidin-based compounds have progressed from animal models of infection, through to Phase I clinical assessment. An analog was shown to be superior to both fluconazole and amphotericin B in mice with systemic candidiasis.[121] Limitations in its spectrum (highly active against *Candida* species, but reduced activity against *Aspergillus*) have been addressed at the preclinical stage by semisynthetic modifications, which have enabled activity against *Aspergillus* through reduction of efflux susceptibility.[118,126]

7.14.6.2 New Antifungal Targets and Leads – Preclinical Discovery

The treatment of fungal infection, in particular systemic mycoses, is based upon the discovery and development of compounds that inhibit a critical mechanism in the pathogen, but not in the host. Given the widespread penetration associated with systemic mycoses, this represents a major challenge to medicinal chemists and drug discovery scientists. The rationale for finding an essential target that is specific for the pathogen has existed for many years following the historic discovery of antibiotic action against bacteria. Fungi poses an additional challenge in that, like their human hosts, they consist of eukaryotic cell(s), and hence share high homology in their genome, and thereby proteome and cellular machinery. The holy grail for a new antifungal drug that addresses the shortcomings of current therapies is a low-dose fungicidal agent with high safety and tolerance that is administered by oral and non-oral routes, and is active against the spectrum of fungi that can cause systemic infections. Some of the approaches and progress toward the goal of finding appropriate new antifungal targets and leads are described below. The combined approaches of synthetic, semisynthetic, and natural-product screening has been exceptionally important for this discovery of

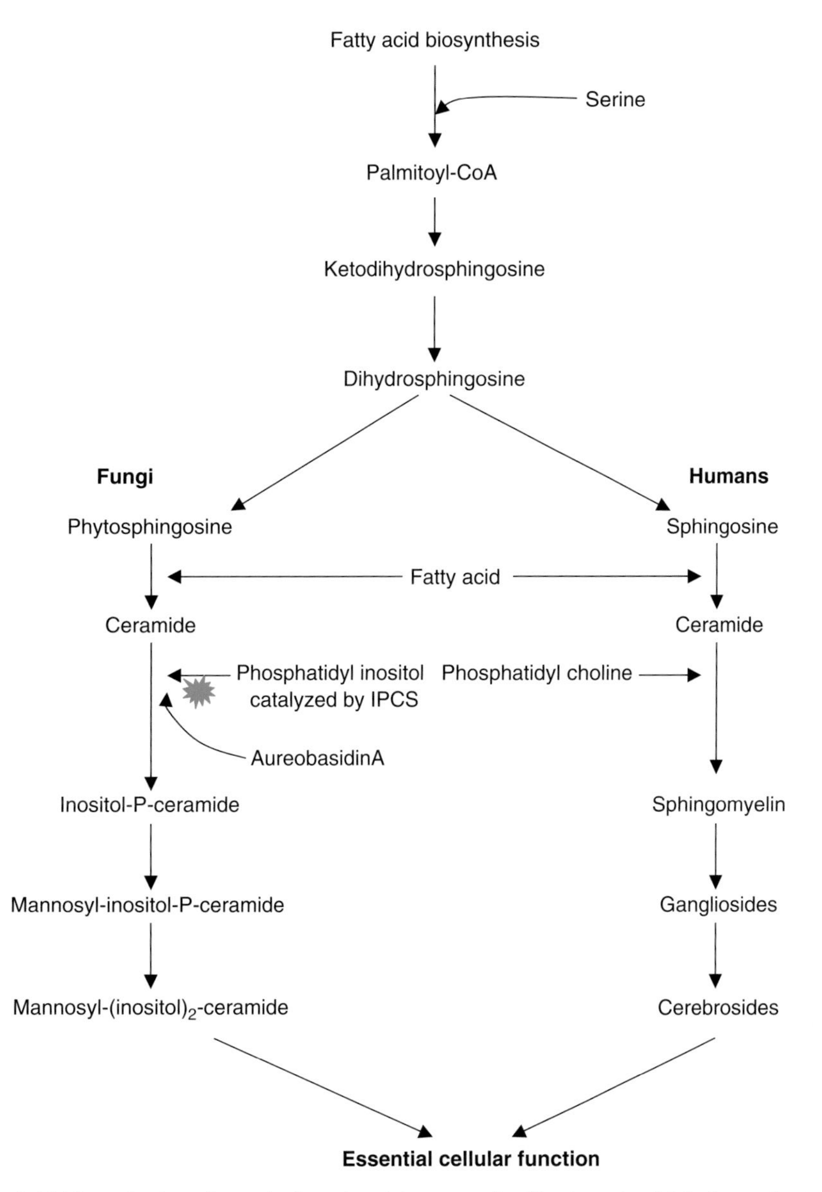

Figure 5 Sphingolipid biosynthesis pathway in fungal and human cells. The branchpoint in the pathway, with separate subsequent enzyme reaction, provides scope for selective inhibition of the fungal pathway. The target for the fungicidal compound aureobasidin A is highlighted.

Figure 6 Aureobasidin A.

antimicrobial agents. This approach has also led to the discovery of new antifungal targets of great potential. An excellent review of natural-product antifungal leads, clinical candidates, and details of their mode of action has been done by Vicente *et al.*[117]

7.14.6.2.1 *N*-Myristoyltransferase (NMT)

NMT catalyzes the co-translational, covalent attachment of myristate to the N-terminal glycine residue of a number of eukaryotic proteins involved in cellular growth and signal transduction, including ADP-ribosylation factors (ARFs). The catalytic cycle and site for known inhibitors is highlighted in **Figure 7**. NMT has been shown to be an essential enzyme in all fungal species tested to date, including *Candida* species,[127,128] and consequently has been targeted for antifungal drug discovery. NMT isoforms have 50–80% amino acid sequence identity between human and fungal pathogens. NMT is also believed to play an essential role in human cell viability and function, which does not make it an obvious antifungal target. Peptide-based and depeptidized inhibitors[129–131] showed NMT inhibition in fungal species, and exerted a measurable inhibition of the function of this enzyme in whole fungal cells via the reduction of ARF myristoylation.[132] This inhibition was selective for *C. albicans* NMT over human, but did not impose sufficient potency for progression as a drug candidate. The leading compounds appeared limited due to their partial peptidic nature, leaving them prone to cleavage and reduced cell permeability. However, high-throughput screening based on scintillation proximity assay technologies run at Pfizer[133] identified the benzothiazole lead CP-123,457 (**Table 7**) as a potent small molecule (IC_{50} *C. albicans* NMT = 1.4 μM), with no measurable potency against the human isoform. A medicinal chemistry campaign on this provided UK-370,485 (IC_{50} *C. albicans* NMT = 40 nM) and UK-362,091 (**Table 7**). The latter compound had dramatically increased potency (IC_{50} *C. albicans* NMT = 10 nM), which was sufficient to drive whole-cell antifungal activity (*C. albicans* MIC = 3.1 μg mL^{-1}). However, this compound and close analogs showed antifungal activity against *C. albicans* alone, with no activity against other pathogens, including *A. fumigatus* and non-*albicans* species of *Candida*. In addition, mechanistic studies indicated that although >90% inhibition of NMT activity in whole *Candida* cells was apparent at its MIC and greater, fungicidality was not achieved. Crystal structures of fungal pathogen NMT have greatly helped in the design of inhibitors.[134–136] The NMT active site residues that were responsible for the high affinity of benzothiazole inhibitors specifically were identified by co-crystallographic studies[133] (**Figure 8**). Bioinformatic studies indicated key residue interactions that were responsible for the narrow spectrum and human selectivity. These were the Phe339 and Ile111 residues in the NMT active site, which formed strong hydrophobic interactions with the benzothiazole moiety of the inhibitor (**Figures 9** and **10**). These interactions did not form in the NMT isoforms from *Aspergillus* and non-*albicans* species of *Candida*, which explained the inactivity and the narrow spectrum of the inhibitors. Similar series to those discovered by Pfizer have since been found with a similar narrow spectrum.[137,138]

7.14.6.2.2 Ribonucleic acid (RNA) polymerase (Pol) III

Traditionally, most agricultural and medicinal antifungal agents have been discovered through whole-cell antifungal testing, with the eventual mode of action elucidation often accomplished later. This approach has been recently implemented using high-throughput screening technologies with follow-up mode-of-action studies to identify new lead compounds and targets for antifungal drug discovery. An example of this includes the discovery of an antifungal RNA Pol III inhibitor.[139] High-throughput screening of a large compound file was undertaken by measuring compound-dependent inhibition of *C. albicans* growth (turbidometric readout). The benzothiophene UK-118,005 (**Table 8**)

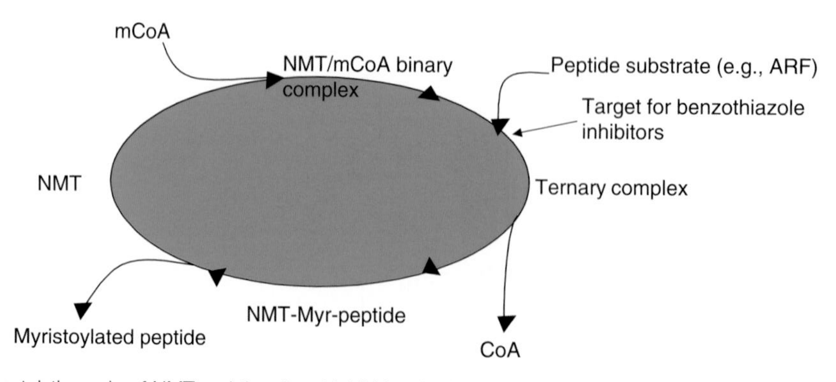

Figure 7 The catalytic cycle of NMT and the site of inhibition for benzothiazole and peptide-based inhibitors.

Table 7 Benzothiazole inhibitors of NMT and antifungal properties

Compound	*Properties*
CP-123,457	*C. albicans* NMT $IC_{50} = 1.4\,\mu M$ Human NMT $IC_{50} > 100\,\mu M$ Antifungal MIC $> 200\,\mu g\,mL^{-1}$ (all species)
UK-370,485	*C. albicans* NMT $IC_{50} = 40\,nM$ Human NMT $IC_{50} > 100\,\mu M$ Antifungal MIC $> 200\,\mu g\,mL^{-1}$ (all species)
UK-362,091	*C. albicans* NMT $IC_{50} = 10\,nM$ Human NMT $IC_{50} > 100\,\mu M$ *C. albicans* MIC $= 3.1\,\mu g\,mL^{-1}$ Non-*albicans* and *Aspergillus* MIC $> 200\,\mu g\,mL^{-1}$

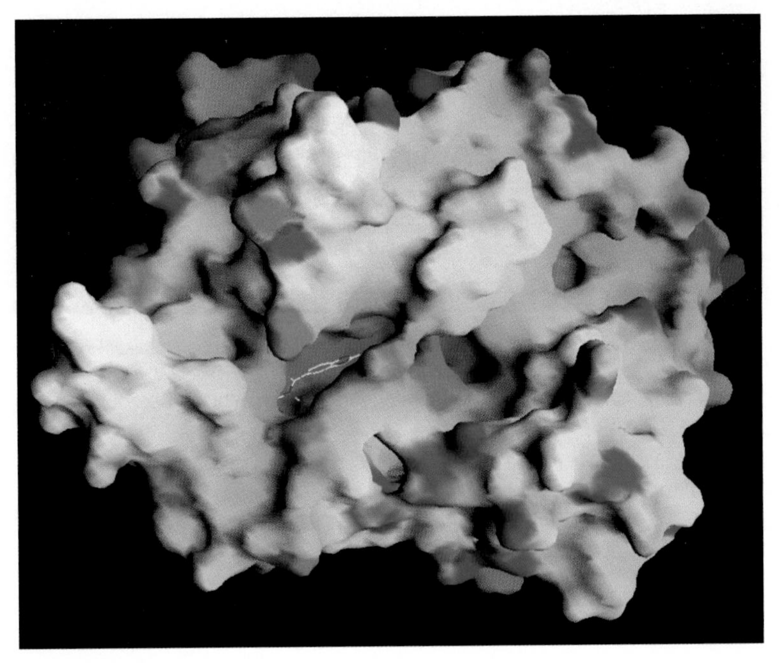

Figure 8 Three-dimensional image of benzothiazole occupation of the *Candida albicans* NMT active site cleft (peptide-binding site) as derived from x-ray diffraction of NMT-inhibitor co-crystals.

was found to have potent antifungal activity against *Candida* (all species) and *Aspergillus*, with a high therapeutic index relative to human cell line cytotoxicity. A genetic approach utilizing the yeast *Saccharomyces cerevisiae* was used to identify the target of antifungal compounds. Three lines of evidence showed that UK-118,005 inhibited cell growth by targeting RNA Pol III in yeast. First, a dominant mutation in the g domain of Rpo31p, the largest subunit of RNA Pol III, conferred resistance to the compound. Second, UK-118,005 rapidly inhibited tRNA synthesis in wild-type cells but

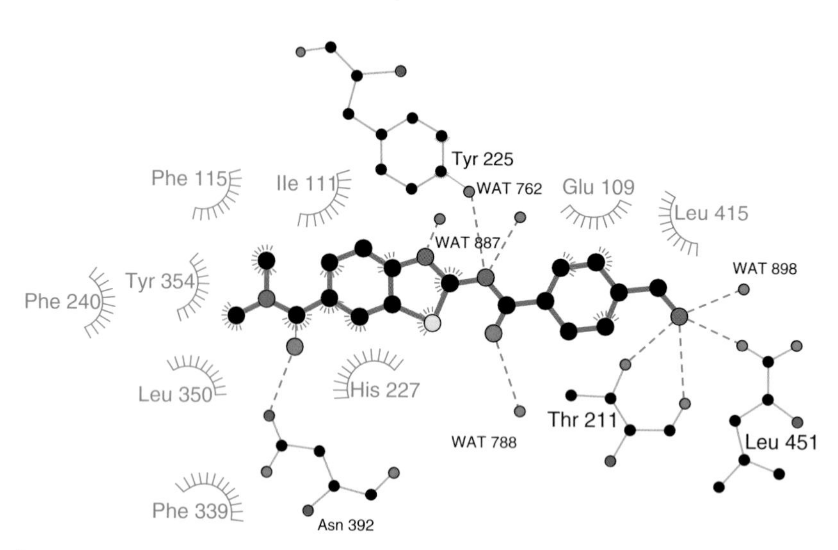

Figure 9 Two-dimensional representation of benzothiazole (UK-370,485) interactions with active site amino acid residues in the *Candida albicans* NMT.

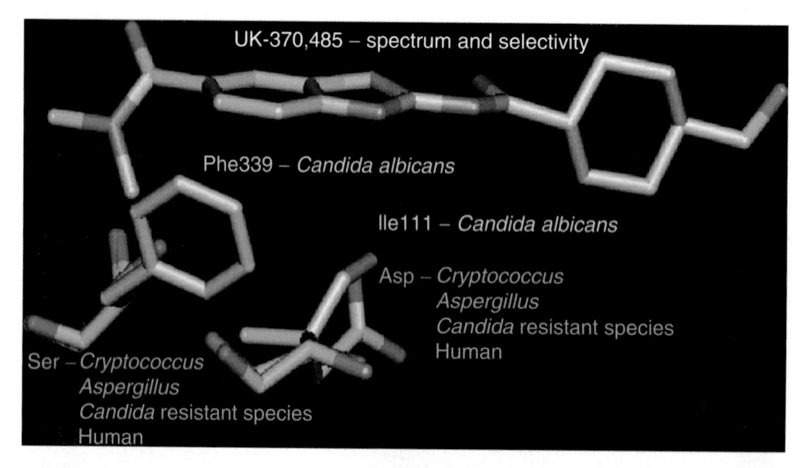

Figure 10 Interaction of key amino acid residues of *Candida albicans* NMT with a benzothiazole highlighting inhibitor. Van der Waals interaction of the benzothiazole moiety with the hydrophobic Ile111 and Phe339 residues of *C. albicans* NMT. The equivalent polar residues (incompatible with respect to forming interactions with the benzothiazole) from other species are highlighted.

Table 8 Antifungal profile of UK 118,005

Structure	*MIC (μg mL⁻¹)*			
	C. albicans	*C. glabrata*	*A. fumaigatus*	*S. cerevisiae*
	6.3	3.1	6.3	0.78

not in UK-118,005-resistant mutants. Third, in biochemical assays, UK-118,005 inhibited tRNA gene transcription in vitro by the wild-type but not the mutant Pol III enzyme. Further examination showed UK-118,005 to inhibit RNA Pol III transcription systems derived from *C. albicans* and human cells. The identification of these inhibitors demonstrates that RNA Pol III can be targeted by small synthetic molecules, to exert a potent antifungal effect. UK-118,005 showed broad-spectrum antifungal activity (MIC = 3.1–12.5 μg mL^{-1} against *Candida* species and *A. fumigatus* (**Table 8**). Unfortunately, this compound showed little selectivity for the fungal/human enzyme in controlled kinetic assays, indicating the need for an extensive medicinal chemistry campaign to generate a fungal selective inhibitor.

7.14.6.2.3 Acetyl-CoA carboxylase (ACCase)

ACCase catalyses the ATP-dependent carboxylation of acetyl-CoA to malonyl-CoA in a multistep reaction (**Figure 11**). This is the first committed step in fatty acid synthesis, is rate-limiting for the pathway, and is tightly regulated. Inhibition of ACCase in fungi results in fatty acid depletion, leading to rapid cell death due to membrane dysfunction.[140,141]

ACCase is a universal enzyme, and has been researched in detail from a number of species. It has been particularly well studied in plants, as ACCase is the target site for numerous highly successful herbicides. Herbicidal ACCase inhibitors show great potency against grass ACCase activity, but do not inhibit mammalian, fungal, or even broadleaf plant ACCase.[142] This underpins their success as nontoxic and selective grassweed killers, and highlights ACCase as having scope for selective inhibition with a lethal effect on the target species. The discovery of soraphen A (**Figure 12**), a potent fungal ACCase inhibitor ($IC_{50} \approx 30$ nM) with broad spectrum and extremely rapid fungicidal activity (MIC against *Candida* species in the sub-microgram per millilter range) triggered considerable interest in ACCase as a target for antifungal drug discovery.[141] It is therefore not surprising that gene knockout studies show ACCase to be an essential enzyme in fungi.[143,144]

In light of the precedent for selective inhibition of ACCase and its association with extremely potent fungicidal activity, high-throughput screening for this and other targets in the lipid pathway to find tractable chemical lead matter are being pursued.[145]

7.14.6.2.4 Fungal protein biosynthesis inhibitors

Protein synthesis has long been considered as an attractive target in the development of antimicrobial agents, in light of the widespread use of antibacterial antibiotics that target the specific areas of this process. However, application of this idea to the field of antifungal therapy is not an easy task, due to the eukaryotic rather than prokaryotic nature of

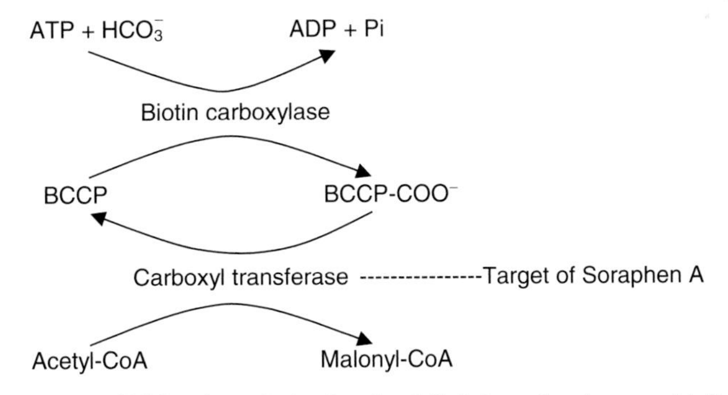

Figure 11 Acetyl-CoA carboxylase (ACCase) catalysis of malonyl-CoA formation from acetyl-CoA. Catalysis is a multistep process, taking place at different catalytic sites of a single multifunctional protein. The reaction proceeds via the ATP-dependent carboxylation of a biotin group at the biotin carboxylase domain of ACCase, with the transfer of this (by a biotin carboxyl carrier peptide – BCCP) to the carboxyl transferase domain of ACCase.

Figure 12 Soraphen A.

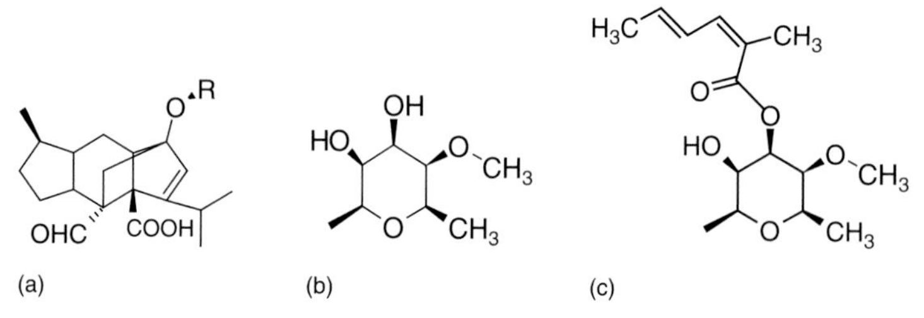

Figure 13 Antifungal sordarins. (a) Core template, (b) parent sordarin (R group), and (c) GR1305402 (R group).

fungi, and therefore the great degree of similarity between the fungal and mammalian protein synthesis machineries. Two soluble elongation factors show some fungal specificity: EF3, a factor that is required by fungal ribosomes only, and EF2, which has been demonstrated to possess at least one functional distinction from its mammalian counterpart. The sordarins are the most important family of antifungal agents acting at the protein synthesis level. Compounds in this class inhibit in vitro translation in *C. albicans*, *C. tropicalis*, *C. kefyr*, and *C. neoformans*, to varying degrees.[117] The lack of activity of the sordarins against *C. krusei*, *C. glabrata*, and *C. parapsilosis*, in comparison with their extremely high levels of potency against *C. albicans*, suggests that these compounds have a highly specific binding site, which may also be the basis for the greater selectivity of these compounds in inhibiting fungal, but not the mammalian, protein synthesis. The most advanced inhibitors of fungal protein biosynthesis are analogs of the natural product lead sordarin lead, GR135402 (see **Figure 13**).[146,147] Its spectrum of activity includes *C. albicans*, *C. tropicalis*, and *C. neoformans*, where impressive antifungal activity has been observed in vitro (MIC $<1\,\mu g\,mL^{-1}$ in many cases), but not in other *Candida* or *Aspergillus* species, which severely restricts its potential as a quality lead. Efficacy has been seen in animal models,[147] although at high dose and predominantly via nonoral routes, reflecting the potential for rapid clearance and limited oral bioavailability.

7.14.7 Further Reading

The textbook edited by Warnock and Richardson[3] and the two volumes edited by San-Blas and Calderone[1,148] are convenient and comprehensive reference points for many aspects of fungal infections and antifungal drugs. In addition, there are a number of useful web resources,[37,149,150] and Chapter 7.15 should be consulted.

References

1. San-Blas, G.; Calderone, R., Eds. *Pathogenic Fungi. Structural Biology and Taxonomy*; Caister Academic Press: Norwich, UK, 2004; Vol. 1.
2. Brandt, M. E.; Warnock, D. W. *J. Chemother.* **2003**, *15*, 36–47.
3. Warnock, D.; Richardson, M. *Fungal Infection Diagnosis and Management*, 3rd ed.; Blackwell: Oxford, 2003.
4. Dutriaux, C.; Saint-Cyr, I.; Desbois, N.; Cales-Quist, D.; Diedhou, A.; Boisseau-Garsaud, A. M. *Ann. Dermatol. Venereol.* **2005**, *132*, 259–262.
5. McGinnis, M. R.; Nordoff, N.; Li, R. K.; Pasarell, L.; Warnock, D. W. *Med. Mycol.* **2001**, *39*, 369–371.
6. Al-Abdely, H. M. *Curr. Opin. Infect. Dis.* **2004**, *17*, 527–532.
7. Allelo, A.; Hay, R. J. Medical Mycology. In *Topley & Wilson's Microbiology and Infectious Infections*, 9th ed.; Edward Arnold: London, 1997; Vol. 4.
8. Gupta, A. K.; Cooper, E. A.; Ryder, J. E.; Nicol, K. A.; Chow, M.; Chaudhry, M. M. *Am. J. Clin. Dermatol.* **2004**, *5*, 225–237.
9. Baran, R.; Gupta, A. K.; Pierard, G. E. *Expert Opin. Pharmacother.* **2005**, *6*, 609–624.
10. Gupta, A. K.; Ryder, J. E.; Lynch, L. E.; Tavakkol, A. *J. Drugs Dermatol.* **2005**, *4*, 302–308.
11. Prabhu, R. M.; Bonnell, M.; Currier, B. L.; Orenstein, R. *Clin. Infect. Dis.* **2004**, *39*, e74–e77.
12. Cortez, K. J.; Walsh, T. J.; Bennett, J. E. *Clin. Infect. Dis.* **2003**, *36*, 1619–1622.
13. Proia, L. A.; Tenorio, A. R. *Antimicrob. Agents Chemother.* **2004**, *48*, 2341.
14. Groll, A. H.; Shah, P. M.; Mentzel, C.; Schneider, M.; Just-Nuebling, G.; Huebner, K. *J. Infect.* **1996**, *33*, 23–32.
15. Bodey, G. *Candidiasis. Pathogenesis, Diagnosis, and Treatment*, 2nd ed.; Raven Press: New York, 1993.
16. Cunha, B. A. *Crit. Care Clin.* **1998**, *14*, 263–282.
17. Selvarangan, R.; Bui, U.; Limaye, A. P.; Cookson, B. T. *J. Clin. Microbiol.* **2003**, *41*, 5660–5664.
18. Pfaller, M. A. *Clin. Infect. Dis.* **1996**, *22*, S89–S94.
19. Hajjeh, R. A.; Sofair, A. N.; Harrison, L. H.; Lyon, G. M.; Arthington-Skaggs, B. A.; Mirza, S. A.; Phelan, M.; Morgan, J.; Lee-Yang, W.; Ciblak, M. A. et al. *J. Clin. Microbiol.* **2004**, *42*, 1519–1527.
20. Hobson, R. P. *J. Hosp. Infect.* **2003**, *55*, 159–168.
21. Fleming, R. V.; Walsh, T. J.; Anaissie, E. J. *Infect. Dis. Clin. North Am.* **2002**, *16*, 915–933.
22. Walsh, T. J.; Groll, A.; Hiemenz, J.; Fleming, R.; Roilides, E.; Anaissie, E. *Clin. Microbiol. Infect.* **2004**, *10*, 48–66.
23. Groll, A. H.; Walsh, T. J. *Clin. Microbiol. Infect.* **2001**, 7, 8–24.
24. Baddley, J. W.; Moser, S. A. *Clin. Lab. Med.* **2004**, *24*, 721–735.
25. Pfaller, M. A.; Diekema, D. J. *J. Clin. Microbiol.* **2004**, *42*, 4419–4431.

26. Garbino, J.; Ondrusova, A.; Baglivo, E.; Lew, D.; Bouchuiguir-Wafa, K.; Rohner, P. *Scand. J. Infect. Dis.* **2002**, *34*, 701–703.
27. Mattei, D.; Mordini, N.; Lo Nigro, C.; Gallamini, A.; Osenda, M.; Pugno, F.; Viscoli, C. *Mycoses* **2003**, *46*, 511–514.
28. Lewis, R. E.; Wiederhold, N. P.; Klepser, M. E. *Antimicrob. Agents Chemother.* **2005**, *49*, 945–951.
29. Schaenman, J. M.; Digiulio, D. B.; Mirels, L. F.; McClenny, N. M.; Berry, G. J.; Fothergill, A. W.; Rinaldi, M. G.; Montoya, J. G. *J. Clin. Microbiol.* **2005**, *43*, 973–977.
30. Husain, S.; Munoz, P.; Forrest, G.; Alexander, B. D.; Somani, J.; Brennan, K.; Wagener, M. M.; Singh, N. *Clin. Infect. Dis.* **2005**, *40*, 89–99.
31. McNeil, M. M.; Nash, S. L.; Hajjeh, R. A.; Phelan, M. A.; Conn, L. A.; Plikaytis, B. D.; Warnock, D. W. *Clin. Infect. Dis.* **2001**, *33*, 641–647.
32. Rees, J. R.; Pinner, R. W.; Hajjeh, R. A.; Brandt, M. E.; Reingold, A. L. *Clin. Infect. Dis.* **1998**, *27*, 1138–1147.
33. Ghannoum, M. A.; Abu-Elteen, K. H. *Mycoses* **1990**, *33*, 265–282.
34. Yang, Y. L. *J. Microbiol. Immunol. Infect.* **2003**, *36*, 223–228.
35. Rinaldi, M. G. Biology and Pathogenicity of *Candida* Species. In *Candidiasis*; Bodey, G., Ed.; Raven Press: New York, 1993, pp 1–20.
36. Bignell, E.; Rogers, T.; Haynes, K. Host Recognition by Fungal Pathogens. In *Pathogenic Fungi: Host Interactions and Emerging Strategies for Control*; San-Blas, G., Calderone, R. A, Eds.; Caister Academic Press: Norwich, 2004; Chapter 1, pp 3–28.
37. Munro, C. A.; Bates, S.; Buurman, E. T.; Hughes, H. B.; Maccallum, D. M.; Bertram, G.; Atrih, A.; Ferguson, M. A.; Bain, J. M.; Brand, A. et al. *J. Biol. Chem.* **2005**, *280*, 1051–1060.
37a. Mycology Online. http://www.mycology.adelaide.edu.au (accessed April 2006).
38. Shen, D. K.; Noodeh, A. D.; Kazemi, A.; Grillot, R.; Robson, G.; Brugere, J. F. *FEMS Microbiol. Lett.* **2004**, *239*, 87–93.
39. Fotedar, R.; Al-Hedaithy, S. S. *Mycoses* **2005**, *48*, 62–67.
40. Dolan, J. W.; Bell, A. C.; Hube, B.; Schaller, M.; Warner, T. F.; Balish, E. *Med. Mycol.* **2004**, *42*, 439–447.
41. Langfelder, K.; Streibel, M.; Jahn, B.; Haase, G.; Brakhage, A. A. *Fungal Genet. Biol.* **2003**, *38*, 143–158.
42. Kogan, T. V.; Jadoun, J.; Mittelman, L.; Hirschberg, K.; Osherov, N. *J. Infect. Dis.* **2004**, *189*, 1965–1973.
43. Varvitarian, S.; Smith, C. B. Pathogenesis, Host Resistance and Predisposing Factors. In *Candidiasis. Pathogenesis, Diagnosis, and Treatment*, 2nd ed.; Bodey, G., Ed.; Raven Press: New York, 1993, pp 59–84.
44. Ashman, R. B.; Farah, C. S.; Wanasaengsakul, S.; Hu, Y.; Pang, G.; Clancy, R. L. *Immunol. Cell Biol.* **2004**, *82*, 196–204.
45. Speth, C.; Rambach, G.; Lass-Florl, C.; Dierich, M. P.; Wurzner, R. *Mycoses* **2004**, *47*, 93–103.
46. Polonelli, L.; Casadevall, A.; Han, Y.; Bernardis, F.; Kirkland, T. N.; Matthews, R. C.; Adriani, D.; Boccanera, M.; Burnie, J. P.; Cassone, A. et al. *Med. Mycol.* **2000**, *38*, 281–292.
47. Clemons, K. V.; Calich, V. L.; Burger, E.; Filler, S. G.; Grazziutti, M.; Murphy, J.; Roilides, E.; Campa, A.; Dias, M. R.; Edwards, J. E., Jr., et al. *Med. Mycol.* **2000**, *38*, 99–111.
48. Morelli, R.; Rosenberg, L. *T. J. Immunol.* **1971**, *107*, 476–480.
49. Romani, L. *Curr. Opin. Microbiol.* **1999**, *2*, 363–367.
50. Romani, L. *J. Leukoc. Biol.* **2000**, *68*, 175–179.
51. Kalfa, V. C.; Roberts, R. L.; Stiehm, E. R. *Ann. Allergy Asthma Immunol.* **2003**, *90*, 259–264.
52. Mencacci, A.; Cenci, E.; Bacci, A.; Montagnoli, C.; Bistoni, F.; Romani, L. *Curr. Pharm. Biotechnol.* **2000**, *1*, 235–251.
53. Standiford, T. J. *Curr. Opin. Pulm. Med.* **1997**, *3*, 81–88.
54. Traynor, T. R.; Huffnagle, G. B. *Med. Mycol.* **2001**, *39*, 41–50.
55. Boktour, M. R.; Kontoyiannis, D. P.; Hanna, H. A.; Hachem, R. Y.; Girgawy, E.; Bodey, G. P.; Raad, I. I. *Cancer* 2004, *101*, 1860–1865.
56. Mansour, M. K.; Levitz, S. M. *Curr. Opin. Microbiol.* **2002**, *5*, 359–365.
57. Bellocchio, S.; Montagnoli, C.; Bozza, S.; Gaziano, R.; Rossi, G.; Mambula, S. S.; Vecchi, A.; Mantovani, A.; Levitz, S. M.; Romani, L. *J. Immunol.* **2004**, *172*, 3059–3069.
58. Groll, A. H.; Walsh, T. J. *Expert Opin. Investig. Drugs* **2001**, *10*, 1545–1558.
59. O'Shaughnessy, E. M.; Shea, Y. M.; Witebsky, F. G. *Infect. Dis. Clinics N. Am.* **2003**, *17*, 135–158.
60. Verweij, P. E.; Meis, J. *Transplant Infect. Dis.* **2000**, *2*, 80–87.
61. McLintock, L. A.; Jones, B. L. *Br. J. Haematol.* **2004**, *126*, 289–297.
62. Kappe, R.; Rimek, D. *Progr. Drug Res.* **2003**, *2*, 39–57.
63. Pazos, C.; Ponton, J.; Del Palacio, A. *J. Clin. Microbiol.* **2005**, *43*, 299–305.
64. Odabasi, Z.; Mattiuzzi, G.; Estey, E.; Kantarjian, H.; Saeki, F.; Ridge, R. J.; Ketchum, P. A.; Finkelman, M. A.; Rex, J. H.; Ostrosky-Zeichner, L. *Clin. Infect. Dis.* **2004**, *39*, 199–205.
65. Viscoli, C.; Machetti, M.; Gazzola, P.; De Maria, A.; Paola, D.; Van Lint, M. T.; Gualandi, F.; Truini, M.; Bacigalupo, A. *J. Clin. Microbiol.* **2002**, *40*, 1496–1499.
66. Sanguinetti, M.; Posteraro, B.; Pagano, L.; Pagliari, G.; Fianchi, L.; Mele, L.; La Sorda, M.; Franco, A.; Fadda, G. *J. Clin. Microbiol.* **2003**, *41*, 3922–3925.
67. Maertens, J.; Glasmacher, A.; Selleslag, D.; Ngai, A.; Ryan, D.; Layton, M.; Taylor, A.; Sable, C.; Kartsonis, N. *Clin. Infect. Dis.* **2005**, *41*, 9–14.
68. Maertens, J.; Verhaegen, J.; Lagrou, K.; Van Eldere, J.; Boogaerts, M. *Blood* **2001**, *97*, 1604–1610.
69. Boutboul, F.; Alberti, C.; Leblanc, T.; Sulahian, A.; Gluckman, E.; Derouin, F.; Ribaud, P. *Clin. Infect. Dis.* **2002**, *34*, 939–943.
70. Weiner, M. H.; Yount, W. J. *J. Clin. Invest.* **1976**, *58*, 1045–1053.
71. Lehmann, P. F.; Reiss, E. *Mycopathologia* **1980**, *70*, 89–93.
72. Yera, H.; Sendid, B.; Francois, N.; Camus, D.; Poulain, D. *Eur. J. Clin. Microbiol. Infect. Dis.* **2001**, *20*, 864–870.
73. Morace, G.; Pagano, L.; Sanguinetti, M.; Posteraro, B.; Mele, L.; Equitani, F.; D'Amore, G.; Leone, G.; Fadda, G. *J. Clin. Microbiol.* **1999**, *37*, 1871–1875.
74. Wahyuningsih, R.; Freisleben, H. J.; Sonntag, H. G.; Schnitzler, P. *J. Clin. Microbiol.* **2000**, *38*, 3016–3021.
75. Chen, S. C.; Halliday, C. L.; Meyer, W. *Med. Mycol.* **2002**, *40*, 333–357.
76. Kawazu, M.; Kanda, Y.; Goyama, S.; Takeshita, M.; Nannya, Y.; Niino, M.; Komeno, Y.; Nakamoto, T.; Kurokawa, M.; Tsujino, S. et al. *Am. J. Hematol.* **2003**, *72*, 27–30.
77. Jordanides, N. E.; Allan, E. K.; McLintock, L. A.; Copland, M.; Devaney, M.; Stewart, K.; Parker, A. N.; Johnson, P. R.; Holyoake, T. L.; Jones, B. L. *Bone Marrow Transplant.* **2005**, *35*, 389–395.
78. Loeffler, J.; Dorn, C.; Hebart, H.; Cox, P.; Magga, S.; Einsele, H. *Diag. Microbiol. Infect. Dis.* **2003**, *45*, 217–220.
79. Huang, D. B.; Ostrosky-Zeichner, L.; Wu, J. J.; Pang, K. R.; Tyring, S. K. *Dermatol. Ther.* **2004**, *17*, 517–522.
80. Petranyi, G.; Meingassner, J. G.; Mieth, H. *Antimicrob. Agents Chemother.* **1987**, *31*, 1558–1561.
81. Favre, B.; Ryder, N. S. *Arch. Biochem. Biophys.* **1997**, *340*, 265–269.

82. Slavin, M. A.; Szer, J.; Grigg, A. P.; Roberts, A. W.; Seymour, J. F.; Sasadeusz, J.; Thursky, K.; Chen, S. C.; Morrissey, C. O.; Heath, C. H. et al. *Intern. Med. J.* **2004**, *34*, 192–200.
83. Pappas, P. G.; Rex, J. H.; Sobel, J. D.; Filler, S. G.; Dismukes, W. E.; Walsh, T. J.; Edwards, J. E. *Clin. Infect. Dis.* **2004**, *38*, 161–189.
84. Rex, J. H.; Walsh, T. J.; Sobel, J. D.; Filler, S. G.; Pappas, P. G.; Dismukes, W. E.; Edwards, J. E. *Clin. Infect. Dis.* **2000**, *30*, 662–678.
85. Stevens, D. A.; Kan, V. L.; Judson, M. A.; Morrison, V. A.; Dummer, S.; Denning, D. W.; Bennett, J. E.; Walsh, T. J.; Patterson, T. F.; Pankey, G. A. *Clin. Infect. Dis.* **2000**, *30*, 696–709.
86. Walsh, T. J.; Teppler, H.; Donowitz, G. R.; Maertens, J. A.; Baden, L. R.; Dmoszynska, A.; Cornely, O. A.; Bourque, M. R.; Lupinacci, R. J.; Sable, C. A. et al. *N. Engl. J. Med.* **2004**, *351*, 1391–1402.
87. Polak, A. *Prog. Drug Res.* 2003, 59–190.
88. Zager, R. A. *Am. J. Kidney Dis.* **2000**, *36*, 238–249.
89. Walsh, T. J.; Melcher, G. P.; Rinaldi, M. G.; Lecciones, J.; McGough, D. A.; Kelly, P.; Lee, J.; Callender, D.; Rubin, M.; Pizzo, P. A. *J. Clin. Microbiol.* **1990**, *28*, 1616–1622.
90. Arikan, S.; Lozano-Chiu, M.; Paetznick, V.; Nangia, S.; Rex, J. H. *J. Clin. Microbiol.* **1999**, *37*, 3946–3951.
91. Jezequel, S. G. *J. Pharm. Pharmacol.* **1994**, *46*, 196–199.
92. Cha, R.; Sobel, J. D. *Expert Rev. Anti Infect. Ther.* **2004**, *2*, 357–366.
93. Perfect, J. R. *Oncology (Huntingt.)* **2004**, *18*, 15–23.
94. Perfect, J. R. *Oncology (Williston Park)* **2004**, *18*, 5–14.
95. Vanden Bossche, H.; Marichal, P.; Gorrens, J.; Coene, M. C. *Br. J. Clin. Pract. Suppl.* **1990**, *71*, 41–46.
96. Richardson, K.; Cooper, K.; Marriott, M. S.; Tarbit, M. H.; Troke, P. F.; Whittle, P. J. *Rev. Infect. Dis.* **1990**, *12*, S267–S271.
97. Donnelly, J. P.; De Pauw, B. E. *Clin. Microbiol. Infect.* **2004**, *10*, 107–117.
98. Herbrecht, R. *Expert Rev. Anti Infect. Ther.* **2004**, *2*, 485–497.
99. Martin, M. V.; Yates, J.; Hitchcock, C. A. *Antimicrob. Agents Chemother.* **1997**, *41*, 13–16.
100. Gallagher, J. C.; Dodds Ashley, E. S.; Drew, R. H.; Perfect, J. R. *Expert Opin. Pharmacother.* **2003**, *4*, 147–164.
101. Purkins, L.; Wood, N.; Greenhalgh, K.; Allen, M. J.; Oliver, S. D. *Br. J. Clin. Pharmacol.* **2003**, *56*, 10–16.
102. Debono, M.; Gordee, R. S. *Annu. Rev. Microbiol.* **1994**, *48*, 471–497.
103. Rafael Sentandreu, M. V. E.; Valentín, E.; Ruiz Herrera, J. The Structure and Composition of the Fungal Cell Wall. In *Pathogenic Fungi: Structural Biology and Taxonomy*; San-Blas, G., Calderone, R. A., Eds.; Caister Academic Press: Norwich, 2004; Chapter 1, pp 3–40.
104. Munro, C.; Gow, N. Chitin Biosynthesis as a Target for Antifungals. In *Antifungal Agents. Discovery and Mode of Action*; G. Dixon, L.; Copping, D., Eds.; Hollowman Bios Scientific Publishers: Oxford, 1995, pp 161–171.
105. Denning, D. W. *Lancet* **2003**, *362*, 1142–1151.
106. Abruzzo, G. K.; Flattery, A. M.; Gill, C. J.; Kong, L.; Smith, J. G.; Pikounis, V. B.; Balkovec, J. M.; Bouffard, A. F.; Dropinski, J. F.; Rosen, H. et al. *Antimicrob. Agents Chemother.* **1997**, *41*, 2333–2338.
107. Maertens, J.; Raad, I.; Petrikkos, G.; Boogaerts, M.; Selleslag, D.; Petersen, F. B.; Sable, C. A.; Kartsonis, N. A.; Ngai, A.; Taylor, A. et al. *Clin. Infect. Dis.* **2004**, *39*, 1563–1571.
108. Schuster, F.; Moelter, C.; Schmid, I.; Graubner, U. B.; Kammer, B.; Belohradsky, B. H.; Fuhrer, M. *Pediatr. Blood Cancer* **2005**, *44*, 682–685.
109. Murdoch, D.; Plosker, G. L. *Drugs* 2004, *64*, 2249–2258, 2259–2260.
110. Onishi, J.; Meinz, M.; Thompson, J.; Curotto, J.; Dreikorn, S.; Rosenbach, M.; Douglas, C.; Abruzzo, G.; Flattery, A.; Kong, L. et al. *Antimicrob. Agents Chemother.* **2000**, *44*, 368–377.
111. Parnham, M. J.; Bogaards, J. J.; Schrander, F.; Schut, M. W.; Oreskovic, K.; Mildner, B. *Biopharm. Drug Dispos.* **2005**, *26*, 27–33.
112. Petraitiene, R.; Petraitis, V.; Kelaher, A. M.; Sarafandi, A. A.; Mickiene, D.; Groll, A. H.; Sein, T.; Bacher, J.; Walsh, T. J. *Antimicrob. Agents Chemother.* **2005**, *49*, 2084–2092.
113. Petraitis, V.; Petraitiene, R.; Kelaher, A. M.; Sarafandi, A. A.; Sein, T.; Mickiene, D.; Bacher, J.; Groll, A. H.; Walsh, T. J. *Antimicrob. Agents Chemother.* **2004**, *48*, 3959–3967.
114. Brockmeyer N, R. M.; Oreskovic, K. *Phase II study of icofungipen (PLD-118) in the treatment of oropharyngeal candidiasis (OPC) in HIV-positive patients*, 44th Interscience Conference of Antimicrobial Agents, 2004.
115. Lipinski, C. A.; Lombardo, F.; Dominy, B. W.; Feeney, P. J. *Adv. Drug Deliv. Rev.* **2001**, *46*, 3–26.
116. Sugimoto, Y.; Sakoh, H.; Yamada, K. *Curr. Drug Targets Infect. Disord.* **2004**, *4*, 311–322.
117. Vicente, M. F.; Basilio, A.; Cabello, A.; Pelaez, F. *Clin. Microbiol. Infect.* **2003**, *9*, 15–32.
118. Zhong, W.; Jeffries, M. W.; Georgopapadakou, N. H. *Antimicrob. Agents Chemother.* **2000**, *44*, 651–653.
119. Nagiec, M. M.; Nagiec, E. E.; Baltisberger, J. A.; Wells, G. B.; Lester, R. L.; Dickson, R. C. *J. Biol. Chem.* **1997**, *272*, 9809–9817.
120. Ikai, K.; Takesako, K.; Shiomi, K.; Moriguchi, M.; Umeda, Y.; Yamamoto, J.; Kato, I.; Naganawa, H. *J. Antibiot. (Tokyo)* **1991**, *44*, 925–933.
121. Takesako, K.; Kuroda, H.; Inoue, T.; Haruna, F.; Yoshikawa, Y.; Kato, I.; Uchida, K.; Hiratani, T.; Yamaguchi, H. *J. Antibiot. (Tokyo)* **1993**, *46*, 1414–1420.
122. Endo, M.; Takesako, K.; Kato, I.; Yamaguchi, H. *Antimicrob. Agents Chemother.* **1997**, *41*, 672–676.
123. Hashida-Okado, T.; Ogawa, A.; Endo, M.; Yasumoto, R.; Takesako, K.; Kato, I. *Mol. Gen. Genet.* **1996**, *251*, 236–244.
124. Kuroda, M.; Hashida-Okado, T.; Yasumoto, R.; Gomi, K.; Kato, I.; Takesako, K. *Mol. Gen. Genet.* **1999**, *261*, 290–296.
125. Heidler, S. A.; Radding, J. A. *Antimicrob. Agents Chemother.* **1995**, *39*, 2765–2769.
126. Kurome, T.; Inoue, T.; Takesako, K.; Kato, I. *J. Antibiot. (Tokyo)* **1998**, *51*, 359–367.
127. Lodge, J. K.; Johnson, R. L.; Weinberg, R. A.; Gordon, J. I. *J. Biol. Chem.* **1994**, *269*, 2996–3009.
128. Weinberg, R. A.; McWherter, C. A.; Freeman, S. K.; Wood, D. C.; Gordon, J. I.; Lee, S. C. *Mol. Microbiol.* **1995**, *16*, 241–250.
129. Devadas, B.; Freeman, S. K.; Zupec, M. E.; Lu, H. F.; Nagarajan, S. R.; Kishore, N. S.; Lodge, J. K.; Kuneman, D. W.; McWherter, C. A.; Vinjamoori, D. V. et al. *J. Med. Chem.* **1997**, *40*, 2609–2625.
130. Sikorski, J. A.; Devadas, B.; Zupec, M. E.; Freeman, S. K.; Brown, D. L.; Lu, H. F.; Nagarajan, S.; Mehta, P. P.; Wade, A. C.; Kishore, N. S. et al. *Biopolymers* **1997**, *43*, 43–71.
131. Nagarajan, S. R.; Devadas, B.; Zupec, M. E.; Freeman, S. K.; Brown, D. L.; Lu, H. F.; Mehta, P. P.; Kishore, N. S.; McWherter, C. A.; Getman, D. P. et al. *J. Med. Chem.* **1997**, *40*, 1422–1438.
132. Lodge, J. K.; Jackson-Machelski, E.; Devadas, B.; Zupec, M. E.; Getman, D. P.; Kishore, N.; Freeman, S. K.; McWherter, C. A.; Sikorski, J. A.; Gordon, J. I. *Microbiology* **1997**, *143*, 357–366.
133. Dorr, P. 3D structure analysis coupled with high throughput screening. Biochemical Society and SGM: Abstract 0950. 148th Ordinary Meeting, March 26–30, 2001.

134. Weston, S. A.; Camble, R.; Colls, J.; Rosenbrock, G.; Taylor, I.; Egerton, M.; Tucker, A. D.; Tunnicliffe, A.; Mistry, A.; Mancia, F. et al. *Nat. Struct. Biol.* **1998**, *5*, 213–221.
135. Sogabe, S.; Masubuchi, M.; Sakata, K.; Fukami, T. A.; Morikami, K.; Shiratori, Y.; Ebiike, H.; Kawasaki, K.; Aoki, Y.; Shimma, N. et al. *Chem. Biol.* **2002**, *9*, 1119–1128.
136. Futterer, K.; Murray, C. L.; Bhatnagar, R. S.; Gokel, G. W.; Gordon, J. I.; Waksman, G. *Acta Crystallogr. D Biol. Crystallogr.* **2001**, *57*, 393–400.
137. Ebara, S.; Naito, H.; Nakazawa, K.; Ishii, F.; Nakamura, M. *Biol. Pharm. Bull.* **2005**, *28*, 591–595.
138. Ebiike, H.; Masubuchi, M.; Liu, P.; Kawasaki, K.; Morikami, K.; Sogabe, S.; Hayase, M.; Fujii, T.; Sakata, K.; Shindoh, H. et al. *Bioorg. Med. Chem. Lett.* **2002**, *12*, 607–610.
139. Wu, L.; Pan, J.; Thoroddsen, V.; Wysong, D. R.; Blackman, R. K.; Bulawa, C. E.; Gould, A. E.; Ocain, T. D.; Dick, L. R.; Errada, P. et al. *Eukaryot. Cell* **2003**, *2*, 256–264.
140. Schneiter, R.; Hitomi, M.; Ivessa, A. S.; Fasch, E. V.; Kohlwein, S. D.; Tartakoff, A. M. *Mol. Cell Biol.* **1996**, *16*, 7161–7172.
141. Pridzun L, S. F.; Reichenbach, H. Inhibition of Fungal ACCase: A Novel Target Discovered with the Myxobacterial Compound Soraphen. In *Antifungal Agents – Discovery and Mode of Action*; Dixon, G. K., Copping, L. G., Holloman, D. W., Eds.; Bios Publ. SCI: Oxford, 1995, pp 99–108.
142. Harwood, J. L. *Biochim. Biophys. Acta* **1996**, *1301*, 7–56.
143. Al-Feel, W.; DeMar, J. C.; Wakil, S. J. *Proc. Natl. Acad. Sci. USA* **2003**, *100*, 3095–3100.
144. Hasslacher, M.; Ivessa, A. S.; Paltauf, F.; Kohlwein, S. D. *J. Biol. Chem.* **1993**, *268*, 10946–10952.
145. Evenson, K. J.; Gronwald, J. W.; Wyse, D. L. *Plant Physiol.* **1994**, *105*, 671–680.
146. Kinsman, O. S.; Chalk, P. A.; Jackson, H. C.; Middleton, R. F.; Shuttleworth, A.; Rudd, B. A.; Jones, C. A.; Noble, H. M.; Wildman, H. G.; Dawson, M. J. et al. *J. Antibiot. (Tokyo)* **1998**, *51*, 41–49.
147. Dominguez, J. M.; Kelly, V. A.; Kinsman, O. S.; Marriott, M. S.; Gomez de las Heras, F.; Martin, J. J. *Antimicrob. Agents Chemother.* **1998**, *42*, 2274–2278.
148. San-Blas, G.; Calderone, R. Eds. *Pathogenic Fungi. Host Interactions and Emerging for Control*; Caister Academc Press: Norwich, 2004, Vol. 2.
149. Doctor Fungus. http://www.doctorfungus.org (accessed April 2006).
150. The Aspergillus Website. http://www.aspergillus.man.ac.uk (accessed April 2006).

Biography

Patrick Dorr gained his BSc in microbiology from Sheffield University, UK, in 1987, and PhD from Kent University, UK, in 1990, following research in bacterial metabolism in Professor Chris Knowles' laboratory. A subsequent move to industry resulted in 5 years at Rhone-Poulenc, researching into new targets for pesticide discovery. In 1995, he moved to Pfizer GRD in Sandwich, as a team leader in antifungals discovery, undertaking research in many aspects of drug discovery, from new target validation and lead seeking to efficacy studies in infection models. He is currently leading the CCR5 and HIV entry preclinical teams for the discovery of new antiviral drugs.

Comprehensive Medicinal Chemistry II
ISBN (set): 0-08-044513-6

ISBN (Volume 7) 0-08-044520-9; pp. 419–443

7.15 Major Antifungal Drugs

A S Bell, Pfizer Global Research and Development, Sandwich, UK

7.15.1 Major Antifungal Drugs

7.15.1.1 The Ideal Antifungal Agent

Over the last 50 years, there has been a sustained effort to discover and develop an ideal treatment for fungal infections of human hosts. When comparing these advances in drug discovery, it is useful to consider what would be the ideal profile for an antifungal agent. Thus an ideal agent would be

- potently active against a fungal-specific target, i.e., one with no close mammalian homologs, which is essential and closely homologous in wide range of fungi
- fungicidal rather than fungistatic due to its mechanism of action
- highly selective against all other mammalian targets, especially metabolising enzymes
- distributed systemically throughout the mammalian host, including into the central nervous system
- soluble enough for formulation by both oral and intravenous route, either as a stand-alone agent or by prodrug/ formulation methodology.

In the following sections, the major classes of antifungal chemotherapy used in the clinic and some possible future directions will be compared to the ideal profile.

7.15.2 Agents Affecting Ergosterol

7.15.2.1 Overview

Ergosterol is the major product of sterol biosynthesis in fungi (and also in some trypanosomes), whereas mammalian systems synthesize cholesterol as the major membrane lipid. While both sterols play a similar role in membrane fluidity, this effect has been shown to be essential for aerobic growth of most fungi. Although cholesterol plays an equally important role in mammalian cells, most cholesterol is obtained through diet, hence inhibition of cholesterol biosynthesis as a side effect is not a major selectivity concern for antifungal agents.

Most of the current antifungal agents interfere with ergosterol function in some way, either through inhibition of various steps in ergosterol biosynthesis (allylamines, azoles, morpholines) or by complexing directly with membrane ergosterol (polyenes).

7.15.2.2 Ergosterol Biosynthesis

Like cholesterol, ergosterol is biosynthetically derived from acetyl-coenzyme A (CoA: **Figure 1**). The nine biosynthetic steps leading to the first sterol intermediate, lanosterol, are identical in both pathways; related enzymes in fungi and mammalian cells catalyze all but one of the remaining steps in either the cholesterol or ergosterol pathways. While there have been reports of compounds inhibiting the fungal-specific step of introduction of the 24-methyl group through sterol Δ24-methyltransferase, to date no drugs have arisen from this approach, hence in all cases selectivity for the fungal target is a key target to avoid toxicological consequences.

The most successful targets have been squalene epoxidase (SE), lanosterol-14α-demethylase and Δ7-8 isomerase, which are discussed in the following sections.

Acetyl CoA → Hydroxymethyl-glutaryl CoA → Mevalonate → Farnesyl pyrophosphate →

SS inhibitors

Squalene → Squalene epoxide → Lanosterol → 14-Desmethyllanosterol →

ALLYLAMINES

AZOLES

→ Zymosterol → Fecosterol → Episterol → Ergosterol

MORPHOLINES

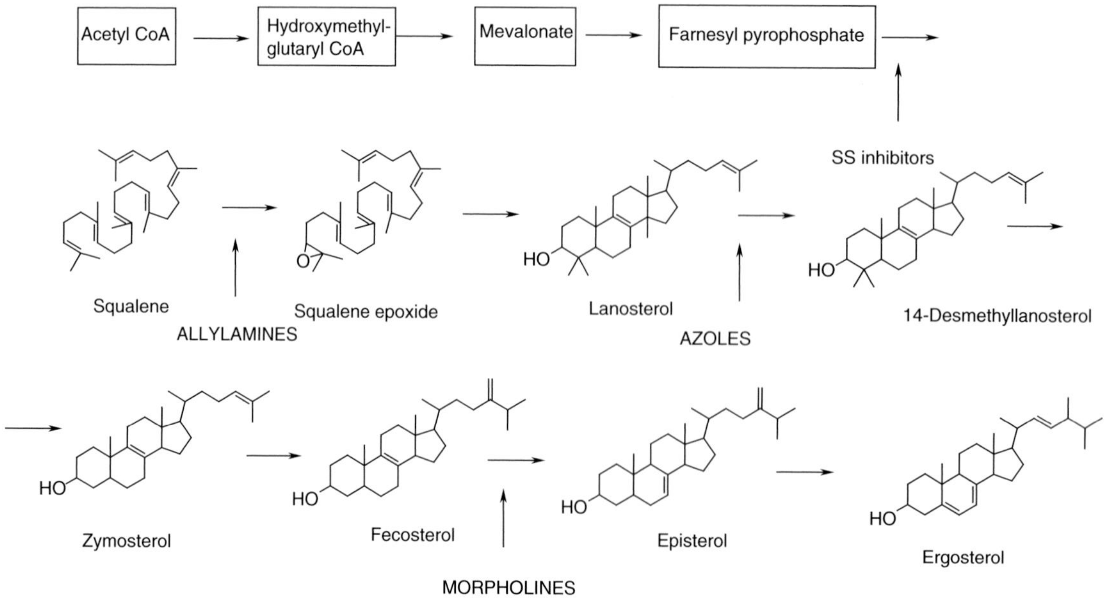

Figure 1 Biosynthetic pathway to ergosterol showing points of inhibition by antifungal drug classes.

7.15.2.3 Squalene Epoxidase Inhibitors

7.15.2.3.1 Medicinal chemistry

SE is a microsomal enzyme catalyzing the conversion of squalene into 2,3-oxidosqualene, which is subsequently cyclized by squalene epoxide cyclase to form lanosterol. Whereas mammalian SE consists of an NADPH-cytochrome *c* reductase activity and a terminal oxidase, the fungal enzymes prefer NADH as the cofactor.[1] The first SE inhibitor, naftifine,[2] was discovered in the late 1970s, as a result of routine whole-cell screening. Based upon the structure of naftifine (**1**), the entire class of SE inhibitors are known as allylamines. Naftifine was found to have activity against a number of human pathogenic fungi, principally those of the dermatophyte class, e.g., *Trichophyton rubrum*. Naftifine was subsequently shown to be a potent blocker of ergosterol biosynthesis[3] by fungal SE inhibition,[1] with excellent selectivity over mammalian homologs (**Table 1**). In *Candida albicans*, naftifine has been shown to have a noncompetitive mode of inhibition.

1

Naftifine's effect on ergosterol biosynthesis results in a fungicidal action against dermatophytes and other filamentous fungi, e.g., *Aspergillus fumigatus*, and against some yeasts, e.g., *C. parapsilosis*. In contrast, its action is fungistatic against most strains of *C. albicans*. The difference is believed to be due to two intrinsic physiological factors in the target fungi; in some cases growth inhibition is achievable at concentrations at which ergosterol biosynthesis is only partly inhibited, whereas in other fungi, total inhibition is necessary. In dermatophytes, only partial inhibition is necessary, which suggests that squalene accumulation may play an important role in causing cell death. This fungicidal action has a slow onset, implying that death is a secondary effect, unlike the polyenes. Ultrastructural studies of cells treated with naftifine show large numbers of lipid bodies, which may represent intracellular squalene accumulation.

Naftifine was also found to be active in vivo in dermatophytosis models such as the guinea-pig trichophytososis model following topical application,[4] though oral activity could only be demonstrated after administration of very high doses that were of no therapeutic relevance.

Initial SAR studies around naftifine[5] demonstrated that its activity was specific since the basic nitrogen atom, the double bond, and the 1-substituted naphthalene ring were all important. In addition, the aromatic rings could not be interchanged. Changes in the distances between the function groups and the aromatic rings could not be altered. However, it was found that rigidification of the structure of naftifine was possible. Incorporation of the basic center into a piperidine ring provided a compound with antifungal activity both in vitro and in vivo (active piperidine analog (**2**)). Activity was restricted to the *R*-enantiomer and this compound was also found to exhibit significantly improved oral activity in the guinea-pig trichophytosis model.

2 **3**

Table 1 Effect of naftifine on biochemistry of different species

Fungus	*MIC* ($\mu g\,mL^{-1}$)	IC_{50} (*ergosterol biosynthesis:* $\mu mol\,L^{-1}$)	IC_{50} (*squalene epoxidase* $\mu mol\,L^{-1}$)
Trichophyton rubrum	0.05	0.005	
Candida parapsilosis	1.6	0.23	0.34
Candida albicans	50	0.35	1.1
Candida glabrata	100	0.34	
Rat liver			144

Data from Ryder, N. S. *Ann. NY Acad. Sci.* **1988**, *544*, 208–220.

Table 2 Effect of terbinafine on biochemistry of different species

Fungus	*MIC* ($\mu g\,mL^{-1}$)	IC_{50} *(ergosterol biosynthesis:* $\mu mol\,L^{-1}$*)*	IC_{50} *(squalene epoxidase:* $\mu mol\,L^{-1}$*)*
Trichophyton rubrum	0.003	0.0005	
Candida parapsilosis	0.4	0.006	0.04
Candida albicans	3.1	0.008	0.03
Candida glabrata	100	0.04	
Rat liver			144

Data from Stütz, A., Petranyi, G. *J. Med. Chem.* **1984**, *27*, 1539–1543.

In subsequent SAR studies, the cinnamyl group of naftifine was replaced by long alkyl chains, with an increasing number of conjugated double bonds. Activity was maximized when two multiple bonds were present but was improved, especially in vivo, when the C4–C5 double bond was replaced by a triple bond. An *E*-configuration of the double and triple bonds was preferred (Terbinafine (**3**); **Table 2**). While the nature of the terminal substituent on the triple bond had relatively little effect on potency, oral activity was optimized by incorporating a *t*-butyl substituent providing terbinafine.[6] Detailed mechanistic studies using terbinafine demonstrated that its mechanism of action was identical to that of naftifine, though terbinafine was significantly more potent and had a superior pharmacokinetic profile.

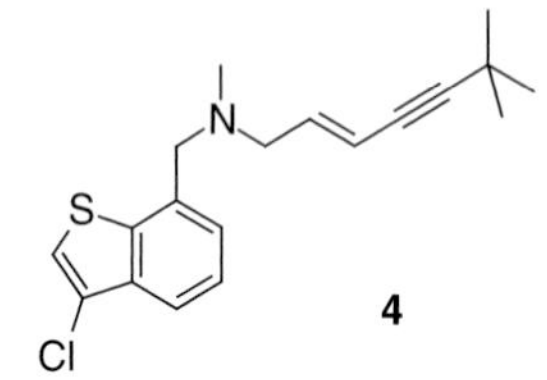

4

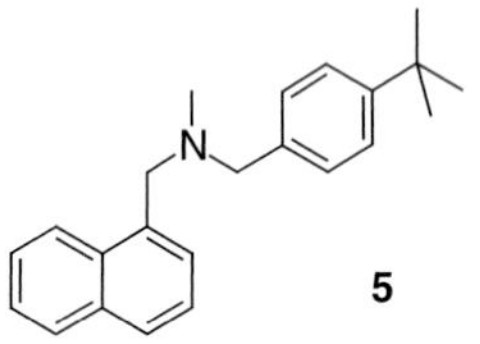

5

Further SAR studies around terbinafine have shown that the naphthalene ring of terbinafine can be replaced by lipophilic heterocycles.[7] Variation of the position of substitution and further substituents on a benzo[b]thiophenyl ring system led to the 3-chloro-derivative (SDZ 87-469 (**4**)), which is the most potent SE inhibitor reported. SDZ 87-469 was also found to be highly effective in in vivo animal infection models.

While optimization of potency led to the pharmaceutically unique enyne fragment, further SAR explorations on the allyl side chain led to the discovery of the homopropargylamines and benzylamines.[8] *Para* substitution on the benzylamine was found to be essential for high antifungal activity and the *t*-butyl derivative, butenafine (**5**),[9] was subsequently developed for topical treatment of dermatophytoses.

7.15.2.3.2 Pharmacokinetic profiles

There have been limited reports on the pharmacokinetics of naftifine. Following topical application in a 1% cream formulation, naftifine penetrates the epidermis in sufficient concentrations of inhibit fungal growth but only 3.8% of the dose was recovered in the urine. Clearance of naftifine was rapid following oral administration, leading to 15 different metabolites.[10] Some 14.3% of the dose was recovered as metabolites, with naphthoic acid predominating. None of the metabolites exhibited significant antifungal activity.

As observed in animal models, and in contrast to naftifine, terbinafine achieves clinically effective concentrations following oral administration. Approximately 70–80% of an orally administered dose is absorbed from the gastrointestinal tract and bioavailability is little affected by food. Following administration of the efficacious dose of 250 mg, a peak plasma concentration of 0.9 mg L^{-1} was achieved within 2 h in healthy volunteers. On multiple dosing to patients, peak and trough concentrations of 3.62 and 1.44 mg L^{-1}[11] were measured at steady state, which is reached after 10–14 days.[12] Terbinafine is a highly lipophilic base and consequently has a very high volume of distribution (13.5 L kg^{-1} at steady state), a result of strong, nonspecific binding to plasma proteins. Terbinafine is distributed to all tissues with particularly high concentrations in the liver, pancreas, and sebum (45 mg kg^{-1}).

Terbinafine is extensively metabolized in humans: the major routes of metabolism are *N*-demethylation, *N*-oxidation, and oxidation of the *t*-butyl group. Plasma elimination occurs with a half-life of 11–16 h in healthy volunteers. However, administration of radiolabeled drug identified an additional elimination phase with a half-life of 90–100 h. The second phase is believed to reflect the slow redistribution of terbinafine from adipose tissue. Total plasma clearance in healthy volunteers was 76 L h^{-1}. Approximately 80% of an oral dose was excreted in urine within 72 h.

7.15.2.3.3 Clinical profiles

Clinical trials[13] with naftifine (Exoril) employed the 1% cream formulation against mycologically confirmed dermatophytosis, e.g. tinea pedis (athlete's foot), for periods of weeks. The predominant infecting organisms were *Trichophyton rubrum* and *T. mentagrophytes,* though activity against cutaneous candidiases has also been observed. Twice-daily application gave the best results, though a once-daily regimen was also effective. Mycological cure rates were high (84%), with similar activities to the first-generation azole, clotrimazole, being observed. Despite its fungicidal mode of action in vitro, modest cure rates (19%) have been observed.

There have been few reports of systemic adverse events following topical application of naftifine. In clinical trials, a few patients developed mild local irritation in the early stages of treatment.

Clinical trials with terbinafine[14] (Lamisil) have shown it to be effective in the treatment of dermatophyte infections following either oral ($250\,mg\,day^{-1}$) or topical (1% cream) administration. Mycological cures have been observed in approximately 90% on patients with tinea pedis, with associated clinical cures in 80% of cases. Oral terbinafine also shows impressive success in the treatment of finger- and toenail infections following oral administration for 3–12 months. Terbinafine is less effective against skin infections caused by *Candida* spp. when given orally, though it is efficacious following topical treatment. There have been few reports of the activity of terbinafine against systemic candidoses.

Terbinafine is well tolerated following oral administration. The predominant side effects are gastrointestinal disturbances and skin reactions, though more unusual reports include tongue discoloration. The potential for drug–drug interaction with terbinafine are low due to its lack of cytochrome P-450 inhibition, though drugs that inhibit or enhance P-450 activity can alter the metabolism of terbinafine.

7.15.2.3.4 Drug resistance

In contrast to the azoles, there are relatively few reports of primary resistance to allylamines, although resistant mutants can be induced through ultraviolet irridation or chemical mutagenesis. In the one primary resistant example,[15] a patient who failed oral terbinafine therapy was found to have a resistant strain of *Trichophyton rubrum,* which was sensitive to azoles but cross-resistant to other SE inhibitors. The remaining examples of terbinafine resistance have been observed in organisms where the multidrug efflux transporter *CDR2* has been induced.[16]

7.15.2.4 Lanosterol-14α-Demethylase Inhibitors

7.15.2.4.1 Enzymology

Demethylation of the 14α-methyl of lanosterol is catalyzed by a specific cytochrome P450 (CYP51 in *Candida albicans*) via a multistep mechanism based on sequential oxidation of the methyl group to the aldehyde oxidation state, followed by an oxidative rearrangement and elimination to leave a double bond. The vast majority of known lanosterol demethylase inhibitors and all of the drugs make use of a mechanism involving coordination to the P450 heme ferric cation core of the enzyme through a metal-chelating group which is part of a structure capable of mimicking the sterol backbone.[17] The azole drugs are selective for CYP51 inhibition over the human CYP3A4, the major metabolizing enzyme.[18]

7.15.2.4.2 Medicinal chemistry

The first inhibitors of lanosterol demethylase were discovered by the Janssen group as part of an effort to synthesize and evaluate the biological activity of a series of imidazole derivatives. The initial leads were 1-phenethylimidazoles bearing a hydroxyl or amino group attached to the benzylic position. Further synthesis around the initial lead demonstrated that a wide range of analogs retained biological activity, including a series of ethers and cyclic acetals derived from intermediate ketones.[19] Among the analogs prepared was the *bis*-(2,4-dichlorophenyl) ether derivative, miconazole (**6**; **Table 3**), which became the first drug of the azole class; subsequent analogs in the same series have led to three further drugs from the Janssen group alone.

The initial studies found that the substitution on the parent primary amine gave moderate activity against dermatophytes but poor activity against yeasts (and bacteria). The ketals were found to have a similar biological profile but in both series activity improved when substituents were introduced on to the aromatic ring. Reduction of several of the ketones and formation of the corresponding benzylethers provided a series with very potent activity against dermatophytes combined with modest activity against *C. albicans* and *Aspergillus fumigatus*. Whereas substitution on the first aryl ring produced a moderate potency increase in the amine series, over a 100-fold increase in anti-*Candida* activity was achieved with 2,4-dichloro aryl examples. Unexpectedly, similar substitution patterns were optimal at both

Table 3 Activity of imidazole antifungals against fungal species

Drug	*MIC (μg mL^{-1}) Trichophyton mentagrophytes*	*MIC (μg mL^{-1}) Candida albicans*
Miconazole	0.001	10
Econazole	0.1	100
Clotrimazole	<1	>100

Data from Godefroi, E. F.; Heeres, J.; Van Cutsem, J.; Janssen, P. A. J. *J. Med. Chem.* **1969**, *12*, 784–791 and Heeres, J.; Backx, L. J. J.; Van Cutsem, J. *J. Med. Chem.* **1976**, *19*, 1148.

positions in miconazole, while substitution at the branch point with a methyl group was not tolerated (**Table 3**). Two other marketed drugs, econazole (**6**) (also Janssen) and tioconazole (**12**)[20] (discovered by Pfizer), are members of the same chemical series.

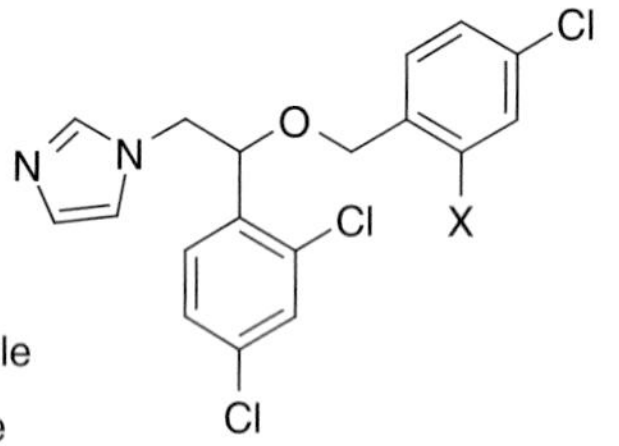

X=Cl Miconazole

X=H Econazole

6 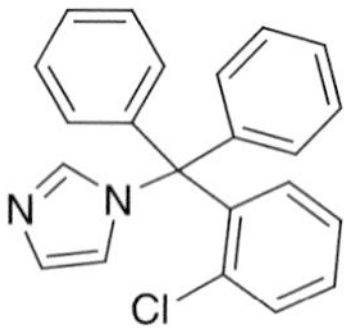 **7**

Follow-on studies around miconazole established that the ether oxygen was not essential for antifungal activity since it was possible to replace the benzylether side chain with alkyl groups.[21] While most of the analogs were highly active against dermatophytes, and some had excellent activity against yeasts, none were active against bacteria. The SAR relationships for aryl substitution were the same as in the miconazole series, while alkyl chains of more than four carbons provided compounds with activity both in vitro and in guinea-pig dermatophyte models.

In parallel with the research that led to miconazole and econazole, scientists at Bayer discovered that the ethyl linker between the imidazole and aryl rings was not essential for biological activity.[22] Instead, trityl substitution on the imidazole ring was found to be sufficient, leading to the 2-chloroderivative, clotrimazole (**7**).

Although the first generation of antifungal azoles were potent in vitro against a wide range of fungi of clinical interest, and were successful in the topical treatment of human infections, their main drawback was the low systemic levels following oral administration. Continuing work at Janssen found that introduction of an additional substituent on the ketal ring of their initial leads was tolerated, leading to ketoconazole.[23] The in vitro activity of ketoconazole (**8**) appeared to be inferior to that of miconazole under standard conditions (Sabouraud's broth) but, in contrast, the activity of ketoconazole was improved by the addition of serum to the antifungal assay while that of miconazole was dramatically reduced. This improvement was mirrored in in vivo experiments in candidosis models, where ketoconazole displayed a dose response in both prophylactic (starting on day of treatment) and curative treatment (starting on day 3 following infection) regimens.

While each of the early azole derivatives makes use of an imidazole ring to coordinate with the heme, the Janssen group found that the triazole analog of ketoconazole (terconazole (**9**)[24]) had a similar in vitro profile (**Table 4**). Surprisingly, in view of later developments, terconazole had an inferior profile compared to ketoconazole in vaginal candidosis models following oral administration.

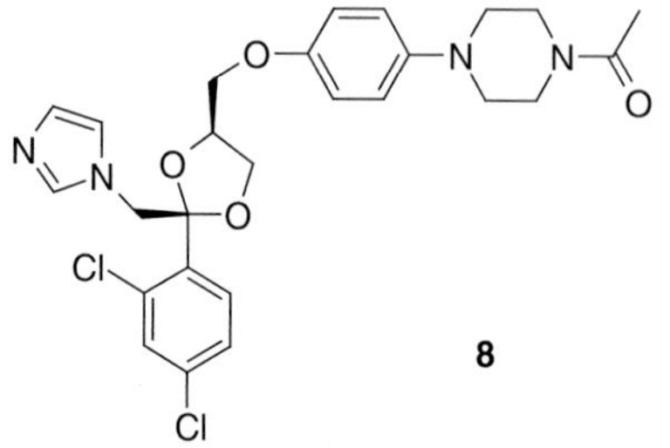

8 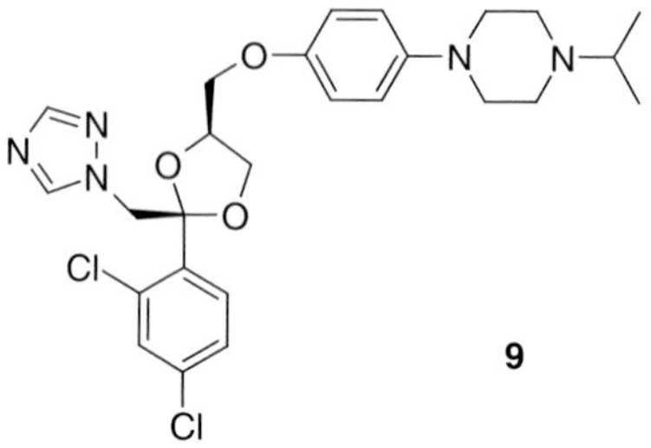 **9**

Table 4 Activity of Janssen antifungal drugs against systemic fungal pathogens

Drug	*MIC ($\mu g\,mL^{-1}$) Candida albicans*	*MIC ($\mu g\,mL^{-1}$) Aspergillus fumigatus*
Miconazole	10 (>10[a])	10 (>10[a])
Ketoconazole	10 (1[a])	100 (10[a])
Terconazole	10 (0.1[a])	>100 (100[a])
Itraconazole	0.1	0.1

Data from Heeres, J.; Backx, L. J. J.; Mostmans, J. H.; Van Cutsem, J. *J. Med. Chem.* **1979**, *22,* 1003–1005, Heeres, J.; Hendrickx, R.; Van Cutsem, J. *J. Med. Chem.* **1983**, *26,* 611–613, and Heeres, J.; Backx, L. J. J.; Van Cutsem, J. *J. Med. Chem.* **1984**, *27,* 894–900.
[a] In the presence of bovine serum.

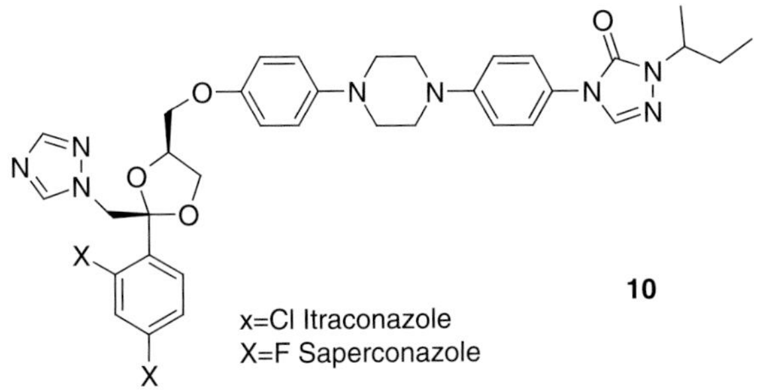

x=Cl Itraconazole
X=F Saperconazole

10

Further elaboration on the terminal piperazine substituent of terconazole led to a series of triazolones[25] with broad-spectrum activity both in vitro and in vivo. *N*-alkylation on the triazolone ring was found to be essential for oral activity, and in this series, the triazole derivatives were found to be superior to the corresponding imidazoles. Extension of the alkyl substituent on the triazolone with straight chains had little effect on the oral activity of the series but introduction of a branching carbon resulted in a dramatic increase in systemic activity with the final marketed drug from this series, itraconazole. The 2,4-difluorophenyl analog of itraconazole (**10**), saperconazole,[26] was taken into development but was found to carcinogenic in long-term toxicology studies.

While itraconazole has potent antifungal activity in its own right, one of the main metabolites, involving hydroxylation of the alkyl side chain on the triazolone ring, is also highly potent. Scientists at Schering incorporated this finding in their closely related series featuring a tetrahydrofuran ring in place of the dioxolane ring of saperconazole. The pentyl side chain of their original preclinical candidate, SCH-51048, was substituted with a hydroxyl group to give posaconazole (SCH-56592: **11**).[27]

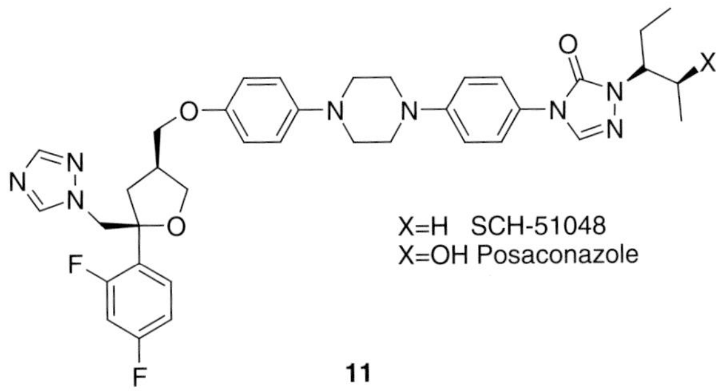

X=H SCH-51048
X=OH Posaconazole

11

Following on from their discovery of tioconazole, researchers at Pfizer adopted an alternative approach to obtaining systemic activity, by aiming for low clearance compounds by reducing lipophilicity.[28] After experimenting with a number of series, they opted for the tertiary alcohol series because this was the most polar series known to have antifungal activity. After extensive effort in this series, they concluded that the imidazole substituent was a metabolic flaw and replaced it with alternative heterocycles, including the 1,2,4-triazole. The triazole UK-46245 (**13**) was found to give improved in vivo activity, despite a sixfold reduction in in vitro potency. Based on the hypothesis that the triazole was more metabolically stable than the imidazole but retained affinity for the target enzyme, they replaced the hexyl chain with another triazole unit. The initial lead, UK-47265 (**13**), was 100 times more potent in a mouse candidosis model than ketoconazole but was found to have modest activity under the standard in vitro assay conditions. Subsequently, it became clear that complex media, especially those containing peptones, antagonized the activity of the more polar azoles, and the assay conditions were modified to a simple medium.

Although UK-47265 had outstanding activity in a wide range of systemic and superficial infection models, it was found to be hepatotoxic in mice and dogs, and teratogenic in rats. An intensive follow-up provided over 100 *bis*-triazole analogs, many with very good in vivo activity. The preferred compound, fluconazole (**14**), has a 2,4-difluorophenyl substituent in place of the dichloroaryl ring of UK-47265. This modification retained the activity of the initial lead in the infection models but was devoid of teratogenicity or hepatoxicity.

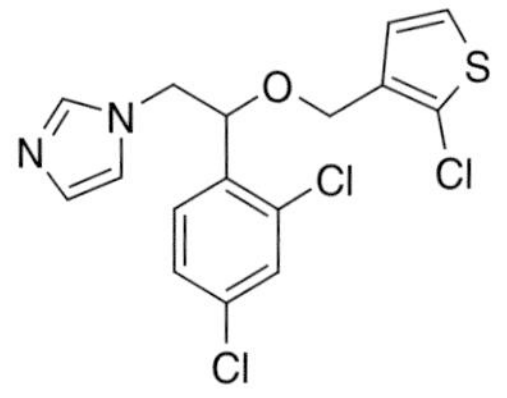

12

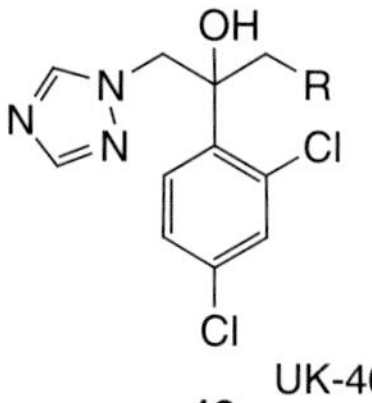

13

UK-46245 R = pentyl

UK-47265 R =

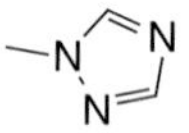

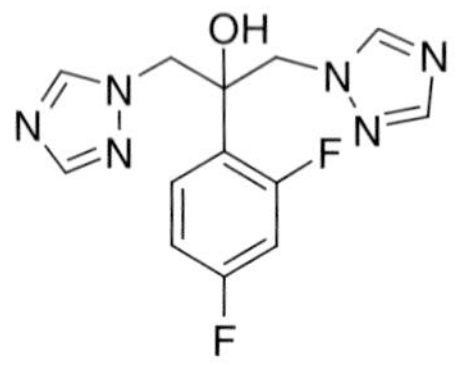

14

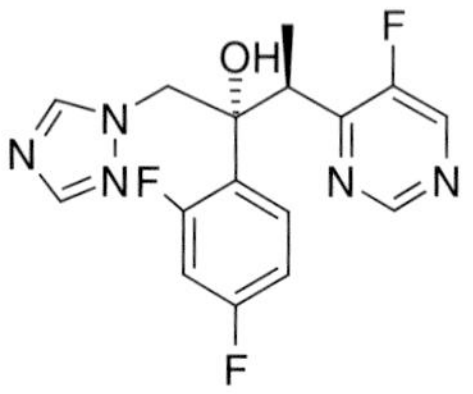

15

During its development, it became clear that fluconazole had superior efficacy to the existing azoles (miconazole, ketoconazole) for the treatment of infections due to *Candida albicans* and *Cryptococcus neoformans*. However, it was poorly effective against *Aspergillus* infections compared to itraconazole. Pfizer scientists found that introducing a methyl group adjacent to one of the triazole rings of fluconazole increased potency against *A. fumigatus*, while retaining potency against the other fungal pathogens. Replacement of the triazole ring with six-membered heterocycles[29] provided compounds with broad-spectrum in vitro activity, and, surprisingly, a fungicidal mechanism of action against *Aspergillus* spp. The optimal compound in the series, voriconazole (**15**), features a 5-fluoro-4-pyrimidinyl substituent. Voriconazole is a single enantiomer whose opposite enantiomer is at least 500-fold less active. The difference in activity of fluconazole and voriconazole against CYP51 in *Aspergillus* has been rationalized as a difference in binding mode.[30]

The beneficial impact of the α-methyl group was also discovered by Sumitomo scientists, who found that one of the triazoles of fluconazole could be replaced by a methyl sulfone substituent (genaconazole: **16**; **Table 5**).[31] The racemate, genaconazole, was co-developed with Schering but was subsequently found to be toxic in long-term safety studies. Resolution of genaconazole[32] demonstrated that the activity resided on one enantiomer, as for voriconazole, but the enantiomer proved to have a similar toxicology profile to the racemate. Interestingly, there is a 60-fold difference in the in vitro potency of genaconazole against *Candida albicans* and *Aspergillus fumigatus*, an effect consistent with the proposed difference in binding modes.[27]

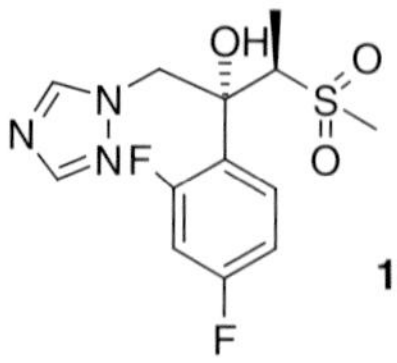

16

Genaconazole

Several other groups have utilized the same motif of a branched alkyl chain attached to a terminal polar group. The most advanced in development is ravuconazole (**17**), a thiazole discovered by Eisai and co-developed with BMS. The Eisai group first worked in the tertiary alcohols series with directly linked heterocycles exemplified by ER-24161

Table 5 Activity of triazole antifungals against systemic pathogens

Compound	*MIC Candida albicans* ($\mu g\, mL^{-1}$)	*MIC Aspergillus fumigatus* ($\mu g\, mL^{-1}$)
Fluconazole	1.0	>50
Voriconazole	0.03	0.09
Genaconazole	0.20	12.5

Data from Dickinson, R. P.; Bell, A. S.; Hitchcock, C. A.; Narayanaswami, S.; Ray, S. J.; Richardson, K.; Troke, P. F. *Biorg. Med. Chem. Lett.* **1996**, *6*, 2031–2036, and Saji, L.; Tamato, K.; Tanio, T.; Okuda, T.; Atsumi, T. *Abstracts of 8th Medicinal Chemistry Symposium*; Osaka, November 1986, p 9.

(**18**).[33] Once preliminary toxicology studies showed significant hepatotoxicity, they introduced an ethyl chain between the quaternary carbon and the thiazole or benzthiazole ring, leading to a series with more potent antifungal activity and a better safety profile.[34] The preferred compound from this series, ravuconazole, features a 4-cyanophenyl substituent on the thiazole ring. Interestingly, there were different SARs for in vitro activity against each of the main fungal pathogens, *Candida albicans*, *C. glabrata*, *Aspergillus fumigatus*, and *Cryptococcus neoformans*, which were best combined in ravuconazole and its 2-methyltetrazol-5-yl analog.

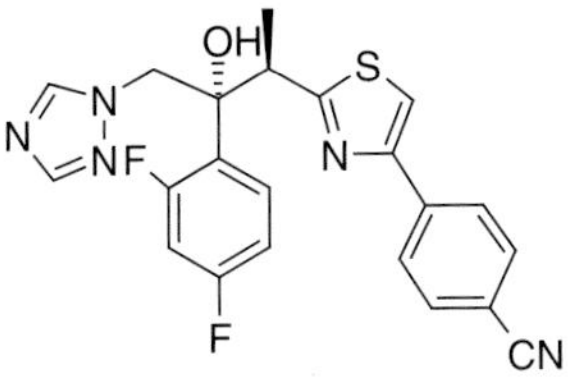

17

Ravuconazole

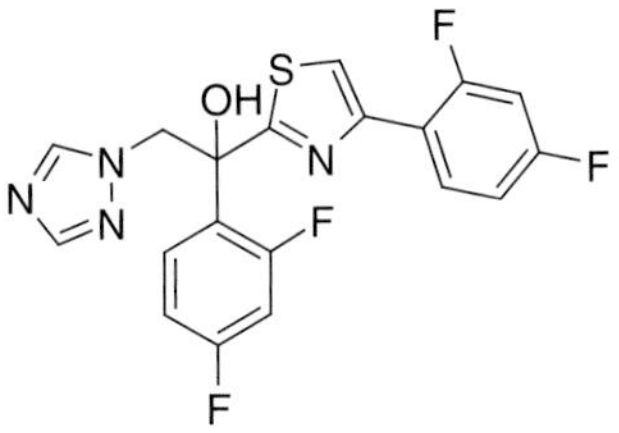

18

ER-24161

While ravuconazole has an excellent activity spectrum, it has poor aqueous solubility (0.6 $\mu g\, mL^{-1}$), which precluded the development of a formulation for intravenous administration. A number of water-soluble prodrugs of ravuconazole have been prepared,[35] the most promising being the phosphonoxymethylether analog, BMS-379224 (**19**).

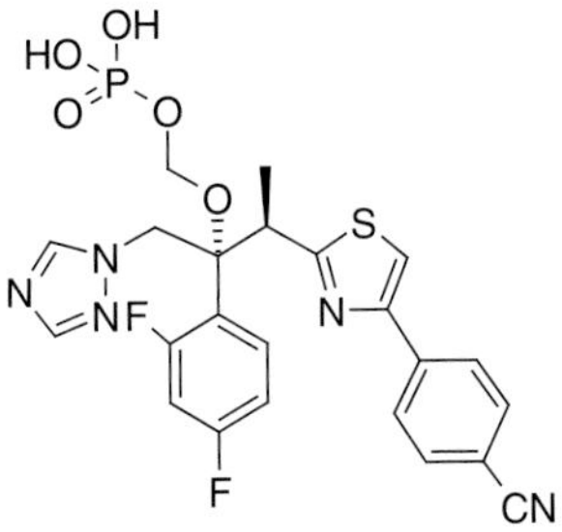

19

BMS-379224

Scientists at Takeda were also pursuing a similar theme in their antifungal discovery program. After initial studies replacing the triazole ring of fluconazole with 1,2,3-triazoles or tetrazoles,[36] they focused on a series of triazolones and tetrazolones. As with the Eisai studies, they found that 4-substituted phenyl substitutents on the azalone rings were optimal. The triazolones were found to be more potent than the corresponding tetrazolones and 4-polyfluoroalkoxy substituents were preferred. As with the other analogous series featuring two adjacent chiral centers, they found that only one of the four possible isomers of their initial lead compound TAK-187 had appreciable antifungal activity (**20**).[37] Subsequently, they found that an imidazolone linker provided a superior profile combined with a terminal tetrazole substituent (TAK-456: **21**).[38] As with ravuconazole, the aqueous solubility of TAK-456 was very low (5 $\mu g\, mL^{-1}$) and a prodrug strategy[39] was required to obtain TAK-457 as an injectable candidate for clinical trials (**22**).

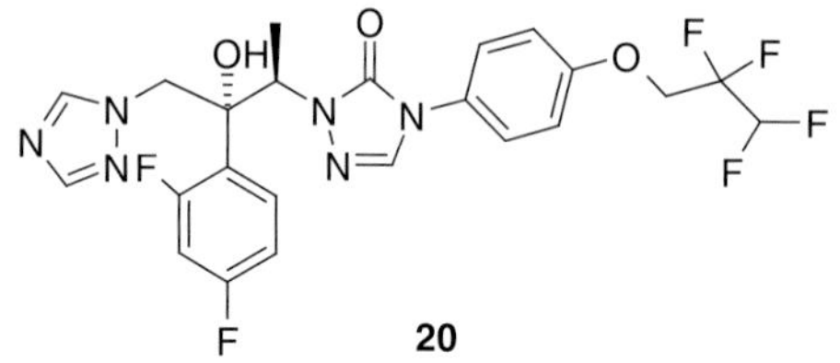

20

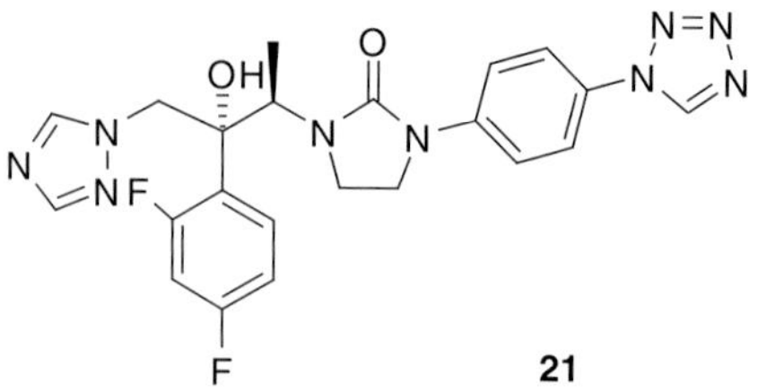

21

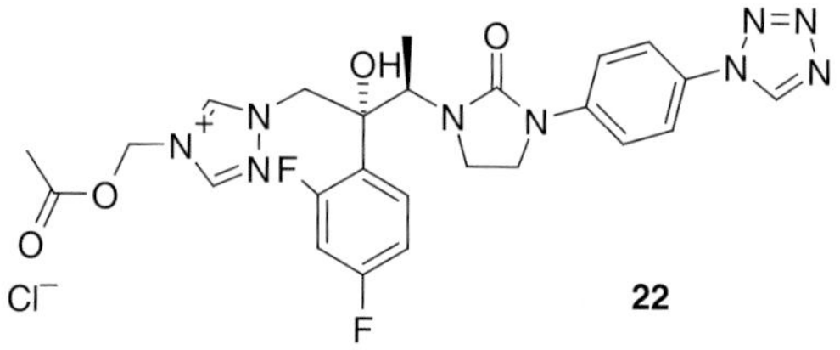

22

7.15.2.4.3 Pharmacokinetic profiles

The first generation of the azoles, which were characterized by poor systemic activity, suffered from a combination of incomplete absorption, high volumes of distribution, and high clearance. For example, the oral bioavailability of miconazole in humans is 25–30%,; its volume of distribution is 21 L kg^{-1}; and less than 1% of the administered dose was excreted unchanged.[40] Consequently, this generation of drug found greater usage by topical or intravaginal application. Thus econazole, administered topically as its nitrate salt, provided high drug levels in the dermal layers but very low systemic levels. Across a number of pharmacokinetic studies, 7% or less of the drug was absorbed.[41] Even when absorbed, econazole undergoes a series of metabolic transformations, resulting in the identification of over 20 urinary metabolites in monkey.

Ketoconazole is better absorbed than the early agents, especially under acidic conditions where the drug is fully dissolved.[42] Co-administration with antacids such as cimetidine reduced blood levels. Although ketoconazole is highly protein-bound (99%) in humans, it is widely distributed, including by penetration into the cerebrospinal fluid. Like the other early azoles, ketoconazole is extensively metabolized, including by oxidation of the imidazole ring and degradation of the piperazine ring. While some of this metabolism occurs on first pass during absorption, ketoconazole is a potent inhibitor of hepatic CYP 3A4, resulting in inhibition of its own metabolism. As a consequence, the half-life of ketoconazole is both dose- and patient-dependent.

As a consequence of its high lipophilicity and low aqueous solubility, gastric acidity is also required for itraconazole absorption.[43] It is best absorbed when administered with food, though there is considerable interpatient variability. The oral bioavailability of itraconazole from a 100 mg solution dose was 55%. Like ketoconazole, bioavailability and half-life are dose-dependent, indicating saturable metabolism. Once absorbed, itraconazole is highly plasma protein-bound (99.8%) and widely distributed (10.7 L kg^{-1}). Although itraconazole is widely metabolized, it does produce an active metabolite by hydroxylation of the triazolone side chain. Following multiple-dose administration of itraconazole for 14 days, concentrations of itraconazole and the hydroxyl metabolite were 1.9 and 3.2 mg mL^{-1}, respectively, at steady state.

In contrast to all previous antifungal azoles, fluconazole is a polar, water-soluble, metabolically stable molecule. Fluconazole[44] is very well absorbed, reaching peak plasma concentrations in 1–2 h and virtually complete bioavailability following oral administration. Neither food nor gastric acidity had any effect. Its volume of distribution is low (0.7–0.8 L kg^{-1}), approximating to body water, with very low plasma protein binding (12%). Fluconazole also crosses membranes easily such that cerebrospinal fluid-to-plasma ratios of 0.5–0.9 are achieved. There is very little evidence for metabolism of fluconazole, which is mainly excreted through the renal elimination route (60–90% of dose). However, there is extensive tubular reabsorption, resulting in a long plasma elimination half-life of 30–40 h, enabling once-daily dosing for systemic candidoses and a single-dose therapy for vaginal candidiasis.

Like fluconazole, voriconazole[45] is polar, with moderate aqueous solubility (0.5 mg mL^{-1}), resulting in rapid absorption (maximum concentration achieved in less than 2 h) and oral bioavailability (96%). Moderate food effects have been observed. Distribution is wide with a steady-state volume of 4.6 L kg^{-1} and moderate binding to plasma proteins (58%). Unlike fluconazole, little (<2%) of orally administered voriconazole is renally eliminated. Instead, voriconazole is metabolized by a range of cytochrome P450s (CYPs), the major route being *N*-oxidation of the pyrimidine ring by CYP2C19 to give an inactive metabolite. The elimination half-life of voriconazole is approximately 6 h.

7.15.2.4.4 Clinical profiles

Econazole can be used to treat dermatophyte infections using a twice-daily regimen for 2–6 weeks, resulting in high cure rates (~90%). In vaginal candidosis, a 3-day treatment regimen using a 15 mg suppository once daily was nearly as effective as a longer treatment (15 days) at a lower dose (50 mg). Both treatments are well tolerated.

The first orally available azole, ketoconazole, is effective in patients with dermatophyte or yeast skin infections, and in systemic infections due to *Candida* and *Histoplasma* spp. For most conditions a 200 mg daily dose is recommended, the exception being vaginal candidosis, where twice-daily dosing for 5 days is required. Administration at higher doses is limited by the ketoconazole's side-effect profile, including elevated liver enzymes and gynecomastia in males.

Due to its wide spectrum of activity, itraconazole is used to treat a wide range of infections. High cure rates were achieved in fingernail and toenail onchomycosis (200 mg day^{-1} for 3 months), dermatophytosis (100 mg day^{-1} for 2–4 weeks), and vaginal candidiasis (400 mg single dose). Unlike ketoconazole, there are few side effects and liver toxicity is rare.

Fluconazole is effective against oropharyngeal and esophageal candidiasis when used orally once daily either as a treatment or prophylactically in patients with acquired immunodeficiency syndrome (AIDS) or undergoing cancer therapy. It is also effective in patients with cryptococcal meningitis, especially as a maintenance therapy. In general a loading dose of twice the daily dose is recommended on the first day of treatment, for example, doses of 200 mg then 100 mg are used for oropharyngeal candidiasis. Higher doses are equally well tolerated and are used for esophagal infections and cryptococcal meningitis. Vaginal candidiasis can be treated using a single 50 mg dose. Fluconazole is available for both oral and intravenous administration, particularly for seriously ill patients, with few side effects.

Owing to its inferior pharmacokinetic profile, voriconazole is dosed twice daily; intravenous doses of 6 mg kg^{-1} are used on the first day, followed by 200 mg orally or continued intravenous dosing at 4 mg kg^{-1}. Voriconazole is recommended for the treatment of adults with invasive aspergillosis and can be used for rare infections caused by *Fusarium* spp. and *Scedosporium apiospermum*, where treatment with other agents has failed. In Europe, it can be used to treat fluconazole-resistant serious *Candida* infections. Its primary use is in immunocompromised patients with progressive, life-threatening infections. Although generally well tolerated, adverse effects such as visual disturbances and skin rash have been observed.

7.15.2.4.5 Drug resistance

Until the late 1980s, resistance to the azole derivatives was rare as they were generally used topically. Since the introduction of the systemically active azoles as the mainstay of antifungal chemotherapy, resistance has become much more widely reported in the literature.[46] A number of different resistance mechanisms have also emerged, ranging from decreased cell wall permeability or increased efflux to increased expression or mutation of the target enzyme, and alteration of the target pathway.

In a related fashion to the way human cells use the adenosine triphosphate-driven pump P-glycoprotein to remove xenobiotics from cancer cells, analogous genes, e.g., *CDR*, are expressed by fungal cells to efflux azoles. In *Candida albicans* another type of protein (MDR) from the major facilitator family or transporters uses membrane potential to efflux azoles, particularly fluconazole.

Changes in the affinity of the target CYP51 is another important mechanism of resistance that has been described in *C. albicans* and *Cryptococcus neoformans*. However, it is important to distinguish between acquired resistance and insensitivity of the specific fungal CYP51, e.g., the *Aspergillus fumigatus* isozyme is much less sensitive to inhibition by fluconazole than the *Candida albicans* CYP51. One clear example was the discovery of four separate mutated CYP51 gene products from resistant *C. albicans* isolates. These genes were expressed in yeast and were shown to have different affinity for different azoles compared to the wild-type enzyme.[47] Overexpression of CYP51 is less common but in one case threefold increased levels were observed in resistant isolates.

Another less common change is an alteration in the sterol biosynthetic pathway. In the presence of azoles, the Δ-5,6-desaturase continues to synthesize 14-methyl sterols from 14α-methylfecosterol, resulting in the formation of a toxic membrane sterol. Mutation of the desaturase then results in the nontoxic methylfecosterol rather than ergosterol, hence the mutant strains are also resistant to polyenes.

7.15.2.5 Other Ergosterol Biosynthesis Inhibitors

Other than the SE inhibitors (allylamines) and lanosterol demethylase inhibitors (azoles), the only other pharmaceutically marketed ergosterol biosynthesis inhibitor is amorolfine (**23**). Amorolfine[48] is a member of the morpholine class of antifungal agent, which includes fenpropiomorph (**24**), a marketed agrofungicide. The mechanism of action of amorolfine is thought to be through inhibition of Δ_7–Δ_8 isomerase (IC_{50} 0.0018 μmol L^{-1}), though amorolfine also inhibits Δ_{14}-reductase (IC_{50} 2.39 μmol L^{-1}), which may result in a synergistic effect.

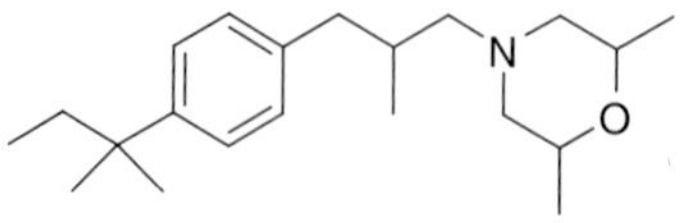

23

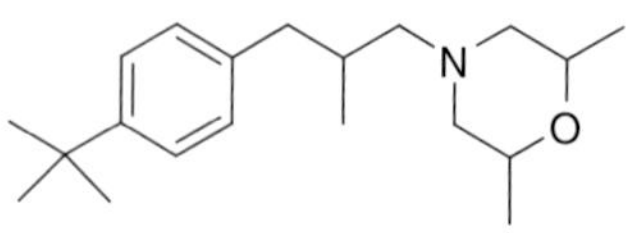

24

Amorolfine possesses a broad antifungal spectrum, including dermatophytes and yeasts; however, it is inactive against *Aspergillus* spp. In most cases, the fungicidal concentration is very close to its minimum inhibitory concentration (MIC). Like the SE inhibitors, this fungicidal effect, particularly, against dermatophytes, may be due to squalene accumulation. Alternatively, electron microscope studies reveal an accumulation of chitin in the cell membrane, which may be the direct effect.

Amorolfine is formulated as a 5% lacquer for the treatment of nail infections. Levels of amorolfine exceed the MICs of most fungi causing nail infections within 24 h of dosing. Penetration into the deepest nail slices at well over the MIC levels are observed within 4 weeks of treatment.

7.15.3 Fungal Cell Wall Disruptors

7.15.3.1 Polyenes

Amphotericin B (**25**), nystatin (**26**), and other polyene antibiotics interact with sterol-containing plasma membranes of sensitive organisms, causing a loss of membrane function, resulting in changes in permeability and disruption of nutrient uptake. Various explanations for the mechanism of action of the polyenes have been proposed, including formation of aqueous pores by rings of polyene molecules spanning the lipid bilayer. However, polyenes also act by interfering with the cellular oxidation–reduction balance. At higher concentrations, the polyenes cause gross membrane damage, a result of binding to membrane sterols affecting the lipid environment of the membrane. Consequently, the polyenes are nonselective in their action, and acute toxicity against the mammalian host is often observed. Over the last 50 years since their discovery, there have been numerous attempts to moderate the toxicity of the polyenes, most recently through liposomal formulations.

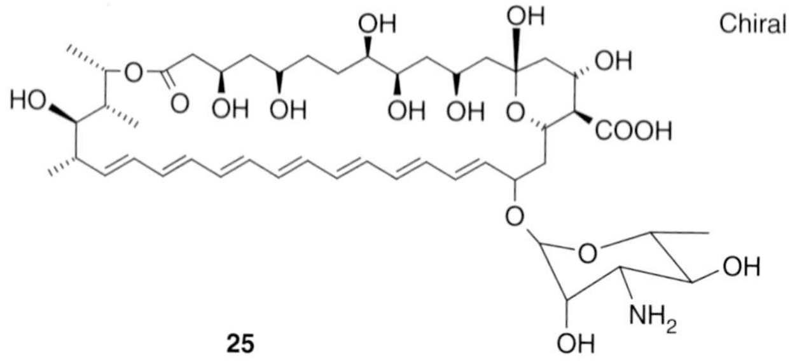

25

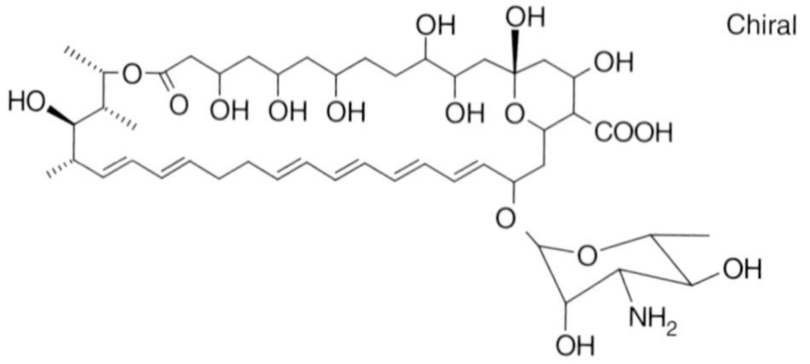

26

7.15.3.1.1 Medicinal chemistry

While the first reference to polyenes dates back as far as 1935, the golden age of this class was the 1950s. The formation of an antifungal antibiotic, fungicidin, by *Streptomyces noursi* was reported in 1951, although pure, crystalline material (subsequently called nystatin) was not obtained until 2 years later, based on submerged culture methods.[49] Amphotericin B, the second polyene, was first isolated in 1955, also from a *Streptomyces* culture. Both molecules are insoluble in water but are soluble in both acidic and basic alcoholic aqueous media, hence the choice of name. The structures of both molecules were not determined for several years, though it was known from ultraviolet spectroscopy that they contained several conjugated double bonds. There have been suggestions[50] that the double bonds are responsible for the ability of these compounds to bind to sterols, but the mechanism of action of the polyenes still remains unclear.

While there have been attempts to prepare semisynthetic analogs of the polyenes,[51] none has progressed far in development and efforts have focused on improvements over the standard desoxycholate solution formulation.

7.15.3.1.2 Clinical profile

Amphotericin B is generally dosed cautiously, starting with 1 mg intravenously to test the severity of the febrile reaction, adding hydrocortisone if necessary. Daily doses are raised to approximately 0.4–0.6 mg kg^{-1}. Hospitalization is required for the prolonged infusions that are necessary for this drug, and nephrotoxicity is a limiting factor. However, it is still used because of its potent fungicidal action. The drug is effective against most of the major fungal pathogens, including candidiasis, blastomycosis, and histomycosis. In cryptococcosis patients, cure rates are lower, usually about 70%, probably a result of low central nervous system penetration. However, response rates in apergillosis sufferers are generally very low.

Nystatin is little used as it is only available in topical formulations.

7.15.3.1.3 Resistance

As a result of its fungicidal action, there are relatively few reports of intrinsic resistance to amphotericin B. In the resistant strains, the most significant feature is an alteration in the lipid composition of the plasma membrane – either the nature of the membrane or the amount of sterol present. Resistance usually results from a lowered affinity of the membrane for amphotericin B.[52]

7.15.4 Fungal Cell Wall Synthesis Inhibitors

7.15.4.1 Overview

Since the fungal cell wall, an essential structure for maintaining cell integrity, is not found in mammalian cells, it offers the opportunity for specific antifungal targets. Although the exact structure of the fungal cell wall is not fully understood, it consists of a complex mixture of proteins and polysaccharides, including glucan, mannans, and chitin. Each of these carbohydrate-based polymers plays an important role in the cell wall via surface binding or structural processes. However, few of the potential targets have been isolated as pure enzymes. Instead, the primary method for identifying new targets has been whole-cell screening, followed by extensive biochemistry to identify the specific target. In addition, due to the polysaccharide nature of many of the enzyme substrates, few of the targets are amenable to small-molecule inhibitors and, consequently, inhibitors are usually natural products or semisynthetic analogs based on the natural products.

7.15.4.2 Glucan Synthase Inhibitors

7.15.4.2.1 Natural product discovery

The first class of glucan synthase inhibitors to be discovered were the echinocandins, which were isolated from fermentation broths of *Aspergillus* cultures in the early 1970s.[53] The echinocandins are a family of closely related lipopeptides, of which echinocandin B is the major component. Structurally, the echinocandins are composed of a complex hexapeptide whose N-terminus is acylated by a long chain carboxylic acid (**27**; **Table 6**). For echinocandin B,

Table 6 Structures of lipopeptide natural products (**27**)

Lipopeptide family	*R*	*R1*	*R2*	*R3*	*R4*	*R5*	*R6*
Echinocandin B	Linoleolyl	OH	OH	Me	H	Me	Me
Mulondocandin	12-Methylmyristoyl	OH	OH	H	H	H	Me
Sporiofungin A	10,12-Dimethylmyristoyl	OH	OH	H	H	CH_2CONH_2	Me
Pneumocandin B	10,12-Dimethylmyristoyl	OH	OH	Me	H	CH_2CONH_2	H
FR-901379	Palmitoyl	OH	OH	Me	OSO_3H	CH_2CONH_2	Me

the cyclic peptide is made up of 4,5-dihydroxyornithine, two threonines, 3-hydroxyproline, 3-hydroxy-4-methylproline and 3,4-dihydroxyhomotryosine.

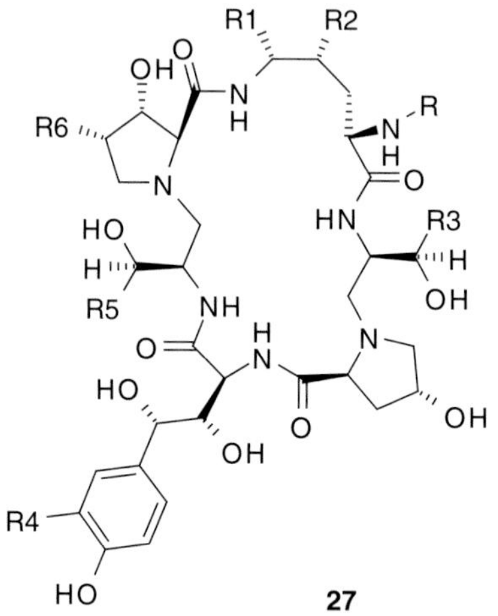

27

Following the discovery of the echinocandins, several related classes of natural product were discovered, including the mulundocandins (featuring a serine residue in place of one threonine), the pneumocandins (where the same threonine is replaced by a 3-hydroxyglutamine), and the sporiofungins (both threonines are substituted by serines). The family of lipopeptides are characterized by excellent in vitro activity against *Candida albicans* (0.1–1 $\mu g\, mL^{-1}$). Several compounds were also protective against candidiasis in animal models following intraperitonal dosing.

Mechanistic studies suggested that the echinocandins and related lipopeptides inhibited cell wall biosynthesis. The compounds were found to inhibit the formation of (1,3)-β-glucan, which is present as the major strucutural component of fungal cell walls. Microfibrils of the glucan are extruded into the interstitial space of the cell, where they are cross-linked by enzymes such as glycosyl transferases to construct the growing cell wall. Interference with the process leads to a weakened wall, cell content leakage, and cell death.

The point of intervention was found to be (1,3)-β-glucan synthase, which is a membrane-bound, multisubunit enzyme, made up of an insoluble catalytic subunit and a soluble regulatory subunit. The active complex binds and polymerizes uridine diphosphate glucose to form glucan. The lipopeptides act via a noncompetitive mechanism but their exact mechanism of action is unknown. However, it has been shown that simultaneous disruption of the genes (*FKS1* and *FKS2*) coding for glucan synthase is lethal and that inhibition with the lipopeptides results in a fungicidal mechanism of action. Since glucan synthesis does not occur in humans, the lipopeptides were thought likely to have the potential for a selective mechanism of action, with low mammalian toxicity. Although this was found to be generally true, unfortunately, several of the early natural products were also found to have a hemolytic action. Subsequent efforts used semisynthesis to search for improvements in the toxicity profile as well as in their antifungal spectrum.

7.15.4.2.2 Medicinal chemistry

The direction of the early medicinal chemistry programs targeting glucan synthase was influenced by the observation that other naturally occurring lipopeptides had different activity and toxicity profiles. These profiles were dependent on the nature of the fatty-acid side chain. Enzymatic removal of the acyl side chain enabled reacylation with a wide range of carboxylic acids. While short-chain acids such as acetic and benzoic acid were inactive, linear chains of around 12 carbon atoms restored activity. Detailed SAR studies were carried out at Lilly with systematic variation of the acyl side chain, including replacement by homologous linear chains (C-12 to C-22), a series of alkoxy- and alkylthio-benzoyl.[54] SARs for in vitro and in vivo anti-*Candida* activity were determined as well as for in vitro hemolytic effects. The three most preferred compounds were progressed to a subchronic toxicity study, which identified the 4-octyloxybenzoyl side chain as optimal, providing the first drug candidate, cilofungin. Unfortunately, cilofungin (**28**) suffered from low aqueous solubility, a characteristic of many members of this class of compound, which required formulation in a co-solvent system. These formulations of the drug led to nephrotoxicity and resulted in the withdrawal of cilofungin from development.[55]

Further investigation of the lipophilic side chain found that whole-cell activity of many molecules from different lipopeptide series correlated with the calculated log P of the lipophilic side chain.[56] The values were used to predict relative changes in activity of the whole-molecule analogs of cilofungin and established that the acyl side chain must have a c log $P>3.5$ and a linear configuration. A large number of side chains containing ether and phenyl linkers were prepared, resulting in the identification of LY-303366 (anidulafungin (**28**)) as the second candidate from this series. Interestingly, both cilofungin and anidulafungin are relatively modest inhibitors of glucan synthase (**28**: **Table 7**).

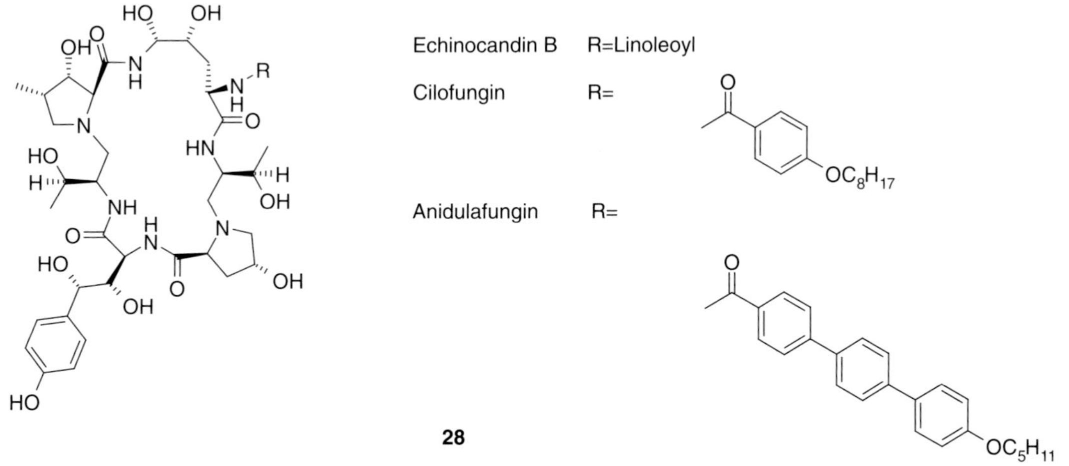

Echinocandin B R=Linoleoyl

Cilofungin R=

Anidulafungin R=

28

In view of its high lipophilicity, it is not surprising that anidulafungin suffers from low aqueous solubility ($<0.1\ \mu g\ mL^{-1}$). Aside from formulation methods to circumvent this issue, a number of water-soluble phosphate prodrugs of anidulafungin and analogs have been prepared.[57] Several analogs combine excellent aqueous solubility with activity in the in vitro and in vivo assays.

In parallel with the work on the echinocandin series, workers at Merck carried out extensive semisynthetic studies on the pneumocandin[58] series they had discovered. Pneumocandin B_0 features a dimethylmyristoyl group as the lipophilic side chain, as well as the β-hydroxyglutamine side chain that is characteristic of the pneumocandins. In a similar finding to the echinocandin SAR, Merck found that in vivo activity could be improved by conversion to phosphate prodrugs[59] and, as for the echinocandins, glucan synthase inhibition by pneumocandins was found not to correlate with whole-cell activity (**Table 8**).[60]

Table 7 Activity of echinocandin antifungals against *Candida albicans*

Compound	*clogP*	*Glucan synthase % inh. at 10 $\mu g\ mL^{-1}$*	*MIC ($\mu g\ mL^{-1}$) versus Candida albicans*	*ED_{50} ($mg\ kg^{-1}$) oral dosing in mouse candidiasis model*
Cilofungin	4.56	76.0	0.312	>400
Anidulafungin	6.47	75.0	0.005	7.8

Data from Debono, M.; Turner W. W.; LaGrandeur, L.; Burkhardt, F. J.; Nissen, J. S.; Nichols, K. K.; Rodriguez, M. J.; Zweifel, M. J.; Zeckner, D. J.; Gordee, R. S. *et al. J. Med. Chem.* **1995**, *38*, 3271–3281.

Table 8 Comparative activity of echinocandin and pneumocandin antifungals against *Candida albicans*

Compound	*Glucan synthase IC_{50} ($\mu mol\ L^{-1}$)*	*MFC ($\mu g\ mL^{-1}$)*	*Candidiasis model ED_{90} ($mg\ kg^{-1}$ per dose)*	*Minimum hemolytic concentration ($\mu g\ mL^{-1}$)*
Echinocandin B	0.25	0.5.	>6	75
Cilofungin	1.0	0.5	3	>400
Pneumocandin B_0	0.07	0.25	3	>400
Phosphate prodrug of pneumocandin	>10	4	3	>400

MFC, Minimum, fungicidal concentration.

Data from Balkovec J. M.; Black, R. M.; Bouffard, F. A.; Dropinski, J. F.; Hammond, M. L. *Proceedings of the XIVth International Symposium on Medicinal Chemistry* **1997**, *28*, 1–13.

Table 9 Activity of pneumocandin derivatives against systemic pathogens

Compound	*Glucan synthase IC_{50} ($\mu mol\,L^{-1}$)*	*MFC ($\mu g\,mL^{-1}$)*	*Candidiasis model ED_{90} ($mg\,kg^{-1}$ per dose)*	*Aspergillosis model ED_{90} ($mg\,kg^{-1}$ per dose)*
Pneumocandin B_o	0.07	0.25	3	>20
Methyl ether	0.1	2	>6	1.8
L-705589	0.01	0.125	0.3	0.06
Aminoethyl thioether	0.006	0.5	0.2	0.52
L-733560	0.001	0.125	0.09	0.03
Caspofungin		0.125		

Instead of relying on changes to the acyl side chain, the Merck group found that activity could be improved by reducing the β-hydroxyglutamine side chain to a β-hydroxyornithine (L-731373) or by replacing the hemiaminal with an aminal ether.[61] An unexpected outcome of these SAR studies was a marked improvement in in vivo potency in an aspergillosis infection model through the change to aminoethyl ethers (L-705589) or thioethers. Combining the change in the glutamine side chain and aminal ether (L-733560) provided unprecedented levels of potency from this series. Finally, replacing the aminal ether with amines provided compounds with an excellent in vivo profile and with superior pharmacokinetics, particularly the aminoethylamino analog (L-743872, caspofungin: **29**; **Table 9**).

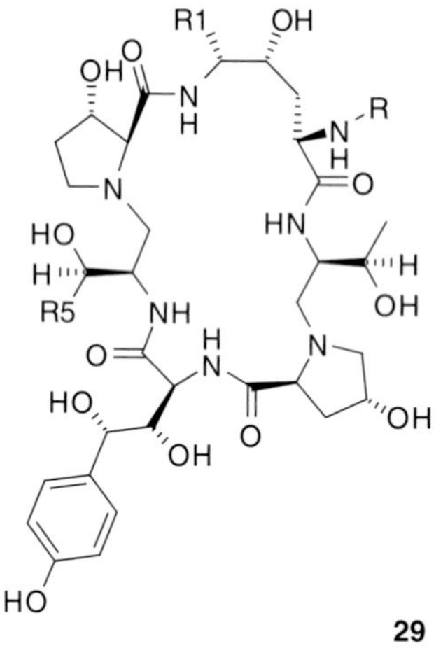

Pneumocandin B_0 R1=OH R5=CH_2CONH_2

L-731373 R1=$OCH_2CH_2NH_2$ R5=CH_2CONH_2

L-705589 R1=OH R5=$CH_2CH_2NH_2$

L-733560 R1=$OCH_2CH_2NH_2$ R5=$CH_2CH_2NH_2$

Caspofungin R1=$NHCH_2CH_2NH_2$ R5=$CH_2CH_2NH_2$

29

A similar series of analogs to the Merck series has been prepared by scientists at Quest based on mulundocandin, which features a serine side chain in place of the β-hydroxyglutamine. Derivatization of the hemiaminal residue with oxygen or sulfur nucleophiles resulted in compounds with similar in vitro activity to the parent but with improved aqueous stability.[62] Subsequent modification of the tyrosine residue using Mannich chemistry provided analogs with improved in vivo activity by both intraperitoneal and oral routes.[63]

Micafungin, the third member of the lipopeptide family to have reached the market, illustrates many of the same themes featuring in the discoveries of anidulafungin and caspofungin. Scientists at Fujisawa identified another natural product (FR-901739)[64] with an attractive activity profile. This natural product, which is analogous to the pneomocandins in bearing a β-hydroxyglutamine side chain, also features a sulfonate adjacent the phenolic hydroxyl present in most candins. Enzymatic cleavage of the natural side chain and replacement by lipophilic groups[65] provided FR-131335 and subsequently, micafungin, a potent glucan synthase inhibitor (*Candida* IC_{50} 0.21 $\mu mol\,L^{-1}$), with excellent whole-cell activity (**30**).

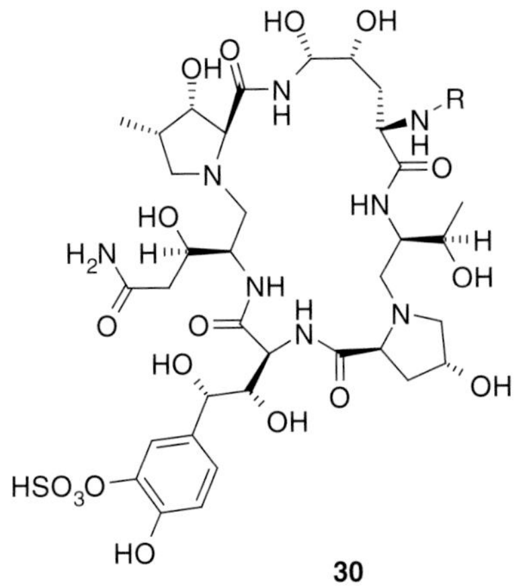

FR-901379 R=palmitoyl

FR-131535 R=

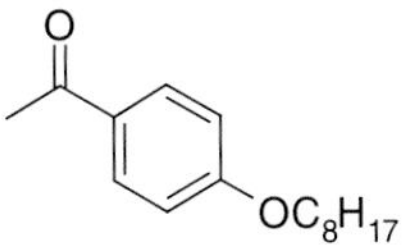

Micafungin R=

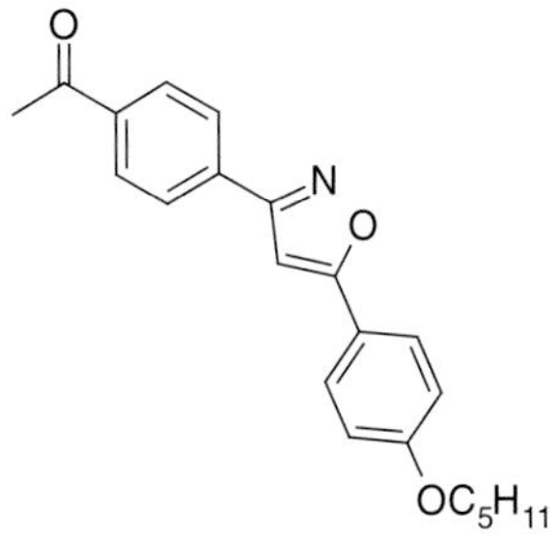

30

In attempts to overcome the limitations of natural product-derived series, alternative total synthesis approaches have been employed for the synthesis of simplified lipopeptide analogs. The Merck group was able to demonstrate the importance of both the 3-hydroxy-4-methylproline and the homotyrosine groups.[66] More recently, workers at Abbott have shown that the disubstituted proline can be replaced by a 3-methylaminoproline or 3-guanidinoproline.[67] Further modifications on the phenolic ring led to the fully synthetic echinocandin analog A-192411 (**31**).[68]

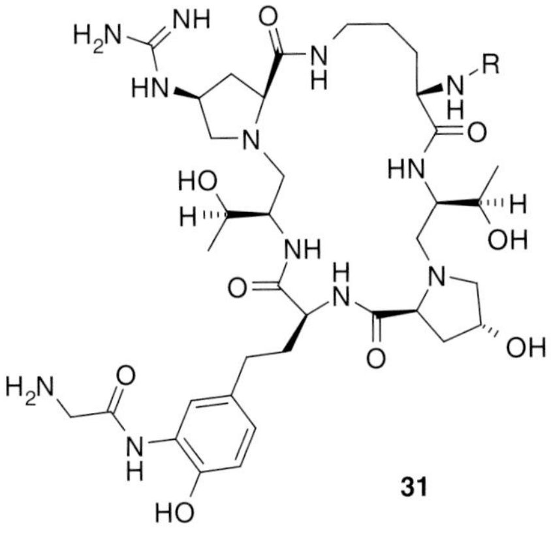

A-192411 R=

31

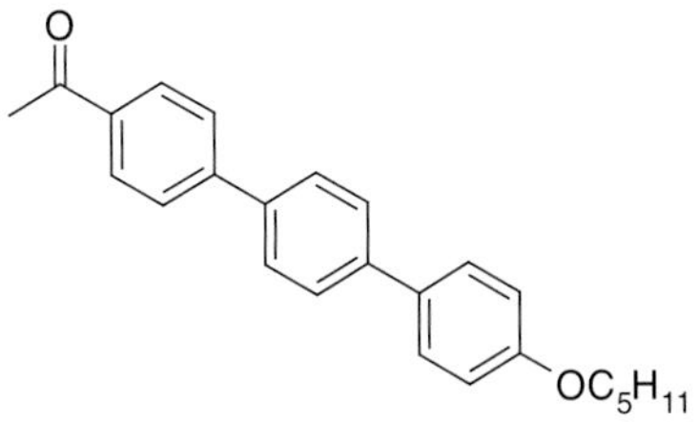

7.15.4.2.3 Pharmacokinetic profiles

The pharmacokinetic profiles of the three launched candin drugs have been well documented. All three agents are administered by intravenous infusion, as a result of low bioavailability, possibly due to their propensity to chemical degradation.

Following intravenous administration of caspofungin,[69] clearance is low in all species (0.26–0.51 $mL\,min^{-1}\,kg^{-1}$), with a modest volume of distribution (0.11–0.27 $L\,kg^{-1}$), resulting in a half-life of 5.2–7.6 h. Plasma protein binding is high (96%) in both mouse and humans. Dose escalation in humans resulted in dose-proportional increase in the area under the curve (AUC) and a modest accumulation was also seen following multiple dosing. Plasma concentration declines in a polyphasic manner in humans, with a short α-phase (1–2 h) followed by an extended β-phase (6–48 h), reflecting distribution into the extracellular fluid ($V_d = 9.7$ l).

Studies with radiolabeled material showed that caspofungin is metabolized by hydrolysis and *N*-acetylation. In addition, caspofungin undergoes spontaneous chemical degradation to an inactive metabolite. Clearance of caspofungin does not appear to involve cytochrome P-450 s, nor does it inhibit P-450 s.

Anidulafungin[70] also has a long terminal half-life in all species (rat: 22.6 h) due to a combination of low clearance and large terminal volume of distribution (33.4 l in humans). For example, in a trial of candidiasis patients receiving 50 $mg\,day^{-1}$ of anidulafungin, clearance was 0.93 $L\,h^{-1}$, resulting in a half-life of 25.6 h. The primary route of degradation of anidulafungin appears to be chemical degradation to ring-opened inactive metabolites. Consequently, like caspofungin, anidulafungin metabolism is little affected by concomitant administration of other agents.

Micafungin[71] is reported to have a similar volume of distribution to the other candins ($0.23\,L\,kg^{-1}$) but, combined with reduced clearance ($0.18\,mL\,min^{-1}\,kg^{-1}$), this results in a long half-life (15 h in humans). The majority of the dose is excreted unchanged in feces, though a small amount of metabolism to a hydroxymetabolite has been observed.[72] No interaction with other drugs has been reported.

7.15.4.2.4 Clinical profiles

Caspofungin has demonstrated efficacy in clinical trials in patients with invasive candidiasis and aspergillosis; patients received a 70 mg loading dose followed by $50\,mg\,day^{-1}$. In a comparative trial with amphotericin B in candidiasis sufferers, caspofungin showed a favorable response in 73% of patients. A similar level of response (72%) was seen in another trial of patients with candidemia. In successfully treated patients, the relapse rates were low (6.4%). There are fewer published studies for patients with aspergillosis infections, although in one, caspofungin demonstrated an effect in 45% of the patients; the reponse was higher in patients with pulmonary infections.

Using a similar dosing regimen to those for caspofungin, the response rate for anidulafungin in esophageal candidiasis was 85%. A lower response rate (72%) was seen in patients with candidemia, though reponse rates were improved at higher doses ($100\,mg\,day^{-1}$). There are no published data on the use of anidulafungin in aspergillosis patients, despite reports of favorable responses in animal infection models.

Micafungin is also effective in the eradication of esophageal candidiasis, with response rates of 77% and 90% observed following 14–21 days' treatment at 100 or $150\,mg\,day^{-1}$, respectively. Similarly high success rates (88%) were observed in patients with newly diagnosed candidemia. Like caspofungin, micafungin is effective against invasive aspergillosis: complete or partial responses were achieved in 71% of cases.

7.15.4.2.5 Drug resistance

Although the candins have been relatively recent introductions to the antifungal drug armamentarium, there are a number of reports of primary resistance to them. While *Candida albicans* is generally sensitive to the candins, some 40% of *C. parapsilosis* strains were found to be insensitive to caspofungin.[73] While this difference in in vitro activity does not always translate to lower clinical success, a multicandin, multiazole-resistant strain has been obtained from a patient with prosthetic valve endocarditis.[74] In another case, isolates from a patient infected with *C. albicans* had increasing MICs, which correlated with in vivo failure following initially successful treatment.[75] Strains of *C. albicans* that are resistant to caspofungin have also been generated in vitro, and, in one key study, this resistance was shown to be due to changes in glucan synthase.[76]

7.15.4.3 Chitin Synthase Inhibitors

As with glucan, chitin is a major and essential component of the fungal cell wall but is absent from mammalian cells. Two classes of natural products, the polyoxins and nikkomycins, are known to be competitive inhibitors of chitin synthase. but neither class has significant antifungal activity.

In one recent report, Roche chemists reported the discovery of a novel chitin synthase inhbitor, RO-41-0986, through high-throughput screening (nikkomycin Z: CaChS $IC_{50} = 3200\,nmol\,L^{-1}$: **32**; RO-41-0986: CaChS $IC_{50} = 11\,300\,nmol\,L^{-1}$: **33**; RO-09-3024: CaChS $IC_{50} = 0.14\,nmol\,L^{-1}$; *Candida albicans* MIC $= 0.10\,mg\,mL^{-1}$: **34** polyoxin D: **35**; aureobasidin A: **36**).[77] Although the initial hits were structurally related to terbinafine, none of the series inhibited SE. Further modifications provided compounds that were potent inhibitors of *Candida* chitin synthase, which translated to whole-cell activity. Unfortunately, the preferred compound, RO-09-3024, was relatively weak against *Aspergillus fumigatus* and was inactive against *Crypotococcus neoformans*.

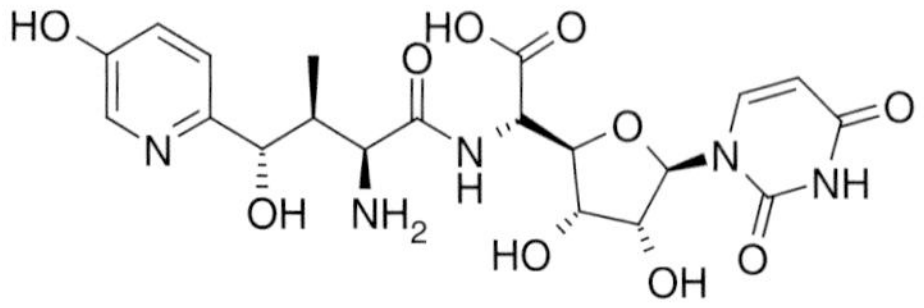

HO HO O O N O N NH N H OH NH2 HO OH O

32

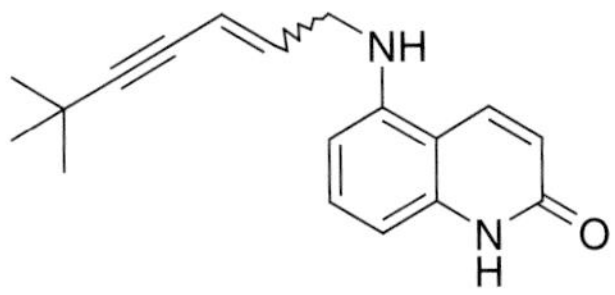

33

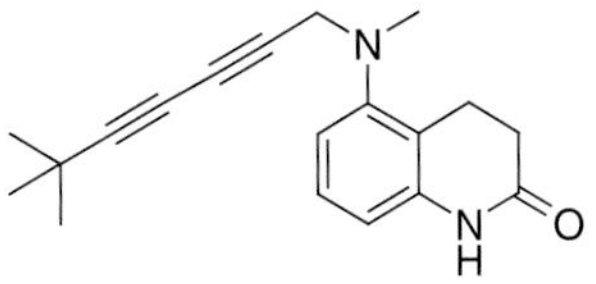

34

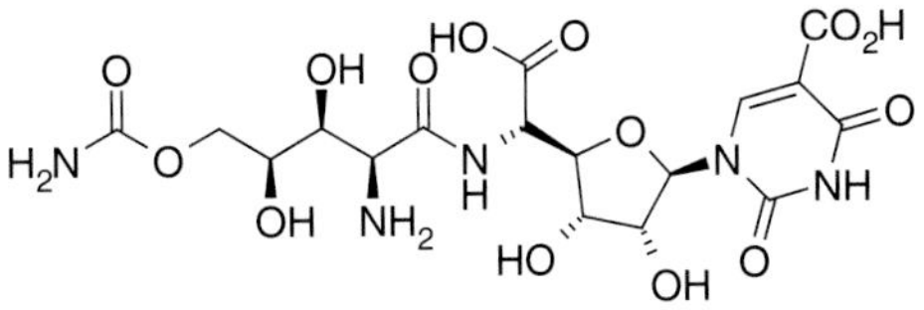

35

7.15.5 New Directions

7.15.5.1 Overview

As the first generation of orally bioavailable systemic antifungal agents reach the end of their patent life, research effort has shifted to a new generation of fungal biological targets. Although some of the new targets have been discovered by the traditional method of whole-cell screening of natural products, other targets have been discovered through genetic approaches based on knowledge of the sequence of the yeast genome. Advances in fungal genetics have also enabled researchers to determine whether a particular target is essential for growth of the target organism, and to clone and express isozymes from the other pathogens to help determine SAR at the target level. In addition, these advances have enabled scientists to identify biological targets with no, or a very structurally distinct, human homolog. However, the challenge of discovering a broad-spectrum antifungal with specificity, or very high selectivity, for the fungi over the human host, combined with good pharmacokinetic profile, remains high. The following sections summarize some of the recent advances in the field, though, as yet, few drug candidates have emerged.

7.15.5.2 Inositol Phosphoryl Ceramide Synthase

Inositol phosphoryl ceramide synthase was identified as an attractive target for antifungals through the traditional method of natural product screening. The aureobasidin class of natural products produced by *Aureobasidium pullulans* are active against many pathogenic fungi, including *Candida albicans, Cryptococcus neoformans*, and *Histoplasma capsulatum*, and are fungicidal against *Candida albicans*.[78] Inositol phosphoryl ceramide synthase is specific to fungi and is essential for fungal growth through its involvement in sphingolipid biosynthesis.[79] The prototypical natural product, aureobasidin A (**36**), was active in a systemic mouse candidiasis model following both oral and intravenous administration but showed little toxicity in mice and dogs.

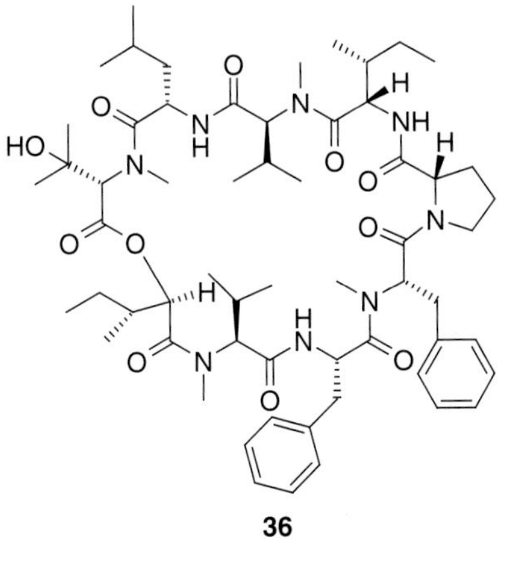

36

Scientists working for Takara Shuzo, who identified the aureobasidins, discovered that they were cyclic depsipeptides made up of a hydroxy acid and eight amino acids. Four of the amino acids are *N*-methylated, resulting in a unique molecular conformation, stabilized by three intramolecular hydrogen bonds.[80]

Several research groups[81] have used total or semisynthetic procedures to prepare aureobasidin A or analogs, mostly in attempts to obtain broad-spectrum activity, since the weak point of the class is poor activity against *Aspergillus fumigatus*. Only the Takara Shuzo group has reported compounds[82] with promising activity against this important human pathogen.

7.15.5.3 *N*-Myristoyl Transferase

Myristoyl coenzyme A (CoA) protein *N*-myristoyl transferase (NMT) is one of the first antifungal targets discovered by a genetic-based approach. NMTs catalyze the transfer of the rare cellular fatty acid, myristate, from myristoyl CoA to the N-terminal glycines of eukaryotic proteins. In *C. albicans*, protein myristoylation is required for the activity of several proteins, including an adenosine diphosphate ribosylation factor. Genetic studies[83] have shown that fungal pathogens require NMT to maintain their viability in vitro and in vivo.

Collaboration between scientists at Searle and Washington University, US, used the sequence of peptide substrates of NMT as a starting point for their drug discovery program. They first discovered that switching the N-terminal glycine of the substrates for alanine resulted in compounds that were inhibitors of *C. albicans* NMT. Secondly, they observed that the identity of residues 2–4 of the peptide were unimportant for recognition as a substrate, whereas serine and lysine residues were preferred at residues 5 and 6 respectively. Replacement of the first four residues of the peptidic inhibitor with an 11-aminoundecanoyl group and simplification of the C-terminal side chain to cyclohexylethylamide provided micromolar inhibitors of both *Candida* and human NMTs. Furthermore, replacement of the N-terminal amine by imidazoles was well tolerated and, surprisingly, introduction of a 2-methyl substituent on the imidazole ring increased both a potency and selectivity over the human enzyme, e.g., NMT inhibitor, SC-58272 (**37**).[84] This inhibitor has also been co-crystallized as a ternary complex with yeast (*Saccharomyces cerivisiae*) NMT and myristoyl CoA, showing that the inhibitor binds in the peptide-binding site.[85]

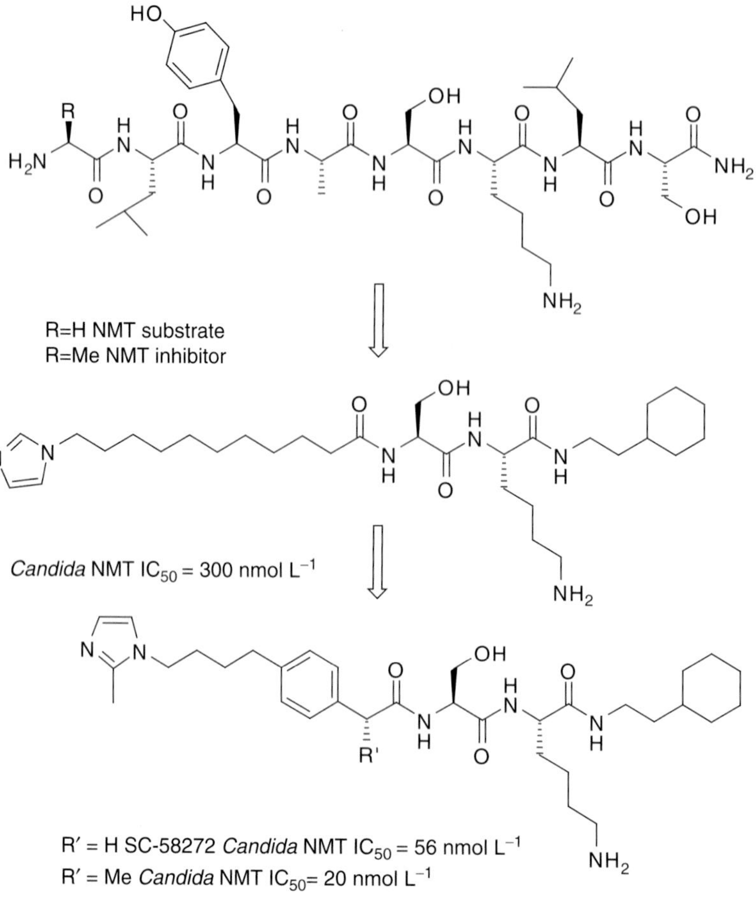

Subsequent SAR studies in the Searle series identified an additional chiral recognition that enhanced potency/selectivity[86] and, secondly, addressed the peptidic nature of the inhibitors by replacing the lysine side chain with heterocycles.[87] Despite extensive effort in the series, antifungal activity was only achieved in some of the weaker, nonselective inhibitors.[88] Consequently other research groups adopted a high-throughput screening approach to the discovery of NMT inhibitors.

The Pfizer group modified a micromolar potency benzothiazole HTS hit, CP-123457, to provide compounds with high potency against *C. albicans* NMT and high selectivity over the human enzyme, e.g. UK-370485[89] and further analogs, e.g., UK-370753[90] with submicromolar potency in a whole-cell assay (**38**). More recently, researchers at SSP used virtual screening to identify novel related benzothiazoles[91] with a similar profile.

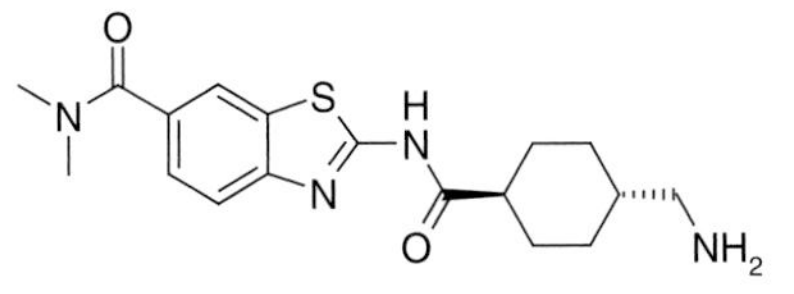

UK-370485 *Candida* NMT IC_{50} = 42 nmol L^{-1}

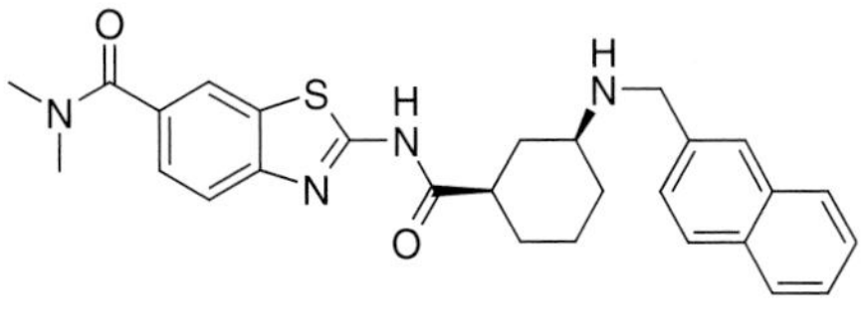

UK-370753 *Candida* NMT IC_{50}= 86 nmol L^{-1}
C. albicans MIC = 0.09 mg mL^{-1}

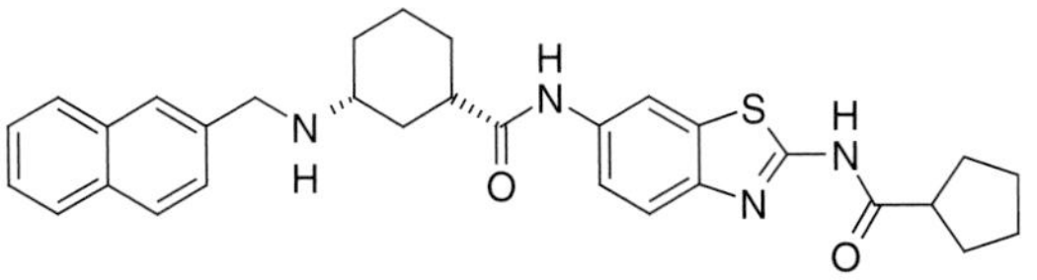

SSP inhbitor *Candida* NMT IC_{50}= 0.49 nmol L^{-1}
C. albicans MIC = 0.78 mg mL^{-1}

38

The Roche group found a benzofuran inhibitor derived from their β-blocker series, which they were able to modify to remove the unwanted activity and increase affinity for the target NMT. Like the Pfizer series, the preferred Roche compound RO-09-4609[92] was highly selective over the human homolog and demonstrated whole-cell activity against *Candida albicans* (**39**). While the initial hits were metabolically vulnerable, subsequent medications identified compounds with activity in animal infection models but with low stability in stimulated gastric medium.[93] Unfortunately, although the Roche team were able to address the stability issues in this series, they were unable to identify compounds with potent activity against both *C. albicans* and *Aspergillus fumigatus* NMTs.[94]

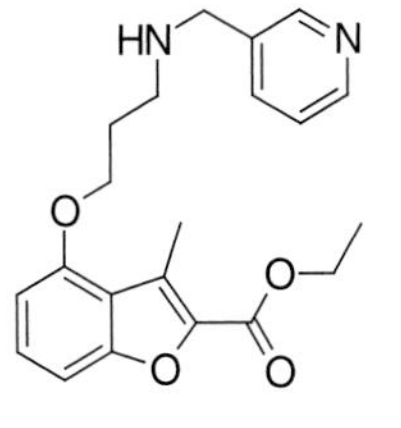

RO-09-4609 *Candida* NMT IC_{50}= 100 nmol L^{-1}
C. albicans MIC IC_{50}= 1.6 mmol L^{-1}

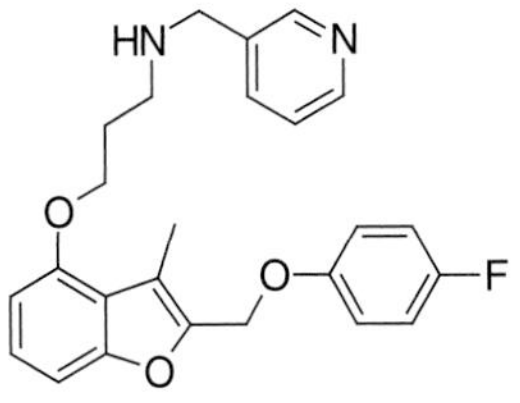

Ref 93 *Candida* NMT IC_{50} = 5.2 nmol L^{-1}
C. albicans MIC = 0.02 mmol L^{-1}

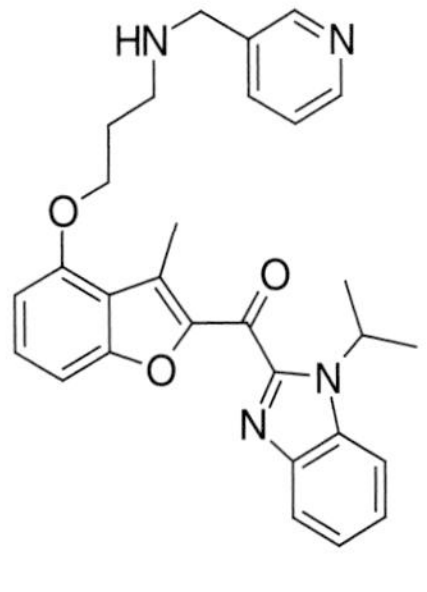

Ref 94 *Candida* NMT IC_{50}= 1.3 nmol L^{-1}
C. albicans MIC = 0.02 mmol L^{-1}
Aspergillus NMT IC_{50}= 480 nmol L^{-1}

39

To date, there are no NMT inhibitor development candidates reported, possibly because of the difficulties in obtaining broad-spectrum antifungal agents with appropriate metabolic stability (*see* 7.14 Fungi and Fungal Disease).

References

1. Ryder, N. S.; Dupont, M.-C. *Biochem. J.* **1985**, *230*, 765–770.
2. Stütz, A. *Angew. Chem. Int. Ed. Engl.* **1987**, *26*, 320–328.
3. Ryder, N. S. *Ann. NY Acad. Sci.* **1988**, *544*, 208–220.
4. Petranyi, G.; Georgopoulos, A.; Mieth, H. *Antimicrob. Agents Chemother.* **1981**, *19*, 390.
5. Stütz, A.; Georgopoulos, A.; Granitzer, W.; Petranyi, G.; Berney, D. *J. Med. Chem.* **1986**, *29*, 112–125.
6. Stütz, A.; Petranyi, G. *J. Med. Chem.* **1984**, *27*, 1539–1543.
7. Nussbaumer, P.; Petranyi, G; Stütz, A. *J. Med. Chem.* **1991**, *34*, 65–73.
8. Nussbaumer, P.; Dorfstatter, G.; Grassberger, M. A.; Leitner, I.; Meingasser, J.; Thirring, K.; Stütz, A. *J. Med. Chem.* **1993**, *36*, 2115–2120.
9. Maeda, T.; Takase, M.; Ishibashi, A.; Yamamoto, T.; Sasaki, K.; Arika, T. *Yakugaku Zasshi* **1991**, *111*, 126.
10. Schatz, F.; Haberl, H.; Battig, F.; Jobstmann, D.; Schulz, G. *Arzneim.-Forsch.* **1986**, *36*, 248.
11. Villars, V.; Jones, T. C. *Clin. Exp. Dermatol. Treat.* **1990**, *1*, 33.
12. Kan, V. L.; Henderson, D. K.; Bennett, J. E. *Antimicrob. Agents Chemother.* **1986**, *30*, 628.
13. Monk, J. P.; Brogden, R. N. *Drugs* **1991**, *42*, 659–672.
14. Balfour, J. A.; Faulds, D. *Drugs* **1992**, *43*, 259–284.
15. Mukherjee, P. K.; Leiditch, S. D.; Isham, N.; Leitner, I.; Ryder, N. S.; Ghannoum, M. A. *Antimicrob. Agents Chemother.* **2003**, *47*, 82–86.
16. Karyotakis, N. C.; Anaissie, E. J.; Hachem, R.; Dignani, M. C.; Samonis, G. *J. Infect. Dis.* **1993**, *168*, 1311–1313.
17. Ji, H.; Zhang, W.; Zhou, Y.; Zhang, M.; Zhu, J.; Song, Y.; Lu, J.; Zhu, J. *J. Med. Chem.* **2000**, *43*, 2493–2505.
18. Lamb, D. C.; Kelly, D. E.; Baldwin, B. C.; Kelly, S. L. *Chemico-Biol. Interact.* **2000**, *125*, 165–175.
19. Godefroi, E. F.; Heeres, J.; Van Cutsem, J.; Janssen, P. A. J. *J. Med. Chem.* **1969**, *12*, 784–791.
20. Jevons, S.; Gymer, G. E.; Brammer, K. W.; Cox, D. A.; Leeming, M. R. G. *Antimicrob. Agents Chemother.* **1979**, *15*, 597.
21. Heeres, J.; Backx, L. J. J.; Van Cutsem, J. *J. Med. Chem.* **1976**, *19*, 1148.
22. Buchel, vK. H.; Draber, W.; Regel, E.; Plempel, M. *Arzneim-Forsch.* **1972**, *22*, 1260–1272.
23. Heeres, J.; Backx, L. J. J.; Mostmans, J. H.; Van Cutsem, J. *J. Med. Chem.* **1979**, *22*, 1003–1005.
24. Heeres, J.; Hendrickx, R.; Van Cutsem, J. *J. Med. Chem.* **1983**, *26*, 611–613.
25. Heeres, J.; Backx, L. J. J.; Van Cutsem, J. *J. Med. Chem.* **1984**, *27*, 894–900.
26. Odds, F. C. *J. Antimicrob. Chemother.* **1989**, *24*, 533–537.
27. Saksena, A. K.; Girijavallabhan, V. M.; Lovey, R. G.; Pike, R. E.; Desai, J. A.; Ganguly, A. K.; Hare, R. S.; Loebenberg, D.; Cacciapouti, A.; Parmegiani, R. M. *Bioorg. Med. Chem. Lett.* **1995**, *5*, 127.
28. Richardson, K. Contemporary Organic Synthesis: The Discovery of Fluconazole, 1996, *3*, 125–132.
29. Dickinson, R. P.; Bell, A. S.; Hitchcock, C. A.; Narayanaswami, S.; Ray, S. J.; Richardson, K.; Troke, P. F. *Biorg. Med. Chem. Lett.* **1996**, *6*, 2031–2036.
30. Gollapudy, R.; Ajmani, S.; Kulkarni, S. A. *Biorg. Med. Chem.* **2004**, *12*, 2937–2950.
31. Saji, I.; Tamato, K.; Tanio, T.; Okuda, T.; Atsumi, T. *Abstracts of 8th Medicinal Chemistry Symposium*; Osaka, November 1986, p 9.
32. Miyauchi, H.; Tanio, T.; Ohashi, N. *Biorg. Med. Chem. Lett.* **1995**, *5*, 933–936.
33. Tsuruoka, A.; Kaku, Y.; Kakinuma, H.; Tsukada, I.; Yanagisawa, M.; Nara, K.; Naito, T. *Chem. Pharm. Bull.* **1997**, *45*, 1169.
34. Tsuruoka, A.; Kaku, Y.; Kakinuma, H.; Tsukada, I.; Yanagisawa, M.; Nara, K.; Naito, T. *Chem. Pharm. Bull.* **1998**, *46*, 623–630.
35. Ueda, Y.; Matiskella, J. D.; Golik, J.; Connolly, T. P.; Hudyma, T. W.; Venkatesh, S.; Dali, M.; Kang, S.-H.; Barbour, N.; Tejwani, R. et al. *Biorg. Med. Chem. Lett.* **2003**, *13*, 3669–3672.
36. Tasaka, A.; Tamura, N.; Matsushita, Y.; Kitazaki, T.; Hayashi, R.; Okonogi, K.; Itoh, K. *Chem. Pharm. Bull.* **1995**, *43*, 432.
37. Tasaka, A.; Kitazaki, T.; Tsuchimori, N.; Matsushita, Y.; Hayashi, R.; Okonogi, K.; Itoh, K. *Chem. Pharm. Bull.* **1997**, *45*, 321–326.
38. Ichikawa, T.; Kitazaki, T.; Matsushita, Y.; Hosono, H.; Yamada, M.; Mizuno, M.; Okonogi, K.; Itoh, K. *Chem. Pharm. Bull.* **2000**, *48*, 1947.
39. Ichikawa, T.; Kitazaki, T.; Matsushita, Y.; Yamada, M.; Hayashi, R.; Yamaguchi, M.; Kiyota, Y.; Okonogi, K; Itoh, K. *Chem. Pharm. Bull.* **2001**, *49*, 1102–1109.
40. Heel, R. C.; Brogden, R. N.; Pakes, G. E.; Speight, T. M.; Avery, G. S. *Drugs* **1980**, *19*, 7–30.
41. Heel, R. C.; Brogden, R. N.; Speight, T. M.; Avery, G. S. *Drugs* **1978**, *16*, 177–201.
42. Heel, R. C.; Brogden, R. N.; Carmine, A.; Morley, P. A.; Speight, T. M.; Avery, G. S. *Drugs* **1982**, *23*, 1–36.
43. Haria, M.; Bryson, H. M.; Goa, K. L. *Drugs* **1996**, *51*, 585–620.
44. Goa, K. L.; Barradell, L. B. *Drugs* **1995**, *50*, 658–690.
45. Muijsers, R. B. R.; Goa, K. L.; Scott, L. J. *Drugs* **2002**, *62*, 2655–2664.
46. Stevens, D. A.; Holmberg, K. *Curr. Opin. Anti-infect. Invest. Drugs* **1999**, *1*, 306–317.
47. Vanden Bossche, H.; Dromer, F.; Impovisi, I.; Lozano-Chiu, M.; Rex, J. H.; Sanglard, D. *Med. Mycol.* **1998**, *36* (suppl. 1), 119–128.
48. Polak, A. *Ann. NY Acad. Sci.* **1988**, *544*, 221–228.
49. Dutcher, J. D.; Boyack, G.; Fox, S. *Antibiotics Ann.* **1953**, *54*, 191–194.
50. Bennett, J. E. *Chemotherapy. Proceedings of the 9th International Congress on Chemotherapy* **1976**, *6*, 105–109.
51. MacLachlan, W. S.; Readshaw, S. A. *Tetrahedron Lett.* **1995**, *36*, 1735.
52. Hitchcock, C. A.; Barrett-Bee, K. J.; Russell, N. J. *J. Med. Vet. Mycol.* **1987**, *25*, 29.
53. Benz, F.; Knusel, F.; Nuesch, J.; Treichler, H.; Voser, W.; Nyfeler, R.; Keller-Schielein, W. *Helvet. Chim. Acta* **1974**, *57*, 2459.
54. Debono, M.; Abbott, B. J.; Fukuda, D. S.; Barnhart, M.; Willard, K. E.; Molloy, R. M.; Michel, K. H.; Turner, J. R.; Butler, T. F.; Hunt, A. H. et al. *Ann. J. Antibiotics.* **1989**, *42*, 389.
55. Turner, W. W.; Rodriguez, M. J. *Curr. Pharm. Design* **1996**, *2*, 209–224.
56. Debono, M.; Turner, W. W.; LaGrandeur, L.; Burkhardt, F. J.; Nissen, J. S.; Nichols, K. K.; Rodriguez, M. J.; Zweifel, M. J.; Zeckner, D. J.; Gordee, R. S. et al. *J. Med. Chem.* **1995**, *38*, 3271–3281.
57. Rodriguez, M. J.; Vasuden, V.; Jamison, J. A.; Borromeo, P. S.; Turner, W. W. *Bioorg. Med. Chem. Lett.* **1999**, *9*, 1863–1868.
58. Schmatz, D. M.; Abruzzo, G.; Powles, M. A.; McFadden, D. C.; Balkovec, J. M.; Black, R. M.; Nollstadt, K.; Bartizal, K. et al. *J. Antibiotics* **1992**, *45*, 1886–1891.
59. Balkovec, J. M.; Black, R. M.; Hammond, M. L.; Heck, J. V.; Zambias, R. A.; Abruzzo, G.; Bartizal, K.; Kropp, H.; Trainor, C.; Schwartz, R. E. et al. *J. Med. Chem.* **1992**, *35*, 194–198.

60. Balkovec, J. M.; Black, R. M.; Bouffard, F. A.; Dropinski, J. F.; Hammond, M. L. *Proceedings of the XIVth International Symposium on Medicinal Chemistry* **1997**, *28*, 1–13.
61. Bouffard, F. A.; Zambias, R. A.; Dropinski, J. F.; Balkovec, J. M.; Hammond, M. L.; Abruzzo, G. K.; Bartizal, K. F.; Marrinan, J. A.; Kurtz, M. B.; McFadden, D. C. et al. *J. Med. Chem.* **1994**, *37*, 222–225.
62. Lal, B.; Gund, V. G.; Gangopadhyay, A. K.; Nadkarni, S. R.; Dikshit, V.; Chatterjee, D. K.; Shirvaikar, R. *Bioorg. Med. Chem.* **2003**, *11*, 5189–5198.
63. Lal, B.; Gund, V. G.; Bhise, N. B.; Gangopadhyay, A. K. *Bioorg. Med. Chem.* **2004**, *12*, 1751–1768.
64. Iwamoto, T.; Fujie, A.; Sakamoto, K.; Tsurumi, Y.; Shigematsu, N.; Yamashita, M.; Hashimoto, S.; Okuhara, M.; Kohsaka, M. et al. *J. Antibiotics* **1994**, *47*, 1084.
65. Fujie, A.; Iwamoto, T.; Sato, B.; Muramatsu, H.; Kasahara, C.; Furuta, T.; Hori, Y.; Hino, M.; Hashimoto, S. et al. *Bioorg. Med. Chem. Lett.* **2001**, *11*, 399–402.
66. Zambias, R. A.; Hammond, M. L.; Heck, J. V.; Bartizal, K.; Trainor, C.; Abruzzo, G.; Schmatz, D. M.; Nollstadt, K. M. et al. *J. Med. Chem.* **1992**, *35*, 2843–2855.
67. Klein, L. L.; Li, L.; Chen, H.-J.; Curty, C. B.; DeGoey, D. A.; Grampovnik, D. J.; Leone, C. L.; Thomas, S. A.; Yeung, C. M.; Funk, K. W. et al. *Bioorg. Med. Chem.* **2000**, *8*, 1677–1696.
68. Wang, W.; Li, Q.; Hasvold, L.; Steiner, B.; Dickman, D. A.; Ding, H.; Clairborne, A.; Chen, H.-J.; Frost, D.; Goldman, R. C. et al. *Bioorg. Med. Chem. Lett.* **2003**, *13*, 489–493.
69. Keating, G. M.; Figgitt, D. P. *Drugs* **2003**, *63*, 2235–2263.
70. Murdoch, D.; Plosker, G. L. *Drugs* **2004**, *64*, 2249–2258.
71. Jarvis, B.; Figgit, D. P.; Scott, L. J. *Drugs* **2004**, *64*, 969–982.
72. Townsend, R.; Terawaka, M.; Kerkesky, I. *J. Clin. Pharmacol.* **2000**, *40*, 1048.
73. Pfaller, M. A.; Diekma, D. J.; Messer, S. A.; Hollis, R. J.; Jones, R. N. *Antimicrob. Agents Chemother.* **2003**, *47*, 1068.
74. Moudgal, V.; Little, T.; Boikov, D.; Vazquez, J. A. *Antimicrob. Agents Chemother.* **2005**, *49*, 767.
75. Hernandez, S.; Lopez-Ribot, J. L.; Najvar, L. K.; McCarthy, D. I.; Bocanegra, R.; Graybill, J. R. *Antimicrob; Agents Chemother.* **2004**, *48*, 1382.
76. Kurtz, M. B.; Abruzzo, G.; Flattery, A.; Bartizal, K.; Marrinan, J. A.; Li, W.; Milligan, J.; Nollstadt, K.; Douglas, C. M. et al. *Infect. Immun.* **1996**, *64*, 3244–3251.
77. Masabuchi, M.; Taniguchi, M.; Umeda, I.; Hattori, K; Suda, H.; Kohchi, Y.; Isshiki, Y.; Sakai, T.; Kohchi, M.; Shirai, M. et al. *Bioorg. Med. Chem Lett.* **2000**, *10*, 1459–1462.
78. Takesako, K.; Kuroda, H.; Inoue, T.; Haruna, F.; Yoshikawa, Y.; Kato, I.; Uchida, K.; Hiratani, T.; Yamaguchi, H. *J. Antibiotics* **1993**, *46*, 1414.
79. Nagiec, M. M.; Nagiec, E. E.; Baltisberger, J. A.; Wells, G. B.; Lester, R. L.; Dickson, R. C. *J. Biol. Chem.* **1997**, *272*, 9809.
80. In, Y.; Ishida, T.; Takesako, K. *J. Peptide Res.* **1999**, *59*, 492.
81. Kurome, T.; Takesako, K. *Curr. Opin. Anti-infect. Invest. Drugs* **2000**, *2*, 375.
82. Kurome, T.; Takesako, K.; Inami, K.; Shiba, T.; Kato, I. *Peptide Sci.* **1999**, *36*, 197–200.
83. Weinberg, R. A.; McWherter, C. A.; Freeman, S. K.; Wood, D. C.; Gordon, J. I.; Lee, S. C. *Mol. Microbiol.* **1995**, *16*, 241.
84. Devadas, B.; Zupec, M. E.; Freeman, S. K.; Brown, D. L.; Nagarajan, S.; Sikorski, J. A.; McWherter, C. A.; Getman, D. P.; Gordon, J. I. *J. Med. Chem.* **1995**, *38*, 1837–1840.
85. Bhatnagar, R. S.; Futterer, K.; Farazi, T. A.; Korolev, S.; Murray, C. L.; Jackson-Machelski, E.; Gokel, G. W.; Gordon, J. I.; Waksman, G. *Nat. Struct. Biol.* **1998**, *5*, 1091–1097.
86. Devadas, B.; Freeman, S. K.; McWherter, C. A.; Kuneman, D. W.; Vinjamoori, D. V.; Sikorski, J. A. *Bioorg. Med. Chem. Lett.* **1996**, *6*, 1977–1982.
87. Brown, D. L.; Devadas, B.; Lu, H-F.; Nagarajan, S.; Zupec, M. E.; Freeman, S. K.; McWherter, C. A.; Getman, D. P.; Sikorski, J. A. *Bioorg Med. Chem. Lett.* **1997**, 7, 379–382.
88. Devadas, B.; Freeman, S. K.; McWherter, C. A.; Kishore, N. S.; Lodge, J. K.; Jackson-Machelski, E.; Brown, D. L.; Gordon, J. I; Sikorski, J. A. *J. Med. Chem.* **1998**, *41*, 996–1000.
89. Armour, D. R.; Bell, A. S.; Kemp, M. I.; Edwards, M. P.; Wood, A. 221st ACS National Meeting 2001 abstract MEDI-349.
90. Bell, A. S.; Armour, D. R.; Edwards, M. P.; Kemp, M. I.; Wood, A. 221st ACS National Meeting 2001 abstract MEDI-350.
91. Yamazaki, K.; Kaneko, Y.; Suwa, K.; Ebara, S.; Nakazawa, K.; Yasuno, K. *Bioorg. Med. Chem.* **2005**, *13*, 2509–2522.
92. Masabuchi, M.; Kawasaki, K.; Ebiike, H.; Ikeda, Y.; Tsujii, S.; Sogabe, S.; Fujii, T.; Sakata, K.; Shiratori, Y.; Aoki, Y. et al. *Bioorg. Med. Chem. Lett.* **2001**, *11*, 1833–1837.
93. Ebiike, H.; Masabuchi, M.; Liu, P.; Kawasaki, K.; Morikami, K.; Sogabe, S.; Hayase, M.; Fujii, T.; Sakata, K.; Shindoh, H. et al. *Bioorg. Med. Chem. Lett.* **2002**, *12*, 607–610.
94. Kawasaki, K.; Masabuchi, M.; Morikami, K.; Sogabe, S.; Aoyama, T.; Ebiike, H.; Niizuma, S.; Hayase, M.; Fujii, T.; Sakata, K. et al. *Bioorg. Med. Chem. Lett.* **2003**, *13*, 87–91.

Biography

Andy S Bell currently works as a Research Fellow at Sandwich Laboratories, Pfizer Global R&D. He joined Pfizer in 1980 following studies at York University, UK. Andy spent the first 9 years of his career working on PDE inhibitors leading to the inotrope/vasodilator (PDE3) candidate, nanterinone, and the PDE5 inhibitor, sildenafil (Viagra, Revatio). Soon after the discovery of sildenafil in 1989, Andy moved to the Antifungals Project, contributing to the candidate nomination of the broad-spectrum agent voriconazole (Vfend), which was launched in the US in 2002.

After working on several exploratory antifungal approaches, Andy moved to a new role in 1999 with responsibility for File Enrichment, as part of Pfizer's collaborations with ArQule and Tripos. He has subsequently been working in the hit-to-lead arena, making use of the File Enrichment investment to generate new lead series for multiple projects.

Comprehensive Medicinal Chemistry II
ISBN (set): 0-08-044513-6

ISBN (Volume 7) 0-08-044520-9; pp. 445–468

7.16 Bacteriology, Major Pathogens, and Diseases

L S Young, California Pacific Medical Center Research Institute, San Francisco, CA, USA

7.16.1 Introduction

The development of drugs to treat infections caused by bacteria must be considered the single major medical advance of the twentieth century. Understanding the significance of this monumental therapeutic development lies in recognizing the relationship between life-sustaining antibacterial therapy to other medical major advances during the past 100 years: these include the advent of modern surgery, the development of life-saving cancer chemotherapy, the advent of organ transplantation, and the ability to treat or palliate connective tissue and inflammatory disorders with immunosuppression. Each of the aforementioned broad categories of intervention leaves the human host at risk for developing life-threatening bacterial sepsis because of breached anatomic barriers, immune suppression, or both. Antimicrobial agents, either given prophylactically or therapeutically, have strikingly reduced deaths from potential pathogens that are often part of normal flora and/or are opportunists, creating disease by taking advantage of impaired host status. Without potent antibacterial agents, many of the advances in modern oncology, surgery, and transplantation would not have been possible.

Historically, there is evidence that empiric treatment approaches, best illustrated by herbal medicines, led to identification of antimicrobial substances long before the germ theory of human disease became accepted. The two best examples appear to be the evolution of quinine as an antimalarial agent and the more recent development of the artemisinin family of drugs. Each derives from ancient empiric uses in which the bark of the cinchona tree (in the case of quinine) or Chinese herbal medicine (artemisinins) became established for the treatment of fever. What was not originally known, however, was the mechanism of activity or the specificity of the antipyretic action: in the case of both quinine and the artemisinins it is now known to be a selective lethal effect directed against the falciparum protozoa.

It was not until the clearly delineated efforts of Paul Ehrlich that modern concepts relating to drug therapy of microbial infections evolved.[1] Ehrlich pioneered the concept of a chemical agent as treatment of a horrendous infectious disease, syphilis, and developed specific chemical therapy (arsenicals) directed against a target within the treponeme. Subsequently, Hitchings and Elion first demonstrated the concept of a specific chemical agent acting upon an enzymatic target that was vital (essential for life) within the infection-causing pathogen.[2] In the latter example, the enzyme dihydrofolate acid reductase (DHFR) is present in both bacteria and mammalian hosts and is an essential biosynthetic pathway. However, pyrimidine inhibitors were found to have significantly different affinities for the bacterial and human host DHFR, and this difference was exploited in the development of trimethoprim as a specific antibacterial therapy.

7.16.2 Definitions

The terms antibiotic, antimicrobial agent, and chemotherapeutic have been used almost interchangeably by many authorities but with understandable confusion (**Table 1**). The concept of an antibiotic ('against life') was originally derived as a therapeutic agent elaborated by living organisms that would act against a living organism causing disease or pathologic changes within the host. As originally promulgated such substances would exert their effect in low concentrations (microgram quantities per milliliter or less) and exert a specificity for a target group of organisms rather than having broad killing activity (such as would be expected with an antiseptic, which is defined as a chemical with broad killing effect but not specificity). Clearly, these were felt to represent small molecules of defined chemical structure. Chemotherapy in its simplest terms refers to treatment with a chemical agent and 'chemotherapy' has been associated with oncology or treatment of tumors. Thus, 'chemotherapy' has at least two medical applications. Some chemotherapeutic agents for cancer were originally isolated by screening for antibiotic activity so there is some blurring of distinctions with applications to treat neoplasm or microbial diseases. Indeed, actinomycin D, bleomycin, daunorubicin, and mitomycin are antitumor antibiotics. Tacrolimus is a macrolide-class antibiotic that is used to prevent organ/graft rejection with immune suppressive activity against T cell mediated host defenses. 'Anti-infective' and 'antimicrobic' have a much more broad meaning and could apply to both natural substances and chemical moieties totally derived by synthetic approaches. The definition of antibacterial agent must now include polypeptide substances that are either naturally isolated defined or synthetically prepared. Just as antibiotics from natural sources have been modified by chemical synthesis, protein chemistry has led to synthetic peptides deliberately produced with the knowledge and insight of an essential target within a disease-causing microorganism. As generally accepted, the term 'antibiotics' can apply to any naturally derived or synthetic chemical compound that kills or restricts the multiplication (growth) of a bacterium either in vitro or in vivo.

7.16.3 Historical Perspective

The single major development in the treatment of bacterial diseases during the past century was the identification of the activity of penicillin.[3] In a parallel track, it was noted that synthetic dyes of the sulfonamide family had antibacterial effects and the evaluation of these compounds led to the development of the sulfonamide family. The most fertile period of new antibacterial drug development came in a relatively short 30-year period between 1940 and 1970 when most of the major classes of drugs that are in use today were identified initially via natural product screening. Thus, the penicillins, cephalosporins, macrolides, tetracyclines, aminoglycosides, and chloramphenicol were all recognized to be of therapeutic benefit between 1940 and 1975. Despite this, no major therapeutic group of agents has become identified through natural product screening in the last quarter of a century. The major developments have come from totally synthetic approaches, which led to the identification of new classes such as the fluoroquinolones and

Table 1 Some definitions

Term	*Definition*
Antimicrobial agent (against microbes)	A broad category of drugs that includes: Antibacterials Antivirals Antiparasitic Antifungals
Antibiotic (against life)	A product of the metabolism of microorganism that posses antibacterial activity at low concentrations and is not toxic to the host
Anti-infective (against infection)	Synonymous with antimicrobial agent
Antiseptic (against infection)	Chemical substance that kills microbes without specificity; for topical rather than systemic use
Chemotherapeutic	Chemical entity used in treatment, possibly synthetic, and includes activity against neoplasms
Antimicrobic or antimicrobial	Used interchangeably with antimicrobial agent

the oxazolidinones. Each of these entails its own intriguing historical drama: the fluoroquinolones, evolving from the isolation of antibacterial fractions from crude quinine preparations, and the oxazolidinones being derived from derivatives of monoamineoxidase inhibitors that upon extended screening displayed antibacterial properties.

7.16.4 Classification of Antimicrobial Agents

The most useful approach to classifying antimicrobial agents probably derives from consideration of the target site of action that underlies the activity of the antimicrobial agent.[1] **Table 2** is a brief simplistic summary of some commonly used compounds and the antibacterial effects or site of action. The clinical basis of effective therapy is that antimicrobial agents will target the disease-causing pathogen with minimal side effects upon the mammalian host. Thus, those that target the cell wall of the bacterium include the penicillins, cephalosporins, and the glycopeptides with good specificity since human cells lack a cell wall. Agents affecting nucleic acid synthesis include those affecting DNA (DHFR inhibitors, fluoroquinolones) or RNA (rifampicin, macrolides, tetracyclines, aminoglycosides, and oxazolidinones). Cell wall structures other than peptidoglycan are targeted by specialized agents such as isoniazid, which appears to act on fatty acid or mycolic acid synthesis within mycobacteria – while again human cells lack these chemical moieties in their cell structures.

7.16.5 Initial Screening and Subsequent Therapeutic Development

Currently used drugs are usually modifications of the first agent of a group of drugs found to have activity. The properties of the group are optimized for stability, pharmacokinetics, and ease of manufacture. Developing improved analogs of existing drugs has been a major activity since the beginning of the antimicrobial era.

The initial discovery process ranges from the selection or screening of natural products to the automated assessment of large chemical libraries that may have initially been developed for other than infectious disease therapeutic purposes (the latter are often referred to as chemical libraries and may include drugs originally developed to treat cancer, central nervous system disorders, and other diseases). The in vitro screening methodology is an essential part of the screening process and it was recognized that experimental or laboratory conditions could widely affect the detection of drug activity. Clearly, the compounds being screened should have activity at physiological conditions within the host.

Table 2 Classification of antimicrobial agents by site of action

Structure/process	*Agent*	*Target*
Cell wall	β-Lactam	Transpeptidase
	Glycopeptide	D-Ala–D-Ala Linkage
	Fosfomycin	Mur A in peptidoglycan synthesis
Cell membrane	Polymyxin B (colistin)	Membrane insertion
	Daptomycin	Block lipoteichoic acid
DNA replication/repair	Quinolones	DNA gyrase A
		Topoisomerase IV
	Novobiacin	Gyrase B
	Pyrimidines	DHFR
RNA polymerization	Rifamycins	RNA polymerase
Protein synthesis	Macrolide	23S ribosomal peptide transferase (Domain V)
	Aminoglycosides	50S ribosomal peptide transferase
	Tetracyclines	Peptidyl transferase
	Oxazolidinones	Initiation phase, protein synthesis
	Chloramphenicol	Peptidyl transferase
	Fusicid acid	Elongation factor G

An in vitro test system could theoretically fail to identify a potentially promising group of compounds that are rapidly metabolized to an active form or show inactivity when the drug is active but only under special conditions, e.g., intracellularly. Simple conditions such as pH or oxygen tension are obvious variables that might affect the screening process. Currently, screening efforts are directed at optimizing sensitivity, but after the initial identification of activity, comparison of activity with known antibacterial compounds and variation in structure–activity relationships are obvious needs.

It has taken decades for fairly uniform criteria to evolve for in vitro antibacterial screening.[13] Clinically important 'indicator organisms' are included in screens. Detection of activity in a screen is but an initial step in the long journey before the identification of therapeutic modalities that can enter into clinical use. In vitro activity is only such and there is no necessary correlate that such in vitro activity will be predictably stable under conditions encountered in the human host, that an agent will not be metabolized rapidly by host enzymatic systems, that a drug will not be adequately distributed due to pharmacologic limitations, or a candidate compound may become inactive at the site of infection because of factors like pH, anaerobiosis, and body fluid osmolarity (among other conditions).

During the initial screening process gravimetric potency against a panel of diverse isolates has obvious appeal. Generally speaking, the lower the concentrations that inhibits select pathogens, the more appealing the candidate. Eventually, the pharmacology of a promising agent requires careful evaluation and from these factors the estimation of the boundaries of susceptibility and resistance (breakpoints). With a totally new class of anti-infectives, all bacteria may appear initially susceptible. The finding that a large population distribution of pathogens is inhibited or killed at concentrations readily achieved in serum would suggest susceptibility. This is the basis for one approach to defining susceptibility and resistance. For most bacteria and candidate therapeutics, defining the quantitative boundaries of susceptibility and resistance is not only essential to development, but important for clinical application. It must be borne in mind that relating activity and susceptibility to drug pharmacokinetics could also be potentially misleading as infections arise in tissues and bloodstream seeding or bacteremia is a subsequent development. Studies of tissue penetration present a more complete picture of the potential of a candidate agent, and demonstration of drug activity in a number of test systems would be further encouraging.

7.16.6 The Drug Evaluation Process

After the in vitro identification of activity, parallel screening of activity and toxicity is usually undertaken by industrial enterprises as well as some preliminary screens in cellular systems. A major concern is toxicity because of the failures among suitable drug candidates during toxicologic screens – the majority of drug candidates are usually excluded in the selection process. In vitro cellular systems may provide surrogate markers for potential problems as well as the all-important potential for inactivation by hepatic or renal metabolism, particularly the cytochrome P450 system, but predicting toxicity or side effects is a science in its infancy. With regard to efficacy, in vitro screens can also identify potential antimicrobial agents that might be concentrated within host tissues such as the phagocytic cells that circulate or those that are fixed of the reticuloendothelial system. Since the latter may be sanctuaries for important intracellular pathogens, screening of activity within macrophage cultures may also identify important biologic potential that is not necessarily revealed by in vitro screens.

7.16.7 Evaluation in Animal Models

Without doubt this is a major hurdle during the process of drug development. The effect of an antimicrobial agent within a living system is, of course, paramount to demonstration of activity prior to considering potential human use. At this point, economic considerations such as the size of the experimental animal the ease of administration of drugs pose important questions in drug development. Unquestionably, the test in a small rodent system, either in mice or in rats, is the most convenient system for evaluating in vivo activity. If studies are promising in small rodents, then the evaluation of infection in larger animal test systems is justified using cats, rabbits, guinea pigs, dogs, or even primates. In each of these circumstances, however, a direct extrapolation to human pharmacokinetics and pharmacodynamics is usually not possible.[4] For instance, it is quite well known that the half-life of drugs in rodents is markedly shorter than in humans, often with a 10-fold difference. There may be species differences in the metabolism of various drugs. If activity is confirmed in a series of animal systems, then clearly such findings are encouraging to proceed with human experimentation. However, there are well-known examples where drugs have not proven to be particularly active in certain animals but have activity in humans. Thus, the effect of the species differences, metabolism, and pharmacokinetic variations should all weigh in the drug evaluation process. In vivo distribution is also an important factor inasmuch as some drugs may be more active in the lung or could even be inactivated within the lung parenchyma. Clearly, drug efficacy at the active site of infection is the key goal in the evaluation of the results of experimental infection. Biologic activity is likely to be the result of the interplay between the intrinsic activity of the tested pharmacologic agent, its stability in the blood, serum, tissue components, its metabolism or distribution, its protein binding, and its stability at the site of infection.

7.16.8 In Vitro Activity: Static or Cidal

In addition to classification by mechanism of action or target site, one of the more useful parameters for assessment is whether a microbe possesses intrinsic killing power or merely serves to limit the growth of an organism. Such distinctions are reflected in the classification of an agent as bactericidal or bacteriostatic. Exposure to agents resulting in a significant decline in viable microbes over a defined time interval are classified as bactericidal, whereas those that merely limit the bacterium's growth are static. Well-known examples of static agents include the sulfonamides. However, some compounds such as chloramphenicol are cidal to some bacteria species but static for others and cidal or static effects can clearly be concentration-dependent. Further blurring distinctions are the rate of killing exhibited by certain agents: those compounds that rapidly kill, e.g., several logs within hours, may be more effective than those in which a 90% or 99% kill is accomplished over a period of half a day or more. It seems intuitive that bactericidal agents would be more effective and desirable than bacteriostatic agents in immunocompromised hosts, the latter being those animals or patients who are immune suppressed or may have some defined impairment to neutrophil function or cellular immunity. Where bactericidal therapy appears called for is in dealing with infections where host defenses are limited or impaired: examples would include in the vegetations of the patient with bacterial endocarditis or in the spinal fluid of the patient with meningitis. In the intact host who has adequate circulating neutrophil function, bacteriostatic drugs may not be notably inferior to those compounds that have bactericidal activity. Equally important may be the tissue penetration of a therapeutic agent – namely, the ability of the therapeutic agent to reach the active site of infection. There is little doubt that with increasing population of immunocompromised patients or individuals with some defect in host response, the impetus has been to develop compounds that have the ability to kill populations of microbes without assistance from host defense.

7.16.9 Clinical Needs

Clinical needs may be viewed from the perspective of the total burden of infectious diseases versus those diseases associated with high morbidity and high mortality. The two are clearly not the same. For instance, acute and chronic urinary infections or acute exacerbations of chronic bronchitis result in a large number of prescribed medications (and thus are economically important) in advanced industrial societies but the impact of drug treatment is in terms of improved clinical well-being rather than mortality. Most urinary tract infections are self-limited processes, although some can be severely symptomatic and a small percentage can result in life-threatenting bacteremia (thus requiring treatment). Chronic exacerbations of bronchitis rarely are associated with mortality but have considerable morbidity and many require patient hospitalization. Some well-designed studies suggest that antimicrobial therapy may hasten the rate of convalescence from exacerbations of bronchitis, decrease hospitalization, and improve quality of life. Thus, the aims of treatment are variable and society may be required to judge the value of intervention. Costs include the monetary costs of treatment and the consequences of drug toxicity and selection of resistant organisms – whose monetary costs are more difficult to appraise.

For advanced industrial societies, categories of major need would appear to be for life-threatening, nosocomial infections and systemic infections caused by multiresistant bacteria.[5] It is clear that at the beginning of the twenty-first century, the resistance that has been noted to develop against almost all major classes of antibacterial agents is the driving force for new drug development (**Table 3**). For the clinical needs in advanced societies, one can thus consider the syndromes of pneumonia, septicemia, and postsurgical infections as those requiring the greatest and most goal-directed novel drug development efforts: bacterial resistance is more likely to be encountered within hospitals and the patients are likely to be more debilitated and immune suppressed. In the quantitatively larger group of less life-threatening infections, there still are unmet needs in developing simpler, safer, more rapidly effective, and less expensive treatments for bacterial respiratory infections, urinary tract infections, and diseases that are complications of an aging population.

Table 3 Clinical needs in the developing world versus advanced industrial societies

Developing world	*Advanced industrial societies*
AIDS (HIV) TBC	Lower respiratory infection
Malaria	Infectious complication of chronic disease
Diarrhea	Diabetes mellitus
Viral	Atherosclerosis
Bacterial	Infections associated with foreign body implants
Protozoal	Nosocomial infections
Upper respiratory infection	Transplant-associated
Lower respiratory infection including tuberculosis	Oncology-associated
Urinary tract infection	Surgery-associated
Zoonoses	Ventilator-associated

In the developing world, the clinical needs are clearly for drugs that reduce mortality. A recent survey indicates the five leading causes of infectious deaths are (in order): lower respiratory infection, human immunodeficiency virus/acquired immune deficiency syndrome (HIV/AIDS), diarrheal diseases, tuberculosis, and malaria.[6] However, mortality from HIV/AIDS is closely associated with tuberculosis so the two are not independent epidemics.[7] Lower respiratory infections are poorly defined from an etiologic perspective in all populations where they occur and do not represent a single nosologic entity. Similarly, diarrheal diseases have been identified as the third leading infectious cause of death on the planet, with total annual mortality exceeding 1.8 million.[14] However, at different age groups the etiology of lethal diarrhea is likely to vary between viruses, bacteria, and parasites and geography plays a major role in etiology. What is clear is that etiologic diagnoses are infrequently made in the developing world and that fact has a major link to morbidity and mortality. The overall incidence of diarrhea is much higher than the mortality figures reveal so that actual numbers of cases cannot be deduced from mortality figures. Additionally, another problem is readily apparent and that is the need for improved diagnostics. Accurate microbiologic diagnosis is likely to be lacking in the developing world. Just distinguishing between a bacterial and protozoal cause of symptoms is likely to be challenging. If therapeutic intervention is likely to be effective in the developing world more readily accessible point of care diagnostics will likely be required. The absence of good, cheap diagnostics is certainly not a problem restricted to the developing world: in advanced industrial societies, the costs of a sputum or urine culture exceed antibiotic prescription costs. This is a driving force in antibiotic overuse for disease processes that are not necessarily caused by bacteria.

Development of improved therapy for tuberculosis is recognized as a major need for the developing world.[7] In areas of high incidence there is often coinfection with HIV, so delivering 'state-of-the-art' antituberculosis therapy is not likely to be successful in the absence of patient access to modern antiretrovirals. In advanced industrial societies, tuberculosis is a curable disease in the vast majority of patients: better than 95% of tuberculosis patients will be cured if they are adherent to a conventional four-drug regimen. Cure rates are much lower with less adherent patients, so the problem of tuberculosis control is not mere availability of drug but treatment follow-up and patient adherence. What would be a major advance and a development that would revolutionize the field would be pharmacologic therapy that is more rapidly effective and sterilizing in its antituberculosis potential such that the treatment course would be less complicated (e.g., contain fewer drugs) and shorter (e.g., half or less than the current half-year duration) with fewer side effects. Thus, while there are more than a dozen licensed drugs active against *Mycobacterium tuberculosis*, improved multidrug treatment regimens are a tremendous unmet need: screening and development efforts for totally new, more rapid antituberculosis treatments have clearcut justification in terms of world needs.

In advanced industrial societies a far larger group (and a growing one) of patients at risk of serious infectious disease are those with prosthetic implants. The implants themselves have been necessitated by the onset of chronic degenerative disease in a progressively aging population or are the added complications of atherosclerosis, obesity, and diabetes mellitus. Millions of individuals now have prosthetic heart valves, vascular grafts, and prosthetic joints: infections of these devices are almost impossible to eradicate with antibacterial therapy alone and effective drugs to treat infection of such devices is a huge unmet need. New effective therapies of these complex syndromes could be a tremendous medical and surgical advance. The latter would include degenerative diseases secondary to atherosclerosis, chronic lung disease complicated by infection, and infections that complicate the insertion of prosthetic devices. As the populations in developing countries grow progressively older, these infectious syndromes of medical progress are becoming more numerous and challenging to treat. However, it is also likely that new and effective agents for treating foreign body infections will require entirely new technologies and a better understanding of the pathogenesis of these infections. Factors leading to microbial adherence to metal, plastic, or synthetic prosthetic materials, mechanisms of microbial persistence in association with such materials, and the selection of drugs that will be active under implant conditions are an entirely new field.

7.16.10 New Antibacterial Drug Development: The Challenges Ahead

An enormous literature has documented the problems with current antibacterial therapy in advanced industrial societies.[8,9] The problem pathogens include an increase in the bacteria of the pneumococcal group which are the most common cause of community-acquired pneumonia and are resistant to both macrolides and penicillins. Where an apparent striking increase in mortality has occurred has been with another group of Gram-positive infections, namely, those due to methicillin-resistant *Staphylococcus aureus* (MRSA). The upsurge in MRSA has occurred quite recently and appears to be a worldwide trend. Staphylococci were initially quite susceptible to the quinolone class of agents but currently no fluoroquinolone is recognized to be a reliable therapeutic to treat staphylococcal disease. It must be recognized that despite the increase in pneumococcal and staphylococcal resistance there are still available therapies: they are costly and many require parenteral administration. New and convenient oral agents to treat MRSA would be a major breakthrough, unless widespread use leads to rapid emergence of resistance. Gram-positive pathogens also include the enterococci which have become resistant not only to penicillins but to vancomycin. While the last 20 years have not been thought to be a major problem with upsurge in Gram-negative pathogens, there are increased, scattered reports of Gram-negative bacteria, such as *Pseudomonas*, *Acinetobacter*, and enteric organisms against which there have been no therapeutic agents that are active, except perhaps colistin or polymyxin-B. This is particularly ominous in that there are no promising drugs for Gram-negative pathogens that appear to be truly novel and nontoxic.

7.16.11 Prospects for New Drug Development

It has been said that only a minority of natural products or sources of natural products have been screened for antibacterial activity. Traditionally, natural product screening has included a variety of geographical sources and has been broadened to a search for drug potentials in organisms that inhabit the murine environment or are native to insects. Recognizing that the potential has barely been comprehensively evaluated, it should be noted that no major group of antimicrobial compounds has come from natural products during the last two decades. From an evolutionary perspective, it is not surprising that microbes would have evolved resistance mechanisms since so many antimicrobial agents are derived from natural sources.[5] However, resistance to totally synthetic agents has been traced to microbes in natural sources that are unlikely to have selected from drug exposure.

In the meanwhile, new drug development has seen considerable progress in the field of antiviral chemotherapy. Although there are problems with the emergence of resistance, there are approved therapeutic interventions for influenza, HIV, and hepatitis. The fundamental principle in any drug development has been the recognition that certain gene products are vital for the metabolism of the virus and chemical moieties can interrupt or target these vital viral processes, ranging from attachment to the virus to a key biosynthetic step. In the revolutionary development of some antiviral medicines, the crucial development has been the identification of a target, whether it be the binding or attachment site of the virus, or an enzyme that is essential for the metabolism or maturation of viral particles. The crystal structure of the active enzymatic site or the crystal structure of the virus attachment site enabled the design of molecular inhibitors that will block viral replication or adherence. Historically, this approach was recognized in bacteria when DHFR was identified as the key target of pyrimidines.[2]

The application of the molecular, structure-based approach to antibacterial chemotherapy has been a vaunted goal, but its pursuit has been very challenging.[10] Bacteria are far more complex microorganisms than viruses. *Streptococcus pneumoniae*, *Staphylococcus aureus*, and *Pseudomonas aeruginosa* have more than 5000 genes as opposed to the 10 or fewer for HIV or the influenza virus.[11] More important, the gene products of these 5000-odd bacterial genes are not necessarily known or their function remains unknown. What is clear is that a far more finite number are 'vital' or essential to the bacterial host as established by genetic knockout techniques. Probably fewer than 100 bacterial genes are the targets for antibacterial chemotherapy after evaluation of the essential function of such genes and the presence of human homologs (which would decrease the likelihood of an effective chemotherapeutic approach). Modeling the inhibitor through crystallography has been accomplished for a number of bacterial genes, yet the identification of a useful therapeutic compound (a specific drug candidate) has stymied further efforts in antibacterial development. In other words, appropriate targets have been identified, such as the amino acid transfer RNA synthetases, yet deriving a small molecule inhibitor that meets all the requirements for a drug (activity, stability, selectivity, pharmacologic properties) is still wanting.[12] Currently, the 'buzz words' in new drug development include not only the genomic approach but such approaches as combinatorial chemistry and enhanced synthetic screens. However, the mere proliferation of chemical moieties fails to take into account the desirable pharmacologic properties of a candidate agent, and a mere elaboration of permutations and combinations around a basic structure may create tremendous problems unless a functional and simple screening process is developed in parallel. Nonetheless, the genomic approach allows those involved in drug development to stay in step with changes in the microbe itself and those that result from mutation and selection.[10] In identifying the active site for macrolide and ketolide drugs, it may be possible to overcome the effects of enzymatic modification such as methylation and define a better inhibitor of ribosomal RNA synthesis. Any property of a pathogen that provides it with survival advantage (e.g., a virulence factor, an efflux system for extruding toxic moieties, a quorum sensing system) is a potential target for chemotherapy. This would appear to be the hope of modern drug development, using the knowledge about mechanisms of resistance to overcome a limitation in current chemotherapy.

References

1. Bryskier, A. *Antimicrobial Agents: Antibacterials and Antifungals*; ASM Press: Romainville, France, 2005.
2. Hitchings, G. H. *Ann. NY Acad. Sci.* **1961**, *23*, 700–708.
3. Walsh, C. *Antibiotics: Actions, Origins, Resistance*; ASM Press: Washington, DC, 2003.
4. Young, L. S.; Bermudez, L. E. *Clin. Infect. Dis.* **2001**, *33*, S221–S226.
5. D'Costa, V. M.; McGrann, K. M.; Hughes, D. W.; Wright, G. D. *Science* **2006**, *311*, 374–377.
6. WHO *World Health Report*; WHO, Geneva, Switzerland, 1994.
7. Young, L. S. *Clin. Infect. Dis.* **1993**, *17*, S436–S441.
8. Craven, D. E.; Shapiro, D. S. *Clin. Infect. Dis.* **2006**, *42*, 179–180.
9. McDonald, L. C. *Clin. Infect. Dis.* **2006**, *42*, S65–S71.
10. Moir, D. T.; Shaw, K. J.; Hare, R. S.; Vovis, G. F. *Antimicrob. Agents Chemother.* **1999**, *43*, 439–446.
11. Stover, C. K.; Pham, X. Q.; Erwin, A. L.; Mizoguchi, S. D.; Warrener, P.; Hickey, M. J.; Brinkman, F. S.; Hufnagle, W. O.; Kowalik, D. J.; Lagrou, M. et al. *Nature* **2000**, *406*, 959–964.
12. Hurdle, J. G.; O'Neill, A. J.; Chopra, I. *Antimicrob. Agents Chemother.* **2005**, *49*, 4821–4833.
13. Clinical and Laboratory Standards Institute, Performance Standards for Antimicrobial Susceptibility Testing; Sixteenth Informational Supplement CSLI document M100-S16.
14. Thielman, N. M.; Guerrant, R. L. Enteric *Escherichia coli* Infections. In *Tropical Infectious Diseases*; Guerrant, R. L., Walker, D. H., Weller, P. F., Eds.; Churchill Livingstone: Philadelphia, 1999; Chapter 21, pp 261–276.

Biography

Lowell S Young is the Director of Kuzell Institute for Arthritis and Infectious Diseases in San Francisco, California, and Clinical Professor of Medicine, University of California, San Francisco. He received his MD degree at Harvard Medical School and postdoctoral training at the Centers for Disease Control and Memorial Sloan-Kettering Cancer Center, New York. His academic interests include infections in immunocompromised hosts (cancer, transplants, AIDS), antimicrobial chemotherapy, and the pathogenesis and treatment of septicemia. He is board certified in internal medicine and infectious diseases. Dr Young has authored over 300 articles in medical journals and contributed to over 60 books and book chapters. He is an editor of *Antimicrobial Agents and Chemotherapy* and serves on the editorial boards of *Clinical Infectious Diseases*. He is a member of the American Society for Clinical Investigation and serves on the clinical guidelines committee of the Infectious Diseases Society of America. Dr Young has received the Langmuir Prize of the Center for Disease Control and the Garrod Medal of the British Society for Antimicrobial Chemotherapy.

Comprehensive Medicinal Chemistry II
ISBN (set): 0-08-044513-6

ISBN (Volume 7) 0-08-044520-9; pp. 469–477

7.17 β-Lactam Antibiotics

C Hubschwerlen, Actelion Pharmaceuticals Ltd, Allschwil, Switzerland

7.17.1 History and Overview

7.17.1.1 Discovery

The β-lactam era began in 1928, when the bacteriologist A Fleming discovered penicillin (**Figure 1**). It was not until 1939 that H Florey and E Chain were able to demonstrate penicillin's ability to kill infectious bacteria. With the increased need for treating wound infections during World War II, huge resources were devoted to elucidate the chemical structure of penicillin and to develop penicillin G as the first modern antibiotic. J Sheehan[1] reported the first total synthesis of penicillin in 1962.

Cephalosporin C (**Figure 1**) was first isolated from cultures of *Cephalosporium acremonium* from a sewer in Sardinia in 1948 by the Italian scientist G Brotzu.[2] In 1960, Eli Lilly launched the first cephalosporin, cefalotin (see **Figure 12**), on the market. Compared with natural penicillins, these early cephalosporins displayed an enhanced β-lactamase stability. R B Woodward[3] reported in 1966 the first total cephalosporin synthesis by ring expansion of penicillin. Both penicillins and cephalosporins were produced by fermentation on a large scale, and further processed to give, respectively, 6-amino-penicillanic acid (APA) and 7-amino-cephalosporanic acid (ACA). These two key intermediates served as basis for chemical programs leading to a large variety of new semisynthetic antibiotics.

Thienamycin (**Figure 1**), a naturally occurring carbapenem, was isolated in the late 1970s from a culture broth of *Streptomyces cattleya*.[4] The carbapenem nucleus had to be synthesized de novo, requiring a vast chemical effort to identify new compounds with improved properties.

Following the discovery of nocardicin A, the first naturally occurring monobactam, sulfazecin (**Figure 1**), was isolated from Gram-negative bacteria (e.g., *Pseudomonas acidophila* and *Acetobacter*) in the early 1980s, simultaneously at Takeda and Squibb.[5–7] Research efforts in this class led to the commercialization of aztreonam and carumonam (see **Figure 11**).

The details of the efforts undertaken during the first 50 years of β-lactam research have been summarized in the first edition of *Comprehensive Medicinal Chemistry*, and in other monographs.[8–10] The aim of the present review is to summarize the scientific activities in this field between 1990 and 2004.

7.17.1.2 Nomenclature

The nomenclature and numbering commonly used for lactams are derived respectively from the penicillin and the cephalosporin nuclei. The denominations α and β refer to substituents below or above the plane of the azetidinone ring, respectively, as depicted in **Figure 2**.

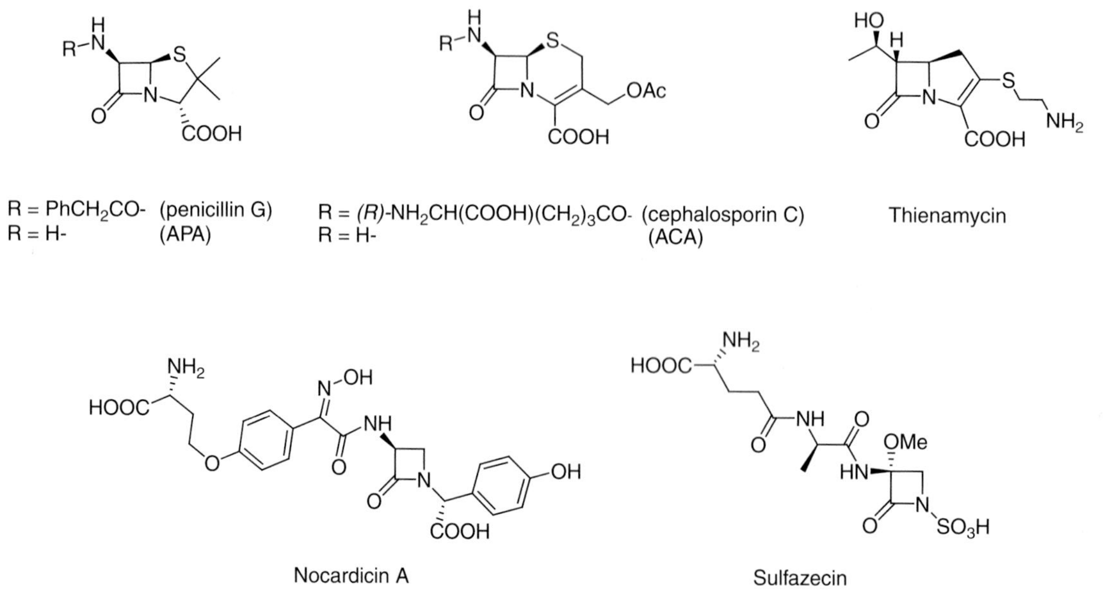

R = $PhCH_2CO$- (penicillin G)
R = H- (APA)

R = (*R*)-$NH_2CH(COOH)(CH_2)_3CO$- (cephalosporin C)
R = H- (ACA)

Thienamycin

Nocardicin A

Sulfazecin

Figure 1 Structures of the natural β-lactam antibiotics penicillin G, cephalosporin C, thienamycin, nocardicin A, and sulfazecin, and the key building blocks APA and ACA.

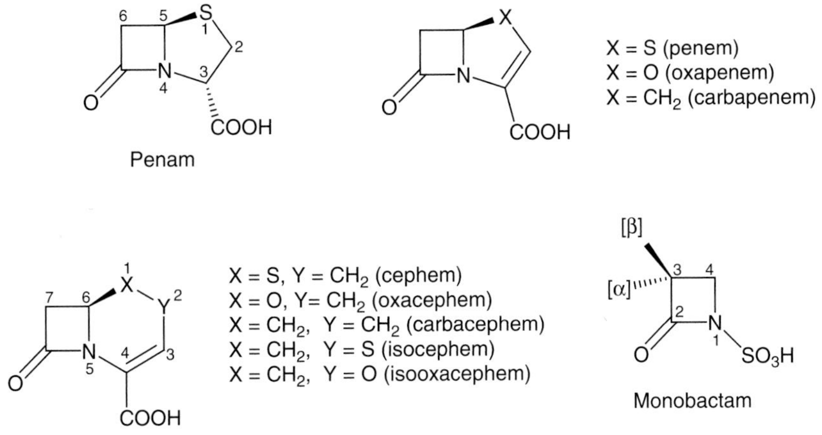

Penam

X = S (penem)
X = O (oxapenem)
X = CH_2 (carbapenem)

X = S, Y = CH_2 (cephem)
X = O, Y = CH_2 (oxacephem)
X = CH_2, Y = CH_2 (carbacephem)
X = CH_2, Y = S (isocephem)
X = CH_2, Y = O (isooxacephem)

Monobactam

Figure 2 Core structures of the major classes of β-lactam antibiotics.

7.17.1.3 Reasons for Success, Trends, and Challenges

β-Lactam antibiotics, especially penicillins and cephalosporins, have emerged as drugs of choice to treat many infections for the following reasons: they cover a broad antibacterial spectrum, they are very safe, and are bactericidal, and some representatives are orally active. However, the tremendous therapeutic potential of these antibiotics has been threatened by the emergence of increasingly resistant bacterial strains as a natural consequence of their use. Factors that exacerbate this phenomenon are misuse and overuse of antimicrobials (notably in the class of extended-spectrum third-generation cephalosporins), an increasing number of immunocompromized hosts, lapses in infection control, the growing use of invasive procedures and devices (such as artificial joints and catheters), and the widespread use of antibiotics in agriculture and animal husbandry. Methicillin-resistant *Staphylococcus aureus* (MRSA) is a significant problem in hospitals, as are vancomycin-resistant enterococci, particularly *Enterococcus faecium*, and multidrug-resistant Gram-negative bacilli. Owing to the high frequency of antibiotic use, an individual is more likely to acquire an antibiotic-resistant infection in an intensive care unit (ICU) than anywhere else. In clinical settings, more than 50% of *Escherichia coli* isolates and more than 90% of *S. aureus* isolates are ampicillin-resistant. Today, several outbreaks of nosocomial infections cannot be efficiently treated with the current armamentarium. In the community, penicillin-resistant *Streptococcus pneumoniae* (PRSP) is of greatest concern, and recent reports also indicate the appearance of community acquired methicillin-resistant *S. aureus* (CAMRSA) infections. Research efforts during the 1990s mainly focused on the discovery of new cephems and carbapenems displaying a useful coverage of MRSA, *Pseudomonas aeruginosa*, and other multiresistant Gram-negative pathogens (e.g., *Klebsiella pneumoniae*).

The global antibacterial drug market, valuated at US $27 billion in 2004, is estimated to remain stable over the next decade. β-Lactam antibiotics correspond to a market share of 45%. The hospital market represents US $8 billion, of which cephalosporins, penicillins, and carbapenems account for US $3.0 billion, $1.5 billion, and $1.0 billion, respectively.[11] During the reporting period (1990–2004), the β-lactam antibiotics share has been reduced in favor of quinolones, macrolides, and glycopeptides. The high price pressure from generic antibiotics (third-generation cephalosporins, quinolones, and macrolides) that are still therapeutically useful has led large pharmaceutical companies to leave the field for more lucrative areas (oncology, life style drugs, etc.). However, medical needs clearly remain present, with bacterial resistance growing inside and outside the hospital.

7.17.2 Mode of Action

7.17.2.1 Cell Wall Cross-Linking

One major difference between eukaryotic and prokaryotic cell envelopes is the presence, in the latter, of a peptidoglycan layer, conferring mechanical strength on the bacterial cell wall. That peptidoglycan is formed from polymeric strands consisting of alternating motifs of *N*-acetylglucosamine and *N*-acetylmuramic acid, appended by a L-Ala-D-Glu-X-D-Ala-D-Ala pentapeptide chain. The nature of X varies from species to species (e.g., *m*-diaminopimelic acid in *E. coli* or L-Lys-$(Gly)_5$ in *S. aureus*). Neighboring peptidoglycan strands are cross-linked by the formation of peptide bridges (**Figure 3**). This final step is catalyzed by a series of transpeptidases that are the targets of β-lactam antibiotics (therefore also called penicillin-binding proteins, PBPs).

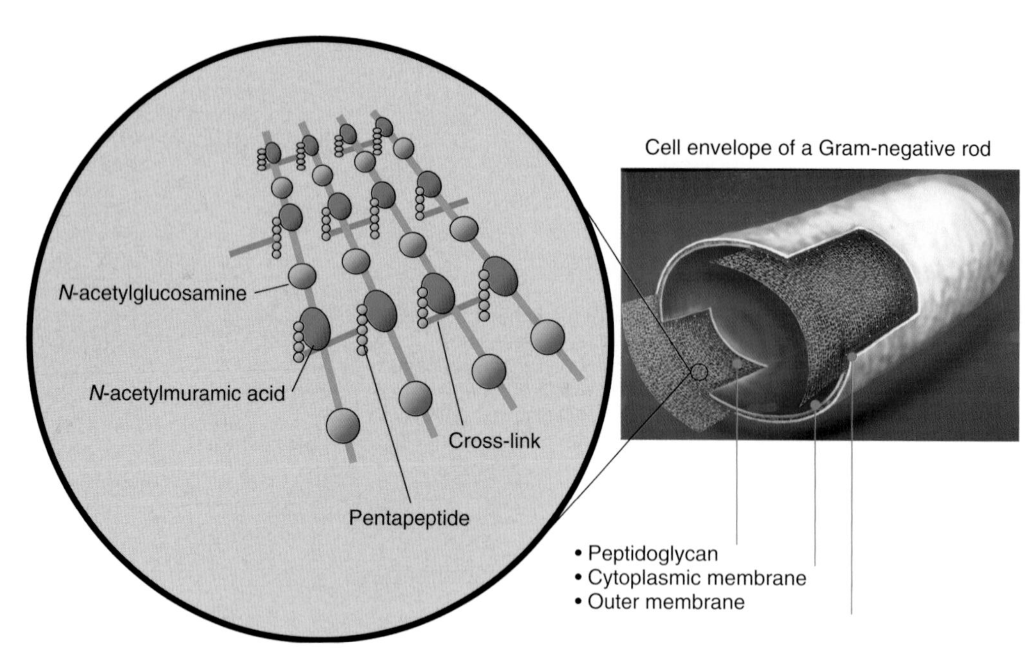

Figure 3 Structure of the bacterial peptidoglycan. (Reproduced from Artist Representation of a Bacterial Cell Envelope.)

Each bacterium possesses a set of PBPs, which have different functions. For instance, PBPs involved in cell duplication (transpeptidases) are essential, whereas those involved in regulation (carboxypeptidases) are nonessential (**Table 1**). All these PBPs, originally numbered according to their increasing molecular weight, have been classified based on sequence homology. The different classes of β-lactam antibiotics display a different pattern of transpeptidase inhibition, resulting in different phenotypes (**Table 1**).

Transpeptidases are serine proteases that cleave the last peptidic bond of the pentapeptide side chain in the nascent peptidoglycan strand, releasing the terminal D-Ala amino acid (**Figures 4** and **5**). The resulting acylated enzymes are attacked by a neighboring terminal amino group of an adjacent peptidoglycan chain, thus achieving the final cross-linking of the peptidoglycan. The integrity of the peptidoglycan network surrounding the bacterium is essential for its survival. The inhibition of essential PBPs leads to an imbalance between the bacterial cell wall autolytic enzymes (autolysins) and the bacterial cell wall synthesizing enzymes (transpeptidases), resulting ultimately in cell lysis mediated by the high internal osmotic pressure.

7.17.2.2 Mode of Action of β-Lactam Antibiotics

β-Lactam antibiotics mimic the terminal D-Ala-D-Ala moiety of the pentapeptide. The reactive β-lactam ring is able to acylate the active serine residue of the transpeptidase, leading to a stable acyl–enzyme intermediate that is still appended with a bulky substituent (the second ring of the β-lactam antibiotics), thus preventing the access of an incoming amino group, required to achieve cross-linking (**Figure 5**). Before the resolution of x-ray structures of β-lactamases and transpeptidases, the mode of action of β-lactam antibiotics was solely explained by the acylating power of the β-lactam ring, as measured by the typical high infrared frequency of the β-lactam carbonyl band ($>1780\,\mathrm{cm}^{-1}$).[12] However, this model did not take into account the enzyme environment. Thus, another model was introduced, based on the distance between the β-lactam carbonyl oxygen atom and the carbon atom of the carboxylate function (3.0–3.9 Å for active versus >4.1 Å for inactive derivatives; **Figure 6**).[13]

The molecular rationalization of the mode of action of β-lactam antibiotics requires the amidic character of the carbonyl group to be reduced, to enable acylation of the active serine residue of the transpeptidases. This is achieved by preventing the overlap of the β-lactam nitrogen π orbital with those of the β-lactam carbonyl group by (1) geometric constraints resulting from the presence of a fused five-membered ring (penams), (2) an inductive effect (monobactams), (3) delocalization of the β-lactam nitrogen electron lone pair into an adjacent unsaturated system (cephems), or (4) by a combination of two effects (penems and carbapenems). In addition, the β-lactam carbonyl group must be located in the oxyanion hole, to provide the correct activation and orientation of the carbonyl group for attack during the acylation and deacylation steps, respectively (see **Figure 8**).

Table 1 PBP numbering, function, and inhibition in some representative bacteria

PBP function	*Organism*			
	S. aureus	*E. faecium*	*S. pneumoniae*	*E. coli*
Transglycosylase plus transpeptidase	2	1a, 1b, 2	1a, 1b, 2a	1a, 1b, 1c
Transpeptidase				
B1	2a	5	No homolog	No homolog
B2	No homolog	No homolog	No homolog	2
B3	No homolog	No homolog	No homolog	3
B4/B5	1, 3	3, 4	2b, 2x	No homolog
Carboxypeptidase				
C1	No homolog	No homolog	No homolog	4
C2	No homolog	No homolog	No homolog	5, 6, 6b, 7
C3	4	6	3	No homolog

In each cell, colored boxes indicate the general level of transpeptidase inhibition by penicillin (first box), third-generation cephalosporins (second box) and carbapenems (third box). Color code: green, strong inhibition; blue, medium inhibition; red, no inhibition.

The unique feature of β-lactam antibiotics is that they target a family of related enzymes (transpeptidases). However, due to the high structural variability of these enzymes in their active site region, it is not possible to obtain high affinity for all of them. Therefore, inhibition relies mainly on the irreversible selective acylation of the active serine residue by the β-lactam ring. Selectivity with regard to other host targets arises from the fact that transpeptidases process an 'unnatural' D-configured substrate. Furthermore, β-lactam antibiotics must remain stable with respect to the action of β-lactamases. These enzymes are closely related to transpeptidases but are able to open the β-lactam ring of the antibiotics.

7.17.3 Mechanisms of Resistance

7.17.3.1 β-Lactamases

Following the early introduction of penicillin G, resistance was rapidly acquired through the production of β-lactamases, which are able to efficiently hydrolyze the β-lactam ring of the antibiotics into the corresponding inactive penicilloic acid derivatives (see **Figure 5**). These enzymes are chromosomal or plasmid encoded. Furthermore, the resistant traits are easily transferred between different types of bacteria, through horizontal plasmid transfer.

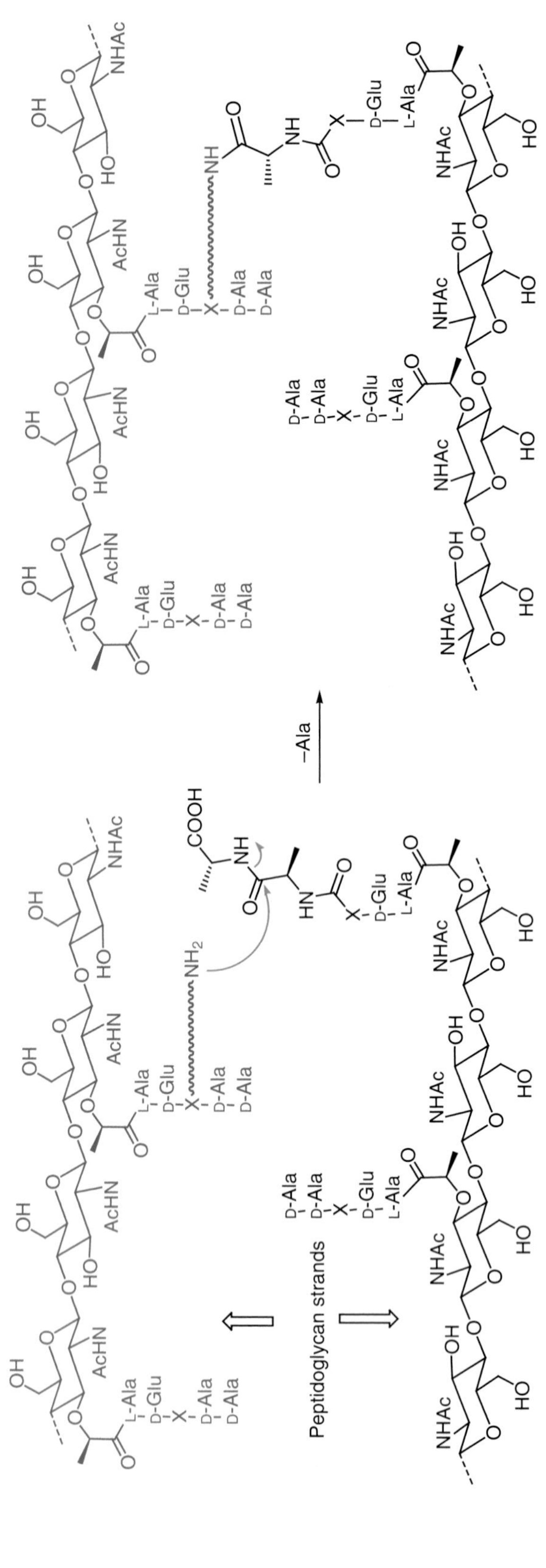

Peptidoglycan strands

–Ala

Figure 4 Cross-linking of the peptidoglycan by transpeptidases.

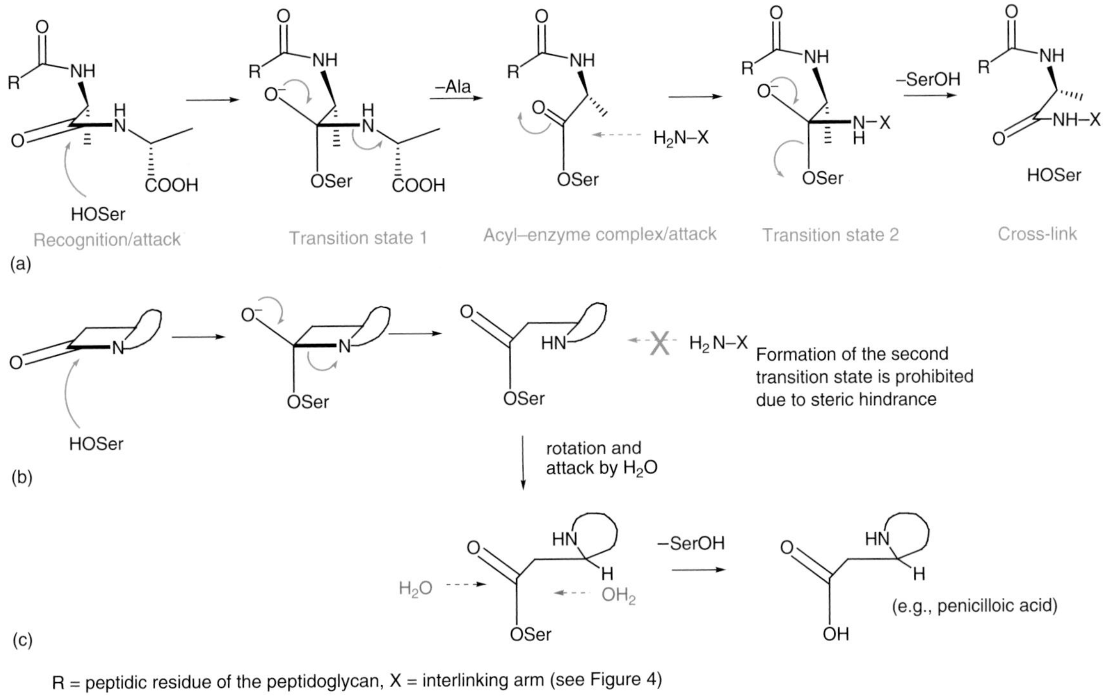

R = peptidic residue of the peptidoglycan, X = interlinking arm (see Figure 4)

Figure 5 (a) Schematic representation of the mechanism of transpeptidation. (b) Transpeptidase inhibition by serine β-lactam antibiotics. (c) Serine β-lactamase-mediated hydrolysis. SerOH represents the active serine residue in both transpeptidases and serine β-lactamases.

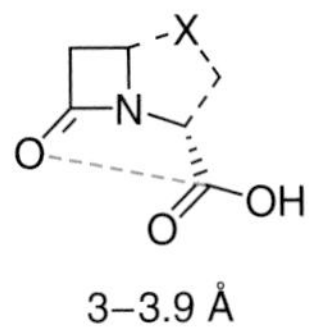

3–3.9 Å

Figure 6 Distance threshold for antibacterial activity.

7.17.3.1.1 Classification

With the different generations of β-lactam antibiotics used in the clinic, different types of β-lactamases appeared, and their number now exceeds 450. Different classifications have been proposed based on their substrate profile or mechanism of action (**Table 2**).[14,15]

There are four broad molecular classes of β-lactamases: A, B, C, and D. Class A, C, and D β-lactamases are serine proteases, sharing common structural and catalytic features with essential transpeptidases, whereas class B β-lactamases are unrelated zinc metallo-β-lactamases (MBLs). The major difference between class A, C, and D β-lactamases lies in the position of the hydrolytic water molecule, which is activated through at least two different mechanisms. In recent years, an increased incidence and prevalence of extended-spectrum β-lactamases (ESBLs) have been observed.[16] These enzymes are able to hydrolyze third-generation cephalosporins, and, therefore, induce resistance. The majority of ESBLs are derived from the widespread broad-spectrum class A β-lactamases TEM-1 and SHV-1. New families of ESBLs, including the CTX-M and OXA-type enzymes, have appeared. Class C β-lactamases (e.g., P99) cause serious therapeutic problems for the treatment of infections related to Enterobacteriaceae, particularly *Enterobacter cloacae*. The influence of class B carbapenemases, belonging to the MBLs, is slowly increasing with the use of carbapenems. Non-metallo-carbapenemases of class A and D seem to be more problematic.[17] Livermore and Williams have described the resistance situation related to the action of β-lactamases.[18]

7.17.3.1.2 Biostructure information

The hydrolysis of β-lactam antibiotics by serine β-lactamases can be described by three major events: (1) substrate recognition, (2) attack of the β-lactam ring by the active serine residue, leading to an acyl–enzyme intermediate, and

Table 2 Classification scheme for β-lactamases

Group	*Enzyme type*	*Molecular class*	*Gene location*[a]	*Examples*
1	Cephalosporinase	C	C/P	*Enterobacter cloacae* P99, AmpC
2a	Penicillinase	A	C/P	*S. aureus*
2b	Broad spectrum	A	C/P	TEM-1, SHV-1
2be	Extended spectrum	A	P	TEM-3, SHV-2
2br	Inhibitor resistant	A	P	TEM-30
2c	Carbenicillinase	A	C/P	PSE-1
2d	Cloxacillinase	D or A	P	OXA-1
2e	Cephalosporinase	A	C	*Proteus vulgaris, Bacteroides fragilis*
2f	Carbapenemase	A	C	*Enterobacter cloacae* IMI-1
3	Metalloenzyme	B	C/P	*Stenotrophomonas maltophilia* L1
4	Penicillinase	ND[b]	C	*Burkholderia cepacia*

[a] C, chromosomal; P, plasmid.
[b] ND, not determined.

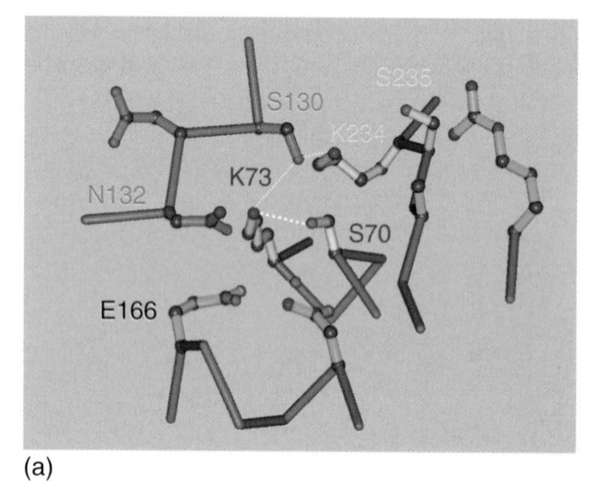

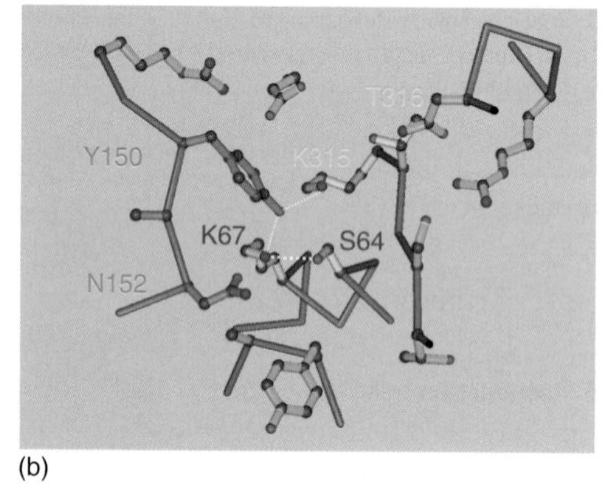

Figure 7 Representation of the active site regions of (a) TEM-1 class A β-lactamase (PDB code 1m40) and (b) *Citrobacter freundii* class C β-lactamase (PDB code 1fr1), showing, despite their low sequence homology, a striking similarity. Conserved residues: KT(S)G (yellow), SXXK (white), and S(Y)XN (magenta).

(3) attack of this latter intermediate by an activated water molecule (see **Figure 5**). The development of new biochemical tools and methodologies allowed quantification of these different steps by determining the affinity, the rates of acylation and deacylation of the active serine residue, and the turnover of the enzyme.

Despite the very low sequence homology (less than 20%) of the different serine β-lactamases, some structural motifs (e.g., SXXK, S(Y)XN, and KT(S)G) are well conserved within class A, C, and D β-lactamases (**Figure 7**). Their three-dimensional structures revealed the location of these conserved residues within their active site region, which, strikingly, displays the same overall architecture. The function of the different domains has been established: SXXK (white) contains the active serine residue, KT(S)G (yellow) is part of a β strand, and forms one side of the active site region, and S(Y)XN (magenta) is responsible for the activation of the active serine residue, and forms the second side of the active cleft. These structures allowed a reasonable understanding of the mechanism of hydrolysis of β-lactam antibiotics.

All these observations also contributed to the elucidation of the mechanism of action of transpeptidases. The gross structure–activity relationships observed in the early days (e.g., the influence of the stereochemistry, positions open to substitution) could be explained as soon as the structures of the transpeptidases became available.

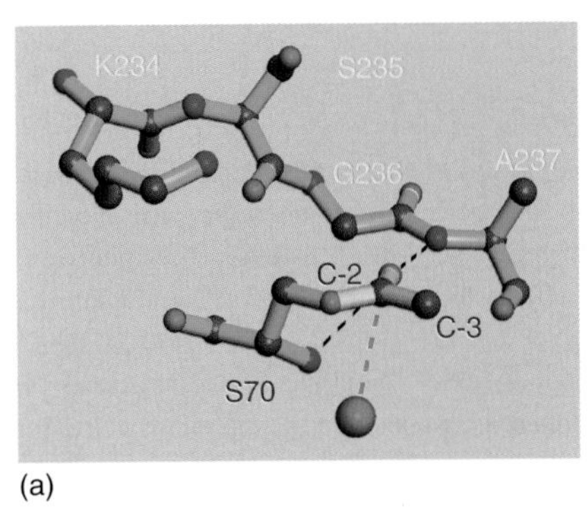

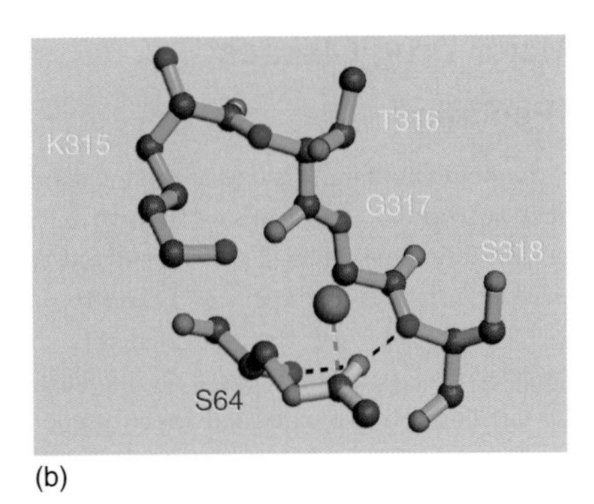

(a) (b)

Figure 8 Representation of the acyl–enzyme complex in (a) TEM-1 class A β-lactamase (PDB code 1m40) and (b) *Citrobacter freundii* class C β-lactamase (PDB code 1fr1), showing the two different positions of the hydrolytic water molecule (red ball). The red dashed line indicates the direction of attack of the hydrolytic water molecule, and the black dashed lines the hydrogen bonds in the oxyanion hole formed by the two backbone NH moieties of S70 and A237 or S64 and S318, respectively. For simplicity, only the active site serine, a portion of the KT(S)G β strand, and two carbon atoms (C-2 and C-3) of the former β-lactam ring are represented.

The difference between transpeptidases and serine β-lactamases results from the presence in the latter of a water molecule in the vicinity of the active site serine and of some conserved amino acids, which activate this water molecule.[19] Indeed, compared with transpeptidases, serine β-lactamases have been re-engineered to accommodate and activate a water molecule flanking the carbonyl group of the acyl–enzyme intermediate. Depending upon the nature of the β-lactamases, the hydrolytic water molecule is located either below (classes A and D) or above (class C) the carbonyl group of the acyl–enzyme complex (**Figure 8**). That water molecule is activated either by a carboxylic acid (Glu166 in class A β-lactamases; see **Figure 7**) or a phenol (Tyr150 in class C β-lactamases; see **Figure 7**).[20,21] Moshabery and co-workers have provided a detailed review on the molecular basis of bacterial resistance, and summarized the current knowledge on the activation of the active serine residue and the hydrolytic water molecule in these enzymes.[22]

7.17.3.2 Transpeptidase Modification

Resistance to β-lactam antibiotics can be caused by structural modifications of the essential PBPs. This is the case in Gram-positive bacteria such as *E. faecium* and *S. pneumoniae* (PBP2x) and in some Gram-negative bacteria. In *S. pneumoniae*, extensive PBP alterations occur by interspecies recombination following transfer of chromosomal DNA fragments from other streptococci, resulting in 'mosaic' genes.[23] Similar mechanisms were detected in other naturally transformable species such as *Neisseria* spp. or *Haemophilus* spp. In contrast, high-level resistance in staphylococci (MRSA) is associated with the production of a unique chromosomally encoded PBP2a (or PBP2′), which is not present in susceptible staphylococci. Ineffective inhibition of PBP2a by β-lactam antibiotics is due to a decreased rate of covalent bond formation during the acylation step as well as a poor fit of the antibiotics in the binding site. Although MRSA was reported in the 1960s, its incidence really started to rise in the mid-1990s, and has been steadily increasing since then. It has become a serious health issue because these strains also acquired resistance to other antibiotics such as quinolones, leaving vancomycin as the drug of last resort. Worryingly, vancomycin resistance has also been recently reported in *S. aureus*.

7.17.3.3 Alteration of the Outer Membrane Permeability of Gram-Negative Bacteria and Active Efflux Systems

In Gram-negative bacteria, antibiotics use porins to penetrate into the periplasmic space. Reduction of the number of porins and the size of pores excludes antibiotics such as cephems and carbapenems. Decreased expression of both the general and some specific porins (e.g., the tonB-dependent iron uptake system or OprD for the passive uptake of basic amino acids in *P. aeruginosa*) has been implicated in clinical resistance. The impaired penetration, combined with nonspecific drug efflux pumps (e.g., overexpression of the MexAB–OprM system or expression of the MexEF–OprN and MexCD–OprJ systems in *P. aeruginosa* and Acr pumps in Enterobacteriaceae), results in a dramatic reduction of the concentration of the antibiotics in the periplasmic space.[24]

7.17.4 Major Drug Classes

7.17.4.1 Penicillins

In primary care, the penicillin family is one of the most valuable groups of antibiotics. Penicillins are bactericidal, highly efficacious against susceptible organisms, well distributed in the body, and some representatives are orally active. Development of synthetic penicillins in the post-World War II period has both broadened the spectrum of activity and enhanced the efficacy of these medications. These research efforts led to the discovery of new groups of penicillins (**Figure 9**).[25]

The anti-staphylococcal penicillins (e.g., methicillin) harbor a bulky lipophilic C-6 substituent, conferring good stability toward penicillinases. The extended-spectrum penicillins (e.g., amoxicillin or ampicillin) contain an amino group in the α-position of the C-6 substituent, which gives both enhanced acid stability and activity against Gram-negative bacteria (e.g., *Haemophilus influenzae* and *E. coli*). The anti-*Pseudomonas* penicillins consist of ureido-penicillins (e.g., piperacillin), which are derivatives of ampicillin, and of penicillins having a carboxy or a sulfoxy group in the α-position of the C-6 substituent (e.g., ticarcillin or sulfocillin). These derivatives are, however, susceptible to β-lactamases. Finally, the amidino-penicillin mecillinam exhibits strong and specific activity against Enterobacteriaceae.[26]

In recent years, the emergence of resistant bacterial strains has limited the usefulness of semisynthetic penicillins. Nonetheless, penicillins remain the drugs of choice for localized soft-tissue infections (SSTs), urinary tract infections (UTIs), and respiratory tract infections (RTIs), and are particularly important in specific situations during pregnancy. Combinations of penicillins with β-lactamase inhibitors such as amoxicillin/clavulanic acid (for RTIs) or piperacillin/tazobactam (for Gram-negative infections) dominate the field. Ampicillin/sulbactam and ticarcillin/clavulanic acid are further examples of marketed combinations. With the exception of research programs aiming at the discovery of new β-lactamase inhibitors containing a penam structure (vide infra), the research effort in this field has been rather limited during the last 20 years.[27] It is, however, worth mentioning the 2-carboxypenams exemplified by T-5575 (**Figure 10**).[28,29] This compound displays an antibacterial spectrum similar to monobactams with almost no activity against Gram-positive bacteria but a Gram-negative spectrum equal or superior to that of ceftazidime or imipenem. As T-5575 is not a good substrate for β-lactamases, it is stable to β-lactamases, with the exception of OXA-1. It binds preferentially to PBP3 of *E. coli*. The synthetic strategy to access T-5575 is described in **Figure 10**.

The key ring closure to form the penam structure is achieved by transformation of the alcohol function of compound **1** into a mesylate or a chloride, and subsequent intramolecular displacement of the leaving group by the anion generated in situ in α-position to the carboxylate group.

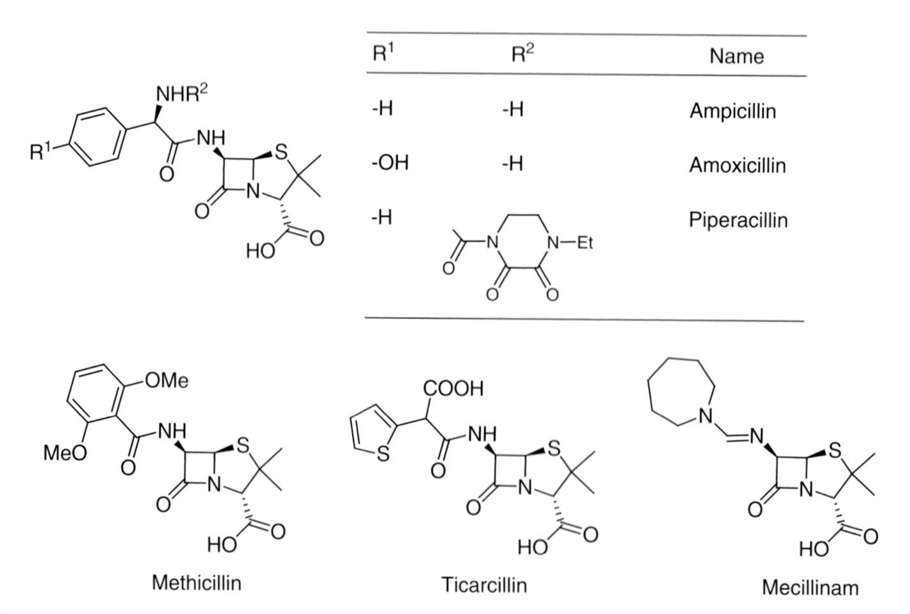

Figure 9 Structures of representative semisynthetic penicillins.

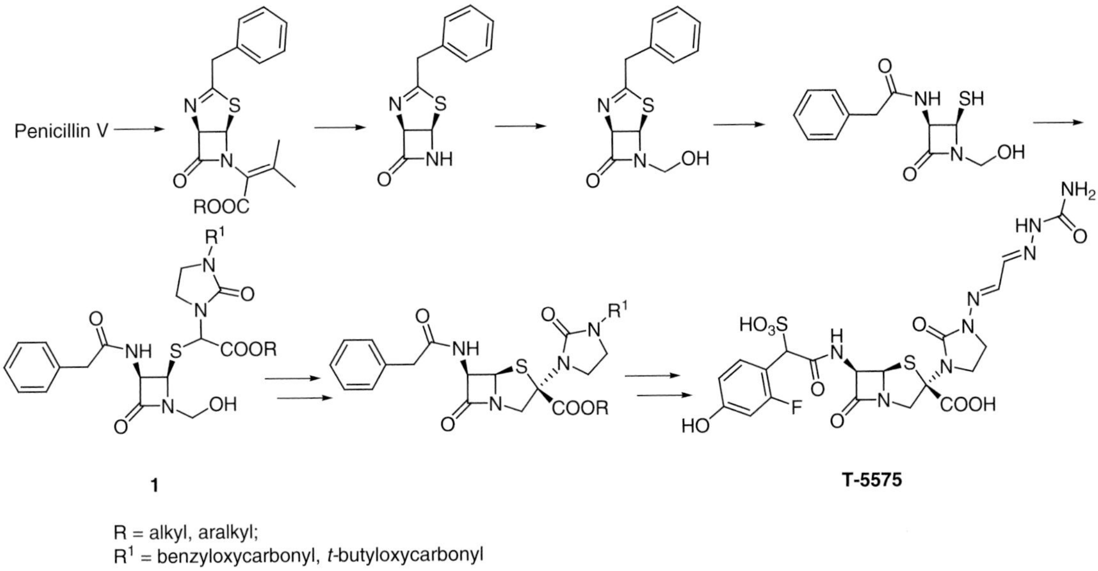

Figure 10 Synthesis of T-5575.

7.17.4.2 Monobactams

The main drawbacks of monobactams are the lack of activity against Gram-positive bacteria, susceptibility to ESBLs of Enterobacteriaceae, and incomplete coverage of *P. aeruginosa*. Aztreonam is mainly prescribed for serious infections caused by Gram-negative bacteria (UTIs, RTIs, and septicemia). However, monobactams might find a niche in the treatment of cystic fibrosis patients.[30,31]

The monobactam nucleus is synthesized following two different routes (**Figure 11**), starting either from L-threonine or L-glyceraldehyde acetonide.[32,33] The key steps for the formation of the azetidinone ring are, respectively, an intramolecular S_N2 displacement or an enantioselective [2 + 2] ketene–imine cycloaddition.

Research into monobactams declined after the introduction of aztreonam and carumonam. Some efforts were made to increase the antibacterial activity against *P. aeruginosa* by exploiting the tonB-dependent iron uptake systems. For this purpose, catechols and other iron-chelating moieties were introduced into the C-3 acylamino side chain (e.g., Syn-2416 or BMS180680; **Figure 11**).[34]

7.17.4.3 Cephalosporins and Their Structural Analogs Oxacephems, Isooxacephems, and Isocephems

Cephalosporin antibiotics have become a major part of the antibiotic formulary for hospitals in affluent countries. They are prescribed for a wide variety of infections, and their popularity stems from lesser allergenic risks compared with those of penicillins, as well as more favorable pharmacokinetic properties and a broader spectrum of activity. This last feature, however, resulted in the selection of microorganisms resistant to these agents. Those organisms not only have pathogenic potential but may also become multiresistant to antibiotics.

7.17.4.3.1 Parenteral cephalosporins

Since 1964, several generations of parenteral cephalosporins have been reported (**Table 3**). All of them were designed to meet medical needs resulting from the appearance of resistance to penicillins as well as, later, to cephalosporins.[35] They differ in their spectra, serum half-lives, penetration into the cerebrospinal fluid, and resistance to β-lactamases.

7.17.4.3.1.1 First- and second-generation cephalosporins

The first-generation, and some second-generation, cephalosporins retain excellent activity against streptococci and methicillin-susceptible *S. aureus* (MSSA). With the exception of the cephamycin-type cephalosporins (e.g., cefoxitin;

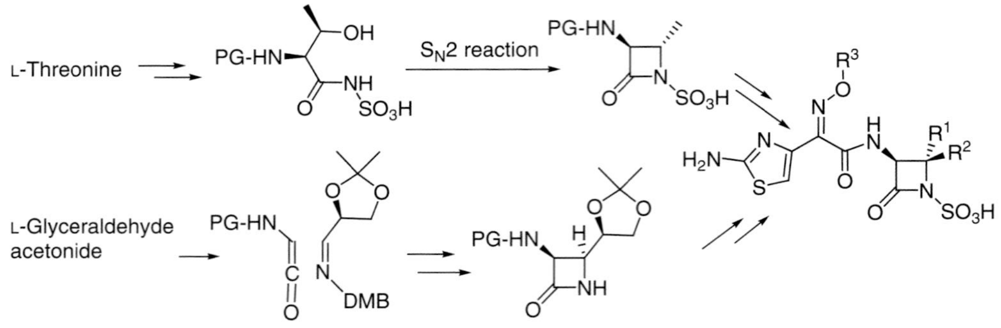

L-Threonine
PG-HN OH NH SO_3H
S_N2 reaction
PG-HN N SO_3H
L-Glyceraldehyde acetonide
PG-HN C O N DMB
PG-HN H NH
H_2N N S N O R^3 H N R^1 R^2 N SO_3H

R^1	R^2	R^3	
-Me	-H	-C(Me)$_2$COOH	Aztreonam
-H	-CH$_2$OCONH$_2$	-CH$_2$COOH	Carumonam
-Me	-H	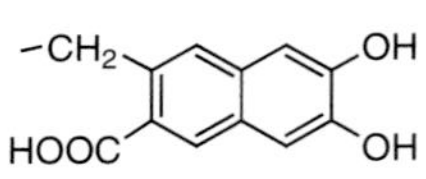 -CH$_2$ (HOOC, OH, OH)	BMS180680/PA1806
-Me	-H	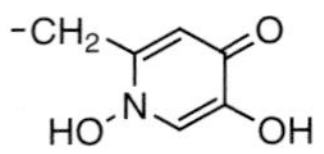 -CH$_2$ (O, HO-N, OH)	Syn-2416/PTX-2416

PG = *t*-butyloxycarbonyl, benzyloxycarbonyl, phthalimido; DMB = 3,4-dimethoxybenzyl

Figure 11 Synthetic routes to monobactams.

Table 3 Overview of marketed parenteral cephalosporins

Cephalosporin generation	*Gram positive (excluding MRSA)*	*Gram negative*	*Examples*
First	+ + +[a]	–	**Cefazolin**[b]
			Cefsulodin[c]
			Cefazaflur
			Cefalotin
			Cefapirin
			Cefradine
			Ceftezole
			Cefaloridine
			Cefacetrile
Second	+ +	+ +	**Cefuroxime**
			Cefoxitin[d]
			Cefmetazole[d]
			Cefbuperazone[d]
			Cefotetan[d]
			Cefamandole
			Cefotiam
			Ceforanide
			Cefonicid
			Cefminox
			Flomoxef[e]
Third	+	+ + +	**Cefotaxime**
			Ceftriaxone
			Ceftazidime
			Cefpimizole
			Ceftizoxime
			Cefoperazone
			Latamoxef[e]
			Cefpiramide
			Ceftiolene
			Cefmenoxime
			Cefodizime (Jp)[f]
			Cefuzonam
Fourth	+	+ + + +	**Cefepime**
			Cefpirome
			Cefoselis (Jp)
			Cefozopran (Jp)

[a] Number of plus signs indicates relative potency.
[b] Compounds highlighted in bold are top-selling drugs in their category.
[c] Exception: developed as an anti-*Pseudomonas* drug.
[d] Cephamycin-type cephalosporin showing enhanced stability toward β-lactamases.
[e] Oxacephem type.
[f] Jp, introduced only in Japan.

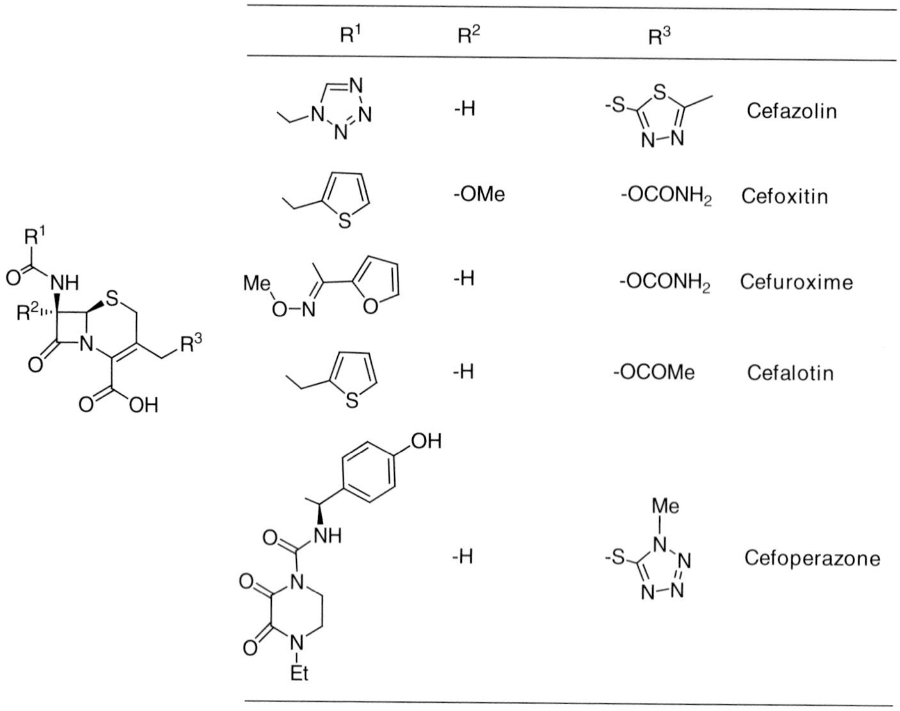

R^1	R^2	R^3	
	-H		Cefazolin
	-OMe	-OCONH$_2$	Cefoxitin
	-H	-OCONH$_2$	Cefuroxime
	-H	-OCOMe	Cefalotin
	-H		Cefoperazone

Figure 12 Representative first- and second-generation parenteral cephalosporins.

Figure 12), containing an additional α-methoxy group at position C-7, which shields the α face of the β-lactam ring, first- and second-generation cephalosporins became very susceptible to β-lactamases, and their use in clinical settings is currently rather restricted.

7.17.4.3.1.2 Third-generation cephalosporins

Third-generation cephalosporins are characterized by a broad spectrum of activity, allowing an expanded Gram-negative coverage. Compared with cephalosporins of the first and second generations, their Gram-positive coverage is weaker, but they possess an increased stability to β-lactamases (**Table 4**). However, the number of reports of resistance to these agents is growing due to their continuous use.

The alkoxyoxyimino-aminothiazolyl moiety in the C-7 acylamino side chain of these cephalosporins is responsible for both their increased resistance to β-lactamase hydrolysis and their ability to penetrate the outer cell envelope of Gram-negative bacilli.[36–39]

The nature of the substituent in position C-3 modulates the reactivity of the β-lactam ring as well as the physicochemical and pharmacokinetic properties of these antibiotics.

A dozen parenteral third-generation cephalosporins have been introduced into clinical use during the 1980s. The most frequently prescribed agents are cefotaxime, ceftriaxone, and ceftazidime (**Figure 13**). They are used in intra-abdominal sepsis, osteomyelitis, UTIs, neonatal sepsis, gonorrhea, and to treat infections in neutropenic patients. Ceftriaxone differs from the two other agents by its long half-life and its dual route of elimination. The use of ceftazidime is mostly restricted to *P. aeruginosa* infections where other agents are contraindicated or ineffective. In this case, it is often co-administered with an aminoglycoside. Cefotaxime and ceftriaxone can be used in nosocomial Gram-negative infections where *P. aeruginosa* is ruled out.

At the beginning of the 1990s, the hospital market was dominated by these three cephalosporins, and the pharmaceutical industry was seeking new cephalosporins with two distinct profiles: either a long-acting ceftazidime or a broad-spectrum cephalosporin that encompasses MSSA and ESBL-producing Gram-negative rods.

7.17.4.3.1.3 Fourth-generation cephalosporins

The first strategy used to improve activity against *P. aeruginosa* was to take advantage of the existence of iron-chelating siderophore channels, in addition to tonB-dependent iron transport systems, by incorporating catechol or catechol

Table 4 MIC_{90} values of selected third- and fourth-generation cephalosporins

Cephalosporin	*MIC_{90} (μg mL^{-1})*						*Ref.*
	MSSA	*S. pneumoniae*	*PRSP*	*Susceptible E. coli*	*Stenotrophomonas maltophilia*	*P. aeruginosa*	
Cefetecol	2	0.5	2	0.25	>64	2	42
RU-59863	0.5	0.03	0.5	0.03	0.25	0.5	43
KP-736	25	0.39	NA	0.1	25	12.5	44
Ro 09-1428	3.1	0.78	NA	0.06	NA	0.39	45
Cefepime	8	0.125	2	0.125	64	8	42
Cefpirome	2	0.125	1	0.125	>64	32	42
Cefoselis	2	0.125	1	0.06	>128	32	46
Cefluprenam	1	0.125	NA	0.06	>128	16	47
Cefozopran	2	0.125	0.39	0.06	64	2	48
Ceftazidime	16	2	>16	0.5	>64	32	42
Ceftriaxone	3.1	0.1	>16	0.06	NA	>32	45

NA, not available.
Marketed drugs are highlighted in bold.

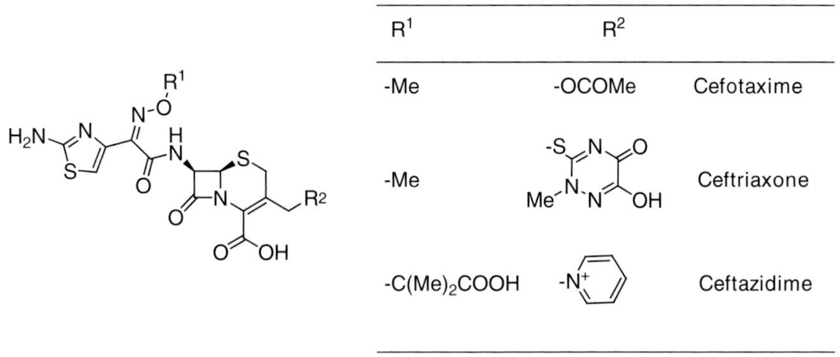

R^1	R^2	
-Me	-OCOMe	Cefotaxime
-Me	(triazinone thio substituent: -S, N, O, Me, N, N, OH)	Ceftriaxone
$-C(Me)_2COOH$	$-N^+$ (pyridinium)	Ceftazidime

Figure 13 Representative third-generation cephalosporins.

bioisostere moieties, mostly at position C-7 on the cephalosporin.[40,41] Cefetecol, RU-59863, KP-736, and Ro 09-1428 are representative examples (**Figure 14**).[42–45]

Most of these compounds display an impressive Gram-negative spectrum, including *Pseudomonas* and ESBL-producing Enterobacteriaceae. In some cases (e.g., RU-59863), enhanced activity against Gram-positive bacteria is observed, albeit without MRSA and enterococci. However, rapid emergence of resistance by *P. aeruginosa* has appeared, arising from downregulation or elimination of the iron uptake mechanisms. Furthermore, the good in vitro activity could not be translated into good in vivo activity. Catechol *O*-methyltransferase (COMT) monomethylates one phenol group of the catechol, leading to much weaker antibiotics. Decreasing the pK_a of the catechol moiety by the introduction of an electron-withdrawing fluorine atom into RU-59863 or replacing the catechol group by the bioisosteric 1,5-dihydroxy-4-pyridone moiety (e.g., KP-736) led to compounds that were more stable toward COMT. However, none of these compounds reached the market for a number of reasons, such as high materials cost, unfavorable pharmacokinetic properties, insufficient in vivo efficacy, and side effects observed during clinical development.

A second strategy, aimed at the optimization of ceftazidime, was followed by further exploiting its zwitterionic character. A series of cephalosporins containing a quaternary nitrogen at C-3 was synthesized, resulting in compounds such as cefepime, cefpirome, cefoselis, cefluprenam, and cefozopran (**Figure 15**).[46–48]

The presence of polar C-3 substituents improved the activity of the aminothiazolyl cephalosporins, especially against staphylococci and *P. aeruginosa*, without loss of activity against members of the Enterobacteriaceae.[49]

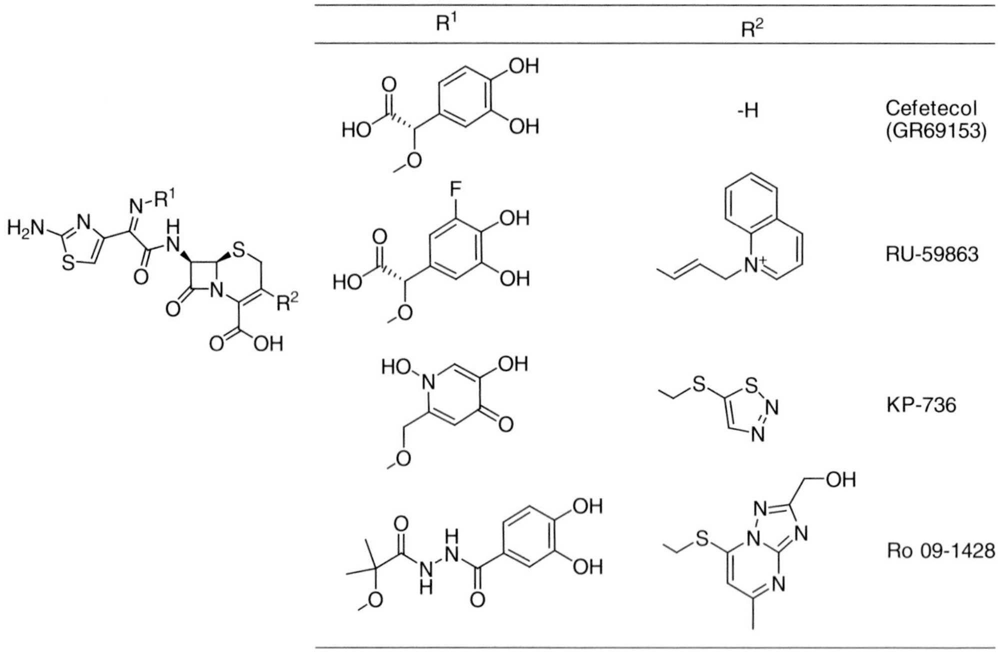

R1	R2	
	-H	Cefetecol (GR69153)
		RU-59863
		KP-736
		Ro 09-1428

Figure 14 Structures of some representative catechol cephalosporins.

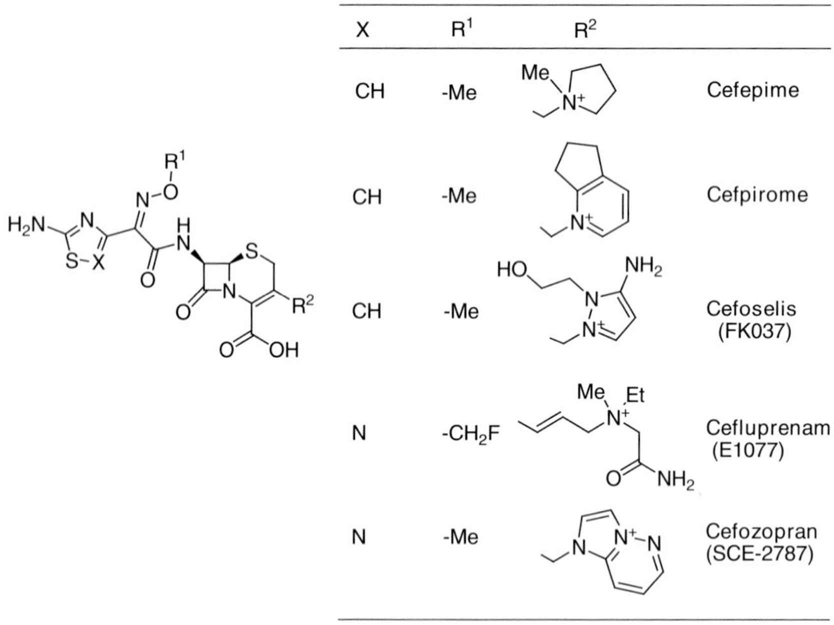

X	R1	R2	
CH	-Me		Cefepime
CH	-Me		Cefpirome
CH	-Me		Cefoselis (FK037)
N	$-CH_2F$		Cefluprenam (E1077)
N	-Me		Cefozopran (SCE-2787)

Figure 15 Structures of some representative noncatechol fourth-generation cephalosporins.

Introduction of an aminothiadiazolyl group in the C-7 side chain of the cephalosporin enhanced antibacterial activity against *P. aeruginosa* by decreasing the affinity for cephalosporinases and by increasing the hydrophilic character, resulting in better penetration through the outer membrane.[50,51] Three compounds, cefepime, cefpirome, and cefoselis, reached the market. Compared with third-generation cephalosporins, they remain active against some, but not all, ceftazidime-resistant Enterobacteriaceae. The anti-*Pseudomonas* activity is generally similar to that of ceftazidime, with the exception of cefepime, which is more active (**Table 4**). These new zwitterionic compounds were more potent than ceftazidime against PRSP and staphylococci, but did not present a full antibacterial coverage due to the lack of activity against MRSA. The absence of significant improvement over ceftazidime against *P. aeruginosa* as well as the insufficient coverage of *S. aureus* explain the currently rather limited commercial success of the fourth-generation cephalosporins. They are mainly used in the treatment of serious nosocomial sepsis in which resistant Gram-negative pathogens are known or suspected to be involved.

7.17.4.3.1.4 Anti-methicillin-resistant *Staphylococcus aureus* Cephalosporins

At the end of the 1990s, the dramatic emergence of MRSA triggered major research efforts to discover new cephalosporins acting against these problem pathogens. Various observations led to incremental improvements toward activity against MRSA. During the optimization of the fourth-generation cephalosporins, it was first observed that some compounds showed noticeable activity against MRSA, albeit insufficient to claim coverage. This was the case for cefoselis and CP6679, where MIC_{90} values of 20 and 2.5 $\mu g\,mL^{-1}$ were observed, respectively (**Figures 15** and **16**).

Enhancing the lipophilic character of the C-7 acylamino residue (e.g., by the introduction of chlorine atoms) improved the activity toward MRSA, as illustrated by RWJ-54428 (for which the prodrug is RWJ-442831; **Figure 16**) and BMS-247243 (**Figure 17**).[52,53]

R^1	X	R^2	R^3	
NH_2	N	$-CH_2F$	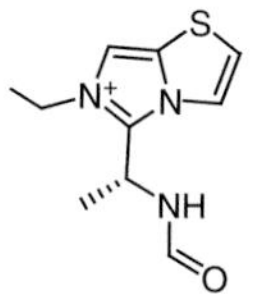	CP6679 (ME1209)
NH_2	N	-Me	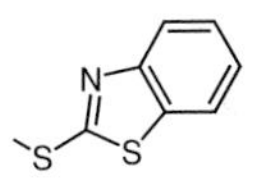	CP-0467
$NHPO_3^{2-}$	N	-Et	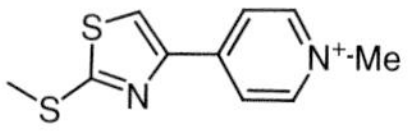	ceftaroline PPI-0903 (TAK599)
NH_2	CCl	-H	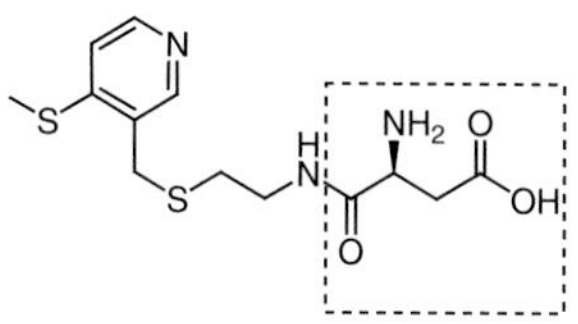	RWJ-442831 (MC-02479)
NH_2	CCl	-H	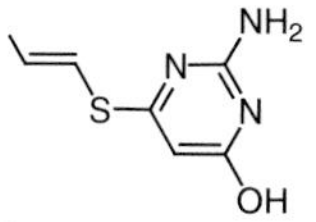	LB11058
NH_2	CH	-H	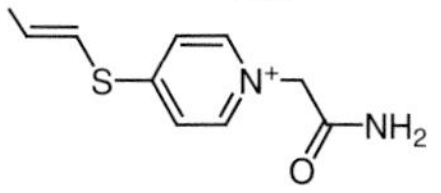	TOC-39
NH_2	N	-H	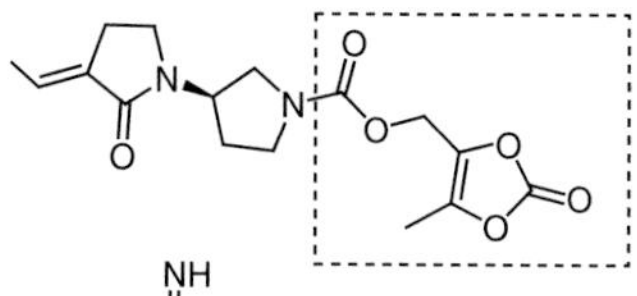	ceftobiprole medocaril (BAL9141)
NH_2	N	$-CH_2F$	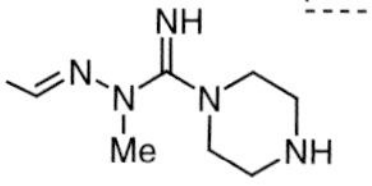	CAB-175 (CB-181963)
NH_2	N	-Et	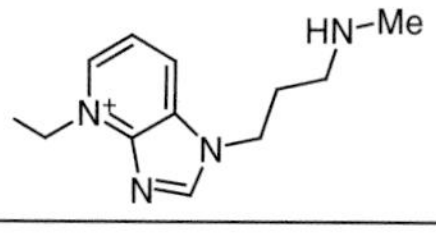	S-3578

Dashed boxes indicate the prodrug moieties.

Figure 16 Selected cephalosporins active against MRSA.

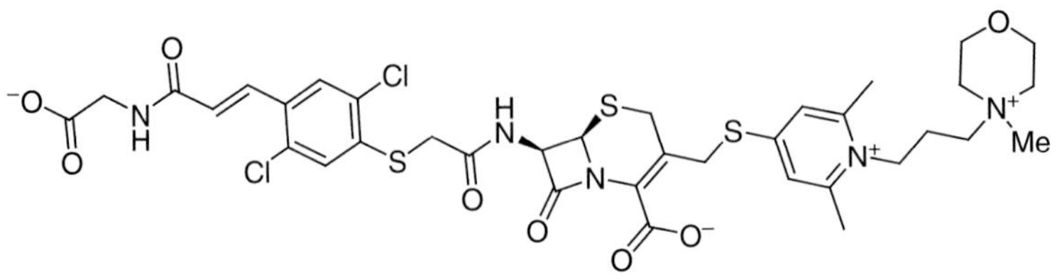

BMS-247243

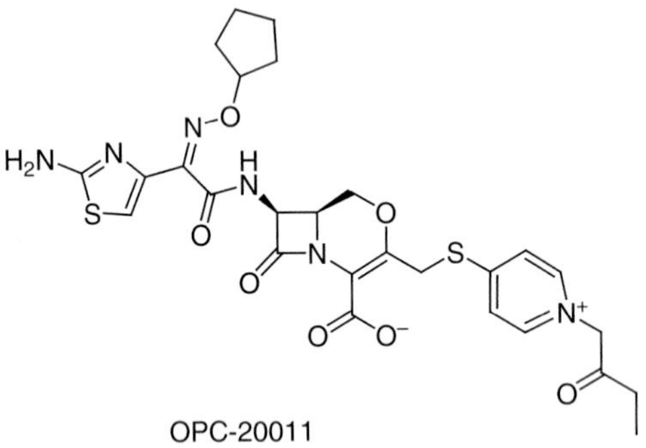

OPC-20011

Figure 17 Structures of additional anti-MRSA cephems.

Table 5 MIC_{90} values for MRSA and inhibition of MRSA PBP2a for some representative anti-MRSA cephalosporins

Cephalosporin	MIC_{90} ($\mu g\,mL^{-1}$)	IC_{50} ($\mu g\,mL^{-1}$)[a]
LB11058	1	0.8
Ceftobiprole[b]	2	0.9
RWJ-54428[b]	2	0.7
T-91825[b]	2	0.9
BMS-247243	2–4	0.7
S-3578[b]	4	4.5
TOC-39	4	NA
CAB-175	4	NA
OPC-20011[c]	6.25	2
CP-0467	6.25	0.5
ME1209 (CP6679)	12.5	5.1

NA, not available.
[a] MRSA PBB2a inhibition.
[b] Compounds still in development in 2004.
[c] Isooxacephem type.

Increasing the lipophilic bulk on the C-3 residue also improved the activity toward MRSA (e.g., 2-mercapto-benzothiazolyl in CP-0467).[54] However, these large substituents, necessary to achieve high affinity toward PBP2a (**Table 5**), precluded broad-spectrum activity, and, in the case of C-7 arylthioacetylamino side chains, reduced the stability toward class A β-lactamases.

A better compromise was found with the 2-aminothiadiazolyl-α-(hydroxyimino)-acetylamino side chain in position C-7, which allows, in some cases, coverage of Gram-negative pathogens.

Enhancing the intrinsic chemical reactivity of the β-lactam ring, by introducing an electron-withdrawing group in position C-3, improved the binding to PBP2a. However, this type of modification compromised the hydrolytic stability, particularly when the C-3 substituent was a good leaving group. Introduction of a noncleavable electron-withdrawing substituent at position C-3 in T-91825 (for which the prodrug is PPI-0903), RWJ-54428 (for which the prodrug is RWJ-442831), LB11058, TOC-39, ceftobiprole, or CAB-175 circumvented this liability (see **Figure 16**).[55–61]

The presence of a large lipophilic residue containing an ammonium group at about 7.5–10 Å from the C-3 connecting atom contributes significantly to the activity against MRSA, as observed in T-91825, RWJ-54428, ceftobiprole, CAB-175, and S-3578.[62]

Overall, these structural requirements represented a challenge to medicinal chemists, to attain drug-likeness criteria and achieve a reasonable materials cost.[63] Indeed, some of these molecules are zwitterionic, resulting in compounds with low aqueous solubility. Prodrug concepts had to be applied to T-91825, RWJ-54428, and ceftobiprole in order to allow intravenous application. High plasma protein binding was another issue, since some of these compounds exceed a level of 98%, resulting in a very low free plasma concentration in the systemic circulation. However, when the physicochemical properties of the molecules were balanced, broad-spectrum activity was observed, including even some strains of Enterobacteriaceae (ceftobiprole and S-3578) and *P. aeruginosa* (ceftobiprole; **Table 6**).

The fact that these molecules contain a noncleavable C-3 substituent required more lengthy syntheses than for the third- and fourth-generation cephalosporins, as illustrated for the synthesis of ceftobiprole (**Figure 18**).[64]

The starting cephem aldehyde **2** had first to be prepared from 7-ACA. Although the geometry of the Wittig olefination was well controlled, the cephem double bond had to be reconjugated with the carboxylate. This equilibration occurred after the transient oxidation of the sulfur to a sulfoxide.

The MIC values against MRSA correlate well with PBP2a inhibition (see **Table 5**). The most promising development compounds exhibit MIC_{90} values between $2\,\mu g\,mL^{-1}$ and $4\,\mu g\,mL^{-1}$ against MRSA.

Compared with the cephems, the isooxacephalosporins and isocephalosporins are less susceptible to the action of β-lactamases, and offer an alternative to noncleavable cephems. However, compounds such as OPC-20011 (see **Figure 17**) suffer from a high materials cost and insufficient activity against MRSA.[65] Its synthetic access is illustrated in **Figure 19**. The core isooxacephem nucleus requires de novo synthesis by a [2+2] ketene–imine cycloaddition, starting from D-threonine.

The good PBP2a inhibition of ceftobiprole has been explained by a combination of enhanced reactivity of the β-lactam ring, higher intrinsic affinity, and a lower rate of de-acylation of the acyl–enzyme complex. The large residue in position C-3 stabilizes the complex by establishing interactions with the enzyme. The pyrrolidinone ring induces a conformational change while the pyrrolidine ring, which is situated partly outside the transpeptidase, participates in the rate acceleration.[66,67]

Although some anti-MRSA cephalosporins display MIC values at the limit of the plasma levels, adequate coverage of MRSA was achieved in vivo. In addition, the positive Phase III results of ceftobiprole medocaril in patients with complicated skin and soft-tissue infections confirmed the potential of β-lactam antibiotics to treat MRSA infections.[68] In addition, their broad-spectrum activity – at least for some of them – allows the coverage of Gram-negative bacteria

Table 6 MIC_{90} values of some relevant anti-MRSA cephalosporins against selected key organisms

Cephalosporin	*MIC_{90} ($\mu g\,mL^{-1}$)*							*Ref.*
	MSSA	*MRSA*	*E. faecalis*	*E. faecium*	*PRSP*	*Non-ESBL producing E. coli*	*P. aeruginosa*	
LB11058	0.25	1	1	>64	0.125	NA	NA	59
Ceftobiprole	0.5	2	4	>32	1	0.06	16	64
RWJ-54428	0.5	2	0.25	8	0.5	>32	>32	61
T-91825	0.5	2	8	>128	0.25	0.25	128	58
S-3578	2	4	>32	NA	1	0.5	64	62
TOC-39	0.39	4	0.78	NA	0.2	0.05	>32	60
Ceftriaxone	4	>32	>32	>32	1	0.125	>32	62

NA, not available.

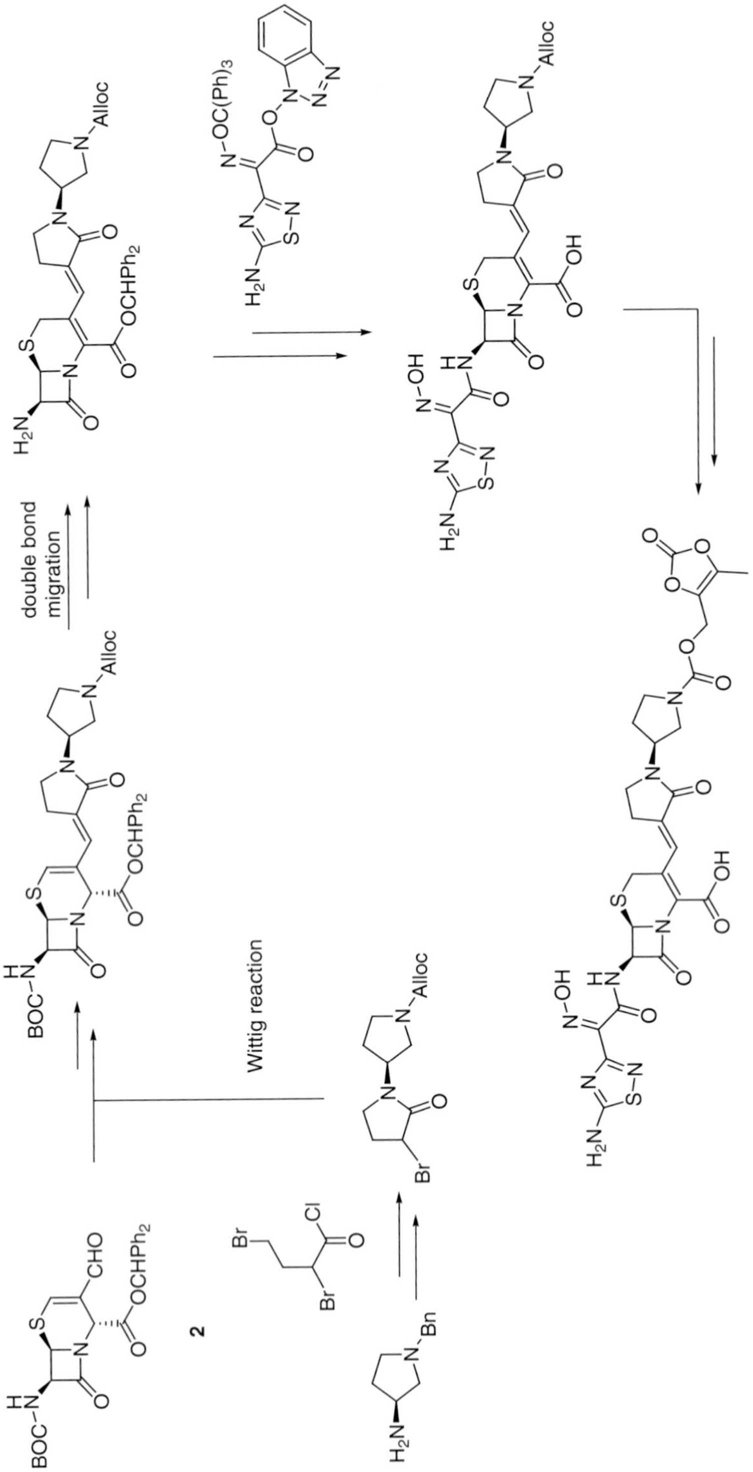

BOC = *t*-butyloxycarbonyl; Alloc = allyloxycarbonyl; Bn = benzyl

Figure 18 Synthesis of ceftobiprole medocaril.

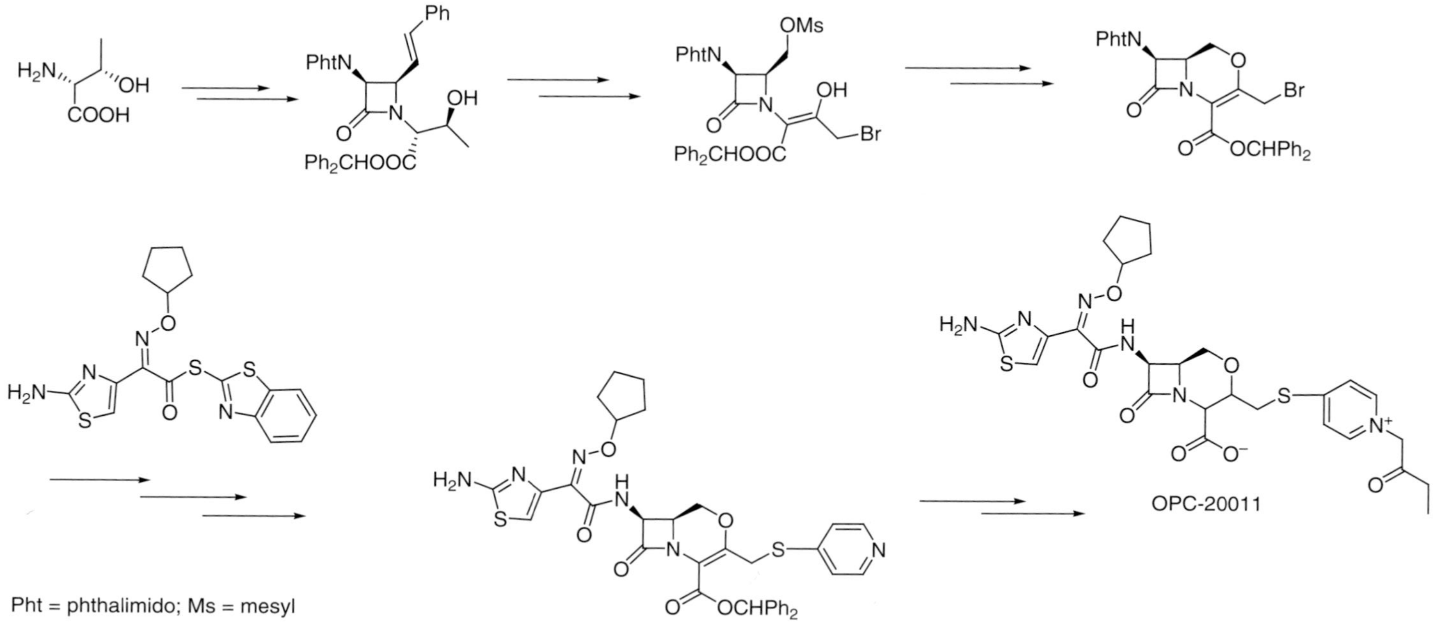

Pht = phthalimido; Ms = mesyl

Figure 19 Synthetic route to OPC-20011.

such as Enterobacteriaceae, and their bactericidal mode of action presents an advantage over that of glycopeptides. However, these molecules remain susceptible to many ESBLs, and do not cover *E. faecium*, which might challenge their broad usage against severe Gram-positive infections in hospital settings.

7.17.4.3.2 Oral cephalosporins

The first oral cephem, cefalexin (**Figure 20**), was introduced in medical practice in 1967. Since then, several cephems have been marketed (**Table 7**).

As penicillins became more susceptible to β-lactamases and less efficacious against *S. pneumoniae* (by horizontal gene transfer and PBP mutations), this created a demand for new oral agents. In addition, in order to foster a parenteral–oral switch, the medical community desired oral third-generation cephalosporins.

From a structural point of view, these cephalosporins can be divided into two groups: those active as free carboxylic acids and those compulsorily administered as ester prodrugs.

The first group includes cephalosporins having a C-7 α-aminoacyl side chain (cefalexin, cefadroxil, cefprozil, cefaclor, and the carbacephem loracarbef; **Figure 20**) or third-generation aminothiazolyl cephalosporins bearing a small substituent at position C-3 (ceftibuten, cefixime, and cefdinir). The only exceptions are FK041 and cefmatilen, which bear larger C-3 residues.[69,70]

The absorption route of this group of cephalosporins was studied, and three active transport systems have been identified: the intestinal proton-dependent peptide transporter PEPT1, a proton-independent dipeptide carrier, and anion exchanger systems. The absorption route of these cephalosporins depends on their physicochemical properties and the nature of the substituents. Cefmatilen, for example, is taken up by a carrier system that recognizes peptides having histidine as the N-terminal amino acid.[71]

The second group consists of parenteral third-generation cephalosporins (e.g., cefcapene, cefetamet, cefditoren, cefpodoxime, and ceftizoxime; **Figure 21** and **Table 8**), displaying adequate activity against RTI pathogens such as *S. pneumoniae*, *Moraxella catarrhalis*, and *Haemophilus influenzae*, and some enteric pathogens.[72–76]

These cephalosporins are administered as esters, and are passively absorbed through the intestinal wall. The cleavage of the ester function occurs under physiological conditions during the passage through the gut. Several

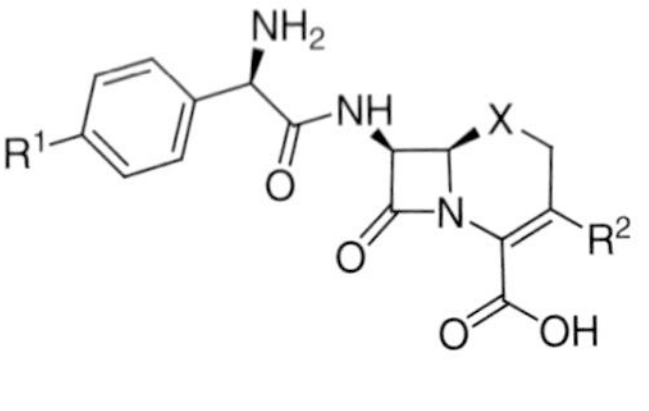

R^1	X	R^2	
-H	S	-Me	Cefalexin
-OH	S	-Me	Cefadroxil
-OH	S	-CH=CHMe (*E*)	Cefprozil
-H	S	-Cl	Cefaclor
-H	CH_2	-Cl	Loracarbef

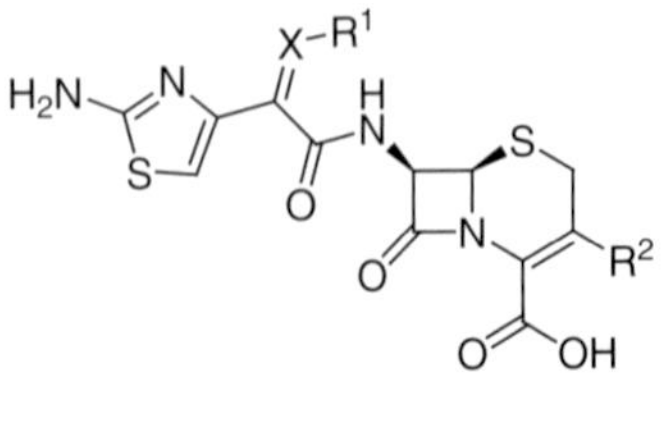

X	R^1	R^2	
CH	$-CH_2CO_2H$	-H	Ceftibuten
N	$-OCH_2CO_2H$	$-CH=CH_2$	Cefixime
N	-OH	$-CH=CH_2$	Cefdinir
N	-OH	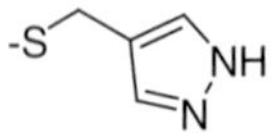	FK041
N	-OH	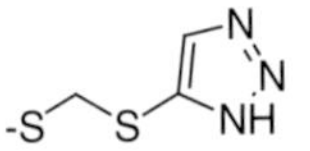	Cefmaliten (S-1090)

Figure 20 Cephalosporins orally active as free acids.

Table 7 Overview of marketed oral cephalosporins

Cephalosporin generation	*Gram positive*	*Gram negative*	*Examples*
First	+ + +[a]	–	**Cefalexin**[b]
			Cefadroxil
			Cefprozil
			Cefatrizine
			Cefradine
			Cefroxadine
Second	+ +	+ +	**Cefaclor**
			Cefuroxime axetil
			Loracarbef[c]
			Cefteram pivoxil
			Cefroxadine
			Cefotiam hexetil
Third	+ +	+ + +	**Cefixime**
			Cefdinir
			Ceftibuten
			Cefcapene pivoxil (S-1108)
			Cefetamet pivoxil
			Cefditoren pivoxil (ME1207)
			Cefpodoxime proxetil
			Ceftizoxime alapivoxil (AS-924)

[a] Number of plus signs indicates relative potency.
[b] Compounds highlighted in bold are top-selling drugs in their category.
[c] Carbacephem type.

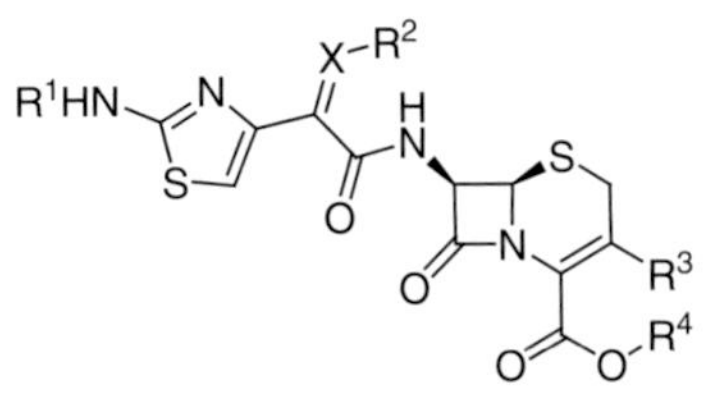

X	R^1	R^2	R^3	R^4	
CH	-H	-Et	$-CH_2OCONH_2$	-PIV	Cefcapene pivoxil
N	-H	-OMe	-Me	-PIV	Cefetamet pivoxil
N	-H	-OMe	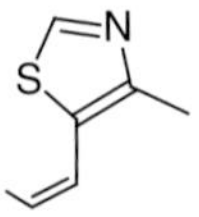	-PIV	Cefditoren pivoxil
N	-H	-OMe	$-CH_2OMe$	-Proxetil	Cefpodoxime proxetil
N	Ala	-OMe	-H	-PIV	Ceftizoxime alapivoxil[a]

[a] Double prodrug

Figure 21 Third-generation cephalosporin esters that have reached the market or are in late-stage development.

Table 8 MIC_{90} values of relevant oral third-generation cephalosporins against some key organisms

Cephalosporin	*MIC_{90} (μg mL^{-1})*						*Ref.*
	MSSA	*S. pneumoniae*	*PRSP*	*H. influenzae*	*M. catarrhalis*	*Non-ESBL producing E. coli*	
Ceftibuten	>64	4	64	0.06	4	0.25	75
Cefixime	32	0.25	64	0.25	0.5	0.5	75
Cefdinir	0.5	0.13	4	0.5	0.5	0.5	75
FK041	0.39	0.2	3.1	0.39	0.78	0.2	70
Cefmatilen	0.25	0.03	4	0.13	2	0.5	69
Cefcapene	1	0.12	2	0.12	0.5	1	108
Cefetamet	>64	1	>64	0.25	0.5	0.5	75
Cefditoren	1.56	0.2	0.78	0.03	0.39	0.78	70
Cefpodoxime	4	0.06	4	0.13	2	0.5	75
Ceftizoxime	6.25	0.5	>32	0.05	0.05	0.05	76
Cefuroxime	2	<0.12	4	1	2	4	75
Cefaclor[a]	16	0.5	>64	16	4	2	75

[a] Reference compound.

Table 9 Representative examples of prodrug moieties

Cephalosporin	*Prodrug moiety*
Pivoxil (PIV)	$-CH_2OCOC(Me)_3$
Acoxil	$-CH_2OCOMe$
Axetil	$-CH(Me)OCOMe$
Pentexil	$-CH(Me)OCOCH(Me)_2$
Hexetil (cilexetil)	$-CH(Me)OCO_2C_6H_{11}$
Proxetil	$-CH(Me)OCO_2CH(Me)_2$
Daloxate (medoxomil)	$-CH_2$ (5-methyl-2-oxo-1,3-dioxol-4-yl structure: O, O, O)
Bopentil	$-CH_2C(=CHEt)CO_2CH_2CH(Me)_2$

prodrug ester moieties have been discovered and evaluated (**Table 9**).[77,78] Among them, the most commonly used prodrug moiety is the pivaloyloxymethyl (PIV) residue. Several prodrugs have reached the market (**Figure 21**).

However, there are issues related to these prodrug esters. Most of them release low-molecular-weight aldehydes (formaldehyde and acetaldehyde), and, in the case of the PIV ester, side effects related to a decreased carnitin concentration have been reported. Furthermore, difficulty in achieving absorption at a sufficiently high level has a negative impact on the cost of materials.

As first- and second-generation oral cephalosporins overcome penicillinase-mediated resistance, they do not show cross-resistance with penicillins in *S. aureus*. Owing to their broader spectrum, third-generation oral cephalosporins offer better protection against fastidious, as well as some Gram-negative, species producing β-lactamases (**Table 8**). They are therefore useful for the treatment of *H. influenzae*, *H. parainfluenzae*, *S. pneumoniae*, and *M. catarrhalis* in acute bacterial exacerbation of chronic bronchitis and in community-acquired pneumonia. They are also effective against strains of *Streptococcus pyogenes* in pharyngitis/tonsillitis, and against MSSA in the treatment of uncomplicated SSTIs.

Despite many attempts in recent years to increase the efficacy and to enlarge the spectrum of these anti-infective agents, improvements with regard to existing compounds have generally been modest. In particular, therapeutically useful activity against PRSP could not be achieved. As these compounds also target the pediatric segment, other properties such as neutral taste are of pivotal importance for the successful development of an acceptable formulation. For example, the PIV esters have an extremely bitter taste, which must be masked by an appropriate galenic formulation. Another important aspect is the disturbance of the intestinal microflora associated with the improved activity against Gram-negative bacteria, leading to an increase in the number of enterococci.[79] Third-generation oral cephalosporins are mainly used in community-acquired infections such as uncomplicated cystitis and upper and lower RTIs.

7.17.4.4 Penems and Oxapenems

7.17.4.4.1 Penems

Woodward[80] designed the penem core structure by combining two different features that activate β-lactam antibiotics: the ring strain introduced by a fused five-membered ring and the delocalization of the β-lactam nitrogen lone pair into a conjugated double bond system. The first prototypes bearing the C-6 acylamino side chain of penicillin V were quite unstable. With the concomitant discovery of the natural carbapenems bearing a C-6 α-hydroxyethyl substituent shielding the β-lactam ring, it was possible, by introducing the same substituent, to synthesize more stable molecules. However, these compounds did not surpass carbapenems in terms of potency against Gram-negative bacteria, especially against *P. aeruginosa*. Furthermore, the compounds were unstable in acidic media, leading to fragmentation and the formation of a thiazole derivative. Faropenem daloxate reached the market in Japan for the treatment of RTIs and SSTIs. Three further compounds are currently in development, MEN 11505 (drug MEN 10700), ritipenem acoxil, and sulopenem (**Figure 22**).[81,82] Bryskier has summarized the activities in this field.[83]

7.17.4.4.2 Oxapenems

Due to the additional strain induced by the shorter C–O bond, the related oxapenems are extremely reactive molecules.[84,85] Only compounds bearing a bulky C-2 substituent (e.g., AM-112; see **Figure 32**) are sufficiently stable. Their antibacterial activity is, however, weak. They present a β-lactamase inhibition potential similar to that of epithienamycin (see **Figure 31**).[86]

7.17.4.5 Carbapenems

7.17.4.5.1 Parenteral anti-*Pseudomonas* carbapenems

The first compounds to reach the market were the C-1 unsubstituted carbapenems imipenem and panipenem (**Figure 23**).[87,88]

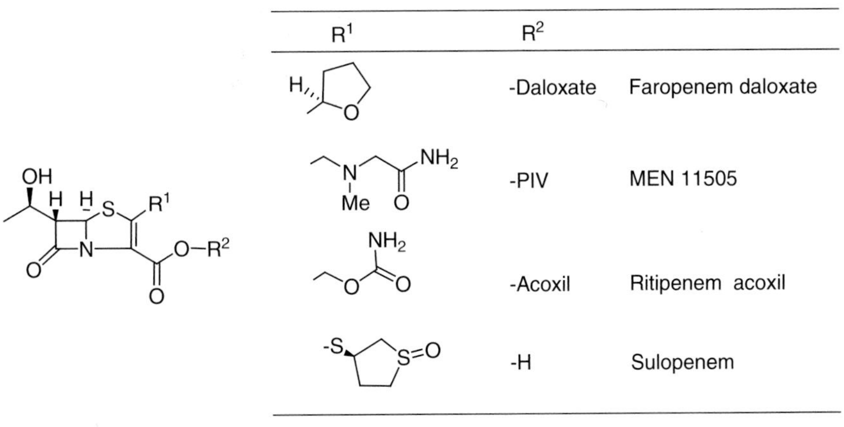

R^1	R^2	
	-Daloxate	Faropenem daloxate
	-PIV	MEN 11505
	-Acoxil	Ritipenem acoxil
	-H	Sulopenem

Figure 22 Chemical structures of penems in development.

R^1	
$-SCH_2CH_2NHCH{=}NH$	Imipenem
[structure]	Panipenem

Figure 23 Structures of imipenem and panipenem.

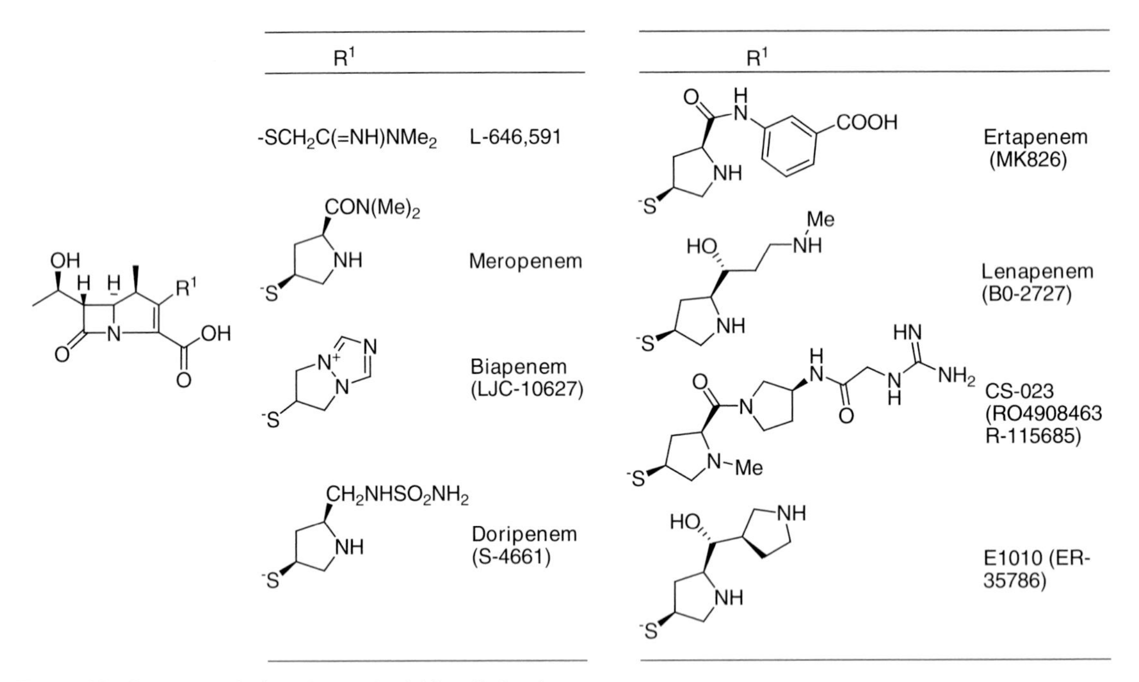

Figure 24 Structures of relevant parenteral 1β-methyl carbapenems.

However, owing to their instability toward renal dehydropeptidase-I (DHP-I) and their renal toxicity, imipenem had to be co-administrated with cilastatin, a DHP-I inhibitor, and panipenem with betamipron, a renal protective agent. Another weakness of these antibiotics is their short half-life, which requires a 6-h dosing interval. Therefore, the main focus of the various research groups has been to identify compounds stable toward DHP-I and presenting more favorable pharmacokinetic and safety profiles. With the emergence of resistance, emphasis was put on improving activity against *P. aeruginosa* and MRSA.

A seminal discovery was made by the Merck group, with the identification of L-646,591 (**Figure 24**), which contains a C-1 β-methyl group, resulting in increased stability toward metabolization by DHP-I. The discovery of effective and highly diastereoselective processes to prepare the key intermediate **3** (**Figure 25**) opened the way for broad synthetic programs.

The carbapenem skeleton is constructed following three different routes. The ring closure is achieved either by rhodium-catalyzed carbene insertion or Dieckmann-type condensation, leading to the 2-keto intermediate **4**. This intermediate is transformed into either its enol-phosphate **5** or its corresponding enol-triflate **6**. The former is further reacted with a nucleophile (e.g., a thiol derivative), and the latter with a halide in transition metal-catalyzed cross-coupling reactions. Alternatively, the 2-thio-substituted carbapenems **7** can be obtained by a thermal intramolecular Wittig–Horner reaction.

The introduction of (3*S*)-pyrrolidinylthio residues at position C-2 ensures sustained antibacterial activity. This effort resulted in the identification of meropenem, which displays roughly the same antibacterial spectrum as imipenem (**Table 10**; weaker activity against Gram-positive aerobes, but enhanced activity against Gram-negative aerobes).

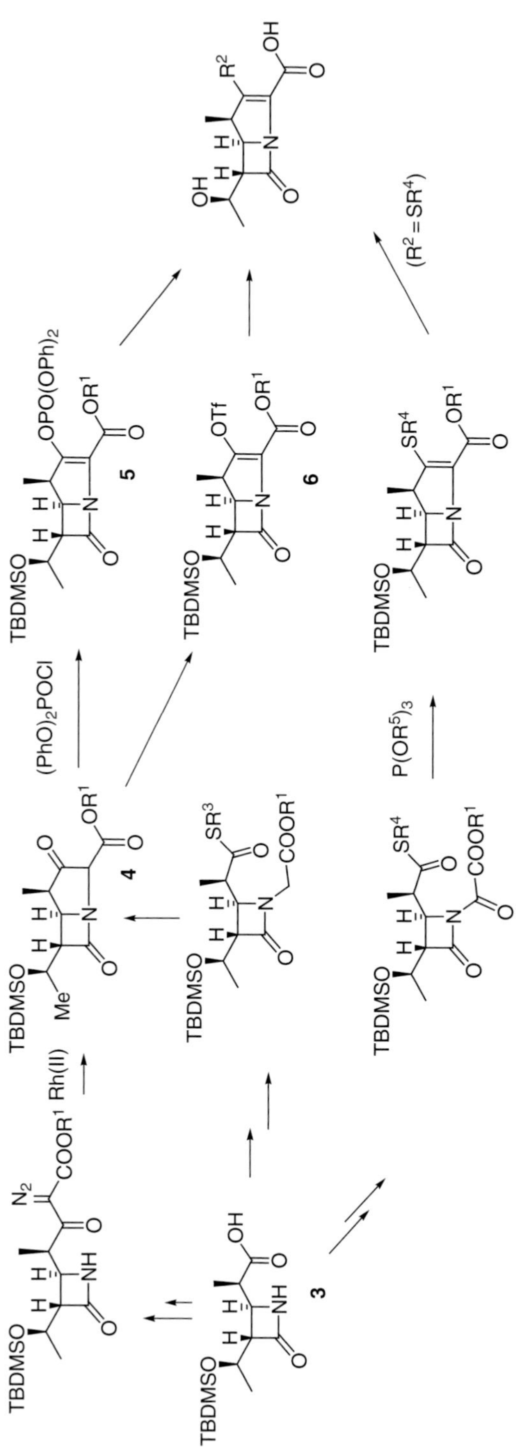

R^1 = allyl, *p*-nitrobenzyl; R^5 = alkyl; TBDMS = *t*-butyldimethylsilyl; Tf = trifluoromethanesulfoxy

Figure 25 General strategy to access carbapenems.

Table 10 MIC_{90} values of relevant parenteral carbapenems against some key organisms

Cephalosporin	*MIC_{90} ($\mu g\,mL^{-1}$)*								*Ref.*
	MSSA	*MRSA*	*Penicillin-susceptible S. pneumoniae*	*PRSP*	*E. coli*	*E. cloacae*	*P. aeruginosa*	*Imipenem-resistant P. aeruginosa*	
Imipenem[a,b]	0.03	>32	<0.008	1	0.5	1	12.5	25	90, 93
Panipenem[a]	0.5	16	<0.015	0.25	0.12	0.25	8	>32	88
Meropenem[a]	0.12	>32	0.008	1	<0.01	4	6.25	25	90
Ertapenem[a]	0.25	>32	0.015	2	<0.01	0.2	12.5	>32	90
Biapenem[a]	0.2	>32	0.015	0.25	0.05	0.125	2	32	96
Doripenem	0.06	>32	0.008	1	0.03	0.2	8	>32	90
Lenapenem	0.1	>16	0.39	NA	0.05	0.2	6.25	12.5	97
CS-023	0.25	4	0.008	1	0.03	0.125	8	NA	94
E1010	0.1	12.5	0.39	NA	0.05	0.125	3.13	6.25	97
ME1036	0.02	2	0.02	0.03	0.12	2	>128	>128	102
SM-232724	0.03	2	0.015	0.25	0.5	16	32	>32	100
J-114,870	0.015	4	0.015	0.25	0.015	1	4	16	96
L-695,256	0.06	1	0.02	0.03	2	16	NA	NA	103

NA, not available.
[a]Marketed compounds.
[b]Compounds highlighted in bold are top-selling drugs in their category.

The most important features of meropenem include enhanced stability toward DHP-I, good solubility allowing bolus injection, and good tolerability.[89]

Meropenem was followed by biapenem (see **Figure 24**) and doripenem, which present only minor improvements over meropenem against *P. aeruginosa*, and by ertapenem which is characterized by a narrower spectrum lacking *P. aeruginosa* but a long half-life, allowing once daily dosing.[90–93]

The prolonged half-life of ertapenem is due to the presence of a carboxylic acid group on the substituent in position C-2, which increases the binding to plasma proteins. With its good activity against ESBL-producing Enterobacteriaceae, ertapenem might be a successor to parenteral third-generation cephalosporins.

Efforts to identify compounds with improved activity against *P. aeruginosa*, while still maintaining the overall broad antibacterial spectrum, have so far been less successful. The most interesting compounds are E1010, lenapenem, CS-023, and J-114,870 (see **Figures 24** and **26**).[94–97] These compounds display a slightly increased activity compared with carbapenem-resistant *P. aeruginosa*, and are stable toward DHP-I. In addition, they have a serum half-life similar or

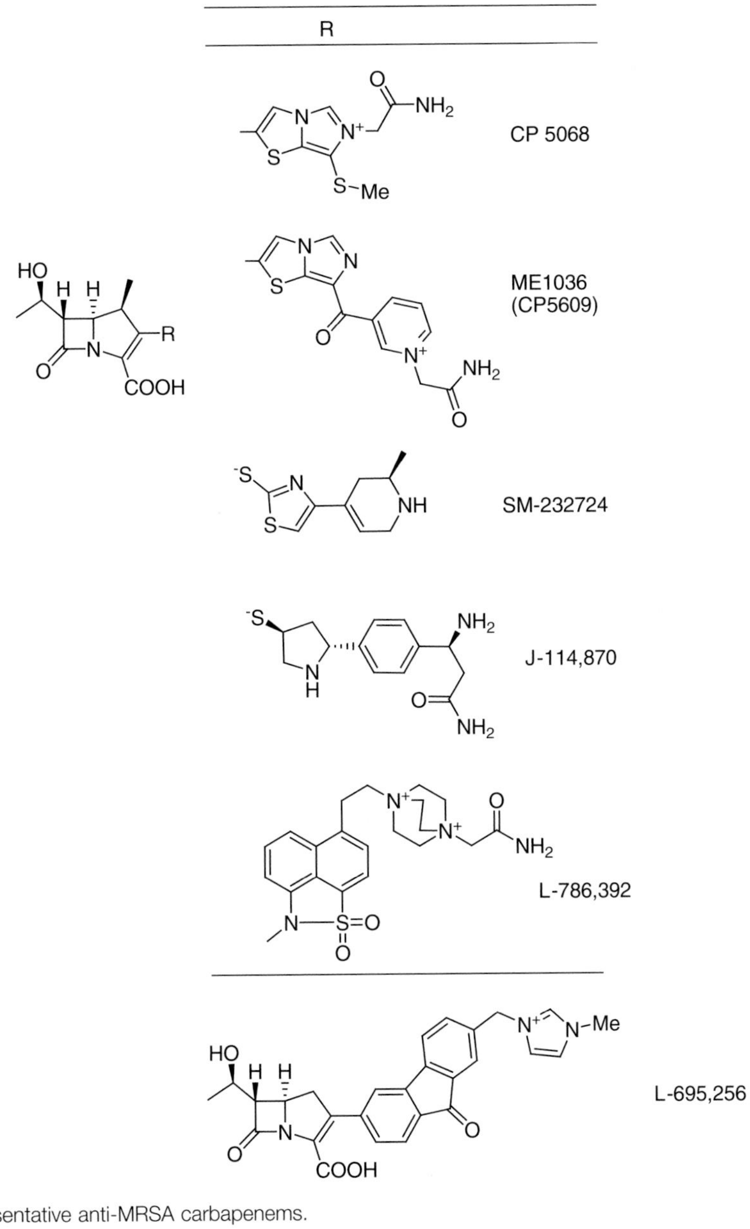

R	
	CP 5068
	ME1036 (CP5609)
	SM-232724
	J-114,870
	L-786,392

L-695,256

Figure 26 Representative anti-MRSA carbapenems.

slightly superior to that of meropenem. The main issues faced in the development of these compounds are the lack of true differentiation factors in comparison to marketed carbapenems and, for some of them, their prohibitive cost. None of these new carbapenems are stable toward hydrolysis by MBLs.

Carbapenems penetrate the periplasmic space of *P. aeruginosa* (and other Gram-negative bacteria) through the general porins. In contrast, imipenem also uses the specific porin OprD2 responsible for the uptake of small hydrophilic peptides. For this reason, carbapenems possessing relevant activity against *P. aeruginosa* are hydrophilic, zwitterionic, and rather small in size. The primary causes of resistance of *P. aeruginosa* are the loss of OprD2 porins (imipenem), the reduced permeability of the water channels (all carbapenems), and the presence of active efflux systems. The overall reduction of the periplasmic concentration of carbapenems, combined with the overproduction of various types of β-lactamases, contributes to the difficulty in inhibiting the growth of *P. aeruginosa*.[98]

7.17.4.5.2 Parenteral anti-methicillin-resistant *Staphylococcus aureus* carbapenems

Some of the broad-spectrum carbapenems displaying activity against *P. aeruginosa* (e.g., E1010, J-114,870, and CS-023; see **Figures 24** and **26**), show moderate activity against MRSA ($MIC_{90} = 4–12\,\mu g\,mL^{-1}$).

A significant level of activity against MRSA ($MIC_{90} < 2\,\mu g\,mL^{-1}$) was obtained with the narrow-spectrum carbapenems CP 5068, SM-232724, ME1036, L-695,256, and L-786,392.[99–104] The last two compounds were not developed further, due to toxicity, notably an immunogenic response associated with the large fluorenonyl group of L-695,256. Like anti-MRSA cephalosporins, the carbapenems active against MRSA harbor a large lipophilic residue containing a positively charged nitrogen atom in position C-2, 9–12 Å from the connecting point.

Among parenteral β-lactam antibiotics, carbapenems exhibit the broadest spectrum and the highest potency. Furthermore, their stability toward hydrolysis by class A and C β-lactamases, especially those from Enterobacteriaceae, gives them a competitive advantage over third-generation cephalosporins. In clinical practice, they are especially used in ICUs to treat bacterial septicemia, febrile neutropenia (neutropenic patients), endocarditis, and pediatric meningitis, as well as SSTIs, lower RTIs, and polymicrobial and intra-abdominal infections. The challenges faced by carbapenems are lack of MRSA coverage, insufficient activity against enterococci, and, for *P. aeruginosa*, incomplete coverage of the spectrum and, in some cases, development of resistance during therapy. Furthermore, carbapenems have selected for difficult-to-treat organisms such as *Burkholderia cepacia*, *Acetinobacter* spp., and *Stenotrophomonas maltophilia*. The identification of carbapenems with a spectrum expansion beyond meropenem/imipenem still remains elusive.

7.17.4.5.3 Oral carbapenems

Oral carbapenems have been developed, since there is a medical need for potent orally active antibiotics. As with oral third-generation cephalosporins, these agents will be used in step-down therapy to discharge hospital settings. They will provide an alternative to quinolones for the control of resistance development. Furthermore, the emergence of PRSP in RTIs and the spread of resistance in UTIs offer additional possible line extensions.

Despite their relatively small size, carbapenems have to be administered as prodrug esters. The most relevant compounds are tacapenem pivoxil (for which the drug is R-95867), DZ-2640 (for which the drug is DU-6681), tebipenem (for which the drug is ebipenem), sanfetrinem cilexetil (for which the drug is GV 104326), and OCA-983 (for which the drug is CL-191121; **Figure 27**).[105–112]

All these compounds display an antibacterial spectrum similar to imipenem, with the exception of *P. aeruginosa*, which is not inhibited (**Table 11**). In particular, they possess good activity against PRSP, fastidious Gram-negative bacteria, and some ESBL-producing Enterobacteriaceae not covered by third-generation cephalosporins. In contrast to imipenem, they are stable toward DHP-I.

Oral carbapenems face two challenges. On one hand, the medical community is reluctant to prescribe potent oral antibiotics in order to control resistance development. On the other hand, the limited stability and bioavailability of these drugs and their complex synthesis result in elevated treatment costs. Furthermore, the gastrointestinal side effects and the spread of resistance among Enterobacteriaceae are even more pronounced than for cephalosporins, and there is a potential for proconvulsive activity (related to binding to the $GABA_A$ receptors). Therefore, research in this class has been directed toward the parenteral hospital market.

The good stability of carbapenems toward β-lactamases-mediated hydrolysis was explained after the resolution of the x-ray structures of the complexes of imipenem and meropenem in class A and C β-lactamases. For class A β-lactamases, the 6-α-hydroxyethyl substituent prevents the attack of the hydrolytic water molecule through both steric and electrostatic factors. In both class A and C β-lactamases, the electrophilic carbonyl center of the acyl–enzyme complex is displaced from the point of hydrolytic attack, thus impairing deacylation.[113,114] **Figure 28** depicts the unproductive acyl–enzyme complex of imipenem in the AmpC β-lactamase (class C).

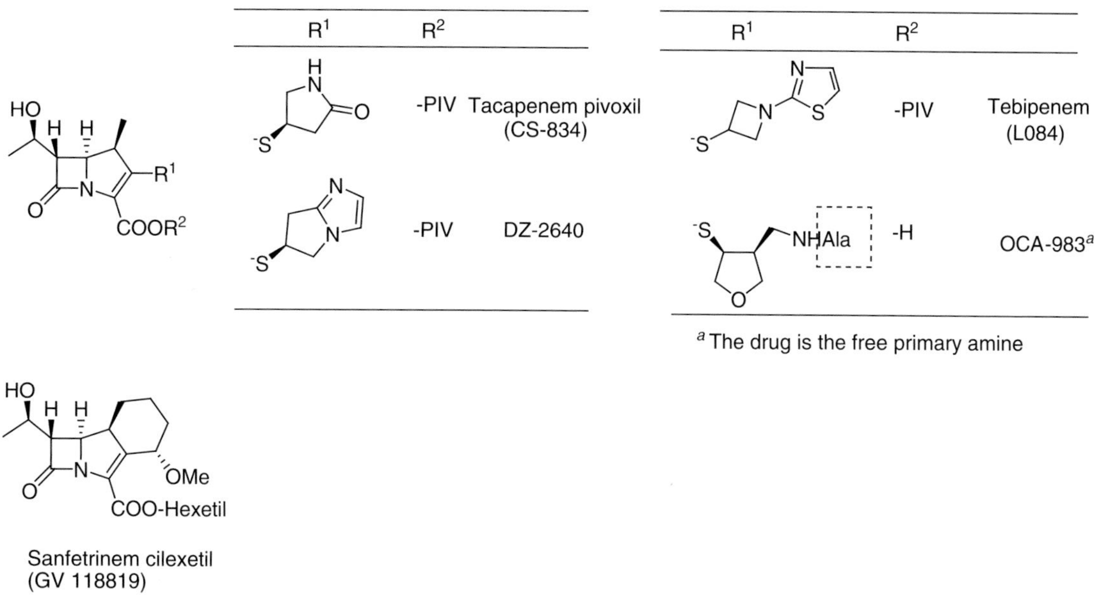

R[1]	R[2]	
(pyrrolidinone thio)	-PIV	Tacapenem pivoxil (CS-834)
(pyrrolo-imidazole thio)	-PIV	DZ-2640

R[1]	R[2]	
(azetidinyl-thiazoline thio)	-PIV	Tebipenem (L084)
(tetrahydrofuranyl thio, NHAla)	-H	OCA-983[a]

[a] The drug is the free primary amine

Sanfetrinem cilexetil
(GV 118819)

Figure 27 Representative oral carbapenems.

Table 11 MIC_{90} values of relevant oral carapenems against some key organisms

Cephalosporin	*MIC₉₀ (μg mL⁻¹)*					*Ref.*
	MSSA	*PRSP*	*M. catarrhalis*	*H. influenzae*	*E. coli*	
Tacapenem	0.2	0.39	0.1	0.39	0.02	105
DU-6681	0.1	0.2	0.1	0.2	0.02	109
Ebipenem	<0.06	0.12	<0.06	0.25	<0.06	107
CL-191,121	0.06	1	0.01	0.5	0.03	112
Sanfetrinem	0.1	0.78	0.05	0.78	0.78	109
Cefdinir[a]	1.56	12.5	0.39	0.78	25	109

[a] Reference compound.

7.17.4.6 Dual-Action Antibiotics

Most of the cephalosporins release their C-3 substituents after acylation of the active serine residue of both transpeptidases and β-lactamases. This feature was used to design dual-action antibiotics (e.g., Ro 23-9424; **Figure 29**), wherein the C-3 substituent contains a second antibiotic such as a quinolone.[115,116] Thus, after acylation, the released quinolone should further penetrate into the cytoplasm, where it reaches its targets (gyrase and topoisomerase IV).

This dual mode of action should have resulted in both an extension of the antibacterial spectrum and overcoming β-lactam antibiotic resistance. However, this approach suffered from three main drawbacks: (1) the dual-action antibiotics, with their large C-3 substituents, were not well recognized by the transpeptidases; (2) the linkage between the two antibiotics was not very stable, leading to the release of the quinolone moiety before the drug reached the pathogens; and (3) a difficult synthesis as well as high cost, since quinolones were not generic compounds. Therefore, the advantages of these drugs over quinolones were limited. Attaching the quinolone via urethane, tertiary amine, or quaternary amine linkages produced compounds with greater stability than those having an ester linkage.[117] Replacement of the cephalosporin by penem or carbapenem increased the potency of these dual-action antibiotics.[118] The concept was further extended, and oxazolidinones were used instead of quinolones.[119]

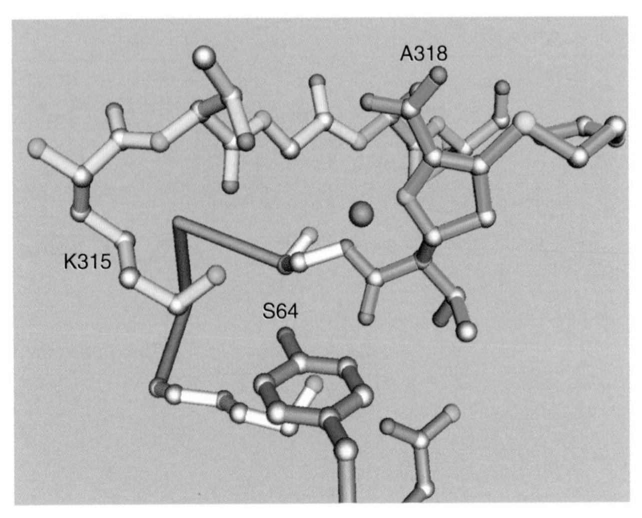

Figure 28 Representation of the acyl–enzyme complex of AmpC and imipenem (PDB code 1LL5) showing that the electrophilic acyl center has rotated away by 180° from the point of hydrolytic attack, marked by a red ball. Conserved residues: KT(S)G on the β strand (yellow), SXXK (white), and YXN (magenta). Imipenem is colored green.

Ro 23-9424

Figure 29 A representative dual-action antibiotic.

7.17.4.7 Non-β-lactam Antibiotics

Many attempts have been made to replace the β-lactam ring by other ring systems. The most interesting results were obtained with bicyclic pyrazolidinone and γ-lactam derivatives (e.g., LY 186826 and compound **8**; **Figure 30**).[120–122]

Using a rational design approach, Lampilas and co-workers synthesized the bridged lactam **9**, which displayed a narrow antibacterial spectrum.[123] The weak antibacterial activity observed with these derivatives could be explained by poor chemical stability, unfavorable steric interactions within the transpeptidase active site, or insufficient activation of the carbonyl function.

It is also worth mentioning the natural product lactivicin, which is a transpeptidase inhibitor. However, its narrow spectrum and its relative chemical instability precluded its use as a therapeutic agent.[124]

7.17.4.8 β-Lactamase Inhibitors

The most common resistance mechanism of bacteria is their ability to hydrolyze β-lactam antibiotics by several types of β-lactamases. An effective counter-measure is to employ a combination product, consisting of a β-lactam antibiotic and a β-lactamase inhibitor. Currently available inhibitors (clavulanic acid, sulbactam, and tazobactam; **Figure 31**), narrowly target class A β-lactamases.

Research was undertaken to understand the molecular basis of the interactions of β-lactam antibiotics and β-lactamase inhibitors with β-lactamases, with the aim of extending the spectrum of inhibition to ESBLs, class C β-lactamases, and class B carbapenemases.

Following the resolution of the structures of various β-lactamases both in native form or complexed with inhibitors, structure- and mechanism-based approaches have been used with success to design new β-lactamase inhibitors. Sandanayaka and Prashad have reviewed this research.[125]

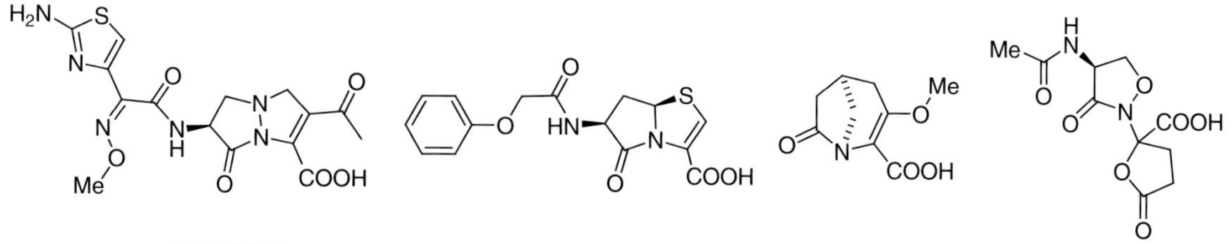

LY186826 8 9 Lactivicin

Figure 30 Structures of non-β-lactam antibiotics.

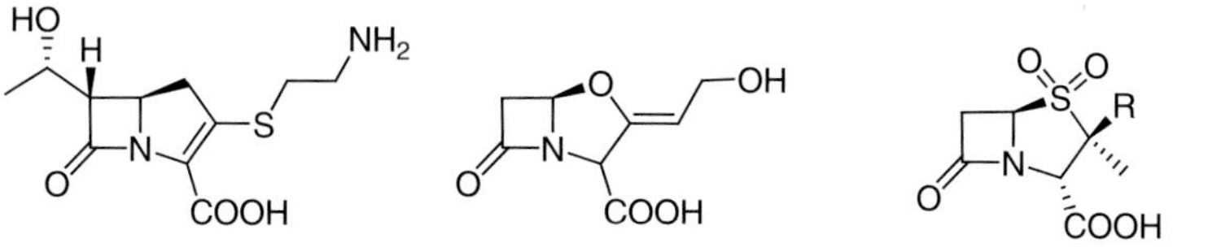

Epithienamycin Clavulanic acid

R = Me- (Sulbactam)
R = 1,2,3-triazol-1-ylmethyl- (Tazobactam)
R = CH=CHCN(*Z*) (Ro 48-1220)

Figure 31 Representatives of different classes of β-lactamase inhibitors.

7.17.4.8.1 Bridged β-Lactams

The information derived from the structure of aztreonam complexed with the class C β-lactamase from *Citrobacter freundii* led to the elucidation of the mechanism of hydrolysis of penicillins and cephalosporins by this type of enzyme. In particular, it was deduced that, at the acyl–enzyme complex stage, prevention of rotation around the C-3–C-4 bond of aztreonam obstructs attack by the hydrolytic water molecule (see **Figure 5c**). Indeed, bridged monobactams (e.g., Ro 47-1317; **Figure 32**) and their corresponding bridged cephems and isooxacephems, which do not allow such a rotation, proved to be potent inhibitors of class C β-lactamases, and showed synergy with third-generation cephalosporins against Enterobacteriaceae.[126]

In addition, some bridged isooxa- and carba-cephems displayed broad-spectrum antibacterial activity, including MSSA and Enterobacteriaceae.[127–129]

7.17.4.8.2 6-Methylidene penams, 6-methylidene penems, and 7-methylidene cephems

Following the early discovery of 6-methylidene penams, which displayed good class A β-lactamase inhibition, the concept was further extended to penam sulfones, penems, and cephem sulfones (e.g., compound **10**; **Figure 32**).[130–133] These compounds inhibit both class A and C β-lactamases. Among them, the penem derivatives (e.g., BRL-42715 and Way-185229) are the most potent β-lactamase inhibitors reported so far, displaying low nanomolar inhibition against both class A and C β-lactamases.[134,135] A hallmark in all those different types of inhibitors is a succession of rearrangements leading to expanded molecules that permanently inactivate the enzyme by forming a hydrolytically more stable acyl–enzyme intermediate (**Figure 33**). Buynak[136] has reviewed the field.

7.17.4.8.3 6-Hydroxyalkyl penams and penam sulfones

The 6α-hydroxyethyl substituent in carbapenem antibiotics is responsible for their good stability toward class A β-lactamases by preventing the attack of the hydrolytic water molecule through both steric and electrostatic factors. The same concept was applied to the penam series, leading to 6α-hydroxyalkyl penam derivatives (e.g., compound **11**; see **Figure 32**), which were found to be potent inhibitors of non-metallo-carbapenemases of class A.[137]

Penam and penam sulfones having a 6β-hydroxyalkyl group in position C-6 (e.g., CP-72,436 and compound **12**) inhibit both class A and C β-lactamases.[138] Lin and co-workers claim that the class C β-lactamase inhibition depends on the steric bulk of the C-6 substituent, which impairs attack by the hydrolytic water molecule.[139] CP-72,436 displays, in addition, antibacterial activity against Gram-positive bacteria.[140]

The 6β-mercaptomethyl penam sulfone derivative **13** represents the first example of an inhibitor able to inhibit metallo β-lactamases by complexing the catalytic zinc atoms of the metallo enzymes as well as serine β-lactamases.[141]

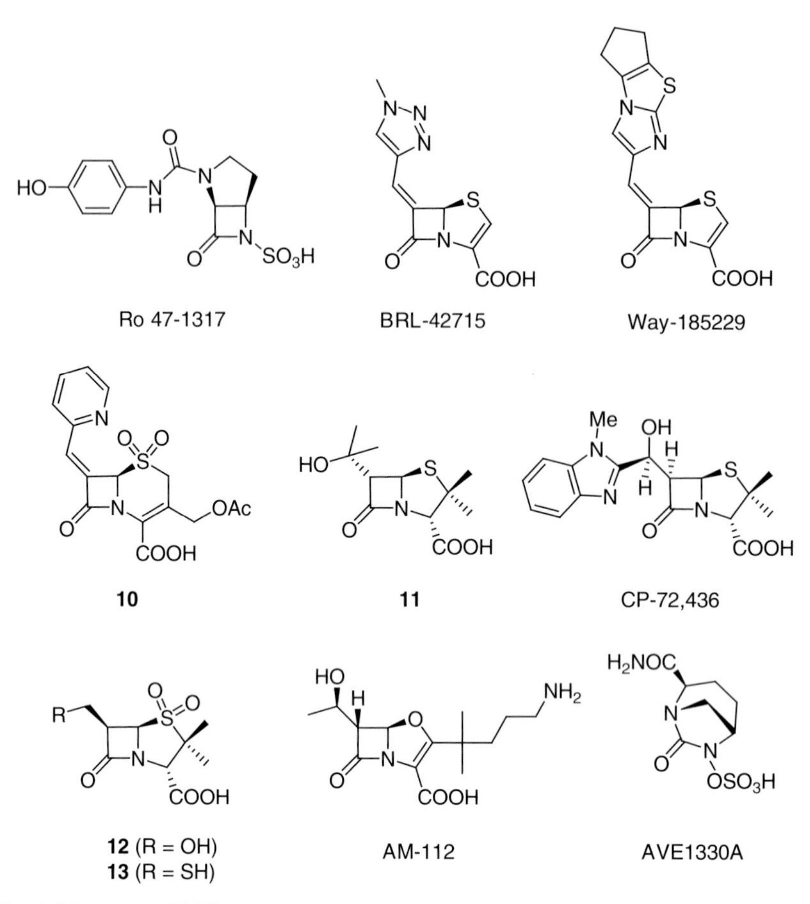

Figure 32 Additional β-lactamase inhibitors.

7.17.4.8.4 2β′-Penam sulfones

The mode of action of these class A β-lactamase inhibitors, illustrated by sulbactam and tazobactam (see **Figure 31**), has been investigated in detail.[142] Following the collapse of the initial transition state arising from the attack of the active serine residue of the class A β-lactamase, the resulting acyl–enzyme intermediate evolves further, leading to an iminium acyl enzyme intermediate. That intermediate undergoes further transformations; for example, transamination with a lysine residue from the enzyme or isomerization into a more stable enaminone derivative (**Figure 33**). The mode of action of inhibitors derived from penams, cephems, oxapenems (e.g., clavulanic acid), and penems follows the same mechanism involving a rearrangement of the central iminium acyl–enzyme intermediate.

Penam sulfones were further studied, leading to Ro 48-1220 (see **Figure 31**).[143] In contrast to tazobactam, this compound displays broad-spectrum β-lactamase inhibition (including both class A and C β-lactamases). The product of the inhibition is a deactivated rearranged acyl–enzyme complex, which occupies a different position in comparison with that of the original putative binding mode. As with most of the penam sulfones, it displays interesting potent activity against *Acinetobacter* species.

7.17.4.8.5 Clavulanic acid and oxapenems

Augmentin (amoxicillin/clavulanic acid) is the best illustration of the commercial success achievable by developing a combination of a β-lactam antibiotic with a class A β-lactamase inhibitor.[144] As with penam sulfones, the inhibition proceeds through different transient intermediates, leading to the formation of a hydrolytically more stable enaminone (**Figure 33**). In contrast to clavulanic acid, oxapenems (e.g., AM-112; see **Figure 32**) display also a significant level of inhibition against class C and D β-lactamases. These compounds, however, are chemically quite reactive.[145]

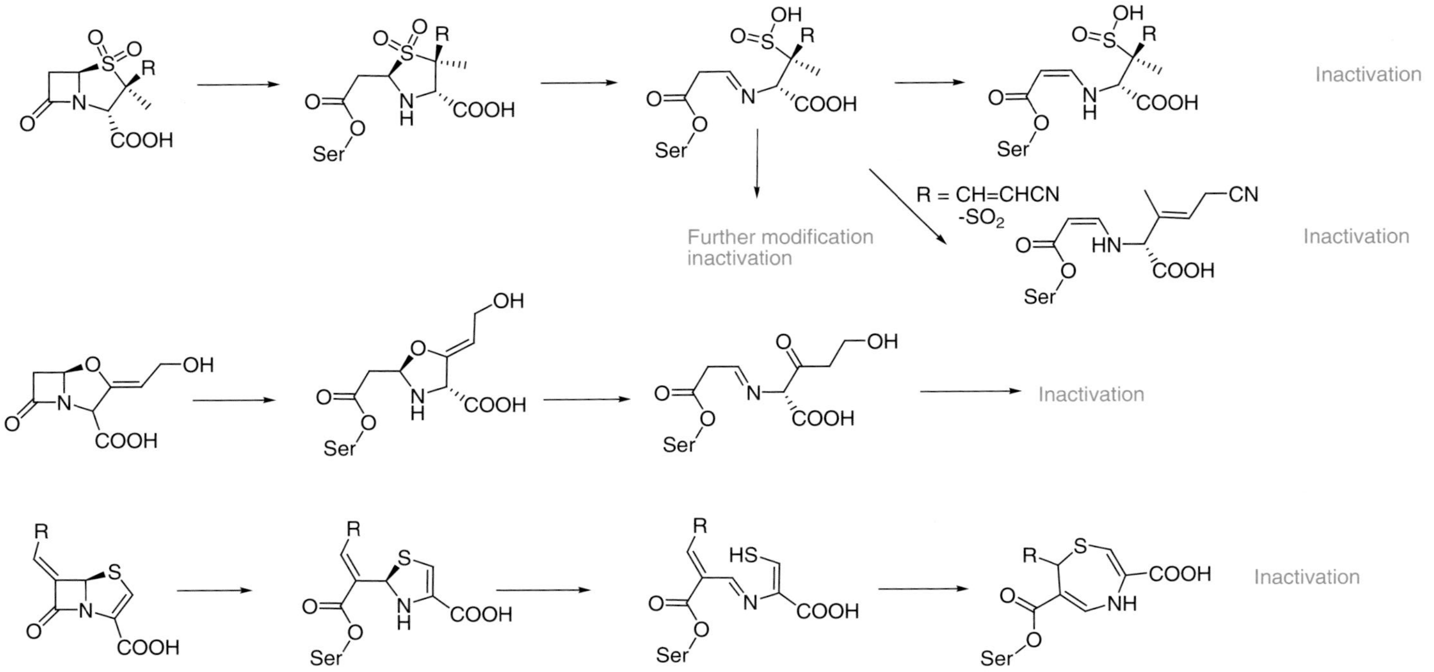

Figure 33 Mechanism of β-lactamase inactivation by various inhibitors.

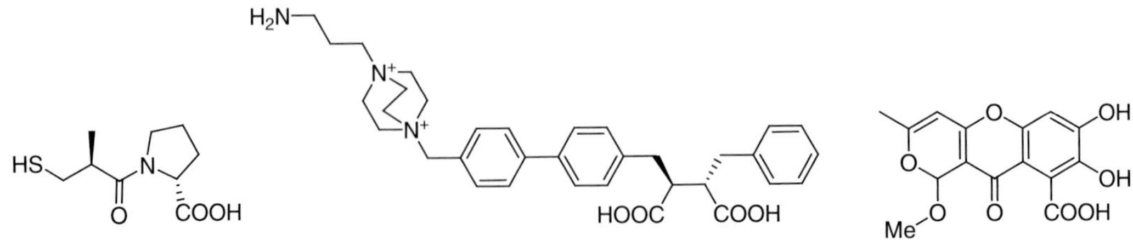

D-Captopril 14 SB238569

Figure 34 Representative MBL inhibitors.

7.17.4.8.6 β-Lactamase inhibitors devoid of a β-lactam ring

Many activated ketone-, boronic acid-, phosphonate-, and phosphonamidate-based transition state analogs have been synthesized as serine β-lactamase inhibitors. Some of them showed potent inhibition, but none have yet made their way to a clinical trial.

An interesting approach was followed at Aventis, leading to the identification of the anti-Bredt bridged lactam AVE1330A (see **Figure 32**), which inhibits both class A and C β-lactamases. In combination with ceftazidime, AVE1330A displays a good coverage of Enterobacteriaceae.[146]

The inhibition of class B β-lactamases was addressed by designing compounds able to coordinate the zinc atom in the active site.[147,148] A series of derivatives containing a thiol or a succinic group has been synthesized (**Figure 34**).

The thiol group of D-captopril and the carboxylic group at the carbon bearing the benzyl residue in compound **14** both displace the catalytic hydroxyl anion in the active site and coordinate two zinc atoms.[149,150]

Screening of natural product extracts against the *Bacillus cereus* II enzyme at GlaxoSmithKline resulted in the identification of phenazines (e.g., SB238569) as MBL inhibitors.[151]

All these inhibitors exhibit good antibacterial synergy with carbapenems against strains producing MBLs.

7.17.5 Conclusions

β-Lactam chemistry will remain a hallmark of the power of medicinal chemistry to improve lead structures of natural origin. Starting from leads, which themselves could not be used in medical practice, it was possible to obtain compounds that were more stable and presented an adequate antibacterial spectrum and a good pharmacokinetic profile. The impressive armamentarium of safe penicillins, cephalosporins, and carbapenems at the disposal of the medical community allows the successful treatment of most infections encountered in the community and in the hospital. However, their success is jeopardized by β-lactamase-mediated inactivation and the continuous emergence of mutated transpeptidases. The major issue addressed in the last 15 years has been the development of resistance. Progress in understanding the mode of action of both transpeptidases and β-lactamases has given new insights for the design of new molecules. However, it will be difficult to enhance further the reactivity of the β-lactam ring without compromising the hydrolytic stability. Thus, the future of this field of chemistry lies in the discovery of new β-lactamase inhibitors that will restore most of the antibacterial activity (with the exception of MRSA and PRSP). Although most of the leading β-lactam antibiotics are coming off-patent, their market share in the next decade will remain over 30% of the pharmaceutical market. With the expected fragmentation of the market, a bright future might also belong to molecules having a tailored narrow antibacterial spectrum.

The use of β-lactams has been extended beyond the scope of the treatment of bacterial infections. They have been used for the selective delivery of oncolytic drugs (using the concept described in Section 7.17.4.6 on dual-action antibiotics), and, in the central nervous system area, cephalosporin-mediated neuroprotection has recently been described.[152–154]

References

1. Sheehan, J. C.; Henry-Logan, K. R. *J. Am. Chem. Soc.* **1962**, *84*, 2983–2990.
2. Brotzu, G. *Lavori Inst. Igiene Cagliari* **1948**, 1–11.
3. Woodward, R. B.; Heusler, K.; Gosteli, J.; Naegeli, P.; Ramage, R.; Ranganathan, S.; Vorbrueggen, H. *J. Am. Chem. Soc.* **1966**, *88*, 852–853.
4. Kahan, J. S.; Kahan, F. M.; Goegelman, R.; Currie, S. A.; Jackson, M.; Stapley, E. O.; Miller, T. W.; Miller, A. K.; Hendlin, D.; Mochales, S. et al. *J. Antibiot.* **1979**, *32*, 1–12.

5. Aoki, H.; Sakai, H.; Kohsaka, Ma.; Konomi, T.; Hosoda, J.; Kubochi, Y.; Iguchi, E.; Imanaka, H. *Antibiot. J.* **1976**, *29*, 492–500.
6. Imada, A.; Kitano, K.; Kintaka, K.; Muroi, M.; Asai, M. *Nature* **1981**, *289*, 590–591.
7. Sykes, R. B.; Cimarusti, C. M.; Bonner, D. P.; Bush, K.; Floyd, D. M.; Georgopapadakou, N. H.; Koster, W. H.; Liu, W. C.; Parker, W. L. *Nature* **1981**, *291*, 489–491.
8. Newall, C.; Hallam, P. β-Lactam Antibiotics: Penicillins and Cephalosporins. In *Comprehensive Medicinal Chemistry*; Hansch, C., Sammes, P. G., Taylor, J. B., Eds.; Pergamon Press: Oxford, UK, 1990; Vol. 5, pp 609–653.
9. Brown, A. G.; Pearson, M. J.; Southgate, R. Other β-Lactam Agents. In *Comprehensive Medicinal Chemistry*; Hansch, C., Sammes, P. G., Taylor, J. B., Eds.; Pergamon Press: Oxford, UK, 1990; Vol. 5, pp 655–703.
10. Page, M. I. *The Chemistry of β-Lactam*; Blackie Academic & Professional: New York, 1992.
11. Anon. Stakeholder insight: the hospital antibacterial market-specialist products drive market growth. *Datamonitor*, 2003.
12. Sweet, R. M. Chemical Biology. In *Cephalosporins and Penicillins*; Flynn, E. H., Ed.; Academic Press: New York, 1972, pp 280–309.
13. Cohen, N. C. *J. Med. Chem.* **1983**, *26*, 259–264.
14. Bush, K.; Jacoby, G. A.; Medeiros, A. A. *Antimicrob. Agents Chemother.* **1995**, *39*, 1211–1233.
15. Ambler, R. P.; Coulson, A. F. W.; Frère, J. M.; Ghuysen, J. M.; Joris, B.; Forsman, M.; Levesque, R. C.; Tiraby, G.; Waley, S. G. *Biochem. J.* **1991**, *276*, 269–272.
16. Jacoby, G. A.; Munoz-Price, L. S. N. *Engl. J. Med.* **2005**, *352*, 380–391.
17. Nordmann, P.; Ronco, E.; Naas, T.; Duport, C.; Michel-Briand, Y.; Labia, R. *Antimicrob. Agents Chemother.* **1993**, *37*, 962–969.
18. Livermore, D. M.; Williams, J. D. β-Lactams: Mode of Action and Mechanisms of Bacterial Resistance. In *Antibiotics in Laboratory Medicine*, 4th ed.; Lorian, V., Ed.; Williams & Wilkins: Baltimore, MD, 1996, pp 502–578.
19. Philippon, A.; Dusart, J.; Joris, B.; Frere, J. M. *Cell Mol. Life Sci.* **1998**, *54*, 341–346.
20. Oefner, C.; D'Arcy, A.; Daly, J. J.; Gubernator, K.; Charnas, R. L.; Heinze, I.; Hubschwerlen, C.; Winkler, F. K. *Nature* **1990**, *343*, 284–288.
21. Nukaga, M.; Mayama, K.; Hujer, A. M.; Bonomo, R. A.; Knox, J. R. *J. Mol. Biol.* **2003**, *328*, 289–301.
22. Fisher, J. F.; Meroueh, S. O.; Mobashery, S. *Chem. Rev.* **2005**, *105*, 395–424.
23. Laible, G.; Spratt, B. G.; Hakenbeck, R. *Mol. Microbiol.* **1991**, *8*, 1993–2002.
24. Mazzariol, A.; Cornaglia, G.; Nikaido, H. *Antimicrob. Agents Chemother.* **2000**, *44*, 1387–1390.
25. Bush, K. β-Lactam Antibiotics: Penicillins. In *Antibiotic and Chemotherapy. Anti-infectives Agents and Their Use in Therapy,*, 8th ed.; Finch, R., Greenwood, D., Norrby, S. R., Whitley, R. J., Eds.; Churchill Livingstone: Edinburgh, UK, 2003, pp 224–258.
26. Tybring, L. *Antimicrob. Agents Chemother.* **1975**, *8*, 266–270.
27. Nathwani, D.; Wood, M. *J. Drugs* **1993**, *45*, 866–894.
28. Matsumura, N.; Minami, S.; Mitsuhashi, S. *J. Antimicrob. Chemother.* **1997**, *39*, 31–34.
29. Ochiai, H.; Yasuo, M.; Fukuda, H.; Yoshino, O.; Minami, S.; Hayashi, T.; Momonoi, K. Bactericidal Penam Derivatives. German Patent 4019960, January 3, 1991.
30. Tiddens, H. *Pediatr. Pulmonol. Suppl.* **2004**, *26*, 92–94.
31. Edlund, C. *Curr. Opin. Anti-Infect. Investig. Drugs* **1999**, *1*, 96–100.
32. Floyd, D. M.; Fritz, A. W.; Cimarusti, C. M. *J. Org. Chem.* **1982**, *47*, 176–178.
33. Hubschwerlen, C.; Schmid, G. *Helv. Chim. Acta.* **1983**, *66*, 2206–2209.
34. Gill, A. E.; Taylor, A.; Lewendon, A.; Unemi, N.; Uji, T.; Nishida, K.; Salama, S.; Singh, R.; Micetich, R. G. 43rd Interscience Conference on Antimicrobial Agents and Chemotherapy, Chicago, IL, 2003; Poster F552, Book of Abstracts, p 236.
35. Bryskier, A. *J. Antibiot.* **2000**, *53*, 1028–1037.
36. Neu, H. C. *Am. J. Med.* **1985**, *79*, 2–13.
37. Durckheimer, W.; Adam, F.; Fischer, G.; Kirrstetter, R. *Adv. Drug Res.* **1988**, *17*, 61–234.
38. Neu, H. C. *Rev. Infect. Dis.* **1983**, *5*, 319–336.
39. Jacoby, G. A.; Medeiros, A. A. *Antimicrob. Agents Chemother.* **1991**, *35*, 1697–1704.
40. Critchley, I. A. *J. Antimicrob. Chemother.* **1990**, *26*, 733–735.
41. Curtis, N. A.; Eisenstadt, R. L.; East, S. J.; Cornford, R. J.; Walker, L. A.; White, A. *J. Antimicrob. Agents Chemother.* **1988**, *32*, 1879–1886.
42. Chin, N. X.; Gu, J. W.; Fang, W.; Neu, H. C. *Antimicrob. Agents Chemother.* **1991**, *35*, 259–266.
43. Erwin, M. E.; Varnam, D.; Jones, R. N. *Diagn. Microbiol. Infect. Dis.* **1997**, *28*, 93–100.
44. Maejima, T.; Inoue, M.; Mitsuhashi, S. *Antimicrob. Agents Chemother.* **1991**, *35*, 104–110.
45. Arisawa, M.; Sekine, Y.; Shimizu, S.; Takano, H.; Angehrn, P.; Then, R. *Antimicrob. Agents Chemother.* **1991**, *35*, 653–659.
46. Fu, K. P.; Foleno, B. D.; Lafredo, S. C.; Lococo, J. M.; Isaacson, D. M. *Antimicrob. Agents Chemother.* **1993**, *37*, 301–307.
47. Toyosawa, T.; Miyazaki, S.; Tsuji, A.; Yamaguchi, K.; Goto, S. *Antimicrob. Agents Chemother.* **1993**, *1*, 60–66.
48. Klein, O.; Chin, N. X.; Huang, H. B.; Neu, H. C. *Antimicrob. Agents Chemother.* **1994**, *38*, 2896–2901.
49. Ida, T.; Tsushima, M.; Ishii, T.; Atsumi, K.; Tamura, A. *J. Infect. Chemother.* **2002**, *8*, 138–144.
50. Watanabe, N.; Sugiyama, I. *J. Antibiot.* **1992**, *45*, 1526–1532.
51. Hiraoka, M.; Inoue, M.; Mitsuhashi, S. *Rev. Infect. Dis.* **1988**, *10*, 746–751.
52. Fung-Tomc, J. C.; Clark, J.; Minassian, B.; Pucci, M.; Tsai, Y. H.; Gradelski, E.; Lamb, L.; Medina, I.; Huczko, E.; Kolek, B. et al. *Antimicrob. Agents Chemother.* **2002**, *46*, 971–976.
53. Glinka, T. W.; Cho, A.; Zhang, Z. J.; Ludwikow, M.; Griffith, D.; Huie, K.; Hecker, S. J.; Dudley, M. N.; Lee, V. J.; Chamberland, S. *J. Antibiot.* **2000**, *53*, 1045–1052.
54. Tsushima, M.; Iwamatsu, K.; Tamura, A.; Shibahara, S. *Bioorg. Med. Chem.* **1998**, *6*, 1009–1017.
55. Hebeisen, P.; Heinze-Krauss, I.; Angehrn, P.; Hohl, P.; Page, M. G.; Then, R. *Antimicrob. Agents Chemother.* **2001**, *45*, 825–836.
56. Jones, R. N.; Deshpande, L. M.; Mutnick, A. H.; Biedenbach, D. J. *J. Antimicrob. Chemother.* **2002**, *50*, 915–932.
57. Huang, V.; Brown, W. J.; Rybak, M. *J. Antimicrob. Agents Chemother.* **2004**, *48*, 2719–2723.
58. Iizawa, Y.; Nagai, J.; Ishikawa, T.; Hashiguchi, S.; Nakao, M.; Miyake, A.; Okonogi, K. *J. Infect. Chemother.* **2004**, *10*, 146–156.
59. Sader, H. S.; Johnson, D. M.; Jones, R. N. *Antimicrob. Agents Chemother.* **2004**, *48*, 53–62.
60. Hanaki, H.; Akagi, H.; Masaru, Y.; Otani, T.; Hyodo, A.; Hiramatsu, K. *Antimicrob. Agents Chemother.* **1995**, *39*, 1120–1126.
61. Chamberland, S.; Blais, J.; Hoang, M.; Dinh, C.; Cotter, D.; Bond, E.; Gannon, C.; Park, C.; Malouin, F.; Dudley, M. N. *Antimicrob. Agents Chemother.* **2001**, *45*, 1422–1430.
62. Fujimura, T.; Yamano, Y.; Yoshida, I.; Shimada, J.; Kuwahara, S. *Antimicrob. Agents Chemother.* **2003**, *47*, 923–931.
63. Glinka, T.; Huie, K.; Cho, A.; Ludwikow, M.; Blais, J.; Griffith, D.; Hecker, S.; Dudley, M. *Bioorg. Med. Chem.* **2003**, *11*, 591–600.

64. Heinze-Krauss, I.; Angehrn, P.; Guerry, P.; Hebeisen, P.; Hubschwerlen, C.; Kompis, I.; Page, M. G.; Richter, H. G.; Runtz, V.; Stalder, H. et al. *J. Med. Chem.* **1996**, *39*, 1864–1871.
65. Matsumoto, M.; Tamaoka, H.; Ishikawa, H.; Kikuchi, M. *Antimicrob. Agents Chemother.* **1998**, *42*, 2943–2949.
66. Page, M.; Bur, D.; Danel, F. 38th Interscience Conference on Antimicrobial Agents and Chemotherapy, San Diego, CA, 1998; Poster F022, Book of Abstracts, p 238.
67. Lim, D.; Strynadka, N. C. *Nat. Struct. Biol.* **2002**, 870–879.
68. Heep, M.; Querner, S.; Harsch, M.; O'Riordan, W. 44th Interscience Conference on Antimicrobial Agents and Chemotherapy, Washington, DC, 2004; Poster L361, Book of Abstracts, p. 380.
69. Tsuji, M.; Ishii, Y.; Ohno, A.; Miyazaki, S.; Yamaguchi, K. *Antimicrob. Agents Chemother.* **1995**, *39*, 2544–2551.
70. Watanabe, Y.; Hatano, K.; Matsumoto, Y.; Tawara, S.; Yamamoto, H.; Kawabata, K.; Takasugi, H.; Matsumoto, F.; Kuwahara, S. *J. Antibiot.* **1999**, *52*, 649–659.
71. Muranushi, N.; Hashimoto, N.; Hirano, K. *Pharm. Res.* **1995**, *12*, 1488–1492.
72. Bryson, H. M.; Brogden, R. N. *Drugs* **1993**, *45*, 589–621.
73. Wellington, K.; Curran, M. P. *Drugs* **2004**, *64*, 2597–2618.
74. Liu, Y. C.; Huang, W. K.; Cheng, D. L. *Chemotherapy* **1997**, *43*, 21–26.
75. Bauernfeind, A.; Jungwirth, R. *Infection* **1991**, *19*, 352–362.
76. Mori, N.; Kodama, T.; Sakai, A.; Suzuki, T.; Sugihara, T.; Yamaguchi, S.; Nishijima, T.; Aoki, A.; Toriya, M.; Kasai, M. et al. *Int. J. Antimicrob. Agents.* **2001**, *18*, 451–461.
77. Mizen, L.; Burton, G. *Pharm. Biotechnol.* **1998**, *11*, 345–365.
78. Hubschwerlem, C.; Charnas, R.; Angehrn, P.; Furlenmeier, A.; Graser, T.; Montavon, M. *J. Antibiot.* **1992**, *45*, 1358–1364.
79. Edlund, C.; Nord, C. E. *J. Antimicrob. Chemother.* **2000**, *46*, 41–48.
80. Woodward, R. B. *Philos. Trans. R. Soc. Lond. B Biol. Sci.* **1980**, *289*, 239–250.
81. Arcamone, F. M.; Altamura, M.; Perrotta, E.; Crea, A.; Manzini, S.; Poma, D.; Salimbeni, A.; Triolo, A.; Maggi, C. A. *J. Antibiot.* **2000**, *53*, 1086–1095.
82. Hoban, D. J. *J. Antimicrob. Chemother.* **1989**, 53–57.
83. Bryskier, A. *Expert Opin. Investig. Drugs.* **1995**, *4*, 705–724.
84. Pfaendler, H. R.; Hendel, W. Z. *Naturforsch. B: Chem. Sci.* **1992**, *47*, 1037–1049.
85. Wild, H.; Metzger, K. G. *Bioorg. Med. Chem. Lett.* **1993**, *3*, 2205–2210.
86. Jamieson, C. E.; Lambert, P. A.; Simpson, I. N. *Antimicrob. Agents Chemother.* **2003**, *47*, 2615–2618.
87. Goa, K. L.; Noble, S. *Drugs* **2003**, *63*, 913–925.
88. Neu, H. C.; Chin, N. X.; Saha, G.; Labthavikul, P. *Antimicrob. Agents Chemother.* **1986**, *30*, 828–834.
89. Wiseman, L. R.; Wagstaff, A. J.; Brogden, R. N.; Bryson, H. M. *Drugs* **1995**, *50*, 73–101.
90. Ge, Y.; Wikler, M. A.; Sahm, D. F.; Blosser-Middleton, R. S.; Karlowsky, J. A. *Antimicrob. Agents Chemother.* **2004**, *48*, 1384–1396.
91. Perry, C. M.; Ibbotson, T. *Drugs* **2002**, *62*, 2221–2234.
92. Cunha, B. A. *Drugs Today* **2002**, *38*, 195–213.
93. Jones, R. N.; Huynh, H. K.; Biedenbach, D. *J. Antimicrob. Agents Chemother.* **2004**, *48*, 3136–3140.
94. Thomson, K. S.; Moland, E. S. *J. Antimicrob. Chemother.* **2004**, *54*, 557–562.
95. Fuchs, P. C.; Barry, A. L.; Brown, S. D. *J. Antimicrob. Chemother.* **2001**, *48*, 23–28.
96. Nagano, R.; Shibata, K.; Adachi, Y.; Imamura, H.; Hashizume, T.; Morishima, H. *Antimicrob. Agents Chemother.* **2000**, *44*, 489–495.
97. Ohba, F.; Nakamura-Kamijo, M.; Watanabe, N.; Katsu, K. *Antimicrob. Agents Chemother.* **1997**, *41*, 298–307.
98. Jacoby, G.; Mills, D. M.; Chow, N. *Antimicrob. Agents Chemother.* **2004**, *48*, 3203–3206.
99. Sunagawa, M.; Itoh, M.; Kubota, K.; Sasaki, A.; Ueda, Y.; Angehrn, P.; Bourson, A.; Goetschi, E.; Hebeisen, P.; Then, R. *J. Antibiot.* **2002**, *55*, 722–757.
100. Ueda, Y.; Sunagawa, M. *Antimicrob. Agents Chemother.* **2003**, *47*, 2471–2480.
101. Shitara, E.; Yamamoto, Y.; Kano, Y.; Maruyama, T.; Kitagasawa, H.; Sasaki, T.; Aihara, K.; Atsumi, K.; Yamada, K.; Tohyama, K. et al. 40th Interscience Conference on Antimicrobial Agents and Chemotherapy, Toronto, Ontario, Canada, 2000; Poster F1236, Book of Abstracts, p 206.
102. Kurazono, M.; Ida, T.; Yamada, K.; Hirai, Y.; Maruyama, T.; Shitara, E.; Yonezawa, M. *Antimicrob. Agents Chemother.* **2004**, *48*, 2831–2837.
103. Rylander, M.; Rollof, J.; Jacobsson, K.; Norrby, S. R. *Antimicrob. Agents Chemother.* **1995**, *39*, 1178–1181.
104. Ratcliffe, R. W.; Wilkening, R. R.; Wildonger, K. J.; Waddell, S. T.; Santorelli, G. M.; Parker, D. L., Jr.; Morgan, J. D.; Blizzard, T. A.; Hammond, M. L.; Heck, J. V. et al. *Bioorg. Med. Chem. Lett.* **1999**, *9*, 679–684.
105. Fukuoka, T.; Ohya, S.; Utsui, Y.; Domon, H.; Takenouchi, T.; Koga, T.; Masuda, N.; Kawada, H.; Kakuta, M.; Kubota, M. et al. *Antimicrob. Agents Chemother.* **1997**, *41*, 2652–2663.
106. Yamaguchi, K.; Domon, H.; Miyazaki, S.; Tateda, K.; Ohno, A.; Ishii, K.; Matsumoto, T.; Furuya, N. *Antimicrob. Agents Chemother.* **1998**, *42*, 555–563.
107. Miyazaki, S.; Hosoyama, T.; Furuya, N.; Ishii, Y.; Matsumoto, T.; Ohno, A.; Tateda, K.; Yamaguchi, K. *Antimicrob. Agents Chemother.* **2001**, *45*, 203–207.
108. Okuda, J.; Otsuki, M.; Oh, T.; Nishino, T. *J. Antimicrob. Chemother.* **2000**, *46*, 101–108.
109. Tanaka, M.; Hohmura, M.; Nishi, T.; Sato, K.; Hayakawa, I. *Antimicrob. Agents Chemother.* **1997**, *41*, 1260–1268.
110. Di Modugno, E.; Erbetti, I.; Ferrari, L.; Galassi, G.; Hammond, S. M.; Xerri, L. *Antimicrob. Agents Chemother.* **1994**, *38*, 2362–2368.
111. Jiraskova, N. *Curr. Opin. Investig. Drugs* **2001**, *2*, 1035–1038.
112. Weiss, W. J.; Petersen, P. J.; Jacobus, N. V.; Lin, Y. I.; Bitha, P.; Testa, R. T. *Antimicrob. Agents Chemother.* **1999**, *43*, 454–459.
113. Maveyraud, L.; Mourey, L.; Kotra, L. P.; Pedelacq, J. D.; Guillet, V.; Mobashery, S.; Samama, J. P. *J. Am. Chem. Soc.* **1998**, *120*, 9748–9752.
114. Beadle, B. M.; Shoichet, B. K. *Antimicrob. Agents Chemother.* **2002**, *46*, 3978–3980.
115. White, R. E.; Demuth, T. P. Jr. Novel Antimicrobial Lactam-Quinolones and Their Preparation and Formulation. European Patent 366189, May 2, 1990.
116. Beskid, G.; Fallat, V.; Lipschitz, E. R.; McGarry, D. H.; Cleeland, R.; Chan, K.; Keith, D. D.; Unowsky, J. *Antimicrob. Agents Chemother.* **1989**, *33*, 1072–1077.
117. Hamilton-Miller, J. M. T. *J. Antimicrob. Chemother.* **1994**, *33*, 197–200.

118. Hershberger, P. M.; Switzer, A. G.; Yelm, K. E.; Coleman, M. C.; Devries, C. I. A.; Rourke, F. J.; Davis, B. W.; Kraft, W. G.; Twinem, T. L.; Koenigs, P. M. et al. Jr. *J. Antibiot.* **1998**, *51*, 857–871.
119. Chung, I. H.; Kim, C. S.; Seo, J. H.; Chung, B. Y. *Arch. Pharmacal. Res.* **1999**, *22*, 579–584.
120. Jungheim, L. N.; Boyd, D. B.; Indelicato, J. M.; Pasini, C. E.; Preston, D. A.; Alborn, W. E., Jr. *J. Med. Chem.* **1991**, *34*, 1732–1739.
121. Baldwin, J. E.; Lynch, G. P.; Pitlik, J. *J. Antibiot.* **1991**, *44*, 1–24.
122. Baldwin, J. E.; Lowe, C.; Schofield, C. J.; Lee, E. *Tetrahedron Lett.* **1986**, *27*, 3461–3464.
123. Aszodi, J.; Rowlands, D. A.; Mauvais, P.; Collette, P.; Bonnefoy, A.; Lampilas, M. *Bioorg. Med. Chem. Lett.* **2004**, *14*, 2489–2492.
124. Nozaki, Y.; Katayama, N.; Harada, S.; Ono, H.; Okazaki, H. *J. Antibiot.* **1989**, *42*, 84–93.
125. Sandanayaka, V. P.; Prashad, A. S. *Curr. Med. Chem.* **2002**, *9*, 1145–1165.
126. Heinze-Krauss, I.; Angehrn, P.; Charnas, R.; Gubernator, K.; Gutknecht, E. M.; Hubschwerlen, C.; Kania, M.; Oefner, C.; Page, M. G. P.; Sogabe, S. et al. *J. Med. Chem.* **1998**, *41*, 3961–3971.
127. Hubschwerlen, C.; Angehrn, P.; Boehringer, M.; Page, M. G. P.; Specklin, J. L. 36th Interscience Conference on Antimicrobial Agents and Chemotherapy, New Orleans, LA, 1996; Poster F157, Book of Abstracts, p 127.
128. Angehrn, P.; Boehringer, M.; Hubschwerlen, C.; Page, M. G. P.; Pflieger, P.; Then, R. 36th Interscience Conference on Antimicrobial Agents and Chemotherapy, New Orleans, LA, 1996; Poster F158, Book of Abstracts, p 127.
129. Pflieger, P.; Angehrn, P.; Boehringer, M.; Hubschwerlen, C.; Page, M. G. P.; Winkler, F. 36th Interscience Conference on Antimicrobial Agents and Chemotherapy, New Orleans, LA, 1996; Poster F159, Book of Abstracts, p 127.
130. Arisawa, M.; Then, R. *J. Antibiot.* **1982**, *35*, 1578–1583.
131. Im, C.; Maiti, S. N.; Micetich, R. G.; Daneshtalab, M.; Atchison, K.; Phillips, O. A.; Kunugita, C. *J. Antibiot.* **1994**, *47*, 1030–1040.
132. Chen, Y. L.; Chang, C. W.; Hedberg, K.; Guarino, K.; Welch, W. M.; Kiessling, L.; Retsema, J. A.; Haskell, S. L.; Anderson, M. *J. Antibiot.* **1987**, *40*, 803–822.
133. Buynak, J. D.; Geng, B.; Bachmann, B.; Hua, L. *Bioorg. Med. Chem. Lett.* **1995**, *5*, 1513–1518.
134. Coleman, K.; Griffin, D. R.; Page, J. W.; Upshon, P. A. *Antimicrob. Agents Chemother.* **1989**, *33*, 1580–1587.
135. Weiss, W. J.; Petersen, P. J.; Murphy, T. M.; Tardio, L.; Yang, Y.; Bradford, P. A.; Venkatesan, A. M.; Abe, T.; Isoda, T.; Mihira, A. et al. *Antimicrob. Agents Chemother.* **2004**, *48*, 4589–4596.
136. Buynak, J. D. *Curr. Med. Chem.* **2004**, *11*, 1951–1964.
137. Mourey, L.; Miyashita, K.; Swaren, P.; Bulychev, A.; Samama, J. P.; Mobashery, S. *J. Am. Chem. Soc.* **1998**, *120*, 9382–9383.
138. Foulds, C. D.; Kosmirak, M.; Sammes, P. G. *J. Chem. Soc. Perkin Trans.* **1985**, *1*, 963–968.
139. Bitha, P.; Li, Z.; Francisco, G. D.; Rasmussen, B. A.; Lin, Y. I. *Bioorg. Med. Chem. Lett.* **1999**, *9*, 991–996.
140. Chen, Y. L.; Hedberg, K.; Guarino, K.; Retsema, J. A.; Anderson, M.; Manousos, K.; Barrett, J. *J. Antibiot.* **1991**, *44*, 870–879.
141. Buynak, J. D.; Chen, H.; Vogeti, L.; Gahachanda, V. K.; Buchanan, C. P. 42nd Interscience Conference on Antimicrobial Agents and Chemotherapy, San Diego, CA, 2002; Poster F344, Book of Abstracts, p 177.
142. Knowles, J. R. *Acc. Chem. Res.* **1985**, *18*, 97–104.
143. Richter, H.; Angehrn, P.; Hubschwerlen, C.; Kania, M.; Page, M. G. P.; Specklin, J. L.; Winkler, F. *J. Med. Chem.* **1996**, *39*, 3712–3722.
144. White, A. R.; Kaye, C.; Poupard, J.; Pypstra, R.; Woodnutt, G.; Wynne, B. *J. Antimicrob. Chemother.* **2004**, *53*, i3–i20.
145. Simpson, I. N.; Urch, C.; Hagen, G.; Albrecht, R.; Sprinkart, B.; Pfaendler, H. R. *J. Antibiot.* **2003**, *56*, 838–847.
146. Bonnefoy, A.; Dupuis-Hamelin, C.; Steier, V.; Delachaume, C.; Seys, C.; Stachyra, T.; Fairley, M.; Guitton, M.; Lampilas, M. *J. Antimicrob. Chemother.* **2004**, *54*, 410–417.
147. Payne, D. J.; Du, W.; Bateson, J. H. *Expert Opin. Investig. Drugs.* **2000**, *9*, 247–261.
148. Di Modugno, E.; Felici, A. *Curr. Opin. Anti-Infect. Investig. Drugs.* **1999**, *1*, 26–39.
149. Garcia-Saez, I.; Hopkins, J.; Papamicael, C.; Franceschini, N.; Amicosante, G.; Rossolini, G. M.; Galleni, M.; Frere, J. M.; Dideberg, O. *J. Biol. Chem.* **2003**, *278*, 23868–23873.
150. Toney, J. H.; Hammond, G. G.; Fitzgerald, P. M.; Sharma, N.; Balkovec, J. M.; Rouen, G. P.; Olson, S. H.; Hammond, M. L.; Greenlee, M. L.; Gao, Y. D. *J. Biol. Chem.* **2001**, *276*, 31913–31918.
151. Payne, D. J.; Hueso-Rodriguez, J. A.; Boyd, H.; Concha, N. O.; Janson, C. A.; Gilpin, M.; Bateson, J. H.; Cheever, C.; Niconovich, N. L.; Pearson, S. et al. *Antimicrob. Agents Chemother.* **2002**, *46*, 1880–1886.
152. Kuhn, D.; Coates, C.; Daniel, K.; Chen, D.; Bhuiyan, M.; Kazi, A.; Turos, E.; Dou, Q. P. *Front. Biosci.* **2004**, *9*, 2605–2617.
153. Veinberg, G.; Shestakova, I.; Vorona, M.; Kanepe, I.; Lukevics, E. *Bioorg. Med. Chem. Lett.* **2004**, *14*, 147–150.
154. Rothstein, J. D.; Patel, S.; Regan, M. R.; Haenggeli, C.; Huang, Y. H.; Bergles, D. E.; Jin, L.; Dykes Hoberg, M.; Vidensky, S.; Chung, D. S. et al. *Nature* **2005**, *433*, 73–77.

Biography

Christian Hubschwerlen received his PhD in chemistry in 1975 from the University of Haute Alsace, France. During his postdoctoral stay at the Woodward Research Institute, in Basel, Switzerland, he worked on penems and γ-lactam antibiotics. In 1980, he joined Hoffmann-La Roche in Basel, Switzerland, where he was involved in various anti-infective projects, targeting cell wall biosynthesis, DNA integrity, and protein synthesis. In particular, he was closely associated with the discovery of the antibiotics carumonam and ceftobiprole medocaril. He acted as deputy head of the anti-infective department. In 2000, he co-founded the Swiss branch of the Munich-based start-up company Morphochem, before joining Actelion Pharmaceuticals Ltd. in 2004, to lead the chemistry effort in that company's novel antibiotic discovery programs.

Comprehensive Medicinal Chemistry II
ISBN (set): 0-08-044513-6

ISBN (Volume 7) 0-08-044520-9; pp. 479–518

7.18 Macrolide Antibiotics

T Kaneko, T J Dougherty, and T V Magee, Pfizer Global Research and Development, Groton, CT, USA

7.18.1 Introduction

Macrolide antibiotics, represented by erythromycin A (**1**), are large polyketide natural products and their semisynthetic derivatives.[1] The naturally occurring macrolide antibiotics are mostly produced by the actinomycete species with structures characterized by a 12–16-membered lactone ring and by the presence of at least one sugar moiety attached to the macrolactone ring (cladinose at C3 and desosamine at C5 in case of erythromycin A). They were discovered in the 'golden years' of natural products, with the disclosure of picromycin (**2**) in 1950, marking the beginning of the macrolide antibiotics era.[2] In 1952, erythromycin A was discovered[3] and was quickly accepted for medical uses due to its safety and spectrum of activity. It continues to be used in the treatment of respiratory tract, skin, soft tissue, and urogenital infections. Although there are other minor congeners (erythromycins B–F, depending on the site of oxygenation and methylation) in the fermentation broth, the term 'erythromycin' usually designates erythromycin A, which is the most abundant and the most potent.

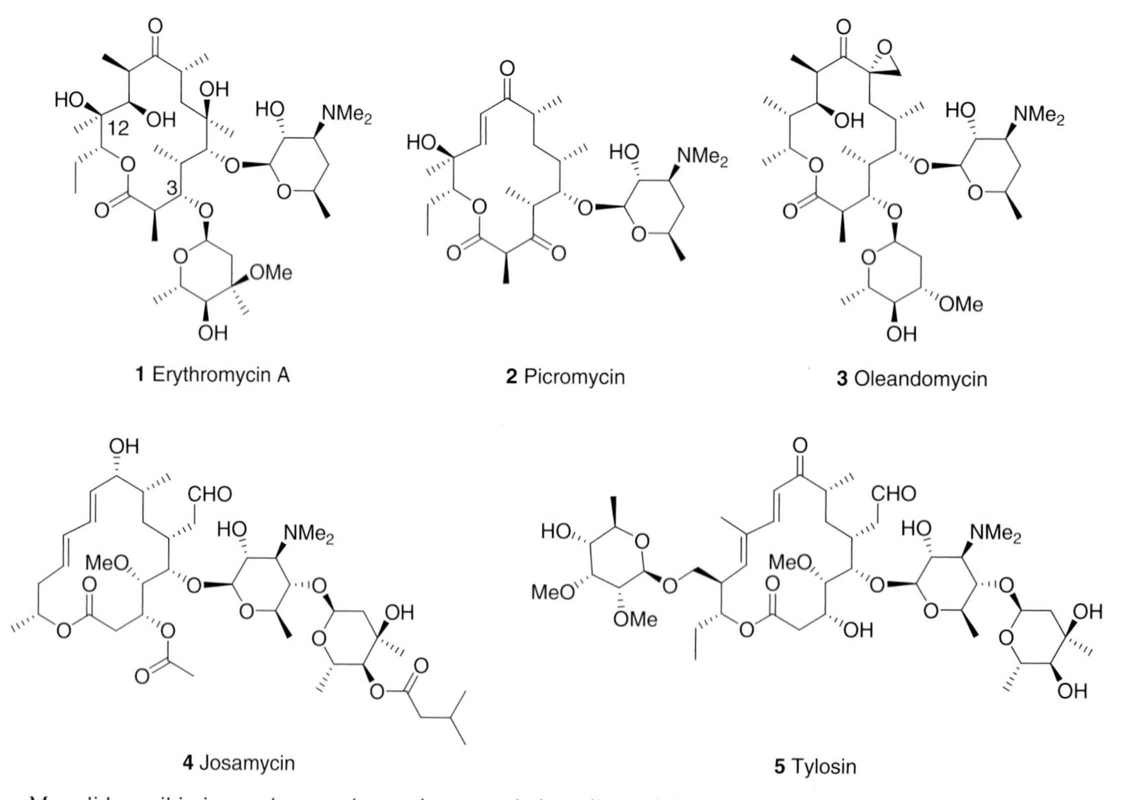

1 Erythromycin A **2** Picromycin **3** Oleandomycin **4** Josamycin **5** Tylosin

Macrolide antibiotics, such as erythromycin, exert their antibacterial activity by binding to the 50S subunit of the bacterial ribosome thereby inhibiting bacterial protein synthesis. Protein synthesis in eukaryotic cells is not inhibited by macrolide antibiotics, and this selectivity contributes to the safety of these agents. Though the discovery of erythromycin was followed by other macrolide antibiotics including oleandomycin (**3**),[4] josamycin (**4**),[5] and tylosin (**5**),[6] erythromycin remains the most widely used macrolide in human medicine. The major drawback of erythromycin is its poor pharmacokinetic properties primarily caused by its acid instability. Toward the end of the 1980s, two semisynthetic derivatives of erythromycin were discovered. These second-generation macrolides, clarithromycin (**6**)[7] and azithromycin (**7**),[8,9] demonstrated improved pharmacokinetic properties and increased spectra of activity, and at present are commercially the most important. In addition, their relative safety makes macrolides the drugs of choice for many respiratory infections in pediatric patients.

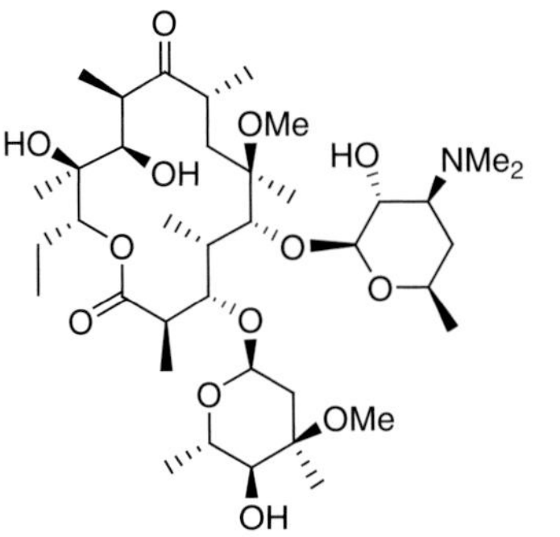

6 Clarithromycin

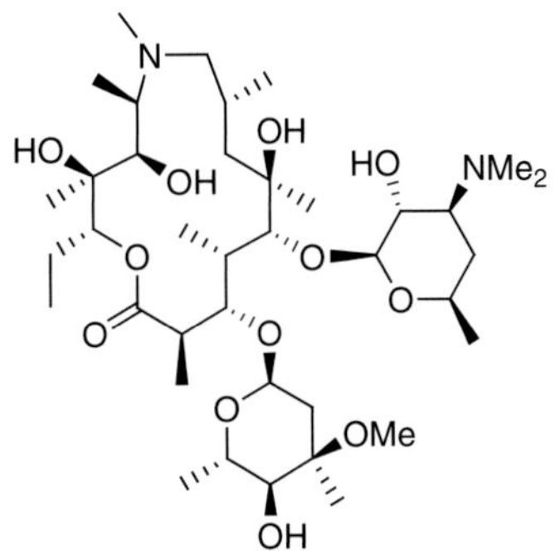

7 Azithromycin

Unfortunately, macrolide-resistant strains of *Streptococcus pneumoniae* were reported[10] almost immediately after the introduction of erythromycin and the proliferation of resistant strains continued in subsequent years. Furthermore, a high percentage of penicillin-resistant streptococci and staphylococci have also developed resistance to macrolides despite their different mechanism of action. The search for new macrolides with activity against these multidrug-resistant pathogens became a focus of intense research in the last two decades. This work culminated in the 1990s with the discovery of a new semisynthetic series of molecules called ketolides, and the recent introduction of telithromycin (Ketek) (**8**)[11] to the market. In ketolides, a ketone functionality is present in place of the cladinose group at C3 of clarithromycin, and a heteroaromatic ring is attached by a tether to the macrolactone ring.[12] It has been hypothesized that the heteroaromatic ring enables additional interactions with the bacterial ribosome, thus overcoming one of the mechanisms of resistance in *S. pneumoniae* and *Streptococcus pyogenes*. The radical structural departure and the newly endowed potency of ketolides resulted in a resurgence of synthetic activity over the last 10 years, as well as recent appearances of patents and publications. In addition, the detailed knowledge of the biosynthetic pathways and genetic engineering of the producing organisms have generated novel macrolides biosynthetically.[13] Although other macrolides played important roles, erythromycin is considered as the predecessor to a continuous medicinal chemistry effort to improve the efficacy of this class of antibiotics. As the macrolide generations progressed, this effort focused initially on pharmacokinetic properties, and, more recently, on spectra of activity.

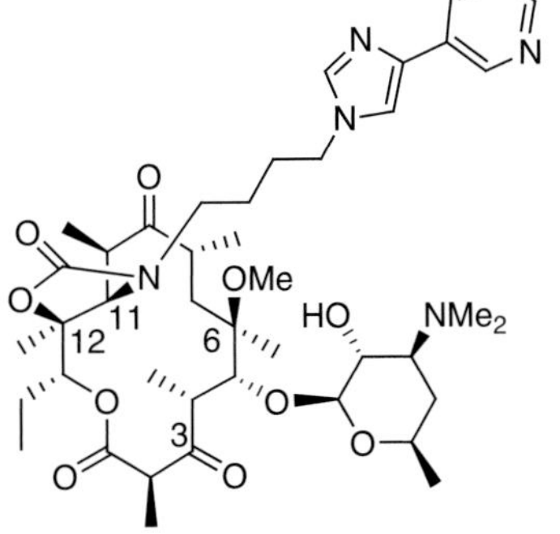

8 Telithromycin

7.18.2 Macrolide Clinical Use

7.18.2.1 Disease States

The macrolide class of antibiotics is used first and foremost in the outpatient treatment of community-acquired respiratory infections.[14] These include: community-acquired pneumonia (CAP),[15] bronchitis,[16] sinusitis,[17] and pharyngtitis.[18] These agents are also used to treat otitis media (middle ear infections) in pediatric and adult populations.[19] The principle pathogens that are associated with these infections and must be covered by the antibiotic are *S. pneumoniae* and *S. pyogenes*, both of which are Gram-positive pathogens, and the Gram-negative organisms *Haemophilus influenzae* and *Moraxella catarrhalis*.

S. pneumoniae represents the most serious pathogen among the group. Pneumococcal pneumonia remains one of the leading causes of death in newborns and the elderly, although this was mitigated to some degree by the introduction of pneumococcal polyvalent vaccines.[20] *S. pneumoniae* remains the leading cause of community-acquired pneumonia in adults, and CAP is a major contributor to morbidity and mortality in elderly patients.[21] Compounding the diagnostic problem in this population is the absence of classic pneumonia symptoms such as fever and cough in a significant proportion of older patients. About 10% of CAP cases become severe enough to require intensive care measures in a hospital setting. Another problem of profound importance with *S. pneumoniae* has been the alarming rate of penicillin resistance and multidrug antibiotic resistance development.[22,23] Macrolide resistance in the pneumococcus has also trended upward over the years, with multiple macrolide resistance mechanisms (see below).[24] In addition to the pneumococcus, *H. influenzae* is also a significant causative agent for community-acquired pneumonia, along with "atypical" organisms such as *Mycoplasma pneumoniae*, *Chlamydia pneumoniae*, and *Legionella pneumophila*. Another type of infection, acute bacterial sinusitis, is an infection of the paranasal sinuses, resulting in facial pain. Organisms identified from these infections are *S. pneumoniae*, *H. influenzae*, *S. pyogenes*, *M. catarrhalis*, and other alpha-hemolytic streptococci.

Yet another syndrome is otitis media, a condition in which there is inflammation in the middle ear and tympanum (eardrum), leading to pain, fever, and hearing impairment. Leading causative agents include *S. pneumoniae*, *H. influenzae*, and *M. catarrhalis*. In addition to infecting the lung, bronchi, sinuses, and middle ear, *S. pneumoniae* and *H. influenzae*, if not treated, can spread from the nasopharynx and cause life-threatening bacteremia and meningitis. In addition, macrolides are often employed in skin and soft tissue infections caused primarily by methicillin- sensitive *Staphylococcus aureus*. Other infections that are treated with the macrolide class of compounds include *Mycobacterium avium* complex infections in acquired immune deficiency syndrome (AIDS) patients. Azithromycin has been used in both the treatment of trachoma (*Chlamydia trachomatis*), which can lead to blindness, and nongonococcal urethritis and cervicitis caused by *C. trachomatis* and *Ureaplasma urealyticum*.

7.18.2.2 Measurement of Antibacterial Activity

The initial determination of an antibiotic's effectiveness against a given strain of microorganism is determined using an in vitro measurement of bacterial growth inhibition. The in vitro potency of an antibiotic is expressed by its minimum inhibitory concentration (MIC). Susceptibilities of individual bacterial isolates to inhibition by an antibiotic are measured in a highly standardized fashion.[25] Following established and approved procedures, the antibiotic is placed in a series of tubes in which twofold serial dilutions of the compound have been made in growth media. The microorganism of interest is then inoculated at a standardized number of cells into the dilutions, and the tubes incubated for a time period (20–22 h). Microbial growth results in cloudiness (turbidity) in the tubes, whereas tubes with higher concentrations of drug are clear, indicative of growth inhibition of the microbe. The concentration of the last tube that is clear in the series (before visible turbid growth is observed) is designated the MIC, usually stated in micrograms per milliliter ($\mu g\,mL^{-1}$). It represents the lowest concentration of drug that inhibits growth.

7.18.3 Mechanism of Action

7.18.3.1 Early Work on the Ribosome

Within a few years of the discovery and deployment of erythromycin as an antibiotic,[26] the first report was published in 1959 on its mechanism of action by the inhibition of bacterial protein synthesis.[27] At the same time, the process of protein synthesis and the role of the ribosome in that process were beginning to emerge. Indeed, erythromycin and other protein synthesis inhibitors, in addition to their therapeutic uses, played an active role as experimental reagents in clarifying ribosomal function. In parallel with the deepening understanding of the mechanism of action studies of the macrolide antibiotics, the understanding of the function of ribosomal components in protein synthesis also underwent drastic revision.[28] However, it was over a period of more than 40 years that a detailed description of ribosome function and the exact interactions of the macrolide antibiotics with the ribosome emerged.

Although a detailed examination of the molecular mechanisms of protein synthesis is beyond the scope of this chapter, some aspects will be covered briefly in the next section to set the stage for the detailed mechanism of action studies. The current view of protein synthesis reveals a complex multistep process with multiple protein factors in addition to the ribosome participation in the process. The process itself occurs in distinct steps (initiation, elongation, termination) that require different factors to proceed. For a more in-depth description and review of protein synthesis, the reader is referred to these publications.[29,30]

7.18.3.2 Overview of Protein Synthesis

The basic tenets of molecular biology emerged during the 1950s and 1960s.[31] During the process of translation, the messenger RNA (mRNA) threads through the ribosome. At the same time, transfer RNAs (tRNAs) with attached amino acids enter the ribosome and, as correct codon–anticodon pairings occur between the mRNA and tRNAs, the process of protein synthesis proceeds. There are multiple sites within the ribosome that are occupied by tRNA molecules during protein synthesis. The incoming tRNA with the attached amino acid enters into a site on the ribosome termed the A (Acceptor) site. Adjacent to the A site is another site termed the P (Peptide) site, where the tRNA has attached to it the growing peptide chain. During the process of protein synthesis, the peptide chain on the P site tRNA is transferred to the amino acid on the tRNA in the A site, in a reaction termed peptidyl transfer. The ribosome translates down the mRNA, and in the process, the tRNA in the A site is moved into the P site. The empty tRNA that was previously in the P site moves into another ribosome site termed the E (Exit) site. At the end of this cycle, the peptide chain is now one amino acid longer, and the A site is empty, ready for the next cognate amino acyl-tRNA.[32] The process has been estimated to have a low error rate of 10^{-3} to 10^{-4} incorrect amino acids incorporated,[33] and the ribosome plays an active role in preserving the fidelity of the translation process.[34]

7.18.3.3 Ribosome Biochemistry

Ribosomes were characterized as large ribonucleoprotein complexes, which sedimented as 70S particles upon ultracentrifugation, and on which the actual process of translation took place. The 70S ribosome could be dissociated into two large subunits, designated as 50S and 30S by their ultracentrifugation sedimentation velocities. The larger 50S subunit consists (in *Escherichia coli*, the reference organism) of two ribosomal RNA (rRNA) species, 23S rRNA and 5S rRNA, along with 31 proteins (designated as L1, L2, etc. for large subunit). The 30S subunit consists of a single RNA species, 16S rRNA, and 21 proteins (S1, S2, etc. for small subunit). Initially, it was unclear as to whether the individual proteins in the two subunits differed in composition from one another, but early experiments with electrophoresis indicated that the individual proteins were unique, confirmed later when DNA sequencing became available. A great deal of detailed experimental work contributed to an increasingly clear picture of the process of protein synthesis programmed by mRNA, and carried out by the complex ribosome in conjunction with other cellular factors.[35] Further work revealed a division of labor between the two subunits that made up the ribosome.

The 30S subunit was the site of the decoding of the mRNA, as the anticodon region of the aminoacylated tRNAs interact with the mRNA codons within the 30S structure. The tRNA with the attached amino acid extends from the 30S into the peptidyl transferase site of the 50S subunit, which catalyzes peptide bond formation in the growing polypeptide chain. There were several models of ribosome function proposed that differed in the specific details of the translocation of the tRNAs.[36,37]

By 1967, it was possible to take the individual ribosome components isolated from the fractionation experiments and reassemble a ribosome that was active in protein synthesis.[38,39] At the same time, hints of complex secondary structure for the rRNA molecules were beginning to accumulate, suggesting that the formation of a significant degree of base-paired duplex structures occurred in the ribosome. Early on, the prevailing view was that the RNA formed a passive scaffolding structure on which the proteins were held in specific configurations. Consistent with this, the ribosomal proteins were believed to be enzymes responsible for the catalytic functions associated with ribosomal protein synthesis. The discovery in the early 1980s that RNA molecules could possess catalytic activities,[40,41] along with accumulating genetic evidence that the rRNA played a role in catalytic functioning of the peptide bond formation by the ribosome, slowly changed the view of ribosome functioning.[42,43] Biochemical techniques were used to strip the protein components from the rRNA, yielding particles that had lost the majority of protein components yet retained peptidyl transferase activity.[44,45] In this picture, the protein components were involved in proper assembly and structural maintenance of the rRNA components, which performed key functions of the ribosome during protein synthesis. The rRNA components had been found to have conserved secondary structures, and the rRNA molecules had areas that were segregated into domains based on the secondary structure.[46–48] It was proposed that the 23S rRNA of the large subunit could be divided into six domains, based on the secondary structure. Increasingly sophisticated genetic manipulations of ribosomal RNA genes identified many mutations in the rRNA coding that resulted in changes in peptidyl transferase, subunit association or decoding.[49] For example, directed mutations in the central loop of domain V of the 23S rRNA demonstrated that changes in key bases (adenine 2060 to cytosine or adenine 2450 to cytosine (*E. coli* numbering)) were dominant lethal events, emphasizing the critical functional importance of these residues. However, a role for ribosomal proteins in peptidyl transfer could not be totally ruled out, since other earlier experiments had implicated L2, L3, and L4 as affecting the reaction.[50,51]

7.18.3.4 Modern View of Ribosome Structure and Function

By the early 1990s, multiple lines of evidence had accumulated for the role of both the 50S and 30S subunit rRNAs as functional entities in protein synthesis.[52] These functions included the participation of rRNA in mRNA selection, tRNA binding, ribosomal subunit association, peptidyl transferase function, and as at least part of the sites of inhibition for antibiotics that inhibited protein synthesis. The role of rRNA in these myriad functions was established primarily through biochemical studies, most notably by affinity labeling and chemical footprinting. Affinity labeling uses chemical groups that can cross-link molecules in close proximity. Experiments were performed that cross-linked the tRNA anticodon with a specific base (cytosine 1400 (*E. coli* numbering)) of the 16S rRNA, implying proximity of these groups.[53] Another set of experiments achieved a cross-link between the 3′ end of the tRNA (where the amino acid is attached) and 23S rRNA, specifically adenine 2451 and 2452 in domain V.[54,55] These results reinforced the orientation of the tRNAs as spanning the two ribosomal subunits. Additionally, cross-links were formed with key bases in the 16S rRNA of the 30S subunit in experiments utilizing photoaffinity mRNA, again implying close interaction between those components. These interactions were also found to be dependent on the presence of tRNA molecules in the ribosome binding sites. Another technique, chemical footprinting, identifies nucleotides that are protected from modification by chemical reagents that react with specific nucleotides. These reagents include kethoxal for modifications of N1 and N2 of guanine, dimethyl sulfate for N1 of adenine, N3 of cytosine, and N7 of guanine, and the carbodiimide CMCT for N3 of uridine and N1 of guanine. Binding of tRNA protects certain rRNA bases from chemical attack, defining interactions of the tRNAs with the ribosomal rRNA. These and other experiments helped determine the number and orientation of the tRNA molecules within the ribosome.

Equally important have been the structural studies of ribosome functions. One of the notable methodologies includes cryoelectron microscopy,[56] which employs images of ribosomes embedded in ice. Multiple images in different orientations are used to obtain a computer reconstruction of a ribosome.[57,58] Among the features identified in this relatively low resolution method (roughly 25 Å) was the channel for mRNA in the small subunit and a channel in the large subunit from which the growing peptide chain emerges from the ribosome. It was also possible to visualize tRNAs in images of 70S ribosomes, using empty and tRNA-loaded ribosomes for comparison.[59] The observed densities correlated with biochemical studies that defined the locations of the A, P, and E sites. More recently, this technique has been used to examine changes in the conformational state of the ribosome during protein synthesis.[60,61] For example, it was possible to discern a ratchet-like motion of the 30S subunit relative to the 50S that plays a role in translocation of the ribosome along the mRNA. Another technique, neutron diffraction, was used to map the relative positions of the proteins within the 30S subunit.[62]

7.18.3.5 Crystallographic Structural Studies

The recent excitement in the ribosome field has been the solution of the high-resolution structure of the ribosome by x-ray crystallography. This work has confirmed many of the prior details of ribosome structure and function, as well as offering new insights into ribosome function and details of antibiotic action and resistance. This work, carried out by several groups, is the culmination of efforts to obtain a crystal structure of the ribosome that started with the first reported crystals of ribosomes in the early 1980s.[63,64] Earlier ribosomal crystallography efforts had centered around ribosomes obtained from *Bacillus stearothermophilus*, *E. coli*, *Thermus thermophilus*, and *Haloarcula marismortui*.[65] Later crystallographic work has focused on *H. marismortui*, *Deinococcus radiodurans*, and *T. thermophilus* ribosome preparations, due to the high quality of data from these crystals. While space permits only a brief summary of these outstanding achievements, several excellent reviews are available that summarize the details to date.[29,66,67]

7.18.3.5.1 Intact ribosome studies

The crystal structure of the intact 70S ribosome from *T. thermophilus* has been resolved to 5.5 Å (**Figure 1**). In addition to the ribosome, the structure also has mRNA and tRNAs in the P and E sites.[68] The interpretation of details at this resolution was possible due to the prior solution of the 30S and 50S structures at higher resolutions (see below). Among

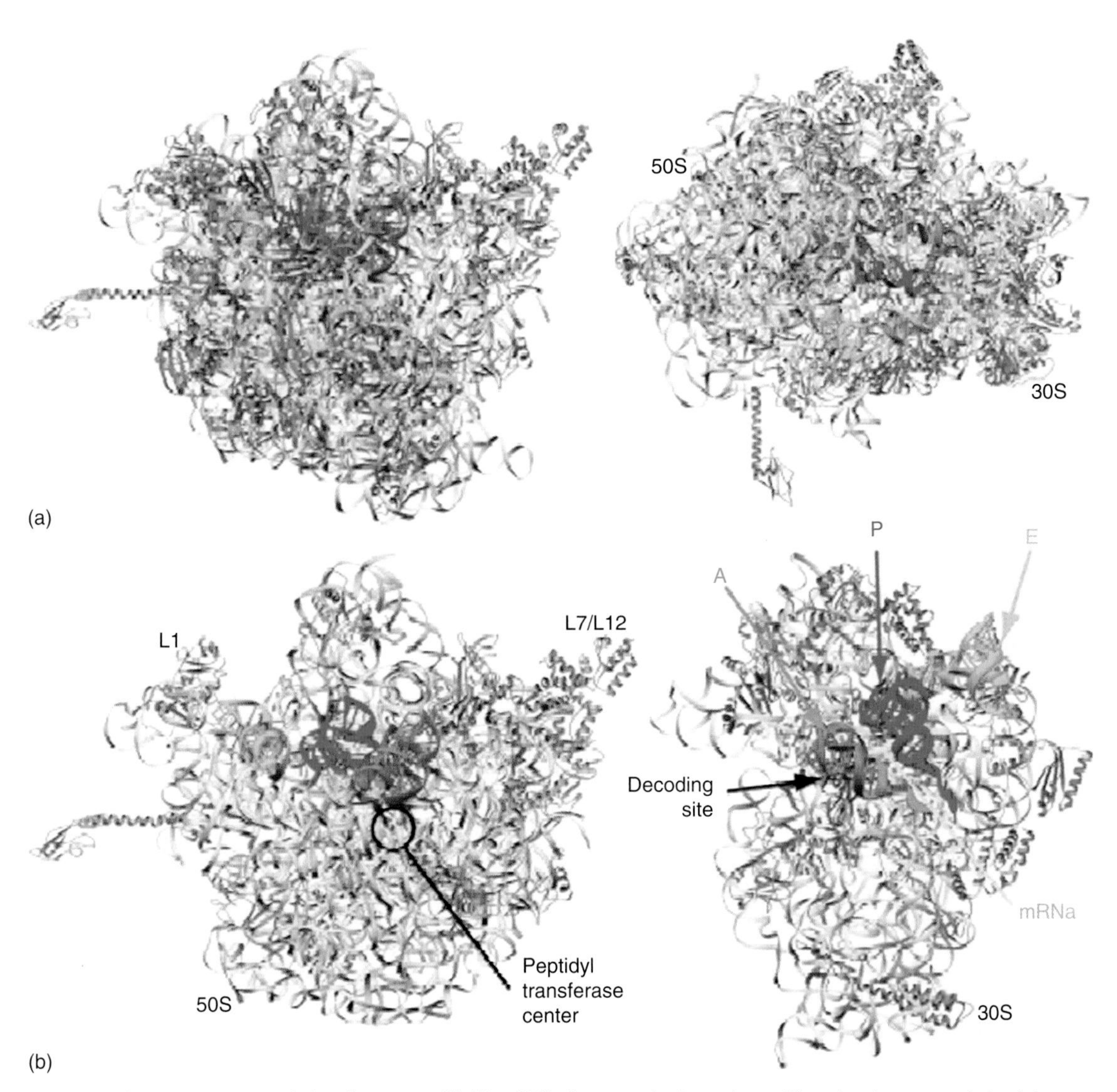

Figure 1 Crystal structure of the ribosome. (A) The 70S ribosome in two views. The structure was derived by x-ray crystallography of the 30S and 50S subunits. The positions of the three tRNAs (red: A site, green: P site, and yellow: E site) are illustrated. (B) The 50S (left) and 30S (right) subunits with tRNAs and the peptidyl transferase site in the 50S and mRNA in the 30S. (Reproduced with permission from Ramakrishnan, V. *Cell* **2002**, *108*, 557–572.)

the findings were that tRNA contact sites with the ribosome are mainly through rRNA and not protein, as predicted by earlier chemical cross-linking and footprinting studies. The tRNA interactions with the ribosome not only stabilize the binding of tRNA to the ribosome, but are also involved in mechanisms that increase the accuracy of the aminoacyl-tRNA recognition for the proper codon, maintain the correct reading frame of the mRNA, move the tRNA within the ribosome, and catalyze formation of the peptide bond. The geometry of the tRNAs within the ribosome was also determined. The anticodon end of the tRNA that interfaces with the mRNA is contained in, and interacts with, elements of the 30S subunit. The rest of the tRNA structure extends into and interacts with the 50S subunit. The ends of the tRNAs with the growing peptide chain (P site) and incoming amino acid (A site) are within 5 Å of each other in the 50S subunit, in an area previously associated by biochemical experiments with the peptidyl transferase activity of the ribosome. Very recently, the two structures of the intact 70S ribosome from *E. coli* were determined to a resolution of 3.5 Å.[69] According to this study, the two structures indicate a fair amount of flexibility within the small subunit. It is hypothesized that the swiveling of the head of the small subunit coupled with the ratchet-like motion is important for the translocation process.

7.18.3.5.2 30S Subunit studies

Subunit structures of both the 30S and 50S subunits at resolutions higher than the intact 70S have been achieved. Two research groups have studied the crystal structure of the 30S subunit.[70,71] The 16S rRNA of the 30S subunit consists of over 50 double-stranded helices connected by irregular single-stranded regions; thus the RNA structure is a three-dimensional arrangement of helices. The packing of the helices determines the overall structure of the four rRNA domains. The ribosomal proteins generally have a globular domain that is on the surface of the 30S, with long extensions that reach into the 30S, making extensive interactions with the rRNA and stabilizing the overall ribosome structure. The A and P tRNA binding sites are each composed of rRNA from two different domains of the 16S rRNA. These two domains have been observed to move relative to one another in cryoelectron microscopy studies, and this motion is almost certainly involved in the motions of tRNAs and mRNA during translation. The three tRNAs bind their anticodon regions in the RNA-rich region that forms a groove in the 30S subunit.[60] Several classes of antibiotics including major clinical classes such as spectinomycin, tetracyclines, and aminoglycosides bind to the 30S as their site of action.[72,73]

7.18.3.5.3 50S Subunit structural studies

The site of action for the macrolide class is the 50S subunit. The large ribosomal subunit structural work on *H. marismortui* improved progressively from 9 Å in 1997 to 5.5 Å in 1999.[74,75] As resolution increased, structural features were identified that corresponded with the earlier electron microscopy studies. In 2000, the *H. marismortui* 50S subunit of the ribosome was solved to the resolution of 2.4 Å.[76] This work established several key facts about the 50S ribosome structure. The individual domains of the rRNA were found to fit together somewhat like a puzzle to form a compact structure with the overall shape of the subunit. The secondary structures, as found previously,[36] consist of rodlike RNA helices joined by irregular domains. These RNA domains interact extensively with each other, forming the large subunit. Most of the 50S ribosomal proteins were found to be on the surface of the ribosome, with one face interacting with the solvent and the other with rRNA domains to stabilize the overall ribosome structure. Twelve of the ribosomal proteins also have extensive "tails," largely of arginine and lysine, which extend into the interior of the ribosome and interact at specific sites with the rRNA over their lengths. The peptidyl transferase activity of the 50S subunit had been associated with nucleotides in the domain V region of the 23S rRNA by footprinting and other biochemical methods. Strikingly, the x-ray structural studies confirmed that there are no proteins within the neighborhood of the peptidyl transferase center, strongly suggesting that the rRNA is indeed involved in catalyzing the transferase reaction.[77] The actual mechanistic details of the peptidyl transferase catalytic reaction remain at this point in some dispute. Precise alignment of the two ends of the tRNAs in the A and P sites is undoubtedly of extreme importance in bringing the incoming amino acid esterified to the tRNA into position for transfer to the peptide on the tRNA in the P site.[78] Specific areas of the 23 rRNA termed the P loop (RNA helix 80) and A loop (RNA helix 92) interact with the three terminal nucleotides (CCA) of the P site and A site tRNAs and position the tRNAs. It had been proposed that the highly conserved adenine 2451 (*E. coli* numbering) was directly involved in the catalysis of peptide bond formation, employing a general acid–base catalysis.[67] Subsequently, however, specific bases (A2451 and G2447) postulated to be involved in chemical catalysis were mutagenized without the loss of peptidyltransferase activity, questioning the significance of these residues in the process.[80] Studies with mutants have suggested that the conserved domain V nucleotides may instead be involved with hydrolysis of the nascent peptide from the ribosome at the termination of synthesis.[81] Another recent study has presented evidence that the rate enhancement of peptide bond formation is exclusively from lowering the entropy of activation, by positioning the substrates and excluding water from the reaction site.[82] An important role for the P site tRNA's terminal 2′-OH in catalysis of peptide bond formation has also been recently proposed.[83] A recent review has summarized the current views on peptide bond formation by the ribosome.[84]

Another feature of the 50S subunit confirmed by the structural studies was the presence of a peptide exit tunnel that leads from the peptidyl transferase center through the large subunit to the exterior. This tunnel is approximately 100 Å long, and between 10 Å and 20 Å in width. The wall of the tunnel is composed primarily of nucleotides from domains I through V of the 23S rRNA, along with nonglobular parts of proteins L4 and L22. It is within this tunnel that the macrolide class of antibiotics bind and have their effects on bacterial protein synthesis.

7.18.3.5.4 Macrolide mechanism of action

Macrolides transport into bacteria by a passive process, independent of energy, with the uncharged form of the drug diffusing across the bacterial cell membrane,[85] and this process is facilitated by the presence of the many intracellular binding sites, the ribosomes, to act as a 'sink.' As described previously, the effect of macrolide antibiotics on the inhibition of protein synthesis was known shortly after erythromycin was introduced.[10] The precise mechanism of

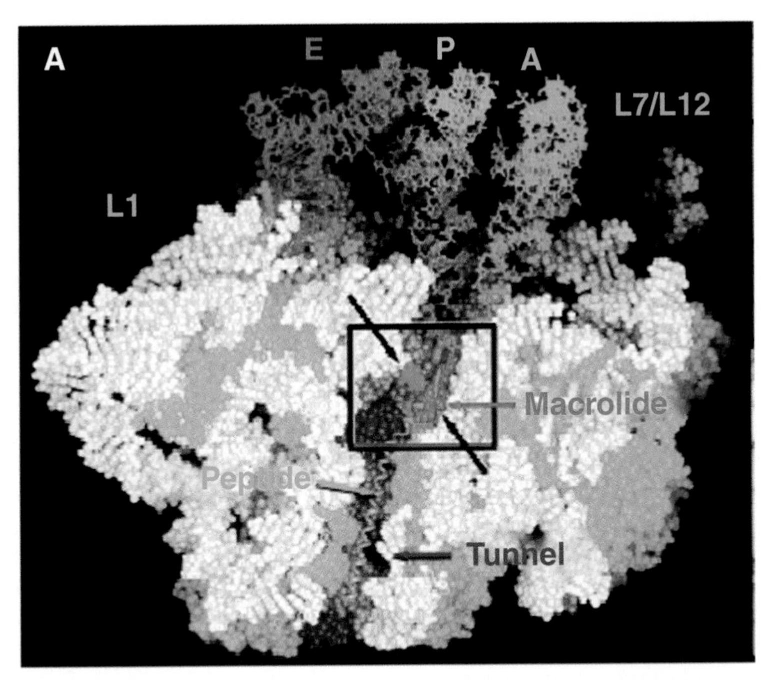

Figure 2 Cross-section of 50S subunit, showing the relative positions of the three tRNA molecules. The upper left diagonal arrow points toward the peptidyl transferase center, the lower right diagonal arrow points toward the macrolide binding site. The peptide exit tunnel (which is occluded by the macrolide binding) is illustrated with a peptide moving downward, toward the outside. (Reproduced with permission from Hansen, J. L.; Ippolito, J. A.; Ban, N.; Nissen, P.; Moore, P. B.; Steitz, T. A. *Mol. Cell* **2002**, *10*, 117–128.)

action, however, was unclear due to the complexity of protein synthesis. It was apparent that macrolides enter bacterial cells and bind to ribosomes reversibly and with high affinity,[86] with K_d values of 10^{-8} to 10^{-10} M. Early studies suggested that macrolides might cause the premature release of peptidyl-tRNA from the ribosome.[87] Footprinting experiments in the presence or absence of drug demonstrated that key nucleotide residues in the 50S rRNA were protected from access to modification by macrolides.[88–90] These results implied, along with genetic mutational studies (e.g., drug resistance mutants) and additional biochemical data, that erythromycin interacted solely with nucleotide residues that were close to the peptidyl transferase center of the 50S subunits.[36,91] Some of the key nucleotide sites involved in binding were identified as A2058, A2059, and A2062 (*E. coli* numbering).

The details of macrolide antibiotics interactions with the ribosome have been clarified by the solution of the high-resolution 50S subunit structure of the ribosome by two research groups (**Figure 2**). The first structure reported was that of the 50S subunit of the halophilic archaeon *H. marismortui*, followed by the structure of the eubacterial *D. radiodurans*.[66,92] These reports were followed by descriptions of the 50S ribosome structures of both organisms with bound macrolide antibiotics.[93,94] Although the general details of the macrolide interactions within the ribosomes of the two organisms are in agreement, the precise interactions and orientations of the drugs in the ribosome differ in these studies. There are significant differences in the two bacterial species used in the studies; for example the *Haloarcula* ribosome has a guanidine at the position equivalent to 2058, which is an adenine that makes a key interaction in macrolide susceptible strains. Other residues implicated in macrolide binding also vary in this organism. As a consequence, *H. marismortui* is not susceptible to macrolide inhibition at concentrations used clinically. In contrast, *Deinococcus* is a eubacterium with 50S rRNA residues that are identical to those found in pathogens, and it is susceptible to erythromycin. Other technical differences between the studies include one group using cocrystallization versus soaking of the drug into the crystals. Other differences may also be attributable to interpretations of the electron densities in the derived structures.

Both sets of results indicate that 14-membered ring macrolides such as erythromycin and clarithromycin, bind largely to nucleotides in domain V of the 23S rRNA, and no direct drug–protein interactions are evident. The 14- and 15-membered (azithromycin) macrolides bind in the ribosomal peptide exit tunnel about 10–15 Å from the peptidyl transferase site. The sugar at C5 (desosamine) extends up the tunnel from the macrocyclic ring toward, but does not reach, the peptidyl transferase center. In the case of the 16-membered macrolides, such as carbomycin and tylosin, the

C5 disaccharide chains (mycaminose-mycarose) extend even farther up the tunnel, and the isobutyrate group on the mycarose of carbomycin extends into the peptidyl transferase center.[93,94] This is in agreement with the previous observation that carbomycin could interfere with binding of the A site substrate, and the isobutyrate group can be seen to be occupying that site in the x-ray structure.[95]

Studies by the Yonath Max Planck group highlight the important hydrogen bond interactions among the desosamine sugar C2′-OH group in the 14- and 15-membered macrolides with adenine 2058 and adenine 2059 equivalents in *Deinococcus* (the 2058 equivalent is a guanidine in *Haloarcula*).[96] Other hydrogen bonds proposed by this group are between three of the hydroxy groups on the macrolactone ring and nucleotides of the 23S rRNA. These bonds are postulated to be between the C6-OH group and the N6 of adenine 2062 (*E. coli* numbering), and the C11-OH and C12-OH are theorized to interact with O4 of adenine 2062 (*E. coli* numbering). The authors, however, remark that it is difficult to explain the high affinity of macrolides for the ribosome on the basis of these interactions alone, and additional hydrophobic effects may be involved in drug binding.

As described previously, the ketolides are a new generation of macrolide derivatives that have been developed to counter macrolide resistance mechanisms (see below). The nature of ketolide binding has been investigated in *Deinococcus*.[96,97] In the case of telithromycin, the aryl-alkyl groups that extend from the cyclic C11/C12 carbamate are believed to make additional contacts with the ribosome, strengthening the compound interaction with the binding site. Another ketolide, cethromycin (ABT-773) also has a cyclic C11/C12 carbamate; however, the quinolinyl-allyl group extends from the C6 position of the lactone ring. Both the aryl-alkyl extension of telithromycin and the quinolyl-allyl of cethromycin have been postulated, based on footprinting experiments, to make additional contacts with domain II of the 23S rRNA, which is folded close to the peptidyltransferase center of domain V in the three-dimensional structure.[98] The x-ray structures from the deinococcal work were interpreted to show the two extensions from telithromycin and cethromycin extending out in divergent directions toward domain II.[97] The published figure for telithromycin seems to place the macrolactone ring in an extended, strained configuration in order to make both the desosamine contacts with the A2058 region and aryl-alkyl contacts with domain II.

In the initial studies by the Steitz and Moore group at Yale with *H. marismortui*, hydrophobic interactions between the lactone ring and the tunnel wall are seen as promoting binding. One face of the macrolactone ring is relatively hydrophobic, and this face is believed to make van der Waals contacts with hydrophobic areas (e.g., the aromatic face of cytosine 2611 (*E. coli* numbering)) on the tunnel wall. The hydrophilic face of the lactone ring is exposed toward the solvent.[79] The 16-membered macrolides have a C20 aldehyde that appears to form a covalent bond with the N6 of 2062 (*E. coli* numbering); this aldehyde moiety is not present in 14- and 15-membered macrolides. The Yale group places the desosamine (or mycaminose) sugar at C5 approximately perpendicular to the plane of the macrolactone ring. All of the macrolides clearly appear to occlude the peptide exit tunnel, effectively blocking the nascent polypeptide from progressing and as a result halting protein synthesis. These observations explain the prior results that 14-membered macrolides allow the ribosome to synthesize short peptides before encountering the block, whereas 16-membered macrolides (such as carbomycin) do not allow formation of peptide bonds.[99]

Subsequent studies by the Yale group have used *H. marismortui* in which the native guanine residue G2058 has been modified by genetic means to an adenine, as found in macrolide susceptible eubacterial species.[100] This modification increased the affinity of the ribosomes for erythromycin by 10^4 over the native form of the *Haloarcula* ribosome, indicative of the importance of this adenine residue in macrolide binding. The orientation of both erythromycin and azithromycin in the modified ribosomes was determined. As in their prior studies with the wild-type ribosome, the erythromycin was found on the floor of the nascent peptide tunnel, with the sugars pointing toward the A and P sites. The hydrophobic side of the macrolactone ring contacts the tunnel wall, whereas the hydrophilic side is oriented toward the tunnel lumen. The drug orientation was basically very similar to that described above for the native G2058 ribosome, and it was found that erythromycin and azithromycin bind in an almost identical fashion. In the telithromycin case, the macrolactone ring was found to be in a virtually identical orientation as that of the erythromycin ring. In contrast to the deinococcal studies, the x-ray studies were interpreted to show the C11/C12 aryl-alkyl side chain folding back above the macrolactone ring with the pyridine group stacking on the C2644 base. The C11/C12 extension was not observed to interact with any bases in the domain II region. While further refinement of the x-ray data will undoubtedly clarify the situation of the specifics of macrolide–ketolide binding, measurement of binding affinities and turnover clearly indicates that ketolides bind tighter and dissociate from the ribosome much more slowly than do earlier macrolides.[101] This longer residence on the ribosome has consequences in maintaining effectiveness against resistant organisms as well as in the induction of resistance, as described below.

Before leaving this section on mechanism of action, it should also be mentioned that a second mechanism for macrolide inhibition of cell growth has been proposed. Work by Champney's group at Quillen College of Medicine has demonstrated that macrolides and ketolides inhibit the assembly of the 50S subunit of the ribosome in a number of pathogenic microorganisms.[102,103] The 50S subunit assembly is believed to be blocked by the prevention of binding of ribosomal proteins by the macrolide, leading to immature particles that are degraded by ribonucleases. The relative significance of this additional mechanism to macrolide action is not clear at this point.

7.18.4 Macrolide Resistance

Macrolide resistance arose in the same year that the erythromycin was introduced to clinical use.[10] Since then, the number of Gram-positive pathogens that are macrolide resistance has continued to rise,[104,105] leading to the impetus for development of newer compounds, such as ketolides, that address resistance. The two most common mechanisms of macrolide resistance are ribosomal modifications and the acquisition of a macrolide efflux pump.[106] The former type of resistance arises either through methylation of 23S rRNA adenine 2058 (*E. coli* numbering) or mutation of adenine 2058 to, most commonly, guanine or uracil. The desosamine sugar in the 14- and 15-membered macrolides as well as the mycaminose of the 16-membered macrolides hydrogen bond with the adenine 2058. Methylation of the C6 amino group of adenine results in steric hindrance resulting in a vastly decreased affinity of the drug for the ribosome. This manifests itself in a great increase in the MIC needed to inhibit the resistant pathogenic bacteria. There is a family of related enzymes responsible for the methylation reaction, designated as Erm methylase enzymes.[107] The genes that encode these enzymes include *ErmA, ErmB, ErmC*, and *ErmAM*, among others. The enzymes use *S*-adenosyl-methionine as a methyl donor, and use 23S rRNA of nascent ribosomes as substrate since the enzymes cannot methylate fully assembled ribosomes.[108] This type of resistance is often designated as MLS_B resistance for macrolide, lincosamide, and streptogramin B resistance, which are structurally unrelated drugs that share part of the same ribosomal binding site, and the binding of which are all adversely affected by the methylation.[109] The genes encoding the Erm methylases are found on mobile genetic elements such as transposons and plasmids, which can readily be transferred among diverse bacterial species.[110]

Mutation of the critical adenine to another base, as well as mutations in other key nucleotides such as A2059, can likewise disrupt the same key interactions as methylation. Many pathogenic bacteria carry multiple copies of extremely similar genes that specify the synthesis of the rRNAs. Mutations of these kinds have been reported in multiple pathogenic species including *S. pneumoniae, S. aureus*, and *H. influenzae*, among others.[111–113] Once the mutation has occurred in one copy of the rRNA genes, it is spread to other copies of the rRNA genes by a nonreciprocal recombinational mechanism similar to gene conversion. Strains found to be resistant usually have more than one copy of the mutated rRNA gene. For example, an *S. aureus* strain that normally carries six copies of the 23S rRNA gene was found to have four identically mutated copies out of six in the resistant strain. An interesting side note is that humans naturally have a guanine at the 2058 equivalent position of the eukaryotic ribosome, and it is one factor in the lack of macrolide inhibitory effects on human cells.

A key aspect of the methylation is that the majority of pathogenic bacteria, which carry the genes encoding the methylase, do not methylate the rRNA in the absence of antibiotic. Only upon encountering an antibiotic, does the methylation begin to take place on the nascent ribosomes. This inducibility of the Erm methylases was first reported in *S. aureus* in 1971.[114] The mechanism of this induction emerged from a number of molecular biology investigations, and has been summarized by Weisblum.[115] The primary control of gene expression occurs in this case at the translational level, as opposed to the more common transcriptional control. In this control system, the coding sequence for the Erm methylase is preceded by a gene for a short 19 amino acid 'leader peptide,' and the two genes are cotranscribed onto a single mRNA. The mRNA also contains several regions of inverted nucleotide repeats, and these regions normally fold into a secondary structure in which two double-stranded RNA loops form, the second of which sequesters the start site (designated the Shine–Dalgarno site) for the translation of the Erm methylase protein, thereby blocking methylase synthesis (**Figure 3**). Under drug-free conditions, only the leader peptide, which serves no known function, is made as the methylase cannot be translated. In this state, the ribosomes are unmethylated, and the organisms would be susceptible to macrolides. In the presence of macrolides, however, the ribosome 'stalls' on the leader peptide coding region, and the ribosome covers one of the RNA regions that forms part of the first double-stranded loop. This allows the other part of the first RNA loop to form an alternative base pair situation with part of the second loop. This new loop formation in turn frees up the Shine–Dalgarno site that immediately precedes the coding region for the Erm methylase, thereby permitting translation of the Erm methylase. The dynamics of the drug–ribosome interaction must be such that although momentary stalling of the leader peptide can occur and unmask the Erm translation start site, there will still be sufficient drug-free ribosomes present to bind to the exposed upstream Shine–Dalgarno start sequence and translate the Erm methylase. This is the case for macrolides such as erythromycin, clarithromycin, and azithromycin. High-level resistance to clarithromycin and azithromycin in clinical isolates of *S. pneumoniae* has been attributed to induction of ErmAM methylase.[116] In contrast, the new ketolide compounds

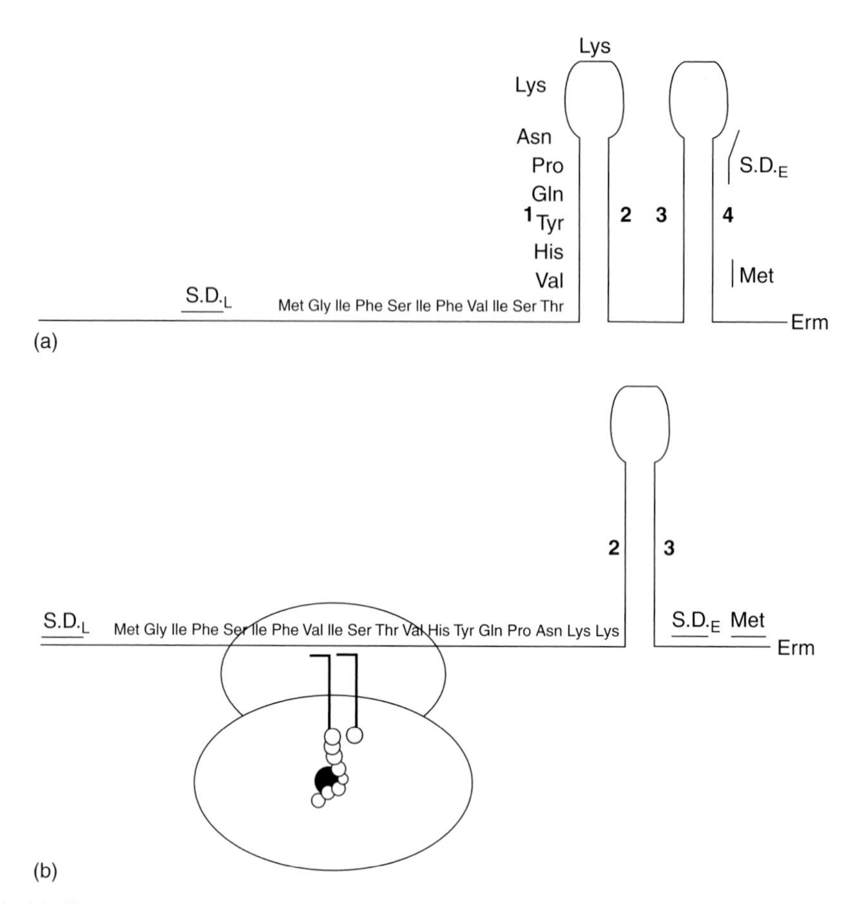

Figure 3 Inducible Erm resistance: the secondary structure of the mRNA that encodes a leader peptide and the Erm methylase gene. In the noninduced state (A), the leader peptide is rapidly synthesized from the first Shine–Dalgarno (S.D.) sequence. The Shine–Dalgarno binding site for Erm methylase remains sequestered and unavailable in the loop 3–4 structure. (B) In the presence of erythromycin, the ribosome translating the leader peptide will stall. This disrupts the loop 1–2 structure and allows the formation of the alternative loop 2–3 structure in the mRNA. This in turn makes the Shine–Dalgarno site available for the translation of the Erm methylase. (Reproduced with permission from Gaynor, M.; and Mankin, A. S. *Curr. Topics Med. Chem.* **2003**, *3*, 949–960.)

were found to be poor inducers of the Erm methylase, and this may be due to their tighter binding and slower dissociation rates from ribosomes, resulting in much lower levels of translation of the Erm methylase.

In some clinical isolates, most notably in *S. aureus*, the expression of the Erm methylase has been found to be constituitive. Most of these strains were found to carry deletions or other mutations in the leader peptide that destabilize the secondary loop structures controlling induction.[117] With currently available macrolides, it is not possible to overcome this constitutively resistant *S. aureus*.

Less commonly, another type of mutational resistance can contribute to macrolide resistance. The L4 and L22 ribosomal proteins assist in maintaining the conformation of the ribosome, primarily through interactions with 23S rRNA domain I, and are important in early assembly.[118] There is also evidence that these proteins interact with the peptidyl transferase region of domain V.[119] Mutations in both L4 and L22 have been reported to confer macrolide resistance (but not resistance to lincosamide), and evidence that they perturb 23S rRNA structural conformation has been presented.[120] These perturbations are believed to disturb the macrolide binding site within the ribosome, which can be visualized by electron microscopy.[121,122] Both clinical and laboratory isolates with various mutations in their L4 or L22, sometimes in conjunction with other macrolide resistance elements have been reported.[123,124]

The other major type of macrolide resistance is due to drug efflux, and is designated as M-type resistance, as it is limited to macrolide class of compounds. In this case, decreased intracellular accumulation of the drug occurs due to an active efflux pump. Several discrete structural families of efflux pumps are found broadly distributed among bacteria.[125] These pumps actively extrude drugs from within bacterial cells, and use either the membrane proton motive force ($\Delta\psi$) or ATP for energy to perform export. Some pumps have relatively narrow substrate range (i.e., only one class of antibiotics) whereas others are broader in exporting diverse antibiotic structuress. In streptococci (e.g., *S. pyogenes*, *S. pneumoniae*) macrolide efflux is mediated by a two-gene cluster, with the first gene designated as *MefA* (originally designated as *MefE* in

S. pneumoniae) and the second as *msrA*.[126,127] The *MefA* pump has 12 transmembrane domains, which span the cytoplasmic membrane, and drug efflux is driven by proton motive force.[128] An additional downstream gene resembles the coding region of the *msrA* efflux pump of *S. aureus*, and has been designated *msrD*. This putative pump is a member of the ATP binding cassette family, presumably using ATP to drive efflux. The efflux genes are transferable among streptococci and reside on a group of mobile genetic elements, a family of related transposons.[129,130] Recently, the role of both the MefA pump and the downstream MsrA-like pump in *S. pneumoniae* has been clarified.[131] The isolated *MefA* gene was found to encode macrolide resistance, and the level of resistance was found not to be influenced by drug induction. The *msrD* gene was found to encode a pump that also conferred macrolide resistance, but appeared to be inducible by drug. This study also confirmed that the two ketolides tested, telithromycin and cethromycin, were not effluxed to any appreciable extent by the macrolide efflux pumps. Whether this effect is due to ketolides being a poorer pump substrate, tighter intracellular ribosome binding by ketolides, or a combination of effects remains unclear at present.[132]

In the case of many Gram-negative bacteria, both outer membrane channels ('porins') and efflux pumps have been shown to play a role in the poor bacterial susceptibility to macrolides.[133,134] In Gram-negative bacteria such as *E. coli*, *Proteus* species, *Enterobacter* species, and others, there is high intrinsic resistance to many antibiotics, including the macrolides, despite the fact that the ribosomes are fully susceptible to these compounds. These Gram-negative organisms have several highly effective efflux pumps coupled with an outer membrane, which differs from the composition of the inner membrane. The outer membrane has lipopolysaccharide on the outer surface, which greatly reduces the penetration of lipophilic compounds and proteins (porins) that form narrow water-filled channels for hydrophilic substrate penetration.[135] This combination of limited drug influx due to the outer membrane barrier properties and highly effective efflux systems results in high intrinsic resistance to macrolides in most Gram-negative organisms. However, in the case of some Gram-negative bacteria, most notably the 'fastidious' species such as *H. influenzae*, these organisms possess pore channel sizes that are larger than those of other Gram-negative bacteria such as *E. coli*. The larger pores, and presumably higher drug influx rates, which may counterbalance efflux to a greater extent, may partially explain the relative sensitivity of Gram-negative bacteria such as *H. influenzae* and *M. catarrhalis* to macrolides.

7.18.5 Major Classes of Macrolides

In the following sections, each class of macrolide antibiotics is highlighted with respect to its in vitro and in vivo activities, pharmacokinetic properties, structure–activity relationships (SARs), and safety properties.

7.18.5.1 Erythromycin

Erythromycin is produced by *Streptomyces erythreus*, a soil organism isolated from the Philippines (recently reclassified as *Saccharopolyspora erythraea*).[136–139] It is a bitter-tasting white powder when used unmodified. Structurally, it is a 14-membered lactone with two sugars, cladinose and desoamine, attached at C3 and C5 positions, respectively (**1**). Due to the dimethylamino moiety of desosamine, erythromycin and its derivatives are slightly basic ($pK_a = 8.8$), which enables purification by acid and base extraction.

Erythromycin's solubility in water is $2\,mg\,mL^{-1}$ and, in spite of its high molecular weight and poly functional groups, it is well suited for chemical manipulations. Erythromycin is soluble in typical organic solvents such as dichloromethane or tetrahydrofuran and, like many of its derivatives, is highly crystalline. The robust lactone linkage is capable of withstanding many synthetic conditions. For example, it is resistant to nucleophilic reactions due to the steric hindrance around the C1 carbonyl group (numbering starts from the lactone carbonyl group and goes counterclockwise around the lactone ring). Furthermore, the five hydroxy groups in the molecule are not uniformly reactive due to the variance in steric environments and neighboring groups. In acetylation reactions, for example, the C2′ hydroxy group is most reactive since the neighboring dimethylamino group acts as an internal base catalyst. Being exposed to the solvent sphere, the next group to be acylated is the C4″ hydroxy group of cladinose. These differences thus make it possible to carry out selective transformations, which can be achieved through protection and deprotection of other functional groups.

Erythromycin is fairly stable in aqueous bases but is somewhat sensitive to acids, which degrade the molecule by cleaving the cladinose sugar as well as forming an internal acetal between the C6 hydroxy and C9 carbonyl groups (see below). This acid-catalyzed cleavage of the sugar is observed with cladinose but not with desosamine since the dimethylamino moiety of desosamine becomes protonated first and the second protonation of the glycosidic linkage is difficult. The x-ray structure and the nuclear magnetic resonance (NMR) studies of erythromycin indicate that most of the polar groups are on one face of the molecule while the other face consists of aliphatic groups.[140,141] This fact may be important in considering the binding of erythromycin to the ribosome as described earlier as well as its penetration into bacterial cells. Erythromycin, besides being an antibiotic in its own right, serves as a rich starting point for new chemistry as shown in the later sections.

7.18.5.1.1 Spectrum of activity

Erythromycin is mainly active against Gram-positive pathogens such as macrolide-susceptible *S. aureus*, *S. pneumoniae*, and *S. pyogenes*. It is used to treat infections of the respiratory tract, skin, soft tissues, and urogenital tract. The increase in the numbers of erythromycin-resistant *S. pneumoniae* and *S. pyogenes* (both Erm and Mef strains), however, makes the use of erythromycin more problematic in the clinic than in the past. Its activity against Gram-negative organisms is limited due to its poor penetration through the outer membrane and active efflux pumps as mentioned before. It is, however, active against some Gram-negative pathogens, including *Neisseria meningitis, N. gonorrhoeae*, and *Bordetella pertussis*, but has relatively little activity against *H. influenzae*, a common Gram-negative pathogen in respiratory tract and middle ear infections. Even though erythromycin's spectrum of activity is not as broad as other major classes of antibiotics such as penicillins, cephalosporins, quinolones, and tetracyclines, one advantage of erythromycin over some of the other agents is its relatively good safety profile as discussed below.

7.18.5.1.2 Structure–activity relationships

Through many studies on structural modifications of erythromycin, there is a wealth of information on its SAR.[142] The macrolactone ring is indispensable for the antibacterial activity, for example, as the hydrolysis of this lactone linkage by an esterase is one way some bacteria cope with this antibiotic.[143] The ring acts as the scaffold for several functional moieties that must be correctly oriented to interact with key regions of the bacterial ribosome. The desosamine sugar also mediates key interactions with the ribosome target. For the antibacterial activity, it is generally considered that the C2′ hydroxy and C3′ dimethylamino groups of desosamine are indispensable. And in fact, acylation or alkylation of the C2′ hydroxy group leads to inactive derivatives as does the removal of the C3′ dimethylamino group or its conversion to N-oxide.[144–146] Similarly, addition of a phosphate group at C2′ hydroxy group is one of the bacterial resistance mechanisms.[147] The significance of these groups for activity can be explained by the recent x-ray structures of macrolide complexed with the bacterial ribosome. The structures show that the dimethylamino group is salt-bridged to the phosphate backbone of the ribosomal RNA, and the C2′ hydroxy group is hydrogen-bonded to the ribosome bases (A2040, A2058, A2059, A2062 (*E. coli* numbering)).[93,94] Mono- or di-demethylation of the dimethylamino group leads to reduced activity, presumably through loss of interaction with the RNA phosphate backbone (**Figure 4**).[148]

In the biosynthesis of erythromycins, C6 and C12 hydroxy groups are incorporated onto the ring late in their synthesis by cytochrome P450 oxidases.[149] A compound without the C12 hydroxy group is designated as erythromycin B (**9**) and has approximately 75–85% of the potency of erythromycin A.[150] A compound without the C6 hydroxy group was prepared by genetic engineering, and also demonstrated reduced activity in comparison to erythromycin A.[151] The C6 hydroxy group, however, can be alkylated with small alkyl groups without eliminating erythromycin's antibacterial activity (see clarithromycin below). Similarly, although the C9 carbonyl group is important for activity, there is significant latitude for modifications to this area of the molecule. X-ray structures reveal that the C9 carbonyl is not involved in strong interactions with the ribosome.[93,94] As a result, the carbonyl group can be converted to an oxime or an amino group (see roxithromycin and dirithromycin respectively below) without limiting antibacterial activity. The freedom to alter this functional group is not absolute, however, as C9 alpha- and beta-alcohols have reduced potency compared with erythromycin.[152] More recently, the replacement of the C13 ethyl group by a methyl, propyl, cyclopropyl, or cyclobutyl group has been achieved by incorporating the corresponding starter units in the biosyntheses by fermentation and the resulting compounds retain most of erythromycin's original antibacterial activity.[153–156] Finally, the C4″ hydroxy group of cladinose can be either acylated or alkylated, with the resulting activity dependent upon the modifying group.[157]

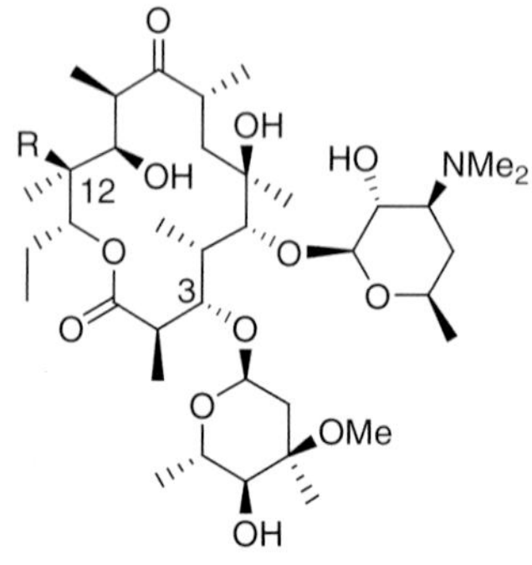

1 Erythromycin A: R = OH
9 Erythromycin B: R = H

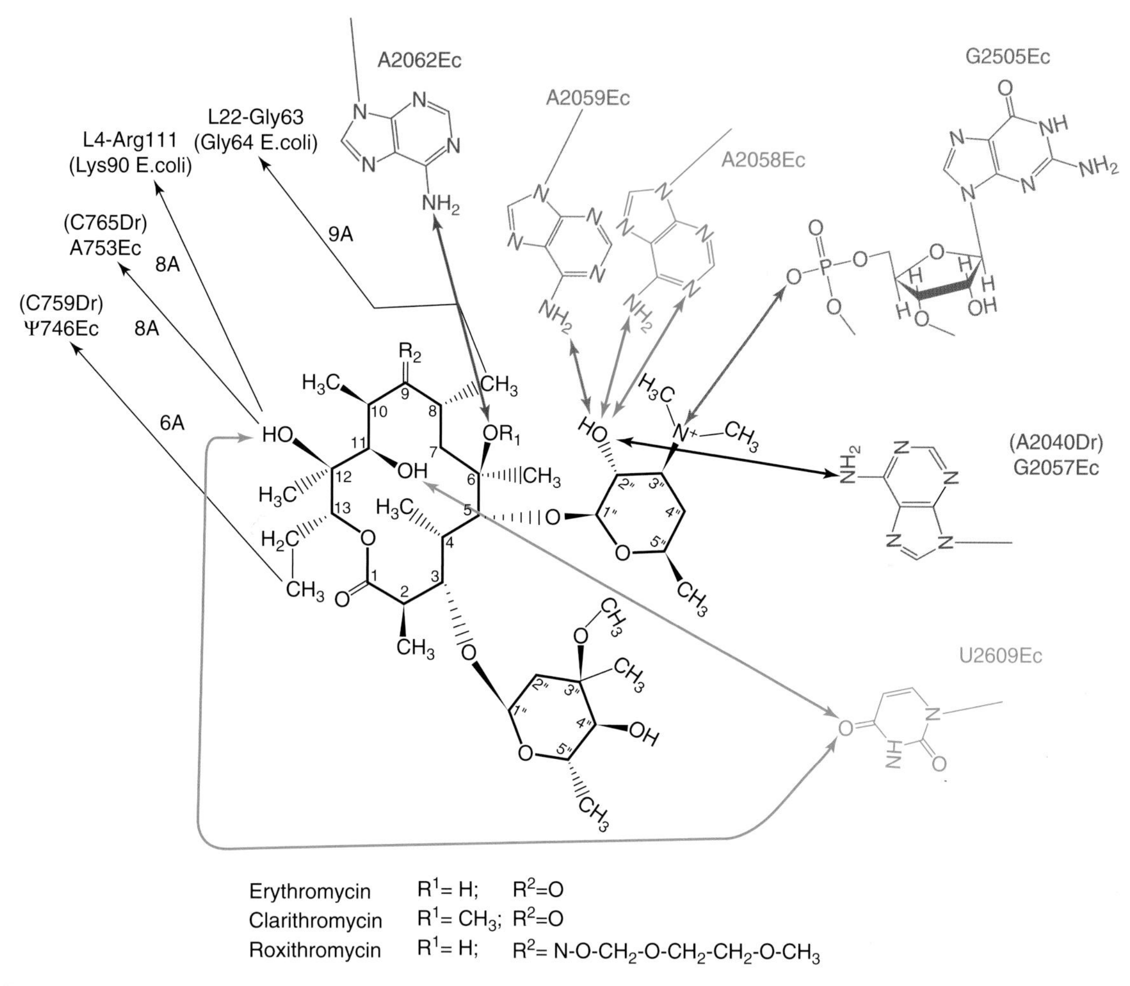

Figure 4 Interaction of macrolides with the peptidyl transferase cavity. (Reproduced with permission from Schlunzen, F. *Nature* **2001**, *413*, 814–821.)

7.18.5.1.3 Pharmacokinetic properties

Erythromycin, like other macrolide antibiotics, is orally administered. Because of its acid instability, however, erythromycin exhibits high interindividual variability in oral bioavailability, with values ranging from 15% to 45%. The maximal plasma concentration after a typical 250–500 mg oral dose is achieved after approximately 4–5 h, and ranges from 0.3 to 3.5 $\mu g\,mL^{-1}$. Erythromycin has a low volume of distribution (0.64 $L\,kg^{-1}$), which contributes to its short half-life (approximately 1.4 h) and necessitates a frequent dosing schedule, typically 250 mg four times a day or 500 mg twice a day. Erythromycin is metabolized in the liver, and its major metabolites are the demethylated product at the dimethylamino group, the N-oxide of the desosamine, and des-cladinose erythromycin, all of which have much reduced antibacterial potency. Erythromycin and its metabolites are eliminated mainly in bile. In terms of tissue distribution, higher concentrations are found in tissues such as the lung and tonsils than in plasma. Like other macrolides, it accumulates in phagocytic cells.[158] Since phagocytic cells (macrophages) migrate to the site of infection and release the macrolide, this is considered to be significant in raising the local concentration of the antibiotic. Erythromycin is 65–90% protein bound with α1-glycoprotein as the main binding partner. The key pharmacokinetic values are summarized in **Table 1**.[159,160] As mentioned earlier, one serious drawback of erythromycin is its instability under acidic conditions such as those encountered in the stomach. With acid catalysis, erythromycin undergoes a series of transformations promoted by the proximity of functional groups and by the entropic factors.[161] **Scheme 1** demonstrates the initial change that occurs, as an internal acetal develops and quickly loses a molecule of water to form a cyclic enol ether. These transformations irreversibly result in microbiologically inactive products. Due to this vulnerability, erythromycin's effectiveness is threatened by both low bioavailability and high interpatient variability. Enteric-coated tablets were developed with some success to protect erythromycin from degradation and to mask its bitter taste.[162]

Table 1 Oral pharmacokinetic properties of first- and second-generation macrolides

Drug	*Typical dose*	*Bioavailability*	C_{max} ($\mu g\, mL^{-1}$)	t_{max} (h)	AUC_{0-24} ($\mu g\, h\, mL^{-1}$)	$t_{1/2}$ a(h)	Vd/F ($L\, kg^{-1}$)	*Plasma protein binding (%)*
Eyrthromycin base	500 mg b.i.d.	15–45%	3.5	2	15	2.5	0.64	65–90
Clarithromycin	500 mg b.i.d.	55%	2.7	2.6	40	4.7	3.1	70
Azithromycin	500 mg qd, day 1 250 mg qd, days 2–5	37%	0.24	3.2	2.1	40–68	31	7–50
Roxithromycin	300 mg qd	78%	10	1.2	130	12	0.87	73–90
Dirithromycin[a]	500 mg qd	10%	0.30	2.7	2.4	17	11	15–30
Flurithromycin	500 mg t.i.d.	ND	2.0	1.5	25	9.0	5.5	70

b.i.d., twice a day; t.i.d., three times a day: qd, once daily; ND, not determined.
[a]Values represent serum levels of erythromycylamine, the active form of dirithromycin.

In attempts to improve the pharmacokinetic properties of erythromycin, various C2′ esters and salts of the amino sugar moiety were generated. For example, erythromycin stearate[163] is a salt formed between the basic desosamine moiety and stearic acid, while erythromycin ethylsuccinate is a prodrug ester formed between the C2′ hydroxy group of desosamine and ethyl succinic acid.[164] Moreover, erythromycin estolate is a propionate ester of the C2′ hydroxy group and also a laurylsulfate salt of desosamine.[165] There are conflicting reports on the extent to which these various preparations improve upon the original bioavailability of erythromycin, and a more successful effort to improve the pharmacokinetic properties can be seen in clarithromycin and azithromycin.

7.18.5.1.4 Safety properties

Erythromycin is well tolerated, with adverse effects usually limited to gastrointestinal discomfort including nausea, diarrhea, and cramps. One area of more recent concern is the potential for drug–drug interactions. It has been known for some time that erythromycin, like many other macrolides, inactivates cytochrome P450 enzymes, with a particularly robust effect on CYP3A4. The inhibitory mechanism involves oxidation of erythromycin's desosamine moiety with cytochrome P450 and inactivation of the heme enzyme by the resulting intermediate (see **Scheme 2**).[166] Given this interaction, erythromycin can interfere with the metabolism of other drugs processed by cytochrome P450 enzymes and influence the pharmacokinetic properties of these molecules. Although most macrolide antibiotics possess desosamine or a similar amino sugar, there is variability of the individual compound's ability to inhibit cytochrome P450 (see azithromycin and 16-membered macrolides). The inactivation appears to depend on the lipophilicity of the molecule, the steric hindrance around the dimethylamino moiety, and the overall shape of the molecule.[167,168] On the other hand, macrolides, including erythromycin, can induce production of cytochrome P450 enzymes, which leads to a reduced level of the second object drug.[169] That is, they can increase the level of cytochrome P450 enzymes in the liver, which accelerates the metabolism and lower the concentration of the second agent. In in vitro studies using primary human hepatocytes, however, an increased level of cytochrome P450 was not observed with erythromycin.[170] At least this potential for drug–drug interactions needs to be recognized when macrolides including erythromycin are coadministered with other drugs.

In terms of the potential for hepatotoxicity, erythromycin itself shows little tendency. Erythromycin estolate, on the other hand, demonstrates an increased tendency to cause cholestasis.[171,172] As is the case with other drugs, cardiovascular adverse effects have recently been investigated more closely. These effects are related to drugs' ability to interact with ion channels in heart muscle, which can lead to changes in contractility termed QTc prolongation. Erythromycin causes prolongation of QTc interval and can induce arrhythmia under certain circumstances.[173] The ranking of this potential among typical macrolides is erythromycin > clarithromycin > azithromycin.[174,175] Combined with its tendency to inhibit cytochrome P450, erythromycin can increase the incidence of arrhythmia, particularly if coadministered with a drug of proarrhythmic potential. It has to be stressed, however, that the incident rates of these adverse effects are low, and that erythromycin is considered to be one of the safer antibiotics.

Thus, erythromycin is a prototype macrolide antibiotic, which has a moderately broad spectrum of activity. Because of the acid instability, its bioavailability is not very high and requires frequent dosing. In order to improve the

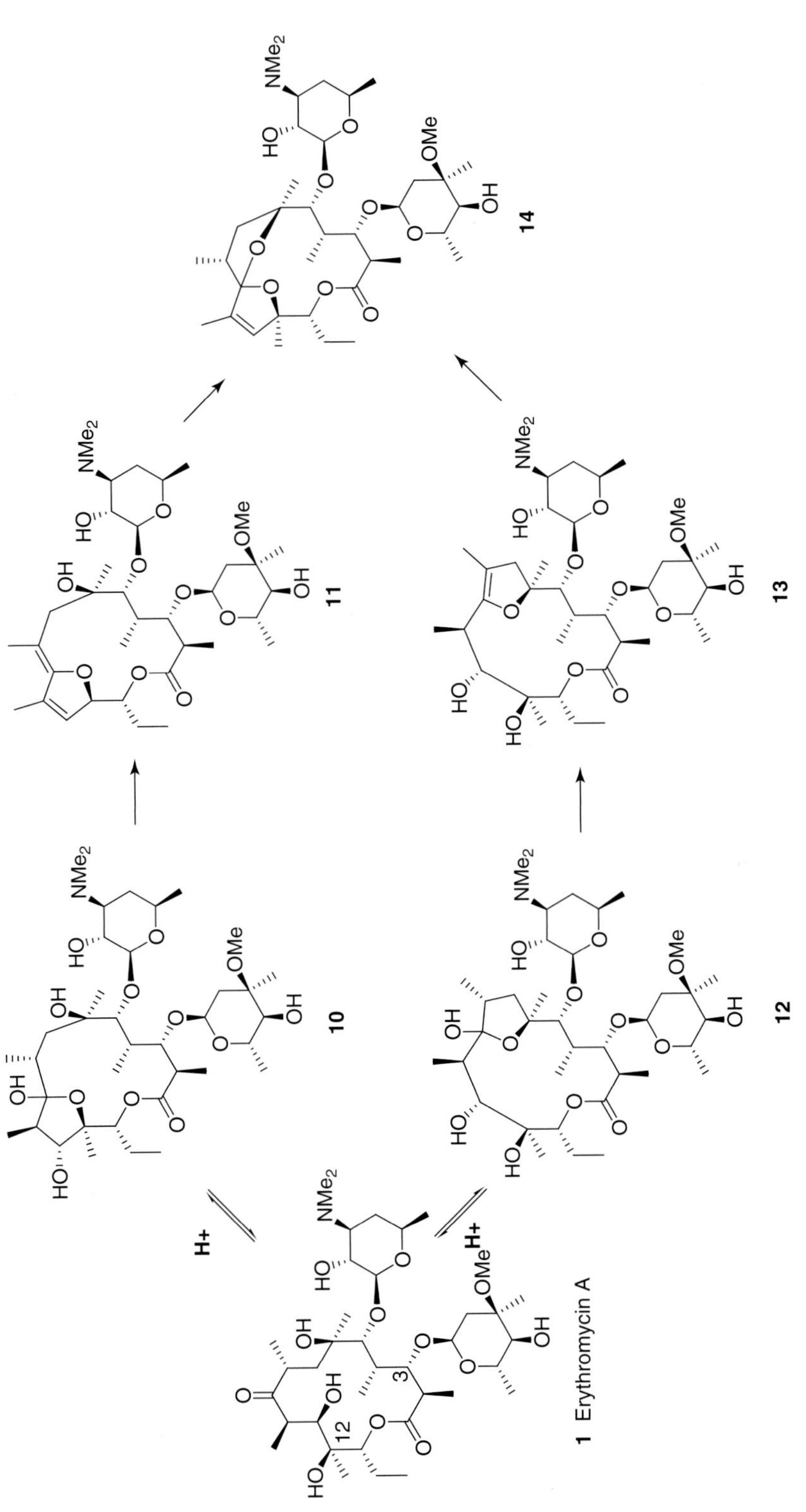

H+

H+

1 Erythromycin A

10

11

12

13

14

Scheme 1

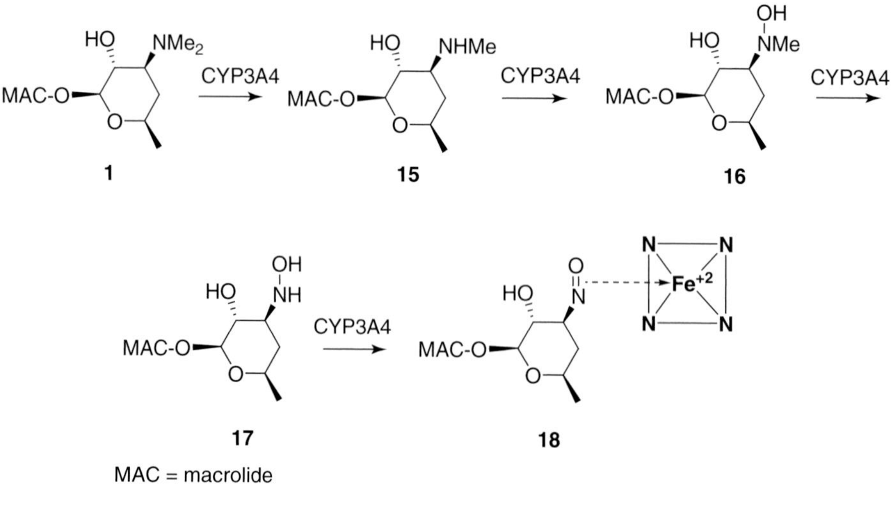

MAC = macrolide

Scheme 2

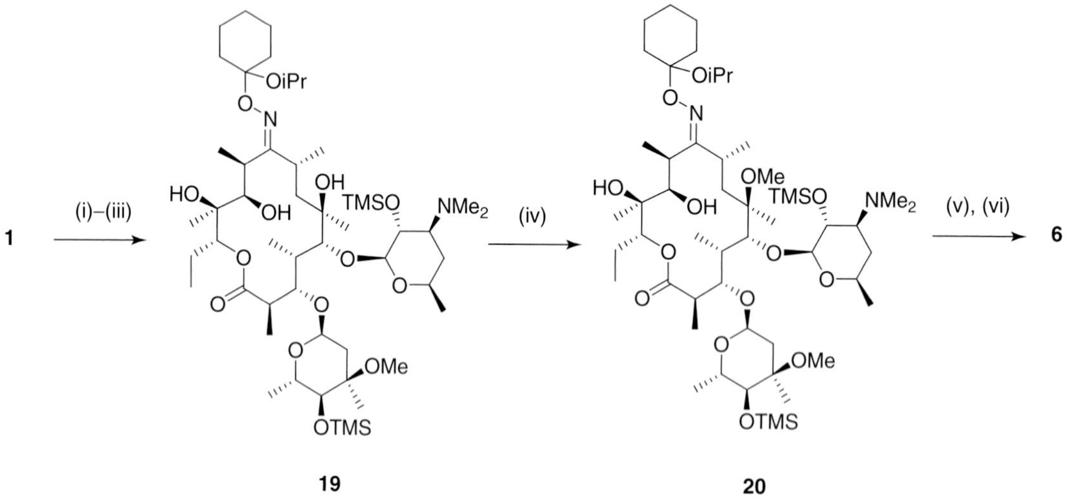

Scheme 3

pharmacokinetic properties and to increase the spectrum of activity a large number of chemical modifications have been carried out. Clarithromycin and azithromycin, compounds belonging to the second generation of macrolides, were discovered from these studies.

7.18.5.2 Clarithromycin

An early effort to avoid the acid-catalyzed inactivation of erythromycin focused on capping any of the hydroxy groups involved in the internal acetal formation. The creation of clarithromycin (**6**) achieved this by converting the C6 hydroxy group into a C6 methyl ether group. Although this alkylation process was not initially selective,[176] considerable effort went into the process chemistry to develop robust selective alkylation. **Scheme 3** demonstrates an improved route in which key steps are: (1) conversion of erythromycin to C9 oxime to avoid the internal acetal formation during the subsequent steps, (2) protection of the C2′ and C4″ hydroxy groups with a trimethylsilyl group to enhance C6 methylation, and (3) protection of this oxime with an easily cleavable O-alkyl group.[177] As mentioned later, clarithromycin is also important as the starting material for the synthesis of ketolides. Without clarithromycin's capped C6 hydroxy group, the C6 hydroxy group would attack the C3 carbonyl group forming a C3–C6 internal acetal and it would be impossible to generate ketolides (see below).

7.18.5.2.1 Spectrum of activity

The spectrum of activity of clarithromycin is similar to that of erythromycin.[178–180] It is active mainly against Gram-positive pathogens such as *S. aureus*, *S. pneumoniae*, and *S. pyogenes*. Clarithromycin appears to be equivalent or slightly more potent than erythromycin against these pathogens. As with erythromycin, it is, however, ineffective against streptococcal strains carrying Erm or Mef resistance mechanisms. Against *M. pneumoniae*, *L. pneumophila*, and *C. pneumoniae*, it is more potent than erythromycin while also active against *M. avium* complex (MAC). Although clarithromycin is only slightly more active than erythromycin against *H. influenzae*, it exhibits clinically useful activity against this pathogen in a synergistic effect with its major human metabolite, C14-hydroxy clarithromycin.[181] It is of interest to note that with the introduction of the C6 methyl ether group, the major metabolite has shifted from the N-demethylation product to the C14 hydroxyation product. Clarithromycin is used clinically to treat respiratory tract infections including community-acquired pneumonia, sinusitis, pharyngitis, bronchitis, and acute exacerbation of chronic bronchitis. Furthermore, it has a high efficacy against MAC infections in human immunodeficiency virus (HIV) patients having a disseminated infection or non-AIDS patients with localized pulmonary disease.[182] Clarithromycin is particularly active against *Helicobacter pylori*, the organism associated with the chronic inflammation of stomach ulcers, and is currently used, along with a proton pump inhibitor and bismuth, in the triple therapy against this pathogen.[183]

7.18.5.2.2 Structure–activity relationships

As the search for an improved macrolide progressed, various alkyl derivatives of erythromycin were investigated. Compounds in which more than two of the five hydroxy groups in erythromycin (or four hydroxy groups in erythromycin B) are methlyated are less potent than erythromycin against *S. aureus*. Similarly, C11 *O*-methyl erythromycin is slightly less potent than erythromycin and the ethyl ethers at C6 or C11 hydroxy group are less potent than the corresponding methyl ethers.[184] There are, however, a number of erythromycin derivatives that proved to be equally effective, including C6,C12-dimethyl erythromycin, which matches erythromycin's activity. The C6 methyl ether is still most potent in vitro among the various ether derivatives. In spite of little differentiation in vitro between erythromycin and clarithromycin, the in vivo studies revealed real advantages of clarithromycin. In mouse systemic infection models, for example, the ED_{50} values (dose of a drug that cures 50% of animals) for clarithromycin given orally were 6–15 times lower than those for erythromycin against *S. aureus*, *S. pyogenes*, and *S. pneumoniae* although the MIC values against these pathogens were essentially identical.[185] It is thus the improved pharmacokinetic properties of clarithromycin that result in its enhanced in vivo activity.

7.18.5.2.3 Pharmacokinetic properties

Clarithromycin's advantage over erythromycin in terms of pharmacokinetic properties is attributable to its improved oral absorption, lower clearance rate, and higher volume of distribution. In mice and rats, the AUC (area under curve) of clarithromycin was three- to fourfold higher than that of erythromycin when both compounds were given orally at $100\,mg\,kg^{-1}$.[186] In humans, clarithromycin has a bioavailability of 50–55% as compared to 15–45% for erythromycin – a difference caused by clarithromycin's increased lipophilicity and its relative stability in acidic environments such as the stomach.[187] A human study in which 200 mg of each drug was administered revealed a C_{max} and AUC of $1.07\,\mu g\,mL^{-1}$ and $7.18\,\mu g \cdot h\,mL^{-1}$ for clarithromycin as opposed to $0.38\,\mu g\,mL^{-1}$ and $1.37\,\mu g \cdot h\,mL^{-1}$ for erythromycin. Clarithromycin's volume of distribution is approximately five times higher than erythromycin, which is at least partially attributable to its increased lipophilicity (ClogP = 2.37 whereas ClogP of erythromycin is 1.61). A considerably higher proportion (23%) is excreted in urine compared to erythromycin (3%).[188] Clarithromycin is metabolized by cytochrome P450 enzymes and, as mentioned earlier, C14-hydroxy clarithromycin is the major metabolite in humans. The peak concentration of C14-hydroxy clarithromycin is achieved within 3 h of oral administration and its level can reach one-third of the parent compound.[189] Clarithromycin's key pharmacokinetic properties are summarized in **Table 1**.[190–193]

7.18.5.2.4 Safety properties

Clarithromycin is well tolerated and its main adverse effects are gastrointestinal effects similar to those of erythromycin. In addition, it inhibits some of the cytochrome P450 enzymes and thus can exhibit drug–drug interactions. Like erythromycin, it can cause QTc prolongation but this tendency appears to be less pronounced than that seen with erythromycin.[176] Clarithromycin thus represents a significant improvement over erythromycin for its in vivo activity and pharmacokinetic properties.

7.18.5.3 Azithromycin

Azithromycin is another second-generation macrolide with improved PK properties and a greater spectrum of activity than erythromycin.[194–196] It belongs to a class of 15-membered macrolides known as azalides due to the additional amino group (aza) in the lactone ring. Derived from the erythromycin C9(*E*)-oxime by a Beckmann rearrangement as shown in **Scheme 4**, azithromycin is a lactone in which erythromycin's problematic C9 carbonyl group has been replaced by an aminomethylene group.[8,9,197] The resulting compound (9-deoxo-9a-aza-9a-homoerythromycin A, **23**) is immune to the acid-catalyzed internal acetal formation observed with erythromycin. Because Beckmann rearrangements are stereospecific, it produces a slightly different ring system, 9-deoxo-8a-aza-8a-homoerythromycin A (**25**) if the starting oxime has the opposite stereochemistry, C9(*Z*) oxime (**24**).[198,199] Azithromycin has an additional basic site ($pK_a = 9.5$)[200] within the lactone ring, which contributes to its improved activity and pharmacokinetic properties such as better activity against fastidious Gram-negative pathogens and higher tissue penetration.

7.18.5.3.1 Spectrum of activity

Azithromycin is significantly more potent than either erythromycin or clarithromycin against the Gram-negative pathogen *H. influenzae*.[201] Like other prominent macrolides, it is highly active against another upper respiratory tract Gram-negative pathogen, *M. catarrhalis*, as well as against an intracellular pathogen, *C. trachomatis*. Clinically, azithromycin is used to combat respiratory tract infections, otitis media, pharyngitis, sinusitis, and trachoma eye infections, and prophylactically against MAC infections in AIDS patients.[202] Its activity, however, against Gram-positive pathogens is slightly diminished compared to erythromycin or clarithromycin.[203,204] Like those two, azithromycin is ineffective against streptococcal strains carrying an Erm of Mef resistance mechanism.

7.18.5.3.2 Structure–activity relationship

The initial product of the Beckmann rearrangement is an unsubstituted amine (**23**). Although **23** itself is quite active in vitro, alkyl groups at N-9a such as methyl, ethyl, or benzyl group at the ring nitrogen increase activity against *S. aureus* and *H. influenzae*.[9] It is of interest that a fairly bulky group such as a benzyl group can be accommodated without losing in vitro potency against these pathogens. Modifications at the C4″ position of cladinose within azalides indicate that it can be an alpha- or beta-alcohol, a ketone, or an alpha-amino group, but the amino derivative has the highest Gram-negative activity.[205] A 9,9a-lactam derivative (**27**) prepared more recently was found to be inactive.[206] The 8a-aza-8a-homoerythromycin derivatives derived from the *Z*-oxime isomer mentioned above show a similar trend to azithromycin analogs, in that the N-alkylated analogues are more potent than unsubstituted analogs.[207] The in vitro

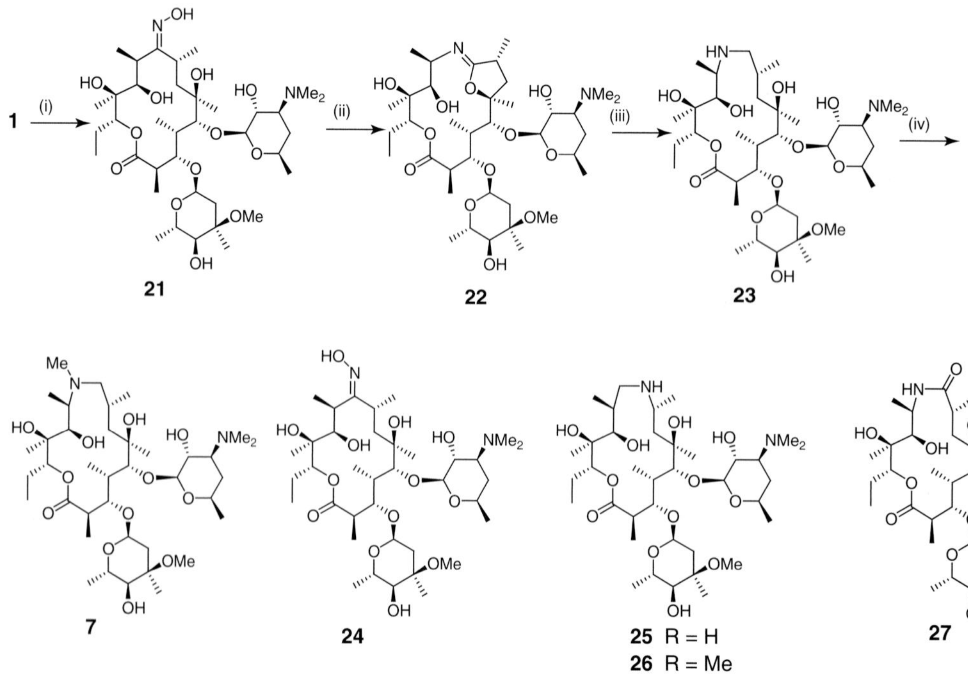

1 (i) → 21 (ii) → 22 (iii) → 23 (iv) → 7

24

25 R = H
26 R = Me

27

Scheme 4

potency of *N*-methyl 8a-aza-8a-homoerythromycin (**26**) was almost identical to that of azithromycin against various Gram-positive and Gram-negative pathogens. Thus, the ring expansion in either direction produced analogs with a spectrum of activity quite distinct from that of erythromycin.

7.18.5.3.3 Pharmacokinetic properties

The presence of a basic center in the macrolactone ring of azithromycin plays an important role in differentiating its pharmacokinetic properties from those of erythromycin and clarithromycin.[203,204] Due to the presence of this second basic center, azithromycin has a very high volume of distribution relative to other macrolides. The oral bioavailability of azithromycin is approximately 37%, and the primary route of elimination is through biliary excretion of unchanged drug. After oral dosing, azithromycin is rapidly distributed into tissues and subsequently undergoes slow release from the tissue compartments. Consequently, the serum levels are lower than observed with erythromycin or clarithromycin, but its tissues concentrations are 10–100-fold higher than the serum.[208] In addition, azithromycin is internalized to a higher degree than erythromycin or clarithromycin by polymorphonuclear leukocytes and macrophages and transported to the infection site, thus increasing the local concentration of the drug.[209] The high volume of distribution also confers a long terminal half-life of 40–68 h, thereby making it possible to limit dosing to once daily. These pharmacokinetic advantages cumulatively mean that in upper respiratory tract infections, a 5-day treatment with azithromycin was found to be as effective as a 10-day treatment with comparator drug.[210,211] Very recently, a new formulation was developed such that only one dosing per course of therapy is necessary.[212,213] The major pharmacokinetic parameters are listed in **Table 1**.[208,214]

7.18.5.3.4 Safety properties

Much like other important macrolides, azithromycin is well tolerated, with the most prominent adverse effect being gastrointestinal disturbance. Unlike erythromycin and clarithromycin, however, azithromycin has little tendency to interact with cytochrome P450 enzymes and studies have shown little potential for drug–drug interactions.[215] In addition, azithromycin's tendency to prolong QTc appears to be lower than that of clarithromycin and erythromycin, and it is therefore less likely to induce arrhythmia.

7.18.5.4 Roxithromycin

Roxithromycin is a derivative of erythromycin in which the C9 carbonyl group is converted to an oxime in order to prevent acid-catalyzed inactivation, more specifically to 9-[*O*-(2-methoxyethoxy)methyl]oxime (**28**).[216–218] It is more lipophillic (ClogP = 3.51) than erythromycin. Since it does not revert to erythromycin under physiological conditions, it is not a prodrug of erythromycin. The in vitro antibacterial spectrum of roxithromycin is similar to that of erythromycin, with activity against macrolide-susceptible *S. aureus*, *S. pyogenes*, and *S. pneumoniae*. It is, however, not particularly potent against *H. influenzae* since its size is considerably larger than erythromycin, which presumably limits penetration through the outer membrane porins. Roxithromycin is characterized by its exceptionally high bioavailability (72–85%), high plasma concentration, and a long half-life in human (see **Table 1**).[160,218] Roxithromycin is not extensively metabolized. In mouse intraperitoneal infection models, roxithromycin was efficacious at lower doses compared with erythromycin, having ED_{50} values 4–14 times lower than those for erythromycin.[217] In human clinical trials, the cure rates in upper or lower respiratory tract infections or genitourinary tract infections were about the same as those of erythromycin. The dosing regimen, however, appeared to be improved. Roxithromycin at 150 mg twice daily or 300 mg once daily produced 83–87% cure rates compared to the 88% cure rate generated with 400 mg of erythromycin ethyl succinate four times a day.[219] Its adverse effects are gastrointestinal, similar to those of other macrolide antibiotics, and it exhibits little cytochrome P450 inhibition.

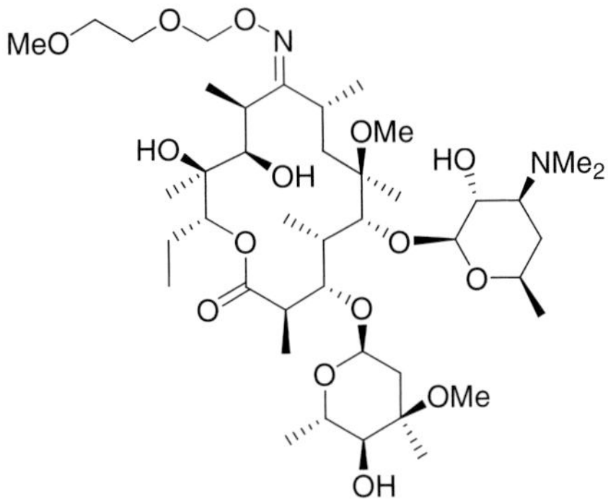

28 Roxithromycin

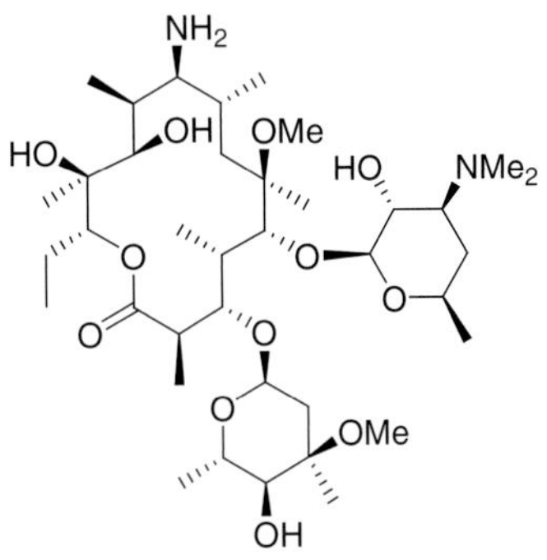

29 Erythromycylamine

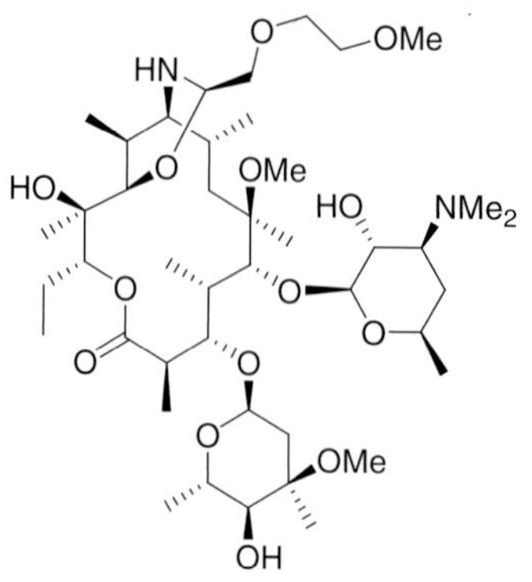

30 Dirithromycin

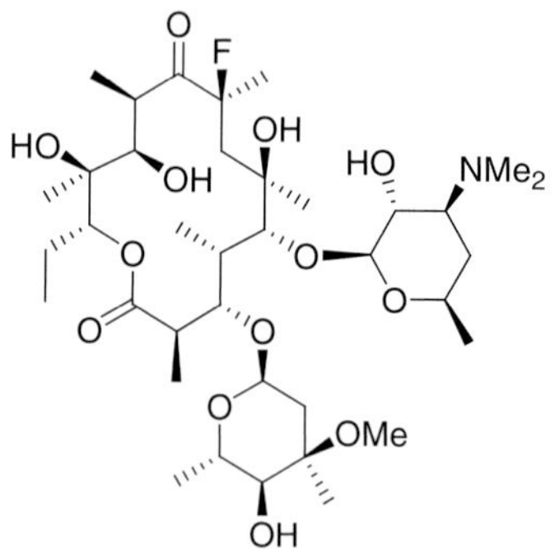

31 Flurithromycin

7.18.5.5 Dirithromycin

Another strategy to prevent the acid-catalyzed inactivation of erythromycin is to convert the C9 carbonyl group into an amine. This is accomplished by the catalytic reduction of C9 oxime or hydrazone. The major isomer, 9(*S*)-erythromycylamine (**29**) has antibacterial activity corresponding to 40–60% of erythromycin, whereas the 9(*R*) isomer is several-fold less potent than erythromycin.[220] Although erythromycylamine retains activity, its absorption is definitely poorer than erythromycin, perhaps due to the presence of a hydrogen bond donor group. In order to improve its absorption, various aldehyde adducts and their reductive amination products were investigated. One of them is an adduct of 2-(2-methoxyethoxy)acetaldehyde (as shown in oxazine structure **30**) in which the C11 hydroxy group reacted with the initial C9 imine or carbinolamine intermediate, known as dirithromycin.[221–223] Under acidic conditions, dirithromycin is rapidly converted to erythromycylamine; thus it is a prodrug of the latter. Dirithromycin possesses an in vitro antibacterial activity similar to that of erythromycin. It provides, however, higher and more prolonged tissue concentrations after oral administration. The pharmacokinetics of dirithromycin are characterized by low oral bioavailability, but relatively large volume of distribution and a long half-life (see **Table 1**).[224–227] It appears that dirithromycin and erythromycylamine have a low potential to interact with cytochrome P450. Little has been published on their cardiovascular effects.

7.18.5.6 Flurithromycin

Flurithromycin is 8(*S*)-fluoro-erythromycin, which was generated by fluorination of 8,9-anhydroerythromycin A 6,9-hemiketal N-oxide[228] or by feeding the fluorinated aglycone to a mutated strain of *S. erythraea*.[229] The spectrum of activity of flurithromycin is almost identical to that of erythromycin but it is more stable to acid as shown by its half-life in artificial gastric juice (38 min) compared with erythromycin (less than 5 min).[230] The increased stability is reflected in the higher serum levels in rat and rabbit compared with a similar dose of erythromycin, and in its lower ED_{50} values in the mouse infection models involving macrolide-sensitive strains of *S. aureus, S. pyogenes*, or *S. pneumoniae*. The pharmacokinetic properties of flurithromycin are also listed in **Table 1**.[231] In a comparative study of flurithromycin ethyl succinate with clarithromycin in infections of lower airways in human, both compounds were efficacious and tolerated to the same extent.[228] Flurithromycin does not appear to inactivate cytochrome P450 enzymes in rat.[232] As mentioned before, inhibition of a ribosome subunit assembly is a potential target of macrolides' mechanism of action. For this inhibitory effect of 30S subunit formation, flurithromycin appears to be more potent than clarithromycin or roxithromycin.[233] These agents – clarithromycin, azithromycin, roxithromycin, dirithromycin, and flurithromycin – are called the second-generation macrolides and they are characterized by their improved spectrum of activity and/or improved pharmacokinetic properties over erythromycin. Unfortunately, their widespread use accelerated the emergence of resistant strains.

7.18.5.7 Third-Generation Macrolides: Ketolides

The monumental success of the second-generation macrolides has brought upon the need to address the rising Gram-positive (streptococcal) resistance to this class (*see* Section 7.18.4). By far the most important development in addressing macrolide resistance is the discovery of the ketolides, so called for their ketone functionality at position C3 of the macrolactone ring. The ketolide class – with particular emphasis on telithromycin (**8**) and cethromycin (**32**) – has been reviewed extensively.[234–237] The advent of the ketolides disproved the long-held notion that the cladinose sugar at C3 was necessary for antibacterial activity, and marked the beginning of the third-generation macrolides. Although the natural product picromycin (**2**) may be formally considered a 'ketolide,' the name is generally applied to semisynthetic derivatives of clarithromycin in which the cladinose sugar has been selectively removed, the resulting C3 hydroxy oxidized to a ketone functionality, and a tethered aryl group has been introduced to provide a novel binding element. In addition to the 3-keto functionality, the tethered aryl moiety is critical to the potency of the ketolides. It is hypothesized that the molecule reaches from the primary point of contact for the macrolides in domain V into a novel region of the ribosome in domain II. This contact with a second domain of the ribosome has been established by footprinting, mutation studies, and more recently by crystallography (*see* Section 7.18.3.5).

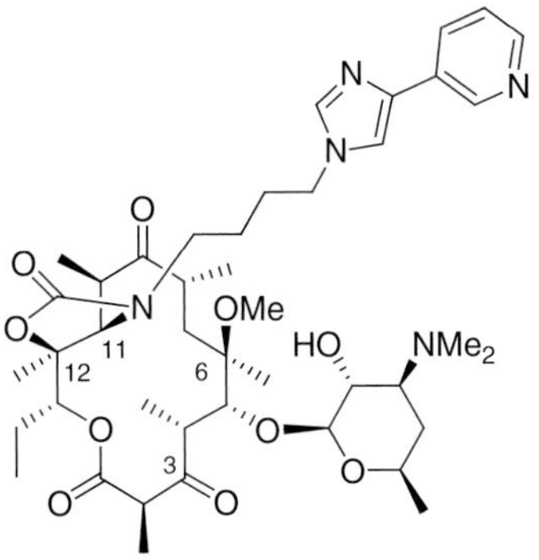

8 Telithromycin

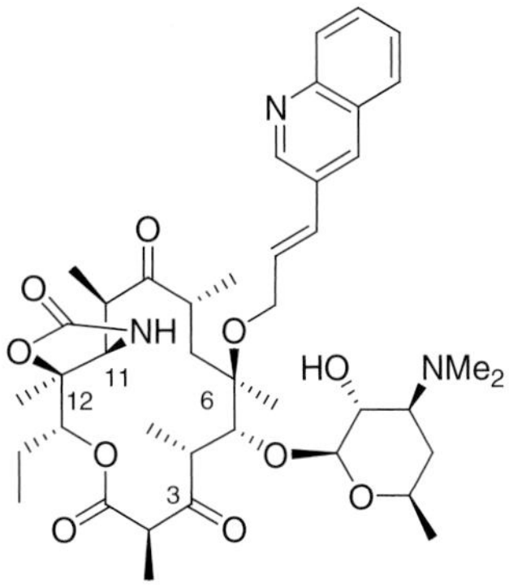

32 Cethromycin

The ketolides offer a number of advantages over the first- and second-generation macrolides. First, they show excellent in vitro activity against both important macrolide-sensitive and -resistant Gram-positive respiratory pathogens, most notably *S. pneumoniae*. The combination of additional binding contacts to the ribosome and the replacement of the cladinose ring with a ketone group yields structures with higher ribosomal affinities, which in turn results in better performance against resistance mechanisms. Second, they have improved potency versus the fastidious Gram-negative respiratory pathogen *H. influenzae* relative to clarithromycin and erythromycin – activity only slightly lower than azithromycin. Finally, they are more stable in acidic media (e.g., as found in the gastrointestinal tract) owing to the lack of a cladinose at C3. A necessary structural feature of the ketolides – in addition to the 3-keto and tethered aryl ring – is the presence of a group other than hydrogen on the C6 hydroxy to avoid hemiacetal formation between C6 hydroxy and C3 carbonyl groups. Another important feature is the presence of a ring fused onto the macrocyclic skeleton, often in the form of a cyclic carbamate spanning C11 and C12. This ring rigidifies the macrocycle and, from the crystallographic data, appears also to participate in ribosome binding. The attachment point of the tethered aryl is used to segregate the ketolides into separate families in this review. First the C11-tethered ketolides, wherein the tether runs through the C11/C12 ring fusion, will be discussed. The first in class and most advanced ketolide, telithromycin (Aventis' Ketek, **8**), belongs to this important family. The second group is the family of C6-tethered ketolides, represented primarily by cethromycin (ABT-773, **32**) from researchers at Abbott Laboratories. Cethromycin, currently still in phase III clinical trials, represents a chemotype that is distinct from telithromycin and the C11 tethered ketolides. Variations on both the C6- and C11-tethered ketolides will be discussed, as will the recently discovered hybrid bridged bicyclo ketolides. In many cases, there is little or no information about the medicinal viability of the variations (i.e., the pharmacokinetics, in vivo efficacy, or the safety) and therefore an emphasis is placed on the chemistry used to generate modified and novel structures and, where possible, the in vitro profile of the resultant analogs.

7.18.5.7.1 C11-Tethered ketolides: telithromycin

Researchers at Abbott Laboratories were the first to develop the chemistry[238] and notice the advantages of C11/C12 cyclic carbamates in overcoming Erm-resistant streptococci. Starting with clarithromycin (**6**), the desosamine and cladinose sugar hydroxy groups were protected with acetyl and benzyloxycarbonyl groups respectively (**Scheme 5**). Treatment with excess carbonyldiimidazole and sodium hexamethyldisilazide in one step resulted in elimination of the C11-hydroxy group and the formation of the O12 acyl imidazole **33**, a key electrophilic intermediate used for the

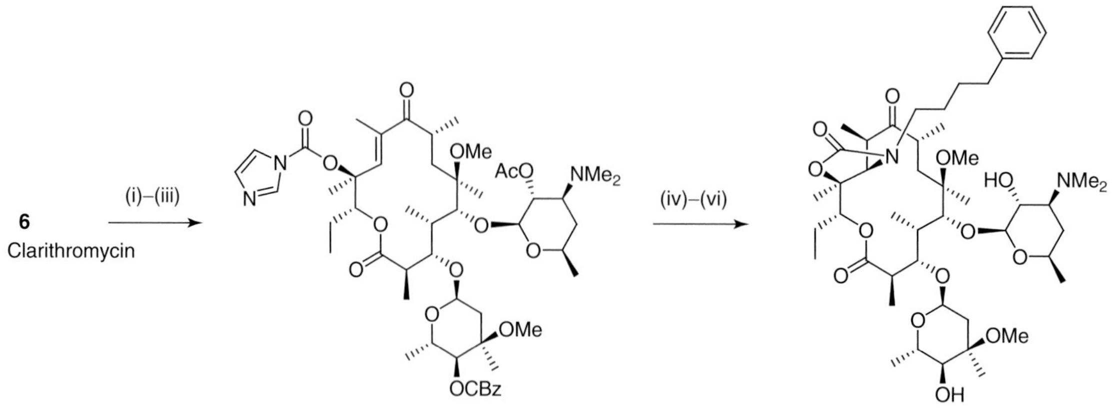

Scheme 5

genesis of C11/C12 carbamates. Upon reacting with a series of amines in aqueous acetonitrile the imidazole of **33** was displaced and the resulting carbamate cyclized in Michael fashion stereoselectively, forming the fused C11/C12 ring and equilibration of the methyl group at C10 to the β-epimer (*R*) predominantly. Under conditions of insufficient equilibration the α-epimer (*S*) was isolated and shown to be less active. Thus, the overall sequence enabled not only formation of the C11/C12 carbamate, but also introduction of a variety of tethered aromatic rings attached to the C11 position through the nitrogen (N11) of the carbamate without disturbing the stereochemistry of the natural system by reintroduction of the C10/C11 stereogenic centers. After removal of the protecting groups on the sugars, analogs with a four-atom tether terminated by a phenyl ring, exemplified by **34**, were found to have in vitro potencies 8–16-fold higher than erythromycin against both the inducible and constitutive MLS *S. pyogenes* strains. Significantly, analogs without the aryl ring at the end of the tether did not show this advantage, indicating the importance of the tethered aryl group in achieving unique interactions with the ribosome.

Although it was later than the Roussel–Uclaf disclosure of what came to be known as the ketolides, Taisho disclosed C11/C12 carbamates on the clarithromycin template with a ketone at C3, in which the cladinose ring was cleaved by acidic hydrolysis and the resulting C3 hydroxy oxidized to a ketone.[239,240] Similar to the Abbott approach described above, the synthetic route involved formation of the acyl imidazole intermediate **35** with excess carbonyldiimidazole and base. When reacted with ethylenediamine, the cyclic carbamate **36** was formed and the primary amine was then condensed in the presence of acetic acid with the ketone at C9 to give the tricyclic intermediate **37** (**Scheme 6**). Significantly, the cladinose sugar was then cleaved leaving the naked hydroxy group at C3 **38**, which was subsequently oxidized under Pfitzner–Moffatt conditions giving, after 2′-acetyl removal, the C3 keto analog TE-802 (**39**). The discovery of **39** was important in that it demonstrated very good activity against the *S. pneumoniae* Mef phenotype, and suggested that the cladinose may be a recognition element for macrolide efflux. However, because it lacked a tethered aryl group, **39** did not exhibit activity against the Erm phenotype and furthermore had relatively poor *H. influenzae* potency.[241]

Researchers at Roussel-Uclaf (subsequently Aventis and currently Sanofi-Aventis) were the first to disclose the hybrid of a N-substituted C11/C12 carbamate with a ketone at C3 to yield compounds that were active against both Mef and Erm *S. pneumoniae*, providing a major breakthrough in the fight to overcome resistance in the macrolide class (**Scheme 7**).[12,242–244] The sequence begins with the removal of the cladinose sugar from clarithromycin (**6**) followed by acetylation of the desosamine 2′-hydroxy group to give **40**. Chemoselective Pfitzner–Moffatt oxidation of the C3 hydroxy then yielded **41**, which was further dehydrated to the enone **42** by selective mesylation of O11 and elimination with DBU (1,8-diazabicyclo[2,2,2]undec-7-ene). After conversion of **42** to the central acyl imidazole intermediate (**43**), different amine-containing fragments were used to displace the imidazole and subsequently form the C11/C12 ring fusion. One variation on the tether came with the C11/C12 carbazates, in which hydrazine hydrate was employed to give intermediate **44**. It should be noted that a 10-fold excess of hydrazine and heating for several hours are both important for the equilibration of the C10 methyl group to the more active *R* configuration.[245] Reductive amination of aldehydes with **44** subsequently yielded a number of analogs, the most promising of which was the quinolin-4-yl ketolide, HMR-3004 (**45**). The in vitro activity of HMR-3004 against macrolide-sensitive and -resistant *S. pneumoniae* and against *H. influenzae* were well within the range of potential therapeutic usefulness. Oral in vivo efficacy was

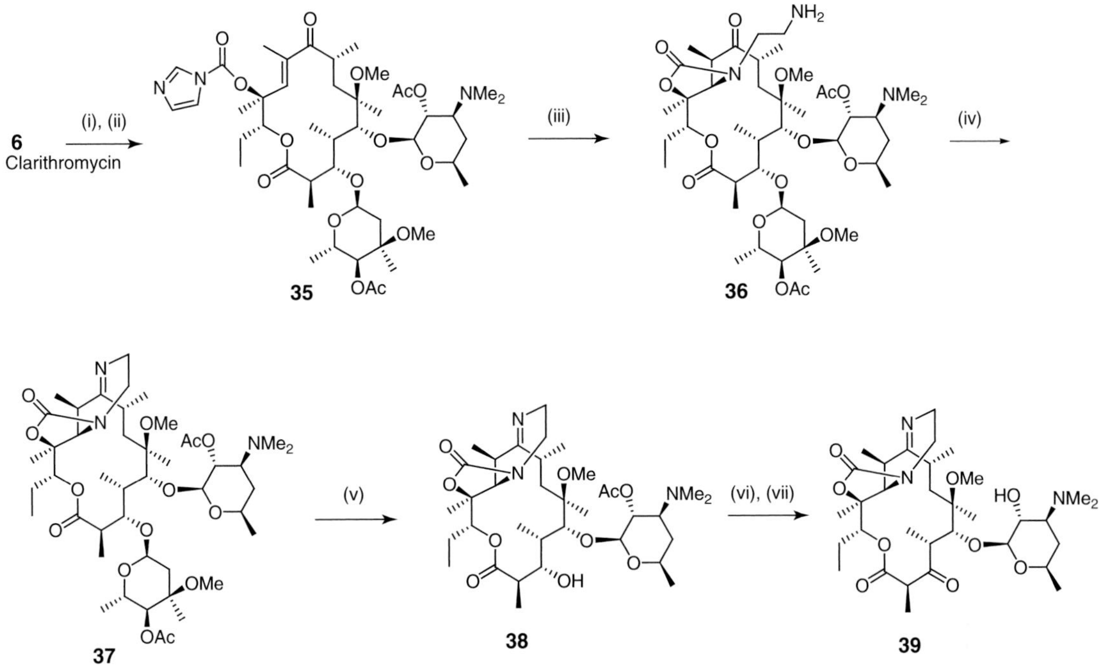

6 Clarithromycin (i), (ii) → 35 (iii) → 36 (iv) → 37 (v) → 38 (vi), (vii) → 39

Scheme 6

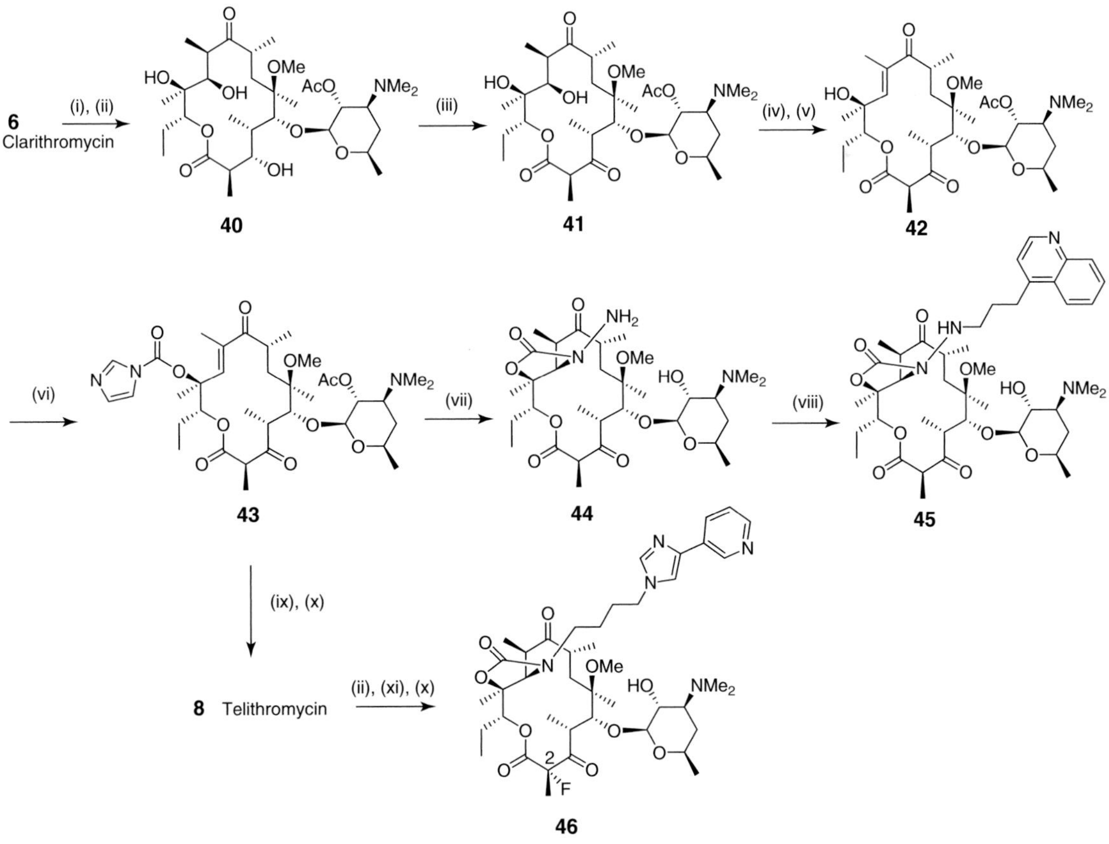

6 Clarithromycin (i), (ii) → 40 (iii) → 41 (iv), (v) → 42 (vi) → 43 (vii) → 44 (viii) → 45

43 (ix), (x) → 8 Telithromycin (ii), (xi), (x) → 46

Scheme 7

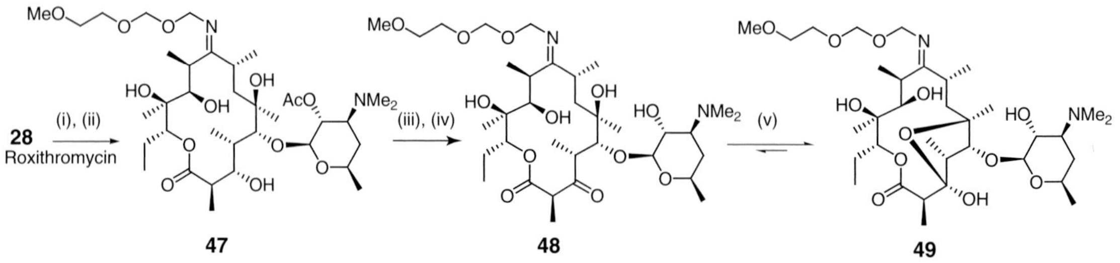

Scheme 8

demonstrated in several intraperitoneal murine models of infection against macrolide resistant pneumococci (50% protective dose (PD_{50}) values ranging from 15 to 42 mg kg^{-1}) where clarithromycin and azithromycin failed (PD_{50} > 100 mg kg^{-1}), and HMR-3004 was subsequently chosen as the first ketolide clinical candidate. The second clinical candidate to emerge and eventually surpass HMR-3004 came out of the carbamate series, in which the *n*-butyl tether is substituted with the 4-(3-pyridinyl)-imidazol-1-yl heteroaryl (telithromycin, HMR-3647, (**8**)).[11,246–248] Telithromycin is the first ketolide approved for clinical use. It is important to note the Aventis chemists also demonstrated that previous reports of C3 keto macrolide derivatives of erythromycin (**1**) (rather than clarithromycin, **6**) were in fact equilibrated to the O6/C3 hemiacetals **49** (**Scheme 8**) and that these compounds were devoid of antibacterial activity. The hemiacetal formation can only be avoided by employing a template with groups other than hydrogen at O6, making this a necessary feature of any ketolide.

7.18.5.7.2 Spectrum of activity

In a study conducted during 2000–2001 of over 10 000 clinical isolates in the USA, telithromycin was potently active against both sensitive and resistant *S. pneumoniae* ($MIC_{90} \leqslant 1\,\mu g\,mL^{-1}$), including macrolide and penicillin resistant strains.[249] Although telithromycin had a very low MIC_{90} ($0.03\,\mu g\,mL^{-1}$) versus *S. pyogenes* in this study, it is worth noting that a certain subset of isolates (most frequently European) have more of the constitutive *Erm* genotype against which telithromycin is considerably less potent ($MIC_{90} = 16\,\mu g\,mL^{-1}$).[250] Whether or not constitutive Erm *S. pyogenes* will increase in significance and become a concern for the empiric treatment of community-acquired respiratory tract infections (RTIs) remains to be seen. Telithromycin also had activity against penicillin-susceptible *S. aureus* ($MIC_{90} \leqslant 0.25\,\mu g\,mL^{-1}$), although not important for respiratory tract infections. Telithromycin had activity comparable to azithromycin against fastidious Gram-negative bacteria, important in community-acquired RTIs, with a *H. influenzae* $MIC_{90} = 4\,\mu g\,mL^{-1}$ and *M. catarrhalis* $MIC_{90} < 1\,\mu g\,mL^{-1}$. It is slightly less potent than azithromycin against these two pathogens. Telithromycin also had very good activity versus atypical and intracellular pathogens such as *M. pneumoniae*, *C. pneumoniae*, and *L. pneumophilia* ($MIC_{90} < 1\,\mu g\,mL^{-1}$ for all).

7.18.5.7.3 Pharmacokinetic properties

The bioavailability of telithromycin for a single 800-mg dose in human was determined to be 57%, achieving a C_{max} of approximately $2\,\mu g\,mL^{-1}$ at 1 h post dose, with an AUC_{0-24} of $12.5\,\mu g\,h\,mL^{-1}$.[244] At steady state after multiple dosing, the terminal elimination half-life was determined to be 9.8 h. The major route of elimination is hepatic metabolism, and at least half of the metabolism is mediated by CYP3A4, which consequently has implications for the drug–drug interaction profile. The volume of distribution of telithromycin after intravenous administration was found to be $2.9\,L\,kg^{-1}$ Distribution to the pulmonary epithelial lining fluid after oral dosing yielded concentrations of $15\,\mu g\,mL^{-1}$ and $3.3\,\mu g\,mL^{-1}$ at 2 h and 12 h post dose, respectively. These levels represent tissue to plasma ratios of 8.6- and 13.8-fold at their respective time points. The protein binding was in the range of 60–70%. The major metabolites result from the loss of the biaryl group at the end of the N11 tether, demethylation at the desoamine nitrogen, and pyridyl N-oxide formation.[251] A review of clinical pharmacokinetics of telithromycin was recently published.[252]

7.18.5.7.4 Safety properties

In clinical trials, the most common adverse effect of telithromycin was gastrointestinal disturbance as observed with other macrolides. Although rare, blurred vision and exacerbation of myasthenia gravis are listed in the label. Telithromycin does not appear to prolong the QTc interval in healthy subjects[253] and does not cause synergistic

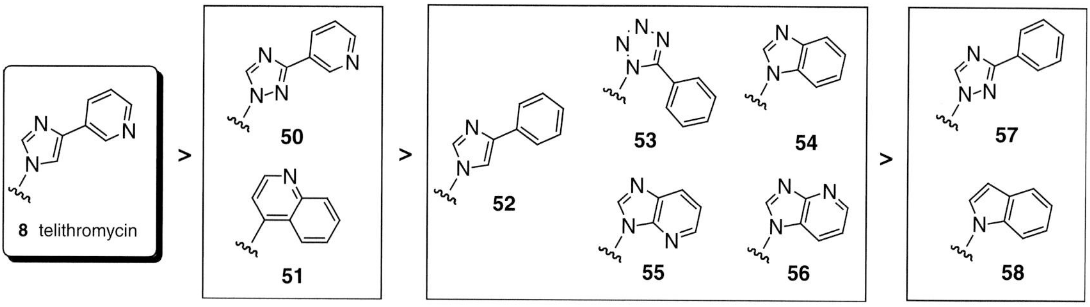

Figure 5 Relative in vitro potencies of telithromycin and other heterocycles.

prolongation when coadministered with sotalol[254] or ketoconazole.[255] Very recently there was a report of three cases of severe drug-induced liver injury, secondary to telithromycin.[79] At the time of this writing the FDA is continuing to evaluate the issue of liver problems in association with use of telithromycin.

7.18.5.7.5 Variations on telithromycin

Concerning the tethers of the N11 side chains, the *n*-propyl carbazate found in HMR-3004 (**45**) and *n*-butyl carbamate of telithromycin (**8**) obviously share a four atom separation between N11 and the aromatic ring but also point to some tolerance for different (possibly heteroatomic) tether constituents. The SARs reported[12] strongly suggested that the four atom tether was preferred in this system. The relative in vitro potencies of telithromycin (**8**), containing the 4-(3-pyridinyl)-imidazol-1-yl heteroaryl group, and other heterocycles are given in **Figure 5**.[11] It should be noted that the activity of telithromycin is negatively affected by the placement of an extra nitrogen in the imidazole ring (triazole, **50**), and also by removal of the pyridyl nitrogen (phenyl imidazole, **52**; also phenyl triazole, **57**). However, the benzimidazole **54** had similar potency to its aza-congeners **55** and **56**. The quinoline analog (**51**), which is the carbon-tether homolog of HMR-3004 (**45**), was less active than telithromycin (**8**). The phenyl tetrazole **53** had moderate potency, while the simple indole **58** was relatively inactive. At the other end of the macrocycle and taking advantage of the acidity of the beta-keto lactone system, position C2 was fluorinated stereoselectively (**Scheme 7**) in the subsequent analogs.[256–258] The sequence involved protection of the 2′-hydroxy group followed by treatment with base and an electrophilic fluoride source to yield, after deprotection, the C2-(*S*)-fluoro analog of telithromycin (HMR-3562, **46**) and a related analog (HMR-3787, **59**). Fluorination at C2 generally improved streptococcal potency, and in some cases improved activity against *H. influenzae*. Chlorination and methylation at C2 on the other hand showed an appreciable drop in antibacterial activity and thus far only fluorination has been productive. Pfizer subsequently disclosed a C11/C12 carbazate with a methyl oxime ether at C9 and a fluorine at C2 (**60**) with excellent potency.[259,260] They also disclosed another C9 oxime derivative (**61**), which contains a methyl substituent in the tether.[261–263]

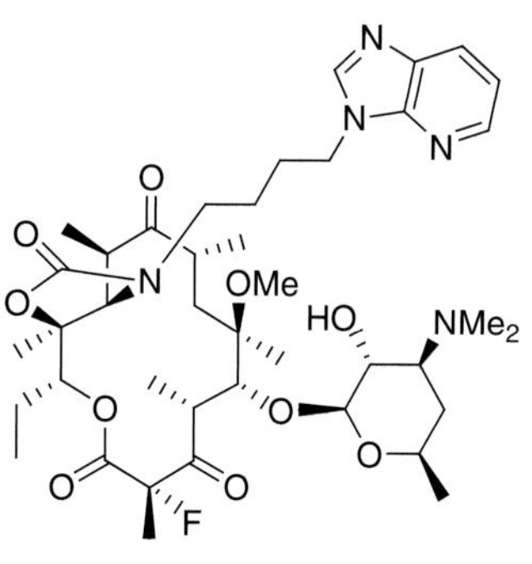

59

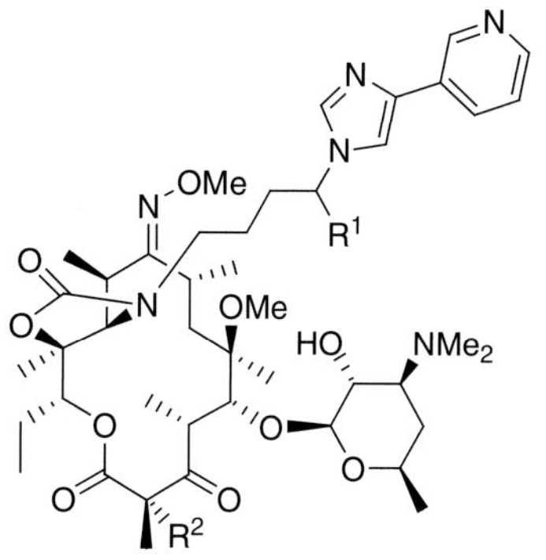

60 R^1 = H, R^2 = F
61 R^1 = (*R*)-Me, R^2 = H

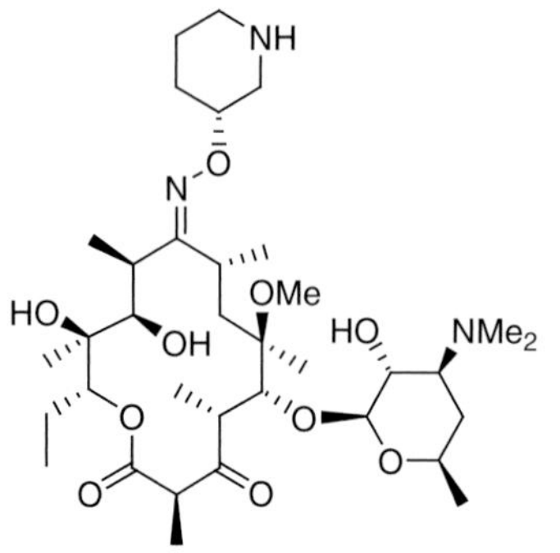

62

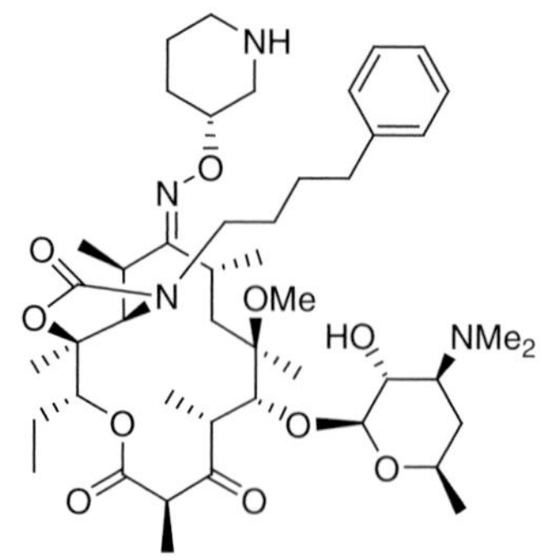

63

Aventis researchers had previously shown[12] that 9-oxime ketolides – specifically piperidinyl oxime ether (**62**) – had activity versus sensitive and resistant *S. pneumoniae* (MIC $\leqslant$0.3 μg mL^{-1}) and against *H. influenzae* (MIC 0.6–2.5 μg mL^{-1}) suggesting the potential for using C9 as a point of attachment in future work. When the piperidinyl oxime ether at C9 was combined with a C11/C12 carbamate substituted with a simple N11 tethered aromatic ring[264] (**63**) the activity versus ErmB-*S. pyogenes* (MIC = 0.6 μg mL^{-1}) was markedly improved over telithromycin (MIC = 5 μg mL^{-1}).

Closer in structure to telithromycin is the C2 fluorinated analog **65** from Optimer Pharmaceuticals, which has good ErmB-*S. pyogenes* potency (MIC $\leqslant$0.5 μg mL^{-1}) (**Scheme 9**).[265] From the C2-fluoro azide intermediate **64** they were able to make a number of 1,2,3-triazole derivatives such as **65** by coupling with aryl alkynes under copper catalysis. This application also disclosed their novel process of desosamine removal by oxidation of the 2′ hydroxy group and subsequent cleavage to generate aglycone (**66**). Modified desosamine sugars were coupled to the O5 hydroxy, leading to the identification of **67**, for example; however, none matched the potency of the parent desosamine analogs, particularly against the streptococcal strains.

Researchers at Chiron Corporation have disclosed various modifications directed at the C12 methyl group, most notably conversion to ethyl and to hydrogen (**68**) and (**69**).[266] The synthetic sequence began with protection of the sugars of clarithromycin, reduction of the C9 ketone, and O9/O11 acetonide formation to form an acetal intermediate.

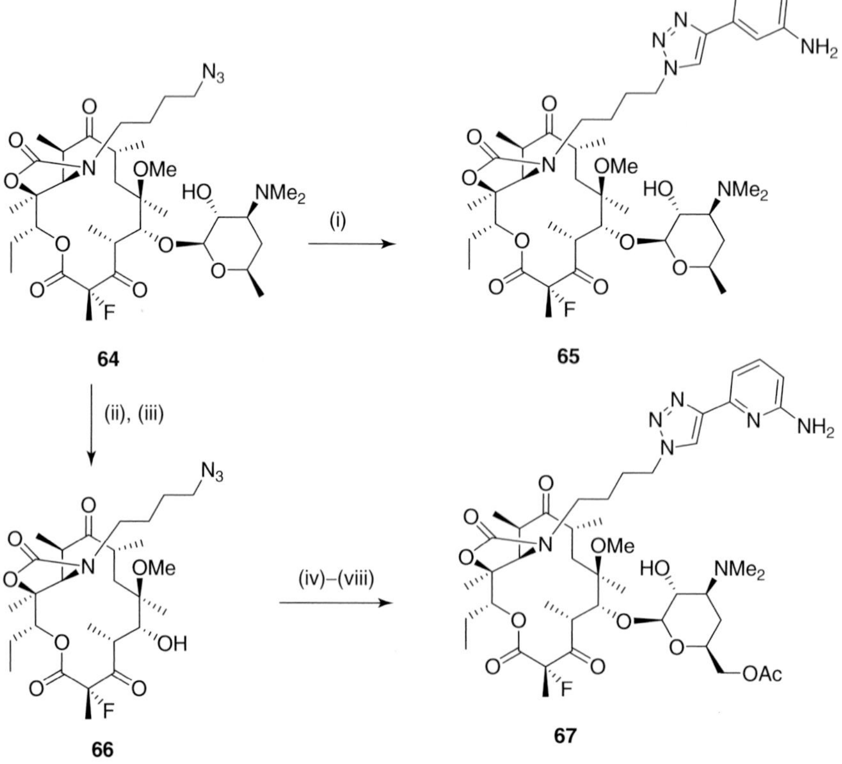

(i)

64

65

(ii), (iii)

(iv)–(viii)

66

67

Scheme 9

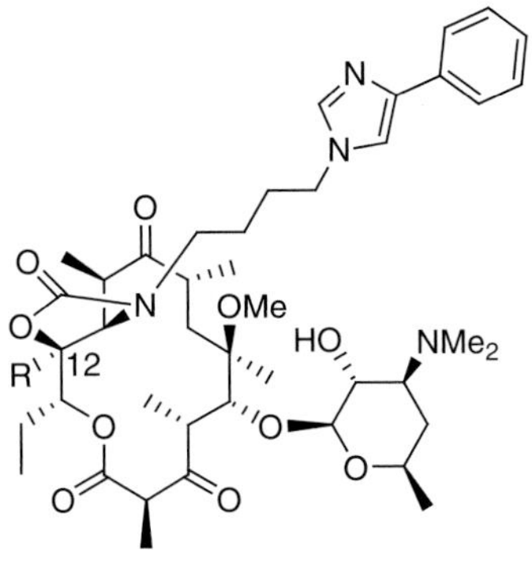

68 R = H
69 R = Et

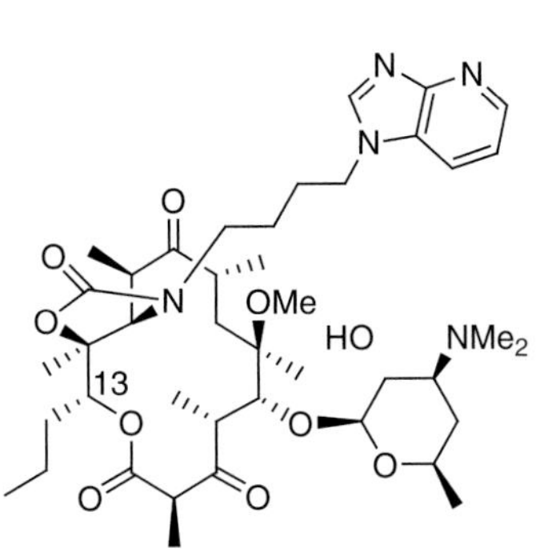

70

The hydroxy at C12 was then eliminated using thionyl chloride and triethylamine to yield the key olefin intermediate (**71**), which was taken in two complementary directions (**Scheme 10**). First, oxidation of **71** with *meta*-chloroperoxybenzoic acid (mCPBA) resulted in stereoselective formation of epoxide (**72**) (after desosamine N-oxide reduction), which was subsequently opened with methyl cuprate to give C12 ethyl intermediate. Acetonide removal gave a triol, which was selectively oxidized with Dess–Martin periodinane at C9 to give **73** and subsequently converted by standard procedures to the C12 ethyl telithromycin homolog (**69**). Second, ozonolysis of olefin **71** yielded ketone **74**, which was then reduced with sodium borohydride and mesylated to give **75**. After acetonide removal and selective oxidation at C9 as before, treatment with DBU gave – after O11 migration and inversion of configuration at C12 – intermediate (**76**), which was converted to the nor-telithromycin analog (**68**). Although telithromycin was not listed as a control, the antibacterial potencies of both **68** and **69** (note the slight modification to the heteroaryl group as well) appeared to be comparable with Mef and Erm *S. pneumoniae* MICs $<1\,\mu g\,mL^{-1}$ (no ErmB *S. pyogenes* reported) and *H. influenzae* MICs of $3–6\,\mu g\,mL^{-1}$. The advantages of C12 modified ketolides over the parent ethyl analogs are not readily apparent from the disclosure, although no doubt they result in physiochemical changes relative to the parent template.

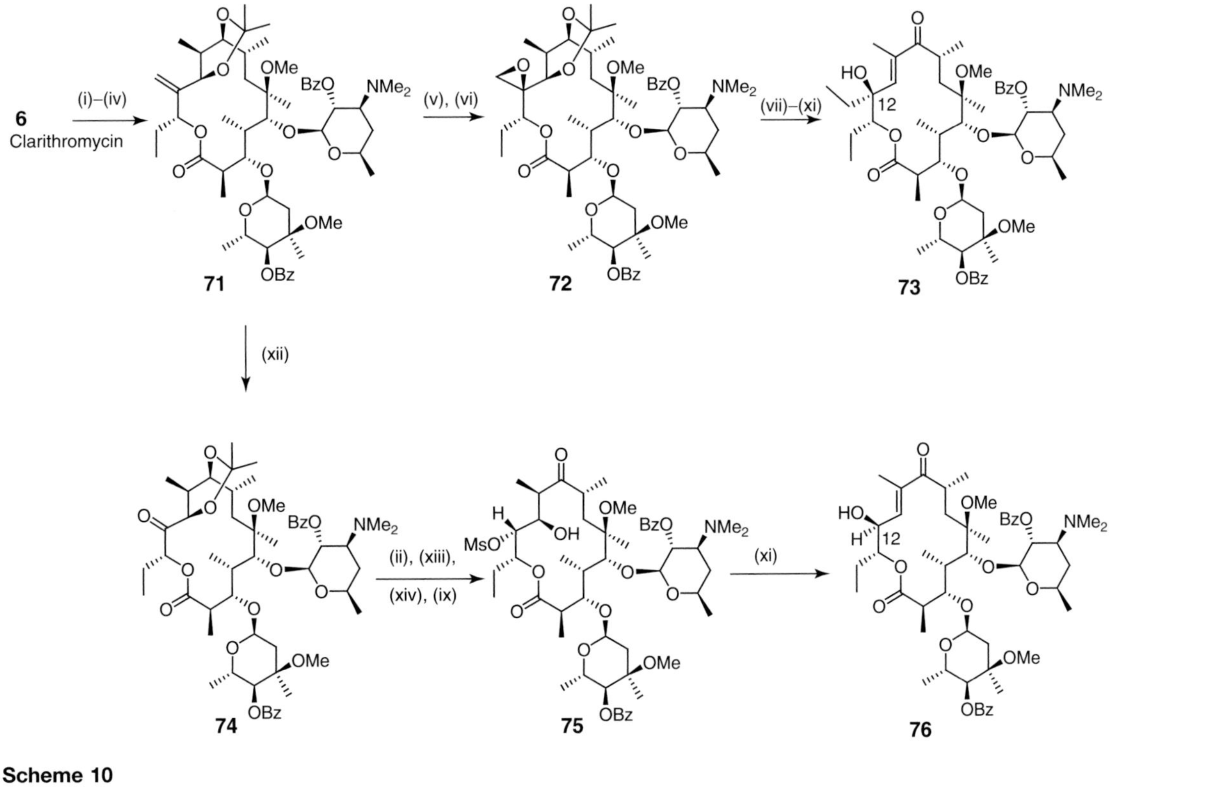

Scheme 10

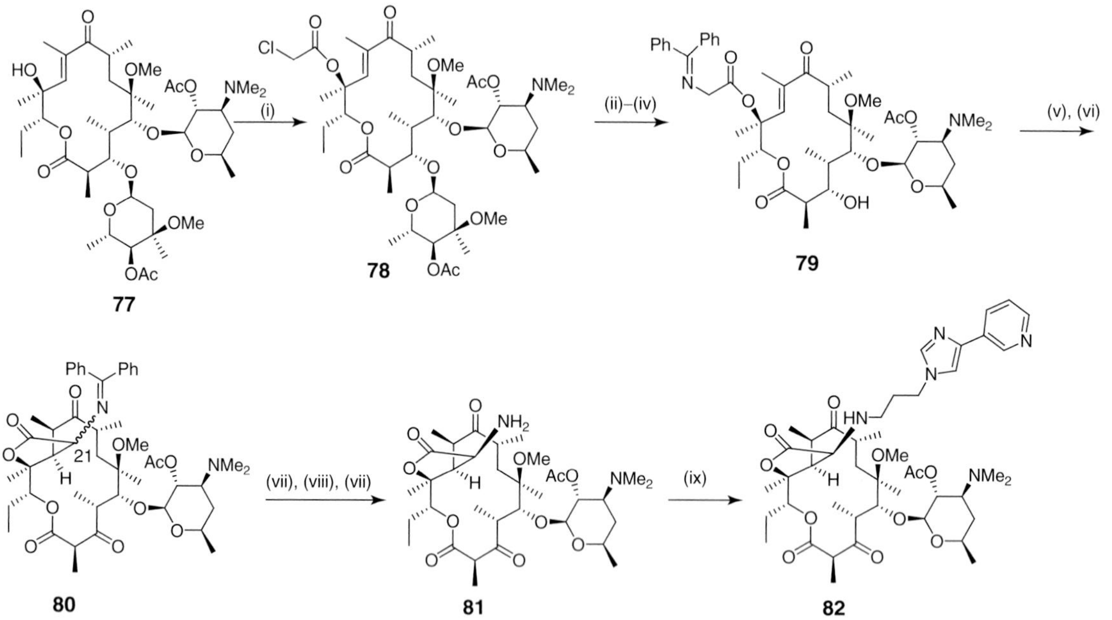

Scheme 11

A collaboration between Kosan Biosciences and Ortho-McNeil Pharmaceuticals has yielded a variation on the macrocyclic ring system through manipulation of the polyketide synthase pathway, focused on changes at the C13 ethyl position.[267–270] The C13 methyl and vinyl analogs were not as potent as the parent ethyl, while propyl analogs could be made as potent as telithromycin upon changing the heteroaromatic headgroup to the imidazo[4,5-*b*]pyridine (**70**).[271] It should be noted that in addition to de novo fermentation, the O6 methyl group must be installed on the clarithromycin homolog before the work of generating the ketolide can begin, imposing an additional cost-of-goods burden to these approaches with no clear medicinal chemistry advantages reported to date.

Researchers at GlaxoSmithKline have reported a C11/C12 lactone ring fused to the macrocycle by attaching a functionalized acetate to O12 followed by Michael addition of the acetate methylene into the alpha,beta-unsaturated enone (**Scheme 11**).[272] A mixture of diastereomers results at the newly generated stereocenter at C21 in intermediate **80** (while controlled at C9 and C10), which can be converted to the more active beta-epimer. This transformation was done by removing the benzhydrylidine protecting group and equilibrating the corresponding benzylidine, which was subsequently removed as well to yield the amine intermediate **81**. This strategy amounts to a transposition of nitrogen and carbon from the C11/C12 carbamate series and introduces the possibility of a basic center in the tether, which could have a profound effect on the physicochemical and pharmacokinetic properties of the compounds. Eleven analogs were identified in the patent application as having MIC $\leq$0.06–8.0 $\mu g\,mL^{-1}$ against erythromycin-resistant *S. pneumoniae* without assigning potency to any compound individually. It would appear from the examples listed that the propylamine (e.g., **82**), the ethylamine (e.g., **83**), and the 4-oxo-butyramide (e.g., **84**) are preferred tethers.

Basilea Pharmaceutica has independently taken a very similar approach using thioether acetates to generate the C11/C12 lactone ring, with stereocontrol over the chiral centers at C11 and the thioether at C21.[273,274] A mixture of diastereomers at C10 were generated in the lactone-formation step; however, these conveniently equilibrated to the more active *R*-isomer simultaneous to the removal of the 2′-acetate protecting group in methanol. The ethylthio ether with a 6-amino-9*H*-purine-9-yl aromatic headgroup **85** matched the Gram-positive potency of telithromycin and was slightly more active against *H. influenzae*. The propyl homolog **86** was almost as potent as **85** and apparently had improved efficacy when dosed both orally (ED_{50} = 7.1 versus 12.0 $mg\,kg^{-1}$) and subcutaneously (ED_{50} < 1.5 versus 1.2 $mg\,kg^{-1}$) in a intraperitoneal murine model against a macrolide-susceptible strain of *S. pyogenes*. No in vivo results were given against a macrolide-resistant strain. Interestingly, the Basilea researchers studied the cytotoxic effects of their analogs on HeLa cells and observed a correlation between cytotoxicity and lipophilicity. For example, compounds with more polar sidechains (**85**, ClogP = 1.19, HeLa IC_{50} = 34.4 μg mL^{-1}) were considerably less toxic than lipophilic ones (**87**, ClogP = 4.40, HeLa IC_{50} = 1.10 μg mL^{-1}) in this system.

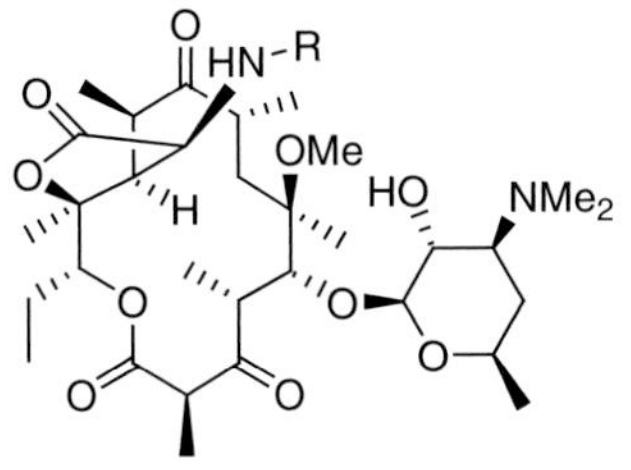

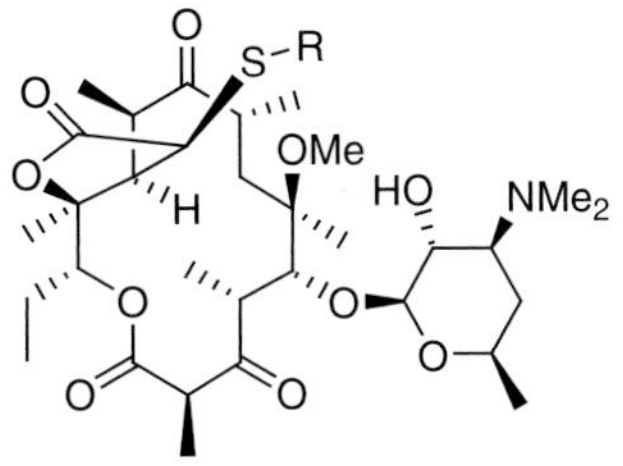

83 R =

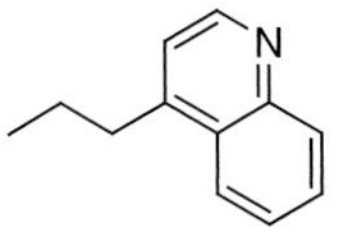

84 R =

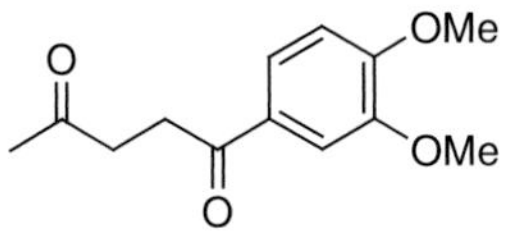

R =

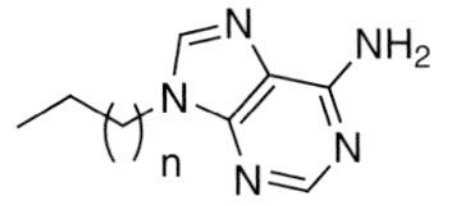

85 n = 1

86 n = 2

R =

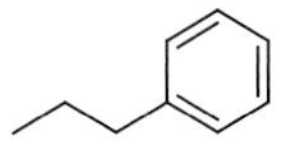

87

Pfizer has disclosed C11/C12 ureas,[275,276] which involved formation of the intermediate ketene acetal **88** from acyl imidazole **43** with a strong base such as DBU, followed by reaction with trimethylsilylazide and reduction to yield amine intermediate **89** (**Scheme 12**). The amine was then converted to the corresponding isocyanate with phosgene and base, which in turn reacted with the side chain amine to give the uncyclized urea. The C11/C12 cyclic urea **90** was subsequently formed by treatment with strong base, and these analogs were found to be slightly less active than their C11/C12 cyclic carbamate counterparts.

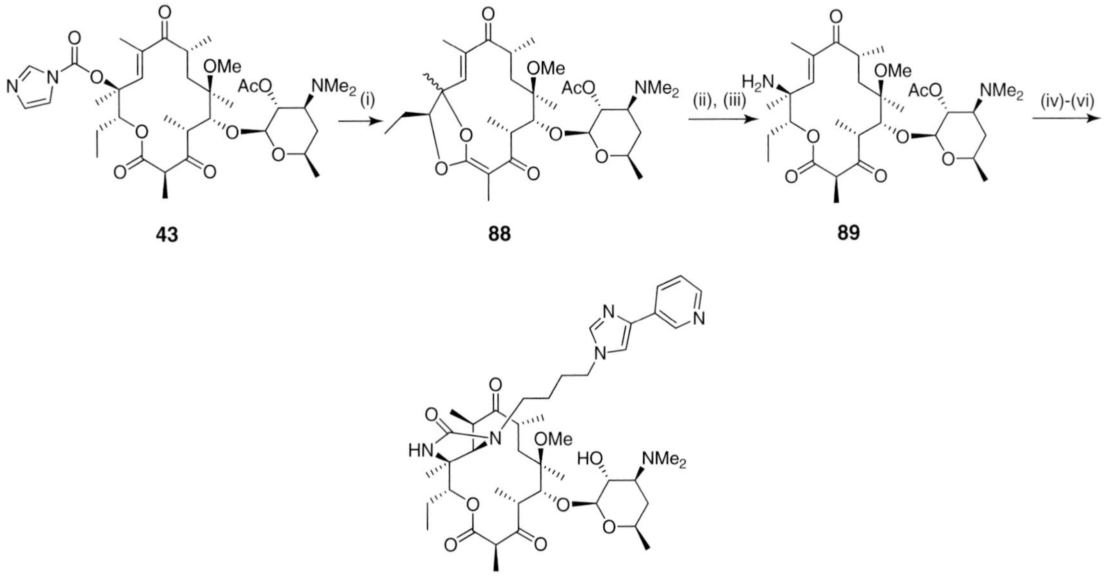

43 (i) **88** (ii), (iii) **89** (iv)-(vi)

90

Scheme 12

7.18.5.8 C6-Tethered Ketolides: Cethromycin

The success of telithromycin has spawned several variants of the C11-tethered ketolides as described earlier, with the clarithromycin template used as the starting material source for these approaches. Researchers at Abbott Laboratories took a different approach by tethering erythromycin A from the C6 position, using extended O6 ethers rather than N11 to reach into domain II.[277–280] It is important to note that the C6 point of tether attachment obviates the problem of hemiacetal formation of the C6 hydroxy with the C3 (and C9) keto group. Following a synthetic scheme analogous to that of clarithromycin, 9-ketaloxime 2′,4″-bis(trimethylsilyl)erythromycin (**19**) was treated with allyl bromide and a hindered base (potassium t-butoxide) to yield O6 allyl ether (**91**) after sequential deprotection (**Scheme 13**).

Removal of the cladinose sugar, reprotection of the 2′-hydroxy group of desosamine and selective oxidation introduced the C3 ketone. Subsequent Heck arylation of the olefin yielded compounds with very promising Mef and Erm *S. pneumoniae* potency. It was found that installation of a C11/C12 carbamate markedly improved activity, and also that stereocontrol of C10 to the *R*-configuration was improved by installing it prior to cladinose removal and C3 oxidation. As in the case with the C11-tethered ketolides, the choice of tether and aryl group were both critical to achieving the desired level of antibacterial potency (variations discussed below). Heck coupling with a variety of aryl halides allowed for efficient analog generation and eventually led to the identification of the 3-quinolinyl allyl compound ABT-773 (**32**), later named cethromycin, for clinical development.[277] Cethromycin has not yet received approval from the US Food and Drug Administration (FDA), being currently in phase III of clinical trials.

7.18.5.8.1 Spectrum of activity

Cethromycin (ABT-773) is active against both macrolide-sensitive and -resistant *S. pneumoniae* and *S. pyogenes*. The activity against Erm-mediated resistant *S. pyogenes* is one of the differentiating factors over telithromycin. Cethromycin appears to be slightly more potent than telithromycin according to MIC_{90} studies against *S. pneumonia* (Erm MIC_{90} 0.015 $\mu g\,mL^{-1}$,

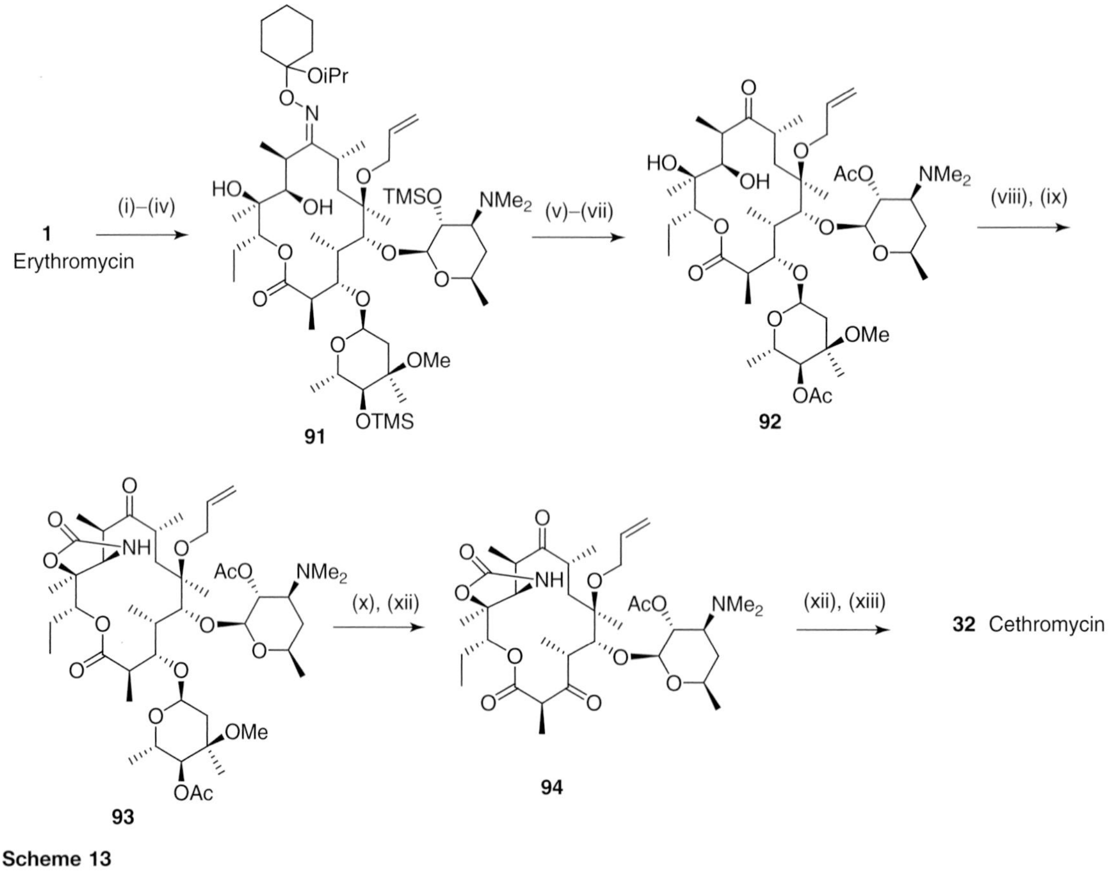

1 Erythromycin —(i)–(iv)→ **91** —(v)–(vii)→ **92** —(viii), (ix)→ **93** —(x), (xii)→ **94** —(xii), (xiii)→ **32** Cethromycin

Scheme 13

Mef MIC_{90} 0.12 μg mL^{-1} versus 0.12 μg mL^{-1} and 1 μg mL^{-1} of telithromycin).[281] Like telithromycin, however, it is not active against constitutively resistant *S. aureus*. Against *H. influenzae* cethromycin is as potent as telithromycin but less than azithromycin. It shows potent activity in vivo against Erm-resistant and Mef-resistant *S. pneumoniae* and the ED_{50} values indicate that it is more potent than telithromycin in those models.[282] The increased potency is most likely due to the tighter binding of cethromycin to the ribosome.[283]

7.18.5.8.2 Pharmacokinetic properties

In mouse, rat, monkey, and dog, cethromycin is slowly absorbed after oral dosing with t_{max} (time that is coincident with C_{max}) appearing at 1.5–6 h. In mice, the volume of distribution was determined to be 1.8 L kg^{-1} and the clearance of 0.81 L h^{-1} kg^{-1} after intravenous dosing of 10 mg kg^{-1}. After oral dosing of 10 mg kg^{-1} a C_{max} of 1.47 μg mL^{-1} AUC∞ of 6.1 μg h mL^{-1} and half-life of 3.9 h were achieved with bioavailability of 49.5%.[284] Lung concentrations of cethromycin were at least 25 times higher than that of plasma in rat after oral dosing of 10 or 30 mg kg^{-1}. Studies involving ^{14}C-labeled cethromycin in rats, dogs, and monkeys indicated that the major metabolites were the N-desmethyl derivative and a product of hydroxylation at the C10 methyl group.[285]

A phase I study of cethromycin indicates that C_{max} ranged from 0.14 to 1.2 μg mL^{-1} after oral dosing of 100 to 1200 mg. The t_{max} after 800 mg of dosing occurred at 3.9 h, and the terminal half-life was 6.7 h at this dose.[286] The main adverse effects were gastrointestinal although there are not many published data available at this time.

7.18.5.8.3 Variations on cethromycin

Initial investigations[268] into the C6-tethered ketolides demonstrated the superiority of the allyl tether over saturated or amine-containing tethers. The propargylic tether (**95**) was subsequently reported to have good potency and the potential for an improved side effect profile relative to telithromycin with respect to gastrointestinal tolerability.[287] Other C6 ketolides containing a propagyl tether have also been disclosed,[288,289] including 9-oxime ether variants such as the 1,5-naphthyridine (**96**) reported to have superior activity against macrolide-resistant *S. pneumoniae* and *S. pyogenes* activity.[290] The C6 carbamate tether (**97**), first reported by researchers at Abbott Laboratories,[291] has been elaborated extensively by chemists at Ortho-McNeil.[292]

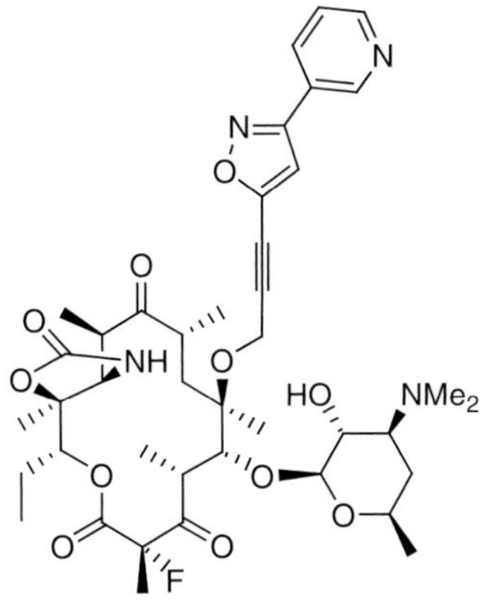

95 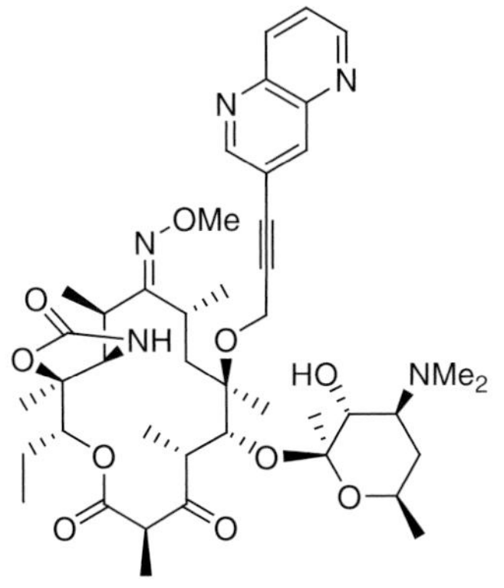 **96** 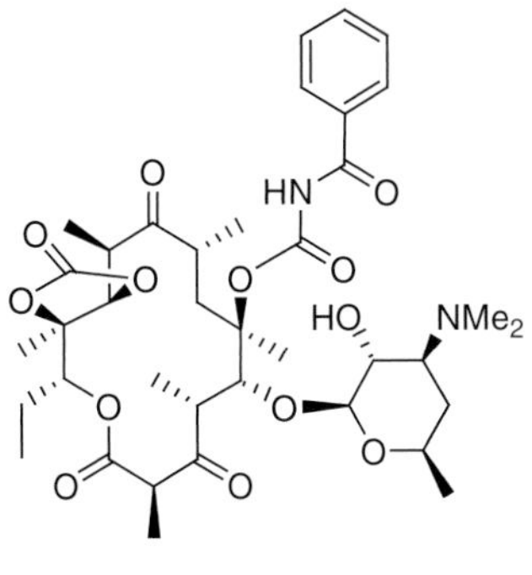 **97**

The carbamate functionality was introduced simultaneously at C6 and C11,C12 following **Scheme 14**. Starting with erythromycin, peracetylation led to the 2′,4″,12-triacetyl intermediate, which was subsequently eliminated to the enone (**102**). Treatment of **102** with trichloroacetylisocyanate yielded **103**, which was cleaved and cyclized to form **104**. The key intermediate **105** was subsequently generated by cladinose removal, C3 oxidation, and deprotection in the usual fashion. Reductive amination of the C6 carbamate nitrogen of **105** under triethylsilane/trifluoroacetic acid conditions with an array of aldehydes allowed for facile synthesis of analogs, several of which (e.g., **98**) had excellent potency against Mef and Erm *S. pneuomoniae* and *H. influenzae* comparable to that of cethromycin and telithromycin.[293] Further elaboration subsequently led to 9-oxime/2-fluoro analogs such as **99** which were similar to telithromycin against Gram-positive strains but appeared less active versus *H. influenzae*.[294] As was the case with telithromycin-related compounds, Ortho-McNeil collaborated with Kosan to synthesize C-13 propyl 2-fluoro analogs of cethromycin such as **100** and **101**.[295,296] These analogs were superior to telithromycin in vitro by two- to fourfold against Mef *S. pnuemoniae* strains, comparable against Erm *S. pneumoniae* and *H. influenzae*, and two- to fourfold less active against 23S rRNA mutants.

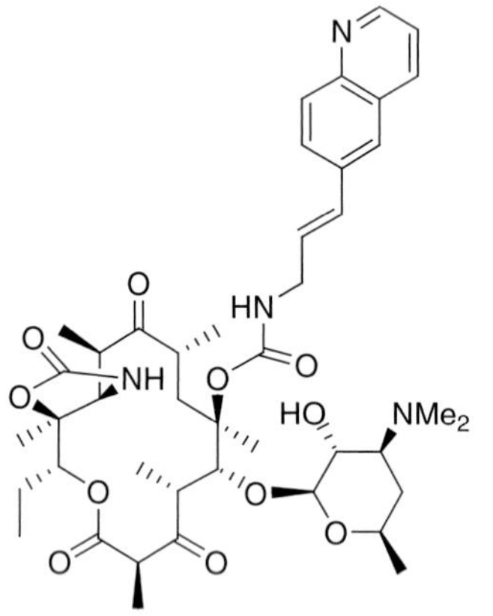

98

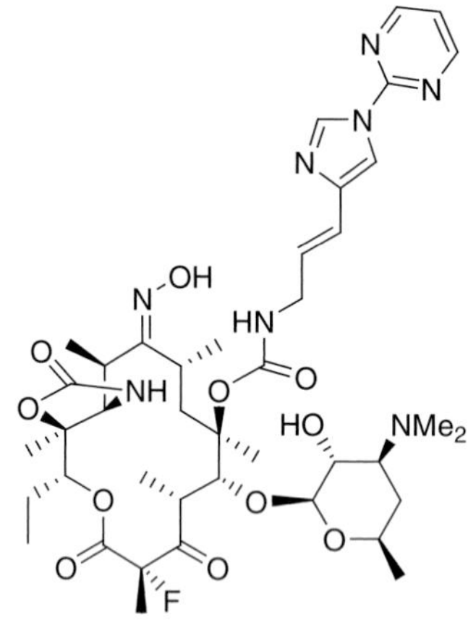

99

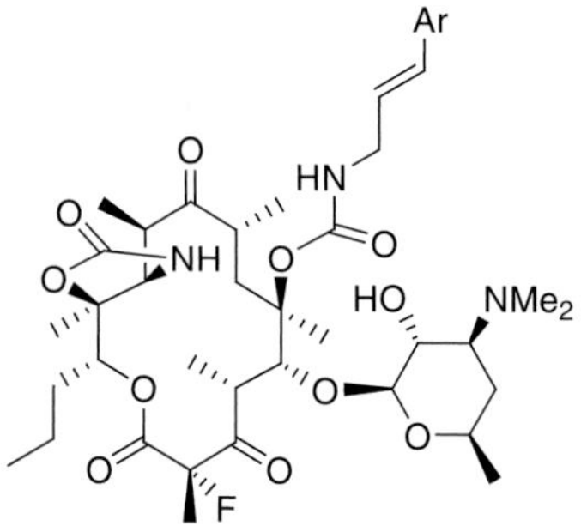

100 Ar =

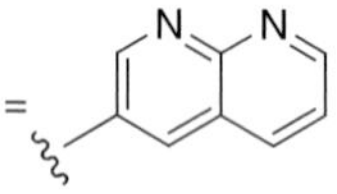

101 Ar =

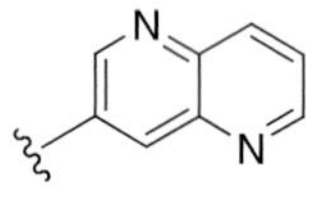

Modifications to the C11/C12 carbamate in the C6-tethered ketolide series were reported by researchers at GlaxoSmithKline. The sequence involved acylation of the O12 position of an O6 allyl intermediate **107** with chloroacetic anhydride after elimination of the C11 hydroxy using phosgene and pyridine to give **107** (**Scheme 15**).

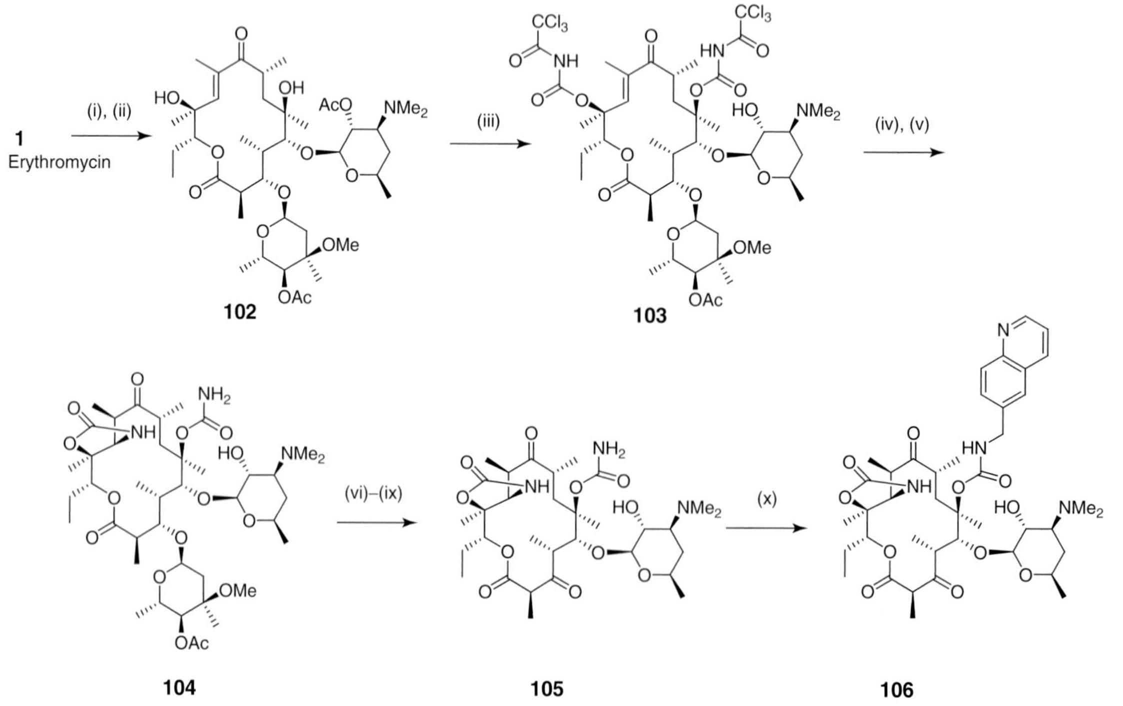

1 Erythromycin (i), (ii) → **102** (iii) → **103** (iv), (v) → **104** (vi)–(ix) → **105** (x) → **106**

Scheme 14

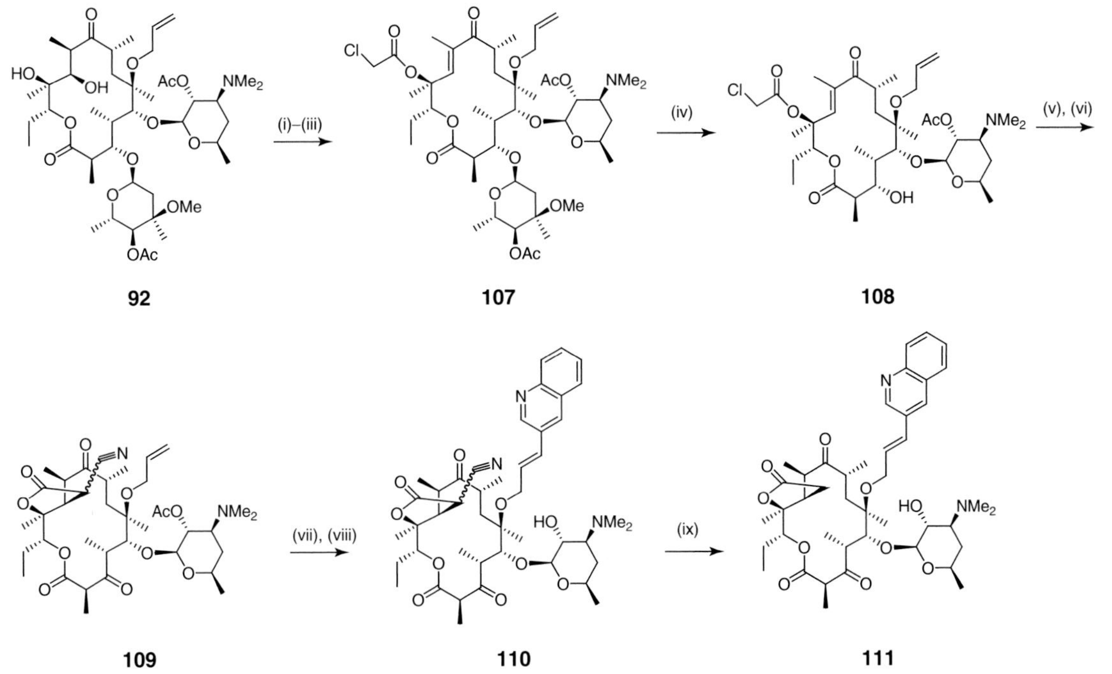

Scheme 15

After removal of the cladinose from **107** to yield **108**, treatment with potassium cyanide resulted in chloride displacement and subsequent Michael addition to give a C11/C12 cyano-substituted lactone. Dess–Martin oxidation of the cyclized adduct gave the 3-keto intermediate **109**, which was further elaborated at the O6 allyl by Heck coupling to brominated heteroaryl rings to give analogs such as the 3-quinolinyl **110**. The cyano group was then removed with activated aluminum oxide to yield the cethromycin homolog **111**, reported to have MICs in the 0.1–64 $\mu g\,mL^{-1}$ range against erythromycin-resistant *S. pneumoniae* strains.[297]

Similarly, researchers at Ortho-McNeil have expanded on their C6 carbamate tethered ketolides by acetylating the O12 of intermediate **112** and formation of the C6 carbamate as before, followed by Michael addition of the O12 acetate methylene using lithium diisopropylamide to give the O6 carbamate C11/C12 lactone **114** (**Scheme 16**).[298] Further elaboration by reductive amination yielded analogs such as **115** and **116**, which were reported to have an excellent in vitro profile, consistent with the C11/C12 carbamates of this series.

7.18.5.9 Acylides

Acylides are macrolide derivatives in which cladinose is removed from the C3 hydroxy group and an acyl group is attached at this site. Although they contain an ester linkage as a result, they are generally stable toward hydrolysis because of steric hindrance around this linkage. The starting point for more recent work is based on the earlier observation that an acylation of the C3 hydroxy group with 3-methoxybenzoyl group restored some antibacterial activity, even though it was only one-fortieth of erythromycin.[299] With different acyl groups, the newer analogs were shown to have activity against (inducible) Erm-resistant *S. aureus* and Mef *S. pneumoniae*. The 4-nitrophenacetyl ester derivative of clarithromycin (TEA 0777, **117**) had such activities, but it lacked potency against *H. influenzae*.[300] When a 3-pyridylacetyl group was introduced in place of the 4-nitrophenacetyl group, the resulting compound gained activity against *H. influenzae*.[301] When the C11 and C12 positions were tied as a cyclic carbonate or a cyclic carbamate (TEA 0929, **118**), the resulting compounds had improved potency against those pathogens. TEA 0929 was as active in vivo as clarithromycin. Introduction of two cyclic carbonate structures in FMA-0122 (**119**) further enhanced the antibacterial activity including the twofold increase against *H. influenzae* compared with azithromycin. This compound, however, was less potent in vivo compared with azithromycin or clarithromycin.[299] This was most likely due to its less favorable

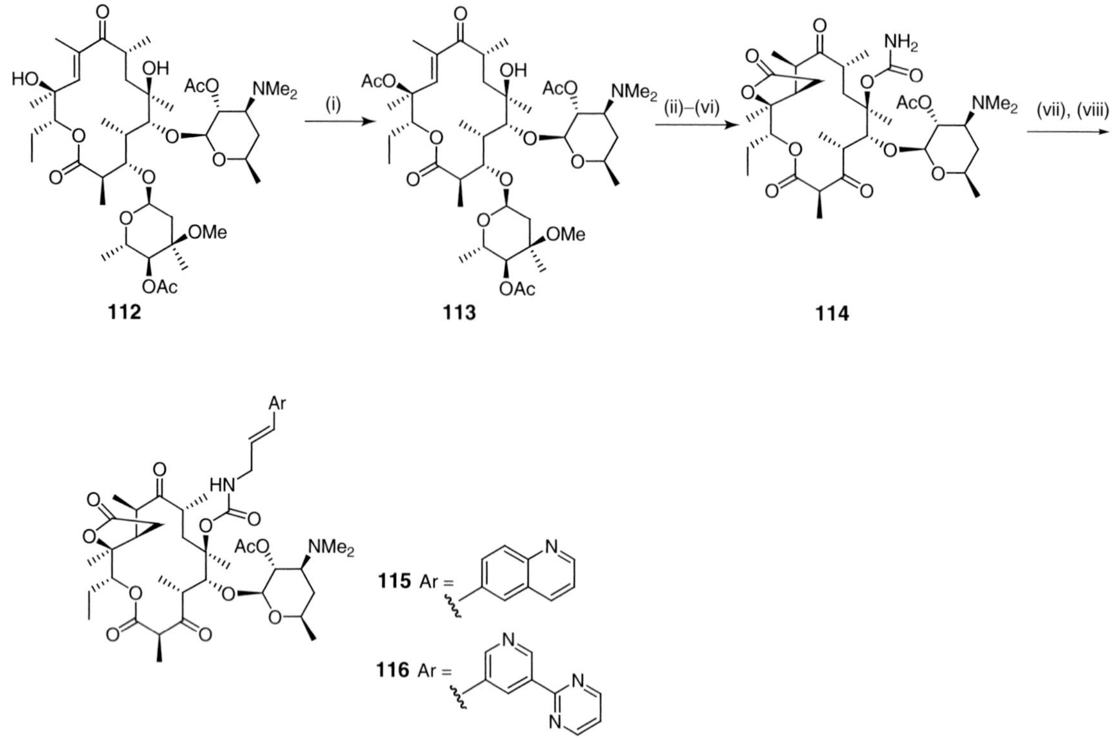

112

113

114

115 Ar =

116 Ar =

Scheme 16

pharmacokinetic properties since both C_{max} and AUC were much less than those of clarithromycin. None of these acylides were active against Erm *S. pneumoniae*. Only when a tethered heterocycle group was attached at the C11/C12 carbamate moiety was activity against Erm-resistant *S. pneumoniae* acquired. FMA-0199 (**120**) and FMA-0481 (**121**) showed potent in vitro activity against Erm- and Mef-resistant *S. pneumoniae*.[302] More recently, FMA-0841 (**122**) and FMA-1082 (**123**) were reported to have good activity against Erm-resistant *S. pneumoniae* and *S. pyogenes*. They also have activity against *H. influenzae* comparable to telithromycin.[303] A new class of acylides represented by structure **124** was reported very recently. They contain a carbamoyl group at C6 as well as a pyridyl-acetate group at C3. This compound has MICs of 0.12, 0.06, and 4 μg mL^{-1} against Erm-resistant *S. pneumoniae*, Mef-resistant *S. pneumoniae*, and *H. influenzae*, respectively.[304]

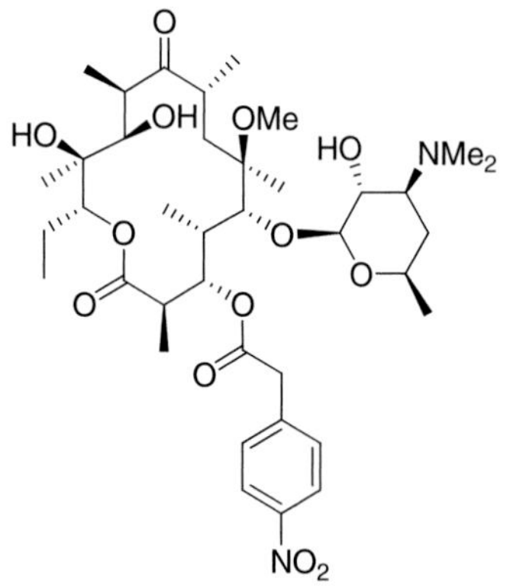

117 TEA-0777

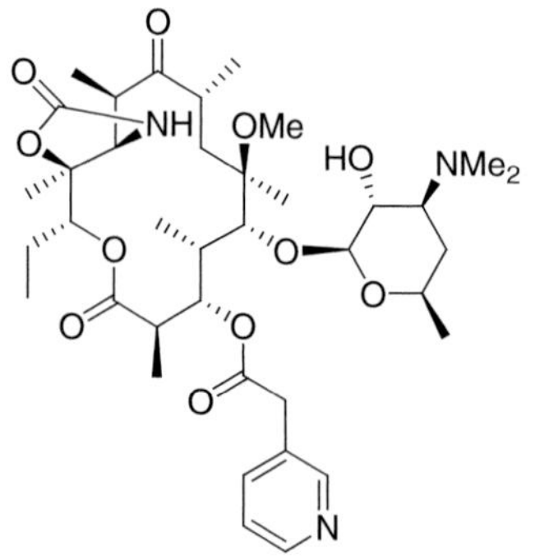

118 TEA 0929

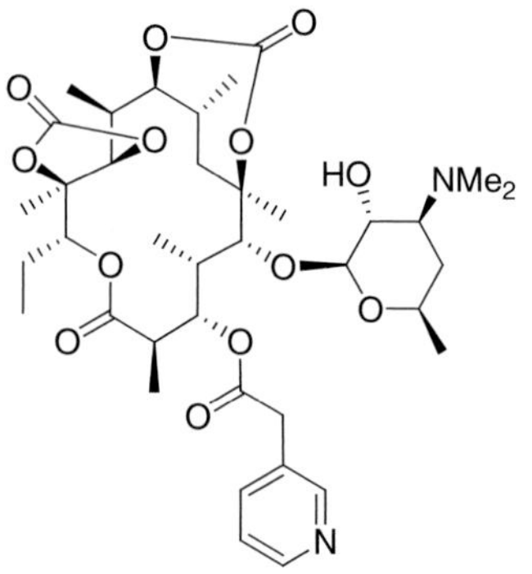

119 FMA-0122

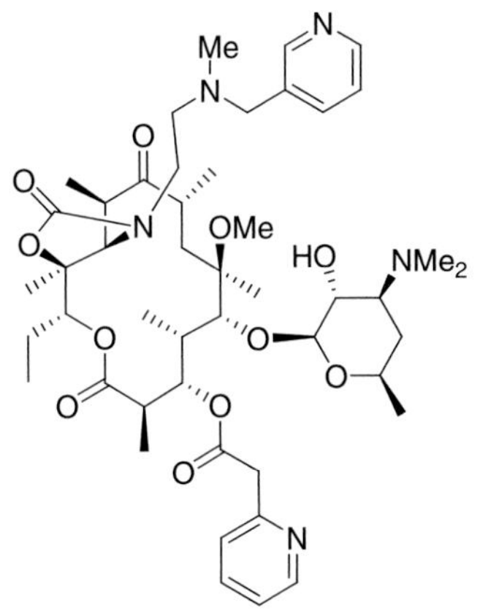

120 FMA-0199

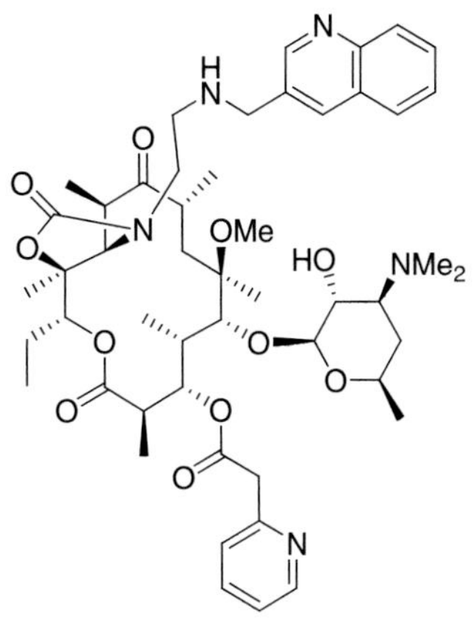

121 FMA-0481

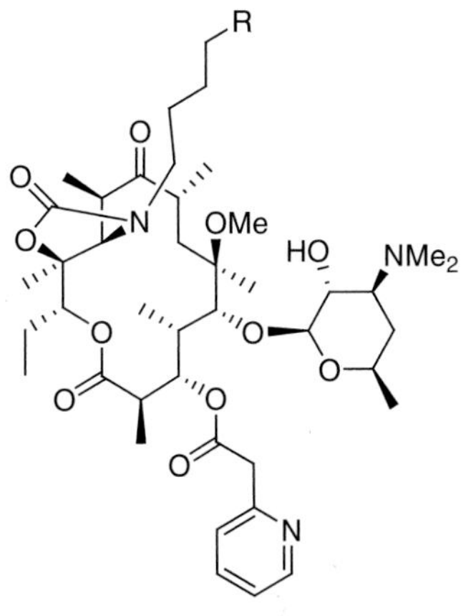

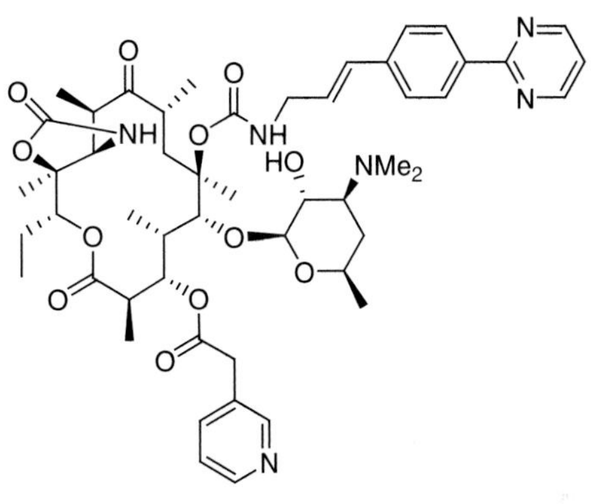

122 FMA-0841 R =

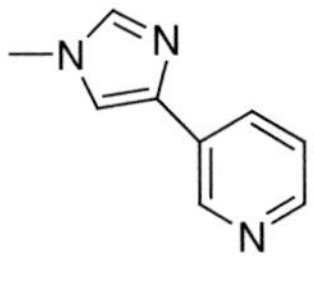

123 FMA1082 R =

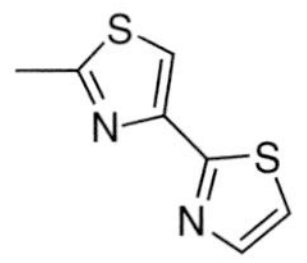

124

7.18.5.10 Bridged Bi-Cyclic Ketolides

Investigators at Enanta Pharmaceuticals recently developed a new class of ketolides in which the C6 and C11 hydroxy groups of erythromycin are tied together (**126**) (**Scheme 17**). As opposed to telithromycin, which is synthesized from clarithromycin, these bi-cyclic ketolides start with erythromycin and the C6,C11 bridge is incorporated by bis-allylic substitutions. After the conversion of the carbon–carbon double bond to a carbonyl group, a heterocycle essential for the secondary interaction (as observed with telithromycin) is attached as an oxime. The acyl-imine moiety at C9 is stable toward hydrolysis and this functional group is retained in the final product. The most advanced bridged bi-cyclic ketolide is known as EP-13420 (**129**), which is quite potent against macrolide-resistant *S. pneumoniae* and

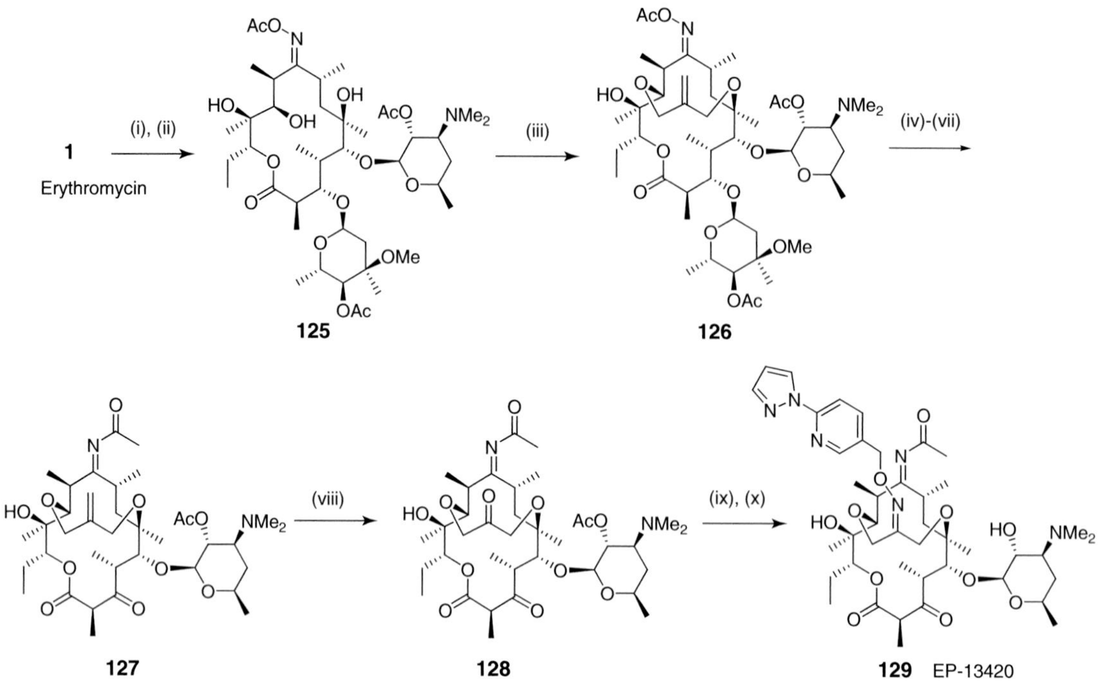

Scheme 17

S. pyogenes.[305] Compared with telithromycin, it has increased potency against macrolide-resistant *S. pyogenes*. Furthermore, it is active against *S. aureus* strains, which are inducibly resistant to macrolides, but not against constitutively resistant *S. aureus*. EP-13420's in vitro potency against *H. influenzae* appears to be slightly less than telithromycin. In mice, C_{max} of 2.5 μg mL^{-1}, AUC_{0-24} 17.0 μg h mL^{-1} and half-life of 3.3 h are achieved after an oral dosing of 15 mg kg^{-1} of EP-13420. The biochemical and genetic studies indicate that it binds in the nascent peptide exit tunnel.[306] It is currently in phase I studies in the USA and Japan while other bridged bi-cyclic ketolides are concurrently in development.

7.18.5.11 16-Membered Macrolide Antibiotics

Almost simultaneous to the discovery of the 14-membered macrolides, 16-membered macrolides were isolated from the fermentation broths of *Streptomyces* and *Micromonospora* species. While a large number of 16-membered macrolides have been identified to date, this class remains underutilized in human therapies perhaps due to its slightly lower potency and narrower spectrum. Two major categories of 16-membered macrolides are represented here by josamycin (**4**) and tylosin (**5**). As demonstrated in these structures, 16-membered macrolides typically contain a diene or an enone system as well as a disaccharide moiety. The first sugar at the C5 position is termed mycaminose. It is identical to desosamine in the 14-membered macrolides, except it contains an additional hydroxy group at the C4′ position, which is glycosylated with a second sugar called mycarose. Although both 14- and 16-membered macrolides are generated by polyketide biosynthetic pathways, the substitution patterns around the macrolactone ring of the 16-membered macrolides are quite different from that of erythromycin. Because of the different arrangement of functional groups, 16-membered macrolides do not undergo the acid-catalyzed internal acetal formation observed with erythromycin. Consequently, they do not suffer from the acid instability that plagues erythromycin. In addition, many of the 16-membered macrolides possess an aldehyde moiety at C20. This moiety is implicated in the formation of a covalent bond with the ribosome protein as seen in the x-ray structure of carbomycin cocrystallized with the ribosome.[93] Like 14-membered macrolides, 16-membered macrolides inhibit bacterial protein synthesis by binding to the 50S subunit of the ribosome. While the binding sites for the different macrolide types overlap, they are not identical. As previously discussed, one striking difference is the case of carbomycin (**130**), which extends the C4″ acyl group toward the peptidyltransferase center whereas the 14-membered macrolides do not possess any substituent that can reach and interact with this center.[93]

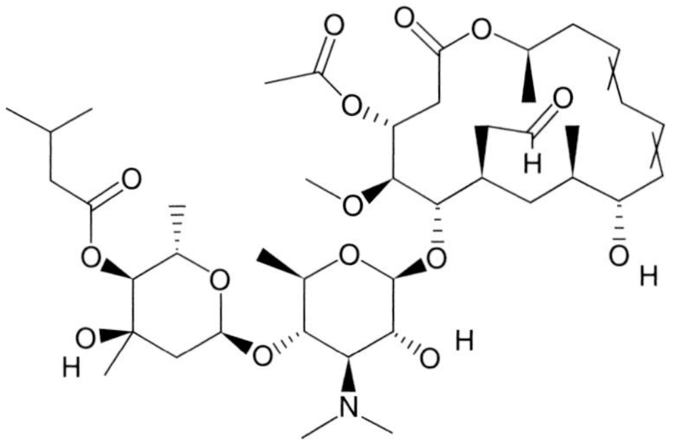

4 Josamycin

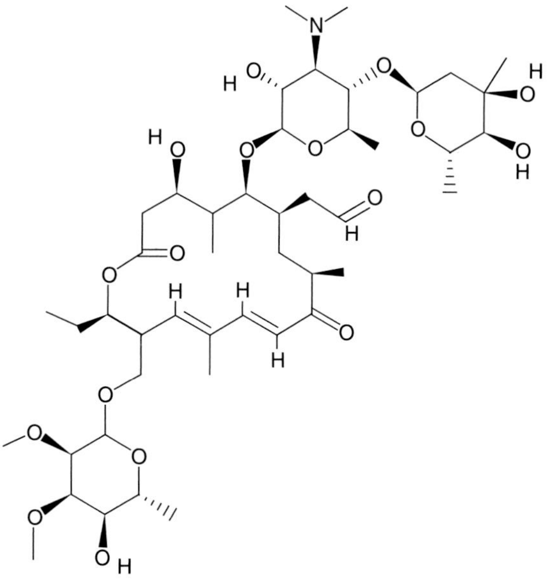

5 Tylosin

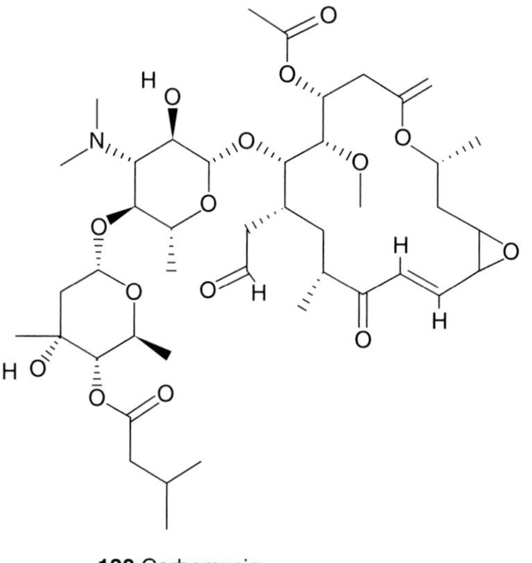

130 Carbomycin

In general, the spectrum of activity of 16-membered macrolides is similar to that of erythromycin, being primarily active against Gram-positive pathogens and mycoplasmas. Their in vitro potency, however, is lower than that of erythromycin, which in the presence of human serum is even further decreased. As a result, relatively few 16-membered macrolides have been developed for commercial uses. Like the first- and second-generation 14- and 15-membered macrolides, they are not effective against constitutive Erm-resistant *S. aureus*. One advantage of 16-membered macrolides, however, is that they are not inducers of Erm resistance. Although not inducers, they are not effective against streptococcal strains if the Erm mechanism is already induced. Conversely, if the Mef efflux systems constitute the resistance mechanism, 16-membered macrolides are effective since they are not substrates of those efflux pumps. Finally, and in contrast to many 14-membered macrolides, 16-membered macrolides are advantageous in that they do not inactivate the cytochrome P450 enzymes.[307]

7.18.5.11.1 Josamycin (leucomycin A3)

Josamycin (4) is produced by *Streptomyces narbonensis* var. *josamyceticus*, and is identical to leucomycin A3, one of the leucomycin components.[308,309] This compound contains an acetoxy group at C3, and an isovaleryoxy group at C4″. There are a large number of derivatives made and the SAR is known[142] to be mainly active against Gram-positive pathogens including macrolide-susceptible *S. aureus*, *S. pneumoniae*, and *S. pyogenes*.[310] Against those bacteria, the MIC_{50} values are approximately twice as high as those of erythromycin. Josamycin is not active against fastidious Gram-negative pathogens such as *H. influenzae* and *M. catarrhalis* due to its inability to penetrate the bacterial outer membrane. It is well absorbed in the gastrointestinal tract and a single dose of 500 mg in human yields peak concentrations of 0.61 to 0.71 $\mu g\,mL^{-1}$.[311] Even though it is well absorbed, josamycin is also rapidly metabolized. The three major metabolites found in human are a hydroxylation product at C14, a hydroxylation product in the isovaleryl group, and the isovaleryl hydrolysis product.[312] Like other macrolides, josamycin tends to penetrate and accumulate in various tissues and body fluids. The acute toxicity of josamycin appears to be quite low, with gastrointestinal disturbance as the most commonly reported side effect. Josamycin is used for the treatment of respiratory and suppurative infections.

7.18.5.11.2 Tylosin

Tylosin (**5**) is produced by *Streptomyces fradiae* and it contains a third sugar unit called mycinose attached at the C23 hydroxy group in addition to mycaminose and mycarose at C5.[313] There is a report suggesting that mycinose increases the ability of the molecule to enter bacterial cells.[314] Tylosin is active against Gram-positive pathogens and mycoplasma, and it is used in veterinary medicine to control respiratory diseases, mastitis, and dysentery in cattle and other farm animals. There have been a large number of derivatives reported and certain SAR is known. For example, desmycosin (**131**), an analog in which mycarose is removed, has an increased potency against *H. influenzae*. Reductive amination products of the C20 aldehyde group of desmycosin have increased oral bioavailability.[315] Among them, Tilmicosin (**132**) is the most widely used commercially against swine respiratory diseases.[316] Attempts to prepare new tylosin derivatives for human respiratory diseases have been recently reported. In these molecules, the mycarose sugar is removed from tylosin and a tethered heterocycle is attached at this position. Compound **133**, for example, has potent *S. pneumoniae* activity regardless of the resistance mechanism in place; however, its *H. influenzae* activity appears to be only modest (4 $\mu g\,mL^{-1}$).[317]

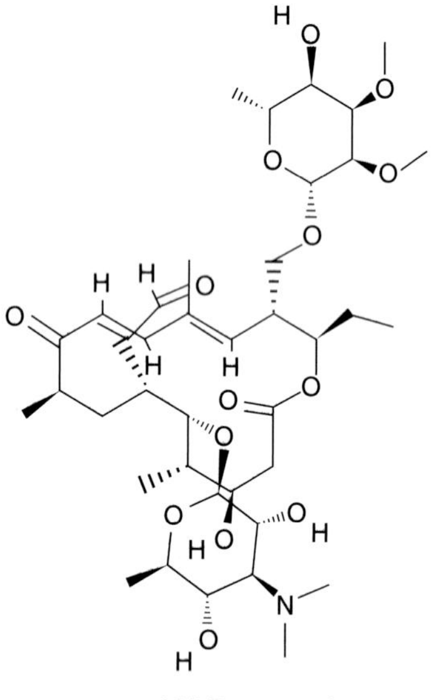

131 Desmycosin

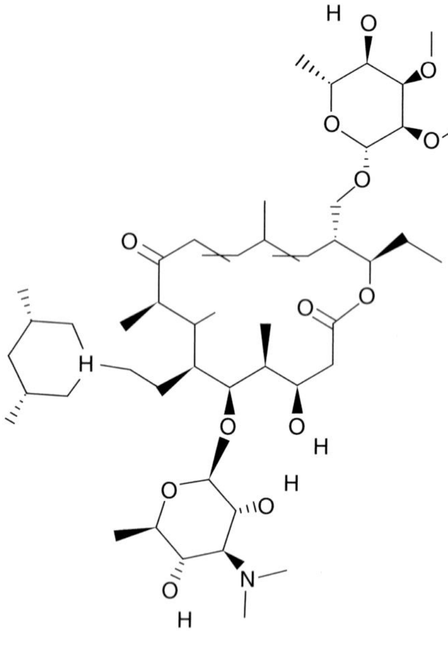

132 Tilmicosin

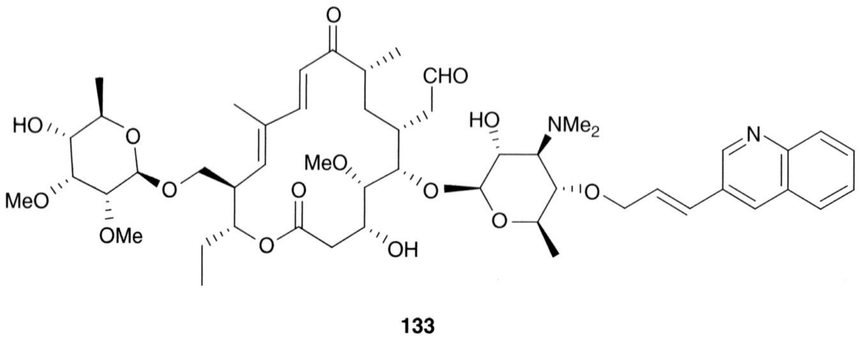

133

7.18.6 Non-Antibacterial Activity of Macrolides

At least some macrolide antibiotics have biological effects in addition to the antibacterial activity already discussed. They include prokinetic, anti-inflammatory, and immunomodulatory effects.[318] For prokinetic effects, it has been known that macrolides such as erythromycin interact with the motilin receptors. This causes some of the gastrointestinal side effects but can also be used in increasing the gastrointestinal motility function. More recently, the advantages of using macrolide antibiotics have been recognized in the treatments of diffuse panbronchiolitis,[319] chronic obstructive pulmonary disease,[320] and cystic fibrosis.[321] Although the mechanism of these additional effects has not been fully elucidated, it appears that macrolides affect transcription factors such as nuclear factor kappa B (NFκB) and consequently affect the level of cytokines. Other potential mechanisms include suppression of mucus production by *Pseudomonas aeruginosa*, which often infect patients with diffuse panbronchiolitis or cystic fibrosis, by interfering with the quorum-sensing mechanism.[322]

7.18.7 Conclusion and Future Directions

Tremendous progress has been made in macrolide chemistry during the last 50 years since the discovery of erythromycin. Although second-generation macrolides such as azithromycin and clarithromycin are the mainstay at present, third-generation macrolides represented by telithromycin are being integrated into the market. This progress notwithstanding, there remains room for improving activity against macrolide-resistant *S. pyogenes* and against *H. influenzae*. As use of third-generation macrolides (ketolides) increases in the future, we will inevitably face ketolide-resistant streptococci and staphylococci. Whether newer ketolides will be able to address this challenge or whether structurally divergent macrolides will be needed is not known at present. Although macrolides are relatively safe, it is, of course, desirable that future macrolides have decreased potentials for drug–drug interactions, cardiac arrhythmia, and hepatotoxicity. While it is likely impossible to satisfy all of these requirements, the challenge for medicinal chemists will be to balance the disparate needs in one molecular entity. Precisely because of the complex molecular structures involved, macrolides represent opportunities for medicinal chemists to design novel structures and explore new chemistries. In addition, with recent advances in the resolution of x-ray crystallography, it is becoming realistic to carry out structure-based drug designs involving the ribosome. As we gain more knowledge of the biochemical pathways by which macrolides are biosynthesized, it is possible to manipulate the enzyme machinery in order to generate novel macrolide templates. Here the major challenge is to obtain high-titer strains to produce enough quantities of templates for the subsequent chemical modifications. It should be also noted that structural modifications of 16-membered macrolides have thus far been limited, and that new breakthroughs may come from this arena. After a half century of clinical uses and chemical manipulations, macrolide antibiotics are thus poised for an exciting era of new growth.

References

1. Schonfeld, W.; Kirst, H. A. *Macrolide Antibiotics*; Birkhauser Verlag: Basel, Switzerland, 2002.
2. Brockmann, H.; Henkel, W. *Naturwissenschaften* **1950**, *37*, 138–139.
3. McGuire, J. M.; Bunch, R. L.; Anderson, R. C.; Boaz, H. E.; Flynn, E. H.; Powell, H. M.; Smith, J. W. *Antibiot. Chemother.* **1952**, *2*, 281–283.
4. Sobin, B. A.; Routin, J. B.; Lees, T. M. Oleandomycin, Its Salts and Production. US Patent 2757123 1956, July 31, 1956.
5. Osono, T.; Oka, Y.; Watanabe, S.; Numazaki, Y.; Moriyama, K.; Ishida, H.; Suzaki, K.; Okami, Y.; Umezawa, H. *J. Antibiot. A* **1967**, *20*, 174–180.
6. McGuire, J. M.; Boniece, W. S.; Higgins, C. E.; Hoehn, M. M.; Stark, W. M.; Westhead, J.; Wolfe, R. N. *Antibiot. Chemother.* **1961**, *11*, 320–327.
7. Morimoto, S.; Takahashi, Y.; Watanabe, W.; Omura, S. *J. Antibiot.* **1984**, *37*, 187–189.
8. Djokic, S.; Kobrehel, G.; Lopotar, N.; Kamenar, B.; Nagl, A.; Mrvos, D. *J. Chem. Res. (S)* **1988**, 152–153.
9. Bright, M.; Nagel, A. A.; Bordner, J.; Deasi, K. A.; Dibrino, J. N.; Nowakowski, J.; Vincent, L.; Watrous, R. M.; Sciavolino, F. C.; English, A. R. et al. *J. Antibiot.* **1988**, *41*, 1029–1047.
10. Finland, M.; Wilcox, C.; Wright, S. S.; Purcell, E. M. *Proc. Soc. Exp. Biol. Med.* **1952**, *81*, 725–729.
11. Denis, A.; Agouridas, C.; Auger, J.-M.; Benedetti, Y.; Bonnefoy, A.; Bretin, F.; Chantot, J.-F.; Dussarat, A.; Fromentin, C.; Gouin Dambrieres, S. L. et al. *Bioorg. Med. Chem. Lett* **1999**, *9*, 3075–3080.
12. Agouridas, C.; Denis, A.; Auger, J.-M.; Benedetti, Y.; Bonnefoy, A.; Bretin, F.; Chantot, J.-F.; Dussarat, A.; Fromentin, C.; Gouin D'Ambrieres, S. L. et al. *J. Med. Chem.* **1998**, *41*, 4080–4100.
13. Katz, L.; Ashley, G. W. *Chem. Rev.* **2005**, *105*, 499–527.
14. Dalhoff, K. *Int. J. Antimicrob. Agents* **2001**, *18*, S39–S44.
15. Dismukes, W. E. Chronic Pneumonia. In *Principles and Practice of Infectious Diseases*, 5th ed.; Mandell, G. R., Bennett, J. E., Dolin, R., Eds.; Churchill Livingstone: New York, 2000, pp 755–767.
16. Reynolds, H. Y. Chronic Bronchitis and Acute Infectious Exacerbations. In *Principles and Practice of Infectious Diseases*, 5th ed.; Mandell, G. R., Bennett, J. E., Dolin, R., Eds.; Churchill Livingstone: New York, 2000, pp 706–710.
17. Gwaltney, J. M., Jr. Sinusitis. In *Principles and Practice of Infectious Diseases*, 5th ed.; Mandell, G. R., Bennett, J. E., Dolin, R, Eds.; Churchill Livingstone: New York, 2000, pp 676–686.

18. Gwaltney, J. R., Jr.; Bisno, A. L. Pharyngitis. In *Principles and Practice of Infectious Diseases*, 5th ed.; Mandell, G. R., Bennett, J. E., Dolin, R., Eds.; Churchill Livingstone: New York, 2000, pp 656–662.
19. Klein, J. O. Otitis Externa, Otitis Media, and Mastioditis. In *Principles and Practice of Infectious Diseases*, 5th ed.; Mandell, G. R., Bennett, J. E., Dolin, R., Eds.; Churchill Livingstone: New York, 2000, pp 669–675.
20. Schrag, S. J.; McGee, L.; Whitney, C. G.; Beall, B.; Craig, A. S.; Choate, M. E.; Jorgensen, J. H.; Facklam, R. R.; Klugman, K. P. Active Bacterial Core Surveillance Team. *Antimicrob. Agents Chemother.* **2004**, *48*, 3016–3023.
21. Neralla, S.; Meyer, K. C. *Drugs Aging* **2004**, *21*, 851–864.
22. Mera, R. M.; Miller, L. A.; Daniels, J. J.; Weil, J. G.; White, A. R. *Diagn. Microb. Infect. Dis.* **2005**, *5*, 195–200.
23. Jacobs, M. R. *Am. J. Med.* **2004**, *117*, 3S–15S.
24. Bozdogan, B.; Bogdanovich, T.; Kosowska, K.; Jacobs, M. R.; Appelbaum, P. C. *Curr. Drug Targets Infect. Disord.* **2004**, *4*, 169–176.
25. Amsterdam, D. *Antibiotics in Laboratory Medicine*, 3rd ed.; Lorian, V., Ed.; Williams & Wilkins: Baltimore, MD 1991, pp 53–105.
26. Haight, T. H.; Finland, M. *Proc. Soc. Exp. Biol. Med.* **1952**, *81*, 175–183.
27. Brock, T. D.; Brock, M. L. *Biochim. Biophys. Acta* **1959**, *33*, 274–275.
28. Echols, H. *Operators and Promoters*; University of California Press: Berkely, CA, 2001, pp 178–198.
29. Ramakrishnan, V. *Cell* **2002**, *108*, 557–572.
30. White, D. *The Physiology and Biochemistry of Prokaryotes*; Oxford University Press: New York, 2000, pp 274–283.
31. Lewin, B. *Gene* Expression; John Wiley & Sons: New York, 1974, pp 117–168.
32. Moat, A. G.; Foster, J. W.; Spector, M. P. In *Microbial Physiology;* 4th ed.; John Wiley & Sons: New York, **2002**, 57–72.
33. Bouadloun, F.; Donner, D.; Kurland, C. G. *EMBO J.* **1983**, *2*, 1351–1356.
34. Ogle, J. M.; Ramakrishnan, V. *Annu. Rev. Biochem.* **2005**, *74*, 129–177.
35. Alberts, B.; Johnson, A.; Lewis, J.; Raff, M.; Roberts, K.; Walter, P. *Molecular Biology of the Cell*, 4th ed.; Garland Publishing Inc.: New York, 2002, pp 335–350.
36. Moazed, D.; Noller, H. F. *Nature* **1989**, *342*, 142–148.
37. Hardesty, B.; Odom, O. W.; Deng, H.-Y. In *The Structure, Function and Genetics of Ribosomes*; Hardesty, B., Kramer, G., Eds.; Springer: New York, 1986, pp 495–508.
38. Staehelin, T.; Raskas, H.; Meselson, M. In *Organizational Biosynthesis*; Vogel, H. J., Lampen, J. O., Bryson, V., Eds.; Academic Press: New York, 1967, pp 443–457.
39. Nomura, M.; Traub, P. In *Organizational Biosynthesis*; Vogel, H. J., Lampen, J. O., Bryson, V., Eds.; Academic Press: New York, 1967, pp 459–476.
40. Cech, T.; Zaug, A.; Grabowski, P. *Cell* **1981**, *27*, 487–496.
41. Guerrier-Takada, C.; Gardiner, K.; Marsh, T.; Pace, N.; Altman, S. *Cell* **1983**, *35*, 849–857.
42. Hampl, H.; Schulze, H.; Nierhaus, K. H. *J. Biol. Chem.* **1981**, *256*, 2284–2288.
43. Franceschi, F. J.; Nierhaus, K. H. *J. Biol. Chem.* **1990**, *265*, 16676–16682.
44. Noller, H. F.; Hoffarth, V.; Zimniak, L. *Science* **1992**, *256*, 1416–1419.
45. Green, R.; Noller, H. *Annu. Rev. Biochem.* **1997**, *66*, 679–716.
46. Woese, C. R.; Magrum, L. J.; Gupta, R.; Siegel, R. B.; Stahl, D. A.; Kop, J.; Crawford, N.; Brosius, J.; Gutell, R.; Hogan, J. J. et al. *Nucleic Acids Res.* **1980**, *8*, 2275–2293.
47. Noller, H. F.; Woese, C. R. *Science* **1981**, *212*, 403–411.
48. Fellner, P.; Ehresmann, C.; Stegler, P.; Ebel, J. P. *Nat. New Biol.* **1972**, *239*, 1–5.
49. Porse, B. T.; Garrett, R. A. *J. Mol. Biol.* **1995**, *249*, 1–10.
50. Schulze, H.; Nierhaus, K. H. *EMBO J.* **1982**, *1*, 609–613.
51. Lotti, M.; Dabbs, E. R.; Hasenbank, R.; Stoffler-Meilicke, M.; Stoffler, G. *Mol. Gen. Genet.* **1983**, *192*, 295–300.
52. Noller, H. F. *Annu. Rev. Biochem.* **1991**, *60*, 191–227.
53. Prince, J. B.; Taylor, B. H.; Thurlow, D. H.; Ofengand, J.; Zimmermann, R. A. *Proc. Natl. Acad. Sci. USA* **1982**, *79*, 5450–5454.
54. Girshovich, A. S.; Bochkareva, E. S.; Kramarov, V. M.; Ovchinnikov, Y. A. *FEBS Lett.* **1974**, *45*, 213–217.
55. Bartha, A.; Kuechler, E.; Steiner, G. In *The Ribosome. Structure, Function and Evolution*; Hill, W. E., Dahlberg, A. E., Garrett, R. A., Moore, P. B., Schlessinger, D., Warner, J. R., Eds.; ASM Press: Washington, DC, 1990, pp 358–365.
56. Frank, J.; Penczek, P.; Grassucci, R.; Srivastava, S. *J. Cell Biol.* **1991**, *115*, 597–605.
57. Stark, H.; Mueller, F.; Orlova, E. V.; Schatz, M.; Dube, P.; Erdemir, T.; Zemlin, F.; Brimacombe, R.; van Heel, M. *Structure* **1995**, *3*, 815–821.
58. Frank, J.; Zhu, J.; Penczek, P.; Li, Y. H.; Srivastava, S.; Verschoor, A.; RadErmacher, M.; Grassucci, R.; Lata, R. K.; Agrawal, R. K. *Nature* **1995**, *376*, 441–444.
59. Agrawal, R. K.; Penczek, P.; Grassucci, R. A.; Li, Y.; Leith, A.; Nierhaus, K. H.; Frank, J. *Science* **1996**, *271*, 1000–1002.
60. Valle, M.; Zavialov, A.; Sengupta, J.; Rawat, U.; Ehrenberg, M.; Frank, J. *Cell* **2003**, *114*, 123–134.
61. Gao, H.; Valle, M.; Ehrenberg, M.; Frank, J. *J. Struct. Biol.* **2004**, *147*, 283–290.
62. Capel, M. S.; Engelman, D. M.; Freeborn, B. R.; Kjeldgaard, M.; Langer, J. A.; Ramakrishnan, V.; Schindler, D. G.; Schneider, D. K.; Schoenborn, B. P.; Sillers, I. Y. et al. *Science* **1987**, *238*, 1403–1406.
63. Yonath, A.; Mussig, J.; Wittmann, H. G. *J. Cell Biochem.* **1982**, *19*, 145–155.
64. Yonath, A.; Bartunik, H. D.; Bartels, K. S.; Wittmann, H. G. *J. Mol. Biol.* **1984**, *177*, 201–206.
65. Glotz, C.; Mussig, J.; Gewitz, H. S.; Makowski, I.; Arad, T.; Yonath, A.; Wittmann, H. G. *Biochem. Int.* **1987**, *15*, 953–960.
66. Jenni, S.; Ban, N. *Curr. Opin. Struct. Biol.* **2003**, *13*, 212–219.
67. Moore, P. B.; Steitz, T. *Nature* **2002**, *418*, 229–235.
68. Yusupov, M. H.; Yusuppov, G. Z.; Baucom, A.; LiebErman, K.; Earnest, T. N.; Cate, J. H. D.; Noller, H. F. *Science* **2001**, *292*, 883–896.
69. Schuwirth, B. S.; Borovinskaya, M. A.; Hau, C. W.; Zhang, W.; Vila-Sanjurjo; Holton, J. H.; Doudna Cata, J. H. *Science* **2005**, *310*, 827–834.
70. Wimberly, B. T.; Brodersen, D. E.; Clemons, W. M.; Morgan-Warner, R. J.; Carter, A. P.; Vonrhein, C.; Hartsch, T.; Ramakrishnan, V. *Nature* **2000**, *407*, 327–339.
71. Schluenzen, F.; Tocijl, A.; Zarivach, R.; Harms, J.; Gluehmann, M.; Janell, D.; Bashan, A.; Bartels, H.; Agmon, I.; Franceschi, F. et al. *Cell* **2000**, *102*, 615–623.
72. Carter, A. P.; Clemons, W. M.; Brodersen, D. E.; Morgan-Warren, R. J.; Wimberly, B. T.; Ramakrishnan, V. *Nature* **2000**, *407*, 340–348.
73. Brodersen, D. E.; Clemons, W. J.; Carter, A. P.; Morgan-Warren, R. J.; Wimberly, B. T.; Ramakrishnan, V. *Cell* **2000**, *103*, 1143–1154.
74. Ban, N.; Freeborn, B.; Nissen, P.; Penczek, P.; Grassuci, R. A.; Sweet, R.; Frank, J.; Moore, P. B.; Steitz, T. A. *Cell* **1998**, *93*, 1105–1115.
75. Ban, N.; Nissen, P.; Hansen, J.; Capel, M.; Moore, P. B.; Steitz, T. A. *Nature* **1999**, *400*, 841–847.

76. Ban, N.; Nissen, P.; Hansen, J.; Moore, P. B.; Steitz, T. A. *Science* **2000**, *289*, 905–920.
77. Nissen, P.; Ban, N.; Hansen, J.; Moore, P. B.; Steitiz, T. A. *Science* **2000**, *289*, 920–930.
78. Gregory, S. T.; Dahlberg, A. E. *Nat. Struct. Mol. Biol.* **2004**, *11*, 586–587.
79. The Involvement of RNA in Ribosome Function. *Ann. Intern. Med.* **2006**, *144*, E1–E6.
80. Polacek, N.; Gaynor, M.; Yassin, A.; Mankin, A. S. *Nature* **2001**, *411*, 498–501.
81. Youngman, E. M.; Brunelle, J. L.; Kochaniak, A. B.; Green, R. *Cell* **2004**, *177*, 589–599.
82. Sievers, A.; Beringer, M.; Rodnina, M. V.; Wolfenden, R. *Proc. Natl. Acad. Sci. USA* **2004**, *101*, 7897–7901.
83. Weinger, J. S.; Parnell, K. M.; Dorner, S.; Green, R.; Strobel, S. A. *Nat. Struct. Mol. Biol.* **2004**, *11*, 1101–1106.
84. Rodnin, M. V.; Beringer, M.; Bieling, P. *Biochem. Soc. Trans.* **2005**, *33*, 493–498.
85. Goldman, R. C.; Scaglione, F. *Curr. Drug Targets* **2004**, *4*, 241–260.
86. Pestka, S. *Antimicrob. Agents Chemother.* **1974**, *6*, 474–479.
87. Menninger, J. R.; Otto, D. P. *Antimicrob. Agents Chemother.* **1982**, *21*, 811–818.
88. Moazed, D.; Noller, H. F. *Biochimie* **1987**, *69*, 879–884.
89. Douthwaite, S. *Nucleic Acids Res.* **1992**, *20*, 4717–4720.
90. Douthwaite, S.; Aagaard, C. J. *J. Mol. Biol.* **1993**, *232*, 725–731.
91. Hansen, L. H.; Mauvais, P.; Douthwaite, S. *Mol. Microbiol.* **1999**, *31*, 623–631.
92. Harms, J.; Schluenzen, F.; Zarivech, R.; Bashan, A.; Gat, S.; Agmon, I.; Bartels, H.; Franceschi, F.; Yonath, A. *Cell* **2001**, *107*, 679–688.
93. Hansen, J.; Ippolito, J. A.; Ban, N.; Nissen, P.; Moore, P. B.; Steitz, T. A. *Mol. Cell* **2002**, *10*, 117–128.
94. Schlunzen, F.; Zarivach, R.; Harms, J.; Bashan, A.; Tocilj, A.; Albrecht, R.; Yonath, A.; Franceschi, F. *Nature* **2001**, *413*, 814–821.
95. Hornig, H.; Woolley, P.; Luhrmann, R. *Biochimie* **1987**, *69*, 803–813.
96. Schluenzen, F.; Harms, J. M.; Franceschi, F.; Hansen, H. A. S.; Bartels, H.; Zarivach, R.; Yonath, A. *Structure* **2003**, *11*, 329–338.
97. Berisio, R.; Harms, J.; Schluenzen, F.; Zarivach, R.; Hansen, H.; Fucini, P.; Yonath, A. *J. Bacteriol.* **2003**, *185*, 4276–4279.
98. Douthwaite, S.; Hansen, L. H.; Mauvais, P. *Mol. Microbiol.* **2000**, *36*, 183–193.
99. Mao, J. C. H.; Robishaw, E. E. *Biochemistry* **1971**, *10*, 2054–2061.
100. Tu, D.; Blaha, G.; Moore, P. B.; Steitz, T. A. *Cell* **2005**, *121*, 257–270.
101. Cao, Z.; Zhong, P.; Ruan, X.; Merta, P.; Capobianco, J. O.; Flamm, K.; Nilus, A. M. *Int. J. Antimicrob. Agents* **2004**, *24*, 362–368.
102. Champney, W. S.; Pelt, J. *Curr. Microbiol.* **2002**, *45*, 328–333.
103. Champney, W. S. *Curr. Drug Targets Infect. Disord.* **2001**, *1*, 19–36.
104. Jacobs, M. R. *Am. J. Med.* **2004**, *117*, 3S–15S.
105. Brown, S. D.; Farrell, D. J. *J. Antimicrob. Chemother.* **2004**, *54*, 123–129.
106. Leclercq, R.; Courvalin, P. *Antimicrob. Agents Chemother.* **2002**, *46*, 2727–2734.
107. Roberts, M. C.; Sutcliffe, J.; Courvalin, P.; Jensen, L. B.; Rood, J.; Seppala, H. *Antimicrob. Agents Chemother.* **1999**, *43*, 2823–2830.
108. Skinner, R.; Cundliffe, E.; Schmidt, F. J. *J. Biol. Chem.* **1983**, *258*, 12702–12706.
109. Docherty, A.; Grandi, G.; Grandi, R.; Gryczan, T. J.; Shivakumar, A. G.; Dubnau, D. *J. Bacteriol.* **1981**, *145*, 129–137.
110. Israeli-Reches, M.; Weinrauch, Y.; Dubnau, D. *Mol. Gen. Genet.* **1984**, *194*, 362–367.
111. Tait-Kamradt, A.; Davies, T.; Cronan, M.; Jacobs, M. R.; Appelbaum, P. C.; Sutcliffe, J. *Antimicrob. Agents Chemother.* **2000**, *44*, 2118–2125.
112. Peric, M.; Bozdogan, B.; Jacobs, M. R.; Appelbaum, P. C. *Antimicrob. Agents Chemother.* **2003**, *47*, 1017–1022.
113. Prunier, A. L.; Malbruny, B.; Tande, D.; Picard, B.; Leclercq, R. *Antimicrob. Agents Chemother.* **2002**, *46*, 3054–3056.
114. Lai, C.-J.; Weisblum, B. *Proc. Natl. Acad. Sci. USA* **1971**, *68*, 856–860.
115. Weisblum, B. *Antimicrob. Agents Chemother.* **1995**, *39*, 797–805.
116. Zhong, P.; Cao, Z.; Hammond, R.; Chen, Y.; Beyer, J.; Shortridge, V. D.; Phan, L. Y.; Pratt, S.; Capobianco, J.; Reich, K. A. et al. *Microb. Drug Resist.* **1999**, *5*, 183–188.
117. Catchpole, I.; Thomas, C.; Davies, A.; Dyke, K. G. *J. Gen. Microbiol.* **1988**, *134*, 697–709.
118. Rohl, R.; Nierhaus, K. H. *Proc. Natl. Acad. Sci. USA* **1982**, *79*, 729–733.
119. Schulze, H.; Nierhaus, K. H. *EMBO J.* **1982**, *1*, 609–613.
120. Gregory, S. T.; Dahlberg, A. E. *J. Mol. Biol.* **1999**, *289*, 827–834.
121. Poehlsgaard, J.; Douthwaite, S. *Curr. Drug Targets* **2002**, *2*, 67–78.
122. Gabashvili, I. S.; Gregory, S. T.; Valle, M.; Grassuci, R.; Worbs, M.; Wahl, M. C.; Dahlberg, A. E.; Frank, J. *Mol. Cell* **2001**, *8*, 181–188.
123. Tait-Kamradt, A.; Davies, T.; Appelbaum, P. C.; Depardieu, F.; Courvalin, P.; Petitpas, J.; Wondrack, L.; Walker, A.; Jacobs, M. R.; Sutcliffe, J. *Antimicrob. Agents Chemother.* **2000**, *44*, 3395–3401.
124. Walsh, F.; Willcock, J.; Amyes, S. *J. Antimicrob. Chemother.* **2003**, *52*, 345–353.
125. Li, X.-Z.; Nikaido, H. *Drugs* **2004**, *64*, 159–204.
126. Sutcliffe, J.; Tait-Kamradt, A.; Wondrack, L. *Antimicrob. Agents Chemother.* **1996**, *40*, 1817–1824.
127. Tait-Kamradt, A.; Clancy, J.; Cronan, M.; Dib-Hajj, F.; Wondrack, L.; Yuan, W.; Sutcliffe, J. *Antimicrob. Agents Chemother.* **1997**, *41*, 2251–2255.
128. Clancy, J.; Petitpas, J.; Dib-Hajj, F.; Yuan, W.; Cronan, M.; Kamath, A. V.; Bergeron, J.; Retsema, J. A. *Mol. Microbiol.* **1996**, *22*, 867–879.
129. Del Grosso, M.; Iannelli, F.; Messina, C.; Santagati, M.; Petrosillo, N.; Stefani, S.; Pozzi, G.; Pantosti, A. *J. Clin. Microbiol.* **2002**, *40*, 774–778.
130. Gay, K.; Stephens, D. S. *J. Infect. Dis.* **2001**, *184*, 56–65.
131. Daly, M. M.; Doktor, S.; Flamm, R.; Shortridge, D. *J. Clin. Microbiol.* **2004**, *42*, 3570–3574.
132. Capobianco, J. O.; Cao, Z.; Shortridge, V. D.; Ma, Z.; Flamm, R. K.; Flamm, R. K.; Zhong, P. *Antimicrob. Agents Chemother.* **2000**, *44*, 1562–1567.
133. Vachon, V.; Kristjanson, D. N.; Coulton, J. W. *Can. J. Microbiol.* **1988**, *34*, 134–140.
134. Sanchez, L.; Pan, W.; Vinas, M.; Nikaido, H. *J. Bacteriol.* **1997**, *179*, 6855–6857.
135. Nikaido, H. *Microbiol. Mol. Biol. Rev.* **2003**, *67*, 593–656.
136. Forfar, J. O.; Maccabe, A. F. *Adv. Antibiot. Chemother.* **1957**, *4*, 115–157.
137. Washington, J. A., II.; Wilson, W. R. *Mayo Clin. Proc.* **1985**, *60*, 189–203.
138. Washington, J. A., II.; Wilson, W. R. *Mayo Clin. Proc.* **1985**, *60*, 271–278.
139. Alvarez-Elcoro, S.; Enzler, M. J. *Mayo Clin. Proc.* **1999**, *74*, 613–634.
140. Egan, R. S.; Perun, T. J.; Martin, J. R.; Mitscher, L. A. *Tetrahedron* **1973**, *29*, 2525–2538.
141. Awan, A.; Brennan, R. J.; Regan, A. C.; Barber, J. *J. Chem. Soc. Perkin Trans.* **2000**, *2*, 1645–1652.
142. Sakakibara, H.; Omura, S. Chemical Modification and Structure–Activity Relationship of Macrolides. In *Macrolide Antibiotics*; Omura, S., Ed.; Academic Press: New York, 1984, pp 85–125.

143. Wondrack, L.; Massa, M.; Yang, B. V.; Sutcliffe, J. *Antimicrob. Agents Chemother.* **1996**, *40*, 992–998.
144. Flynn, E. H.; Sigal, M. V., Jr.; Wiely, P. F.; Gerzon, K. *J. Am. Chem. Soc.* **1954**, *76*, 3121–3131.
145. Jones, P. H.; Rowley, E. K. *J. Org. Chem.* **1968**, *32*, 665–670.
146. Martin, Y. C.; Jones, P. H.; Perun, T. J.; Grundy, W. E.; Bell, S.; Bower, R. R.; Shipkowitz, N. L. *J. Med. Chem.* **1972**, *15*, 635–638.
147. Noguchi, N.; Emuma, A.; Matsuyam, H.; O'Hara, K.; Sasatsu, M.; Kono, M. *Antimicrob. Agents Chemother.* **1995**, *39*, 2359–2363.
148. Higgins, C.; Pittenger, R. C.; McGuire, J. M. *Antibiot. Chemother.* **1953**, *3*, 50.
149. Shafiee, A.; Hutchinson, C. R. *Biochemistry* **1987**, *26*, 6204–6210.
150. Kibwage, I. O.; Hoogmartens, J.; Roets, E.; Vanderhaeghe, H.; Verbist, L.; Dubost, M.; Pascal, C.; Petitjean, P.; Levol, G. *Antimicrob. Agents Chemother.* **1985**, *28*, 630–633.
151. Weber, J. M.; Leung, J. O.; Swanson, S. J.; Idler, K. B.; McAlpine, J. B. *Science* **1991**, *252*, 114–117.
152. Tadanier, J.; Martin, J. R.; Goldstein, A. W.; Hirner, E. A. *J. Org. Chem.* **1978**, *43*, 2351–2501.
153. Paecy, M. S.; Dirlam, J. P.; Geldart, R. W.; Leadlay, P. F.; McArthur, M. A. I.; McCormick, E. L.; Monday, R. A.; O'Connell, T. N.; Staunton, J.; Winchester, T. J. *J. Antibiot.* **1998**, *51*, 1029–1034.
154. Wilkinson, B.; Kendrew, S. G.; Sheridan, R. M.; Leadlay, P. F. *Expert Opin. Therapeut. Patents* **2003**, *13*, 1579–1606.
155. Jacobsen, J. R.; Hutchinson, C. R.; Cane, D. E.; Khosla, C. *Science* **1997**, *227*, 367–369.
156. Katz, L.; Ashley, G. W. *Chem. Rev.* **2005**, *105*, 499–528.
157. Fernandes, P. B.; Baker, W. R.; Freiberg, L. A.; Hardy, D. J.; McDonald, E. J. *Antimicrob. Agents Chemother.* **1989**, *33*, 78–81.
158. Pocidalo, J.-J.; Albert, F.; Desnottes, J. F.; Kerbaum, S. *J. Antimicrob. Chemother.* **1985**, *16*, 165–174.
159. Periti, P.; Mazzei, T.; Mini, E.; Novelli, A. *Clin. Pharmacokin.* **1989**, *16*, 193–214.
160. Jain, R.; Danziger, L. H. *Curr. Pharma. Des.* **2004**, *10*, 3045–3053.
161. Kurath, P.; Jones, P. H.; Egan, R. S.; Perun, T. J. *Experientia* **1971**, *27*, 362.
162. Fraser, D. G. *Am. J. Hosp. Pharm.* **1980**, *37*, 467–474.
163. Booth, R.; Dale, J. K. Erythromycin Esters. US Patent 2,862,921, Dec. 2, 1958.
164. Steffansen, B.; Bundgaard, H. *Int. J. Pharm.* **1989**, *56*, 159–168.
165. Bray, M. D.; Stephens, V. C. Sulfate Salt of Erythromycin Monoester. US Patent 3,000,874, Sept. 19, 1961.
166. Bensoussan, C.; Delaforge, M.; Mansuy, D. *Biochem. Pharmacol.* **1995**, *49*, 591–602.
167. Milberg, P.; Eckardt, L.; Bruns, H.-J.; Biertz, J.; Ramtin, S.; Reinsch, N.; Fleischer, D.; Kirchhof, P.; Fabritz, L.; Breithardt, G. et al. *Pharmcol. Exp. Therapeut.* **2002**, *303*, 218–225.
168. Periti, P.; Mazzei, T.; Mini, E.; Novelli, A. *Clin. Pharmacokinet.* **1992**, *23*, 106–131.
169. Mansuy, D. *Pharmacol. Ther.* **1987**, *33*, 41–45.
170. Ledirac, N.; De Sousa, G.; Fontaine, F.; Agouridas, C.; Gugenheim, J.; Lorenzon, G.; Rahmani, R. *Drug Metab. Dispos.* **2000**, *28*, 1391–1393.
171. Viluksela, M.; Hanhijarvi, H.; Husband, R. F.; Kosma, V.-M.; Collan, Y.; Mannisto, P. T. *J. Antimicrob. Chemother.* **1988**, *21*, 9–27.
172. Tolman, K. G.; Sannella, J. J.; Freston, J. S. *Ann. Intern. Med.* **1974**, *81*, 58–60.
173. Iannini, P. B. *Expert Opin. Drug Saf.* **2002**, *1*, 121–128.
174. Milberg, P.; Eckardt, L.; Bruns, H.-J.; Biertz, J.; Ramtin, S.; Reinsch, N.; Fleischer, D.; Kirchhof, P.; Fabritz, L.; Breinhardt, G. et al. *J. Pharmacol. Exp. Ther.* **2002**, *303*, 218–225.
175. Ohtani, H.; Taninaka, C.; Hanada, E.; Kotaki, H.; Sato, H.; Sawada, Y.; Iga, T. *Antimicrob. Agents Chemother.* **2000**, *44*, 2630–2637.
176. Morimoto, S.; Takahashi, Y.; Watanabe, W.; Omura, S. *J. Antibiot.* **1984**, *37*, 187–189.
177. Morimoto, S.; Adachi, T.; Matsunaga, T.; Kashimura, M.; Asaka, T.; Watanabe, Y.; Sota, K.; Sekiuchi, K. Erythromycin A Derivatives. US Patent 4,990,602, Feb. 5, 1991.
178. Alvarez-Elcoro, S.; Enzler, M. J. *Mayo Clin. Proc.* **1999**, *74*, 613–634.
179. Sturgill, M. G.; Rapp, R. P. *Ann. Pharmacother.* **1992**, *26*, 1099–1108.
180. Peters, D. H.; Clissold, S. P. *Drugs* **1992**, *44*, 117–164.
181. Hardy, D. J.; Swanson, R. N.; Rode, R. A.; Marsh, K.; Shipkowitz, N. L.; Clement, J. J. *Antimicrob. Agents Chemother.* **1990**, *34*, 1407–1413.
182. Heifets, L. *Tuberc. Lung Dis.* **1996**, *77*, 19–26.
183. Ulmer, H.-J.; Beckering, A.; Gatz, G. *Helicobacter* **2003**, *8*, 95–104.
184. Morimoto, S.; Misawa, Y.; Adachi, T.; Nagate, T.; Watanabe, Y.; Omura, S. *J. Antibiot.* **1990**, *43*, 286–294.
185. Morimoto, S.; Nagate, T.; Sugita, K.; Ono, T.; Numata, K.; Miyachi, J.; Misawa, Y.; Yamada, K.; Omura, S. *J. Antibiot.* **1990**, *43*, 295–305.
186. Hardy, D. J.; Swanson, R. N.; Rode, R. A.; Marsh, K.; Shipkowitz, N. L.; Clement, J. J. *Antimicrob. Agents Chemother.* **1990**, *34*, 1407–1413.
187. Chu, S. Y.; Deaton, R.; Cavanaugh, J. *Antimicrob. Agents Chemother.* **1992**, *36*, 1147–1150.
188. Suwa, T.; Yoshida, H.; Fukushima, K.; Nagate, T. *Chemother.* **1989**, *36*, 198–204.
189. Ferrero, J. L.; Bopp, B. A.; Marsh, K. C. *Drug Metab. Dispos.* **1990**, *18*, 441–446.
190. Biaxin® product label.
191. Chu, S.-Y.; Deaton, R.; Cavanaugh, J. *Antimicrob. Agents Chemother.* **1992**, *36*, 1147–1150.
192. Chu, S.-Y.; Wilson, D. S.; Deaton, R. L.; Machenthun, A. V.; Eason, C. N.; Cavanaugh, J. H. *J. Clin. Pharmacol.* **1993**, *33*, 719–726.
193. Rodvold, K. A. *Clin. Pharmacokinet.* **1999**, *5*, 385–398.
194. Neu, H. C. *Am. J. Med.* **1991**, *91*, 12S–18S.
195. Peters, D. H.; Friedel, H. A.; Mc Tavish, D. *Drugs* **1992**, *44*, 750–799.
196. Duran, J. M.; Amsden, G. W. *Expert Opin. Pharmacother.* **2000**, *1*, 489–505.
197. Djokic, S.; Kobrehel, G. Nouveaux derives de l'erythromycine. Belgium Patent 892,357, July 1, 1982.
198. Wilkening, R.; Ratcliffe, R. W.; Doss, G. A.; Bartizal, K. F.; Graham, A. C.; Herbert, C. M. *Bioorg. Med. Chem. Lett.* **1993**, *3*, 1287–1292.
199. Yang, B.; Goldsmith, M.; Rizzi, J. P. *Tetrahedron Lett.* **1994**, *35*, 3025–3028.
200. McFarland, J. W.; Berger, C. M.; Froshauer, S. A.; Hayashi, S. F.; Hecker, S. J.; Jaynes, B. H.; Jefson, M. R.; Kamicker, B. J.; Lipinski, C. A.; Lundy, K. M. et al. *J. Med. Chem.* **1997**, *40*, 1340–1346.
201. Williams, J. D. *Eur. J. Clin. Microbiol. Infect. Dis.* **1991**, *10*, 813–820.
202. Dunne, M.; Foulds, G.; Retsema, J. A. *Am. J. Med.* **1997**, *102*, 37–49.
203. Lode, H.; Boner, K.; Koeppe, P.; Schaberg, T. *J. Antimicrob. Chemother.* **1996**, *37*, 1–8.
204. Dunn, C. J.; Barradell, L. B. *Drugs* **1996**, *51*, 483–505.
205. Bronk, B. S.; Letavic, M. A.; Bertsche, C. D.; George, D. M.; Hayashi, S. F.; Kamicker, B. J.; Kolosko, N. L.; Norcia, L. J.; Rushing, M. A.; Santoro, S. L. et al. *Bioorg. Med. Chem. Lett.* **2003**, *13*, 1955–1958.

206. Fattori, R.; Pelacini, F.; Romagnano, S.; Fronza, G.; Rallo, R. *J. Antibiot.* **1996**, *49*, 938–940.
207. Wilkening, R.; Ratcliffe, R. W.; Doss, G. A.; Bartizal, K. F.; Graham, A. C.; Herbert, C. M. *Bioorg. Med. Chem. Lett.* **1993**, *3*, 1287–1292.
208. Foulds, G.; Shepard, R. M.; Johnson, R. B. *J. Antimicrob. Chemother.* **1990**, *25*, 73–82.
209. Gladue, R. P.; Bright, M.; Issacson, R.; Newborg, M. F. *Antimicrob. Agents Chemother.* **1989**, *33*, 277–282.
210. Dark, D. *Am. J. Med.* **1991**, *91*, 31S–35S.
211. Kinasewitz, G.; Wood, R. G. *Eur. J. Clin. Microbiol. Infect. Dis.* **1991**, *10*, 872–877.
212. Murry, J. J.; Empranza, P.; Lesinskas, E.; Tawadrous, M.; Breen, J. D. *Otolaryngol. – Head Neck Surg.* **2005**, *133*, 194–201.
213. D'Ignazio, J.; Camere, M. A.; Lewis, D. E.; Jorgensen, D.; Breen, J. D. *Antimicrob. Agents Chemother.* **2005**, *49*, 4035–4041.
214. Zithromax® product label.
215. Felstead, S. *Path. Biol.* **1995**, *43*, 512–514.
216. Puri, S. K.; Lassman, H. B. *J. Antimicrob. Chemother.* **1987**, *20*, 89–100.
217. Young, R. A.; Gonzalez, J. P.; Sorkin, E. M. *Drugs* **1989**, *37*, 8–41.
218. Paulsen, O. *Drugs Today* **1991**, *27*, 193–222.
219. Herron, J. M. *J. Antimicrob. Chemother.* **1987**, *20*, 139–144.
220. Massey, M. A.; Kitchell, B. S.; Martin, L. D.; Gerzon, K. *J. Med. Chem.* **1974**, *17*, 105–107.
221. Counter, F. T.; Ensminger, P. W.; Preston, D. A.; Wu, C.-Y. E.; Greene, J. M.; Felty-Duckworth, A. M.; Paschal, J. W.; Kirst, H. A. *Antimicrob. Agent Chemother.* **1991**, *35*, 1116–1126.
222. Brogden, R. N.; Peters, D. H. *Drugs* **1994**, *48*, 599–616.
223. Derriennic, M. *Drugs Today* **1995**, *31*, 111–115.
224. Dynabac® product label.
225. Geerdes-Fenge, H. F.; Goetschi, B.; Rau, M.; Borner, K.; Koeppe, P.; Wettich, K.; Lode, H. *Eur. J. Clin. Pharmacol.* **1997**, *53*, 127–133.
226. Sides, G. D.; Cerimele, B. J.; Black, H. R.; Busch, U.; DeSante, L. A. *J. Antimicrob. Chemother.* **1993**, *31*, 65–75.
227. Bergegne-Berezin, E. *J. Antimicrob. Chemother.* **1993**, *31*, 77–87.
228. Zanuso, G.; Toscano, L.; Pezzali, R.; Seghetti, E. *Proc. Int. Congr. Chemother.* **1983**, 107/45–107/48.
229. Toscano, L.; Fioriello, G.; Spagnoli, R.; Cappelletti, L.; Zanuso, G. *J. Antibiot.* **1983**, *36*, 1439–1440.
230. Vagliasindi, M. *Int. J. Clin. Pharmacol. Ther.* **1997**, *35*, 245–249.
231. Benoni, G.; Cuzzolin, L.; Leone, R.; Consolo, U.; Ferronato, G.; Bertrand, C.; Puchetti, V.; Fracasso, M. E. *Antimicrob. Agents Chemother.* **1988**, *32*, 1875–1878.
232. Villa, P.; Corti, F.; Guaitani, A.; Bartosek, I.; Casacci, F.; De Marchi, F.; Pacei, E. *J. Antibiot.* **1986**, *39*, 463–468.
233. Mabe, S.; Eller, J.; Champney, W. S. *Curr. Microbiol.* **2004**, *49*, 248–254.
234. Zhanel, G. G.; Walters, M.; Noreddin, A.; Vercaigne, L. M.; Wierzbowski, A.; Embil, J. M.; Gin, A. S.; Douthwaite, S.; Hoban, D. J. *Drugs* **2002**, *62*, 1771–1804.
235. Zhanel, G. G.; Hisanaga, T.; Nichol, K.; Wierzbowski, A.; Hoban, D. J. *Exp. Opin. Emerg. Drugs* **2003**, *8*, 297–321.
236. Zhanel, G. G.; Hoban, D. J. *Exp. Opin. Pharmacother.* **2002**, *3*, 277–297.
237. Nilius, A. M.; Ma, Z. *Curr. Opin. Pharmacol.* **2002**, *2*, 493–500.
238. Baker, W. R.; Clark, J. D.; Stephens, R. L.; Kim, K. H. *J. Org. Chem.* **1988**, *53*, 2340–2345.
239. Asaka, T.; Kashimura, M.; Misawa, Y.; Morimoto, S.; Hatayama, K. (Taisho Pharmaceutical Co.). Preparation of 5-O-Desosaminylerythronolide A Derivative as Antibacterial Agent. World Patent Appl. 21199 A1, 1993.
240. Asaka, T.; Kashimura, M.; Misawa, Y.; Morimoto, S.; Hatayama, K. (Taisho Pharmaceutical Co.) Preparation of 5-O-Desosaminylerythronolide A Derivatives as Antibacterial Agents. World Patents. Appl. 21200 A1, 1993.
241. Kashimura, M.; Asaka, T.; Misawa, Y.; Matsumoto, K.; Morimoto, S. *J. Antibiot.* **2001**, *54*, 664–678.
242. Agouridas, C.; Benedetti, Y.; Denis, A.; Fromentin, C.; Gouin D'Ambrieres, S.; Le Martret, O.; Chantot, J.-F. Presented at 34th Interscience Conference on Antimicrobial Agents and Chemotherapy, 1994, Abstr. F-164.
243. Agouridas, C.; Bonnefoy, A.; Chantot, J.-F.; Le Martret, O.; Denis, A. (Roussel-Uclaf). European Patent Appl. 0596802 A1, 1994.
244. Agouridas, C.; Benedetti, Y.; Denis, A.; Le Martret, O.; Chantot, J.-F. Presented at 35th Interscience Conference on Antimicrobial Agents and Chemotherapy, 1995, Abstr. F-157.
245. Griesgraber, G.; Or, Y. S.; Chu, D. T. W.; Nilius, A. M.; Johnson, P. M.; Flamm, R. K.; Henry, R. F. *J. Antibiot.* **1996**, *49*, 465–477.
246. Wellington, K.; Noble, S. *Drugs* **2004**, *64*, 1683–1694.
247. Bryskier, A. *Clin. Microbial. Infect.* **2000**, *6*, 661–669.
248. Bryskier, A.; Agouridas, C.; Chantot, J.-F. *Infect. Dis. Ther.* **2000**, *23*, 79–102.
249. Doern, G. V.; Brown, S. D. *J. Infect.* **2004**, *48*, 56–65.
250. Giovanetti, E.; Montanari, M. P.; Marchetti, F.; Varaldo, P. E. *J. Antimicrob. Chemother.* **2000**, *46*, 905–908.
251. FDA Briefing Package, New Drug Application (NDA) 21-144, Ketek™ (telithromycin), Apr. 26, 2001.
252. Shi, J.; Montay, G.; Bhargava, V. O. *Clin. Pharmacokin.* **2005**, *44*, 915–934.
253. Demolis, J.-L.; Vacheron, F.; Cardus, S.; Funk-Brentano, C. *Clin. Pharmacol. Ther.* **2003**, *73*, 242–252.
254. Demolis, J.-L.; Strabach, S.; Vacheron, F.; Funck-Brentano, C. *Br. J. Clin. Pharmacol.* **2005**, *60*, 120–127.
255. Shi, J.; Chapel, S.; Montay, G.; Hardy, P.; Barrett, J. S.; Sica, D.; Swan, S. K.; Noveck, R.; Leroy, B.; Bhargava, V. O. *Int. J. Clin. Pharmacol. Ther.* **2005**, *43*, 123–133.
256. Agouridas, C.; Bretin, F.; Chantot, J.-F. French Patent Appl. 2742757 A1, 1997.
257. Denis, A.; Bretin, F.; Fromentin, C.; Bonnet, A.; Piltan, G.; Bonnefoy, A.; Agouridas, C. *Bioorg. Med. Chem. Lett.* **2000**, *10*, 2019–2022.
258. Denis, A.; Bonnefoy, A. *Drugs Future* **2001**, *26*, 975–984.
259. Kaneko, T.; McMillen, W. T. (Pfizer, Inc.). Ketolide Antibiotics. World Patent Appl. 44761 A2, 2000.
260. Kaneko, T.; McMillen, W.; Sutcliffe, J.; Duignan, J. Presented at 40th Interscience Conference on Antimicrobial Agents and Chemotherapy, 2000, Abstr. F-1815.
261. Su, W.; Smyth, K.; Rainville, J.; Wu, Y.; Wons, R.; Durkin, D.; Finegan, S.; Cimochowski, C.; Girard, D.; Brennan, L. et al. Presented at 41st Interscience Conference on Antimicrobial Agents and Chemotherapy, 2001, Abstr. F-1168.
262. Dirard, D.; Cimochowski, C. R.; Finegan, S. M.; Su, W. G. Presented at 41st Interscience Conference on Antimicrobial Agents and Chemoherapy, 2001, Abstr. F-1169.
263. Girard, D.; Finegan, S. M.; Su. W. G.; Zhang, D.; Raunig, D. L. Presented at 41st Interscience Conference on Antimicrobial Agents and Chemotherapy, 2001, Abstr. F-1170.

264. Denis, A.; Pejac, J.-M.; Bretin, F.; Bonnefoy, A. *Bioorg. Med. Chem. Lett.* **2003**, *11*, 2389–2394.
265. Liang, C.-H.; Duffield, J.; Romero, A.; Chiu, Y.-H.; Rabuka, D.; Yao, S.; Sucheck, S.; Marby, K.; Shue, Y.-K. (Optimer Pharmaceuticals). Novel Antibacterial Agents. World Patent Appl. 080391 A2, 2004.
266. Chu, D.; Burger, M.; Lin, X.; Roll, G. L.; Plattner, J.; Rico, A. (Chiron Corp.). C12 Modified Erythromycin Macrolides and Ketolides Having Antibacterial Activity. World Patent Appl. 004509 A2, 2003.
267. Hlasta, D.; Henninger, T.; Grant, E.; Khosla, C.; Chu, D. (Ortho-McNeil Pharmaceutical, Inc.). Ketolide Antibacterials. World Patent Appl. 62783 A2, 2000.
268. Chu, D. T. W.; Ashley, G. W. (Kosan Biosciences). Macrolide Antiinfective Agents. World Patent Appl. 63224 A2, 2000.
269. Chu, D. T. W. (Kosan Biosciences). Macrolide Antiinfective Agents. World Patent Appl. 63225 A2, 2000.
270. Hlasta, D.; Henninger, T. C.; Grant, E. B.; Khosla, C.; Chu, D. T. W.; Ashley, G. (Ortho-McNeil Pharmaceutical, Inc.). Ketolide Antibacterials. World Patent. Appl. 32918 A2, 2002.
271. Macielag, M.; Abbanat, D.; Ashley, G.; Baum, E.; Fardis, M.; Foleno, B.; Fu, H.; Grant, E.; Henninger, T.; Hilliard, J.; et al. Presented at 41st Interscience Conference on Antimicrobial Agents and Chemotherapy, 2001, Abstr. F-1174.
272. Andreotti, D.; Arista, L.; Biondi, S.; Cardullo, F.; Damiani, F.; Lociuro, S.; Marchioro, C.; Merlo, G.; Mingardi, A.; Noccolai, D. et al. (Glaxo Group Ltd.). Macrolide Antibiotics. World Patent Appl. 50091 A1, 2002.
273. Guerry, P.; Kellenberger, J. L.; Blanchard, S. (Basilea Pharmaceutica A.-G.). Macrolides with Antibacterial Activity. World Patent Appl. 072588 A1, 2003.
274. Hunziker, D.; Wyss, P.-C.; Angehrn, P.; Mueller, A.; Marty, H.-P.; Halm, R.; Kellenberger, L.; Bitsch, V.; Biringer, G.; Arnold, W. et al. *Bioorg. Med. Chem.* **2004**, *12*, 3505–3519.
275. Kaneko, T. (Pfizer, Inc.). Novel Ketolide Antibiotics. World Patent Appl. 26224 A2, 2000.
276. Kaneko, T.; McMillen, W. T. (Pfizer, Inc.). Macrolide Antibiotics. European Patent Appl. 1167376 A1, 2002.
277. Or, Y. S.; Clark, R. F.; Wang, S.; Chu, D. T. W.; Nilius, A. M.; Flamm, R. K.; Mitten, M.; Ewing, P.; Alder, J.; Ma, Z. *J. Med. Chem.* **2000**, *43*, 1045–1049.
278. Clark, R. F.; Ma, Z.; Wang, S.; Griesgraber, G.; Tufano, Ml.; Yong, H.; Li, L.; Zhang, X.; Nilius, A. M.; Chu, D. T. W. et al. *Bioorg. Med. Chem. Lett.* **2000**, *10*, 815–816.
279. Ma, Z.; Clark, R. F.; Brazzale, A.; Wang, S.; Rupp, M. J.; Li, L.; Griesgraber, G.; Zhang, S.; Yong, H.; Phan, L. T. et al. *J. Med. Chem.* **2001**, *44*, 4137–4156.
280. Morimoto, S.; Adachi, T.; Matsunaga, T.; Kashimura, M.; Asaka, T.; Watanabe, Y.; Sota, K.; Sekiushi, K. (Taisho Pharmaceutical Co. Ltd.). Erythomycin A Derivatives. US Patent Appl. 4,990,602, Feb. 5, 1991.
281. Shortridge, V. D.; Ramer, N. C.; Beyer, J.; Ma, Z.; Or, Y. Presented at 39th Interscience Conference on Antimicrobial Agents and Chemotherapy, 1999, Abstr. F-2136.
282. Or, Y. S. *J. Med. Chem.* **2000**, *43*, 1045–1049.
283. Cao, Z.; Zhong, P.; Ruan, X.; Merta, P.; Capobianco, J. O.; Flamm, R. K.; Nilius, A. M. *Int. J. Antimicrob. Agents* **2004**, *24*, 362–368.
284. Hernandez, L.; Sadrzadeh, N.; Krill, S.; Ma, Z.; Marsh, K. Presented at 39th Interscience Conference on Antimicrobial Agents and Chemotherapy, 1999, Abstr. F 2148.
285. Guan, Z.; Javanti, V.; Johnson, M.; Nequest, G.; Reisch, T.; Roberts, E.; Schmidt, J.; Rotert, G.; Surber, B.; Thomas, S. et al. Presented at 39th Interscience Conference on Antimicrobial Agents and Chemotherapy, 1999, Abstr. F 2149.
286. Pradhan, R. S.; Gustavson, L. E.; Londo, D. D.; Zhang, Y.; Zhang, J.; Paris, M. M. Presented at 40th Interscience Conference on Antimicrobial Agents and Chemotherapy, Abstr. F-2135.
287. Ma, Z.; Pham, L. T.; Zhang, S.; Djuric, S. (Abbott Laboratories). 6-O Substituted Erythromycin Derivatives Having Improved Gastrointestinal Tolerance. World Patent Appl. 32919 A2, 2002.
288. Phan, L. T.; Clark, R. F.; Rupp, M.; Or, Y. S.; Chu, D. T. W.; Ma, Z. *Org. Lett.* **2000**, *2*, 2951–2954.
289. Keyes, R. F.; Carter, J. J.; Englund, E. E.; Daly, M. M.; Stone, G. G.; Nilius, A. M.; Ma, Z. *J. Med. Chem.* **2003**, *46*, 1795–1798.
290. Beebe, X.; Yang, F.; Bui, M. H.; Mitten, M. J.; Ma, Z.; Nilius, A. M.; Djuric, S. W. *Bioorg. Med. Chem. Lett.* **2004**, *14*, 2417–2421.
291. Phan, L. T.; Or, Y. S.; Ma, Z.; Chen, Y. (Abbott Laboratories). 6-O-Carbamate Ketolide Derivatives. World Patent Appl. 75156 A1, 2000.
292. Henninger, T. C.; Xu, X. (Ortho-McNeil Pharmaceutical, Inc.). 6-O-Carbamoyl Ketolide Derivatives of Erythromycin Useful As Antibacterials. World Patent Appl. 46204 A1, 2002.
293. Henninger, T.; Abbanat, D.; Baum, E.; Foleno, B.; Wira, E.; Xu, X.; Bush, K.; Macielag, M. Presented at 42nd Interscience Conference on Antimicrobial Agents and Chemotherapy, 2002, Abstr. F-1661.
294. Xu, X.; Henninger, T.; Abbanat, D.; Bush, K.; Foleno, B.; Hilliard, J.; Macielag, M. *Bioorg. Med. Chem. Lett.* **2005**, *15*, 883–887.
295. Abbanat, D. R.; Ashley, G.; Carney, J.; Fardis, M.; Foleno, B.; Hilliard, J.; Li, Y.; Licari, P.; Loeloff, M.; Macielag, M. et al. Presented at 41st Interscience Conference on Antimicrobial Agents and Chemotherapy, 2001, Abstr. F-1173.
296. Abbanat, D.; Webb, G.; Foleno, B.; Li, Y.; Macielag, M.; Montenegro, D.; Wira, E.; Bush, K. *Antimicrob. Agents Chemother.* **2005**, *49*, 309–315.
297. Andreotti, D.; Biondi, S.; Lociuro, S. (Glaxo Group Ltd). Macrolide Antibiotics. World Patent Appl. 50092 A1, 2002.
298. Grant, E. B.; III., Henninger, T. C.; Macielag, M. J.; Guiadeen, D. (Ortho-McNeil Pharmaceutical, Inc.). 6-O-Carbamate-11,12-lacto-ketolide Antimicrobials. World Patent Appl. 024986 A1, 2003.
299. LeMahieu, R. A.; Carson, M.; Kierstead, R. W.; Pestka, S. *J. Med. Chem.* **1975**, *18*, 849–851.
300. Tanikawa, T.; Asaka, T.; Kashimura, M.; Misawa, Y.; Suzuki, K.; Sato, M.; Kameo, K.; Morimoto, S.; Nishida, A. *J. Med. Chem.* **2001**, *44*, 4027–4030.
301. Tanikawa, T.; Asaka, T.; Kashima, M.; Suzuki, K.; Sugiyama, H.; Sato, M.; Kameo, K.; Morimoto, S.; Nishida, A. *J. Med. Chem.* **2003**, *46*, 2706–2715.
302. Asaka, T.; Kashimura, M.; Manaka, A.; Tanikawa, T.; Ishii, T.; Sugimoto, T.; Suzuki, K.; Sugiyama, H.; Akashi, T.; Saito, H. et al., Presented at 39th Interscience Conference on Antimicrobial Agents and Chemotherapy, 1999, Abstr. F-2159.
303. Tanikawa, T.; Manaka, A.; Sugimoto, T.; Suzuki, K.; Sugiyama, H.; Nakaumi, K.; Yamasaki, Y.; Kato, T.; Asaka, T. Presented at 43rd Interscience Conference on Antimicrobial Agents and Chemotherapy, 2003, Abstr. F-1188.
304. Zhu, B.; Marinelli, B.; Abbabat. D.; Folono, B.; Henninger, T.; Wira, E. Presented at 227th Natl. Meeting of the American Chemical Society, 2004.
305. Wang, G.; Niu, D.; Qiu, Y.-L.; Phan, L. T.; Chen, Z.; Polemeropoulos, A.; Or, Y. S. *Org. Lett.* **2004**, *6*, 4455–4458.
306. Xiong, L.; Korkhim, Y.; Mankin, A. S. *Antimicrob. Agent Chemother.* **2005**, *49*, 281–288.
307. Pessayre, D.; Larrey, D.; Funck-Brentano, C.; Benhamou, J. P. *J. Antimicrob. Chemother.* **1985**, *16*, 181–194.
308. Nakayama, I. In *Macrolide Antibiot*; Omura, S., Eds., 1984, 261–300.

309. Omura, S.; Hironaka, Y.; Hata, T. *J. Antibiot.* **1970**, *23*, 511–513.
310. Osono, T. *Drug Acti. Drug Resist. Bacteria* **1971**, *1*, 41–120.
311. Periti, P.; Mazzei, T.; Mini, E.; Novelli, A. *Clin. Pharmacokin.* **1989**, *16*, 193–214.
312. Osono, T.; Umezawa, H. *J. Antimicrob. Chemother.* **1985**, *6*, 151–166.
313. Gray, P. P.; Bhuwapathanapun, S. *Drugs Pharmaceut. Sci.* **1984**, *22*, 743–757.
314. Omura, S.; Inokoshi, J.; Matsubara, H.; Tanaka, H. *J. Antibiotics* **1983**, *36*, 1709–1712.
315. Kirst, H. A.; Toth, J. E.; Debono, M.; Willard, K. E.; Truedell, B. A.; Ott, J. L.; Counter, F. T.; Felty-Duckworth, A. M.; Pekarek, R. S. *J. Med. Chem.* **1988**, *31*, 1631–1641.
316. Debono, M.; Willard, K. E.; Kirst, H. A.; Wind, J. A.; Crouse, G. D.; Tao, E. V.; Vicenzi, J. T.; Counter, F. T.; Ott, J. L.; Ose, E. E. *J. Antibiot.* **1989**, *42*, 1253–1267.
317. Ly, T. P.; Jian, T.; Chen, Z.; Qiu, Y.-L.; Wang, Z.; Beach, T.; Polemeropoulos, A.; Or, Y. S. *J. Med. Chem.* **2004**, *47*, 2965–2968.
318. Labro, M.-T. *Curr. Pharm. Des.* **2004**, *10*, 3067–3080.
319. Keicho, N.; Kudoh, S. *Am. J. Respir. Med.* **2002**, *1*, 119–131.
320. Gotfried, M. H. *Chest* **2004**, *125*, 52S–61S.
321. Prescott, W. A., Jr.; Johnson, C. E. *Pharmacotherapy* **2005**, *25*, 555–573.
322. Imamura, Y.; Yanagihara, K.; Mizuta, Y.; Seki, M.; Ohno, H.; Higashiyama, Y.; Miyazaki, Y.; Tsukamoto, K.; Hirakata, Y.; Tomono, K. et al. *Antimicrob. Agents Chemother.* **2004**, *48*, 3457–3461.

Biographies

Takushi Kaneko, born in Hiroshima, Japan, studied at the University of Missouri, where he obtained a BSc in 1970. He obtained his PhD in 1974, at the University of Michigan, under the direction of Prof J P Marino, working in the area of sulfur ylides. He then carried out a postdoctoral research at Harvard University under Prof Yoshito Kishi, working on the total synthesis of saxitoxin. Since 1977, he worked in Bristol-Myers Research and Development in the area of cancer chemotherapy. In 1989, he moved to Pfizer, Inc. and has since worked in the areas of cancer chemotherapy, natural product discovery, and antibacterials. He is currently a Research Fellow in Pfizer Global R&D. His interests lie in medicinal chemistry, antibacterial agents, new reaction methods, and natural product isolation.

Thomas J Dougherty received his BSc in Microbiology with honors in 1973 from The University of the Sciences in Philadelphia (formerly the Philadelphia College of Pharmacy & Science). In 1978, he obtained his PhD in Microbiology from Thomas Jefferson University (Philadelphia, PA). He then joined the laboratory of Alexander Tomasz at the

Rockefeller University (New York, NY) as a postdoctoral fellow, and was subsequently promoted to Assistant Professor. While at Rockefeller University, he conducted studies on the biochemistry and genetics of penicillin resistance in pathogenic bacteria. From 1986–2001, he was in the Department of Microbiology at Bristol-Myers Squibb Co., working on various early and late stage antibiotic programs. In 2001, he joined Pfizer, Inc., and is currently a Research Fellow in the Antibacterial Department at Groton, CT. Dr Dougherty has published over 50 research articles and reviews to date.

Thomas V Magee received a BSc in chemistry from the University of Wisconsin-Madison in 1985, subsequently obtained a doctorate with emphasis on the total synthesis of natural products under the direction of Prof Gilbert Stork at Columbia University in 1991. After a postdoctoral fellowship at Memorial Sloan-Kettering Cancer Center with Prof Samuel Danishefsky, he started working at Pfizer in 1993 and since that time has worked on allergy and respiratory diseases, antivirals, and gastrointestinal disorders. He is currently in focused on the antibacterials field and retains an interest in natural product classes such as the macrolides.

Comprehensive Medicinal Chemistry II
ISBN (set): 0-08-044513-6

ISBN (Volume 7) 0-08-044520-9; pp. 519–566

7.19 Quinolone Antibacterial Agents

A S Wagman, Novartis Institutes for BioMedical Research, Inc., Emeryville, CA, USA
M P Wentland, Rensselaer Polytechnic Institute, Troy, NY, USA

7.19.1 Introduction

7.19.1.1 Overview of Quinolones

Quinolone antimicrobial agents meet a vital need in the arsenal used to fight both community-acquired and serious hospital-acquired infections. While early representatives of the quinolone class, such as nalidixic acid (**1**), oxolinic acid (**2**), and cinoxacin (**3**), were mainly used for the treatment of urinary tract infections (UTI), modern quinolone agents, having evolved over the course of 30 years of medicinal chemistry research, play a central role in the management of respiratory tract infections (RTI), UTI, sexually transmitted diseases (STD), gastrointestinal and abdominal infections, skin and soft tissue infections, and infections of the bone and joints among many other uses.[1] Of the modern quinolones, ciprofloxacin (**4**), introduced into the US market in 1986, is one of the most prescribed of the class and continues to exhibit one of the best in vitro spectrums of activity against Gram-negative pathogens, especially the difficult to treat *Pseudomonas aeruginosa*. Newer quinolone agents, such as gatifloxacin (**5**), gemifloxacin (**6**), levofloxacin (**7**), and moxifloxacin (**8**), show improved activity toward important RTI bacteria including Gram-positive and anaerobic species.[2] Additionally, these new agents have improved bioavailability and plasma half-lives compared to ciprofloxacin, which allows for more convenient once daily oral dosing.

Adverse effects of the quinolone class are typically minimal, including gastrointestinal disturbances, dizziness, headache, and skin rashes. However, serious adverse effects, such as phototoxicity, QTc interval prolongation leading to torsades de pointes, and hepatoxicity seen with use of trovafloxacin (**9**),[3] have prompted extensive phase 4, or postmarketing, surveillance of safety data to define the safety profile of the new generation of quinolones.[4] Few drug–drug interactions have been noted with quinolone use, although all quinolones chelate metal ions such as zinc, calcium, and magnesium, which can interfere with antacid treatments. With advantages such as broad-spectrum activity, effectiveness against important resistant strains (e.g., penicillin and macrolide resistant *Streptococcus pneumoniae* for RTI), good pharmacokinetic parameters, and excellent clinical efficacy the application of new-generation quinolones will continue to grow for the treatment of community-acquired pneumonia (CAP), acute exacerbations of chronic bronchitis (AECB), acute sinusitis, UTI, and in many more situations where an oral agent would be favorable over an intravenous antibacterial.[2]

7.19.1.2 Brief History of Quinolone Agents

The family of quinolone antibiotics sprang from the discovery of nalidixic acid (**1**) by George Lesher and coworkers in 1962.[5] The original lead compound, a 7-chloro-4-quinolone (**10**), was a by-product isolated from a recrystallization

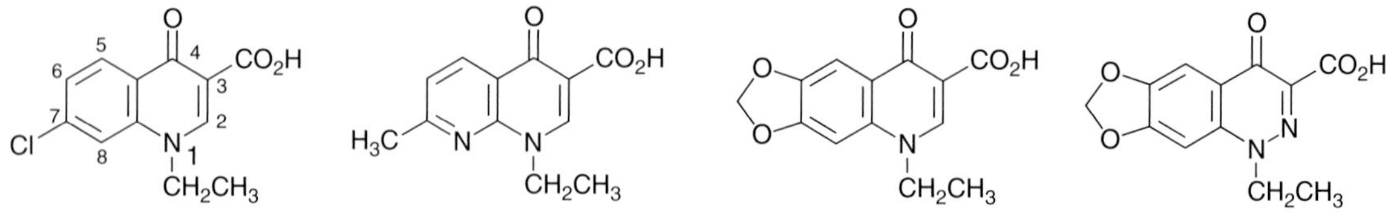

Original Lead 4-Quinolone (**10**) — Nalidixic acid (**1**) — Oxolinic acid (**2**) — Cinoxacin (**3**)

Figure 1 Early lead quinolone agents.

mother liquor during the synthesis of chloroquine. Thus, modern fluoroquinolones can trace their linage back to a serendipitous offshoot of antimalarial research. In the clinic, nalidixic acid (**1**), which is excreted in urine, was primarily used for the treatment of uncomplicated UTI caused by Gram-negative microbes such as *Escherichia coli*, *Klebsiella*, and *Proteus* spp. Unfortunately due to its poor activity against Gram-positive organisms, lack of potency against *P. aeruginosa*, inadequate serum concentrations, poor tissue distribution, and frequent incidence of adverse effects, nalidixic acid's utility and benefits were limited.[6] However, the discovery of nalidixic acid inspired a rush of innovative research and development leading to a new class of antibiotic.

Introduced into the clinic in the 1970s, compounds such as oxolinic acid,[7] cinoxacin[8] and pipemidic acid gave marginal improvement over nalidixic acid (**Figure 1**). These agents still lacked broad-spectrum activity and had modest serum levels, but could be dosed twice daily to treat UTI. The discovery of the first fluoroquinolone, norfloxacin (**11**), led to a marked improvement in the activity against Gram-negative species, including *P. aeruginosa*. Norfloxacin was also active against problematic bacteria such as gentamicin-resistant *P. aeruginosa*, penicillin-resistant *Neisseria gonorrhoeae*, Enterobacteriaceae, and methicillin-resistant *Staphylococcus aureus*. However, unlike with *S. aureus*, norfloxacin had less activity against most other aerobic, Gram-positive organisms and very little effect on anaerobic bacteria.[9] While the enhanced Gram-negative activity due to the C-6 fluoro group was a breakthrough, norfloxacin, with high kidney concentrations and renal excretion as the major route of elimination, was still mostly limited to the treatment of UTI. However, the UTI cure rate using norfloxacin was improved over nalidixic acid, oxolinic acid, and cinoxacin.[10] While high-sustained serum concentrations were difficult to attain, the drug concentrated in prostatic tissues and bile leading to some utility in treating bacterial prostatitis and gastroenteritis.[11]

A subtle change from the N-1 ethyl of norfloxacin to an N-1 cyclopropyl group led to the discovery of ciprofloxacin in the 1980s. To date, this fluoroquinolone maintains the most potent spectrum of activity against Gram-negative pathogens and an improved profile of activity against Gram-positive bacteria.[12,13] Combining its extended spectrum activity and improved pharmacokinetic profile (e.g., improved C_{max}, volume of distribution and bioavailability), ciprofloxacin became the first quinolone used for treating infections outside of UTI.[14] The favorable pharmacokinetics and extended half-life allowed for twice-daily dosing while maintaining a minimal potential for adverse side effects.[13] Because of its impressive Gram-negative spectrum, ciprofloxacin, typically in combination with a Gram-positive agent, continues to find wide application in the treatment of hospital-acquired and serious resistant infections, including RTI and UTI. However, due to relatively modest activity against strains of *S. pneumoniae*, ciprofloxacin has limited utility for treatment of common RTIs such as CAP.[15]

7.19.1.3 Current Marketed Drugs

With the goal of identifying new fluoroquinolones with efficacy in the major market of community-acquired RTIs, pharmaceutical research focused on extending the antimicrobial spectrum and enhancing the pharmacokinetic profiles of new molecules. Addressing the limitations of ciprofloxacin and other older marketed fluoroquinolones (see **Table 1** and **Figure 2**), namely activity against Gram-positive cocci (*S. pneumoniae*), distribution of drug to lung tissue, longer plasma half-life, decreased frequency of resistant mutant selection, and improved safety profiles, modern fluoroquinolones (see **Table 2** and **Figure 3**), such as gatifloxacin, moxifloxacin, and gemifloxacin, are finding increasing clinical utility in indications including lower-RTI, UTI, intra-abdominal infections, sexually transmitted diseases, and skin and soft tissue infections.[2] The efficacy, success rates, and safety of quinolones, e.g., ciprofloxacin, levofloxacin, moxifloxacin, gatifloxacin, and trovafloxacin, in clinical studies for CAP and AECB have been recently summarized.[16] These recent additions to the fluoroquinolone family are typically prescribed for once daily dosing, are more efficacious for CAP, and tend to exhibit an improved safety profile (e.g., lower potential for drug–drug interactions).[17] However, with increased spectrum of activity, which includes Gram-positive, anaerobic and atypical

Table 1 Older fluoroquinolones marketed for human use

Generic name	*Structure*	*Brand name*	*First intro.*	*US*	*Company*	*Limited use due to*[17]:
Norfloxacin	(**11**)	Noroxin	1983	yes	Kyorin/Merck	
Pefloxacin	(**13**)	Peflacine	1985	no	Roger Bellon	Tendopathies, phototoxicity
Ofloxacin	(**14**)	Floxin	1985	yes	Daiichi/Ortho	
Ciprofloxacin	(**4**)	Cipro	1986	yes	Bayer	
Enoxacin	(**12**)	Penetrex	1986	yes	Dainippon/RPR	CYP inhibition
Lomafloxacin	(**15**)	Maxaquin	1989	yes	Hokuriku/Unimed	Phototoxicity
Tosufloxacin	(**16**)	Tosuxacin	1990	no	Abbott/Toyama	Thrombocytopenia, nephritis
Temafloxacin[a]	(**17**)	Omniflox	1992	yes	Abbott	Hemolytic uremic syndrome
Fleroxacin	(**18**)	Quinodic	1992	no	Roche	Phototoxicity, CNS effects
Nadifloxacin	(**19**)	Acutim	1992	no	Otsuka	
Rufloxacin	(**20**)	Qari	1992	no	Mediolanum	
General characteristics of this class: – Best versus Gram-(−) bacilli – Less versus Gram-(+) cocci – Little versus anaerobes						Indications broadened (from UTI) to include: – Gastrointestinal – Genital – Bone – Joint

[a]Withdrawn from the market.

pathogens, modern fluoroquinolones have been developed at the cost of *P. aeruginosa* activity, which limits their utility in some hospital-acquired and chronic infections.

Improvements in the potency and antibacterial spectrum of the newer fluoroquinolones stem from subtle modifications of the ciprofloxacin (**4**) structure. Incorporation of the enoxacin (**12**) naphthyridine ring nitrogen at position 8 gave compounds such as trovafloxacin (**9**) and gemifloxacin (**6**) with greatly enhanced activity against Gram-positive respiratory pathogens including *S. pneumoniae*, *Streptococcus pyogenes*, *S. aureus*, and methicillin-resistant *Staphylococcus aureus* (MRSA) and ciprofloxacin/norfloxacin-resistant strains.[2,18,19] Alternately, use of a small C-8 substituent, such as a halide or methoxy group, increases anaerobic activity and water solubility.[20] These groups also impart an advantage in oral bioavailability over older drugs giving high serum concentrations and tissue concentrations, which are typically several fold higher than serum. The enhanced bioavailability allow for simple sequential therapy switching from intravenous to oral dosing, thus decreasing patient's hospital time. The C-8 methoxy group has the advantage over C-8 halogens in reducing the incidence of phototoxicity. Moxifloxacin (**8**) and gatifloxacin (**5**) have not shown phototoxicity, hepatotoxicity, or significant cardiac events.[21] However, even the newer fluoroquinolones exhibit QT prolongation, which is an inherent adverse effect of this class of antimicrobials, but has not been associated with substantial cardiac manifestations.[21] A further attractive feature of moxifloxacin and gatifloxacin is the observation that in vitro these C-8 methoxy agents show a lower tendency to select for resistance and appear to be less affected by mutations in either DNA gyrase or topoisomerase IV.[22] These studies could indicate that moxifloxacin and gatifloxacin may evade the emergence of resistant strains and maintain their effectiveness in the clinic longer than previous fluoroquinolones. With enhanced activity against clinically important Gram-positive and atypical pathogens while maintaining a broad Gram-negative spectrum, the modern fluoroquinolones are highly effective in treating UTI, AECB, and CAP. These agents have superior efficacy versus older antibacterials against resistant strains, but are deficient in activity against *P. aeruginosa*, MRSA, and vancomycin-resistant enterococci (VRE). With once-daily dosing, improved safety profiles, and decreased tendency to select for resistance during therapy, moxifloxacin, gatifloxacin, and gemifloxacin mark an important advancement in treatment of infectious disease with the added potential clinical value of prolonged service in the face of growing bacterial resistance.

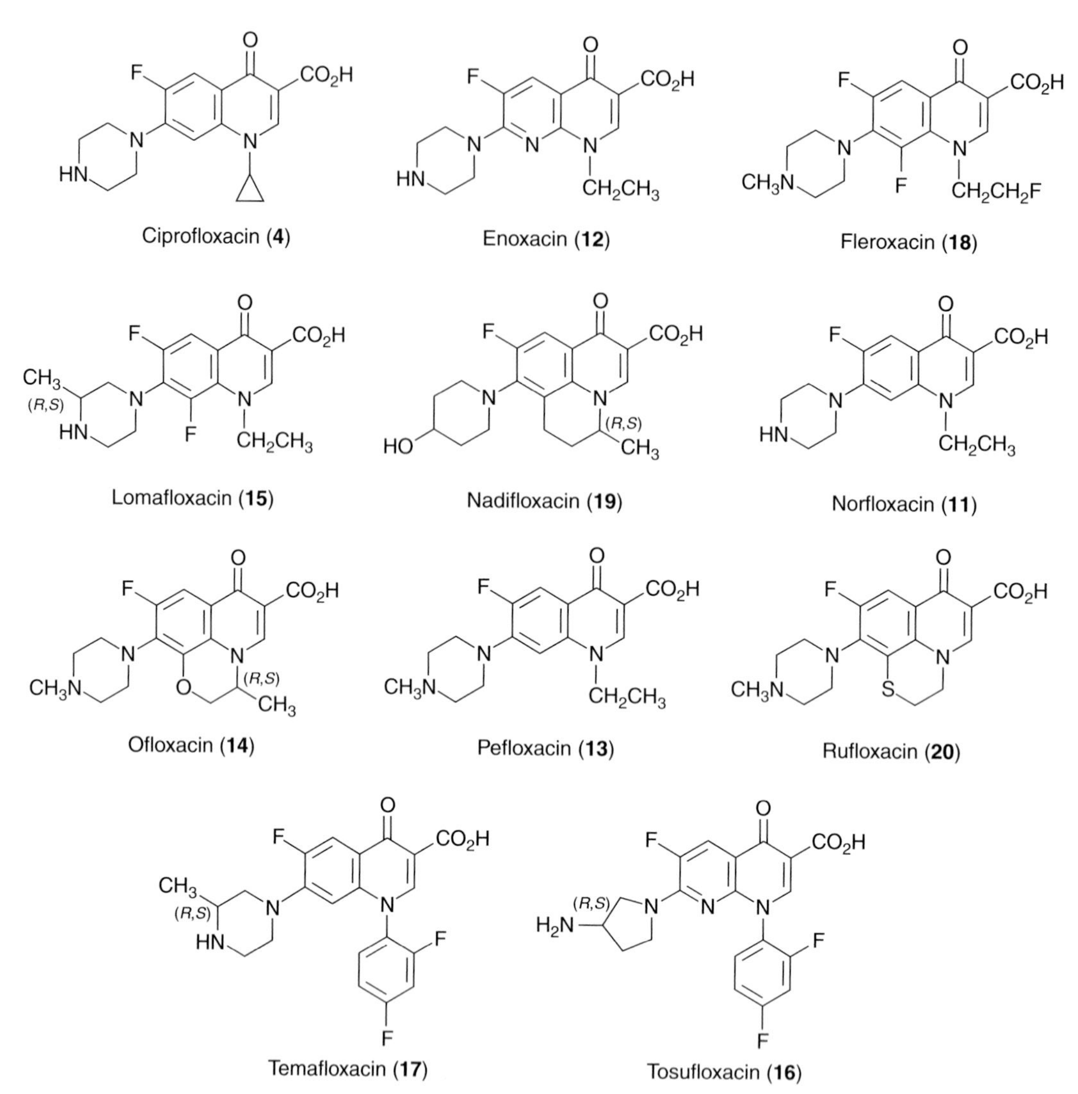

Figure 2 Older marketed fluoroquinolones.

7.19.2 Mechanism of Antibacterial Action

Fluoroquinolones disrupt bacterial DNA synthesis leading to rapid bacterial cell death. The specific targets, DNA gyrase and topoisomerase IV, are members of the topoisomerase class of enzymes.[23] Discovered long after clinical use of fluoroquinolones, the targets of antimicrobial action are DNA-binding enzymes that control the topology of DNA that are required for cell growth and division. Since DNA gyrase and topoisomerase IV are intracellular enzymes, quinolones must penetrate the bacterial membrane and accumulate in the bacterial cell to a high enough concentration to produce antibacterial activity. Simple diffusion across the bacterial membrane leads to antimicrobial activity in *E. coli* and *S. aureus*.[24] However, in other Gram-negative species, quinolone accumulation and antibacterial action has been linked to active uptake by porins in the outer membrane.[25] Defense mechanisms, such as small molecule pumps, may also play a role in reducing the accumulation of quinolones. The defensive systems can drastically reduce the effectiveness of quinolones in some organisms leading to resistance, which will be discussed in a later section.

There are two groups of topoisomerases within the class, which have distinct biochemical mechanisms and physiological roles, although both groups regulate the superhelical state of bacterial DNA. The nature or tightness of the superhelicity controls the initiation of DNA replication and the resulting transcription of many bacterial genes. Type I topoisomerases, which include topoisomerase I and topoisomerase III, are not inhibited by quinolones or other antibiotics. Type I topoisomerases catalyze the cleavage of single strands of DNA causing breaks, single strand passage,

Table 2 Modern fluoroquinolones marketed for human use

Generic name	*Structure*	*Brand name*	*First intro.*	*US*	*Company*	*Limited use due to*[17]:
Levofloxacin	(**7**)	Levaquin	1993	yes	Daiichi/Ortho	
Sparfloxacin	(**21**)	Zagam	1993	yes	Dainippon/RPR	QTc elongation
Grepafloxacin[a]	(**22**)	Raxar	1997	yes	Otsuka/GW	QTc elongation, phototoxicity
Trovafloxacin[a]	(**9**)	Trovan	1997	yes	Pfizer	Hepatotoxicity, CNS effects
Gatifloxacin	(**5**)	Tequin	1999	yes	Kyorin/BMS	
Moxifloxacin	(**8**)	Avelox	1999	yes	Bayer	
Gemifloxacin	(**6**)	Factive	2003	yes	GeneSoft/LG Life Sciences	

General characteristics of this class:
- Improved Gram-(+) activity
- Improved anaerobe activity

Indications broadened to include:
- Respiratory

[a] Withdrawn from the market.

and religation to relieve superhelical twists and decatenate interlocked circles of DNA.[26] Type I topoisomerases remove negative superhelical twists in DNA, an action that opposes the activity of DNA gyrase. The antimicrobial activity of fluoroquinolones is manifest mostly due to inhibition of DNA gyrase and topoisomerase IV, which are both type II topoisomerases. Type II topoisomerases catalyze cleavage, passage, and religation of double-stranded DNA.

DNA gyrase, which was the first target of quinolones to be identified, is an essential (required for life) enzyme in bacteria.[27] Bacterial DNA is normally in equilibrium between a closed circular double-stranded state (relaxed) and a highly negatively supercoiled state (**Figure 4**). DNA gyrase controls bacterial DNA topology and chromosome function by maintaining DNA negative supercoiling (superhelical twists). Crucial for DNA replication, DNA gyrase is responsible for relieving the positive superhelical twist that accumulates in advance of an active DNA replication fork.[28] Additionally, DNA gyrase is involved in bending and folding DNA and removing knots (**Figure 4**). The active enzyme is comprised of four subunits, two A and two B subunits, which are the products of the *gyrA* and *gyrB* genes, respectively.[29] Subunit A is the quinolone binding site and requires an active site Tyr-122 for catalytic activity. X-ray structures have been obtained for: (1) fragment of N-terminal domain ('breakage-union' and quinolone resistance-determining region (QRDR) domains) of A subunit[30]; and (2) yeast topoisomerase II.[31] The ATP-binding domain is found in the B subunit, which mediates the ATPase activity of DNA gyrase. A crystal structure is available for the 43-kDa N-terminal domain (ATPase and DNA capture domains) of subunit B.[32]

Based on both biochemical and structural evidence, the catalytic cycle of DNA gyrase can be expressed as the opening and closing of a series of 'molecular gates' coupled to ATP hydrolysis.[33] DNA gyrase makes contact with and wraps a 120 bp segment of DNA into a positive supercoil resulting in a noncovalent enzyme – DNA binding complex (**Figure 5**). The next step is the formation of a cleavable complex in which Tyr-122 of A subunits covalently links to DNA via phosphodiester cleavage in the typical 4-base stagger configuration. At this stage, strand passage is possible. Strand passage occurs when ATP binding induces a structural change with concomitant transport of DNA through transiently cleaved DNA bound in the enzyme complex. Finally, the DNA ends are religated (resealed) resulting in formation of a negative supercoil. ATP hydrolysis triggers regeneration of prestrand passage conformation and dissociation of ADP regenerates the active enzyme completing the enzymatic cycle.

The second target of fluoroquinolones, topoisomerase IV, is responsible for separating the catenated (interlinked) circular DNA daughter molecules, which are the product of DNA replication.[34] Once the two newly replicated DNA molecules are separated by the cleavage, passage and religation process, the bacterial cell is primed to divide isolating one copy of the DNA in each of the newly formed daughter cells. Thus, while the catalytic activity of DNA gyrase and topoisomerase IV are similar mechanically, they play distinct and essential roles in DNA replication and cell proliferation.[35] Topoisomerase IV, like DNA gyrase, is a tetramer comprised of two pairs of two subunits encoded by the genes *parC* and *parE*. Being closely related to DNA gyrase in structure and function, the genes and products show close homology; *parC* for *gyrA* and *parE* for *gyrB*.[36] The primary difference between topoisomerase IV and DNA gyrase is how

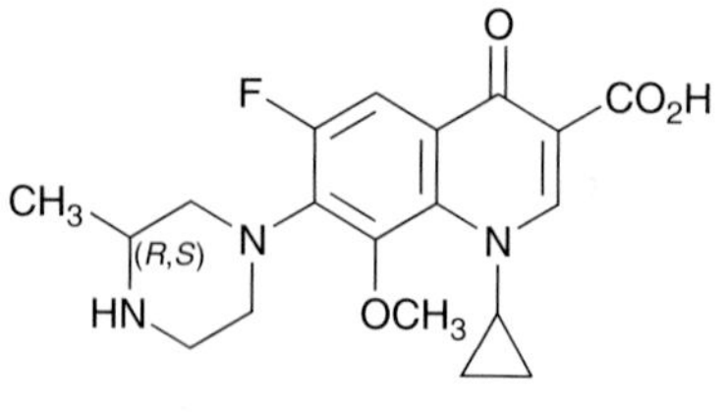

Gatifloxacin (**5**)

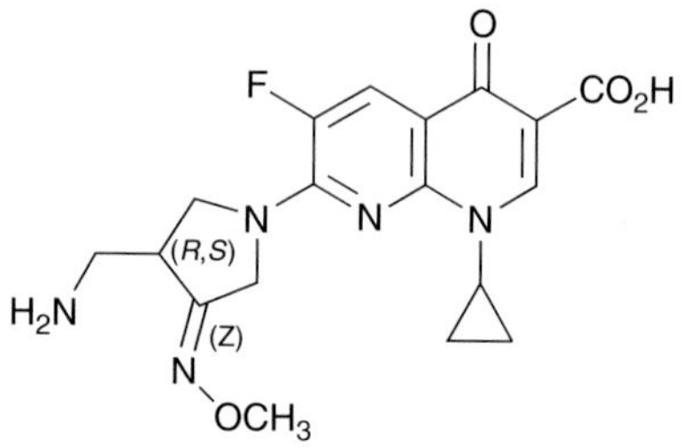

Gemifloxacin (**6**)

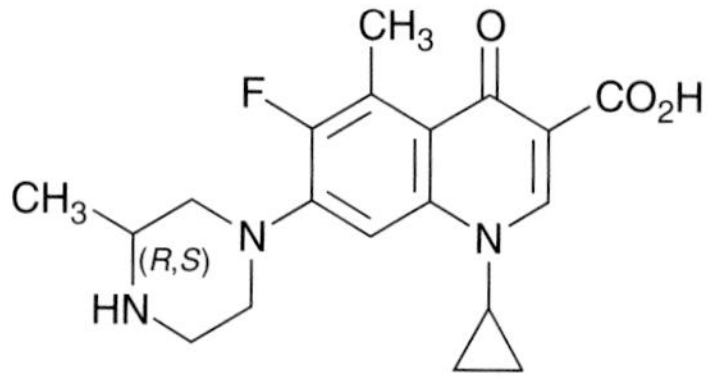

Grepafloxacin (**22**)

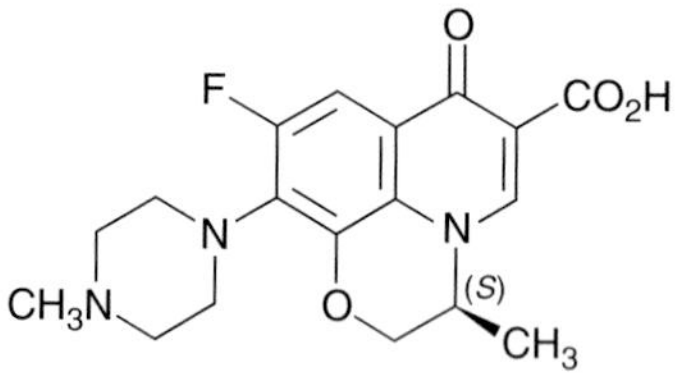

Levofloxacin (**7**)

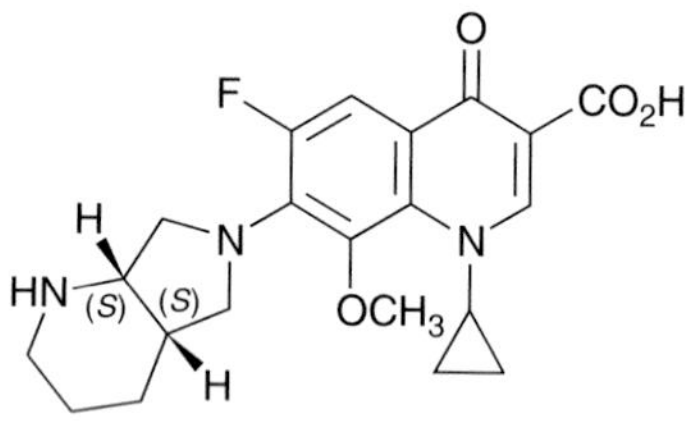

Moxifloxacin (**8**)

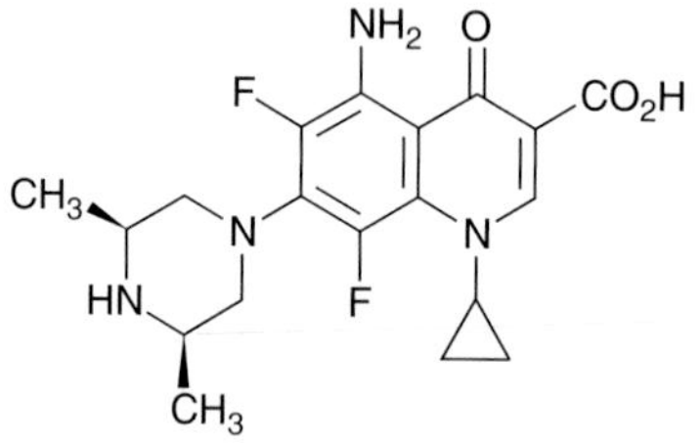

Sparfloxacin (**21**)

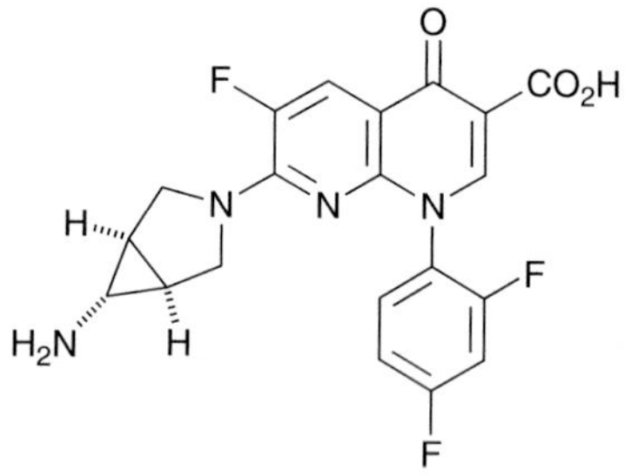

Trovafloxacin (**9**)

Figure 3 Modern marketed fluoroquinolones.

DNA associates with the enzymes. In DNA gyrase, noncovalent bonding interactions between DNA and the enzyme supports a close association or 'wrapping' of the DNA around the protein. The 'wrapping' favors 'intramolecular' strand passage. Alternatively, topoisomerase IV does not display DNA'wrapping' and thus favors 'intermolecular' strand passage. Deletion of a portion of the C-terminal domain ('wrapping' part) of GyrA converts DNA gyrase into an enzyme that cannot catalyze supercoiling, but has strong topoisomerase IV-like decatenation activity.[37]

Numerous studies have examined the molecular target specificity of quinolones in a wide spectrum of bacteria. In Gram-negative species such as *E. coli*, the primary target appears to be DNA gyrase.[38] Evidence suggests that quinolones bind to the enzyme–DNA complex after strand cleavage thus stabilizing the complex and inhibiting the strand religation step. The stabilized complex creates a DNA break that the cell is poorly equipped to repair. The stabilized or 'trapped' complex is proposed to act as a cellular poison, which may help explain the disparity between quinolone potency at enzyme and cellular levels (frequently MICs are much lower than IC_{50} or CC_{50} values).[39] Formation of the stabilized ternary complex results in inhibition of DNA synthesis and cell growth by blockage of replication fork. Lethal DNA breakdown (double strand breaks) then occurs, which requires RNA and protein synthesis to repair the damage. As a result, quinolones as a class are bactericidal. In response to the cellular damage, bacteria trigger the SOS pathway; expression of a set of genes that involve DNA repair, DNA recombination (e.g., nonviable cell), and mutagenesis (e.g., potential resistance). However, inhibition of DNA synthesis may be uncoupled with bactericidal activity in some cases, thus opening the possibility of alternative pathways or additional factors involved in the mechanism of action.[40]

In Gram-negative species such as *E. coli*, quinolone treatment generates resistant strains that have specific single mutations in *gyrA* and *gyrB*. These findings indicate that stabilization of DNA gyrase is the main mode of action in

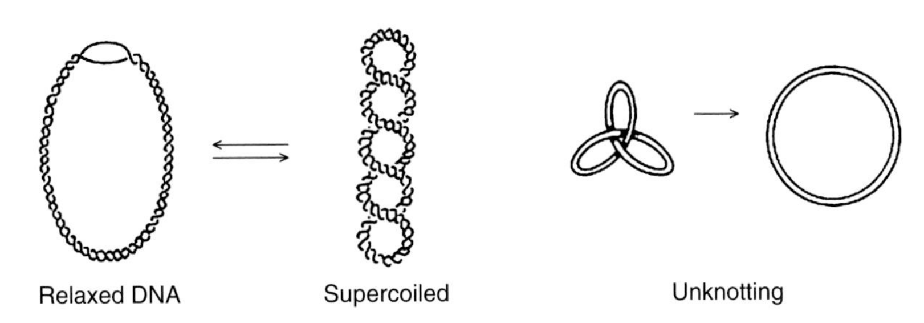

Figure 4 DNA supercoiling and knotting.

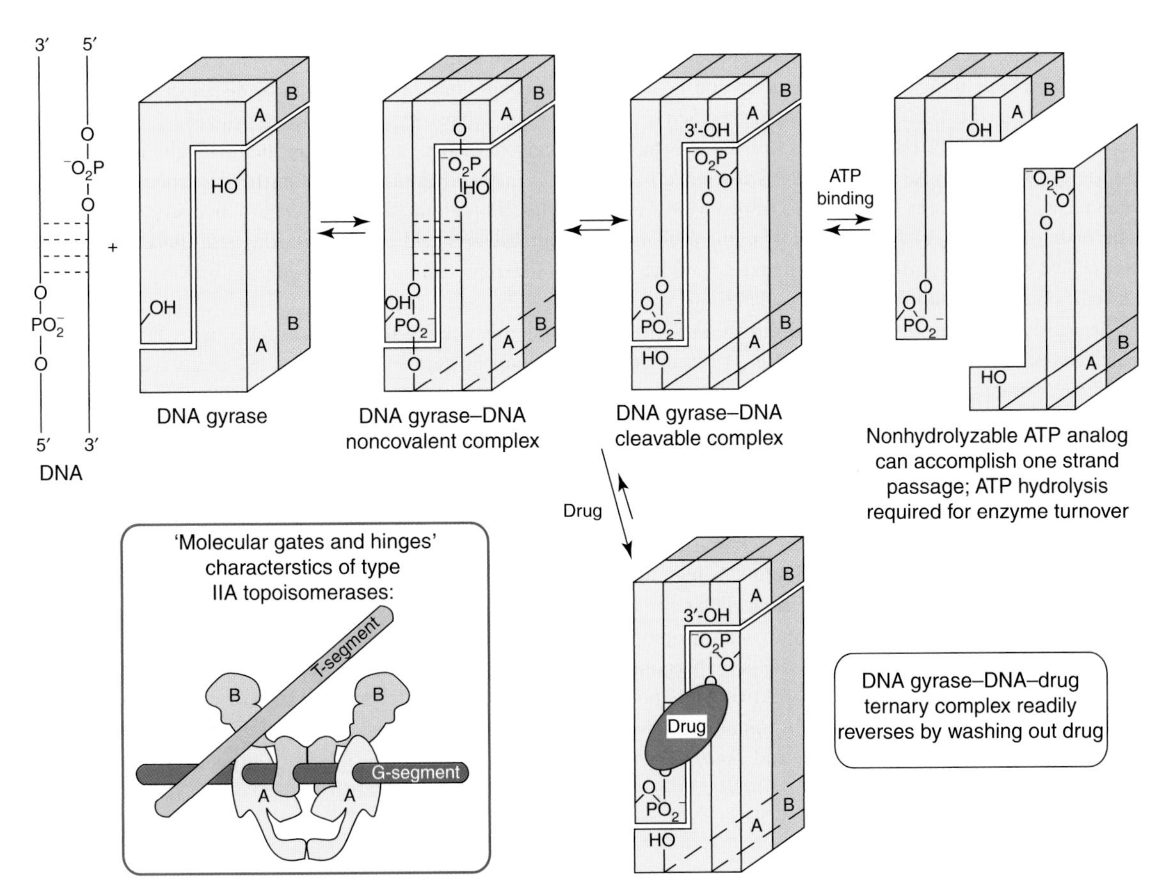

Figure 5 DNA gyrase: cleavable complex formation, strand passage, and quinolone binding.

E. coli. These mutations confer resistance by lowering the binding affinity of DNA gyrase for the quinolone.[41] In contrast to Gram-negative bacteria, quinolone resistance in Gram-positives, such as *S. pneumoniae* and *S. aureus*, tends to result from single mutations in the topoisomerase IV genes, *parC* and *parE* in *S. pneumoniae*, suggesting that topoisomerase IV is the primary quinolone target in the Gram-positive species.[38,42] In a study of inhibition of enzyme catalysis by a battery of quinolones in *E. coli* and *S. aureus*, the most sensitive topoisomerase in each species followed the expected trend: in *E. coli* DNA gyrase was favored over topoisomerase IV by a ratio of 15- to 27-fold while in *S. aureus* topoisomerase IV was favored by 2- to 21-fold.[43] However, the structure of the quinolone plays a role in differentiating between the topoisomerase targets. Recent studies of newer quinolones have demonstrated that the gyrA subunit of DNA gyrase is the primary target for moxifloxacin and gatifloxacin in *S. pneumoniae*.[44] Based on these observations, changes in substitution on the quinolone C-7 or C-8 positions can play a significant role in the target preference. Although there is a general trend for target specificity between Gram-positive and Gram-negative species, the primary topoisomerase target for a given bacteria is the enzyme most sensitive to a specific quinolone.

7.19.3 Structure–Activity Relationships of Quinolone Antibiotics

The history, development, and structure–activity relationships (SARs) of the quinolones have been extensively reviewed.[45] In this section, the general quinolone SAR and that of newer generations of quinolones will be examined. SARs will be discussed in terms of target specificity; target activity in vitro (e.g., binding and functional assays), and antimicrobial activity with comments on species susceptibility (e.g., minimum inhibitory concentration (MIC) data). Notes on structure–toxicology relationships will also be made. The description is not meant to be exhaustive, but rather an overview of commonly observed trends. Before considering the SAR, the synthesis of quinolones should be reviewed to get some perspective on the structural variations and available points of substitution.

The classic approach uses the Gould–Jacobs synthesis (**Scheme 1**) in which an aniline is reacted with diethyl 2-(ethoxymethylene)malonate.[46] Under high heat conditions, the enamine forms an intermediate ketene, which annulates to yield the quinolone. The nitrogen at position 1 can easily be alkylated with S_N2-amenable alkyl halides. The final acid is revealed by saponification. This process is very amenable for scale-up with high yields and the starting material anilines are also widely available. However, the Gould–Jacobs synthesis has the disadvantage that it can lead to regioisomers depending on the aniline used and 1-position substituents must be brought in by displacement reactions.

Overcoming the problems with regiospecificity, an alternative cycloaracylation process utilizes an ortho-chloro benzoic acid or ketone (**Scheme 2**).[47] In this synthesis, the regiochemistry can be set by the ortho-chloro group and the yields are typically high. This route is amenable to scale-up. A major advantage of this synthetic sequence is the use of an enamine, which can be generated from a wide array of amines. This allows for easy access to diversity at position 1. The main disadvantage of the synthesis is the availability of suitable aryl acid chloride starting materials. Of note are several fast analoging syntheses, which take advantage of this well explored chemistry to produce libraries of quinolones on solid phase or in solution.[48]

At position 1, a substituent on the nitrogen other than hydrogen is needed for antibacterial activity. Historically, an ethyl group, as seen in nalidixic acid (**1**), is the optimal alkyl group in quinolones and has led to good antimicrobial activity.[45] Through the 1970s and 1980s, bioisosteric replacement strategies were quite effective. Utilizing the SYBYL molecular modeling software, activity was found to correlate well with STERIMOL (length and width approximations) and molar refractivity (MR) as a description of substituent bulk.[49] Simple bioisosteric replacements led to several compounds that entered into clinical development, e.g., fleroxacin (**18**) (1-CH_2CH_2F) and amifloxacin (**24**) (1-$NHCH_3$). In a modern series of quinolones, the ethyl group at N-1 was similarly potent in a gyrase assay (**Table 3**).[49] However, the ethyl group has improved antibacterial activity in *E. coli* among the bioisosteric replacements with *i*-Pr showing the highest MIC in the group. These differences may be due to the substituent's effect on the compound's membrane permeability and ability to reach the intracellular site of action.

An improvement over the ethyl group is a cyclopropyl group at N-1 as seen in ciprofloxacin With the compact cyclic constraint of the cyclopropyl group, gyrase activity is much improved over either ethyl or *i*-Pr (**Table 4**).[49] Moreover, there is a notable improvement in Gram-negative MIC potency in most species. Tested against 410 clinical isolates, the minimum inhibitory concentrations of ciprofloxacin for 90% of Enterobacteriaceae, *P. aeruginosa*, *Haemophilus influenzae*, *N. gonorrhoeae*, streptococci, *S. aureus*, and *Bacteroides fragilis* strains were 0.008–2 $\mu g\,mL^{-1}$,[50] although some strains of

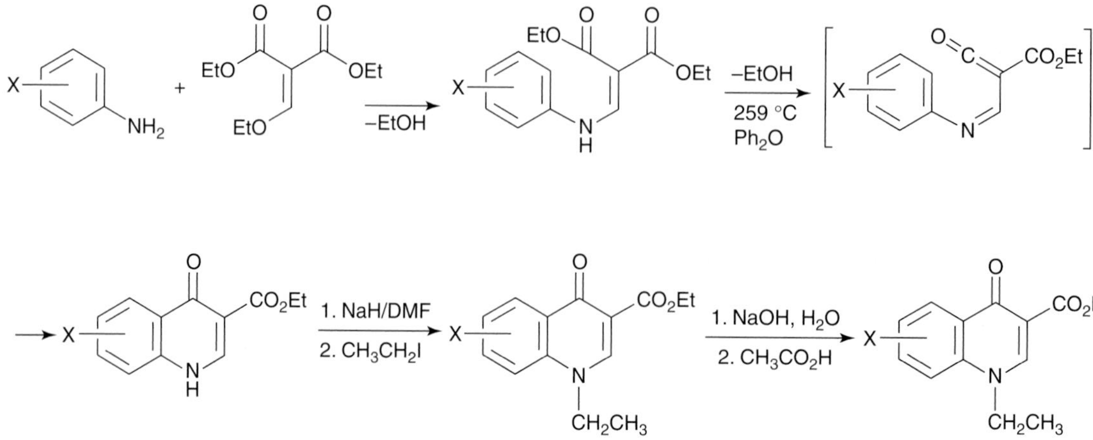

Scheme 1

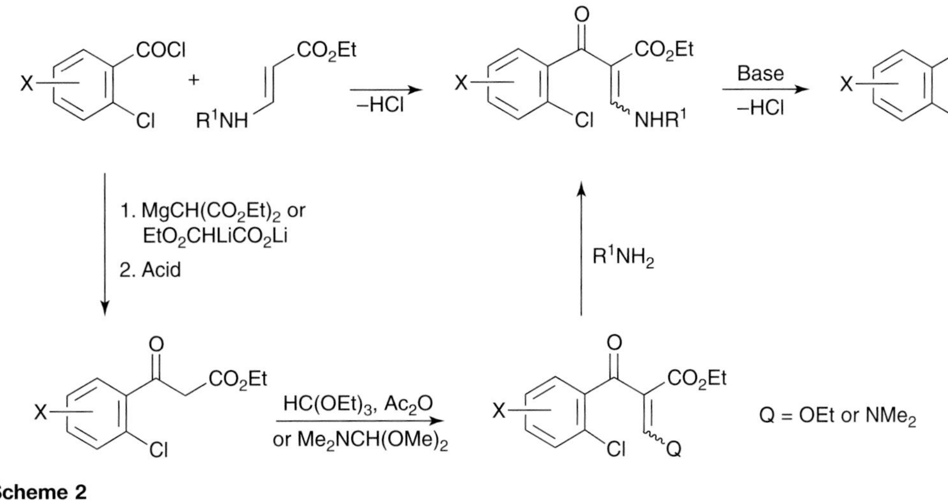

Scheme 2

Table 3 Position 1 SAR and Gram-negative activity

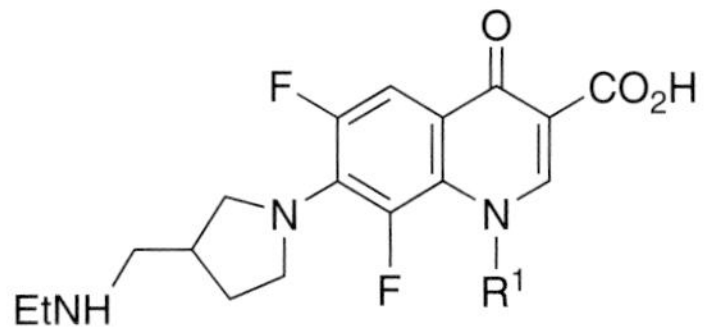

R^1	*Gyrase cleavage*[a] *(μg mL*$^{-1}$*)*	*Gram-negative*[b] *MIC (μg mL*$^{-1}$*)*
CH_3	5.0	0.53
CH_2CH_3	2.5	0.17
$CH(CH_3)_2$	5.0	1.05
$CH=CH_2$	0.8	0.30
CH_2CH_2F	2.5	0.23
OCH_3	2.5	0.46
$NHCH_3$	2.5	0.30

[a] Lowest (drug) necessary to induce *E. coli* DNA gyrase-mediated cleavage of DNA.
[b] Geometric mean calculated for Gram-negative Enterobacteriaceae.

S. aureus are not sensitive to ciprofloxacin and the cyclopropyl compound **25** in **Table 4**. Unfortunately, certain compounds with a cyclopropyl on N-1, (hetero)aryl at C-7 and fluorine at C-8 have yielded potent mammalian topoisomerase II inhibitors.[49] Inhibition of the mammalian topoisomerase could lead to adverse side effects in animals and humans.

In addition to the N-1 ethyl substituent as in norfloxacin (NOR) or the cyclopropyl group in ciprofloxacin (CIP), a phenyl group at N-1 is also potent as a DNA gyrase inhibitor and has an improved Gram-negative spectrum of activity (**Table 5**).[51] The spectrum of activity can be augmented with substituents around the N-1 phenyl group. In the series shown, the MIC potency was best with a para -OH group or fluoro group and was worse with larger groups like bromo or methoxy in the relative order: 4′-OH > F > H > Me = Cl > Br ≫ OMe (highest to lowest in vitro potency).[51] In particular, the 2,4-difluorophenyl group improved Gram-positive activity as seen in the improved *S. aureus* MICs. The benefit in Gram-positive activity and broad Gram-negative spectrum was also apparent in the 1,8-naphthyridines such as difloxacin (**27**), tosufloxacin (**16**), and trovafloxacin (**9**).[52] An added benefit was the excellent oral efficacy displayed by these compounds.

Table 4 Position 1 SAR of cycloalkyl groups

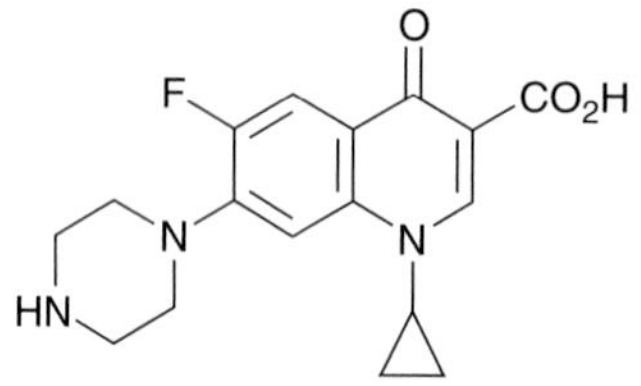

Ciprofloxacin (**4**)

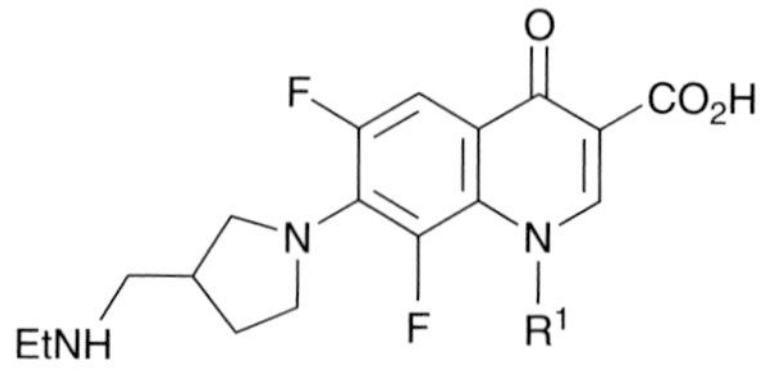

25

Cmpd/R[1]	*Gyrase cleavage[a] (μg mL^{-1})*	*Gram-negative[b] MIC (μg mL^{-1})*
CIP (**4**)	0.5	0.06
CH_2CH_3	2.5	0.17
c-C_3H_5	0.25	0.06
$CH(CH_3)_2$	5.0	1.05
CH_2-c-C_3H_5	10	2.09
c-C_4H_7	2.5	0.40
c-C_5H_9	5.0	3.60

[a] Lowest (drug) necessary to induce *E. coli* DNA gyrase-mediated cleavage of DNA.
[b] Geometric mean calculated for Gram-negative Enterobacteriaceae.

Table 5 Position 1 SAR of aryl groups

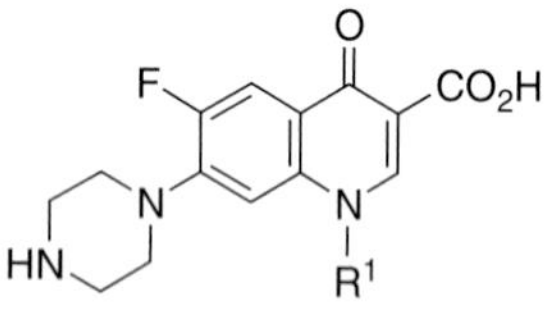

26

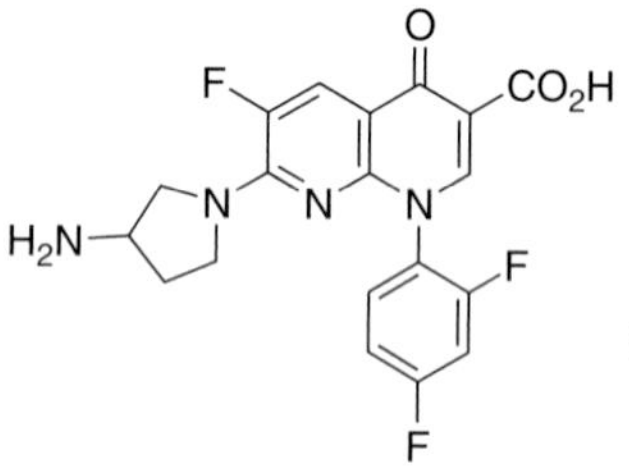

Tosufloxacin (**16**)

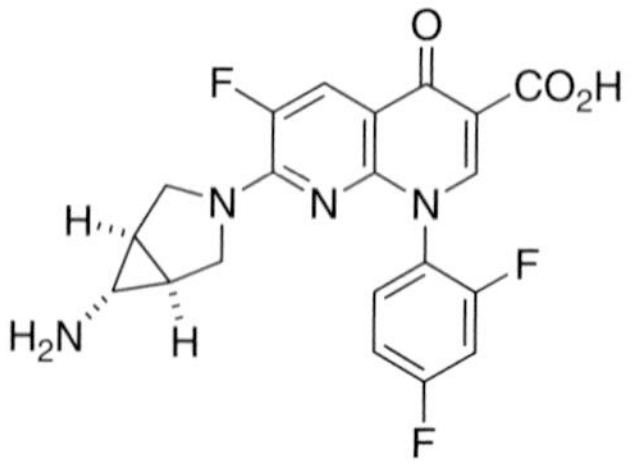

Trovafloxacin (**9**)

Compound	*R[1] in* **26**	*MIC (μg mL^{-1})*			
		Ec[a]	*Pa[a]*	*Sa[a]*	*Sp[a]*
NOR (**11**)	CH_2CH_3	0.1	0.39	0.78	3.10
CIP (**4**)	c-C_3H_5	0.02	0.1	0.2	0.39
	C_6H_5	0.2	0.78	0.39	1.56
Difloxacin (**27**)	4′-FC_6H_4 (MePip)	0.05	0.39	0.2	0.78
	2′,4′-FC_6H_3 (MePip)	0.2	1.56	0.1	0.78

[a] *E. coli* (Ec), *P. aeruginosa* (Pa), *S. aureus* (Sa), *S. pneumoniae* (Sp).

Increasing the bulk of the N-1 substituent from ethyl or cyclopropyl to a *tert*-butyl greatly improves the Gram-positive antibacterial activity, especially the activity against *S. aureus* (**Table 6**).[53] However, moving the bulky group further out by one methylene (1–CH_2–c–C_3H_5) or adding a methyl to the cyclopropyl group at any position is deleterious for in vitro potency particularly against the Gram-negative bacteria. The improvement in Gram-positive activity with an N-1 *tert*-butyl group holds true in the 1,8-naphthyridines, unfortunately at the expense of *P. aeruginosa* activity. The *tert*-butyl group also imparts improved ADME (absorption, distribution, metabolism, and

Table 6 Position 1 SAR of bulky alkyl groups

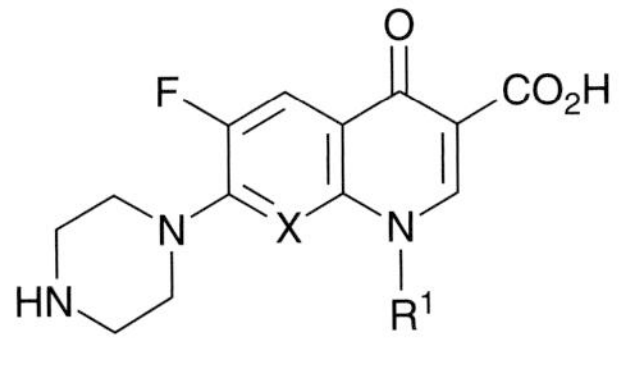

28

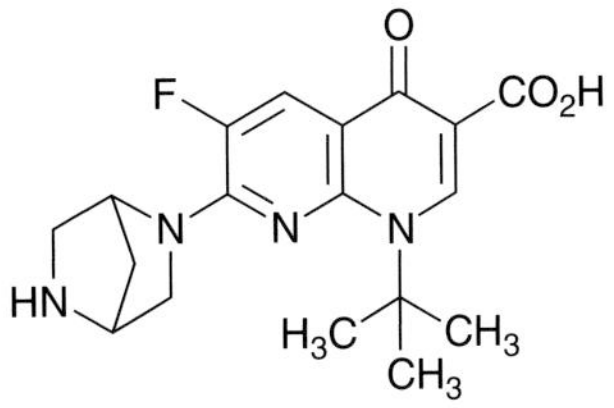

BMY 40062 (**29**)

Compound	*R[1] in 28*	*X*	*MIC (μg mL^{-1})*			*PD$_{50}$ (mg kg^{-1} p.o.)*
			Ec[a]	*Pa[a]*	*Sa[a]*	*versus Sa*
NOR (**11**)	CH_2CH_3	CH	0.13	0.5	0.25	10
CIP (**4**)	c-C_3H_5	CH	0.03	0.13	0.13	4
	$C(CH_3)_3$	CH	0.06	0.5	0.06	8
	1-CH_2-c-C_3H_5	CH	0.25	0.5	0.25	na
	$C(CH_3)_3$	N	0.015	1.0	0.06	1.1

[a] *E. coli* (Ec), *P. aeruginosa* (Pa), *S. aureus* (Sa).

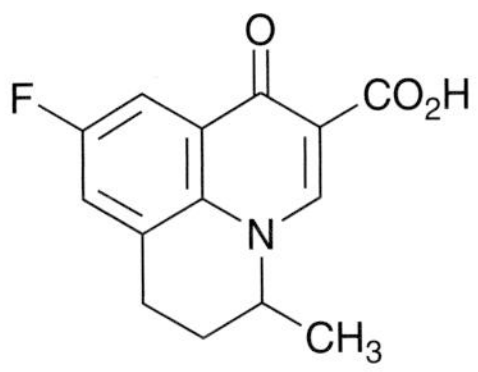

Flumequine (**30**)

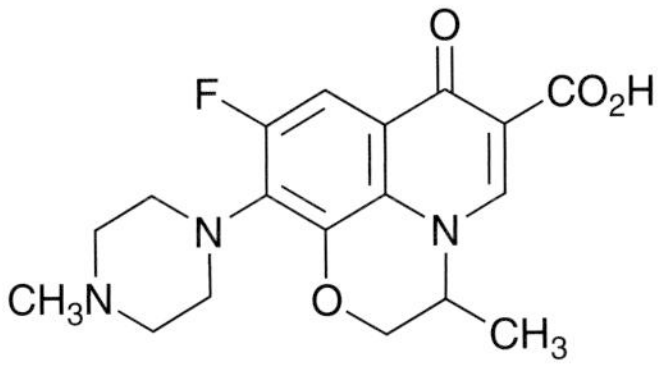

Ofloxacin (**14**)

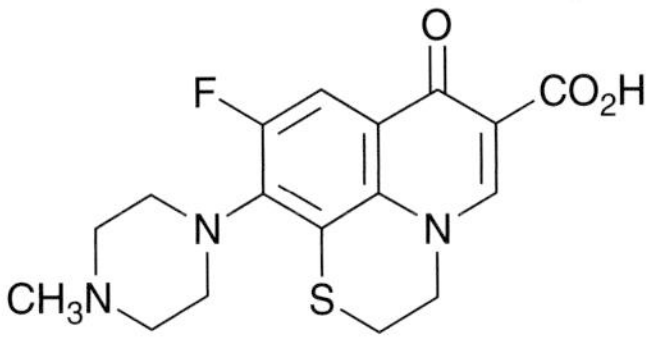

Rufloxacin (**20**)

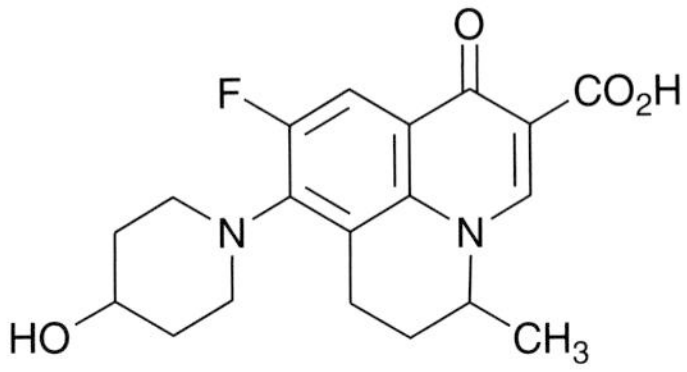

Nadifloxacin (**19**)

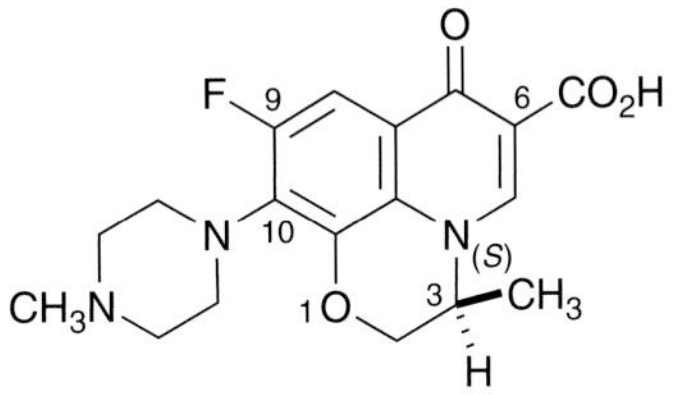

Levofloxacin (**7**)

Figure 6 Marketed 1,8-bridged quinolones.

excretion) properties leading to greater maximum concentration (C_{max}), area under the plasma concentration curve (AUC), and half-life ($t_{1/2}$). The superior pharmacokinetic (PK) profile and excellent Gram-positive MICs of this series yield outstanding efficacy against *S. aureus* in the mouse systemic infection model. With optimized in vitro and in vivo microbiological activity, better bioavailability than ciprofloxacin and low toxicity, BMY 40062 (**29**) is an advanced representative of this series that was chosen for clinical evaluation.[54]

Further exploration in the 1-position has led to preparation of tricyclic quinolone ring systems in which a new saturated ring is formed to give the 1,8-bridged quinolones.[45] These novel analogs at first opened up new intellectual property space and led to several marketed drugs (**Figure 6**), but did not intrinsically provide better antibacterial activity over the nonbridged quinolones. However, this situation changed as methyl groups were installed on the 3 position of the bridging ring to mimic the *tert*-butyl and cyclopropyl groups at N-1 in the traditional quinolones.[55] While the exact structural and conformational requirements of substituents in the N-1 position with respect to their role in DNA gyrase activity are not completely understood, there is a distinct stereochemical preference for the methyl at the bridging 3-position. For example, levofloxacin (**7**) ((−)-ofloxacin (**14**)) is 8–128 times more active than

Table 7 1,8-Bridged Quinolones: SAR at position-3

Ofloxacin (**14**)/levofloxacin (**7**) **31** **32**

R	*R*[1]	*MIC (μg mL*$^{-1}$*)*	*IC*$_{50}$ *(μg mL*$^{-1}$*)*	
		Ec[a]	*DNA gyrase*	*TOP2*
H/CH_3	H/CH_3 (**14**)	0.05	0.76	1870
CH_3	H (**7**)	0.025	0.38	1380
H	CH_3	0.78	4.7	2550
H	H	0.10	3.1	178
Exocylic methylene (**31**)		0.05	0.70	64
Spirocyclopropyl (**32**)/(**14**)		0.125 (0.008)	1.0 (1.0)	

[a]*E. coli* (Ec).

HC(OEt)$_3$, Ac_2O

KOH, H_2O, THF 70%

MePip

Levofloxacin (**7**)

Scheme 3

(+)-ofloxacin against both Gram-positive and Gram-negative bacteria, and is twice as active as the racemate (**Table 7**).[56] The synthetic scheme to make these 1,8-bridged quinolones gives regio- and stereospecific products and high yields, but is limited to the availability of chiral amino alcohols (**Scheme 3**).[57]

A stereospecific methyl group plays an important role in DNA gyrase activity. The levofloxacin (**7**) stereochemistry gives approximately a 12-fold increase in DNA gyrase activity (**Table 7**).[58] However, the eudismic ratio for the MICs (~32) is much greater than expected from the 12-fold ratio against DNA gyrase in the ofloxacin core. A gem-dimethyl group at position 3 on the bridge is not very active by MIC, and the spirocyclic cyclopropane is approximately 16-fold poorer than ofloxacin.[59] Interestingly, the spirocyclic cyclopropane compound gives the same DNA gyrase activity when tested side-by-side leading to the observation that the spirocyclic variant's MIC cannot be predicted by its IC_{50}[59] Breaking the plane of the 1,8-bridge with a methyl is also important in maintaining selectivity against mammalian topoisomerase 2.[58] A group in the plane of the bridge at position 3, such as the exocylic methylene compound (**31**), displays strong mammalian topoisomerase 2 (TOP2) activity.

Historically, the 2-H, 3-CO_2H, and 4-oxo substituents are required for activity.[45] However, in the 1980s, some novel groups were explored, some of which gave potent activity in vitro. While a linear thiomethyl group at the 2-position

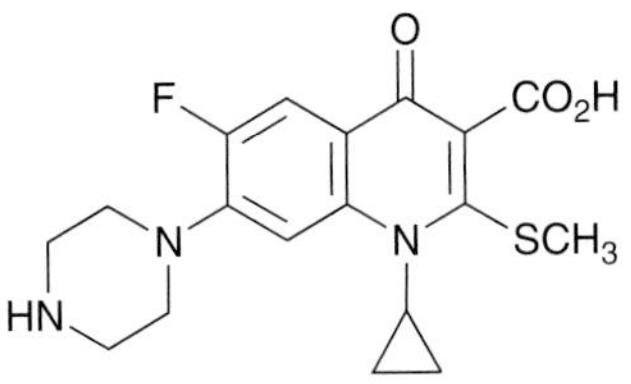

33

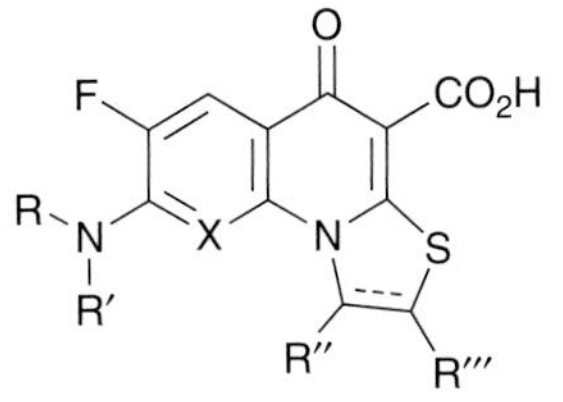

34

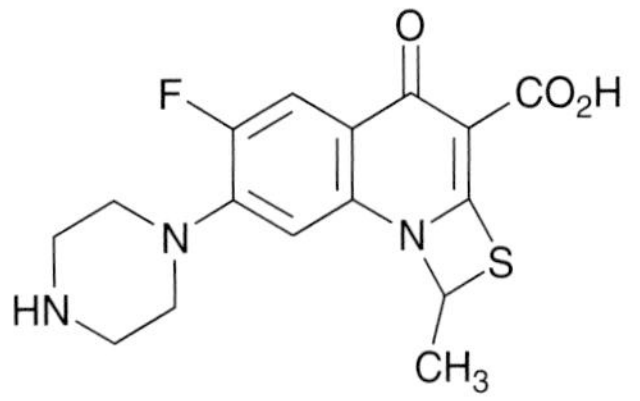

NM 394 (**35**)

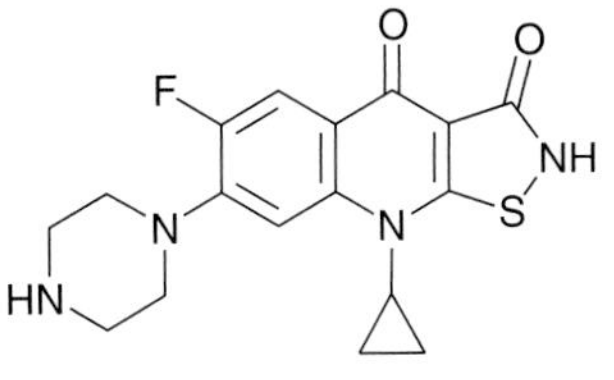

36

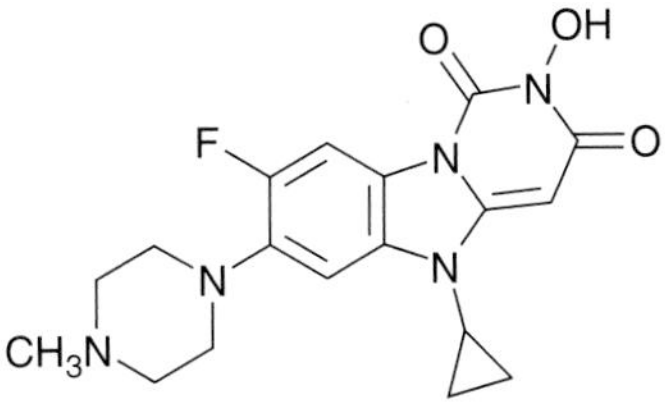

37

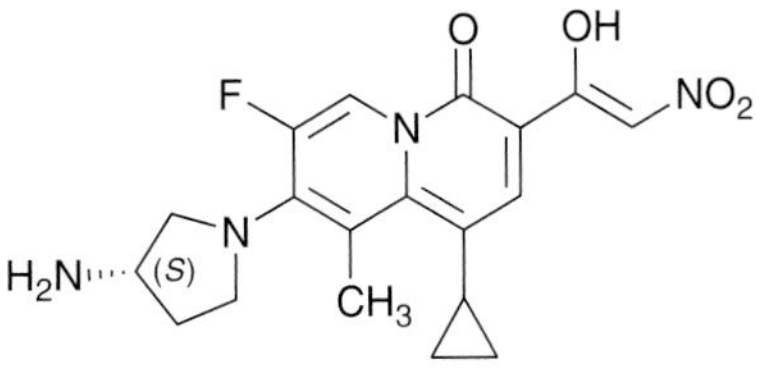

38

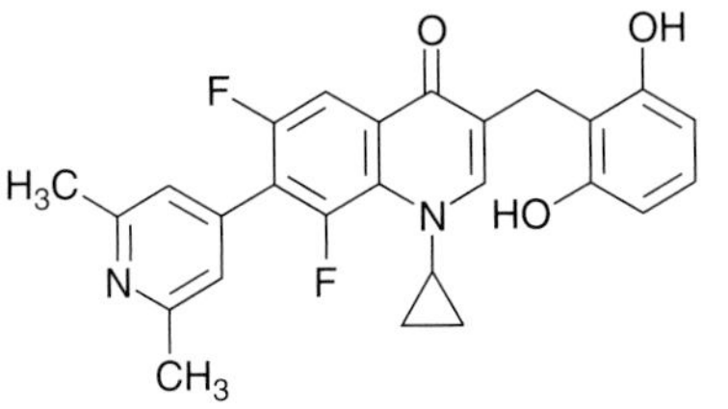

39

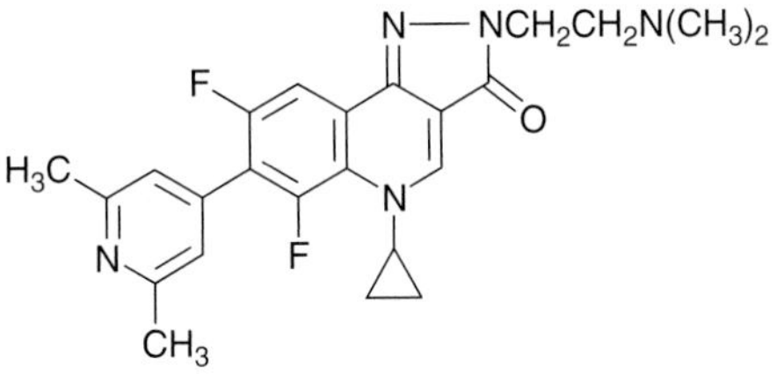

40

Figure 7 2-, 3-, and 4-Position SARs.

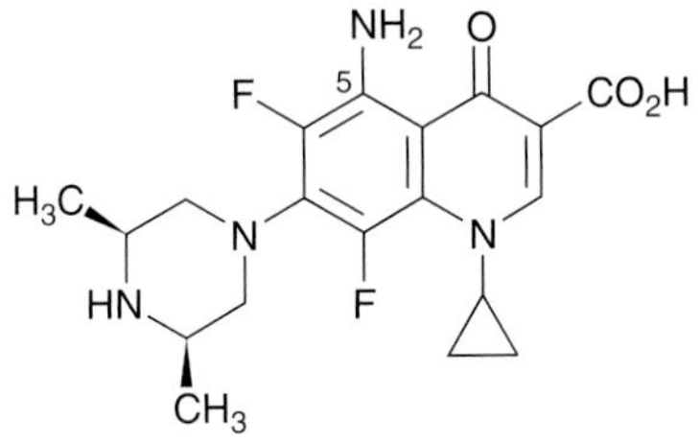

Sparfloxacin (**21**)

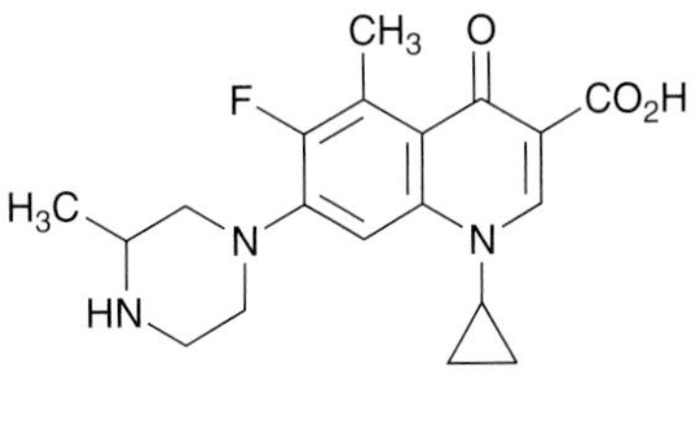

Grepafloxacin (**22**)

Figure 8 5-Position SAR and the fluoroquinolones.

(**33**) was not active, a 2-position sulfur was tolerated when cyclized to the 1- or 3-positions (**Figure 7**).[60] Numerous active 1,2-fused compounds (**34**) have been reported with the highly active compound NM 394 (**35**) among them.[61] In the 3-position, certain acidic groups can replace the carboxylate. The 2- and 3-positions can be fused into a ring (**36**), which is acidic and active against both DNA gyrase and mammalian topoisomerase 2.[62] Other acidic group arrangements (**37**, **38**) also show potent activity against DNA gyrase.[63,64] The 3-position analogs (**39**, **40**) of WIN 57294 while active against DNA gyrase were also potent inhibitors of mammalian topoisomerase 2.[65] To date, no useful replacements for the 4-position carbonyl group have been identified.

Few substituents are fruitful in the 5-position. Thus far, 5-NH_2 (sparfloxacin **21**) and 5-CH_3 (grepafloxacin **22**) are the most beneficial for in vitro activity (**Figure 8**). Larger amine or alkyl groups are less effective. The effects of these groups are dependent on the nature of other substituents. For example, the 5-NH_2 increases activity (including Gram-positive activity) with 1-cyclopropyl, decreases activity with 1-CH_2CH_3 or 1-(2,4-F_2)-C_6H_3, and is most effective when combined with 8-fluoro groups.[66] The 5-position methyl group improves the ADME properties including tissue penetration, C_{max}, AUC, and $t_{1/2}$.[67] Both groups increase the lipophilicity as seen in increased log$P$ values.

Table 8 6-Position SAR – regioisomeric fluoroquinolones

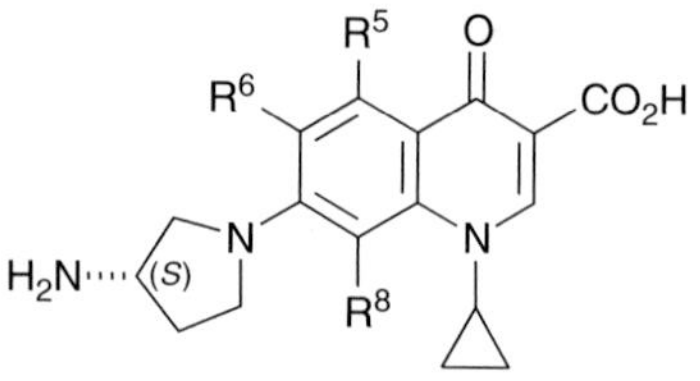

Compound			*MIC* ($\mu g\,mL^{-1}$)		
R^5	R^6	R^8	*Ec*[a]	*Pa*[a]	*Sa*[a]
F	H	H	0.13	4.0	1.0
H	F	H	0.004	0.25	0.03
H	H	F	0.008	0.25	0.03
H	F	F	0.008	0.13	0.008

[a] *E. coli* (Ec), *P. aeruginosa* (Pa), *S. aureus* (Sa).

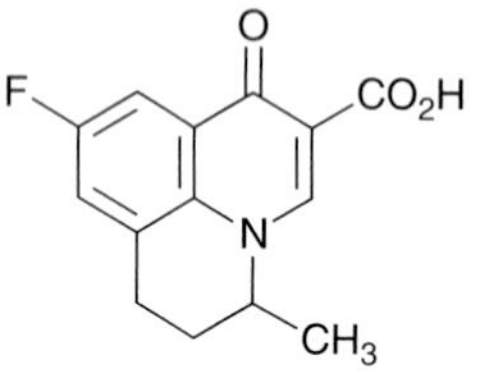

Flumequine (**30**)

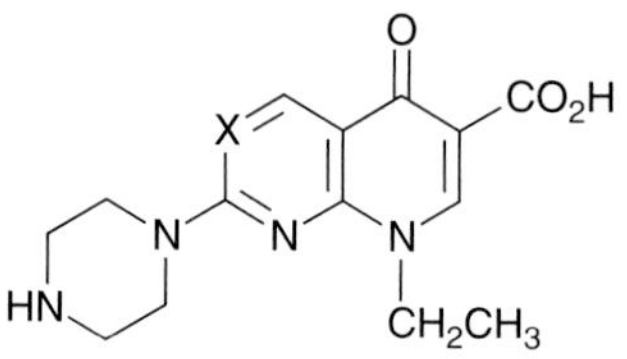

Pipemidic acid; X = N (**41**)

Figure 9 6-Position SAR.

Early exploration of substitution in the 5-position led to compounds like flumequine (**30**), which has a 6-fluoro group.[68] In many cases as in norfloxacin, a 6-fluoro group enhances the DNA gyrase inhibitory activity and MIC potency, which led to the group of 6-fluoro compounds being called the fluoroquinolones.[69] For example, norfloxacin with a 6-fluoro group is 18-fold more potent against *E. coli* DNA gyrase and 63-fold more potent in *E. coli* by MIC than the 6-H analog.[70] Additionally with this arrangement of substituents, 6-fluoro is 10-fold more potent by MIC than its 8-fluoro isomer. In the 1,8-bridged quinolones, levofloxacin is 10-fold more potent against *E. coli* DNA gyrase and 500-fold more potent in *E. coli* by MIC than the des-F compound.[71] In the pipemidic acid core (**41**) X=N (see **Figure 9**), the 6-fluoro analog (X=CF) is active whereas pipemidic acid (**41**) X=N and the X=CH compounds have very poor activity against *E. coli* DNA gyrase and MIC.[70] Alternatively, a 6-NH_2 can provide useful activity depending on other substitutions around the quinolone.[72] Of significance, the antimicrobial activity afforded by quinolone substituents is often dependent on the nature of the substitution pattern as a whole, not simply driven by a single functional group.[67] For example, in contrast to the trends seen in compounds with a C-7 piperazine such as norfloxacin, the 6-fluoro can be repositioned to the 8-position without a loss of activity when the 7-position is an (*S*)-pyrrolidin-3-amine (**Table 8**).[73] From this study, the conclusion that can be drawn is that the effect of the 8-fluoro is dependent on the 1- and 7-substituents.

While the fluoroquinolones imparted greater potency and DNA gyrase activity, the non-fluoroquinolones (NFQs) (des-6-fluoro) (e.g., PGE 9262932 (**42**), PGE 4175997 (**43**), and PGE 9509924 (**44**)) display less susceptibility than other quinolones to existing mechanisms of resistance in the important Gram-positive species *S. aureus* (**Table 9**).[74,75] For example, PGE 9509924 shows much better activity against quinolone-resistant *S. aureus* species EN1252aSp, which has a double mutant; Ser84 to Leu in *gyrA* and Ser80 to Phe in *grlA*. This may be related to the propensity for the NFQs to exploit both targets (DNA gyrase and topoisomerase IV) better than their FQ counterparts, such as ciprofloxacin or trovafloxacin. A clear example can be seen in the comparison of BMS-284756 (**45**) to its 6-fluoro analog BMS-340280 (**46**) in **Table 10**.[76] While both compounds exhibit excellent activity in vitro against wild-type and quinolone-resistant Gram-positive and Gram-negative bacteria, the 6-H analog has improved topoisomerase IV (topo IV) potency and mammalian topoisomerase 2 selectivity. There was little enantiopreference observed in this series. In general, the NFQs demonstrate broad-spectrum activity in vitro and in vivo especially versus Gram-positive organisms, and tend to

Table 9 Nonfluorinated quinolones (NFQs)

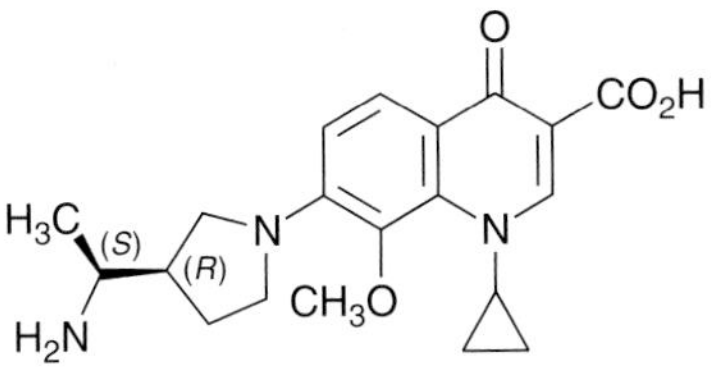

PGE 9262932 (**42**) 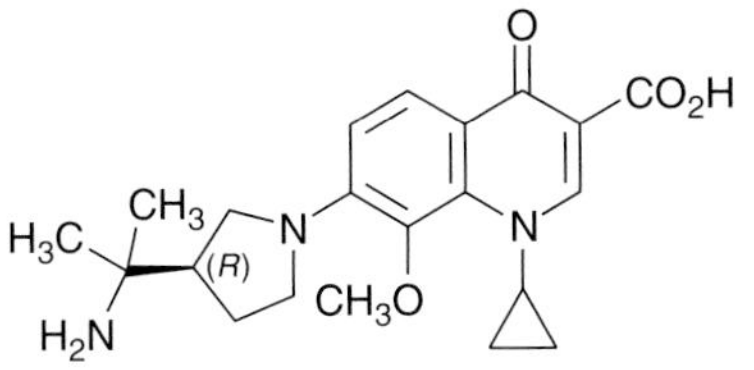 PGE 4175997 (**43**) 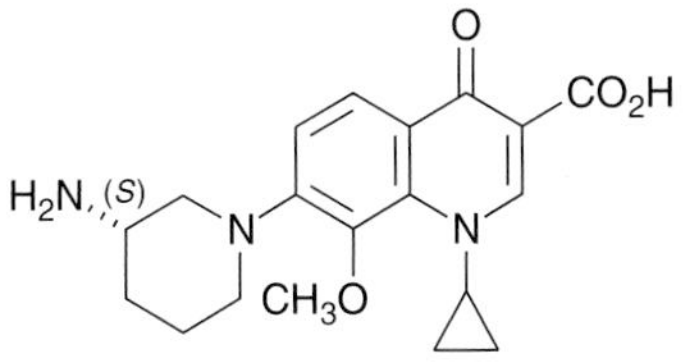 PGE 9509924 (**44**)

Compound	*S. aureus whole-cell target inhibition assay – MIC (μg mL⁻¹)*	
	ISP794[a]	*EN1252aSp*[b]
PGE 9509924 (**44**)	0.063	0.5
CIP (**4**)	0.25	32
Trovafloxacin (**9**)	0.063	4.0

[a] ISP794, quinolone-sensitive parent strain.
[b] EN1252aSp, double mutant; Ser84 to Leu in *gyrA* and Ser80 to Phe in *grlA*.

Table 10 Nonfluorinated quinlones (NFQs) topoisomerase activity

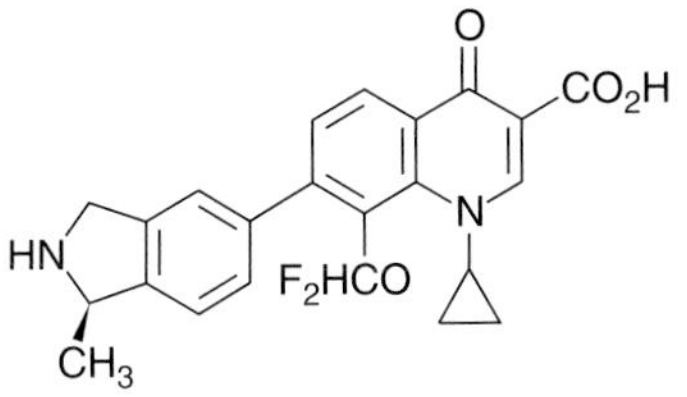

BMS-284756 (**45**) 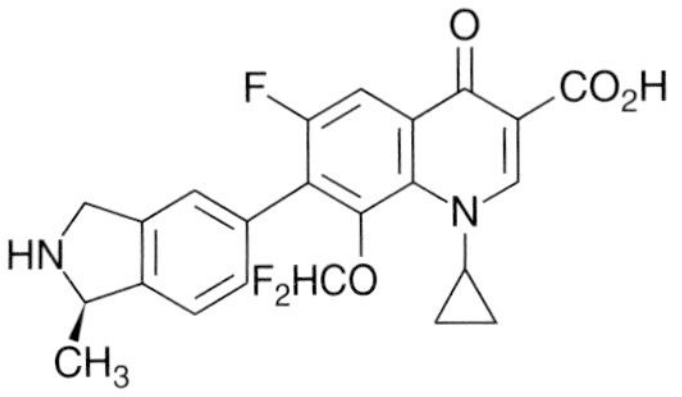 BMS-340280 (**46**) 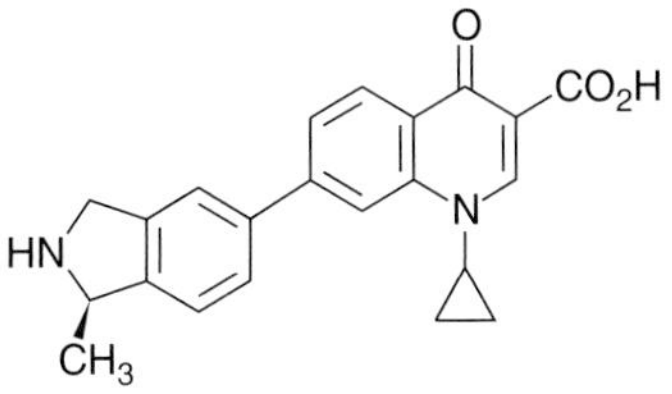 BMS-340278 (**47**)

Compound	*IC_{50} (μg mL⁻¹)*			*MIC (μg mL⁻¹)*	
	Ec gyrase	*Sa topo IV*	*TOP2*	*Ec*	*Sa*
BMS-284756 (**45**)	0.17	2.19	510	0.015	0.007
BMS-340280 (**46**)	0.16	4.60	128	0.015	0.007
BMS-340278 (**47**)	0.19	2.04	89	0.007	0.015

E. coli (Ec), *S. aureus* (Sa).

have lower genotoxicity than their fluorinated counterparts. However, it must be noted that there are exceptions and the contribution of the 6-fluoro group depends on the nature of the 1-, 7-, and 8-substituents as seen in BMS-340278 (**47**), which has less eroded mammalian topoisomerase 2 selectivity (**Table 10**).[76]

A wide variety of groups can be installed at the 7-position and are typically used to modulate physicochemical and ADME properties.[45c,45e] A piperazine contributes to intrinsic microbiological activity; however, a distal nitrogen is not required for activity. In general, a basic nitrogen group in the distal position of a 7-position piperazine group improves ADME and in vivo performance. Alkylation of the 4′-N of the piperazine (e.g., pefloxacin (**13**); R = CH_3) (**Figure 10**) improved PD_{50} values (especially p.o.), but are metabolized to NH yielding the active metabolite norfloxacin (**11**). In a 5-membered ring, 3-(alkyl)aminomethyl- (CI-934 (**48**)) and 3-aminopyrrolidine (tosufloxacin (**16**)) groups led to enhanced Gram-positive activity.[45c,45e,77] Variation of small alkyl groups can improve AMDE and Gram-positive potency. For example, the 3′-methyl of temafloxacin (**17**) improves the half-life over norfloxacin and the 5′-methyl analog of tosufloxacin gives a fivefold increase in AUC.

While the 7-position pyrrolidines tend to enhance Gram-positive activity over the piperazine of ciprofloxacin, the Gram-negative and DNA gyrase activity are not necessarily improved (**Table 11**).[78] Further, DNA gyrase and in vitro potency are not greatly affected by the chirality of the groups appended to the pyrrolidine. Chirality of distal

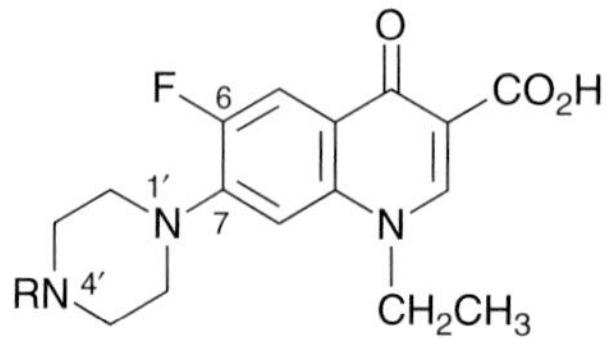

Norfloxacin (**11**); R = H
Pefloxacin (**13**); R = CH_3

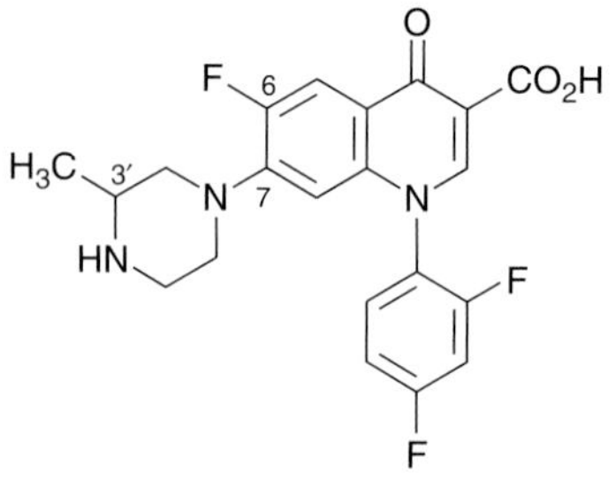

Temafloxacin (**17**)

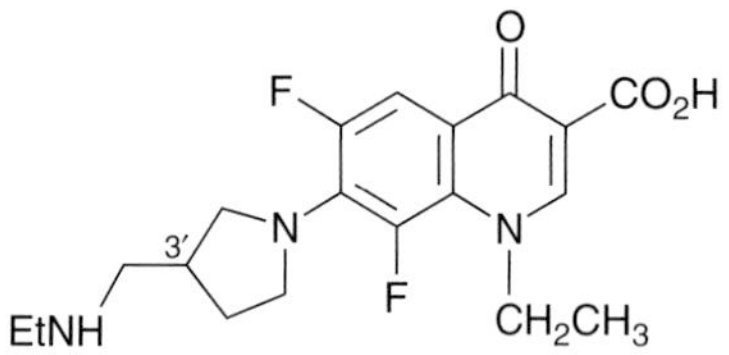

CI-934 (**48**)

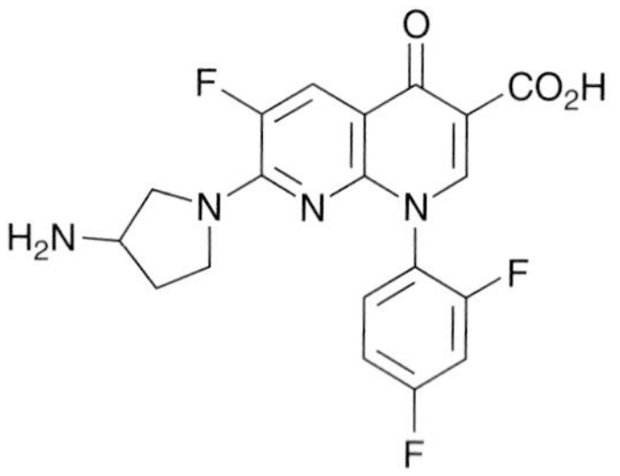

Tosufloxacin (**16**)

Figure 10 Alkylation of the 4′-N of the piperazine.

Table 11 7-Position SAR – pyrrolidines

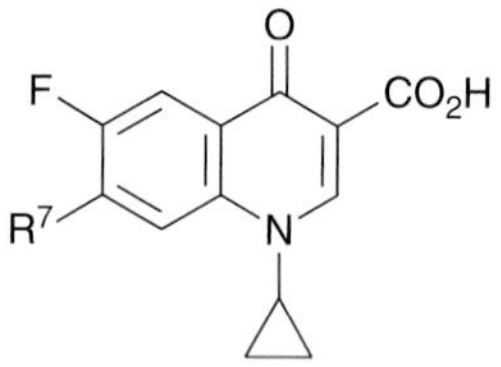

Compound R^7	*Gyrase cleavage IC_{50} ($\mu g\,mL^{-1}$)*	*MIC ($\mu g\,mL^{-1}$ mean) Gram-(−)*	*MIC ($\mu g\,mL^{-1}$ mean) Gram-(+)*
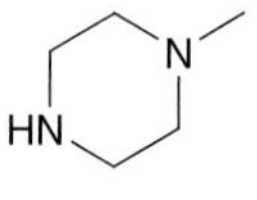	0.5	0.057	0.91
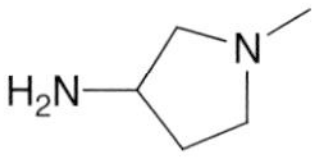	0.1	0.033	0.066
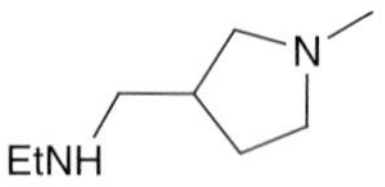	1.0	0.23	0.10

stereogenic centers on 7-position groups generally has little effect on gyrase and/or in vitro activity, which can be seen in early studies of CI-934,[79] temafloxacin[80] and BMS-284756 (**45**).[76] For stereogenic centers on the 7-position pyrrolidine moiety closer to the quinolone ring, chirality had a significant impact. For example, the difference between (3*S*,5*S*)-5-methylpyrrolidin-3-amine and n (3*S*,5*R*)-5-methylpyrrolidin-3-amine is 0.6 versus 7.5 $\mu g\,mL^{-1}$ in DNA gyrase cleavage IC_{50} and 20-fold in *E. coli* MIC.[81] Recent studies for 7-[3-(1-aminomethyl)-1-pyrrolidinyl] groups also showed significant differences.[82] In a continuing search for novel proprietary structures and enhanced spectrum of activity, the bicyclic 7-position systems of trovafloxacin (**9**) and moxifloxacin (**8**) provide excellent Gram-positive potency and improved ADME properties including reduced metabolism, better bioavailability, and greater AUC.[67]

Table 12 7-Position SAR – (hetero)aromatic groups

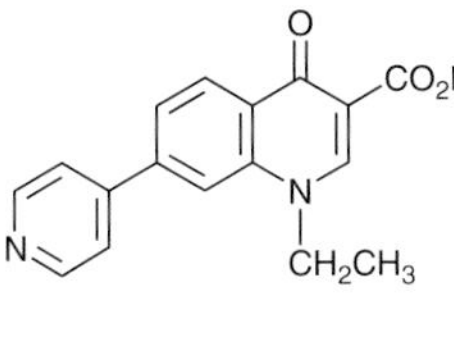

50 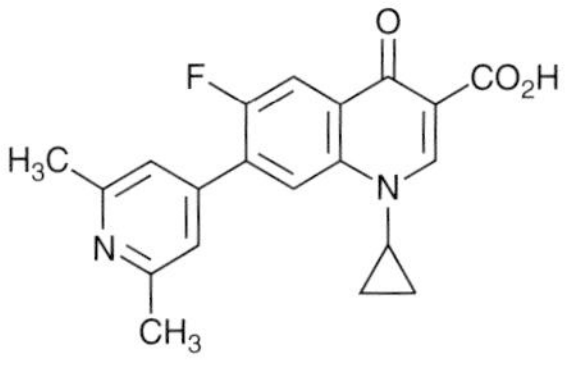 | WIN 57273 (**51**) 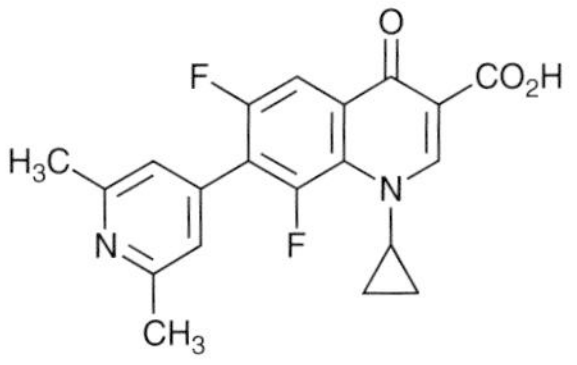 | WIN 57294 (**52**) 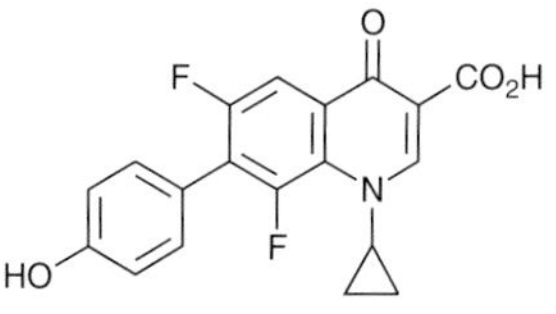 | CP-115,953 (**53**)

Compound	*MIC (μg mL⁻¹)*			
	Ec[a]	*Pa*[a]	*Sa*[a]	*Sp*[a]
Pyridine (**50**)	0.03	2.0	0.06	0.25
WIN 57273 (**51**)	0.03	2.0	0.008	0.03
WIN 57294 (**52**)	0.06	4.0	0.008	0.016

[a] *E. coli* (Ec), *P. aeruginosa* (Pa), *S. aureus* (Sa), *S. pneumoniae* (Sp).

Aromatic and heteroaromatic rings are also useful groups at the 7-position. Consistent with other SARs at the 7-position, addition of alkyl groups to the 7-substituent (e.g., pyridine 50 versus WIN 57273 (**51**)) enhances Gram-positive activity (**Table 12**).[83] WIN 57273 (51) exhibits outstanding in vivo activity versus Gram-positive pathogens and anaerobes. However, the addition of a fluoro group at the 8-position of 7-aryl quinolones tends to greatly enhance mammalian topoisomerase 2 while not significantly modifying the antibacterial activity, e.g., 52 and CP-115,953 (**53**).[84]

The 8-position of the quinolones is amenable to some variation and is used to improve ADME properties and reduce toxicity.[45c,45g] Primarily, the 8-position substituent can improve anaerobic activity, improve efficacy, and modulate adverse effects. The potential for the 8-position to vary the in vitro activity is highly dependent on the organism and substitution at the 1- and 7-positions. However, the change from 8-H in norfloxacin to a ring nitrogen as in enoxacin or an 8-F group can improve PK properties, especially bioavailability.[45g] Hydrophilic groups (e.g., NH_2, OH) tend to diminish gyrase/in vitro activity while small hydrophobic groups can show improvement. While 8-F and 8-Cl can show improved spectrum, these groups are associated with a serious class adverse effect of photosensitivity in animals/humans.[45g] These groups can also increase mammalian topoisomerase 2 activity, which can be ameliorated with a 2′-fluoro group on the 1-position cyclopropy in sitafloxacin (DU-6859a) (**54**).[45g,85] For sitafloxacin, the 8-Cl group increased activity versus Gram-positive and quinolone-resistant organisms. The 8-$OCHF_2$ of BMS-284756 decreases human topoisomerase 2 liability of 1-cyclopropyl, 7-heteroaryl groups.[76] The –OCH_3 of moxifloxacin increased activity for Gram-positive, quinolone- and methicillin-resistant organisms, and reduced the phototoxicity potential.[45g,87] The potential for phototoxicity is related to radicals generated from the 8-position halides; this adverse side effect has not been noted in the clinical use of other 8-OMe compounds such as gatifloxacin (**5**).[88] Incorporation of an 8-CH_3 has proven useful in the case of 2-pyridones, but has been of limited utility in classical 4-quinolones.[89]

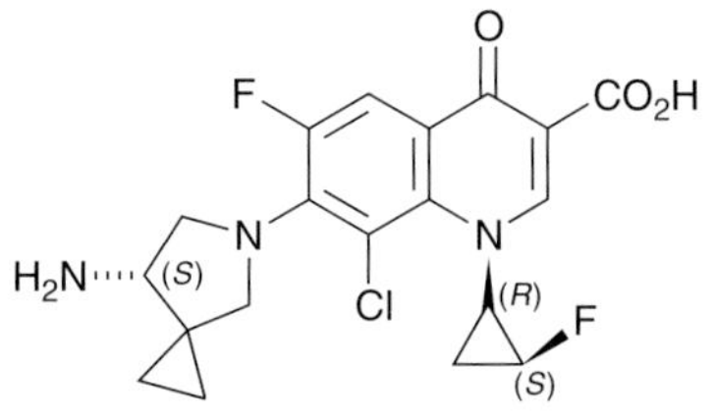

Sitafloxacin (**54**)

Interestingly, the 8-OMe group as seen in gatifloxacin has dual-targeting properties with a preference for bacterial topoisomerase IV, which is revealed in Gram-positive activity. In studies of the role of 8-OMe in *S. pneumoniae*, the 8-OMe group showed enhanced activity versus both targets, although the effect is greater for DNA gyrase than topoisomerase IV, and led to improved MIC potency (**Table 13**).[90] In a survey of 8-position SAR, an 8-H or halide tended to give better DNA gyrase and MIC activity while the PK properties were driven to a greater extent by the 7-position group with piperazine (PIP) giving better oral potency than the amino pyrrolidine (**Table 14**).[91] The ADME effect of changing from an 8-H to the 8-ring nitrogen of the 1,8-naphthyridine of trovafloxacin (**9**) can be clearly seen in

Table 13 Role of 8-methoxy in *S. pneumoniae* activity

Gatifloxacin (**5**); $R^8 = OCH_3$ Ciprofloxacin (**4**); $R^8 = H$

Compound	*IC_{50} (μg mL^{-1})*		*MIC (μg mL^{-1})*	
	Sp gyrase	*Sp topo IV*	*IC_{50} ratio*	*Sp HD553*
$R^8 = OCH_3$; (**5**)	25.9	7.3	3.6	0.25
$R^8 = H$	181	11.7	16	0.5
$R^8 = OCH_3$	19.4	4.1	4.8	0.25
$R^8 = H$; (**4**)	138	6.9	20	0.5

S. pneumoniae (Sp).

Table 14 8-Position SAR

Compound		*Gyrase cleavage IC_{50} (μg mL^{-1})*	*MIC in (μg mL^{-1}) (mean)*		*Ec PD_{50} (mg kg^{-1})*	
X	*R^7*		*Gram-(−)*	*Gram-(+)*	*p.o.*	*s.c.*
N	Pip	1.0	0.088	1.2	0.7	0.4
CH	Pip	0.5	0.057	0.91	1	0.3
CF	Pip	0.5	0.066	0.41	0.5	0.3
CCl	Pip	0.5	0.038	0.13	0.8	0.4
N	AP	1.0	0.013	0.12	2	0.6
CH	AP	0.1	0.025	0.066	3	0.5
CF	AP	0.1	0.013	0.044	1	0.2
CCl	AP	0.5	0.013	0.038	3	0.6

E. coli (Ec); amino pyrrolidine (AP).

the enhanced PK properties ($t_{1/2}$, AUC, C_{max}) in orally dosed monkeys (**Table 15**).[92] An overview of the general SAR and STR trends of the quinolones are summarized in **Figure 11**.[45g] Of significance, the degree of benefit of functional group(s) is frequently dependent on the nature of other substituents.[20]

At least brief mention should be made concerning the interesting nonclassical quinolone series, the 2-pyridones. The 2-pyridines are isoelectronic and isosteric analogs of the typical 4-quinolones. While this series, typified by

Table 15 Quinoline versus 1,8-naphthyridine – effect on ADME properties

X	*R[1]*	*20 mg kg^{-1} oral dose in monkeys*		
		$t_{1/2}$ *(h)*	*AUC (μg h^{-1} mL^{-1})*	C_{max} *(μg mL^{-1})*
CH	c-C_3H_5	3.7	22	2.3
N	c-C_3H_5	5.4	74	6.0
CH	2′,4′-F_2-C_6H_3	1.8	4	1.1
N	2′,4′-F_2-C_6H_3 (**23**) (Trov)	8.8	42	4.1

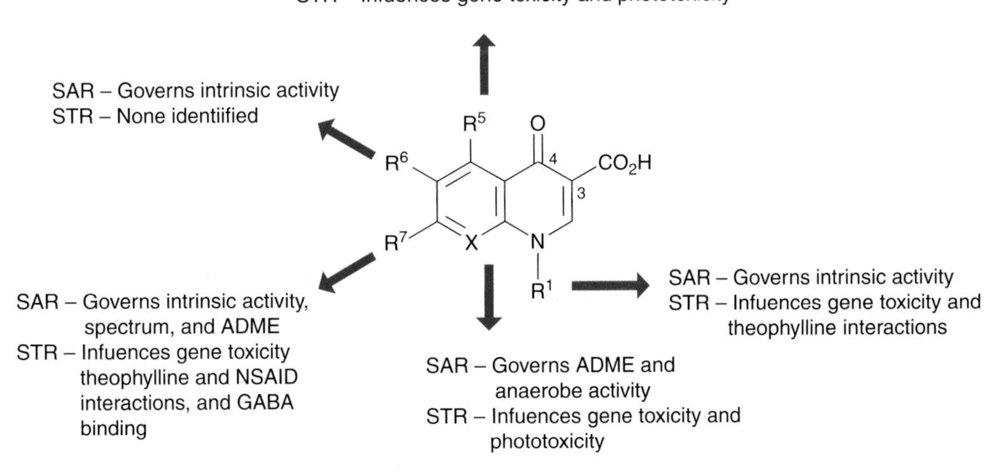

Figure 11 Structure–activity and structure–toxicity relationship summary.

ABT-719 (**55**), shares many of the broad SAR trends, there are some unique activity relations specific to the 2-pyridones. In **Scheme 4** where X=N, this pyrido[1,2-a]pyrimidine subseries shows potent DNA gyrase and antibacterial activity comparable to ciprofloxacin[93] However, the pyrido[1,2-a]pyrimidine subseries (X=N) is less potent in vitro than the corresponding 4-oxoquinolizine subseries (X=CR). For the 4-oxoquinolizine subseries (X=CR), the SAR parallels the classical quinolones with the following exceptions: (1) X=CF, CCl, or COMe shows inferior in vitro potency versus corresponding quinolones; (2) X=CMe (e.g., ABT-719 (**55**)) is far superior to corresponding quinolones; and (3) for the series, gyrase inhibition and MICs do not correlate as well as seen in the quinolones.[93] In an advanced example of the 2-pyridones, ABT-719, chirality does play a role as ABT-719 is fivefold more potent against DNA gyrase than its enantiomer. While both enantiomers are nearly equivalent in potency in vitro, ABT-719 (the (*S*)-isomer) displays enhanced ADME with *c.* threefold higher C_{max}, improved oral bioavailability, and aqueous solubility. ABT-719 shows an excellent spectrum of in vitro activity including Gram-positive, quinolone-resistant, vancomycin-resistant, and anaerobic bacteria. ABT-719 and analogs tend to exhibit superior in vivo efficacy (oral antibacterial spectrum) and ADME properties. Nitrogen juxtaposition seems transferable to other quinolone cores (e.g., levofloxacin (**7**)).[93]

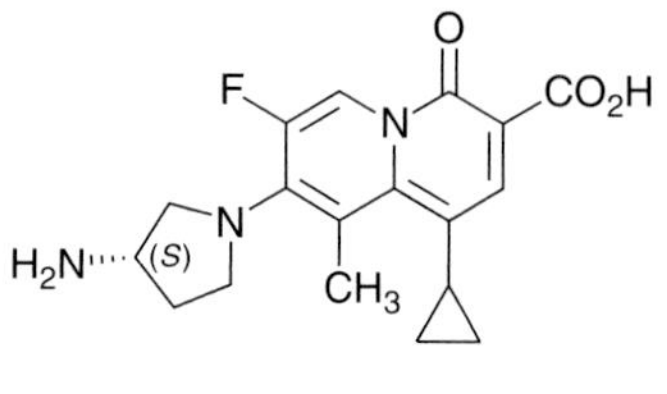

ABT-719 (**55**)

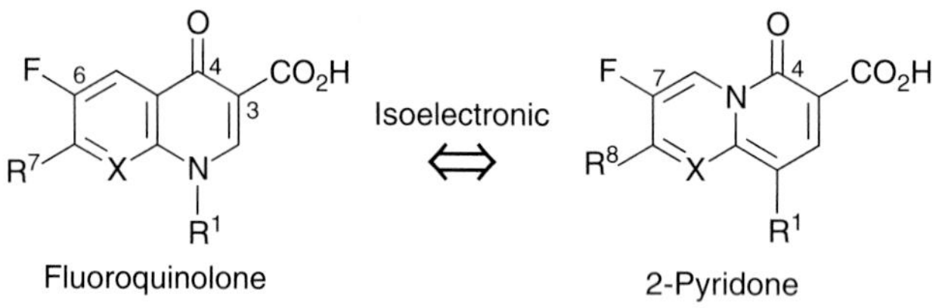

Fluoroquinolone Isoelectronic 2-Pyridone

Scheme 4

7.19.4 Antibacterial Resistance Mechanisms

Bacterial resistance to quinolones arises from specific chromosomal mutations, which alter the targets of activity (DNA gyrase and topoisomerase IV) or from nonspecific mechanisms of resistance including upregulation of drug efflux pumps and alterations in the bacterial membrane, which reduce the drug's permeation into the cell.[94] From genetic assessment of resistance in *S. pneumoniae*, mutations in particular regions of the genes *gyrA* and *parC*, the QRDR, are most often associated with quinolone target-specific resistance. In a large study of *S. pneumoniae*, the most common mutations that led to resistance were Ser-81/Phe or Tyr in *gyrA* and Ser-79/Tyr in *parC*,[95] although Ser-79/Phe in *parC* has also been seen.[96] Similar *gyrA* and *parC* mutations have been observed to lead to quinolone resistance in *E. coli* and other bacteria.[94a] While single mutations in one gene led to low-level resistance, high-level resistance arises from at least two sequential mutations.[97] The most resistant strains of pneumocci have been demonstrated to have double mutations in both *gyrA* and *parC*.[98] Only a few reports of mutations in gyrB or parE/grlB in any bacteria have been described.[94a]

Efflux pumps can have a profound effect on the ability of a quinolone (or in fact any small molecule antibiotic) to accumulate intercellularly and thus cause resistance.[99] Efflux pumps are a primary mechanism of resistance in Gram-negative species, such as *P. aeruginosa*.[100] After fluoroquinolone therapy, isolated *P. aeruginosa* clones frequently overexpress at least one of the multiple efflux pumps. While in the Gram-positive species *S. pneumoniae*, the effect of efflux pumps on resistance seems to be limited and quinolone specific.[101] Analysis of ciprofloxacin resistance in *S. pneumoniae* with the efflux inhibitor reserpine indicates that *c*. 50% of these clinical isolates may have enhanced efflux and could be clinically relevant. Efflux systems have been identified for many species. The norfloxacin efflux pump has been identified for *S. aureus* (NorA) and for *S. pneumoniae* (PmrA). In *P. aeruginosa*, several efflux pump systems have been identified with the major system encoded by mex (Multiple EffluX) genes, e.g., mexAmexB-oprM, mexCD-OprJ system, and mexEF-oprN. In *E. coli* the efflux pump is the acrAB-tolC system.[94a] The extent of efflux action is dependent on the bacterial species, the efflux pumps expressed in species or strains and the nature of the particular quinolone under study. However, there is a general trend that lipophilic compounds, such as gatifloxacin (**5**) and grepafloxacin (**22**), are poorer pump substrates than the more hydrophilic molecules, such as ciprofloxacin (**4**) and sparfloxacin (**21**), as determined in *S. pneumoniae* and *S. aureus*.[102] From the observation of reduced efflux of moxifloxacin (**8**) in *S. pneumoniae*, it has been proposed that larger, sterically bulky groups at the 7-position can confer some pump avoidance.[103]

Adding to the arsenal of defense mechanisms, Gram-negative outer membranes limit small molecule permeability and reduce their susceptibility to many antimicrobial agents. Changes in drug susceptibility can be traced to mutations that result in alterations in the outer membrane component lipopolysaccharide (LPS) and porins.[104] These outer membrane permeability changes along with active forms of efflux combine synergistically leading to multidrug resistance in Gram-negative bacteria. Transferable plasmid-born resistance is well known for other classes of antibiotic agent, however, the recent discovery of a quinolone resistance plasmid, named qnr, has altered this view. Qnr-bearing strains of Gram-negative bacteria generate quinolone resistant mutants at a much higher rate than wild-type organisms, and also raise the possibility of rapid dissemination of quinolone resistance.[105] Thus far, unlike β-lactam antibiotics, no quinolone-degrading enzymes or inactivating proteins have been identified in bacterial species, although enrofloxacin has been found to be metabolized by a fungus.[106]

Reducing the spread and incidence of antibiotic resistance is of prime concern in both the clinical and research arenas.[107] For example, blocking the SOS pathway by inhibiting the protease LexA in *E. coli* renders the bacteria unable to evolve resistance to ciprofloxacin (**4**) in vivo.[108] An alternative approach for reducing the rate of resistance mutation is based on the fact that high-level resistance in quinolones arises from double mutations (one each in DNA gyrase and topoisomerase IV), which is a low frequency event. This may in part explain why quinolone resistance has been relatively slow to develop and spread. While treating bacteria at concentrations below the MIC would tend to select for resistance mutants, higher concentrations of quinolones above the MIC (termed the mutant prevention concentration (MPC)) can reduce the possibility for selection of antibacterial resistance mutations in vitro, and may prevent mutant selection.[109] The MPC is set at the concentration (typically 2- to 10-fold above the MIC), which will inhibit the growth of most or all susceptible and moderately resistant bacteria, those with a single-step mutation. Treating bacteria at high concentrations, such as at or above the MPC, rapidly reduces the number of bacterial cells, both susceptible and moderately resistant.[110] Owing to the infrequency of quinolone chromosomal resistance mutations (10^{-7} to 10^{-10} for quinolones in *S. pneumoniae*), statistically there are too few growing cells to quickly select for resistance mutations as observed during in vitro experiments. Although administering doses of quinolones to humans at or above the MPC has not been specifically investigated in clinical trials, treatment of an infection at or above the MPC is expected to reduce the propensity for development of antibacterial resistance during acute or chronic treatment of infections.[111] Clinical reports indicate that higher doses of quinolones do result in improved treatment outcomes and suppression of bacterial resistance. For example, in a recent clinical study of critically ill patients with nosocomial lower respiratory tract infections caused by Gram-negative bacilli (e.g., *P. aeruginosa*), data indicates that treatment using ciprofloxacin at C_{max}/MIC ratios of 10:1 and $AUIC_{24}$ ratios of 100–125 maximizes bacterial eradication, thereby preventing the emergence of resistant bacteria during treatment.[112]

7.19.5 Pharmacokinetics and Pharmacodynamics

The pharmacokinetics, pharmacodynamics, ADME properties, microbiology, and clinical use of modern quinolones has been expertly reviewed.[113] The clinical utility of quinolones for specific applications, such as upper and lower respiratory infections,[114] urinary tract infections,[114,115] ocular infections,[116] gastrointestinal infections,[117] skin infections,[118] and mycobacterial infections,[119] has also been recently reviewed. Current clinical perspectives, efficacy, and adverse effects with ciprofloxacin,[120] gatifloxacin,[121] gemifloxacin,[122] levofloxacin,[123] and moxifloxacin[124] have been reported. As a class, quinolones are typically highly bioavailable (70–99%) with maximal concentrations (C_{max}) occurring between 1 and 3 h post oral dosing and half-lives between 1 and 4 h. The volumes of distribution tend (V_d) to be relatively high ($1.1-7.7\,L\,kg^{-1}$), in many cases exceeding body volume. The large V_d is an indication that quinolones can distribute systemically to various tissues and can accumulate in some tissues. Taking into consideration the fact that quinolones have low serum protein binding (30–60%) with the exception of trovafloxacin, the free fraction of quinolone in tissue or the unbound percentage of drug is relatively high and available to act on bacterial infection.[113c,113d] Quinolones bind strongly to cations, which can lower their absorption. Coadministration of quinolones with aluminum-, magnesium-, or calcium-containing antacids can have a pronounced effect on bioavailability.[125] Clearance of quinolones can be through renal and nonrenal mechanisms and depends on the specific agent. While clinafloxacin (**56**), gatifloxacin, levofloxacin, and sitafloxacin are primarily excreted intact as parent (>50%) in the urine (which is a positive tract for treating urinary tract infections), gemifloxacin, grepafloxacin, moxifloxacin, and sparfloxacin are excreted by nonrenal routes.[113d] The identification of metabolites and some SARs for quinolone metabolism routes has been examined,[126] as has modes of quinolone intestinal transport.[127]

A variety of factors are taken into consideration when describing the pharmacodynamics of quinolones. Most simply, pharmacodynamics is a relationship between a drug's concentration in serum (or tissue) and the resulting pharmacological and toxicological effects. Typical parameters used to characterize the pharmacodynamic effect of a quinolone are MIC for a specific bacteria or a species (MIC^{90}) as a measure of drug potency and a measurement of drug exposure in vivo over 24 h, which is usually the AUC_{24} (a relationship between the C_{max} and time of exposure). However, there are further critical factors to bear in mind, such as the two types of bacterial killing kinetics: concentration-dependent or time-dependent killing and the free fraction of drug, e.g., the non-protein- or tissue-bound fraction of drug.[113d] Overall, quinolones are considered to act with concentration-dependent action and a close correlation to bacterial eradication and clinical effectiveness is a ratio of either AUC_{24}/MIC or C_{max}/MIC[128] A target ratio of >10 for C_{max}/MIC has been associated with successful clinical outcomes and prevention of resistance mutations.[113d] However, this ratio is highly dependent on the bacterial species, quinolone agent, and infection with many outcome breakpoints indicating much higher ratios. A further important attribute of quinolones is the phenomenon of persistent antibacterial activity or postantibiotic effect (PAE) in which bacterial growth will continue to be suppressed significantly after the drug concentration falls below the MIC.[129]

7.19.6 Adverse Effects and Drug Interactions

With safety as a major concern, e.g., the discontinuation of development or restricted label of grepafloxacin, clinafloxacin, temafloxacin, and trovafloxacin,[130] new quinolones, such as moxifloxicin, are being scrutinized for adverse effects.[21a] Thus, improved safety, as well as resistance profiles, are driving current research. The latest developments in DNA gyrase inhibitors addressing these issues has been published with a focus on the C-8 methoxy fluoroquinolones moxifloxacin and gatifloxacin and new advanced research compounds.[131] Quinolones with C-8 methoxy groups have shown a lower propensity to induce resistance in *S. pneumoniae*.[132] And thus far after treatment of more than 3 million patients, moxifloxacin has shown an improved safety profile with a notable lack of phototoxicity, hepatoxicity, and significant cardiac events. Overall, the safety profile of quinolones as a class is quite good and the agents are considered well tolerated. However, to protect patient health, adverse events are monitored during clinical phase trials and after approval during postmarketing surveillance. The most common adverse effects in the quinolone class are gastrointestinal tract disturbance (nausea and diarrhea) and central nervous system (CNS) effects (headache and dizziness).[133] These side effects of treatment are typically mild and do not require cessation of therapy. Importantly, there are rare, but potentially serious quinolone-related adverse reactions that involve the cardiovascular system (QT interval prolongation), musculoskeletal system (tendinitis and tendon rupture), endocrine system (glucose homeostasis dysregulation), renal system (crystalluria, interstitial nephritis, and acute renal failure), the CNS (seizures), allergic reactions (rash), and phototoxicity.[113d,133]

To date, the incidence of quinolone-induced cardiotoxicity (torsades de pointes) is rare with only a few cases reported per 10 million prescriptions for ciprofloxacin, levofloxacin, and moxifloxacin.[134] A significant incidence of torsades de pointes is seen with gatifloxacin with 27 cases reported per 10 million prescriptions. Torsades de pointes is a serious adverse effect and can lead to ventricular tachycardia and fibrillation due to prolongation of cardiac repolarization. The cardiotoxicity and details of HERG channel blockage leading to QTc prolongation was reviewed for both quinolones and macrolides.[134] Early reports of cartilage toxicity in young animals has been the cause for caution in the pediatric use of quinolones. A critical discussion of the clinical use of quinolones in pediatric infections reveals an excellent record of efficacy and safety profile.[135] However, while tendinopathy is a rare adverse event, it is none the less a serious and debilitating adverse reaction. With over 3500 cases of tendon toxicity associated with quinolone use, very little is known about the mechanism or prediction of the toxicity. Risk factors in patients that seem to be related to tendon toxicity are renal disease or concurrent corticosteroid use.[136] The safety profiles of newer quinolones have recently been published. Gemifloxacin was found to be well tolerated by the 'at risk' elderly population having renal or hepatic impairment and when coadministered with omeprazole, digoxin, theophylline, and warfarin.[137] With more than 50 000 patients treated, moxifloxacin is also well tolerated in elderly patients with renal dysfunction, or with mild to moderate hepatic impairment. No adverse drug–drug interactions were noted with commonly prescribed drugs.[138] For further reference and SAR, a recent review covering the most commonly prescribed fluoroquinolones examines the mechanisms that lead to adverse side effects, such as cardiotoxicity, phototoxicity, and CNS effects.[139]

7.19.7 Current Development Programs

This section will highlight the current status of leading quinolone antibiotics; those compound expected to enter development shortly, those that are in development currently, and those that have recently been dropped from clinical studies. As previously mentioned, safety is a major concern in clinical evaluation of quinolones. In response to clinical surveillance, several quinolones have been discontinued from development (grepafloxacin, clinafloxacin (PD 127391)) or have restricted labels (trovafloxacin). Several new agents under investigation have progressed into clinical trials or been approved. Balofloxacin (Q-35, Q-roxin) (**57**), an orally active quinolone was approved by the Korean FDA in December 2001 for UTI. Balofloxacin (Q-35, Q-roxin) has shown excellent activity against antibiotic-resistant Gram-positive bacteria and decreased phototoxicity in comparison with members of the same class.[140] Having completed phase III studies, an NDA filing is expected for garenoxacin (BMS-284756, T3811) (**58**) in 2005.[141] Garenoxacin demonstrated outstanding Gram-positive coverage including *S. pneumoniae*, 4- to 16-fold greater anti-mycoplasma activity, and a favorable side effect profile compared to current lead quinolones.[142]

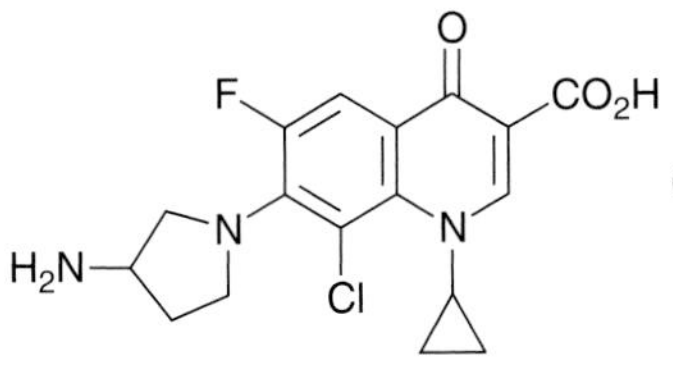

Clinafloxacin (**56**) 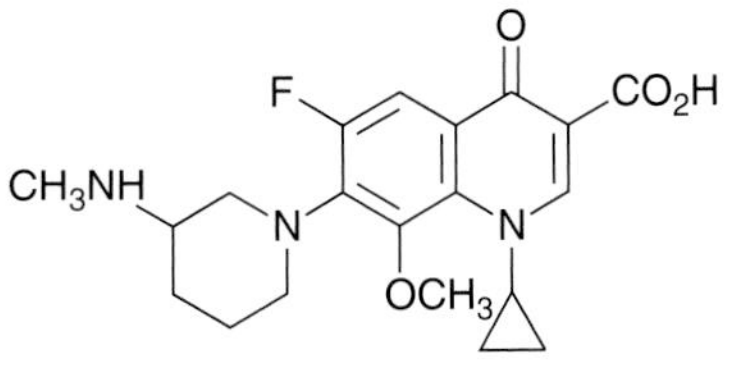 Balofloxacin (**57**) 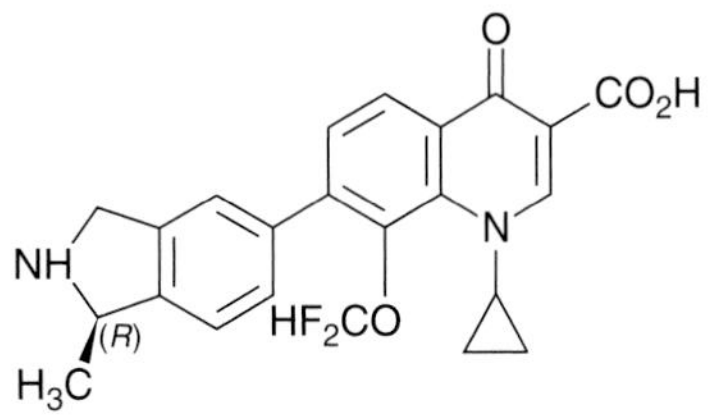 Garenoxacin (**58**)

Pazufloxacin (T-3761) (**59**), an injectable synthetic quinolone, was launched in Japan in September 2002.[143] Phase III studies of pazufloxacin revealed efficacy and overall safety similar to tosufloxacin.[144] Premafloxacin (U 95376) (**60**), which was no longer under development in 2003,[145] was highly active against resistant *S. aureus*.[146] Prulifloxacin (NM441) (**61**), a prodrug of ulifloxacin (NM394), shows good activity against *P. aeruginosa*,[147] and was launched in Japan in 2002. Prulifloxacin (NM441) entered US phase III trials as OPT-99 and is expected to reach the market in 2008.[148] Sitafloxacin (DU-6859a) was in phase III trials in the US and Japan for severe and drug-resistant Gram-negative bacterial infections.[149] The antibacterial activity in vitro and the PK in advanced animal models have been reported.[150,151] No development has been reported for ABT-255 (**62**) or ABT-719 (**63**) for several years.[152,153] Notably, ABT-255 shows a high in vitro and in vivo activity versus rifampin-sensitive/resistant and ethambutol-sensitive/resistant *Mycobacterium tuberculosis*.[154] Discontinued from Phase I, ecenofloxacin (CFC-222) (**64**) showed good PK and ADME properties, but did not exhibit significant improvement in activity over other marketed quinolones.[155] Antibacterial activity was similarly undistinguishing for cadrofloxacin (CS-940) (**65**) which was discontinued from phase II trials in Japan.[156,157]

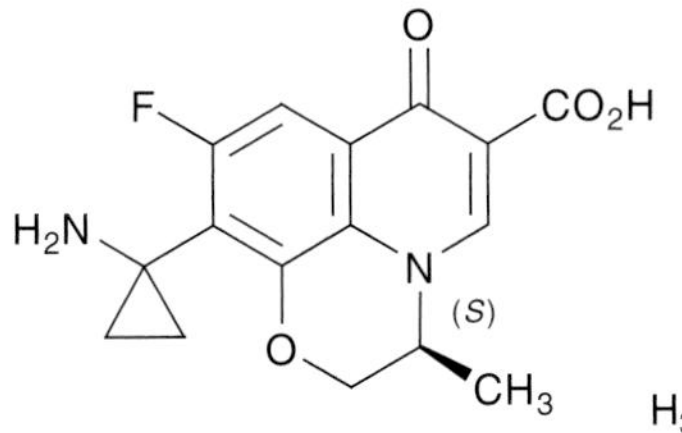

Pazufloxacin (**59**) 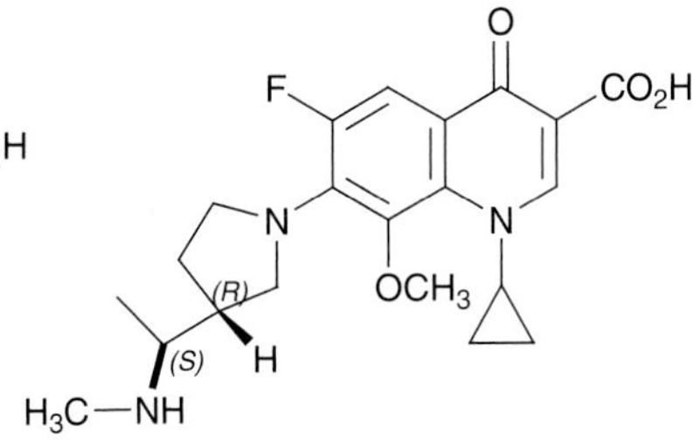 Premafloxacin (**60**)

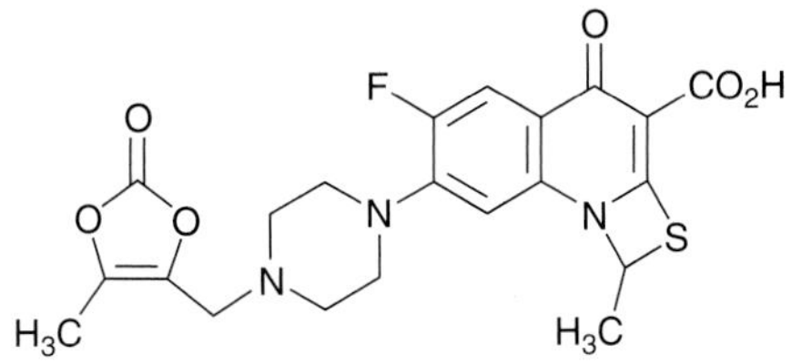

Prulifloxacin (**61**) 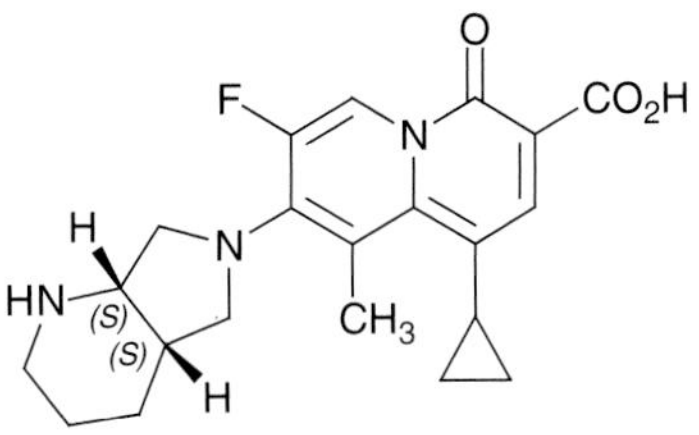 ABT-255 (**62**)

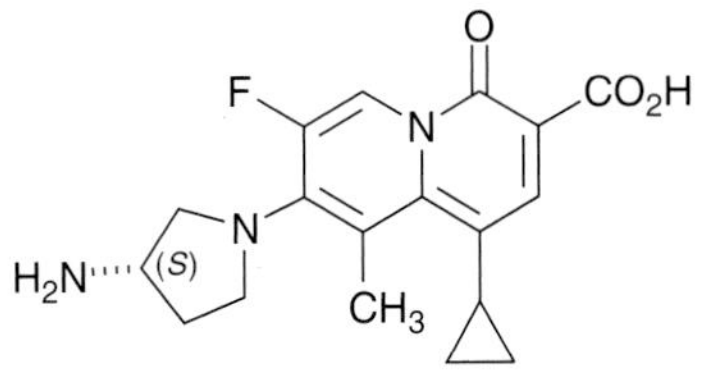

ABT-719 (**63**) 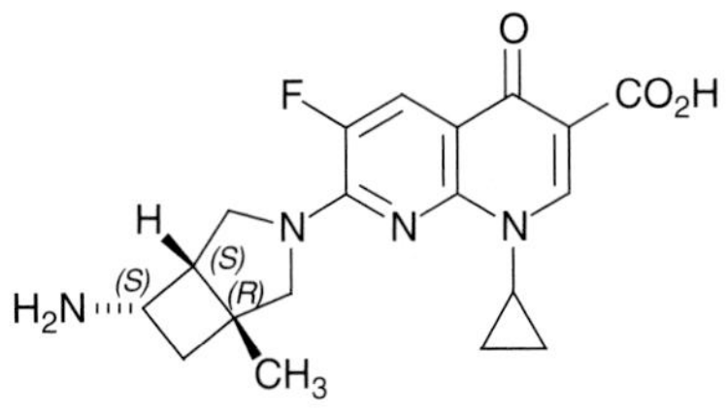 Ecenofloxacin (**64**) 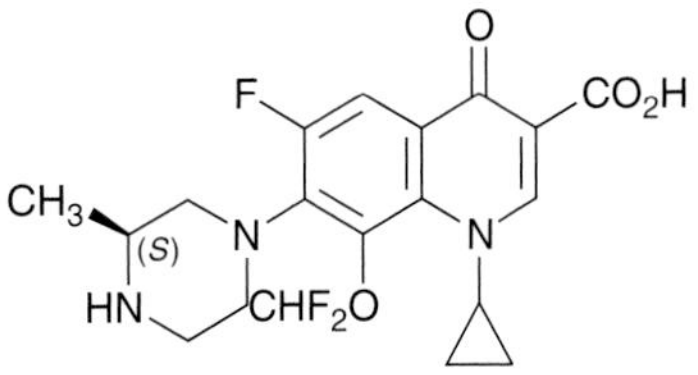 Cadrofloxacin (**65**)

After analysis of phase I data in Japan, DK-507k (**66**) was pulled from further development.[158] However, DK-507k was highly active against *Mycoplasma pneumoniae* and resistant strains of *H. influenzae*.[159] DQ-113 (**67**) has shown excellent efficacy in murine models of pneumonia caused by penicillin-resistant *S. pneumoniae* (PRSP) and hematogenous pulmonary infection with MRSA and vancomycin-insensitive *Staphylococcu aureus* (VISA), but no development activity reported since 2002.[160] Having been evaluated in phase I trials in Japan in 1995,[161] DV-7751a (**68**) displayed good MIC activity against *S. aureus*, *Staphylococcus epidermidis*, *S. pyogenes*, and *Enterococcus faecalis*.[162] No further development has been reported and the toxicity profile in rats, monkeys, and dogs has been disclosed.[163] Fandofloxacin (DW-116) (**69**) had been in phase II trials for UTI in 2000, but no development has been recently communicated.[164] Fandofloxacin, while displaying good antimicrobial activity and PK,[165,166] showed liver toxicity and developmental toxicity in rats.[167] DW-286 (**70**) was found to be active against quinolone-resistant strains of *S. aureus*, and compared well with marketed quinolones against a panel of Gram-positive strains (*S. epidermidis*, *S. pneumoniae*, and *E. faecalis*).[168,169] DW-286 was reported to begin phase I trials in 2005.[170] Olamufloxacin (HSR-903) (**71**) was in phase III trials up to 2001.[171] Olamufloxacin was active against a spectrum of Gram-positive and Gram-negative bacteria in vitro and in vivo, most notably against *Legionella* spp., and *P. aeruginosa*.[172]

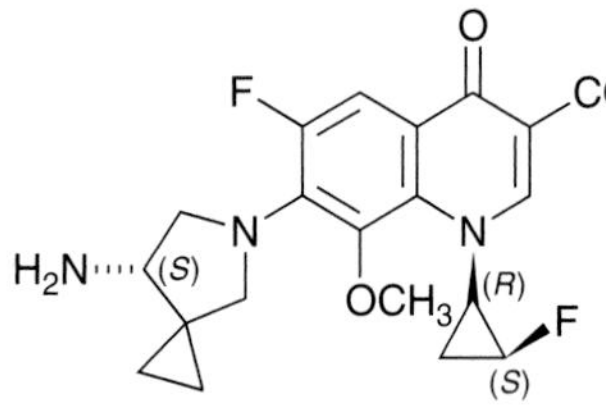

DK-507k (**66**)

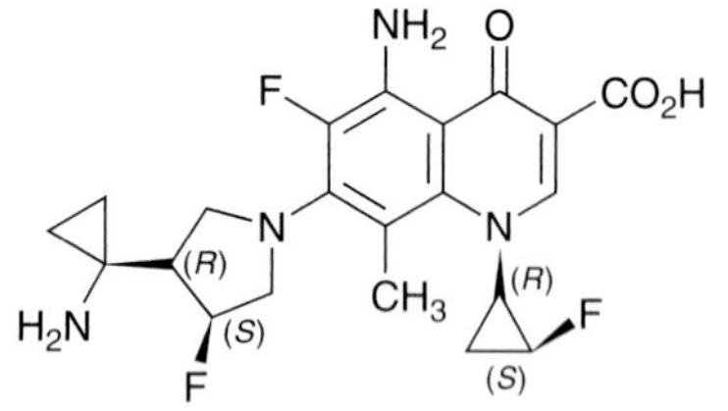

DQ-113 (**67**)

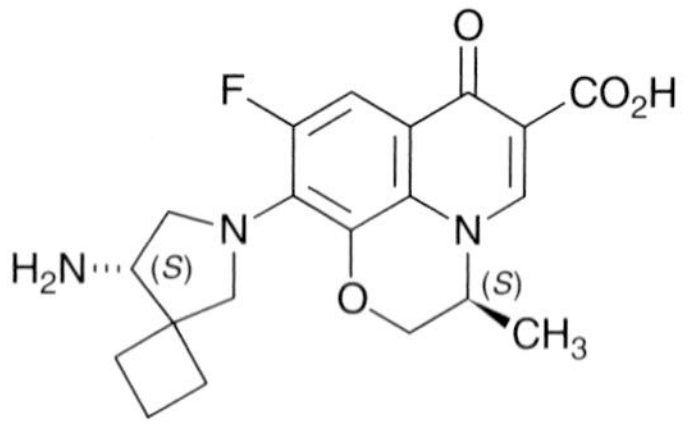

DV-7751a (**68**)

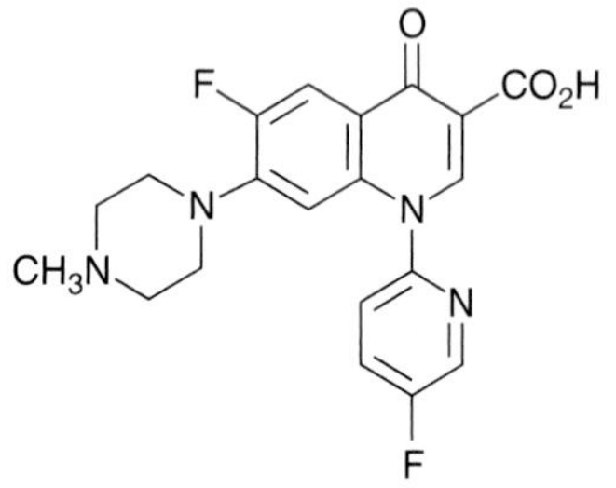

Fandofloxacin (**69**)

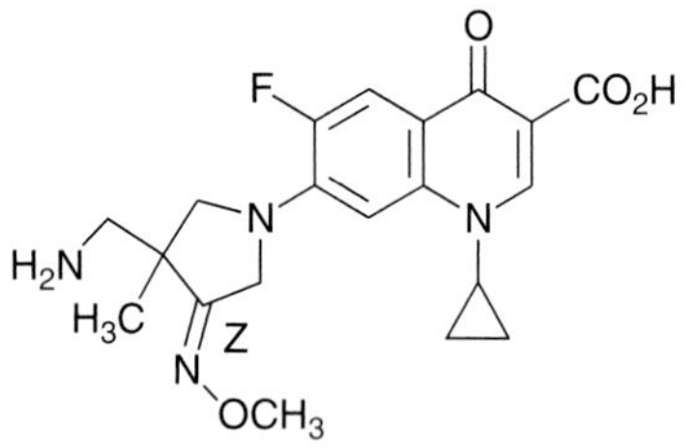

DW-286 (**70**)

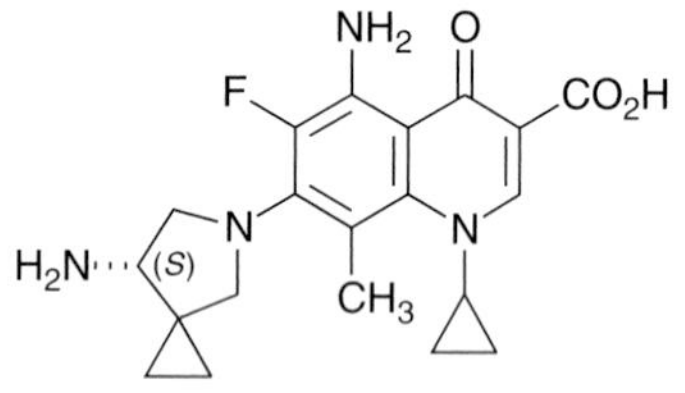

Olamufloxacin (**71**)

TG-873870 (presumed to be one of a series of nonfluorinated quinolones, e.g., PGE-9602021 or PGE-9509924 (**72**)) was reported to be moving to phase II clinical trials in 2006.[173] PGE-9509924 has broad spectrum in vitro and in vivo activity especially versus Gram-positive organisms. Interestingly, the post-antibiotic effect (PAE) was found to be longer for PGE-9509924 than other quinolones.[174] Nonfluorinated quinolones, such as PGE-9509924, are less susceptible than other quinolone classes to existing mechanisms of resistance in *S. aureus* and seemingly have lower genotoxicity potential.[175] T-3912 (**73**), a preclinical compound, was being examined for topical applications, e.g., dermatology, otorhinolaryngology, and ophthalmology. T-3912 has demonstrated good activity against microbes causing topical infections, e.g., *S. aureus*, *S. epidermidis*, and *Propionibacterium acnes*.[176] WQ-3034 (ABT-492) (**74**) was expected to enter phase II clinical trials for UTI and RTI in 2001, however, no development has been reported to date.[177] WQ-3034 showed excellent activity in vitro compared with marketed antibacterials,[178] and particularly good activity against respiratory tract pathogens including quinolone-resistant staphylococci and streptococci, vancomycin-resistant enterococci, anaerobic bacteria, *P. aeruginosa*, enterobacteriaceae, *Legionella* spp., and mycobacteria.[179] Y-688 (**75**) showed enhanced activity against Gram-positive species and reduced propensity for inducing photosensitivity.[180] Even with excellent MIC potency, Y-688 failed to show efficacy against ciprofloxacin-resistant staphylococci in a rat model of endocarditis and quickly selected for resistance in in vitro studies.[181]

DX-619 (**76**) was in phase I in 2004 for drug-resistant Gram-positive bacterial infections.[182] This compound had excellent activity against quinolone-resistant staphylococci and compared favorably with marketed agents against a large panel of clinical isolates, including Gram-negative species.[183,184] DX-619 was efficacious against VISA in the hematogenous pulmonary infection model, endocarditis due to MRSA, and had an excellent safety profile in rodents, dogs and monkeys.[185] Preclinical reports on DW-224a (**77**) show excellent MIC potency against respiratory bacteria *M. pneumoniae*, *M. hominis* and *L. pneumophilia*, *C. pneumoniae* and quinolone-resistant pathogens,[186] and a longer PAE than other lead quinolones.[187] The safety pharmacology, toxicity, and toxicokinetics of DW-224a have been published showing no adverse effects at a level of 10 mg kg^{-1} for both sexes in dog.[188]

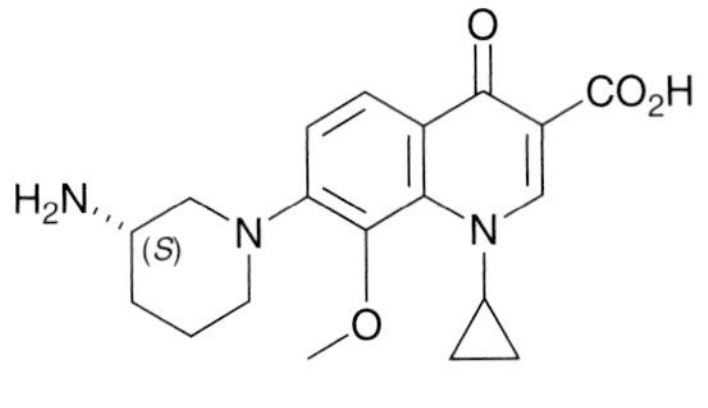

PGE-9509924 (**72**) 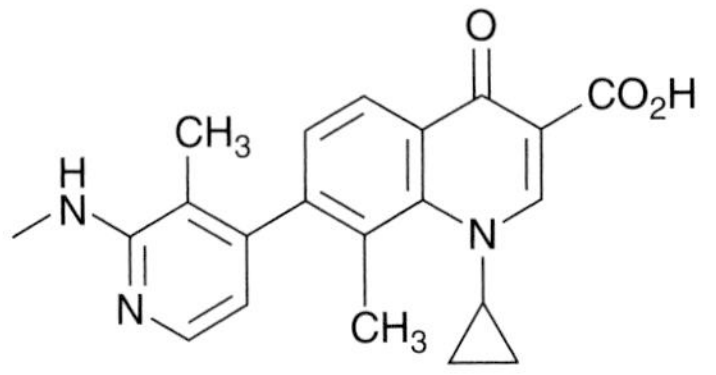 T-3912 (**73**) 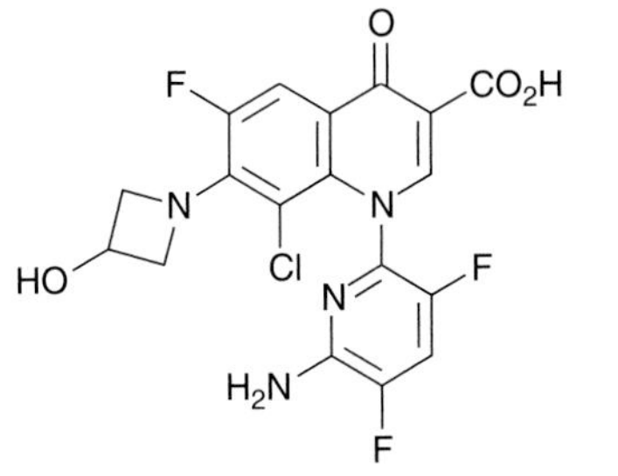 WQ-3034 (**74**)

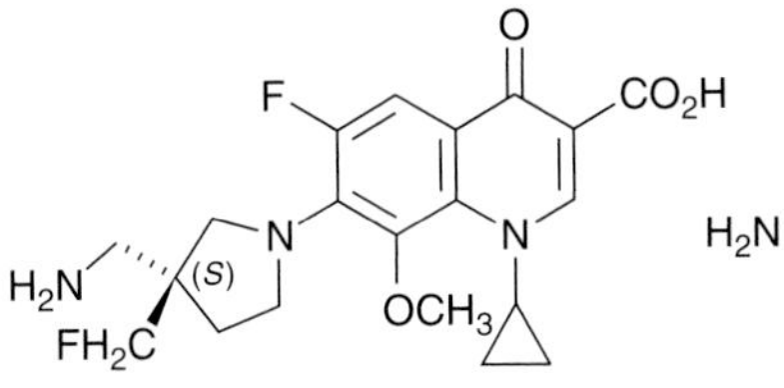

Y-688 (**75**) 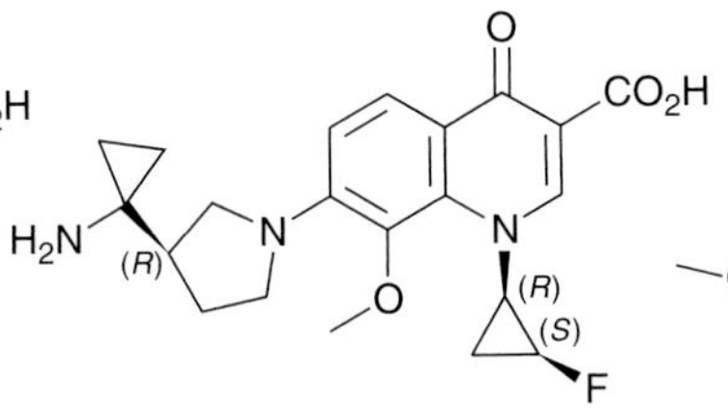 DX-619 (**76**) 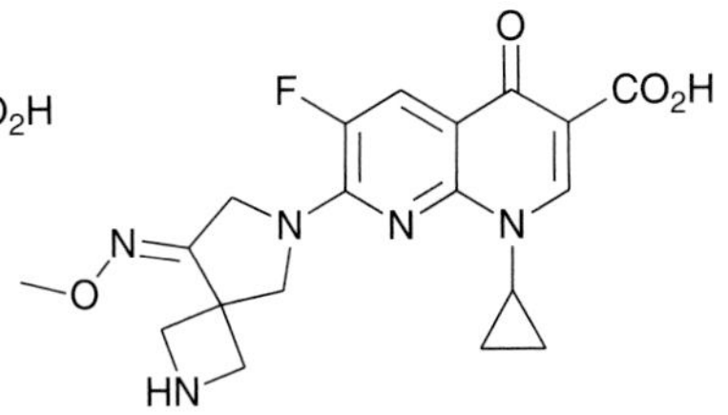 DW-224a (**77**)

7.19.8 Conclusion

The quinolone class of antibiotics has enjoyed a rich history of medicinal chemistry applied toward improving bacterial spectrum, pharmacokinetics, efficacy, and adverse side effect profiles. While great advances have been made concerning SARs and structure–toxicity relationships, there is still room for improved quinolone agents, especially agents with superior safety profiles for the community and pediatric markets and enhanced spectrum of activity particularly for treatment of serious nosocomial and chronic infections. Although bacterial resistance to quinolone compounds has not yet become widespread in the community setting due to the dual mechanism of action against DNA gyrase and topoisomerase IV, resistance is currently a concern in nosocomial infections. In the near future, quinolone resistance might be stemmed by treating bacteria above their mutant prevention concentration (MPC). However, as the use of quinolones becomes more routine replacing those antimicrobial agents that become obsolete due to spreading resistance, quinolone-resistant strains will become more prevalent. Consequently, research and discovery of novel quinolones and inhibitors of DNA gyrase and topoisomerase IV will be necessary to meet current and future unmet clinical needs.

References

1. For general reviews see: (a) Albrecht, R. *Prog. Drug Res.* **1977**, *21*, 9–104; (b) Rosen, T. *Prog. Med. Chem.* **1990**, *27*, 235–295; (c) Siporin, C.; Heifetz, C.; Domagala, J., Eds.; *The New Generation of Quinolones*; Marcel Dekker, Inc.: New York, 1990; (d) Wolfson, J. S.; Hooper, D. C., Eds.; *Quinolone Antibacterial Agents*, 2nd ed.; American Society for Microbiology: Washington, DC, 1993; (e) Andriole, V. T., Ed.; *The Quinolones*, 3rd ed.; Academic Press, Inc.: New York, 2000; (f) Hooper, D. C.; Rubinstein, E., Eds.; *Quinolone Antibacterial Agents*, 3rd ed.; American Society for Microbiology: Washington, DC, 2003.
2. Blondeau, J. M. *Expert Opin. Investig. Drugs* **2001**, *10*, 213–237.
3. Carbon, C. *Chemotherapy* **2001**, *47*, 9–14.
4. Rubinstein, E. *Chemotherapy* **2001**, *47*, 3–8.
5. Lesher, G. Y.; Froelich, E. J.; Gruett, M. D.; Bailey, J. H.; Brundage, R. P. *J. Med. Chem.* **1962**, *5*, 1063–1065.
6. (a) Buchbinder, M.; Webb, J. C.; Anderson, L.; McCabe, W. R. *Antimicrob. Agents Chemother.* **1962**, *6*, 308–317; (b) Ronald, A. R.; Turck, M.; Petersdorf, R. G. *New Eng. J. Med.* **1966**, *275*, 1081–1089; (c) Blondeau, J. M. *J. Antimicrob. Chemother.* **1999**, *43*, 1–11.
7. (a) Guyer, B. M.; Whitford, G. M. *Curr. Med. Res. Opin.* **1974–1975**, *2*, 636–640; (b) Gleckman, R.; Alvarez, S.; Joubert, D. W.; Matthews, S. J. *Am. J. Hosp. Pharm.* **1979**, *36*, 1077–1079.
8. (a) Guay, D. R. *Drug Intel. Clin. Pharm.* **1982**, *16*, 916–921; (b) Paulson, D. F. *Urology* **1982**, *20*, 138–140; (c) Scavone, J. M.; Gleckman, R. A.; Fraser, D. G. *Pharmacotherapy* **1982**, *2*, 266–272; (d) Sisca, T. S.; Heel, R. C.; Romankiewicz, J. A. *Drugs* **1983**, *25*, 544–569.
9. Rowen, R. C; Michel, D. J.; Thompson, J. C. *Pharmacotherapy* **1987**, *7*, 92–110.
10. Sabbour, M. S.; El Bokl, M. A.; Osman, L. M. *Infection* **1984**, *12*, 377–380.
11. (a) Wise, R. *J. Antimicrob. Chemother.* **1984**, *13*, 59–64; (b) Holmes, B.; Brogden, R. N.; Richards, D. M. *Drugs* **1985**, *30*, 482–513; (c) Goldstein, E. J. C. *Am. J. Med.* **1987**, *82*, 17–33; (d) Wolfson, J. S.; Hooper, D. C. *Ann. Intern. Med.* **1988**, *108*, 238–251.

12. (a) Wise, R.; Andrews, J. M.; Edwards, L. J. *Antimicrob. Agents Chemother.* **1983**, *23*, 559–564; (b) Fass, R. J. *Antimicrob. Agents Chemother.* **1983**, *24*, 568–574; (c) Eliopoulos, G. M.; Gardella, A.; Moellering, R. C., Jr. *Antimicrob. Agents Chemother.* **1984**, *25*, 331–335.
13. Davis, R.; Markham, A.; Balfour, J. A. *Drugs* **1996**, *51*, 1019–1074.
14. Goldstein, E. J. C. *Clin. Infect. Diseases* **1996**, *23*, S25–S30.
15. Weiss, K.; Laverdiere, M.; Restieri, C. *J. Antimicrob. Chemother.* **1998**, *42*, 523–525.
16. (a) Garcia-Rodriguez, J. A.; Munoz Bellido, J. L. *Int. J. Antimicrob. Agents* **2000**, *16*, 281–285; (b) Guthrie, R. *Chest* **2001**, *120*, 2021–2034; (c) Stahlmann, R. *Toxicol. Lett.* **2002**, *127*, 269–277; (d) Blondeau, J. M. *Expert Opin. Investig. Drugs* **2000**, *9*, 383–413; (e) Breen, J.; Skuba, K.; Grasela, D. *J. Respir. Dis.* **1999**, *20*, S70–S76; (f) Ball, P.; Mandell, L.; Niki, Y.; Tillotson, G. *Drug Safety: Int. J. Med. Toxicol. Drug Experience* **1999**, *21*, 407–421.
17. (a) Stahlmann, R. *Toxicol. Lett.* **2002**, *127*, 269–277; (b) Blondeau, J. M. *Expert Opin. Investig. Drugs* **2000**, *9*, 383–413; (c) Breen, J.; Skuba, K.; Grasela, D. *J. Respir. Dis.* **1999**, *20*, S70–S76; (d) Ball, P.; Mandell, L.; Niki, Y.; Tillotson, G. *Drug Safety: Int. J. Med. Toxicol. Drug Experience* **1999**, *21*, 407–421.
18. Blondeau, J. M.; Hansen, G.; Metzler, K. L.; Borsos, S.; Irvine, L. B.; Blanco, L. *Int. J. Antimicrob. Agents* **2003**, *22*, 147–154.
19. Blondeau, J. M.; Missaghi, B. *Exp. Opin. Pharmacother.* **2004**, *5*, 1117–1152.
20. Domagala, J. M.; Domagala, J. M. *J. Antimicrob. Chemother.* **1994**, *33*, 685–706.
21. (a) Blondeau, J. M.; Hansen, G. T. *Exp. Opin. Pharmacother.* **2001**, *2*, 317–335; (b) Perry, C. M.; Ormrod, D.; Hurst, M.; Onrust, S. V. *Drugs* **2002**, *62*, 169–207.
22. (a) Lu, T.; Zhao, X.; Li, X.; Drlica-Wagner, A.; Wang, J.-Y.; Domagala, J.; Drlica, K. *Antimicrob. Agents Chemother.* **2001**, *45*, 2703–2709; (b) Ince, D.; Aras, R.; Hooper, D. C. *Drugs* **1999**, *58*, 132–133; (c) Ince, D.; Aras, R.; Hooper, D. C. *Drugs* **1999**, *58*, 134–135; (d) Zhao, B. Y.; Pine, R.; Domagala, J.; Drlica, K. *Antimicrob. Agents Chemother.* **1999**, *43*, 661–666.
23. (a) Hooper, D. C. *Drugs* **1995**, *49*, 10–15; (b) Shen, L. L.; Baranowski, J.; Pernet, A. G. *Biochemistry* **1989**, *28*, 3879–3885; (c) Shen, L. L.; Mitscher, L. A.; Sharma, P. N.; O'Donnell, T. J.; Chu, D. W.; Cooper, C. S.; Rosen, T.; Pernet, A. G. *Biochemistry* **1989**, *28*, 3886–3894.
24. Piddock, L. J. V.; Jin, Y.-F.; Ricci, V.; Asuquo, A. E. *J. Antimicrob. Chemother.* **1999**, *43*, 61–70.
25. Bryan, L. E.; Bedard, J.; Wong, S.; Chamberland, S. *Clin. Invest. Med.* **1989**, *12*, 14–19.
26. Wang, J. C. *J. Mol. Biol.* **1971**, *55*, 523–533.
27. Gellert, M.; O'Dea, M. H.; Mizuuchi, K.; Nash, H. *Proc. Natl. Acad. Sci. USA* **1976**, *73*, 3872–3876.
28. Reece, R. J.; Maxwell, A. *Crit. Rev. Biochem. Mol. Biol.* **1991**, *26*, 335–375.
29. Wang, J. C. *Ann. Rev. Biochem.* **1996**, *65*, 635–692.
30. Morais Cabral, J. H.; Jackson, A. P.; Smith, C. V.; Shikotra, N.; Maxwell, A.; Liddington, R. C. *Nature* **1997**, *388*, 903–906.
31. Fass, D.; Bogden, C. E.; Berger, J. M. *Nat. Struct. Biol.* **1999**, *6*, 322–326.
32. Wigley, D. B.; Davies, G. J.; Dodson, E. J.; Maxwell, A.; Dodson, G. *Nature* **1991**, *351*, 624–629.
33. (a) Champoux, J. J. *Annu. Rev. Biochem.* **2001**, *70*, 369–413; (b) Heddle, J. G.; Barnard, F. M.; Wentzell, L. M.; Maxwell, A. *Nucleosides Nucleotides Nucleic Acids* **2000**, *19*, 1249–1264.
34. Kato, J.; Nishimura, Y.; Imamura, R.; Niki, H.; Hiraga, S.; Suzuki, H. *Cell* **1990**, *63*, 393–404.
35. Ullsperger, C.; Cozzarelli, N. R. *J. Biol. Chem.* **1996**, *271*, 31549–31555.
36. Kato, J.; Nishimura, Y.; Imamura, R.; Niki, H.; Hiraga, S.; Suzuki, H. *Cell* **1990**, *63*, 393–404.
37. Kampranis, S. C.; Maxwell, A. *Proc. Natl. Acad. Sci. USA* **1996**, *93*, 14416–14421.
38. (a) Hooper, D. C.; Hooper, D. C. *Drugs* **1995**, *49*, 10–15; (b) Hoshino, K.; Kitamura, A.; Morrissey, I.; Sato, K.; Kato, J.-I.; Ikeda, H. *Antimicrob. Agents Chemother.* **1994**, *38*, 2623–2627.
39. (a) Drlica, K.; Zhao, X. *Microbiol. Mol. Biol. Rev.* **1997**, *61*, 377–392; (b) Shen, L. L.; Kohlbrenner, W. E.; Weigl, D.; Baranowski, J. *J. Biol. Chem.* **1989**, *264*, 2973–2978.
40. Deitz, W. H.; Cook, T. M; Goss, W. A. *J. Bacteriol.* **1966**, *91*, 768–773.
41. Willmott, C. J. R.; Maxwell, A. *Antimicrob. Agents Chemother.* **1993**, *37*, 126–127.
42. (a) Ng, E. Y.; Trucksis, M.; Hooper, D. C. *Antimicrob. Agents Chemother.* **1996**, *40*, 1881–1888; (b) Saiki, A. Y. C.; Shen, L. L.; Chen, C.-M.; Baranowski, J.; Lerner, C. G. *Antimicrob. Agents Chemother.* **1999**, *43*, 1574–1577; (c) Blanche, F.; Cameron, B.; Bernard, F.-X.; Maton, L.; Manse, B.; Ferrero, L.; Ratet, N.; Lecoq, C.; Goniot, A.; Bisch, D. et al. *Antimicrob. Agents Chemother.* **1996**, *40*, 2714–2720.
43. Tanaka, M.; Onodera, Y.; Uchida, Y.; Sato, K.; Hayakawa, I. *Antimicrob. Agents Chemother.* **1997**, *41*, 2362–2366.
44. (a) Pestova, E.; Millichap, J. J.; Noskin, G. A.; Peterson, L. R. *J. Antimicrob. Chemother.* **2000**, *45*, 583–590; (b) Fukuda, H.; Hiramatsu, K. *Antimicrob. Agents Chemother.* **1999**, *43*, 410–412.
45. (a) Albrecht, R. *Prog. Drug Res.* **1977**, *21*, 9–104; (b) Siporin, C.; Heifetz, C.; Domagala, J., Eds.; *The New Generation of Quinolones*; Marcel Dekker, Inc.: New York, 1990; (c) Wentland, M. P. Structure–Activity Relationships of Fluoroquinolones. In *The New Generation of Quinolones*; Siporin, C., Heifetz, C., Domagala, J., Eds.; Marcel Dekker, Inc.: New York, 1990, pp 1–43; (d) Chu, D. T. W.; Fernandes, P. B. *Adv. Drug Res.* **1991**, *21*, 39–144; (e) Mitscher, L. A.; Devasthale, P.; Zavod, R. Structure–Activity Relationships. In *Quinolone Antibacterial Agents*, 2nd ed.; Wolfson, J. S., Hooper, D. C., Eds.; American Society for Microbiology: Washington, DC, 1993, pp 3–52; (f) Wolfson, J. S.; Hooper, D. C., Eds.; *Quinolone Antibacterial Agents*, 2nd ed.; American Society for Microbiology: Washington, DC, 1993; (g) Domagala, J. M., *J. Antimicrob. Chemother.* **1994**, *33*, 685–706; (h) Andriole, V. T., Ed.; *The Quinolones*, 3rd ed.; Academic Press, Inc.: New York, 2000; (i) Mitscher, L. A.; Ma, Z. Structure–Activity Relationships of Quinolones. In *Fluoroquinolone Antibiotics*; Ronald, A. R., Low, D. E., Eds.; Birkhauser Verlag: Boston, MA, 2003, pp 11–48.
46. Gould, R. G., Jr.; Jacobs, W. A. *J. Am. Chem. Soc.* **1939**, *61*, 2890–2895.
47. (a) Grohe, K.; Heitzer, H. *Liebigs Ann. Chem.* **1987**, *10*, 29–37; (b) Grohe, K.; Heitzer, H. *Liebigs Ann. Chem.* **1987**, 871–879.
48. (a) MacDonald, A. A.; Dewitt, S. H.; Ramage, R. *Chimia* **1996**, *50*, 266–270; (b) Frank, K. E.; Jung, M.; Mitscher, L. A. *Comb. Chem. High Throughput Screen.* **1998**, *1*, 73–87; (c) Frank, K. E.; Devasthale, P. V.; Gentry, E. J.; Ravikumar, V. T.; Keschavarz-Shokri, A.; Mitscher, L. A.; Nilius, A.; Shen, L. L.; Shawar, R.; Baker, W. R. *Comb. Chem. High Throughput Screen.* **1998**, *1*, 89–99; (d) Srivastava, S. K.; Haq, W.; Murthy, P. K.; Chauhan, P. M. S. *Bioorg. Med. Chem. Lett.* **1999**, *9*, 1885–1888.
49. Domagala, J. M.; Heifetz, C. L.; Hutt, M. P.; Mich, T. F.; Nichols, J. B.; Solomon, M.; Worth, D. F. *J. Med. Chem.* **1988**, *31*, 991–1001.
50. Wise, R.; Andrews, J. M.; Edwards, L. J. *Antimicrob. Agents Chemother.* **1983**, *23*, 559.
51. Chu, D. T. W.; Fernandes, P. B.; Claiborne, A. K.; Pihuleac, E.; Nordeen, C. W.; Maleczka, R. E., Jr.; Pernet, A. G. *J. Med. Chem.* **1985**, *28*, 1558–1584.
52. Chu, D. T. W.; Fernandes, P. B.; Claiborne, A. K.; Gracey, E. H.; Pernet, A. G. *J. Med. Chem.* **1986**, *29*, 2363–2369.
53. Brouzard, D.; Di Cesare, P.; Essiz, M.; Jacquet, J. P.; Remuzon, P.; Weber, A.; Oki, T.; Masuyoshi, M. *J. Med. Chem.* **1989**, *32*, 537–542.

54. (a) Fung-Tomc, J.; Desiderio, J. V.; Tsai, Y. H.; Warr, G.; Kessler, R. E. *Antimicrob. Agents Chemother.* **1989**, *33*, 906–914; (b) Bouzard, D.; Di Cesare, P.; Essiz, M.; Jacquet, J. P.; Kiechel, J. R.; Remuzon, P.; Weber, A.; Oki, T.; Masuyoshi, M.; Kessler, R. E. *J. Med. Chem.* **1990**, *33*, 1344–1352.
55. Mitscher, L. A.; Sharma, P. N.; Chu, D. T. W.; Shen, L. L.; Pernet, A. G. *J. Med. Chem.* **1987**, *30*, 2283–2286.
56. Hayakawa, I.; Atarashi, S.; Yokohama, S.; Imamura, M.; Sakano, K.; Furukawa, M. *Antimicrob. Agents Chemother.* **1985**, *29*, 163.
57. Mitscher, L. A.; Sharma, P. N.; Chu, D. T. W.; Shen, L. L.; Pernet, A. G. *J. Med. Chem.* **1987**, *30*, 2283–2286.
58. Hoshino, K.; Sato, K.; Akahane, K.; Yoshida, A.; Hayakawa, I.; Sato, M.; Une, T.; Osada, Y. *Antimicrob. Agents Chemother.* **1991**, *35*, 309–312.
59. Wentland, M. P.; Perni, R. B.; Dorff, P. H.; Rake, J. B. *J. Med. Chem.* **1988**, *31*, 1694–1697.
60. Chu, D. T. W.; Lico, I. M.; Claiborne, A. K.; Plattner, J. J.; Pernet, A. G. *Drugs Exptl. Clin. Res.* **1990**, *16*, 215–224.
61. Segawa, J.; Kitano, M.; Kazuno, K.; Matsuoka, M.; Shirahase, I.; Ozaki, M.; Matsuda, M.; Tomii, Y.; Kise, M. *J. Med. Chem.* **1992**, *35*, 4727–4738.
62. Chu, D. T. W. *Drugs Future* **1992**, *17*, 1101–1109.
63. Hubschwerlen, C.; Pflieger, P.; Specklin, J.-L.; Gubernator, K.; Gmünder, H.; Angehrn, P.; Kompis, I. *J. Med. Chem.* **1992**, *35*, 1385–1392.
64. Li, Q.; Chu, D. T. W.; Claiborne, A.; Cooper, C. S.; Lee, C. M.; Raye, K.; Berst, K. B.; Donner, P.; Wang, W.; Hasvold, L. et al. *J. Med. Chem.* **1996**, *39*, 3070–3088.
65. (a) Wentland, M. P.; Lesher, G. Y.; Reuman, M.; Gruett, M. D.; Singh, B.; Aldous, S. C.; Dorff, P. H.; Rake, J. B.; Coughlin, S. A. *J. Med. Chem.* **1993**, *36*, 2801–2809; (b) Eissenstat, M. A.; Kuo, G.-H.; Weaver, J. D., III; Wentland, M. P.; Robinson, R. G.; Klingbeil, K.; Danz, D. W.; Corbett, T. H.; Coughlin, S. A. *Bioorgan. Med. Chem. Lett.* **1995**, *5*, 1021–1026; (c) Wentland, M. P.; Aldous, S. C.; Gruett, M. D.; Perni, R. B.; Powles, R. G.; Danz, D. W.; Klingbeil, K. M.; Peverly, A. D.; Robinson, R. G.; Corbett, T. H. et al. *Bioorgan. Med. Chem. Lett.* **1995**, *5*, 405–410.
66. Domagala, J. M.; Bridges, A. J.; Culbertson, T. P.; Gambino, L.; Hagen, S. E.; Karrick, G.; Porter, K.; Sanchez, J. P.; Sesnie, J. A.; Spense, F. et al. *J. Med. Chem.* **1991**, *34*, 1142–1154.
67. Gootz, T. D.; Brighty, K. E. *Med. Res. Rev.* **1996**, *16*, 433–486.
68. Stilwell, G.; Holmes, K.; Turck, M. *Antimicrob. Agents Chemother.* **1975**, *7*, 483–485.
69. Koga, H.; Itoh, A.; Murayama, S.; Suzue, S.; Irikura, T. *J. Med. Chem.* **1980**, *23*, 1358–1363.
70. Domagala, J. M.; Hanna, L. D.; Heifetz, C. L.; Hutt, M. P.; Mich, T. F.; Sanchez, J. P.; Solomon, M. *J. Med. Chem.* **1986**, *29*, 394–404.
71. Gray, J. L.; Almstead, J.-I. K.; Gallagher, C. P.; Hu, X. E.; Kim, N. K.; Taylor, C. J.; Twinem, T. L.; Wallace, C. D.; Ledoussal, B. *Bioorgan. Med. Chem.* **2003**, *11*, 2373–2375.
72. Cecchetti, V.; Filipponi, E.; Fravolini, A.; Tabarrini, O.; Bonelli, D.; Clementi, M.; Cruciani, G.; Clemeti, S. *J. Med. Chem.* **1997**, *40*, 1698–1706.
73. Ledoussal, B.; Bouzard, D.; Coroneos, E. *J. Med. Chem.* **1992**, *35*, 198–200.
74. Ledoussal, B.; Almstead, J.-I. K.; Gray, J. L.; Hu, E. X.; Roychoudhury, S. *Curr. Med. Chem.: Anti-Infect. Agents* **2003**, *2*, 13–25.
75. Roychoudhury, S.; Catrenich, C. E.; McIntosh, E. J.; McKeever, H. D.; Makin, K. M.; Koenigs, P. M.; Ledoussal, B. *Antimicrob. Agents Chemother.* **2001**, *45*, 1115–1120.
76. Lawrence, L. E.; Wu, P.; Fan, L.; Gouveia, K. E.; Card, A.; Casperson, M.; Denbleyker, K.; Barrett, J. F. *J. Antimicrob. Chemother.* **2001**, *48*, 195–201.
77. Mandell, W.; Neu, H. C. *Antimicrob. Agents Chemother.* **1986**, *29*, 852–857.
78. Sanchez, J. P.; Domagala, J. M.; Hagen, S. E.; Heifetz, C. L.; Hutt, M. P.; Nichols, J. B.; Trehan, A. K. *J. Med. Chem.* **1988**, *31*, 983–991.
79. Ulbertson, T. P. C.; Domagala, J. M.; Nichols, J. B.; Priebe, S.; Skeean, R. W. *J. Med. Chem.* **1987**, *30*, 1711–1715.
80. Chu, D. T. W.; Nordeen, C. W.; Hardy, D. J.; Swanson, R. N.; Giardina, W. J.; Pernet, A. G.; Plattner, J. J. *J. Med. Chem.* **1991**, *34*, 168–174.
81. (a) Rosen, T.; Chu, D. T. W.; Lico, I.; Fernandes, P. B.; Marsh, K.; Shen, L.; Cepa, V. G.; Pernet, A. G. *J. Med. Chem.* **1988**, *31*, 1598–1611; (b) Rosen, T.; Chu, D. T. W.; Lico, I; Fernandes, P. B.; Shen, L.; Borodkin, S.; Pernet, A. G. *J. Med. Chem.* **1988**, *31*, 1586–1590.
82. Domagala, J. M.; Hagen, S. E.; Joannides, T.; Kiely, J. S.; Laborde, E.; Schroeder, M. C.; Sesnie, J. A.; Shapiro, M. A.; Suto, M. J.; Vanderroest, S. *J. Med. Chem.* **1993**, *36*, 871–882.
83. Reuman, M.; Daum, S.; Singh, B.; Wentland, M. P.; Perni, R. B.; Pennock, P.; Carabateas, P. M.; Gruett, M. D.; Saindane, M. T.; Dorff, P. H. *J. Med. Chem.* **1995**, *38*, 2531–2540.
84. Robinson, M. J.; Martin, B. A.; Gootz, T. D.; McGuirk, P. R.; Moynihan, M.; Sutcliffe, J. A.; Osheroff, N. *J. Biol. Chem.* **1991**, *266*, 14585–14592.
85. (a) Sato, K.; Hoshino, K.; Tanaka, M.; Hayakawa, I.; Osada, Y.; Nishino, T.; Mitsuhashi, S. Antibacterial Activity of New Quinolones and Their Modes of Action. In *Molecular Biology of DNA Topoisomerases and Its Application to Chemotherapy*; Andow, T., Ikeda, H., Oguro, M., Eds.; CRC Press, Inc.: Boca Raton, FL, 1993, pp 167–176; (b) Sato, K.; Hoshino, K.; Tanaka, M.; Hayakawa, I.; Osada, Y. *J. Antimicrob. Chemother.* **1992**, *36*, 1491–1498.
87. Blondeau, J. M.; Hansen, G. T. *Exp. Opin. Pharmacother.* **2001**, *2*, 317–335.
88. Perry, C. M.; Ormrod, D.; Hurst, M.; Onrust, S. V. *Drugs* **2002**, *62*, 169–207.
89. Cecchetti, V.; Fravolini, A.; Palumbo, M.; Sissi, C.; Tabarrini, O.; Terni, P.; Xin, T. *J. Med. Chem.* **1996**, *39*, 4952–4957.
90. Kishii, R.; Takei, M.; Fukuda, H.; Hayashi, K.; Hosaka, M. *Antimicrob. Agents Chemother.* **2003**, *47*, 77–81.
91. Sanchez, J. P.; Domagala, J. M.; Hagen, S. E.; Heifetz, C. L.; Hutt, M. P.; Nichols, J. B.; Trehan, A. K. *J. Med. Chem.* **1988**, *31*, 983–991.
92. Brighty, K. E.; Gootz, T. D. *J. Antimicrob. Chemother.* **1997**, *39*, 1–14.
93. Li, Q.; Mitscher, L. L; Shen, L. L. *Med. Res. Rev.* **2000**, *20*, 231–293.
94. (a) Piddock, L. J. V. *Drugs* **1999**, *58*, 1118; (b) Hooper, D. C. *Drug Resist. Updates* **1999**, *2*, 3855; (c) Piddock, L. J. V. *Drugs* **1995**, *49*, 2935; (d) Hooper, D. C.; Wolfson, J. S. Mechanisms of Bacterial Resistance to Quinolones. In *Quinolone Antimicrobial Agents*, 2nd ed.; Hooper, D. C., Wolfson, J. S., Eds.; American Society for Microbiology: Washington, DC, 1993, pp 97–118.
95. Jones, M. E.; Sahm, D. F.; Martin, N.; Scheuring, S.; Heisig, P.; Thornsberry, C.; Kohrer, K.; Schmitz, F.-J. *Antimicrob. Agents Chemother.* **2000**, *44*, 462–466.
96. Broskey, J.; Coleman, K.; Gwynn, M. N.; McCloskey, L.; Traini, C.; Voelker, L.; Warren, R. *J. Antimicrob. Chemother.* **2000**, *45*, 95–99.
97. (a) Janoir, C.; Zeller, V.; Kitzis, M.-D.; Moreau, N. J.; Gutmann, L. *Antimicrob. Agents Chemother.* **1996**, *40*, 2760–2764; (b) Pestova, E.; Beyer, R; Cianciotto, N. P.; Noskin, G. A.; Peterson, L. R. *Antimicrob. Agents Chemother.* **1999**, *43*, 2000–2004.
98. Varon, E.; Janoir, C.; Kitzis, M.-D.; Gutmann, L. *Antimicrob. Agemts Chemother.* **1999**, *43*, 302–306.
99. (a) Kaatz, G. W. *Frontiers Antimicrob. Resist.* **2005**, 275–285; (b) Li, X.-Z.; Nikaido, H. *Drugs* **2004**, *64*, 159–204.
100. (a) Nikaido, H. *Frontiers Antimicrob. Resist.* 2005, 261–274; (b) Poole, K. *Antimicrob. Agents Chemother.* **2000**, *44*, 2233–2241.
101. Zhanel, G. G.; Hoban, D. J.; Schurek, K.; Karlowsky, J. A. *Int. J. Antimicrob. Agents* **2004**, *24*, 529–535.
102. (a) Brenwald, N. P.; Gill, M. J.; Wise, R. *Antimicrob. Agents Chemother.* **1998**, *42*, 2032–2035; (b) Zeller, V.; Janoir, C.; Kitzis, M.-D.; Gutmann, L.; Moreau, N. J. *Antimicrob. Agents Chemother.* **1997**, *41*, 1973–1978; (c) Takenouchi, T.; Tabata, F.; Iwata, Y.; Hanzawa, H.; Sugawara, M.; Ohya, S. *Antimicrob. Agents Chemother.* **1996**, *40*, 1835–1842.

103. Beyer, R.; Pestova, E.; Millichap, J. J.; Stosor, V.; Noskin, G. A.; Peterson, L. R. *Antimicrob. Agents Chemother.* **2000**, *44*, 798–801.
104. Poole, K. *Curr. Pharm. Biotech.* **2002**, *3*, 77–98.
105. Li, X.-Z. *Int. J. Antimicrob. Agents* **2005**, *25*, 453–463.
106. Wetzstein, H.-G.; Schmeer, N.; Karl, W. *Appl. Environ. Microb.* **1997**, *63*, 4272–4281.
107. (a) Hawkey, P. M. *J. Antimicrob. Chemother.* **2003**, *51*, 2935; (b) Dougherty, T. J.; Beaulieu, D.; Barrett, J. F. *Drug Disc. Today* **2001**, *6*.
108. (a) Cirz, R. T.; Chin, J. K.; Andes, D. R.; de Crecy-Lagard, V.; Craig, W. A.; Romesberg, F. E. *PLoS Biology* **2005**, *3*, 1024–1033; (b) Cirz, R. T.; Chin, J. K.; Andes, D. R.; de Crecy-Lagard, V.; Craig, W. A. *PLoS Biology* **2005**, *3*, 936.
109. Blondeau, J. M.; Zhao, X.; Hansen, G.; Drlica, K. *Antimicrob. Agents Chemother.* **2001**, *45*, 433–438.
110. Hermsen, E. D.; Hovde, L. B.; Konstantinides, G. N.; Rotschafer, J. C. *Antimicrob. Agents Chemother.* **2005**, *49*, 1633–1635.
111. (a) Hansen, G. T.; Blondeau, J. M. *Therapy* **2005**, *2*, 6166; (b) Blondeau, J. M.; Hansen, G.; Metzler, K.; Hedlin, P. *J. Chemother.* **2004**, *16*, 1–19.
112. Forrest, A.; Nix, D. E.; Ballow, C. H.; Goss, T. F.; Birmingham, M. C.; Schentag, J. J. *Antimicrob. Agents Chemother.* **1993**, *37*, 1073–1081.
113. (a) Wispelwey, B. *Clinical Infect. Dis.* **2005**, *41*, S127–S135; (b) Zhanel, G. G.; Noreddin, A. M. *Fluoroquinolone Antibiotics* **2003**, 87–105; (c) Lode, H.; Allewelt, M. *J. Antimicrob. Chemother.* **2002**, *49*, 709–712; (d) Zhanel, G. G.; Ennis, K.; Vercaigne, L.; Walkty, A.; Gin, A. S.; Embil, J.; Smith, H.; Hoban, D. J. *Drugs* **2002**, *62*, 13–59.
114. (a) Kuhnke, A.; Lode, H. *Eur Respir. Monogr.* **2004**, *9*, 94112; (b) Liu, H. H.; Liu, H. H. *Curr. Therap. Res.* **2004**, *65*, 225–238; (c) Leibovitz, E.; Dagan, R. *Fluoroquinolone Antibiotics* **2003**, 167–176; (d) Low, D. E. *Fluoroquinolone Antibiotics* **2003**, 189–203; (e) Marrie, T. J. *Fluoroquinolone Antibiotics* 2003, 177–188; (f) Blondeau, J. M. *Expert Opin. Investig. Drugs* **2001**, *10*, 213–237.
115. (a) Carson, C.; Naber, K. G. *Drugs* **2004**, *64*, 1359–1373; (b) Johnson, J. R. Fluoroquinolones in Urinary Tract Infection. In *Fluoroquinolone Antibiotics*; Ronald, A. R., Low, D. E., Eds.; Birkhauser Verlag: Boston, 2003, pp 107–119.
116. Kowalski, R. P.; Dhaliwal, D. K. *Exp. Rev. Anti Infect. Ther.* **2005**, *3*, 131–139.
117. Chow, A. W. *Fluoroquinolone Antibiotics* **2003**, 137–166.
118. Martin, S. J.; Zeigler, D. G. *Exp. Opin. Pharmacother.* **2004**, *5*, 237–246.
119. Jacobs, M. R. *Curr. Pharm. Des.* **2004**, *10*, 3213–3220.
120. Al-Omar, M. A. *Profiles Drug Substances, Excipients, Related Methodology* **2004**, *31*, 209–214.
121. Keam, S. J.; Croom, K. F.; Keating, G. M. *Drugs* **2005**, *65*, 695–724.
122. (a) Bhavnani, S. M.; Andes, D. R. *Pharmacotherapy* **2005**, *25*, 717–740; (b) Blondeau, J. M.; Missaghi, B. *Exp. Opin. Pharmacother.* **2004**, *5*, 1117–1152; (c) File, T. M., Jr.; Tillotson, G. S. *Exp. Rev. Anti Infect. Ther.* **2004**, *2*, 831–843; (d) Yoo, B. K.; Triller, D. M.; Yong, C.-S.; Lodise, T. P. *Ann. Pharmacother.* **2004**, *38*, 1226–1235.
123. Blasi, F. J. *Chemother* **2004**, *16*, 11–14.
124. (a) Miravitlles, M. *Exp. Opin. Pharmacother* **2005**, *6*, 283–293; (b) Keating, G. M.; Scott, L. J. *Drugs* **2004**, *64*, 2347–2377; (c) Blondeau, J. M.; Hansen, G. T. *Exp. Opin. Pharmacother.* **2001**, *2*, 317–335; (d) Talan, D. A. *Clin. Infect. Dis.* **2001**, *32*, S64–S71.
125. Radandt, J. M.; Marchbanks, C. R.; Dudley, M. N. *Clin. Infect. Dis.* **1992**, *14*, 272–284.
126. (a) Lode, H.; Hoeffken, G.; Boeckk, M.; Deppermann, N.; Borner, K.; Koeppe, P. *J. Antimicrob. Chemother.* **1990**, *26*, 41–49; (b) Outman, W. R.; Nightingale, C. H. *Am. J. Med.* **1989**, *87*, 3S7–42S; (c) Soergel, F. *Rev. Infect. Dis.* **1989**, *11*, S1119–S1129.
127. (a) Yamaguchi, H.; Yano, I.; Saito, H.; Inui, K. *Eur. J. Pharmacol.* **2001**, *431*, 297–303; (b) Cormet, E.; Huneau, J. F.; Bouras, M.; Carbon, C.; Rubinstein, E.; Tome, D. *J. Pharm. Sci.* **1997**, *86*, 33–36.
128. Hyatt, J. M.; McKinnon, P. S.; Zimmer, G. S.; Schentag, J. J. *Clin. Pharmacokinetics* **1995**, *28*, 143–160.
129. Zhanel, G. G.; Noreddin, A. M. *Cur. Opin. Pharmacol.* **2001**, *1*, 459–463.
130. FDC Reports. *The Pink Sheet* **1999**, *61*, 11.
131. Kim, O. K.; Barrett, J. F.; Ohemeng, K. *Exp. Opin. Invest. Drugs* **2001**, *10*, 199–212.
132. Schmitz, F.-J.; Boos, M.; Mayer, S.; Hafner, D.; Jagusch, H.; Verhoef, J.; Fluit, A. *Antimicrob. Agents Chemother.* **2001**, *45*, 2666–2667.
133. (a) Owens, R. C., Jr.; Ambrose, P. G. *Clin. Infect. Dis.* **2005**, *41*, S144–S157; (b) Mandell, L.A. *Fluoroquinolone Antibiotics* **2003**, 73–85.
134. Iannini, P. B.; Iannini, P. B. *Expert Opin. Drug Saf.* **2002**, *1*, 121–128.
135. Cuzzolin, L.; Fanos, V. *Expert. Opin. Drug Saf.* **2002**, *1*, 319–324.
136. (a) Melhus, A. *Expert. Opin. Drug Saf.* **2005**, *4*, 299–309; (b) Khaliq, Y.; Zhanel, G. G. *Clin. Infect. Dis.* **2003**, *36*, 1404–1410.
137. Ball, P.; Mandell, L.; Patou, G.; Dankner, W.; Tillotson, G. *Int. J. Antimicrob. Agents* **2004**, *23*, 421–429.
138. Ball, P.; Stahlmann, R.; Kubin, R.; Choudhri, S.; Owens, R. *Clin. Ther.* **2004**, *26*, 940–950.
139. De Sarro, A.; De Sarro, G. *Curr. Med. Chem.* **2001**, *8*, 371–384.
140. Domagala, J. M.; Domagala, J. M. *J. Antimicrob. Chemother.* **1994**, *33*, 685–706.
141. Schering-Plough Corp Schering licenses Toyama's garenoxacin antibiotic. Press Release, June 22, 2004.
142. (a) Zhao, X.; Eisner, W.; Perl-Rosenthal, N.; Kreiswirth, B.; Drlica, K. *Antimicrob. Agents Chemother.* **2003**, *47*, 1023–1027; (b) Bronson, J. J.; Barrett, J. F. *Curr. Med. Chem.* **2001**, *8*, 1775–1793; (c) Frechette, R.; Frechette, R. *Curr. Opin. Invest. Drugs* **2001**, *2*, 1706–1711.
143. Mitsubishi Pharma Corp. Japan's first domestically originated new quinolone antibacterial agent, Press Release, September 2, 2002
144. Johnson, A. P. *Curr. Opin. Investig. Drugs* **2000**, *1*, 52–57.
145. Fleck, T. J.; McWhorter, W. W., Jr.; DeKam, R. N.; Pearlman, B. A. *J. Org. Chem.* **2003**, *68*, 9612–9617.
146. Ince, D.; Hooper, D. C. *Antimicrob. Agents Chemother.* **2000**, *44*, 3344–3350.
147. Keam, S. J.; Perry, C. M. *Drugs* **2004**, *64*, 2221–2234.
148. Optimer Pharmaceuticals. Product pipeline. Optimer Pharmaceuticals Inc. Company Website (www.optimerpharma.com, accessed on November 3, 2004 and October 31, 2005).
149. World-class people can find world-class drugs: Mr Morita of Daiichi. *Pharma Japan* **2003**, February 17, *1832*, 1; (a) Hoffman-Roberts, H. L.; Babcock, E. C.; Mitropoulos, I. F. *Expert Opin. Invest. Drugs* **2005**, *14*, 973–995.
150. Bogdanovich, T.; Esel, D.; Kelly, L. M.; Bozdogan, B.; Credito, K.; Lin, G.; Smith, K.; Ednie, L. M.; Hoellman, D. B.; Appelbaum, P. C. *Antimicrob. Agents Chemother.* **2005**, *49*, 3325–3333.
151. Tachibana, M.; Tanaka, M.; Mitsugi, K.; Jin, Y.; Takaichi, M.; Okazaki, O. *Arzneim-Forsch* **2004**, *54*, 898–905.
152. Oleksijew, A.; Meulbroek, J.; Ewing, P.; Jarvis, K.; Mitten, M.; Paige, L.; Tovcimak, A.; Nukkula, M.; Chu, D.; Alder, J. D. *Antimicrob. Agents Chemother.* **1998**, *42*, 2674–2677.
153. Saiki, A. Y. C.; Shen, L. L.; Chen, C.-M.; Baranowski, J.; Lerner, C. G. *Antimicrob. Agents Chemother.* **1999**, *43*, 1574–1577.
154. Oleksijew, A.; Meulbroek, J.; Ewing, P.; Jarvis, K.; Mitten, M.; Paige, L.; Tovcimak, A.; Nukkula, M.; Chu, D.; Alder, J. D. *Antimicrob. Agents Chemother.* **1998**, *42*, 2674–2677.

155. Jung, Y.H.; Lee, K.S.; Bhang, W.Y.; Hwang, M.Y.; Yeon, K.J. Kim, J.W.; Yoon, Y.H.; Lee, K.H.; Park, K. H. *Pharmacokinetic Study of CFC-222, a New Potent Quinolone, Using Radiolabelled Compound*, 35th Interscience Conference on Antimicrobial Agents and Chemotherapy, San Francisco, 1995; Abs. F201.
156. Masuda, N.; Takahashi, Y.; Otsuki, M.; Ibuki, E.; Miyoshi, H.; Nishino, T. *Antimicrob. Agents Chemother.* **1996**, *40*, 1201–1207.
157. Sankyo Co Ltd. R&D pipeline Company website (www.sankyo.co.jp/english/, accessed May 31, 2002).
158. Pfizer Inc. Pfizer to discontinue development of anti-infective compound DK-507k, Press Release, February 19, 2004.
159. Ubukata, K.; Hasegawa, K.; Morozumi, M.; Murayama, S. *Antibacterial Activity of DK-507k, a Novel Fluoroquinolone, against Clinical Isolates in Pediatric Infections*, 43rd Interscience Conference on Antimicrobial Agents and Chemotherapy, Chicago, 2003; Abstr. F-426.
160. (a) Inagaki, H.; Miyauchi, S.; Miyauchi, R. N.; Kawato, H. C.; Ohki, H.; Matsuhashi, N.; Kawakami, K.; Takahashi, H.; Takemura, M. *J. Med. Chem.* **2003**, *46*, 1005–1015; (b) Yanagihara, K.; Otsu, Y.; Kaneko, Y.; Miyazaki, Y.; Hirakata, Y.; Tomono, K.; Tashiro, T.; Kohno, S. *Efficacy of DQ-113 on Penicillin-Resistant* Streptococcus pneumoniae (PRSP) *in a Mouse Model of Pneumonia*, 42nd Interscience Conference on Antimicrobial Agents and Chemotherapy, San Diego, 2002; Abs. F576; (c) Kaneko, Y.; Yanagihara, K.; Imamura, Y.; Miyazaki, Y.; Hirakata, Y.; Tomono, K.; Tashiro, T.; Kohno, S. *Efficacy of DQ-113 on MRSA and Vancomycin-Insensitive* Staphylococcus aureus (VISA) *Hematogenous Pulmonary Infection Model*, 42nd Interscience Conference on Antimicrobial Agents and Chemotherapy, San Diego, 2002; Abs. F577.
161. Tanaka, M.; Tamura, K.; Atarashi, S.; Kubo, Y.; Oliver, S. D.; Bentley, M.; Hakusui, H. *Xenobiotica* **1995**, *25*, 1119–1125.
162. Kawakami, K.; Atarashi, S.; Kimura, S.; Takemura, M.; Hayakawa, I. *Chem. Pharm. Bull.* **1998**, *46*, 1710–1715.
163. (a) Shimoda, K.; Akahane, K.; Nomura, M.; Kato, M. *Arzneim.-Forsch* **1996**, *46*, 625–628; (b) Sugawara, T.; Yoshida, M.; Shimoda, K.; Takada, S.; Miyamoto, M.; Nomura, M.; Kato, M. *Arzneim-Forsch* **1996**, *46*, 705–710; (c) Yoshida, K.; Matsubayashi, K.; Sekiguchi, M.; Hayakawa, I. *Chemotherapy* **1997**, *43*, 332–339.
164. Dong Wha Pharmaceutical Industry Co. Ltd., Company website (http://www.dong-wha.co.kr/english/, accessed on April 13, 2000).
165. Choi, K.-H.; Hong, J.-S.; Kim, S.-K.; Lee, D.-K.; Yoon, S.-J.; Choi, E.-C. *J. Antimicrob. Chemother.* **1997**, *39*, 509–514.
166. Lee, D.-K. *Drugs* **1995**, *49*, 323–325.
167. (a) Kim, J.-C.; Shin, D.-H.; Kim, S.-H.; Oh, K.-S.; Jung, Y.-H.; Kwon, H.-J.; Yun, H.-I.; Shin, H.-C.; Chung, M.-K. *Food Chem. Toxicol.* **2004**, *42*, 389–395; (b) Kim, J.-C.; Shin, D.-H.; Ahn, T.-H.; Kang, S.-S.; Song, S.-W.; Han, J.; Kim, C.-Y.; Ha, C.-S.; Chung, M.-K. *Food Chem. Toxicol.* **2003**, *41*, 637–645; (c) Kim, J.-C.; Yun, H.-I.; Shin, H.-C.; Hur, J.-D.; Lee, J.-H.; Chung, M.-K. *J. Appl. Pharmacol.* **2002**, *10*, 43–49.
168. Yun, H.-J.; Min, Y.-H.; Jo, Y.-W.; Shim, M-J.; Choi, E.-C. *Int. J. Antimicrob. Agents* **2005**, *25*, 334–337.
169. Yun, H.-J.; Min, Y.-H.; Lim, J.-A.; Kang, J.-W.; Kim, S.-Y.; Kim, M.-J.; Jeong, J.-H.; Choi, Y.-J.; Kwon, H.-J.; Jung, Y.-H. et al. *Antimicrob. Agents Chemother.* **2002**, *46*, 3071–3074.
170. Market information. *Pharma Koreana* **2005**, *15* (2).
171. White M. D. Abbott Laboratories. Drug development pipeline Merrill Lynch Global Healthcare Conference, February 6–8, 2001, Abbott Laboratories Company Website (www.abbott.com, accessed on February 8, 2001).
172. (a) Higa, F.; Arakaki, N.; Tateyama, M.; Koide, M.; Shinzato, T.; Kawakami, K.; Saito, A. *J. Antimicrob. Chemother.* **2003**, *52*, 920–924; (b) Yoshizumi, S.; Takahashi, Y.; Murata, M.; Domon, H.; Furuya, N.; Ishii, Y.; Matsumoto, T.; Ohno, A.; Tateda, K.; Miyazaki, S. et al. *J. Antimicrob. Chemother.* **2001**, *48*, 137–140.
173. TaiGen Biotechnology Co. Ltd. TaiGen begins Phase Ib trial of quinolone antibiotic, Press Release, June 18, 2005.
174. (a) Odenholt, I.; Loewdin, E.; Cars, O. *Antimicrob. Agents Chemother.* **2003**, *47*, 3352–3356; (b) Jones, M. E.; Critchley, I. A.; Karlowsky, J. A.; Blosser-Middleton, R. S.; Schmitz, F.-J.; Thornsberry, C.; Sahm, D. F. *Antimicrob. Agents Chemother.* **2002**, *46*, 1651–1657.
175. Hu, X. E.; Kim, N. K.; Gray, J. L.; Almstead, J.-I. K.; Seibel, W. L.; Ledoussal, B. *J. Med. Chem.* **2003**, *46*, 3655–3661.
176. (a) Yamakawa, T.; Mitsuyama, J.; Hayashi, K. *J. Antimicrob. Chemother.* **2002**, *49*, 455–465; (b) Toyama Chemical Co. Ltd. Synthetic antibacterial agent T-3912 licensed out in Japan and overseas, Toyama Chemical Co. Ltd., Press Release, November 6, 2001.
177. Abbott Laboratories. Credit Suisse First Boston Health Care Conference, Abbott Laboratories, Company Presentation, November 14, 2001.
178. (a) Sillerstroem, E.; Wahlund, E.; Nord, C. E. *J. Chemother.* **2004**, *16*, 227–229; (b) Almer, L. S.; Hoffrage, J. B.; Keller, E. L.; Flamm, R. K.; Shortridge, V. D. *Antimicrob. Agents Chemother.* **2004**, *48*, 2771–2777; (c) Hammerschlag, M. R.; Roblin, P. M. *J. Antimicrob. Chemother.* **2004**, *54*, 281–282.
179. Mealy, N. E.; Castaner, J. *Drugs Future* **2002**, *27*, 1033–1038.
180. (a) Kitani, H.; Kuroda, T.; Moriguchi, A.; Hikida, K.; Ao, H.; Yokoyama, Y.; Hirayama, F.; Ikeda, Y. *Novel 7-Substituted-Fluoroquinolones as Potent Antibacterial Agents: Synthesis and Structure-Activity Relationships*, 35th Interscience Conference on Antimicrobial Agents and Chemotherapy, San Francisco, 1995; Abs. F190; (b) Kitani, H.; Kuroda, T.; Moriguchi, A.; Ao, H.; Hirayama, F.; Ikeda, Y.; Kawakita, T. *Bioorg. Med. Chem. Lett.* **1997**, 7, 515–520; (c) Macgowan, A. P.; Bowker, K. E.; Wootton, M.; Holt, H. A.; Reeves, D. S. *Antimicrob. Agents Chemother.* **1998**, *42*, 419–424.
181. Entenza, J. M.; Marchetti, O.; Glauser, M. P.; Moreillon, P. *Antimicrob. Agents Chemother.* **1998**, *42*, 1889–1894.
182. Satoh, O. Daiichi Pharmaceutical Co., Ltd., Japan. Updated Research Strategies of Daiichi to Improve Candidate Selection from Streamlined Processes, IBC Annual Asia-Pacific Congress – Drug Discovery and Technology, October 18–20, 2004.
183. Bogdanovich, T.; Esel, D.; Kelly, L. M.; Bozdogan, B.; Credito, K.; Lin, G.; Smith, K.; Ednie, L. M.; Hoellman, D. B.; Appelbaum, P. C. *Antimicrob. Agents Chemother.* **2005**, *49*, 3325–3333.
184. (a) Fujikawa, K.; Chiba, M.; Tanaka, M.; Sato, K. *Antimicrob. Agents Chemother.* **2005**, *49*, 3040–3045; (b) Nishino, T.; Otsuki, M. *In vitro Comparison of DX-619, a Novel Des-F(6)-Quinolone, with Other Quinolones, Linezolid and Vancomycin Tested against Recent Clinical Isolates*, 44th Interscience Conference on Antimicrobial Agents and Chemotherapy, Washington, DC, 2004; Abs. F-1936.
185. (a) Yanagihara, K.; Tashiro, M.; Okada, M.; Ohno, H.; Miyazaki, Y.; Hirakata, Y.; Tashiro, T.; Kohno, S. *Efficacy of DX-619 in MRSA and Vancomycin-Insensitive* Staphylococcus aureus *(VISA) Hematogenous Pulmonary Infection Model*, 43rd Interscience Conference on Antimicrobial Agents and Chemotherapy, Chicago, 2003; Abs. F-1057; (b) Kurosaka, Y.; Nishida, S.; Ishii, C.; Sawada, Y.; Namba, K.; Otani, T.; Sato, K. *DX-619, a Novel Des-F(6)-Quinolone: Pharmacodynamics (PD) Activity and Therapeutic Efficacy in Animal Infection Models*, 43rd Interscience Conference on Antimicrobial Agents and Chemotherapy, Chicago, 2003; Abs. F-1061; (c) Tsuchiya, Y.; Goto, K.; Igarashi, M.; Jindo, T.; Furuhama, K. *DX-619, a Novel Des-F(6)-Quinolone: Safety Evaluation in Preclinical Studies*, 43rd Interscience Conference on Antimicrobial Agents and Chemotherapy, Chicago, 2003; Abs. F-1062.
186. Williams, L.; Shackcloth, J.; Jung, Y.; Morrissey, I. *Activity of DW-224a and Comparators against Atypical Bacterial Respiratory Pathogens*, 44th Interscience Conference on Antimicrobial Agents and Chemotherapy, Washington, DC, 2004; Abs. F-1946.
187. Kwak, J.; Seol, M.; Kim, H.; Park, H.; Choi, D.; Jung, Y. *Bactericidal Activities of DW-224a, a New Fluoroquinolone*, 44th Interscience Conference on Antimicrobial Agents and Chemotherapy, Washington, DC, 2004; Abs. F-1947.
188. (a) Kim, E.-J.; Shin, W.-H.; Kim, K.-S.; Han, S.-S. *Drug Chem. Toxicol.* **2004**, *27*, 295–307; (b) Han, J.; Kim, J.-C.; Chung, M.-K.; Kim, B.; Choi, D.-R. *Biol. Pharma. Bull.* **2003**, *26*, 832–839.

Biographies

Allan S Wagman started his medicinal chemistry career in 1996 when he joined the fledgling Small Molecule Drug Discovery group at Chiron Corporation. He has held various positions in the group, leading teams in projects spanning kinase inhibitors, antivirals, and antibiotics. His work has led to several advanced lead compounds and development compounds with the most relevant being novel inhibitors of Gram-negative LpxC and new inhaled antibiotics to expand the TOBI (inhaled tobramycin) franchise. Previously, Allan worked on the total synthesis of ginkgolide B during a postdoctoral fellowship with Prof Michael T Crimmins at the University of North Carolina at Chapel Hill. He earned his PhD in the laboratory of Prof Stephen F Martin at the University of Texas at Austin in 1995. Allan currently works at the Emeryville, CA site of the Novartis Institutes for BioMedical Research, Inc.

Mark P Wentland began his career in drug discovery in 1970 when he joined the medicinal chemistry department at Sterling Winthrop Inc. He earned his PhD in synthetic organic chemistry in 1970 at Rice University. During his 24 years at Sterling Winthrop, he held various positions of scientific and administrative responsibility with his last positions being Sterling Winthrop Fellow and Oncology Discovery Co-Chair. In 1994, he joined the chemistry faculty at Rensselaer Polytechnic Institute in Troy, NY. At Rensselaer, he maintains a federally funded research program in medicinal chemistry aimed at identifying novel, long-acting oral agents to treat cocaine and heroin addiction in humans. He regularly lectures at pharmaceutical and biotech companies on applications of physicochemical properties of organic molecules to lead identification and development. During his career, he has made significant contributions to the discovery of nine drug candidates, six of which have been advanced to clinical trials. His homepage can be found at: http://www.rpi.edu/~wentmp/.

Comprehensive Medicinal Chemistry II
ISBN (set): 0-08-044513-6

ISBN (Volume 7) 0-08-044520-9; pp. 567–596

7.20 The Antibiotic and Nonantibiotic Tetracyclines

M L Nelson and M Y Ismail, Paratek Pharmaceuticals, Inc., Boston, MA, USA

7.20.1 Introduction and the Tetracycline Record

The tetracyclines are a family of polyketide natural products or their semisynthetic derivatives based upon the naphthacene tetracycline ring system (**Figure 1**) which have been used widely for over 50 years in human healthcare, veterinary medicine, and agriculture. The scientific literature on the tetracyclines began in 1948 with a publication by Duggar describing the fermentation, isolation, and preliminary chemical characteristics of chlortetracycline **1**, tradenamed Aureomycin, which was the first antibiotic to have extraordinary activity against a broad spectrum of bacteria.[1]

Other tetracyclines were soon discovered and the number of citations and scientific front grew quickly (**Figure 2**), covering the antibacterial characteristics of newer compounds and analogs and their usefulness as antibiotics. By the

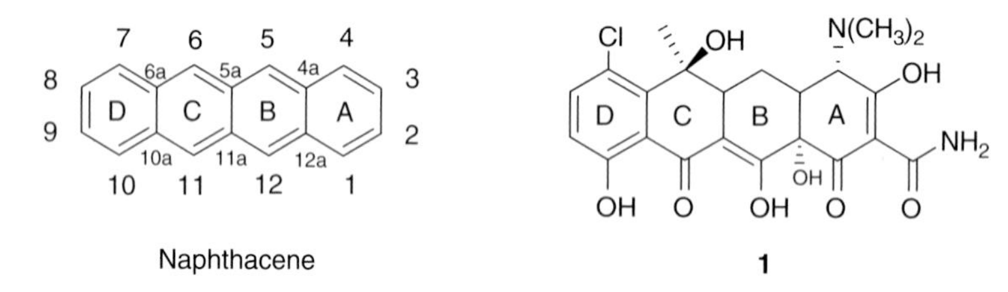

Figure 1 The chemical of structure of the naphthacene ABCD ring system and structural locants of chlortetracycline (**1**), the first natural product tetracycline discovered.

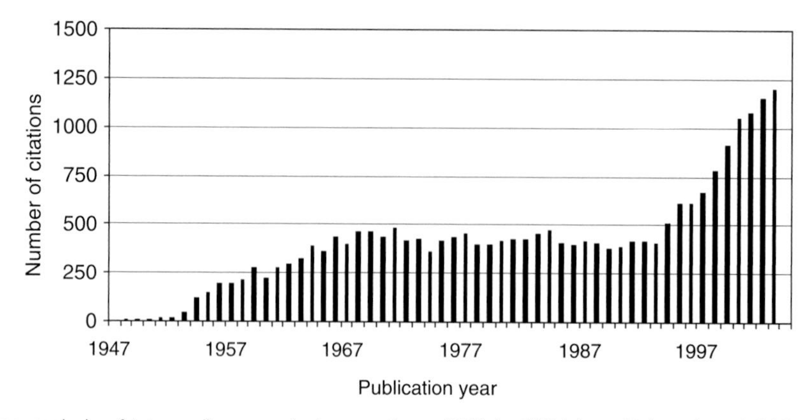

Figure 2 Citation analysis of tetracycline records by year from 1947 to 2004 from Pubmed and CAS scientific literature databases.

mid-1960s, several hundred papers per year were published concerning the tetracyclines, describing the synthesis of newer analogs, their use in chemotherapy, and the rise of antibiotic resistance in antibacterial chemotherapy. The field had reached a plateau in these areas that lasted over 25 years.

In the mid-1980s, newer uses and nonantibiotic applications of the tetracyclines related to inflammation and their effect modulating mammalian disease states began to be reported and have added to the growth of the tetracycline scientific front. Close inspection and citation analysis shows scientific clustering of areas of interest to medicinal chemistry in microbiology, nonantibiotic uses related to inflammation, neurodegeneration, and other interrelated diseases, and in genetic engineering, where tetracycline-regulated gene expression systems are generating new tools for molecular biology and for the study of the function of biological targets.

Given the history and the promising future of the tetracyclines in chemistry, biology, and medicine, the chemical and biological properties and structure–activity relationships (SARs) related to these topics will be covered and salient features of each scientific front described.

7.20.2 Tetracycline Generations and Origins

The tetracyclines derived as first-generation natural products (**Figure 3**) are chlortetracycline **1**, oxytetracycline **2**, and tetracycline **3**, while the semisynthetic second-generation derivatives minocycline **4**, doxycycline **5**, and demeclocycline **6** (a natural product derived from strain mutation) have been studied and used widely. Only one third-generation tetracycline in clinical use has emerged in the last 30 years, tigecycline **7** (Tygacil),[2] while more recent novel tetracyclines such as PTK 0796[3] **8** have reported favorable preclinical bioactivity.

Chlortetracycline **1** was discovered through extensive screening of soil samples bioprospecting for antibiotic producing organisms, and is the product of aerobic fermentation from the soil actinomycete *Streptomyces aureofaciens* A377 (NRRL 2209).[4] The next compound described was oxytetracycline **2**,[5] designated Terramycin, isolated from *Streptomyces rimosus* S-3279 (NRRL 2234). Catalytic hydrogenation of chlortetracycline led to the antibiotic tetracycline **3**,[6] designated Tetracyn, and is the core structure for which the entire family of tetracyclines derives its name. Tetracyn was the first semisynthetic tetracycline used clinically, although now it is produced solely via fermentation by biochemical mutants of *Streptomyces* and/or the manipulation of the fermentation medium.

First generation tetracyclines

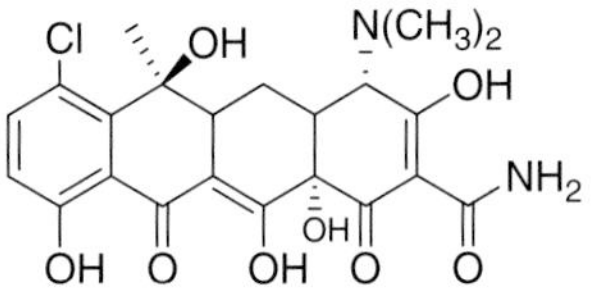

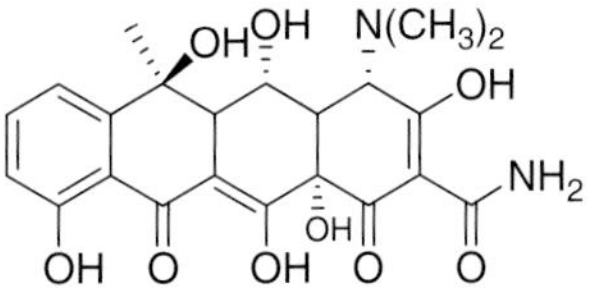

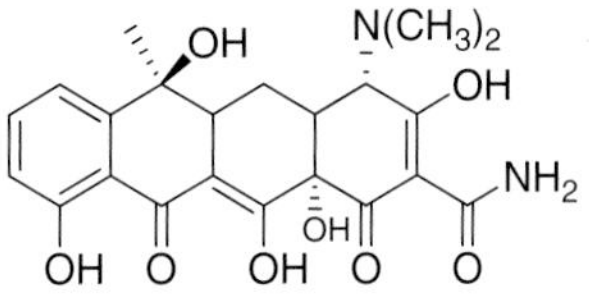

Second generation tetracyclines

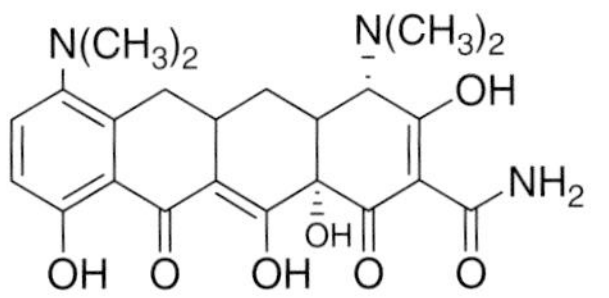

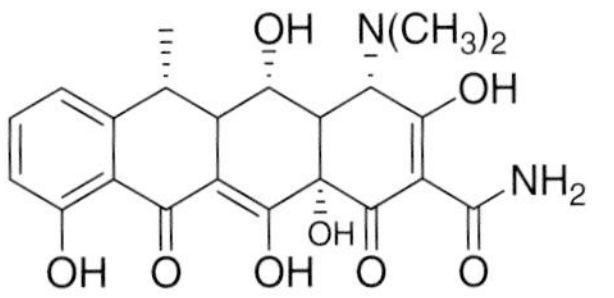

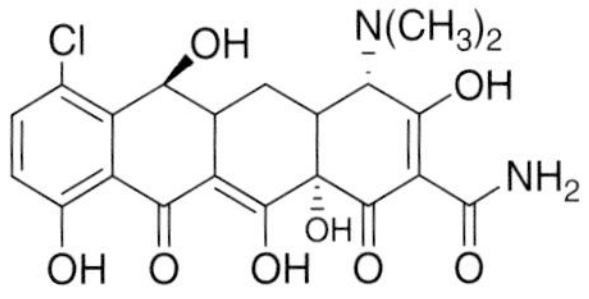

Third generation tetracyclines

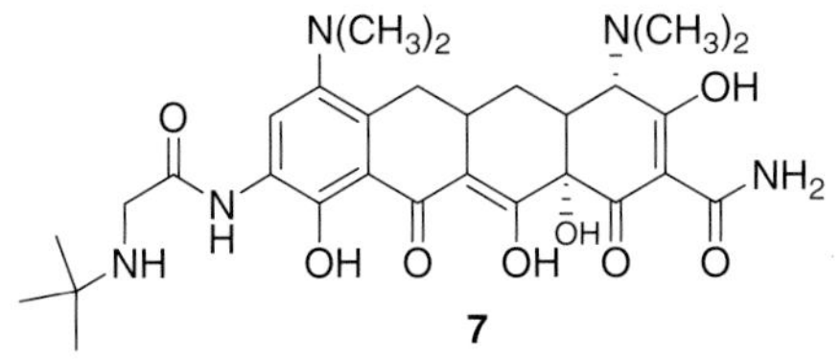

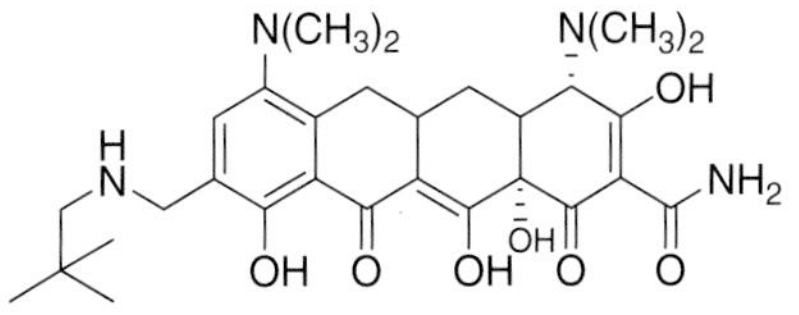

Figure 3 The major first, second, and third generation tetracyclines.

7.20.3 Tetracycline Biosynthesis

The biosynthesis of the tetracyclines was studied primarily in *Streptomyces* spp. and the pathways of oxytetracycline and chlortetracycline production determined by chemically or biologically manipulating producing organisms, which modified their metabolism while charting the pathways and enzymology leading to their production.[7–9] The sequence of reactions that direct tetracycline biosynthesis show that the polyketide backbone, ring foldings and closures, and exocyclic modifications that produce the naphthacene ring scaffold along with its numerous chemical functional groups are conserved between species and can also occur in other genera, including *Nocardia, Dactylosporangium, Actinomadura,* and *Penicillium* spp., among other aerobic-fermentative microorganisms.

While not every biosynthesis intermediate has been isolated, radiolabel incorporation experiments with malonate and acetate subunits showed that the tetracycline ring system is formed through addition of malonylamyl-CoA and acetyl-CoA units (**Figure 4, I**) to form a linear nona- or decaketide precursor **II**, followed by a concerted series of enzyme-mediated foldings, and ring closures and stereochemical transformations by type II polyketide synthases, forming the basic naphthacene ring system.[10] Position C6 methylation yields 6-methylpretetramide **9**, a stable and isolable precursor, which is further modified to the critical intermediate 4-hydroxy-6-methylpretetramide **10**.[11] Pretetramides are substrates for hydroxylase and ketoreductase enzymes, yielding 4-hydroxy-12-deoxyanhydrotetracycline **11**. The introduction of the C12a-hydroxyl group by hydroxylases and ketoreductase modification changes the structurally planar pretetramide molecule to the pharmacophore pattern associated with bioactive tetracyclines, where the A ring is now right-angular compared to the BCD rings, and the three-dimensional antibacterial shape is created **12**. Bioamination of position C4 forms 4-aminotetracycline **13**, followed by methylation resulting in 4-dimethylaminoanhydrotetracycline **14**, an intermediate that has never been isolated within the pathway and is toxic to the bearing

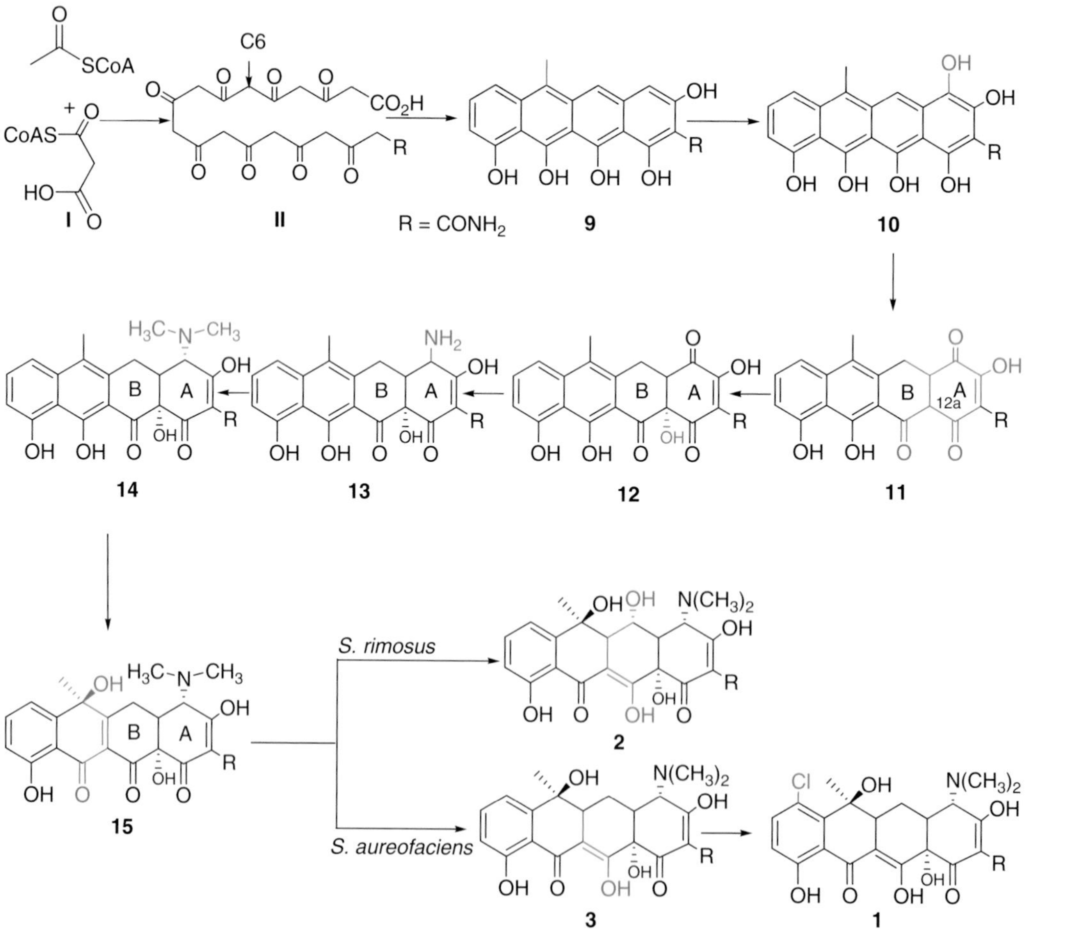

Figure 4 Biosynthetic pathway for chlortetracycline, oxytetracycline, and tetracycline in *Streptomyces* species. Structural features in red represent chemical changes that occur within the pathway step.

organism.[12] Hydroxylation at position C6 and the formation of an unsaturated center between positions C5a and C11a produces a pivotal intermediate, 6-hydroxy-5a(11a)-dehydrotetracycline **15**. Using the producer strain *S. rimosus*, the unsaturated substrate is hydroxylated at position C5, producing oxytetracycline **2**, while using *S. aureofaciens* hydroxylation does not occur, forming tetracycline **3**. Further halogenation via haloperoxidases produces C7 chlortetracycline **1**. Tetracycline **3** can also be produced directly by *S. aureofaciens* mutants that show depressed chlorination ability, or by the addition of heterocyclic chlorination inhibitors.[13]

7.20.3.1 Tetracyclines Derived from Other Actinomycetes and Microorganisms

Mutant strains of *Streptomyces* are now used to produce commercially manufactured tetracyclines, while less common natural tetracyclines such as demeclocycline **6** and demecycline **16** (**Figure 5**), are produced via induced biochemical mutations and the modification of key steps within its biosynthetic pathway.[14] Demecycline and demeclocycline are produced by C6 methylation blocked mutants, and demeclocycline, which is used clinically, is obtained by biological conversion using the C6 methylation inhibitor sulfaguanidine. The presence of the C6 hydroxyl group and the C7 chlorine in demeclocycline has also made this compound a useful synthetic intermediate in the semisynthesis of second-generation tetracyclines related to sancycline **17**, the chemically simplest tetracycline that still maintains antibacterial activity, and minocycline **4**.

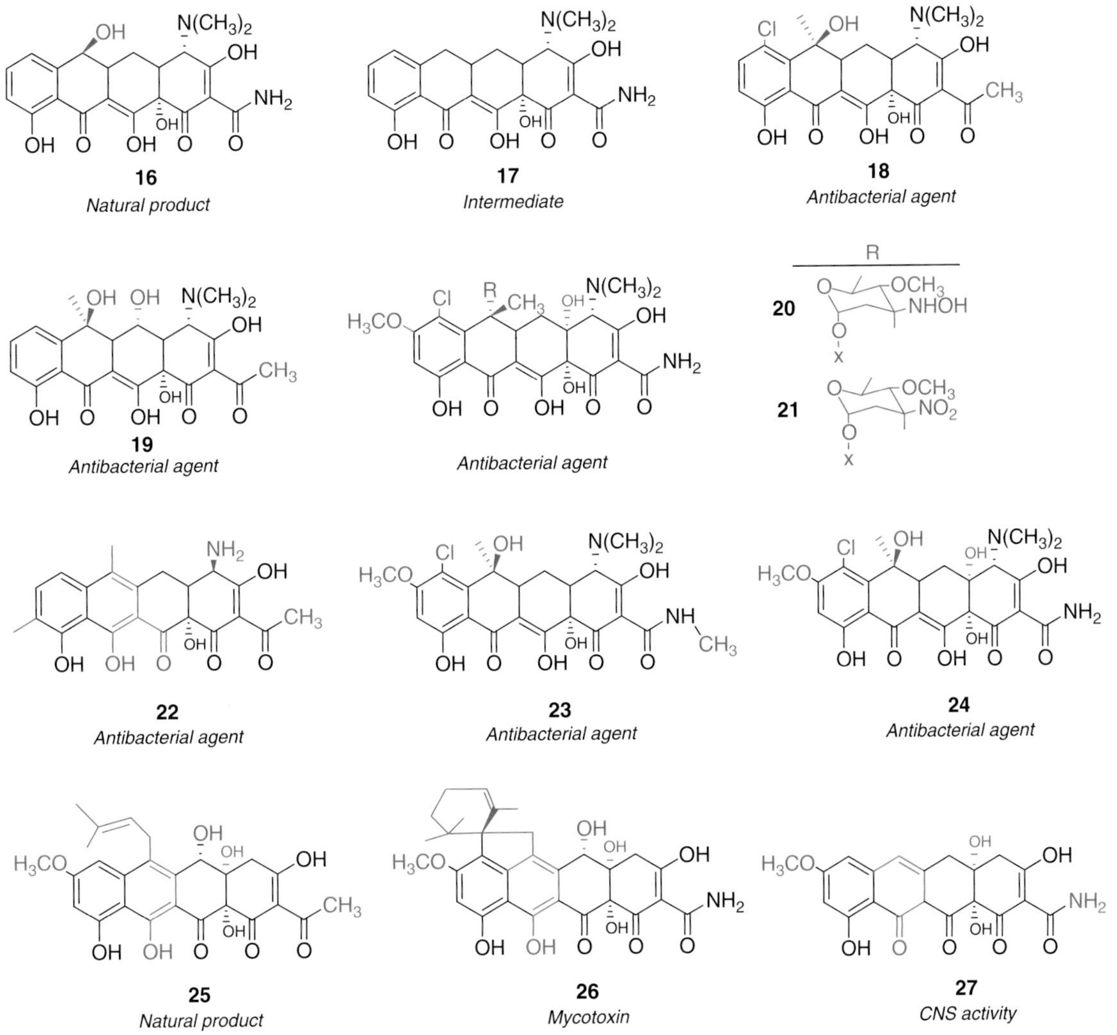

16 *Natural product*

17 *Intermediate*

18 *Antibacterial agent*

19 *Antibacterial agent*

20, **21** *Antibacterial agent*

22 *Antibacterial agent*

23 *Antibacterial agent*

24 *Antibacterial agent*

25 *Natural product*

26 *Mycotoxin*

27 *CNS activity*

Figure 5 Less common natural product tetracyclines derived from *Streptomyces* species and their semisynthetic intermediates and natural product tetracyclines from other Actinomycetes and other microorganisms. Structural features in red represent chemical changes and substituents differing from the minimum pharmacophore for antibacterial activity, sancycline. Uses or activity detailed in italic print below the structure.

The soil microorganisms of the class Actinomycetes are composed of over 3000 separate species in 40 distinct genera, many of which have been found to produce tetracycline natural products. Some *Streptomyces* species can produce novel and active tetracyclines during biosynthesis by malonyl-CoA pathway changes, substituting the usual C2 carboxamide group with an acetyl moiety to yield 2-acetylchlortetracycline **18** and Terramycin X **19**, structurally unusual tetracyclines with antibacterial activity.[15,16]

Dactylocycline A **20** and B **21** are produced by *Dactylosporangium* sp. SC14051, and are chemically unique, possessing a β-aminoglycoside substructure at position C6 and a C4a hydroxyl group, compared to the natural tetracyclines.[17] Chelocardin **22**, produced by *Nocardia sulphurea* (NRRL 2822), is composed of aromatized C and D rings and an acetyl group at position C2 with an diastereomeric C4 β-amino group, compared to the usual C4 α-amino group in the natural tetracyclines.[18] All of these compounds were found to have antibacterial activity primarily against Gram-positive bacteria, while chelocardin maintained Gram-negative antibacterial activity. The tetracycline Sch 33256 **23** was isolated from *Actinomadura brunnea*, and resembled chlortetracycline in addition to a C2 *N*-methylcarboxamide and a C8 position methoxy group.[19] Other 8-methoxylated tetracyclines, such as Sch 34164 **24**, possess a C4a hydroxyl group with the methyl substituent at the carboxamide absent.[20]

Tetracyclines are also produced by fungi, where hypomycetin **25** has been isolated from the mycophilic fungus *Hypomyces aurantius*,[21] a compound structurally similar to a product isolated from *Penicillium viridicatum*, the mycotoxin viridicatumtoxin **26**.[22] A more recent tetracycline found to be structurally and biologically unique was isolated during the fermentation of *Aspergillus niger*, BMS-192548 **27**, a tetracycline devoid of the C4 dimethylamino group and with a C4a hydroxyl group. This compound was inactive against bacteria but had unexpected central nervous system (CNS) activity, binding to neuropeptide Y receptors in mammalian cells with the potential to modify numerous neural and physiological processes.[23]

7.20.4 The Antibiotic Tetracyclines

All antibiotic tetracyclines have in common a linearly arranged naphthacene ring (**Figure 6**) with oxygen and nitrogen containing functional groups along the designated lower peripheral region, the 2N region and the C3–C4 region. The lower peripheral region outlines a pharmacophore pattern of hydroxyl, keto-enol, and carbonyl groups across the C1–C10 positions, and its spatial arrangement is responsible for the potency and range of biological activities in bacteria, where chemical modification eliminates antibacterial activity. Chemical modifications along the upper periphery, positions C5–C9, can also influence activity, depending upon position and nature of the chemical substitution, and derivatives at these positions have produced clinically significant antibiotics.

The position C2-exocyclic carbonyl and the C3-keto-enolate group are also required for antimicrobial activity, while broad-spectrum activity is dependent on the presence and stereochemical orientation of the C4 dimethylamino group. The 2N amide nitrogen can also be chemically modified, producing compounds of variable antibacterial activity.

The tetracyclines have been semisynthetically modified by disjunctive or conjunctive approaches, by removal or subtraction of functional groups, or by addition of substituents at modifiable positions, and many are reported

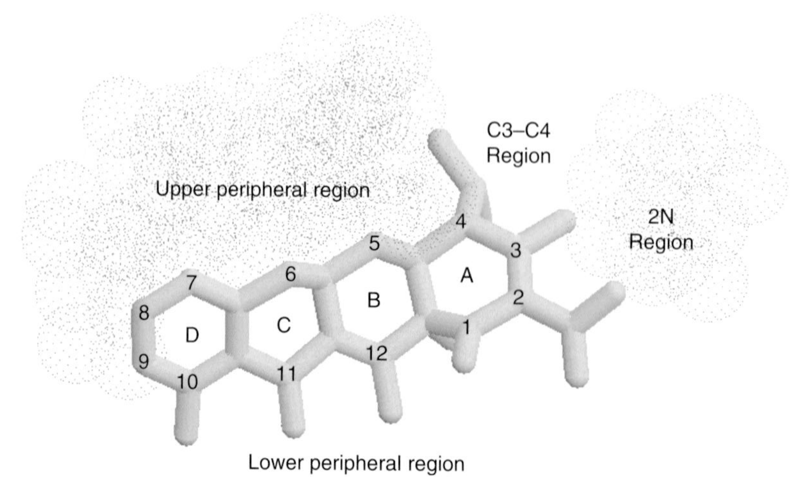

Figure 6 The tetracycline molecule minimum pharmacophore and the designated upper periphery, lower periphery, 2N and C3–C4 regions. (Drawn using Alchemy 2000 software with PM3 and MM3 geometry optimization, Tripos, Inc., St. Louis, MO.)

synthesized. Both methods of modification either alone or combined have led to improvements in tetracycline bioactivity against bacteria and the production of therapeutically and commercially valuable compounds. Chemical changes along the lower periphery hinder the formation of tautomeric and metal binding substructures needed for antibacterial activity, while their SARs as antibacterial agents relies primarily on chemically modifiable positions along the upper peripheral region.

7.20.4.1 Chemistry of the A-Ring and Antibacterial Activity

The A-ring of the natural tetracyclines possesses five different functional groups simultaneously (**Figure 7**). Position C1 possesses a carbonyl moiety, C2 an unsubstituted carboxamide group, C3 a keto-enol group, and C4 a dimethylamino group naturally alpha (below) to the plane of the ring system. At the C12a ring juncture a tertiary hydroxyl group is located alpha to the ring plane and one the most crucial functional groups needed for maintaining antibacterial activity. Semisynthetic removal of the C12a hydroxyl group results in a total loss of activity.

The C1–C2 exocyclic carbonyl-C3 position tricarbonylmethane substructure within the A-ring is crucial for antibacterial activity, producing an acidic proton with an approximate pK_a of 3.1–3.5,[24] and is ionized at physiological pH producing a highly tautomerized structure. Chemical modifications that change the tricarbonylmethane substructure drastically alters proton dissociation properties and abolishes antibacterial activity.

7.20.4.2 C2 Modifications and 2N Prodrugs of the Tetracyclines

Modification of the C2-exocyclic carbonyl to a nitrile functional group **28** eliminates antibacterial activity, at least in whole cells, while chelocardin **22**, possessing a C2 acetyl group, remains active, indicative of the importance of the exocyclic carbonyl group in maintaining activity.[25] The C2 carboxamide nitrogen may be modified, and is synthetically reactive with aldehydes, ketones and C- and N-Mannich base reagents, where the latter has been used to produce 2N aminoalkylated prodrug derivatives of the tetracyclines.[26] Reaction with formaldehyde and amines with the C2 amide N forms tetracyclines with increased aqueous solubility of two or more orders of magnitude, with altered bioavailability and decreased toxicity. Changes in the amine substituent can modify the solvolytic half-life of the prodrug back to the parent molecule both in vitro and in vivo,[26,27] and two have found clinical utility, pyrrolidinylmethyl **29** (Rolitetracycline) and lysinylmethyl **30** (Lymecycline), derivatives of tetracycline with hydrolysis half-lives of 45 and 62 min, respectively. Both antibiotics are approved for use in Europe for an array of bacterial infections, but not in the USA.

7.20.4.3 C4 Derivatives and Epimerization of the Dimethylamino Group

The naturally oriented alpha-C4 dimethylamino group functions as a base (**Figure 8**), forming a conjugate acid with a weak pK_a of approximately 9.4. The dimethylamino group is important for maintaining a broad spectrum of activity against bacteria where the alpha-epimer (*S*) is much more potent against Gram-negative bacteria, while the beta-epimer (*R*) activity is decreased by almost 95% for both Gram-negative and Gram-positive bacteria.[28] Alpha to beta epimerization is induced by acidic pH ranges (2–6), basic pH ranges (>7.5), and other factors, including the presence of chelating metals, neighboring substituents, solvents, and buffering systems.[29] Epimerization also occurs in vivo, where 4-β-epitetracyclines have been detected in the muscle and organ tissues of animals.[29,30]

The kinetics of epimerization have been determined to be a first-order reversible reaction with an activation energy of 20 kcal $mole^{-1}$, whose equilibrium concentration and rate are dependent upon pH and the presence of cations.[31]

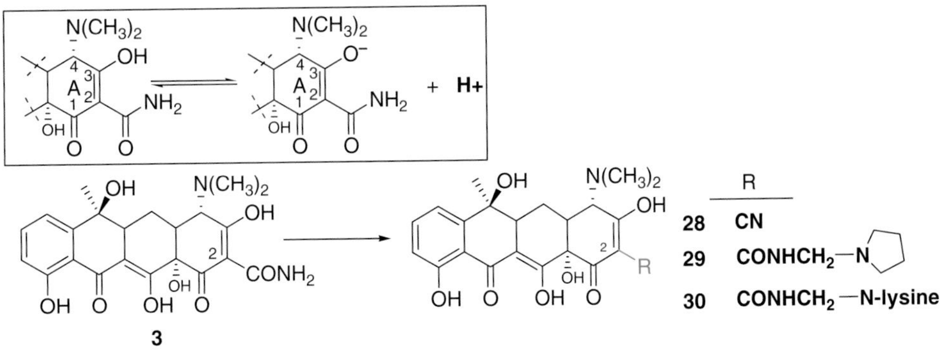

Figure 7 Proton dissociation of the A-ring (boxed area) and C2 derivatives of the tetracyclines.

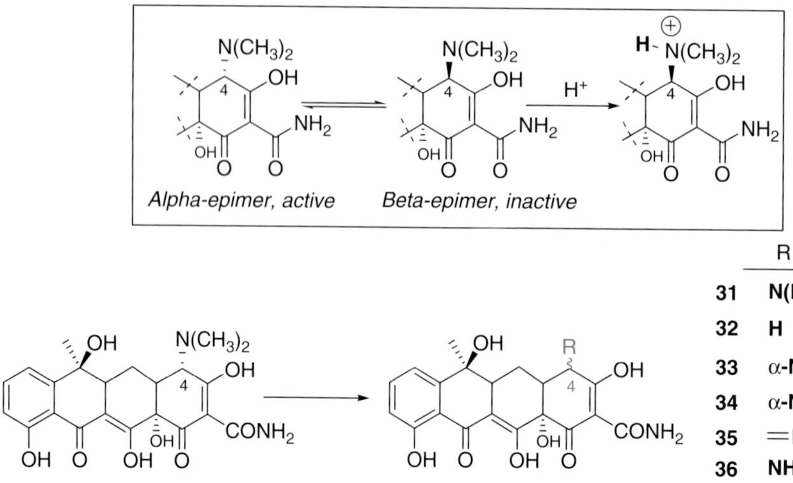

Alpha-epimer, active

Beta-epimer, inactive

	R	
31	$N(R^1R^2R^3)^+ X^-$	
32	H	(X = halogen)
33	α-NH_2	
34	α-NR^1R^2, R^1 = Me, R^2 = Et	
35	=NOH	
36	NHNHR	R = H,Me,Ph
37	α or β-OH	

Figure 8 Protonation and epimerization of the dimethylamino functional group (boxed area) and C4 derivatives of the tetracyclines.

Semisynthesis of more novel antibacterial derivatives of the tetracyclines depends upon using chemical reagents and conditions that minimize the alpha to beta epimerization process.[32]

Chemical conversion of the C4-dimethylamino group to quaternary ammonium betaine analogs eliminates antibacterial activity **31**.[33] Removal of the C4 methyl betaine by reagents or reductive electrolysis affords C4-dedimethylamino tetracyclines **32**,[34] compounds that lose antibacterial activity, at least against Gram-negative bacteria. The C4-dedimethylamino series of compounds produced from the natural and semisynthetic tetracyclines are collectively known as the chemically modified tetracyclines, sometimes referenced as the CMTs. These compounds are active only in vitro against Gram-positive bacteria with decreased potency compared to their parent structures with minimum inhibitory concentration (MIC) values between 0.8 and 1.5 $\mu g\,mL^{-1}$.[35] Their lack of in vitro Gram-negative activity may be explained by decreased passage across Gram-negative cellular membranes and in vivo activity by their low aqueous solubility and high serum binding properties. Unsubstituted C4 amino derivatives of tetracyclines **33** are inactive, while increasing the amino carbon chain length **34** or substitution to a C4 oximino **35**, hydrazone **36**, or hydroxyl group **37**, produces compounds of lesser or no activity.[36,37] C4 derivatives **33–37** are only synthetically feasible using the first-generation tetracyclines or those possessing a 6-hydroxyl group.

7.20.4.4 Chemistry of the B-Ring and Antibacterial Activity

The B-ring lower periphery harbors a C12a hydroxyl group and a keto-enol substructure within the C-ring with tautomerism across carbons C11, C11a, and C12 (**Figure 9**). This is a major pharmacophore region capable of deprotonation, generating a third macroscopic pK_a value of 7.5. Chemical modification of the inherent beta-diketone substructure and any chemical changes to it will eliminate all antibacterial activity.

The 11,12-pyrazolo product of the reaction of tetracycline with hydrazine **38**,[38] is completely inactive, demonstrating the relationship of the beta-diketone substructure, metal binding, and chelation potential between positions C11 and C12 and bioactivity. Further changes in the B-ring lower periphery, by reductive removal of the C12a hydroxyl group **39**,[39] results in a loss of bioactivity, presumably due to major structural and steric changes. The binding of metals along the lower periphery in this region and the conformation due to the 12a hydroxyl group create a major pharmacophore responsible for ligand formation, cation binding, and the three-dimensional structure needed at the drug–ribosome interface.

There are relatively few descriptions of upper periphery B-ring modified tetracyclines, although 5-hydroxy-6-deoxytetracyclines such as doxycycline have been reported to form esters at the C5 hydroxyl group **40** with nonpolar carboxylic acids in anhydrous HF or methanesulfonic acid.[40] Position C5-esters of doxycycline have activity against Gram-positive bacteria, both in vitro and in vivo, although a limited number of congeners were studied.

7.20.4.5 Chemistry of the C-Ring and Antibacterial Activity

The natural products chlortetracycline, tetracycline, and oxytetracycline all possess a chemically unstable C6-hydroxyl group (**Figure 10**), and form C-ring dehydrated and aromatized anhydrotetracyclines **41**, in vitro and in vivo,

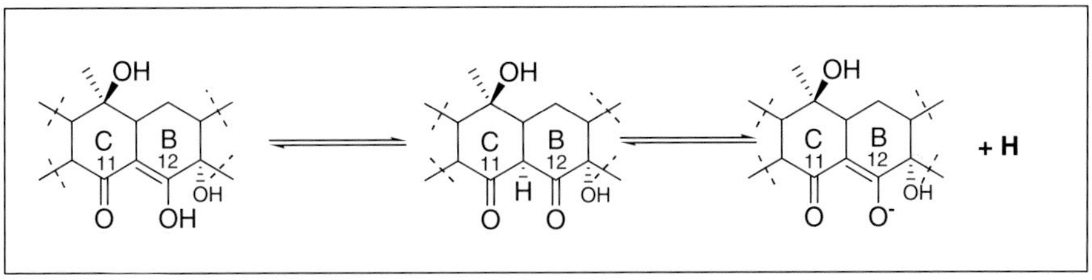

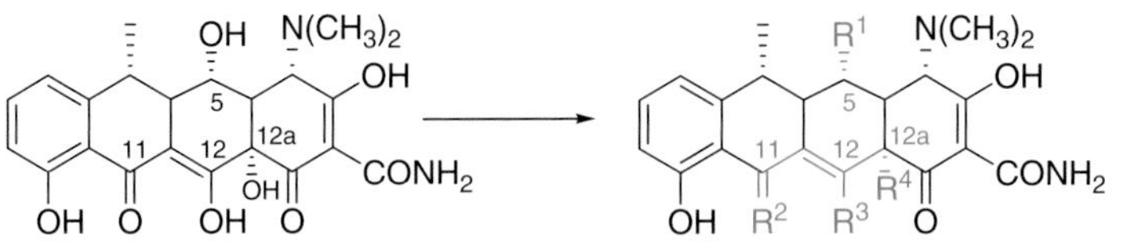

	R^1	R^2	R^3	R^4
38	OH	N—NH		OH
39	OH	O	OH	H
40	O_2CR	O	OH	OH

R = alkyl, phenyl

Figure 9 Proton dissociation of the C–D ring (boxed area) and C5 and C11–C12 derivatives of the tetracyclines.

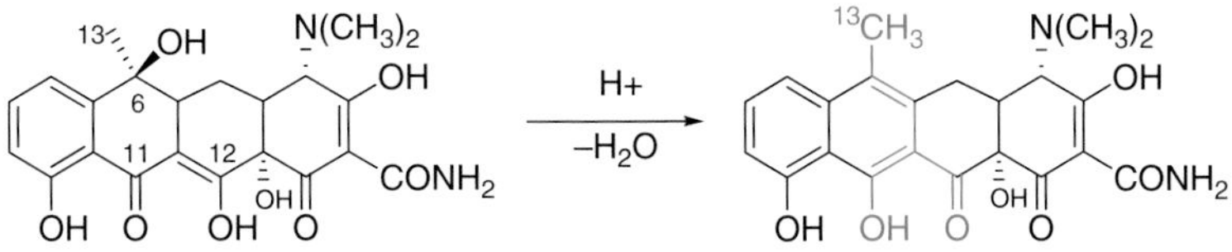

Figure 10 Formation of anhydrotetracyclines from 6-OH tetracyclines.

particularly in the presence of acids.[41] These by-products are common in commercial samples of tetracyclines possessing a reactive C6-hydroxyl group, unless extra measures are taken to assure compound purity. Furthermore, anhydrotetracyclines can have numerous deleterious biological effects, such as increased toxicity and phototoxicity while exhibiting decreased antibacterial activity.[42,43] Because chemical modifications and harsh reagents can readily form anhydrotetracyclines, breakthroughs in tetracycline semi synthesis were few until derivatives lacking the C6-OH functional group were discovered.

7.20.4.5.1 Second generation tetracyclines: the synthesis and antibacterial activity of methacycline

The degradation of tetracyclines to C-ring anhydrotetracyclines can be avoided by protection of the C11a position with chlorine or bromine (**Figure 11**), stabilizing the C-ring and forming 11a-Cl-6,12-hemiketal oxytetracycline **42**.[44] Protection of position C11a allows further chemical transformations in acid and under more harsh chemical conditions. Oxytetracycline 6,12-hemiketal **42** in anhydrous HF undergoes an exocyclic dehydration forming a C6-C13 double bond **43**, which upon removal of the 11a halogen with reducing reagents yields methacycline **44**, tradenamed Rondomycin, an antibiotic that possesses activity against a broad spectrum of bacteria and is used clinically in countries other than the USA. Methacycline is also used in veterinary and agricultural applications and as a valuable synthetic intermediate for the synthesis of doxycycline **5**, a clinically useful and potent antibiotic.

7.20.4.6 Position C13 Bacterial Efflux Protein Inhibitors

Derivatives of methacycline **44** based on its radical catalyzed reaction with mercaptans have yielded a potent inhibitor of bacterial efflux transport proteins related to the Tet family of proteins, commonly found in resistant Gram-negative and Gram-positive bacteria.[45–47] Tet proteins belong to the major-facilitator superfamily of drug efflux proteins, and are partly responsible for tetracycline resistance in bacteria, by specifically removing tetracyclines from within the cytoplasm, decreasing intracellular concentrations, and rendering them ineffective at the level of the ribosome.

The derivative 13-cyclopentyl-5-hydroxytetracycline **45** (13-CPTC) (**Figure 12**) was found to competitively inhibit the Tet(B) efflux protein in everted membrane vesicles and in whole cells,[46] allowing the intracellular accumulation of clinically used doxycycline **5**, thereby reversing antibiotic resistance due to drug efflux. 13-CPTC had no effect on bacterial membranes by perturbation, and was active in synergy with doxycycline against Gram-negative resistant

Figure 11 Synthetic pathway resulting in methacycline.

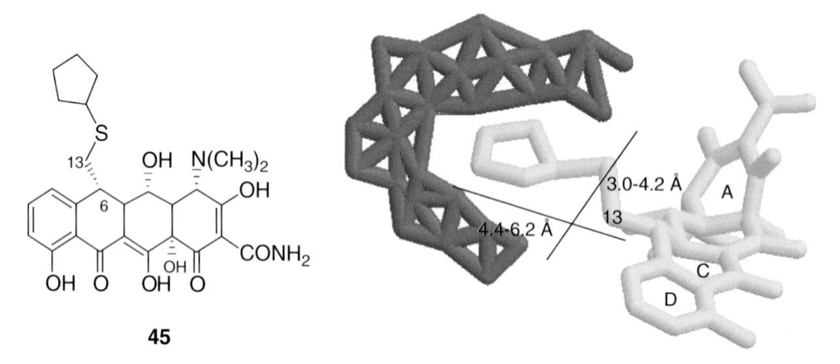

Figure 12 The Tet efflux protein inhibitor 13-cyclopentylthio-5-hydroxy tetracycline (13-CPTC) and structural delineation of the hydrophobic pocket (axes X and Y in angstroms) within the Tet(B) efflux protein described by 13-alkylthio-5-hydroxy tetracyclines. (Drawn using Alchemy 2000 software with PM3 and MM3 geometry optimization, Tripos, Inc., St. Louis, MO.)

Escherichia coli cells expressing the Tet(B) efflux protein. 13-CPTC also was active against tetracycline-resistant Gram-positive bacteria, including bacteria harboring efflux and ribosomal protection mechanisms.

Structure–activity studies using the Tet(B) protein in the vesicle assay and an extensive series of C13-thiol analogs demonstrated that optimal efflux inhibition was attained when the C13 substituent was confined between steric length values of 4.4 to 6.2 Å, with width constraints between 3.0 and 4.2 Å.[47] The most potent compounds with similar steric limitations defined a hydrophobic pocket within the domains of the efflux protein, where a defined range of lipophilicity π values between 1.0 and 2.5 were the most active, while polar compounds of similar size where inactive as efflux protein inhibitors. The 13-thiol derivative studies showed that efflux-mediated tetracycline resistance in bacteria could be readily reversed using a substrate-based drug design approach.

7.20.4.7 The Synthesis and Antibacterial Activity of Doxycycline

Catalytic reduction of methacycline **44** affords a diastereomeric mixture of 6-α-methyl-6-deoxytetracycline **5** and 6-β-methyl-6-deoxytetracycline **46**, the former of which, when isolated from its β-methyl isomer, was found to be a potent broad-spectrum antibiotic (**Figure 13**).[48] Process development for the production of commercial quantities of the 6-α-methyl isomer relied on the reaction of benzyl mercaptan with methacycline **47**, followed by reduction to yield doxycycline **5**, tradenamed Vibramycin, a clinically used and commercially successful tetracycline. Today doxycycline may be synthesized by this pathway or by using stereospecific hydrogenation catalysts that reduce methacycline directly to doxycycline,[49] although when using these methods a small percentage of the β-isomer (<5%) is also produced.

Doxycycline is used clinically for a broad spectrum of community-acquired bacterial infections and is exceptionally active and indicated for *Bacillus anthracis*, or anthrax infections.[50] More commonly, doxycycline is active against the spirochete *Borrelia burgdorfii*, the causative agent in Lyme's disease, where a single dose (100 mg), can prevent the emergence of a *Borrelia* infection following a tick bite.[51]

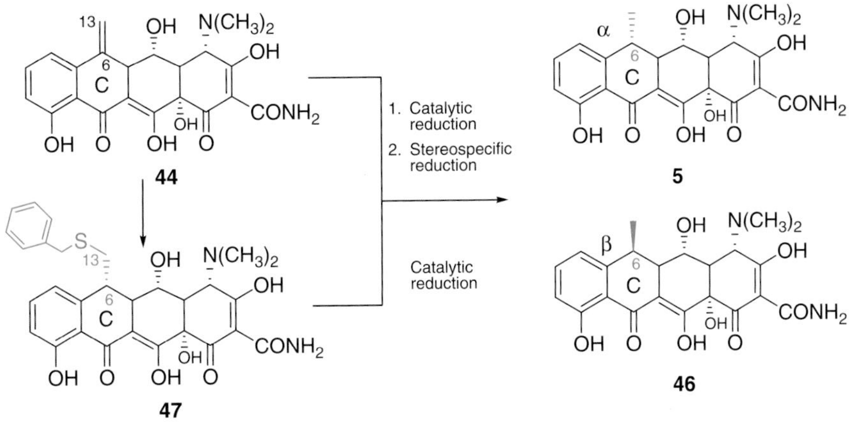

Figure 13 Synthesis of doxycycline via 13-benzylthio-5-hydroxy intermediate or by stereospecific reduction.

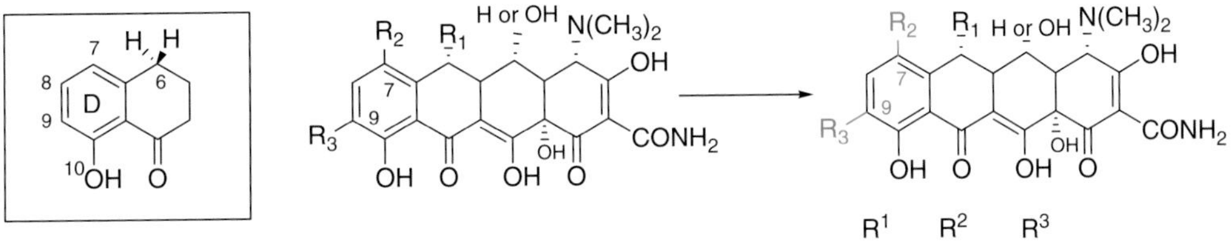

	R^1	R^2	R^3
48	CH_3	NO_2	H
49	CH_3	H	NO_2
50	H	NO_2	H
51	H	H	NO_2

	R^1	R^2	R^3
52	CH_3	NH_2	H
53	CH_3	H	NH_2
54	H	NH_2	H
55	H	H	NH_2

	R^1	R^2	R^3
56	H	Cl,Br,I	H
57	H	H	Cl,Br,I
58	H	N-imide	H
59	H	H	N-imide

Figure 14 Description of the D-ring positions (boxed area) and C7 and C9 derivatives of doxycycline and sancycline.

Doxycycline also is approved and indicated for use in periodontal disease under the tradename Periostat, and is the first tetracycline marketed as a low-dose formulation (20 mg day^{-1}) and matrix metalloproteinase inhibitor in human tissues and is active independent of its antibiotic activity.[52]

7.20.4.8 Chemistry of the D-Ring and Antibacterial Activity

The aromatic D-ring of the tetracyclines, comprising positions C7, C8, and C9, and an electron-donating C10 phenol group, can be chemically modified via electrophilic aromatic substitution (EAS) reactions, provided the tetracycline does not possess an acid-labile C6 hydroxyl group (**Figure 14**). Electron density calculations indicate nucleophilic potential at positions C7 and C9, while position C8 is electron deficient and chemically unreactive to EAS reagents. Doxycycline **5** and sancycline **17** readily undergo nitration in acids forming the corresponding C7 and C9 nitro derivatives, which are catalytically reduced to the C7 and C9 amino compounds **48–55**.[53,54] The C7 and C9 aromatic positions are also halogenated nonregiospecifically **56**, **57**, or regiospecifically at C7 **56**,[32] and can react with electrophilic species such as *N*-hydroxymethylimide reagents to form the C7 and C9 *N*-methylimide derivatives of 6-deoxytetracyclines **58**, **59**.[3,55] All direct C7 and C9 substitutions are generally inactive as antibacterial agents, but have been used as reactive intermediates for the synthesis of several of the most potent and clinically relevant tetracycline antibacterials, including those now in or entering clinical use.

7.20.4.9 Synthesis of Minocycline

Sancycline **17**[56] is produced industrially, which upon further regiospecific nitration and reduction yields 7-aminosancycline **54** (**Figure 15**), an inactive compound against bacteria. Reductive alkylation with formaldehyde affords 7-dimethylamino sancycline,[57] known as minocycline **4**, and tradenamed Minocin, a tetracycline possessing superior broad-spectrum

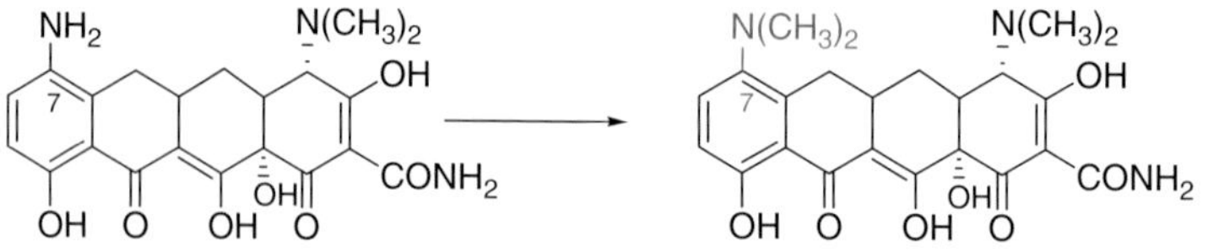

Figure 15 Synthesis of minocycline by reductive alkylation.

antibacterial activity. As the last tetracycline approved for clinical use in 1971, minocycline is active against tetracycline-susceptible and some tetracycline-resistant strains of *Staphylococcus* and vancomycin-resistant Gram-positive bacteria. Reductive alkylation using other aldehydes and increasing the alkyl chain length, or forming alicyclic C7 amino derivatives showed decreased antibiotic activity,[58] as did the C9 dimethylamino derivative of sancycline.

Quantitative SAR studies of C7 derivatives suggests that electron-withdrawing groups are more active due to increased ionization along the lower peripheral region and increased affinity for cations subsequent to bacterial ribosomal binding.[59] The C7 dimethylamino functional group of minocycline, as a base, would therefore be expected to be inactive; however, protonation in cellular microregions of low pH can induce a temporary electron-withdrawing effect, explaining its excellent potency. 7-NH_2 sancycline **54**, however, lacks antibacterial activity due to its weaker activity as a base and less tendency to protonate and form electron-withdrawing species. Minocycline is also one of the most lipophilic tetracyclines, and both its electronic properties and lipophilicity may account for its enhanced activity both in vitro and in vivo.

7.20.4.10 Third-Generation Tetracyclines: Synthesis of Minocycline, Sancycline, and Doxycycline Derivatives

Within the past 15 years, reports in the synthesis and microbiology of several new tetracyclines have signaled a renewed interest in this family of molecules, representing a new third-generation of derivatives. Semisynthetic derivatives of the second-generation tetracyclines, minocycline **4**, sancycline **17**, and doxycycline **5** (**Figures 16–18**), represent compounds that are currently being studied in human clinical trials or preclinical investigations or are at earlier stages of development. New third-generation tetracyclines were synthesized by a conjunctive approach, derivatizing core scaffolds that are clinically used or are chemically stable, followed by optimization of their activity, especially against tetracycline- and antibiotic-resistant bacteria. This approach has led to the development of the glycylcyclines and tigecycline, and approved for clinical use, and PTK 0796, a novel 9-aminomethyltetracycline recently entering Phase I clinical studies.

Other recent approaches to tetracycline semisynthesis have utilized the application of more modern reactions to tetracycline scaffolds, and the use of transition-metal-based chemistries has increased chemical diversity at the C7, C8, and C9 positions of the naphthacene ring system, and is a versatile and facile way of synthesizing numerous new classes of tetracyclines.

7.20.4.11 The Synthesis of the Glycylcyclines and Tigecycline

Minocycline **4** and sancycline **17** undergo nitration and reduction to form 9-NH_2 minocycline **60** and 9-NH_2 sancycline **55** (**Figure 16**).[2] The aniline functional group is reactive with acid chlorides in protic solvents, where bromoacetylbromide forms an amide bond and a reactive halogen center for nitrogen nucleophiles (**61**, **62**). Further reaction by amine substitution results in the synthesis of a glycine subunit, and is the genesis of the new class of tetracyclines, the glycylcyclines (**63–65**). Using a series of amines, substituted derivatives were prepared and the SARs determined, affording excellent activity against a broad spectrum of bacteria,[60] including tetracycline-resistant bacteria bearing efflux and ribosomal protection mechanisms, with the dimethylglycine derivatives **63** and **64** showing the most activity. Further explorations using the minocycline scaffold led to synthesis of tigecycline, formerly known as GAR-936 **65**, where t-butylamine substitution produced an antibiotic of exceptional potency.[61,62] Tigecycline is the first third-generation tetracycline to reach the clinic as a hospital-based injectable antibiotic in over 30 years, possessing a broad spectrum of activity against tetracycline-susceptible and -resistant bacteria.[63–65] Similar reactions applied to the doxycycline scaffold produced glycylaminodoxycycline derivatives such as compound **66**, which were less active than their minocycline counterparts, while amino acids and dipeptides **67** covalently coupled with standard reagents with the 9-anilino group of doxycycline.[53]

	R^1	R^2	R^3
60, 61, 63	$N(CH_3)_2$	CH_3	CH_3
55, 62, 64	H	CH_3	CH_3

	R^1		R^1
66	$COCH_2N(CH_3)_2$	**67**	AA; AA-AA

A A = amino acid

Figure 16 Synthesis of the glycylcyclines, tigecycline, and 9-amino derivatives of doxycycline.

Figure 17 Synthesis of 9-aminomethylcyclines and PTK 0796 from minocycline.

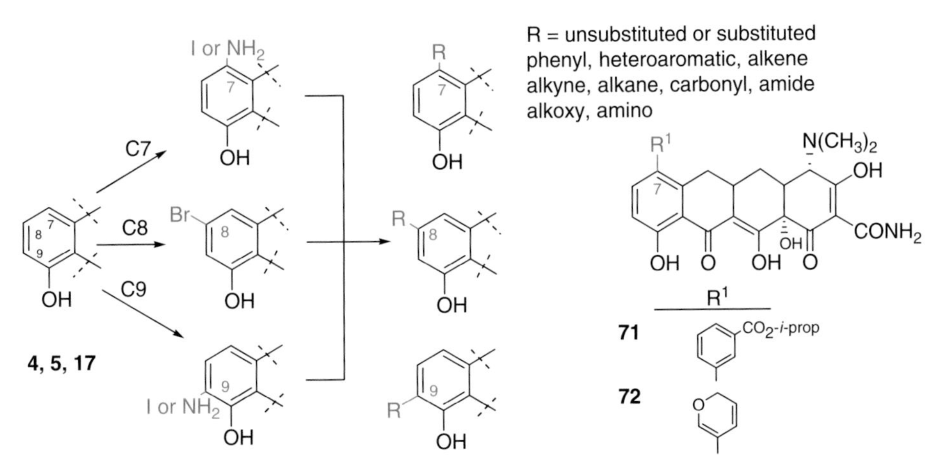

Figure 18 Synthesis of C7, C8, and C9 tetracyclines derivatives via transition metal catalyzed reactions.

7.20.4.12 The 9-Aminomethylcyclines and PTK 0796

Reaction of minocycline **4** with *N*-hydroxymethylphthalimide forms 9-methylphthalimide minocycline **68** in strong acids (**Figure 17**), which upon treatment with dimethylamine yields 9-aminomethyl minocycline **69**.[3] The primary amine readily undergoes reductive alkylation with aldehydes forming C9 alkylaminomethylminocyclines, products of which are active against a broad spectrum of bacteria both in vitro[66,67] and in in vivo animal models.[68] PTK 0796 **8**, the pivalaldehyde reaction reductive alkylation product, was the most potent of an extensive series studied, and was also active against tetracycline-resistant bacteria and vancomycin-resistant enterococci,[69] and was efficacious in soft tissue infections as an intravenously administered antibiotic.

7.20.4.13 Transition Metal-Catalyzed Reactions of the Tetracyclines

More recently it was reported that D-ring aniline or halogenated tetracyclines can undergo palladium-catalyzed reactions with a variety of reagents forming Heck, Suzuki, and other coupling products at positions C7, C8, C9, and C13, greatly expanding the chemical diversity at these positions by forming carbon–carbon bonds with alkenes, alkynes, phenylboronic acids, and carbonylation reagents (**Figure 18**).[32] Minocycline **4**, sancycline **17**, and doxycycline **5** D-ring reactive intermediates and methacycline **45** were used to form a large array of chemically diverse C- and D-ring derivatives, creating numerous new classes of tetracyclines for biological evaluation. Many of these compounds are still in the exploration stage, where their spectrum and use as biological agents are being assessed. Preliminary reports of the derivatives demonstrate in vitro potency as antibacterial agents, particularly against tetracycline-resistant bacteria, where 7-phenyl sancycline derivatives **70** or heteroaryl compounds **71** were active against resistant Gram-positive bacteria.[3] Against the parasite *Cryptosporidium parvum* numerous C9 alkynyl and alkenyl derivatives of doxycycline and sancycline displayed potent antiparasitic activity, in a limited number of congeners studied.[70]

7.20.5 Antibacterial and Antimicrobial Uses of the Tetracyclines

Clinically, the tetracyclines are best known for treating microbial infections (**Figure 19**), due to their bacteriostatic activity against a wide range of pathogens among the Gram-negative and Gram-positive bacteria and atypical intracellular pathogens, such as parasites, the rickettsias, and closely related α-proteobacteria. They may also act as bactericidal agents, at higher concentrations, and by mechanisms that are unclear.[71] Tetracyclines are active against infections caused by the mycoplasmas, chlamydia, and protozoans such as *Giardia, Cryptosporidium,* and *Toxoplasmosis* species.

Tetracyclines are also used prophylactically and in combination therapy with other chemotherapeutic agents against chloroquine-resistant malaria due to *Plasmodium falciparum*. More recently, tetracyclines have also been found to be active against α-proteobacteria *Wolbachia* species from the order Rickettsiales, endosymbiotic bacteria that are commensal in insects and with filarial nematodes, furthering the potential of the tetracyclines for use against a wide array of tropical diseases and agricultural pests.

7.20.5.1 Antibacterial Uses of the Tetracyclines

The tetracyclines have differences in antibacterial spectrum of action against Gram-positive and Gram-negative bacteria, and since resistance and cross-resistance to these agents can occur, susceptibility testing of clinical tetracyclines against resistant bacteria is recommended. Currently doxycycline, minocycline, and demeclocycline are approved and indicated for oral administration and systemic use in the USA, while oxytetracycline is indicated for the treatment of ophthalmic and urinary tract infections (**Figure 19**).[72]

Of the first-generation tetracyclines, oxytetracycline and tetracycline were the most potent oral agents against a wide spectrum of aerobic and anaerobic bacteria, with less toxic side effects compared to chlortetracycline.[73] The second generation brought increases in potency, spectrum, oral bioavailability, and safety, where doxycycline and minocycline in particular, demonstrated activity against resistant strains in both Gram-positive and Gram-negative bacteria and against some strains of *Mycobacterium tuberculosis*. Pharmacological studies with minocycline show that an oral dose results in higher peak levels, a longer serum half-life, and less urinary excretion compared to other tetracyclines.[74]

The third generation furthers the spectrum of activity against bacteria, and both tigecycline and PTK 0796 have potent activity against Gram-positive, Gram-negative, and atypical bacteria, especially resistant strains possessing antibiotic efflux or ribosomal protection mechanisms.[66,67,75] Both compounds are the latest tetracyclines to show desirable activity against bacteria and are being developed as intravenously administered antibiotics.

First generation

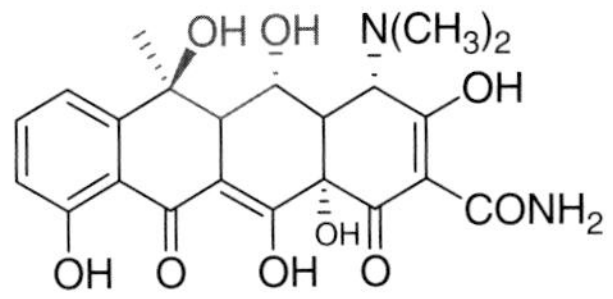

Oxytetracycline
Terramycin

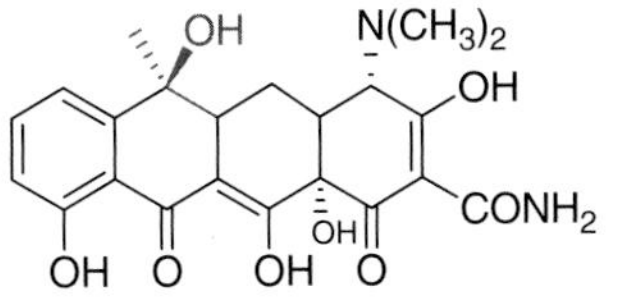

Tetracycline
Achromycin

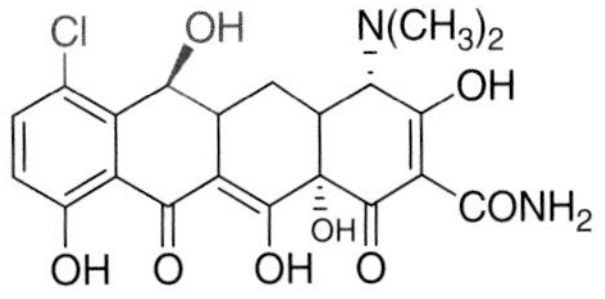

Demeclocycline
Declomycin

Second generation

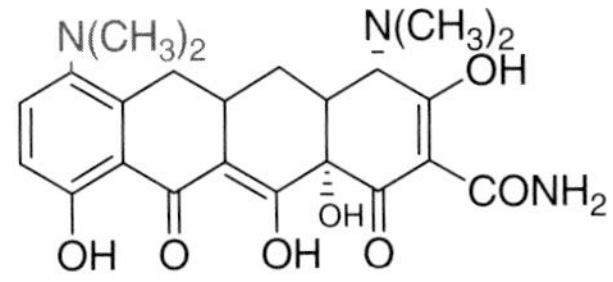

Minocycline
Minocin

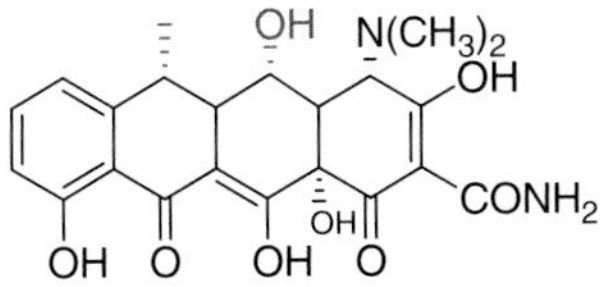

Doxycycline
Vibramycin Periostat

Primary indications	Secondary indications[a]
Bacillus anthracis	*Streptococcus* species[b]
Bacteroides species	Amebiasis[c]
Borrelia recurrentis and other species	Severe acne[d]
Chlamydia species	*Escherichia coli*[b]
Clostridium species	*Enterobacter aerogenes*[b]
Hemophilis ducreyi	*Shigella* species[b]
Mycoplasma pneumoniae	*Klebsiella* species[b]
Malaria prophylaxis–*Plasmodium species*	
Neisseria species	
Psittacosis and ornithosis	
Rickettsiae	
Treponema species	
Vibrio cholera and other species	
Yersinia pestis	

[a] When beta-lactams are contraindicated.
[b] Susceptible strains.
[c] Adjunct to amebicides.
[d] Adjunct to standard therapy.

Representative minimum inhibitory concentration values (μg mL^{-1})

	E. coli $Tc^{sensitive}$	*E. coli* $Tc^{resistant\,a}$	*S. aureus* $Tc^{sensitive}$	*S. aureus* $Tc^{resistant\,a}$
Tetracycline	0.25	64	0.06	64
Doxycycline	0.5	32	0.06	4
Minocycline	0.5	8	0.5	1

[a]Efflux-mediated resistance.

Figure 19 Approved tetracyclines and their uses against bacterial and microbial infections along with representative MIC values of tetracycline, doxycycline, and minocycline against Gram-negative and Gram-positive bacteria.

7.20.5.2 Specific Antimicrobial Uses of the Tetracyclines

7.20.5.2.1 Acne

The tetracyclines are commonly used to treat moderate to severe acne vulgaris, a dermatological disease with both bacterial and inflammatory components. Oxytetracycline has been found to reduce the volume and area of inflammation in an animal model of acne, decreasing the proportion of polymophonuclear leukocytes,[76] while minocycline and tetracycline augmented the immune response of patients with severe acne, increasing levels of interleukin-1 (IL1) in comedone lesions. Increased levels of cytokines, such as epidermal IL1, promote wound healing and resolve inflammation, independent of the presence of bacterial flora such as *Staphylococcus* and *Propionibacterium acnes*.[77] Minocycline can also inhibit enzymes responsible for cellular growth in *Propionibacterium* species, where lipase production is inhibited at much lower concentrations than needed to inhibit its growth.[78] Lymecycline, also demonstrates anti-acne activity in patients, where 300 mg once daily is effective and well tolerated, with a mechanism of action presumed similar to oxytetracycline.[79]

7.20.5.2.2 Tetracyclines and malaria

Tetracyclines have clinical utility against malaria parasites caused by *Plasmodium* species, where tetracycline, minocycline, and doxycycline are active. Doxycycline has been used clinically for malarial prophylaxis in travelers and in geographical areas where resistance to first choice antimalarial agents such as chloroquine is widespread, and is used in conjunction with quinine for malaria treatment.[80–82] The tetracyclines show activity against cultured parasites[83] and against *Plasmodium berghei* in murine models,[84] while recent reports have described more novel tetracyclines with potent in vitro and in vivo activity.[85]

Tetracyclines affect the parasite life cycle at the later trophozoite stage, where daughter parasites are found incapable of maturation and further growth is inhibited. During this stage, energy production by malarial mitochondria may be compromised by the tetracyclines,[86] where electron-transport proteins related to mitochondrial metabolite biosynthesis are depressed,[87] mitochondrial protein synthesis inhibited in a dose-dependent manner,[88] and plastid activity decreased.[86] Tetracyclines have no effect on mitochondrial membrane potential as compared to other mitochondria inhibitors capable of damaging membrane function.[89]

7.20.5.2.3 Tetracyclines and *Chlamydia*

Infections caused by intracellular obligate parasites from the genera *Chlamydia* are an important cause of human disease and are treatable using tetracycline,[90] minocycline,[91] and doxycycline,[92] all found to be active in vitro against clinical isolates of *C. pneumoniae*, *C. trachomatis*, and the veterinary pathogen *C. pecorum*. In the clinic, *C. trachomatis* is treated using doxycycline or tetracycline, which are the drugs of choice for this common infection.[93]

Tetracyclines encapsulated in neutral liposomes may increase the delivery of tetracyclines intracellularly to mammalian cells infected by *Chlamydia* spp., as compared to free tetracyclines,[94] augmenting their bacteriocidal activity. Unfortunately, newer strains of *Chlamydia* harbor resistance genes causing resistance to the tetracyclines, particularly in clinical[95] and veterinary isolates,[96] prompting changes in the chemotherapy of chlamydial infections.

7.20.5.2.4 Tetracyclines, rickettsias, and obligate intracellular parasites

Rocky Mountain Spotted Fever (RMSF), erlichiosis and Q-fever are caused by α-proteobacteria intracellular obligate parasites, some of the most common insect vector-transmitted diseases found in North America. The natural product tetracyclines demonstrated potent activity against RMSF both in vitro and in vivo, giving clinicians for the first time a treatment for this disease that had a high incidence of mortality.[97] Early mechanism of action studies suggested that they affected protein synthesis in whole bacterial cells.[98] Given the close phylogenetic relationships of the α-proteobacteria and endosymbiotic mitochondria in eukaryotes, the effects of tetracyclines show common traits as cellular growth inhibitors and as mitochondrial protein synthesis inhibitors.[99]

7.20.5.2.5 Tetracyclines and mycoplasmas

Both the natural product tetracyclines and semisynthetic derivatives possess potent activity against *Mycoplasma* spp. and *Ureaplasma* spp., and are indicated for clinical use against these organisms,[100] provided tetracycline resistance is not a complicating factor.[101] Tetracyclines also have the added effect of decreasing the release of mycoplasma neurotoxins,[102] and a long in vitro postantibiotic effect against *Mycoplasma pneumoniae*, even in the presence of serum, furthering their usefulness against *Mycoplasma* community-transmissible respiratory pathogens.

7.20.5.2.6 Tetracyclines and other endosymbiotic bacteria

Bacteria from the genus *Wolbachia* are found as endosymbionts with filarial nematodes and numerous insect species, where *Wolbachia* are passed maternally to offspring and modulate the host reproductive biology and metabolic fitness.[103] *Wolbachia* also cause sex ratio distortion in certain insects, changing their population dynamics by the feminization of males in various insect species.[104]

Oxytetracycline is active against *Wolbachia* bacteria and *Onchocerca ochengi* filarial worms, a surrogate laboratory strain used for the study of the causative agent of river blindness, *Onchocerca volvulus*, a common endemic in developing countries.[103] It is hypothesized that tetracyclines disrupt the growth and metabolism of *Wolbachia*, leading to the arrest of hostgrowth ultimately leading to macrofilaricidal activity against the *Onchocerca* nematode.

7.20.5.2.7 Tetracyclines and other microbes

Lipophilic tetracyclines are more active against a wide variety of parasites including *Giardia lamblia*,[105] *Entamoeba histolytica*,[106] and *Cryptosporidium parvum*.[70] Cell-free extracts of *Giardia* showed that all tetracyclines retain similar protein synthesis inhibition activity, but differences occur among the tetracyclines due to intracellular accumulation of the drug. Against acute toxoplasmosis caused by *T. gondii* in a murine model of infection, minocycline showed increased potency; however, less lipophilic tetracyclines were far less effective in reducing brain cysts, correlating with the ability of minocycline to cross the blood–brain barrier.[107]

7.20.5.2.8 Antibacterial quantitative structure–activity studies

Quantitative structure–activity relationship (QSAR) studies of the tetracyclines suggest that antibacterial activity depends upon inherent perturbation energy and its interaction with the ribosome, and that the electronic properties of the lower periphery C10 to C12 positions are influenced by the C6 and C7 upper periphery substituents, where subtle changes can modulate potency.[59] In silico approaches to tetracycline drug design indicate that potency is even more complex, dependent on molecular descriptors characterizing lipophilicity and inter- and intramolecular H-bonding, where changes may predictively produce tetracyclines of differing potency.[108] Furthermore, oral bioavailability and activity may be dependent upon the total number of rotatable bonds, the total H bond count, and the polar surface area,[109] in addition to lipophilicity.

7.20.6 Antibacterial and Chemical Properties of the Tetracyclines

7.20.6.1 Tetracyclines, Bacterial Uptake, and Membrane Activity

In fractionated Gram-negative *E. coli* cell membranes, passage and uptake of tetracyclines occurs by passive diffusion, with no differences between membranes prepared from tetracycline-susceptible and -resistant cells.[110] Partitioning experiments show that tetracycline transfer through lipid phases is governed by two tetracycline conformers, a zwitterionic (A) and a neutral form (B) (**Figure 20**), the latter responsible for crossing the lipoidal membrane, although the concentration of the neutral species in an aqueous phase is small.[111]

In Gram-negative bacteria, increasing the hydrophobicity of a tetracycline decreases its ability to cross the outer cell membrane, where OmpC and OmpF channel proteins allow pore mediated diffusion via a tetracycline–Mg complex, attracted by the Donnan electrochemical potential established by the periplasmic membrane.[112] Tetracyclines freely diffuse

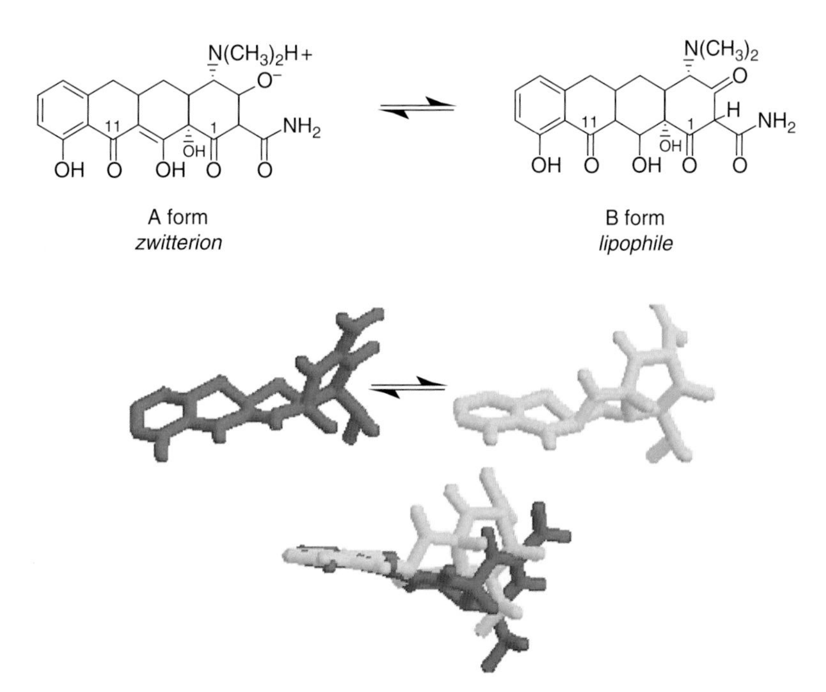

Figure 20 The zwitterionic A form and neutral and lipophilic B form of the tetracyclines and their structural differences shown by superimposition. (Drawn using Alchemy 2000 software with PM3 and MM3 geometry optimization, Tripos, Inc., St. Louis, MO.)

through the periplasm, crossing the bilayer of the cytoplasmic membrane in both Gram-negative and Gram-positive bacteria, as presumably electroneutral species in an energy independent process through apolar segments of the cell membrane. Transport systems driven by energy and pH-dependent translocating proteins remain elusive, although uptake of tetracycline into the cytosol of all bacterial cells is found to be dependent on the total cellular energization and proton symport.[113]

The uptake of tetracycline, 2*N*-tetracycline nitrile, and a glycylcycline analog of sancycline were studied using liposomes and *E. coli* cells both possessing the Tet repressor protein, capable of sensing tetracyclines, coupled to a fluorescent protein reporter. This system allows observation of the membrane uptake kinetics and the trafficking of tetracyclines, readily measured as fluorescence changes, and the permeation coefficients calculated.[114] Tetracycline and the glycylcycline analog showed similar permeation properties in both liposomes and whole cells, while the 2*N*-tetracycline nitrile derivative was at least 400-fold less permeant, perhaps explaining its lack of antibacterial activity.

The molecular weight barrier of *E. coli* cells for antibacterial agents is approximately 600 atomic mass units, where agents exceeding this limit penetrate poorly and are inactive. Within the tetracycline family, penetrative ability with *E. coli* show parabolic relationships comparing lipophilicity to activity, with the outer *E. coli* envelope preferentially excluding more hydrophobic tetracyclines.[115] Mutant cells lacking the polysaccharide envelope are routinely used for assessing the activity of hydrophobic tetracyclines such as the 13-alkylthiotetracyclines, where once the outer membrane is removed they regain Gram-negative activity.[45]

7.20.6.2 Typical and Atypical Tetracyclines

From studies of the different tetracyclines and their effects on Gram-negative bacterial membranes the tetracyclines can be further divided into two classes based upon mechanism: the typical tetracyclines, which act as classical protein synthesis inhibitors with no adverse effects on membranes, and the atypical tetracyclines, those that can perturb membranes modifying macromolecular synthesis and cellular pathways, leading to a bactericidal mode of action and lytic release of cellular constituents from the cell.[116] Atypical tetracyclines possessing lipophilic character, such as chelocardin and anhydrotetracycline, inhibit protein synthesis at much higher concentrations, but decrease radiolabel precursor incorporation into protein, DNA, and RNA, suggesting membrane perturbation activity and a loss of precursor uptake capability.[117] Lipophilic tetracyclines were also found to change cytoplasmic membrane shape and cell morphology, while lysis of spheroplasts derived from *E. coli* cells was spared with some atypical tetracyclines.

Perturbation activity of the atypical tetracyclines may arise from an increased retention time within the membrane, where the equilibrium between the zwitterionic (**Figure 20**) and lipophilic species is altered, favoring the B form and affecting energy-dependent enzymes and biosynthetic reactions crucial to membrane and cell wall synthesis and integrity. Some tetracyclines, such as sancycline, fall into neither category and are capable of potent protein synthesis inhibition while affecting macromolecular synthesis and membrane function.

7.20.6.3 Mechanism of Action and Antibacterial Activity

The primary antibacterial mechanism of the typical tetracyclines is attributed to the inhibition of protein synthesis at the ribosome, as demonstrated by inhibition in whole cells[118] and in cell-free translation systems in both the Gram-positive and Gram-negative bacteria.[119,120] However, other macromolecular synthesis pathways (DNA, RNA, and cell wall synthesis) may appear affected due to the downregulation of crucial enzymes in their biosynthesis. Protein synthesis inhibition can also affect membrane synthesis and lipid turnover, changing envelope proteins, cellular morphology, and the uptake of DNA and RNA precursors.[121]

Tetracyclines bind to isolated *E. coli* ribosomes with strong affinity and specificity to the 30S subunit, delineated as the head region,[122] with multiple low-affinity binding sites located on the 50S subunit, accompanying the S proteins, namely, S5, S7, S13, S14, and S18 subunit proteins.[123]

Radiolabeled tetracycline, ^{3}H-Tc, used as a photoaffinity label to specifically probe the tetracycline–*E. coli* ribosome interaction, showed a high degree of specificity and affinity covalently binding to the 30S and S7 subunits.[124] Tetracycline labeling and ribosomal component mapping studies uses the inherent photolability of ^{3}H-Tc to form free radical species,[125] whose specificity is enhanced by the use of radical quenching reagents, although competing side-reactions and tetracycline photoadducts may occur.

Once reversibly bound to the ribosome, tetracycline interferes with the binding of codon-specific aminoacyl transfer RNA at the A site via a noncompetitive allosteric effect, stopping the aminoacyl transfer reaction, backing up amino acid–tRNA complexes within the cytosol, and inhibiting protein synthesis.[126] Biochemical footprinting studies suggest that the S7 and 16S rRNA bases G693, A892, U1052, C1054, G1300, and G1338 are involved in the tetracycline–ribosome interaction,[127] although direct binding and other conformational changes may occur in 16S rRNA.

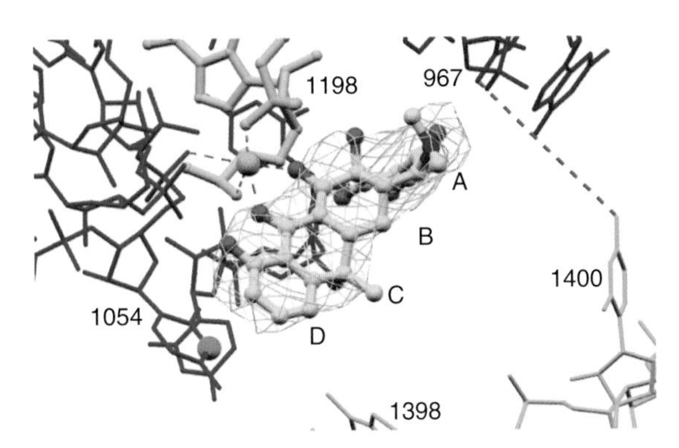

Figure 21 X-ray crystal structure of tetracycline binding to the 30S portion of the ribosome. (The graphic was provided by Dr. V. Ramakrishnan, MRC, Laboratory of Molecular Biology, Cambridge, UK, author of the study.)

The structural basis for tetracycline binding to prokaryotic ribosomes has been studied via x-ray crystallography using ribosomes from the thermophilic bacterium *Thermus thermophilus*, where 30S ribosomal subunits were crystallized and soaked or cocrystallized with tetracycline and the x-ray structures determined diffracting with a resolution of 3.4 Å (**Figure 21**).[128] The findings confirm a primary tetracycline binding site near the acceptor region for tRNA forming a binding pocket 20 Å wide and 7 Å deep with an irregular groove of helix 34 and interacting with its carbohydrate–phosphate scaffold. The hydrophilic lower peripheral region of tetracycline makes numerous H-bonds with phosphate oxygens at G1197–G1198, while a lone magnesium ion forms salt bridges with the lower peripheral region, which is present even in the absence of tetracycline, maintaining the 30S tertiary structure.[129]

Overall, the effect of tetracycline binding to the A-site is to hinder rotation of the tRNA after ribosomal binding via steric interactions and the dislodging of the complex from the ribosome. This structural binding model is consistent with tetracycline biochemical footprinting studies and other factors related to peptidyl transfer and protein synthesis, and represents the current understanding of the mode of action of tetracyclines and the structural biology of ribosomal inhibition.

7.20.7 Antibacterial Resistance to the Tetracyclines

Antibacterial resistance to the tetracyclines occurred after their mass introduction into clinical and agricultural practices, as evidenced by the low incidence of tetracycline resistance in coliforms obtained from cored glacial ice pre-dating 1947.[130] Clinically, widespread tetracycline use has resulted in a loss of activity against numerous bacterial pathogens, manifest by three main mechanisms or phenotypes of tetracycline resistance: antibiotic efflux via transport proteins, ribosomal protection, and tetracycline inactivation mechanisms. Tetracycline resistance mechanisms and the genetic determinants responsible have been studied extensively, and are found to be the result of mobile genes within the environment that move freely between bacterial species. Their genetic origins have been systematically studied and labeled, given a letter or number designation in order of their discovery, spanning many different clinically relevant Gram-positive and Gram-negative bacteria (**Table 1**).[131,132]

7.20.7.1 The Tet Repressor Resistance System

The expression of resistance mechanisms to the tetracyclines are regulated by intricate cellular mechanisms, where the Tet repressor protein, one of the best-studied systems, regulates transcription of drug efflux proteins in Gram-negative bacteria. In the absence of tetracycline, Tet repressor homodimers bind to *tet* operators, blocking transcription of genes coding for the repressor and an efflux protein. Binding of a tetracycline–Mg complex in nanomolar concentrations to the repressor decreases affinity to the DNA operator region, and transcription of the tetracycline efflux protein and a repressor protein occurs.[136] Different tetracyclines also bind with different affinity to the repressor, causing increases or decreases in transcription induction ability. Anhydrotetracycline showed the most potent binding with the repressor, 35-fold better than tetracycline, while other derivatives modified at C4 through C7 showed considerably less activity.[137]

X-ray crystallography studies of the repressor show a binding pocket for tetracycline that also accommodates a divalent cation along the lower peripheral region between C11 and C12, along with water molecules forming a water zipper with an extensive H-bonding network.[138]

Table 1 Tetracycline resistance determinants and their phenotype mechanism of action. (The table was provided by Dr. Laura McMurry, The Center for Adaptation Genetics and Drug Resistance, Tufts University School of Medicine, Boston, MA)

Determinant (class)	*Mechanism*
Tet A, B, C, D, E, F, G, H	Efflux, group 1
I, J, Y, Z, 30, 31, 33	Efflux, group 1
Tet K, 38	Efflux, group 2
Tet L (plasmid)	Efflux, group 2
Tet L (chromosome)	Efflux, group 2
Tet M, O, 32, tet	Ribosomal protection
Tet P (protein A)	Efflux, group 4
Tet P (protein B)	Ribosomal protection
Tet Q, S, T, W, 36	Ribosomal protection
Tet V	Efflux, group 5
Tet X, 37	Degradation
Tet 34	Unknown
Tet 35	Efflux, group 7
OtrA	Ribosomal protection
otrB, tcr3 (tcrC)	Efflux, group 3
OtrC	Efflux, group 6
no designation	16S rRNA mutation

See [133–135].

7.20.7.2 Tetracycline Resistance by Efflux

The Tet repressor system occurs mainly in Gram-negative bacteria, where expression leads to the production of membrane proteins that remove tetracyclines from within the cell with differing affinity for tetracycline substrates.[48] Minocycline, the glycylcyclines, and the aminomethylcyclines are active against efflux mediated resistant bacteria, and are not subject to efflux, while C13-substituted tetracyclines can act to inhibit tetracycline efflux proteins.[46] The major tetracycline efflux proteins are further subdivided into seven groups based on sequence homology, substrate specificity, and genetic origins, and are labeled Groups 1–7 (**Table 2**).

Group 1 include mostly Gram-negative bacterial proteins, of which Tet(A) and Tet(B) are the most studied. This group is composed of 12-membrane spanning helices with protruding cytoplasmic and periplasmic loops, exchanging a proton for tetracycline during antiport, although the binding site has not been precisely determined.[139]

Group 2 describes efflux proteins found primarily in Gram-positive bacteria with 14 transmembrane helices which are found in bacteria that are usually sensitive to second and third-generation tetracyclines.

Groups, 3–7 cover efflux proteins that are found in *Streptomyces* tetracycline producer strains, and in *Clostridium, Mycobacterium,* and *Corynebacterium* species, and are not as well characterized.

7.20.7.3 Tetracycline Resistance by Ribosomal Protection

Gram-negative and Gram-positive bacteria can also shield ribosomes from the action of tetracyclines via ribosomal protection proteins, the most studied of which are Tet(O) and Tet(M), cytoplasmic proteins which mediate resistance by dislodging tetracycline from the ribosome. Both Tet(O) and Tet(M) are dependent upon GTP as an energy source in order to function.[140–142] Tetracyclines prevent the binding of amino acid–tRNA, and induce a conformational change that results in a nonproductive protein elongation cycle. Tet(O) binds to the ribosome, forcing tetracycline release and the continuation of the protein elongation cycle. Whether the Tet(O) is bound with higher affinity after tetracycline binding or blocks an open A-site is still open to question, although ribosomal protection mechanisms are increasingly important in antibacterial chemotherapy.

The third-generation tetracyclines are active against bacteria harboring ribosomal protection mechanisms including Tet(O) and Tet(M) proteins, giving hope that compounds effective in inhibiting or bypassing efflux mechanisms can also have extended activity on multiple resistance mechanisms and phenotypes.

7.20.7.4 Tetracycline Resistance by Inactivation

Only one substantiated case where tetracycline is chemically modified by enzymatic processes has been reported. The *tet*(X) gene encodes a flavin-monooxygenase, a 44 kDa protein which inserts a hydroxyl group at position C11a that further degrades the tetracycline scaffold, although it is not a common mechanism of resistance, found primarily in anaerobic bacteria.[143]

In the future, in order to produce newer and more active antibacterial tetracyclines resistance mechanisms and their SARs must be further understood. The clinical utility of newer tetracycline rests on the ability to inhibit the growth of bacteria, those susceptible and those resistant to the tetracyclines by the above underlying mechanisms.

7.20.8 Chemical Properties Related to Biological Activity

7.20.8.1 Structure

Spectroscopic techniques have been used to study the conformational structures possible with the tetracyclines under different conditions and in the biological milieu, and show that the A-ring along the C4a–C12a axis is twisted at lower pH values, gradually unfolding as the pH is raised, forming an extended conformation at neutral pH and greater.[144] Such conformations are also sensitive to changes in metal binding and the C4 dimethylamino group configuration, where epimerization also leads to conformational changes.[145] Circular dichroism spectra of oxytetracycline in aqueous or organic solvents can distinguish between two major conformers, where oxytetracycline base shifts between a zwitterionic form, predominant in aqueous environments, and the lipophilic species prevalent in organic solvents (**Figure 19**).[146] The interconversion between the two forms and their conformational adaptability is a significant factor in their overall antibacterial activity.

X-ray crystal structures of the tetracyclines have also distinguished between the two forms, and other crystal structures have been determined for several of the natural product tetracyclines and semisynthesis products, and the results corroborate the studies performed by circular dichroism.[147] In the natural tetracyclines, the x-ray structures show extensive intramolecular H bonding along the lower peripheral region and the C2 amide nitrogen to the C3 enolate proton, while chemically modified tetracyclines, particularly those structurally changed along the lower periphery, show drastic changes in chemical conformation.

Tautomerism, hydrogen shifts due to deprotonations and the formation of keto-enolate groups, also plays a major role in the chemical behavior and biological activity of tetracyclines. With four deprotonation sites, tetracyclines can form up to 64 different tautomeric structures. Computational studies show that the extended form is the most stable, although six different tautomers have been calculated to exist within 10 kcal mol^{-1} of each other.[148,149] Nuclear magnetic resonance (NMR) spectroscopy and ^{13}C shifts have also been used to distinguish between the major conformers, based upon the shift values along the upper peripheral region.

The ability of tetracyclines to change conformations and tautomerize in different solvents, molecular environments, and biological matrices is just one of reasons the tetracyclines may harbor bioactivity in both prokaryotes and eukaryotes, the ability to interact with many structurally different receptors by conformational adaptation.

7.20.9 Chemical Properties of the Tetracyclines

7.20.9.1 Metal Binding

One of the prominent chemical features of the lower periphery intact tetracyclines is their ability to form complexes with mono-, di-, and trivalent cations, resulting in different chemical conformations and stabilized structures via coulombic interactions. Solvent forces, pH, and the protonation state of the tetracyclines all direct the nature and degree of metal binding, and differences in affinity are directed by the inherent chemical structure of the tetracycline and its lower periphery functional groups. Spectroscopic studies have shown that the C11–C12 keto-enol functional group within the BCD chromophore is the primary binding site for calcium and magnesium ions at low pH, while other sites are operable at higher pH values.[150,151] The tetracycline to metal binding ratio for calcium and magnesium are 1:2, with differences in avidity, dependent upon the binding ion and pH, while bimetallic complexes are evident at higher pH values. Calcium forms a primary complex first at positions C11–C12, followed by a second at C12–C12a–C1. Magnesium binds first to C11 and C12 and secondarily at the C3–C4 functional groups.[152] Tetracyclines that are not lower periphery intact, those chemically modified at C11 and C12, lose metal avidity, forming 1:1 complexes through other probable tetracycline-metal binding sites at the C12a and C4 positions.[153]

Most metals bind avidly with tetracyclines and include the Group I alkali ions, Li, Na, and K, in order of decreasing affinity, with association constants around 50 mM.[154] Tetracyclines bind to Group II and III metals and transition metals, including biologically relevant Fe, Co, Mn, and Zn, among others, with similar stoichiometries and binding sites within the tetracycline scaffold. Their relative order of decreasing affinity for metals is $Fe^{3+} > Al^{3+} = Cu^{2+} > Co^{2+} = Fe^{2+} > Zn^{2+} > Mn^{2+} > Mg^{2+} > Ca^{2+}$, depending upon the solvent and the pH system used, where the conditional stability constants range from a high $\log K_1$ with Fe^{3+} (9.9) to a low of $\log K_1$ value of 3.0 for Ca^{2+} binding.[155]

Tetracyclines that are modified along the upper periphery also display changes in metal binding capacity. The C4 dedimethylamino tetracyclines show slightly altered binding characteristics in the presence of Ca^{2+} and Mg^{2+}, with a ligand:metal binding ratio of 1:2 for Ca and 1:1 for Mg binding, indicating the role that the C4 functional group may play in metal binding, at least at higher pH values.[156] Methacycline is incapable of forming binuclear complexes with both Ca^{2+} and Mg^{2+}, changing the ligand:metal coordination complex and ultimately influencing bioactivity.[157]

Once chelated to metals, tetracyclines can act as ionophores, transporting metals through lipophilic phases and cellular membranes into intracellular compartments and changing divalent ion dynamics in situ.[158] While only chlortetracycline has been thoroughly studied, it remains to be determined if other tetracyclines can act as ionophores, as it is assumed that tetracyclines are transported in vivo as Ca^{2+} or Mg^{2+} chelates. Complexation characteristics are related to biological properties, with >99% of a tetracycline complexed and bound and <1% in the free uncomplexed state.[159]

Tetracycline–metal bridge complexes can also form with DNA[160] and serum albumin,[161] further influencing their tertiary structure and activity in biological systems. Antibacterial activity of the tetracyclines is modulated in the presence of certain calcium and magnesium salts both in vitro and in vivo. Some cations and their counterions may slightly increase activity against bacteria, while others may decrease or abolish their activity altogether.[162]

Iron can also affect the bioactivity of the tetracyclines, where binding can decrease activity, which is reversible upon the addition of iron chelation agents.[163] The accumulation of tetracyclines in bacteria is also sensitive to iron concentrations, where high levels can block uptake and activity, while bacteria in iron-depleted media are more susceptible to growth suppression.[164]

Tetracyclines can also complex with zinc, a finding that may have an impact on the study of their activity against zinc-dependent matrix metalloproteinases (MMPs) involved in the maintenance of cell and tissue growth, structural morphogenesis and tissue support. MMPs possess multiple zinc and calcium catalytic and binding domains, especially in MMP-7 or matrilysin, a protein linked to a number of pathological processes related to inflammation. The inhibition of matrilysin by doxycycline shows lack of binding to the zinc catalytic site, but binding sites are found adjacent to the zinc and calcium atom binding domains in the drug–protein complex. Proximal binding of doxycycline with these structural centers induces conformational changes leading to matrilysin destabilization and degradation.[165] These results are consistent with other studies showing a mechanism of action by inhibition of MMP activity via the chelation of calcium, or co-factor inhibition.[166]

7.20.9.2 Tetracycline Photophysics, Phototoxicity, and Antioxidant Activity

The natural tetracyclines, B- or C-ring modified, or interrupted analogs and clinically used compounds, can sensitize both Type I photoreactions, those that can produce radical or reactive oxygen species (ROS), or Type II reactions, those that can generate biologically damaging singlet oxygen, $^1O_2$, via energy transfer from the photosensitizer to molecular oxygen.

Type I reactions can occur in hydroxide solutions,[167] while aerated solutions absorb oxygen when irradiated by ultraviolet (UV) light,[168] forming tetracycline A-ring quinone substructures.[169] Other C4 photoproducts and ROS have been detected with tetracycline UV irradiation and oxygenation, where 4-dedimethylamino tetracycline was found produce a radical species at C4 that was long-lived before C4 quinone formation,[170] while photolysis of tetracycline forms anhydrotetracycline with reducing agents.[171]

Superoxide anion radical (O_{2-}) formation by UV irradiation of different tetracyclines decreased in the order chlortetracycline > oxytetracycline > demeclocycline ≫ doxycycline = tetracycline = minocycline. These findings trend with clinical reports that chlortetracycline and oxytetracycline are potent photosensitizers, while minocycline lacks phototoxicity, suggesting that O_{2-} production may be involved in tetracycline-induced phototoxicity.[172] Using *E. coli* cells, photoilluminated tetracyclines were found to produce O_{2-}, H_2O_2, and ·OH radicals,[173] while ·OH generation by the tetracyclines caused damage to DNA, lipids and carbohydrates in the presence of iron and copper salts.[174]

Type II reactions of the tetracyclines yields $^1O_2$ following the same trend in ability, where demeclocycline > tetracycline > minocycline, and correlating with clinical observations, suggesting that $^1O_2$ generation occurs present during photosensitization.[175] Studies using polymorphonuclear leukocytes also showed photosensitization caused by tetracyclines following the same order of compound activity which was inhibited by azide and enhanced by deuterium oxide, again implicating $^1O_2$.[176]

Screening of tetracyclines for phototoxic potential in vitro uses the 3T3 neutral red uptake assay, where fibroblast cells at different concentrations of drug are exposed to UV-A or UV-A plus UV-B light followed by dye uptake measurement, a measure of cell viability.[177] In fibroblasts, the order of cytotoxic potential of tetracyclines was found to parallel that of susceptibility to Type I and II reactions, where 4-dedimethylamino sancycline ≫ doxycycline > tetracycline > minocycline.[178]

Metal binding of the tetracyclines may decrease the formation of Type I and II photoproducts and photosensitization,[179] while autooxidative processes can release both product types, even in the dark and in the presence of cations.[180]

In patients, photoreaction to the tetracyclines, especially doxycycline, is dose-dependent, where phototoxicity is observed in 3%, 20%, or 40% of patients receiving 100, 150, or 200 mg day^{-1}, respectively.[181] The order of phototoxicity ranked in order of most to least clinically phototoxic is: demeclocycline > doxycycline > tetracycline > methacycline, while minocycline is seldom phototoxic in humans. The order of phototoxicity is closely related to the lipophilicity of the tetracyclines, with the exception of minocycline, which is the most lipophilic tetracycline followed by demeclocycline. This anomaly suggests that drug compartmentalization may be responsible in part for phototoxicity, perturbing and damaging cell membranes and cellular organelles such as mitochondria, where doxycycline has been found to localize.[182]

Conversely, tetracyclines also act as antioxidants, able to scavenge free radical species and the secondary messengers NO and peroxynitrite (ONOO–). Chemically, tetracyclines are also potent scavengers of the ·OH,[183] OCl,[184] and ONOO–,[185] while in polymophonuclear cells, intracellular calcium movement triggers the oxidative burst and chemotaxis that is readily inhibited by tetracyclines.[186] Oxytetracycline alters the morphology of inflammation filtrates, decreasing the number of polymorphonuclear cells present while doxycycline and C4 dedimethyl tetracyclines inhibit NO production in lipopolysaccharide-stimulated macrophage cells, presumably through posttranscriptional regulation of inducible nitric oxide synthase mRNA.[187]

7.20.9.3 Physicochemical Properties, Pharmacokinetics, and Formulations

As solids, tetracyclines are yellow polymorphic powders, dependent upon the method of manufacturing and isolation of the final product and their salts, as evidenced by the two different crystalline forms of oxytetracycline obtained in organic or aqueous solvents.[188] The water content of the powders varies between 3% and 18%, depending upon the final solvents used in its manufacturing process, and are hygroscopic powders as HCl salts. The microcrystalline state can affect dissolution, where differences are notable for tetracycline free base and HCl salt forms, dependent on hydration state, which can affect bioavailability.[189] The aqueous solubility of the tetracyclines changes at different pH values, as the relative proportion of zwitterionic forms of the tetracyclines changes[190] according to the species generated, the solvent, and the pH. The accompanying anion also affects aqueous solubility. Solubility of tetracycline is approximately 2 mg mL^{-1}, the HCl salt 11 mg mL^{-1}, and the salt of phosphoric acid 15 mg mL^{-1}, all subject to the common ion effect, where the Cl^- anion presence in vivo from gastric HCl can decrease aqueous solubility.[191] Between pH values of 5.0 and 6.5 tetracyclines are the most lipophilic, a trend that applies to most of the tetracyclines with a relative order of partitioning in octanol and water at pH 7.5 of minocycline > doxycycline > methacycline > demeclocycline > tetracycline > oxytetracycline.[192] The lipophilicity also affects oral absorbtion, where it is inversely related to mean plasma concentration after oral dosing, and the ability to penetrate the blood–brain barrier, where only minocycline and doxycycline cross to any measurable extent.

As a tetracycline becomes more lipophilic it also becomes more serum protein bound, changing overall bioavailability and the maximal concentration attainable and the tetracycline half-life. The relative order of serum protein binding in plasma follows the order for lipophilicity: minocycline > doxycycline > methacycline > demeclocycline > tetracycline > oxytetracycline.

Calcium-rich food and dairy products, particularly milk, can decrease the oral absorbtion of tetracycline, while minocycline is less affected,[193] while both compounds show decreased absorbtion in the presence of ferrous iron and magnesium polyvalent cations and sodium-based antacids. A combination of bismuth subsalicylate, tetracycline salts, and metronidazole is used widely for the treatment of stomach ulcers caused by *Helicobacter pylori* infections, although some formulations may adversely affect tetracycline bioavailability.[194]

7.20.10 Nonantibiotic Effects of the Tetracyclines

More recent studies have demonstrated that minocycline can affect the physiology of neurodegenerative processes in Alzheimer's disease,[195] cerebral ischemia,[196] Parkinson's disease,[197] multiple sclerosis,[198] and amyotrophic lateral sclerosis,[199] presumably via the interference of mitochondrial biochemistry and other cellular processes. Minocycline was also neuroprotective in acute spinal cord injury, where it stopped the permeability transition-mediated release of mitochondrial cytochrome *c*, a secondary injury mechanism in spinal cord trauma.[200]

7.20.10.1 Tetracyclines and Eukaryotic Cell Membranes

The effect of tetracyclines on eukaryotic cell membranes is less well known, and it was thought that the selectivity of tetracyclines as antibiotics was due to their inability to cross mammalian cell membranes. That notion was dispelled when tetracyclines were found to inhibit the growth of certain mammalian cell lines at concentrations similar to the MIC values needed to inhibit bacterial growth. Oral epithelial cells were found to allow minocycline to accumulate 40-fold compared to the surrounding medium, influenced by pH, organic and inorganic ions, and competitive inhibitors, such as organic cations, particularly alkaloids, and by doxycycline,[201] while human polymorphonuclear leukocytes show significant uptake of tetracycline with a half-life of uptake of approx. 15 min and an apparent activation energy of cellular uptake of 52.2 kJ mol^{-1}.[202]

Oral absorption of the tetracyclines in the intestine may be dependent on carrier proteins, as demonstrated in an animal model, where at low concentrations transport was linear and saturable.[203] However, oral absorption can also be impeded by di- and trivalent metals in the gastrointestinal tract, a factor taken into account where calcium- and magnesium-containing foodstuffs can decrease therapeutic efficacy. Metal binding limitations have also restricted the development of a Caco-2 transport model for the tetracyclines, where correlating transepithelial transport rates with tetracycline structures has not yet been reported due to ionic interference of apical-to-basolateral flux due to cell culture media, although a minimum calcium model has been developed.[204]

The cytotoxicity of tetracyclines on cultured mammalian cells shows a dose-dependent effect at concentrations higher than those expected to be reached in vivo. Toxicity against kidney and liver cell lines is absent, consistent with the finding that tetracyclines are not hepatotoxic or nephrotoxic in vivo.[205] Gingival cultured human fibroblasts showed no indication of chromosomal aberrations for the major tetracyclines studied, even at high concentrations and in the presence of exogenous metabolic activation by mitochondrial supernatant, indicating the lack of clastogenic activity for the parent tetracyclines and possible oxidative metabolites.[206]

Tetracyclines lacking a C4 dimethylamino group have been shown to induce apoptosis at the level of the caspase cascade and on Bcl-2 and c-myc mRNA expression in the J774 macrophage cell line, inducing programmed cell death in a dose-dependent manner.[207] Dedimethylamino tetracyclines show that some members may be potential anticancer agents, triggering apoptosis, activating the proteolytic caspases and modulating physiology resulting in cell death.

Against cultured articular chondrocytes doxycycline demonstrated activity in inhibiting the expression of a type X collagen epitope, and was shown to decrease collagenase and gelatinase activities and thus matrix degradation.[208] In a series of tetracyclines examined, doxycycline was the most potent, while minocycline was found to inhibit the production of inducible nitric oxide synthase by inhibition of mRNA expression[209] substantiating their beneficial effects in the treatment of rheumatoid arthritis.

7.20.10.2 Tetracyclines and Hematological Effects

Both granulocytes and agranulocytes have been shown to be biologically affected by tetracyclines. In alveolar macrophage, tetracycline can accumulate with a ratio of cellular to extracellular concentration of around 0.5, indicating efficient intracellular uptake.[210] Phagocytosis was inhibited only at high concentrations of tetracycline, while there is some evidence to suggest that phagocytic activity may be increased at therapeutic levels of drug, at least in vitro.[211] Doxycycline may also influence binding of mononuclear leukocytes and neutrophils to hematological macromolecules, inhibiting cation-dependent adherence at therapeutic concentrations, while higher concentrations enhanced neutrophil binding.[212] Doxycycline can also decrease lipopolysaccharide-stimulated inducible nitric oxide synthase activity, suggesting that modulation of NO levels may result in anti-inflammatory activity.[213]

Tetracycline and minocycline also suppress the mitotic response of lymphocytes stimulated with interleukin-1B, with minocycline exhibiting more potent activity at supratherapeutic levels[214] while doxycycline reduced DNA synthesis in vitro in mitogen-stimulated lymphocytes derived from normal patients.[215] Doxycycline modulates human natural killer cell-mediated cytotoxicity against target cells, where inhibition is not correlated with direct cytotoxicity but is the result of immunosuppresion activity,[216] while other tetracyclines showed no effect. At therapeutic concentrations doxycycline can also suppress immunoglobulin secretion by activated B cells, having significant implications for tetracycline therapy in immunoglobulin-mediated autoimmune or allergic diseases.[217]

Activated T cells and their proliferation, which are responsible in part for the inflammation caused by rheumatoid arthritis, are inhibited by minocycline and restored by the addition of exogenous calcium, indicating that modulation of T-cell activity by minocycline is chelation dependent.[218]

In leukocytes, and at therapeutic levels, tetracyclines can modulate chemotaxis and migration, while leaving response to human complement unaffected. This may be one component of the effect of tetracyclines in the treatment of acne vulgaris, the abatement of white blood cell infiltration during inflammation.[219]

7.20.10.3 Tetracyclines and Effect on Matrix Metalloproteinases and Cytokines

In 2001, doxycycline was the first tetracycline approved for the inhibition of MMPs, and is now marketed under the tradename Periostat, a low-dose purportedly nonantibiotic formulation. Minocycline was the first tetracycline found to reduce collagenase activity[220] in a number of different clinically relevant models of tissue destruction, and could also change prostaglandin and cytokine levels in vitro,[221] affecting tissue remodeling systems and resorbtion processes. Tetracyclines can affect angiogenesis,[222] modulate cell invasion and migration,[223] and alter cellular metabolism in fibroblasts,[224] all through modulating levels of MMP activity. The tetracyclines can also scavenge peroxynitrite, affecting inflammation processes and decreasing secondary effects of tissue damage.[225]

Since the initial description of their anti-MMP activity the tetracyclines have found activity in a broad variety of cellular enzymes implicated in inflammation and tissue degradation diseases. A partial list of the MMP and inflammatory activities reported of minocycline, doxycycline, and the 4-dedimethylamino tetracyclines is shown in **Table 2**, and represents a selective compilation relevant to chemotherapeutic areas and medicinal chemistry.

Tetracyclines have been shown to decrease levels of cellular cytokines, where minocycline lowered tumor necrosis factor (TNF)-α and (TNF)-α mRNA levels by 40% in human peripheral monocytes,[237] decreased levels of phospholipase A_2,[238] and modulated α-melanocyte-stimulating hormone production by keratinocytes in vitro, an anti-inflammatory peptide produced in response to minocycline.[239]

Because of their exceptional activity against MMPs and cytokines, and their use as antibacterial agents, doxycycline and minocycline have been studied in animal models and in humans for several MMP-related pathologies, and have shown activity in preventing MMP-related coronary syndromes,[240] modulation of ventricular remodeling in myocardial infarction,[241] and reducing MMP-9 activity in abdominal aortic aneurysms in animal models[242] and in humans.[243]

Minocycline, doxycycline, and the 4-dedimethyl tetracyclines are the most studied tetracyclines as MMP inhibitors, and one common feature is their similar lipophilic character, in the order: 4-dedimethyl sancycline (also known as CMT-3) > minocycline > doxycycline. While the SARs of other position-modified tetracyclines need to be explored to derive correlations of activity and structure, the limited number of compounds mentioned account for a majority of the literature on their effects on MMPs and cytokine modulation.

Table 2 Minocycline, doxycycline and 4-dedimethyltetracycline and their effects on matrix-metalloproteinases in vitro and in vivo

Minocycline	Gelatinase B inhibition[226]
	Angiogenesis inhibition[227]
Doxycycline	MMP-9 and angiogenesis[228]
	Neutrophil collagenase[229]
	Anticollagenase activity[230]
	MMPs:arthritic and diabetic rates[231]
4-Dedimethyl Tcs	MMPs and metastasis[232]
	NO synthesis and cytokines[233]
	Prevention of ARDS[234]
	NO synthases and NO[235]
	Chondroprotective effect[236]

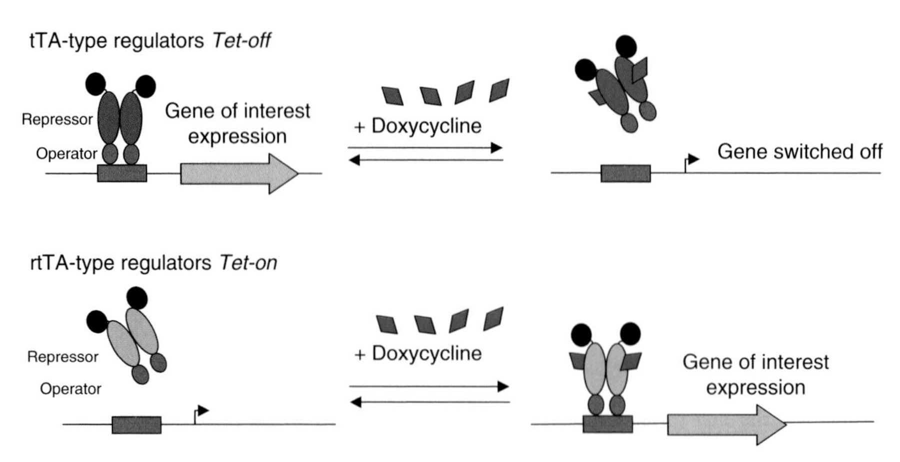

Figure 22 The Tet-Off and Tet-On genetic constructs and their regulation of gene expression by doxycycline. (The graphic was provided by Prof. Dr. Wolfgang Hillen, Lehrstuhl für Mikrobiologie, Institut für Mikrobiologie, Biochemie und Genetik, Friedrich-Alexander Universität, Erlangen, Germany.)

7.20.11 Other Tetracycline-Based Technologies: The Tet-On–Tet-Off Operons and Gene Expression Systems

One of the major scientific fronts based on the tetracyclines that has emerged over the past decade is the Tet-On–Tet-Off gene expression system initially developed by Gossen and Bujard,[244] a genetically engineered operon based on the Tet repressor protein-transcriptional transactivator construct that can be precisely controlled by a binding tetracycline. The construct allows the binding of a tetracycline to the repressor resulting in the repression or expression of a chosen gene (**Figure 22**). The tetracycline-controlled regulatory system has become a widely used system in a variety of prokaryotic,[244] eukaryotic,[245] and transgenic applications,[246] where Tet systems provide insights into gene expression, biological development, and the fundamentals and phenotypes responsible for disease processes.

In its most basic description, the Tet repressor is fused with a viral VP16 transcription factor, binding to *tetO* and controlling downstream expression of cDNA. The technology is based upon two control mechanisms that in the presence of a low dose ($\mathrm{ng\,mL^{-1}}$) of a tetracycline turn the selected gene off (tTA dependent) or on (rtTA dependent), resulting in the loss or gain, respectively, of a desired protein and phenotype. In both systems gene expression is regulated from an exogenously supplied tetracycline, allowing strict reduction or induction control over the gene of interest. The tTA system responds primarily to tetracycline, doxycycline, and anhydrotetracycline, while the rtTA system responds to doxycycline and anhydrotetracycline.[247] Both systems use the Tet repressor protein, although the rtTA system has point mutations within the repressor that expresses genes in the presence of doxycycline, the effector molecule of choice in tissue cultures and organisms.

Tet system based regulation has been the subject of reviews[248] concerning the molecular biology, transgenics, and its many applications described for the structure–function of gene expression and pathways in a wide variety of transgenic organisms. The Tet systems have been used to control gene function in yeast and in transgenic insects,[249] and have also been used to study gene function in hundreds of other applications.

The selective and conditional expression of genes and their resulting phenotypes and roles in pathogenesis can now be efficiently probed, generating specific biological targets for pharmacological study both in vitro and in vivo.

7.20.12 Conclusions and the Future of the Tetracyclines

Tetracyclines, as dynamic molecules, possess unique and potent biological activities against a wide array of prokaryotic and eukaryotic targets, and in some cases the SARs and chemical prerequisites for activity are known, especially against bacteria, while against other targets, they are virtually unknown. It has been shown that they have agonist, antagonist, and physiological properties against a wide array of proteins, enzymes, and macromolecular targets, and are effective in mammalian models of chemotherapy against inflammation and eukaryotic disease states, and hold promise for treatment in humans.

From inspection of the literature it becomes clearly evident that the tetracyclines are chemically and biologically 'promiscuous,' able to adapt to different chemical and biological environments, changing conformations interchangeably and readily. While discovered and initially developed as antibiotics, they also hold promise as nonantibiotic compounds for future study and use.

The scientific fronts that are immediately recognizable for the future fall into three main categories, namely, newer and more potent tetracyclines and antibiotic resistance, the nonantibiotic uses of tetracyclines targeted toward inflammation and tissue destructive diseases, and the use of the tetracyclines in the Tet repressor controlled gene switch, which has a promising future all to itself in target identification, transgenics, and gene therapy. In view of the widespread uses of the tetracyclines and their value as chemotherapeutic agents, the further study and application of medicinal chemistry principles is warranted.

References

1. Duggar, B. M. *Ann. NY Acad. Sci.* **1948**, *51*, 177–181.
2. Sum, P. E.; Lee, V. J.; Testa, R. T.; Hlavka, J. J.; Ellestad, G. A.; Bloom, J. D.; Gluzman, Y.; Tally, F. P. *J. Med. Chem.* **1994**, *37*, 184–188.
3. Bhatia, D.; Bowser, T.; Chen, J.; Ismail, M.; McIntyre, L.;Mechiche, M.; Nelson, M.; Ohemeng, K.; Verma, A. *43rd Interscience Conference on Antimicrobial Agents and Chemotherapy,* Chicago, IL, **2003**, Abs 2420.
4. Duggar, B. M. *Ann. NY Acad. Sci.* **1948**, *51*, 1171–1181.
5. Findley, A. C.; Hobby, G. L.; Pan, S. Y.; Regna, J. B.; Routien, D. B.; Seeley, D. B.; Shull, G. M.; Sobin, B. A.; Solomens, I. A.; Vinson, J. W. et al. *Science* **1950**, *111*, 85.
6. Conover, L. H.; Moreland, W. T.; English, A. R.; Stephens, C. R.; Pilgrim, F. J. *J. Am. Chem. Soc.* **1953**, *75*, 5455.
7. McCormick, J. R. D.; Joachim, U. H.; Jensen, E. R.; Johnson, S.; Sjolander, N. O. *J. Am. Chem. Soc.* **1965**, *87*, 1793–1794.
8. Bu'lock, J. D. In *Comprehensive Organic Chemistry.*; Haslam, E., Ed.; Pergamon Press: Oxford, UK, 1979; Vol. 5, pp 927–987.
9. Hostalek, Z.; Tinterova, M.; Jechova, V.; Blumauerova, M.; Suchy, J.; Vanek, Z. *Biotechnol. Bioeng.* **1969**, *11*, 539–548.
10. Behal, V.; Cudlin, J.; Vanek, Z. *Folia Microbiol.* **1969**, *14*, 117–120.
11. McCormick, J. R. D.; Jensen, E. R. *J. Am. Chem. Soc.* **1969**, *91*, 206.
12. Goodman, J. J.; Marishin, M.; Backus, E. J. *J. Bacteriol.* **1955**, *69*, 70–72.
13. Goodman, J. J.; Matrishin, M. *Nature* **1968**, *219*, 291–292.
14. McCormick, J. R. D.; Sjoander, N. O.; Hirsch, U.; Jensen, E. R.; Doershuk, A. P. *J. Am. Chem. Soc.* **1957**, *79*, 4561–4563.
15. Hochstein, F. A.; Schach von Wittenau, M.; Tanner, F. W.; Murai, K. *J. Am. Chem. Soc.* **1960**, 5934-5937.
16. Miller, M. W.; Hochstein, F. A. *J. Org. Chem.* **1962**, *27*, 2525–2528.
17. Wells, J. S.; O'Sullivan, J.; Aklonis, C.; Ax, H. A.; Tymiak, A. A.; Kirsch, D. R.; Trejo, W. H.; Principe, P. *J. Antibiotics* **1992**, *45*, 1892–1898.
18. Mitscher, L. A.; Rosenbrook, W., Jr.,; Andres, W. W.; Egan, R. S.; Schenck, J.; Juvarkar, J. V. *Antimicrob. Agents Chemother.* **1970**, *10*, 38–41.
19. Patel, M.; Gullo, V. P.; Hegde, V. R.; Horan, A. C.; Gentile, F.; Marquez, J. A.; Miller, G. H.; Puar, M. S.; Waitz, J. A. *J. Antibiotics* **1987**, *40*, 1408–1413.
20. Patel, M.; Gullo, V. P.; Hegde, V. R.; Horan, A. C.; Marquez, J. A.; Vaughan, R.; Puar, M. S.; Miller, G. H. *J. Antibiotics* **1987**, *40*, 1414–1418.
21. Breinholt, J.; Jensen, G. W.; Kjaer, A.; Olsen, C. E.; Rosendahl, C. N. *Acta Chem. Scand.* **1997**, *51*, 855–860.
22. Kabutao, C.; Silverton, J. V.; Akiyama, T.; Sankawa, V.; Hutchinson, R. D.; Steyn, P. S.; Vleggar, R. *J. Chem. Soc. Chem. Comm.* **1976**, 728-729.
23. Shu, Y. Z.; Cutrone, J. Q.; Klohr, S. E.; Huang, S. *J. Antibiot.* **1995**, *48*, 1060–1065.
24. Schneider, S. In *Tetracyclines in Biology, Chemistry and Medicine*; Nelson, M. L., Hillen, W., Greenwald, R. A., Eds.; Birkhauser: Basel, 2001, pp 65–104.
25. Garmaise, D. L.; Chu, D. T.; Bernstein, E.; Inaba, M.; Stamm, J. M. *J. Med. Chem.* **1979**, *22*, 559–564.
26. Gottstein, W. J.; Minor, W. F.; Cheney, L. C. *J. Am. Chem. Soc.* **1958**, *81*, 1198–1201.
27. Hughes, D.; Wilson, W. L.; Buttefield, A. G.; Pound, N. J. *J. Pharm. Pharmacol.* **1974**, *26*, 79–80.
28. Doershuk, A. P.; Bitler, B. A.; McCormick, J. R. D. *J. Am. Chem. Soc.* **1955**, 77, 4687.
29. Yuen, P. H.; Sokoloski, T. D. *J. Pharm. Sci.* **1977**, *66*, 1648–1650.
30. Blanchflower, J. W.; McCracken, R. J.; Haggan, A. S.; Kennedy, G. D. *J. Chromatogr. B, Biomed. Sci. Appl.* **1997**, *692*, 351–360.
31. Remmers, E. G.; Sieger, G. M.; Doerschuk, A. P. *J. Pharm. Sci.* **1963**, *52*, 752–756.
32. Nelson, M. L.; Ismail, M. Y.; McIntyre, L.; Bhatia, B.; Viski, P.; Rennie, G.; Andorsky, D.; Messersmith, D.; Stapleton, K.; Dumornay, J. et al. *J. Org. Chem.* **2003**, *68*, 5838–5851.
33. Boothe, J. H.; Bonvicino, G. E.; Waller, C. Y.; Petisi, J. P.; Wilkinson, R. W.; Broshard, R. B. *J. Am. Chem. Soc.* **1958**, *80*, 1654–1657.
34. Stephens, C. R.; Conover, L. H.; Pasternack, R.; Hochstein, F. A.; Moreland, W. T.; Regna, P. P.; Pilgrim, F. J.; Brunings, J. K.; Woodward, R. B. *J. Am. Chem. Soc.* **1954**, *76*, 3568–3575.
35. McCormick, J. R. D.; Fox, S. M.; Smith, L. L.; Bitler, B. A.; Reichenthal, J.; Origoni, V. E.; Muller, W. H.; Winterbottom, R.; Doerschuk, A. P. *J. Am. Chem. Soc.* **1957**, *79*, 2849–2858.
36. Esse, R. C.; Lowery, J. A.; Tamorria, C. R.; Sieger, G. M. *J. Am. Chem. Soc.* **1964**, *86*, 3875–3877.
37. Valcalvi, U.; Campanella, G.; Pacini, N. *Gazz. Chim. Ital.* **1963**, *93*, 916–928.
38. Green, A.; Boothe, J. H. *J. Am. Chem. Soc.* **1960**, *82*, 3949–3953.
39. Hochstein, F. A.; Stephens, C. R.; Conover, L. H.; Regna, P. P.; Pasternack, R.; Brunings, K. J.; Woodward, R. B. *J. Am. Chem. Soc.* **1953**, *75*, 5455–5475.
40. Bernardi, L.; DeCastiglione, R.; Colonna, V.; Masi, P.; Mazzoleni, R. *Il farmaco-Ed. Sci.* **1975**, *29*, 902–909.
41. Walton, V. C.; Howlett, M. R.; Selzer, G. B. *J. Pharm. Sci.* **1970**, *59*, 1160–1164.
42. Cullen, S. I.; Crounse, R. G. *J. Invest. Dermatol.* **1965**, *45*, 263–268.
43. Blackwood, R. K.; Beereboom, J. J.; Rennhard, H. H.; Schach von Wittenau, M.; Stephens, C. R. *J. Am. Chem. Soc.* **1961**, *83*, 2773–2775.
44. Blackwood, R. K.; Beereboom, J. J.; Rennhard, H. H.; Schach von Wittenau, M.; Stephens, C. R., Jr. *J. Am. Chem. Soc.* **1963**, *85*, 3943–3953.
45. Nelson, M. L.; Park, B. H.; Andrews, J. S.; Georgian, V. A.; Thomas, R. C.; Levy, S. B. *J. Med. Chem.* **1993**, *36*, 370–377.
46. Nelson, M. L.; Levy, S. B. *Antimicrob. Agents Chemother.* **1999**, *43*, 1719–1724.
47. Nelson, M. L.; Park, B. H.; Levy, S. B. *J. Med. Chem.* **1994**, *37*, 1355–1361.

48. Stephens, C. R.; Beereboom, J. J.; Rennhard, H. H.; Gordon, P. N.; Murai, K.; Blackwood, R. K.; Schach von Wittenau, M. *J. Am. Chem. Soc.* **1963**, *85*, 2643–2652.
49. Pirotte, B.; Felekidis, A.; Fontaine, M.; Demonceau, A.; Noels, A. F.; Delarge, J.; Chizhevsky, L. T.; Zinevich, T. V.; Pisareva, I. V.; Bregadze, V. I. *Tetrahedron Lett.* **1993**, *34*, 1471–1474.
50. Centers for Disease Control, *Morb. Mortal. Wkly Rep.* **2001**, *50*, 909–919.
51. Nadelman, R. B.; Nowakowski, J.; Fish, D.; Falco, R. C.; Freeman, K.; McKenna, D.; Welch, P.; Marcus, R.; Aguero-Rosenfeld, M. E.; Dennis, D. T. et al. *N. Engl. J. Med.* **2001**, *345*, 79–84.
52. Thomas, J. G.; Metheny, R. J.; Karakiozis, J. M.; Wetzel, J. M.; Krout, R. *J. Adv. Dent. Res.* **1998**, *12*, 32–39.
53. Barden, T. C.; Buckwalter, B. L.; Testa, R. T.; Petersen, P. J.; Lee, V. J. *J. Med. Chem.* **1994**, *37*, 3205–3211.
54. Hlavka, J. J.; Schneller, A.; Krazinski, H.; Boothe, J. H. *J. Am. Chem. Soc.* **1962**, *84*, 1426–1430.
55. Martell, M. J., Jr.; Ross, A. S.; Boothe, J. H. *J. Med. Chem.* **1967**, *10*, 359–363.
56. McCormick, J. R. D.; Jensen, E. R.; Miller, P. A.; Doershuk, A. P. *J. Am. Chem. Soc.* **1960**, *82*, 3381–3386.
57. Church, R.; Schaub, R. E.; Weiss, M. J. *J. Org. Chem.* **1971**, *36*, 723–725.
58. Martell, M. J., Jr.; Boothe, J. H. *J. Med. Chem.* **1967**, *10*, 44–46.
59. Peradejordi, F.; Martin, A. N.; Cammarata, A. *J. Pharm. Sci.* **1971**, *60*, 576–582.
60. Johnson, D. M.; Jones, R. N. *Diag. Microbiol. Infect. Dis.* **1996**, *24*, 53–57.
61. Sum, P. E.; Petersen, P. J.; Jacobus, N. V.; Testa, R. T.; Lang, S. A. *212th ACS National Meeting*, Orlando, FL, **1996**, Abs MEDI-188.
62. Sum, P. E.; Petersen, P. *Bioorg. Med. Chem. Lett.* **1999**, *9*, 1459–1462.
63. Projan, S. J. *Pharmacotherapy* **2000**, *20*, 219S–223S.
64. Chopra, I. *Expert Opin. Invest. Drugs* **1994**, *3*, 191–193.
65. Zhanel, G. G.; Homenuik, K.; Nichol, K.; Noreddin, A.; Vercaigne, L.; Embil, J.; Gin, A.; Karlowsky, J. A.; Hoban, D. J. *Drugs* **2004**, *64*, 63–88.
66. Traczewski, M. M.; Brown, S. D. *41st Interscience Conference on Antimicrobial Agents and Chemotherapy*, Chicago, IL, **2003**, Abs 2463.
67. Macone, A.; Donatelli, J.; Dumont, T.; Levy, S. B.; Tanaka, S. K. *41st Interscience Conference on Antimicrobial Agents and Chemotherapy*, Chicago, IL, **2003**, Abs 2439.
68. McKenney, D.; Quinn, J. M.; Jackson, C. L.; Guilmet, J. L.; Landry, J. A.; Tanaka, S. K.; Cannon, E. P. *41st Interscience Conference on Antimicrobial Agents and Chemotherapy*, Chicago, IL, **2003**, Abs 2627.
69. Endermann, R.; Ladel, C. H.; Broetz-Oesterhelt, H.; Labischinski, H. *14th European Congress of Clinical Microbiology and Infectious Diseases*. Prague, **2004**, Abs P929.
70. Levy, S. B.; Nelson, M. L. Tetracycline compounds for the treatment of *Cryptosporidium parvum* related disorders. U.S. Patent 6,833,365, Dec 21, **2004**.
71. Daschner, F. *Zentralblatt Bakteriol. Parasit. Infektionskrankheit. Hygiene Medi. Mikrobiol. Parasitol.* **1977**, *239*, 527–534.
72. *Physicians Desk Reference* 59th ed.; Thomson PDR: **2005**.
73. Kylin, O.; Wall, E. U. *Excerpta Medica Physiol. Biochem. Pharmacol.* **1955**, *8*, 1114–1115.
74. Williams, J. D.; Farrell, W.; Wood, M. *Proceedings 9th Int. Congr. Chemother.* **1976**, 357-362.
75. Gales, A. C.; Jones, R. N. *Diagn. Microbiol. Infecti. Dis.* **2000**, *36*, 19–36.
76. Dalziel, K.; Dykes, P. J.; Marks, R. *Br. J. Exp. Pathol.* **1987**, *68*, 67–70.
77. Eady, E. A.; Ingham, E.; Walters, C. E.; Cove, J. H.; Cunliffe, W. J. *J. Invest. Dermatol.* **1993**, *101*, 86–91.
78. Akamatsu, H.; Horio, T. *Jap. Pharmacol. Ther.* **2000**, *28*, 1107–1109.
79. Dubertret, L.; Alirezai, M.; Rostain, G.; Lahfa, M.; Forsea, D.; Dimitrie, A.; Niculae, B.; Simola, M.; Horvath, A.; Mizzi, F. *Eur. J. Dermatol.* **2003**, *13*, 44–48.
80. Colwell, E. J. *SE Asian J. Trop. Med. Public Health* **1972**, *3*, 190–197.
81. Rieckmann, K. H.; Powell, R. D.; McNamara, J. V.; Willerson, D., Jr.; Lass, L.; Frischer, H.; Carson, P. E. *Am. J. Trop. Med. Hyg.* **1971**, *20*, 811–815.
82. Pang, L. W.; Limsomwong, N.; Boudreau, E. F.; Singharaj, P. *Lancet* **1987**, *1*, 1161–1164.
83. Divo, A. A.; Geary, T. G.; Jensen, J. B. *Antimicrob. Agents Chemother.* **1985**, *27*, 21–27.
84. Jacobs, R. L.; Koontz, L. C. *Exp. Parasitol.* **1976**, *40*, 116–123.
85. Draper, M. P.; Ohemeng, K.; Jones, G.; Bhatia, B.; Assefa, H.; Honeyman, L.; Chen, J.; Garrity-Ryan, L.; Nelson, M.; Rosenthal, P. J. et al. *Interscience Conference on Antimicrobial Agents and Chemotherapy*, New Orleans, LA, **2005**, Abs P423.
86. Lin, Q.; Katakura, K.; Suzuki, M. *FEBS Lett.* **2002**, *515*, 71–74.
87. Prapunwattana, P.; O'Sullivan, W. J.; Yuthavong, Y. *Mol. Biochem. Parasitol.* **1988**, *27*, 119–124.
88. Kiatfuengfoo, R.; Suthiphongchai, T.; Prapunwattana, P.; Yuthavong, Y. *Mol. Biochem. Parasitol.* **1989**, *34*, 109–115.
89. Srivastava, I. K.; Rottenberg, H.; Vaidya, A. B. *J. Biol. Chem.* **1997**, *272*, 3961–3966.
90. Chirgwin, K.; Roblin, P. M.; Hammerschlag, M. R. *Antimicrob. Agents Chemother.* **1989**, *33*, 1634–1635.
91. Pudjiatmoko; Fukushi, H.; Ochiai, Y.; Yamaguchi, T.; Hirai, K. *Microbiol. Immunol.* **1998**, *42*, 61–63.
92. Bowie, W. R. *Proceedings 6th Int. Symp. Human Chlamydial Infect*, 524–527, **1986**.
93. Samra, Z.; Rosenberg, S.; Soffer, Y.; Dan, M. *Diag. Microbiol. Infect. Dis.* **2001**, *39*, 177–179.
94. Al-Awadhi, H.; Stokes, G. V.; Reich, M. *J. Antimicrob. Chemother.* **1992**, *30*, 303–311.
95. Lefevre, J. C.; Lepargneur, J. P.; Guion, D.; Bei, S. *Pathol. Biol.* **1997**, *45*, 376–378.
96. Lenart, J.; Andersen, A. A.; Rockey, D. D. *Antimicrob. Agents Chemother.* **2001**, *45*, 2198–2203.
97. Pearson, M. *The Million Dollar Bugs*; Putnam: New York, 1969.
98. Kohno, S.; Natsume, Y.; Shishido, A. *Jap. J. Med. Sci. Biol.* **1961**, *14*, 213–222.
99. Andersson, S. G. E.; Zomorodipour, A.; Andersson, J. O.; Sicheritz-Ponten, T.; Alsmark, U. C. M.; Podowski, R. M.; Naslund, A. K.; Eriksson, A. S.; Winkler, H. H.; Kurland, C. G. *Nature* **1998**, *396*, 133–140.
100. Bonissol, C.; Stoiljkovic, B.; Kona, P.; Fleury, J. *Pathol. Biol.* **1983**, *31*, 492–496.
101. Cummings, M. C.; McCormack, W. M. *Antimicrob. Agents Chemother.* **1990**, *34*, 2297–2299.
102. Thomas, L.; Bitensky, M. W. *J. Exp. Med.* **1966**, *124*, 1089–1098.
103. Langworthy, N. G.; Renz, A.; Mackenstedt, U.; Henkle-Duhrsen, K.; Bronsvoort, M. B. D. C.; Tanya, V. N.; Donnelly, M. J.; Trees, A. J. *Proc. R. Soc. Lond. B* **2000**, *267*, 1063–1069.
104. Holden, P. R.; Jones, P.; Brookfield, J. F. Y. *Genet. Res.* **1993**, *62*, 23–29.
105. Edlind, T. D. *Antimicrob. Agents Chemother.* **1989**, *33*, 2144–2145.

106. Katiyar, S. K.; Edlind, T. D. *Antimicrob. Agents Chemother.* **1991**, *35*, 2198–2202.
107. Chang, H. R.; Comte, R.; Piguet, P. F.; Pechere, J. C. *Antimicrob. Agents Chemother.* **1991**, *27*, 639–645.
108. Cronin, M. T. D.; Aptula, A. O.; Dearden, J. C.; Duffy, J. C.; Netzeva, T. I.; Patel, H.; Rowe, P. H.; Schultz, T. W.; Worth, A. P.; Voutzoulidis, K. et al. *J. Chem. Inf. Comput. Sci.* **2002**, *42*, 869–878.
109. Veber, D. F.; Johnson, S. R.; Cheng, H. Y.; Smith, B. R.; Ward, K. W.; Kopple, K. D. *J. Med. Chem.* **2002**, *45*, 2615–2623.
110. Franklin, T. J. *Biochem. J.* **1971**, *123*, 267–273.
111. Terada, H.; Inagi, T. *Chem. Pharm. Bull.* **1975**, *23*, 1960–1968.
112. Nikaido, H.; Thanassi, D. G. *Antimicrob. Agents Chemother.* **1993**, *37*, 1393–1399.
113. Munske, G. R.; Lindley, E. V.; Magnuson, J. A. *J. Bacteriol.* **1984**, *158*, 49–54.
114. Sigler, A.; Schubert, P.; Hillen, W.; Niederweis, M. *Eur. J. Biochem.* **2000**, *267*, 527–534.
115. Clark, D. *Microbios* **1984**, *40*, 107–115.
116. Oliva, B.; Guay, G.; McNicholas, P.; Ellestad, G.; Chopra, I. *Antimicrob. Agents Chemother.* **1992**, *36*, 913–919.
117. Rasmussen, B.; Noller, H. F.; Daubresse, G.; Oliva, B.; Misulovin, Z.; Rothstein, D. M.; Gluzman, Y.; Tally, F. P.; Chopra, I. *Antimicrob. Agents Chemother.* **1991**, *35*, 2306–2311.
118. Gale, E. F.; Folkes, J. P. *Biochemistry* **1953**, *53*, 493–498.
119. Day, L. E. *J. Bacteriol.* **1966**, *91*, 1917–1923.
120. Laskin, A. I.; May Chan, W. *Biochem. Biophys. Res. Commun.* **1964**, *14*, 137–142.
121. Hirashima, A.; Childs, G.; Inouye, M. *J. Mol. Biol* **1973**, *79*, 373–389.
122. Tritton, T. R. *Biochemistry* **1977**, *16*, 4133–4138.
123. Last, J. A. *Biochim. Biophys. Acta* **1969**, *195*, 506–514.
124. Goldman, R. A.; Hasan, T.; Hall, C. C.; Strycharz, W. A.; Cooperman, B. S. *Biochemistry* **1983**, *22*, 359–368.
125. Cooperman, B. S. *Ann. NY Acad. Sci.* **1980**, *346*, 302–323.
126. Semenkov, Y. P.; Makarov, E. M.; Makhno, V. I.; Kirillov, S. V. *FEBS Lett.* **1982**, *144*, 125–129.
127. Moazed, D.; Noller, H. F. *Nature* **1987**, *327*, 389–394.
128. Brodersen, D. E.; Clemons, W. M., Jr.; Carter, A. P.; Morgan-Warren, R. J.; Wimberly, B. T.; Ramakrishnan, V. *Cell* **2000**, *103*, 1143–1154.
129. Carter, A. P.; Clemons, W. M.; Brodersen, D. E.; Morgan-Warren, R. J.; Wimberly, B. T.; Ramakrishnan, V. *Nature* **2000**, *407*, 340–348.
130. Dancer, S. J.; Shears, P.; Platt, D. J. *J. Appl. Microbiol.* **1997**, *82*, 597–609.
131. Levy, S. B. *J. Antimicrob. Chemother.* **1989**, *24*, 1–3.
132. Levy, S. B.; McMurry, L. M.; Burdett, V.; Courvalin, P.; Hillen, W.; Roberts, M. C.; Taylor, D. E. *Antimicrob. Agents Chemother.* **1989**, *33*, 1373–1374.
133. Levy, S. B.; McMurry, L. M.; Barbosa, T. M.; Burdett, V.; Courvalin, P.; Hillen, W.; Roberts, M. C.; Rood, J. I.; Taylor, D. E. *Antimicrob. Agents Chemother.* **1999**, *43*, 1523–1524.
134. McMurry, L. M.; Levy, S. B. Tetracycline Resistance in Gram-Positive Bacteria. In *Gram-Positive Pathogens*; Fischetti, V. A., Novick, R. P., Feretti, J. J., Portnoy, D. A., Roberts, M. C., Rood, J. I., Taylor, D. E., Eds.; ASM Press: Washington, DC, 2000, pp 660–677.
135. Chopra, I.; Roberts, M. *Microbiol. Mol. Biol. Revi.* **2001**, *65*, 232–260.
136. Hillen, W.; Wissmann, A. *Top. Mol. Struct. Biol.* **1989**, *10*, 143–162.
137. Degenkolb, J.; Takahashi, M.; Ellestad, G. A.; Hillen, W. *Antimicrob. Agents Chemother.* **1991**, *35*, 1591–1595.
138. Kisker, C.; Hinrichs, W.; Tovar, K.; Hillen, W.; Saenger, W. *J. Mol. Biol.* **1995**, *247*, 260–280.
139. Yamaguchi, A.; Udagawa, T.; Sawai, T. *J. Biol. Chem.* **1990**, *265*, 4809–4813.
140. Spahn, C. M. T.; Blaha, G.; Agrawal, R. K.; Penczek, P.; Grassucci, R. A.; Trieber, C. A.; Connell, S. R.; Taylor, D. E.; Nierhaus, K. H.; Frank, J. *Mol. Cell* **2001**, *7*, 1037–1045.
141. Burdett, V. *J. Bacteriol.* **1993**, *175*, 7209–7215.
142. Burdett, V. *J. Bacteriol.* **1996**, *178*, 3246–3251.
143. Yang, W.; Morre, I. F.; Koteva, K. P.; Bareich, D. C.; Hughes, D. W.; Wright, G. D. *J. Biol. Chem.* **2004**, *279*, 52346–52352.
144. Grosheva, V. I. *Antibiot. Khimioter.* **1992**, *37*, 11–14.
145. Mitscher, L. A.; Bonacci, A. C.; Sokoloski, T. D. *Antimicrob. Agents Chemother.* **1968**, *8*, 78–86.
146. Hughes, L. J.; Stezowski, J. J.; Hughes, R. E. *J. Am. Chem. Soc.* **1979**, *101*, 7655–7656.
147. Stezowski, J. J. *J. Am. Chem. Soc.* **1976**, *98*, 6012–6018.
148. Othersen, O. G.; Beierlein, F.; Lanig, H.; Clark, T. *J. Phys. Chem. B* **2003**, *107*, 13743–13749.
149. Duarte, H. A.; Carvalho, S.; Paniago, E. B.; Simas, A. M. *J. Pharm. Sci.* **1999**, *88*, 111–120.
150. Mitscher, L. A.; Bonacci, A. C.; Sokoloski, T. D. *Tetrahedron Lett.* **1968**, 5361–5364.
151. Mitscher, L. A.; Bonacci, A. C.; Slater, B. J.; Hacker, A. K.; Sokoloski, T. D. *Antimicrob. Agents Chemother.* **1969**, *9*, 111–115.
152. Schmitt, M. O.; Schneider, S. *Phys. Chem. Comm.* **2000**, *9*, 42–55.
153. Doluisio, J. T.; Martin, A. N. *J. Med. Chem.* **1963**, *6*, 16–20.
154. Coibion, C.; Laszlo, P. *Biochem. Pharmacol.* **1979**, *28*, 1367–1372.
155. Martin, R. B. *Met. Ions Biol. Syst.* **1985**, *19*, 19–52.
156. Newman, E. C.; Frank, C. W. *J. Pharm. Sci.* **1976**, *65*, 1728–1732.
157. Lambs, L.; Berthon, G. *Inorgan. Chim. Acta* **1988**, *151*, 33–43.
158. White, J. R.; Pearce, F. F. *Biochemistry* **1982**, *21*, 6309–6312.
159. Berthon, G.; Brion, M.; Lambs, L. *J. Inorg. Biochem.* **1983**, *19*, 1–18.
160. Kohn, K. W. *Nature* **1961**, *191*, 1156–1158.
161. Doluisio, J. T. *J. Med. Chem.* **1963**, *6*, 20–23.
162. Gupta, R. P.; Yadav, B. N.; Tiwari, O. P.; Srivastava, A. K. *Inorg. Chim. Acta* **1979**, *32*, L95–L96.
163. Al-Bakieh, N. H.; Weinberg, E. D. *Microbios Lett.* **1981**, *18*, 13–16.
164. Avery, A. M.; Goddard, H. J.; Sumner, E. R.; Avery, S. V. *Antimicrob. Agents Chemother.* **2004**, *48*, 1892–1894.
165. Garcia, R. A.; Pantazatos, D. P.; Gessner, C. R.; Go, K. V.; Woods, V. L.; Villareal, F. J. *Mol. Pharmacol.* **2005**, *67*, 1128–1136.
166. Smith, J. G. N.; Mickler, E. A.; Hasty, K. A. *Arthritis Rheum.* **1999**, *42*, 1140–1146.
167. Lagercrantz, C.; Yhland, M. *Acta Chem. Scand.* **1963**, *17*, 2568–2570.
168. Wiebe, J. A.; Moore, D. E. *J. Pharm. Sci.* **1977**, *66*, 186–189.
169. Davies, A. K.; McKellar, J. F.; Phillips, G. O.; Reid, A. G. *J. Chem. Soc., Perkin Trans. 2, Phys. Orga. Chem.* **1979**, 369–375.

170. Moore, D. E.; Fallon, M. P.; Burt, C. D. *Int. J. Pharm.* **1983**, *14*, 133–142.
171. Hasan, T.; Allen, M.; Cooperman, B. S. *J. Org. Chem.* **1985**, *50*, 1755–1757.
172. Li, A. S.; Roethling, H. P.; Cummings, K. B.; Chignell, C. F. *Biochem. Biophys. Res. Commun.* **1987**, *146*, 1191–1195.
173. Martin, J. P., Jr.; Colina, K.; Logsdon, N. *J. Bacteriol.* **1987**, *169*, 2516–2522.
174. Quinlan, G. J.; Gutteridge, J. M. *Free Radical Biol. Med.* **1988**, *5*, 341–348.
175. Glette, J.; Sandberg, S. *Biochem. Pharmacol.* **1986**, *35*, 2883–2885.
176. Glette, J.; Sandberg, S.; Haneberg, B.; Solberg, C. O. *Antimicrob. Agents Chemother.* **1984**, *26*, 489–492.
177. Duffy, P. A.; Bennett, A.; Roberts, M.; Flint, O. P. *Mol. Toxicol.* **1988**, *1*, 579–587.
178. Zerler, B.; Roemer, E.; Raabe, H.; Sizemore, A.; Reeves, A.; Harbell, J. Evaluation of the Phototoxic Potential of Chemically Modified Tetracyclines with the 3T3 Neutral Red Uptake Photoxicicity Test. In *Progress in the Reduction, Refinement and Replacement of Animal Experimentation*; Balls, M., van Zeller, A. M., Halder, M. E., Eds.; Elsevier Science: Amsterdam, 2000, pp 545–554.
179. Riaz, M.; Pilpel, N. *J. Pharm. Pharmacol.* **1984**, *36*, 153–156.
180. Kruk, I.; Lichszteld, K.; Michalsak, T.; Nizinkiewicz, K.; Wronska, J. *J. Photochem. Photobiol. B* **1992**, *14*, 329–343.
181. Layton, A. M.; Cunliffe, W. J. *Clin. Exp. Dermatol.* **1993**, *18*, 425–427.
182. Salet, C.; Moreno, G. *J. Photochem. Photobiol. B* **1990**, *5*, 133–150.
183. van Barr, H. M.; van de Kerkhof, P. C.; Mier, P. D.; Happle, R. *Br. J. Dermatol.* **1987**, *117*, 131–132.
184. Wasil, M.; Halliwell, B.; Moorhouse, C. P. *Biochem. Pharmacol.* **1988**, *37*, 775–778.
185. Whiteman, M.; Kaur, H.; Halliwell, B. *Ann. Rheum. Dis.* **1996**, *55*, 383–387.
186. Takeshige, K.; Matsumoto, T.; Nabi, Z. F.; Minakami, S. *Adv. Exp. Med. Biol.* **1982**, *141*, 453–461.
187. Amin, A. R.; Patel, R. N.; Thakker, G. D.; Lowenstein, C. J.; Attur, M. G.; Abramson, S. B. *FEBS Lett.* **1997**, *410*, 259–264.
188. Grakovskaia, L. K.; Nesterova, L. *Antibiotiki* **1982**, *27*, 815–820.
189. Miyazaki, S.; Nakano, M.; Arita, T. *Chem. Pharm. Bull.* **1975**, *23*, 552–558.
190. Stephens, C. R.; Murai, K.; Brunings, K. J.; Woodward, R. B. *J. Am. Chem. Soc.* **1956**, *78*, 4155–4158.
191. Bogardus, J. B.; Blackwood, R. K., Jr. *J. Pharm. Sci.* **1979**, *68*, 1183–1184.
192. Colaizzi, J. L.; Klink, P. R. *J. Pharm. Sci.* **1969**, *58*, 1184–1189.
193. Leyden, J. J. *Int. Congr. Symp. Series Roy. Soc. Med.* **1985**, *95*, 87–92.
194. Healy, D. P.; Dansereau, R. J.; Dunn, A. B.; Clendening, C. E.; Mounts, A. W.; Deepe, G. S., Jr. *Ann. Pharmacother.* **1997**, *31*, 1460–1464.
195. Sopher, B. L.; Fukuchi, K. I.; Kavanagh, T. J.; Furlong, C. E.; Martin, G. M. *Mol. Chem. Neuropathol.* **1996**, *29*, 153–168.
196. Arvin, K. L.; Han, B. H.; Du, Y.; Lin, S. Z.; Paul, S. M.; Holtzman, D. M. *Ann. Neurol.* **2002**, *52*, 54–61.
197. Hirsch, E. C.; Breidert, T.; Rousselet, E.; Hunot, S.; Hartmann, A.; Michel, P. P. *Ann. NY Acad. Sci.* **2003**, *991*, 214–228.
198. Giuliani, F.; Metz, L. M.; Wilson, T.; Fan, Y.; Bar-Or, A.; Yong, V. W. *J. Neuroimmunol.* **2005**, *158*, 213–221.
199. Kriz, J.; Nguyen, M. D.; Julien, J. P. *Neurobiol. Dis.* **2002**, *10*, 268–278.
200. Teng, Y. D.; Choi, H.; Onario, R. C.; Zhu, S.; Desilets, F. C.; Lan, S.; Woodard, E. J.; Snyder, E. Y.; Eichler, M. E.; Friedlander, R. M. *Proc. Natl. Acad. Sci. USA* **2004**, *101*, 3071–3076.
201. Brayton, J. J.; Yang, Q.; Nakkula, R. J.; Walters, J. D. *J. Periodontol.* **2002**, *73*, 1267–1272.
202. Laufen, H.; Widfeuer, A. *Arzneimittel-Forschung* **1989**, *39*, 233–235.
203. Meshali, M. M.; Attia, I. E. *Pharmazie* **1978**, *33*, 107–108.
204. Nicklin, P. L.; Irwin, W. J.; Hassan, I.; Mackay, M. *Int. J. Pharmaceut.* **1995**, *123*, 187–197.
205. Bacon, J. A.; Linseman, D. A.; Raczniak, T. J. *Toxicol. In Vitro* **1990**, *4*, 384–388.
206. Rokukawa, A.; Tsutsui, T. *Nippon Shishubyo Gakkai Kaishi* **1997**, *39*, 338–347.
207. D'Agostino, P.; Ferlazzo, V.; Milano, S.; La Rosa, M.; Di Bella, G.; Caruso, R.; Barbera, C.; Grimaudo, S.; Tolomeo, M.; Feo, S.; Cillari, E. *Int. Immunopharmacol.* **2003**, *3*, 63–73.
208. Davies, S. R.; Cole, A. A.; Schmid, T. M. *J. Biol. Chem.* **1996**, *271*, 25966–25970.
209. Sadowski, T.; Steinmeyer, J. *J. Rheumatol.* **2001**, *28*, 336–340.
210. Johnson, J. D.; Hand, W. L.; Francis, J. B.; King-Thompson, N.; Corwin, R. W. *J. Lab. Clin. Med.* **1980**, *95*, 429–439.
211. Cifarelli, A.; Forte, N.; Lombardi, L.; Pepe, G.; Paradisi, F. *J. Infect.* **1982**, *5*, 183–188.
212. Gabler, W. L.; Tsukuda, N. *Res. Commun. Chem. Pathol. Pharmacol.* **1991**, *74*, 131–140.
213. D'Agostino, P.; Arcoleo, F.; Barbera, C.; Di Bella, G.; La Rosa, M.; Misiano, G.; Milano, S.; Brai, M.; Cammarata, G.; Feo, S.; Cillari, E. *Eur. J. Pharmacol.* **1998**, *346*, 283–290.
214. Ingham, E.; Turnbull, L.; Kearney, J. N. *J. Antimicrob. Chemother.* **1991**, *27*, 607–617.
215. Potts, R. C.; Hassan, H. A.; Brown, R. A.; MacConnachie, A.; Gibbs, J. H.; Robertson, A. J.; Beck, J. S. *Clin. Exp. Immunol.* **1983**, *53*, 458–464.
216. Goh, D. H. B.; Ferrante, A. *Int. J. Immunopharmacol.* **1984**, *6*, 51–54.
217. Kuzin, I. I.; Snyder, J. E.; Ugine, G. D.; Wu, D.; Lee, S.; Bushnell, T., Jr.; Insel, R. A.; Young, F. M.; Bottaro, A. *Int. Immunol.* **2001**, *13*, 921–931.
218. Kloppenburg, M.; Verweij, C. L.; Miltenburg, A. M. M.; Verhoeven, A. J.; Daha, M. R.; Dijkmans, B. A. C.; Breedveld, F. C. *Clin. Exp. Immunol.* **1995**, *102*, 635–641.
219. Esterly, N. B.; Furey, N. L.; Flanagan, L. E. *J. Invest. Dermatol.* **1978**, *70*, 51–55.
220. Golub, L. M.; Lee, H. M.; Lehrer, G.; Nemiroff, A.; McNamara, T. F.; Kaplan, R.; Ramamurthy, N. S. *J. Periodont. Res.* **1983**, *18*, 516–526.
221. Elattor, T. M.; Lin, H. S.; Schulz, R. *J. Periodont. Res.* **1988**, *23*, 285–286.
222. Tamargo, R.; Bok, R.; Brem, H. *Cancer Chemother. Pharmacol.* **1995**, *36*, 418–424.
223. Sotomayer, E. A.; Teicher, B. A.; Schwartz, G. N.; Holden, S. A.; Menon, K.; Herman, T. S.; Frei, E. *Cancer Chemother. Pharmacol.* **1992**, *30*, 377–394.
224. Soory, M.; Tilakaratne, A. *Arch. Oral Biol.* **2000**, *45*, 257–265.
225. Whiteman, M.; Halliwell, B. *Free Radical Res.* **1997**, *26*, 49–56.
226. Paemen, L.; Martens, E.; Norga, K.; Masure, S.; Roets, E.; Hoogmartens, J.; Opdenakker, G. *Biochem. Pharmacol.* **1996**, *52*, 105–111.
227. Gilbertson-Bealding, S.; Powers, E. A.; Stamp-Cole, M.; Scott, P. S.; Wallace, T. L.; Copeland, J.; Petzold, G.; Mitchell, M.; Ledbetter, S. *Cancer Chemother. Pharmacol.* **1995**, *36*, 418–424.
228. Lee, C. Z.; Xu, B.; Hashimoto, T.; McCulloch, C. E.; Yang, G.-Y.; Young, W. L. *Stroke* **2004**, *35*, 1715–1719.
229. Sorsa, T.; Konttinen, Y. T.; Lindy, O.; Suomalainen, K.; Ingham, T.; Saari, H.; Halinen, S.; Lee, H. M.; Golub, L. M. *Agents Actions Suppl.* **1993**, *39*, 225–229.

230. Suomalainen, K.; Sorsa, T.; Golub, L. M.; Ramamurthy, N.; Lee, H. M.; Uitto, V. J.; Saari, H.; Konttinen, Y. T. *Antimicrob. Agents Chemother.* **1992**, *36*, 227–229.
231. Golub, L.; Greenwald, R.; Ramamurthy, N.; Zucker, S.; Ramsammy, L.; McNamara, T. *Proceedings Matrix Metalloproteinase Conf.*, **1992**, 315–316.
232. Acharya, M. R.; Venitz, J.; Figg, W. D.; Sparreboom, A. *Drug Resist. Updates* **2004**, *7*, 195–208.
233. D'Agostino, P.; Ferlazzo, V.; Milano, S.; La Rosa, M.; Di Bella, G.; Caruso, R.; Barbera, C.; Grimaudo, S.; Tolomeo, M.; Feo, S.; Cillari, E. *Int. Immunopharmacol.* **2001**, *1*, 1765–1776.
234. Carney, D. E.; McCann, U. G.; Schiller, H. J.; Gatto, L. A.; Steinberg, J.; Picone, A. L.; Nieman, G. F. *J. Surg. Res.* **2001**, *99*, 245–252.
235. Trachtman, H.; Futterweit, S.; Greenwald, R.; Moak, S.; Singhal, P.; Franki, N.; Amin, A. R. *Biochem. Biophys. Res. Commun.* **1996**, *229*, 243–248.
236. Amin, A. R.; Attur, M. G.; Thakker, G. D.; Patel, P. D.; Vyas, P. R.; Patel, R. N.; Patel, I. R.; Abramson, S. B. *Proc. Natl. Acad. Sci. USA* **1996**, *93*, 14014–14019.
237. Kloppenburg, M.; Brinkman, B. M. N.; de Rooij-Dijk, H. H.; Miltenburg, A. M. M.; Daha, M. R.; Breedveld, F. C.; Dijkmans, B. A. C.; Verweij, C. L. *Antimicrob. Agents Chemother.* **1996**, *40*, 934–940.
238. Pruzanski, W.; Greenwald, R. A.; Street, I. P.; Laliberte, F.; Stefanski, E.; Vadas, P. *Biochem. Pharmacol.* **1992**, *44*, 1165–1170.
239. Sainte-Marie, I.; Tenaud, I.; Jumbou, O.; Dreno, B. *Acta Dermatol. Venereol.* **1999**, *79*, 265–267.
240. Brown, D. L.; Desai, K. K.; Vakili, B. A.; Nouneh, C.; Lee, H. M.; Golub, L. M. *Arterioscler. Thromb. Vasc. Biol.* **2004**, *24*, 733–738.
241. Villareal, F. J.; Griffin, M.; Omens, J.; Dillman, W.; Nguyen, J.; Covell, J. *Circulation* **2003**, *12*, 1482–1487.
242. Kaito, K.; Urayama, H.; Watanabe, G. *Surgery Today* **2003**, *33*, 426–433.
243. Thompson, R. W.; Baxter, B. T. *Ann. NY Acad. Sci.* **1999**, *878*, 159–178.
244. Gossen, M.; Bujard, H. *Proc. Natl. Acad. Sci. USA* **1992**, *89*, 471–475.
245. Gossen, M.; Freundlieb, S.; Bender, G.; Mueller, G.; Hillen, W.; Bujard, H. *Science* **1995**, *268*, 1766–1769.
246. Furth, P. A.; St. Onge, L.; Boeger, H.; Gruss, P.; Gossen, M.; Kistner, A.; Bujard, H.; Henninghausen, L. *Proc. Natl. Acad. Sci. USA* **1994**, *91*, 9302–9306.
247. Baron, U.; Schnappinger, D.; Helbl, V.; Gossen, M.; Hillen, W.; Bujard, H. *Proc. Natl. Acad. Sci. USA* **1999**, *96*, 1013–1018.
248. Gossen, M.; Bujard, H. In *Tetracyclines in Biology, Chemistry and Medicine*; Nelson, M., Hillen, W., Greenwald, R. A., Eds.; Birkhauser: Basel, Switzerland, 2001, pp 139–159.
249. Alphey, L. Autoregulatory Insect Expression System, Oxitec Limited, GB, Br. Patent Appl. 2404382 **2005**.

Biographies

Mark L Nelson, PhD, has a BSc degree in chemistry and microbiology from Gannon University, Erie, PA, and a doctoral degree in medicinal chemistry from Temple University, Philadelphia, PA. Postdoctoral research was conducted at Tufts University School of Medicine, the Center for Adaptation Genetics and Drug Resistance with Stuart B Levy, MD, studying the SAR of bacterial efflux protein inhibitors and the synthesis of novel tetracycline antibiotics and nonantibiotic compounds. His research with Dr Levy led to the formation of Paratek Pharmaceuticals, Inc., Boston, MA, where he is currently Senior Director of Chemistry. He has acted as Chair of several conferences devoted to new antibiotics and nonantibiotic properties of the tetracyclines, and has numerous publications and patents devoted to the subject. He has been a Fulbright Distinguished Lecturer and received the Rho Chi Pharmaceutical Society Kallelis–Lynch Lectures Award for research on bacterial resistance mechanisms and novel chemistries applied to antibiotics.

Mohamed Y Ismail, PhD, received a BSc degree summa cum laude in chemistry from Somali National University, Mogadishu, Somalia. After spending a year as an assistant lecturer, he won the prestigious Fulbright Scholarship to study at the University of Massachusetts, Amherst, under the direction of Professor Louis Carpino where he finished his doctoral work in peptide chemistry. Postdoctoral research was conducted at the Center for Adaptation Genetics and Drug Resistance, Tufts University School of Medicine, with Stuart B Levy, MD. He is currently principal scientist at Paratek Pharmaceuticals, Inc., Boston, MA where his work with Dr Nelson in the development of tetracyclines led to the novel compound PTK 0796, a clinical candidate for drug-resistant infections.

Comprehensive Medicinal Chemistry II
ISBN (set): 0-08-044513-6

ISBN (Volume 7) 0-08-044520-9; pp. 597–628

7.21 Aminoglycosides Antibiotics

H A Kirst, Consultant, Indianapolis, IN, USA
N E Allen, Indiana University School of Medicine, Indianapolis, IN, USA

7.21.1 Fermentation-Derived Aminoglycosides

Aminoglycosides are an older but still important class of antibiotics. Their gross structures comprise an aminocyclitol onto which are attached one or more amino- and/or neutral sugars. The class is also known by the more definitive name of aminocyclitol antibiotics. Aminoglycosides are natural products derived from fermentation of soil microbes or semisynthetic derivatives of those secondary metabolites. Their structures contain multiple amino and hydroxyl groups, making them highly polar, strongly basic, water-soluble molecules. Several books and reviews detail the extensive structural diversity of this class.[1–6]

The first aminoglycosides were quickly recognized as valuable antibiotics to treat Gram-negative bacterial infections and tuberculosis, but problems were encountered, such as development of antibiotic resistance, ototoxicity, and nephrotoxicity. The emergence of aminoglycosides into their current role required decades to learn and apply details of microbial resistance, drug administration, pharmacodynamics, and toxicology. Despite past success, other antibiotics have been replacing aminoglycosides and challenges for improvements in this class await modern medicinal chemistry.

7.21.1.1 Streptomycin

The paradigm of low-molecular-weight antibiotic substances of microbial origin is generally considered to begin with the discovery of penicillin. Once its medical value was recognized, the search was begun for other antibiotics produced by other microbes.

Streptomycin, from *Streptomyces griseus*, was the first reported aminoglycoside.[7] This early work further established the feasibility of microbial products as sources of new antibiotics that continued for several decades.[8] Streptomycin is a pseudotrisaccharide having a monosubstituted aminocyclitol to which a disaccharide is attached (**Figure 1**).[9–11] In contrast to penicillin, streptomycin inhibited Gram-positive and Gram-negative bacteria and *Mycobacterium tuberculosis*. Rapid development of resistance soon limited its utility as a single agent, so its principal current use is in combinations with other agents to treat tuberculosis.[12]

7.21.1.2 4,5-Disubstituted-2-Deoxystreptamine Family

In contrast to streptomycin, most important aminoglycosides contain 2-deoxystreptamine (2-DOS) as the aminocyclitol. They are classified by differences in number, type, and position of substituents on 2-DOS. The neomycin complex from *S. fradiae* was the first series having the 4,5-disubstituted-2-DOS motif, typified by neomycin B (**Figure 2**).[13–15] Although neomycin's broad spectrum was originally encouraging, systemic toxicity subsequently limited it to localized uses such as topical applications.

The paromomycin complex was found and given different names by different researchers in the 1950s.[16] Paromomycin I, isolated from *S. rimosus*, differs from neomycin B in the 4-*O*-aminosugar on 2-DOS (**Figure 2**).[14–17] This single change improved antiparasitic activity such that paromomycin has been used to treat certain intestinal parasites.[18] Lividomycin B, isolated from *S. lividus*, lacks the 3′-hydroxyl group of paromomycin I, rendering it nonsusceptible to 3′-*O*-phosphorylating enzymes.[4,19] Ribostamycin, from *S. ribosidificus*, is a pseudotrisaccharide having only one aminosugar.[20] Lividomycin and ribostamycin have been used in Japan to treat infections of Gram-negative

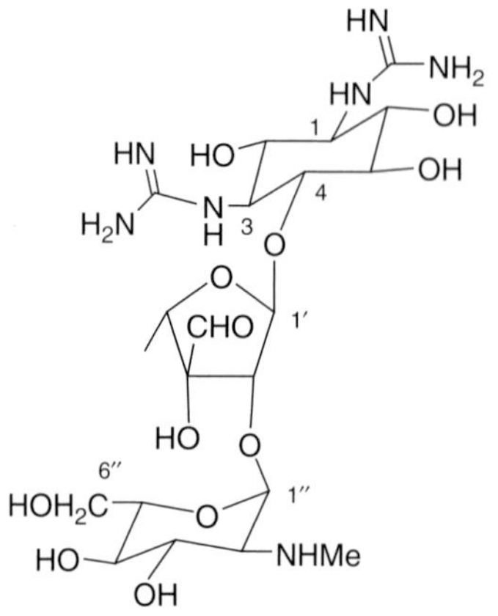

HN
NH2
HN
1
OH
HN HO
OH
N
4
H2N
H 3
O
O
CHO
1′
HO
O
6″
1″
HOH2C
O
HO
NHMe
OH

Figure 1 Structure of streptomycin.

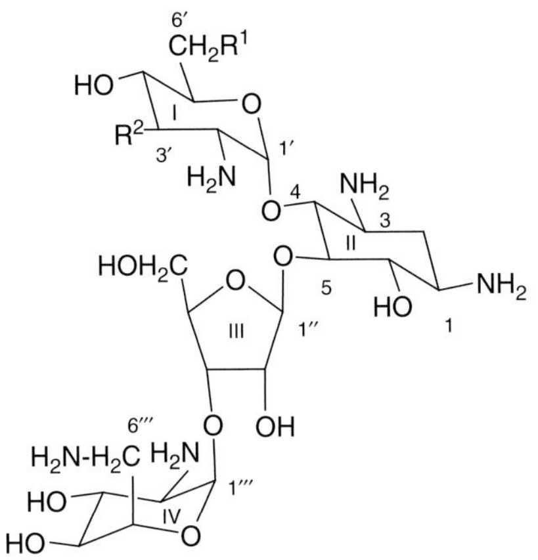

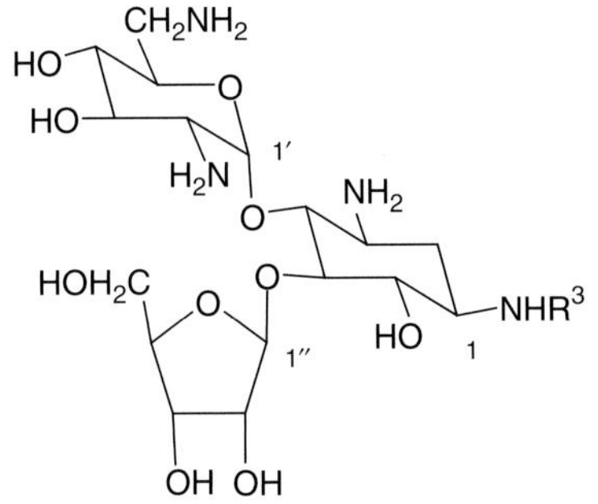

Neomycin B: $R^1 = NH_2$, $R^2 = OH$
Paromomycin I: $R^1 = OH$, $R^2 = OH$
Lividomycin B: $R^1 = OH$, $R^2 = H$

Ribostamycin: $R^3 = H$
Butirosin B: R^3 = 4-amino-2(*S*)-hydroxybutyryl

Figure 2 Representative 4,5-di-*O*-substituted-2-DOS aminoglycosides.

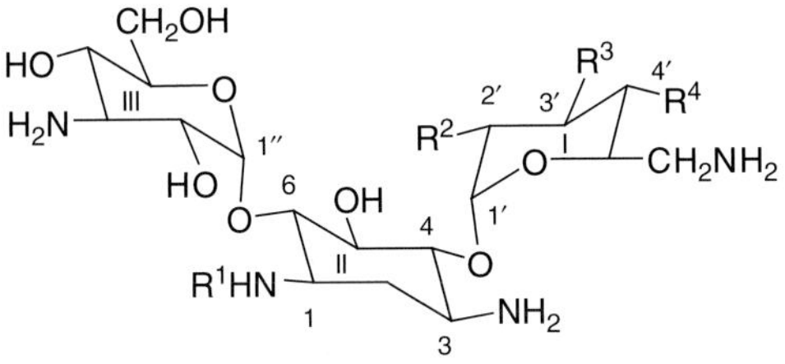

Kanamycin A: $R^1 = H$, $R^2 = R^3 = R^4 = OH$
Kanamycin B: $R^1 = H$, $R^2 = NH_2$, $R^3 = R^4 = OH$
Tobramycin: $R^1 = H$, $R^2 = NH_2$, $R^3 = H$, $R^4 = OH$
Dibekacin: $R^1 = H$, $R^2 = NH_2$, $R^3 = R^4 = H$
Amikacin: R^1 = 4-amino-2(*S*)-hydroxybutyryl, $R^2 = R^3 = R^4 = OH$
Habekacin: R^1 = 4-amino-2(*S*)-hydroxybutyryl, $R^2 = NH_2$, $R^3 = R^4 = H$

Figure 3 Representatives of the kanamycin–tobramycin group.

bacteria.[2] The butirosins, produced by *Bacillus circulans*, have the unusual 4-amino-2(*S*)-hydroxybutyryl (AHBA) group on the 1-amino group of 2-DOS (**Figure 2**).[21] This unique amide provided the critical lead that chemists extended to other aminoglycosides to create analogs active against many aminoglycoside-resistant bacteria. However, despite many 4,5-disubstituted-2-DOS structures, the 4,6-disubstituted-2-DOS group became the more important aminoglycoside family.

7.21.1.3 4,6-Disubstituted-2-Deoxystreptamine Family

4,6-Disubstituted-2-DOS aminoglycosides are divided into the kanamycin–tobramycin and gentamicin–sisomicin series. Kanamycin A, produced by *S. kanamyceticus*, was the first member of this family, while kanamycin B differs in having a 4-*O*-diaminosugar on 2-DOS (**Figure 3**).[22–24] This single change makes the latter approximately twice as active in vitro.[4] Both kanamycins exhibit a broad spectrum of activity and have been important commercial antibiotics in many countries.[2]

In the early 1960s at Schering Corp., the search for antibiotics was directed toward microbes other than *Streptomyces*. *Micromonospora* species yielded the gentamicin complex whose members differ from the kanamycins in the aminosugars attached to 2-DOS (**Figure 4**).[25–27] The commercial product, gentamicin, containing predominantly factors C_1, C_{1a},

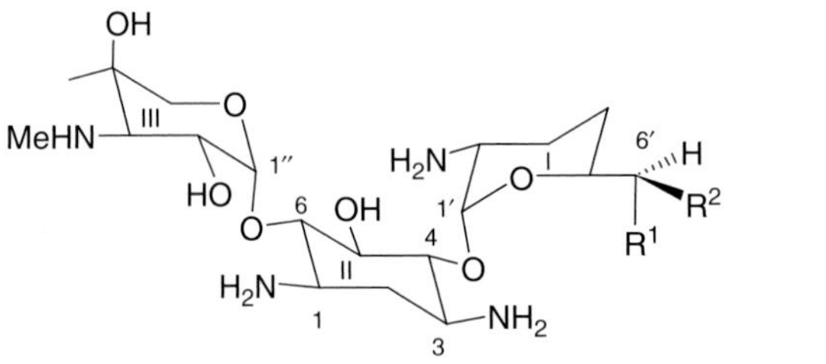

Gentamicin C_1: R^1 = NHMe, R^2 = Me
Gentamicin C_{1a}: R^1 = NH_2, R^2 = H
Gentamicin C_2: R^1 = NH_2, R^2 = Me
Gentamicin C_{2a}: R^1 = Me, R^2 = NH_2
Gentamicin C_{2b} (Micronomicin): R^1 = NHMe, R^2 = H

Figure 4 Structures of gentamicin C factors.

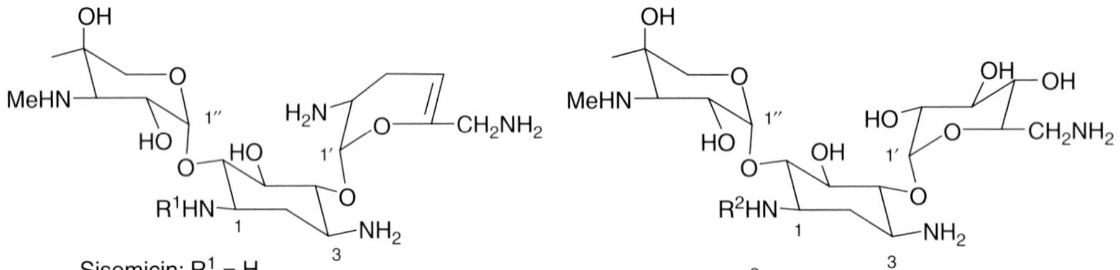

Sisomicin: R^1 = H
Netilmicin: R^1 = Et

Gentamicin B: R^2 = H
Isepamicin: R^2 = 3-amino-2(*S*)-hydroxypropionyl

Figure 5 Semisynthetic derivatives of the gentamicin–sisomicin group.

and C_2, pioneered the medical expansion of aminoglycoside therapy against Gram-negative bacteria. Its long clinical history and low cost made it the most widely used aminoglycoside when resistance was not an issue.[3,28–31]

Tobramycin is the hydrolysis product of nebramycin factor 5′, one component of the nebramycin complex from culture broths of *S. tenebrarius*.[32] Tobramycin is 3′-deoxykanamycin B (**Figure 3**).[33] This one change renders it more potent than the kanamycins against Gram-negative bacteria, including certain resistant strains.[4] Like gentamicin, tobramycin is widely used and often selected for its activity against *Pseudomonas aeruginosa*.[28–30,34–36] Later studies of *Micromonospora* yielded sisomicin from *M. inyoensis*, whose distinctive feature is an unsaturated aminosugar (**Figure 5**).[37] Sisomicin is more potent than gentamicin and became a commercial product in certain countries, but it never broadly supplanted gentamicin or tobramycin.[3] Sagamicin, obtained from *M. sagamiensis*, is gentamicin C_{2b} (**Figure 4**), which was commercialized in Japan as micronomicin.[38,39]

7.21.1.4 Additional Diaminocyclitol Antibiotics

Spectinomycin is an older antibiotic from *S. spectabilis* whose structure contains an unusual tricyclic ring (**Figure 6**).[40] Apramycin (**Figure 7**) is a unique member of the nebramycins and is used to treat Gram-negative infections in veterinary medicine.[41,42] Hygromycin B (**Figure 8**), produced by *S. hygroscopicus*, exhibits activities against both prokaryotic and eukaryotic cells.[43,44] During the 1970s, the search for new antibiotics yielded several series whose structures contain a 1,4-diaminocyclitol. The first series was fortimicin, produced by *M. olivoasterospora*. Activity of fortimicin A (**Figure 9**) was comparable to that of kanamycin, but included some aminoglycoside-resistant strains.[45,46] Related series were isolated by other groups and named sporaricin, dactimicin, sannamycin, or istamycin.[1,3,5] Fortimicin A was developed in Japan as astromicin and its success offers encouragement that other new structures might be found beyond the traditional structural motifs.[3,39]

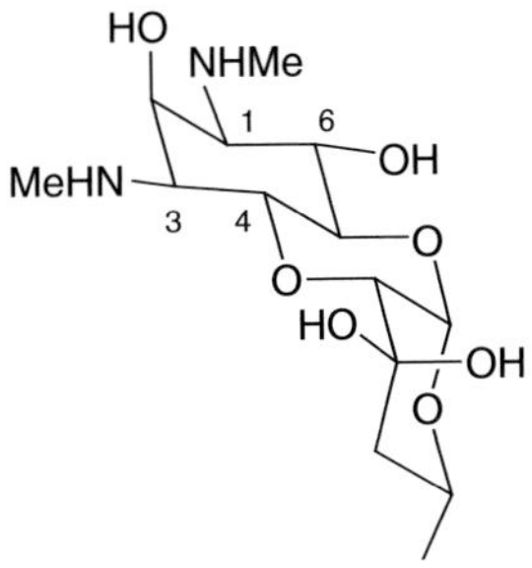

Figure 6 Structure of spectinomycin hydrate.

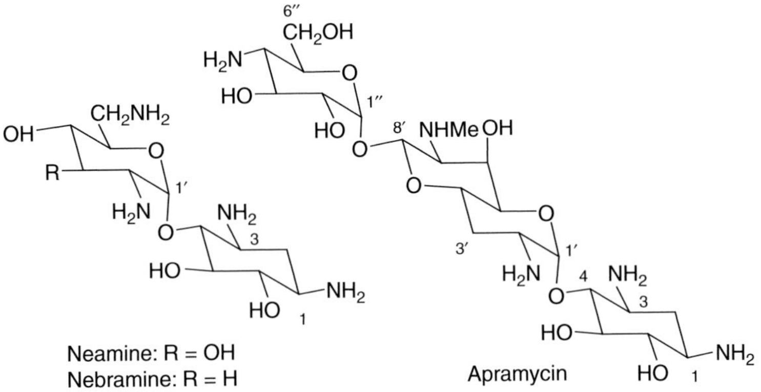

Figure 7 Neamine, nebramine, and apramycin.

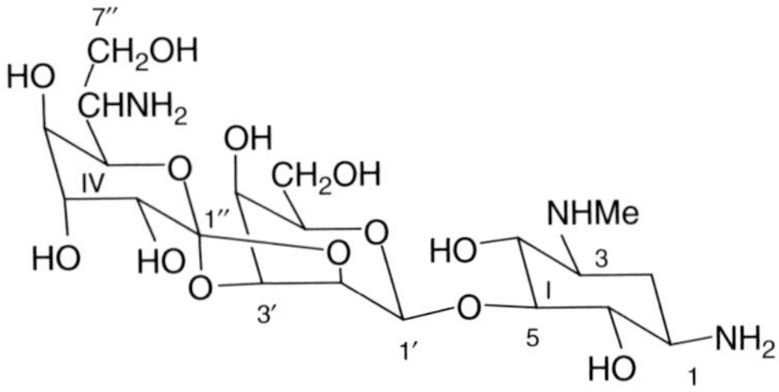

Figure 8 Structure of hygromycin B.

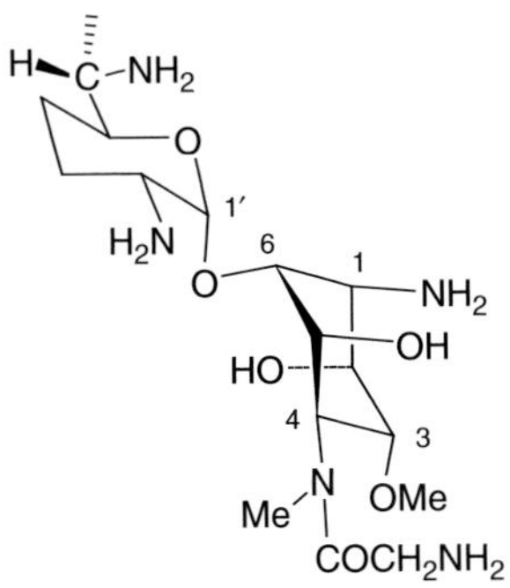

Figure 9 Structure of fortimicin A.

7.21.1.5 Other Natural Products

Fermentation screening programs began to decline in the 1980s and novel aminoglycosides were generally absent. Reinvestigation of older work has yielded unique poly-aminosaccharides named saccharomicin from an uncommon *Saccharothrix* species.[47,48] No aminocyclitols have been isolated from the abundant nitrogenous compounds formed by marine plants or animals, but istamycin was formed by a marine actinomycete, *Streptomyces tenjimariensis*, so future searches of marine sources might yield new molecules.[49–52] Natural products will probably remain a part of future screening programs.[53] However, research has become oriented toward other antibiotic classes to the extent that recent surveys of corporate pipelines list no aminoglycosides in development.[54–56] Uncertainties of screening and a current focus on new antibiotic classes that circumvent microbial resistance make future efforts unpredictable for discovery of new aminoglycoside structures.

7.21.2 Semisynthetic Aminoglycosides

Dihydrostreptomycin was prepared in 1946 by reducing the parent's aldehyde group, but clinical use was halted due to irreversible ototoxicity.[57,58] As more aminoglycosides were found, directions for modification were seen by comparing in vitro activity among natural structures, which provided clues that were applied to synthesis of analogs. Valuable guiding principles arose from mechanism of resistance studies that revealed the chemistry of inactivating enzymes. Furthermore, total synthesis was extended to create analogs, biosynthetic pathways were altered, and bioconversions led to structural changes. Hundreds of derivatives were prepared in the search for improved efficacy coupled with fewer and less severe side effects.[4,59–61] The most successful types of modification were deoxygenation of the 3′,4′-hydroxyl groups and substitution of the 1-amino group.

7.21.2.1 Dibekacin

By the early 1970s, the critical role of the amino and hydroxyl groups on in vitro activity was known and structure correlations with mechanisms of resistance had emerged.[62–65] One significant revelation was 3′-deoxyaminoglycosides were more active than their 3′-hydroxy analogs, especially against some resistant strains. Discovery of inactivation of kanamycin A via 3′-*O*-phosphorylation further suggested removing that hydroxyl group might improve activity.[66] While preparing new deoxy derivatives, a facile synthesis of 3′,4′-dideoxykanamycin B (**Figure 3**) was achieved.[67] Named dibekacin, it showed strong activity against Gram-negative bacteria, including certain resistant strains, and became an important antibiotic in Japan and other countries.[3,29,68]

7.21.2.2 Amikacin, Netilmicin, Isepamicin, and Habekacin

The unexpected activity against aminoglycoside-resistant strains conferred by the novel AHBA amide of butirosin prompted syntheses of many analogs from other aminoglycosides.[4] Acylation of kanamycin A was extensively studied and its 1-*N*-AHBA derivative was synthesized.[69] Named amikacin (**Figure 3**), it inhibited bacteria that enzymatically inactivated aminoglycosides by a variety of mechanisms.[4,68,70] Due to its strong activity that included many resistant strains, amikacin became the most widely used semisynthetic aminoglycoside.[2,30,31,71]

Due to amikacin's prominence, numerous analogs were prepared in the search for new compounds that might have even more advantageous properties, but only a few analogs became products. The scope of AHBA-type substitutions was logically expanded to *N*-alkylations, from which 1-*N*-ethylsisomicin was found to display activity against certain resistant bacteria and lower potential for chronic toxicity.[72] Under the name of netilmicin (**Figure 5**), it became an important clinical antibiotic.[3,28–30,73] Derivatization of other gentamicin factors led to selection of 1-*N*-(3-amino-2(*S*)-hydroxypropionyl)gentamicin B (**Figure 5**) for further evaluation.[74] Named isepamicin, it is another more recent semisynthetic aminoglycoside that has activity against certain resistant bacteria.[3,28–30,68,75] Subsequent modification of the semisynthetic dibekacin to its 1-*N*-AHBA derivative produced broad activity that included certain aminoglycoside-resistant bacteria (**Figure 3**).[76] Under the name habekacin (arbekacin sulfate), it was successfully commercialized in Japan and other countries.[29,68,77]

7.21.2.3 Recent Trends

Despite an absence of new structures, derivatization of old aminoglycosides has had a renaissance.[78] While some work extended earlier structure–activity relationship (SAR) directions, the resurgence was also driven by expanding knowledge about mechanisms of microbiological and pharmacological actions and microbial resistance. Binding of aminoglycosides to RNA and ribosomes is a particular research focus.[79–82] Advances in technologies such as combinatorial chemistry, high-throughput screening, x-ray crystallography, nuclear magnetic resonance (NMR) spectroscopy, and molecular modeling contribute to the renewed effort. New compounds provide additional means to probe details of antibiotic activity or uncover new biological activities. Efficacious aminoglycosides that circumvent antibiotic resistance and dissociate antimicrobial activity from patient toxicity remain the overarching objective.

7.21.2.4 *N*-Acylation

The AHBA ligand was converted to *N*-amidino or *N*-guanidino derivatives or replaced by a 4-amino-2(*S*)-fluorobutyryl group, but with little change in activity profile.[83,84] Replacing the 2(*S*)-hydroxyl group in AHBA by its *N*-hydroxyurea isostere also retained activity.[85] Other *N*-acyl derivatives were prepared for binding studies.[86–89] Acylation of aminoglycosides with deacetylcephalothin created hybrids designed to target the site of aminoglycoside release upon cleavage by beta-lactamase.[90] Amikacin was polymerized via *N*-acylation of its AHBA amino group, but this particular multiple antibiotic linkage was inactive.[91] Although much of the scope of the AHBA group has been defined, the prominence of amino groups will likely ensure further attempts to improve activities via *N*-acylation strategies, perhaps aided by RNA binding studies.

7.21.2.5 Hydroxyl Group Replacement

Modifications of hydroxyl groups continue to probe the scope of their role in activity and resistance. The 6″-hydroxyl group in kanamycin B was systematically replaced, but no structure–activity correlations were found.[92,93] Activity was reduced in several 6′-oxidized or homologated derivatives of paromamine.[94] Oxidation of the 3′-hydroxyl group of kanamycin formed the 3′-ketone in equilibrium with its hydrate that regenerated parent when enzymatically phosphorylated.[95] A report of 5,4″-bis-epi-habekacin (TS2037) extends other attempts to find broad spectrum, less toxic compounds via 5-epi or 5-fluoro derivatives.[96–98] In contrast, halogenation of other hydroxyl groups produced more limited effects,[97] but replacement of a 4′-hydroxyl group by fluorine lowered susceptibility to 3′-phosphotransferases.[99] Although some derivatives show interesting features, no successful clinical candidate has yet emerged from these efforts.

7.21.2.6 Aminoalcohol Moieties

A critical structural motif is the multiplicity of aminoalcohol moieties, a focus of recent interest as essential aminoglycoside pharmacophores.[78,79,100–102] In one early approach, an axial hydroxymethyl group, added into gentamicin C_2 at C1 of 2-DOS, conferred activity against resistant bacteria and lowered nephrotoxicity, but a series of kanamycin B analogs unexpectedly did not show similar features.[103] Introducing such C-substituents to create new unnatural aminoalcohol functions has apparently not been further pursued, although aminoalcohols are incorporated in newer *O*-alkyl substituents (*see* Section 7.21.2.8). Chelates that readily form between aminoalcohols and certain metal cations have long been used as temporary selective protecting groups during synthesis.[104,105] Although such complexes cause biological effects due to oxidation or cleavage reactions, these effects might not occur under physiological conditions.[106,107]

7.21.2.7 *O*-Glycosylation

The carbohydrate nature of aminoglycosides suggests creating new antibiotics by altering saccharides on core moieties.[108] The pseudodisaccharide neamine (**Figure 7**) and analogs constitute a minimum structure for antibacterial activity.[78,79,109,110] Surprisingly, activity was not substantially dependent on D- versus L-enantiomers, and comparable data should be generated about toxicity.[109,110] Glycosylation of three analogs of 2-DOS produced pseudodisaccharides related to neamine, but no activity data were reported.[111] Future trends in glycosylation are directed toward developing combinatorial chemistry to efficiently generate and assay larger series of derivatives.[112–115]

Suitable combinatorial methods were found to selectively glycosylate neamine at its 5- or 6-hydroxyl group and tobramycin analogs were synthesized from a protected intermediate.[116,117] Using a process termed glyco-optimization (OPopS), another group prepared other analogs of tobramycin for further study.[118] Another process called glycodiversification was developed to guide SAR studies of pyranmycins, a series of 5-*O*-pyranosyl derivatives of neamine having activity comparable to neomycin plus greater acid stability.[119–121] This approach was also extended to 6-*O*-pyranosyl derivatives.[122] However, some 6-*O*-disaccharide derivatives of nebramine had reduced activity attributed to steric problems.[118] Analogs of neomycin were prepared while a group of even larger pseudopentasaccharide 5″-*O*-glycosyl derivatives of neomycin B suggested activity was retained in a wide array of *O*-glycosyl derivatives.[123–126] The SAR of glycosyl derivatives is still incomplete and combinatorial methods to prepare and assay more diverse libraries should assist efforts to understand SAR principles.

7.21.2.8 Glycosyl Mimetics

Several groups have sought mimetics of glycosyl substituents as a strategy to create new derivatives. Replacing the aminosugar of neamine by changes at C4 of 2-DOS generally lowered activity.[127–129] Using neamine or analogs as substrates, several 5-*O*- or 6-*O*- substitutions did not enhance potency even if inhibition of bacterial translation was improved.[86,130,131] In contrast, a small group of doubly modified 1-*N*-acyl-6-*O*-alkyl derivatives of neamine displayed good activity.[87,132] Starting from larger cores, activity was retained when aminoalkyl groups replaced the fourth ring of neomycin.[123] Active hybrid structures were created by alkylating pseudotrisaccharide hydroxyl groups that are not naturally glycosylated.[120,133,134] These varying effects of structure on activity exemplify the incomplete state of SAR knowledge and the consequent difficulties in designing modified aminoglycosides.

7.21.2.9 Replacement of 2-Deoxystreptamine

Attempts have been made to replace 2-DOS as the central scaffold of aminoglycoside structures. Among heterocyclic replacements, a designed group of azepane glycosides was more active than piperidine glycosides.[135,136] Replacing 2-DOS by acyclic moieties and aminosugar dimers without any linker between sugars have generally showed lesser activity.[137–141] Antibiotic potency in these series was typically low, so it is not clear if further investigations might raise it toward more useful levels.

7.21.2.10 Other Derivatives

Dimers of 2-DOS via 4,4-di-*O*-alkyl linkers strengthened binding to RNA while neamine or nebramine dimerized through 5,5-di-*O*-alkyl linkages gave compounds with Gram-negative activity.[142–144] A small library of derivatives for screening was prepared by solution-phase 6′-*N*-alkylation of neamine.[145] A systematic set of deamino derivatives of neamine and kanamycin were poor substrates of 3′-*O*-phosphorylating enzymes.[146] Inhibitors of this enzyme were also prepared by tethering neamine to adenosine via 3′-*O*-methylene bridges of variable length.[147] These examples reflect some innovative approaches to modification that will likely expand in the future.

7.21.2.11 Products from Altered Biosynthesis

Many pathways and transformations by which microbes form these structures have been deduced.[148–150] In an early strategy (mutasynthesis) for preparing new aminocyclitols, mutants of producing organisms were created by blocking formation of endogenous 2-DOS; these strains were then fed analogs of 2-DOS to be incorporated into new structures.[151,152] This process was applied to several aminoglycosides, and some analogs received extended evaluation, but none progressed to commercial status.

Progress was slower than for some antibiotic classes, but additional details are emerging about the genetics and biochemistry of aminoglycoside biosynthesis, including gene clusters from more producing organisms.[153–155] As methodology develops to obtain genetic material from traditionally unculturable microbes, this metagenomic source may contribute new aminoglycoside biosynthetic genes.[156–158] Additional gene clusters and enzymes may open more opportunities for manipulating and altering those genes and pathways to create modified structures.[159–161] Interchange of genetic sequences between different gene clusters (i.e., combinatorial biosynthesis) may become possible for generating new compounds, an approach already established for polyketide synthases and nonribosomal peptide synthetases.[162] Application of these and future technologies in the postgenomic era may provide means to create novel aminoglycosides.

7.21.3 Potency, Spectrum, and Related Properties of Aminoglycosides

The spectrum of in vitro activity of aminoglycosides includes aerobic and facultative Gram-negative bacteria.[163] Aminoglycosides are most often used as empiric therapy to treat serious infections caused by aerobic Gram-negative bacilli including but not limited to *Escherichia coli, Klebsiella* spp., *Enterobacter* spp., *Hemophilus influenzae, Serratia* spp.,

Table 1 Bacterial susceptibility to aminoglycoside antibiotics

Family or Species	*Strains*	*% Susceptible*									*Reference*
		Gm	*Tob*	*Amk*	*Isp*	*Abk*	*Net*	*Kan*	*Sm*	*Neo*	
Enterobacteriaceae	30634	87	85								395
	48440	91	90	97							396
E. coli	>3800	93	94	99							397
K. pneumoniae	>1900	91	91	98							397
P. aeruginosa	716	67	80	85	81						398
	52637	73	86	91							399
	488	81	90	94							400
	56	38	38	39	41		38				401
S. aureus											
MSSA	243	72	71	88		99	99	67	55	91	402
MRSA	439	5	2	10		87	82	2	69	37	402

Gm, gentamicin; Tob, tobramycin; Amk, amikacin; Isp, isepamicin; Net, netilmicin; Abk, arbekacin; Kan, kanamycin; Sm, streptomycin; Neo, neomycin; MSSA, methicillin-sensitive *S. aureus*; MRSA, methicillin-resistant *S. aureus*.

Providencia spp., and *Pseudomonas aeruginosa*. These agents also have in vitro activity against *Staphylococcus aureus*, streptococci, and enterococci but uptake by these bacteria, especially enterococci, is often poor.[28] Combinations are often used to augment activity against Gram-positive bacteria. Aminoglycosides in the kanamycin–gentamicin series receive the most extensive use in human medicine and the choice of agent to treat a susceptible infection is often decided by local resistance issues or the need to avoid toxicities. Large surveys that monitor bacterial resistance reveal a high degree of susceptibility still exists to gentamicin and tobramycin, and susceptibility may often be higher for newer derivatives that overcome certain resistance mechanisms (**Table 1**). However, small or localized populations of bacteria may exhibit different susceptibility patterns (**Table 1**).

Activity of streptomycin against *Mycobacterium tuberculosis* and its role in treating tuberculosis are well known while amikacin and kanamycin are second-line antituberculosis drugs. Streptomycin is also indicated for treatment of plague (*Yersinia pestis*) and tularemia (*Francisella tularensis*).[163] Some aminoglycosides (e.g., paromomycin, geneticin,[164] and hygromycin B) are active against eukaryotic organisms and paromomycin is used to treat intestinal amebiasis (*Entamoeba histolytica*).[163] Hygromycin B is a veterinary anthelmintic used to treat ascaris infections.[165] Use of spectinomycin is limited to treating infections caused by *Neisseria gonorrhoeae*.[163]

Combinations of an aminoglycoside plus a penicillin or cephalosporin are synergistic against a variety of bacterial pathogens.[28,163] Combinations having demonstrated synergistic activity include: tobramycin plus a beta-lactam against a variety of Gram-negative clinical isolates,[166] gentamicin plus penicillin against *Enterococcus faecium* including high-level penicillin-resistant strains,[167] and gentamicin plus a beta-lactam against aminoglycoside-resistant *P. aeruginosa*.[168] A combination of gentamicin and vancomycin was synergistic against penicillin-resistant pneumococci in a study in which inclusion of the cell wall inhibitor increased penetration of the aminoglycoside.[169] Beta-lactams and glycopeptides, all cell wall inhibitors, are thought to potentiate accumulation of aminoglycosides by disruption of a cell wall barrier.[169] The clinical use of combinations enhances bactericidal activity and can avoid persistence of resistant subpopulations.[163]

With the exception of spectinomycin, aminoglycosides are bactericidal[28,163] and show concentration-dependent killing.[170] This feature makes aminoglycosides attractive for treating diseases such as endocarditis or meningitis because treatment failures can result from using bacteriostatic antibiotics.[171] The mechanism of lethality has not been fully explained and remains unusual because most agents that act on the ribosome are bacteriostatic. Lethality could be related to miscoding and production of mistranslated proteins. Inhibition by aminoglycosides leads to permeabilization of the cytoplasmic membrane.[172,173] Loss of cell viability could result from accumulation of truncated proteins in the cytoplasmic membrane forming nonspecific channels permeable to small molecules.[173] However, the rationale for this notion has been questioned.[174] Rather than a direct cause of lethality, membrane damage may be required to assure sufficient aminoglycoside uptake to effect an irreversible inhibition of ribosome function,[173,175] ribosome assembly,[176] or another critical cellular process.

Aminoglycosides exhibit a postantibiotic effect (PAE) whereby suppression of bacterial growth persists after the agent has been cleared by drug metabolism and elimination.[28,177,178] The PAE for aminoglycosides was demonstrated in animal infection[178] and in vitro[177] models. During treatment with aminoglycosides, the PAE is prolonged by the activity of host leukocytes due to aminoglycoside-enhanced phagocytic activity.[177,179] The combination of PAE and rapid, concentration-dependent killing allows for flexible dosing schedules including daily dosing of aminoglycosides.[179] Aminoglycosides in combination with other agents that demonstrate a PAE exhibit a prolonged PAE compared to use of aminoglycoside alone.

7.21.4 Determinants of Activity

As a class, aminoglycosides bind to the small (30S) subunit of the bacterial ribosome and inhibit protein biosynthesis.[180,181] Some aminoglycosides bind to both prokaryotic and eukaryotic ribosomes and inhibit bacterial and mammalian protein biosynthesis.[97] Streptomycin has been extensively studied as a model for aminoglycoside–ribosome interactions, and early studies[182] suggested that the antibacterial effects derived from direct binding of this agent to 30S ribosomal proteins.[182,183] This seemed a reasonable explanation given that numerous spontaneous high-level streptomycin-resistance mutations could be traced to an alteration of 30S ribosomal protein S12.[182] Early studies also tended to focus on what appeared to be the pleiotropic behavior and effects of aminoglycosides, i.e., irreversible uptake, membrane damage, misreading, and inhibition of protein elongation.[175] However, it is now largely recognized that aminoglycosides, including streptomycin, inhibit by binding to 16S rRNA of the 30S subunit. Although ribosomal proteins are important for the structure and function of the ribosome, the catalytic properties of the ribosome derive from rRNA and define the ribosome as a ribozyme.[184,185]

Inhibition of ribosome function requires that inhibitors cross the cytoplasmic membrane and accumulate in the cytoplasm. For aminoglycosides, this process is self-promoted and consists of three phases.[186–188] An initial ionic binding to the cell surface is concentration-dependent but energy-independent. Being highly charged, cationic, hydrophilic molecules, aminoglycosides efficiently penetrate the outer membrane of Gram-negative bacteria, most likely via hydrophilic porin proteins.[188–190] Initial ionic binding is followed by two energy-dependent phases (EDP-I and EDP-II).[188,190] EDP-I uptake is slow and concentration-dependent, and occurs prior to loss of viability. EDP-II is a rapid, linear transport across the cytoplasmic membrane and may require interaction with the ribosome as it coincides with aminoglycoside-induced lethality.[187,188] In aerobic bacteria, EDP-II is dependent on a membrane potential and can be blocked by inhibitors of electron transport and uncouplers of oxidative phosphorylation.[188]

Deciphering the interactions between aminoglycosides and the ribosome has relied on use of a combination of mutagenic, biochemical, and structural techniques. The earliest attempts to identify the target of aminoglycosides included analysis of resistance mutations that identified the ribosome, and the decoding step(s) in protein biosynthesis as the primary target.[182] Biochemical methods such as chemical footprinting and cross-linking have been used to identify the sites and specific rRNA nucleotides with which the aminoglycosides interact. Chemical footprinting methods use RNA modifying reagents (e.g., dimethylsulfate) to create a footprint identifying the sites where rRNA nucleotides are protected from modification due to binding of the antibiotic. Biophysical methods such as electron microscopy, NMR spectroscopy, and x-ray crystallography have had a huge impact on how the ribosome and its interactions with aminoglycosides are perceived at the molecular and atomic levels.[180] These techniques have enabled development of three-dimensional structures of oligonucleotide analogs of rRNA, as well as the entire 30S subunit (containing >99% of the 16S rRNA and most of the ribosomal proteins) complexed with aminoglycosides. Carter *et al.*[191] have presented the highest-resolution crystal structures to date of 30S subunits complexed with aminoglycosides.

Aminoglycosides bind to 16S rRNA of the 30S subunit and cause miscoding (misreading of mRNA) and/or inhibition of the translocation step.[192–194] Protein synthesis occurs on the ribosome in three main stages: chain initiation, elongation, and termination.[195–197] These stages involve sequential interactions and movements of mRNA, aminoacyl-tRNAs, and peptidyl-tRNAs on the ribosome in order to synthesize a polypeptide of defined sequence. Following an initiation stage wherein a start codon is recognized by formylmethionyl-tRNA, each additional amino acid is brought to the A (acceptor) site of the ribosome as an aminoacyl-tRNA. Specific codon–anticodon interactions facilitate binding of each aminoacyl-tRNA in the A site according to the sequence encoded in mRNA. This process, called decoding, requires the participation of an elongation factor (EF-Tu) and guanosine triphosphate (GTP). The elongating nascent peptide is located in the adjacent P (peptidyl) site as peptidyl-tRNA. Transpeptidation forms a peptide bond between the new amino acid and the amino acid at the C-terminus of the elongating peptide. This transfers the elongating peptide to the tRNA in the A site. Translocation, requiring another elongation factor (EF-G) and GTP, moves the peptidyl-tRNA from the A site into the P site making room for the next aminoacyl-tRNA. Chain termination occurs when release factors recognize a termination codon in mRNA leading to release of the polypeptide from the ribosome. The inhibitory effects of aminoglycosides on codon recognition and translocation derive from their binding to the decoding A site region of 16S rRNA and related sites that affect translocation.[196,198]

7.21.4.1 Streptomycin

Accuracy of the decoding process in translation is determined at the initial codon–anticodon recognition stage and a subsequent proofreading stage.[199,200] Miscoding induced by streptomycin or other aminoglycosides results in the incorporation of near-cognate amino acids into a polypeptide that prematurely dissociates from the ribosome.[201,202] This can be discerned on both the whole cell [203] and ribosome [204] levels. On the whole-cell level, amino acid auxotrophs of *E. coli* can grow in the absence of a required amino acid when the growth medium is supplemented with sublethal concentrations of streptomycin.[182] This is called phenotypic suppression. Streptomycin can also suppress the expression of certain nonsense and missense mutations. On the ribosome level, streptomycin induces translation errors in amino acid incorporation by isolated ribosomes in a cell-free system.[205] Streptomycin induces miscoding by stabilizing the ribosome in a state of ribosomal ambiguity (*ram*).[191,201] Streptomycin increases affinity of the A site for tRNA which enhances the opportunity for near-cognate recognition.[206–208] Certain streptomycin-resistance mutations in ribosomal proteins S4 and S5 affect codon–anticodon recognition and have the effect of decreasing translational fidelity.[200] Proofreading occurs subsequent to codon–anticodon recognition but prior to peptide bond formation.[209] Streptomycin-dependent mutations (restrictive, hyperaccurate phenotype) in ribosomal protein S12 affect proof-reading by decreasing stability of tRNA–ribosome interactions which has the effect of enhancing fidelity.[210,211] Growth of these strains requires the presence of streptomycin to stabilize the ribosome and allow translation to proceed. With the exception of spectinomycin, which blocks translocation, miscoding appears to be a property of the aminoglycoside class.[205,212]

Streptomycin binds tightly to nucleotides located in five different helices in four separate domains of 16S rRNA.[191,198] These areas are near the decoding region and are distinct from where several other aminoglycosides bind.[180] Streptomycin also interacts directly with a ribosomal protein side chain, the only aminoglycoside that does so; it contacts a lysine residue in ribosomal protein S12.[191] The effects of streptomycin on codon–anticodon recognition and proofreading derive from binding to these sites on the ribosome. Ribosomal decoding requires a conformational switch in 16S rRNA to ensure fidelity of translation.[201,213] Streptomycin appears to interfere with this conformational switch by stabilizing the RNA near the decoding site in a *ram* state that favors increased affinity for tRNA. This leads to increased near-cognate recognition and interference with the proofreading process. Protein S12 participates in this process as streptomycin resistance mutations in S12 can reverse the streptomycin-conferred stabilization thereby lessening miscoding and enhancing proofreading.[214]

7.21.4.2 Spectinomycin

The unusual fused ring structure of spectinomycin binds to nucleotides in the minor groove of 16S rRNA at the upper part of helix 34 outside the decoding region.[191,198] Mutations in this region of 16S rRNA confer resistance to spectinomycin.[215,216] Spectinomycin does not cause miscoding nor the conformational changes in rRNA typical of miscoding agents.[217] Rather, spectinomycin inhibits the elongation factor-dependent translocation of A site peptidyl tRNA to the P site.[206,215,218] Binding of spectinomycin may stabilize the structure of helix 34 preventing a transient conformational change in the helix necessary for movement of tRNAs from the A to P site during translocation.[219] A cytosine-to-uracil transition at nucleotide position 1192 in helix 34 confers resistance to spectinomycin.[220] The same nucleotide position is a cross-link site to ribosomal protein S5.[221] Mutations in S5 confer resistance to spectinomycin supporting the notion that S5 may affect helix 34 and participate in the translocation process.[221,222] Resistance could be due to an alteration in the structure of S5 that prevents access of spectinomycin to its binding site in helix 34.

7.21.4.3 2-Deoxystreptamine Aminoglycosides

The disubstituted 2-DOS-containing aminoglycosides bind in the decoding region of the A site of 16S rRNA where codon–anticodon recognition occurs.[183,198,223,224] The decoding region is located near the 3′ end of 16S rRNA (nucleotides numbered 1400–1500). No ribosomal proteins have been identified that bind specifically in this region.[225] Protein S12 is the only ribosomal protein located near enough to the decoding region that could conceivably play a role in the decoding process[191] which may explain how mutations in S12 confer aminoglycoside resistance. The 4,5- and 4,6-disubstituted-2-DOS aminoglycosides share conserved functional groups on their aminocyclitol and ring I sugars (**Figures 2–5**), no doubt accounting for similarities in binding sites and activities of the members of these families. The 2-DOS moiety of all 4,5- and 4,6-disubstituted aminoglycosides interacts directly with two consecutive nucleotide base pairs located in the decoding region of the A site, C1407-G1494 and U1406-U1495.[226–228] These base pairs are conserved in all prokaryotic and eukaryotic ribosomes.[226]

Paromomycin (4,5-disubstituted) binds as a 1:1 complex in the major groove of 16S rRNA near an internal loop in the decoding region.[191] Nucleotides in the A site critical for paromomycin–RNA interactions have been identified by NMR using a 27-membered oligonucleotide mimetic of the decoding region complexed with the aminoglycoside.[229] The NMR data indicate that an asymmetric internal loop is required for binding and suggest that binding to the A site induces local conformational changes in the RNA. When bound, the antibiotic assumes an L-shaped conformation with ring I positioned in a pocket formed by A1492 and the A1408–A1493 base pair with the other rings positioned at a right angle to ring I.[226] Specific nucleotides contacted by functional groups on each ring have been identified.[191,226] Binding of paromomycin induces conformational changes in the decoding region facilitating the binding of near-cognate tRNAs accounting for the miscoding caused by this agent.[201,230] Binding is dependent on base sequence since mutations in this region result in decreased binding.[229,231–233] Other 4,5-disubstituted-2-DOS aminoglycosides (neomycin, neamine, and ribostamycin) share very similar binding patterns with paromomycin.[229]

Although the specificity of paromomycin activity is toward prokaryotes, paromomycin also interacts with the decoding region of eukaryotic ribosomes and inhibits ribosomal function.[234–237] An explanation for specificity resides in the nucleotide at position 1408 of the decoding region.[238,239] In wild-type *E. coli* and other prokaryotic organisms 1408 is always adenosine; it is always guanosine in eukaryotes.[240] Paromomycin binds to the decoding region of the A site of the 80S ribosome, yet binding is noticeably reduced compared to that in prokaryotes.[229] Whereas A1408 along with A1492 and A1493 facilitate a conformational change conducive to paromomycin binding on bacterial ribosomes, no such conformational change occurs when guanosine replaces adenosine at this position.[238] Mutations introducing a guanosine at position 1408 in *E. coli* confer low-level resistance to paromomycin but high-level resistance to neomycin.[241] Paromomycin and neomycin differ only in the functional group at C6′. Aminoglycosides having 6′-hydroxy (paromomycin) are more active on G1408-containing 16S rRNA than are those having 6′-ammonium (neomycin).[241] The extensive eukaryotic activity of G418 (geneticin)[164,242,243] can be at least partially explained on the basis of its having a 6′-hydroxyl group.[239]

The 4,6-disubstituted-2-DOS aminoglycosides (e.g., kanamycins and gentamicins) share similar functional groups on their aminocyclitol and 4-*O*-aminosugars with paromomycin and neomycin (**Figures 2–5**). Like the 4,5-disubstituted aminoglycosides, members of this family interact with 16S rRNA by binding to the decoding region of the A site and cause miscoding.[227] Each of the three components of the gentamicin C complex binds to the same site on the 30S subunit but with different affinities. NMR structure studies[227] show that gentamicin C_{1a} binds as a 1:1 complex in the major groove within the internal loop of the A site. The binding geometry of rings I and II of gentamicin C_{1a} and paromomycin are similar and rings I and II of both agents bind into the same RNA pocket of the decoding region. The two aminoglycosides differ, however, in the manner in which the remaining rings bind. Ring III of gentamicin C_{1a} makes sequence-specific contacts with RNA of the upper stem of the decoding region. Although rings III and IV of paromomycin contribute to binding affinity,[244] these rings make less significant contacts with RNA of the lower stem. This may help explain why the gentamicin–kanamycin-type aminoglycosides have been more useful clinically than the neomycin–paromomycin-types. Mutations in the rRNA of the upper stem completely prevent binding of gentamicin C_{1a} but have no effect on binding by paromomycin.[227] Rings I and II of neomycin, neamine, and ribostamycin likewise are sufficient to confer binding specificity.[244] The importance of rings I and II for the 4,6-disubstituted aminoglycosides is further supported by similar findings from an x-ray crystallographic examination of tobramycin bound to the decoding region.[228]

Apramycin, a unique 4-*O*-monosubstituted-2-DOS aminoglycoside, binds in the decoding region and interacts with some of the same nucleotides with which the 4,5- and 4,6-disubstituted agents interact.[245,246] However, unlike paromomycins, kanamycins, and gentamicins, apramycin inhibits translocation and causes only limited miscoding effects.[247] Unique interactions in the decoding region involving the bicyclic and terminal sugars of apramycin may account for the specific effects of this agent on translocation.[246]

7.21.4.4 Hygromycin B

Hygromycin B interacts with 16S rRNA in a manner different from that of paromomycin, streptomycin, or spectinomycin.[248] Hygromycin B binds close to the A site near the top of helix 44 of 16S rRNA,[191,198,228] with one ring (ring IV) binding close to where the P site mRNA codon is located.[248] Hygromycin B causes miscoding but to a lesser extent than some of the other aminoglycosides.[242,249] Its major effect is inhibition of translocation[242,250,251] which is reflected by the binding site. Inhibition of translocation sequesters peptidyl-tRNA in the A site and increases affinity of the A site for aminoacyl-tRNA[249,252] which provides an explanation for the miscoding effects. Nucleotides located in helix 44 close to where hygromycin B binds have been shown to affect fidelity of translation.[213] Although hygromycin B binds to eukaryotic ribosomes and inhibits polypeptide polymerization[242,250,251] chemical reactivity experiments[198] indicate that this agent does not protect A1408, but rather enhances its reactivity most likely by an indirect effect.[248]

7.21.5 Nontraditional Pharmacologic Activities

As agents that interfere with ribosomal RNA function and induce miscoding, aminoglycosides have demonstrated potential for a variety of nontraditional applications.[253,254] Cystic fibrosis (CF) is an inherited disorder due to deleterious mutations in the CF transmembrane conductance regulator (CFTR) protein responsible for chloride and sodium transport. Some cases of CF are due to the presence of stop mutations in the CFTR gene. Gentamicin and G418 suppress the expression of these stop mutations in cell culture resulting in production of full-length CFTR protein and restoration of cAMP-activated chloride channel activity.[255,256] In provocative studies on CF individuals carrying a defective stop mutation in their CFTR gene,[257,258] topical administration of gentamicin to nasal epithelia restored CFTR function. Detection of full-length CFTR in these individuals is consistent with 'read through' of the stop codon promoted by gentamicin. Gentamicin had no effect on CF patients carrying the ΔF508 deletion mutation in their CFTR gene.[258] Aminoglycosides could be used to correct or circumvent other inherited disorders due to stop mutations, e.g., Duchenne muscular dystrophy. In *mdx* mice there is a stop mutation in the dystrophin gene, and these mice fail to produce dystrophin protein. Cultured muscle cells from *mdx* mice expressed dystrophin protein and localized it to the cell membrane when exposed to gentamicin.[259]

Aminoglycosides inhibit activity of numerous ribozymes (catalytically active RNA molecules), including self-splicing of group I introns.[260] Streptomycin inhibits self-splicing of group I introns including the thymidylate synthase intron of phage T4[261,262] and the sun Y intron from phage T4.[261] Chemical probing of the sun Y intron[263] showed that neomycin, 5-epi-sisomicin, and streptomycin were bound to sun Y RNA under conditions wherein splicing was completely inhibited. All three agents gave essentially the same footprint on the intron RNA. Kanamycin A did not inhibit splicing and had no effect on the pattern of chemical modification of RNA. Neomycin also inhibits the cleavage reaction of hammerhead ribozymes.[264] Paromomycin is tenfold less effective than neomycin, suggesting that the amino groups of neomycin are important for the antibiotic–hammerhead interaction.

Rev and Tat are essential, viral, regulatory proteins encoded by the human immunodeficiency virus (HIV-1). Both proteins bind to separate, specific sites (designated RRE and TAR, respectively) on viral RNA and function in viral replication. Aminoglycosides antagonize replication by interfering with interaction of Rev [253,265,266] and Tat [253,267] with their binding sites. Neomycin interferes with functioning of the Rev–RRE complex.[266] It binds competitively to RNA in the RRE site but also forms a ternary complex with Rev and RRE.[266] Neomycin also binds to the Tat–TAR complex in a noncompetitive manner and effects a conformational change that increases dissociation of Tat and TAR.[267] Affinity of neomycin for binding TAR correlates directly with inhibitory potency.[268] Streptomycin has a similar effect on TAR but potency of inhibition was less than with neomycin. There may be at least two neomycin binding sites in TAR since binding of a dimer of neomycin was more than tenfold stronger than the monomer.[269] There may be an RNA structural motif rather than a specific nucleotide sequence common to RNAs that bind neomycin.[270] Another potential target for aminoglycosides in HIV-1 is the dimerization initiation site in genomic RNA. Model complexes involving this site resemble the prokaryotic 16S ribosomal A site and bind neomycin and paromomycin.[271]

7.21.6 Antimicrobial Resistance

7.21.6.1 Transport Alterations

Because aminoglycoside transport is energy-dependent, mutations in electron transport or the lack of same (e.g., anaerobes) confer reduced susceptibility or resistance to aminoglycosides.[186] Poor permeability to aminoglycosides is particularly common among clinical isolates of *P. aeruginosa*.[272] Organisms deemed to be impermeable to aminoglycosides tend to show reduced susceptibility to multiple members of this class.[273] So-called adaptive resistance to aminoglycosides, especially in *Pseudomonas*,[274] is thought due to accumulated effects on membrane structure and function that negatively impact the energy-dependent phases of uptake.[275] Multidrug active efflux transporters have been identified in Gram-negative bacteria that extrude aminoglycosides and can confer low-level resistance to numerous aminoglycosides.[276–279]

7.21.6.2 Ribosomal Alterations

High-level, single-step resistance is readily induced to streptomycin or spectinomycin.[182] This resistance is due to alteration of ribosomal proteins that are indirectly involved in the interaction of these agents with the 30S subunit. Alterations in protein S12 confer streptomycin resistance and alterations in protein S5 confer resistance to spectinomycin. Low-level resistance to some of the 2-DOS-containing aminoglycosides due to protein alterations has been described but the mutations are not single-step.[280]

Posttranslational modification of 16S rRNA by enzymatic methylation of nucleotide bases confers resistance to aminoglycosides and is a common mechanism found in aminoglycoside-producing microorganisms.[223,281] This mechanism has been described in the gentamicin,[282,283] kanamycin,[281] tobramycin, [284] and istamycin [284] producers. The sites methylated are in the decoding region where the aminoglycosides bind, and the particular nucleotides methylated determine the specificity of resistance.[281] For example, a methylase from the gentamicin-producer (*Micromonospora purpurea*) methylates nucleotide 1405 in the decoding region and confers resistance to both kanamycin and gentamicin.[281] This mechanism of resistance also has been identified in nonaminoglycoside-producing, pathogenic organisms. A *Klebsiella pneumoniae* isolate has been described producing a methyltransferase that methylates 16S rRNA and confers high-level resistance to all aminoglycosides currently used in human medicine except streptomycin.[285] Similar methyltransferases, conferring high-level resistance to numerous aminoglycosides, have been found in *Serratia marcescens* and *P. aeruginosa*.[286]

7.21.6.3 Enzymatic Modification of Aminoglycosides

The most common and clinically significant mechanism of resistance to aminoglycosides is antibiotic modification by enzymes that *N*-acetylate (acetyl coenzyme A-dependent acetyltransferases), *O*-nucleotidylate (nucleotide triphosphate-dependent nucleotidyltransferases) or *O*-phosphorylate (ATP-dependent phosphotransferases) aminoglycosides.[193,287] Modifying enzymes are found in both Gram-negative and Gram-positive bacteria, can be plasmid or chromosomally encoded,[288] and are closely related to enzymes from aminoglycoside-producing organisms.[289,290] Modification has a deleterious effect on binding to rRNA accounting for resistance.[291,292] The enzymes within each of the three classes differ with respect to their regiospecificities of action and spectra of resistance conferred.[289,292] Some enzymes modify at more than one site, and a widespread, bifunctional enzyme acetylates the 6′ and phosphorylates the 2″ positions of several aminoglycosides.[289,293] Specific rules of nomenclature and classification have been adopted for these enzymes to ensure consistent terminology.[289] Crystal structures of modifying enzymes representing each of the three classes complexed with an aminoglycoside substrate have been reported revealing interesting similarities between binding to resistance enzymes and binding to A site rRNA.[192–194]

The resistance level conferred by aminoglycoside modifying enzymes is determined, in part, by kinetic parameters. For most enzymes, V/K (V_{max}/K_m) appears to be the best predictor of resistance level conferred as measured by in vitro minimal inhibitory concentrations.[193,294,295] Many of the modifying enzymes are periplasmic and the extent of resistance conferred is influenced by a combination of decreased antibiotic transport and inadequate interactions with rRNA.[296] Whether or not the presence of a modifying enzyme confers clinically significant resistance depends on both extent of enzyme-catalyzed modification and how readily the antibiotic is transported. Effective transport of an aminoglycoside requires EDP II. The inability of the modified antibiotic to interact with the ribosome precludes induction of EDP II, effectively slowing transport and allowing more aminoglycoside to be modified.[188]

Adenylyltransferases transfer an adenosine monophosphate from Mg-ATP to a hydroxyl substituent to form an adenylylated aminoglycoside plus magnesium pyrophosphate.[193] The three-dimensional crystal structure of a kanamycin adenylyltransferase from *S. aureus* complexed with substrates has been reported.[297,298] The enzyme forms a dimeric structure with each monomer binding ATP and kanamycin.

Phosphotransferases transfer the gamma-phosphoryl group of Mg-ATP to a hydroxyl substituent forming an *O*-phosphoryl aminoglycoside plus Mg-ADP.[193] Examination of an aminoglycoside phosphotransferase from *Enterococcus* showed that the enzyme forms a dimer covalently linked through two cysteine bridges.[299,300] The enzyme is active as a monomer or dimer, and its three-dimensional structure reveals how it modifies a wide range of structurally distinct aminoglycosides. The aminoglycoside–enzyme complex bears a structural similarity to aminoglycosides bound into the A site of the 30S ribosome.[301] The overall structure of this enzyme also bears a striking resemblance to eukaryotic protein kinases indicating relatedness despite a lack of sequence homology.

Acetyltransferases transfer an acetyl group from acetyl coenzyme A to an amino group of the aminoglycoside forming an *N*-acetyl derivative plus coenzyme A.[193] Three-dimensional structures of acetyltransferases with different regiospecificities from *Mycobacterium tuberculosis*,[302] *Serratia marcescens*,[303] *Salmonella enterica*,[304] and *Enterococcus faecium*[305,306] indicate all form dimers. Closer examination of the enzyme from *E. faecium*, including comparison with a mutant enzyme which does not form a dimer, revealed two nonequivalent binding sites suggesting cooperative binding of the aminoglycoside substrate to the dimer.[306] Unlike these acetyltransferases, the bifunctional acetyltransferase–phosphotransferase purifies as a monomer.[307]

7.21.6.4 Circumvention of Resistance

As previously discussed in this chapter, efforts to identify novel aminoglycosides or new derivatives have focused on molecules that circumvent inactivation by aminoglycoside modifying enzymes. However, other efforts to overcome resistance due to enzymatic modification also have been tried.[287] Williams and Northrop[308] described in 1979 an acetylgentamicin-CoA analog that inhibited 3-*N*-gentamicin acetyltransferase. Although the inhibitor bound tightly to the enzyme in vitro, there was no potentiation of antibiotic activity against resistant bacteria. Several inhibitors of eukaryotic protein kinases inhibit aminoglycoside phosphotransferases.[309,310] Isoquinolinesulfonamides are particularly effective as competitive inhibitors, with respect to ATP, of aminoglycoside phosphorylation.[309] 7-Hydroxytropolone, isolated as a natural product, is a competitive inhibitor, with respect to ATP, of 2″-*O*-adenylyltransferase.[311,312] Combinations of 7-hydroxytropolone plus aminoglycoside substrates for the enzyme were active against strains harboring the 2″-*O*-adenylyltransferase. A variety of cationic antimicrobial peptides inhibit aminoglycoside acetyl- and phosphotransferases.[313] In particular, the bovine peptide indolicidin and analogs were effective inhibitors of both modifying enzyme classes.

1-(Bromomethyl)phenanthrine and related compounds covalently inactivate the acetyltransferase portion of the bifunctional acetyltransferase–phosphotransferase enzyme.[314] Several neamine dimers were active against bacteria producing the same bifunctional modifying enzyme.[142] These compounds bind to A site rRNA inhibiting bacterial growth, are not substrates for the acetyltransferase portion of the modifying enzyme, and are potent competitive inhibitors of the phosphotransferase portion. Other approaches to circumventing resistance include mechanism-based aminoglycoside structures that irreversibly inactivate aminoglycoside-3′-phosphotransferases,[146] and a 3′-keto derivative of kanamycin A which, when phosphorylated, spontaneously hydrolyzes to regenerate the 3′-keto derivative.[95,292] Rather than circumvent resistance, this approach would confer immunity to the inactivating mechanism.

Pursuing a much different approach, DeNap *et al.*[315,316] used apramycin as an RNA mimic to interfere with the process of plasmid replication and eliminate plasmids that carry resistance determinants. In their experiments, apramycin caused a dose-dependent loss of the resistance plasmid. This approach is based on mechanisms of plasmid incompatibility which limit the extent to which plasmids are replicated and passed on to subsequent generations. With elimination of the plasmid carrying the resistance determinant, antibiotic susceptibility is restored, obviating the need for agents to block the activity of gene products of a resistance determinant.

7.21.7 Clinical Pharmacology

Aminoglycosides are poorly absorbed from the gastrointestinal tract and are administered parenterally.[28] Their high water solubility and low protein-binding facilitate distribution.[28,163] Intracellular concentrations tend to be low except in the proximal renal tubule; aminoglycosides are eliminated nonmetabolized by the kidneys, primarily through glomerular filtration. Multiple dosing of aminoglycosides can lead to accumulation in the renal cortex resulting in nephrotoxicity.

Concentration-dependent killing and extended PAE by aminoglycosides allow for once-daily dosing.[179,317,318] The efficacy of this regimen relies on achieving a peak serum concentration (C_{max}) of antibiotic above the minimum inhibitory concentration (MIC) that results in bactericidal activity for several hours after dosing.[163] Excretion of antibiotic subsequently lowers its serum concentration to a trough level near or even below the MIC, but does not impair continued bacterial killing due to the long PAE, while the amount of potentially toxic antibiotic is lowered before the next day's dose.[317] Furthermore, the extent of the PAE increases with increasing C_{max}. Thus, a single daily dosing regimen can reduce aminoglycoside accumulation in the renal cortex, lessening the risk or severity of nephrotoxicity while maintaining bactericidal effectiveness.[319]

7.21.8 Formulation Chemistry and Drug Delivery Technology

Physicochemical features of drugs influence their delivery to bodily sites of action.[320,321] Aminoglycosides form acid-addition salts for parenteral administration in aqueous media. For localized therapy, aminoglycosides are formulated for placement at sites not treated as well systemically. Aminoglycoside-impregnated polymethylmethacrylate (PMMA) beads reduce infection when implanted at surgical bone and soft tissue sites, but they must later be surgically removed.[322] Biodegradable carriers are preferable, such as collagen sponges, bioceramics, composites with other materials, or microspheres prepared from hydroxyacid-polymers and copolymers.[322–330] Many of these systems are still in research stages. Local drug delivery is also important for sites other than the musculoskeletal system. Neomycin is commonly used as a topical antibiotic.[331] Tobramycin's activity against *Pseudomonas* is useful for treating burns, so it was studied in a skin wound treatment matrix.[332] Aminoglycosides are important for treating certain ophthalmic infections and new delivery methods are being studied.[333–337] Some aminoglycosides are used for gastrointestinal sterilization since they are not well absorbed orally in the absence of gastrointestinal disorders.[338] Lipophilic surfactant excipients have been studied as a means to enhance oral absorption.[339,340]

A different approach involves drug carriers that transport an antibiotic to its intended site while reducing exposure for systemic toxicity, such as liposome encapsulation.[321,341–343] Nanotechnology will likely find important applications such as development of nanoparticles as colloidal drug carriers.[321] Gentamicin was embedded in bacterial membrane vesicles to penetrate the Gram-negative outer membrane and deliver antibiotic with a bactericidal enzyme.[344] Derivatives were synthesized as prodrugs to provide extended or controlled drug release.[345,346] Polymeric aminoglycoside prodrugs have been considered.[91] While new drug delivery matrices and techniques may be experimentally promising, benefits will eventually need to be proven in clinical trials.

7.21.9 Analytical and Bioanalytical Chemistry

Many analytical methods for aminoglycosides are available to screen samples, identify compounds, monitor separations, test quality or purity of bulk or formulated drugs, or assay concentrations, degradation products, metabolites, or residues from diverse material.[347–351] To minimize patient toxicity with optimum efficacy, therapeutic drug monitoring was developed to guide dosing by correlating aminoglycoside concentrations with indicators of toxicity.[352–356] Commonly used analytical methods include different microbiological, radiological, immunological, and chromatographic assays. Microbiological assays are still used for many applications since they are readily performed without expensive specialized equipment. Automated immunoassays are popular, but an assay for each test compound is required. Chromatographic methods must separate and quantify nonvolatile, polar, water-soluble, basic molecules lacking a natural ultraviolet or fluorescent chromophore, which require alternative detection systems.[357–360] Development of new analytical methods for aminoglycosides should continue in order to meet needs for rapid, accurate analysis of large numbers of samples.

7.21.10 Toxicology

The toxicological liabilities of aminoglycosides, particularly nephro- and ototoxicity, are well known.[338,361–363] Side effects are monitored, but predictive markers only reveal onset of observable patient toxicity.[352,363–365] Animal model and in vitro studies help to define toxic events while many strategies and agents to ameliorate toxicity have been proposed. However, despite decades of research, primary molecular causes of toxicity remain to be proven. Without such information, important clues are missing to rationally find aminoglycosides having reduced toxicity, and progress is dependent on resource-intensive testing of low numbers of compounds and serendipity.

Nephrotoxicity begins after aminoglycosides enter kidney proximal tubules, probably facilitated by initial binding to acidic membrane phospholipids.[363,366] Computational analysis of binding to model lipids was used to seek structure–toxicity correlations, but such binding may not itself cause toxicity.[88,367] Other mechanisms proposed for toxicity include uptake by the endocytic receptor megalin,[366] interaction with renal NMDA receptors,[368] generation of reactive oxygen species,[369,370] disruption of chaperone action,[371,372] induction of apoptosis,[373] or inhibition of renal enzymes.[88] Microbiologically-inactive 6′-*N*-aminoacyl derivatives of netilmicin and kanamycin inhibited membrane-embedded phospholipase, an indicator of potential toxicity, suggesting dissociation between aminoglycoside antibacterial activity and some toxicity parameters may be possible.[88] Separating the causative origin(s) of toxicity from the many secondary downstream effects remain key to progress in this area.

Proposed mechanisms for the origin of ototoxicity exhibit similar themes, although irreversibility of hearing loss makes cochlear toxicity much more critical to avoid.[362,374–377] The initial step appears to be aminoglycoside binding to acidic membrane phospholipids and glycosaminoglycans in the ear followed by intracellular uptake and a subsequent cascade of events. Other proposed mechanisms for ototoxicity include generation of reactive oxygen species,[375,378] activation of NMDA receptors,[379] interference with F-actin,[380] and mitochondrial mutations.[378,381] Some apramycin derivatives showed no correlation between loss of antibiotic activity and NMDA-receptor activation, again suggesting aminoglycoside antibacterial activity and some toxicity parameters may be dissociated.[382] Although aminoglycosides may lose antibacterial activity and retain activity in assays indicative of potential toxicity, the reverse is needed, i.e., retain antibiotic activity with reduced toxicity. Application of modern screening and genomic technologies such as microarray analysis and gene expression profiling may provide new mechanistic insights into aminoglycoside toxicity and novel ways to prevent it.[383,384]

7.21.11 Trends and Advances in Aminoglycoside Medicinal Chemistry

Two trends in aminoglycoside chemistry are: (1) rational design of derivatives based on insights into mechanisms of action and resistance, and (2) high-throughput methods to synthesize and assay more compounds. The former is aided by advances in x-ray crystallography, NMR spectroscopy, and physicochemical analysis that yield more molecular detail for an increasing number of aminoglycoside complexes with RNA and protein.[181,194,385,386] These data build a basis for initiating computational chemistry models such as aminoglycoside docking into RNA and may eventually allow applications such as screening virtual compound libraries in silico.[387–389] Studies to find small aminoglycoside mimetics have already been reported.[390–393] New synthetic methods to more selectively manipulate specific functional groups will also apply to creating larger libraries for screening. New approaches such as principles from click chemistry might be applied to find new binding ligands.[394] A systems-biology approach may resolve the complexities of aminoglycoside toxicity. If mechanistic information coupled with imaginative synthesis produces useful new antibiotics, the aminoglycoside renaissance will have succeeded.

References

1. Hooper, I. R. The Naturally Occurring Aminoglycoside Antibiotics. In *Aminoglycoside Antibiotics*; Umezawa, H., Hooper, I. R., Eds.; Springer-Verlag: Berlin, 1982, pp 1–35.
2. Daniels, P. J. L. Antibiotics (Aminoglycosides). In *Kirk-Othmer Encyclopedia of Chemical Technology*, 3rd ed.; Grayson, M., Ed.; Wiley-Interscience: New York, **1978**; Vol. 2, pp 819–852.
3. McGregor, D. Antibiotics (Aminoglycosides). In *Kirk-Othmer Encyclopedia of Chemical Technology*, 4th ed.; Howe-Grant, M., Ed.; John Wiley: New York, **1992**; Vol. 2, pp 904–926.
4. Price, K. E.; Godfrey, J. C.; Kawaguchi, H. Effect of Structural Modifications on the Biological Properties of Aminoglycoside Antibiotics Containing 2-Deoxystreptamine. In *Structure–Activity Relationships among the Semisynthetic Antibiotics*; Perlman, D., Ed.; Academic Press: New York, 1977, pp 239–395.
5. Kirst, H. A. Aminoglycoside, Macrolide, Glycopeptide and Miscellaneous Antibacterial Antibiotics. In *Burger's Medicinal Chemistry and Drug Discovery*, 5th ed.; Wolff, M. E., Ed.; John Wiley: New York, **1996**; Vol. 2, pp 463–525.
6. Umezawa, S. *Adv. Carbohyd. Chem. Biochem.* **1974**, *30*, 111–182.
7. Schatz, A.; Bugie, E.; Waksman, S. A. *Proc. Soc. Exp. Biol. Med.* **1944**, *55*, 66–69.
8. Waksman, S. A. *Science* **1953**, *118*, 259–266.
9. Lemieux, R. U.; Wolfrom, M. L. *Adv. Carbohyd. Chem.* **1948**, *3*, 337–384.
10. Kuehl, F. A., Jr.; Peck, R. L.; Hoffhine, C. E., Jr.; Folkers, K. *J. Am. Chem. Soc.* **1948**, *70*, 2325–2330.
11. Neidle, S.; Rogers, D.; Hursthouse, M. B. *Tetrahedron Lett.* **1968**, *9*, 4725–4728.
12. Alford, R. H.; Wallace, R. J., Jr. Antimycobacterial Agents. In *Principles and Practice of Infectious Diseases*, 6th ed.; Mandell, G. L., Bennett, J. E., Dolin, R., Eds.; Churchill Livingstone: Philadelphia, PA, **2005**; Vol. 1, pp 489–501.
13. Waksman, S. A.; Lechevalier, H. A. *Science* **1949**, *109*, 305–307.
14. Hichens, M.; Rinehart, K. L., Jr. *J. Am. Chem. Soc.* **1963**, *85*, 1547–1548.
15. Rinehart, K. L., Jr. *The Neomycins and Related Antibiotics*; John Wiley: New York, 1964.
16. Schillings, R. T.; Schaffner, C. P. *Antimicrob. Agents Chemother.* **1961**, 274–285.
17. Haskell, T. H.; French, J. C.; Bartz, Q. R. *J. Am. Chem. Soc.* **1959**, *81*, 3482–3483.
18. Jernigan, J. A.; Pearson, R. D. Agents Active against Parasites and Pneumocystis. In *Principles and Practice of Infectious Diseases*, 6th ed.; Mandell, G. L., Bennett, J. E., Dolin, R., Eds.; Churchill Livingston: Philadelphia, PA, **2005**; Vol. 1, pp 568–603.
19. Mori, T.; Kyotani, Y.; Watanabe, I.; Oda, T. *J. Antibiot.* **1972**, *25*, 149–150.
20. Akita, E.; Tsuruoka, T.; Ezaki, N.; Niida, T. *J. Antibiot.* **1970**, *23*, 173–183.
21. Woo, P. W. K.; Dion, H. W.; Bartz, Q. R. *Tetrahedron Lett.* **1971**, *12*, 2625–2628.
22. Umezawa, H.; Ueda, M.; Maeda, K.; Yagashita, K.; Kondo, S.; Okami, Y.; Utahara, R.; Osato, Y.; Nitta, K.; Takeuchi, T. *J. Antibiot.* **1957**, *A10*, 181–188.
23. Koyama, G.; Iitaka, Y.; Maeda, K.; Umezawa, H. *Tetrahedron Lett.* **1968**, *9*, 1875–1879.
24. Ito, T.; Nishio, M.; Ogawa, H. *J. Antibiot.* **1964**, *A17*, 189–193.
25. Cooper, D. J.; Daniels, P. J. L.; Yudis, M. D.; Marigliano, H. M.; Guthrie, R. D.; Bukhari, S. T. K. *J. Chem. Soc. (C)* **1971**, 3126–3129.
26. Cooper, D. J. *Pure Appl. Chem.* **1971**, *28*, 455–467.
27. Berdy, J.; Pauncz, J. K.; Vajna, Z. M.; Horvath, G.; Gyimesi, J.; Koczka, I. *J. Antibiot.* **1977**, *30*, 945–954.

28. Gonzalez, L. S., III; Spencer, J. P. *Am. Family Phys.* **1998**, *58*, 1811–1820.
29. Zembower, T. R.; Noskin, G. A.; Postelnick, M. J.; Nguyen, C.; Peterson, L. R. *Int. J. Antimicrob. Agents* **1998**, *10*, 95–105.
30. Kumana, C. R.; Yuen, K. Y. *Drugs* **1994**, *47*, 902–913.
31. Edson, R. S.; Terrell, C. L. *Mayo Clin. Proc.* **1999**, *74*, 519–528.
32. Koch, K. F.; Davis, F. A.; Rhoades, J. A. *J. Antibiot.* **1973**, *26*, 745–751.
33. Koch, K. F.; Rhoades, J. A. *Antimicrob. Agents Chemother.* **1971**, 309–313.
34. Gialdroni Grassi, G. *J. Chemother.* **1995**, *7*, 344–354.
35. Singh, M. P. *Curr. Opin. Investig. Drugs* **2001**, *2*, 755–765.
36. Cheer, S. M.; Waugh, J.; Noble, S. *Drugs* **2003**, *63*, 2501–2520.
37. Reimann, H.; Cooper, D. J.; Mallams, A. K.; Jaret, R. S.; Yehaskel, A.; Kugelman, M.; Vernay, H. F.; Schumacher, D. *J. Org. Chem.* **1974**, *39*, 1451–1457.
38. Fukuda, M.; Sasaki, K. *Jpn. J. Ophthalmol.* **2002**, *46*, 384–390.
39. Hotta, K.; Sunada, A.; Ikeda, Y.; Kondo, S. *J. Antibiot.* **2000**, *53*, 1168–1174.
40. Cochran, T. G.; Abraham, D. J.; Martin, L. L. *J. Chem. Soc. Chem. Commun.* **1972**, 494–495.
41. O'Connor, S.; Lam, L. K. T.; Jones, N. D.; Chaney, M. O. *J. Org. Chem.* **1976**, *41*, 2087–2092.
42. Prescott, J. F. Aminoglycosides and Aminocyclitols. In *Antimicrobial Therapy in Veterinary Medicine*, 3rd ed.; Prescott, J. F., Baggot, J. D., Walker, R. D., Eds.; Iowa State University Press: Ames, IA, 2000, pp 191–228.
43. Neuss, N.; Koch, K. F.; Molloy, B. B.; Day, W.; Huckstep, L. L.; Dorman, D. E.; Roberts, J. D. *Helv. Chim. Acta* **1970**, *53*, 2314–2319.
44. Inouye, S.; Shomura, T.; Watanabe, H.; Totsugawa, K.; Niida, T. *J. Antibiot.* **1973**, *26*, 374–385.
45. Egan, R. S.; Stanaszek, R. S.; Cirovic, M.; Mueller, S. L.; Tadanier, J.; Martin, J. R.; Collum, P.; Goldstein, A. W.; DeVault, R. L.; Sinclair, A. C. et al. *J. Antibiot.* **1977**, *30*, 552–563.
46. Girolami, R. L.; Stamm, J. M. *J. Antibiot.* **1977**, *30*, 564–570.
47. Kong, F.; Zhao, N.; Siegel, M. M.; Janota, K.; Ashcroft, J. S.; Koehn, F. E.; Borders, D. B.; Carter, G. T. *J. Am. Chem. Soc.* **1998**, *120*, 13301–13311.
48. Singh, M. P.; Peterson, P. J.; Weiss, W. J.; Kong, F.; Greenstein, M. *Antimicrob. Agents Chemother.* **2000**, *44*, 2154–2159.
49. Blunt, J. W.; Copp, B. R.; Munro, M. H. G.; Northcote, P. T.; Prinsep, M. R. *Nat. Prod. Rep.* **2003**, *20*, 1–48.
50. Hotta, K.; Yoshida, M.; Hamada, M.; Okami, Y. *J. Antibiot.* **1980**, *33*, 1515–1520.
51. Slattery, M.; Rajbhandari, I.; Wesson, K. *Microb. Ecol.* **2001**, *41*, 90–96.
52. Burgess, J. G.; Jordan, E. M.; Bregu, M.; Mearns-Spragg, A.; Boyd, K. G. *J. Biotechnol.* **1999**, *70*, 27–32.
53. Donadio, S.; Carrano, L.; Brandi, L.; Serina, S.; Soffientini, A.; Raimondi, E.; Montanini, N.; Sosio, M.; Gualerzi, C. O. *J. Biotechnol.* **2002**, *99*, 175–185.
54. Bush, K.; Macielag, M.; Weidner-Wells, M. *Curr. Opin. Microbiol.* **2004**, *7*, 466–476.
55. Bronson, J. J.; Barrett, J. F. *Curr. Med. Chem.* **2001**, *8*, 1775–1793.
56. Bax, R.; Mullan, N.; Verhoef, J. *Int. J. Antimicrob. Agents* **2000**, *16*, 51–59.
57. Peck, R. L.; Hoffhine, C. E., Jr.; Folkers, K. *J. Am. Chem. Soc.* **1946**, *68*, 1390–1391.
58. Waksman, S. A.; Lechevalier, H. A. *The Actinomycetes*; Williams & Wilkins: Baltimore, MD, 1962; Vol. 3, pp 256–257.
59. Umezawa, H.; Tsuchiya, T. Total Synthesis and Chemical Modification of the Aminoglycoside Antibiotics. In *Aminoglycoside Antibiotics*; Umezawa, H., Hooper, I. R., Eds.; Springer-Verlag: Berlin, 1982, pp 37–110.
60. Cox, D. A.; Richardson, K.; Ross, B. C. The Aminoglycosides. In *Topics in Antibiotic Chemistry*; Sammes, P., Ed.; Ellis Horwood: Chichester, UK, 1977, pp 5–90.
61. Reden, J.; Durckheimer, W. *Topics Curr. Chem.* **1979**, *83*, 105–170.
62. Benveniste, R.; Davies, J. *Antimicrob. Agents Chemother.* **1973**, *4*, 402–409.
63. Benveniste, R.; Davies, J. *Annu. Rev. Biochem.* **1973**, *42*, 471–506.
64. Umezawa, H. *Adv. Carbohyd. Chem. Biochem.* **1974**, *30*, 183–225.
65. Mitsuhashi, S. *Drug Action and Drug Resistance in Bacteria, Vol. 2, Aminoglycoside Antibiotics*; University Park Press: Baltimore, MD, 1975.
66. Umezawa, H.; Okanishi, M.; Kondo, S.; Hamana, K.; Utahara, R.; Maeda, K.; Mitsuhashi, S. *Science* **1967**, *157*, 1559–1561.
67. Umezawa, H.; Umezawa, S.; Tsuchiya, T.; Okazaki, Y. *J. Antibiot.* **1971**, *24*, 485–487.
68. Kondo, S.; Hotta, K. *J. Infect. Chemother.* **1999**, *5*, 1–9.
69. Kawaguchi, H.; Naito, T.; Nakagawa, S.; Fujisawa, K. *J. Antibiot.* **1972**, *25*, 695–708.
70. Kawaguchi, H.; Naito, T. In *Chronicles of Drug Discovery.*; Bindra, J., Lednicer, D., Eds.; Wiley-Interscience: New York, **1983**; Vol. 2, pp 207–234.
71. Cunha, B. A. *Sem. Respir. Infect.* **2002**, *17*, 231–239.
72. Wright, J. J. *J. Chem. Soc. Chem. Commun.* **1976**, 206–208.
73. Campoli-Richards, D. M.; Chaplin, S.; Sayee, R. H.; Goa, K. L. *Drugs* **1989**, *38*, 703–756.
74. Nagabhushan, T. L.; Cooper, A. B.; Tsai, H.; Daniels, P. J. L.; Miller, G. H. *J. Antibiot.* **1978**, *31*, 681–687.
75. Tod, M.; Padoin, C.; Petitjean, O. *Clin. Pharmacokinet.* **2000**, *38*, 205–223.
76. Kondo, S.; Iinuma, K.; Yamamoto, H.; Maeda, K.; Umezawa, H. *J. Antibiot.* **1973**, *26*, 412–415.
77. Suzuki, K. *Pediatr. Int.* **2003**, *45*, 301–306.
78. Kotra, L. P.; Mobashery, S. *Curr. Org. Chem.* **2001**, *5*, 193–205.
79. Tok, J. B.; Bi, L. *Curr. Top. Med. Chem.* **2003**, *3*, 1001–1019.
80. Tor, Y. *ChemBioChem* **2003**, *4*, 998–1007.
81. Vicens, Q.; Westhof, E. *ChemBioChem* **2003**, *4*, 1018–1023.
82. Hermann, T. *Biopolymers* **2003**, *70*, 4–18.
83. Yamasaki, T.; Narita, Y.; Hoshi, H.; Aburaki, S.; Kamei, H.; Naito, T.; Kawaguchi, H. *J. Antibiot.* **1991**, *44*, 646–658.
84. Hoshi, H.; Aburaki, S.; Iimura, S.; Yamasaki, T.; Naito, T.; Kawaguchi, H. *J. Antibiot.* **1990**, *43*, 858–872.
85. Hanessian, S.; Kornienko, A.; Swayze, E. E. *Tetrahedron* **2003**, *59*, 995–1007.
86. Vourloumis, D.; Winters, G. C.; Simonsen, K. B.; Takahashi, M.; Ayida, B. K.; Shandrick, S.; Zhao, Q.; Han, Q.; Hermann, T. *ChemBioChem* **2005**, *6*, 58–65.
87. Haddad, J.; Kotra, L. P.; Llano-Sotelo, B.; Kim, C.; Azucena, E. F., Jr.; Liu, M.; Vakulenko, S. B.; Chow, C. S.; Mobashery, S. *J. Am. Chem. Soc.* **2002**, *124*, 3229–3237.

88. Kotretsou, S.; Mingeot-Leclercq, M. P.; Constantinou-Kokotou, V.; Brasseur, R.; Georgiadis, M. P.; Tulkens, P. M. *J. Med. Chem.* **1995**, *38*, 4710–4719.
89. Wang, Y.; Rando, R. R. *Chem. Biol.* **1995**, *2*, 281–290.
90. Grapsas, I.; Lerner, S. A.; Mobashery, S. *Arch. Pharm.* **2001**, *334*, 295–301.
91. Tanaka, H.; Nishida, Y.; Furuta, Y.; Kobayashi, K. *Bioorg. Med. Chem. Lett.* **2002**, *12*, 1723–1726.
92. Van Schepdael, A.; Delcourt, J.; Mulier, M.; Busson, R.; Verbist, L.; Vanderhaeghe, H. J.; Mingeot-Leclercq, M. P.; Tulkens, P. M.; Claes, P. J. *J. Med. Chem* **1991**, *34*, 1468–1475.
93. Mingeot-Leclercq, M. P.; Van Schepdael, A.; Brasseur, R.; Busson, R.; Vanderhaeghe, H. J.; Claes, P. J.; Tulkens, P. M. *J. Med. Chem.* **1991**, *34*, 1476–1482.
94. Simonsen, K. B.; Ayida, B. K.; Vourloumis, D.; Takahashi, M.; Winters, G. C.; Barluenga, S.; Qamar, S.; Shandrick, S.; Zhao, Q.; Hermann, T. *ChemBioChem* **2002**, *3*, 1223–1228.
95. Haddad, J.; Vakulenko, S.; Mobashery, S. *J. Am. Chem. Soc.* **1999**, *121*, 11922–11923.
96. Maebashi, K.; Usui, T.; Hiraiwa, Y.; Akiyama, Y.; Otsuka, K.; Ozaki, S.; Murakami, S.; Nakano, Y.; Yamamoto, M.; Seki, A. et al. *Abstracts of the 44th Interscience Conference on Antimicrobial Agents and Chemotherapy*, Oct 30–Nov 2, 2004; Washington, DC, F-716.
97. Mingeot-Leclercq, M. P.; Glupczynski, Y.; Tulkens, P. M. *Antimicrob. Agents Chemother.* **1999**, *43*, 727–737.
98. Shitara, T.; Umemura, E.; Tsuchiya, T.; Matsuno, T. *Carbohydr. Res.* **1995**, *276*, 75–89.
99. Kim, C.; Haddad, J.; Vakulenko, S. B.; Meroueh, S. O.; Wu, Y.; Yan, H.; Mobashery, S. *Biochemistry* **2004**, *43*, 2373–2383.
100. Tok, J. B.-H.; Rando, R. R. *J. Am. Chem. Soc.* **1998**, *120*, 8279–8280.
101. Hendrix, M.; Alper, P. B.; Priestley, E. S.; Wong, C.-H. *Angew. Chem. Int. Ed. Engl.* **1997**, *36*, 95–98.
102. Hermann, T. *Biochemie* **2002**, *84*, 869–875.
103. Van Schepdael, A.; Busson, R.; Vanderhaeghe, H. J.; Claes, P. J.; Verbist, L.; Mingeot-Leclercq, M. P.; Brasseur, R.; Tulkens, P. M. *J. Med. Chem* **1991**, *34*, 1483–1492.
104. Lee, S. H.; Cheong, C. S. *Tetrahedron* **2001**, *57*, 4801–4815.
105. Grapsas, I.; Massova, I.; Mobashery, S. *Tetrahedron* **1998**, *54*, 7705–7720.
106. Lesniak, W.; Harris, W. R.; Kravitz, J. Y.; Schacht, J.; Pecoraro, V. L. *Inorg. Chem.* **2003**, *42*, 1420–1429.
107. Szczepanik, W.; Dworniczek, E.; Ciesiolka, J.; Wrzesinski, J.; Skala, J.; Jezowska-Bojczuk, M. *J. Inorg. Biochem.* **2003**, *94*, 355–364.
108. Ritter, T. K.; Wong, C.-H. *Angew. Chem. Int. Ed. Engl.* **2001**, *40*, 3508–3533.
109. Ryu, D. H.; Litovchick, A.; Rando, R. R. *Biochemistry* **2002**, *41*, 10499–10509.
110. Ryu, D. H.; Tan, C.-H.; Rando, R. R. *Bioorg. Med. Chem. Lett.* **2003**, *13*, 901–903.
111. Verhelst, S. H. L.; Magnee, L.; Wennekes, T.; Wiedenhof, W.; van der Marel, G. A.; Overkleeft, H. S.; van Boeckel, C. A. A.; van Boom, J. H. *Eur. J. Org. Chem.* **2004**, 2404–2410.
112. Sucheck, S. J.; Shue, Y.-K. *Curr. Opin. Drug Disc. Dev.* **2001**, *4*, 462–470.
113. Bryan, M. C.; Wong, C.-H. *Tetrahedron Lett.* **2004**, *45*, 3639–3642.
114. Disney, M. D.; Seeberger, P. H. *Chem. Eur. J.* **2004**, *10*, 3308–3314.
115. Disney, M. D.; Magnet, S.; Blanchard, J. S.; Seeberger, P. H. *Angew. Chem. Int. Ed. Engl.* **2004**, *43*, 1591–1594.
116. Chou, C.-H.; Wu, C.-S.; Chen, C.-H.; Lu, L.-D.; Kulkarni, S. S.; Wong, C.-H.; Hung, S.-C. *Org. Lett.* **2004**, *6*, 585–588.
117. Liang, F.-S.; Wang, S.-K.; Nakatani, T.; Wong, C.-H. *Angew. Chem. Int. Ed. Engl.* **2004**, *43*, 6496–6500.
118. Yao, S.; Sgarbi, P. W. M.; Marby, K. A.; Rabucka, D.; O'Hare, S. M.; Cheng, M. L.; Bairi, M.; Hu, C.; Hwang, S.-B.; Hwang, C.-K. et al. *Bioorg. Med. Chem. Lett.* **2004**, *14*, 3733–3738.
119. Elchert, B.; Li, J.; Wang, J.; Hui, Y.; Rai, R.; Ptak, R.; Ward, P.; Takemoto, J. Y.; Bensaci, M.; Chang, C.-W. T. *J. Org. Chem.* **2004**, *69*, 1513–1523.
120. Li, J.; Wang, J.; Hui, Y.; Chang, C.-W. T. *Org. Lett.* **2003**, *5*, 431–434.
121. Wang, J.; Li, J.; Czyryca, P. G.; Chang, H.; Kao, J.; Chang, C.-W. T. *Bioorg. Med. Chem. Lett.* **2004**, *14*, 4389–4393.
122. Li, J.; Wang, J.; Czyryca, P. G.; Chang, H.; Orsak, T. W.; Evanson, R.; Chang, C.-W. T. *Org. Lett.* **2004**, *6*, 1381–1384.
123. Alper, P. B.; Hendrix, M.; Sears, P.; Wong, C.-H. *J. Am. Chem. Soc.* **1998**, *120*, 1965–1978.
124. Seeberger, P. H.; Baumann, M.; Zhang, G.; Kanemitsu, T.; Swayze, E. E.; Hofstadler, S. A.; Griffey, R. H. *Synlett* **2003**, 1323–1326.
125. Ding, Y.; Swayze, E. E.; Hofstadler, S. A.; Griffey, R. H. *Tetrahedron Lett.* **2000**, *41*, 4049–4052.
126. Fridman, M.; Belakhov, V.; Yaron, S.; Baasov, T. *Org. Lett.* **2003**, *5*, 3575–3578.
127. Ding, Y.; Hofstadler, S. A.; Swayze, E. E.; Griffey, R. H. *Org. Lett.* **2001**, *3*, 1621–1623.
128. Vourloumis, D.; Takahashi, M.; Winters, G. C.; Simonsen, K. B.; Ayida, B. K.; Barluenga, S.; Qamar, S.; Shandrick, S.; Zhao, Q.; Hermann, T. *Bioorg. Med. Chem. Lett.* **2002**, *12*, 3367–3372.
129. Ding, Y.; Hofstadler, S. A.; Swayze, E. E.; Risen, L.; Griffey, R. H. *Angew. Chem. Int. Ed. Engl.* **2003**, *42*, 3409–3412.
130. Greenberg, W. A.; Priestley, E. S.; Sears, P. S.; Alper, P. B.; Rosenbohm, C.; Hendrix, M.; Hung, S.-C.; Wong, C.-H. *J. Am. Chem. Soc.* **1999**, *121*, 6527–6541.
131. Hanessian, S.; Tremblay, M.; Kornienko, A.; Moitessier, N. *Tetrahedron* **2001**, *57*, 3255–3265.
132. Russell, R. J. M.; Murray, J. B.; Lentzen, G.; Haddad, J.; Mobashery, S. *J. Am. Chem. Soc.* **2003**, *125*, 3410–3411.
133. Hanessian, S.; Tremblay, M.; Swayze, E. E. *Tetrahedron* **2003**, *59*, 983–993.
134. François, B.; Szychowski, J.; Adhikari, S. S.; Pachamuthu, K.; Swayze, E. E.; Griffey, R. H.; Migawa, M. T.; Westhof, E.; Hanessian, S. *Angew. Chem. Int. Ed. Engl.* **2004**, *43*, 6735–6738.
135. Barluenga, S.; Simonsen, K. B.; Littlefield, E. S.; Ayida, B. K.; Vourloumis, D.; Winters, G. C.; Takahashi, M.; Shandrick, S.; Zhao, Q.; Han, Q. et al. *Bioorg. Med. Chem. Lett.* **2004**, *14*, 713–718.
136. Simonsen, K. B.; Ayida, B. K.; Vourloumis, D.; Winters, G. C.; Takahashi, M.; Shandrick, S.; Zhao, Q.; Hermann, T. *ChemBioChem* **2003**, *4*, 886–890.
137. Wong, C.-H.; Hendrix, M.; Manning, D. D.; Rosenbohm, C.; Greenberg, W. A. *J. Am. Chem. Soc.* **1998**, *120*, 8319–8327.
138. Vourloumis, D.; Winters, G. C.; Takahashi, M.; Simonsen, K. B.; Ayida, B. K.; Shandrick, S.; Zhao, Q.; Hermann, T. *ChemBioChem* **2003**, *4*, 879–885.
139. Wu, B.; Yang, J.; Robinson, D.; Hofstadler, S.; Griffey, R.; Swayze, E. E.; He, Y. *Bioorg. Med. Chem. Lett.* **2003**, *13*, 3915–3918.
140. Venot, A.; Swayze, E. E.; Griffey, R. H.; Boons, G.-J. *ChemBioChem* **2004**, *5*, 1228–1236.
141. Sofia, M. J.; Allanson, N.; Hatzenbuhler, N. T.; Jain, R.; Kakarla, R.; Kogan, N.; Liang, R.; Liu, D.; Silva, D. J.; Wang, H. et al. *J. Med. Chem.* **1999**, *42*, 3193–3198.
142. Sucheck, S. J.; Wong, A. L.; Koeller, K. M.; Boehr, D. D.; Draker, K.; Sears, P.; Wright, G. D.; Wong, C.-H. *J. Am. Chem. Soc.* **2000**, *122*, 5230–5231.

143. Liu, X.; Thomas, J. R.; Hergenrother, P. J. *J. Am. Chem. Soc.* **2004**, *126*, 9196–9197.
144. Agnelli, F.; Sucheck, S. J.; Marby, K. A.; Rabuka, D.; Yao, S.-L.; Sears, P. S.; Liang, F.-S.; Wong, C.-H. *Angew. Chem. Int. Ed. Engl.* **2004**, *43*, 1562–1566.
145. Nunns, C. L.; Spence, L. A.; Slater, M. J.; Berrisford, D. J. *Tetrahedron Lett.* **1999**, *40*, 9341–9345.
146. Roestamadji, J.; Grapsas, I.; Mobashery, S. *J. Am. Chem. Soc.* **1995**, *117*, 11060–11069.
147. Liu, M.; Haddad, J.; Azucena, E.; Kotra, L. P.; Kirzhner, M.; Mobashery, S. *J. Org. Chem.* **2000**, *65*, 7422–7431.
148. Okuda, T.; Ito, Y. Biosynthesis and Mutasynthesis of Aminoglycoside Antibiotics. In *Aminoglycoside Antibiotics*; Umezawa, H., Hooper, I. R., Eds.; Springer-Verlag: Berlin, 1982, pp 111–203.
149. Piepersberg, W.; Distler, J. Aminoglycosides and Sugar Components in Other Secondary Metabolites. In *Biotechnology*; 2nd ed.; Kleinkauf, H., von Doehren, H., Eds.; VCH: Weinheim, **1997**; Vol. 7, pp. 397–488.
150. Piepersberg, W. Molecular Biology, Biochemistry and Fermentation of Aminoglycoside Antibiotics. In *Biotechnology of Antibiotics*; Strohl, W. R., Ed.; Marcel Dekker: New York, 1997, pp 81–164.
151. Rinehart, K. L., Jr.; Fang, J.; Jin, W.; Pearce, C. J.; Tadano, K.; Toyokuni, K. *Dev. Indust. Microbiol.* **1985**, *26*, 117–128.
152. Rinehart, K. L., Jr. Biosynthesis and Mutasynthesis of Aminocyclitol Antibiotics. In *Aminocyclitol Antibiotics*; Rinehart, K. L., Jr., Suami, T., Eds.; American Chemical Society: Washington, DC, 1980, pp 335–370.
153. Kharel, M. K.; Subba, B.; Basnet, D. B.; Woo, J. S.; Lee, H. C.; Kiou, K.; Sohng, J. K. *Arch. Biochem. Biophys.* **2004**, *429*, 204–214.
154. Unwin, J.; Standage, S.; Alexander, D.; Hosted, T., Jr.; Horan, A. C.; Wellington, E. M. *J. Antibiot.* **2004**, *57*, 436–445.
155. Du, Y.; Li, T.; Wang, Y. G.; Xia, H. *Curr. Microbiol.* **2004**, *49*, 99–107.
156. Daniel, R. *Curr. Opin. Biotechnol.* **2004**, *15*, 199–204.
157. Handelsman, J. *Microbiol. Mol. Biol. Rev.* **2004**, *68*, 669–685.
158. Cowan, D. A. *Trends Biotechnol.* **2000**, *18*, 14–16.
159. Walsh, C.; Freel Meyers, C. L.; Losey, H. C. *J. Med. Chem.* **2003**, *46*, 3425–3436.
160. Hu, Y.; Walker, S. *Chem. Biol.* **2002**, *9*, 1287–1296.
161. Nedal, A.; Zotchev, S. B. *Appl. Microbiol. Biotechnol.* **2004**, *64*, 7–15.
162. Hutchinson, C. R. *Biotechnology* **1994**, *12*, 375–380.
163. Gilbert, D. N. Aminoglycosides. In *Principles and Practice of Infectious Diseases*, 6th ed.; Mandell, G. L., Bennett, J. E., Dolin, R., Eds.; Churchill Livingston: Philadelphia, PA, **2005**; Vol. 1, pp 328–356.
164. Bar-Nun, S.; Shneyour, Y.; Beckman, J. S. *Biochim. Biophys. Acta* **1983**, *741*, 123–127.
165. Lindquist, W. D. *J. Am. Vet. Med. Assoc.* **1958**, *132*, 72–75.
166. Owens, R. C., Jr.; Banevicius, M. A.; Nicolau, D. P.; Nightingale, C. H.; Quintiliani, R. *Antimicrob. Agents Chemother.* **1997**, *41*, 2586–2588.
167. Torres, C.; Tenorio, C.; Lantero, M.; Gastanares, M. J.; Baquero, F. *Antimicrob. Agents Chemother.* **1993**, *37*, 2427–2431.
168. Baltch, A. L.; Bassey, C.; Hammer, M. C.; Smith, R. P.; Conroy, J. V.; Michelson, P. B. *J. Antimicrob. Chemother.* **1991**, *27*, 801–808.
169. Cottagnoud, P.; Cottagnoud, M.; Tauber, M. G. *Antimicrob. Agents Chemother.* **2003**, *47*, 144–147.
170. MacArthur, R. D.; Lolans, V. T.; Zar, F. A.; Jackson, G. G. *J. Infect. Dis.* **1984**, *150*, 778–779.
171. Finberg, R. W.; Moellering, R. C.; Tally, F. P.; Craig, W. A.; Pankey, G. A.; Dellinger, E. P.; West, M. A.; Joshi, M.; Linden, P. K.; Rolston, K. V. et al. *Clin. Infect. Dis.* **2004**, *39*, 1314–1320.
172. Busse, H. J.; Wostmann, C.; Bakker, E. P. *J. Gen. Microbiol.* **1992**, *138*, 551–561.
173. Davis, B. D.; Chen, L.; Tai, P. C. *Proc. Natl. Acad. Sci. USA* **1986**, *83*, 6164–6168.
174. Nichols, W. W. *J. Antimicrob. Chemother.* **1989**, *23*, 673–676.
175. Davis, B. D. *Microbiol. Rev.* **1987**, *51*, 341–350.
176. Mehta, R.; Champney, W. S. *Antimicrob. Agents Chemother.* **2002**, *46*, 1546–1549.
177. Craig, W. A. *J. Antimicrob. Chemother.* **1993**, *31*, 149–158.
178. Gudmundsson, S.; Einarsson, S.; Erlendsdottir, H.; Moffat, J.; Bayer, W.; Craig, W. A. *J. Antimicrob. Chemother.* **1993**, *31*, 177–191.
179. Novelli, A.; Mazzei, T.; Fallani, S.; Cassetta, M. I.; Conti, S. *J. Chemother.* **1995**, *7*, 355–362.
180. Puglisi, J. D.; Blanchard, J. S.; Green, R. *Nat. Struct. Biol.* **2000**, *7*, 855–861.
181. Puglisi, J. D.; Blanchard, S. C.; Dahlquist, K. D.; Eason, R. G.; Fourmy, D.; Lynch, S. R.; Recht, M. I.; Yoshizawa, S. Aminoglycoside Antibiotics and Decoding. In *The Ribosome: Structure, Function, Antibiotics, and Cellular Interactions*; Garrett, R. A., Douthwaite, S. R., Liljas, A., Matheson, A. T., Moore, P. B., Noller, H. F., Eds.; American Society for Microbiology Press: Washington, DC, 2000, pp 419–429.
182. Gale, E. F.; Cundliffe, E.; Reynolds, P. E.; Richmond, M. H.; Waring, M. J. *The Molecular Basis of Antibiotic Action*, 2nd ed.; John Wiley: New York, 1981.
183. Cundliffe, E. Recognition Sites for Antibiotics within rRNA. In *The Ribosome: Structure, Function and Evolution*; Hill, W. E., Dahlberg, A., Garrett, R. A., Moore, P. B., Schlessinger, R. A., Warner, J. R., Eds.; American Society for Microbiology Press: Washington, DC, 1990, pp 479–490.
184. Steitz, T. A.; Moore, P. B. *Trends Biochem.* **2003**, *28*, 411–418.
185. Lilley, D. M. *ChemBioChem.* **2001**, *2*, 31–35.
186. Hancock, R. E. W. *J. Antimicrob. Chemother.* **1981**, *8*, 249–276.
187. Hancock, R. E. W. *J. Antimicrob. Chemother.* **1981**, *8*, 429–445.
188. Taber, H. W.; Mueller, J. P.; Miller, P. F.; Arrow, A. S. *Microbiol. Rev.* **1987**, *51*, 439–457.
189. Nakae, R.; Nakae, T. *Antimicrob. Agents Chemother.* **1982**, *22*, 554–559.
190. Bryan, L. E.; Kwan, S. *Antimicrob. Agents Chemother.* **1983**, *23*, 835–845.
191. Carter, A. P.; Clemons, W. M., Jr.; Broderson, D. E.; Morgan-Warren, R. J.; Wimberly, B. T.; Ramakrishnan, V. *Nature* **2000**, *407*, 340–348.
192. Wong, C. H.; Hendrix, M.; Priestley, E. S.; Greenberg, W. A. *Chem. Biol.* **1998**, *5*, 397–406.
193. Magnet, S.; Blanchard, J. S. *Chem. Rev.* **2005**, *105*, 477–497.
194. Vicens, Q.; Westhof, E. *Biopolymers* **2003**, *70*, 42–57.
195. Green, R.; Noller, H. F. *Annu. Rev. Biochem.* **1997**, *66*, 679–716.
196. Spahn, C. M. T.; Prescott, C. D. *J. Mol. Biol.* **1996**, *74*, 423–439.
197. Ramakrishnan, V. *Cell* **2002**, *108*, 557–572.
198. Moazed, D.; Noller, H. F. *Nature* **1987**, *327*, 389–394.
199. Pape, T.; Wintermeyer, W.; Rodnina, M. V. *EMBO J.* **1999**, *18*, 3800–3807.
200. Rodnina, M. V.; Wintermeyer, W. *Trends Biochem. Sci.* **2001**, *26*, 124–130.
201. Pape, T.; Wintermeyer, W.; Rodnina, M. V. *Nat. Struct. Biol.* **2000**, *7*, 104–107.

202. Davies, J.; Gorini, L.; Davis, B. D. *Mol. Pharmacol.* **1965**, *1*, 93–106.
203. Gorini, L.; Gunderson, W.; Burger, M. *Cold Spring Harbor Symp. Quant. Biol.* **1961**, *26*, 173–182.
204. Davies, J. E. *Proc. Natl. Acad. Sci. USA* **1964**, *51*, 659–664.
205. Ruusala, T.; Kurland, C. G. *Mol. Gen. Genet.* **1984**, *198*, 100–104.
206. Peske, F.; Savelsbergh, A.; Katunin, V. I.; Wintermeyer, W. *J. Mol. Biol.* **2004**, *343*, 1183–1194.
207. Noller, H. F. *Annu. Rev. Biochem.* **1991**, *60*, 191–227.
208. Powers, T.; Noller, H. F. *J. Mol. Biol.* **1994**, *235*, 156–172.
209. Ruusala, T.; Ehrenberg, M.; Kurland, C. G. *EMBO J.* **1982**, *1*, 741–745.
210. Kirimi, R.; Ehrenberg, M. *Eur. J. Biochem.* **1994**, *226*, 355–360.
211. Bilgin, N.; Claesens, F.; Pahverk, H.; Ehrenberg, M. *J. Mol. Biol.* **1992**, *224*, 1011–1027.
212. Jelenc, P. C.; Kurland, C. G. *Mol. Gen. Genet.* **1984**, *194*, 195–199.
213. Lodmell, J. S.; Dahlberg, A. *Science* **1997**, *277*, 1262–1267.
214. Ogle, J. M.; Carter, A. P.; Ramakrishnan, V. *Trends Biochem. Sci.* **2003**, *28*, 259–266.
215. Bilgin, N.; Richter, A. A.; Ehrenberg, M.; Dahlberg, A. E.; Kurland, C. G. *EMBO J.* **1990**, *9*, 735–739.
216. Brink, M. F.; Brink, G.; Verbeet, M. P.; de Boer, H. A. *Nucleic Acids Res.* **1994**, *22*, 325–331.
217. Jerinic, O.; Joseph, S. *J. Mol. Biol.* **2000**, *304*, 707–713.
218. Wallace, B. J.; Tai, P.-C.; Davis, B. D. *Proc. Natl. Acad. Sci. USA* **1974**, *71*, 1634–1638.
219. Brimacombe, R.; Atmadja, J.; Stiege, W.; Schuler, D. *J. Mol. Biol.* **1988**, *199*, 115–136.
220. Sigmund, C. D.; Ettayebi, M.; Morgan, E. A. *Nucleic Acids Res.* **1984**, *12*, 4653–4663.
221. Ramakrishnan, V.; White, S. W. *Nature* **1992**, *358*, 768–771.
222. Davies, C.; Bussiere, D. E.; Golden, B. L.; Porter, S. J.; Ramakrishnan, V.; White, S. W. *J. Mol. Biol.* **1998**, *279*, 873–888.
223. Beauclerk, A. A. D.; Cundliffe, E. *J. Mol. Biol.* **1987**, *193*, 661–671.
224. Purohit, P.; Stern, P. *Nature* **1994**, *370*, 659–662.
225. Powers, T.; Noller, H. F. *RNA* **1995**, *1*, 194–209.
226. Fourmy, D.; Recht, M. I.; Blanchard, S. C.; Puglisi, J. D. *Science* **1996**, *274*, 1367–1371.
227. Yoshizawa, S.; Fourmy, D.; Puglisi, J. D. *EMBO J.* **1998**, *17*, 6437–6448.
228. Vicens, Q.; Westhof, E. *Chem. Biol.* **2002**, *9*, 747–755.
229. Recht, M. I.; Fourmy, D.; Blanchard, S. C.; Dahlquist, K. D. *J. Mol. Biol.* **1996**, *262*, 421–436.
230. Ogle, J. M.; Broderson, D. E.; Clemons, W. M., Jr.; Tarry, M. J.; Carter, A. P.; Ramakrishnan, V. *Science* **2001**, *292*, 897–902.
231. DeStasio, E. A.; Dahlberg, A. E. *J. Mol. Biol.* **1990**, *212*, 127–133.
232. DeStasio, E. A.; Moazed, D.; Noller, H. F.; Dahlberg, A. E. *EMBO J.* **1989**, *8*, 1213–1216.
233. Recht, M. I.; Douthwaite, S.; Dahlquist, K. D.; Puglisi, J. D. *J. Mol. Biol.* **1999**, *286*, 33–43.
234. Li, M.; Tzagoloff, A.; Underbrink-Lyon, K.; Martin, N. *J . Biol. Chem,* **1982**, *257*, 5921–5928.
235. Spangler, E. A.; Blackburn, E. H. *J. Biol. Chem.* **1985**, *260*, 6334–6340.
236. Edlind, T. D. *Antimicrob. Agents Chemother.* **1989**, *33*, 484–485.
237. Palmer, E.; Wilhelm, J. M. *Cell* **1978**, *13*, 329–334.
238. Lynch, S. R.; Puglisi, J. D. *J. Mol. Biol.* **2001**, *306*, 1037–1058.
239. Recht, M. I.; Douthwaite, S.; Puglisi, J. D. *EMBO J.* **1999**, *18*, 3133–3138.
240. Van de Peer, Y.; Van den Broeck, I.; De Rijk, P.; De Wachter, R. *Nucleic Acids Res.* **1994**, *22*, 3488–3494.
241. Pfister, P.; Hobbie, S.; Vicens, Q.; Bottger, E. C.; Westhof, E. *ChemBioChem.* **2003**, *4*, 1078–1088.
242. Eustice, D. C.; Wilhelm, J. M. *Antimicrob. Agents Chemother.* **1984**, *26*, 53–60.
243. Waitz, J. A.; Sabatelli, F.; Menzel, F.; Moss, E. L., Jr. *Antimicrob. Agents Chemother.* **1974**, *6*, 579–581.
244. Fourmy, D.; Recht, M. I.; Puglisi, J. D. *J. Mol. Biol.* **1998**, *277*, 347–362.
245. Woodcock, J.; Moazed, D.; Cannon, M.; Davies, J.; Noller, H. F. *EMBO J.* **1991**, *10*, 3099–3103.
246. Han, Q.; Zhao, Q.; Fish, S.; Simonsen, K. B.; Vourloumis, D.; Froelich, J. M.; Wall, D.; Hermann, T. *Angew. Chem. Int. Ed. Engl.* **2005**, *44*, 2694–2700.
247. Perzynski, S.; Cannon, M.; Cundliffe, E.; Chahwala, S. B.; Davies, J. *Eur. J. Biochem.* **1979**, *99*, 623–628.
248. Broderson, D. E.; Clemons, W. M., Jr.; Carter, A. P.; Morgan-Warren, R. J.; Wimberly, B. T.; Ramakrishnan, V. *Cell* **2000**, *103*, 1143–1154.
249. Eustice, D. C.; Wilhelm, J. M. *Biochemistry* **1984**, *23*, 1462–1467.
250. Gonzales, A.; Jimenez, A.; Vasquez, D.; Davies, J.; Schindler, D. *Biochim. Biophys. Acta* **1978**, *521*, 459–469.
251. Cabanas, M. J.; Vazquez, D.; Modelell, J. *Biochem. Biophys. Res. Commun.* **1978**, *83*, 991–997.
252. Cabanas, M. J.; Vazquez, D.; Modelell, J. *Eur. J. Biochem.* **1978**, *87*, 21–27.
253. Hermann, T.; Westhof, E. *Curr. Opin. Biotechnol.* **1998**, *9*, 66–73.
254. Schroeder, R.; Waldisch, C.; Wank, H. *EMBO J.* **2000**, *19*, 1–19.
255. Howard, M.; Frizzell, R. A.; Bedwell, D. M. *Nat. Med.* **1996**, *2*, 467–469.
256. Lukacs, G. L.; Durie, P. R. *N. Engl. J. Med.* **2005**, *349*, 1401–1404.
257. Wilschanski, M.; Famini, C.; Blau, H.; Rivlin, J.; Augarten, A.; Avital, A.; Kerem, B.; Kerem, E. *Am. J. Resp. Crit. Care Med.* **2000**, *161*, 860–865.
258. Wilshanski, M.; Yahav, Y.; Yaacov, Y.; Blau, H.; Bentur, L.; Rivlin, J.; Aviram, M.; Bdolah-Abram, T.; Bebok, Z.; Shushi, L. et al. *N. Engl. J. Med.* **2003**, *349*, 1433–1441.
259. Barton-Davis, E. R.; Cordier, L.; Shorturuma, D.; Leland, S. E.; Sweeney, H. L. *J. Clin. Invest.* **1999**, *104*, 375–381.
260. von Ahsen, U.; Davies, J.; Schroeder, R. *J. Mol. Biol.* **1992**, *226*, 935–941.
261. von Ahsen, U.; Davies, J.; Schroeder, R. *Nature* **1991**, *353*, 368–370.
262. von Ahsen, U.; Schroeder, R. *Nucleic Acids Res.* **1991**, *19*, 2261–2265.
263. von Ahsen, U.; Noller, H. F. *Science* **1993**, *260*, 1500–1503.
264. Stage, T. K.; Hertel, K. J.; Uhlenbeck, O. C. *RNA* **1995**, *1*, 95–101.
265. Zapp, M. L.; Stern, S.; Green, M. R. *Cell* **1993**, *74*, 969–978.
266. Lacourciére, K. A.; Stivers, J. T.; Marino, J. P. *Biochemistry* **2000**, *39*, 5630–5641.
267. Wang, S.; Huber, P. W.; Cui, M.; Czarnik, A. W.; Mei, H. Y. *Biochemistry* **1998**, *37*, 5549–5557.
268. Tassew, N.; Thompson, M. *Org. Biomol. Chem.* **2003**, *1*, 3268–3270.
269. Tok, J. B.; Dunn, L. J.; Des Jean, R. C. *Bioorg. Med. Chem. Lett.* **2001**, *11*, 1127–1131.

270. Wallis, M. G.; von Ahsen, U.; Schroeder, R.; Famulok, M. *Chem. Biol.* **1995**, *2*, 543–552.
271. Ennifar, E.; Paillart, J.-C.; Marquet, R.; Ehresmann, B.; Ehresmann, C.; Dumas, P.; Walter, P. *J. Biol. Chem.* **2003**, *278*, 2723–2730.
272. Poole, K. *Antimicrob. Agents Chemother.* **2005**, *49*, 479–487.
273. Phillips, I.; Shannon, K. *Br. Med. Bull.* **1984**, *40*, 28–35.
274. Karlowsky, J. A.; Zelenitsky, S. A.; Zhanel, G. G. *Pharmacother.* **1997**, *17*, 549–555.
275. Daikos, G. L.; Jackson, G. G.; Lolans, V. T.; Livermore, D. M. *J. Infect. Dis.* **1990**, *162*, 414–420.
276. Aires, J. R.; Kohler, T.; Nikaido, H.; Plesiat, P. *Antimicrob. Agents Chemother.* **1999**, *43*, 1298–1300.
277. Magnet, S. P.; Courvalin, P.; Lambert, T. *Antimicrob. Agents Chemother.* **2001**, *45*, 3375–3380.
278. Moore, R. A.; De Shazer, D.; Reckseidler, S.; Weissman, A.; Woods, D. E. *Antimicrob. Agents Chemother.* **1999**, *43*, 465–470.
279. Rosenberg, E. Y.; Ma, D.; Nikaido, H. *J. Bacteriol.* **2000**, *182*, 1754–1756.
280. Allen, N. E. *Prog. Med. Chem.* **1995**, *32*, 157–238.
281. Cundliffe, E. *Annu. Rev. Microbiol.* **1989**, *43*, 207–233.
282. Cundliffe, E. *Gene* **1992**, *115*, 75–84.
283. Thompson, J.; Skeggs, P. A.; Cundliffe, E. *Mol. Gen. Genet.* **1985**, *201*, 168–173.
284. Piendl, W.; Bock, A.; Cundliffe, E. *Mol. Gen. Genet.* **1984**, *197*, 24–29.
285. Galimand, M.; Courvalin, P.; Lambert, T. *Antimicrob. Agents Chemother.* **2003**, *47*, 2565–2571.
286. Doi, Y.; Yokoyama, K.; Yamane, K.; Wachino, J.; Shibata, N.; Yagi, T.; Shibayama, K. *Antimicrob. Agents Chemother.* **2004**, *48*, 491–496.
287. Vakulenko, S.; Mobashery, S. *Clin. Microbiol. Rev.* **2003**, *16*, 430–450.
288. Rather, P. N. *Drug Resis. Updates* **1998**, *1*, 285–291.
289. Shaw, K. J.; Rather, P. N.; Hare, R. S.; Miller, G. H. *Microbiol. Rev.* **1993**, *57*, 138–163.
290. Cundliffe, E. Antibiotic Biosynthesis: Some Thoughts on "Why?" and "How?". In *The Ribosome: Structure, Function, Antibiotics, and Cellular Interactions*; Garrett, R. A., Douthwaite, S. R., Liljas, A., Matheson, A. T., Moore, P. B., Noller, H. F., Eds.; American Society for Microbiology Press: Washington, DC, 2000, pp 409–417.
291. Llano-Sotelo, B.; Azucena, E. F., Jr.; Kotra, L. P.; Mobashery, S.; Chow, C. S. *Chem. Biol.* **2002**, *9*, 455–463.
292. Kotra, L. P.; Haddad, J.; Mobashery, S. *Antimicrob. Agents Chemother.* **2000**, *44*, 3249–3256.
293. Ferretti, J. J.; Gilmore, K. S.; Courvalin, P. *J. Bacteriol.* **1986**, *167*, 431–438.
294. Williams, J. W.; Northrop, D. B. *J. Biol. Chem.* **1978**, *253*, 5908–5914.
295. Radika, K.; Northrop, D. B. *Antimicrob. Agents Chemother.* **1984**, *25*, 479–482.
296. Dickie, P.; Bryan, L. E.; Pickard, M. A. *Antimicrob. Agents Chemother.* **1978**, *14*, 569–580.
297. Sakon, J.; Liao, H. H.; Kanikula, A. M.; Benning, M. M.; Rayment, I.; Holden, H. H. *Biochemistry* **1993**, *32*, 11977–11984.
298. Pederson, L. C.; Benning, M. M.; Holden, H. M. *Biochemistry* **1995**, *34*, 13305–13311.
299. McKay, G. A.; Thompson, P. R.; Wright, G. D. *Biochemistry* **1994**, *33*, 6936–6944.
300. Hon, W.-C.; McKay, G. A.; Thompson, P. R.; Sweet, R. M.; Yang, D. S. C.; Wright, G. D.; Berghuis, A. M. *Cell* **1997**, *89*, 887–895.
301. Fong, D. H.; Berghuis, A. M. *EMBO J.* **2002**, *21*, 2323–2331.
302. Vetting, M. W.; Hegde, S. S.; Javid-Majd, F.; Blanchard, J. S.; Roderick, S. L. *Nat. Struct. Biol.* **2002**, *9*, 653–658.
303. Wolf, E.; Vassilev, A.; Makino, Y.; Sali, A.; Nakatain, Y.; Burley, S. K. *Cell* **1998**, *94*, 439–449.
304. Vetting, M. W.; Magnet, S.; Nieves, E.; Roderick, S. L.; Blanchard, J. S. *Chem. Biol.* **2004**, *11*, 565–573.
305. Wright, G. D.; Ladak, P. *Antimicrob. Agents Chemother.* **1997**, *41*, 956–960.
306. Draker, K. A.; Northrup, D. B.; Wright, G. D. *Biochemistry* **2003**, *42*, 6565–6574.
307. Daigle, D. M.; Hughes, D. W.; Wright, G. D. *Chem. Biol.* **1999**, *6*, 99–110.
308. Williams, J. W.; Northrop, D. B. *J. Antibiot.* **1979**, *32*, 1147–1154.
309. Daigle, D. M.; McKay, G. A.; Wright, G. D. *J. Biol. Chem.* **1997**, *272*, 24755–24758.
310. Burk, D. L.; Berghuis, A. M. *Pharmacol. Ther.* **2002**, *93*, 283–292.
311. Allen, N. E.; Alborn, W. E., Jr.; Hobbs, J. N., Jr.; Kirst, H. A. *Antimicrob. Agents Chemother.* **1982**, *22*, 824–831.
312. Kirst, H. A.; Marconi, G. G.; Counter, F. T.; Ensminger, P. W.; Jones, N. D.; Chaney, M. O.; Toth, J. E.; Allen, N. E. *J. Antibiot.* **1982**, *35*, 1651–1657.
313. Boehr, D. D.; Draker, K. A.; Koteva, K.; Bains, M.; Hancock, R. E. W.; Wright, G. D. *Chem. Biol.* **2003**, *10*, 189–196.
314. Boehr, D. D.; Jenkins, S. L.; Wright, G. D. *J. Biol. Chem.* **2003**, *278*, 12873–12880.
315. DeNap, J. C. B.; Thomas, J. R.; Musk, D. J.; Hergenrother, P. J. *J. Am. Chem. Soc.* **2004**, *126*, 15402–15404.
316. DeNap, J. C. B.; Hergenrother, P. J. *Org. Biomol. Chem.* **2005**, *3*, 959–966.
317. Gilbert, D. N. *Antimicrob. Agents Chemother.* **1991**, *35*, 399–405.
318. McLean, A. J.; IonnidesDemos, L. L.; Li, S. C.; Bastone, E. B.; Spicer, W. J. *J. Antimicrob. Chemother.* **1993**, *32*, 301–305.
319. Giuliano, R. A.; Verpooten, G. A.; DeBroe, M. E. *Am. J. Kidney Dis.* **1986**, *8*, 297–303.
320. Gardner, C. R.; Walsh, C. T.; Almarsson, O. *Nat. Rev. Drug Disc.* **2004**, *3*, 926–934.
321. Barratt, G. *Cell Mol. Life Sci.* **2003**, *60*, 21–37.
322. Kanellakopoulou, K.; Giamarellos-Bourboulis, E. J. *Drugs* **2000**, *59*, 1223–1232.
323. Ruszczak, Z.; Friess, W. *Adv. Drug Deliv. Rev.* **2003**, *55*, 1679–1698.
324. Sivakumar, M.; Rao, K. P. *J. Biomed. Mat. Res. A* **2003**, *65*, 222–228.
325. El-Ghannam, A.; Ahmed, K.; Omran, M. *J. Biomed. Mat. Res. B* **2005**, *73*, 277–284.
326. Lucke, M.; Schmidmaier, G.; Sadoni, S.; Wildemann, B.; Schiller, R.; Haas, N. P.; Raschke, M. *Bone* **2003**, *32*, 521–531.
327. Rossi, S.; Azghani, A. O.; Omri, A. *J. Antimicrob. Chemother.* **2004**, *54*, 1013–1018.
328. Li, L. C.; Deng, J.; Stephens, D. *Adv. Drug Deliv. Rev.* **2002**, *54*, 963–986.
329. Liu, S.-J.; Tsai, Y.-E.; Ueng, S. W.-N.; Chen, E.-C. *Biomaterials* **2005**, *26*, 4662–4669.
330. Prior, S.; Gander, B.; Lecaroz, C.; Irache, J. M.; Gamazo, C. *J. Antimicrob. Chemother.* **2004**, *53*, 981–988.
331. Spann, C. T.; Taylor, S. C.; Weinberg, J. M. *Clin. Dermatol.* **2003**, *21*, 70–77.
332. Park, S.-N.; Kim, J. K.; Suh, H. *Biomaterials* **2004**, *25*, 3689–3698.
333. Robert, P.-Y.; Adenis, J.-P. *Drugs* **2001**, *61*, 175–185.
334. Leeming, J. P. *Clin. Pharmacokinet.* **1999**, *37*, 351–360.
335. Eljarrat-Binstock, E.; Raiskup, F.; Stepensky, D.; Domb, A. J.; Frucht-Pery, J. *Invest. Ophthalmol. Vis. Sci.* **2004**, *45*, 2543–2548.
336. Changez, M.; Koul, V.; Krishna, B.; Dinda, A. K.; Choudhary, V. *Biomaterials* **2004**, *25*, 139–146.

337. Khopade, A. J.; Arulsudar, N.; Khopade, S. A.; Hartmann, J. *Biomacromolecules* **2005**, *6*, 229–234.
338. Chambers, H. F. The Aminoglycosides. In *The Pharmacological Basis of Therapeutics*, 10th ed.; Hardman, J. G., Limbird, L. E., Gilman, A. G., Eds.; McGraw-Hill: New York, 2001, pp 1219–1238.
339. Hu, Z.; Tawa, R.; Konishi, T.; Shibata, N.; Takada, K. *Life Sci.* **2001**, *69*, 2899–2910.
340. Ross, B. P.; DeCruz, S. E.; Lynch, T. B.; Davis-Goff, K.; Toth, I. *J. Med. Chem.* **2004**, *47*, 1251–1258.
341. Schiffelers, R.; Storm, G.; Bakker-Woudenberg, I. *J. Antimicrob. Chemother.* **2001**, *48*, 333–344.
342. Yanagihara, K. *Curr. Pharm. Des.* **2002**, *8*, 475–482.
343. Howell, S. B. *Cancer J.* **2001**, 7, 219–227.
344. Allan, N. D.; Beveridge, T. J. *Antimicrob. Agents Chemother.* **2003**, *47*, 2962–2965.
345. Schechter, Y.; Tsubery, H.; Fridkin, M. *J. Med. Chem.* **2002**, *45*, 4264–4270.
346. Greenwald, R. B.; Zhao, H.; Yang, K.; Reddy, P.; Martinez, A. *J. Med. Chem.* **2004**, *47*, 726–734.
347. Stead, D. A. *J. Chromatogr., B* **2000**, *747*, 69–93.
348. Flurer, C. L. *Electrophoresis* **2003**, *24*, 4116–4127.
349. Soltes, L. *Biomed. Chromatogr.* **1999**, *13*, 3–10.
350. Isoherranen, N.; Soback, S. *J. Assoc. Offic. Anal. Chem.* **1999**, *82*, 1017–1047.
351. Tawa, R.; Matsunaga, H.; Fujimoto, T. *J. Chromatogr., A* **1998**, *812*, 141–150.
352. Matthews, I.; Kirkpatrick, C.; Holford, N. *Br. J. Clin. Pharmacol.* **2004**, *58*, 8–19.
353. Turnidge, J. *Infect. Dis. Clin. North Am.* **2003**, *17*, 503–528.
354. Bartal, C.; Danon, A.; Schlaeffer, F.; Reisenberg, K.; Alkan, M.; Smoliakov, R.; Sidi, A.; Almog, Y. *Am. J. Med.* **2003**, *114*, 194–198.
355. Touw, D. J.; Neef, C.; Thomson, A. H.; Vinks, A. A. *Ther. Drug Monit.* **2005**, *27*, 10–17.
356. Hammett-Stabler, C. A.; Johns, T. *Clin. Chem.* **1998**, *44*, 1129–1140.
357. Oertel, R.; Neumeister, V.; Kirch, W. *J. Chromatogr., A* **2004**, *1058*, 197–201.
358. Clarot, I.; Chaimbault, P.; Hasdenteufel, F.; Netter, P.; Nicolas, A. *J. Chromatogr., A* **2004**, *1031*, 281–287.
359. Kim, B.-H.; Lee, S. C.; Lee, H. J.; Ok, J. H. *Biomed. Chromatogr.* **2003**, *17*, 396–403.
360. Yang, H.-H.; Zhu, Q.-Z.; Qu, H.-Y.; Chen, X.-L.; Ding, M.-T.; Xu, J.-G. *Anal. Biochem.* **2002**, *308*, 71–76.
361. Whelton, A.; Neu, H. C.; *The Aminoglycosides: Microbiology, Clinical Use and Toxicology*; Marcel Dekker: New York, 1982.
362. *Ototoxicity: Basic Science and Clinical Applications*; Henderson, D., Salvi, R. J., Quaranta, A., McFadden, S. L., Burkard, R. F., Eds.; Academy of Sciences: New York, 1999; Vol. 884.
363. Mingeot-Leclercq, M. P.; Tulkens, P. M. *Antimicrob. Agents Chemother.* **1999**, *43*, 1003–1012.
364. Rougier, R.; Claude, D.; Maurin, M.; Sedoglavic, A.; Ducher, M.; Corvaisier, S.; Jelliffe, R.; Maire, P. *Antimicrob. Agents Chemother.* **2003**, *47*, 1010–1016.
365. Kirkpatrick, C. M. J.; Duffull, S. B.; Begg, E. J.; Frampton, C. *Ther. Drug Monit.* **2003**, *25*, 623–630.
366. Nagai, J.; Takano, M. *Drug Metab. Pharmacokin.* **2004**, *19*, 159–170.
367. Tulkens, P. M.; Mingeot-Leclercq, M. P.; Laurent, G.; Brasseur, R. Conformational and Biochemical Analysis of the Interactions between Phospholipids and Aminoglycoside Antibiotics in Relation to Toxicity. In *Molecular Description of Biological Membranes by Computer Aided Conformational Analysis*; Brasseur, R., Ed.; CRC Press: Boca Raton, FL, 1990, pp 63–93.
368. Leung, J. C.; Marphis, T.; Craver, R. D.; Silverstein, D. M. *Kidney Int.* **2004**, *66*, 167–176.
369. Baliga, R.; Ueda, N.; Walker, P. D.; Shah, S. V. *Drug Metab. Rev.* **1999**, *31*, 971–997.
370. Ali, B. H. *Food Chem. Toxicol.* **2003**, *41*, 1447–1452.
371. Horibe, T.; Matsui, H.; Tanaka, M.; Nagai, H.; Yamaguchi, Y.; Kato, K.; Kikuchi, M. *Biochem. Biophys. Res. Commun.* **2004**, *323*, 281–287.
372. Miyazaki, T.; Sagawa, R.; Honma, T.; Noguchi, S.; Harada, T.; Komatsuda, A.; Ohtani, H.; Wakui, H.; Sawada, K.; Otaka, M. et al. *J. Biol. Chem.* **2004**, *279*, 17295–17300.
373. El Mouedden, M.; Laurent, G.; Mingeot-Leclercq, M.-P.; Taper, H. S.; Cumps, J.; Tulkens, P. M. *Antimicrob. Agents Chemother.* **2000**, *44*, 665–675.
374. Nakashima, T.; Teranishi, M.; Hibi, T.; Kobayashi, M.; Umemura, M. *Acta Otolaryngol.* **2000**, *120*, 904–911.
375. Wu, W. J.; Sha, S. H.; Schacht, J. *Audiol. Neurootol.* **2002**, 7, 171–174.
376. Black, F. O.; Pesznecker, S.; Stallings, V. *Otol. Neurootol.* **2004**, *25*, 559–569.
377. Halsey, K.; Skjonsberg, S.; Ulfendahl, M.; Dolan, D. F. *Hearing Res.* **2005**, *201*, 99–108.
378. Bates, D. E. *Drugs Today* **2003**, *39*, 277–285.
379. Basile, A. S.; Huang, J.-M.; Xie, C.; Webster, D.; Berlin, C.; Skolnick, P. *Nat. Med.* **1996**, *2*, 1338–1343.
380. Kopaczynska, M.; Lauer, M.; Schulz, A.; Wang, T.; Schaefer, A.; Fuhrhop, J.-H. *Langmuir* **2004**, *20*, 9270–9275.
381. Hutchin, T. P.; Cortopassi, G. A. *Cell. Mol. Life Sci.* **2000**, *57*, 1927–1937.
382. Harvey, S. C.; Li, X.; Skolnick, P.; Kirst, H. A. *Eur. J. Pharmacol.* **2000**, *387*, 1–7.
383. Kramer, J. A.; Pettit, S. D.; Amin, R. P.; Bertram, T. A.; Car, B.; Cunningham, M.; Curtiss, S. W.; Davis, J. W.; Kind, C.; Lawton, M. et al. *Environ. Health Perspect.* **2004**, *112*, 460–464.
384. Amin, R. P.; Vickers, A. E.; Sistare, F.; Thompson, K. L.; Roman, R. J.; Lawton, M.; Kramer, J.; Hamadeh, H. K.; Collins, J.; Grissom, S. et al. *Environ. Health Perspect.* **2004**, *112*, 465–479.
385. Kaul, M.; Barbieri, C. M.; Kerrigan, J. E.; Pilch, D. S. *J. Mol. Biol.* **2003**, *326*, 1373–1387.
386. Pilch, D. S.; Kaul, M.; Barbieri, C. M.; Kerrigan, J. E. *Biopolymers* **2003**, *70*, 58–79.
387. Detering, C.; Varani, G. *J. Med. Chem.* **2004**, *47*, 4188–4201.
388. Hermann, T.; Westhof, E. *Comb. Chem. High Throughput Screen.* **2000**, *3*, 219–234.
389. Hermann, T.; Westhof, E. *J. Med. Chem.* **1999**, *42*, 1250–1261.
390. Foloppe, N.; Chen, I.-J.; Davis, B.; Hold, A.; Morley, D.; Howes, R. *Bioorg. Med. Chem.* **2004**, *12*, 935–947.
391. Maddaford, S. P.; Motamed, M.; Turner, K. B.; Choi, M. S. K.; Ramnauth, J.; Rakhit, S.; Hudgins, R. R.; Fabris, D.; Johnson, P. E. *Bioorg. Med. Chem. Lett.* **2004**, *14*, 5987–5990.
392. Yu, L.; Oost, T. K.; Schkeryantz, J. M.; Yang, J.; Janowick, D.; Fesik, S. W. *J. Am. Chem. Soc.* **2003**, *125*, 4444–4450.
393. Davis, P. W.; Osgood, S. A.; Hebert, N.; Sprankle, K. G.; Swayze, E. E. *Biotechnol. Bioeng. (Comb. Chem.)* **1998/1999**, *61*, 143–154.
394. Manetsch, R.; Krasinski, A.; Radic, Z.; Raushel, J.; Taylor, P.; Sharpless, K. B. *J. Am. Chem. Soc.* **2004**, *126*, 12809–12818.
395. Turner, P. J. *Diagn. Microbiol. Infect. Dis.* **2004**, *50*, 291–293.
396. Sader, H. S.; Biedenbach, D. J.; Jones, R. N. *Diagn. Microbiol. Infect. Dis.* **2003**, *47*, 361–364.

397. Karlowsky, J. A.; Jones, M. E.; Draghi, D. C.; Thornsberry, C.; Dahm, D. F.; Volturo, G. A. *Ann. Clin. Microbiol. Antimicrob.* **2004**, *3*, 7.
398. Van Eldere, J. *J. Antimicrob. Chemother.* **2003**, *51*, 347–352.
399. Flamm, R. K.; Weaver, M. K.; Thornsberry, C.; Jones, M. E.; Karlowsky, J. A.; Sahm, D. F. *Antimicrob. Agents Chemother.* **2004**, *48*, 2431–2436.
400. Jones, R. N. *Chest* **2001**, *119*, S397–S404.
401. Biswas, S. K.; Kelkar, R. S. *Indian J. Cancer* **2002**, *39*, 135–138.
402. Kim, H. B.; Jang, H.-C.; Nam, H. J.; Lee, Y. S.; Kim, B. S.; Park, W. B.; Lee, K. D.; Choi, Y. J.; Park, S. W.; Oh, M.-D. et al. *Antimicrob. Agents Chemother.* **2004**, *48*, 1124–1127.

Biographies

Herbert A Kirst received his PhD degree in organic chemistry from Harvard University in 1971. After a postdoctoral fellowship at the California Institute of Technology, he joined the fermentation products research division of Eli Lilly and Company in 1973 where he worked on the chemistry of many natural products and antibiotics. He transferred to Lilly's Elanco Animal Health Division in 1994 and retired from Lilly in 2003. He is currently an independent consultant in natural products and antibiotic research.

Norris E Allen received a PhD in microbiology from the University of Maryland in 1969. Following postdoctoral studies at the University of Pennsylvania in Philadelphia, he joined Eli Lilly and Company in Indianapolis and was involved in antimicrobial discovery and development for more than 30 years. He retired from Lilly in 2001 and is currently Adjunct Professor in the Department of Microbiology and Immunology at the Indiana University School of Medicine.

Comprehensive Medicinal Chemistry II
ISBN (set): 0-08-044513-6

ISBN (Volume 7) 0-08-044520-9; pp. 629–652

7.22 Anti-Gram Positive Agents of Natural Product Origins

V J Lee, Adesis, Inc., New Castle, DE, USA

7.22.1 Introduction

With the exception of prontosil, a synthetic antibacterial, the earlier antibiotics for human use originated as natural products produced by nonpathogenic bacteria or fungi. These are secondary metabolites, produced as defenses by the producing organisms against other organisms.[1–3,3a,3b] The more common first-generation natural products (beta-lactams, tetracyclines, macrolides), by virtue of their intrinsic activity against both Gram-positive and Gram-negative screening organisms, prompted pharmaceutical company decision makers to commit significant chemistry resources to enhance their drug properties. Through the power of medicinal chemistry, a number of products with broad antibacterial or antifungal spectrum were introduced that satisfied numerous clinical needs.

Natural products, active only against Gram-positive pathogens, have historically not engendered enthusiasm for commercialization. Many of them were discovered during the 1970–80s, when broad antibacterial spectrum was a requisite for marketing. However, the advent of the multidrug-resistant Gram-positive pathogens prompted renewed interest in them. Recently, they have become key sources for newer experimental agents that show exquisite activity against the problematic Gram-positive pathogens. This chapter focuses on initiatives with natural products that have afforded products with primary activity against multidrug-resistant Gram-positive pathogens (*Staphylococcus aureus*, *Streptococcus pneumoniae*, *Enterococcus faecalis*, *Enterococcus faecium*, and related pathogens). While there are others that should be mentioned, in many cases there is insufficient information to provide a comprehensive analysis.

7.22.2 Background

Resistance to penicillin in *S. aureus* was observed shortly after the introduction of penicillin in the 1950s[4]; this resistance was due exclusively to beta-lactamase inactivation of penicillin. The extensive clinical use of the lactamase-stable penicillins (methicillin, oxacillin) and broader spectrum beta-lactamase stable cephalosporins (aminothiazolyl cephalosporins) resulted in proliferation of the low-level multidrug-resistant Gram-positive pathogens. Historically, such epidemiological trends occur subsequent to the introduction of each new class of anti-infectives. The recognition that pathogens could possess multiple resistance mechanisms, which would make them insensitive to

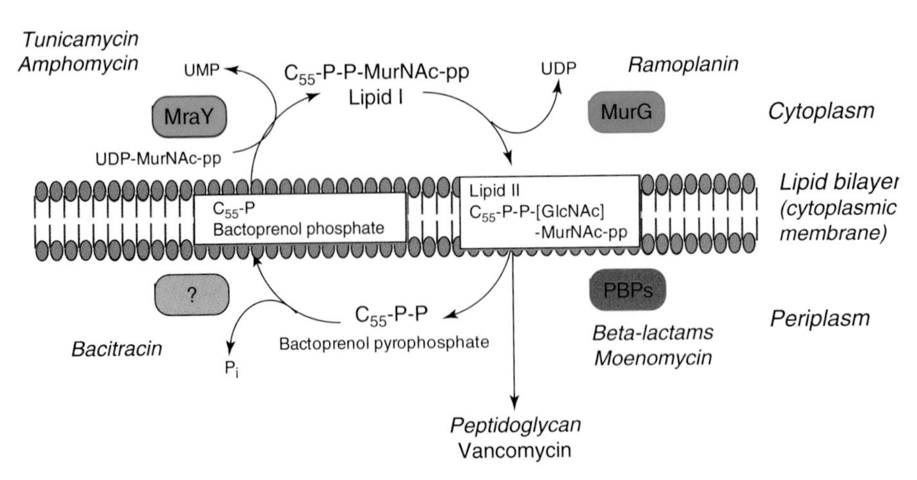

Figure 1 Cell wall (peptidoglycan) biosynthesis and site of inhibition by antibiotics.

multiple classes of antibacterials, emphasizes the importance of identifying new agents for problematic Gram-positive pathogens.

One notable example is the clinical interplay of beta-lactams and vancomycin. Beta-lactam resistance in methicillin-resistant *S. aureus* (MRSA) is conferred by the *mec*A gene, which encodes for the expression of a modified penicillin-binding protein PBP2a that has low affinity for most beta-lactams.[5,6] Similarly, insensitivity to beta-lactams by enterococci[7,7a] and pneumococci[8,8a] is also due to poor affinity to relevant PBPs.

In both diagnosed and suspected clinical cases of MRSA or enterococcal infections, vancomycin is the drug of choice. While the mechanism of resistance to vancomycin in enterococci is distinctly different from that of beta-lactam resistance, the laboratory observation that glycopeptide resistance was transferred from enterococci to *S. aureus* inferred that vancomycin could lose its clinical utility. A few years later, vancomycin-resistant enterococci were reported from clinical strains.[6,9,9a,9b]

Introducing new antibacterials possessing exquisite anti-MRSA activity, as well as having anti-enteroccocal activity, would provide alternatives to the use of vancomycin as the antibiotic of last recourse for Gram-positive infections. In contrast to the typical community-acquired infections, where the therapeutic indices for antibiotics are substantial, there is greater acceptability of narrow therapeutic indices for treatments of infections due to problematic Gram-positive pathogens in a hospital setting.

Many anti-Gram-positive natural products have a high molecular weight and are highly lipophilic, the same properties that hinder their ability to penetrate the cell membrane of Gram-negative organisms. Thus, many of them exert their antibacterial action on Gram-positive organisms through interactions on biochemical targets involved in the biosynthesis of the peptidoglycan building blocks for the cell wall. These targets are intimately connected to the cytoplasm, the lipid bilayer, periplasmic space, or the cell wall (**Figure 1**). While inhibiting a single biochemical target that has antibacterial effect is a minimal requisite, inhibition of several biochemical targets in close physical proximity minimizes rapid formation of resistance. Some of the antibiotics discussed in this chapter highlight these efforts.

7.22.3 Glycopeptides

Vancomycin (**1**), produced by *Streptomyces orientalis*, is a glycopeptide antibiotic that complexes with the D-Ala-D-Ala terminus of peptidoglycan precursors, which inhibits the peptidoglycan coupling and transpeptidation reactions (**Figure 2**). Its characteristic antibacterial spectrum against Gram-positive bacteria such as multidrug-resistant *E. faecalis*, *E. faecium*, *Clostridium difficile*, and MRSA, with negligible anti-Gram-negative activity, complements the antibacterial spectrum of the more popular third-generation cephalosporins and carbapenems. As empiric use of combination therapy with vancomycin increased,[10] it was feared that the incidence of glycopeptide-intermediate-sensitive *S. aureus* (GISA) (also called vancomycin-intermediate VISA strains) would arise.[11,11a,11b] Thus, research into glycopeptides has focused on: (1) enhancing activity against the GISA and VISA strains, including vancomycin-resistant enteroccocci (VRE); and (2) improving pharmacological properties.

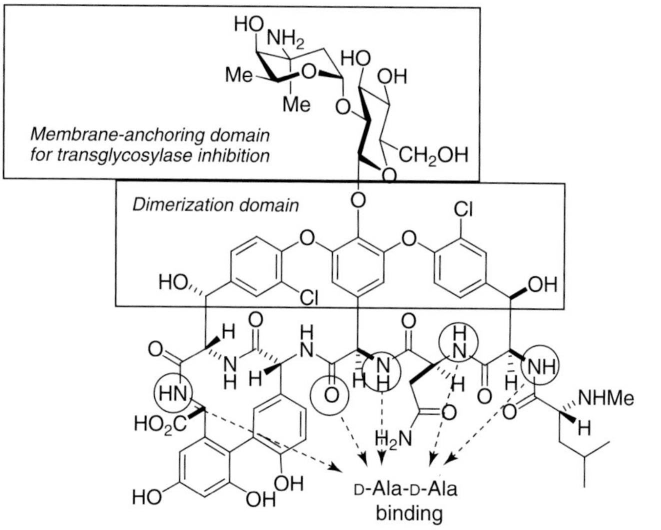

Membrane-anchoring domain for transglycosylase inhibition

Dimerization domain

D-Ala-D-Ala binding

Figure 2 Glycopeptide multifunctionality.

VRE have recently been identified; their resistance is due to the presence of gene clusters (vanHAX or vanHBBXB) that encode for the synthesis of abnormal peptidoglycan precursors with either the D-Ala-D-Lac or D-Ala-D-Ser terminus, which have low affinity for vancomycin.[12] The VanA phenotype enterococci are resistant to both vancomycin (**1**) and teicoplanin (**2**), while the vancomycin-resistant VanB phenotype isolates respond to teicoplanin.

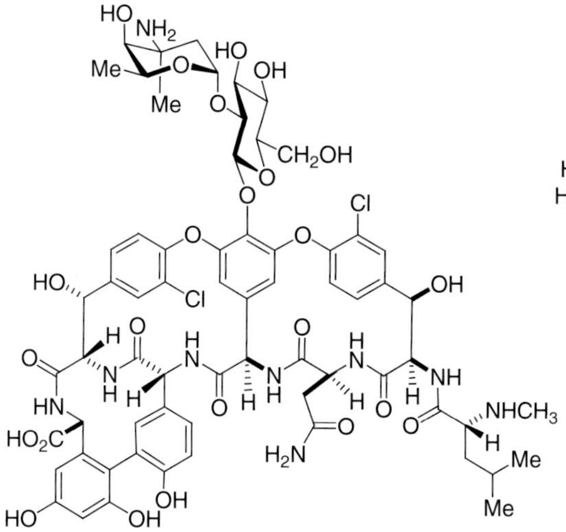

1 Vancomycin

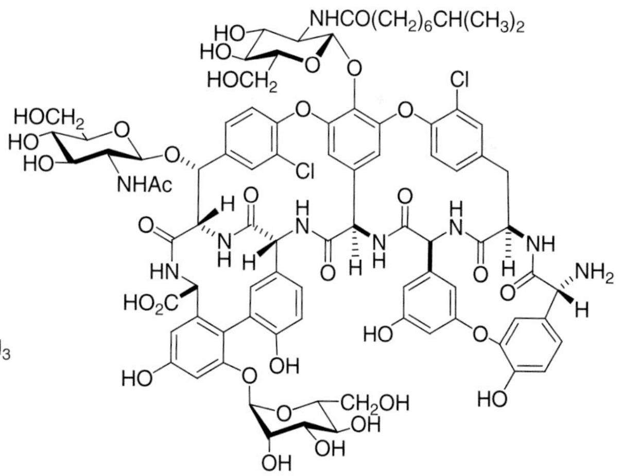

2 Teicoplanin

Epidemiological studies of glycopeptide resistance showed significant geographic variability, with higher incidences in Europe compared with the US. This coincides with the introduction in 1985 of avoparcin (**3**), a related glycopeptide, as a growth promoter for pigs and poultry in Europe but never approved for use in the US.[13] Growth promoters function to facilitate more efficient usage of animal feed stocks for meat production through modification of the intestinal flora. *Enteroccoccus faecium* in the animals develops resistance factors to (**3**), and by default to other glycopeptides; the resistant bacteria are transferred to humans either through consumption of partially cooked poultry or meat products or animal carriage in feedlots.[14] The implications of dual use of antibiotics for human and animal applications is discussed below.

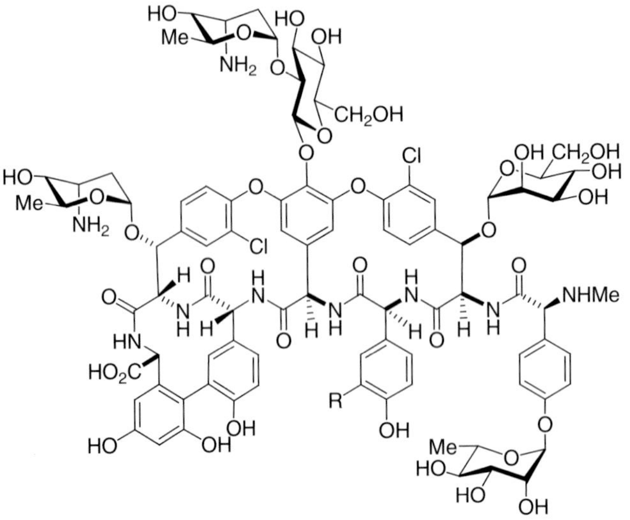

3 Avoparcin (growth promoter)
R = H,Cl

The recent generation of glycopeptides, dalbavancin (**5**) and oritavacin (**7**), are noted for their activity against the subset of problematic Gram-positives that are resistant to specific glycopeptides (VanA–C resistance enterococci), and longer half-lives. In contrast, telavancin (**8**) is active against some of the glycopeptide-resistant strains, but its half-life is comparable to vancomycin.

The common structural modifications are the incorporation of a lipophilic moiety on the Northern carbohydrate unit of the various glycopeptides, which enhances antibacterial activity against VRE and retains anti-MRSA activity. The presence of this unit facilitates anchoring of the antibiotic to the bacterial membrane, which is a prerequisite for inhibiting transglycosylase function. Specifically (1) A-40926 (**4**) is converted to dalbavancin (**5**),[15] (2) chloroeremomycin (**6**) (A82846B; LY264826) is converted to oritavancin (**7**),[16] and (3) vancomycin (**1**) is converted to telavancin (**8**).[17] These modifications significantly impact their mode-of-action and their bactericidal activity.[18] In contrast to vancomycin, which is a time-dependent bactericidal agent with *S. aureus*, both dalbavancin (**5**) and oritavancin (**7**) are concentration-dependent bactericidal agents against both *S. aureus* and enterococci.[19] Telavancin (**8**) shows rapid bactericidal activity.

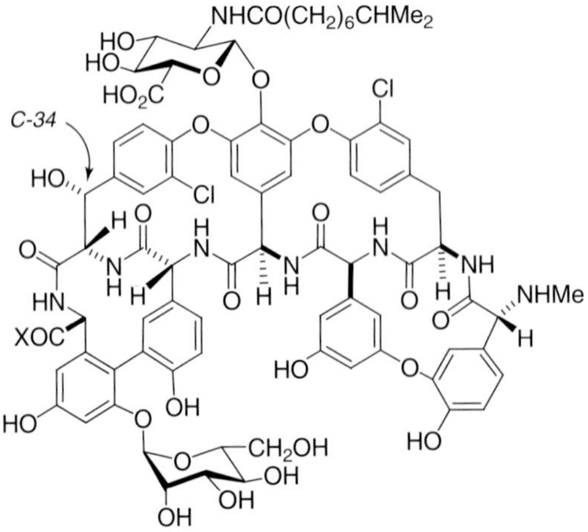

4 X = OH (A40926 complex)
5 X = $NHCH_2CH_2NMe_2$
(BI-397; MDL 633397) dalbavancin

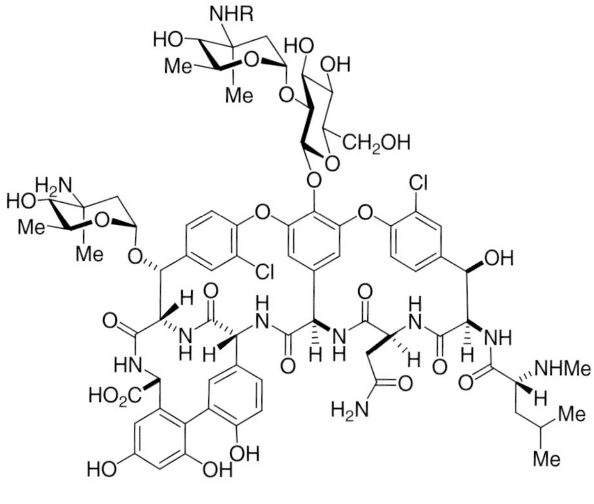

6 R = H chloroeremomycin (A82846B)
7 R = 4-Cl-biphenyl-CH_2 oritavancin

In the natural products arena, one of the challenges for medicinal chemists is to have a cogent biology-driven rationale for selecting specific scaffolds for modification. The technical events that led to the discovery of dalbavancin and telavancin warrant discussion. The original natural product A40926 (**4**) was identified via a D-ala-D-ala affinity screen; it was found subsequently to have comparable in vitro activity to vancomycin and teicoplanin against aerobic and anaerobic Gram-positive bacteria, including some VRE. However, it showed activity against *Neisseria gonorrhoeae*, a Gram-negative bacteria, which is not observed with other glycopeptides.[20] While modifications of A40926 (**4**) did not afford analogs with additional potency for Gram-negative bacteria, they did maintain the characteristic anti-Gram-positive activity of glycopeptides.

A pivotal aspect was the discovery that basic amides (**10**) of 34-des(acetylglucosaminyl)-34-deoxyteicoplanin (**9**) were active in vitro against vancomycin-resistant and teicoplanin-resistant strains of *E. faecalis* and *E. faecium*[21] However, the 34-hydroxy group, as found in the A40926 scaffold, provides significant stabilization to the Western half of the macrocycle. Similar modifications on A40926 showed similar in vitro advantages with enhanced pharmacokinetic advantages.

Other researchers, understanding the multiple roles of the transglycosylase binding domain and self-association domains, modified vancomycin (**1**) or chloroeremomycin (**6**) accordingly with lipophilic moieties.[22–24] However, such modifications can impart undesirable PK (pharmacokinetic)-ADME (absorption, distribution, metabolism, and excretion) properties that can limit their clinical usage.[25] *N*-(Decylaminoethyl)vancomycin (**11**), with a compelling in vitro profile, is poorly excreted and distributed preferentially to the liver and kidney. However, with telavancin (**8**), the high lipophilicity of *N*-(decylaminoethyl)vancomycin (**11**) is balanced by placement of hydrophilic functionalities at distal sites, which do not interfere with the key recognition sites for antibacterial function.[17,26]

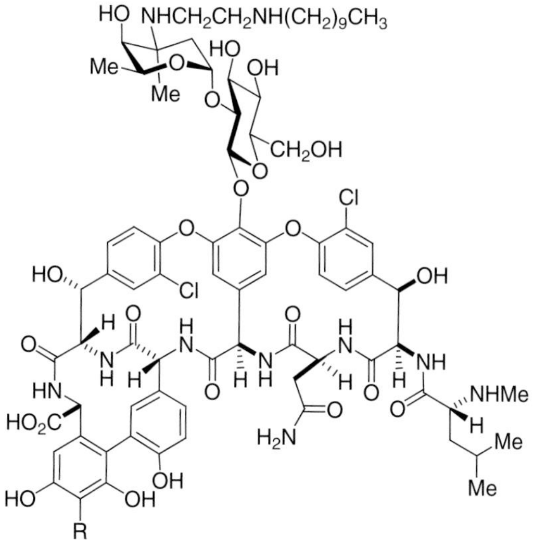

8 R = $CH_2NHCH_2PO(OH)_2$ televancin
11 R = H

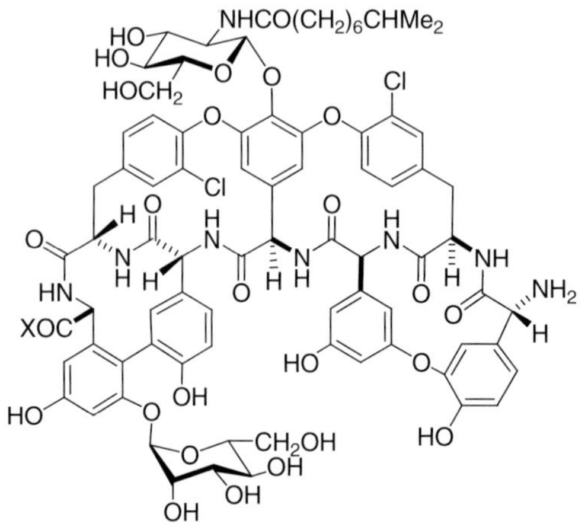

9 X=OH 34-des(acetylglucosaminyl)-34-deoxyteicoplanin
10 X=NH$(CH_2)_n$NR^{1}R^2 (n = 2–3)
34-des(acetylglucosaminyl)-34-deoxyteicoplanin amides

A comparative tabulation of the in vitro characteristics of vancomycin, oritavancin, dalbavancin, and telavancin is presented in **Tables 1** and **2**[27,28] with key pharmacokinetic properties of glycopeptides **1**, **2**, **5**, **7**, and **8** showing significant variations in **Table 3**.[29] It is expected that both dalbavancin (**5**) and oritavancin (**7**) will be approved for treatment of problematic Gram-positives, for once-a-week and twice-a-week dosing, respectively. The dosing regimen for televancin should be once-a-day versus vancomycin, which is thrice daily.

7.22.4 Pseudomonic Acid

Mupirocin (pseudomonic acid A) **12**, produced by *Pseudomonas fluorescens*,[30] is bactericidal against *S. aureus* and *S. epidermidis*, including methicillin-resistant and beta-lactamase producing strains, and some *Streptococcus* species, including *S. pyogenes*, *S. agalactiae*, and *S. viridans*[31] The in vitro minimum inhibitory concentrations (MICs) against methicillin-sensitive and -resistant *S. aureus* are $<1\,\mu g\,mL^{-1}$. However, it is inactive against anaerobic streptococci, enterococci, Gram-negative organisms, anaerobes, and fungi. Furthermore, it has minimal activity against normal skin flora such as *Micrococcus*, *Corynebacterium*, and *Propionibacterium* species (acne indication) **Table 4**.[30]

It is used for prophylaxis and treatment of primary and secondary infections of the skin, skin appendages, and mucous membranes due to *S. aureus* and *S. epidermidis*, including methicillin-resistant (MRSA, MRSE) and beta-lactamase-producing strains.

Mupirocin inhibits RNA and protein synthesis by selectively binding to the bacterial isoleucyl-tRNA synthetase (IleS), preventing the formation of isoleucyl-tRNA, which, in turn, inhibits incorporation of isoleucine into the nascent polypeptide chain.[32] This bactericidal mechanism of action is unique to mupirocin, and cross-resistance between mupirocin and other classes of antibiotics has not been reported. When mupirocin resistance does occur, it is due to a modified isoleucyl-tRNA synthetase.[33] The other resistance mechanism, high-level plasmid-mediated resistance, is found occasionally in some strains of *S. aureus* and coagulase-negative staphylococci (MIC $1024\,\mu g\,mL^{-1}$).[34]

Owing to metabolic hydrolytic instability, which produces the inactive monic acid, mupirocin is administered topically. Since it has been shown previously that modifications on the tetrahydropyranyl unit resulted in loss of activity, only modifications to replace the hydrolytically labile ester moiety are a prerequisite for developing an oral agent. Simple substitution with a ketone, ether, or thioether linkage yielded less potent analogs,[35] while substitution with the isosteric oxazole motif **13–14** resulted in comparable potency.[36–38]

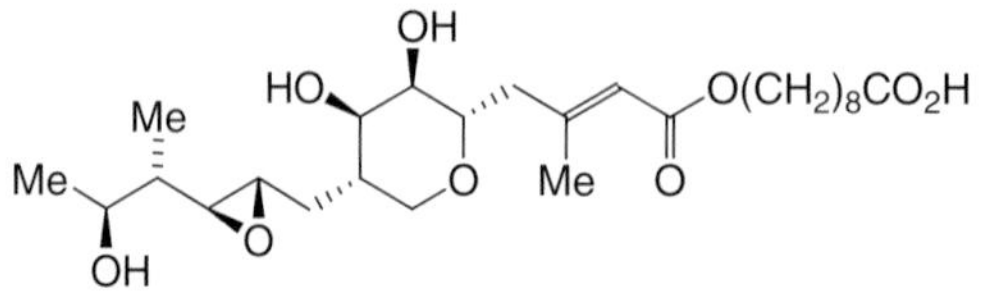

12

Table 1 Comparative in vitro profile of dalbavancin (**5**), oritavancin (**7**), teicoplanin (**2**), and vancomycin (**1**) against streptococci and enterococci

Organism	*Drug*	*N*	*MIC* ($\mu g\, mL^{-1}$)		
			Range	MIC_{50}	MIC_{90}
S. aureus (Meth^S)	Vancomycin	10	0.125–1	1	1
	Teicoplanin	10	0.25–8	0.5	4
	Dalbavancin	10	≤0.03–0.5	0.06	0.125
	Oritavancin	10	0.125–1	0.5	1
S. aureus (Meth^R)	Vancomycin	23	0.5–4	1	4
	Teicoplanin	23	0.13–8	0.5	8
	Dalbavancin	23	0.06–1	0.13	0.25
	Oritavancin	23	0.13–4	0.5	2
S. epidermidis (Meth^S)	Vancomycin	13	0.125–1	1	1
	Teicoplanin	13	0.25–16	4	8
	Dalbavancin	13	≤0.03–0.25	0.06	0.25
	Oritavancin	13	0.25–1	0.5	2
S. epidermidis (Meth^R)	Vancomycin	12	1–4	2	4
	Teicoplanin	12	1–16	4	16
	Dalbavancin	12	≤0.03–1	0.06	0.25
	Oritavancin	12	0.25–4	0.5	1
S. haemolyticus (Meth^S)	Vancomycin	10	1–4	1	2
	Teicoplanin	10	1–32	4	32
	Dalbavancin	10	≤0.03–0.25	≤0.03	0.125
	Oritavancin	10	0.06–1	0.25	1
S. haemolyticus (Meth^R)	Vancomycin	12	0.5–8	2	4
	Teicoplanin	12	2–128	16	32
	Dalbavancin	12	≤0.03–4	0.13	0.5
	Oritavancin	12	0.125–1	0.5	1
S. pneumoniae (Pen^S)	Vancomycin	12	0.125–0.5	0.5	0.5
	Teicoplanin	12	0.008–0.06	0.03	0.06
	Dalbavancin	12	0.016–0.125	0.03	0.06
	Oritavancin	12	≤0.002–0.06	≤0.002	0.008
S. pneumoniae (Pen^R)	Vancomycin	5	0.25–2	0.25[a]	–
	Teicoplanin	5	0.016–0.13	0.03[a]	–
	Dalbavancin	5	0.008–0.125	0.03[a]	–
	Oritavancin	5	≤0.002–0.06	≤0.002[a]	–

continued

Table 1 Continued

Organism	*Drug*	*N*	*MIC (μg mL⁻¹)*		
			Range	*MIC₅₀*	*MIC₉₀*
Enterococcus spp. (Van^S^)	Vancomycin	6	0.25–4	0.5[a]	–
	Teicoplanin	6	0.13–0.5	0.13[a]	–
	Dalbavancin	6	0.06–0.125	0.125[a]	–
	Oritavancin	6	0.06–0.25	0.06[a]	–
Enterococcus spp. (VanA)	Vancomycin	21	>128	>128	>128
	Teicoplanin	21	64–>128	>128	>128
	Dalbavancin	21	0.5–>128	32	>128
	Oritavancin	21	0.06–1	0.25	1
Enterococcus spp. (VanB)	Vancomycin	10	8–128	128	128
	Teicoplanin	10	0.13–8	1	2
	Dalbavancin	10	0.02–2	0.13	1
	Oritavancin	10	≤0.03–0.125	0.03	0.125

[a] Owing to inadequate strain population size, MIC_{50} is provided.

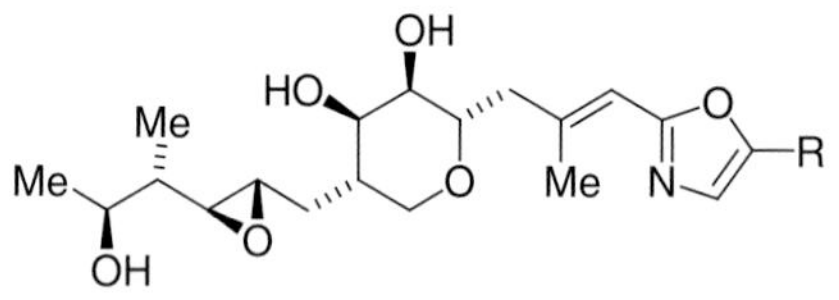

Me Me OH HO OH O O Me N O R

13 a–e

Me Me OH HO OH O O Me N O R_1

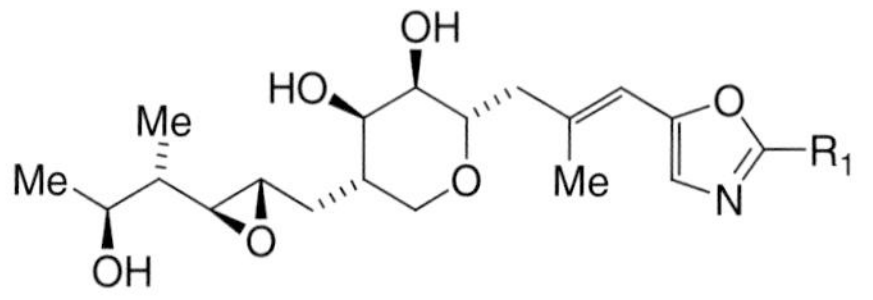

14

7.22.5 Streptogramins

The streptogramin class of antibiotics typically comprises a mixture of two structural classes (groups A and B) found in a 70:30 mix, respectively; group A compounds are polyunsaturated macrolactams, while group B compounds are cyclic hexadepsipeptides. The individual components display moderate bacteriostatic activity in vitro and in vivo. As a mixture, they function synergistically to enhance bacteriostatic activity by binding at two sites on the bacterial ribosome to disrupt the translation of mRNA into protein. Both group A and group B compounds bind to the peptidyltransferase domain of the bacterial ribosome. The group A compounds interfere with the polypeptide elongation sequence by preventing the binding of aa-tRNA to the ribosome, while the group B compounds destabilize the peptidyl–tRNA complex. The synergy between the group A and group B compounds results from an enhanced binding of the group B compounds to the ribosome, which is further enhanced by the induced conformational change on binding of the group A agent. The first-generation streptogramin (**15**), a 70:30 mixture of pristinamycin II_A (**15a**) and pristinamycin I_A (**15b**), had a compelling antibacterial profile but extremely low solubility.

Table 2 Comparative in vitro profile of vancomycin (**1**), teicoplanin (**2**), and telavancin (**8**) against aerobic Gram-positive bacteria

Organism	*Drug*	*N*	*MIC* ($\mu g\, mL^{-1}$)		
			Range	MIC_{50}	MIC_{90}
S. aureus (Meth^S)	Vancomycin	30	1–2	2	2
	Teicoplanin	30	0.5–2	1	2
	Telavancin	30	0.25–1	0.5	0.5
S. aureus (Meth^R)	Vancomycin	30	1–2	1	1
	Teicoplanin	30	0.5–16	1	2
	Telavancin	30	0.125–1	0.25	0.5
S. epidermidis (Meth^S)	Vancomycin	30	0.5–4	2	2
	Teicoplanin	30	0.25–16	1	4
	Telavancin	30	0.125–1	0.5	0.5
S. epidermidis (Meth^R)	Vancomycin	30	1–4	2	2
	Teicoplanin	30	0.5–8	2	8
	Telavancin	30	0.25–2	0.25	1
S. haemolyticus	Vancomycin	25	1–2	2	2
	Teicoplanin	25	0.5–16	4	8
	Telavancin	25	0.25–1	0.5	0.5
S. pneumoniae	Vancomycin	50	0.25–0.5	0.5	0.5
	Teicoplanin	50	0.03–0.125	0.06	0.125
	Telavancin	50	0.008–0.03	0.016	0.016
E. faecalis (Van^S)	Vancomycin	29	0.5–2	1	2
	Teicoplanin	29	0.03–0.5	0.125	0.25
	Telavancin	29	0.06–1	0.5	1
E. faecalis (Van^R)	Vancomycin	21	4–>256	128	>256
	Teicoplanin	21	0.06–32	4	32
	Telavancin	21	0.25–4	2	4
E. faecium (Van^S)	Vancomycin	28	0.5–2	1	1
	Teicoplanin	28	0.5–2	0.5	1
	Telavancin	28	0.06–0.5	0.25	0.5
E. faecium (Van^R)	Vancomycin	22	64–>256	256	>256
	Teicoplanin	22	4–64	8	32
	Telavancin	22	0.5–8	2	4

Table 3 Comparative pharmacokinetic profile, of dalbavancin (**5**), oritavancin (**7**), telavancin (**8**), teicoplanin (**2**), and vancomycin (**1**)

	Dalbavancin	*Oritavancin*	*Telavancin*	*Vancomycin*	*Teicoplanin*
Terminal half-life (h) (man)	257	<360	8	4–8	83–168
Clearance ($L h^{-1} kg^{-1}$) (man)	0.0006	0.008–033	0.012 (man)	0.058	0.011
V_{ss} (L) (man)	9.16	0.65–1.92	0.15	0.7	
Renal excretion (%)	42	<5	40–45	80–90	80
Protein binding (%) (human)	98	85.7–89.9	94–96	10–55	90

Table 4 In vitro profile of hydrolytically stable mupirocin analogs vis-à-vis mupirocin

Substituent or reference drug	*MIC ($\mu g mL^{-1}$)*			
	S. aureus NCTC 6571	*S. pyogenes CN10*	*S. pneumoniae PU7*	*H. influenzae Q1*
12 Mupirocin	<0.06	<0.06	<0.06	<0.06
13a R = $(CH_2)_3CH_3$	1	0.5	4	1
13b R = $(CH_2)_3CO_2Me$	1	0.5	4	1
13c R = 1,3-dithiane	1	1	4	1
13d R = 2-methyl-5-nitro-imidazol-2-yl	<0.06	<0.06	<0.06	<0.06
13e R = 5-nitrofur-2-yl	<0.06	<0.06	<0.06	0.06

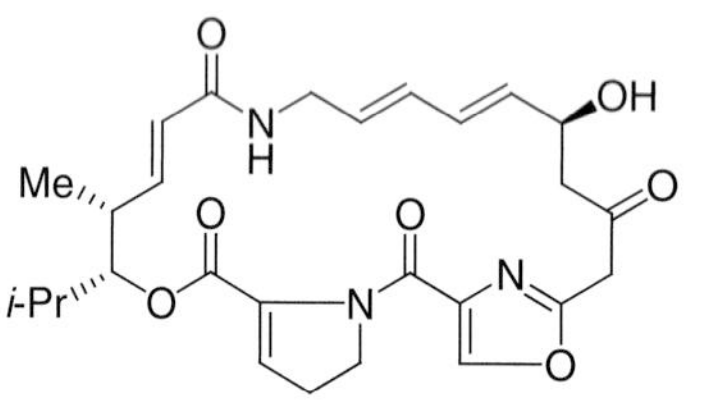

15a Pristinamycin II_A (group A)

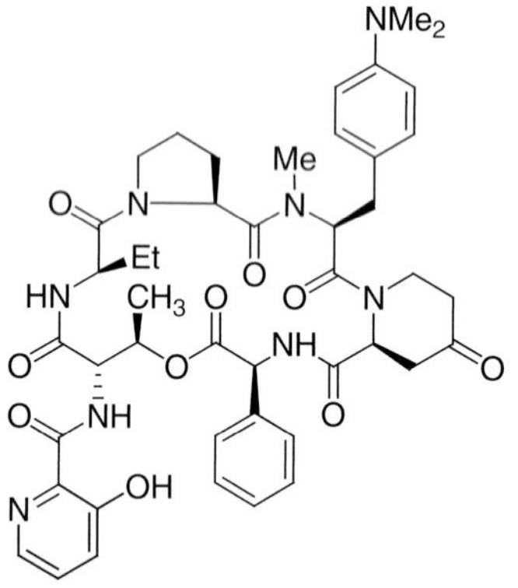

15b Pristinamycin I_A (group B)

Synercid (**17**), a 70:30 mixture of dalfopristin (RP 54476) (**17a**) and quinopristin (RP 57669) (**17b**), was introduced in the mid-1980s for treatment of vancomycin-resistant enterococci and MRSA.[39] However, its clinical usage is limited by the necessity for administration via a central cathether due to poor solubility. In parallel, RPR106972 (**18**), a mixture of pristinamycin II_B (**18a**) and pristinamycin I_B (**18b**), was found to have oral bioavailability. Note the structural subtleties between **15a** and **18a** and **15b** and **18b**, which provide adequate bioavailability for **18**. While **18** is less active then **17** against staphylococci, it is more potent against streptococci and enterococci.[40,41]

The orally bioavailable second-generation complex NXL103 (**19**), consisting of RPR132552 (**19a**) and RPR202868 (**19b**), has significantly improved potency against the problematic Gram-positive pathogens including multidrug-resistant *S. pneumoniae* (**Table 5**).

Table 5 Comparative in vitro profile of NXL103 (**19**), PRP132552 (**19a**), RPR202868 (**19b**), and synercid (**17**) against Gram-positive bacteria

Organism	*Drug*	*N*	*MIC* ($\mu g\,mL^{-1}$)		
			Range	MIC_{50}	MIC_{90}
S. aureus (Oxacillin^S)	XRP2868	30	0.06–0.25	0.125	0.125
	RPR202868	30	16–>128	16	32
	PRP132552	30	0.25–0.50	0.5	0.5
	Synercid	30	0.12–0.5	0.25	0.5
S. aureus (Oxacillin^R)	XRP2868	45	0.06–0.5	0.25	0.5
	RPR202868	45	16–>128	>128	>128
	PRP132552	45	0.25–2	0.5	1.0
	Synercid	45	0.12–1.0	0.5	1.0
S. epidermidis	XRP2868	14	0.06–1	0.06	0.5
	RPR202868	14	8–>128	16	>128
	PRP132552	14	0.125–4	0.125	4
	Synercid	14	≤0.06–0.5	0.125	0.25
S. pyogenes	XRP2868	27	≤0.06–0.5	0.5	0.5
	RPR202868	27	1.0–4	2	4
	PRP132552	27	≤0.06–0.25	<0.06	0.12
	Synercid	27	0.25–1	0.25	0.5
E. faecalis	XRP2868	24	0.125–2	1	2
	RPR202868	24	4–>128	16	>128
	PRP132552	24	8–>128	128	>128
	Synercid	24	1–32	8	16
E. faecalis (VanA)	XRP2868	10	1	1	1
	RPR202868	10	16–64	32	64
	PRP132552	10	64–>128	128	>128
	Synercid	10	4–8	8	8
E. faecalis (VanB)	XRP2868	23	0.5–4	2	2
	RPR202868	23	4–>128	>128	>128
	PRP132552	23	64–>128	128	>128
	Synercid	23	4–32	16	32
E. faecium (VanA)	XRP2868	32	≤0.06–8	0.12	1
	RPR202868	32	2–>128	>128	>128
	PRP132552	32	0.25–128	0.25	64
	Synercid	32	0.5–32	0.5	8
E. faecium (VanB)	XRP2868	31	≤0.06–2	0.12	1
	RPR202868	31	2–>128	>128	>128
	PRP132552	31	≤0.06–64	0.5	16
	Synercid	31	0.25–16	0.5	4

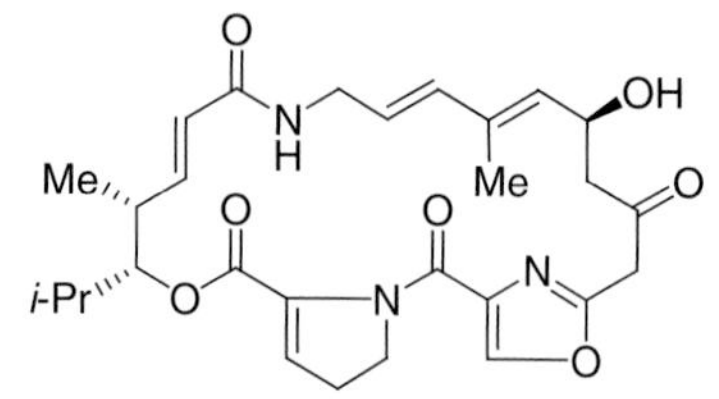

16a Virginiamycin M_1 (group A)

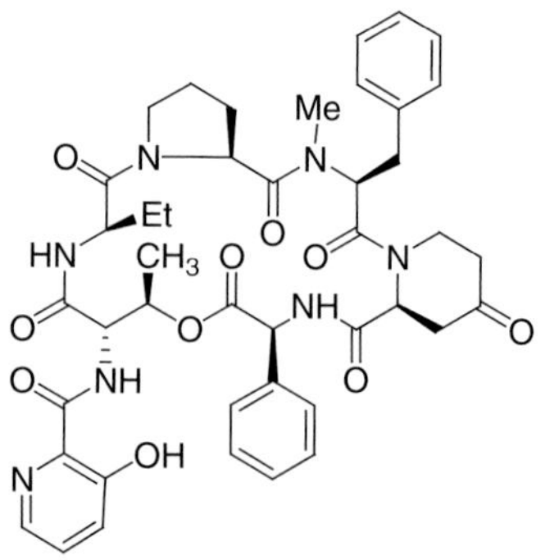

16b Virginiamycin S_1 (group B)

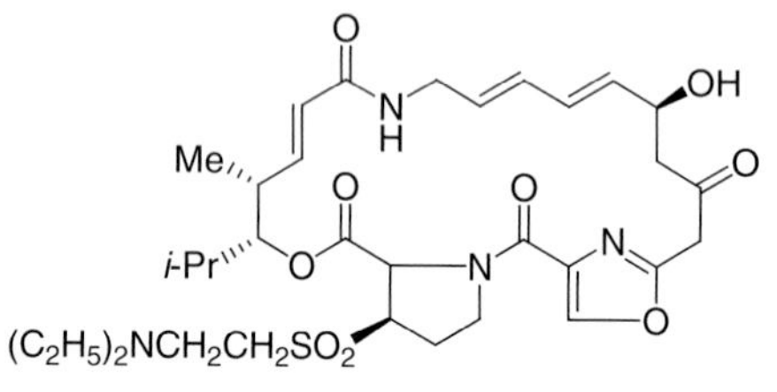

17a Dalfopristin (RP 54476)

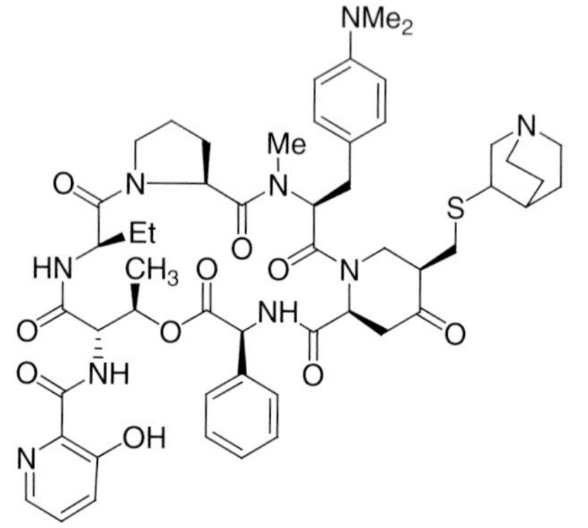

17b Quinopristin (RP 57669)

17 Synercid (RP-59500)

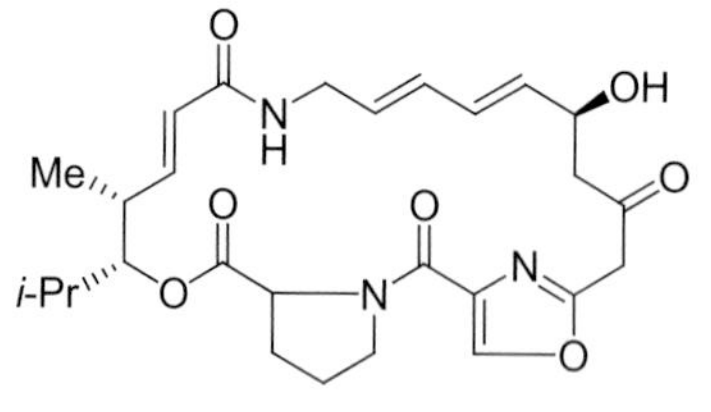

18a Pristinamycin II_B (RPR106950)

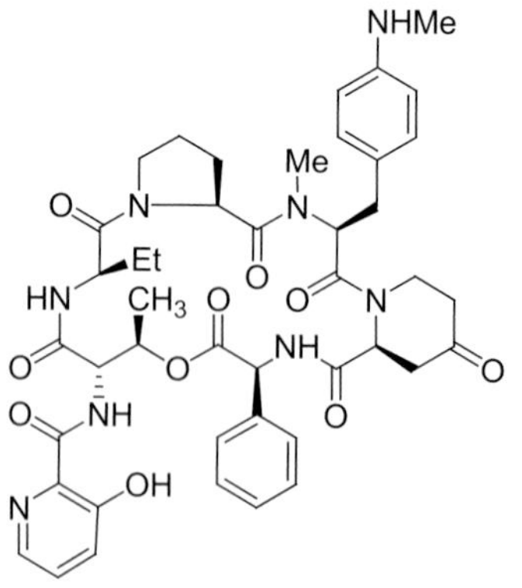

18b Pristinamycin I_B (RPR112808)

18 RPR106972

Soon after the introduction of Synercid (**17**), epidemiological studies on resistance factors in *E. faecium* showed origins from the animal kingdom. This was due to the extensive use of the related antibiotic virginiamycin (**16**) in animal husbandry as growth promoter prior to the development of streptogramins for human use. This created a reservoir of resistance genes in animal pathogens (*E. faecium*). Bacterial resistance to the streptogramins occurs through (1) ribosomal modification, (2) substrate inactivation processes, or (3) efflux (only coagulase-negative staphylococci); the major source of resistance is due to the MLS_B resistance component.

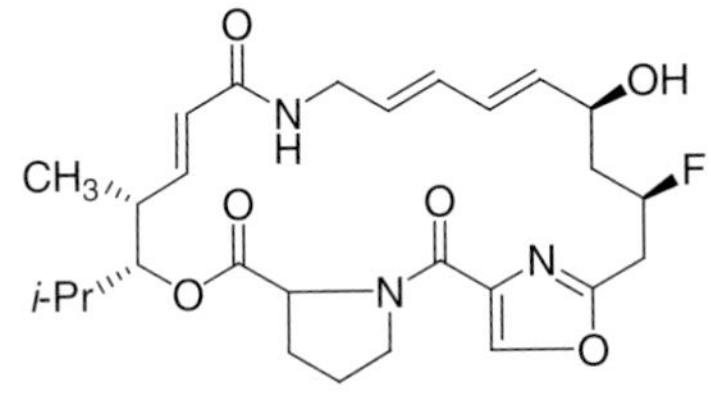

19a RPR132552
(group A)

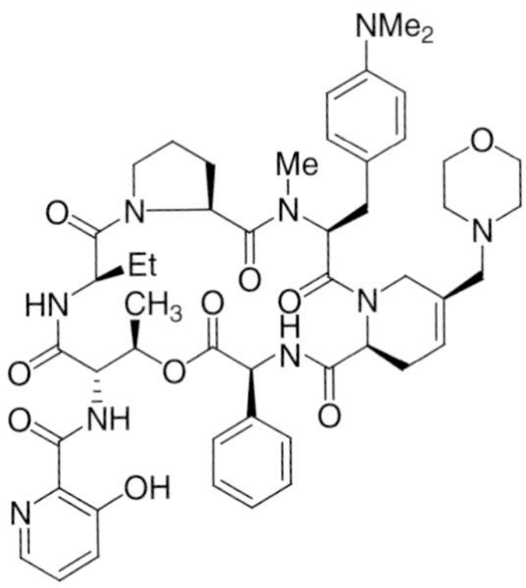

19b RPR202868
(group B)

19 NXL103 (XRP 2868)

7.22.6 Antibiotics from Nontraditional Structural Classes

The rising incidence of infections due to problematic Gram-positive pathogens has prompted reanalysis of natural product screening leads that had been de-emphasized earlier due to their limited antibacterial spectrum.

7.22.6.1 Lipopeptides

7.22.6.1.1 Daptomycin

Daptomycin (**20**), produced by *Streptomyces roseosporus*,[42,43] was identified in the late 1970s due to its exquisite activity against methicillin-sensitive *S. aureus* (MSSA) and methicillin-resistant *S. aureus* (MRSA), with rapid bactericidal action. Additional in vitro studies showed it to be effective against vancomycin-intermediate *S. aureus* (VISA) and vancomycin-resistant *S. aureus* (VRSA), *S. pneumoniae* and other streptococci, vancomycin-sensitive and vancomycin-resistant *E. faecium* and *E. faecalis*.[44–46] Its lack of activity against Gram-negative bacteria is due to its inability to cross the outer membrane.

Other analogs, which vary on the lipophilic moiety, are concomitantly produced in the fermentations. While they demonstrated similar antibacterial spectrum, they have different levels of untoward effects in animal models.[47] Daptomycin appears to have an adequately balanced therapeutic profile.

Data on the mode of action of daptomycin points to its antimicrobial activity due to calcium-dependent binding to the cytoplasmic membrane. On binding, the lipophilic moiety (i.e., *N*-decanoyl) perturbs the cellular membrane, which results in rapid depolarization and subsequent bacterial cell death, but not cellular lysis.[48]

Initial clinical studies were suspended due to concerns regarding skeletal muscle toxicity, manifested by muscle weakness and elevated levels of creatine phosphokinase (CPK). As a result of understanding the pharmacodynamic and pharmacokinetic properties of daptomycin, a revised dosing regimen was used, which minimized muscle toxicity so that it was adequate for clinical studies.[49] In 2003, daptomycin was approved for complicated skin and skin structure infections (abscesses, diabetic ulcers, and wound infections). Subsequent clinical trials are ongoing for endocarditis due to *S. aureus*.

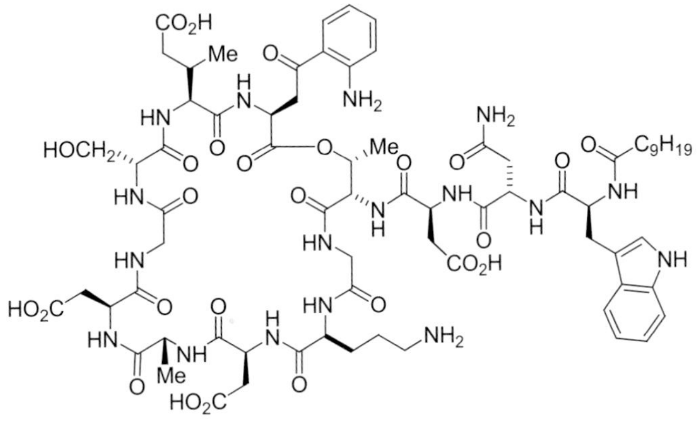

20 Daptomycin

7.22.6.1.2 Amphomycin analogs

The narrow therapeutic profile of daptomycin prompted consideration of other lipopeptides with intrinsically better therapeutic indices. Presumably, structural modifications on such a scaffold would improve potency against problematic Gram-positive bacteria, while retaining the better therapeutic indices. Amphomycin (**21**), a lipocyclodecapeptide produced by *Streptomyces griseus*,[50,51] with better intrinsic therapeutic indices, was studied as a scaffold for structural modifications. The amphomycin class antibiotics inhibit peptidoglycan biosynthesis in Gram-positive bacteria. Specifically, they block transfer of phospho-*N*-acetylmuramyl pentapeptide from the UDP derivative to undecaprenylmonophosphate by interaction with dolichylmono-phosphate.[52]

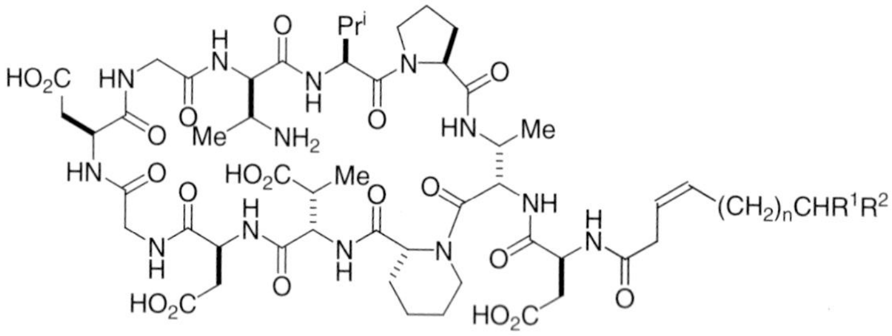

21 n = 5, R^1 = Me, R^2 = Et (Amphomycin)
22 n = 6, R^1 = R^2 = Me (Friulimicin)

In contrast to daptomycin, which was obtained directly from mutasynthesis fermentation, the modifications on the amphomycin scaffold are more challenging because they were performed at multiple sites on intrinsically unstable intermediates.[53] M-1396 (**23**) has shown comparable activity to daptomycin, including the calcium-dependent effect on MICs.[54] Other presentations[55,56] and patent applications[57,58] show analogs **24** with additional sites of modification.

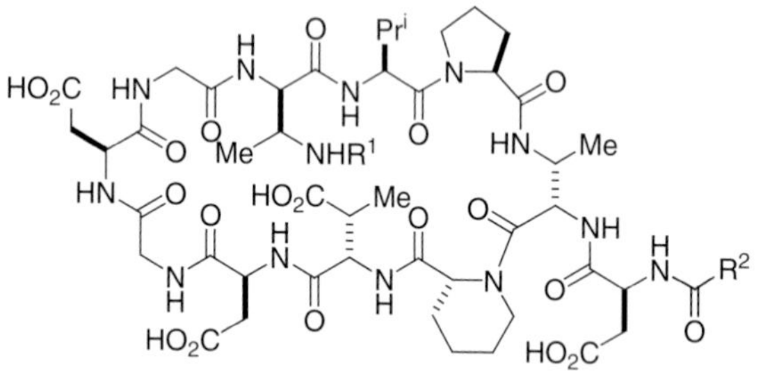

23 M-1396
R^1 = $COCH_2CH_2NH_2$, R^2 = $(CH_2)_{13}CH_3$

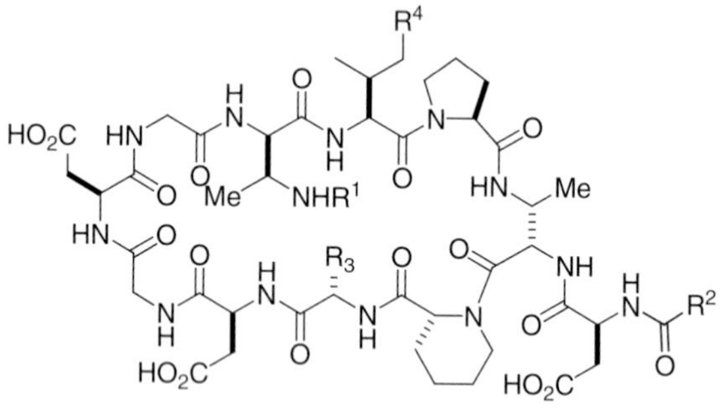

24

7.22.6.2 Ramoplanin

Ramoplanin (**25**), a glycolipodepsipeptide produced by *Actinoplanes* spp. ATCC 33076,[59,59a] is highly active against Gram-positive aerobic and anaerobic bacteria.[60,61] Ramoplanin blocks bacterial cell wall biosynthesis by inhibition of bacterial transglycosylases by binding tightly to lipid II, the substrate for the enzyme.[62] It displays rapid bactericidal mode of action against both vancomycin-sensitive and vancomycin-resistant *E. faecium* and *E. faecalis*[63,64,64a] While its low solubility and toxic profile precludes systemic applications, ramoplanin is currently in phase III clinical trials for the eradication of vancomycin-resistant *E. faecium* (VREF) and/or *C. difficile* in the gastrointestinal tract.[65]

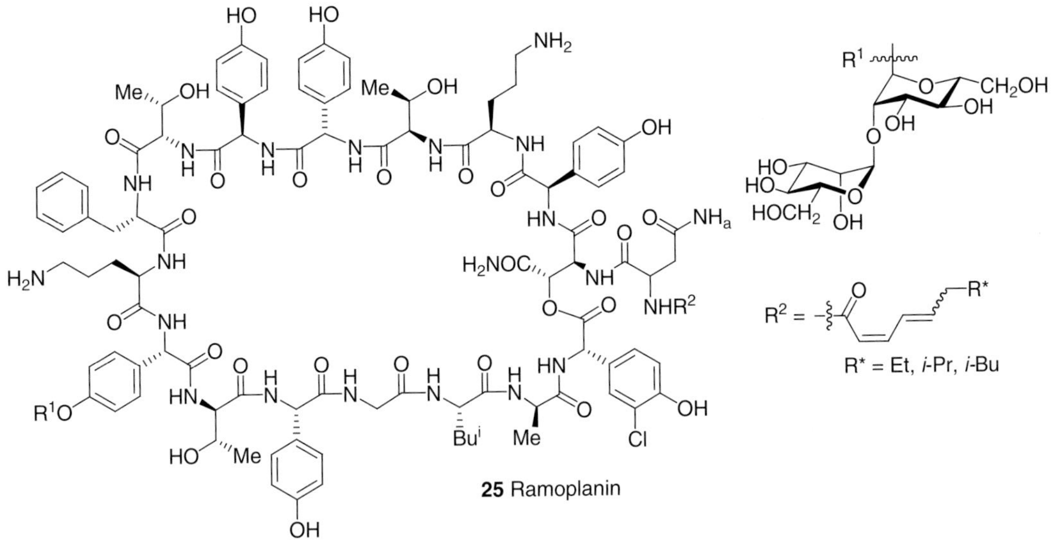

25 Ramoplanin

R* = Et, *i*-Pr, *i*-Bu

7.22.6.3 Orthosomycins

These are either septasaccharide or octasaccharides characterized by the presence of two ortho-ester moieties. Everninomicin D (**26**), produced by *Micromonospora carbonacea*,[66] is highly active against Gram-positive bacteria and was studied clinically for problematic Gram-positive infections.[67] However, its solubility is very low; attempts to increase solubility in conjunction with other structure–activity modifications yielded evernimicin (SCH-27899, **27**), which is also effective against MRSA and VRE.[68] In spite of the cessation of clinical trials due to untoward activities (nephrotoxicity and neurotoxicity), the antibacterial properties of **27** (**Table 6**) and mode of action are compelling. Mechanistically, **27** binds exclusively to the 50S ribosomal subunit and inhibits translation.[69]

As in the case of the streptogramins, the promise of the everninomicins as a human antibacterial was weakened by the presence of animal reservoirs of resistance in *E. faecalis*, which were imparted from the use of avilamycin (**28**) as a feed additive in Europe.[14]

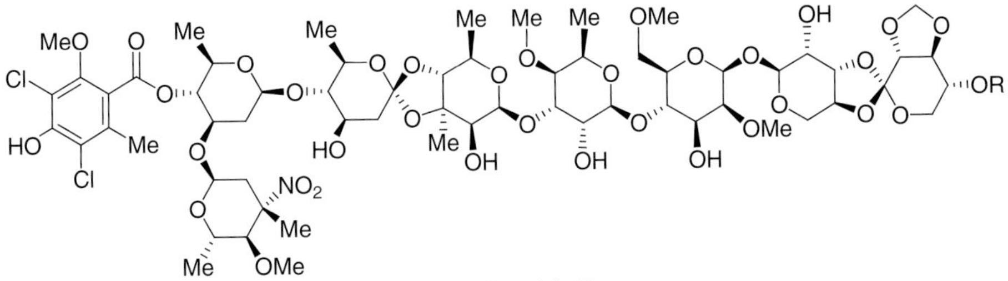

26 R=H everninomicin D
27 R=2,4-dihydroxy-6-benzoyl
(SCH-27899, evernimicin)

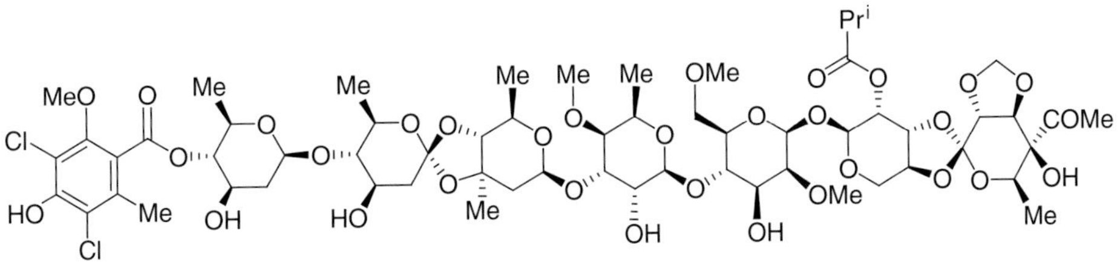

28 Avilamycin (LY-048740)

Table 6 Comparative in vitro profile of evernimicin (**27**) and comparator agents against aerobic Gram-positive bacteria

Organism	*Drug*	*N*	*Etest MIC* ($\mu g\, mL^{-1}$)	
			MIC_{50}	MIC_{90}
S. pneumoniae (Pen^S)	Evernimicin	971	0.023	0.047
	Synercid	971	0.5	1.0
	Vancomycin	971	0.5	0.75
	Ceftriaxone	971	0.023	0.047
	Erythromycin	971	0.125	0.38
S. pneumoniae (Pen^I)	Evernimicin	364	0.032	0.094
	Synercid	364	0.5	1.0
	Vancomycin	364	0.5	0.75
	Ceftriaxone	364	0.38	1.0
	Erythromycin	364	0.25	>256
S. pneumoniae (Pen^R)	Evernimicin	117	0.032	0.064
	Synercid	117	0.5	1.0
	Vancomycin	117	0.5	1.0
	Ceftriaxone	117	1.0	4.0
	Erythromycin	117	4.0	>256
Enterococcus spp. (Van^S)	Evernimicin	495	0.38	1.0
	Synercid	496	16	>32
	Vancomycin	484	2	3
	Ampicillin	495	1	2
Enterococcus spp. (Van^R)	Evernimicin	28	0.19	0.5
	Synercid	28	1	4
	Vancomycin	28	>256	>256
	Ampicillin	27	64	>256
E. faecium (Van^S)	Evernimicin	157	0.25	0.75
	Synercid	157	1.5	8.0
	Vancomycin	157	1.0	2
	Ampicillin	155	24	>256
E. faecium (Van^R)	Evernimicin	107	0.25	1.0
	Synercid	107	1.0	8.0
	Vancomycin	107	>256	>256
	Ampicillin	107	>256	>256

Table 6 Continued

Organism	*Drug*	*N*	*Etest MIC* ($\mu g\,mL^{-1}$)	
			MIC$_{50}$	*MIC*$_{90}$
E. faecalis (Van^S)	Evernimicin	726	0.38	1.0
	Synercid	726	16	>32
	Vancomycin	721	2.0	3.0
	Ampicillin	721	1.0	1.5
E. faecalis (Van^R)	Evernimicin	3	0.5[a]	
	Synercid	3	16[a]	
	Vancomycin	3	6.0[a]	
	Ampicillin	3	0.75[a]	

Erythromycin is comparator agent for *S. pneumoniae*; Ampicillin is comparator agent for enterococcal infections.
[a] Owing to inadequate strain population size, MIC$_{50}$ is provided.

7.22.7 Summary

The advances in antibacterial therapies in the twentieth century are being weakened by the formation of resistance in many pathogens. This occurs in a period when corporate support for research in acute therapeutics has shifted to life-style drugs or chronic therapeutics. Simultaneously, new treatment modalities that require various levels of immunomodulation have resulted in increases in nosocomial infections. More incidents of infections that were formerly found in nosocomial settings have appeared in the community setting (e.g., MRSA). Thus, new antibiotics, either from existing structural classes or new classes, are needed. In some clinical circumstances, especially for critically ill patients, creative applications of pharmacodynamic and pharmacokinetic strategies will be needed to fully realize the potential of new antibacterials.

References

1. Williams, D. H.; Stone, M. J.; Hauck, P. R.; Rahman, S. K. *J. Nat. Prod.* **1989**, *52*, 1189–1208.
2. Vining, L. C. *Annu. Rev. Microbiol.* **1990**, *44*, 395–427.
3. Woodruff, H. B.; Hernandez, S.; Stapley, E. D. *Hindustan Antibiot. Bull.* **1979**, *21*, 71–84.

3a. Verdine, G. L. *Nature* **1996**, *384*, 11–13.

3b. Singh, M. P.; Greenstein, M. *Curr. Opin. Drug Disc. Dev.* **2000**, *3*, 167–176.

4. Kirby, W. M. M. *Science* **1944**, *99*, 452–453.
5. Chambers, H. F. *Clin. Microbiol. Rev.* **1997**, *10*, 789–791.
6. Noble, W. C.; Virani, Z.; Cree, R. G. *FEMS Microbiol. Lett.* **1992**, *72*, 195–198.
7. Fontana, R.; Cerini, R.; Longoni, P.; Grossato, A.; Canepari, P. *J. Bacteriol.* **1983**, *155*, 1343–1350.

7a. Williamson, R.; le Bouguenec, C.; Gutmann, L.; Horaud, T. *J. Gen. Microbiol.* **1985**, *131*, 1933–1940.

8. Reichmann, P.; Konig, A.; Marton, A.; Hakenbeck, R. *Microb. Drug Resist.* **1996**, *2*, 177–181.

8a. Hakenbeck, R.; Konig, A.; Kern, I.; van der Linden, M.; Keck, W.; Billot-Klein, D.; Legrand, R.; Schoo, B.; Gutmann, L. *J. Bacteriol.* **1998**, *180*, 1831–1840.

9. Smith, T. L.; Pearson, M. L.; Wilcox, K. R.; Cruz, C.; Lancaster, M. V.; Robinson-Dunn, B.; Tenover, F. C.; Zervos, M. J.; Band, J. D.; White, E. et al. *N. Engl. J. Med.* **1999**, *340*, 493–501.

9a. Sieradzki, K.; Roberts, R. B.; Haber, S. W.; Tomasz, A. *N. Engl. J. Med.* **1999**, *340*, 517–523.

9b. Chang, S.; Sievert, D.; Hageman, J.; Boulton, M. L.; Tenover, F. C.; Downes, F. P.; Shah, S.; Rudrik, J. T.; Pupp, G. R.; Brown, W. J.; et al. *N. Engl. J. Med.* **2003**, *348*, 1342–1347.

10. Kirst, H. A.; Thompson, D. G.; Nicas, T. I. *Antimicrob. Agents Chemother.* **1998**, *42*, 1303–1304.
11. Moellering, R. C., Jr. *J. Antimicrob. Chemother.* **1991**, *28*, 1–12.

11a. Frieden, T. R.; Munsiff, S. S.; Low, D. E.; Willey, B. M.; Williams, G.; Faur, Y.; Eisner, W.; Warren, S.; Kreiswirth, B. *Lancet* **1993**, *342*, 76–79.

11b. Boyce, J. M. *Infect. Dis. Clin. North Am.* **1997**, *11*, 367–384.

12. Walsh, C. T.; Fisher, S. L.; Park, I. S.; Prahalad, M.; Wu, Z. *Chem. Biol.* **1996**, *3*, 21–28.
13. Bager, F.; Madsen, M.; Christensen, J.; Aarestrup, F. M. *Prev. Vet. Med.* **1997**, *31*, 95–112.
14. Wegener, H. C.; Aarestrup, F. M.; Jensen, L. B.; Hammerum, A. M.; Bager, F. *Emerg. Infect. Dis.* **1999**, *5*, 329–335.

15. Hermann, R.; Ripamonti, F.; Romano, G.; Restelli, E.; Ferrari, P.; Goldstein, B. P.; Berti, M.; Ciabatti, R. J. *J. Antibiot. (Tokyo)* **1996**, *49*, 1236–1248.
16. Cooper, R. D.; Snyder, N. J.; Zweifel, M. J.; Staszak, M. A.; Wilkie, S. C.; Nicas, T. I.; Mullen, D. L.; Butler, T. F.; Rodriguez, M. J.; Huff, B. E. et al. *J. Antibiot. (Tokyo)* **1996**, *49*, 575–581.
17. Leadbetter, M. R.; Adams, S. M.; Bazzini, B.; Fatheree, P. R.; Karr, D. E.; Krause, K. M.; Lam, B. M. T.; Linsell, M. S.; Nodwell, M. B.; Pace, J. L. et al. *J. Antibiot. (Tokyo)* **2004**, *57*, 326–336.
18. Van Bambeke, F.; Laethem, Y. V.; Courvalin, P.; Tulkens, P. M. *Drugs* **2004**, *64*, 913–936.
19. Zelenitsky, S. A.; Karlowsky, J. A.; Zhanel, G. G.; Hoban, D. J.; Nicas, T. *Antimicrob. Agents Chemother.* **1997**, *41*, 1407–1408.
20. Goldstein, B. P.; Selva, E.; Gastaldo, L.; Berti, M.; Pallanza, R.; Ripamonti, F.; Ferrari, P.; Denaro, M.; Arioli, V.; Cassani, G. *Antimicrob. Agents Chemother.* **1987**, *31*, 1961–1966.
21. Malabarba, A.; Ciabatti, R.; Kettenring, J.; Ferrari, P.; Scotti, R.; Goldstein, B. P.; Denaro, M. *J. Antibiot. (Tokyo)* **1994**, *47*, 1493–1506.
22. Mackey, J. P.; Gerhard, U.; Beauregard, D. A.; Williams, D. H.; Westwell, M. S.; Searle, M. S. *J. Am. Chem. Soc.* **1994**, *116*, 4581–4590.
23. Beauregard, D. A.; Williams, D. H.; Gwynn, M. N.; Knowles, D. *J. Antimicrob. Agents Chemother.* **1995**, *39*, 781–785.
24. Printsevskaya, S. S.; Pavlov, A. Y.; Olsufyeva, E. N.; Mirchink, E. P.; Preobrazhenskaya, M. N. *J. Med. Chem.* **2003**, *46*, 1204–1209.
25. Van Bambeke, F.; Saffran, J.; Mingeot-Leclercq, M.-P.; Tulkens, P. M. *Antimicrob. Agents Chemother.* **2005**, *49*, 1695–1700.
26. Pace, J. L.; Krause, K.; Johnston, D.; Debabov, D.; Wu, T.; Farrington, L.; Higgins, D. L. L.; Christensen, B.; Judice, J. K.; Koniga, K. *Antimicrob. Agents Chemother.* **2003**, *47*, 3602–3604.
27. Candiani, G.; Abbondi, M.; Borgonovi, M.; Roman, G.; Parenti, F. *J. Antimicrob. Chemother.* **1999**, *44*, 179–192.
28. King, A.; Phillips, I.; Kaniga, K. *J. Antimicrob. Chemother.* **2004**, *53*, 797–803.
29. Dorr, M. B.; Jabes, D.; Cavaleri, M.; Dowell, J.; Mosconi, G.; Malabarba, A.; White, R. J.; Henkel, T. J. *J. Antimicrob. Chemother.* **2005**, *55*, ii25–ii30.
30. Fuller, A. T.; Mellows, G.; Woolford, M.; Banks, G. T.; Barrow, K. D.; Chain, E. B. *Nature* **1971**, *234*, 416–417.
31. Slocombe, B.; Perry, C. *J. Hosp. Infect.* **1991** (Suppl B) 19–25.
32. Silvian, L. F.; Wang, J.; Steitz, T. A. *Science* **1999**, *285*, 1074–1077.
33. Gilbart, J.; Perry, C. R.; Slocombe, B. *Antimicrob. Agents Chemother.* **1993**, *37*, 32–38.
34. Udo, E. E.; Farook, V. S.; Eiman, M.; Mokadas, E. M.; Jacob, L. E.; Sanyal, S. C. *Int. J. Infect. Dis.* **1999**, *3*, 82–87.
35. Klein, L. L.; Yeung, C. M.; Kurath, P.; Mao, J. C.; Fernandes, P. B.; Lartey, P. A.; Pernet, A. G. *J. Med. Chem.* **1989**, *32*, 151–160.
36. Brown, P.; Davies, D. T.; O'Hanlon, P. J.; Wilson, J. M. *J. Med. Chem.* **1996**, *39*, 446–457.
37. Broom, N. J. P.; Cassels, R.; Cheng, H.-Y.; Elder, J. S.; Hannan, P. C. T.; Masson, N.; O'Hanlon, P. J.; Pope, A.; Wilson, J. M. *J. Med. Chem.* **1996**, *39*, 3596–3600.
38. Brown, P.; Best, D.; Broom, N. J. P.; Cassels, R.; Mitchell, T. J.; Osborne, N. F.; Wilson, J. M. *J. Med. Chem.* **1997**, *40*, 2563–2570.
39. Barriere, J. C.; Berthaud, N.; Beyer, D.; Dutka-Malen, S.; Paris, J. M.; Desnottes, J. F. *Curr. Pharm. Des* **1998**, *4*, 155–180.
40. Berthaud, N.; Charles, Y.; Gouin, A.-M.; Mereau, B.; Rousseau, J.; Desnottes, J.-F.; Macgowan, A.P. *RPR 106972, A New Oral Streptogramin: In Vitro Antibacterial Activity.* 35th Interscience Conference on Antimicrobial Agents and Chemotherapy, Washington, DC, 1995; Abs F112.
41. Spangler, S. K.; Jacobs, M. R.; Appelbaum, P. C. *Antimicrob. Agents Chemother.* **1996**, *40*, 481–484.
42. Debono, M.; Barnhart, M.; Carrell, C. B.; Hoffman, J. A.; Occolowitz, J. L.; Abbott, B. J.; Fukuda, D. S.; Hamill, R. L.; Biemann, K.; Herlihy, W. C. *J. Antibiot. (Tokyo)* **1987**, *40*, 761–777.
43. Baltz, R. H.; Miao, V.; Wrigley, S. K. *Nat. Prod. Rep.* **2005**, *22*, 717–741.
44. Howe, R. A.; Noel, A. R.; Tomaselli, S.; Bowker, K. E.; Walsh, T. R.; Macgowan, A. P. *Killing Activity of Daptomycin against Vancomycin Intermediate Staphylococcus aureus (VISA) and Hetero-VISA.* 43rd Interscience Conference on Antimicrobial Agents and Chemotherapy, Chicago, IL, Sept 14–17, 2003; Abs C1-1641.
45. Rybak, M. J.; Hershberger, E.; Moldovan, T.; Grucz, R. G. *Antimicrob. Agents Chemother.* **2000**, *44*, 1062–1066.
46. Streit, J. M.; Jones, R. N.; Sader, H. S. *J. Antimicrob. Chemother.* **2004**, *53*, 669–674.
47. Debono, M.; Abbott, B. J.; Molloy, R. M.; Fukuda, D. S.; Hunt, A. H.; Daupert, V. M.; Counter, F. T.; Ott, J. L.; Carrell, C. B.; Howard, L. C. et al. *J. Antibiot. (Tokyo)* **1988**, *41*, 1093–1105.
48. Silverman, J. A.; Perlmutter, N. G.; Shapiro, H. M. *Antimicrob. Agents Chemother.* **2003**, *47*, 2538–2544.
49. Carpenter, C. F.; Chambers, H. F. *Clin. Infect. Dis.* **2004**, *38*, 994–1000.
50. Heinemann, B.; Kaplan, M. A.; Muir, R. D.; Hooper, I. R. *Antibiot. Chemother.* **1953**, *3*, 1239–1242.
51. Vertesy, L.; Ehlers, E.; Kogler, H.; Kurz, M.; Meiwes, J.; Seibert, G.; Vogel, M.; Hammann, P. *J. Antibiot. (Tokyo)* **2000**, *53*, 816–827.
52. Banerjee, D. K. *J. Biol. Chem.* **1989**, *264*, 2024–2028.
53. Cameron, D. R.; Chen, Y.; Dugourd, D.; Sun, J.; Wang, L.; Borders, D. B.; Curran, W. V.; Leese, R. A. *New Laspartomycin-Based Semi-Synthetic Lipopeptide Antibiotics.* 229th American Chemical Society Meeting, San Diego, CA, Mar 13–17, 2005; Abs 358.
54. Wacowich-Sgarbi, S. A.; Boyd, V. A.; Cameron, D. R.; Chen, Y.; Dugourd, D.; Jia, Q.; Nodwell, M.; Sgarbi, P. W. M.; Sun, J.; Wang, L. et al. *Synthesis and Structure–Activity Relationship (SAR) Studies on Dab-9 Substitutions of the Lipopeptide Antibiotic Amphomycin.* 229th American Chemical Society Meeting, San Diego, CA, Mar 13–17, 2005; Abs 367.
55. Sgarbi, P. W. M.; Boyd, V. A.; Cameron, D. R.; Chen, Y.; Jia, Q.; Nodwell, M.; Siu, R.; Sun, J.; Wacowich-Sgarbi, S. A.; Wang, L. et al. *Synthesis and Structure–Activity Relationship (SAR) Studies on the Lipophilic Tail of the Lipopeptide Antibiotic Amphomycin.* 229th American Chemical Society Meeting, San Diego, CA, Mar 13–17, 2005; Abs 368.
56. Rubinchik, E.; Pasetka, C. J.; Liu, B.; Omilusik, K. D.; Scott, A.; Cameron, D. R.; Boyd, V. A.; Borders, D. B.; Leese, R. A.; Curran, W. V. et al. *Novel Antimicrobial Lipopeptides with Long In vivo Half-Lives.* 105th General Meeting of American Society of Microbiology, Atlanta, GA, June 5–9, 2005; Abs A-056.
57. Cameron, D. R.; Boyd, V. A.; Leese, R. A.; Curran, W. V.; Francis, N. D.; Jarolmen, H.; Borders, D. B.; Sgarbi, P. W. M.; Wacowich-Sgarbi, S. A.; Nodwell, M. et al. Compositions of Lipopeptide Antibiotics and Methods of Use Thereof, WO 05/000878; US 2005/153876, Filed June 28, 2004.
58. Fardis, M.; Cameron, D. R.; Boyd, V. A. Dab9 Derivatives of Lipopeptide Antibiotics and Methods of Making and Using the Same, WO 03/057724; US 2004/138107, Filed Jan 3, 2003.
59. Cavalleri, B.; Pagani, H.; Volpe, G.; Selva, E.; Parenti, F. *J. Antibiot. (Tokyo)* **1984**, *37*, 309–317.

59a. Pallanza, R.; Berti, M.; Scotti, R.; Randisi, E.; Arioli, V. *J. Antibiot. (Tokyo)* **1984**, *37*, 318–324.

60. Jiang, W.; Wanner, J.; Lee, R. J.; Bounaud, P.-Y.; Boger, D. L. *J. Am. Chem. Soc.* **2002**, *124*, 5288–5290.
61. Montecalvo, M. *J. Antimicrob. Chem.* **2003**, *51*, iii31–iii35.

62. Hu, Y.; Helm, J. S.; Chen, L.; Ye, X.-Y.; Walker, S. *J. Am. Chem. Soc.* **2003**, *125*, 8736–8737.
63. Collins, L. A.; Eliopoulos, G. M.; Wennersten, C. B.; Ferraro, M. J.; Moellering, R. C. *Antimicrob. Agent Chemother.* **1993**, *37*, 1364–1366.
64. Mobarakai, N.; Quile, J. M.; Landman, D. *Antimicrob. Agents Chemother.* **1994**, *38*, 385–387.
64a. Johnson, C. C.; Taylor, S.; Pitsakis, P.; May, P.; Levinson, M. E. *Antimicrob. Agents Chemother.* **1992**, *36*, 2342–2345.
65. Wong, M. T.; Kauffman, C. A.; Standiford, H. C.; Linden, P.; Fort, G.; Fuchs, H. J.; Porter, S. B.; Wenzel, R. P.; Ramoplanin VRE Clinical Study Group. *Clin. Infect. Dis.* **2001**, *33*, 1476–1482.
66. Weinstein, M. J.; Luedemann, G. M.; Oden, E. M.; Wagman, G. H. *Antimicrob. Agents Chemother.* **1964**, *10*, 24–32.
67. Ganguly, A. K.; Sarre, O. Z.; Greeves, D.; Morton, J. *J. Am. Chem. Soc.* **1975**, *97*, 1982–1985.
68. Jones, R. N.; Hare, R. S.; Sabatelli, F. J. Ziracin Susceptibility Testing Group *J. Antimicrob. Chemother.* **2001**, *47*, 15–25.
69. McNicholas, P. M.; Najarian, D. J.; Mann, P. A.; Hesk, D.; Hare, R. S.; Shaw, K. J.; Black, T. A. *Antimicrob. Agents Chemother.* **2000**, *44*, 1121–1126.

Biography

Ving J Lee received his BA and PhD from Ohio State University and the University of Illinois-Urbana-Champaign, respectively. After postdoctoral appointments at the University of Illinois and Harvard, he has held positions of increasing responsibility at Lederle Laboratories (Cyanamid), Microcide Pharmaceuticals, Anacor Pharmaceuticals, and Adesis Inc. (Chief Executive Officer and Chief Scientific Officer).

His interests encompass the following: new synthetic methodologies; systems biology for drug discovery and development; ex vivo methods for preclinical development and investigational toxicology; new uses for natural products and existing pharmaceuticals; new anti-infectives; and new therapeutic opportunities.

Comprehensive Medicinal Chemistry II
ISBN (set): 0-08-044513-6

ISBN (Volume 7) 0-08-044520-9; pp. 653–671

7.23 Oxazolidinone Antibiotics

S J Brickner, Pfizer Inc., Groton, CT, USA

7.23.1 Introduction

In the late 1980s, the medical community began witnessing the first clear signs of what was to become a frightening escalation in the incidence of multidrug resistance in Gram-positive bacteria, pathogens capable of causing some of the most serious infections in humans. In a time when ostensibly no technological challenge is unsolvable, it is at first thought difficult to accept the incongruous fact that many of the revered antibiotics discovered just a few decades ago have now become completely ineffective against an increasing number of specific strains of bacterial pathogens. Each year, nearly 2 million people, or about 6% of all hospitalized patients in the US, get a hospital-acquired (or nosocomial) infection, of which the Centers for Disease Control and Prevention (CDC) has reported 70% are resistant to one or more antibiotics.[1] Multidrug resistance has become a dire problem, and it supplies the motive and has created the genuine need to continue the search for completely new, effective antibiotics that will have different mechanisms of action (MOAs) not shared with older antibiotic classes.

This review chronicles the discovery of the oxazolidinone antibacterial agents,[2–4] an entirely new class of drugs having potent activity against important susceptible and multidrug resistant (MDR) Gram-positive pathogens. Linezolid (**1**, Zyvox),[5] the first, and thus far, only member of the oxazolidinones to be approved for use in humans, was discovered at the Upjohn Company, and developed and marketed by the Pharmacia Corporation (now Pfizer Inc.) for the treatment of serious Gram-positive bacterial infections. As such, the oxazolidinones represent the first completely new class of antibiotics with a novel MOA to gain regulatory approval in over 35 years,[6] since the early 1960s when nalidixic acid[7] (the progenitor of the synthetic fluoroquinolones[8]) was approved.

Following the announcement of successful clinical trials with the early oxazolidinones, to date more than 25 companies and numerous academic and clinical research laboratories from around the world have been reporting an extensive collective body of research findings on the oxazolidinones and related isosteric analogs. This intensely competitive activity is a testament to both the serious medical need presented by MDR Gram-positive infections and the desire to discover second-generation oxazolidinones with improvement over linezolid. Given the rapid expansion and massive volume of research activities published in this area, particularly within the past decade, by necessity there has been a need to limit the overall scope of this review chapter. The primary focus here centers on key attributes and aspects of the three oxazolidinones for which there are published clinical trial data: linezolid (**1**, U-100766, PNU-100766),[5] eperezolid (**2**, U-100592, PNU-100592, Pharmacia & Upjohn Co.),[5] and ranbezolid (**3**, RBx-7644, Ranbaxy Research Laboratories),[9] and drawing from the available pre-clinical data on the first oxazolidinone clinical candidates, DuP-105 (**4**) and DuP-721 (**5**, DuPont).[10] The secondary emphasis concerns what are considered the key advances in the structure–activity relationships (SARs) developed beyond these five compounds. This review provides an overview of the medical need, key attributes, discovery and development history, SAR, safety profile, MOA, issues, and directions for future research into this important new class of oxazolidinone antibiotics. The reader is directed to other excellent reviews on this rapidly moving area of research,[2,4,9,11–19] as well as the companion review in Volume 8 of this publication, *see* 8.13 Zyvox.

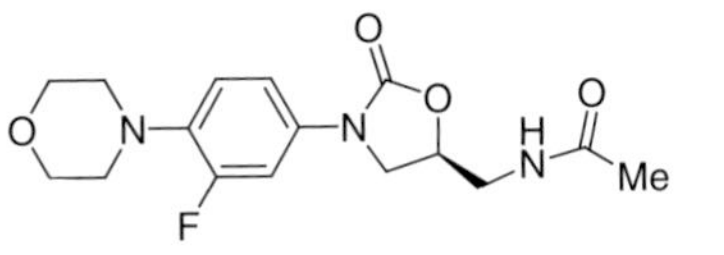

1

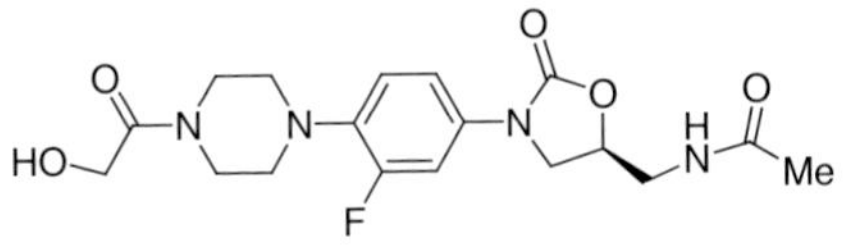

2

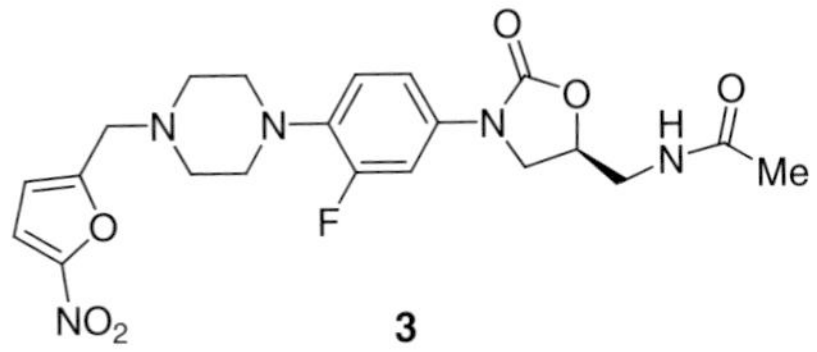

3

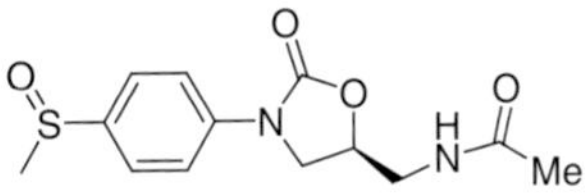

4

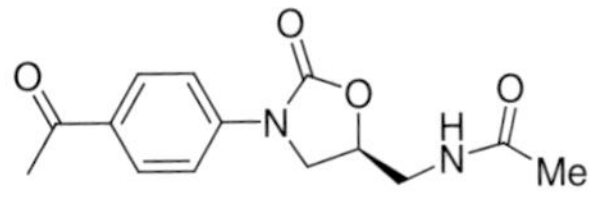

5

7.23.2 Disease State

Gram-positive bacteria differ most fundamentally from Gram-negative bacteria in that the latter have a second outer membrane that presents a potential penetration barrier to antibiotics, in addition to the cytoplasmic membrane found within all bacteria. Given this, one could presume that the goal of killing or arresting the growth of a Gram-positive bacterium with any given antibiotic should therefore be an easier objective than that required for a Gram-negative bacterium. The reality is that, over the past two decades, an increasing number of strains of certain serious Gram-positive pathogens have become adept at developing or acquiring the means to circumvent the antibacterial action of virtually all previously effective antibiotics. These MDR Gram-positive staphylococcal, enterococcal, and streptococcal strains are becoming increasingly predominant among isolates collected from patients with hospital-acquired (nosocomial) infections around the world, and they cause some of the most difficult-to-treat infections encountered in the last 15 years.[20] Associated with such antibiotic-resistant infections are higher medical costs, increased severity of infection, and elevated mortality relative to similar infections caused by antibiotic-sensitive strains.[21–23] While MDR Gram-negative bacteria have also become problematic in hospitals, the collective incidence of those pathogens in nosocomial infections is considerably lower than those of the Gram-positives.

Amongst the Gram-positives, *Streptococcus pneumoniae* is considered a major bacterial pathogen, capable of causing considerable human morbidity and mortality. These microorganisms are implicated in many cases of community-acquired pneumonia, respiratory tract and middle-ear infections, and, far less commonly, meningitis. Mortality rates for elderly patients with bacteremia, or infection of the bloodstream, are typically about 20%.[24a,24b]

The virulent pathogen *Staphylococcus aureus* is the leading cause of hospital-acquired pneumonia, and the second most common cause of bacteremia.[25] It is commonly implicated in infections of skin and soft tissues, postoperative surgical sites, bone and joint, diabetic foot, and the urinary tract, and causes septicemia and endocarditis. At present, there is a high prevalence of methicillin-resistant *S. aureus* (MRSA) strains that are isolated in hospitals throughout the world[26]; these strains have been shown to be just as virulent as the methicillin-susceptible *S. aureus* (MSSA) strains.[27] While bacteremia with MRSA or MSSA can rapidly reach a critical stage with ensuing shock and death due to the production of multiple staphylococcal exoprotein toxins and virulence factors,[28,29a,29b] higher mortality rates have been reported due to MRSA versus MSSA bacteremias.[30] Factors that have been associated with acquiring bacteremia with MRSA versus MSSA include old age, prior antibiotic treatment, prolonged hospitalization, urinary catherization, or placement of a nasogastric tube.[31] Diabetic foot infections are the leading cause of lower-extremity amputations and diabetes-related hospitalization in the US,[32] and a rising prevalence of MRSA in foot ulcers of diabetic patients has been reported.[33a,33b] Strains of methicillin-resistant *Staphylococcus epidermidis* (MRSE) are also very highly prevalent in the US.

Recently there has been a surprising rise in the incidence of isolates of community-acquired MRSA (CA-MRSA) strains, which are genetically differentiated from the hospital-acquired MRSA (HA-MRSA)[34]; it is now apparent that

HA-MRSA strains are not the source of these CA-MRSA infections. Previously rare, CA-MRSA strains have been implicated in a number of infectious outbreaks in the US among various groups of healthy children and adults, including some sports teams and prison inmates. They are most often associated with skin or skin structure infections,[35a,35b] and CA-MRSA are now being recognized as common, highly virulent, and serious pathogens.[36] With invasive CA-MRSA infections there have been fatal outcomes, including those with severe necrotizing pneumonia.[37] These strains typically harbor multiple virulence genes that encode for a number of enterotoxins, including the highly prevalent Panton-Valentine leukocidin,[26,38] which causes dermal necrosis.[39]

Strains of *Enterococcus faecium* and *E. faecalis*, which were previously considered typical constituents of the human gastrointestinal tract and not a serious medical problem, have now acquired high-level resistance to the glycopeptide vancomycin. These strains are now very problematic, particularly in large hospitals primarily within the US, but also in other countries. Vancomycin-resistant enterococci (VRE) have become the second most common cause of bloodstream infections in intensive care units (ICUs) in the US. VRE bacteremias correlate to significantly higher rates of infection-related mortality than vancomycin-susceptible enterococcal infections,[40–44] and to high rates of severe sepsis and septic shock.[44] One study found liver transplant, acquired immunodeficiency syndrome (AIDS) or human immunodeficiency virus (HIV)-positive status, drug abuse, or prior exposure to vancomycin to be the most predictive factors associated with VRE bacteremia.[43]

Mycobacterium tuberculosis (TB) is an insidious scourge that has afflicted humans for thousands of years, and still remains one of the leading causes of death in the world, now estimated at 1.8 million per year.[45] TB infects over one-third of the world's population, and may persist for years in asymptomatic latent TB-infected individuals, where it hides in macrophages. About 10% of people carrying the latent bacterium will eventually convert to active TB infection.[46] Outbreaks of MDR strains of TB (MDR-TB) have been very problematic in prison populations of republics of the former Soviet Union,[47] and there were several disturbing hospital outbreaks in New York City during the 1990s.[48]

7.23.3 Disease Basis

The introduction of pathogenic bacteria into the normally aseptic bloodstream results in bacteremia, with the likelihood of impending septicemia and systemic infection. As *S. aureus* and *S. epidermidis* organisms reside on the normal human skin, they can be introduced into the bloodstream by, among other means, injury, surgery, illegal drug injection, or an underlying illness such as pneumonia. *S. aureus* is capable of producing multiple toxins, and without swift and effective antibiotic treatment, a systemic staphylococcal infection can rapidly turn fatal.

7.23.3.1 Hospital-Acquired Methicillin-Resistant *Staphylococcus aureus* and *Staphylococcus epidermidis*

MRSA (now defined as strains with an oxacillin minimum inhibitory concentration (MIC) $\geq 4\,\mu g\,mL^{-1}$) are resistant to not only methicillin and oxacillin, but virtually all β-lactams, aminoglycosides, tetracyclines, chloramphenicol, lincosamides, macrolides, trimethoprim/sulfamethoxazole, and fluoroquinolone antibiotics.[49] MRSA remain susceptible to vancomycin (with the exception of those strains noted below), quinupristin/dalfopristin (QD), and linezolid.[26] According to the CDC, the incidence of HA-MRSA in the US during 2003 was 59.5% of the *S. aureus* isolates collected from ICU patients, representing an 11% increase over the mean rate of resistance for the 5 prior years.[50] Most current data on MRSA incidence is 60% in Taiwan,[49] 74% in Hong Kong, 72% in Japan,[51] and 20% in Europe.[52] There are at least five types of a mobile staphylococcal cassette chromosome *mecA* (SCC*mec*) gene that confers β-lactam resistance,[53] by virtue of encoding an altered penicillin-binding protein 2a (PBP2a); the SCC*mec* types I–III are associated with HA-MRSA. MRSE are now extremely common, as ICU surveillance by the CDC showed an 89% incidence in the US in 2003,[50] while a report from Taiwan in 2000 showed the incidence there was 84%.[49]

7.23.3.2 Community-Acquired Methicillin-Resistant *Staphylococcus aureus*

CA-MRSA are characterized as strains of SCC*mec* types IV and V,[26] which typically do not have the genetic elements that encode resistance to other antibiotic classes like the HA-MRSA, and are thus primarily resistant only to β-lactams.[54] The overall prevalence of CA-MRSA strains in the community is uncertain, but is apparently on the rise.[34] A recent report found the incidences in the two US cities of Baltimore, MD, and Atlanta, GA, to be 18 and 26 cases per 100 000 residents, respectively.[36] Recommended effective treatment regimens for skin infections with CA-MRSA following incision and drainage are linezolid, vancomycin, or trimethoprim/sulfamethoxazole plus rifampin.[55]

7.23.3.3 Vancomycin-Resistant Enterococci

The first reports of VRE,[56] encoded by the *vanA* gene,[57a,57b] appeared in 1988. US CDC surveillance efforts indicate that, from January 1989 to early 1993, there was a 20-fold increase in the percentage of VRE in hospital-acquired infections in ICUs, and that many of these VRE were resistant to all available antibiotics.[20,40,58] As of December 2003, the incidence of VRE was 28.5%, up 12% over the mean rate of resistance found in the prior 5 years.[50]

7.23.3.4 Vancomycin-Resistant *Staphylococcus aureus* (VRSA)

The horizontal transfer of the *vanA* gene from VRE to MRSA was experimentally demonstrated by Nobel in 1992.[59] As of this writing, there have been three confirmed clinical isolates of high-level vancomycin-resistant *S. aureus* (VRSA, MIC $\geq 32\,\mu g\,mL^{-1}$).[60,61a–61c] All three were confirmed to contain the *vanA* gene, indicating acquisition of this resistance determinant via interspecies transfer from VRE.[61b,62] These VRSA strains were found to be MDR, yet still susceptible to linezolid, ranbezolid, chloramphenicol, trimethoprim/sulfamethoxazole, minocycline, and QD.[61c,63,64] *S. aureus* strains that have intermediate resistance to vancomycin (VISA: vancomycin MIC = 8–16 $\mu g\,mL^{-1}$), or intermediate resistance to both of the marketed glycopeptides, vancomycin and teicoplanin (GISA), have been reported. Thus far, indications are that these strains remain uncommon.[65] The oxazolidinones linezolid and ranbezolid retain excellent in vitro activity against these strains.[17,64]

7.23.3.5 Penicillin-Resistant *Streptococcus pneumoniae* (PRSP)

For nearly two decades following the introduction of penicillin in 1943, nearly all strains of *S. pneumoniae* were fully susceptible to this β-lactam, and until 1988, the incidence of PRSP was <0.1%. By the early 1990s, high-level PRSP strains with altered, low-affinity PBP targets[66] were becoming increasingly prevalent around the world.[67] US survey results[68] tracking the winter of 2002–03 *S. pneumoniae* community-acquired respiratory tract infection isolates show that 34% were penicillin-resistant (including intermediate resistance) and 22% of these PRSP isolates were MDR. All were susceptible to linezolid, vancomycin, and the ketolide telithromycin. In Taiwan, the incidence of nonsusceptible *S. pneumoniae* was 69%,[49] and 90% of the PRSP are now resistant to three or more classes of antibiotics.[69] Rates of resistance in *S. pneumoniae* in the US now appear to be leveling off or declining, with the exception of resistance to fluoroquinolones.[68]

7.23.3.6 *Mycobacterium tuberculosis*

Treatment of TB is challenging, due to the multitude of drugs required and protracted length of treatment, and poor patient compliance (often requiring direct observation of therapy programs). World Health Organization guidelines for recommended therapy is an initial four-drug regimen of isoniazid, rifampin, pyrazinamide, and ethambutal or streptomycin. MDR-TB presents even more difficult treatment challenges, as 1–2 years of drug therapy is required.[46]

7.23.3.7 Fastidious Gram-Negatives

Haemophilus influenzae and *Moraxella catarrhalis* are two fastidious Gram-negative organisms involved in respiratory tract infections. A goal of many working in the oxazolidinone field has been to increase the generally weak potency of the oxazolidinones against these species, for potential use in the community infection market. The oxazolidinones have poor activity against Gram-negative Enterobacteriaceae, like *Escherichia coli* or *Pseudomonas aeruginosa*. Buysse *et al.*[70] demonstrated that this is due to drug export from the cell by AcrAB transmembrane efflux pumps. It should be noted that the oxazolidinones have considerable in vitro activity against a number of other species that will not be covered in detail, including other aerobes, Gram-negative and Gram-positive anaerobes,[71] *Nocardia*,[72] actinomycetes, *Mycoplasma*, and *Chlamydia* species.[71,73,74]

7.23.4 Experimental Disease Models

There are excellent in vitro assays of antibacterial activity and in vivo efficacy models of infection available that allow for rapid cycles of SAR progression, and highly reliable predictions of efficacy in humans. The fundamental assay of antibacterial activity is the MIC, which assesses the amount of drug in $\mu g\,mL^{-1}$ required to prevent the growth of bacteria. Target-based in vitro assays are also of use to the medicinal chemist, usually in providing binding affinity information (assessed as an IC_{50} $\mu mol\,L^{-1}$ value) to assist the progression of SAR, and providing an indication of whether cell wall penetration problems may explain poor MICs. Structure-based drug design is amenable when high-resolution x-ray crystal structures of the molecular target are available with bound co-crystallized inhibitors. In the absence of that information, the research program leading to linezolid was conducted using MICs and IC_{50} data from a cell-free coupled transcription-translation assay.[5] Subsequently, various groups have examined in vitro assays for assessing safety (see below). Those compounds with good in vitro potency and pharmacokinetic (PK) properties are typically progressed into in vivo efficacy evaluations. There are many well-precedented, reliable rodent models of infection.[15] The systemic (bacteremia) and pulmonary infection models most typically involve administration of drug by oral (p.o.), subcutaneous (s.c.), or intravenous (i.v.) routes to groups of mice, each given a dosage level within an escalating range of doses, following experimental infection through inoculation in the peritoneum or nasal cavity with a lethal dose of bacteria. After several days, the number of surviving mice at each dosage level is determined, and an ED_{50} value in $mg\,kg^{-1}$ is calculated, being the effective dose required to protect 50% of the infected animals from death. Tissue infection models (mouse thigh or lung abscess, or gerbil middle ear) do not monitor survival as an endpoint, but rather reduction of colony-forming units (CFU) in harvested infected tissue, by drug treatment, relative to untreated control animals. Over many decades of research, these models have established an excellent track record of great utility in a predictive sense of the human efficacy for compounds demonstrating low ED_{50} or reductions of CFU, provided that the human exposure and pharmacokinetic–pharmacodynamic (PK/PD) parameters are similar to those for the species used in these models. A preclinical drug candidate will be extensively evaluated for in vivo activity against multiple strains of key pathogens to demonstrate activity against susceptible and resistant bacteria in support of a projected product profile.

7.23.5 Clinical Trial Issues

DuP-721 and DuP-105 were discontinued from clinical trials.[3] Eperezolid[5] and ranbezolid[9] have completed phase I clinical trials. As of this writing, no additional information concerning the current development status of ranbezolid has become available. Eperezolid was not advanced further on the basis of the superior human PK properties of linezolid. Pfizer Inc. now markets linezolid, which progressed through phase IV clinical trials and gained regulatory approval in many countries.

The oxazolidinones have excellent antibacterial activity against many of the MDR Gram-positive bacteria that cause severe, life-threatening infections. While the prevalence of susceptible Gram-positive infections has increased dramatically worldwide[75,76] (see above), there have been cases where recruitment of patients in a particular clinical trial with infections of certain resistant strains may be difficult, given the scarcity of those strains at that particular location.

During clinical trials with linezolid for VRE, there were no US Food and Drug Administration (FDA)-approved comparator drugs available that were effective against VRE. Pharmacia & Upjohn worked with the FDA to establish an acceptable clinical protocol for assessment of efficacy against VRE in a dose comparison trial. VRE-infected patients were treated with either 200 or 600 mg b.i.d. linezolid (i.v.). The demonstrated differential clinical cure rates of 73.7% and 88.6%, respectively, was sufficient to establish the effectiveness of linezolid against this serious pathogen.

Challenges for new oxazolidinone agents include demonstrating a significant advantage over the existing approved agent linezolid, while retaining most of its compelling attributes, such as the excellent oral bioavailability and efficacy, and lack of interaction with cytochrome P450 (CYP) enzymes. Targeted improvements could be once-a-day dosing, improved microbiological spectrum to include fastidious Gram-negatives, and reduced monoamine oxidase A (MAO-A) inhibition.

7.23.6 Current Treatment

7.23.6.1 Oxazolidinones and Competing Agents

Linezolid is the only member of the oxazolidinones thus far to have been approved for use in humans for the treatment of Gram-positive infections. A brief description follows of currently approved agents that compete with it for the treatment of these serious Gram-positive infections.

7.23.6.1.1 Vancomycin

Vancomycin[77] has been widely used for the treatment and surgical prophylaxis of serious Gram-positive infections, in particular for MDR infections such as MRSA. As vancomycin lacks oral bioavailability, it must be given by i.v. administration. Significant adverse effects that can be encountered and may result in discontinuation of use include rash, red-man syndrome, and nephrotoxicity.[78]

7.23.6.1.2 Teichoplanin

Teichoplanin[79] is currently the only other glycopeptide used in Europe and elsewhere, but is not approved for use in the US. Like vancomycin, it is only available for i.v. administration.

7.23.6.1.3 Quinupristin/Dalfopristin

Q/D (Synercid, King Pharmaceuticals)[80,81] is a synergistic, bactericidal combination of two modified streptogramins. Q/D was approved in the US for the treatment of VRE bacteremia with *E. faecium*, and complicated skin and skin structure infections with MSSA and *Streptococcus pyogenes*[82]; it is not effective against *E. faecalis*, and cases of superinfection with *E. faecalis* while on Q/D therapy have been reported.[83] The drug has encountered resistance (10% in Taiwan, 7% in Europe, and 1–2% in the US) in strains of *E. faecium*.[49,84] Q/D must be administered parenterally via a central venous line to avoid irritation to peripheral veins; notable are a number of significant adverse effects, including sometimes severe arthralgia and myalgia.[85,86]

7.23.6.1.4 Daptomycin

Daptomycin (Cubicin, Cubist Pharmaceuticals)[87] is a member of a bactericidal, novel class of cyclic lipopeptides, available for i.v. administration only, with a once-daily dose of $4\,mg\,kg^{-1}$. Daptomycin was discovered at Eli Lilly, and their early clinical trials were halted when toleration issues were seen with some patients dosed i.v. at $4\,mg\,kg^{-1}$ twice daily, with skeletal muscle symptoms reported.[87] Thus far, it was approved by the FDA in 2003 only for the treatment of Gram-positive infections associated with complicated skin and skin structure infections, for MSSA, MRSA, *E. faecalis* (vancomycin-susceptible only) and certain strains of β-hemolytic streptococci. Its MOA involves inserting into the bacterial cytoplasmic membrane, depolarizing the membrane potential, and shutting down adenosine triphosphate synthesis.[88,89] Development of a daptomycin-resistant MRSA strain during 27 days of therapy for MRSA bacteremia has been reported.[90]

7.23.6.1.5 Tigecycline

Tigecycline (Tygacil, Wyeth)[91] is a modified tetracycline (glycylcycline) approved in June 2005 in the US for the treatment of complicated skin and abdominal infections. Available for i.v. administration only, it has potent broad-spectrum in vitro activity against the MDR Gram-positive and Gram-negative organisms.

7.23.6.2 Origin of the Oxazolidinones - Early Structure–Activity Relationships Leading to Clinical Candidates

The oxazolidinone SAR has been extensively reviewed, covering periods of up to 1996,[2] 1999,[92] 2000,[93] and 2004.[12,13,94a,94b] The most salient SAR features leading to the clinical candidates are summarized here; more recent SAR advances are covered in the following section.

7.23.6.3 Structure–Activity Relationships Leading to DuP-721 and DuP-105

The earliest described 3-aryl-2-oxazolidinones with antibacterial activity were reported by E. I. du Pont de Nemours & Company (DuPont) in 1978.[95] Screening efforts at DuPont uncovered the lead compound, a 3-aryl 5-chloromethyl-2-oxazolidinone (6), which had activity against plant bacterial and fungal pathogens. In an account of the history of the oxazolidinones, Ranger[3] has reported that Fugitt and Luckenbaugh,[96] based on some prior work of Gregory with chloramphenicol analogs, had prepared this screening lead. Gregory[97] proceeded to explore the SAR, preparing other 5-halomethyl-3-aryl-2-oxazolidinones, some with oral activity against *Staphylococcus aureus* and *Escherichia coli* in animal models of infection. S-6123 (7) is a 5-hydroxymethyl 3-aryl-2-oxazolidinone[97] that demonstrated marginal in vitro activity, yet demonstrated good in vivo efficacy against *E. coli* in rats.[98] A series of papers by Gregory *et al.*[99–101] nicely illuminated their SAR explorations in three particular areas of the molecule that resulted in defining the optimal 5-(*S*)-acetamidomethyl-3-phenyl-2-oxazolidinone pharmacophores. This explicitly concerned the 5′-methyl substituent,[101] where acetamide was optimal, and demonstrated that only the 5-(*S*)-enantiomer has antibacterial activity; the 4-phenyl substituent,[99] where the acetyl was most active. Study of poly-substitution on the phenyl by Park *et al.*[100] showed that, while 2,4-, 2,5-, or 3,5-disubstituted aryl and 3,4,5- or 2,4,6-trisubstituted aryl analogs had poor activity at best, various 3,4-disubstituted (including annulated) phenyl compounds had good activity, provided the 3-substituent was smaller than bromine.

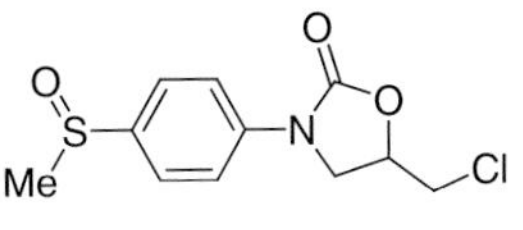

O S Me N O O Cl

6

O O=S H₂N N O O OH

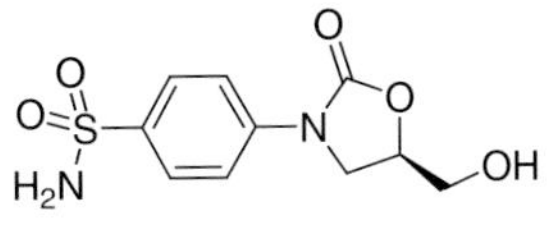

7

The DuPont work culminated with the identification of DuP-721 and DuP-105, which were disclosed in 1987.[10] DuP-721 and DuP-105 demonstrated good oral absorption in the rat,[101,102] and potent in vitro activity against important Gram-positive bacteria, anaerobes, and *M. tuberculosis*. Of the two, DuP-721 was about 4–16-fold more potent, with MICs against 90% of the tested strains (MIC_{90}s) of $\leq 4\,\mu g\,mL^{-1}$ versus important staphylococci, enterococci, and streptococci.[103] In animal models of infection with Gram-positive pathogens (including MRSA), DuP-721 was essentially equipotent with vancomycin. As synthetic compounds, these oxazolidinones presumably would not encounter preexisting resistance, of the manner associated with natural products, where resistance genes from the producing microorganism have been shared with other species. Both compounds were reported to have entered phase I clinical trials[104,105]; however, no data have been published from those clinical studies. Shortly after the 1987 presentation of these compounds, DuPont terminated its oxazolidinone program,[106a,106b] reportedly due to observation of liver toxicity in rats treated with DuP-721.[3]

7.23.6.4 Structure–Activity Relationships Leading to Eperezolid and Linezolid

Following the disclosure of DuPont's oxazolidinone clinical candidates, the author initiated a medicinal chemistry exploratory project on the oxazolidinones in late 1987 at what was then the Upjohn Company. The laboratory's initial focus turned to the examination of various rigidified compounds designed as restricted rotation analogs of DuP-721. To expedite the progress of this small burgeoning project,[2] a decision was made to work initially in the racemic series. Five- and six-membered aryl cyclic ketone analogs of DuP-721 proved to have interesting activity, and one compound would come to play a pivotal role in the Upjohn program. The indanone (±)-U-82965 (**8**, subsequently reported by DuPont)[100,107] and the tetralone (±)-U-85055 (**9**)[108] both demonstrated in vitro and in vivo activity on a par with (±)-DuP-721. (±)-U-82965 and (±)-DuP-721 were evaluated and shown to have similar PK profiles in the rat.[2] Substitution of a methyl group α to the U-82965 indanone carbonyl resulted in two- to fourfold enhancement of potency, giving a compound with comparable in vivo activity to (±)-DuP-721.[108]

In response to the receipt of fragmentary information that DuPont had terminated its oxazolidinone program apparently due to toxicology findings,[2,3] a side-by-side comparative toxicology study of (±)-DuP-721 and (±)-U-82965 was conducted by R. C. Piper *et al.* (unpublished data) at Upjohn. The protocol involved twice-daily oral dosing of three rats per sex at $100\,mg\,kg^{-1}$, for 30 days. (±)-DuP-721 demonstrated lethal toxicity in one rat, and two rats were sacrificed in a moribund state; bone marrow atrophy and severe progressive weight loss were observed.[15] In contrast, (±)-U-82965 was well tolerated, with no signs of drug-related toxicity.[2] Based on the close structural similarity of these compounds, the differentiated toleration profile of U-82965 played a very critical role in the decision to progress the Upjohn program, by demonstrating the existence of a structure–toxicity relationship (STR) for the oxazolidinones.

7.23.6.5 5′-Indolinyl Oxazolidinones

Supporting the decision to advance the Upjohn program based on the outcome with U-82965 were results from another novel series that substantially added to the definition of the STR, the 5′-indolinyl oxazolidinones. The goal in preparing the 5′-indolines was to explore the benefit of replacing the U-82965 ketone with an amide moiety.[109] The *N*-hydroxyacetyl-5′-indoline oxazolidinone (±)-U-85112 ((±)-**10**) demonstrated good potency and oral efficacy, but was slightly less active than (±)-DuP-721 and vancomycin (dosed s.c.). (±)-U-85112 proved to have an excellent safety profile in 30-day rat toxicology studies, being one of the best tolerated oxazolidinones examined.[15] Subsequently the active 5-(*S*)-enantiomer U-97456 (**10**, PNU-97546) was prepared, and it was also very well tolerated in the 30-day protocol in the rat, dosed at $50\,mg\,kg^{-1}$ (p.o., b.i.d). As would later be shown by Ciske *et al.*,[110] similar to the earlier result with α-Me substitution on U-82965, addition of an (*R*)-methyl group α to the indoline nitrogen in U-97456 also provided a twofold potency increase.

The toxicology findings with U-97456 were the first to establish clearly the utility, from a toleration standpoint, of the substitution of a nitrogen atom directly attached to the *para* position of the phenyl oxazolidinone. The indanone, 5′-indoline and other toleration studies provided a strong foundation for the Upjohn oxazolidinone progression strategy, which became based on the early toxicological examination of each new promising lead series. This unconventional strategy allowed the Upjohn team to establish and use SAR and STR to advance the program to drug candidate identification. Others have reviewed the toxicological aspects of the oxazolidinones in more detail.[13,111,112]

7.23.6.6 Troponyl Oxazolidinones

A potent series extensively explored by Barbachyn *et al.*[113] was the 3-(4-troponyl-phenyl)-2-oxazolidinones, where fluorinated phenyl compound (**11**) demonstrated MICs $\leq 1\,\mu g\,mL^{-1}$ for a panel of Gram-positives and oral in vivo activity against MSSA comparable to vancomycin; however, low aqueous solubility and poor PK properties proved to be undesirable characteristics for the advancement of this series.[11] Nonetheless, this series established the utility of a 3-fluorine substitution on the phenyl ring in increasing potency.

7.23.6.7 3-(4-Piperazinyl-3-fluorophenyl)-2-Oxazolidinones

Two significant advances in the Upjohn SAR[5] were the identification by Hutchinson *et al.*[2,114] of the piperazine subclass, and the finding noted above for the tropones, by Barbachyn *et al.*,[2,11] that substituting one or two fluorine atoms on the phenyl positions flanking the piperazine increased potency about twofold. The latter finding was similar to SAR in the fluoroquinolone class of antibiotics, where fluorine substitution brought about enhanced potency and improvement of PK properties.[115] In contrast, Brittelli and co-workers[100] reported that, for DuP-721, fluorine substitution at the same 3-phenyl position had reduced potency twofold.

The decision to install the piperazine ring had been influenced by the demonstrated advantage of a nitrogen atom substituted at the 4-phenyl position, as noted with the 5′-indoline (U-97456), and recognition, again from the fluoroquinolone class, that installation of a piperazine on that template provided beneficial properties, as exemplified by ciprofloxacin (**12**). Weighing the individual attributes of each of these lead series, a decision was made to focus the search for a drug candidate within the piperazinyl oxazolidinones.[5,15]

7.23.6.8 Eperezolid (U-100592)

Following that decision, the Upjohn team aggressively explored a wide variety of substituents on the piperazine distal nitrogen,[2,11,114,116a,116b] and eventually found the hydroxyacetyl moiety to be the most optimal. The hydroxyacetyl of U-100592 was the same optimal nitrogen substituent previously identified with U-97456 in the 5′-indolines. It is notable that others have subsequently reported additional oxazolidinone series wherein the hydroxyacetyl nitrogen substituent has also been found to afford optimal activity.[117–120] Considerable exploration of the C-5 side-chain SAR beyond the breadth of the prior study[101] was examined by the Upjohn group in the piperazinyl series.[121] Ultimately, while more potent amides were identified in this series (e.g., the dichloroacetamidomethyl was two- to fourfold more potent in vitro than the acetamidomethyl), the overall optimal properties of the 5-acetamidomethyl side chain, as identified by DuPont,[101] also proved to be advantageous for the two Upjohn drug candidates.

On the basis of the following attributes, eperezolid was selected as the first drug candidate for clinical evaluation by Upjohn[5]: (1) its potent in vitro antibacterial activity (see below)[122]; (2) its excellent oral and parenteral efficacy in numerous in vivo mouse models of Gram-positive infection, including MRSA and *E. faecalis*, with ED_{50}s essentially equivalent to vancomycin (s.c.); (3) its excellent efficacy in mice versus vancomycin-resistant *E. faecium*, with an ED_{50} of $12.5\,mg\,kg^{-1}$;[15,123] (4) its excellent aqueous solubility ($4.2\,mg\,mL^{-1}$); and (5) its good PK properties in the rat and dog.[2,126] Additionally, eperezolid was very well tolerated in 30-day toxicology studies in the rat and dog, dosed orally, with the no-observed-adverse-effect-level (NOAEL) of $25\,mg\,kg^{-1}day^{-1}$, and only minimal adverse effects at $80\,mg\,kg^{-1}day^{-1}$ for both species.[124]

7.23.6.9 Linezolid (Zyvox, U-100766, PNU-100766)

Examination of morpholine as a bioisostere for the piperazine ring provided linezolid, a compound that proved to have exceptional PK properties in animal models[125] and in humans, compared to eperezolid. In the rat oral PK studies at $25\,mg\,kg^{-1}$, significant differences were noted in the profile of linezolid and eperezolid, where oral bioavailabilities were 109% and 56%, and area under the curve (AUC) values were 42.6 and $9.2\,\mu g\,mL^{-1}\,h^{-1}$, respectively.[126] Linezolid undergoes renal tubular reabsorption.[125] In the dog PK studies, at $25\,mg\,kg^{-1}$, linezolid and eperezolid had oral bioavailabilities of 97%[125] and 100%,[126] respectively. As found with the piperazinyl and troponyl oxazolidinone series, substitution of a 3-F on the phenyl group increased the in vitro activity by twofold over the des-fluorophenyl morpholine congener.[127] Linezolid was only slightly less potent in vitro than eperezolid (see below),[122] but, like eperezolid, compared very favorably with vancomycin in demonstrating excellent efficacy in multiple mouse models of Gram-positive infection, including PRSP, MRSE, MRSA, and *E. faecalis*[123]; while in a mouse model of vancomycin-resistant *E. faecium* infection, linezolid demonstrated an ED_{50} of $24\,mg\,kg^{-1}$.[123] Linezolid displayed excellent aqueous solubility ($3.7\,mg\,mL^{-1}$), which greatly facilitated the i.v. formulation development. Linezolid was very well tolerated in 30-day toxicology studies in the rat and dog, dosed orally, with a NOAEL of $20\,mg\,kg^{-1}\,day^{-1}$ found for both species; additionally, doses of 50 and $40\,mg\,kg^{-1}\,day^{-1}$ were well tolerated, with only mild adverse effects for the rat and dog, respectively.[128] On the basis of these data, linezolid was selected as the second drug candidate for clinical evaluation by Upjohn.[5]

7.23.6.10 Thiomorpholine U-100480 (PNU-100480)

Contemporaneous with the identification of linezolid was another compound of interest, U-100480 (**13**), that resulted from substitution of thiomorpholine for the piperazinyl ring. This compound demonstrated very potent antimycobacterial activity,[129] having excellent in vitro activity against several MDR-TB strains, and was found as active as isoniazid in an in vivo model of *M. tuberculosis* infection.[130] U-100480 was very well tolerated in a 30-day rat toxicology study at $50\,mg\,kg^{-1}$ (p.o., b.i.d). The compound undergoes metabolism in the rat to two metabolites that also possessed good antimycobacterial activity – the corresponding sulfoxide and sulfone.[129]

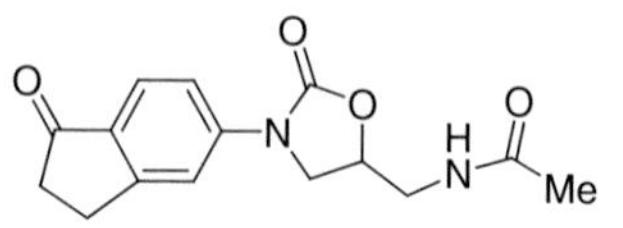

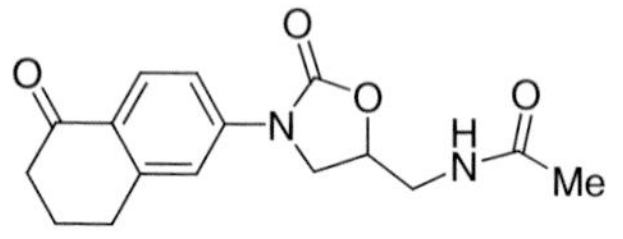

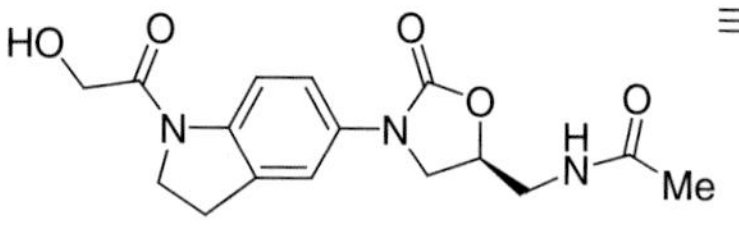

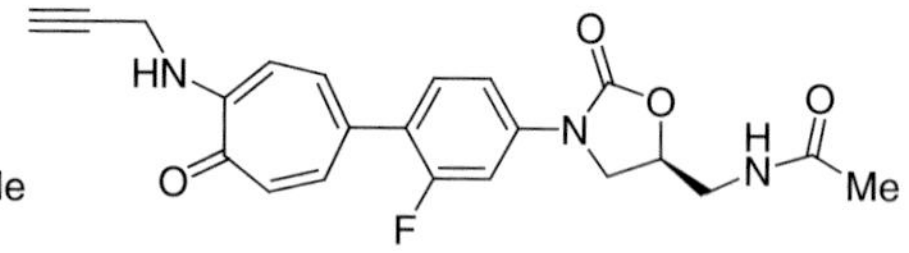

7.23.6.11 Tricyclic-Fused Oxazolidinone Analogs of DuP-721 and Linezolid

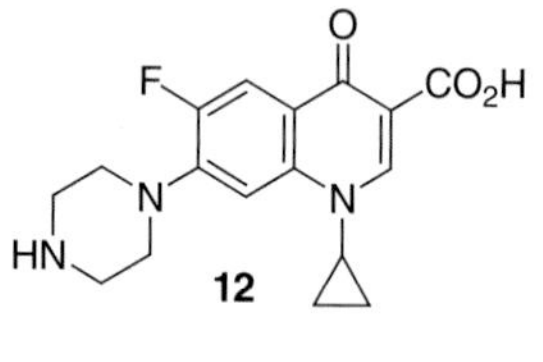

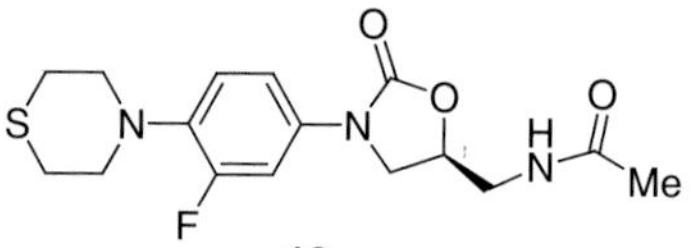

An area of early interest at Upjohn was to examine the effect of restricted rotation about the aryl oxazolidinone C–N bond with rigid tricyclic-fused oxazolidinones (**14**), wherein a one- or two-carbon bridge was installed between the *ortho*-phenyl and C-4 oxazolidinone positions.[2,127] Only those [6,5,5]-tricyclic-fused oxazolidinones (**14a**)[127] having a *trans* relationship for the C-4 and C-5 substituents proved active. Contrary to computational considerations predicting superior activity of [6,6,5]-tricyclic analogs (**14b**),[131] on the basis of torsional angles between the phenyl and oxazolidinone rings being most similar to those in DuP-721, rather it was the [6,5,5]-tricyclics that were found to be significantly more active than the [6,6,5] series.[2,127] For the purpose of illustrating the full range of the STR understandings, racemic pyridyl tricyclic (**15**), a highly potent in vivo orally active pyridyl-substituted [6,5,5]-tricyclic with in vivo activity equal to vancomycin,[2] was only poorly tolerated in the rat toxicology study.

An enantioselective (98.8% ee) synthesis[132] provided the chiral [6,5,5]-tricyclic analog of DuP-721 (**16**), which had MICs half those of DuP-721. The racemic des-fluoro [6,5,5]-tricyclic version of linezolid (**17**) was most notable for its poor in vitro activity.[127] Rajagopalan and co-workers[133] prepared a different series of conformationally constrained analogs of linezolid, where the morpholine ring was fused on to the 3-fluorophenyl ring at the 4- and 5-positions, and the nitrogen at the 4-position was also encompassed by a fused pyrrolidine ring. For compound **18** with the 5-acetamidomethyl group, near identical in vitro potency with linezolid was observed, while structure **19** with the corresponding methyl thiocarbamate side chain had twofold better in vitro activity, and was equipotent with linezolid in vivo.

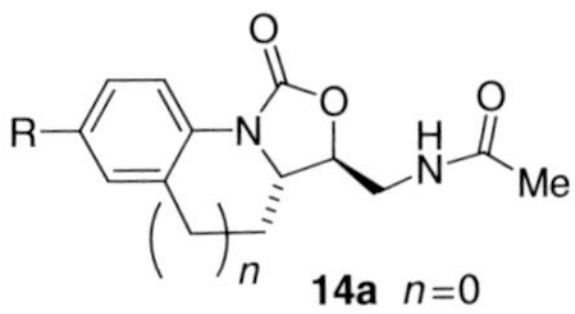

14a $n=0$
14b $n=1$

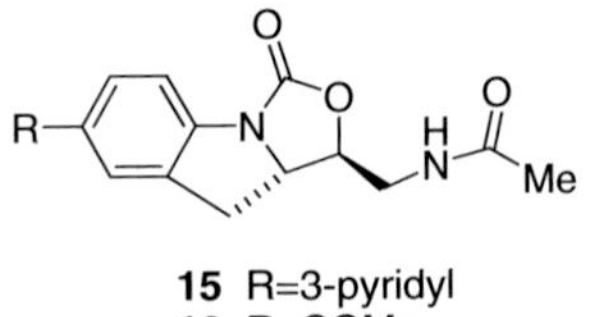

15 R=3-pyridyl
16 R=COMe
17 R=morpholinyl

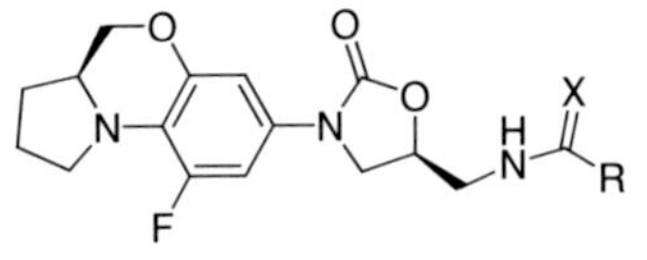

18 X=O R=Me
19 X=S R=OMe

7.23.6.12 Ranbezolid (RBx-7644)

Rattan[9] from Ranbaxy Research Labs reported in 2003 that ranbezolid completed phase I clinical trials conducted in the UK. The ranbezolid structure differs from eperezolid in that eperezolid's hydroxyacetyl group is replaced by a 2-(5-nitrofuranyl)methyl moiety, an appendage found in an older class of antibiotics, the nitrofurans. That class of nitrofuran antibiotics is thought to function by in vivo reduction of the nitro group to reactive intermediates, which cause DNA damage to the bacterium.[134–136] In the rat, oral bioavailability of ranbezolid was 30%.[9] In rat and dog i.v. PK studies, ranbezolid demonstrated an absolute bioavailability of 107% and 66%, and volumes of distribution of 2.3 and 3.9 $L\,kg^{-1}$, respectively. Ranbezolid was well tolerated in 28-day rat and dog b.i.d. toxicology studies, with a NOAEL of 50 and 40 $mg\,kg^{-1}$, respectively. The compound was reported to have no genetoxicity in in vitro and in vivo assays.[9,137]

7.23.6.13 Recent Structure–Activity Relationship Advances

Given the large volume of recent research activity in the oxazolidinone field, only those SAR advances providing highly potent activity and deemed as most significant extensions of the prior established SAR (leading up to linezolid) are covered in the following section. Several reviews have been published detailing the oxazolidinone SAR advances in a more comprehensive fashion.[11–13]

Dong-A Pharmaceuticals reported on preclinical studies with DA-7867 (**20**),[138] where an *N*-methyltetrazole was substituted on a pyridyl-3-fluorophenyl oxazolidinone. DA-7867 is highly potent in vitro, with approximately four- to eightfold improved MIC_{90}s versus linezolid for MSSA and MRSA, enterococci (including VRE), and *M. tuberculosis*, and significantly improved activity against fastidious Gram-negatives and certain anaerobic organisms.[138] DA-7867 demonstrated improved efficacy versus MRSA in a mouse model of infection versus linezolid ($ED_{50}=2.6$ versus 8.3 $mg\,kg^{-1}$).[140] In the rat, DA-7867 had an oral bioavailability of 71%, and demonstrated a long oral half-life (12.4 h). With a postantibiotic effect 1.5 times longer than linezolid, the potential for once-daily dosing was projected.[141]

Genin *et al.*[142] at Pharmacia described a series of various N–C-linked azolylphenyl oxazolidinones, where the morpholine of linezolid was replaced with heterocyclic amines; the 3-cyanopyrrole (**21**) demonstrated *c.* fourfold improved in vitro activity over linezolid, against the fastidious Gram-negative and Gram-positive pathogens, and excellent oral in vivo activity versus *S. aureus* ($ED_{50}=1.9\,mg\,kg^{-1}$) versus eperezolid ($ED_{50}=4.0\,mg\,kg^{-1}$).

7.23.6.13.1 5-Acetamidomethyl replacements

Gravestock *et al.*[120] at AstraZeneca highlighted AZD-2563 (**22**) as a potential clinical candidate; however, subsequent information from the company stated that work with AZD-2563 had been discontinued.[143] AZD-2563 appeared to represent a very significant SAR breakthrough in the replacement of the 5-(*S*)-methyl's acetamido side chain with an amide isostere – an ether-linked isoxazole, which was the most optimal of a series of O-linked heterocycles.[120] AZD-2563 versus linezolid demonstrated the same or twofold better in vitro activity for Gram-positive aerobes[144] and equipotent activity against anaerobes.[145] The compound has PK properties such that once-a-day dosing in humans had been projected.[146] Additional acetamide replacements with good in vitro potency identified by AstraZeneca were the ether-linked thiadiazoles, aminothiadiazoles, and the 1,2,3-triazoles (**23**); with some compounds, there was an eight- to 10-fold improvement in MICs versus linezolid for the Gram-positives. The 1,2,3-triazoles (**23a–d**) had MICs of 2 $\mu g\,mL^{-1}$ versus *H. influenzae*, and (**23b, c**) demonstrated a three- or fourfold reduction in inhibitory activity versus MAO-A compared with the acetamide; for structure **23d**, a more than 30-fold reduction was seen.[147] Phillips *et al.*[148] from Kuwait University also reported on a triazole analog of linezolid, PH-027 (**24**), which had MICs equivalent to or twofold better than linezolid against Gram-positive pathogens.

Groups at Bayer[149] and Pharmacia[150] independently were the first to establish that substituting a thiocarbonyl on the oxazolidinone 5-methylamine in place of the acetamide leads to very significant improvements in in vitro potency, a finding that has been widely exploited by many other researchers. The SAR of this group has been reviewed[13,151]; however, it does not appear that any of these compounds has as yet progressed to the clinic. In some reported cases, in vivo efficacy of the thioacetamide at a given dose range was poor or not observed, when the corresponding acetamide congener gave good efficacy when dosed in the same range.[152,153] Thomas *et al.*[154,155] at Pfizer, in collaboration with Vicuron researchers, reported the interesting series of oxazolidinones, butenolides, and isoxazolinones with a reversed carboxamide side chain directly attached at the oxazolidinone C-5 position. Dihydropyrrolopyridine (**25**) demonstrated an MIC versus *S. aureus* of 0.5 $\mu g\,mL^{-1}$, while for the compound (**26**) corresponding to linezolid, the MIC was 4 $\mu g\,mL^{-1}$. This compound was obtained by alkylation of lithiated carbamate (**39**) with potassium (2*R*)-glycidate, followed by amidation of the carboxylic acid.

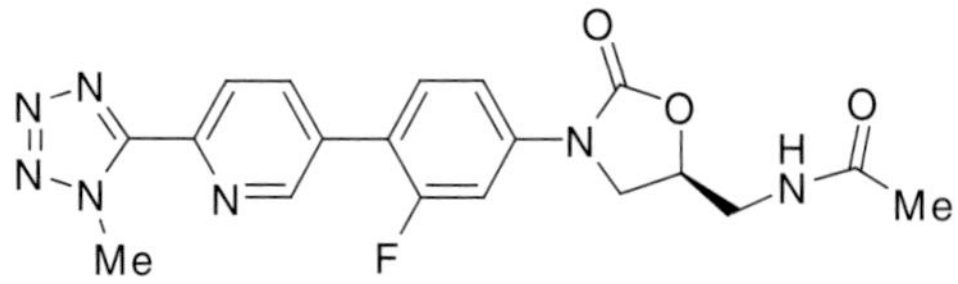

20

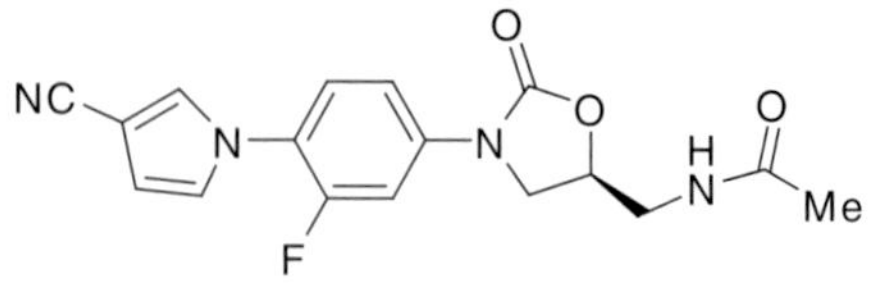

21

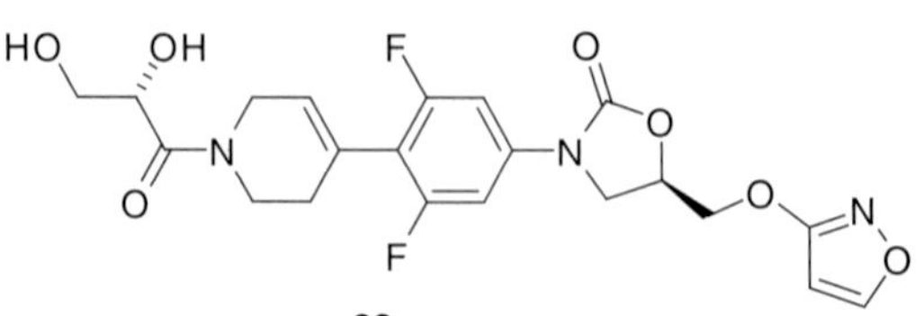

22

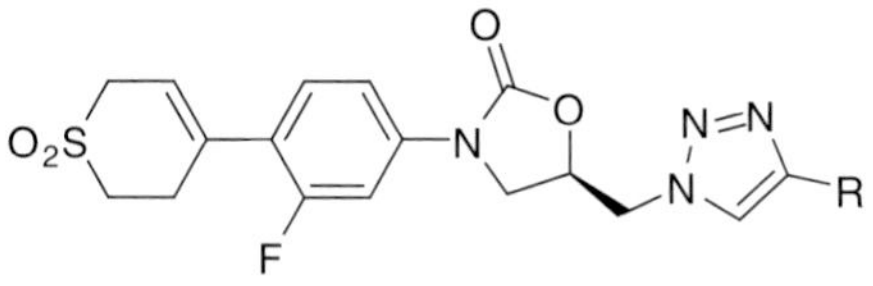

23a R= H
23b R= Me
23c R= Br
23d R= CCH

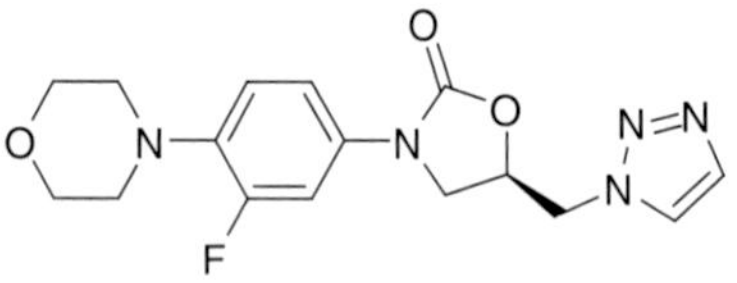

24

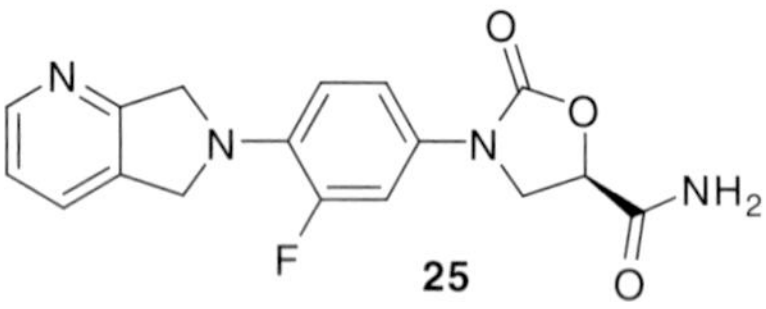

25

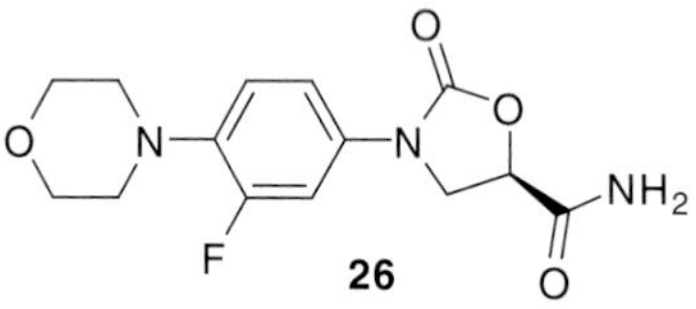

26

Paget *et al.*[156] at Johnson & Johnson also reported that replacement of the morpholine of linezolid with a dihydropyrrolopyridine gave the 5-acetamidomethyl compound (**27**) with fourfold improved MICs over linezolid, including against linezolid-resistant *S. aureus*, and comparable oral in vivo efficacy versus *S. aureus*. Thomasco *et al.* at Pharmacia[153] described a series of 1,3,4-thiadiazole fluorophenyl oxazolidinones. The methylthiadiazole analog (**28**) demonstrated MICs of 2 $\mu g\,mL^{-1}$ against the fastidious Gram-negatives, and two- to 10-fold improvement over linezolid against the Gram-positives. The compound had an oral ED_{50} twofold higher than that of linezolid in a mouse model of MSSA infection. Tucker *et al.*[116a,116b] at Pharmacia & Upjohn found that a cyanopyridine (e.g., **29**), a pyrimidine, or a pyridazene on the distal piperazine nitrogen demonstrated in vitro potency and in vivo efficacy against *S. aureus* that was comparable to linezolid.

Sciotti *et al.*[157] at Abbott described a series of halostilbene and heterostilbene oxazolidinones. Compound **30** demonstrated 65% oral bioavailability in the mouse, MICs equal or twofold better than linezolid, and a fivefold relative reduction in inhibition of MAO-B. Lohray *et al.*[158] reported on replacements of the hydroxyacetyl of eperezolid with a series of substituted cinnamoyl groups, the most potent being the *para*-hydroxyl **31a**, which was generally twofold more active in vitro than linezolid. Substitution of the acetamide of **31a** with the thioacetamide **31b** further increased in vitro activity generally fourfold.

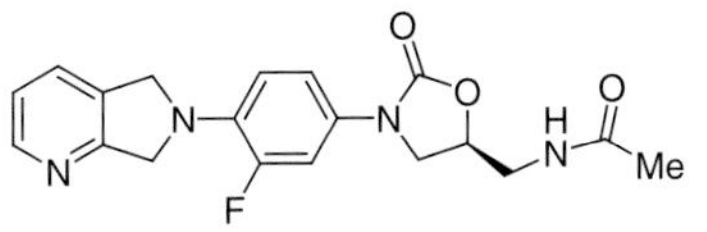

27

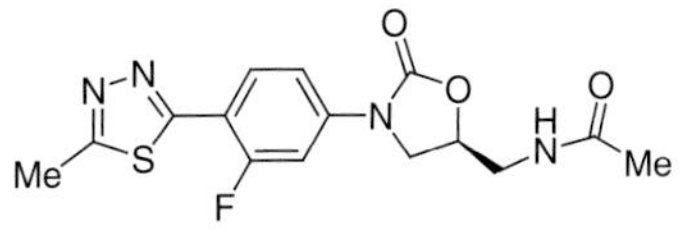

28

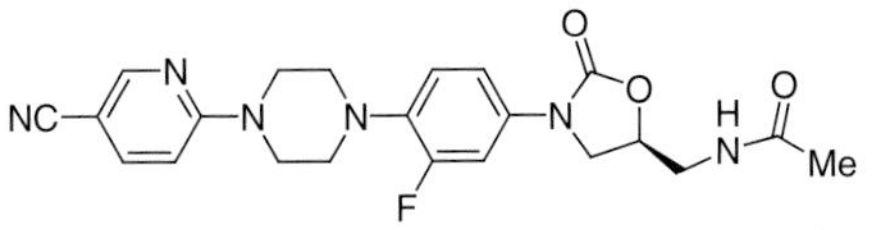

29

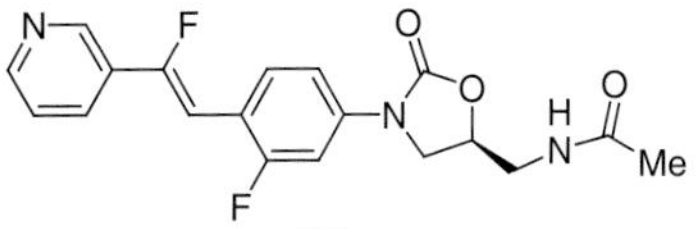

30

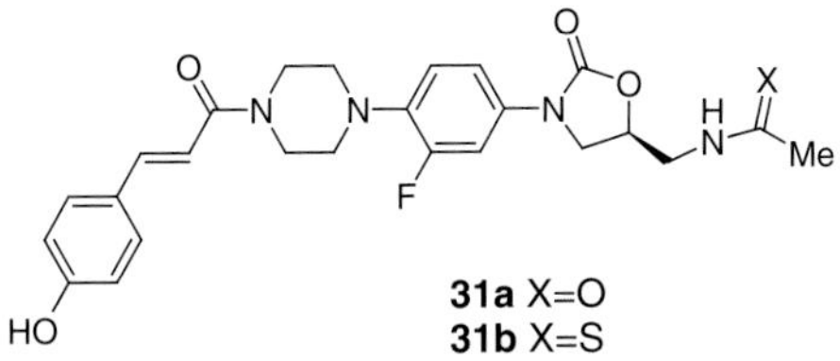

31a X=O
31b X=S

7.23.6.13.2 Oxazolidinone core replacements

Following the early work on replacement of the oxazolidinone ring of DuP-721 with a butenolide ring (**32**) as reported by Denis and Villette,[2,159] other groups exploring this surrogate ring have also found butenolides with good in vitro activity, by replacing the phenyl's acetyl moiety of **32** with either a troponyl,[160] 2-cyano-1-methylethenyl,[161] or a 4-thiomorpholinyl-3-fluoro[162] moiety. Barbachyn *et al.*[118] at Pharmacia were the first to report on a series of phenylisoxazolines with the 5-(*R*)-acetamidomethyl side chain. The compound corresponding to eperezolid with an additional fluorine (**33**) demonstrated in vitro and in vivo activity comparable to linezolid, while the morpholinyl congener directly corresponding to linezolid was about twofold less active in vitro and in vivo. Another oxazolidinone replacement identified at Bristol-Myers Squibb group[163] is the isoxazolinone ring. This finding is interesting in that there is no chiral center in these compounds at what was C-5 of the oxazolidinone. The most potent compound was **34**, with in vitro activity comparable to linezolid, but in the presence of 50% calf serum, the activity was reduced 4–16-fold due to a substantial protein-binding effect.

7.23.6.13.3 Dual-action agents

Various oxazolidinone hybrids in conjunction with other antibiotics have been explored as a means of acquiring a broadened spectrum of activity, including Gram-negatives, and increased potency. Agents that have been chemically linked with oxazolidinones include streptogramins, erythromycin, and a lincosamide (Theravance),[164] azithromycin (Rib-X Pharmaceuticals),[165] and aminoglycoside neomycin (Korea Institute of Science and Technology).[166] Several groups have filed patents claiming a compound (**35**), wherein the fluoroquinolone ciprofloxacin (X=CH) is linked through a shared piperazine ring with the core found in eperezolid.[167–169] This hybrid compound demonstrated very potent activity against the Gram-positive strains of interest, good in vitro activity against the fastidious Gram-negatives, but only weak activity versus the Gram-negatives aerobes. Good to modest activity (MICs of 4 and 8 μg mL^{-1}) against linezolid-resistant *E. faecalis* and linezolid-resistant *S. aureus* was obtained, respectively.[170] A variety of other nonaromatic heterocycles have been investigated by Hubschwerlen *et al.*[168] at Morphochem as spacers linking these two antibacterial templates. The best compound (**36**) showed very potent activity, and demonstrated in general a 10-fold improvement in MICs versus linezolid, and bactericidal activity. Gordeev *et al.*[167] at Vicuron in collaboration with Pharmacia researchers also described a biaryl oxazolidinone–quinolone hybrid (i.e., without a spacer ring), which displayed an antibacterial spectrum more characteristic of a quinolone.

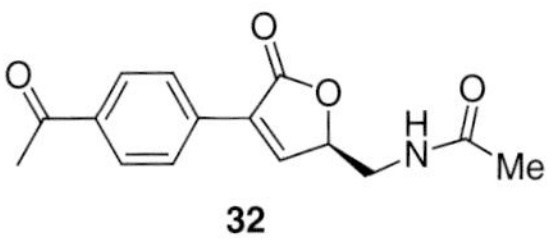

32

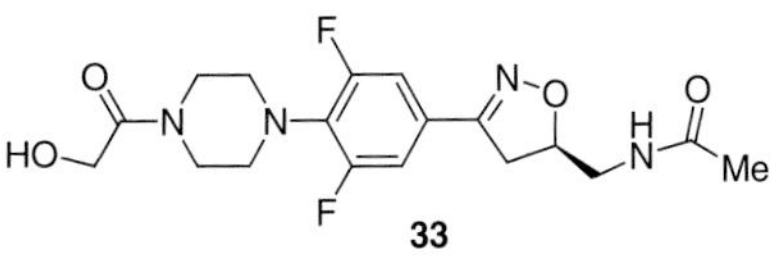

33

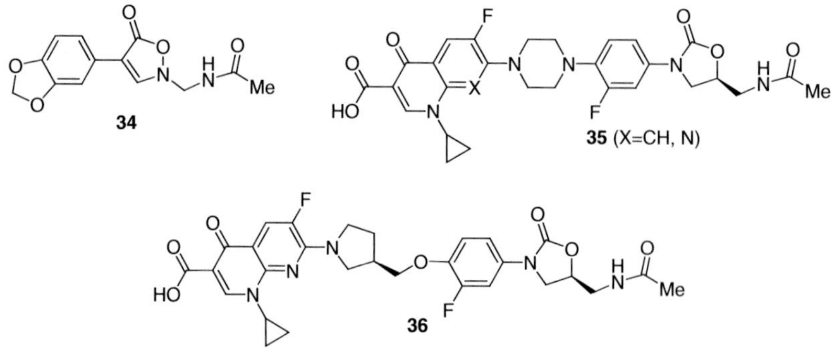

34

35 (X=CH, N)

36

7.23.6.14 Synthesis of Enantiomerically Enriched Oxazolidinones

Critical to the success of an oxazolidinone preclinical and clinical development program was the identification of viable and efficient synthetic routes to enantiomerically enriched oxazolidinones. Early preparation of 5-(*S*)-acetamido-methyl-2-oxazolidinones at DuPont relied on a chiral resolution.[101] Wang *et al.*[171] subsequently employed the isocyanate-epoxide cyclization[172,173] using (*R*)-glycidyl butyrate (**37**), providing ester **38**, and after saponification, the requisite 5-(*R*)-hydroxymethyl alcohol intermediate to DuP-721.[100] The narrow scope of commercially available isocyanates can be a limitation on the breadth of diversity of oxazolidinone templates readily accessible by this method, primarily as synthesis of isocyanates from anilines typically employs phosgene, which in the discovery laboratory setting introduces considerable safety concerns.

An alternate approach developed at Upjohn (referred to as the Manninen reaction)[5,174,175] relies on alkylation of a readily derived *N*-lithiated aryl carbamate (derived from **39**) with **37**, to give directly the key (*R*)-5-hydroxymethyl-3-aryl-2-oxazolidinone intermediate (**40**) in high yield and extremely high % ee. The lithium counter ion was demonstrated to be critical to the success of this reaction, as, with use of sodium or potassium bases, significant untoward side reactions are encountered.[174] The Manninen reaction was carried out on a 100-kg scale for synthesis of early clinical bulk supplies at Upjohn. A variation was subsequently developed at Pharmacia for use in a large-scale, concise synthesis of linezolid[176,177] that employs chloropropane diol (**41**)[11] or 1-chloro-2-acetoxy-3-acetamidopropane (**42**) in place of the epoxide **37**. Of particular note is the highly convergent, one-step synthesis of linezolid from **39** and **42**.[177] Another recently reported approach by Madar *et al.*[178] at Abbott utilizes a Buchwald/Hartwig palladium-catalyzed coupling of aryl bromides with 3-unsubstituted 5-(*S*)-oxazolidinones, and was applied to the synthesis of DuP-721.

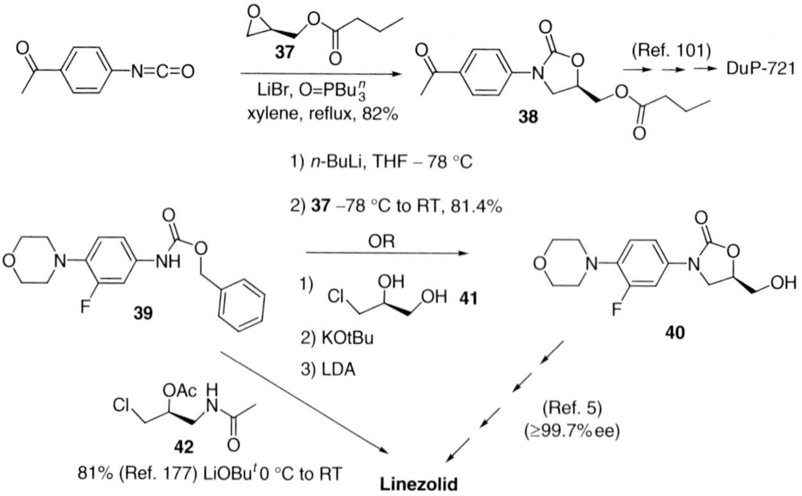

N=C=O

37

LiBr, O=PBu$^n_3$

xylene, reflux, 82%

38

(Ref. 101)

DuP-721

1) *n*-BuLi, THF – 78 °C

2) **37** –78 °C to RT, 81.4%

OR

1) Cl OH OH **41**

2) KOtBu

3) LDA

39

40

42

81% (Ref. 177) LiOBu^t 0 °C to RT

(Ref. 5)

(≥99.7% ee)

Linezolid

7.23.6.15 Extent of Use

Following priority review, the FDA approved linezolid on 17 April, 2000 for an initial set of indications included below; subsequently additional indications were approved, and those that are currently approved in the US are the following. Linezolid is approved for the treatment of hospital- and community-acquired pneumonia caused by *Staphylococcus aureus*

(MSSA or MRSA) or *Streptococcus pneumoniae* (penicillin-susceptible or MDR strains), and VRE *E. faecium* (including concurrent bacteremias). Linezolid is the first oral agent ever approved for the treatment of VRE infections,[179,180] and is the only approved therapy for hospital-acquired MDR *S. pneumoniae*. Linezolid is also approved for the treatment of complicated skin and skin structure infections, including diabetic foot infections without concomitant osteomyelitis, which are caused by MSSA and MRSA, *S. pyogenes*, or *S. agalactiae*. Linezolid has been approved for use in children and newborns against Gram-positive infections.

The prescribed duration of linezolid therapy for complicated skin and skin structure infections is typically 7–14 days; for VRE, 14–28 days; for nosocomial pneumonia, 7–21 days; and for diabetic foot infections, 7–28 days.[181] The typical dose for adults is 600 mg q12 h, except for uncomplicated skin and skin structure infections, for which the recommended dose is 400 mg q12 h. The availability of i.v., oral tablets and oral suspension formulations, coupled with the 100% oral bioavailability of linezolid, has advantages over competing agents in allowing for easy switch from i.v. to oral therapy. The cost benefits of this aspect, in allowing earlier discharge of patients from hospital to continue therapy with the oral form, have been noted.[78]

Details of the linezolid phase III clinical trials that supported the approved indications have been extensively reviewed elsewhere[15,3,74,182]; only a brief summary is presented here. For nosocomial Gram-positive pneumonia, cure rates for linezolid versus vancomycin were 66.4% and 68.1%, respectively.[183] For community-acquired pneumonia versus either i.v. ceftriaxone or oral cefpodoxime proxetil, clinical cure rates were 83.0% (linezolid) versus 76.4% (for the cephalosporin comparators), and for the patients with bacteremia with *S. pneumoniae*, the respective clinical cure rates were 93.1% versus 68.2%.[184] For complicated skin and skin structure infections, respective clinical cure rates for linezolid versus oxacillin followed by dicloxacin were 88.6% and 85.8%.[185]

Schentag and co-workers[186] reported results of the linezolid compassionate-use program in treating MDR Gram-positive infections, where they found linezolid gave high rates of clinical cure in the study set of patients with complicated infections (46% were bacteremias, 11% endocarditis, and 31% line-related infections). While at this writing, linezolid has not been approved for treating endocarditis, there have been additional reports of successful linezolid treatment of infective endocarditis, including one case with VRE.[187,188]

7.23.6.16 Limitations: Adverse Effects

Linezolid is considered to be well tolerated.[78] The most common side effects observed in seven phase III clinical trials with linezolid were (% incidence) diarrhea (2.8–11%), nausea (3.4–9.6%), and headache (0.5–11.3%).[181] There is an association of reversible myelosuppression (anemia, leucopenia, or pancytopenia) with longer-term linezolid usage, particularly for courses of therapy over 2 weeks[181]; weekly monitoring of complete blood count status is recommended. Thrombocytopenia (defined as a decrease in platelet count below 75% of normal) was observed in 2.4% of linezolid-treated patients in phase III clinical trials and 1.5% of those treated with comparator compounds. In the compassionate-use basis trials, tolerance of linezolid was deemed to be very good, with 7.4% of the patients experiencing thrombocytopenia.[186] Reports based on postapproval marketing experience have shown higher rates of patients with reduction in platelet counts.[189a,189b] Lactic acidosis has been reported in some patients treated with linezolid.[190,191] There have been reports of neuropathies, both peripheral and optic,[192,193] primarily with patients treated with linezolid well beyond the recommended 28 days; the causal relationship with drug treatment has not been established.[194] Pharmacia & Upjohn researchers reported on the utility of vitamin B (B_2, B_6, B_{12}, or folic acid) co-therapy in reducing side effects with oxazolidinones, including neuropathy and hematological effects.[195] Spellberg *et al.*[196] reported that vitamin B_6 treatment reversed cytopenia seen in two patients on linezolid treatment, and concluded that this treatment may have utility in preventing or modifying the course of cytopenias associated with long-term linezolid therapy.

7.23.6.17 Sites and Mechanism of Action

7.23.6.17.1 Overview

The oxazolidinones have a unique MOA, interfering with the bacterial protein synthesis initiation process, versus elongation steps (as do chloramphenicol, erythromycin, and clindamycin), or the termination process.[197] For an excellent discussion of steps in the initiation process, refer to Bozdogan and Appelbaum.[18] As a consequence, there is no cross-resistance of the oxazolidinones with other classes of antibiotics.[198] There has been a considerable amount of continuing research effort to understand more detailed aspects of the mechanism, which is still not entirely elucidated. An early report concluded that the action of DuP-721 was due to inhibition of a step preceding the interaction of the 30S ribosome subunit and *N*-formylmethionyl-tRNA (fMet-tRNA) with the initiator codon,[199] such as synthesis of fMet-tRNA; however, work by Shinabarger *et al.*[200] with linezolid and eperezolid ruled out this mechanism. Stockman and co-workers[201] at Pharmacia studied the binding of eperezolid to *E. coli* bacterial ribosomes by ^{1}H nuclear magnetic resonance transferred nuclear Overhauser enhancement, and demonstrated that it bound only to the 50S subunit, not the 30S.

Some of the strongest evidence supporting the current MOA understanding stems from studies at Pharmacia by Colca *et al.*,[202] where a radioactive photo affinity probe attached to the eperezolid hydroxyacetyl moiety (**43**) was used to map the site of action. Photolysis in actively growing *Staphylococcus aureus* cells initiated cross-linking of compound **43** to the universally conserved nucleotide at position A-2602 (*E. coli* numbering) in the peptidyl transferase center of 23S rRNA; it also cross-linked to tRNA, ribosomal protein L27, and LepA (although this was not considered to be a direct target). These results suggest that the oxazolidinone-binding site is in the vicinity of the ribosomal peptidyl transferase center near the P site, which is generally consistent with the studies examining sites of oxazolidinone resistance mutations in various bacteria, that cluster about the peptidyl transferase catalytic site, in the central loop of domain V of 23S rRNA.[203] Böttger and co-workers[204] have reported that the spectrum of the resistance mutations for linezolid exhibits a strong bias for species specificity, which they characterized as a unique aspect of this drug.

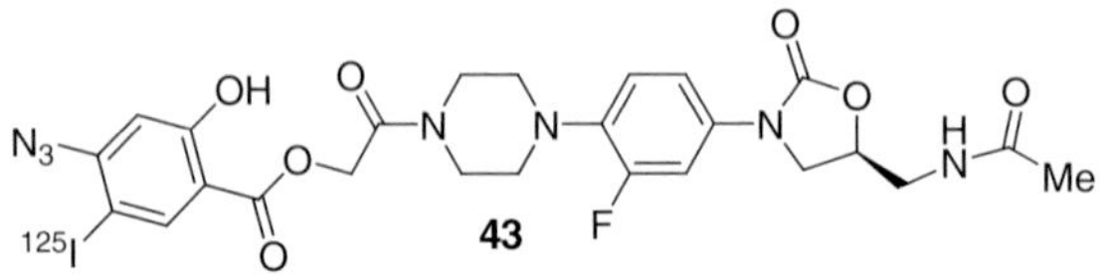

N_3 OH O O N N N O O H N O Me ^{125}I O **43** F

While the oxazolidinone-binding site overlaps with binding sites of the peptidyl transferase inhibitors lincomycin and chloramphenicol,[203,205] they have a distinctly different mode of action, as linezolid is a poor inhibitor of the peptidyl transferase reaction.[203,206] Oxazolidinone binding may prevent the fMet-tRNA from binding at the P site at the initiation stage, and thereby inhibit formation of the first peptide bond.[207a,207b] Overall, the consequence of oxazolidinone binding appears to be to block the assembly of the ternary 70S initiation complex from the preassociated mRNA-30S ribosomal subunit, the 50S subunit, and fMet-tRNA.[200,208] Of particular note, a high-resolution x-ray crystal structure of linezolid bound to the large ribosomal subunit of *Haloarcula marismortui* has been obtained by Rib-X Pharmaceuticals.[209]

Dahlberg and co-workers[210] reported the intriguing finding that, in a transformed *E. coli* strain, linezolid causes significant frameshifting, and promotes stop codon read-through (or nonsense suppression) at concentrations below the MIC, but does not cause misincorporation of amino acids, as occurs with aminoglycosides. They concluded that the oxazolidinone mechanism might involve an effect on translational fidelity.

7.23.6.18 Bacteriostatic Activity and Suppression of Virulence Factors

The oxazolidinones are generally bacteriostatic antibacterial agents, with the exception of some strains of *Streptococcus pneumoniae*, for which linezolid and eperezolid have been shown to be bactericidal[122]; ranbezolid is reported to be bacteriostatic against pneumococci.[211] The conventional microbiological definition of bactericidality is the ability of an antibiotic to kill $\geqslant 10^3$ bacteria over a 24-h incubation period, under strict laboratory conditions. Most bacteriostatic agents are able to kill bacteria, but at a lesser extent in that time course than is required by the definition of bactericidality. As noted by Pankey and Sabath,[212] consideration of the relevance of the distinction of bactericidal versus bacteriostatic MOA for predicting outcomes in the treatment of Gram-positive infections must be combined with PK/PD data. They note there may be some advantages of bacteriostatic mechanisms, citing clindamycin's inhibition of production of toxic shock syndrome toxin.

Linezolid significantly inhibits the expression of several bacterial virulence factors involved in infections with *Staphylococcus aureus* and *Streptococcus pyogenes*, decreasing levels of various exoproteins produced in these pathogens, such as hemolytic enzymes and toxins that cause tissue destruction, even at low concentrations representing only fractional levels of the MIC value.[28,213] When *S. pyogenes* was grown in sub-MIC concentrations of linezolid, the extent of phagocytic ingestion by human neutrophils was substantially increased.[213] It has been proposed that linezolid's high clinical cure rate in complicated skin and soft-tissue infections may be partially due to this capacity to inhibit the production of such damaging substances, in addition to its high concentration in the skin.[185] Pagano *et al.*[215] demonstrated linezolid has potent in vitro antiadhesion activity against clinical isolates of *Staphylococcus epidermidis*, which is associated with infections of indwelling devices, showing a superior sustained effect over vancomycin when concentrations fell below the MIC or were delayed by 4–6 h.

7.23.6.19 Oxazolidinone Relative Potency and Breadth of Spectrum

Linezolid has MIC_{90} values against staphylococci (MSSA and MRSA) of 4 μg mL^{-1}, coagulase-negative staphylococci (oxacillin-susceptible and resistant) of 2 μg mL^{-1}, streptococci (penicillin-resistant and susceptible) of 1 μg mL^{-1}, and enterococci (vancomycin-susceptible and VRE) of 2 μg mL^{-1}.[16,122] Eperezolid and ranbezolid both demonstrate a slight potency advantage over linezolid against some Gram-positive strains. Comparative MIC_{90} ratios relating relative activity for ranbezolid to linezolid against: (1) pneumococci; (2) penicillin-intermediate-resistant pneumococci; (3) MSSA; (4) MRSA; and (5) *E. faecium* are: 2, 4, 2, 1, and 0.5.[9] In general, eperezolid in vitro is comparable to or twofold more active than linezolid[2,122]; the MIC_{90} ratios comparing the relative activity of eperezolid to linezolid for the organisms above are: 2, 2, 1, 1, and 2. Against a wide panel of Gram-negative and Gram-positive anaerobes, ranbezolid was eightfold more active than linezolid, with ranbezolid, linezolid, and clindamycin MIC_{90}s of 0.5, 4, and 8 μg mL^{-1}, respectively. [86,216] In panels of Gram-positive clinical isolates, DuP-721 was generally fourfold more active than DuP-105,[198,217] and had equal or slightly less activity than linezolid.[2] DuP-105, DuP-721 and eperezolid have MIC_{90}s of *c.* 16 μg mL^{-1} versus anaerobe *Bacteroides fragilis*.[2]

In a CDC study of VISA and *S. aureus* strains with reduced susceptibility to vancomycin (SA-RVS, MIC ≥ 4 μg mL^{-1}) collected in the US, linezolid was reported to be the only antibacterial agent of 13 tested wherein 100% of the isolates ($n = 18$) of VISA and SA-RVS were fully susceptible to drug.[218] For the fastidious Gram-negatives, for *Haemophilus influenzae*, linezolid and ranbezolid have MIC_{90}s of 32 μg mL^{-1},[9,219] representing only modest activity, while for *Moraxella catarrhalis*, ranbezolid demonstrated improved activity over linezolid, where MIC_{90}s were 2 and 8 μg mL^{-1}, respectively.[216] Fritsche and co-workers[220] concluded that linezolid had the broadest spectrum of in vitro activity amongst all antibiotics that target the most common ICU pathogens, in examining 1321 North American Gram-positive isolates collected in 2001 from 25 hospital ICUs.

7.23.6.20 Advantages: Linezolid Pharmacokinetic Profile

The clinical data collected with linezolid have been extensively reviewed.[4,17,182,221] The superior phase I human PK profile of linezolid led to its selection over eperezolid for advancement into phase II clinical studies.[221] Most notably, linezolid has an oral bioavailability of approximately 100%, and can be administered with or without food. The drug is rapidly absorbed, with a T_{max} of 1–2 h. Dosing of 625 mg oral linezolid q12 h resulted in average peak serum levels of 18.8 μg mL^{-1} and average C_{min} levels of 8 μg mL^{-1}. For the 625 mg i.v. dose, steady-state PK C_{max} and C_{min} values were 15.7 and 3.8 μg mL^{-1}, respectively. As linezolid is cleared more rapidly in children younger than 12 years, the recommended dose is 10 mg kg^{-1} every 8 h (not to exceed 600 mg every 12 h) to approximate similar drug exposure to adults.[221]

Linezolid dosed i.v. or orally gives mean plasma concentrations ≥ MIC_{90} for the entire 12-h dosing period, for the target Gram-positive pathogens. Linezolid has a postantibiotic effect of 3.6–3.9 h, which, along with an elimination half-life of 5–7 h, supports the twice-daily dosing (q12 h). This PK profile allows for the ready switch from i.v. therapy to oral administration without the need for dosage adjustments. This can provide advantages over vancomycin[222] and potentially other competing therapies (QD, daptomycin, and tigecycline) that lack an oral formulation, by reducing the length of stay and the cost of hospitalization.

The human plasma protein binding of linezolid is 31%, and its volume of distribution at steady state is 40–50 L; hence, it demonstrates good tissue penetration.[221] Studies supporting the q12 h dosing regimen for skin infections and pneumonia included PK studies in healthy volunteers, where drug concentration in skin blister fluid was shown to equal or exceed steady-state plasma concentrations.[223] In similar human studies examining lung epithelial lining fluid (ELF), ELF concentrations of linezolid exceeded fourfold and twofold the plasma levels at 4 and 12 h post last dose (plasma = 15.5 ± 4.9 and 10.2 ± 2.3 μg mL^{-1}, respectively).[224] As the MIC_{90} levels for staphylococci, streptococci, and enterococci are all ≤ 4 μg mL^{-1}, linezolid plasma, skin blister fluid, and ELF levels exceeded the MIC_{90}s for the dosing interval. The high levels of linezolid in the ELF could factor into the clinical success of patients with hospital-acquired pneumonia due to MRSA,[225a,225b] where both higher survival rates (80% versus 63.5%) and clinical cure rates (59% versus 35.5%) were reported for linezolid versus vancomycin, respectively. Linezolid was shown to penetrate rapidly into human muscle, bone, and fat tissues (assessed during hip replacement surgery); drug concentrations in those tissues were 90%, 50%, and 30% of plasma, respectively.[226] Linezolid rapidly achieves a concentration of *c.* 5 μg mL^{-1} in the aqueous humor of the noninflamed human eye after a 600 mg i.v. dose.[227]

7.23.6.20.1 Pharmacokinetic/pharmacodynamic projections

In in vivo mouse thigh infection models, PK/PD studies demonstrated that the antibacterial efficacy of linezolid correlated best with the time the plasma level exceeds the MIC.[228] Craig[229] demonstrated that the free drug serum concentration needed to be above the MIC concentration for at least 40–50% of the dosing interval to achieve efficacy. Similar PK/PD parameters were reported for ranbezolid.[9,230]

7.23.6.20.2 Human metabolism and clearance profile

Linezolid is not metabolized by CYP isozymes, nor does it inhibit or induce major CYP isoforms.[231] Hence no CYP-induced drug interactions are observed. The metabolism in humans is nonenzymatically mediated, and produces two metabolites, an *N*-(hydroxyethyl) glycine (**44**) and an aminoethoxy acetic acid (**45**), derived from oxidative cleavage of the morpholine ring. Both metabolites lack antibacterial activity. The investigation of the mechanism of metabolism points to a nonenzymatic oxidation process.[221,231] The linezolid human clearance pathway is a combination of both renal and nonrenal clearance, the latter accounting for ~65% of the total clearance. As renal clearance of linezolid is low, the PK of linezolid is not altered in renally impaired patients, but there is evidence the metabolites could accumulate in such patients.[232] Percentage of dose excreted in urine is 30% parent drug, 10% (**44**), and 40% (**45**). Less than 10% of the dose is found in the feces as the two major metabolites.[181,221]

Ranbezolid is metabolized by human liver microsomal CYP enzymes, primarily CYP3A4, and partially inhibits CYP2E1.[9] In phase I clinical studies where healthy male volunteers were dosed orally with one dose of ranbezolid, the T_{max} was ~1 h, and the mean terminal half-life was 1.6 ± 0.1 h. After an 800 mg dose, the plasma level exceeded $2\ \mu g\ mL^{-1}$ for *c.* 5 h, representing a substantially lower level of exposure than demonstrated by linezolid (see above). Ranbaxy reported the maximum tolerated dose was 600 mg, and that the compound was well tolerated.[9]

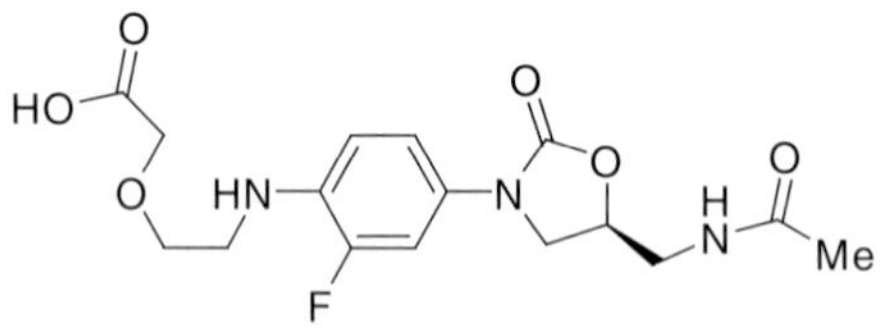

7.23.7 Unmet Medical Needs

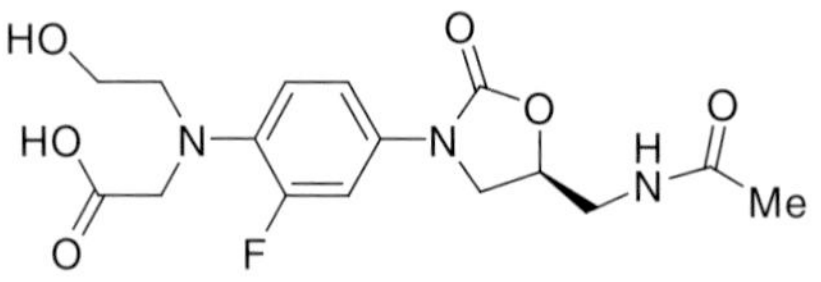

7.23.7.1 *Mycobacterium tuberculosis*

Ashtekar *et al.*[233–235] first described the activity of oxazolidinones against *M. tuberculosis*. Against 25 *M. tuberculosis* clinical isolates, DuP-721 had MICs of $0.3–1.25\ \mu g\ mL^{-1}$, comparable to rifampin, and was not cross-resistant with existing anti-TB drugs. In a murine in vivo model of *M. tuberculosis*, (±)-DuP-721 demonstrated some activity when administered for 17 days, but the increase in survival time was considerably inferior to that seen with isoniazid or rifampin dosed for only 2 days.[234] Other oxazolidinones have been reported with excellent in vitro potency against MDR-TB and efficacy in vivo in animal models of TB infection, most notably the thiomorpholine U-100480.[129] Linezolid was found to have MICs $\leq 1\ \mu g\ mL^{-1}$ versus 117 isolates of *M. tuberculosis*,[236] but is not approved for the treatment of TB. Challenges to using oxazolidinones for TB therapy have included identifying compounds that will not raise resistance to other bacteria over the protracted course of therapy, and finding oxazolidinones with reduced myelosuppression for long-term treatment.

7.23.7.2 Resistance

The eventual acquisition of resistance by bacteria to any antibiotic used in medicine is inevitable. One of the attractive attributes of the oxazolidinones is that it has generally been very difficult to demonstrate spontaneous resistance in vitro to the Gram-positive pathogens of interest. The rate of spontaneous resistance development to concentrations twice the MIC of eperezolid or linezolid in *S. aureus* and *S. epidermidis* strains was reported to be below detectable limits, indicating a spontaneous mutation frequency of $< 10^{-9}$ to 8×10^{-11}.[122,237] One aspect that lends to slow resistance development with the oxazolidinones is that there is a gene dosage effect that correlates with the extent of resistance.[238] This means that point mutations in Gram-positive organisms like *S. aureus* must occur in more than one of the multiple copies of 23S rRNA genes, in order to engender a significant degree of resistance.[197,239] In *S. aureus*, linezolid resistance is conferred by point mutations in bases located within the domain V peptidyl transferase region of 23S rRNA; this results in substitution of a guanine by a uracil at position 2576 (G2576U) or a G2447U mutation. Similarly, in *Enterococcus faecium* and *E. faecalis*, a G2576U mutation in clinical isolates has been seen.[197,240] There has been a report of two linezolid-resistant (MICs of 4 μg mL^{-1}) *Streptococcus pneumoniae* clinical isolates in a study of 7746 isolates.[241]

Bacterial resistance to linezolid in clinical trials has been reviewed.[18,239] Resistance to linezolid developed in 18 patients infected with *E. faecium* and one patient infected with *E. faecalis*. No resistance was seen in patients infected with staphylococci or streptococci in the clinical studies. Postapproval, there have been only a few reports of documented clinical isolates of linezolid-resistant *Staphylococcus aureus*. A clinical isolate of MRSA resistant to linezolid has been reported, with a G2576U mutation.[242] A broad surveillance program examining 9833 Gram-positive isolates collected from January 2001 to July 2002 found that 0.08% were resistant to linezolid.[243] There have been several reported cases where linezolid-resistant isolates of VRE have been isolated from patients undergoing longer-term (typically >4 weeks) treatment[244,245]; in one case, there was nosocomial spread of a linezolid-resistant VRE strain to several patients.[246] Risk factors for linezolid resistance include long duration of therapy over 14 days; treatment of difficult-to-cure infections, e.g., those caused by VRE and associated with an indwelling device or undrained abscess, where it may be necessary to treat with protracted linezolid therapy[186]; or use of inappropriately low dosage.[181]

7.23.7.3 Monoamine Oxidase Inhibition

The ability of oxazolidinones to inhibit MAO has been reviewed.[2,247,248] Marketed members of a class of reversible and competitive MAO inhibitors include the antidepressant 5-hydroxymethyl-3-aryl-2-oxazolidinone toloxatone (Humoryl),[249] and cimoxatone,[250] a 5-methoxymethyl-3-aryl-2-oxazolidinone. Other 3-aryl-2-oxazolidinones may therefore have undesirable side effects if there is significant inhibition of MAO-A or MAO-B enzymes. As such, there is the potential for interaction with serotonergic and adrenergic agents, leading to pressor responses, and serotonin syndrome resulting from hyperstimulation of the serotonin receptor $5HT_{1A}$. Linezolid ($K_i = 55$ μmol L^{-1})[251] and eperezolid are weak, reversible, nonselective inhibitors of MAO-A, an enzyme that metabolizes monoamine neurotransmitters, such as serotonin.[252] It is reported that ranbezolid has a similar effect to linezolid on MAO-A.[9] The 10 published reports of serotonin syndrome-like clinical events in those using linezolid and a selective serotonin reuptake inhibitor have been reviewed.[253,254a,254b] A clinical study demonstrated linezolid was similar to moclobemide in decreasing the oral clearance of tyramine relative to placebo.[255] While moclobemide has no food restrictions,[255] patients taking linezolid have been advised to avoid large quantities of foods (e.g., aged cheese) or beverages (e.g., red wine, beer) with a high tyramine content.[181]

7.23.7.4 Reversible Myelosuppression

Myelosuppression in the form of thrombocytopenia and anemia has been reported infrequently in patients treated with linezolid in the long term (>2 weeks). The reduction in platelets and red blood cells is generally rapidly reversible upon cessation of treatment, and no cases have involved aplastic anemia. It is recommended that complete blood counts be monitored weekly, and linezolid therapy discontinuation be considered if myelosuppression develops.[181] In the first year following FDA approval of linezolid, there were 72 reports of hematological abnormalities (32 of which were thrombocytopenia)[256] out of >55 000 treated patients, representing a total overall incidence of 1 in 750; nearly all were in patients on linezolid >4 weeks' duration. This rate was considered comparable to other antibiotic therapies; the hematological effects of other antimicrobials have been reviewed.[256] In the first prospective observational study of myelosuppression in long-term treatment with either vancomycin or linezolid of orthopedic (bone or prosthetic joint) Gram-positive bacterial infections, Yu and co-workers[257] concluded that the incidence of hematological effects was not higher in the linezolid-treated group than the vancomycin-treated group (thrombocytopenia rates of 3.3 versus 5.1 cases per 1000 days of therapy, respectively). They did note a higher incidence of thrombocytopenia in patients treated with vancomycin prior to the linezolid therapy. Similarly, Nasraway *et al.*[258] in analyzing 686 patients with nosocomial pneumonia randomized for ≥5 days treatment with linezolid or vancomycin, did not find a statistical difference between the two drug treatments in the risk for thrombocytopenia.

The mechanism of reversible myelosuppression with linezolid is not known. Several groups have published studies pointing to a possible role of mitochondrial protein synthesis (MPS) inhibition. Bartel *et al.*[259] at Bayer published use of a murine fibroblast in vitro assay for assessment of toxicity of oxazolidinones. Sciotti *et al.*[157] at Abbott reported use of an in vitro MPS inhibition assay as a screen for toxicity of a series of halostilbene oxazolidinones, showing these agents inhibit ^{35}S-methionine incorporation into mitochondrial protein. They examined MPS inhibition as a possible cause of the hematological effects of the oxazolidinones, testing linezolid and chloramphenicol in an in vitro rat heart MPS assay, and reported IC_{50}s of 12 and $<1\ \mu mol\ L^{-1}$, respectively, while their fluorostilbene had an $IC_{50} > 300\ \mu mol\ L^{-1}$. Leach and co-workers[214] at Pfizer reported that eperezolid inhibits proliferation of mammalian cells as a result of inhibition of MPS, and found that linezolid has an MPS IC_{50} of $16 \pm 2\ \mu mol\ L^{-1}$. Palenzuela *et al.*[191] have hypothesized that the rare cases of lactic acidosis, as well as potentially other toxicities seen in protracted treatment with linezolid, may be due to inhibition of MPS, and propose that mitochondrial polymorphism in the 16S rRNA gene might impart genetic susceptibility to linezolid toxicity, based on precedent[139] from aminoglycoside-associated hearing loss. Peripheral and optic neuropathy has been reported in some patients treated with linezolid, primarily those undergoing long-term treatment, greater than the maximum recommended duration of 28 days, but there are also cases of visual blurring that were reported in some patients who received a shorter course of therapy.[181]

7.23.8 New Research Areas

Crystallization of a ribosome from a bacterium with close homology to a human pathogen could allow potential x-ray structural determination of co-crystallized oxazolidinone bound to the ribosome, and aid structure-based drug design. In the ideal, a detailed understanding of the differential binding of oxazolidinones to mitochondrial ribosomes could potentially advance the search for compounds having improved toleration.

7.23.8.1 Areas for Improvement in Future Oxazolidinones

While linezolid remains the only oxazolidinone thus far marketed, the oxazolidinone field remains a highly competitive area of research at many companies. Future areas for improvement over linezolid include enhanced potency, cidality, once-a-day dosing, expansion of the spectrum of activity to include fastidious Gram-negative organisms, longer-term tolerability, and reduced MAO-A inhibition. AstraZeneca researchers have developed a computational binding model for designing oxazolidinone analogs with reduced MAO-A inhibition.[147] Ranbezolid appears to have some in vitro potency advantages over linezolid, although it appears the human PK profile diverges significantly from that of linezolid, with lower oral exposure and its interaction with CYP enzymes.

With a call from many for the development of narrow-spectrum antibiotics as a means of reducing the incidence of multidrug resistance, efforts to address the need for improved, fast, inexpensive, and reliable diagnostics must also advance, to assist physicians in the rapid identification of the offending microbe. An oxazolidinone compound with the capacity for long-term use without hematological effects may significantly expand the utility of oxazolidinones for use in those indications where extended therapy is often necessary, including osteomyelitis, endocarditis, and potentially TB. For the use of an oxazolidinone for TB, it would be desirable to separate the antimycobacterial activity from the Gram-positive activity, to reduce the potential spread of broad oxazolidinone resistance. Critical objectives for consideration of second-generation oxazolidinones include enhanced fastidious Gram-negative activity, and improvement of the activity against linezolid-resistant strains. It will also be of interest to monitor the clinical utility of combination therapies of other drugs with oxazolidinones, either as physical combinations or chemically linked in nature.

With the observation of what seems to be an ever-increasing rise in the incidence of multidrug resistance, it is most highly likely that the need for completely new antibiotic classes functioning by new mechanisms will continue to grow. Such new antibiotics classes, like the oxazolidinones and lipopeptides, will have potential hope for reducing the frequency and speed of the inevitable development of resistance. Linezolid has established the oxazolidinones as a very important class of antibiotics useful for treating susceptible and MDR Gram-positive infections. Opportunities exist to improve upon linezolid, and advances in the understanding of the finer details of the mechanism and specific site of the oxazolidinone interaction with the ribosome (both bacterial and mitochondrial) will undoubtedly assist in the design of more potent compounds with potential improvement in the toleration profile.

References

1. Panlilio, A. L.; Culver, D. H.; Gaynes, R. P.; Banerjee, S.; Henderson, T. S.; Tolson, J. S.; Martone, W. J. *Infect. Control Hosp. Epidemiol.* **1992**, *13*, 582–586.
2. Brickner, S. J. *Curr. Pharm. Des.* **1996**, *2*, 175–194.
3. Ranger, L. Linezolid and the Oxazolidinones – A New Class of Antibiotics. In *Creation of a Novel Class: The Oxazolidinone Antibiotics;* Batts, D. H., Kollef, M. H., Lipsky, B. A., Nicolau, D. P., Weigelt, J. A., Eds.; Innova Institute for Medical Education: Tampa, Florida, 2004; Chapter 3, p 30.
4. Zurenko, G. E.; Ford, C. W.; Hutchinson, D. K.; Brickner, S. J.; Barbachyn, M. R. *Exp. Opin. Invest. Drugs* **1997**, *6*, 151–158.
5. Brickner, S. J.; Hutchinson, D. K.; Barbachyn, M. R.; Manninen, P. R.; Ulanowicz, D. A.; Garmon, S. A.; Grega, K. C.; Hendges, S. K.; Toops, D. S.; Ford, C. W. et al. *J. Med. Chem.* **1996**, *39*, 673–679.

6. Walsh, C.; Wright, G. *Chem. Rev.* **2005**, *105*, 391–393.
7. Lesher, G. Y.; Froelich, E. J.; Gruett, M. D.; Bailey, J. H.; Brundage, R. P. *J. Med. Pharm. Chem.* **1962**, *5*, 1063–1068.
8. Mitscher, L. A. *Chem. Rev.* **2005**, *105*, 559–592.
9. Rattan, A. *Drugs Future* **2003**, *28*, 1070–1077.
10. Slee, A. M.; Wuonola, M. A.; McRipley, R. J.; Zajac, I.; Zawada, M. J.; Bartholomew, P. T.; Gregory, W. A.; Forbes, M. *Abstracts of Papers*; Proceedings of the 27th Interscience Conference Antimicrobial Agents and Chemotherapy, 1987; American Society for Microbiology: Washington, DC; Abs 244.
11. Barbachyn, M. R.; Ford, C. W. *Angewandte Chemie, Int. Ed.* **2003**, *42*, 2010–2023.
12. Hutchinson, D. K. *Exp. Opin. Ther. Patents* **2004**, *14*, 1309–1328.
13. Hutchinson, D. K. *Curr. Topics Med. Chem.* **2003**, *3*, 1021–1042.
14. Xiong, Y.-Q.; Yeaman, M. R.; Bayer, A. S. *Drugs Today* **2000**, *36*, 631–639.
15. Ford, C. W.; Zurenko, G. E.; Barbachyn, M. R. *Curr Drug Targets – Infect. Disord.* **2001**, *1*, 181–199.
16a. Fung, H. B.; Kirschenbaum, H. L; Ojofeitimi, B. O. *Clin. Ther.* **2001**, *23*, 356–391.
16b. Diekema, D. J.; Jones, R. N. *Drugs* **2000**, *59*, 7–16.
16c. Clemett, D.; Markham, A. *Drugs* **2000**, *59*, 815–827.
16d. Zabransky, R. J. *Clin. Microbiol. Newslett.* **2002**, *24*, 25–30.
16e. Livermore, D. M. *Int. J. Antimicrob. Agents* **2000**, *16*, S3–S10.
16f. Zurenko, G. E.; Gibson, J. K.; Shinabarger, D. L.; Aristoff, P. A.; Ford, C. W.; Tarpley, W. G. *Curr. Opin. Pharmacol.* **2001**, *1*, 470–476.
17. Perry, C. M.; Jarvis, B. *Drugs* **2001**, *61*, 525–551.
18. Bozdogan, B.; Appelbaum, P. C. *Int. J. Antimicrob. Agents* **2004**, *23*, 113–119.
19. Stevens, D. L.; Dotter, B.; Madaras-Kelly, K. *Exp. Rev. Anti-Infect. Ther.* **2004**, *2*, 51–59.
20. Livermore, D. M. *Clin. Infect. Dis.* **2003**, *36*, S11–S23.
21. Linden, P. K.; Pasculle, A. W.; Manez, R.; Kramer, D. J.; Fung, J. J.; Pinna, A. D.; Kusne, S. *Clin. Infect. Dis.* **1996**, *22*, 663–670.
22a. Rubin, R. J.; Harrington, C. A.; Poon, A.; Dietrich, K.; Greene, J. A.; Moiduddin, A. *Emerg. Infect. Dis.* **1999**, *5*, 9–17.
22b. Paladino, J. A. *Am. J. Health Syst. Pharm.* **2000**, *57*, S10–S12.
23. Feikin, D. R.; Schuchat, A.; Kolczak, M.; Hadler, J.; Harrison, L. H.; Lefkowitz, L.; McGeer, A.; Farley, M. M.; Vugia, D. J.; Lexau, C. et al. *Am. J. Public Health* **2000**, *90*, 223–229.
24a. Turett, G. S.; Blum, S.; Telzak, E. E. *Arch. Intern. Med.* **2001**, *161*, 2141–2144.
24b. Raz, R.; Elhanan, G.; Shimoni, Z.; Kitzes, R.; Rudnicki, C.; Igra, Y.; Yinnon, A. *Clin. Infect. Dis.* **1997**, *24*, 1164–1168.
25. Centers for Disease Control and Prevention. *National Nosocomial Infections Surveillance (NNIS) System Report. Data Summary from January 1999–May 1999.* Centers for Disease Control and Prevention: Atlanta, GA, 1999.
26. Deresinski, S. *Clin. Infect. Dis.* **2005**, *40*, 562–573.
27. Muder, R. R.; Brennen, C.; Wagener, M. W.; Vickers, R. M.; Rihs, J. D.; Hancock, G. A.; Yee, Y. C.; Miller, J. M.; Yu, V. L. *Ann. Intern. Med.* **1991**, *114*, 107–112.
28. Bernardo, K.; Pakulat, N.; Fleer, S.; Schnaith, A.; Utermöhlen, O; Krut, O.; Müller, S.; Krönke, M. *Antimicrob. Agents Chemother.* **2004**, *48*, 546–555.
29a. Soriano, A.; Martínez, J. A.; Mensa, J.; Marco, F.; Almela, M.; Moreno-Martínez, A.; Sánchez, F.; Muñoz, I.; Jiménez de Anta, M. T.; Soriano, E. *Clin. Infect. Dis.* **2000**, *30*, 368–373.
29b. Melzer, M.; Eykyn, S. J.; Gransden, W. R.; Chinn, S. *Clin. Infect. Dis.* **2003**, *37*, 1453–1460.
30. Cosgrove, S. E.; Sakoulas, G.; Perencevich, E. N.; Schwaber, M. J.; Karchmer, A. W.; Carmeli, Y. *Clin. Infect. Dis.* **2003**, *36*, 53–59.
31. Romero-Vivas, J.; Rubio, M.; Fernandez, C.; Picazo, J. J. *Clin. Infect. Dis.* **1995**, *21*, 1417–1423.
32. Lipsky, B. A.; Itani, K.; Norden, C. *Clin. Infect. Dis.* **2004**, *38*, 17–24.
33a. Tentolouris, N.; Jude, E. B.; Smirnof, I.; Knowles, E. A.; Boulton, A. J. M. *Diabetic Med.* **1999**, *16*, 767–771.
33b. Dang, C. N.; Prasad, Y. D. M.; Boulton, A. J. M.; Jude, E. B. *Diabetic Med.* **2003**, *20*, 159–161.
34. Chambers, H. F. *Emerg. Infect. Dis.* **2001**, 7, 178–182.
35a. Centers for Disease Control and Prevention. *Morb. Mortal. Wkly Rep.* **1999**, *48*, 707–710.
35b. Centers for Disease Control and Prevention. *Morb. Mortal. Wkly Rep.* **2003**, *52*, 88.
36. Fridkin, S. K; Hageman, J. C.; Morrison, M.; Thomson Sanza, L.; Como-Sabetti, K.; Jernigan, J. A.; Harriman, K.; Harrison, L. H.; Lynfield, R.; Farley, M. M. *N. Engl. J. Med.* **2005**, *352*, 1436–1444.
37. Francis, J. S.; Doherty, M. C.; Lopatin, U.; Johnston, C. P.; Sinha, G.; Ross, T.; Cai, M.; Hansel, N. N.; Perl, T.; Ticehurst, J. R. et al. *Clin. Infect. Dis.* **2005**, *40*, 100–107.
38. Panton, P. N.; Valentine, F. C. O. *Lancet* **1932**, *222*, 506–508.
39. Ward, P. D.; Turner, W. H. *Infect. Immun.* **1980**, *28*, 393–397.
40. Centers for Disease Control and Prevention. *Morb. Mortal. Wkly Rep.* **1993**, *42*, 597–599.
41. Lodise, T. P.; McKinnon, P. S.; Tam, V. H.; Rybak, M. J. *Clin. Infect. Dis.* **2002**, *34*, 922–929.
42. Stosor, V.; Peterson, L. R.; Postelnick, M.; Noskin, G. A. *Arch. Intern. Med.* **1998**, *158*, 522–527.
43. Bhavnani, S. M.; Drake, J. A.; Forrest, A.; Deinhart, J. A.; Jones, R. N.; Biedenbach, D. J.; Ballow, C. H. *Diagn. Microbiol. Infect. Dis.* **2000**, *36*, 145–158.
44. Edmond, M. B.; Ober, J. F.; Dawson, J. D.; Weinbaum, D. L.; Wenzel, R. P. *Clin. Infect. Dis.* **1996**, *23*, 1234–1239.
45. Corbett, E. L.; Watt, C. J.; Walker, N.; Maher, D.; Williams, B. G.; Raviglione, M. C.; Dye, C. *Arch. Intern. Med.* **2003**, *163*, 1009–1021.
46. Weir, E.; Fisman, D. N. *Can. Med. Assoc. J.* **2003**, *169*, 937–938.
47. Coninx, R; Mathieu, C.; Debacker, M.; Mirzoev, F.; Ismaelov, A.; de Haller, R.; Meddings, D. R. *Lancet* **1999**, *353*, 969–973.
48. Munsiff, S. S.; Nivin, B.; Sacajiu, G.; Mathema, B.; Bifani, P.; Kreiswirth, B. N. *J. Infect. Dis.* **2003**, *188*, 356–363.
49. McDonald, L. C.; Lauderdale, T.-L. Shiau Y.-R.; Chen, P.-C.; Lai, J.-F.; Wang, H.-Y.; Ho, M. *Int. J. Antimicrob. Agents* **2004**, *23*, 362–370.
50. Cardo, D.; Horan, T.; Andrus, M.; Dembinski, M.; Edwards, J.; Peavy, G.; Tolson, J.; Wagner, D. *Am. J. Infect. Control.* **2004**, *32*, 470–485.
51. Diekema, D. J.; Pfaller, M. A.; Schmitz, F. J.; Smayevsky, J.; Bell, J.; Jones, R. N.; Beach, M. *Clin. Infect. Dis.* **2001**, *32*, S114–S132.
52. Tiemersma, E. W.; Bronzwaer, S. L. A. M.; Lyytikäinen, O.; Degener, J. E.; Schrijnemakers, P.; Bruinsma, N.; Monen, J.; Witte, W.; Grundmann, H. *Emerg. Infect. Dis.* **2004**, *10*, 1627–1634.
53. Fisher, J. F.; Merouch, S. O.; Mobashery, S. *Chem. Rev.* **2005**, *105*, 395–424.
54. Naimi, T. S.; LeDell, K. H.; Como-Sabetti, K.; Borchardt, S. M.; Boxrud, D. J.; Etienne, J.; Johnson, S. K.; Vandenesch, F.; Fridkin, S.; O'Boyle, C. et al. *J. A. M. A.* **2003**, *290*, 2976–2984.

55. Iyer, S.; Jones, D. H. *J. Am. Acad. Dermatol.* **2004**, *50*, 854–858.
56. Kahne, D.; Leimkuhler, C.; Lu, W.; Walsh, C. *Chem. Rev.* **2005**, *105*, 425–448.
57a. Uttley, A. H.; Collins, C. H.; Naidoo, J.; George, R. C. *Lancet* **1988**, *i*, 57–58.
57b. Leclercq, R.; Derlot, E.; Duval, J.; Courvalin, P. *N. Engl. J. Med.* **1988**, *319*, 157–161.
58. Woodford, N. *J. Med. Microbiol.* **1998**, *47*, 849–862.
59. Noble, W. C.; Virani, Z.; Cree, R. G. A. *FEMS Microbiol. Lett.* **1992**, *93*, 195–198.
60. Weigel, L. M.; Clewell, D. B.; Gill, S. R.; Clark, N. C.; McDougal, L. K.; Flannagan, S. E.; Kolonay, J. F.; Shetty, J.; Killgore, G. E.; Tenover, F. C. *Science* **2003**, *302*, 1569–1571.
61a. Centers for Disease Control and Prevention. *Morb. Mortal. Wkly Rep.* **2002**, *51*, 565–567.
61b. Centers for Disease Control and Prevention. *Morb. Mortal. Wkly Rep.* **2002**, *51*, 902.
61c. Centers for Disease Control and Prevention. *Morb. Mortal. Wkly Rep.* **2004**, *53*, 322–323.
62. Willems, R. J. L.; Top, J.; van Santen, M.; Robinson, D. A.; Coque, T. M.; Baquero, F.; Grundmann, H.; Bonten, M. J. M. *Emerg. Infect. Dis.* **2005**, *11*, 821–828.
63. Tenover, F. C.; Weigel, L. M.; Appelbaum, P. C.; McDougal, L. K.; Chaitram, J.; McAllister, S.; Clark, N.; Killgore, G.; O'Hara, C. M.; Jevitt, L. et al. *Antimicrob. Agents Chemother.* **2004**, *48*, 275–280.
64. Bozdogan, B.; Esel, D.; Whitener, C.; Browne, F. A; Appelbaum, P. C. *J. Antimicrob. Chemother.* **2003**, *52*, 864–868.
65. Kaatz, G. W.; Rybak, M. J. *Emerg. Drugs* **2001**, *6*, 43–55.
66. Laible, G.; Spratt, B. G.; Hakenbeck, R. *Mol. Microbiol.* **1991**, *5*, 1993–2002.
67. Klugman, K. P.; Koornhof, H. J. *Lancet*, **1989**, *ii*, 444.
68. Doern, G. V.; Richter, S. S.; Miller, A.; Miller, N.; Rice, C.; Heilmann, K.; Beekmann, S. *Clin. Infect. Dis.* **2005**, *41*, 139–148.
69. Hsueh, P. R.; Teng, L. J.; Lee, L. N.; Yang, P. C.; Ho, S. W.; Luh, K. T. *J. Clin. Microbiol.* **1999**, *37*, 897–901.
70. Buysse, J. M.; Demyan, W. F.; Dunyak, D. S.; Stapert, D.; Hamel, J. C.; Ford, C. W. *Abstracts of Papers*; Proceedings of the 36th Interscience Conference on Antimicrobial Agents and Chemotherapy, New Orleans; American Society for Microbiology: Washington, DC, 1996; Abs C-42.
71. Yagi, B. H.; Zurenko, G. E. *Anaerobe* **1997**, *3*, 301–306.
72. Moylett, E. H.; Pacheco, S. E.; Brown-Elliott, B. A.; Perry, T. R.; Buescher, E. S.; Birmingham, M. C.; Schentag, J. J.; Gimbel, J. F.; Apodaca, A.; Schwartz, M. A. et al. *Clin. Infect. Dis.* **2003**, *36*, 313–318.
73. Zurenko, G. E.; Yagi, B. H.; Schaadt, R. D.; Allison, J. W.; Kilburn, J. O.; Glickman, S. E.; Hutchinson, D. K.; Barbachyn, M. R.; Brickner, S. J. *Antimicrob. Agents Chemother.* **1996**, *40*, 839–845.
74. Diekema, D. J.; Jones, R. N. *Lancet* **2001**, *358*, 1975–1982.
75. Pfaller, M. A.; Jones, R. N.; Doern, G. V.; Kugler, K. *Antimicrob. Agents Chemother.* **1998**, *42*, 1762–1770.
76. Jones, R. N. *Am. J. Med.* **1996**, *100*, 3S–12S.
77. Yao, R. C.; Crandall, L. W. *Drugs Pharm. Sci.* **1994**, *63*, 1–27.
78. Wood, M. J. *J. Antimicrob. Chemother.* **1996**, *37*, 209–222.
79. Cassani, G.; Lancini, G. C.; Cavalleri, B.; Parenti, F. *Chim. Oggi* **1987**, *5*, 11–13.
80. Leclercq, R.; Nantas, L.; Soussy, C. J.; Duval, J. *J. Antimicrob. Chemother.* **1992**, *30*, 67–75.
81. Eliopoulos, G. M. *Clin. Infect. Dis.* **2003**, *36*, 473–481.
82. Lamb, H. M.; Figgilt, D. P.; Faulds, D. *Drugs* **1999**, *58*, 1061–1097.
83. Chow, J. W.; Davidson, A.; Sanford, E.; Zervos, M. J. *Clin. Infect. Dis.* **1997**, *24*, 91–92.
84. Hershberger, E.; Donabedian, S.; Konstantinou, K.; Zervos, M. J. *Clin. Infect. Dis.* **2004**, *38*, 92–98.
85. Livermore, D. *J. Antimicrob. Chemother.* **2000**, *46*, 347–350.
86. Olsen, K. M.; Rebuck, J. A.; Rupp, M. E. *Clin. Infect. Dis.* **2001**, *32*, e83–e86.
87. Alder, J. D. *Drugs Today* **2005**, *41*, 81–90.
88. Tally, F. P.; DeBruin, M. F. *J. Antimicrob. Chemother.* **2000**, *46*, 523–526.
89. Raja, A.; LaBonte, J.; Lebbos, J.; Kirkpatrick, P. *Nat. Rev.* **2003**, *2*, 943–944.
90. Mangili, A.; Bica, I.; Snydman, R.; Hamer, D. H. *Clin. Infect. Dis.* **2005**, *40*, 1058–1060.
91. Johnson, A. P. *Curr. Opin. Anti-Infect. Invest. Drugs* **2000**, *2*, 164–170.
92. Riedl, B.; Endermann, R. *Exp. Opin. Ther. Patents* **1999**, *9*, 625–633.
93. Genin, M. J. *Exp, Opin. Ther. Patents* **2000**, *10*, 1405–1414.
94a. Katritzky, A. R.; Fara, D. C.; Karelson, M. *Bioorg. Med. Chem.* **2004**, *12*, 3027–3035.
94b. Karki, R. G.; Kulkarni, V. M. *Bioorg. Med. Chem.* **2001**, *9*, 3153–3160.
95. Fugitt, R. B.; Luckenbaugh, R. W. 5-Halomethyl-3-phenyl-2-oxazolidinones; US Patent 4,128,654, December 5, 1978.
96. Fugitt, R. B.; Luckenbaugh, R. W. 3-(*p*-Alkylsulfonylphenyl)oxazolidinone Derivatives as Antibacterial Agents; US Patent 4,340,606, July 20, 1982.
97. Gregory, W. A. *p*-Oxooxazolidinylbenzene Compounds as Antibacterial Agents; US Patent no. 4,461,773, July 24, 1984.
98. Daly, J. S.; Eliopoulos, G. M.; Willey, S.; Moellering, R. C., Jr. *Antimicrob. Agents Chemother.* **1988**, *32*, 1341.
99. Gregory, W. A.; Brittelli, D. R.; Wang, C.-L. J.; Kezar, H. S.; Carlson, R. K.; Park, C.-H.; Corless, P. F.; Miller, S. J.; Rajagopalan, P.; Wuonola, M. A. et al. *J. Med. Chem.* **1990**, *33*, 2569–2578.
100. Park, C. H.; Brittelli, D. R.; Wang, C. L.-J.; Marsh, F. D.; Gregory, W. A.; Wuonola, M. A.; McRipley, R. J.; Eberly, V. S.; Slee, A. M.; Forbes, M. *J. Med. Chem.* **1992**, *35*, 1156–1165.
101. Gregory, W. A.; Brittelli, D. R.; Wang, C.-L. J; Wuonola, M. A.; McRipley, R. J.; Eustice, D. C.; Eberly, V. S.; Bartholomew, P. T.; Slee, A. M.; Forbes, M. *J. Med. Chem.* **1989**, *32*, 1673–1681.
102. Zajac, I.; Lam, G. N.; Hoffman, H. E.; Slee, A. M. *Abstracts of Papers*; Proceedings of the 27th Interscientific Conference on Antimicrobial Agents and Chemotherapy; American Society for Microbiology: Washington, DC, 1987; Abs no. 247.
103. Slee, A. M.; Wuonola, M. A.; McRipley, R. J.; Zajac, I.; Zawada, M. J.; Bartholomew, P. T.; Gregory, W. A.; Forbes, M. *Antimicrob. Agents Chemother.* **1987**, *31*, 1791–1797.
104. *Pharmaprojects*; PJB Publications: Richmond, Surrey, UK, April 12, 1995.
105. *Scrip*, October 13, **1987**, *1250*, p 25.
106a. *Pharmcast-Int.* **1995**, February, 7-1-487.
106b. *Pharmcast-Int.* **1995**, February, 7-1-484.
107. Wang, C.-L. J.; Wuonola, M. A. Preparation of 5-(Aminomethyl)-3-phenyl-2-oxazolidinone Derivatives and Antibacterial Pharmaceuticals Containing them; US Patent 4,801,600, January 31, 1989.

108. Brickner, S. J. Antibacterial 3-(Fused-ring substituted)phenyl-5.beta-amidomethyloxazolidin-2-ones; US Patent 5,225,565, July 6, 1993.
109. Brickner, S. J. 5′Indolinyl-5.beta.-amidomethyloxazolidin-2-ones; US Patent 5,164,510, November 17, 1992.
110. Ciske, F. L.; Barbachyn, M. R.; Genin, M. J.; Grega, K. C.; Lee, C. S.; Dolak, L. A.; Seest, E. P.; Watt, W.; Adams, W. J.; Friis, J. M. et al. *Bioorg. Med. Chem. Lett.* **2003**, *13*, 4235–4239.
111. French, G. *J. Antimicrob. Chemother.* **2003**, *51*, ii45–ii53.
112. Harwood, P. J.; Giannoudis, P. V. *Exp. Opin. Drug Safety* **2004**, *3*, 405–414.
113. Barbachyn, M. R.; Toops, D. S.; Ulanowicz, D. A.; Grega, K. C.; Brickner, S. J.; Ford, C. W.; Zurenko, G. E.; Hamel, J. C.; Schaadt, R. D.; Stapert, D. et al. *Bioorg. Med. Chem. Lett.* **1996**, *6*, 1003–1008.
114. Hutchinson, D. K.; Barbachyn, M. R.; Brickner, S. J.; Buysse, J. M.; Demyan, W. F.; Ford, C. W.; Garmon, S. A.; Grega, K. C.; Hendges, S. K. et al. *212th ACS National Meeting*, Orlando, FL, August 25–29, 1996; MEDI-192.
115. Chu, D. T. W.; Fernandes, P. B. *Antimicrob. Agents Chemother.* **1989**, *33*, 131–135.
116a. Tucker, J. A.; Brickner, S. J.; Ulanowicz, D. A. Oxazolidinone Antibacterial Agents having a Six-membered Heteroaromatic Ring; US Patent 5,719,154, February 17, 1998.
116b. Tucker, J. A.; Allwine, D. A.; Grega, K. C.; Barbachyn, M. R.; Klock, J. L.; Adamski, J. L.; Brickner, S. J.; Hutchinson, D. K.; Ford, C. W.; Zurenko, G. E. et al. *J. Med. Chem.* **1998**, *41*, 3727–3735.
117. Weidner-Wells, M. A.; Boggs, C. M.; Foleno, B. D.; Melton, J.; Bush, K.; Goldschmidt, R. M.; Hlasta, D. J. *Bioorg. Med. Chem.* **2002**, *10*, 2345–2351.
118. Barbachyn, M. R.; Cleek, G. J.; Dolak, L. A.; Garmon, S. A.; Morris, J.; Seest, E. P.; Thomas, R. C.; Toops, D. S.; Watt, W.; Wishka, D. G. et al. *J. Med. Chem.* **2003**, *46*, 284–302.
119. Johnson, P. D.; Aristoff, P. A.; Zurenko, G. E.; Schaadt, R. D.; Yagi, B. H.; Ford, C. W.; Hamel, J. C.; Stapert, D.; Moerman, J. K. *Bioorg. Med. Chem. Lett.* **2003**, *13*, 4197–4200.
120. Gravestock, M. B.; Acton, D. G.; Betts, M. J.; Dennis, M.; Hatter, G.; McGregor, A.; Swain, M. L.; Wilson, R. G.; Woods, L.; Wookey, A. *Bioorg. Med. Chem. Lett.* **2003**, *13*, 4179–4186.
121. Ulanowicz, D. A.; Brickner, S. J.; Ford, C. W.; Zurenko, G. E. *Abstracts of Papers*; Proceedings of the 37th Interscience Conference on Antimicrobial Agents and Chemotherapy; American Society for Microbiology: Washington, DC, 1997; Abs F-21.
122. Zurenko, G. E.; Yagi, B. H.; Schaadt, R. D.; Allison, J. W.; Kilburn, J. O.; Glickman, S. E.; Hutchinson, D. K.; Barbachyn, M. R.; Brickner, S. J. *Antimicrob. Agents Chemother.* **1996**, *40*, 839–845.
123. Ford, C. W.; Hamel, J. C.; Wilson, D. M.; Moerman, J. K.; Stapert, D.; Yancey, R. J., Jr.; Hutchinson, D. K.; Barbachyn, M. R.; Brickner, S. J. *Antimicrob. Agents Chemother.* **1996**, *40*, 1508–1513.
124. Piper, R. C.; Lund, J. E.; Denlinger, R. H.; Platte, T. F.; Brown, W. P.; Brown, P. K.; Palmer, J. R. *Abstracts of Papers*; Proceedings of the 35th Interscience Conference on Antimicrobial Agents and Chemotherapy, American Society for Microbiology: Washington, DC, 1995; Paper F223.
125. Slatter, J. G.; Adams, L. A.; Bush, E. C.; Chiba, K.; Daley-Yates, P. T.; Feenstra, K. L.; Koike, S.; Ozawa, N.; Peng, G. W.; Sams, J. P. et al. *Xenobiotica* **2002**, *32*, 907–924.
126. Martin, I. J.; Daley-Yates, P. T. *Abstracts of Papers*; Proceedings of the 35th Interscience Conference on Antimicrobial Agents and Chemotherapy; American Society for Microbiology: Washington, DC, 1995; Paper F222.
127. Gleave, D. M.; Brickner, S. J.; Manninen, P. R.; Allwine, D. A.; Lovasz, K. D.; Rohrer, D. C.; Tucker, J. A.; Zurenko, G. E.; Ford, C. W. *Bioorg. Med. Chem. Lett.* **1998**, *8*, 1231–1236.
128. Koike, S.; Miura, H.; Nakamura, R.; Chiba, K.; Moe, J. B. *Abstracts of Papers*; Proceedings of the 35th Interscience Conference on Antimicrobial Agents and Chemotherapy; American Society for Microbiology: Washington, DC, 1995; Paper F224.
129. Barbachyn, M. R.; Hutchinson, D. K.; Brickner, S. J.; Cynamon, M. H.; Kilburn, J. O.; Klemens, S. P.; Glickman, S. E.; Grega, K. C.; Hendges, S. K.; Toops, D. S. et al. *J. Med. Chem.* **1996**, *39*, 680–685.
130. Cynamon, M. H.; Klemens, S. P.; Sharpe, C. A.; Chase, S. *Antimicrob. Agents Chemother.* **1999**, *43*, 1189–1191.
131. Brickner, S. J. Tricyclic [6,6,5]-fused Oxazolidinone Antibacterial Agents; US Patent 5,247,090, September 21, 1993.
132. Gleave, D. M.; Brickner, S. J. *J. Org. Chem.* **1996**, *61*, 6470–6474.
133. Selvakumar, N.; Srinivas, D.; Kumar Khera, M.; Sitaram Kumar, M.; Mamidi, R. N. V. S.; Sarnaik, H.; Charavaryamath, C.; Srinivasa Rao, B.; Raheem, M. A.; Das, J. et al. *J. Med. Chem.* **2002**, *45*, 3953–3962.
134a. Pires, J. R.; Saito, C.; Gomes, S. L.; Giesbrecht, A. M.; Amaral, A. T. *J. Med. Chem.* **2001**, *44*, 3673–3681.
134b. Edwards, D. I. *J. Antimicrob. Chemother.* **1993**, *31*, 9–20.
135. Hof, H.; Chakraborty, T.; Royer, R.; Buisson, J. P. *Drugs Exp. Clin. Res.* **1987**, *13*, 635–639.
136. Chatterjee, S. N. *Proc. Ind. Natl. Sci. Acad. B: Biol. Sci.* **1997**, *63*, 477–499.
137. Shingatgeri, V. M.; Venkatesha, U.; Ranvir, R. K.; Chaudhari, G. R.; Vachaspati, P.; Sharma, N.; Kakkar, S.; Saini, G. S.; Manisha, T.; Shahnaz, A. et al. *Abstracts of Papers*; Proceedings of the 42nd Interscience Conference on Antimicrobial Agents and Chemotherapy, American Society for Microbiology: Washington, DC, 2002; Abs F-1300.
138. Yong, D.; Yum, J. H.; Lee, K.; Chong, Y.; Choi, S. H.; Rhee, J. K. *Antimicrob. Agents Chemother.* **2004**, *48*, 352–357.
139. Hutchin, T.; Cortopassi, G. *Antimicrob. Agents Chemother.* **1994**, *38*, 2517–2520.
140. Lee, T. H.; Kim, D. H.; Cho, J. H.; Choi, S. H.; Im, W. B.; Rhee, J. K. *Abstracts of Papers*; Proceedings of the 42nd Interscience Conference on Antimicrobial Agents and Chemotherapy; American Society for Microbiology: Washington, DC, 2002; Abs F-1314.
141. Bae, S. K.; Chung, W.-S.; Kim, E. J.; Rhee, J. K.; Kwon, J. W.; Kim, W. B.; Lee, M. G. *Antimicrob. Agents Chemother.* **2004**, *48*, 659–662.
142. Genin, M. J.; Allwine, D. A.; Anderson, D. J.; Barbachyn, M. R.; Emmert, D. E.; Garmon, S. A.; Graber, D. R.; Grega, K. C.; Hester, J. E.; Hutchinson, D. K. et al. *J. Med. Chem.* **2000**, *43*, 953–970.
143. AstraZeneca R&D Pipeline: Discontinued Projects 25 July 2002. http://www.astrazeneca.com (accessed Aug 2006).
144. Johnson, A. P.; Warner, M.; Livermore, D. M. *J. Antimicrob. Chemother.* **2002**, *50*, 89–93.
145. Ednie, L. M.; Jacobs, M. R.; Appelbaum, P. C. *J. Antimicrob. Chemother.* **2002**, *50*, 101–105.
146. Arundel, P. A. *Abstracts of Papers,* Proceedings of the 41st Interscience Conference on Antimicrobial Agents and Chemotherapy; American Society for Microbiology: Washington, DC, 2001; abstract F-21, Paper 1039.
147. Reck, F.; Zhou, F.; Girardot, M.; Kern, G.; Eyermann, C. J.; Hales, N. J.; Ramsay, R. R.; Gravestock, M. B. *J. Med. Chem.* **2005**, *48*, 499–506.
148. Phillips, O. A.; Udo, E. E.; Ali, A. A. M.; Al-Hassawi, N. *Bioorg. Med. Chem.* **2003**, *11*, 35–41.
149. Riedl, B.; Haebich, D.; Stolle, A.; Ruppelt, M.; Bartel, S.; Guarrieri, W.; Endermann, R.; Kroll, H.-P. Preparation of Bactericidal Pyridothienyl- and Pyridofuryloxazolidinones. Ger. Offen. Patent 19601264, 1997; *Chem. Abstr.* **1997**, *127*, 149139a.

150. Hester, J. B.; Nidy, E. G.; Perricone, S. C.; Poel, T.-J. (Pharmacia & Upjohn Co.) Preparation of Thiocarbonyloxazolidinones as Antibacterial Agents. 1998 WO 9854161 A1. *Chem. Abstr.* **1998**, *130*, 38373.
151. Tokuyama, R.; Takahashi, Y.; Tomita, Y.; Suzuki, T.; Yoshida, T.; Iwasaki, N.; Kado, N.; Okezaki, E.; Nagata, O. *Chem. Pharm. Bull.* **2001**, *49*, 347–352.
152. Lee, C. S.; Allwine, D. A.; Barbachyn, M. R.; Grega, K. C.; Dolak, L. A.; Ford, C. W.; Jensen, R. M.; Seest, E. P.; Hamel, J. C.; Schaadt, R. D. et al. *Bioorg. Med. Chem.* **2001**, *9*, 3243–3253.
153. Thomasco, L. M.; Gadwood, R. C.; Weaver, E. A.; Ochoada, J. M.; Ford, C. W.; Zurenko, G. E.; Hamel, J. C.; Stapert, D.; Moerman, J. K.; Schaadt, R. D. et al. *Bioorg. Med. Chem. Lett.* **2003**, *13*, 4193–4196.
154. Thomas, R. C.; Poel, T.-J.; Barbachyn, M. R.; Gordeev, M. F.; Luehr, G. W.; Renslo, A.; Singh, U. (Pharmacia & Upjohn) *N*-Aryl-2-oxazolidinone-5-carboxamides and Their Derivatives; and Their Use as Antibacterials. 2003 WO 03072553. *Chem. Abstr.* **2003**, *139*, 214457.
155. Thomas, R. C.; Poel, T.-J.; Barbachyn, M. R.; Gordeev, M. F.; Luehr, G. W.; Renslo, A.; Singh, U.; Josyula, V. P. V. N. (Pfizer) *N*-Aryl-2-oxazolidinone-5-carboxamides and Their Derivatives; US Patent Application 147760 A1, July 29, 2004.
156. Paget, S. D.; Foleno, B. D.; Boggs, C. M.; Goldschmidt, R. M.; Hlasta, D. H.; Weidner-Wells, M. A.; Werblood, H. M.; Wira, E.; Bush, K.; Macielag, M. J. *Bioorg. Med. Chem. Lett.* **2003**, *13*, 4173–4177.
157. Sciotti, R. J.; Pliushchev, M.; Wiedeman, P. E.; Balli, D.; Flamm, R.; Nilius, A. M.; Marsh, K.; Stolarik, D.; Jolly, R.; Ulrich, R. et al. *Bioorg. Med. Chem. Lett.* **2002**, *12*, 2121–2123.
158. Lohray, B. B.; Lohray, V. B.; Srivastava, B. K.; Gupta, S.; Solanki, M.; Kapadnis, P.; Takale, V.; Pandya, P. *Bioorg. Med. Chem. Lett.* **2004**, *14*, 3139–3142.
159. Denis, A.; Villette, T. *Bioorg. Med. Chem. Lett.* **1994**, *4*, 1925–1930.
160. Hester, J. B., Jr.; Brickner, S. J.; Barbachyn, M. R.; Hutchinson, D. K.; Toops, D. S., Preparation of 5-Amidomethyl-3-arylbutyrolactones as Antibacterial Agents; US Patent 5,708,169, January 13, 1998.
161. Borthwick, A. D.; Biggadike, K.; Rocherolle, V.; Cox, D. M.; Chung, G. A. C. *Med. Chem. Res.* **1996**, *6*, 22–27.
162. Gravestock, M. B. 5-(Acetamidomethyl)-3-aryldihydrofuran-2-one and tetrahydrofuran-2-one Derivatives with Antibiotic Activity. WO 9714690, 24 April, 1997. *Chem. Abstr.* **1997**, *126*, 343482.
163. Snyder, L. B.; Meng, Z.; Mate, R.; D'Andrea, S. V.; Marinier, A.; Quesnelle, C. A.; Gill, P.; DenBleyker, K. L.; Fung-Tomc, J. C.; Frosco, M. et al. *Bioorg. Med. Chem. Lett.* **2004**, *14*, 4735–4739.
164. Griffin, J. H.; Pace, J. L. (Theravance) Multivalent Macrolide Antibiotics; US Patent Application 0176670, September 18, 2003.
165. Farmer, J. J.; Sutcliffe, J. A.; Bhattacharjee, A. (Rib-X Pharmaceuticals, Inc.) Preparation of Bifunctional Heterocyclic Azithromycin Compounds Useful as Anti-infective, Anti-proliferative, Anti-inflammatory, and Prokinetic Agents; World Patent Application WO 078770, 16 September, 2004. *Chem. Abstr.* **2004**, *141*, 260999.
166. Yu, J.; Lee, J.; Kwon, M.; Pae, A.; Koh, H. Heterodimeric Conjugates of Neomycin-oxazolidinone, Their Preparation and Their Use. PCT International Application 2003, Korean Institute of Science and Technology, WO 066648 A1. *Chem. Abstr.* **2003** *139*, 173775.
167. Gordeev, M. F.; Hackbarth, C.; Barbachyn, M. R.; Banitt, L. S.; Gage, J. R.; Luehr, G. W.; Gomez, M.; Trias, J.; Morin, S. E.; Zurenko, G. E. et al. *Bioorg. Med. Chem. Lett.* **2003**, *13*, 4213–4216.
168. Hubschwerlen, C.; Specklin, J. L.; Baeschlin, D. K.; Borer, Y.; Haefeli, S.; Sigwalt, C.; Schroeder, S.; Locher, H. H. *Bioorg. Med. Chem. Lett.* **2003**, *13*, 4229–4233.
169. Mourelle Mancini, M. L.; Huguet Clotet, J.; Hidalgo Rodriguez, J.; Del Castillo, J. C. Preparation of Fluoroquinolonyl Derivatives of Oxazolidinones as Antibacterial Agents. PCT International Application 2003 Vita-Invest SA, WO 002560 A1. *Chem. Abstr.* **2003**, *138*, 89799.
170. Hubschwerlen, C.; Specklin, J.-L.; Sigwalt, C.; Schroeder, S.; Locher, H. H. *Bioorg. Med. Chem.* **2003**, *11*, 2313–2319.
171. Wang, C.-L. J.; Gregory, W. A.; Wuonola, M. A. *Tetrahedron* **1989**, *45*, 1323–1326.
172. Herweh, J. E.; Kauffman, W. J. *Tetrahedron Lett.* **1971**, 809–812.
173. Speranza, G. P.; Peppel, W. J. *J. Org. Chem.* **1958**, *23*, 1922–1924.
174. Manninen, P. R.; Little, H. A.; Brickner, S. J. *Book of Abstracts*; 212th ACS National Meeting, Orlando, FL, August 25–29, 1996. American Chemical Society: Washington, DC, 1996; ORGN-389.
175. Manninen, P. R.; Brickner, S. J. *Org. Syn.* **2004**, *81*, 112–120.
176. Pearlman, B. A.; Perrault, W. R.; Barbachyn, M. R.; Manninen, P. R.; Toops, D. S.; Houser, D.; Fleck, T. J. Process to Prepare Oxazolidinones, US Patent 5837870, November 17, 1998.
177. Perrault, W. R.; Pearlman, B. A.; Godrej, D. B.; Jeganathan, A.; Yamagata, K.; Chen, J. J.; Lu, C. V.; Herrinton, P. M.; Gadwood, R. C.; Chan, L. et al. *Org. Process Res. Dev.* **2003**, *7*, 533–546.
178. Madar, D. J.; Kopecka, H.; Pireh, D.; Pease, J.; Pliushchev, M.; Sciotti, R. J.; Wiedeman, P. E.; Djuric, S. W. *Tetrahedron Lett.* **2001**, *42*, 3681–3684.
179. Bain, K. T.; Wittbrodt, E. T. *Ann. Pharmacother.* **2001**, *35*, 566–575.
180. Burleson, B. S.; Ritchie, D. J.; Micek, S. T.; Dunne, W. M. *Pharmacotherapy* **2004**, *24*, 1225–1231.
181. Zyvox^{TM} (linezolid) [package insert]; Kalamazoo, MI; Pharmacia & Upjohn 2003.
182. Moellering, R. C., Jr. *Ann. Intern. Med.* **2003**, *138*, 135–142.
183. Rubinstein, E.; Cammarata, S. K.; Oliphant, T. H.; Wunderink, R. G. *Clin. Infect. Dis.* **2001**, *32*, 402–412.
184. San Pedro, G. S.; Cammarata, S. K.; Oliphant, T. H.; Todisco, T. *Scand. J. Infect. Dis.* **2002**, *34*, 720–728.
185. Stevens, D. L.; Smith, L. G.; Bruss, J. B.; McConnell-Martin, M. A.; Duvall, S. E.; Todd, W. M.; Hafkin, B. *Antimicrob. Agents Chemother.* **2000**, *44*, 3408–3413.
186. Birmingham, M. C.; Rayner, C. R.; Meagher, A. K.; Flavin, S. M.; Batts, D. H.; Schentag, J. *J. Clin. Infect. Dis.* **2003**, *36*, 159–167.
187. Dresser, L. D.; Birmingham, M. C.; Karchmer, A. W.; Rayner, C. R.; Flavin, S. M.; Schentag, J. J. *Abstracts of Papers,* Proceedings of the 40th Interscience Conference on Antimicrobial Agents and Chemotherapy; American Society for Microbiology: Washington, DC, 2000; Abs 2239.
188. Babcock, H. M.; Ritchie, D. J.; Christiansen, E.; Starlin, R.; Little, R.; Stanley, S. *Clin. Infect. Dis.* **2001**, *32*, 1373–1375.
189a. Attasi, K.; Hershberger, E.; Alam, R.; Zervos, M. J. *Clin. Infect. Dis.* **2002**, *34*, 695–698.
189b. Orrick, J. J.; Johns, T.; Janelle, J.; Ramphal, R. *Clin. Infect. Dis.* **2002**, *35*, 348–349.
190. Apodaca, A. A.; Rakita, R. M. *N. Engl. J. Med.* **2003**, *348*, 86–87.
191. Palenzuela, L.; Hahn, N. M.; Nelson, R. P., Jr.; Arno, J. N.; Schobert, C.; Bethel, R.; Ostrowski, L. A.; Sharma, M. R.; Datta, P. P.; Agrawal, R. K. et al. *Clin. Infect. Dis.* **2005**, *40*, e113–e116.
192. Bressler, A. M.; Zimmer, S. M.; Gilmore, J. L.; Somani, J. *Lancet Infect. Dis.* **2004**, *4*, 528–531.
193. Zivkovic, S. A.; Lacomis, D. *Neurology* **2005**, *64*, 926–927.

194. Lee, E.; Burger, S.; Shah, J.; Melton, C.; Mullen, M.; Warren, F.; Press, R. *Clin. Infect. Dis.* **2003**, *37*, 1389–1391.
195. Martin, J. P., Jr.; Dupuis, M. J.; Herberg, J. T. (Pharmacia & Upjohn Co.) Co-therapy with an Oxazolidinone and a Vitamin B to Prevent Oxazolidinone-Associated Side Effects, 2003; WO 2003 063862 A1. *Chem. Abstr.* **2003**, *139*, 144006.
196. Spellberg, B.; Yoo, T.; Bayer, A. S. *J. Antimicrob. Chemother.* **2004**, *54*, 832–835.
197. Shinabarger, D. L. *Exp. Opin. Invest. Drugs* **1999**, *8*, 1195–1202.
198. Daly, J. S.; Eliopoulos, G. M.; Reiszner, E.; Moellering, R. C., Jr. *J. Antimicrob. Chemother.* **1988**, *21*, 721–730.
199. Eustice, D. C.; Feldman, P. A.; Zajac, I.; Slee, A. M. *Antimicrob. Agents Chemother.* **1988**, *32*, 1218–1222.
200. Shinabarger, D. L.; Marotti, K. R.; Murray, R. W.; Lin, A. H.; Melchior, E. P.; Swaney, S.; Dunyak, D. S.; Demyan, W. F.; Buysse, J. M. *Antimicrob. Agents Chemother.* **1997**, *41*, 2132–2136.
201. Zhou, C. C.; Swaney, S. M.; Shinabarger, D. L.; Stockman, B. J. *Antimicrob. Agents Chemother.* **2002**, *46*, 625–629.
202. Colca, J. R.; McDonald, W. G.; Waldon, D. J.; Thomasco, L. M.; Gadwood, R. C.; Lund, E. T.; Cavey, G. S.; Mathews, W. R.; Adams, L. D.; Cecil, E. T. et al. *J. Biol. Chem.* **2003**, *278*, 21972–21979.
203. Kloss, P.; Xiong, L.; Shinabarger, D. L.; Mankin, A. S. *J. Mol. Biol.* **1999**, *294*, 93–101.
204. Sander, P.; Belova, L.; Kidan, Y. G.; Pfister, P.; Mankin, A. S.; Böttger, E. C. *Mol. Microbiol.* **2002**, *46*, 1295–1304.
205. Lin, A. H.; Murray, R. W.; Vidmar, T. J.; Marotti, K. R. *Antimicrob. Agents Chemother.* **1997**, *41*, 2127–2131.
206. Xiong, L.; Kloss, P.; Douthwaite, S.; Andersen, N. M.; Swaney, S.; Shinabarger, D. L.; Mankin, A. S. *J. Bacteriol.* **2000**, *182*, 5325–5331.
207a. Aoki, H.; Ke, L.; Poppe, S. M.; Poel, T.-J.; Weaver, E. A.; Gadwood, R. C.; Thomas, R. C.; Shinabarger, D. L.; Ganoza, M. C. *Antimicrob. Agents Chemother.* **2002**, *46*, 1080–1085.
207b. Patel, U.; Yan, Y. P.; Hobbs, F. W., Jr.; Kaczmarczyk, J.; Slee, A. M.; Pompliano, D. L.; Kurilla, M. G.; Bobkova, E. V. *J. Biol. Chem.* **2001**, *276*, 37199–37205.
208. Swaney, S. M.; Aoki, H.; Ganoza, M. C.; Shinabarger, D. L. *Antimicrob. Agents Chemother.* **1998**, *42*, 3251–3255.
209. Steitz, T. A.; Moore, P. B.; Ban, N.; Nissen, P.; Hansen, J.; Sutcliffe, J. A.; Oyelere, A. K.; Ippolito, J. A. Crystal Structures of Ribosome 50S Subunit and Its Complexes with Protein Synthesis Inhibitors and Use for Homology Modeling and Rational Antibiotic Design. European Patent Application 2003, EP 1308457 A1 20030507 *Chem. Abstr.* **2003**, *138*, 350275.
210. Thompson, J.; O'Connor, M.; Mills, J. A.; Dahlberg, A. E. *J. Mol. Biol.* **2002**, *322*, 273–279.
211. Hoellman, D. B.; Lin, G.; Ednie, L. M.; Rattan, A.; Jacobs, M. R.; Appelbaum, P. C. *Antimicrob. Agents Chemother.* **2003**, *47*, 1148–1150.
212. Pankey, G. A.; Sabath, L. D. *Clin. Infect. Dis.* **2004**, *38*, 864–870.
213. Gemmell, C. G.; Ford, C. W. *J. Antimicrob. Chemother.* **2002**, *50*, 665–672.
214. Nagiec, E. E.; Wu, L.; Swaney, S. M.; Chosay, J. G.; Ross, D. E.; Brieland, J. K.; Leach, K., L. *Antimicrob. Agents Chemother.* **2005**, *49*, 3896–3902.
215. Pagano, P. J.; Buchanan, L. V.; Dailey, C. F.; Haas, J. V.; Van Enk, R. A.; Gibson, J. K. *Int. J. Antimicrob. Agents* **2004**, *23*, 226–234.
216. Rattan, A.; Mehta, A.; Das, B.; Pandya, M.; Bhateja, P.; Mathur, T.; Singhal, S.; Sood, R.; Malhotra, S.; Yadav, A. et al. *Abstracts of Papers*; Proceedings of the 42nd Interscience Conference on Antimicrobial Agents and Chemotherapy; American Society for Microbiology, Washington, DC, 2002; Abs F-1291.
217. Maple, P. A. C.; Hamilton-Miller, J. M. T.; Brumfitt, W. *J. Antimicrob. Chemother.* **1989**, *23*, 517–525.
218. Fridkin, S. K.; Hageman, J.; McDougal, L. K.; Mohammed, J.; Jarvis, W. R.; Perl, T. M.; Tenover, F. C. *Clin. Infect. Dis.* **2003**, *36*, 429–439.
219. Biedenbach, D. J.; Jones, R. N. *Diag. Microbiol. Infect. Dis.* **2001**, *39*, 49–53.
220. Streit, J. M.; Jones, R. N.; Sader, H. S.; Fritsche, T. R. *Int. J. Antimicrob. Agents* **2004**, *24*, 111–118.
221. Stalker, D. J.; Jungbluth, G. L. *Clin. Pharmacokinet.* **2003**, *42*, 1129–1140.
222. Li, J. Z.; Willke, R. J.; Rittenhouse, B. E.; Glick, H. A. *Pharmacotherapy* **2002**, *22*, 45S–54S.
223. Gee, T.; Ellis, R.; Marshall, G.; Andrews, J.; Ashby, J.; Wise, R. *Antimicrob. Agents Chemother.* **2001**, *45*, 1843–1846.
224. Conte, J. E., Jr.; Golden, J. A.; Kipps, J.; Zurlinden, E. *Antimicrob. Agents Chemother.* **2002**, *46*, 1475–1480.
225a. Wunderink, R. G.; Rello, J.; Cammarata, S. K.; Croos-Dabrera, R. V.; Kollef, M. H. *Chest* **2003**, *124*, 1789–1797.
225b. Kollef, M. H.; Rello, J.; Cammarata, S. K.; Croos-Dabrera, R. V.; Wunderink, R. G. *Intens. Care Med.* **2004**, *30*, 388–394.
226. Lovering, A. M.; Zhang, J.; Bannister, G. C.; Lankester, B. J. A.; Brown, J. H. M.; Narendra, G.; MacGowan, A. P. *J. Antimicrob. Chemother.* **2002**, *50*, 73–77.
227. Vázquez, E. G.; Mensa, J.; López, Y.; Couchard, P. D.; Soy, D.; Fontenla, J. R.; Sarasa, M.; Carné, X.; Montull, E. *Antimicrob. Agents Chemother.* **2004**, *48*, 670–672.
228. Andes, D.; van Ogtrop, M. L.; Peng, J.; Craig, W. A. *Antimicrob. Agents Chemother.* **2002**, *46*, 3484–3489.
229. Craig, W. A. *Clin. Infect. Dis.* **2001**, *33*, S233–S237.
230. Malhotra, S.; Gautam, R.; Bhadauriya, T.; Malhotra, S.; Kakkar, S.; Paliwal, J. K; Rattan A. *23rd International Congress Chemotherapy* (June 7–9, Durban), 2003; Abs MO 157.
231. Wienkers, L. C. *Drug Metab. Dispos.* **2000**, *28*, 1014–1017.
232. Brier, M. E.; Stalker, D. J.; Aronoff, G. R.; Batts, D. H.; Ryan, K. K.; O'Grady, M.; Hopkins, N.; Jungbluth, G. L. *Antimicrob. Agents Chemother.* **2003**, *47*, 2775–2780.
233. Ashtekar, D. R.; Costa-Periera, R.; Nagarajan, K.; Vishvanathan, N.; Bhatt, A. D.; Rittel, W. *Antimicrob. Agents Chemother.* **1993**, *37*, 183–186.
234. Ashtekar, D. R.; Costa-Periera, R.; Shrinivasan, T.; Iyyer, R.; Vshvanathan, N.; Rittel, W. *Diagn. Microbiol. Infect. Dis.* **1991**, *14*, 465–471.
235. Ashtekar, D. R.; Costa-Periera, R.; Ayyer, R.; Shrinivasan, T.; Vishvanathan, N.; Nagarajan, K. *Abstracts of Papers*; Proceedings of the 29th Interscience Conference on Antimicrobial Agents and Chemotherapy; American Society for Microbiology: Washington, DC, 1989: paper 889.
236. Alcala, L.; Ruiz-Serrano, M. J.; Turegano, C. P.-F.; De Viedma, D. G.; Diaz-Infantes, M.; Marin-Arriaza, M.; Bouza, E. *Antimicrob. Agents Chemother.* **2003**, *47*, 416–417.
237. Kaatz, G. W.; Seo, S. M. *Antimicrob. Agents Chemother.* **1996**, *40*, 799–801.
238. Marshall, S. H.; Donskey, C. J.; Hutton-Thomas, R.; Salata, R. A.; Rice, L. B. *Antimicrob. Agents Chemother.* **2002**, *46*, 3334–3336.
239. Meka, V. G.; Gold, H. S. *Clin. Infect. Dis.* **2004**, *39*, 1010–1015.
240. Prystrowsky, J.; Siddiqui, F.; Chosay, J.; Shinabarger, D. L.; Millichap, J.; Peterson, L. R.; Noskin, G. A. *Antimicrob. Agents Chemother.* **2001**, *45*, 2154–2156.
241. Farrell, D. J.; Morrissey, I.; Bakker, S.; Buckridge, S.; Felmingham, D. *Antimicrob. Agents Chemother.* **2004**, *48*, 3169–3171.
242. Tsiodras, S.; Gold, H. S.; Sakoulas, G.; Eliopoulos, G. M.; Wennersten, C.; Venkataraman, L.; Moellering, R. C., Jr.; Ferraro, M. J. *Lancet* **2001**, *358*, 207–208.
243. Mutnick, A. H.; Enne, V.; Jones, R. N. *Ann. Pharmacother.* **2003**, *37*, 769–774.
244. Gonzales, R. D.; Schreckenberger, P. C.; Graham, M. B.; Kelkar, S.; Den Besten, K.; Quinn, J. P. *Lancet* **2001**, *357*, 1179.

245. Min, S. S.; Weber, D. J.; Donovan, B. J.; Gilligan, P. H.; Cairns, B. A.; Doern, G. V.; Rutala, W. A. *Clin. Infect. Dis.* **2003**, *36*, 1210–1211.
246. Herrero, I. A.; Issa, N. C.; Patel, R. *N. Engl. J. Med.* **2002**, *346*, 867–869.
247. Strolin Benedetti, M.; Dostert, P. *J. Neural Transmiss. Suppl.* **1987**, *23*, 103–119.
248. Ramsay, R. R.; Gravestock, M. B. *Mini-Rev. Med. Chem.* **2003**, *3*, 129–136.
249. Moureau, F.; Wouters, J.; Vercauteren, D. P.; Collin, S.; Evrard, G.; Durant, F.; Ducrey, F.; Koenig, J. J.; Jarreau, F. X. *Eur. J. Med. Chem.* **1992**, *27*, 939–948.
250. Fowler, C. J.; Strolin Benedetti, M. *J. Neurochem.* **1983**, *40*, 510–513.
251. Humphrey, S. J.; Curry, J. T.; Turman, C. N.; Stryd, R. P. *J. Cardiovasc. Pharmacol.* **2001**, *37*, 548–563.
252. Martin, J. P.; Herberg. J. T.; Slatter, J. G.; Depuis, M. J. *Abstracts of Papers*; Proceedings of the 38th Interscience Conference on Antimicrobial Agents and Chemotherapy, American Society for Microbiology: Washington, DC, 1998; Paper A-85.
253. Bergeron, L.; Boulé, M.; Perreault, S. *Ann. Pharmacother.* **2005**, *39*, 956–961.
254a. Wigen, C. L.; Goetz, M. B. *Clin. Infect. Dis.* **2002**, *34*, 1651–1652.
254b. Gillman, P. K. *Clin. Infect. Dis.* **2003**, *37*, 1274–1275.
255. Antal, E. J.; Hendershot, P. E.; Batts, D. H.; Sheu, W. P.; Hopkins, N. K.; Donaldson, K. M. *J. Clin. Pharmacol.* **2001**, *41*, 552–562.
256. Kuter, D. J.; Tillotson, G. S. *Pharmacotherapy* **2001**, *21*, 1010–1013.
257. Rao, N.; Ziran, B. H.; Wagener, M. M.; Santa, E. R.; Yu, V. L. *Clin. Infect. Dis.* **2004**, *38*, 1058–1064.
258. Nasraway, S. A.; Shorr, A. F.; Kuter, D. J.; O'Grady, N.; Le, V. H.; Cammarata, S. K. *Clin. Infect. Dis.* **2003**, *37*, 1609–1616.
259. Bartel, S.; Endermann, R.; Guarnieri, W.; Haebich, D.; Haerter, M.; Kroll, H.; Raddatz, S.; Riedl, B.; Rosentreter, U.; Ruppelt, M. et al. *Abstracts of Papers*; Proceedings of the 40th Interscience Conference on Antimicrobial Agents and Chemotherapy; American Society for Microbiology: Washington, DC, 2000; Abs F-1823.

Biography

Steven J Brickner, PhD, is Research Fellow, Antibacterials Chemistry, at Pfizer Global Research and Development in Groton, CT. He graduated from Miami University (Ohio) with a BS in Chemistry with Honors in 1976. He received his MS and PhD degrees in Organic Chemistry from Cornell University and was an NIH Postdoctoral Research Fellow at the University of Wisconsin-Madison. Dr Brickner is a medicinal chemist with over 20 years of research experience in the pharmaceutical industry, all focused on the discovery and development of novel antibacterial agents. He is an inventor/co-inventor on 21 US patents, and has published numerous scientific papers within the areas of oxazolidinones and novel azetidinones. Since 1997, Dr Brickner has been a member of the Forum on Microbial Threats at the Institute of Medicine (National Academy of Sciences), and is a member of the Editorial Advisory Board for *Current Pharmaceutical Design*. He was named the 2002–03 Outstanding Alumni Lecturer, College of Arts and Science, Miami University (Ohio), and was a co-recipient of the 2003 American Chemical Society's 31st Northeast Regional Industrial Innovation Award. Prior to joining Pfizer in 1996, he led a team at Pharmacia and Upjohn that discovered and developed Zyvox (linezolid).

Comprehensive Medicinal Chemistry II
ISBN (set): 0-08-044513-6

ISBN (Volume 7) 0-08-044520-9; pp. 673–698

7.24 Antimycobacterium Agents

Z Ma, A M Ginsberg, and M Spigelman, Global Alliance for TB Drug Development, New York, NY, USA

7.24.1 Introduction

The mycobacteria are responsible for more human disease than any other bacterial genus. There are 115 species of mycobacteria currently recognized,[1] over 30 of which are capable of causing human disease. Mycobacteria are nonmotile, acid-fast, weakly Gram-positive bacilli, in the shape of slender, straight, or slightly curved rods. They are subdivided into facultative and obligate human pathogens, and further classified based on their growth rate in culture into rapid- and slow-growing species. The most pathogenic members of the genus are *Mycobacterium tuberculosis*, the species responsible for tuberculosis (TB), *M. leprae*, the agent that causes leprosy, and *M. ulcerans*, the cause of Buruli ulcer, a devastating, necrotizing infection of the skin, highly prevalent in many tropical countries, most prominently in West Africa.

With the explosion of the AIDS epidemic over the past several decades, the global TB epidemic has grown and spread. Disease due to mycobacteria other than TB (MOTT; also known as nontuberculous mycobacteria, atypical mycobacteria, or environmental mycobacteria) has also become significantly more prevalent, as human immunodeficiency virus (HIV)-infected individuals are particularly susceptible to a number of mycobacterial infections, including, for example, infections due to *M. avium intracellulare* complex (MAC). This chapter will focus primarily on drug therapy for TB, due to the exceptional global morbidity and mortality burden of the disease – both current drugs and novel compounds under development.

7.24.2 Disease State

Worldwide, TB is the second-leading cause of death due to a single infectious agent, after AIDS. In 2003, there were an estimated 1.7 million deaths due to TB and 8.8 million new cases, representing a 1% annual growth rate globally.[2] Without modern anti-TB chemotherapy, which began with the discovery of streptomycin by Schatz, Bugie, and Waksman in 1944,[3] death rates due to TB would be significantly higher. TB appears to have been present, although relatively rare, in the prehistoric era in mammals and humans, and *M. tuberculosis* DNA has been identified in pre-Columbian mummies from Peru.[4] By the 1500s, a TB epidemic known as 'The Great White Plague' had spread through Europe. TB continued to spread and increase in incidence throughout Western Europe and North America from the 1600s to early 1800s, when approximately 25% of all deaths were due to TB. It remained a major cause of mortality throughout the first half of the twentieth century in these parts of the world, causing an estimated 110 000 deaths per year in the USA in the early 1900s, before its incidence began to decline. This decline is most convincingly attributed to a combination of improved living and sanitation standards, and perhaps the survival of a more TB-resistant human population.[5] Subsequently, the TB epidemic spread to Africa and Asia, primarily in the latter half of the twentieth century. Today, the highest morbidity and mortality burden due to TB is found in sub-Saharan Africa, parts of Asia, and the Russian Federation.[2]

Currently, one-third of the world's population, approximately 2 billion people, is estimated to be infected with *M. tuberculosis*. Although the majority (over 90%) of *M. tuberculosis*-infected individuals remain latently infected throughout their lives without experiencing any clinical symptoms or being infectious to others, this population represents an enormous reservoir of future cases of active, transmissible disease. Among these latently infected individuals, those with intact immune systems have a 5–10% lifetime risk of developing active (clinically symptomatic) TB and becoming capable of transmitting the infection to others; those who are HIV-infected or otherwise immunocompromised have an estimated 8% per year probability of developing active TB. Latent TB infection (LTBI) is typically curable, but requires a lengthy treatment with drugs having significant rates of toxicity,[6] and therefore treatment often is not provided to these clinically asymptomatic individuals.

7.24.2.1 Human Immunodeficiency Virus–Tuberculosis

The HIV and TB epidemics not only co-exist in most parts of the world today, but are synergistic. In individuals latently infected with *M. tuberculosis*, HIV co-infection is estimated to increase the risk of progression to active TB by 50-fold on a yearly basis compared with those who are not infected with HIV. The risk in these co-infected individuals of activating their *M. tuberculosis* infections appears to correlate with their degree of suppression of cellular immunity as measured by CD4+ cell counts. Presentation of TB also correlates with CD4+ cell counts: patients with higher counts present with typical pulmonary TB, whereas those with low CD4+ cell counts tend to present with extrapulmonary TB, and more often have sputum smears and cultures negative for *M. tuberculosis*, making their TB more difficult to diagnose. Approximately, 12 million individuals are currently co-infected with HIV and *M. tuberculosis* (HIV–TB co-infected), and approximately 15% of AIDS patients globally die of TB.[7]

Treatment of TB in AIDS patients is complicated by drug–drug interactions between antiretroviral agents and anti-TB agents, particularly HIV protease inhibitors and rifamycin derivatives, especially rifampicin (rifampin in the USA). Rifampicin induces hepatic microsomal enzymes, specifically cytochromes CYP1A2, CYP2C9, CYP2C19, and CYP3A4 and P-glycoprotein (P-gp), as well as weakly inducing 2D6. Rifampicin is also a substrate for P-gp. HIV protease inhibitors such as indinavir and nelfinavir are substrates for and inhibitors of CYP3A4 and P-gp. The most significant result of these interactions is that rifampicin decreases the serum half-lives of these antiretroviral agents. Rifabutin, a rifampicin analog, can be prescribed instead to patients on indinavir or nelfinavir, but rifabutin is not entirely free of these interactions. HIV-infected patients also tend to have decreased exposures to standard TB drugs, most likely from poor absorption. As a result, effective and safe co-administration of TB treatment and HAART (highly active antiretroviral therapy) requires careful monitoring of drug serum levels and is impractical in many endemic country settings, further compromising effective treatment of HIV–TB co-infected patients.

7.24.2.2 Multidrug Resistant-Tuberculosis (MDR-TB)

Multidrug resistant TB (MDR-TB) is increasing in prevalence, and threatens the ability of standard control measures to contain the global TB epidemic.[8] In TB, drug resistance is mostly a human-made problem, resulting from inappropriate prescribing, poor treatment adherence, irregular drug supply, and/or poor drug quality. The most recent report of the World Health Organization (WHO) Global Project on Anti-Tuberculosis Drug Resistance Surveillance found drug resistance in 74 of 77 settings tested from 1999 to 2002 and in all regions of the world. The highest levels of MDR-TB among newly diagnosed cases were found in countries of the former Soviet Union, China, Ecuador, and Israel. Globally, a median of 1.1% of all newly diagnosed cases were found to be MDR (range: 0–14.2%); in previously treated cases, a median of 7% were MDR (range: 0–58%), and a median of 18.4% demonstrated resistance to at least one of the first-line TB drugs (range: 0–82%).[2] Treatment of MDR-TB is significantly more costly, toxic, and complex than treatment of drug-sensitive TB, and relies on a battery of second-line drugs.

7.24.3 Disease Basis

TB can take many forms in the human host, but the most common is pulmonary disease, in which the bacilli are inhaled from aerosols generated, for example, by coughing, forceful breathing, or sneezing by a patient with active disease, and are deposited on the alveolar surfaces of the lung in the terminal air sacs. There they are taken up by and replicate within macrophages, where they reside within a membrane-bound vacuole and inhibit maturation of the vacuole.[9–12]

The typical pattern of disease has been described as occurring in four stages.[13] The first is the implantation of inhaled bacteria in alveoli, which occurs 3–8 weeks postinhalation. It is followed by dissemination through the lymphatic circulation to the regional lymph nodes in the lung, forming the primary (or Ghon) complex. In the second stage, which occurs over the ensuing approximate 3 months, the bacteria circulate through the bloodstream to other organs. Some patients suffer fatal disease during this stage, known as 'primary' disease. The third stage can occur at any time up to 2 years following infection, and is typically marked by inflammation of the pleural surfaces (pleurisy) and severe chest pain.[14] The fourth and final stage consists of resolution of the primary complex, and can take several years. In some cases, extrapulmonary disease can become manifest during this time. Before the HIV epidemic, approximately 5–10% of newly diagnosed cases were extrapulmonary.[15] In HIV-positive individuals, however, over 50% of cases are extrapulmonary. Because extrapulmonary disease is significantly more difficult to diagnose than pulmonary TB, these figures are estimations. As noted above, most immunocompetent, infected persons do not exhibit clinical disease, and remain latently infected throughout their lifetimes. But postprimary or 'reactivation' disease occurs in approximately 10% of these individuals at some point during their lives.

As noted above, the mycobacteria are intracellular pathogens, residing primarily within host macrophages. They have a complex and unique cell wall, which has been elucidated through research extending back to the 1960s. This research has been based on a combination of classic biochemical techniques with more modern genomics, nuclear magnetic resonance, and mass spectral analysis. The cell wall structure is composed of a peptidoglycan core covalently joined through a linker (L-Rha-D-GlcNAc-P) to galactofuran, which itself is connected to highly branched arabinofurans, which are in turn attached to mycolic acids. The last form a lipid barricade, key to many aspects of TB pathogenesis.[16] The metabolic pathways involved in synthesis of these cell wall lipids have the potential to be attractive drug targets once elucidated.

Other targets that will presumably be crucial to the development of drugs to shorten current therapy are those involved in the pathways essential to bacterial survival during the persistent state. Persistence is operationally defined here as the phenotypic bacterial state(s) that enable(s) *M. tuberculosis* to evade chemotherapy for prolonged periods, and thus is responsible for the extended duration of current TB treatment regimens. Genomic approaches have been used to identify genes and pathways transcriptionally active during persistence.[17] The pathologic hallmark of TB is the caseating granuloma. The host immune response and molecular mechanisms responsible for the formation of the caseating granuloma are not yet fully defined, but are likely to be the key, ultimately, to understanding *M. tuberculosis* persistence and latency.

7.24.4 Experimental Disease Models

Experimental disease models of TB, as in other disease areas, play a critical role in the development of effective diagnostics, vaccines, and therapeutic agents. Since the identification of *M. tuberculosis* as the etiological agent of TB,[18] many forms of in vitro and in vivo models have been developed. TB is a complicated disease involving many disease forms and stages, and the mechanism by which *M. tuberculosis* becomes persistent is still largely unknown. The validation of many of the disease models is difficult if not impossible, and the predictive value of these models continues to be the subject of debate. This section will focus on in vitro and in vivo models that are relevant to and commonly used in TB drug discovery research.

7.24.4.1 In Vitro Drug Susceptibility Models

The most commonly used in vitro model is a drug susceptibility test that measures the minimum inhibitory concentration (MIC) of a given drug or drug combination against *M. tuberculosis* in its exponential growth phase.[19] This model is performed in a rich, highly oxygenated culture medium. There have been several new variations of the susceptibility test described for rapid, high-throughput screening of drug susceptibilities.[20] Because of the rapid, high-throughput nature, the susceptibility tests are primarily used for confirming biochemical leads, developing structure–activity relationships (SARs), and evaluating the microbial susceptibility or resistance to a given drug. Organisms are generally more susceptible to drugs during the exponential phase of growth under oxygen- and nutrient-rich conditions. In an infected host, pathogens may adopt different physiological growth states, depending on the local growth conditions. In certain loci such as inside a lung granuloma, pathogens adopt a slowly replicating or nonreplicating state, and are extremely hard to eradicate. Therefore, susceptibility tests performed under nutrient- and oxygen-rich conditions may have limited value for predicting the efficacy against pathogens in an infected host.

To address the deficiency of the conventional susceptibility tests, many in vitro persistence models have been introduced.[21] The Wayne model is perhaps the most commonly used and cited persistence model; in this model, the persistent state is induced by slow oxygen depletion in capped tubes.[22] The Wayne model appears to be most useful in predicting in vivo bacterial sterilizing activity against TB. However, this model has clear limitations as a drug discovery tool due to its low throughput, long testing duration, and requirement for large amount of test compounds.

Another widely used susceptibility model is the ex vivo intracellular macrophage model.[23] This model is supported by a body of evidence demonstrating that macrophages are the predominantly infected cell type in TB[24] and may play an important role in drug persistence and latent infections.

7.24.4.2 In Vivo Models for Acute, Chronic, and Latent Infections

Animal models have served an important function in the development of therapeutic agents against TB.[25] The models vary significantly depending on a number of parameters, including animal species, bacterial strains, and routes and stages of infection. It is crucial to use models that are relevant to the issues being addressed. For a lead optimization program, the predictability, speed, throughput, and sample size are important factors to be considered. Mouse models have become the primary choice at this stage, in order to quickly screen a large number of compounds. After a drug candidate is selected, models with the best predictive value for treatment results in human disease are the best choices for studying proper dose regimens and drug combinations. In this instance, guinea pig, rabbit, or even monkey models could be considered. However, to date, the best substantiated model for this purpose is the aerosol-infected mouse model.

Various mouse models have been developed to mimic *M. tuberculosis* infections in the acute, chronic, or latent stage.[26] Compared with models in other animal species, the mouse models are better characterized and have clear advantages in the lead optimization stage of a drug discovery program. The acute infection model[27,28] can be used for quickly screening a large number of compounds and assessing some of the important pharmacokinetic/pharmacodynamic parameters, such as drug oral availability and penetration into the infection locus. Compounds with good efficacy in the acute model can then be moved into the chronic infection model,[29] to assess efficacy in the chronic phase of infection. Mouse models have played a key role in developing the current anti-TB agents. However, due to differences in the host immune response to *M. tuberculosis* infection, mouse models are not expected to reproduce the human disease in all aspects. Most notably, mice do not form caseating granulomas in response to *M. tuberculosis* infection.

The guinea pig and rabbit have also been used to develop TB models. These models may have some specific advantages over the mouse model; in particular, the lung pathology from infected guinea pigs and rabbits resembles more closely that from TB patients. Rabbits are the only nonprimate known to form caseating granulomas. These species may therefore be better than mouse models for studying persistent and latent infections. However, additional studies are clearly needed to validate this hypothesis. Both the guinea pig and rabbit models are more expensive and require a larger amount of test compounds than the mouse models. The monkey model may mimic human TB most closely of all the animal models.[30] However, economic and ethical concerns limit its value in drug development.

7.24.5 Clinical Trial Issues

The design of clinical trials to evaluate new TB drugs faces a number of key challenges. These include, first, that efficacy trials are of long duration – due to both the ability of the bacteria to evade at least the current chemotherapeutic agents for a prolonged period (i.e., 6-month current treatment duration) and the lack of surrogate markers that could substitute reliably for the need to measure long-term relapse rates as a primary efficacy endpoint. The current state-of-the-art is the assessment of the efficacy of a treatment based on relapse rates in the first year or more following completion of therapy. Although the percentage of patients converting their sputum to bacteriologic negativity after 2 months of treatment is used in early clinical trials as an indicator of the ability of a regimen to shorten treatment duration, this endpoint is neither very sensitive nor rigorously validated at this time. A second major challenge for the design of TB treatment trials is that TB therapy must consist of multiple drugs used in combination to prevent development of resistance. Therefore, conventionally, one new drug under development is added to or substituted into the current regimen at a time, necessitating 6–8 years to evaluate each new drug in sequence, and therefore potentially two or more decades to test an entirely new multidrug regimen.

The design of clinical trials for new MDR-TB treatments will be even more challenging, as each patient should in theory receive an individualized treatment regimen based on the drug susceptibilities of that patient's own strain of *M. tuberculosis*. Consequently, it is difficult to devise appropriate control groups for these trials, and therefore difficult to accurately assess the efficacy of a new drug or regimen.

Lastly, trials of new treatments for LTBI require relatively large patient numbers and long duration times, as the endpoint is development of active disease, which can occur, as noted previously, in a small number of infected patients as well as decades after the time of infection.[31–34] Further complicating the design of these trials has been the lack of a sufficiently specific diagnostic tool. The tuberculin skin test, the standard tool for diagnosing infection with *M. tuberculosis*, can be positive not only as a result of infection with *M. tuberculosis*, but also from bacillus Calmette–Guérin (BCG) vaccination or exposure to mycobacteria other than TB. New-generation diagnostics recently introduced and based on T cell responses to *M. tuberculosis*-specific antigens should be an improvement in this latter regard.[35–37] Another approach taken recently by some investigators to streamline the clinical evaluation of new regimens for the treatment of LTBI is to study populations at high risk of developing active disease, such as HIV-positive patients.[38–41] This approach limits the duration of follow-up needed to observe an adequate number of clinical events for achievement of statistical power.

7.24.6 Current Treatment

Currently, active TB is treated by combination therapies that consist of three or more drugs (four, most typically) selected from more than a dozen known anti-TB agents. Directly observed treatment, short course (DOTS), is considered the regimen and treatment approach of choice for active TB, and has been recommended and promulgated globally by the WHO. During the treatment, patients with active TB are typically administered isoniazid, rifampicin, pyrazinamide, and ethambutol for 2 months (the intensive phase), followed by isoniazid and rifampicin for an additional 4 months (the continuation phase). The initial intention of a combination therapy was to minimize the development of resistance to streptomycin after that drug was first introduced. More recently, it has come to be believed by many in the field that various drugs in the standard regimen act orchestrally against different populations of *M. tuberculosis*[42] Isoniazid, a cell wall synthesis inhibitor, kills actively growing bacteria rapidly, and plays a key role in eradicating the replicating population. Rifampicin, an inhibitor of RNA synthesis, is active against both replicating and slowly or nonreplicating bacteria. Pyrazinamide, presumably an inhibitor of proton motive force, appears to be active only under acidic conditions during the first 2 months of therapy. Rifampicin and pyrazinamide played a major role in shortening the duration of therapy, from more than 24 months to 6 months currently. It is reasonable to believe that the mechanism of action of each individual agent dictates the role of this agent in TB therapy. The more than one dozen anti-TB agents presently in the arsenal for the treatment and prevention of TB can be divided into six groups, based on their mechanisms of action, as show in **Figure 1**. The mechanisms of action for some agents are not totally defined, and therefore what is indicated in **Figure 1** for these agents must be treated as hypothetical at this stage.

Streptomycin was initially used intramuscularly in combination with first-line drugs to treat active disease, but due to the inconvenience of parenteral delivery and concern about the transmission of HIV, it was later replaced by orally available ethambutol in the recommended regimen. LTBI is currently best treated by daily isoniazid for 9 months.

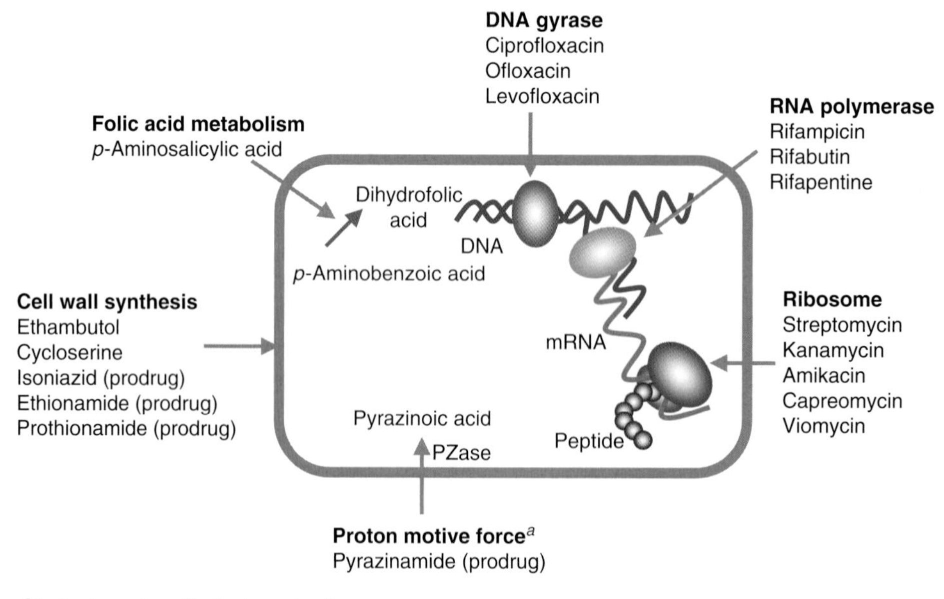

Figure 1 Schematic illustration of the sites of action for the available anti-TB agents.

Table 1 The current anti-TB agents, their available dose form, year of discovery, source, and mechanism of action

Agent	*Dose form*[a]	*Discovery*[b]	*Source*	*Mechanism*
Streptomycin	IM	1944	*Streptomyces griseus*	Inhibits ribosome/protein synthesis
p-Aminosalicyclic acid	PO	1946[d]	Synthetic	Inhibits folic acid synthesis and iron metabolism
Viomycin	IM	1951	*Streptomyces puniceus*	Inhibits ribosome/protein synthesis
Isoniazid[c]	PO/IM	1952[d]	Synthetic	Inhibits cell wall macolic acid synthesis
Pyrazinamide[c]	PO	1952[d]	Synthetic	Inhibits proton motive force and/or cell wall synthesis
Cycloserin	PO	1955	*Streptomyces orchidaceus*	Inhibits cell wall peptidoglycan synthesis
Ethionamide[c]	PO	1956[d]	Synthetic	Inhibits cell wall mycolic acid synthesis
Prothionamide[c]	PO	1956[d]	Synthetic	Inhibits cell wall mycolic acid synthesis
Kanamycin	IM/IV	1957	*Streptomyces kanamyceticus*	Inhibits ribosome/protein synthesis
Ethambutol	PO	1961	Synthetic	Inhibits arabinosyl transferase/cell wall synthesis
Capreomycin	IM	1961	*Streptomyces capreolus*	Inhibits ribosome/protein synthesis
Rifampicin	PO/IV	1966	Semisynthetic	Inhibits RNA polymerase/RNA synthesis
Amikacin	IM/IV	1972	Semisynthetic	Inhibits ribosome/protein synthesis
Rifapentine	PO	1976	Semisynthetic	Inhibits RNA polymerase/RNA synthesis
Rifabutin	PO	1979	Semisynthetic	Inhibits RNA polymerase/RNA synthesis
Ofloxacin	PO/IV	1982	Synthetic	Inhibits DNA gyrase/translation and transcription
Ciprofloxacin	PO/IV	1983	Synthetic	Inhibits DNA gyrase/translation and transcription
Levofloxacin	PO/IV	1987	Synthetic	Inhibits DNA gyrase/translation and transcription

[a] IM, intramuscular; IV, intravascular; PO, peroral.
[b] See *Merck Index*, 13th edition.
[c] Parent compound is a prodrug.
[d] Listed as year of the first introduction for TB use.

The remaining agents are used ad hoc in combinations to treat MDR-TB and infections that have failed to respond to first-line drugs. The treatment of MDR-TB infections typically requires at least 18 months of therapy. The majority of the known anti-TB agents were introduced during the antibiotic golden era between 1940 and 1960. The origins, available dose forms, years of discovery, and mechanisms of action of these anti-TB agents are summarized in **Table 1**.

7.24.6.1 Rifamycin Class

Rifamycins are broad-spectrum agents having a unique ansa structure (**Figure 2**). They were initially isolated from *Streptomyces mediteranei* in 1957.[43] The early members of the family are generally undesirable as therapeutic agents, due primarily to poor potency, low solubility, poor bioavailability, or short half-life. Structural modification of the natural rifamycins led to several derivatives that are highly potent and orally available. Currently, three semisynthetic compounds, rifampicin, rifapentine, and rifabutin, are in clinical use. Rifampicin has become the cornerstone of the current therapy, mainly responsible for reducing the treatment duration from 12 months to the current 6 months. The newer members, rifapentine and rifabutin, have demonstrated some advantages over rifampicin, including a longer half-life, a reduced potential for drug–drug interactions, and/or activity against some rifampicin-resistant strains.

7.24.6.1.1 Sites and mechanisms of action

Rifamycins are potent RNA polymerase (RNAP) inhibitors. The detailed interactions between rifampicin and RNAP have been elucidated by high-resolution crystal structure studies of the *Thermus aquaticus* core enzyme complexed with rifampicin.[44] Rifampicin binds to a deep pocket of the β subunit within the RNA channel that is about 12 Å away from the active center. The drug works by blocking the path of the elongating RNA when the transcript reaches two to three

	R^1	R^2
Rifamycin B	–OH	$-OCH_2COOH$
Rifamycin O	=O	=(1,3-dioxolan-4-on)-2-yl
Rifamycin S	=O	=O
Rifamycin SV	–OH	–OH

Figure 2 Structures of some early naturally occurring rifamycins and the rifamycin numbering system.

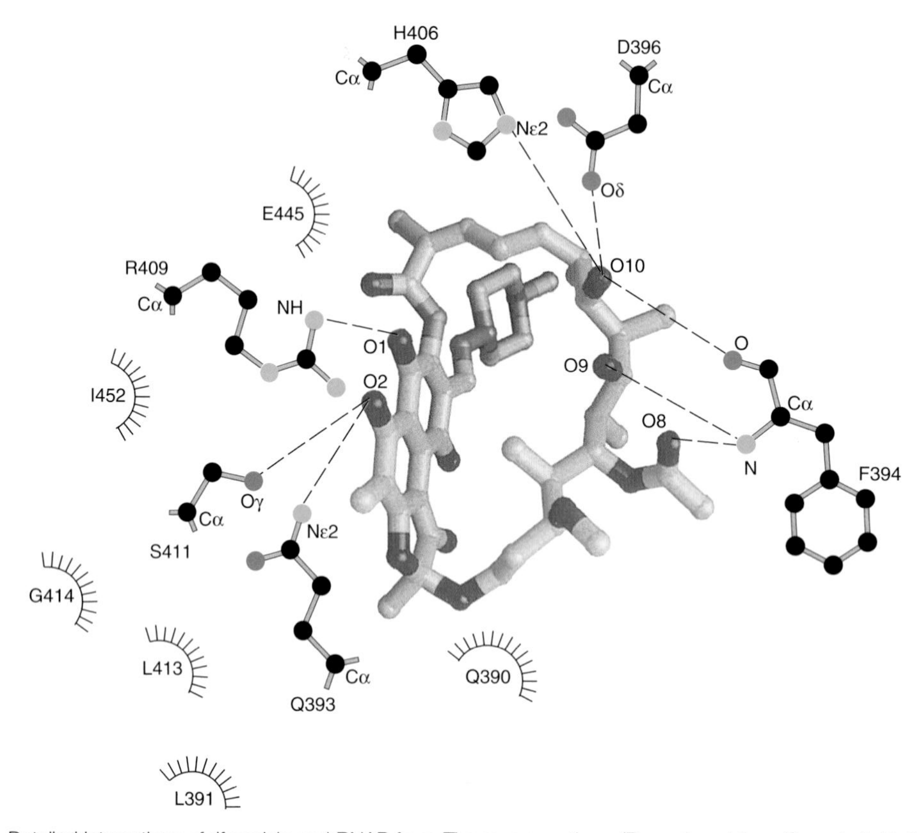

Figure 3 Detailed interactions of rifampicin and RNAP from *Thermus aquaticus*. (Reproduced from Campbell, E. A.; Korzheva, N.; Mustaev, A.; Murakami, K.; Nair, S.; Goldfarb, A.; Darst, S. A. *Cell* **2001**, *104*, 901–912, Copyright (2001), with permission from Elsevier.)

nucleotides in length. Resistance to rifamycins can occur frequently, and is mainly due to point mutations in the rifamycin-binding region of the β subunit of RNAP. A single-step mutation of one of the key residues in the binding region generally leads to high-level resistance (**Figure 3**). The most commonly observed mutations among *M. tuberculosis* clinical isolates are Q432, F433, H445, S450, and L452 residues (corresponding to Q393, F394, H406, S411, and L413 positions of *Thermus aquaticus*, respectively). The S450 and H445 mutations lead to high-level rifampicin resistance.[45]

7.24.6.1.2 Structure–activity relationships

Chemical modifications of the natural products have produced several clinically important semisynthetic rifamycins with improved pharmacological profiles. The SARs accumulated to date are highlighted in **Figure 4**.[46]

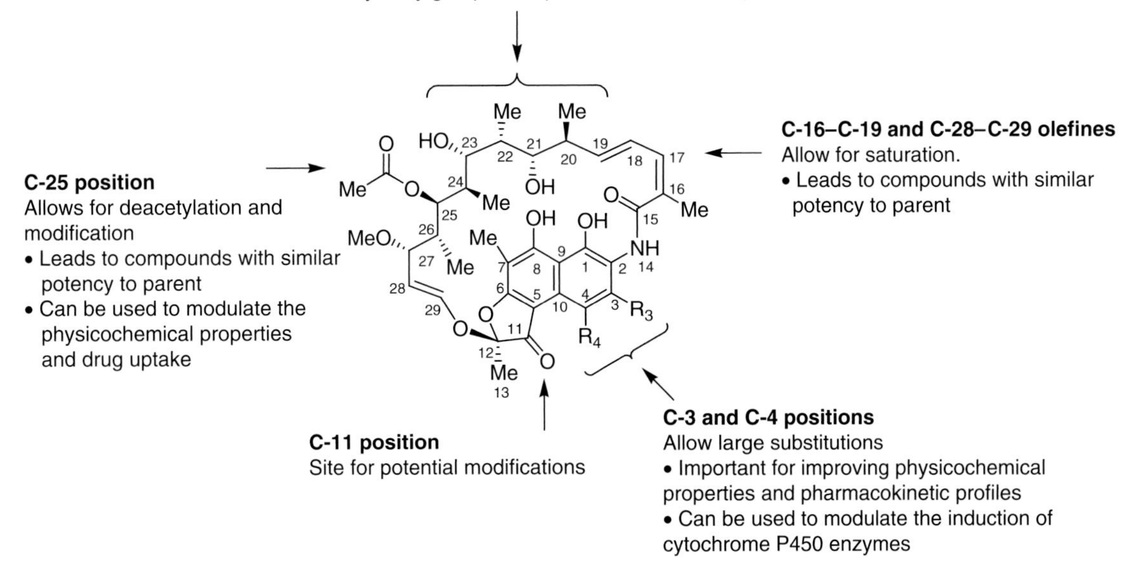
C-1, C-8, C-21, and C-23 hydroxy groups
Essential for activity
• Modification of these groups leads to inactive compounds, with exception of C-1 –OH to =O conversion
• Modification of other positions that causes conformational changes to these hydroxy groups also produce inactive compounds
C-25 position
Allows for deacetylation and modification
• Leads to compounds with similar potency to parent
• Can be used to modulate the physicochemical properties and drug uptake
C-16–C-19 and C-28–C-29 olefines
Allow for saturation.
• Leads to compounds with similar potency to parent
C-11 position
Site for potential modifications
C-3 and C-4 positions
Allow large substitutions
• Important for improving physicochemical properties and pharmacokinetic profiles
• Can be used to modulate the induction of cytochrome P450 enzymes

Figure 4 Important SARs of the rifamycin class.

The four hydroxy groups located at the C-1, C-8, C-21, and C-23 positions of the rifamycin scaffold are essential for antibacterial activity. Any modifications to these hydroxy groups, with the exception of the C-1 –OH to =O conversion, led to compounds with reduced activity. Other modifications that changed the conformation of these hydroxy groups also led to inactive compounds. Based on the co-crystal structure of rifampicin–RNAP, these hydroxy groups directly interact with RNAP through hydrogen bonding. Any modification that interferes with such interactions would lead to compounds with reduced binding affinity.

Saturation of one or more of the double bonds located at the C-16/C-17, C-18/C-19, and C-28/C-29 positions generally leads to compounds with equivalent or slightly reduced activity compared with their parent compounds. These modifications are not significant enough to have any major impact on potency, nor provide any advantage over the parent compounds. The C-25 deacetylation products are often observed as the metabolites of rifamycins. These C-25 hydroxy metabolites are generally as potent as their parent. Other modifications of the acetyl group also led to potent compounds with different physicochemical properties. However, any modifications to this position could be 'temporary' because the ester groups attached to this position are readily cleaved in vivo. The liability of the C-25 ester linkage has been exploited as an advantage to develop a rifamycin-based 'Trojan Horse' that brings other impermeable drug molecules into the cells.[47] Siderophore scaffolds were also attached to this position, and successfully transport rifamycins into Gram-negative bacteria.[48]

The most important rifamycin derivatives are those with modifications at the C-3 and C-4 positions. Synthetically, these positions are very accessible and easy to modify. Structurally, these positions point to the open space of the RNAP-binding pocket, and allow for significant modifications. Modification at the C-3 and C-4 positions often yields compounds with significantly improved physicochemical properties and pharmacokinetic profiles.[49] Structural modification of 3-formyl rifamycin SV led to rifampicin, a more potent and orally active compound that has become the cornerstone of modern anti-TB therapy. Further modification of the rifampicin series produced rifapentine, a cyclopentyl analog of rifampicin. Rifapentine has a longer half-life than rifamycin, and therefore can be used as intermittent therapy. Rifabutin was produced by bridging the C-3 and C-4 positions, to form a spiro structure. This molecule has the advantages relative to rifampicin of improved tissue penetration and reduced cytochrome P450 enzyme induction. Rifalazil is a newer member of the rifamycin family currently in clinical development for other indications. This compound has a benzoxazino group fused to the C3 and C4 positions, and has a long half-life of 61 h, no substantial cytochrome P450 induction, and activity against certain rifampicin-resistant bacteria.[50]

The co-crystal structure of rifampicin–RNAP provides some insights into potential positions for future modification.[44] The C-11 carbonyl group points to the open space of the rifampicin-binding pocket, and is a potential site for modulating physicochemical properties and pharmacokinetics of the drug class.

7.24.6.1.3 Comparisons of available agents within the class

Currently, four agents in the rifamycin class, rifampicin, rifapentine, rifabutin, and rifalazil (**Figure 5**), are in clinical use or in clinical development. Despite the structural similarities, these agents possess very distinct characteristics. The main advantages and disadvantages of these analogs are compared in **Table 2**.

7.24.6.1.4 Limitations and future directions

Rifamycins are one of the few drug classes active against bacteria in both replicating and nonreplicating states, and are mainly responsible for shortening TB treatment from over 12 months to 6 months. However, the current members of the rifamycin family are optimized for activity against bacteria in the replicating state, which is not a good predictor of their efficacy for shortening therapy. By utilizing proper assays against bacteria in the drug-persistent state, it is anticipated that a better rifamycin with optimized activity against persistent *M. tuberculosis* and greater potential for shortening therapy could be developed.

In addition to several of the intrinsic safety issues associated with the rifamycin class, two specific limitations are particularly important for modern TB therapy. The first limitation is the high prevalence of rifampicin resistance among clinical isolates of *M. tuberculosis*. Although rifabutin and rifalazil are active against some rifampicin-resistant *M. tuberculosis*, cross-resistance is expected. New agents that overcome rifampicin resistance would be advantageous. Second, rifampicin is a strong inducer of several cytochrome P450 enzymes, including CYP3A4, which leads to drug–drug interactions when co-administered with antiretroviral therapy, particularly HIV protease inhibitors and non-nucleoside reverse transcriptase inhibitors. This causes difficulties for the effective management of TB-HIV co-infections. Some progress has been made in this area, with the introduction of rifabutin and rifalazil.[51] New agents that are free of drug–drug interactions would be highly desirable.

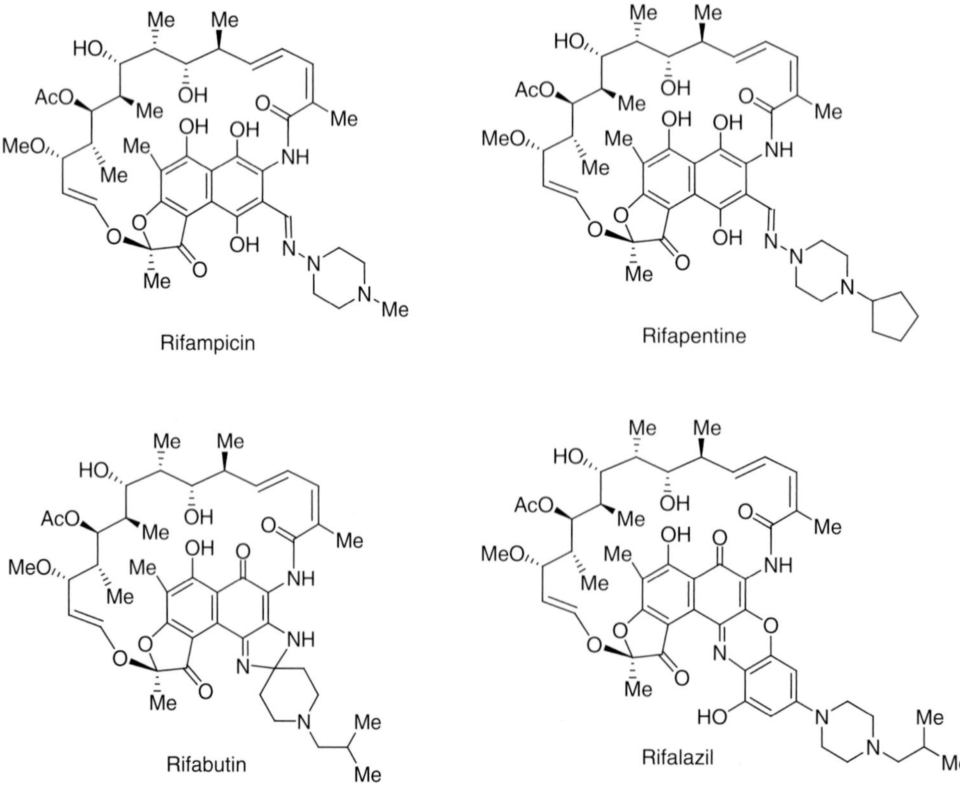

Figure 5 Structures of rifampicin, rifapentine, rifabutin, and rifalazil.

Table 2 Comparisons of rifampicin, rifapentine, rifabutin, and rifalazil as anti-TB agents

Compound (introduced)	*Key advantage(s)*	*Key disadvantage(s)*	*Current/potential use*
Rifampicin (1966)	Mainly responsible for shortening TB therapy from 12 months to 6 months	Drug–drug interactions, and drug resistance	Key component of the first-line regimen
Rifapentine (1976)	Longer half-life than rifampicin, slightly reduced cytochrome P450 induction	Cross-resistance with rifampicin, drug–drug interaction, food effects	Potential agent for intermittent therapy
Rifabutin (1979)	Higher tissue level, reduced cytochrome P450 induction and active against some rifampicin-resistant strains	Partial cross-resistance with rifampicin, low oral availability	Treatment of MAC. Potential use in TB-HIV co-infections
Rifalazil (1990)	Highly potent, no cytochrome P450 induction, longer half-life than rifapentine, active against some rifampicin-resistant strains	Partial cross-resistance with rifampicin, skin discoloration, flu-like symptoms	Potential use for intermittent therapy and TB-HIV co-infections

7.24.6.2 Isoniazid and Related Compounds

Isoniazid, also called isonicotinic acid hydrazide, is a synthetic compound first prepared in 1912. The anti-TB activity of this compound was first reported in 1952 during the search for more potent derivatives of nicotinamide, an early lead against TB. Subsequent studies also identified ethionamide and prothionamide as having potent anti-TB activities.

Isoniazid is one of the most widely used anti-TB agents, and one of the key components of first-line therapy for active disease. A 9-month isoniazid monotherapy is used for the treatment of latent infections. Isoniazid is highly effective against replicating *M. tuberculosis*, and is mainly responsible for the early reduction of the bacterial load in the initial phase of therapy. Ethionamide and prothionamide have limited applications, and are mainly used as second-line agents to treat patients who have failed to respond to first-line therapy.

7.24.6.2.1 Sites and mechanisms of action

Isoniazid, ethionamide, and prothionamide are all prodrugs. Isoniazid is activated by a catalase peroxidase enzyme (KatG). Mutations in KatG account for the majority of isoniazid resistance. The active species of isoniazid appears to have multiple cellular targets.[52] The primary target of isoniazid is believed to be InhA, a NADH-dependent enoyl acyl carrier protein reductase involved in the synthesis of mycolic acid.[53,54] The activated species, presumably an isonicotinic acyl radical, forms an adduct with the NAD radical. The resulting isonicotinic acyl-NADH adduct (INA) binds to InhA, and inhibits the synthesis of mycolic acid, an essential cell wall component of *M. tuberculosis* (**Figure 6**).[55]

The crystal structure formed between InhA isolated from *M. tuberculosis* and INA indicated that INA binds to InhA in a competitive fashion with NADH (**Figure 7**), and mutations within the NADH-binding region of InhA confer isoniazid resistance among clinical isolates.

The primary target of ethionamide is also believed to be InhA. However, this agent is activated by a different enzyme, EthA. Prothionamide is presumed to have the same mechanism of action as ethionamide due to their structural similarity. Resistance to isoniazid, ethionamide, and prothionamide is mainly due to either mutation of the activation enzymes (KatG or EthA) or their molecular target (InhA).[56] Therefore, cross-resistance between isoniazid and ethionamide/prothionamide is expected, but not common.

7.24.6.2.2 Structure–activity relationships

The discovery of isoniazid was a serendipitous event. Isoniazid was identified by pursuing derivatives of nicotinamide, an earlier anti-TB lead. Optimization of nicotinamide directed by in vitro and in vivo anti-TB assays produced two distinct drugs with very different characteristics (**Figure 8**), isoniazid and pyrazinamide.[57] While pyrazinamide exhibits a similar biological profile and shows cross-resistance with nicotinamide, isoniazid is biologically and mechanistically distinct from its parent. The discovery of isoniazid and pyrazinamide fully illustrated the advantages and potential risks

Figure 6 Activation of isoniazid and formation of isonicotinic acyl–NADH.

Figure 7 Structures of INA and the active site of InhA reveal key interactions based on the x-ray crystal structure (distances in angstroms). (Reprinted with permission from Rozwarski, D. A.; Grant, G. A.; Barton, D. H. R.; Jacobs, W. R.; Sacchettini, J. C. *Science* **1998**, *279*, 98–102. Copyright 1998 AAAS.)

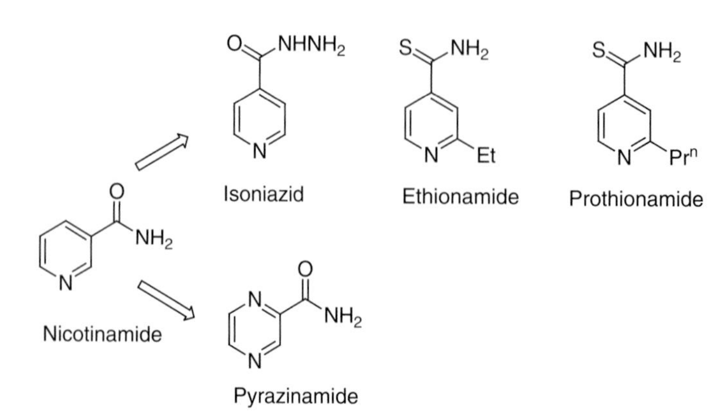

Figure 8 Structures of the early lead nicotinamide and the follow-up compounds isoniazid, ethionamide, prothionamide, and pyrazinamide.

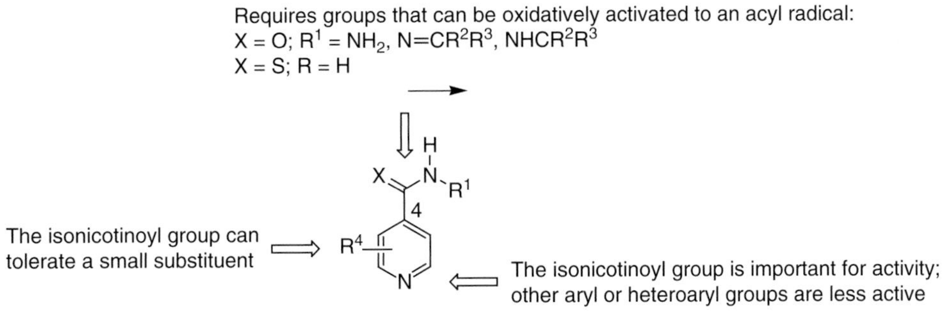

Requires groups that can be oxidatively activated to an acyl radical:
X = O; $R^1 = NH_2$, $N{=}CR^2R^3$, $NHCR^2R^3$
X = S; R = H

The isonicotinoyl group can tolerate a small substituent

The isonicotinoyl group is important for activity; other aryl or heteroaryl groups are less active

Figure 9 Highlights of the SARs of the isoniazid series.

associated with the whole-cell-based approach for lead optimization. Lead optimization solely directed by whole-cell activity can lead to compounds with 'off-target' mechanisms. In most cases, the off-target activities are nonselective and unwanted. However, in certain instances, the off-target activities can lead to serendipitous discovery of useful drugs, such as isoniazid.

Most of the medicinal chemistry on the isoniazid series was completed in the 1950s. There are two independent tracks of SARs for this drug series, one governing the activation of the prodrug and another governing the interaction of the active species with InhA. Many groups at the 4-position can be activated inside *M. tuberculosis*, and are possible prodrug structures (**Figure 9**).[58] After activation, the structural requirements for InhA binding appear to be very stringent. Only the isonicotinoyl core structure is substantially active among many aryl and heteroaryl groups explored.[59] The isonicotinoyl group could be substituted with lower alkyl groups while maintaining excellent activity.

7.24.6.2.3 Comparisons of available agents within the class

As a first-line drug, isoniazid is by far the most commonly used agent within the class. Due to its relatively good efficacy and safety, isoniazid has become one of the key components of modern anti-TB therapy. Ethionamide and prothionamide are mainly used as second-line drugs to treat MDR-TB and patients who cannot tolerate first-line agents. Recently, isoniazid resistance has become a major problem in many parts of the world.[8] The majority of isoniazid-resistant clinical isolates have mutations in the activating enzyme, KatG. Since ethionamide and prothionamide are activated by a different enzyme, EthA, these agents show little cross-resistance with isoniazid. Both ethionamide and prothionamide are less tolerable compared with isoniazid.[60]

7.24.6.2.4 Limitations and future directions

There are two main limitations associated with the isoniazid class: the high prevalence of drug resistance and the low efficacy against drug-persistent bacilli. In addition, the toxicity of this drug class, particularly ethionamide and prothionamide, is a significant concern.[52]

Recent work in this area mainly focused on overcoming isoniazid resistance. A new agent that directly targets InhA without the need for KatG or EthA activation could potentially be effective against the majority of isoniazid-resistant strains. InhA inhibitors that bind to a different binding site than isoniazid may be able to overcome the remaining cause of resistance – InhA mutations.

A new member of the isoniazid family that is effective against isoniazid-resistant strains and is safer than isoniazid will likely find utility in the treatment of MDR-TB. However, it is unclear whether an agent that targets cell wall biosynthesis can play a major role in shortening therapy. The definite answer to this question will not be available, of course, until such an InhA inhibitor is identified and its efficacy against persistent *M. tuberculosis* is tested in in vitro and in vivo models, and eventually by clinical trials.

7.24.6.3 Pyrazinamide and Related Compounds

Pyrazinamide is a close analog of nicotinamide, and shares the same root with isoniazid (see **Figure 8**). Pyrazinamide is another first-line agent, and one of the two agents that played a significant role in shortening the duration of therapy from 12 months to 6 months.[61] Pyrazinamide appears to be active only under acidic conditions. The drug is more efficacious in vivo than would be predicted by its in vitro potency.

7.24.6.3.1 Sites and mechanisms of action

Pyrazinamide is a prodrug that requires activation by pyrazinamidase (PZase). The activation product is pyrazinoic acid, which is believed to be the active species. Mutations in PZase lead to loss of PZase activity, and are mainly responsible for the development of pyrazinamide resistance. The same mutations show cross-resistance to nicotinamide, suggesting that pyrazinamide and nicotinamide are activated by the same enzyme. Pyrazinoic acid, the active form of pyrazinamide, remains sensitive to these mutants. No pyrazinoic acid-resistant mutants have been identified, and the exact target of pyrazinoic acid is unclear. Although evidence based on an experiment done with 5-chloropyrazinamide, a close analog of pyrazinamide, on *M. smegmatis* suggested that fatty acid biosynthesis I (FAS-I) is the potential target of pyrazinoic acid,[62] this hypothesis was later questioned.[63] Recently, a new mechanism of action for pyrazinamide was proposed, suggesting that disruption of the proton motive force and energy production is the basis of its antibacterial activity.[64] According to this hypothesis, pyrazinoic anion, under acidic conditions, serves as a proton carrier that transports protons from the outer membrane to the intracellular space. This process effectively depletes the proton motive force and impacts energy production. This hypothesis helps to explain some of the unusual properties of pyrazinamide, including its requirement for acidic conditions and its activity against persistent bacilli.

7.24.6.3.2 Structure–activity relationships

Optimization of pyrazinamide was performed mainly in the 1950s as a follow-up to work on isonicotinamide. The SARs of this series have been extremely hard to elucidate since the series is essentially inactive under normal culture conditions. Even under acidic conditions at pH 5.5, the potency of pyrazinamide is in the range of 16 $\mu g\,mL^{-1}$ or higher. Therefore, the SARs for pyrazinamide are mainly derived through in vivo animal studies, and the data that are available for SAR analysis are very limited. One should take extra precautions when performing a SAR analysis for a drug series whose mechanism of action is not well defined. As an example, 5-chloropyrazinamide is a structural analog of pyrazinamide, but appears to have a different mechanism of action.[63] Therefore, 5-cloropyrazinamide should not be included in the SAR analysis of the pyrazinamide series. Instead, this compound could serve as a potential lead for a new series.

As prodrugs, the pyrazinamide series also follows two independent SAR tracks, one for PZase activation and another for target interaction by the resulting acid (**Figure 10**). The R^1 group serves two important functions. First, the active species, pyrazinoic acid, is usually in its anionic form at physiological pH, and is not permeable or bioavailable. The R^1 group masks the acid group, and allows the drug to pass through cell membranes. Second, the killing effect of the active species is unlikely to be a selective process. The selective removal of the R^1 group by *M. tuberculosis* and other bacterial species provides the desired selectivity. In principle, any structures that can be selectively removed by mycobacteria would be a good choice for the R^1 group. Esters, in addition to amides, have therefore been explored as prodrugs.[65] The esters appeared to have a broader spectrum of activity than amides, and did not show cross-resistance with the corresponding amide, indicating that they are activated by different enzymes.[66] Information regarding the structural requirements for the aromatic group is limited, and most of the relevant work was performed many years ago, in the 1950s. In general, structures that lead to acids with a relatively lower pK_a are more active than those that produce acids with a higher pK_a. Thus, pyrazinoic acid has better activity than nicotinic acid. The pK_a of the active species can be influenced by both the structures of the aromatic system and its substitution.

7.24.6.3.3 Comparisons of available agents within the class

Pyrazinamide is the only agent currently used clinically for the treatment of TB in this class. The initial lead, nicotinamide, was abandoned before reaching clinical development due to potential antagonism with isoniazid. This

The best R1 group is amide, a potential R1 group is an ester that could be activated by bacterial esterase
Substitutions are generally not tolerated
R2
X
O
R1
N
Pyrazinoyl (X = N) is better than nicotinoyl (X = CH); other aryl or herteroaryl groups are less active

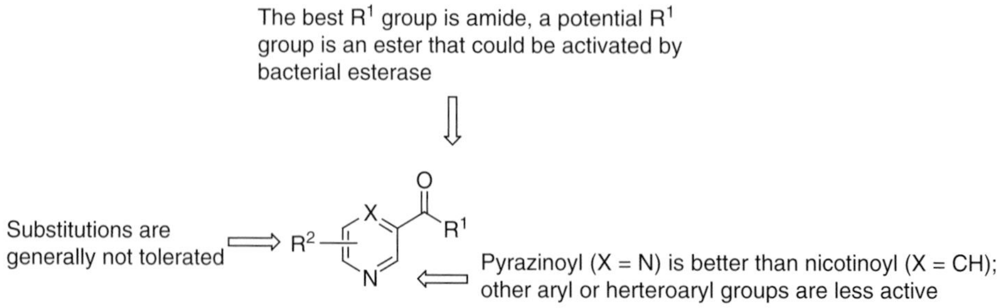

Figure 10 Highlights of the available SARs for the pyrazinamide series.

compound resurfaced recently as a potential agent that might have dual activities against both TB and HIV.[57] Pyrazinamide is more active both in vitro and in vivo than nicotinamide against *M. tuberculosis*. In addition, nicotinamide appears to show antagonism with isoniziad, while pyrazinamide does not.

7.24.6.3.4 Limitations and future directions

The most obvious limitation of pyrazinamide is, in addition to its marginal safety profile, its narrow window of activity. This agent is active only during the first 2 months of treatment and under acidic conditions. No evidence suggests that pyrizinamide has any therapeutic benefit during the continuation phase of treatment after the initial 2 months of therapy. Presumably, because pyrazinamide is only active under acidic conditions, it plays a limited role in preventing the development of resistance to co-administered agents.[67]

Due to its unique mechanism, pyrazinamide may serve as a useful tool for understanding the biology of persistence. Conversely, a better understanding of the mechanism of action of this agent may lead to new strategies to overcome persistence.

To overcome the limitations of pyrazinamide, a safer compound active against both replicating and nonreplicating bacterial populations is highly desirable. One can envision that a pyrazinamide analog with a dual mechanism of action could achieve this goal. 5-Chloropyrazinamide, a compound that appears to target both FAS-I and the proton motive force, might serve as one interesting lead.[68]

7.24.6.4 Aminoglycosides and Polypeptides

Aminoglycosides are widely used antibiotics with broad-spectrum activity. The first agent in the class, streptomycin, was isolated in the 1940s from *Streptomyces griseus*. Many additional aminoglycosides were later isolated or synthesized. Currently, three agents in the class, streptomycin, kanamycin, and amikacin, are used for the treatment of TB (**Figure 11**). Among these agents, streptomycin and kanamycin are natural products, and amikacin is a semisynthetic compound derived from kanamycin. In 1948, the first recorded, randomized, placebo-controlled trial was conducted, its purpose being to evaluate the efficacy of streptomycin as an antitubercular agent.[42] Aminoglycosides are not orally active, and have limited intracellular activity. In addition, nephrotoxicity, ototoxicity, and other adverse reactions limit their use. Aminoglycosides are used widely for the treatment of other infections, as discussed in other chapters. In this section, we will focus on the anti-TB applications of this drug class.

Polypeptides, capreomycin, and viomycin are structurally related compounds that possess similar mechanisms of action, pharmacokinetic, potency and toxicity profiles, to aminoglycosides (**Figure 12**). Capreomycin was isolated from *Streptomyces capreolus* in 1961, and viomycin, which is produced by various *Streptomyces* species, was first reported in 1951. Both compounds are parenteral agents, and possess significant nephrotoxicity and ototoxicity.

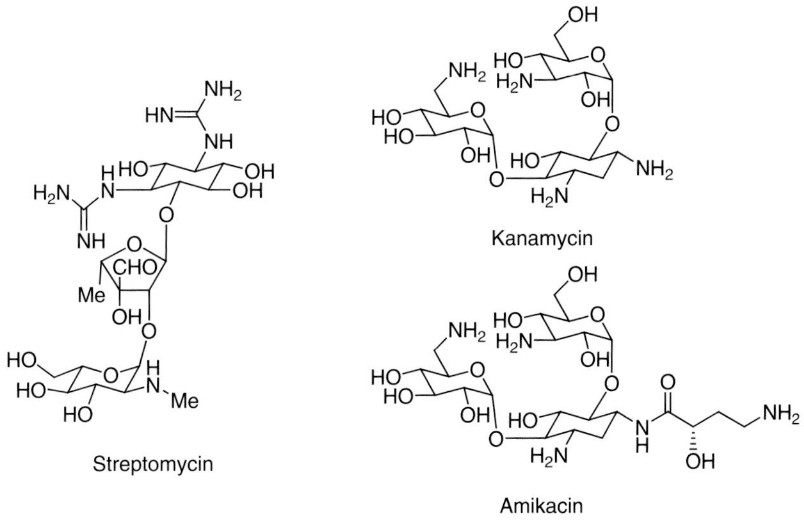

Streptomycin

Kanamycin

Amikacin

Figure 11 Structures of aminoglycosides that have been used for the treatment of TB. Amikacin is a semisynthetic compound derived from kanamycin.

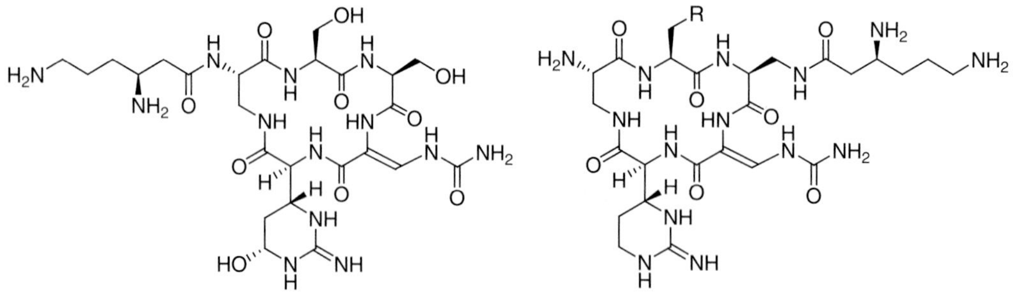

Viomycin

Capreomycin 1A: R = OH
Capreomycin 1B: R = H

Figure 12 Polypeptide anti-TB agents.

7.24.6.4.1 Sites and mechanisms of action

At physiological pH, aminoglycosides are charged molecules. They have limited permeability across the lipid bilayers of cellular membranes. Aminoglycosides are transported into the cell by an energy-dependent drug transport system that utilizes the proton gradient as its driving force.[69] The proton gradient decreases in an anaerobic environment, and, as a consequence, the uptake of aminoglycosides is reduced, leading to a reduction in antibacterial activity.[70] The aminoglycoside class, therefore, may have limited activity against drug-persistent cell populations. Once across the membrane, the drug is trapped and accumulates inside the bacteria, resulting in bactericidal events. Aminoglycosides inhibit protein synthesis by binding to the small subunit (30S) of the bacterial ribosome. High-resolution crystal structures of two aminoglycosides, streptomycin and paromomycin, complexed with the 30S ribosomal subunit from *Thermus thermophilus*, have been elucidated.[71] Based on the crystal structure, aminoglycosides bind to an area adjacent to the decoding site in the 30S subunit of the ribosome, and cause decoding errors. Streptomycin binds to four nucleotides of 16S ribosomal RNA and a single amino acid of ribosomal protein S12. Paromomycin appears to interact with a slightly different binding site in the same general region. Mutations of the S12 protein and 16S ribosomal RNA account for the majority of aminoglycoside resistance in *M. tuberculosis*. Interestingly, aminoglycoside-modifying enzymes commonly found in other bacterial species have not been found in *M. tuberculosis*.

Capreomycin and viomycin are also protein synthesis inhibitors, and appear to bind at the interface between the 30S and 50S subunits.[72] Viomycin and capreomycin resistance are largely due to methylation or mutation of ribosomes. Amikacin, kanamycin, capreomycin, and viomycin do not generally exhibit cross-resistance with streptomycin, and are currently being used for the treatment of MDR-TB.[73] There is significant genotypic overlap among the mutations responsible for resistance to amikacin, kanamycin, capreomycin, and viomycin.[74] Cross-resistance among them is therefore expected and is commonly observed.

7.24.6.4.2 Structure–activity relationships

Currently, there are approximately a dozen naturally occurring and semisynthetic aminoglycosides available for the treatment of various bacterial infections. Among them, streptomycin possesses a unique streptidine group; all other aminoglycosides possess a 2-deoxystreptamine moiety. The SARs for this drug class are insufficiently defined, particularly against *M. tuberculosis*. As suggested by the cross-resistance data and the crystal structures of ribosomes and aminoglycosides, each aminoglycoside appears to have a slightly different binding site and mechanism of action. The promiscuous binding of this drug class has increased the difficulty of developing generalized SARs. Among all the aminoglycosides, amikacin is slightly more potent against *M. tuberculosis* than streptomycin and kanamycin. Other members of the class appear to have insufficient activity against *M. tuberculosis*. Knowledge of the SARs for the polypeptide family is totally lacking.

7.24.6.4.3 Comparisons of available agents within the class

All anti-TB agents within the aminoglycoside and polypeptide families have similar potency, pharmacokinetic and toxicity profiles, with little variation between them. Streptomycin is used as a first-line agent, and resistance to it is common among clinical isolates. Streptomycin is the least nephrotoxic aminoglycoside; however, it is highly ototoxic. Cross-resistance between streptomycin and other aminoglycosides and polypeptides is not common, and, therefore, kanamycin, amikicin, capreomycin, and viomycin can be used as second-line agents to treat MDR-TB. One of the major differences between capreomycin and aminoglycosides is their anaerobic activity against *M. tuberculosis*. Capreomycin is active against *M. tuberculosis* under anaerobic conditions, while aminoglycosides show limited activity against this cell population.[75] Differences in drug transport may explain this difference – aminoglycosides require active uptake, which slows down substantially under anaerobic conditions, while capreomycin may be transported via a different mechanism that is still active under these conditions.

7.24.6.4.4 Limitations and future directions

There are a number of limitations for the aminoglycoside and polypeptide classes. Lack of oral availability has limited the use of these agents. Nephrotoxicity and ototoxicity are significant among members of these classes, and extra care must be taken when these drugs are administered. In addition, resistance to streptomycin is common, and further limits its use. Cross-resistance between the second-line aminoglycosides and polypeptides is common, and combinations of these agents are not recommended.

Aminoglycosides also lack activity against intracellular mycobacteria and mycobacteria in their nonreplicating state. Therefore, these agents have little role in eradicating mycobacteria after the initial phase of treatment and in shortening the duration of therapy. In this regard, the anaerobic activity of capreomycin is interesting and worth further investigation. Understanding the mechanism of the anaerobic activity of capreomycin may give us some indication how to further improve this important property.

7.24.6.5 Quinolones

Quinolones belong to one of the few classes of antimicrobial agents that are totally synthetic in origin.[76] The first quinolone, nalidixic acid, was introduced in the 1960s, and is a narrow-spectrum agent against Gram-negative organisms. The coverage of this drug class was expanded significantly by the introduction of fluoroquinolones, as second-generation agents, in the 1980s. Certain second-generation fluoroquinolones, such as ciprofloxacin and ofloxacin, are active against *M. tuberculosis*, and widely used for the treatment of MDR-TB. The anti-TB activity of the quinolone class was further improved by the introduction of third-generation agents, exemplified by moxifloxacin and gatifloxacin. The third-generation quinolones demonstrate potent sterilizing activity against *M. tuberculosis* in both the replicating and nonreplicating states. Both moxifloxacin and gatifloxacin are currently being evaluated in clinical trials as potential therapies to treat TB. A brief history of the quinolone class is illustrated in **Figure 13**.

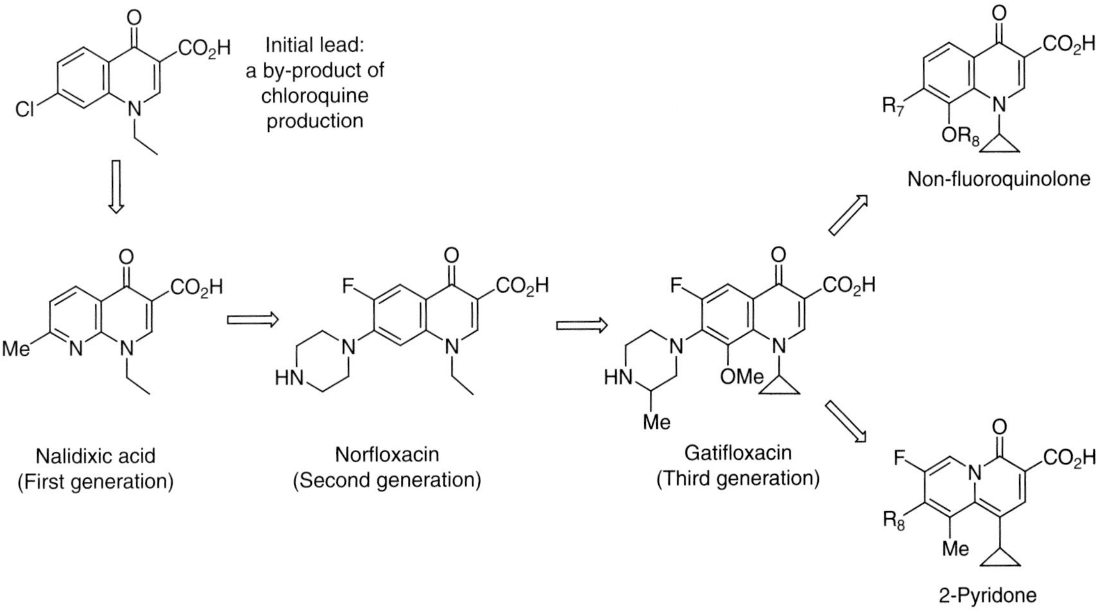

Figure 13 A summary of the quinolone class.

The safety of this drug class for long-term use has not been well defined. In addition, quinolone resistance has become a serious problem among many pathogens, such as *Escherichia coli* and *Staphylococcus aureus*, and there is concern whether relatively long-duration treatment with quinolones for TB will lead to development of resistance in commensal organisms. If quinolones become commonly used for TB, it will be important to monitor and evaluate the potential for such resistance to become a clinical problem.

7.24.6.5.1 Sites and mechanisms of action

The molecular targets of the quinolone class are DNA topoisomerases, both topoisomerase II, also known as DNA gyrase, and topoisomerase IV. DNA gyrase is essential for DNA replication, transcription, and repair, and topoisomerase IV is involved in the partitioning of chromosomal DNA during cell division. Therefore, DNA gyrase is thought to be a more important target during the nonreplicating state. Quinolones are dual-action agents against organisms that require both DNA gyrase and topoisomerase IV for viability. In *M. tuberculosis*, DNA gyrase appears to be the only type II topoisomerase present, based on genetic studies, and is likely the sole target for the quinolone class.[77] Because of the difference in molecular targets, as well as cell wall structures, quinolones (**Figure 14**) exhibit different SARs against *M. tuberculosis* compared with other organisms (**Table 3**).[78]

Quinolone resistance among *M. tuberculosis* strains is not common, and is relatively less well defined. The major resistance mechanisms, as in other bacterial species, are DNA gyrase mutations and drug efflux.

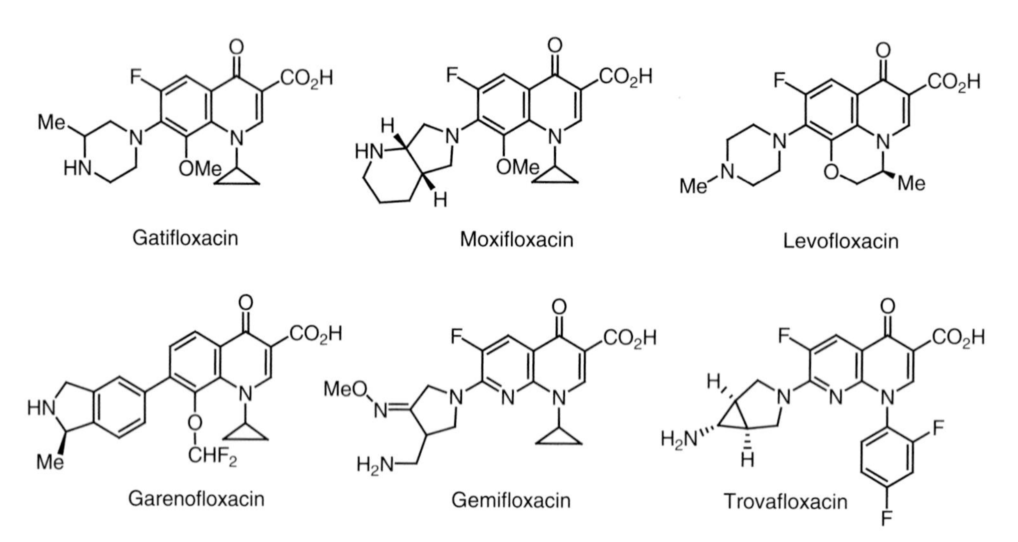

Figure 14 Structures of selected reference quinolones.

Table 3 In vitro activities of selected quinolones against *S. aureus*, *E. coli*, and *M. tuberculosis* and inhibitory (IC_{50}) and DNA cleavage (CC_{50}) activities against *M. tuberculosis* DNA gyrase

Quinolones	*S. aureus*	*E. coli*	*M. tuberculosis*	*M. tuberculosis DNA gyrase*	
	MIC ($\mu g\,mL^{-1}$)	*MIC* ($\mu g\,mL^{-1}$)	*MIC* ($\mu g\,mL^{-1}$)	IC_{50} ($\mu g\,mL^{-1}$)[a]	CC_{50} ($\mu g\,mL^{-1}$)[b]
Gatifloxacin	0.05	0.02	0.12	3	4
Moxifloxacin	0.03	0.015	0.5	4.5	4
Levofloxacin	0.12	0.008	0.5	5	12
Garenofloxacin	0.01	0.015	2	13	15
Gemifloxacin	0.01	0.13	4	11	6
Trovafloxacin	0.03	0.02	16	15	25

[a] IC_{50}: drug concentration that inhibits DNA supercoiling by 50% compared with no drug control.
[b] CC_{50}: drug concentration that induces 50% of the maximum DNA cleavage.

7.24.6.5.2 Structure–activity relationships

Quinolones have been extensively optimized against many common pathogens, and a significant amount of SAR knowledge against these pathogens is available.[76] The general SARs and structure–toxicity relationships (STRs) of the quinolone class are summarized in **Figure 15**.[79]

The SARs against mycobacteria are much less well defined, and most of these studies are primarily focused on activity against *M. avium*, *M. fortuitum*, or *M. smegmatis*. Positions that have significant impact on antimycobacterial activity are the N-1, C-7, and C-8 positions. At the N-1 position, the order of potency from high to low is *t*-butyl, cyclopropyl, 2,4-difluorophenyl, ethyl/cyclobutyl, and isopropyl. At the C-7 position, piperizine and pyrrolidine appear to have similar activity.[80] The contribution of the C-8 group is dependent on the structure at the N-1 position. When the N-1 substituent is a cyclopropyl group, the order of potency for the C-8 group from high to low is C–OMe, C–Br, C–Cl, C–F/C–H/C–OEt, N, and C–CF3. When the N-1 group is *t*-butyl, N is better than C–H.[81] For *M. tuberculosis*, the SAR information is extremely limited, but indicates that the potency of the quinolone class against *M. tuberculosis* is mainly driven by DNA gyrase interactions, as evidenced by a good correlation between DNA gyrase inhibitory activity and MICs.[78,82,83] Similarly to other antimycobacterial SARs, structures at the N-1, C-7, and C-8 positions are important for anti-TB activity. A potent anti-*M. tuberculosis* quinolone generally has the following structural characteristics: a cyclopropyl group at the N-1 position, a pyrrolidine or piperazine group at the C-7 position, and an F, Cl, or OMe group at the C-8 position.

7.24.6.5.3 Comparisons of available agents within the class

The second generation of quinolones, ciprofloxacin, ofloxacin, and levofloxacin, are currently in use as second-line TB drugs. These agents are slightly less active than rifampicin and isoniazid against *M. tuberculosis*. Resistance to fluoroquinolones is rare among clinical isolates of *M. tuberculosis*, so these agents are often useful for the treatment of MDR-TB.

Third-generation fluoroquinolones, moxifloxacin and gatifloxacin, are more potent than the second-generation agents. As more potent DNA gyrase inhibitors, moxifloxacin and gatifloxacin demonstrate significantly better sterilizing activity against persistent and rifampicin-tolerant *M. tuberculosis*.[84] Preclinical and limited clinical data indicate that regimens containing these agents could potentially shorten the current 6-month therapy.[85–87] Currently, clinical studies are ongoing to evaluate the potential of moxifloxacin and gatifloxacin as components of first-line anti-TB therapy.

7.24.6.5.4 Limitations and future directions

Despite widespread use of quinolones in second-line treatment of TB, none of the available agents are truly optimized for activity against *M. tuberculosis*. The SAR divergence against *M. tuberculosis* and other pathogens (**Table 3**) illustrates

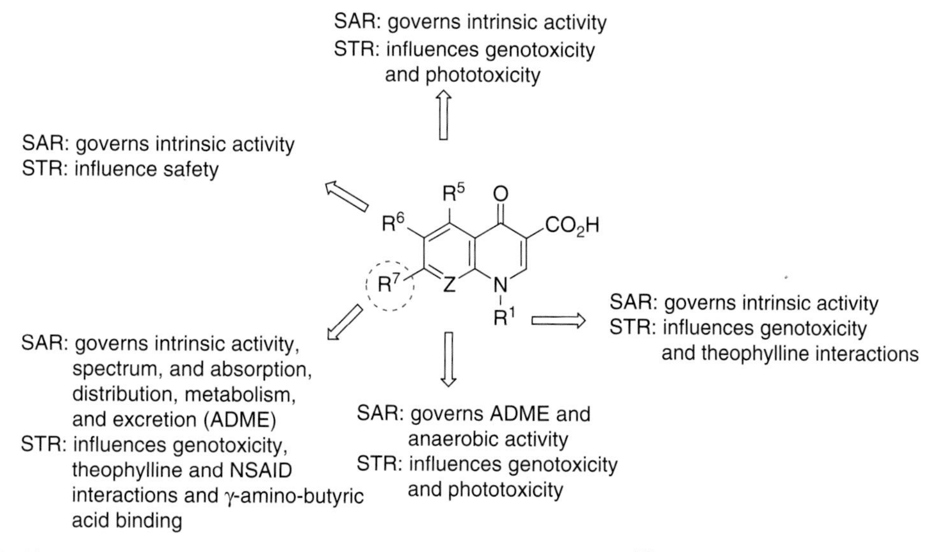

Figure 15 Highlights of the SAR and STR relationships for the quinolone class.[79]

the need for a dedicated lead optimization program focusing on activity against *M. tuberculosis*. Quinolones optimized against *M. tuberculosis* would hold great potential for shortening the duration of therapy. For the treatment of TB, quinolones are generally given for a long period and in combination with other drugs. The long-term safety and potential for drug–drug interactions of this class have not been well characterized. Development of resistance to quinolones is a concern, as discussed in Section 7.24.6.5 above, due to the high incidence of quinolone resistance among many other pathogens, although the frequency in *M. tuberculosis* is still quite low.

There are two new series of DNA gyrase inhibitors on the horizon: non-fluoroquinolones (NFQs) and 2-pyridones or quinolizinones. Members of the NFQ series were advanced to clinical development recently for other indications. This series has a potency profile similar to that of the third-generation agents, and possesses a better safety/tolerability profile. The 2-pyridones are potent DNA gyrase inhibitors, and more potent against Gram-positive and anaerobic organisms than the third-generation compounds. Both the NFQ and 2-pyridone series are potential leads for further optimization against *M. tuberculosis*.

7.24.6.6 Miscellaneous Agents: Ethambutol, Cycloserine, and *p*-Aminosalicylic Acid (PAS)

7.24.6.6.1 Ethambutol

Ethambutol (*N*,*N*′-bis(1-hydroxymethylpropyl)ethylenediamine) is a derivative of *N*,*N*′-diisopropylethylenediamine, the initial lead identified by random screening.[88] Ethambutol is a narrow-spectrum bacteriostatic agent against *M. tuberculosis*, and has low activity against nonreplicating organisms. It contributes little, if any, to the shortening of TB therapy. The main function of ethambutol is to prevent the emergence of resistance to other agents in the combination therapy.

The precise mechanisms of action and resistance to ethambutol are not fully defined. The primary targets of ethambutol appear to be the arabinosyltransferase enzymes encoded by the *emb*A and *emb*B genes, which are involved in cell wall assembly.[89]

The interaction of ethambutol with its molecular target appears to be very stereospecific – only one of the four enatiomers, (*S*,*S*)-ethambutol, is active against *M. tuberculosis* (**Figure 16**). Further optimization of this series by combinatorial synthesis produced a new compound, SQ-109, a highly lipophilic compound that appears to have a different mechanism of action from ethambutol.[90]

The major limitation of ethambutol and the ethylene diamine series with respect to TB treatment is their limited role in shortening therapy. This limitation could be mechanism based, and further optimization of potency may not address the problem. Another cautionary note is that optimization of the series should be done with careful monitoring of whole-cell activity in parallel to enzymatic activity, to avoid unwelcome off-target effects.

7.24.6.6.2 Cycloserine

Cycloserine is a natural product, initially isolated from *Streptomyces orchidaceus* in the 1950s.[91] This compound is a broad-spectrum agent, active against both Gram-positive and Gram-negative organisms. Cycloserine is only marginally active against *M. tuberculosis*, and is primarily used for the treatment of MDR-TB. The use of cycloserine is limited due to its weak activity and frequent adverse reactions.

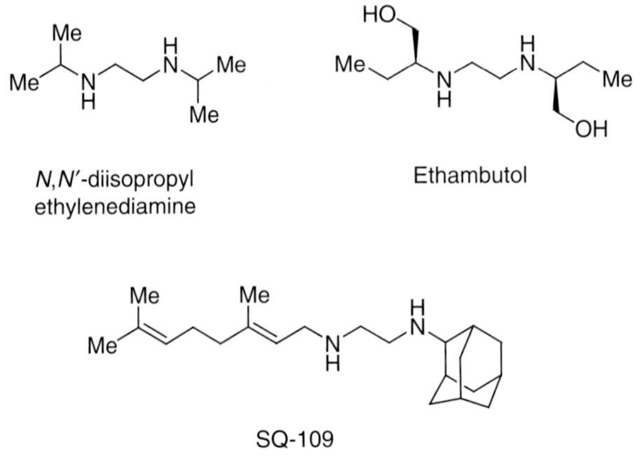

Figure 16 Structures of key ethylenediamine derivatives.

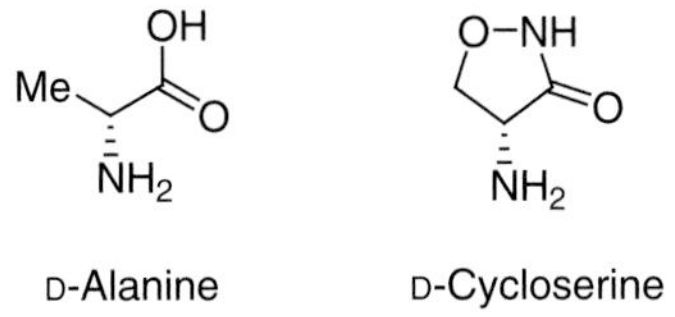

Figure 17 Structures of D-alanine and D-cycloserine.

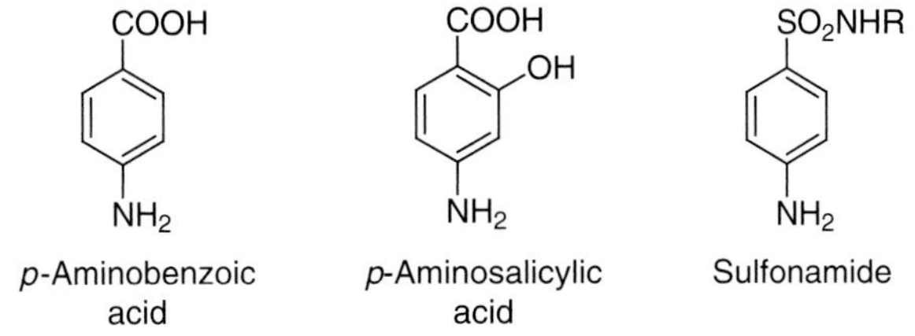

Figure 18 Structures of *p*-aminobenzoic acid, PAS, and sulfonamides.

The primary target of cycloserine is D-alanine racemase, an enzyme responsible for the interconversion of alanine enantiomers, and essential for the synthesis of the bacterial cell wall.[92] Cycloserine is believed to act as a structural analog of D-alanine, and covalently binds to the reactive center of the enzyme (**Figure 17**). The high-resolution crystal structure of D-alanine racemase from *M. tuberculosis,* which recently became available, should help to define further the mechanism of action of this agent.[93]

The main limitation of cycloserine, besides its relatively weak activity, is its toxicity profile. In addition, there is little evidence to suggest that this agent would play any role in eradicating drug-persistent bacterial populations. Without significant improvement in potency and safety, this series will eventually be replaced by other agents that are more potent and safer to use.

7.24.6.6.3 *p*-Aminosalicylic acid

p-Aminosalicylic acid (PAS), a synthetic compound, has been known for more than 100 years. Its anti-TB activity was first reported in the 1940s during the early era of antibiotic discovery.[94] The use of PAS has declined significantly because more potent and safer agents have become available. Currently, PAS is used only for the treatment of MDR-TB, when susceptibility to this agent is known or expected.

PAS is considered a structural analog of *p*-aminobenzoic acid, and inhibits the synthesis of folic acid (**Figure 18**). Different from sulfonamides in spectrum, PAS is only active against *M. tuberculosis.*

PAS has very limited utility for the treatment of TB due to its weak efficacy and low tolerability. Inhibition of folic acid synthesis generally results in a bacteriostatic effect that subsides during the persistent phase of growth. Although a safer and more efficacious agent in this class could be useful as second-line therapy, there is little evidence to suggest that even more potent inhibition of folic acid biosynthesis would have significant effect on the duration of TB treatment.

7.24.7 Unmet Medical Needs

Although the current regimen for treating active TB (2 months of isoniazid, rifampicin, pyrazinamide, and ethambutol, followed by 4 months of isoniazid and rifampicin) is efficacious when properly prescribed and adhered to, its long duration and complexity directly contribute to the development of drug resistance and hamper the global public health community's ability to effectively control the TB epidemic, particularly in settings of high HIV prevalence. Key priorities for TB drug development are: (1) shorter, simpler regimens for effective treatment of active TB; (2) regimens that can be easily and safely administered simultaneously with highly active antiretroviral therapy (i.e., that do not demonstrate significant drug–drug interactions with commonly used antiretroviral agents); (3) regimens that are safe and efficacious against MDR strains of *M. tuberculosis*; and (4), a safe, short prophylactic regimen for use against LTBI. This last priority, in conjunction with an effective vaccine to prevent infection, would ultimately have the greatest impact in eliminating TB as a public health problem – by removing the currently vast reservoir of future patients with active disease. Improved treatment for LTBI also poses the greatest inherent challenge – to understand the underlying mechanisms of latent infection well enough to identify key molecular targets for new drugs and to identify surrogate markers to streamline clinical trials of novel preventive therapies.

7.24.8 New Research Areas

The last new drug to be incorporated in the current standard anti-TB regimen, rifampicin, was introduced about 40 years ago. Since then, very few new agents for TB have been developed because of the lack of market opportunities. Recently, however, we have observed a surge of research and development activities in the TB therapeutic area due to heightened attention to and increased resources for this urgent, global, public health need. There are several new chemical series that are currently being optimized against *M. tuberculosis*. These new series are generally from one of three sources: (1) existing anti-TB agents, (2) known antibiotic classes not yet approved for TB treatment, and (3) novel chemical series. Examples of these three groups are shown in **Table 4**. Key factors to be considered when prioritizing the development of new agents for TB are their ability to overcome drug persistence (to shorten therapy), their potential effectiveness in treating MDR-TB, and their appropriateness (ease of use and safety) for HIV co-infected patients.

Figure 19 illustrates the molecular targets of the new drug series that are currently under investigation. Different from the existing agents, the target pathways of the new series are believed to be more essential mechanistically to the

Table 4 Examples of new drug series that are currently under investigation

Group 1: existing anti-TB class	*Group 2: known antibiotic class not yet approved for TB*	*Group 3: novel drug class*
Rifamycins	Macrolides/ketolides	Nitroimidazoles
Ethambutol series	Pleuromutilins	Diarylquinolines
	Oxazolidinones	Novel InhA inhibitors
	Quinolones	Novel ICL inhibitors
	PDF inhibitors	Novel gyrase inhibitors

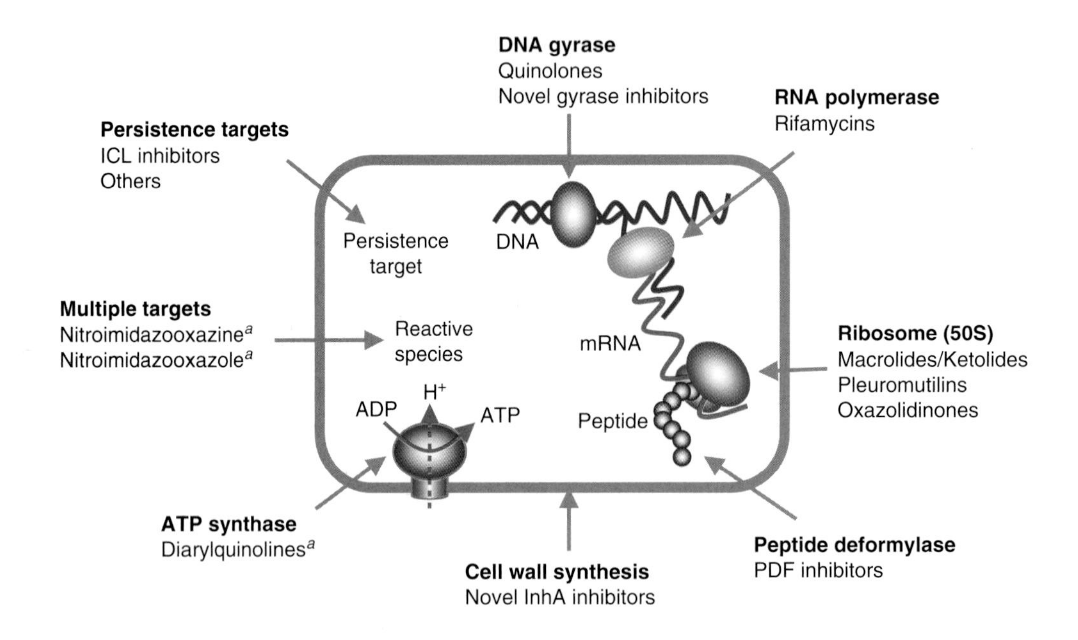

Figure 19 Schematic illustration of the sites of action of potential new anti-TB agents currently under investigation.

nonreplicating state, and therefore have a better chance to shorten treatment duration. Compounds derived from the existing anti-TB agents and known antibiotic classes have been discussed elsewhere. This section will focus primarily on two novel compound series, nitroimidazoles and diarylquinolines. Other novel series will be discussed briefly.

7.24.8.1 Nitroimidazole Class: PA-824 and OPC-67683

The structural and mechanistic ancestor of the current nitroimidazoles is a natural product called azomycin (2-nitroimidazole), isolated from streptomycete in the 1950s (**Figure 20**). Structural modification of azomycin led to the introduction of metronidazole and several other first-generation compounds.[95] Metronidazole is widely used for the treatment of protozoan and anaerobic infections. Although active against *M. tuberculosis* under anaerobic conditions, metronidazole is inactive against *M. tuberculosis* under aerobic conditions. The compound appeared to be mutagenic based on Ames tests, and carcinogenic in mice and rats. CGI-17341 is a second-generation nitroimidazole with a 4-nitroimidazooxazole structure. This compound showed potent activity against *M. tuberculosis* under both anaerobic and aerobic conditions. However, the development of CGI-17341 was abandoned due to the mutagenic potential of the compound.

Further development of the nitroimidzole series led to the discovery of PA-824, a 4-nitroimidazooxazine compound with a larger substituent attached to the oxazine ring.[96] PA-824 possesses all the beneficial attributes of CGI-17341, with activity against *M. tuberculosis* under both aerobic and anaerobic conditions. This compound is active against MDR-TB strains in vitro, indicating a novel mechanism of action. More importantly, PA-824 is not mutagenic, based on various in vitro and in vivo studies. OPC-67683 is a more recent analog of CGI-17341, with the same 4-nitroimidazooxazole scaffold as CGI-17341 but with a larger substituent linked to a quaternary carbon of the oxazole ring.[97] OPC-67683 showed even better in vitro and in vivo potency than PA-824 against *M. tuberculosis*. This compound also appears not to be mutagenic.[98]

7.24.8.1.1 Sites and mechanisms of action

Nitroimidazoles are prodrugs requiring bioreduction of the nitro group for antimycobacterial activity; however, the exact targets of the bioreductive species are unknown. Most researchers believe that the bioreductive species, which are highly reactive, damage various intracellular targets, leading to cell death. This killing mechanism is nonselective and does not discriminate between eukaryotic and prokaryotic cells. The demonstrated selectivity of the nitroimidazole class must therefore come from a selective drug activation process. It is known that *M. tuberculosis* and many other anaerobic organisms possess certain electron transport systems with a reduction potential that is low enough to reduce 4-nitroimidazoles.

As noted previously, PA-824 and OPC-67683 are active under both aerobic and anaerobic conditions. The activation mechanism of these agents may be different under these different conditions. Evidence suggests that under aerobic conditions, PA-824 is activated by an F420-dependent glucose-6-phosphate dehydrogenase (Fgd). Mutations in the gene encoding the F420 enzyme (*fgd*) are responsible for some instances of PA-824 resistance identified in vitro. Under anaerobic conditions, however, nitroimidazoles are potentially activated by a different reduction system. This phenomenon has been observed in other bacterial systems. For example, *Helicobacter pylori* mutants that are resistant to metronidazole under microaerophilic conditions demonstrate restored drug sensitivity when they are tested under anaerobic conditions.[99] Nitrofuran-resistant mutants of *E. coli* also have restored drug sensitivity under anaerobic

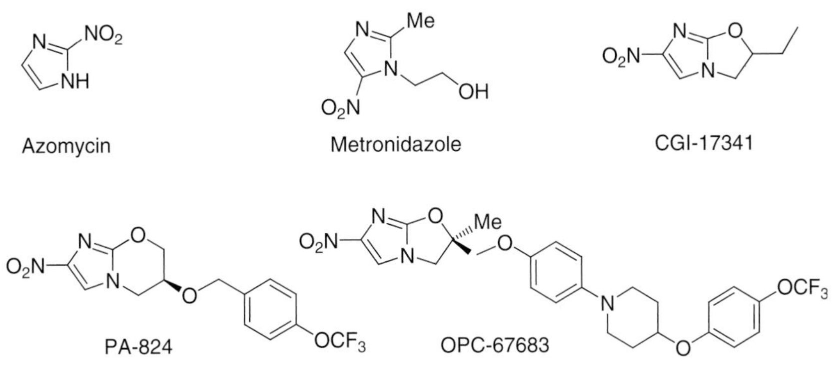

Azomycin Metronidazole CGI-17341 PA-824 OPC-67683

Figure 20 Structures of important nitroimidazoles.

conditions.[100] These studies suggest that the activation systems utilized under anaerobic conditions differ substantially from those utilized under aerobic conditions.

Preliminary studies suggested that PA-824 inhibits both protein and lipid synthesis but does not affect nucleic acid synthesis. Cell treatment with PA-824 results in accumulation of hydroxymycolic acid, with a concomitant reduction in ketomycolic acids, suggesting inhibition of an enzyme responsible for the oxidation of hydroxymycolate to ketomycolate. PA-824 did not exhibit cross-resistance with other therapeutic agents when tested against MDR clinical isolates. The frequency of resistance development is relatively low (10^{-7}), and appears to be due mainly to mutation of drug activation enzymes.

7.24.8.1.2 Structure–activity relationships

As to be expected for such a new series, the SARs for the nitroimidazole class are insufficient and incomplete. **Figure 21** summarizes the preliminary SARs and STRs for this series. It appears that the 4-nitroimidazole core is essential for antibacterial activity, which is consistent with the mechanism of action discussed above. The nitroaromatic group is also believed to be the source for mutagenicity. The substitution on the oxazine or oxazole ring is critical to many properties, including potency, physicochemical properties, pharmacokinetic profiles, and mutagenicity. Substitution with a larger group generally reduces or totally removes the potential for mutagenicity. The absolute stereochemistry of the substituent group is important for the 4-nitroimidazo-oxazine series (the (*S*) configuration is more potent than the (*R*) configuration) but has no effect on the 4-nitroimidazo-oxazole series.

7.24.8.1.3 Comparisons of compounds within the class

There are two compounds, PA-824 and OPC-67683, within the nitroimidazole class that are currently under investigation as anti-TB agents. Preclinical data indicate that OPC-67683 is the more active compound both in vitro and in vivo. However, therapeutic indices based on human clinical trials are not yet available for these agents.

7.24.8.1.3.1 PA-824

PA-824 is a potent compound against a variety of drug-sensitive and MDR-TB isolates, with MICs in the range of $\leqslant 0.015$ to $0.25\ \mu g\,mL^{-1}$ The activity is highly selective, with potent activity only against BCG and *M. tuberculosis* among the mycobacterial species tested, and without significant activity against a broad range of Gram-positive and Gram-negative bacteria (with the exception of *Helicbacter pylori* and some anaerobes). Perhaps of even greater significance is the finding in anaerobic culture that PA-824 has activity against nonreplicating bacilli, indicating its potential for activity against persisting organisms and therefore for shortening the treatment duration. Longer term mouse studies with PA-824 at $50\ mg\,kg^{-1}\,day^{-1}$ demonstrated a reduction in the bacillary lung burden similar to that of isoniazid at $25\ mg\,kg^{-1}\,day^{-1}$ with all PA-824-treated mice surviving while all untreated control animals died by day 35. In a guinea pig aerosol infection model, daily oral administration of PA-824 at $37\ mg\,kg^{-1}\,day^{-1}$ for 35 days also produced reductions of *M. tuberculosis* counts in lungs and spleens comparable to those produced by isoniazid.

When tested in the Ames assay, both with and without S9 activation, PA-824, unlike CGI-17341, demonstrated no evidence of mutagenicity. Chromosomal aberration, mouse micronucleus, and mouse lymphoma tests have all been negative, demonstrating lack of genotoxic potential. PA-824 neither inhibits nor is metabolized by major cytochrome

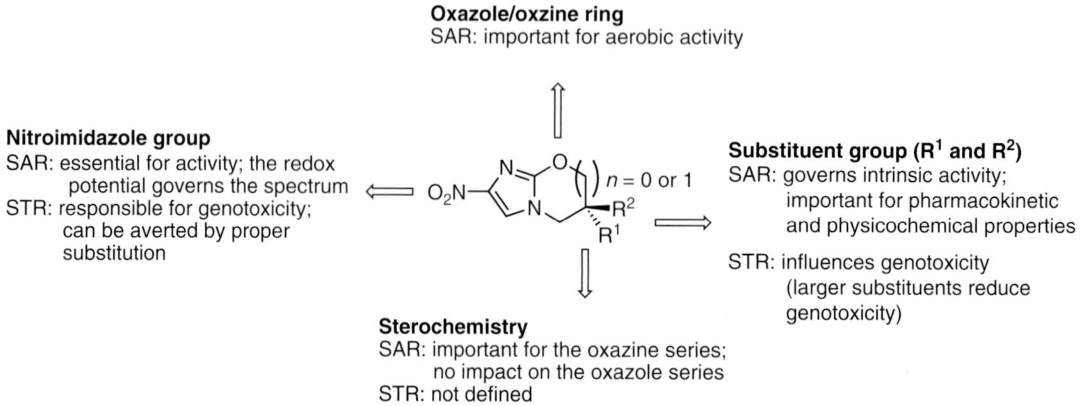

Figure 21 Highlights of preliminary SAR and STR relationships of the nitroimidazole class.

P450 enzyme isoforms in vitro, importantly indicating a low potential for drug–drug interactions, including with presently used AIDS antiretrovirals. Pharmacokinetic studies of PA-824 in the rat indicate excellent tissue penetration. Total exposure in various tissues as measured by AUC (area under the curve) is three- to eightfold higher than that in plasma.

7.24.8.1.3.2 OPC-67683

The 4-nitroimidazo-oxazole OPC-67683 has potent in vitro antimicrobial activity against *M. tuberculosis*. MICs against multiple clinically isolated *M. tuberculosis* strains range from 0.006 to 0.024 $\mu g\,mL^{-1}$. As with PA-824, OPC-67683 shows no cross-resistance with any of the currently used first-line TB drugs. There are also no indications from preclinical testing of mutagenicity or potential for cytochrome P450 enzyme-mediated drug–drug interactions. Based on their relatively similar chemical structure, it is likely that the mechanisms of action of PA-824 and OPC-67683 will prove to be the same. In a chronic infection mouse model, the efficacy of OPC-67683 is superior to that of the currently used TB drugs. In these experiments, the effective plasma concentration was 0.100 $\mu g\,mL^{-1}$, which was achieved with an oral dose of 0.625 $mg\,kg^{-1}$ confirming the remarkable in vivo potency of this compound. In nonclinical in vitro and in vivo studies, OPC-67683 in combination with various first-line TB drugs shows synergistic, additive, or no appreciable interaction, but does not demonstrate any evidence of antagonistic activity.

7.24.8.1.4 Limitations and future directions

The nitroimidazole class holds great potential for addressing several key issues in TB therapy. For the reasons stated earlier, these agents could have a major impact on TB treatment by shortening therapy, and safely and efficaciously treating both MDR-TB and HIV–TB co-infections. However, the 4-nitroimidazole series appears to have divergent SARs for aerobic and anaerobic activities. Neither PA-824 nor OPC-67683 has been optimized against persistent *M. tuberculosis* under anaerobic conditions, and SARs against this important population of *M. tuberculosis* are totally lacking. Therefore, optimization of this drug class against nonreplicating *M. tuberculosis* is an important future direction. It will also be important to gain a better understanding of the mechanisms underlying the mutagenicity of some members and their relationship to chemical structures. The risk of a nitroimidazole for inducing mutagenicity needs to be further addressed.

7.24.8.2 Diarylquinolines

Diarylquinolines belong to a novel class of anti-TB agents initially identified by a whole-cell-based screen against *M. smegmatis*. Structural modification of the initial lead compound led to the discovery of TMC-207 (previously known as R207910, **Figure 22**), which is currently under clinical development.[101] TMC-207 possesses a novel mechanism of action, and is highly potent against both drug-susceptible and MDR *M. tuberculosis*. Preclinical animal studies indicate that the compound has the potential to shorten the duration of TB therapy.

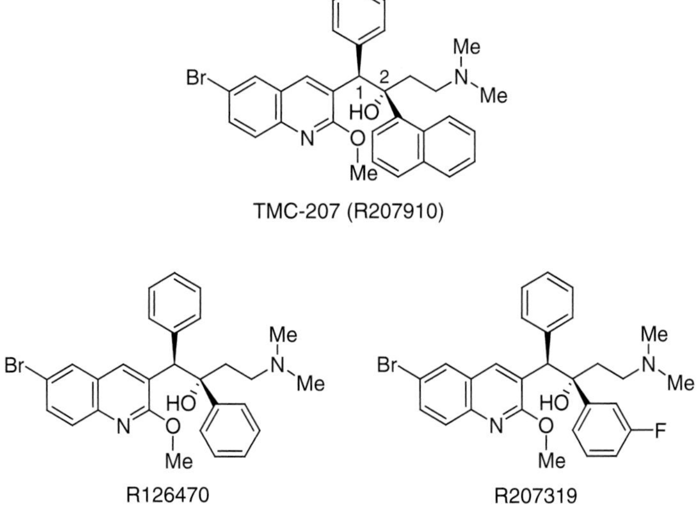

Br N O Me HO 1 2 N Me Me

TMC-207 (R207910)

R126470

R207319

Figure 22 Structures of TMC-207 and analogs.

7.24.8.2.1 Sites and mechanisms of action

Based on gene sequences of drug-resistant mutants, the mechanism of action of TMC-207 has been postulated to be inhibition of the proton pump of ATP synthase. Point mutations that conferred resistance to TMC-207 were identified in both *M. tuberculosis* and *M. smegmatis*. In three independent mutants, the only gene commonly affected encoded AtpE, a part of the F0 subunit of ATP synthase. Further transformation studies confirmed the importance of the *atpE* gene in the mechanistic pathway of TMC-207. Consistent with having a novel mechanism of action, TMC-207 showed potent activity against *M. tuberculosis* isolates resistant to a variety of anti-TB agents.

7.24.8.2.2 Structure–activity relationships

Information regarding the SARs for this new drug class is very limited. The diarylquinioline series has two chiral centers and four enantiomers. It appears that only one enantiomer, (1*R*,2*S*), is active against *M. tuberculosis*, clearly indicating that the drug-binding site on the molecular target (likely ATP synthase) is highly stereospecific.

R126470 and R207319 (**Figure 22**) are two close analogs of TMC-207 that have anti-TB activity in vivo.[101] R126470 has the same configuration as TMC-207, and a phenyl moiety, instead of a naphthyl group, at the C-2 position. This compound showed a bacteriostatic effect in a mouse model. R207319 has a 3-fluorophenyl group at the C-2 position, and showed weak bactericidal activity in the mouse model. TMC-207 was significantly more efficacious than the other two analogs in the same mouse model. These results clearly indicate the importance of the C-2 moiety.

7.24.8.2.3 Comparisons of compounds within the class

TMC-207 is the only agent in the class that is currently under clinical development. This compound is active in vitro against both drug-susceptible and drug-resistant strains of *M. tuberculosis* (MIC = 0.06 μg mL^{-1}). Strains tested included those resistant to a wide variety of commonly used drugs, including isoniazid, rifampicin, streptomycin, ethambutol, pyrazinamide, and the fluoroquinolones. While TMC-207 is active in vitro against other mycobacteria, including *M. smegmatis*, *M. kansasii*, *M. bovis*, *M. avium*, and *M. fortuitum*, the compound is not active against a variety of Gram-positive and Gram-negative organisms such as *Nocardia asteroides*, *E. coli*, *S. aureus*, *Enterococcus faecium*, or *Haemophilus influenzae*. Of note, two resistant *M. smegmatis* isolates were not cross-resistant to a wide range of antibiotics, including the fluoroquinolones.

Pharmacokinetic studies in mice have shown rapid absorption, with extensive tissue distribution in liver, kidney, heart, spleen, and lung. The half-lives ranged from 28.1 to 92 h in tissues and from 43.7 to 64 h in plasma. One contribution to the relatively long half-life appears to be slow redistribution from tissue compartments.

TMC-207 has also demonstrated significant in vivo activity in mouse models of both established and nonestablished infections. In the nonestablished disease model, mice were treated for 4 weeks, beginning the day after inoculation. In this setting, a once-weekly dose of 12.5 mg kg^{-1} was almost as efficacious as 6.5 mg kg^{-1} given five times per week. At 12.5 and 25 mg kg^{-1}, TMC-207 was more efficacious than isoniazid at 25 mg kg^{-1}. In the established disease model, treatment was begun 12–14 days after inoculation. In combination studies in the mouse, the substitution of TMC-207 for any of the three commonly used drugs (isoniazid, rifampicin, or pyrazinamide) had greater efficacy than the standard regimen of isoniazid, rifampicin, and pyrazinamide. The combination of either TMC-207, isoniazid, and pyrazinamide, or TMC-207, rifampicin, and pyrazinamide, resulted in negative spleen and lung cultures after 8 weeks of therapy, while the standard therapy (rifampicin, isoniazid, and pyrazinamide) led to positive cultures both in lungs (0.97 log CFU) and spleen (1.91 log CFU).

TMC-207 has also been tested in Phase I pharmacokinetic and safety studies in healthy volunteers. Single oral administration of TMC-207 at doses ranging from 10 to 700 mg revealed the drug to be well absorbed, with peak serum concentrations at approximately 5 h post-dose. The pharmacokinetics were dose-proportional over the range studied. A multiple ascending dose study (once-daily doses of TMC-207 at 50, 150, and 400 mg day^{-1} for 14 days) was then performed in healthy volunteers. Accumulation was observed with a doubling of the AUC on the 14th day, compared with day 1. Of note, the average observed AUCs were greater than those that achieved optimal activity in an established infection in the mouse. Safety evaluations revealed only mild or moderate adverse events, with the majority considered only possibly related to the study drug.

7.24.8.2.4 Limitations and future directions

The diarylquinoline TMC-207 represents a novel drug class with excellent potential for the treatment of TB, particularly in shortening therapy and treating MDR-TB. As with any other new series, the safety and efficacy of the compound need to be evaluated thoroughly in clinical trials. A back-up program that focuses on addressing potential safety issues identified during clinical development is important. The diarylquinoline was identified and optimized based on activity against replicating bacteria. SARs for activity against bacteria in the nonreplicating state need to be further elucidated.

7.24.8.3 Other Novel Classes: Pyrroles, Oxazolidinones, and Macrolides

7.24.8.3.1 Pyrroles

Another class of compounds being investigated for TB therapy is the pyrroles. First described in 1998 as having good antimycobacterial activity, the most potent compound was designated BM212 (**Figure 23**).[102] MICs for BM212 ranged between 0.7 and 1.5 $\mu g\,mL^{-1}$ against several strains of *M. tuberculosis*. The MICs for strains resistant to the commonly used antitubercular drugs were similar to those for sensitive strains, indicating the compound most likely has a novel mechanism of action. However, no mechanism of action has yet been elucidated for this class of compound. Some non-TB mycobacterial strains also appeared to be sensitive to BM212, albeit with MICs higher than those for *M. tuberculosis*.

A novel pyrrole compound, LL3858, is currently in Phase I clinical development for TB in India.[103] This compound has sub-micromolar MICs, and has been reported to have significant efficacy in a mouse model of TB. In combination with currently used anti-TB drugs, LL3858 sterilized lungs and spleens of infected animals in a shorter timeframe than conventional therapy.

7.24.8.3.2 Macrolides

Macrolides, a well-known antibiotic class, are initially isolated from *Streptomyces erythreus* in the 1950s. Macrolides are potent inhibitors of protein synthesis, via binding to the 50S ribosomal subunit of bacteria at the peptidyl transferase center formed by 23S rRNA. Studies have suggested they also block the formation of the 50S subunit in growing cells. Macrolides could therefore add a novel mechanism of action to TB combination therapy, and thereby also hold out the promise of being equally effective against MDR-TB and drug-sensitive TB. The macrolides, known to be orally active, have also proven to be safe and well tolerated when used for non-TB indications. Key for TB treatment, the macrolides tend to exhibit high levels of intracellular activity and extensive distribution into the lungs. Macrolides have already proven to be clinically useful in the treatment of other mycobacterial diseases including MAC and leprosy.

The major challenge for this class is its weak activity against *M. tuberculosis*, which needs to be further optimized. In addition, several members of the macrolide class are known inhibitors of cytochrome P450 enzymes associated with drug–drug interactions. This class of antibiotics is currently undergoing optimization for potential TB therapy.[104]

7.24.8.3.3 Oxazolidinones

The oxazolidinones, a relatively new class of antimicrobial agents, exert their antimicrobial effect by inhibiting protein synthesis through binding to the 70S ribosomal initiation complex. They have a relatively broad spectrum of activity, including anaerobic and Gram-positive aerobic bacteria as well as mycobacteria.[105] The first and thus far only oxazolidinone to be approved is linezolid. Although not approved for use in TB, linezolid has in vitro activity against *M. tuberculosis*. Of the oxazolidinones that have been evaluated for their in vitro activity, the most active compound appears to be PNU-100480 (**Figure 24**), which demonstrates potency similar to that of isoniazid or rifampicin.[105]

Because of the clinical availability of linezolid, it has been used anecdotally in patients with MDR-TB, and demonstrates biological activity, as evidenced by sputum culture conversion.[106] However, with relatively long-term use in MDR-TB patients, there are emerging reports of peripheral and optic neuropathy.[107]

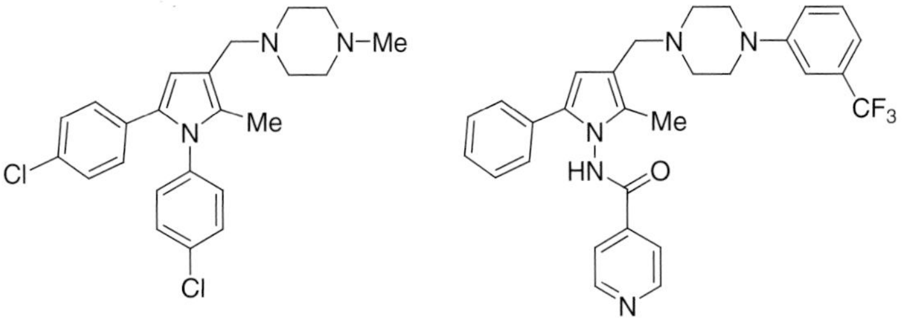

Figure 23 Structures of important pyrrole compounds.

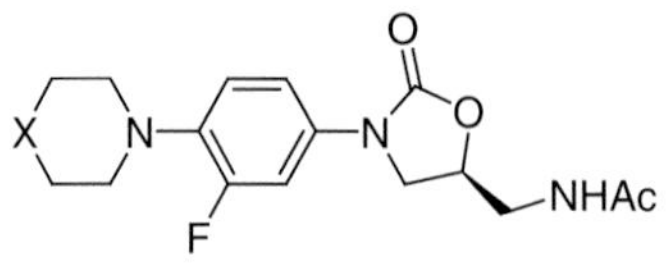

Figure 24 Structures of important oxazolidinones.

Even though the oxazolidinones have the potential to be a useful addition to the armamentarium of anti-TB drugs, there has never been a truly concerted effort to optimize their activity against *M. tuberculosis*. In the meantime, the neuropathic side effects of linezolid, emerging with its long-term use, will require careful monitoring as this drug becomes more commonly used in the treatment of MDR-TB. A lead optimization program focusing on improving efficacy and safety of this drug class would be an important development.

7.24.8.4 Novel Drug Targets for Persistence

All the current anti-TB agents were identified by their ability to kill or inhibit *M. tuberculosis* in the exponential phase of growth in vitro, with the exception of pyrazinamide, whose antimycobacterial activity was discovered directly in vivo in a mouse model. Most of these agents are highly potent against replicating *M. tuberculosis* but have limited activity against *M. tuberculosis* in anaerobic or nutrient-depleted conditions in their nonreplicating state. This phenomenon is largely due to the fact that in the nonreplicating state, the drug targets of these agents are inactive, and inhibiting theses enzymes therefore has limited impact on the viability of the bacilli. Recently, significant progress has been made in identifying *M. tuberculosis* factors believed to play an essential role in the ability of *M. tuberculosis* to adopt and/or maintain a persistent state. Some of these factors are believed to be suitable drug targets for therapeutic interventions.[108,109] Drugs targeting these enzymes are likely to have a significant impact on cell viability in the persistent state, and therefore may shorten the duration of therapy.

There is significant disagreement, however, regarding how persistence is induced and maintained in the host. Various in vitro conditions have been used to mimic the host environment and to identify putative persistence factors, including oxygen depletion, nutrient depletion, and nitric oxide stress.[110] Whole-genome microarray experiments conducted under various stress conditions suggest that gene expression patterns under these various conditions have little overlap.[111] The lack of correlation among gene expression patterns under different in vitro stress conditions further illustrates the high risks involved with this approach. Without a clear understanding of the clinical relevance of these models, research should first focus on the small numbers of genes that are commonly expressed under different conditions. This group of genes may provide a better chance to identify true persistence factors. Another caveat relevant to this approach is that these putative persistence targets are often not essential under the normal growth conditions used to assay drug susceptibility. Compounds that inhibit such targets will have to be tested under special conditions or directly in vivo. This requirement adds significantly to the challenge of the lead optimization process. Currently, most programs focused on agents active against persistent *M. tuberculosis* are in the target validation or lead identification stages, the very early steps of a drug discovery program.

7.24.9 Conclusion

While current, standard, TB drugs can be effective, they necessitate lengthy and complex treatment regimens, lessening rates of patient adherence. This lack of compliance in turn has created a significant drug resistance problem and hampered control of the global TB epidemic. The lack of strong market opportunities has hindered the development of new TB drugs until very recently. Currently, increased resources and public attention have led to a resurgence in TB drug research and development activities, leading to optimism that shorter, simpler, and safer treatment regimens will be developed and registered in the foreseeable future. Medicinal chemistry is playing, and will continue to play, a key role in this endeavor.

References

1. Euzeby, J. P. List of bacterial names with standing in nomenclature – Genus *Mycobacterium*. http://www.bacterio.cict.fr/m/mycobacterium.html (accessed Aug 2006).
2. WHO *Global Tuberculosis Control: Surveillance, Planning, Financing*; World Health Organization: Geneva, Switzerland, 2005.
3. Schatz, A.; Bugie, E.; Waksman, S. A. *Proc. Soc. Exp. Biol. Med.* **1944**, *55*, 66–69.

4. Donoghue, H. D.; Spigelman, M.; Greenblatt, C. L.; Lev-Maor, G.; Bar-Gal, G. K.; Matheson, C.; Vernon, K.; Nerlich, A. G.; Zink, A. R. *Lancet Infect. Dis.* **2004**, *4*, 584–592.
5. Stead, W. W. *Ann. Intern. Med.* **1992**, *116*, 937–941.
6. American Thoracic Society. *Am. J. Respir. Crit. Care Med.* **2000**, *161*, S221–S247.
7. WHO *State of the Art of New Vaccines: Research & Development*; World Health Organization: Geneva, Switzerland, 2003.
8. Aziz, M. A.; Wright, A. *Clin. Infect. Dis.* **2005**, *15*, S258–S262.
9. Armstrong, J. A.; Hart, P. D. *J. Exp. Med.* **1971**, *134*, 713–740.
10. Clemens, D. L.; Horwitz, M. A. *J. Exp. Med.* **1995**, *181*, 257–270.
11. Via, L. E.; Deretic, D.; Ulmer, R. J.; Hibler, N. S.; Huber, L. A.; Deretic, V. *J. Biol. Chem.* **1997**, *272*, 13326–13331.
12. Glickman, M. S.; Jacobs, W. R., Jr. *Cell* **2001**, *104*, 477–485.
13. Wallgren, A. *Tubercle* **1948**, *29*, 245–251.
14. Kamholz, S. L. Pleural tuberculosis. In *Tuberculosis*; Rom, W. N., Garay, S., Eds.; Little, Brown and Co: Boston, MA, 1996, pp 483–491.
15. Talavera, W.; Miranda, R.; Lessnau, K.; Klapholz, L. Extrapulmonary tuberculosis. In *Tuberculosis: Current Concepts and Treatment*, 2nd ed.; Friedman, L. N., Ed.; CRC Press: Boca Raton, FL, 2001, pp 139–190.
16. Brennan, P. J. *Tuberculosis (Edinb.)* **2003**, *83*, 91–97.
17. Sassetti, C. M.; Rubin, E. J. *Proc. Natl. Acad. Sci. USA* **2003**, *100*, 12989–12994.
18. Koch, R. *Berlin. Klin. Wochenschr.* **1882**, *15*, 221–230.
19. Heifets, L. B. Drug Susceptibility Tests in the Management of Chemotherapy of Tuberculosis. In *Drug susceptibility in the Chemotherapy of Mycobacterial Infections*; Heifets, L. B., Ed.; CRC Press: Boca Raton, FL, 1991, pp 89–122.
20. Collins, L.; Franzblau, S. G. *Antimicrob. Agents Chemother.* **1997**, *41*, 1004–1009.
21. Gomez, J. E.; McKinney, J. D. Persistence and Drug Tolerance. In *Tuberculosis*, 2nd ed.; Rom, W. N., Garay, S. M., Eds.; Lippincott Williams & Wlkins: Philadelphia, PA, 2004, pp 101–114.
22. Wayne, L. G.; Hayes, L. G. *Infect. Immun.* **1996**, *64*, 2062–2069.
23. Nathan, C. F.; Ehrt, S. Nitric Oxide in Tuberculosis. In *Tuberculosis*, 2nd ed.; Rom, W. N., Garay, S. M., Eds.; Lippincott Williams & Wlkins: Philadelphia, PA, 2004, pp 215–235.
24. Rook, G. A. W. *Immunol. Ser.* **1994**, *60*, 249–261.
25. Flynn, J. L.; Chan, J. Animal Models of Tuberculosis. In *Tuberculosis*, 2nd ed.; Rom, W. N., Garay, S. M., Eds.; Lippincott Williams & Wlkins: Philadelphia, PA, 2004, pp 237–250.
26. Ome, I. M.; Collins, F. M. Mouse Model of Tuberculosis. In *Tuberculosis: Pathogenesis, Protection, and Control*; Bloom, B. R., Ed.; American Society for Microbiology: Washington, DC, 1994, pp 113–134.
27. Lenaerts, A. J.; Gruppo, V.; Brooks, J. V.; Orme, I. M. *Antimicrob. Agents Chemother.* **2003**, *47*, 783–785.
28. Nikonenko, B. V.; Samala, R.; Einck, L.; Nacy, C. A. *Antimicrob. Agents Chemother.* **2004**, *48*, 4550–4555.
29. Adams, L. B.; Sinha, I.; Franzblau, S. G.; Krahenbuhl, J. L.; Mehta, R. T. *Antimicrob. Agents Chemother.* **1999**, *43*, 1638–1643.
30. Capuano, S. V., III; Croix, D. A.; Pawar, S.; Zinovik, A.; Myers, A.; Lin, P. L.; Bissel, S.; Fuhrman, C.; Klein, E.; Flynn, J. L. *Infect. Immun.* **2003**, *71*, 5831–5844.
31. Ferebee, S. H.; Mount, F. W. *Am. Rev. Respir. Dis.* **1962**, *85*, 490–510.
32. Comstock, G. W.; Ferebee, S. H.; Hammes, L. M. *Am. Rev. Respir. Dis.* **1967**, *95*, 935–943.
33. Zuber, P. L. F.; McKenna, M. T.; Binkin, N. J.; Onorato, I. M.; Castro, K. G. *JAMA* **1997**, *278*, 304–307.
34. International Union Against Tuberculosis Committee on Prophylaxis. *Bull. World Health Organ.* **1982**, *60*, 555–564.
35. Streeton, J. A.; Desem, N.; Jones, S. L. *Int. J. Tuberc. Lung Dis.* **1998**, *2*, 443–450.
36. Pai, M.; Gokhale, K.; Joshi, R.; Dogra, S.; Kalantri, S.; Mendiratta, D. K.; Narang, P.; Daley, C. L.; Granich, R. M.; Mazurek, G. H. et al. *JAMA* **2005**, *293*, 2746–2755.
37. Kang, Y. A.; Lee, H. W.; Yoon, H. I.; Cho, B.; Han, S. K.; Shim, Y. S.; Yim, J. J. *JAMA* **2005**, *293*, 2756–2761.
38. Mwinga, A.; Hosp, M.; Godrey-Faussett, P.; Quigley, M.; Mwaba, P.; Mugala, B. N.; Nyirenda, O.; Luo, N.; Pobee, J.; Elliott, A. M. et al. *AIDS* **1998**, *12*, 2447–2457.
39. Gordin, F.; Chaisson, R. E.; Matts, J. P.; Miller, C.; de Lourdes Garcia, M.; Hafner, R.; Valdespino, J. L.; Coberly, J.; Schechter, M.; Klukowicz, A. J. et al. *JAMA* **2000**, *283*, 1445–1450.
40. Halsey, N. A.; Coberly, J. S.; Desormeaux, J. *Lancet* **1998**, *351*, 786–792.
41. Hawken, M. P.; Meme, H. K.; Elliott, L. C.; Chakaya, J. M.; Morris, J. S.; Githui, W. A.; Juma, E. S.; Odhiambo, J. A.; Thiong'o, L. N.; Kimari, J. N. et al. *AIDS* **1997**, *11*, 875–882.
42. Mitchison, D. A. *Am. J. Respir. Crit. Care Med.* **2005**, *171*, 699–706.
43. Sensi, P. *Rev. Infect. Dis.* **1983**, *5*, S402–S406.
44. Campbell, E. A.; Korzheva, N.; Mustaev, A.; Murakami, K.; Nair, S.; Goldfarb, A.; Darst, S. A. *Cell* **2001**, *104*, 901–912.
45. Mariam, D. H.; Mengistu, Y.; Hoffner, S. E.; Andersson, D. I. *Antimicrob. Agents Chemother.* **2004**, *48*, 1289–1294.
46. Bacchi, A.; Pelizzi, G.; Nebuloni, M.; Ferrari, P. *J. Med. Chem.* **1998**, *41*, 2319–2332.
47. Michaelis, A. F.; Maulding, H. V.; Sayada, C.; Zha, C. Rifamycin derivatives for drug targeting. International Patent WO 2003045319, 2003.
48. Ferguson, A. D.; Kodding, J.; Walker, G.; Bos, C.; Coulton, J. W.; Diederichs, K.; Braun, V.; Welte, W. *Structure* **2001**, *9*, 707–716.
49. Floss, H. G.; Yu, T.-W. *Chem. Rev.* **2005**, *105*, 621–632.
50. Rothstein, D. M.; Hartman, A. D.; Cynamon, M. H.; Eisenstein, B. I. *Expert Opin. Investig. Drugs* **2003**, *12*, 255–271.
51. Li, A. P.; Reith, M. K.; Rasmussen, A.; Gorski, J. C.; Hall, S. D.; Xu, L.; Kaminski, D. L.; Cheng, L. K. *Chem. Biol. Interact.* **1997**, *107*, 17–30.
52. Zhang, Y. Isoniazid. In *Tuberculosis*, 2nd ed.; Rom, W. N., Garay, S. M., Eds.; Lippincott Williams & Wlkins: Philadelphia, PA, 2004, pp 739–758.
53. Larsen, M. H.; Vilcheze, C.; Kremer, L.; Besra, G. S.; Parsons, L.; Salfinger, M.; Heifets, L.; Hazbon, M. H.; Alland, D.; Sacchettini, J. C. et al. *Mol. Microbiol.* **2002**, *46*, 453–466.
54. Kremer, L.; Dover, L. G.; Morbidoni, H. R.; Vilcheze, C.; Maughan, W. N.; Baulard, A.; Tu, S.-C.; Honore, N.; Deretic, V.; Sacchettini, J. C. et al. *J. Biol. Chem.* **2003**, *278*, 20547–20554.
55. Rozwarski, D. A.; Grant, G. A.; Barton, D. H. R.; Jacobs, W. R.; Sacchettini, J. C. *Science* **1998**, *279*, 98–102.
56. Morlock, G. P.; Metchock, B.; Sikes, D.; Crawford, J. T.; Cooksey, R. C. *Antimicrob. Agents Chemother.* **2003**, *47*, 3799–3805.
57. Murray, M. F. *Clin. Inf. Dis.* **2003**, *36*, 453–460.
58. Maccari, R.; Ottana, R.; Vigorita, M. G. *Bioorg. Med. Chem. Lett.* **2005**, *15*, 2509–2513.
59. Pasqualoto, K. F. M.; Ferreira, E. I.; Santos-Filho, O. A.; Hopfinger, A. J. *J. Med. Chem.* **2004**, *47*, 3755–3764.

60. Schwartz, W. S. *Am. Rev. Respirat. Dis.* **1966**, *93*, 685–692.
61. Wade, M. M.; Zhang, Y. *Frontiers Biosci.* **2004**, *9*, 975–994.
62. Zimhony, O.; Cox, J. S.; Welch, J. T.; Vilcheze, C.; Jacobs, W. R. *Nat. Med.* **2000**, *6*, 1043–1047.
63. Boshoff, H. I.; Mizrahi, V.; Barry, C. E., III *J. Bacteriol.* **2002**, *184*, 2167–2172.
64. Zhang, Y.; Wade, M. M.; Scorpio, A.; Zhang, H.; Sun, Z. *J. Antimicrob. Chemother.* **2003**, *52*, 790–795.
65. Cynamon, M. H.; Gimi, R.; Gyenes, F.; Sharpe, C. A.; Bergmann, K. E.; Han, H. J.; Gregor, L. B.; Rapolu, R.; Luciano, G.; Welch, J. T. *J. Med. Chem.* **1995**, *38*, 3902–3907.
66. Speirs, R. J.; Welch, J. T.; Cynamon, M. H. *Antimicrob. Agents Chemother.* **1995**, *39*, 1269–1271.
67. Mitchison, D. A. *Chest* **1979**, *76S*, 771–781.
68. Cynamon, M. H.; Speirs, R. J.; Welch, J. T. *Antimicrob. Agents Chemother.* **1998**, *42*, 462–463.
69. Bryan, L. E.; Kawan, S. *Antimicrob. Agents Chemother.* **1983**, *23*, 835–845.
70. Mates, S. M.; Patel, L.; Kaback, H. R.; Miller, M. H. *Antimicrob. Agents Chemother.* **1983**, *23*, 526–530.
71. Carter, A. P.; Clemons, W. M.; Brodersen, D. E.; Morgan-Warren, R. J.; Wimberly, B. T.; Ramakrishnan, V. *Nature* **2000**, *407*, 340–348.
72. Maus, C. E.; Plikaytis, B. B.; Shinnick, T. M. *Antimicrob. Agents Chemother.* **2005**, *49*, 571–577.
73. Alangaden, G. J.; Kreiswirth, B. N.; Aouad, A.; Khetarpal, M.; Igno, F. R.; Moghazeh, S. L.; Manavathu, E. K.; Lerner, S. A. *Antimicrob. Agents Chemother.* **1998**, *42*, 1295–1297.
74. Maus, C. E.; Plikaytis, B. B.; Shinnick, T. M. *Antimicrob. Agents Chemother.* **2005**, *49*, 3192–3197.
75. Heifets, L.; Simon, J.; Pham, V. *Ann. Clin. Microbiol. Antimicrob.* **2005**, *4*, 1–7.
76. Mitscher, L. A. *Chem. Rev.* **2005**, *105*, 559–592.
77. Cole, S. T.; Brosch, R.; Parkhill, J.; Garnier, T.; Churcher, C.; Harris, D.; Gordon, S. V.; Eiglmeier, K.; Gas, S.; Barry, C. E., III et al. *Nature* **1998**, *393*, 537–544.
78. Aubry, A.; Pan, X.-S.; Fisher, L. M.; Jarlier, V.; Cambau, E. *Antimicrob. Agents Chemother.* **2004**, *48*, 1281–1288.
79. Domagala, J. M. *J. Antimicrob. Chemother.* **1994**, *33*, 685–706.
80. Renau, T. E.; Sanchez, J. P.; Gage, J. W.; Dever, J. A.; Shapiro, M. A.; Gracheck, S. J.; Domagala, J. M. *J. Med. Chem.* **1996**, *39*, 729–735.
81. Renau, T. E.; Gage, J. W.; Dever, J. A.; Roland, G. E.; Joannides, E. T.; Shapiro, M. A.; Sanchez, J. P.; Gracheck, S. J.; Domagala, J. M.; Jacobs, M. R. et al. *Antimicrob. Agents Chemother.* **1996**, *40*, 2363–2368.
82. Onodera, Y.; Tanaka, M.; Sato, K. *J. Antimicrob. Chemother.* **2001**, *47*, 447–450.
83. Kawakami, K.; Namba, K.; Tanaka, M.; Matsuhashi, N.; Sato, K.; Takemura, M. *Antimicrob. Agents Chemother.* **2000**, *44*, 2126–2129.
84. Hu, Y.; Coates, A. R.; Mitchison, D. A. *Antimicrob. Agents Chemother.* **2003**, *47*, 653–657.
85. Tuberculosis Research Centre (Indian Council of Medical Research). *Ind. J. Tub.* **2002**, *49*, 27–38.
86. Nuermberger, E. L.; Yoshimatsu, T.; Tyagi, S.; Williams, K.; Rosenthal, I.; O'Brien, R. J.; Vernon, A. A.; Chaisson, R. E.; Bishai, W. R.; Grosset, J. H. *Am. J. Respir. Crit. Care Med.* **2004**, *170*, 1131–1134.
87. Nuermberger, E. L.; Yoshimatsu, T.; Tyagi, S.; O'Brien, R. J.; Vernon, A. N.; Chaisson, R. E.; Bishai, W. R.; Grosset, J. H. *Am. J. Respir. Crit. Care Med.* **2004**, *169*, 421–426.
88. Shepherd, R. G.; Baughn, C.; Cantrall, M. L.; Goodstein, B.; Thomas, J. P.; Wilkinson, R. G. *Ann. NY Acad. Sci.* **1966**, *135*, 686–710.
89. Telenti, A.; Philipp, W. J.; Sreevatsan, S.; Bernasconi, C.; Stockbauer, K. E.; Wieles, B.; Musser, J. M.; Jacobs, W. R. *Nat. Med.* **1997**, *3*, 567–570.
90. Jia, L.; Tomaszewski, J. E.; Hanrahan, C.; Coward, L.; Noker, P.; Gorman, G.; Nikonenko, B.; Protopopova, M. *Br. J. Pharmacol.* **2005**, *144*, 80–87.
91. Harned, R. L.; Hidy, P. H.; La Baw, E. K. *Antibiot. Chemother.* **1955**, *5*, 204–205.
92. Feng, Z.; Barletta, R. G. *Antimicrob. Agents Chemother.* **2003**, *47*, 283–291.
93. LeMagueres, P.; Im, H.; Ebalunode, J.; Strych, U.; Benedik, M. J.; Briggs, J. M.; Kohn, H.; Krause, K. L. *Biochemistry* **2005**, *44*, 1471–1481.
94. Lehmann, J. *Lancet* **1946**, *250*, 15–16.
95. Barry, C. E., 3rd; Boshoff, H. I.; Dowd, C. S. *Curr. Pharm. Des.* **2004**, *10*, 3239–3262.
96. Stover, C. K.; Warrener, P.; VanDevanter, D. R.; Sherman, D. R.; Arain, T. M.; Langhorne, M. H.; Anderson, S. W.; Towell, J. A.; Yuan, Y.; McMurray, D. N. et al. *Nature* **2000**, *405*, 962–966.
97. Tsubouchi, H.; Sasaki, H.; Haraguchi, Y.; Itotani, M.; Kuroda, H.; Miyamura, S.; Hasegawa, T.; Kuroda, T.; Goto, F. Synthesis and antituberculous activity of a novel series of optically active 6-nitro-2,3-dihydroimidazo[2,1-*b*]oxazoles. 45th Interscience Conference on Antimicrobial Agents and Chemotherapy, Washington, DC, Dec 16–19, 2005, Poster F1473.
98. Hashizume, H.; Ohguro, K.; Itoh, T.; Shiragiku, T.; Ohara, Y.; Awogi, T.; Matsumoto, M. *OPC-67683, A Novel Imidazo-Oxazole Antituberculous Compound, Free from Mutagenicity: Structure–Activity–Mutagenesity Relationships*. 45th Interscience Conference on Antimicrobial Agents and Chemotherapy, Washington, DC, Dec 16–19, 2005, Poster F1464.
99. Smith, M. A.; Edwards, D. I. *J. Antimicrob. Chemother.* **1995**, *36*, 453–461.
100. Asnis, R. E.; Cohen, F. B.; Gots, J. S. *Antibiot. Chemother.* **1952**, *2*, 123–129.
101. Andries, K.; Verhasselt, P.; Guillemont, J.; Gohlmann, H. W.; Neefs, J. M.; Winkler, H.; Van Gestel, J.; Timmerman, P.; Zhu, M.; Lee, E. et al. *Science* **2005**, *307*, 223–227.
102. Diedda, D.; Lampis, G.; Fioravanti, R.; Biava, M.; Porretta, G. C.; Zanetti, S.; Pompei, R. *Antimicrob. Agents Chemother.* **1998**, *42*, 3035–3037.
103. Arora, S. K.; Sinha, N.; Sinha, R.; Bateja, R.; Sharma, S.; Upadhayaya, R. S. *Design, Synthesis, Modeling and Activity of Novel Anti-Tubercular Compounds*. Abstracts of Papers, 227th ACS National Meeting, Anaheim, CA, March 28–April 1, 2004.
104. Falzari, K.; Zhu, Z.; Pan, D.; Liu, H.; Hongmanee, P.; Franzblau, S. G. *Antimicrob. Agents Chemother.* **2005**, *49*, 1447–1454.
105. Cynamon, M. H.; Kelmens, S. P.; Sharpe, C. A.; Chase, S. *Antimicrob. Agents Chemother.* **1999**, *43*, 1189–1191.
106. Fortun, J.; Martin-Davila, P.; Navas, E.; Perez-Elias, M. J.; Cobo, J.; Tato, M.; De la Pedrosa, E. G.-G.; Gomez-Mampaso, E.; Moreno, S. *J. Antimicrob. Chemother.* **2005**, *56*, 180–185.
107. Bressler, A. M.; Zimmer, S. M.; Gilmore, J. L.; Somani, J. *Lancet Infect. Dis.* **2004**, *4*, 528–531.
108. McKinney, J. D. *Nat. Med.* **2000**, *6*, 1330–1333.
109. Gomez, J. E.; McKinney, J. D. *Tuberculosis (Edinb.)* **2004**, *84*, 29–44.
110. Zhang, Y. *Frontiers Biosci.* **2004**, *9*, 1136–1156.
111. Kendall, S. L.; Rison, S. C.; Movahedzadeh, F.; Frita, R.; Stoker, N. G. *Trends Microbiol.* **2004**, *12*, 537–544.

Biographies

Zhenkun Ma is the head of research at the Global Alliance for TB Drug Development. An experienced researcher in drug discovery, he was formerly the director of chemistry at Cumbre Inc., a Dallas-based biotechnology firm specializing in antibacterial drug discovery research. Prior to his experience at Cumbre, he worked at Abbott Laboratories for 8 years, where, as a member of the prestigious Volwiler Society, he led antibacterial programs and oversaw the medicinal chemistry for macrolide and other antibacterial programs. In 1997, he received the Chairman's Award for the discovery of cethromycin, a novel ketolide antibiotic currently in late-stage clinical development. A native of northeastern China, he studied chemistry and organic chemistry at Beijing University, and holds a PhD in organic chemistry from the University of Connecticut, Storrs. The holder of more than 40 US patents and patent applications, Ma has also authored and co-authored more than 80 peer-reviewed articles and meeting presentations.

Ann M Ginsberg has been the head of clinical development for the Global Alliance for TB Drug Development since June 2004. A highly regarded tuberculosis expert, and a former director for project management at Merck & Co., Inc., she brings 15 years of experience at the US National Institutes of Health, starting in the National Cancer Institute as a medical staff fellow and resident in anatomic pathology. She subsequently joined the National Institute of Diabetes, Digestive and Kidney Diseases as a senior staff fellow in the Laboratory of Cellular and Developmental Biology. In 1995, she joined the National Institute for Allergy and Infectious Diseases as a program officer for tuberculosis, leprosy, and other mycobacterial diseases, and was appointed chief of the respiratory diseases branch in 2000. Trained as a molecular biologist and Board Certified Anatomic Pathologist, Ginsberg holds a BA from Harvard University, an MD from Columbia University, and a PhD from Washington University. She has authored numerous scientific publications, and has received several prominent awards, including the US Department of Health and Human Services Secretary's Award for Distinguished Service in 2000. Currently a member of the Board of Directors of the Aeras Global TB Vaccine Foundation, she has served on multiple global health committees.

Melvin Spigelman is the director of research and development for the Global Alliance for TB Drug Development. A highly regarded expert in domestic and international drug research and development, Spigelman spent a decade managing drug R&D at Knoll Pharmaceuticals (a division of BASF Pharma). As the vice president of R&D at Knoll, Spigelman directed clinical development and supervised R&D activities from basic discovery to regulatory approval. As part of Knoll's senior R&D management team, he established global R&D processes, oversaw a marked increase in US regulatory filings and approvals, and supervised joint R&D programs with pharmaceutical companies. Starting as a director of oncology and immunology in 1989, Spigelman became the vice president of R&D at Knoll until its acquisition by Abbott Laboratories in 2001. He received his undergraduate degree from Brown University and his medical degree from the Mt. Sinai School of Medicine, where he specialized in internal medicine and neoplastic diseases. Spigelman holds board certifications from the American Board of Internal Medicine, the American Board's Subspecialty Board of Medical Oncology, and the American Board of Preventive Medicine, and was the recipient of the American Cancer Society Clinical Oncology Career Development Award (1985–88).

Comprehensive Medicinal Chemistry II
ISBN (set): 0-08-044513-6

ISBN (Volume 7) 0-08-044520-9; pp. 699–730

7.25 Impact of Genomics-Emerging Targets for Antibacterial Therapy

J F Barrett,[*] Merck Research Laboratories, Rahway, NJ, USA

7.25.1 Impact of Genomics-Emerging Targets for Antibacterial Therapy

Any chapter describing the impact of genomics-emerging targets on antibacterial therapy would be a relatively short one if the impact was measured purely by the number of genomics targets on the market, or even in clinical development. There are simply none to date. But the discipline of 'genomics' and its inevitable impact on antibacterial therapy can be measured by the advancement of the discipline of antibacterial discovery, as genomics has enabled an understanding of bacteria growth, replication, pathogenicity, and infection that is unprecedented in its scope and detail.

[*]Deceased.

The discipline of genomics has made impact in the following key areas of antibacterial discovery research: understanding the bacterial infection; understanding the nature and use of targets; the multiple subdisciplines of genomics when considered as a 'tool'; the interface of genomes and chemical space in drug discovery; the nature, measurement, and assessment of resistance; and the almost unlimited applications of genomics in the drug discovery process of identifying antibacterial agents. Within this synopsis, each of these areas of impact by genomics will be reviewed.

7.25.2 Introduction: The Disease State of Bacterial Infection

The need for novel antimicrobial agents to combat evolving resistance among many human bacterial pathogens worldwide is clear and imminent,[1–4] if for no other reason we need only to reflect back on the preantibiotic era.[5] The introduction of sulphanilamide into clinical use for leprosy in 1936 and the industrial production of penicillin in 1941 as the first commercially available antibiotics for human use was followed by the release of several new classes of antibiotics in the 1940s, 1950s, and 1960s. These 'miracle' compounds (mostly penicillins, streptomycin, tetracyclines, sulfonamides, macrolides, semisynthetic penicillins, cephalosporins, and glycopeptides) were all natural products isolated from soil microbes, and were discovered by simple bacterial growth inhibition screens.[6,7] And whereas antibiotic resistance was initially just a tool used as a drug marker[8] in the laboratory, and infrequent drug resistance in the clinic failed to raise any concern, today the antibiotic resistance problem has grown to include all the serious pathogens and all classes of antimicrobial compounds.[9,10] The drop in efficacy of many classical antibacterial agents (e.g., penicillins, cephalosporins, tetracyclines, glycopeptides, and macrolides) as well as the newer antibacterial agents (quinolones, carbapenems, oxazolidinones) is a direct measurement of the loss of susceptibility and increases in resistance – and these in vitro measurements of inhibition of bacteria are usually indicative of in vivo efficacy (or lack of). Almost in parallel to the rise in antibiotic resistance in the 1960s, efforts by the pharmaceutical industry to support basic drug discovery efforts to identify new classes of antimicrobial agents dropped precipitously due to a number of concerns such as medical need, commercial viability, and superceding priorities within other therapeutic areas[11,12] as the boom of chronic disease research areas opened up opportunity for the pharmaceutical industry to invest in. Only within the last decade has there been the recognition of the increasingly serious nature of resistance,[9–13] and a consensus has emerged that as part of an overall strategy to control emerging and drug-resistant pathogens, novel antibiotic classes must be an essential part of the solution.[14–17]

Although there have been efforts (i.e., restricted use, cycling antibiotics, etc.) to reduce resistance, there can be no arguing that the identification of the so-called 'novel' bacterial targets, and exploitation of these targets by the identification of inhibitors of the(se) target(s) matched with whole cell antibacterial activity is the way to combat resistance to existing drugs. The existing classes of antibacterials and antibiotics with broad-based clinical utility are relatively few, and these target a small list of essential bacterial processes.[16,17] While it can be argued that not every gene product identified as essential in bacteria is an ideal antibacterial target, it is clear that cataloging and characterizing the total set of essential bacterial gene products could lead to selection of the 'best' targets to which novel inhibitors could be identified to work against.[14,18–20]

To this end, both academic and industrial laboratories ushered in a wave of 'genomics' approaches to the identification of novel bacterial targets soon after the publication of the first two whole-genome sequences of bacterial pathogens – *Haemophilus influenzae* and *Mycoplasma genitalium* – in 1995.[21,22] In a matter of just 10 years, we have moved from just two true pathogen whole-genome sequences to over 500 sequenced genomes in 2005.[23–26] To date, however, no genome-associated or genome-sourced target has been successfully exploited to the point of having a genomic target inhibitor advanced to the marketplace, but the measurement of success or failure must be tempered to the recognition of the historic time-line of this genomics targets era (i.e., whole-genome sequence available to the research community).[14,27,28]

7.25.3 Targets

To begin any assessment of choosing the best target(s), one must understand what properties determine the quality of a target. The preferred antibacterial target should possess the following properties: (1) broad spectrum, such that the target should be present in all relevant pathogenic bacteria (i.e., proteins of high sequence similarity that are distributed among microbes, termed orthologs, are considered high interest targets); (2) the gene(s) would encode an essential function whose reduction, absence, or inhibition would result in a bacteriostatic or bactericidal effect on the pathogen; (3) any inhibitor of such a target would be a novel chemotype lacking preexisting resistance with currently used antibacterial agents; and (4) selectivity for bacterial cells such that a similar gene in mammalian cells reduce the likelihood of adverse effects if given to humans to treat a bacterial infection.

There are at least two philosophically different approaches to the identification of novel targets in bacteria. One approach is to identify the bacterial pathogen's genes that are essential for growth (which is typically measured in vitro on artificial microbiological media) and for which the 'novel' genes are identified by any of several methods (some of which are described below in this overview). There is a historical precedence that the inhibition of an in vitro essential gene product usually makes an excellent antibacterial candidate. In fact, virtually all the antibiotic classes presently in use to treat bacterial infections in the clinic inhibit bacterial growth in vitro; by the nature of this inhibition in vitro the paradigm of the clinical microbiology laboratory determining minimum inhibitory concentrations (MICs) usually provides the susceptibility report for which physicians dispense one antibiotic over the other, and this process forms the basis for clinical antimicrobial susceptibility testing.[29,30]

In stark contrast, a different, novel strategy is to identify targets that are only expressed or only essential in the context of the infected host.[31,32] Inhibitors of these 'in vivo essential' processes would presumably interfere with continued survival of the microorganism within infected tissues, and would constitute a previously undiscovered class of antimicrobial inhibitors.[31,32] However, the major difficulties of this approach is the inability to determine the susceptibility of an agent whose inhibitory action is only manifest in the infection state of the host, and to design the clinical trials to allow the licensing of such an agent.

Essential processes that are candidates as antibacterial targets can be identified by at least two different strategies.[33,34] In the first approach, essential genes may be identified in the laboratory setting by the selection of mutant strains of bacteria from a large population of bacteria subject to harsh conditions (such as mutagens) by which mutant strains are isolated and the genetic defect determined by genetic, biochemical, or molecular approaches.[33] With breakthrough technologies, this mutagenesis can now be accomplished by any of a number of techniques (chemical and genetic) described below in detail. In contrast, an alternate strategy consists of the preselection of specific gene targets/products and are subsequently disrupted for essentiality in a directed, target-specific fashion.[34] Both approaches work, and both have their advantages and disadvantages.

7.25.3.1 Targets: What Are They?

A 'target' in antibacterial research is typically defined as the obligatory essential gene product(s) in the bacteria that a researcher will seek to inhibit with the intention of attaining antimicrobial activity in the growing bacterium. Since the early 1950s when most antibacterial research biology was empirical, including the microbiological assessment of antibacterial activity, the targets of the first wave of antibacterial agents has been successfully identified by a combination of microbiology, biochemistry, and genetics (i.e., premolecular biology era).[35–37] Initially, the research community dealt with just a handful of defined cell wall targets, (e.g., carboxypeptidases, transpeptidases (cross-linking), transglycosylation (disaccharide-peptide chain elongation), and D-alanine racemace (alanine isomer conversion), D-alanyl-D-alanine synthetase (D-alanine bonding to D-alanine), and these led to multiple generations of lead compound scaffolds to which all the major pharmaceutical companies worked on, resulting in over 200 penicillins and cephalosporins in just 30 years against the two cell wall targets they inhibit.[35] Likewise, in the 1960s and 1970s, the cell wall targets became the model for similar efforts with protein synthesis inhibitors (primarily chloramphenicols, tetracyclines, macrolides, and aminoglycosides), ribonucleic acid (RNA) synthesis inhibitors (rifamycins), and deoxyribonucleic acid (DNA) synthesis inhibitors.[37,38] In contrast to these >30 years of empirical discovery, the genomics era opened the door for investigators to attack any target in the cell by enabling the identification, cloning, characterization, and screening-against any essential target in the pathogen.

7.25.3.2 Targets: Where Do We Find Them?

If this question had been answered some 40 years ago, one would most likely hear "in the cell wall or protein biosynthesis machinery" as these were the two major areas of antimicrobial research for many years. Within these two target arenas one can find D-ala-D-ala-ligase, transpeptidases, carboxypeptidases, and transglycosylases as the most likely to be mentioned targets.[35,37,38] However, in use, the exact molecular target in this most complex assembly processes of DNA, RNA, and protein synthesis was not initially known and accessible for target-specific screening. Now with the arena of genomics upon us, virtually any gene product encoded on the bacterial chromosome or plasmid element within the bacteria cell can be targeted by just having the ability to sequence the DNA of whole bacteria genomes, as the determination of sequence is simply a tool in the 'tool chest' of antibacterial discovery scientists so there is no longer a limit to the number of targets or access to these specific targets – but the real key to success is picking the 'right' target(s).

7.25.3.3 Targets: How Do We Find Them?

There are many options for finding a target(s) in bacteria, and there are probably more than a hundred research platforms among the many pharmaceutical companies and biotechnology organizations that provide a defined paradigm for target identification, selection, and exploitation. In the simplest answer, targets can be identified in several ways: (1) use existing targets; (2) use a priori knowledge to identify or select a target; (3) select at random by identifying a microbial gene product, characterizing it, mapping it back to the particular gene, and then ascertaining as to whether it may be a viable target; (4) use a reverse mapping of empirically identified antimicrobial activity back to its putative target (so-called 'reverse chemical genomics'); or (5) employ the discipline of 'genomics' to find the target desired.

The identification and selection of the 'right' target may ultimately be a $5 billion to $20 billion question if the right call is made and the blockbuster antibacterial agent emerges (e.g., Biaxin, Zithromax, Cipro, Levaquin, etc.), where peak sales may exceed $1 billion. Alternatively, the downside of picking the 'wrong' target can be the loss of the program, the therapeutic area in a company, or the financial future of a small organization as the costs have now reached $800 million for development of a single antibacterial agent.[39]

7.25.3.4 Targets: An Essential Part, but Just a Part, of the Success Equation

The selection of the target is a major undertaking, a major responsibility and risk point, but the selection of the 'right' target is just a part of the overall equation for successful antibacterial drug discovery. The downstream effort to identify an active inhibitor against the selected target, optimize or even build in antimicrobial activity, maintain selectivity over the host, and optimize all of the pharmacological feature that makes a compound a drug candidate is an immense undertaking that may very well be the major shortfall in the lack of sustained success in the R&D of antibacterial agents over the past 20 years.[40,41]

Somewhere along the way, the science of 'genomics' migrated from being a 'tool' to become an applied goal in itself – an endpoint in many organizations to measure success. The goal was to identify and characterize all of the bacterial genes, and many organizations both in the large pharmaceutical industry and smaller biotech companies set forward a plan to sequence the genome of bacterial pathogens, identify all of the gene products of interest, advance them to high-throughput screens (HTSs)[42] with the intent of identifying a novel scaffold as a starting point for medicinal chemistry optimization, or rational drug design efforts to design inhibitors that could be synthesized in the laboratory as antibacterials.[43,44] In many cases, the output was an outstanding scientific achievement – whether a complete genome sequenced and 'owned' by a company's patent, whole genomes characterized for their top broad-spectrum targets based on sequence homology and similarity, or the identification of a novel gene product by taking an open reading frame (ORF) and identifying the protein's function in the whole cell bacterium.

In fact, with the identification and easily available cataloging of all essential gene products through the use of microbial genomics, it is now possible to establish in vitro screens of high throughput and ultrasensitivity screening to detect inhibitors of bacteria in both compound libraries and natural product sources.[42] Through multidisciplinary collaborations using these advancements in molecular biology, genetics, and protein chemistry, it has been possible to conduct detailed structural biology analyses on purified bacterial targets with the goal of using computer-aided drug design matched with historical knowledge of the medicinal chemistry of specific inhibitors to assist the medicinal chemist in tweaking the design of an optimized lead inhibitor.[43,44] More lore than fact, the structural-based design of inhibitors uses models of preexisting chemical and structural behaviors to model presumptive binding of potential inhibitors – but it takes true biochemical characterization to actually assign the 'quality' stamp to a virtual lead.

Ji provides a most explicit and through review of the successful application of genomics for the discovery of novel antibacterial targets for antibiotic therapy.[45] Citing three significant uses of their program (essential gene identification, evaluation of targets for selectivity and spectrum, and mode of action determination), Ji also provides an 'industrialized' perspective on many of the technologies described within this manuscript in the applied setting.[45]

While this breakthrough in technology provided fantastic results in producing millions of gene sequencing data points, dozens of models of comparative genomics, numerous bioinformatics platforms designed to achieve the 'best' prediction of the homology-driven selection of bacterial targets, the fundamental point is that there really never was a shortage of targets, but rather a shortage of quality, novel 'lead' compounds against any target. And after 10 years in this post-1995 release of the first two microbial genomes,[21,22] we have just an empirically identified novel oxazolidinones class of antibacterial agent, and a number of minor, but incrementally modified scaffolds that show promise (e.g., ketolides, novel macrolides) – but no new drugs from this genomics effort.

7.25.4 Genomics as a Scientific Discipline

7.25.4.1 Genomics Defined

As defined in multiple dictionaries and on multiple scientific websites, genomics is the study of "the branch of genetics that studies organisms in terms of their genomes (their full DNA sequences)." This discipline has emerged from the roots of molecular biology in need of sophisticated analysis tools with the birth of whole-genome sequencing requiring the simultaneously analysis of hundreds of millions of data points in the search for novel antibacterial targets.[19,20,45–59] The underlying assumption of employing this technology is that microbial genomics will provide a number of essential, validated targets as possible candidates for the discovery of new antimicrobials, and genomics has not disappointed.

7.25.4.2 Genomics: Multiple Views of Accomplishment and Failure

The role of genomics in the drug discovery process may be regarded as either the most underappreciated or overemphasized process of target analyses in the early drug discovery phase. Introduced as an analysis tool subsequent to the availability of whole-genome sequences from bacteria, users of these analysis tools were able to draw sequence–function linkages within a single bacterial species, and expand the comparison to other bacterial species.[18,19,45–59] Specific sequences, attributed to coding for a specific gene, became the sequence template to which one could search for the same gene in other pathogenic species, and if it was found this would provide assurances that the particular gene being examined, was in fact, a broad-spectrum target. Thus target selection based on conservation of sequence could be as simple as completing a particular sequence comparison between two or more bacteria – and simple conservation would be indicative of gene conservation. But organizations involved in target selection soon found that the presence of a sequence, its degree of homology and/or similarity, devoid of high human sequence homology, and a list of up to another dozen qualifiers required to pick a 'good' target was just the beginning of the process, and that genomics was just part of the overall drug discovery tool chest.

There has been a great deal of success in the selection and work-up of novel genomics targets,[45–59] but none has progressed to the point of being in late stage development or on the market. Examples of genomics-derived targets include: aminoacyl-tRNA synthases,[48–50,57] fatty acid biosynthesis targets,[55,56] peptide deformylase,[49,54] and two-component system targets (of which a subset are essential genes).[19,51,52]

The use of genomics as a target selection tools has been visibly documented by publication and patents by many industrial organizations, noteworthy among them GlaxoSmithKline, Bristol-Myers Squibb, AstraZeneca, Schering-Plough, Wyeth, Johnson & Johnson, Genome Therapeutics (GTC, now named Oscient), and multiple biotechnology organizations formed around their own proprietary version of a genomics-facilitated discovery platform over the past 10 years. Building on both proprietary and nonproprietary whole-genome microbial DNA sequences, researchers at these organizations have assembled a selection paradigm for their own favorite novel antibacterial targets. All intended to use this paradigm as the first step to identify novel inhibitors against the novel targets, assuming that the use of totally unexploited targets as 'catch-basins' for inhibitors would undoubtedly identify novel inhibitors as antibacterials not subject to preexisting resistance.

Using part of the GlaxoSmithKline experience as a model system comparing more than 60 bacteria, the fatty acid biosynthesis pathway was one of numerous pathway targets dissected and pushed into their drug discovery paradigm in search of these novel antibacterials against pathogens of clinical importance. Among the targets pursued by GlaxoSmithKline were FabF/B, FabI/K/L, FabA/Z, FabH, and FabG.[56] Output from their efforts has been to identify the putative target of the diazaborine antibacterials as FabI antibacterials, a clarification of cerulenin as an inhibitor characterizing the prokaryotic versus eukaryotic homolog binding, and confirmation of the targets of triclosan and thiolactomycin.[56] In addition, McDevitt and Rosenberg[18] and Payne and colleagues[55] in reviewing GlaxoSmithKline-examined targets, have reported several lead series of compounds against multiple targets through their genomics program, including inhibitors against t-RNA synthases, MurA, the chorismate biosynthetic pathway, the isoprenoid biosynthesis pathway, multiple cell division pathway targets, and two-component systems.[55] Within these programs are perhaps the best published, applied example of the successful application of genomics in applying the tool of genomics to drug discovery – but it also demonstrates the difficulty of identifying novel progressible leads as part of any genomics program, as no candidate compound has emerged from this most impressive discovery program.

7.25.4.3 Genomics: Multiple Categories of Research

There are multiple categories of genomics-based targets to consider in thoroughly addressing this topic, among them: (1) true genomics identified and validated targets; (2) genomics identified targets; and (3) genomics validated targets.

A fourth category, with no representation, could be genomics-based targets leading to drugs – and this topic is briefly mentioned below.

The process of drug discovery into which 'genomics' approaches/technology are applied is shown below in **Figure 1**. The generalized process of genomics is just the starting point of a prolonged process taking between 7 and 10 years. Within this front-ended effort in which a target is selected, the importance and impact of selecting the right target is immense – in that the wrong choice of a target that is not conserved, or not fully validated as essential in all relevant pathogens, can be devastating some 7–10 years after the target selection (if the drug candidate fails to match the medical-need product profile); thus target selection needs to be 100% accurate to sustain commercial interest in the antimicrobial drug discovery discipline.

To understand the impact of genomics on the discovery process, one needs to understand the options in the genomics 'tool chest.' Among these tools are wet-biology tools such as whole-genome sequencing, computer-driven analyses of whole-genome sequencing data allowing the design and usage of knockouts to test essentiality, wet-biology proteomics, wet-biology essentiality determination protocols, virtual pathway analyses tools, structure elucidation tools, protein domain motif comparison tools, microarrays, and sequence concordance analyses (to mention just a few). The integrated use of these 'tools' is usually at the beginning of the discovery process for any given target ultimately selected for advancement into the identification of a lead process, but in reality it is often a continuous process of selecting the next target to progress through the often one-target-at-a-time format for screening, as virtually all mature industrial programs have a portfolio of different stage, targeted programs progressing at any one time.

Perhaps the most important consideration is the complementary nature of these genomics tools, for no individual tool in itself serves to address all techniques, issues, and solutions. There are bacterial species variations in the

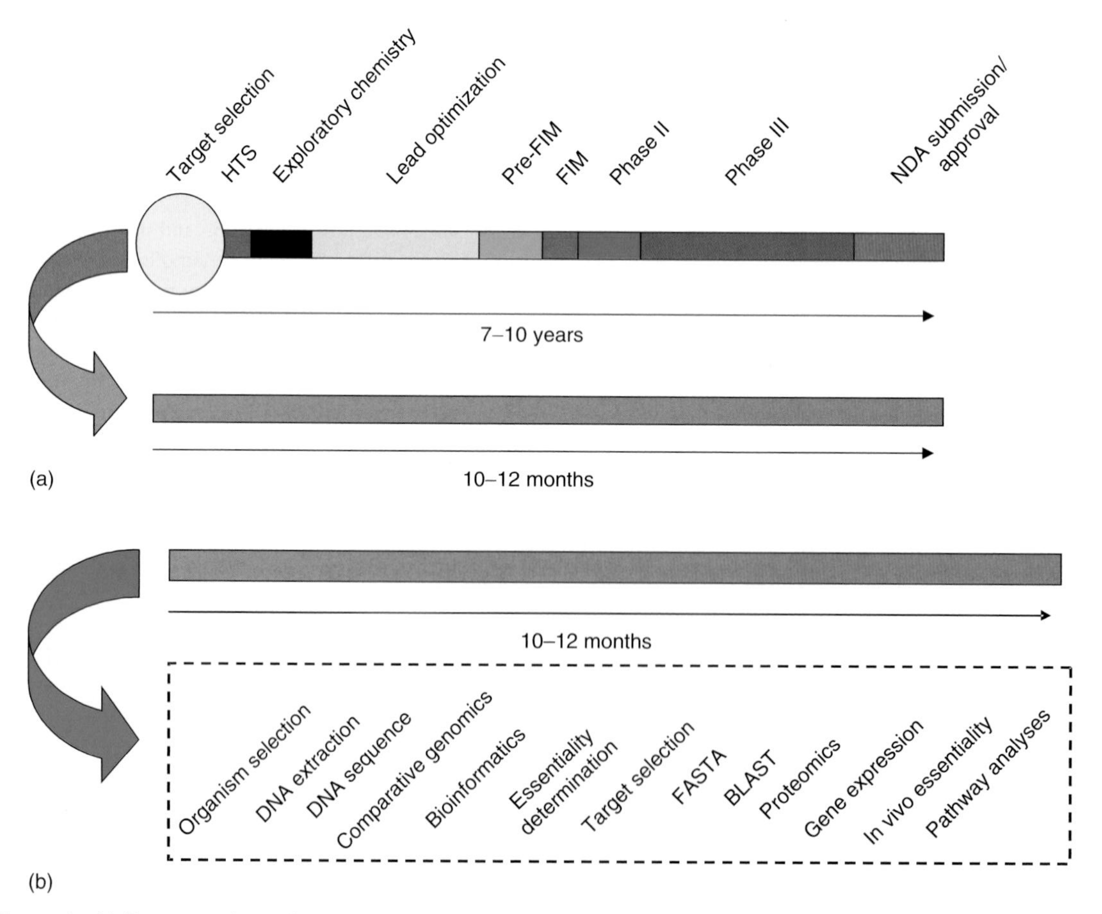

Figure 1 (a) Representative antibacterial discovery time-line with approximately 7–10 years from target selection to drug approval. (b) An expanded time-line of target selection; representative protocols/processes involving genomics are listed as occurring in this early stage discovery process.

robustness for any particular genetic tools providing the genomics data/knowledge, as well as exceptions to every rule of thumb predicted by the secondary analyses of raw sequencing data. In the end, virtually every target in any bacterium must be individually scrutinized by the classical measurements of fitness for target selection described above.

7.25.4.4 Genomics: What has it Provided Us?

With the arrival of genomics in 1995,[21,22] there has been a dramatic release of bacterial genome DNA sequences resulting in the documentation of virtually all of the potential antibacterial targets to which a function can be assigned by either a priori knowledge or a combination of biochemical, molecular, microbiological, genetic, structural, microbial physiology, or other techniques. In general, this is approximately half of the theoretical ORFs in any bacterium, with the other 50% of the unknown ORFs yet to be identified. Perhaps nowhere has there been greater impact of bacterial genomics than in the realm of microbial research,[18,45–59] in particular, bacterial research. Since the first bacterial genome sequence was published in 1995,[21] and the second genome (from *Mycoplasma genitalium*) in the same year[22] the number of bacterial genome sequences numbers in the hundreds and continues to expand as the technology advances and more bacterial species are added to the growing list of genomic databases. In fact, there are 395 bacterial genomes available in ERGO,[23] 252 bacterial genomes available in the NCBI database, 179 available at TIGR,[25] and 104 available bacterial genomes available through the Joint Genome Program/Microbial Genome Program[26] at the time of submission of this manuscript.

Beyond the hundreds of bacterial whole-genome DNA sequences now available, the genomics era has also led to the design, implementation, and improvement in a number of genomics and bioinformatics tools that can help researchers better characterize targets and make batter choices in selecting the target(s) they pursue.[60,61] As shown below in **Table 1**, there is a short list of top criteria required for choosing the best targets for antibacterial drug discovery, and among the top two criteria must be essentiality and spectrum. In fact, of the ten criteria listed in usefulness in selecting a target in **Table 1**, half of them are direct readouts from the application of genomics to the drug discovery process.

Table 1 Criteria for best quality antibacterial target selection

Target property	*Preferred property*	*Comment*
Spectrum	Broad-spectrum inhibition drives empirical therapy; highly desirable	A narrower spectrum may be desired, but commercially limited
Essentiality	Inhibition leads to bacterial death	Standard of care is to kill (or at least cripple) the bacterium; virulence inhibition may be effective, but is not currently accepted in clinical practice
Selectivity	Inhibitor less likely to be mechanism-based toxic to humans	Marketed drugs exist against targets with homology to eukaryotic (incl. human) targets
Multiplicity of action by inhibitor	Dual action	Serendipitously most successful 'classical' targets inibit two targets
Bactericidal	Killing bacteria is preferred over bacteriostatic inhibition	Bacteriostatic drugs on the market demonstrate efficacy (e.g., macrolides)
HTS available	Mechanism-based whole cell screening preferred	There are alternative methods to identifying inhibitors/leads
Control drug available	Advantage in gauging 'active' in HTS	Alternative approaches available
Structural data available	Cloned, expressed, purified, and crystallized protein	Enhances the opportunity to 'design' the chemical chemotype and/or further undesatnd the interaction of a compound in the active site
Druggability	Suitable for targeting inhibition	Subjective labeling of a chemotype based on its chemical nature, reactivity, size, ability to be modified, and a priori medicinal chemistry knowledge
Do-ability	Setting aside goals and theory, the ability to conduct all aspects of the drug discovery process in working with a target is a criteria in itself	Practical matters such as technical success and intellectual property position enters into the do-ability equation

7.25.5 Genomics Tools

There are many 'deliverables' from the effort of genomics, and although not a single antibacterial drug has made it to the marketplace from the genomics era[58] – that may be asking a little too much for just 10 years into the program, but the breakthroughs in understanding targets has been immensely impressive.[19,45–59] The underappreciated problem of discovery is that one may pick the best target meeting all quality criteria, arrange for the most sensitive, most thorough screening of synthetic compound libraries or natural product libraries – and there is no guarantee that a quality, progressible lead compound may be found, and moreover there is no guarantee that it will meet the minimum criteria for a starting point that can be optimized to a candidate compound. To this end, the more through a discovery program is in placing key filters for quality upfront, the sooner one can make the call in triaging hits identified in HTS.

7.25.5.1 Comparative Genomics

There may be dozens of different ways to approach the technical issues of identifying and characterizing novel antibacterial targets, and none is necessarily better than the next. There are, however, differences in leadership opinion as to what constitutes a valid discovery target for a novel antibacterial agent. One routine approach in finding new antibacterial targets is to identify genes whose products are essential for bacterial survival (under a standard environment in the laboratory). To this goal, a number of genomic techniques have been developed to identify this subset of genes in vitro in the test tube laboratory setting.

All marketed antibacterial compounds permeate the cell wall or membrane of the bacterium, and interact with a target(s) that inhibits or kills bacteria that are grown on defined, artificial laboratory media. In fact, the basis of standardization across clinical microbiology laboratory susceptibility testing is reliant on the standardization of media testing so as to assure consistent performance of susceptibility testing.[29,30] This in vitro determination of growth inhibition is the 'gold standard' of antibacterial therapy, but it is not the only approach. Some research teams have sought to identify essential targets in bacterial growth, attachment, invasion, infection, and virulence in the host.[62] These virulence-related targets are usually expressed only within the environment of growing in the host[62] or within host-simulatory laboratory setting (i.e., tissue or organ culture). It may be hypothesized that antivirulence or pathogenicity targets may be less likely to select for resistance with the belief that the less exogenously insulted pathogen will not be in the stress mode to select for a mutant strain without selective pressure usually found with antibacterial treatment (where it's a matter of death or selection of a less-optimal mutant strain in the population if such a mutant exists).

7.25.5.2 Genome Comparisions by Computer Tools

Prior to 'genomics,' microbial or molecular geneticists had been routinely using basic computer tools to assist in their cloning (e.g., design of cloning or restriction sites) and DNA sequencing experiments (reading/conversion of their sequence bands to sequence data), but this was exclusively done in dealing with a single bacterium and a single gene, that is, to examine a specific gene (or operon), and analyze this very limited DNA region on the bacterial genome after purification of a smaller fragment. Few truly accurate genome maps were available, and those that did exist were simple representations of gene order, linkages, and distances between genes ascertained by classical genetic manipulation and/or mapping techniques.[63]

The age of genomics ushered in a new wave of technology based on need. The need was simple: how to cross-analyze two or more whole-genome DNA sequences, not to mention the need to analyze all of the individual bacterium's genome.[64] Beyond the simple genome-to-genome comparison, the desire to cross-check the sequence of multiple pathogens[64] simultaneously to address strategic issues of 'spectrum' of a potential target required the ability to analyze both the sequence of DNA and the residue-by-residue comparison of proteins. Thus the need for computer capacity, speed, and analyses tools drove the design of a computer 'tool chest' to assist the biologist with their analyses of genomes. Key programs were designed and customized to be used to assemble the genome(s) themselves from the various sequence techniques such as random shotgun sequencing,[65] where the assembling of massive numbers of sequence files into larger contiguous sequences (contigs), and ultimately handling megabase sequence size range information such as with the TIGR Assembler tool for assembly of large shotgun cloning sequencing projects.[66] Quick to follow this DNA sequence analyses was genome-wide annotation, in which programs such as BLAST[67] and FASTA[68] were brought online to handle genome-wide similarity searches to identify and annotate genes and proteins.[67–69]

A breakthrough approach to identifying genes in a whole genomes without reliance on DNA sequence is based on the Markov models (GeneMark).[70] This model determines the differences in coding and noncoding area based on features such as three-base periodicity. Additional design tools such as GLIMMER[71] and build-off of the Markov models to add flexibility in sequence analyses based on varying sequence length in its analyses, gene overlaps, and other sophisticated analysis tools too detailed for this review which enable operon detection based on sequence analysis.[71–74] A excellent example of use of the Markov models was reported by Borodovsky and Colleagues[75] in which an analyses of three classes of *Escherichia coli* genes employing a phased Markov model for a nucleotide sequence of each gene class was developed for predicting genes using the GeneMark program.

In terms of the 'deliverables' from this melding of molecular biology sequencing and computer-aided analysis tools is the ability to undertake massive computational analyses and comparisons of two or more pathogen genomes simultaneously to determine both DNA and protein sequence homology and similarity in targets.[76,77] For example, a genome-by-genome cross-comparison of protein sequences using FASTA to build tables of gene(s) similarities that can be accessed via a web-based interface[76] has become the norm. One such example is the website featuring 123-Genomics[77] billed as a 'Genomics, Proteomics and Bioinformatics Knowledge Base'; this is an open data analyses site enabling anyone to build on the knowledge of others in the determination of function and structure from gene sequence. For example, the HOBACGEN (homologous bacterial genes) database can be accessed for comparative genomic analyses by selecting gene families and homology common to particular bacterium in the database.[77] Such a cross-analyses enables the investigator to predict broad-spectrum antibacterial targets,[77–79] which can subsequently be verified by 'essentiality' checking through wet-biology experimentation in the laboratory. A most sobering fact is that despite the sophistication and advancement of the genomics and bioinformatic tools used in analyzing genomic information generated by multiple bacterial genomic sequences, in the best-characterized organism, *E. coli*, just under 40% of the ORFs identified by computational methods encode proteins of unknown function.[80–82]

One example of the application of this complex discipline for target selection was described in a microbial genome concordance analysis.[83] This concordance analysis performed a FASTA comparison of multiple genomes (the *E. coli* genome against *Bacillus subtilis*, *H. influenzae*, *Mycobacterium tuberculosis*, and *Helicobacter pylori*) at the amino acid level, and assembles tables for subsequent access by a web-based interface. The retrieval of data sequences, both raw and analyzed, can be made based upon customized criteria set forth by the end user, which is in contrast to other comparison tools such as COG, HOBACGEN, and others that use fixed default constraints,[82,84] and subtracted out sequences representative of close homology to the yeast *Saccharomyces cerevisiae*.

Use of the programs mentioned above as well as others[85–87] define analyses sets of microbial genes that can be tested for 'quality' as novel antimicrobial targets. As discussed previously, criteria for ideal targets include broad distribution of the target (gene) among pathogenic bacteria, and a minimal homology or similarity to eukaryotic genomes, especially the human genome. Applied examples include *E. coli* and human dihydrofolate reductase, the antibacterial target for trimethoprim (28% identical at the amino acid level) and *E. coli* and human topoisomerase II, the target for quinolone antibacterials (20% identical). In both of these cases, the bioinformatics analyses enabled an understanding of the selectivity of these two somewhat related genes in prokaryotes and eukaryotes, with an explanation as to why selectivity could be attained (i.e., selectivity despite similarities to human orthologs even though the level of overall identity of proteins among various bacterial species appears low).

7.25.5.3 The Target Space Meets Chemical Space

Within the gene 'space' of a given pathogen exists a repertoire of all of a typical pathogen's genes, both essential and nonessential (**Figure 2**). In comparison to another unrelated pathogen, there is a subset of the genes conserved between the two pathogens being compared, and within any *n* number of pathogens, there lies a smaller subset of conserved genes between all *n* pathogens (**Figure 2**). And out of these conserved genes is a smaller subset of essential genes. The approach of the investigator using genomics enables the ability to assess the anticipated spectrum to be obtained by successfully targeting a particular gene found to be overlapping in the target space in the designed organisms. For example, for modern day respiratory tract infection (RTI) oral antibacterial agents, it is required that the antibacterial agent have intrinsic activity against *Streptococcus pneumoniae*, *H. influenzae*, *Streptococcus pyogenes*, *Moraxella catarrhalis*, and three atypical pathogens (*Mycoplasma pneumoniae*, *Chlamydia pneumoniae*, and *Legionella pneumophila*). With today's tools, the analysis of common genes within these 10 pathogens is simple (as compared to just 10 years ago with the tools described above), but this is merely one part of a very complex paradigm whereby the 'target space' defined as the bioinformatics target for an antibacterial must now be matched with the equally limited 'inhibitor/chemical space' (**Figure 3**).

The quality of a lead may be the greatest Achilles' heel in any antibacterial drug discovery program. Always a point of controversy and uncertainty, unless a drug makes it to the market, there is no assurance of its 'quality' among discovery investigators, as the subjective evaluation of a lead's quality is full of opinion, personal experience, a priori knowledge,

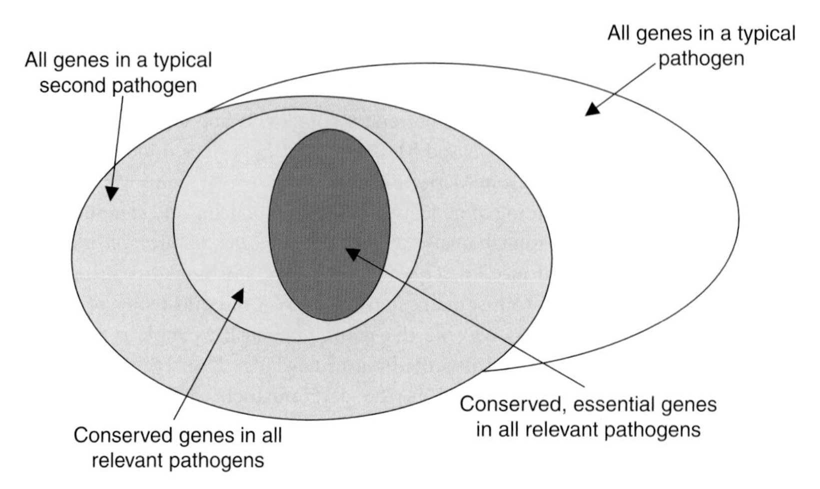

Figure 2 The ever-shrinking 'space' of genomics targets through the process of comparative genomics in which the sequences of genomes are compared against one another, seeking the common genes in all of the relevant genomes (purple overlap) which is significantly smaller than the overall subset of essential genes in any given pathogen.

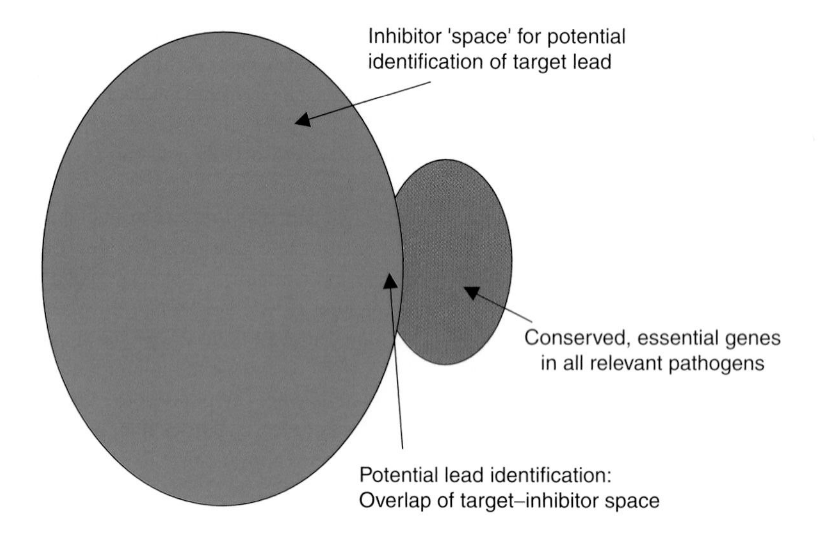

Figure 3 Intercept of the 'target space' and 'inhibitor/chemical space' for any given drug spectrum analysis.

technical know-how, bias – and occasionally fact. There are both biological and chemistry criteria for quality of a lead, with all assessment made based primarily on pre-existing knowledge from experience – with no real logical manner to assess quality in a novel chemical series. What one can do, however, is set a series of target criteria for a candidate, and work backwards toward the minimally acceptable criteria for a lead. For example, typically investigators may screen for antibacterial or target activity either to the limit of solubility or $\sim$40–100 μM, but the assessment for a lead may have a minimal threshold of an IC_{50} of 10 μM, and/or an MIC of $<8\,\mu g\,mL^{-1}$. A lead must target a gene product conserved across all RTI pathogens to be considered a lead for a community RTI drug, but only against key community skin pathogens (mostly Gram-positive pathogens) to be considered as a community skin and soft tissue infection agent. An inhibitor with no selectivity over the mammalian homolog target, or overtly cytotoxic based on its chemical structure (i.e., not mechanism-based toxicity) could not be considered a quality lead. The lesson learned from historical experience, both success and failure, is that the sooner one can make the assessment of the quality of a lead, the more certain one can be in progressing the lead to the candidate stage. And the one last lesson from experience is that there is no guarantee that a quality inhibitor will be found against all targets.

7.25.6 Experimental Proof of Essentiality and 'Quality' of Targets

There have been numerous approaches[33,34,88–92] or the demonstration of essentiality in bacteria including targeted gene disruptions (the so-called knockouts),[33,34,81,88,89] conditional lethals,[90] and random mutagenesis (including transposon mutagenesis,[91] cassette mutagenesis,[83] and shotgun antisense).[92]

7.25.6.1 Essentiality Determination by Random Mutagenesis: Conditional Lethals

One of the earliest examples of identification of essential genes in bacteria was the isolation of temperature-sensitive mutants.[80,93–96] These mutants are 'conditional' in that they have a phenotype that allows for growth at a temperature varying from wild-type growth, while not supporting growth of the parent (nonmutant) strain. These are mostly high-temperature-restricted mutants, although cold-sensitive mutants have also been isolated and characterized. The design strategy is such that an essential gene that is mutated such that its gene product is temperature sensitive, and the target gene functions with a lower efficiency than the wild-type gene.

One specific application to the antibacterial drug discovery process is to use large numbers of temperature-sensitive mutants as screening tools to identify any inhibitors against the same essential gene, but which due to its thermal lability is a more sensitive target and thus should provide for a more sensitive screen for antibacterials.[95] This design allows for use in HTS assays, but the assumption is that the thermal labile protein is similar enough to the wild-type that an inhibitor identified against the temperature-sensitive protein will act as an inhibitor against the wild-type target too.

7.25.6.2 Gene Knockout Systems: Transposons

Another method to identify genes that encode essential products is by gene knockouts.[33,34,80,88,89] There are at least two fundamentally different approaches that can be employed. In one general case, transposons can be utilized to knock out genes by random insertion into the total genome. The transposons usually possess some selective marker (e.g., antibiotic resistance), and cells with an integrated transposon are identified by the selection process. Used in this simple format, the process identifies essential genes by virtue of the absence of isolation of transposons in these genes, i.e., a gene with a transposon disruption is not an essential gene.[97,98] Problems limit the usefulness in that mapping of transposon inserts by saturation is very laborious, and transposon knockouts suffer from polarity effects that can lead to false conclusions (by virtue of polarity raising some ambiguity as to what gene is actually essential where there are multiple genes on an operon). One method used to resolve such issues is the use of balanced lethals, in which plasmids with the wild-type allele are employed with the transposon knockout.[99,100] Additional problems to deal with are that transposons are generally not random in their sites of insertion,[101] and distal insertions into any given gene may not truly inactivate it.[102,103] In response to these problems, variations of transposon mutagenesis have been implemented such as GAMBIT (genomic analysis and mapping by in vivo transposition).[102,103] In this technique, a variant of the eukaryotic mariner transposon, which is essentially a random insertion transposon, has been modified with antibiotic resistance genes[102] removing the ambiguity of any analyses.

7.25.7 Genes to Screen Approach

7.25.7.1 From Genes to Screens

The identification of the genes that represent potential novel targets is a fundamental early step along the path to the discovery of novel compounds. This is the first of many subsequent steps to progress a candidate gene to the point where the encoded protein can serve as a tool to identify an inhibitor against the in situ target gene in the bacterium. In the case in which the similarity of a chosen essential gene (e.g., DNA gyrase, *gyr*A/*gyr*B) has such significant similarities and identities to genes in other pathogens, its is easy to understand the path forward for HTS to identify a novel DNA gyrase inhibitor; unfortunately the design of an HTS assay for this target is not so easy. Where gene sequence and structure of an ORF has such similarity that the function can be ascertained and tested in a functional biochemical assay, the path forward is equally clear. And if the target protein catalyzes the hypothesized reaction, it is usually straightforward to establish an HTS based on the functional activity. In the case of conserved ORF genes, several additional technologies may be used to discern function and/or screen for inhibitors. Potential roles of unknown function genes may be explored experimentally by DNA microarrays, proteomics, regulatory pathway analyses, physiological responses, and structural tools (nuclear magnetic resonance imaging (NMR), x-ray crystallography, and NMR-mass spectrometry).[104–112]

One of the many creative data-driven analyses that has emerged from the genomics paradigm is the ability to overlay the catalog of genes (containing ORFs) with a combination of microbial physiology, biochemistry, microbiology, and the preexisting understanding of pathways to ascertain the identification and/or function of unknown genes.[104,105] One excellent example is the EcoCyc database which describes the genome and gene products of *E. coli*, including the major biosynthetic pathways of DNA, RNA, cell wall and protein biosynthesis, but also includes the network of regulatory pathways (i.e., metabolism and signal transduction).[104,105] The EcoCys database describes over 4000 genes of *E. coli*, and with all known *E. coli* enzymes and regulatory networks, assembles these metabolic activities into almost 130 metabolic pathways in a fully annotated 'pathway analysis.' Similar analyses in other organisms provide the opportunity to match necessary metabolic functions to ORFs in search of function.

7.25.8 Genomics Tool Chest

7.25.8.1 Gene Expression Levels

The measurement of transcription of gene expression, whether individual or operon, can be controlled in any of several regulatory systems. With the birth of whole-genome sequencing, multiple new technologies have been designed to enable the monitoring of genome-wide transcription so as to fully understand the regulatory control network. There are essentially two distinct approaches to this technology: (1) DNA-based microarrays for large-scale monitoring of mRNA levels; and (2) the Affymetrix GeneChip arrays. Whereas both technologies provide for the synthesis of DNA-based microarrays on a large scale, allowing for the monitoring of changes in mRNA levels, the technical details and outputs are quite distinct.

7.25.8.1.1 Deoxyribonucleic acid-based microarrays

In the first type protocol, single- or double-stranded DNA (several hundred base pairs in length) is spotted or 'printed' onto glass slides.[113–114] mRNA is extracted and prepared from two comparative sources, conditions, drug treatments, etc., and the qualitative and/or quantitative differences in expression are determined (i.e., isolated and reverse-transcribed). For example, a comparison may be done in a before and after drug treatment, or a healthy versus diseased state tissue. To distinguish between the two environmental conditions, the two reverse transcription reactions employ different fluorophores to 'label' the two samples.[115] These fluorophore-labeled cDNA samples are mixed and hybridized to the glass slide DNA array.[116] After processing, the relative amounts of fluorescence bound to each spot are determined, and the data can be analyzed for quantitative differences.[116] An example of this in antibacterial research may be found in an experiment designed to identify the putative target of protein synthesis. If a pathogen is grown in defined conditions and treated in one or two parallel protocols (e.g., no drug treatment and treatment with a protein synthesis inhibitor), and both samples are processed for the mRNA at specific time point – then one may be able to ascertain the target(s) of the inhibitor. Alternatively, the same analyses may be done with a proteomic analyses of drug-treated cultures (as compared to the untreated control cells).

7.25.8.1.2 Affymetrix GeneChip arrays

In the second approach, single-stranded oligonucleotides are synthesized on a proprietary substrate using photolithography methods. Any single gene is represented by unique 25mer oligonucleotides derived from defined genome sequences. After hybridization, the GeneChip is scanned in the proprietary Affymetrix chip reader, and the levels of gene transcripts are quantitated.[117,118] Both techniques enable the investigator to measure the differential expression of mRNA under defined conditions, thus enabling a genome-wide determination of gene expression under different conditions.[119] An example of the technology may be to identify and follow the up- and downregulation of gene expression in response to any insult to the given bacterial culture. In this case, a more unambiguous outcome may arise from the detection of specific oligonucleotide sequence tags.

7.25.8.1.3 Regulated gene expression

Equally exciting is the advanced understanding that the industry has gained on the use of genetic systems based on regulated gene expression as a 'tool' for advancing discovery research. Within virtually every R&D program, the use of regulated expression of genes plays some role in the success of their programs.[4,18,19,45–59] Among the practical uses of these regulatable gene expression systems are constructs for: the demonstration of essentiality; the selection of targets; the determination of function of targets; mode of action studies for inhibitors; the overexpression of target proteins; and the constructs designed for use in whole cell screening toward identifying target inhibitors.

7.25.8.2 Proteomics

Another technology enabled by genomics is the discipline of proteomics.[120–123] Proteomics has been defined as "the measurement of the total expressed protein complement of a cell under defined conditions."[124] Proteins are usually separated on one-dimensional gels for overexpression work in molecular biology, but for the more sophisticated mapping and quantitation of protein profiles in the whole organism, gels are usually two-dimensional (isoelectric point versus molecular weight), intent on the selection of protein 'spots,' which are subsequently followed by proteolytic digestion and matrix-assisted laser desorption ionization-time of flight mass spectrometry to identify the individual peptides.[125] These proteolytic spots can then be correlated to a gene product by referencing back to a genomic database and predicting a protein.[120–123] For the measurement of an antibacterial's effect on a pathogen, the pathogen is grown under different concentrations of an antibiotic and the cultures harvested and processed as described above (with a no-drug control culture in parallel), and the entire protein complement of the pathogen is resolved on two separate gels and the 'proteomic' profiles compared.[120,121,123] If done in a time-based study, one can see the effects of a drug on a pathogen over its generation time, and draw specific conclusions as to its primary and secondary (and indirect) mode of inhibition. Low-abundance proteins and membrane-bound proteins may pose a problem at present, but advanced methodology is addressing these technical issues.[126,127] The successful application of proteomics to microbial genomes has been illustrated in several hallmark papers.[128–130] All three of these studies are examples of the use of proteomics together with transcriptional profiling to identify the function of ORFs, which subsequently enablies the design of inhibitor screens, and to profile the bacterial cell responses to a novel antibacterial.[128–130]

7.25.8.3 In Vivo Essentiality Determination

A number of researchers hypothesize that novel antibacterial targets may be found in the inhibition of gene products that are expressed by the pathogen in the infected host.[102] These virulence target genes would only be expressed within the environment in which their expression is required for the successful infection or maintenance of infection in the host, and which are activated only under these conditions.[131–134] The massive collection of genomic data has provided new insights into the molecular organization of bacterial pathogens, revealing the basic genetic and metabolic structures that support the viability of the microbes in their pathogenic state in the host. Identifying novel compounds that interfere with the expression or function of one or more of these factors would provide the ability to abort the infection.[135–137] This is not a bactericidal approach per se, but it does portend to prevent infection. Two specific approaches to the in vivo essentiality techniques are described below.

7.25.8.3.1 Signature-tagged mutagenesis

This was one of the first industrialized scale effort undertaken on a comprehensive genome wide system to search for in vivo 'essential' genes that were specifically involved in virulence.[138,139] The basic idea is to incorporate a 'label' or 'tag,' which in this case is a random oligonucleotide sequence of ~40 bp, flanked by defined, constant DNA sequencing primer. These random 40-bp tags are ligated into the specific transposon to be used for mutagenesis, providing the individual transposon with a unique 'tag.'[134,135] These transposon libraries are then used to mutate a bacterial species, and individual antibiotic resistant colonies, each with a unique transposon inserted into a different gene, are recovered.[140] Pools of the transposon mutated cells are used to infect individual animals, and surviving bacteria are subsequently recovered from the infected animals by virtue of the presence of the transposon knocking out the in vivo-requiring 'essential' virulence gene. The unique tag are amplified out by polymerize chain reaction (PCR), and used as probes against colony blots of the original input organisms in the inoculum pool. Loss of a signal indicates that the transposon had inserted into a gene that was essential for survival in the infected animal, hence the ability to identify in vivo 'essential' genes.[138–140]

7.25.8.3.2 In vivo expression technology

A variation of this theme of in vivo determination of essentiality is the use of a genetic system composed of an ampicillin-resistant suicide delivery plasmid that contains randomly inserted small DNA fragments introduced into defined a cloning site (usually upstream of a promoterless essential bacterial gene transcriptionally fused to a β-galactosidase gene).[141,142] This in vivo expression technology (IVET) delivers a plasmid into the chromosome of a recipient bacterial strain (either Δ*pur*A or Δ*thy*A) and recombined into the chromosome using the inserted DNA fragment for homology. Recombinant cells, deficient in their ability to produce either purine or thymidine, are selected by the addition of thymidine or purine – and these deficient cells are used to infect a host animal (typically a mouse). During the in vivo growth of the recombinant cells in the host, only bacteria with promoters in the inserted DNA turned on in the animal host would survive, as the induced promoter would drive the *pur*A or *thy*A gene, essential for survival. Surviving bacteria are collected from the host animals and screened on laboratory media for β-galactosidase activity. Clones that contained promoters turned on only in the host would show little activity on the β-galactosidase indicator media in vitro, hence the selection of in vivo 'essential' genes. Other variants of IVET have used green fluorescent protein and sucrose sensitivity as selections.[143] One problem with this IVET system is that genes that are transiently expressed during infection process may not be identified.

7.25.9 Conclusions

This chapter has attempted to outline some examples of the provided by the 'deliverables' genomics revolution ultimately in support of novel antibacterials R&D. Whereas the output from 10 years of prokaryotic genomics effort is devoid of any marketed antibacterial, there has been a major contribution to the tool set that researchers commonly use in everyday laboratory practices. Among these 'deliverables' are:

1. Genomics offers a unique opportunity to identify novel, essential, validated targets.
2. Genomics offers inroads to technological advancements utilizing molecular techniques of DNA characterization and computer-aided bioinformatics analyses of homology, similarity, and motif to provide comparative genome analyses of drug spectrum.
3. Genomics has offered advancements in the understanding of the internal physiology of the pathogen's growth and virulence.
4. Genomics has prompted the 'invention' of data analyses tools to handle millions of data points simultaneously.
5. Genomics has enabled a better understanding of the mechanism of action of antibacterial agents.
6. Genomics has enabled an understanding of the pathogen-wide metabolic changes under conditions of insult or stress (e.g., antibiotic pressure).
7. Genomics has opened the door to improved therapeutic option for the treatment of bacterial infections; these include both novel antibacterials and advances in nonantibitoic therapy (antivirulence).

The availability of genomic sequence data for nearly all clinically relevant bacterial pathogens provides an important resource for antibacterial drug discovery research. It provides the know-how for use of a comparative genomic analyses in the identification of potential new broad-spectrum targets with low human homology (providing some assurance against mammalian toxicity), or more selective, narrow-spectrum antibiotics. This same technology enables the determination of essentiality by design and use of so-called knockout gene strategies, while providing the validation of the target as a drug-inhibitable essential gene. And the genomics technology provides the foundation for the complementary use of molecular biology and biochemistry to allow for the construction of expression systems for production of native or recombinant proteins and the development of high-throughput assays for the detection of inhibitors of the target.

There have been dozens of organizations that have delved into this genomics arena, and many papers have been published and companies launched (and failed) on individual genomics platforms. In the end, with genomics recognized as an efficient tool for probing the whole bacterial genome, there have been multiple reports of the top candidate targets recommended for use in HTS to identify novel lead antibacterial agents, with the assumption that unique classes of molecules not yet discovered would be identified as antibacterials.[144] In addition, and not the topic of this review, genomics has opened the door for a new approach to vaccines and diagnostics.[145]

Pharmaceutical and biotech organizations were quick to recognize that this sudden massive release of microbial DNA sequence could be mined for the identification of novel targets, including the potential to target ORFs as antimicrobial targets (even in the absence of known function). Although a massive effort was unleashed toward the complete capture of relevant pathogens DNA sequence in an attempt to fully understand the basis for 'good' targets and broad-spectrum agents, it appears that in many cases the process of genomics determination of novel targets became an end in itself, and the commercial organizations that built a genomics-only platform found that continued capital support would decrease without a candidate drug in the near future. What became the boom business of prokaryotic genomics research in the 1990s has all but gone away in 2005. However, in its effort to identify significant numbers of novel bacterial targets, the microbial genome efforts serves as a model for the development of computational tools and experimental genomic strategies such as the microarrays, proteomics, the study of gene regulation, and the complex nature of DNA sequence.

Microbial genomics may go far beyond the identification of novel targets; it may improve the discovery process by identifying bacterial gene products that are involved in antibiotic permeability and transport, as well as efflux. Genomic studies may also clarify how resistance to novel classes appears, is expressed and regulated, and is transferred between bacteria. Genomic studies may also suggest novel approaches to reducing the emergence of resistance in the first place. One such example is the mutant prevention concept (MPC) in fluoroquinolones[146,147] or oxazolidinones.[148] This concept offers guidance, based on the molecular understanding of drug-target interactions, toward reducing the emergence of quinolone- or oxazolidinones-resistant population of bacterial pathogens.

The success of genomics is not to be measured solely by the ambitious goal of identifying novel gene targets to subsequently select chemical scaffolds for lead optimization and eventually progression to candidate compounds, it is to be measured by the total output of the genomics technology that has advanced numerous processes that has contributed to the advancement of overall antibacterial drug discovery as has been described above. With the onslaught of antibacterial resistance,[149–151] bacterial genomics continues to offer hope for future novel antibacterial targets and agents, but its pay-off is already here in terms of its overall contributions to enabling technologies of overall drug discovery.

References

1. Chopra, I. *Expert Rev. Anti-infect. Ther.* **2003**, *1*, 45–55.
2. Livermore, D. M. *Ann. Med.* **2003**, *35*, 226–234.
3. Eur., J. *Clin. Milcrob. Inf. Dis.* **2005**, *24*, 83–99.
4. Ray, A.; Rice, L. B. *Therapy* **2004**, *1*, 1–5.
5. Lederberg, J. *Science* **2000**, *288*, 287–293.
6. Silver, L.; Bostian, K. *Eur. J. Clin. Microbiol. Infect. Dis.* **1990**, *9*, 455–461.
7. Davies, J. *Science* **1994**, *264*, 375–382.
8. Hamilton-Miller, J. M. T. *Int. J. Antimicrob. Agents* **2004**, *23*, 209–218.
9. Neu, H. C. *Science* **1992**, *257*, 1064–1073.
10. Levy, S. B. *N. Engl. J. Med.* **1998**, *338*, 1376–1378.
11. Bush, K. *ASM News* **2004**, *70*, 282–287.
12. Shlaes, D. *Curr. Opin. Pharmacol.* **2003**, *3*, 470–473.
13. Barrett, C.; Barrett, J. F. *Curr. Opin. Biotechnol.* **2003**, *14*, 1–6.
14. Moir, D. T.; Shaw, K. J.; Hare, R. S. *Antimicrob. Agents Chemother.* **1999**, *43*, 439–446.
15. Walsh, C. *Nature* **2000**, *406*, 775–781.
16. Walsh, C. *Nat. Rev. Microbiol.* **2003**, *1*, 65–70.
17. Hall, B. G. *Nat. Rev. Microbiol.* **2004**, *2*, 430–435.
18. McDevitt, D.; Rosenberg, M. *Trends Microbiol.* **2001**, *9*, 611–617.
19. Chan, P. F.; Macarron, R.; Payne, D. J. *J. Curr. Drug Targets, Infect. Disorders* **2002**, *2*, 291–308.
20. Loferer, H. *Mol. Med. Today* **2000**, *6*, 470–474.
21. Fleischmann, R. D.; Adams, M. D. *Science* **1995**, *269*, 496–512.
22. Fraser, C. M.; Gocayne, J. D.; White, O. *Science* **1995**, *270*, 397–403.
23. http://igweb.integratedgenomics.com/ ERGO/supplement/genomes.html (accessed Aug 2006).
24. http://www.ncbi.nlm.nih.gov/ (accessed Aug 2006).
25. http://www.tigr.org/tigr-scripts/CMR2 /CMRGenomes.spl (accessed Aug 2006).
26. http://microbialgenome.org/ (accessed Aug 2006).
27. Dougherty, T. J.; Barrett, J. F. *Ann. Rep. Med. Chem.* **2002**, *37*, 95–104.
28. Perrière, G.; Duret, L.; Gouy, M. *Genome Res.* **2000**, *10*, 379–385.
29. Jorgenson, J. H.; Ferraro, M. J. *Clin. Infect. Dis.* **1998**, *26*, 973–980.
30. Woods, G. L. *Infect. Dis. Clin. North Am.* **1995**, *9*, 463–481.
31. Mahan, M. J.; Tobias, J. W.; Slauch, J. M.; Hanna, P. C.; Collier, R. J.; Mekalanos, J. J. *Proc. Natl. Acad. Sci. USA* **1995**, *92*, 669–673.
32. Slauch, J. M.; Mahan, M. J.; Mekalanos, J. J. *Methods Enzymol.* **1994**, *235*, 481–492.
33. Thanassi, J. A.; Hartman-Neumann, S. L.; Dougherty, T. J. *Nucleic Acids Res.* **2002**, *30*, 3152–3162.
34. Herring, C. D.; Blattner, F. R. *J. Bacteriol.* **2004**, *186*, 2673–2681.
35. Silver, L. L. *Curr. Opin. Microbiol.* **2003**, *6*, 431–438.
36. Demain, A. J. *Nat. Biotechnol.* **2002**, *20*, 331.
37. Moat, A. G.; Foster, J. W. *Microbial Physiology*; Wiley-Liss, Inc.: New York, 1995.
38. Franklin, T. J.; Snow, G. A. *Biochemistry of Antimicrobial Action*; Chapman & Hall: New York, 1991.
39. Workman, P. *Curr. Pharm. Des.* **2003**, *9*, 891–902.
40. Overbye, K.; Barrett, J. F. *Drug Disc. Today* **2005**, *10*, 45–52.
41. Ng, J. H.; Ilag, L. L. *Drug Disc. Today* **2004**, *9*, 59–60.
42. Butler, M. S. *J. Nat. Prod.* **2004**, *67*, 2141–2153.
43. Hu, X.; Nguyen, K. T.; Jiang, V. C.; Lofland, D.; Moser, H. E.; Pei, D. *J. Med. Chem.* **2004**, *47*, 4941–4949.
44. Schuffenhauer, A.; Popov, M.; Schopfer, U.; Acklin, P.; Stanek, J.; Jacoby, E. *Comb. Chem. H-T Screen.* **2004**, 7, 771–781.
45. Ji, Y. *Pharmacogenomics* **2002**, *3*, 315–323.
46. Lerner, C. G.; Beutel, B. A. *Curr. Drug Targets Infect. Dis.* **2002**, *2*, 109–119.
47. Rosamond, J.; Allsop, A. *Science* **2000**, *287*, 1973–1976.
48. Fritz, B.; Raczniak, G. A. *Bio. Drugs* **2002**, *16*, 331–337.
49. Mills, S. J. *Antimicrob. Chemother.* **2003**, *51*, 749–752.
50. Payne, D. J.; Wallis, N. G.; Gentry, D. R. *Curr. Opin. Drug Disc. Dev.* **2000**, *3*, 177–190.
51. Rodrigue, A.; Quentin, Y.; Lazdunski, A. *Trends Microbiol.* **2000**, *8*, 498–504.

52. Throup, J. P.; Koretke, K. K.; Bryant, A. P. *Mol. Microbiol.* **2000**, *35*, 566–576.
53. Dougherty, T. J.; Barrett, J. F.; Pucci, M. J. *Curr. Pharmacol. Des.* **2002**, *8*, 1119–1135.
54. Kreusch, A.; Spraggon, G.; Lee, C. C. *J. Mol. Biol.* **2003**, *330*, 309–321.
55. Payne, D. J.; Holmes, D. J.; Rosenberg, M. *Curr. Opin. Investig. Drugs* **2001**, *2*, 1028–1034.
56. Payne, D. J.; Warren, P. V.; Holmes, D. J. *Drug Disc. Today* **2001**, *6*, 537–544.
57. Kim, S.; Lee, S. W.; Choi, E. C. *Appl. Microbiol. Biotechnol.* **2003**, *61*, 278–288.
58. Rogers, B. L. *Curr. Opin. Drug Discov Dev.* **2004**, 7, 211–222.
59. Dougherty, T. J.; Barrett, J. F.; Pucci, M. J. In *Microbial Genomics and Drug Discovery*; Dougherty, T. J., Projan, S. J., Eds.; Marcel Dekker: New York, 2003, pp 71–96.
60. Chalker, A. F.; Lunsford, R. D. *Pharmacol. Therapeut.* **2002**, *95*, 1–20.
61. Haney, S. A.; Alkane, L. E.; Dunman, P. M. *Curr. Pharmaceut. Design* **2002**, *8*, 1099–1118.
62. Isaacson, R. E. *Expert Opin. Investig. Drugs* **1997**, *6*, 1009–1018.
63. Bachmann, B. J. *Microbiol Rev.* **1990**, *54*, 130–197.
64. Searls, D. B. *Drug Disc. Today* **2000**, *5*, 135–143.
65. Frangeul, L.; Nelson, K. E.; Buchrieser, C. *Microbiology* **1999**, *145*, 2625–2634.
66. Sutton, G.; White, O.; Adams, M.; Kerlauge, A. *Genome Sci. Technol.* **1995**, *1*, 9–19.
67. Altschul, S. F.; Gish, W.; Miller, W. *J. Mol. Biol.* **1990**, *215*, 403–410.
68. Pearson, W. R.; Lipman, D. J. *Proc. Natl. Acad. Sci. USA* **1988**, *85*, 2444–2448.
69. Brenner, S. E.; Chothia, C.; Hubbard, T. J. P. *Proc. Natl. Acad. Sci. USA* **1998**, *95*, 6073–6078.
70. Borodovsky, M.; McIninch, J. D. *Comput. Chem.* **1993**, *17*, 123–133.
71. Salzberg, S. L.; Delcher, A. L.; Kasif, S. *Nucleic Acids Res.* **1998**, *26*, 544–548.
72. Delcher, A. L.; Harmon, D.; Kasif, S. *Nucleic Acids Res.* **1999**, *27*, 4636–4641.
73. Shmatkov, A. M.; Melikyan, A. A.; Chernouska, F. L. *Bioinformatics* **1999**, *15*, 874–876.
74. Yada, T.; Nakao, M.; Totoki, Y.; Nakai, K. *Bioinformatics* **1999**, *15*, 987–993.
75. Borodovsky, M.; McIninch, J. D.; Koonin, E. V. *Nucl. Acid. Res.* **1995**, *23*, 3554–3562.
76. Brucolleri, R. E.; Dougherty, T. J.; Davison, D. B. *Nucleic Acids Res.* **1998**, *26*, 4482–4486.
77. Perrière, G.; Duret, L.; G ouy, M. *Genome Res.* **2000**, *10*, 379–385. http://www.123genomics.com/files/databases.html (accessed Aug 2006).
78. Miller, J. H.; Reznikoff, W. S. *The Operon*; Cold Spring Harbor Laboratory Press: Plainview, NY, 1980.
79. Overbeek, R.; Fonstein, M.; D'Souza, M. *Proc. Natl. Acad. Sci. USA.* **1999**, *96*, 2896–2901.
80. Fraser, C. M.; Eisen, J.; Fleischmann, R. D. *Emerg. Infect Dis.* **2000**, *6*, 505–512.
81. Freiberg, C.; Wieland, B.; Splatmann, F. *J. Mol. Biotech.* **2001**, *3*, 483–489.
82. Graves, J. F.; Biswas, G. D.; Sparling, P. F. *J. Bacteriol.* **1982**, *152*, 1071–1077.
83. Dougherty, B. A.; Smith, H. O. *Microbiology* **1999**, *145*, 401–409.
84. Danner, D. B.; Deich, R. A.; Sisco, K. L. *Gene* **1980**, *11*, 311–318.
85. Murphy, K. C. *J. Bacteriol.* **1998**, *180*, 2063–2071.
86. Murphy, K. C.; Campellone, K. G.; Poteete, A. R. *Gene* **2000**, *246*, 321–330.
87. Yu, D.; Ellis, H. M.; Lee, E. C. *Proc Natl. Acad. Sci. USA* **2000**, *97*, 5978–5983.
88. Jana, M.; Luong, T. T.; Komatsuzawa, H. *Plasmid* **2000**, *44*, 100–104.
89. Link, A. J.; Phillips, D.; Church, G. M. *J. Bacteriol.* **1997**, *179*, 6228–6237.
90. Schmid, M. B. In *Antibiotic Development and Resistance*; Hughes, D., Anderson, D. I., Eds.; Taylor & Francis: New York, 2001, pp 197–208.
91. Akerley, B. J.; Rubin, E. J.; Camilli, A. *Proc. Natl. Acad. Sci. USA* **1998**, *95*, 8927–8932.
92. Forsyth, R. A.; Hasselbeck, R. J.; Ohlsen, K. L. *Mol. Microbiol.* **2002**, *43*, 1387–1400.
93. Hirota, Y.; Ryter, A.; Jacob, F. *Cold Spring Harbor Symp. Q. Biol.* **1968**, *33*, 677–693.
94. Schmid, M. B.; Kapur, N.; Isaacson, D. R. *Genetics* **1989**, *123*, 625–633.
95. Sturgeon, J.; Ingram, L. O. *J. Bacteriol.* **1978**, *133*, 256–264.
96. Overbeek, R.; Fonstein, M.; D'Souza, M. *Proc. Natl. Acad. Sci. USA* **1999**, *96*, 2896–2901.
97. Hutchison, C. A.; Peterson, S. N.; Gill, S. R. *Science* **1999**, *286*, 2165–2169.
98. Takiff, H. E.; Baker, T.; Copeland, T. *J. Bacteriol.* **1992**, *174*, 1544–1553.
99. Murphy, C. K.; Stewart, E. J.; Beckwith, J. *Gene* **1995**, *155*, 1–7.
100. Gaiano, N.; Amsterdam, A.; Kawakami, K. *Nature* **1996**, *383*, 829–832.
101. Bender, J.; Kleckner, N. *Proc. Natl. Acad. Sci. USA* **1992**, *89*, 7996–8000.
102. Lehoux, D. E.; Sanschagrin, F.; Levesque, R. C. *Curr. Opin. Microbiol.* **2001**, *4*, 515–519.
103. Reich, K. A.; Chovan, L.; Hessler, P. *J. Bacteriol.* **1999**, *181*, 4961–4968.
104. Karp, P. D.; Riley, M.; Saier, M.; Paulson, I. T.; Paley, S. M.; Pellegrini-Toole, A. *Nucleic Acids Res.* **2000**, *28*, 56–59.
105. Maranas, C. D.; Borgard, A. P. *Metabol. Engin.* **2001**, *3*, 98–99.
106. Ouzounis, C. A.; Karp, P. D. *Genome Res.* **2000**, *10*, 568–571.
107. Ogata, H.; Goto, S.; Fujibuchi, W. *Biosystems* **1998**, *47*, 119–128.
108. Overbeek, R.; Larsen, N.; Pusch, G. D. *Nucleic Acids Res.* **2000**, *28*, 123–125.
109. Blattner, F. R.; Plunkett, G.; Bloch, C. A. *Science* **1997**, *277*, 1453–1474.
110. Betz, S. F.; Baxter, S. M.; Fetrow, J. S. *Drug Disc. Today* **2002**, 7, 865–871.
111. Sanishvili, R.; Yakunin, A. F.; Laskowski, R. A. *J. Biol. Chem.* **2003**, *278*, 26039–26045.
112. Kimber, M. S.; Vallee, F.; Houston, S. *Proteins: Struc. Funct. Genet.* **2003**, *51*, 562–568.
113. Eisen, M. B.; Brown, P. O. *Methods Enzymol.* **1999**, *303*, 179–205.
114. DeRisi, J. L.; Iyer, V. R.; Brown, P. O. *Science* **1997**, *278*, 680–686.
115. Schena, M.; Shalon, D.; Davis, R. W. *Science* **1995**, *270*, 467–470.
116. Toranon, P.; Kolehmainen, M.; Wong, G. *FEBS Letts.* **1999**, *451*, 142–146.
117. Chee, M.; Yang, R.; Hubbell, E. *Science* **1996**, *274*, 610–614.
118. Harrington, C. A.; Rosenow, C.; Retief, J. *Curr. Opin. Microbiol.* **2000**, *3*, 285–291.
119. de Saizieu, A.; Certa, U.; Warrington, J. *J. Nature Biotech.* **1998**, *16*, 45–48.
120. Freiberg, C.; Brotz-Oesterhelt, H.; Labischinski, H. *Curr. Opin. Microbiol.* **2004**, 7, 451–459.

121. Yakunin, A. F.; Yee, A. A.; Savchenko, A. *Curr. Opin. Chem. Biol.* **2004**, *8*, 42–48.
122. Yee, A.; Pardee, K.; Christendat, D. *Accounts Chem. Res.* **2003**, *3*, 183–189.
123. Haley-Vicente, D.; Edwards, D. J. *Curr. Opin. Drug Disc. Dev.* **2003**, *6*, 322–332.
124. Pandey, A.; Mann, M. *Nature.* **2000**, *405*, 837–846.
125. Washburn, M. P.; Yates, J. R. *Curr. Opin. Microbiol.* **2000**, *3*, 292–297.
126. Fountoulakis, M.; Takacs, M.-F.; Takacs, B. *J. Chromatorg. A* **1999**, *833*, 157–168.
127. Molloy, M.; Herbert, B.; Williams, K. *Electrophoresis* **1999**, *20*, 701–704.
128. Gmuender, H.; Kuralti, K.; Di Padova, X. Y. *Genome Res.* **2001**, *11*, 28–42.
129. Langen, H.; Takacs, B.; Evers, S. *Electrophoresis* **2000**, *21*, 411–429.
130. McAtee, C. P.; Hoffman, P. S.; Berg, D. E. *Proteomics* **2001**, *1*, 516–521.
131. Deiwick, J.; Hensel, M. *Electrophoresis* **1999**, *20*, 813–817.
132. Rakeman, J. L.; Miller, S. I. *Trends Microbiol.* **1999**, *7*, 221–223.
133. Garcia Vescovi, E.; Soncini, F. C.; Groisman, E. A. *Res Microbiol.* **1994**, *145*, 473–480.
134. Foster, J. W.; Park, Y. K.; Bang, I. S. *Microbiology* **1994**, *140*, 341–352.
135. Alksne, L. E.; Projan, S. J. *Curr. Opin. Biotechnol.* **2000**, *11*, 625–636.
136. Heithoff, D. M.; Sinsheimer, R. L.; Low, D. A. *Phil. Trans. R. Soc. Lond. B* **2000**, *355*, 633–642.
137. Buysse, J. M. *Frontiers Med. Chem.* **2004**, *1*, 529–542.
138. Mei, J. M.; Nourbakhsh, F.; Ford, C. W. *Mol. Microbiol.* **1997**, *26*, 399–407.
139. Unsworth, K. E.; Holden, D. W. *Phil. Trans. R. Soc. Lond. B* **2000**, *355*, 613–642.
140. Shea, J. E.; Santangelo, J. D.; Feldman, R. G. *Curr. Opin. Microbiol.* **2000**, *3*, 451–458.
141. Mahan, M. J.; Slauch, J. M.; Mekalanos, J. J. *Science.* **1993**, *259*, 686–688.
142. Heithoff, D. M.; Conner, C. P.; Hanna, P. C. *Proc. Natl. Acad. Sci. USA* **1997**, *94*, 934–939.
143. Camilli, A.; Mekalanos, J. J. *Mol. Microbiol.* **1995**, *18*, 671–683.
144. Allsop, A. E. *Curr. Opin. Microbiol.* **1998**, *1*, 530–534.
145. Raczniak, G.; Ibba, M.; Soll, D. *Toxicology* **2001**, *160*, 181–189.
146. Hermsen, E. D.; Hovde, B.; Konstantinides, G. N.; Rotschafer, J. C. *Antimicrob. Agents Chemother.* **2005**, *49*, 1633–1635.
147. Blondeau, J. M.; Hansen, G.; Metzler, K.; Hedlin, P. *J. Chemother.* **2004**, *16*, 1–19.
148. Rodriquez, J. C.; Cebrian, L.; Lopez, M.; Ruiz, M.; Jimenez, I.; Royo, G. *J. Antimicrob. Chemother.* **2004**, *53*, 441–444.
149. Barrett, J. F. *Expert Opin. Pharmacother.* **2001**, *2*, 201–204.
150. Ray, A.; Rice, L. B. *Therapy* **2004**, *1*, 1–5.
151. Pirnay, J.-P.; Vos, D.; Zizi, M.; Heyman, P. *Expert Opin. Anti-Infective Ther.* **2003**, *1*, 523–525.

Biography

John F Barrett was a Senior Director of Infectious Diseases Research at Merck Research Laboratories. With 20 years service in the pharmaceutical industry, he gained experience in a wide breadth of antibacterial and antifungal discovery and development while efforts at Pfizer, Johnson & Johnson's RWJPRI, and the Bristol-Myers Squibb Pharmaceutical Research Institute (BMSPRI) before joining Merck in 2001. John received his bachelor's degree in biochemistry from Temple University (Philadelphia, PA), and his doctorate in microbiology and microbial physiology from the Temple University School of Medicine. He completed a postdoctoral fellowship in genetic engineering with Roy Curtiss III, PhD (UAB and Washington University), and joined Pfizer in 1985, RWJPRI in 1989, and BMSPRI in 1997, and Merck in 2001. His research efforts have included β-lactams, penams, penems, β-lactamase inhibitors, quinolones, novel gyrase inhibitors, HIV gp120/CD4 antagonists, cephems, oxazolidinones, thiazolyl peptides, two-component system signal transduction inhibitors, virulence factors, novel antibacterial targets, and mammalian topoisomerase II inhibitors. John was a member of numerous editorial boards and author of over 250 publications. He was a former Division A ASM Chairperson, and was named a Fellow in the American Academy of Microbiology in 2000. He led a group of research scientists seeking the identification and development of novel antimicrobials through a combination of classical discovery strategies and techniques, and proprietary genomics efforts. He had been involved with the development of several antibacterials currently on the market including sulbactam, ofloxacin, levofloxacin, and gatifloxacin, and in the late-stage development candidate, the des-quinolone garenoxacin (BMS-284756/T-3811).

Comprehensive Medicinal Chemistry II
ISBN (set): 0-08-044513-6

ISBN (Volume 7) 0-08-044520-9; pp. 731–748

7.26 Overview of Parasitic Infections

P J Rosenthal, University of California, San Francisco, CA, USA

7.26.1 Introduction

Parasitic organisms are responsible for a huge number of illnesses in humans (for detailed discussions of parasitic infections and the biology of human parasites, see [1]). Although all infectious pathogens can be considered human parasites, the term parasite is generally restricted to two large groups of eukaryotic organisms that inflict human disease: unicellular protozoans and multicellular helminths. Among these are species responsible for many of the most widespread and devastating human infections. In general, protozoan parasites reproduce in human hosts, leading to the generation of large numbers of parasites in the body, and serious illnesses. Malaria, leishmaniasis, African trypanosomiasis, American trypanosomiasis, and amebiasis are among the most deadly human infections. Effective treatment of protozoan infections generally requires complete eradication of infecting parasites. Most helminths do not reproduce in humans, so disease is the consequence of infection with large numbers of organisms. Some helminths are quite benign, only occasionally causing disease, but the very high prevalence of infection with, for example, a number of roundworms, leads to important morbidities, even though most infections are asymptomatic. Other helminths are less prevalent, but nonetheless infect millions, and are responsible for significant human morbidity and mortality. Treatment of some helminth infections requires parasite eradication, but for a number of infections antiparasitic drugs are most widely used to control infections both in individuals and populations, without necessarily fully eradicating infecting worms.

It so happens that most, although not all, parasitic infections disproportionately affect the most economically disadvantaged populations in the world. Many parasites that are rarely, if ever, encountered in the developed world are extremely common in developing countries. The consequence of this fact is that, for most parasitic diseases, the incentive for drug discovery is low, as it is difficult for pharmaceutical companies to generate profits from drugs for developing world populations. For this reason, our ability to treat parasitic infections is remarkably inadequate. Many of the earliest anti-infective therapies were directed against parasites, for example quinine, which was developed in the early nineteenth century as an antimalarial drug.[2] Yet, in the early twenty-first century we lack highly effective therapies for many parasitic infections. For some diseases, such as malaria, effective therapies are available, but for complex reasons, including the high cost of newer agents, most infected individuals still receive older drugs that are suboptimal due to drug resistance and toxicity. For other diseases, including African and American trypanosomiasis, available therapies are poor, due to limited efficacy, resistance, and unacceptable toxicity. New drugs are greatly needed for many human parasitic diseases, but development efforts have been slow. Recently, some progress has been seen on this front, with increased funding for antiparasitic drug discovery from industry and public–private partnerships, and the future for antiparasitic drug development has become a bit more promising.

Another population that is disproportionately affected by parasitic infections is immunocompromised individuals. A number of protozoan parasites cause most illnesses in the immunocompromised. For example, toxoplasmosis is usually a harmless chronic infection and cryptosporidial infections are generally mild in most healthy individuals, but these and other parasites cause a range of serious and at times deadly manifestations in immunocompromised hosts. Another important reason for antiparasitic drug discovery is the need to improve therapies for infections in immunocompromised individuals in both developed and developing countries.

7.26.2 Disease Burden from Parasitic Diseases

The burden of parasitic diseases of humans is enormous (**Table 1**). It is quite difficult to ascertain accurate epidemiologic data, as most infections occur in underserved populations with limited healthcare infrastructures.

Table 1 Summary of human parasitic diseases

Disease	*Key species*	*Vectors*	*Geographic distribution*	*Prevalence*[a]	*Annual incidence*[a]	*Annual mortality*[a]
Malaria	*Plasmodium falciparum, P. vivax, P. ovale, P. malariae*	Anopheline mosquitoes	Most of tropics	2 777 000	213 743 000	856 000
Leishmaniasis	*Leishmania* spp.	Sand flies	Much of tropics	1 278 000	414 000	50 000
African trypanosomiasis	*Trypanosoma brucei*	Tsetse fly	Sub-Saharan Africa	267 000	60 000	47 000
American trypanosomiasis	*Trypanosoma cruzi*	Triatome insects	Latin America	15 785 000	728 000	19 000
Amebiasis	*Entamoeba histolytica*		Worldwide, especially tropics			
Toxoplasmosis	*Toxoplasma gondii*		Worldwide			
Giardiasis	*Giardia lamblia*		Worldwide, especially tropics			
Trichomoniasis	*Trichomonas vaginalis*		Worldwide			
Babesiosis	*Babesia* spp.		Scattered locations			
Cryptosporidiosis	*Cryptosporidum* spp.		Worldwide, especially tropics			
Isosporiasis	*Isospora belli*		Worldwide, especially tropics			
Hookworm	*Ancylostoma duodenale, Necator americanus*		Worldwide, especially tropics	1 298 000 000	159 000 000	65 000
Ascariasis	*Ascaris lumbricoides*		Worldwide, especially tropics	1 472 000 000	335 000 000	60 000
Trichuriasis	*Trichuris trichuria*		Worldwide, especially tropics	1 049 000 000	220 000 000	10 000
Strongyloidiasis	*Strongyloides stercoralis*		Worldwide, especially tropics	70 000 000		
Trichinellosis	*Trichinella spiralis*		Worldwide			
Drancunculiasis	*Dracunculus medinensis*	Copepods	Africa	80 000		
Filariasis	*Wuchereria bancrofti, Brugia malayi, Brugia timori*	Mosquitoes	Worldwide in tropics	120 000 000	44 000 000	
Loiasis	*Loa loa*	Tabanid flies	Africa	13 000 000		
Onchocerciasis	*Onchocerca volvulus*	Blackflies	Africa, Latin America	17 660 000	270 000	45 000
Schistosomiasis	*Schistosoma* species		Worldwide in tropics	201 380 000	20 000 000	20 000
Liver flukes	*Clonorchis sinensis, Opisthorchis* species, *Fasciola hepatica*		Worldwide, especially Asia	19 740 000		
Intestinal flukes	*Fasciolopsis buski* and others		Worldwide, especially Asia	1 110 000		
Lung flukes	*Paragonimus westermani* and others		Worldwide, especially Asia	20 680 000		
Fish tapeworm	*Diphyllobothrium latum*		Worldwide	9 000 000		
Beef tapeworm	*Taenia saginata*		Worldwide	77 000 000		
Pork tapeworm	*Taenia solium*		Worldwide	10 000 000		
Dwarf tapeworm	*Hymenolepsis nana*		Worldwide	75 000 000		
Echinococcosis	*Echinococcus* spp.		Scattered areas worldwide	2 700 000		

[a] Estimates of prevalence, incidence, and mortality vary quite widely. The estimates provided here are from[27] for protozoan infections in 1990 and[46] for helminth infections.

Nonetheless, significant effort has gone into estimating the impact of the most important parasitic infections. In many cases, most notably for protozoan infections such as malaria, the burden consists of acute infections, with extensive acute morbidity and mortality. In other cases, in particular a number of helminth infections such as schistosomiasis, chronic disease ensues after infection with large numbers of organisms. In addition, many asymptomatic individuals are infected with helminths, serving as potential future sufferers of chronic diseases and as reservoirs for the transmission of infections. In some cases, such as malaria, leishmaniasis, and trypanosomiasis, case-mortality is high. In other cases, despite limited mortality, morbidity may be extensive, such as with many helminth infections. In addition, for infections where prevalence is extremely high, even quite low case-mortality levels can lead to important morbidity and mortality. The data on infection and disease prevalence, incidence, and mortality in **Table 1** are from a number of sources, and of varied reliability, but nonetheless they provide insight into the profound importance of parasitic infections of humans. Some data are quite old, but sadly this fact is of little importance, as progress against most parasitic diseases has been very limited in recent years. Additional epidemiologic data for diseases of particular importance are provided in sections on individual infections below.

7.26.3 Protozoan Infections

Protozoan infections are the leading causes of morbidity and mortality among parasitic infections of humans. The most important human protozoan infections are summarized in **Tables 1** and **2**. Additional information is provided on infections of greatest importance in drug discovery efforts.

7.26.3.1 Malaria

7.26.3.1.1 Biology of malaria parasites

Human malaria is caused by four species of apicomplexan parasites, *Plasmodium falciparum*, *P. vivax*, *P. ovale*, and *P. malariae*. *P. falciparum* is by far the most medically important parasite, as it is the only malaria parasite that commonly causes severe disease and death. The virulence of *P. falciparum* is due to a number of biological factors, including its ability to infect erythrocytes of all ages and the binding of *P. falciparum*-infected erythrocytes to human endothelial cells, which mediates various disease processes.[3] *P. vivax* is about as common as *P. falciparum*, but is generally not responsible for severe disease. The other two plasmodial species are quite uncommon.

All malaria parasites have a similar life cycle.[3] Human infection is initiated by inoculation of sporozoites during the bite of an anopheline mosquito. Circulating sporozoites rapidly invade liver cells, and exoerythrocytic stage tissue schizonts mature in the liver. Merozoites are subsequently released from the liver and invade erythrocytes, where parasites divide asexually. Repeated cycles of erythrocyte infection can lead to the infection of many erythrocytes and serious disease. Sexual stage gametocytes also develop in erythrocytes before being taken up by mosquitoes, where they develop into infective sporozoites.

Clinical illness is caused only by the erythrocytic cycle of malaria parasites. Antimalarial therapy focuses on eradicating parasites at this stage of their life cycle.[4] Most drugs are not effective against other stages, but some offer activity against liver stages, to facilitate chemoprophylaxis, or against gametocytes, to offer benefit in diminishing parasite transmission to others.

In *P. falciparum* and *P. malariae* infections, only one cycle of liver cell invasion and multiplication occurs, and liver infection ceases spontaneously in less than 4 weeks. Thus, treatment that eliminates erythrocytic parasites will cure these infections. In *P. vivax* and *P. ovale* infections, a dormant hepatic stage, the hypnozoite, is not eradicated by most drugs, and subsequent relapses can therefore occur after therapy directed against erythrocytic parasites. Eradication of both erythrocytic and hepatic parasites is required to cure these infections.

7.26.3.1.2 Epidemiology of malaria

Malaria is the most deadly protozoan infection of humans. Hundreds of millions of cases of malaria occur annually, and infections with *P. falciparum*, the most virulent human malaria parasite, probably lead to over a million deaths each year.[5,6] Indeed, a recent analysis suggested that prior estimates of incidence were low, and that 515 million episodes of clinical *P. falciparum* infection occurred in 2002.[7] Most severe disease and deaths from malaria are in children; it has been estimated to cause 10.7% of all deaths in children under 5 years of age.[8] Despite extensive control efforts, the incidence of malaria is not decreasing in most endemic areas of the world, and in some areas it is clearly increasing.[9] A major reason for the persistence of the severe malaria problem is the increasing resistance of parasites to available

Table 2 Available drugs and key drug needs for parasitic infections

Disease	*Available drugs*	*Key drug needs*
Malaria	Chloroquine, amodiaquine, quinine, mefloquine, primaquine, sulfadoxine–pyrimethamine and other antifolates, artemisinins (including combination regimens), atovaquone, doxycycline and other antibiotics, halofantrine	Inexpensive, orally bioavailable, safe compounds for combination therapy for uncomplicated malaria, including that caused by highly drug-resistant parasites; improved drugs for chemoprophylaxis
Leishmaniasis	Pentavalent antimonials, amphotericin B, pentamidine, miltefosine	Inexpensive, nontoxic drugs active against drug-resistant disease
African trypanosomiasis	Pentamidine, suramin, melarsoprol, eflornithine	Drugs with efficacy against resistant parasites and decreased toxicity compared to current agents
American trypanosomiasis	Nifurtimox, benznidazole	Drugs with ability to eradicate fully early infections and to provide benefit when used in chronic and advanced infections
Amebiasis	Metronidazole or tinidazole	
Toxoplasmosis	Pyrimethamine plus sulfadiazine or clindamycin	Therapies with decreased toxicity
Giardiasis	Tinidazole or metronidazole	Therapies with improved efficacy
Trichomoniasis	Metronidazole	Therapy for metronidazole-resistant infections
Babesiosis	Quinine plus clindamycin	
Cryptosporidiosis	Nitazoxanide	Therapies with improved efficacy
Isosporiasis	Trimethoprim–sulfamethoxazole	
Hookworm	Albendazole or mebendazole	
Ascariasis	Albendazole or mebendazole or pyrantel pamoate	
Trichuriasis	Albendazole or mebendazole	
Strongyloidiasis	Ivermectin or thiabendazole or albendazole	
Trichinellosis	Mebendazole or albendazole (uncertain benefit)	
Drancunculiasis	Thiabendazole or metronidazole and worm removal	
Filariasis	Diethylcarbamazine or ivermectin	
Loiasis	Diethylcarbamazine or ivermectin	
Onchocerciasis	Ivermectin or diethylcarbamazine	
Schistosomiasis	Praziquantel	
Liver flukes	Praziquantel or albendazole or triclabendazole	
Intestinal flukes	Praziquantel	
Lung flukes	Praziquantel	
Fish tapeworm	Praziquantel or niclosamide	
Beef tapeworm	Praziquantel or niclosamide	
Pork tapeworm	Praziquantel or niclosamide or albendazole (cysticercosis)	
Dwarf tapeworm	Praziquantel or niclosamide	
Echinococcosis	Albendazole and cyst removal	

chemotherapeutic agents. Resistance to chloroquine, the most widely used antimalarial over the last half-century, is now very common, and other available antimalarials are limited by resistance, high cost, and toxicity.[10]

Malaria is endemic throughout most of the tropics. However, its epidemiology varies greatly in different endemic regions. Malaria is most highly endemic in sub-Saharan Africa, which suffers the bulk of worldwide malarial morbidity and mortality. The particularly high prevalence of malaria in Africa is probably due to the high transmission efficiency of local anopheline vectors and other environmental factors. The level of transmission of malaria in much of sub-Saharan Africa is truly remarkable; in many areas individuals are infected, on the average, more than once each day. With increasing transmission comes increasing levels of antimalarial immunity. Immunity is never complete, but in areas with high transmission malaria is primarily a disease of young children, with older children and adults relatively protected against severe disease. A second reason for the particular severity of malaria in Africa is that in this area, the disease is nearly all due to *P. falciparum*, the most virulent parasite; in most other endemic areas *P. vivax* is as common or more common than *P. falciparum* as a cause of disease. Nonetheless, although numbers are lower than in Africa, falciparum malaria causes large numbers of illnesses and deaths in South America, South Asia, and parts of Oceania.

A key factor in understanding the current epidemiology of malaria is the epidemiology of drug resistance. Drug resistance is primarily a problem with *P. falciparum*. Most important are resistance to chloroquine, the mainstay for antimalarial therapy from about 1950 until recently, and to antifolates, the most common replacements for chloroquine. Chloroquine resistance was first seen in the late 1950s in South America and Southeast Asia, spread quite slowly around the world, but is now deeply seated in nearly all malarious areas.[10] Resistance to antifolates is also now a large and increasing problem.

7.26.3.1.3 Clinical manifestations of malaria infection

Although it is the most common cause of death from parasitic infection, malaria most commonly presents as a relatively mild febrile illness. Millions of episodes of malaria are treated with self-administered drugs, drugs provided by health workers or shopkeepers with limited medical training, or in clinics without laboratory diagnostic capabilities. On the other hand, a subset of infections, particularly those caused by *P. falciparum*, will progress to severe disease. The progression to severe illness is much more likely in an individual without prior antimalarial immunity, such as a traveler to the tropics from a nonendemic area. The most common presentation of malaria is fever accompanied by nonspecific symptoms including headache, weakness, sweats, chills, arthralgias, and myalgias. Common manifestations of severe malaria are profound anemia, most commonly in children in endemic areas, cerebral malaria, with progressive neurological dysfunction, and severe pulmonary disease, each of which can commonly progress to death. Many organ systems can be affected by malaria; heart failure secondary to anemia, renal failure, respiratory failure, and hepatic dysfunction are all common. However, prompt therapy of severe malaria is often followed by excellent therapeutic outcomes.

7.26.3.1.4 Drugs to treat malaria

Numerous drugs are available to treat malaria.[4,11,12] However, many drugs suffer from one or more of three main concerns: diminishing efficacy due to the spread of drug-resistant parasites, toxicity, and high cost that limits use in developing countries. It is important to consider antimalarial drugs based on three quite different indications. The greatest need is for safe, inexpensive, oral drugs for the therapy of uncomplicated malaria in developing countries. For this indication drugs must be very inexpensive and safe enough for use in unsupervised settings. Considering the challenges of assuring patient compliance in settings with limited healthcare infrastructures, they should also ideally be effective when used over brief courses (generally up to 3 days) with once-daily dosing. A second need is for rapidly acting drugs to treat severe malaria; these drugs are typically given intravenously, but other routes (e.g., per rectal) are of interest in some settings with limited technology.[13] Cost is less of a concern for the treatment of severe malaria, as the number of doses needed is much lower than for uncomplicated disease. A third need is for chemoprophylaxis against infection, most commonly used in travelers from nonendemic to endemic countries. In this case, much higher costs can be tolerated. However, drugs used for chemoprophylaxis should be extremely safe and should be effective when used with weekly or once-daily dosing.

7.26.3.1.4.1 Drugs to treat uncomplicated malaria

Most difficulties in the treatment of malaria are with *P. falciparum*. Infections with other species are generally successfully treated with chloroquine, although increasing resistance of *P. vivax* to chloroquine has been observed.[14] For *P. vivax* and *P. ovale* infections, primaquine must also be given to eradicate liver hypnozoites and prevent subsequent relapse of infection.

For many years chloroquine was an outstanding drug for uncomplicated malaria, offering a very cheap, very safe, rapidly acting, and highly effective therapy for falciparum malaria. However, chloroquine resistance has gradually spread around the world, such that the use of this drug is now appropriate in only a few areas. Despite decreasing efficacy, due to its low cost and the familiarity of patients and clinicians with this drug, it is still widely used, especially in Africa. However, there is a growing consensus that in nearly all areas chloroquine should be replaced by other drugs for the routine treatment of malaria. Related drugs, including amodiaquine and mefloquine, offer much better efficacy in most areas, although resistance to these drugs is also a problem in some areas. Quinine retains excellent efficacy in most areas, but it is fairly expensive and toxic, relegating its use for uncomplicated malaria primarily to second-line therapy after the failure of another regimen. Antifolates offer potent antimalarial activity, and sulfadoxine–pyrimethamine (Fansidar), which inhibits dihydrofolate reductase and dihydropteroate synthase, has been used as an inexpensive replacement for chloroquine. However, resistance to sulfadoxine–pyrimethamine has been seen to develop rapidly after widespread use.[10] In Southeast Asia, where resistance to chloroquine and sulfadoxine–pyrimethamine was seen earlier than in Africa, other drugs are now routinely used. In Thailand, the use of mefloquine was followed by the identification of parasites resistant to this drug, but the use of a combination of mefloquine and artesunate has shown excellent efficacy against highly resistant parasite.[15] Artesunate is one of a growing class of antimalarials derived from artemisinin, a natural product developed in China.[16,17] These drugs offer very rapid antimalarial activity and are not yet limited by drug resistance. However, they have very short half-lives, so must be used in combination with longer-acting drugs to avoid unacceptable levels of late recrudescences. Artemisinin-based combination therapy (ACT) is rapidly becoming the standard for the therapy of uncomplicated malaria.[14,18] Although mefloquine–artesunate has been successful in Thailand, it is probably too toxic and too expensive for widespread use in more disadvantaged populations, such as Africa. Other ACT regimens currently under study (and increasingly used) are amodiaquine–artesunate, lumefantrine–artemether, and piperaquine–dihydroartemisinin.[19] However, ACTs remain much more expensive than older drugs, are of uncertain safety in pregnant women, and have to date been relatively little tested in Africa, where the need for new therapies is greatest.[20] Particularly in Africa, most therapies for uncomplicated malaria will be used outside of medical supervision, meaning that many doses of antimalarials may be used when the true cause of fever is not malaria, and it is unclear if ACT regimens are appropriate or affordable enough for this widespread use. Other non-ACT combination regimens can play a role in some settings. In particular, amodiaquine–sulfadoxine–pyrimethamine has shown surprisingly good efficacy, even in areas with moderate resistance to the individual components of the combination.[21] Additional antimalarial drugs that show efficacy but are of limited utility in developing countries due to high cost and/or toxicity concerns are halofantrine and the combination of atovaquone and proguanil. The field of antimalarial drug discovery is now increasingly active,[22] and it is anticipated that a number of artemisin-based and other new antimalarial drugs will be undergoing clinical testing in the near future.

7.26.3.1.4.2 Drugs for complicated malaria

The standard therapy for complicated malaria has for many years been parenteral quinine. Courses of quinine, which is quite poorly tolerated, can be shortened by adding doxycycline or clindamycin; these antibiotics are slow acting, but effective in combination. Recent studies have shown a number of artemisinins to offer equivalent efficacy to quinine for severe malaria, and it is likely that these new regimens, if affordable, will play important roles in the treatment of severe disease.

7.26.3.1.4.3 Drugs for chemoprophylaxis against malaria

It has been standard practice for many years for travelers from nonendemic to endemic regions to receive medication to prevent malaria.[23] Weekly dosing of chloroquine was the standard for this indication, but resistance to this drug in nearly all malarious areas now mandates the use of other drugs for most travelers. Most commonly used at present are mefloquine, which is effective but relatively poorly tolerated, atovaquone–proguanil, a new quite expensive regimen, and doxycycline, which is limited by photosensitivity and gastrointestinal toxicity. Improved chemoprophylactic agents are needed. Ironically, although this need is tiny compared to that for therapies for the millions infected with drug-resistant malaria parasites, this is the only indication for which antimalarial drug development may be profitable, and so the development of drugs for chemoprophylaxis may drive the identification of new therapies.

Chemoprophylaxis for malaria is not deemed appropriate for widespread use in endemic countries due to high costs, potential toxicities, difficulties of assuring compliance, and potential to select for resistant parasites. However, another use of chemoprophylaxis is for endemic populations at particular risk for severe malaria, notably young children and pregnant women. Intermittent preventive therapy of pregnant women with chloroquine was used previously, and the use of sulfadoxine–pyrimethamine for this indication has offered benefit.[24] More recently, intermittent treatment of infants with sulfadoxine–pyrimethamine diminished malarial morbidity.[25] However, benefits of chloroquine and sulfadoxine–pyrimethamine for these indications may in large part be due to the unusually long half-lives of these drugs, and with increasing resistance, it is not clear which other drugs can fill this role.

7.26.3.2 Leishmaniasis

7.26.3.2.1 Biology of leishmanial parasites

Leishmanial parasites cause three quite different clinical syndromes, depending on the infecting species. Visceral leishmaniasis (kala azar) is the most important, as it is a severe disseminated disease that is commonly fatal. Cutaneous leishmaniasis is a common cause of chronic skin ulcers. Mucocutaneous leishmaniasis is an uncommon syndrome involving mucosal lesions of the respiratory tract after prior cutaneous disease. Leishmaniasis is transmitted by sand flies of the genus *Lutzomyia* or *Phlebotomus*. The flies transmit promastigotes to humans, which develop into amastigotes in human macrophages. In visceral disease, infected macrophages disseminate, leading to infection of the entire reticuloendothelial system. In cutaneous disease infected macrophages cause only local skin disease, except for some species which can occasionally progress to distant mucocutaneous involvement.

7.26.3.2.2 Epidemiology of leishmaniasis

Visceral leishmaniasis is endemic in the Indian subcontinent, other regions of central Asia, and parts of Africa, the Middle East, and Latin America, a total of 62 countries. Over 90% of cases occur in five countries, India, Bangladesh, Nepal, Sudan, and Brazil. Recent estimates are that 200 million people are at risk, and that there are 500 000 episodes each year, leading to over 40 000 recorded and likely many additional unreported deaths.[26] Importantly, the incidence of visceral leishmaniasis has been increasing in recent years. Cutaneous leishmaniasis is endemic in scattered areas around the world, with an estimated worldwide incidence of 1.5 million in 1990.[27] It is particularly important in regions of the Middle East, the Mediterranean, Africa, India, Asia, and Central and South America. Mucocutaneous leishmaniasis occurs only Central and South America.

Drug resistance is an important concern for visceral leishmaniasis. Notably, 40% of cases in India are reportedly now resistant to pentavalent antimony, although resistance in other areas is uncommon.[28]

7.26.3.2.3 Clinical manifestations of leishmanial infection

Visceral leishmaniasis causes a febrile wasting illness with progressive hepatosplenomegaly. Untreated, the disease progresses slowly until death from anemia, malnutrition, bleeding, or secondary infections. Cutaneous leishmaniasis presents with large painless skin ulcers that heal slowly, over months to over a year, often with residual scarring. Mucocutaneous disease presents with destructive mucosal lesions of the mouth, nose, larynx, or trachea.

7.26.3.2.4 Drugs for leishmaniasis

Standard drugs for all forms of leishmaniasis are pentavalent antimonials (sodium stibogluconate and meglumine antimonate). These drugs generally have good efficacy, but they require a long course of parenteral therapy and entail significant toxicity. In addition, resistance is an increasing problem, as noted above.[28,29] Alternative therapies for visceral leishmaniasis include amphotericin B, another parenteral therapy with significant toxicity. Lipid-associated formulations of amphotericin B decrease toxicity and allow a shortened treatment course, but costs are prohibitive for developing countries.[30] Pentamidine is a third parenteral therapy with major toxicities, and its use is limited also by increasing drug resistance. Two important new drugs for leishmaniasis are miltefosine and paromomycin. Miltefosine, which was registered in India in 2002, is an effective oral agent for visceral leishmaniasis, although concerns with this drug include the rapid selection of resistant parasites and genotoxicity, and its overall role is not yet clear.[30] Paromomycin appears to be promising as an alternative parenteral therapy, but funding constraints have delayed development of the drug for this indication. For cutaneous leishmaniasis, pentavalent antimonials are generally used, although in some cases no therapy may be needed. Miltefosine may offer a simpler therapy. Multiple new approaches to antileishmanial chemotherapy are under investigation.[29,30]

7.26.3.3 African Trypanosomiasis

7.26.3.3.1 Biology of African trypanosomes

African trypanosomiasis is caused by two subspecies of *Trypanosoma brucei*. The parasites are transmitted by tsetse flies. The insects inoculate trypomastigotes during a blood meal. Disease is caused by parasites that live and multiply extracellularly in the bloodstream and other spaces. African trypanosomes are thus unusual for human protozoan parasites in successfully surviving as free forms in the bloodstream despite an immune response against the parasite.

7.26.3.3.2 Epidemiology of African trypanosomiasis

Two similar syndromes, both referred to as sleeping sickness, are caused by African trypanosomes. West African trypanosomiasis is caused by *T. brucei gambiense* and transmitted in tropical rain forests of West and Central Africa. East African trypanosomiasis is caused by *T. brucei rhodesiense* and transmitted in wooded and savanna regions of Central and East Africa. Overall, approximately 50 million Africans are at risk of sleeping sickness, with a recent estimate of 100 000 cases and 66 000 deaths each year.[31] Of concern, the incidence of African trypanosomiasis appears to be increasing.[32]

7.26.3.3.3 Clinical manifestations of African trypanosomiasis

West African trypanosomiasis is characterized by a chronic febrile illness, which may not begin until months after the acquisition of infection. Meningoencephalitic disease may ensue months to years after the onset of infection, with the development of irritability, personality changes, somnolence, and headaches. Progressive neurological abnormalities end in death. East African trypanosomiasis has a much more acute course, with the illness typically presenting a few days after an infectious bite. The disease progresses more rapidly, but with similar findings to West African trypanosomiasis, ending in death from severe neurological dysfunction.

7.26.3.3.4 Drugs for African trypanosomiasis

Available treatments for African trypanosomiasis are remarkable in that they are mostly very old and very toxic. The limited armamentarium for this disease is a major concern.[32] Disease that has not yet involved the central nervous system responds fairly well to therapy. Pentamidine (used since 1940) is quite effective for West African disease, although the therapy is parenteral and quite toxic. Suramin (used since the 1920s) is effective against early East African disease, but it is also quite toxic. Three drugs are available for advanced central nervous system disease. Melarsoprol (used since 1949) can be used for both subspecies, and is the only reliable drug for East African disease. It is very toxic, and in addition suffers from decreasing efficacy in some areas, probably due to parasite drug resistance.[32] An important relatively new drug is eflornithine, which has been available since the 1990s. Eflornithine is an effective therapy for advanced West African, but not East African trypanosomiasis. Although clearly less toxic than melarsoprol, it does have some important safety concerns. It requires parenteral dosing, although an oral formulation is under development.

7.26.3.4 American Trypanosomiasis (Chagas' Disease)

7.26.3.4.1 Biology of American trypanosomes

Trypanosoma cruzi, the cause of American trypanosomiasis, is transmitted by a number of species of triatome insects (kissing bugs). Transmission occurs when mucous membranes or skin breaks are contaminated with insect feces containing infective trypomastigotes. In the human host parasites invade multiple cell types, transform into amastigotes, and multiply intracellularly. The parasite can also be transmitted by blood transfusions. The most heavily parasitized tissue is muscle, with most disease manifestations due to infection of cardiac and smooth muscle. Chronic infection leads to gradual destruction of cardiac and gastrointestinal smooth muscle over many years.

7.26.3.4.2 Epidemiology of Chagas' disease

Chagas' disease is a zoonosis that occurs through most of Latin America. The parasite infects many mammalian species, and humans are an incidental host. The disease is transmitted principally in areas where primitive housing is used, allowing cohabitation of humans and disease vectors. Thus, Chagas' disease is primarily a problem of poor rural inhabitants of Latin America. Recent extensive efforts to control the disease have had success, and transmission has decreased markedly in much of South America. Despite these advances, Chagas' disease was recently estimated to be responsible for 10–12 million infections and up to 45 000 deaths each year.[33]

7.26.3.4.3 Clinical manifestations of Chagas' disease

Acute infections with *T. cruzi* are commonly asymptomatic, but may include swelling at the site of inoculation followed by fever, malaise, other nonspecific findings and, rarely, serious neurological or cardiac disease. The major public health importance of Chagas' disease is chronic infection. Disease manifestations often present many years after primary infection. Chronic infection progresses to serious clinical problems, including severe heart failure, arrhythmias, and extreme dilatation of the esophagus or colon.

7.26.3.4.4 Drugs for Chagas' disease

Available therapies for Chagas' disease consist of just two drugs, and these are clearly inadequate.[34] Nifurtimox, an oral agent, reduces the severity and duration of acute Chagas' disease, but long courses are required, adverse events are common, and parasitological cure is not achieved in 30% of patients. Benznidazole, another oral agent, has similar efficacy and limitations. It is not clear whether it is beneficial to treat chronic Chagas' disease with either nifurtimox or benznidazole.[34] New therapeutic approaches for Chagas' disease are under investigation.[34]

7.26.3.5 Gastrointestinal and Genitourinary Protozoan Infections

7.26.3.5.1 Biology of gastrointestinal and genitourinary protozoan infections

Infections with *Entamoeba histolytica, Giardia lamblia, Cryptosporidium* parasites, *Isospora belli*, and other gastrointestinal protozoans are initiated by ingestion of infectious cysts or oocysts. Amebiasis causes an invasive, inflammatory enteritis, but the other pathogens cause noninflammatory diarrhea. *Entamoeba* can also invade other tissues, most commonly the liver. *Trichomonas vaginalis* is primarily transmitted sexually and causes genitourinary disease.

7.26.3.5.2 Epidemiology of gastrointestinal and genitourinary protozoan infections

Amebiasis is most common in developing countries. It causes approximately 50 million cases and 100 000 deaths each year.[35] Giardiasis and cryptosporidiosis are unusual among protozoans, in causing significant disease, mostly related to contaminated water, in both developed and developing countries. Trichomoniasis is among the most common sexually transmitted diseases in women and a fairly common cause of urethritis in men.

7.26.3.5.3 Clinical manifestations of gastrointestinal and genitourinary protozoan infections

Amebiasis most commonly presents with colitis and diarrhea, and can progress to severe dysentery. Amebiasis can also cause extraintestinal disease, most notably liver abscesses. Other intestinal protozoans cause diarrhea, which can be severe in immunocompromised individuals infected with *Cryptosporidium*, *Isospora*, and some other parasites. Trichomoniasis causes vaginitis in women and urethritis in men.

7.26.3.5.4 Drugs for gastrointestinal and genitourinary protozoan infections

Amebiasis, giardiasis, and trichomoniasis are usually treated with metronidazole or tinidazole[36] Drug resistance is of uncertain significance, but appears to be an increasing problem with *Trichomonas*.[37] Cryptosporidiosis is usually self-limited in those with normal immunity, but the treatment of severe disease in immunocompromised hosts has been very challenging. Nitazoxanide was recently approved for the treatment of cryptosporidiosis and giardiasis, although efficacy against severe cryptosporidiosis in AIDS patients has not clearly been demonstrated.[38] *Isospora* infection is treated with trimethoprim–sulfamethoxazole, which is quite effective, even in AIDS patients.

7.26.3.6 Toxoplasmosis

7.26.3.6.1 Biology of *Toxoplasma gondii*

Toxoplasma gondii infects many species of mammals and birds. The only definitive hosts are various species of cats; all others, including humans, are intermediate hosts. Humans are infected by ingesting undercooked meat containing tissue cysts, by ingestion of water or food contaminated with oocysts from cat feces, or by congenital transmission. In the intestines, sporozoites released from oocysts or bradyzoites released from tissue cysts invade and multiply within epithelial cells as tachyzoites, and spread to other locations via the lymphatics and bloodstream. Tachyzoites subsequently invade many cell types, leading to rapid replication and cell destruction. Some tachyzoites differentiate to bradyzoites, which replicate slowly and form tissue cysts. In immunocompetent individuals, infection is followed by a chronic subclinical infection. With immune compromise, latent infections become active, and rapid proliferation of tachyzoites leads to destructive lesions, most commonly in the brain, eyes, heart, and lungs.

7.26.3.6.2 Epidemiology of toxoplasmosis

Toxoplasmosis is a worldwide zoonosis. In different regions the relative importance of transmission due to the ingestion of undercooked meat and the ingestion of food contaminated by cat feces may vary. Toxoplasmosis is most important as a cause of disease after reactivation in immunocompromised individuals. Presentations differ with different types of immune deficiency. With AIDS, encephalitis due to toxoplasmosis is one of the most important opportunistic infections, and may occur in up to half of patients who do not receive antiretroviral or antiparasitic therapy.[39] Involvement of other organs, in particular the lungs and heart, is more common in patients immunocompromised due to organ transplantation. Congenital toxoplasmosis occurs after acute infection during pregnancy and is an important cause of both neonatal disease and long-term sequellae of infection.

7.26.3.6.3 Clinical manifestations of toxoplasmosis

Acute toxoplasmosis is usually asymptomatic, but may present with lymphadenopathy or a mild febrile illness in immunocompetent hosts. In patients with AIDS, toxoplasmosis usually presents with a severe encephalopathy. In patients with other forms of immunodeficiency, encephalopathy, pneumonitis, and myocarditis are common.

7.26.3.6.4 Drugs for toxoplasmosis

Available drugs do not eradicate encysted bradyzoites, but act against tachyzoites to control active disease. The standard therapy is pyrimethamine plus either sulfadiazine or clindamycin. Trimethoprim–sulfamethoxazole, which is used for prophylaxis against a number of infections in immunocompromised individuals, is effective for the prevention of toxoplasmosis.[40]

7.26.4 Helminth Infections

A great many different worm species, including nematodes, trematodes, and cestodes, are important human pathogens. In general, drug needs are less for helmintic infections, as most of the infections are usually less serious than the most important protozoan infections and some excellent drugs active against multiple helminth species are already available. However, effective drugs can be of great importance for both the treatment and control of helminth infections. Helminths of importance for drug discovery are detailed in **Tables 1** and **2**, and important infections are described in more detail below.

7.26.4.1 Intestinal Nematode (Round Worm) Infections

7.26.4.1.1 Biology of nematodes

Nematodes infect humans after the ingestion of eggs (ascariasis and trichuriasis) or the passage of infective larvae through the skin (hookworm infections and strongyloidiasis). Worms migrate through the gut or other tissues to the intestines. Infections can persist for years, and can be particularly prolonged in strongyloidiasis.

7.26.4.1.2 Epidemiology of nematode infections

Intestinal nematode infections are very common in developing countries. Over 80% of populations may be infected with *Ascaris*.[41] For all of the nematodes, low-level infections are very common, but probably not of clinical consequence. Heavy infections are much less common, but lead to clinical disease. Deaths are uncommon, but can result from abdominal complications. In addition, it is estimated that severe anemia secondary to hookworm infection is responsible for about 60 000 deaths per year.[42]

7.26.4.1.3 Clinical manifestations of nematode infections

Ascariasis and trichuriasis are typically asymptomatic, but *Ascaris* can cause pulmonary symptoms during passage of larvae through the lungs. Heavy infections can lead to gastrointestinal symptoms and occasional obstructive processes. Although serious manifestations are very uncommon, these infections are so prevalent that they are a common cause of surgical admission for abdominal disease in the tropics. Hookworm can cause gastrointestinal symptoms; the major consequence of infection is chronic anemia, which can be severe. Strongyloidiasis can also cause acute pulmonary symptoms. Chronic infection is usually asymptomatic, but can include intestinal symptoms due to adult worms and nonspecific symptoms due to migrating larvae. Uniquely among the nematodes, *Strongyloides* can replicate in the human host; this process is of particular relevance in immunodeficient individuals, in whom hyperinfection syndromes with multisystem involvement, bacterial superinfection, and high levels of mortality can be seen.[43]

7.26.4.1.4 Drugs for nematode infections

A number of effective drugs are available for nematode infections. Indications vary for the different parasites. Important agents include albendazole, mebendazole, thiabendazole, pyrantel pamoate, and ivermectin. Short courses are generally highly effective, although *Strongyloides* hyperinfection requires prolonged therapy.

7.26.4.2 Tissue Nematode Infections

7.26.4.2.1 Biology of tissue nematodes

Trichinosis results from ingestion of *Trichinella* larvae in undercooked meat. The larvae circulate in the bloodstream and invade skeletal muscle. Dracunculiasis occurs after drinking water contaminated with small crustaceans infected with *Dracunculus medinensis*. Larvae are released in the stomach and penetrate the intestinal mucosa. Worms subsequently develop and mate in the retroperitoneum, followed by migration to subcutaneous tissues, protrusion of worms, and release of larvae upon contact with water. Filariasis is transmitted by mosquitoes. Infective larvae develop in lymph nodes into adult worms, which can live for years, releasing microfilariae into the bloodstream. Loiasis is transmitted by tabanid flies. Larvae develop into adult worms, which migrate through subcutaneous tissues, causing painless soft tissue swellings. Onchocerciasis is transmitted by black flies. Larvae develop into adult worms, which become surrounded by an inflammatory reaction in connective tissues, and which release microfilariae into the circulation.

7.26.4.2.2 Epidemiology of tissue nematodes

Trichinosis occurs worldwide, but uncommonly. Dracunculiasis occurs in sub-Saharan Africa; the incidence of the disease has decreased markedly due to successful control efforts. Different species of filarial worms have different distributions, but overall filarial infections are seen in most of the tropics. Overall prevalence of these parasites is estimated at 120 million. Loiasis occurs in parts of West and Central Africa. Onchocerciasis occurs in West and Central Africa and parts of the Middle East and Latin America. The disease is believed to be prevalent in millions of Africans. Most importantly, it is the second leading cause of preventable blindness in the world, with an estimated prevalence of 500 000 for visual impairment and 270 000 for blindness.[44]

7.26.4.2.3 Clinical manifestations of tissue nematode infections

Trichinosis is usually subclinical, but can present with fever, periorbital edema, myositis, and many nonspecific symptoms. Drancunculiasis causes a painful ulcer, usually on the legs, often with a visible protruding worm. Filarial disease presents most commonly as lymphangitis progressing to chronic manifestations of lymphatic obstruction, including elephantiasis. Loiasis causes transient subcutaneous swellings; adult worms may be noticed passing through subconjunctival tissue in the eye. Early onchocerciasis presents most commonly with a pruritic rash. Chronic infection progresses to inflammatory eye disease, and eventually blindness.

7.26.4.2.4 Drugs for tissue nematodes

No therapy effectively eradicates *Trichinella* larvae in tissue, although mebendazole or albendazole may offer some benefit. Dracunculiasis is treated with thiabendazole or metronidazole, followed by physical removal of infected worms. Treatment of filariasis, loiasis, and onchocerciasis with diethylcarbamazine or ivermectin reduces circulating microfilariae, but does not reliably kill adult worms, and can be complicated by serious acute inflammatory processes. For onchocerciasis, ivermectin is the safer therapy, and offers benefit with repeated therapy to destroy microfilariae and gradually reduce numbers of adult worms.[44] Intermittent therapy with ivermectin has also been used for the control of onchocerciasis, with good success.[44] An interesting new approach now under investigation is to use antibiotics, in particular doxycycline, to eradicate the bacterial symbiont *Wolbachia*, and thereby control onchocerciasis.[44]

7.26.4.3 Trematode Infections

7.26.4.3.1 Biology of trematode infections

Schistosomiasis occurs after invasion of human skin by cercariae released from infected fresh-water snails. Schistosomulae subsequently migrate to the lungs and liver, mature to adult worms, and migrate to mesenteric (or, for *S. hematobium*, urinary tract) veins. Disease is caused primarily by inflammatory responses to eggs and larvae near the location of adult worms. Liver flukes are transmitted by ingestion of undercooked fish (*Clonorchis*, *Opisthorchis*) or aquatic vegetation (*Fasciola*), followed by maturation in the biliary tract, where worms can live for decades. Intestinal flukes (*Fasciolopsis* and other species) are acquired after ingestion of contaminated plants; adult flukes inhabit the small intestine. *Paragonimus*, the lung fluke, is acquired from ingestion of undercooked fresh-water crayfish and crabs; adults migrate to the lungs.

7.26.4.3.2 Epidemiology of trematodes

The most important trematode infection is schistosomiasis. Different species of schistosomes are distributed throughout the tropics. The worldwide prevalence of infection was estimated at 208 million, leading to 8000 deaths in 1990.[27] Other fluke infections have varied geographic distributions; in particular, many flukes are prevalent in Asia.

7.26.4.3.3 Clinical manifestations of trematode infections

Schistosomiasis generally presents with disease in the liver or urinary tract after chronic infection with large numbers of flukes. Important late manifestations are portal vein fibrosis, gastrointestinal bleeding, liver disease, and genitourinary disease. Liver fluke infections lead to biliary tract inflammation and obstruction. Lung flukes cause chronic pulmonary disease, often with nodular or cystic disease, productive cough, and other nonspecific pulmonary symptoms. For many fluke infections, acute disease from small numbers of migrating worms can also be seen, and disease at sites distant from the usual locations of individual species can occur. In some cases, inflammatory processes linked to chronic infections have been associated with increased risks of malignancy.

7.26.4.3.4 Drugs for trematode infections

Schistosomiasis, intestinal flukes, and lung flukes are generally treated with praziquantel. Liver flukes are treated with praziquantel or albendazole, except for fascioliasis, which is treated with triclabendazole.

7.26.4.4 Cestode (Tapeworm) Infections

7.26.4.4.1 Biology of cestode infections

Tapeworm infections are all acquired by ingesting worm cysts or eggs. The most common infections result from undercooked fish (*Diphyllobothrium latum*), beef (*Taenia saginata*), and pork (*Taenia solium*). Other tapeworms can be spread person-to-person (*Hymenolepsis nana*) or with contamination of food by feces from infected dogs (*Echinococcus* species). Mature worms reside in the gut, releasing large numbers of eggs, but usually causing little disease.

7.26.4.4.2 Epidemiology of cestodes

Tapeworm infections are common, particularly in the tropics, but clinical consequences of infection are much less common. However, cysticercosis, a consequence of *T. solium* infection, is an important cause of seizures and other neurological disease in certain areas, including parts of Latin America and Southeast Asia.

7.26.4.4.3 Clinical manifestations of cestode infections

Infections with the fish, beef, and pork tapeworms are usually asymptomatic. *Diphyllobothrium latum* infection can lead to vitamin B_{12} deficiency. An important syndrome related to *T. solium* infection is cysticercosis. This syndrome is due to tissue infection with parasite cysts, most often in the brain, following ingestion of food contaminated with parasite eggs from pig feces. Neurocysticercosis causes inflammatory brain lesions, with seizures, meningitis, and other neurological sequellae. Echinococcosis causes large cysts (hydatid disease), most commonly in the liver and lung.

7.26.4.4.4 Drugs for cestode infections

Tapeworm infections are generally treated with niclosamide or praziquantel. Neurocysticercosis can be treated with praziquantel or albendazole. However, the killing of active cysts can be accompanied by inflammatory changes, with worsening of neurological symptoms, and therefore the advisability of treating neurocysticercosis remains unclear. With antiparasitic treatment, corticosteroids are often coadministered to limit inflammation. Hydatid disease is treated by careful surgical resection of cysts, with perioperative administration of albendazole.

7.26.5 The Future of Drug Discovery and Development for Parasitic Infections

Of tremendous importance in considerations of drug development for parasitic diseases is the general lack of a traditional market for drugs for these indications. Huge amounts of drugs are used for these common diseases, but the vast majority of this usage is in developing countries, so opportunities for profits are minimal. Contrasts between healthcare spending in developing and developed countries are enormous, leading to striking, at times perhaps obscene parallels. For example, eflornithine is a promising and potentially life-saving new drug for African trypanosomiasis, but for some years it was only available as an expensive topical cream for removal of unwanted facial hair. However, despite continued disparities in wealth, there are some reasons for optimism regarding drug development for parasitic diseases.

First, advances in medical science and chemistry promise the possibility of more cheaply discovering, developing, and producing drugs. Second, transportation and communication advances continue to lead to increased contact between all world populations, and it is increasingly difficult for those in the developed world to ignore health problems of poor countries. Some of these health problems present in developed countries and are at risk to increase there, many are risks to travelers from developed to developing countries, and all impact on attempts to improve health and social circumstances in the poorest countries. Thus, programs to attack major infectious diseases of developing countries are increasingly promoted by governments of developed countries and nongovernmental organizations. Third, academic interest in research on parasitic diseases and drug targets remains high. Academicians realize that diseases which are of relatively little interest among industry groups offer niches for productive research. Fourth, some companies are increasingly interested in research on diseases of underserved populations, perhaps to ease public perceptions that their business practices are not adequately serving the public. Finally, the concept of public–private partnerships has emerged, whereby major drug discovery and development efforts are supported by a combination of public funding, industrial partners, and dedicated funding agencies.[45] Despite the reasons for cautious optimism, there is no call for complacency. Massive disparities in healthcare spending persist, spending for 'modern' diseases such as heart disease and cancer still far outweighs that for parasitic diseases that affect millions, and the incidence of many of the most important parasitic diseases is not decreasing. More than ever, aggressive drug discovery efforts directed against leading parasitic diseases are needed.

References

1. Bogitsch, B. J.; Carter, C. E.; Oeltmann, T. N. *Human Parasitology*; Elsevier/Academic Press: Amsterdam, the Netherlands, 2005.
2. Meshnick, S. R.; Dobson, M. J. The History of Antimalarial Drugs. In *Antimalarial Chemotherapy: Mechanisms of Action, Resistance, and New Directions in Drug Discovery*; Rosenthal, P. J., Ed.; Humana Press: Totowa, NJ, 2001, pp 15–25.
3. Miller, L. H.; Baruch, D. I.; Marsh, K.; Doumbo, O. K. *Nature* **2002**, *415*, 673–679.
4. Rosenthal, P. J. Antiprotozoal Drugs. In *Basic and Clinical Pharmacology*; Katzung, B. G., Ed.; Lange Medical Books/McGraw-Hill: New York, 2004, pp 864–885.
5. Breman, J. G. *Am. J. Trop. Med. Hyg.* **2001**, *64*, 1–11.
6. Greenwood, B. M.; Bojang, K.; Whitty, C. J.; Targett, G. A. *Lancet* **2005**, *365*, 1487–1498.
7. Snow, R. W.; Guerra, C. A.; Noor, A. M.; Myint, H. Y.; Hay, S. I. *Nature* **2005**, *434*, 214–217.
8. Stein, C. E.; Inoue, M.; Fat, D. M. *Semin. Pediatr. Infect. Dis.* **2004**, *15*, 125–129.
9. Guerin, P. J.; Olliaro, P.; Nosten, F.; Druilhe, P.; Laxminarayan, R.; Binka, F.; Kilama, W. L.; Ford, N.; White, N. J. *Lancet Infect. Dis.* **2002**, *2*, 564–573.
10. Wongsrichanalai, C.; Pickard, A. L.; Wernsdorfer, W. H.; Meshnick, S. R. *Lancet Infect. Dis.* **2002**, *2*, 209–218.
11. Maitland, K.; Makanga, M.; Williams, T. N. *Curr. Opin. Infect. Dis.* **2004**, *17*, 405–412.
12. Baird, J. K. *N. Engl. J. Med.* **2005**, *352*, 1565–1577.
13. Barnes, K. I.; Mwenechanya, J.; Tembo, M.; McIlleron, H.; Folb, P. I.; Ribeiro, I.; Little, F.; Gomes, M.; Molyneux, M. E. *Lancet* **2004**, *363*, 1598–1605.
14. White, N. J. *J. Clin. Invest.* **2004**, *113*, 1084–1092.
15. Nosten, F.; van Vugt, M.; Price, R.; Luxemburger, C.; Thway, K. L.; Brockman, A.; McGready, R.; ter Kuile, F.; Looareesuwan, S.; White, N. J. *Lancet* **2000**, *356*, 297–302.
16. Haynes, R. K. *Curr. Opin. Infect. Dis.* **2001**, *14*, 719–726.
17. Meshnick, S. R. Artemisinin and its Derivatives. In *Antimalarial Chemotherapy: Mechanisms of Action, Resistance, and New Directions in Drug Discovery*; Rosenthal, P. J., Ed.; Humana Press: Totowa, NJ, 2001, pp 191–201.
18. Adjuik, M.; Babiker, A.; Garner, P.; Olliaro, P.; Taylor, W.; White, N. *Lancet* **2004**, *363*, 9–17.
19. Olliaro, P. L.; Taylor, W. R. *J. Postgrad. Med.* **2004**, *50*, 40–44.
20. Mutabingwa, T. K.; Anthony, D.; Heller, A.; Hallett, R.; Ahmed, J.; Drakeley, C.; Greenwood, B. M.; Whitty, C. J. *Lancet* **2005**, *365*, 1474–1480.
21. Staedke, S. G.; Mpimbaza, A.; Kamya, M. R.; Nzarubara, B. K.; Dorsey, G.; Rosenthal, P. J. *Lancet* **2004**, *364*, 1950–1957.
22. Rosenthal, P. J. *J. Exp. Biol.* **2003**, *206*, 3735–3744.
23. Ryan, E. T.; Kain, K. C. *N. Engl. J. Med.* **2000**, *342*, 1716–1725.
24. Greenwood, B. *Am. J. Trop. Med. Hyg.* **2004**, *70*, 1–7.
25. Rosen, J. B.; Breman, J. G. *Lancet* **2004**, *363*, 1386–1388.
26. Guerin, P. J.; Olliaro, P.; Sundar, S.; Boelaert, M.; Croft, S. L.; Desjeux, P.; Wasunna, M. K.; Bryceson, A. D. *Lancet Infect. Dis.* **2002**, *2*, 494–501.
27. Murray, C. J. L.; Lopez, A. D. *Global Health Statistics*; Harvard School of Public Health: Cambridge, MA, 1996.
28. Sundar, S. *Trop Med. Int. Health* **2001**, *6*, 849–854.
29. Murray, H. W. *Antimicrob. Agents Chemother.* **2001**, *45*, 2185–2197.
30. Murray, H. W. *Am. J. Trop. Med. Hyg.* **2004**, *71*, 787–794.
31. Pepin, J.; Meda, H. A. *Adv. Parasitol.* **2001**, *49*, 71–132.
32. Legros, D.; Ollivier, G.; Gastellu-Etchegorry, M.; Paquet, C.; Burri, C.; Jannin, J.; Buscher, P. *Lancet Infect. Dis.* **2002**, *2*, 437–440.
33. Kirchhoff, L. V. Trypanosoma Species (American Trypanosomiasis, Chagas' Disease): Biology of Trypanosomes. In *Principles and Practice of Infectious Diseases*; Mandell, G. L., Bennett, J. E., Dolin, R., Eds.; Elsevier/Churchill Livingstone: Philadelphia, PA, 2005, pp 3156–3164.
34. Urbina, J. A.; Docampo, R. *Trends Parasitol.* **2003**, *19*, 495–501.
35. Ravdin, J. I.; Stauffer, W. M. *Entamoeba histolytica* (amebiasis). In *Principles and Practice of Infectious Diseases*; Mandell, G. L., Bennett, J. E., Dolin, R., Eds.; Elsevier/Churchill Livingstone: Philadelphia, PA, 2005, pp 3097–3111.
36. Haque, R.; Huston, C. D.; Hughes, M.; Houpt, E.; Petri, W. A., Jr. *N. Engl. J. Med.* **2003**, *348*, 1565–1573.

37. Cudmore, S. L.; Delgaty, K. L.; Hayward-McClelland, S. F.; Petrin, D. P.; Garber, G. E. *Clin. Microbiol. Rev.* **2004**, *17*, 783–793.
38. Fox, L. M.; Saravolatz, L. D. *Clin. Infect. Dis.* **2005**, *40*, 1173–1180.
39. Luft, B. J.; Remington, J. S. *Clin. Infect. Dis.* **1992**, *15*, 211–222.
40. Podzamczer, D.; Salazar, A.; Jimenez, J.; Consiglio, E.; Santin, M.; Casanova, A.; Rufi, G.; Gudiol, F. *Ann. Intern. Med.* **1995**, *122*, 755–761.
41. Crompton, D. W. *Adv. Parasitol.* **2001**, *48*, 285–375.
42. Maguire, J. H. Intestinal Nematodes (Roundworms). In *Principles and Practice of Infectious Diseases*; Mandell, G. L., Bennett, J. E., Dolin, R., Eds.; Elsevier/Churchill Livingstone: Philadelphia, PA, 2005, pp 3260–3267.
43. Grove, D. I. *Adv. Parasitol.* **1996**, *38*, 251–309.
44. Hoerauf, A.; Buttner, D. W.; Adjei, O.; Pearlman, E. *BMJ* **2003**, *326*, 207–210.
45. Nwaka, S.; Ridley, R. G. *Nat. Rev. Drug Disc.* **2003**, *2*, 919–928.
46. Crompton, D. W. *J. Parasitol.* **1999**, *85*, 397–403.

Biography

Philip J Rosenthal, MD, is a professor in the Department of Medicine at the University of California, San Francisco. He received a BS in Biochemistry from the State University of New York at Stony Brook, an MD from New York University, training in Internal Medicine at the University of Michigan, and then training in Infectious Diseases at the University of California, San Francisco. His research interests include the biochemistry of malaria parasites, antimalarial drug discovery, and translational studies of antimalarial drug efficacy and resistance.

Comprehensive Medicinal Chemistry II
ISBN (set): 0-08-044513-6
ISBN (Volume 7) 0-08-044520-9; pp. 749–763

7.27 Advances in the Discovery of New Antimalarials

K M Muraleedharan and M A Avery, University of Mississippi, University, MS, USA

7.27.1 Disease State

7.27.1.1 Introduction

Malaria remains one of the deadliest diseases on this planet, which accounts for 300–500 million clinical cases and up to 2.7 million deaths each year. About 90% of these casualties occur in tropical Africa – the great majority being children under the age of 5.[1] Apart from Africa, malaria is prevalent in Asia, Central and South America, the Middle East, the island of New Guinea, Haiti, the Dominican Republic, and the Pacific Islands. It has been estimated that about 40% of the world's population live in areas where malaria is common, and the threat from this disease has increased in recent years due to changing global climate, resistance developed by the parasite to traditional drugs, and increasing international travel which exposes nonimmune populations to malaria parasites. The fact that at least one child dies of malaria every 40 seconds shows the devastating effect of this disease.[2]

7.27.1.2 Infection and Clinical Symptoms

Malaria is caused by protozoan parasites of the genus *Plasmodium*. There are over 100 species of malaria parasites, capable of infecting a wide variety of hosts, including reptiles, birds, rodents, and primates. However, only four of these species can cause human malaria, namely *Plasmodium falciparum*, *P. vivax*, *P. ovale*, and *P. malariae*. *Plasmodium falciparum* is the most virulent and is responsible for high infant mortality, especially in Africa. *Plasmodium vivax* can exist in temperate as well as tropical climates and is the second largest contributor to clinical malaria cases. *Plasmodium malariae* and *P. ovale* infections are less common and are generally not life-threatening. The parasites are transmitted to human beings through the bite of infected female *Anopheles* mosquitoes. There are approximately 422 species of *Anopheles* mosquitoes worldwide, of which around 70 are capable of transmitting malaria. The parasites go through a series of developmental stages during their life cycle, which involve a sexual phase within the insect vector and an asexual phase in a vertebrate host.

The sporozoites, after entering the bloodstream of the host, reach liver hepatocytes and undergo exoerythrocytic schizogony, resulting in the formation of merozoites. These merozoites, which are released into the bloodstream, rapidly invade erythrocytes and go through ring, early, and late trophozoite stages and then undergo multiple rounds of nuclear divisions without cytokinesis to form schizonts. Mature schizonts burst and release large numbers of fresh merozoites to repeat the cycle. Some of the blood stage parasites can differentiate into sexual forms called gametocytes, which when taken up by mosquitoes transform into micro- and macrogametes. These sexual forms fuse to form zygotes and subsequently transform into motile ookinetes. The ookinetes penetrate the gut epithelial cells and develop into oocysts, which undergo mitotic divisions to produce sporozoites. Mature oocysts rupture and release sporozoites which migrate and invade the salivary glands, from where they are injected into the host during the next blood meal.[3]

The clinical symptoms of malaria usually appear 7–9 days after a bite by an infected mosquito. The parasitized erythrocytes in general can initiate three principal pathogenic features – hemodynamic, immunodynamic, and metabolic perturbations. The overall outcome of the infection depends on the interactions between these pathways.[4] Fever, shaking chills, headaches, and backaches are the common symptoms, and the body temperature can rise up to 40–41 °C. Other characteristics of this disease include vomiting, diarrhea, coughing, and yellowing (jaundice) of the skin and the whites of the eyes.[5,6] In the case of patients with low immunity levels, *Plasmodium* infection can lead to a condition known as severe malaria. This is usually observed with *P. falciparum* infections and can manifest in different ways such as (1) cerebral malaria, characterized by abnormal behavior, seizures, coma, or other neurological abnormalities; (2) severe anemia; (3) hemoglobinuria; (4) pulmonary edema or acute respiratory distress syndrome; (5) abnormalities in blood coagulation and thrombocytopenia (decrease in blood platelets); (6) enlargement of liver and spleen; and (7) cardiovascular collapse and shock. It is important to note that the severity and range of symptoms depend on various factors such as the immunity level and genetic characteristics of the host and the type of parasite involved.[7] In some instances, the parasite can stay dormant inside the host for up to 5 years and then recur as in the cases of *P. vivax* and *P. ovale*, having stages such as hypnozoites.[8,9] Failure in early detection or the lack of proper medication can result in death. There are several methods available for malaria diagnosis, which include (1) microscopic examination of Giemsa-stained thick and thin blood smears, (2) methods based on fluorescent microscopy or the detection of nucleic acids, and (3) tests based on immunoassays.[10,11]

7.27.1.3 Ligand–Receptor Interactions during Host Cell Invasion

As described above, sporozoites, merozoites, and ookinetes are the three invasive forms of the parasite observed during the *Plasmodium* life cycle.[12] Soon after they are released into the bloodstream of the vertebrate host, sporozoites find their way into liver hepatocytes after crossing the sinusoidal cell layer, which is composed of highly fenestrated endothelial and Kupffer cells. The high speed and selectivity with which this invasion occurs suggest specific ligand–receptor recognition events involving parasite and host cell surface molecules. Even though the complete molecular details of this process remain to be unraveled, studies have shown that circumsporozoite proteins (CSP), which cover the entire surface of sporozoites, bind with high affinity to specific hepatocyte heparan sulfate proteoglycans (HSPGs) that extend from the Disse space to the sinusoidal lumen through the endothelial fenestrae, and facilitate the initial attachment.[13–17] Parasites may then cross the endothelial layer directly or traverse through Kupffer cells and reach the target. A series of studies by Ishino *et al.* highlight the involvement of a protein named 'sporozoites microneme protein essential for cell traversal' in the cell traversal process.[18–20] Another protein, thrombospondin-related anonymous protein (TRAP), otherwise known as sporozoite surface protein 2, is also considered to play an important role in hepatocyte invasion and parasite motility.[21,22] Recent studies by Silvie *et al.* have also demonstrated the involvement of a microneme derived protein, known as apical membrane antigen 1 (AMA-1) in the invasion process.[23]

All the clinical symptoms of malaria occur during the erythrocytic phase of the *Plasmodium* life cycle. Erythrocyte invasion of merozoites is a complex process and can be viewed as taking place in four distinct phases.[24–27] Initial

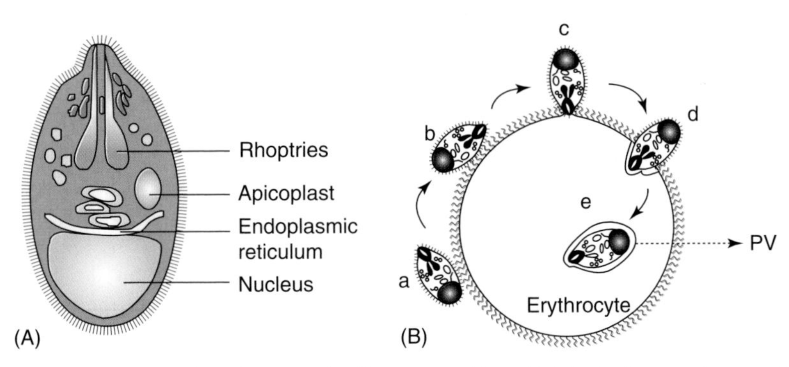

Figure 1 (A) Diagrammatic cross-section of merozoite. (Reproduced from Ralph, S. A.; van Dooren, G. G.; Waller, R. F.; Crawford, M. J.; Fraunholz, M. J.; Foth, B. J.; Tonkin, C. J.; Roos, D. S.; McFadden, G. I. *Nat. Rev. Microbiol.* **2004**, *2*, 203–216, with permission from Nature Reviews Drug Discovery. Copyright (2005) Macmillan Magazines Ltd.) (B) (a–d) Erythrocyte invasion of merozoites, (e) intraerythrocytic parasite inside the parasitophorous vacuole (PV).

binding with the target erythrocyte is followed by a reorientation step whereby the parasite brings its apical end in contact with the host cell membrane. This is followed by junction formation and subsequently parasite entry (**Figure 1**). Several merozoite proteins have been identified and implicated in the erythrocyte invasion process which include merozoite surface proteins (MSP), rhoptry-associated proteins, and those derived from micronemes.[25]

A number of proteins belonging to the MSP class have been identified and biochemically characterized from various *Plasmodium* species over the last two decades. They are important not only because of their involvement in initial erythrocyte binding but also due to their ability to induce immune response in host cells that make them attractive candidates for vaccine development. These include MSP-1,[29,30] MSP-2,[31,32] MSP-3,[33–35] MSP-4,[36] MSP-5,[37,38] MSP-6,[39,40] MSP-7,[41] MSP-8,[42] MSP-9,[43] and MSP-10,[44,45] from *P. falciparum*, and MSP-1,[46–50] MSP-3 (α, β, and γ),[51–53] MSP-4 and MSP-5,[54] MSP-8,[55] and MSP-9[43,56,57] from *P. vivax*. Similar proteins have also been identified from other species of *Plasmodium* such as *P. knowlesi* [58,59] and *P. cynomolgi*.[60] Many of these MSPs are anchored to the parasite surface by glycophosphatidylinositol (GPI) moiety.[61]

Erythrocyte-binding antigen 175 (EBA-175) is a microneme-derived protein involved in the binding of merozoites to the erythrocyte membrane through sialic acid residues on glycophorin A, the major sialoglycoprotein on the surface of erythrocytes.[62–64] A recent report by Gruner *et al.* has shown that EBA-175 is expressed not only in blood stage parasites, but also on the surface of sporozoites in hepatocytes.[65] The facts that structural analogs of sialic acid can interfere with the binding of EBA-175 to glycophorin A, and high-affinity binding peptides corresponding to protein segments in EBA protein family can block merozoite entry of erythrocytes show the importance of this family of proteins as chemotherapeutic targets and vaccine candidates.[66,67]

Extensive studies aimed at understanding the host cell preferences of malaria parasites have shown the involvement of another protein known as reticulocyte binding protein (RBP) in the host cell invasion of merozoites.[68] As the name implies, this class of proteins are known to bind to human reticulocytes with high affinities. Two proteins belonging to this class, namely PvRBP-1 and PvRBP-2, with molecular masses of 325 kDa and 330 kDa, respectively, have been identified from *P. vivax*, and they possess hydrophobic transmembrane domains at the C-termini with relatively shorter cytoplasmic domains.[69–71] PvRBP-1 is a dimeric protein, linked via disulfide bonds, which forms a complex with RBP-2 through noncovalent interactions. Even though genes homologous to *P. vivax* RBP-1 and RBP-2 are present in *P. falciparum*, there is no indication of their specific affinity toward reticulocytes.[72,73] Studies by Triglia *et al.* recently demonstrated the involvement of *P. falciparum* reticulocyte binding homolog 1 (PfRh-1) in the sialic acid dependent invasion of human erythrocytes.[74] It is interesting to note that PvRBP-2 shares noticeable homology to a *P. yoelii* 235 kDa protein which is known to bind to erythrocytes in general, without any specificity to reticulocytes.[75–77]

Another well-studied protein from merozoites that aid in erythrocyte invasion is AMA-1. This class of proteins has been identified from a number of plasmodium species such as *P. falciparum*,[78] *P. vivax*,[79] *P. knowlesi*,[80] *P. fragile*,[81] and *P. chabaudi*.[82] They are initially localized in the 'neck' of each rhoptry, which, during the time of release from erythrocytes, get translocated to the entire surface of merozoites. Recently, a gene that codes for a protein having similar features to AMA-1 and those in the Duffy binding protein (DBP) family was identified from *P. yoelii* and *P. berghei* and named MAEBL.[83,84] Subsequent studies have indicated the presence of its analog in *P. falciparum*, which localizes with rhoptry proteins.[85] In an interesting study by Kariu *et al.*, sporozoites in the oocyst were also found to carry MAEBL, which was shown to have an important role in their invasion of mosquito salivary glands.[86]

Two noncovalently bound protein complexes located in rhoptries have also been characterized. Of these, the higher molecular weight complex, known as the Rhop complex, includes three distinct proteins of molecular weight 140, 130, and 110/105 kDa each.[87–89] The second low molecular weight complex consists of three members, namely RAP-1, RAP-2, and RAP-3, of 80, 42, and 37 kDa, respectively.[90–98] Studies using blood stage *P. falciparum* parasites with a truncated RAP-1 gene showed that this protein is required to localize RAP-2 to rhoptries, supporting the hypothesis that the secretory pathways in parasites are required for rhoptry biogenesis.[99,100] Even though further studies are needed to get a complete picture on the roles of these proteins in erythrocyte invasion, recent efforts have shown their presence in membrane structures of infected erythrocytes.[101,102]

It should be noted that the above-mentioned classes of proteins do not represent the entire list of molecules involved in host cell recognition and invasion. However, as mentioned above, many of them have importance in the development of vaccine candidates.

7.27.2 Disease Basis

7.27.2.1 Epidemiology

Malarial epidemiology deals with the study of incidence, distribution, determinants, and control measures related to the infection. The complex interactions between hosts, parasites, and vectors, coupled with the emergence of this disease in areas from where it was once eliminated or suppressed, make the study elaborate and difficult. Malaria is considered endemic in regions where there is constant and steady prevalence of the disease over a period of many successive years. The term epidemic is used to address situations when there is periodic or occasional sharp increase in malaria cases. In addition, there is a more general classification, such as 'stable malaria' and 'unstable malaria', to address the state of this disease. The former means that there is a high transmission rate without any noticeable fluctuation over many years, whereas the latter refers to fluctuation in transmission with the possibility of epidemics.

7.27.2.2 Geographical Distribution

Due to the lack of a proper reporting system for infectious diseases in many regions such as Africa, it is difficult to get a complete and accurate picture on annual malaria incidences and mortalities. The disease is prevalent in tropical regions of the world where there is a suitable climate for vector survival and parasite transmission. Even though there have been discrepancies in reporting annual malaria incidents, studies over the last several decades show that the disease is geographically restricted and mainly affects the poorest populations. The economic growth in many countries has been hampered by continuous episodes of malaria and other infectious diseases. In the year 2000, malaria is estimated to have claimed the lives of approximately 803 000 children under the age of 5 in sub-Saharan Africa. Malaria during pregnancy leads to the birth of infants with low birth weights, which contributes to infant mortality. The increase in malaria-related deaths in the 1980s and 1990s has been attributed to growing resistance to existing drugs such as chloroquine. Recent studies using a combination of epidemiological, geographical, and demographical data have shown that in 2002 alone, almost 2.2 billion people were exposed to infections by *P. falciparum*, which resulted in ~515 million clinical cases. The great majority of these clinical events occurred in Africa (70%) and South East Asia (25%).[103] An estimation of regional contribution to the global malaria burden, principal vectors and parasites involved, and the geographical distribution of this disease from the *World Malaria Report 2005* are presented in **Table 1** and **Figure 2**, respectively.[1]

Infections due to *P. vivax* account for 70–80 million clinical cases annually and are responsible for more than 50% of malaria cases outside Africa, especially in the Middle East, Asia, the Western Pacific, and Central and South America.[104] Its relatively low prevalence in West Africa has been attributed to the presence of Duffy-negative blood group variants, which hinders the erythrocytic invasion of parasites.[7,105] Out of about 1.06 million malaria cases reported in the Americas in 1997, almost 393 000 (37%) were in Brazil, where malaria transmission is considered hypoendemic as per World Health Organization (WHO) standards.[106] A recent study conducted by TropNetEurope surveillance network showed a large number of imported *P. vivax* malaria cases among European travelers and immigrants.[107,108] In many parts of South East Asia, infections due to *P. vivax* have increased in recent years. *Plasmodium vivax* was prevalent in South Korea until the 1960s, after which it was controlled, and thought to have been eradicated by 1984. However, it has re-emerged, and annual incidents have shown a steady increase since 1993. Polymorphism of the AMA-1 gene was studied recently to obtain information on the genetic characteristics of the parasites involved in these infections, which showed that this emerging *P. vivax* possesses two genotypes of AMA-1 having similarity to Chinese genotypes.[109,110]

Table 1 Regional distribution of malaria

	Africa	*Asia*	*The Americas*
Plasmodium species	*P. falciparum* (93%)	*P. falciparum* (35%)	*P. falciparum* (18%)
	P. vivax or its mixed infections with *P. falciparum* (7%)	*P. vivax*	*P. vivax* (72%)
			P. malariae
Principal vectors	*A. gambiae,*	*A. culicifacies,*	*A. albimanus* (Central America)
	A. funestus	*A. minimus*	*A. darlingi* (Amazon Basin)
		A. annularis,	
		A. dirus, A. Fluviatilis,	
		A. maculipennis,	
		A. sacharovi,	
		A. superpictus,	
		A. farauti	
Population at risk	66%	49%	14%
Contribution to the global burden of clinical malaria cases	59%	38%	3%
Contribution to the global burden of *P. falciparum* cases	74%	25%	1%
Contribution to global malaria mortality burden	89%	10%	<1%

7.27.2.3 Factors Affecting the Distribution and Clinical Manifestation of Malaria

Dynamics of malaria transmission depend on several factors, such as the density of vectors, susceptibility of these vectors to transmit different parasites, frequency with which the vector takes the blood meal from the host, duration of sporogony,[111–113] parasite diversity,[114,115] interaction between species,[116] and the genetic characteristics of the hosts.

Of the four *Plasmodium* species that affect humans, *P. falciparum* is the most extensively studied, due in part to the availability of its cloned cell lines. Studies have demonstrated considerable variability in isoenzymes, drug resistance, adhesion, and antigenic and genotypic characteristics of these cloned cell lines.[117,118] *Plasmodium vivax* also shows extensive diversity within the species.[114,119,120] In malaria endemic areas, co-occurrence of more than one parasitic species within the same human host or vector is a common observation.[121,122] When two genetically distinct clones of a species get cotransmitted from human host to the vector during the obligatory sexual phase in their life cycle, chances of outcrossing between them increase, leading to the formation of new genotypes. Such types of interactions may affect parameters such as pathology and infection dynamics. According to a recent study conducted in Antula, the association of *P. falciparum*/*P. malariae* was found common in humans. At the same time, analysis of mid guts and salivary glands of the mosquitoes in the same area indicated a combination of *P. falciparum*/*P. ovale*.[116]

Blood slide surveys in areas of high malaria transmission have shown that the density of *P. falciparum* gametocytes decrease in an age-specific manner.[123] One of the theories put forward to explain this observation is based on the fact that early gametocytes and trophozoites express the same repertoire of *P. falciparum* erythrocyte membrane protein 1 (PfEMP-1) on the infected cell membrane. The early maturation of trophozoites confers naturally acquired immunity on PfEMP-1, which would subsequently hinder the development of gametocytes.[124,125]

One of the important characteristics of *P. falciparum* is its ability to switch surface antigens to evade the host immune system.[126–130] A multigene family, known as 'var genes,' which includes a group of 50 different genes is responsible for

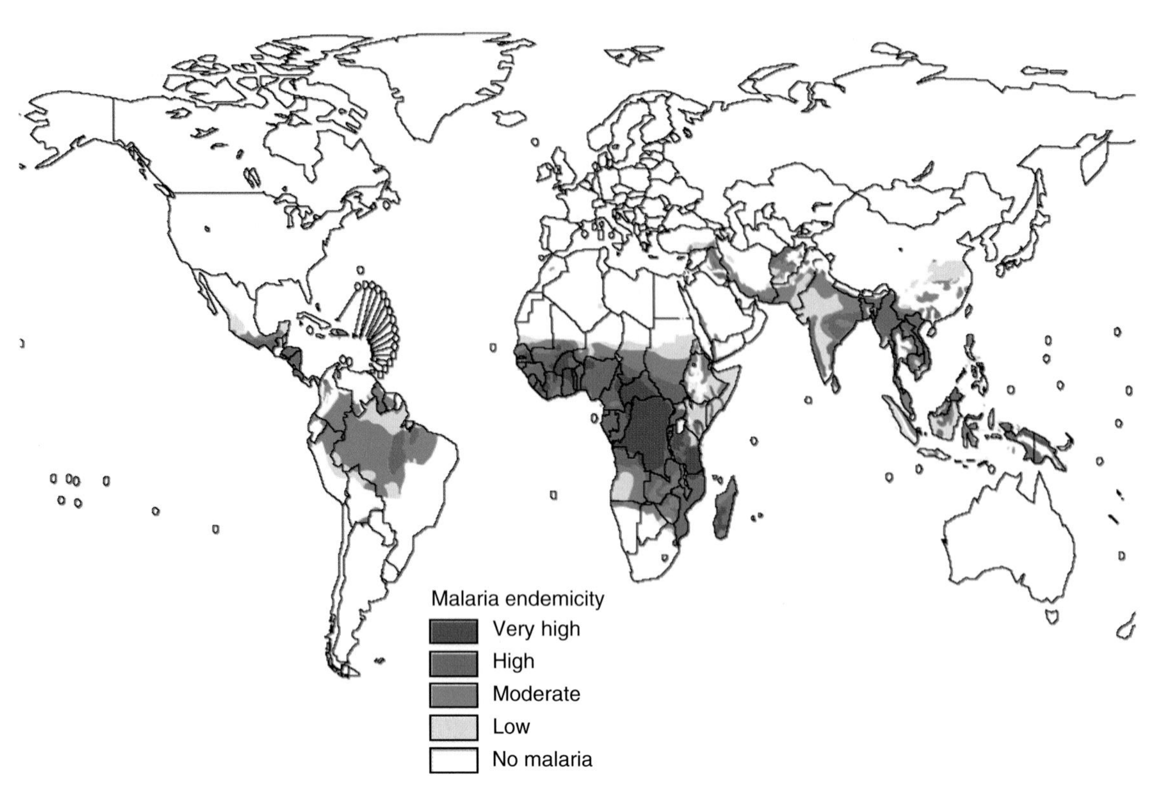

Figure 2 Map showing the prevalence of malaria. (Reproduced from *World Malaria Report*; World Health Organization: Geneva, Switzerland, 2005, with permission from WHO/UNICEF.)

this.[131,132] Differential expressions of individual genes from this group enable the parasite to change its surface coat, which help the parasites to escape the host immune response. This is a strategy to prolong the survival of the parasites within human hosts to ensure their transmission to vectors.

Host genetic factors have profound influence on the progression of malaria.[133] A careful analysis of the high gene frequency of thalassemia in certain Mediterranean populations led Haldane (1948) to propose that this disease might have come under intense selection and prevailed, as it gave a heterozygote advantage against malaria.[134,135] This hypothesis was strengthened by subsequent studies involving people with sickle-cell trait who showed reduced susceptibility to malaria compared to normal individuals.[136–138] In addition, studies have demonstrated that red blood characteristics such as Duffy blood group determinant $F^{ya}F^{yb}$ [105,139–141] and deficiency in glucose 6-phosphate dehydrogenase (G6PD) can also confer innate resistance to malaria infection.[142]

Recent reports indicate that genetic variability as the result of selection is not just restricted to abnormalities associated with erythrocytes. Studies conducted by Hill *et al.* involving West African children have shown that a human leukocyte class I antigen (HLA-Bw53) and a HLA class II haplotype (DRB1*1302-DQB1*0501) are associated with protection from severe malaria in the region.[135,143] Further, a large case control study conducted among Gambian children showed that those who were homozygous for tumor necrosis factor 2 (TNF-2) allele, a variant of the TNF-α gene promoter region, had a sevenfold increased risk of dying from cerebral malaria.[144,145]

7.27.3 Control Measures

Preventive measures against malaria include (1) vector control; (2) use of bed nets, insecticides, or insecticide-treated bed nets; (3) early diagnosis; (4) increasing the availability and effectiveness of therapeutic agents; and (5) development of vaccines. Resistance developed by parasites and mosquitoes against commonly used chemopreventive agents has become a significant obstacle in controlling malaria. However, current knowledge about the genome sequences of various *Plasmodium* species, humans, and *Anopheles* vectors would be helpful in understanding the biology of parasite–host–vector interactions and in the identification of new drugs and vaccine candidates.[146–148]

One of the limitations with commonly used insecticides, such as dichlorodiphenyl trichloroethane (DDT), apart from resistance developed by vectors, is the possible environmental pollution.[149] Bed nets treated with insecticide such as pyrethroid have proven to be effective, but there are concerns about the development of resistance after extended exposures.[150–152] Another strategy being pursued to control malaria transmission is the development of genetically modified mosquitoes which are either sterile[153] or incapable of transmitting the parasite.[154–157] Arguments against this include the possibility of these transgenic mosquitoes being vectors for other diseases and resistance development.[158] Another related approach is to use entomopathogenic fungi, such as *Beauveria bassiana* or *Metarhizium anisopliae*, which can either kill the vectors or reduce their capacity to transmit the parasite.[159,160]

Even though limitations exist due to the spread of disease in vast geographical areas lacking hospital facilities and drug availability, the last decade has witnessed an increased coordination of healthcare activities, both nationally and internationally, initiated by groups like the 'Roll Back Malaria' partnership, launched in 1988 by WHO, UNICEF, UNDP, and the World Bank.[1]

7.27.4 Experimental Disease Models

The feasibility of culturing *P. falciparum* in human erythrocytes under experimental conditions has helped in developing protocols for evaluating antimalarial drugs in vitro.[161] As one progresses toward studies in animal models, such as rodents or primates, the species specificity of malaria parasites becomes an obstacle. However, murine plasmodium models, especially those involving *P. berghei*, *P. yoelii*, *P. chabaudi*, and *P. vinckei*, are useful in the preliminary evaluation of antimalarial drug candidates. Of these, *P. berghei*, and to a lesser extent *P. chabaudi*, are generally used in efficacy studies. The method involves comparing the parasitemia and survival times in treated and untreated mice, after four daily doses of various drug candidates.[161] While studying the efficacy of chemotherapeutic agents, it is important to note that biological characteristics of parasites, such as the duration of the schizogonic cycle, time of schizogony, and synchronous or asynchronous nature of their development, can influence drug sensitivity.[133] Some biological characteristics of plasmodium species used in rodent models are given in **Table 2**.[133]

If the drug acts at a particular stage in the life cycle of the parasite, its overall sensitivity would depend on the parasitic stages present at the time of the treatment. So, *P. chabaudi* and *P. vinckei*, which develop synchronously in blood, are generally more sensitive compared to those like *P. yoelii* and *P. berghei*, which develop asynchronously. In chronotherapy, the time of drug administration is adjusted such that its concentration in blood reaches maximum when the parasites of their sensitive stage are present in circulation. Studies have shown that antimalarial drugs, such as chloroquine, quinine, and mefloquine (see below) exert maximum effect on the midterm trophozoite stage. Similarly, parasites at the ring stage or young trophozoite stage are more sensitive to arteether, whereas dividing schizonts are sensitive toward pyrimethamine.[162,163]

Table 2 Biological characteristics of various parasite species used in rodent models

	P. yoelii	*P. berghei*	*P. vinckei*	*P. chabaudi*
Subspecies/strains	*P. yoelii yoelii* *P. yoelii nigeriensis* *P. yoelii killicki*	K173 ANKA NK65 SP11	*P. vinckei vinckei* *P. vinckei petteri* *P. vinckei lentum* *P. vinckei bruce chwatti*	*P. chabaudi chabaudi* *P. chabaudi adami*
Pre-erythocytic schizogony	~45 h	~50 h	~60 h	~54 h
Schizogony	~18 h, asynchronous	~21 h, asynchronous	~24 h, synchronous	~24 h, synchronous
Species and chloroquine sensitivity index	*P. y. killicki 194ZZ* (10) *P. y. yoelii 265BY* (10) *P. y. nigeriensis* (25)	*P. b. ANKA* (3) *P. b. NK65* (5)	*P. v. petteri 106HW* (1) *P. v. lentum 194 ZZL* (3) *P. v. vinckei* (3)	*P. c. adami 887KA* (1) *P. c. chabaudi 864VD* (3)
Host cell preference	*P. y.* shows affinity of reticulocytes	ANKA reticulocytes; NK 65 all cells	*P.v.v* and *P.v.p* normocytes	Normocytes and reticulocytes

Aotus and *Saimiri* monkeys have emerged as suitable experimental models to study human malaria and are useful in the preclinical evaluation of antimalarial drugs and vaccine candidates.[133,164] Some of the early developments in this area include the induction of *P. falciparum* infections in *Aotus* monkeys,[165] infection of *Aotus trivirgatus* and splenectomized *Saimiri sciureus* with *P. vivax*,[166,167] and the infection of *Aotus* monkeys with *P. malariae*.[168] Although various plasmodium species exhibit comparable life cycles in humans and other phylogenetically related primates, it should be noted that these experimental models cannot reproduce all the clinical features of human malaria. So, one should be careful in drawing conclusions from such studies, as the disease characteristics are dependent on host karyotypes, phenotypes, and parasitic strains. However, nonhuman primate models have proven to be useful in understanding disease characteristics, such as malaria-induced anemia[169] and pathogenesis of cerebral malaria,[170,171] and immune response.[172–175]

7.27.5 Current Treatment

7.27.5.1 Commonly Used Antimalarial Agents

Antimalarial agents can be broadly classified into three groups based on the stage of plasmodium life cycle where they act: (1) tissue schizonticides which prevent the development of liver stage parasites, (2) blood schizonticides which act at the intraerythrocytic stage, and (3) gametocides that can inhibit the development of sexual forms of the parasite in blood and prevent their transmission to mosquitoes. It is possible that the parasite may be sensitive toward a drug at more than one stage during its development.[176] The following sections describe commonly used antimalarial drugs, their limitations, and advantages. New lead compounds identified based on existing drugs or by methods such as high throughput screening or computational design are also mentioned at appropriate places.

7.27.5.1.1 Quinolines

Quinine (**1**), an alkaloid first isolated from the bark of the cinchona tree in Peru, has been in use as an antimalarial drug since the eighteenth century.[177] Even though the widespread use of this drug is limited due to toxicity, bitter taste, and adverse effects, such as nausea, tinnitus, and deafness, quinine served as a starting point to design new drugs with a better pharmacological profile.[178,179] These include (1) compounds in the 4-aminoquinoline series, such as chloroquine (**2**) and amodiaquine, (**3**) and the acridine derivative pyronaridine (**4**); (2) aryl-amino alcohols related to quinine, like mefloquine, halofantrine, and lumefantrine; and (3) 8-aminoquinolines exemplified by primaquine and tafenoquine.

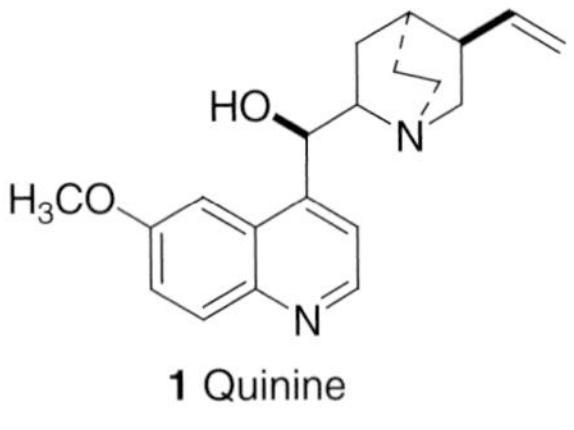

1 Quinine

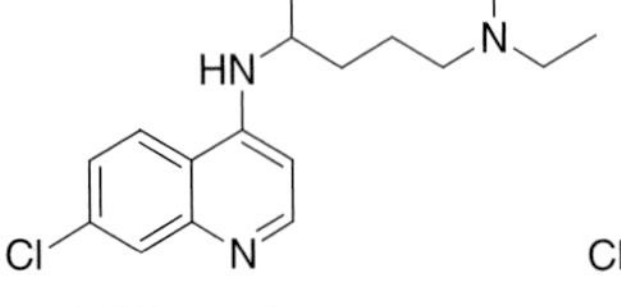

2 Chloroquine

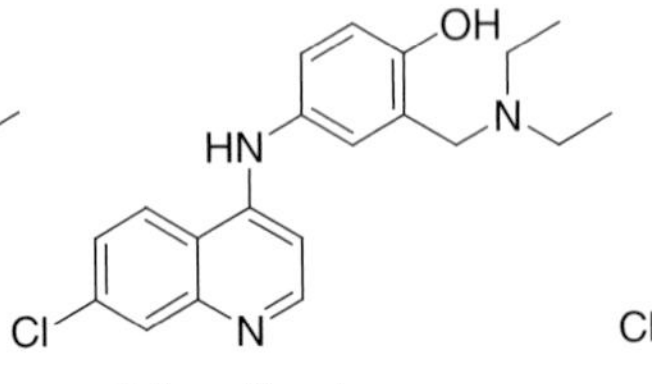

3 Amodiaquine

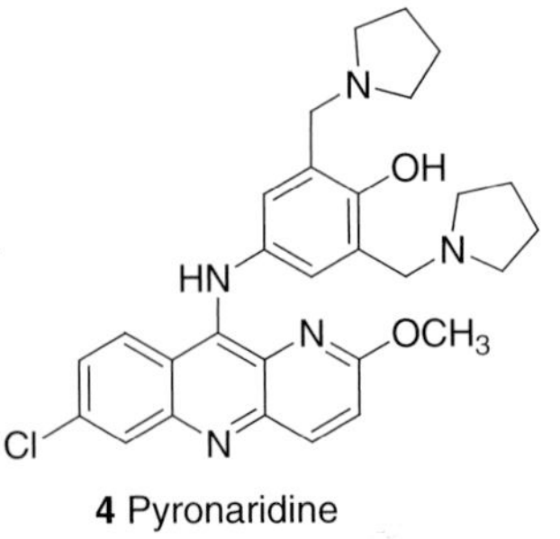

4 Pyronaridine

7.27.5.1.1.1 4-Aminoquinolines

Chloroquine (CQ, **2**) was developed as a result of intense antimalarial drug development efforts in the USA during World War II, but the compound was familiar to Germans as early as 1934 under the name resochin.[176] The safety, efficacy, and low cost brought chloroquine to the front lines to treat malaria, and it was used extensively for almost two decades after its first introduction in 1944–45 – until the parasites developed resistance in the 1960s. Amodiaquine (AQ, **3**) is structurally related to CQ and is active against drug-resistant strains of *P. falciparum*.[180,181] Even though it is more effective in parasite clearance than CQ, the clinical use of amodiaquine has been limited due to hepatotoxicity, agranulocytosis, and cross-resistance with CQ.[182] Pyronaridine (**4**), an acridine derivative having resemblance to CQ and AQ, was first developed in China in 1970 and has proven to be very effective against all four *Plasmodium* species affecting humans, including drug-resistant strains.[183–185]

Members of the quinoline family in general exert their effect during the intraerythrocytic phase of the *Plasmodium* life cycle where the parasites show tremendous increase in metabolic activities and make use of host cell constituents for their biosynthetic needs.[186] Hemoglobin catabolism, which occurs within the digestive food vacuoles, is one of the important pathways by which these parasites acquire amino acids. The involvement of three classes of enzymes, namely plasmepsins, falcipains, and falcilysin, has been implicated in this process, and each has gained attention as important chemotherapeutic targets (see below). As the redox-active heme moieties generated during hemoglobin degradation are toxic, the parasites biomineralize them to nontoxic hemozoin (malaria pigment). The ability of chloroquine to inhibit hemozoin formation suggests that this and related compounds may be interfering with the heme-detoxification process, making the parasites susceptible to oxidative stress by heme.[187,188] The exact molecular details of this interference have been the subject of much discussion, and studies over the last several years tend to show that inhibition of hemozoin formation may either be due to the direct complexation of quinolines with hematin (hydroxyferriprotoporphyrin IX), an autooxidation product of heme, or due to a capping effect whereby the drug binds to the growing face of the hemozoin crystal, thus preventing its growth.[189] The ability of members of this class to interfere with heme binding to histidine rich protein II (HRP-II), a protein involved in hemozoin formation,[190] and a recent report showing chloroquine binding with lactate dehydrogenase[191] point toward the possible existence of additional biological targets.

After the emergence of parasites that are resistant to chloroquine, a number of structure–activity relationship (SAR) studies were initiated to understand the stereoelectronic factors that are essential for the observed antiplasmodial action and those characteristics that contribute to parasitic resistance.[186,192] The following general conclusions could be derived from these studies.

1. The weak base property of chloroquine allows it is diffuse through plasma as well as vacuolar membranes. Its protonation under the acidic conditions of the food vacuole traps the molecule inside, leading to accumulation.
2. The 4-amino quinoline nucleus is essential for complexation with hematin; but this alone is not sufficient for the inhibition of hemozoin formation. Simple quinoline or its 3-, 5-, 6-, or 8-amino derivatives do not form noticeable complexes with hematin, whereas its 2- and 4-amino substituted analogs do, and the major component of their stability arises from π–π interactions with the porphyrin system.[193]
3. The aminoalkyl side chain in chloroquine helps in the accumulation of the drug inside the food vacuole and assists in the complexation of the quinoline nucleus with the porphyrin system. Modification of this side chain either by varying its length or attaching new chemical groups can circumvent chloroquine resistance.[186,194] Although this is not a permanent solution to deal with resistance, such modifications have provided a number of interesting compounds with favorable therapeutic profiles, some of which are presented in **Table 3**.
4. The presence of chlorine at the 7-position is essential for the inhibition of hemozoin formation and its replacement with other halogens, such as iodine or bromine, do not significantly alter the biological activities of these compounds. Substitution of hydrogens in the quinoline ring with other groups influence the pK_a of the ring as well as the side chain nitrogens and may indirectly affect the stability of the hematin–drug complex.[197,201,202]

An overview of various factors described above is pictorially presented in **Figure 3**.

Since toxic side effects limited the use of amodiaquine, there have been several attempts to understand the molecular basis for this toxicity and to develop better candidates devoid of adverse effects. Available evidences indicate that the quinone-imine intermediate **9**, formed as a result of metabolism of AQ in liver, alkylates various biological targets and is responsible for the toxicity.[203]

Table 3 Various chloroquine analogs having improved activities against resistant strains of the parasite

Compounds	*Biological characteristics*
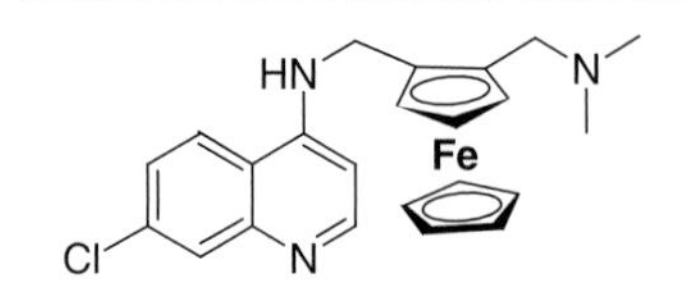 **5** Ferroquine (as tartrate)	Ferroquine (**5**) is ∼22 times more potent than chloroquine against resistant strain of *P. falciparum* in vitro. After a 4-day in vivo test in mice infected with *P. berghei* (NS), only 20% showed recrudescence when observed for 60 days, whereas all mice treated with CQ showed recrudescence.[195,196]
$NH(CH_2)_nNEt_2$ Cl N **6**	Analogs such as **6**, with shorter side chains ($n=2$–3) or larger chains ($n=10$–12), are almost 10 times more potent than CQ in vitro against resistant strains.[197]
HN N Cl N **7**	Compound **7** showed an in vitro IC_{50} value of 49 ± 14 nM (compared to 315 ± 82 nM for CQ) against resistant strains of *P. falciparum*. However, this and related analogs with shorter side chains in general showed low in vivo efficacy and cross resistance with CQ.[198]
HN R NH Cl N N Cl **8** Bis-quinoline R=*trans* 1, 2-cyclohexyl	Bis-quinoline (**8**) showed an IC_{50} value of 1.4 nM against W2 clone of *P. falciparum*, (relative to 100 nM for CQ) and 100% cure when tested in vivo against *P. berghei* at 320 mg kg^{-1} dose.[199,200]

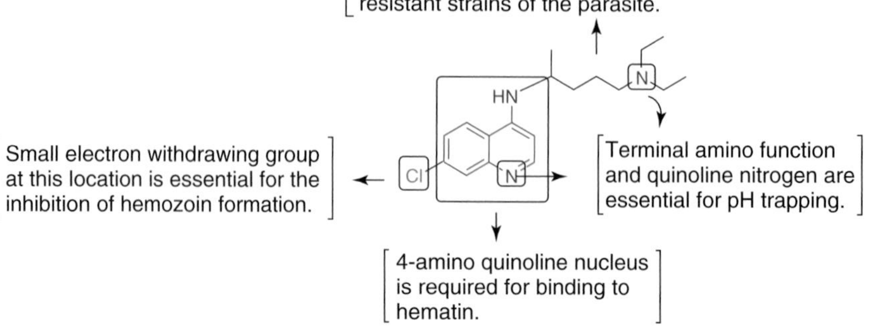

Figure 3 Structural features of chloroquine that contribute to its biological activity.

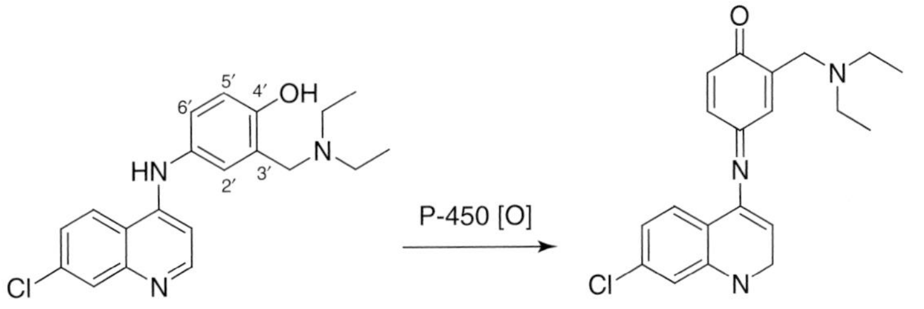

P-450 [O]

3 Amodiaquine

9 Amodiaquine quinone imine

Introduction of various groups at the 3′ and 5′ positions of the amodiaquine side chain was initially considered as a strategy to increase the lipophilicity of drugs and to reduce the cross-resistance which normally arises after side chain metabolism.[204,205] Several compounds in this series have been synthesized and analyzed (e.g., **10–12**).[206] Even though these compounds are more potent than AQ in vitro and in vivo, toxicity remains a problem.[207] In an elegant approach by O'Neill *et al.*, a number of AQ analogs were synthesized by interchanging the position of hydroxy and diethyl-aminomethyl groups and were evaluated for antimalarial potencies and toxicities.[208] It was assumed that the formation of the quinone-imine intermediate is electronically unfeasible in such systems, which, at the same time, possess necessary groups to interact with a biological target. This strategy has given very promising results in initial studies. Thus, compound **13** (isoquine) was found to be more potent than CQ and AQ without any signs of toxicity.

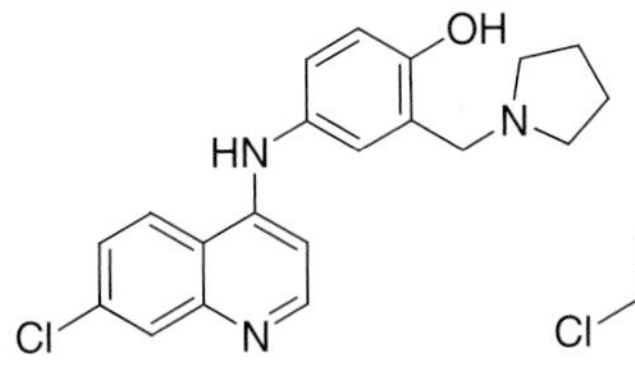

10 Amopyroquine

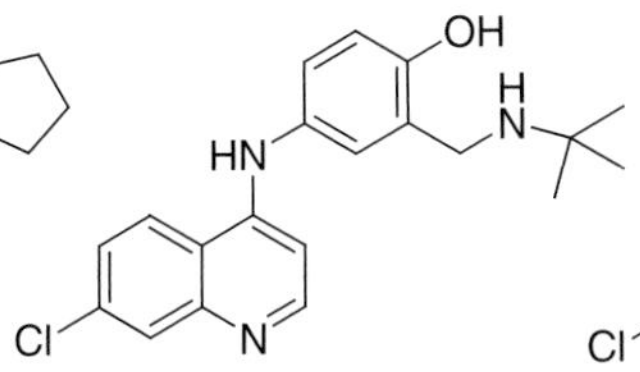

11 tert-Butylamodiaquine

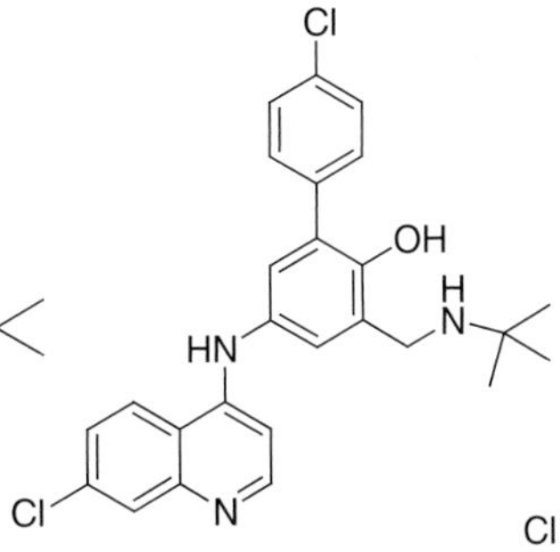

12 Tebuquine

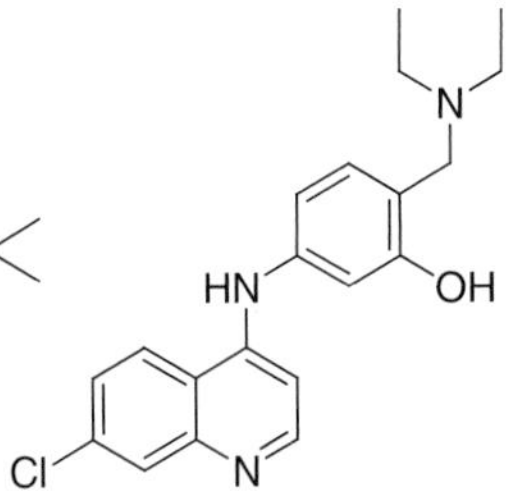

13 Isoquine

7.27.5.1.1.2 Quinoline-methanols and aryl alcohols

Mefloquine (**14**), commercially known as Lariam, is a structural analog of quinine and was developed by the Walter Reed Army Institute in the 1960s.[209] Although mefloquine is useful prophylatically, its long half-life (2–3 weeks) has contributed to resistance. Because of reports indicating possible side effects, such as insomnia, anxiety, vivid dreams, and visual disturbances, mefloquine is not recommended for people suffering from depression, anxiety, and other major psychological disorders.[210]

Halofantrine (**15**) is another therapeutic agent developed by the Walter Reed Army Institute which is active against chloroquine-resistant *Plasmodium* parasites.[211,212] However, cardiotoxicity has limited its usefulness as a drug.[213] Recent structure–toxicity relationship (STR) studies using halofantrine and its metabolite have shown that its N-desbutyl derivative is equally potent and is devoid of cardiotoxicity.[213,214] A structurally related molecule, lumefantrine (**16**), originally synthesized by the Institute of Military Medical Sciences in Beijing in the 1970s is a potent antimalarial agent and has been shown to be capable of acting synergistically with artemether (see below). Their combination, known as riamet or coartemether (Coartem), has been introduced for the treatment of uncomplicated *P. falciparum* malaria in a number of countries.[215] Although lumefantnine is safe, there are concerns about its possible cross-resistance with mefloquine, which needs to be addressed to maintain the therapeutic potential of this combination.[216] Like halofantrine, the N-desbutyl derivative of lumefantrine, is ~4 times more potent compared to the parent compound.[217]

14 Mefloquine **15** Halofantrine **16** Lumefantrine

7.27.5.1.1.3 8-Aminoquinolines

The first member in this class, pamaquine (**17**), was developed by German researchers in 1925 and was given special attention because of its efficiency against liver stage parasites.[218] This was followed by the synthesis of primaquine (**18**), an N,N-deethylated analog of pamaquine, capable of targeting pre-erythrocytic stages of *P. falciparum*, *P. vivax*, and *P. ovale*, including hypnozoites, with excellent prophylactic activity.[219,220] Even though it is not active against blood stage parasites, primaquine has been shown to inhibit the maturation of fertile gametocytes.[221] Because of toxicological concern, this compound was subjected to further structural optimization, which led to the development of tafenoquine (**19**) with a longer plasma half-life (2–3 weeks), less toxicity, and improved activities against both liver and blood stage parasites.[222,223] Its mode of action against erythrocytic parasites is believed to be similar to that of 4-aminoquinolines,[224] but an alternate mechanism involving interference with mitochondrial function could be involved in its activity against liver stage parasites and gametocytes.[225]

17 Pamaquine **18** Primaquine **19** Tafenoquine

The main drawbacks of primaquine include its oxidative conversion to carboxy primaquine (**20**)[226] and toxic effects such as hemolytic anemia in patients with G6PD deficiency.[227–229] A recent report by Jain *et al.* shows that introduction of bulky substituents at the 2-position of primaquine can lead to drug candidates with less toxicity, improved metabolic stability, and good blood schizonticidal activity (e.g., **21**).[230] In another study, a double prodrug approach was used by Moreira and co-workers to improve the metabolic stability of primaquine. Thus, by masking the terminal nitrogen by making it part of an imidazolidin-4-one system, the authors were able to identify new candidates (e.g., **22**) with improved gametocidal activities.[231]

20 **21** **22**

7.27.5.1.1.4 The resistance to quinolines

The resistance developed by *P. falciparum* to quinolines in general and chloroquine in particular has greatly increased the mortality rates in malaria endemic areas.[232] Even though *P. vivax* and *P. malariae* remained sensitive to CQ for a very long time, the past decade has witnessed the emergence of resistant strains which could adversely affect malaria control efforts in this new century.[233,234] One of the main characteristics of resistant strains is the reduced accumulation of chloroquine inside the food vacuole compared to that in sensitive strains.[235] Based on genetic and biochemical studies, a number of theories have been put forward to explain the molecular basis of chloroquine resistance, which include (1) the involvement of an altered Na^{+}/H^{+} exchanger which causes an increase in cytoplasmic pH and decrease in CQ uptake,[236] (2) amplification of the *Pfmdr* 1 gene which results in the overexpression of Pgh-1 protein, causing drug efflux, similar to that operative in multidrug resistant cancer cells, and (3) mutations in a food vacuole membrane transporter known as *P. falciparum* chloroquine resistance transporter (PfCRT). Even though more evidence is needed to support the Na^{+}/H^{+} exchanger theory, recent studies by Reed *et al.* have shown that mutations in Pgh-1 can not only contribute to CQ resistance by decreasing the drug accumulation, but can also modulate resistance to other dugs such as mefloquine, halofantrine, and quinine.[237] However, it has been suggested that mutations in Pgh-1 alone may not be sufficient to confer CQ resistance, but may also require mutations in other genes. More recently, studies by Sidhu *et al.* have clearly demonstrated the association of mutations in PfCRT with CQ resistance.[238] This protein is localized in the digestive vacuolar membrane and is believed to contain 10 transmembrane domains. Mutations in this protein, particularly a Lys76→Thr (K76T) change in the first transmembrane domain, has shown to be associated with CQ resistance. This change essentially results in the loss of a positive charge from the putative transporter pore, which allows the diprotonated CQ to escape from the food vacuole, thereby reducing its concentration inside.

7.27.5.1.2 **Artemisinin and its analogs**

Artemisinin (**23**), a sesquiterpene endoperoxide isolated from the plant *Artemisia annua,* is exceptional because of its activity against drug-resistant strains of *P. falciparum*. The medicinal value of herbal extracts from this plant was known to the Chinese for centuries and is well documented in various books such as *Zhou Hou Bei fi Fang* (The Handbook of Prescriptions for Emergencies) written by Ge Hong (AD 281–340) and *Ben Cao Gang Mu* (Compendium of Materia Medica) compiled in 1596.[239] After its structure elucidation in 1979,[240] a large number of synthetic efforts were initiated to improve the potency of artemisinin and address some of its drawbacks such as short plasma half-life, limited bioavailability, and poor solubility in oil and water. Consequently, reduction of artemisinin gave dihydroartemisinin (**24**), which is about seven times more potent than the parent compound in vitro.[241] Simple chemical modifications of this lactol led to the development of clinically useful first-generation artemisinin derivatives such as artemether (**25**),[242] arteether (**26**),[243,244] and artesunate (**27**).[245,246]

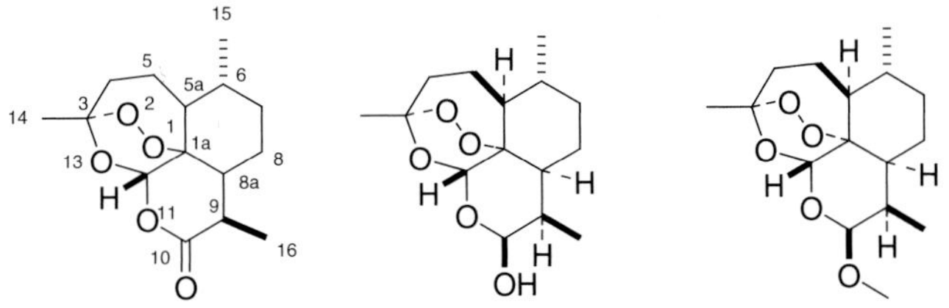

23 Artemisinin **24** Dihydroartemisinin **25** Artemether

Artemether and arteether are oil-soluble drugs and are well absorbed when administered intramuscularly, whereas artesunate is water-soluble and generally given through intravenous route.[247] These analogs are safe, fast-acting, and effective against uncomplicated *P. falciparum* malaria, severe malaria, and blood stage *P. vivax* infections, and have gametocidal activities. Since recrudescence is a problem when used as monotherapy, these compounds are generally given in combination with other long-acting antimalarial agents (see below).[248] Efforts to improve the hydrolytic stability of artesunate led to the identification of relatively more stable and less toxic artelinic acid (**28**), which is currently in the developmental stage.[249] Related analogs of artesunate with aliphatic carboxylic acids linked at C10 through an ether linkage, however, were not as active as artesunate or artelinic acid.[250]

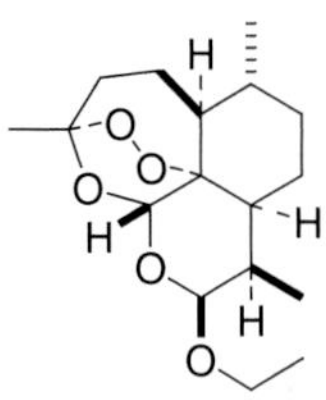

26 Arteether 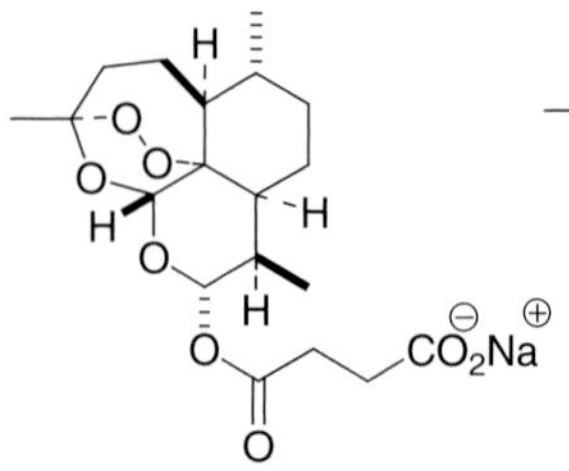 **27** Sodium artesunate 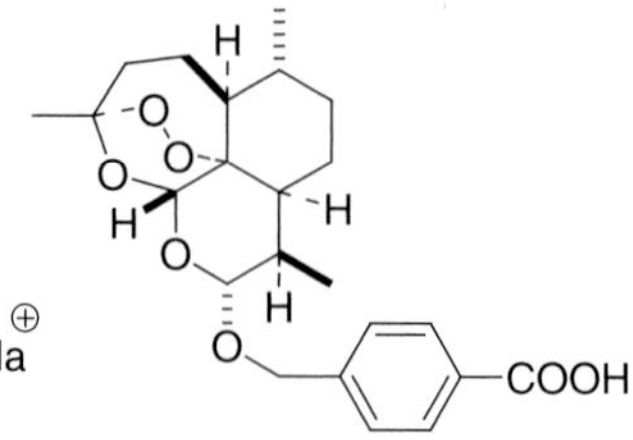 **28** Artelinic acid

Many of these artemisinin analogs, although potent, possess acid-labile acetal functions and have short biological half-lives.[251] In addition, they are metabolized to dihydroartemisinin, which is known to induce neurotoxicity in animal models.[252,253] Even though similar toxicity has not yet been substantiated in humans,[254] the main focus during the last decade has been on development of new artemisinin analogs, either by introducing chemical groups that confer stability to the acetal functionality or by using a nonacetal type linkage at C10 which would resist metabolism to dihydroartemisinin. Representative examples from these studies are presented in **Table 4**. In addition to derivatization at C10, there have been several other significant efforts to understand the effect of chemical modification at positions in artemisinin such as C3, C11, and C9 on its antimalarial potency. These analogs were prepared either by total synthesis or by semisynthetic methods. Studies in this direction have not only led to the identification of new series of artemisinin analogs with outstanding activities, but have also helped to conduct quantitative SAR (QSAR) studies, which would be helpful in the design of next-generation artemisinin analogs. Even though antimalarial potency relative to existing drugs is a major criterion during the selection of new analogs, it is important to consider their cost of production, since affordability remains one of the major factors which determine the success of malaria treatment in underdeveloped countries.

Although various structural modifications of artemisinin and their effects on antimalarial potencies have been investigated,[266,267] derivatization at positions C4 to C8 remained a challenge to chemists because of the difficulty in the introduction of functional groups by usual synthetic operations. However, fungal fermentation techniques have recently emerged as a promising alternative to generate hitherto inaccessible derivatives (see structures **41–46**).[268–270] Preliminary studies from our laboratory have shown that many of these hydroxylated analogs can be chemically modified to afford a new series of highly potent antimalarial drug candidates.

The mechanism of action of artemisinin is believed to involve an initial interaction of the endoperoxide group with heme (generated as a result of hemoglobin degradation), which leads to the formation of free radical intermediates capable of alkylating various biological targets.[271,272] Various steps involved in this pathway are schematically represented in **Figure 4**.

Initial efforts to identify the biological target of artemisinin using radiolabeled dihydroartemisinin have shown that this class of compounds, on activation, get covalently attached to proteins such as translationally controlled tumor protein.[273] Related studies have also shown that artemisinin can interfere with the heme polymerization process by complexing with heme[274] and is also capable of forming covalent adducts with reduced glutathione under experimental conditions.[275] More recently, Eckstein-Ludwig and co-workers have demonstrated that PfATP6, a SERCA type Ca^{2+} ATPase in *P. falciparum*, is an important target of artemisinin and its analogs, and that the peroxide group in the molecule and an iron source is important for its activity against PfATP6.[276] While the possible existence of multiple targets explains the delay in resistance development, it is logical to think that the overall antiparasitic activities of the artemisinin class of compounds could be influenced by various factors including their uptake, availability at the target sites, and activation.

Based on results from the mechanistic investigations with artemisinin analogs, particularly those showing the importance of the peroxide group in their antimalarial potency, a number of studies were simultaneously pursued to see whether simple synthetic peroxides could show similar antiplasmodial action. Various developments in this area have recently been reviewed by Tang *et al.*[277] Important prototypes belonging to this class include 1,2,4-trioxanes, 1,2,3-trioxanes, and tetraoxanes, representative examples from which are presented in **Table 5**.

7.27.5.1.3 Protease inhibitors

Proteases represent an important class of biomolecules which take part in a myriad of functions including digestion, growth, differentiation, cell signaling, wound healing, immunological responses, and apoptosis.[284] As mentioned earlier, malaria parasites depend on hemoglobin catabolism for the supply of amino acids. This process takes place within the digestive food vacuole of the parasite and is brought about by the action of a number of proteases, which belong to three major classes, namely aspartic, cysteine, and metalloproteases. The following section endeavors to summarize recent progress in the design and development of drug candidates that can target enzymes involved in the hemoglobin degradation pathway.

Table 4 Artemisinin analogs

Compound	*Biological characteristics*
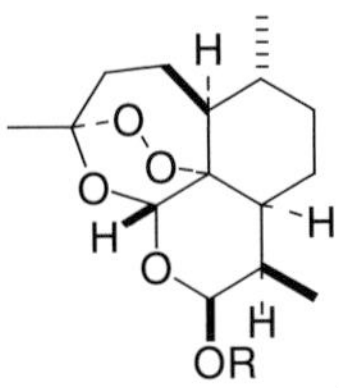 **29** R 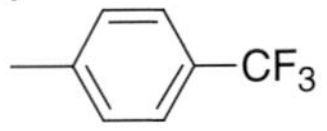 = —(C$_6$H$_4$)—CF$_3$	**29** is orally active with an ED_{50} (*P. berghei*) value of 2.7 mg kg^{-1} (cf. sodium artesunate = 4.0 mg kg^{-1}). There was no sign of metabolism to dihydroartemisinin.[255]
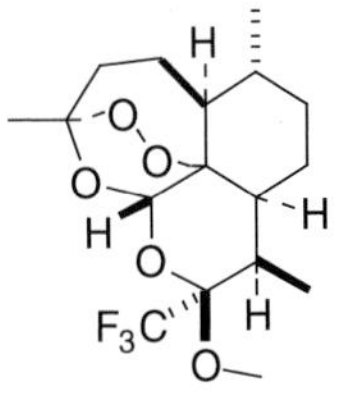 **30**	**30** is ~35 times more stable than artemether. ED_{50} (*P. berghei*, intraperitoneal route) = 1.25 mg kg^{-1} (cf. artemether = 2.5 mg kg^{-1}). There was complete parasite clearance after a 4-day treatment in mice infected with *P. berghei*, with no recrudescence up to 25 days.[256]
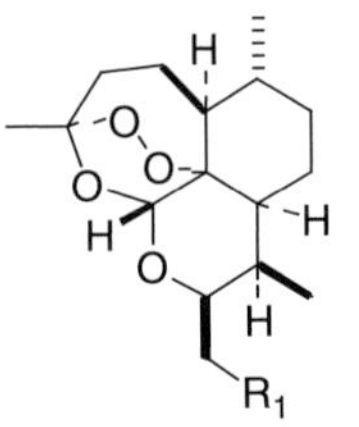 **31** $R_1 = -CH_2OC(O)CH_2CH_2-CO_2H$	**31** showed an IC_{50} (*P. falciparum*, W2 clone) value of 0.4 nM (cf. artemisinin = 10 nM).[257]
32 R_1 = 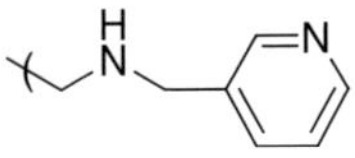	**32** showed an IC_{50} (*P. falciparum*, W2 clone) value of 1.0 nM (cf. artemisinin = 10 nM).[257]
33 R_1 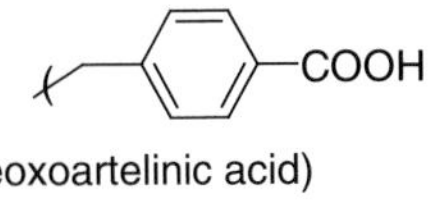 = —COOH (Deoxoartelinic acid)	**33** is ~23 times more stable than arteether. Its ED_{50} (*P. falciparum*, K1 strain) = 0.6 ng mL^{-1} (cf. artelinic acid = 1.4 ng mL^{-1}) and showed ~98.7% suppression of parasitemia in *P. chabaudi* infected mice (10 mg kg^{-1} day^{-1}, intraperitoneally × 4), with 100% survival rate, compared to 60% for arteether.[258]
34 R_1 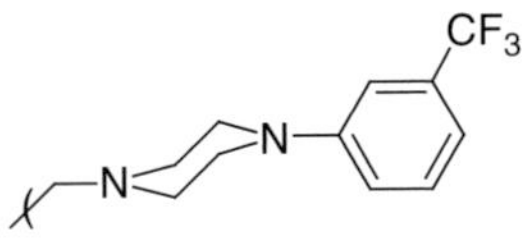 = —CF$_3$	Compound **34** showed an ED_{50} (*P. berghei*, oral route) value of 3.12 mg kg^{-1} (cf. artemether = 6.02 mg kg^{-1}).[259]
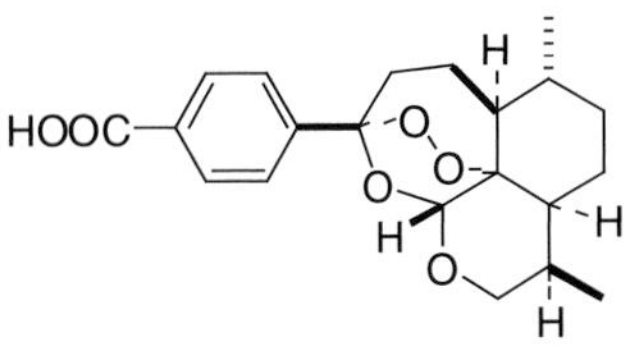 **35**	**35** is thermally stable and showed very good aqueous solubility. ED_{50} (*P. berghei*, oral route) = 15 mg kg^{-1} (cf. artelinic acid = 9.6 mg kg^{-1}).[260]

continued

Table 4 Continued

Compound	*Biological characteristics*
36	Compound **36** is ~21 times more potent than artemisinin (**23**) when tested in vitro against W2 clones of *P. falciparum*. Groups such as Et or ^{n}Pr at C3 or C16 positions of **23** seem to improve the activity; however, sterically demanding groups at these positions led to an overall reduction in potency.[261]
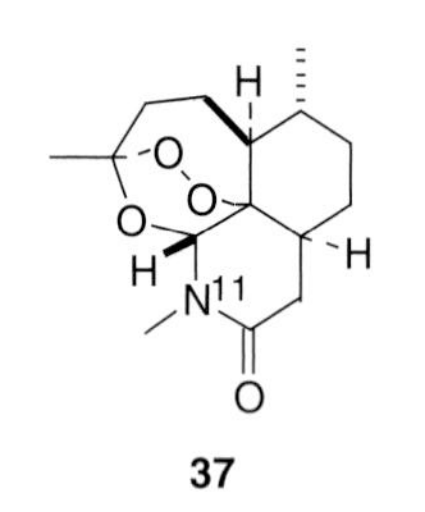 **37**	This compound (**37**) is ~5 times more potent than artemisinin (in vitro) against W2 clones of *P. falciparum*. Short aliphatic carbon chains (C1–C3) at N11 seem to slightly improve or retain the activity of artemisinin, but no significant improvement in the potency could be seen on C11 derivatization.[262]
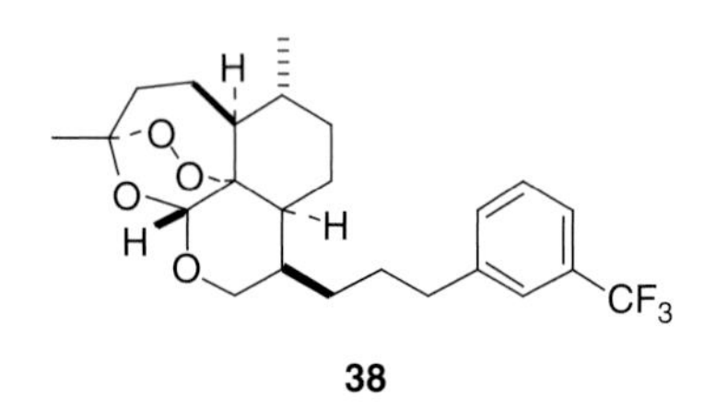 **38**	**38** showed ED_{50} (*P. berghei* N; oral or subcutaneous routes) values of 1.25 or 0.4 $mg\,kg^{-1}$ (cf. sodium artesunate = 2.4 or 0.8 $mg\,kg^{-1}$).[263]
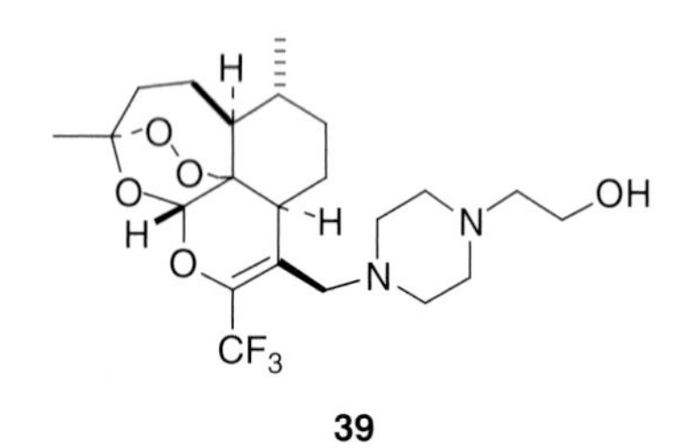 **39**	**39** led to 100% reduction in parasitemia by day 4, on oral as well as subcutaneous administration in *P. berghei* infected mice ($ED_{90} = <10\,mg\,kg^{-1}$). It is more potent than artesunate and showed ~25 times more oral bioavailability than artemether.[264]
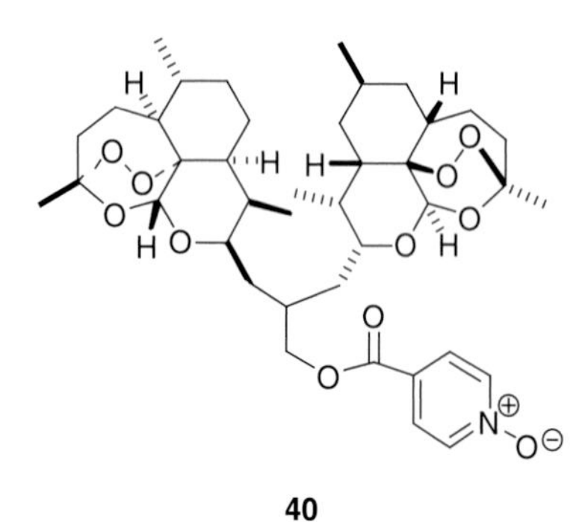 **40**	The EC_{50} of **40** against CQ sensitive strain of *P. falciparum* (NF54) = 0.53 nM (cf. artesunate = 1.5 nM). The compound showed good in vivo efficacy both via intravenous and oral routes with ED_{50} ($mg\,kg^{-1}\,day^{-1} \times 4$) = 0.8 (intravenous) and 2.0 (oral), relative to 5.5 (intravenous) and 8.0 (oral) for artesunate.[265]

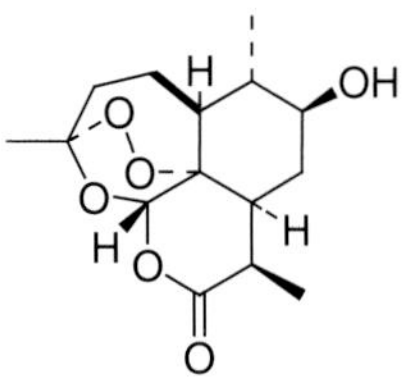

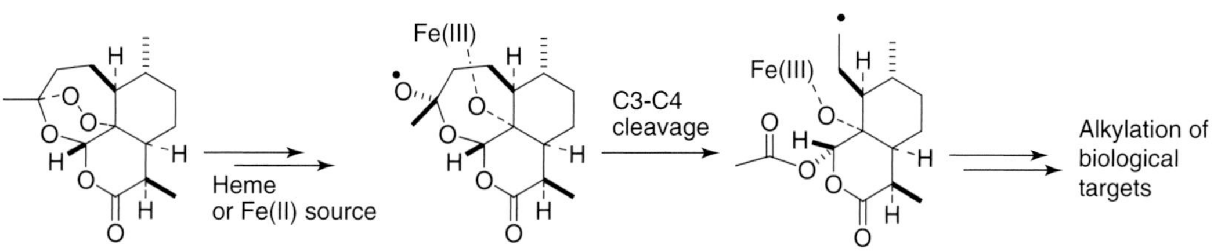

Figure 4 Mode of action of artemisinin.

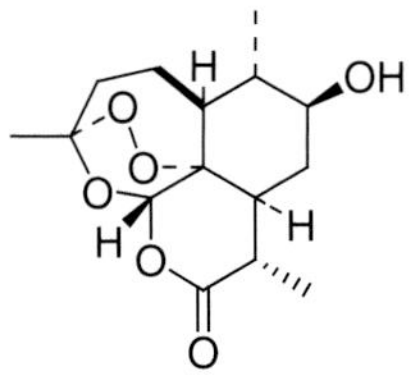

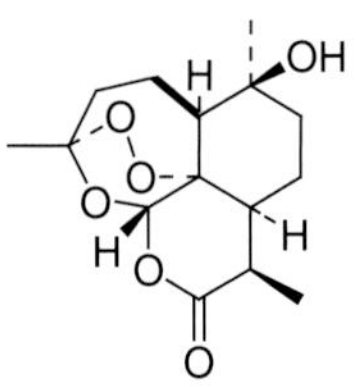

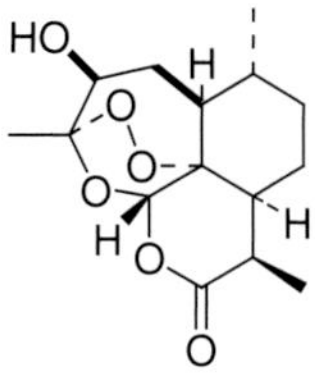

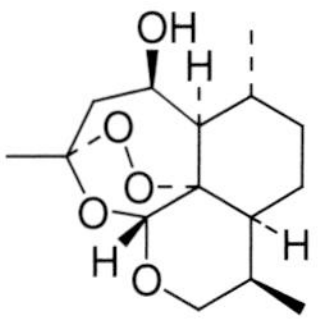

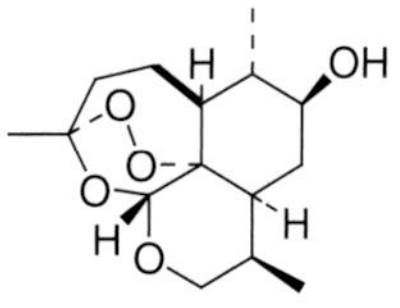

7.27.5.1.3.1 Aspartic proteases: the plasmepsins

Available data indicate that *P. falciparum* genome codes for at least 10 aspartic proteases, of which four, associated with the food vacuole, have been identified and biochemically characterized.[285,286] These enzymes, collectively known as plasmepsins, take part in the initial breakdown of hemoglobin into smaller fragments. The first two members, plasmepsins I and II (Pfpm I and Pfpm II), share 73% sequence identity and mediate the first cleavage of the Phe33-Leu34 peptide bond in the hinge region of the alpha chain, making the molecule susceptible for further degradation.[287] Plasmepsin IV and a histo-aspartic protease (HAP) are the next members in this series and share about 60% sequence identity with the former two. Plasmepsins I, II, and IV possess two aspartates in their active sites, whereas one of these aspartates is replaced by a histidine residue in the case of HAP. Sequence analysis has shown that cathepsin D is the closest human analog related to the plasmepsins and has a 35% sequence identity with Pfpm II.[288] Even though all these members are needed for the breakdown of hemoglobin, recent gene knockout studies involving *P. falciparum* show considerable functional redundancy among these enzymes.[285]

Of the four enzymes, PfpmI and II are the most studied as candidates for target-based inhibitor design. Crystal structures of Pfpm II and IV in complexation with pepstatin A, a known inhibitor of cathepsin D,[289,290] and a homology model of HAP[291] are available and have helped in understanding the active sites and binding modes of these enzymes (**Figure 5**). In addition, recent studies with combinatorial libraries of chromogenic peptide substrates have given very valuable information regarding the active site specificity of plasmepsins.[292] These studies indicate that hydrophobic groups at P1 and P1′ sites, such as phenylalanine or alanine, would be optimal for effective hydrolysis. Based on this and similar studies, a number of inhibitors have been designed, synthesized, and evaluated to target plasmepsins, which are presented in **Table 6**. While structural similarity among these enzymes would help in the design of novel drug candidates that can target two or more members of this family, it is important to incorporate structural features in the design that would improve their selectivity over cathepsin D.

7.27.5.1.3.2 Cysteine proteases: the falcipains

Three papain-like cysteine proteases, known as the falcipains, have been identified from *P. falciparum*, and studies have shown that at least two of them have important roles in hemoglobin degradation.[298,299] They appear to act downstream of aspartic proteases and are new targets for malaria chemotherapy. Of these, falcipain-1, encoded on chromosome 14, was identified in erythrocytic parasites, but studies with this were limited due to low abundance and difficulty in

Table 5 Synthetic peroxides related to artemisinin

Compound	*Biological characteristics*
47 Fenozan BO7	**47** is equally active both orally and subcutaneously ($ED_{50} = 2.5\,mg\,kg^{-1}$) against *P. berghei* in mice model.[278]
48 Arteflene	Arteflene (**48**) showed an ED_{50} value of $2.7\,mg\,kg^{-1}$ subcutaneously and $10\,mg\,kg^{-1}$ orally, and exhibited longer half-life compared to arteether or artesunic acid.[279]
49	**49** was orally efficacious and showed an ED_{50} (subcutaneous route) of $0.43\,mg\,kg^{-1}$ against *P. berghei* (N), relative to $0.5\,mg\,kg^{-1}$ for artemether and $0.9\,mg\,kg^{-1}$ for arteether. No significant toxicity.[280]
50	**50** showed an IC_{50} value of 6.2 nM (cf. artemisinin = 10 nM) against CQ resistant strain of *P. falciparum*. This was inactive orally.[281]
51	**51** was orally active when tested in mice infected with *P. berghei* ($ED_{50} = 20\,mg\,kg^{-1}$ versus $13\,mg\,kg^{-1}$ for artemisinin), with no significant toxicity.[282]
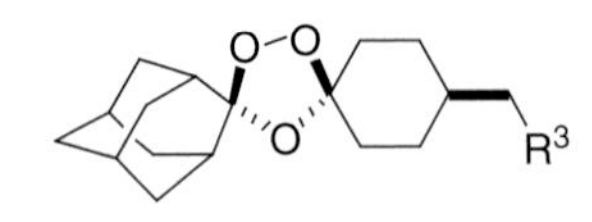 $R^3 = C(O)NHCH_2C(CH_3)_2NH_2$ **52**	**52** showed better potency, oral bioavailability, and prolonged duration of action compared to traditional drugs such as artesunate and artemether and is under clinical development. The compound exhibited prophylactic activity when $R^3 = NH_2$.[283]

Table 6 Plasmepsin inhibitors

Compound	*Biological characteristics*
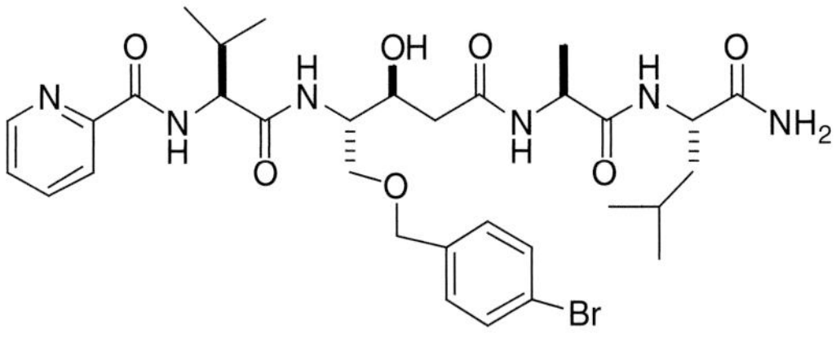 **53**	**53** is one of the lead compounds from studies on statin-like inhibitors. This compound exhibited K_i values of 0.5 and 2.2 nM, respectively, against Pfpm I and II, and caused 51% growth inhibition of *P. falciparum* at 5 μM concentration.[293] The crystal structure of this compound with Pfpm II is presented in **Figure 5**. A number of promising leads related to this molecule were reported recently.[294]
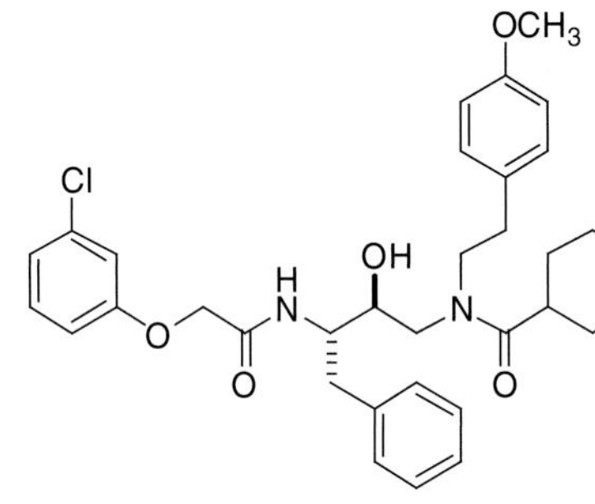 **54**	Iterative optimization using small molecule libraries led to the identification of **54**, which showed a K_i value of 4.3 ± 0.1 nM against Pfm II, relative to 63 ± 12 nM for human cathepsin D (hCatD). However, its in vitro IC_{50} value against *P. falciparum* growth was only 1–2 μM.[288]
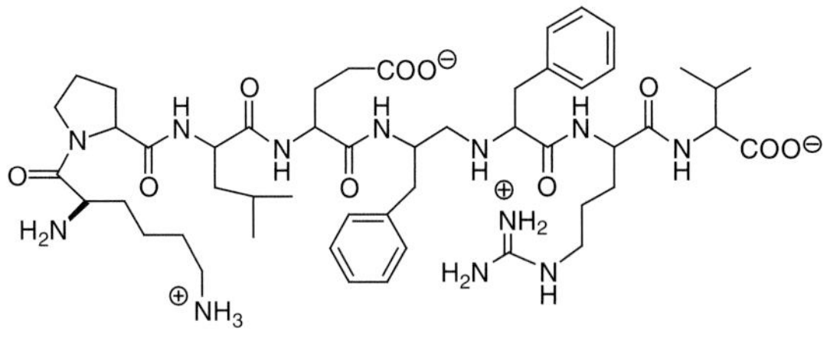 **55**	The K_i of **55** against Pfpm IV = 85 pM and that against Pfpm II = 4.3 μM. The relative value for hCatD is 4.7 nM. In vitro assay against *P. falciparum* showed an IC_{50} value of 58 μM.[292]
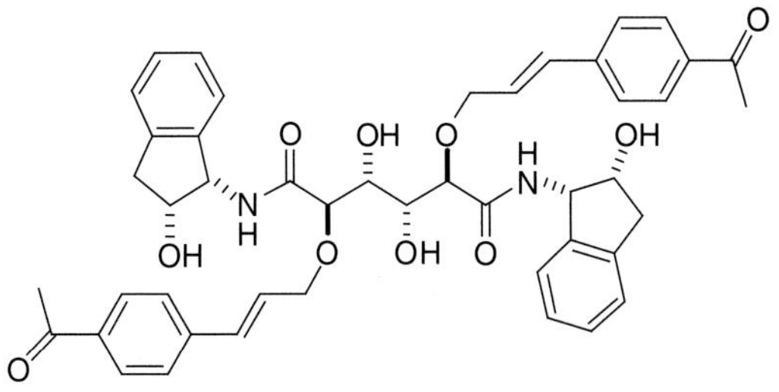 **56**	Compound **56** showed K_i values of 0.8 and 6 nM, respectively, against Pfpm I and II, with no measurable affinity toward hCatD. It caused a growth inhibition of 78% at 5 μM concentration when tested in vitro against *P. falciparum*.[295]
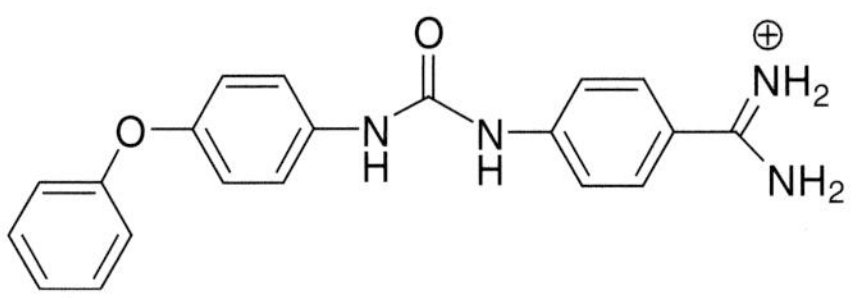 **57**	Compound **57** was identified by library screening at the Walter Reed Army Institute. This inhibited plasmepsins with $K_i = 1–6$ μM, relative to a value of > 280 μM against hCatD, and showed an IC_{50} value of 0.03–0.16 μg mL^{-1} against *P. falciparum* growth in vitro.[296,297]

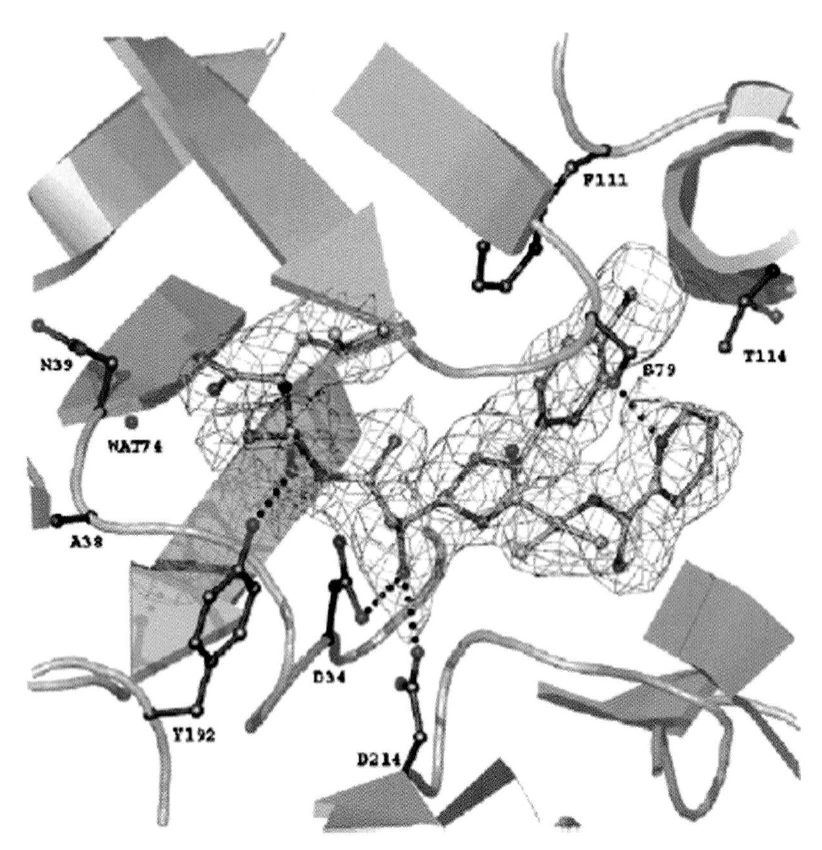

Figure 5 Ribbon diagram showing the binding of compound **53** in the cavity of Pfpm II. Specific hydrogen bonds with side chains of D34, D214, Y192, and S79 are clearly indicated. (Reprinted with permission from Johansson, P.-O.; Lindberg, J.; Blackman, M. J.; Kvarnstroem, I.; Vrang, L.; Hamelink, E.; Hallberg, A.; Rosenquist, S. A.; Samuelsson, B. *J. Med. Chem.* **2005**, *48*, 4400–4409. Copyright (2005) American Chemical Society.)

developing heterologous expression systems. Recent studies by Sijwali *et al.* using recombinant parasites with disrupted *falcipain-1* gene have shown that this enzyme is not essential during erythrocytic development of the parasite.[300] However, similar studies have shown its importance in the development of oocysts in the mosquito midgut.[301] The parasite's genome also codes for two identical copies of falcipain-2 (*falcipain-2* and *falcipain-2'*)[302] and a single copy of *falcipain-3*,[303] of which *falcipain-2* and *falcipain-3* have been isolated, characterized, and proven to be essential for hemoglobin degradation. In addition, their *P. vivax* orthologs, named *vivipain-2* and *vivipain-3*, have also been cloned and expressed in *E. coli* and are found to share some of the structural and functional characteristics with their falcipain counterparts.[304,305] A number of inhibitors against falcipains have been studied over the last few years, and several of them, especially those in the vinyl sulfone and fluoromethylketone series, have given promising results.[306,307] Even though the crystal structures of falcipains are not currently available, their sequence similarity with other members of the family has helped to develop homology models of falcipain-2 and falcipain-3.[308,309] A virtual database screening using these homology models has helped us to identify a number of potential leads with interesting activities against falcipains (**Figure 6**).[310] Some of the important examples from this and similar studies are presented in **Table 7**.

Chalcones are another interesting class of compounds that have shown activity against cysteine proteases (see below). Even though resistance development could be a problem, recent studies by Rosenthal and co-workers using a vinyl sulfone inhibitor suggest that the selection of resistant strains under laboratory conditions is associated with elevations of cysteine protease activity, *falcipain-2* and *falcipain-3* copy numbers, and transcription of the falcipain gene, and not by mutations in these proteins.[311]

7.27.5.1.3.3 Falcilysin

Small stretches of peptide fragments generated by the action of cysteine proteases are then acted upon by a zinc metalloprotease known as falcilysin. This is a relatively large, 1193 amino acid protein belonging to the M16 family of

Table 7 Cysteine protease inhibitors

Compound	*Biological characteristics*
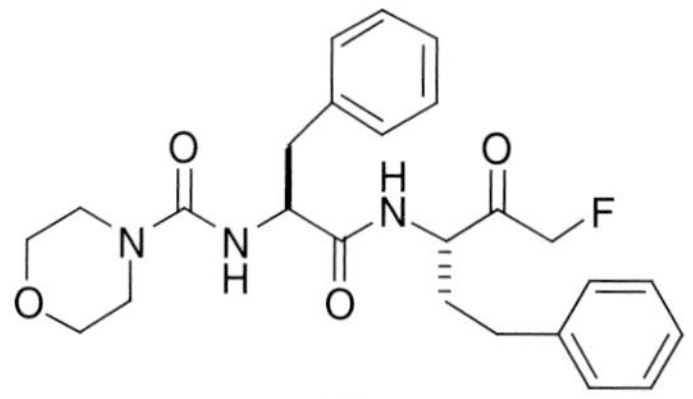 **58**	IC_{50} (*P. falciparum* culture) = 4.0 nM; IC_{50} (falcipain) = 3.0 nM; Compound **58** was able to prolong the survival time of mice infected with *P. vinckei* when studied in vivo.[312]
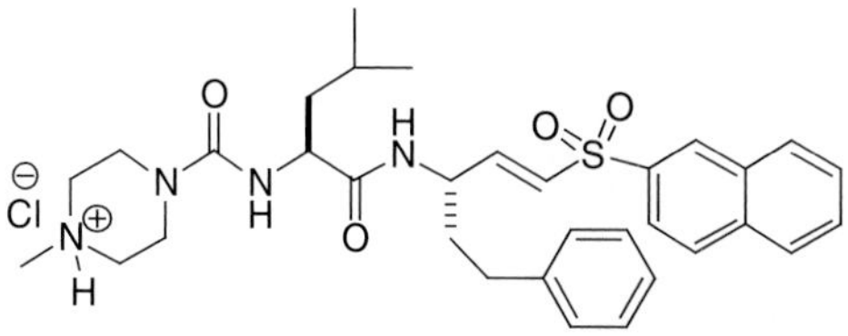 **59**	IC_{50} (*P. falciparum* culture) = 0.4 nM; IC_{50} (falcipain) = 2 nM; when studied in vivo, **59** delayed the progression of murine malaria and cured about 40% of treated animals.[312]
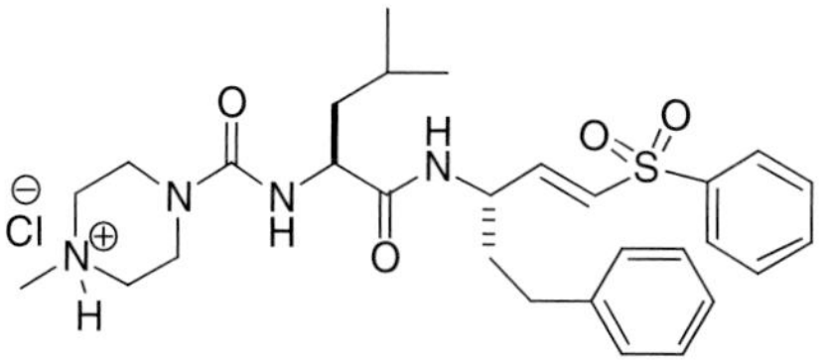 **60**	IC_{50} (*P. falciparum* culture) = 2.0 nM; IC_{50} (falcipain) = 5 nM.[312]
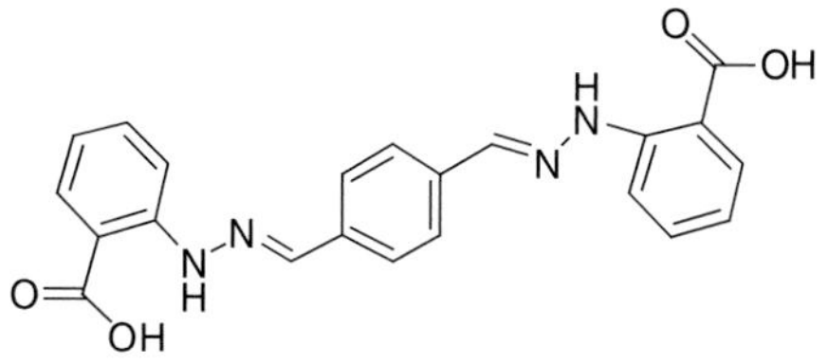 **61**	Compound **61** was identified by virtual screening of a library of ~241 000 compounds from ChemBridge database. This showed IC_{50} values of 1.0 and 4.9 μM, respectively, against falcipain-2 and 3, and exhibited an in vitro EC_{50} value of 41 μM against W2 clones of *P. falciparum*. The docking profile of this in the active site of falcipain II is shown in **Figure 6**.[310]

clan ME zinc metalloproteases and is characterized by an inverted active site motif of HXXEH in comparison with HEXXH found in other members of the family.[313] Studies have shown that falcilysin is active at acidic pH, but shows a pH-dependent substrate specificity. Even though immunofluorescence studies show the localization of this protein inside the food vacuole, there are evidences suggesting its presence in vesicular structures outside this organelle.[314] In-depth structural and functional details of this protein are currently limited, and future studies in this direction are expected to give information useful to develop new drug candidates.

7.27.5.1.4 Inhibitors of the glycolytic pathway

A functional citric acid cycle is absent in plasmodium parasites, and their main energy source is considered to be the ATPs generated during anaerobic fermentation of glucose. To meet the high energy requirement of the parasites, the infected erythrocytes consume 30–100 times more glucose than uninfected ones, and glycolysis is essential for parasite

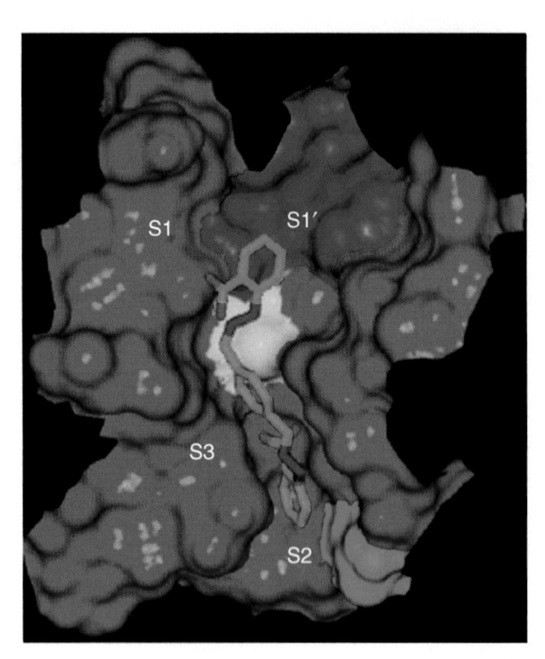

Figure 6 Binding of compound **61** in the active site of falcipain II. The surface of the catalytic cysteine is colored yellow. (Reprinted with permission from Desai, P. V.; Patny, A.; Sabnis, Y.; Tekwani, B.; Gut, J.; Rosenthal, P.; Srivastava, A.; Avery, M. *J. Med. Chem.* **2004**, *47*, 6609–6615. Copyright (2005) American Chemical Society.)

survival.[315] Almost all enzymes involved in this process have been identified from *P. falciparum*, and the crystal structures of three of them, triosephosphate isomerase (PfTIM),[316] lactate dehydrogenase (PfLDH),[317] and fructose-1,6-bisphosphate aldolase (PfALDO),[318] are currently available. Since a number of key differences exist between these enzymes and their human counterparts, many of them have gained attention as possible targets for therapeutic intervention.

During the second stage in glycolysis, fructose-1,6-bisphosphate aldolase catalyzes the conversion of fructose-1,6-bisphosphate to glyceraldehyde-3-phosphate and dihydroxyacetone phosphate. Comparison of the crystal structures of PfALDO and human aldolase A has indicated a few key features that could be useful in drug design. One of these is the presence of a 'pocket' formed by residues in the loop region 290–300 and those from two nearby loops, which is hydrophobic and more constricted in PfALDO than that in human aldolase A.[318,319] Even though reports on small molecule inhibitors of PfALDO are limited, homologous peptides of human band-3 protein and those derived from *P. falciparum* alpha-tubulin have been shown to inhibit PfALDO.[320] In another approach, antisense oligonucleotide technique was used to target PfALDO messenger RNA (mRNA), which subsequently led to the growth inhibition of erythrocytic *P. falciparum*.[321]

Triosephosphate isomerase is involved in the isomerization of dihydroxyacetone phosphate to glyceraldehyde-3-phosphate, which is essential to maintain the efficiency of glycolysis. Screening a database of compounds against the crystal structure of PfTIM led to the identification of several anionic dyes with good docking scores and activities, and one of the compounds investigated, congo red (**62**), showed an IC_{50} value of 30.4 μM against the enzyme.[322] In another interesting study by Singh *et al.*, peptide fragments corresponding to loops 1 and 3 of PfTIM were used to disrupt the dimerization of TIM which is essential for its enzymatic activity.[323] One of these peptides, which contained residues 68–79 of loop 3, was able to inhibit the enzyme with submillimolar IC_{50} values. Even though a number of studies pertaining to the conformational details of PfTIM are available,[324] reports on rational drug design based on this enzyme are currently very limited. Future exploration in this direction could give new lead compounds useful in drug development against malaria.

Lactate dehydrogenase (LDH) is the last enzyme in the glycolytic pathway and is responsible for the conversion of pyruvate to lactate with concomitant formation of NAD^+ from NADH. Since a constant supply of NADH is essential for glycolysis, LDH also assists in the regeneration of NADH from NAD^+. Even though the inhibition of this enzyme could kill the parasite, it is important to develop inhibitors which can differentiate it from human LDH isoforms. Important structural and kinetic differences that are useful in studies with this enzyme are (1) a change in the positioning of NADH in PfLDH compared to that in the human enzyme, as indicated by a displacement of the nicotinamide ring by about 1.2 Å due to sequence changes in the cofactor binding region, (2) changes in the sequence and secondary structure of a loop adjacent to the active site, (3) larger size of the active site cleft in PfLDH compared to its human counterpart, (4) unlike human LDH, the malarial enzyme is not inhibited by high concentrations of pyruvate, and (5) malarial LDH is very active with a synthetic coenzyme, 3-acetylpyridine adenine dinucleotide (APAD), which makes it useful as a probe to test parasitemia in human blood.[319,325,326]

Over the years, a number of investigations have been conducted to identify new drug candidates that can inhibit PfLDH. One of the important compounds identified was gossypol (**63**), a polyphenolic binaphthyl disesquiterpene isolated from cotton seeds. It is competitive for NADH and showed LDH inhibition with a submicromolar IC_{50} value (0.7 μM). Although this compound retained similar activity when tested in vitro against malaria parasites ($IC_{50} = 7$ μM), its cytotoxicity, arising from the aldehyde groups, and poor selectivity against human LDHs were limiting factors in drug development efforts.[327] A synthetic analog based on the structure of gossypol, 7-*p*-trifluoromethylbenzyl-8-deoxyhemigossylic acid (**64**), was later developed, which showed K_i values of 13, 81, and 0.2 μM, respectively, against human muscle, heart, and PfLDHs.[328] More recently, high-throughput screening of a library of compounds led to the discovery of a number of azole-based small molecules having good selectivities and activities against PfLDH. Thus, compound **65** showed an IC_{50} value of 0.65 μM against PfLDH (relative value against human LDH = 72.05 μM) and exhibited promising in vitro ($IC_{50} = 18.6$ μM against K1 strain of *P. falciparum*) and in vivo (41% reduction in parasitemia after a 4-day test) activities against malaria parasites.[329] Crystallographic studies showed that these analogs bind in a similar fashion to that of pyruvate in the active site of the enzyme, and suggest the possibility of developing new antimalarial agents based on such scaffolds. More studies are needed in this area to identify better drug candidates having activities ideal for therapeutic use, with good selectivities over human enzymes.

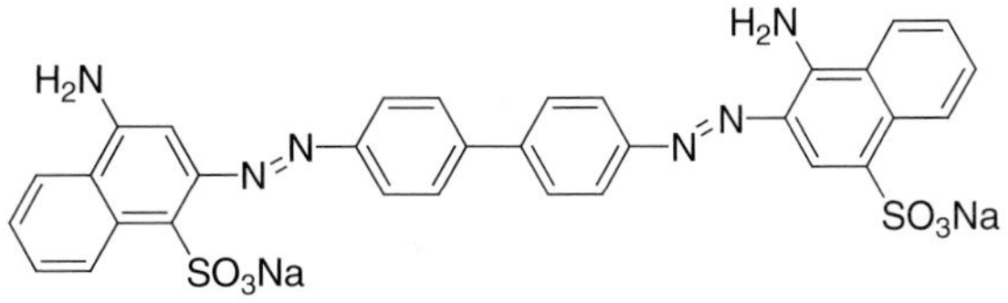

62 Congo red

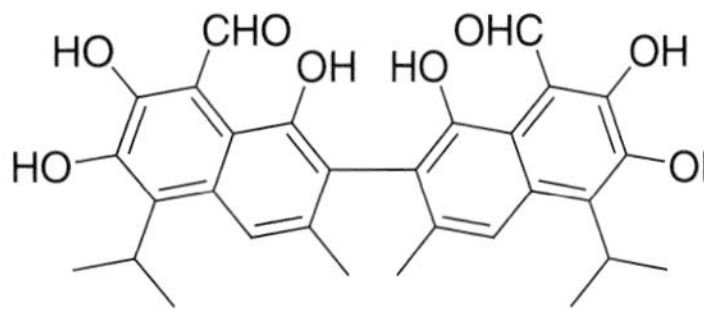

63 Gossypol

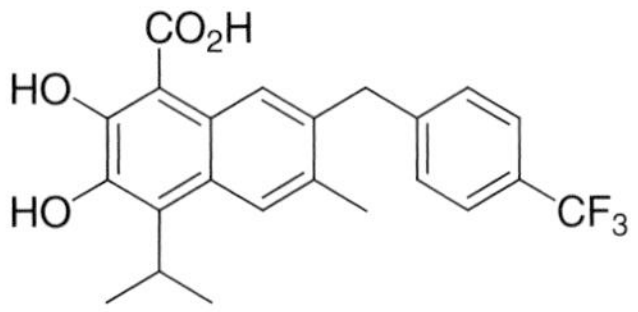

64

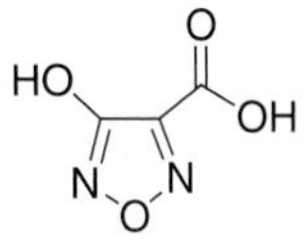

65

7.27.5.1.5 Drugs targeting nucleotide metabolism

The rapidly growing and dividing intraerythrocytic parasites require ready supply of nucleotides for maintaining their genetic makeup and to carry out other vital cellular functions. Since host erythrocytes lack the nucleotide biosynthetic machinery, the parasites have evolved their own pathways to acquire these nucleotides. All evidences indicate that the parasites derive purines through a salvage pathway and pyrimidines by de novo biosynthetic route.[330] Many enzymes involved in these pathways have been under scrutiny as drug targets,[331] and the following section would highlight important developments in this area.

7.27.5.1.5.1 Purine salvage pathway

The uptake of purine is believed to be mediated by a saturable adenosine transporter system which localizes to the parasite's plasma membrane.[332] Other possible transport pathways include the operation of a nonsaturable anion-selective ion channel or the parasite's tubovesicular system.[333] Of these, only the saturable adenosine transporter is characterized at the molecular level and is unique because of its ability to transport L-nucleosides in addition to physiological D-nucleosides.[332] It could be possible to exploit this by designing L-nucleoside analogs to target some of the downstream targets, as demonstrated by the inhibition of adenosine deaminase by the L-nucleoside analog of coformycin.[334]

One of the most extensively studied plasmodial enzymes in the purine salvage pathway is hypoxanthine-guanine-xanthine phosphoribosyltransferase (HGXPRTase), which catalyzes the transfer of a 5-phosphoribosyl group from

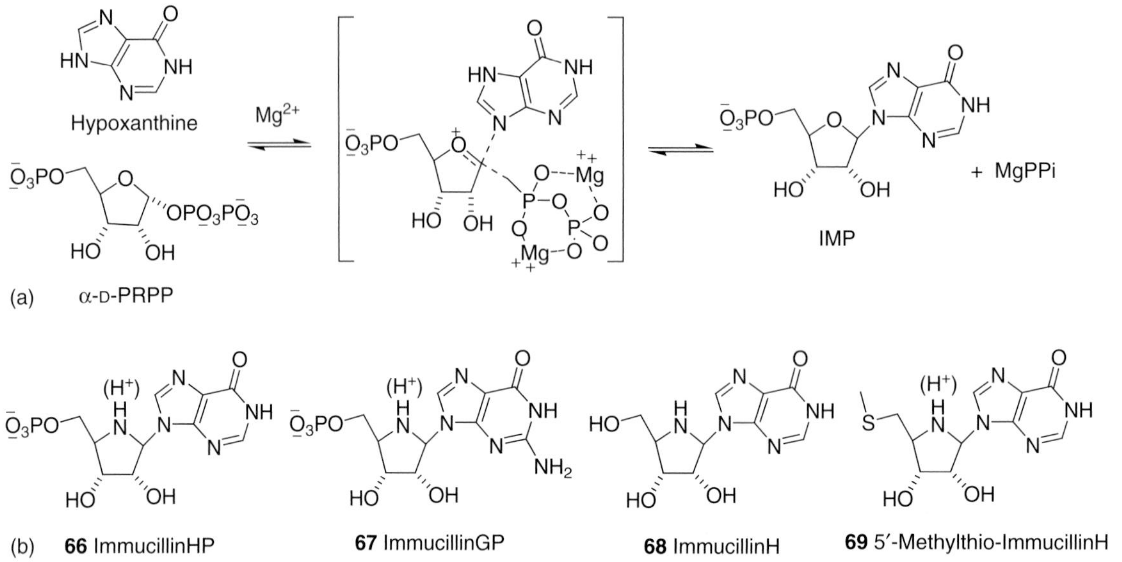

Figure 7 (a) Reaction and transition state involved during catalysis by purine phosphoribosyl transferase. (b) Structures of various transition-state analogs.

5-phospho-α-D-ribofuranosyl-1-pyrophosphate (α-D-PRPP) to hypoxanthine, guanine, or xanthine and form intracellular inosine monophosphate (IMP), guanine monophosphate, or xanthine monophosphate (XMP), respectively (**Figure 7**). IMP, formed in this way, can serve as a precursor to AMP and guanine monophosphate. Based on amino acid sequence, human HGPRTase is 44% identical to HGXPRTase from *P. falciparum*, and the crystal structures of these enzymes bound to the transition state analogs immucillinGP (**67**) and immucillinHP (**66**), respectively, together with pyrophosphate and Mg^{2+} are available.[335,336] Even though these structures are largely identical, one difference between these two enzymes is the inability of human HGPRTase to use xanthine as the substrate.[337] A recent report by Makobongo *et al.* has shown the antigenic role of PfHGXPRTase in developing cell-mediated immunity against malaria. In this study, recombinant PfHGXPRTase was found to activate T cells in vitro, and immunization with it reduced the parasite growth in all mice studied.[338] Another important enzyme involved in purine salvage pathway is purine nucleoside phosphorylase (PNP), which catalyzes the phosphorolysis of inosine to hypoxanthine. Although the inhibition of *P. falciparum* PNP by a transition state analog such as immucillinH (**68**) was shown to kill the parasite, it was necessary to inhibit human PNP to completely block the supply of hypoxanthine.[339,340] Based on information from the crystal structure of PfPNP in complexation with immucillinH.SO_4, Schramm and co-workers recently identified 5′-methylthio-immucillinH (**69**) as a selective inhibitor of PfPNP with 112-fold selectivity over human analog.[341] Considering the structural and mechanistic similarities between host and parasite enzymes in the purine salvage pathway, selective targeting of one or more enzymes seems to be essential for the development of new drug candidates.

7.27.5.1.5.2 De novo pyrimidine synthesis

Unlike the salvage of purines, plasmodium parasites derive pyrimidines through a de novo biosynthetic route.[330] A number of enzymes involved in this pathway, which is linked to folate biosynthesis, have been identified and clinically exploited as antimalarial drug targets, which include dihydrofolate reductase-thymidylate synthase (DHFR-TS),[342] dihydropteroate synthase (DHPS),[343] and dihydroorotate dehydrogenase (DHODH).[344] The dihydrofolate reductase domain of DHFR-TS is responsible for the reduction of dihydrofolate to tetrahydrofolate with concomitant oxidation of the NADPH cofactor. Pyrimethamine (**70**) and proguanil (**71**) are the best-known antimalarial drugs that target DHFR; the latter exerting effects through its active metabolite cycloguanil (**72**), produced in vivo by the action of cytochrome P450.[345] Additionally, chloroproguanil (**73**),[346] and compounds in the phenoxypropoxy-biguanide series such as PS-15 (**74**) and PS-26 (**75**),[347,348] are the new drugs in developmental stages which act through active metabolites analogous to proguanil. Even though the co-occurrence of DHFR and TS domains in the same polypeptide chain is a typical feature of the *P. falciparum* enzyme, the latter is highly conserved, which makes selective targeting difficult. Compounds such as 5-fluoroorotate have been shown to kill malaria parasites by inhibiting the TS domain, but causes significant toxicity to the host, due to the incorporation of 5-fluoropyrimidine into the nucleotide pool. Interestingly, combining an anti-TS inhibitor 1843U89 (**76**) with thymidine resulted in selective inhibition of parasite growth with greatly reduced host toxicity, showing the feasibility of targeting this enzyme for therapeutic use.[349] In addition, crystallographic studies of PfDHFR-TS have shown the importance of a junction region in this bifunctional enzyme ideal for drug targeting.[350]

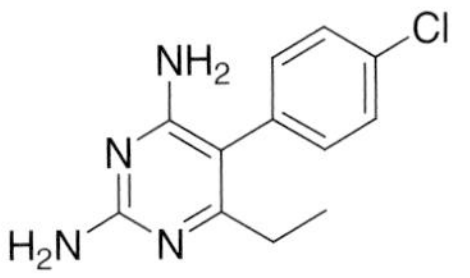

70 Pyrimethamine

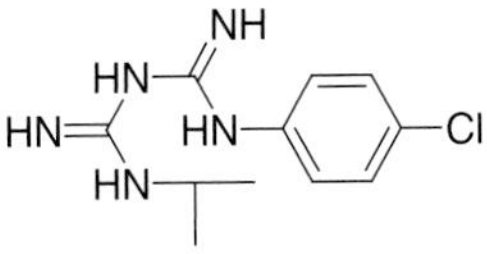

71 Proguanil

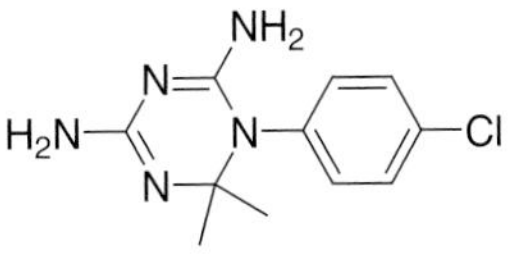

72 Cycloguanil

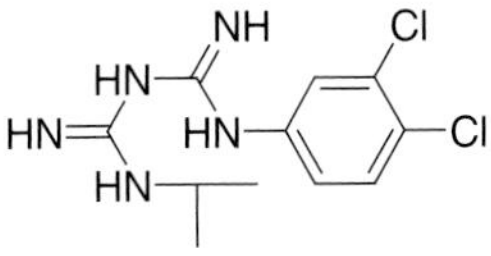

73 Chloroproguanil

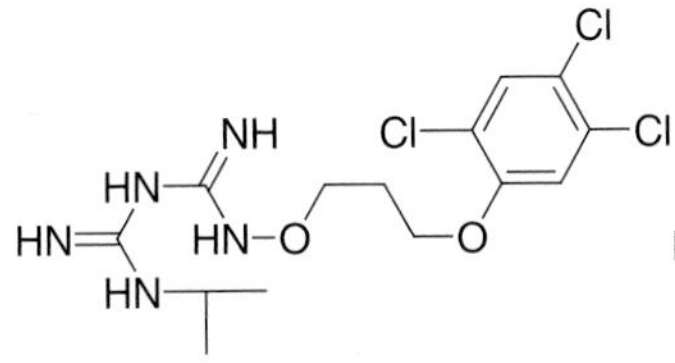

74 PS-15

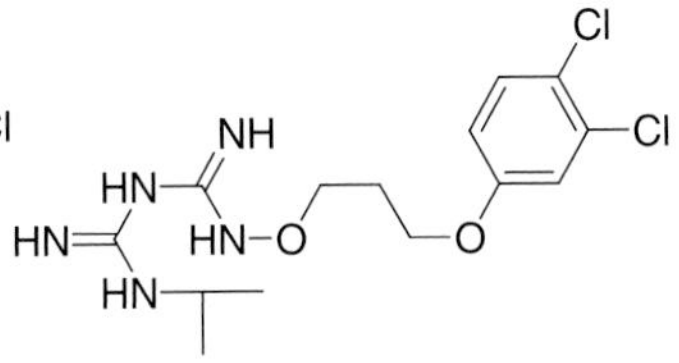

75 PS-26

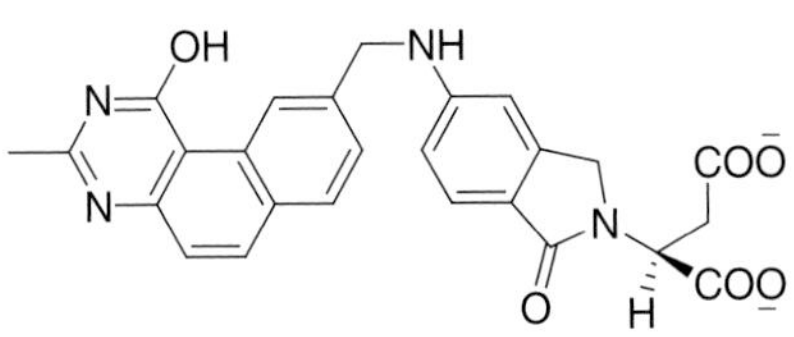

76 1843U89

Even though pyrimethamine and cycloguanil were effective against malaria, resistance developed rather quickly by mutations on DHFR at various positions such as 16, 50, 51, 59, 108, and 164.[351,352] Studies involving crystal structures and homology models of DHFR-TS have highlighted a number of features concerning the structural basis of drug resistance, the most important one being the steric clash between *p*-substituted chlorine atoms in pyrimethamine (**70**) and cycloguanil (**72**) with side chain of Asn108 in mutant enzymes.[350,353] In the case of cycloguanil, interaction between one of the 2,2-dimethyl groups and the side chain of Val16 has also been implicated in drug resistance.[354] In light of these observations, a number of investigations were carried out to develop new antifolates based on pyrimethamine and cycloguanil that can overcome resistance. The results from these studies suggest that the resistance can be reversed by (1) aromatic ring modification, either by using an unsubstituted or *m*-chloro- or 3,4-dichloro- derivative of pyrimethamine or cycloguanil,[355] (2) use of a flexible side chain which provides conformational freedom to fit in the active site,[352] and (3) removal of one of the alkyl groups in cyclogunanil, which avoids steric clash with Val16.[354] In addition to the modifications on existing drugs, docking and database screening approach recently identified a number of new compounds with promising anti-DHFR activities.[356] Some representative examples of lead compounds that emerged from these studies are presented in Table 8.

Sulfa drugs structurally mimic *p*-aminobenzoic acid and compete for the active site of dihydropteroate synthase which catalyzes the formation of dihydropteroate from hydroxymethyldihydropterin. Sulfadoxine (**81**) is the best known antimalarial agent in this category and is usually used in combination with pyrimethamine.[357] As with DHFR inhibitors, resistance to sulfadoxine is a common problem in many parts of the world due to mutations in DHPS.[358] Another drug, dapsone (**82**), together with chloroproguanil (**73**), is under development as a possible replacement for sulfadoxine–pyrimethamine.[346]

Malarial DHODH is a mitochondrial enzyme which catalyzes the oxidation of dihydroorotate to orotate, an important step in pyrimidine biosynthesis.[359] The activity of this enzyme is linked to the mitochondrial electron transport system which makes it an indirect target of atovaquone, one of the antimalarial drugs in clinical use (see below).[360] Even though inhibitors of human DHODH such as A77 1726 (**83**) did not significantly inhibit PfDHODH, most likely due to the variation of amino acid residues in the active site,[344] high-throughput screening of ~220 000 compounds recently identified some selective PfDHODH inhibitors, exemplified by compound **84** which showed an IC_{50} value of 16 nM against the *P. falciparum* enzyme with ~12 500-fold selectivity over its human counterpart.[361] Thus, it seems possible to exploit the difference in active site residues of human and parasite DHODHs in the design of new antiplasmodial agents.

Table 8 New DHFR inhibitors

Compound	*Biological characteristics*
77	**77** showed a K_i value of 17.8±0.8 nM against mutant (A16V+S108T) DHFR and an IC_{50} value of 19±8 nM against mutant *P. falciparum* clones. It showed ~73 times more binding affinity toward mutant DHFR and is ~120 times more potent in vitro against the mutant clone, compared to cycloguanil.[354]
78	The K_i value of **78** against PfDHFR mutants (N51I+C59R+S108N+I164L) = 1.7±0.2 nM, compared to a value of 385±163 nM for pyrimethamine. Its in vitro IC_{50} value against *P. falciparum* clone with the same quadruple mutation = 4.6±0.2 μM, relative to >100 μM for pyrimethamine.[355]
79	**79** showed IC_{50} values of 3.57±0.52 μM and 3.02±0.29 μM, respectively, against wild-type and mutant (quadruple) clones of *P. falciparum*. The flexible phenyl substitution was used to overcome steric restrictions at the active site of mutant enzyme.[352]
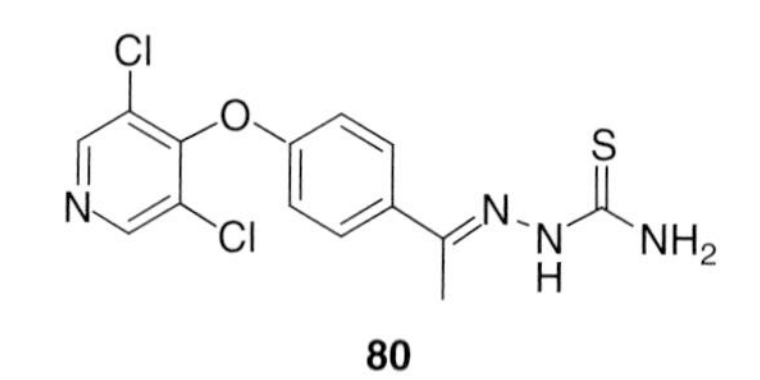 **80**	**80** is one of the compounds identified after screening a three-dimensional database of druglike molecules against the homology model of PfDHFR. Its K_i values against wild-type and mutated (N51I+C59R+S108N+I164L) PfDHFR were 0.9±0.08 μM and 1.50±0.03 μM, respectively.[356]

81 Sulfadoxine

82 Dapsone

83 A77 1726

84

7.27.5.1.6 Atovaquone: targeting mitochondrial membrane potential

As mentioned, catalysis by DHODH is closely linked to the mitochondrial electron transport chain, where the electrons are funneled to the cytochrome complex through ubiquinone. Atovaquone (**85**) is a structural analog of ubiquinone (**86**) which can bind competitively with the cytochrome bc_1 complex leading to impaired electron flow and mitochondrial membrane potential.[362] This drug, in combination with proguanil (malarone), has been recommended for the treatment and prophylaxis of uncomplicated *P. falciparum* malaria.[363,364] When atovaquone was used alone, resistance developed rather quickly due to mutation on the cytochrome *b* gene.[365,366] However, in combination with proguanil, a synergistic interaction and an improved potency none observed.[367] Studies have shown that it is proguanil itself and not its metabolite cycloguanil that is responsible for the biological effect.[368] Even though this combination has been effective in treating malaria, there is growing concern about the long-term efficacy of this combination due to decreased synergy and treatment failures once resistance has developed.[369,370]

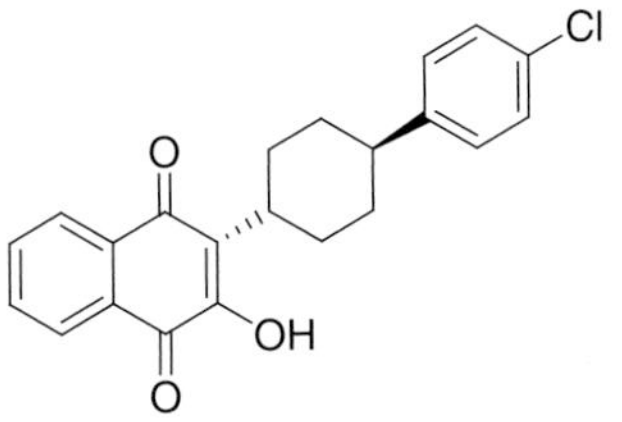

85 Atovaquone

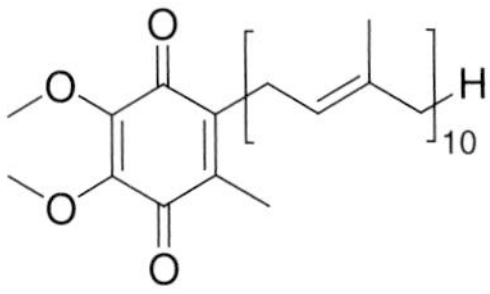

86 Ubiquinone

7.27.5.1.7 Farnesyltransferase inhibitors

Farnesylation is one of the important steps in the posttranslational modification of proteins associated with intracellular signal transduction.[371] The process is catalyzed by a heterodimeric zinc protein known as farnesyltransferase, and involves the transfer of a farnesyl group from farnesylpyrophosphate to the C-terminal cysteine sulfur of the target proteins such as Ras.[372] These protein substrates, in general, have characteristic C-terminal consensus sequences of the type CAAX (C, cysteine; A, aliphatic amino acid; X, serine or methionine) and prenylation is important for their translocation to the membrane where they form part of the signaling network. Since constitutive activation of Ras is a major contributory factor in a number of malignant human tumors, its inactivation by interfering with the farnesylation step has been extensively studied as a strategy to develop new anticancer agents.[373,374]

After the identification of farnesyltransferase from *P. falciparum* (PfFTase),[375,376] a number of studies were conducted using the available repertoire of farnesyltransferase inhibitors (PfFTIs), which showed that inhibition of PfFTase is lethal to the parasite. Thus, compounds **87** and **88** inhibited PfFTase with IC_{50} values of 8.0 and 0.9 nM, respectively, and showed ED_{50} values of 180 nM and 5.0 nM, respectively, against *P. falciparum* growth in vitro.[377] When **88** was tested in mice infected with *P. berghei* with 200 mg kg^{-1} day^{-1}dose for 7 days, 60% of the mice showed complete recovery, with no sign of infection on observation for 60 days. Another compound (**89**), based on a benzophenone scaffold, was recently developed by Schlitzer and co-workers, which showed an IC_{50} value of 210 ± 21 nM against PfFTase, with an ED_{50} of 21 mg kg^{-1} when studied in mice infected with *P. vinckei*.[378]

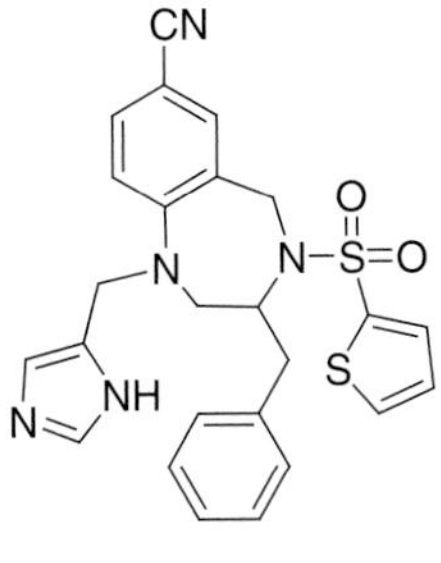

87 BMS-214662

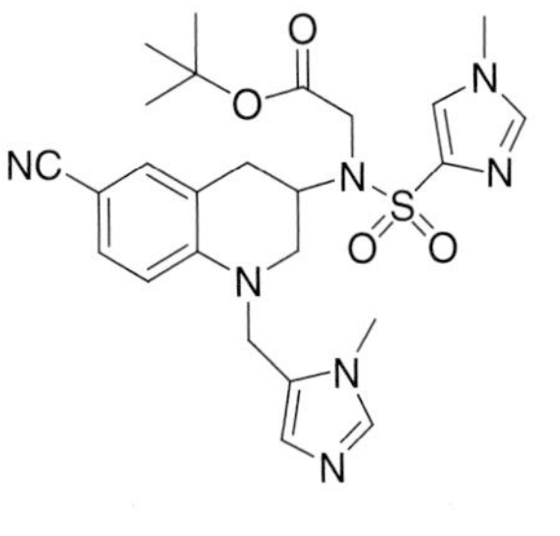

88 BMS-386914

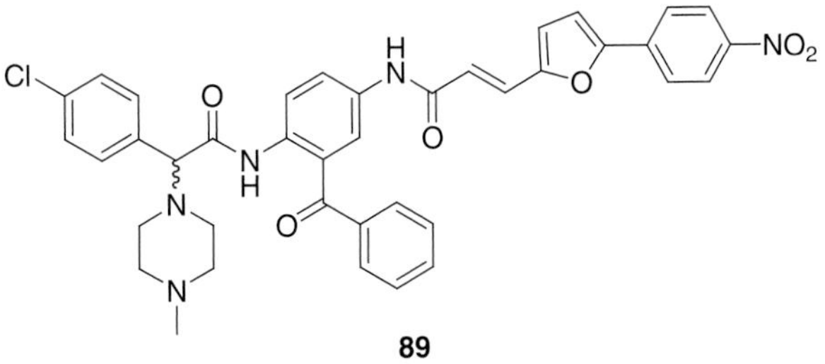

89

Even though the development of effective PfFTIs has been hampered due to the difficulty in developing heterologous expression systems and the lack of a crystal structure, similarity in the active site residues of PfFTase with its mammalian analogs of known crystal structures would be useful in understanding ligand receptor interactions and drug design.[378] While future work would undoubtedly lead to better PfFTase-based antimalarial agents, early studies have indicated the possibility of resistance development when the inhibitors are used alone, and recommend the use of a combination of PfFTIs with other active antimalarials for therapeutic applications.[379]

7.27.5.1.8 Apicoplast and new drug targets

Apicoplast is a plastid-like organelle found in members of Phylum Apicomplexa, which includes *Plasmodium* spp. and *Toxoplasma gondii*.[380] It is believed that these parasites, during early stages of their evolution, acquired this organelle through secondary endosymbiosis that involved initial engulfment of a cyanobacterium by an alga and subsequently the engulfment of the latter by the apicomplexan ancestor.[381,382] Even though the function of this organelle remained unknown for a long time, this turned out to be something of a gold mine for researchers looking for new biosynthetic pathways for drug development.[383,384] Apicoplast contains a 35 kb circular DNA, which is translationally active.[385,386] A number of nuclear encoded proteins have also been identified, which contain N-terminal bipartite signal sequences to direct them to apicoplast through the secretory pathway.[387,388] Fatty acid and isoprenoid biosyntheses are the two important biosynthetic pathways associated with this organelle and are distinct from that of mammals, making the enzymes involved in them important targets for drug development.[28]

7.27.5.1.8.1 Fatty acid biosynthesis

Fatty acid biosynthesis is important for cell growth, differentiation, and homoeostasis. Unlike in animals and fungi, where a single multifunctional enzyme known as type I fatty acid synthase catalyzes all, the steps in fatty acid biosynthesis, bacteria, plants, and *Plasmodium* spp. use a type II fatty acid biosynthetic pathway (**Figure 8**) in which each step is catalyzed by separate enzymes.[389]

Table 9 lists some of the compounds identified thus far that inhibit various enzymes involved during fatty acid biosynthesis. It is interesting to note that one of these compounds, triclosan (**94**), a known inhibitor of FabI and a widely used antibacterial agent,[390] inhibited *P. falciparum* growth in vitro with an IC_{50} value of 0.7 μM. When tested in vivo in mice infected with *P. berghei* at a 3.0 mg kg^{-1} dose subcutaneously, this compound inhibited parasitemia by 75% within 24 h of drug administration.[391] Compounds Genz-8575 (**97**) and Genz-10850 (**98**) were identified by high-throughput screening of a library of compounds against enoyl acyl carrier protein reductase of *Mycobacterium tuberculosis* (inhA). In addition to inhA, they also inhibited the corresponding *Plasmodium* enzyme (PfENR) with IC_{50} values of

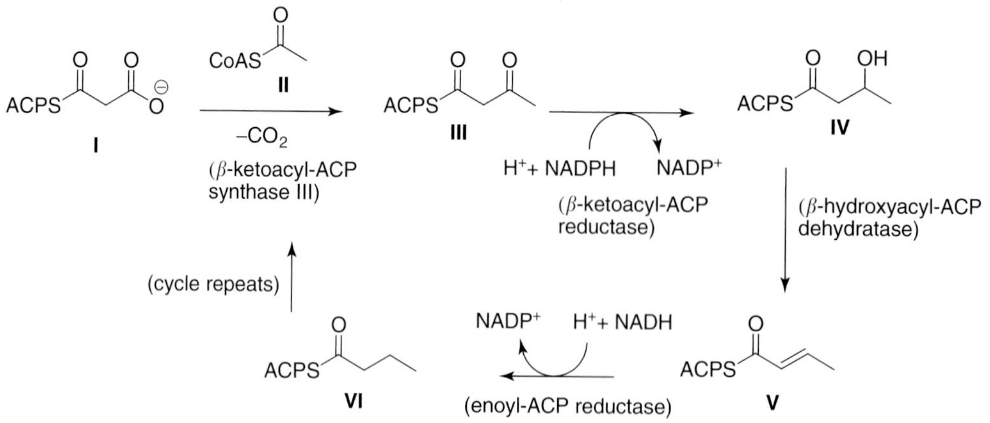

Figure 8 Steps involved in type II fatty acid biosynthesis. The process involves (1) the conversion of acetyl-CoA to malonyl-CoA (catalyzed by acetyl-CoA carboxylase (ACC)) and then to malonyl-ACP (I) (catalyzed by FabD), (2) the condensation of malonyl-ACP with another molecule of acetyl-CoA (II) to form β-ketoacyl-ACP (III) (catalyzed by β-ketoacyl-ACP synthase III or FabH), (3) reduction of β-ketoacyl-ACP to β-hydroxyacyl-ACP (IV) by β-oxoacyl-ACP reductase (or FabG) and its dehydration by β-hydroxyacyl-ACP dehydratases (FabZ/A) to enoyl-ACP (V), and (4) reduction of V by enoyl-ACP reductase (FabI) to form butyryl-ACP (VI), which then reenters the FAS cycle (catalyzed by FabB/F), and gets elongated by two carbon atoms per cycle. Enoyl-ACP can also be a precursor for the synthesis of unsaturated fatty acids through catalysis by FabA and FabB.

Table 9 Compounds and their targets involved in the fatty acid synthesis in malaria parasites

Compound	*Proposed target(s)*	*P. falciparum growth inhibition (IC_{50})*	*Reference*
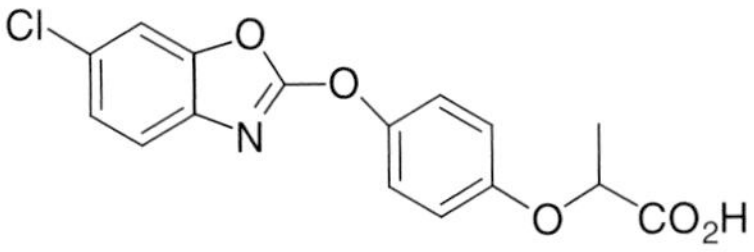 **90** Fenoxaprop	ACC	$144 \pm 22\,\mu M$	Waller *et al.*[393]
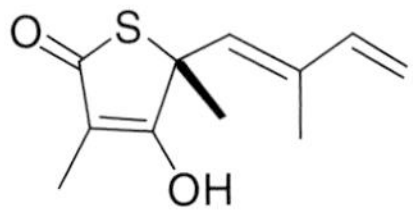 **91** Thiolactomycin	Fab B/F/H	$49 \pm 8\,\mu M$	Waller *et al.*[393]
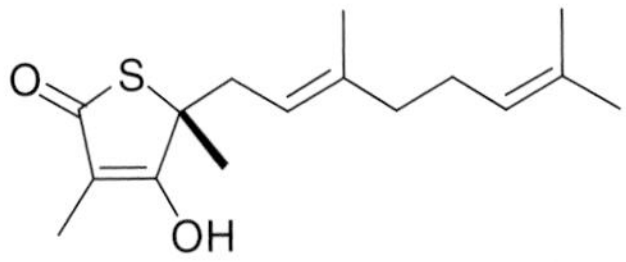 **92**	Fab B/F/H	$8 \pm 0.2\,\mu M$	Waller *et al.*[393]
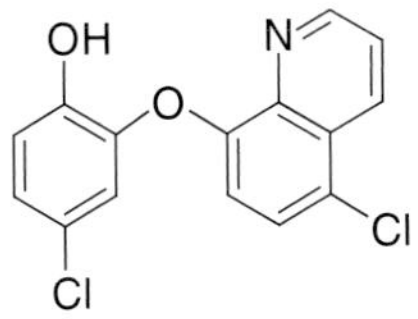 **93** NAS-91	FabZ	7.4 μM	Sharma *et al.*[394]
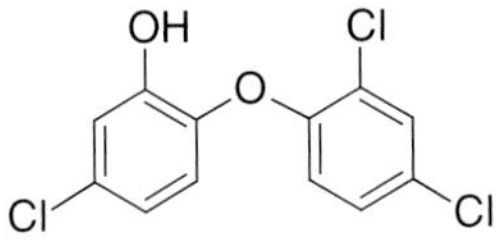 **94** Triclosan	FabI	0.7 μM	Surolia *et al.*[391]
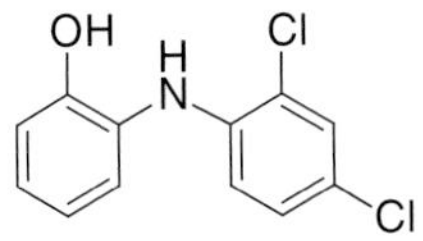 **95**	FabI	1.8–6.0 μM	Perozzo *et al.*[395]
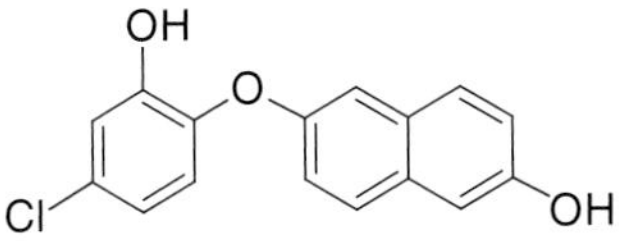 **96**	FabI	5.9–19.5 μM	Perozzo *et al.*[395]

continued

Table 9 Continued

Compound	*Proposed target(s)*	*P. falciparum growth inhibition (IC_{50})*	*Reference*
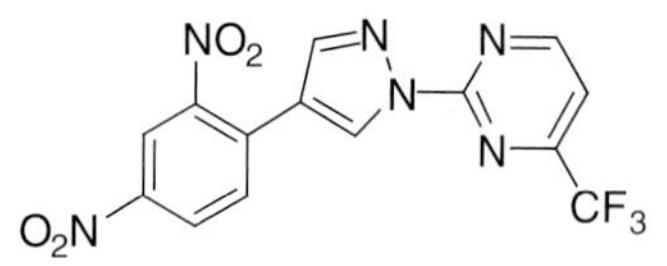 **97** Genz-8575	FabI	10–16 μM	Kuo *et al.*[392]
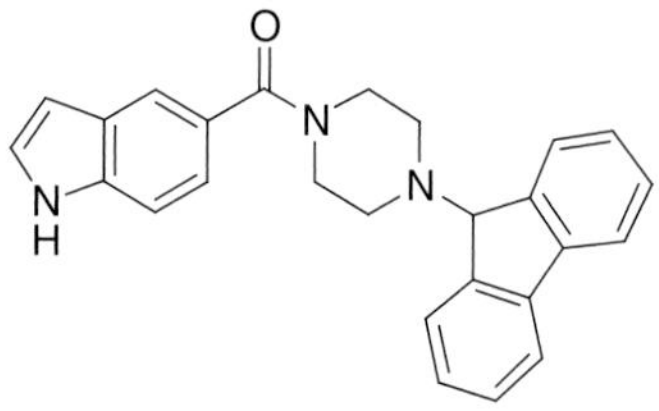 **98** Genz-10850	FabI	14–32 μM	Kuo *et al.*[392]

32 μM and 18 μM, respectively, and were active against *P. falciparum* growth in vitro (**Table 9**).[392] Structural optimization on various compounds identified already, and future design based on the active site details of enzymes involved in fatty acid synthesis are expected to give better candidates for medicinal use.

7.27.5.1.8.2 Isoprenoid biosynthesis

Like other apicomplexan parasites, *Plasmodium* has also been shown to use a distinct pathway called the 1-deoxy-D-xylulose-5-phosphate (DOXP) pathway or a nonmevalonate pathway for the preparation of isopentenyl pyrophosphate (IPP) and dimethylallyl pyrophosphate (DMAPP), which are the precursors for isoprenoids (**Figure 9**).[396,397] The first step in this pathway involves the condensation of pyruvate and D-glyceraldehyde-3-phosphate (G3P) to produce DOXP. In the second step, DOXP-reductoisomerase (DXR) catalyzes the conversion of DOXP to 2C-methyl-D-erythritol-4-phosphate (MEP), which is followed by the transfer of this erythritol derivative to a nucleotide, resulting in 4-diphosphocytidyl-2C-methyl-D-erythrytol (CDP-ME). This is followed by its phosphorylation by CDP-ME kinase (CMK) to produce 4-diphosphocytidyl-2C-methyl-D-erythritol-2-phosphate (CDP-MEP). It is then converted to 2C-methyl-D-erythritol-2,4-cyclodiphosphate (ME-cPP), which in the next two enzyme catalyzed steps undergoes reduction and elimination to 1-hydroxy-2-methyl-2-(*E*)-butenyl-4-diphosphate (HMBPP) and then to IPP and DMAPP.

After the original report by Hassan *et al.* describing the presence of DXS and DOXP genes in *P. falciparum*,[398] a series of studies during the last few years have indicated the presence of other downstream enzymes in the malaria parasite.[28] Fosmidomycin (**99**), a known antibacterial agent that targets DOXP-reductoisomerase, has been shown to inhibit the plasmodial enzyme in a dose-dependent manner. This compound also inhibited the growth of *P. falciparum* in culture, and in vivo studies in mice have shown that it is well tolerated even at 300 mg kg^{-1} dose, and the parasitemia reduced to <1% after a 5 mg kg^{-1} day^{-1} × 4 intraperitoneal administration.[398] Even though early studies on humans with uncomplicated *P. falciparum* malaria showed very good clinical efficacy and tolerability, fosmidomycin as a monotherapy suffered from high rates of recrudescence.[399] Further drug interaction studies by Wiesner *et al.* showed a synergistic interaction of fosmidomycin with clindamycin (see below),[400] and recent trials among African children indicate that a 5-day twice daily regimen of this combination is well tolerated and effective in parasite clearance.[401,402] While its activity against resistant strains of the parasite and tolerability remain an advantage, fosmidomycin has a short plasma half-life and moderate absorption rates, which may be a limitation.[403,404] Another drawback is the potential liver toxicity observed for the fosmidomycin/clindamycin combination during human trials, which needs to be addressed to improve the therapeutic potential.[402]

Three fosmidomycin analogs have shown improved potency compared to the parent compound. Of these, compound **100** (FR900098) was approximately twice as active as fosmidomycin when tested in vivo in mice infected with *P. vinckei*.[398] Similarly, the prodrug esters **101** and **102** showed improved oral bioavailability when tested in mice.[405,406]

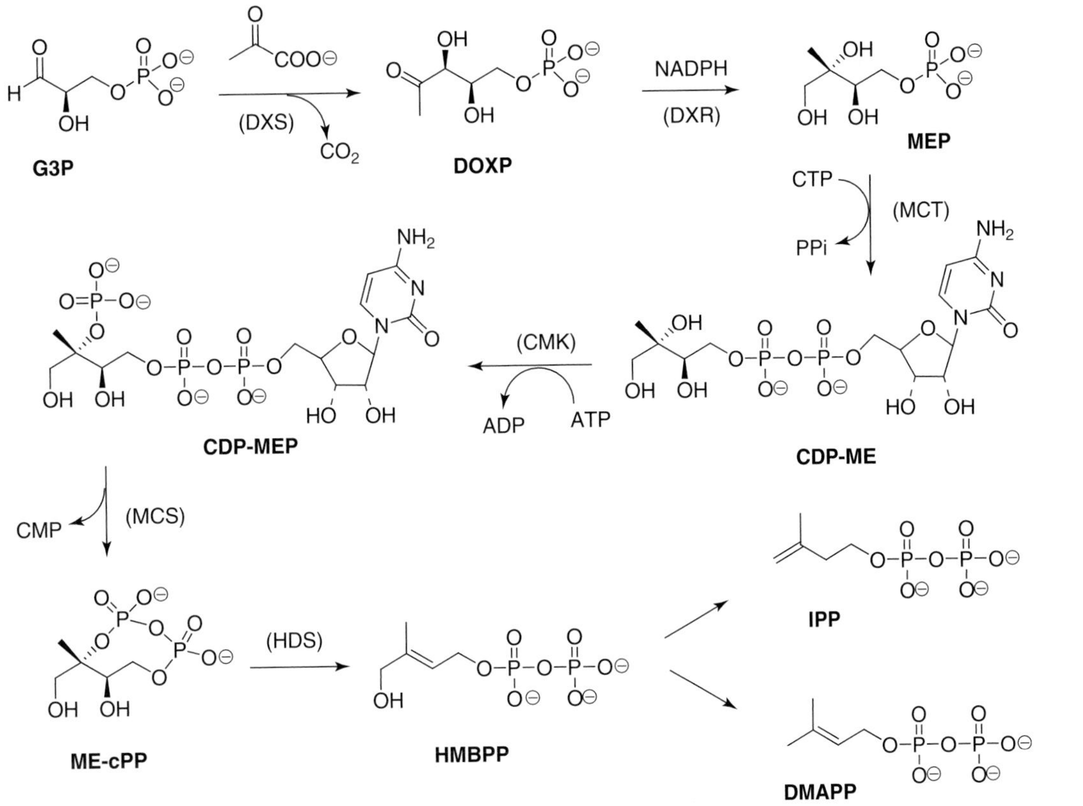

Figure 9 Nonmevalonate pathway in isoprenoid synthesis. DXS, DOXP synthase; DXR, DOXP reductoisomerase; MCT, MEP cytidyltransferase; CMK, CDP-ME kinase; MCS, ME-cPP synthase; HDS, HMBPP synthase.

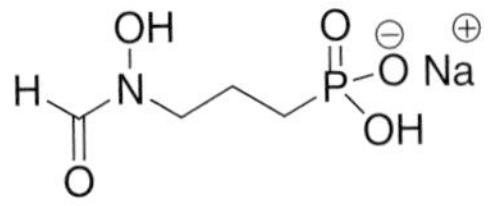

99 Fosmidomycin

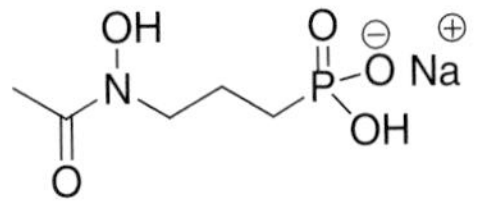

100 FR900098

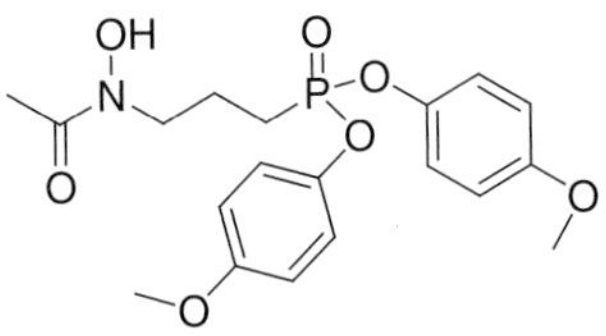

101 FR900098-prodrug

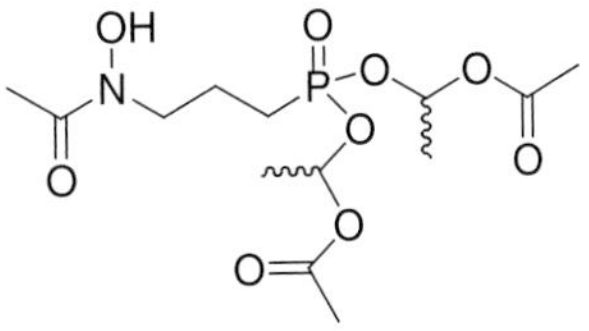

102 FR900098-prodrug

7.27.5.1.9 Antibiotics

Analysis of the effect of various drugs on apicoplast functions has shown that the compounds which target biosynthetic pathways beneficial for the entire organism (e.g., fatty acid biosynthesis) show their effect very rapidly, whereas those which affect only the plastid functions act relatively slowly, leading to a delayed death phenotype.[380] The apicoplast DNA mainly codes for transfer RNAs (tRNAs), ribosomal RNAs (rRNAs), three subunits of RNA polymerase, elongation factor Tu, and ribosomal proteins, which are principally associated with housekeeping functions.[383,407] The antimalarial activities of a number of antibiotics have already been established,[408,409] and studies aimed at understanding their individual targets have gained momentum in recent years. Since slow acting antibiotics are effective either prophylatically or when used in combination with other potent antimalarials.

Ciprofloxacin (**103**) has been shown to block the apicoplast DNA replication in *P. falciparum*, most probably by inhibiting DNA gyrase, a prokaryotic type II topoisomerase, without inhibiting nuclear replication.[410,411] Similarly, the antimalarial activity of rifampicin (**104**), which is known to target bacterial RNA polymerase, suggests that this drug may be targeting a similar apicoplast derived protein in *Plasmodium*.[412,413] Even though there was no evidence for translation in apicoplast, studies by Sushma *et al.* recently demonstrated the translation of the *tufA* gene encoded by the *P. falciparum* apicoplast genome, which suggests that the plastid DNA is translationally active.[385] Antibiotics such as lincosamides (lincomycin and clindamycin (**105**)) and macrolides (erythromycin and azithromycin (**106**)), which block bacterial protein synthesis by interacting with the peptidyl transferase domain of 23S rRNA, are capable of inhibiting the growth of *P. falciparum*, suggesting a similar mechanism operative in the case of the malaria parasite.[414] The antimalarial activity of doxycycline (**107**), which targets ribosomal protein synthesis,[415] and similar activity shown by thiostrepton (which blocks bacterial translation by interfering with GTPase activity on the 23S rRNA),[416–419] are other examples which show the potential of apicoplast translational systems as possible targets for chemotherapy.

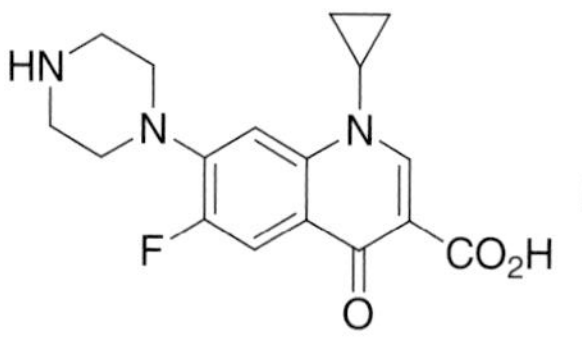

103 Ciprofloxacin

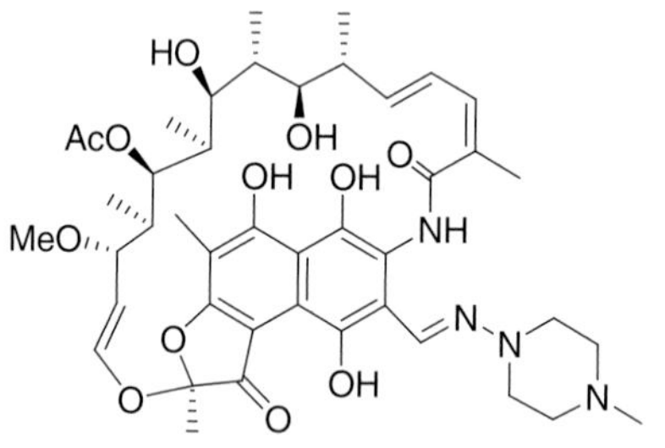

104 Rifampicin

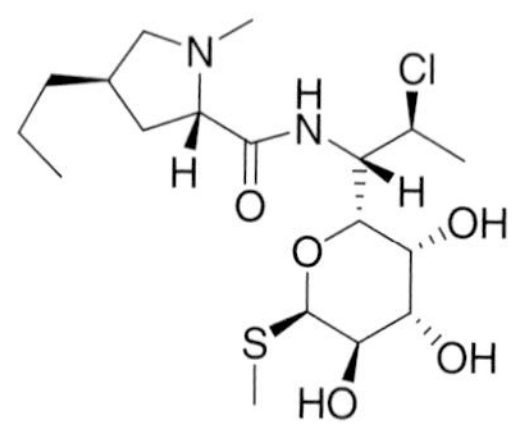

105 Clindamycin

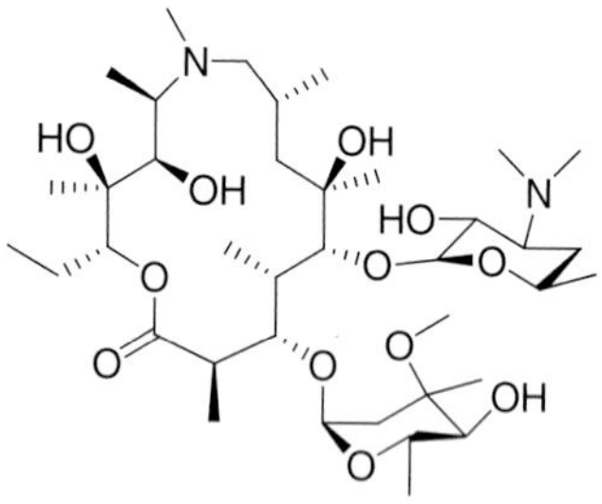

106 Azithromycin

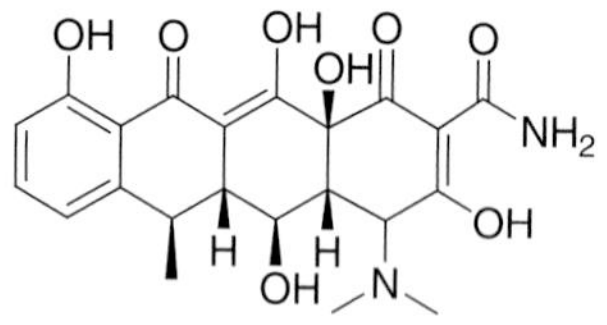

107 Doxycycline

Doxycycline, which has been recommended as a prophylactic agent, is also effective in combination with chloroquine to treat CQ-resistant *P. falciparum* malaria. This combination, however, is only modestly effective against *P. vivax* infections.[420,421] Since doxycycline has photosensitizing effects, it is contraindicated in small children and pregnant women, which is a limitation to its wide acceptability.[422,423] Clindamycin/chloroquine is another useful combination that has been shown to be effective in areas with high rates of chloroquine resistance. Even though clindamycin is generally well tolerated, its limitations are the high cost and the possible development of *Clostridium difficile* associated diarrhea.[424] The prophylactic activity of azithromycin has also been studied, and the results indicate that this drug can confer excellent protection against *P. vivax* infections, with modest prophylactic effects against *P. falciparum*.[425] However, a recent study involving a combination of azithromycin with chloroquine shows that this combination is effective in treating *P. falciparum* malaria.[426]

7.27.5.1.10 Peptide deformylase (PDF) as a therapeutic target

Prokaryotic translation begins with N-formylmethionine, and the resulting proteins undergo N-terminal modification to become functionally mature. This involves stepwise removal of the N-formyl group catalyzed by PDF, and then the methionine residue through the action of methionine aminopeptidase.[427] Studies have shown that the proteins which fail to undergo this modification are inactive, and in recent years, there has been a surge in scientific efforts to develop new antibacterial agents by targeting PDF.[428–430] A nuclear encoded PDF with a signal sequence, indicating its localization in apicoplast, was recently identified from *P. falciparum*,[431] and its crystal structure is available at 2.2 Å resolution.[432,433] Its sensitivity toward a number of standard PDF inhibitors has highlighted its importance as a drug target.[431,434]

Although the PDF from human mitochondria was originally believed to be nonfunctional, recent studies have shown that this enzyme is active and sensitive to a number of actinonin-based PDF inhibitors. These compounds exhibited antiproliferative activities on tumor cells by affecting the mitochondrial membrane potential.[435] Another factor which may come as an obstacle in drug design based on PfPDF is the possibility of resistance development as has been demonstrated recently in studies using *Escherichia coli*.[436]

7.27.5.1.11 Inhibitors of phospholipid metabolism

Phospholipids (PLs) are important constituents of biological membranes. The PL content in erythrocytes increases by approximately 500% after *Plasmodium* infection.[437,438] Phosphatidylcholine and phosphatidylethanolamine (PE) comprise about 85% of the total PL in infected erythrocytes and are synthesized by a de novo route. During the synthesis of phosphatidylcholine, steps involving cholinephosphate cytidyltransferase and choline uptake are rate limiting and likely regulatory, whereas in the case of PE, the ethanolaminephosphate cytidyltransferase step is believed to be rate limiting. Due to the importance of PL biosynthesis in the development of parasites, coupled with its absence in mature human erythrocytes, this pathway has been targeted for antimalarial drug development.[439,440] Generally, compounds that can interfere with the incorporation of polar head groups and impair the PL biosynthesis are lethal to the parasite.[441]

Systematic studies involving a series of choline analogs with mono- and bis-quaternary ammonium salts have identified a series of lead compounds with promising antiplasmodial activities.[442] Important observations from SAR studies could be discerned from these derivatives: (1) for monoquaternary ammonium salts, increased lipophilicity around the terminal nitrogen (tris-propylammonium group is optimal) and a 12 carbon alkyl chain are ideal for activity, and (2) for bis-quaternary ammonium salts, lipophilicity around nitrogen and an alkyl spacer up to 21 carbon atoms long are found to increase the potency.

Monoquaternary salts (E10) **108** and (E13) **109** inhibited *P. falciparum* growth in vitro with IC_{50} values of 64 nM and 33 , respectively. Comparatively, the bis-quaternary ammonium salts were superior in their potency – exemplified by (G25) **110** and (G19) **111**, which showed IC_{50} values of 0.64 nM and 3.0 pM, respectively.[442] In vivo studies in *Aotus* monkeys (against *P. falciparum*) showed that at a dose as low as 0.03 mg kg^{-1} (twice daily), G25 was able to clear the parasitemia by day 4, without recrudescence for up to 60 days.[443] Studies using G25 in rhesus monkeys infected with *P. cynomolgi* (a species close to *P. vivax*) showed a similar trend, where, with a dose of 0.15 mg kg^{-1} twice daily for 8 days (intramuscularly), there was total parasite clearance by day 4, without recrudescence for up to 35 days. Even though compounds such as G25 studied here have shown good selectivity (~1000-fold) over human cell lines, possible interaction of this class of compounds with the human cholinergic system may be a potential limitation.[443]

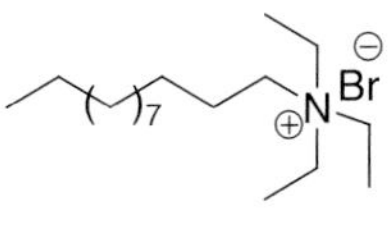

108 E10 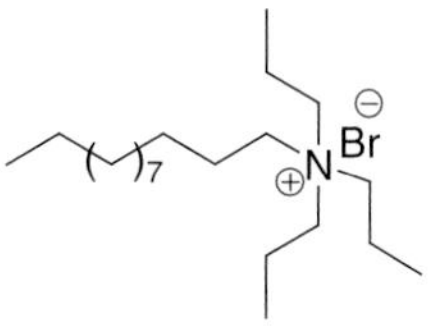 **109** E13 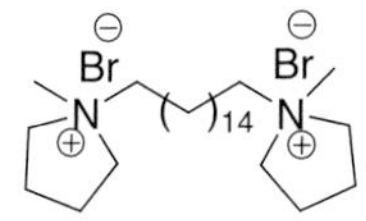 **110** G25

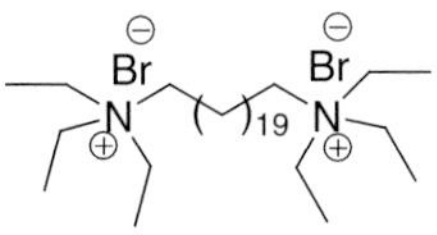

111 G19 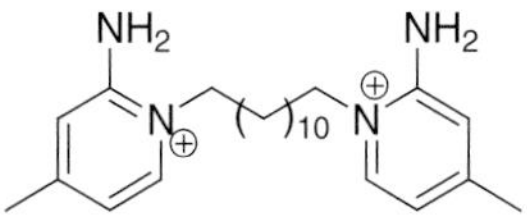 **112** MS1

A related compound, (MS1) **112**, where the bis-quaternary ammonium function is replaced by aromatic amidine functions, is another lead in this area which is reported to have improved oral bioavailability.[178] Another interesting development is the identification of a series of bis-thioester derivatives – prodrugs of bis-thiazolium salts, which were designed to improve oral bioavailability.[444] One of the compounds studied, (TE3) **113**, showed an ED_{50} of 5 mg kg^{-1} when tested orally in mice infected with *P. vinckei*. This compound exhibited excellent potency when evaluated in rhesus monkeys infected with *P. cynomolgi*, with complete clearance of parasitemia without recrudescence after four daily oral doses of 3 mg kg^{-1}.

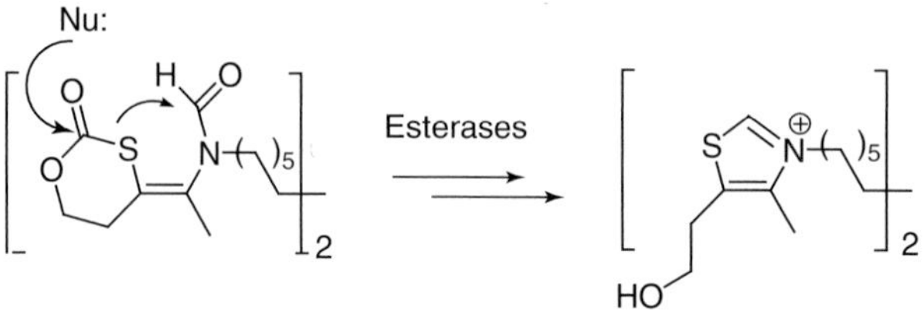

Nu: H O O S N ( )5 O 2 Esterases S N ( )5 2 HO

113 TE3 Bisthiazolium salt

7.27.5.1.12 Chalcones

Interest in chalcones as antimalarial drug candidates started after the discovery of antimalarial activity associated with licochalcone A (**114**), a natural product isolated from Chinese licorice root.[445] In an unrelated study based on the structure of compound **115**, one of the lead compounds identified after virtual database screening of a library of compounds against the homology model of malarial cysteine protease, various chalcones were synthesized and an SAR study was carried out to understand the stereoelectronic parameters that confer the highest potency.[446] The results from this study suggest the following trends: (1) the presence of electron withdrawing substituents on ring A or a quinoline ring at this location favors good activity, (2) in general, the presence of electron donating groups on ring B is associated with better antimalarial activity, and (3) an α,β-unsaturated ketone bridge is essential for activity. The most potent compound identified from this study was chalcone **116** with an IC_{50} value of $0.23 \pm 0.01\,\mu M$ against a drug-resistant strain of *P. falciparum*.

Similar studies using libraries of alkoxylated and hydroxylated chalcones further confirmed the association of good antimalarial activities with the presence of a 3-quinolinyl group at ring A, and led to the identification of compound **117** which inhibited *P. falciparum* growth with an IC_{50} value of $2.0\,\mu M$.[447] Another chalcone, **118**, from this study, which showed an in vitro activity of $2.4\,\mu M$ against *P. falciparum*, was as good as chloroquine in prolonging the lifespan of mice infected with *P. berghei*. However, there are examples from this and related reports,[448] where some of the chalcones which showed poor in vitro activities turned out to be potent when tested in vivo, as exemplified by compound **119** ($IC_{50} = 20\,\mu M$), which was comparable to chloroguine in increasing the survival time of *P. berghei* infected mice.[447]

In a recent report by Domínguez *et al.*, a series of phenylurenyl chalcone derivatives were synthesized and their antimalarial potencies compared with their ability to inhibit (1) malarial cysteine protease (falcipain-2), (2) globin hydrolysis, and (3) hemozoin formation.[449] The results from this study did not show a direct correlation between these factors and indicate that chalcones may be exerting their antimalarial activities via multiple mechanisms. The most potent compound (**120**) from this study inhibited *P. falciparum* growth in vitro with an IC_{50} of $1.76\,\mu M$ but showed only modest activity when tested in vivo. Another report by Ziegler *et al.* involving licochalcone A seemed to shed light on the uncertainties regarding the correlation between in vitro and in vivo data observed for various chalcones.[450] The authors found that this natural product is membrane active and can transform normal erythrocytes into echinocytes, and the observed antiparasitic activity was due to an indirect effect on the host cell membrane. Similar studies with previously reported chalcones may give a better understanding about the mode of action of these compounds.

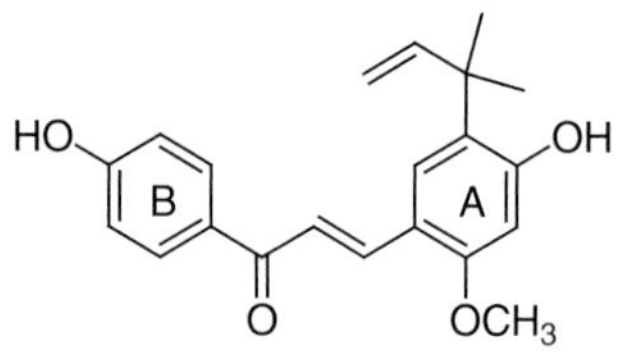

HO B O A OH OCH3

114 Licochalcone A

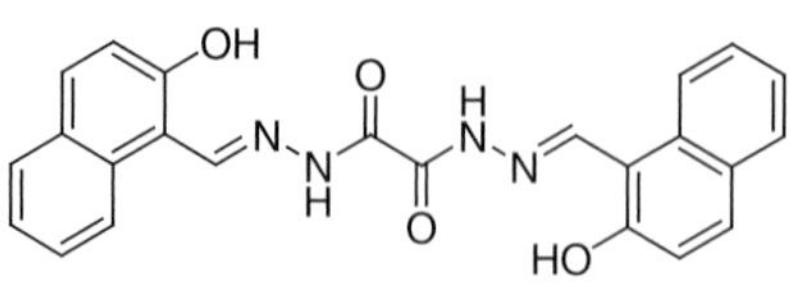

OH N N H O O H N N HO

115

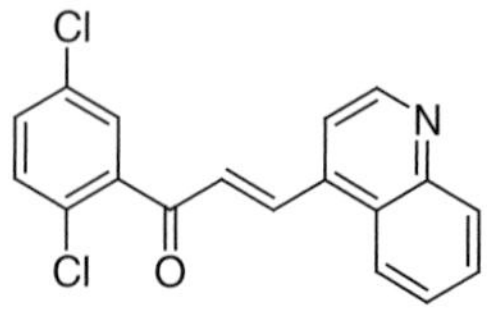

Cl Cl O N

116

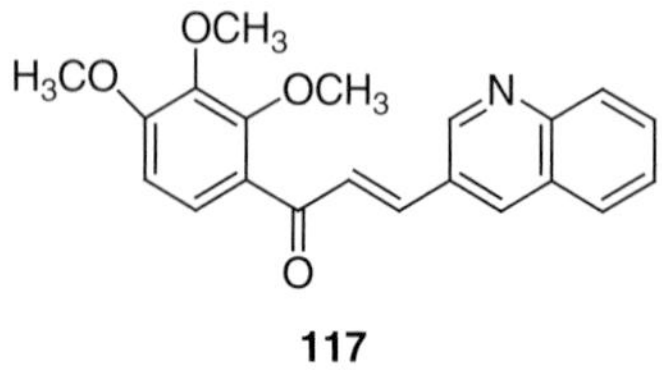

OCH3 H3CO OCH3 O N

117

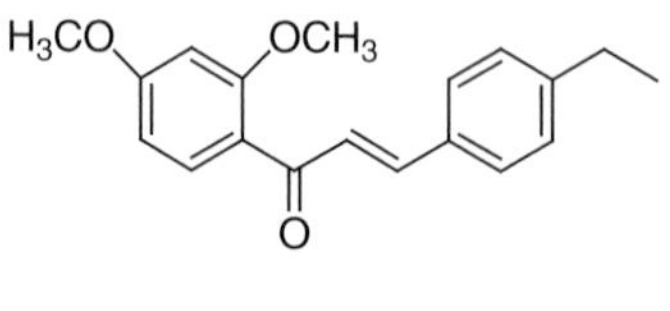

H3CO OCH3 O

118

HO OH O

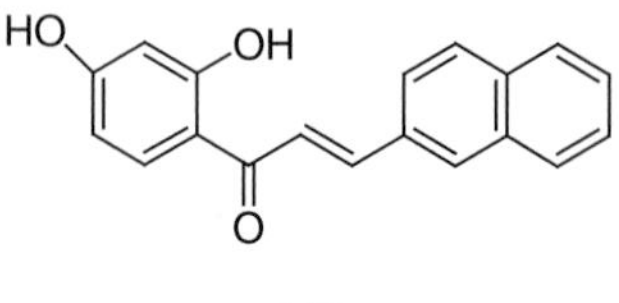

119

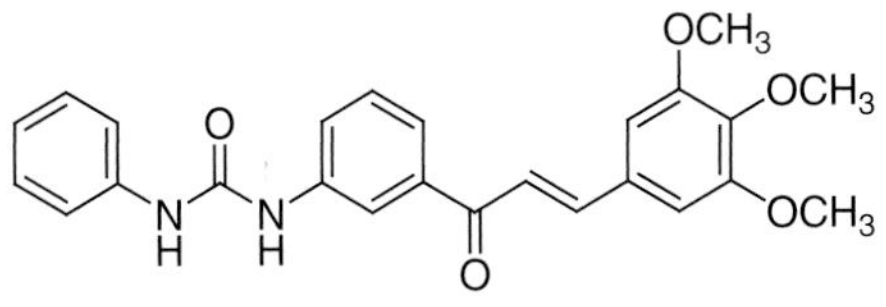

120

7.27.5.1.13 Miscellaneous compounds

In addition to various antimalarial drug classes described above, a number of interesting compounds of synthetic or natural product origin with promising antiparasitic activities have been reported over the last several years and are presented in **Table 10**.

7.27.5.2 Combination Therapies

Since parasites have developed resistance to most of the existing drugs, new therapeutic strategies need to be considered and implemented to manage the morbidities and mortalities due to *Plasmodium* infections. As continuous exposure of parasites to subtherapeutic levels of a drug is one of the reasons for development of resistance, combination of two or more antimalarial drugs having different biological targets, with additive or synergistic interactions, has been identified as a viable strategy to deal with this situation. While choosing a combination of drugs, it is necessary to make sure that there is no harmful interaction between them in vivo. The final goal is "effective parasite clearance in the shortest duration of time," and biological characteristics of individual components in a drug combination could show profound influence on the effectiveness of the treatment. Blood elimination half-life is an important factor in the cases of erythrocytic schizonticides since early elimination of one of the components may lead to selection pressure from the other. However, if the component with shorter half-life can clear the majority of parasites (e.g., artemisinin analogs), the second drug with the longer half-life (e.g., mefloquine, lumefantrine, or amodiaquine) would face only a few parasites, with low risk of resistance development.[471]

Sulfadoxine/pyrimethamine (SP) is a useful combination that is available as tablets or as an injectable solution. Even though SP was highly effective against *P. falciparum*, the longer half-lives of the components have resulted in resistance development in areas of high transmission such as East Africa.[358] Studies showing the synergistic interaction between atovaquone and proguanil have led to the development of a combination involving these two drugs (Malarone), which is highly effective against *P. falciparum*, including its chloroquine- and mefloquine-resistant strains. Chloroproguanil plus dapsone, known as Lapdap, is another combination currently undergoing advanced clinical trials. In areas of *P. falciparum* infections, where sensitivity to quinine is declining, a combination of quinine with doxycycline is another interesting option.[248,472]

Unique properties of artemisinin and its analogs, such as rapid parasite clearance, activity against multidrug-resistant parasites, limited toxicity, reduction of gametocyte carriage, and rapid resolution of clinical symptoms, have made them important ingredients in a number of drug combinations. A coformulation of artemether and lumefantrine, known as Coartem, is highly effective against multidrug-resistant strains of *P. falciparum* and is the most viable artemisinin-based combination now. Although no serious adverse effects have been documented, Coartem is not recommended for pregnant or lactating women, as safety in these groups remain to be established. Other interesting combinations based on artemisinin under development are (1) artesunate + mefloquine, (2) artesunate + amodiaquine, (3) Artesunate + SP, (4) artesunate + pyronaridine, and (5) chloroproguanil + dapsone + artesunate (CDA or Lapdap plus).[248,472]

7.27.5.3 Development of Malaria Vaccine

Nearly all stages in the *Plasmodium* life cycle can be viewed as targets for vaccine development. While various laboratory-based experiments and field observations (such as the development of clinical immunity after continuous exposure to parasites in the case of individuals living in endemic areas) give hope for vaccine(s) against malaria, the main challenge comes from the complex life cycle of the parasites, having a number of stages expressing a wide variety of surface antigens.[473] This, coupled with the lack of proper in vitro and in vivo surrogate assays, slows down the entire process. One of the most promising results pertaining to vaccines against sporozoite-stage parasites came from studies using irradiated sporozoites which conferred protection against infections in mice, nonhuman primates, and humans.[474,475] The main limitation with this approach is the difficulty in having a large amount of parasites for vaccine development. CSP, which is expressed in sporozoites and liver stage parasites, is one of the pre-erythrocytic vaccine candidates that have undergone detailed investigation.[476] Since sporozoites remain in blood only for a very short duration, the vaccines targeting them should be able to elicit high levels of antibody response continuously, which is a major limitation.

Table 10 New antimalarial lead candidates

Compound	*Proposed targets/sites of action*	*Comments*
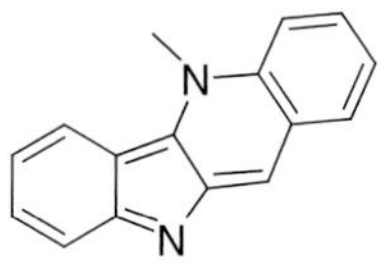 **121** Cryptolepine	DNA intercalation and stabilization of DNA topoisomerase II complex/ the inhibition of hemozoin formation	Isolated from West African climbing herb *Cryptolepis sanguinolenta*; IC_{50} (*P. falciparum*, K1 strain) = 0.44±0.22 μM; although cytotoxicity limits its use as antimalarial agent, suitable derivatization which modulates its interactions with DNA and topoisomerase II may improve its therapeutic profile.[451,452]
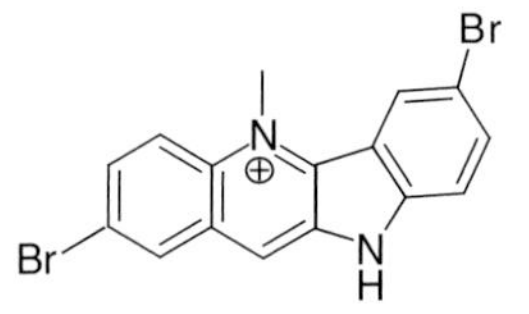 **122** 2,7-Dibromocryptolepine	Same as for **121**	**122** showed an ED_{50} of 6.92 mg kg^{-1} when administered intraperitoneally in mice infected with *P. berghei*.[451] At a 25 mg kg^{-1} dose, parasitemia was suppressed by 91.4% with no significant toxicity.
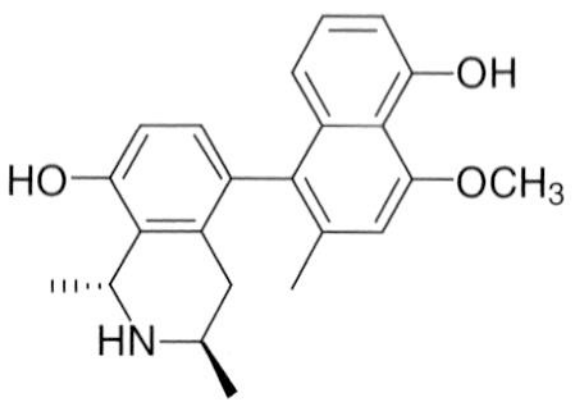 **123** Dioncophylline C	—	In vivo studies in *P. berghei* infected mice at 50 mg kg^{-1} day^{-1} × 4 dose of **123** (orally) caused complete clearance of parasitemia without noticeable toxicity; ED_{50} (orally) = 10.71 mg kg^{-1}.[452,453]
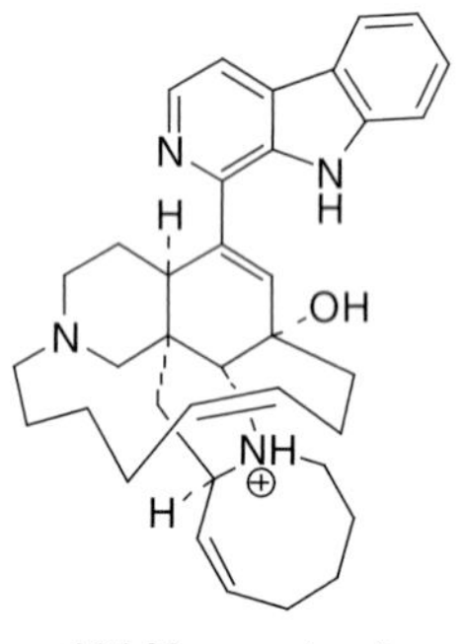 **124** Manzamine A	—	Manzamine A is a beta-carboline alkaloid isolated from a marine sponge species. A single intrapentoneal administration of **124** at 50 μmol kg^{-1} dose prolonged the survival time of *P. berghei* infected mice for more than 10 days.[454] It has a narrow therapeutic index, but it should be possible to develop new analogs with better therapeutic profile based on existing literature reports.[455,456]
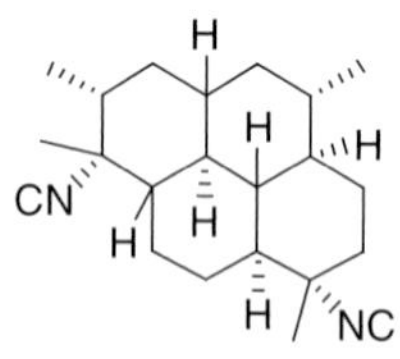 **125** Diisocyanoadociane	Heme	IC_{50} (*P. falciparum*, D6) = 14.48 nM.[457]
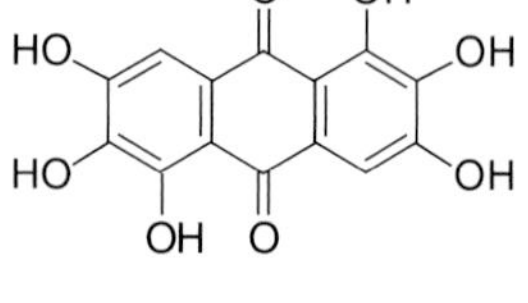 **126** Ruffigallol	Heme polymerization	IC_{50} = 35 nM (*P. falciparum*, D6)[458]; compound **126** showed considerable synergy when combined with exifone (**127**), through the formation of an active xanthone intermediate.[459]

Table 10 Continued

Compound	*Proposed targets/sites of action*	*Comments*
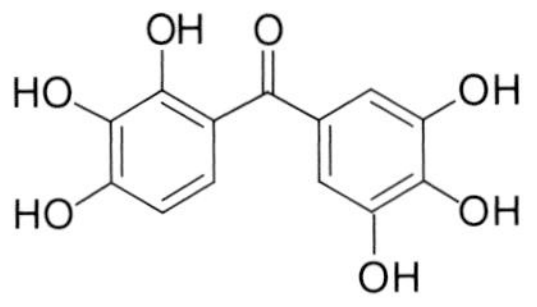 **127** Exifone	Heme polymerization	$IC_{50} = 4.1\ \mu M$ (*P. falciparum*, D6).[452]
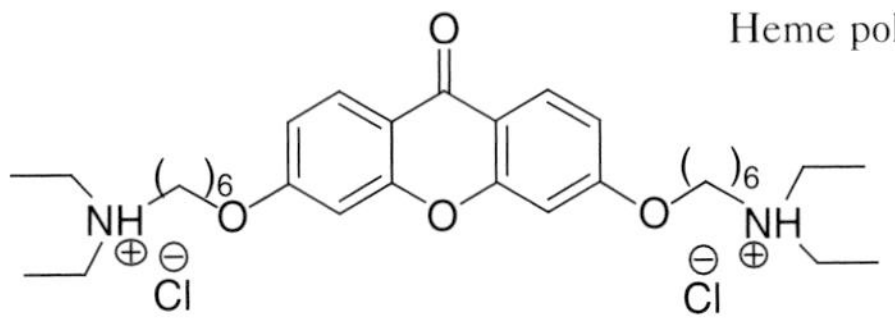 **128**	Heme polymerization	$IC_{50} = 0.07\ \mu M$ (*P. falciparum*, W2).[460]
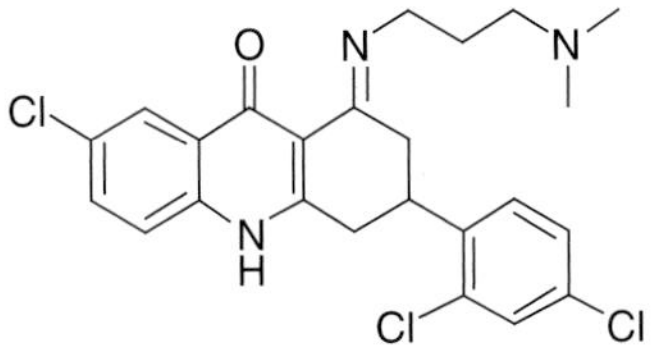 **129** WR243251	Heme/cellular respiration[461,462]	With a dose of $16\ mg\ kg^{-1}$ of **129** for 3 days, all *Aotus* monkeys studied (infected with *P. falciparum*, CQ resistant strain) were cured.[463]
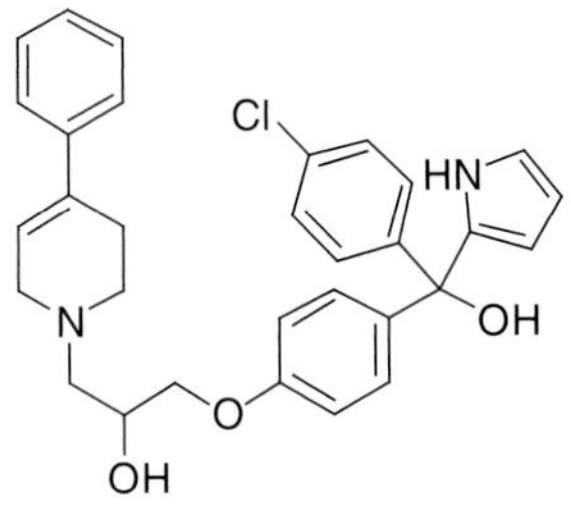 **130** Ro 06-9075	Heme polymerization	**130** was identified by high throughput screening; IC_{50} (*P. falciparum*, K1 strain) $= 0.97\ \mu M$; it displayed oral activity in mice infected with *P. berghei*, with an ED_{90} of about $100\ mg\ kg^{-1}$.[464]
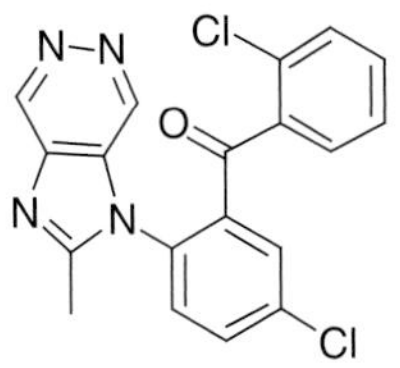 **131** Ro 22-8014	Heme polymerization	IC_{50} (*P. falciparum*, K1 strain) $= 1.32\ \mu M$.[464]
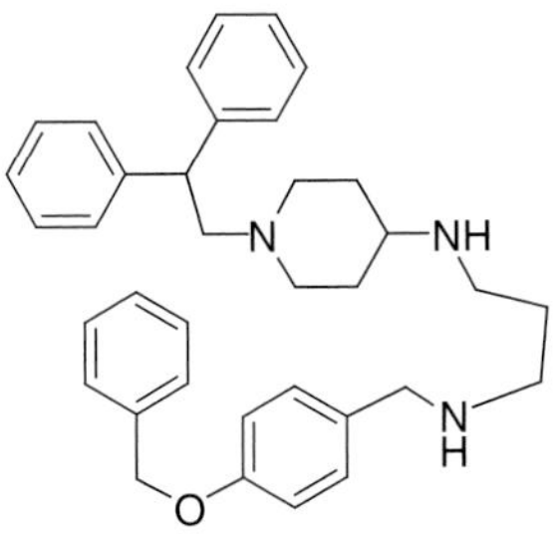 **132**	Heme polymerization	$IC_{50} = 11\ nM$ and $70\ nM$, respectively, against CQ sensitive and resistant strains.[190]

continued

Table 10 Continued

Compound	*Proposed targets/sites of action*	*Comments*
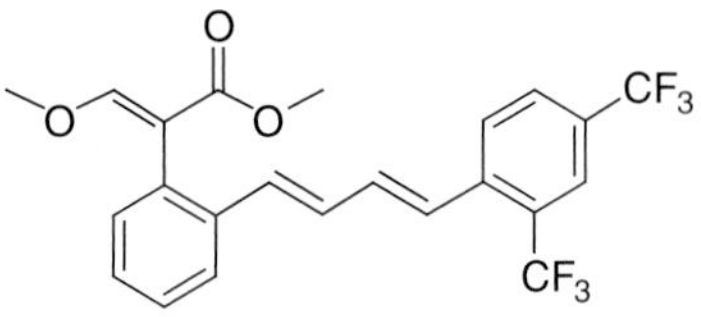 **133**	Mitochondrial electron transport	**133** showed IC_{50} values of 0.06 nM and 0.13 nM, respectively, against CQ sensitive and resistant strains of *P. falciparum*; $ED_{50/90}$ (*P. berghei*) = 0.42/1.44 mg kg^{-1} (orally).[465]
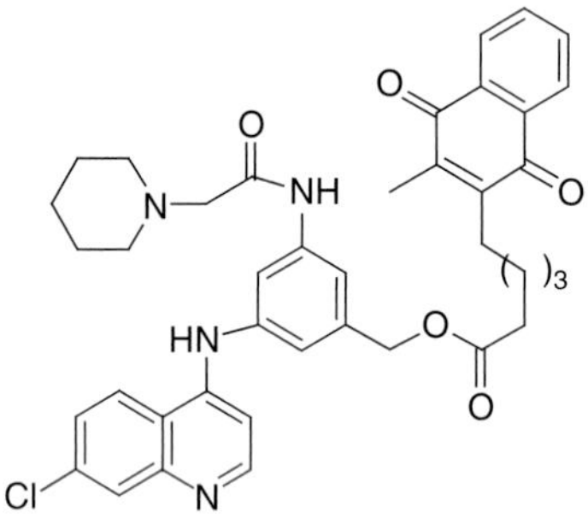 **134**	Heme/glutathione reductase	IC_{50} (*P. falciparum*, FcB1R) = 23.1 ± 6.9 nM; a 40 mg kg^{-1} day^{-1} × 4 dose (orally) of **134** significantly prolonged the survival time of mice infected with *P. berghei*.[466]
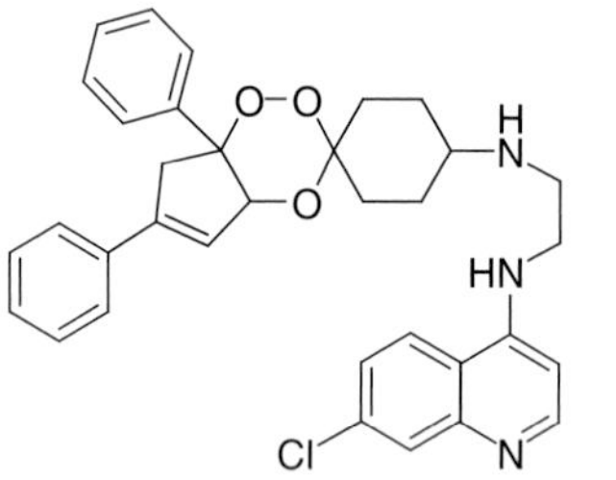 **135** DU-1102	—	Trioxaquine **135** has the structural features of peroxidic and 4-aminoquinoline classes of antimalarials. It showed a mean IC_{50} value of 227 nM when tested against 19 CQ resistant *P. falciparum* isolates from Cameroon.[467]
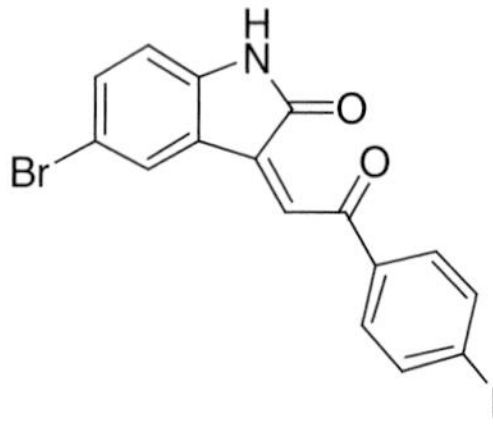 **136**	Pfmrk, PfPK5	IC_{50}: Pfmrk = 1.40 μM, PfPK5 = 190 μM, and CDK1 = 29 μM. This compound did not show noticeable antimalarial activity in vitro.[468]
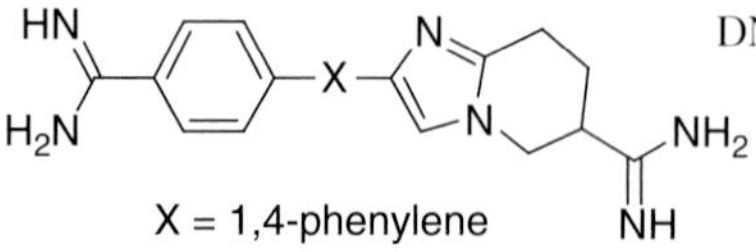 X = 1,4-phenylene **137**	DNA binding	IC_{50} (*P. falciparum*) = 14 nM.[469]
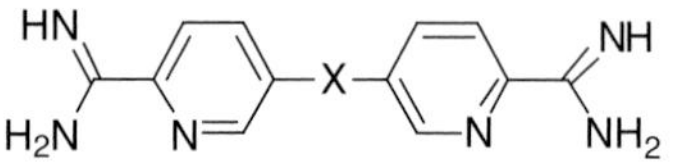 X = 2,5-furanyl **138**	DNA binding	IC_{50} (*P. falciparum*) = 3.9 nM.[470]

Another vaccine, RTS,S/SBAS2, with portions containing repeats and the C-terminal regions of CSP fused to the hepatitis B virus surface antigen (HBsAg), is another important candidate which has shown protection in six out of seven human volunteers.[477,478] However, immunity was short-lived, which was a disadvantage.[479] Of the other liver stage antigens, LSA-1 and LSA-3 have received special attention in recent years and are undergoing preclinical evaluation.[480–482]

MSPs, such as MSP-3 and MSP-1_{19} (derived from MSP-1),[483–486] and those associated with rhoptries, such as AMA-1,[487,488] RAP-1, and RAP-2,[489] are the important erythrocyte-stage vaccine candidates undergoing detailed examination. Ookinete surface proteins P25 and P28, another set of proteins, are attractive transmission-blocking vaccine candidates which can prevent the development of oocysts in mosquitoes.[490,491] Based on the notion that a combination of antigens from different parasitic stages can offer better protection, a number of 'multiantigen' systems have been developed and studied over the last few years. One of the important candidates from this is SPf66, having three blood stage antigens linked together by the NANP peptide derived from CSP.[475,492,493] Even though this showed high protective effects in trials conducted in Colombia, the same could not be confirmed by subsequent studies.[475] Another approach involved using viral vectors carrying genes coding for proteins expressed during various stages in the parasite's life cycle. One of the candidates from this group, NYVAC-Pf7, is a pox-vectored vaccine candidate that can express proteins from sporozoite (CSP and TRAP), liver (LSA-1), erythrocytic stage (MSP-1, SERP, AMA-1) and sexual stages (Pfs25), but in a phase I/IIa trial, only one out of 35 volunteers showed protection from infection.[494] Other multiantigen vaccine candidates under development are: (1) CDC/NII MAL VAC-1 (containing 21 B and T cell epitopes of pre-erythrocytic, erythrocytic, and sexual stages of the parasite, such as CSP, LSA-1, MSP-1, TRAP, MSP-2, AMA-1, RAP-1, EBA-175, and Pfg27),[475,495] (2) FALVAC-2 (containing MSP-1_{19}, Pfs25, region II of EBA-175 and 30 B cell epitopes and 25 T cell epitopes from 13 stage-specific antigens),[475] and (3) recombinant FP19 and MVA vaccines encoding the pre-erythrocytic stage malaria antigen TRAP and multiepitope string.[496,497] Although there are encouraging results from these studies, limitations associated with the multiantigen approach are the difficulty in identifying all the antigens from different parasitic stages, their delivery and the necessity to elicit the required level of immune response from them.

DNA vaccine strategy, which is based on the feasibility of expressing foreign DNA in a mammalian cell to give the corresponding protein, is another important method useful in malaria vaccine development. Early studies in mice and monkey models, with DNA plasmids encoding various malaria antigens have given promising results, but the relatively low immunogenicity of DNA vaccine candidates is a challenge. Since it is well suited for the multiantigen approach, there exists much hope in this area for the development of new vaccine candidates carrying diverse antigenic sites.[498–501] Even though there are encouraging results, considering the complex nature of the parasites, their interactions with each other and with hosts, it may take several more decades to develop an effective and widely applicable malaria vaccine. Strong partnership involving developing countries, creative strategies, and leadership from developed nations are critical in achieving this goal faster.[502]

7.27.6 Summary and Future Prospects

Morbidities and mortalities associated with malaria continue to be a major challenge of the twenty-first century. The resistance developed by plasmodium parasites to conventional drugs, such as chloroquine and mefloquine, and the emergence of infections in new areas of the world are the important factors that hinder malaria control programs. Despite millions of people being at risk, malaria remains economically unattractive to pharmaceutical industries, as the total turnover from investment could be rather low. Since affordability of drugs is a major criterion that determines success in underdeveloped nations, all drugs, whether alone or in combination, should be accessible to the public at a reasonable price. As the prevalence of parasite species is different in different parts of the globe, the selection of drugs (or their combination) should be based on their effectiveness against the particular species or strains. Some of the ongoing drug development efforts include (1) the structural modification of existing drugs to improve their potency or to overcome resistance, (2) identifying the combination of two or more drugs with additive or synergistic interactions, (3) identification of new lead candidates by high-throughput screening of natural products or synthetic molecules, and (4) structure-based approaches including virtual database screening or rational design based on individual targets.

Of all the available drugs, artemisinin and its derivatives are the only class against which parasites have not yet developed resistance. Even though the number of new drugs at advanced stages of preclinical or clinical developments is limited, the structural information available about various enzymes associated with metabolic pathways, such as fatty acid biosynthesis, isoprenoid biosynthesis, glycolytic pathway, and choline uptake, should greatly assist the drug development efforts of the next decade. Of these pathways, those which have minimum interactions with that in humans are of particular importance. Most of the existing drugs target erythrocytic parasites, and the development of new candidates that target liver stage parasites, or those which can reduce gametocyte carriage should also reduce infection and transmission rates. In order to improve the efficacy of existing drugs, it is important to understand their mechanisms of action so that the three-dimensional structural information about the target can be used for their structure optimization using computational methods. Similarly, information pertaining to the resistance development would indeed provide ideas to introduce new structural features that would help overcome the resistance. Since 'prevention is better than cure,' it is important to implement new strategies or improve the existing ones for effective vector control. Identification of new prophylactic agents is another area that could reduce the number of infections. Even though the development of vaccines against malaria still remains a major challenge, the information available from *Plasmodium* genome data and the feasibility of expressing multiple antigens through various biochemical tools would be helpful in identifying and studying new antigens, alone or in combination. Strong international collaboration, such as that envisioned by the Roll Back Malaria partnership, is going to be important for the success of malaria control programs of the next decade.

Acknowledgment

We thank the Centers for Disease Control and Prevention for financial support (CDC cooperative agreements 1U01 CI000211-02 and U50/CCU423310-03).

References

1. *World Malaria Report*; World Health Organization: Geneva, Switzerland, 2005.
2. Sachs, J.; Malaney, P. *Nature* **2002**, *415*, 680–685.
3. Phillips, R. S. *Clin. Microbiol. Rev.* **2001**, *14*, 208–226.
4. Barsoum, R. S. *J. Am. Soc. Nephrol.* **2000**, *11*, 2147–2154.
5. Kallander, K.; Nsungwa-Sabiiti, J.; Peterson, S. *Acta Trop.* **2004**, *90*, 211–214.
6. Nsungwa-Sabiiti, J.; Kallander, K.; Nsabagasani, X.; Namusisi, K.; Pariyo, G.; Johansson, A.; Tomson, G.; Peterson, S. *Trop. Med. Int. Health* **2004**, *9*, 1191–1199.
7. Miller, L. H.; Baruch, D. I.; Marsh, K.; Doumbo, O. K. *Nature* **2002**, *415*, 673–679.
8. White, N. J. *N. Engl. J. Med.* **1996**, *335*, 800–806.
9. Krotoski, W. A. *Prog. Clin. Parasitol.* **1989**, *1*, 1–19.
10. Makler, M. T.; Palmer, C. J.; Ager, A. L. *Ann. Trop. Med. Parasitol.* **1998**, *92*, 419–433.
11. Makler, M. T.; Piper, R. C.; Milhous, W. K. *Parasitol. Today* **1998**, *14*, 376–377.
12. Soldati, D.; Foth, B. J.; Cowman, A. F. *Trends Parasitol.* **2004**, *20*, 567–574.
13. Tewari, R.; Rathore, D.; Crisanti, A. *Cell. Microbiol.* **2005**, *7*, 699–707.
14. Ancsin, J. B.; Kisilevsky, R. *J. Biol. Chem.* **2004**, *279*, 21824–21832.
15. Rathore, D.; McCutchan, T. F.; Garboczi, D. N.; Toida, T.; Hernaiz, M. J.; LeBrun, L. A.; Lang, S. C.; Linhardt, R. J. *Biochemistry* **2001**, *40*, 11518–11524.
16. Rathore, D.; Hrstka, S. C. L.; Sacci, J. B., Jr.; De la Vega, P.; Linhardt, R. J.; Kumar, S.; McCutchan, T. F. *J. Biol. Chem.* **2003**, *278*, 40905–40910.
17. Rathore, D.; Sacci, J. B.; De la Vega, P.; McCutchan, T. F. *J. Biol. Chem.* **2002**, *277*, 7092–7098.
18. Masao, Y.; Tomoko, I. *Cell. Microbiol.* **2004**, *6*, 1119–1125.
19. Ishino, T.; Chinzei, Y.; Yuda, M. *Cell. Microbiol.* **2005**, *7*, 199–208.
20. Ishino, T.; Yano, K.; Chinzei, Y.; Yuda, M. *PLoS Biol.* **2004**, *2*, 77–84.
21. Sultan, A. A. *Int. Microbiol.* **1999**, *2*, 155–160.
22. Sultan, A. A.; Thathy, V.; Frevert, U.; Robson, K. J. H.; Crisanti, A.; Nussenzweig, V.; Nussenzweig, R. S.; Menard, R. *Cell* **1997**, *90*, 511–522.
23. Silvie, O.; Franetich, J.-F.; Charrin, S.; Mueller, M. S.; Siau, A.; Bodescot, M.; Rubinstein, E.; Hannoun, L.; Charoenvit, Y.; Kocken, C. H. et al. *J. Biol. Chem.* **2004**, *279*, 9490–9496.
24. Chitnis, C. E. *Curr. Opin. Hematol.* **2001**, *8*, 85–91.
25. Chitnis, C. E.; Blackman, M. J. *Parasitol. Today* **2000**, *16*, 411–415.
26. Sim, B. K. L.; Chitnis, C. E.; Wasniowska, K.; Hadley, T. J.; Miller, L. H. *Science* **1994**, *264*, 1941–1944.
27. Gaur, D.; Mayer, D. C. G.; Miller, L. H. *Int. J. Parasitol.* **2004**, *34*, 1413–1429.
28. Ralph, S. A.; van Dooren, G. G.; Waller, R. F.; Crawford, M. J.; Fraunholz, M. J.; Foth, B. J.; Tonkin, C. J.; Roos, D. S.; McFadden, G. I. *Nat. Rev. Microbiol.* **2004**, *2*, 203–216.
29. Diggs, C. L.; Ballou, W. R.; Miller, L. H. *Parasitol. Today* **1993**, *9*, 300–302.
30. Miller, L. H.; Roberts, T.; Shahabuddin, M.; McCutchan, T. F. *Mol. Biochem. Parasitol.* **1993**, *59*, 1–14.
31. Smythe, J. A.; Coppel, R. L.; Brown, G. V.; Ramasamy, R.; Kemp, D. J.; Anders, R. F. *Proc. Natl. Acad. Sci. USA* **1988**, *85*, 5195–5199.
32. Smythe, J. A.; Coppel, R. L.; Day, K. P.; Martin, R. K.; Oduola, A. M.; Kemp, D. J.; Anders, R. F. *Proc. Natl. Acad. Sci. USA* **1991**, *88*, 1751–1755.
33. Oeuvray, C.; Bouharoun-Tayoun, H.; Grass-Masse, H.; Lepers, J. P.; Ralamboranto, L.; Tartar, A.; Druilhe, P. *Mem. Inst. Oswaldo Cruz* **1994**, *89*, 77–80.
34. McColl, D. J.; Silva, A.; Foley, M.; Kun, J. F.; Favaloro, J. M.; Thompson, J. K.; Marshall, V. M.; Coppel, R. L.; Kemp, D. J.; Anders, R. F. *Mol. Biochem. Parasitol.* **1994**, *68*, 53–67.
35. McColl, D. J.; Anders, R. F. *Mol. Biochem. Parasitol.* **1997**, *90*, 21–31.
36. Marshall, V. M.; Silva, A.; Foley, M.; Cranmer, S.; Wang, L.; McColl, D. J.; Kemp, D. J.; Coppel, R. L. *Infect. Immun.* **1997**, *65*, 4460–4467.
37. Wu, T.; Black, C. G.; Wang, L.; Hibbs, A. R.; Coppel, R. L. *Mol. Biochem. Parasitol.* **1999**, *103*, 243–250.
38. Kedzierski, L.; Black, C. G.; Goschnick, M. W.; Stowers, A. W.; Coppel, R. L. *Infect. Immun.* **2002**, *70*, 6606–6613.

39. Singh, S.; Soe, S.; Roussilhon, C.; Corradin, G.; Druilhe, P. *Infect. Immun.* **2005**, *73*, 1235–1238.
40. Pearce, J. A.; Triglia, T.; Hodder, A. N.; Jackson, D. C.; Cowman, A. F.; Anders, R. F. *Infect. Immun.* **2004**, *72*, 2321–2328.
41. Pachebat, J. A.; Ling, I. T.; Grainger, M.; Trucco, C.; Howell, S.; Fernandez-Reyes, D.; Gunaratne, R.; Holder, A. A. *Mol. Biochem. Parasitol.* **2001**, *117*, 83–89.
42. Puentes, A.; Garcia, J.; Ocampo, M.; Rodriguez, L.; Vera, R.; Curtidor, H.; Lopez, R.; Suarez, J.; Valbuena, J.; Vanegas, M. et al. *Peptides* **2003**, *24*, 1015–1023.
43. Vargas-Serrato, E.; Barnwell, J. W.; Ingravallo, P.; Perler, F. B.; Galinski, M. R. *Mol. Biochem. Parasitol.* **2002**, *120*, 41–52.
44. Puentes, A.; Ocampo, M.; Rodriguez, L. E.; Vera, R.; Valbuena, J.; Curtidor, H.; Garcia, J.; Lopez, R.; Tovar, D.; Cortes, J. et al. *Biochimie* **2005**, *87*, 461–472.
45. Black, C. G.; Wang, L.; Wu, T.; Coppel, R. L. *Mol. Biochem. Parasitol.* **2003**, *127*, 59–68.
46. Ak, M.; Babaoglu, A.; Dagci, H.; Tuerk, M.; Bayram, S.; Ertabaklar, H.; Oezcel, M. A.; Uener, A.; Charoenvit, Y.; Kumar, S. et al. *Hybrid. Hybridomics* **2004**, *23*, 133–136.
47. Rodriguez, L. E.; Urquiza, M.; Ocampo, M.; Curtidor, H.; Suarez, J.; Garcia, J.; Vera, R.; Puentes, A.; Lopez, R.; Pinto, M. et al. *Vaccine* **2002**, *20*, 1331–1339.
48. De Oliveira, C. I.; Wunderlich, G.; Levitus, G.; Soares, I. S.; Rodrigues, M. M.; Tsuji, M.; Del Portillo, H. A. *Vaccine* **1999**, *17*, 2959–2968.
49. del Portillo, H. A.; Longacre, S.; Khouri, E.; David, P. H. *Proc. Natl. Acad. Sci. USA* **1991**, *88*, 4030–4034.
50. Gibson, H. L.; Tucker, J. E.; Kaslow, D. C.; Krettli, A. U.; Collins, W. E.; Kiefer, M. C.; Bathurst, I. C.; Barr, P. J. *Mol. Biochem. Parasitol.* **1992**, *50*, 325–333.
51. Rayner, J. C.; Corredor, V.; Feldman, D.; Ingravallo, P.; Iderabdullah, F.; Galinski, M. R.; Barnwell, J. W. *Parasitology* **2002**, *125*, 393–405.
52. Galinski, M. R.; Corredor-Medina, C.; Povoa, M.; Crosby, J.; Ingravallo, P.; Barnwell, J. W. *Mol. Biochem. Parasitol.* **1999**, *101*, 131–147.
53. Galinski, M. R.; Ingravallo, P.; Corredor-Medina, C.; Al-Khedery, B.; Povoa, M.; Barnwell, J. W. *Mol. Biochem. Parasitol.* **2001**, *115*, 41–53.
54. Black, C. G.; Barnwell, J. W.; Huber, C. S.; Galinski, M. R.; Coppel, R. L. *Mol. Biochem. Parasitol.* **2002**, *120*, 215–224.
55. Perez-Leal, O.; Sierra, A. Y.; Barrero, C. A.; Moncada, C.; Martinez, P.; Cortes, J.; Lopez, Y.; Torres, E.; Salazar, L. M.; Patarroyo, M. A. *Biochem. Biophys. Res. Commun.* **2004**, *324*, 1393–1399.
56. Oliveira-Ferreira, J.; Vargas-Serrato, E.; Barnwell, J. W.; Moreno, A.; Galinski, M. R. *Vaccine* **2004**, *22*, 2023–2030.
57. Vargas-Serrato, E.; Corredor, V.; Galinski, M. R. *J. Mol. Epidemiol. Evol. Genet. Infect. Diseases* **2003**, *3*, 67–73.
58. Garman, S. C.; Simcoke, W. N.; Stowers, A. W.; Garboczi, D. N. *J. Biol. Chem.* **2003**, *278*, 7264–7269.
59. Blackman, M. J.; Dennis, E. D.; Hirst, E. M. A.; Kocken, C. H.; Scott-Finnigan, T. J.; Thomas, A. W. *Exp. Parasitol.* **1996**, *83*, 229–239.
60. Longacre, S. *Mol. Biochem. Parasitol.* **1995**, *74*, 105–111.
61. Gerold, P.; Schofield, L.; Blackman, M. J.; Holder, A. A.; Schwarz, R. T. *Mol. Biochem. Parasitol.* **1996**, *75*, 131–143.
62. Sim, B. K. L. *Parasitol. Today* **1995**, *11*, 213–217.
63. Sim, B. K. L.; Toyoshima, T.; Haynes, J. D.; Aikawa, M. *Mol. Biochem. Parasitol.* **1992**, *51*, 157–159.
64. Duraisingh, M. T.; Maier, A. G.; Triglia, T.; Cowman, A. F. *Proc. Natl. Acad. Sci. USA* **2003**, *100*, 4796–4801.
65. Gruner, A. C.; Brahimi, K.; Letourneur, F.; Renia, L.; Eling, W.; Snounou, G.; Druilhe, P. *J. Infect. Dis.* **2001**, *184*, 892–897.
66. Bharara, R.; Singh, S.; Pattnaik, P.; Chitnis, C. E.; Sharma, A. *Mol. Biochem. Parasitol.* **2004**, *138*, 123–129.
67. Valbuena, J. J.; Bravo, R. V.; Ocampo, M.; Lopez, R.; Rodriguez, L. E.; Curtidor, H.; Puentes, A.; Garcia, J. E.; Tovar, D.; Gomez, J. et al. *Biochem. Biophys. Res. Commun.* **2004**, *321*, 835–844.
68. Mons, B. *Blood Cells* **1990**, *16*, 299–312.
69. Galinski, M. R.; Medina, C. C.; Ingravallo, P.; Barnwell, J. W. *Cell* **1992**, *69*, 1213–1226.
70. Galinski, M. R.; Xu, M.; Barnwell, J. W. *Mol. Biochem. Parasitol.* **2000**, *108*, 257–262.
71. Galinski, M. R.; Barnwell, J. W. *Parasitol. Today* **1996**, *12*, 20–29.
72. Triglia, T.; Thompson, J.; Caruana, S. R.; Delorenzi, M.; Speed, T.; Cowman, A. F. *Infect. Immun.* **2001**, *69*, 1084–1092.
73. Ocampo, M.; Vera, R.; Rodriguez, L. E.; Curtidor, H.; Suarez, J.; Garcia, J.; Puentes, A.; Lopez, R.; Valbuena, J.; Tovar, D. et al. *Parasitol. Int.* **2004**, *53*, 77–88.
74. Triglia, T.; Duraisingh, M. T.; Good, R. T.; Cowman, A. F. *Mol. Microbiol.* **2005**, *55*, 162–174.
75. Khan, S. M.; Jarra, W.; Preiser, P. R. *Mol. Biochem. Parasitol.* **2001**, *117*, 1–10.
76. Keen, J. K.; Sinha, K. A.; Brown, K. N.; Holder, A. A. *Mol. Biochem. Parasitol.* **1994**, *65*, 171–177.
77. Ogun, S. A.; Holder, A. A. *Mol. Biochem. Parasitol.* **1996**, *76*, 321–324.
78. Peterson, M. G.; Marshall, V. M.; Smythe, J. A.; Crewther, P. E.; Lew, A.; Silva, A.; Anders, R. F.; Kemp, D. J. *Mol. Cell. Biol.* **1989**, *9*, 3151–3154.
79. Cheng, Q.; Saul, A. *Mol. Biochem. Parasitol.* **1994**, *65*, 183–187.
80. Waters, A. P.; Thomas, A. W.; Deans, J. A.; Mitchell, G. H.; Hudson, D. E.; Miller, L. H.; McCutchan, T. F.; Cohen, S. *J. Biol. Chem.* **1990**, *265*, 17974–17979.
81. Peterson, M. G.; Nguyen-Dinh, P.; Marshall, V. M.; Elliott, J. F.; Collins, W. E.; Anders, R. F.; Kemp, D. J. *Mol. Biochem. Parasitol.* **1990**, *39*, 279–283.
82. Marshall, V. M.; Peterson, M. G.; Lew, A. M.; Kemp, D. J. *Mol. Biochem. Parasitol.* **1989**, *37*, 281–283.
83. Kappe, S. H.; Noe, A. R.; Fraser, T. S.; Blair, P. L.; Adams, J. H. *Proc. Natl. Acad. Sci. USA* **1998**, *95*, 1230–1235.
84. Noe, A. R.; Adams, J. H. *Mol. Biochem. Parasitol.* **1998**, *96*, 27–35.
85. Blair, P. L.; Kappe, S. H. I.; Maciel, J. E.; Balu, B.; Adams, J. H. *Mol. Biochem. Parasitol.* **2002**, *122*, 35–44.
86. Kariu, T.; Yuda, M.; Yano, K.; Chinzei, Y. *J. Exp. Med.* **2002**, *195*, 1317–1323.
87. Cooper, J. A.; Ingram, L. T.; Bushell, G. R.; Fardoulys, C. A.; Stenzel, D.; Schofield, L.; Saul, A. J. *Mol. Biochem. Parasitol.* **1988**, *29*, 251–260.
88. Kaneko, O.; Tsuboi, T.; Ling, I. T.; Howell, S.; Shirano, M.; Tachibana, M.; Cao, Y. M.; Holder, A. A.; Torii, M. *Mol. Biochem. Parasitol.* **2001**, *118*, 223–231.
89. Holder, A. A.; Freeman, R. R.; Uni, S.; Aikawa, M. *Mol. Biochem. Parasitol.* **1985**, *14*, 293–303.
90. Bushell, G. R.; Ingram, L. T.; Fardoulys, C. A.; Cooper, J. A. *Mol. Biochem. Parasitol.* **1988**, *28*, 105–112.
91. Clark, J. T.; Anand, R.; Akoglu, T.; McBride, J. S. *Parasitol. Res.* **1987**, *73*, 425–434.
92. Howard, R. F.; Stanley, H. A.; Campbell, G. H.; Reese, R. T. *Am. J. Trop. Med. Hyg.* **1984**, *33*, 1055–1059.
93. Jakobsen, P. H.; Kurtzhals, J. A.; Riley, E. M.; Hviid, L.; Theander, T. G.; Morris-Jones, S.; Jensen, J. B.; Bayoumi, R. A.; Ridley, R. G.; Greenwood, B. M. *Parasit. Immunol.* **1997**, *19*, 387–393.
94. Ridley, R. G.; Lahm, H. W.; Takacs, B.; Scaife, J. G. *Mol. Biochem. Parasitol.* **1991**, *47*, 245–246.
95. Ridley, R. G.; Takacs, B.; Etlinger, H.; Scaife, J. G. *Parasitology* **1990**, *101* (Pt 2), 187–192.

96. Ridley, R. G.; Takacs, B.; Lahm, H. W.; Delves, C. J.; Goman, M.; Certa, U.; Matile, H.; Woollett, G. R.; Scaife, J. G. *Mol. Biochem. Parasitol.* **1990**, *41*, 125–134.
97. Schofield, L.; Bushell, G. R.; Cooper, J. A.; Saul, A. J.; Upcroft, J. A.; Kidson, C. *Mol. Biochem. Parasitol.* **1986**, *18*, 183–195.
98. Howard, R. F.; Reese, R. T. *Exp. Parasitol.* **1990**, *71*, 330–342.
99. Baldi, D. L.; Andrews, K. T.; Waller, R. F.; Roos, D. S.; Howard, R. F.; Crabb, B. S.; Cowman, A. F. *EMBO J.* **2000**, *19*, 2435–2443.
100. Howard, R. F.; Schmidt, C. M. *Mol. Biochem. Parasitol.* **1995**, *74*, 43–54.
101. Lustigman, S.; Anders, R. F.; Brown, G. V.; Coppel, R. L. *Mol. Biochem. Parasitol.* **1988**, *30*, 217–224.
102. Sam-Yellowe, T. Y.; Shio, H.; Perkins, M. E. *J. Cell Biol.* **1988**, *106*, 1507–1513.
103. Snow, R. W.; Guerra, C. A.; Noor, A. M.; Myint, H. Y.; Hay, S. I. *Nature* **2005**, *434*, 214–217.
104. Mendis, K.; Sina, B. J.; Marchesini, P.; Carter, R. *Am. J. Trop. Med. Hyg.* **2001**, *64*, 97–106.
105. Martinez, P.; Suarez, C. F.; Cardenas, P. P.; Patarroyo, M. A. *Parasitology* **2004**, *128*, 353–366.
106. Duarte, E. C.; Gyorkos, T. W.; Pang, L.; Abrahamowicz, M. *Am. J. Trop. Med. Hyg.* **2004**, *70*, 229–237.
107. Muhlberger, N.; Jelinek, T.; Gascon, J.; Probst, M.; Zoller, T.; Schunk, M.; Beran, J.; Gjorup, I.; Behrens, R. H.; Clerinx, J. et al. *Malaria J.* **2004**, *3:5* (8 March 2004).
108. Jelinek, T.; Schulte, C.; Behrens, R.; Grobusch, M. P.; Coulaud, J. P.; Bisoffi, Z.; Matteelli, A.; Clerinx, J.; Corachan, M.; Puente, S. et al. *Clin. Infect. Dis.* **2002**, *34*, 572–576.
109. Han, E.-T.; Park, J.-H.; Shin, E.-H.; Choi, M.-H.; Oh, M.-D.; Chai, J.-Y. *Korean J. Parasitol.* **2002**, *40*, 157–162.
110. Park, J.-W.; Klein Terry, A.; Lee, H.-C.; Pacha Laura, A.; Ryu, S.-H.; Yeom, J.-S.; Moon, S.-H.; Kim, T.-S.; Chai, J.-Y.; Oh, M.-D. et al. *Am. J. Trop. Med. Hyg.* **2003**, *69*, 159–167.
111. Montalvo Alvarez, A. M.; Landau, I.; Baccam, D. *Rev. Inst. Med. Trop. São Paulo* **1991**, *33*, 421–426.
112. Kissinger, J. C.; Collins, W. E.; Li, J.; McCutchan, T. F. *J. Parasitol.* **1998**, *84*, 278–282.
113. Teklehaimanot Hailay, D.; Lipsitch, M.; Teklehaimanot, A.; Schwartz, J. *Malaria J.* **2004**, *3:41* (12 November 2004).
114. Cui, L.; Escalante, A. A.; Imwong, M.; Snounou, G. *Trends Parasitol.* **2003**, *19*, 220–226.
115. Paul, R. E. L.; Packer, M. J.; Walmsley, M.; Lagog, M.; Ranford-Cartwright, L. C.; Paru, R.; Day, K. P. *Science* **1995**, *269*, 1709–1711.
116. Arez, A. P.; Pinto, J.; Palsson, K.; Snounou, G.; Jaenson Thomas, G. T.; do Rosario, V. E. *Am. J. Trop. Med. Hyg.* **2003**, *68*, 161–168.
117. Triglia, T.; Wellems, T. E.; Kemp, D. J. *Parasitol. Today* **1992**, *8*, 225–229.
118. Kemp, D. J.; Cowman, A. F.; Walliker, D. *Adv. Parasitol.* **1990**, *29*, 75–149.
119. Joshi, H.; Subbarao, S. K.; Adak, T.; Nanda, N.; Ghosh, S. K.; Carter, R.; Sharma, V. P. *Trans. R. Soc. Trop. Med. Hyg.* **1997**, *91*, 231–235.
120. Joshi, H.; Subbarao, S. K.; Raghavendra, K.; Sharma, V. P. *Trans. R. Soc. Trop. Med. Hyg.* **1989**, *83*, 179–181.
121. Arez, A. P.; Palsson, K.; Pinto, J.; Franco, A. S.; Dinis, J.; Jaenson, T. G.; Snounou, G.; do Rosario, V. E. *Parassitologia* **1997**, *39*, 65–70.
122. Pinto, J.; Arez, A. P.; Franco, S.; do Rosario, V. E.; Palsson, K.; Jaenson, T. G.; Snounou, G. *Ann. Trop. Med. Parasitol.* **1997**, *91*, 217–219.
123. Wahlgren, M.; Perlmann, P. *Malaria: Molecular and Clinical Aspects*; Harwood: Chichester, UK, 1999.
124. Day, K. P.; Hayward, R. E.; Dyer, M. *Parasitology* **1998**, *116*, S95–S109.
125. Piper, K. P.; Hayward, R. E.; Cox, M. J.; Day, K. P. *Infect. Immun.* **1999**, *67*, 6369–6374.
126. Calderwood, M. S.; Gannoun-Zaki, L.; Wellems, T. E.; Deitsch, K. W. *J. Biol. Chem.* **2003**, *278*, 34125–34132.
127. Duraisingh, M. T.; Voss, T. S.; Marty, A. J.; Duffy, M. F.; Good, R. T.; Thompson, J. K.; Freitas-Junior, L. H.; Schert, A.; Crabb, B. S.; Cowman, A. F. *Cell* **2005**, *121*, 13–24.
128. Jensen, A. T. R.; Magistrado, P.; Sharp, S.; Joergensen, L.; Lavstsen, T.; Chiucchiuini, A.; Salanti, A.; Vestergaard, L. S.; Lusingu, J. P.; Hermsen, R. et al. *J. Exp. Med.* **2004**, *199*, 1179–1190.
129. Khattab, A.; Kremsner, P. G.; Klinkert, M.-Q. *J. Infect. Dis.* **2003**, *187*, 477–483.
130. Rowe, J. A.; Kyes, S. A.; Rogerson, S. J.; Babiker, H. A.; Raza, A. *J. Infect. Dis.* **2002**, *185*, 1207–1211.
131. Su, X. Z.; Heatwole, V. M.; Wertheimer, S. P.; Guinet, F.; Herrfeldt, J. A.; Peterson, D. S.; Ravetch, J. A.; Wellems, T. E. *Cell* **1995**, *82*, 89–100.
132. Smith, J. D.; Chitnis, C. E.; Craig, A. G.; Roberts, D. J.; Hudson-Taylor, D. E.; Peterson, D. S.; Pinches, R.; Newbold, C. I.; Miller, L. H. *Cell* **1995**, *82*, 101–110.
133. Sherman, I. W. *Malaria: Parasite Biology, Pathogenesis, and Protection*; American Society Microbiology Press: Washington, DC, 1998.
134. Haldane, J. B. S. *Hereditas* **1948**, *35*, 267–273.
135. Hill, A. V.; Elvin, J.; Willis, A. C.; Aidoo, M.; Allsopp, C. E.; Gotch, F. M.; Gao, X. M.; Takiguchi, M.; Greenwood, B. M.; Townsend, A. R. *Nature* **1992**, *360*, 434–439.
136. Olumese, P. E.; Adeyemo, A. A.; Ademowo, O. G.; Gbadegesin, R. A.; Sodeinde, O.; Walker, O. *Ann. Trop. Paediatr.* **1997**, *17*, 141–145.
137. Abu-Zeid, Y. A.; Abdulhadi, N. H.; Hviid, L.; Theander, T. G.; Saeed, B. O.; Jepsen, S.; Jensen, J. B.; Bayoumi, R. A. *Scand. J. Immunol.* **1991**, *34*, 237–242.
138. Abu-Zeid, Y. A.; Theander, T. G.; Abdulhadi, N. H.; Hviid, L.; Saeed, B. O.; Jepsen, S.; Jensen, J. B.; Bayoumi, R. A. *Clin. Exp. Immunol.* **1992**, *88*, 112–118.
139. Chitnis, C. E.; Chaudhuri, A.; Horuk, R.; Pogo, A. O.; Miller, L. H. *J. Exp. Med.* **1996**, *184*, 1531–1536.
140. Miller, L. H.; Mason, S. J.; Clyde, D. F.; McGinniss, M. H. *N. Engl. J. Med.* **1976**, *295*, 302–304.
141. Miller, L. H.; McGinniss, M. H.; Holland, P. V.; Sigmon, P. *Am. J. Trop. Med. Hyg.* **1978**, *27*, 1069–1072.
142. Ruwende, C.; Hill, A. *J. Mol. Med.* **1998**, *76*, 581–588.
143. Hill, A. V.; Allsopp, C. E.; Kwiatkowski, D.; Anstey, N. M.; Twumasi, P.; Rowe, P. A.; Bennett, S.; Brewster, D.; McMichael, A. J.; Greenwood, B. M. *Nature* **1991**, *352*, 595–600.
144. McGuire, W.; Hill, A. V.; Allsopp, C. E.; Greenwood, B. M.; Kwiatkowski, D. *Nature* **1994**, *371*, 508–510.
145. McGuire, W.; Knight, J. C.; Hill, A. V.; Allsopp, C. E.; Greenwood, B. M.; Kwiatkowski, D. *J. Infect. Dis.* **1999**, *179*, 287–290.
146. Toure, Y. T.; Oduola, A. M. J.; Morel, C. M. *Trends Parasitol.* **2004**, *20*, 142–149.
147. Holt, R. A.; Mani Subramanian, G.; Halpern, A.; Sutton, G. G.; Charlab, R.; Nusskern, D. R.; Wincker, P.; Clark, A. G.; Ribeiro, J. M. C.; Wides, R. et al. *Science* **2002**, *298*, 129–149.
148. Hoffman Stephen, L.; Subramanian, G. M.; Collins Frank, H.; Venter, J. C. *Nature* **2002**, *415*, 702–709.
149. Greenwood, B.; Mutabingwa, T. *Nature* **2002**, *415*, 670–672.
150. Najera, J. A.; Zaim, M. Malaria Vector Control: Decision Making Criteria and Procedures for Judicious Use of Insecticides, World Health Organization, 2002; WHO/CDS/WHOPES/2002.5/Rev.1.
151. Weill, M.; Lutfalia, G.; Mogensen, K.; Chandre, F.; Berthomieu, A.; Berticat, C.; Pasteur, N.; Philips, A.; Fort, P.; Raymond, M. *Nature* **2004**, *429*, 262.

152. Chandre, F.; Darrier, F.; Manga, L.; Akogbeto, M.; Faye, O.; Mouchet, J.; Guillet, P. *Bull. World Health Org.* **1999**, *77*, 230–234.
153. Alphey, L.; Beard, C. B.; Billingsley, P.; Coetzee, M.; Crisanti, A.; Curtis, C.; Eggleston, P.; Godfray, C.; Hemingway, J.; Jacobs-Lorena, M. et al. *Science* **2002**, *298*, 119–121.
154. Ghosh, A. K.; Moreira, L. A.; Jacobs-Lorena, M. *Insect Biochem.* **2002**, *32*, 1325–1331.
155. Ito, J.; Ghosh, A.; Moreira, L. A.; Wimmer, E. A.; Jacobs-Lorena, M. *Nature* **2002**, *417*, 452–455.
156. Moreira, L. A.; Ghosh, A. K.; Abraham, E. G.; Jacobs-Lorena, M. *Int. J. Parasitol.* **2002**, *32*, 1599–1605.
157. Osta, M. A.; Christophides, G. K.; Kafatos, F. C. *Science* **2004**, *303*, 2030–2033.
158. Riehle, M. A.; Srinivasan, P.; Moreira, C. K.; Jacobs-Lorena, M. *J. Exp. Biol.* **2003**, *206*, 3809–3816.
159. Scholte, E.-J.; Ng'habi, K.; Kihonda, J.; Takken, W.; Paaijmans, K.; Abdulla, S.; Killeen, G. F.; Knols, B. G. J. *Science* **2005**, *308*, 1641–1642.
160. Blanford, S.; Chan, B. H. K.; Jenkins, N.; Sim, D.; Turner, R. J.; Read, A. F.; Thomas, M. B. *Science* **2005**, *308*, 1638–1641.
161. Fidock, D. A.; Rosenthal, P. J.; Croft, S. L.; Brun, R.; Nwaka, S. *Nat. Rev. Drug Disc.* **2004**, *3*, 509–520.
162. Rieckmann, K.; Suebsaeng, L.; Rooney, W. *Am. J. Trop. Med. Hyg.* **1987**, *37*, 211–216.
163. Caillard, V.; Beaute-Lafitte, A.; Chabaud, A.; Ginsburg, H.; Landau, I. *J. Parasitol.* **1995**, *81*, 295–301.
164. Kaneko, O.; Soubes, S. C.; Miller, L. H. *Exp. Parasitol.* **1999**, *93*, 116–119.
165. Geiman, Q. M.; Meagher, M. J. *Nature* **1967**, *215*, 437–439.
166. Young, M. D.; Baerg, D. C.; Rossan, R. N. *Trans. R. Soc. Trop. Med. Hyg.* **1971**, *65*, 835–836.
167. Young, M. D.; Porter, J. A., Jr.; Johnson, C. M. *Science* **1966**, *153*, 1006–1007.
168. Geiman, Q. M.; Siddiqui, W. A. *Am. J. Trop. Med. Hyg.* **1969**, *18*, 351–354.
169. Egan, A. F.; Fabucci, M. E.; Saul, A.; Kaslow, D. C.; Miller, L. H. *Blood* **2002**, *99*, 3863–3866.
170. Gay, F.; Robert, C.; Pouvelle, B.; Peyrol, S.; Scherf, A.; Gysin, J. *J. Immunol. Methods* **1995**, *184*, 15–28.
171. Fujioka, H.; Millet, P.; Maeno, Y.; Nakazawa, S.; Ito, Y.; Howard, R. J.; Collins, W. E.; Aikawa, M. *Exp. Parasitol.* **1994**, *78*, 371–376.
172. Cubillos, M.; Alba, M. P.; Bermudez, A.; Trujillo, M.; Patarroyo, M. E. *Biochimie* **2003**, *85*, 651–657.
173. Arevalo-Herrera, M.; Herrera, S. *Mol. Immunol.* **2001**, *38*, 443–455.
174. Herrera, S.; Perlaza Blanca, L.; Bonelo, A.; Arevalo-Herrera, M. *Int. J. Parasitol.* **2002**, *32*, 1625–1635.
175. Collins, W. E. *Methods Mol. Med.* **2002**, *72*, 85–92.
176. Frederich, M.; Dogne, J.-M.; Angenot, L.; De Mol, P. *Curr. Med. Chem.* **2002**, *9*, 1435–1456.
177. Kaufman, T. S.; Ruveda, E. A. *Angew. Chem. Int. Ed. Engl.* **2005**, *44*, 854–885.
178. Wiesner, J.; Ortmann, R.; Jomaa, H.; Schlitzer, M. *Angew. Chem. Int. Ed. Engl.* **2003**, *42*, 5274–5293.
179. Ridley, R. G. *Nature* **2002**, *415*, 686–693.
180. Thomas, F.; Erhart, A.; D'Alessandro, U. *Lancet Infect. Dis.* **2004**, *4*, 235–239.
181. Ndounga, M.; Basco, L. K. *Acta Trop.* **2003**, *88*, 27–32.
182. Neftel, K. A.; Woodtly, W.; Schmid, M.; Frick, P. G.; Fehr, J. *Br. Med. J. (Clin. Res. Ed).* **1986**, *292*, 721–723.
183. Ringwald, P.; Bickii, J.; Basco, L. *Lancet* **1996**, *347*, 24–28.
184. Ringwald, P.; Bickii, J.; Basco, L. K. *Clin. Infect. Dis.* **1998**, *26*, 946–953.
185. Ringwald, P.; Bickii, J.; Same-Ekobo, A.; Basco, L. K. *Antimicrob. Agents Chemother.* **1997**, *41*, 317–2319.
186. Egan, T. J. *Drug Des. Rev.* **2004**, *1*, 93–110.
187. Vippagunta, S. R.; Dorn, A.; Matile, H.; Bhattacharjee, A. K.; Karle, J. M.; Ellis, W. Y.; Ridley, R. G.; Vennerstrom, J. L. *J. Med. Chem.* **1999**, *42*, 4630–4639.
188. Vippagunta, S. R.; Dorn, A.; Ridley, R. G.; Vennerstrom, J. L. *Biochim. Biophys. Acta* **2000**, *1475*, 133–140.
189. Ziegler, J.; Linck, R.; Wright, D. W. *Curr. Med. Chem.* **2001**, *8*, 171–189.
190. Choi, C. Y. H.; Schneider, E. L.; Kim, J. M.; Gluzman, I. Y.; Goldberg, D. E.; Ellman, J. A.; Marletta, M. A. *Chem. Biol.* **2002**, *9*, 881–889.
191. Menting, J. G. T.; Tilley, L.; Deady, L. W.; Ng, K.; Simpson, R. J.; Cowman, A. F.; Foley, M. *Mol. Biochem. Parasitol.* **1997**, *88*, 215–224.
192. Egan, T. J. *Mini-Rev. Med. Chem.* **2001**, *1*, 113–123.
193. Egan, T. J.; Hunter, R.; Kaschula, C. H.; Marques, H. M.; Misplon, A.; Walden, J. *J. Med. Chem.* **2000**, *43*, 283–291.
194. Stocks, P. A.; Raynes, K. J.; Bray, P. G.; Park, B. K.; O'Neill, P. M.; Ward, S. A. *J. Med. Chem.* **2002**, *45*, 4975–4983.
195. Biot, C.; Glorian, G.; Maciejewski, L. A.; Brocard, J. S.; Domarle, O.; Blampain, G.; Millet, P.; Georges, A. J.; Abessolo, H.; Dive, D. et al. *J. Med. Chem.* **1997**, *40*, 3715–3718.
196. Biot, C.; Taramelli, D.; Forfar-Bares, I.; Maciejewski, L. A.; Boyce, M.; Nowogrocki, G.; Brocard, J. S.; Basilico, N.; Olliaro, P.; Egan, T. J. *Mol. Pharmaceut.* **2005**, *2*, 185–193.
197. De, D.; Krogstad, F. M.; Byers, L. D.; Krogstad, D. J. *J. Med. Chem.* **1998**, *41*, 4918–4926.
198. Ridley, R. G.; Hofheinz, W.; Matile, H.; Jaquet, C.; Dorn, A.; Masciadri, R.; Jolidon, S.; Richter, W. F.; Guenzi, A.; Girometta, M. A. et al. *Antimicrob. Agents Chemother.* **1996**, *40*, 1846–1854.
199. Vennerstrom, J. L.; Ellis, W. Y.; Ager, A. L., Jr.; Andersen, S. L.; Gerena, L.; Milhous, W. K. *J. Med. Chem.* **1992**, *35*, 2129–2134.
200. Raynes, K. *Int. J. Parasitol.* **1999**, *29*, 367–379.
201. Kaschula, C. H.; Egan, T. J.; Hunter, R.; Basilico, N.; Parapini, S.; Taramelli, D.; Pasini, E.; Monti, D. *J. Med. Chem.* **2002**, *45*, 3531–3539.
202. Madrid, P. B.; Sherrill, J.; Liou, A. P.; Weisman, J. L.; DeRisi, J. L.; Guy, R. K. *Bioorg. Med. Chem. Lett.* **2005**, *15*, 1015–1018.
203. Maggs, J. L.; Tingle, M. D.; Kitteringham, N. R.; Park, B. K. *Biochem. Pharmacol.* **1988**, *37*, 303–311.
204. Hawley, S. R.; Bray, P. G.; O'Neill, P. M.; Naisbitt, D. J.; Park, B. K.; Ward, S. A. *Antimicrob. Agents Chemother.* **1996**, *40*, 2345–2349.
205. Naisbitt, D. J.; Williams, D. P.; O'Neill, P. M.; Maggs, J. L.; Willock, D. J.; Pirmohamed, M.; Park, B. K. *Chem. Res. Toxicol.* **1998**, *11*, 1586–1595.
206. Raynes, K. J.; Stocks, P. A.; O'Neill, P. M.; Park, B. K.; Ward, S. A. *J. Med. Chem.* **1999**, *42*, 2747–2751.
207. Naisbitt, D. J.; Ruscoe, J. E.; Williams, D.; O'Neill, P. M.; Pirmohamed, M.; Park, B. K. *J. Pharmacol. Exp. Ther.* **1997**, *280*, 884–893.
208. O'Neill, P. M.; Mukhtar, A.; Stocks, P. A.; Randle, L. E.; Hindley, S.; Ward, S. A.; Storr, R. C.; Bickley, J. F.; O'Neil, I. A.; Maggs, J. L. et al. *J. Med. Chem.* **2003**, *46*, 4933–4945.
209. Palmer, K. J.; Holliday, S. M.; Brogden, R. N. *Drugs* **1993**, *45*, 430–475.
210. van Riemsdijk, M. M.; Sturkenboom, M. C. J. M.; Pepplinkhuizen, L.; Stricker, B. H. C. *J. Clin. Psychiatry* **2005**, *66*, 199–204.
211. Cosgriff, T. M.; Boudreau, E. F.; Pamplin, C. L.; Doberstyn, E. B.; Desjardins, R. E.; Canfield, C. J. *Am. J. Trop. Med. Hyg.* **1982**, *31*, 1075–1079.
212. Horton, R. J. *Parasitol. Today* **1988**, *4*, 238–239.
213. Wesche, D. L.; Schuster, B. G.; Wang, W.-X.; Woosley, R. L. *Clin. Pharmacol. Therapeut.* **2000**, *67*, 521–529.
214. Basco, L. K.; Peytavin, G.; Gimenez, F.; Genissel, B.; Farinotti, R.; Bras, J. L. *Trop. Med. Parasitol.* **1994**, *45*, 45–46.
215. Falade, C.; Makanga, M.; Premji, Z.; Ortmann, C.-E.; Stockmeyer, M.; Ibarra de Palacios, P. *Trans. R. Soc. Trop. Med. Hyg.* **2005**, *99*, 459–467.

216. Sisowath, C.; Stroemberg, J.; Martensson, A.; Msellem, M.; Obondo, C.; Bjoerkman, A.; Gil, J. P. *J. Infect. Dis.* **2005**, *191*, 1014–1017.
217. Noedl, H.; Allmendinger, T.; Prajakwong, S.; Wernsdorfer, G.; Wernsdorfer, W. H. *Antimicrob. Agents Chemother.* **2001**, *45*, 2106–2109.
218. Greenwood, D. *J. Antimicrob. Chemother.* **1995**, *36*, 857–872.
219. Baird, J. K.; Fryauff, D. J.; Hoffman, S. L. *Clin. Infect. Dis.* **2003**, *37*, 1659–1667.
220. Schwartz, E.; Regev-Yochay, G. *Clin. Infect. Dis.* **1999**, *29*, 1502–1506.
221. Kamtekar, K. D.; Gogtay, N. J.; Dalvi, S. S.; Karnad, D. R.; Chogle, A. R.; Aigal, U.; Kshirsagar, N. A. *Ann. Trop. Med. Parasitol.* **2004**, *98*, 453–458.
222. Walsh, D. S.; Eamsila, C.; Sasiprapha, T.; Sangkharomya, S.; Khaewsathien, P.; Supakalin, P.; Tang, D. B.; Jarasrumgsichol, P.; Cherdchu, C.; Edstein, M. D. et al. *J. Infect. Dis.* **2004**, *190*, 1456–1463.
223. Lell, B.; Faucher, J.-F.; Missinou, M. A.; Borrmann, S.; Dangelmaier, O.; Horton, J.; Kremsner, P. G. *Lancet* **2000**, *355*, 2041–2045.
224. Vennerstrom, J. L.; Nuzum, E. O.; Miller, R. E.; Dorn, A.; Gerena, L.; Dande, P. A.; Ellis, W. Y.; Ridley, R. G.; Milhous, W. K. *Antimicrob. Agents Chemother.* **1999**, *43*, 598–602.
225. Olenick, J. G.; Hahn, F. E. *Antimicrob. Agents Chemother.* **1972**, *1*, 259–262.
226. Constantino, L.; Paixao, P.; Moreira, R.; Portela, M. J.; do Rosario, V. E.; Iley, J. *Exp. Toxicol. Pathol.* **1999**, *51*, 299–303.
227. Bolchoz, L. J. C.; Gelasco, A. K.; Jollow, D. J.; McMillan, D. C. *J. Pharmacol. Exp. Ther.* **2002**, *303*, 1121–1129.
228. Silva Monica, C. M.; Santos Eliane, B.; Costal Elenild, G.; Filho Manoel, G. S.; Guerreiro Joao, F.; Povoa Marinete, M. *Rev. Soc. Bras. Med. Trop.* **2004**, *37*, 215–217.
229. Fletcher, K. A.; Barton, P. F.; Kelly, J. A. *Biochem. Pharmacol.* **1988**, *37*, 2683–2690.
230. Jain, M.; Vangapandu, S.; Sachdeva, S.; Singh, S.; Singh, P. P.; Jena, G. B.; Tikoo, K.; Ramarao, P.; Kaul, C. L.; Jain, R. *J. Med. Chem.* **2004**, *47*, 285–287.
231. Araujo, M. J.; Bom, J.; Capela, R.; Casimiro, C.; Chambel, P.; Gomes, P.; Iley, J.; Lopes, F.; Morais, J.; Moreira, R. et al. *J. Med. Chem.* **2005**, *48*, 888–892.
232. Trape, J. F. *Am. J. Trop. Med. Hyg.* **2001**, *64*, 12–17.
233. Maguire, J. D.; Sumawinata, I. W.; Masbar, S.; Laksana, B.; Prodjodipuro, P.; Susanti, I.; Sismadi, P.; Mahmud, N.; Bangs, M. J.; Baird, J. K. *Lancet* **2002**, *360*, 58–60.
234. Baird, J. K. *Antimicrob. Agents Chemother.* **2004**, *48*, 4075–4083.
235. Sanchez, C. P.; McLean, J. E.; Stein, W.; Lanzer, M. *Biochemistry* **2004**, *43*, 16365–16373.
236. Sanchez, C. P.; Wunsch, S.; Lanzer, M. *J. Biol. Chem.* **1997**, *272*, 2652–2658.
237. Reed, M. B.; Saliba, K. J.; Caruana, S. R.; Kirk, K.; Cowman, A. F. *Nature* **2000**, *403*, 906–909.
238. Sidhu, A. B. S.; Verdier-Pinard, D.; Fidock, D. A. *Science* **2002**, *298*, 210–213.
239. O'Neill, P. M.; Posner, G. H. *J. Med. Chem.* **2004**, *47*, 2945–2964.
240. Butler, A. R.; Conforti, L.; Hulme, P.; Renton, L. M.; Rutherford, T. J. *J. Chem. Soc. Perkin Trans.* **1999**, *2*, 2089–2092.
241. Janse, C. J.; Waters, A. P.; Kos, J.; Lugt, C. B. *Int. J. Parasitol.* **1994**, *24*, 589–594.
242. Sowunmi, A.; Oduola, A. M. J. *Acta Trop.* **1996**, *61*, 57–63.
243. Brossi, A.; Venugopalan, B.; Dominguez Gerpe, L.; Yeh, H. J. C.; Flippen-Anderson, J. L.; Buchs, P.; Luo, X. D.; Milhous, W.; Peters, W. *J. Med. Chem.* **1988**, *31*, 645–650.
244. Mandal, P. K.; Sarkar, N.; Pal, A. *Indian J. Med. Res.* **2004**, *119*, 28–32.
245. Price, R.; Van Vugt, M.; Nosten, F.; Luxemburger, C.; Brockman, A.; Phaipun, L.; Chongsuphajaisiddhi, T.; White, N. *Am. J. Trop. Med. Hyg.* **1998**, *59*, 883–888.
246. Barradell, L. B.; Fitton, A. *Drugs* **1995**, *50*, 714–741.
247. Balint, G. A. *Pharmacol. Ther.* **2001**, *90*, 261–265.
248. *Antimalarial Drug Combination Therapy: Report of a Technical Consultation*; World Health Organization: Geneva, Switzerland, 2001.
249. Lin, A. J.; Klayman, D. L.; Milhous, W. K. *J. Med. Chem.* **1987**, *30*, 2147–2150.
250. Lin, A. J.; Lee, M.; Klayman, D. L. *J. Med. Chem.* **1989**, *32*, 1249–1252.
251. Baker, J. K.; McChesney, J. D.; Chi, H. T. *Pharm. Res.* **1993**, *10*, 662–666.
252. Brewer, T. G.; Grate, S. J.; Peggins, J. O.; Weina, P. J.; Petras, J. M.; Levine, B. S.; Heiffer, M. H.; Schuster, B. G. *Am. J. Trop. Med. Hyg.* **1994**, *51*, 251–259.
253. Brewer, T. G.; Peggins, J. O.; Grate, S. J.; Petras, J. M.; Levine, B. S.; Weina, P. J.; Swearengen, J.; Heiffer, M. H.; Schuster, B. G. *Trans. R. Soc. Trop. Med. Hyg.* **1994**, *88*, 33–36.
254. Gordi, T.; Lepist, E.-I. *Toxicol. Lett.* **2004**, *147*, 99–107.
255. O'Neill, P. M.; Miller, A.; Bishop, L. P. D.; Hindley, S.; Maggs, J. L.; Ward, S. A.; Roberts, S. M.; Scheinmann, F.; Stachulski, A. V.; Posner, G. H. et al. *J. Med. Chem.* **2001**, *44*, 58–68.
256. Magueur, G.; Crousse, B.; Charneau, S.; Grellier, P.; Begue, J.-P.; Bonnet-Delpon, D. *J. Med. Chem.* **2004**, *47*, 2694–2699.
257. Khac, V. T.; Nguyen, V. T.; Tran, V. S. *Bioorg. Med. Chem. Lett.* **2005**, *15*, 2629–2631.
258. Jung, M.; Lee, K.; Kendrick, H.; Robinson, B. L.; Croft, S. L. *J. Med. Chem.* **2002**, *45*, 4940–4944.
259. Hindley, S.; Ward, S. A.; Storr, R. C.; Searle, N. L.; Bray, P. G.; Park, B. K.; Davies, J.; O'Neill, P. M. *J. Med. Chem.* **2002**, *45*, 1052–1063.
260. Posner, G. H.; Jeon, H. B.; Ploypradith, P.; Paik, I.-H.; Borstnik, K.; Xie, S.; Shapiro, T. A. *J. Med. Chem.* **2002**, *45*, 3824–3828.
261. Avery, M. A.; Mehrotra, S.; Bonk, J. D.; Vroman, J. A.; Goins, D. K.; Miller, R. *J. Med. Chem.* **1996**, *39*, 2900–2906.
262. Avery, M. A.; Bonk, J. D.; Chong, W. K.; Mehrotra, S.; Miller, R.; Milhous, W.; Goins, D. K.; Venkatesan, S.; Wyandt, C.; Khan, I. *J. Med. Chem.* **1995**, *38*, 5038–5044.
263. Avery, M. A.; Alvim-Gaston, M.; Vroman, J. A.; Wu, B.; Ager, A.; Peters, W.; Robinson, B. L.; Charman, W. *J. Med. Chem.* **2002**, *45*, 4321–4335.
264. Grellepois, F.; Chorki, F.; Ourevitch, M.; Charneau, S.; Grellier, P.; McIntosh, K. A.; Charman, W. N.; Pradines, B.; Crousse, B.; Bonnet-Delpon, D. et al. *J. Med. Chem.* **2004**, *47*, 1423–1433.
265. Posner, G. H.; McRiner, A. J.; Paik, I.-H.; Sur, S.; Borstnik, K.; Xie, S.; Shapiro, T. A.; Alagbala, A.; Foster, B. *J. Med. Chem.* **2004**, *47*, 1299–1301.
266. Avery, M. A.; Alvim-Gaston, M.; Rodrigues, C. R.; Barreiro, E. J.; Cohen, F. E.; Sabnis, Y. A.; Woolfrey, J. R. *J. Med. Chem.* **2002**, *45*, 292–303.
267. Avery, M. A.; Muraleedharan, K. M.; Desai, P. V.; Bandyopadhyaya, A. K.; Furtado, M. M.; Tekwani, B. L. *J. Med. Chem.* **2003**, *46*, 4244–4258.
268. Parshikov, I. A.; Muraleedharan, K. M.; Miriyala, B.; Avery, M. A.; Williamson, J. S. *J. Nat. Prod.* **2004**, *67*, 1595–1597.
269. Parshikov, I. A.; Muraleedharan, K. M.; Avery, M. A.; Williamson, J. S. *Appl. Microbiol. Biotechnol.* **2004**, *64*, 782–786.
270. Zhan, J.-X.; Zhang, Y.-X.; Guo, H.-Z.; Han, J.; Ning, L.-L.; Guo, D.-A. *J. Nat. Prod.* **2002**, *65*, 1693–1695.

271. Selmeczi, K.; Robert, A.; Claparols, C.; Meunier, B. *FEBS Lett.* **2003**, *556*, 245–248.
272. Robert, A.; Dechy-Cabaret, O.; Cazelles, J.; Meunier, B. *Acc. Chem. Res.* **2002**, *35*, 167–174.
273. Bhisutthibhan, J.; Pan, X.-Q.; Hossler, P. A.; Walker, D. J.; Yowell, C. A.; Carlton, J.; Dame, J. B.; Meshnick, S. R. *J. Biol. Chem.* **1998**, *273*, 16192–16198.
274. Kannan, R.; Sahal, D.; Chauhan, V. S. *Chem. Biol.* **2002**, *9*, 321–332.
275. Wang, D.-Y.; Wu, Y.-L. *Chem. Commun.* **2000**, 2193-2194.
276. Eckstein-Ludwig, U.; Webb, R. J.; van Goethem, I. D. A.; East, J. M.; Lee, A. G.; Kimura, M.; O'Neill, P. M.; Bray, P. G.; Ward, S. A.; Krishna, S. *Nature* **2003**, *424*, 957–961.
277. Tang, Y.; Dong, Y.; Vennerstrom, J. L. *Med. Res. Rev.* **2004**, *24*, 425–448.
278. Peters, W.; Robinson, B. L.; Rossiter, J. C.; Misra, D.; Jefford, C. W. *Ann. Trop. Med. Parasitol.* **1993**, *87*, 9–16.
279. Hofheinz, W.; Burgin, H.; Gocke, E.; Jaquet, C.; Masciadri, R.; Schmid, G.; Stohler, H.; Urwyler, H. *Trop. Med. Parasitol.* **1994**, *45*, 261–265.
280. Bachi, M. D.; Korshin, E. E.; Hoos, R.; Szpilman, A. M.; Ploypradith, P.; Xie, S.; Shapiro, T. A.; Posner, G. H. *J. Med. Chem.* **2003**, *46*, 2516–2533.
281. Dong, Y. *Mini-Revi. Med. Chem.* **2002**, *2*, 113–123.
282. Kim, H.-S.; Nagai, Y.; Ono, K.; Begum, K.; Wataya, Y.; Hamada, Y.; Tsuchiya, K.; Masuyama, A.; Nojima, M.; McCullough, K. J. *J. Med. Chem.* **2001**, *44*, 2357–2361.
283. Vennerstrom, J. L.; Arbe-Barnes, S.; Brun, R.; Charman, S. A.; Chiu, F. C. K.; Chollet, J.; Dong, Y.; Dorn, A.; Hunziker, D.; Matile, H. et al. *Nature* **2004**, *430*, 900–904.
284. Leung, D.; Abbenante, G.; Fairlie, D. P. *J. Med. Chem.* **2000**, *43*, 305–341.
285. Omara-Opyene, A. L.; Moura, P. A.; Sulsona, C. R.; Bonilla, J. A.; Yowell, C. A.; Fujioka, H.; Fidock, D. A.; Dame, J. B. *J. Biol. Chem.* **2004**, *279*, 54088–54096.
286. Banerjee, R.; Liu, J.; Beatty, W.; Pelosof, L.; Klemba, M.; Goldberg, D. E. *Proc. Natl. Acad. Sci. USA* **2002**, *99*, 990–995.
287. Brinkworth, R. I.; Prociv, P.; Loukas, A.; Brindley, P. J. *J. Biol. Chem.* **2001**, *276*, 38844–38851.
288. Haque, T. S.; Skillman, A. G.; Lee, C. E.; Habashita, H.; Gluzman, I. Y.; Ewing, T. J.; Goldberg, D. E.; Kuntz, I. D.; Ellman, J. A. *J. Med. Chem.* **1999**, *42*, 1428–1440.
289. Silva, A. M.; Lee, A. Y.; Gulnik, S. V.; Maier, P.; Collins, J.; Bhat, T. N.; Collins, P. J.; Cachau, R. E.; Luker, K. E.; Gluzman, I. Y. et al. *Proc. Natl. Acad. Sci. USA* **1996**, *93*, 10034–10039.
290. Bernstein, N. K.; Cherney, M. M.; Yowell, C. A.; Dame, J. B.; James, M. N. G. *J. Mol. Biol.* **2003**, *329*, 505–524.
291. Bjelic, S.; Aaqvist, J. *Biochemistry* **2004**, *43*, 14521–14528.
292. Beyer, B. B.; Johnson, J. V.; Chung, A. Y.; Li, T.; Madabushi, A.; Agbandje-McKenna, M.; McKenna, R.; Dame, J. B.; Dunn, B. M. *Biochemistry* **2005**, *44*, 1768–1779.
293. Johansson, P.-O.; Chen, Y.; Belfrage, A. K.; Blackman, M. J.; Kvarnstroem, I.; Jansson, K.; Vrang, L.; Hamelink, E.; Hallberg, A.; Rosenquist, A. et al. *J. Med. Chem.* **2004**, *47*, 3353–3366.
294. Johansson, P.-O.; Lindberg, J.; Blackman, M. J.; Kvarnstroem, I.; Vrang, L.; Hamelink, E.; Hallberg, A.; Rosenquist, S. A.; Samuelsson, B. *J. Med. Chem.* **2005**, *48*, 4400–4409.
295. Ersmark, K.; Feierberg, I.; Bjelic, S.; Hamelink, E.; Hackett, F.; Blackman, M. J.; Hulten, J.; Samuelsson, B.; Qvist, J.; Hallberg, A. *J. Med. Chem.* **2004**, *47*, 110–122.
296. Bhattacharya, G.; Gerena, L.; Jiang, S.; Werbovetz, K. A. *Lett. Drug Des. Disc.* **2005**, *2*, 162–164.
297. Jiang, S.; Prigge, S. T.; Wei, L.; Gao, Y.-E.; Hudson, T. H.; Gerena, L.; Dame, J. B.; Kyle, D. E. *Antimicrob. Agents Chemother.* **2001**, *45*, 2577–2584.
298. Rosenthal, P. J.; McKerrow, J. H.; Aikawa, M.; Nagasawa, H.; Leech, J. H. *J. Clin. Invest.* **1988**, *82*, 1560–1566.
299. Rosenthal, P. J. *Exp. Parasitol.* **1995**, *80*, 272–281.
300. Sijwali, P. S.; Kato, K.; Seydel, K. B.; Gut, J.; Lehman, J.; Klemba, M.; Goldberg, D. E.; Miller, L. H.; Rosenthal, P. J. *Proc. Natl. Acad. Sci. USA* **2004**, *101*, 8721–8726.
301. Eksi, S.; Czesny, B.; Greenbaum, D. C.; Bogyo, M.; Williamson, K. C. *Mol. Microbiol.* **2004**, *53*, 243–250.
302. Sijwali, P. S.; Rosenthal, P. J. *Proc. Natl. Acad. Sci. USA* **2004**, *101*, 4384–4389.
303. Sijwali, P. S.; Shenai, B. R.; Gut, J.; Singh, A.; Rosenthal, P. J. *Biochem. J.* **2001**, *360*, 481–489.
304. Na, B.-K.; Shenai, B. R.; Sijwali, P. S.; Choe, Y.; Pandey, K. C.; Singh, A.; Craik, C. S.; Rosenthal, P. J. *Biochem. J.* **2004**, *378*, 529–538.
305. Desai, P. V.; Avery, M. A. *J. Biomol. Struct. Dyn.* **2003**, *21*, 781–790.
306. Singh, A.; Rosenthal, P. J. *Antimicrob. Agents Chemother.* **2001**, *45*, 949–951.
307. Shenai, B. R.; Lee, B. J.; Alvarez-Hernandez, A.; Chong, P. Y.; Emal, C. D.; Neitz, R. J.; Roush, W. R.; Rosenthal, P. J. *Antimicrob. Agents Chemother.* **2003**, *47*, 154–160.
308. Sabnis, Y.; Rosenthal, P. J.; Desai, P.; Avery, M. A. *J. Biomol. Struct. Dyn.* **2002**, *19*, 765–774.
309. Sabnis, Y. A.; Desai, P. V.; Rosenthal, P. J.; Avery, M. A. *Protein Sci.* **2003**, *12*, 501–509.
310. Desai, P. V.; Patny, A.; Sabnis, Y.; Tekwani, B.; Gut, J.; Rosenthal, P.; Srivastava, A.; Avery, M. *J. Med. Chem.* **2004**, *47*, 6609–6615.
311. Singh, A.; Rosenthal, P. J. *J. Biol. Chem.* **2004**, *279*, 35236–35241.
312. Olson, J. E.; Lee, G. K.; Semenov, A.; Rosenthal, P. J. *Bioorg. Med. Chem.* **1999**, *7*, 633–638.
313. Eggleson, K. K.; Duffin, K. L.; Goldberg, D. E. *J. Biol. Chem.* **1999**, *274*, 32411–32417.
314. Murata, C. E.; Goldberg, D. E. *J. Biol. Chem.* **2003**, *278*, 38022–38028.
315. Roth, E., Jr. *Blood Cells* **1990**, *16*, 453–460, discussion 461–456.
316. Velanker, S. S.; Ray, S. S.; Gokhale, R. S.; Suma, S.; Balaram, H.; Balaram, P.; Murthy, M. R. N. *Structure* **1997**, *5*, 751–761.
317. Dunn, C. R.; Banfield, M. J.; Barker, J. J.; Higham, C. W.; Moreton, K. M.; Turgut-Balik, D.; Brady, R. L.; Holbrook, J. J. *Nat. Struct. Biol.* **1996**, *3*, 912–915.
318. Kim, H.; Certa, U.; Dobeli, H.; Jakob, P.; Hol, W. G. *Biochemistry* **1998**, *37*, 4388–4396.
319. Brady, R. L.; Cameron, A. *Curr. Drug Targets* **2004**, *5*, 137–149.
320. Itin, C.; Burki, Y.; Certa, U.; Dobeli, H. *Mol. Biochem. Parasitol.* **1993**, *58*, 135–143.
321. Wanidworanun, C.; Nagel, R. L.; Shear, H. L. *Mol. Biochem. Parasitol.* **1999**, *102*, 91–101.
322. Joubert, F.; Neitz, A. W. H.; Louw, A. I. *Proteins: Struct. Funct., Genet.* **2001**, *45*, 136–143.
323. Singh, S. K.; Maithal, K.; Balaram, H.; Balaram, P. *FEBS Lett.* **2001**, *501*, 19–23.
324. Eaazhisai, K.; Balaram, H.; Balaram, P.; Murthy, M. R. N. *J. Mol. Biol.* **2004**, *343*, 671–684.

325. Brown, W. M.; Yowell, C. A.; Hoard, A.; Vander Jagt, T. A.; Hunsaker, L. A.; Deck, L. M.; Royer, R. E.; Piper, R. C.; Dame, J. B.; Makler, M. T. et al. *Biochemistry* **2004**, *43*, 6219–6229.
326. Sessions, R. B.; Dewar, V.; Clarke, A. R.; Holbrook, J. J. *Protein Eng.* **1997**, *10*, 301–306.
327. Jagt, D. L. V.; Deck, L. M.; Royer, R. E. *Curr. Med. Chem.* **2000**, 7, 479–498.
328. Deck, L. M.; Royer, R. E.; Chamblee, B. B.; Hernandez, V. M.; Malone, R. R.; Torres, J. E.; Hunsaker, L. A.; Piper, R. C.; Makler, M. T.; Vander Jagt, D. L. *J. Med. Chem.* **1998**, *41*, 3879–3887.
329. Cameron, A.; Read, J.; Tranter, R.; Winter, V. J.; Sessions, R. B.; Brady, R. L.; Vivas, L.; Easton, A.; Kendrick, H.; Croft, S. L. et al. *J. Biol. Chem.* **2004**, *279*, 31429–31439.
330. Raman, J.; Balaram, H. *Med. Chem. Rev.* **2004**, *1*, 465–473.
331. Queen, S. A.; Vander Jagt, D. L.; Reyes, P. *Antimicrob. Agents Chemother.* **1990**, *34*, 1393–1398.
332. Rager, N.; Mamoun, C. B.; Carter, N. S.; Goldberg, D. E.; Ullman, B. *J. Biol. Chem.* **2001**, *276*, 41095–41099.
333. Carter, N. S.; Ben Mamoun, C.; Liu, W.; Silva, E. O.; Landfear, S. M.; Goldberg, D. E.; Ullman, B. *J. Biol. Chem.* **2000**, *275*, 10683–10691.
334. Brown, D. M.; Netting, A. G.; Chun, B. K.; Choi, Y.; Chu, C. K.; Gero, A. M. *Nucleosides Nucleotides* **1999**, *18*, 2521–2532.
335. Shi, W.; Li, C. M.; Tyler, P. C.; Furneaux, R. H.; Cahill, S. M.; Girvin, M. E.; Grubmeyer, C.; Schramm, V. L.; Almo, S. C. *Biochemistry* **1999**, *38*, 9872–9880.
336. Shi, W.; Li, C. M.; Tyler, P. C.; Furneaux, R. H.; Grubmeyer, C.; Schramm, V. L.; Almo, S. C. *Nat. Struct. Biol.* **1999**, *6*, 588–593.
337. Queen, S. A.; Vander Jagt, D.; Reyes, P. *Mol. Biochem. Parasitol.* **1988**, *30*, 123–133.
338. Makobongo, M. O.; Riding, G.; Xu, H.; Hirunpetcharat, C.; Keough, D.; de Jersey, J.; Willadsen, P.; Good, M. F. *Proc. Natl. Acad. Sci. USA* **2003**, *100*, 2628–2633.
339. Kicska, G. A.; Tyler, P. C.; Evans, G. B.; Furneaux, R. H.; Schramm, V. L.; Kim, K. *J. Biol. Chem.* **2002**, *277*, 3226–3231.
340. Kicska, G. A.; Tyler, P. C.; Evans, G. B.; Furneaux, R. H.; Kim, K.; Schramm, V. L. *J. Biol. Chem.* **2002**, *277*, 3219–3225.
341. Shi, W.; Ting, L.-M.; Kicska, G. A.; Lewandowicz, A.; Tyler, P. C.; Evans, G. B.; Furneaux, R. H.; Kim, K.; Almo, S. C.; Schramm, V. L. *J. Biol. Chem.* **2004**, *279*, 18103–18106.
342. Yuthavong, Y.; Yuvaniyama, J.; Chitnumsub, P.; Vanichtanankul, J.; Chusacultanachai, S.; Tarnchompoo, B.; Vilaivan, T.; Kamchonwongpaisan, S. *Parasitology* **2005**, *130*, 249–259.
343. Triglia, T.; Menting, J. G. T.; Wilson, C.; Cowman, A. F. *Proc. Natl. Acad. Sci. USA* **1997**, *94*, 13944–13949.
344. Baldwin, J.; Farajallah, A. M.; Malmquist, N. A.; Rathod, P. K.; Phillips, M. A. *J. Biol. Chem.* **2002**, 277, 41827–41834.
345. Eke, F. U.; Anochie, I. *Curr. Ther. Res.* **2003**, *64*, 616–625.
346. Winstanley, P. *Trop. Med. Int. Health* **2001**, *6*, 952–954.
347. Shearer, T. W.; Kozar, M. P.; O'Neil, M. T.; Smith, P. L.; Schiehser, G. A.; Jacobus, D. P.; Diaz, D. S.; Yang, Y.-S.; Milhous, W. K.; Skillman, D. R. *J. Med. Chem.* **2005**, *48*, 2805–2813.
348. Jensen, N. P.; Ager, A. L.; Bliss, R. A.; Canfield, C. J.; Kotecka, B. M.; Rieckmann, K. H.; Terpinski, J.; Jacobus, D. P. *J. Med. Chem.* **2001**, *44*, 3925–3931.
349. Jiang, L.; Lee, P.-C.; White, J.; Rathod, P. K. *Antimicrob. Agents Chemother.* **2000**, *44*, 1047–1050.
350. Yuvaniyama, J.; Chitnumsub, P.; Kamchonwongpaisan, S.; Vanichtanankul, J.; Sirawaraporn, W.; Taylor, P.; Walkinshaw, M. D.; Yuthavong, Y. *Nat. Struct. Biol.* **2003**, *10*, 357–365.
351. Gregson, A.; Plowe, C. V. *Pharmacol. Rev.* **2005**, *57*, 117–145.
352. Sirichaiwat, C.; Intaraudom, C.; Kamchonwongpaisan, S.; Vanichtanankul, J.; Thebtaranonth, Y.; Yuthavong, Y. *J. Med. Chem.* **2004**, *47*, 345–354.
353. Chitnumsub, P.; Yavaniyama, J.; Vanichtanankul, J.; Kamchonwongpaisan, S.; Walkinshaw, M. D.; Yuthavong, Y. *Acta Crystallogr. Sect. D* **2004**, *60*, 780–783.
354. Yuthavong, Y.; Vilaivan, T.; Chareonsethakul, N.; Kamchonwongpaisan, S.; Sirawaraporn, W.; Quarrell, R.; Lowe, G. *J. Med. Chem.* **2000**, *43*, 2738–2744.
355. Kamchonwongpaisan, S.; Quarrell, R.; Charoensetakul, N.; Ponsinet, R.; Vilaivan, T.; Vanichtanankul, J.; Tarnchompoo, B.; Sirawaraporn, W.; Lowe, G.; Yuthavong, Y. *J. Med. Chem.* **2004**, *47*, 673–680.
356. Rastelli, G.; Pacchioni, S.; Sirawaraporn, W.; Sirawaraporn, R.; Parenti, M. D.; Ferrari, A. M. *J. Med. Chem.* **2003**, *46*, 2834–2845.
357. Mabuza, A.; Govere, J.; la Grange, K.; Mngomezulu, N.; Allen, E.; Zitha, A.; Mbokazi, F.; Durrheim, D.; Barnes, K. *S. Afr. Med. J.* **2005**, *95*, 346–349.
358. Gatton, M. L.; Martin, L. B.; Cheng, Q. *Antimicrob. Agents Chemother.* **2004**, *48*, 2116–2123.
359. Krungkrai, J. *Biochim. Biophys. Acta* **1995**, *1243*, 351–360.
360. Ittarat, I.; Asawamahasakda, W.; Meshnick, S. R. *Exp. Parasitol.* **1994**, *79*, 50–56.
361. Baldwin, J.; Michnoff, C. H.; Malmquist, N. A.; White, J.; Roth, M. G.; Rathod, P. K.; Phillips, M. A. *J. Biol. Chem.* **2005**, *280*, 21847–21853.
362. Baggish, A. L.; Hill, D. R. *Antimicrob. Agents Chemother.* **2002**, *46*, 1163–1173.
363. Sabchareon, A.; Attanath, P.; Phanuaksook, P.; Chanthavanich, P.; Poonpanich, Y.; Mookmanee, D.; Chongsuphajaisiddhi, T.; Sadler, B. M.; Hussein, Z.; Canfield, C. J. et al. *Trans. R. Soc. Trop. Med. Hyg.* **1998**, *92*, 201–206.
364. McKeage, K.; Scott, L. J. *Drugs* **2003**, *63*, 597–623.
365. Mather, M. W.; Darrouzet, E.; Valkova-Valchanova, M.; Cooley, J. W.; McIntosh, M. T.; Daldal, F.; Vaidya, A. B. *J. Biol. Chem.* **2005**, *280*, 27458–27465.
366. Kessl, J. J.; Ha, K. H.; Merritt, A. K.; Lange, B. B.; Hill, P.; Meunier, B.; Meshnick, S. R.; Trumpower, B. L. *J. Biol. Chem.* **2005**, *280*, 17142–17148.
367. Canfield, C. J.; Pudney, M.; Gutteridge, W. E. *Exp. Parasitol.* **1995**, *80*, 373–381.
368. Srivastava, I. K.; Vaidya, A. B. *Antimicrob. Agents Chemother.* **1999**, *43*, 1334–1339.
369. Fivelman, Q. L.; Adagu, I. S.; Warhurst, D. C. *Antimicrob. Agents Chemother.* **2004**, *48*, 4097–4102.
370. Fivelman, Q. L.; Butcher, G. A.; Adagu, I. S.; Warhurst, D. C.; Pasvol, G. *Malaria J.* **2002**, *1:1* (8 February 2002).
371. Hinterding, K.; Alonso-Diaz, D.; Waldmann, H. *Angew. Chem. Int. Ed. Engl.* **1998**, *37*, 688–749.
372. Anura, S. H. *J. Natl. Cancer Inst.* **2001**, *93*, 1062–1074.
373. Reuter, C. W. M.; Morgan, M. A.; Bergmann, L. *Blood* **2000**, *96*, 1655–1669.
374. Alsina, M.; Fonseca, R.; Wilson, E. F.; Belle, A. N.; Gerbino, E.; Price-Troska, T.; Overton, R. M.; Ahmann, G.; Bruzek, L. M.; Adjei, A. A. et al. *Blood* **2004**, *103*, 3271–3277.
375. Chakrabarti, D.; Azam, T.; DelVecchio, C.; Qiu, L.; Park, Y.-I.; Allen, C. M. *Mol. Biochem. Parasitol.* **1998**, *94*, 175–184.

376. Chakrabarti, D.; Da Silva, T.; Barger, J.; Paquette, S.; Patel, H.; Patterson, S.; Allen, C. M. *J. Biol. Chem.* **2002**, *277*, 42066–42073.
377. Nallan, L.; Bauer, K. D.; Bendale, P.; Rivas, K.; Yokoyama, K.; Horney, C. P.; Rao Pendyala, P.; Floyd, D.; Lombardo, L. J.; Williams, D. K. et al. *J. Med. Chem.* **2005**, *48*, 3704–3713.
378. Wiesner, J.; Kettler, K.; Sakowski, J.; Ortmann, R.; Katzin, A. M.; Kimura, E. A.; Silber, K.; Klebe, G.; Jomaa, H.; Schlitzer, M. *Angew. Chem. Int. Ed. Engl.* **2003**, *43*, 251–254.
379. Eastman, R. T.; White, J.; Hucke, O.; Bauer, K.; Yokoyama, K.; Nallan, L.; Chakrabarti, D.; Verlinde, C. L. M. J.; Gelb, M. H.; Rathod, P. K. et al. *J. Biol. Chem.* **2005**, *280*, 13554–13559.
380. Foth, B. J.; McFadden, G. I. *Int. Rev. Cytol.* **2003**, *224*, 57–110.
381. Koehler, S.; Delwiche, C. F.; Denny, P. W.; Tilney, L. G.; Webster, P.; Wilson, R. J. M.; Palmer, J. D.; Roos, D. S. *Science* **1997**, *275*, 1485–1489.
382. Fast, N. M.; Kissinger, J. C.; Roos, D. S.; Keeling, P. J. *Mol. Biol. Evol.* **2001**, *18*, 418–426.
383. Roos, D. S.; Crawford, M. J.; Donald, R. G. K.; Fraunholz, M.; Harb, O. S.; He, C. Y.; Kissinger, J. C.; Shaw, M. K.; Striepen, B. *Phil. Trans. R. Soc. London, Ser. B* **2002**, *357*, 35–46.
384. Fichera, M. E.; Roos, D. S. *Nature* **1997**, *390*, 407–409.
385. Chaubey, S.; Kumar, A.; Singh, D.; Habib, S. *Mol. Microbiol.* **2005**, *56*, 81–89.
386. Singh, D.; Chaubey, S.; Habib, S. *Mol. Biochem. Parasitol.* **2003**, *126*, 9–14.
387. Waller, R. F.; Reed, M. B.; Cowman, A. F.; McFadden, G. I. *EMBO J.* **2000**, *19*, 1794–1802.
388. Waller, R. F.; Keeling, P. J.; Donald, R. G. K.; Striepen, B.; Handman, E.; Lang-Unnasch, N.; Cowman, A. F.; Besra, G. S.; Roos, D. S.; McFadden, G. I. *Proc. Natl. Acad. Sci. USA* **1998**, *95*, 12352–12357.
389. Surolia, A.; Ramya, T. N. C.; Ramya, V.; Surolia, N. *Biochem. J.* **2004**, *383*, 401–412.
390. Sivaraman, S.; Sullivan, T. J.; Johnson, F.; Novichenok, P.; Cui, G.; Simmerling, C.; Tonge, P. J. *J. Med. Chem.* **2004**, *47*, 509–518.
391. Surolia, N.; Surolia, A. *Nat. Med.* **2001**, *7*, 167–173.
392. Kuo, M. R.; Morbidoni, H. R.; Alland, D.; Sneddon, S. F.; Gourlie, B. B.; Staveski, M. M.; Leonard, M.; Gregory, J. S.; Janjigian, A. D.; Yee, C. et al. *J. Biol. Chem.* **2003**, *278*, 20851–20859.
393. Waller, R. F.; Ralph, S. A.; Reed, M. B.; Su, V.; Douglas, J. D.; Minnikin, D. E.; Cowman, A. F.; Besra, G. S.; McFadden, G. I. *Antimicrob. Agents Chemother.* **2003**, *47*, 297–301.
394. Sharma, S. K.; Kapoor, M.; Ramya, T. N. C.; Kumar, S.; Kumar, G.; Modak, R.; Sharma, S.; Surolia, N.; Surolia, A. *J. Biol. Chem.* **2003**, *278*, 45661–45671.
395. Perozzo, R.; Kuo, M.; Sidhu, A. B. S.; Valiyaveettil, J. T.; Bittman, R.; Jacobs, W. R., Jr.; Fidock, D. A.; Sacchettini, J. C. *J. Biol. Chem.* **2002**, *277*, 13106–13114.
396. Hunter, W. N.; Bond, C. S.; Gabrielsen, M.; Kemp, L. E. *Biochem. Soc. Trans.* **2003**, *31*, 537–542.
397. Rodriguez-Concepcion, M.; Boronat, A. *Plant Physiol.* **2002**, *130*, 1079–1089.
398. Jomaa, H.; Wiesner, J.; Sanderbrand, S.; Altincicek, B.; Weidemeyer, C.; Hintz, M.; Turbachova, I.; Eberl, M.; Zeidler, J.; Lichtenthaler, H. K. et al. *Science* **1999**, *285*, 1573–1576.
399. Lell, B.; Ruangweerayut, R.; Wiesner, J.; Missinou, M. A.; Schindler, A.; Baranek, T.; Hintz, M.; Hutchinson, D.; Jomaa, H.; Kremsner, P. G. *Antimicrob. Agents Chemother.* **2003**, *47*, 735–738.
400. Wiesner, J.; Henschker, D.; Hutchinson, D. B.; Beck, E.; Jomaa, H. *Antimicrob. Agents Chemother.* **2002**, *46*, 2889–2894.
401. Borrmann, S.; Issifou, S.; Esser, G.; Adegnika, A. A.; Ramharter, M.; Matsiegui, P.-B.; Oyakhirome, S.; Mawili-Mboumba, D. P.; Missinou, M. A.; Kun, J. F. J. et al. *J. Infect. Dis.* **2004**, *190*, 1534–1540.
402. Borrmann, S.; Adegnika, A. A.; Matsiegui, P.-B.; Issifou, S.; Schindler, A.; Mawili-Mboumba, D. P.; Baranek, T.; Wiesner, J.; Jomaa, H.; Kremsner, P. G. *J. Infect. Dis.* **2004**, *189*, 901–908.
403. Kuemmerle, H. P.; Murakawa, T.; Sakamoto, H.; Sato, N.; Konishi, T.; De Santis, F. *Int. J. Clin. Pharmacol.* **1985**, *23*, 521–528.
404. Tsuchiya, T.; Ishibashi, K.; Terakawa, M.; Nishiyama, M.; Itoh, N.; Noguchi, H. *Eur. J. Drug Metab. Pharmacokinet.* **1982**, *7*, 59–64.
405. Reichenberg, A.; Wiesner, J.; Weidemeyer, C.; Dreiseidler, E.; Sanderbrand, S.; Altincicek, B.; Beck, E.; Schlitzer, M.; Jomaa, H. *Bioorg. Med. Chem. Lett.* **2001**, *11*, 833–835.
406. Ortmann, R.; Wiesner, J.; Reichenberg, A.; Henschker, D.; Beck, E.; Jomaa, H.; Schlitzer, M. *Bioorg. Med. Chem. Lett.* **2003**, *13*, 2163–2166.
407. Wilson, R. J. M. *J. Mol. Biol.* **2002**, *319*, 257–274.
408. Pradines, B.; Rogier, C.; Fusai, T.; Mosnier, J.; Daries, W.; Barret, E.; Parzy, D. *Antimicrob. Agents Chemother.* **2001**, *45*, 1746–1750.
409. Pradines, B.; Ramiandrasoa, F.; Rolain, J. M.; Rogier, C.; Mosnier, J.; Daries, W.; Fusai, T.; Kunesch, G.; Le Bras, J.; Parzy, D. *Antimicrob. Agents Chemother.* **2002**, *46*, 225–228.
410. Weissig, V.; Vetro-Widenhouse, T. S.; Rowe, T. C. *DNA Cell Biol.* **1997**, *16*, 1483–1492.
411. Williamson, D. H.; Preiser, P. R.; Moore, P. W.; McCready, S.; Strath, M.; Wilson, R. J. M. *Mol. Microbiol.* **2002**, *45*, 533–542.
412. Strath, M.; Scott-Finnigan, T.; Gardner, M.; Williamson, D.; Wilson, I. *Trans. R. Soc. Trop. Med. Hyg.* **1993**, *87*, 211–216.
413. Gardner, M. J.; Williamson, D. H.; Wilson, R. J. M. *Mol. Biochem. Parasitol.* **1991**, *44*, 115–123.
414. Ralph, S. A.; D'Ombrain, M. C.; McFadden, G. I. *Drug Resist. Updates* **2001**, *4*, 145–151.
415. Budimulja, A. S.; Syafruddin, X.; Tapchaisri, P.; Wilairat, P.; Marzuki, S. *Mol. Biochem. Parasitol.* **1997**, *84*, 137–141.
416. McConkey, G. A.; Rogers, M. J.; McCutchan, T. F. *J. Biol. Chem.* **1997**, *272*, 2046–2049.
417. Rogers, M. J.; Bukhman, Y. V.; McCutchan, T. F.; Draper, D. E. *RNA* **1997**, *3*, 815–820.
418. Rodnina, M. V.; Savelsbergh, A.; Matassova, N. B.; Katunin, V. I.; Semenkov, Y. P.; Wintermeyer, W. *Proc. Natl. Acad. Sci. USA* **1999**, *96*, 9586–9590.
419. Clough, B.; Strath, M.; Preiser, P.; Denny, P.; Wilson, I. *FEBS Lett.* **1997**, *406*, 123–125.
420. Taylor, W. R. J.; Widjaja, H.; Richie, T. L.; Basri, H.; Ohrt, C.; Tjitra, E.; Taufik, E.; Jones, T. R.; Kain, K. C.; Hoffman, S. L. *Am. J. Trop. Med. Hyg.* **2001**, *64*, 223–228.
421. Kain, K. C.; Shanks, G. D.; Keystone, J. S. *Clin. Infect. Dis.* **2001**, *33*, 226–234.
422. Yong, C. K.; Prendiville, J.; Peacock, D. L.; Wong, L. T.; Davidson, A. G. *Pediatrics* **2000**, *106*, E13.
423. Shea, C. R.; Hefetz, Y.; Gillies, R.; Wimberly, J.; Dalickas, G.; Hasan, T. *J. Biol. Chem.* **1990**, *265*, 5977–5982.
424. Lell, B.; Kremsner, P. G. *Antimicrob. Agents Chemother.* **2002**, *46*, 2315–2320.
425. Taylor, W. R. J.; Richie, T. L.; Fryauff, D. J.; Picarima, H.; Ohrt, C.; Tang, D.; Braitman, D.; Murphy, G. S.; Widjaja, H.; Tjitra, E. et al. *Clin. Infect. Dis.* **1999**, *28*, 74–81.
426. Dunne, M. W.; Singh, N.; Shukle, M.; Valecha, N.; Bhattacharyya, P. C.; Dev, V.; Patel, K.; Mohapatra, M. K.; Lakhani, J.; Benner, R. et al. *J. Infect. Dis.* **2005**, *191*, 1582–1588.

427. Vaughan, M. D.; Sampson, P. B.; Honek, J. F. *Curr. Med. Chem.* **2002**, *9*, 385–409.
428. Clements, J. M.; Beckett, R. P.; Brown, A.; Catlin, G.; Lobell, M.; Palan, S.; Thomas, W.; Whittaker, M.; Wood, S.; Salama, S. et al. *Antimicrob. Agents Chemother.* **2001**, *45*, 563–570.
429. Chen, D. Z.; Patel, D. V.; Hackbarth, C. J.; Wang, W.; Dreyer, G.; Young, D. C.; Margolis, P. S.; Wu, C.; Ni, Z.-J.; Trias, J. et al. *Biochemistry* **2000**, *39*, 1256–1262.
430. Hackbarth, C. J.; Chen, D. Z.; Lewis, J. G.; Clark, K.; Mangold, J. B.; Cramer, J. A.; Margolis, P. S.; Wang, W.; Koehn, J.; Wu, C. et al. *Antimicrob. Agents Chemother.* **2002**, *46*, 2752–2764.
431. Bracchi-Ricard, V.; Nguyen, K. T.; Zhou, Y.; Rajagopalan, P. T. R.; Chakrabarti, D.; Pei, D. *Arch. Biochem. Biophys.* **2001**, *396*, 162–170.
432. Kumar, A.; Nguyen, K. T.; Srivathsan, S.; Ornstein, B.; Turley, S.; Hirsh, I.; Pei, D.; Hol, W. G. J. *Structure* **2002**, *10*, 357–367.
433. Robien, M. A.; Nguyen, K. T.; Kumar, A.; Hirsh, I.; Turley, S.; Pei, D.; Hol, W. G. J. *Protein Sci.* **2004**, *13*, 1155–1163.
434. Wiesner, J.; Sanderbrand, S.; Altincicek, B.; Beck, E.; Jomaa, H. *Trends Parasitol.* **2001**, *17*, 7–8.
435. Lee, M. D.; She, Y.; Soskis, M. J.; Borella, C. P.; Gardner, J. R.; Hayes, P. A.; Dy, B. M.; Heaney, M. L.; Philips, M. R.; Bornmann, W. G. et al. *J. Clin. Invest.* **2004**, *114*, 1107–1116.
436. Apfel, C. M.; Locher, H.; Evers, S.; Takacs, B.; Hubschwerlen, C.; Pirson, W.; Page, M. G. P.; Keck, W. *Antimicrob. Agents Chemother.* **2001**, *45*, 1058–1064.
437. Vial, H. J.; Ancelin, M. L.; Philippot, J. R.; Thuet, M. J. *Blood Cells* **1990**, *16*, 531–555, discussion 556–561.
438. Ancelin, M. L.; Calas, M.; Bompart, J.; Cordina, G.; Martin, D.; Ben Bari, M.; Jei, T.; Druilhe, P.; Vial, H. J. *Blood* **1998**, *91*, 1426–1437.
439. Biagini, G. A.; Pasini, E. M.; Hughes, R.; De Koning, H. P.; Vial, H. J.; O'Neill, P. M.; Ward, S. A.; Bray, P. G. *Blood* **2004**, *104*, 3372–3377.
440. Biagini, G. A.; Ward, S. A.; Bray, P. G. *Trends Parasitol.* **2005**, *21*, 299–301.
441. Ancelin, M. L.; Calas, M.; Vidal-Sailhan, V.; Herbute, S.; Ringwald, P.; Vial, H. J. *Antimicrob. Agents Chemother.* **2003**, *47*, 2590–2597.
442. Calas, M.; Ancelin, M. L.; Cordina, G.; Portefaix, P.; Piquet, G.; Vidal-Sailhan, V.; Vial, H. *J. Med. Chem.* **2000**, *43*, 505–516.
443. Wengelnik, K.; Vidal, V.; Ancelin, M. L.; Cathiard, A.-M.; Morgat, J. L.; Kocken, C. H.; Calas, M.; Herrera, S.; Thomas, A. W.; Vial, H. J. *Science* **2002**, *295*, 1311–1314.
444. Vial, H. J.; Wein, S.; Farenc, C.; Kocken, C.; Nicolas, O.; Ancelin, M. L.; Bressolle, F.; Thomas, A.; Calas, M. *Proc. Natl. Acad. Sci. USA* **2004**, *101*, 15458–15463.
445. Chen, M.; Theander, T. G.; Christensen, S. B.; Hviid, L.; Zhai, L.; Kharazmi, A. *Antimicrob. Agents Chemother.* **1994**, *38*, 1470–1475.
446. Li, R.; Chen, X.; Gong, B.; Dominguez, J. N.; Davidson, E.; Kurzban, G.; Miller, R. E.; Nuzum, E. O.; Rosenthal, P. J. et al. *J. Med. Chem.* **1995**, *38*, 5031–5037.
447. Liu, M.; Wilairat, P.; Go, M.-L. *J. Med. Chem.* **2001**, *44*, 4443–4452.
448. Dominguez, J. N.; Charris, J. E.; Lobo, G.; Gamboa de Dominguez, N.; Moreno, M. M.; Riggione, F.; Sanchez, E.; Olson, J.; Rosenthal, P. J. *Eur. J. Med. Chem.* **2001**, *36*, 555–560.
449. Dominguez, J. N.; Leon, C.; Rodrigues, J.; Gamboa de Dominguez, N.; Gut, J.; Rosenthal, P. J. *J. Med. Chem.* **2005**, *48*, 3654–3658.
450. Ziegler, H. L.; Hansen, H. S.; Staerk, D.; Christensen, S. B.; Haegerstrand, H.; Jaroszewski, J. W. *Antimicrob. Agents Chemother.* **2004**, *48*, 4067–4071.
451. Onyeibor, O.; Croft, S. L.; Dodson, H. I.; Feiz-Haddad, M.; Kendrick, H.; Millington, N. J.; Parapini, S.; Phillips, R. M.; Seville, S.; Shnyder, S. D. et al. *J. Med. Chem.* **2005**, *48*, 2701–2709.
452. Go, M.-L. *Med. Res. Rev.* **2003**, *23*, 456–487.
453. Francois, G.; Timperman, G.; Eling, W.; Assi, L. A.; Holenz, J.; Bringmann, G. *Antimicrob. Agents Chemother.* **1997**, *41*, 2533–2539.
454. Ang, K. K. H.; Holmes, M. J.; Higa, T.; Hamann, M. T.; Kara, U. A. K. *Antimicrob. Agents Chemother.* **2000**, *44*, 1645–1649.
455. Humphrey, J. M.; Liao, Y.; Ali, A.; Rein, T.; Wong, Y.-L.; Chen, H.-J.; Courtney, A. K.; Martin, S. F. *J. Am. Chem. Soc.* **2002**, *124*, 8584–8592.
456. Winkler, J. D.; Axten, J. M. *J. Am. Chem. Soc.* **1998**, *120*, 6425–6426.
457. Wright, A. D.; Wang, H.; Gurrath, M.; Koenig, G. M.; Kocak, G.; Neumann, G.; Loria, P.; Foley, M.; Tilley, L. *J. Med. Chem.* **2001**, *44*, 873–885.
458. Winter, R. W.; Cornell, K. A.; Johnson, L. L.; Isabelle, L. M.; Hinrichs, D. J.; Riscoe, M. K. *Bioorg. Med. Chem. Lett.* **1995**, *5*, 1927–1932.
459. Winter, R. W.; Cornell, K. A.; Johnson, L. L.; Ignatushchemko, M.; Hinrichs, D. J.; Riscoe, M. K. *Antimicrob. Agents Chemother.* **1996**, *40*, 1408–1411.
460. Kelly, J. X.; Winter, R.; Peyton, D. H.; Hinrichs, D. J.; Riscoe, M. *Antimicrob. Agents Chemother.* **2002**, *46*, 144–150.
461. Suswam, E.; Kyle, D.; Lang-Unnasch, N. *Exp. Parasitol.* **2001**, *98*, 180–187.
462. Dorn, A.; Scovill, J. P.; Ellis, W. Y.; Matile, H.; Ridley, R. G.; Vennerstrom, J. L. *Am. J. Trop. Med. Hyg.* **2001**, *65*, 19–20.
463. Berman, J.; Brown, L.; Miller, R.; Andersen, S. L.; McGreevy, P.; Schuster, B. G.; Ellisk, W.; Ager, A.; Rossan, R. *Antimicrob. Agents Chemother.* **1994**, *38*, 1753–1756.
464. Kurosawa, Y.; Dorn, A.; Kitsuji-Shirane, M.; Shimada, H.; Satoh, T.; Matile, H.; Hofheinz, W.; Masciadri, R.; Kansy, M.; Ridley, R. G. *Antimicrob. Agents Chemother.* **2000**, *44*, 2638–2644.
465. Alzeer, J.; Chollet, J.; Heinze-Krauss, I.; Hubschwerlen, C.; Matile, H.; Ridley, R. G. *J. Med. Chem.* **2000**, *43*, 560–568.
466. Davioud-Charvet, E.; Delarue, S.; Biot, C.; Schwoebel, B.; Boehme, C. C.; Muessigbrodt, A.; Maes, L.; Sergheraert, C.; Grellier, P.; Schirmer, R. H. et al. *J. Med. Chem.* **2001**, *44*, 4268–4276.
467. Basco, L. K.; Dechy-Cabaret, O.; Ndounga, M.; Meche, F. S.; Robert, A.; Meunier, B. *Antimicrob. Agents Chemother.* **2001**, *45*, 1886–1888.
468. Woodard, C. L.; Li, Z.; Kathcart, A. K.; Terrell, J.; Gerena, L.; Lopez-Sanchez, M.; Kyle, D. E.; Bhattacharjee, A. K.; Nichols, D. A.; Ellis, W. et al. *J. Med. Chem.* **2003**, *46*, 3877–3882.
469. Ismail, M. A.; Brun, R.; Wenzler, T.; Tanious, F. A.; Wilson, W. D.; Boykin, D. W. *J. Med. Chem.* **2004**, *47*, 3658–3664.
470. Ismail, M. A.; Brun, R.; Easterbrook, J. D.; Tanious, F. A.; Wilson, W. D.; Boykin, D. W. *J. Med. Chem.* **2003**, *46*, 4761–4769.
471. Danis, M.; Bricaire, F. *Fundam. Clin. Pharmacol.* **2003**, *17*, 155–160.
472. *The Use of Antimalarial Drugs: Report of an Informal Consultation*; World Health Organization: Geneva, Switzerland, 2001.
473. Holder, A. A. *Proc. Natl. Acad. Sci. USA* **1999**, *96*, 1167–1169.
474. Waters, A. P.; Mota, M. M.; van Dijk, M. R.; Janse, C. J. *Science* **2005**, *307*, 528–530.
475. Carvalho, L. J. M.; Daniel-Ribeiro, C. T.; Goto, H. *Scand. J. Immunol.* **2002**, *56*, 327–343.
476. Khusmith, S.; Charoenvit, Y.; Kumar, S.; Sedegah, M.; Beaudoin, R. L.; Hoffman, S. L. *Science* **1991**, *252*, 715–718.
477. Doherty, J. F.; Pinder, M.; Tornieporth, N.; Carton, C.; Vigneron, L.; Milligan, P.; Ballou, W. R.; Holland, C. A.; Kester, K. E.; Voss, G. et al. *Am. J. Trop. Med. Hyg.* **1999**, *61*, 865–868.

478. Stoute, J. A.; Slaoui, M.; Heppner, D. G.; Momin, P.; Kester, K. E.; Desmons, P.; Wellde, B. T.; Garcon, N.; Krzych, U.; Marchand, M. *N. Engl. J. Med.* **1997**, *336*, 86–91.
479. Bojang, K. A.; Milligan, P. J.; Pinder, M.; Vigneron, L.; Alloueche, A.; Kester, K. E.; Ballou, W. R.; Conway, D. J.; Reece, W. H.; Gothard, P. et al. *Lancet* **2001**, *358*, 1927–1934.
480. Migot-Nabias, F.; Deloron, P.; Ringwald, P.; Dubois, B.; Mayombo, J.; Minh, T. N.; Fievet, N.; Millet, P.; Luty, A. *Trans. R. Soc. Trop. Med. Hyg.* **2000**, *94*, 557–562.
481. Daubersies, P.; Thomas, A. W.; Millet, P.; Brahimi, K.; Langermans, J. A. M.; Ollomo, B.; Ben Mohamed, L.; Slierendregt, B.; Eling, W.; Van Belkum, A. et al. *Nat. Med.* **2000**, *6*, 1258–1263.
482. Joshi, S. K.; Bharadwaj, A.; Chatterjee, S.; Chauhan, V. S. *Infect. Immun.* **2000**, *68*, 141–150.
483. Cavanagh, D. R.; Dodoo, D.; Hviid, L.; Kurtzhals, J. A. L.; Theander, T. G.; Akanmori, B. D.; Polley, S.; Conway, D. J.; Koram, K.; McBride, J. S. *Infect. Immun.* **2004**, *72*, 6492–6502.
484. Wang, L.; Goschnick, M. W.; Coppel, R. L. *Infect. Immun.* **2004**, *72*, 6172–6175.
485. John, C. C.; O'Donnell, R. A.; Sumba, P. O.; Moormann, A. M.; De Koning-Ward, T. F.; King, C. L.; Kazura, J. W.; Crabb, B. S. *J. Immunol.* **2004**, *173*, 666–672.
486. Hisaeda, H.; Saul, A.; Reece, J. J.; Kennedy, M. C.; Long, C. A.; Miller, L. H.; Stowers, A. W. *J. Infect. Dis.* **2002**, *185*, 657–664.
487. Malkin, E. M.; Diemert, D. J.; McArthur, J. H.; Perreault, J. R.; Miles, A. P.; Giersing, B. K.; Mullen, G. E.; Orcutt, A.; Muratova, O.; Awkal, M. et al. *Infect. Immun.* **2005**, *73*, 3677–3685.
488. Pizarro, J. C.; Vulliez-Le Normand, B.; Chesne-Seck, M.-L.; Collins, C. R.; Withers-Martinez, C.; Hackett, F.; Blackman, M. J.; Faber, B. W.; Remarque, E. J.; Kocken, C. H. M. et al. *Science* **2005**, *308*, 408–411.
489. Collins, W. E.; Walduck, A.; Sullivan, J. S.; Andrews, K.; Stowers, A.; Morris, C. L.; Jennings, V.; Yang, C.; Kendall, J.; Lin, Q. et al. *Am. J. Trop. Med. Hyg.* **2000**, *62*, 466–479.
490. Malkin, E. M.; Durbin, A. P.; Diemert, D. J.; Sattabongkot, J.; Wu, Y.; Miura, K.; Long, C. A.; Lambert, L.; Miles, A. P.; Wang, J. et al. *Vaccine* **2005**, *23*, 3131–3138.
491. Tomas, A. M.; Margos, G.; Dimopoulos, G.; Van Lin, L. H. M.; De Koning-Ward, T. F.; Sinha, R.; Lupetti, P.; Beetsma, A. L.; Rodriguez, M. C.; Karras, M. et al. *EMBO J.* **2001**, *23*, 3975–3983.
492. Valero, M. V.; Amador, R.; Aponte, J. J.; Narvaez, A.; Galindo, C.; Silva, Y.; Rosas, J.; Guzman, F.; Patarroyo, M. E. *Vaccine* **1996**, *14*, 1466–1470.
493. Patarroyo, M. E.; Romero, P.; Torres, M. L.; Clavijo, P.; Moreno, A.; Martinez, A.; Rodriguez, R.; Guzman, F.; Cabezas, E. *Nature* **1987**, *328*, 629–632.
494. Ockenhouse, C. F.; Sun, P. F.; Lanar, D. E.; Wellde, B. T.; Hall, B. T.; Kester, K.; Stoute, J. A.; Magill, A.; Krzych, U.; Farley, L. et al. *J. Infect. Dis.* **1998**, *177*, 1664–1673.
495. Shi, Y. P.; Das, P.; Holloway, B.; Udhayakumar, V.; Tongren, J. E.; Candal, F.; Biswas, S.; Ahmad, R.; Hasnain, S. E.; Lal, A. A. *Vaccine* **2000**, *18*, 2902–2914.
496. Moorthy, V. S.; Imoukhuede, E. B.; Milligan, P.; Bojang, K.; Keating, S.; Kaye, P.; Pinder, M.; Gilbert, S. C.; Walraven, G.; Greenwood, B. M. et al. *PLoS Med.* **2004**, *1*, 128–136.
497. Webster, D. P.; Dunachie, S.; Vuola, J. M.; Berthoud, T.; Keating, S.; Laidlaw, S. M.; McConkey, S. J.; Poulton, I.; Andrews, L.; Andersen, R. F. et al. *Proc. Natl. Acad. Sci. USA* **2005**, *102*, 4836–4841.
498. Rainczuk, A.; Scorza, T.; Spithill, T. W.; Smooker, P. M. *Infect. Immun.* **2004**, *72*, 5565–5573.
499. Scorza, T.; Grubb, K.; Smooker, P.; Rainczuk, A.; Proll, D.; Spithill, T. W. *Infect. Immun.* **2005**, *73*, 2974–2985.
500. Kongkasuriyachai, D.; Bartels-Andrews, L.; Stowers, A.; Collins, W. E.; Sullivan, J.; Sattabongkot, J.; Torii, M.; Tsuboi, T.; Kumar, N. *Vaccine* **2004**, *22*, 3205–3213.
501. Richie, T. L.; Saul, A. *Nature* **2002**, *415*, 694–701.
502. Ballou, W. R. *Exp. Opin. Emerg. Drugs* **2005**, *10*, 489–503.

Biographies

Kannoth M Muraleedharan received his MSc in chemistry from the Mahatma Gandhi University, Kerala, India. He completed his PhD degree in 1999 under the supervision of Prof S Ranganathan at the Regional Research Laboratory, Trivandrum, where he worked on the synthesis of model chemical systems to understand molecular assembly and disassembly, the key phenomena involved in every biological function. In 2001, he joined the research group of Prof Mitchell A Avery at the University of Mississippi for postdoctoral studies and at present is an Assistant Professor at the Department of Chemistry, Indian Institute of Technology Madras, India. His current research interests include the synthesis of broad-spectrum anti-infective agents, mechanistic investigations using molecular probes, and studies on protein–small molecule interactions.

Mitchell A Avery received his BS in chemistry in 1975 from Oregon State University and his PhD in organic chemistry in 1979 from the University of California–Santa Cruz. After working as a postdoctoral researcher from 1979 to 1981 at the University of California–San Diego and Oregon State University, he became a research associate in the Life Sciences Division of SRI International, Menlo Park, CA, eventually becoming Director of the Steroids and Natural Products Program. Rejoining academia in 1990, Professor Avery was an Associate Professor in the Department of Chemistry at the University of North Dakota in Grand Forks until he joined the Department of Medicinal Chemistry at University of Mississippi in 1994. He has risen through the ranks to become the Interim Chair of the Department of Medicinal Chemistry, a Professor of Medicinal Chemistry, Chemistry, and Biochemistry, and Research Professor through the Research Institute of Pharmaceutical Sciences and National Center for Natural Products Research at the University of Mississippi. He is also the director of the Laboratory of Applied Drug Design and Synthesis at the University of Mississippi, which is funded through the Centers for Disease Control and Prevention to further his work in developing new drugs for combating malaria and other infectious diseases worldwide. During his research career, Dr Avery's work has focused on a number of rational methods for discovery and/or development of new bioactive substances for human or veterinary use.

Comprehensive Medicinal Chemistry II
ISBN (set): 0-08-044513-6

ISBN (Volume 7) 0-08-044520-9; pp. 765–814

7.28 Antiprotozoal Agents (African Trypanosomiasis, Chagas Disease, and Leishmaniasis)

P M Woster, Wayne State University, Detroit, MI, USA

7.28.1 Introduction

A variety of parasitic protozoa are capable of producing human infection that can, in some cases, be life threatening. The majority of these organisms are considered minor health threats, either because they have a low level of occurrence, or because they can be effectively treated with existing agents. Among these infections are amebiasis (caused by various species of *Entamoeba*), *Giardia lamblia*, trichomoniasis (caused by species related to *Trichomonas vaginalis*), and infection caused by a variety of more obscure protozoans. A subset of protozoa are causative agents in important opportunistic infections in immunocompromised patients. Among these are infections by *Microsporidia*, such as *Encephalitozoon cuniculi*, *Enterocytozoon bieneusi*, and other protozoa, such as *Cryptosporidium parvum*, *Toxoplasma gondii*, and the fungal-related *Pneumocystis carinii*. These diseases are either self-limiting or can be effectively treated in patients with normal immune function. However, in immunocompromised patients, these infections can be difficult to manage with currently available agents. There is currently no effective therapy for infections caused by *Microsporidia* or *Cryptosporidia*, and the treatment of *Toxoplasma gondii* and *Pneumocystis carinii* produces variable and unsatisfactory results. Because they occur in developed nations, these parasitic diseases are being studied in a variety of laboratories, and new agents to treat these infections are under development.

The World Health Organization (WHO) has designated certain parasitic organisms as serious disease threats to global health, all of which are major infectious diseases that account for a large portion of global morbidity. Most prevalent among these infections is malaria, which is caused by four strains of the *Plasmodium* family. Recent developments in medicinal chemistry approaches to this disease are discussed in Chapter 7.27 of this volume. Infectious diseases caused by protozoa in the *kinetoplastid* family *Trypanosomatidae*, including human African trypanosomiasis (HAT), Chagas disease (also known as American trypanosomiasis), and leishmaniasis represent major threats to human health, particularly in underdeveloped countries. Medicinal chemistry approaches to the treatment of these three diseases are the focus of this chapter. Current therapies for parasitic infection by trypanosomatids are inadequate, especially in light of the emergence of drug-resistant parasitic strains. Many of the currently used drugs are toxic or sporadically efficacious, and there are no effective treatments for some late-stage parasitic diseases. Most of the currently used drugs are old and difficult to administer in the poor conditions found in areas where these diseases are endemic. Drug discovery efforts against the diseases mentioned above are limited, either because infected persons in underdeveloped areas cannot afford even a single course of therapy, or because the infected population is too small to justify the required research expenditures. Efforts to fight parasitic diseases in Third World Nations are often hampered by economic issues and political turmoil. This virtually assures that the world's most impoverished people will continue to bear the major burden of these parasitic diseases. Clearly, there is a need for new antiinfective agents that are potent, nontoxic, and inexpensive to manufacture. The recent publication of the complete genomes for *T. brucei*, *T. cruzi*, and *L. major*,[1–4] the so-called hat trick of antiparasitic research,[5] presents a unique opportunity for the discovery, characterization, and validation of new targets for chemotherapy aimed at the trypanosomatids. This report describes recent progress toward identifying suitable agents to treat parasitic diseases caused by organisms of the *Trypanosoma* and *Leishmania* families.

7.28.2 Human African Trypanosomiasis

7.28.2.1 Epidemiology

HAT is caused by two members of the *kinetoplastid* family *Trypanosoma brucei brucei* including *T. brucei gambiense* and *T. brucei rhodesiense*.[6] *Tb rhodesiense* produces an acute form of the disease known as East African trypanosomiasis, while *Tb gambiense* produces a chronic disease called West African trypanosomiasis. *Tb brucei* does not infect humans, but is a threat to livestock, and it has been suggested that these animals can act as a reservoir for *Tb gambiense* and *Tb rhodesiense*. Sleeping sickness is a daily threat to over 60 million people in 36 countries of sub-Saharan Africa (**Figure 1**). Of that number, only 3 to 4 million people are regularly monitored by health centers that can provide screening. Detection of the disease calls for major human and material resources, such as well-equipped health centers and qualified staff. Because medical resources are lacking in rural areas of the Third World, most people with sleeping sickness die prior to diagnosis. Nearly 45 000 new cases are reported annually, although the actual number may be as much as 10 times higher. WHO now estimates that more than 48 000 people died from HAT in 2004, and that 300 000–500 000 new cases occur each year.[7] Major epidemics of the disease occurred between 1896 and 1906, in 1920, and a third epidemic began in 1970 that persists even today. Following the 1920 epidemic, surveillance by mobile teams nearly eradicated the disease, and it had nearly disappeared between 1960 and 1965. Subsequently, relaxation of surveillance, combined with political unrest and destruction of the health care infrastructure, allowed an alarming resurgence of HAT. Today, in some areas of Angola, the Democratic Republic of the Congo and southern Sudan, the incidence of HAT is between 20% and 50%, and sleeping sickness has become the first or second greatest cause of mortality, ahead of HIV/AIDS.

7.28.2.2 Life Cycle of *Trypanosoma brucei* in Human African Trypanosomiasis

Tb gambiense and *Tb rhodesiense* are transmitted by the bite of the tsetse flies *Glossina fuscipes*, *Glossina palpalis*, or *Glossina morsitans*.[6] During a blood meal from an infected host, the tsetse fly ingests trypanosomes in the short, stumpy morphological form. This form is a growth-arrested version of the parasite that has moderate mitochondrial activity and is adapted for transmission to the insect vector. Following ingestion of trypanosomes, a complex life cycle begins (**Figure 2**) with the formation of the trypanosomal procyclic (promastigote) form in the tsetse midgut.[6,8,9] Procyclic trypanosomes cease proliferation and move to the salivary gland, where they are converted into epimastigote forms that attach through their flagella. Both the promastigote and epimastigote morphological forms are proliferative. Subsequently, the organism develops a variable surface glycoprotein (VSG) coat to avoid detection by the mammalian host immune system, and is converted into the nonproliferative metacyclic form for transmission. During a second tsetse blood meal, metacyclic organisms are injected into the mammalian host through the insect salivary gland, and the flagellated, morphologically

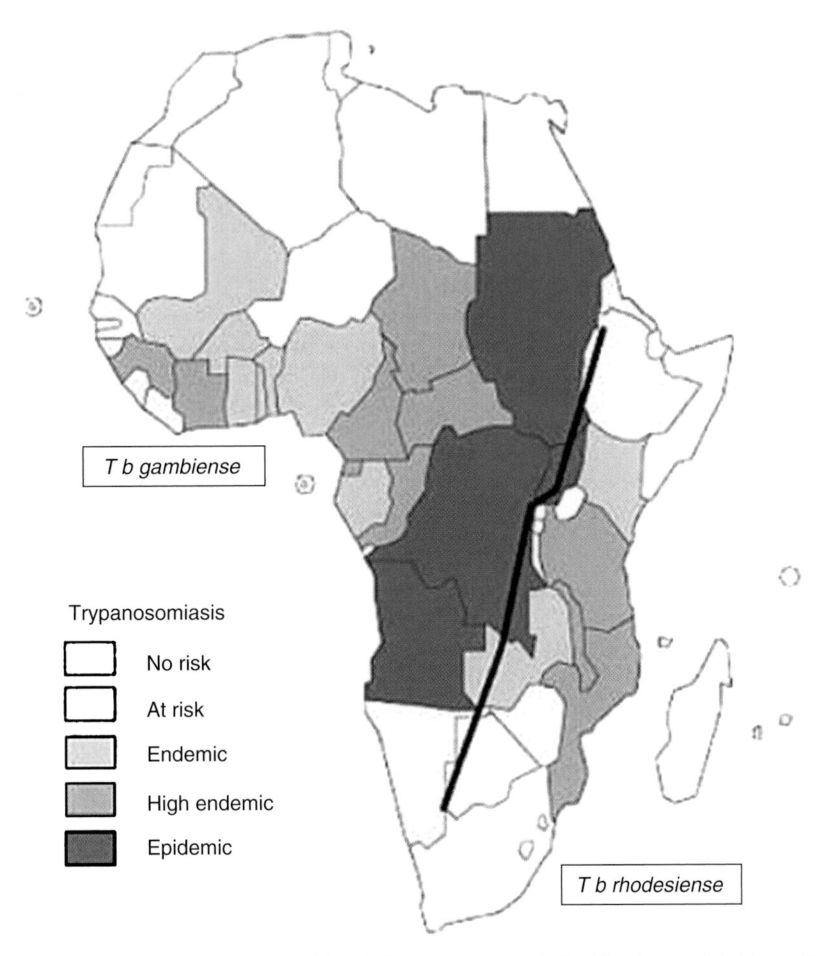

Figure 1 Geographical occurrence of East and West African trypanosomiasis. Map by the World Health Organization via http://www.medicalecology.org/diseases/d_african_trypano.htm.

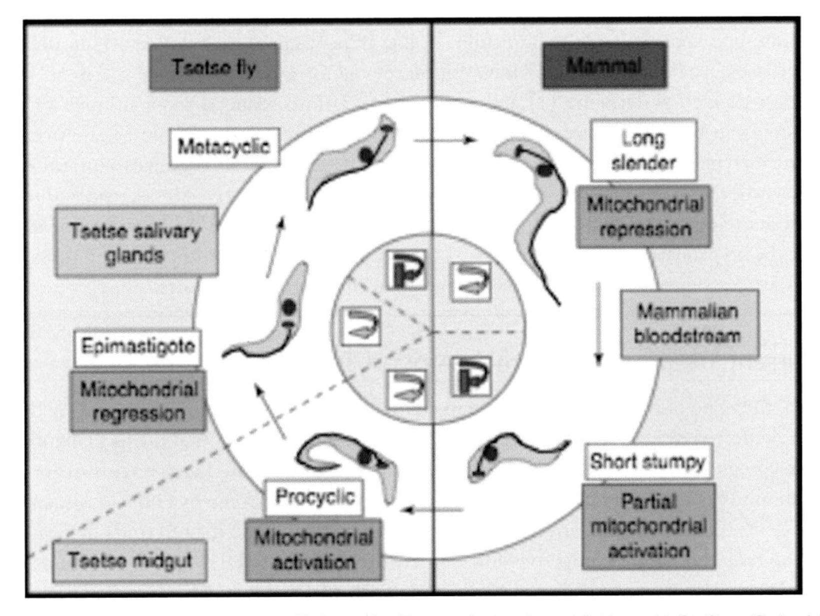

Figure 2 Life cycle of the African trypanosome. (Adapted with permission from McKean, P. G. *Curr. Opin. Microbiol.* **2003**, *6*, 600–607.)

slender trypomastigote bloodstream form of the organism develops in the blood and lymphatic system of the host. The trypomastigote bloodstream form expresses the bloodstream-stage-specific VSG coat, and is proliferative. In this form, the kinetoplast (the mitochondrial genome of the parasite) is located at the posterior end of the cell and mitochondrial activity is relatively repressed. As the parasite burden increases in the bloodstream, differentiation to morphologically stumpy forms occurs, and the cycle begins anew. The first signs of Stage 1 disease appear in the lymph nodes, and in most instances a nodule or ulcer forms at the site of infection.[6] Symptoms include malaise, headache, and an undulating fever, making differential diagnosis of the disease difficult. More serious manifestations of Stage 1 disease, such as pericardial and pulmonary edema, have been noted in the more acute *Tb rhodesiense* form of the infection. Stage 2 disease develops in a few weeks from *Tb rhodesiense* infection, or in a few months or even years in the case of *Tb gambiense*. This stage is initiated by parasitic invasion of internal organs and the central nervous system (CNS), causing the variations in diurnal and nocturnal sleep patterns that led to the coining of the phrase 'sleeping sickness.' Stage 2 disease can only be confirmed by lumbar puncture; however, trypanosomes are difficult to detect in cerebrospinal fluid, and diagnosis is often made by detecting elevated levels of IgM and lymphocytes.

7.28.2.3 Trypanosomal Biochemistry

Like many parasitic organisms, the cellular biochemistry of trypanosomes has not been fully elucidated, but a number of pathways that are parasite-specific have been identified as potential targets for chemotherapy. Trypanosomes are single-celled, flagellated protozoan parasites that express a number of organelles (nucleus, kinetoplast–mitochondrion, lysosome, endosome, flagellar pocket, Golgi, and glycosome) that have specific cellular functions.[9,10] Among these, the glycosome is a cellular microbody that is associated with trypanosomal peroxisomes, and which contains a variety of metabolic enzymes, including the first seven steps of the glycolytic pathway and the pentose phosphate shunt.[10,11] Thus, the glycolytic pathway in trypanosomes is distinct from mammalian glycolysis, in that it does not occur in the cytoplasm. Parasite glycolysis has therefore been studied as a potential drug target by designing inhibitors that exploit differences in the structure of individual glycolytic enzymes, or that interrupt transport of metabolites into the glycosome.[12] Interestingly, when in the insect vector, the parasite develops a conventional cytochrome chain and tricarboxylic acid (TCA) cycle. However, in the vertebrate host, trypanosomes depend entirely upon glucose for energy and are highly aerobic, despite the fact that the kinetoplast–mitochondrion completely lacks cytochromes. Instead, mitochondrial oxygen consumption is based on an alternative oxidase that most likely does not produce ATP. It has been shown that trypanosomal mitochondria can also adapt to low oxygen levels in the host through the use of unique anaerobic pathways that employ enzymes such as trypanosome alternative oxidase (TAO). This method of respiration, and in particular the enzyme TAO, is distinct from those of the host,[13,14] and as such represents a viable target for chemotherapy.

Polyamine metabolism in African trypanosomes is similar to mammalian polyamine metabolism, in that the organism synthesizes putrescine and spermidine from ornithine,[15] and trypanosomal forms of ornithine decarboxylase (ODC) and spermidine synthase have been identified. However, these organisms do not produce spermine, but instead convert two molecules of host-derived glutathione (**1**) and spermidine (**2**) into reduced trypanothione (**4**), which is used to protect the organism against oxidative stress. The formation of reduced trypanothione (**4**, **Figure 3**) is mediated by two ATP-dependent enzymes: glutathionylspermidine synthetase (GSpS) that produces glutathionylspermidine (**3**), and trypanothione synthetase (TS) that produces **4**.[16] In the presence of oxidative stress, oxidized trypanothione (**5**) is formed and must be recycled to the reduced form **4** by a third enzyme unique to the parasite, trypanothione reductase (TR). Importantly, the trypanothione pathway contains three enzymes that are unique to the parasite, and thus these enzymes may be viewed as targets for the rational design of antitrypanosomal agents.

7.28.2.4 Current Treatment of Human African Trypanosomiasis

At the present time, there are only four drugs that are in common use for the treatment of early- and late-stage HAT, as shown in **Figure 4**. Unfortunately, diagnosis of early-stage trypanosomiasis is difficult, especially in rural areas, and as such many patients progress to late-stage disease before seeking treatment. Early-stage disease is usually treated effectively with suramin (**6**) or pentamidine (**7**). Suramin enters the trypanosome by endocytosis, and is more than 75% bound to serum proteins following administration,[17,18] a fact that appears to be related to its mechanism of action. The drug may hamper the required parasitic uptake of low-density lipoproteins (LDL) by receptor-mediated endocytosis, it may associate with cytosolic enzymes inside the parasite, or with the highly positively charged glycolytic enzymes in the glycosome.[17] Pentamidine is thought to act by inhibiting the P2 adenosine uptake system in the parasite.[19–21] This system will be described in more detail later in this chapter. Trypanosomal strains that are resistant to suramin and

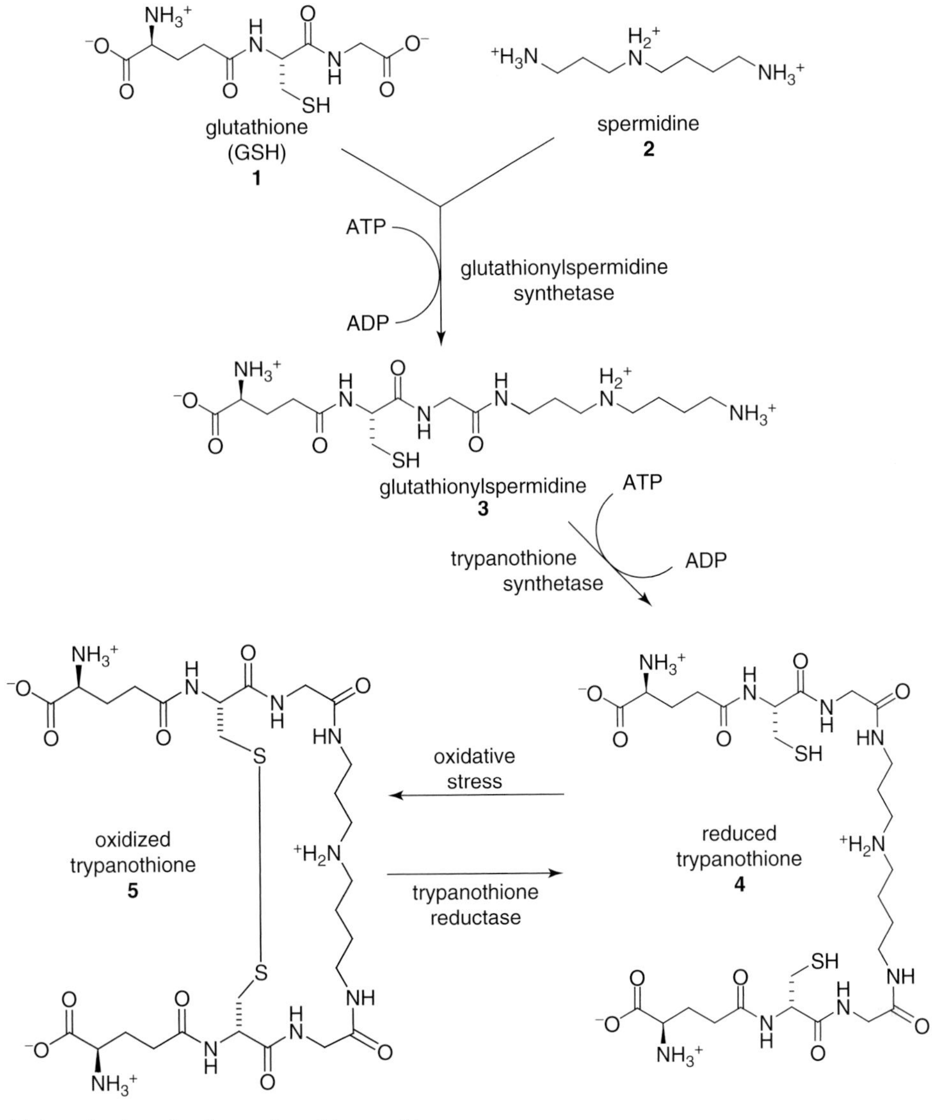

Figure 3 The synthesis and redox cycling of trypanothione.

pentamidine have been detected. Suramin resistance may be due to development of a drug extrusion complex,[21] while pentamidine resistance is likely due to mutations in trypanosomal transporter proteins, or to the ability of the parasite to use alternate transporters for adenosine.[21] There are only two effective treatments for late-stage trypanosomiasis, and these agents must penetrate the CNS in order to be effective. End-stage trypanosomiasis is treated with melarsoprol (Mel B, **8**), which was discovered in 1949. Following administration, melarsoprol is converted in vivo into its active metabolite, melarsen oxide (Mel Ox, **9**),[22] which has a plasma half-life of about 30 min. This arsenical drug produces its antitrypanosomal effect by forming a covalent complex with trypanothione known as Mel T, thus inactivating it and exposing the organism to oxidative damage. The treatment regimen for HAT with melarsoprol is outdated and complicated, and patients require hospitalization and monitoring during therapy. Patients are usually given three series of four intravenous injections with interval of 10 days between each series.[23] This schedule may not be the most effective and could possibly account, in part, for frequently reported side effects,[24,25] including a 10% incidence of reactive encephalopathy which is fatal in 3–5% of patients.[26] Melarsoprol is only soluble in propylene glycol, and is marketed as a 3.6% solution. As soon as a vial is opened, the drug must be used immediately, as it begins to deteriorate. In addition, administration is painful, and thrombophlebitis at the injection site is common.[17] Many patients who survive melarsoprol treatment suffer from serious neurological sequelae, and additional adverse effects arise following the covalent binding of melarsoprol to native biomolecules that subsequently become antigenic. The situation is further complicated by the emergence of arsenic-resistant strains of *Tb gambiense* and *Tb rhodesiense*, a group which is now comprised of 30% of all trypanosomes.[24] It is likely that this resistance is mediated by mutations in the P2 transporter, which is necessary for the import of melarsoprol into the organism.[19]

The ODC inhibitor eflornithine (**10**, **Figure 4**) has been shown to be curative in end-stage infections caused by *Tb gambiense*, but it is ineffective against late-stage *Tb rhodesiense* infection.[27–29] Eflornithine, also known as difluoromethylornithine (DFMO), is the only new molecule approved for the treatment of HAT over the last 50 years, and is considered a second-line agent for the treatment of arsenic (and hence melarsoprol) resistant *Tb gambiense*. The drug is most commonly dosed at 100 mg kg^{-1} of body weight as a short infusion at intervals of 6 h for 14 days (150 mg kg^{-1} in children). Eflornithine has limited efficacy against *Tb rhodesiense*, most likely because this parasite exhibits a much higher ODC turnover. Adverse reactions to eflornithine are reversible at the end of the treatment, and include convulsions (7%); gastrointestinal symptoms such as nausea, vomiting, and diarrhea (10–39%); bone marrow toxicity leading to anemia, leukopenia, and thrombocytopenia (25–50%); hearing impairment (5% in cancer patients); and alopecia (5–10%). Because it is trypanostatic rather than trypanocidal, it is a rather slow-acting drug. Eflornithine acts as an irreversible, enzyme-activated inactivator of ODC, as shown in **Figure 5**, for the mouse form of the enzyme.[30,31] Like the substrate ornithine, eflornithine forms a Schiff base with the pyridoxal phosphate that is tightly bound to the enzyme through LYS-89.[31] Decarboxylation occurs through the normal mechanism, and the resulting flow of electrons results in the elimination of a fluorine on the α-methyl group. The resulting adduct then acts as a Michael acceptor, and has been shown to bind covalently to CYS-360 in the catalytic site of the mouse enzyme, leading to irreversible inactivation. Although the drug is an effective inactivator, new ODC is rapidly produced through compensatory protein synthesis, and thus eflornithine must be given in large doses over extended periods to be effective. Because the synthesis of eflornithine involves a difluorination on the industrial scale, it is expensive to produce. Although it is marketed in the Unites States for a lifestyle disease, its availability in impoverished nations is limited, a problem that WHO is trying to address. Comprehensive reviews of current agents used to treat trypanosomiasis have recently been published.[32,33]

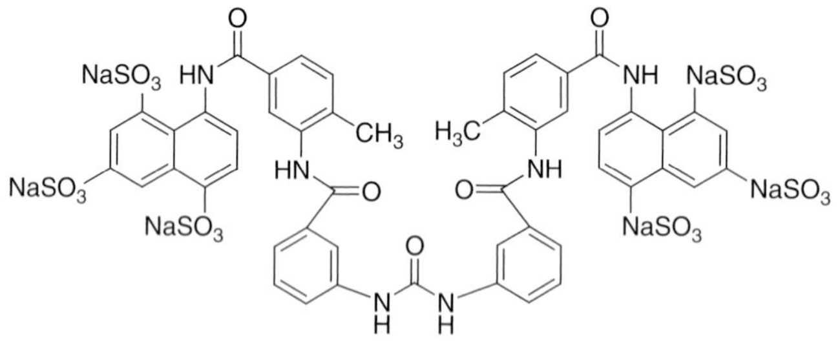

6

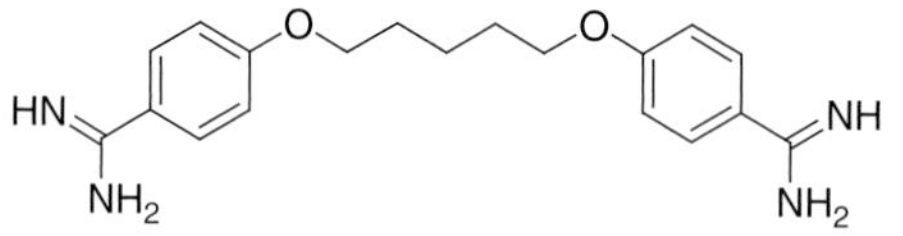

7

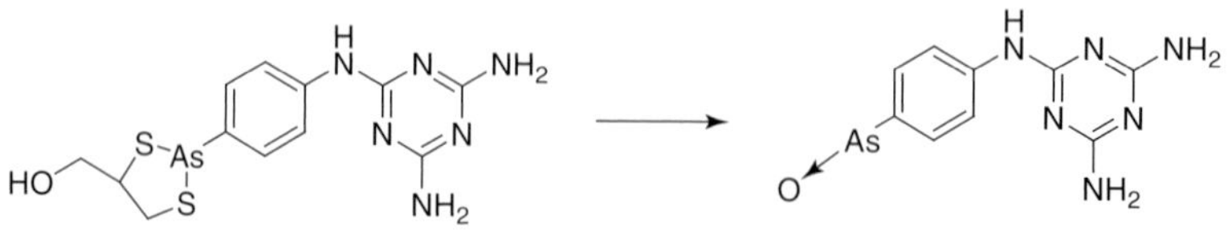

8

9

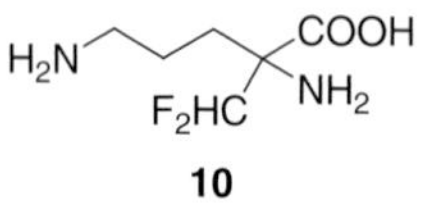

10

Figure 4 Structures of suramin (**6**), pentamidine (**7**), melarsoprol (**9**), melarsen oxide (**9**), and eflornithine (**10**).

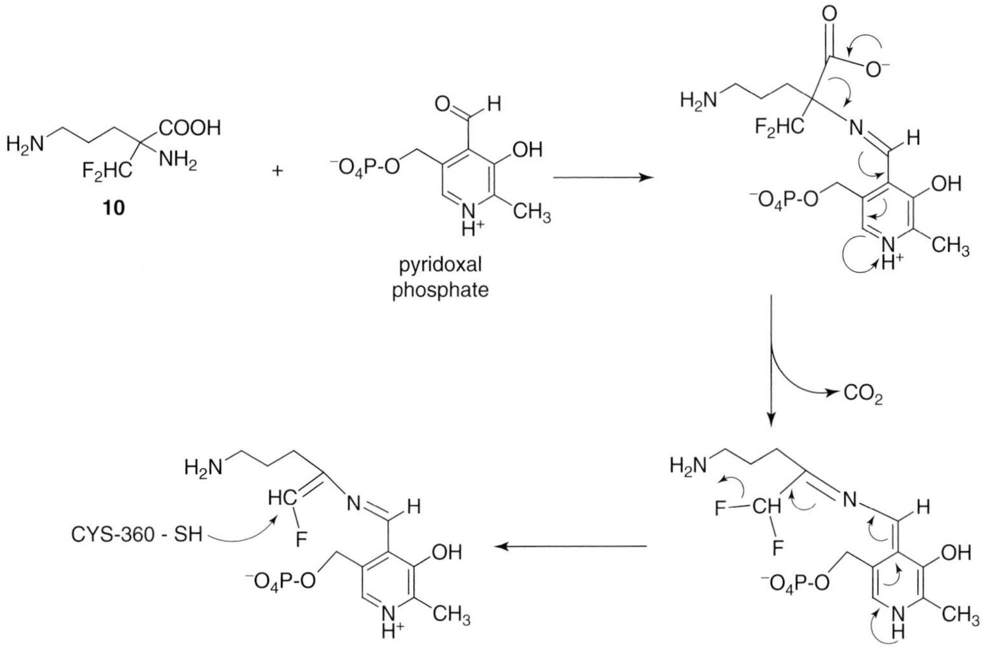

Figure 5 Inactivation of mouse ornithine decarboxylase by eflornithine, **10**.

7.28.3 Chagas Disease

7.28.3.1 Epidemiology and Clinical Manifestation

American trypanosomiasis, known as Chagas disease, was first characterized by Carlos Chagas in Brazil in 1909, and continues to be among the most important diseases in tropical and subtropical Mexico, Central America, and South America.[34,35] Chagas disease is caused by the trypanosomatid species *T. cruzi*, which is similar but distinct from the species that cause HAT. According to WHO, 25% of the total population in Central and South America is at risk (see **Figure 6**). Currently, there are between 16 and 18 million people infected, with 6 million cases advancing to clinically significant disease and more than 45 000 deaths annually. The disease is transmitted through the bite of several species of triatomine bugs (also referred to as reduviid bugs or 'kissing bugs'), including *Triatoma infestans*, *Triatoma dimidiata*, and *Rhodnius prolixa*, all of which live in dry, forested areas. These insects hide during the day in dark crevices or behind objects that are abundant in the type of housing used in endemic areas, as well as in animal nests and thatched roofs. At night, the insects emerge and feed on the blood of a variety of mammals, including humans. During a blood meal, the disease is not transmitted by the bite of the insect, but rather through its feces, which contains the organism. The parasite enters the host when the wound is scratched, or through the conjunctiva of the eye or the mucosa of the nose or mouth, and invades a variety of cell types including macrophages, smooth and striated muscle, and fibroblasts. When the site of infection is near the eye, acute swelling of one eyelid is observed, a phenomenon known as Romaña's sign. After a 1–2 week incubation period, the disease progresses through three phases, termed as acute, indeterminate, and chronic. In the acute phase, patients have high levels of the parasite in blood and tissues, and symptoms are generally mild (high fever and edema). Immunosuppressed patients and children can develop a more severe form of infection, with cardiac involvement and encephalomyelitis. Following the acute phase, the parasite burden in blood and tissues decreases dramatically, although low levels of the parasite are still detectable in certain tissues. Patients then experience an asymptomatic lag period of anywhere from 10 to 30 years, which is known as the indeterminate form of Chagas disease. A percentage of these patients advance to chronic Chagas disease, and develop moderate to severe clinical symptoms including cardiomyopathy, heart failure, and digestive tract abnormalities such as megacolon and megaoesophagus. If severe enough, these manifestations of the disease are the main determinants of morbidity. Because tissue damage can be extensive even when parasite levels are low in the affected tissues, it has been postulated that morbidity occurs as a result of autoimmune responses.[36,37]

Figure 6 Distribution of chagas disease (from the World Health Organization, http://www.who.int/ctd/chagas/geo.htm).

7.28.3.2 Life Cycle of *Trypanosoma cruzi* in Chagas Disease[34,35]

Triatomine bugs ingest both the trypomastigote and amastigote form of the organism during a blood meal from an infected host, and these forms of the organism pass into the midgut of the insect. In the midgut, trypomastigotes exist in slender and broad (or short, stumpy) forms, and both forms are converted into amastigotes, which are able to replicate. Flagella develop and begin to function (spheromastigote form), followed by transition to the epimastigote stage, which is also replicative. The epimastigotes attach to the cuticle area of the hindgut through hydrophobic interactions, and undergo transformation to their metacyclic form, which detaches and is excreted in feces. During a second blood meal, the organism enters the host through the wound, as described above. The organism enters the host cell via a unique, energy-dependent mechanism[38] that appears to depend on several trypomastigote surface recognition glycoproteins. These mucin-like glycoproteins are threonine-rich, *O*-glycosylated molecules that function as sialic acid acceptors during the infective stage. Inside the host cell, trypomastigotes are contained within a so-called parasitophorous vacuole, but subsequently escape, differentiate into amastigotes and replicate freely in the cytosol. After nine cycles of binary division, amastigotes differentiate back into highly motile trypomastigotes, which are released upon host-cell rupture, causing the acute form of Chagas disease. Parasites invade muscle and other nonphagocytic cells as an immune evasion strategy during the chronic stage of the disease, but it is unknown how this invasion is accomplished (**Figure 7**).

7.28.3.3 Current Treatment of Chagas Disease

T. cruzi is much more difficult to treat than HAT, since this trypanosomatid parasite is intracellular, and drugs used for the disease must pass through mammalian and parasite cell membranes to be effective. Laboratory and clinical studies conducted since 1969 have demonstrated that nifurtimox (**11**, **Figure 8**) and benznidazole (**12**, **Figure 8**) are the best agents for treating human *T. cruzi* infection, although they are far from being ideal drugs.[39] Nifurtimox and benznidazole are indicated in the acute phase of the infection, the congenital form, reactivation of disease associated with immunosuppression, and in transfusions and organ transplants involving infected individuals. Nifurtimox is administered at 8–10 $mg\,kg^{-1}\,day^{-1}$ in adults and $<15\,mg\,kg^{-1}\,day^{-1}$ in children for a period of 60–90 days, while benznidazole is given at $5\,mg\,kg^{-1}\,day^{-1}$ in adults and $<10\,mg\,kg^{-1}\,day^{-1}$ in children for 60 days. Both drugs are taken orally and must be given in 2–3 fractions after meals. They are generally well tolerated by children, particularly in

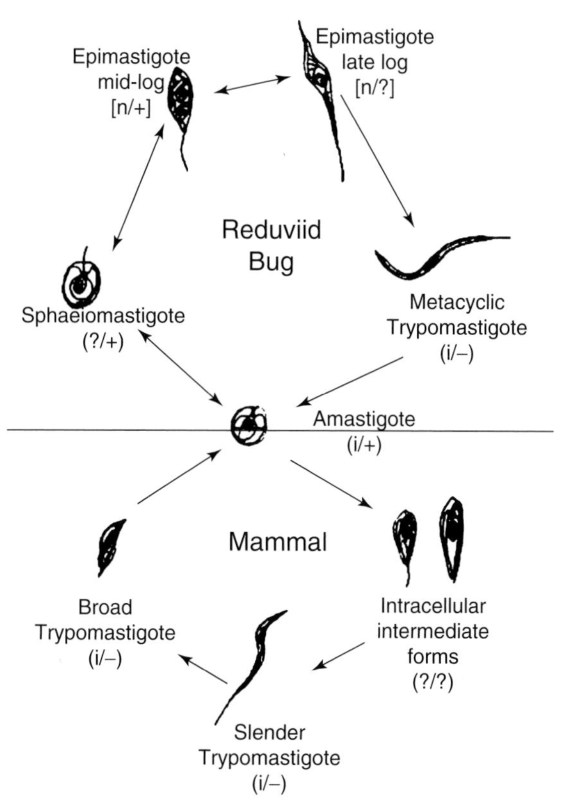

Figure 7 Life cycle of *Trypanosoma cruzi*. n, non-infective; i, infective; +, proliferative; –, nonproliferative. (Reproduced with permission from Tyler, K. M.; Engman, D. M. *Int. J. Parasitol.* **2001**, *31*, 472.)

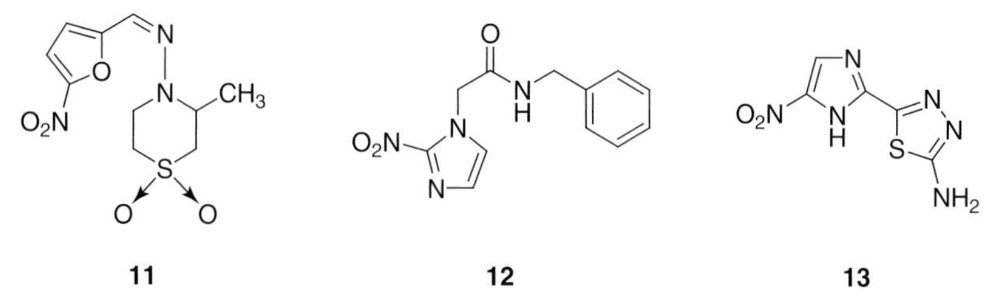

Figure 8 Structures of nifurtimox **11**, benznidazole **12**, and megazol **13**.

the acute phase of the disease, but relatively frequent and severe gastrointestinal or dermatological adverse reactions may be observed. Recurrence of the disease is a significant problem, and as such these drugs are considered generally ineffective. The main limitations of both drugs are their long courses of administration and the occurrence of adverse side effects. The related compound megazol (**13**, **Figure 8**) has also been used for Chagas disease, but its use was discontinued because of severe mutagenic and cytotoxic effects.[40]

7.28.4 Leishmaniasis

7.28.4.1 Epidemiology and Clinical Manifestations

Leishmaniasis is caused by one of 20 strains of the trypanosomatid parasite *Leishmania*, and currently threatens 350 million men, women, and children in 88 countries around the world. Health statistics are only maintained in 32 of the 88 countries affected by leishmaniasis, and as such, a substantial number of cases are never reported. It is estimated by WHO that 2 million new cases (1.5 million cutaneous and 500 000 visceral, see below) occur annually, with an estimated 12 million people infected worldwide. It is transmitted through an insect vector following the bite of a female sandfly from one of 30 species of the genus *Phlebotomus*. The disease can be divided into three categories (cutaneous, mucocutaneous, and visceral leishmaniasis), and is further categorized as Old or New World leishmaniasis, depending on the geographic location of the various parasite species causing the disease. Cutaneous and mucocutaneous leishmaniasis are diseases that are generally not life threatening, but the resulting disfigurement can cause social stigma that has a significant impact on lifestyle. Visceral leishmaniasis, also called kala-azar, produces life-threatening systemic infection if left untreated.[41] Although leishmaniasis is endemic to rural South America and Africa, it is encroaching on urban areas with substandard hygienic practices, and is also a risk for travelers in affected areas (e.g., Gulf War veterans). In addition, leishmania as an opportunistic infection in HIV patients has become a significant health threat in some areas.[42]

Cutaneous leishmaniasis manifests itself in several forms based on the species of parasite involved in the infection.[33,42–44] Old World cutaneous leishmaniasis is primarily caused by the species *L. major*, *L. aethiopica*, and *L. tropica*, and is transmitted by phlebotomine sandflies *P. papatasi* and *P. sergentii*.[42–44] The recidivans (lupoid) form may appear as a complication of cutaneous infection with *L. tropica*, and presents as erythematous papules that develop near the scars of previously healed lesions.[45] Old World cutaneous disease has been detected in rats, but dogs and humans are considered the major reservoirs for this form of leishmaniasis. The vast majority of cases of Old World cutaneous leishmaniasis occur in the Sudan, Afghanistan, Iran, Saudi Arabia, and Syria, and normally produce skin ulcers on the exposed parts of the body such as the face, arms, and legs. The disease can produce as many as 200 lesions, causing serious disability and permanent scarring. In most cases, the disease is self-curing, although the healing process can take a year or more. New World cutaneous leishmaniasis extends from Texas to central South America, and has been found in the Caribbean, but occurs primarily in Brazil and Peru.[42–44] The disease is transmitted by female sandflies of the genus *Lutzomyia*, and infections are caused by a variety of strains, including *L. mexicana*, *L. garnhami*, *L. lainsoni*, *L. venezuelensis*, *L. peruviana*, *L. colombiensis*, *L. guyanensis*, *L. amazonensis*, *L. panamensis*, and *L. pifanoi*. Unlike Old World disease, New World cutaneous leishmaniasis is passed between various mammals (dogs, rodents, opossum, and anteaters), as well as among human hosts. The symptoms of New World cutaneous leishmaniasis are similar to those of Old World disease, but a wide variation in the degree of tissue destruction occurs, depending on the strain of the infection. Treatment with ketoconazole or benznidazole is generally given, especially if the infection involves the soft cartilage of the ear.

The trypanosomatid parasite *L. braziliensis* is the main causative organism for mucocutaneous leishmaniasis, and more than 90% of this form of leishmaniasis occurs in Bolivia, Brazil, and Peru.[42–44] The initial lesions are similar to those seen in both forms of cutaneous leishmaniasis, but develop into more severe lesions of the mucosa in about 80% of untreated cases. These secondary lesions can lead to partial or total destruction of the mucous membranes of the nose, mouth, and throat, and in severe cases the resulting disfigurement can result in victims being humiliated and cast out from society.

Visceral leishmaniasis, also known as kala-azar in India or dumdum fever in Africa, is caused by organisms of the *Leishmania donovani* complex (*L. donovani*, *L. infantum*, and *L. chagasi*).[42–44,46] More than 90% of visceral leishmaniasis cases occur in Bangladesh, Brazil, India, Nepal, and Sudan. *L. donovani* is the primary cause of visceral leishmaniasis in India and East Africa. The disease is also caused by *L. infantum* in Mediterranean countries and *L. chagasi* in the New World. Human beings are the only known reservoir of *L. donovani*, but canines, especially domestic and stray dogs, provide a reservoir for *L. infantum* and *L. chagasi*. These differences have a major impact on the control of the disease and the emergence of drug resistance. Typically, patients with visceral leishmaniasis present with irregular bouts of fever, cough, abdominal pain, diarrhea, epistaxis, splenomegaly, hepatomegaly, substantial weight loss, and moderate to severe anemia. If left untreated, the fatality rate in developing countries can be as high as 100% within 2 years. As the organism develops in the skin cells, spleen, liver, or bone marrow of the host, patients experience anemia and cachexia, and ultimately succumb to the parasite. There were more than 41 000 recorded deaths owing to visceral leishmaniasis in 2000.[46]

7.28.4.2 Life Cycle of *Leishmania*

All leishmania strains develop through very similar life-cycle stages, as summarized in **Figure 9**. *Leishmania* have two distinct phases in their life cycle, one in the mammalian host and one in the insect vector.[43,47] The insect phase begins when the appropriate sandfly ingests *leishmania* amastigotes during a blood meal from a mammalian carrier. The amastigote form is either round or oval in shape, 3–5 μm in size, and contains a distinct nucleus, a kinetoplast, and a short, intracytoplasmic flagellum. The amastigotes then transform into promastigotes, which are slender, motile organisms, 10–15 μm in length, with a single anterior flagellum. The promastigote form migrates to the midgut of the phlebotomine sandfly, multiplies by longitudinal fission, and then attaches to the wall of the insect hypostome

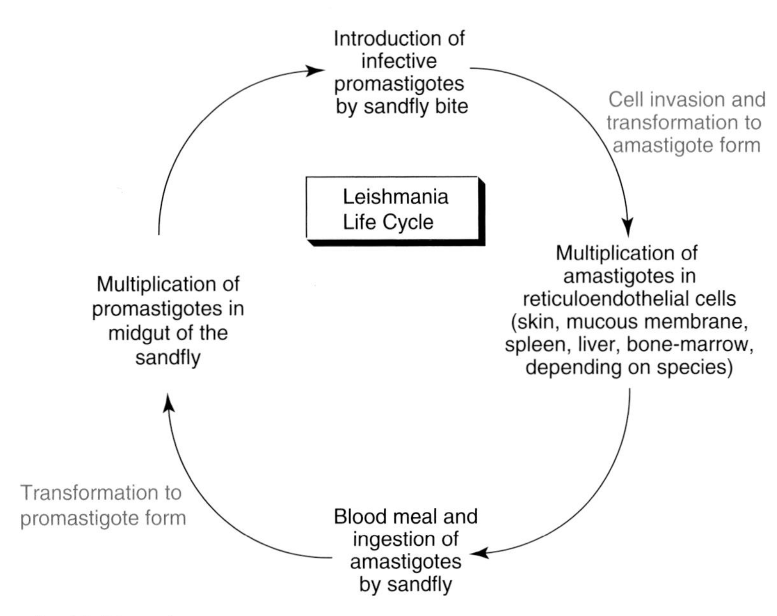

Figure 9 The life cycle of *Leishmania* spp.

through their flagella. When the insect takes a second blood meal, promastigotes are transferred into the mammalian host, initiating the mammalian phase of the life cycle, and they are engulfed by reticuloendothelial cells and converted into amastigotes. Depending on the species, the affected reticuloendothelial cells can be subcutaneous (cutaneous leishmaniasis), mucosal (mucocutaneous leishmaniasis), or in other areas of the body such as bone, liver, and spleen (visceral leishmaniasis). The amastigote form reproduces by binary fission until the reticuloendothelial cell is destroyed, and the liberated parasites are then phagocytized by other reticuloendothelial cells of the same type.

7.28.4.3 Current Treatment of Leishmaniasis

Because leishmania are biochemically similar to trypanosomes, many of the antitrypanosomals mentioned above are also effective against *L. donovani*. This is not universally true, however, due to the differences in tissue distribution between the parasites. In general, the intracellular amastigote form is considered the target for chemotherapy. The standard treatment for visceral leishmaniasis is sodium stibogluconate **14** or meglumine antimoniate **15**, moderately toxic antimonials for which resistance is increasing.[33,48,49] These essentially equivalent compounds are actually complex mixtures of carbohydrate/antimony complexes, which are dose-based on their antimony(V) content. The mechanism of the antileishmanial activity of antimony remains to be elucidated, but it is known that **14** and **15** suppress glycolysis and fatty acid metabolism in the parasitic glycosome. They may also interact with the leishmanial form of trypanothione.[50] Antimonials are administered intravenously at 20–40 mg kg^{-1} day for 20–40 days. In addition to high cost and a difficult administration schedule, there are numerous resistant strains, and antimonial toxicity is frequent when HIV coinfection is present.[46] In addition, there are quality control problems and batch-to-batch variability for both branded and generic antimonial preparations. The poor quality of some generic formulations of the drug in India has led to serious toxicity.[51] Pentamidine (**7**, **Figure 4**) is also marginally effective in the visceral form of leishmaniasis when given at 2–4 mg kg^{-1} by intramuscular injection, either daily or every other day, for a course of 12–15 injections. Owing to toxicity and resistance, especially in India, pentamidine is being abandoned as a second-line treatment for leishmaniasis.[46] The macrocyclic antibiotic amphotericin B (**16**) is used as a second-line treatment, and is administered at 7–20 mg kg^{-1} total dose intravenously for up to 20 days. Cure rates have been reported as high as 97%, and there have been no reports of resistance, but the drug is nonideal owing to dose-limiting toxicity.[46] Lipid-associated formulations of amphotericin B are highly effective against visceral leishmaniasis and better tolerated than the conventional preparation. Liposomal amphotericin B was studied in India, Kenya, and Brazil,[52] and minimum doses of 6, 14, and 21 mg kg^{-1}, respectively, were necessary to provide 95% cure rates.[53] However, such products are prohibitively expensive, and as such their utility in the Third World is severely restricted (**Figure 10**).

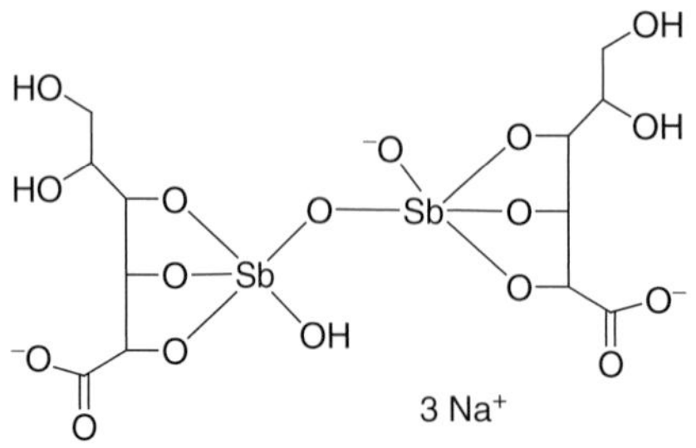

Figure 10 The structures of sodium stibogluconate **14**, meglumine antimoniate **15**, and amphotericin B **16**.

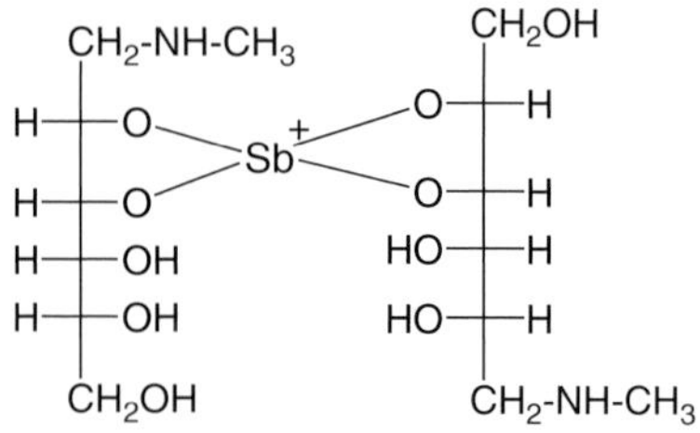

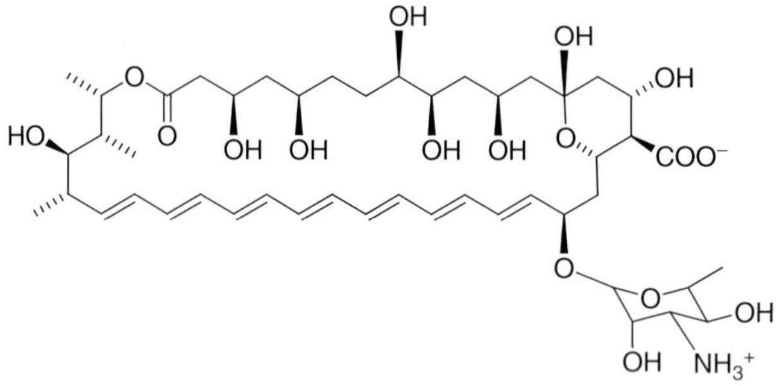

7.28.5 Progress in Drug Research for Human African Trypanosomiasis, Chagas Disease, and Leishmaniasis

7.28.5.1 Potential New Therapies for *Trypanosoma brucei* and *Trypanosoma cruzi*

7.28.5.1.1 Agents targeted to the polyamine pathway

A number of novel agents have been developed which target the trypanosomal polyamine biosynthetic pathway, including the associated trypanothione redox system. The phosphate-based transition state analogs **17** and **18** were designed to mimic the tetrahedral transition state of GSpS.[54–56] These analogs inhibited GSpS from *Escherichia coli* with K_i values of 6.0 and 3.2 μM, respectively. In addition, **18** was shown to be a slow-binding inhibitor of the enzyme, and formed an E–I* complex with a 410-fold higher affinity than the collisional E–I complex. Despite these promising early results, compounds **17** and **18** have not been developed further as antiparasitic agents. A number of *N*-(3-phenylpropyl)-substituted spermine analogs have been reported that act as potent inhibitors of TR.[57] Compounds **19** and **20** were the most potent inhibitors, with K_i values of 0.61 and 0.15 μM against TR from *T. cruzi*. Compound **19** showed significant activity against African trypanosomes in vitro, with IC_{50} values between 0.12 and 0.16 μM against four strains of trypanosomal clinical isolates, including the arsenic-resistant K 243 As-10-3 variant of *Tb rhodesiense*.

It has been suggested that alkyl- and aralkyl-substituted polyamine analogs disrupt polyamine metabolism because the nitrogen pK_a values have been altered, leading to reductions in their degree of protonation at physiological pH.[58] A variety of these analogs have been synthesized that selectively enter proliferating cells using the polyamine transport system, where they downregulate the biosynthetic enzymes ODC and *S*-adenosylmethionine decarboxylase (AdoMet-DC). However, these agents do not substitute for the natural polyamines in terms of their cell growth and survival functions,[16] and ultimately produce polyamine depletion and cell death. An extended series of polyamine analogs based on the lead structure MDL 27695 (**21**) has recently been reported. Compound **21** was shown to possess significant antimalarial and antileishmanial activity in vitro and in vivo, and produced cures in murine models of malaria and leishmania.[59–62] Subsequently, **21** was found to possess antitrypanosomal activity as well,[63] prompting the synthesis of the series of analogs mentioned above. Compounds **22** and **23** were both effective in vitro against *Tb brucei* LAB 110 as well as the *Tb rhodesiense* isolates K 243, K 269, and K 243 As-10-3.[16,64] The most impressive activity was found in the case of **23**, which inhibited the growth of K 243 with an IC_{50} of 40 nM, the same IC_{50} as melarsen oxide. Further, **23** inhibited growth in the K 243 As-10-3 arsenic-resistant strain ($IC_{50} = 165$ nM), against which melarsoprol is inactive. These initial studies indicated that analogs with a 3-3-3 carbon backbone had poor activity as antiparasitic agents, but were effective antitumor compounds. By contrast, expanding the center section of the molecule to afford a 3-7-3 backbone produced analogs that had little antitumor activity, but were effective antiparasitic agents.[16,63,64] Subsequent structure–activity studies involving a library of over 200 alkyl- and aralkylpolyamine analogs revealed that both symmetrically and unsymmetrically substituted analogs possess potent antitrypanosomal activity, and that the central chain length of these derivatives can vary between 3 and 7 carbons, as shown in **Figure 12**. Thus, compound **24**, which has a 3-7-3 carbon backbone and unsymmetrical alkyl substituents on the terminal nitrogens, inhibits the growth of *Tb brucei* Lab 110 and *Tb rhodesiense* K 243 with IC_{50} values of 0.06 and 0.07 μM, respectively. Compound **28**, which differs from **23** (**Figure 11**) only in the length of the intermediate carbon chain, inhibits the same strains of *Tb brucei* with IC_{50} values of 1.6 and 0.9 μM, respectively. It is significant to note that both **24** and **28** were effective against the K 243 As-10-3 arsenic-resistant strain of *Tb rhodesiense* with IC_{50} values of 0.185 and 6.25 μM, respectively. Analogs **24–28** were found to be effective inhibitors of TR, while having no activity against the human form of glutathione reductase (GR) (unpublished observations). These analogs are currently being evaluated for antitrypanosomal activity in vivo (**Figure 12**).

As outlined above, alkylpolyamines are able to disrupt cellular metabolism because the pK_a values for their internal and terminal nitrogen moieties have been altered, leading to changes in the percent ionization of each nitrogen. The natural polyamines exert their effects at various cellular sites, many of which have not been identified, and as such, the charge recognition of specific polyamines at these sites is critical to normal cellular function.[58] This strategy has been used to develop a number of successful antitumor agents related to bis(ethyl)norspermine, some of which have entered human clinical trials. All of these analogs appear to take advantage of the specific polyamine transport system to enter

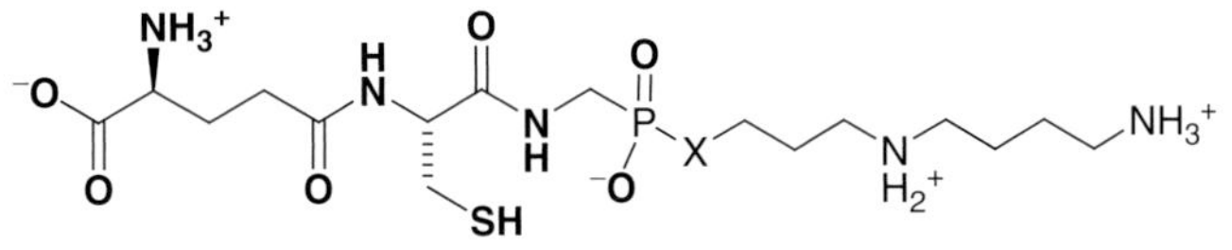

17 (X = O)
18 ($X = CH_2$)

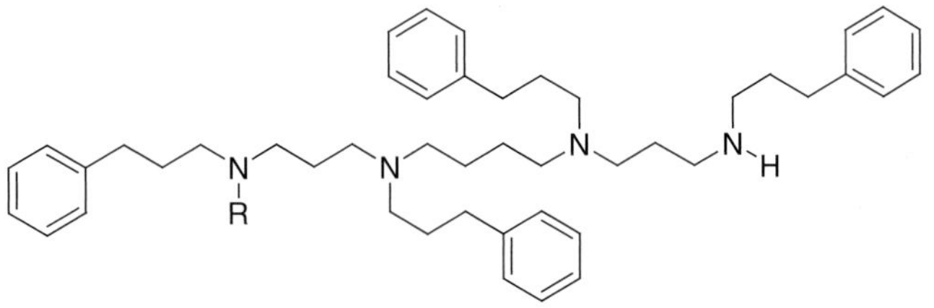

19 (R = H)
20 ($R = CH_2CH_2CH_2$-Ph)

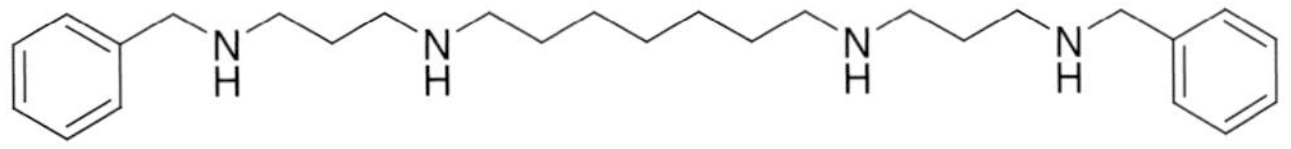

21

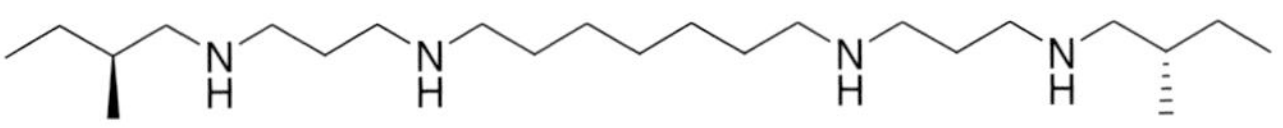

22

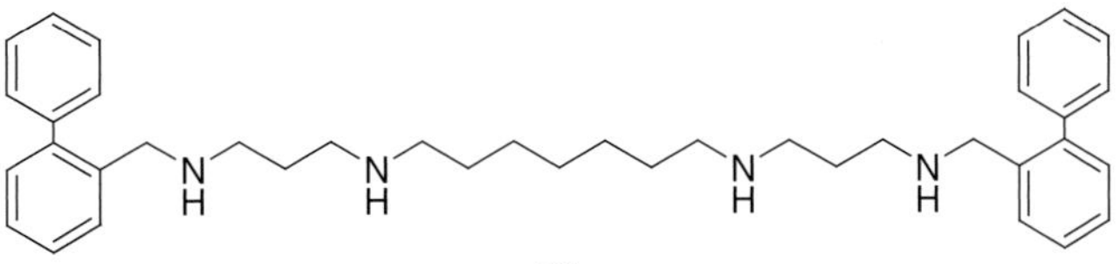

23

Figure 11 Polyamine-based antitrypanosomal agents **17–23**.

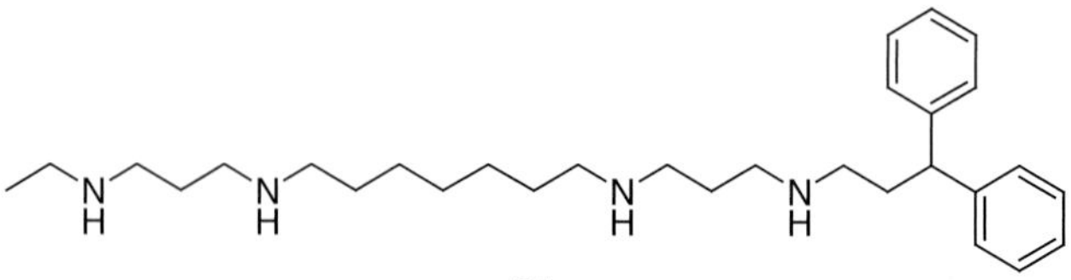

Figure 12 Structures of antitrypanosomal polyamine analogs **24–28**.

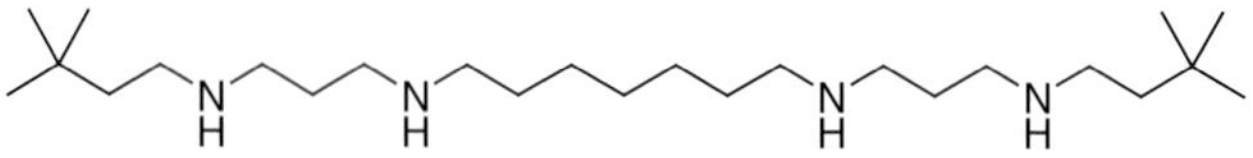

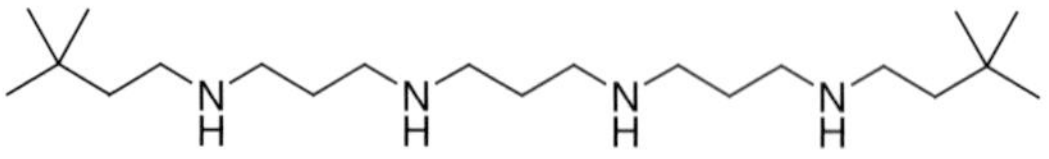

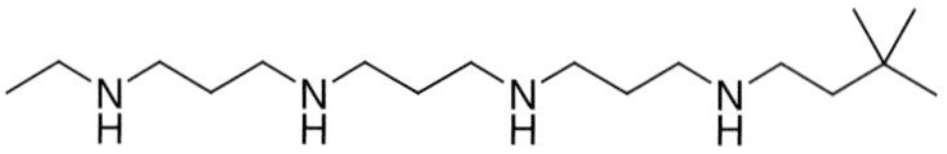

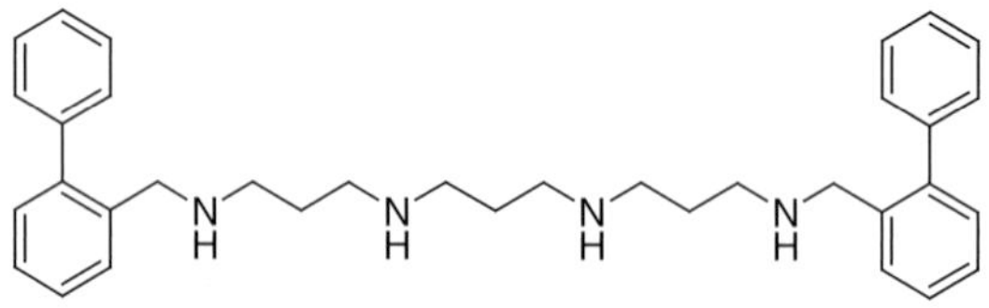

cells. It was reasoned that 3-7-3 polyamine analogs containing substituted terminal guanidines would have similar effects on polyamine metabolism, since guanidine moieties would produce even greater charge perturbations when compared with the natural polyamines. A series of alkyl- and aralkyl-substituted polyaminoguanidines and polyaminobiguanides represented by compounds **29–33** (**Figure 13**) were thus synthesized and evaluated as inhibitors of TR, and as antitrypanosomal agents in vitro. These analogs are potent inhibitors of TR (IC_{50} values for **29**: 2.24 μM; **30**: 4.72 μM; **31**: 69.5 μM; **32**: 4.16 μM; **33**: 2.96 μM) but have no significant effects on GR ($IC_{50} \gg 100$ μM) in an in vitro purified enzyme assay system. In addition, they were found to have IC_{50} values as low as 90 nM against cultured bloodforms of *Tb brucei* and *Tb rhodesiense* KETRI 243 at comparable levels (e.g., **29**, $IC_{50} = 2.05$ μM; **30**, $IC_{50} = 0.85$ μM; **31**, $IC_{50} = 0.39$ μM; **32**, $IC_{50} = 1.95$ μM; **33**, $IC_{50} = 0.22$ μM). These analogs are also being examined in animal models of trypanosomiasis.

Molecular modeling studies suggested that various tricyclic antidepressant compounds could have affinity for TR,[65] and these observations were confirmed by studies involving purified TR from *T. cruzi*.[66] The most active of these analogs was clomipramine, **34** (**Figure 14**) , with a K_i of 6.5 μM against TR from *T. cruzi*. Subsequent studies indicated that phenothiazines such as chlorpromazine **35** were also effective TR inhibitors.[67] Attempts to reduce the neuroleptic side effects of these analogs by forming diphenylsulfide derivatives related to **34** and **35**, either alone or as dimers linked by spermidine moieties, did not produce any analogs that were more inhibitory than **34**.[68,69] Related analogs such as fluorenone **36** and acridine **37**, which also included guanidine moieties in their structure (*see* Section 7.28.5.1.2) were effective inhibitors of trypanosomal growth with IC_{50} values of 7.3 and 66.9 μM, respectively, against *Tb rhodesiense* in culture.[70] *N*-Alkylated fluorene analogs such as **38** ($IC_{50} = 16.9$ μM) were also effective against *Tb rhodesiense* in vitro. Although they were not particularly potent against the organism in vitro, analogs **36–38** produced cures in murine models of *Tb rhodesiense* infections in 60–75% of treated animals. The ability of compounds **36–38** and their homologues to inhibit TR was never determined.

7.28.5.1.2 Trypanocides that utilize unique parasitic nucleotide transporters

It has recently been shown that trypanosomes do not synthesize purines, but instead import them using two unusual adenosine transporters termed P1 and P2.[19,71,72] The P1 transporter is responsible for importing adenosine and inosine,

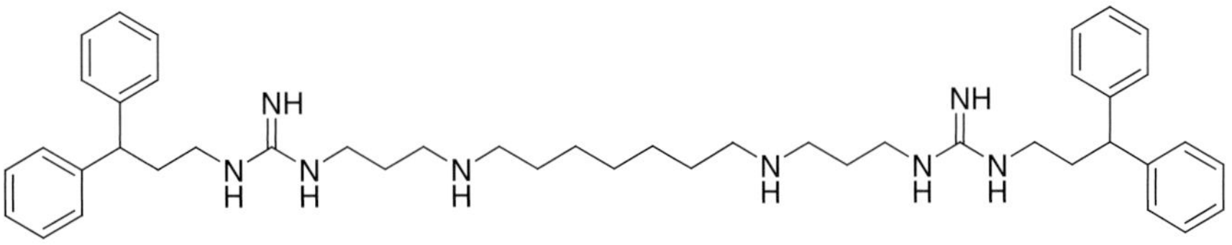

29

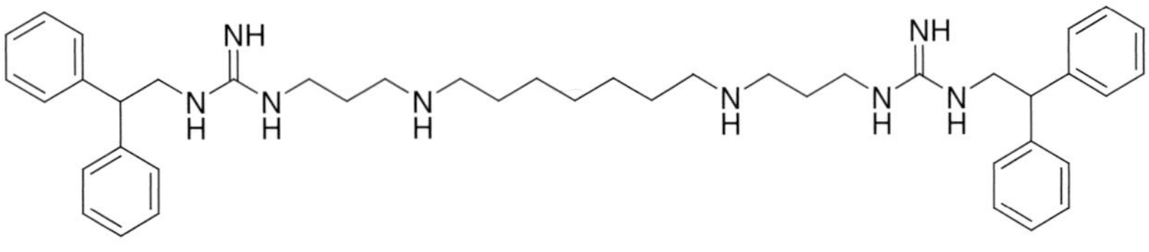

30

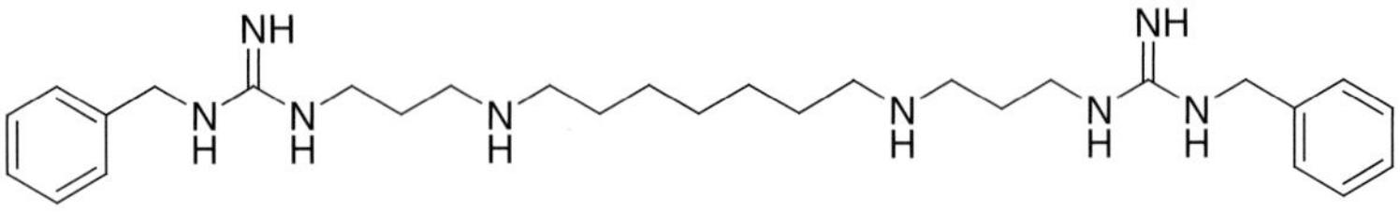

31

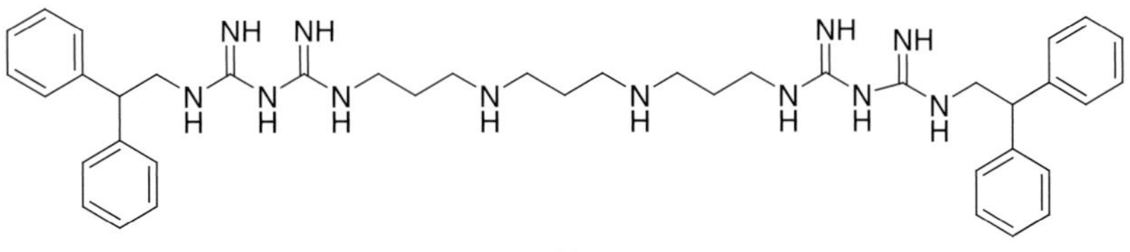

32

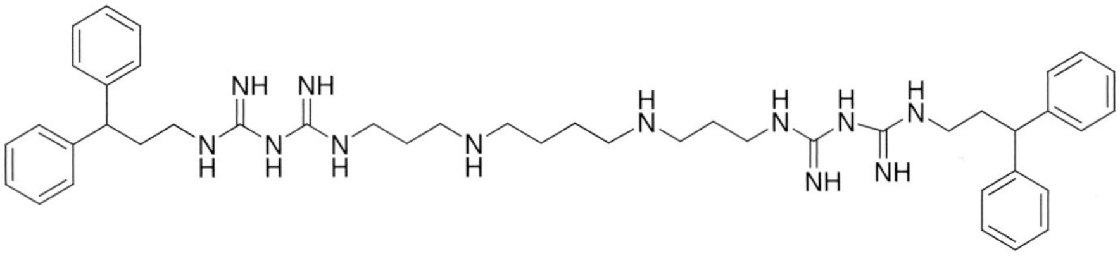

33

Figure 13 Structures of polyaminoguanidines **29–31** and polyaminobiguanides **32–33**.

while the P2 receptor accepts a number of substrates, including adenine, adenosine, suramin **6**, pentamidine **7**, and arsenical drugs such as melarsoprol, **8**. The P2 transporter is missing in arsenic-resistant strains.[71,72] A number of nucleoside analogs have been synthesized that appear to act as substrates for the P1 or P2 transporter, and these derivatives act by as yet unknown mechanisms. The *trans*-1*S*,4*S* isomer of **39** is an irreversible inactivator of mammalian and bacterial AdoMet-DC, and also acts as a trypanocide with an IC_{50} of 0.9 μM.[73] Another inhibitor of this enzyme, AbeAdo **40**, was found to be a substrate for the P2 transporter in *Tb rhodesiense*, and exhibited a nanomolar IC_{50} value.[74] It was also curative for *Tb rhodesiense* infections in mice.[75–77] AbeAdo acts as an irreversible, enzyme-activated inactivator of AdoMet-DC, and thus effectively disrupts polyamine biosynthesis in the parasite, causing accumulation of *S*-adenosylmethionine in trypanosomes. However, this drug has never been developed for therapeutic use. Nucleoside analogs related to the methylthioadenosine analog hydroxyethylthioadenosine (HETA), **41** enter trypanosomes via the P2 and *S*-adenosylmethionine transport systems.[76] HETA is concentrated 10–80-fold in trypanosomes, and kills the parasite by reducing protein methylation, leading to in vivo in cures against a number of strains.[78] The (+)-isomer of 7-deaza-noraristeromycin, **42**, appears to utilize the trypanosomal P1 system, and shows activity against *Tb brucei* (IC_{50} = 0.16 μM) and the arsenic-resistant K 243 As-10-3 strain (IC_{50} = 5.3 μM) (**Figure 15**).[79]

There has been a recent resurgence in interest in the use of amidines and guanidines as antitrypanosomal agents, many of which are amidine analogs (**Figure 16**) related to pentamidine (**7**). These compounds were synthesized to

Figure 14 Structures of tricyclic amines and guanidines with antitrypanosomal activity.

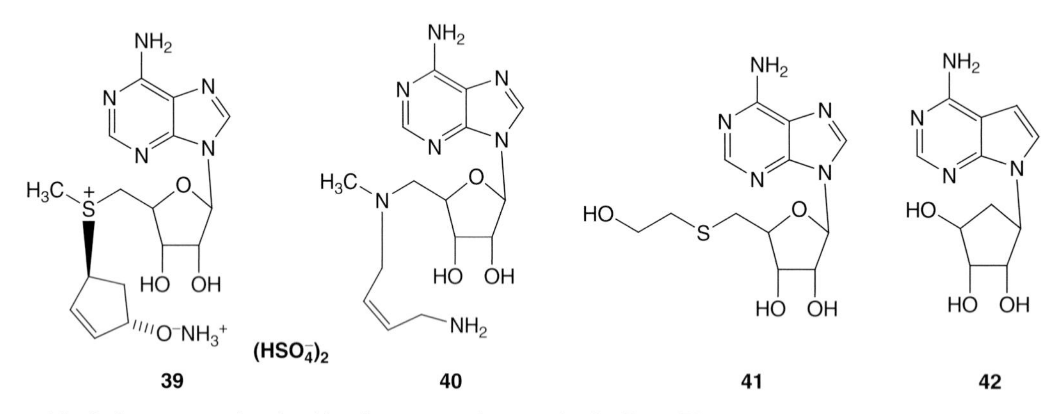

Figure 15 Antitrypanosomal nucleosides that act as substrates for the P1 or P2 transport system.

eliminate the significant toxicity caused by the administration of pentamidine. In addition to pentamidine **7** and the closely related amidine berenil **43**, conformationally restricted methylglyoxal (bis)guanylhydrazone (MGBG) analogs such as CGP 40215 (**44**) have shown significant antitrypanosomal activity.[80] Nearly all of these analogs are aromatic amidines that are structurally similar to pentamidine. Linear analogs such as the (bis)amidine undecanediamidine (**45**) and the (bis)guanidine synthalin (**46**) have also shown promising antitrypanosomal activity.[80] Substituted (bis)guanide analogs related to synthalin were recently identified by virtual screening, and found to inhibit TR from *T. cruzi* with K_i values between 2 and 47 μM.[81] The in vivo activity of these analogs has not yet been assessed. Using the structure–activity data for the P2 transport system, a series of triazine-substituted polyamines related to **47** and **48** were synthesized and evaluated. Compound **48** was an excellent substrate for the P2 transport system as determined by an adenosine uptake competition assay, and exhibited a 0.18 μM IC_{50} value against *Tb rhodesiense*.[82] However, these analogs were not effective in vivo.

Pentamidine and its analogs are primarily useful in the treatment of early-stage trypanosomiasis. Attempts to design conformationally restricted analogs of pentamidine have resulted in the discovery of agents with moderately improved activity (**Figure 17**). Dicationic amidines with unsaturated internal carbon chains, such as **49** and **50**, exhibited IC_{50} values between 2 and 9 μM against *Tb brucei* and three strains of *Tb rhodesiense* (KETRI 243, 269, and 243 As-10-3).[83] Compound **50** was more potent in vitro against all strains of *Tb rhodesiense* tested. Compounds **49** and **50** produced cures

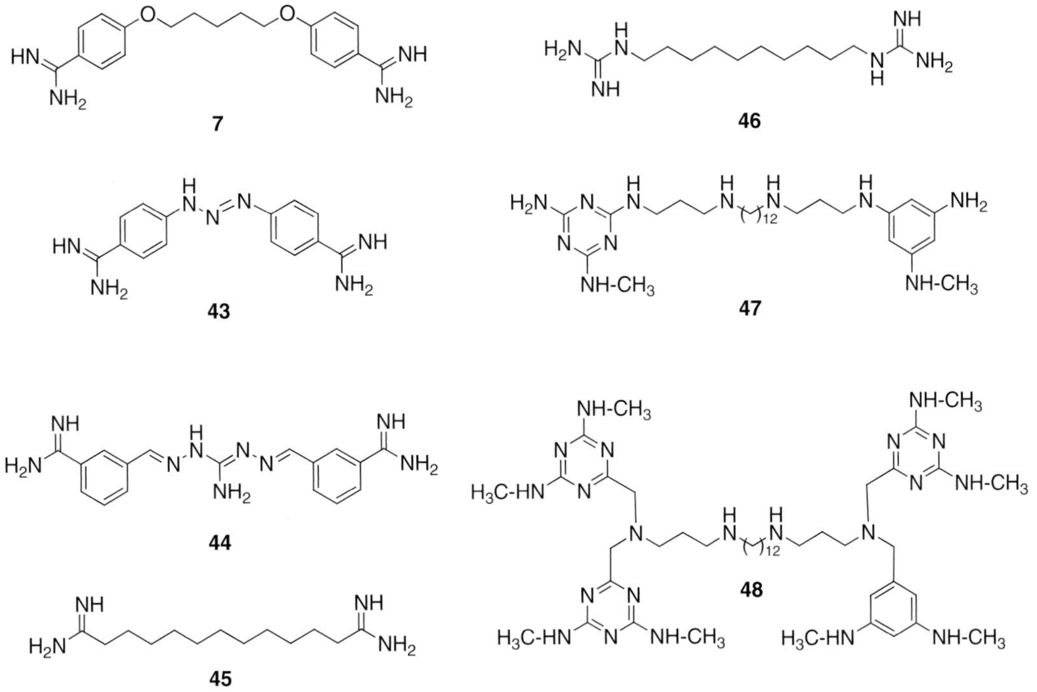

Figure 16 Antitrypanosomal amidines pentamidine **7** and related analogs **43–48**.

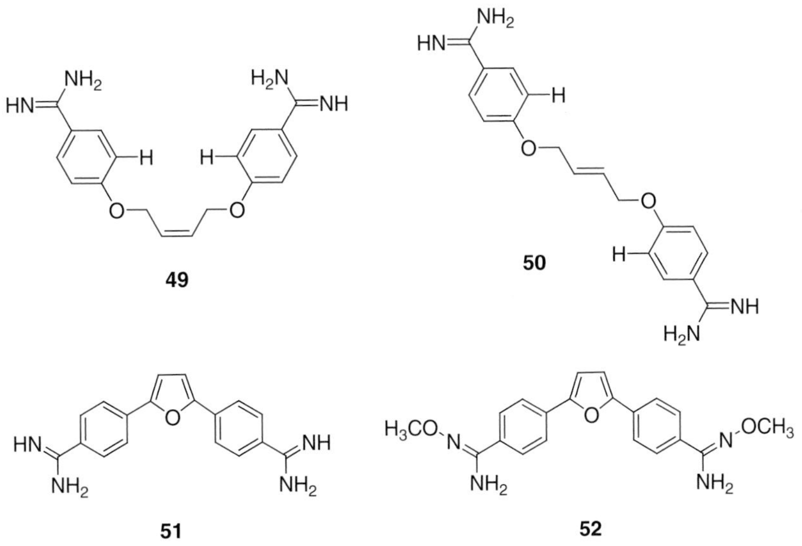

Figure 17 Conformationally restricted analogs of pentamidine with antitrypanosomal activity.

in murine models of trypanosomal infection involving *Tb brucei* Lab 110 and the clinical isolates *Tb rhodesiense* KETRI 269, 2002, and 2538, although with IC_{50} values 2–3-fold higher than pentamidine. There has been considerable interest in the conformationally restricted pentamidine analogs DB75 (**51**) and DB289 (**52**), which are currently undergoing clinical trials.[84] DB75 (**51**), also known as furamidine, is a potent trypanocide in vitro and in vivo, but does not exhibit significant oral bioavailability. In addition, it is of little value in the treatment of late-stage trypanosomiasis, either by oral or intravenous administration, because it does not readily penetrate into the CNS. To circumvent these problems, the (bis)-*N*-methoxy analog DB289 (**52**) was synthesized as a prodrug form of **51**.[84] DB289 was found to have good oral bioavailability, leading to current clinical trials for early-stage trypanosomiasis in Central Africa. However, the drug does not seem to be effective in late-stage disease, regardless of route of administration.[84] Brain levels of DB75 and DB289 were low in treated animals, and the drug appeared to be sequestered in brain parenchyma (DB75) or poorly transported across the blood–brain barrier (DB289). It has recently been shown that DB289 is metabolized to DB75 in freshly isolated rat hepatocytes.[85] Thus, the activity of DB289 likely depends on conversion into the active constituent, DB75, an event that appears to occur exclusively in the periphery. The somewhat disappointing activity of DB289 in late-stage trypanosomiasis underscores the fact that there is a continuing need to discover new, more effective agents for the treatment of late-stage trypanosomiasis.

7.28.5.1.3 Inhibitors of parasitic glucose metabolism and glycosomal transport

Glycolysis plays a critical role in energy metabolism in bloodstream form trypanosomes.[86] In this form, oxidative metabolism involving mitochondrial Krebs cycle enzymes and oxidative phosphorylation is largely repressed, and the organism is entirely dependent on the conversion of glucose into pyruvate for its ATP supply (**Figure 18**).[87,88] Thus, glycolysis is regarded as a potential target for antitrypanosomal agents. All of the glycolytic enzymes of *T. brucei* have been purified directly or by recombinant expression in *E. coli*, and their kinetic properties have been determined.[89] As described above, the majority of trypanosomal glycolytic enzymes are sequestered inside the glycosome. Compartmentalization of the enzymes involved in glucose metabolism also plays an important role in trypanosomal glucose oxidation and energy production.[90,91] Experiments involving siRNA knockdown of the trypanosomal forms of

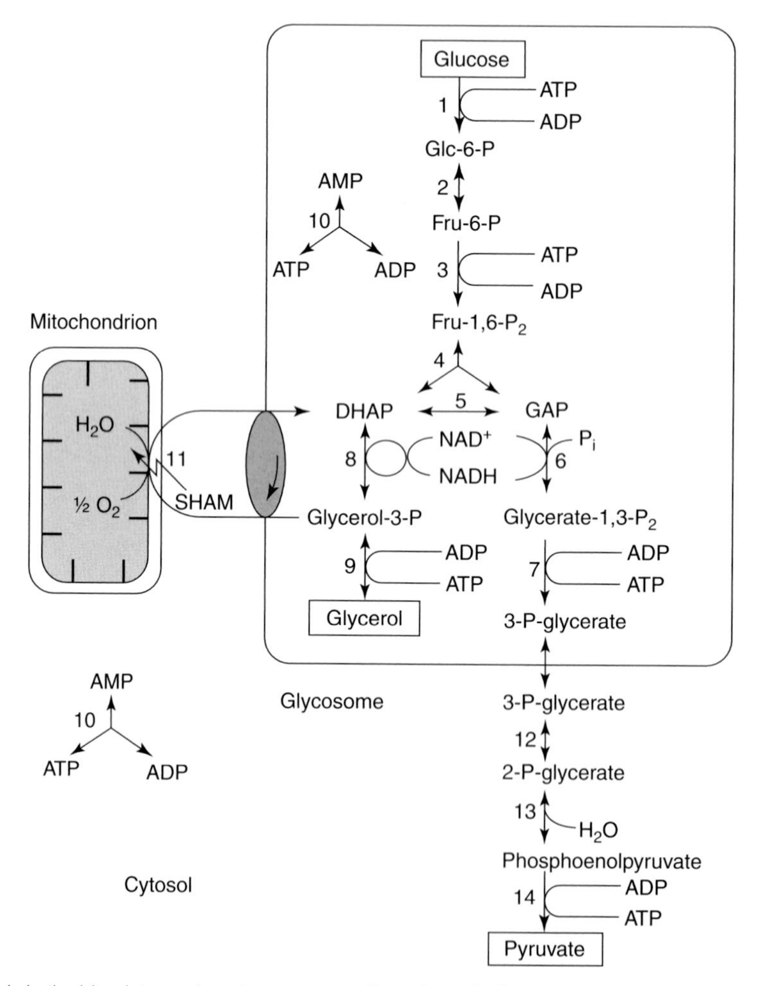

Figure 18 Glycolysis in the bloodstream form trypanosome. Boxed metabolites are nutrients or end-products of metabolism. Enzymes: 1, hexokinase; 2, phosphoglucose isomerase; 3, phosphofructokinase; 4, aldolase; 5, triose-phosphate isomerase; 6, glyceraldehyde-3-phosphate dehydrogenase; 7, phosphoglycerate kinase; 8, glycerol-3-phosphate dehydrogenase; 9, glycerol kinase; 10, adenylate kinase; 11, glycerol-3-phosphate oxidase; 12, phosphoglycerate mutase; 13, enolase; 14, pyruvate kinase. (Reproduced with permission from Opperdoes, F. R.; Michels, P. A. M. *Int. J. Parasitol.* **2001**, *31*, 481.)

hexokinase, phosphofructokinase, phosphoglycerate mutase, enolase, and pyruvate kinase suggest that there are unique and novel control mechanisms for parasitic glycolysis, and that hexokinase, phosphofructokinase, and pyruvate kinase are present in excess.[92] Interestingly, depletion of phosphofructokinase and enolase had an effect on the activity, but not the expression, of other glycolytic enzymes. The first seven enzymes of the glycolytic pathway that are sequestered in the glycosome, as well as associated cytosol enzymes, were affected.

In addition to glycolysis, other glucose metabolizing pathways have been described in trypanosomatids. A pentose phosphate pathway has been described in *T. cruzi*, which can be detected in all morphological forms.[93] The procyclic stage of *T. brucei* expresses a soluble nicotinamide adenine dinucleotide (NADH)-dependent fumarate reductase in the glycosomes that is responsible for the production of about 70% of the excreted succinate, the major end product of glucose metabolism in this form of the parasite.[94] Procyclic *T. brucei* uses parts of the Krebs cycle for purposes other than degradation of mitochondrial substrates. Citrate synthase, pyruvate dehydrogenase, and malate dehydrogenase are used to transport acetyl-CoA units from the mitochondrion to the cytosol for fatty acid biosynthesis, while α-ketoglutarate dehydrogenase and succinyl-CoA synthetase are used for the conversion of proline and glutamate into succinate. In addition, succinate dehydrogenase and fumarase are most likely used for conversion of succinate into malate for use in gluconeogenesis.[95]

In addition to the metabolic enzymes involved in trypanosomal glucose oxidation, glycosomal transport is a critical facet of energy production which is absolutely required for growth.[96] *T. brucei* bloodstream forms replicate extracellularly under conditions where plasma glucose concentrations are maintained between 3.5 and 5 mM. The rate of glucose flux is high (170 nmol $min^{-1} mg^{-1}$ of cellular protein) and the parasite is thus absolutely dependent on a continuous supply of exogenous glucose to maintain ATP levels. It has been shown that *T. brucei* long slender bloodstream forms take up glucose by facilitated transport with a K_m of approximately 1 mM.[97,98] Glucose transport in the bloodstream form of the protozoan parasite *T. brucei* was characterized by monitoring D-glucose uptake, and proved to be a concentration-dependent, saturable process typical of a carrier-mediated transport system, with an apparent K_m of 0.49 mM.[99] Seven genes encoding proteins involved in glucose transport have been identified and cloned from various kinetoplastid species. All these transporters belong to the same glucose transporter superfamily as the mammalian erythrocyte transporter GLUT1. Although these proteins are members of the same superfamily, there appear to be significant structural and kinetic differences between the mammalian and trypanosomal forms of the glucose transporter.[100] Some species (e.g., *Tb brucei*) are exposed to drastically different glucose concentrations in the mammalian bloodstream and insect midgut, and have thus evolved two distinct transporters to accommodate this fluctuation. By contrast, *T. cruzi* completes its life cycle predominantly under conditions of relative glucose deprivation, both inside mammalian host cells and within the reduviid bug midgut, and have a single, relatively high-affinity type, transporter.[101] Like the pathways of glucose metabolism, the glucose transport system represents a viable target for chemotherapy. In addition, this transporter may be able to selectively import substances that are toxic to the parasite.[102]

Although parasitic glucose metabolism has been identified as a viable drug target,[103] relatively little research to discover inhibitors has been undertaken. The adenosine analog tubercidin (**53, Figure 19**) has been shown to disrupt glucose metabolism in procyclic *T. brucei*.[104] Studies involving the RNAi library indicated that **53** is an inhibitor of trypanosomal hexokinase, and also appears to inhibit the hexose transporter. Adenosine itself was found to inhibit trypanosomatid glycolysis, but has a poor IC_{50} value (50 mM). However, structure-based design of a series of adenosine analogs[105] as inhibitors of glyceraldehyde-3-phosphate dehydrogenase resulted in the identification of compounds with submicromolar affinity for trypanosomal glyceraldehyde-3-phosphate dehydrogenase, but little affinity for the human form of the enzyme. These compounds, typified by **54** (**Figure 19**), were effective inhibitors of the purified enzyme, and also inhibited the growth of *T. brucei* and *T. cruzi* with IC_{50} values between 3 and 50 μM.[106] A series of glucosamine analogs were found to act as competitive inhibitors of trypanosomal hexokinase, with a K_i of 2.8 μM.[107] Interestingly, these analogs had no effect on the corresponding yeast enzyme in vitro. The most active analog, compound **55** (**Figure 19**), also inhibited trypanosomal growth in culture, but with a relatively high LD_{50} (3.5 mM). The high LD_{50} value was attributed to the fact that the inhibitor had to compete for the enzyme-binding site with endogenous glucose in high concentration (5 mM). 2,5-Anhydro-D-mannitols substituted at position 1 with various arylamino groups have been studied as inhibitors of trypanosomal phosphofructokinase.[108] These compounds most likely bind to the ATP-binding site rather than the fructose 6-phosphate site with affinity constants of approximately 100 mM. A compound with an electrophilic isothiocyanate group, **56**, displayed an irreversible inactivation pattern with K_i and k_{inact} values of 130 μM and 0.26 min^{-1}, respectively. The residue involved in this specific inactivation of the parasite enzyme was identified by site-directed mutagenesis as Lys 227. Hydroxynaphthaldehyde phosphate derivatives such as **57** have recently been shown to be irreversible inactivators of fructose-1,6-bisphosphate aldolase from rabbit muscle.[109] It has been proposed that these analogs may be effective inhibitors of the trypanosomal form of the enzyme, and these studies are underway.

These compounds appeared to bind at the ATP-binding site, instead of the fructose 6-phosphate site, with affinity constants of approximately 100 mM.

The glucose transport system represents an important point for control of glycolytic flux in trypanosomatids.[110] The specificity of glucose transport was investigated by inhibitor studies, and glucose uptake was shown to be sodium-independent and unaffected by the H^+-ATPase inhibitor *N,N'*-dicyclohexylcarbodiimide or the uncoupler carbonyl-cyanide-4-(trifluoromethoxy)phenylhydrazone.[99] However, highly significant inhibition was obtained with both phloretin (**58**, $K_i = 64\,\mu M$) and cytochalasin B (**59**, $K_i = 0.44\,\mu M$). These analogs are shown in **Figure 20**. In each case, inhibition was noncompetitive, and it was partially reversible for phloretin and completely reversible for cytochalasin B. Interestingly, glucosamine derivatives related to compound **55** (**Figure 19**)[107] were effective inhibitors

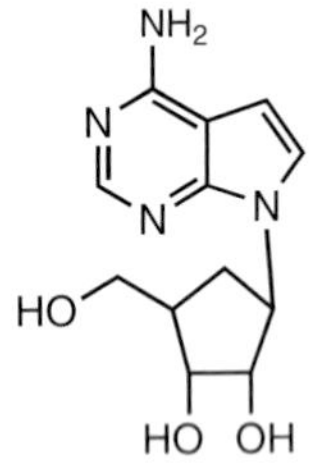

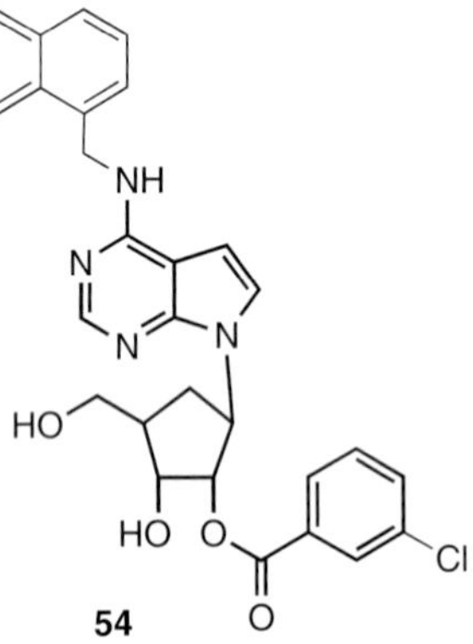

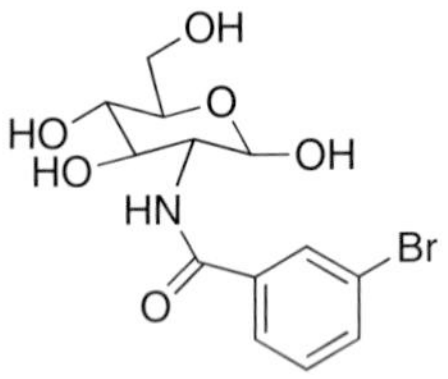

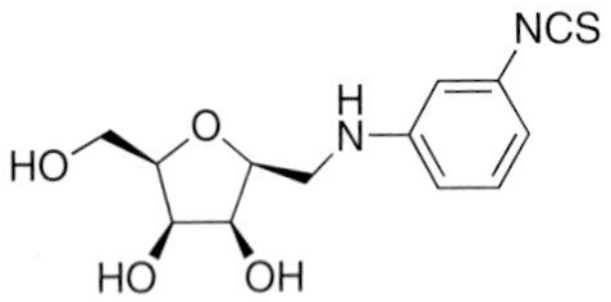

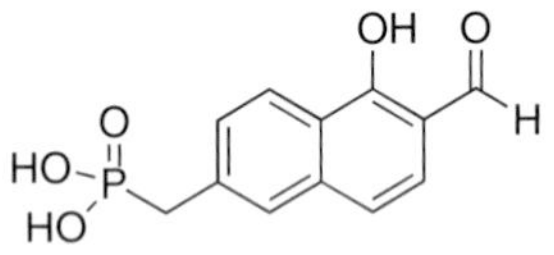

Figure 19 Synthetic inhibitors of trypanosomal glucose metabolism.

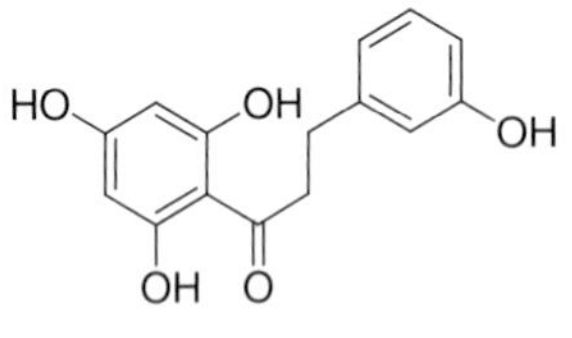

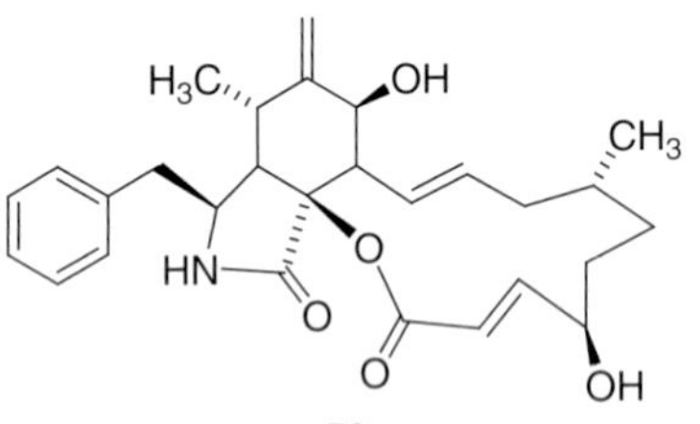

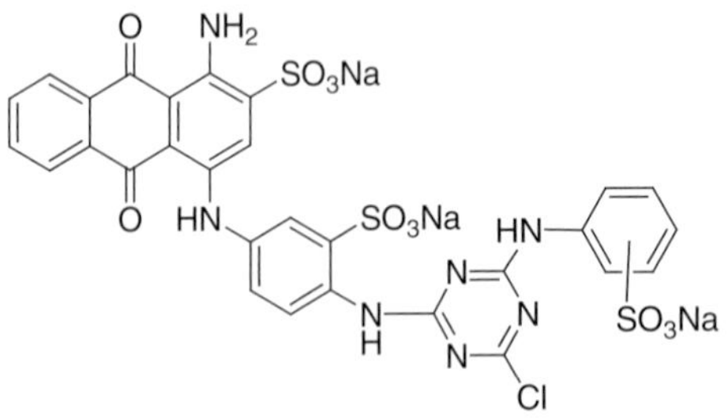

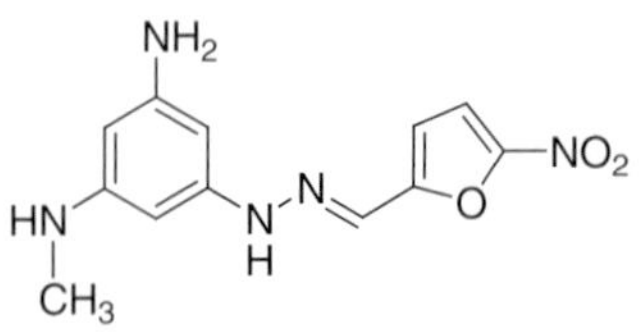

Figure 20 Inhibitors of the trypanosomal glucose transport system.

of trypanosomal glucose transport, with K_i values as low as 200 μM.[111] A number of triazine dyes have been shown to inhibit glucose transport in mammalian cell lines, and were subsequently found to be effective inhibitors of trypanosomal glucose transport.[100] The most active of these analogs, cibacron blue (**60**, **Figure 20**) produced an 89% inhibition of glucose transport activity in *T. brucei* at a concentration of 100 μM. Unfortunately, the effect of these triazines on growth was not determined. Other triazines have been described recently that exhibit antitrypanosomal activity, such as the melamine-based nitroheterocycle **61**.[112] All of the analogs that contained the melamine core exhibited affinity for the parasitic P2 transport system, most likely because they had structural features in common with known P2 substrates. Melamine-based analogs with a nitrofuran moiety were most active against *Tb rhodesiense* in vitro, with IC_{50} values as low as 18 nM. Alterations to the nitroheterocycle, such as isosteric replacement of the ring nitrogen or removal of the nitro group, resulted in dramatic decrease in activity. Compound **61** was also curative for *Tb brucei* infections in a murine model at 20 mg kg^{-1} for 4 days.[113] However, it was not determined if these analogs inhibited glucose transport as well.

7.28.5.1.4 Inhibitors of parasitic lipid metabolism

Carbohydrate-based inhibitors of glycosyl phosphatidylinositol have been studied as trypanocides, since they interfere with the remodeling of surface recognition and anchoring glycoproteins.[114] It has been shown that trypanosomatids require large amounts of myristate to support remodeling of their glycosyl phosphatidylinositol glycoprotein anchors. However, the parasite was considered to be unable to synthesize fatty acids, and myristate does not occur in sufficient amounts in the host bloodstream to support this requirement. It has recently been found that trypanosomes actually do synthesize fatty acids using a synthetic pathway in which myristate is preferentially incorporated into glycosyl phosphatidylinositols, but not into other lipids. The antibiotic thiolactomycin, **62**, inhibits parasitic myristate synthesis, and is an effective antitrypanosomal agent, suggesting that the myristate pathway is a potential chemotherapeutic target.[115] Analogs of thiolactomycin have been described that are being evaluated as antiparasitic agents.[116] Recently, the trypanosomal sterol biosynthetic pathway has become a target of interest in the design of antiparasitic agents. The human protein farnesyl transferase inhibitor tipifarnib (**63**, **Figure 21**) effectively inhibits protein farnesyl transferase from *T. cruzi* with an IC_{50} of 75 nM. Unexpectedly, the IC_{50} against *T. cruzi* amastigotes in

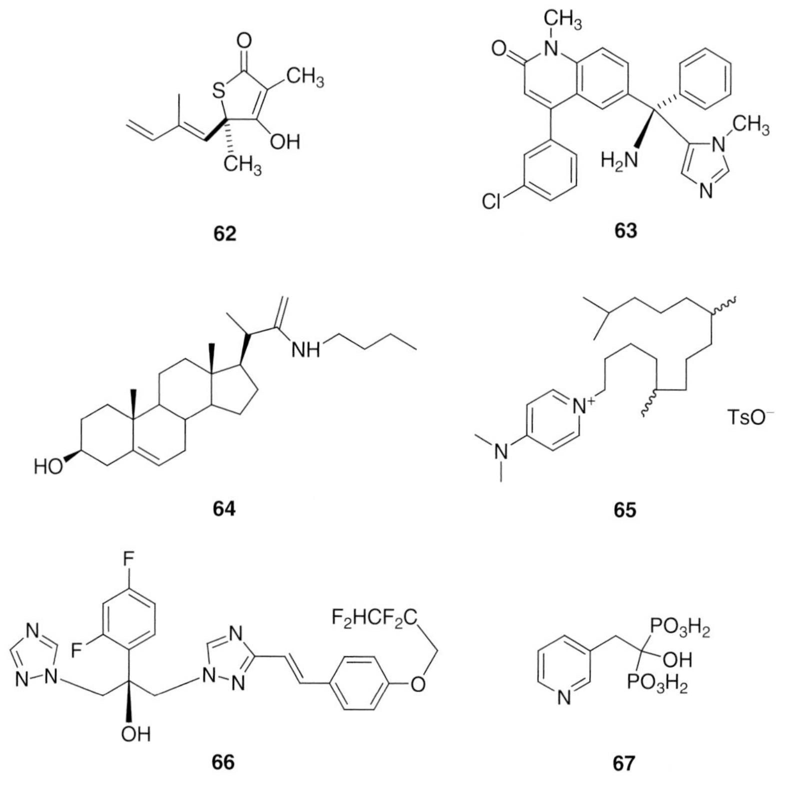

Figure 21 Inhibitors of trypanosomal lipid metabolism.

culture was considerably lower (4 nM). This observation suggested that tipifarnib acts by a second, distinct mechanism of action, and it was subsequently determined that tipifarnib also inhibits trypanosomal cytochrome P450 sterol 14-demethylase.[117] Sterols related to **64** were found to be excellent growth inhibitors in *L. donovani*, *L. major*, and *T. cruzi* with relatively low toxicity to host cells.[118,119] Compounds related to **64** act as transition state analog inhibitors of 24-sterol methyltransferase from *L. major*.[119] Some analogs in the series were also effective against *Tb rhodesiense*, although they appeared to act at a site other than 24-sterol methyltransferase. Inhibitors of oxidosqualene cyclase have been shown to possess antitrypanosomal activity when evaluated against four strains of *T. cruzi* in vitro.[120] Analog **65** inhibited trypanosomal growth at concentrations between 2 and 6 nM, but was significantly more toxic than ketoconazole or benznidazole in this system.[120] The bis-triazole D0870, **66**, inhibits the growth of *T. cruzi* epimastigotes ($IC_{50} = 0.1\ \mu M$), and induces radical cures in murine models of the infection.[121] A number of related triazoles, such as the antifungal agents posaconazole and ravuconazole, are also being evaluated as antitrypanosomal agents in preclinical trials.[33] Risedronate, **67**, and a series of related bis-phosphonates have also been shown to be effective against *T. brucei* and *T. cruzi*, and also showed low micromolar activity against *L. donovani*, *P. falciparum*, and *Toxoplasma gondii*.[122] Both **66** and **67** appear to act by disrupting parasitic sterol biosynthesis.

7.28.5.1.5 Inhibitors of parasitic proteases

Proteolytic activity in trypanosomatids was first identified as a viable drug target in 1983.[123] The protozoan parasites *T. cruzi*, *T. brucei*, and *Leishmania* sp. produce large quantities of papain-like lysosomal proteases. In plants and mammals, analogous proteases are regulated by endogenous inhibitors of the cystatin family.[124] A search for similar regulators of parasite cysteine proteases resulted in the discovery of chagasin, a *T. cruzi* inhibitor of the endogenous cysteine protease cruzipain.[125] Subsequently, homologues were identified in the genomes of all parasites mentioned above, as well as in bacteria.[126] Some of the identified proteins were recombinantly expressed and exhibited inhibitory activity against mammalian papain-like cysteine proteases.[127] In African trypanosomes, three major cysteine proteases have been studied: rhodesain from *Tb rhodesiense* and its equivalent form in cattle, congopain from *T. congolense* and brucipain from *Tb brucei* (also known as trypanopain-Tb). A *T. cruzi* cysteine protease known as cruzain or cruzipain has also been identified, and has been the subject of efforts at structure-based drug design.[128] All of these enzymes, which exhibit amino acid sequence homology to mammalian cathepsin L, promote lysosomal activity and have been identified in all life cycle stages of the protozoa, especially during the infective stage of parasite development.[129] Inhibition of the cysteine protease by the diazomethylketone inhibitor Z-Phe-Ala-CHN_2 (**68**, **Figure 22**) was shown to kill *T. brucei* in culture and in a murine model,[130] thus validating trypanosomal cysteine proteases as therapeutic drug targets. Since then, numerous inhibitors of cysteine protease activity have been developed as potential antiparasitic agents. Epoxysuccinate analogs related to the protease inhibitor E-64[131] have been shown to be effective irreversible inhibitors of the cysteine protease cruzain, isolated from *T. cruzi*. One such analog, compound **69**, inhibits the enzyme with an IC_{50} of < 10 nM, and has little effect on other cysteine proteases such as cathepsin B or papain.[132] Compound **70**, also known as K777, is also a potent inhibitor of cruzain and is currently being studied in preclinical trials.[33] Related second-generation vinyl sulfonamides such as **71** are also effective inactivators of cruzain and are effective against *T. cruzi* in cultured J744 macrophages.[133] Semicarbazones such as **72** are effective inhibitors of cruzain ($IC_{50} = 60$ nM) and rhodesain, a related cysteine protease isolated from *Tb rhodesiense* ($IC_{50} = 50$ nM).[134,135] Longer peptides, such as the 50-mer congopain inhibitor Pcp27, act as inhibitors of this protease, but are rapidly hydrolyzed by the enzyme.[136] A number of submicromolar inhibitors for cruzain have been identified from libraries of ketone-based cysteine protease inhibitors through solid-phase parallel synthesis.[137] Compound **73** inhibited cruzain with a K_i of 0.9 nM and had almost no activity against the related cysteine proteases, cathepsin B and L. A complete description of cysteine protease inhibitors and their use as potential therapeutic agents is beyond the scope of this chapter, but this topic has been recently reviewed.[138,139] Despite numerous efforts to identify inhibitors of parasitic cysteine proteases as antiparasitic agents, to date, none of these analogs have advanced to human clinical trials.

7.28.5.2 Potential New Therapies for Leishmaniasis

Many of the approaches to therapy of trypanosomiasis mentioned above also apply to the treatment of leishmania because of the similarity between leishmania and the organisms that cause trypanosomiasis. Emerging treatment modalities for leishmaniasis have recently been reviewed.[46,48,49,140] Much attention has been paid to the agents paromomycin (**74**) and miltefosine (**75**) (**Figure 23**), which are currently undergoing clinical trials.[141] The aminoglycoside antibiotic paromomycin (**74**) was found to possess antileishmanial activity in the 1960s, against both the visceral and cutaneous forms. Paromomycin is a drug with poor oral bioavailability and must be administered parenterally. Development of the parenteral formulation for visceral leishmaniasis has been slow, but Phase 2 clinical

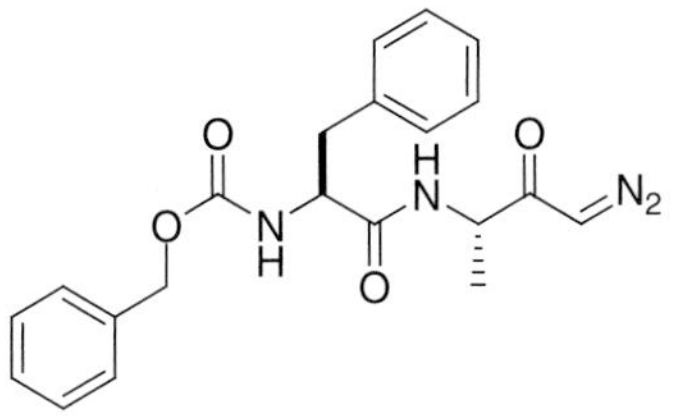

Figure 22 Representative inhibitors of parasitic cysteine proteases.

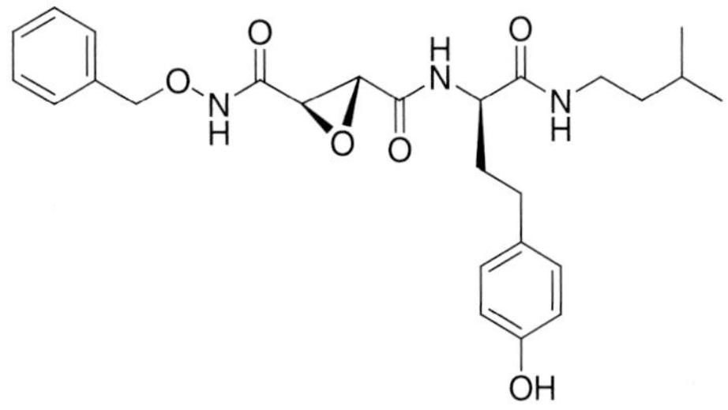

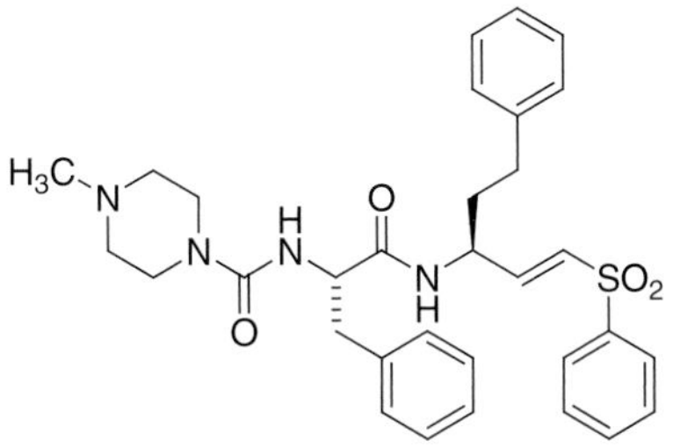

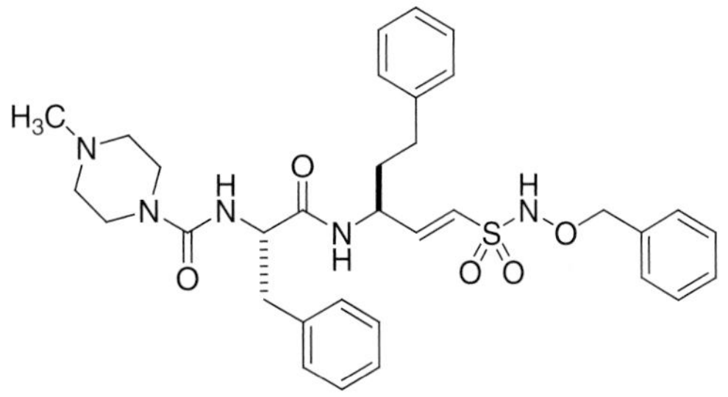

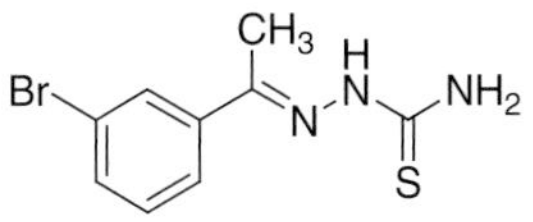

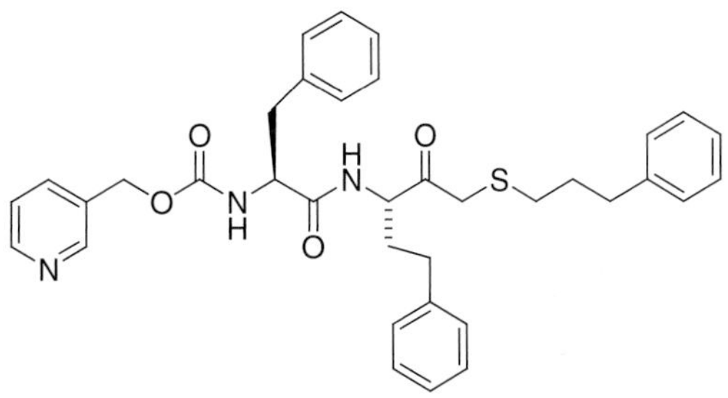

trials in India and Kenya have been promising, with 90% of patients cured of visceral leishmaniasis following treatment with 15 mg kg^{-1} daily for 20 days.[142] The drug is effective against antimony-refractory cases, and can also be used topically in the treatment of cutaneous leishmaniasis. The mechanism of antileishmanial activity of paromomycin is unknown, but may relate to inhibition of RNA synthesis.[143] The antileishmanial activity of miltefosine (**75**) was initially discovered in the mid-1980s, and it was shown to be efficacious in several experimental models.[144] In the mid-1990s, clinical trials and co-development of **75** were accomplished through a partnership between Zentaris and WHO. In a resulting Phase 3 clinical trial, 94% of visceral leishmaniasis patients were cured with an oral dose of 2.5 mg kg^{-1} of miltefosine daily for 28 days,[145] and as a result miltefosine was registered in India in March 2002 for oral treatment of visceral leishmaniasis. Although the mechanism of action of miltefosine is unknown, it may involve inhibition of protein kinase C or of phosphatidylcholine biosynthesis. A major limitation of miltefosine is significant teratogenicity that precludes its use in women of childbearing age. The orally acting 8-aminoquinoline sitamaquine (**76**) has demonstrated activity against visceral leishmaniasis. The antileishmanial activity of this compound was first identified in the 1970s at the Walter Reed Army Institute of Research, and the drug is currently in development by GlaxoSmithKline.[146] Phase 1 and 2 clinical trials have been completed with varying levels of success. For example, 67% of patients were cured of visceral *L. chagasi* infections in Brazil when treated with 2 mg kg^{-1} daily for 28 days, while in Kenya, 92% were cured of visceral leishmaniasis when treated with 1.7 mg kg^{-1} daily for 28 days. Sitamaquine is rapidly metabolized, forming the corresponding desethyl and 4-hydroxymethyl derivatives, which are likely active metabolites that are responsible for the activity of **76**. Toxicity appears to be relatively mild and limited to methemoglobinaemia. Recent studies indicate that cure of leishmaniasis appears to be dependent upon the development of an effective immune response that

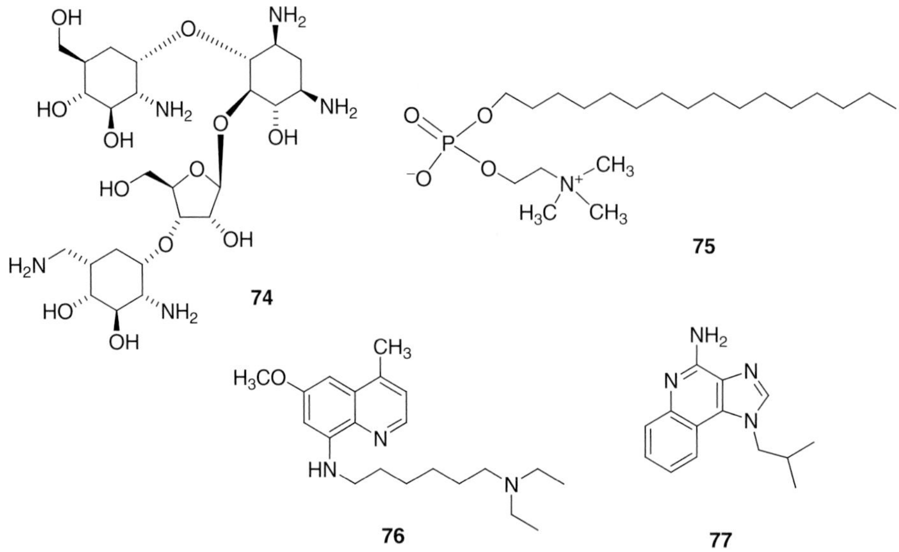

Figure 23 Agents in clinical trials for the treatment of visceral or cutaneous leishmania.

activates macrophages to produce toxic nitrogen and oxygen metabolites to kill the intracellular amastigotes.[6,33,34,147,148] Recently discovered immunopotentiating drugs have shown potential for leishmaniasis treatment. The imidazoquinoline imiquimod (**77**) induces nitric oxide (NO) production in macrophages, and was shown to have antileishmanial activity via macrophage activation in experimental models[149] and in clinical studies on cutaneous leishmania in combination with antimonials against cutaneous leishmaniasis.[36,150] Treatment with 5 mg kg^{-1} orally for 5 days resulted in a 60% reduction in the number of *L. donovani* liver amastigotes in mice.[151]

As was mentioned above, the (bis)benzylpolyamine **21** has been shown to be an effective agent in vitro and in vivo,[59–62] but the development of this drug for use in leishmaniasis was discontinued in the early 1990s. This compound, as well as the associated polyamine analogs **22–28**, is now being evaluated in vitro and in vivo against leishmania. A number of the antitrypanosomal bis-phosphonates such as **67** (**Figure 21**) are also effective against *L. donovani* in vitro.[122] Novel approaches include the specific dihydrofolate reductase inhibitor **78** ($IC_{50} = 6$ mM),[43,152] and (terpyridine)-platinum(II) complexes such as **79**, which are effective against *L. donovani* (100% inhibition at 1 μM), *T. cruzi* (65% inhibition at 1 μM), and *Tb brucei* (100% inhibition at 30 nM).[153] The triphenyltin complex **80** also has antileishmanial activity, and reduces the number of amastigotes in cultured hamster macrophages to undetectable levels at a concentration of 10 mg L^{-1}.[154] A wide variety of natural products have been isolated that exhibit antileishmanial activity, but relatively few have been the subject of systematic structure – activity relationship (SAR) studies.[155] A series of dihydro-β-agarofuran sesquiterpenes were isolated and a series of semisynthetic analogs were synthesized for evaluation as antileishmanial agents. Compound **81** was particularly effective, producing a 96.7% growth inhibition of multidrug-resistant *L. tropica* at 15 μM.[156] Finally, antimitotic agents have been identified that disrupt tubulin assembly in both *L. donovani* and *T. brucei* in vitro.[157] The substituted 3,5-dinitrosulfanilamide **82** exhibited IC_{50} values of 2.6 and 0.2 μM, respectively, against *L. donovani* amastigotes and *Tb brucei* in culture, with little effect on the J774 macrophage and PC3 prostate cell lines. This analog produced a 100% inhibition of tubulin assembly in *L. donovani* at a concentration of 10 mM, but was much less effective in vivo (**Figure 24**).

Advances in the genetics of *Leishmania*, coupled with genomic[1] and proteomic[158] investigations, are beginning to identify new therapeutic targets for the treatment of leishmania. Characterization of leishmanial membrane transporters[159] should result in identification of new targets for selective drug design. The leishmania surface coat is exquisitely regulated during the life cycle of the organism, and as such should provide new drug targeting strategies.[160] In addition, cell trafficking in leishmania involves unique methods of protein targeting that could present new targets for rational drug design.[161] As parasitic biochemical processes are better understood, and as genomics and proteomics aid in the understanding of leishmanial transcription, new targets for drug design will emerge. *L. donovani* has a nucleotide importer analogous to the trypanosomal P1 protein, but no analogs have been targeted to the transporter to date. The parasite also contains a trypanothione pathway that may be involved in the mechanism of activity of antimonials,[50] but this assembly has not been specifically targeted for drug design. Efforts are also underway to identify

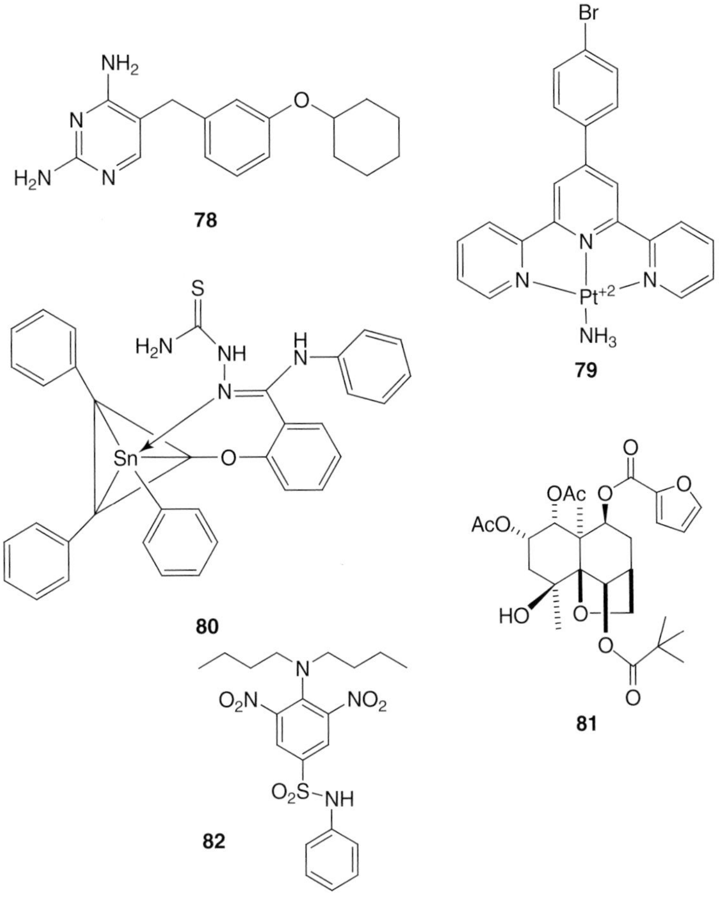

Figure 24 Antileishmanial agents designed for new targets for drug design.

an effective vaccine against leishmania, but these studies have not yielded a marketable vaccine to date.[162] Although hampered by geographical and economic factors, there is hope that effective, nontoxic cures for parasitic infection will be found.

7.28.6 Conclusion

The agents that are currently available for the treatment of infection caused by the trypanosomatids *T. brucei*, *T. cruzi*, and *Leishmania* are inadequate, and there is a need to develop new agents that are effective and inexpensive to produce. It is unfortunate that, despite much research, the standard of therapy for these diseases is essentially the same as it was 25 years ago. The emergence of resistant strains of these organisms presents a serious problem because there are no effective agents to treat these infections. Thus, parasitic diseases remain a major cause of morbidity in the Third World. Drug discovery efforts for diseases caused by trypanosomatids in the developed nations are limited for a variety of reasons, not the least of which is that the affected population is not regarded as a viable market for the development of new drug entities. In addition, these diseases are not common in the northern hemisphere where the bulk of drug discovery efforts take place. However, many of the efforts to introduce new agents into the clinic are fueled by partnerships between WHO and the pharmaceutical industry.

Drug discovery efforts for parasitic diseases will be greatly aided by the recent publication of complete genomes for the trypanosomatids mentioned above. Resulting genomic and proteomic studies will result in the identification of new, parasite-specific targets. In addition, the drug discovery efforts outlined above have resulted in promising leads, and compounds have been identified with activities that rival currently used agents, but with less toxicity and greater activity against resistant strains. These efforts, combined with new methods for vector control and higher levels of awareness in the affected population, could result in the eradication of these diseases in the near future.

References

1. Ivens, A. C.; Peacock, C. S.; Worthey, E. A.; Murphy, L.; Aggarwal, G.; Berriman, M.; Sisk, E.; Rajandream, M. A.; Adlem, E.; Aert, R. et al. *Science* **2005**, *309*, 436–442.
2. El-Sayed, N. M.; Myler, P. J.; Blandin, G.; Berriman, M.; Crabtree, J.; Aggarwal, G.; Caler, E.; Renauld, H.; Worthey, E. A.; Hertz-Fowler, C. et al. *Science* **2005**, *309*, 404–409.
3. El-Sayed, N. M.; Myler, P. J.; Bartholomeu, D. C.; Nilsson, D.; Aggarwal, G.; Tran, A.-N.; Ghedin, E.; Worthey, E. A.; Delcher, A. L.; Blandin, G. et al. *Science* **2005**, *309*, 409–415.
4. Berriman, M.; Ghedin, E.; Hertz-Fowler, C.; Blandin, G.; Renauld, H.; Bartholomeu, D. C.; Lennard, N. J.; Caler, E.; Hamlin, N. E.; Haas, B. et al. *Science* **2005**, *309*, 416–422.
5. Butler, D. *Nature* **2005**, *436*, 337.
6. Barrett, M. P.; Burchmore, R. J. S.; Stich, A.; Lazzari, J. O.; Frasch, A. C.; Cazzulo, J. J.; Krishna, S. *Lancet* **2003**, *362*, 1469.
7. World Health Organization Report, 2004.
8. Matthews, K. R. *J. Cell. Sci.* **2005**, *118*, 283–290.
9. McKean, P. G. *Curr. Opin. Microbiol.* **2003**, *6*, 600–607.
10. Guerra-Giraldez, C.; Quijada, L.; Clayton, C. E. *J. Cell. Sci.* **2002**, *115*, 2651–2658.
11. Clayton, C. E.; Michels, P. *Parasitol. Today* **1996**, *12*, 465.
12. Verlinde, C. L. M. J.; Hannaert, V.; Blonski, C.; Willson, M.; Perie, J. J.; Fothergill-Gilmore, L. A.; Opperdoes, F. R.; Gelb, M. H.; Hol, W. G. J.; Michels, P. A. M. *Drug Resist. Updat.* **2001**, *4*, 50.
13. Kita, K.; Nihei, C.; Tomitsuka, E. *Curr. Med. Chem.* **2003**, *10*, 2535–2548.
14. Nihei, C.; Fukai, Y.; Kita, K. *Biochimi. Biophys. Acta (BBA) – Mol. Basis Dis.* **2002**, *1587*, 234.
15. Bacchi, C. J.; Yarlett, N.; Goldberg, B.; Bitonti, A. J.; McCann, P. P. *Biochemical Protozoology*, 1st ed; Taylor and Francis: Washington, DC, 1991, pp 469–481.
16. Casero, R. A., Jr.; Woster, P. M. *J. Med. Chem.* **2001**, *44*, 1–26.
17. Nok, A. *J. Parasitol. Res.* **2003**, *90*, 71.
18. Vansterkenburg, E. L.; Coppens, I.; Wilting, J.; Bos, O. J.; Fischer, M. J.; Janssen, L. H.; Opperdoes, F. R. *Acta Trop.* **1993**, *54*, 237–250.
19. Carter, N. S.; Berger, B. J.; Fairlamb, A. H. *J. Biol. Chem.* **1995**, *270*, 28153–28157.
20. Berger, B. J.; Carter, N. S.; Fairlamb, A. H. *Mol. Biochem. Parasitol.* **1995**, *69*, 289.
21. de Koning, H. P. *Int. J. Parasitol.* **2001**, *31*, 511.
22. World Health Organization TDR News **1992**, *39*, 1–2.
23. Friedham, E. A. H. *Am. J. Trop. Med. Hyg.* **1949**, *29*, 173–180.
24. Barrett, M. P.; Fairlamb, A. H. *Parasitol. Today* **1999**, *15*, 136–140.
25. Kuzoe, F. A. *Acta Trop.* **1993**, *54*, 153–162.
26. Pepin, J.; Milord, F. *Adv. Parasitol.* **1994**, *33*, 1–47.
27. Bacchi, C. J.; Nathan, H. C.; Hutner, S. H.; McCann, P. P.; Sjoerdsma, A. *Science* **1980**, *210*, 332–334.
28. Schechter, P. J.; Sjoerdsma, A. *Parisitol. Today* **1984**, *2*, 223–224.
29. Burri, C.; Brun, R. *Parasitol. Res.* **2003**, *90*, S49–S52.
30. Bey, P.; Gerhart, F.; Van Dorsselaer, V.; Danzin, C. *J. Med. Chem.* **1983**, *26*, 1551–1556.
31. Poulin, R.; Lu, L.; Ackermann, B.; Bey, P.; Pegg, A. E. *J. Biol. Chem.* **1992**, *267*, 150–158.
32. Werbovetz, K. A. *Curr. Med. Chem.* **2000**, 7, 835–860.
33. Croft, S. L.; Barrett, M. P.; Urbina, J. A. *Trends Parasitol.* **2005**, *21*, 508.
34. Andrade, L. O.; Andrews, N. W. *Nat. Rev. Microbiol.* **2005**, *3*, 819.
35. Tyler, K. M.; Engman, D. M. *Int. J. Parasitol.* **2001**, *31*, 472.
36. Engman, D. M.; Leon, J. S. *Acta Trop.* **2002**, *81*, 123.
37. Kalil, J.; Cunha-Neto, E. *Parasitol. Today* **1996**, *12*, 396.
38. Burleigh, B. A.; Andrews, N. W. *Annu. Rev. Microbiol.* **1995**, *49*, 175–200.
39. Paulinoa, M.; Iribarnea, F.; Dubinb, M.; Aguilera-Moralesc, S.; Tapiad, O.; Stoppani, A. O. *Mini Rev. Med. Chem.* **2005**, *5*, 499–519.
40. Poli, P.; Aline de Mello, M.; Buschini, A.; Mortara, R. A.; Northfleet de Albuquerque, C.; da Silva, S.; Rossi, C.; Zucchi, T. M. A. D. *Biochem. Pharmacol.* **2002**, *64*, 1617.
41. Herwaldt, B. L. *Lancet* **1999**, *354*, 1191–1199.
42. Pasquau, F.; Ena, J.; Sanchez, R.; Cuadrado, J. M.; Amador, C.; Flores, J.; Benito, C.; Redondo, C.; Lacruz, J.; Abril, V. et al. *Eur. J. Clin. Microbiol. Infect. Dis.* **2005**, *24*, 411–418.
43. Garcia, L. S.; Bruckner, D. A. *Diagnostic Medical Parisitology*, 2nd ed; American Society of Microbiology: Washington, DC, 1993, pp 139, 158.
44. Davies, C. R.; Kaye, P.; Croft, S. L.; Sundar, S. *BMJ* **2003**, *326*, 377–382.
45. Puig, L.; Pradinaud, R. *Ann. Trop. Med. Parasitol.* **2003**, *97*, 107–114.
46. Guerin, P. J.; Olliaro, P.; Sundar, S.; Boelaert, M.; Croft, S. L.; Desjeux, P.; Wasunna, M. K.; Bryceson, A. D. M. *Lancet Infect. Dis.* **2002**, *2*, 494.
47. Hepburn, N. C. *Clin. Exp. Dermatol.* **2000**, *25*, 363–370.
48. Croft, S. L.; Coombs, G. H. *Trends Parasitol.* **2003**, *19*, 502.
49. Werbovetz, K. A. *Expert Opin. Ther. Targets* **2002**, *6*, 407–422.
50. Ouellette, M.; Drummelsmith, J.; Papadopoulou, B. *Drug Resist. Updat.* **2004**, 7, 257.
51. Sundar, S.; Sinha, P. R.; Agrawal, N. K.; Srivastava, R.; Rainey, P. M.; Berman, J. D.; Murray, H. W.; Singh, V. P. *Am. J. Trop. Med. Hyg.* **1998**, *59*, 139–143.
52. Sundar, S.; Goyal, A. K.; More, D. K.; Singh, M. K.; Murray, H. W. *Ann. Trop. Med. Parasitol.* **1998**, *92*, 755–764.
53. Berman, J. D.; Badaro, R.; Thakur, C. P.; Wasunna, K. M.; Behbehani, K.; Davidson, R.; Kuzoe, F.; Pang, L.; Weerasuriya, K.; Bryceson, A. D. *Bull. World Health Organ.* **1998**, *76*, 25–32.
54. Lin, C. H.; Chen, S.; Kwon, D. S.; Coward, J. K.; Walsh, C. T. *Chem. Biol.* **1997**, *4*, 859–866.
55. Chen, S.; Lin, C. H.; Kwon, D. S.; Walsh, C. T.; Coward, J. K. *J. Med. Chem.* **1997**, *40*, 3842–3850.
56. Kwon, D. S.; Lin, C. H.; Chen, S.; Coward, J. K.; Walsh, C. T.; Bollinger, J. M., Jr. *J. Biol. Chem.* **1997**, *272*, 2429–2436.
57. Li, Z.; Fennie, M. W.; Ganem, B.; Hancock, M. T.; Kobaslija, M.; Rattendi, D.; Bacchi, C. J.; O'Sullivan, M. C. *Bioorg. Med. Chem. Lett.* **2001**, *11*, 251.

58. Bergeron, R. J.; McManis, J. S.; Weimar, W. R.; Schreier, K. M.; Gao, F.; Wu, Q.; Ortiz-Ocasio, J.; Luchetta, G. R.; Porter, C.; Vinson, J. R. *J. Med. Chem.* **1995**, *38*, 2278–2285.
59. Baumann, R. J.; Hanson, W. L.; McCann, P. P.; Sjoerdsma, A.; Bitonti, A. *J. Antimicrob. Agents Chemother.* **1990**, *34*, 722–727.
60. Baumann, R. J.; McCann, P. P.; Bitonti, A. *J. Antimicrob. Agents Chemother.* **1991**, *35*, 1403–1407.
61. Bitonti, A. J.; Bush, T. L.; McCann, P. P. *Biochem. J.* **1989**, *257*, 769–774.
62. Bitonti, A. J.; Dumont, J. A.; Bush, T. L.; Edwards, M. L.; Stemerick, D. M.; McCann, P. P.; Sjoerdsma, A. *Proc. Natl. Acad. Sci. USA* **1989**, *86*, 651–655.
63. Bellevue, F. H., III.; Boahbedason, M.; Wu, R.; Woster, P. M.; Casero, J. R. A.; Rattendi, D.; Lane, S.; Bacchi, C. J. *Bioorg. Med. Chem. Lett.* **1996**, *6*, 2765.
64. Zou, Y.; Wu, Z.; Sirisoma, N.; Woster, P. M.; Casero, R. A., Jr.; Weiss, L. M.; Rattendi, D.; Lane, S.; Bacchi, C. *J. Bioorg. Med. Chem. Lett.* **2001**, *11*, 1613–1617.
65. Benson, T. J.; McKie, J. H.; Garforth, J.; Borges, A.; Fairlamb, A. H.; Douglas, K. T. *Biochem. J.* **1992**, *286*, 9–11.
66. Garforth, J.; Yin, H.; McKie, J. H.; Douglas, K. T.; Fairlamb, A. H. *J. Enzyme Inhib.* **1997**, *12*, 161–173.
67. Chan, C.; Yin, H.; Garforth, J.; McKie, J. H.; Jaouhari, R.; Speers, P.; Douglas, K. T.; Rock, P. J.; Yardley, V.; Croft, S. L. et al. *J. Med. Chem.* **1998**, *41*, 148–156.
68. Bonnet, B.; Soullez, D.; Davioud-Charvet, E.; Landry, V.; Horvath, D.; Sergheraert, C. *Bioorg. Med. Chem.* **1997**, *5*, 1249.
69. Girault, S.; Davioud-Charvet, E.; Salmon, L.; Berecibar, A.; Debreu, M. A.; Sergheraert, C. *Bioorg. Med. Chem. Lett.* **1998**, *8*, 1175–1180.
70. Arafa, R. K.; Brun, R.; Wenzler, T.; Tanious, F. A.; Wilson, W. D.; Stephens, C. E.; Boykin, D. W. *J. Med. Chem.* **2005**, *48*, 5480–5488.
71. Carter, N. S.; Fairlamb, A. H. *Nature* **1993**, *361*, 173–176.
72. de Koning, H. P.; Bridges, D. J.; Burchmore, R. J. S. *FEMS Microbiol. Rev.* **2005**, *29*, 987.
73. Guo, J.; Wu, Y. Q.; Rattendi, D.; Bacchi, C. J.; Woster, P. M. *J. Med. Chem.* **1995**, *38*, 1770–1777.
74. Goldberg, B.; Yarlett, N.; Sufrin, J.; Lloyd, D.; Bacchi, C. *J. FASEB J.* **1997**, *11*, 256–260.
75. Bitonti, A. J.; Byers, T. L.; Bush, T. L.; Casara, P. J.; Bacchi, C. J.; Clarkson, A. B., Jr.; McCann, P. P.; Sjoerdsma, A. *Antimicrob. Agents Chemother.* **1990**, *34*, 1485–1490.
76. Byers, T. L.; Casara, P.; Bitonti, A. *J. Biochem. J.* **1992**, *283*, 755–758.
77. Bacchi, C. J.; Nathan, H. C.; Yarlett, N.; Goldberg, B.; McCann, P. P.; Bitonti, A. J.; Sjoerdsma, A. *Antimicrob. Agents Chemother.* **1992**, *36*, 2736–2740.
78. Bacchi, C. J.; Goldberg, B.; Rattendi, D.; Gorrell, T. E.; Spiess, A. J.; Sufrin, J. R. *Biochem. Pharmacol.* **1999**, *57*, 89.
79. Seley, K. L.; Schneller, S. W.; Rattendi, D.; Lane, S.; Bacchi, C. *J. Antimicrob. Agents Chemother.* **1997**, *41*, 1658–1661.
80. Dardonville, C.; Brun, R. *J. Med. Chem.* **2004**, *47*, 2296–2307.
81. Meiering, S.; Inhoff, O.; Mies, J.; Vincek, A.; Garcia, G.; Kramer, B.; Dormeyer, M.; Krauth-Siegel, R. L. *J. Med. Chem.* **2005**, *48*, 4793–4802.
82. Klenke, B.; Stewart, M.; Barrett, M. P.; Brun, R.; Gilbert, I. H. *J. Med. Chem.* **2001**, *44*, 3440–3452.
83. Donkor, I. O.; Assefa, H.; Rattendi, D.; Lane, S.; Vargas, M.; Goldberg, B.; Bacchi, C. *Eur. J. Med. Chem.* **2001**, *36*, 531.
84. Sturk, L. M.; Brock, J. L.; Bagnell, C. R.; Hall, J. E.; Tidwell, R. R. *Acta Trop.* **2004**, *91*, 131–143.
85. Zhou, L.; Thakker, D. R.; Voyksner, R. D.; Anbazhagan, M.; Boykin, D. W.; Hall, J. E.; Tidwell, R. R. *J. Mass Spectrom.* **2004**, *39*, 351–360.
86. Besteiro, S.; Barrett, M. P.; Riviere, L.; Bringaud, F. *Trends Parasitol.* **2005**, *21*, 185.
87. Coustou, V.; Besteiro, S.; Biran, M.; Diolez, P.; Bouchaud, V.; Voisin, P.; Michels, P. A. M.; Canioni, P.; Baltz, T.; Bringaud, F. *J. Biol. Chem.* **2003**, *278*, 49625–49635.
88. van Weelden, S. W. H.; Fast, B.; Vogt, A.; van der Meer, P.; Saas, J.; van Hellemond, J. J.; Tielens, A. G. M.; Boshart, M. *J. Biol. Chem.* **2003**, *278*, 12854–12863.
89. Opperdoes, F. R.; Michels, P. A. M. *Int. J. Parasitol.* **2001**, *31*, 481.
90. Kessler, P. S.; Parsons, M. *J. Biol. Chem.* **2005**, *280*, 9030–9036.
91. Bakker, B. M.; Mensonides, F. I. C.; Teusink, B.; van Hoek, P.; Michels, P. A. M.; Westerhoff, H. V. *Proc. Natl. Acad. Sci.* **2000**, *97*, 2087–2092.
92. Albert, M.-A.; Haanstra, J. R.; Hannaert, V.; Van Roy, J.; Opperdoes, F. R.; Bakker, B. M.; Michels, P. A. M. *J. Biol. Chem.* **2005**, *280*, 28306–28315.
93. Cazzulo, J. J. *J. Bioenerg. Biomembr. (Historical Archive)* **1994**, *26*, 157.
94. Coustou, V.; Besteiro, S.; Riviere, L.; Biran, M.; Biteau, N.; Franconi, J.-M.; Boshart, M.; Baltz, T.; Bringaud, F. *J. Biol. Chem.* **2005**, *280*, 16559–16570.
95. van Weelden, S. W. H.; van Hellemond, J. J.; Opperdoes, F. R.; Tielens, A. G. M. *J. Biol. Chem.* **2005**, *280*, 12451–12460.
96. Furuya, T.; Kessler, P.; Jardim, A.; Schnaufer, A.; Crudder, C.; Parsons, M. *Proc. Natl. Acad. Sci.* **2002**, *99*, 14177–14182.
97. Bakker, B. M.; Walsh, M. C.; ter Kuile, B. H.; Mensonides, F. I.; Michels, P. A.; Opperdoes, F. R.; Westerhoff, H. V. *Proc. Natl. Acad. Sci. USA* **1999**, *96*, 10098.
98. Wille, U.; Seyfang, A.; Duszenko, M. *Eur. J. Biochem.* **1996**, *236*, 228–233.
99. Seyfang, A.; Duszenko, M. *Eur. J. Biochem.* **1991**, *202*, 191–196.
100. Bayele, H. *Parasitol. Res.* **2001**, *87*, 911.
101. Tetaud, E.; Barrett, M. P.; Bringaud, F.; Baltz, T. *Biochem. J.* **1997**, *325*, 569–580.
102. Barnard, J. P.; Reynafarje, B.; Pedersen, P. L. *J. Biol. Chem.* **1993**, *268*, 3654–3661.
103. Perie, J.; Riviere-Alric, I.; Blonski, C.; Gefflaut, T.; de Viguerie, N. L.; Trinquier, M.; Willson, M.; Opperdoes, F. R.; Callens, M. *Pharmacol. Ther.* **1993**, *60*, 347.
104. Drew, M. E.; Morris, J. C.; Wang, Z.; Wells, L.; Sanchez, M.; Landfear, S. M.; Englund, P. T. *J. Biol. Chem.* **2003**, *278*, 46596–46600.
105. Bressi, J. C.; Verlinde, C. L. M. J.; Aronov, A. M.; Shaw, M. L.; Shin, S. S.; Nguyen, L. N.; Suresh, S.; Buckner, F. S.; Van Voorhis, W. C.; Kuntz, I. D. et al. *J. Med. Chem.* **2001**, *44*, 2080–2093.
106. Aronov, A. M.; Suresh, S.; Buckner, F. S.; Van Voorhis, W. C.; Verlinde, C. L. M. J.; Opperdoes, F. R.; Hol, W. G. J.; Gelb, M. H. *Proc. Natl. Acad. Sci. USA* **1999**, *96*, 4273–4278.
107. Willson, M.; Sanejouand, Y.-H.; Perie, J.; Hannaert, V.; Opperdoes, F. *Chem. Biol.* **2002**, *9*, 839.
108. Claustre, S.; Denier, C.; Lakhdar-Ghazal, F.; Lougare, A.; Lopez, C.; Chevalier, N.; Michels, P. A. M.; Perie, J.; Willson, M. *Biochemistry* **2002**, *41*, 10183–10193.
109. Dax, C.; Coincon, M.; Sygusch, J.; Blonski, C. *Biochemistry* **2005**, *44*, 5430–5443.
110. Bakker, B. M.; Walsh, M. C.; ter Kuile, B. H.; Mensonides, F. I. C.; Michels, P. A. M.; Opperdoes, F. R.; Westerhoff, H. V. *Proc. Natl. Acad. Sci. USA* **1999**, *96*, 10098–10103.
111. Claustre, S.; Bringaud, F.; Azema, L.; Baron, R.; Perie, J.; Willson, M. *Carbohydr. Res.* **1999**, *315*, 339–344.

112. Baliani, A.; Bueno, G. J.; Stewart, M. L.; Yardley, V.; Brun, R.; Barrett, M. P.; Gilbert, I. H. *J. Med. Chem.* **2005**, *48*, 5570–5579.
113. Stewart, M. L.; Bueno, G. J.; Baliani, A.; Klenke, B.; Brun, R.; Brock, J. M.; Gilbert, I. H.; Barrett, M. P. *Antimicrob. Agents Chemother.* **2004**, *48*, 1733–1738.
114. Smith, T. K.; Sharma, D. K.; Crossman, A.; Brimacombe, J. S.; Ferguson, M. A. *EMBO J.* **1999**, *18*, 5922–5930.
115. Morita, Y. S.; Paul, K. S.; Englund, P. T. *Science* **2000**, *288*, 140–143.
116. McFadden, J. M.; Medghalchi, S. M.; Thupari, J. N.; Pinn, M. L.; Vadlamudi, A.; Miller, K. I.; Kuhajda, F. P.; Townsend, C. A. *J. Med. Chem.* **2005**, *48*, 946–961.
117. Hucke, O.; Gelb, M. H.; Verlinde, C. L. M. J.; Buckner, F. S. *J. Med. Chem.* **2005**, *48*, 5415–5418.
118. Orenes Lorente, S.; Rodrigues, J. C. F.; Jimenez Jimenez, C.; Joyce-Menekse, M.; Rodrigues, C.; Croft, S. L.; Yardley, V.; de Luca-Fradley, K.; Ruiz-Perez, L. M.; Urbina, J. et al. *Antimicrob. Agents Chemother.* **2004**, *48*, 2937–2950.
119. Lorente, S. O.; Jimenez, C. J.; Gros, L.; Yardley, V.; de Luca-Fradley, K.; Croft, S. L. A.; Urbina, J.; Ruiz-Perez, L. M.; Pacanowska, D. G.; Gilbert, I. H. *Bioorg. Med. Chem.* **2005**, *13*, 5435.
120. Buckner, F. S.; Griffin, J. H.; Wilson, A. J.; Van Voorhis, W. C. *Antimicrob. Agents Chemother.* **2001**, *45*, 1210–1215.
121. Liendo, A.; Lazardi, K.; Urbina, J. A. *J. Antimicrob. Chemother.* **1998**, *41*, 197–205.
122. Martin, M. B.; Grimley, J. S.; Lewis, J. C.; Heath, H. T.; Bailey, B. N.; Kendrick, H.; Yardley, V.; Caldera, A.; Lira, R.; Urbina, J. A. et al. *J. Med. Chem.* **2001**, *44*, 909–916.
123. North, M. J.; Coombs, G. H.; Barry, J. D. *Mol. Biochem. Parasitol.* **1983**, *9*, 161–180.
124. McKerrow, J. H.; Engel, J. C.; Caffrey, C. R. *Bioorg. Med. Chem.* **1999**, *7*, 639.
125. Monteiro, A. C. S.; Abrahamson, M.; Lima, A. P. C. A.; Vannier-Santos, M. A.; Scharfstein, J. *J. Cell. Sci.* **2001**, *114*, 3933–3942.
126. Rigden, D. J.; Mosolov, V. V.; Galperin, M. Y. *Protein Sci.* **2002**, *11*, 1971–1977.
127. Sanderson, S. J.; Westrop, G. D.; Scharfstein, J.; Mottram, J. C.; Coombs, G. H. *FEBS Lett.* **2003**, *542*, 12.
128. McKerrow, J. H.; McGrath, M. E.; Engel, J. C. *Parasitol. Today* **1995**, *11*, 279.
129. Lecaille, F.; Kaleta, J.; Bromme, D. *Chem. Rev.* **2002**, *102*, 4459–4488.
130. Scory, S.; Caffrey, C. R.; Stierhof, Y.-D.; Ruppel, A.; Steverding, D. *Experimental Parasitol.* **1999**, *91*, 327.
131. Barrett, A. J.; Kembhavi, A. A.; Brown, M. A.; Kirschke, H.; Knight, C. G.; Tamai, M.; Hanada, K. *Biochem. J.* **1982**, *201*, 189–198.
132. Roush, W. R.; Hernandez, A. A.; McKerrow, J. H.; Selzer, P. M.; Hansell, E.; Engel, J. C. *Tetrahedron* **2000**, *56*, 9747.
133. Roush, W. R.; Cheng, J.; Knapp-Reed, B.; Alvarez-Hernandez, A.; McKerrow, J. H.; Hansell, E.; Engel, J. C. *Bioorg. Med. Chem. Lett.* **2001**, *11*, 2759.
134. Du, X.; Guo, C.; Hansell, E.; Doyle, P. S.; Caffrey, C. R.; Holler, T. P.; McKerrow, J. H.; Cohen, F. E. *J. Med. Chem.* **2002**, *45*, 2695–2707.
135. Greenbaum, D. C.; Mackey, Z.; Hansell, E.; Doyle, P.; Gut, J.; Caffrey, C. R.; Lehrman, J.; Rosenthal, P. J.; McKerrow, J. H.; Chibale, K. *J. Med. Chem.* **2004**, *47*, 3212–3219.
136. Godat, E.; Chowdhury, S.; Lecaille, F.; Belghazi, M.; Purisima, E. O.; Lalmanach, G. *Biochemistry* **2005**, *44*, 10486–10493.
137. Huang, L.; Lee, A.; Ellman, J. A. *J. Med. Chem.* **2002**, *45*, 676–684.
138. Dubin, G. *Cell. Mol. Life Sci.* **2005**, *62*, 653–669.
139. Powers, J. C.; Asgian, J. L.; Ekici, O. D.; James, K. E. *Chem. Rev.* **2002**, *102*, 4639–4750.
140. Sundar, S.; Rai, M. *Expert Opin. Pharmacother.* **2005**, *6*, 2821–2829.
141. Berman, J. *Expert Opin. Investig. Drugs* **2005**, *14*, 1337–1346.
142. Thakur, C. P.; Kanyok, T. P.; Pandey, A. K.; Sinha, G. P.; Messick, C.; Olliaro, P. *Trans. R. Soc. Trop. Med. Hyg.* **2000**, *94*, 432–433.
143. Maarouf, M.; Lawrence, Fo.; Brown, S.; Robert-Gero, M. *Parasitol. Res.* **1997**, *83*, 198.
144. Croft, S. L.; Seifert, K.; Duchene, M. *Mol. Biochem. Parasitol.* **2003**, *126*, 165.
145. Sundar, S.; Jha, T. K.; Thakur, C. P.; Engel, J.; Sindermann, H.; Fischer, C.; Junge, K.; Bryceson, A.; Berman, J. *N. Engl. J. Med.* **2002**, *347*, 1739–1746.
146. Yeates, C. *Curr. Opin. Investig. Drugs* **2002**, *3*, 1446–1452.
147. Berhe, N.; Wolday, D.; Hailu, A.; Abraham, Y.; Ali, A.; Gebre-Michael, T.; Desjeux, P.; Sonnerborg, A.; Akuffo, H.; Britton, S. *AIDS* **1999**, *13*, 1921–1925.
148. Alvar, J.; Canavate, C.; Gutierrez-Solar, B.; Jimenez, M.; Laguna, F.; Lopez-Velez, R.; Molina, R.; Moreno, J. *Clin. Microbiol. Rev.* **1997**, *10*, 298–319.
149. Buates, S.; Matlashewski, G. *J. Infect. Dis.* **1999**, *179*, 1485–1494.
150. Arevalo, I.; Ward, B.; Miller, R.; Meng, T. C.; Najar, E.; Alvarez, E.; Matlashewski, G.; Llanos-Cuentas, A. *Clin. Infect. Dis.* **2001**, *33*, 1847–1851.
151. Smith, A. C.; Yardley, V.; Rhodes, J.; Croft, S. L. *Antimicrob. Agents Chemother.* **2000**, *44*, 1494–1498.
152. Chowdhury, S. F.; Villamor, V. B.; Guerrero, R. H.; Leal, I.; Brun, R.; Croft, S. L.; Goodman, J. M.; Maes, L.; Ruiz-Perez, L. M.; Pacanowska, D. G.; Gilbert, I. H. *J. Med. Chem.* **1999**, *42*, 4300–4312.
153. Lowe, G.; Droz, A. S.; Vilaivan, T.; Weaver, G. W.; Tweedale, L.; Pratt, J. M.; Rock, P.; Yardley, V.; Croft, S. L. *J. Med. Chem.* **1999**, *42*, 999–1006.
154. Raychaudhury, B.; Banerjee, S.; Gupta, S.; Singh, R. V.; Datta, S. C. *Acta Trop.* **2005**, *95*, 1.
155. Rocha, L. G.; Almeida, J. R.; Macedo, R. O.; Barbosa-Filho, J. M. *Phytomedicine* **2005**, *12*, 514–535.
156. Cortes-Selva, F.; Campillo, M.; Reyes, C. P.; Jimenez, I. A.; Castanys, S.; Bazzocchi, I. L.; Pardo, L.; Gamarro, F.; Ravelo, A. G. *J. Med. Chem.* **2004**, *47*, 576–587.
157. Bhattacharya, G.; Herman, J.; Delfin, D.; Salem, M. M.; Barszcz, T.; Mollet, M.; Riccio, G.; Brun, R.; Werbovetz, K. A. *J. Med. Chem.* **2004**, *47*, 1823–1832.
158. Drummelsmith, J.; Brochu, V.; Girard, I.; Messier, N.; Ouellette, M. *Mol. Cell. Proteomics: MCP* **2003**, *2*, 146.
159. Landfear, S. M. *Curr. Opin. Microbiol.* **2000**, *3*, 417.
160. Ilgoutz, S. C.; McConville, M. J. *Int. J. Parasitol.* **2001**, *31*, 899.
161. Costa-Pinto, D.; Trindade, L. S.; McMahon-Pratt, D.; Traub-Cseko, Y. M. *Int. J. Parasitol.* **2001**, *31*, 536.
162. Ghosh, M.; Bandyopadhyay, S. *Mol. Cell. Biochem.* **2003**, *253*, 199–205.

Biography

Patrick M Woster is Professor of Medicinal Chemistry at the Eugene Applebaum College of Pharmacy and Health Sciences at Wayne State University. He received a BS in Pharmacy from the University of Nebraska Medical Center in 1978, and a PhD in Medicinal Chemistry from the University of Nebraska – Lincoln in 1986. Following postdoctoral work in chemistry at Rensselaer Polytechnic Institute, and in medicinal chemistry at the University of Michigan, he joined the WSU Faculty in 1988. Professor Woster teaches undergraduate biochemistry and medicinal chemistry courses, graduate level courses in medicinal and bioorganic chemistry, and has developed a series of WWW-based tutorials for students in medicinal chemistry. He has been voted Teacher of the Year eight times, and received the WSU President's Award for Excellence in Teaching in 1993. He was also awarded a Wayne State University Career Development Chair in 1997.

Professor Woster maintains an active research program that has been funded by several agencies, including the National Institutes of Health and the World Health Association. Ongoing projects in the Woster laboratories include the synthesis of alkylpolyamines as antitumor or antiparasitic agents, synthesis of novel inhibitors of histone deacetylase, solution- and solid-phase synthesis of peptide-based inhibitors of plasmepsin, synthesis of *S*-adenosylmethionine analogs as mechanism-based enzyme inhibitors and synthesis of furanocoumarins as inhibitors of cytochrome P450. Dr Woster has directed a number of PhD dissertations and Master's theses, and has mentored five postdoctoral associates. To date, he has authored more than 65 articles in peer-reviewed research journals, and more than 100 research abstracts and invited presentations. Dr Woster has also served as a member of two NIH study sections, and on editorial boards or as a reviewer for numerous scientific journals.

Professor Woster has been a member of ACS and the Division of Medicinal Chemistry since 1979, and is also a member of the Organic Chemistry and Biological Chemistry divisions. He assumed an active service role in the Division of Medicinal Chemistry in 1994 as a member of the Membership Committee and Chair of the Electronic Communications Committee. Dr Woster was a member of the Division Long Range Planning Committee (1998–2000), and organized 4 symposia for National ACS meetings. He has served as Secretary and Public Relations Chair for the Division of Medicinal Chemistry since 1998. Dr Woster is an active member of the American Association of Colleges of Pharmacy, where he is a Past Chair of the Section of Teachers of Chemistry. During 2004–05, he was a member of the AACP Executive Board of Directors, and Chair of the Academic Sections Coordinating Committee.

Comprehensive Medicinal Chemistry II
ISBN (set): 0-08-044513-6

ISBN (Volume 7) 0-08-044520-9; pp. 815–843

7.29 Recent Advances in Inflammatory and Immunological Diseases: Focus on Arthritis Therapy

R Magolda, Wyeth Research, Princeton, NJ, USA
T Kelly, Boehringer-Ingelheim Inc., Ridgefield, CT, USA
R Newton, Incyte Corporation, Wilmington, DE, USA
J S Skotnicki, Wyeth Research, Pearl River, NY, USA
J Trzaskos, Bristol-Myers Squibb Corporation, Lawrenceville, NJ, USA

7.29.1 Introduction

Many disorders fall under the umbrella of inflammatory and immunological diseases, including some that affect a significant number of individuals (e.g., rheumatoid arthritis (RA), osteoarthritis (OA), asthma, psoriasis, inflammatory bowel disease), and others that are more rare (e.g., systemic lupus erythematosus (SLE) or multiple sclerosis). Despite affecting different tissues and organs, inflammatory and immunological diseases have several common features. All of these diseases involve the activation and recruitment of bone marrow derived white cells (leukocytes) in response to ill-defined stimuli. Normally, cells involved in such processes work to restore the normal state of host tissue while ridding the body of foreign pathogens (e.g., bacteria). Leukocytes arrive at the site of tissue injury and release inflammatory mediators and proteinases that first remove the infectious source and then start the repair process to remove damaged proteins involved, initiate wound repair, and ultimately return the tissue to normal structure and function. These white cells or leukocytes include neutrophils, monocytes, macrophages, T cells, B cells, mast cells, eosinophils, and basophils. A disease state begins to arise when an imbalance occurs between host defense and repair mechanisms leading to excessive tissue degradation or aberrant repair. Drug discovery strategies have been aimed at controlling this imbalance. This chapter will briefly describe the function of these cells in several inflammatory and immunological diseases. This review's focus is on new approaches employed since 1990 that seek to re-establish the defense–repair balance and afford new treatment options for patients with various autoimmune/inflammatory diseases.

7.29.2 Role of White Cells in Normal and Disease States

Bone marrow derived pluripotent hematopoietic stem cells are precursors to the family of myeloid and lymphoid cells involved in both cellular and humoral immunity.[1] The innate immune response, which involves primarily cells of the myeloid lineage, is triggered by nonantigen-specific detection of invading foreign pathogens. As a first-line defense, leukocytes are marshaled to the site of insults to engulf and digest foreign intruders, enabling the host to ward off infection. In these instances, detection can occur through protein opsonins that associate with complement activation or coagulation cascade byproducts as well as through pattern recognition receptors also known as toll-like receptors (TLRs) that are expressed on the cell surface and mediate cell activation. Adaptive immunity involves the assembly of antigen-specific cells along with antibodies associated with activation of primarily cells of the lymphoid lineage that are the product of a defense system requiring the display of at least some part of the foreign intruder. Digestion of a pathogen protein generates unique peptide fragments that are ultimately displayed on the surface of an antigen-presenting cell. The antigen is presented on the surface as a unique signature peptide or epitope in association with major histocompatibility complex (MHC) class II proteins and, when the presenting cell associates with a T lymphocyte, this engagement leads to the activation of only T lymphocytes that recognize the specific epitope. The simultaneous association of various costimulatory interactions that occur between the T lymphocyte and the antigen-presenting cell enhances this activation. The effector mechanisms of adaptive immunity can present as the subsequent production of antigen specific antibodies, antigen specific effector T cells or coordinated recruitment of cell members of the innate immune system. This cascade of lymphocyte activation also leaves the host with residual T cells that represent the 'memory' of this response (hence memory T cells) that are available for a subsequent amplified response to presentation of the same antigens.

One way to illustrate these repair and tissue destruction concepts is to compare two diseases of the joint that present with an inflammatory component, RA and OA. Both RA and OA involve an erosion of cartilage leading to a compromise in joint integrity, joint function, and pain. However, the predisposing factors and disease progression follow quite different paths as revealed by tissue histopathology in diseased joints. Early RA is marked by the relatively acute onset of swelling and pain and the joint histology is characterized by an infiltration of macrophages and B lymphocytes. As the inflammatory imbalance proceeds, there is continued influx of cells with ongoing tissue destruction. Due to the continued destruction of tissue, the repair processes that normally occur also become dysfunctional, leading to excessive proliferation of synovial tissue and the production of a cellular mass known as pannus located adjacent to cartilage. As the pannus continues to expand it invades the joint space, eroding cartilage and eventually bone, leading to a significant compromise in joint instability. While the etiology of RA is unknown, genetic susceptibility to RA is associated with immune-related proteins such as tumor necrosis factor alpha (TNF-α). Successful treatment of RA involves targeting the immune system, and it is interesting that these immune proteins are also involved in the immune response to infections such as *Mycobacterium tuberculosis*. This suggests that RA, along with other autoimmune diseases, is the result of selective genetic pressure exerted by epidemics of the recent past. Alternately, it has been suggested that RA is triggered by infection with a pathogen, possibly a virus, which expresses epitopes that mimic host proteins and leads to the expansion of a reaction to self that escapes normal regulation. Further support for this notion was the emergence of RA in Europe during the 1500–1600s. It has been speculated that artists like Reuben and others captured this new disease in their paintings.[2,3]

Osteoarthritis has been around for millennia as documented by x-rays of Egyptian mummies.[4] In contrast to RA, susceptibility to OA is associated with joint overuse, injury, and obesity rather than genetics associated with the immune system as seen in RA. Osteoarthritis is characterized by a gradual onset of swelling and pain that is reflective of worn out rather than inflamed tissue. In OA, cartilage is first lost due to normal wear and tear or as a result of joint injury (e.g., anterior crucial ligament (ACL) tear). The resultant cartilage erosion leads to a grinding of the bone surfaces and stiffness that becomes characteristic of OA. The advancing joint instability leads to osteophyte formation and progressive joint pain. Initially, local cells attempt to repair this condition. At some point, damage exceeds repair and the joint becomes swollen and inflamed.[5] This can trigger the influx of leukocytes that can release inflammatory mediators and accelerate the joint destruction. Depending on the disease severity, patients cycle from a highly painful inflamed joint to a quiescent or less inflamed joint. In both RA and OA, joints reach a point where joint destruction outpaces repair. Ultimately, the patient loses joint mobility, so that a joint replacement may be required.

7.29.3 Recent Advances in Research on Inflammatory and Immunological Diseases

7.29.3.1 Arthritis

Arthritis is estimated to afflict about 1% of the world's population.[6] Ultimately, joint immobility drastically impacts quality of life and significantly impacts worker productivity and the economy.[7] Prior to and during the 1960s, therapies were targeted at managing the symptoms of arthritis and were focused on pain relief and reduction in swelling. The use of aspirin, steroids, and later nonsteroidal anti-inflammatory drugs (NSAIDs) were the main treatment options for those who suffered from RA and OA. Likewise, clinical studies focused on pain and inflammation. The American College of Rheumatology created a clinical scoring system that incorporated over 30 clinical parameters to assess disease level and progression measurements.[8] Endpoints for approval by the US Food and Drug Administration (FDA) in clinical trials include relief from clinical signs and symptoms (e.g., pain), gain in functionality, quality of life, and objective inflammatory markers like C-reactive protein (CRP).

In the early 1970–80s, treatment options for RA expanded. In an attempt to move beyond symptom management, many investigators began to accept the view that the synovial pannus represented a form of unregulated growth, not dissimilar to many cancers. One notable consequence was the introduction of an agent used for cancer treatment, namely the antiproliferative drug methotrexate (MTX). While on the surface the hypothesis seemed intriguing, the actual dose of MTX that was effective in RA was significantly lower than could be rationalized by its antiproliferative activity. Nevertheless, MTX and similar compounds with undefined mechanisms of action (gold salts, sulfasalazine, antimalarials, and penicillamine) were considered the preferred therapy to be used once the disease had progressed beyond the mild stage and the joint began to show progressive damage. Combinations of agents targeting both the symptoms and the underlying disease progression became the major components of the therapeutic strategy – the 'pyramid of therapy' approach (**Figure 1**). Based upon disease severity, patients were initially treated with NSAIDs, steroids were then added on as a second line and finally an antiproliferative like MTX was brought in as the third line of treatment.[9] Side effects were managed by replacing agents within each class.

Over the past decade, cellular pathways that may alter disease progression have emerged as targets for RA (e.g., anticytokine) therapy. Clear endpoints have been defined to monitor the progression of joint destruction over time periods ranging from 1 to 2 years. Indeed, many of the newer agents demonstrate a significant reduction in disease progression over this time period. In 1999, the FDA added joint structural integrity as an approvable clinical endpoint that could lead to drug registration. Agents that can be documented to retard or even halt disease progression can now be identified and, ultimately, treatment options beyond the traditional pain management therapies are available for RA. A number of drugs are now on the market (e.g., leflunomide, enbrel, remicade, and humira, as well as the former 'gold standard' MTX) that have been shown to block disease progression in RA.

While clinical markers are available to identify early RA patients (e.g., CRA, human leukocyte antigen (HLA)) and clear endpoints for monitoring disease are available, identifying early stage OA patients and monitoring their disease progression is significantly more challenging. Risk factors (e.g., family history, obesity, unilateral OA) have been identified yet correlation with the onset of OA remains elusive. The advent of pharmacogenetics and additional diagnostic tools may assist identifying patients at earlier stages of disease. The lack of overt inflammation as a primary driver of disease has made addressing therapy of OA quite challenging from a scientific standpoint, and therapies for OA primarily target symptom management. The major challenge in OA still occurs in designing clinical studies for agents

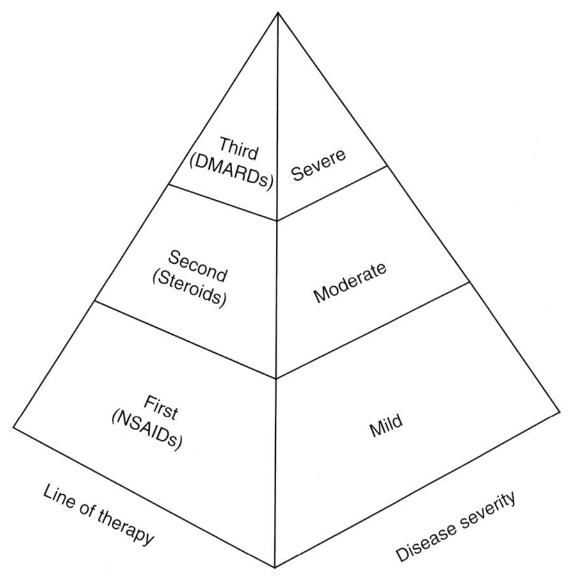

Figure 1 Classical 'pyramid of therapy' for rheumatoid arthritis.

that seek to blunt or halt disease progression. Due to the nature and progression of the disease, changes in joint physiology and architecture can be small and difficult to monitor over a reasonable timeframe in the clinic. The FDA has worked with academic and industrial institutions for over a decade to provide guidance and tools for clinical investigators.[10] For example, imaging methods like magnetic resonance imaging (MRI) or x-ray provide physicians with tools that can measure and therefore monitor changes of the disease over time. As the cartilage in a joint erodes in OA patients, erosions in the cartilage surface appear and the joint space is compressed. This results in a narrowing of joint space measurable by x-ray. On the other hand, MRI can reveal viability and quality of joint cartilage. Both methods can reliably measure significant joint changes within 6-month intervals. Both tools play a major role in defining disease progression, although the value of these measures in the absence of a standard therapy for OA is still controversial. Unfortunately, in OA there are very few compounds that show dramatic and significant effects in blocking progression of joint space narrowing. To date, only glucosamine and chondroitin sulfate have been shown to be useful and then only in the treatment of mild to moderate OA. Based on the available data both therapies are useful when combined with other modalities such as weight loss and exercise. Both agents appear to relieve pain and improve range of the joint motion, and thus may also display mild anti-inflammatory effects, but many hypothesize that their primary mode of action may be in providing critical nutritional supplements to cartilage, similar to calcium supplementation in osteoporosis.

7.29.3.2 Respiratory Diseases

Asthma and chronic obstructive pulmonary disease (COPD) are two major respiratory diseases include affecting a significant portion of the population. Asthma is characterized by inflammatory cell infiltration into the airway tissue leading to airway hyperresponsiveness and producing the clinical signs of wheezing, coughing, and tightness of the chest. About 4% of the world's population suffers from asthma.[11] Asthma may result from a variety of causes from genetic predisposition to environmental triggers.[12] For example, it has been shown that communities exposed to airborne industrial waste have an increased incidence of asthma. However, there is a growing incidence of asthma (and inflammatory bowel disease) in communities not exposed to airborne industrial particulates populated by individuals that practice good personal hygiene. These findings have prompted a working hypothesis that individuals not exposed to antigens in their youth lack the appropriate humeral immunity to ward off airborne pathogens later in life. This suggests an immunological pathogenesis for at least some cases of asthma. Epidemiology studies are under way to help better understand the growth of this disease.

Glucocorticoid steroids and β2-agonists are commonly prescribed treatments for asthmatics. Steroids suppress leukocyte activation and infiltration into lung tissue while β2-agonists serve to arrest vasoconstriction of lung airways. An aerosol device or inhaler can administer both agents to patients. During an asthmatic attack, immediate relief is needed and this mode of delivery is most effective. Recently, orally administered leukotriene receptor antagonists or biosynthesis inhibitors have been used to address both vasoconstriction and inflammation. These agents stabilize the asthmatic airway, reduce attacks, and improve the overall quality of life. Patients still use the steroid and β2-agonist aerosol therapy to offer relief from acute attacks. Steroid resistance is observed in a number of asthmatics. These individuals are believed to have a mutant form of the glucocorticoid receptor (beta form) that lowers the effectiveness of steroid therapy leaving the asthmatic with limited therapeutic options. Investigation of agents that may treat steroid resistant asthma is a focus for several academic and industrial research groups.

Chronic obstructive pulmonary disease is caused largely by environmental factors, particularly smoking. The incidence of COPD ranges between 2% and 4% of the world population. It is considered the fourth most prominent cause of death in the world with projections that it will reach the third leading cause by the end of the decade.[12] Inhaled smoke irritates the lung airways recruiting inflammatory cells. These cells release proteinases initiating lung tissue remodeling that leads to a reduction in airway plasticity. These changes attract inflammatory cells leading to a proinflammatory state. The progression of COPD is very slow, unlike that of asthma which tends to be very rapid. Cessation of smoking retards further progression. Recovery to a normal lung function depends on the degree of insult caused by smoking.

Much like asthma therapy, the most commonly prescribed agents for COPD are steroids used to dampen the inflammatory response and beta-blockers to stop vasoconstriction of the airways. New treatment approaches for COPD have targeted neutrophils, eosinophils, and mast cell activation or migration into lung tissue. Several clinical studies are under way to investigate novel therapies to treat this growing condition.

7.29.3.3 Other Inflammatory and Immunological Diseases

The worldwide prevalence of multiple sclerosis is estimated at over 2 million individuals.[13] As an inflammatory disease, multiple sclerosis attacks nerve tissue in the brain and spinal cord. Inflammatory cells are believed to enter the central nervous system erroneously targeting host epitopes for destruction. Subsequent plaque formation in the brain progressively reduces voluntary and involuntary control of bodily functions. This disease generally displays clinical signs when individuals reach the third decade of life and is found more often in women than in men. The pathogenesis of the disease is poorly understood. Initial treatment approaches typically employ steroids during disease flares to address symptoms. As the disease progresses, biologics like beta-interferon are employed that can alter the reactivity of a subset of involved T cells. Unfortunately only about 40–55% of the treated individuals are relapse-free after 2 years of therapy.[14] Other therapies target T cell proliferation but carry a significant side effect burden. Several approaches to regulating T cell activation or leukocyte migration are currently under investigation clinically. To date, an effective small-molecule treatment for multiple sclerosis with good efficacy and therapeutic index remains elusive.

SLE is an autoimmune disease that impacts about 0.02% of the world's population.[15] The disease causes an overproduction of antibodies that recognize a subset of membrane phospholipids. Immune complexes ultimately overload the kidney, the main organ responsible for removing antibody complexes from the bloodstream. Ultimately the patient experiences renal failure and requires a kidney transplant. The origin of this condition is poorly understood. Treatment involves high doses of anti-inflammatory steroids along with plasma dialysis to deplete the patient of antibody complexes. As in the case of multiple sclerosis, several clinical trials are being conducted to investigate the use of immunosuppressants and other immunological therapies.[16]

Organ transplantation would not be possible without organ donors and immunosuppressant therapy. In 2005, the National Organ Bank in the USA tracked in 2005 about 21 000 transplants that were conducted from January through September leaving about 90 000 individuals on a transplant waiting list. Shortage of organ donors limits the number of transplants each year. Once an organ has been transplanted into the recipient, the new organ is recognized as foreign by the host causing activation of the immune system. Unless the activated T cells are suppressed, the host would reject the transplanted organ. The current treatment employs a cocktail of medicines that reduce T cell activity. Combining anti-inflammatory drugs (e.g., glucocorticoids) with immunosuppressants (e.g., CsA, FK506, rapamycin) leads to effective therapy. The challenge is to maintain a balance between the integrity of the transplanted organ while leaving sufficient immunological capacity to reduce infection. While the overall prognosis for such transplant recipients is good, the search for far more effective therapies that reduce side effects or possibly induce foreign organ tolerance is under way in several laboratories.

7.29.4 Recent Therapeutic Advances for Arthritis

We now turn our attention to describe how treatment options have changed for patients with arthritis over the past 15 years. Given the breadth of such a topic, our intent is not to cover every therapy. Rather, we have elected to focus on new therapies, how these new agents were discovered, plus how these results have directed research efforts to seek new therapies. These new treatment options and strategies offer clinicians and patients greater flexibility to provide efficacy while improving safety. Each section builds on the 'pyramid of therapy'[17] and shows how these new therapies have impacted how patients are treated.

7.29.4.1 Arachidonic Acid Cascade

For over five decades, drug discovery approaches have aimed at regulating the production and biotransformation of the fatty acid arachidonic acid. Arachidonic acid is a polyunsaturated, 20 carbon fatty acid (a tetra-unsaturated eicosinoid) that is transformed by cyclooxygenases into an endoperoxide (PGH2) and ultimately prostaglandins or thromboxanes. These lipid hormones have inflammatory and aniti-inflammatory properties that affect a number of biological systems. The collective complex of enzymes and eicosinoids is known as the arachidonic acid cascade (**Figure 2**). Phospholipase A2 (PLA2) liberates arachidonic acid from membrane phospholipids and starts the biotransformation into prostaglandins and thromboxanes via the cyclooxygenase pathway or via the lipoxygenases to leukotrienes. For several decades the focus was on producing NSAIDs (e.g., ibuprofen, naproxen, diclofenac) by blocking cyclooxygenase. These inhibitors limited the production of prostaglandins and thromboxanes. Such agents are highly effective for the treating the 'heat of inflammation,' pain, swelling, and fever. Yet upon chronic use, NSAIDs are known to cause dose-limiting side effects like gastric ulcers and renal abnormalities. There is some evidence to indicate that the gastric ulcers are caused by an increase in lipoxygenase products.[18] Blocking cycloxygenase allows more arachidonic acid to be directed through the lipoxygenase pathway. Thus a safe NSAID has been long sought for treating arthritis. Since 1990, the main focus of this search has been on highly selective inhibitors of phospholipase A2, cyclooxygenases and lipoxygenases.

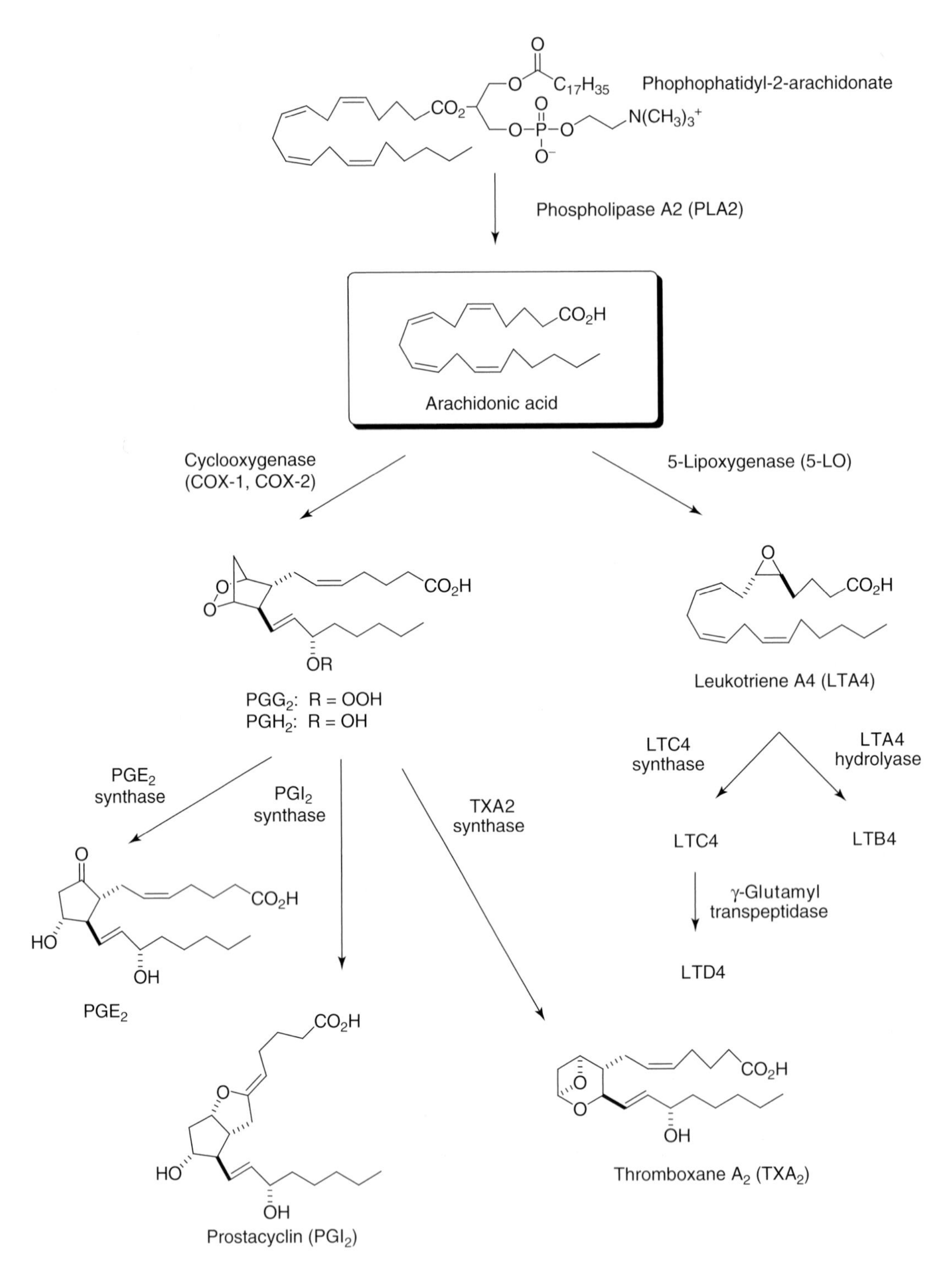

Figure 2 The arachidonic acid cascade.

7.29.4.1.1 Phospholipase A2

Phospholipase A2 (PLA2) is a serine hydrolase that liberates arachidonic acid from membrane phospholipids. Reducing arachidonic acid liberation deprives both cycloxygenase and lipoxygenase pathways of substrate. If increased leukotriene production is responsible for causing gastric distress,[18] then a PLA2 inhibitor should block both pathways to offer the efficacy of an NSAID without the gastric ulcer risk. Over the last 25 years, many groups have been pursing

PLA2 inhibitors with very few agents reaching the clinic. There are several sources of phospholipases that are considered more relevant than others. There are 10 secretory forms of PLA2 (sPLA2) with sPLA2-IIa being viewed as the most abundant in RA synovial tissue.[19] The other form viewed as being relevant for RA is one of the three cytosolic forms known as cPLA2-α.[20] Besides the complexity with the identification of the most suitable form of PLA2 to inhibit, measuring enzyme inhibition is also challenging with this target. The enzyme assays used were confounded by the aggregating nature of phospholipids leading to misleading results. Several investigators have developed assays that allow for assessing PLA2 inhibition under a variety of conditions.[21]

Despite these hurdles several PLA2 inhibitors have been identified in the last 15 years and listed below are some noteworthy examples that were felt to advance the discovery of PLA2 inhibitors for the treatment of arthritis. The range of examples includes substrate-based inhibitors, natural products, and most recently a series of highly selective inhibitors of different PLA2 families. Since PLA2 uses phospholipids substrates, active site inhibitors will have lipophilic features. The challenge for the medicinal chemist is to identify an agent that would balance lipophilic requirements of PLA2 active site with druglike properties.

Some substrate-based inhibitors were identified that served as tools for cellular investigation. The recent discovery of GIVA-PLA2 inhibitors has been reported and has employed a transition-state inhibitor design approach. The designed agent AX006 (**1**) rapidly and reversibly inhibits cPLA2 with good potency and selectivity verses other PLA2 family members. It has comparable analgesic efficacy when dosed orally to that of the NSAID, indomethacin.[22] The use of the diketo functionality has been exploited by several groups and incorporated into traditional druglike scaffolds. Natural products have also offered a source of PLA2 inhibitors with the discovery of manoalide.[23] Researchers have shown manoalide to irreversibly and covalently attach to the catalytic serine in synovial fluid cPLA2. Recently a related marine natural product, cacospongionolide (**2**), was also shown to be selective for synovial fluid cPLA2 along with anti-inflammatory activity in animal models both topically and systematically.[24]

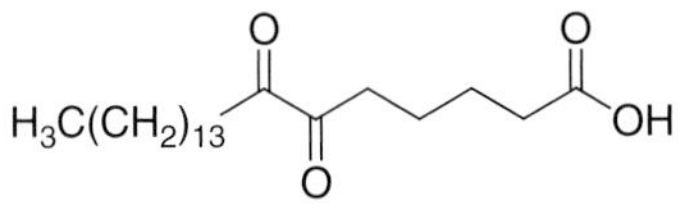

1 AX006

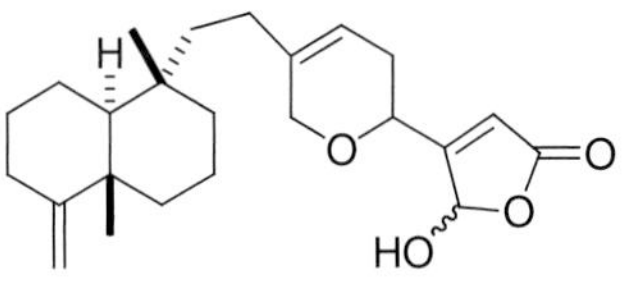

2 Cacospongionolide

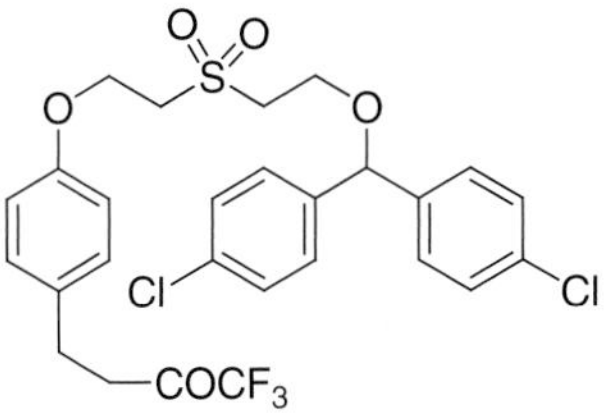

3 BMS-229724

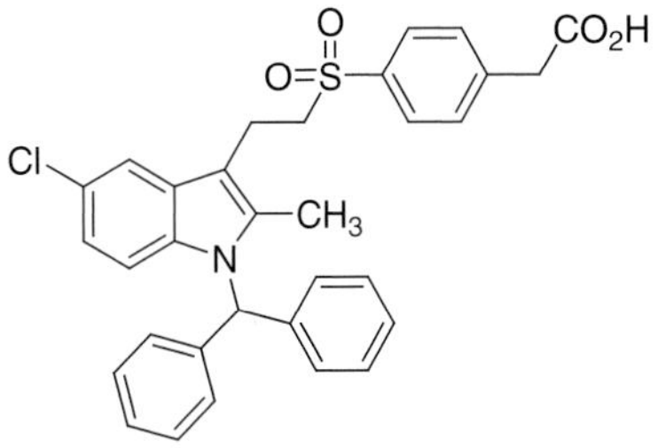

4 (McKew *et al.*)[25]

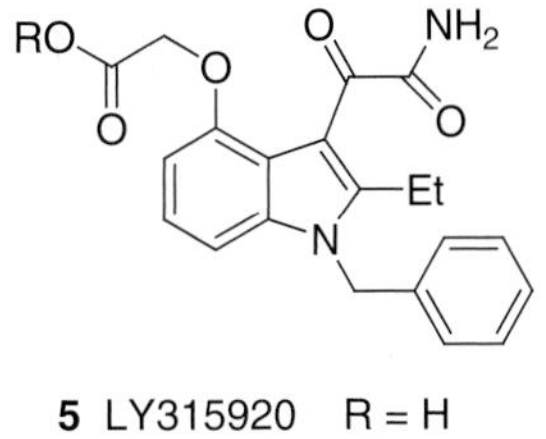

5 LY315920 R = H
6 LY333013 R = Me

Inhibitors of cPLA2-α have been reported. Two examples, BMS-229724 (**3**) and **4**, provide evidence that selective cPLA2-α inhibitors can be obtained with druglike properties suitable for further investigation.[25] A report of a clinical study using a selective sPLA2 inhibitor has provided mixed results in RA. Based upon preliminary efficacy observed in an RA clinical trial with parentally dosed LY-315920 (**5**), a follow-up clinical study was conducted. A randomized, double blind placebo controlled 12-week study in moderate RA using LY-333013 (**6**), the methyl ester prodrug of **5**, showed very little efficacy.[26] The authors feel adequate active drug was delivered. This study raises questions on the role of sPLA2 inhibition in RA and the level of selectivity needed. Additional studies will be required to determine the future of sPLA2 inhibition in RA. Along with the clinical data that may become available for cPLA2 inhibitors, the promise of efficacy and safety may be realized for RA patients with this challenging target for drug discovery.

7.29.4.1.2 Cyclooxygenase

Cyclooxygenase inhibitors continue to be the standard of care for treating RA and OA pain. With the NSAID safety risk of gastrointestinal (GI) ulcers or lesions, several approaches have been employed by physicians to minimize this risk while providing relief to the arthritis suffer.[27] One approach was to limit upper GI exposure by the use of enterocoated tablets or slow release capsules. Depending on the patient history, physicians will coprescribe a PGE1 analog (miniprosol)[28] or a proton pump inhibitor[29] with an NSAID. This combination affords the pain relief while providing ulcer generation protection. Another option is to test patients for *Helicobacter pylori*, a gut bacterium. Some individuals are more prone to NSAID-induced GI lesions when this gut bacterium is present. In these cases, physicians prescribe an antibacterial agent along with the NSAID.[30] These work-around approaches minimized the NSAID GI side effect while offering patients the desired pain relief.

An alternative approach emerged in the 1990s with the discovery of a second form of cyclooxygenase. Cyclooxygenase-1 (PGHS-1, COX-1) is present in most tissues and has been the traditional target for most NSAIDs. Levels of COX-1 itself remain relatively stable suggesting a role in arachidonic acid cascade homeostasis. The second form of cyclooxygenase called cyclooxygenase-2 (PGHS-2, COX-2) is present in inflammatory cells.[31] Activated leukocytes can express COX-2 at levels that far exceed COX-1. At the site of inflammation, these activated leukocytes would produce elevated levels of arachidonic acid cascade products. The two enzymes share 60% gene sequence identify and have similar enzymatic properties and kinetics. The approached adopted by several groups was to prepare selective COX-2 inhibitors. These agents were designed to localize to the site of inflammation where leukocytes were activated and bypass COX-1 located in the gut. This approach led to a new generation of NSAIDs with an improved GI safety profile.

Comparison of COX-1 versus COX-2 inhibition profiles revealed that most known NSAIDs blocked both enzymes with most being more effective at blocking COX-1 than COX-2. Yet there were examples that showed greater COX-2 potency suggesting that selectivity was attainable. Groups started looking at compounds with improved GI or renal profiles to obtain a structural insight into selectivity. One of the early leads was DuP 697 (**7**), an agent shown to afford improved renal safety in contrast to most known NSAIDs.[32] With an extended half-life in animal and humans (unpublished results), the GI safety margin was never realized. It was later found that **7** was a very selective inhibitor (70-fold) for COX-2 over COX-1 and proved to be a structural lead for many COX-2 selective inhibitors.[33] Another feature was the time-dependent enzymatic inhibition of this class of compound, which proved to be very significant in providing COX-2 selectivity.[34]

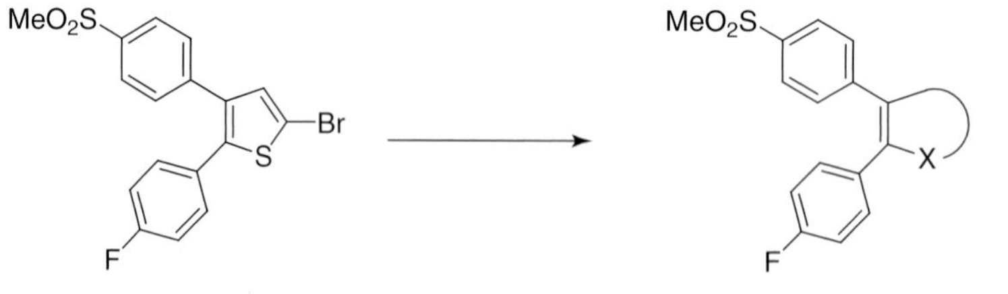

7 DuP 697

8 Diaryl heterocyclic scaffold

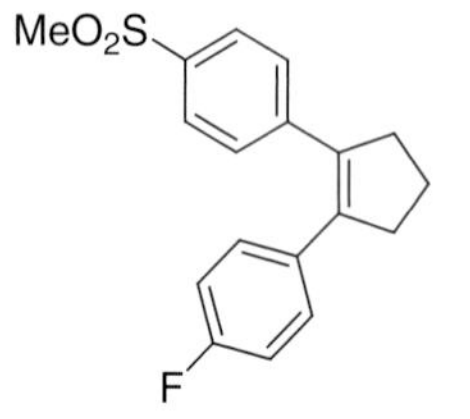

9 SC-57666

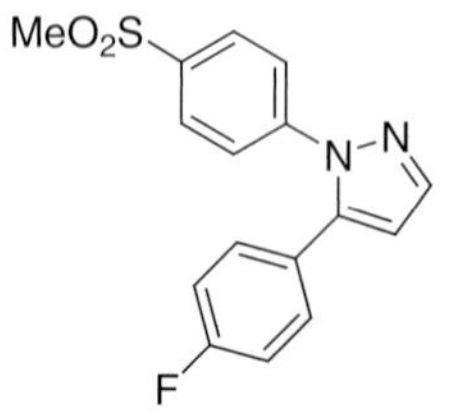

10 SC-58125

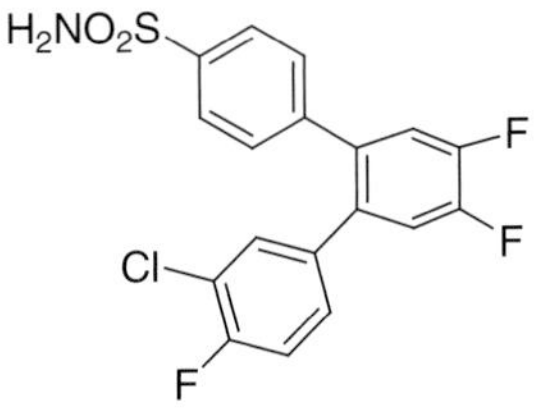

11 (Li *et al.*)[36]

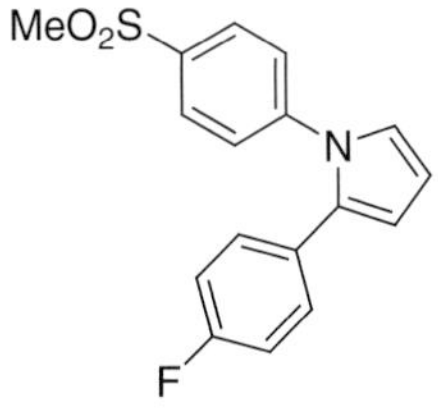

12 (Khanna *et al.*)[36]

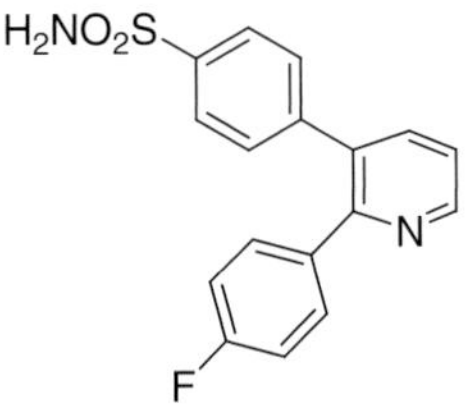

13 (Pinto *et al.*)[36]

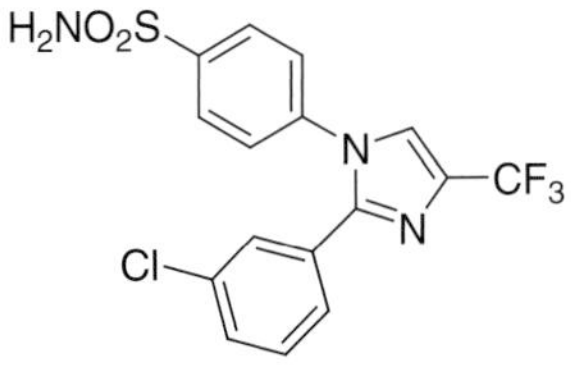

14 (Khanna *et al.*)[36]

Starting with DuP 697 several groups systematically probed this structural class to reveal a common set of substituents displayed on this diaryl heterocyclic scaffold (**8**). The Searle group had identified SC-57666 (**9**) and SC-58125 (**10**) as some early selective COX-2 inhibitors.[35] Expansion into larger rings or into other heterocycles (**11–14**) was investigated by a number of groups.[36] In all cases, a common motif emerged with the electronically deficient aryl rings affording the desired profile.

Based upon this foundation, below are listed the current collection of highly selective COX-2 inhibitors that have been approved: celecoxib (**15**),[37] rofecoxib (**16**),[38] and valdecoxib (**17**),[39] or are pending registrations etoricoxib (**18**),[40] and lumiracoxib (**19**).[41] A modestly selective COX-2 inhibitor, meloxicam (**20**),[42] has recently entered the market. Each of these drugs achieve the desired potency and display an improved GI safety profile over most NSAIDs to warrant the advancement for the treatment of arthritis.

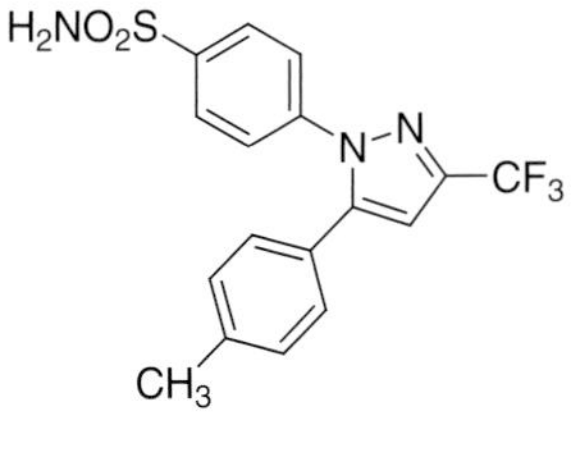

15 Celecoxib

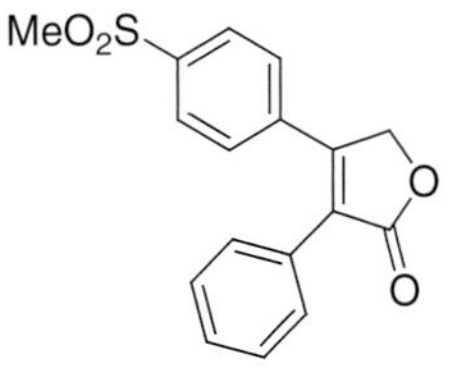

16 Rofecoxib

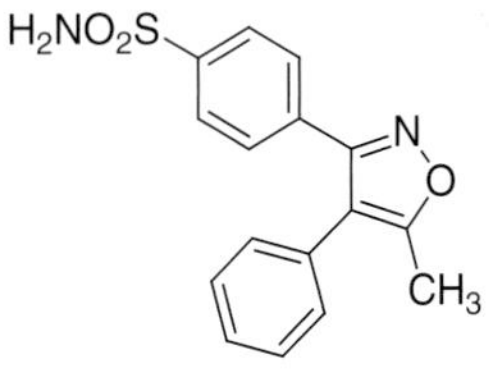

17 Valdecoxib

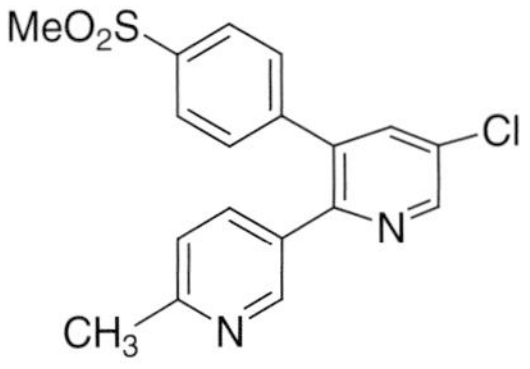

18 Etoricoxib

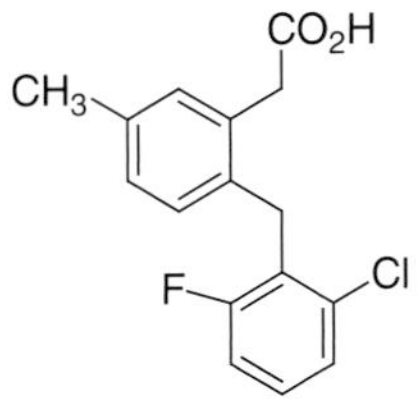

19 Lumiracoxib

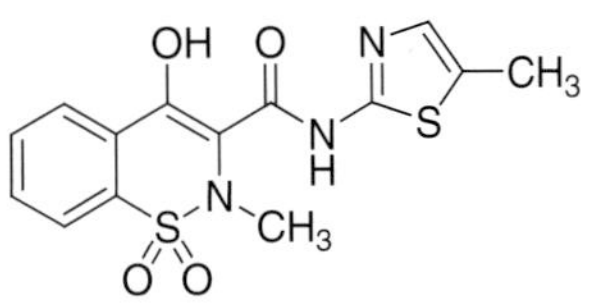

20 Meloxicam

Since 2000, several papers have appeared to suggest that highly selective COX-2 inhibitors display adverse cardiovascular events.[43] The scientific basis for the adverse events has stimulated significant research by investigators. With the emergence of this new risk, one of the COX-2 inhibitors has been removed from the market (rofecoxib), while the others are under additional review with the FDA to further define the risks associated with these new modalities.

7.29.4.1.3 Leukotrienes

Lipoxygenase (LO) is an enzyme that converts arachidonic acid into leukotrienes.[44] As shown in **Figure 1**, arachidonic acid is first expoxidized by 5-lipoxygenase (5-LO) followed by the addition of cystiene peptides to afford a new family of leukotrienes (e.g., LTC4, LTD4). All of these leukotrienes have the ability to activate leukocytes and increase inflammation.[45] Drug discovery efforts were undertaken to intervene in this pathway. One of the key approaches was to block the initial lipoxygenase 5-LO. This resulted in the discovery of the drug zileuton (**21**).[46] Another approach was to target the LTD4 receptor leading to drugs like montelukast (**22**),[47] zafirlukast (**23**),[48] and pranilukast (**24**).[49] While very effective in asthma, where infiltrating leukocytes drive the disease, these leukotriene pathway blockers were only modestly effective in treating RA.[50]

Some groups have explored the idea of coupling a COX inhibitor with a LO inhibitor.[51] Known as CO-LO inhibitors, this approach offers a way to achieve the efficacy of an NSAID balanced by the GI safety offered by blocking the LO pathway.[18] Although they have been explored for over 25 years, none of these agents has reached the market. A recent dual inhibitor, licofelone (**25**), has entered phase III trials for OA and the initial reports are encouraging.[52] Additional clinical studies are needed to see if this approach will be effective alternative for arthritis patients.

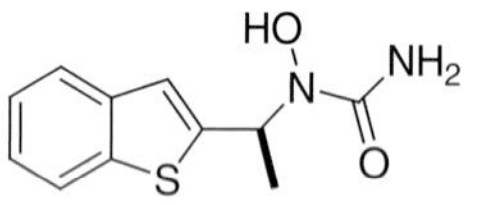

21 Zileuton

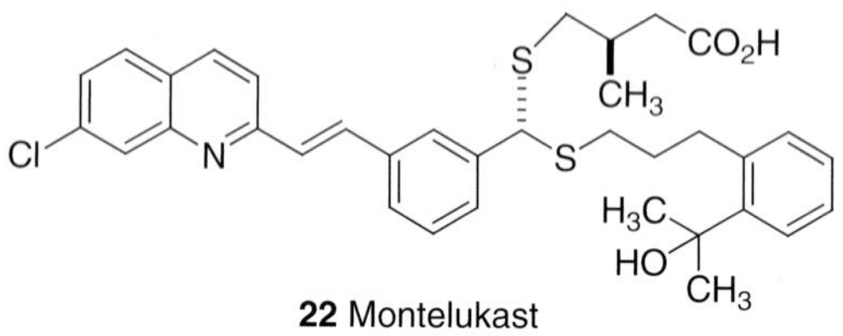

22 Montelukast

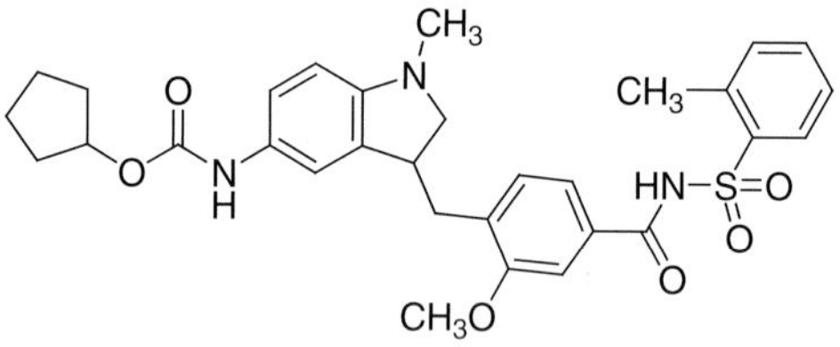

23 Zafirlukast

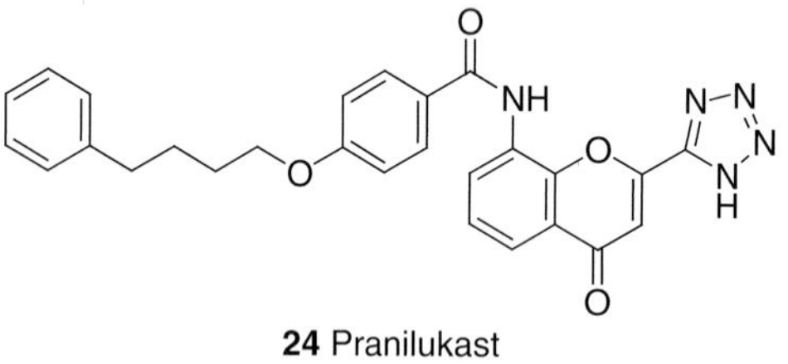

24 Pranilukast

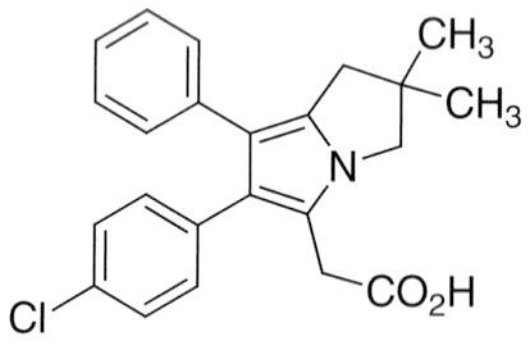

25 Licofelone

7.29.4.2 Steroids

Corticosteroids have been used to treat inflammatory and immunological diseases systemically (e.g., RA) and topically (e.g., psoriasis, asthma). Prior to NSAIDs, steroids were the primary therapy for arthritis patients. Now such agents are confined to the second tier of the 'pyramid of therapy.' Synthetic modifications of corticosteroids have improved potency and drug delivery issues. Some of the most commonly prescribed steroids are methylprednisolone or hydrocortisone acetate (RA, organ transplantation) and beclomethasone dipropionanate (asthma). While highly effective these medicines have safety limitations with chronic dosing leading to significant bone and cartilage loss along with a number of adverse effects (e.g., electrolyte imbalance, hyperglycemia, osteoporosis).[53] Several research groups are seeking a safer steroid that retains the potent anti-inflammatory properties.

Anti-inflammatory steroids (glucocorticoids) bind to the glucocorticoid receptor (GR), a member of the nuclear hormone receptor (NHR) family. Once steroid binds, the receptor–ligand complex translocates from the cytoplasm to the nucleus of the cell. In the nucleus the complex acts as a transcriptional factor to either transactivate or transrepress various target genes. There are data to support the notion that steroid-mediated transactivation of specific genes through glucocorticoid receptor elements (GREs) contributes to the side effects (e.g., diabetes, osteoporosis) of steroid use, while transrepression which entails interaction of the steroid–receptor complex with other transcriptional factors such as activating protein 1 (AP-1) and nuclear factor kappa B (NFκB) is responsible for the beneficial or anti-inflammatory properties of steroids. This hypothesis[54] is driven by the observation that many 'inflammatory' genes such as proteinases (e.g., matrix metalloproteinases (MMPs)) and cytokines (e.g., TNF-α) are positively regulated by the transcription factors AP-1 and NFκB, whose action if countered by the transrepressive properties of steroids. Suppression of the inflammatory response by steroids re-establishes the balance between joint repair and destruction. Since this mechanism requires gene regulation, the onset of steroid action or relief to patients takes 24–48 h, contrasting with NSAIDs where symptom improvement occurs within minutes to an hour of drug delivery. The search for the safe steroid has remains elusive. The prospect of achieving such a feat requires a ligand-bound GR complex to adopt a conformation that transrepresses AP-1 while not transactivating GREs. While this is a challenging area, several promising results have recently appeared in the literature.

Several examples of novel anti-inflammatory compounds that exploit interaction with the GR have appeared in the patent literature.[55] Compound **26** represents an example of compounds described in a patent by Schering AG that have been pursued as nonsteroidal anti-inflammatory agents that act through a GR-mediated mechanism. Additional compounds that exploit the central quaternary center with the hydroxyl and trifluoromethyl groups have been described in applications by Boehringer-Ingelheim (**27**)[56] and by GlaxoSmithKline (**28**).[57] Both groups disclose compounds that are dissociated in cellular systems that measure transrepression versus a cellular system that measures transactivation, thus demonstrating a cellular proof of concept.

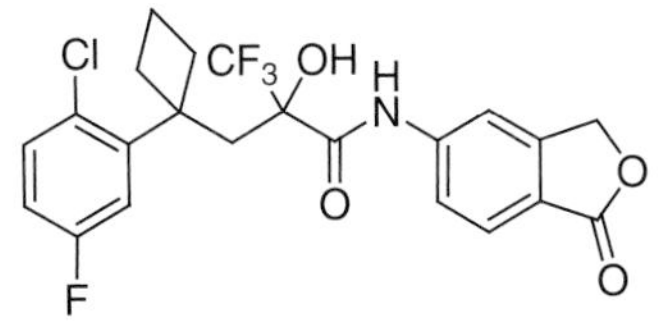

26 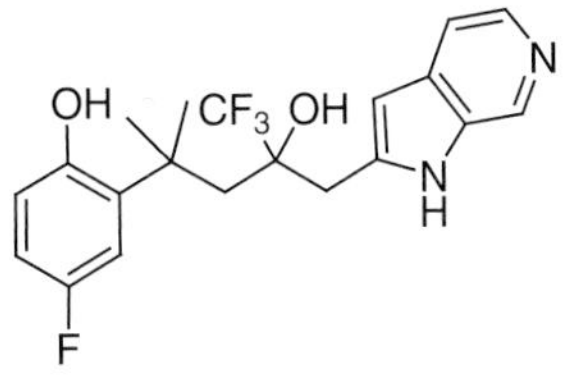 **27**

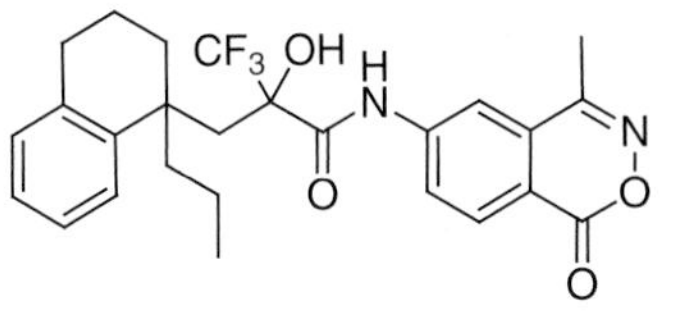

28

Additional structural classes have also started to emerge as effective glucocorticoid mimetics.[58] Compounds **29–31** highlight three examples. Compound **29** from GlaxoSmithKline demonstrated that the esterification of the steroid nucleus leads to compounds that are dissociated in cellular systems. Compound **30** from UCSF is also reported to have dexamethasone-like efficacy and a cellularly dissociated profile.[59] The modified glucocorticoid core structure compound **31** is also reported in a US patent by Pfizer to be dissociated in an osteoblast assay that could be indicative of a bone-sparing glucocorticoid.[60]

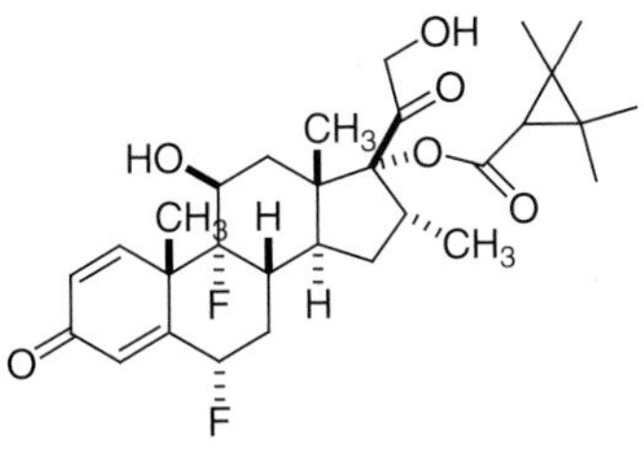

29 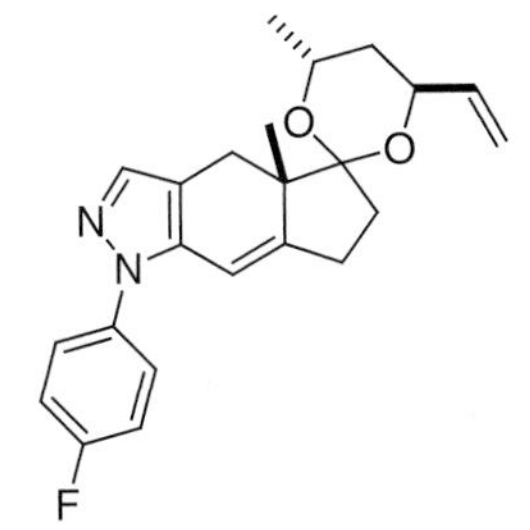 **30** 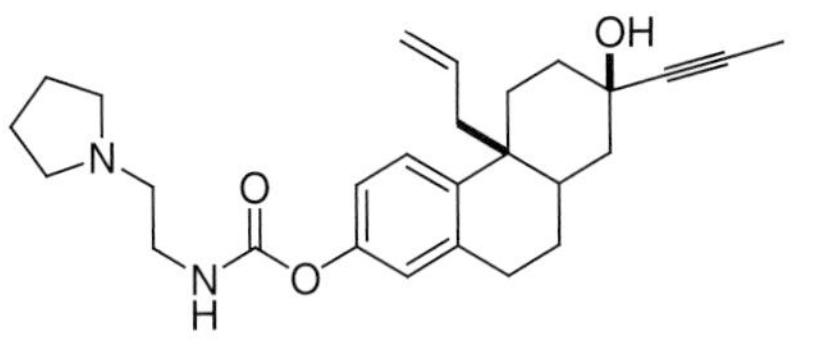 **31**

Another nuclear hormone receptor that has recently emerged as a potential target to treat immunological diseases is the estrogen receptor. Besides the observed gender bias for females over males for RA or multiple sclerosis, these diseases seemed to be blunted during pregnancy when estrogen levels are high.[61] Several studies have speculated that populations of T cells are altered while others have suggested that estrogen has suppressive activity toward the NFκB pathway. Unfortunately, use of estrogen directly has limitations due to other hormonal effects. Fortunately, tools have recently emerged that target one of the two forms of the estrogen receptor (estrogen receptor-β versus estrogen receptor-α). Results show that selective estrogen receptor-β agonists – ERB-041 (**32**), ERB-196 (**33**), and 8β-VE2, (**34**) – lack many of the uterine modulating properties associated with estrogen.[62] More importantly these estrogen receptor-β selective agents have been shown to be effective preclinically in animal models of arthritis and inflammatory bowel disease. These agents are currently awaiting clinical evaluation for RA. Another tool compound is the estrogen receptor ligand that suppresses NFκB activation without activating the estrogen receptor-mediated uterine activities. One of these molecules (SIM-916, **35**) has been shown to be effective in animal models of RA and is entering clinical trials for RA.[63]

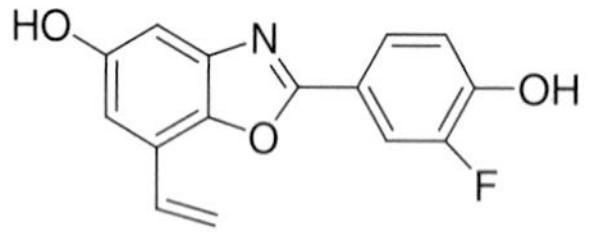

32 ERB-041

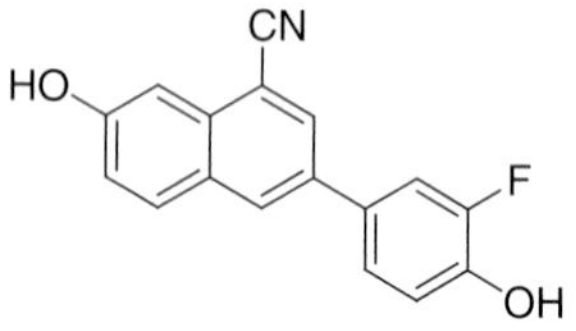

33 ERB-196

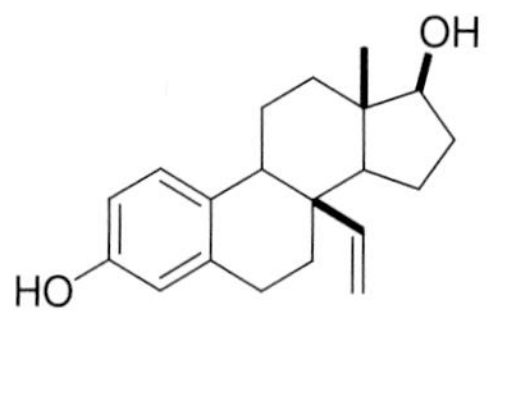

34 8β-VE2

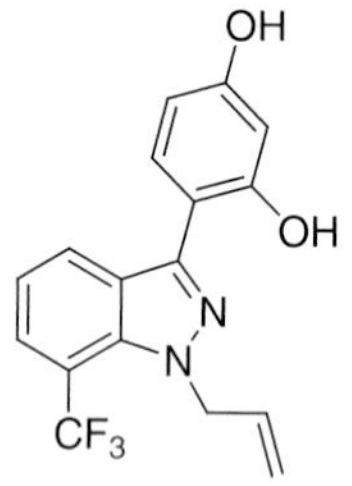

35 SIM-916

Overall the nuclear hormone receptor family offers novel opportunities for further study. The challenge will be the identification of selective agents that retain the steroid efficacy while improving the side effect profile.

7.29.4.3 Disease-Modifying Anti-Rheumatic Drugs

The standard of care for moderate to severe RA is MTX (MTX, **36**), developed during the mid 1980s.[64] MTX occupies the top level in the pyramid of therapy. A dihydrofolate reductase inhibitor, MTX blocks cell division. Originally developed for cancer, MTX has been very effective in dampening rapid T cell proliferation in RA patients. Based upon blocking disease progression, this class of agent became known as disease-modifying anti-rheumatic drugs (DMARDs). Administered with an NSAID to control pain and a steroid as a broad anti-inflammatory, MTX offers the added ability to suppress T cell activity and pannus proliferation for the moderate to severe RA patient. Hepatotoxicity can limit the use of MTX and some patients require routine liver function tests to avoid overdosing.[65]

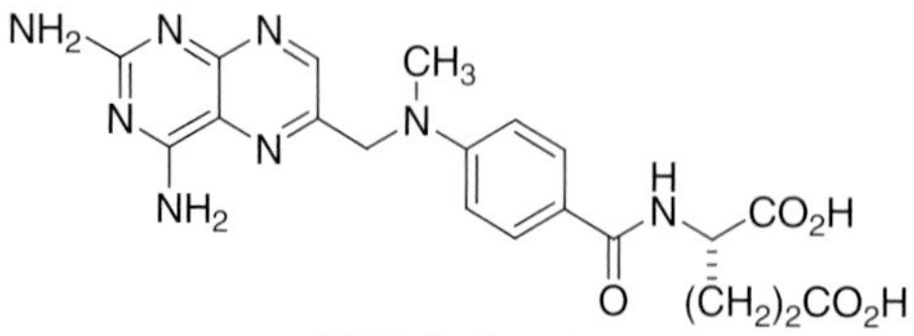

36 Methotrexate

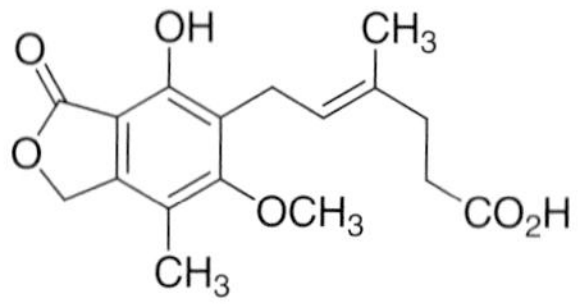

37 Mycophenolic acid MPA

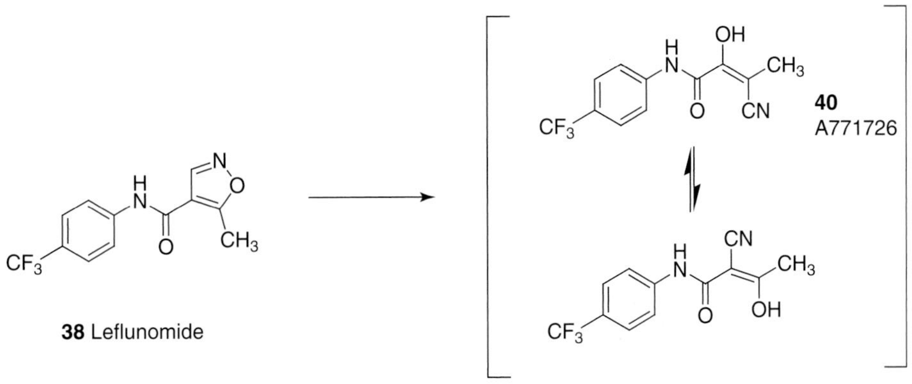

38 Leflunomide

40 A771726

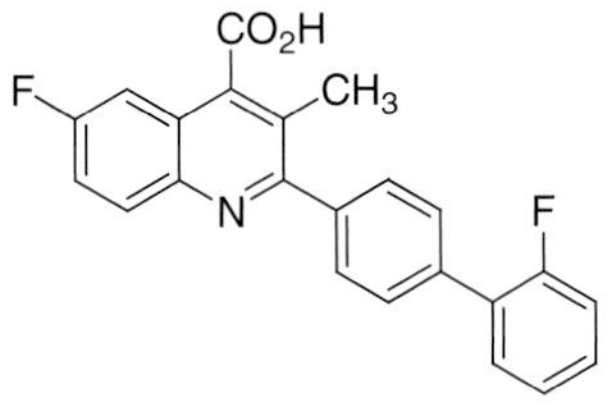

39 Brequinar

Several other agents known to affect T cell proliferation have been investigated in the clinic. Most of these approaches exploited either purine (e.g., mycophenolic acid, **37**) or pyrimidine (e.g., leflunomide, **38**; brequinar, **39**) biosynthesis.[66] Again rapidly dividing cells were most impacted by such therapies leading to side effects found primarily in the GI tract. These agents have advanced into clinical studies but have not all advanced to approval. Leflunomide has been approved for RA and serves as an alternative to MTX. It is a prodrug that converts to the active species, A771726 (**40**).[67] The mechanism of action for arthritis is unclear since the active species can inhibit dihydroorotate dehydrogenase (DHODase) and some kinases.[68]

While an area of intense study during the early 1990s, this DMARD approach to controlling arthritis has diminished. With the advent of highly specific agents like anticyctokine therapies (e.g., anti-TNF) patients have a number of disease modifying alternatives. By combining an anti-TNF agent with MTX one can lower the dose of both drugs. This reduces potential side effects of both agents while providing patients a powerful and effective combination to treat RA.

7.29.4.4 Structure-Modifying Anti-Inflammatory Drugs

Thus far we have described arthritis therapies that deal with pain or disease proliferation. Seldom do these therapies delay disease-induced structural changes of the joint as measured with x-ray (joint space narrowing) or MRI (cartilage viability). Structure-modifying anti-inflammatory drugs (SMADs) offer a new frontier of therapy. Cartilage protectant agents could offer a treatment option that might stabilize the joint and improve the quality of life for arthritis patients. Such therapies would also delay the need for joint replacements. The approaches described below may offer patients these medical options.

7.29.4.4.1 Blocking cytokines

Cyctokines are proteins secreted by or found on the surface of leukocytes that serve to regulate cellular activity. These proteins are transiently expressed upon activation with some being pro-inflammatory (e.g., TNF-α, interleukin-1 beta (IL1β)) while others display anti-inflammatory activity (e.g., IL10). Temporally, some cyctokines appear earlier than others and serve to activate others cytokines (e.g., TNF-α before IL1β). This 'hierarchy of cytokine production' suggested that targeting the upstream cytokine would afford more effective therapy.[69] The proinflammatory cytokines activate cells to produce members of the arachidonic acid cascade that contribute to pain and inflammation. In addition, a variety of proteinases (e.g., MMPs) are secreted, leading to cartilage and bone destruction in the affected joint. The pioneering work of Feldmann[69] highlighted the importance of blocking TNF-α in arthritis by showing that first RA synovial cells were very sensitive to TNF-α inhibition, then blocking TNF-α is highly effective in animal models, and finally very high levels of TNF-α are found in RA joints especially in joints undergoing significant cartilage destruction. Based upon these findings, blocking TNF-α became a focal point for drug therapy in the 1990s.

One TNF-α blocking approach was the use of antibodies. Two FDA approved antibody therapies for RA are remicade[70] and humira.[71] Remicade is a chimeric neutralizing antibody of mouse fragment variable domain (Fv) and human immunoglotulin G1 (IgG1). In clinical trials, remicade was used with MTX and shown to be very effective in treating pain and inflammation in patients. Antiremicade antibodies where minimized by this combination therapy permitting extension of dosing such that joint space narrowing was significantly reduced when patients were treated by remicade infusion over extended periods. This improvement of joint structure was a new and promising benefit with anti-TNF therapy. The latest anti-TNF-α agent approved was humira, a humanized anti-TNF-α antibody. This molecule also showed significant improvement in symptoms as well as joint structure verses placebo.

An alternative biopharmaceutical to remicade or humira is enbrel,[72] a fusion protein of two p75 TNF-α receptor extracellular domains linked to the fragment constant domain (Fc) fragment of an IgG. Developed at about the same time as remicade, enbrel can be administered with MTX to patients and shows very similar efficacy results on joint inflammation and integrity. Even bone erosion was notably improved. The discovery of these anti-TNF therapies truly has led to a significant improvement in the arsenal of RA therapy. The ability to affect inflammation and joint integrity really was a significant step forward for patients.

Blocking TNF-α production spawned a number of small-molecule approaches. One approach is blocking a TNF-α converting enzyme (TACE, ADAM-17). TACE is a type I transmembrane, multidomain, zinc containing metalloproteinase and unique member of the adamalysin family.[73] It is the primary enzyme responsible for cleaving membrane-bound 26 kDa proTNF-α to generate soluble 17 kDa TNF-α. This proteolytic processing of 26 kDa proTNF-α occurs at its physiological site (between Ala76 and Val77) to yield the biologically active cytokine as a soluble, noncovalently bound trimer. Blocking the production of soluble TNF-α by small-molecule active site inhibitors could address a variety of pathological conditions including RA, Crohn's disease, multiple sclerosis, ulcerative colitis, and congestive heart failure. Inhibition of TACE provides an alternative strategy both to the use biologics (infliximab, remicade; etanercept, enbrel; adalimumab, humira) or other small-molecule approaches (receptor antagonists, signal transduction, kinases).[74] Recent reviews concerning the biology of TACE inhibition[75] and the medicinal chemistry strategies[76] for the design of inhibitors have appeared. Immunex (**41**)[77] and Glaxo (**42**)[78] have reported on the isolation, purification, and cloning of TACE.

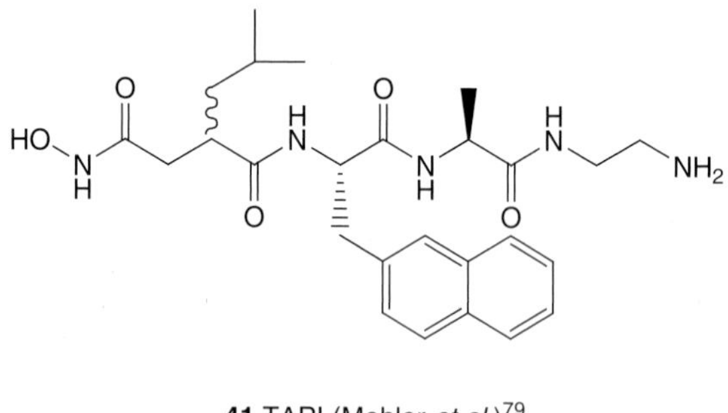

41 TAPI (Mohler *et al.*)[79]

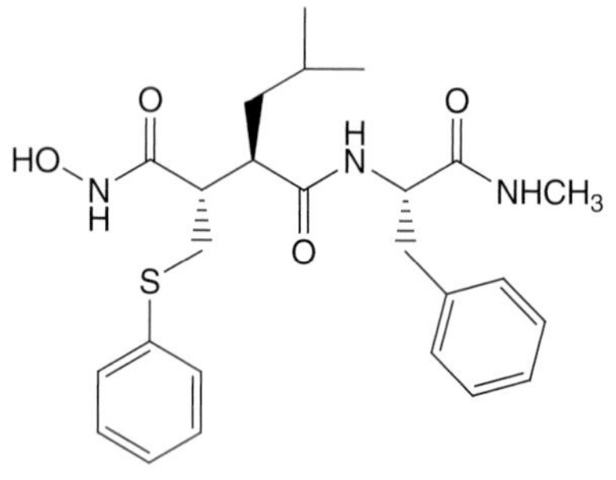

42 GI 129471 (McGeehan *et al.*)[79]

TACE is an attractive medicinal chemistry target for lead generation and optimization. The rich literature and broad experience in the design of the matrix metalloproteinase inhibitors (MMPIs), coupled with knowledge of the substrate and mechanism of action, has made this a fertile area of research. Early generation MMPIs have been shown to block the release of TNF-α from cells.[79] Solution of the crystal structure of the catalytic domain of TACE has been reported.[80] Though sharing some structural features in the active site with the MMPs, the angular and deep S1′–S3′ pocket of TACE is a distinguishing characteristic. Differential affinity for this unique S1′–S3′ pocket of TACE and the S1′ pocket of the MMPs offers design opportunities to enhance potency against TACE and optimize selectivity versus these closely related enzymes (see compounds **43–47**). Consideration of these structural aspects has been reviewed.[81]

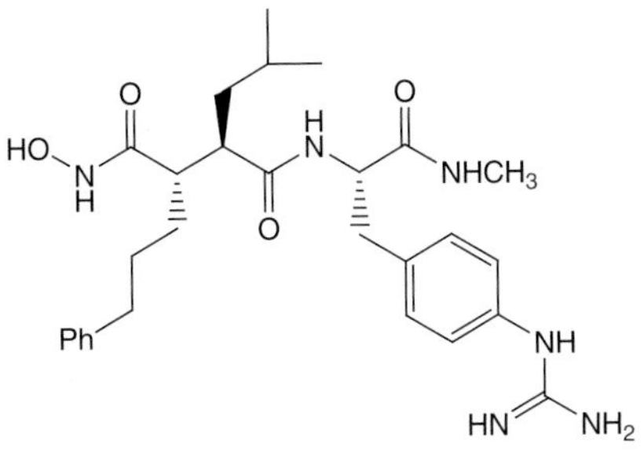

43 (Fujisawa *et al.*)[82]

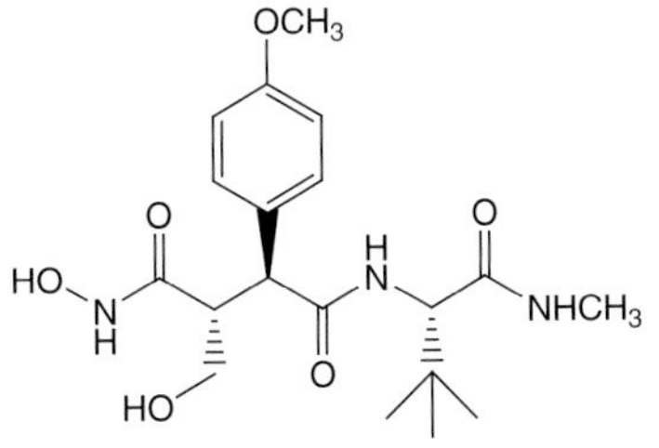

44 (Kottirsch *et al.*)[82]

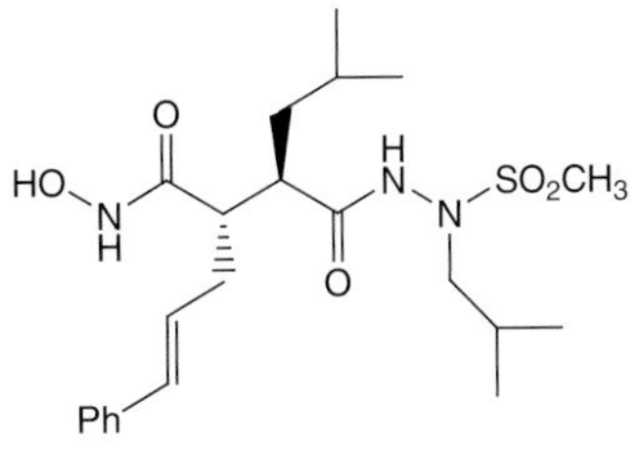

45 Ro 32-7315 (Beck *et al.*)[82]

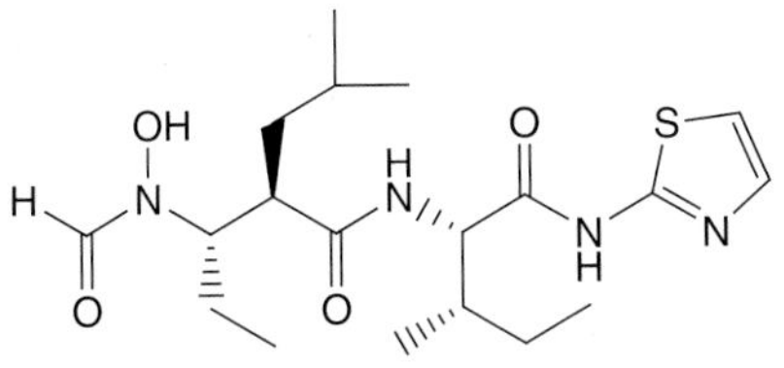

46 (Musso *et al.*)[82]

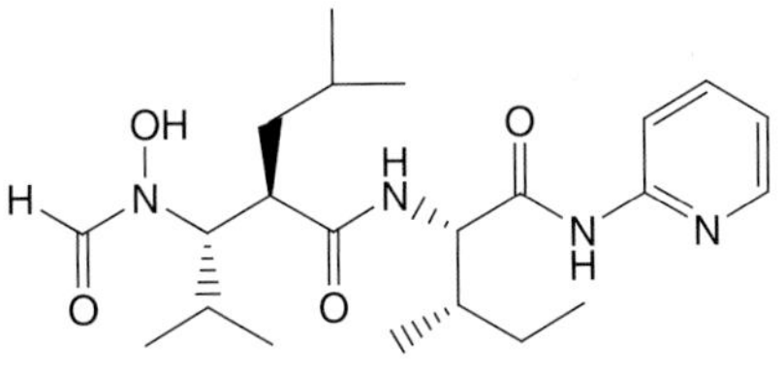

47 GW3333 (Musso *et al.*)[82]

Recognition of key structural features of TACE in the context of inhibitor design provides a framework for the generation of compounds to study the local polarity of the active site. In a number of cases, the design paradigm extended and adapted the tactics employed if the MMPI arena which involved the introduction of a zinc chelator and an appropriate enzyme affinity substituent (P1′ group) to a suitable scaffold. The scaffolds are both peptidic and nonpeptidic, and generally the zinc chelator is a hydroxamic acid. Implementation of these concepts by combining the principles of medicinal chemistry, x-ray crystallography, and molecular modeling has led to the discovery of unique classes of potent TACE inhibitors. Representative examples of TACE inhibitors are shown (compounds **48–55**). Systematic identification of new scaffolds and their modification afforded improved in vivo profiles, physical characteristics, and pharmacokinetic/pharmacodynamic properties. Introduction of novel P1′ groups and further functionalization have yielded enhanced potency and selectivity. Significant advances in inhibitor design and lead optimization have been realized. These have evolved with the development of structure–activity relationship (SAR) studies and as the knowledge of the structure features of the enzyme were refined. Nonetheless, direct translation of potent enzyme inhibition to cellular activity remains elusive. Musculoskeletal side effects are still an issue and the definition of the requisite enzyme inhibitory profile or combination is still awaits solution. From these efforts, two compounds have advanced to clinical studies, DPC333 (BMS-561392, **54**) (Bristol-Myers Squibb) and TMI-005 (**55**), (Wyeth). Numerous other molecules have proceeded to advanced preclinical lead status.[82]

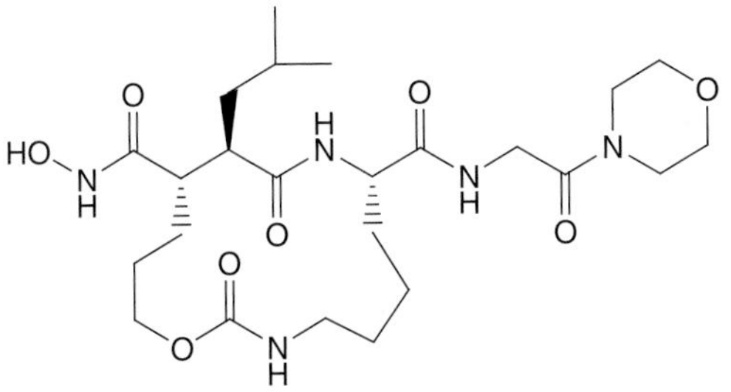

48 (Xue *et al.*)[82]

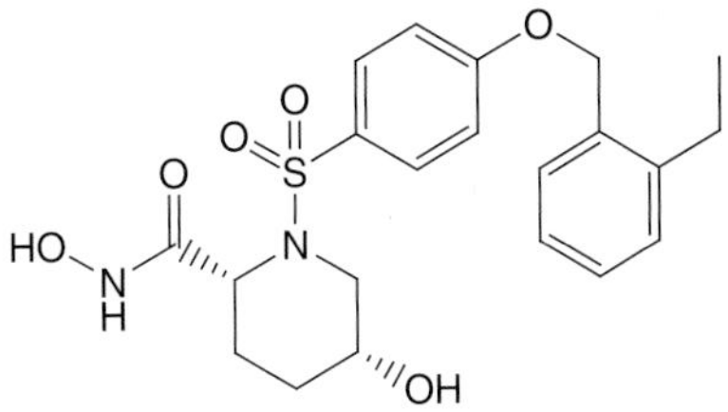

49 (Letavic *et al.*)[82]

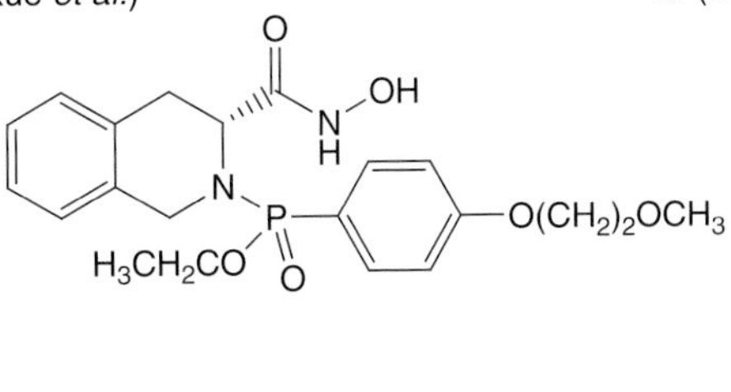

50 (Sawa *et al.*)[82]

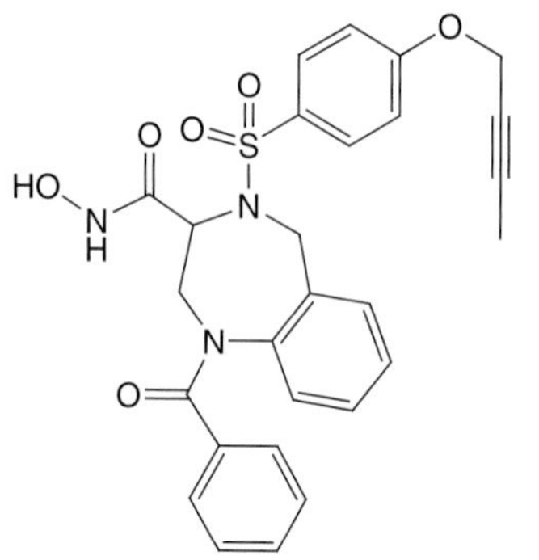

51 (Levin *et al.*)[82]

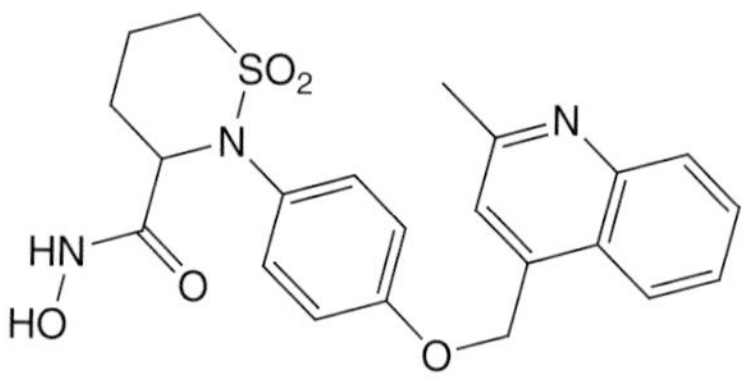

52 (Cherney *et al.*)[82]

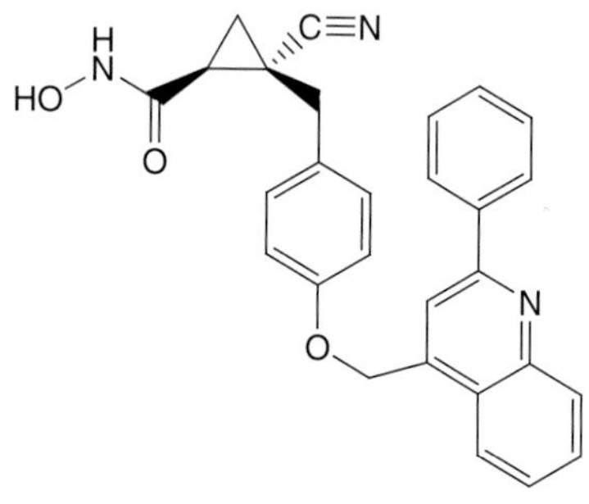

53 (Zhu *et al.*)[82]

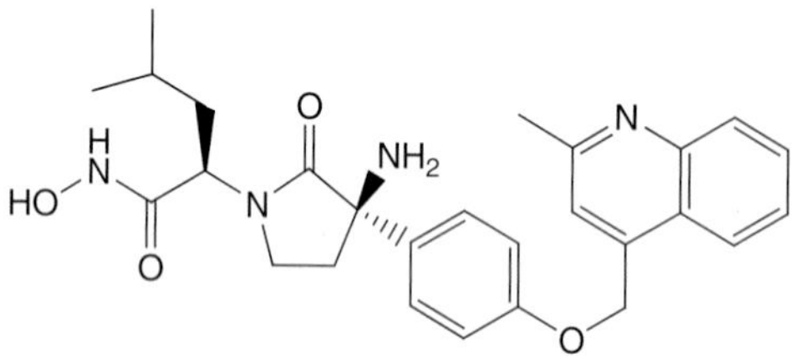

54 DPC-333 (BMS-561392; Liu *et al.*)[82]

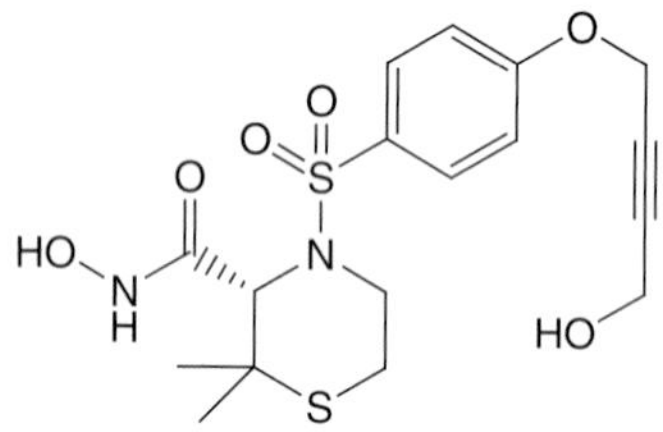

55 TMI-005 (Levin *et al.*)[82]

A recent concern has been raised that TACE inhibition leads to the build-up of membrane TNF on the surface of cells. Membrane TNF-α can bind and then activate cells expressing the TNF-α receptor on their cell surface. We now know that TACE is also involved in the liberation of TNF-α receptors from the cell surface by a process referred to as shedding. The implications of the combined build-up of membrane associated TNF-α and its cell surface receptor are poorly understood. Studies were conducted in vitro to show that membrane TNF-α is internalized then digested by the cell. What is not clear is the impact of membrane TNF-α after chronic treatment in vivo. Some preliminary reports[83] indicate that liver toxicity in dogs was observed after chronic exposure (52 weeks) with TACE inhibitors. Additional studies will need to be conducted to characterize these findings and how they relate to the use of TACE inhibitors in RA patients.

Interleukin-1β (IL1β) is another cyctokine that groups targeted for blocking.[84] Activated downstream of TNF-α, IL1β activates other cyctokines (e.g., IL6) and proteinases (e.g., MMPs). In particular, IL1β is very effective in degrading cartilage by activating proteinases (e.g., aggrecanase, MMPs) along with suppressing the biosynthesis of tissue inhibitor of metalloproteinases (TIMPs), an endogenous inhibitor of MMPs.[85] Thus IL1β becomes a target to block since it shifts the balance of cartilage homestasis to a stage of accelerated cartilage destruction.[86] In RA patients, elevated levels of the native IL1β receptor antagonist are observed (IL1ra). The levels of IL1ra appear to be inadequate to re-equilibrate joint homeostasis.[87] By increasing levels of IL1ra in the affected joint, the degree of joint damage should be minimized. This served as the basis to advance into the clinic anakinra, a recombinant methionyl version of the IL1β receptor antagonist that binds to the IL1β receptor. Anakinra demonstrated efficacy in animals and later in RA patients preventing symptoms (pain, edema) and slowing progressive joint destruction.[88] Currently approved for RA, anakinra offers an additional tool for treating RA.

With the clinical proof of concept, groups have been exploring small-molecule approaches to block IL1β production. Like TNF-α, IL1β is posttranslationally modified from a proform to the active cytokine. The proteinase involved is called IL1β converting enzyme (ICE-1).[89] Several groups have been exploring ICE inhibitors with the most advanced agent in phase II trials (Pralnacasan, **56**).[90]

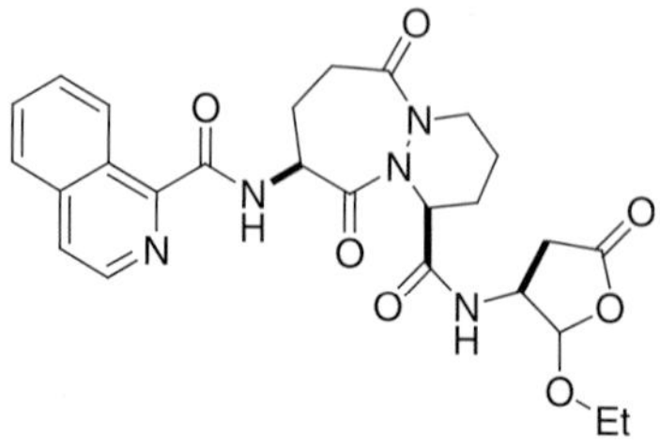

56 Pralnacasan

Approaches to block TNF-α and IL1β have also emerged that do not employ biologics or converting enzyme inhibitors. Several candidate targets are being investigated but two of the most intensely and recently studied targets are the kinases p38 and I-kappa-β-kinase (IKKβ). Both of these kinases are involved with the gene activation of multiple cytokines along with other proinflammatory genes. Some elegant kinase inhibitor design has resulted from the pursuit of these two kinase targets.

The role kinase p38 in regulating cytokine production was first established by the Lee *et al.*[91] This group originally observed a diaryl heterocyclic series of molecules that were able to regulate both TNF-α and IL1β (called CSAIDs). The biological mechanism was later found to be inhibition of p38 kinase. The compounds, exemplified by the clinical compound SB242235 (**57**), were very effective anti-inflammatory agents in animal studies but were stopped in clinical trials due to toxicity.[92] The demonstration of the potential for selective kinase inhibitors opened the door for extensive SAR on multiple structural classes of p38 inhibitor.[93] Inhibitors of kinases such as p38 traditionally target the ATP binding site. Since this region is fairly conserved across the kinases gene family, the identification of increasingly selective p38 inhibitors became the quest of several research groups. One interesting mode of inhibition arose from the discovery of unique binding modes of certain p38 inhibitors. The group at Boehringer-Ingelheim group led this advancement with the discovery of BIRB796 (**58**).[94] This p38 inhibitor bound partially in a allosteric pocket and induced a conformation of p38 that prevents ATP binding. In a phase I clinical trial, BIRB796 demonstrated good inhibition of lipopolysaccharide (LPS)-induced TNF-α production but the compound was stopped due to unacceptable liver function test (LFT) elevations at targeted doses.

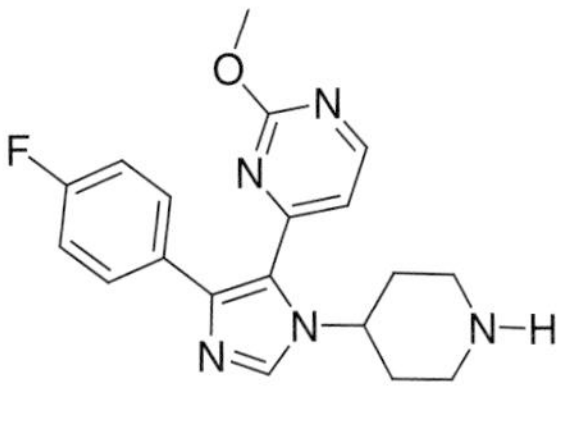

57 SB242235

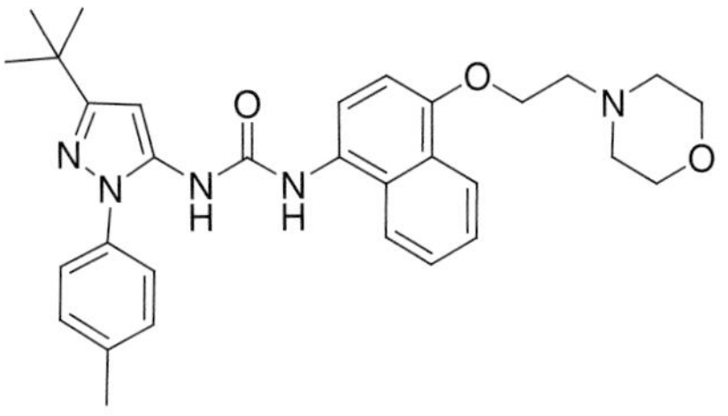

58 BIRB796

Several additional groups with p38 inhibitors are conducting clinical trials. A recent review[95] covers clinical data on many compounds, including AMG-548 (**59**). In single- and multidose trials, AMG-548 reached concentrations that were sufficient for nearly complete inhibition of inflammatory cytokine production upon ex vivo LPS challenge. However, the development of the compound was suspended due to elevated LFTs.

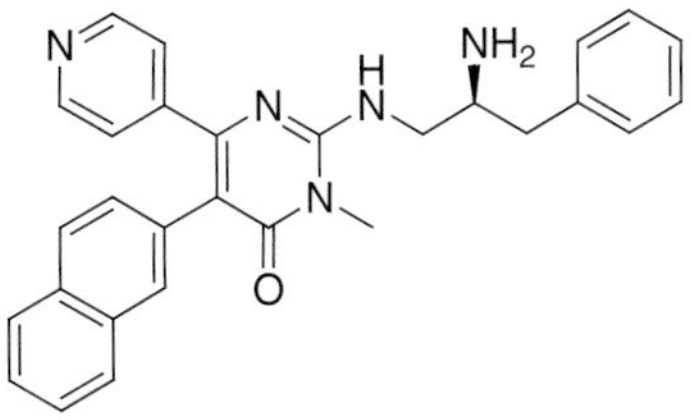

59 AMG-548

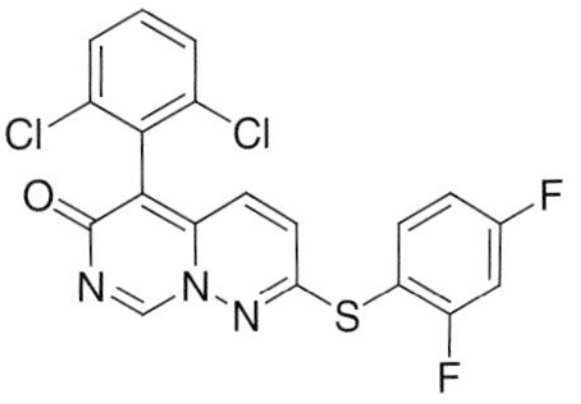

60 VX-745

Vertex designed VX-745 (**60**) based on crystallographic data and progressed it into phase II clinical trials. The compound was stopped however due to elevated LFTs in the trials as well as central nervous system side effects seen in preclinical studies.[96] A second-generation compound that does not penetrate the blood–brain barrier (VX-702) is reported to be in clinical development. Due to the appearance of adverse events in clinical trials with compounds from several structural classes, there is a growing concern about the long-term viability of p38 inhibitors in the treatment of chronic diseases. However, compounds continue to advance carefully into clinical trials. A compound from Scios (Scio-469) (the structure has not yet been published) is reported in phase II for RA.[97] This compound did not demonstrate dose-limiting toxicity in the phase I trials.

The second kinase that has attracted significant attention is IKKβ. This kinase phosphorylates I-κB and activates signals downstream of NFκB. Blocking the release of NFκB is expected to prevent the transcription of numerous proinflammatory cytokines, chemokines, adhesion molecules, and inducible effector enzymes. Additionally, NFκB regulates the transcription of antiapoptotic factors so IKK inhibitors are also expected to be of potential benefit in the treatment of various cancers. There has been much preclinical activity in this area but no reports of compounds that have entered clinical trials. This may be due to concerns over potential toxicities arising from the central role that IKK plays in cell death and development. The central role of IKKβ in regulating both proinflammatory responses as well as early development has been supported by genetic experiments in knockout mice.

Several structural classes have been shown to inhibit IKKβ. A recent review covers many structural classes.[98] These compounds have typical hydrogen bond donor–acceptor pairs that are predicted to bind to the hinge region of the ATP binding pocket in various modes. The quinazoline series from Signal was one of the first series to be reported as IKKβ selective inhibitors with an IC_{50} of 62 nM for the advanced candidate SPC-839 (**61**). SPC-839 also has reported activity in rat LPS challenge and paw edema models. Aventis and Millenium have had a joint program to discover and develop anti-inflammatory agents. Two series have emerged from this work exemplified by the beta-carboline series.[99] Both series are reported to have values of IC_{50} against IKKβ in the 5–200 nM range as well as activity in cellular assays of NFκB translation. PS-1145 (**62**) was reported to lower TNF-α production in LPS-challenged mice along with compound **63**. A number of companies have developed SAR around an apparent similar hit, the aminothiophenecarboxamide motif present in compound **64**.[100] The base compound is a fairly weak IKK inhibitor; however, modifications such as the incorporation of a urea group compound **65**, or the fusion of a second aromatic ring compound **66** preserves the hydrogen bonding motif and provides access to additional polar and nonpolar interactions producing more potent compounds.

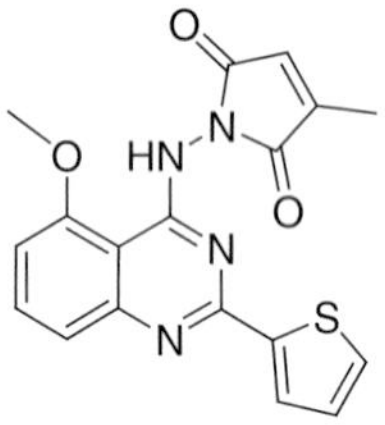

61 SPC-839 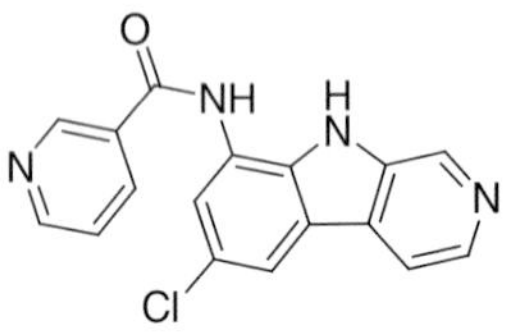 **62** PS-1145 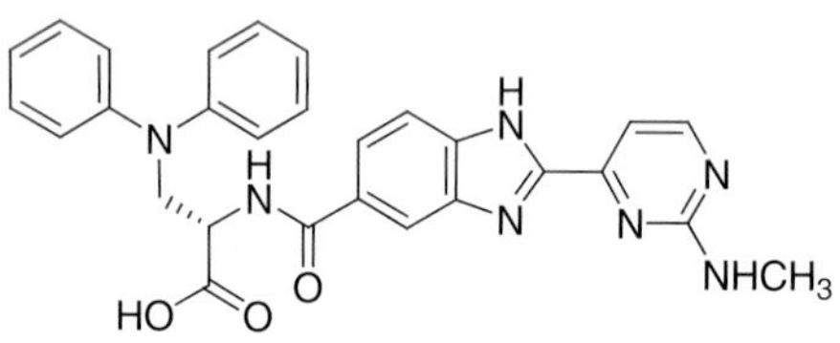 **63**

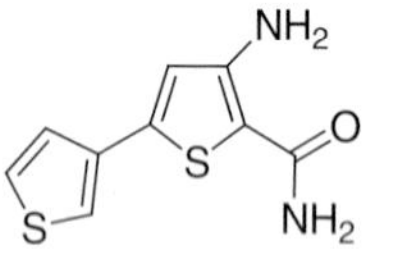

64 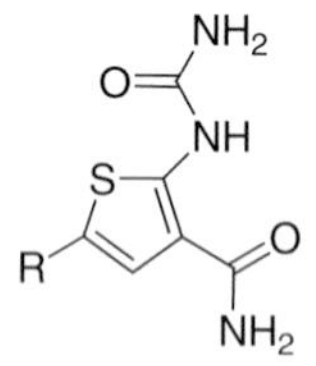 **65** 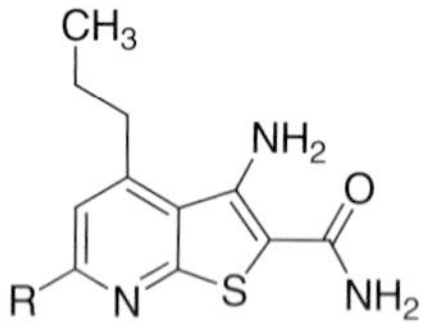 **66**

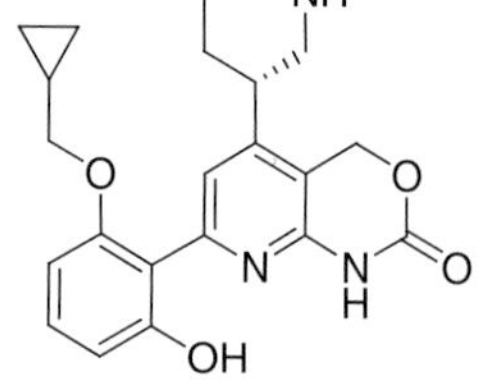

67 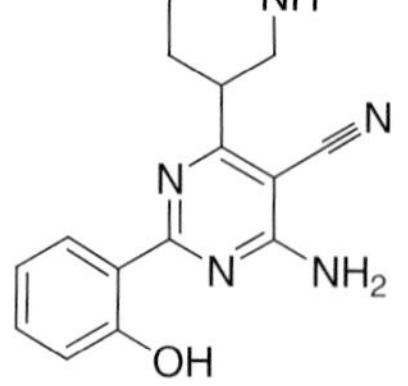 **68** 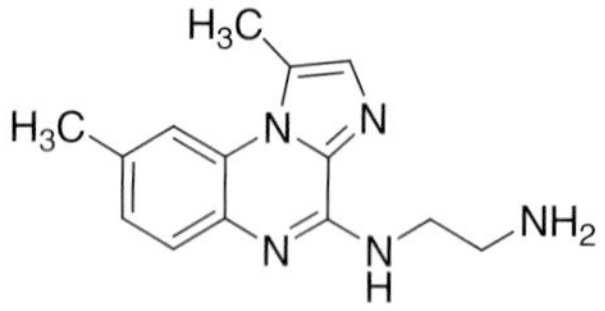 **69** BMS-345541

No crystal structure of IKKβ has been published which may have also hampered the progress in this field. However, the low nM inhibitor **67** shows additional means of filling more of the putative ATP binding pocket than the weaker but related compound **68**. Compound **67** has been reported to be effective in a rat airway inflammation model.[101]

BMS-345541 (**69**) is a unique compound among reported IKK inhibitors as it is an allosteric modulator of the enzyme. The compound has moderate potency in the enzyme assay (300 nM) but has shown activity in both murine arthritis and LPS challenge models.[102]

The biologics have provided clinical proof of concept (e.g., enbrel, remicade, humira, anakinra). Blocking the cytokines TNF-α and/or IL1β by small-molecule approaches remains the focal point of several discovery groups. The prospect of providing relief of symptoms while retarding joint space narrowing makes these approaches very exciting. Hopefully, these approaches will afford new medicines for arthritis patients.

7.29.4.4.2 Joint space narrowing therapies

Several groups have been seeking agents that protect joints from both cartilage and bone loss. Such agents may not offer relief from the symptoms of arthritis (e.g., pain) but may maintain or even improve joint vitality. There are several treatment options for arthritis patients but very few modalities to maintain joint integrity. Investigations into joint physiology in arthritis patients reveal an imbalance between joint destruction verses homeostasis. Typically, an overproduction of MMPs is a feature of both RA and OA. Some MMPs have been found to be involved in many tissues while others have been shown to be specific joint components like cartilage or bone. Research over the past 15 years has eliminated a number of MMPs from consideration as therapeutic targets. Two of the remaining and most attractive candidates for arthritis remain MMP-13 and aggrecanse.

The MMPs are members of a family of more than 20 zinc-requiring enzymes, regulated by cytokines and growth factors that are intimately involved in the normal maintenance of the extracellular matrix. Endogenous inhibitors of MMPs such as α_2-macroglobulin and TIMPS modulate the proteolytic activity locally in the tissue. Disruption of this subtle physiological balance between the MMPs and these endogenous inhibitors is associated with a number of pathological conditions including OA and RA, tumor metastasis and angiogenesis, corneal ulceration, multiple sclerosis, periodontal disease, and atherosclerosis. Hypotheses concerning the specific MMPs related disease target(s) have evolved. It is now widely believed that MMP-13 plays a key role in the connective tissue destruction in OA. The fundamental constituent of articular cartilage, type II collagen, is the primary substrate for MMP-13 and this enzyme is more efficient in the degradation of type II collagen than other MMPs. Also, increased concentration of MMP-13 is found in the synovial fluid of OA patients.

Inhibition of MMP-13 has emerged as a critical goal for the treatment of OA. Though modulation is possible at several biochemical sites, direct inhibition of enzyme action by small molecules that bind to the active site of the enzyme is a particularly attractive medicinal chemistry targets for therapeutic intervention. Comprehensive reviews of MMP and MMP inhibition have appeared.[103]

Early inhibitors in this class were peptide-based molecules with structural similarity to the protein substrate. These inhibitors contained a zinc-chelating group (hydroxamate, carboxylate, phosphate) and side chain substituent (P1′ group) for effective interaction with enzyme subsite (S1′ pocket) proximal to the scissile bond. Affinity for the S1′ subsite is integral for enzyme potency and selectivity against other MMPs and related enzymes. Marimastat (**70**)[104] is a well-known example of broad spectrum peptide-based MMPI. As the research evolved, a class of nonpeptide MMPIs was discovered, represented by sulfonamide CGS-27023A (**71**).[105] Compound **71** embraced the same functional features, a zinc chelator, a P1′ group, but with a nonpeptide scaffold which could afford substantial opportunities for new design strategies, improved physical and pharmaceutical properties and orally bioavailability – all limitations with the peptide inhibitors. As the field continued to mature, issues of selectivity, in vivo activity, and pharmacokinetic/pharmacodynamic hurdles needed to be addressed. The intricate three-dimensional interaction between enzyme and inhibitor were examined and exploited to expand the understanding of the enzyme–inhibitor interactions. Medicinal chemistry strategies utilizing structure-based design or pharmacophore-based design were executed in compound optimization with respect to potency and selectivity. Nonpeptide scaffolds were designed to enhanced interactions with the enzyme backbone and effective positioning of the chelating group and P1′ moiety for optimal interactions in the active site. This knowledge base coupled with medicinal chemistry instinct and experience, helped to define allowable, productive, and optimal sites of lead compound modification to improve physicochemical, pharmacological, pharmaceutical, and pharmacokinetic properties as well. Reviews of structural and computational aspects of the MMPs and MMPIs have appeared.[106]

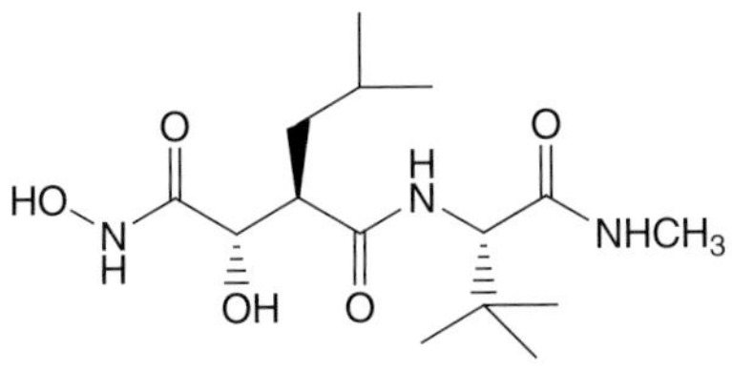

70 Marimastat[104]

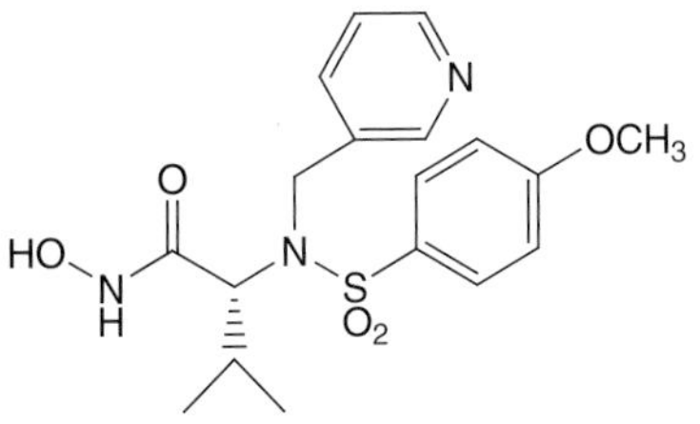

71 CGS-27023A[105]

Despite these successes in the identification of potent compounds, many MMPIs exhibit musculoskeletal syndrome (MSS: fibroplasia, tendonitis, etc.) as a dose-limiting side effect. Whether MSS is related to inhibition of specific MMPs, combinations, and/or related enzymes has not been determined unequivocally.

The design and synthesis of MMP-13 inhibitors remains an active and important area of research. Several compounds have advanced to the clinic for OA, oncology, or late stage preclinical studies. Representative classes of MMP-13 inhibitors are shown (see compounds **72–85**).[107–120]

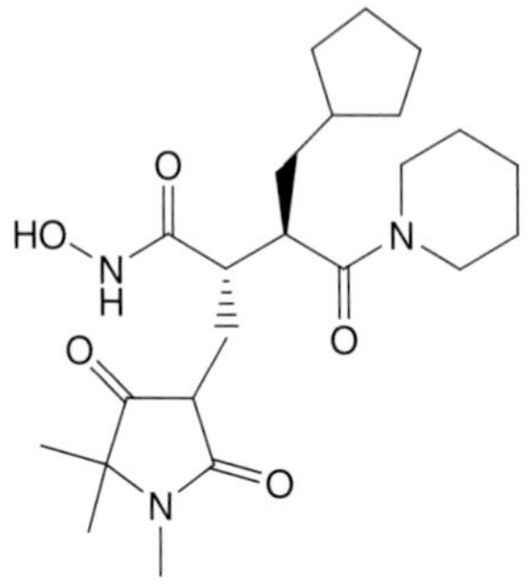

72 Trocade[107]

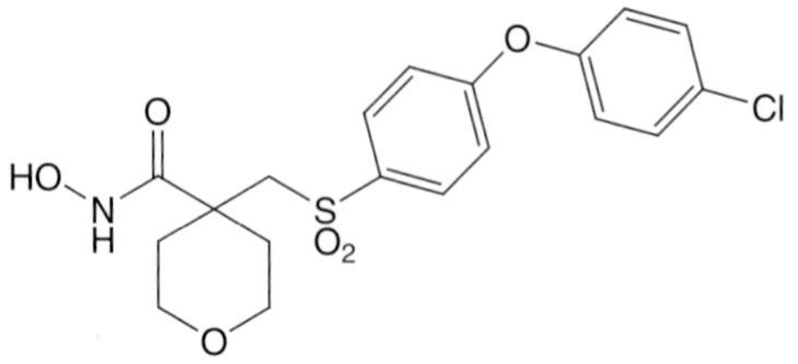

73 RS-130, 830[108]

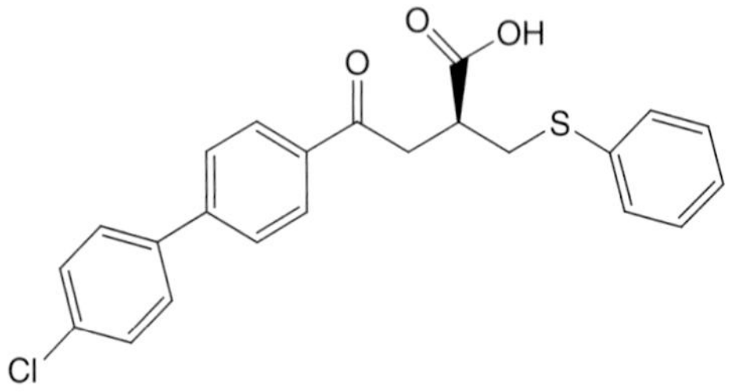

74 Bay-12-9566[109]

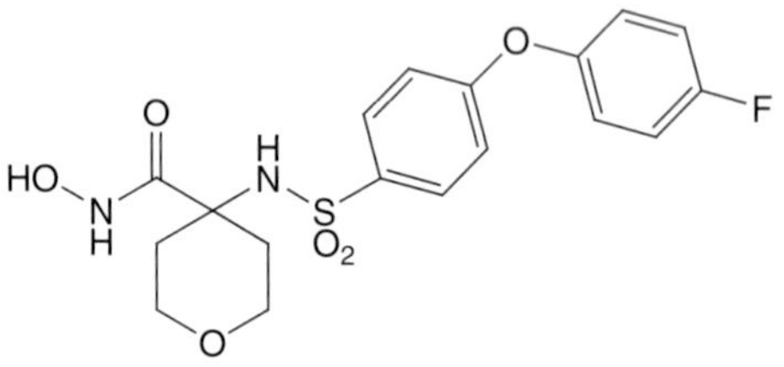

75 (Reiter *et al.*)[110]

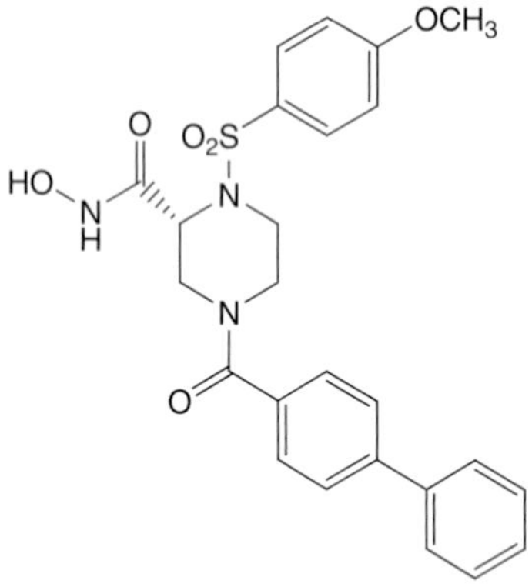

76 (Cheng *et al.*)[111]

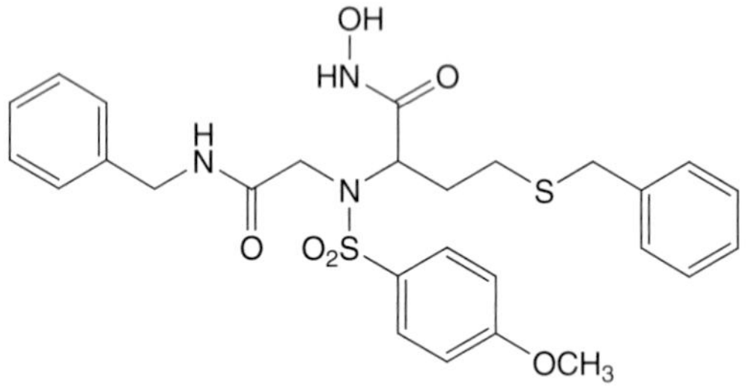

77 (Hanessian *et al.*)[112]

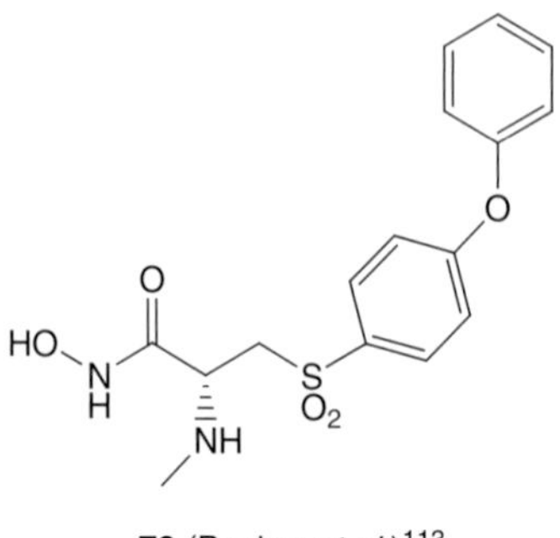

78 (Becker *et al.*)[113]

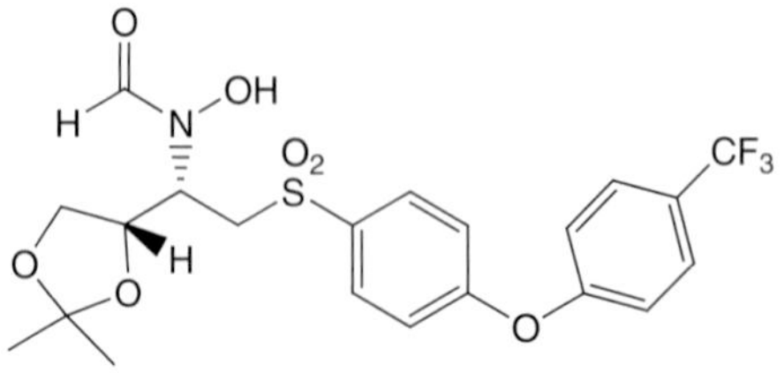

79 (Wada *et al.*)[114]

80 (O'Brien *et al.*)[115]

81 (Aranapakam *et al.*)[116]

82 (Picard *et al.*)[117]

83 (Barvian *et al.*)[118]

84 (Engel *et al.*)[119]

85 FR255031[120]

The other matrix component of cartilage is proteoglycan. This matrix protein is primarily digested by aggrecanase, a member of the ADAM family of MMPs (ADAM-TS-3, ADAM-TS-4).[121] Aggrecanase appears to be very specific for proteoglycans leaving fragments found in high concentration in the joints of OA patients. While structurally related to MMPs, these ADAM-TS family members contain a specific proteoglycan recognition motif that aligns the MMP machinery to recognize and digest a specific epitope found in proteoglycans. By blocking this enzyme, studies have shown that proteoglycans in cartilage are preserved in cartilage explants after IL1β treatment. Such an inhibitor would be a chondroprotective agent and may offer arthritis patients a disease-modifying and joint-preserving agent. Also, the recent studies with aggrecanase knockout animals have provided some preclinical proof of concept for this unique target.[122] Recent comprehensive reviews of aggrecanases and inhibitors have appeared.[123] Some selective inhibitors (outlined below) have been used as tools to study the role of aggrecanse in arthritis (see compounds **86–89**).[124–127]

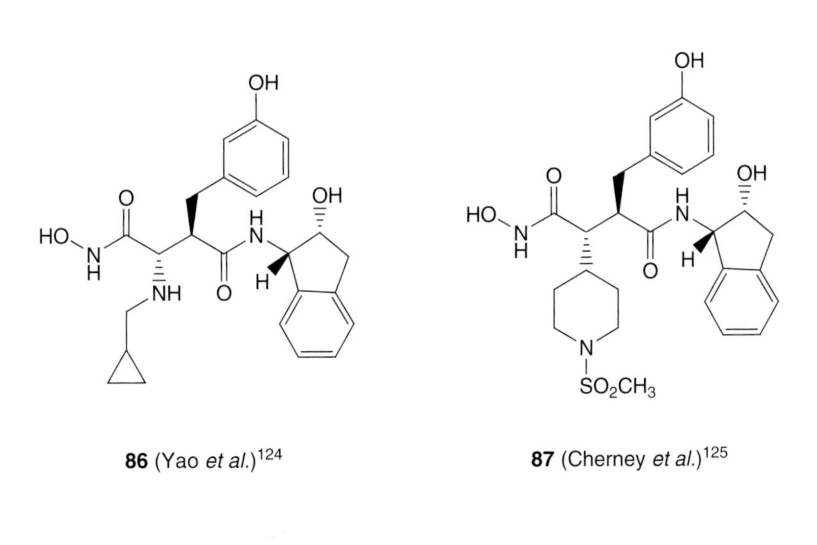

86 (Yao *et al.*)[124]

87 (Cherney *et al.*)[125]

88 (Noe *et al.*)[126]

89 (Xiang *et al.*)[127]

Despite the advances of selective MMP-13 or aggrecanase inhibitors, the paucity of validated animal models of OA remains a real challenge for this field of study. As noted above, cartilage loss in arthritis patients takes several years. In some cases, inflammation can dominate the joint environment releasing other factors that accelerate the joint destruction rendering a selective MMP-13 or aggrecanase inhibitor less effective. Taken together these technical hurdles highlight the complexities of finding suitable therapies to treat OA. Investigators are systematically investigating these new tools in old and new animal models of disease. The development of biomarkers will further aid the development in this area. These efforts should result in the generation of data that would help the scientific community reach a decision on the feasibility of such agents to treat joint destruction in arthritis patents.

7.29.5 Conclusions and Future Directions

Over the past 15 years a number of new treatment options have been provided for arthritis patients. More drugs for arthritis have been approved for arthritis in the last 15 years than the previous 15 years. Two existing therapeutic categories (NSAIDs, antiproliferatives) added new therapies with some significant advantages as opposed to modest improvements to known NSAIDs. With the discovery and FDA approval of COX-2 inhibitors clinicians have new clinical tools. While the future of these therapies remains a bit clouded today, the prospect of additional studies may help identify an appropriate treatment paradigm that may exploit the advantages of COX-2 inhibitors (if any) for patients. The second category in the 'pyramid of therapy' is steroids. New approaches are being pursued but have not yielded any new drugs. Several candidates are being studied preclinically and clinically. Further expansion of this target type (e.g., NHR) has afforded new therapy options from agents traditionally involved with women's health. The addition of a third arm drug, leflunomide, has provided much needed alternatives to MTX. Finally, two new categories are evolving that may be added to the 'pyramid of therapy.' One entails blocking cytokines and the introduction of four drug approvals for RA. Small-molecule approaches remain elusive but studies are under way to find the most appropriate alternative (e.g., TACE, ICE-1, p38, or IKKβ inhibitors). The second approach entails direct blockade of MMPs or ADAM-TS proteinases preserving cartilage in RA and OA. These last two categories truly constitute a movement from treating disease symptoms to impacting the disease progression and ultimately joint health. The anticytokine approaches offer real hope with clinically showing significant improvement in joint integrity. Expectations are that over time these cartilage protective agents from the MMP and ADAM-TS family of inhibitors may offer patients some disease altering therapy. Only time will tell if these new and exciting approaches will work in patients.

References

1. Janeway, C. A.; Travers, P.; Walport, M.; Capra, J. D. In *Immuno Biology: The Immune system in Health and Disease*; 4 ed.; Garland: New York, 1999.
2. Kirwan, J. R. *Arthrit. Rheum.* **2004**, *50*, 1–4.
3. Appelbloom, T. *Rheumatology* **2005**, *44*, 681–683; Appelboom, T.; De Boelpaepe, C.; Famaey, J. P.; Ehrich, H. *JAMA* **1981**, *245*, 483–486.
4. Rogers, J.; Watt, I.; Dieppe, P. *Ann. Rheum. Dis.* **1985**, *44*, 113–120.
5. Lajeunesse, D.; Reboul, P. *Curr. Opin. Rheumatol.* **2003**, *15*, 628–633; Blanco, F. J.; Guitian, R.; Vazquez-Martul, E.; de Toro, F. J.; Galdo, F. *Arthrit. Rheum.* **1998**, *41*, 284–289.
6. Symmons, D. P. M. *Best Prac. Res. Clin. Rheumatol.* **2002**, *16*, 707–722.
7. Kvien, T. K. *Pharmacoeconomics* **2004**, *22*, 1–12.
8. Felson, D. T.; Anderson, J. J.; Boers, M.; Bombardier, C.; Furst, D.; Goldsmith, C.; Katz, L. M.; Lightfoot, R., Jr.; Paulus, H.; Strand, V. *Arthrit. Rheum.* **1995**, *38*, 727–735.
9. Weinblatt, M. E.; Coblyn, J. S.; Fox, D. A.; Fraser, P. A.; Holdsworth, D. E.; Glass, D. N.; Trentham, D. E. *N. Engl. J. Med.* **1985**, *312*, 818–822.
10. *Guidance for Industry: Clinical Development Programs for Drugs, Devices, and Biological Products for the Treatment of Rheumatoid Arthritis (RA)*; Food and Drug Administration: Rockville, MD, 1999. http://www.fda.gov (accessed Aug 2006).
11. Magnussen, H.; Richter, K.; Taube, C. *Clin. Exp. Allergy* **1998**, *28*, 187–194.

12. Decramer, M.; Selroos, O. *Int. J. Clin. Pract.* **2005**, *59*, 385–398.
13. Kurtzke, J. F. *Acta Neurnol. Scand.* **1995**, *161*, 23–33.
14. Wingerchuk, D. M.; Noseworthy, J. H. *Neurology* **2002**, *58*, S40–S48; OWIMS Study Group. *Neurology* **1999**, *53*, 679–686.
15. Tebbe, B.; Orfanos, C. E. *Lupus* **1997**, *6*, 96–104.
16. Urowitz, M. B.; Feletar, M.; Bruce, I. N.; Ibanez, D.; Gladman, D. D. *J. Rheumatol.* **2005**, *32*, 1467–1472.
17. Bensen, W. G.; Adachi, J. D.; Tugwell, P. X. *J. Rheumatol.* **1990**, *17*, 987–999.
18. Rainsford, K. D. *Agent Actions* **1987**, *21*, 316–325; Wallace, J. L. *Can. J. Physiol. Pharm.* **1993**, *71*, 98–105.
19. Masuda, S.; Murakami, M.; Komiyama, K.; Ishihara, M.; Ishikawa, Y.; Ishii, T.; Kudo, I. *FEBS J.* **2005**, *272*, 655–672; Pruzanski, W. *J. Rheumatol.* **2005**, *32*, 399–402.
20. Uozumi, N.; Kume, K.; Nagase, T.; Nakatani, N.; Ishi, S.; Tashiro, F.; Komagata, Y.; Maki, K.; Ikuta, K.; Ouchi, Y. et al. *Nature* **1997**, *390*, 618–622; Bonventre, J.; Huang, Z.; Taheri, M. R.; O'Leary, E.; Li, E.; Moskowitz, M. A.; Sapirstein, A. *Nature* **1997**, *390*, 622–625; Hegen, M.; Sun, L.; Kume, K.; Goad, M. E.; Nickerson-Nutter, C. I.; Shimizu, T.; Clark, J. D. *J. Exp. Med.* **2003**, *197*, 1297–1302.
21. Davidson, F. F.; Lister, M. D.; Dennis, E. A. *J. Biol. Chem.* **1990**, *265*, 5602–5609; Yang, H. C.; Mosior, M.; Johnson, C. A.; Chen, Y.; Dennis, E. A. *Anal. Biochem.* **1999**, *269*, 278–288; Cajal, Y.; Berg, O. G.; Jain, M. K. *Biochemistry* **2004**, *43*, 9256–9264; Berg, O. G.; Yu, B. Z.; Chang, C.; Koehler, K. A.; Jain, M. K. *Biochemistry* **2004**, *43*, 7999–8013; Leslie, C. C.; Gelb, M. H. *Methods Mol. Biol.* **2004**, *284*, 229–242.
22. Kokotos, G.; Six, D. A.; Loukas, V.; Smith, T.; Constantinou-Kokotous, V.; Hadjipavlou-Litina, D.; Kotsovolou, S.; Chiou, A.; Beltzner, C. C.; Dennis, E. A. *J. Med. Chem.* **2004**, *47*, 3615–3628.
23. Jacobson, P. B.; Marshall, L. A.; Sung, A.; Jacobs, R. S. *Biochem. Pharmacol.* **1990**, *39*, 1557–1564.
24. Pastor, P. G.; De Rosa, S.; De Giulio, A.; Paya, M.; Alcaraz, M. J. *Br. J. Pharmacol.* **1999**, *126*, 301–311.
25. Bruke, J. R.; Davern, L. B.; Stanley, P. L.; Gregor, K. R.; Banville, J.; Remillard, R.; Russell, J. W.; Brassil, P. J.; Witmer, M. R.; Johnson, G. et al. *J. Pharm. Exp. Therap.* **2001**, *298*, 376–385; McKew, J. C.; Foley, M. A.; Thakker, P.; Behnke, M. L.; Lovering, F. E.; Sum, F. W.; Tam, S.; Wu, K.; Shen, W. H.; Zhang, W. et al. *J. Med. Chem.* **2006**, *49*, 135–158.
26. Bradley, J. D.; Dmitrienko, A. A.; Kivitz, A. J.; Gluck, O. S.; Weaver, A. L.; Wiesenhutter, C.; Myers, S. L.; Sides, G. D. *J. Rheumatol.* **2005**, *32*, 417–423; Snyder, D. W.; Bach, N. J.; Dillard, R. D.; Draheim, D.; Carlsom, D. G.; Fox, N.; Roehm, N. W.; Armstrong, C. T.; Chang, C. H.; Hartley, L. W. et al. *J. Exp. Therap.* **1999**, *288*, 1117–1124.
27. Laine, L. *Gastroenterology* **2001**, *120*, 594–606.
28. Graham, D. Y.; Agrawal, N. M.; Campbell, D. R.; Haber, M. M.; Collis, C.; Luskasik, N. L.; Huang, B.; *Arch. Intern. Med.* **2002**, 169–175; Chong, Y. S.; Su, L. L.; Arulkumaran, S. *Obstet. Gynecol. Surv.* **2004**, *59*, 128–140.
29. Wolfe, M. M.; Lichtenstein, D. R.; Singh, G. *N. Engl. J. Med.* **1999**, *340*, 1888–1899.
30. Chan, F. K. L.; Chung, S. C. S.; Suen, B. Y.; Lee, Y. T.; Leung, W. K.; Leung, V. K. S.; Wu, J. C. Y.; Lau, J. Y. W.; Hui, Y.; Lai, M. S. et al. *N. Engl. J. Med.* **2001**, *344*, 967–973.
31. Xie, W. L.; Chipman, J. G.; Robertson, D. L.; Erikson, R. L.; Simmons, D. L. *Proc. Natl. Acad. Sci. USA* **1991**, *88*, 2692–2696; O'Banion, M. K.; Sadowski, H. B.; Winn, V.; Young, D. A. *J. Biol. Chem.* **1991**, *266*, 23261–23267; Kujubu, D. A.; Fletcher, B. S.; Varnum, B. C.; Lim, R. W.; Herschman, H. R. *J. Biol. Chem.* **1991**, *266*, 12866–12872.
32. Gans, K. R.; Galbraith, W.; Roman, R. J.; Haber, S. B.; Kerr, J. S.; Schmidt, W. K.; Smith, C.; Hewes, W. E.; Ackerman, N. R. *J. Pharmacol. Exp. Therapeut.* **1990**, *254*, 180–187.
33. Reitz, D. B.; Seibert, K. *Ann. Rep. Med. Chem.* **1995**, *30*, 179–188.
34. Copeland, R. A.; Williams, J. M.; Giannaras, J.; Nurnberg, S.; Covington, M.; Pinto, D.; Pick, S.; Trzaskos, J. *Proc. Natl. Acad. Sci. USA* **1994**, *91*, 11202–11206; Lanzo, C. A.; Sutin, J.; Rowlinson, S.; Talley, J.; Marnett, L. *J. Biochemistry* **2000**, *39*, 6228–6234.
35. Li, J.; Anderson, G. D.; Burton, E. G.; Cogburn, J. N.; Collins, J. T.; Garland, D. J.; Gregory, S. A.; Huang, H. C.; Isakson, P. C.; Koboldt, C. M. et al. *J. Med. Chem.* **1995**, *38*, 4570–4578; Seibert, K.; Zhang, Y.; Leahy, K.; Hauser, S.; Masferrer, J.; Perkins, W.; Lee, L.; Isakson, P. C. *Proc. Natl. Acad. Sci. USA* **1994**, *91*, 12013–12017.
36. Li, J. J.; Norton, M. B.; Reinhard, E. J.; Anderson, G. D.; Gregory, S. A.; Koboldt, C. M.; Masferrer, J. L.; Perkins, W. E.; Seibert, K.; Zhang, Y. et al. *J. Med. Chem.* **1996**, *39*, 1846–1856; Khanna, I. K.; Weier, R. M.; Yu, Y.; Collins, P. W.; Miyashiro, J. M.; Koboldt, C. M.; Veenhuizen, A. W.; Currie, J. L.; Seibert, K.; Isakson, P. C. *J. Med. Chem.* **1997**, *40*, 1619–1633; Khanna, I. K.; Weier, R. M.; Yu, Y.; Xu, X.; Koszyk, F. J.; Collins, P. W.; Koboldt, C. M.; Veenhuizen, A. W.; Perkins, W. E.; Casler, J. et al. *J. Med. Chem.* **1997**, *40*, 1634–1647; Pinto, D. J.; Batt, D. G.; Pitts, W. J.; Petraitis, J. J.; Orwat, M. J.; Wang, S.; Jetter, J. W.; Sherk, S. R.; Houghton, G.; Copeland, R. A. et al. *Bioorg. Med. Chem. Lett.* **1999**, *9*, 191–924.
37. Penning, T. D.; Talley, J. T.; Bertenshaw, S. R.; Carter, J. S.; Collins, P. W.; Docter, S.; Graneto, M. J.; Lee, L. F.; Malecha, J. W.; Miyashiro, J. M. et al. *J. Med. Chem.* **1997**, *40*, 1347–1365.
38. Prasit, P.; Wang, Z.; Brideau, C.; Chan, C. C.; Charleson, S.; Cromlish, W.; Ethier, D.; Evans, J. F.; Ford-Hutchinson, A. W.; Gauthier, J. Y. et al. *Bioorg. Med. Chem. Lett.* **1999**, *9*, 1773–1778.
39. Talley, J. J.; Brown, D. L.; Carter, J. S.; Graneto, M. J.; Koboldt, C. M.; Masferrer, J. L.; Perkins, W. E.; Rogers, R. S.; Shaffer, A. F.; Zhang, Y. Y. et al. *J. Med. Chem.* **2000**, *43*, 775–777.
40. Riendeau, D.; Percival, M. D.; Brideau, C.; Dube, D.; Ethier, D.; Falgueyret, J. P.; Friesen, R. W.; Gordon, R.; Greig, G.; Guay, J. et al. *J. Pharmacol. Exp. Ther.* **2001**, *296*, 558–566.
41. Esser, R.; Berry, J. C.; Du, Z.; Dawson-King, J.; Fox, A.; Fujimoto, R. A.; Haston, W.; Kimble, E.; Koehler, J.; Quadros, J. et al. *Br. J. Pharmacol.* **2005**, *144*, 538–550.
42. Ogino, K.; Hatanaka, K.; Kawamura, M.; Ohno, T.; Harada, Y. *Pharmacology* **2000**, *61*, 244–250.
43. Wong, D.; Wang, M.; Cheng, Y.; Fitzgerald, G. A. *Curr. Opin. Pharmacol.* **2005**, *5*, 204–210; Mukherjee, D.; Nissen, S. E.; Topol, E. J. *JAMA* **2001**, *286*, 954–959; Solomon, D. H.; Schneeweiss, S.; Glynn, R. L.; Kiyota, Y.; Levin, R.; Mogun, H.; Avord, J. *Circulation* **2004**, *44*, 140–145.
44. Borgeat, P.; Samuelson, B. *J. Biol. Chem.* **1979**, *254*, 2643–2646.
45. Penrose, J. F.; Austen, K. F.; Larn, B. K. Leukotrienes: Biosynthetic Pathways, Release and Receptor-Mediated Actions with Relevance to Disease States. In *Inflammation: Basic Principles and Clinical Correlates*; Gallin, J. L., Snyderman, R., Eds.; Lippincott, Williams & Wilkins: Philadelphia, PA, 1999, pp 361–372.
46. Brooks, C. D. W.; Stewart, A. O.; Basha, A.; Bhatia, P.; Ratajczyk, J. D.; Martin, J. G.; Craig, R. A.; Kolasa, T.; Bouska, J. B.; Lanni, C. et al. *J. Med. Chem.* **1995**, *38*, 4768–4775.
47. Noonan, M. J.; Chervinsky, P.; Brandon, M.; Zhang, J.; Kundu, S.; McBurney, J.; Reiss, T. F. *Eur. Respir. J.* **1998**, *11*, 1232–1239.
48. Turner, C. R.; Smith, W. B.; Andresen, C. J.; Swindell, A. C.; Watson, J. W. *Pulmon. Pharmacol.* **1994**, 7, 49–58.
49. Nabe, T.; Kohno, S.; Tanpo, T.; Saeki, Y.; Yarnamura, H.; Horiba, M.; Ohata, K. *Prostagland. Leukotri. Ess. Fatty Acids* **1994**, *51*, 163–171.

50. Harris, R. R.; Carter, G. W.; Bell, R. L.; Moore, J. L.; Brooks, D. W. *Int. J. Immunopharmacol.* **1995**, *17*, 147–156.
51. Celotti, F.; Durand, T. *Prostaglandi. Lipid Medi.* **2003**, *71*, 147–162; Parente, L. *J. Rheumatology* **2002**, *28*, 2375–2382.
52. Alvaro-Gracia, J. M. *Rheumatology* **2004**, *43*, 121–125; Rotondo, S.; Krauze-Brzosko, K.; Evangelista, V.; Cerletti, C. *Eur. J. Pharm.* **2004**, *488*, 79–83; Charlier, C.; Michaux, C. *Eur. J. Med. Chem.* **2003**, *38*, 645–659.
53. Schimmer, B. P.; Parker, K. L. In *Goodman & Gilman's The Pharmacological Basis of Therapeutics*; Hardman, J. G., Limbird, L. E., Eds.; McGraw-Hill: New York, 2001, pp 1649–1678.
54. Mangelsdorf, D. J.; Thummel, C.; Beato, M.; Herrlich, P.; Schultz, G.; Umesono, K.; Blumberg, B.; Kastner, P.; Mark, M.; Chambon, P. et al. *Cell* **1995**, *83*, 835–839; McKay, L. I.; Cidlowski, J. A. *Endocrinol. Rev.* **1999**, *20*, 435–459.
55. Jaroch, S.; Lehmann, M.; Schmees, N.; Buchmann, B.; Rehwinkel, H.; Droescher, P.; Skuballa, W.; Krolikiewicz, K.; Hennekes, H.; Schaecke, H. et al. World Patent Application WO2002/10143, 2002.
56. Bekkali, Y.; Betageri, R.; Gilmore, T.; Cardozo, M.; Kirrane, T.; Kuzmich, D.; Proudfoot, J.; Takahashi, H.; Thomson, D.; Wang, J. et al. World Patent Application WO2003/082280, 2003.
57. Barker, M.; Demaine, D.; House, D.; Inglis, G.; Johnston, M.; Jones, H.; Macdonald, S.; Mclay, I.; Shananhan, S.; Skone, P. et al. World Patent Application WO2004/071389, 2004.
58. Biggadike, K.; Morton, G.; Needham, D. World Patent Application WO2003/072592, 2003.
59. Scanlan, T. S.; Shah, N. World Patent Application WO203/061651, 2003.
60. Buckbinder, L. US Patent Application 2003/0224349, 2003.
61. Lang, T. J. *Clin. Immunol.* **2004**, *113*, 224–230; Salem, M. L. *Curr. Drug Targets Inflamm. Allergy* **2004**, *3*, 97–104.
62. Malamas, M. S.; Manas, E. S.; McDevitt, R. E.; Gunawan, I.; Zhang, X. B.; Collini, M. D.; Miler, C. P.; Dinh, T.; Henderson, R. A.; Keith, J. C., Jr. et al. *J. Med. Chem.* **2004**, *47*, 5021–5040; Mewshaw, R. E.; Edsall, R. J.; Yang, C.; Manas, E. S.; Xu, Z. B.; Henderson, R. A.; Keith, J. C.; Harris, H. A. *J. Med. Chem.* **2005**, *48*, 3953–3979; Harris, H.; Albert, L. M.; Leathurby, Y.; Malamas, M. S.; Mewshaw, R. E.; Miller, C. P.; Kharode, Y. P.; Marzolf, J.; Komm, B. S.; Winneker, R. C. et al. *Endocrinology* **2003**, *144*, 4241–4249; Hegele-Hartung, C.; Siebel, P.; Peters, O.; Muller, G.; Hillisch, A.; Walter, A.; Kraetzschmar, J.; Fritzemeier, K-H. *Proc. Natl. Acad. Sci. USA* **2004**, *101*, 5129–5134.
63. Steffan, R. J.; Matelan, E.; Ashwell, M. A.; Moore, W. J.; Solvibile, W. R.; Trybulski, E.; Chadwick, C. C.; Chippari, S.; Kenney, T.; Eckert, A. et al. *J. Med. Chem.* **2004**, *47*, 6435–6438; Chadwick, C. C.; Chippari, S.; Matelan, E.; Borges-Marcucci, L.; Eckert, A. M.; Keith, J. C., Jr.; Albert, L. M.; Leathurby, Y.; Harris, H. A.; Bhat, R. A. et al. *Proc. Natl. Acad. Sci. USA* **2005**, *102*, 2543–2548.
64. O'Dell, J. R.; Haire, C. E.; Erikson, N.; Drymalski, W.; Palmer, W.; Eckhoff, P. J.; Garwood, V.; Maloley, P.; Klassen, L. W.; Wees, S. et al. *N. Engl. J. Med.* **1996**, *334*, 1287–1291; Weinblatt, M. E.; Kremer, J. M.; Coblyn, J. S.; Maier, A. L.; Helfgott, S. M.; Morrell, M.; Byrne, V. M.; Kaymakcian, M. V.; Strand, V. *Arthrit. Rheum.* **1999**, *42*, 1322–1328.
65. Van Ede, A. E.; Laan, R. F.; Blom, H. J.; de Abreu, R. A.; van de Putte, L. B. *Sem. Arthrit. Rheum.* **1998**, *27*, 277–292.
66. Pankiewicz, C. *Exert. Opin. Ther. Patents* **1999**, *9*, 55–65; Batt, D.G., *Expert Opin. Ther. Patents* **1999**, 41–54.
67. Martin, K.; Bentaberry, F.; Dumoulin, C.; Dehais, J.; Haramburu, F.; Begaud, B.; Schaeverbeke, T. *Clin. Exp. Rheumatol.* **2005**, *23*, 80–84.
68. Cherwinski, H. M.; Byars, N.; Ballaron, S. J.; nakano, G. M.; Young, J. M.; Ransom, J. T. *Inflamm. Res.* **1995**, *44*, 317–322; Davis, J. P.; Copeland, R. A.; Pitts, W. J.; Magolda, R. L. *Biochemistry* **1996**, *35*, 1270–1273; Papageorgiou, C.; Zurini, M.; Weber, H. P.; Borer, X. *Bioorg. Chem.* **1997**, *25*, 233–238.
69. Elliott, M. J.; Feldmann, M.; Maini, R. N. *Int. J. Immunopharmacol.* **1995**, *17*, 141–145; Feldmann, M.; Brennan, F. M.; Williams, R. O.; Woody, J. N.; Maini, R. N. *Best Pract. Res. Clin. Rheumatol.* **2004**, *18*, 59–80; Vilcek, J.; Feldmann, M. *Trends Pharmacol. Sci.* **2004**, *25*, 201–209.
70. Braun, J.; Sieper, J. *Exp. Opin. Biol. Ther.* **2003**, *3*, 141–168.
71. Furst, D. E.; Schiff, M. H.; Fleischmann, R. M.; Strand, V.; Birbara, C. A. *J. Rheumatol.* **2003**, *30*, 2563–2571.
72. Baumgartner, S. W.; Fleischmann, R. M.; Moreland, L. W.; Schiff, M. H.; Markenson, J.; Whitmore, J. B. *J. Rheumatol.* **2004**, *31*, 1532–1537.
73. Black, R. A. *J. Biochem. Cell Biol.* **2002**, *34*, 1–5; Moss, M.; Becherer, J. D.; Milla, M.; Pahel, G.; Lambert, M.; Andrews, R.; Frye, S.; Haffner, C.; Cowan, D.; Maloney, P. et al. In *Metalloproteinases as Targets for Anti-Inflammatory drugs*; Bottomley, K. M., Bradshaw, D., Nixon, J. S., Eds.; Birkhauser Verlag: Basel, Germany, 1999, pp 187–203; Killar, L.; White, J.; Black, R.; Peschon, J. *Ann. NY Acad. Sci.* **1999**, *878*, 442–452; Moss, M. L.; White, J. M.; Lambert, M. H.; Andrews, R. C. *Drug Disc. Today* **2001**, *6*, 417–426; Black, R. A.; White, J. M. *Curr. Opin. Cell. Biol.* **1998**, *10*, 654–659.
74. Newton, R. C.; Decicco, C. P. *J. Med. Chem.* **1999**, *42*, 2295–2314; Moreland, L. M. *J. Rheumatol.* **1999**, *26*, 7–15; Palladino, M. A.; Bahjat, F. R.; Theodorakis, E. A.; Moldawer, L. L. *Nat. Rev. Drug Disc.* **2003**, *2*, 736–746; Reimold, A. M. *Curr. Drug Targets Inflamm. Allergy* **2002**, *1*, 377–392; Wagner, G.; Laufer, S. *Med. Res. Rev.* **2006**, *26*, 1–62.
75. Newton, R. C.; Solomon, K. A.; Covington, M. B.; Decicco, C. P.; Haley, P. J.; Friedman, S. M.; Vaddi, K. *Ann. Rheum. Dis.* **2001**, *60*, iii25–iii32.
76. Skotnicki, J. S.; Levin, J. I. *Ann. Rep. Med. Chem.* **2003**, *38*, 153–162; Nelson, F. C.; Zask, A. *Expert Opin. Investig. Drugs* **1999**, *8*, 383–392; Le, G. T.; Abbenante, G. *Curr. Med. Chem.* **2005**, *12*, 2963–2977.
77. Black, R.; Rauch, C.; Kozlosky, C.; Peschon, J.; Slack, J.; Wolfson, M.; Castner, B.; Stocking, K.; Reddy, P.; Srivivasan, S. et al. *Nature* **1997**, *385*, 729–733.
78. Moss, M.; Jin, S.; Milla, M.; Burkhart, W.; Carter, L.; Chen, W.; Clay, W.; Didsbury, J.; Hassler, D.; Hoffman, C. et al. *Nature* **1997**, *385*, 733–736.
79. Mohler, K.; Sleath, P.; Fitzner, J.; Cerretti, D.; Alderson, M.; Kewar, S.; Torrance, D.; Otten-Evans, C.; Greenstreet, T.; Weerawarna, K. et al. *Nature* **1994**, *370*, 218–220; McGeehan, G.; Becherer, D.; Bast, R.; Boyer, C.; Champion, B.; Connolly, K.; Conway, J.; Furdon, P.; Karp, S.; Kidao, S. et al. *Nature* **1994**, *370*, 558–561; Gearing, A.; Beckett, P.; Christodoulou, M.; Churchill, M.; Clemments, J.; Davidson, A.; Drummond, A.; Galloway, W.; Gilbert, R.; Gordon, J. et al. *Nature* **1994**, *370*, 555–557.
80. Maskos, K.; Fernandez-Catalan, C.; Huber, R.; Bourenkov, G. P.; Bartunik, H.; Ellestad, G. A.; Reddy, P.; Wolfson, M. F.; Rauch, C. T.; Castner, B. J. et al. *Proc. Natl. Acad. Sci. USA* **1998**, *95*, 3408–3412.
81. Wasserman, Z. R.; Duan, J. J.; Voss, M. E.; Xue, C. B.; Cherney, R. J.; Nelson, D. J.; Hardman, K. D.; Decicco, C. P. *Chem. Biol.* **2003**, *10*, 215–223; Rush, T. S.; Powers, R. *Curr. Top. Med. Chem.* **2004**, *4*, 1311–1327; Solomon, A.; Rosenblum, G.; Gonzalez, P. E.; Leonard, J. D.; Mobashery, S.; Milla, M. E.; Sagi, I. *J. Biol. Chem.* **2004**, *279*, 31646–31654; Lukacova, V.; Zhang, Y.; Kroll, D. M.; Raha, S.; Comez, D.; Balaz, S. *J. Med. Chem.* **2005**, *48*, 2361–2370.
82. Fujisawa, T.; Igeta, K.; Odake, S.; Morita, Y.; Yasuda, J.; Morikawa, T. *Bioorg. Med. Chem.* **2002**, *10*, 2569–2581; Kottirsch, G.; Koch, G.; Feifel, R.; Neumann, U. *J. Med. Chem.* **2002**, *45*, 2289–2293; Trifilieff, A.; Walker, C.; Keller, T.; Kottirsch, G.; Neumann, U. *Br. J. Pharmacol.* **2002**, *135*, 1655–1664; Beck, G.; Bottomley, G.; Bradshaw, D.; Brewster, M.; Broadhurst, M.; Devos, R.; Hill, C.; Johnson, W.; Kim, H. J.;

Kirtland, S. et al. *J. Pharmacol. Exp. Ther.* **2002**, *302*, 390–396; Musso, D.; Moss, M.; Anderson, M.; Andrews, R.; Austin, R.; Beaudet, E.; Becherer, D.; Bubacz, D.; Bickett, M.; Chan, J. et al. *Bioorg. Med. Chem. Lett.* **2001**, *11*, 2147–2151; Conway, J. G.; Andrews, R. C.; Beaudet, B.; Bickett, D. M.; Boncek, V.; Brodie, T. A.; Clark, R. L.; Crumrine, R. C.; Leenitzer, M. A.; McDougald, D. L. et al. *J. Pharmacol. Exp. Ther.* **2001**, *298*, 900–908; Xue, C.; Voss, M.; Nelson, D.; Duan, J.; Cherney, R.; Jacobson, I.; He, X.; Roderick, J.; Chen, L. et al. *J. Med. Chem.* **2001**, *44*, 2636–2660; Xue, C.; He, X.; Corbett, R.; Roderick, J.; Wasserman, Z.; Liu, R.; Jaffee, B.; Covington, M.; Qian, M.; Trzaskos, J. et al. *J. Med. Chem.* **2001**, *44*, 3351–3354; Holms, J.; Mast, K.; Marcotte, P.; Elmore, I.; Li, J.; Pease, L.; Glaser, K.; Morgan, D.; Michaelides, M.; Davidsen, S. *Bioorg. Med. Chem. Lett.* **2001**, *11*, 2907–2910; Letavic, M. A.; Axt, M. Z.; Barberia, J. T.; Carty, T. J.; Danley, D. E.; Geoghegan, K. F.; Halim, N. S.; Hoth, L. R.; Kamath, A. V.; Laird, E. R. et al. *Bioorg. Med. Chem. Lett.* **2002**, *12*, 1387–1390; Sawa, M.; Kiyio, T.; Kurokawa, K.; Kumihara, H.; Yamamoto, M.; Miyasaka, T.; Ito, Y.; Hirayama, R.; Inoue, T.; Kirii, Y. et al. *J. Med. Chem.* **2002**, *45*, 919–929; Levin, J. I.; Nelson, F. C.; Delos Santos, E.; Du, M. T.; MacEwan, G.; Chen, J. C.; Ayral-Kaloustian, S.; Xu, J.; Jin, G.; Cummons, T. et al. *Bioorg. Med. Chem. Lett.* **2002**, *14*, 4147–4151; Levin, J. I.; Chen, J. M.; Laakso, L.; Du, M; Schmid, J.; Xu, W.; Cummons, T.; Xu, J.; Zhang, Y.; Jin, G. et al. *Bioorg. Med. Chem. Lett.* **2006**, *16*, 1605–1609; Hegen, M.; Zhang, Y.; Levin, J.; Xu, J.; Cummons, T.; Harding, K.; Albert, L.; Leathurby, Y.; Sheppard, B. J.; Leach, M. W. et al. *Ann. Rheum. Dis.* **2005**, *64*, 154; Zhang, Y.; Xu, J.; Udata, C.; McDevitt, J.; Cummons, T.; Sun, L.; Zhu, Y.; Li, G.; Rao, V.; Wang, Q. et al. *Ann. Rheum. Dis.* **2005**, *64*, 462–463; Liu, R., Magolda, R., Duan, J., Vaddi, K., Maduskuie, T., Qian, M., Collins, R., Taylor, T., Giannaris, J. *Inflamm. Res.* **2002**, S125; Vaddi, K. G.; Magolda, R. L.; Haley, P. J.; Collins, R. J.; Taylor, T. L.; Maduskuie, T. P.; Decicco, C. P.; Newton, R. C.; Friedman, S. M. *65th Ann. Mtg. Of the Amer. Col. Rheum.* 2001, San Francisco, Abstr. 255; Grootveld, M.; McDermott, M. F. *Curr. Opin. Invest. Drugs* **2003**, *4*, 598–602; Cherney, R. J.; King, B. W.; Gilmore, J. L.; Liu, R.-Q.; Covington, M. B.; Duan, J. J.; Decicco, C. P. *Bioorg. Med. Chem. Lett.* **2006**, *16*, 1028–1031; Zhu, R.; Mazzola, R.; Sinning, L.; Lavey, B.; Zhou, G.; Spitler, J.; Wong, S.-C.; Orth, P.; Guo, Z.; Kong, J. et al. *230th National Meeting of the Amer. Chem. Soc.*, 2005, Washington, DC; Abstr. Medi 293.

83. Car, B. D. Presented at PharmaDiscovery Conference 2005; Washington DC (May 2005).
84. Dinarello, C. A. *Blood* **1996**, *87*, 2095–2147.
85. Arend, W. P.; Dayer, J. M. *Arthrit. Rheum.* **1995**, *38*, 151–160.
86. Van den Berg, W. B. *Curr. Opin. Rheumatol.* **1997**, *9*, 221–228.
87. Dinarello, C. A. *N. Engl. J. Med.* **2000**, *343*, 732–734.
88. Nuki, G.; Bresnihan, B.; Bear, M. B.; McCabe, D. *Arthrit. Rheum.* **2002**, *46*, 2838–2846.
89. Randle, J. C.; Harding, M. W.; Ku, G.; Schonharting, M.; Kurrle, R. *Exp. Opin. Invest. Drugs* **2001**, *10*, 1207–1209.
90. Rudolphi, K.; Gerwin, N.; Verzijl, N.; van der Kraan, P.; van der Berg, W. *Osteoarthrit. Cartil.* **2003**, *11*, 738–746.
91. Lee, J. C.; Laydon, J. T.; McDonnell, P. C.; Gallagher, T. F.; Kumar, S.; Green, D.; McNutty, D.; Blumenthal, M. J.; Heys, J. R.; Landvatter, S. W. et al. *Nature* **1994**, *372*, 739–746.
92. Adams, J. L.; Badger, A. M.; Kumar, S.; Lee, J. C. *Prog. Med. Chem.* **2001**, *38*, 1–60.
93. Wagner, G.; Laufer, S. *Med. Res. Rev.* **2006**, *26*, 1–62; Cirillo, P. F.; Pargellis, C.; Regan, J. *Curr. Top. Med. Chem.* **2002**, *2*, 1019–1033.
94. Regan, J.; Breitfelder, S.; Cirillo, P.; Gilmore, T.; Graham, A. G.; Hickey, E.; Klaus, B.; Madwed, J.; Moriak, M.; Moss, M. et al. *J. Med. Chem.* **2002**, *45*, 2994–3008; Pargellis, C. A.; Tong, L.; Churchill, L.; Cirillo, P.; Gilmore, T.; Graham, A. G.; Grob, P. M.; Hickey, E. R.; Moss, N.; Pav, S. et al. *Nat. Struct. Biol.* **2002**, *9*, 268–272.
95. Dominguez, C.; Powers, D. A.; Tamayo, N. *Curr. Opi. Drug Disc. Dev.* **2005**, *8*, 421–430.
96. Fitzgerald, C. E.; Patel, S. B.; Becker, J. W.; Cameron, P. M.; Zaller, D.; Pikounas, V. B.; O'Keefe, S. J.; Scapin, G. *Nat. Struct. Biol.* **2003**, *10*, 764–769.
97. Nikas, S. N.; Drosos, A. A. *Curr. Opin. Invest. Drugs* **2004**, *5*, 1205–1212.
98. Karin, M.; Yamamoto, Y.; Wang, Q. M. *Nat. Rev. Drug Disc.* **2004**, *3*, 17–26; Palanki, M.; Suto, M. World Patent Application 99/01441, 1999. Palanki, M. S. S.; Gayo-Fung, L. M.; Shevlin, G. I.; Erdman, P.; Sato, M.; Goldman, M.; Ransone, L. J.; Spooner, C. *Bioorg. Med. Chem. Lett.* **2002**, *12*, 2573–2577.
99. Ritzeler, O.; Castro, A.; Grenier, L.; Soucy, F.; Hancock, W. W.; Mazdiyasni, H.; Palombella, V.; Adams, J. World Patent Application 01/68648, 2001; Haddad, E-B.; Ritzeler, O.; Aldous, D. J.; Cox, P. J. World Patent Application 2005/113544, 2005. Castro, A. C.; Dang, L. C.; Soucy, F.; Grenier, L.; Mazdiyasni, H.; Hottelet, M.; Parent, L.; Pien, C.; Palombella, V.; Adams, J. *Bioorg. Med. Chem. Lett.* **2003**, *13*, 2419–2422.
100. Kishore, N.; Sommers, C.; Mathialagan, S.; Guzova, J.; Yao, M.; Hauser, S.; Huynh, K.; Bonar, S.; Mielke, C.; Albee, L. et al. *J. Biol. Chem.* **2003**, *278*, 32861–32871; Baxter, A.; Brough, S.; Cooper, A.; Floettmann, E.; Foster, S.; Harding, C.; Kettle, J.; McInally, T.; Martin, C.; Mobbs, M. et al. *Bioorg. Med. Lett.* **2004**, *14*, 2817–2822; Cywin, C. L.; Chen, Z.; Fleck, R. W.; Hao, M. H.; Hickey, E.; Liu, W.; Marshall, D. R.; Nemoto, P.; Sorcek, R. J.; Sun, S. et al. US Patent 6,964,956, 2005.
101. Murata, T.; Sasaki, S.; Yoshino, T.; Sato, H.; Koriyama, Y.; Nunami, N.; Yamauchi, M.; Fukushima, K.; Grosser, R.; Fuchikami, K. et al. World Patent Application WO2003/076447, 2003; Clare, M.; Hagen, T. J.; Houdek, S. C.; Lennon, P.; Weier, R, M.; Xu, X. World Patent Application, 2005/040133, 2005; Ziegelbauer, K.; Gantner, F.; Lukacs, N. W.; Berlin, A.; Fuchikami, K.; Niki, T.; Sakai, K.; Inbe, H.; Takeshita, K.; Ishimosi, M. et al. *Br. J. Pharmacol.* **2005**, *145*, 178–192.
102. Burke, J. R.; Pattoli, M. A.; Gregor, K. R.; Brassil, P. J.; MacMaster, J. F.; McIntyre, K. W.; Yang, X.; Iotzova, V. S.; Clarke, W.; Strnad, J. et al. *J. Biol. Chem.* **2003**, *278*, 1450–1456.
103. Brown, S.; Merouch, S. O.; Fridman, R.; Mobashery, S. *Curr. Top. Med. Chem.* **2004**, *4*, 1227–1238; Hanessian, S.; Moitessier, N. *Curr. Top. Med. Chem.* **2004**, *4*, 1269–1287; Wada, C. K. *Curr. Top. Med. Chem.* **2004**, *4*, 1255–1267; Levin, J. I. *Curr. Top. Med. Chem.* **2004**, *4*, 1289–1310; Skiles, J. W.; Gonnella, N. C.; Jeng, A. Y. *Curr. Med. Chem.* **2004**, *11*, 2911–2977; Skotnicki, J. S.; Levin, J. I.; Zask, A.; Killar, L. M. In *Metalloproteinases as Targets for Anti-Inflammatory drugs*; Bottomley, K. M., Bradshaw, D., Nixon, J. S., Eds.; Birkhauser Verlag: Basel, Germany, 1999, pp 17–57.
104. Rasmussen, H. S.; McCann, P. P. *Pharmacol. Ther.* **1997**, *75*, 69–75.
105. MacPherson, L. J.; Bayburt, E. K.; Capparelli, M. P.; Carroll, B. J.; Goldstein, R.; Justice, M. R.; Zhu, L.; Hu, S.; Melton, R. A.; Fryer, L. et al. *J. Med. Chem.* **1997**, *40*, 2525–2532.
106. Bode, W.; Fernandez-Catalan, C.; Tschesche, H.; Grams, F.; Nagase, H.; Maskos, K. *Cell Mol. Life Sci.* **1999**, *55*, 639–652; Borkakoti, N. In *Metalloproteinases as Targets for Anti-Inflammatory Drugs*; Bottomley, K. M., Bradshaw, D., Nixon, J. S., Eds.; Birkhauser Verlag: Basel, Germany, 1999, pp 1–16; Nagase, H. In *Matrix Metalloproteinases in Cancer Therapy*; Clendeninn, N. J., Appelt, K., Eds.; Humana Press: Totowa, NJ, USA, 2001, pp 39–66; Rao, B. G. *Curr. Pharm. Design.* **2005**, *11*, 295–322.
107. Lorenz, H.-M. *Curr. Opin. Anti-Inflamm. Immunomod. Invest. Drugs* **2000**, *2*, 47–52.
108. Lovejoy, B.; Welsh, A. R.; Carr, S.; Luong, C.; Broka, C.; Hendricks, R. T.; Campbell, J. A.; Walker, K. A. M.; Martin, R.; Van Wart, H. et al. *Nat. Struct. Biol.* **1999**, *6*, 217–221.

109. Leff, R. L.; Elias, I.; Ionescu, M.; Reiner, A.; Poole, A. R. *J. Rheumatol.* **2003**, *30*, 544–549.
110. Reiter, L. A.; Robinson, R. P.; McClure, K. F.; Jones, C. S.; Reese, M. R.; Mitchell, P. G.; Otterness, I. G.; Bliven, M. L.; Liras, J.; Cortina, S. R. et al. *Bioorg. Med. Chem. Lett.* **2004**, *14*, 3389–3395.
111. Cheng, M.; De, B.; Pikul, S.; Almstead, N. G.; Natchus, M. G.; Anastasio, M. V.; McPhail, S. J.; Snider, C. E.; Taiwo, Y. O.; Chen, L. et al. *J. Med. Chem.* **2000**, *43*, 369–380.
112. Hanessian, S.; Moitessier, N.; Gauchet, C.; Viau, M. *J. Med. Chem.* **2001**, *44*, 3066–3073.
113. Becker, D. P.; Barta, T. E.; Bedell, L.; DeCrescenzo, G.; Freskos, J.; Getman, D. P.; Hockerman, S. L.; Li, M.; Mehta, P.; Mischke, B. et al. *Bioorg. Med. Chem. Lett.* **2001**, *11*, 2719–2722.
114. Wada, C. K.; Holms, J. H.; Curtin, M. L.; Dai, Y.; Florjancic, A. S.; Garland, R. B.; Guo, Y.; Heyman, H. R.; Stacey, J. R.; Steinman, D. H. et al. *J. Med. Chem.* **2002**, *45*, 219–232.
115. O'Brien, P. M.; Ortwine, D. F.; Pavlovsky, A. G.; Picard, J. A.; Sliskovic, D. R.; Roth, B. D.; Dyer, R. D.; Johnson, L. L.; Man, C. F.; Hallak, H. *J. Med. Chem.* **2000**, *43*, 156–166.
116. Aranapakam, V.; Davis, J. M.; Grosu, G. T.; Baker, J.; Ellingboe, J.; Zask, A.; Levin, J. I.; Sandanayaka, V. P.; Du, M.; Skotnicki, J. S. et al. *J. Med. Chem.* **2003**, *46*, 2376–2396.
117. Picard, J. A.; Wilson, M. W.; Lumb, J. T. World Patent Application. WO2002/064578, 2002.
118. Barvian, N. C.; Patt, W. C.; World Patent Application WO2002/064571, 2002.
119. Engel, C. K.; Pirard, B.; Schimanski, S.; Kirsch, R.; Habermann, J.; Klingler, O.; Schlotte, V.; Weithmann, K. U.; Wendt, K. U. *Chem. Biol.* **2005**, *12*, 181–189.
120. Ishikawa, T.; Nishigaki, F.; Miyata, S.; Hirayama, Y.; Minoura, K.; Imanishi, J.; Neya, M.; Mizutabi, T.; Imamura, Y.; Naritomi, Y. et al. *Br. J. Pharmacol.* **2005**, *144*, 133–143.
121. Tortorella, M. D.; Burn, T. C.; Pratta, M. A.; Abbaszade, I.; Hollis, J. M.; Liu, R.; Rosenfeld, S. A.; Copeland, R. A.; Decicco, C. P.; Wynn, R. et al. *Science* **1999**, *284*, 1664–1666; Tortorella, M. D.; Malfait, A. M.; Decicco, C.; Arner, E. A. *Osteoarthrit. Cartilage* **2001**, *9*, 539–552.
122. Glasson, S. S.; Askew, R.; Sheppard, B.; Carito, B.; Blanchet, T.; Ma, H.-L.; Flannery, C. R.; Peluso, D.; Kanki, K.; Yang, Z. et al. *Nature* **2005**, *434*, 644–648; Stanton, H.; Rogerson, F. M.; East, C. J.; Golub, S. B.; Lawlor, K. E.; Meeker, C. T.; Little, C. B.; Last, K.; Farmer, P. J.; Campbell, I. K. et al. *Nature* **2005**, *434*, 648–652.
123. Liu, R. Q.; Trzaskos, J. M. *Curr. Med. Chem. Anti-Inflamm. Anti-Allergy Agents* **2005**, *4*, 251–264; Nagase, H.; Kashiwagi, M. *Arthrit. Res. Ther.* **2003**, *5*, 94–103.
124. Yao, W.; Wasserman, Z. R.; Chao, M.; Reddy, G.; Shi, E.; Liu, R.-Q.; Covington, M. B.; Arner, E. C.; Pratta, M. A.; Tortorella, M. et al. *J. Med. Chem.* **2001**, *44*, 3347–3350.
125. Cherney, R. J.; Mo, R.; Meyer, D. T.; Wang, L.; Yao, W.; Wasserman, Z. R.; Liu, R.-Q.; Covington, M. B.; Tortorella, M. D.; Arner, E. C. et al. *Bioorg. Med. Chem. Lett.* **2003**, *13*, 1297–1300.
126. Noe, M. C.; Natarajan, V.; Snow, S. L.; Mitchell, P. G.; Lopresti-Morrow, L.; Reeves, L. M.; Yocum, S. A.; Carty, T. J.; Barberia, J. T.; Sweeney, F. J. et al. *Bioorg. Med. Chem. Lett.* **2005**, *15*, 2808–2811.
127. Xiang, J. S.; Hu, Y.; Rush, T. S.; Thomason, J. R.; Ipek, M.; Sum, P.-E.; Abrous, L.; Sabatini, J. J.; Georgiadis, K.; Reifenberg, E. et al. *Bioorg. Med. Chem. Lett.* **2006**, *16*, 311–316.

Biographies

Ron Magolda received a BS in Chemistry and a BSG in general science from Villanova University in 1976. In 1980, he obtained a PhD in organic chemistry from the University of Pennsylvania. Dr Magolda began his industrial career at DuPont Central Research, where he conducted medicinal chemistry research in the fields of inflammation and cardiovascular disease. During his 18-year career at DuPont, Dr Magolda led several different medicinal chemistry groups. Ultimately, he directed the Inflammatory Diseases Therapeutic Area, a group that encompassed both chemists and biologists. In 1998, he moved to Boehringer-Ingelheim (Ridgefield, CT) where he headed a chemistry department that included a synthetic effort focused on inflammatory research (medicinal chemistry and combichem) as well as technology groups working on structural research (protein crystallography) and ultra-high-throughput screening. In

2002, Dr Magolda joined Wyeth Research as the Vice President of Medicinal Chemistry with responsibilities in Princeton, NJ and Collegeville, PA. His therapeutic focus has now expanded to include neurological diseases and woman's health indications. Dr Magolda continues to apply his efforts to the discovery and development of novel therapies that can treat major disease indications, especially where available therapies are ineffective.

Terence Kelly received his chemistry education at Rensselaer Polytechnic Institute (BS, 1982) and the University of Texas at Austin (PhD, 1988). After postdoctoral studies at Yale University in the laboratory of Prof Harry Wasserman, Dr Kelly joined Boehringer-Ingelheim in Ridgefield, CT in 1990. During this period, Dr Kelly has worked primarily in the areas of virology and immunological diseases. In particular, he has been active in the area of leukocyte cell adhesion, especially the discovery and profiling of LFA-1 antagonists. His current position at Boehringer-Ingelheim is Vice President of Chemistry.

Robert Newton received his BS degree in biology from Niagara University and his PhD in molecular, cellular, and developmental biology from Iowa State University. He joined DuPont Central Research and then moved to DuPont Medical Products as a Group Leader in Immunology and Inflammatory Diseases. He subsequently joined the DuPont Merck joint venture and remained with the company when it became DuPont Pharmaceuticals, eventually heading the Inflammatory Diseases Department. Shortly after Bristol Myers Squibb acquired DuPont Pharma he joined Incyte Corporation where he is currently a Vice President in Biology Research. Dr Newton's research interest continues to be in understanding basic biological concepts and using this information to identify new therapeutic opportunities, particularly for small molecules. He has also conducted research in the area of macrophages and macrophage biology that has led to applications in inflammation, cardiovascular, metabolic, and cancer biology.

Jerauld S Skotnicki received his BA in chemistry from the College of the Holy Cross, MA in chemistry from Dartmouth College, and PhD from Princeton University. Prior to his studies at Princeton, Jerry was a staff chemist at Lederle Laboratories for 3 years. Following his PhD, he joined the Biochemicals Department at DuPont (1.5 years) as Research Chemist. In 1982, Dr Skotnicki moved to Wyeth where he has been a Medicinal Chemist at Wyeth at three sites: in Radnor, PA; Princeton, NJ; and later Pearl River, NY. During this period Dr Skotnicki rose to the level of Director. Since 2004, his current position is Senior Director of Chemical Sciences Interface, Chemical and Screening Sciences at Wyeth. He has responsibility for research activities at the junction of Discovery and Development, as well as management of external alliances in Chemical and Screening Sciences. His research experience has been in the areas of immunoinflammaory diseases, infectious diseases, musculoskeletal diseases, oncology, and transplantation. Dr Skotnicki was the recipient (2004) of the Thomas Alva Edison Patent Award in 'Emerging Technologies' for his contributions to the discovery of Temsirolimus. He is currently an editor of *Inflammation Research* and has served on the Board of Directors of the *Inflammation Research Association* (2000–04).

Jim Trzaskos received his BS in chemistry from St Lawrence University (1973), and his PhD in biochemistry from the University of Vermont (1979). Following his postdoctoral training in lipid biochemistry with Prof James L Gaylor at the University of Missouri, Columbia, Jim joined the DuPont Company in 1981. At DuPont, Dr Trzaskos focused his initial research on cholesterol biosynthesis with emphasis on the enzymatic conversion of lanosterol to cholesterol in the latter stages of the biosynthetic pathway. He continued to work in lipid metabolism when he joined the Inflammatory Diseases group at DuPont Pharmaceuticals where he worked on prostaglandin synthesis and COX-2 inhibitor design. He also was involved in the enzymology of various metalloproteinases including TACE and aggrecanase, cell signaling programs including MEK and IKK, and the study of dissociated steroid drug discovery. Dr Trzaskos now works at Bristol-Myers Squibb as a Research Fellow where he has responsibility for the chemokine biology programs.

Comprehensive Medicinal Chemistry II
ISBN (set): 0-08-044513-6

ISBN (Volume 7) 0-08-044520-9; pp. 845–872

7.30 Asthma and Chronic Obstructive Pulmonary Disease

G P Roth, Abbott Bioresearch Center, Worcester, MA, USA
D W Green, Amgen Inc., Cambridge, MA, USA

7.30.1 Disease State

Asthma and chronic obstructive pulmonary disease (COPD) are serious respiratory diseases that not only compromise the health and daily activities of those affected, but also impose an economic burden on countries around the globe. As communities become more industrialized, the incidence of asthma and COPD increases. It is estimated the global incidence of asthma is 300 million people and accounts for one in every 250 deaths.[1] Global estimates for COPD predict it will rank fifth as a worldwide health burden by the year 2020.[2] In 1996 COPD was accountable for over 100 000 deaths in the US, putting it behind heart disease, cancer, and stroke as the major mortality diseases.[3] Age, sex, race, and socioeconomic status are more important factors in the epidemiology of COPD than asthma. Incidence and mortality of COPD incidence increase with age and prevalence is higher among males than females (although the incidence in females is rising), whites than nonwhites, and blue-collar than white-collar workers.[4] Asthma occurs across all demographics, but its escalating incidence in children is particularly alarming. It is the leading cause of childhood hospitalization and school absenteeism in the US. In 1998, it was estimated that more than 15 million people in the US have asthma, leading to 500 000 hospitalizations and 5 000 deaths annually.[3] At least 16 million people in the US have COPD, which is divided between 14 million with chronic bronchitis and 2 million with emphysema.[5] The total annual costs in the US for asthma[6] and COPD[7] have been estimated at over $12 billion and $14 billion, respectively. A comparison of the epidemiology for asthma and COPD is summarized in **Table 1**.

Asthma and COPD are both characterized by airway obstruction that leads to significant breathlessness (dyspnea). Several diagnostic indicators can be used to distinguish the diseases from one another[8] as summarized and compared in **Table 2**. Asthma typically involves a reversible airflow limitation, the onset usually occurs in childhood, the symptoms vary from day to day, and a family history is common. Airflow limitation in COPD is not fully reversible, the onset occurs

Table 1 Comparison of asthma and COPD epidemiology in the USA

	Asthma	*COPD*
Occurrence	15 million	16 million
Estimated annual costs	$12 billion (2001)	$14 billion (2000)
Deaths (Center for Disease Control (CDC) Reports)	4261 (2002)	119 054 (2000)
Common age of onset	Childhood	Over 50 years
Incidence by sex	Male = female	Male > female
Incidence by race	No difference	White > nonwhite
Incidence by socioeconomic status	No difference	Blue-collar > white-collar

Table 2 Similarities and differences between asthma and COPD

	Asthma	*COPD*
Stimulus	Allergens	Environmental pollutant
	Viral infections	
	Changes in temperature or humidity	
	Exercise	
	Environmental pollutants	
	Strong emotional reactions	
Extent of airway inflammation	All airways (except parenchyma)	Peripheral
Airway obstruction and dyspnea	Yes; reversible	Yes; not reversible
Cellular activation	T cell and B cells	Macrophages
	IgE production	Cytokines
	Mast cells and basophils	Proteases
	Leukotrienes	Prostaglandins
	Histamine	Leukotrienes
	Cytokines	Growth factors
	Proteases	Reactive O_2
	Growth factors	
Primary T cells	$CD4^+$	$CD8^+$
Infiltrating cells	Eosinophils	Neutrophils
Destruction of parenchyma	No	Yes

in mid-life and is usually correlated with a long history of cigarette smoking, and the symptoms are slowly progressive. Severe cases of asthma can be more difficult to distinguish from COPD, especially if the reversibility of airway obstruction is diminished.

In asthma the frequency and severity of symptoms can vary greatly in the population as the disease progresses in an individual.[9] An asthma attack can be triggered by a number of different factors including allergens (animal dander, dust mites, etc.), viral infections, changes in temperature or humidity, exercise, environmental pollutants, or strong emotional reactions (anxiety, crying, laughing, etc.).[10–14] Bronchoconstriction and airway inflammation contribute to airway hyperreactivity in asthma, which typically results in coughing and wheezing. During the attack airway flow is decreased by the constriction of the bronchi and can be exacerbated by inflammation and increased mucus secretion. Tachypnea and tachycardia are usually observed and in more severe cases patients have difficulty speaking and symptoms of cyanosis (blue skin) can appear as the attack worsens.

Pulmonary function tests are used to evaluate the progression of asthma and the response to therapeutic agents. The most common test is spirometry, which measures the maximum airflow during expiration (peak expiratory flow (PEF) or forced expiratory lung volume (FEV_1)). The diagnosis can also be confirmed by bronchial hyperresponsiveness to histamine or methacholine.[14] Asthma severity is classified into four stages, which are defined by the frequency of attacks and results of spirometry tests.[9] In the least severe stage, mild intermittent asthma, attacks occur at a frequency of less than two per week, night-time episodes are infrequent, and the effect on PEF or FEV_1 is less than 20% relative to healthy subjects. The next most severe stage, mild persistent asthma, is also characterized by a change in PEF or FEV_1 of less than 20% but the frequency of attacks is three to six episodes per week and more than two night-time attacks per month. In the third stage, moderate persistent asthma, the frequency of attacks (daily with more than five night-time episodes per month) and the effect on lung function (20–40% change in PEF or FEV_1) both increase. In the most critical stage, severe persistent asthma, attacks are continuous and the effect on lung function is greater than a 40% change in PEF or FEV_1.

In contrast to asthma, COPD can almost always be linked to exposure to preventable environmental irritants. In developed countries it is estimated that cigarette smoking accounts for more than 95% of COPD cases. Other risk factors include exposure to air pollution such as indoor burning fuels, occupational exposure, and poor diet.[15] The onset of symptoms is usually observed in people aged 50–60 who have a long history of cigarette smoking. Most patients have symptoms of chronic bronchitis and emphysema as well as mucus plugging but may differ in the degree of emphysema and chronic bronchitis.[16] Individuals with an α_1-antitrypsin deficiency associated with a homozygous Z variant develop emphysema earlier in life and this is worsened by smoking. Although this indicates a genetic predisposition to COPD, fewer than 1% of COPD patients have an α_1-antitrypsin deficiency and other variants that result in decreased serum levels of α_1-antitrypsin have not been linked to COPD.[17]

COPD and asthma can typically be distinguished from one another based upon the extent of obstruction, inflammation, and tissue damage in the airways.[15] Asthma usually involves inflammation of all airways whereas inflammation in COPD is usually restricted to the bronchioles (peripheral or 'small' airways). Inflammation in asthma results in airway hyperresponsiveness, whereas inflammation in COPD results in alveolar destruction. In both diseases inflammation of the epithelial cells lining the airways (asthma) or bronchioles (COPD) results in enlargement of mucus-secreting glands and the number of goblet cells, causing hypersecretion of mucus. Chronic inflammation in COPD and severe asthma results in repeated cycles of injury and repair, causing fibrosis and narrowing of the small airways. These processes ultimately lead to dilation and destruction of the bronchioles in COPD followed by the formation of large distended air sacs (bullae) in the lung. The symptoms associated with this COPD pathology are a progressive dyspnea along with chronic cough and sputum production.[8]

The severity of COPD has been classified into three stages distinguished by symptoms and spirometry scores relative to normal pulmonary function.[8] Individuals with a chronic cough and sputum production but no change in spirometry scores are considered at Stage 0 and at risk. Stage I (mild COPD) is characterized by mild airflow limitation resulting in an FEV_1 score decreased by up to 20% that is usually, but not always, associated with chronic cough and sputum production. At Stage II (moderate COPD) patients usually seek medical attention due to dyspnea and increased cough and sputum production. Pulmonary function is decreased by more than 20% at this stage and can be further subdivided into Stage IIA or IIb depending upon whether FEV_1 has been effected more than 50% or not, respectively. In Stage III (severe) COPD pulmonary function has decreased more than 70% and/or clinical symptoms of respiratory failure or right heart failure are observed. Quality of life is significantly affected and exacerbations of the disease can be life threatening.

7.30.2 Disease Basis

The cellular pathology of both asthma and COPD involves several types of immune cells and mediators, and the cellular progression of disease follows a general sequence of leukocyte activation, trafficking, and airway remodeling. Both diseases will be reviewed in the context of cellular pathology and therapeutic targets involved in each step.

7.30.2.1 Asthma: Leukocyte Activation and Trafficking

Current models of asthma pathology divide the disease into an early and late immune response.[18] The early immune response refers to the events that lead up to the production of immunoglobulin E (IgE) (**Figure 1**) and cause acute bronchospasm, edema, and airflow obstruction that is reversible.

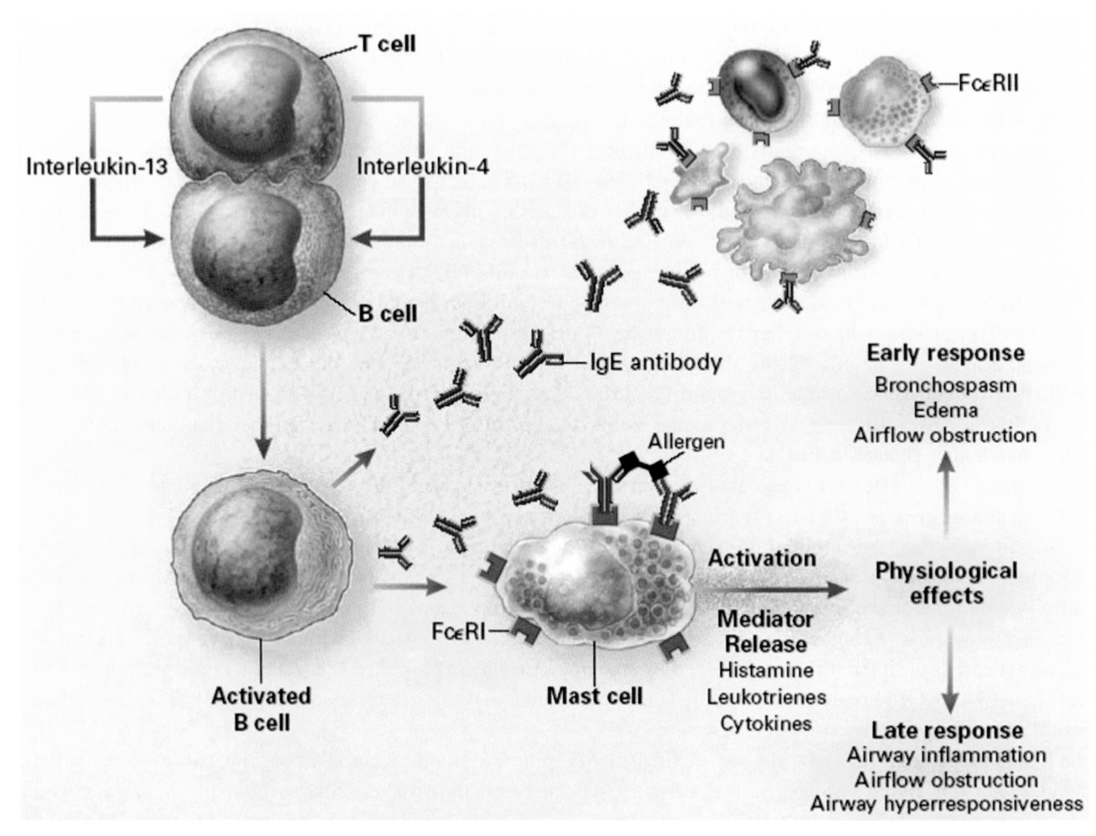

Figure 1 Production of IgE and its consequences in the early asthma immune response. (Reproduced with permission from Busse, W. W.; Lemanske, R. F. *N. Engl. J. Med.* **2001**, *344*, 350–362.)

The acute process is initiated when a dendritic cell in the airway lining encounters an antigen and is stimulated to migrate to the lymph node. There the antigen is presented to T and B cells, and the interaction between these cells (referred to as costimulation) activates the B cell to produce IgE. Costimulation and the subsequent activation of B cells is mediated by the interaction of two pairs of receptors between T and B cells. The first pair involves interleukin-4 (IL4) or IL13 secreted from T cells and the respective receptor on the B cell. The second pair involves CD40 on the surface of B cells and its receptor on the surface of T cells. Costimulation results in the activation of intracellular signaling pathways that ultimately converge on the nucleus and lead to the expression and production of IgE. The antibodies circulate and bind to high-affinity receptors (FcεRI) on mast cells and basophils and to low-affinity receptors (FcεRII) on lymphocytes, eosinophils, platelets, and macrophages. The cross-linking of IgE-bound FcεRI receptors on mast cells activates intracellular signaling pathways that result in degranulation of histamine and leukotrienes into the mucosal and submucosal sites of the airway. The cysteinyl leukotrienes (LTC_4, LTD_4, LTE_4) are peptide-conjugated lipids that arise from arachadonic acid metabolism, and along with histamine, cause contraction of bronchial smooth muscle.[19]

Mast cells also play a role in the late asthma response that leads to airway hyperresponsiveness (**Figure 2**). Activated mast cells also release cytokines such as the interleukins 1 through 5 (IL1, IL2, IL3, IL4, IL5), interferon-γ (IFN-γ), and tumor necrosis factor-α (TNF-α) as well as granulocyte–macrophage colony-stimulating factor (GM-CSF).[18] GM-CSF, IL3, and in particular IL5 are known to stimulate production of eosinophils in the bone marrow.[20] Eosinophils released into the circulation constitutively express proteins such as the $\alpha_4\beta_1$ integrin VLA-4 (very late antigen-4), the β_2 integrin LFA-1 (lymphocyte function-associated antigen-1), L selectin, and P selectin glycoprotein ligand (PSGL)-1. These proteins selectively bind to the adhesion molecules and selectins on the endothelial cells.[21,22] Cytokines such as IL4, IL13, and other products of inflammation induce the expression of proteins on the surface of endothelial cells lining the vasculature such as the intracellular and vascular cell-adhesion molecules ICAM-1 and VCAM-1, respectively, and the E and P selectins.[23] The interaction between eosinophil and endothelial cell surface proteins is regulated by the interaction of specific chemoattractant cytokines known as

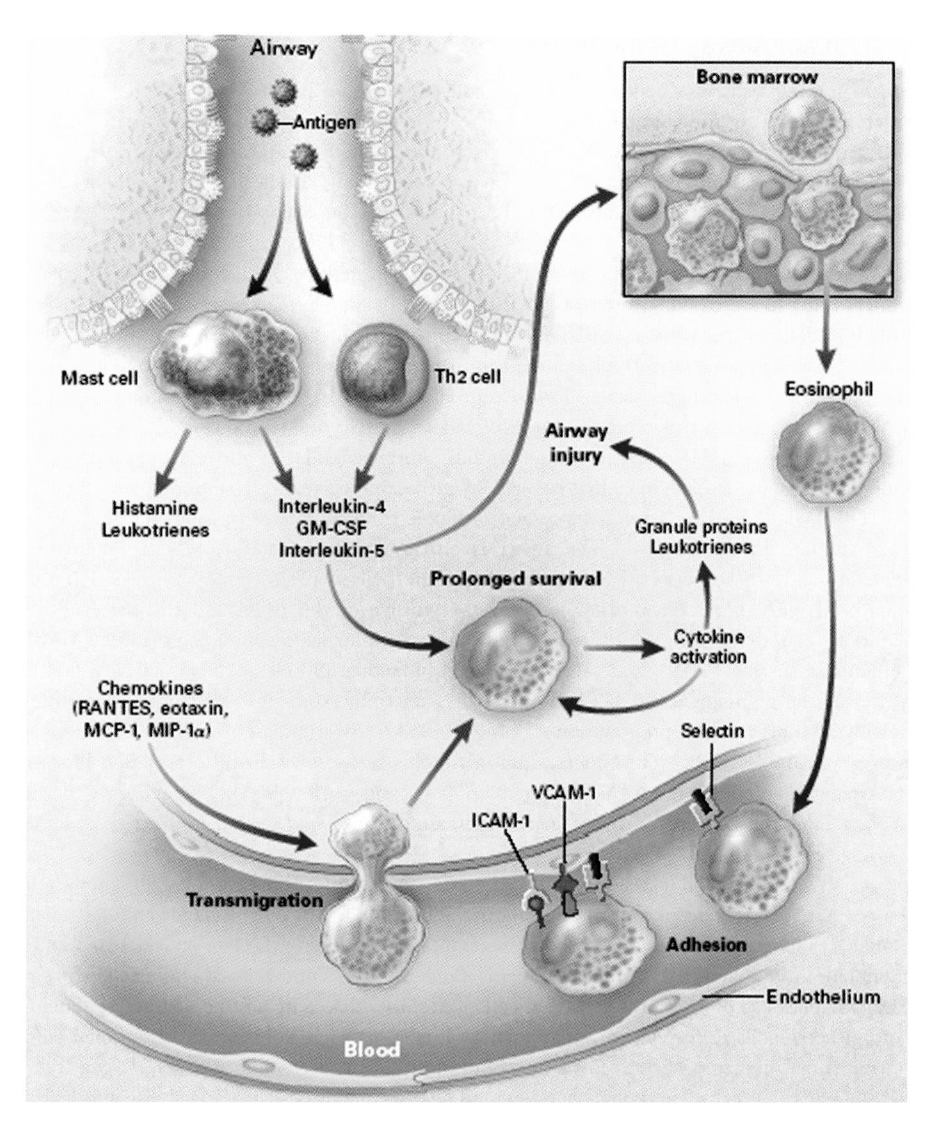

Figure 2 Mast cell, eosinophils, and Th2 cells in the late asthma immune response. (Reproduced with permission from Busse, W. W.; Lemanske, R. F. *N. Engl. J. Med.* **2001**, *344*, 350–362.) For definitions see text.

chemokines with specific receptors on eosinophils. In particular, the chemokine eotaxin (CCL11) produced by endothelial cells in response to IL4 or IL13[24] binds to the chemokine receptor CCR3 on the surface of eosinophils and can influence the affinity of the cell surface interactions between endothelial cells and eosinophils.[25] Eosinophils can also migrate in response to chemoattractants such as prostaglandin D_2 through the receptors DP (D_2 prostaglandin) and $CRTH_2$ (chemoattractant receptor-homologous molecule expressed on Th2 cells).[26] The result of all of these interactions is the rolling of eosinophils across the endothelial lining of the vasculature, which enables trafficking to inflammation sites in airways.

Antigen stimulation of naive $CD4^+$ T cells not only results in costimulation of B cells to produce IgE, but also in the differentiation of T cells into Th2 cells.[27] Naive $CD4^+$ T cells can differentiate into either Th1 or Th2 cells. These two types of T cells can be distinguished from one another by the types of cytokines they produce upon stimulation. Th1 cells play a role in protective immunity and produce IL2 and IFN-γ. Th2 cells predominate in the inflamed tissue of asthmatics along with mast cells produce cytokines that mediate allergic inflammation (IL4, IL5, IL6, IL9, and IL13). Th2 cell migration to inflammation sites from the lymph nodes is essentially mediated by the same selectins, integrins, chemokines, and chemoattractants involved in eosinophil trafficking.[28] The proliferation of Th2 cells at inflammatory sites can be stimulated by neighboring eosinophils as the latter can produce lymphokines such as IL4.[29]

7.30.2.2 Asthma: Airway Remodeling

The convergence of mast cells, eosinophils, and Th2 cells in the airways can lead to thickening of the airway walls, which is referred to as airway remodeling.[13,30] The individual abnormalities involved in airway remodeling include smooth muscle hypertrophy and hyperplasia, extracellular matrix deposition, alterations in epithelial and goblet cells, and vascular remodeling, which all contribute to increasing airflow resistance. Mast cells, eosinophils, and Th2 cells at inflammatory sites produce cytokines, growth factors, and other mediators that contribute to these processes.

In severe cases of asthma increased smooth muscle mass can reduce the airway lumen three- to fourfold, significantly contributing to increased airway resistance. Mast cell degranulation releases mitogens such as histamine, IL1β, LTD_4, platelet-derived growth factor (PDGF) and tryptase which can induce smooth muscle cell hypertrophy and hyperplasia.[31,32] Based upon in vitro results under serum-free conditions, direct contact of Th2 cells with smooth muscle cells is believed to also stimulate smooth muscle proliferation.[33] In addition to inflammatory cells, stimulated endothelial cells can produce growth factors that induce smooth muscle proliferation.[34]

The extracellular matrix (ECM) is composed of fibrous proteins (collagen and elastin) and adhesive proteins (fibronectin and laminin) embedded in a hydrous gel of glycosaminoglycans (hyaluronic acid) and proteoglycans (versican, etc.). This hydrated mix of carbohydrates and insoluble fibrous proteins provides resilient scaffolding which supports tissue structure in the airway.[13] The ECM is also a dynamic structure where a homeostatic balance between degradation and synthesis is critical, and fibroblasts are cells that govern this process. Fibroblasts produce the components of the ECM as well as the proteases that turn over the ECM and the protease inhibitors that regulate the proteolytic activity. These protease/inhibitor pairs include matrix metalloproteases (MMP) and tissue inhibitors of metalloproteases (TIMP), and there are several individual protein members of both families.[35] During inflammation, transforming growth factor β (TGF-β) is released from eosinophils and drives the differentiation of fibroblasts into myofibroblasts. The proliferation of lung fibroblasts is stimulated by mast cell tryptase through a protease-activated receptor, PAR-2.[36] The proliferation of fibroblasts and/or their differentiation into myofibroblasts contributes to excessive production of ECM components and disruption of the homeostatic balance between proteases and their inhibitors, which can result in the increased fibrosis and decreased elasticity of the airway wall observed in asthma.[37]

Thickening of epithelial cells and increased mucus secretion are part of asthma pathology. The epithelium provides a protective barrier between the airway and the vasculature and is lined with mucus glands and epithelial goblet cells that produce mucus. The epithelium is composed of several layers including the lamina reticularis and basal lamina (basement membrane). In asthma the lamina reticularis thickens but not the basal lamina.[13] Deposition of ECM components by fibroblasts and myofibroblasts can contribute to thickening of the lamina reticularis as well as proliferation of epithelial cells in response to inflammatory mediators such as histamine. Epithelial damage can also result from inflammatory mediators such as histamine from mast cells and granule proteins, oxygen free radicals, and TNF-α from eosinophils that cause epithelial shedding and lead to microvascular permeability and leakage of plasma into the airways. This contributes to restricting the airway as plasma proteins can mix with mucus and inflammatory cells to form viscid luminal plugs. Hypersecretion of mucus can be stimulated by mast cell histamine, PGD_2 and cysteinyl leukotrienes, and mucus glands and goblet cells can proliferate and enlarge in response to cytokines as well as proinflammatory and toxic mediators released by eosinophils.

As a consequence of the hypertrophy and hyperplasia of smooth muscle and epithelial cells, blood vessels enlarge and new blood vessels are produced (angiogenesis).[38,39] Blood vessel enlargement, sometimes referred to as nonsprouting angiogenesis, is commonly proportionally to the proliferation of endothelial cells. There are a number of agents that can stimulate blood vessel enlargement and angiogenesis, and they are directly or indirectly attributable to mast cells, eosinophils or Th2 cells in asthma. IL15 is a known angiogenic mitogen and can is produced by activated Th2 cells and fibroblasts. The TNF-α produced by eosinophils is also an angiogenic factor. Epithelial cells and fibroblasts are sources of the angiogenic stimuli epidermal growth factor (EGF), fibroblast growth factor (FGF), and platelet-derived growth factor (PDGF).[34]

7.30.2.3 Chronic Obstructive Pulmonary Disease: Leukocyte Activation and Trafficking

Airway remodeling, leukocyte activation, and trafficking of leukocytes are also important to the progression of disease in COPD and each process has common and distinct features from asthma. Even though COPD is roughly equivalent to asthma with respect to global incidence, mortality, and socioeconomic impact, little is known about the disease mechanism relative to asthma.

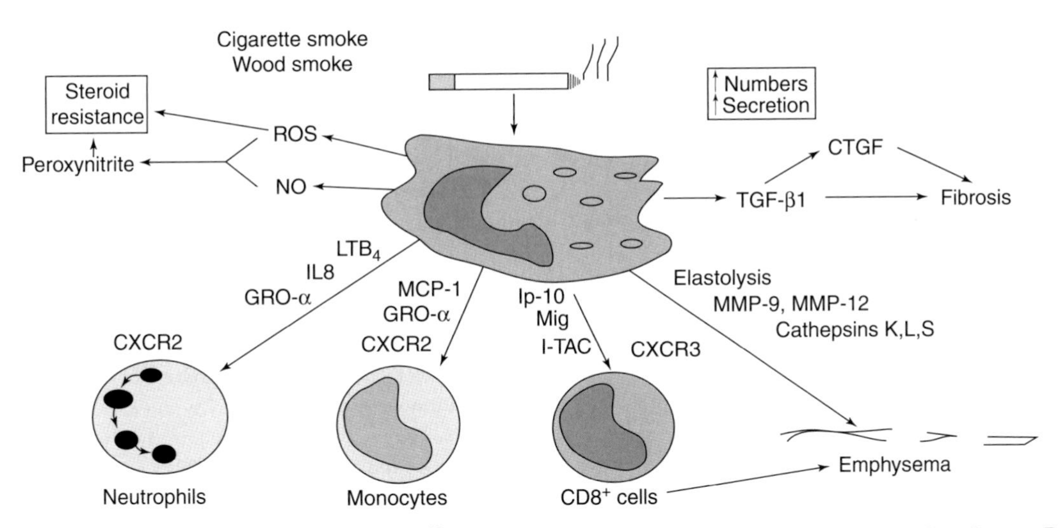

Figure 3 Cellular mechanisms involved in COPD.[15] For definitions see text. (Reproduced with permission from Barnes, P. J.; Shapiro, S. D., Pauwels, R. A. *Eur. Respir. J.* **2003**, *22*, 672–688 © European Respiratory Society Journals Ltd.)

In contrast to the wide variety of molecular antigens and other factors that are asthma triggers, COPD is associated with an abnormal inflammatory response of the lungs to noxious particles and gases.[40] Also in contrast to asthma, the sequence of molecular and cellular events involved in the progression of the disease is less well defined in COPD. Some of the cellular mechanisms involved in the progression of COPD are depicted in **Figure 3**. As reactive oxygen species are known to activate immune cells, it is likely that components of exogenous pollutants such a tobacco smoke trigger inflammation in COPD. Dendritic cells are abundant in the lung and increased in the airways of smokers, so they are likely to be involved in initiating the COPD inflammatory response.[41] However, there is no evidence to link dendritic cell activation to COPD inflammation. It is very likely that macrophages are involved in the pathogenesis of COPD as they are dramatically increased in the lung tissue, airways, bronchiolar lavage, and sputum of COPD patients[42] and can be localized to sites of alveolar damage.[43] Macrophages can be activated by cigarette smoke, which induces the production of cytokines (IL8, TNF-α), elastolytic proteases (MMPs, cathepsins), lipid mediators such as prostaglandin E_2 (PGE_2) and leukotriene B_4 (LTB_4), and GM-CSF, which releases neutrophils from bone marrow. Interestingly, corticosteroids are not effective in treating inflammation in COPD patients and cytokine and protease production by macrophages from COPD patients is refractory to corticosteroids, but the production of GM-CSF is not.[44] Cigarette smoke can also induce the production of GM-CSF and the cytokines TNF-α, IL1β, and IL8 in lung epithelial cells. Neutrophils are also plentiful in the sputum and bronchiolar lavage, but not in the lung tissue and airways of COPD patients.[45] It is possible they are relatively transient in lung tissue and airways relative to macrophages. T cells are increased in the lung tissue and peripheral and central airways of COPD patients, but in contrast to asthma, $CD8^+$ T_c1 cells predominate over $CD4^+$ Th2 cells.[45] There is no increase in eosinophils, mast cells, or the bronchoconstrictive agents histamine and cysteinyl leukotrienes in COPD.[46]

Macrophages are resident in the lungs of COPD patients, but neutrophils and T cells must migrate from the bone marrow and lymph node, respectively, to the lung. Like eosinophils and $CD4^+$ T cells in asthma, neutrophils and $CD8^+$ T cells traffic to inflammatory sites through the interaction of VLA-4 and LFA-1 integrins and PSGL-1 with endothelial adhesion molecules and selectins, respectively. Chemotaxis of $CD8^+$ T cells is mediated through interactions between the chemokine receptor CXCR3 and its ligands IP-10, I-TAC, and Mig.[40] In contrast to eosinophils, neutrophil chemoattractants include the cytokine/chemokine IL8 and the leukotriene LTB_4. IL8 can bind to either CXCR1 or CXCR2 chemokine receptors on neutrophils. Binding to CXCR1 is low affinity relative to the high-affinity binding to CXCR2, and the former is thought to mediate effector functions such as protease and mediator release while the latter is responsible for chemotaxis.[40] The neutrophil receptor for LTB_4 is BLT_1, but this receptor can also be found on Th1 and Th2 cells.[47]

7.30.2.4 Chronic Obstructive Pulmonary Disease: Remodeling

There are multiple hypotheses that address the mechanism of airway obstruction and alveolar damage in COPD, and all of these mechanisms probably contribute to the pathology of the disease. As mentioned earlier, reactive oxygen species can activate immune cells. Reactive oxygen species such as superoxide anions and hydroxyl radicals are generated as metabolic by-products by healthy cells but are quickly converted to less toxic molecules by enzymes such as catalase and superoxide dismutase or nonenzymatic antioxidants such as glutathione and ascorbic acid. In COPD the production of reactive oxygen species by cigarette smoke and activated neutrophils and macrophages can overwhelm antioxidants to generate oxidative stress in the lungs.[40] Increased levels of reactive oxygen species cause increased mucus production, constriction of airway smooth muscle, and plasma exudation in the airway. Epithelial cells can also be stimulated to produce TGF-β, which can cause fibrosis through proliferation of fibroblasts and differentiation into myofibroblasts.

An imbalance between proteases and their endogenous inhibitors can cause increased mucus production and tissue destruction. Serine proteases produced by neutrophils such as elastase and cathepsin G stimulate mucus secretion from goblet epithelial cells and mucus glands.[48] Neutrophil elastase can also destroy elastin, which contributes to a loss of elasticity in the lung tissue. Neutrophil elastase is usually kept in check by α_1-antyrypsin inhibitor, but oxidation of the inhibitor by cigarette smoke or reactive oxygen species can result in increased elastase activity in COPD.[40] A genetic deficiency of α_1-antyrypsin is known to exacerbate COPD. The activity of matrix metalloproteases and cathepsins produced by neutrophils and macrophages can overwhelm endogenous inhibitors such as TIMPs and cystatins. The importance of MMP-12 to the development of COPD was demonstrated with MMP-12$^{-/-}$ mice exposed to tobacco smoke as wild-type mice developed emphysema but knockout mice did not.[49]

The cellular mechanisms involved in the pathology of asthma and COPD provide opportunities for therapeutic intervention in these debilitating diseases, and the similarities and differences between these diseases are summarized in **Table 2**. Blocking the activation of leukocytes has been approached by interfering with intracellular signaling pathways through inhibition of phosphodiesterase type 4 (PDE4) or by blocking the inflammatory mediators such as leukotrienes by inhibiting their biosynthesis (5-lipoxygenase) or their interaction with receptors (BLT_1 receptor antagonists). Interfering with the recruitment and trafficking of eosinophils in asthma has focused on blocking the action of IL5 or antagonizing the CCR3 and CRTH2 receptors, while blocking the trafficking of neutrophils in COPD has been focused on chemokine receptors such as CXCR2. The degranulation of leukocytes has also been targeted through inhibition of the proteases that are released in this process including tryptase, elastase, cathepsins and matrix metalloproteases.

New approaches to identifying disease targets and treating asthma and COPD are also being explored. Genomic linkage studies have led to the identification of new asthma targets such as the metalloprotease ADAM33,[50] a dipeptidyl protease (DPP10) involved in cytokine processing,[51] and a putative transcription factor (PHF11).[52] Although genomic linkage has not identified potential COPD therapeutic targets, a deficiency in α_1-antitrypsin inhibitor has been shown to increase the susceptibility as mentioned in a previous section. In addition to several antibody approaches that will be mentioned throughout the chapter, other nonconventional approaches have been considered as therapeutic strategies. Given the importance of reactive oxygen species in COPD, antioxidant approaches such as *N*-acetylcysteine have been initiated but antioxidant therapy with good bioavailability and sufficient potency to be effective in vivo has not been achieved.[53]

7.30.3 Experimental Disease Models

In any therapeutic area of drug discovery, animal models have limits with respect to mimicking human disease. However, several asthma and COPD animal models have been developed that capture particular aspects of the human disease. Most asthma models involve immunization and subsequent challenge with protein antigens such as ovalbumin or immunogenic proteins from the nematode *Ascaris suum*[54] or cockroaches.[55] A wider variety of exogenous agents are used to create emphysema in animals such as proteases (elastase, papain), chemicals (nitrogen dioxide, ozone), particulates (coal dust, silica), and cigarette smoke.[56] The effect of these agents in both disease models is measured by pulmonary function, cellular infiltrate (bronchiolar lavage), and histopathology. Most clinical therapeutic agents are effective in mammalian asthma and COPD models. Montelukast, a clinical cysteinyl leukotriene-1 receptor antagonist, is efficacious in ovalbumin mouse[57] and *A. suum* sheep[58] asthma models. Zileuton, a 5-liopxygenase inhibitor, has also been shown to be efficacious in the *A. suum* sheep asthma model.[59] The adverse effects of β_2-agonist enantiomers such as (*S*)-albuterol were characterized in mouse asthma models and could influence therapeutic approaches with respect to using racemic mixtures or single enantiomers.[60] Thus, animal models can be predictive of the human disease and have proven to be a valuable tool in the development of therapeutic agents for asthma and COPD.

The majority of asthma and COPD animal models have been developed in rodents such as mice, rats, and guinea pigs. In addition to the conventional model development described above, mice have afforded a genetic approach to disease mechanisms in asthma and COPD through the development of transgenic and knockout mice. Genetic manipulation of mice can be used for target identification in conventional models. It was mentioned in the previous section that the importance of MMP-12 in the development of COPD was evaluated in MMP-12$^{-/-}$ mice exposed to tobacco smoke.[49] Recent studies using BLT_1 knockout mice have demonstrated that the LTB_4/BLT_1 axis is required for the Th2 response in an ovalbumin-induced asthma model.[61] The generation of PDE4D knockout mice and subsequent testing in an ovalbumin asthma model demonstrated that this PDE4 subtype is responsible for bronchoconstriction in response to cholinergic stimulation but did not differ from wild-type mice with respect to Th2-driven inflammation.[62] This has important implications for the strategic design of PDE4 inhibitors that target single or multiple PDE4 isoforms. The role of IFN-γ in the pathology of COPD has been evaluated in transgenic mice overexpressing IFN-γ in the lung, which caused alveolar enlargement, enhanced lung volumes and pulmonary compliance, and macrophage and neutrophil-enriched inflammation. Induction of matrix metalloproteases and cathepsins were observed with a concomitant decrease in secreted protease inhibitors.[63] Constitutive expression of IL13 in the lungs of transgenic mice causes an infiltration of eosinophils, lymphocytes, and macrophages into the airways and hyperresponsiveness resembling asthma, and has been used to identify therapeutic targets associated with the IL13 mechanism such as adenosine receptors.[64] Transgenic mice devoid of eosinophils have been created by lineage-specific expression of a dyptheria toxin A protein that is selectively expressed in eosinophils (PHIL)[65] or mutation of a transcription factor essential for eosinophil differentiation (Δdbl GATA).[66] Although both approaches resulted in mice devoid of eosinophils, the PHIL mice were protected but the Δdbl GATA mice were not from airway hyperresponsiveness and excess mucus production after immunization with ovalbumin. The Δdbl GATA mice did demonstrate attenuated airway remodeling, which was not measured in PHIL mice. The results observed with the Δdbl GATA mice more closely mirror the lack of effects observed in human clinical trials with an anti-IL5 antibody, which also target eosinophils. The differences between the two transgenic mouse studies has been attributed to the different strains that were used to create the PHIL (B6) and Δdbl GATA (Balb/c) mice, which coincidentally match the phenotypic results of IL5 knockout mice in B6 or Balb/c strains. These studies illustrate the challenge of modeling multigenic diseases such as asthma in animals.[67]

In addition to enabling genetic approaches for animal model development, mice have several advantages over other species for animal models with respect to reproductive turnover, short lifespans, and relatively inexpensive breeding, housing, and maintenance costs. However, measuring pulmonary function in mice can be a challenge due to their size and the obligate nasal breathing which filters tobacco smoke inefficiently in COPD models. There are also physiological differences from humans such as an absence of goblet cells and few submucosal glands in the trachea, and there is less branching of the bronchial tree without respiratory bronchioles.[68] Significant variation between species and strains with respect to modeling COPD has been reported.[69] At a molecular level, mice can diverge from humans with respect to drug target homology and cellular mechanisms. For example, there is only 70% amino acid sequence identity between mouse and human CXCR2 receptors and mice do not have an orthologous protein for the human CXCR2 ligand IL8. Nonhuman primate models offer the advantages of genetic, anatomical, and physiological similarity to humans and relative ease in measuring pulmonary function.[70] Although the major drawback to using nonhuman primate models is the cost relative to rodents, the genetic similarity to humans can be critical for testing biological therapeutic agents such as antibodies against IL5[71] and differences in cellular mechanisms between species preclude the use of rodents as with CXCR2 and IL8.[72]

7.30.4 Clinical Trial Issues

Over the past decade several new classes of drugs targeting asthma and/or COPD have advanced into clinical trials. It is perhaps not surprising that as asthma and COPD are complicated, multifactorial diseases clinical studies must be designed carefully to evaluate the desired outcome. For example, it was mentioned in a previous section that spirometry is the traditional method for assessing pulmonary function and drug efficacy in asthma and COPD. This is a very facicle way to measure the efficacy of bronchodilators and other agents that target the acute bronchospasm and airway flow obstruction phases of the diseases. However, it is only an indirect measurement on the efficacy of agents that target the inflammatory aspects of the disease. As inflammation is now more widely recognized as a critical part of the asthma and COPD disease mechanisms, measuring inflammation biomarkers as well as pulmonary function has become a more routine aspect of clinical trials.[53,73] The effect of anti-inflammatory agents has been addressed by invasive techniques such as bronchoscopy with biopsy, bronchiolar lavage, or endobronchial brushings and by less invasive techniques such as induced sputum, exhaled breath condensate, or gas analysis.[74] A trial measuring the effect of the leukotriene receptor antagonist montelukast examined the effect on serum concentrations of the soluble IL2 receptor, IL4, soluble ICAM-1, and peripheral blood eosinophils in children with asthma. Significant decreases in each of these parameters correlated with improved pulmonary function.[75] Other inflammatory biomarker assays have been developed such as measuring levels of carboxyhemoglobin in patients with COPD.[76] It must be noted that the measurement of biomarkers can be complicated by existing therapies. Asthmatic patients are known to have increased eosinophils in sputum and exhale higher concentrations of nitric oxide. As measuring exhaled nitric oxide would be a noninvasive, there was interest in developing it as a clinical biomarker. However, dissociation between exhaled nitric oxide and sputum eosinophil counts can result from inhaled corticosteroids. Therefore, exhaled nitric oxide is not viewed as a reliable biomarker given the prevalence of inhaled corticosteroid use.[77]

Other challenges in clinical trials include specific issues around the targets and therapies being developed. A significant challenge around the development of PDE4 inhibitors involved eliminating emetic effects associated with the parent molecule rolipram. Although the exact mechanisms involved in the emetic effect have been debated between high- and low-affinity interactions and PDE4 subclasses, several PDE4 inhibitors have apparently overcome this obstacle and have advanced through clinical trials. Cilomilast from GlaxoSmithKline and rofumilast from Altana and Pfizer are in preregistration for COPD, and tetomilast from Otsuka has advanced into phase II trials for COPD.[78]

The development of therapeutic agents targeting eosinophils such as CCR3 antagonists has been called into question based upon negative results in a phase II asthma trial for mepolizumab (SB-240563), a humanized antibody against IL5.[79] Although treatment with the antibody significantly decreased blood and airway eosinophils, no significant effect was observed in improving late asthmatic response or histamine-mediated airway responsiveness. However, the design and implementation of the clinical trial has been questioned such that it cannot be readily concluded that IL5 and eosinophils are not involved in the late asthmatic response or airway hyperreactivity.[80] Although it was not efficacious in asthma trials, mepolizumab has advanced into phase III clinical trials for hypereosinophilic syndrome.[81] Despite the controversy around IL5 and eosinophils in asthma, at least two CCR3 antagonists have advanced into clinical trials for asthma; GSK-766994 is in phase II and BMS-639623 (DPC-168) is in phase I.[82] GlaxoSmithKline has also advanced a CXCR2 antagonist (SB-322235) in clinical trials to target neutrophil trafficking in COPD.[83]

As proteases have proven to be tractable and successful drug targets in other therapeutic areas (cardiovascular, antivirals, etc.), parallel approaches have been attempted in asthma and COPD around neutrophil elastase, tryptase, cathepsins, and matrix metalloproteases. ONO Pharmaceuticals has developed and launched the neutrophil elastase inhibitor Sivelestat (ONO-5046) in Japan for acute lung injury[84] and it is anticipated that these inhibitors could be efficacious in COPD.[85] Several pharmaceutical companies have been active in developing inhibitors of mast cell β-tryptase, but few have reached clinical trials.[86] Axys/Celera advanced APC-366 into clinical trials for asthma despite poor selectivity against other serine proteases such as trypsin and thrombin, but despite indications of clinical safety and efficacy trials were stopped due to formulation issues. Johnson & Johnson have also advanced a tryptase inhibitor into phase II trials for asthma (RWJ-56423), but poor bioavailability has limited its administration to aerosol delivery. Inhibitors of cathepsins or matrix metalloproteases have not progressed into clinical trials for asthma or COPD, but are being evaluated for other inflammatory diseases. Serono and Vernalis (formerly British Biotech) are reported to have advanced an MMP-12 inhibitor with a primary indication of multiple sclerosis into phase I clinical trials and a cathepsin S inhibitor with a psoriasis indication has been advanced into phase I trials by Celera. Efficacy in these inflammatory diseases could lead to clinical evaluation in asthma and COPD.

Intensive efforts to develop therapeutic agents that interfere with the biosynthesis and activity of leukotrienes have resulted in clinical success against asthma and more modest success against COPD.[87] Zileuton (Zyflo), a 5-lipoxygenase inhibitor,[88] and the leukotriene receptor antagonists zapfirlukast (Accolate)[89] and montelukast (Singulair)[90], were some of the first drugs to demonstrate clinical success but also highlight some of the issues in clinical development of these drug classes. The clearance of zileuton from the circulation is relatively fast (half-life of 4 h) and requires dosing at four times per day. The liver cytochrome P450 enzymes CYP1A2, CYP2C9, and CYP3A4, mediate zileuton clearance, and inhibition of CYP1A2 can limit the use of zileuton as it appears to interfere with the clearance of other drugs such as propanolol, theophyline, and warafin.[91] Leukotriene receptor antagonists such as zapfirlukast and montelukast are more effective when used in combination with other anti-inflammatory drugs such as inhaled corticosteroids rather than as a monotherapy.[92] However, caution must be taken when the use of inhaled corticosteroids is tapered as incidences of Chung–Strauss syndrome, an eosinophilic infiltrative disorder, has been reported in combination with zapfirlukast[93] or montelukast.[94] The incidence of Chung–Strauss syndrome associated with either leukotriene receptor antagonist is relatively low and is thought to predate the antagonist therapy and is unmasked when inhaled corticosteroid is reduced. Thus, the risks of either 5-lipoxygenase inhibitors or leukotriene receptor antagonists are relatively small when compared to their therapeutic benefit.

7.30.5 Current Treatment

Therapeutic approaches against asthma and COPD have been tailored to the severity of disease.[8,9] In all stages of asthma, inhaled short-acting β_2-agonists such as albuterol are recommended for quick relief from bronchospasms on an as-needed basis (**Table 3**). Oral steroids such as triamcinolone may also be prescribed for acute relief. This regimen is usually sufficient to treat mild intermittent asthma. In mild, moderate, and severe persistent stages, inhaled corticosteroids (beclametasone) and long-acting β-agonists (salmeterol, formterol) are commonly prescribed for daily use. Alternative approaches for mild and moderate persistent asthma include theophylline or leukotriene modifiers such as montelukast or zileuton. In some severe cases, systemic use of steroids such as prednisone has been shown to be effective. For childhood and exercise-induced asthma, mast cell stabilizers such as nedocromil are prescribed. An antibody against IgE, omalizumab (Xolair), has been demonstrated to be clinically effective on asthma exacerbation rates, use of corticosteroids, and symptoms in both adults and children and was launched for use in the USA in 2003.[95]

In COPD smoking cessation is a primary objective for treatment as a long history of smoking is almost always linked with developing the disease. The recommended therapy for those at Stage 0 (and also healthy individuals) is to avoid risk factors such as tobacco smoke and other environmental pollutants (**Table 4**). Bronchodilators such as β2-agonists are prescribed to patients in Stages I to III for symptomatic management of the disease. For Stage I bronchodilators are recommended on an as-needed basis and for more advanced stages of COPD they are used regularly. Corticosteroids, either inhaled or delivered orally, are recommended to patients with a spirometric response to the agent and/or are in

Table 3 Asthma severity and treatment

Severity	*Symptoms*	*Pulmonary function (PEF or FEV_1)*	*Treatment*
Mild, intermittent	Attacks <1 per week, night-time episodes infrequent	Decreased <20%	Inhaled β_2-agonist as needed
Mild, persistent	Attacks 3–6 per week but <1 per day, >2 night-time episodes per month	Decreased <20%	Inhaled β_2-agonist as needed Daily inhaled corticosteroid or long-acting β_2-agonist
Moderate, persistent	Daily attacks, >5 night-time episodes per month	Decreased 20–40%	Inhaled β_2-agonist as needed Daily inhaled corticosteroid or long-acting β_2-agonist
Severe, persistent	Continuous attacks	Decreased >40%	Inhaled β_2-agonist as needed Daily inhaled corticosteroid or long-acting β_2-agonist

Table 4 COPD severity and treatment

Severity (stage)	*Pulmonary function (PEF or FEV_1)*	*Treatment*
At risk (0)	Normal spirometry but chronic cough and sputum production	Avoid risk factors
Mild (I)	Decreased <20%	Inhaled β_2-agonist as needed
Moderate (IIA)	Decreased at least 20% but not more than 50%	Inhaled β_2-agonist regularly; inhaled corticosteroid if effective
Moderate (IIB)	Decreased at least 50% but not more than 70%	
Severe (III)	Decreased >70%	Inhaled β_2-agonist regularly Inhaled corticosteroid if effective; oxygen therapy; surgery

Stage II or III of the disease. However, corticosteroids are not effective in suppressing inflammation in COPD patients.[40] In severe (Stage III) COPD, long-term oxygen exposure to patients with respiratory failure is beneficial but surgical treatments such as bullectomy, lung volume reduction, or lung transplantation may have to be considered.

7.30.6 Unmet Medical Needs

As indicated in the previous section, current therapies for asthma and COPD are primarily directed at treating the symptoms (e.g., β_2-agonists for bronchoconstriction) rather than the disease. Treating symptoms has been generally effective in the management of asthma, but not for a progressive disease like COPD. Even in asthma the use of corticosteroids, especially in children, has risks associated with this general immunosuppressive agent. New therapeutic approaches that target the immune mechanisms associated with the disease have shown clinical success (zileuton, montekulast, etc.) and have initiated the search for new therapies targeting immune mechanisms in COPD and asthma.

7.30.7 New Research Areas

7.30.7.1 Chemokine Receptor CCR3

The chemotactic cytokine (chemokine) eotaxin (CCL11), and its 7-transmembrane G protein-coupled receptor (GPCR) CCR3 have been shown to be involved in modulating the trafficking of eosinophils to areas of allergen challenge. While the ligand CCL11 is specific to its receptor target CCR3 has been shown to be quite promiscuous in that at least 11 other endogenous ligands or their isoforms also exhibit affinity. Of the characterized 'allergy related' chemokine receptors,[96–98] CCR3 has received the most attention due to a strong clinical observation that lung disfunction correlates with the number of eosinophils that accumulate in the bronchioalveolar compartments, specifically in cases of bronchial asthma.[99,100]

Although CCR3 is highly expressed on eosinophils (estimated to be 40 000–50 000 sites per cell) it is also expressed on basophils as well as a subset of T lymphocytes with Th2 properties. Additional supportive, beneficial aspects of modulating CCR3 pharmacology have been validated by a number of mechanisms. In the murine knockout models, $CCR3^{-/-}$ animals have been shown to exhibit greatly attenuated recruitment of eosinophils to skin patches after repeated sensitization with ovalbumin and exhibit decreased airway hyperresponsiveness to methacholine with 90% reduction of cellular influx into the bronchioaveloar lavage after a single ovalbumin challenge.[101] Similar experiments with $CCL11^{-/-}$ mice show a 70% reduction of relevant cellularity after challenge.[102] While several small-molecule approaches will be reviewed within this section, it is worth mentioning that biologic approaches including one involving monoclonal antibodies targeted at blocking ligand–receptor interactions have also been investigated. An example of a murine antibody against CCL11 showed a reduction of in vivo eosinophil recruitment to the lung in response to ovalbumin by 50%.[103] Again, much like the knockout studies, the corresponding receptor antibody 7B11 is able to block second messenger Ca^{2+} flux and chemotaxis in human eosinophils in vitro.[104]

Several small-molecule medicinal chemistry examples that demonstrate the relevance of CCR3 antagonism in asthma models are presented. In addition to these illustrative examples, a recent, more comprehensive, chemotype survey of the patent literature can be found in the review by Naya and Saeki.[105]

High-throughput screening of corporate compound collections has lead to a variety of published pharmacophores that effectively antagonize the CCR3 receptor and subsequent downstream signaling cascade. A nonselective CCR1/CCR3 antagonist (**1**, CCR1 binding IC_{50} = 0.9 μM versus CCR3 binding IC_{50} = 0.6 μM) was identified and a strategy based on a binding hypothesis to generate a focused combinatorial library that would incorporate cross receptor selectivity SAR as developed.[106–108]

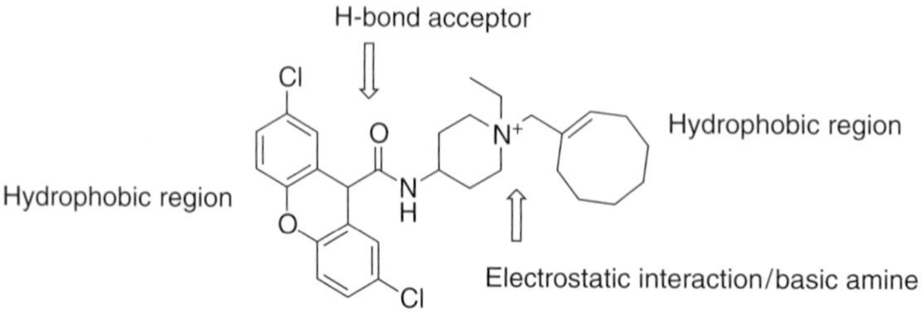

H-bond acceptor
Hydrophobic region
Hydrophobic region
Electrostatic interaction/basic amine

1

This exercise subsequently generated a 770-member focused library that delivered the 2-(benzthiazole)thioacetamide (**2**) with good potency and >800-fold improved selectivity. It is worth noting that **2** is also the first reported noncompetitive ligand described for CCR3 since binding IC_{50} values did not increase predictably despite a 100-fold increase in $[^{125}I]$eotaxin concentration.

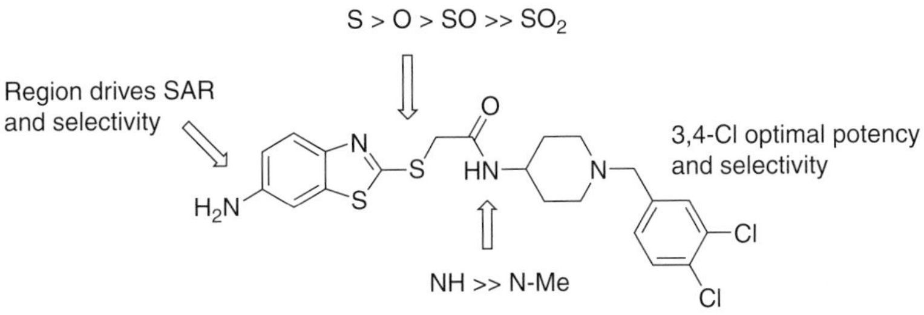

$S > O > SO >> SO_2$

Region drives SAR and selectivity

3,4-Cl optimal potency and selectivity

NH >> N-Me

2 Binding CCR3 IC_{50} = 2.3 nM
Ca^{2+} flux IC_{50} = 27 nM

A variation of this chemotype advanced to rodent and nonhuman primate models of asthma. Analog **3** was evaluated at 40 mg kg^{-1} (subcutaneous) in *Cynomologus* monkeys challenged with inhaled *Ascaris* pathogen resulting in a 50% reduction of BAL eosinophils when compared to vehicle-treated animals.[109]

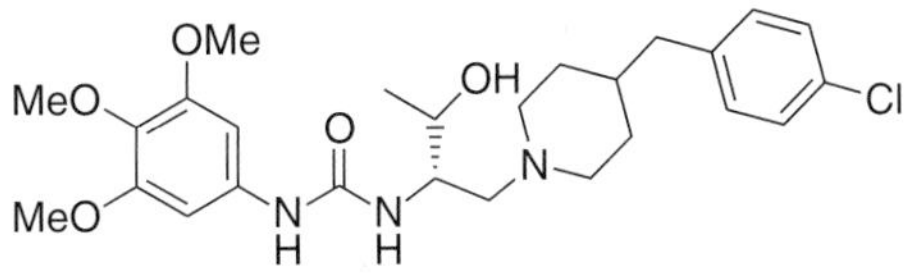

3 Binding IC_{50} = 4 nM
Eosinophil chemotaxis IC_{50} = 1.8 nM

In one lead optimization example, the structure–activity relationship (SAR) was explored with a synthesis strategy focused on exploiting the fact that constrained variants of the piperidine ring can alter the spatial presentation of the basic nitrogen which may interact with a key conserved glutamic acid residue located within the TM7 regional loop. The bridged bicyclic analog **4** ultimately offered no potency increase over the parent piperidine derivative.[110]

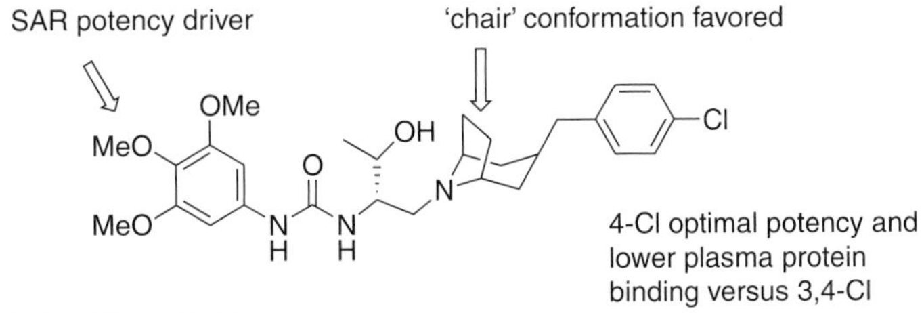

SAR potency driver

'chair' conformation favored

4-Cl optimal potency and lower plasma protein binding versus 3,4-Cl

4 Binding IC_{50} = 8 nM
Chemotaxis IC_{50} = 2.4 nM
Poor in vivo PK profile

The poor pharmacokinetic profile of **4** is driven by a combination of lipophilicity and a basic nitrogen that resulted in a high volume of distribution and high clearance rate. This example also illustrates another important strategy typically used in medicinal chemistry programs. Conformational restriction of rings or chains can often maintain target receptor potency while assisting in improving selectivity over other off-target 7TM receptors, typically those in the central and peripheral nervous system.

A common benzyl piperidine motif has emerged within various chemotypes reported to be selective antagonists of CCR3. The bispiperidine series **5** was optimized using a systematic, iterative process coupled with library array expansion where chemically feasible (amide formation).

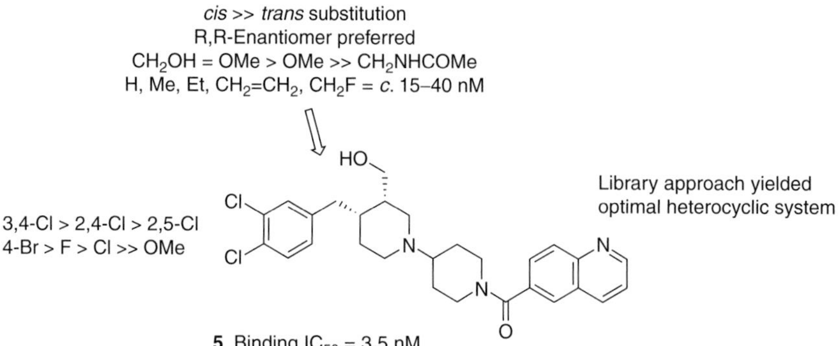

cis >> *trans* substitution
R,R-Enantiomer preferred
CH_2OH = OMe > OMe >> $CH_2NHCOMe$
H, Me, Et, CH_2=CH_2, CH_2F = *c.* 15–40 nM

3,4-Cl > 2,4-Cl > 2,5-Cl
4-Br > F > Cl >> OMe

Library approach yielded optimal heterocyclic system

5 Binding IC_{50} = 3.5 nM
Ca^{2+} flux IC_{50} = 9 nM
Eosinophil chemotaxis IC_{50} = 160 nM

Overall the series showed good chemokine receptor cross-selectivity with only 11% inhibition and 8% inhibition of CCR4 and CCR8 at 1 μM, respectively. It is important to note that while **5** maintained good affinity for the human and monkey CCR3 receptors (K_i = 3.3 and 7.2 nM) much lower affinity for the rodent receptors (rat K_i = 2981 nM; mouse K_i = 606 nM) is seen. This problem of species selectivity is not atypical and often hinders many GPCR optimization efforts providing significant challenges with in vivo profiling. As a final point, this series was also hindered by an undesired 85% (1 μM) hERG inhibition as measured in a voltage clamp assay.[111]

Within the lead optimization strategy for this chemical series the orientation of the bispiperidine ring system was reversed as shown below (**6**). The SAR indicated that the central pharmacophore could indeed be reversed and reasonable CCR3 affinity could be maintained. Movement of the amide moiety to the 3-position of the reversed piperidine followed by repositioning of the amide atom array provided an initial 60 nM inhibitor as measured in the binding assay. The requirement for a 3,4-dichloro substituted benzyl group was confirmed and, based on a survey of various CCR3 active chemotypes, this motif appears to provide an optimal binding interaction with a putative lipophilic region of the receptor. Further SAR exploration yielded an unexpected surprise, that is, while competitive inhibition of CCL11 binding was maintained, the functional activity (using a [^{35}S]GTPγS binding assay)[112] reversed to that of an agonist as illustrated by analogs **7** versus **8**.

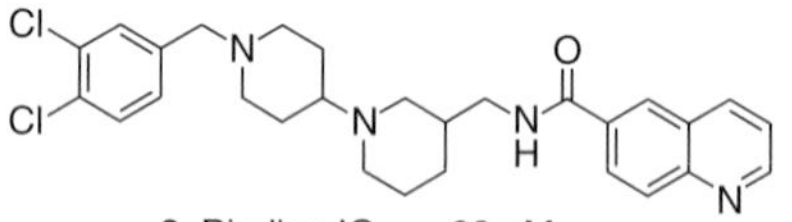

6 Binding IC_{50} = 23 nM
Ca^{2+} flux IC_{50} = 215 nM
BAF3 chemotaxis IC_{50} = 136
Poor rat PK (AUC iv)

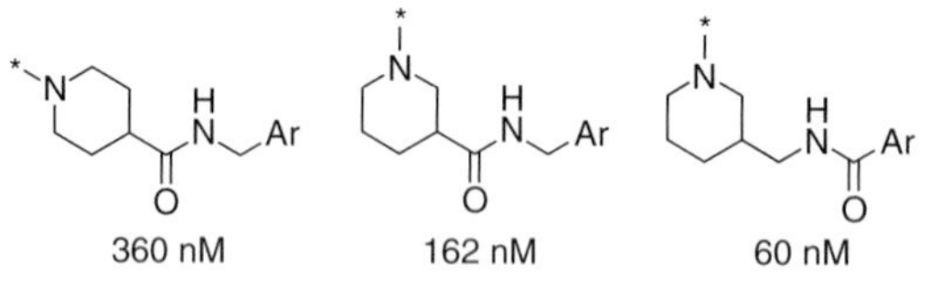

360 nM 162 nM 60 nM

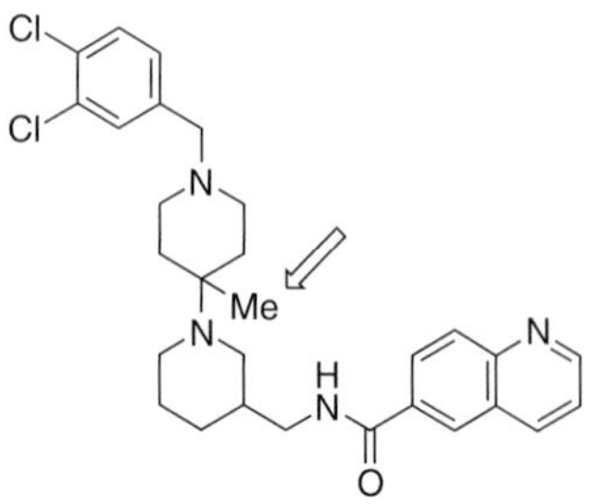

7 Antagonist activity
Binding K_i = 50 nM
E_{max}%(GTPγ S) = 8%

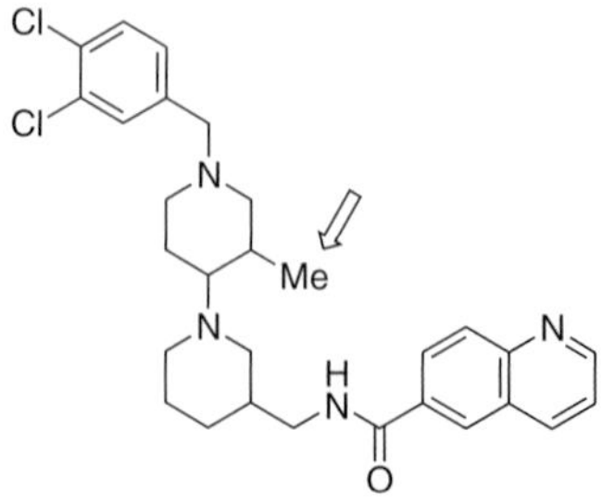

8 Agonist activity
Binding K_i = 7.3 nM
E_{max}%(GTPγ S) = 96%

Similar reversal in functional activity has also been seen in another chemotype. During the investigation and optimization of a series of pyrrolidinohydroquinazolines, analog **9** was identified as an antagonist with promising potency. Further exploration furnished analog **10** with improved binding; however, agonist activity was conferred.[113]

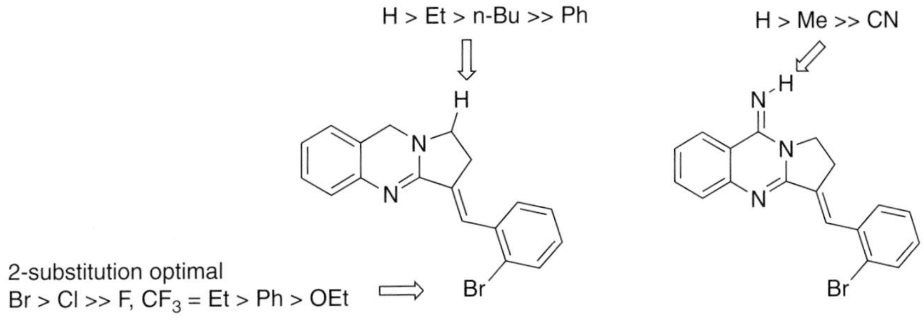

H > Et > n-Bu >> Ph

H > Me >> CN

2-substitution optimal
Br > Cl >> F, CF_3 = Et > Ph > OEt

9 Functional antagonist
Binding K_i = 90 nM

10 Functional agonist
Binding K_i = 28 nM

Another series of potent analogs containing benzylpiperidine and urea moieties illustrate clever SAR analog strategies that allowed for the optimization of CCR3 potency, target selectivity over neuroreceptors, and pharmacokinetic properties.[114–117] In an iterative sequence the HTS hits represented by **11a, b** were converted to the corresponding urea linked analog **12** based upon a competitive analysis of the chemotype/target literature. Subsequent recognition of a latent benzylpiperidine (bold lines in **11a**) furnished the lead analog **13**.

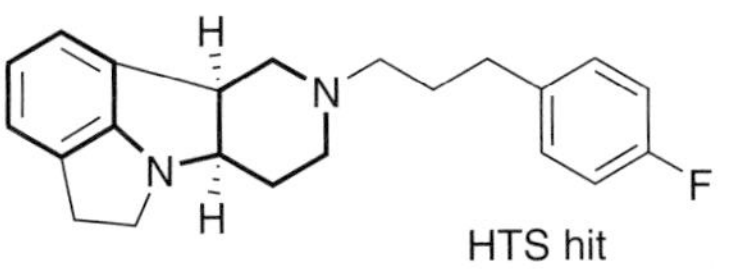

HTS hit
Binding IC_{50} = 500 nM

11a

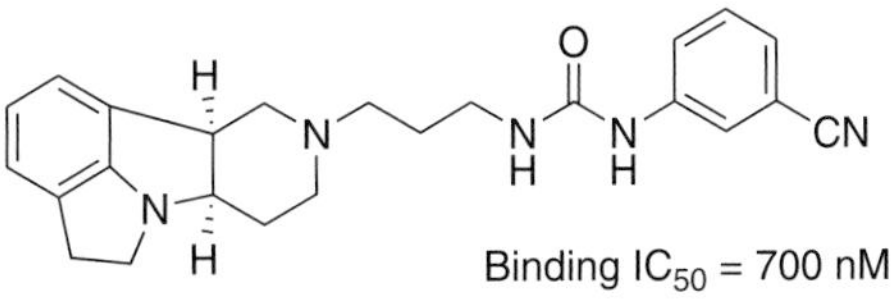

Binding IC_{50} = 700 nM

12

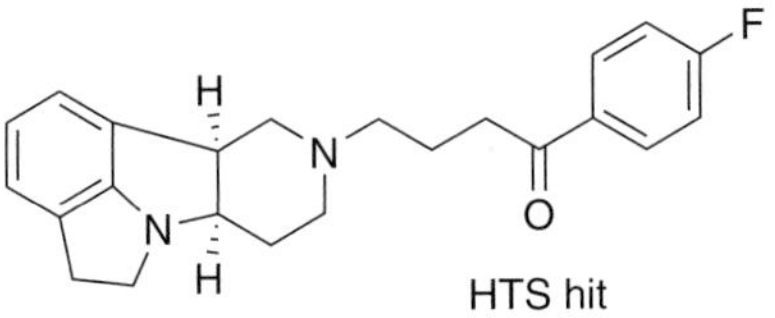

HTS hit
Binding IC_{50} = 1μM

11b

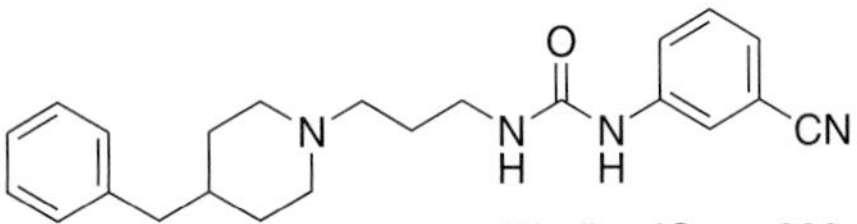

Binding IC_{50} = 200 nM

13

Structural changes in the piperidine functionality and continued exploration of aromatic substitution improved the selectivity of the antagonists for CCR3 and drove potency to the picomolar range (**14**).

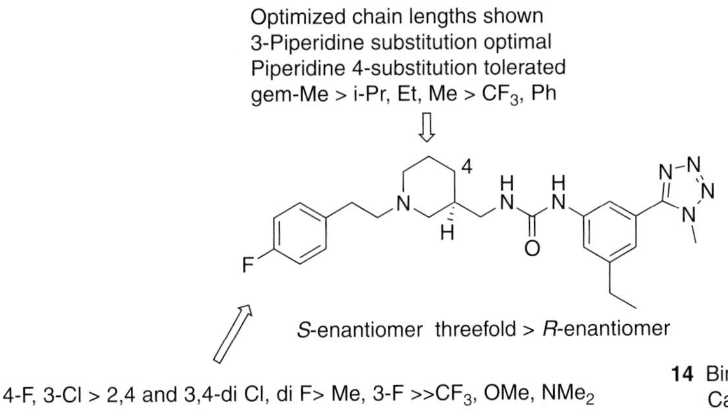

Optimized chain lengths shown
3-Piperidine substitution optimal
Piperidine 4-substitution tolerated
gem-Me > i-Pr, Et, Me > CF_3, Ph

S-enantiomer threefold > *R*-enantiomer

4-F, 3-Cl > 2,4 and 3,4-di Cl, di F> Me, 3-F >>CF_3, OMe, NMe_2

14 Binding IC_{50} = 0.7 nM
Ca^{2+} flux IC_{50} = 27 nM
Chemotaxis = 51% I at 30 nM

The search for improved receptor selectivity continued with SAR exploration of the linker region, chain lengths, and aromatic substitution, as well as identification of the optimal diastereomeric isomer which ultimately lead to analog **15** which proceeded to phase I clinical studies.

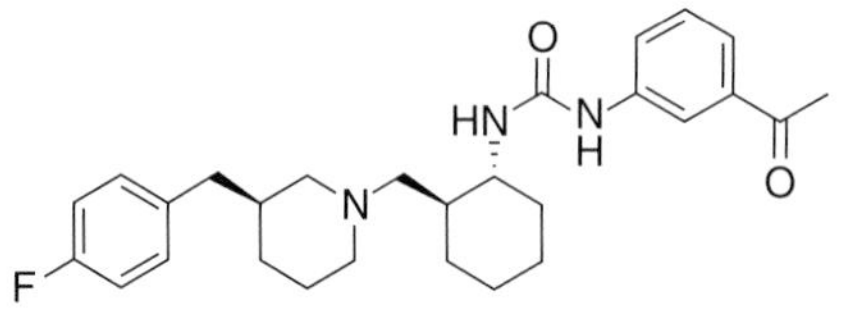

15 DPC-168 / BMS-639623

Binding CHO IC_{50}=2 nM
Binding hEos IC_{50} = 800 pM
Ca^{2+} flux IC_{50} = 8 nM
Eosinophil chemotaxis IC_{50} = 34 pM
Selective over D2 and 5HT2a,2c

Murine F% = 20
Cyno F% = 8
Chimp F% = 22
$t_{1/2}$ = 2–5h
CL = 1.2–2 (L/h)/kg
Protein binding = 96%

OVA challenge: 50% 30 mpk, 86% 100 mpk po

Although a detailed account of the discovery and SAR surrounding a second clinical candidate has not yet been disclosed, the pharmacophore elements of the urea linkage and the 3,4-dichloro substituted benzylpiperidine has been translated to the novel morpholine analog **16** as disclosed in a series of nine patent applications.[118]

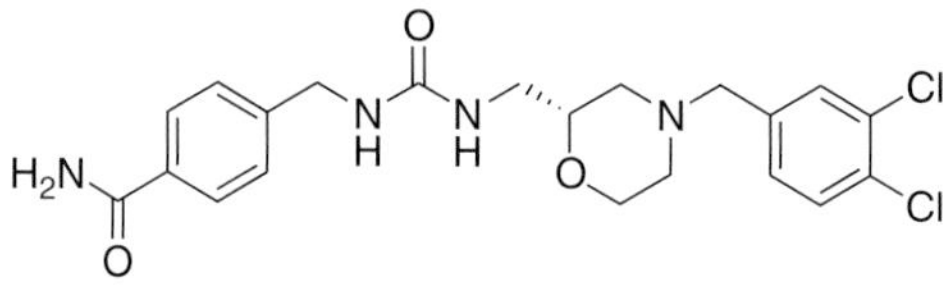

16 GSK-766994
Phase II clinical candidate for asthma and allergic rhinitis

In conclusion, there is strong supportive evidence linking eosinophil recruitment and activation in the lung tissue with asthma. Recently CCR3 has become a popular target and the landscape is very competitive. Given the lack of crystal-based structural information to guide SAR development, computational approaches to ligand design are emerging as exemplified by a five-dimensional quantitative SAR (5D-QSAR) model.[119] Several excellent reviews exist that more fully survey the breadth of patent literature.[105] It is also known that eosinophil trafficking is not an exclusively CCR3 mediated event and the clinical trials will determine whether CCR3 antagonists will achieve therapeutic potential.

7.30.7.2 CRTH2 and the DP-1 Receptor

The prostaglandin D_2 (PGD_2) is the predominant prostanoid produced by allergen activated mast cells and it mediates its effects as an agonist ligand for two different 7TM-GPCR receptors, DP-1 (45 nM) and CRTH2 (61 nM; chemoattractant receptor-homologous molecule expressed on T cells, has also been named DP-2) (see **Figure 4**).[120] PGD_2, along with a host of arachadonic acid derived prostanoid agonists, are inflammatory mediators responsible for a host of events including bronchoconstriction mediated by contraction of smooth muscle. CRTH2 is expressed on and mediates chemotactic responses of eosinophils, basophils, and Th2 cells and plays a role in inducing eosinophil degranulation thus, it stands to reason that activation of this receptor by PGD_2 may play a key role in modulating inflammatory response (**Figure 4**). CRTH2 is coupled with a Gαi-type G protein and is thought to be involved in post-PGD_2 stimulatory activities such as induction of cell migration and upregulation of adhesion molecules.[121,122] Research targeting small-molecule mediators of the CRTH2 receptor is a relatively new field. Although no clinical therapeutics have yet emerged a number of intriguing small-molecule probes have been discovered. Recently, it has been shown that the nonsteroidal anti-inflammatory indomethacin (**17**) has affinity for the receptor. In the course of identifying PGD_2 as the sole CRTH2 ligand produced by mast cells, indomethacin (a lipoxygenase inhibitor) was used to suppress PGD_2 production. It was unexpectedly found that **17** exhibited an agonistic effect thus activating the cells toward migration.[123] This observation is clinically significant because it demonstrates that a widely utilized COX inhibitor can stimulate rather than inhibit a chemotactic receptor leading to undesired migration of leukocytes at therapeutic blood levels.

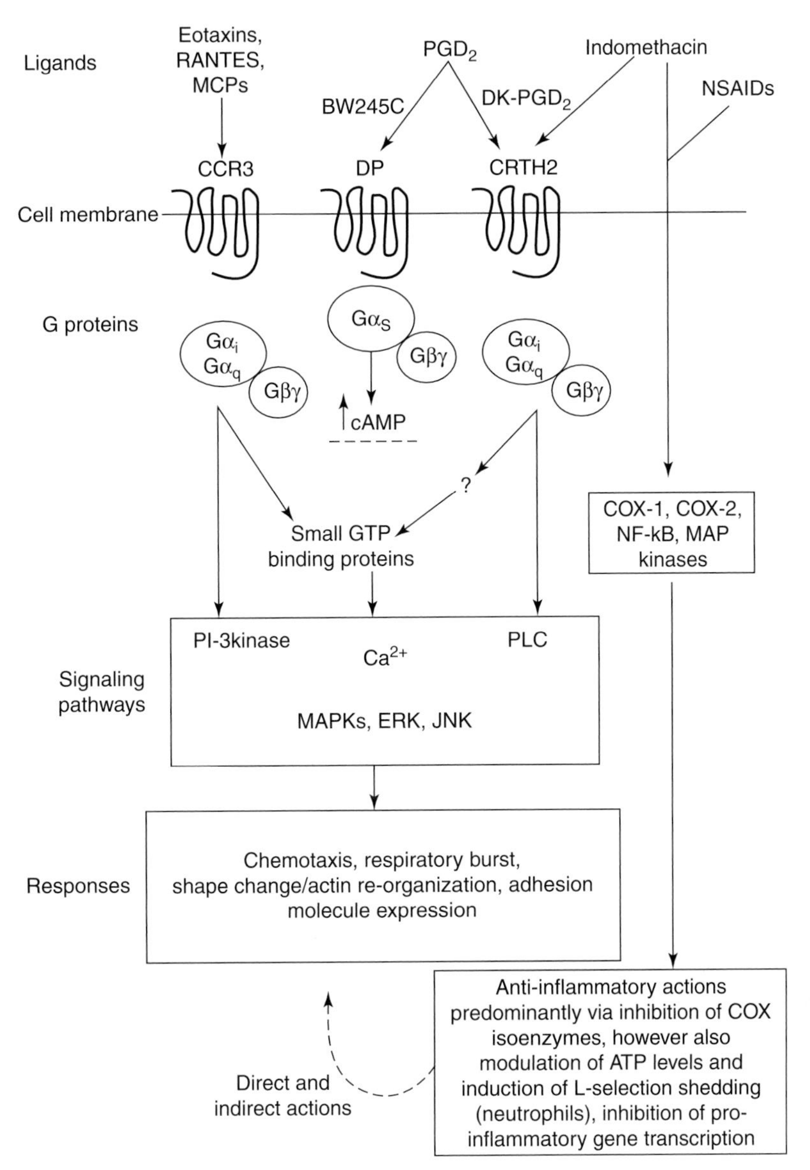

Figure 4 Proposed signaling pathways for the CCR3 and CRTH2 receptors. (Reproduced with permission from Stubbs, V. E. L. *et al. J. Biol. Chem.* **2002**, *277*, 26012 © American Society for Biochemistry & Molecular Biology.)

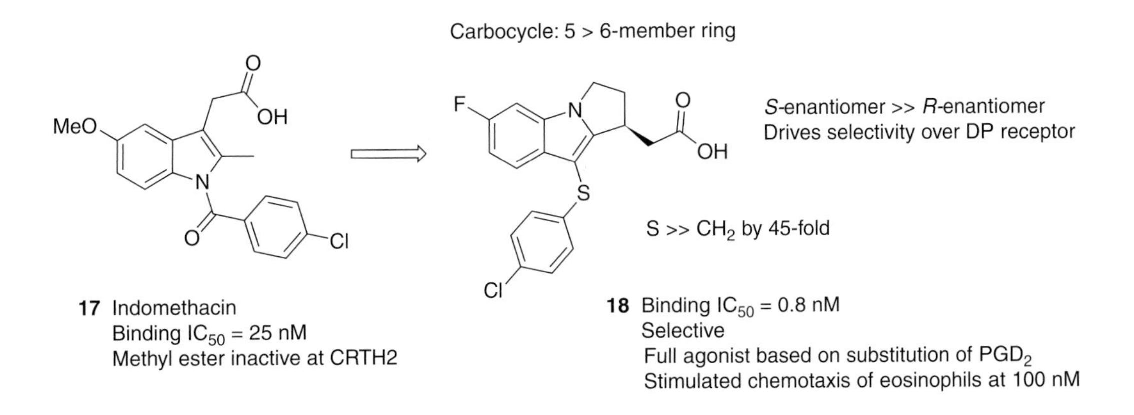

In order to understand the in vivo role of this receptor, a potent and highly selective agonist ligand for CRTH2 was derived from manipulation of the core indole acetic acid moiety of indomethacin.[124,125] Analog **18** exhibited nearly complete selectivity for CRTH2 over all other prostaniod receptors and the (*S*)-configuration at the chiral center was pivotal for selectivity over DP-1, in which activity is conferred with the corresponding (*R*)-enantiomer. The analog has excellent pharmacokinetic properties with bioavailability of 48% and should prove useful for in vivo mechanistic studies.

The structurally related analog ramatroban (**19**) is one of the first reported CRTH2 antagonists. It is currently marketed in Japan for treatment of allergic rhinitis and has been characterized as a selective TP antagonist (thromboxane A_2 receptor).[126] The drug candidate is currently in phase III clinical trials for treatment of asthma.[127,128]

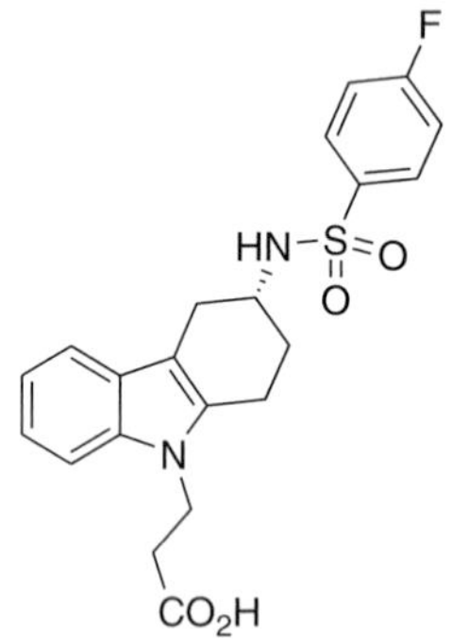

19 ramatroban
Antagonist activity
CRTH2 binding IC_{50} = 100 nM
Ca^{2+} flux IC_{50} = 30 nM

During further mechanistic characterization it was discovered that **19** is also an antagonist of CRTH2 (measured against ^{3}H-PGD_2) suggesting its clinical efficacy may in part be through its modulation of chemotaxis. Further synthesis and optimization of the ramatroban scaffold has furnished a series of isosteric analogs highlighted by **20**. Reversal of the scaffold substituents and subsequent optimization of both aromatic substitution and chain length between the acid functional group and the indole nitrogen provided an antagonist 400-fold more selective for CRTH2 versus the TP receptor.[129]

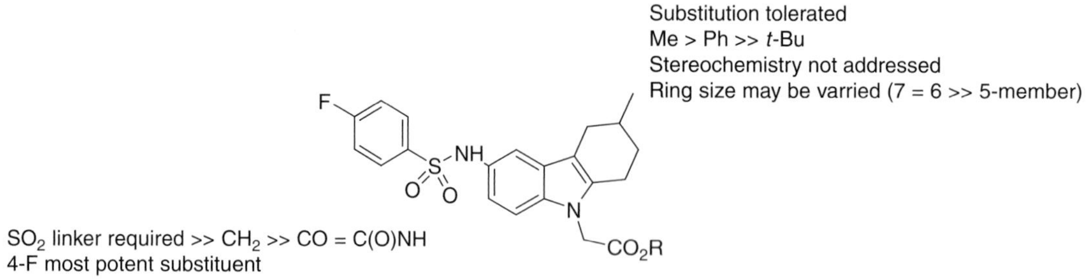

Substitution tolerated
Me > Ph >> *t*-Bu
Stereochemistry not addressed
Ring size may be varried (7 = 6 >> 5-member)

SO_2 linker required >> CH_2 >> CO = C(O)NH
4-F most potent substituent

Acetate > propionate >> benzoate
Alkyl substitution ablates activity

20 Binding HEK293 K_i = 13 nM
Ca^{2+} flux IC_{50} = 9.7 nM
Selective over TP

It has also been shown that subtle manipulation of the ramatroban scaffold can greatly shift the selectivity profile leading to complete loss of activity at the TP and DP-1 receptors. *N*-Methylation of the sulfonamide in combination with shortening the acidic side chain furnished **21** as a highly potent antagonist. The antagonist profile included an assessment of inositol phosphate, cAMP signaling second messenger functional activity and inhibition of agonist mediated β-arrestin translocation.

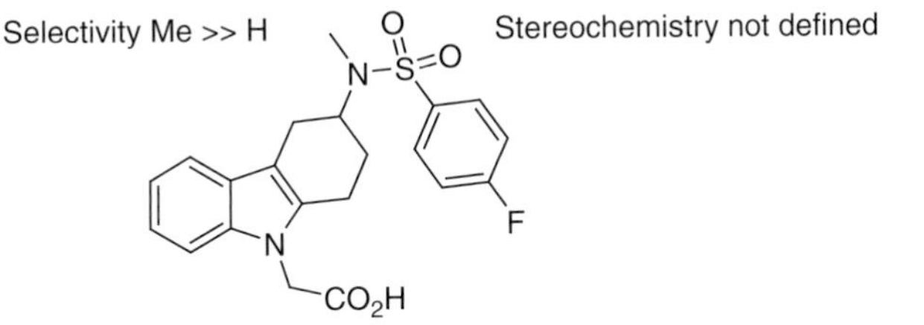

Selectivity Me >> H

Stereochemistry not defined

21 hCRTH2 binding IC_{50} = 600 pM
Inositol phosphate IC_{50} = 1.2 nM
β-arrestin *trans* IC_{50} = 3 nM

Chain length alters selectivity and potency(one methylene optimal)

The analogs were prepared as racemates and resolution was not accomplished since the intent was to use **21** as a mechanistic biochemical tool.[130] Additional chemotypes centering on indole, quinoline, and carbazole derivatives have been reported in the patent literature[131] indicating intense competition in seeking a clinically relevant CRTH2 antagonist within a narrow privileged chemotype.

Given the PGD_2 ligand relationship between CRTH2 and DP-1, it is appropriate to bin discussion of these two receptors together even though they do not share significant sequence homology. It is important to remember that even though these 7TM-GPCR receptors share a common ligand, DP-1 belongs to the prostaniod receptor family ($G_{\alpha s}$-coupled signaling) and CRTH2 is a member of the chemokine receptor family thus differentiating physiological functional roles. As well, DP-1 is primarily expressed on smooth muscle and epithelial cells while CRTH2 is expressed primarily on T cells, basophils, and eosinophils. The development of novel antagonists in this area has been slow and only few reports on the efficacy of DP-1 receptor antagonists in asthma models or against human disease are available.[132] Most frequently reviewed is the hydantoin prostaniod mimetic **22** which exhibits moderate competitive binding but is shown to be a partial agonist in functional assays.[133]

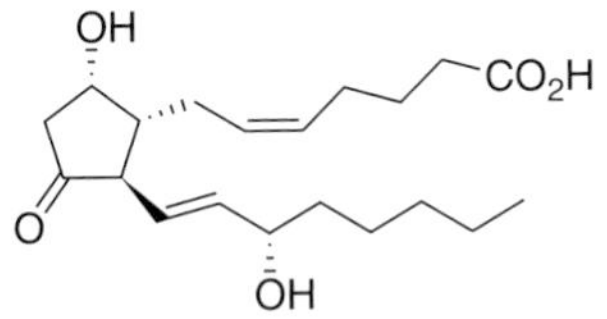

PGD2 DP-receptor binding K_i = 21 nM

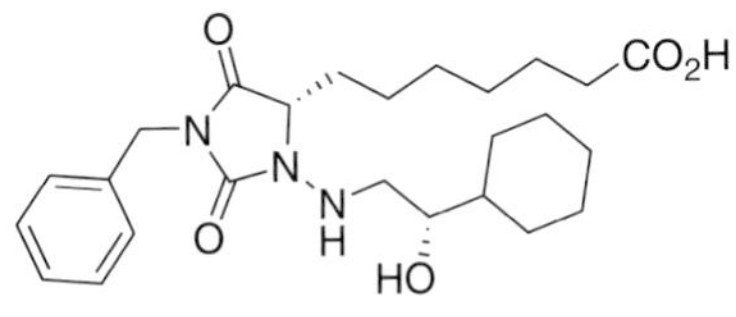

22 BW A868C
Binding K_i = 220 nM

One approach to analog generation has been to exploit receptor class cross-reactivity and develop new chemical entities from prostaglandin scaffolds. In this case, the bicyclo[2.2.1]heptane ring system,[134] as shown in analog **23**, and the bicyclo[3.1.1]heptane system[135] **24** has provided a new class of antagonists originating from a thromboxane A_2 hit in high-throughput screening. Based on observations in past programs, both series of analogs were optimized with a focus on the ω-chain that seemed most important for PGD_2 activity (versus other family receptors). This included the incorporation and extensive SAR of aromatic moieties linked with a proper spacer. Ultimately the dibenzofuran derivative **23** was selected as the (+) – enantiomer because of its selectivity profile.

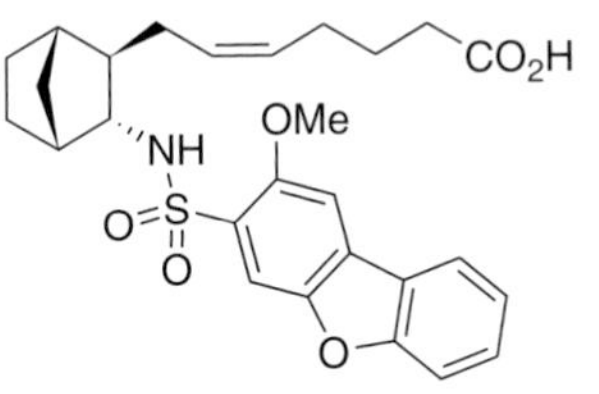

23 Binding IC_{50} = 24 nM
Functional cAMP IC_{50} = 52 nM

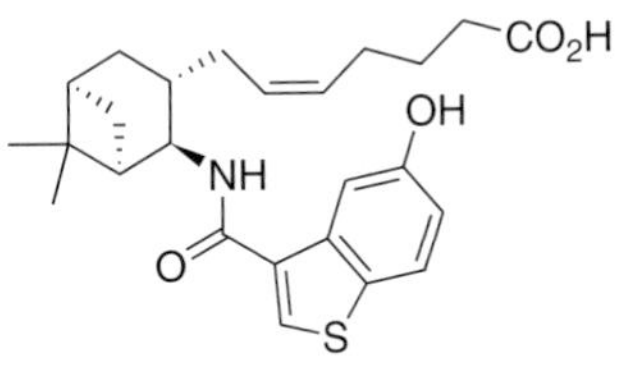

24 (*S*-5751) Binding IC_{50} = 1.9 nM
Functional cAMP IC_{50} = 0.9 nM

In general, the sulfonamide moiety was observed to be critical and it was suggested that the sulfonamide NH mimicked the C15 hydroxyl proton of PGD_2. Analog **23** was subsequently taken in vivo and exhibited 42% inhibition, based on airway resistance, in the guinea pig asthma model at 10 mg kg^{-1} (orally). In a similar manner an alternate ring structure was optimized to provide **24** after a stereochemical SAR study indicated the (1*R*,2*R*,3*S*,5*S*)-isomer exhibited optimal potency. Next, the sulfonamide of **23** was substituted with the simpler amide linker. Subsequent optimization of the heterocycle and substituents furnished an analog with good in vivo efficacy in the guinea pig asthma model (70% inhibition at $10\,mg\,kg^{-1}$, orally). This analog is currently considered a promising clinical candidate, now in development.

Several other pharmacophore cores have also emerged. In a hit-to-lead exercise, chemical manipulation of the anti-inflammatory indomethacin has furnished an indole-based chemotype that is currently under investigation. Iterative synthesis and SAR at each substituent on indomethacin lead increased affinity for the DP-1 receptor with selectivity over the EP subset. Potency seemed to have peaked until the acetic acid moiety was transferred from the indole C3 position to C4 providing analog **25** with an optimal substitution pattern.[136]

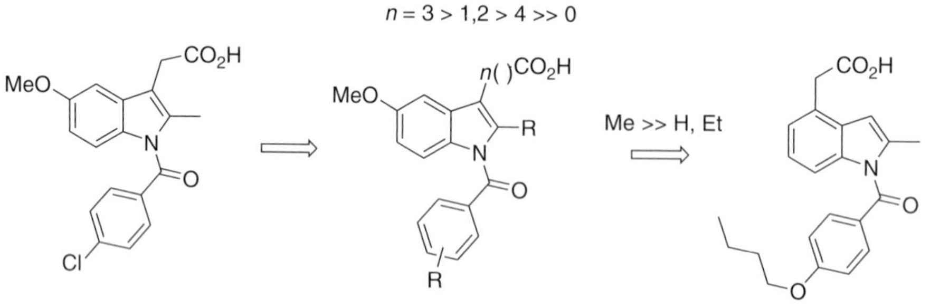

n = 3 > 1,2 > 4 >> 0

Me >> H, Et

Indomethacin
DP binding IC_{50} = >10 μM

O-Bu: *para* > *meta* >> *ortho*

25 Binding IC_{50} = 300 nM

With the lead structure **25** identified, optimization continued through exploration of the indole nitrogen substituent.[137] Employing a resin-based parallel synthesis approach, the *para*-alkoxy substitution was evaluated. A series of chain and aryl ring substituents were prepared and the SAR data lead to the *N*-methylbenzomorpholine **26**, specifically the (*S*)-stereoisomer.

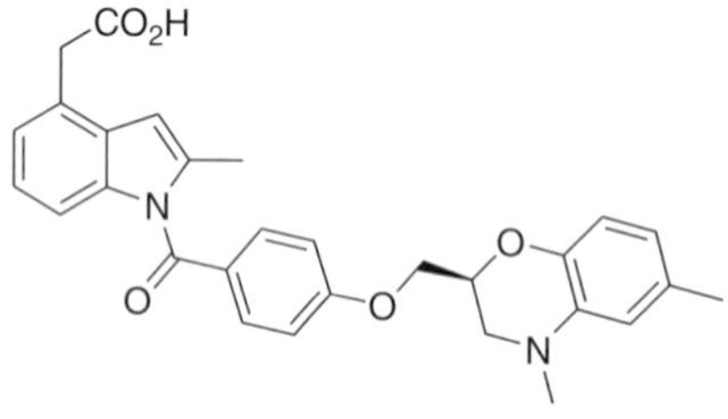

26 Human DP binding IC_{50} = 2.1 nM
Murine DP binding IC_{50} = 18 nM

A pharmacokinetic study demonstrated that **26** had good oral bioavailability (48% at $10\,mg\,kg^{-1}$), long $t_{1/2}$ (8 h), and moderate clearance with good tissue distribution. Efficacy in an asthma model has not been reported however suppression of allergic inflammatory response was demonstrated in an ovalbumin (OVA)-induced increase of vascular permeability in guinea pig conjunctiva.

In another chemotype derived from ramatroban, the development candidate **27** demonstrated exquisite PGD_2 binding inhibition potency at the DP-1 receptor.[138] The analog is reported to be relatively nonselective inhibiting TXA_2 at 11 nM and CRTH2 at 910 nM (note similarity to analog **18**).

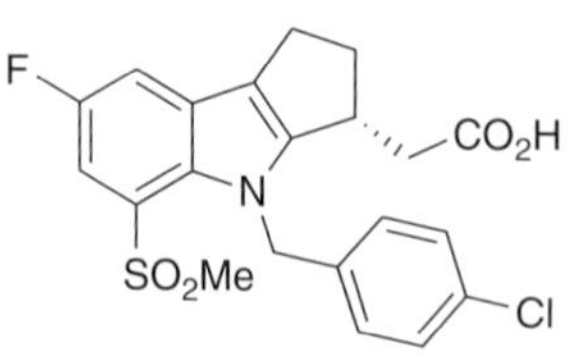

27 Merck L-888839
DP binding K_i = 0.07 nM
CRTH2 binding K_i = 910 nM

In the sheep asthma model, **27** demonstrated efficacy with 33% inhibition of early phase and 74% inhibition of late phase bronchoconstriction at 3 $mg\,kg^{-1}$. Also discovered as a result of a high-throughput screening campaign was the structurally related *N*-benzyl benzimidazole analog **28**. To explore the SAR the key 4-chloro-*N*-benzyl moiety was retained and a systematic survey of aromatic substitution and linkers at the 2-position were probed, ultimately furnishing the selective lead **29**.[139]

4-substitution optimal
CO_2Me = methyl oxadiazole > COMe >
$CONMe_2$ > SO_2NH_2 > CO_2H > H

CH_2 > S > S=O > NH

28 DP binding K_i = 73 nM
TP binding K_i = 9 μM

29 DP binding K_i = 38 nM
TP binding K_i = >75 μM

In summary, it is clear that the anti-inflammatory indomethacin (as a CRTH2 agonist) and ramatroban (a TXA_2 antagonist with similar activity in CRTH2) have played a pivotal role in defining a general pharmacophoric elements for two functionally different GPCRs sharing a common, prostaniod ligand. As well, structurally diverse prostaglandin mimetics provide an additional pharmacophore apparently unique to the DP-1 receptor. A number of interesting small-molecule probes with both agonist and antagonist activity have been developed. This may lead to future discovery of yet unknown prostaniod receptors that have evolved with differential functional activity.

7.30.7.3 Phosphodiesterase Type 4

The development of PDE4 inhibitors has been an active area of research for over 10 years now. These inhibitors act by increasing intracellular concentrations of cyclic AMP (by inhibiting degradation) resulting in a broad range of anti-inflammatory effects, driven by reduction of TNF-α production, on various effector cells involved in both asthma and COPD. As well, increased cAMP levels have also been shown to relax airway smooth muscle. Although the phosphodiesterase family consists of 11 subtypes they differ in primary sequence, substrate specificity, and cofactor requirements with PDE4 being the isozyme abundantly expressed across lymphocytes. During past decade, and especially in the past 5 years, a series of first- and second-generation compounds have progressed to the clinic.[140,141] Development has been slowed due to basic pharmacology issues that lead to a narrow therapeutic index over side effects. The first in class analog, rolipram (**30**) has a propensity to induce nausea and vomiting at therapeutic doses.[142] Rolipram has proven to be an excellent research tool and further hypotheses of refining the therapeutic window over side effects have been generated. Initial attention focused on the rolipram binding site, which, besides being a catalytic binding site, is believed to exist in two non- or slowly interconvertible conformations, exhibiting either a high and low binding affinity state for **30**.[143] Modulation of $PDE4_{high}$, which is believed to be generally expressed in the central nervous system, produces the adverse effect of emesis, while inhibition of the $PDE4_{low}$ affinity state has been associated with favorable anti-inflammatory activity. With this hypothesis, the design of the second-generation inhibitors such as rofumilast (Daxas, **31**)[144] and cilomilast (Ariflo, **32**)[145,146] have shown better efficacy and therapeutic potential in the clinic.

To date, the sole hypothesis of high versus low affinity receptor states has come under question especially since cDNA has been identified for isoforms of PDE4 leading to four distinct PDE4 subtypes. All four PDE4 sybtypes have the same general structure, especially in the catalytic domain. Regions N-terminal to the conserved catalytic domain show the subtype variability and are believed to be involved in regulation of the catalytic activity and intracellular targeting. A therapeutic hypothesis has emerged that is founded on subtype selectivity. The data is still controversial but the common belief is that selectivity for either subtype-PDE4b or PDE4d can lead to an improved clinical profile over side effects.[147,148] It has been shown that analog **32** which has 10-fold selectivity for PDE4d is better tolerated and this may support the hypothesis that selectivity for subtype-PDE4d can contribute to improved clinical tolerance.

Whether the strategy of specifically targeting subtype-PDE4b over PDE4d, or in combination with selectivity toward low versus high affinity states, will help achieve an ideal therapeutic ratio for future analogs remains to be seen. There is also an emerging third strategic component, that is, dual inhibition of a second PDE gene family isozyme along with PDE4. Preliminary investigations indicate that addition of a PDE3 inhibitor, motapizone (in vitro) provides functional synergy as measured by TNF-α inhibition in macrophages.[149]

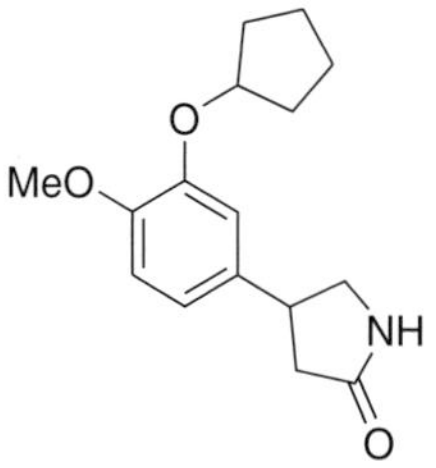

30 Rolipram (Schering)
PDE4$_{lo}$ IC$_{50}$ = 300 nM
PDE4$_{hi}$ IC$_{50}$ = 5 nM
PDE4a IC$_{50}$ = 67 nM
PDE4b IC$_{50}$ = 220 nM
PDE4c IC$_{50}$ = 1250 nM
PDE4d IC$_{50}$ = 79 nM

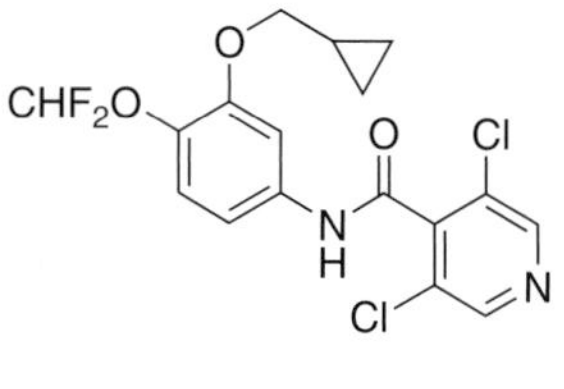

31 Roflumilast (Altana)
PDE4 IC$_{50}$ = 0.8 nM
10-fold more potent in PDE4c
Reported to exhibit no other subtype selectivity

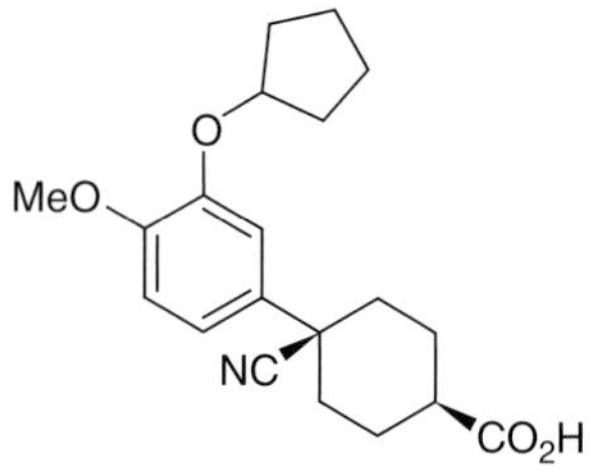

32 Cilomilast (GSK)
PDE4$_{lo}$ IC$_{50}$ = 95 nM
PDE4$_{hi}$ IC$_{50}$ = 120 nM
PDE4a IC$_{50}$ = 160 nM
PDE4b IC$_{50}$ = 120 nM
PDE4c IC$_{50}$ = 8700 nM
PDE4d IC$_{50}$ = 13 nM

There has also been an ongoing and aggressive research effort which has identified a number of new chemotypes still under investigation. To emphasize the point over 450 patents applications targeting PDE4 have been published in just under the past 10 years. An analysis of the medicinal chemistry literature reveals several medicinal chemistry synthesis strategies that have provided fruitful analog classes. A wealth of rolipram-like carbachol derivatives have been reported. An excellent example is analog **33**,[150] which was dropped from development due to lack of efficacy at a dose that did not result in emesis. Good to potent catalytic site PDE4 inhibition is seen with **34** as well as a conformationally constrained oxindole variant **35**[151] which are selective for the low-affinity binding state. It is interesting to note that removal of the oxindole fused phenyl ring in **35** increases potency (PDE4$_{lo}$ IC$_{50}$ = 6.3 nM); however, selectivity is lost for the high-affinity binding site (PDE4$_{hi}$ IC$_{50}$ = 7.5 nM).

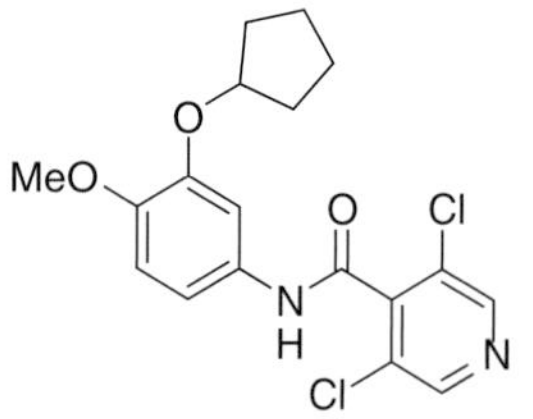

33 Piclamilast (RPR/Aventis)
PDE4$_{lo}$ IC$_{50}$ = 1 nM

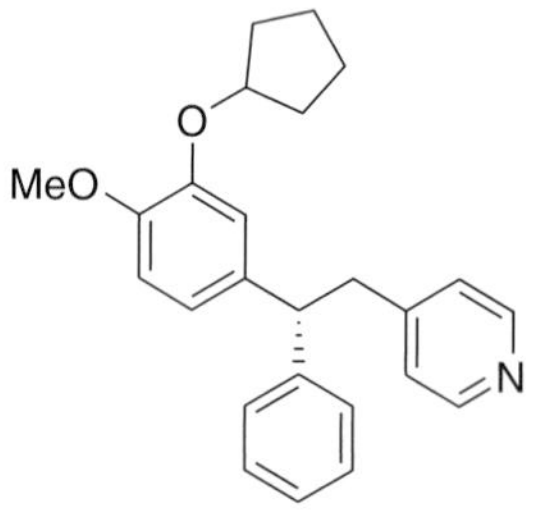

34 CDP 840 (Celltech)
PDE4$_{lo}$ IC$_{50}$ = 4.5 nM

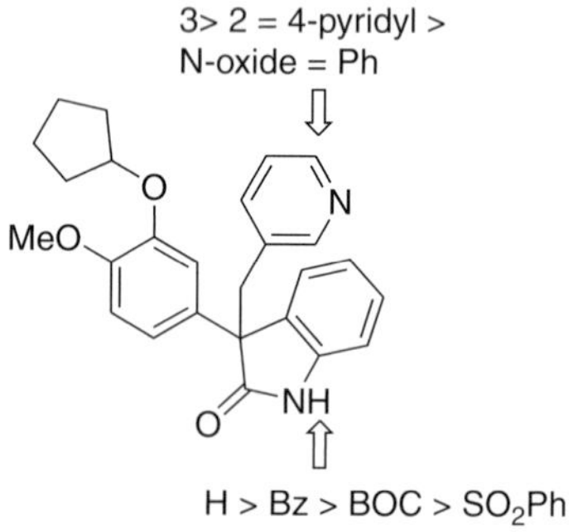

3> 2 = 4-pyridyl > N-oxide = Ph
H > Bz > BOC > SO$_2$Ph
35 PDE4$_{lo}$ IC$_{50}$ = 400 nM

In the quest to reduce compound affinity for the [^{3}H]rolipram high-affinity binding site, another series of oxindoles were derived. Analog **36** also strategically substitutes the rolipram pyrrolidone ring with a similar oxindole motif followed by SAR exploration of the phenyl substitution.[152] Extensive SAR did not allow for any correlation between the high- and low-affinity sites and it was found that electron withdrawing substituents in the 5-position of the oxindole phenyl ring were favored along with the *Z*-configuration at the alkene linker. Nothing larger then *N*-methyl was tolerated.

In another variant, a benzimidazole ring analog **37** was used to provide a proposed key hydrogen-binding motif at the catalytic site.[153] This substitution allowed for inhibition potencies in the nanomolar range. Considerable potency was afforded through the installation of an indanyl group and acceptable in vivo activity was seen in an antigen induces airway obstruction model (78% inhibition at 10 mg kg^{-1}).

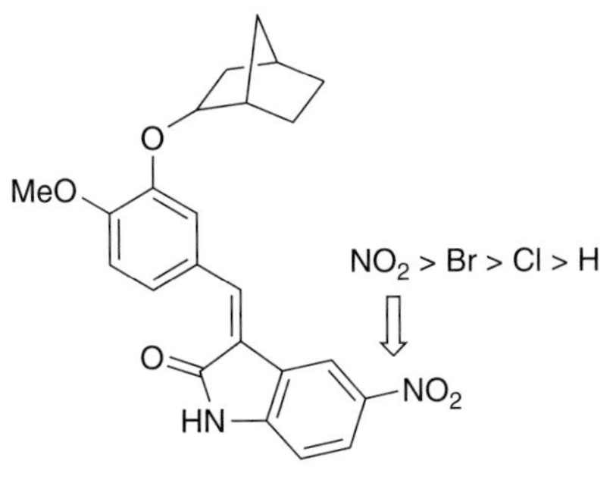

NO$_2$ > Br > Cl > H

36 PDE4$_{lo}$ IC$_{50}$ = 1.2 μM
PDE4$_{hi}$ IC$_{50}$ = 62 μM

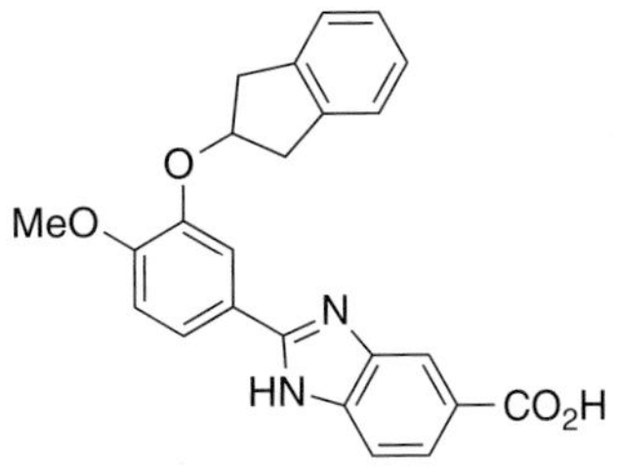

37 PDE4$_{lo}$ IC$_{50}$ = 2 nM
PDE4$_{hi}$ IC$_{50}$ = 32 nM

Using CDP-840 (**34**) as a scaffold, continued SAR exploration to improve the overall profile has identified analog series **38** as potential development candidates.[154–156] Initial analogs with the ideal in vitro profile exhibited excessive in vivo half-life due to the metabolically resistant bis(trifluoromethyl) phenylcarbinol moiety (**38a**).

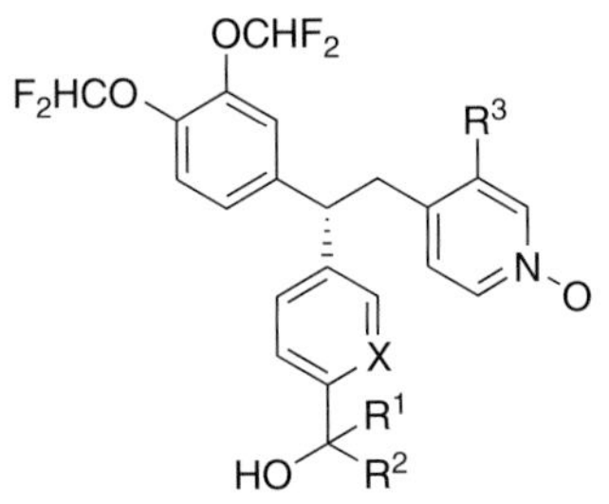

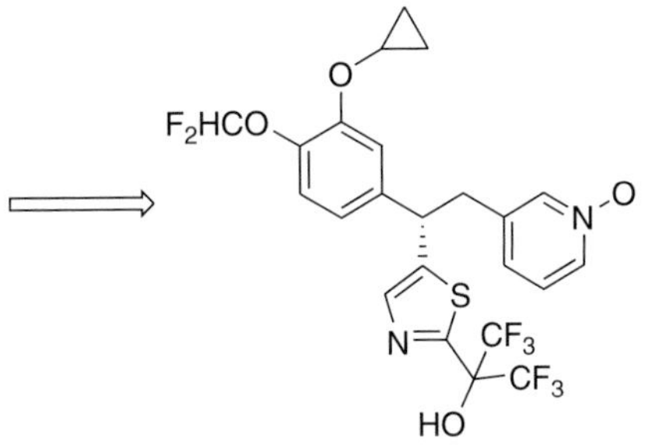

38a L-791943: R^1, R^2 = CF$_3$, R^3 = H, X = CH
PDE4A IC$_{50}$ = 4.2 nM
Rat $t_{1/2}$ = > 24h

38b L-826141: R^1, R^2 = CF$_3$, R^3 = Me, X = CH
PDE4A IC$_{50}$ = 1.3 nM
Rat $t_{1/2}$ = 10 h

38c R^1 = Me, R^2 = Ph, R^3 = H, X = N
PDE4A IC$_{50}$ = 2 nM
Rat $t_{1/2}$ = 1.5 h

39 L-869298
PDE4A IC$_{50}$ = 0.5 nM

Further optimization improved the pharmacokinetic properties to provide **38b**; however, when cardiovascular pharmacology studies were conducted on a related analog (**38c**), a significant QTc interval prolongation was observed at 3 mg kg^{-1} in dogs. A second round of refinement ultimately furnished **39**, which not only further improved potency but exhibited a reduced binding to the hERG channel (K_i = 61 μM).[157] The analog exhibited human whole blood inhibition of TNF-α potency of 90 nM and was efficacious in the *Acaris*-induced bronchoconstriction model in sheep (68%/92% early/late at 0.5 mg kg^{-1}).

Additional novel analogs have been produced through the evolution of the rolipram pyrrolidone moiety. Although the substituted catechol fragment has been retained, analogs **40**, **41**, and **42** illustrate the expanded scope of hetero- and carbocyclic fragments that provide varied activity profiles.

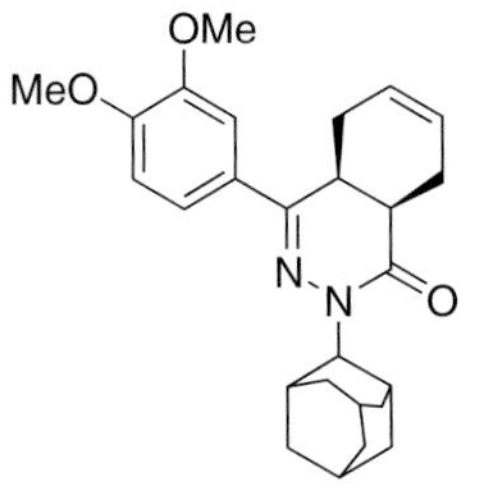

40 PDE4 IC$_{50}$ = 5 nM

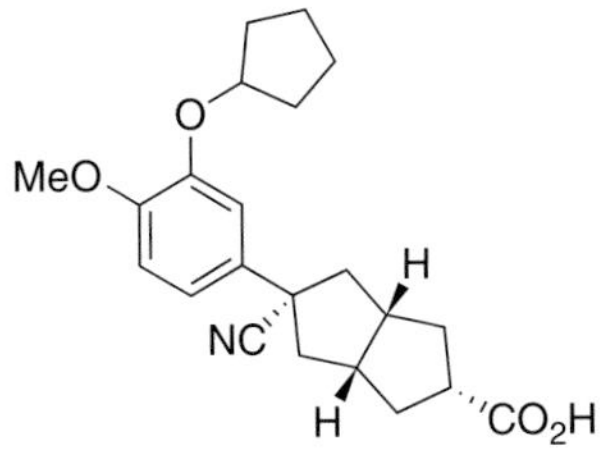

41 PDE4$_{lo}$ IC$_{50}$ = 9 nM

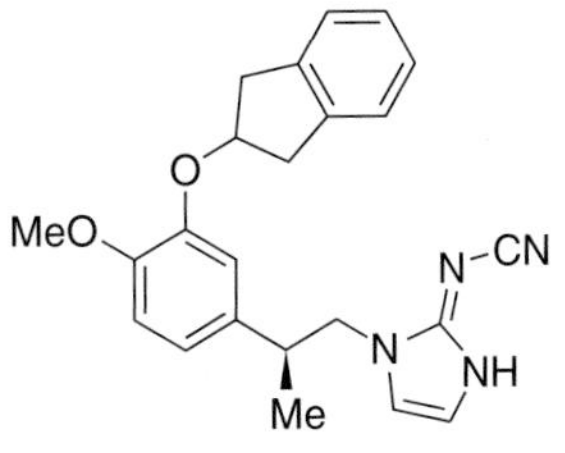

42 PDE4B IC$_{50}$ = 19 nM

Besides potent enzyme inhibition the racemic variant of the hexahydrophthalazinone **40** also demonstrated in vivo efficacy.[158,159] Further study and diastereomeric resolution identified the (+)-isomer as the desired isomer offering 250-fold increase in potency. A second strategy was investigated with this scaffold using a combinatorial library/docking approach. The report is informative as it captures the modern concept of technology-driven synthesis using a scaffold-linker-functional group approach to lead finding.[160] Building on the scaffold of cilomilast (**32**), further exploration of the spatial arrangement of the key pharmacophores (carboxy, nitrile, and carbachol moieties) furnished **41** which was equipotent with **32** but demonstrated enhanced inhibition of TNF-α (0.4 versus 1.7 mg kg^{-1}, orally, respectively).[161] A novel series of imidazol-2-one based derivatives such as **42** have also been described.[162] Potent PDE4 inhibitors were found and the corresponding 2-cyanoimino moiety of the (*S*)-isomer was shown to greatly inhibit gastrointestinal side effects and lower affinity for $PDE4_{hi}$ over rolipram.

The 3-methoxy-4-cyclopentoxycarbachol motif has gradually given way to additional heterocyclic variants. The indole isostere **43** is one of the first reported carbachol variants.[163–165] The initial SAR was analogous to that for rolipram, and an earlier analog **33**, as exhibited by the trend for lipophilic substituents on the heterocyclic system. Another variant, **44**, was nearly equipotent for both the high- and low-affinity binding states supporting the pharmacophore relationship to the parent rolipram.[166]

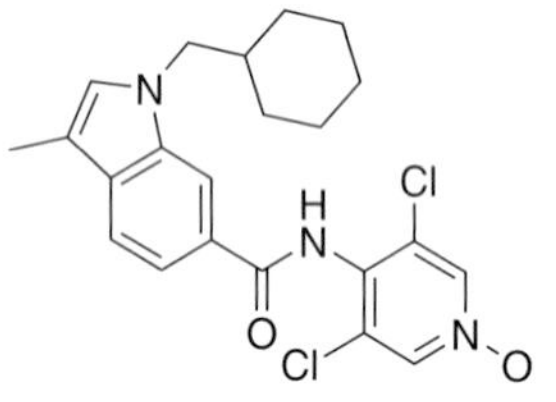

Lipophilic SAR

NH favored over N-Me

43 $PDE4_{lo}$ IC_{50} = 30 nM
$PDE4_{hi}$ K_i = 86 nM
ED_{50} in vivo TNF-α release = 7 mg kg^{-1}

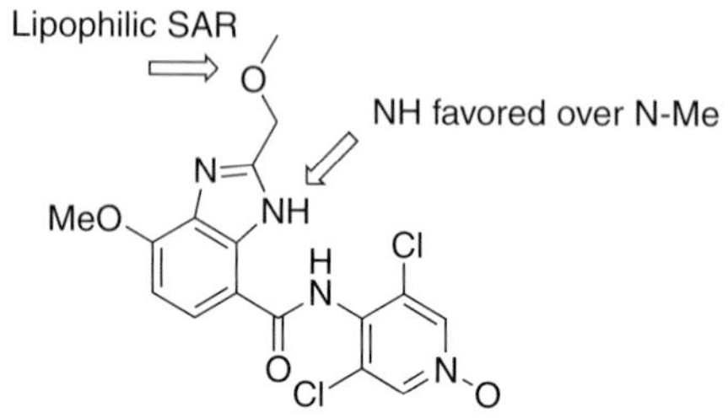

44 $PDE4_{lo}$ IC_{50} = 45 nM
$PDE4_{hi}$ K_i = 45 nM

In another example, phthalazine analog **45** further challenges the carbachol substitution by taking advantage of the 4-position of the phthalazine ring.[167–169] The ring constrained design strategy was based on replacement of the carbonyl functionality with a π-bond of an aromatic ring. A new structural series has been identified based on the benzodiazepine ring system.[170–172] Analog **46** demonstrated submicromolar activity with good selectivity as well as in vivo inhibition of TNF-α production.

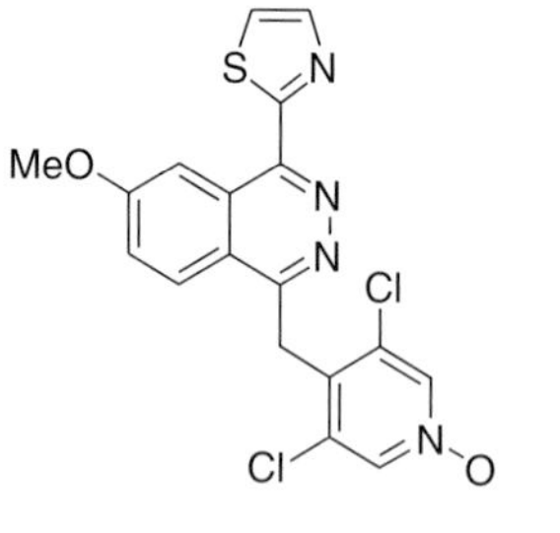

45 $PDE4_{lo}$ IC_{50} = 4 nM
$PDE4_{hi}$ K_i = 17 nM
TNF-α release IC_{50} = 3 nM

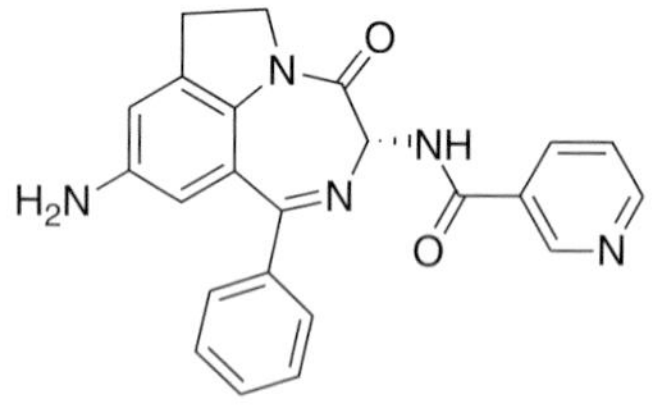

46 PDE4 U937 cell IC_{50} = 270 nM
PDE 1,3,5 IC_{50} = > 100 μM

In summary, a large scope of structurally diverse compounds have been identified that demonstrate acceptable activity in preclinical animal asthma models. A void in the medicinal chemistry literature exists for analogs based on the weak, nonselective inhibitor theophylline with only a few new micromolar leads being reported. The primary basis for design has been the prototypical inhibitor rolipram where many strategies for modifying the pyrrolidone, carbachol, or both fragments have proven fruitful in the quest for a selective inhibitor with attenuated (but not eliminated) side effects. The development of a third generation of analogs using these structural clues[173,174] will be coupled with a greater structural understanding of enzyme isoforms and affinity states although the role and selectivity for PDE4b versus PDE4d remains to be resolved. Given the high value of the target, new screening strategies are being employed such as fragment based x-ray identification of alternate pharmacophores.[175] The ultimate clinical outlook is encouraging as demonstrated by several advanced phase II and III candidates with improved therapeutic windows over undesired emesis as a side effect.

7.30.7.4 Mast Cell Tryptase and Chymase

Mast cells are highly granulated cells found in connective tissues at maturity and are important modulators of various inflammatory conditions, most importantly, allergic reactions that trigger asthma. A major component of mast cells are serine proteases that comprises up to 20–30% of the total protein mass. Specific protease activity has been clearly demonstrated to play a key role in pulmonary inflammation and allergic disease. One target of recent interest is β-tryptase which is classified as a serine protease which is found almost exclusively in mast cells.[176] Several additional isoforms have been identified resulting in a family size of five to date, with β-tryptase being in high relative abundance in mast cells. The crystal structure has been solved at high resolution (1.77 Å) revealing a ringlike homotetrameric species with the active site facing a central pore.[177,178] Loss off enzymatic activity is believed to occur through a spontaneous inactivation/dissociation process and there has been some controversy as to the physiological conditions that cause this and the role of heparin in stabilizing the tetrameric form.[179,180] It is worth noting that each subunit is believed to have catalytic activity and these active sites are located within the central cavity and have limited steric accessibility. This structural feature apparently limits the substrate selectivity toward small peptides (or peptidomimetics) or peptide loops/chains that be projected to the catalytic site. The intense interest to develop a therapeutic agent is driven primarily by a decade worth of data derived from in vitro and in vivo experiments that suggest a proinflammatory role for this enzyme specifically in the pathophysiology of asthma.[181] Consistent with mast cell degranulation products exhibiting a causative effect, several therapeutic approaches to inhibit mast cell activation have already demonstrated clinical utility. For example, mast cell stabilizers such as sodium cromoglycate and anti-IgE therapy have been effective.

Release of tryptase upon mast cell degranulation, in parallel with histamine, PGD_2, and LTC_4, triggers airway bronchoconstriction and hyperresponsivness and may have a role in fibrosis and thus airway remodeling. This is validated in animal models; tryptase increases airway hyperresponsivness and induces a cellular inflammatory infiltrate characteristic to human asthma.[182] Tryptase released from mast cells after allergen cross-linking can stimulate neighboring mast cells resulting in amplification of the degranulation process. Cellular activation is then mediated through proteolytic cleavage of the N-terminus chain of the 7TM-GPCR receptor PAR-2, located on smooth muscle, epithelial cells, and leukocytes. PAR-2 activation is dependant on this proteolysis since it unmasks the N-terminal tethered peptide ligand sequence (SLIGKV) which binds to the extracellular loop 2 initiating signal transduction. Evidence for PAR-2's role in inflammation is supported by the mouse knockout that was shown to have reduced airway inflammation following antigen challenge.[183]

The strategy for designing therapeutic inhibitors of tryptase centers on three approaches, which are classified as monofunctional or bifunctional inhibitors (these classes are also referred to as monobasic and dibasic inhibitors) and tetramer disruption. The clinical candidate **47** represents the concept of a monofunctional inhibitor.[184] This analog proceeded to phase II trials at which point it was discontinued most likely due to off-target effects based on lack of compound selectivity against thrombin and trypsin.[185]

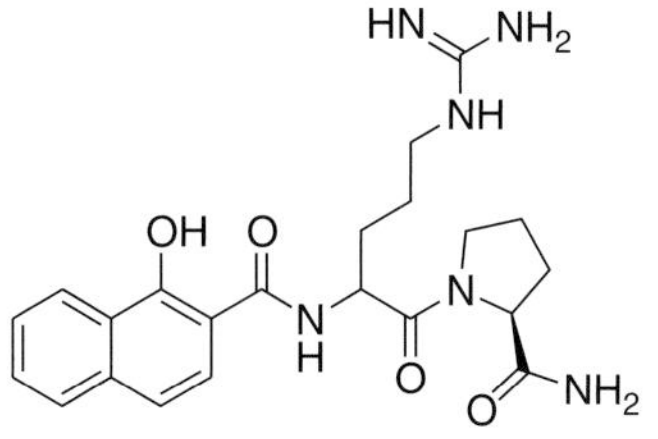

47 Clinical candidate APC-366
Tryptase K_i = 330 nM
Trypsin K_i = 162 nM

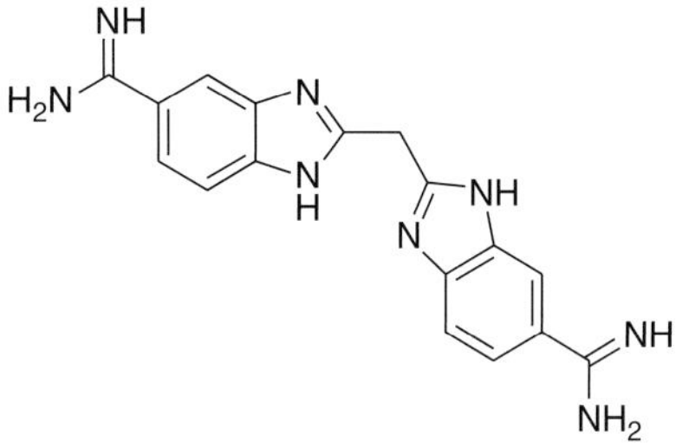

48 BABIM
Tryptase K_i = 2 nM
Trypsin K_i = 17 nM

The peptide analog was demonstrated to bind to each enzyme subunit in a 4:1 stoichiometry using a key salt bridge between the arginine side chain and Asp 189 at the base of the subunits P1 pocket. This is accomplished in an irreversible manner believed to be driven by serine addition to an activated form of the hydroxyl naphthalene moiety. Another prototypical monobasic inhibitor is represented by analog **48**. Again, the amidine moiety plays a key role as a guanidine isostere at the P1 pocket and although this potent molecule behaves in a reversible manner, it is not selective for the target.[186] The guanidine/amidine motif is prevalent across a wide variety of inhibitor series as is the peptidic nature of the scaffold. Analog **49** illustrates these features and is an example of a chemotype with greatly improved selectivity over that of BABIM. Analogs **50** and **51** further illustrate scaffolds that adequately present an essential lysine or arginine-like residue to the enzyme. In particular, **51** demonstrated efficacy in vivo in a sheep model on antigen-induced asthmatic response with 70–75% blockade of the early response and complete ablation of the late response and airway hyperresponsiveness.[187,188]

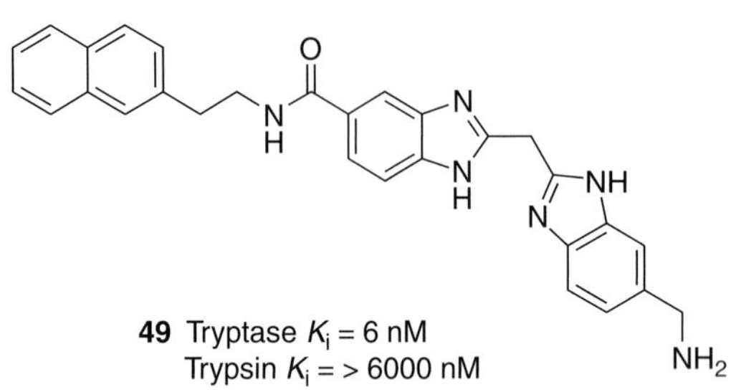

49 Tryptase K_i = 6 nM
Trypsin K_i = > 6000 nM

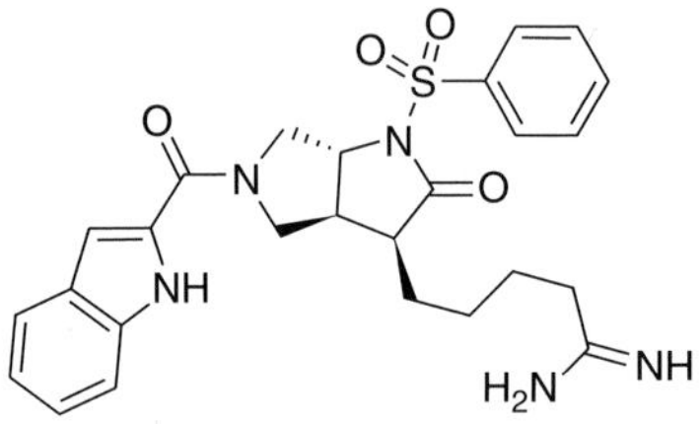

50 Tryptase K_i = 6 nM

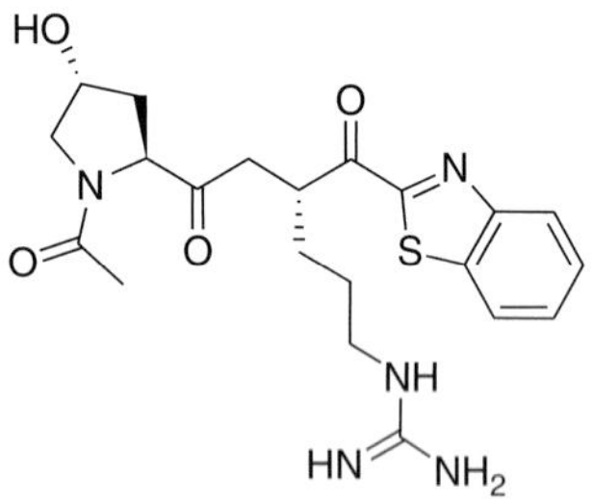

51 Tryptase K_i = 10 nM

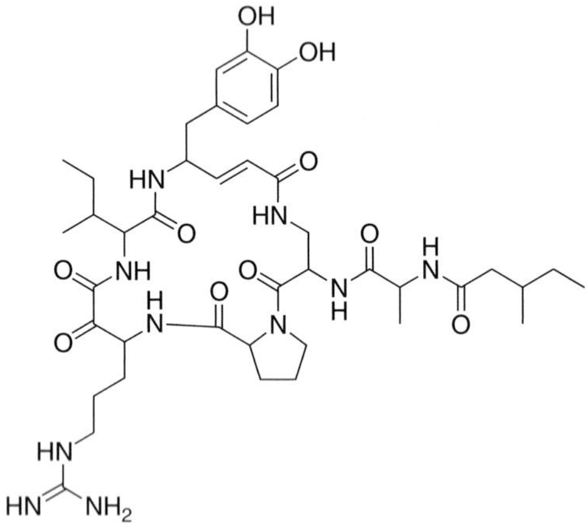

52 Tryptase K_i = 5 nM

The screening of natural product extracts has provided a macrocyclic peptide from the marine sponge of the genus *Ircinia*. Cyclotheonamide E4 (**52**) is potent inhibitor of tryptase with marginal selectivity over trypsin and thrombin.[189] A proven drug design strategy is illustrated in the atypical (nonbasic) analog **53** where the electrophilic saccharin moiety is targeted in a time-dependent manner by an active site serine, thus becoming covalently bound to the enzyme in a reversible manner.[190]

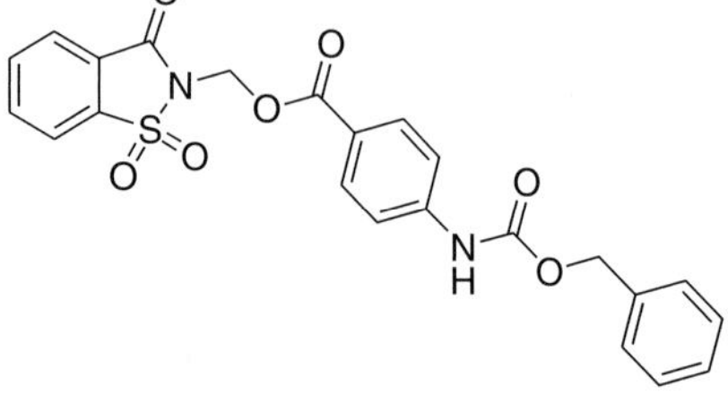

53 Tryptase IC_{50} = 64 nM

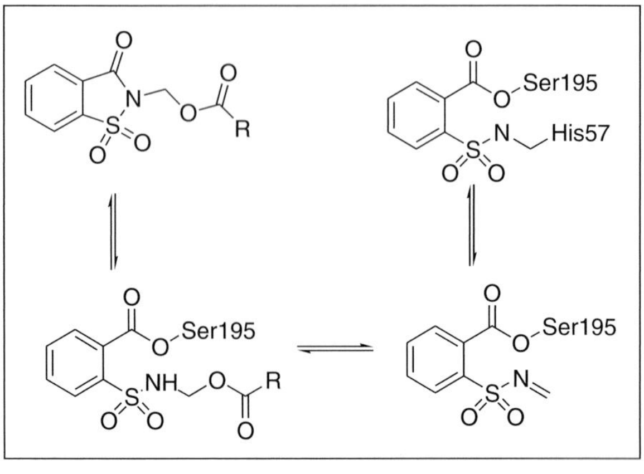

Two additional examples of novel scaffolds displaying either the arginine (**54**) or lysine (**55**) residues are shown. The β-lactam derivative **54** [191] exhibited excellent potency compared to early lead compounds but still did not offer good selectivity against trypsin. Further refinement of substituents lead to the identification of additional analogs exhibiting >3000-fold selectivity against trypsin as well selectivity against a broader panel of other proteases.

54 Tryptase IC_{50} = 6 nM

55 Tryptase IC_{50} = 68 nM
Selectivity >10 000-fold

By leveraging past SAR work on factor Xa and the fact that there is high homology between the factor Xa and tryptase S1 pockets, a strategy was employed where 2500 proprietary compounds from the factor Xa program were screened against tryptase.[192,193] The resulting hits were then optimized using a rational approach based on crystallography that resulted in potent, selective inhibitors exemplified by analog **55**.

In order to take advantage of the tetrameric enzyme structure a first generation series of bifunctional inhibitors have been rationally designed. **Figure 5** illustrates both the tetrameric nature of β-tryptase with a conceptualized bifunctional molecule interacting with two subunits. The general design strategy is to develop compounds that possess an extended length of about 20–30 Å in a manner that bridges two monomeric active sites. The inhibitors reported to date typically display a symmetrical motif with the following design characteristics: head–linker–scaffold–linker–head. As with the monofunctional analogs they also display an arginine or lysine-like functional group attached to a heterocyclic core. Analogs **56**, **57**, and **58** represent such designs.[194]

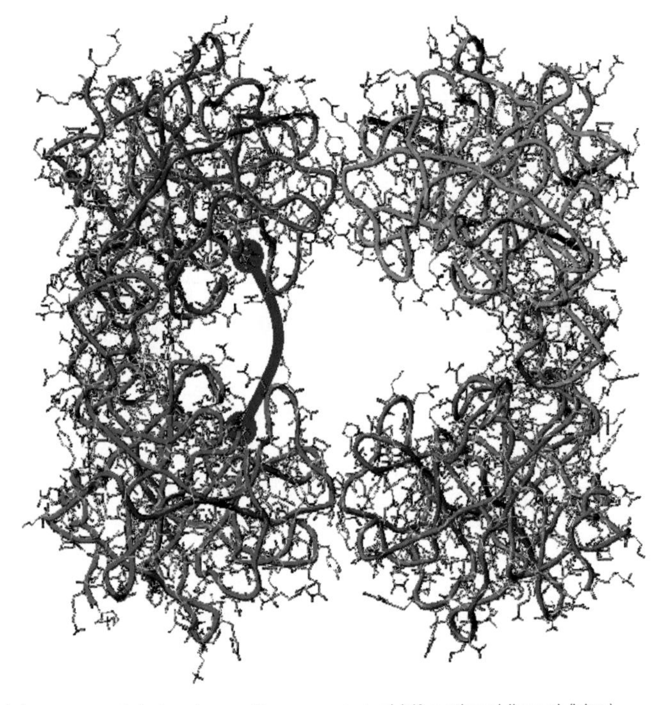

Figure 5 Beta-tryptase tetramer crystal structure with a conceptual bifunctional ligand (blue).

56 Tryptase $K_i = 1.3$ nM

57 Tryptase $K_i = 7.5$ nM

58 Tryptase $K_i = < 1$ nM

An analysis of analog **58** illustrates the role played by the amidine moieties play in building the adjacent S1 pocket interactions. It is subnanomolar in potency and a single change such as exchange of an amidine for a nitrile results in analogs 2000-fold less potent. The analog possesses excellent selectivity (1 μM, trypsin; >100 μM, thrombin; 4 μM, plasmin) and has been shown to be efficacious when delivered intratracheally at 1 mg kg^{-1} in a guinea pig airway hyperresponsiveness screen. This is an important point as the bifunctional inhibitors are limited to powder inhalation formulations given their generally poor pharmacokinetic properties. Another way to exploit the bifunctional concept toward inhibitor design is illustrated by analog **59**. In this example the smaller monofunctional monomer can form a dimeric species through hydrogen bonding of the primary amide group. The SAR shows that a charged acid species is essentially inactive while the corresponding ethyl ester exhibited only modest potency (K_i = 88 and 59 μM, respectively).[195] When compared to the dimers linked by a covalent bond, the pseudodimeric associative approach may offer some advantages in terms of molecules with lower molecular weight, better solubility, and improved pharmacokinetic properties.

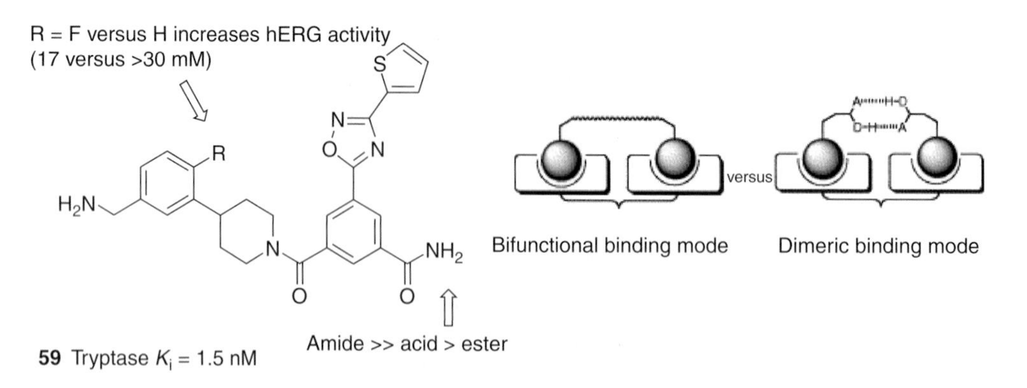

59 Tryptase $K_i = 1.5$ nM

The last strategy involves the concept of tetramer disruption. Although controversial, it is believed that dissociation of the heparin-stabilized tryptase tetramer into monomeric units results in irreversible inactivation of the enzyme. It has been shown that the heparin antagonists polybrene (IC_{50} = 3.6 nM) and protamine (IC_{50} = 65 nM) are both potent inhibitors of human lung tryptase either competitively or noncompetitively.[196] Protamine is an arginine-rich polycationic protein which is involved in DNA binding in spermatozoa, and it has been used clinically as a heparin antagonist to reverse the effect of injected heparin during surgery. Other examples of inhibitors are lactoferrin, which is a cationic, neutrophil-derived protein of 78 kDa which possesses unique heparin-binding domains, as well as the similar cationic protein myeloperoxidase (118 kDa).[197] Chymase is classified as a chymotrypsin-like serine protease. Human chymase is restricted to mast cell populations and is abundant in the skin and submucosal tissue of the respiratory and gastrointestinal tracts.[198] It is stored along with tryptase in secretory granules but is released as a macromolecular complex with carboxypeptidase and proteoglycans in response to immunological stimuli. The enzyme complex has an extended substrate binding site that prefers a phenylalanine residue at pocket P1 and hydrophobic residues at P2 and P3 in the catalytic cleft. This confers a somewhat broad spectrum of activity including degradation of extracellular matrix proteins (tissue remodeling) and the synthesis of angiotensin-2. These features suggest that inhibition of chymase could also provide therapeutic benefit in cardiovascular disease.[199] The design of early inhibitors used a traditional strategy targeting inhibition through substrate mimicry. These early peptides and peptidomimetics were designed to be recognized by the enzyme and contained a activated functional substituent such as a carbonyl group made susceptible to nucleophilic addition by the active site Ser195 hydroxyl group.

Several typical examples illustrate the different types or reactive moieties utilized. In general, α-haloketones, aldehydes, boronic acids, and phosphonate esters can all provide the electrophilic component necessary for reversible covalent interaction. The reader is referred to the appropriate references and reviews since this is a general, well-established protease inhibition strategy.[200,201]

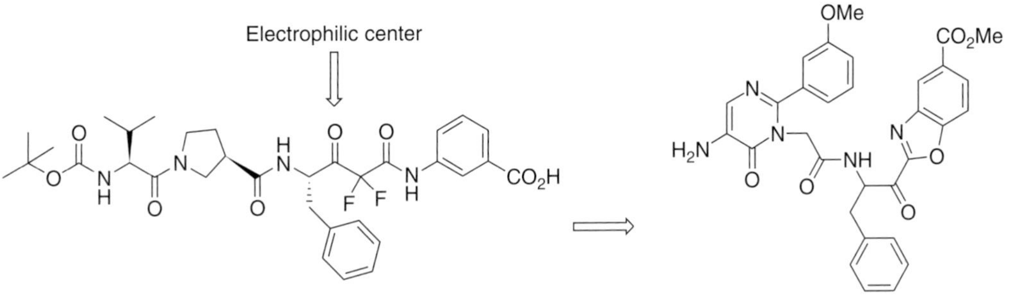

Electrophilic center

60 Example peptide-'warhead' inhibitor chymase K_i = 5.6 nM

61 Example peptidomimetic chymase K_i = 4.8 nM

Despite excellent potency, the therapeutic use of peptide-based inhibitors is still limited due to poor bioavailability. Reports on nonpeptidic inhibitors are somewhat limited and this is currently an active area of research. An example is illustrated by the continued refinement of analog **60** to furnish **61**, which demonstrates that the combination of an aryl–pyrimidone scaffold representing P2–P3 interactions and the use of a heterocyclic ketone can rival potency while offering good PK, selectivity, and bioavailability.[202,203] The aryl–pyrimidone scaffold has spawned a host of inhibitors whereby researchers have mixed and matched pocket SAR information from peptide programs with a variety of electrophilic 'warheads.'[204,205] The β-lactams **62** and **63** also diverge from the traditional peptide SAR.

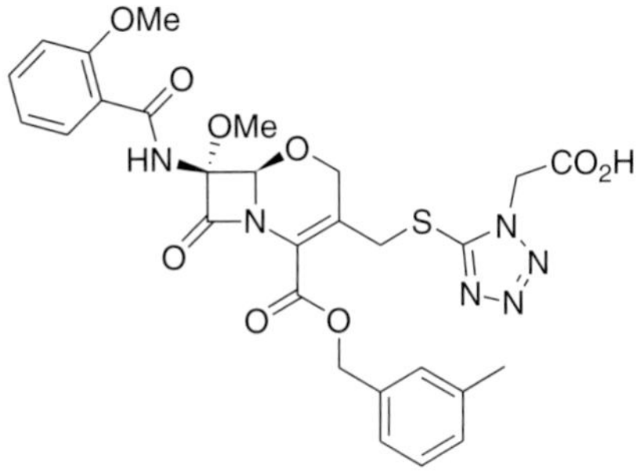

62 Chymase K_i = 6 nM

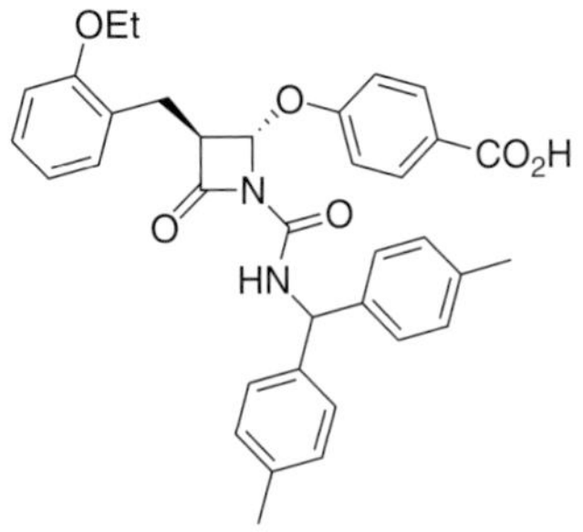

63 Shionogi 'BCEAB' Chymase K_i = 4 nM Cat G IC_{50} = 35 nM

Development of a high-throughput screening hit originating from an antibacterial program lead to the discovery of **62**, which offers high potency and is somewhat selective over a panel of proteases.[206] A related analog, **63**, falls within this compound class and, as expected, the active site Ser195 hydroxyl group is thought to attack the β-lactam carbonyl leading to ring cleavage and acylation of the enzyme.[207]

Analog **64** represents an emerging scaffold with a novel ketophosphonate moiety that offers potency for chymase and also has been shown to inhibit another mast cell serine protease, cathepsin G (which plays a role in the activation of PAR-4 leading to platelet activation).[200] Worth noting is the observation that the β-lactam **63** is also reported to have a similar dual-target specificity demonstrating a cathepsin G in vitro potency of 35 nM. It has been shown that dual inhibition or selectivity for either serine protease can be modulated through SAR modifications at a related specificity pocket in the active site.

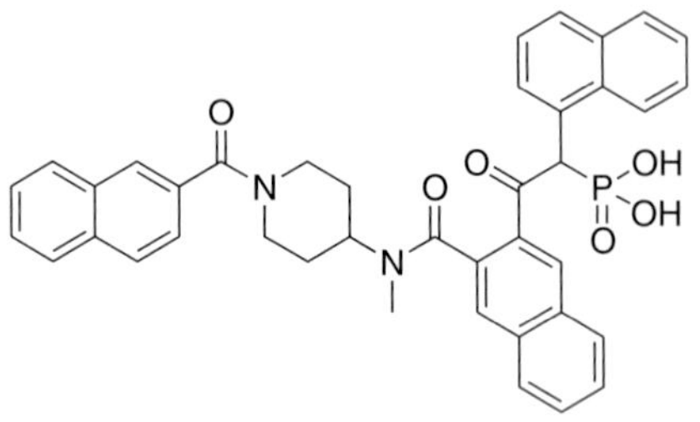

64 Johnson & Johnson
Chymase K_i = 2.3 nM
Cat G K_i = 38 nM

Recent reports indicate that **64** is advancing to the clinic and has demonstrated in vivo efficacy in preventing both early- and late-stage airway hyperreactivity in the sheep model at 0.1 mg kg^{-1} (twice daily aerosol dosing). For completeness, a brief survey of additional classes of chymase inhibitors is illustrated by structures **65–67** below.[205] Each exhibits a unique selectivity profile over related proteases and their optimization has been driven by structure-based design.

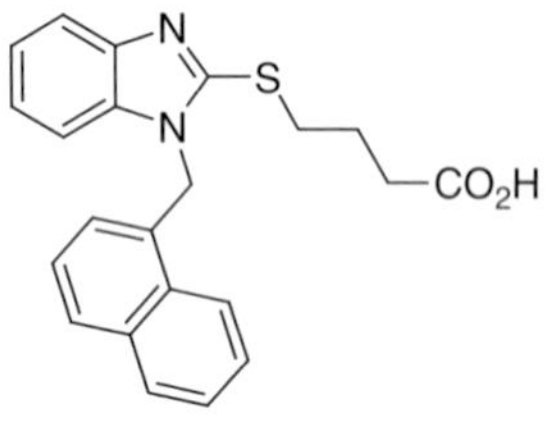

65 Chymase IC_{50} = 1 nM

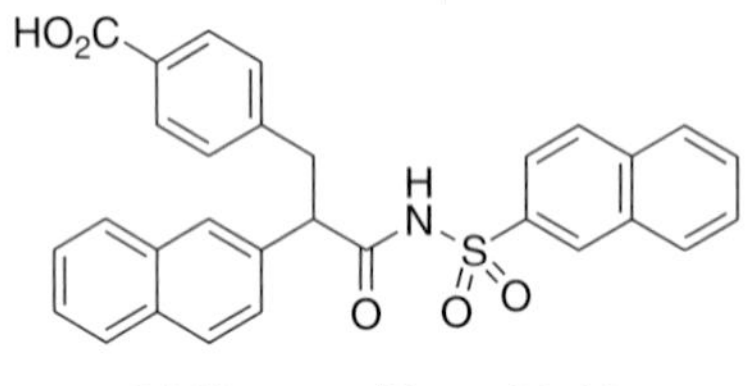

66 Chymase IC_{50} = 17 nM

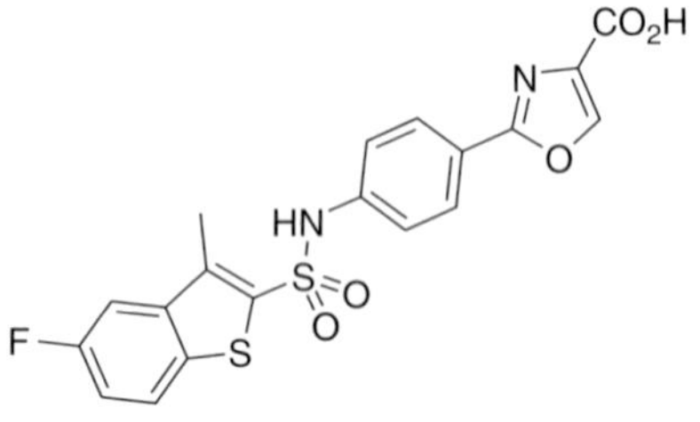

67 Chymase IC_{50} = 2 nM

Overall, mast cells and mast cell degranulation products play a critical role in modulation of the immune system. Based on a variety of biological studies, β-tryptase appears to be best associated with a variety of allergic disease states including asthma. Recent advances in the design of effective inhibitors has been driven by a solid understanding of enzyme structure and the monofunctional series of analogs seems to hold the most promise. The bifunctional series is the result of clever exploitation of the enzymes tetrameric structure, however, these molecules may suffer from downstream development issues based on physicochemical properties. The development of chymase inhibitors has also taken a structure-guided approach, since many analogs are peptide or peptidomimetic in nature thus taking advantage of active site catalytic residues in order to covalently inactivate the enzyme. The physiological and pathological roles of chymase are still to be fully elucidated and are an active area of research. Priority efforts are aligned to this need to establish validation for the involvement of chymase in both cardiovascular and autoimmune diseases. In general, the role of mast cell proteases is still under active investigation and the greatest challenge is to design and deliver a selective, orally available inhibitor with robust efficacy.

7.30.7.5 Interleukin-5 (IL5) Receptor Inhibitors

Pronounced eosinophilic infiltration is believed to play a crucial role in the pathogenesis of asthma. The cytokine IL5 is produced by activated T cells, mast cells, and eosinophils, and functions as a promoter for the proliferation of bone marrow precursor cells and their subsequent differentiation into mature basophils and eosinophils.[208] IL5 also has been shown to promote eosinophil survival and activation during allergic inflammation. The high affinity receptor for IL5 consists of two subunits (α- and β-chain) with the α-chain being the specific or unique IL5 binding subunit in low affinity and the β-chain assisting in the formation of a high-affinity state receptor through interaction with the α-chain. In this arrangement, the β-chain is shared by two other receptors, GM-CSF and IL3.[211]

As a method of target validation, it has been shown that wild-type mice treated with an anti-IL5 neutralizing antibody or IL5$^{-/-}$ genetic knockouts[209] are able to suppress eosinophil recruitment to the lung and inhibit airway hyperresponsivness to antigen challenge. While animal studies using the antibody approach were encouraging, human clinical trials with mild atopic asthmatics did not show consistent effects on lung function despite a marked suppression of circulating eosinophils in the blood. Currently this study design and resulting data has sparked a vigorous debate on the role of both IL5 and eosinophils, in general, in the pathogenesis of asthmatic disease.[210] Nevertheless, this level of target validation has generated the interest of medicinal chemistry and the search for small molecule inhibitors of the IL5 receptor has received some attention. Analog **68** was shown to inhibit binding of IL5 to its receptor on peripheral human eosinophils.[212,213] Given the lack of published, advanced data and modest potency, this compound may have simply been used as a research tool to validate the targets signaling pathway through the JAK kinase manifold.

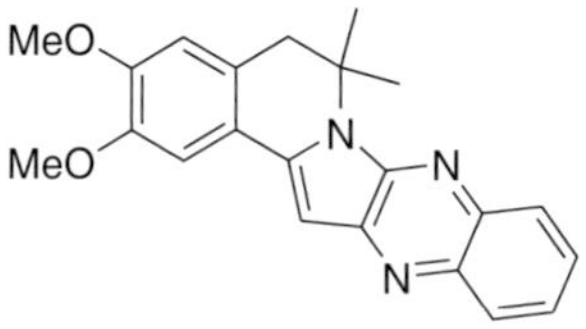

Receptor Cys66-SH (Domain 1)

68 YM-90709
Receptor binding IC_{50} = 1 μM

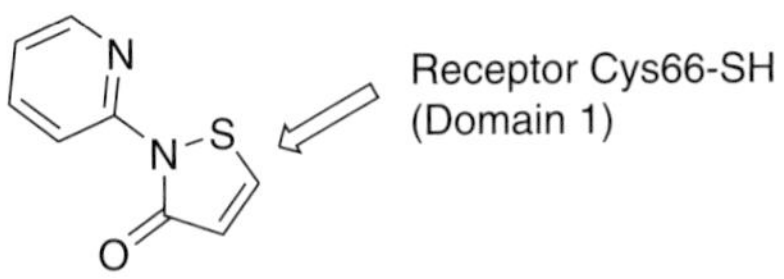

69 Isothiazolone chemical probe
Receptor binding IC_{50} = ~20 μM

Another early exploratory approach has identified a class of compounds shown to covalently modify the receptor. A series of isothiazolones (**69**) were shown to specifically inhibit the IL5/IL5R interaction at the α-chain. This was verified through the generation of a Cys66 alanine mutant that still bound IL5, which was not disrupted in the presence of the probe molecule.[214] While not therapeutic drugs, these molecules illustrate that inhibition of the receptor is chemically tractable. High-throughput screening has also uncovered several other compound classes that can modulate IL5 binding. Several other irreversible ligands have been identified employing a surface plasmon resonance/mass spectrometry affinity experiment. Analogs **70** and **71** are moderately active at the IL5 receptor and were cross-reactive with the IL3 and GM-CSF receptors.[215]

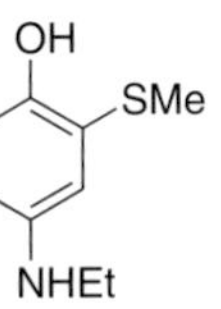

70 Chemical probe
Receptor binding IC_{50} = ~8 μM

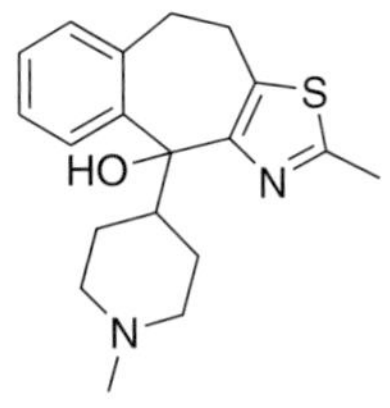

71 Chemical probe
Receptor binding IC_{50} = ~11 μM

Again, druglike molecules were not identified in this study. The probes furthered the understanding of receptor function with regards to a shared β-chain. Included in this section is a series of recent compounds that have been demonstrated to inhibit the production of IL5, rather than specifically target the receptor. The azauracil derivative **72** has demonstrated in vitro potency by inhibition of IL5 protein in activated human whole blood and in isolated human peripheral blood monocytes as well as mouse splenocytes.[216] Message levels for other cytokines remained unchanged.

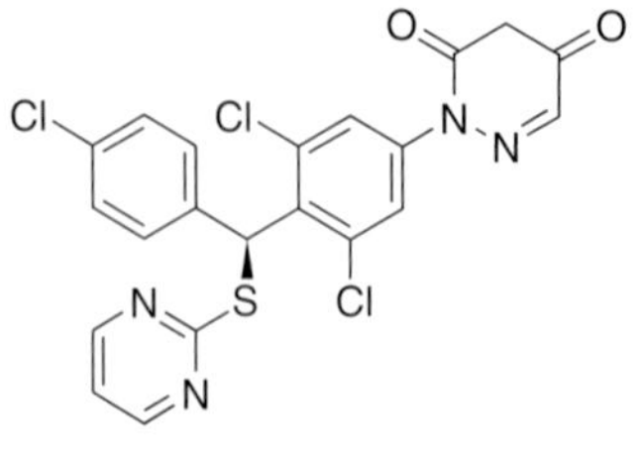

72 R146225
hWhole blood IC_{50} = 34 nM
hPBMC IC_{50} = 24 nM
Murine splenocytes IC_{50} = 6 nM

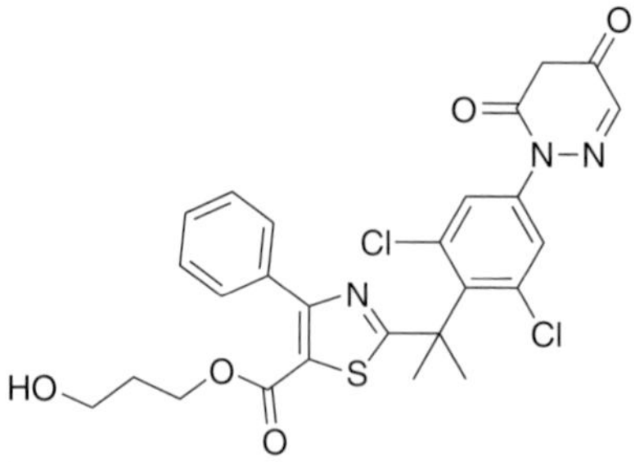

73 Whole blood IC_{50} = 78 nM
MCP-1 IC_{50} = 220 nM
MCP-2 IC_{50} = 580 nM
MCP-3 IC_{50} = 80 nM

Since the observed potency was sufficient to test in vivo, the compound showed efficacy at 0.6–2.5 mg kg^{-1} orally in a mouse model measuring pulmonary accumulation of eosinophils. It was found that chronic dosing in pregnant rats and rabbits induced signs of teratogenesis and further development was halted. This triggered further SAR activity with an objective to eliminate this adverse effect through the design of an inhaled prodrug as illustrated by analog **73**. After a survey of functionality, it was found that the hydroxypropyl ester offered high metabolic stability in lung tissue while it was rapidly metabolized to the inactive carboxylic acid by human liver preparations.[217] This effectively translated to low and short-lived plasma levels of **73**. The activity profile suggested that the analogs might target Th2 cells given the reduced levels of the chemotactic ligands MCP-1,2,3 as well as IL4. In vivo activity was demonstrated by intratracheal administration of an aerosol in the *Ascaris* sheep model. Efficacy was seen whether the compound was dosed pre- or post-antigen challenge with complete abrogation of the late phase response as well as reduced bronchial hyperreactivity to inhaled carbachol.

Several approaches targeting the cytokine IL5 have been investigated using a variety of strategies in the past years.[218,219] These include antibody approaches toward the receptor, small-molecule probes to inhibit ligand binding to the receptor, and agents that prevent production of the cytokine itself. Given the dearth of small-molecule research in the area, perhaps spawned by the failure of antibodies in the clinical setting, may lead one to conclude that given asthma has a multifactorial pathogenesis, inhibition of a single signaling agent may not be sufficient for therapy.

7.30.7.6 Inhibition of Leukotriene Biosynthesis

The past decade has witnessed an explosion of activity in the field of inhibition leukotrienes and outstanding, comprehensive reviews are available to the reader.[88] Several intersections for intervention in the arachidonic acid pathway have resulted in a rich history of chemotypes and strategies that an entire volume could be devoted to the discussion. The following targets best represent these intersections: inhibitors of 5-lipoxygenase (5-LO), inhibitors of the redox protein, five lipoxygenase activating protein (FLAP), the leukotriene receptor LTD_4 (CysLT1), and LTB_4. There has been interest on other pathway enzymes with limited reports of small molecule research targeting LTA_4 hydrolase.

5-Lipoxygenase is a dioxygenase that incorporates molecular oxygen into arachidonic acid giving rise to the metastable epoxide LTA_4. Further processing by either LTA_4 hydrolase (neutrophils and monocytes) or LTC_4 synthetase (mast cells and eosinophils) generated either LTB_4 or the cysteinyl leukotrienes LTC_4, LTD_4, and LTE_4 (**Figure 6**).

Intervention of 5-LO at the start of the cascade has been an intense area of research since its discovery in 1983.[220,221] Although an attractive target, there has been some concern about possible unwanted side effects especially since an entire metabolic process would be interfered with. Some genetic knockout information is available and the mice appear to develop normally.[222] In the design of inhibitors, many researchers targeted the active site, nonheme iron atom and built in a weak chelator such as a hydroxamic group as best illustrated by Zyflo (**74**, zileuton, Abbott).[223]

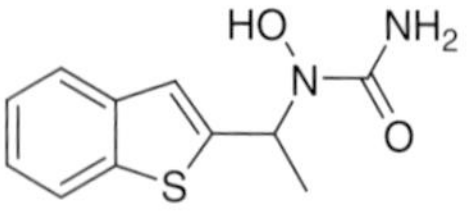

74 Zileuton (Abbott)
hWhole blood IC_{50} = 700 nM

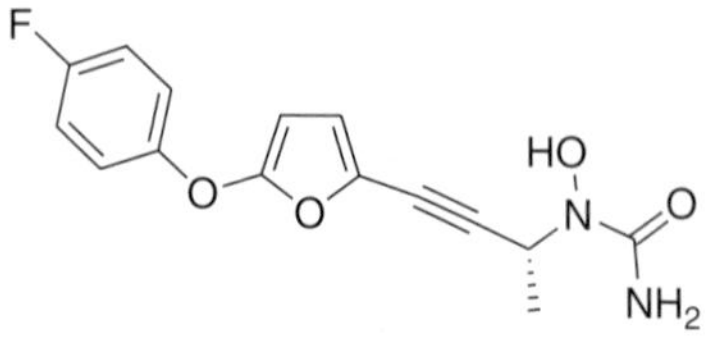

75 hWhole blood IC_{50} = 80 nM

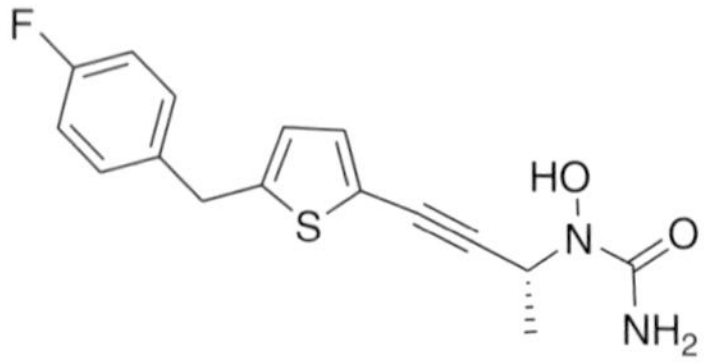

76 Atreleuton (Abbott)
hWhole blood IC_{50} = 150 nM

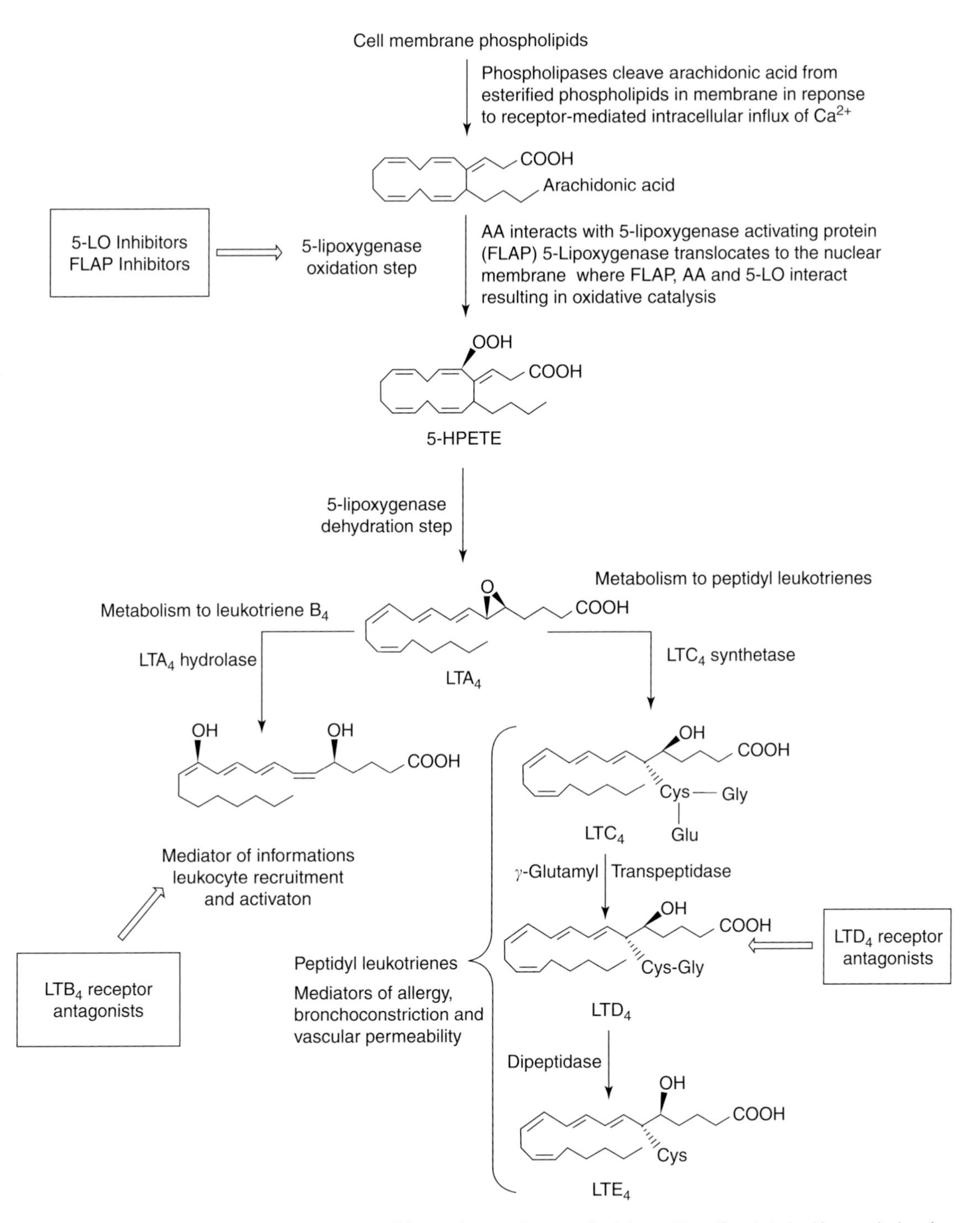

Figure 6 Leukotriene inflammatory pathway and key points of therapeutic intervention. (Reprinted with permission from Brooks, C. D. W.; Summers, J. B. *J. Med. Chem.* **1996**, *39*, 2629 © American Chemical Society.)

Further structural modifications following this design strategy generated the second-generation iron ligand inhibitors **75** and **76**.[224–226] The objective of the advanced analogs was to improve potency and to improve duration of action by inhibition of glucuronidation rate.[224] Analog **76**, as the (*R*)-isomer, provided sustained ex vivo duration of 5-LO inhibition (as measure in a human whole blood assay) and progressed to the clinic with discontinuation after phase III trials due to elevated liver enzymes. It was about fivefold more potent that zileuton in animal models of bronchospasm and had an oral half-life of ~16 h.

A unique approach is represented by analog **77** which combines two concepts that could prove beneficial in allergy and asthma.[227] A series of *N*-hydroxycarbamates containing a histamine H1 antagonist pharmacophore was constructed and exhibited the desired dual activity profile. Preclinical work in the guinea pig OVA model showed **77** to have a better overall profile that a traditional 5-LO analog such as **74**. Other dual target strategies have been employed including COX-/5-LO, PAF/5-LO, and lipid and sugar metabolism enhancers/5-LO, although little is reported in the medicinal chemistry literature. Beyond this approach, the set strategy has been to explore lipophilic groups and linkers stemming from a terminal hydroxyurea moiety and many examples exist in the review articles. Several isosters of hydroxyureas have been devised (e.g., analog **78**)[133,228] offering modest potency; however, second-generation substrate inhibitors emerged that were 'designed' to fit the 5-LO active site without the need for an iron-chelating group. Analogs **79** and **80** represent ongoing discovery efforts in this area.[229–231]

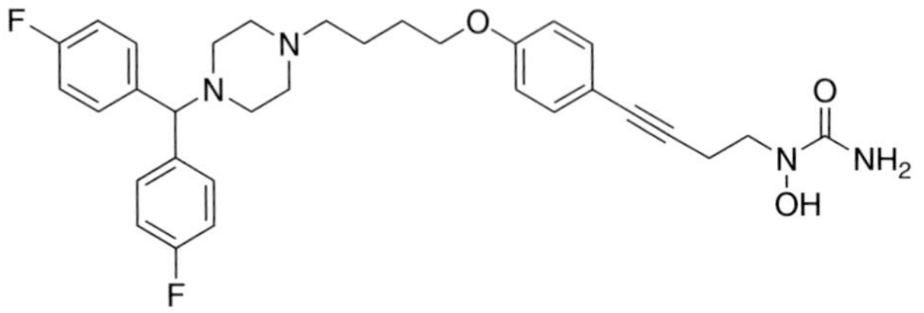

Iron chelation

77 Dual H1/5-LO inhibitor
H1 binding IC_{50} = 3.6 nM
hWhole blood 5LO IC_{50} = 89 nM

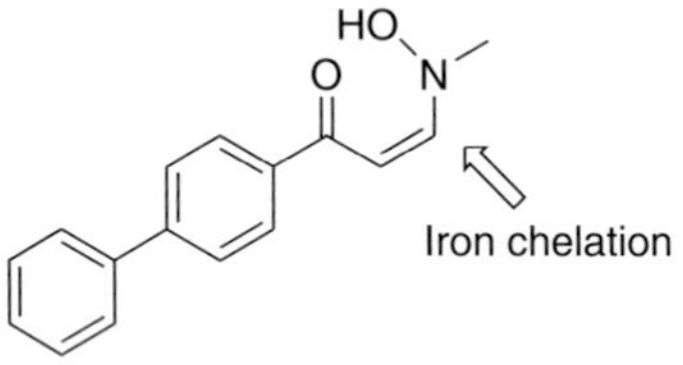

78 Rat basophil leukemia cell IC_{50} = 60 nM

Representative 5-LO substrate inhibitors

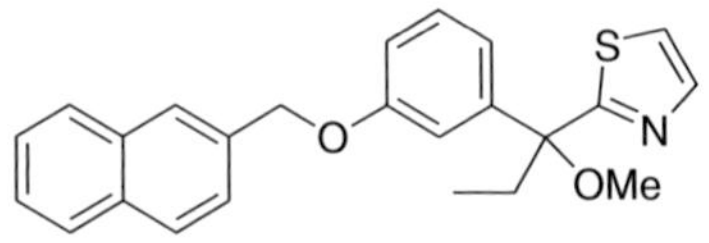

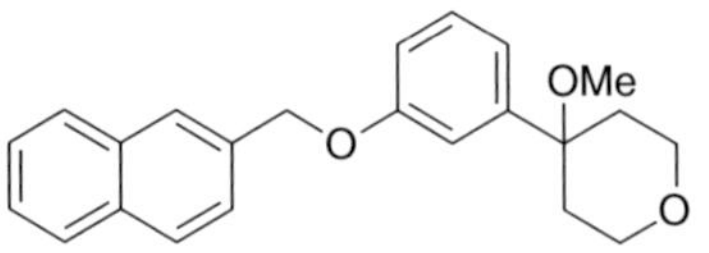

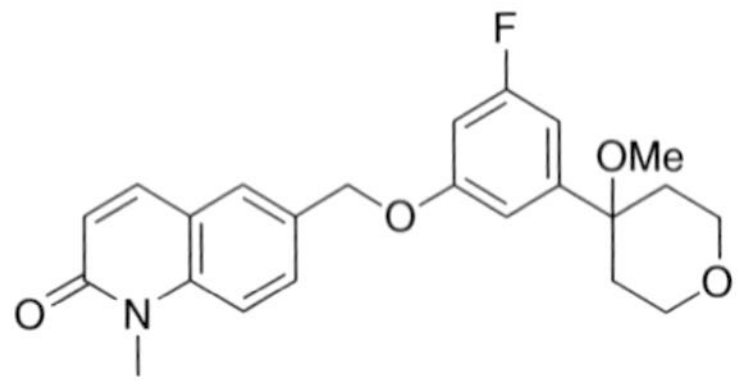

79 ZD-2138 (Zeneca)
hWhole blood LTB4 IC_{50} = 24 nM

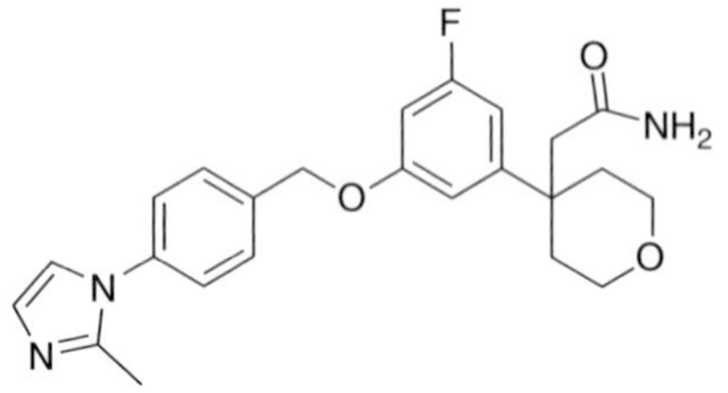

80 CJ-13454 (Pfizer)
hWhole blood LTB4 IC_{50} = 34 nM

The first to be evaluated in the clinic was analog **79** in a 1-month safety study. A single dose (350 mg orally) was shown to completely inhibit LTB_4 production in ex vivo blood samples. Phase II studies were designed as allergen challenge in asthmatics using **79** at the same does. No effect was seen on both early and late stage response, however identical ex vivo efficacy was seen. After several additional studies development was ultimately stopped. As a follow-on, the imidazole analog **80** has illustrated improved pharmacokinetic and toxicology characteristics though clinical efficacy remains to be seen.

FLAP inhibitors comprise a class of compounds that can inhibit leukotriene biosynthesis without direct inhibition of 5-LO. The indomethacin derivative **81** proved indispensable in the discovery of FLAP and understanding the differential activity observed in intact versus lysed cell results on leukotriene biosynthesis.[232] While the precise mechanism of action is not clearly understood, FLAP is hypothesized to be essential for activation of 5-LO through assisting in either the membrane translocation process or as a vehicle for presentation of arachdonic acid. A series of first-generation (**81, 83**) and second-generation (**82, 84**) inhibitors were discovered and developed in the early to mid-1990s.[233] By combining pharmacophoric regions of the first-generation lead compounds researchers were able to further optimize analog potency. A key observation was the conserved quinoline moiety which turned out to be a prominent binding motif in many FLAP SAR programs.

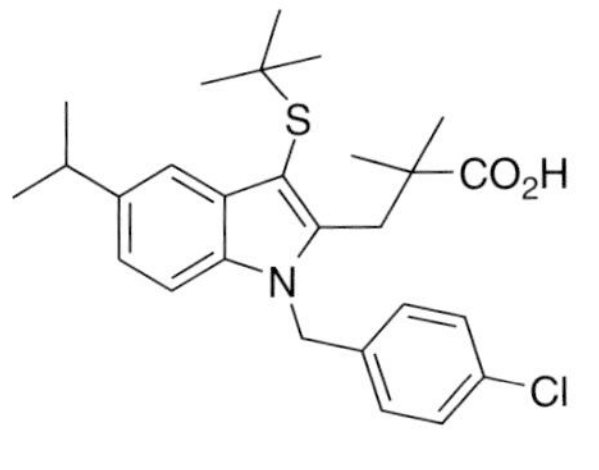

81 MK-886
Binding IC_{50} = 23 nM

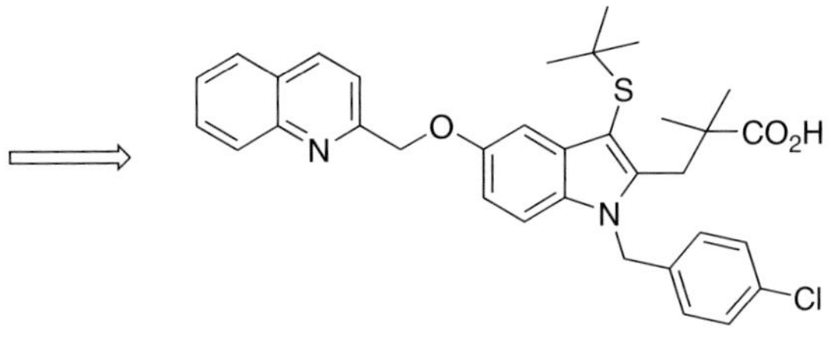

82 MK-591
Binding IC_{50} = 2 nM

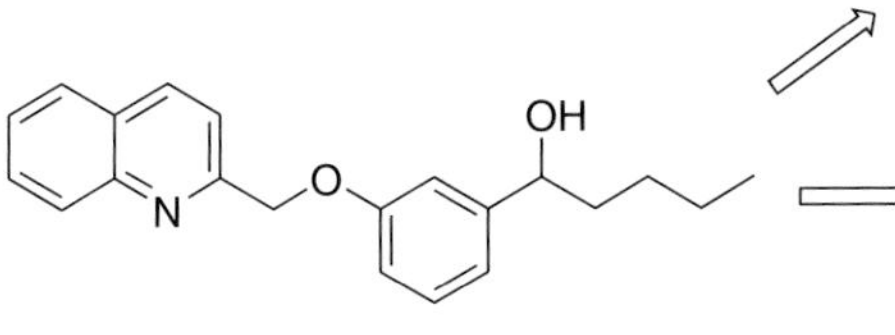

83 REV-5901

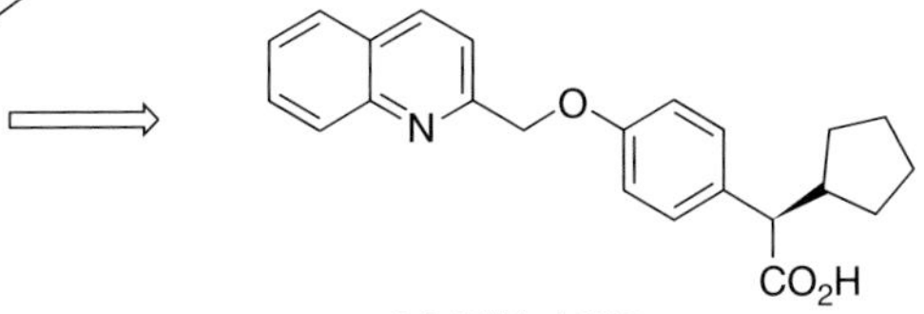

84 BAY-x1005

A number of clinical trials were initiated with varied results. Compound **81**, when dosed at 500 mg orally in asthmatic patients prior to antigen challenge improved both early and late response by 58% and 44%, respectively, but was discontinued based on the modest leukotriene inhibition measured ex vivo.[234,235] The second-generation analog **82** demonstrated better early and modest late phase response (79% and 39%, respectively) and the ex vivo inhibition of leukotriene biosynthesis was 98% on ionophore challenge of whole blood up to 24 h. This demonstrated the clinical efficacy of FLAP inhibition; however, larger studies showed that patient respiratory improvement was not aligned with in vitro compound potency.[233] To date, no clinical candidate from these programs has emerged to the market. From a medicinal chemistry point of view, optimization strategies including 'oxime insertion' proved fruitful as illustrated by analog **85** which provided the desired improvement in solubility and absorption properties.[236]

84 BAY-x1005

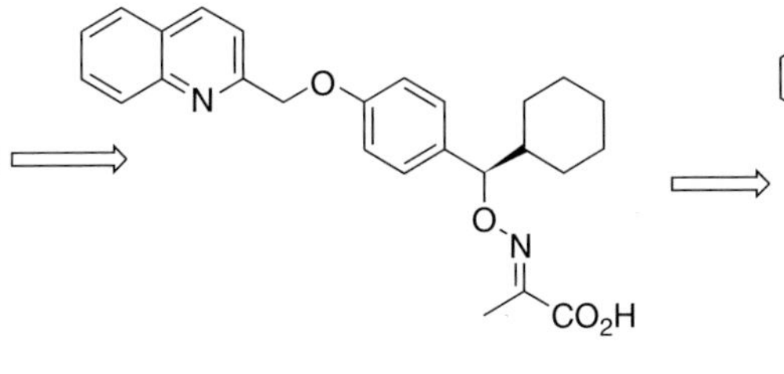

R = H or Me

85 Oxime insertion analog
hPMNL LTB_4 inhibition IC_{50} = 15 nM
In vivo ED_{50} = 0.5 mg kg^{-1}

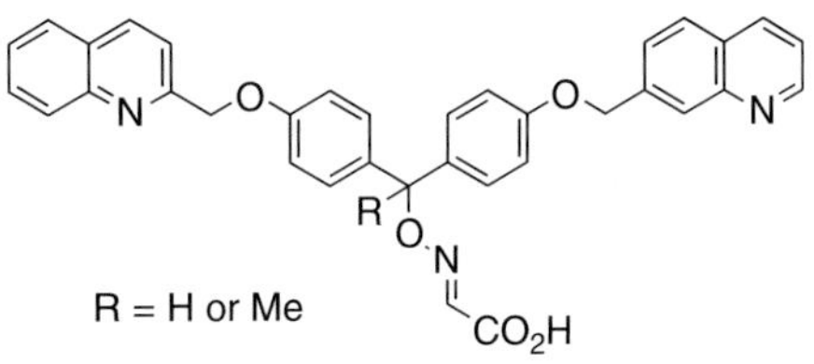

86 Symmetrical core analog
hPMNL LTB_4 inhibition IC_{50} = 20 nM
In vivo ED_{50} = 1.5 mg kg^{-1}

Further refinement, as demonstrated by the symmetrical core analog **86**, removed the stereocenter (which proved too costly to develop) yet maintained superior potency in inhibiting LTB_4 formation in human neutrophils challenged with a calcium ionophore. Given the efficacy in rodent asthma models, analog **86** progressed to phase I clinical trials.[237,238]

The best-represented compounds in the area of LTD_4 (CysLT1) inhibition are the marketed drugs Singulair (montelukast, Merck),[90,239] Accolate (zafirlukast, AstraZeneca),[89,240] and Ultair (pranlukast, Ono).[241] The history of the development of these drugs is a fascinating story and stems from SAR and privileged pharmacophores identified in 5-LO and FLAP research programs. Most importantly it was the seminal discovery that the active components of slow-reacting substance of anaphylaxis (SRS-A) were the CysLTs, specifically LTD_4, LTC_4, and LTE_4.[242,243] This, coupled with the strong evidence that SRS-A played a pivotal role in bronchoconstriction launched intense research efforts. The resulting analogs generally comprise combined lipid, acid, and peptide mimetics representing the three key regions and stereochemistry of LTD_4.

Montelukast (**88**) represents a second-generation inhibitor stemming from its congener **87** which was designed by combining the FLAP active quinioline pharmacophore with a thioacetal unit found in early high-throughput screening hits.[244,245]

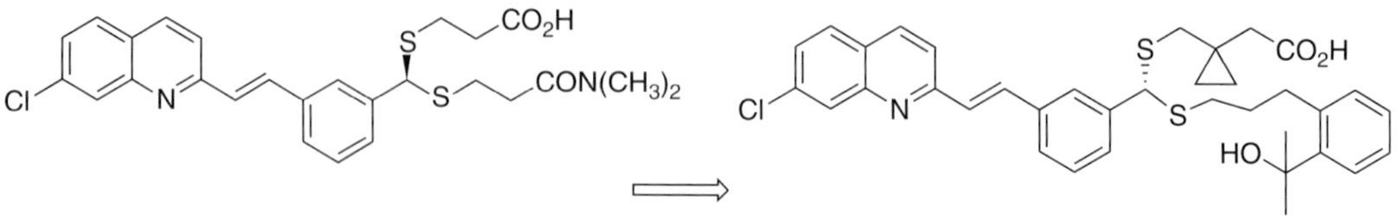

87 Verlukast (Merck)
Binding IC_{50} = 0.8 nM

88 Singulair (montelukast, Merck)
Binding IC_{50} = 0.5 nM

Verlukast was in early clinical trials and only demonstrated about a 13% improvement in lung function following a single oral or aerosol dose. The compound was ultimately withdrawn due to liver function abnormality. Replacement of one of the thioacetal side chains with an arylalkyl moiety and installation of the geminal cyclopropane substituent furnished the potent analog **88** which was devoid of undesired side effects especially peroxisomal enzyme induction which had plagued the first generation series. In moderate asthmatics, a single 100 mg dose produced an immediate 10% improvement in base line FEV_1 which stayed constant during a 9-h study. Pranlukast (**89**) and zafirlukast (**90**) were developed using similar strategies.

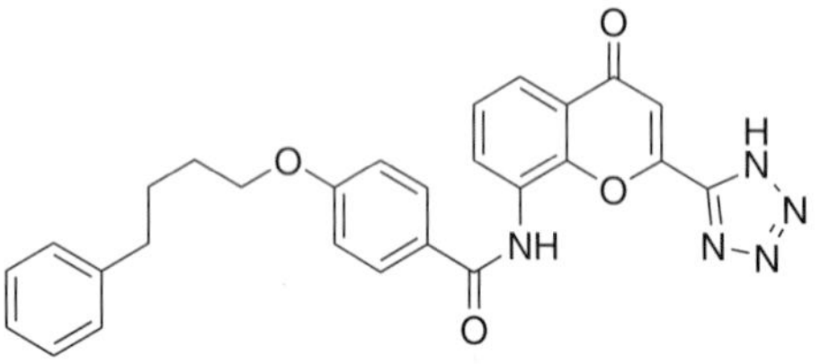

89 Pranlukast (Ono)
Binding K_i = 46 nM

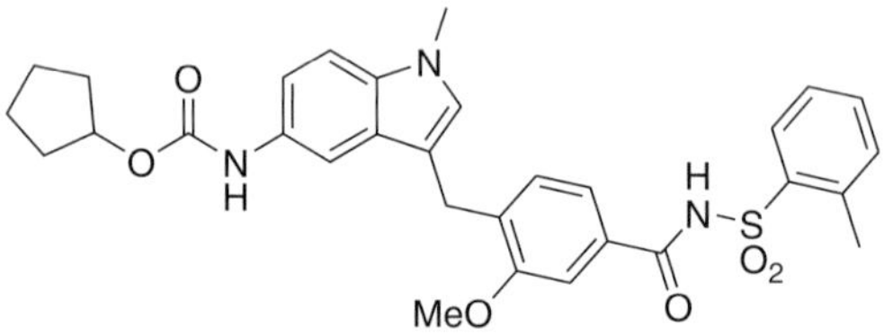

90 Zafirlukast (AstraZeneca)
Binding K_i = 0.5 nM

Analog **89** was identified from a weak high-throughput screening lead and was developed with a strategy that involved replacement of a benzoic acid moiety with the chromone carboxylate core which increased potency 100-fold. Subsequent replacement of the carboxylate with a tetrazole isostere furnished both in vitro and in vivo potency increases. Finally, optimization of the lipophilic tail resulted in the clinical candidate that was approved in Japan in 1995 and was the first LTD_4 (CysLT1) antagonist to be commercially available.[246,247]

The discovery of zafirlukast (**90**) is the subject of several excellent SAR review articles.[248,249] The indole-based inhibitor was essentially developed by a hybridization between a leukotriene pharmacophore based on LTD_4 itself and a hydroxyacetophenone series of antagonists found in a screening exercise. Early analogs suffered from pharmacokinetic issues and poor bioavailability ($<1\%$). Continued optimization of the SAR furnished the clinical candidate that demonstrated a 117-fold shift (within 2 h) of the dose–response curve using the LTD_4 induced bronchoconstriction challenge in asthmatics and was effective in blocking a variety of environmental asthmatic triggers. Zafirlukast was the first leukotriene modulator approved in the USA, which was then followed by montelukast.

Currently research has slowed in this area and it appears no new agents are being developed. This most likely is because the market is saturated and these agents are still used in combination therapy with steroids and β-agonists as a primary care standard for asthma. There has been some controversy over the effectiveness of these agents in the general population and that still remains to be resolved. Current clinical work is focused on their use for other modes of inflammatory disease such as atopic dermatitis, aspirin- and exercise-induced asthma, as well as seasonal allergic rhinitis.

The discovery of LTB_4 antagonists has been slower that of the LTD_4 (CysLT1) antagonists. Pharmacological interest has focused on the proinflammatory properties of this lipid mediator that is produced by several cell types including neutrophils and macrophages. LTB_4 has also been shown to stimulate degranulation and chemotaxis of inflammatory cells through a receptor-based mechanism. Recently the distribution of LTB_4 receptors in human hematopoietic cells has been studied and it was shown that it is widely distributed and also plays a role in endothelial cells but not platelets.[250] A variety of diverse chemotypes have been discovered in the past decade with few new reports since the mid-1990s. Two fundamental research strategies toward drug design have emerged, one based on the discovery of a key fragment and the other based on mimicking the structure of LTB_4.

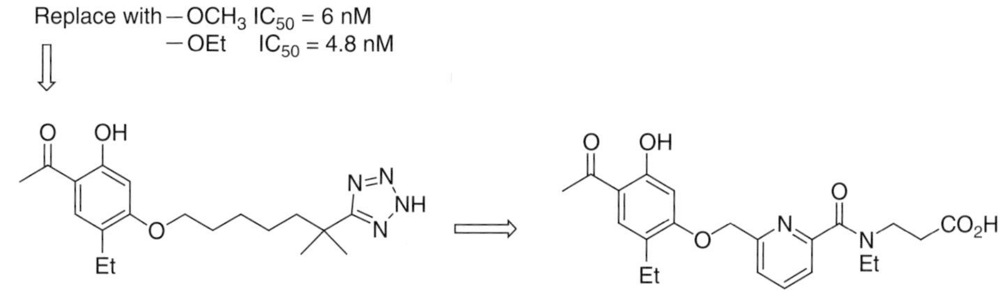

Replace with $-OCH_3$ IC_{50} = 6 nM
$-OEt$ IC_{50} = 4.8 nM

91 Binding IC_{50} = 85 nM (human neutrophils)

92 Binding IC_{50} = 300 nM (human neutrophils)

A key pharmacophore was found, along with some fundamental SAR resulting in various derivatives based on the 5-ethyl-2,4-dihydroxyacetophone moiety which was hypothesized to be a mimetic of the triene moiety in LTB_4. Two such examples are illustrated by analogs **91** and **92**.[251–253] The location of the alkyl chain of **91** proved critical to selectivity over LTD_4. Further SAR work optimizing chain length, carboxy isosteres, and aryl SAR ultimately drove binding potency down to 4.8 nM and afforded a potent analog that was active in blocking LTB_4-induced bronchoconstriction in vivo. Several reviews have been published that are devoted to analogs derived from the parent leukotriene structure.[254]

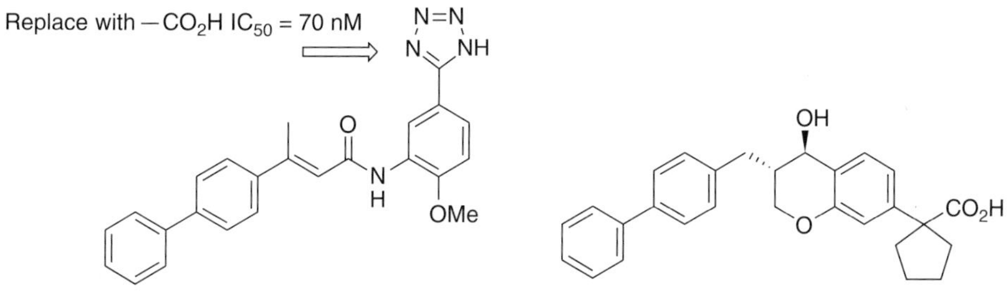

Replace with $-CO_2H$ IC_{50} = 70 nM

93 Binding IC_{50} = 17 nM (human neutrophils)

94 Binding IC_{50} = 6.4 nM (human neutrophils)

Two recent leukotriene-like examples are illustrated by analogs **93** and **94**.[255,256] Both the biaryl and carboxy (or isostere) functionality are required for potency and the design concept is based on a rigid platform with key functionality appropriately displayed. Analog **94** demonstrated high affinity for the LTB_4 receptor and for related cell responses. Oral administration in the guinea pig at 3 mg kg^{-1} blocked 75% of neutrophil response to 12-R-HETE challenge. Phase I studies used ex vivo inhibition of LTB_4-induced CD11b in healthy subjects. The analog produced a 10-fold shift in the dose response curve and was well tolerated up to 640 mg kg^{-1}. Due to its exceptionally long half-life ($t_{1/2}$ = 420 h) it was discontinued for development.[257] Overall, clinical success has yet to be achieved with this class of antagonists. It is not clear whether the receptors response to LTB_4 is exclusive or subject to stimulation by other more potent chemoattractants or whether the primary role of LTB_4 may be coupled with binding to the BLTR2 receptor.

A key enzyme in the biosynthesis of LTB_4 is LTA_4 hydrolase, which is a zinc-containing metalloenzyme and is involved in the rate-limiting step for the production of LTB_4 through the hydrolysis of the epoxide moiety. Many early analogs were dipeptidic in nature as illustrated by the kelatorphan analog **95**.[258]

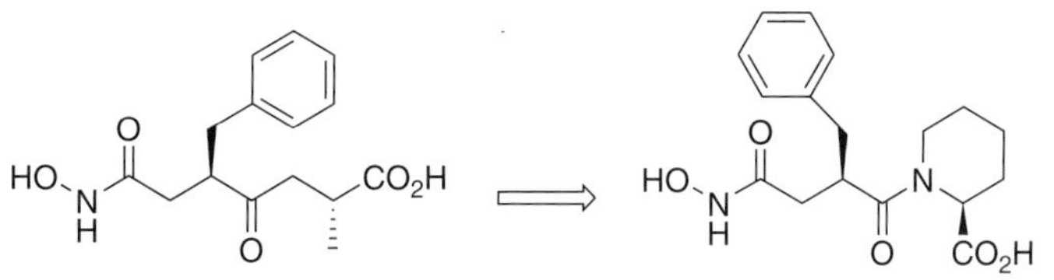

Kelatorphan
LTA_4 hydrolase IC_{50} = 5 nM
Aminopeptidase IC_{50} = 7 nM
NEP IC_{50} = 46 nM
ACE IC_{50} = > 10 μM

95
LTA_4 hydrolase IC_{50} = 8 nM
Aminopeptidase IC_{50} = 10 nM
NEP IC_{50} = 17 μM
ACE IC_{50} = 40 μM

A systematic study of the peptide and substituent SAR demonstrated that potent compounds could be derived with selectivity against NEP and ACE could be gained. Overall, the analogs retained their aminopetdidase activity. In all cases the hydroxamic moiety was required to retain inhibitory activity.

More recently a series of nonpeptide nitrogen heterocycles have been described that are effective inhibitors of LTA_4 hydrolase. They are unique in that they do not appear to bind zinc as the hydroxamic acid inhibitors do and unlike the peptide, they demonstrate good oral bioavailability. Analogs **96**, **97**, and **98** illustrate the SAR of this class.[259]

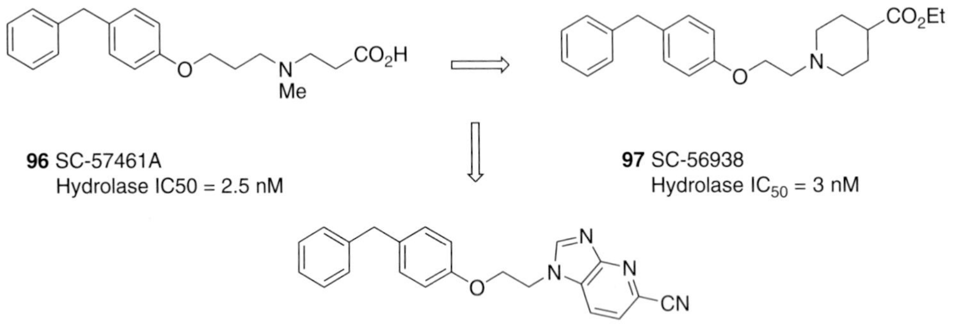

96 SC-57461A
Hydrolase IC50 = 2.5 nM

97 SC-56938
Hydrolase IC_{50} = 3 nM

98 Hydrolase IC_{50} = 0.8 nM

As can be seen in the examples, increased potency can be obtained through optimization of the amine or heterocyclic component. The series produced two clinical candidates that were chosen based on ex vivo versus in vitro potency using blood samples. The primary candidate was **96** that was designed through a series of iterative steps exploring chain length and linker substitution as well as amine substitution. Besides potency in the enzyme assay good whole blood LTB_4 production inhibition was seen (IC_{50} = 49 nM). This analog demonstrated excellent oral bioavailability in the mouse ex vivo assay with an ED_{90} in the 1–3 mg kg^{-1} range. The clinical backup was analog **97** (mouse ex vivo LTB_4 production ED_{90} = 97% Inhibition at 3 mg kg^{-1}) which was selected based on a rhesus monkey assay and showed excellent potency when dosed orally, inhibiting the production of LTB_4 with an ED_{90} < 10 mg kg^{-1}. Unfortunately, these compounds failed preclinical safety studies due to the accumulation of a long-lived metabolite in adipose tissue and mild hepatic toxicity.[260]

To date, leukotriene modulation has been approached from several different avenues resulting in few marketed products. Most clinical trials have focused and concluded for asthma and only recently has the trend shifted toward other inflammation targets.

References

1. Masoli, M.; Fabian, D.; Holt, S.; Beasley, R. *Allergy* **2004**, *59*, 469–478.
2. Lopez, A. D.; Murray, C. C. J. L. *Nat. Med.* **1998**, *4*, 1241–1243.
3. Mannino, D. M.; Homa, D. M.; Pertowski, C. A.; Ashizawa, A.; Nixon, L.; Johnson, C. A. *Morb. Mortal. Wkly Rep.* **1998**, *47*, 1–28.
4. Strassels, S. A.; Smith, D. H.; Sullivan, S. D.; Mahajan, P. S. *Chest* **2001**, *119*, 344–352.
5. Tinkelman, D.; Corsello, P. *Am. J. Manag. Care* **2003**, *9*, 767–771.
6. Weiss, K. B.; Sullivan, S. D. *J. Allergy Clin. Immunol.* **2001**, *107*, 3–8.
7. Hurd, S. *Chest* **2000**, *117*, 1–4.
8. Pauwels, R. A.; Buist, A. S.; Calverley, P. M. A.; Jenkins, C. R.; Hurd, S. S. *Am. J. Respir. Crit. Care Med.* **2001**, *163*, 1256–1276.
9. Williams, S. G.; Schmidt, D. K; Redd, S. C.; Storms, W. *Morb. Mortal. Wkly Rep.* **2003**, *52*, 1–8.
10. McFadden, E. R.; Gilbert, I. A. *N. Engl. J. Med.* **1994**, *330*, 1362–1367.
11. Martinez, F. M. *Am. J. Respir. Crit. Care Med.* **1995**, *151*, 1644–1647.
12. Platts-Mills, T. A. E.; Carter, M. C. *N. Engl. J. Med.* **1997**, *336*, 1382–1384.
13. Bousquet, J.; Jeffery, P. K.; Busse, W. W.; Johnson, M.; Vignola, A. M. *Am. J. Respir. Crit. Care Med.* **2000**, *161*, 1720–1745.
14. Stirling, R. G.; Chung, K. F. *Allergy* **2001**, *56*, 825–840.
15. Barnes, P. J.; Shapiro, S. D.; Pauwels, R. A. *Eur. Respir. J.* **2003**, *22*, 672–688.
16. Barnes, P. J. *N. Engl. J. Med.* **2000**, *343*, 269–280.
17. Mahadeva, R.; Lomas, D. A. *Thorax* **1998**, *53*, 501–505.
18. Busse, W. W.; Lemanske, R. F. *N. Engl. J. Med.* **2001**, *344*, 350–362.
19. Kanaoka, Y.; Boyce, J. A. *J. Immunol.* **2004**, *173*, 1503–1510.
20. Sanderson, C. J. *Blood* **1992**, *79*, 3101–3109.
21. Sriramarao, P.; DiScipio, R. G.; Cobb, R. R.; Cybulsky, M.; Stachnick, G.; Castaneda, D.; Elices, M.; Broide, D. H. *Blood* **2000**, *95*, 592–601.

22. Larbi, K. Y.; Dangerfield, J. P.; Culley, F. J.; Marshall, D.; Haskard, D. O.; Jose, P. J.; Williams, T. J.; Nourshargh, S. *J. Leukoc. Biol.* **2003**, *73*, 65–73.
23. Carlos, T. M.; Harlan, J. M. *Blood* **1994**, *84*, 2068–2101.
24. Terada, N.; Hamano, N.; Nomura, T.; Numata, T.; Hirai, K.; Nakajima, T.; Yamada, H.; Yoshie, O.; Ikeda-Ito, T.; Konno, A. *Clin. Exp. Allergy* **2000**, *30*, 348–355.
25. Tachimoto, H.; Burdick, M. M.; Hudson, S. A.; Kikiuchi, M.; Konstantopolous, K.; Bochner, B. S. *J. Immunol.* **2000**, *165*, 2748–2754.
26. Hata, A. N.; Zent, R.; Breyer, M. D.; Breyer, R. M. *J. Pharmacol. Exp. Therapeut.* **2003**, *306*, 463–470.
27. El Biaze, M.; Boniface, S.; Koscher, V.; Mamessier, E.; Dupuy, P.; Milhe, F.; Ramadour, M.; Vervloet, D.; Magnan, A. *Allergy* **2003**, *58*, 844–853.
28. Clissi, B.; D'Ambrosio, D.; Geginat, J.; Colantonio, L.; Morrot, A.; Freshney, N. W.; Downward, J.; Sinigaglia, F.; Pardi, R. *J. Immunol.* **2000**, *164*, 3292–3300.
29. Shi, H.-Z.; Humables, A.; Gerard, C.; Jin, Z.; Weller, P. F. *J. Clin. Invest.* **2000**, *105*, 945–953.
30. Lazaar, A. L.; Panettieri, R. A. *Am. J. Med.* **2003**, *115*, 652–659.
31. Hawker, K. M.; Johnson, P. R. A.; Hughes, J. M.; Black, J. L. *Am. J. Physiol. Lung Cell Mol. Physiol.* **1998**, *275*, L469–L477.
32. Page, S.; Ammit, A. J.; Black, J. L.; Armour, C. L. *Am. J. Physiol. Lung Cell Mol. Physiol.* **2001**, *281*, L1313–L1323.
33. Chung, K. F. *Eur. Respir. J.* **2000**, *15*, 961–968.
34. Barnes, P. J.; Chung, K. F.; Page, C. P. *Pharmacol. Rev.* **1998**, *50*, 515–596.
35. Parks, W. C.; Shapiro, S. D. *Respir. Res.* **2001**, *2*, 10–19.
36. Akers, I. A.; Parsons, M.; Hill, M. R.; Hollenberg, M. D.; Sanjar, S.; Laurent, G. J.; McAnulty, R. J. *Am. J. Physiol. Lung Cell Mol. Physiol.* **2000**, *278*, L193–L201.
37. Vignola, A. M.; Mirabella, F.; Costanzo, G.; Di Giorgi, R.; Gjomarkaj, M.; Bellia, V.; Bonsignore, G. *Chest* **2003**, *123*, 417–423.
38. Busse, W.; Elias, J.; Sheppard, D.; Banks-Schlegel, S. *Am. J. Respir. Crit. Care Med.* **1999**, *160*, 1035–1042.
39. McDonald, D. M. *Am. J. Respir. Crit. Care Med.* **2001**, *164*, S39–S45.
40. Barnes, P. J. *Pharmacol. Rev.* **2004**, *56*, 515–548.
41. Soler, P.; Moreau, A.; Basset, F.; Hance, A. J. *Am. Rev. Respir. Dis.* **1989**, *139*, 1112–1117.
42. Shapiro, S. *Am. J. Respir. Crit. Care Med.* **1999**, *160*, S29–S32.
43. Meshi, B.; Vitalis, T. Z.; Ionescu, D.; Elliott, W. M.; Liu, C.; Wang, X.-D.; Hayashi, S.; Hogg, J. C. *Am. J. Respir. Mol. Cell. Biol.* **2002**, *26*, 52–57.
44. Keatings, V. M.; Barnes, P. J. *Am. J. Respir. Crit. Care Med.* **1997**, *155*, 453–459.
45. Finkelstein, R.; Fraser, R. S.; Ghezzo, H.; Cosio, M. G. *Am. J. Respir. Crit. Care Med.* **1995**, *152*, 1666–1672.
46. Csoma, Z.; Kharitonov, S. A.; Balint, B.; Bush, A.; Wilson, N. M.; Barnes, P. J. *Am. J. Respir. Crit. Care Med.* **2002**, *166*, 1345–1349.
47. Tager, A. M.; Bromley, S. K.; Medoff, B. D.; Islam, S. A.; Bercury, S. D.; Friedrich, E. B.; Carafone, A. D.; Gerstzen, R. E.; Luster, A. D. *Nat. Immunol.* **2003**, *4*, 982–990.
48. Sommerhoff, C. P.; Nadel, J. A.; Basbaum, C. B.; Caughey, G. H. *J. Clin. Invest.* **1990**, *85*, 682–689.
49. Hautamaki, R. D.; Kobayashi, D. K.; Senior, R. M.; Shapiro, S. D. *Science* **1997**, *277*, 2002–2004.
50. Van Eerdewegh, P.; Little, R. D.; Dupuis, J.; Del Mastro, R. G.; Falls, K.; Simon, J.; Torrey, D.; Pandit, S.; McKenny, J.; Braunschweiger, K. et al. *Nature* **2002**, *418*, 426–430.
51. Allen, M.; Heinzmann, A.; Noguchi, E.; Abecasis, G.; Broxholme, J.; Ponting, C. P.; Bhattacharyya, S.; Tinsley, J.; Zhang, Y.; Holt, R. et al. *Nat. Genet.* **2003**, *35*, 258–263.
52. Zhang, Y.; Leaves, N. I.; Anderson, G. G.; Ponting, C. P.; Broxholme, J.; Holt, R.; Edser, P.; Bhattacharyya, S.; Dunham, A.; Adcock, I. M. et al. *Nat. Genet.* **2003**, *34*, 181–186.
53. Buhl, R.; Farmer, S. G. *Proc. Am. Thorac. Soc.* **2005**, *2*, 83–93.
54. Ambler, J.; Miller, J. N.; Orr, T. S. C. *Int. Arch. Allergy Appl. Immunol.* **1994**, *46*, 427–437.
55. Campbell, E. M.; Kunkel, S. L.; Strieter, R. M.; Lukacs, N. W. *J. Immunol.* **1998**, *161*, 7047–7051.
56. Mahadeva, R.; Shapiro, S. D. *Thorax* **2002**, *57*, 908–914.
57. Pizzichini, E.; Leff, J. A.; Reiss, T. F.; Hendeles, L.; Boulet, L.-P.; Wei, L. X.; Efthimiadis, A. E.; Zhang, J.; Hargreave, F. E. *Eur. Respir. J.* **1999**, *14*, 12–18.
58. Sabater, J. R.; Wanner, A.; Abraham, W. M. *Am. J. Respir. Crit. Care Med.* **2002**, *166*, 1457–1460.
59. Abraham, W. L.; Ahmed, A.; Cortes, A.; Sielczak, M. W.; Hinz, W.; Bouska, J.; Lanni, C.; Bell, R. L. *Eur. J. Pharmacol.* **1992**, *217*, 119–126.
60. Henderson, W. R.; Banerjee, E. R; Chi, E. Y. *J. Allergy Clin. Immunol.* **2005**, *116*, 332–340.
61. Terawaki, K.; Yokomizo, T.; Nagase, T.; Toda, A.; Taniguchi, M.; Hasizume, K.; Yagi, T.; Shimizu, T. *J. Immunol.* **2005**, *175*, 4217–4225.
62. Hansen, G.; Jin, S. L. C.; Umetsu, D. T.; Conti, M. *Proc. Natl. Acad. Sci. USA* **2000**, *97*, 6751–6756.
63. Wang, Z.; Zheng, T.; Zhu, Z.; Homer, R. J.; Riese, R. J.; Chapman, H. A.; Shapiro, S. D.; Elias, J. A. *J. Exp. Med.* **2000**, *192*, 1587–1599.
64. Elias, J. A.; Zheng, T.; Lee, C. G.; Homer, R. J.; Chen, Q.; Ma, B.; Blackburn, M.; Zhu, Z. *Chest* **2003**, *123*, 339–345.
65. Lee, J. J.; Dimina, D.; Macias, M. P.; Ochkur, S. I.; McGarry, M. P.; O'Neill, K. R.; Protheroe, C.; Pero, R.; Nguyen, T.; Cormier, S. A. et al. *Science* **2004**, *305*, 1773–1776.
66. Humbles, A. A.; Lloyd, C. M.; McMillan, S. J.; Friend, D. S.; Xanthou, G.; McKenna, E. E.; Ghiran, S.; Gerard, N. P.; Yu, C.; Orkin, S. H. et al. *Science* **2004**, *305*, 1776–1779.
67. Wills-Karp, M.; Karp, C. L. *Science* **2004**, *305*, 1726–1729.
68. Dawkins, P. A.; Stockley, R. A. *Thorax* **2001**, *56*, 972–977.
69. Wright, J. L.; Chung, A. *Chest* **2002**, *122*, 301–306.
70. Coffman, R. L.; Hessel, E. M. *J. Exp. Med.* **2005**, *201*, 1875–1879.
71. Mauser, P. J.; Pitman, A. M.; Fernandez, X.; Feran, S. K.; Adams, G. K.; Kreutner, W.; Egan, R. W.; Chapman, R. W. *Am. J. Respir. Crit. Care Med.* **1995**, *152*, 467–472.
72. Chang, M. M.-J.; Wu, R.; Plopper, C. G.; Hyde, D. M. *Am. J. Physiol.* **1998**, *275*, L524–L532.
73. Frew, A. J. *J. Allergy Clin. Immunol.* **2002**, *109*, 210–213.
74. Renard, S. I. *Am. Proc. Am. Thorac. Soc.* **2004**, *1*, 282–287.
75. Stelmach, I.; Jerzymska, J.; Kuna, P. *J. Allergy Clin. Immunol.* **2002**, *109*, 257–263.
76. Yasuda, H.; Yarnaya, M.; Nakayma, K.; Ebihara, S.; Sasaki, T.; Okinaga, S.; Inoue, D.; Asada, M.; Nemoto, M.; Saski, H. *Am. J. Respir. Crit. Care Med.* **2005**, *171*, 1246–1251.
77. Berlyne, G. S.; Parameswaran, K.; Kamada, D.; Efthimiadis, A.; Hargreave, F. E. *J. Allergy Clin. Immunol.* **2000**, *106*, 638–644.
78. Odingo, J. O. *Expert Opin. Ther. Patents* **2005**, *15*, 773–787.

79. Lecklie, M. J.; ten Brinke, A.; Khan, J.; Diamant, Z.; O'Connor, B. J.; Walls, C. M.; Mathur, A. K.; Cowley, H. C.; Chung, K. F.; Djukanovic, R. et al. *Lancet* **2000**, *356*, 2144–2148.
80. O'Byrne, P. M.; Inman, M. D.; Parameswaran, K. J. *Allergy Clin. Immunol.* **2001**, *108*, 503–508.
81. Plötz, S.-G.; Uwe-Simon, H.; Darsow, U.; Simon, D.; Vassina, E.; Yousefi, S.; Hein, R.; Smith, T.; Behrendt, H.; Ring, J. *N. Engl. J. Med.* **2003**, *349*, 2334–2339.
82. Naya, A.; Saeki, T. *Expert Opin. Ther. Patents* **2004**, *14*, 7–16.
83. Traves, S. L.; Donnelly, L. E. *Curr. Respir. Med. Rev.* **2005**, *1*, 15–32.
84. Hoshi, K.; Kurosawa, S.; Kato, M.; Andoh, K.; Satoh, D.; Kaise, A. *Tohuku J. Exp. Med.* **2005**, *207*, 143–148.
85. Barnes, P. J. *Eur. Respir. Rev.* **2005**, *14*, 2–11.
86. Cairns, J. A. *Pulmon. Pharmacol. Ther.* **2005**, *18*, 55–66.
87. Donohue, J. F. *Chest* **2004**, *126*, 125S–137S.
88. Brooks, C. D. W.; Summers, J. B. *J. Med. Chem.* **1996**, *39*, 2629–2654.
89. Kelloway, J. S. *Ann. Pharmacother.* **1997**, *31*, 1012–1021.
90. Noonan, M. J.; Chervinsky, P.; Brandon, M.; Zhang, J.; Kundu, S.; McBurney, J.; Reiss, T. F. *Eur. Respir. J.* **1998**, *11*, 1232–1239.
91. Lu, P.; Schrag, M. L.; Slaughter, D. E.; Raab, C. E.; Shou, M.; Rodrigues, A. D. *Drug Metab. Dispos.* **2003**, *31*, 1352–1360.
92. Currie, G. P.; Srivastava, P.; Dempsey, O. J.; Lee, D. K. C. *Q. J. Med.* **2005**, *98*, 171–182.
93. Wechsler, M. E.; Garpestad, E.; Flier, S. R.; Kocher, O.; Weiland, D. A.; Polito, A. J.; Klinek, M. M.; Bigby, T. D.; Wong, G. A.; Helmers, R. A. et al. *JAMA* **1998**, *279*, 455–457.
94. Solans, R.; Bosch, J. A.; Selva, A.; Orriols, R.; Vilardell, M. *Thorax* **2000**, *57*, 183–185.
95. Djukanovic, R.; Wilson, S. J.; Kraft, M.; Jarjour, N. N.; Steel, M.; Chung, K. F.; Bao, W.; Fowler-Taylor, A.; Matthews, J.; Busse, W. W. et al. *Am. J. Respir. Crit. Care Med.* **2004**, *170*, 583–593.
96. Bisset, L. R.; Schmid-Grendelmeier, P. *Curr. Opin. Pulmon. Med.* **2004**, *11*, 35–42.
97. Rot, A.; VonAdrian, U. H. *Ann. Rev. Immunol.* **2004**, *22*, 891–928.
98. Carter, P. H. *Curr. Opin. Chem. Biol.* **2002**, *6*, 510–525.
99. Wardlaw, A. J.; Dunnette, S.; Gleich, G. J.; Collins, J. V.; Kay, A. B. *Am. Rev. Respir. Dis.* **1988**, *137*, 62–69.
100. Corrigan, C. *Curr. Opin. Invest. Drugs* **2000**, *1*, 321–328.
101. Humbles, A. A.; Lu, B.; Friend, D. S.; Okinaga, S.; Logarawi, J.; Martin, T. R.; Gerard, N. P.; Gerard, C. *Proc. Natl. Acad. Sci. USA* **2002**, *99*, 1479–1484.
102. Rothenberg, M. E.; MacLean, J. A.; Pearlman, E.; Luster, A. D.; Leder, P. *J. Exp. Med.* **1997**, *185*, 785–790.
103. Gonzalo, J. A.; Lloyd, C. M.; Kremer, L. *J. Clin. Invest.* **1996**, *98*, 2332–2345.
104. Heath, H.; Qin, S.; Rao, P.; LaRosa, G.; Kassam, N.; Ponath, P. D.; Mackay, C. R. *J. Clin. Invest.* **1997**, *99*, 178–184.
105. Naya, A.; Saeki, T. *Expert Opin. Ther. Patents* **2004**, *14*, 7–16.
106. Naya, A.; Kobayashi, K.; Ishikawa, M.; Ohwaki, K.; Saeki, T.; Noguchi, K.; Ohtake, N. *Bioorg. Med. Chem. Lett.* **2001**, *11*, 1219–1223.
107. Saeki, T.; Ohwaki, K.; Naya, A.; Kobayashi, K.; Ishikawa, M.; Ohtake, N.; Noguchi, K. *Biochem. Biophys. Res. Commun.* **2001**, *281*, 779–782.
108. Naya, A.; Kobayashi, K.; Ishikawa, M.; Ohwaki, K.; Saeki, T.; Noguchi, K.; Ohtake, N. *Chem. Pharm. Bull.* **2003**, *51*, 697–701.
109. Bryan, S. A.; Jose, P. J.; Topping, J. R.; Wilhelm, R.; Soderberg, C.; Kertesz, D.; Barnes, P. J.; Williams, T. J.; Hansell, T. T.; Sobroe, I. *Am. J. Resp. Crit. Care Med.* **2002**, *165*, 1602.
110. Gong, L.; Hogg, J. H.; Collier, J.; Wilhelm, R. S.; Soderberg, C. *Bioorg. Med. Chem. Lett.* **2003**, *13*, 3597–3600.
111. Ting, P. C.; Lee, J. F.; Wu, J.; Umland, S. P.; Aslanian, R.; Cao, Z. J.; Dong, Y.; Garlisi, C. G.; Gilbert, E. J.; Huang, Y. et al. *Bioorg. Med. Chem. Lett.* **2005**, *15*, 1375–1378.
112. Wan, Y.; Jakway, J. P.; Qiu, H.; Shah, H.; Garlisi, C. G.; Tian, F.; Ting, P.; Hesk, D.; Egan, R. W.; Billah, M. M.; Umland, S. P. *Eur. J. Pharmacol.* **2002**, *456*, 1–10.
113. Anderskewitz, R.; Bauer, R.; Bodenbach, G.; Gester, D.; Gramlich, B.; Morschhauser, G.; Birke, F. W. *Bioorg. Med. Chem. Lett.* **2005**, *15*, 669–673.
114. Wacker, D. A.; Santella, J. B., III; Gardner, D. S.; Varnes, J. G.; Estrella, M.; DeLucca, G. V.; Ko, S. S.; Tanabe, K.; Watson, P. S.; Welch, P. K. et al. *Bioorg. Med. Chem. Lett.* **2002**, *12*, 1785–1789.
115. Batt, D. G.; Houghton, G. C.; Roderick, J. R.; Santella, J. B., III; Wacker, D. A.; Welch, P. K.; Orlovsky, Y. I.; Wadman, E. A.; Trzaskos, J. M.; Davies, P. et al. *Bioorg. Med. Chem. Lett.* **2005**, *15*, 787–791.
116. Varnes, J. G.; Gardner, D. S.; Santella, J. B., III; Duncia, J. V.; Estrella, M.; Watson, P. S.; Clark, C. M.; Ko, S. S.; Welch, P.; Covington, M. et al. *Bioorg. Med. Chem. Lett.* **2004**, *14*, 1645–1649.
117. DeLucca, G. V.; Kim, U. T.; Vargo, B. J.; Duncia, J. V.; Santella, J. B., III; Gardner, D. S.; Zheng, C.; Liauw, A.; Wang, Z.; Emmett, G. et al. *J. Med. Chem.* **2005**, *48*, 2194–2211.
118. *Expert Opin. Ther. Patents* **2004**, *14*, 577-582.
119. Vedani, A.; Dobler, M.; Dillinger, H.; Hasselbach, K. M.; Birke, F.; Lill, M. A. *J. Med. Chem.* **2005**, *48*, 1515–1527.
120. Hirai, H.; Tanaka, K.; Yoshie, O.; Ogawa, K.; Kenmotsu, K.; Takamori, Y.; Ichimasa, M.; Sugamura, K.; Nakamura, M.; Takano, S.; Nagata, K. *J. Exp. Med.* **2001**, *193*, 255–261.
121. Liu, F.; Gonzalo, J. A.; Manning, S.; O'Connell, L. E.; Fedyk, E. R.; Burke, K. E.; Elder, A. M.; Pulido, J. C.; Cao, W.; Tayber, O. et al. *Prostaglandins Leukot. Essent. Fatty Acids* **2005**, *76*, 133–147.
122. Sinigaglia, F.; Bordignon, P. P.; d'Ambrosio, D. *Clin. Exp. Allergy Rev.* **2004**, *4*, 167–170.
123. Hirai, H.; Tanaka, K.; Takano, S.; Ichimasa, M.; Nakamura, M.; Nagata, K. *J. Immunol.* **2002**, *165*, 981–985.
124. Gervais, F. G.; Morello, J.-P.; Beaulieu, C.; Sawyer, N.; Denis, D.; Greig, G.; Malebranche, D.; O'Neill, G. P. *Mol. Pharmacol.* **2005**, *67*, 1834–1839.
125. Hata, A. N.; Lybrand, T. P.; Marnett, L. J.; Breyer, R. M. *Mol. Pharmacol.* **2005**, *67*, 640–647.
126. Sugimoto, H.; Shichijo, M.; Iino, T.; Manabe, Y.; Watanabe, A.; Shimazaki, M.; Gantner, F.; Bacon, K. B. *J. Pharmacol. Exp. Ther.* **2003**, *305*, 347–352.
127. Ohkubo, K.; Gotoh, M. *Allergol. Int.* **2003**, *52*, 131–138.
128. Aizawa, H.; Shigyo, M.; Nogami, H.; Hirose, T.; Hara, N. *Chest* **1996**, *109*, 338–342.
129. Robarge, M. J.; Bom, D. C.; Tumey, L. N.; Varga, N.; Gleason, E.; Silver, D.; Song, J.; Murphy, S. M.; Ekema, G.; Doucette, C. et al. *Bioorg. Med. Chem. Lett.* **2005**, *15*, 1749–1753.
130. Ulven, T.; Kostenis, E. *J. Med. Chem.* **2005**, *48*, 897–900.

131. *Expert Opin. Ther. Patents* **2005**, *15*, 115-117.
132. Kabashima, K.; Narumiya, S. *Prostaglandins Leukot. Essent. Fatty Acids* **2003**, *69*, 187–194.
133. Coleman, R. A.; Kennedy, I.; Humphrey, P. P. A.; Bunce, K.; Lumley, P. In *Comprehensive Medicinal Chemistry*; Emmet, J. C., Ed.; Pergamon Press: Oxford, UK, 1989; Vol. 3, pp 643–714.
134. Mitsumori, S.; Tsuri, T.; Honma, T.; Hiramatsu, Y.; Okada, T.; Hashizume, H.; Inagaki, M.; Arimura, A.; Yasui, K.; Asanuma, F. et al. *J. Med. Chem.* **2003**, *46*, 2436–2445.
135. Mitsumori, S.; Tsuri, T.; Honma, T.; Hiramatsu, Y.; Okada, T.; Hashizume, H.; Kida, S.; Inagaki, M.; Arimura, A.; Yasui, K. et al. *J. Med. Chem.* **2003**, *46*, 2446–2455.
136. Torisu, K.; Kobayashi, K.; Iwahashi, M.; Egashira, H.; Nakai, Y.; Okada, Y.; Nanbu, F.; Ohuchida, H.; Nakai, H.; Toda, M. *Bioorg. Med. Chem. Lett.* **2004**, *14*, 4557–4562.
137. Torisu, K.; Kobayashi, K.; Iwahashi, M.; Nakai, Y.; Onoda, T.; Nagase, T.; Sugimoto, I.; Okada, Y.; Matsumoto, R.; Nanbu, F. et al. *Bioorg. Med. Chem.* **2004**, *12*, 5361–5378.
138. *O'Neill, G. The Prostanoid DP1 Receptor as a Target for Respiratory Disease*. Proceedings of the 12th International Conference of the Inflammation Research Association, Bolton Landing, NY, Oct 3–7, 2004.
139. Beaulieu, C.; Wang, Z.; Denis, D.; Greig, G.; Lamontagne, S.; O'Neill, G.; Slipetz, D.; Wang, J. *Bioorg. Med. Chem. Lett.* **2004**, *14*, 3195–3199.
140. Giembycz, M. A. *Monaldi Arch. Chest. Dis.* **2002**, *57*, 48–64.
141. Souness, J. E.; Alsous, D.; Sargent, C. *Immunopharmacology* **2000**, *47*, 127–162.
142. Giembycz, M. A. *Curr. Opin. Pharmacol.* **2005**, *5*, 238–244.
143. Souness, J. E.; Rao, S. *Cell. Signal.* **1997**, *9*, 227–236.
144. Reid, P. *Curr. Opin. Investig. Drugs* **2002**, *3*, 1165–1170.
145. Giembycz, M. A. *Expert Opin. Investig. Drugs* **2001**, *10*, 1361–1379.
146. Christensen, S. B.; Guider, A.; Forster, C. J.; Gleason, J. G.; Bender, P. E.; Karpinski, J. M.; DeWolf, W. E., Jr.; Barnette, M. S.; Underwood, D. C.; Griswold, D. E. et al. *J. Med. Chem.* **1998**, *41*, 821–835.
147. Lipworth, B. J. *Lancet* **2005**, *365*, 167–175.
148. MacKenzie, S. J. *Allergol. Int.* **2004**, *53*, 101–110.
149. Odingo, J. O. *Expert Opin. Ther. Patents* **2005**, *15*, 773–787.
150. Ashton, M. J.; Cook, D. C.; Fenton, G.; Karlsson, J. A.; Palfreyman, M. N.; Raeburn, D.; Ratcliffe, A. J.; Souness, J. E.; Thurairatnam, S.; Vicker, N. *J. Med. Chem.* **1994**, *37*, 1696–1703.
151. Hulme, C.; Poli, G. B.; Huang, F. C.; Souness, J. E.; Djuric, S. W. *Bioorg. Med. Chem. Lett.* **1998**, *8*, 175–178.
152. Masamune, H.; Cheng, J. B.; Cooper, K.; Eggler, J. F.; Marfat, A.; Marshall, S. C.; Shirley, J. T.; Tickner, J. E.; Umland, J. P.; Vazquez, E. *Bioorg. Med. Chem. Lett.* **1995**, *5*, 1965–1968.
153. Cheng, J. B.; Cooper, K.; Duplantier, A. J.; Eggler, J. F.; Kraus, K. G.; Marshall, S. C.; Marfat, A.; Masamune, H.; Shirley, J. T.; Tickner, J. E. et al. *Bioorg. Med. Chem. Lett.* **1995**, *5*, 1969–1972.
154. Guay, D.; Hamel, P.; Blouin, M.; Brideau, C.; Chan, C. C.; Chauret, N.; Ducharme, Y.; Huang, Z.; Girard, M.; Jones, T. R. et al. *Bioorg. Med. Chem. Lett.* **2002**, *12*, 1457–1461.
155. Frenette, R.; Blouin, M.; Brideau, C.; Chauret, N.; Ducharme, Y.; Friesen, R. W.; Hamel, P.; Jones, T. R.; Laliberté, F.; Li, C. et al. *Bioorg. Med. Chem. Lett.* **2002**, *12*, 3009–3013.
156. Ducharme, Y.; Friesen, R. W.; Blouin, M.; Côté, B.; Dubé, D.; Ethier, D.; Frenette, R.; Laliberté, F.; Mancini, J.; Masson, P. et al. *Bioorg. Med. Chem. Lett.* **2003**, *13*, 1923–1926.
157. Friesen, R. W.; Ducharme, Y.; Ball, R. G.; Blouin, M.; Boulet, L.; Côté, B.; Frenette, R.; Girard, M.; Guay, D.; Huang, Z. et al. *J. Med. Chem.* **2003**, *46*, 2413–2426.
158. Van der Mey, M.; Boss, H.; Couwenberg, D.; Hatzelmann, A.; Sterk, G. J.; Goubitz, K.; Schenk, H.; Timmerman, H. *J. Med. Chem.* **2002**, *45*, 2526–2533.
159. Van der Mey, M.; Boss, H.; Hatzelmann, A.; Van der Laan, I. J.; Sterk, G. J.; Timmerman, H. *J. Med. Chem.* **2002**, *45*, 2520–2525.
160. Krier, M.; de Araújo-Júnior, J. X.; Schmitt, M.; Duranton, J.; Justiano-Basaran, H.; Lugnier, C.; Bourguignon, J. J.; Rogna Van der Mey, M.; Boss, H.; Couwenberg, D. et al. *J. Med. Chem.* **2005**, *48*, 3816–3822.
161. Ochiai, H.; Ohtani, T.; Ishida, A.; Kishikawa, K.; Obata, T.; Nakai, H.; Toda, M. *Bioorg. Med. Chem. Lett.* **2004**, *14*, 1323–1327.
162. Andrés, J. I.; Alonso, J. M.; Díaz, A.; Fernández, J.; Iturrino, L.; Martínez, P.; Matesanz, E.; Freyne, E. J.; Deroose, F.; Boeckx, G. et al. *Bioorg. Med. Chem. Lett.* **2002**, *12*, 653–658.
163. Stafford, J. A.; Feldman, P. L.; Marron, B. E.; Schoenen, F. J.; Valvano, N. L.; Unwalla, R. J.; Domanico, P. L.; Brawley, E. S.; Leesnitzer, M. A.; Rose, D. A. et al. *Bioorg. Med. Chem. Lett.* **1994**, *4*, 1855–1860.
164. Hulme, C.; Mathew, R.; Moriarty, K.; Miller, B.; Ramanjulu, M.; Cox, P.; Souness, J.; Page, K. M.; Uhl, J.; Travis, J. et al. *Bioorg. Med. Chem. Lett.* **1998**, *8*, 3053–3058.
165. Hulme, C.; Moriarty, K.; Miller, B.; Mathew, R.; Ramanjulu, M.; Cox, P.; Souness, J.; Page, K. M.; Uhl, J.; Travis, J. et al. *Bioorg. Med. Chem. Lett.* **1998**, *8*, 1867–1872.
166. Regan, J.; Bruno, J.; McGarry, D.; Poli, G.; Hanney, B.; Bower, S.; Travis, J.; Sweeney, D.; Miller, B.; Souness, J. et al. *Bioorg. Med. Chem. Lett.* **1998**, *8*, 2737–2742.
167. Napoletano, M.; Norcini, G.; Pellacini, F.; Marchini, F.; Morazzoni, G.; Ferlenga, P.; Pradella, L. *Bioorg. Med. Chem. Lett.* **2000**, *10*, 2235–2238.
168. Napoletano, M.; Norcini, G.; Pellacini, F.; Marchini, F.; Morazzoni, G.; Ferlenga, P.; Pradella, L. *Bioorg. Med. Chem. Lett.* **2001**, *11*, 33–37.
169. Napoletano, M.; Norcini, G.; Pellacini, F.; Marchini, F.; Morazzoni, G.; Fattori, R.; Ferlenga, P.; Pradella, L. *Bioorg. Med. Chem. Lett.* **2002**, *12*, 5–8.
170. Pascal, Y.; Andrianjara, C. R.; Auclair, E.; Avenel, N.; Bertin, B.; Calvet, A.; Feru, F.; Lardon, S.; Moodley, I.; Ouagued, M. et al. *Bioorg. Med. Chem. Lett.* **2000**, *10*, 35–38.
171. Devillers, I.; Pevet, I.; Jacobelli, H.; Durand, C.; Fasquelle, V.; Puaud, J.; Gaudilliere, B.; Idrissi, M.; Moreau, F.; Wrigglesworth, R. *Bioorg. Med. Chem. Lett.* **2004**, *14*, 3303–3306.
172. Burnouf, C.; Auclair, E.; Avenel, N.; Bertin, B.; Bigot, C.; Calvet, A.; Chan, K.; Durand, C.; Fasquelle, V.; Feru, F. et al. *J. Med. Chem.* **2000**, *43*, 4850–4867.
173. Manallack, D. T.; Hughes, R. A.; Thompson, P. E. *J. Med. Chem.* **2005**, *48*, 3449–3462.
174. Conti, M. *Nat. Struct. Mol. Biol.* **2004**, *11*, 809–810.
175. Card, G. L.; Blasdel, L.; England, B. P.; Zhang, C.; Suzuki, Y.; Gillette, S.; Fong, D.; Ibrahim, P. N.; Artis, D. R.; Bollag, G. et al. *Nat. Biotechnol.* **2005**, *23*, 201–207.

176. Schwartz, L. B.; Irani, A. M. A.; Roller, K.; Castells, M. C.; Schechter, N. M. *J. Immunol.* **1987**, *138*, 2611–2615.
177. Pereira, P. J. B.; Bergner, A.; Macedo-Riberiro, S.; Huber, G.; Matschiner, G.; Fritz, H.; Sommerhoff, C. P.; Bode, W. *Nature* **1998**, *392*, 306.
178. Sommerhoff, C. P.; Bode, W.; Matschiner, G.; Bergner, A.; Fritz, H. *Biochim. Biophys. Acta* **2000**, *1477*, 75–89.
179. Schwartz, L. B.; Bradford, T. R. *J. Biol. Chem.* **1986**, *261*, 7372–7379.
180. Selwood, T.; Smolensky, H.; McCaslin, D. R.; Schechter, N. M. *Biochemistry* **2005**, *44*, 3580–3590.
181. Burgess, L. E. *Drug News Perspect.* **2000**, *13*, 147–157.
182. Gangloff, A. R. *Curr. Opin. Invest. Drugs* **2000**, *1*, 79–85.
183. Carins, J. A. *Pulm. Pharmacol. Ther.* **2005**, *18*, 55–66.
184. Rice, K. D.; Sprengeler, P. A. *Curr. Opin. Drug Disc. Dev.* **1999**, *2*, 463–474.
185. Clark, J. M.; Moore, W. R.; Tanaka, R. D. *Drugs Future* **1996**, *21*, 811–816.
186. Caughey, G. H.; Raymond, W. W.; Bacci, E.; Lombardy, R. J.; Tidwell, R. R. *J. Pharmacol. Exp. Ther.* **1993**, *264*, 676–682.
187. Combrink, K. D.; Gulgeze, H. B.; Meanwell, N. A.; Pearce, B. C.; Zulan, P.; Bisacchi, G. S.; Roberts, D. G. M.; Stanley, P.; Seiler, S. M. *J. Med. Chem.* **1998**, *41*, 4854–4860.
188. Costanzo, M. J.; Yabut, S. C.; Almond, A. R.; Andrade-Gordon, P.; Corcoran, T. W.; deGaravilla, L.; Kauffman, J. A.; Abraham, W. M.; Recacha, R.; Chattopadhyay, D. et al. *J. Med. Chem.* **2003**, *46*, 3865–3876.
189. Murakami, Y.; Takei, M.; Shindo, K.; Kitazume, J.; Tanaka, J.; Higa, T.; Fukamachi, H. *J. Nat. Prod.* **2002**, *65*, 259–261.
190. Yu, K. L.; Civiello, R.; Roberts, D. G. M.; Seiler, S. M.; Meanwell, N. A. *Bioorg. Med. Chem. Lett.* **1999**, *9*, 663–666.
191. Qian, X.; Zheng, B.; Burke, B.; Saindane, M. T.; Kronenthal, D. R. *J. Org. Chem.* **2002**, *67*, 3595-3600 (and references cited therein).
192. Levell, J.; Astles, P.; Eastwood, P.; Carins, J.; Houille, O.; Aldous, S.; Merriman, G.; Whiteley, B.; Pribish, J.; Czekaj, M. et al. *Bioorg. Med. Chem.* **2005**, *13*, 2859–2872.
193. Hopkins, C. R.; Czekaj, M.; Kaye, S. S.; Gao, Z.; Pribish, J.; Pauls, H.; Liang, G.; Sides, K.; Cramer, D.; Carins, J. et al. *Bioorg. Med. Chem. Lett.* **2005**, *15*, 2734–2737.
194. Martin, T. J. Fifth International Electronic Conference on Synthetic Organic Chemistry (ECSOC-5), Sep 2001, C0011. http://www.mdpi.org/ (accessed April 2006).
195. Vaz, R. J.; Gao, Z.; Pribish, J.; Chen, X.; Levell, J.; Davis, L.; Albert, E.; Brollo, M.; Ugolini, A.; Cramer, D. M. et al. *Bioorg. Med. Chem. Lett.* **2004**, *14*, 6053–6056.
196. Hallgren, J.; Estrada, S.; Karlson, U.; Alving, K.; Pejler, G. *Biochemistry* **2001**, *40*, 7342–7349.
197. Cregar, L.; Elrod, K. C.; Putnam, D.; Moore, W. R. *Arch. Biochem. Biophys.* **1999**, *366*, 125–130.
198. Miller, H. R. P.; Pemberton, A. D. *Immunology* **2002**, *105*, 375–390.
199. Dell'Italia, L. J.; Husain, A. *Curr. Opin. Cardiol.* **2002**, *17*, 374–379.
200. Maryanoff, B. E. *J. Med. Chem.* **2004**, *47*, 769–787.
201. Doggrell, S. A.; Wanstall, J. C. *Expert Opin. Invest. Drugs* **2003**, *12*, 1429–1432.
202. Akahoshi, F.; Ashimori, A.; Sakashita, H.; Yoshimura, T.; Eda, M.; Imada, T.; Nakajima, M.; Mitsutomi, N.; Kuwahara, S.; Ohtsuka, T. et al. *J. Med. Chem.* **2001**, *44*, 1297–1304.
203. Akahoshi, F.; Ashimori, A.; Sakashita, H.; Yoshimura, T.; Imada, T.; Nakajima, M.; Mitsutomi, N.; Kuwahara, S.; Ohtsuka, T.; Fukaya, C. et al. *J. Med. Chem.* **2001**, *44*, 1285–1296.
204. Akahoshi, F. *Drugs Future* **2002**, *27*, 765–770.
205. Muto, T.; Fukami, H. *IDrugs* **2002**, *5*, 1141–1150.
206. Aoyama, Y.; Uenaka, M.; Konoike, T.; Iso, Y.; Nishitani, Y.; Kanda, A.; Naya, N.; Nakajima, M. *Bioorg. Med. Chem. Lett.* **2000**, *10*, 2397–2401.
207. Aoyama, Y.; Uenaka, M.; Kii, M.; Tanaka, M.; Konoike, T.; Hayasaki-Kajiwara, Y.; Naya, N.; Nakajima, M. *Bioorg. Med. Chem.* **2001**, *9*, 3065–3075.
208. Greenfeder, S.; Umland, S. P.; Cuss, F. M.; Chapman, R. W.; Egan, R. W. *Respir. Res.* **2001**, *2*, 71–79.
209. Uings, I.; McKinnon, M. *Curr. Pharma. Des.* **2002**, *8*, 1837–1844.
210. Minnicozzi, M. *Expert Opin. Ther. Patents* **1999**, *9*, 147–156.
211. Miyajima, A.; Kitamura, T.; Harada, N.; Yokota, T.; Arai, K. *Annu. Rev. Immunol.* **1992**, *10*, 295–331.
212. Morokata, T.; Suzuki, K.; Ida, K.; Yamada, T. *Immunol. Lett.* **2005**, *98*, 161–165.
213. Morokata, T.; Ida, K.; Yamada, T. *Int. Immunol.* **2002**, *2*, 1693–1702.
214. Devos, R.; Guisez, Y.; Plaetinck, G.; Cornelis, S.; Tavernier, J.; Van der Heuden, J.; Foley, L. H.; Scheffler, G. *Eur. J. Biochem.* **1994**, *225*, 635–640.
215. Wiekowski, M.; Prosser, D.; Taremi, S.; Tsarbopoulos, A.; Jenh, C. H.; Chou, C. C.; Lundell, D.; Zavodny, P.; Narula, S. *Eur. J. Biochem.* **1997**, *246*, 625–632.
216. Van Wauwe, J.; Aerts, F.; Cools, M.; Deroose, F.; Freyne, E.; Goossens, J.; Hermans, B.; Lacrampe, J.; Van Genechten, H.; Van Gerven, F. et al. *J. Pharmacol. Exp. Ther.* **2000**, *295*, 655–661.
217. Freyne, E. J.; Lacrampe, J. F.; Deroose, F.; Boeckx, G. M.; Willems, M.; Embrechts, W.; Coesemans, E.; Willems, J. J.; Fortin, J. M.; Ligney, Y. et al. *J. Med. Chem.* **2005**, *48*, 2167–2175.
218. Williams, T. J. *J. Clin. Invest.* **2004**, *113*, 507–509.
219. O'Byrne, P. M.; Inman, M. D.; Parameswaran, K. *J. Allergy Clin. Immunol.* **2001**, 503-508.
220. Samuelson, B. *Science* **1983**, *220*, 568.
221. Werz, O.; Steinhilber, D. *Expert Opin. Ther. Patents* **2005**, *15*, 505–519.
222. Chen, X. S.; Naumann, T. A.; Kurre, U.; Jenkins, N. A.; Copeland, N. G.; Funk, C. D. *J. Biol. Chem.* **1995**, *270*, 17993–17999 (and references cited therein).
223. Carter, G. W.; Young, P. R.; Albert, D. H.; Bouska, J. B.; Dyer, R. D.; Bell, R. L.; Summers, J. B.; Brooks, D. W.; Rubin, P. R.; Kesterson, J. In *Leukotrienes and Prostanoids in Health and Disease, New Trends in Lipid Mediators Research*; Zor, U., Naor, Z., Danon, A., Eds.; Karger: Basel, Switzerland, 1989; Vol. 3, pp 50–55.
224. Brooks, C. D. W.; Stewart, A. O.; Kolasa, T.; Basha, A.; Bhatia, P.; Ratajczyk, J. D.; Craig, R. A.; Gunn, D.; Harris, R. A.; Bouska, J. B. et al. *Pure Appl. Chem.* **1998**, *70*, 271–274.
225. Stewart, A. O.; Bhatia, P. A.; Martin, J. G.; Summers, J. B.; Rodriques, K. E.; Martin, M. B.; Holms, J. H.; Moore, J. L.; Craig, R. A.; Kolasa, T. et al. *J. Med. Chem.* **1997**, *40*, 1955–1968.
226. Brooks, C. D. W.; Stewart, A. O.; Basha, A.; Bhatia, P.; Ratajczyk, J. D.; Martin, J. G.; Craig, R. A.; Kolasa, T.; Bouska, J. B.; Lanni, C. et al. *J. Med. Chem.* **1995**, *38*, 4768–4775.

227. Lewis, T. A.; Bayless, L.; DiPesa, A. J.; Eckman, J. B.; Gillard, M.; Libertine, L.; Scannell, R. T.; Wypij, D. M.; Young, M. A. *Bioorg. Med. Chem. Lett.* **2005**, *15*, 1083–1085.
228. Wright, S. W.; Pinto, D. J.; Sherk, S. R.; Green, A. M.; Magolda, R. L.; *Bioorg. Med. Chem. Lett.* **1992**, *2*, 1079–1084.
229. Mano, T.; Stevens, R. W.; Ando, K.; Nakao, K.; Okumura, Y.; Sakakibara, M.; Okumura, T.; Tamura, T.; Miyamoto, K. *Bioorg. Med. Chem.* **2003**, *11*, 3879–3887.
230. Mano, T.; Stevens, R. W.; Okumura, Y.; Kawai, M.; Okumura, T.; Sakaibara, M. *Bioorg. Med. Chem. Lett.* **2005**, *15*, 2611–2615.
231. Mano, T.; Okumura, Y.; Sakaibara, M.; Okumura, T.; Tamura, T.; Miyamoto, K.; Stevens, R. W. *J. Med. Chem.* **2004**, *47*, 720–725.
232. Young, R. N.; Gillard, J. W.; Hutchinson, J. H.; Leger, S.; Prasit, P. *J. Lipid Mediators* **1993**, *6*, 233–238.
233. Prasit, P.; Belley, M.; Brideau, C.; Chan, C.; Charleson, S.; Evans, J. F.; Fortin, R.; Ford-Hutchinson, A. W.; Gillard, J. W.; Guay, J. et al. *Bioorg. Med. Chem. Lett.* **1992**, *2*, 1395–1398.
234. O'Byrne, P. M.; Israel, E.; Deazen, J. M. *Ann. Int. Med.* **1997**, *127*, 472–480.
235. Friedman, B. S.; Bel, E. H.; Buntinx, A.; Tanaka, W.; Han, Y.; Shingo, S.; Spector, R.; Sterk, P. *Am. Rev. Respir. Dis.* **1993**, *147*, 839–844.
236. Bel, E. H.; Tanaka, W.; Spector, R.; Freidman, B.; VondeVeen, J. H.; Dijkman, J. H.; Sterk, P. J. *Am. Rev. Respir. Dis.* **1990**, *141*, A31.
237. Kolasa, T.; Gunn, D. E.; Bhatia, P.; Woods, K. W.; Gane, T.; Stewart, A. O.; Bouska, J. B.; Harris, R. R.; Hulkower, K. I.; Malo, P. E. et al. *J. Med. Chem.* **2000**, *43*, 690–705.
238. Kolasa, T.; Gunn, D. E.; Bhatia, P.; Basha, A.; Craig, R. A.; Stewart, A. O.; Bouska, J. B.; Harris, R. R.; Hulkower, K. I.; Malo, P. E. et al. *J. Med. Chem.* **2000**, *43*, 3322–3334.
239. Nayak, A. *Expert Opin. Pharmacother.* **2004**, *5*, 679–686.
240. Turner, C. R.; Smith, W. B.; Andresen, C. J.; Swindell, A. C.; Watson, J. W. *Pulmon. Pharmacol.* **1994**, *7*, 49–58.
241. Nabe, T.; Kohno, S.; Tanpo, T.; Saeki, Y.; Yarnamura, H.; Horiba, M.; Ohata, K. *Prostaglandins Leukot. Essent. Fatty Acids* **1994**, *51*, 163–171.
242. Dahlen, S.; Hedqvist, P.; Hammarstrom, S.; Samuelsson, S. *Nature* **1980**, *4*, 484–486.
243. Murphy, R.; Hammarstrom, C. S.; Samuelsson, B. *Proc. Natl. Acad. Sci. USA* **1979**, *76*, 4275–4279.
244. Zamboni, R.; Belley, M.; Champion, E.; Charette, L.; DeHaven, R.; Frenette, R.; Gauthier, J. Y.; Jones, T. R.; Leger, S.; Masson, P. et al. *J. Med. Chem.* **1992**, *35*, 3832–3844.
245. Labelle, M.; Belley, M.; Gareau, Y.; Gauthier, J. Y.; Guay, D.; Gordon, R.; Grossman, S. G.; Jones, T. R.; Leblanc, Y.; McAuliffe, M. et al. *Bioorg. Med. Chem. Lett.* **1995**, *5*, 283–288.
246. Nakagawa, T.; Mizushima, Y.; Ishii, A.; Nambu, F.; Motoishi, M.; Yui, Y.; Shida, T.; Miyamoto, T. *Adv. Prostaglandin Thromboxane Leukot. Res.* **1991**, *21a*, 465.
247. Fujimura, M.; Sakahoto, S.; Kamis, Y.; Matsuda, T. *Respir. Med.* **1993**, *87*, 133.
248. Matassa, V. G.; Maduskuie, T. P., Jr.; Shapiro, H. S.; Hesp, B.; Snyder, D. W.; Aharony, D.; Krell, R. D.; Keith, R. A. *J. Med. Chem.* **1990**, *33*, 1781–1790.
249. Brown, F. J.; Yee, Y. K.; Cronk, L. A.; Hebbel, K. C.; Krell, R. D.; Snyder, D. W. *J. Med. Chem.* **1990**, *33*, 1771–1781.
250. Dasari, V. R.; Jin, J.; Kunapuli, S. P. *Immunopharmacology* **2000**, *48*, 157–163.
251. Sawyer, J. S.; Baldwin, R. F.; Froelich, L. L.; Saussy, D. L., Jr.; Jackson, W. T. *Bioorg. Med. Chem. Lett.* **1993**, *3*, 1981–1984.
252. Sofia, M. J.; Jackson, W. T.; Saussy, D. L., Jr.; Silbaugh, S. A.; Froelich, L. L.; Cockerham, S. L.; Stengel, P. W. *Bioorg. Med. Chem. Lett.* **1992**, *2*, 1669–1674.
253. Kawakami, H.; Ohmi, N.; Nagata, H. *Bioorg. Med. Chem. Lett.* **1994**, *4*, 1461–1466.
254. Koch, K.; Melvin, L. S.; Reiter, L. A.; Biggers, M. S.; Showell, H. J.; Griffiths, R. J.; Pettipher, E. R.; Cheng, J. B.; Milici, A. J.; Breslow, R. et al. *J. Med. Chem.* **1994**, *37*, 3197–3199 (and references cited therein).
255. Greenspan, P. D.; Main, A. J.; Bhagwat, S. S.; Barsky, L. I.; Doti, R. A.; Engle, A. R.; Frey, L. M.; Zhou, H.; Lipson, K. E.; Chin, M. H. et al. *Bioorg. Med. Chem. Lett.* **1997**, *7*, 949–954.
256. Reiter, L. A.; Melvin, L. S., Jr.; Crean, G. L.; Showell, H. J.; Koch, K.; Biggers, M. S.; Cheng, J. B.; Breslow, R.; Conklyn, M. J.; Farrell, C. A. et al. *Bioorg. Med. Chem. Lett.* **1997**, *7*, 2307–2312.
257. Koch, K.; Melvin, L. S.; Reiter, L. A.; Biggers, M. S.; Showell, H. J.; Pettipher, E. R.; Hackman, B.; Cheng, J. B.; Milici, A. J.; Breslow, R. et al. In 7th International Conference of the Inflammation Research Association, White Haven, PA, 1994, W6.
258. Penning, T. D.; Askonas, L. J.; Djuric, S. W.; Haack, R. A.; Yu, S. S.; Michener, M. L.; Krivi, G. G.; Pyla, E. Y. *Bioorg. Med. Chem. Lett.* **1995**, *5*, 2517–2522.
259. Penning, T. D. *Curr. Pharmac. Des.* 2001, 7, 163–179 (and references cited therein).
260. Penning, T. D.; Chandrakumar, N. S.; Desai, B. N.; Djuric, S. W.; Gasiecki, A. F.; Liang, C. D.; Miyashiro, J. M.; Russell, M. A.; Askonas, L. J.; Gierse, J. K. et al. *Bioorg. Med. Chem. Lett.* **2002**, *12*, 3383–3386.

Biographies

Gregory P Roth, PhD, is currently Associate Director of Medicinal Chemistry at the Abbott Bioresearch Center located in Worcester, MA. Greg obtained his doctorate degree in asymmetric synthesis under the supervision of Prof Albert I Meyers at Colorado State University in 1988. Prior to joining Abbott Laboratories, Greg contributed to drug discovery efforts at Bristol-Myers Squibb starting in 1988 and then joined Boehringer Ingelheim Pharmaceuticals in 1994 where he was involved in process research, combinatorial chemistry, and immunoscience-based drug discovery programs.

David W Green, PhD, is Senior Associate Director of High Throughput Screening and Molecular Phamacology at Amgen, Inc. in Cambridge, MA. He received his graduate training in mechanistic enzymology at the University of Iowa with Prof Bryce V Plapp (PhD, 1988) and as a postdoctoral fellow in Monsanto Corporate Research. He joined SmithKline Beecham in 1989 and contributed to anti-inflammatory and antiviral drug discovery teams. David went on to lead cardiovascular drug discovery programs at Bristol-Myers Squibb before managing enzymology and HTS groups involved in structure-based drug design at 3-Dimensional and Cubist Pharmaceuticals. Before joining Amgen, he was Associate Director of Molecular and Cellular Biology at Abbott Bioresearch Center where his responsibilities included leading projects targeting cellular receptors for the development of anti-inflammatory drugs.

Comprehensive Medicinal Chemistry II
ISBN (set): 0-08-044513-6

ISBN (Volume 7) 0-08-044520-9; pp. 873–916

7.31 Treatment of Transplantation Rejection and Multiple Sclerosis

J S Skotnicki, Wyeth Research, Pearl River, NY, USA
D M Huryn, University of Pennsylvania, Philadelphia, PA, USA

7.31.1 Introduction

The immune system is critical in the regulation of many essential functions. Disruption or aberration of these processes is manifested in a number of autoimmune diseases[1] including rheumatoid arthritis (RA), multiple sclerosis, inflammatory bowel disease (IBD), cystic fibrosis (CF), type 1 diabetes, systemic lupus erythematosus (SLE), and asthma. Additionally, the activated immune response is intrinsically involved in transplantation rejection. Accordingly, modulation of the immune system at one or more junctures in the cascade provides an inviting target for therapy. However, due to the importance and complexity in maintaining a responsive immune system, absolute immunosuppression is not a viable option as a therapeutic strategy. Nonetheless, in debilitating diseases and where productive alternatives are not feasible, attenuation of immune processes is an effective strategy.

Because of the commonality of mechanistic principles, therapeutic strategies for the treatment of specific autoimmune diseases are often extended beyond the original target. Compound or compound class optimization by the tenets of medicinal chemistry and pharmacology are developed to yield lead compounds for the original target; further analyses and creative thought elucidate that these lead compounds exhibit properties that are superior or appropriate for other pathologies. Thus, significant therapeutic advances have been made without fidelity to a sole target or malady.

Based on space considerations and since many of the aforementioned diseases are covered elsewhere in this compendium, this chapter will focus on two of these treatment modalities, transplantation rejection and multiple sclerosis. In both cases, there are examples of approved treatments and potential treatments, and in numerous examples these drugs or drug candidates have multiple mechanisms. In each case, there are examples of the drug and treatment preceding the mechanistic understanding and thus the drugs provided fundamental tools to explore the science. With this surge in understanding, novel molecular targets and newer agents to affect the immune response have been identified. Described within is the focus on important agents for the prevention of rejection in transplantation and treatment of multiple sclerosis regardless of pharmacological ancestry.

7.31.2 Transplantation Rejection

7.31.2.1 Background

For the treatment of end-stage organ failure of the kidneys, lung, heart, liver, pancreas, inter alia, a standard and effective approach involves organ transplantation. This approach is direct in concept to replace the diseased organ with the ultimate resumption in function and health. In addition to a variety of concerns and factors associated with the surgical process, a fundamental issue is allograft rejection. The immune system employs a complicated and interconnected array of macromolecules to provide a matrix of complementary and compensatory feedback mechanisms. In this way, the immune system is engineered to serve as the primary guardian of the body to foreign invasion. By this capacity for the body to recognize and address the difference between self and alien, the transplanted organ is susceptible to self-defense in the form of rejection.

7.31.2.2 Therapeutic Need

An approach to address graft rejection is to modulate or suppress the immune system. From the early stages of transplantation research and therapy, scientists have developed and tested hypotheses to affect the correct balance between prevention of rejection and untoward immunosuppressive effects using agents that inhibit one or more components of the immune response. As the basic and clinical research evolved, improvements have been noted and, importantly, the understanding of these complex issues has provided new opportunities for intervention. The tactics employed for the use of immunosuppressive drugs involve therapies related to induction, maintenance, and reversal of rejection. Recent reviews concerning the identification and characterization of immunosuppressants for use in the control of transplantation rejection have appeared.[2–12]

Nonetheless, some critical problems attendant to molecular intrusion in the immune system are not anticipated and remain unsolved. Immunosuppressive agents often possess inherent drug toxicity to the transplanted organs (e.g., kidney) or display other pathologies, including nephrotoxicity, hepatotoxicity, neurotoxicity, cardiovascular liabilities, and gastrointestinal complications. These are often characterized by very tight therapeutic indices requiring substantial toxicodynamic dose monitoring. Another inherent attribute in the strategy is the possibility of nonspecific immunosuppression, wherein infection and malignancy are significant therapeutic compromises. General reviews concerning side effects have been published.[13–18] The widely used agents for the prevention of transplantation rejection fall into a number of mechanistic categories. Some convenient characterizations are protein therapeutics, antimetabolites, glucocorticoids, inhibitors of calcineurin, mTOR, nucleotide synthesis, specific tyrosine kinases 3-hydroxy-3-methylglutaryl coenzyme A (HMG-CoA) reductase, and G protein-coupled receptor (GPCR) antagonists. Members of each mechanistic class possess specific therapeutic advantages but also carry potential for side effects.

Careful examination of this dichotomy and continued experimentation have been used for therapeutic optimization. Combination therapy offers opportunities for systematic discrimination of mechanism-based side effects. Often times the effects for each drug are additive to allow regulation of dose of individual drugs. In some instances, the effects are synergistic taking advantage of upstream or downstream interruptions in a particular immune system cascade. Individual protocols take into account several diagnostic attributes (organ system, characteristics of the patient, other pathology, other pharmacology). These therapeutic regimens are complex.

7.31.2.3 Current Treatment Options

Herein are depicted select compounds used to prevent transplantation rejection. The diversity of both the molecular targets and the structures of the agents are noteworthy. Details of the clinical or preclinical attributes are beyond the scope of this chapter and are discussed in the publications listed in the References. Protein approaches[19–21] and glucocorticoids[22–26] (*see* Section 7.31.3.4.3) have been thoroughly reviewed and will not be considered further.

7.31.2.3.1 Calcineurin inhibitors

Cyclosporin (**1**)[27–30] is a cyclic peptide and a prominent immunosuppressive agent approved for use in organ transplantation. It forms a molecular complex with cyclophilin and this complex blocks calcineurin with resulting inhibition of T cell proliferation via inhibition of interleukin-2 (IL2) production. Neoral is an improved microemulsion form of **1**. A semisynthetic derivative of cyclosporin, ISAtx-247, is reported in development for a number of autoimmune diseases.[31] Tacrolimus (**2**)[32–36] is a macrolide antibiotic with immunosuppressive activity and is structurally unrelated to cyclosporin. Tacrolimus forms a molecular complex with a distinct immunophilin, FKBP12, and this complex also binds calcineurin, but with greater molar potency than the cyclosporin–cyclophilin complex. The blocking of calcineurin activation is one basis for the observed toxicities of these molecules.

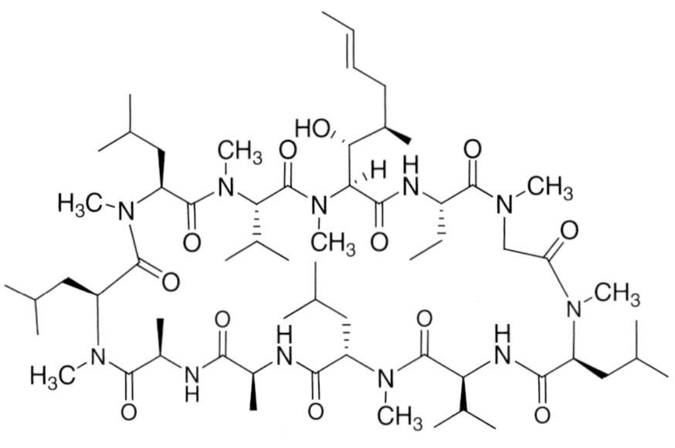

1 Cyclosporin

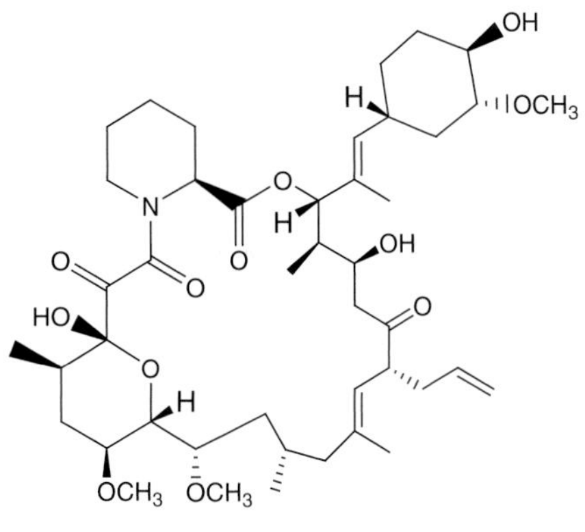

2 Tacrolimus

7.31.2.3.2 mTOR inhibitors

Sirolimus (rapamycin, **3**)[37–41] is a macrolide antibiotic as well as an important immunosuppressive agent, first approved for use in kidney transplantation in 1999. Though sharing some structural similarities with tacrolimus, sirolimus interferes at a different stage of the T cell activation cascade yielding a unique immunosuppressive mechanism. Sirolimus likewise binds to FKBP12, and this complex in turn blocks mTOR resulting in the arrest of cell-cycle progression at the G1/S phase. Because of this distinct mechanism, sirolimus is able to act synergistically with cyclosporin. Everolimus (**4**),[42,43] an ether analog of sirolimus that operates by the same mechanism, is an effective immunosuppressant. Early and extensive mechanistic studies using cyclosporin, tacrolimus, and sirolomus as pharmacology probes contributed significantly to the understanding of intracellular signal transduction pathways.[44–49]

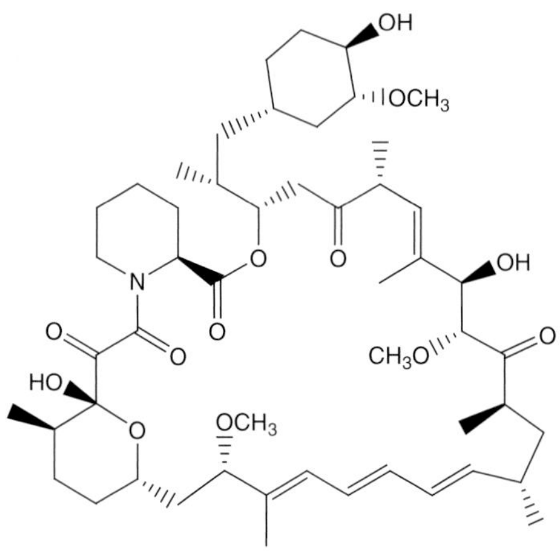

3 Sirolimus

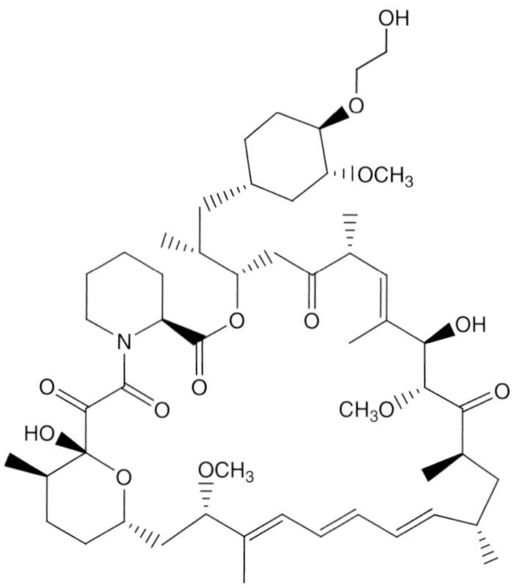

4 Evirolimus

7.31.2.3.3 Inhibitors of nucleotide synthesis

Mycophenolate mofetil (**5**),[50–54] an ester prodrug of mycophenolic acid, is a noncompetitive inhibitor of inosine 5′-monophosphate dehydrogenase, IMPDH. By this activity, mycophenolic acid blocks de novo synthesis of guanosine nucleotides, thus inhibiting lymphocyte proliferation and cell-mediated immune responses. Following solution of the structure of IMPDH, other inhibitors such as VX-497 (**6**)[55,56] were identified via structure-based design.

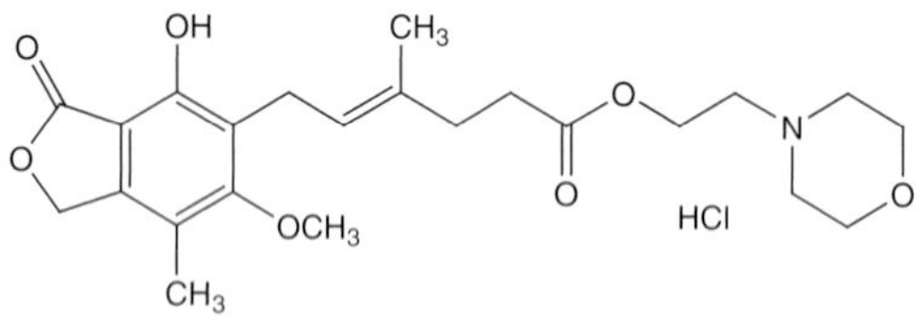

5 Mycophenolate mofetil

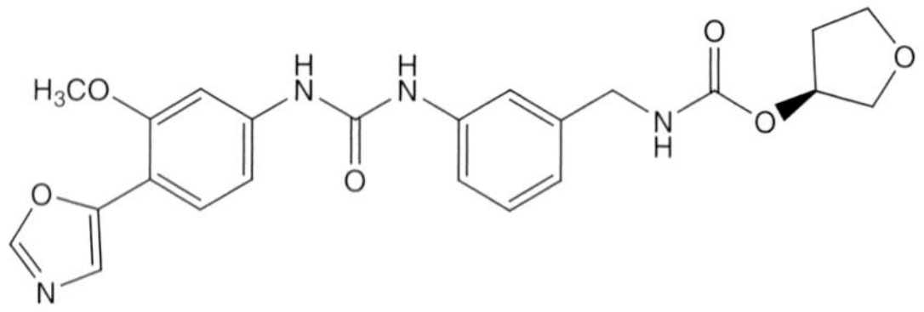

6 VX-497

Leflunomide (**7**)[57,58] is an inhibitor of dihydroorotate dehydrogenase (DHODH). It suppresses de novo biosynthesis of pyrimidines, blocking the proliferation of lymphocytes. Leflunomide has been approved for RA and has limited use in transplantation due in part to its long $t_{1/2}$ in humans (15 days). FK-778 (**8**)[59,60] is a structurally related, newer generation DHODH inhibitor that exhibits more favorable pharmacokinetic properties.

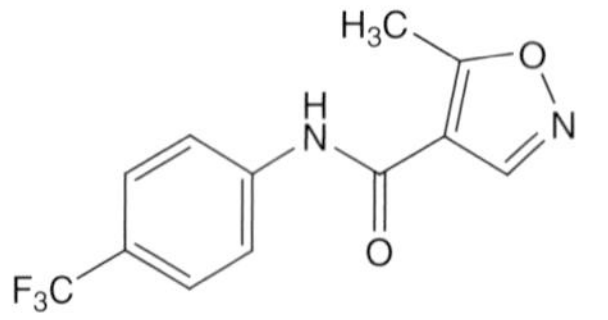

7 Leflunomide

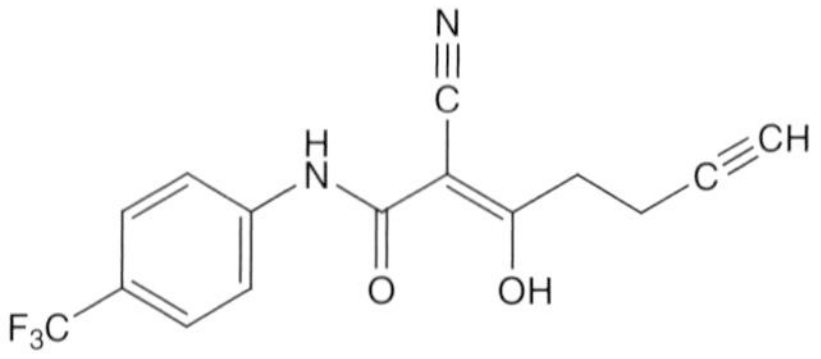

8 FK-778

7.31.2.3.4 Statins

The statins, exemplified by pravastatin (**9**) are inhibitors of HMG-CoA.[61–63] These are used clinically for lipid-lowering capacity, but have also been shown to have anti-inflammatory and immunomodulating properties. The statins have been examined for cardiovascular benefits in the transplantation patient, but there is evidence reported suggesting that statins may be involved in control of rejection.[64–66] The statins are further discussed in Section 7.31.3.4.11.

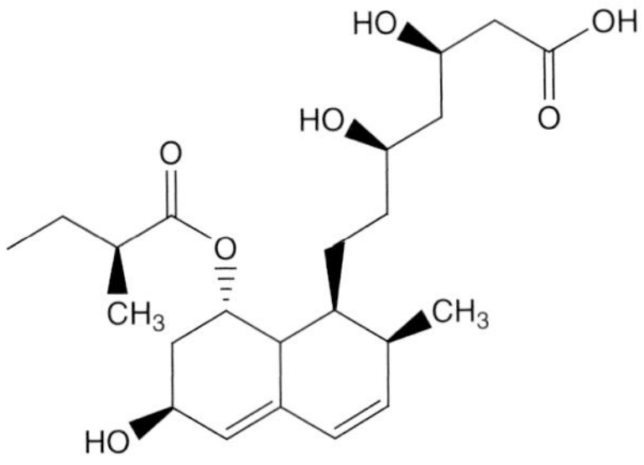

9 Pravastatin

7.31.2.3.5 Chemokine receptor antagonists

The chemokines[67–69] are a family of structurally related proteins that bind to GPCRs, and are intimately involved in the activation and recruitment of leukocytes. They display pleiotropic biological effects and are implicated in a variety of pathological conditions. The multiple chemokine receptor pathways are complex. SAR studies have been challenging and have been directed mainly to autoimmune diseases. Nonetheless, the involvement of specific chemokine receptors in allograft rejection has been discussed.[70–74] BX-471 (**10**),[75,76] a potent and selective CCR1 antagonist that is reported to be active in both multiple sclerosis and transplantation models, is a good example.

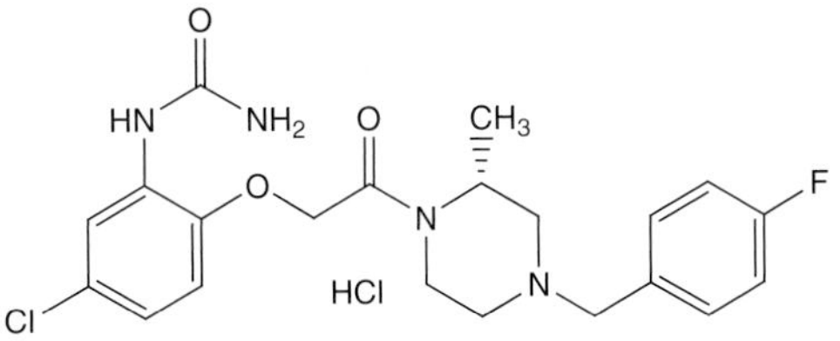

10 BX-471

7.31.2.3.6 Lysophospholipid receptor agonists

Lysophospholipids are a family of simple phospholipids that signal through GPCRs and are involved in a broad range of biological processes. The potential of lysophospholipid receptors as targets for immunomodulation in transplant therapy has been reviewed.[77–79] One member of this family is sphingosine 1-phosphate (S1P), which is involved in lymphocyte development and B and T cell recirculation. FTY720 (**11**)[80–83] is a nonselective sphingosine 1-phosphate receptor (S1P-R) agonist that is in development for kidney transplantation. FTY720 operates via a unique mechanism of action involving the reduction of peripheral lymphocytes by their sequestering to lymph nodes.

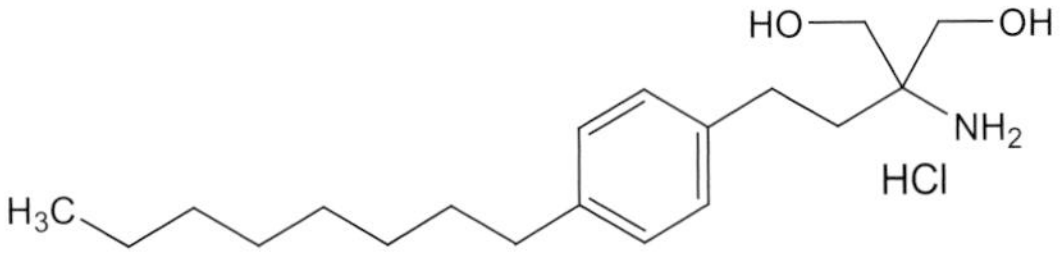

11 FTY-720

7.31.2.3.7 Tyrosine kinase inhibitors

The tyrosine receptor kinases are ubiquitous members of a family that are key in signal transduction processes. They are involved in a number of pathways implicated in oncology and inflammation. Lck and JAK3 are members of this family, expressed on T lymphocytes and involved in T cell receptor signaling. These kinases have been studied as potential targets for the prevention of transplantation rejection. Following SAR optimization, lck inhibitor A-420983 (**12**)[84,85] and JAK-3 inhibitor CP-690,550 (**13**)[86] have been identified and are illustrative of this concept.

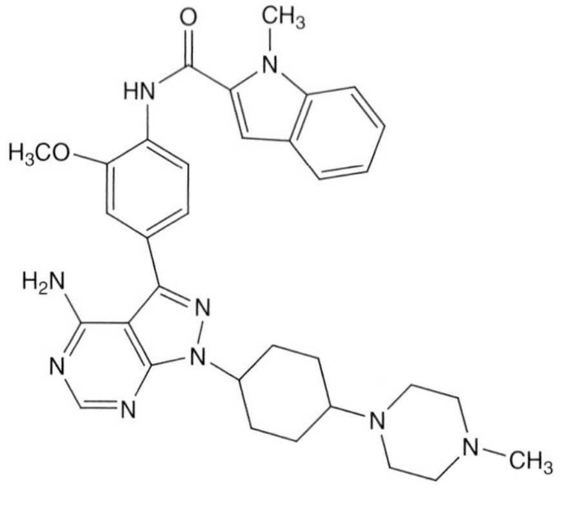

12 A-420983

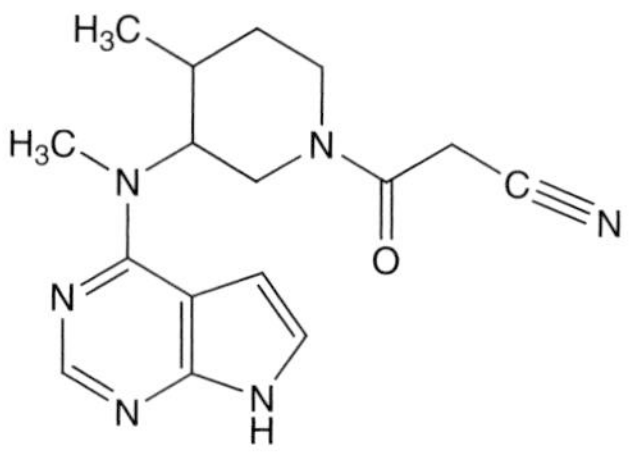

13 CP-690,550

7.31.2.4 Emerging Research Areas

The continued, and hopefully increased, availability of organs for transplantation will encourage further developments in the search for effective and safe therapeutics.[7,87–99] Undoubtedly, medicinal chemistry efforts will continue to pursue improved agents that act through proven mechanisms (e.g., mTOR, calcineurin, nucleotide biosynthesis inhibition, etc.) Emerging areas include tolerance induction, whereby the specific immune response to allograft antigens is prevented. Specific approaches to induce tolerance include targeting the costimulatory pathways (e.g., B7/CD28, anti-CD52, major histocompatibility complex (MHC) peptides, anti-CD40 ligand), clonal deletion (e.g., T cell depleting antibodies and immunotoxin conjugates, CTLA-4-Ig, targeting Fas, TNF and Trance pathways, bone marrow micro-chimerism), Toll-like receptors antagonism, and inhibition of the complement cascade.

The control of graft-versus-host disease is another promising avenue in the area of transplantation, as is the use of antisense approaches. Finally, pharmacogenomic approaches will allow personalized therapy and therapeutic regimes that are highly specific to the individual patient and the transplanted organ.

7.31.2.5 Summary

Evolution of the science of immunosuppression has provided significant breakthrough achievements in the control of rejection. Cyclosporin, tacrolimus, sirolimus, mycophenolate mofetil, as well as antibody approaches (e.g., OKT-3, basiliximab, daclizumab) are effective, but have limitations. The balance between prevention of organ rejection and intolerable immunosuppressive properties has yet to be achieved universally. Future antirejection drugs will need to optimize the immunosuppressive profile to ensure organ protection at the expense of dose and schedule limited side effects. In this way, toxicity to the organ and/or the patient, and susceptibility to infection and malignancy will be minimized or eliminated.

7.31.3 Multiple Sclerosis

7.31.3.1 Background

Multiple sclerosis is considered to be the most common disorder of the CNS, affecting approximately one million people worldwide, primarily women, and often spanning a course of 30–40 years. While its etiology is unknown, multiple sclerosis is believed to result from an autoimmune response that targets components of the CNS. It is a complex disease characterized by heterogeneity – in symptoms, pathology, disease course, and treatment efficacy. Symptoms of multiple sclerosis vary among patients, are usually episodic, and include fatigue, limb weakness, clumsiness, vision disturbances, and bowel and bladder symptoms. Over the long course of the disease, progressive weakening is common, and within 15 years of onset, 50% of patients will develop disabilities severe enough to require assistance in walking.[100–102]

The heterogeneity of the disease has prompted standardized definitions of four subtypes of multiple sclerosis.[103] The relapsing-remitting (RRMS) form of the disease is the most common, occurring in about 80% of cases. Disease onset typically occurs during young adulthood (20s–30s), and affects mainly woman (2:1). It is characterized by acute episodes of neurological symptoms, typically developing over days, followed by remission, which is often complete. In 50% of RRMS patients, the disease course changes and neurological deterioration gradually continues with or without relapses. This form of the disease is termed secondary progressive multiple sclerosis (SPMS). Primary progressive multiple sclerosis (PPMS) is a form of the disease that affects approximately 20% of patients. Symptom severity gradually and continuously worsens from disease onset, with an absence of acute attacks. The incidence of PPMS among men and women is similar, and disease onset is typically later (40–60 years) than RRMS. Progressive relapsing multiple sclerosis (PRMS) is an uncommon form of the disease, and is characterized by gradual deterioration of neurological function with acute episodes superimposed.[100]

7.31.3.2 Disease Basis

While the event that initiates multiple sclerosis is unknown, it has been proposed that T cells recognizing self-antigen, such as myelin, become activated and initiate a proinflammatory response. Cytokines and tumor necrosis factor (TNF) are released, further amplifying the immune response. Immune cells cross the blood–brain barrier (BBB) and enter the CNS, where proinflammatory cytokines and chemokines are released and induce demyelination. Damage to myelin is believed to cause the early symptoms of multiple sclerosis, however, the molecular mechanisms responsible for myelin degradation are under debate. Several scenarios have been suggested including: damage from microglia and macrophages, cytokine- induced injury, direct antibody binding to myelin, and complement-mediated injury. Regardless of the cause, damaged myelin results in axons being exposed, the inefficient or even loss of nerve conductance, and onset of neurological symptoms. Proinflammatory and proimmune mediators, such as cytokines, chemokines, complement and proteases, may do further damage by acting on the exposed axon, resulting in irreversible injury and neurological deterioration. The mechanism by which remission and resolution of symptoms occurs is also poorly understood. Current hypotheses focus on the release of immunosuppressive cytokines and resolution of the inflammatory response and spontaneous re-myelination, redistribution of sodium channels to restore conduction, and the action of oligodendrocyte precursor cells to regenerate myelin.[101,102]

The observation that multiple sclerosis relapses are noted to occur after viral infection supports a hypothesis of a viral etiology for multiple sclerosis.[101] The molecular mimicry theory applied to multiple sclerosis[104] suggests that viral peptide sequences mimic sequences in proteins, such as myelin basic protein (MBP), proteolipid protein (PLP), and myelin oligodendrocyte glycoprotein (MOG), that constitute the myelin sheath. After viral infection, antigen presenting cells (APCs) and activated T cells mistakenly recognize self-antigens in the CNS, leading to their destruction, and ultimately myelin degradation.

Evidence supporting an environmental factor as a cause of multiple sclerosis includes a correlation between disease prevalence and geographic location, several examples of 'clustering' of cases, and the change in risk after migration. Disease incidence is highest in Northern Europe, the middle part of North America, and southern Australia, with the demarcation line considered to be above 40° latitude. Migration from an area of lower incidence to an area of higher incidence changes an individual's risk for developing multiple sclerosis, and age at the time of migration appears to affect how the increased risk is manifested. While these epidemiological characteristics suggest that environmental factors play a role in susceptibility to multiple sclerosis, other data confound this theory, and the precise nature of a putative environmental factor is unknown.[101]

While an environmental cause of multiple sclerosis is intriguing, the role of genetics in susceptibility is unequivocal. Caucasians are twice as likely to be afflicted with multiple sclerosis than other ethnic groups, and women are more likely to develop the disease than men. First-degree relatives of multiple sclerosis patients have a 20–40 times higher risk of developing the disease than the general population. Presence of the HLA-DR2 allele increases the risk of developing multiple sclerosis considerably, and those populations that have a high frequency of this allele have the highest risk of disease development. The progression and severity of the disease may also rely on a genetic component. Despite the strong genetic component to the disease, most cases are sporadic, and defining the specific genes involved and how they are transmitted is still under investigation.[101]

Owing to the heterogeneous nature of the disease, it has been suggested that multiple sclerosis may in fact be a series of different diseases, resulting from different exposures (pathogenic, environmental, genetic), and having different pathologies.[101] The lack of understanding of the initiating events and the specific pathology of the disease make drug discovery efforts in targeting multiple sclerosis particularly difficult.

7.31.3.3 Experimental Disease Models

Experimental autoimmune encephalomyelitis (EAE) is the animal model most commonly used to mimic human multiple sclerosis, and has been a valuable model to study the disease process itself. It is induced in animals, often rodents such as rat or guinea pig, but also primates, by inoculation with a variety of antigens such as MBP, or even whole brain homogenates. Induction by passive transfer of activated myelin-specific T cells is also common. EAE, like multiple sclerosis, is characterized by infiltration of the CNS by $CD4^+$ T cells and macrophages resulting in damage to myelin, and eventually impaired nerve conductance and paralysis. By the appropriate choice of species, antigen, and protocol, disease progression in EAE can be tailored to mimic either an acute illness or a chronic relapsing form of the disease. While the similarities between multiple sclerosis and EAE, particularly its clinical presentation, pattern of genetic susceptibility, and observed pathology, are strong, some notable differences are observed in the response to specific therapies, with a trend toward effective therapies in EAE being ineffective in multiple sclerosis.[100,105,106] Viral models whose pathology mimic human multiple sclerosis have also been used, albeit less frequently. Infection of mice with Theiler's murine encephalomyelitis virus (TMEV) causes chronic progressive immune-mediated demyelination that can be ameliorated by immunosuppressive treatment.[107]

7.31.3.4 Current Treatment Options

Application of conventional drug discovery strategies to multiple sclerosis is difficult due to the lack of understanding of the clinico-pathologic events in the disease process. Current treatment strategies primarily rely on reducing and/or ameliorating the putative inappropriate immune response, with few therapies available to repair damage. During relapses, additional therapies are often prescribed. Therapeutic approaches can be broadly divided into drugs that produce immunosuppression and those that modulate the immune system. Neither approach is ideal. Immunosuppressants typically work through a cytotoxic effect on immune cells. These have broad effects, target several different pathways, and are often very effective. Unfortunately, the likelihood of significant toxicity (cardiotoxicity, malignancies, risk of infection) is also high, and this risk must be balanced with the therapeutic benefit. Immunomodulators act through shifting the immune response to an anti-inflammatory response, and are generally perceived to be less toxic, but also may be less effective in a wide range of patients. In reality, many multiple sclerosis drugs appear to work through several different mechanisms producing both immunosuppressive and immunomodulatory effects.[102,106] For the purposes of this review, drugs currently approved to treat multiple sclerosis and those approved to treat other disorders, but with potential to treat multiple sclerosis, will be covered in this subsection. Other agents that show promise in early stages of drug discovery or clinical trials will be covered in Section 7.31.3.5.

7.31.3.4.1 Interferon beta

Patients with the relapsing–remitting form of multiple sclerosis are frequently treated with interferon beta. Two forms of the recombinant protein are prescribed: interferon β-1a, identical to natural interferon beta; or interferon β-1b, a nonglycosylated form containing a single amino acid substitution. Clinical trials of these agents demonstrated a significant reduction in disease progression as evidenced by MRI.[100] Interferon beta acts through a number of different mechanisms in both the CNS and periphery.[102] By activating the interferon receptor, its administration is believed to cause a series of events that result in antiproliferative, immunomodulatory, and (perhaps) antiviral effects.[102,108] Interferon beta may be most beneficial at the very early stages of disease, but whether it has a significant effect over the course of the disease is unknown.

7.31.3.4.2 Glatiramer acetate

Glatiramer acetate is a mixture of copolymers of four amino acids (L-alanine, L-glutamic acid, L-lysine, and L-tyrosine) in specific stoichiometry that ranges in length from 40 to 90 amino acids. The polymers, originally designed to mimic sequences of MBP, are rapidly degraded to the individual amino acids upon administration. In animal models of EAE, glatiramer acetate suppresses disease induced by MBP, as well as PLP and MOG.[100,102,108,109] The mechanism by which glatiramer acetate exerts its beneficial effect has been the subject of considerable debate and study. Recent reports describe it acting at different pathways of the immune response, such as blocking MHC, antagonizing T cell receptors, and inducing regulatory T cells.[109,110] The broad immune-modulating activity of glatiramer acetate has also shown benefit in other immune disorders, such as IBD, and during graft rejection.[109]

7.31.3.4.3 Corticosteroids

To treat the underlying immune and inflammatory response, corticosteroids (e.g., methylprednisolone, **14**) are prescribed to treat acute relapses. Glucocorticosteriod effects are broad based, and involve immunomodulatory as well as anti-inflammatory effects.[23–27,108] By reducing the inflammatory events during a relapse, edema is reduced, the leaky blood–brain barrier is restored, and axonal conductance is improved. The mechanism by which these agents work in multiple sclerosis has been recently reviewed.[111,112]

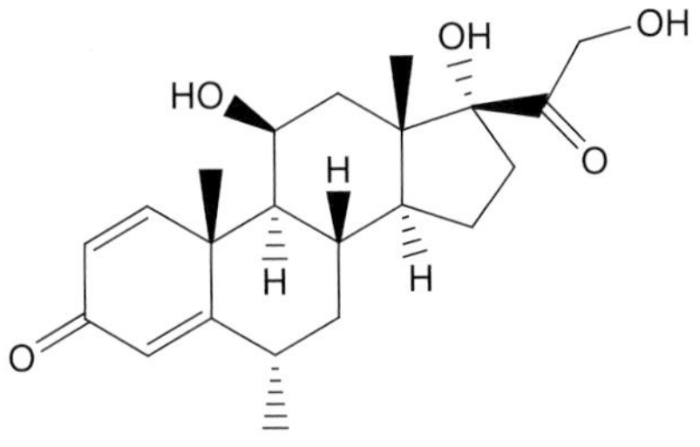

14 Methylprednisolone

7.31.3.4.4 Azathioprine

Azathioprine (**15**) is an immunosuppressant, once widely used to treat patients who had undergone organ transplantation and as a second-line therapy for those diagnosed with autoimmune diseases.[100,108] Initially, it was believed to act as a prodrug of 6-mercaptopurine (**16**, **Scheme 1**), which, through its incorporation into RNA and DNA, targets activation, proliferation, and differentiation of T and B cells. More recent data indicates that azathioprine's mode of action may be more complex. It is extensively metabolized (although the extent of metabolism varies among patients) to other purine analogs, which themselves may exert immunosuppressant activity. Some have also suggested that the imidazole product **17**, generated during the formation of 6-mercaptopurine, is responsible for some of the drug's biological effects.[113] A number of reviews on this topic have appeared.[114,115]

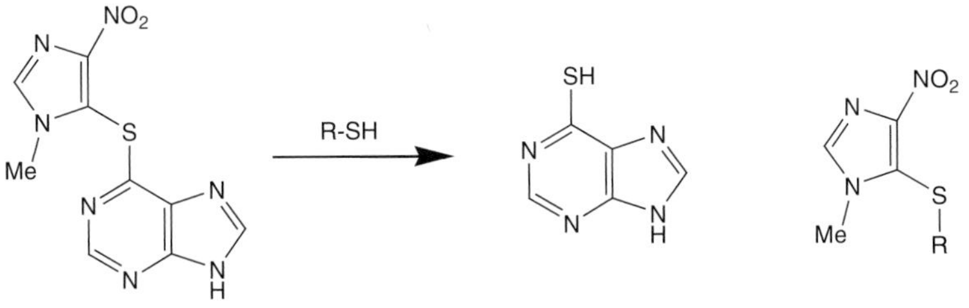

R-SH

15 Azathioprine **16** 6-Mercaptopurine **17**

Scheme 1

7.31.3.4.5 Methotrexate

Methotrexate (**18**) is a viable treatment option for patients with progressive forms of the disease. It inhibits the enzyme dihydrofolate reductase (DHFR), essential for the synthesis of nucleotides, thereby preventing cell proliferation. This inhibition results in broad anti-inflammatory and immunosuppressive effects. Methotrexate is widely used in anticancer therapy, in addition to its use in other autoimmune diseases such as rheumatoid arthritis.[100]

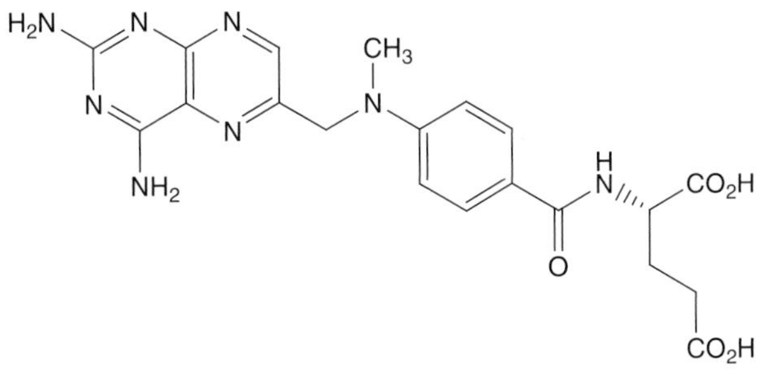

18 Methotrexate

7.31.3.4.6 Cyclophosphamide

Owing to its considerable side effects, cyclophosphamide (**19**, **Scheme 2**) is typically prescribed for patients with severe and rapidly progressing disease. Similar to the sulfur mustard used as a chemical weapon in World War I, it is an alkylating agent with considerable cytotoxic and immunosuppressive effects, most frequently used to treat cancers and autoimmune diseases. Cyclophosphamide's mechanism of action relies on metabolic activation that first produces aldophosphamide (**20**), which in turn generates phorphoramide mustard (**21**) and acrolein (**22**). Phosphoramide mustard, through an aziridine intermediate, reacts with a number of targets including nucleic acids and enzymes, and displays some selectivity toward lymphocytes, although the reasons behind the selectivity are poorly understood.[116] Weiner and Cohen[117] have recently reviewed the multiple sclerosis-specific effects of cyclophosphamide.

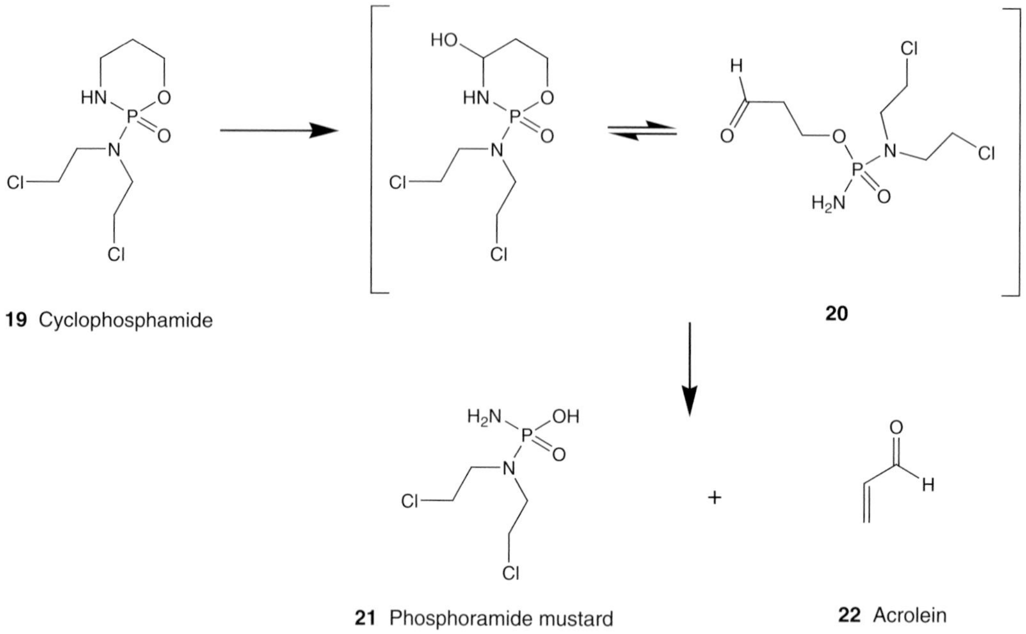

19 Cyclophosphamide

20

21 Phosphoramide mustard

22 Acrolein

Scheme 2

7.31.3.4.7 Mitoxantrone

By virtue of its inhibition of topoisomerase II, an enzyme essential for DNA/RNA synthesis, mitoxantrone (**23**) interferes with cell proliferation and promotes cell death.[102] It has been used historically as an anticancer agent, and is believed to work by suppressing T and B cells. Recently, it has been suggested that its efficacy in multiple sclerosis is through induction of apoptosis in antigen-presenting cells, and deactivation of macrophages, the cells primarily responsible for demyelination. However, owing to its side effect profile, the drug is most useful in treating patients with very active disease or those refractive to other therapies.[102,108,118]

7.31.3.4.8 Anti-VLA therapies

Natalizumab is a humanized antibody that targets the alpha four subunit of the integrin, alpha-4 beta-1 (also known as VLA-4). By targeting this integrin's interaction with vascular cell adhesion molecule-1 (VCAM-1), natalizumab prevents entry of lymphocytes into the CNS, and is one of the few therapies that targets multiple sclerosis-selective events, rather than general immunosuppression or immunomodulation. After very promising results in EAE models, and in clinical trials, natalizumab was approved to treat patients with RRMS. Soon after its approval, three cases of progressive multifocal leukoencephalopathy, a serious demyelinating disease of viral etiology, were seen in patients, and the drug was voluntarily withdrawn from the market. After additional review, in June 2006 the FDA approved the resumption of marketing based on a restricted distribution program. Recent data from animal models identify the potential for issues, such as the rapid return of paralytic disease after cessation of treatment (rebound) and worsening of disease if the antibody was administered at the peak of relapsing disease. A number of very thorough reviews of this area have appeared.[119–122]

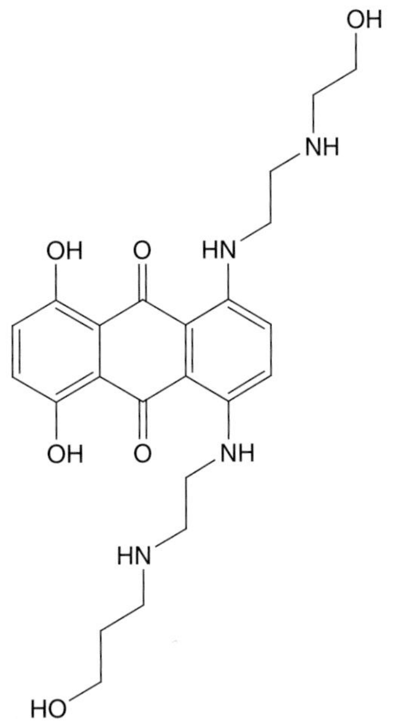

23 Mitoxantrone

The initially promising results from antibody studies prompted significant efforts toward the design of nonpeptide, small molecule inhibitors of VLA-4.[123,124] While much of the focus of the work has been in the asthma arena, several reports of small molecules, such as **24–26**, that exhibit efficacy in EAE models have been reported.[125–129] However, the potential for rebound and exacerbation of disease after treatment with VLA-antagonists,[130] as well as issues of possible developmental disorders,[131,132] raises concerns about the potential of anti-VLA-4 and anti-adhesion therapies in large numbers of multiple sclerosis patients.

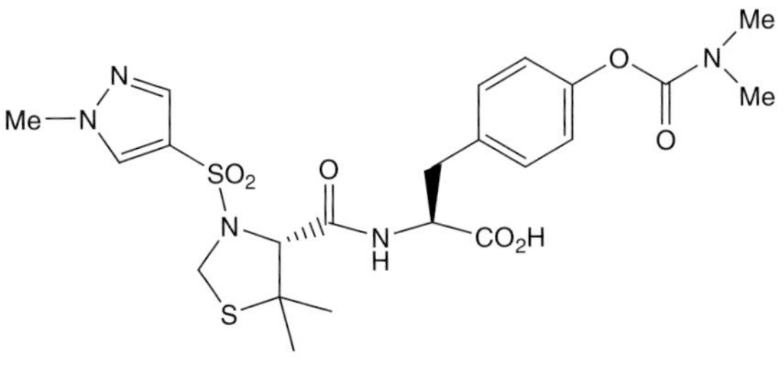

24

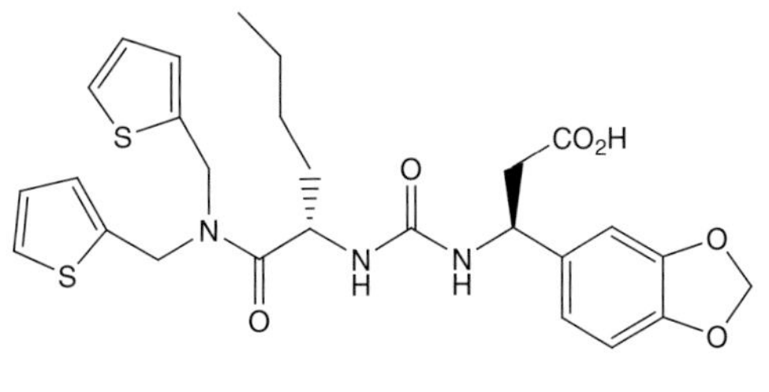

25

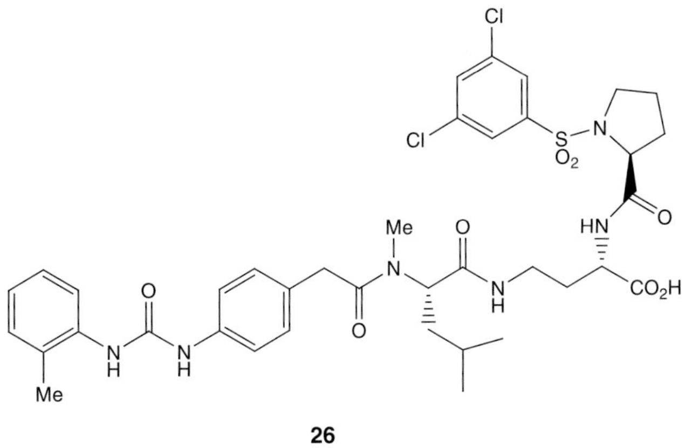

26

7.31.3.4.9 Intravenous immunoglobulins

The mechanism by which intravenous immunoglobulins (IVIgs) provide a beneficial effect in multiple sclerosis patients is complex and multifactorial. Effects on the immune system may be mediated through: anti-idiotype interactions; downregulation and/or neutralization of cytokines and complement-mediated effects; and inhibiting Fc receptors. Studies in animal models of multiple sclerosis (TMEV) showed the possibility of IVIgs promoting re-myelination; however, clinical studies have not yet supported those findings, and the treatment remains a valuable second-line therapy for RRMS.[133–136]

7.31.3.4.10 Cyclosporin

Cyclosporin (**1**) has been used in multiple sclerosis patients based on its potent immunosuppressive effects; however, its significant side effects reduce its utility in this patient population.[100] The use of this drug in transplantation is discussed in Section 7.31.2.3.1.

7.31.3.4.11 Statins

Statins, currently used as lipid-lowering drugs, have recently been shown to exhibit certain anti-inflammatory and immunomodulatory effects in vitro and in models of EAE. Their mechanism of action in multiple sclerosis patients is not understood, and may involve inhibition of HMG-CoA reductase or direct effects on the immune system. By inhibition of HMG-CoA reductase, statins may prevent posttranslational modifications of key proteins (e.g., Rho) involved in the immune response. Alternatively, statins have recently been reported to antagonize the adhesion molecule LFA-1 through binding to an allosteric site, and thereby preventing adhesion and costimulation of lymphocytes mediated by LFA-1. In a small study in RRMS patients, simvastatin (**27**) treatment resulted in a significant decrease in the number of new lesions as measured by MRI. The relative safety of this class of drugs and their ease of administration are significant benefits, but whether statins will exacerbate multiple sclerosis symptoms, as some have proposed, and whether they can become a mainstay of multiple sclerosis therapy remains to be determined.[61,64,108,137–141]

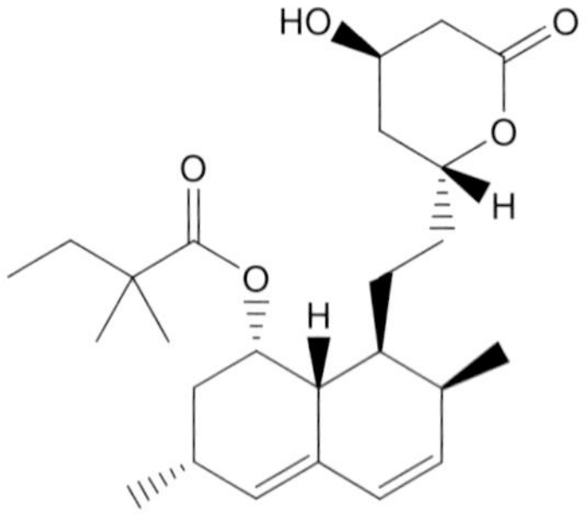

HO O O H O O H

27 Simvastatin

7.31.3.4.12 Steroid hormones

A number of observations have suggested a beneficial role for steroid hormones in multiple sclerosis patients. While woman are twice as likely as men to be diagnosed with the disease, progression is typically slower in females than in males. Furthermore, during pregnancy multiple sclerosis symptoms and relapses decline significantly, followed by a dramatic increase post-partum. A number of studies in EAE have demonstrated that steroid hormones (e.g., estriol, **28**) can delay progression, and in some cases, prevent disease. The mechanism by which estrogens work is unclear and may involve not only effects on the immune system, but also effects on enhancement and promotion of re-myelination. Small clinical trials have shown promising results, and larger studies are underway. The potential of estrogens and selective estrogen receptor modulators (SERMs) in multiple sclerosis has been reviewed.[142–144]

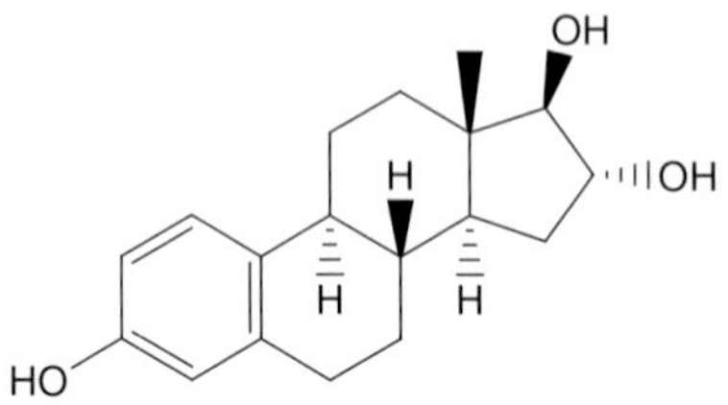

OH H OH H H HO

28 Estriol

7.31.3.5 Emerging Research Areas

According to the National Multiple Sclerosis Society,[173] over one hundred clinical trials in multiple sclerosis patients were conducted in 2005. Many of the drugs currently in clinical trials were approved for other therapies, and their utility in the multiple sclerosis patient population is being evaluated. In earlier stages of the drug discovery effort are a number of novel approaches (see a recent review by Wiendl and Kieseier [145]). For the purpose of this chapter, those describing peptides as antagonists or elicitors of immune responses and new GPCR targets will be highlighted.

The unexpected activity of peptide copolymers, such as glatiramer acetate, in modulating an immune response, as well as the relative safety of the drug, has promoted new efforts aimed at identifying other peptides that may act similarly. A number of approaches focus on interfering with the binding of the T cells to the putative multiple sclerosis auto-antigen, MBP. Peptide 15-mers designed to mimic the MBP-85-99 sequence inhibited the binding of MBP-85-99 to HLA-DR2 and were effective in EAE models.[146] Similar strategies using constrained cyclic peptides, peptide mimetics,[147] and small molecules[148] have also been described, and cite preliminary in vitro results. Ideally, drugs working through this mechanism may be more specific, and therefore more effective, than current therapies, without resulting in the broad immunosuppression which produces significant side effects.[149] Peptides based on the sequence of MBP have also been used to induce tolerance to EAE, by inducing anergy in specific T cell populations.[150]

Several different groups have reported the potential of targeting GPCRs as an approach to immune modulation in multiple sclerosis patients. One promising approach involves the exploitation of lysophospholipid GPCR receptors' role in mediating immunological effects. FTY720 (**11**) is phosphorylated in vivo, and that molecule acts as a broad agonist at S1P receptors and is effective in EAE models. Current efforts focus on identifying molecules with S1P receptor subtype specificity.[77,151–155] Other GPCR targets for small-molecule multiple sclerosis drug discovery include cannabinoid,[156–159] histamine,[160] and chemokine (*see* Section 7.31.2.3.5) receptors.[161–165]

A wide range of other approaches has also been reported, as have molecules with intriguing activities, however, results are very preliminary. These approaches include cytokine (IL2) suppression,[166] mimicking the natural immunosuppressive effects of tryptophan catabolites,[167] inhibition of potassium channel Kv1.3,[168] and targeting transcription factors involved in the immune response.[169] This last hypothesis is supported by the findings that glitazones provide an attenuation of symptoms in EAE models.

In addition to small-molecule drug discovery efforts, protein therapeutic and vaccine strategies are also being pursued.[102,108,118] Promising new strategies to treat multiple sclerosis patients include the use of neurotrophic factors, neuroprotective agents, T cell targeting therapies (e.g., CTLA-4 Ig),[170] bone marrow and glial cell transplantation, leukocyte-targeted antibodies, and antisense[99] and siRNA approaches.[171]

7.31.3.6 Summary

The availability of interferon beta and glatiramer acetate has had a profound effect on the treatment of multiple sclerosis patients. While previous therapies involved immunosuppression, typically with cytotoxic agents, these two biological products have a more subtle effect and work by shifting the immune response away from a destructive response to a beneficial one. Despite these advances, preventative therapies, and those that can elicit repair mechanisms, are lacking, as is a cure. Current treatments typically focus on attenuating the immune response and preventing further damage to myelin and axons. Therapeutic strategies to promote re-myelination and repair damage are areas for future advances.[172] Furthermore, research on multiple sclerosis-specific targets remains an area ripe for future efforts. Further advances in understanding its etiology, diverse pathophysiology, and disease course are required before significant gains in new therapies for multiple sclerosis can be realized.

7.31.4 Future Considerations

While graft rejection and multiple sclerosis are very different conditions, both rely on an autoimmune response, and treatment requires a delicate balance between modulating the immune system to prevent symptoms, while at the same time maintaining an appropriate immune response. The availability of drugs that suppress the immune system has ushered in a new era in medicine by enabling organ transplants to become a reality. The use of these same agents and others to treat autoimmune diseases such as multiple sclerosis has allowed a greater understanding of these diseases. As advances in understanding the immune response, its role in disease, and its compensatory mechanisms are made, effective, yet safe, approaches to its modulation will be forthcoming, and ultimately more effective and safe treatments will become available to transplant and multiple sclerosis patients.

References

1. Davidson, A.; Diamond, B. *N. Engl. J. Med.* **2001**, *345*, 340–350.
2. Calne, R. *Transplant. Proc.* **2005**, *37*, 1979–1983.
3. Hoerbelt, R.; Muniappan, A.; Madsen, J. C.; Allan, J. S. *Curr. Opin. Investig. Drugs* **2004**, *5*, 489–498.
4. Halloran, P. F. *N. Engl. J. Med.* **2004**, *351*, 2715–2729.
5. First, M. R. *Transplant. Proc.* **2002**, *34*, 1369–1371.
6. Kahan, B. D. *Transplant. Proc.* **2001**, *33*, 3035–3037.
7. Dumont, F. J. *Curr. Opin. Investig. Drugs* **2001**, *2*, 357–363.
8. Helderman, J. H. *Transplant. Rev.* **2000**, *14*, 177–182.
9. Kahan, B. D.; Kirken, R. A.; Stepkowski, S. M. *Transplant. Proc.* **2003**, *35*, 1621–1623.
10. Kahan, B. D.; Keown, P.; Levy, G. A.; Johnston, A. *Clin. Ther.* **2002**, *24*, 330–350.
11. Al-khaldi, A.; Robbins, R. C. *Ann. Rev. Med.* **2006**, *57*, 455–471.
12. Perry, I.; Neuberger, J. *Clin. Exp. Immunol.* **2005**, *139*, 2–10.
13. Serkova, N.; Christians, U. *Curr. Opin. Investig. Drugs* **2003**, *4*, 1287–1296.
14. Danovitch, G. M. *Transplant. Rev.* **2000**, *14*, 65–81.
15. Kasiske, B. L.; Ballantyne, C. M. *Transplant. Rev.* **2002**, *16*, 1–21.
16. Helderman, J. H. *Transplant. Proc.* **2001**, *33*, 2S–3S.
17. Paul, L. C. *Transplant. Proc.* **2001**, *33*, 2089–2091.
18. Grinyo, J. M. *Transplant. Proc.* **2001**, *33*, 4S–6S.
19. Colvin, R. B.; Smith, R. N. *Nat. Rev. Immunol.* **2005**, *5*, 807–817.
20. Pascher, A.; Klupp, J. *Biodrugs* **2005**, *19*, 211–231.
21. Chapman, T. M.; Keating, G. M. *Drugs* **2003**, *63*, 2803–2835.
22. Rosen, J.; Miner, J. N. *Curr. Med. Chem.: Immunol., Endocr. Metab. Agents* **2002**, *2*, 11–22.
23. Rhen, T.; Cidlowski, J. A. *N. Eng. J. Med.* **2005**, *353*, 1711–1723.
24. Reichardt, H. M. *Curr. Pharm. Des.* **2004**, *10*, 2797–2805.
25. Coghlan, M. J.; Elmore, S. W.; Kym, P. R.; Kort, M. E. *Ann. Rep. Med. Chem.* **2002**, *37*, 167–176.
26. Ponticelli, C. *Transplant. Proc.* **2005**, *37*, 3597–3599.
27. Schrem, H.; Luck, R.; Becker, T.; Nashan, B.; Klempnauer, J. *Transplant. Proc.* **2004**, *36*, 2525–2531.
28. Ponticelli, C.; Tarantino, A.; Campise, M.; Montagnino, G.; Aroldi, A.; Passerini, P. *Transplant. Proc.* **2004**, *36*, 557S–560S.
29. Hamawy, M. M.; Knechtle, S. J. *Transplant. Rev.* **2003**, *17*, 165–171.
30. Matsuda, S.; Koyasu, S. *Immunopharmacology* **2000**, *47*, 119–125.
31. Dumont, F. J. *Curr. Opin. Investig. Drugs* **2004**, *5*, 542–550.
32. Gerwirtz, A. T.; Sitaraman, S. V. *Curr. Opin. Investig. Drugs* **2002**, *3*, 1307–1311.
33. Shapiro, R. *Transplant. Proc.* **2001**, *33*, 3158–3160.
34. Dumont, F. J. *Curr. Med. Chem.* **2000**, *7*, 731–748.
35. Siekierka, J. J.; Staruch, M. J.; Hung, S. H. Y.; Sigal, N. H. *J. Immunol.* **1989**, *143*, 1580–1583.
36. Brazelton, T. R.; Morris, R. *Curr. Opin. Immunol.* **1996**, *8*, 710–720.
37. Sehgal, S. N. *Transplant. Proc.* **2003**, *35*, 7S–14S.
38. Watson, C. J. E. *Transplant. Rev.* **2001**, *15*, 165–177.
39. Calne, R. *Transplant. Proc.* **2003**, *35*, 15S–17S.
40. Camardo, J. *Transplant. Proc.* **2003**, *35*, 18S–24S.
41. Abraham, R. T. *Curr. Opin. Immunol.* **1998**, *10*, 330–336.
42. Lehmkuhl, H.; Ross, H.; Eisen, H.; Valantine, H. *Transplant. Proc.* **2005**, *37*, 4145–4149.
43. Dumont, F. J. *Curr. Opin. Investig. Drugs* **2001**, *2*, 1220–1234.
44. Schreiber, S. L. *Science* **1991**, *251*, 283–287.
45. Crabtree, G. R. *Science* **1991**, *251*, 355–361.
46. Sigal, N. H.; Dumont, F. J. *Ann. Rev. Immunol.* **1992**, *10*, 519–560.
47. Schreiber, S. L.; Crabtree, G. R. *Immunol. Today* **1992**, *13*, 136–142.
48. Liu, J. *Immunol. Today* **1993**, *14*, 290–295.
49. Bierer, B. E.; Hollander, G.; Fruman, D.; Burakoff, S. J. *Curr. Opin. Immunol.* **1993**, *5*, 763–773.
50. Allison, A. C.; Fugui, E. M. *Transplantation* **2005**, *80*, S181–S190.
51. Hesselink, D. A.; van Gelder, T. *Transplant. Rev.* **2003**, *17*, 158–163.
52. Allison, A. C.; Eugui, E. M. *Immunopharmacology* **2000**, *47*, 85–118.
53. Land, W.; Schneeberger, H.; Weiss, M.; Ege, T.; Stumpfig, L. *Transplant. Proc.* **2001**, *33*, 29S–35S.
54. Becker, B. N. *Transplant. Proc.* **1999**, *31*, 2777–2778.
55. Decker, C. J.; Heiser, A. D.; Chaturvedi, P. R.; Faust, T. J.; Ku, G.; Moseley, S.; Nimmesgern, E. *Drugs Exp. Clin. Res.* **2001**, *27*, 89–95.
56. Jain, J.; Almquist, S. J.; Shlyakhter, D.; Harding, M. W. *J. Pharm. Sci.* **2001**, *90*, 625–637.
57. Williams, J. W.; Mital, D.; Chong, A.; Kottayil, A.; Millis, M.; Longstreth, J.; Huang, W.; Brady, L.; Jensik, S. *Transplantation* **2002**, *73*, 358–366.
58. Hardinger, K. L.; Wang, C. D.; Schnitzler, M. A.; Miller, B. W.; Jendrisak, M. D.; Shenoy, S.; Lowell, J. A.; Brennan, D. C. *Am. J. Transplant.* **2002**, *2*, 867–871.
59. Kaplan, M. *Curr. Opin. Invest. Drugs* **2005**, *6*, 526–536.
60. Pan, F.; Ebbs, A.; Wynn, C.; Erickson, L.; Jang, M.-S.; Crews, G.; Fisniku, O.; Kobayashi, M.; Paul, L. C.; Benediktsson, H. et al., *Transplantation* **2003**, *75*, 1110–1114.
61. Jain, M. K.; Ridker, P. M. *Nat. Rev.* **2005**, *4*, 977–987.
62. Fellstrom, B.; Holdaas, H.; Jardine, A. *Transplant. Rev.* **2004**, *18*, 122–128.
63. Weitz-Schmidt, G. *Trends Pharmacol. Sci.* **2002**, *23*, 482–486.
64. Kwak, B.; Mulhaupt, F.; Myit, S.; Mach, F. *Nat. Med.* **2000**, *6*, 1399–1402.
65. Kobashigawa, J. A. *Am. J. Transplant.* **2004**, *4*, 1013–1018.

66. Kobashigawa, J. A.; Katznelson, S.; Laks, H.; Johnson, J. A.; Yeatman, L.; Wang, X. M.; Chia, D.; Terasaki, P. I.; Sabad, A.; Cogert, G. A. et al. *N. Engl. J. Med.* **1995**, *333*, 621–627.
67. Onuffer, J. J.; Horuk, R. *Trends Pharmacol. Sci.* **2002**, *23*, 459–467.
68. von Andrian, U. H.; Mackay, C. R. *N. Engl. J. Med.* **2000**, *343*, 1020–1034.
69. Medina, J. C.; Johnson, M. G.; Collins, T. L. *Ann. Rep. Med. Chem.* **2005**, *40*, 215–225.
70. Panzer, U.; Reinking, R.; Steinmetz, O. M.; Zahner, G.; Sudbeck, U.; Fehr, S.; Pfalzer, B.; Schneider, A.; Thaiss, F.; Mack, M. et al. *Transplantation* **2004**, *78*, 1341–1350.
71. Fahmy, N. M.; Yamani, M. H.; Starling, R. C.; Ratliff, N. B.; Young, J. B.; McCarthy, P. M.; Feng, J.; Novick, N. C.; Fairchild, R. L. *Transplantation* **2003**, *75*, 72–78.
72. Hancock, W. W.; Wang, L.; Ye, Q.; Han, R.; Lee, I. *Curr. Opin. Immunol.* **2003**, *15*, 479–486.
73. Hancock, W. W.; Lu, B.; Gao, W.; Csizmadia, V.; Faia, K.; King, J. A.; Smiley, S. T.; Ling, M.; Gerard, N. P.; Gerard, C. *J. Exp. Med.* **2000**, *192*, 1515–1519.
74. Hancock, W. W.; Gao, W.; Faia, K. L.; Csizmadia, V. *Curr. Opin. Immunol.* **2000**, *12*, 511–516.
75. Elices, M. J. *Curr. Opin. Invest. Drugs* **2002**, *3*, 865–869.
76. Horuk, R.; Clayberger, C.; Krensky, A. M.; Wang, Z.; Grone, H.-J.; Weber, C.; Weber, K. S. C.; Nelson, P. J.; May, K.; Rosser, M. et al. *J. Biol. Chem.* **2001**, *276*, 4199–4204.
77. Chun, J.; Rosen, H. *Curr. Pharm. Des.* **2006**, *12*, 161–171.
78. Gardell, S. E.; Dubi, A. E.; Chun, J. *Trends Mol. Med.* **2006**, *12*, 65–75.
79. Goetzl, E. J.; Rosen, H. *J. Clin. Invest.* **2004**, *114*, 1531–1537.
80. Suzuki, S. *Transplant. Proc.* **1999**, *31*, 2779–2782.
81. Kahan, B. D. *Transplant. Proc.* **2001**, *33*, 3081–3083.
82. Ferguson, R. *Transplant. Proc.* **2004**, *36*, 549S–553S.
83. Elices, M. J. *Curr. Opin. Investig. Drugs* **2004**, *5*, 1137–1140.
84. Borhani, D. W.; Calderwood, D. J.; Friedman, M. M.; Hirst, G. C.; Li, B.; Leung, A. K.; McRae, B.; Ratnofsky, S.; Ritter, K.; Waegell, W. *Bioorg. Med. Chem. Lett.* **2004**, *14*, 2613–2616.
85. Kamens, J. S.; Ratnofsky, S. E.; Hirst, G. C. *Curr. Opin. Investig. Drugs* **2001**, *2*, 1213–1219.
86. Changelian, P. S.; Flanagan, M. E.; Ball, D. J.; Kent, C. R.; Magnuson, K. S.; Martin, W. H.; Rizzuti, B. J.; Sawyer, P. S.; Perry, B. D.; Brissette, W. H. et al. *Science* **2003**, *302*, 875–878.
87. Pomfret, E. A.; Feng, S.; Hale, D. A.; Magee, J. C.; Mulligan, M.; Knechtle, S. J. *Am. J. Transplant.* **2006**, *6*, 275–290.
88. Newell, K. A.; Larsen, C. P.; Kirk, A. D. *Transplantation* **2006**, *81*, 1–6.
89. Jiang, S.; Herrera, O.; Lechler, R. I. *Curr. Opin. Immunol.* **2004**, *16*, 550–557.
90. Rotrosen, D.; Matthews, J. B.; Bluestone, J. A. *J. Allergy Clin. Immunol.* **2002**, *110*, 17–23.
91. Sacks, S. H.; Chowdhury, P.; Zhou, W. *Curr. Opin. Immunol.* **2003**, *15*, 487–492.
92. Andrade, C. F.; Waddell, T. K.; Keshavjee, S.; Liu, M. *Am. J. Transplant.* **2005**, *5*, 969–975.
93. Heeger, P. S. *Am. J. Transplant.* **2003**, *3*, 525–533.
94. Cattaneo, D.; Perico, N.; Remuzzi, G. *Am. J. Transplant.* **2004**, *4*, 299–310.
95. Dumont, F. J. *Curr. Opin. Investig. Drugs* **2002**, *3*, 1453–1467.
96. Levy, R. *Curr. Med. Chem.: Immunol., Endocr. Metab. Agents* **2005**, *5*, 585–597.
97. Ferrara, J. L. M.; Mineishi, S. *Curr. Med. Chem.: Immunol., Endocr. Metab. Agents* **2005**, *5*, 539–545.
98. Iwasaki, T. *Curr. Med. Chem.: Immunol., Endocr. Metab. Agents* **2005**, *5*, 565–573.
99. Mourich, D. V.; Marshall, N. B. *Curr. Opin. Pharmacol.* **2005**, *5*, 508–512.
100. Rudick, R. A.; Cohen, J. A.; Weinstock-Guttman, B.; Kinkel, R. P.; Ransohoff, R. M. *N. Engl. J. Med.* **1997**, *337*, 1604–1611.
101. Noseworthy, J. H.; Lucchinetti, C.; Rodriguez, M.; Weinshenker, B. G. *N. Engl. J. Med.* **2000**, *343*, 938–952.
102. Chofflon, M. *Biodrugs* **2005**, *19*, 299–308.
103. Lublin, F. D.; Reingold, S. C. *Neurology* **1996**, *46*, 907–911.
104. Wekerle, M.; Hohlfeld, R. *N. Engl. J. Med.* **2003**, *349*, 185–186.
105. Kanwar, J. R. *Curr. Med. Chem.* **2005**, *12*, 2947–2962.
106. Steinman, L. *Neuron* **1999**, *24*, 511–514.
107. Lipton, H. L.; Dal Canto, M. C. *Science* **1976**, *192*, 62–64.
108. Neuhaus, O.; Archelos, J. J.; Hartung, H.-P. *Trends Pharmacol. Sci.* **2003**, *24*, 131–138.
109. Arnon, R.; Aharoni, R. *Proc. Natl. Acad. Sci. USA* **2004**, *101*, 14593–14598.
110. Farina, C.; Weber, M. S.; Meinl, E.; Wekerle, H.; Hohlfeld, R. *Lancet Neurol.* **2005**, *4*, 567–575.
111. Gold, R.; Buttgereit, F.; Toyka, K. V. *J. Immunol.* **2001**, *117*, 1–8.
112. Sloka, J. S.; Stefanelli, M. *Mult. Scler.* **2005**, *11*, 425–432.
113. Crawford, D. J. K.; Maddocks, J. L.; Jones, D. N.; Szawlowski, P. *J. Med. Chem.* **1996**, *39*, 2690–2695.
114. Aarbakea, J.; Janka-Schaub, G.; Elion, G. B. *Trends Pharmacol. Sci.* **1997**, *18*, 3–7.
115. El-Azhary, R. A. *Int. J. Dermatol.* **2003**, *42*, 335–341.
116. Bischoff, P. L.; Holl, V.; Coelho, D.; Dufour, P.; Luu, B.; Weltin, D. *Curr. Med. Chem.* **2000**, 7, 693–713.
117. Weiner, H. L.; Cohen, J. A. *Mult. Scler.* **2002**, *8*, 142–154.
118. Polman, C. H.; Uitdehaag, B. M. J. *Lancet Neurol.* **2003**, *2*, 563–566.
119. Steinman, L. *Nat. Rev. Drug Disc.* **2005**, *4*, 510–519.
120. Rice, G. P. A.; Hartung, H.-P.; Calabresi, P. A. *Neurology* **2005**, *64*, 1336–1342.
121. Sheremata, W. A.; Minagar, A.; Alexander, J. S.; Vollmer, T. *CNS Drugs* **2005**, *18*, 909–922.
122. Yusuf-Makagiansar, H.; Anderson, M. E.; Yakovleva, T. V.; Murray, J. S.; Siahaan, T. J. *Med. Res. Rev.* **2002**, *22*, 146–167.
123. Tilley, J. W. *Expert Opin. Ther. Patents* **2002**, *12*, 991–1008.
124. Yang, G. X.; Hagmann, W. K. *Med. Res. Rev.* **2003**, *23*, 369–392.
125. Piraino, P. S.; Yednock, T. A.; Freedman, S. B.; Messersmith, E. K.; Pleiss, M. A.; Vandevert, C.; Thorsett, E. D.; Karlik, S. J. *J. Neuroimmunol.* **2002**, *131*, 147–159.
126. Cannella, B.; Gaupp, S.; Tilton, R. G.; Raine, C. S. *J. Neurosci. Res.* **2003**, *71*, 407–416.

127. Leone, D. R.; Giza, K.; Gill, A.; Dolinski, B. M.; Yang, W.; Perper, S.; Scott, D. M.; Lee, W.-C.; Cornebise, M.; Wortham, K. et al. *J. Pharm. Exp. Ther.* **2003**, *305*, 1150–1162.
128. Huryn, D. M.; Konradi, A. W.; Ashwell, S.; Freedman, S. B.; Lombardo, L. J.; Pleiss, M. A.; Thorsett, E. D.; Yednock, T.; Kennedy, J. D. *Curr. Top. Med. Chem.* **2004**, *4*, 1473–1484.
129. Pepinsky, R. B.; Lee, W. C.; Cornebise, M.; Gill, A.; Wortham, K.; Chen, L.; Leone, D. R.; Giza, K.; Doninsky, B. M.; Perper, S. et al. *J. Pharm. Exp. Ther.* **2005**, *312*, 742–750.
130. Theien, B. E.; Vanderlugt, C. L.; Nickerson-Nutter, C.; Cornebise, M.; Scott, D. M.; Perper, S. J.; Whalley, E. T.; Miller, S. D. *Blood* **2003**, *102*, 4464–4471.
131. Spence, S.; Vetter, C.; Hagmann, W. K.; Van Riper, G.; Williams, H.; Mumford, R. A.; Lanza, T. J.; Lin, L. S.; Schmidt, J. A. *Teratology* **2002**, *65*, 26–37.
132. Crofts, F.; Pino, M.; De Lise, B.; Guittin, P.; Barbellion, S.; Brunel, P.; Potdevin, S.; Bergmann, B.; Hofmann, T.; Lerman, S. et al. *Birth Defects Res. Part B Dev. Reprod. Toxicol.* **2004**, *71*, 55–68.
133. Trebst, C.; Stangel, M. *Curr. Pharm. Des.* **2006**, *12*, 241–249.
134. Stangel, M.; Hartung, H.-P. *J. Neurol. Neurosurg. Psychiatry* **2002**, *72*, 1–4.
135. Wiles, C. M.; Brown, P.; Chapel, H.; Guerrini, R.; Hughes, R. A. C.; Martin, T. D.; McCrone, P.; Newsom-Davis, J.; Palace, J.; Rees, J. H. et al. *J. Neurol. Neurosurg. Psychiatry* **2002**, *72*, 440–448.
136. Jorgensen, S. H.; Sorensen, P. S. *J. Neurol. Sci.* **2005**, *233*, 61–65.
137. Weitz-Schmidt, G.; Welzenbach, K.; Brinkmann, V.; Kamata, T.; Kallen, J.; Bruns, C.; Cottens, S.; Takada, Y.; Hommel, U. *Nat. Med.* **2001**, *7*, 687–692.
138. Youssef, S.; Stuve, O.; Patarroyo, J. C.; Ruiz, P. J.; Radosevich, J. L.; Hur, E. M.; Bravo, M.; Mitchell, D. J.; Sobel, R. A.; Steinman, L. et al. *Nature* **2002**, *420*, 78–84.
139. Neuhaus, P.; Archelos, J. J.; Hartung, H.-P. *Mult. Scler.* **2003**, *9*, 429–430.
140. Neuhaus, O.; Stuve, O.; Zamvil, S. S.; Hartung, S.-P. *CNS Drugs* **2005**, *19*, 833–841.
141. Darlington, C. *Curr. Opin. Investig. Drugs* **2005**, *6*, 667–671.
142. Darlington, C. *Curr. Opin. Investig. Drugs* **2002**, *3*, 911–914.
143. El-Etr, M.; Vukusic, S.; Gignoux, L.; Durand-Dubief, F.; Achiti, I.; Baulieu, E. E.; Confavreux, C. *J. Neurol. Sci.* **2005**, *233*, 49–54.
144. Nalbandian, G.; Kovats, S. *Curr. Med. Chem.: Immunol., Endocr. Metab. Agents* **2005**, *5*, 85–91.
145. Wiendl, H.; Kieseier, B. C. *Expert Opin. Investig. Drugs* **2003**, *12*, 704–712.
146. Stern, J. N. H.; Illes, Z.; Reddy, J.; Keskin, D. B.; Fridkis-Hareli, M.; Kuchroo, V. K.; Strominger, J. L. *PNAS* **2005**, *102*, 1620–1625.
147. Matsoukas, J.; Apostolopoulos, V.; Mavromoustakos, T. *Mini Rev. Med. Chem.* **2001**, *1*, 273–282.
148. Koehler, N. K.; Yang, C.-Y.; Varady, J.; Lu, Y.; Wu, X.-W.; Liu, M.; Yin, D.; Bartels, M.; Xy, B.-Y.; Roller, P. P. et al. *J. Med. Chem.* **2004**, *47*, 4989–4997.
149. Hohlfeld, R.; Wekerle, H. *Proc. Natl. Acad. Sci. USA* **2004**, *101*, 14599–14606.
150. Gaur, A.; Wiers, B.; Liu, A.; Rothbard, J.; Fathman, C. G. *Science* **1992**, *258*, 1491–1494.
151. Brinkmann, V.; Davis, M. D.; Heise, C. E.; Albert, R.; Cottens, S.; Hof, R.; Bruns, C.; Prieschl, E.; Baumruker, T.; Hiestand, P. et al. *J. Biol. Chem.* **2002**, *277*, 21453–21457.
152. Mandala, S.; Hajdu, R.; Bergstrom, J.; Quackenbush, E.; Xie, J.; Milligan, J.; Thornton, R.; Shie, G.-J.; Card, D.; Keohane, C. et al. *Science* **2002**, *296*, 346–349.
153. Fujino, M.; Funeshima, N.; Kitazawa, Y.; Kimura, H.; Amemiya, H.; Suzuki, S.; Li, X.-K. *J. Pharm. Exp. Ther.* **2003**, *305*, 70–77.
154. Hale, J. J.; Lynch, C. L.; Neway, W.; Mills, S. G.; Hajdu, R.; Keohane, C. A.; Rosenbach, M. J.; Milligan, J. A.; Shei, G. J.; Parent, S. A. et al. *J. Med. Chem.* **2004**, *47*, 6662–6665.
155. Li, Z.; Chen, W.; Hale, J. J.; Lynch, C. L.; Mills, S. G.; Hajdu, R.; Keohane, C. A.; Rosenbach, M. J.; Mulligan, J. A.; Shei, G.-J. et al. *J. Med. Chem.* **2005**, *48*, 6169–6172.
156. Smith, P. F. *Curr. Opin. Investig. Drugs* **2004**, *5*, 748–754.
157. Maresz, K.; Carrier, E. J.; Ponomarev, E. D.; Hillard, C. J.; Dittel, B. N. *J. Neurochem.* **2005**, *95*, 437–445.
158. Malfitano, A. M.; Matarese, G.; Pisanti, S.; Grimaldi, C.; Laezza, C.; Bisogno, T.; DiMarzo, V.; Lechler, R. I.; Bifulco, M. *J. Neuroimmunol.* **2005**, *171*, 110–119.
159. Mestre, L.; Correa, F.; Arevalo-Martin, A.; Moline-Holgado, E.; Valenti, M.; Ortar, G.; Di Marzo, V.; Guaza, C. *J. Neurochem.* **2005**, *92*, 1327–1339.
160. Pedotti, R.; Steinman, L. *Curr. Med. Chem.-Anti-Inflamm. Anti-Allergy Agents* **2005**, *4*, 637–643.
161. White, F. A.; Bhangoo, S. K.; Miller, R. J. *Nat. Rev. Drug Disc.* **2005**, *4*, 834–844.
162. Glabinski, A. R.; Ransohoff, R. M. *Curr. Opin. Investig. Drugs* **2001**, *2*, 1712–1719.
163. Hendriks, J. J. A.; de Vries, H. E.; van der Pol, S. M. A.; van den Berg, T. K.; van Tol, E. A. F.; Dijkstra, C. D. *Biochem. Pharmacol.* **2003**, *65*, 877–885.
164. Chen, L.; Pei, G.; Zhang, W. *Curr. Pharm. Des.* **2004**, *10*, 1045–1055.
165. Brodmerkel, C. M.; Huber, R.; Covington, M.; Diamond, S.; Hall, L.; Collins, R.; Leffet, L.; Gallagher, K.; Feldman, P.; Collier, P. et al. *J. Immunol.* **2005**, *175*, 5370–7378.
166. Bouerat, L.; Fensholdt, J.; Liang, X.; Havez, S.; Nielsen, S. F.; Hansen, J. R.; Bolvig, S.; Andersson, C. *J. Med. Chem.* **2005**, *48*, 5412–5414.
167. Platten, M.; Ho, P. P.; Youssef, S.; Fontoura, P.; Garren, H.; Hur, E. M.; Gupta, R.; Lee, L. Y.; Kidd, B. A.; Robinson, W. H. et al. *Science* **2005**, *310*, 850–855.
168. Beeton, C.; Change, K. G. *Neuroscientist* **2005**, *11*, 550–562.
169. Eggert, M.; Kluter, A.; Zettl, U. K.; Neeck, G. *Curr. Pharm. Des.* **2004**, *10*, 2787–2796.
170. Alegre, M.-L.; Fallarino, F. *Curr. Pharm. Des.* **2006**, *12*, 149–160.
171. Lovett-Racke, A. E.; Cravens, P. D.; Gocke, A. R.; Racke, M. K.; Stuve, O. *Arch. Neurol.* **2005**, *62*, 1810–1813.
172. Zhao, C.; Fancy, S. P. J.; Kotter, M. R.; Li, W.-W.; Franklin, R. J. M. *J. Neurol. Sci.* **2005**, *233*, 87–91.
173. National Multiple Sclerosis Society. http://www.nationalmssociety.org (accessed May 2006).

Biographies

Jerauld S Skotnicki was born in Niagara Falls, NY. He received a BA in Chemistry from the College of the Holy Cross, an MA in Chemistry from Dartmouth College (under Prof G W Gribble), and a PhD from Princeton University (under Prof E C Taylor). Prior to his studies at Princeton, Dr Skotnicki was staff chemist at Lederle Laboratories for 3 years. Following his PhD, he joined the Biochemicals Department at DuPont as Research Chemist. In 1982, Dr Skotnicki joined Wyeth as a Medicinal Chemist at Wyeth Laboratories in Radnor, PA and Wyeth-Ayerst Research in Princeton, NJ, and he became a Director of Medicinal Chemistry at Wyeth in Pearl River, NY. His research experience has been in the areas of immunoinflammatory diseases, infectious diseases, musculoskeletal diseases, oncology, and transplantation. Since 2004, in his current position as Senior Director of Chemical Sciences Interface, Chemical and Screening Sciences at Wyeth he has responsibility for research activities at the junction of discovery and development, as well as management of external alliances in Chemical and Screening Sciences. Dr Skotnicki was the recipient (2004) of the Thomas Alva Edison Patent Award in 'Emerging Technologies' for his contributions to the discovery of temsirolimus. He is currently an editor of *Inflammation Research* and serves as a manuscript reviewer for the *Journal of Medicinal Chemistry*, *Journal of Organic Chemistry*, *Bioorganic and Medicinal Chemistry Letters*, *Tetrahedron*, *Chemistry & Biology*, and *Medicinal Chemistry*. He has served on the Board of Directors of the Inflammation Research Association (2000–04).

Donna M Huryn received her BA degree from Cornell University and her PhD in Organic Chemistry from the University of Pennsylvania. She joined the Chemistry Department at Hoffmann-La Roche and worked in the areas of anticancer, anti-inflammatory, immunonology, and anti-viral agents. In 1993, she moved to Wyeth Research's Chemical Sciences Department where she led the CNS Medicinal Chemistry Department in projects targeting multiple sclerosis, Alzheimer's disease, and depression, among others, and then led the Chemical Sciences Interface Department. Dr Huryn is currently Associate Director of the Chemistry Core at the Penn Center for Molecular Discovery (University of Pennsylvania), Scientific Advisor to the University of Pittsburgh Center for Chemical Methodology and Library Design, Senior Scientific Fellow at the Pittsburgh Molecular Library Screening Center, and

Adjunct Professor of Pharmaceutical Sciences at the University of Pittsburgh. In addition, she is a consultant to several technology and biotech companies. She has served on a number of local and national committees of the American Chemical Society, and is a member of the Editorial Advisory Board of Organic Letters. Dr Huryn was a member of the NIH Medicinal Chemistry Study Section from 1996 to 2000, and is a frequent ad hoc member of the Drug Discovery, Development and Delivery SBIR Study Section.

Comprehensive Medicinal Chemistry II
ISBN (set): 0-08-044513-6

ISBN (Volume 7) 0-08-044520-9; pp. 917–934

7.32 Overview of Dermatological Diseases

H R Jalian, S Takahashi, and J Kim, David Geffen School of Medicine at UCLA, Los Angeles, CA, USA

7.32.1 Acne

Acne vulgaris is a near-universal cutaneous disorder of the pilosebaceous follicle. More than 45 million people in the US are affected with acne and over 85% of individuals 12–25 years old suffer from acne. From 1996 to 1998, US consumers spent over one billion dollars for systemic acne medications alone and millions more for topical and over-the-counter treatments. Acne has a significant effect on a patient's psyche and self-esteem and, if not properly treated, can result in pitted or hypertrophic scarring. Although the pathogenesis is not entirely understood, it is thought to be multifactorial, involving abnormal hyperkeratinization, increased sebum production, hormones, cutaneous microbes, and the host immune response. In order to devise effective treatments, understanding the underlying pathogenesis is essential to identify potential therapeutic targets. This review will highlight recent discoveries in the pathogenesis of acne, as well as discuss both established therapies and future trends in treatment.

7.32.1.1 Pathogenesis

The pathogenesis of acne is multifactorial and includes follicular hyperkeratinization, increased sebum production, hormones, colonization with *Propionibacterium acnes*, and immunologic influences (**Table 1**). Acne develops in the pilosebaceous unit, which consists of epidermal cells lining the hair follicle and the sebaceous gland. Alteration in the pattern of keratinization within the follicle contributes to sebaceous follicular obstruction, causing the formation of microcomedones, which are the precursors of acne lesions. Sebum production is important for the development of acne and is regulated by a number of different factors, including local androgen production and neuroendocrine influences. In addition, *P. acnes*, a Gram-positive anaerobe, plays an important role in inflammation by secreting lipases, proteases, hyaluronidases, and chemotactic factors. *P. acnes* also directly activates immune cells, inducing an inflammatory response in the host. New insights into the pathogenesis of acne will be reviewed.

7.32.1.1.1 Follicular hyperkeratinization

Normally, follicular keratin is loosely organized. In acne, ultrastructural changes in keratin are initially observed in the lower portion of the follicular infundibulum and later in the entire follicle. The keratinous material becomes dense, intracellular lipid droplets accumulate, the number and size of keratohyaline granules increase, and lamellar granules decrease, leading to the formation of the comedone.[1] Also, proliferation of keratinocytes in comedones is higher compared to normal follicles[2] and infundibular keratinocytes are more cohesive. Local deficiencies in essential fatty acids, specifically linoleic acid, may play a role in follicular hyperkeratinization.[3] Because acne patients have increased sebum secretion, it is theorized that this results in decreased linoleic acid within the follicular epithelium. Consequently this is thought to induce follicular hyperkeratosis and decreased epithelial function. In addition, the proinflammatory cytokine interleukin (IL)-1α, produced by lymphocytes and keratinocytes, has been shown to play a role in pathogenesis. IL-1α induces hyperkeratinization of infundibular keratinocytes, which can be blocked by the addition of an IL1 receptor antagonist, and may contribute to hyperkeratinization within the follicle.[4]

7.32.1.1.2 Sebum production

The increased production of sebum is a critical component in the pathogenesis of acne. Sebaceous glands in humans develop in the 13th to 15th week of gestation and are associated with hair follicles over the entire body. The largest

Table 1 Summary of the pathogenesis of acne

Etiology	*Mechanism*	*Treatment*
Abnormal Hyperkeratinization	Higher rate of keratinocyte proliferation in comedones Interleukin-1α produced by keratinocytes contributes to hyperkeratinization	Retinoids
Sebum production	Higher rate of sebum secretion in acne patients Sebum production required for acne	Retinoids Laser Antiandrogens
Androgens	Androgens required for sebum production Increased 5α-reductase in biopsy of acne lesions	Antiandrogens
Propionibacterium acnes	Secretes enzymes that lead to comedone formation and rupture Activates monocytes, contributing to inflammation Produces neutrophil chemotactant factor	Topical antibiotics Oral antibiotics Benzoyl peroxide Azelaic acid Light therapy

glands and maximum gland density are found on the face and scalp. At birth, sebaceous gland activity is high but begins to decline into childhood. At around age 7, sebum production increases and continues to rise into adolescence.[5] The maturation of sebaceous glands and production of sebum are required for the development of acne. In addition, the average rate of sebum secretion is higher in individuals with acne than in those without.[6] The exact mechanism underlying the regulation of human sebum production has not been fully defined but is likely regulated by androgens, neuroendocrine factors, and other factors regulating lipid metabolism.

7.32.1.1.3 Androgens

Androgens are required to produce significant quantities of sebum. In fact, castrated males and individuals with genetic errors in the production or function of androgen receptors do not develop acne.[7] In the majority of patients with acne serum androgens are normal.[8] It is likely the local production of androgens within the skin that correlates more directly with the development of acne.[9] Dehydroepiandrosterone sulfate (DHEAS) may regulate sebaceous gland activity through its conversion into more potent androgens in the epidermis. The skin possesses each of the enzymes for the conversion of DHEAS into potent androgens, including the enzyme 5α-reductase.[10] This enzyme is required for the conversion of testosterone into higher affinity dihydrotestosterone (DHT), and exists in two isoforms. The type I isozyme is present on the scalp, chest, and sebaceous glands. Type II is mostly present on genitourinary tissue, dermal papillae, and outer-root sheath of hair follicles.[10–12] In areas susceptible to acne, higher activity of type I 5α-reductase has been observed.[13] In addition, skin biopsies in men and women with acne have increased expression of 5α-reductase.[14] Local conversion of androgen precursors may play a role in the pathogenesis of acne and could be a potential target for therapeutic intervention.

7.32.1.1.4 Neuroendocrine factors

It is well accepted that stress can aggravate acne, although the mechanism of how this occurs is unknown. Recently, cutaneous neurogenic factors, notably substance P (SP), have been implicated in the pathogenesis of acne.[15] Nerve fibers expressing SP were detected in close apposition to sebaceous glands in skin from acne patients but not in normal skin. SP is known to promote the proliferation and differentiation of sebaceous glands in vitro.[16] Also, SP was shown to increase expression of neutral endopeptidase, which has the capability of degrading various neuropeptides and thus decreasing inflammation.

Melanocortins, specifically α-melanocyte-stimulating hormone (α-MSH) and corticotropin-releasing hormone (CRH), are known to increase sebum production in rodents. Recent studies of sebocyte cell lines and immunohistological studies on human sebaceous glands revealed the expression of melanocortin-1 receptor (MC-1R).[17] In addition, the knockout of the melanocortin-5 receptor results in hypoplastic sebaceous glands and reduced sebum production in murine models.[18] α-MSH may also influence IL8 secretion by sebocytes and promote inflammation in acne.

CRH is the most proximal element of the hypothalamic–pituitary–adrenal axis and coordinates the body's response to stress. CRH, CRH-binding protein, and CRH receptors (CRH-R1 and 2) have been detected in a sebocyte cell line in vitro.[19] CRH induces an increase in the synthesis of sebaceous lipids and also upregulates mRNA levels of enzymes important for testosterone synthesis in sebocytes.

Peroxisome proliferator-activated receptors (PPARs) are a family of nuclear receptors known to form heterodimers with the retinoid × receptor.[20] This results in the transcription of genes involved in a variety of processes, including lipid metabolism, cellular proliferation, and differentiation. PPARs are strongly expressed in human sebaceous glands where they may play some role in regulating sebum production.[21] Moreover, agonists of PPAR-γ in rats have been shown to increase lipid accumulation.[22] Further experiments are needed to determine the role, if any, PPARs have in the pathogenesis of acne.

7.32.1.1.5 *Propionibacterium acnes*

P. acnes colonizes sebaceous follicles and is found in the fairly anaerobic, lipid-rich microcomedones. Colonization with *P. acnes* is thought to occur at some time during adolescence. In 11–15-year-olds, virtually no *P. acnes* is found in unaffected individuals.[23] However, teenagers with acne have significantly higher *P. acnes* counts compared to age-matched controls.[24] It is a point of interest to note that, in older individuals, the number of organisms is the same in those with or without acne.[23] *P. acnes* produces an extracellular lipase that hydrolyzes sebum triglycerides to glycerol and free fatty acids, which may contribute to comedone formation. In addition, enzymes produced by *P. acnes* lead to the rupture of the comedonal wall, the impetus for the larger inflammatory lesions.

P. acnes has also been shown to induce monocyte activation through nuclear factor kappa B (NFκB).[25,26] In addition, monocyte cytokine production in response to *P. acnes* occurs through a Toll-like receptor (TLR)2-dependent mechanism.[26] Kim *et al.* demonstrated that macrophages surrounding the pilosebaceous follicles in acne lesions expressed TLR2, providing evidence for a TLR-mediated mechanism occurring at the site of disease.[25]

P. acnes may further contribute to the inflammatory process by producing neutrophil chemotactic factor (NCF). Neutrophils attracted by bacterial chemoattractants release inflammatory mediators such as lysosomal enzymes, resulting in the formation of free radicals and other reactive oxygen species (ROS).[27]

7.32.1.2 Treatment

There are a variety of agents available for the treatment of acne. Used alone or in combination, these treatments exert their effect on one or more of the pathogenic factors. Currently, the main therapeutic targets are aimed at: (1) correcting the altered pattern of follicular keratinization; (2) decreasing sebaceous gland activity and sebum production; (3) decreasing the follicular *P. acnes* population, and subsequent inflammatory mediators; and (4) producing an antiinflammatory effect. Topical agents are often used first-line in mild to moderate acne, and systemic agents are used in more severe cases. In acne that is severe, likely to result in scarring, or with clear endocrine or hormonal etiology, systemic agents are often used on initial presentation (**Figure 1**). In the following section, traditional acne treatments, as well as new developments in light and laser therapy for the treatment of acne, will be discussed.

7.32.1.2.1 Antibacterial agents

Benzoyl peroxide is a topical antibacterial agent that has been a mainstay in the treatment of acne. One advantage to the use of benzoyl peroxide over other antibacterial agents is that bacterial resistance does not develop. In addition

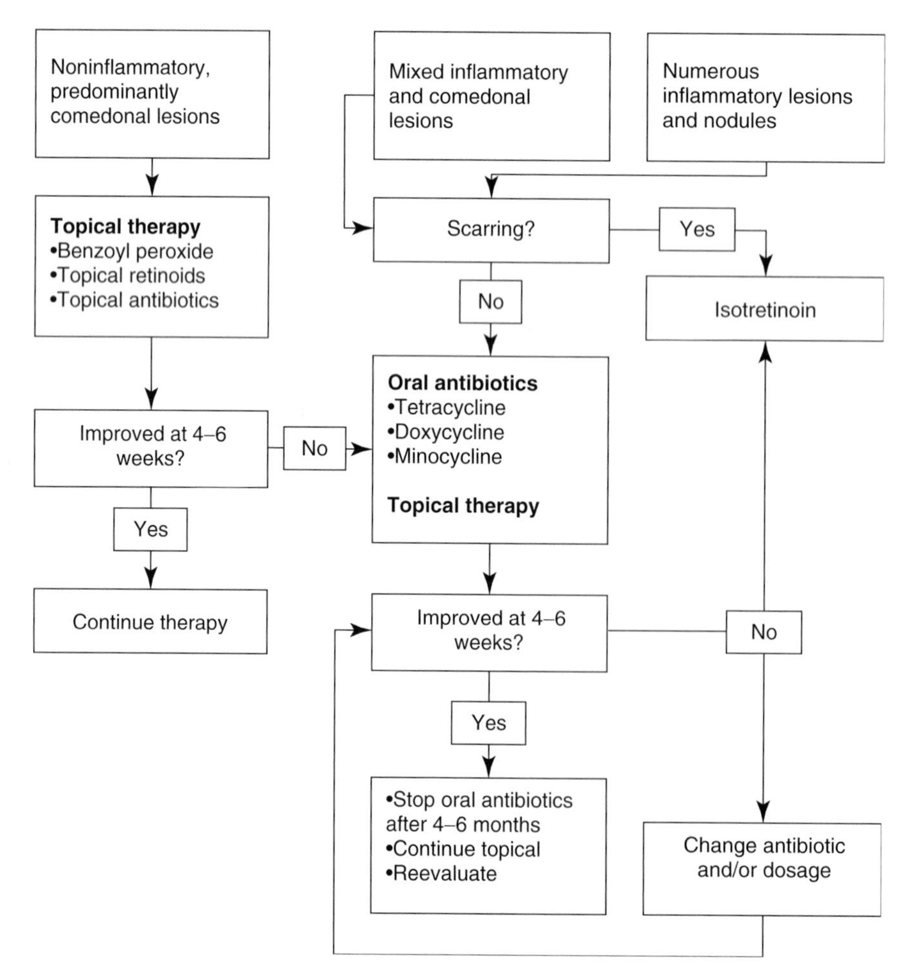

Figure 1 Treatment algorithm for acne.

to its antibacterial effect, benzoyl peroxide may also have antiinflammatory actions by blocking ROS production in neutrophils in a dose-dependent manner.[28] Used alone, or in combination with other agents, benzoyl peroxide has remained an important treatment option. One study measured bacterial counts from treated individuals, and found that 6% benzoyl peroxide gel is superior in suppressing *P. acnes* when compared to 1% clindamycin gel.[29] Multiple studies have shown that, when combined with topical clindamycin, benzoyl peroxide treatment resulted in decreased *P. acnes* counts and improvement in number of lesions.[30,31] In addition, with the concomitant use of benzoyl peroxide and clindamycin, the number of clindamycin-resistant bacteria did not increase in comparison to clindamycin use alone.

Inflammatory acne is traditionally treated with topical and oral antibiotics. Tetracyclines and macrolide antibiotics have been shown to suppress directly the number of *P. acnes*. A major concern in the use of antibiotics is the selection for antibiotic resistance *P. acnes*. Over a 10-year period, the prevalence of skin colonization by propionibacteria resistant to commonly used antiacne antibiotics in patients rose steadily.[32] Resistance to erythromycin was most common, with the majority of these strains being cross-resistant to clindamycin. Tetracycline resistance was encountered less often and did not show a dramatic increase over time. A second major concern is the possible link between chronic antibiotic use and the increased risk of developing breast cancer.[33] Although this association is not clear, there is much concern from the public and further studies are needed at this time to clarify this issue.

In addition to their antibacterial activity, macrolides may also possess some antiinflammatory activity. Akamatsu *et al.* demonstrated that the macrolide antibiotic, roxithromycin, inhibits the bacterial lipase and the production of neutrophil chemotactic factor produced by *P. acnes*.[34]

Another agent that has shown some antibacterial activity is azelaic acid (AZA). AZA is a naturally occurring dicarboxylic acid that has been shown to be beneficial in treating acne and has demonstrated significant antibacterial activity.[35] One study found a 30-fold decrease in *Propionibacterium* species, demonstrating similar efficacy to oral tetracycline treatment.[36] AZA has also been shown to decrease *P. acnes* protein synthesis and may be related to disruption in bacterial cell transmembrane pH gradient. In addition, AZA has a depigmenting effect by inhibiting tyrosinase, an enzyme necessary for melanogenesis, and may be beneficial in patients with the tendency for hyperpigmented scarring.

7.32.1.2.2 Topical retinoids

Topical retinoids are critical in the treatment of acne, as they treat the primary lesions in acne, the comedone. The mechanism of action is through the retinoic acid receptor (RAR), which affects genes involved in cell proliferation, cell differentiation, and inflammation. Numerous topical retinoids are available, including tretinoin, adapalene, and tazarotene, all with similar efficacies.[37–39] Liu *et al.* have recently shown that the antiinflammatory response of all-*trans* retinoic acid (ATRA) was through direct effects on monocytes.[40] In vitro studies show that ATRA has its anti-inflammatory effect in acne by downregulating TLR2/1 and CD14 expression, and subsequently decreasing the release of proinflammatory cytokines from monocytes.

7.32.1.2.3 Oral retinoids

Isotretinion is one of the most effective and definitive treatments for certain types of acne and has traditionally been reserved for severe cases due to the risk of side effects. Common side effects include cheilitis, dryness of mucous membranes, xerosis, conjunctivitis, and pruritus. More serious side effects include hypertriglyceridemia, hepatotoxicity, visual changes, and the known teratogenicity. Unfortunately the number of pregnancies during isotretinoin use has not declined, despite the carefully implemented program to control isotretinoin use. One debated adverse effect of isotretinoin is depression and the increased risk for suicides. Ng *et al.* found no significant relationship between isotretinoin and depression.[41] Chia *et al.* followed a cohort of 101 subjects with moderate to severe acne being treated with isotretinoin over a 3–4-month period.[42] This study found that there was no increased evidence of clinically significant depression in the treatment group versus the control group. Although individual cases have been reported, no studies have shown a clear relationship between isotretinoin use and depression. Clinicians should be cautioned to monitor patients for signs of depression and assess suicide risk if appropriate.

7.32.1.2.4 Hormonal agents

Agents that lower androgens are used to treat acne in females. Ortho Tri-Cyclen and Estrostep are oral contraceptives (OCP) that have been approved by the US Food and Drug Administration (FDA) for the treatment of acne. OCPs containing 20 μg ethinylestradiol and 100 μg levonorgestrel decrease both inflammatory and noninflammatory lesions.[43] Also, a decrease in the serum levels of free androgens has been observed in patients taking OCPs. Other agents used include spironolactone, a potassium-sparing diuretic, which also has weak antiandrogenic properties.

No serious side effects were found with long-term use of spironolactone and it has been shown to be efficacious in the treatment of acne.[44,45] Other agents that can potentially be used include gonadotropin-releasing hormone (GnRH) agonists, although the side-effect profile limits its use.

Peripheral enzyme inhibitors may be an effective therapy in disrupting local androgen influence on sebaceous glands. Finasteride, a 5α-reductase inhibitor, decreases local production of DHT in genitourinary tissue, but does not adequately target the isozyme present in the skin. In addition, this drug has been shown to have no effect on local sebum production in the skin. In the future, specific inhibitors of the type I isozyme, present in skin, may be effective in the treatment of acne.

7.32.1.2.5 Light therapy

Light, laser, and radiofrequency (RF) therapy are increasingly being used in the management of acne. Given the rise in antibiotic resistance, and unknown consequences of long-term retinoid and antibiotic use, photomedicine has become an attractive option that may prove more convenient, safe, and efficacious. Current modalities in photomedicine for the treatment of acne are summarized in **Table 2**.

Light is currently being used as an alternative approach for decreasing the follicular *P. acnes* population. Photoactivating bacterial porphyrins with visible or ultraviolet (UV) light causes the formation of free radicals and ROS, inducing membrane damage and subsequent bacterial cell death.[46–48] This is independent of antibiotic resistance, and does not affect the remainder of the epithelium or follicle, as these cells do not contain porphyrins.

Either noncoherent, narrow-band light or noncoherent, broad spectrum light is used in the treatment of acne. Narrow-band blue light has known efficacy in the treatment of acne and also has minimal associated side effects. Several studies have demonstrated reductions in lesions using narrow-band blue light (400–420 nm). Sigurdsson *et al.* demonstrated improvement in acne lesions on the face and trunk without significant side effects after treatment with broad-spectrum green and violet light.[49] Other studies have shown that red light therapy limits *P. acnes* viability, alters macrophage activity, and may penetrate the epidermis deeper than blue light.[50] Several commercially available devices are available and FDA-approved for the treatment of acne, including Acne Lamp (narrow-band blue and red light), Clearlight (narrow-band blue light), and Omnilux Blue (narrow-band blue light).

Longer wavelengths than blue light have been postulated to provide more porphyrin activation by being able to penetrate deeper into the follicle. One such device, the Intense pulsed light, administers broad-spectrum, noncoherent visible light via a flashlamp. Several studies demonstrate the efficacy of this device with up to 68–87% clearance of lesions in patients.[51] The studies involving both broadband and narrow-band light have shown promise in the treatment of acne.

Photodynamic therapy involves the use of photosensitizing agents that generate free radicals upon stimulation with light. These agents can be administered topically or orally, and are preferentially absorbed by the pilosebaceous unit. Initial studies involving the use of aminolevulinic acid (ALA) as a photosensitizer have shown promise in the treatment of acne. One study found that ALA–red light combination decreased sebum production. Although long-term reductions in lesions were seen in treatment groups, most patients suffer an acneiform eruption, pain, and worsening of symptoms upon initiation of therapy.[52] ALA–blue light combination, however, has similar efficacy to red light, but does not have the adverse side effects associated with ALA–red light treatment.[53]

Table 2 Light, laser, and radiofrequency for the treatment of acne

Treatment	*Mechanism of action*
Narrow-spectrum light	Activates bacterial porphyrins, generating reactive oxygen species
Broad-spectrum light	Activates bacterial porphyrins
	Deeper dermal penetration than narrow-spectrum light
Photodynamic therapy	Photosensitizing agent localizes to pilosebaceous follicle, producing free radicals upon stimulation with light
Lasers	Decreases *Propionibacterium acnes* counts
	Thermoablation of sebaceous glands
	Photothermolysis of dilated vasculature
Radiofrequency	Thermally inhibits sebaceous glands

7.32.1.2.6 Lasers

Several different lasers have been used for the treatment of acne. The potassium-titanyl-phosphate laser (KTP: Aura, Laserscope) which emits green light at 532 nm, decreases sebum production, but has not shown significant effect on *P. acnes* counts.[54] Other lasers, such as the pulsed dye laser (PDL), have shown mixed benefit in the short-term improvement in inflammatory acne, although efficacy is dependent on laser parameters.[55] One study found no significant improvement in facial acne after two treatments with PDL.[56] In contrast to the KTP and PDL lasers, the 1450 nm diode laser (Smoothbeam, Candela Corp.), and the neodymium:yttrium aluminum garnet (Nd:YAG) laser (Cooltouch CT3) have been shown to penetrate to the level of the sebaceous gland and exert their effect directly on sebocytes.[57] It is unknown whether treatments aimed at reducing the *P. acnes* population or those that focus on sebaceous gland ablation are more effective in treating acne. Prospective studies and head-to-head comparisons are needed to evaluate the true efficacy of these treatment modalities.

7.32.1.2.7 Radiofrequency

Finally, RF has emerged as one of the newest modalities used to treat acne. One such device, Thermacool TC (Thermage, Inc.), utilizes RF energy (6–250 MHz) to heat the dermis, while a cooling device prevents damage to the overlying epidermis.[58] Preliminary studies have reported a significant reduction of inflammatory acne in 82% of 22 patients treated in one study.[59] In addition, improvement of acne scarring was seen. Although initial studies are promising, large-scale RF studies are pending and definitive evidence of the efficacy of this treatment modality is currently not available. A combination of RF and light therapy has been reported to be effective in the treatment of inflammatory acne. Studies exploring the possible combinations and applications are needed to optimize treatment regimens.

7.32.1.3 Conclusion

Acne is one of the most common skin diseases treated by dermatologists, pediatricians, and family medicine physicians. Despite its near-universal prevalence, the pathogenesis of acne is only partially understood. Although numerous topical therapies are available for acne, they are usually only effective in mild to moderate acne. In severe cases, oral antibacterial agents, antihormonal agents, and systemic retinoid therapy can be used to treat the lesions, but these treatments have concerning side effects. Finally, light, laser, and RF therapy have shown promise, but well-controlled studies are not available and further investigation is needed in order to optimize these technologies. Despite the variety of treatments available, acne can still be difficult to treat, and even with treatment acne may result in scarring and irreparable damage to patients' self-esteem and psychological well-being. In light of this, more research is needed to better understand the pathogenesis of acne in order to develop novel, targeted therapeutic modalities. With the increase of antibiotic resistance and many clinicians' reservations of widespread use of systemic retinoids, new treatment options are needed to treat acne effectively.

7.32.2 Psoriasis

Psoriasis is one of the most common chronic inflammatory skin conditions. It affects 1–3% of the population in the US.[60] Age of onset of psoriasis is most commonly during adulthood in the third decade; however early onset during childhood is not uncommon.[61] Psoriasis is inherited as a polygenic trait and one-third of patients with psoriasis have a positive family history of psoriasis.[62] The cause of psoriasis remains unknown; however common trigger factors which induce or exacerbate psoriasis have been identified. These include: physical trauma (the Koebner phenomenon), infection (acute streptococcal infection, human immunodeficiency virus (HIV)), stress, and medications (withdrawal of systemic steroids, lithium, antimalarials, beta-blockers, interferon). Although the exact mechanism of development of psoriasis and psoriatic arthritis (PsA) remain unknown, there have been recent advances in our understanding of the immunopathogenesis of psoriasis. T cells have been found to play an essential role in the pathogenesis of psoriasis.[63] Tumor necrosis factor-α (TNF-α), a proinflammatory cytokine produced by activated T cells, keratinocytes, and dendritic cells, also plays an important role in the pathogenesis of psoriasis.[64] Increased understanding of the immunopathogenesis of psoriasis and PsA has led to the development of new biological therapies.

7.32.2.1 Pathogenesis

Although initially thought to be a primary keratinization disorder, the pathogenesis of psoriasis is now believed to be a multifactorial process, with T cells and adaptive immunity taking a central role in the immunopathogenesis model. In addition to the importance of type 1 T cells (Th1), innate immune defenses within the skin may also contribute to disease progression.

7.32.2.1.1 T-cell response and cytokines

Psoriasis is now widely considered to be a T-cell-mediated inflammatory disease.[65] This is supported by the fact that the majority of inflammatory cells within the dermis in skin lesions are CD4+ and CD8+ T cells.[66] In addition, these cells were shown to express CD45RO, suggesting their effector and memory status.[66] Also, compounds known to target T cells, such as ciclosporin, were found to be effective in the treatment of psoriasis, lending credence to the central role of T cells.[67] The association between psoriasis with certain major histocompatibility complex (MHC) alleles, notably human leukocyte antigen (HLA)-B13, HLA-B17, further supports the pathogenic role of T cells.[68]

Cytokine dysregulation as a result of aberrant T-cell activation may partially account for the psoriatic phenotype. A predominance of Th1 cytokines, including IL2, interferon-gamma (IFN-γ), and TNF-α, are seen in activated T cells within psoriatic lesions.[69,70] The production of proinflammatory cytokines increases the expression of adhesion molecules on both endothelial cells and keratinocytes, allowing for extravasation and migration, respectively, of activated immune effector cells.[71,72] IFN-γ, long considered to play a central role in psoriasis, has been shown to cause epidermal thickening and hyperkeratinization when injected into nonlesional skin of psoriasis patients.[73]

The emergence of TNF-α as a therapeutic target has suggested a more central role for the cytokine in the pathogenesis of psoriasis. TNF-α is increased in lesional skin, and also within the synovium of patients with PsA.[74] Moreover, TNF-α is known to induce expression of adhesion molecules, such as intercellular adhesion molecule (ICAM)-1,[75] and vascular endothelial growth factor (VEGF),[76] mediating extravasation of leukocytes and angiogenesis, respectively. Finally, TNF-α is able to induce numerous other proinflammatory cytokines, including IFN-γ, IL8,[77] a potent chemokine, and also can potentiate its own production through indirect pathways.

7.32.2.1.2 Innate immunity

Although the role of T cells and adaptive immunity has been well established in the pathogenesis of psoriasis, less is known about cellular elements of the innate immune system, namely natural killer (NK) cells, dendritic cells (DC), neutrophils, and keratinocytes. NK cells, a separate lineage of CD3+T cells recognized to be part of the innate immune system, are capable of rapid antigen response and subsequent production of cytokines, including IFN-γ.[78] A large number of NK cells are present in lesional skin versus nonlesional skin or skin from normal patients.[79] Activated DCs, professional antigen-presenting cells, are increased within the dermis of lesional skin in psoriasis patients[80] and may be an important source of TNF-α within the lesion.[77] Neutrophils are present in both a perivascular distribution and also within the epidermis of psoriatic lesions, and are important in the pathogenesis of psoriasis.[81] Moreover, drug-induced agranulocytosis results in remission of psoriasis.[82] Finally, the interaction of NK, DCs, and neutrophils with keratinocytes may result in increased production of proinflammatory cytokines, including TNF-α and IL6. In addition, keratinocytes are capable of producing IL8, a potent chemokine, capable of attracting leukocytes to the lesion.[83]

Noncellular elements of the innate immune system, such as antimicrobial peptides, and TLRs, have gained recent attention in psoriasis. Antimicrobial peptides, such as human beta defensins and cathelicidin (LL-37), are known to be produced by keratinocytes. In psoriatic lesions, there is increased expression of human beta defensin-2, human beta defensin-3, and LL-37, compared to normal skin.[84,85] The presence of increased concentrations of antimicrobial peptides may account for the relative resistance of psoriatic plaques to superinfection. Also, upregulation may indicate an increase in the local innate immune response. TLRs, a group of pattern recognition receptors of the innate immune system capable of recognizing conserved microbial motifs, may play a role in the inflammatory response in psoriasis. TLR1, 2, and 5 are expressed by keratinocytes in psoriasis, and the localization of these TLRs is different in the psoriatic epidermis versus normal epidermis.[86] Miller *et al.* demonstrated in psoriatic epidermis that the expression of TLR5 and TLR9 is more diffuse than in normal skin, indicating a possible role for TLRs in the activation of the innate immune system in psoriasis.[87] Further studies are needed to investigate the role of antimicrobial peptides, and TLRs in the pathogenesis of psoriasis.

7.32.2.2 Clinical Findings

The classic skin findings in psoriasis are well-demarcated, erythematous papules, often coalescing into plaques, with silvery (micaceous) scales. Removal of scales result in punctate hemorrhages, known as Auspitz' sign. Although lesions may occur anywhere on the body, sites of predilection include extensor surfaces of the extremities (e.g., elbows and knees), and the scalp. Fifty percent of patients with psoriasis have fingernail and 35% have toenail involvement,[60] including pitting, onycholysis (spontaneous separation of the nail plate starting at the distal free margin and progressing proximally) subungual hyperkeratosis, and the so-called oil spot, a yellowish-brown discoloration under the nail plate, pathognomonic for psoriasis.

There are several clinical variants of psoriasis, differing in both the nature of the primary lesion and localization. Guttate psoriasis is characterized by numerous disseminated small papules and plaques, with a majority of the lesions occurring on the trunk. This type of psoriasis often occurs in adolescents and young adults and is usually associated with group A streptococcal infection. Often, treatment with antibiotics may improve or clear the psoriasis. Palmoplantar psoriasis, as the name implies, predominantly affects the palms and soles, with typical lesions being scaly, and erythematous, similar to lesions elsewhere in psoriasis. There is also marked keratoderma, thickening, and scaling of the skin. Pustules may also occur, a condition known as palmoplantar pustulosis. Generalized pustular psoriasis (von Zumbusch type) is a severe variant characterized by disseminated sterile pustules which can coalesce into lakes of pus. These patients may suffer from fever, elevated sedimentation rate, leukocytosis, and arthropathy. Von Zumbusch psoriasis is often triggered by withdrawal from systemic corticosteroids, and may also be provoked by pregnancy and infections, although most cases remain idiopathic, with no known trigger identified. It is often recalcitrant to treatment and poses a challenge to both patients and clinicians. Erythrodermic psoriasis, characterized by a scaling, pruritic, inflammatory skin eruption that involves all or almost all of the body surface, usually occurs in the setting of known worsening or unstable psoriasis but may uncommonly be the first presentation of psoriasis. Systemic illness and excessive use of potent corticosteroids are known precipitating factors. Patients are at increased risk for bacterial infections and septicemia, secondary to the compromised skin barrier.

7.32.2.2.1 Psoriatic arthritis

PsA is a seronegative spondyloarthropathy characterized by distal interphalangeal joint involvement, enthesitis (inflammation of the tendon at point of insertion into the bone), and dactylitis (inflammation of the entire digit). The prevalence of PsA within the population is unknown and ranges from 0.04% to 0.1%.[88,89] In patients with psoriasis, the prevalence ranges from 25% to 35%.[90,91] The condition is distinct from rheumatoid arthritis given the absence of rheumatoid factor,[92] the asymmetric distribution, and marked involvement of the axial skeleton. There is a known association of PsA with the major histocompatibility type HLA-B27, and this has led to classification of PsA with other seronegative spondyloarthropathies such as ankylosing spondylitis.

Clinically, PsA presents as an inflammatory arthritis, with patients often noting pain and tenderness over affected joints, and associated swelling. Patients also suffer from tendinitis, which may affect the finger joints, but can also be more disseminated, affecting the plantar fascia and Achilles tendon. On radiologic examination of the digits, the prototypic finding is the 'pencil in a cup,' marginal erosions at the interphalangeal joints. The sacroiliac joint may be involved, presenting with inflammatory back pain and marked morning stiffness. Most PsA occurs in patients with psoriasis vulgaris, although it occurs less frequently in patients with pustular and guttate psoriasis. Patients with PsA have higher incidence of nail lesions than in psoriasis patients without arthropathy. No correlation between severity of psoriasis and the presence of PsA has been observed.

7.32.2.3 Laboratory Findings

Although there are no serologic tests specific to psoriasis, elevated antistreptolysin-O titers can be seen in patients with guttate psoriasis. In addition, there is an association with psoriasis and HIV, and patients with risk factors for HIV who present with psoriasis should have HIV serology. In 50% of patients with psoriasis, an increase of serum uric acid is observed.

Diagnosis is mainly based on clinical presentation and skin biopsy. Histologically, psoriasis is characterized by acanthosis, with elongation of the rete ridges. Hyperkeratosis is commonly seen. Marked parakaratosis, retention of nuclei by keratinocytes in the stratum corneum, is invariably present and is reflective of the increased proliferation rate of keratinocytes. A mixed inflammatory infiltrate can often be seen on biopsy. Pockets of neutrophils within the epidermis, Munro microabscesses, can be identified.

7.32.2.4 Treatment

Historically, effective treatments for psoriasis have been available for almost 100 years. The choice of treatment is often decided by the severity of the psoriasis, taking into account the patient's psoriasis area and severity index (PASI) (**Table 3**), the body surface area involved (**Table 4**), the patient's quality of life, and whether the patient's psoriasis is recalcitrant to treatment. Topical treatments are often used in mild psoriasis, with systemic agents being reserved for moderate to severe cases. The emergence of biologic agents, many of which are designed to interfere with the aberrant immune response at the molecular level, has provided a successful treatment option for severe cases. In addition, light therapy has proven to be effective in a subset of patients. Treatment options for psoriasis will briefly be reviewed.

Table 3 Psoriasis area and severity index (PASI)

Four sections: head (H: 10%), trunk (T: 30%), legs (L: 40%), arms (U: 20%)

Area of involvement for each section: 0–6 (*A*)

0	1	2	3	4	5	6
0%	<10%	10–29%	30–49%	50–69%	70–89%	>89%

Erythema, infiltration, desquamation: 0–4 (*E*, *I*, *D*)

0	1	2	3	4
None	Minimal	Moderate	Severe	Maximum

PASI $= 0.1\ (E_H + I_H + D_H) \times A_H + 0.3\ (E_T + I_T + D_T) \times A_T + 0.4\ (E_L + I_L + D_L) \times A_L + 0.2\ (E_U + I_U + D_U) \times A_U$

Table 4 Body surface area

Part of the body	%
Head	9
Chest and abdomen	18
Back and buttocks	18
Arms (including axilla)	18
Hand	1[a]
Legs	36
Genitals	1

[a] Included in the surface area measure for arms.

7.32.2.4.1 Topical treatments

Topical treatment for psoriasis began over 100 years ago with the use of anthralin (dithranol). Anthralin has moderate efficacy, but its use has waned because of staining caused by the compound, and skin irritation. Although the mechanism of action is not completely understood, in vitro studies have demonstrated that anthralin inhibits monocyte secretion of IL6, IL8, and TNF-α.[93] Anthralin was also found to activate NFκB in a murine keratinocyte model,[94] and decreased expression of epidermal growth factor receptors in vitro.[95] Coal tar has also been used for many years, but similar to anthralin, its use is limited by staining and an offensive odor. Although the mechanism of action is largely unknown, it appears to have antiproliferative and antiinflammatory effects, and is clinically efficacious.[96] Coal tar is effective alone in the treatment of mild to moderate psoriasis, but is also useful as an adjunct treatment to UVB in the treatment of refractory cases.

Keratolytic agents such as salicylic acid (SA) and urea also have an established role in the treatment of psoriasis. SA is useful as an adjunctive treatment for psoriasis. For example, when combined with topical corticosteroids, SA enhances skin penetration. SA and urea are also useful in the treatment of thick plaques with hyperkeratotic scales[97] and are effective in the treatment of palmoplantar psoriasis and keratoderma.

Topical corticosteroids, in various potencies and formulations, remain the most popular treatment in psoriasis. These range in strength from over-the-counter 1% hydrocortisone (low potency), to more robust formulations such as 0.1% triamcinolone (mid potency) and 0.05% fluocinonide (high potency), and the ultra-high potency 0.05% clobetasol and 0.05% betamethasone. Different delivery vehicles are also available for use on different body sites and time of application: foams, solutions, and gels can be used for scalp treatment, creams are best for daytime use, and ointments, which are often more effective but messy, for nighttime use. Although steroids are an effective treatment for localized lesions, side effects include cutaneous atrophy, telangiectasias, striae, and, if large quanitities are used, suppression of the hypothalamic–pituitary–adrenal axis. Care should be taken to use lower potency steroids on the face and intertriginous areas that are more prone to steroid side effects.

Vitamin D analogs such as calcipotriene have been available for the treatment of psoriasis since the 1990s. Calcipotriene is a vitamin D_3 analog that binds the vitamin D receptor, and has been shown to inhibit hyperproliferation and abberant differentiation of keratinocytes.[98–100] Calcipotriene is an effective nonsteroidal therapy in psoriasis.

In addition, when combined with steroids, it is an effective adjunct in the treatment of psoriasis.[101] Generally, vitamin D analogs are well tolerated, with irritation being the most common side effect. Caution should be taken, however, because large quantities of calcipotriene in excess of 120 g may result in hypercalcemia.[102]

Tazarotene, a topical retinoid available in 0.05% and 0.1% gels and cream formulations, is effective in the treatment of plaque psoriasis both as monotherapy and in combination with other agents.[103] Tazarotene binds RAR and regulates the transcription of genes, presumably involved in the the pathogenesis of psoriasis.[104,105] Generally, adverse reactions are limited to local cutaneous irritation, similar to other topical retinoids. In combination with corticosteroids, tazarotene has been shown to increase the efficacy of coritcosteroid therapy,[106–108] decrease poststeroid treatment flare, and also reduce cutaneous atrophy.[109] Tazarotene has also shown promise with the concomitant use of UVB, showing significant improvement in plaque and scaling, as well as reducing median UVB exposure versus UVB alone.[110,111]

7.32.2.4.2 Phototherapy

Light therapy dates back to the 1920s when William Goeckerman used UVB light with topical application of tar to treat psoriasis (now known as the Goeckerman regimen).[112] UVB is thought to inhibit DNA synthesis and keratinocyte hyperproliferation. This leads to local immunosuppressive effects and the production of antiinflammatory cytokines.[113] Broadband UVB is safe and effective for psoriasis, but requires at least three treatments per week for several months, making patient compliance difficult. The most effective wavelengths for the treatment of psoriasis are between 300 and 313 nm, and this led to the development and use of narrow-band UVB in the mid-1980s for the treatment of psoriasis.[114] One study found that 63–80% of patients treated with narrow-band UVB will achieve clearance,[113] but that up to 20 treatments are required to see a 50% improvement.[115] Short-term side effects of UVB include erythema, which, if severe enough, may be treated with topical corticosteroids. Long-term side effects include the potential increased risk for skin cancer, although the incidence of skin cancer in patients treated with either broadband or narrow-band UVB has not been accurately quantified.[116] More studies are needed to evaluate the efficacy and the long-term safety of UVB therapy.

Photochemotherapy, the use of a photosensitizing agent (psoralen) in conjunction with exposure to UVA light, is currently approved by the FDA for the treatment of severe psoriasis. The use of systemic or topical psoralens with long-wavelength UVA irradiation (PUVA) is effective, with partial or complete remission seen in 70–90% of patients.[117] Remission rates up to 40% have been reported in patients 1 year posttreatment, even after a single course.[118] However, like other forms of light therapy, PUVA carries an increased risk of skin cancer, with premature aging of skin, and an increased incidence of actinic keratoses.[119] Prospective studies have demonstrated a clear relationship with PUVA and the increased risk of skin cancer, with an approximately 11-fold increase in risk of cutaneous squamous cell carcinoma 10 years posttreatment.[120] There may also be an increased risk of basal cell carcinoma[120] and melanoma,[121] although the evidence is less conclusive.

The 308 nm excimer laser has had encouraging results in the treatment of psoriasis.[122] A recent multicenter clinical trial of 124 patients reported significant improvement of mild to moderate psoriasis in adults, with high patient satisfaction rates.[123] Advantages of the excimer laser over narrow-band UVB include the ability to deliver focused UVB to affected skin, sparing normal skin, and preventing some of the adverse effects of conventional light therapy. In addition, because a higher and more focused dose of UVB can be delivered per treatment, patients treated with the excimer laser have quicker response, and require less treatment for remission.[124]

7.32.2.4.3 Systemic therapy

Traditionally, systemic therapy has been used in patients with a large body surface area of involvement, and also in cases that have been refractory to more conservative therapy. For years, methotrexate (MTX), ciclosporin, and oral retinoids were the mainstay in the systemic treatment of psoriasis. With the advent of the newer biologic agents, the systemic treatment of psoriasis has changed significantly, offering more options for those who fail conventional therapies.

MTX has been used for the treatment of psoriasis since the 1960s, and remains one of the first lines of systemic therapy in many patients. Despite its widespread use, the mechanism of action has not clearly been elucidated. Much of the beneficial effect of MTX is thought to stem from its role as an immunosuppressant.[125] Unfortunately, there are few well-designed, randomized controlled trials evaluating the efficacy of MTX in the treatment of psoriasis. One literature review, based mainly on case reports and retrospective analysis, reported that MTX can reduce disease severity by at least half in 75% of patients.[126] Serious side effects are associated with the medication, including hepatotoxicity and bone marrow suppression. Clinicians should routinely monitor the patient's complete blood count and liver function tests to survey for adverse drug reaction. In addition, liver biopsy is done when total MTX dose reaches 1.5 g.

Ciclosporin was inadvertently used for the treatment of psoriasis in 1979. The agent, which is a potent immunosuppressive, helped shed light on the importance of the immune system and T cells in the pathogenesis of psoriasis. Clinical trials have demonstrated that ciclosporin is effective in inducing remission when used daily for a short period, and also for maintaining remission.[126] A recent randomized controlled trial comparing ciclosporin with MTX found no statistically significant difference in efficacy and duration of remission.[127] Nephrotoxicity, hypertension, susceptibility to infection secondary to immunosuppression, and increased risk of malignancy are the main side effects of ciclosporin therapy. Routine laboratory studies, including serum creatinine, are recommended biweekly during the first 2 months of treatment and monthly thereafter to monitor for nephrotoxicity.

Oral retinoids were introduced approximately 20 years ago for the treatment of psoriasis. Acitretin, an oral retinoid that has been FDA-approved for the treatment of psoriasis, exerts its effect through the RAR. Side effects are typical of systemic retinoids and include cheilitis, hypertriglyceridemia, muscle aches and pains, hair loss, and thinning of the nail plates. One of the most serious side effects is its teratogenic effects on the fetus, and acitretin should be used with caution in females of child-bearing age. Acitretin is also effective when combined with phototherapy. Low-dose acitretin in combination with either UVB or PUVA is highly effective in inducing remission with minimal side effects.[128]

7.32.2.4.4 Biologic agents

Advances in understanding the role that cytokines and the immune system play in the pathogenesis of psoriasis have led to the development of medications aimed at specifically reducing inflammation in psoriasis (**Table 5**). Alefacept was the first biologic agent approved for the treatment of chronic plaque psoriasis by the FDA. Alefacept is a fusion protein consisting of lymphocyte function-associated antigen 3 (LFA-3), fused with the Fc portion of human immunoglobulin (Ig)G1. The soluble LFA-3 binds CD2 which is required for the costimulation of T cells, preventing effective T-cell activation.[129] The drug also has the advantage of inducing apoptosis in T cells, via interaction with NK cells and macrophages through its Fc portion.[129] A phase III clinical trial investigating the efficacy of intramuscular alefacept demonstrated that 56% and 70% of patients showed greater than 50% improvement in PASI score after one or two courses of treatment, respectively. In addition, the effects of alefacept were found to be long-lasting, without evidence of rebound or flare posttreatment.[130] Generally, alefacept is well tolerated, with injection site reactions and chills postinjection being the two most common complaints. No increase in infection or malignancy has been observed, although additional studies are needed to assess fully the true effects of long-term immunosuppression with alefacept.

Etanercept is a fully soluble, recombinant TNF-α receptor administered subcutaneously, which binds TNF and prevents physiologic signaling through the cellular membrane receptor. Etanercept is approved by the FDA for the treatment of psoriasis and PsA. Multiple clinical trials have shown the efficacy of etanercept, with one phase II clinical trial showing a mean percentage improvement in PASI scores of 67% after 24 weeks of treatment.[131] In addition to improvement in PASI scores, a marked increase in quality of life was seen after 24 weeks of treatment.[131] A larger phase III clinical trial demonstrated the efficacy of twice a week etanercept, with significant differences in the mean percentage improvement of PASI score evident at week 2 of treatment.[132] Etanercept is considered safe, with injection site reactions being the most cited complaint. Other rarer but more serious side effects include opportunistic infections and perhaps an increased risk in lymphoma.[133]

Table 5 Summary of biologic agents approved for the treatment of psoriasis

	Structure	*Mechanism of action*	*Route*	*Adverse effects*
Alefacept	LFA-IgG1 fusion protein	Binds CD2, preventing T-cell activation	IM	Injection site reaction Chills postinjection
Etanercept	Soluble, recombinant TNF-α receptor	Binds TNF-α preventing physiologic signaling	SC	Injection site reaction Increased opportunistic infections? Increased risk of lymphoma
Efalizumab	Monoclonal antibody to CD11a	Binds CD11a, preventing migration of T cells	SC	Flu-like symptoms Thrombocytopenia
Infliximab	Monoclonal antibody to TNF-α	Binds TNF-α preventing physiologic signaling	IV	Infusion reactions Increased opportunistic infections Lupus-like syndrome

Efalizumab, a monoclonal antibody, targets CD11a on the surface of T cells.[134] CD11a is the alpha subunit of LFA-1, involved in the activation and migration of T cells into peripheral tissues.[135] Four phase III clinical trials have characterized the efficacy of efalizumab administered subcutaneously, measuring the PASI-75, the dose needed for 75% improvement in the PASI score. In all four trials, 22–37% of patients achieved a PASI-75, a statistically significant improvement when compared to placebo.[136,137] In addition to the improvement in PASI, quality of life, as measured by the Dermatology Life Quality Index (DLQI), was also significantly improved in all four trials. Efalizumab is generally well tolerated, with flu-like symptoms being the most common complaint. No significant risk of infection while on therapy or malignancy has been associated with the drug. Several patients developed thrombocytopenia during therapy, therefore checking platelets monthly for the first 3 months of treatment is recommended.

Infliximab is a recombinant monoclonal antibody that targets TNF-α. By binding TNF-α, in both membrane-bound and soluble forms, infliximab prevents TNF from binding to its natural cellular receptor, decreasing subsequent inflammation[138] and inducing cell-mediated cytotoxicity in cells expressing TNF-α on their surface.[139] Currently, infliximab is approved for the treatment of both psoriasis and PsA. Results of randomized clinical trials of infliximab for the treatment of psoriasis have demonstrated good efficacy, with one phase II trial showing 88% of patients achieving PASI-75 compared to 5% in the placebo group.[140] Adverse reactions include infusion reactions in 16% of patients, infections, and lupus-like syndrome. The long-term risk of malignancy in patients treated with infliximab is unknown.

7.32.3 Atopic Dermatitis

Atopic dermatitis is an inherited chronic inflammatory condition of the skin characterized by the hallmark symptom of pruritus. It is often associated with a personal or family history of the atopic triad of atopic dermatitis, allergic rhinitis, and/or asthma. The mode of inheritance of atopic dermatitis is not clear, but it appears to be inherited as a polygenic trait.[141] Atopic dermatitis affects over 10% of the overall population, with increased prevalence in industrialized countries.[142] Most patients (50–60%) are diagnosed with atopic dermatitis within the first year of life. In more than 85% of patients, the disease is present by 5 years of age.[143] Prevalence of disease decreases with age; however, recurrent or persistent disease is commonly seen.

7.32.3.1 Pathogenesis

Although the pathogenesis of atopic dermatitis remains unclear, it appears to be an interaction between immune dysregulation and epidermal barrier dysfunction. An abnormal local immune response, by CD4+ (Th2) helper T lymphocytes and their interaction with dermal dendritic cells and epidermal Langerhans cells, seems to play an important role in the immunopathogenesis. The reactivity may be due to presentation of self-peptides, presentation of processed foreign antigen, or binding of microbial superantigen.[144] Proposed mechanisms of immunopathogenesis in addition to this excessive T-cell activation include abnormal antigen-presenting Langerhans cells, increased IgE synthesis, excessive phosphodiesterase dysregulation, and monocyte hyperreactivity.[143]

In the acute phase of disease, there is a prominent production of cytokines IL4 and IL13, indicating the role of Th2 cells in initiation of disease.[145] It appears that allergens or antigens such as staphylococcal superantigen processed by Langerhans cells induce activation of the Th2 cells during the acute phase. Consequently, and most likely due to IL10 production by keratinocytes and macrophages, there is a differentiation of T-helper precursor cells to Th2 cells. Th2 cell production of IL13, IL4, and IL5 induces B-cell production of IgE, which promotes mast cell degranulation.

However, chronic inflammation in atopic disease is associated with increased levels of expression of cytokines IFN-γ and IL5, indicating the predominant role of Th1 cells in this stage of disease.[146] The chronic inflammatory response is dominated by a Th1 type of response driven by the infiltration of IL12 expressing eosinophils and macrophages.

IgE is involved in the pathogenesis of atopic dermatitis by several mechanisms.[147] Elevated levels of circulating IgE are seen in up to 80% of patients with atopic dermatitis.[142] IgE autoreactivity may play a role in the pathogenesis of atopic dermatitis. Most patients with atopic dermatitis have circulating IgE antibodies directed against human proteins.[148] IgE is also found on the cell surfaces of Langerhans cells and macrophages within the inflammatory infiltrate of skin lesions in atopic dermatitis.[149] However, over 20% of patients with atopic dermatitis have normal levels of IgE, suggesting that IgE is not a requirement in the production of clinical disease.

The role of food and inhaled allergens in atopic dermatitis is an area of great interest. It is estimated that 10–20% of those with atopic dermatitis have food allergies that are clinically relevant.[150] The most common reaction to a food allergen is an urticarial or morbilliform eruption, which has an onset within 30–120 min of food consumption and a duration of 30–120 min. Allergies to eggs, peanuts, milk, wheat, fish, and soy comprise 90% of food allergies in children, and most children are allergic to one or two foods.[151] While food allergen avoidance may be helpful in preventing urticarial eruptions and anaphylactic reactions, there is still inconclusive evidence that food allergens play a role in the development of eczematous lesions. The role of inhaled allergens such as plant allergens and housedust mites remains unclear as well.

Since over 90% of patients with atopic dermatitis are colonized with *Staphylococcus aureus* in skin lesions,[152] its potential role in dermatitis has been of widespread interest. More than half of patients with atopic dermatitis have *S. aureus* isolates that produce superantigens such as staphylococcal enterotoxins (SE) A, SEB, SEC, SED, SEE, toxic shock syndrome toxin-1, and exfoliative toxin A and B.[153] Most patients with atopic dermatitis make IgE antibodies directed against staphylococcal toxins.[154] Superantigens stimulate marked activation of T cells and macrophages and augment the initial inflammatory response in the skin of patients with atopic dermatitis.[155]

7.32.3.2 Clinical Presentation

Atopic dermatitis is usually the first manifestation of the atopic triad of eczema, allergic rhinitis, and asthma. It is a chronic disease characterized by periods of exacerbation and remission. Clinical disease has been characterized by three phases: (1) the infantile phase up to 2 years of age; (2) the childhood phase from 2 years of age to puberty; and (3) the adult phase from puberty onward.

7.32.3.2.1 Infantile phase

Infantile atopic dermatitis usually develops between 2 and 6 months of age. Sixty percent develop clinical disease by the age of 1.[156] During this phase, common sites of involvement include the face, scalp, and extensor surfaces of the lower extremities and may later involve the antecubital and popliteal fossae. The diaper area is usually spared. Characteristic lesions are ill-defined, scaly, erythematous papules and patches with or without weeping and crusting. Infants affected by atopic dermatitis suffer from pruritus, which may disrupt their sleeping and eating patterns. Generalized xerosis often accompanies the inflammatory lesions.

7.32.3.2.2 Childhood phase

During the childhood phase, lesions are distributed primarily on flexural surfaces such as the antecubital and popliteal fossae, neck, flexural wrists and ankles, and umbilicus. There is less facial involvement. Lesions are ill-defined, erythematous, scaly papules, patches and plaques with or without crusting and excoriations. Due to frequent scratching, fingernails may appear smooth and shiny with a buffed appearance. If facial pruritus is prominent, rubbing may also produce sparse eyebrows with broken hairs.

7.32.3.2.3 Adult phase

Atopic dermatitis spontaneously resolves in most after the age of 20, but some have persistent disease. During the adult phase, after puberty, clinical lesions are usually diffuse, involving the face, neck, trunk, and extremities. Characteristic lesions include erythematous, scaly, lichenified patches and plaques on a background of xerosis. The face of adults with atopic dermatitis often has a characteristic central pallor.

7.32.3.3 Associated Clinical Findings

Ichthyosis vulgaris is characterized by generalized xerosis and mild generalized hyperkeratosis. The lower legs are primarily affected. Xerosis is also a result of physiologic changes in atopic skin, resulting in increased transepidermal water loss.[157] Keratosis pilaris, perifollicular hyperkeratosis that is most prominent on the posterior surface of the upper extremities, but may involve thighs, buttocks, and face, is also commonly seen in those with atopic dermatitis. Hyperlinear palms and soles are seen in one-third to one-half of patients with atopic dermatitis and are often seen in those with ichthyosis. These features tend to be persistent despite improvement of atopic disease. A crease under the lower eyelid, the Dennie–Morgan fold, is found in patients with atopic dermatitis. It may be present at birth or develop in infancy. This fold may be persistent throughout life.

There are characteristic eye findings in patients with atopic dermatitis, including keratoconjunctivitis, photophobia, and eyelid pruritus.[158] Cataracts have been seen in up to 13–21% of patients with severe atopic dermatitis.[159,160] Most are posterior subcapsular. Pityriasis alba is a common condition which occurs in the presence or absence of atopic dermatitis. It is characterized by ill-defined hypopigmented patches on the cheeks and upper trunk of children and becomes more noticeable after sun exposure (e.g., at the end of summer). It most likely represents postinflammatory hypopigmentation following a subclinical dermatitis. The hypopigmentation resolves spontaneously. Symptomatic treatment with a low-potency topical steroid may be given. Otherwise emollients and sun protection are recommended.

Dyshidrotic eczema involving the palms, soles, and lateral digits commonly affects those with atopic dermatitis, but may be seen independently as well. Clinically it is characterized by intensely pruritic, deep-seated vesicles resembling tapioca pudding, erythema, scaling, and crusting.

7.32.3.4 Complications

Secondary bacterial and viral infections commonly occur in atopic dermatitis. This may be related to decreased cell-mediated immunity and reduced chemotaxis.

Patients with atopic dermatitis have higher colonization rates of *S. aureus* on both unaffected and affected skin surfaces.[161] Interestingly, despite these high colonization rates and disruptions in skin integrity due to erosions and excoriations, severe systemic infections are uncommon. Secondary impetiginization often occurs, especially in infants and children, and is characterized by the development of oozing and honey-colored crusting. Treatment with topical and/or systemic antibiotics is recommended.

Eczema herpeticum, or Kaposi's varicelliform eruption, is a secondary infection of atopic dermatitis with herpes simplex virus. Following exposure to the virus and an incubation period of 5–10 days, an eruption of pruritic vesicles develops within areas of preexisting dermatitis. The vesicles rupture, leaving painful hemorrhagic crusts and punched-out erosions. It is believed that children with atopic dermatitis may be more susceptible to herpes simplex infection due to decreased numbers of NK cell and IL2 receptors.[162] Systemic antiviral therapy for 7–10 days is recommended. Treatment with Burow's solution compresses is helpful in weeping, eroded areas. Coverage with systemic antibiotics for secondary bacterial infection may be needed.

Other cutaneous viral infections, verruca and molluscum contagiosum, may be more common or more widespread in the setting of atopic dermatitis.[163,164] Molluscum contagiosum infections tend to be widespread and more difficult to treat in patients with atopic dermatitis. Additionally, a pruritic eczematous dermatitis often develops around active lesions, often prior to involution of the molluscum lesion.

7.32.3.5 Histopathology

The histopathologic findings of atopic dermatitis are nonspecific, nondiagnostic, and dependent upon the type of clinical lesion present. In acute dermatitis, there is marked spongiosis (intercellular edema) with possible formation of intraepidermal vesicles and/or bullae. There is an associated perivascular lymphocytic infiltrate in the upper dermis. In subacute lesions, there is spongiosis and vesiculation, but to a lesser degree than is present in acute dermatitis. Acanthosis with elongation of the rete ridges and patchy parakeratosis is present as well. In chronic dermatitis, there is hyperkeratosis, acanthosis with a psoriasiform pattern, and minimal spongiosis. Within the epidermis are increased numbers of IgE-bearing Langerhans cells.[165] Within the dermis is a mononuclear infiltrate with many macrophages.

7.32.3.6 Treatment

Management of atopic dermatitis due to its chronic, relapsing nature requires good communication between doctor and patient. Those affected by atopic dermatitis have increased transepidermal water loss and impaired skin barrier function due to reduced ceramide levels in the stratum corneum.[166] Therefore skin hydration is essential in obtaining and maintaining control of atopic disease. Current recommendations for bathing are once- or twice-daily bathing for 5–15 min in lukewarm water followed by application of an emollient over damp skin within 3 min of getting out of the bath.[167] Emollients which are in the formulation of a thick cream or ointment maintain skin hydration better than lotions. White petrolatum and mineral oil are inexpensive and frequently used emollients. Use of emollients containing urea, lactic acid, glycolic acid, and propylene glycol may cause stinging and irritation to inflamed skin and should be avoided in such areas. Such ingredients may, however, be helpful in dry, lichenified areas of involvement.

Topical corticosteroids have been the mainstay of therapy in atopic dermatitis. The potency of the topical corticosteroid should be selected by what is safe and effective for the age and body surface area affected in the patient. As a general rule, low-potency steroids are used for areas with thin skin, such as the face, neck, and intertriginous areas. Medium-potency steroids are used for the trunk and extremities. High-potency steroids may be needed in areas of thick skin such as the palms and soles or in patients with long-standing disease with more lichenification. Atopic patients may prefer to use ointments during acute stages of their disease because creams may cause stinging when applied to inflamed skin. However, ointments are generally less cosmetically elegant for patients, especially in hot, humid climates. Creams, lotions, or gels may be preferable to some patients, especially during the maintenance phase.

Although there is widespread concern about using topical steroids, they can be used safely and effectively with proper instruction.[168] This is often particularly concerning to parents of infants and children with atopic dermatitis. Use of low- to medium-potency topical corticosteroids over many years in children and adolescents with moderate to severe disease rarely causes hypothalamic–pituitary axis suppression.[169] Use of moderate-potency steroids for prolonged periods of time has also demonstrated decreased growth velocity in prepubertal children.[170]

Local side effects from topical corticosteroid use may occur, particularly with the inappropriate use of high- and ultrapotent corticosteroids in areas such as the face, neck, and intertiginous areas or of prolonged duration. These local side effects include atrophy, striae, perioral dermatitis, and local hypertrichosis.

Systemic corticosteroid use is reserved for the short-term treatment of severe, acute flares. Systemic corticosteroids are not indicated for long-term therapy of disease due to multiple systemic adverse effects with chronic use. For severe, chronic disease which is uncontrollable with topical therapy alone, alternative adjuvant therapy such as phototherapy or other systemic therapies may be considered.

Topical macrolides, tacrolimus and pimecrolimus, were recent introductions to our therapeutic regime for atopic dermatitis. They act by competitive inhibition of calcineurin, an intracellular phosphatase necessary for the transcription of multiple T-cell inflammatory cytokines.[142]

Tacrolimus (FK 506) systemically was developed to prevent transplant rejection. Topical formulations of tacrolimus in concentrations of 0.03% and 0.1% are commercially available. Multicenter randomized, double-blind safety and efficacy studies in adults and in children down to 2 years of age have demonstrated efficacy and safety as compared to vehicle.[171–173] Twice-daily application of tacrolimus ointment is recommended. Local side effects include mild to moderate but transient pruritus and burning and erythema at application sites. No tachyphylaxis, contact allergy, photosensitivity, pigmentary changes, atrophy, or increased incidence of infection has been seen.

Pimecrolimus is an ascomycin derivative which was developed for the treatment of inflammatory skin diseases. Twice-daily application of pimecrolimus cream has also been shown to be safe and effective in the treatment of mild to moderate atopic dermatitis in children and adults.[174,175] Similar to tacrolimus, no tachyphylaxis, contact allergy, photosensitivity, atrophy, or increased risk of infection was seen. Long-term studies to evaluate the safety and risk of development of malignancy in children and adults using these topical macrolides are necessary and underway to determine the long-term risks.

Oral antihistamines have been used for the treatment of pruritus associated with atopic dermatitis. Their efficacy in relieving pruritus is unclear. However, sedating antihistamines may be useful in children and adults to promote sleep which is often disrupted by pruritus associated with flares of dermatitis.

Use of oral antibiotics should be instituted for any signs or symptoms of secondary impetiginization and may be considered for use in an acute flare. Most commonly used antibiotics are those which cover *S. aureus*, such as dicloxacillin, cephalexin, or erythromycin. Obtaining a bacterial culture with antibiotic sensitivities may be indicated for unresponsive cases to rule out antibiotic resistance. For patients with recurrent infections, nasal swabs for bacterial culture should be done to rule out nasal staphylococcal carriage. Nasal carriage may be treated with intranasal mupirocin to the anterior nares twice daily for 10 days. Treatment with oral antiviral agents such as aciclovir, famciclovir, or valaciclovir should be given for secondary infection with herpes simplex.

Phototherapy may be useful in treating patients with moderate to severe atopic dermatitis that is recalcitrant to topical therapy. UVB $\pm$ UVA, UVA_1, narrow-band UVB and PUVA have all been used in the treatment of atopic dermatitis. Although the exact mechanism of action is unclear, UV therapy affects the local immune system in the skin and has been shown to inhibit antigen presentation by Langerhans cells and T-cell activation, and alters cytokine production by keratinocytes.[176] Initiation of phototherapy is not recommended during an acute flare.

In severe cases of atopic dermatitis which is recalcitrant to other therapies, systemic immuosuppressive therapy with agents such as ciclosporin, azathioprine, MTX, and IFN-γ may be considered. However, systemic risks and side effects must be strongly considered and monitored when using these medications.

References

1. Knutson, D. D. *J. Invest. Dermatol.* **1974**, *62*, 288–307.
2. Knaggs, H. E.; Holland, D. B.; Morris, C.; Wood, E. J.; Cunliffe, W. J. *J. Invest. Dermatol.* **1994**, *102*, 89–92.
3. Downing, D. T.; Stewart, M. E.; Wertz, P. W.; Strauss, J. S. *J. Am. Acad. Dermatol.* **1986**, *14*, 221–225.
4. Guy, R.; Green, M. R.; Kealey, T. *J. Invest. Dermatol.* **1996**, *106*, 176–182.
5. Stewart, M. E.; Steele, W. A.; Downing, D. T. *J. Invest. Dermatol.* **1989**, *92*, 371–378.
6. Harris, H. H.; Downing, D. T.; Stewart, M. E.; Strauss, J. S. *J. Am. Acad. Dermatol.* **1983**, *8*, 200–203.
7. Shaw, J. C. *Clin. Dermatol. Am. J. Clin. Dermatol.* **2002**, *3*, 571–578.
8. Levell, M. J.; Cawood, M. L.; Burke, B.; Cunliffe, W. J. *Br. J. Dermatol.* **1989**, *120*, 649–654.

9. Lookingbill, D. P.; Horton, R.; Demers, L. M.; Egan, N.; Marks, J. G., Jr.; Santen, R. J. *J. Am. Acad. Dermatol.* **1985**, *12*, 481–487.
10. Chen, W.; Thiboutot, D.; Zouboulis, C. C. *J. Invest. Dermatol.* **2002**, *119*, 992–1007.
11. Thiboutot, D. *J. Am. Acad. Dermatol.* **2002**, *47*, 109–117.
12. Strauss, J. S.; Thiboutot, D. M. Disease of the Sebaceous Glands. In *Fitzpatrick's Dermatology in General Medicine*; Freedberg, I. M., Eisen, A. Z., Wolff, K., Eds.; McGraw-Hill Health Professions Division: New York, 1999, pp 769–784.
13. Thiboutot, D.; Harris, G.; Iles, V.; Cimis, G.; Gilliland, K.; Hagari, S. *J. Invest. Dermatol.* **1995**, *105*, 209–214.
14. Sansone, G.; Reisner, R. M. *J. Invest. Dermatol.* **1971**, *56*, 366–372.
15. Toyoda, M.; Nakamura, M.; Morohashi, M. *Eur. J. Dermatol.* **2002**, *12*, 422–427.
16. Toyoda, M.; Nakamura, M.; Makino, T. *Exp. Dermatol.* **2002**, *11*, 241–247.
17. Bohm, M.; Schiller, M.; Stander, S.; Seltmann, H.; Li, Z.; Brzoska, T.; Metze, D.; Schioth, H. B.; Skottner, A.; Seiffert, K. et al. *J. Invest. Dermatol.* **2002**, *118*, 533–539.
18. Chen, W.; Kelly, M. A.; Opitz-Araya, X.; Thomas, R. E.; Low, M. J.; Cone, R. D. *Cell* **1997**, *12*, 789–798.
19. Zouboulis, C. C.; Seltmann, H.; Hiroi, N.; Chen, W.; Young, M.; Oeff, M.; Scherbaum, W. A.; Orfanos, C. E.; McCann, S. M.; Bornstein, S. R. *Proc. Natl. Acad. Sci. USA* **2002**, *99*, 7148–7153.
20. Issemann, I.; Prince, R. A.; Tugwood, J. D.; Green, S. *J. Mol. Endocrinol.* **1993**, *11*, 37–47.
21. Rosenfield, R. L.; Deplewski, D; Greene, M. E. *Horm. Res.* **2000**, *54*, 269–274.
22. Rosenfield, R. L.; Kentsis, A.; Deplewski, D.; Ciletti, N. *J. Invest. Dermatol.* **1999**, *112*, 226–232.
23. Leyden, J. J.; McGinley, K. J.; Mills, O. H.; Kligman, A. M. *J. Invest. Dermatol.* **1975**, *65*, 382–384.
24. Oberemok, S. S.; Shalita, A. R. *Cutis* **2002**, *70*, 101–105.
25. Chen, Q.; Koga, T.; Uchi, H.; Hara, H.; Terao, H.; Moroi, Y.; Urabe, K.; Furue, M. *J. Dermatol. Sci.* **2002**, *29*, 97–103.
26. Kim, J.; Ochoa, M. T.; Krutzik, S. R.; Takeuchi, O.; Uematsu, S.; Legaspi, A. J.; Brightbill, H. D.; Holland, D.; Cunliffe, W. J.; Akira, S. et al. *J. Immunol.* **2002**, *169*, 1535–1541.
27. Jain, A.; Sangal, L.; Basal, E.; Kaushal, G. P.; Agarwal, S. K. *Dermatol. Online J.* **2002**, *8*, 2.
28. Hegemann, L.; Toso, S. M.; Kitay, K.; Webster, G. F. *Br. J. Dermatol.* **1994**, *130*, 569–575.
29. Gans, E. H.; Kligman, A. M. *J. Dermatol. Treat.* **2002**, *13*, 107–110.
30. Cunliffe, W. J.; Holland, K. T.; Bojar, R.; Levy, S. F. *Clin. Ther.* **2002**, *24*, 1117–1133.
31. Leyden, J. J. *Cutis* **2002**, *69*, 475–480.
32. Coates, P.; Vyakrnam, S.; Eady, E. A.; Jones, C. E.; Cove, J. H.; Cunliffe, W. J. *Br. J. Dermatol.* **2002**, *146*, 840–848.
33. Velicer, C. M.; Heckbert, S. R.; Lampe, J. W.; Potter, J. D.; Robertson, C. A.; Taplin, S. H. *JAMA* **2004**, *291*, 827–835.
34. Akamatsu, H.; Tomita, T.; Horio, T. *Dermatology* **2002**, *204*, 277–280.
35. Nazzaro-Porro, M.; Passi, S.; Picardo, M.; Breathnach, A.; Clayton, R.; Zina, G. *Br. J. Dermatol.* **1983**, *109*, 45–48.
36. Bladon, P. T.; Burke, B. M.; Cunliffe, W. J.; Forster, R. A.; Holland, K. T.; King, K. *Br. J. Dermatol.* **1986**, *114*, 493–499.
37. Ioannides, D.; Rigopoulos, D.; Katsambas, A. *Br. J. Dermatol.* **2002**, *147*, 523–527.
38. Cunliffe, W. J.; Danby, F. W.; Dunlap, F.; Gold, M. H.; Gratton, D.; Greenspan, A. *Eur. J. Dermatol.* **2002**, *12*, 350–354.
39. Leyden, J. J.; Tanghetti, E. A.; Miller, B.; Ung, M.; Berson, D.; Lee, J. *Cutis* **2002**, *69*, 12–19.
40. Liu, P. T.; Krutzik, S. R.; Kim, J.; Modlin, R. L. *J. Immunol.* **2005**, *174*, 2467–2470.
41. Ng, C. H.; Tam, M. M.; Celi, E.; Tate, B.; Schweitzer, I. *Aust. J. Dermatol.* **2002**, *43*, 262–268.
42. Chia, C. Y.; Lane, W.; Chibnall, J.; Allen, A.; Siegfried, E. *Arch. Dermatol.* **2005**, *141*, 557–560.
43. Leyden, J.; Shalita, A.; Hordinsky, M.; Swinyer, L.; Stanczyk, F. Z.; Weber, M. E. *J. Am. Acad. Dermatol.* **2002**, *47*, 399–409.
44. Shaw, J. C.; White, L. E. *J. Cutan. Med. Surg.* **2002**, *6*, 541–545.
45. Yemisci, A.; Gorgulu, A.; Piskin, S. J. *Eur. Acad. Dermatol. Venereol.* **2005**, *19*, 163–166.
46. Huang, X.; Nakanishi, K.; Berova, N. *Chirality* **2000**, *12*, 237–255.
47. Meffert, H.; Scherf, H. P.; Sonnichsen, N. *Dermatol. Monatsschr.* **1987**, *173*, 678–679.
48. Arakane, K.; Ryu, A.; Hayashi, C.; Masunaga, T.; Shinmoto, K.; Mashiko, S.; Nagano, T.; Hirobe, M. *Biochem. Biophys. Res. Commun.* **1996**, *25*, 578–582.
49. Sigurdsson, V.; Knulst, A. C.; van Weelden, H. *Dermatology* **1997**, *194*, 256–260.
50. Konig, K.; Teschke, M.; Sigusch, B.; Glockmann, E.; Eick, S.; Pfister, W. *Cell Mol. Biol.* **2000**, *46*, 1297–1303.
51. Elman, M.; Lebzelter, J. *Dermatol. Surg.* **2004**, *30*, 139–146.
52. Hongcharu, W.; Taylor, C. R.; Chang, Y.; Aghassi, D.; Suthamjariya, K.; Anderson, R. R. *J. Invest. Dermatol.* **2000**, *115*, 183–192.
53. Goldman, M. P.; Boyce, S. M. *J. Drugs Dermatol.* **2003**, *2*, 393–396.
54. Bowes, L. E.; Manstein, D.; Anderson, R. R. *Lasers Surg. Med.* **2003**, *18*, S6–S7.
55. Seaton, E. D.; Charakida, A.; Mouser, P. E.; Grace, I.; Clement, R. M.; Chu, A. C. *Lancet* **2003**, *362*, 1347–1352.
56. Orringer, J. S.; Kang, S.; Hamilton, T.; Schumacher, W.; Cho, S.; Hammerberg, C.; Fisher, G. J.; Karimipour, D. J.; Johnson, T. M.; Voorhees, J. J. *JAMA* **2004**, *291*, 2834–2839.
57. Paithankar, D. Y.; Ross, E. V.; Saleh, B. A.; Blair, M. A.; Graham, B. S. *Lasers Surg. Med.* **2002**, *31*, 106–114.
58. Zelickson, B. D.; Kist, D.; Bernstein, E.; Brown, D. B.; Ksenzenko, S.; Burns, J.; Kilmer, S.; Mehregan, D.; Pope, K. *Arch. Dermatol.* **2004**, *140*, 204–209.
59. Ruiz-Esparza, J.; Gomez, J. B. *Dermatol. Surg.* **2003**, *29*, 333–339.
60. Farber, E. M.; Nall, M. L. *Dermatologica* **1974**, *148*, 118.
61. Hoede, K. *Hautarzt* **1957**, *8*, 433.
62. Andersen, C.; Henseler, T. *Hautarzt* **1982**, *33*, 214.
63. Bata-Csorgo, Z.; Hammerberg, C.; Voorhees, J. J.; Cooper, K. D. *J. Invest. Dermatol.* **1995**, *105*, 89S–94S.
64. Terajima, S.; Higaki, M.; Igarashi, Y.; Nogita, T.; Kawashima, M. *Arch. Dermatol. Res.* **1998**, *290*, 246–250.
65. Cristophers, E. *Int. Arch. Allergy Immunol.* **1996**, *110*, 199–206.
66. Bos, J. D.; De Rie, M. A. *Immunol. Today* **1999**, *50*, 40–46.
67. Mueller, W.; Herrmann, B. *N. Engl. J. Med.* **1979**, *301*, 555.
68. Christophers, E. *Clin. Exp. Dermatol.* **2001**, *26*, 314–320.
69. Uyemura, K.; Yamamura, M.; Fivenson, D. F.; Modlin, R. L.; Nickoloff, B. J. *J. Invest. Dermatol.* **1993**, *101*, 701–705.
70. Schlaak, J. F.; Buslau, M.; Jochum, W.; Hermann, E.; Girndt, M.; Gallati, H.; Meyer zum Buschenfelde, K. H.; Fleischer, B. *J. Invest. Dermatol.* **1994**, *102*, 145–149.

71. Griffiths, C. E. M.; Voorhees, J. J.; Nickoloff, B. J. *J. Am. Acad. Dermatol.* **1989**, *20*, 617–629.
72. Barker, J. N. W. N.; Sarma, V.; Mitra, R. S.; Dixit, V. M.; Nickoloff, B. J. *J. Clin. Invest.* **1990**, *85*, 605–608.
73. Fierlbeck, G.; Rassner, G.; Muller, C. *Arch. Dermatol.* **1990**, *83*, 416–420.
74. Ritchlin, C.; Haas-Smith, S. A.; Hicks, D.; Cappuccio, J.; Osterland, C. K.; Looney, R. J. *J. Rheumatol.* **1998**, *25*, 1544–1552.
75. Terajima, S.; Higaki, M.; Igarashi, Y.; Nogita, T.; Kawashima, M. *Arch. Dermatol. Res.* **1998**, *290*, 246–252.
76. Detmar, M.; Yeo, K. T.; Nagy, J. A.; Van de Water, L.; Brown, L. F.; Berse, B.; Elicker, B. M.; Ledbetter, S.; Dvorak, H. F. *J. Invest. Dermatol.* **1995**, *105*, 44–50.
77. Nickoloff, B. J.; Karabin, G. D.; Barker, J. N.; Griffiths, C. E.; Sarma, V.; Mitra, R. S.; Elder, J. T.; Kunkel, S. L.; Dixit, V. M. *Am. J. Pathol.* **1991**, *138*, 129–140.
78. Cerwenka, A.; Lanier, L. L. *Nat. Rev. Immunol.* **2001**, *1*, 41–49.
79. Cameron, A. L.; Kirby, B.; Fei, W.; Griffiths, C. E. M. *Arch. Dermatol. Res.* **2002**, *294*, 363–369.
80. Teunissen, M. B. M. Langerhans Cells and Other Skin Dendritic Cells. In *Skin Immune System: Cutaneous Immunology and Clinical Immunodermatology*, 3rd ed.; CRC Press: Boca Raton, FL, 2005, pp 123–182.
81. Jablonka, S. Immunology Mechanisms in Psoriasis: Role of Polymorphonuclear Leukocytes. In *Psoriasis*; Farber, E. M., Nall, L. M., Morhenn, V., Jacobs, P. H., Eds.; Elsevier: New York, 1987, pp 131–137.
82. Toichi, E.; Tachibana, T.; Furukawa, F. *J. Am. Acad. Dermatol.* **2000**, *43*, 391–395.
83. Chang, E. Y.; Hammerberg, C.; Fisher, G.; Baadsgaard, O.; Ellis, C. N.; Voorhees, J. J.; Cooper, K. D. *Arch. Dermatol.* **1992**, *128*, 1479–1485.
84. Ong, P. Y.; Ohtake, T.; Brandt, C.; Strickland, I.; Boguniewicz, M.; Ganz, T.; Gallo, R. L.; Leung, D. Y. *N. Engl. J. Med.* **2002**, *347*, 1151–1160.
85. Nomura, I.; Goleva, E.; Howell, M. D.; Hamid, Q. A.; Ong, P. Y.; Hall, C. F.; Darst, M. A.; Gao, B.; Boguniewicz, M.; Travers, J. B. et al. *J. Immunol.* **2003**, *171*, 3262–3269.
86. Baker, B. S.; Ovigne, J. M.; Powles, A. V.; Corcoran, S.; Fry, L. *Br. J. Dermatol.* **2003**, *148*, 670–679.
87. Miller, L. S.; Sorensen, O. E.; Liu, P. T.; Jalian, H. R.; Eshtiaghpour, D.; Behmanesh, B. E.; Chung, W.; Starner, T. D.; Kim, J.; Sieling, P. A. et al. *J. Immunol.* **2005**, *174*, 6137–6143.
88. Gladman, D. D. Psoriatic Arthritis. In *Classification and Assessment of Rheumatic Disease: Part 1*. Bailliere's Clinical Rheumatology; Silman, A. J., Symmons, D. P. M., Eds.; International Practice and Research: Bailliere Tindall, London, 1995, pp 319–329.
89. Shbeeb, M.; Uramoto, K. M.; Gibson, L. E.; O'Fallon, W. M.; Gabriel, S. E. *J. Rheumatol.* **2000**, *27*, 1247–1250.
90. Scarpa, R.; Oriente, P.; Pulino, A.; Torella, M.; Vignone, L.; Riccio, A.; Biondi Oriente, C. *Br J. Rheumatol.* **1984**, *23*, 246–250.
91. Zachariae, H. *Clin. Dermatol. Am. J. Clin. Dermatol.* **2003**, *4*, 441–447.
92. Wright, V.; Moll, J. M. H. Psoriatic Arthritis. In *Seronegative Polyarthritis*; Wright, V., Moll, J. M. H., Eds.; North-Holland Publishing: Amsterdam, 1976, pp 169–223.
93. Mrowietz, U.; Jessat, H.; Schwarz, A.; Schwarz, T. *Br. J. Dermatol.* **1997**, *136*, 542–547.
94. Schmidt, K. N.; Podda, M.; Packer, L.; Baeuerle, P. A. *J. Immunol.* **1996**, *156*, 4514–4519.
95. Kemeny, L.; Michel, G.; Arenberger, P.; Ruzicka, T. *Acta Dermatol. Venereol.* **1993**, *73*, 37–40.
96. Dodd, W. A. *Dermatol. Clin.* **1993**, *11*, 131–135.
97. Krochmal, L.; Wang, J. C.; Patel, B.; Rodgers, J. *J. Am. Acad. Dermatol.* **1989**, *21*, 979–984.
98. Nagpal, S.; Lu, J.; Boehm, M. F. *Curr. Med. Chem.* **2001**, *8*, 1661–1679.
99. Jensen, A. M.; Llado, M. B.; Skov, L.; Hansen, E. R.; Larsen, J. K.; Baadsgaard, O. *Br. J. Dermatol.* **1998**, *139*, 984–991.
100. Stewart, D. G.; Lewis, H. M. *J. Clin. Pharm. Ther.* **1996**, *21*, 143–148.
101. Fenton, C.; Plosker, G. L. *Clin. Dermatol. Am. J. Clin. Dermatol.* **2004**, *5*, 463–478.
102. Scott, L. J.; Dunn, C. J.; Goa, K. L. *Clin. Dermatol. Am. J. Clin. Dermatol.* **1998**, *38*, 1010–1011.
103. Dando, T. M. *Clin. Dermatol. Am. J. Clin. Dermatol.* **2005**, *6*, 255–272.
104. Chandraratna, R. A. *J. Am. Acad. Dermatol.* **1997**, *37*, S12–S17.
105. Chandraratna, R. A. *J. Am. Acad. Dermatol.* **1998**, *39*, S124–S128.
106. Tanghetti, E. A. *Cutis* **2000**, *66*, 4–11.
107. Poulin, Y. P. *Cutis* **1999**, *63*, 41–48.
108. Lebwohl, M. G.; Breneman, D. L.; Goffe, B. S.; Grossman, J. R.; Ling, M. R.; Milbauer, J.; Pincus, S. H.; Sibbald, R. G.; Swinyer, L. J.; Weinstein, G. D. et al. *J. Am. Acad. Dermatol.* **1998**, *39*, 590–596.
109. Kaidbey, K.; Kopper, S. C.; Sefton, J.; Gibson, J. R. *Int. J. Dermatol.* **2001**, *40*, 468–471.
110. Koo, J. Y. *J. Am. Acad. Dermatol.* **1998**, *39*, S144–S148.
111. Koo, J. Y.; Lowe, N. J.; Lew-Kaya, D. A.; Vasilopoulos, A. I.; Lue, J. C.; Sefton, J.; Gibson, J. R. *J. Am. Acad. Dermatol.* **2000**, *43*, 821–828.
112. Perry, H. O.; Soderstrom, C. W.; Schulze, R. W. *Arch. Dermatol.* **1968**, *98*, 178–182.
113. Ibbotson, S. H.; Bilsland, D.; Cox, N. H.; Dawe, R. S.; Diffey, B.; Edwards, C.; Farr, P. M.; Ferguson, J.; Hart, G.; Hawk, J. et al. *Br. J. Dermatol.* **2004**, *151*, 283–297.
114. Coven, T. R.; Burack, L. H.; Gilleaudeau, R.; Keogh, M.; Ozawa, M.; Krueger, J. G. *Arch. Dermatol.* **1997**, *133*, 1514–1522.
115. Zanolli, M. *J. Am. Acad. Dermatol.* **2003**, *49*, S78–S86.
116. Weischer, M.; Blum, A.; Eberhard, F.; Rocken, M.; Berneburg, M. *Acta Dermatol. Venereol.* **2004**, *84*, 370–374.
117. Lauharanta, J. *Clin. Dermatol.* **1997**, *15*, 769–780.
118. Koo, J.; Lebwohl, M. *J. Am. Acad. Dermatol.* **1999**, *41*, 51–59.
119. Lowe, N. J.; Chizhevsky, V.; Gabriel, H. *Clin. Dermatol.* **1997**, *15*, 745–752.
120. Stern, R. S.; Lange, R. *J. Invest. Dermatol.* **1988**, *91*, 120–124.
121. Stern, R. S. *J. Am. Acad. Dermatol.* **2001**, *44*, 755–761.
122. Asawanonda, P.; Anderson, R. R.; Chang, Y.; Taylor, C. R. *Arch. Dermatol.* **2000**, *136*, 619–624.
123. Feldman, S. R.; Mellen, B. G.; Housman, T. S.; Fitzpatrick, R. E.; Geronemus, R. G.; Friedman, P. M.; Vasily, D. B.; Morison, W. L. *J. Am. Acad. Dermatol.* **2002**, *46*, 900–906.
124. Trehan, M.; Taylor, C. R. *J. Am. Acad. Dermatol.* **2002**, *47*, 701–708.
125. McClure, S. L.; Valentine, J.; Gordon, K. B. *Drug Saf.* **2002**, *25*, 913–927.
126. Griffiths, C. E.; Clark, C. M.; Chalmers, R. J.; Li Wan Po, A.; Williams, H. C. *Health Technol. Assess.* **2000**, *4*, 1–125.

127. Heydendael, V. M.; Spuls, P. I.; Opmeer, B. C.; de Borgie, C. A.; Reitsma, J. B.; Goldschmidt, W. F.; Bossuyt, P. M.; Bos, J. D.; de Rie, M. A. *N. Engl. J. Med.* **2003**, *349*, 658–665.
128. Lebwohl, M.; Drake, L.; Menter, A.; Koo, J.; Gottlieb, A. B.; Zanolli, M.; Young, M.; McClelland, P. *J. Am. Acad. Dermatol.* **2001**, *45*, 544–553.
129. Majeau, G. R.; Meier, W.; Jimmo, B.; Kioussis, D.; Hochman, P. S. *J. Immunol.* **1994**, *152*, 2753–2767.
130. Lebwohl, M.; Christophers, E.; Langley, R.; Ortonne, J. P.; Roberts, J.; Griffiths, C. E. Alefacept Clinical Study Group. *Arch. Dermatol.* **2003**, *139*, 719–727.
131. Gottlieb, A. B.; Matheson, R. T.; Lowe, N.; Krueger, G. G.; Kang, S.; Goffe, B. S.; Gaspari, A. A.; Ling, M.; Weinstein, G. D.; Nayak, A. et al. *Arch. Dermatol.* **2003**, *139*, 1627–1632.
132. Leonardi, C. L.; Powers, J. L.; Matheson, R. T.; Goffe, B. S.; Zitnik, R.; Wang, A.; Gottlieb, A. B. Etanercept Psoriasis Study Group. *N. Engl. J. Med.* **2003**, *349*, 2014–2022.
133. Mellemkjaer, L.; Linet, M. S.; Gridley, G.; Frisch, M.; Moller, H.; Olsen, J. H. *Eur. J. Cancer* **1996**, *32A*, 1753–1757.
134. Werther, W. A.; Gonzalez, T. N.; O'Connor, S. J.; McCabe, S.; Chan, B.; Hotaling, T.; Champe, M.; Fox, J. A.; Jardieu, P. M.; Berman, P. W. et al. *J. Immunol.* **1996**, *157*, 4986–4995.
135. Dedrick, R. L.; Walicke, P.; Garovoy, M. *Transpl. Immunol.* **2002**, *9*, 181–186.
136. Gordon, K. B.; Papp, K. A.; Hamilton, T. K.; Walicke, P. A.; Dummer, W.; Li, N.; Bresnahan, B. W.; Menter, A. Efalizumab Study Group. *JAMA* **2004**, *290*, 3073–3080.
137. Lebwohl, M.; Tyring, S. K.; Hamilton, T. K.; Toth, D.; Glazer, S.; Tawfik, N. H.; Walicke, P.; Dummer, W.; Wang, X.; Garovoy, M. R. et al. *N. Engl. J. Med.* **2003**, *349*, 2004–2013.
138. Knight, D. M.; Trinh, H.; Le, J.; Siegel, S.; Shealy, D.; McDonough, M.; Scallon, B.; Moore, M. A.; Vilcek, J.; Daddona, P. et al. *Mol. Immunol.* **1993**, *30*, 1443–1453.
139. Scallon, B. J.; Moore, M. A.; Trinh, H.; Knight, D. M.; Grayeb, J. *Cytokine* **1995**, 7, 251–259.
140. Gottlieb, A. B.; Evans, R.; Li, S.; Dooley, L. T.; Guzzo, C. A.; Baker, D.; Bala, M.; Marano, C. W.; Menter, A. *J. Am. Acad. Dermatol.* **2004**, *51*, 534–542.
141. Cookson, W. O. C. M. *Ann. Med.* **1994**, *26*, 351–353.
142. Leung, D. Y. M. *J. Allergy Clin. Immunol.* **2000**, *105*, 860–876.
143. Leung, D. Y. M. *Clin. Exp. Immunol.* **1997**, *107*, 25–30.
144. Cooper, K. D. *J. Invest. Dermatol.* **1994**, *102*, 128–137.
145. Hamid, Q.; Boguniewicz, M.; Leung, D. Y. M. *J. Clin. Invest.* **1994**, *94*, 870–876.
146. Hanifin, J. M.; Chan, S. *J. Am. Acad. Dermatol.* **1999**, *41*, 72–77.
147. Leung, D. Y. M.; Soter, N. A. *J. Am. Acad. Dermatol.* **2001**, *44*, S1–S12.
148. Valenta, R.; Maurer, D.; Steiner, R. *J. Invest. Dermatol.* **1996**, *107*, 203–208.
149. Leung, D. Y. M.; Schneeberger, E. E.; Siraganina, R. P.; Geha, R. S.; Bhan, A. K. *J. Invest. Dermatol.* **1997**, *108*, 336–342.
150. Hanifin, J. M. *Pediatr. Dermatol.* **1986**, *3*, 161–174.
151. Sampson, H. A.; McCaskill, C. C. *J. Pediatr.* **1988**, *107*, 669–675.
152. Leyden, J. E.; Marples, R. R.; Kligman, A. M. *Br. J. Dermatol.* **1974**, *90*, 525–530.
153. Bunikowski, R.; Mielke, M.; Skarabis, H.; Herz, U.; Bergmann, R. L.; Wahn, U.; Renz, H. *J. Allergy Clin. Immunol.* **1999**, *103*, 119–124.
154. Leung, D. Y.; Harbeck, R.; Bina, P.; Reiser, R. F.; Yang, E.; Norris, D. A.; Hanifin, J. M.; Sampson, H. A. *J. Clin. Invest.* **1993**, *92*, 1374–1380.
155. Leung, D. Y. M. *J. Am. Acad. Dermatol.* **2001**, *45*, S13–S16.
156. Williams, H. C.; Wuthrich, B. The Natural History of Atopic Dermatitis. In *Atopic Dermatitis: The Epidemiology, Causes, and Prevention of Atopic Eczema*; Williams, H. C., Eds.; Cambridge University Press: New York, 2000, p 43.
157. Linde, Y. W. *Acta Dermatol Venereol. Suppl.* **1992**, *177S*, 9–13.
158. Gelmetti, C. *Pediatr. Dermatol.* **1992**, *9*, 380–382.
159. Garrity, J. A.; Liesegang, T. J. *Can. J. Ophthalmol.* **1984**, *19*, 21–24.
160. Katavisto, M. *Acta Ophthalmol.* **1949**, *27*, 581.
161. Aly, R. *Acta Dermatol. Venereol.* **1980**, *92S*, 16.
162. Goodyear, H. M.; McLeish, P.; Randall, S. *Br. J. Dermatol.* **1996**, *134*, 85–93.
163. Currie, J. M.; Wright, R. C.; Miller, O. G. *Cutis* **1971**, *8*, 243.
164. Solomon, L. M.; Telner, P. *Can. Med. Assoc. J.* **1966**, *95*, 978–979.
165. Barker, J. N. W. M.; Alegre, V. A.; MacDonlad, D. M. *J. Invest. Dermatol.* **1988**, *90*, 117.
166. Hara, J.; Higuchi, K.; Okamoto, R.; Kawashima, M.; Imokawa, G. *J. Invest. Dermatol.* **2000**, *115*, 406–413.
167. Hanifin, J. M.; Tofte, S. J. *J. Allergy Clin. Immunol.* **1999**, *104*, S123–S125.
168. Akers, W. *Arch. Dermatol.* **1980**, *116*, 786–788.
169. Ellison, J. A.; Patel, L.; Ray, D. W.; David, T. J.; Clayton, P. E. *Pediatrics* **2000**, *105*, 794–799.
170. Patel, L.; Clayton, P. E.; Addison, G. M.; Price, D. A.; David, T. J. *Arch. Dis. Child.* **1998**, *79*, 169–172.
171. Hanifin, J. M.; Ling, M. R.; Langley, R.; Beneman, D.; Rafal, E. *J. Am. Acad. Dermatol.* **2001**, *44*, S28–S38.
172. Soter, N. A.; Fleischer, A. B., Jr.; Webster, G. F.; Monroe, E.; Lawrence, I. *J. Am. Acad. Dermatol.* **2001**, *44*, S39–S46.
173. Paller, A.; Eichenfield, L. F.; Leung, D. Y.; Stewart, D.; Appell, M. *J. Am. Acad. Dermatol.* **2001**, *44*, S47–S57.
174. Eichenfield, L. F.; Lucky, A. W.; Boguniewicz, M.; Langley, R. G.; Cherill, R.; Marshall, K.; Bush, C.; Graeber, M. *J. Am. Acad. Dermatol.* **2002**, *46*, 495–504.
175. Van Leent, E. J.; Graber, M.; Thurston, M.; Wagenaar, A.; Spuls, P. I.; Bos, J. D. *Arch. Dermatol.* **1998**, *134*, 806–809.
176. Cooper, K. D. *Clin. Rev. Allergy* **1993**, *11*, 543–557.

Biographies

H Ray Jalian completed his Bachelors of Science in Microbiology, Immunology, and Molecular Genetics at the University of California, Los Angeles. He is currently a fourth-year medical student at the David Geffen School of Medicine at UCLA. Current research interests include identifying cell wall components of *Propionibacterium acnes* that may contribute to the inflammatory process in acne.

Stefani Takahashi MD attended medical school at the University of Southern California. After a pediatric internship at Children's Hospital of Los Angeles, she completed her dermatology residency at UCLA. Dr Takahashi is Clinical Assistant Professor of Dermatology and Medicine at UCLA. Her clinical interests are general dermatology and pediatric dermatology and she is co-director of pediatric dermatology at UCLA. She is a member of the American Academy of Dermatology and Women's Dermatologic Society and is a board member of the Los Angeles Metropolitan Dermatological Society.

Jenny Kim MD, PhD completed her medical degree, dermatology residency, and Mohs micrographic surgery fellowship at the UCLA. She received her PhD from UCLA in immunology and is an Assistant Professor of Dermatology and

Medicine at UCLA. Dr Kim's clinical interest is in skin cancer surgery, lasers, and aesthetic dermatology surgery. Her research focuses on the immunologic mechanisms of host responses in cutaneous inflammatory diseases and neoplasm, including acne, psoriasis, and skin cancers. Dr Kim also conducts translational medicine, bringing scientific results from the bench to the bedside. Dr Kim's research has contributed to the field of dermatology in recent years and she has been recognized with many awards from the American Academy of Dermatology, the Society for Investigative Dermatology, and the American Society for Dermatologic Surgery. Dr Kim speaks at both national and international meetings and reviews many articles for peer-reviewed journals in both dermatology and immunology. She has consulted for various pharmaceutical and cosmeceutical companies.

Comprehensive Medicinal Chemistry II
ISBN (set): 0-08-044513-6

ISBN (Volume 7) 0-08-044520-9; pp. 935–955

7.33 Advances in the Discovery of Acne and Rosacea Treatments

S J Baker, Anacor Pharmaceuticals, Palo Alto, CA, USA

7.33.1 Introduction

There are two forms of acne: the most common known form is acne vulgaris (acne), which mostly affects teenagers but can continue to manifest symptoms into the early 20s. The second form is acne rosacea (rosacea), which mostly affects 30- to 60-year-olds. Although neither form is life-threatening, the quality of life for sufferers can be greatly diminished. The obvious facial blemishes caused by these conditions inflict a psychosocial impact on the individual and can cause embarrassment in the workplace or at social events. Acne is particularly troublesome to young people in their teens to 20s, when socializing with friends and courting are important. Young people in this age group want to look their best and acne directly opposes this, which often leads to depressed emotions. In addition, acne can be painful and in severe cases permanent facial scarring may occur. Like teenagers, rosacea sufferers are also concerned about their appearance. However, those who suffer from moderate-to-severe rosacea show symptoms that are commonly confused with excess alcohol consumption, which can lead to further embarrassment, especially in the workplace. For these reasons and more, acne and rosacea sufferers actively want to be rid of their condition and will often seek medication. This chapter will review these acne types, the pathogenesis of the disease, and current treatments. For further reading, see reviews by Fulton[1] and Thiboutot.[2]

7.33.2 Acne Vulgaris

Acne vulgaris is a complex skin disease that only occurs in humans in their pilosebaceous follicles (**Figure 1**). The pilosebaceous follicles are vellus hair follicles that contain a sebaceous gland and occur most prevalently on the head

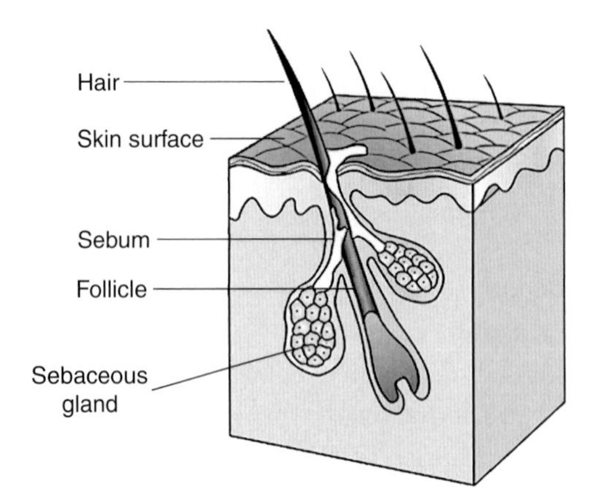

Figure 1 Pilosebaceous follicle. (Reproduced from the National Institute of Arthritis and Musculoskeletal and Skin Diseases (NIAMS), National Institutes of Health booklet *Questions and Answers About Acne,* published on its website: http://www.niams.nih.gov/hi/topics/acne/acne.htm.)

Androsterone Testosterone

Figure 2 Androgen hormones that cause an increase in sebum levels.

and upper trunk.[3] The sebaceous glands produce sebum, an oily secretion that helps to preserve and moisturize skin and hair.

Acne can be categorized as either noninflammatory or inflammatory.[3,4] Three major biological events must occur in order for noninflammatory acne to occur: (1) overproduction of sebum; (2) hyperkeratinization and blockage of the sebaceous follicle; and (3) proliferation of *Propionibacterium acnes*. Events (1) and (2) do not need to occur in order. At this stage, a fourth event, an inflammatory response, can be triggered causing the condition to progress from noninflammatory acne to inflammatory acne.[5] Removing any one of these factors will lead to the resolution of the condition. The events are described below.

7.33.2.1 Event 1: Overproduction of Sebum

Events (1) and (2) likely occur in either order but both must be functional to move to event (3). For acne to occur, sebum must build up within the pilosebaceous follicle. Normal sebum production is not sufficient to cause acne. Sebum production is linked to androgen hormones such as testosterone and androsterone (**Figure 2**). During puberty or pregnancy, androgen levels are increased and, in turn, this increases sebum levels. Adolescents and pregnant women are prone to acne until hormone levels stabilize.

7.33.2.2 Event 2: Hyperkeratinization and Blockage of the Sebaceous Follicle

The second event leading to acne is hyperkeratinization, which is caused by increased production of keratinocytes (epidermal skin cells that produce keratin, the structural protein of skin, hair and nails) and a buildup of other follicular debris, including melanin. This process results in blockage of the pilosebaceous follicle, preventing the natural cleansing of the follicle by sebum, also being overproduced, resulting in a buildup of sebum. These events culminate in the formation of a microcomedo (**Figure 3**).

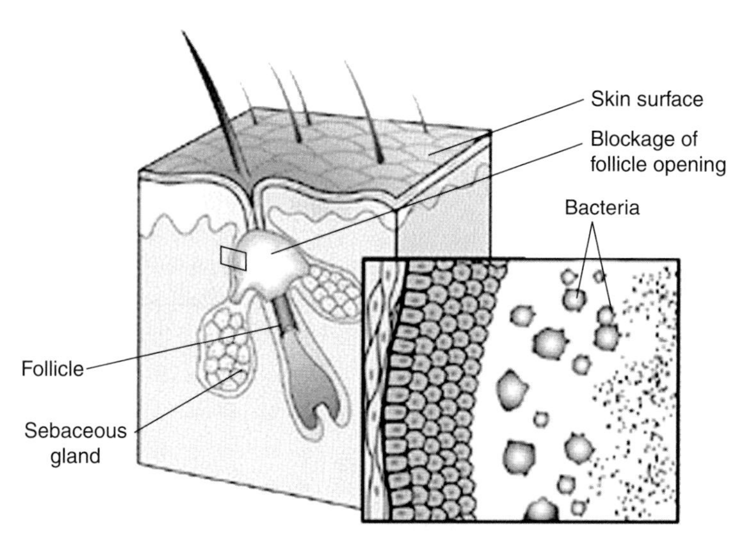

Figure 3 A microcomedo formed by blockage of the follicle and buildup of sebum. This microcomedo provides a favorable environment for *Propionibacterium acnes* bacteria. (Reproduced from the National Institute of Arthritis and Musculoskeletal and Skin Diseases (NIAMS), National Institutes of Health booklet *Questions and Answers About Acne,* published on its website: http://www.niams.nih.gov/hi/topics/acne/acne.htm.)

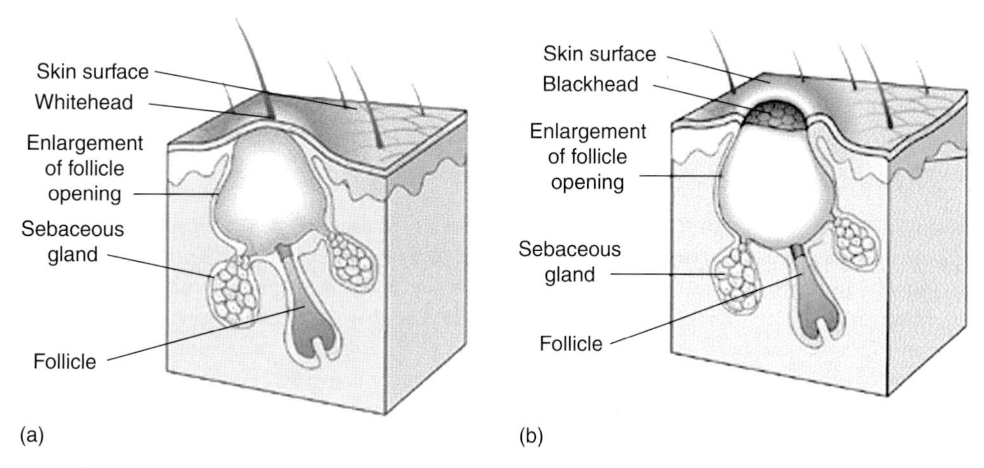

Figure 4 (a) Closed comedo (whitehead); (b) open comedo (blackhead), blocked by melanin and other follicular debris. (Reproduced from the National Institute of Arthritis and Musculoskeletal and Skin Diseases (NIAMS), National Institutes of Health booklet *Questions and Answers About Acne,* published on its website: http://www.niams.nih.gov/hi/topics/acne/acne.htm.)

7.33.2.3 Event 3: Proliferation of *P. acnes*

The third component of acne is the growth of the anaerobic bacterium *P. acnes*, from which the condition is named. *P. acnes* is part of the natural skin flora and resides within the pilosebaceous follicles in all skin and skin types. Since *P. acnes* is a commensal organism and is present on the skin, its presence alone is not sufficient to induce the formation of acne. However, the blocked follicle and sebum buildup provide a lipid-rich, anaerobic environment within which the bacterium *P. acnes* thrives. This proliferation causes a further buildup of pressure within the follicle, leading to a closed comedo, or whitehead (**Figure 4a**). This closed comedo has one of two fates: firstly, in noninflammatory acne, the pressure breaks at the skin surface and the closed comedo develops into an open comedo, or blackhead (**Figure 4b**). The black color of the comedo is caused by the presence of melanin blocking the follicle internally and not because of dirt particles lodged there from external sources (despite the popular myth that acne is caused through lack of hygiene). In this case, the blackhead is eventually washed away and the follicle returns to normal. A second fate of the closed comedo is inflammatory acne. This condition is caused when increased pressure within the comedo breaks not at the surface, but in the opposite direction, toward the dermis.[3]

7.33.2.4 Event 4: Inflammation

Spillage of the bacteria *P. acnes* into the dermis triggers an inflammatory response leading to a papule (inflamed red pimples). Accumulation of inflammatory cells, including white blood cells combined with other cellular debris (commonly known as pus), results in the formation of pustules (pimples containing pus). If these papules and pustules are untreated and grow in size, they can develop into nodules and cysts, respectively. At this stage, the condition is quite painful.

The severity of acne can be classified into four categories[6,7]: (1) mild; (2) mild-to-moderate; (3) moderate-to-severe; and (4) severe. Mild acne is noninflammatory comedonal acne (low numbers of whiteheads and blackheads). Mild-to-moderate acne is inflammatory acne with low-to-moderate numbers of papules and pustules. Moderate-to-severe acne has a larger presence of papules and pustules, which are likely to result in eventual scarring. Severe acne involves the presence of papules and pustules larger than 1 cm in diameter, termed nodules and cysts, respectively, which are painful and result in permanent scarring.

7.33.3 Overview of Acne Treatments[5,7–12]

The integrated nature of the pathogenesis of this disease allows for effective medical intervention. There are several events in the pathogenesis pathway of acne (**Figure 5**) that can be targeted in order to develop treatments for acne. Targeting one or more of these events will lead to the collapse of this pathway and resolve the condition. Current treatments can be classified into three main classes : (1) antibacterial agents that target *P. acnes*, thereby removing the bacterial component of the condition; (2) retinoids, that target hyperkeratinization, sebum production, and inflammation; and (3) steroids, that target androgen production and inflammation. A less popular method of treatment is phototherapy; however, this has increased in use in recent years. Drugs used in the treatment of acne typically have one main biological activity (e.g., antibacterial), but many also have a secondary activity (e.g., anti-inflammatory). Acne is treated in several ways depending on the severity of the disease and treatments are prescribed alone or in combination.

7.33.4 Acne Rosacea[13,14]

Rosacea mostly affects adults aged 30–60 and is usually limited to the face, especially areas that flush, such as the cheeks and nose. It has a vascular and inflammatory component, commonly characterized by the reddening of the skin (erythema), but can also appear as telangiectasias (deep red blotches), papules and pustules, and sebaceous gland hypertrophy. However, rosacea is differentiated from acne as it does not have the central comedo associated with acne.

The intensity of rosacea has been categorized into four stages.[15,16] Stage 1 is characterized by frequent blushing with easy irritation of the skin. Stage 2 occurs when a vascular component is present, leading to persistent blushing and early telangiectasias. Stage 3 involves a deeper erythema with telangiectasias increasing and papules and pustules

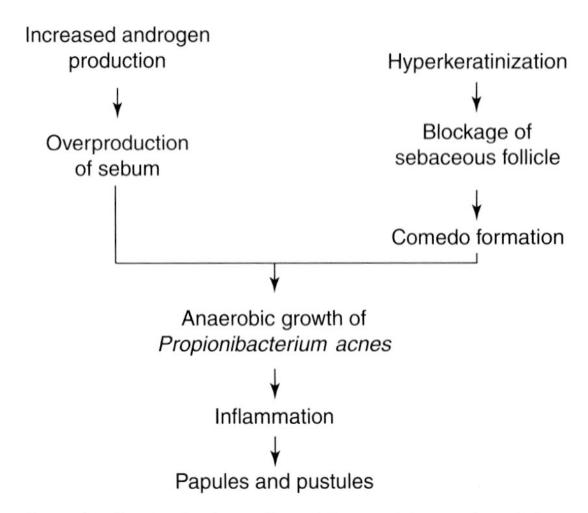

Figure 5 The pathogenesis pathway leading to the formation of the pastules and pustules commonly associated with acne.

appearing. Stage 4 occurs when nasal edema, called rhinophyma, develops, and ocular inflammation begins. In severe cases, rosacea can progress to irreversible disfigurement and loss of vision.[17]

The cause of rosacea is unknown. Erythema arises because of dilation of facial blood vessels, while edema is caused by increased blood flow and accumulation of fluids. Several hypotheses have been suggested as to the causative factors of rosacea, including infectious, immunological, genetic, climatic, and psychological. However, no firm conclusions have been drawn. Unlike acne, the involvement of *P. acnes* is not widely considered to be an important factor. On the other hand, the bacteria *Helicobactor pylori*, which has been linked with stomach ulcers, has been implicated as a factor. However, antibacterial eradication of *H. pylori* in one clinical trial had no effect on the condition.[18] The current lack of understanding of the pathogenesis of rosacea challenges healthcare providers and physicians who treat this disease.

7.33.5 Overview of Rosacea Treatments[19]

The current paradigm for the successful treatment of rosacea involves managing rather than eradicating the disease. There are a number of trigger factors involved at the onset of rosacea that sufferers can avoid,[4] including sun exposure, irritating facial cosmetics, and spicy food. Although alcohol does not act as a trigger factor, it can exacerbate the condition. Since the cause of rosacea is poorly understood, the development of medicines against specific targets has not been pursued. Instead, treatments for other closely related diseases are usually prescribed. Similar to acne, rosacea is treated with antibacterial agents and oral retinoids. However, rosacea is also treated with topical corticosteroids, which target inflammation, and antihypertensives, to reduce the facial flushing.

7.33.6 Antibacterial Treatments for Acne and Rosacea

7.33.6.1 Topical Antibiotics[5–11]

The first modern treatment for acne was benzoyl peroxide (BPO, **1**) applied topically. BPO was first utilized in the 1930s and is still widely used today in concentrations typically ranging from 2.5% to 10%. BPO is effective at killing *P. acnes* resident on the skin and in the follicles preventing the formation of papules and pustules. BPO's mechanism of action is thought to involve the degradation of bacterial proteins by a free-radical mechanism, leading to bacterial death. BPO is known to cause local skin irritation and can be painful to apply, especially on a daily basis. A current major area of research in dermatology-based pharmaceutical companies involves the development of novel formulations containing BPO that produce decreased irritation. Some of these are already available on the market.

In addition to local skin irritation, BPO causes other side effects, including dry skin, contact allergy, and bleaching of skin and fabrics. Despite these adverse effects, BPO has stood the test of time and remains the gold standard for topical antibacterial treatment of acne.

The success of BPO validated the use of antibacterial agents as a treatment method for acne and has given rise to the widespread use of two other antibacterial medications: erythromycin (**2**) and clindamycin (**3**). Erythromycin is a macrolide antibiotic and clindamycin is a lincosamide antibiotic. Both exert a similar mechanism of action through binding to the bacterial 50S ribosomal subunit, causing dissociation of peptidyl tRNA from the ribosome and thus shutting down bacterial protein synthesis. Side effects of these antibacterial agents include local skin irritation, hypersensitivity, and bacterial resistance.

Another topical antibacterial agent used to treat acne is sodium sulfacetamide (**4**), which mimics *p*-aminobenzoic acid (PABA) and inhibits the bacterial enzyme dihydropteroate synthase (DHPS), for which PABA is a substrate. This activity stops the biosynthesis of tetrahydrofolic acid and therefore stops the biosynthesis of DNA.

Since DHPS is a classic target for antibacterial therapy, it comes as no surprise that others have followed this lead. The most recent drug to be approved for acne is dapsone (**5**),[20,21] which also targets DHPS. Daspone was first synthesized in 1908 as an azo-dye but was not tested for antibacterial activity until the late 1930s. Dapsone also possesses anti-inflammatory activity, although the mechanism of action for this activity is unknown. The dual activity of dapsone makes it an ideal treatment for acne by eliminating the causative bacteria and reducing inflammation and redness. The combined effect produces a resolution of the infected follicle and leads to an improved appearance of the skin.

Another topical antibacterial agent used for the treatment of acne is nadifloxacin (**6**),[22] which is a member of the fluoroquinolone family of antibacterial agents. These compounds kill bacteria by inhibiting the bacterial enzymes DNA gyrase and topoisomerase, stopping DNA synthesis. Nadifloxacin has also been reported to have antioxidant[23] and anti-inflammatory properties.[24]

Topical antibacterial agents for the treatment of rosacea also include BPO (**1**) and sodium sulfacetamide (**4**). However, metronidazole (**7**) is more commonly used and is considered to be first-line treatment for early to mid-stage rosacea.[15] Metronidazole is active against anaerobic bacteria and parasites, although its mechanism of action remains unknown.

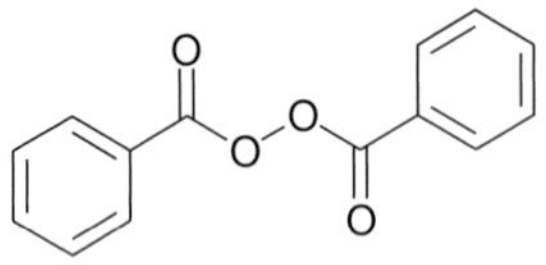

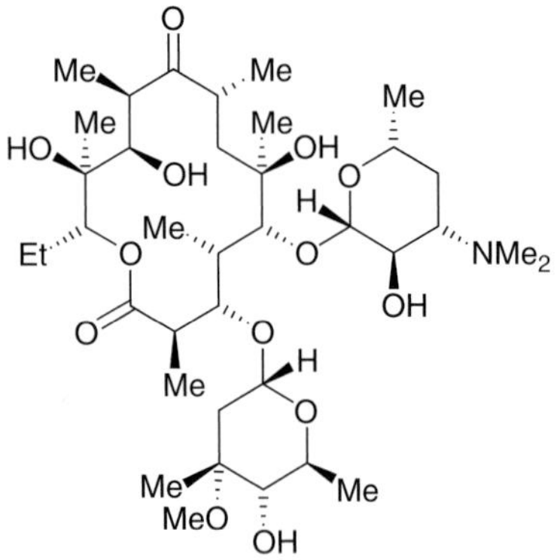

7.33.6.2 Systemic Treatments[12]

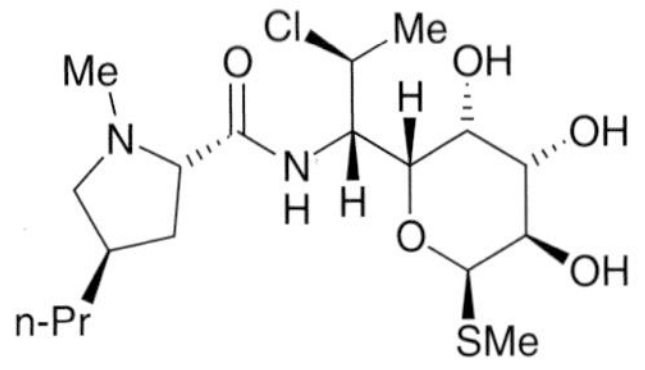

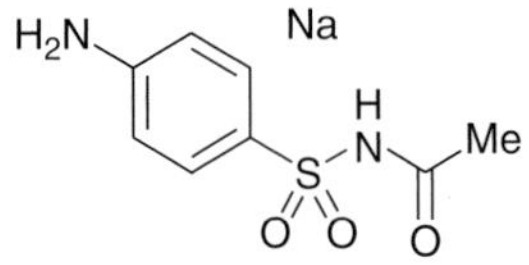

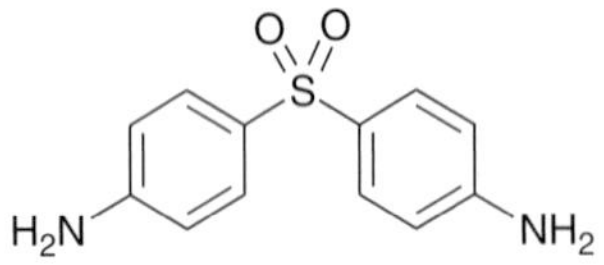

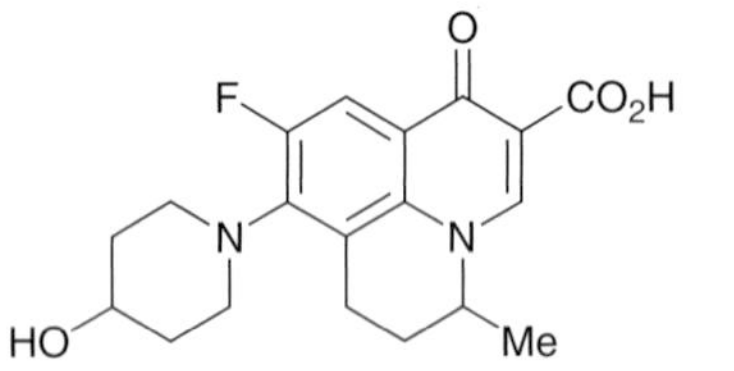

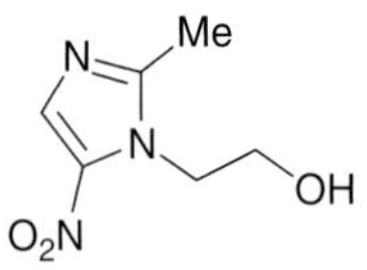

Systemic antibacterial treatment is usually prescribed for moderate-to-severe acne, especially when potential scarring is a risk. Erythromycin (**2**) and clindamycin (**3**), which are used in topical formulations to treat acne, are also used systemically. Erythromycin (**2**) and the tetracycline class of antibacterial agents, including tetracycline (**8**), doxycycline (**9**), and minocycline (**10**), are most often used to treat moderate-to-severe acne. The tetracycline class has a similar mechanism of action to erythromycin, that is, through binding to the bacterial ribosome and inhibiting protein biosynthesis. However, tetracyclines bind to the 30S subunit and inhibit the codon–anticodon interaction as opposed to binding to the 50S subunit and causing dissociation of the peptidyl tRNA. Other antibacterial agents used to treat acne include the macrolide azithromycin (**11**), trimethoprim (**12**), which inhibits dihydrofolate reductase and therefore DNA biosynthesis, and sulfamethoxazole (**13**), which inhibits DHPS and also affects DNA biosynthesis.

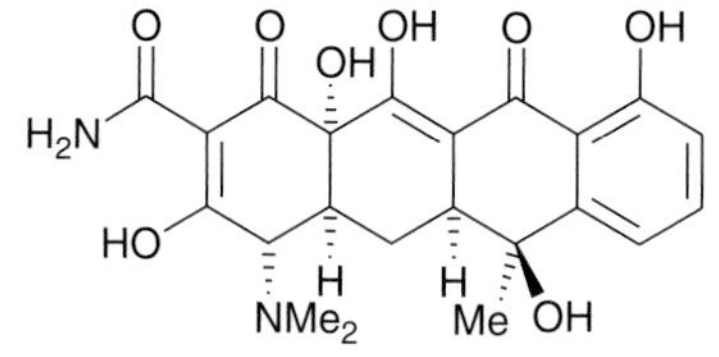

8

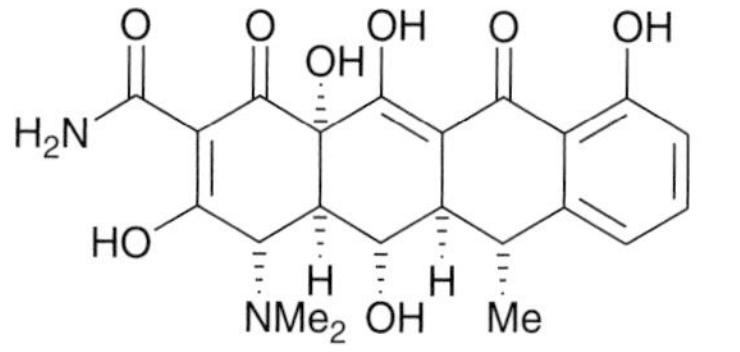

9

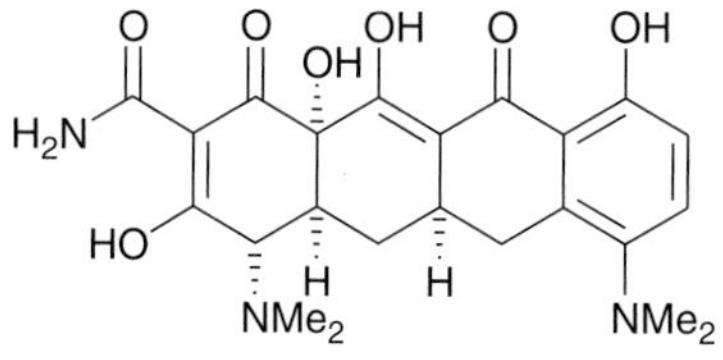

10

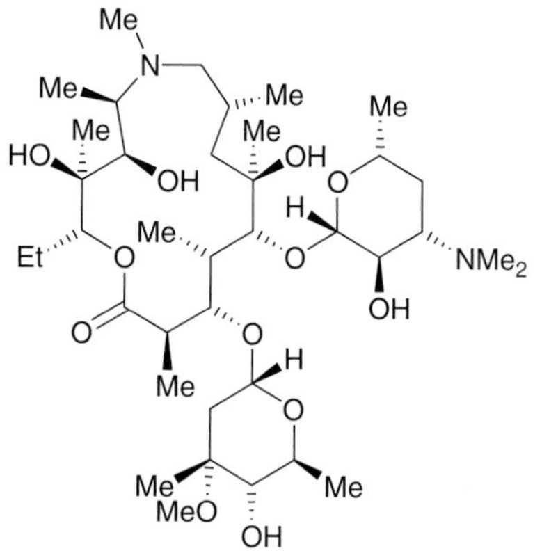

11

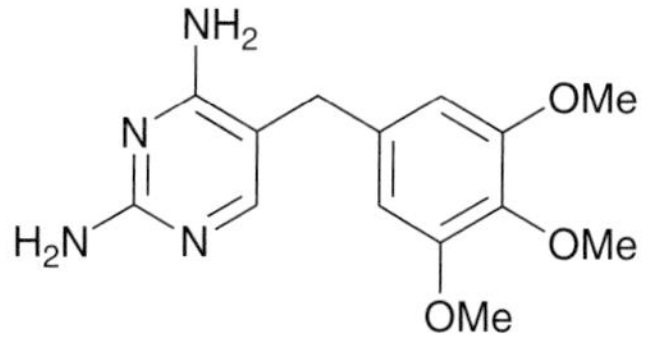

12

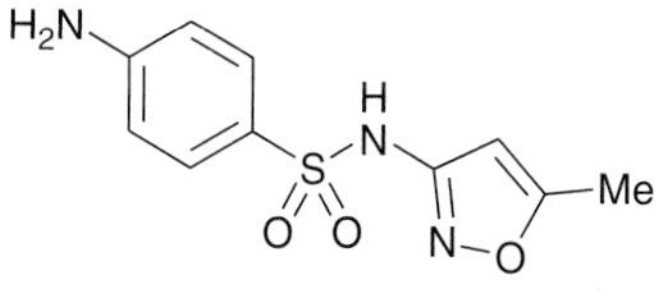

13

Rosacea responds well to systemic antibacterial therapy[15] Tetracycline (**8**) is the treatment of choice, although doxycycline (**9**), minocycline (**10**), erythromycin (**2**), and metronidazole (**7**) are also used.

7.33.7 Retinoids[5–12,25]

The term retinoid is used to describe natural or synthetic analogs of retinol, or vitamin A (**14**). Retinol is a fat-soluble vitamin found in the skin and is required to maintain healthy skin and hair. It is also important for vision, growth and development, immune functions, and reproduction. The first-generation and possibly most important retinoids were all-*trans* retinoic acid, tretinoin (**15**), and its *cis* isomer isotretinoin (**16**). Tretinoin was first used topically to treat acne in 1969 and has remained widely used.[26] Tretinoin has been shown to be far more potent and less toxic than vitamin A, although it does cause skin irritation when applied topically. In fact, early hypotheses suggested that the irritation was the mechanism of action for these compounds. This hypothesis has since been shown to be incorrect and the elimination of this unpleasant side effect has been a focus of medicinal chemistry efforts. Isotretinoin (**16**) revolutionized acne treatment when it was introduced as an oral treatment in 1979 and remains one of the most effective treatments for acne. It is widely regarded as the last line of acne treatment and is usually reserved for the most severe cases. More recently, isotretinoin has been a major focus of both medical and media attention because of potentially serious side effects, including birth defects and a possible connection with teenage suicides leading the US Food and Drug Administration (FDA) to begin monitoring patients on this drug treatment.[27]

Retinoids have multiple mechanisms of action: they normalize keratinization by modulating keratinocyte proliferation and inducing orthokeratosis, they have anti-inflammatory properties, and they are comedolytic. Isotretinoin also inhibits sebaceous secretion.

Retinoids elicit their activity by binding to nuclear retinoid receptors, of which there are six: retinoic acid receptors (RARs) alpha, beta, and gamma, and retinoid X receptors (RXRs) alpha, beta, and gamma. RARs are activated by tretinoin (**15**) and RXRs are activated by 9-*cis* retinoic acid (**17**). Retinoid receptors must dimerize in order to elicit their biological activity. Combinations of RAR–RXR and RXR–RXR dimers exist, but there are no reports of RAR–RAR dimers. Retinoid receptors are primarily located in the nucleus and different combinations of the hetero- or homodimers bind to different and specific DNA sequences, termed retinoic acid response elements (RAREs). Binding of retinoids to these receptor–RARE complexes initiates the expression of many different genes, leading to different cellular effects.[28]

Retinoid receptors have been well studied and their molecular structures have been solved. They belong to the nuclear receptor superfamily, which also includes steroid and thyroid receptors.[29] Biological and structural information of the retinoid receptors has helped to understand receptor specificity and hence determine the individual function of specific receptors.

Given that different retinoid structures activate different retinoid receptors and that first-generation compounds bind nonspecifically to several retinoid receptors, these early retinoids cause many side effects. Therefore, medicinal chemistry efforts[30,31] have focused on elucidating retinoid receptor structure–activity relationships (SARs) in order to find retinoids that bind single receptors and thus produce a single cellular function without side effects.[30]

The importance of the carboxylic acid group of retinoids was determined when studies showed that retinoic acid was more active than retinol.[31] In addition, retinoic acid did not accumulate in the liver, as retinol did. This carboxylic acid group has remained a key pharmacophore in retinoids whether as the free acid, the prodrug ester, or some other bioisostere.

The lipophilic cyclohexyl ring was also found to be an important feature and extensive SAR has been performed in this region, including conversion to a heteroaromatic ring, to the adamantyl group, or to other groups with increased lipophilic bulk.

The acid and cyclohexyl groups are bound by an unsaturated linking region and this has also been the focus of SAR studies. Like many other receptors, conformational restriction has led to increased affinity for the retinoid receptors and various conformations elicit different receptor selectivity. This conformational restriction includes replacing alkenes with alkynes and conjugated olefins with aromatic carbocyclic or heterocyclic rings.

This classical medicinal chemistry has led to the development of the latest generation of retinoids: adapalene (**18**) and tazarotene (**19**). Tazarotene was developed to treat the skin condition psoriasis, which responds well to retinoid therapy. Tazarotene is also used to treat acne.

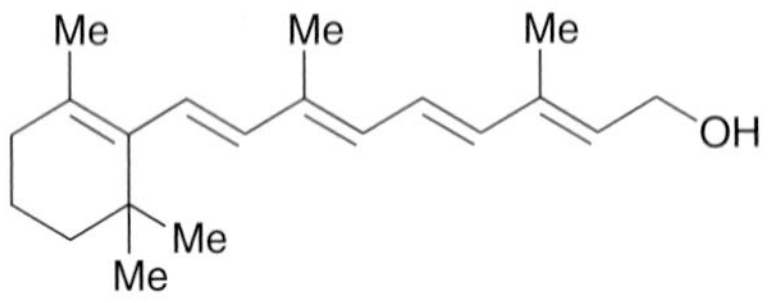

Me Me Me OH Me Me

14

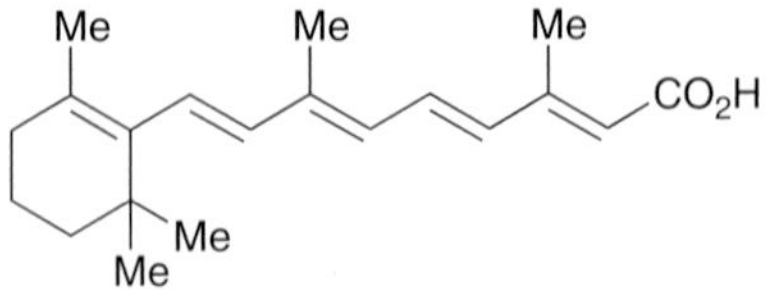

Me Me Me CO_2H Me Me

15

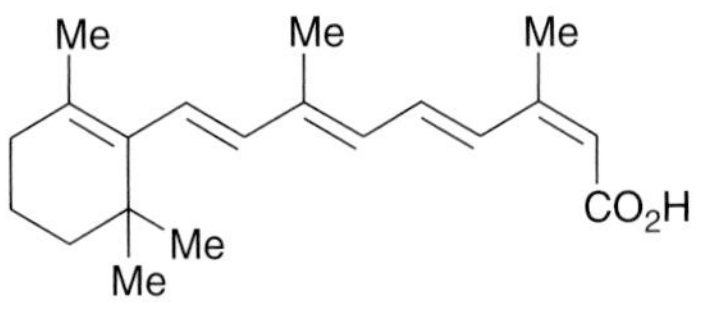

Me Me Me CO_2H Me Me

16

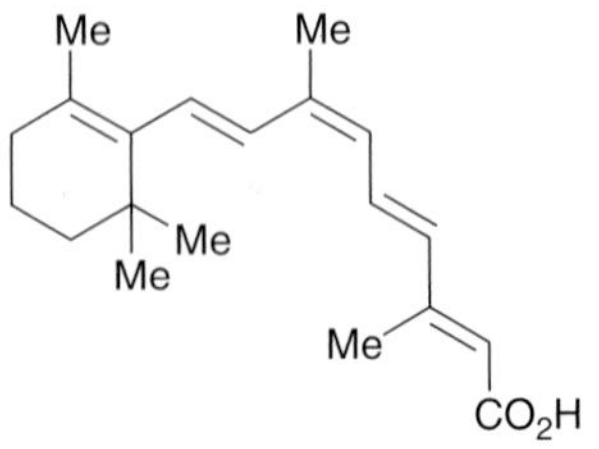

Me Me Me Me Me CO_2H

17

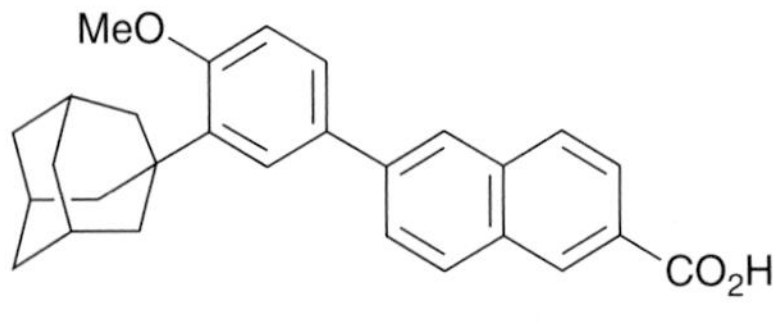

MeO CO_2H

18

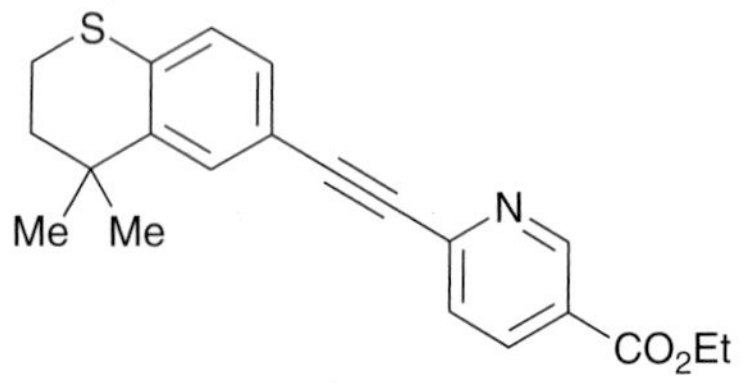

S Me Me N CO_2Et

19

Tretinoin (**15**) is applied topically in concentrations ranging from 0.025% to 0.1% and formulations have been developed in order to reduce skin irritation, but this side effect remains a problem. The newer retinoids, adapalene and tazarotene, also cause skin irritation.

Retinoid therapy has been very successful for acne treatment and, because of its mechanism of action, combination therapy with an antibiotic is an attractive proposition. However, retinoids are chemically incompatible with BPO so it has been challenging to develop a topical formulation that combines these two active agents.

7.33.8 Steroids

One of the causative factors leading to acne is increased androgen levels. Androgen stimulates androgen receptors, which, among other effects, leads to in increased sebum levels. Androgen receptors are activated by androgens such as testosterone; however, dihydrotestosterone (DHT) has a much greater affinity for this receptor. DHT is biosynthesized from testosterone by 5-α-reductase (5αR: **Scheme 1**). 5αR is a membrane-bound NADPH-dependent enzyme, of which there are two subtypes: 5αR1, found predominantly in the skin and liver, and 5αR2, found predominantly in the prostate, genital skin fibroblasts, and liver. In order to treat acne, methods of reducing androgen levels and normalizing their effect represent an obvious target. Current therapeutic approaches involve inhibiting 5αR to prevent the biosynthesis of DHT, or by directly competing with DHT for the androgen receptor with an antagonist.[32,33]

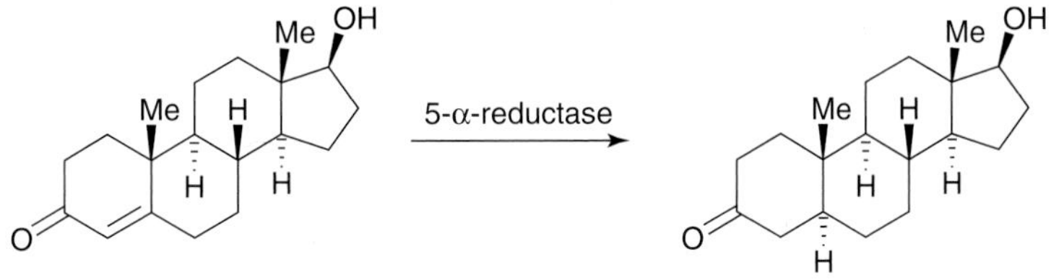

Testosterone — 5-α-reductase → Dihydrotestosterone (DHT)

Scheme 1

Finasteride (**20**) and epristeride (**21**) are systemic treatments that inhibit 5αR. Finasteride has been reported to work by forming a covalent adduct with the reduced form of nicotinamide adenine dinucleotide phosphate (NADPH) in the active site of 5αR, leading to an estimated K_i of 0.3 pmol L^{-1}. It is more selective (~100×) for the 5αR2 enzyme than the 5αR1. Epristeride is more selective for 5αR2 (~400×). It inhibits the enzyme in a reversible manner and has a K_i of 1 nmol L^{-1}. However, there are many disorders that are androgen-mediated so these drugs have been primarily developed for prostrate hyperplasia. Clinical trials for other indications, including acne, rosacea, alopecia, and hirsutism, have been secondary to trials in benign prostrate hyperplasia and prostate cancer.

In other steroid treatments for acne, androgen levels in women can be lowered using spironolactone (**22**) and desogestrel (**23**). These compounds compete with DHT for the androgen receptor.[5–10]

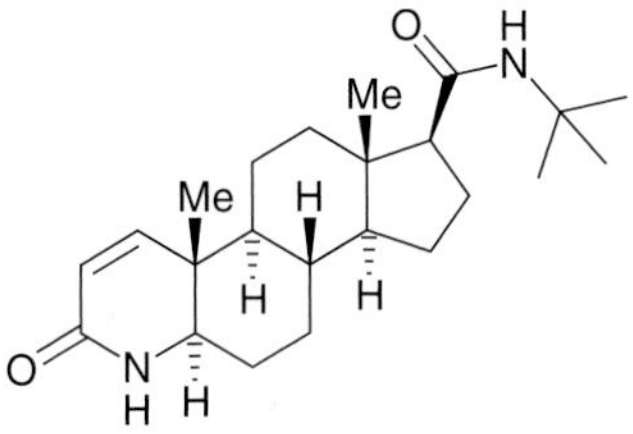

20 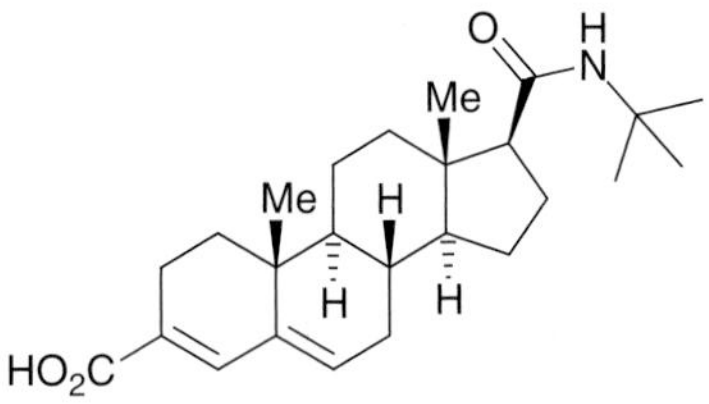 **21**

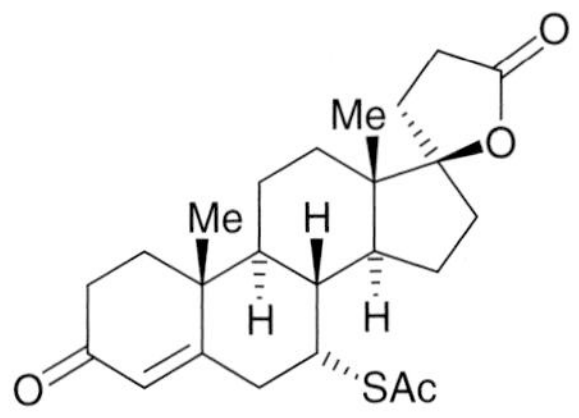

22 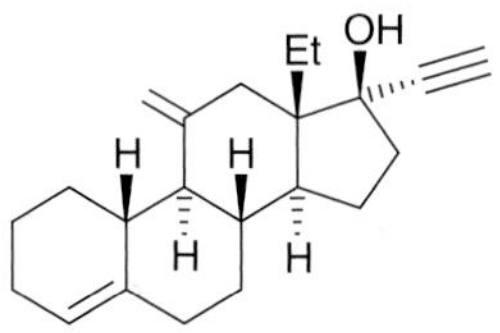 **23**

7.33.9 Other Treatments[5–12]

Over the years, there have been many treatments for acne, some with scientific merit, while others are more speculative. Among the more modern treatments, salicylic acid (**24**) has been commonly used to treat acne due to its keratolytic activity. Azelaic acid (**25**), a naturally occurring dicarboxylic acid that has antimicrobial activity and decreases hyperketatosis, has also been used for the treatment of acne. α-Hydroxy acids such as glycolic acid (**26**) or lactic acid (**27**) are used topically to treat acne and work through the desquamation of stratum corneum. This causes acceleration of the skin's cellular 'conveyor belt' to replace the lost cells quickly. Sun exposure, ultraviolet radiation, and phototherapy have been used to treat acne and, to some extent, have provided marginal improvement. However, it is unknown whether this treatment has directly improved the condition or merely deepened the color of the skin and thereby masking the appearance of acne. The obvious risks associated with phototherapy, namely early skin aging and skin cancer, demonstrate the extent of treatment some people will seek in exchange for a few years of improved appearance.

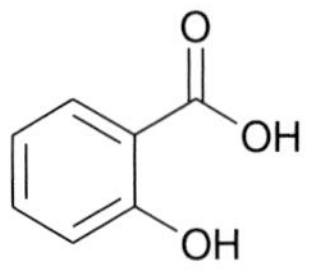

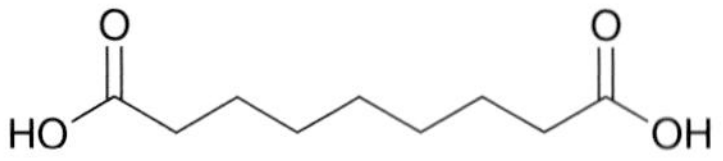

7.33.10 Combination Treatments

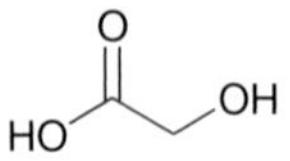

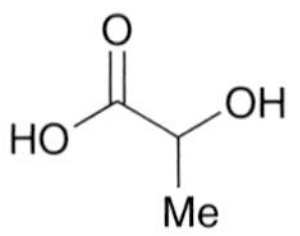

Given that the pathogenesis of acne is complex and there are several ways to block the formation of papules and pustules, it is hardly surprising that clinicians regularly resort to prescribing multiple treatments for acne.[7] For instance, the combination of an antibactetial agent with a retinoid would remove offending bacteria, normalize keratinization, and reduce inflammation, thus attacking the disease from multiple angles. Due to the common practice of prescribing multiple treatments, one might expect drug companies to package combination treatments in a single container. However, for the most part, this is not the case. The potent and broad-spectrum antibacterial activity of BPO makes it a treatment of choice in combination therapy and, indeed, BPO has been combined with the antibacterial agents erythromycin and clindamycin. The reason for these combinations is that although BPO is extremely effective, it causes unpleasant skin irritation. By combining BPO with another antibacterial agent, the concentration of BPO can be reduced, thus reducing the level of skin irritation without significantly reducing antibacterial activity. However, BPO is a reactive oxidizing agent and so the cotreatment must be stable to these conditions.

7.33.11 Current Clinical Pipeline

There are currently several drugs in clinical trials for acne and rosacea, most of which are 'me-too' drugs, including clindamycin, doxycyclin, and azithromycin, that have been reformulated, or retinoids modified as prodrug esters or modified at the hydrophobic region. 5αR inhibitors, which are structural modifications of finasteride (**20**) and epristeride (**21**), in development for benign prostate hyperplasia, are also in trials for acne as a secondary indication.

Two novel treatments in clinical trials for acne include XMP.629 (**28**)[34,35] and PSK-3841 (**29**).[36,37] XMP.629 is a peptidomimetic of the amino acid sequence 90–98 from the human defense protein bactericidal/permeability-increasing protein (BP1). XMP.629 has potent and rapid in vitro activity against *P. acnes*, including antibiotic-resistant strains. This has been developed in a topical formulation, which overcomes the absorption, distribution, metabolism, and excretion (ADME) problems peptidemimetics usually face when administered systemically. In preliminary phase II trials, the vehicle response was higher than anticipated, so no statistical difference was observed between the vehicle and treatment groups.[38]

PSK-3841 (**29**) is a nonsteroid androgen receptor antagonist, which is in phase II clinical trials for alopecia and a preliminary phase II clinical trial for acne.

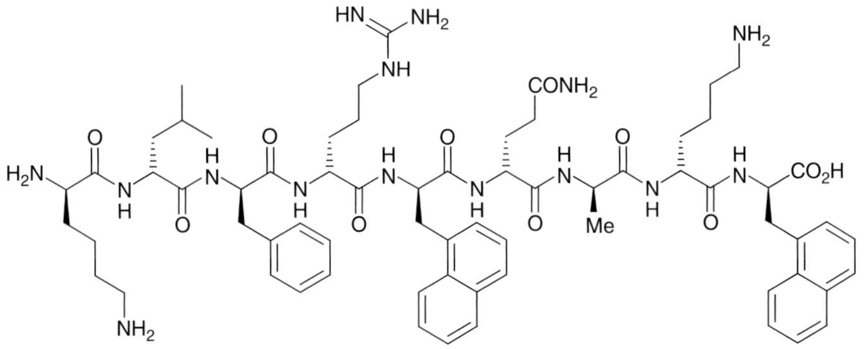

28

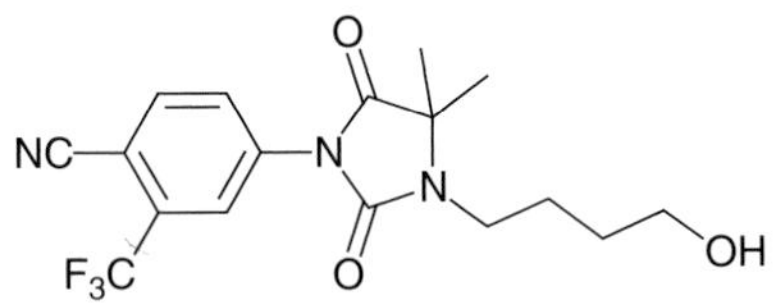

29

7.33.12 Summary

Acne vulgaris and acne rosacea are not life-threatening diseases and it can be argued that they are merely an episodic cosmetic disorder. However, worldwide sales of prescription medications are currently $3.3 billion, demonstrating a clear desire to find effective treatments for these sometimes painful and socially embarrassing conditions. There are many treatments available for acne and sufferers will commonly try new treatments. In addition, dermatologists tend to prescribe multiple treatments and frequently change medications. Taken together, these factors do not attract large pharmaceutical companies to invest sizable budgets in dedicated antiacne research programs. Recently, studies into the pathophysiology of acne have led to a greater understanding of the disease and have allowed dermatologists to prescribe treatments with existing drugs previously approved for other diseases that are, in fact, useful for acne. Unfortunately, the same is not true for rosacea and many rosacea treatments are 'hand-me-downs' from acne in the hope they will also succeed with rosacea. It is likely that, in the near future, increased research, and therefore understanding, into the pathogenesis of rosacea will lead to improved treatment regimes for this condition. Current treatments for acne and rosacea include antibacterial agents to remove pathogenic bacteria, retinoids to normalize the skin, steroids to normalize hormone levels, and other miscellaneous skin-modifying treatments. These treatments have significantly improved these diseases and combination therapy is widespread. However, many of these common treatments cause undesirable side effects. Optimization of these treatments through medicinal chemistry and the development of novel agents will provide new treatment regimens with minimal side effects and improve the armamentarium of the dermatologist.

References

1. Fulton, J. E. *Acne Rx. What Acne Really is and How to Eliminate its Devastating Effects!* James E. Fulton: USA, 2001.
2. Thiboutot, D. Acne 1991–2001. *J. Am. Acad. Dermatol.* **2002**, *47*, 109–117.
3. Toyoda, M.; Morohashi, M. *Med. Electron Microsc.* **2001**, *34*, 29–40.
4. Webster, G. F. *Br. Med. J.* **2002**, *325*, 472–478.
5. Haider, A.; Shaw, J. C. *JAMA* **2004**, *292*, 726–735.
6. James, W. D. *N. Engl. J. Med.* **2005**, *352*, 1463–1472.
7. Leyden, J. J. *J. Am. Acad. Dermatol.* **2003**, *49*, S200–S210.
8. Thiboutot, D. *Arch. Fam. Med.* **2000**, *9*, 179–187.
9. Lee, D. J.; Van Dyke, G. S.; Kim, J. *Curr. Opin. Pediatr.* **2003**, *15*, 405–410.
10. Bershad, S. V. *Mt Sinai J. Med.* **2001**, *68*, 279–286.
11. Akhavan, A.; Bershad, S. *Am. J. Clin. Dermatol.* **2003**, *4*, 473–492.
12. Kunynetz, R. *Skin Ther. Lett.* **2002**, 7, 1–7.
13. Macsai, M. S.; Mannis, M. J.; Huntley, A. C.; *Acne Rosacea. Eye and Skin Diseases*; Lippincott-Williams & Wilkins: Baltimore, USA, 1996, pp. 335–345.
14. Bamford, J. T. M. *Semin. Cutan. Med. Surg.* **2001**, *20*, 199–206.
15. Cohen, A. F.; Tiemstra, J. D. *JABFP* **2002**, *15*, 214–217.

16. Zuber, T. J. *Primary Care* **2000**, *27*, 309–318.
17. Kligman, A. M. *Arch. Dermatol.* **1997**, *133*, 89–90.
18. Bamford, J. T.; Tilden, R.; Blankush, J.; Gangeness, D. E. *Arch. Dermatol.* **1999**, *135*, 659–663.
19. Webster, G. F. *Semin. Cutan. Med. Surg.* **2001**, *20*, 207–208.
20. Wolf, R.; Matz, H.; Orion, E.; Tuzun, B.; Tuzun. Y. *Dermatol. Online J.* **2002**, *8*, article 2. Avilable online at: http://dermatology.cdlib.org/DOJvol8num1/ (accessed Aug 2006).
21. Aczone gel 5% package insert. Available online at: http://www.fda.gov/cder/foi/label/2005/021794lbl.pdf (accessed Aug 2006).
22. Gollnick, H.; Schramm, M. *Dermatology* **1998**, *196*, 119–125.
23. Akamatsu, H.; Sasaki, H.; Kurokawa, I.; Nishijima, S.; Asada, Y.; Niwa, Y. *J. Int. Med. Res.* **1995**, *23*, 19–26.
24. Kuwahara, K.; Keiichi, T.; Kitagaki, H.; Tsukamoto, T.; Kikuchi, M. *J. Dermatol. Sci.* **2005**, *38*, 47–55.
25. Bershad, S. *Semin. Cutan. Med. Surg.* **2001**, *20*, 154–161.
26. Kligman, A. M.; Fulton, J. E.; Plegwig, G. *Arch. Dermatol.* **1969**, *99*, 469–476.
27. http://www.fda.gov/cder/drug/ (accessed Aug 2006).
28. Bastien, J.; Rochette-Egly, C. *Gene* **2004**, *328*, 1–16.
29. Mangelsdorf, D. J.; Umesono, K.; Evans, R. M. The Retinoid Receptors. In *Retinoids*, 2nd ed.; Sporn, M. B.; Roberts, A. B.; Goodman, D. S., Eds.; Raven Press: New York, 1994, pp 319–349.
30. Thacher, S. M.; Vasudevan, J.; Chandraratner, R. A. S. *Curr. Pharm. Des.* **2000**, *6*, 25–58.
31. Dawson, M. I.; Zhang, X. *Curr. Med. Chem.* **2002**, *9*, 623–637.
32. Harris, G. S.; Kozarich, J. W. *Curr. Opin. Chem. Biol.* **1997**, *1*, 254–259.
33. Bratoeff, E.; Cabeza, M.; Ramirez, E.; Heuze, Y.; Flores, E. *Curr. Med. Chem.* **2005**, *12*, 927–943.
34. XMP.629 Profile. Investigational antimicrobial compound currently under development for mild-to-moderate acne vulgaris. Available online at: http://www.xoma.com/ (accessed Aug 2006).
35. Cano, R. J.; Lambert, L. H. Jr.; Wrighton, K.; Haak-Frendscho, M. In Vitro Efficacy of XMP.629, a Novel Antimicrobial Peptide for Acne Indication. Poster A-065, American Society for Microbiology General Meeting, May 23–27, 2004, New Orleans, LA, USA.
36. Topical anti-androgen PSK3841. Available online at: http://www.proskelia.com/rd_32.htm (accessed Aug 2006).
37. Battmann, T.; Bonfils, A.; Branche, C.; Humbert, J.; Goubet, F.; Teutsch, G.; Philibert, D. *J. Steroid Biochem. Mol. Biol.* **1994**, *48*, 55–60.
38. Product development: XMP.629 for acne. Available online at: http://www.xoma.com (accessed Aug 2006).

Biography

Stephen J Baker received a BSc in Medicinal Chemistry from the University of Sussex, UK. He was awarded a DPhil in organic chemistry, also at the University of Sussex, under the supervision of Prof Douglas W Young. His research involved studying enzymes involved in the biosynthesis of pteridine cofactors. Dr Baker then moved to the Pennsylvania State University, USA, to conduct postdoctoral studies under the guidance of Prof Stephen J Benkovic. During this time he worked as part of a team in collaboration with Prof Lucy Shapiro at Stanford University, which led to the discovery of a new class of boron-containing antibacterial agents and subsequently, the foundation of a new company, Anacor Pharmaceuticals, Inc. Following his postdoctoral studies, Dr Baker moved to the University of Leicester, UK, as a lecturer in organic chemistry and began research designing and developing inhibitors of protein–protein interactions. In 2003, Dr Baker moved to Anacor Pharmaceuticals, Inc., CA, USA, as a Senior Scientist in Medicinal Chemistry. He currently holds the position of Manager, Medicinal Chemistry. Dr Baker has been involved in the discovery and development of two compounds currently in clinical trials, one for acne, rosacea, and atopic dermatitis and the second for onychomycosis.

Comprehensive Medicinal Chemistry II
ISBN (set): 0-08-044513-6

ISBN (Volume 7) 0-08-044520-9; pp. 957–968

7.34 New Treatments for Psoriasis and Atopic Dermatitis

M J Elices, PharmaMar USA, Cambridge, MA, USA
T Arrhenius, Del Mar, CA, USA

7.34.1 Introduction

The cliché 'beauty is skin deep' exposes a deeply seated image in modern Western societies: physical appearance is very important socially, especially to our own psyche. Along with the fact that the skin is the body's largest organ, this makes dermatological pathologies a potential source of anguish and social ostracism. Take the chronic autoimmune skin disease psoriasis; it may not be a generally accepted life-threatening condition, but it does give rise to stress, poor quality of life, morbidity, and disability in afflicted individuals.

The skin and immune system are two major barriers that operate jointly to confer protection upon higher organisms. By and large, the former constitutes an engineering safety barrier while the latter utilizes both innate and adaptive strategies to carry out its role under a constant state of body surveillance. Thus, it stands to reason that most types of dysfunction at the crossroads of these two major systems should result in untoward pathologies that, by their own nature, are multifactorial because many players do participate. Also, one should not ignore a genetic component as in other autoimmune conditions, even though it may not follow a straightforward Mendelian pattern.

Two inflammatory skin diseases, namely, atopic dermatitis (AD) or eczema and psoriasis, whose prevalence is relatively high and rising in industrialized societies, have several commonalties. They both are: (1) chronic, (2) relapsing, (3) possibly of an autoimmune nature, (4) exhibiting defects in skin barrier function (AD in terms of an impaired innate immune system, and psoriasis in terms of hypertrophy of the epidermal layer and parakeratosis), and (5) of unknown etiology. Yet they have very different manifestations and particularly natural histories. Of note, AD typically develops during infancy as part of the so-called 'atopic march,' an ominous sounding term that includes asthma and allergic rhinitis,[1] while initial flares of psoriasis tend to occur in young adults. Of course, these are sweeping generalizations because both AD and psoriasis can actually occur at any age.

AD and psoriasis are also clinically very different.[1–3] Whereas there is no objective test to diagnose AD, clinicians can usually identify psoriasis because patients present characteristically well demarcated, often symmetrical, erythematous skin plaques with silvery scales that tend to slough off. From an immunologic standpoint, high serum titers of immunoglobulin E (IgE) antibodies (Abs) frequently characterize AD and at least in its initial acute phase it exhibits a T_{H2} cytokine phenotype (i.e., high expression of the interleukins of IL4 and IL13). In contrast, psoriasis is generally considered a disease exhibiting a T_{H1} cytokine phenotype since interferon gamma (IFN-γ) plays a central role. Undoubtedly, genetic predisposition in terms of an individual's major histocompatibility complex haplotype provides a founding seed so that either of these disorders may evolve in time as a full-fledged pathology.

Within the last five years or so, many comprehensive and authoritative articles have reviewed psoriasis and AD in terms of immunopathology[1–6] and clinical management,[7–13] even though understanding of many issues is still woefully lacking. Since our intent is not to regurgitate widely available information, this chapter will focus primarily on future trends in drug development to address psoriasis and AD. Thus, recently developed animal models, novel biological targets, and emerging therapeutics will instead be discussed.

7.34.2 Animal Models of Skin Inflammation

Selecting drugs for clinical studies and commercialization involves winnowing out a single lead compound and a handful of potential backups from a large number of chemicals, perhaps even hundreds of thousands. Consequently, the emphasis typically rests on quantitative elimination and speed in the early stages of the process. By design, in vitro systems are best suited to meet these objectives and thus should probably be used first. This approach also bypasses obvious ethical issues regarding the use of animals in drug development. Ultimately, a limited cohort of drug candidates must, however, be tested in an in vivo system. Therein lies the challenge since animal models highly predictive of human disease are not always available.

Key technological advances that took place in the latter part of the twentieth century, such as the development of immunodeficient, transgenic, and so-called 'knockout' animal systems, have facilitated a more specific pharmacologic approach to human disease. This has particularly paid fruitful dividends for skin inflammation. Nevertheless, one cannot say with certainty that there is a 'gold standard' model of skin inflammation for either AD or psoriasis. Thus, we will review some systems that are currently available to scientists in the field of dermatology.

7.34.2.1 The Classic Skin Inflammation Model

The classic studies on guinea pig contact and delayed-type hypersensitivity reactions by Benacerraf and co-workers provided the background for much-employed animal models for skin inflammatory conditions.[14,15] In fact, cutaneous contact hypersensitivity (CHS) became the workhorse of skin inflammation research for the last 40 years of the twentieth century, even though technically it may be best described in the pharmacology parlance as an 'intrinsic activity' model.

Pathologically, CHS more closely resembles AD. It consists of an inducing or sensitizing phase in which a small-molecule haptenic chemical such as dinitrofluorobenzene (DNFB), oxazolone (OXA), 12-*O*-tetradecanoylphorbol-13-acetate, or trinitrochlorobenzene (TNCB) is formulated in an appropriate vehicle and applied epicutaneously on the shaved abdomen of a naive animal, typically a mouse. This process will make treated animals immunized, or in other words, render them sensitive to the aforementioned hapten. After 1 or 2 weeks, the ear thickness of mice is quantitated with a caliper to obtain a baseline measurement. Then, they are challenged with the identical substance as above (i.e., DNFB, OXA, 12-*O*-tetradecanoylphorbol-13-acetate, or TNCB, respectively) on one ear while the contralateral ear is simply painted with vehicle. This design constitutes the so-called active form of CHS.

Alternatively, spleens from animals immunized as described in the previous paragraph are taken and T lymphocytes are purified from corresponding splenocytes using suitable techniques (e.g., chromatography on a glasswool column). The resulting hapten-sensitive T cells are injected intravenously into naive recipient mice and then challenged as above. This experimental protocol represents the adoptive transfer modality or so-called passive form of CHS.

Ultimately, the endpoint is again determined with a caliper, roughly 24 h after challenge and is illustrated as net increase in ear thickness. Parenthetically, it should be noted that each animal serves as its own control by virtue of the experimental design in either its active or passive formats (i.e., lateral ear challenged with sensitizing chemical and contralateral ear challenged with vehicle alone).

As mentioned earlier, several immunogenic haptenic chemicals have been used for CHS, namely, DNFB, OXA, 12-*O*-tetradecanoylphorbol-13-acetate, or TNCB.[16–22]

OXA has been reported to be a particularly convenient sensitizing immunogen.[23–26] It is relatively simple to formulate and use, is capable of generating the endpoint of ear swelling with a relatively fast turnaround for the standards of animal work (thus, a 'medium throughput' system), and gives rise to quantitative and reproducible data, particularly in its adoptive transfer modality.

In the category of classic delayed-type hypersensitivity (DTH) models, a transgenic system was recently developed that overexpressed vascular endothelial growth factor (VEGF)-A in the epidermis.[27] The authors followed the standard monitoring of ear thickness as endpoint.

7.34.2.2 Immunodeficient Models of Skin Inflammation

Since the CHS approach (see previous paragraph) does not address specific pathologic features of human psoriasis, a pharmacologic 'disease model' resembling this human condition may be recapitulated with xenotransplantation systems in immunodeficient animals. Thus, a clever animal model utilizes transplantation of human psoriatic plaques on severe combined immunodeficient (*scid/scid*, or SCID) mice recipients.[28] In this system, a graft bed is created on the back of a shaved SCID mouse making sure that an intact vessel plexus is maintained on the fascia of back muscles. Subsequently, a partial-thickness human skin biopsy is orthotopically xenografted following appropriate consent from the corresponding human psoriatic donor. A clear advantage of this system is that when corresponding punch biopsies are snap-frozen, it becomes amenable to using well-known histopathology endpoints of the psoriatic skin, such as the length of fingerlike rete ridges, in order to assess the potential efficacy of candidate drugs. Alternatively, other endpoints such as epidermal thickness, grade of parakeratosis, and numbers of inflammatory cells and cycling keratinocytes may be determined by using paraffin-embedded punch biopsies.[29]

Another transplantation model uses symptomless prepsoriatic human skin grafted onto immunodeficient homozygous mice lacking the recombination activating gene (*RAG*)-*2* and also engineered to be deficient in both type I and II interferon (IFN) receptors.[30] In these studies, histopathology indices associated with direct human clinical relevance such as acanthosis and papillomatosis, as defined by Fraki *et al.*[31] were monitored as endpoints.

7.34.2.3 Genetically Engineered Models (GEM) of Skin Inflammation

A particularly timely model of psoriasis is one utilizing mice with a CD18 hypomorphic mutation that results in approximately 10% of the CD18 expression present on wild-type littermates.[32] Indeed, reduced expression but not complete abrogation of CD18 appears to be necessary for the development of skin manifestations similar to human psoriasis in mice of the appropriate genetic background.[33]

Homozygous transgenic mice overexpressing VEGF-A are particularly amenable to using the standard CHS design as indicated above. The significant advantage of this system is that animals develop skin inflammatory lesions characterized by epidermal hyperplasia and dense T lymphocyte infiltration, which make them more aligned with psoriasis than classic CHS.[27] Nonetheless, it should be pointed out that homozygous transgenic mice, overexpressing VEGF-A, spontaneously develop a form of psoriasis after 6 months of age, while heterozygous mice do not.[34]

Males of the wild-type mice strain DBA/1 spontaneously develop a disease when they age (i.e., about 24 weeks and older) that is pathologically comparable to human psoriatic arthritis in its involvement of chronic inflammation and joint destruction.[35] Characteristically, ankylosing enthesitis and dactylitis are symptoms that can be used as endpoints in the aging male DBA/1 mice model albeit disease incidence is highly variable, sometimes not exceeding 50%.[35]

A transgenic mouse model containing the psoriasis-associated susceptibility gene *HCR* has recently been developed, which allows one to monitor specific genetic signatures by using microarray technology.[36] Consequently, potential comparison with those found in biopsies from human psoriasis patients may thus be feasible and suggests that suitable surrogate markers may be followed upon drug treatment both in preclinical studies and later on during clinical trials.

Lastly, targeted ablation of the epiregulin gene results in mice with the $EP^{-/-}$ mutation that spontaneously develop chronic dermatitis.[37]

A particularly timely recent report[38] promises to delve into the etiology of psoriasis. A corresponding animal model was thus developed using mice that carried conditional inducible deletions of the *JunB* and *c-Jun* genes specifically in basal layer keratinocytes (i.e., $JunB^{-/-}/c\text{-}Jun^{-/-}$ double knockout animals). The phenotype of these mice exhibited scaly plaques in ears, paws, and tails, and histologically, they showed thickened epidermis with characteristic fingerlike rete ridges, hyperkeratosis, and parakeratosis, as well as skin infiltration by T lymphocytes, neutrophils, and macrophages. Further characterization of the phenotype of these epidermal-specific $Jun^{-/-}/c\text{-}Jun^{-/-}$ double mutant mice reinforced their resemblance to human psoriasis in that mRNA expression for cytokines IL1-α, IL1β, IFN-γ, and tumor necrosis factor alpha (TNF-α) as well as chemokines providing strong neutrophil and T cell chemoattraction such as macrophage inflammatory protein-2 (MIP-2) (IL8 in humans), and MIP-1-α, IP-10, and monocyte chemoattractant protein-1 (MCP-1), respectively, was upregulated in skin lesions. Interestingly, the presence of T cells did not appear to be required for the appearance of psoriasis symptoms, but merely had an exacerbating effect. Taken together, these data suggest that dysregulation of *JunB* and *c-Jun* are likely early events in psoriasis and a potentially relevant animal model of the disease may now be available to discover and test novel therapies for psoriasis.

7.34.2.4 Conclusions on Animal Models

The plethora of systems developed around the time of the 'human genome revolution' is likely to pay off regarding the discovery of new targeted therapies for skin inflammatory conditions, especially psoriasis. Undoubtedly, this will take some time to come to fruition because these models are relatively new and thus have been subjected to a fairly limited amount of experimental testing. In addition, the issue of predictability immediately springs to mind for any of these systems in terms of discovering and developing brand-new drugs with putative novel mechanisms of action. For instance, mice with CD18 hypomorphic mutations are very much aligned with therapies recently approved by the FDA that target CD18 (e.g., efalizumab). However, one wonders if the use of this system will bias drug discovery toward substances that solely operate along this signaling pathway.

Since a wide consensus on the 'gold standard' animal model for skin inflammation does not yet exist, we recommend the time-honored pharmacologic approach of first using an 'intrinsic activity model' to select a handful of drug candidates and then follow up with a more narrowly defined 'disease model' that identifies a lead candidate and backups.

7.34.3 Present and Future Approaches to the Treatment of Skin Inflammation

7.34.3.1 T Cell Activation and Signaling

T cell activation and signaling may be considered the newest targeted therapies based on an immunologic understanding of skin inflammation.[7] Two novel biologics have recently received FDA approval for psoriasis, i.e., alefacept (an inhibitor of CD2 binding to lymphocyte function-associated antigen-3 (LFA-3)) and efalizumab, a monoclonal antibody (MAb) blocking interaction between LFA-1 and intercellular adhesion molecule 1 (ICAM-1). LFA-3 and LFA-1 are completely unrelated molecules since LFA-3 is a monomeric member of the immunoglobulin gene superfamily whereas LFA-1 is a heterodimeric member of the β2 integrin subfamily. Shared features include their expression on T lymphocytes, general involvement in T cell activation, and the fact that they can both be loosely described as cell adhesion receptors.

Alefacept is a recombinant humanized LFA-3/IgG1 fusion protein which binds to skin T lymphocytes, but not dendritic cells, thus resulting in a muted characteristic T_{H1}-type inflammatory response with a particularly reduced production of the cytokine IFN-γ.[39] Administration of alefacept (15 mg) once weekly for 12 weeks by the intramuscular route resulted in a statistically significant diminution in the endpoint psoriasis area and severity index (PASI) score 2 weeks after termination of treatment compared to the placebo-treated cohort, and this effect was also apparent when alefacept (7.5 mg) was dosed intravenously using the same regimen.[40] Importantly, phase III trials demonstrated that the drug was generally well tolerated.[40]

Efalizumab, in turn, is a humanized MAb that binds to the alpha subunit of LFA-1 (α_L or CD11a) and in so doing it prevents T cell activation. The mechanism of action appears to involve blocking localization of the $CD8^+$ memory T cell subset to the psoriatic skin in a reversible fashion.[41] In double-blind trials, patients with moderate to severe plaque psoriasis experienced a statistically significant reduction in their PASI score compared to placebo-treated psoriatics when receiving once weekly subcutaneous administrations of efalizumab at a $1.0\,mg\,kg^{-1}$ dose for 12 weeks.[42,43]

The investigational drug CTLA4Ig (abatacept, Bristol-Myers Squibb) is a fusion protein that inhibits the CD28–B7 (CD80/CD86) molecular interaction and thus prevents T cell costimulation.[44] Interestingly, CTLA4Ig possesses the ability to inhibit production of antibodies to double-stranded DNA and thus it may be of utility to treat human systemic lupus erythematosus, as shown in a murine model of nephritis.[44] Moreover, it appears that CTLA4Ig may enhance the relative percentage of CD4/CD25 regulatory T cells and thus induce tolerance.[45] Notwithstanding this data, CTLA4Ig given by intravenous infusions was also clinically efficacious in patients with psoriasis vulgaris in an open label, dose escalation phase I study, and most importantly was well tolerated.[46] The mechanism of action of CTLA4Ig appeared to involve impaired activation of lesional T cells, keratinocytes, and mature dendritic cells particularly in the epidermis.[46] Therefore, CTLA4Ig may hold therapeutic promise for the treatment of psoriasis.

The calcineurin signaling pathway contains well-known targets that are addressed by immunosuppressive therapies, namely, cyclosporine A (Cs A) and the topical immunosuppressants (TIMs) (see below). However, well-characterized systemic adverse events are also known to arise from the use of these drugs, chiefly kidney toxicity. In fact, it is generally believed that one cannot separate the anti-inflammatory properties of these agents from their untoward side effects. Nonetheless, the Cs A analog ISAtx-247 may challenge this widely held notion.[47] In Phase II studies with moderate to severe psoriasis patients, oral administration of ISAtx-247 (0.25–0.75 $mg\,kg^{-1}$) twice a day for 12 weeks caused statistically significant reduction in 74% and 18% of psoriatics in the high- and low-dose cohort, respectively, 6 weeks after treatment cessation relative to the placebo control group.[47] Importantly, ISAtx-247 administration in this study did not give rise to any increases in serum creatinine levels outside of the normal range that could then be indicative of potential nephrotoxicity.[47] Thus, a soon to be initiated phase III trial in psoriasis patients may settle the issue of whether ISAtx-247 indeed holds promise as a safer and apparently more potent alternative to Cs A and thereby becomes a bona fide second-generation drug as an antipsoriatic therapy.

7.34.3.2 Topical Immunosuppressants

Two other relatively new agents (**Table 1**) are tacrolimus (FK506) and pimecrolimus (33-epi-chloro-33-desoxyascomycin), which are antagonists of T cell signaling and are recommended for AD therapy.[48–50] AD is a source of parents' anguish particularly in very young infants, especially since AD in children within modern Western societies has a relatively high 10–20% incidence.[51] Tacrolimus and pimecrolimus are also known collectively as topical immunosuppressants (TIMs) and operate via inhibition of the calcineurin pathway.

In 1987, the macrolide antibiotic FK506 (the active principle of tacrolimus) was isolated and characterized by fermentation from the bacterium *Streptomyces tsukubaensis*.[52] FK506 is capable of binding to the cytoplasmic FK506 binding protein (FKBP) FKBP12, and this ensuing complex inhibits calcineurin, a phosphatase whose activity is linked to signal transduction pathways initiated by engagement of the T cell receptor (TCR) by its cognate antigen (principally involving c-jun NH_2-terminal (JNK) and p38 kinases).

From a structure–activity relationship (SAR) standpoint, FK506 (**Table 1**) is related to a family of streptomycete-derived compounds that differ slightly in terms of their in vitro potency.[53] Separately, the structurally similar macrolide rapamycin, which was first isolated from a soil fungus found on Easter Island (Rapa Nui), behaves antagonistically with FK506,[54] thus suggesting that they share binding contacts to their common FKBP12 target. Yet, they must sharply differ in their binding to downstream targets since the FKBP12–rapamycin complex acts via mTOR (mammalian target of rapamycin) along the TCR signaling pathway in terms of its immunosuppressive action.[55]

Data on the relative clinical superiority of either TIM above are mixed and difficult to sort out,[51] although it should be pointed out that tacrolimus and pimecrolimus were first indicated for moderate-to-severe and mild-to-moderate AD, respectively.[48] Based on the eczema area severity index (EASI) endpoint, a recent review suggested that topical tacrolimus had a better efficacy performance and faster onset than pimecrolimus cream in adults, children with moderate to severe disease, and children with mild disease 1 week after termination of treatment.[56] On the safety side, tacrolimus use was associated with a greater number of application site reactions on day 1, but there was no statistical difference with pimecrolimus thereafter.[56] On the other hand, mild to moderate symptoms of irritation (burning, redness, and itching) occurred in 20% and 10% of tacrolimus- and pimecrolimus-treated children, respectively, even though they typically subsided after a few days.[51] In summary, TIMs appear to be efficacious agents for AD therapy, even in young children. However, only limited clinical experience currently exists with their long-term use, and potential adverse events may include nephrotoxicity, neurotoxicity, and diabetogenicity.

Table 1 Topical therapies for AD

Name (date of FDA approval)	*Structure*	*Molecular Target*	*Indication*
Tacrolimus (FK506) (December 2000)	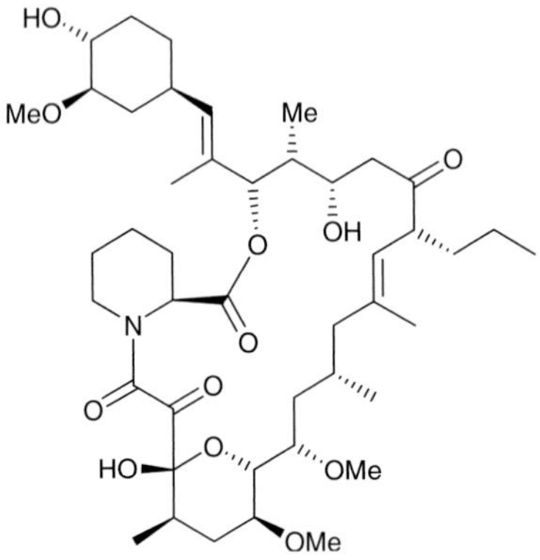	Calcineurin (via FKBP)	Moderate to severe AD
Pimecrolimus (July 2002)	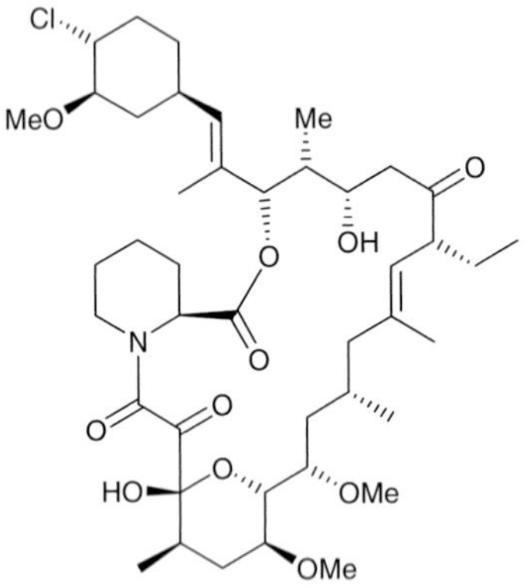	Calcineurin (via macrophilin-12 (FKBP12))	Mild to moderate AD
Prednicarbate (June 2002)	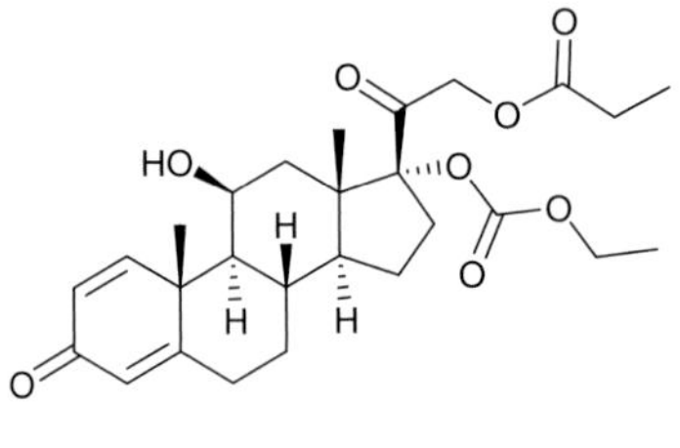	Glucocorticoid receptor	Acute and chronic corticosteroid-responsive dermatoses
Fluticasone (April 2004, first generic)	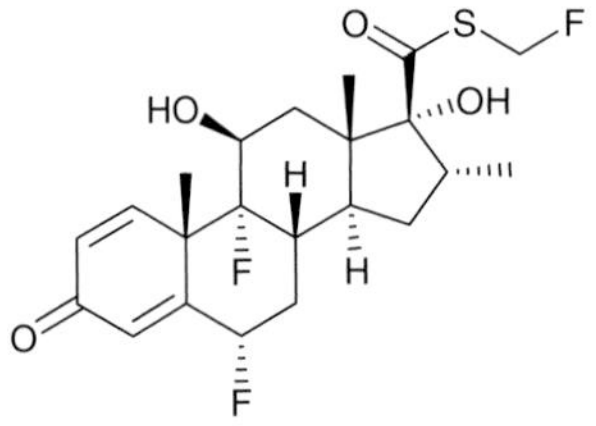	Glucocorticoid receptor	Corticosteroid-responsive dermatoses

7.34.3.3 Cytokines

In addition to targeting TNF-α, e.g., with the fusion protein etanercept and the humanized blocking MAb infliximab, other cytokines prominently expressed in psoriasis patients have been pursued as potential therapeutic agents and are reviewed below.

7.34.3.3.1 Etanercept

Etanercept is a recombinant TNF-α antagonist which received FDA approval as a subcutaneous monotherapy in adult patients with moderate to severe psoriasis and is also indicated in psoriatic arthritis. In patients with moderate to severe psoriasis, either 25 or 50 mg etanercept therapy administered twice weekly for 12 weeks caused a statistically significantly reduction in the PASI score of treated patients relative to placebo.[57] In a 1-year nonblinded extension study, short-term treatment with 25 mg etanercept twice weekly appeared to slow down joint damage significantly in patients with psoriatic arthritis as judged by radiographic analysis, and this was applicable either alone or in combination with methotrexate.[58] Since etanercept therapy showed a good profile of tolerability in these trials, this medication provides a promising new alternative for patients with chronic moderate to severe plaque psoriasis or with psoriatic arthritis.

7.34.3.3.2 Infliximab

Recently, the efficacy of infliximab was evaluated in patients with active psoriatic arthritis in the IMPACT-1[59] and -2[60] clinical trials. In the IMPACT-1 placebo-controlled study, the objective was to address articular and dermatologic manifestations of active psoriatic arthritis, and patients were treated with 5 mg kg^{-1} infliximab via intravenous infusions at weeks 0, 2, 6, 14; and after week 16, all patients received infliximab every 8 weeks until week 50.[59] Until the end of the trial on week 50, sustained infliximab treatment gave rise to significant reductions in PASI scores and improvement in articular and dermatologic manifestations of psoriatic arthritis.[59] The IMPACT-2 phase III study, in turn, enrolled 200 patients receiving 5 mg kg^{-1} infusions of infliximab at weeks 0, 2, 6, 14, and 22.[60] The results were fairly dramatic in that psoriatic arthritis patients exhibiting disease involvement in at least 3% of their body surface area at baseline experienced a highly significant reduction in their PASI scores relative to placebo at week 14, and these improvements continued through week 24.[60] Significant effects were also seen in dactylitis and enthesitis at week 14 and the MAb was well tolerated.[60] Thus, a relatively short-term infliximab therapy at a dose of 5 mg kg^{-1} appears to improve psoriatic arthritis symptoms significantly while it displays an acceptable tolerability profile.

7.34.3.3.3 Adalimumab

Lastly, adalimumab is a purely human TNF-α MAb, which is not yet approved for psoriasis, but has been approved for rheumatoid arthritis; yet, a phase II trial in psoriasis provides encouraging hints of efficacy.[61]

7.34.3.3.4 Interleukin-15

Blockade of interleukin (IL)-15 may be beneficial since IL15 triggers upregulation of the prototype T_{H1} inflammatory cytokine IFN-γ, as well as TNF-α and IL17 in tissue from psoriatic patients. In a xenotransplantation model using SCID mice, three weekly injections of a MAb to IL15 on days 1, 8, and 15 caused statistically significant amelioration of human psoriasis lesions, namely, length of epidermal thickness (from the stratum corneum to the deepest end of the rete ridges), number of infiltrating mononuclear inflammatory cells in the upper dermis, and number of cycling keratinocytes, when compared to vehicle-treated control animals.[29] Moreover, the antipsoriatic effect of 20 mg kg^{-1} IL15 MAb given intraperitoneally appeared to be superior to that of 10 mg kg^{-1} Cs A given intraperitoneally, an accepted standard treatment for human psoriasis, on these same endpoints when this drug was administered every other day for 15 days.[29] Thus, inhibition of IL15 shows promise as a potential therapy for psoriasis based on preclinical data with a blocking MAb.

7.34.3.3.5 Interleukin-18

IL18 is a cytokine that exhibits proinflammatory features some of which operate via activation of the nuclear factor kappa B (NFκB) pathway.[62] In particular, generation of interferon IFN-γ is upregulated by IL18 and this suggests that antagonists of IL18 may exert anti-inflammatory effects in conditions in which a T_{H1} inflammatory phenotype dominates, such as rheumatoid arthritis,[63,64] Crohn's disease, and psoriasis. In addition, animal models appear to bear out this hypothesis.[65] Nevertheless, IL18-deficient mice exhibited chronic dermatitis that developed spontaneously.[66] This suggests, in turn, that IL18 blockade may be efficacious in human AD. Actually, whether IL18 gives rise to T_{H1}- or T_{H2}-type inflammation may depend on concomitant exposure to IL12.[67] In conclusion, IL18-dependent therapies may

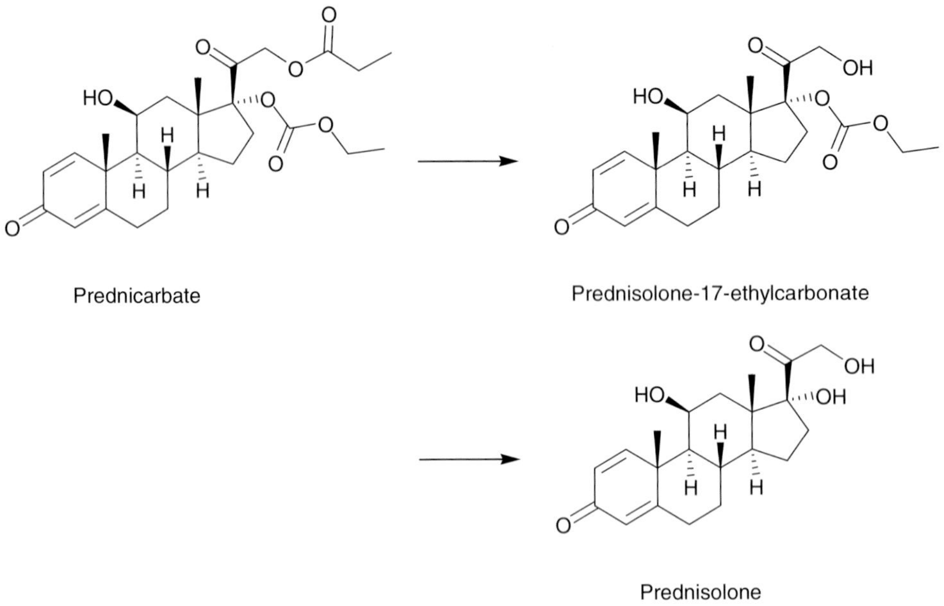

Prednicarbate

Prednisolone-17-ethylcarbonate

Prednisolone

Scheme 1

be useful in the treatment of skin inflammatory conditions, be it psoriasis (mostly T_{H1}-driven) or AD (mostly T_{H2}-driven), but this awaits further elucidation.

Parenthetically, a promising therapeutic approach that also targets IL18-dependent inflammation, particularly psoriasis, involves using the naturally occurring decoy receptor IL18 binding protein.[65]

7.34.3.4 Glucocorticoid Receptor Antagonists

Corticosteroid administration has proven to be effective in controlling the symptoms of many inflammatory diseases including AD. However, the frequently severe side effects of these drugs, particularly when given systemically, has limited their long-term use and stimulated the search for topical glucocorticoid receptor antagonists with an optimized benefit to risk ratio. From these efforts, a new class of nonhalogenated double-ester-type glucocorticosteroids has emerged, typified by prednicarbate (**Table 1**). Prednicarbate is rapidly metabolized in the skin to the more active metabolite, prednisolone-17-ethylcarbonate, which is then slowly converted to the less active prednisolone during permeation (**Scheme 1**). As a result of these transformations, that lead to favorable pharmacokinetic properties, prednicarbate has proven to be a relatively effective and safe agent for the treatment of corticosteroid-responsive dermatoses. The pharmacological properties and therapeutic use of this compound have been thoroughly reviewed elsewhere.[68]

7.34.3.5 Vitamin D_3 Analogs

This category of compounds[69] has been in use as topical ointments since the 1990s, e.g., calcitriol, calcipotriol (Denmark, 1990), and tacalcitol (Japan, 1993 marketed by Teijin) (**Table 2**). Based on data from a DTH mouse model, tacalcitol (100 ng) appears to be effective at reducing cutaneous inflammatory infiltrates and mast cell degranulation, which in turn may modulate keratinocyte proliferation.[18] In fact, these agents are indicated for topical administration to adult patients and appear to be suitable for inflammatory keratotic dermatoses such as mild to moderate psoriasis. In terms of their safety profile,[70] they do not seem to induce skin atrophy like corticosteroids, but do increase photosensitivity and cause hypercalcemia.[71,72] However, they appear to be superior when combined with corticosteroids[73,74] and ultraviolet (UV) treatment,[75] thus suggesting their compatibility with rotational therapy schemes (**Table 3**).

7.34.3.6 Antagonists of Folic Acid Metabolism

The standard-bearer in this category is methotrexate, a classic drug commonly used as an anticancer agent. In the rotational therapy regimen for psoriasis (**Table 3**), methotrexate is an often used drug along with Cs A, even though a

Table 2 Topical therapies for psoriasis

Name	*Structure*	*Indication*
Coal tar	Hydrocarbon	Rotational therapy
Calcitriol	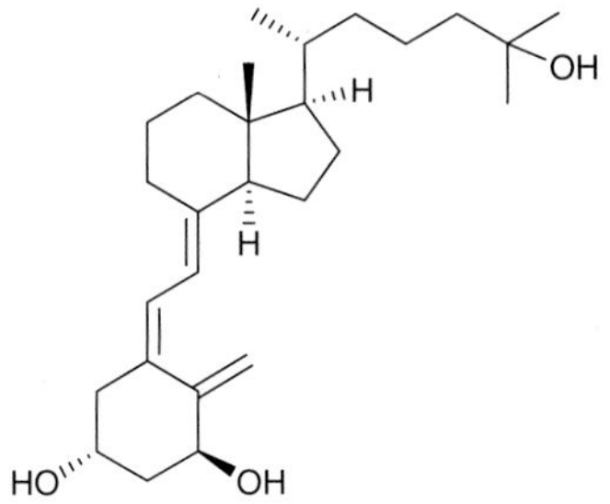	Mild to moderate psoriasis
Calcipotriol	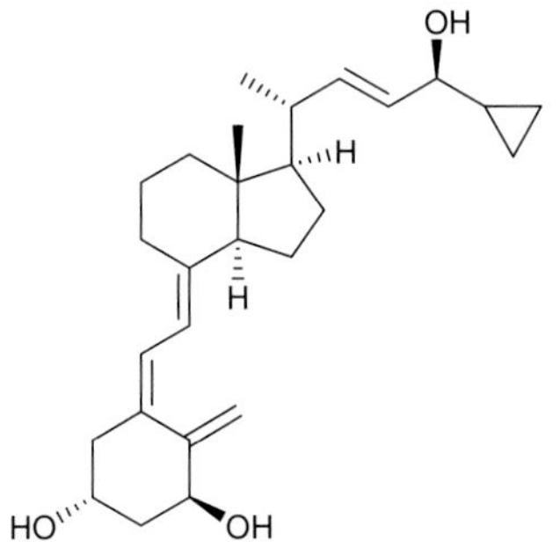	Mild to moderate psoriasis
Tacalcitol	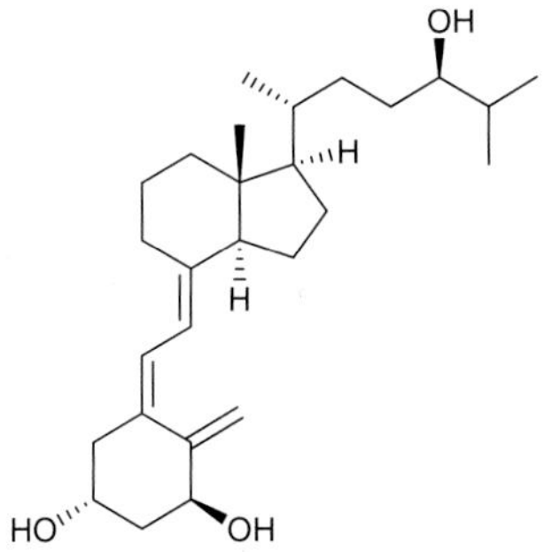	Mild to moderate psoriasis
Dithranol	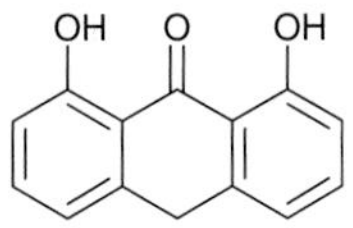	Mild psoriasis

relatively small 2003 clinical trial study ($n = 85$ patients) showed that there was no statistically significant difference between the two drug treatments after 16 weeks of therapy both in terms of the classic clinical endpoints PASI score and physicians' global assessments.[76] The preliminary conclusion from this clinical study is that individualized safety may be a prioritizing criterion, rather than efficacy, for either of the two drugs mentioned above when designing a treatment plan that fits a specific psoriasis patient.

Table 3 Rotational therapies for psoriasis

Name	*Structure*	*Indication*
UV B	N/A	Moderate psoriasis
Psoralen plus UV A (PUVA)	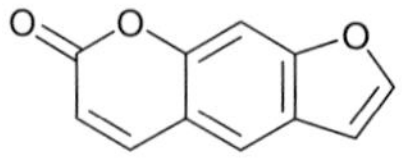 + UV A	Severe psoriasis
Methotrexate	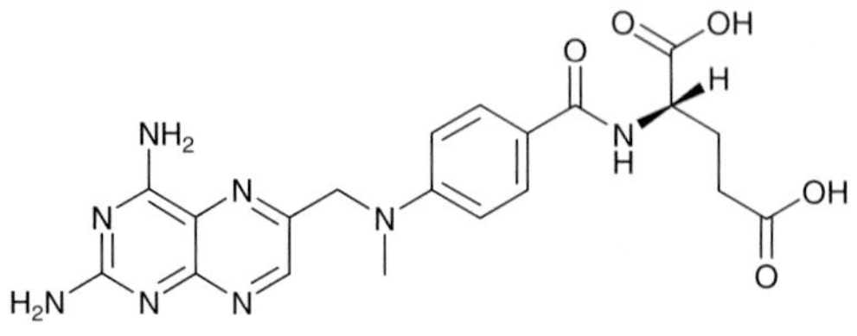	Severe psoriasis
Acitretin	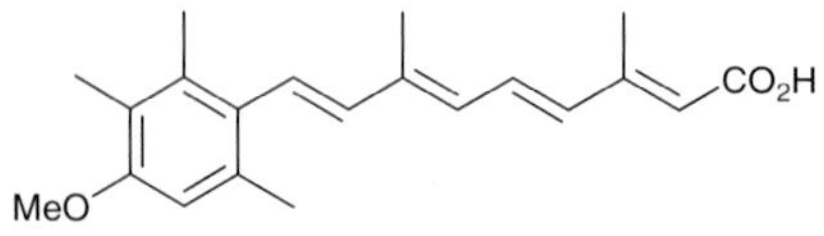 CO_2H, MeO	Severe psoriasis
Etretinate	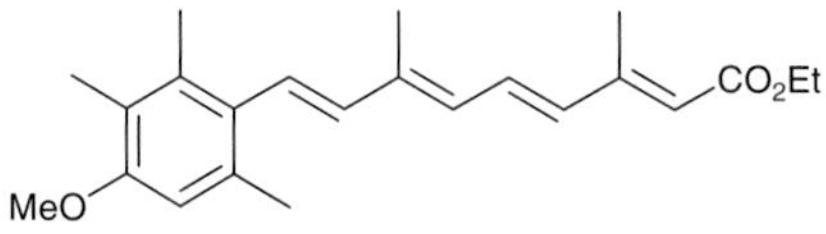 CO_2Et, MeO	Severe psoriasis
Cyclosporine A	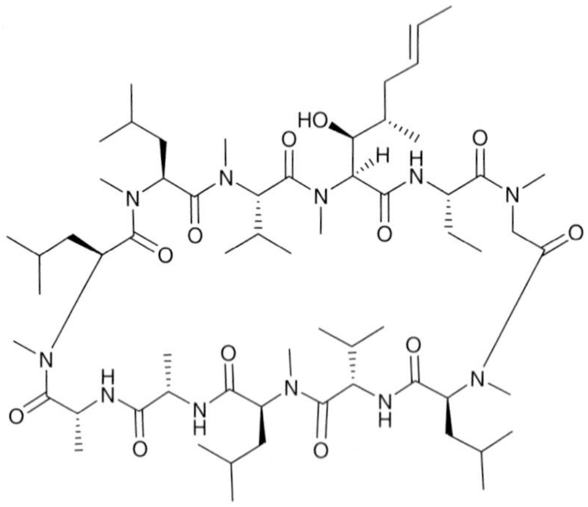	Severe psoriasis

7.34.3.7 Angiogenesis

Blood vessel sprouting occurs in the adult when certain specific growth factors stimulate endothelial cell proliferation thus giving rise to angiogenesis. Angiogenesis is not only critical to tumor spread in carcinogenesis, but is also fairly common in autoimmune inflammatory conditions, e.g., rheumatoid arthritis and psoriasis. VEGF-A is an angiogenic molecule that exerts its biological action by interacting with its receptors VEGFR-1 (Flt-1) and VEGFR-2 (Flk-1) found on endothelial cells. In a classic DTH model using OXA, transgenic mice specifically overexpressing a VEGF-A isoform in their skin developed psoriasis-like chronic inflammatory lesions compared to wild-type animals that solely exhibited the prototype acute phase in ear thickness, which peaked at around 24 h and subsided by 5 days after OXA challenge.[27] Histologically, the skin of VEGF-A-overexpressing mice demonstrated hyperplasia, parakeratosis, fingerlike protrusions into the dermis, and dense inflammatory infiltrates, particularly $CD4^+$ cells in the dermis and $CD8^+$ cells in the epidermis, even 1 month after OXA challenge.[27] In sharp contrast, wild-type animals essentially showed no signs of inflammation and a normal-looking skin architecture at 1 month post challenge.[27]

To demonstrate the therapeutic effect of blocking VEGF-A systemically, MAbs to VEGFR-1 and VEGFR-2 administered together, but not either one alone or an isotype-matched control, significantly inhibited lymphatic vessel enlargement and skin infiltration by $CD11b^+$ leukocytes in a classic OXA-induced DTH model after OXA challenge using wild-type mice.[27] Thus, antagonizing VEGF-A may be a novel biological strategy to treat psoriasis with the caveat that it presumably may have to affect both VEGF-A receptors, VEGFR-1 and VEGFR-2, in order to be clinically effective.

7.34.3.8 Phosphodiesterase 4 (PDE4) Antagonists

The phosphodiesterase 4 (PDE4) family is an attractive therapeutic target for the treatment of AD. Leukocytes from AD patients have reduced cyclic AMP (cAMP) levels, resulting from abnormally high PDE4 activity.[77,78] Increased intracellular cAMP levels, resulting from the inhibition of PDE4, have pronounced anti-inflammatory effects. The recruitment of immune cells and release of proinflammatory mediators are blocked by the inhibition of PDE4. Topical administration of the PDE4 selective inhibitor, CP80633 (Pfizer, Inc.) (**Table 4**), has been demonstrated to result in reduced skin inflammation in AD patients.[79] These observations have contributed to a long-standing interest in developing selective PDE4 inhibitors as therapeutic agents. However, the clinical utility of systemic administered compounds continues to be limited by unpleasant side effects, most notably nausea and vomiting, and no PDE4 selective inhibitor is yet available on the market. The causes of the side effects remain poorly understood. Alternative approaches to the development of PDE4-targeted drugs, with diminished side effects, are being considered.[80] These include the development of PDE4 subfamily specific inhibitors, the targeting of noncatalytic sites of the enzyme, controlling the phosphorylation state of PDE4, and modulating the interaction of PDE4 with other regulatory proteins.

7.34.3.9 Antileukotriene Drugs

The leukotrienes (LTs) are derived from arachidonic acid metabolism via the enzymatic activity of 5-lipoxygenase and they are produced in response to a variety of cell-activating stimuli. Two classes of LTs exist: (1) the so-called cysteinyl LTs, LTC4, LTD4, and LTE4, which all contain a thioether-linked peptide; and (2) LTB4, characterized by the absence of a thioether linkage.

Enhanced production of both cysteinyl and noncysteinyl LTs has been implicated in the pathogenesis of AD. Although the results of in vitro studies on LT production in AD have been inconsistent,[81] elevated LT levels have been reported in leukocytes[82,83] and in suction blister fluid[84–87] from AD patients. In addition, urine concentrations of LTE4 have been shown to correlate with AD disease severity.[88–90] Small clinical and case studies with the LTD4 receptor antagonists, montelukast[91–93] (**Table 4**) and zafirlukast[94] (**Table 4**) further support the involvement of LTs in AD. Thus, these agents are reported to be a safe and effective alternative when considering steroid-sparing therapies for the management of patients with AD.

7.34.3.10 Mitogen-Activated Protein Kinase p38 (MAPK p38) Inhibitors

The mitogen-activated protein kinase p38 (MAPK p38) has emerged as an attractive target for the treatment of inflammatory disorders. This kinase plays a key role in the regulation of inflammatory cytokine production since MAPK p38 inhibitors have been demonstrated to reduce levels of the T_{H1}-type cytokines TNF-α and IL1, both in vitro and in vivo.[95–99] Consequently, the development of MAPK p38 inhibitors has experienced a flurry of attention for the treatment of T_{H1}-driven inflammation.[100–108] Structurally diverse classes of MAPK p38 inhibitors have been discovered. Several of these classes are exemplified in **Figure 1**.[97,108–110]

In addition to serving as a general regulator of inflammatory cytokine production, the p38d isoform of MAPK p38 has specifically been shown to play an important role in inducing keratinocyte differentiation.[111] This observation, prompted a recent investigation into the expression of MAPK p38, as well as extracellular signal regulated kinase

Table 4 Other chemical structures

Name	*Structure*	*Comment*
CP80633	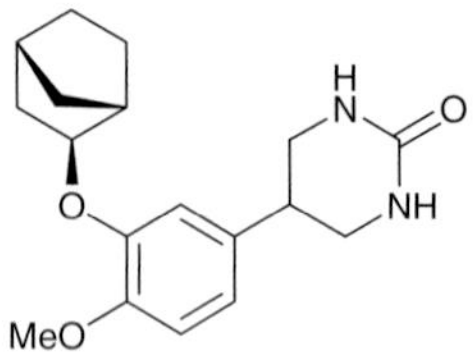	PDE4 antagonist
CP99994	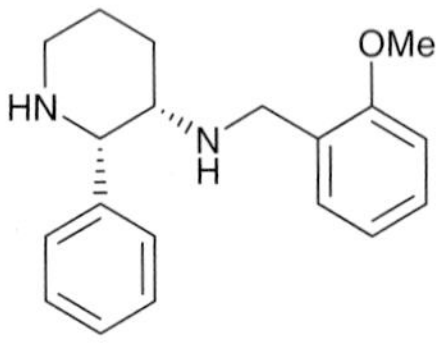	NK-1 antagonist
FTY-720	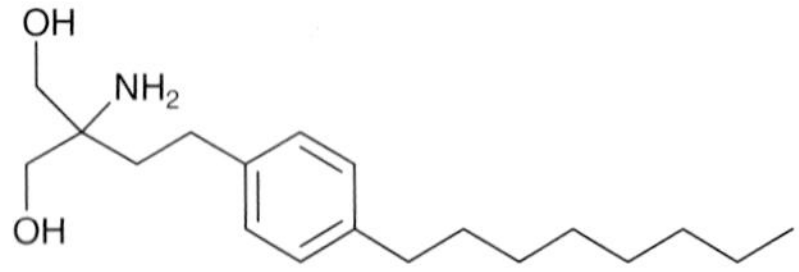	$S1P_1$ agonist
Kahalalide F	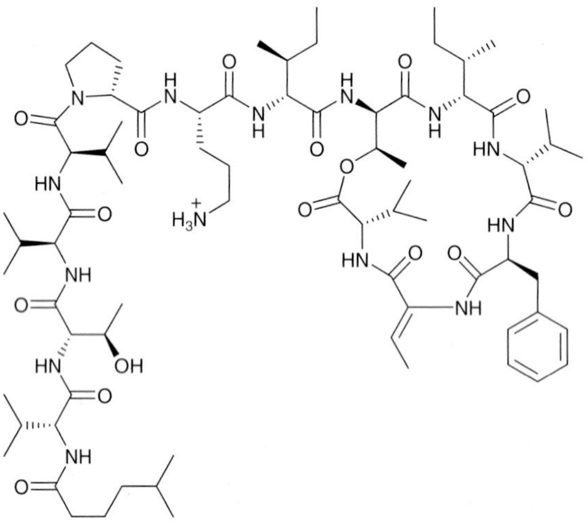	Target currently unknown
Montelukast	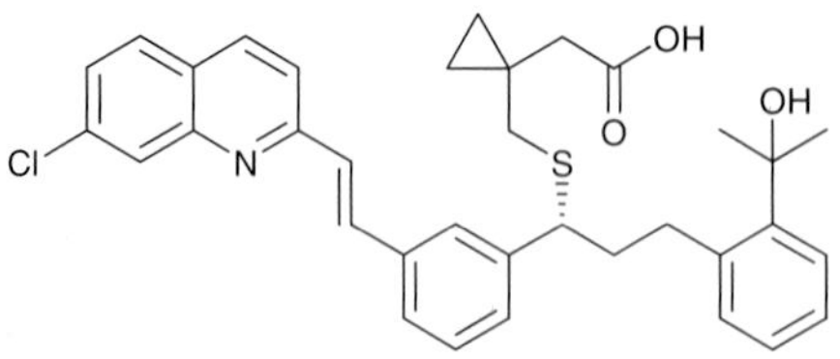	LT receptor antagonist
Zafirlukast	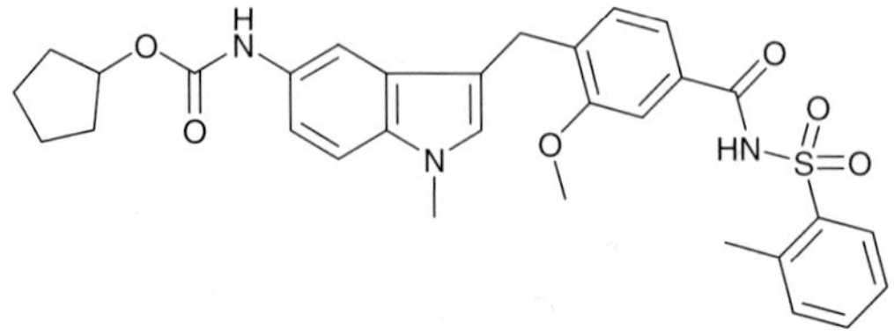	LT receptor antagonist

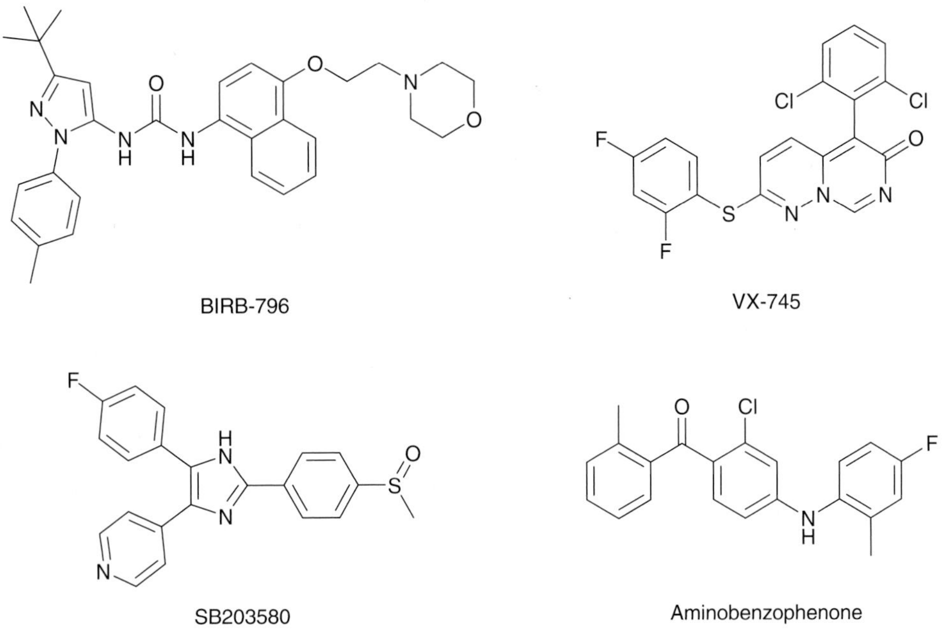

BIRB-796

VX-745

SB203580

Aminobenzophenone

Figure 1 Representative examples of MAPK p38 inhibitors.

(ERK) and JNK in psoriatic skin.[112] In fact, the phosphorylated forms of MAPK p38 and ERK were upregulated at lesional sites in psoriasis, and as a result, MAPK p38 activity was increased in psoriatic skin tissues. Furthermore, clearance of the psoriatic lesions, induced by climatotherapy at the Dead Sea for 4 weeks, led to a normalization in the MAPK p38 and ERK activities. These data provide suggestive evidence to support the validity of these two kinases as targets for the treatment of psoriasis. In particular, the development of topical agents may be appealing in light of potential toxicities often associated with systemic administration of kinase inhibitors.

7.34.3.11 Chemokine Receptor 3 (CCR3) Inhibitors

The chemokine receptor, CCR3, a member of the seven transmembrane domain G protein-coupled receptor (GPCR) family, has been the subject of considerable investigation. The potential for the use of antagonists of CCR3 as therapeutic agents for the treatment of allergic diseases is the subject of a recent, excellent review[113] and is briefly summarized below, in the context of AD.

The infiltration of eosinophils is a hallmark of allergic skin diseases, such as AD. Chemokines are involved in the recruitment of these eosinophils, and the chemokine receptor CCR3 is regarded as the principal mediator of eosinophil chemotaxis. This observation has triggered the search for compounds that inhibit the binding of CCR3 to its ligands. The result of these efforts is the discovery of CCR3 blocking antibodies, as well as peptide and small-molecule CCR3 antagonists, some of which have entered into clinical trials.

As expected, these CCR3 inhibitors interfere with eosinophil chemotaxis in vitro, and additionally, they prevent calcium fluxes. However, the ability of CCR3 antagonists to block eosinophil recruitment in in vivo models of AD has not yet been reported and no data on the activities of these inhibitors in humans have been published. The results of ongoing studies, on this exciting class of compounds, are awaited with much anticipation.

7.34.3.12 Neurogenic Inflammation

There is evidence that neurogenic inflammation, the response that is triggered within the peripheral nervous system, plays a role in exacerbating several human inflammatory conditions including psoriasis.[28] Specifically, keratinocytes demonstrate augmented levels of nerve growth factor (NGF) and also there is upregulation of NGF receptor (NGF-R) in the nerves surrounding psoriatic plaques.[28] This suggests that targeting NGF or the high-affinity NGF-R with either specific neutralizing antibodies or antagonists, respectively, may offer therapeutic benefit to psoriasis patients.

The NGF-R antagonist K252a (Alexis Biochemical) at a dose of 50 μg kg^{-1} twice a day for 14 days was delivered by intralesional injection on psoriatic plaques in the human/SCID mouse xenotransplantation model.[114] By way of preamble, grafted psoriatic lesions showed marked upregulation of NGF expression in keratinocyte areas and enhanced regeneration of NGF-R-positive nerve fibers as early as 4 weeks; however, a 2-week treatment with K252a blunted the latter proliferative response.[114] Moreover, K252a dosing brought about a statistically significant reduction in the length of rete ridges in psoriatic skin as compared to placebo-treated controls.[114] Other endpoints such as number of dermal papillary nerves also demonstrated a significant decrease in K252a-dosed animals relative to either the beginning of treatment or those found in control mice at the same time point.[114] In conclusion, antagonizing NGF-R with K252a results in preclinical efficacy and supports further development of this compound as a putative agent for psoriasis.

By their nature, skin diseases such as psoriasis possess a stress-induced inflammatory component and thus neurogenic inflammation plays a role. A key agonist in this system is substance P (SP), which binds to neurokinin-1 (NK-1) receptors, and skin mast cells express NK-1 receptors thus suggesting that stress may induce cutaneous mast cell degranulation. Recently, injection of CP99994 (**Table 4**), a specific nonpeptide NK-1 antagonist, inhibited stress-induced cutaneous mast cell degranulation in a rat preclinical model.[115] These data gives support to the notion that CP99994 may be used therapeutically to treat psoriasis.

7.34.3.13 Other Targets

Kahalalide F (KF) is a marine-derived cyclic depsipeptide that is being studied as an antitumor agent and has entered phase I clinical trials for psoriasis in Spain (PharmaMar) (**Table 4**). Mechanism of action data suggest that KF-treated cells undergo severe cytoplasmic swelling and vacuolization, and other types of intracellular damage,[116] some of which are reminiscent of type II programmed cell death or autophagy.[117]

Based on the argument that mRNA transcripts for several cytokines of the T_{H1}-type such as IL2, IL6, and IFN-γ were downregulated[118,119] by the compound FTY-720 (Novartis), (**Table 4**) one may hypothesize that this compound or a suitable analog may also have utility in the T_{H1}-driven disorder psoriasis. FTY-720 is a sphingosine-1-phosphate receptor-1 ($S1P_1$) agonist (recently reviewed by Elices[120]), and as such it represents a novel mechanism of action for skin inflammation. Nonetheless, its potential clinical utility in diseases like psoriasis remains speculative at this juncture.

7.34.4 Final Considerations

Clinically, rotational therapy is commonly utilized in modern psoriasis practice (**Table 3**), and it is based on the use of approved drugs with significant systemic toxicities. Some of those drugs are also given to cancer patients, thus restricting their applicability to time-limited courses of treatment, typically no more than two years. By definition, this disease management strategy makes psoriasis a permanently underserved medical need. Furthermore, it points to at least two areas in which advances are needed relative to already approved therapies: new drugs with different mechanisms of action and medications with an enhanced tolerability profiles, or both.

Notwithstanding this mainstay of clinical practice, the last half a decade has been prodigious in bearing fruit as newly approved therapeutic modalities to skin inflammatory disorders have appeared on the market. In psoriasis, alefacept and efalizumab have ushered in the era of targeted immunotherapy especially for the adult population. For pediatric patients, particularly young infants, the TIMs tacrolimus and pimecrolimus appear to be reasonably safe. They indeed offer a ray of hope and relief for many parents. In summary, we anticipate, with a modicum of optimism, that the momentum that has been gained in the development of new therapies for the treatment of skin inflammatory disorders will continue unabated over the next decade.

References

1. Leung, D. Y.; Boguniewicz, M.; Howell, M. D.; Nomura, I.; Hamid, Q. A. *J. Clin. Invest.* **2004**, *113*, 651–657.
2. Nickoloff, B. J.; Nestle, F. O. *J. Clin. Invest.* **2004**, *113*, 1664–1675.
3. Schön, M.; Boehncke, W.-H. *N. Eng. J. Med.* **2005**, *352*, 1899–1912.
4. Lowes, M. A.; Lew, W.; Krueger, J. G. *Dermatol. Clin.* **2004**, *22*, 349–369.
5. Wollenberg, A.; Kraft, S.; Oppel, T.; Bieber, T. *Clin. Exp. Dermatol.* **2000**, *25*, 530–534.
6. Barker, J. N. *Lancet* **1991**, *338*, 227–230.
7. Kupper, T. S. *N. Engl. J. Med.* **2003**, *349*, 1987–1990.
8. Novak, N.; Kwiek, B.; Bieber, T. *Clin. Exp. Dermatol.* **2005**, *30*, 160–164.
9. Roos, T. C.; Geuer, S.; Roos, S.; Brost, H. *Drugs* **2004**, *64*, 2639–2666.
10. Sidbury, R.; Hanifin, J. M. *Clin. Exp. Dermatol.* **2000**, *25*, 559–566.
11. Thestrup-Pedersen, K. *Clin. Exp. Dermatol.* **2000**, *25*, 535–543.

12. Williams, H. C. *Clin. Exp. Dermatol.* **2000**, *25*, 522–529.
13. Krueger, J. G. *J. Am. Acad. Dermatol.* **2002**, *46*, 1–26.
14. Benacerraf, B.; Gell, P. G. *Immunology* **1959**, *2*, 219–229.
15. Gell, P. G.; Benacerraf, B. *J. Exp. Med.* **1961**, *113*, 571–585.
16. Gorbachev, A. V.; Fairchild, R. L. *J. Immunol.* **2004**, *172*, 2286–2295.
17. Mascia, F.; Mariani, V.; Girolomoni, G.; Pastore, S. *Am. J. Pathol.* **2003**, *163*, 303–312.
18. Sato, H.; Nakayama, Y.; Yamashita, C.; Uno, H. *J. Dermatol.* **2004**, *31*, 200–217.
19. Schwarz, A.; Maeda, A.; Wild, M. K.; Kernebeck, K.; Gross, N.; Aragane, Y.; Beissert, S.; Vestweber, D.; Schwarz, T. *J. Immunol.* **2004**, *172*, 1036–1043.
20. Varona, R.; Villares, R.; Carramolino, L.; Goya, I.; Zaballos, A.; Gutierrez, J.; Torres, M.; Martinez, A. C.; Marquez, G. *J. Clin. Invest.* **2001**, *107*, R37–R45.
21. Gautam, S.; Battisto, J.; Major, J. A.; Armstrong, D.; Stoler, M.; Hamilton, T. A. *J. Leukoc. Biol.* **1994**, *55*, 452–460.
22. Ferguson, T. A.; Kupper, T. S. *J. Immunol.* **1993**, *150*, 1172–1182.
23. Arrhenius, T.; Chiem, A.; Elices, M.; He, Y.-B.; Jia, L.; Maewal, A.; Muller, D.; Gaeta, F. In *Peptides: Chemistry, Structure and Biology*; Kaumaya, P. T. P., Hodges, R. S., Eds.; Mayflower Scientific: Columbus, OH, 1995, pp 337–339.
24. Elices, M. J. *CIBA Found. Symp.* **1995**, *189*, 79-85; discussion 85-90, 174–176.
25. Elices, M. J.; Tamraz, S.; Tollefson, V.; Vollger, L. W. *Clin. Exp. Rheumatol.* **1993**, *11*, S77–S80.
26. Tamraz, S.; Arrhenius, T.; Chiem, A.; Forrest, M. J.; Gaeta, F. C.; He, Y. B.; Lei, J.; Maewal, A.; Phillips, M. L.; Vollger, L. W. et al. *Springer Semin. Immunopathol.* **1995**, *16*, 437–441.
27. Kunstfeld, R.; Hirakawa, S.; Hong, Y. K.; Schacht, V.; Lange-Asschenfeldt, B.; Velasco, P.; Lin, C.; Fiebiger, E.; Wei, X.; Wu, Y. et al. *Blood* **2004**, *104*, 1048–1057.
28. Raychaudhuri, S. P.; Raychaudhuri, S. K. *Prog. Brain Res.* **2004**, *146*, 433–437.
29. Villadsen, L. S.; Schuurman, J.; Beurskens, F.; Dam, T. N.; Dagnaes-Hansen, F.; Skov, L.; Rygaard, J.; Voorhorst-Ogink, M. M.; Gerritsen, A. F.; van Dijk, M. A. et al. *J. Clin. Invest.* **2003**, *112*, 1571–1580.
30. Boyman, O.; Hefti, H. P.; Conrad, C.; Nickoloff, B. J.; Suter, M.; Nestle, F. O. *J. Exp. Med.* **2004**, *199*, 731–736.
31. Fraki, J. E.; Briggaman, R. A.; Lazarus, G. S. *J. Invest. Dermatol.* **1983**, *80*, 31s–35s.
32. Kess, D.; Peters, T.; Zamek, J.; Wickenhauser, C.; Tawadros, S.; Loser, K.; Varga, G.; Grabbe, S.; Nischt, R.; Sunderkotter, C. et al. *J. Immunol.* **2003**, *171*, 5697–5706.
33. Barlow, S. C.; Collins, R. G.; Ball, N. J.; Weaver, C. T.; Schoeb, T. R.; Bullard, D. C. *Am. J. Pathol.* **2003**, *163*, 197–202.
34. Xia, Y. P.; Li, B.; Hylton, D.; Detmar, M.; Yancopoulos, G. D.; Rudge, J. S. *Blood* **2003**, *102*, 161–168.
35. Lories, R. J.; Matthys, P.; de Vlam, K.; Derese, I.; Luyten, F. P. *Ann. Rheum. Dis.* **2004**, *63*, 595–598.
36. Elomaa, O.; Majuri, I.; Suomela, S.; Asumalahti, K.; Jiao, H.; Mirzaei, Z.; Rozell, B.; Dahlman-Wright, K.; Pispa, J.; Kere, J. et al. *Hum. Mol. Genet.* **2004**, *13*, 1551–1561.
37. Shirasawa, S.; Sugiyama, S.; Baba, I.; Inokuchi, J.; Sekine, S.; Ogino, K.; Kawamura, Y.; Dohi, T.; Fujimoto, M.; Sasazuki, T. *Proc. Natl. Acad. Sci. USA* **2004**, *101*, 13921–13926.
38. Zenz, R.; Eferl, R.; Kenner, L.; Florin, L.; Hummerich, L.; Mehic, D.; Scheuch, H.; Angel, P.; Tschachler, E.; Wagner, E. F. *Nature* **2005**, *437*, 369–375.
39. Chamian, F.; Lowes, M. A.; Lin, S. L.; Lee, E.; Kikuchi, T.; Gilleaudeau, P.; Sullivan-Whalen, M.; Cardinale, I.; Khatcherian, A.; Novitskaya, I. et al. *Proc. Natl. Acad. Sci. USA* **2005**, *102*, 2075–2080.
40. Weinberg, J. M. *Clin. Ther.* **2003**, *25*, 2487–2505.
41. Vugmeyster, Y.; Kikuchi, T.; Lowes, M. A.; Chamian, F.; Kagen, M.; Gilleaudeau, P.; Lee, E.; Howell, K.; Bodary, S.; Dummer, W.; Krueger, J. G. *Clin. Immunol.* **2004**, *113*, 38–46.
42. Wellington, K.; Perry, C. M. *Am. J. Clin. Dermatol.* **2005**, *6*, 113-118; discussion 119–120.
43. Lebwohl, M.; Tyring, S. K.; Hamilton, T. K.; Toth, D.; Glazer, S.; Tawfik, N. H.; Walicke, P.; Dummer, W.; Wang, X.; Garovoy, M. R. et al. *N. Engl. J. Med.* **2003**, *349*, 2004–2013.
44. Dall'Era, M.; Davis, J. *Lupus* **2004**, *13*, 372–376.
45. Miao, G.; Ito, T.; Uchikoshi, F.; Akamaru, Y.; Kiyomoto, T.; Komodo, H.; Song, J.; Nozawa, M.; Matsuda, H. *Transplantation* **2004**, *78*, 59–64.
46. Abrams, J. R.; Kelley, S. L.; Hayes, E.; Kikuchi, T.; Brown, M. J.; Kang, S.; Lebwohl, M. G.; Guzzo, C. A.; Jegasothy, B. V. et al. *J. Exp. Med.* **2000**, *192*, 681–694.
47. Dumont, F. J. *Curr. Opin. Invest. Drugs* **2004**, *5*, 542–550.
48. Weinberg, J. M. *J. Manag. Care Pharm.* **2005**, *11*, 56–64.
49. Wellington, K.; Noble, S. *Am. J. Clin. Dermatol.* **2004**, *5*, 479–495.
50. Wolff, K. *Dermatol. Clin.* **2004**, *22*, 461–465, ix–x.
51. Breuer, K.; Werfel, T.; Kapp, A. *Am. J. Clin. Dermatol.* **2005**, *6*, 65–77.
52. Kino, T.; Hatanaka, H.; Hashimoto, M.; Nishiyama, M.; Goto, T.; Okuhara, M.; Kohsaka, M.; Aoki, H.; Imanaka, H. *J. Antibiot. (Tokyo)* **1987**, *40*, 1249–1255.
53. Hatanaka, H.; Kino, T.; Miyata, S.; Inamura, N.; Kuroda, A.; Goto, T.; Tanaka, H.; Okuhara, M. *J. Antibiot. (Tokyo)* **1988**, *41*, 1592–1601.
54. Dumont, F. J.; Melino, M. R.; Staruch, M. J.; Koprak, S. L.; Fischer, P. A.; Sigal, N. H. *J. Immunol.* **1990**, *144*, 1418–1424.
55. Hong, J. C.; Kahan, B. D. *Semin. Nephrol.* **2000**, *20*, 108–125.
56. Paller, A. S.; Lebwohl, M.; Fleischer, A. B., Jr.; Antaya, R.; Langley, R. G.; Kirsner, R. S.; Blum, R. R.; Rico, M. J.; Jaracz, E.; Crowe, A. et al. *J. Am. Acad. Dermatol.* **2005**, *52*, 810–822.
57. Leonardi, C. L.; Powers, J. L.; Matheson, R. T.; Goffe, B. S.; Zitnik, R.; Wang, A.; Gottlieb, A. B. *N. Engl. J. Med.* **2003**, *349*, 2014–2022.
58. Goldsmith, D. R.; Wagstaff, A. J. *Am. J. Clin. Dermatol.* **2005**, *6*, 121–136.
59. Antoni, C. E.; Kavanaugh, A.; Kirkham, B.; Tutuncu, Z.; Burmester, G. R.; Schneider, U.; Furst, D. E.; Molitor, J.; Keystone, E.; Gladman, D. et al. *Arthritis Rheum.* **2005**, *52*, 1227–1236.
60. Antoni, C.; Krueger, G. G.; de Vlam, K.; Birbara, C.; Beutler, A.; Guzzo, C.; Zhou, B.; Dooley, L. T.; Kavanaugh, A. *Ann. Rheum. Dis.* **2005**, *52*, 1227–1236.
61. Scheinfeld, N. *J. Dermatol. Treat.* **2004**, *15*, 348–352.
62. Suk, K.; Yeou Kim, S.; Kim, H. *Immunol. Lett.* **2001**, 77, 79–85.

63. McInnes, I. B.; Liew, F. Y.; Gracie, J. A. *Arthritis Res. Ther.* **2005**, 7, 38–41.
64. Berenson, L. S.; Ota, N.; Murphy, K. M. *Immunol. Rev.* **2004**, *202*, 157–174.
65. Muhl, H.; Pfeilschifter, J. *Eur. J. Pharmacol.* **2004**, *500*, 63–71.
66. Tsutsui, H.; Yoshimoto, T.; Hayashi, N.; Mizutani, H.; Nakanishi, K. *Immunol. Rev.* **2004**, *202*, 115–138.
67. Nakanishi, K.; Yoshimoto, T.; Tsutsui, H.; Okamura, H. *Cytokine Growth Factor Rev.* **2001**, *12*, 53–72.
68. Spencer, C. M.; Wagstaff, A. J. *BioDrugs* **1998**, *9*, 61–86.
69. Uhoda, I.; Quatresooz, P.; Hermanns-Le, T.; Pierard-Franchimont, C.; Arrese, J. E.; Pierard, G. E. *Dermatology* **2003**, *206*, 366–369.
70. Bruner, C. R.; Feldman, S. R.; Ventrapragada, M.; Fleischer, A. B., Jr. *Dermatol. Online J.* **2003**, *9*, 2.
71. Kawaguchi, M.; Mitsuhashi, Y.; Kondo, S. *J. Dermatol.* **2003**, *30*, 801–804.
72. Wietrzyk, J.; Pelczynska, M.; Madej, J.; Dzimira, S.; Kusnierczyk, H.; Kutner, A.; Szelejewski, W.; Opolski, A. *Steroids* **2004**, *69*, 629–635.
73. Ortonne, J. P.; Kaufmann, R.; Lecha, M.; Goodfield, M. *Dermatology* **2004**, *209*, 308–313.
74. Guenther, L. C. *Am. J. Clin. Dermatol.* **2004**, *5*, 71–77.
75. Katayama, I.; Ashida, M.; Maeda, A.; Eishi, K.; Murota, H.; Bae, S. J. *Eur. J. Dermatol.* **2003**, *13*, 372–376.
76. Heydendael, V. M.; Spuls, P. I.; Opmeer, B. C.; de Borgie, C. A.; Reitsma, J. B.; Goldschmidt, W. F.; Bossuyt, P. M.; Bos, J. D.; de Rie, M. A. *N. Engl. J. Med.* **2003**, *349*, 658–665.
77. Grewe, S. R.; Chan, S. C.; Hanifin, J. M. *J. Allergy Clin. Immunol.* **1982**, *70*, 452–457.
78. Safko, M. J.; Chan, S. C.; Cooper, K. D.; Hanifin, J. M. *J. Allergy Clin. Immunol.* **1981**, *68*, 218–225.
79. Hanifin, J. M.; Chan, S. C.; Cheng, J. B.; Tofte, S. J.; Henderson, W. R., Jr.,; Kirby, D. S.; Weiner, E. S. *J. Invest. Dermatol.* **1996**, *107*, 51–56.
80. MacKenzie, S. J. *Allergol. Int.* **2004**, *53*, 101–110.
81. Wedi, B.; Kapp, A. *BioDrugs* **2001**, *15*, 729–743.
82. Ikai, K.; Imamura, S. *Prostaglandins Leukot. Essent. Fatty Acids* **1993**, *48*, 409–416.
83. Neuber, K.; Hilger, R. A.; Konig, W. *Immunology* **1991**, *73*, 83–87.
84. Fogh, K.; Herlin, T.; Kragballe, K. *J. Allergy Clin. Immunol.* **1989**, *83*, 450–455.
85. Thorsen, S.; Fogh, K.; Broby-Johansen, U.; Sondergaard, J. *Allergy* **1990**, *45*, 457–463.
86. Ruzicka, T.; Simmet, T.; Peskar, B. A.; Braun-Falco, O. *Lancet* **1984**, *1*, 222–223.
87. Ruzicka, T.; Simmet, T.; Peskar, B. A.; Ring, J. *J. Invest. Dermatol.* **1986**, *86*, 105–108.
88. Fauler, J.; Neumann, C.; Tsikas, D.; Frolich, J. *Br. J. Dermatol.* **1993**, *128*, 627–630.
89. Oymar, K.; Aksnes, L. *Allergy* **2005**, *60*, 86–89.
90. Hon, K. L.; Leung, T. F.; Ma, K. C.; Li, A. M.; Wong, Y.; Li, C. Y.; Chan, I. H.; Fok, T. F. *Clin. Exp. Dermatol.* **2004**, *29*, 277–281.
91. Rackal, J. M.; Vender, R. B. *Skin Therapy Lett.* **2004**, *9*, 1–5.
92. Concha del Rio, L. E.; Arroyave, C. M. *Rev. Alerg. Mex.* **2003**, *50*, 187–191.
93. Yanase, D. J.; David-Bajar, K. *J. Am. Acad. Dermatol.* **2001**, *44*, 89–93.
94. Carucci, J. A.; Washenik, K.; Weinstein, A.; Shupack, J.; Cohen, D. E. *Arch. Dermatol.* **1998**, *134*, 785–786.
95. Lee, J. C.; Laydon, J. T.; McDonnell, P. C.; Gallagher, T. F.; Kumar, S.; Green, D.; McNulty, D.; Blumenthal, M. J.; Heys, J. R.; Landvatter, S. W. et al. *Nature* **1994**, *372*, 739–746.
96. Young, P.; McDonnell, P.; Dunnington, D.; Hand, A.; Laydon, J.; Lee, J. *Agents Actions* **1993**, *39* (Spec No.), C67–C69.
97. Gallagher, T. F.; Seibel, G. L.; Kassis, S.; Laydon, J. T.; Blumenthal, M. J.; Lee, J. C.; Lee, D.; Boehm, J. C.; Fier-Thompson, S. M.; Abt, J. W. et al. *Bioorg. Med. Chem.* **1997**, *5*, 49–64.
98. Badger, A. M.; Bradbeer, J. N.; Votta, B.; Lee, J. C.; Adams, J. L.; Griswold, D. E. *J. Pharmacol. Exp. Ther.* **1996**, *279*, 1453–1461.
99. Jackson, J. R.; Bolognese, B.; Hillegass, L.; Kassis, S.; Adams, J.; Griswold, D. E.; Winkler, J. D. *J. Pharmacol. Exp. Ther.* **1998**, *284*, 687–692.
100. Herlaar, E.; Brown, Z. *Mol. Med. Today* **1999**, *5*, 439–447.
101. Salituro, F. G.; Germann, U. A.; Wilson, K. P.; Bemis, G. W.; Fox, T.; Su, M. S. *Curr. Med. Chem.* **1999**, *6*, 807–823.
102. Dumas, J.; Sibley, R.; Riedl, B.; Monahan, M. K.; Lee, W.; Lowinger, T. B.; Redman, A. M.; Johnson, J. S.; Kingery-Wood, J.; Scott, W. J. et al. *Bioorg. Med. Chem. Lett.* **2000**, *10*, 2047–2050.
103. Redman, A. M.; Johnson, J. S.; Dally, R.; Swartz, S.; Wild, H.; Paulsen, H.; Caringal, Y.; Gunn, D.; Renick, J.; Osterhout, M. et al. *Bioorg. Med. Chem. Lett.* **2001**, *11*, 9–12.
104. Hanson, G. H. *Exp. Opin. Ther. Patents* **1997**, 7, 729–733.
105. Boehm, J. C.; Adams, J. L. *Exp. Opin. Ther. Patents* **2000**, *10*, 25–37.
106. Anonymous *Exp. Opin. Ther. Patents* **1999**, *9*, 975–979.
107. Anonymous *Exp. Opin. Ther. Patents* **2000**, *10*, 1151–1154.
108. Ottosen, E. R.; Sorensen, M. D.; Bjorkling, F.; Skak-Nielsen, T.; Fjording, M. S.; Aaes, H.; Binderup, L. *J. Med. Chem.* **2003**, *46*, 5651–5662.
109. Pargellis, C.; Tong, L.; Churchill, L.; Cirillo, P. F.; Gilmore, T.; Graham, A. G.; Grob, P. M.; Hickey, E. R.; Moss, N.; Pav, S. et al. *Nat. Struct. Biol.* **2002**, *9*, 268–272.
110. Bemis, G. W.; Salituro, F. G.; Duffy, J. P.; Cochran, J. E.; Harrington, E. M.; Murcko, M. A.; Wilson, K. P.; Su, M.; Galullo, V. P. Patent WO 199827098, 1998.
111. Eckert, R. L.; Efimova, T.; Balasubramanian, S.; Crish, J. F.; Bone, F.; Dashti, S. *J. Invest. Dermatol.* **2003**, *120*, 823–828.
112. Johansen, C.; Kragballe, K.; Westergaard, M.; Henningsen, J.; Kristiansen, K.; Iversen, L. *Br. J. Dermatol.* **2005**, *152*, 37–42.
113. Elsner, J.; Escher, S. E.; Forssmann, U. *Allergy* **2004**, *59*, 1243–1258.
114. Raychaudhuri, S. P.; Sanyal, M.; Weltman, H.; Kundu-Raychaudhuri, S. *J. Invest. Dermatol.* **2004**, *122*, 812–819.
115. Erin, N.; Ersoy, Y.; Ercan, F.; Akici, A.; Oktay, S. *Clin. Exp. Dermatol.* **2004**, *29*, 644–648.
116. Suarez, Y.; Gonzalez, L.; Cuadrado, A.; Berciano, M.; Lafarga, M.; Munoz, A. *Mol. Cancer Ther.* **2003**, *2*, 863–872.
117. Shintani, T.; Klionsky, D. J. *Science* **2004**, *306*, 990–995.
118. Fujino, M.; Funeshima, N.; Kitazawa, Y.; Kimura, H.; Amemiya, H.; Suzuki, S.; Li, X. K. *J. Pharmacol. Exp. Ther.* **2003**, *305*, 70–77.
119. Webb, M.; Tham, C. S.; Lin, F. F.; Lariosa-Willingham, K.; Yu, N.; Hale, J.; Mandala, S.; Chun, J.; Rao, T. S. *J. Neuroimmunol.* **2004**, *153*, 108–121.
120. Elices, M. J. *Curr. Opin. Investig. Drugs* **2004**, *5*, 1137–1140.

Biographies

Mariano J Elices obtained a PhD in Biological Chemistry from the University of Michigan Medical School in Ann Arbor. He then took a postdoctoral fellowship at the Harvard Medical School where he was involved in groundbreaking research at the Dana Farber Cancer Institute on the cell adhesion receptors known as integrins. The involvement of integrins in inflammation led him to a career in the pharmaceutical industry in Southern California. Today, he works at PharmaMar USA in Cambridge, the American subsidiary of the Spanish biopharmaceutical company PharmaMar, which specializes in oncology applications of novel chemical entities derived from marine sources.

Thomas Arrhenius is a consultant, based in San Diego, California. For the past 15 years he has been involved in drug discovery research in the biotech and pharmaceutical industries.

Comprehensive Medicinal Chemistry II
ISBN (set): 0-08-044513-6

ISBN (Volume 7) 0-08-044520-9; pp. 969–985

INDEX FOR VOLUME 7

Notes

Abbreviations

COPD – chronic obstructive pulmonary disease
SAR – structure–activity relationship

Cross-reference terms in italics are general cross-references, or refer to subentry terms within the main entry (the main entry is not repeated to save space). Readers are also advised to refer to the end of each article for additional cross-references – not all of these cross-references have been included in the index cross-references.

The index is arranged in set-out style with a maximum of three levels of heading. Major discussion of a subject is indicated by bold page numbers. Page numbers suffixed by T and F refer to Tables and Figures respectively. vs. indicates a comparison.

This index is in letter-by-letter order, whereby hyphens and spaces within index headings are ignored in the alphabetization. Prefixes and terms in parentheses are excluded from the initial alphabetization.

Any method, model or other subject, associated with the name of the developer (e.g. name's model) does NOT imply that Elsevier, nor the indexers, have assumed the right to name models/methods after the authors of the papers in which they are described. This is merely a succinct phrase to refer to a model/method developed/described by the relevant author, so that the subentry could be alphabetized under the most pertinent name.

A

B

C

D

E

F

G

H

I

M

O

P

Q

R

S

U

V

W

X

Y

Z